W9-BUV-327

THE NEW
*Fowler's*
**Modern English Usage**

THE NEW

*Fowler's*

# Modern
# English
# Usage

*first edited by* H. W. Fowler

THIRD EDITION
*edited by* R. W. Burchfield

CLARENDON PRESS    OXFORD
1996

Oxford University Press, Great Clarendon Street, Oxford OX2 6DP

Oxford New York

Athens Auckland Bangkok Bogota Bombay Buenos Aires
Calcutta Cape Town Dar es Salaam Delhi Florence Hong Kong
Istanbul Karachi Kuala Lumpur Madras Madrid Melbourne
Mexico City Nairobi Paris Singapore Taipei Tokyo Toronto
and associated companies in
Berlin Ibadan

Oxford is a trade mark of Oxford University Press

© Oxford University Press 1965, 1996

First edition 1926
Second edition 1965
Third edition 1996

All rights reserved. No part of this publication may be reproduced,
stored in a retrieval system, or transmitted, in any form or by any means,
without the prior permission in writing of Oxford University Press.
Within the UK, exceptions are allowed in respect of any fair dealing for the
purpose of research or private study, or criticism or review, as permitted
under the Copyright, Designs and Patents Act, 1988, or in the case of
reprographic reproduction in accordance with the terms of the licences
issued by the Copyright Licensing Agency. Enquiries concerning
reproduction outside these terms and in other countries should be
sent to the Rights Department, Oxford University Press,
at the address above.

This book is sold subject to the condition that it shall not, by way
of trade or otherwise, be lent, re-sold, hired out or otherwise circulated
without the publisher's prior consent in any form of binding or cover
other than that in which it is published and without a similar condition
including this condition being imposed on the subsequent purchaser

British Library Cataloguing in Publication Data
The New Fowler's Modern English Usage—3rd ed.
English language—Dictionaries
Burchfield, R.W.

Library of Congress Cataloging in Publication Data
The New Fowler's Modern English Usage—3rd ed./edited by R.W. Burchfield
1. English language—Dictionaries.   I. Burchfield, R.W. (Robert William).
II. Fowler, H.W. (Henry Watson), 1858-1933.
ISBN 0-19-869126-2

10 9 8 7 6 5 4 3 2 1

Data-capture by Jayvee Ltd., Trivandrum, India
Typeset in Swift and Meta by Latimer Trend Ltd., Plymouth
Printed in the United States of America
on acid-free paper

For my beloved wife

*Elizabeth Austen Burchfield*

## CONFLICTING VIEWS

Ours is a Copious Language, and Trying to Strangers.

Mr Podsnap in Dickens's *Our Mutual Friend*, 1865

Grammar is like walking. You have to think about it when you start but if you have to go on thinking about it you fall over. It should come as second nature.

Alice Thomas Ellis in *The Spectator*, 1989

Was she becoming, like the century, illiterate?

a character in Iris Murdoch's *The Book and the Brotherhood*, 1987

'How charming. Now, "Luney". How do you spell that?' Swayed by the drawing of her breath, the [Haitian] girl took a moment to dream, then said with a far-off resonance, 'You don' spell dat, ma'am, you sez it.'

Barbara Neil, *The Possession of Delia Sutherland*, 1993

## DISLIKES

Comments by members of a Usage Panel on the use of *hopefully* as a sentence adverb meaning 'it is to be hoped', as reported in the *Harper Dictionary of Contemporary Usage* (2nd edn, 1985):

I have fought this for some years, will fight it till I die. It is barbaric, illiterate, offensive, damnable, and inexcusable.

I don't like chalk squeaking on blackboards either.

'Hopefully' is useful or it would not be used so universally.

'Grounded' meaning a withdrawal of privileges is a word I dislike. It's off the television (*Roseanne* notably) but now in common use. (I just heard it on *Emmerdale Farm*, where they probably think it's dialect). I would almost prefer 'gated', deriving from Forties public school stories in *Hotspur* and *Wizard*. Other current dislikes: 'Brits'; 'for starters'; 'sorted'; and (when used intransitively) 'hurting'.

Alan Bennett in *London Review of Books*, 4 Jan. 1996

Henry Watson Fowler[1] (1858–1933) is a legendary figure and his *Dictionary of Modern English Usage* (*MEU*), first published in 1926, is one of the most celebrated reference books of the twentieth century. It was the work of a private scholar writing in virtual seclusion in the island of Guernsey; later, after the 1914–18 war, he lived mostly in the village of Hinton St George in Somerset. His background was typical of that of hundreds of middle-class young men of the second half of the Victorian period: educated at Rugby School and Balliol College, Oxford (where he read Classics), he went on to spend seventeen years (1882–99) teaching Classics and English at Sedburgh School in north-west Yorkshire (now Cumbria). There followed a four-year period in London as a freelance essayist, after which he joined his younger brother, Francis George Fowler, in Guernsey in 1903. In two separate granite cottages, fifty yards apart, the brothers embarked on and completed three ambitious projects. First, they translated the Greek works of Lucian of Samosata (1905); they then wrote *The King's English* (1906), the precursor of *MEU*, and compiled *The Concise Oxford Dictionary* (1911). After an adventurous interlude in the army in France in 1915–16, and after the death of his brother in 1918, Fowler finished the *Pocket Oxford Dictionary* in 1924, and *MEU* in 1926, by which time he was 68.

What I want to stress is the isolation of Fowler from the mainstream of the linguistic scholarship of his day, and his heavy dependence on school-masterly textbooks in which the rules of grammar, rhetoric, punctuation, spelling, and so on, were set down in a quite basic manner. For him, the ancient Greek and Latin classics (including the metrical conventions of the poets), the best-known works of Renaissance and post-Renaissance English literature, and the language used in them formed part of a three-coloured flag. This linguistic flag was to be saluted and revered, and, as far as possible, everything it represented was to be preserved intact.

The book that emerged in 1926, *Modern English Usage*, was aimed at a domestic audience. Fowler disclaimed any knowledge of American English and by implication, the varieties of English spoken and written in Canada, Australia, New Zealand, South Africa, and elsewhere. In a letter written to his publishers in 1911 he drew attention to further limits of his horizon:

> We have our eyes not on the foreigner, but on the half-educated Englishman of literary proclivities who wants to know Can I say so-&-so? . . . Is this use English? . . . Not but what we may be of some use to the foreigner who knows English pretty well; but the foreigner as such we must leave out of consideration.

For his illustrative examples Fowler often turned to the *OED* and drew on them to support his arguments. Above all, however, he turned to newspapers

---

[1] An affectionate biographical sketch of Fowler by his friend G. G. Coulton was published in 1935 as Tract xliii of the Society for Pure English.

(though he seldom specifies which ones) because they reflected and revealed the solecistic waywardness of 'the half-educated' general public in a much more dramatic fashion than did works of English literature. As any lexicographer or grammarian knows, newspapers, by the very nature of the circumstances in which they are prepared, inevitably contain a higher proportion of deviations from standard language, misprints, and solecisms than works such as novels that are thoroughly copy-edited by professional editors in publishing houses.

Perhaps as a hangover from Fowler's days as a schoolmaster, his scholarship needed to be enlivened by a veneer of idiosyncrasy and humour. The King's English (1906) had a trail of conventional articles on alliteration, archaism, negatives, omission of relatives, the split infinitive, and so on; but it also had more unexpected, indeed opaque, titles to articles, for example, 'airs and graces', 'between two stools', 'false scent', 'unequal yokefellows', and 'wens and hypertrophied members'. Most of these amusing headwords were retained in MEU, and were joined by others, for example, 'battered ornaments', 'out of the frying-pan', 'pairs and snares', and 'swapping horses'. They have endeared the book to Fowler's devotees, but no longer have their interest or appeal and are not preserved in this new edition. The material in them has been redistributed under much more transparent heads.

Before embarking on the preparation of the third edition I carefully analysed the contents of MEU 1926, and the emphasis turned out to be a little unexpected. The largest contingent of entries were those under the general heading 'differentiation', though the actual entries were deposited at their correct alphabetical place. There were scores of entries distinguishing related or like-sounding words, admission/admittance, affect/effect, childish/childlike, continual/continuous, and so on. Many were gems of conciseness (or concision), with the distinctions clearly brought out. Others were quirky, opinionated, and based on inadequate evidence. MEU 1926 was also much concerned with the plurals of words of foreign origin, especially those ending in -o (adagio, cargo, concerto, potato, etc.), -um (asylum, curriculum, memorandum, etc.), and -us (apparatus, corpus, virus, etc.). These were usually cross-referenced to neat articles where the various types were discussed as groups, e.g. -O(E)S; -UM; and -US, with further details supplied s.v. LATIN PLURALS. All these entries have been preserved and expanded in the present edition.

High in Fowler's order of priority were prosodic and other poetical terms derived from classical literature and used, often with modifications, by English poets: alcaics, alexandrine, anacrusis, arsis, etc. The chalk-lined hand of the classics master at Sedburgh is most clearly observed in this group of words. I decided, on balance, that these articles, with minor modifications, should be retained in the third edition. As was customary at the time, Fowler used a respelling system when discussing the pronunciation of individual words. In the third edition this system has been replaced by the symbols of the International Phonetic Alphabet to bring the book into accord with the practice of other Oxford dictionaries (except those prepared for schoolchildren). For the convenience of readers a table of the IPA symbols is provided on p. xv. Somewhat surprising is the relative lack of space given in the first edition to disputed usage as such. Of course there are articles, many of them classics of their kind, on matters such as *aggravate* (= annoy), *all right* (as against *alright*), the choice between *between* and *among*, *under the circumstances* (as against *in the circumstances*), 'preposition at end', and so on, but they are by no means the most prominent articles in the book.

The mystery remains: why has this schoolmasterly, quixotic, idiosyncratic, and somewhat vulnerable book, in a form only lightly revised once, in 1965, by Ernest Gowers, retained its hold on the imagination of all but professional linguistic scholars for just on seventy years? It sold very well on publication, and has remained in print ever since. People of all kinds continue to tell me that they use it 'all the time', and that 'it never lets them down'. In the space of three weeks a judge, a colonel, and a retired curator of Greek and Roman antiquities at the British Museum told me on separate social occasions that they have the book close at hand at all times. They all looked anxious when I mentioned a few changes that I have made in the new edition, including the placing of twentieth-century changes in their historical dimension and the introduction of the International Phonetic Alphabet. 'I wish you hadn't told me that,' one commented. The slightly haunted looks they gave me were those of passengers fearing that they were going to miss their connection.

From the outset it was obvious to me that a standard work on English usage needed to be based on satisfactory modern evidence and that a great deal of this evidence could be obtained and classified by electronic means. In September 1986, after the completion of the *Supplement to the OED* and the *New Zealand Pocket Oxford Dictionary*, and coinciding with other work, including the editing of a volume of essays called *Studies in Lexicography*, I obtained a personal computer and began to establish a database consisting of ten independent fields corresponding to obvious categories of grammar and usage. The ten fields were adjectives, adverbs, concord, gerunds, infinitives, nouns and articles, ordinaries (a convenient term for points of disputed usage), passives, pronouns, and subjunctives. The fields that I created enabled me to assign specific numbers to the various types of gerunds, passives, subjunctives, etc., and these types soon began to multiply as my reading of sources continued. The numbering system enabled me to retrieve and print out all examples of a specified type, e.g. gerunds 3 = possessive with gerund: *I was proud of his being accepted at such a good school—New Yorker*, 1986; and gerunds 4 = possessive not used with gerund: *how could she think of the baby being born in the house—A. S. Byatt*, 1985. In the end my gerunds field contained examples of more than 100 types of gerundial constructions, and, like all my fields, it is infinitely extendible. Some of the fields, and especially the one containing examples of constructions in which infinitives occur, are much larger. The ordinaries field contains, for example, a formidable array of controversial uses of *due to*, *like* used as a conjunction, *of* used by children and poorly educated people to mean 'have', unattached participles, irregular or unstable past tenses of verbs, e.g. *hove/heaved, sneaked/snuck, spelled/spelt*, and numerous other types including *try and* (used beside *try to*) followed by an infinitive.

The database was programmed in such a way that I could retrieve from it all examples of specified words that randomly occurred in the sentences keyed in for other purposes—words such as *about, better, if, more, though, too*, and also specified parts of words, e.g. all words in the database that happened to end in *-edly* (*allegedly, markedly*, etc.). This database is small by the standards of the great university- and business-based corpora. But its value lies in the fact that it contains material from sources that I have selected myself, and examples that I have chosen and keyed in myself—in computer terminology, it contains no garbage. It is a private, personalized database of English uses and constructions of the 1980s and 1990s. By 1990 the time-consuming process of collecting and classifying evidence needed to be modified, as the writing of articles for the book, which I had begun in August 1987, had only reached the end of the letter C by October 1990. From then onward I continued to build up my paper-slip files, but began to rely much more than hitherto on the evidence available in the OED Department's electronic and paper-slip files (see the Acknowledgements on p. xiii).

A usage manual of the *MEU* kind reflects its sources. The bulk of the material in this book has been obtained from a systematic reading of British and American newspapers, periodicals, and fiction of the 1980s and 1990s in approximately equal proportions. Thus British sources drawn on include national newspapers like *The Times*, the *Sunday Times*, and the *Observer*; periodicals like the *Spectator, Encounter* (until it folded in 1990), the *London Review of Books, The Face*, and a number of others; journals like the *Bodleian Library Record* and *English*; and fiction by a very large number of writers including Kingsley Amis, Martin Amis, Julian Barnes, William Boyd, Anita Brookner, Penelope Lively, Iris Murdoch, and Nigel Williams. An equivalent amount of material has been drawn from American newspapers, especially in the Chicago area (where I have a regular correspondent), periodicals such as the *New Yorker*, the *New York Review of Books*, the *Bulletin of the American Academy of Arts and Sciences, Dædalus*; and fiction by a wide range of writers including Saul Bellow, Garrison Keillor, Philip Roth, John Updike, and Tom Wolfe.

I have also collected a more limited range of material from other English-speaking countries, especially Australia, Canada, New Zealand, and South Africa, e.g. the work of Peter Carey and Thomas Keneally (Australia), Alice Munro and Robertson Davies (Canada), Maurice Gee and Maurice Shadbolt (New Zealand), and Menán du Plessis and André Brink (South Africa). I have drawn too on material derived from learned journals, including *English Studies* (Amsterdam), *English World-Wide* (Amsterdam), *World Englishes* (Oxford), and *The International Journal of Lexicography* (Oxford).

I should mention that between 1988 and 1992 I wrote regular (at first fortnightly, later monthly) *MEU*-type articles in the *Sunday Times*. The more important of these were collected and published in my book *Points of View* (OUP, 1992). This exercise provided me with a considerable amount of feedback from readers. Other exploratory essays that I have written on aspects of modern English usage have appeared (*a*) in my book *Unlocking the English Language* (Faber & Faber, 1989), including a description of the controversial migration of some personal pronouns to and from their traditional positions; (*b*) as an essay on grammatical concord in *The English Reference Grammar*, edited by Gerhard Leitner (Max Niemeyer, 1986); (*c*) as an article illustrating differences of attitude to traditional grammar as shown in the novels of Jeffrey Archer and Anita Brookner (in *The State of the Language*, edited by Christopher Ricks and Leonard Michaels (University of California Press, 1990)); and (*d*) an outline account of my policy for this book in *Aspects of English and Italian Lexicology and Lexicography*, edited by David Hart (Bagatto Libri, 1993).

Anyone who has spent nearly thirty years, as I did, editing a major dictionary on historical principles is bound to prefer an historical approach to English usage to one that is limitedly descriptive. Judgements based on the distribution of competing constructions or pronunciations are intrinsically fragile and diminished in value if the constructions are not also examined historically. This third edition of *MEU* provides essential details of how and when new usages occurred whenever it is relevant and interesting to do so. Examples may be found on a great many pages, including the following articles: (competing meanings) MUTUAL; REFUTE; (rise and fall of certain suffixes) -ESS; -ETTE; (semantic change) GAY; HECTIC; HORRID; (20C. changes of pronunciation) HERCULEAN; LEGEND; MYTH; PARIAH; PROTEIN (and several other words ending in -*ein(e)*).

I judged it to be essential to retain the traditional terminology of English grammar: there are no tree diagrams, no epistemic modality (except to explain what the term means), no generative grammar. The indefinite article *a/an* is called the indefinite article, not a central determiner. Adverbs are not complicatedly divided into adjuncts, conjuncts, disjuncts, and subjuncts: standard speakers can communicate well enough without having to analyse their adverbs into four substantially overlapping types.

Fowler's name remains on the title-page, even though his book has been largely rewritten in this third edition. I hope that a way will be found to keep the 1926 masterpiece in print for at least another seventy years. It shows what it was like to be linguistically aware before a new race of synchronic linguistic giants appeared, and before the advent of new electronic technology made it possible to scrutinize standard varieties of English in many countries throughout the world with minute thoroughness. It is not, of course, as antiquated as Ælfric's Grammar nor yet as those of Ben Jonson or Robert Lowth. But it is a fossil all the same, and an enduring monument to all that was linguistically acceptable in the standard English of the southern counties of England in the first quarter of the twentieth century.

The pages that follow attempt, with the aid of quotational evidence drawn from identified sources, to guide readers to make sensible choices in linguistically controversial areas of words, meanings, grammatical constructions, and pronunciations. Several articles stress the desirability of removing gobbledegook or officialese from public documents and letters. Political correctness gets its full share of attention, as do linguistic aspects of the powerful feminist movement in the twentieth century. It is written at a time when there are many varieties of standard English, all making different choices from the material notionally available to them. It is also a time when pessimists are writing gloomily about declining standards, the loss of valuable distinctions in meaning, the introduction of unappetising vogue words and slang. But I refuse to be a pessimist. I am sure that the English language is not collapsing—more severe changes have come about in past centuries than any that have occurred in the twentieth century—and in the English language, used well, we still have, and will continue to have, a tool of extraordinary strength and flexibility.

## Acknowledgements

It gives me great pleasure to set down my obligation to the many people who have contributed in one way or another to the preparation of this edition. First and foremost I owe an immeasurable debt to my former colleagues in the OED Department, who allowed me unrestricted access to their rich electronic and paper-slip quotation files and to the electronic databases (e.g. NEXIS) to which they themselves have access. Once it had been decided to identify the sources of the quotational evidence rather than to rely on unattributed illustrative examples or merely invented examples, the book could never have been assembled without such privileged access, even in the nine years it has taken to write it. Major contributors included a retired American lexicographer, Mr Frank G. Pickel (of Evanston, Illinois), a diplomat (now retired), Sir Brian L. Barder, and a library researcher, Mr George Chowdharay-Best.

Indispensable help of various kinds—suggestions for new entries, criticism of existing articles, judgements about particular words or constructions, and so on—have come from the following people: Mr David Annett, Mr Don Barton, Mr P. R. Bonnett, Sir James Craig, Mr G. Crawford, Dr Robert D. Eagleson, Dr J. A. D. Ewart, Mr Bryan A. Garner, Dr Valerie Grundy, Mr William E. Hutchins, Mr Kenneth R. Lake, Professor Geoffrey Lewis, Mr E. W. Noll, Dr Stefania Nuccorini, Mr Jim Powell, Professor James Sutherland, Mr Ernest Trehern, Mr B. Verity, Mr F. R. le P. Warner, Professor Emer. Hugh E. Wilkinson, Mr C. F. Wright.

I am also greatly indebted to Sarah Barrett, who brought her considerable copy-editing skills to bear on the complexities of this book.

## Dedication, 1926

*To the memory of my brother*

FRANCIS GEORGE FOWLER, M.A. CANTAB.

*who shared with me the planning of this book,*
*but did not live to share the writing.*

I think of it as it should have been, with its prolixities docked, its dullnesses enlivened, its fads eliminated, its truths multiplied. He had a nimbler wit, a better sense of proportion, and a more open mind, than his twelve-year-older partner; and it is a matter of regret that we had not, at a certain point, arranged our undertakings otherwise than we did.

In 1911 we started work simultaneously on the *Pocket Oxford Dictionary* and this book; living close together, we could, and did, compare notes; but each was to get one book into shape by writing its first quarter or half; and so much only had been done before the war. The one in which, as the less mechanical, his ideas and contributions would have had much the greater value had been assigned, by ill chance, to me. In 1918 he died, aged 47, of tuberculosis contracted during service with the B.E.F. in 1915–16.

The present book accordingly contains none of his actual writing; but, having been designed in consultation with him, it is the last fruit of a partnership that began in 1903 with our translation of Lucian.

H. W. F.

The pronunciation system is that of the International Phonetic Alphabet (IPA) and, except where otherwise specified, is based on the pronunciation, widely called 'Received Pronunciation' or RP, of educated people in southern England. The necessary adjustments have been made when standard American English pronunciations are given.

The symbols used, with typical examples, are as follows:

## Consonants

*b, d, f, h, k, l, m, n, p, r, s, t, v, w,* and *z* have their usual English values. Other symbols are used as follows:

| | | | | | |
|---|---|---|---|---|---|
| g | (get) | ŋ | (ring) | ʒ | (decision) |
| tʃ | (chip) | θ | (thin) | j | (yes) |
| dʒ | (jar) | ð | (this) | | |
| x | (loch) | ʃ | (she) | | |

## Vowels

| *Short vowels* | | *Long vowels* | | *Diphthongs* | |
|---|---|---|---|---|---|
| æ | (cat) | ɑː | (arm) | eɪ | (day) |
| e | (bed) | iː | (see) | aɪ | (my) |
| ə | (ago) | ɔː | (saw) | ɔɪ | (boy) |
| ɪ | (sit) | ɜː | (her) | əʊ | (no) |
| ɒ | (hot) | uː | (too) | aʊ | (how) |
| ʌ | (run) | | | ɪə | (near) |
| ʊ | (put) | | | eə | (hair) |
| | | | | ʊə | (poor) |
| | | | | aɪə | (fire) |
| | | | | aʊə | (sour) |

The main or primary stress of a word is shown by a superior ' placed immediately before the relevant syllable. When a secondary stress is called for this is indicated by an inferior ˌ placed immediately before the relevant syllable.

The mark ˜ (called a tilde) indicates a nasalized sound, as in the following sounds that are not natural in English:

| | | | | | |
|---|---|---|---|---|---|
| æ̃ | (timbre) | ɑ̃ | (élan) | ɔ̃ | (garçon) |

## Abbreviations and Symbols

| | |
|---|---|
| † | obsolete |
| → | becomes |
| * | unacceptable construction, spelling, etc. |
| * | precedes a reconstructed etymological formation |
| ~ | varies freely with; by comparison with |

SMALL CAPITALS refer the reader to the article so indicated, for further information.

| | | | |
|---|---|---|---|
| *a, ante* | before, not later than | e.g. | *exempli gratia,* 'for example' |
| abbrev. | abbreviation, abbreviated as | ELT | English Language Teaching |
| abl. | ablative | Eng. | English |
| acc. | accusative | esp. | especially |
| adj. | adjective | et al. | *et alii,* 'and others' |
| adv. | adverb | exc. | except |
| advbl | adverbial | | |
| advt | advertisement | f. | from |
| AmE | American English | fem. | feminine |
| arch. | archaic | fig. | figuratively |
| attrib. | attributive(ly) | Fr. | French |
| Aust. | Australian | | |
| aux. | auxiliary | Ger. | German |
| AV | Authorized Version (of the Bible) | gen. | genitive |
| | | Gk | Greek |
| BEV | Black English Vernacular (US) | hist. | historical, with historical reference |
| BrE | British English | | |
| | | ibid. | *ibidem,* 'in the same book or passage' |
| *c* | *circa* | idem | 'the same' |
| c. | century, centuries | i.e. | *id est,* 'that is' |
| Canad. | Canadian | IE | Indo-European |
| cf. | compare | indic. | indicative |
| colloq. | colloquial | infin. | infinitive |
| compar. | comparative | intr. | intransitive |
| conj. | conjunction | Ir. | Irish |
| const. | construed (with) | irreg. | irregular(ly) |
| | | It. | Italian |
| d. | died | ITV | Independent Television (UK) |
| dat. | dative | | |
| det. | determiner | | |
| dial. | dialect, -al | L, Lat. | Latin |
| Du. | Dutch | lit. | literally |
| EC | European Community | | |

| | |
|---|---|
| masc. | masculine |
| math. | mathematical |
| MDu. | Middle Dutch |
| ME | Middle English |
| medL | medieval Latin |
| MLG | Middle Low German |
| mod. | modern |
| modE | modern English |
| modF | modern French |
| modL | modern Latin |
| mus. | music |
| | |
| n., n.pl. | noun, plural noun |
| NAmer. | North American |
| naut. | nautical |
| neut. | neuter |
| NEXIS | proprietary name of a large electronic database |
| NIr. | Northern Irish |
| nom. | nominative |
| NT | New Testament |
| NZ(E) | New Zealand (English) |
| | |
| obs. | obsolete |
| obsolesc. | obsolescent |
| occas. | occasional(ly) |
| OE | Old English |
| OF | Old French |
| OFris. | Old Frisian |
| ON | Old Norse |
| OProvençal | Old Provençal |
| orig. | originally |
| OUP | Oxford University Press |
| | |
| pa.t. | past tense |
| pa.pple | past participle |
| perh. | perhaps |
| pers. | person |
| Pg. | Portuguese |
| phr. | phrase |
| pl. | plural |
| poet. | poetic |
| popL | popular Latin |
| ppl | participial |

| | |
|---|---|
| pple | participle |
| pref. | prefix |
| prep. | preposition, prepositional |
| pres. | present |
| prob. | probably |
| pron. | pronoun |
| pronunc. | pronunciation |
| | |
| quot. | quotation |
| | |
| RC | Roman Catholic |
| refl. | reflexive |
| RP | Received Pronunciation (in BrE) |
| | |
| SAfr. | South African |
| SAmer. | South American |
| sb. | substantive |
| Sc. | Scottish |
| sc. | *scilicet*, 'understand' or 'supply' |
| sing. | singular |
| Sp. | Spanish |
| SPE | Society for Pure English |
| spec. | specifically |
| superl. | superlative |
| s.v. | *sub voce*, 'under the word' |
| | |
| t. | tense |
| theol. | theological |
| tr. | translation (of) |
| trans. | transitive |
| | |
| UK | United Kingdom |
| ult. | ultimately |
| UN | United Nations |
| US | United States |
| usu. | usually |
| | |
| v., vs. | verb, verbs |
| vbl | verbal |
| vol. | volume |
| | |
| WGmc | West Germanic |

## Bibliographical Abbreviations

| | |
|---|---|
| Alford | Henry Alford, *The Queen's English*, 1864 |
| *Amer. N. & Q.* | *American Notes & Queries* |
| *Amer. Speech* | *American Speech* |
| *Ann.* | *Annual* |
| | |
| Baldick | C. Baldick, *The Concise Oxford Dictionary of Literary Terms*, 1990 |
| *BMJ* | *British Medical Journal* |
| *Bodl. Libr. Rec.* | *Bodleian Library Record* |
| *Bull. Amer. Acad. Arts & Sci.* | *Bulletin of the American Academy of Arts and Sciences* |
| Burchfield | R. Burchfield, *The Spoken Word: a BBC Guide*, 1981 |
| | |
| *CGEL* | *A Comprehensive Grammar of the English Language*, ed. R. Quirk et al., 1985 |
| *Chr. Sci. Monitor* | *Christian Science Monitor* |
| *Chron.* | *Chronicle* |
| *COD* | *The Concise Oxford Dictionary*, 8th edn, 1990; 9th edn, 1995 |
| *Conc. Scots Dict.* | *Concise Scots Dictionary* |
| Crystal | D. Crystal, *A First Dictionary of Linguistics and Phonetics*, 1980 |
| | |
| *DARE* | *Dictionary of American Regional English*, ed. F. G. Cassidy et al., 2 vols. (A–H), 1985, 1991 |
| *Dict.* | *Dictionary (of )* |
| *Dict. Eng. Usage* | *Dictionary of English Usage* |
| | |
| *Eccles. Hist.* | *Ecclesiastical History* |
| *Encycl.* | *Encyclop(a)edia* |
| *European Sociol. Rev.* | *European Sociological Review* |
| | |
| Fowler | H. W. and F. G. Fowler, *The King's English*, 1906 |
| | |
| Garner | Bryan A. Garner, *A Dictionary of Modern Legal Usage*, 1987 |
| *Gaz.* | *Gazette* |
| Gimson | A. C. Gimson, *An Introduction to the Pronunciation of English*, 3rd edn, 1980 |
| | |
| Hartmann and Stork | R. R. K. Hartmann and F. C. Stork, *Dictionary of Language and Linguistics*, 1973 |
| *Hart's Rules* | *Hart's Rules for Compositors and Readers at the University Press, Oxford*, 39th edn., 1983 |
| | |
| *Internat.* | *International* |
| | |
| Jespersen | Otto Jespersen, *A Modern English Grammar on Historical Principles*, i–vii, 1909–49 |
| Jones | Daniel Jones, *An English Pronouncing Dictionary*, 1917 |
| *Jrnl* | *Journal (of )* |
| *Jrnl RSA* | *Journal of the Royal Society of Arts* |

| | |
|---|---|
| *London Rev. Bks* | *London Review of Books* |
| Lyons | J. Lyons, *Semantics*, 2 vols., 1977 |
| | |
| *Mag.* | *Magazine* |
| Mitchell | Bruce Mitchell, *Old English Syntax*, 2 vols., 1985 |
| | |
| *N&Q* | *Notes & Queries* |
| *NEB* | *New English Bible* |
| *New SOED* | *The New Shorter Oxford English Dictionary*, 2 vols., 1993 |
| *NY Rev. Bks* | *New York Review of Books* |
| | |
| *OCELang.* | *The Oxford Companion to the English Language*, ed. Tom McArthur, 1992 |
| *OCELit.* | *The Oxford Companion to English Literature*, ed. Margaret Drabble, 5th edn, 1985 |
| *ODCIE* | *Oxford Dictionary of Current Idiomatic English*, ed. A. P. Cowie et al., 2 vols, 1975, 1983 |
| *ODEE* | *Oxford Dictionary of English Etymology*, 1966 |
| *ODWE* | *Oxford Dictionary for Writers and Editors*, 1981 |
| *OED* | *The Oxford English Dictionary*, issued in parts 1884–1928; as 12 vols., 1933 |
| *OED 2* | *Oxford English Dictionary*, 2nd edn, 20 vols., 1989 |
| *OEDS* | *A Supplement to the Oxford English Dictionary*, 4 vols., 1972–86 |
| *OGEU* | *The Oxford Guide to English Usage*, 2nd edn, 1993 |
| *OMEU* | *The Oxford Miniguide to English Usage*, 1983 |
| *Oxf. Dict. Eng. Gramm.* | *The Oxford Dictionary of English Grammar*, 1994 |
| | |
| *Parl. Aff.* | *Parliamentary Affairs* |
| *POD* | *The Pocket Oxford Dictionary* |
| Poutsma | Hendrik Poutsma, *A Grammar of Late Modern English*, 5 vols., 1926–9 |
| *Publ. Amer. Dial. Soc.* | *Publications of the American Dialect Society* |
| | |
| Q | *Quarterly* |
| Quirk | R. Quirk et al., *A Grammar of Contemporary English*, 1972 |
| | |
| *Reg.* | *Register* |
| *Rep.* | *Report* |
| *Rev.* | *Review (of )* |
| | |
| *Sci. Amer.* | *Scientific American* |
| Smith | Egerton Smith, *The Principles of English Metre*, 1923 |
| | |
| *Tel.* | *Telegraph* |
| *TES* | *The Times Educational Supplement* |
| *THES* | *The Times Higher Educational Supplement* |
| *TLS* | *The Times Literary Supplement* |
| *Trans. Philol. Soc.* | *Transactions of the Philological Society* |
| Tulloch | S. Tulloch, *The Oxford Dictionary of New Words*, 1991 |
| | |
| Visser | F. Th. Visser, *An Historical Syntax of the English Language*, parts i–iii, 4 vols., 1963–73 |

| | |
|---|---|
| Wales | Katie Wales, *A Dictionary of Stylistics*, 1989 |
| Walker | John Walker, *A Critical Pronouncing Dictionary and Expositor of the English Language*, 1791; 4th edn 1806 |
| *WDEU* | *Webster's Dictionary of English Usage*, 1989 |
| Webster's Third | *Webster's Third New International Dictionary*, 1961 |
| Wells | J. C. Wells, *Longman Pronunciation Dictionary*, 1990 |

For convenience, the dates assigned to the works of Shakespeare are those given in the Bibliography of the *OED*. The texts of the individual works are cited from the original-spelling edition of Stanley Wells and Gary Taylor, OUP, 1986.

All examples from the Bible are cited from the Authorized Version of 1611 (quoted from the 'exact reprint' with an introduction by Alfred W. Pollard, OUP, 1985).

# Modern English Usage

# Aa

**a-¹,** a prefix of privation or negation, represents Greek ἀ- before a consonant and becomes *an-* (Greek ἀν-) before a vowel. It occurs (*a*) in words representing Greek compounds, mostly adopted through French or Latin, in which the prefix is wholly or partially obscured, as *abyss, adamant, amethyst, amorphous, anarchy, anomalous*; (*b*) in terms of the arts or sciences, having Greek bases, but coming into English through medieval or modern Latin, as *abranchiate* (having no gills), *anaesthesia* (absence of sensation), *anorexia* (want of appetite), *apetalous* (without petals), *aphasia* (loss of speech); (*c*) in words formed in the 19c. on Greek elements, as *agnostic, aseptic*. In the last century or so, privative *a-*, most commonly with the pronunciation /eɪ/, has come to be attached to a limited number of adjectives in general use, as *ahistoric* (1937), *ahistorical* (1957), *amoral* (1882), *apolitical* (1952), *asocial* (1883), and *atypical* (1885), and to their derivatives (*amorality*, etc.). But it is far from being a free-forming prefix.

**a-².** In origin a preposition (OED *a* prep.¹ 13), it is recorded from 1523 onward in many literary works, (*a*) with *be*: engaged in (*She is a taking of her last farewell*—Bunyan; *'twas the Bishops or Judges a coming*—Leigh Hunt); (*b*) with a verb of motion: to, into (*Simon Peter sayde vnto them: I goo a fysshynge*—Tyndale; *We were able to set the loan a going again*—Jefferson). Now interpreted as a prefix joined to an *-ing* word, it seems to be having a limited revival, imparting an element of informality or whimsicality to the present participle or gerund that it precedes. Examples: *Maybe you'll be a-basking in the sun this afternoon*—BBC Radio 4, 1988; *the basement is going a-begging*—P. Bailey, 1986; *Kris Kirk … plans to return, a-brandishing the manuscript* [of his book] *by the end of March*—Melody Maker, 1988; *there's trouble a-brewing down the line between Michael Dukakis and Jesse Jackson*—Chicago Sun-Times, 1988; *'Time's a-wasting,' said Ben*—A. Lejeune, 1986. The best-known use of the

prefix, perhaps, is in *The times they are a-changing* (a line in a 1960s song by the songwriter Bob Dylan).

**-a.** Now being printed more and more to represent the sound that replaces *of* in rapid (esp. demotic) speech, as in *kinda* ( = kind of), *loadsa, sorta*. Examples: *always slippin in and out a faces*—M. Doane, 1988 (US); *I'se a real credit, she said, loadsa times*—N. Virtue, 1988 (NZ).

Cf. also *-a = have, 've* (as a hypercorrection) in *coulda* ( = could have), *shoulda, woulda*; and *-a = to* in *gonna* ( = going to), *gotta* ( = got to), and *wanna* ( = want to).

**a, an.** **1** In origin, *a* (and its by-form *an*), which is usually called the indefinite article, but now, by many grammarians, an indefinite or central determiner, is a version of OE *ān* 'one'. It largely split off in function from the numeral about 1150, though in some circumstances the two are still interchangeable (e.g. *the Smiths have two daughters and a son*) or are hardly distinguishable in meaning (*he gave me a glass of water*).

**2** In most circumstances *a* is pronounced /ə/ and *an* is pronounced /ən/, but, when emphasized, as in slow dictation, /eɪ/ and /æn/ respectively. Public speakers, including broadcasters, often use the emphatic form /eɪ/ of *a* when there is no call for it: *she has a* (pause) *difficult task ahead of her*. It is only a short step from this to the unacceptable *she has a* (pause) *embarrassing task ahead of her*.

**3** Before all normal vowels or diphthongs *an* is obligatory (*an actor, an eagle, an illness, an Old Master, an uncle*). Before a syllable beginning in its written form with a vowel but pronounced with a consonantal sound, *a* is used (*a eulogy, a unit, a use; a one, a once-only*). Before all consonants except silent *h, a* is customary: *a book, a history, a home, a household name, a memorial service, a puddle, a young man*; but, with silent *h, an hour, an honour*.

Opinion is divided over the form to use before *h*-words in which the first syllable is unstressed: the thoroughly modern thing to do is to use *a* (never *an*) together with an aspirated *h* (*a habitual, a heroic, a historical, a Homeric, a hypothesis*), but not to demur if others use *an* with minimal or nil aspiration given to the following *h* (*an historic* /ən (h)ɪsˈtɒrɪk/, *an horrific* /ən (h)ɒˈrɪfɪk/, etc.). Three special cases: *an hotel* (with no aspiration in the second word) is now old-fashioned (E. Waugh and 1930s), but by no means extinct (*Encounter*, 1987; A. Brink, 1988); in *humble*, the *h* was originally mute and the pronunciation /ˈʌmb(ə)l/ prevailed until the 19c. but is now obsolete: it should therefore be preceded by *a*, not *an*; AmE *herb*, being pronounced with silent *h*, is always preceded by *an*, but the same word in BrE, being pronounced with an aspirated *h*, by *a*.

At the present time, especially in written English, there is abundant evidence for the use of *an* before *habitual, historian, historic(al), horrific*, and *horrendous*, but the choice of form remains open.

**4** With single letters and groups of letters *pronounced as letters*, be guided by the pronunciation: *a B road, a CFC refrigerant, a KLM flight, a TUC leader*; but *an A road, an FA cup final, an MCC ruling, an OUP book, an SAS unit*, on the assumption that these will not be mentally expanded to 'a Football Association cup final', 'a Marylebone Cricket Club ruling', etc. Acronyms beginning with a consonant are preceded by *a* not *an*: *a NATO* (pronounced /ˈneɪtəʊ/) *conference*.

**5** The indefinite article normally precedes the word or words it determines (*a popular history*). However, it follows the adjectives *many, such,* and *what* (*many a year, such a family, what an awful nuisance!*). It also follows any adjective preceded by *as* or *how* (*Iris Murdoch is as good a writer as Virginia Woolf*; *he did not know how tiresome a person she would be*), and often an adjective preceded by *so* (*so bold a move deserved success*), but *such a bold move* is more usual. In some circumstances the positioning is optional: either before or after an adjective preceded by *too* (*too strict a regime*, or *a too strict regime*), and before or after the adverbs *quite* and *rather* (*at quite an early hour* or *at a quite early hour*; *it's rather a hard puzzle* or *it's a*

*rather hard puzzle*). By contrast, when *quite* is used with *few*, the only possible ordering is *quite a few*.

**6** The phrase *a good few* in the sense 'a considerable number' (*a good few didn't turn up at all*), once rather restricted in use, has moved markedly toward the unnoticed neutral area of the language. So has *a fair few* (*a fair few of them were on the cellphone*—A. Coren, 1989). *Quite a few* ( = a considerable number (of)), which was originally US, is also now in general use.

**7** The indefinite article has a number of other idiomatic uses: *his duties as* (*a*) *judge* (optionally omitted); *he owns a Van Gogh* (a painting by Van Gogh); *an eleventh-century Aldhelm* (an 11c. MS of one of Aldhelm's works); *one of my older records, a Smetana* (a composition by Smetana); *a village Bradman* (a Bradman-like cricketer); *Jon broke a rib* (used with parts of the body when there is more than one); *half an hour, once a fortnight, fifty miles an hour, £15,000 a year* (in measure phrases); *I once knew a Lucy Jones* (a person called Lucy Jones); *a Mr Armitage called while you were out* (a person called Mr Armitage unknown to the speaker). In fixed phrases like *a knife, fork, and spoon*, the indefinite article is not repeated; but if emphasis is required, or if the sequence requires *an* as well as *a* (*a minute, an hour, or a day*) omission is not desirable.

**8** *Unacceptable* a *in the standard language but what of the future?* There are signs that *a* is intruding into the proper territory of *an* in AmE of various kinds. Two scholars recently presented substantial evidence of this phenomenon in both the unscripted and scripted speech of American politicians, entertainers, etc., pronounced both as /eɪ/ and as /ə/ (*a apple, a interesting, a ultimate,* etc.). Also in representations of the speech of American blacks (e.g. *He had a old Ford somebody gave him*—M. Golden, 1989; *My old dad lost one of his legs, had it bit off by a alligator this time he's fishing the rim canal*—E. Leonard, 1994). This is not the same as the emphatic *a* of 2 above.

**9** *A grammatical oddity.* The indefinite article has been used with nouns of multitude (*a thousand selected men, a dozen eggs*) for centuries. What seems to be a fairly recent extension of this use is the type indef. article + adj. + pl. noun (usu.

preceded by a large number): *The police found themselves confronted by an estimated two hundred youths*; *It* [sc. a stone dyke] *was an astonishing 30 feet wide, six feet high and 500 yards long.*

**abacus.** Pl. *abacuses* or, pedantically, *abaci* /-saɪ/. See -US 1.

**abbreviations. 1** *Curtailed words.* The practice of curtailing ordinary words was roundly condemned in the 18c. Thus Addison in the *Spectator* (1711): 'It is perhaps this Humour of speaking no more than we needs must which has so miserably curtailed some of our Words ... as in *mob., rep., pos., incog.* and the like.' In his *Proposal for Correcting, Improving, and Ascertaining the English Tongue* (1712), Swift proposed the publication of an annual *index expurgatorius* 'to ... condemn those barbarous mutilations of vowels and syllables [including the curtailments mentioned by Addison]'. In more recent times hostility to this method of word-formation has abated. Some of the words formerly condemned have survived and thrived, while others have not. Examples (showing the date of first record of the shortened form in the *OED*):

(a) *Now obsolete or obsolescent*
    *brig*(antine) 1720
    *coster*(monger) 1851
    *phiz* (physiognomy) 1688
    *rep*(utation) 1705
    *spats* (spatterdashes) 1802

(b) *In current use but with varying degrees of informality*
    *ad*(vertisement) 1841
    *bike* (bicycle) 1882
    *bra*(ssière) 1936
    (omni)*bus* 1832
    *cab*(riolet) 1827
    (violon)*cello* 1876
    *cinema*(tograph) 1909
    *cox*(swain) 1869
    *deb*(utante) 1920
    *dynamo*(-electric machine) 1882
    (in)*flu*(enza) 1839
    *fridge* (refrigerator) 1935
    *gym*(nasium, -nastics) 1871
    *incog*(nito) 1700
    (women's, etc.) *lib*(eration) 1970
    *mike* (microphone) 1927
    *mob*(ile vulgus) 1688
    *pants* (pantaloons) 1840
    (tele)*phone* 1884
    *photo*(graph) 1860
    *pi*(ous) *c.*1870
    *polio*(myelitis) 1931
    *pram* (perambulator) 1884

    *prefab*(ricated building) 1937
    *prep*(aration) 1862
    *pro*(fessional) 1866
    *pub*(lic house) 1859
    *quad*(rangle) 1820
    *quotes* (quotation marks, quotations) 1888
    *rep*(resentative n.) 1896
    *rhino*(ceros) 1884
    *spec*(ulation) 1794
    *specs* (spectacles) 1826
    *Strad*(ivarius) 1884
    *stylo*(graph pen) 1890
    *taxi*(meter cab) 1907
    (de)*tec*(tive) 1879
    *telly* (television) 1942
    *turps* (turpentine) 1823
    *typo*(grapher) 1816
    *vet*(erinary surgeon) 1862
    *viva* (voce) 1891
    *zoo*(logical garden) 1847

**2** *Abbreviations and contractions.* For the circumstances in which a full point is still used, see FULL STOP 2.

**abdomen.** Standard speakers normally place the main stress on the first syllable.

**abductee.** A recent word (first recorded 1975) for 'a person who has been abducted' (*Mozambique, which was short of food, had ordered recruits and abductees back across the border—Daily Tel.*, 1979). See -EE[1] 2.

**abduction.** In the sense 'The act of illegally carrying off or leading away anyone, such as a wife, child, ward ... Applied to any leading away of a minor under the age of sixteen, without the consent of the parent or guardian; and the *forcible* carrying off of anyone above that age' (*OED*), the word entered the language in the 18c. It was distinguished from *kidnapping* which was at first (late 17c.) applied to the snatching of children or others in order to provide servants or labourers for the American plantations. The two words are now partially divided in law and custom. *Abduction* commonly implies a leading away of a minor (with or without the minor's consent) for marriage or seduction or the breaking of a legal custodial arrangement for the children of divorced parents. There is some overlap of meaning with *kidnapping* (double headline in *Times*, 1989: *Midnight abduction by gunmen may be linked to smuggling inquiry. Armed gang kidnaps businessman at pub*), but *kidnapping* now tends most often to be applied to the seizure and carrying off of a person for the purpose of demanding a ransom from his or her family or employers. *Hijacking* is the illicit commandeering of a vehicle, especially an aeroplane, and forcing it to be taken to a new destination, usu. in pursuit of a political aim. All three words were applied to the seizure and detention of political hostages in the Middle East (esp. in Lebanon) in the 1980s, and all continue to be used.

**abductor.** Thus spelt, not *-er*. See *-OR.*

**abetter, -or.** In non-legal contexts 'one who abets' is now normally an *abetter*, but *abettor* 'is the constant form of the word as a legal term' (*OED*). See *-OR.*

**abide.** A verb in retreat does not always show its full plumage. Except in a few senses, *abide* has fallen somewhat into disuse and its conjugational forms have become reduced. The once regular past tense *abode* has given way to *abided* (*except for the odd pilchard he's never abided fish*— E. J. Howard, 1965; contrast the archaic use in *Ulmo ... abode not in Valinor*—J. R. R. Tolkien, *ante* 1973), esp. in the phrasal verb *abide by* 'to stand firm by' (*he abided by our decision*), or is replaced by another verb (cf. Luke (AV) 8: 27 *and ware no clothes, neither abode in any house*; (NEB) *he had neither worn clothes nor lived in a house*). When used in its dominant sense 'to bear, tolerate', *abide* is most frequently used with a preceding modal auxiliary in negative contexts (*those ordinary Aryan Australian girls whose coarse complexions and lumpy features he could not abide*—H. Jacobson, 1986). In its other senses, it tends to be used mostly in the present tense, most famously as an imperative in a hymn (*Abide with me; fast falls the eventide*—H. F. Lyte, *ante* 1845), or as a participial adjective (*I accept this award with an abiding faith in America*—M. L. King, 1964, accepting the Nobel Peace prize).

**abjure.** This verb, from Fr. *abjurer*, L *abjūrāre* 'to deny on oath', should be carefully distinguished from *adjure*, from Fr. *adjurer*, L *adjūrāre* 'to swear to (something)'. *Abjure* means principally 'to renounce on oath' (*he had abjured, he thought, all superstition*—I. Murdoch, 1985) or 'to swear perpetual absence from (one's country, etc.)' (*to abjure the realm, town, commonwealth*, etc. = to swear to abandon it for ever). By contrast, *adjure*

means 'to charge (a person) under oath (to do something)', or, more frequently, 'to request earnestly'. It is usually followed by a direct object and a *to*-infinitive (*they were all talking at once, adjuring each other to have fresh cups of tea*—L. A. G. Strong, 1948). Anita Brookner used the wrong word of the two in *A Friend from England* (1987): *He flapped a hand at us both, abjured us to behave ourselves, and was gone.*

**-able, -ible.**

| | |
|---|---|
| **1** | Origin. |
| **2** | Alterations to the stem of the source-word. |
| **3** | Both endings in use. |
| **4** | Forms differ in meaning. |
| **5** | Cognate forms. |
| **6** | List of the main words ending in *-able*. |
| **7** | List of the main words ending in *-ible*. |
| **8** | Negative forms of adjectives ending in *-able* or *-ible*. |

**1** *Origin.* To begin with, words ending in *-able* owed their form to the L termination *-ābilis* or to the OF *-able* (or both) by the addition of *-bilis* to verbs in *-āre*, Fr. *-er*, as *amāre, amābilis* ('to love, lovable'), *mūtāre, mūtābilis* ('to change, changeable'), but this method of formation was extended to verbs with other stems, e.g. *capere, capābilis*→*capable*, and to nouns, as *favor, favōrābilis*→*favourable*. The suffix was later perceived as a living element that could be added to English or French roots (whether nouns or verbs) without regard to the existence of a Latin word in *-ābilis*: *actionable* (16c.), *bearable* (16c.), *clubbable* (Dr Johnson), *forgivable* (16c.), *objectionable* (18c.), even *unget-at-able* (19c.).

Words ending in *-ible* reflect Fr. *-ible*, L *-ībilis*, from *-i-, -ī-*, a connective or stem-vowel of L verbs in *-ēre, -ere, -īre* + *-bilis*, e.g. Fr. *admissible*, late L *admissibilis* (from *admiss-*, ppl stem of *admittere*)→*admissible*; medL *audībilis* (from L *audīre* 'to hear')→*audible*.

**2** *Alterations to the stem of the source-word.* The suffix *-able* is a living one, and may be appended to any transitive verb to make an adjective with the senses 'able', or 'liable', or 'allowed', or 'worthy', or 'requiring', or 'bound' *to be—ed* or 'capable of being—*ed*', e.g. *conceive*→ *conceivable* that can be (mentally) conceived. The suffix, in other words, normally has a modal meaning.

Other -*able* words are formed from nouns (e.g. *actionable*). There are also many common words in which the sense of -*able* either is or seems to be not passive but active, e.g. *agreeable* (willing to agree), *comfortable* (having the power to obviate hardship), *viable* (able to live). It is also worth noting that many common words containing the suffixes -*able* or -*ible* have no matching verb or noun, e.g. *affable*, *amenable*, *delectable*, *feasible*, and *plausible*. It should also be borne in mind that the sequences of letters -*able* and -*ible* in some words do not constitute suffixes. The words *parable*, *syllable*, and *vegetable* and the words *bible*, *crucible*, and *thurible* are formations of a different kind, and, of course, have no power to generate related forms in -*ability*, -*ibility*.

*Alterations*. The stem of the source-word undergoes alteration in various circumstances: (i) With some exceptions, words ending in silent -*e* lose the *e* when -*able* is added, e.g. *adorable*, *excusable*, *removable*, *usable*. By custom, in the house style of OUP, a few words retain the *e* when its loss could lead to ambiguity or excessive disguise of the root form: e.g. *giveable*, *hireable*, *likeable*, *liveable*, *nameable*, *rateable*, *saleable* (see 6 below). But this custom is widely abandoned now, esp. in the US and Australia. (ii) Polysyllabic words ending in -*y* (as a simple syllable, not -*ay*, -*oy*, etc.) replace *y* by *i*, e.g. *dutiable*, *rectifiable*, *undeniable* (but *buyable*, *employable*, *flyable*, *payable*). (iii) In words ending in -*ce* or -*ge*, the *e* is retained as an indication of the 'soft' sound of *c* or *g*, e.g. *bridgeable*, *changeable*, *chargeable*, *noticeable*, *peaceable*, *serviceable* (iv) Words ending in -*ee* retain both letters: *agreeable*, *foreseeable*. (v) Many verbs of more than two syllables ending in -*ate* lose this ending in the corresponding adjectives in -*able*: *alienable* (not *alienatable*), *appreciable*, *calculable*, *demonstrable*, *estimable*, *penetrable*, *tolerable* (but, when the verb is disyllabic, *creatable*, *debatable*, *dictatable*, *locatable*). (vi) In words of English formation, a final consonant is usually doubled before -*able* when it is also doubled in the present participle: *biddable*, *discussable*, *forgettable*, *regrettable*. Words ending in -*fer* constitute a problem: *confer* (*conferrable*), *defer* (*deferrable*), *differ* (no form in -*able*), *infer* (*inferable*), *offer* (*offerable*), *prefer* (*preferable*), *proffer* (*profferable*), *refer* (*referable*), *suffer* (*sufferable*), *transfer* (*transferable*). But

*inferrable* is widely used and could not be said to be wrong.

**3** A number of words are recorded both with -*able* and with -*ible*. In the following list the preferred form is placed first:

| | |
|---|---|
| collapsible | collapsable |
| collectable | collectible (common in NAmer.) |
| condensable | condensible |
| confusable | confusible |
| connectible | connectable |
| deductible | deductable |
| discussible | discussable |
| extendible | extendable |
| extractable | extractible |
| gullible | gullable (19c. only) |
| ignitable | ignitible |
| indetectable | indetectible |
| preventable | preventible |

**4** Some adjectives differ in meaning in their -*able* and -*ible* forms:

*contractable*: liable to be contracted or acquired, as a disease or a habit
*contractible*: capable of contracting or drawing together, contractile
*conversable*: with whom one can converse
*conversible*: capable or being converted or transposed (? *obs*.)
*forceable*: capable of being forced open
*forcible*: effected by means of force (*forcible entry*), having logical force
*impartable*: capable of being imparted (*rare*)
*impartible*: not subject to partition
*impassable*: that cannot be traversed, closed (of a mountain pass, etc.)
*impassible*: incapable of feeling or emotion; incapable of suffering injury; (theol.) not subject to suffering
*passable*: that can pass muster; that can be traversed (of a mountain pass, etc.)
*passible* (theol.) capable of feeling or suffering (from late L *passibilis*, from L *pati*, *pass*- to suffer)

See also INFUSABLE, INFUSIBLE.

**5** Some words in -*able* stand alongside cognate words of similar formation: *destroyable* is occasionally felt to be as idiomatic as *destructible* in the sense 'capable of being destroyed' (*This has been a month for the world to remind itself that terrorism in the air is neither omnipotent nor wholly destroyable*—Economist, 1985; *garlic can make red blood cells more frugile and destructible*—Chicago Tribune, 1989); *eatable* is optionally available instead of the more frequently used *edible*; *solvable* may be preferred in some contexts because *soluble* is frequently used to mean 'that can be dissolved'; *submersible* is competing with *submergible* in the language of the oil industry; a mistake is often called *uncorrectable* (*this poignant exploration of life in late middle age is a harrowing examination of uncorrectable mistakes—Boston Globe*, 1989), because *incorrigible* has become ethical in sense.

**6** For convenience of reference a reasonably full list of words in *-able* follows.
\* signifies 'in OUP house style'; many printing houses in the UK and abroad omit the medial *-e-*.

abominable, acceptable, accountable, actionable, adaptable, adjustable, administrable, admittable, adorable, advisable, agreeable, alienable, allocable, allowable, amenable, amiable, amicable, analysable, applicable, appreciable, approachable, arable, arguable, arrestable, ascribable, assessable, attainable, attributable, available

bankable, bearable, believable, biddable, \*blameable, breakable, bribable, bridgeable

calculable, capable, certifiable, changeable, chargeable, clubbable, collectable (see 3), comfortable, committable, conceivable, condensable (see 3), conductable, conferrable, confinable, comformable, confusable (see 3), connectable, consolable, contestable, contractable (see 4), conversable (see 4), convictable, correctable, creatable, creditable, culpable, curable

datable, debatable, declinable, deferrable, definable, demonstrable, dependable, deplorable, desirable, despicable, destroyable (see 5), detachable, detectable, detestable, developable, dilatable, dispensable, disposable, disputable, dissolvable, drivable, durable, dutiable

eatable (see 5), educable, endorsable, enforceable, equable, evadable, excisable, excitable, excludable, excusable, expandable, expendable, expiable, extractable (see 3)

fashionable, finable, flammable, forceable (see 4), foreseeable, forgettable, forgivable, framable

gettable, \*giveable

\*hireable

ignitable (see 3), illimitable, imaginable, imitable, immovable, immutable, impalpable, impartable (see 4), impassable (see 4), impeccable,

imperturbable, implacable, impregnable, impressionable, improvable, incapable, incurable, indefatigable, indescribable, indetectable (see 3), indictable, indispensable, indistinguishable, indubitable, inferable, inflammable, inflatable, inimitable, inseparable, insufferable, intolerable, intractable, irreconcilable, irrefutable, irreplaceable

justifiable

knowledgeable

laughable, leviable, \*likeable, \*liveable, losable, lovable

machinable, malleable, manageable, manoeuvrable, marriageable, measurable, movable, mutable

\*nameable, noticeable

objectionable, obtainable, offerable, operable, opposable

palatable, passable (see 4), payable, peaceable, penetrable, perishable, permeable, persuadable, pleasurable, preferable, prescribable, presentable, preventable (see 3), processable, profferable, pronounceable, provable

\*rateable, readable, receivable, recognizable, reconcilable, rectifiable, refusable, registrable, regrettable, reliable, removable, reputable, retractable

\*saleable, serviceable, \*sizeable, solvable (see 5), statable, storable, sufferable, suitable, superannuable

teachable, \*timeable, tolerable, traceable, tradable, transferable, tuneable

unconscionable, uncorrectable (see 5), undeniable, unexceptionable, unflappable, unget-at-able, unknowable, unmistakable, unpronounceable, unscalable, \*unshakeable, unspeakable, untouchable, usable

viable

washable

**7** A list of some of the most frequently used words in *-ible*:

accessible, adducible, admissible, audible, avertible

collapsible (see 3), collectible (see 3), combustible, compatible, comprehensible, connectible (see 3), contemptible, contractible (see 4), conversible (see 4), convertible, corrigible, credible

deducible, deductible (see 3), defensible, destructible (see 5), diffusible, digestible, dirigible, discernible, discussible (see 3), dismissible, divisible edible, educible, eligible, exhaustible, expansible, expressible, extendible (see 3), extensible

fallible, feasible, flexible, forcible (see 4)

gullible (see 3)

horrible

impartible (see 4), impassible (see 4), implausible, inaccessible, incompatible, inadmissible, incomprehensible, incorrigible (see 5), incredible, indelible, indestructible, indigestible, indivisible, ineligible, infallible, inflexible, intangible, intelligible, invincible, invisible, irascible, irreducible, irrepressible, irresistible, irresponsible, irreversible

legible

negligible

omissible, ostensible

passible (see 4), perceptible, perfectible, permissible, persuasible, plausible, possible

reducible, reprehensible, reproducible, resistible, responsible, reversible, risible

sensible, submergible (see 5), submersible (see 5), suggestible, suppressible, susceptible

terrible, transfusible

vendible, visible

**8** *Negative forms of adjectives in '-able' or '-ible'.* Such adjectives are used with especial frequency in negative contexts, and complex rules govern the choice of *in-* or *un-* or some other element. A simple contrast is shown in the pairs *conceivable/inconceivable* and *bearable/unbearable*. In other negative formations the matter was settled in ancient times by the assimilation of L *in-* to the initial letter of the stem of the word. Thus the negative forms of *illimitable, implacable, incredible, incurable, irascible,* and *irrefutable,* for example, were settled in Latin long before these words were adopted in English.

**able to.** Used with the verb *to be,* and followed by a *to-*infinitive in the active voice, this semi-auxiliary is a natural part of the language, whether with an animate or, to some extent, with an inanimate subject (*I have not been able to write for several days*—G. Vidal, 1955; *a yarn about not being able to keep a job*—K. Amis, 1978; *By his proceeding to the beach ... the next phase of the attack was able to proceed*—New Yorker, 1986). The attribution of ability to inanimate subjects, however,

is not always advisable and seems altogether too forced when the following infinitive is in the passive voice (*No evidence that an air rifle was able to be fired*—Times (heading of Law Report), 1988). In general it is better to use the modal verb *can/could* with passive infinitives, or express the sentence actively.

**ablution.** From the 16c. onward *ablution(s)* has been used for the process of washing the body as a religious rite, and also, from the 18c., in liturgical ritual, as a name given to 'the wine and water used to rinse the chalice, and wash the fingers of the celebrant after the communion'. The word was drawn into lay use in the 18c. as a somewhat light-hearted term for 'the washing or cleansing of one's person'. John Buchan in *The Thirty-Nine Steps* (1915) speaks of *Mr. Turnbull's Sunday ablutions,* and Booth Tarkington has a character in *The Magnificent Ambersons* (1918) who *stood with a towel in the doorway, concluding some sketchy ablutions before going downstairs.* The word continues to be used from time to time: *Lewis's tiny room was put at the services of his ablutions and the writing up of his notes*—A. Brookner, 1989; *Morse waited with keen anticipation until his morning ablutions were complete*—C. Dexter, 1989; *I got out of bed and felt very heavy in the limbs, but I did my ablutions and dressed*—E. L. Doctorow, 1989. For most of the century, however, its dominant serious use has been in the armed services in the UK, where *ablutions* is the customary word for the process of washing one's person or for the building (also called the *ablution(s) block*) set aside for this purpose. Fowler classified the word as an example of 'pedantic humour', and it is still labelled as 'jocular' or 'humorous' in several desk dictionaries. It is especially common in the phrase *to perform one's ablutions.*

**abode.** In the sense 'a dwelling-place, a place of ordinary habitation', the word is tending to fall into disuse except in the expression (*of*) *no fixed abode,* used of a person who has no settled place of habitation. But it has not entirely gone. Examples: *The house, standing at the edge of a fair-sized tract of woodland and once, perhaps, the abode of gamekeepers*—K. Amis, 1974; *God was a tramp. He had no fixed abode*—B. Rubens, 1987; *at the Snowgoose Lodge where Rita had booked them as there* *was no room in her humble abode*—B. Anderson, 1991. It was much used in recent times in the phrase *right of abode*, esp. as applied to the appeal by the citizens of Hong Kong for Britain to grant them an automatic right to settle in Britain.

**abolishment.** This word and *abolition* both entered the language in the 16c. and for two centuries they coexisted as synonyms (used of sins, faith, guilt, and other abstract concepts) with approximately equal currency. The choice of *abolition* by those working for the abandonment, first of slavery, and then of capital punishment, ensured the relative supremacy of this member of the pair. Contextual needs, however, still leave room for *abolishment* to be used with fair frequency: *Whitlam favored abolishment of Australian appeals to the Privy Council—Facts on File Inc.* (NEXIS), 1975; *The deregulation of financial markets and abolishment of fixed commission rates had encouraged many firms . . . to prepare for* [etc.]—*Institutional Investor Inc.* (NEXIS), 1989; *It's a negation of him, an abolishment of him, like ripping a medal off his chest*—M. Atwood in *New Yorker*, 1990.

**aborigines.** This Latinate word (specifically applied since the 16c. to the inhabitants of a country *ab origine*, i.e. from the beginning) has largely given way to *aboriginals* in the plural. For the singular, the etymologically indefensible form *Aborigine* has become firmly established in Australia (first recorded there in 1829). In Australian contexts, an initial capital should be used for all forms of the word: thus *Aboriginal(s)*, *Aborigine(s)*. The abbreviated form *Abo*, used with varying degrees of affection and hostility, is common in Australia, both as a noun and as an adjective.

**aborticide.** In the great modern debates about abortion and the 'right to life' (the expression is applied attrib., with hyphens, to a person or group opposed to the premature ending of a normal pregnancy), *aborticide*, meaning (1) the destruction of the foetus in the uterus, and (2) a drug or other agent that causes abortion, is tending to displace *foeticide* (US *feticide*). The word is so recent that it does not appear in *OED* 2, whereas *foeticide* is recorded from 1844 onward.

**abortive.** The daring but unsuccessful attempt in 1979 to free American hostages held in Iran was widely described as 'abortive'. The American journalist William Safire condemned the use as 'absurd' on the grounds that the suffix *-ive* implies continuity, and insisted that the right word was 'aborted'. In fact the notion of past or present continuity is not a necessary component of the word. The central meaning of *abortive* since the 16c. has been 'coming to nought, fruitless, unsuccessful'. It can be applied to attempts, efforts, missions, coups, proposals, etc., indeed to any action that proves to be unsuccessful (e.g. *an abortive attempt to do the* Times *crossword*—A. N. Wilson, 1982).

**abound.** This verb can be used with propriety in the senses 'to be plentiful', 'to be rich *in*', and 'to teem *with*': *Mulberry trees abound in Oxford*—J. Morris, 1978; *a few years since this country abounded in wild animals*—A. Moorehead, 1963; *the book is badly written and abounds with mistakes.* As the *OED* expresses it: 'A place abounds *with* all those things, which abound *in* it; it abounds *in* those things only which by their abundance give it a character, or add to its resources.'

**about.** 1 *As adverb.* With a following *to*-infinitive (transitive, intransitive, or passive), *be about* has the force of a future participle (cf. L *scripturus* about to write, on the point of writing): *she was about to mash the potatoes*; *direct negotiations are about to resume*; *Macbeth's tyranny was about to be opposed.* By contrast, in negative contexts, and esp. in the US, the implication is often one of not intending to do something or of determination not to do something: *Peace is the mission of the American people and we are not about to be deterred*—attrib. to L. B. Johnson; *I'm not about to foist something on the general public just for the sake of releasing something*—Record Mirror, 1982.

2 *As preposition.* In the senses 'near in number, scale, degree, time, etc.; approximately at, approximately equal to' (*about half, fifty, nine o'clock, ten per cent, about 186,000 miles a second*), *about* is the traditional word in the UK, but is now in competition with the US *around* (*the convention adjourned around four o'clock*; *around fifty people came to the party*, etc.).

**3** In the sense 'concerning', *about* can be used as a simple preposition (*quarrels about territory*) or as a conjunction introducing a *which-, how-*, etc., clause (*a great deal of discussion about which versions should be used*; *your campaign address about how the Tories had watched New Zealand slide into depression*). Since the 1930s, the phrase *to be (all) about* has also been frequently used in the sense 'to be primarily concerned with' (*Love and war were about winning, not fair play*—A. Price, 1982). Cf. also the phrase *what it's all about (really)*: *That's what love is all about*—song, 1937; *They like the feeling that they have had to fight other men for possession. That is what it is all about, really*—A. Brookner 1984.

**4** *Part of speech.* In contexts of the type *he studied Russian for about a year*, it should be noted (as *CGEL* points out) that 'about' is not a preposition but an intensifying predeterminer as part of the noun phrase, as can be seen from the fact that it is itself preceded by a preposition. In *she is about twenty*, 'about' is an adverb since it can be replaced by other adverbs (*roughly, approximately*) in the same construction.

**5** *Tending to replace 'of'.* Dwight Bolinger (*World Englishes*, 1988, vii. 238 f.) records many instances of a tendency in informal English for *about* to be used to replace *of*: *We're more aware about it*; *the Vietnamese are disdainful about Chinese cooking*; *the issue about how such things are monitored.* These uses, brought about by a process called 'reiconization', are still lurking at the edge of acceptability.

**above.** **1** For about a century, cautionary notes have been written in usage guides about the undesirability of using *above* as an adjective and as a noun. In the course of the 18c., by ellipsis of a following past participle, as *said, written, mentioned*, 'above' came to be used attributively = 'the above explanation'; when the following noun was also suppressed, 'the above' came to be used absolutely (*OED*). Thus *the above list, decision, words*, etc.; *pictures like the above.* Both uses have become part of the natural machinery of the language: *During the above speech French is becoming increasingly agitated*—T. Stoppard, 1976; *Following from the above, the Japanese will not be rushed into making decisions*—Randle and Watanabe, 1985.

**2** Old grumblings about the use of *above* to mean 'more than' are shown to be out of order by the continuous record of the use since the 16c.: e.g. *It was neuer Acted: or if it was, not aboue once*—Shakespeare *Hamlet*; *it is above a week since I saw Miss Crawford*—Austen, 1814; *he doesn't look a day above forty.* Nevertheless either *more than* or *over* is the more usual and natural expression in such contexts.

**abridgement.** The better form, rather than *abridgment*. See MUTE E. See also -MENT.

**abrogate** means 'to repeal, annul, cancel' (*to abrogate laws, rules, a treaty*, etc.), whereas *arrogate* means 'to lay claim to without justification'. Examples: (abrogate) *the Cabinet clung stubbornly to the belief that the mere signing of the agreement itself abrogated imperial preferential tariffs*—D. Acheson, 1969; *He abrogated at once the Penal Code*—W. H. Auden, 1969; *Such behaviour abrogated the ceremoniousness of feeding*—J. Carey, 1977; (arrogate) *That sort of writing which has arrogated to itself the epithet 'creative'*—D. J. Enright, 1966; *The illegal but effective authority which the Assembly of the United Nations seemed now to have arrogated to itself*—H. Macmillan, 1971.

**absent.** The emergence of this word used as a preposition meaning 'in the absence of, without' underlines the pleasing fecundity of American English: *Absent such an appeal, the constitutional issues were conclusively determined against Ender*—NY Law Jrnl, 1972; *Finally, absent an agreement, one must ask what the likely terms of political debate would be in 1988*—Dædalus, 1987; *Absent any major change in government programs, many companies have no clear idea how they will pay for them*—Chicago Tribune, 1989. It is first recorded in 1944 in legal writing, and by the 1970s had branched out into more general use in the US.

**absolute adjectives.** See ADJECTIVE 4.

**absolute comparative.** *CGEL* 7.85 draws attention to the existence of comparatives used where no explicit comparison is made though it may be implied. Thus *senior* retains its comparative force in *A is senior to B*, but cannot be used in a *than-* clause. We can say *A is older than B*, but we cannot say *\*A is senior than B*.

There is a wide range of adjectives with a comparative form that can be used in contexts where the comparison is not made explicit: the *better-* class hotels, the *greater* London area, *higher* education, the *major* political parties, an *older* man, a *prior* claim, of *superior* quality, the *younger* generation. Such implied comparisons are a feature of advertising language: *higher* mileage, a *smoother* finish, etc.

**absolute construction.** **1** Such a construction stands out of explicit accord with a following clause. It may be verbless, but usu. contains a verb in its present or past participial form. Examples of the simplest types: Given Didi's condition, *he performed an heroic feat in Israel*—D. Athill, 1986; that done, *they drove the animal through a side gate*—E. O'Brien, 1988; The washing up finished, *Jennifer called through the echoey building*—F. Weldon, 1988; That said, *Edgar concluded the missive by reminding his employees that* [etc.]—*Chicago Tribune*, 1989; Our business done, *we were now kinder to each other*—*New Yorker*, 1991. Stereotyped or formulaic absolute phrases include *all told, all things considered, God willing, other things being equal, present company excepted, putting it mildly, roughly speaking, to say the least, weather/time permitting.* They do not pose any threat of disunion to the following clause.

**2** Fowler (1926) and Gowers (1965) objected forcibly to the placing of a comma between the noun and the participle in the absolute use: e.g. *The King, having read his speech from the throne, their Majesties retired*; *Bath King of Arms, having bowed first to those Knights Grand Cross who have been installed ... they thereupon sit in the seats assigned to them.* Their warnings seem to have been heeded. I have not encountered any such aberrant commas myself.

**3** A formally similar construction occurs when an absolute phrase containing a participle in *-ing* or *-ed* refers directly to the subject of the attached clause: *Looking at Jim, I remembered the first time I had seen him*—Encounter, 1988; *Located in the Smith homestead, Fairweather prepared for his first patrol*—M. Shadbolt, 1986 (NZ). (It was 'I' who was looking at Jim, and 'Fairweather' who was located in the Smith homestead.) Occasionally the connection need not be exact: *As a speaker, I*

*thought him excellent* lies just within the borders of acceptability. By contrast, the following examples lie outside the acceptable area: Picking up my Bible, *the hill seemed the only place to go just then*—J. Winterson, 1985; Packing to leave, *her fingertips had felt numb on contact with her belongings*—M. Duckworth, 1986. For further examples see UNATTACHED PARTICIPLES.

**4** *CGEL* 15.59 points out that in formal scientific writing, unrelated constructions have become institutionalized 'where the implied subject is to be identified with the *I, we,* and *you* of the writer(s) or reader(s)': When treating patients with language retardation ... *the therapy consists of* [etc.]; To check on the reliability of the first experiment, *the experiment was replicated with a second set of subjects.*

**absolutely.** This word has a string of important meanings in the broad area 'in an absolute position, manner, or degree' (independently; arbitrarily; in grammar, without the usual construction; unconditionally; etc.). It has also come to be used as a mere intensive (*absolutely awful, dreadful, essential, improbable, out of the question, shattered, terrible*). In such contexts, the word if anything tends to lessen the power of the following adjective or adjectival phrase. In conversation, *absolutely* is a pleasingly old-world variation of 'yes, quite so': *I trust that we are still brothers-in-arms?—Absolutely. Pals*—R. Stout, 1937; *I said, 'Art Floyd?' He smiled. 'Absolutely,' he said*—R. B. Parker, 1986. With *not*, it is often used in speech as an emphatic refusal or denial: *'Are you going to the office party this year?'—'Absolutely not.'*; *'Did you exceed the speed limit?'—'Absolutely not.'*

**absolute possessives.** Under this term are included the possessive pronouns *hers, his, its, ours, theirs,* and *yours,* and also (except in the archaic adjectival use, as *mine/thine eyes*) *mine* and *thine.* None of the *-s* forms takes an apostrophe. The ordinary uses are straightforward (except for *its,* see below): *the house is hers, his, ours, mine,* etc.; *I met a friend of yours, hers,* etc. Matters become more complicated when two or more possessives refer to a single noun that follows the last of them. In such cases the *-s* and *-ne* forms are incorrect. The correct forms are shown in *your and our and his efforts* (not *yours* and *ours*); *either my or your informant must have lied* (not *mine*); *her and his strong contempt* (not *hers*). Rearrangement of the pronouns removes any risk of error: thus *his efforts and yours and ours, either your informant or mine must have lied*, etc. *Its* is the only pronoun in the series that normally cannot be used predicatively or in the double possessive construction: thus *its tail is red*, but not *\*this tail is its* nor *\*a mate of its. CGEL* 6.29n. points out, however, that independent *its* is occasionally found, e.g. *History has its lessons and fiction its; She knew the accident was either her husband's fault or the car's: it turned out to be not his but its*. In such cases strong emphasis is placed on the contrasted pronouns.

**absolute superlative.** Like absolute comparatives, absolute superlatives, i.e. superlatives used merely to express a very high degree of the quality or attribute, without definite comparison with other objects, occur occasionally in informal language: *she is most peculiar, your letter is most kind* ( = extremely kind). Forms in *-est* can also be used in an absolute manner: *she is the strangest woman, it is the sweetest hat, he is the happiest of babies.*

**absolve.** Formerly pronounced either /æb'z-/ or /æb's-/, but the pronunciation with /-z-/ now seems to be dominant, presumably under the influence of *resolve*. Its normal constructions are of three kinds: (not common) followed by a phrase or clause led by *for*; (both common) followed by a phrase or clause led by either *from* or *of*. Examples: (*a*) *One's conscience may be pretty well absolved for not admiring this man*—Miss Mitford, 1817; *We may perhaps absolve Ford for the language of the article*—R. Burlingame, 1949 (*WDEU*); (*b*) *absolve me from all spot of sin*—J. Agee, 1950; *Having thus absolved himself from the duty of making the essential discriminations*—F. R. Leavis, 1952; (*c*) *it absolved him of all responsibility*—L. A. G. Strong, 1948; *Dollar was absolved of personal liability for the line's debts*—*Time*, 1950 (*WDEU*).

**absorbedly.** Four syllables. See -EDLY.

**abstract nouns.** Gowers (1965), as part of his drive for the use of plain English in official documents, pointed to a marked tendency (which he termed *abstractitis*) in modern writing for abstract nouns to be used in such a way that the meaning is obscured. *Participation by the men in the control of the industry is non-existent*, he said, should be rewritten as *The men have no part in the control of the industry*. He gave some other (unattributed) examples of official pomposity, e.g. *Strangeness of samples has been shown to lead to relative rejection of products in the comparative absence of clues to a frame of reference within which judgement may take place* ... The complaint was not new. In *The King's English* (1906), the Fowler brothers attacked 'the far-fetched, the abstract, the periphrastic, the long, and the Romance'. They cited numerous examples of injudiciously chosen diction, including this one from *The Times*: *The signs of the times point to the necessity of the modification of the system of administration* (rewrite as *It is becoming clear that the administrative system must be modified*).

In the bureaucratic offices of our society inflated sentences are still being written, but the drive towards the use of everyday English in national and local government documents, in legal writing, and elsewhere is gathering momentum. Gowers's *Plain Words* (1948), and its successive editions, are vigorous and healthy tracts which have helped to eradicate some of the verbosity, superfluity, and opaqueness lurking in official documents and memoranda.

**absurd.** The standard pronunciation is /əb'sɜːd/, but the less frequent pronunciation with medial /z/ is used by some speakers (e.g. by N. Chomsky in a lecture in Oxford in Nov. 1992).

**abusage.** An obsolete 16c. and 17c. word partially revived by Eric Partridge in his *Usage and Abusage: A Guide to Good English* (1942).

**abysmal, abyssal.** Of the respective base-nouns, *abysm* 'bottomless gulf, deep immeasurable space' has retreated into obsolescence except in reflections of Shakespeare's line in *The Tempest* (1612), *What seest thou els in the dark-backward and Abisme of Time? Abyss*, on the other hand, is a customary word, whether literal or figurative, for '[a place, etc., of] immeasurable depth'. A person facing defeat in an election or humiliation of some other kind, for example, is often said to be 'staring into the abyss'.

The derivatives *abysmal* and *abyssal* have moved in opposite directions. Residual uses of *abysmal* applied to gorges, precipices, deep space, etc., can be found (*the abysmal depths of interstellar space*— P. W. Merrill, 1938; *the abysmal depths of the ocean*—T. Barbour, 1944 (*WDEU*)), but the word is mostly used in figurative contexts: (*a*) = bottomless, fathomless (*Contemplate with despair the abysmal depths of your incapacity*—A. Burgess, 1987); (*b*) = extremely bad (*Guatemala's abysmal human rights record*—NY Times, 1984; *the abysmal quality of what is produced* [in Romania]—NY Rev. Bks, 1989).

*Abyssal*, once also used figuratively, is now only in restricted use as a technical term in oceanography, 'belonging to one of the deepest levels of the ocean' (*the ocean basin floor ... is called the abyssal floor*—A. C. and A. Duxbury, 1984).

**Academe.** In ancient Greece, ἀκαδημία was the proper name of a garden near Athens sacred to the hero Academus. In the 4c. BC it was the site of a philosophical school established there by Plato. Shakespeare modified the Greek name to the trisyllabic *Achademe* in *Love's Labour's Lost* and used it to mean 'a place of learning'. Milton's line *The olive-grove of Academe, Plato's retirement* (*Paradise Regained*, IV. 244) also used the trisyllabic form of the word, but applied it specifically to Plato's philosophical school. Later writers followed Shakespeare's lead, and, particularly since the publication of Thackeray's *Pendennis* in 1849 (*the least snugly sheltered arbour among the groves of Academe*), the word *Academe* (frequently but not invariably with an initial capital) has been modishly used by many writers to mean 'the academic community, the world of university scholarship', notably in the phrase *the groves of Academe* (cf. *Atque inter silvas Academi quaerere verum*— Horace, *Ep.* 2. 2. 45). Mary McCarthy's novel *The Groves of Academe* (1952) helped to establish the currency of the expression. It is a tangled story, but it is clear that *Academe* (or *academe*) has now slipped away from its Greek original, and has passed into general use as an acceptable expression (*One of the most remarkable pieces of hostile reviewing ever seen in academe*—Jonathan Clark, 1989). More recently (first recorded in 1946), it has been

joined by *Academia* (or *academia*) in the same sense (*Businessmen liked to adopt the language of academia*—A. Sampson, 1971).

**academic.** The serious uses of this word (first recorded in the late 16c.) remain firm, but a little more than a century ago (first noted 1886) it developed a depreciatory range of meanings as well, 'unpractical, merely theoretical, having no practical applications', e.g. *All the discussion, Sirs, is—academic. The war has begun already*—H. G. Wells, 1929; *The strike vote ... was dismissed as 'largely academic' by Merseyside Health Authority*—Times, 1990.

**Academy.** *The Academy, the Garden, the Lyceum, the Porch,* and *the Tub* are names used for the five chief schools of Greek philosophy, their founders, adherents, and doctrines: *the Academy,* Plato, the Platonists, and Platonism; *the Garden,* Epicurus, the Epicureans, and Epicureanism; *the Lyceum,* Aristotle, the Aristotelians, and Aristotelianism; *the Porch* (or *the Painted Porch,* Gk στοὰ ποικίλη), Zeno, the Stoics, and Stoicism; *the Tub,* Diogenes (who lived in extreme poverty, legend has it, in a tub), the Cynics, and Cynicism.

**acatalectic.** Not wanting a syllable in the last metrical foot; complete in its syllables: a technical term in prosody. *Háppy | field or | móssy | cávern* (Keats) and *Cóme, you | prétty | fálse-eyed | wánton* (Campion) are trochaic (—˘) tetrameters in which the last foot is acatalectic. Cf. CATALECTIC.

**accent.** 1 The noun is pronounced on the first syllable and the verb on the second. See NOUN AND ADJECTIVE ACCENT; NOUN AND VERB ACCENT.

2 In general use, an accent is 'an individual, local, or national mode of pronunciation' (*COD*), as 'he has a slight accent, a strong provincial accent, an Irish, Scottish, etc., accent'. Examples: *She had ... the accent of a good finishing school*—J. Braine, 1957; *RP is, within England, a non-regional accent; other educated accents all have local characteristics*—D. Abercrombie, 1964; *My mother came from Nashville, Tennessee, and try as she would she couldn't obliterate her accent*—Lord Hailsham, 1981. It is also used to mean the position of the stress in a word; *You must*

*pronounce this all as one word with the accent on the first syllable*—C. S. Lewis, 1955.

**accent, accentuate.** In figurative senses ( = to emphasize, heighten, make conspicuous, intensify, etc.), *accentuate* is now much the commoner of the two verbs, but *accent* is by no means extinct. Examples: (accent) *The great piers ... are accented at the cardinal points by shafts of dark lias*—R. J. King, 1877; *Blouse woven of Dacron® polyester is accented with a big, self-tie bow*—Sears Catalog *1985*, 1984; (accentuate) *I observed a severe grey skirt, the waist accentuated by a leather belt*—W. Golding, 1967; *The villagey feeling was accentuated by the use of tile-hanging*—M. Girouard, 1977; *Collingwood also has a rather learned look, accentuated by steel spectacles*—R. Cobb, 1985. In the literal sense 'to place an accent on a syllable or word', the only word used (of the two) is *accent* (e.g. *'begin'* is accented on the second syllable).

**acceptance, acceptation.** The first of these, though coming later into the language, has gradually driven out most of the everyday uses of the second during the last four centuries, except that *acceptation* is still the more usual of the two words in the sense 'the particular sense, or the generally accepted meaning, of a word or phrase'. So we find *acceptance* used in contexts of receiving or accepting gifts, payments, aid, pleasure, duty, invitations, persons (into a group), hardship, and all manner of other things; but *No up-to-date language-user could unselfconsciously speak or write the words* gay *or* queer *in their old acceptations*—Verbatim, 1986; *This endlessly fascinating anthology of ancient Hebrew literature was ... against all plausible acceptations of the word, on its way to becoming a book*—R. Alter, 1987.

**accepter, acceptor.** In general contexts the word is normally spelt *accepter* (*he is no accepter of conventions*), but in special senses in law and science *acceptor*: *the Bill of Exchange ... is an order written by the drawer and addressed to the acceptor*—J. E. T. Rogers, 1868; *the ionization energy of donors is less than that of acceptors*—Physical Review, 1949; *another possible electron acceptor such as sulphate or perhaps even nitrate may be reduced*—Forestry, 1986. See -ER AND -OR; -OR.

**accept, except.** No educated person would confuse the two in meaning or function, since all they have in common is that in fast speech they both approximate to /ək'sept/. As a result poor spellers may occasionally set down the wrong word. David Crystal reports in his book *Who Cares About English Usage?* (1984) that several of twenty English undergraduates before whom he set the pair of sentences *Shall we accept/except his invitation to dinner?* chose *except* not *accept*.

**accept of.** Partly no doubt because of its use in the Bible (*they shall accept of the punishment of their iniquitie*—Lev. (AV) 26: 43) in 1611, *accept of* in the sense 'to accept (something offered, esp. something unwelcome)' survived from the 16c. until the first half of the 20c., but it appears to be obsolete now.

**access, accession.** 1 As nouns. There are very few contexts in which one of these can be substituted for the other without the meaning's being modified. In contexts of arriving, *accession* means arrival, admission, *access* opportunity of arriving or of admission. Accordingly, *accession to the throne* means becoming sovereign, *access to the throne* opportunity of petitioning the sovereign. The idea of increase or augmentation, often present in *accession* (*recent accessions to the library are not yet catalogued*; *an accession of strength*), is foreign to *access*; *an access of fever, fury, joy, despair, paralysis, pessimism*, etc., is a fit or sudden attack of it, which may occur whatever the previous state of mind may have been, *accession* would not be used in such a manner; *people without readers' tickets have no access to the library*, i.e. are not permitted to enter, is a normal use; *accession* cannot be so used.

2 As verbs. Since the 1890s, *accession* has established itself as a regular verb meaning 'to enter in the accessions register of a library' (*the new books have been promptly accessioned*—G. M. Jones, 1892). In more recent times (first recorded 1975), *access* (perhaps as a back-formation) has occasionally been used in the same sense (*the Lowestoft Hoard had to be accessed*—TLS, 1978). Since the early 1960s, *access* has been used in computer language in the sense 'to gain access to (data, etc., held in a computer or computer-based system, or the system itself)' (*OED*): *The library's*

*statistical section uses its Polis terminal to access various statistical databases—Times,* 1983. It is part of the everyday vocabulary of the subject.

**accessary, accessory.** 1 These two words were given separate entries in the *OED* because their derivations differ. Moreover, in meaning, both as nouns and as adjectives, they mostly ran on separate tracks until about 1900. The *-ary* form traditionally tended to imply complicity in an offence, acting as (or one who acts as) a subordinate in a crime, while the *-ory* form was mostly reserved for general contexts where someone or something contributed in a subordinate way. In the course of the 20c. *accessory* has come to prevail in all contexts both as a noun and as an adjective. *ODWE* s.v. *accessory* (n. and adj.) says uncompromisingly 'use now in all senses, not *-ary*'. Examples: *As the one person who knew of their illegalities I felt that I was becoming an accessory after the fact*—S. Unwin, 1960; *if he buried the captain, as he says, he's an accessory*—R. Macdonald, 1971; *accessory ideas associated with the principal idea*—M. Cohen, 1977; *accessory glands in the internal male and female genitalia of insects*—*Zool. Jrnl Linnean Soc.,* 1988. As a noun, *accessory* has also become widely used (usu. in the pl.) in the 20c. for the smaller articles of dress (gloves, handbag, etc.) or the extras in a motor vehicle (foglights, radio, etc.). Examples: *Accessories . . . may be considered essential to an outfit*—A. Lurie, 1981; *auto accessories.*

2 Both words were frequently pronounced with the main stress on the first syllable until the early part of the present century, but now /æk'sesərɪ/ is the standard pronunciation. Some uneducated people pronounce *accessory* with initial /æs-/ as if it were spelt 'assessory'.

**access, excess.** The *OED* (in a section published in 1884) no doubt correctly said that '*access* is frequently found written for *excess* sb., chiefly by phonetic confusion; but the senses also approach in 8 ['addition, increase'] above'. Murray would not have made such a statement without substantial evidence. It is just possible still to imagine circumstances in which uncertainty might arise. The distinction between (say) *an access of energy* (outburst of) and *an excess of energy* (too much), and between *an access of rage* (emotional outburst) and *an excess of rage* (something judged to be 'over the top') seems to be observed in most printed work. The danger area is in the choice of word to indicate a surge, or alternatively a surfeit, of emotions. There is a marked gradation in moving from an *access* of loyalty, wickedness, willingness, etc., to an *excess* of the same.

**accessorize.** *A hankie to accessorize a costume,* reported the linguistic journal *American Speech* in 1939, when launching what looked like being one of the least-lovable words of the 20c. But fashion writers have adopted the word in *Vogue* and elsewhere, and it has settled into acceptability esp. in the language of fashion and of interior decoration. Examples: *In pre-glasnost days she would have accessorized her gear with a small party badge in the lapel of a sombre suit*—*Times,* 1988; *The Sultan has accessorized himself with epaulettes, medals and several strings of pearls*—*Times,* 1989; *The ground floor reception rooms are resplendent with new brushbox flooring . . . accessorized with traditional rugs*—*Belle* (Aust.), 1983; *Earth tone nylon covers of seats in a car easily accessorize with any hue*—*Toronto Star,* 1984.

**accidence.** It is perhaps as well to set down here (since the terminology of traditional grammar is less well known than it once was) that accidence is 'that part of Grammar which treats of the Accidents or inflections of words' (*OED*). The word is first recorded in a context of 1509. For example, the endings of the parts of regular English verbs fall under the heading of accidence: *talk* (base), *talking* (-*ing* participle), *talks* (3rd person singular present indicative), and *talked* (past tense or past participle). Accidence also covers *inter alia* the inflections of adjectives (*tall, taller, tallest*) and of nouns (*girl, girls, girl's, girls'*).

**accidently.** This non-standard form, used instead of *accidentally,* is recorded as early as 1611 and is still encountered occasionally in print: e.g. *When I have accidently collided with it*—B. Rubens, 1985; *Julie Kohler (Moreau) loses her groom on the steps of the church when he is accidently shot by a group of men fooling about with a* *gun—Listener*, 1987. But it is very much a minority form.

**acclimate, acclimatize.** Both words mean 'to habituate to a new climate (lit. or fig.)'. The slightly older *acclimate* (first recorded 1792) has totally given way to *acclimatize* (1836) in the UK, whether as transitive or intransitive verbs, and partially in the US, though American examples of *acclimate* and derivatives are not difficult to find: *'I'll drop in after dinner,' Dick promised. 'First I must get acclimated.'*—F. Scott Fitzgerald, 1934; *but as they are acclimated and toughened to the native condition they suffer far less than we do*—F. Lloyd Wright, 1954; *Harry, by now almost acclimated to Lela, her Olympic highs and her subterranean lows*—M. Doane, 1988. *Acclimatize* (and derivatives) is the normal word, for instance, in the works of Chris Bonington (UK mountaineer), Paul Scott (UK novelist), and Carl Sagan (US scientist).

**accommodate, accommodation.** Two *c*s, two *m*s. The verb *accommodate* is construed with *to* when it means 'to adapt' (*He would easily have accommodated his body to the rule of never turning his head on his shoulders*—R. Graves, 1938; *his eyes quickly accommodated to the gloom*—mod.; *the second noun accommodated in form to the native cognate*—E. S. Olszewska, 1962; *Fussell's own vision of human (well, American) society and how to accommodate to it*—Listener, 1984); and with *with* when it means 'to equip, supply, oblige' (*Major Kent was accommodated with a hammock chair*—G. A. Birmingham, 1908; *Can you accommodate me with cash for a cheque?*—OED). *Accommodations* (in the plural), once common in the UK for 'lodgings, living premises', is now more or less restricted to AmE (*overnight accommodations*).

**accompanist.** Now the standard form of the word for 'a person who plays a musical accompaniment'. The by-form *accompanyist* (also first recorded in the 19c.), used e.g. by Dickens in *Oliver Twist*, has fallen out of favour in BrE but is occasionally encountered in AmE.

**accomplice, accomplish.** The standard pronunciation of both words is now with /-kʌm-/ not /-kɒm-/, though the first thirteen editions of Daniel Jones's *English* 

*Pronouncing Dictionary* (1917–67), presumably correctly reflecting the mood of that time, gave precedence to the second of these two pronunciations.

**accord.** *Of his owne accord he went vnto you* (2 Cor. (AV) 8: 17) and *She had no idea of giving up Felix of her own accord* (Trollope, 1862) are earlier examples of the still customary idiom 'of one's own accord': *her face, of its own accord, folded into a false, obedient smile*—E. O'Brien, 1989. *On mine owne accord, Ile off* (Shakespeare, *Winter's Tale*, 1611) illustrates the now obsolete use of *on* as the head of this phrase.

**according.** **1** *according as.* Fowler's long warning (1926) against repetition of the phrase now seems dated and unnecessary. It is now regarded as a subordinating conjunction meaning 'depending on whether, to the extent to which' (*everyone contributes according as he or she is able*); and, as the *OED* says, it is normally confined to contexts where the accordance is with one of two or more alternatives (*Llanabba Castle presents two quite different aspects, according as you approach it from the Bangor or the coast road*—E. Waugh, 1928).

**2** *according to*, used as a complex preposition, means (*a*) in a manner that is consistent with (something) (*everything went according to plan*); (*b*) as stated by (someone) or in (something): (as a factual statement) *according to our records your current account is in credit*; *The Gospel according to St Matthew*; *according to the new regulations*; *I have acted according to my conscience*—A Blunt, 1979; (with an element of uncertainty or disbelief) *according to you, you were at school this morning, but according to the teacher you were not*; (*c*) in a manner or degree that is in proportion to (something): *salary according to experience*; *arrange the blocks according to size and colour*; *My price varied from twenty to fifty pounds according to the neighbourhood and the customer*—G. Greene, 1966.

**account.** The slightly formal prepositional phrase *on account of* in Virginia Woolf's *Diary* (1915) (*Rhubarb was forbidden him, on account of its acidity*), and in Anita Brookner's *Latecomers* (1988) (*He remained miserable and ashamed, largely on account of his appetite which continued to torment him*) means 'by reason of, because of'. It contrasts sharply with (a) the non-standard *on accounts of* shown in *they giv' it the name, on accounts of Old Harmon living solitary there* (Dickens, *Our Mutual Friend*, Victorian working-class context); and (b) the 20c. slang conjunctional phrases *on account* and *account of* as illustrated in the following examples: *I was feeling kind of down, on account that tooth of mine was giving me the devil*—P. G. Wodehouse, 1936; *Fred's five foot ten ... but I tell him he's still a shrimp, account of I'm so tall*—*Horizon*, 1942; *Take your three days off, Mr. Barlow, only don't expect to be paid for them on account you're thinking up some fancy ideas*—E. Waugh, 1948; *Account of you think you're tough you're going up to State Prison where you'll have to prove it*—E. Leonard, 1994 (US).

**accountable.** Most frequently used of persons being *accountable to* (another person or persons) *for* (something), i.e. liable to be called to account: *Labour MPs ... would at least have the authority to keep a Labour government accountable to them for what they say they would do and are required to do*—T. Benn, 1979. It was formerly often used in the sense 'explicable', i.e. as an antonym of *unaccountable* (*By George—it was a very accountable obstinacy*—G. Eliot, 1876), but this use is not often encountered nowadays.

**accoutrement.** Thus spelt in BrE, but also as *accouterment* in AmE.

**accumulative, cumulative.** These two adjectives of approximately the same meaning ('arising from accumulation') have been contextually competing for roles since the 17c. Nowadays one is much more likely to encounter *cumulative*, both in general contexts (*cumulative arguments, effect, evidence, force*, etc.) and in more specialized areas of legal sentencing, dividends, reference library indexes, errors, voting, etc. If a sentence is tending in the direction of needing an adjective meaning 'increasing in amount, force, intensity, number, etc., over a period of time', and the context is general not technical, *cumulative* rather than *accumulative* is the word that is more likely to be chosen (*the cumulative effects of smoking, cumulative proof of a conspiracy*). But the choice of word remains fairly open, e.g. *accumulative scraps of evidence*—M. Wheeler, 1972: *The draw weight*

*on a woman's bow is about 34 lb, which makes for an accumulative weight of about five tons in the course of the competition*—*Times*, 1988.

**accuse.** A person is accused *of* (a crime, misbehaviour, etc.) in modern English: e.g. *you ... accused me of affectation last time*—M. Drabble, 1963; *he accused the sound technicians of sabotaging the record*—M. Puzo, 1969; *People jumped up and accused her of making common cause with the Nazis*—D. May, 1988. In former times other constructions were available and acceptable: e.g. *he is accused for too much conniving at the factious disturbers thereof*—Thomas Fuller, 1655; *The Romanists accuse the Protestants for their indifference*—Southey, 1809. But the construction with *of* has prevailed.

**accused.** *The accused*, 'he or she who is accused in a court of justice', is an everyday use. *The* or *an accused man, person, banker, teacher*, etc., are also routinely acceptable (only the sex or the occupation of the accused person being given). It is inadvisable, however, to use expressions like *the accused thief* or *the accused rapist*, i.e. to indicate the nature of the alleged crime.

**acerb, acerbic.** Of these two competing adjectives, *acerb* (first recorded in 1657) appears to be retreating, while *acerbic* (1865) is advancing. But the battle is far from over. Both words are used in the literal sense 'sour, bitter' (*acerb(ic) apples, lemon juice*, etc.), but are found much more frequently in figurative contexts. Examples: (acerb) *Many of his acerb remarks about wives and marriage may or may not apply to his own*—D. M. Frame, 1964; *The acerb after effects of the tragic act of love*—J. Updike, 1978; *Ricks's Victorians tend to be more like himself than they are like Q: acerb, amusing, self-aware*—*TLS*, 1987; (acerbic) *The fury he aroused in the acerbic breast of Karl Marx*—*TLS*, 1971; *Barney Frank ... agile, acerbic and ferociously intelligent, the kind of Democrat who struck fear in the hearts of Republicans*—*Newsweek*, 1989; *'Mastergate' by Larry Gelbart, which just opened on Broadway, is full of acerbic political satire*—*Chicago Sun-Times*, 1989.

The erroneous spelling *ascerbic* turns up in print from time to time—perhaps modelled on *ascetic*—but has no validity.

The corresponding adverbial form is *acerbically*, not *\*acerbly*.

**Achilles' heel.** For many years, in the house style of OUP, an apostrophe was insisted on for *Achilles' heel* but not for *Achilles tendon*. *This state of affairs has been rectified in the COD* (1990), where both phrases are allotted an apostrophe. It must be stressed, however, that the presence or absence of an apostrophe in these two expressions is not a matter of rule, but of custom. Consistency in a given publication is desirable, whichever form is chosen.

**acid.** Since the 1960s, when *acid* was first used to mean 'the hallucinogenic drug LSD', the word has developed all the sinister connotations of a widely despised sub-culture. Persons taking such drugs came to be called *acid heads* or *acid freaks*; their way of life came to depend upon going on *acid trips* at *acid parties* or elsewhere, in the presence of *acid rock* (loud music). More recently (1988), the unrelated term *Acid House* (or *House*) has been applied to a style of music and dancing imported to Britain from Chicago, and characterized by the widespread use by participants of a designer drug called Ecstasy. These uses are striking examples of the transformation of the primary sense of a basic term in a technical subject, namely chemistry.

**acid rain.** A stock example of a phrase that was in use for a long period (first recorded 1859) before becoming part of International English. Environmentalists brought the matter of the pollution of forests by acid rain, i.e. rain with significantly increased acidity as a result of atmospheric pollution, to the UN in 1972, and the term has now become part of the 'Green' vocabulary known to us all (along with *CFC*, *greenhouse effect*, *ozone layer*, etc.).

**acid test.** See POPULARIZED TECHNICALITIES.

**acknowledgement.** In most printed work this is the preferred spelling in BrE, whereas the form without medial *-e- (-ledgment)* is more usual in AmE. The choice is a matter of convention not of correctness or error.

**acoustic.** 1 *Pronunciation*. In 1917 Daniel Jones gave priority to /-ˈkaʊst-/ in his *English Pronouncing Dictionary*. By 1963 (12th edn.) this pronunciation was labelled old-fashioned, and preference was given to /-ˈkuːst-/, which remains the normal pronunciation. H. W. Fowler's forecast in 1926 ('If the word came into popular use, it would probably be with -ow-') was based upon traditional assumptions about the English pronunciation of Greek.

2 *Acoustics* is construed as a singular noun when used to mean 'the science of sound' (e.g. *Acoustics is a branch of physics*), and as a plural when used to mean 'the acoustic properties (of a building)' (e.g. *the acoustics of the church hall are not impressive*).

**acqu-.** See AQU-, ACQU-.

**acquaintanceship.** The logical progress of ideas in a sentence occasionally allows this word to slip into print, sometimes when *acquaintance* would have served instead. Examples: *Aunt Kate of Hungerford or some other member of her wide acquaintanceship of old ladies*—P. Norman, 1979; *visits to and by members of the other group with acquaintanceship with immigrants*—European Sociol. Rev., 1986; *we'd somehow crossed the barrier between acquaintanceship and friendship*—Truckin' Life (Aust.), 1986.

**acronym.** 1 Since 1943, words formed from the initial letters of other words have been known as *acronyms* (the word itself formed from *acro-*, Gk ἀκρο-, combining form of ἄκρος 'topmost' + *-onym* after *homonym*). A word was clearly needed for this prolific method of modern word-formation. The test of a true acronym is often assumed to be that it should be pronounceable as a word within the normal word patterns of English. By such a reckoning, *BBC* is not an acronym but an abbreviation; whereas *Nato* ( = North Atlantic Treaty Organization), being pronounceable like *Cato*, is an acronym. Examples of familiar acronyms include: *Aids* (acquired immune deficiency syndrome), *Anzac* (Australian and New Zealand Army Corps), *ASH* (Action on Smoking and Health), *SALT* (Strategic Arms Limitation Talks), *SAM* (surface-to-air missile), *Unesco* (United Nations Educational, Scientific, and Cultural Organization), and *WASP* (White Anglo-Saxon Protestant). Many words of this kind, especially those that are the names of organizations, begin by being written with uniform capitals and full stops, and only gradually attain the status and shape of ordinary words after constant use: thus U.N.E.S.C.O. → UNESCO → Unesco. Others remain written with uniform capitals (ASH, SALT, SAM, WASP, above) but without full stops. Still others were written with uniform lower-case letters virtually from the beginning: *laser* (light amplification by stimulated emission of radiation), *radar* (radio detection and ranging), and the US word *snafu* (situation normal—all fouled up).

**2** The limitations of the term being not widely known to the general public, *acronym* is also often applied to abbreviations that are familiar but are not pronounceable as words. Thus *EC* (European Community), *FBI* (Federal Bureau of Investigation), and *VCR* (videocassette recorder). Such terms are also called *initialisms*.

**act, action.** **1** The distinction between the two words is not always absolute: *we are judged by our acts* or *by our actions*. In some of its senses, *act* (derived in the 14c. from OF *acte*) refers directly to L *actus* 'a doing' and in others to L *actum* 'a thing done'. In general, *action* means 'the doing of (something)' and has tended over the centuries to prevent *act* from being used in the more abstract senses. We can speak only of the *action*, not the *act*, of a machine, when we mean the way it acts; and *action* alone has a kind of collective sense, as in *his action throughout* (i.e. his acts or actions as a whole) *was correct, he took decisive action* (freq. involving a series of separate acts). The *actions* of a person are usually viewed as occupying some time in doing, in other words are the habitual or ordinary deeds of a person, the sum of which constitutes his or her *conduct. Act*, by contrast, normally means something brought about at a stroke or something of short duration. As such it is frequently followed by *of* and a noun (*an act of God, an act of cruelty, folly, madness, mercy*, etc.).

**2** *Action* is freely used in the attributive position (*action committee, painting, photography, replay*, etc.). *Act* cannot be so used.

**3** Both words form part of fixed phrases, idioms, or proverbs: (act) *my act and deed, Act I* (of a play), *a variety act* (at a circus, etc.); *to put on an act, caught in the act, clean up one's act, get one's act together*; (action) *killed in action, out of action, to take action, actions speak louder than words*; (colloq.) *where the action is* (the centre of activity), *a piece* (or *share*) *of the action* (implying participation in some activity).

**action** (as verb). In its modern use in the sense 'to take action on (a request, etc.)', the word is best left at present to the tight-lipped language of business managers, e.g. *Dismissal will be actioned when the balance of probabilities suggests that an employee has committed a criminal act—Daily Tel., 1981.*

**activate, actuate.** These two verbs were on a collision course in general contexts in the 19c. and *activate* became obsolete (and was so labelled in the *OED*) for a while. Its substantial use in chemistry and physics in the 20c. has brought *activate* back into prominence. At the present time, *activate* is the term used when the context requires 'to render active' (of carbon, molecules, etc.), 'to make radio-active' (*Most of the elements situated between boron and calcium have been activated under the influence of α-rays*—R. W. Lawson, 1938), 'to aerate (sewage) as a means of purification', or some other technical or scientific sense. It is also widely applied to much less technical items like burglar alarms, traffic lights, flight plans, and, less commonly, to behaviour that is motivated by some set of circumstances (*Are they activated by concern for public morality?*—*Essays & Studies*, 1961). *Actuate*, by contrast, is much less often encountered in technical contexts, though in practice devices, diaphragms, forces, pinions, pistons, and so on, as shown in 18c. and 19c. examples in the *OED*, are still *actuated* by this or that instrument or agent. Abstract qualities like anger, greed, jealousy, malice, etc. are only grudgingly *activated* and more commonly *actuated*. There is no guarantee that this state of affairs is permanent. Examples of *actuate*:

*His opposition was actuated by a different and more compelling motive than that of her other relatives*—D. Cecil, 1948; *Peirce was actuated by the analogy with science, not by a vision*—J. Barzun, 1983; *the union campaign ... was actuated by political rather than industrial considerations*—*New Statesman,* 1985.

**actuality.** A century ago *actuality* stood beside *actualness* as virtual synonyms in the sense 'the quality or state of being actual'. *Actuality* also meant 'the state of being real, reality', and, in the plural, 'actual existing conditions or circumstances' (*his words were directed to the actualities of the case*). *Actualness* has dropped by the wayside, whereas *actuality* has retained its older senses while acquiring a new concrete one, namely 'a film record or radio or television broadcast of an event as it actually occurs' (*actuality film, material, programme*).

**actually.** One of a number of adverbs (*definitely, really, surely,* etc.) that at present tend to be overused as emphasizers. The traditional use of the word to mean 'in fact, in reality' is shown in sentences like *Often it wasn't actually a railway station but a special stopping place in the middle of nowhere*—*New Yorker,* 1987. Emphasizing uses of the word are not new: the OED (sense 5) lists examples from Goldsmith, Ruskin, and others in which *actually* is added 'to vouch for statements which seem surprising, incredible, or exaggerated', e.g. *I had some dispositions to be a scholar and had actually learned my letters*—Goldsmith, 1762. In many modern contexts, however, the degree of surprise, incredibility, or exaggeration being vouched for is often a little less easy to discern, e.g. (used as a sentence adverb) *But I'd like to see those scrap books again, actually*—Lee Smith, 1983; '*It was a fairly rough night, actually, sir. One way or another.*'—A. Price, 1987; *except that instead of going into the technological future I had to go back in time. Actually, I don't really mean that, back in time*—J. Barnes, 1989; (in a normal adverbial position) '*And we don't actually have the money, either,*' *my mother said*—A. Munro, 1987. In each case the writers concerned appropriately used this low-key device as part of the natural language of their characters. The problem is how to bring about a reduction in the low-key devices that litter the language, particularly in spoken English. Further examples: (part of an assertion or counter-assertion) *We think it's quite all right, actually*—D. Heffron, 1976; (adding slight emphasis) *I am not actually in a position to make a decision*—H. Secombe, 1981; (reinforcing a negative statement) '*I told you, I've got problems at work.*' '*Actually, you didn't.*'—S. Mackay, 1981; (when giving advice) *Actually, it might be a good idea not to travel from Gatwick in the summer this year*—mod. Clearly a useful, but an overused word.

**actuate.** See ACTIVATE.

**acuity, acuteness.** *Acuity* (first recorded 1543) has long vied with *acuteness* (1627) in lit. and fig. contexts. In practice the former word has retreated somewhat: nowadays it collocates principally with hearing, understanding, vision, wit, and certain diseases. Examples: *People that I know who like to read poetry with acuity and intelligence*—A. Hecht, 1981; *almost drowsy, but with no loss of mental acuity*—I. Asimov, 1982; *being able to see 6/12 on a visual acuity chart*—*Nursing Times,* 1986. For nearly all other areas broadly signifying 'sharpness, perceptiveness', *acuteness* is the more usual word. The two words have not, however, drifted into separate semantic bins in the manner of numerous other *-ty*/*-ness* pairs (*casualty/casualness, ingenuity/ingenuousness,* etc.).

**acumen.** The Victorian (and therefore the OED's) pronunciation was /əˈkjuːmən/, stressed on the second syllable. This still seems to be the dominant pronunciation in AmE, but it has given way to /ˈækjʊmən/, stressed on the first syllable, in BrE.

**ad,** a frequent colloquial shortened form of *advertisement* or *advertising* (so *small ad, adman,* etc.), first recorded in 1841 and now in very widespread informal use.

**AD** (*anno Domini,* in the year of the Lord) should always be placed (in small capitals when printed) before the numerals: AD 900. It is customary, however, to write 'the third century AD' after the model of 'the third century BC'. See BC.

**adagio.** (Mus.) slow; as noun, pl. *adagios.* See -O(E)S 4.

**adamant.** Derived from L *adamantem* (nom. *adamāns, adamās* (earlier in Gk), it was brought into English before the Conquest as a noun meaning a hard rock or mineral ('as to which vague, contradictory, and fabulous notions long prevailed', *OED*). In modern use it is 'only a poetical or rhetorical name for the embodiment of surpassing hardness' (*OED*). But since the 1930s it has passed into use as an adjective (*His appointment had met with the adamant opposition of almost all the Fellows*—T. Sharpe, 1974), from which the adverb *adamantly* is derived (*When she mentions him at all in her diary, it is in adamantly negative terms*—S. Quinn, 1988). Cf. ANATHEMA 2.

**adapter, adaptor.** In an ideal world, an *adapter* is a person who adapts (something or to something), whereas an *adaptor* is a device which adapts (something). The distinction is not an absolute one, however, and any rule-stating goes against the evidence. One is likely to encounter technical examples of *adapters* (e.g. *the UG-176 adapter, the T1 TMS 380 L.A.N adapter chipset*) in any English-speaking area; similarly *adaptors* ( = persons) may be found in any form of English. See -ER AND -OR; -OR.

**adaption.** For Swift and Dickens, *adaption* was an acceptable formation meaning 'adaptation, the action of adapting'. At no time, however, has it presented a serious challenge to the regular form *adaptation*: in any substantial corpus of evidence *adaptation* is at least ten times more likely to occur than *adaption*. Nevertheless it should be noted that the shorter form turns up from time to time in good sources: e.g. *The Russian adaption of foreign models*—TLS, 1983; *The Quadruplex system is still used today by broadcasters, but was quite unsuitable for adaption to domestic use*—New Scientist, 1983; *But I daresay you'll display your usual power of adaption*—P. Lively, 1987. There is considerable variation in usage in the derivatives: *adaptionist* (adj.) is frequent, but so are *adaptationism, -ist, adaptational, -alism, -alist*.

***ad captandum,*** a L phrase (in full *ad captandum vulgus*) meaning '(calculated) to take the fancy of (the crowd)'. Used in English since the 18c. of unsound or specious arguments or opinions (*an ad captandum presentation of the facts, an ad captandum sentimentality*).

**addenda.** 1 This is a plural form: it should be used only when listing a number of items; if there is only one, the heading should be *Addendum*.

2 *Addenda* is occasionally (and regrettably) used with a singular construction when it means 'a list of additional items' (*a new edition with an invaluable addenda*).

**addicted to.** This should be followed by an ordinary noun (*is addicted to whisky*) or by a verbal noun in *-ing* (*is addicted to reading science fiction*). Formerly (until some point in the 19c.) it could also be followed by an infinitive: *I have been so little addicted to take my opinions from my uncle that ...*—J. Austen, 1814.

**addle, addled.** The ancient noun *addle* (OE *adela* 'stinking urine, mire') came to be applied as an adj. to eggs in the 13c. (*addle egg* was equated with medL *ovum ūrīnae* 'egg of urine' or 'putrid liquid'), and by the 16c. to virtually anything seen as decomposed, muddled, unsound, or idle. Since then it has gradually become restricted in use to describe (rotten) eggs and muddled brains. The forms in current use are *addle* (verb) (both eggs and brains are capable of 'addling'); *addle-brain(ed), addle-head(ed), addle-pate(d)*, and the corresponding participial adj. *addled*.

**addresses.** It is now customary to use as little punctuation as possible in addresses. Thus (omitting commas at the end of lines):

The College Secretary
St Peter's College
Oxford OX1 2DL

and (omitting the comma after a street name):

44 High Street

**-ade.** Nearly all disyllabic or polysyllabic words ending in *-ade* are derived from French (some being drawn into that language from other Romance languages). The majority have passed through a stage of being pronounced with final /-ɑːd/, but are now normally pronounced with /-eɪd/, e.g. *accolade, arcade, balustrade, brigade, brocade, cascade, cavalcade, crusade, lemonade, marmalade, masquerade, palisade, parade, serenade*, and *tirade*. A small

group, e.g. *esplanade, fanfaronnade, fusil-
lade, glissade, pomade,* and *rodomontade,*
are still commonly pronounced with
either /-eɪd/ or /-ɑːd/; and *aubade, ballade,
charade, façade,* and *promenade* always
have /-ɑːd/.

**adequate.** 1 In the sense 'proportionate
(to the requirements), sufficient', *adequate* is most commonly used without
a complement (*the interest on his investment
was small but adequate; there is an adequate
supply of food in the flooded area*). When it
has one, the complement is now just as
commonly introduced by *for* as by *to*: cf.
*very reliable and adequate to our purposes;
she has adequate grounds for divorce; their
earnings are adequate for/to their needs. Adequate* is also frequently followed by a *to*-
infinitive: *Is language adequate to describe
it?*

2 It is notionally not subject to comparison (adequate, it can be argued, is
adequate, neither more nor less), but
absoluteness is not one of the more
characteristic features of our language.
The level of adequacy is variously graded
by the use of *more/most, less/least,* or *just,
very,* etc.: *a more adequate return; the most
adequate description yet released of the horror
of the hijacking; you can get along quite
comfortably if you are just adequate; some
very adequate salaries; a slapdash but sufficiently adequate manner.*

3 *Adequate* is also idiomatically used
to mean 'barely sufficient': *The standard
rapidly sinks to a level which is, at best,
adequate but at worst incompetent.*

**adherence, adhesion.** These two
nouns, both adopted from French in the
17c., go back ultimately to two Latin
nouns, each of which is derived from
the L verb *adhaerēre, adhaes-* 'to adhere'.
*Adherence* is now mostly used in figurative
senses (adherence to a political party, a
point of view, one's beliefs, usu. implying
a continuing allegiance, and often in
spite of difficulties). *Adhesion* has tended
more and more to imply physical contact
between surfaces, e.g. the grip of wheels
on road or rail, the sticking or gluing
together of two surfaces. *Adhesion* also
has a technical sense in medicine, 'a
mass of fibrous connective tissue joining
two surfaces that are normally separate'

(*OED* 2). Nevertheless there is an overlapping area in which *adherence* is used literally (e.g. adherence of petals and sepals)
and *adhesion* figuratively (e.g. *The present
humble writer, who has declared his adhesion
to Macaulay's dictum that Rymer was the
worst critic who ever lived*—G. Saintsbury,
1916; *He is in fact more rigid in his adhesion
to his old doctrines*—B. Webb, 1952; *Others
. . . fell under his control through the adhesion
to France of their ruler, the Prince-Bishop of
Liège*—W. S. Churchill, 1957).

***ad hoc,*** L, lit. 'to this', has been recorded
in English since the 17c. in the sense 'for
this specific purpose, to this end' (*a sum
not far off two millions will have to be provided* ad hoc *by the Chancellor of the Exchequer*), but more particularly as a quasi-
adj. meaning '(arranged) for some particular purpose' (*an* ad hoc *committee is
to be set up*) or '(in a manner) not planned
in advance' (*a remark of the* ad hoc *kind*).
It should normally be printed in italic,
and in strict use is not capable of being
qualified by an intensifying word like
*very* or a down-toning word like *fairly*. In
the 20c. it has become very commonly
used and has also generated an array
of startlingly un-latinate derivatives, e.g.
*ad-hoc-ery, ad-hoc-ism, ad-hoc-ness* (sometimes with medial hyphens, sometimes
without).

***ad idem,*** = L, lit. 'to the same (thing)',
has continued to appear sporadically in
English printed work since the 16c., used
as an adverbial phrase meaning 'on the
same point, making direct reference to
the matter in hand, à propos' (*We think
you would have a strong case for claiming
that there is no contract on the ground that
there was no consensus* ad idem—*Financial
Times,* 1983). It should normally be
printed in italic, and used sparingly because it is likely to be not known except
by someone with knowledge of Latin.

**adieu.** In the house style of OUP, the
plural is *adieus* not *adieux,* pronounced
/-z/. See -x.

**adjacent** means 'lying near', but a definition of nearness or proximity, as the
*OED* says s.v. *adjacent,* by no means pre-
cludes the possibility that two *adjacent*
things can actually *touch* each other. An
*adjacent inn* is normally one within a
hundred yards or so, but it can be an

adjoining property; *adjacent angles* in a triangle are separated by the length of one side of the triangle; *adjacent tables* are next to each other, but with a space between. Recent examples: *a little field adjacent to his avenue: Thaw and his sister slept in adjacent rooms*; *the shop, adjacent to the Palace, but divided from it by the breadth of a steep narrow street*. Because of the elasticity of the concept of nearness, it is permissible (though the need will seldom arise) to speak of someone or something being *very adjacent*.

**adjectivally.** Until the 19c. the normal adverb corresponding to the word *adjective* was *adjectively*, but it has now been entirely replaced by *adjectivally* on the model of *adverbially, pronominally, verbally*, etc. (not *adverbly, *pronounly, etc.).

**adjective.**

 1  As name of a part of speech.
 2  Attributive and predicative.
 3  Comparison.
 4  'Absolute' adjectives.
 5  Position of adjectives.
 6  Hyphenation.
 7  Compound adjectives.
 8  Adjectives used as adverbs.
 9  Adjectives as nouns.
10  Transferred epithets.

**1** *As name of a part of speech.* The italicized words in 'a *black* cat' and 'a body *politic*', used as an addition to the name of a thing to describe the thing more fully or definitely (as the *OED* defines it), were usually called *noun adjectives* from the 15c. to the 18c. The term *noun adjective* (as distinguished from *noun substantive*) was not recognized as being one of the primary parts of speech. Joseph Priestley, in *The Rudiments of English Grammar* (1761), was perhaps the first English grammarian to recognize the *adjective* as an independent part of speech, though some earlier writers had used the term in this way.

Since the mid-19c., some writers have used the word *modifier* to signify 'a word, phrase, or clause which modifies another' (*OED*). Within this terminology, the italicized words in 'the *black* cat', 'a body *politic*', and 'the *city* council' are all modifiers. In traditional grammar, *home* in *home counties* and *city* in *city council* are called attributive uses of the nouns *home* and *city*.

An adjective has three forms, traditionally called a positive (*hot*), a comparative (*hotter*), and a superlative (*hottest*). In some modern grammars (e.g. *CGEL*), the base form is called the *absolute*, not the *positive*, form.

In this book I retain the terminology of traditional grammar.

**2** *Attributive and predicative.* Most adjectives can be used both attributively (a *black* cat, a *gloomy* outlook) and predicatively (the cat is *black*, the outlook is *gloomy*, he found the door *shut*), that is, can within limits be placed after the noun to which it refers. Some adjectives, however, are normally restricted to the predicative position, e.g. *afraid* (*he is afraid* but not *the afraid boy*, though *the somewhat afraid boy* is admissible), *answerable* (*he is* answerable *to his superiors*), *rife* (*speculation was* rife *that . . .*), *tantamount* (*his action is* tantamount *to treason*), etc. Conversely, numerous adjectives (if the meaning of the word is to remain unchanged) must be used only in the attributive position: e.g. *he is a big eater* not *as an eater he is big*; *the sheer richness of his material* not *the richness of his material is sheer*. So also *mere repetition, my old self, pure fabrication, on the stout side, a tall order, the whole occasion*, etc.

**3** *Comparison.* Monosyllabic and disyllabic adjectives normally form their comparative and superlative forms by adding *-er* and *-est* (*soft, softer, softest*). Polysyllabic adjectives are more comfortably preceded by *more* and *most* (*more frightening, a most remarkable woman*). For special effect, a polysyllabic adjective is sometimes used in an unexpected *-er* or *-est* form: e.g. *'Curiouser and curiouser!' cried Alice*—L. Carroll, 1865; *one of the generousest creatures alive*—Thackeray, 1847/8; *Texas A&M's Shelby Metcalf, the winningest coach in Southwest Conference basketball history, was relieved of his duties Monday*—*Chicago Tribune*, 1990.

**4** *'Absolute' adjectives.* Certain adjectives are normally incapable of modification by adverbs like *largely, more, quite, too*, or *very*: e.g. *absolute, complete, equal, excellent, impossible, infinite, perfect, possible, supreme, total, unique, utter*; (see also ADEQUATE, ADJACENT). But English is not a language of unbreakable rules, and contextual needs often bring theoretically unconventional uses into being, e.g.

The ... ghosts ... made the place absolutely impossible—Harper's Mag., 1884; All animals are equal but some animals are more equal than others—G. Orwell, 1945; his profile is—oh Aunt Frances—most utterly perfect—J. Gardam, 1985.

**5** *Position of adjectives.* In numerous fixed expressions, an adjective is placed immediately after the noun it governs: e.g. *attorney-general, body politic, court martial, fee simple, heir apparent, notary public, poet laureate, postmaster-general, president elect, situations vacant, vice-chancellor designate, the village proper.* These are to be distinguished from cases in which an adjective just happens to follow the noun it governs (e.g. *The waiter ... picked up our dirty glasses in his fingertips, his eyes impassive—Encounter,* 1987; *1992, that hailed watershed for the SM between matters past and matters future—Linguist,* 1991), or when the natural order is reversed for rhetorical effect (e.g. *And goats don't have it* [sc. self-consciousness], *they live in a light perpetual—Maurice Gee,* 1990; *before the loving hands of the Almighty cradled him in bliss eternal—N. Williams,* 1992).

**6** *Hyphenation.* There is an increasing and undesirable tendency at present to insert a hyphen in the type *a highly competitive market, a newly adopted constituency* (thus *a highly-competitive market,* etc.), i.e. where an adverb in -ly governs an adjective which is immediately followed by a noun. Printers and writers are sharply divided in the matter. Examples from my files: (*a*) *unwontedly clean clothes, a comparably enormous step, a genuinely wintry blackness, those handsomely engraved certificates, a statistically significant relationship;* but also (*b*) *the abundant recently-published material, lawfully-elected prime ministers, fiercely-illuminated buildings, professionally-inclined Eastern European Jews, their scarcely-filled baskets of food.* It is to be hoped that the hyphenless type will prevail.

**7** *Compound adjectives.* Compound adjectives of the types noun + adjective, noun + past participle, and noun + participle in -ing have proliferated in the 20c. Examples: (*a*) *accident-prone* (1926), *acid-free* (1930), *child-proof* (1956), *computer-literate* (1976), *host-specific* (1969), *machine-readable* (1961), *sentence-final* (1949), *sentence-initial* (1964), *water-insoluble* (1946), *word-final* (1918), *word-initial* (1918).

(*b*) *computer-aided* (1962), *custom-built* (1925), *hand-operated* (1936). (*c*) *data-handling* (1964), *pressure-reducing* (1934), *stress-relieving* (1938).

A new kind of compound adjective emerging in technical and scientific work is the type *landscape ecological principles* ( = the principles of landscape ecology). From the starting-point *landscape ecology* (the name of an academic subject), some writers are unwisely tempted into converting the second noun into an adjective to produce *landscape ecological principles.* Similarly from *physical geography* (name of subject) emerge such phrases as *physical geographical studies.* In all such cases it is better to use an *of-* or *in-* construction: thus *studies in physical geography, the rules of diachronic linguistics, research in environmental psychology, students of historical geography,* etc.

**8** *Adjectives used as adverbs.* In formal written work, adjectives are not often used as adverbs, but such uses are common enough and mostly unobjectionable in informal speech (e.g. come *clean,* come *quick,* drive *slow,* hold *tight*). To these may be added *real* and *sure,* which in the UK are often taken to be tokens of informal NAmer. speech (*that was* real *nice, I* sure *liked seeing you*), but are playfully used in other regions, including the UK, as well.

It is important to recognize that an adjective and an adverb sometimes have the same form: *He left in the* late *afternoon* (adjective); *He left* late *in the afternoon* (adverb). The adverbial form *lately* has a different meaning: *Have you been to Oxford* lately? ( = recently). Other examples in which the adverb and the adjective have the same form are (*CGEL* 7.8): *clean, close, deep, fine, light, straight,* and *wide;* (with -ly) *early, likely, monthly, nightly,* etc. Adverbs without -ly and those with -ly often occur in close proximity: *'I play straight, I choose wisely, Harry,' he assured me*—J. le Carré, 1989.

**9** *Adjectives as nouns.* For many centuries English adjectives have been put to service as nouns while remaining in use as adjectives. Thus (*a*), all of which can be used as count nouns:

|  | first recorded as adj. | first recorded as noun |
|---|---|---|
| ancient | 1490 | 1541 |
| classic | 1613 | 1711 |

|            | first recorded as adj. | first recorded as noun |
|------------|------------------------|------------------------|
| classified | 1889                   | 1961                   |
| daily      | 1470                   | 1832 (newspaper)       |
| explosive  | 1667                   | 1874                   |
| intellectual | 1398                 | 1599                   |

(b) adjectives preceded by the and used as non-count nouns to indicate 'that which is—' or 'those who are—':

| beautiful 1526 | the beautiful 1756 |
|----------------|--------------------|
| poor 1200      | the (deserving, etc.) poor 1230 |
| sublime 1604   | the sublime 1679   |
| unemployed 1600 | the unemployed 1782 |

**10** *Transferred epithets.* A curiosity of our language is the way in which an adjective can be made to operate in such a way that it has merely an oblique relevance to the noun it immediately qualifies. Examples: *'It's not your stupid place,'* she says. *'It's anyone's place.'*—P. Lively, 1987 (the person addressed, not the place, is stupid); *the possibility of somebody getting killed is only balanced by the improbability of either side adequately policing these* melancholy *waters* [sc. the sea area round the Falklands]—*Times,* 1987 (the waters themselves are not melancholy); *I will be sitting quietly at the kitchen table stirring an* absent-minded *cup of coffee*—*Chicago Tribune,* 1989 (the person, not the coffee, is absent-minded). The traditional name for this phenomenon is 'transferred epithet' or 'hypallage'.

On other aspects of the behaviour of adjs, see ABSOLUTE COMPARATIVE; ABSOLUTE SUPERLATIVE; ADVERB; DOUBLE COMPARISON; PARTS OF SPEECH. See also FUN and MAGIC, used in the 20c. as quasi-adjs.

**adjudicator.** Regularly spelt with final -or.

**adjunct.** This grammatical term is used in various ways. **1** 'Any word or words expanding the essential parts of a sentence; an amplification or "enlargement" of the subject, predicate, etc.' (OED).

**2** 'In Jespersen's terminology, a word or group of words of the second rank of importance in a phrase or sentence (contrasted with a *primary* and a *subjunct*)' (OEDS).

**3** 'A term used in grammatical theory to refer to an optional or secondary element in a construction; an adjunct may

be removed without the structural identity of the rest of the construction being affected', (D. Crystal, 1980).

**4** 'There are four broad categories of grammatical function [in adverbials]: *adjunct, subjunct, disjunct,* and *conjunct*' (CGEL 8.24).

These four definitions underline the care with which the terminology of modern English grammar must be approached. Unless otherwise specified, I use *adjunct* in this book in the manner defined by the OED (and by Crystal).

**adjure.** See ABJURE.

**adjust.** Two intransitive uses of this verb have entered the language in the 20c.: **1** To adapt oneself *to* (something): e.g. *She seemed to have adjusted to her new status with little difficulty*—L. Niven, 1983; (used absolutely) *He may try to adjust by staying with people of the same group as his family*—Listener, 1962; *She needs time to adjust*—S. King, 1979.

**2** (intransitive for passive) To be capable of being adjusted: e.g. *The barrel ... can adjust right up to the neck of the lamp*—Habitat Catalogue, 1982. These are minor examples of the way in which English silently colonizes new territory all the time.

**3** Among the smoke-signals of our market-driven economy is a new transitive use. Economic statements, figures, etc., are now often described as being adjusted *for* inflation, etc.: e.g. *Lenders vary as to when they adjust your repayments for tax relief*—What Mortgage, 1986.

**administer.** Some functions of the verb are now under threat from *administrate* (see 1, 3), while *administer* itself is now being increasingly used in medical contexts in place of *minister* (see 4). **1** For many centuries, administrators have *administered* their establishments, institutions, departments, etc.: (a modern example) *The Rezzoris were minor Austrian gentry administering the outposts of empire*—London Rev. Bks, 1990. In recent decades, there is evidence that the longer form *administrate* (first recorded in the 17c.) is increasingly being used as a kind of newly invented back-formation from *administration* and is now awkwardly challenging *administer* in its traditional sense. Examples: *the machinery of such aid* is still primed by administrators eager to go out and administrate—Times, 1981; The Sports Council has begun a major investigation into discovering new ways to administrate a drug-detecting system—Times, 1988.

2 An older generation did not believe that it was right, except in a humorous context, to administer blows, a rebuke, etc., other verbs being deemed to be more appropriate. Such uses now seem to have moved into the uncontroversial core vocabulary of the language. Examples: the corporal chastisement ... was administered with birch rods—W. S. Churchill, 1956; two others held her feet while the headmaster administered the cane—B. Emecheta, 1974.

3 Administer is still routinely used to mean 'to give (medicine, an injection, etc.' (I was brimming with alcohol—administered to loosen my tongue—A. Price, 1982; If my profession had a false white coat, this would be the moment when I administered the wicked injection—J. le Carré, 1989); and 'to celebrate (a religious rite)' (a preacher called in to administer last rites—J. T. Story, 1969). Even in the first of these senses, administrate is sometimes (unwisely) being used (In 1947 it had not yet been realized that vaccination against smallpox administrated in the early months of pregnancy is exceedingly dangerous for the foetus—Author, 1994).

4 In medical contexts administer to is now being increasingly (and erroneously?) used instead of minister to (an injured person, etc.). Examples: the fact that Ranjit is alive today is a tribute to the ambulance attendants who administered to him at the scene—Oxford Times, 1977; Trevor Proudfoot, the supremo of the statuary workshop, administers to a wild boar from Clarement—National Trust Magazine, 1989; American doctors, being vastly rich, have better things to do with their leisure time than administer to patients at weekends—Times, 1994.

**admission, admittance.** Like so many other doublets, these two words have competed with each other for several centuries (admission first recorded in 1494, admittance in 1589) without ever establishing totally separate roles. The base meaning of both is 'the action of admitting, letting in, to a place', often, but not always, applied to the person

being admitted as well as to the person (a doorman, porter, etc.) granting admittance. In most circumstances admission seems to be dominant now. It is the only one of the pair to develop attributive uses (admission fee, money, ticket) and elliptical uses thereof (Admission £1). The tutor, secretary, etc., dealing with admissions to colleges and universities is called the admissions tutor, secretary, officer, etc. Admission is the customary word for the action of being received into an office or position (he gained admission to the Society of Antiquaries). Admission is also the word used for the admitting of a crime, guilt, etc., a confession. But admittance hangs on determinedly, esp. in the phrase used on doors and gates, No admittance except on business, and in sentences like this one from A. L. Barker's The Gooseboy (1987): When admittance [to a fan club] was denied them, the maturer members tried to climb over. Also, occasionally, in general contexts where admission is the more usual word: That is an admittance that there is variation, but an assertion that there is nothing more to be said—Word, 1984; The DTI's lack of admittance of negligence in this affair is a travesty of justice—Times, 1988.

**admit.** 1 This verb is most commonly used with a direct object (he admitted it) or with a following clause led by that (he admitted that he had no money). Both as a transitive verb and as an intransitive one, admit was once freely interchangeable in most senses with the phrasal verb admit of. Nowadays admit of is restricted to the senses 'present an opening' and 'leave room for', e.g. a hypothesis admits by its nature of being opposable; his truthfulness admits of no exceptions; it seems to admit of so many interpretations; the circumstances will not admit of delay. Even in these senses, the construction seems old-fashioned.

2 The phrase admit to ( = confess to, acknowledge) is a relatively recent addition to the language: Senior Ministry of Defence officials yesterday admitted to a catalogue of errors in the development of a new light anti-tank weapon—Times, 1989. The OED's earliest example is one of 1936.

**-ado.** The older trisyllabic and polysyllabic English words ending in -ado are often assumed to be direct adoptions of Spanish or Portuguese words. Many of them are in fact either (a) affected refashionings of French words in -ade (thus *ambuscado*, found in Shakespeare and in many other writers, is a refashioning of *ambuscade*, from Fr. *embuscade*); or (b) adaptations (-*ado* being 'felt' to be a Spanish or Portuguese ending) of -*ada* in the original language (e.g. *tornado* from Sp. *tronada* thunderstorm). A number of -*ado* words remain firmly in the language: (normally pronounced with /-ɑːdəʊ/: *aficionado, amontillado, avocado, bravado, desperado, incommunicado*; (with /-eɪdəʊ/) *bastinado, tornado*. Others are archaic or obsolete (if pronounced, they would normally have /-ɑːdəʊ/, e.g. *ambuscado, barricado, grenado, scalado. Mikado*, always with /-ɑːdəʊ/, is unconnected, being from Japanese (from *mi* 'august, honourable' + *kado* 'door').

**adopted, adoptive.** When parents adopt a child, the child, traditionally, is said to be *adopted*, and those taking the child into their family *adoptive*. Usage has partly eroded this useful distinction and the two words are now occasionally used interchangeably. In wider use, *adopted* is the customary word applied to a new area, country, town, etc., that a person has chosen to live in, but the *OED* and other dictionaries cite examples where *adoptive* is used instead (*adoptive country, home, his adoptive England*, etc.). The traditional distinction between *adopted* and *adoptive* appears to be crumbling.

**adrift.** This 17c. adverb (in origin a phrase) = 'in a drifting condition' has acquired a new range of senses in the language of late 20c. sports commentators, namely 'short or wide of (a target), behind (a competitor)': *their score had reached a wonderful 276 for 4, only 52 runs adrift*—J. Laker, 1977; *West Bromwich are 2 points adrift of City*—BBC Radio 4, 1978. The use has spread to more general contexts: *Reed ... was already £16m adrift at the pretax level at the nine-month stage*—*Times*, 1983.

**adult.** **1** Now usually pronounced /ˈædʌlt/ in RP and /əˈdʌlt/ in AmE, but the distribution is uneven among educated speakers throughout the English-speaking world. It is worth recalling that the *OED* (in 1884) and Daniel Jones (in 1917)

gave priority to /əˈdʌlt/. See NOUN AND ADJECTIVE ACCENT.

**2** Since the late 1950s, changing social attitudes have caused the word *adult* to be used euphemistically in the sense 'sexually explicit', applied to certain categories of films, magazines, shows, etc.

**adumbrate.** This Latinate word (L *adumbrāre* 'to overshadow'; cf. L *umbra* 'shadow') first enriched the language in the 16c. It can be most acceptably used in the sense 'to represent in outline, to foreshadow' when the surrounding text suggests that the writer or speaker has a good command of such vocabulary. It is out of place in office memoranda and the handouts of bureaucratic officials.

**advance.** **1** *advance, advanced. Advance* is one of a class of nouns that has been used so frequently as a qualifier before another noun (*advance copy, guard, payment*, etc.) that it has joined the category of words which the *OED* labels as '*attrib. passing into adj.*'. This quasi-adjectival use is recorded in the *OED* from c1910 onward: e.g. *advance warning signs should not be provided unless owing to a turn in the road ... A* Ministry of Transport Report, 1933; *often managing to arrive without advance notice*—E. Roosevelt, 1962; *advance information about the layout of the plant*—A. Hailey, 1979.

For the most part it elbows its way clear of *advanced*, which means 'far on in development': e.g. *an advanced young woman* (progressive in outlook), *an advanced degree* (more senior than an MA), *at his advanced age* (he is elderly).

**2** *advance, advancement. Advance* is much the commoner word of the two: e.g. *a new advance on the capital city* (approach to), *the advance of knowledge* (progress), *an advance of £10* (paid before the due time), *the advance of old age* (onset), *seats booked in advance* (before the game or performance), and so on. *Advancement* is far from extinct: e.g. *1985/6 was another year of great advancement for Glaxo Inc.* (progress); *the structure of the department allows for speedy advancement* (promotion), *the advancement of the main aims of the EC* (furtherance). But there are not many contexts in which *advancement* can be used in the general sense of progress, except in *of*-phrases of a type that call to mind Francis Bacon's title *The Advancement of Learning* (1605).

**advantage, vantage.** Etymologically both derived from OF *avantage*, these two words have gradually prised themselves apart over the centuries to the point that *vantage* effectively survives in general use only in the phrase *vantage point* (occasionally *vantage ground*) = (*a*) a good position from which to observe troop movements, etc., (*b*) a point of view having some observable advantage. When reading slightly older books, though, one often encounters (and is not troubled or puzzled by) expressions like *coign of vantage*, *vantage of ground*, *have* (someone) *at vantage*; and it is only in the course of the 20c. that the formula *advantage Becker* has taken the place in tennis of the former terms *vantage-in* (when the point after 'Deuce' has been called has been won by the server) and *vantage-out* (when the server has lost the point).

**adventurous, venturesome, adventuresome, venturous.** First recorded in *c*.1374, 1661, 1731, and 1565 respectively, only *adventurous* and *venturesome* have wide currency in modern English. The other two are occasionally used in a somewhat self-conscious manner: *Adventuresome readers can extend QWING to simulate the effects of an innovation introduced on banks more than a decade ago—Scientific American*, 1985; *The very venturous Huxley who was said to have become a Muslim in order to divorce her husband quickly—J. Huxley*, 1986.

**adverb.**

1 Main types.
2 Sentence adverbs.
3 Derivational adverbs.
4 Adverbs of time ending in -*s*.
5 Position of adverbs.
6 Same spelling as adjectives.
7 The type *to* + adverb + infinitive.

**1** *Main types.* One of the largest of all word-classes, the *adverb* is here to be taken as including the several hundred of the simplest kind like *else*, *how*, *just*, *out*, *over*, *quite*, *soon*, and *though*, most of which also have roles as prepositions, adjectives, conjunctions, or other parts of speech; adverbs formed from adjectives by the addition of -*ly* (*sad/sadly*); those beginning with *a*- (*apart*) or ending, sometimes optionally, with -*s* (*always*,

*backward*(*s*)); compound adverbs like *downstairs*, *herewith*, *outside*, *underfoot*, *whereupon*; adverbial phrases or adverbials (e.g. *for a time*); and conjunctive adverbs or conjuncts (e.g. *so* in *He has a large salary, so he must pay a lot of tax*).

Readers are referred to the extensive treatment of the adverb in *CGEL* (7.46–8.135), but should bear in mind that *CGEL*'s fourfold classification of adverbials into adjuncts, subjuncts, disjuncts, and conjuncts has not been adopted here, as being beyond the needs of anyone except professional grammarians.

**2** *Sentence adverbs.* See HOPEFULLY; SENTENCE ADVERB.

**3** *Derivational adverbs.* Large classes of adverbs and adverbial phrases are formed by adding certain suffixes to a preceding noun or adjective. The main types include -*fashion* (a use first recorded in 1633 in the compound adverb *broad-angle-fashion*); -*style* (e.g. *cowboy-style*); -*ward*(*s*) (e.g. *afterward*(*s*), *backward*(*s*), *downward*(*s*)); -*way*(*s*) (e.g. *anyway*, *halfway*, *partway*, *sideways*); -*wise* (e.g. *clockwise*, *likewise*, *lengthwise*). Of these, the only group that has run into criticism in the 20c. is -*wise*. *Standard examples. Mr Jones walked crabwise along the path—M. Wesley*, 1983; *Debbi can't even get a word in edgewise—Lee Smith*, 1983; *I wasn't good enough, quality-wise—A. L. Barker*, 1987. Somewhat more extreme examples are listed in the *OED* (*wise* sb.[1] 3b), most of them from earlier centuries: *anthem-wise*, *hind-foot-wise*, *lozenge-wise*, *Rossetti-wise*, *Saint Joseph-wise*. Unless used semi-jocularly, such formations often seem obtrusive if they are used in formal contexts.

**4** *Adverbs of time ending in -s.* Adverbs of time are of various kinds: (simple) *ever*, *now*, *often*, *once*, *soon*, etc.; (in -*ly*) *early*, *eventually*, *occasionally*, *recently*, *usually*, etc.; (in -*s*) *always*, *days*, *Mondays*, *nights*, *nowadays*, etc. Of these only the last group calls for comment. *Alway*, familiar from AV (*I would not liue* alway—Job 7: 16), where it alternates with *alwayes*, has receded into archaism. *Nowadays* is used throughout the English-speaking world, and has generated the potentially jocular word *nowanights* (used, for example, by Thoreau, Beerbohm, and Joyce). *Nights* ( = at night, each night), in origin an adverbial genitive (found already in *Beowulf*), has had a continuous history in English since OE, but in some contexts is beginning to be disused in the UK though it is retained elsewhere: *I lay awake nights for a week sweating about this four hundred and fifty dollars—New Yorker,* 1987.

There is a range of uses in which the context requires the specification of a day or days of the week or a season: *You know I'm completely stuffed* Fridays–R. Hall, 1978 (NZ); *We nip off to her place* afternoons–R. Hall, 1978 (NZ); *So from now on gentlemen,* Tuesdays *and* Thursdays ... *you're going to learn to think like white men*–V. O'Sullivan, 1985 (NZ); *Tuesday night, the board approved the addition of a new subsection*–Chicago Tribune, 1987; *Benn, childless, was free* on weekends–S. Bellow, 1987; *I was to be offered an option of taking her to live with me* summers–S. Bellow, 1987; *Noriega ... said* Monday *the U.S. Southern Command in Panama ... threatens the Central American nation*–USA Today, 1988. None of these seven examples, all taken from NZ and US sources, would pass muster in a UK publication. The same notions would be expressed in a different manner, e.g. *on Fridays, in the afternoons, on Tuesdays and Thursdays, On Tuesday night, on Monday, at weekends, every summer.*

**5** *Position of adverbs.* Leaving aside the more obvious constructions, e.g. those in which an adverb modifies an adjective (*very* angry), a noun (we stayed there for *quite* a while), another adverb (*too* modestly), or heads a clause (*perhaps* I may speak now), there are various accepted placings of adverbs within phrases or clauses. (*a*) (between auxiliaries) *a car dealer who could* certainly *have afforded to hire somebody;* (between an auxiliary and a main verb) *Roosevelt's financial policy was* roundly *criticized in 1933; he had* inadvertently *joined a lonely-hearts club.* But there are many reasons for altering the position of the adverb (change of meaning, emphasis, rhythm, etc.): e.g. *there is little chance that the student will* function *effectively after he returns to China.* (*b*) (adverb does not separate a verb from its object) *Gradually the Chinese Communists* abandoned *the Soviet methods; he dutifully observes all its quaint rules; They aim to set each subject briefly* into context; *Did he hear her* correctly?; *This alarmed me* greatly; *a dichotomy which simply* highlighted *Owen's own guilt-feelings.*

**6** *Same spelling as an adjective.* See ADJECTIVE 8.

**7** For the type *to* + adverb + infinitive, see SPLIT INFINITIVE.

**adversary.** Stress on the first syllable.

**adverse, averse.** The two words are close in meaning but not identical. *Adverse* is most commonly used attributively in three main senses: 'unfavourable' (*an adverse balance of trade, adverse circumstances, adverse weather conditions*); 'hostile' (*adverse criticism, an adverse reaction*); 'harmful' (*the adverse effects of drugs*). When the context requires a construction of the type 'unfavourable + *to* (a person)', the adjective selected is very rarely *adverse.* The latest example in the *OED* (*The presidential election of 1860 was adverse to the cause of slavery,* 1867) and a 1942 use by Winston Churchill (*the whole Parliamentary tradition as built up in this country ... is adverse to it*) both have a tinge of archaism about them. The type *he is not adverse to an occasional cigar* should be avoided: use *averse to* or express the idea in another way.

**advert.** A frequent colloquial abbreviation of *advertisement* with a continuous history in the language (though it has never been common in AmE) since the second half of the 19c.: *Reading the adverts for deodorants, hair-transplants, and staff agencies*–G. Swift, 1981; *if you examine the adverts for personal computers ... you'll find that almost nowhere do the ads promise you any kind of concrete benefit*–Your Computer, 1984.

**advert** (verb). The absence of this word from two of the main ELT dictionaries (those of OUP and Collins) adds weight to the view that the word may be slipping slowly into obsolescence. When it is used—most often in formal contexts—it normally means 'to mention, to pay heed (to), to call attention (to)': *the Duke adverted to his own position*–L. Strachey, 1921; *I walked down to the centre of the town without adverting to my surroundings*–'F. O'Brien', 1939; *His criticism had worried her a good deal and she adverts to it frequently in her diary*–Q. Bell, 1972. But *refer* and *allude* are now more frequently used in such contexts.

**advertise.** Spelt thus, not -*ize.* See -ISE 1.

**advertisement.** Always pronounced in RP with the main stress on the second syllable, but, as Walter Nash reminds us in *English Usage* (1986), the stress is placed on the third syllable in many regional varieties of English.

**advertising, language of.** All linguistic means of persuasion have been explored to the limit by advertising copywriters in the 20c. The linguistic strategies involved are numerous and are often complex. Some of the main techniques are listed and illustrated here.

(*a*) Enticement by the use of vague promises, claims, and undertakings: *Performance counts*; *The experience you seek*; *Because quality matters*; *Happiness is a cigar called Hamlet*; (after showing on the screen a feat of derring-do) *All because the lady loves Milk Tray.*

(*b*) Exaggeration verging on deception: *30% more cleaning power than any other detergent*; *Nails need the unique fortification of Clinique Daily Nail Saver.*

(*c*) Word-play: (portable air conditioners) *The Heat Beaters*; *The Lofa Sofa*; (a Landrover type of vehicle) *The Isuku Trooper . . . Guns Across the Moor. Shoots Down the Motorway . . . will allow you nothing to grouse about*; (advertising a T-shirt with a bull-terrier design stamped on the front) *Put on the Dog*; (a shampoo) *because we're dedicated to the medicated*; (disc-shaped sweet) *Who put the art into Smarties?*; *If you want to get ahead get a hat*; *My mate. Marmite*; *Slam in the lamb*; (water privatization) *You too could be an H₂ Owner* (ITV, 9 Sept. 1989).

(*d*) Slogans, catchphrases, and semi-proverbs: (Inter-Continental Hotels) *It's where you go when you've arrived*; *A Strong Company for a Strong Country*; (Bovril) *that sinking feeling*; (a dishwasher) *Life's too short for doing dishes*; (Doan's Backache Kidney Pills) *Every picture tells a story.*

(*e*) Appeals to expert opinion: *That's why more dentists and hygienists recommend Search 4 than any other toothbrush.*

(*f*) Use of comfortable-sounding adjectives: *custom-dyed, hand-crafted, fully crafted, ozone-friendly, ultra-pure*, etc. With widely varying punctuation: e.g. *hand crafted, hand-crafted, Hand Crafted, HAND CRAFTED*, etc.

(*g*) Expressed as a (usually mild) command: (Esso) *Put a tiger in your tank*;

(Mazda) *Don't let anything come between you and seventh heaven.*

(*h*) Expressed as a question: *Have You Heard About Our New Trade Secret?*; (a home perm) *Which twin has the Tony?*

(*i*) Use of slang: (Foster's beer) *The amber nectar.*

(*j*) Leading in with a negative statement: (Toyota) *Why we'll never make a perfect car.*

(*k*) Alteration of spelling: *Beanz Meanz Heinz*; (Schweppes) *Sch . . . you know who*; *Drink a pinta milka day.*

(*l*) Use of foreign expressions not necessarily known to the reader or viewer: (preceded by a conversation in German between two executives) *Vorsprung durch Technik.*

**advise, advice. 1** It should not be necessary to say that *advise* is a verb meaning 'to offer advice (to), to recommend' (*he advises the Japanese about nutritional matters*; *they advised her not to pursue the matter further*), and that *advice* is a noun meaning principally 'an opinion given or offered as to future action' (*to take advice from an accountant*; *an advice note*), but apparently the two words (or spellings) are sometimes confused.

**2** The words have related meanings in certain contexts, esp. in business language: the verb in the sense 'to inform (officially)' (*please advise us when the goods have been dispatched*), and the noun in the sense 'a formal notice of transaction' (*the firm sent advice that the goods had been dispatched*).

**advisedly. 1** Pronounce as four syllables. See -EDLY.

**2** The only surviving sense (from the four listed in the OED) is 'after careful thought, as a result of deliberation': *most of his sayings must have been almost unintelligible to English country folk—I use the word 'folk' advisedly as suggesting a certain mental solidity*—G. A. Birmingham, 1949.

**adviser, advisor.** The OED makes it plain that both forms occur with equal frequency throughout the English-speaking world. A decade or two ago one's impression was that *adviser* was the more usual spelling in BrE and *advisor* in AmE, but impressions are one thing and hard evidence another. My preference is for *adviser.* Examples: (adviser) *An adviser to*

*Charles V, Fosca seeks to advise the world*—L. Appignanesi, 1988; *the Service would never forgive me a mucky divorce, dear—not its legal adviser, it couldn't*—J. le Carré, 1989; (advisor) *The goose ... shuffled off to see if she could find some advisors*—J. Winterson, 1985; *He was working as an advisor on governmental agricultural schemes*—L. Ingalls, 1986 (US).

**advocate** (verb). **1** In a letter written in 1798 Benjamin Franklin asked Noah Webster to use his authority to 'reprobate' this verb, which he imagined had been recently formed from the noun. In fact the verb had been in use for at least a century and a half (first recorded in 1641, though in the literal sense 'to act as advocate').

**2** Fowler's view (1926) that 'unlike *recommend, propose, urge,* and other verbs, *advocate* is not idiomatically followed by a *that*-clause, but only by an ordinary noun or a verbal noun' has proved to be unsound. All three constructions are found: (noun) *He had been expelled by the National Executive for continuing to advocate a political alliance with Communists*—George Brown, 1971; (verbal noun) *I would advocate the keeping of animals at school*—A. S. Neill, 1915; (*that*-clause) *the UN envoy advocated that sanctions be imposed on South Africa.*

**-ae, -as,** as plurals of noun in *-a*. Most English nouns in *-a* are from Latin (or latinized Greek) nominative feminine singular nouns, which have in Latin the plural ending *-ae*. But some have a different origin: e.g. *indaba* is from Zulu, *sofa* from Arabic, and *swastika* from Sanskrit; *subpoena* is not nominative; *comma* and *drama* are neuter; *addenda, data,* and *stamina* are plural: with all such words a pl. form in *-ae* is impossible.

Of the majority, some retain the *-ae* for all uses (*algae, larvae*); some alternate with *-as* (usu. *antennae* in BrE, but *antennas* = aerials; *formulas* in general work, *formulae* in scientific; *lacunae* or *lacunas; nebulae* or *nebulas*); and some always have *-as* (*areas, ideas, villas*). See LATIN PLURALS.

**æ, œ. 1** Ligatured vowels, which once graced the pages of every well-set book, have substantially fallen out of use. The 39th edn of *Hart's Rules* (1983) ruled that the combinations æ and œ should each

be printed as two letters, (*a*) in the names of botanical and bacteriological families and orders ending in *-aceae* and in the names of zoological families and subfamilies ending in *-idae* and *-inae* respectively; (*b*) in Latin and Greek words, e.g. *Aeneid, Aeschylus, Caesar, Oedipus, Phoenicia*; and also in English words that are derived from Latin and Greek, e.g. *formulae, phoenix.*

Ligatures should still be printed in OE words (*Ælfric, Cædmon*), and in French words (*manœuvre, œillade, œuvre, trompe-l'œil*). Also in Danish, Icelandic, and Norwegian words and in the relevant vowels of the IPA system.

**2** *Spelling.* Since the late 19c. there has been a marked drift away from *ae* and *oe* to simple *e* in words derived from Greek and Latin. The movement has been relatively slow in BrE: *oecology* and *oeconomy* have become *ecology* and *economy*, and *mediaeval* is now mostly printed as *medieval*; but *oe* is retained in e.g. *Oedipus* (or *Œdipus*), *oesophagus*, and the 20c. word *oestrogen*. In AmE the substitution has proceeded apace, resulting in the dominance of e.g. *esophagus* and *estrogen* (but *Oedipus*). Other common examples (showing the unevenness of the process):

| BrE | AmE |
| --- | --- |
| aegis | aegis (also *egis*) |
| aeon | aeon (also *eon*) |
| aesthetics | aesthetics (also *esthetics*) |
| anaemic | anemic (also *anaemic*) |
| anaesthetic | anesthetic |
| archaeology | archaeology (also *archeology*) |
| Caesarean, -ian | Cesarean, -ian |
| diarrhoea | diarrhea |
| ecumenical | ecumenical |
| encyclopaedia (also *-pedia*) | encyclopedia |
| faeces | feces (also *faeces*) |
| gynaecology | gynecology |
| homoeopathy | homeopathy |
| larvae (pl.) | larvae |
| oenology | enology |
| paediatrics | pediatrics |
| paean (song of praise) | paean (also *pean*) |
| palaeography | paleography |
| pedagogy (i.e. not *paed-*) | pedagogy |

See also CERULEAN; FETID; FOETUS.

**aegis, aeon** are the spellings recommended (except in AmE, in which *egis*

and *eon* are frequently used). See Æ, Œ.
For more on *aeon* see TIME.

**aerate.** Formerly pronounced /ˈeɪəreɪt/
with three syllables, but now always
/ˈeəreɪt/ with two. The old spelling with
a diaeresis (*aërate*) has been abandoned.

**aerial.** Formerly pronounced /eɪˈɪərɪəl/
with four syllables and often printed as
*aërial*, but now always /ˈeərɪəl/ with three
syllables, printed without a diaeresis.

**aerie, aery** (nest of bird of prey). Use
*eyrie*.

**aero-, air-. 1** At some point in the first
half of the 20c. the pronunciation of
*aero-* as /ˈeɪərəʊ/ with three syllables was
abandoned and /ˈeərəʊ/ with two used
instead.

**2** Most of the words formed with
the prefix *aero-* have remained stable
(e.g. *aerobiology, aerodynamics, aeroelasti-
city, aerofoil, aeronautics, aerosol,* and *aerospace*). But *aerodrome* has largely yielded
to *airport*, while *aeroplane* remains the
usual term in BrE as against *airplane* in
AmE. Combinations beginning with *air-*
are customary both in BrE and in AmE
for the language of aviation: *airborne,
aircraft, aircrew, airlift, airmail, airman, air
traffic controller,* etc.

**aesthete, aesthetic.** Normally pronounced /ˈiːsθiːt/ and /iːsˈθetɪk/ in RP, but,
with the spellings *esthete, esthetic,* with
initial /es-/ in AmE.

**aet., aetat.** (with full points). Abbreviations of L *aetatis* 'of or at the age of',
used occasionally in English (*Cor pulmon-
ale, aet. 60, Audenshaw, to be admitted*—
Diary of H. Selbourne, 1960–3), esp. in
captions to photographs.

**affect, effect 1** *As verbs.* These verbs are
not synonyms requiring differentiation,
but words of totally different meaning
neither of which can ever be substituted
for the other. *Affect,* apart from other
senses in which it is not liable to be
confused with *effect,* means 'have an influence on, produce an effect on, concern, effect a change in'. *Effect* means
'bring about, cause, produce, result in,
have as result, accomplish'. Examples:
(a) *These measures chiefly affect* (i.e. are
aimed at) *drug-pushers. It does not affect*
(i.e. concern) *me. It may seriously affect* (i.e.

have an adverse effect on) *his health. This
will not affect* (i.e. change, have any bearing on) *his chances of promotion.* (b) *A single
glass of brandy may effect* (i.e. bring about)
*his recovery. This will not effect* (i.e. secure)
*his promotion. The prisoner effected* (i.e.
made good) *his escape.*

*Affect* in the sense 'assume (a character), pretend to have or feel, etc.' (*As he
reached the pick-up point, he should affect to
slow down as if hunting for a car*—J. le Carré,
1989) is a quite separate verb of different
origin.

**2** *As nouns. Affect* survives only as a
technical term in psychology meaning
'an emotion, a feeling, or a desire, esp.
as leading to action' (*COD*), whereas *effect*
has a broad range of senses (result, consequence; impression produced on a
spectator, hearer, etc.) and idiomatic
uses (*to good effect, to take effect, personal
effects,* i.e. property, *sound effects,* etc.).

**3** Examples of erroneous uses: *A
further disadvantage that can* effect *both
print quality and runnability is the shive
content*—A Designer's Guide to Paper & Board,
1989 (read *affect); the judge postponed sentence because he was angered by the victim's
injuries and did not want emotion to* effect
*his decision*—Times, 1989 (read *affect); if
Hogg and Richardsons* [sic] *apprehensions
about the* affect *of yesterdays* [sic] *unemployment numbers are on line ... it is definitely
goodbye Paul Keating*—Austral. Financial
Rev., 1993 (read *effect*).

**affinity.** When the word implies a mutual relationship or attraction, it is normally followed by *between* or *with: the
affinity between Britain and most of her former colonies; Beckett ... stresses that he wrote
the little book on order, not out of any deep
affinity with Proust*—M. Esslin, 1980. If the
feeling, etc., is one-sided, other words
(sympathy, affection, feeling, etc.) are
normally called on and are followed by
*for. Affinity for* (art, Christianity, controversy, politics, etc.), i.e. a liking for such
things, is a more reluctant choice but is
occasionally found (*Shevek felt a kind of
affinity for the man*—U. Le Guin, 1974).
It is more common in scientific work:
*Aluminum, ferromanganese or ferrosilicon is
added to the molten metal ... to provide
material with a higher affinity for oxygen
than the carbon*—E. P. Degarmo, 1984.

**affirmative.** From its use in US signalling and radio communication where, for the sake of clarity, it is used as a quasi-adv. instead of 'yes', *affirmative* has made a cautious entrance in ordinary prose. Thus (military) '*Roger, you say a boat full of bad guys and machine guns. Is this for real? Over.' 'That's affirmative, son.'*—T. Clancy, 1987; and (non-military) '*You awake over there?' 'Affirmative.'*—T. O'Brien, 1976. Cf. NEGATIVE.

**affirmative action.** An American phrase brought into being in the 1930s to signify positive action taken by employers to ensure that minority ethnic groups are not discriminated against during recruitment or employment. It is sometimes regarded by the majority groups as, in some circumstances, giving an unfair advantage to members of the minority groups.

**affix.** A grammatical term that includes prefixes like *post-* in *postscript*, suffixes like *-ward* in *windward*, and the italicized elements in *un-*touch-*able*. The term is also used for infixes like Eliza Doolittle's abso-*blooming*-lutely.

**afflict.** See INFLICT.

**à fond** means 'to the bottom, i.e. thoroughly', whereas *au fond* means 'at bottom, basically'. Contrast *The Comte de la Roche knows one subject* à fond: *Women*—A. Christie, 1928; and *Au fond every man is a hunter, a cave-man*—M. Lutyens, 1933. Use ital. for both expressions.

**afore-.** A combining form used to form somewhat ponderous words like *aforementioned* and *aforesaid* (both of which are frequently encountered in legal documents). It is also used in the neat fixed expression *malice aforethought*.

**a fortiori.** Lit. 'from yet firmer (an argument)', a term used first in logic and then generally, introducing a fact that, if another fact already accepted is true, must also and still more obviously be true, i.e. with yet stronger reason, more conclusively (*It could not have been finished in a week*; a fortiori *not in a day*). Pronounced /eɪ fɔːtrˈɔːraɪ/.

**African-American.** The favoured word in the US at present (1995) for a black person seems to be *African-American*, though *black* (or *Black*) is still widely used. Also used as a compound adj.

**after.** Used in Anglo-Irish with the verb *to be* + a present participle in a manner unknown elsewhere: (*a*) to be in the act of; to be on the point of (doing something): *If it wasn't the turnips it was the pigs were after breaking loose*—P. Kavanagh, 1938; *The cork's after breakin' on me, said Bimbo*—R. Doyle, 1991. (*b*) to have just (done something): *I would have enclosed the thirty-five dollars I owe you, only I'm already after sealing the envelope*—Lore and Language, 1979.

**aftermath.** This agricultural word meaning 'a second or later mowing; the crop of grass which springs up after the first mowing in early summer' (*Winnal Moors on the Itchen, managed by the Hampshire and Isle of Wight Trust for hay and aftermath grazing*—Natural World, 1984–5) developed a figurative meaning in the 17c., 'a state or condition left by a (usually unpleasant) event, or some further occurrence arising from it': *Depression is sometimes an immediate aftermath of completing a piece of work*—A. Storr, 1979; *The aftermath of the wedding seemed to mean different things to different people. Princess Anne confessed … to having 'a slight hangover from a very enjoyable wedding'*—Times, 1981.

**afterward(s)** (adv.). The geographical distribution is fairly clear-cut. *Afterward* is restricted to North America, whereas *afterwards* is the customary form in the UK, Australia, NZ, and South Africa, and an optional by-form in North America. Contrast (*a*) *Afterward, he had a long and satisfying career with the city Welfare Department*—New Yorker, 1987; *Afterward they were enormously and finally sick of each other*—A. Munro, 1987 (Canad.); (*b*) *not long afterwards Menelaus inherited the throne of Sparta*—K. Lines, 1973 (UK); '*But afterwards,' he added, turning solemn again, 'you wash your hands with soap and hot water.'*—M. Richler, 1980 (Canad.); *It's only afterwards … that meaning frees itself and becomes a symbol*—M. du Plessis, 1983 (SAfr.); *it seems to me bad business to pay afterwards for a favour which has already been freely done*—T. Keneally, 1985 (Aust.); *and afterwards he has to drag us backstage to say hello to one of the actresses*—J. McInerny, 1988 (US); *Often afterwards she wondered about the beating of* *those hands on the door*—S. Koea, 1989 (NZ).

**age.** 1 *I haven't seen her in ages*—J. McInerny, 1988. *In ages* is the AmE equivalent of BrE *for ages.*

2 *It all started when he got diphtheria, at age eighteen*—New Yorker, 1991. The BrE equivalent of AmE *at age eighteen* is *at the age of eighteen.*

3 For near synonyms of *age*, see TIME.

**aged.** *Aged 66, the house had aged well,* etc., pronounced /eɪdʒd/; *an aged man,* etc., pronounced /'eɪdʒɪd/.

**ageing.** For preference use this spelling, not *aging*, though both spellings are in common use, with *aging* esp. common in AmE.

**agenda.** 1 The essential plurality of this word ( = things to be done) has become virtually extinct. Its dominant sense now is 'a list of items of business to be considered at a meeting, etc.' (*Mrs Walton said she hadn't a spare moment. She had a busy agenda*—B. Bainbridge, 1975; *There is a feeling that we have got to draw up a new agenda now*—Marxism Today, 1986) and it has produced a plural in *-s* (*our students' ideals and agendas*—Dædalus, 1988). The former singular form *agendum* (i.e. just one item to be discussed) is used principally in the out-of-term meetings of Oxbridge governing bodies and other such admirable institutions: *The Estates Bursar was called upon to introduce Agenda Item 3. (They mean Agendum 3, thought Jake.)*—K. Amis, 1978.

2 The phrase *hidden agenda* (first recorded 1971) has somewhat sinister connotations. It occurs most frequently in descriptions of the undisclosed motives of an opposing group or groups: *There's family politics, sure, but ... when we get into disagreements there's no hidden agenda*—Cambridge (Mass.) Chronicle, 1986: *Sex was the hidden agenda at these discussions*—M. Atwood, 1987.

**aggrandize.** At some point between 1917 (when Daniel Jones's *English Pronouncing Dictionary* was first published) and 1963 (12th edn), the main stress moved to the second syllable (doubtless under the influence of *aggrandizement* /ə'grændɪzmənt/) from its traditional placing on the first.

**aggravate.** A half a century ago, schoolmasters were insisting that *aggravate* could be properly used only in the sense 'to increase the gravity of, to make more grievous or burdensome' (applied to crime, grief, illness, misery, terror, etc.), a use first recorded in the late 16c. The meaning 'to exasperate, incense, make (someone) angry', despite the fact that it had existed in good sources since the early 17c., 'should be left to the uneducated. It is for the most part a feminine or childish colloquialism, but intrudes occasionally into the newspapers' (Fowler, 1926). The two senses now stand side by side in a relatively unthreatening manner. Thus: (older sense) *England is very uncomfortable & everything is being done by the bureaucrats to aggravate the discomfort*—E. Waugh, 1941; *But these misfortunes were greatly aggravated by the policies of the English Government*—W. S. Churchill, 1958; *the monetarists were able to claim that incomes policy only aggravated the malady*—London Rev. Bks, 1981; (later sense) *Don't aggravate yourself. I'll be no trouble*—A. Lurie, 1965; *Do not aggravate them, be quiet, smile nicely*—P. Carey, 1982; *It was aggravating that he had to do so many little jobs himself*—M. Wesley, 1983; *Jane Fairfax aggravates her in all sorts of ways*—T. Tanner, 1986.

**aggravation.** The 20c. has witnessed an increase in the harassment of appointed or elected officials and of other people in positions of authority, e.g. schoolteachers. The word most commonly applied to such aggressive behaviour is *aggravation* (first recorded in this sense in 1939); its colloquial equivalent is *aggro* (1969). *Aggravation*, though, still retains its older sense 'a making graver or more heinous'. Representative examples: *The second was the aggravation (i.e making worse) of the difficulties of the poor countries*—Encounter, 1981; *in retreat from such dangers and aggravations (i.e. acts of aggression) as seemed bound to interfere with normal life in Hove*—P. Lively, 1981; *A certain amount of agricultural aggro is a regular part of the French way of public life*—Times, 1984.

**aggressive.** The commercial world of the 20c. has added the senses 'self-assertive; energetic, enterprising' to the word when it is applied to the techniques of marketing and salesmanship or to the people marketing or selling goods or services: *We are seeking an aggressive senior level manager with excellent business acumen*— ad. in *Times*, 1985; *Sales People. For an aggressive home improvement company. Experience not necessary as we prefer to train*—ad. in *Cambridge* (Mass.) *Chron.*, 1986.

**aggressor,** not *aggresser.*

**agitator,** not *agitater.*

**ago.** If *ago* is used, and the event to be dated is given by a clause, it must normally be one beginning with *that* or *when* and not *since*. The right forms are: *He died 20 years ago* (no clause); *it is 20 years since he died* (not *ago*); *it was 20 years ago that he died*; *That was years ago, when he was a young man.* The tautologous construction *it was 20 years ago since he died* should not be used.

**agree.** Alongside the commonest uses of the verb with a prepositional complement (*agree on, about, to, upon, with*), the *OED* records a transitive use ('to arrange, concert, or settle (a thing in which various interests are concerned)') which has played a part in the language since the 16c. (first recorded in 1523). Older examples, which seem to have caused no concern, can be consulted in the *OED*. In the 20c. a not dissimilar transitive use has become common but somewhat controversial in Britain but remains rare or non-existent in America. Examples: (an accountant's letter) *the tax inspector has agreed your allowances*; *The Russians have agreed a wide list of categories*— *Bookseller*, 1959; *the difficulty of agreeing a definition of mysticism*—*Listener*, 1963; *There is nothing unusual about agreeing the end but disputing the means*—*Times*, 1986.

**agreement** (in grammar).

1 Definition.
2 Normal concord.
3 Nouns joined by *and*.
4 Nouns joined by quasi-coordinators.
5 Collective nouns.
6 Indefinite pronouns.
7 Subject–complement agreement.
8 Attraction.
9 *one of those who*.
10 Titles.
11 Subjects.
12 Names of illnesses.

1 *Definition.* Also called (*grammatical*)

*concord*, it signifies 'formal agreement between words as parts of speech, expressing the relation of fact between things and their attributes or predicates' (*OED*). This formal agreement, as the *OED* says, 'consists in the words concerned being put in the same case, number, gender, and person, as far as the inflexional structure of the language provides for this, or as other considerations (in respect to gender and number) do not forbid it'.

2 *Normal concord.* In present-day English, agreement in number between subject and verb is overwhelmingly normal: The climate was *not brilliant* (in standard English one just cannot say *The climate were not brilliant*). This is true even if another phrase containing a noun in the plural intervenes: The wooden platform *between the pillars* was *green and rotten*; a line *of caravans* lumbers *northward*. If the subject is plural, the verb must also be plural: The supplies were *all laid out*; *the doorknobs gleam.*

Similarly, pronouns must agree in gender and number with their antecedents: he had never been close enough to *a girl* to consider making *her* his wife; *she adored her* little Renault ... *She drove it fast*; *Madeleine took charge of* the guests, *leading* them *through the hall.*

3 *Nouns joined by 'and'.* Two nouns joined by *and* normally form a plural subject and require a plural verb: Unreason and inevitability *go hand in hand*; Historians and dramatists approach *the truth by different routes.* Except that a composite subject may be thought of as a single theme and then is followed by a singular verb: Tarring and feathering was *too good for Meakin as far as I was concerned*; *A certain* cynicism and resignation comes *along with the poverty of Italian comedy*; *the* usefulness and credibility *of such an arms control agreement* hinges *on the reliability of seismic technology.* (Such constructions existed in OE and ME: examples are listed in Mitchell 1985 and Visser 1963.) Naturally there are borderline cases in which judgements will differ: The Service, and the country, owes *him much more than may be generally recognized* (?remove the commas, *owe*); The celebration and mirroring of this union ... *is the marriage of Apollonian lyric and*

*Dionysian music* (influenced by the singular complement); *The hurt and disbelief of patients' friends and families* is *already quite real* (*hurt and disbelief* not 'a single theme'); *The extent and severity of drug use in the United States* has *been a shock to Dr* ——, *the medical director of Lifeline* (*extent and severity* not 'a single theme').

**4** *Nouns joined by quasi-coordinators.* Nouns joined by other linking words or quasi-coordinators (e.g. *accompanied by, as well as, not to mention, together with,* etc.) are followed by a singular verb if the first noun or noun phrase is singular: *A very profitable company such as British Telecom, along with many other companies in the UK,* is *not prepared to pay a reasonable amount*; *Daddy had on the hairy tweed jacket with leather elbow patches which, together with his pipe,* was *his trade mark.* So too when the linking words are *or* or *nor*: *Every run-out or lbw* is *similarly analysed*; *neither our mourning nor our rejoicing* is *as the world's* is (T. S. Eliot, 1935).

**5** *Collective nouns.* The idea of notional agreement affects this group and also those dealt with in **6** and **7**. In BrE, collective nouns may be correctly followed by either a singular or a plural verb: (sing.) *Each succeeding generation of gallery visitors* finds *it easier to recognize Cubist subject-matter*; *A group of four young men, in denim overalls,* was *standing close to him*; *the Oxford University Press* publishes *many scholarly monographs each year*; (pl.) *The jury* retired *at five minutes past five o'clock to consider their verdict*; *Let us hope that the Ministry of Defence* are *on our side this time.*

In AmE it is customary for a singular verb to be used with collective nouns: *a fleet of helicopters* was *flying low* (*New Yorker*, 1986); *the government routinely* imposes *differential taxes on hotels, bars ... and the like* (*Bull. Amer. Acad. Arts & Sci.*, 1987); *the KGB* uses *blackmail and intimidation* (*Chr. Sci. Monitor*, 1987). But some collective nouns in AmE, esp. those of the type 'a + noun + of + plural noun', optionally govern a plural verb: *a couple of them* rattle *on amusingly* (*New Yorker*, 1986); *a handful of bathers* were *bobbing about in the waves* (P. Roth, 1987); *you think of the States as a country where the majority of all the shareholdings* are *in the hands of women* (*Dædalus*, 1987); *a rich and detailed picture of a world in which a*

multitude *of elements* were *intertwined* (*NY. Rev. Bks*, 1989).

**6** *Indefinite pronouns.* In many constructions, these (*each, either, every, everybody, neither, nobody, none, no one,* etc.) govern a singular verb, but sometimes contextual considerations lead to the use of a plural verb: (sing.) *Neither of these figures* illuminates *the case against Trident* (D. Steel, 1985); *none of her features* is *particularly striking* (D. Lodge, 1962); (pl.) *Neither the Government nor the tribunal, surely,* want *to bear responsibility for ...* (*Daily Tel.*, 1987); *None of our fundamental problems* have *been solved* (*London Rev. Bks*, 1987).

Sometimes there is a clash of agreements within a given sentence: *Everyone* was *in their shirt-sleeves* (F. Tuohy, 1984); *No one in their senses* wants *to create instability* (D. Healey, 1985); *I really resent it when I call somebody who's not home and they don't have an answering machine* (*Chicago Tribune*, 1988).

Constructions in which an indefinite pronoun is followed by a plural pronoun often result from a desire to avoid using the combination 'his or her' (or equivalent) or from a fear of being thought of as a male chauvinist: *Why would anyone plan their own funeral?*; *we don't want anyone to hurt themselves*; *It must have been someone who wanted to clear their conscience*; *Nobody wants to return to the car park and find that their car has been clamped.*

**7** *Subject–complement agreement.* When a subject and a complement of different number are separated by the verb *to be*, the verb should agree with the number of the subject: (sing.) *the only traffic* is *ox-carts and bicycles*; *Gustave* is *other animals as well*; *The problem* is *the windows*; *The result* was *these awkward declarations*; *The view it obscured* was *pipes, fire escapes, a sooty-walled well*; *Our main problem* is *the older pesticides*; (pl.) *These socials* were *a big deal to her*; *The house and garden* were *a powerful cauldron of heat and light*; *the March events in Poland* were *a natural stage in the evolution of communism.*

In certain limited circumstances the verb correctly agrees with an object-complement: *Forty droschen's* [i.e. is] *the price*; *More nurses* is *the next item on the agenda* (i.e. the subject of *more nurses*). But in the following examples from a major daily newspaper *is* would have been better than *are*: *What is remembered are the warmth of his personality and his capacity for giving and receiving friendship*; *Another legacy are injuries which could keep both* [players] *out of the big match.* Cf. BE 1.

**8** *Attraction.* Absence of agreement often arises from the phenomenon known as 'attraction' or 'proximity', i.e. the verb is forced out of agreement with the true subject by the proximity of an intervening element in a different number from that of the subject. Older literature abounds with examples: *The outside of her garments were of lawn* (Marlowe); *The posture of your blows are yet unknown* (Shakespeare). Modern examples, all of them reprehensible: *Copyright of Vivienne's papers are in the keeping of the Haigh-Wood family* (*Literary Review*, 1985); (headline) *Cost of hospital AIDS tests vary widely in state, survey finds* (*Chicago Tribune*, 1988); *the spread of nuclear weapons and technology are likely to make the true picture very different* (*Dædalus*, 1991); *At least one in two churches are likely to be burgled next year* (*Times*, 1992).

**9** *one of those who.* When the emphasis is placed on the immediate antecedent (as it normally is), the construction requires a plural verb in the following clause: *Lily had been ... one of those numerous people who are simply famous for being famous* (I. Murdoch, 1987); *That's one of those propositions that become harder to sustain the further they're explored* (K. Amis, 1988). Exceptions occur when the writer or speaker presumably regards *one* as governing the verb in the subordinate clause: *Perhaps you were one of those fellows who sees tricks everywhere* (P. Carey, 1985); *I am one of those people who wants others to do what I think they should* (J. Bakewell, 1988). See also ONE 4.

**10** *Titles.* Titles of books, films, etc., and names of firms that are plural in form are followed by a singular verb: *Great Expectations is an account of development of identity*; *Star Wars has diverted some six billion dollars from the federal treasury*; *Television's 'One Man and His Dog' is with us again*; *Harrods has become as much a logo as a shop.*

**11** *Subjects.* The names of subjects that are plural in form govern singular verbs:

*the physics involved* is *sophisticated*; *international relations* has *become an arena in which defeat tends to outweigh success*; *by a process whose mathematics* has *always eluded me*; *politics* is *the art of the possible.* But a plural verb is required when such words are used generally and not as the name of a subject: *the acoustics of the room* are *dreadful*; *your politics* (i.e. views on political matters or political parties) are *different from mine.* See -ICS 2.

**12** *Names of illnesses.* Those that are plural in form (as well as those that are singular) govern a singular verb: *Aids* has *merely hastened the process*; *measles* is *normally a childhood disease*; *mumps often* occurs *in adults.*

Clearly, strict agreement is one of the most desirable aims in well-constructed English sentences, but there is a brittleness at the edge of the concept of grammatical agreement which is an unmistakable component of the language, in its guarded as well as in its unguarded moments.

See also COLLECTIVE NOUN; EITHER 4; MANY A; NEITHER 4; THERE IS; WHEREABOUTS; and the entries for many personal and indefinite pronouns.

**agricultur(al)ist.** See -IST 2.

**ahold.** Apart from an isolated use in Shakespeare's *The Tempest* (1610), not recorded as an adverb before 1872; now used colloquially (alternating with *aholt*), esp. in AmE, with verbs like *catch*, *get*, *lay*, *take*. (*A kid gets ahold of it* [*sc.* a firearm], *you have an immediate situation*—D. DeLillo, 1985; '*Come on, grab ahold*,' *he said, slapping hard on the side of the pumpkin*—*New Yorker*, 1988). But the more usual idioms are *to catch*, *get*, etc., *hold* (of something), with no *a-* prefixed.

**-aholic.** A suffix (also *-(o)holic*) abstracted from *alcoholic* and used to form nouns suggesting that a person is addicted to the object, activity, etc., given in the first element. First recorded in 1965, this suffix has swiftly moved into general use, esp. in the word *workaholic*, but also in such semi-jocular formations as *bookaholic*, *chocoholic*, *computerholic*, *golfaholic*, *newsaholic*, *spendaholic.*

**aid.** The noun was first recorded in the sense 'material (esp. economic) help given by one country to another' in 1940

(Christian aid, foreign aid, Marshall aid, etc.), but became even more widely prevalent in 1984 and later years as the second element in the names of events, etc., organized to raise money for charitable causes (Band Aid, Fashion Aid, Live Aid, etc.). Such formations were triggered by Band Aid, the name of a rock music group formed by Bob Geldof in 1984 to raise money for the relief of famine in Ethiopia.

**aide.** A person who helps is occasionally called an aid. An aide, by contrast, is (a) short for an aide-de-camp, a high-ranking officer in the armed services (e.g. Brigadier Monson summoned his five closest aides for a working lunch—N. Barber, 1984); (b) a person employed as an assistant or an ancillary worker, esp. in a hospital or as a visitor to the home of an ill or elderly person (e.g. There is domiciliary care ... offered by 200 home helps, 18 aides and their organisers—Daily Tel., 1984; Just before he died a nurse's aide brought his dinner tray into the room—E. L. Doctorow, 1989).

**aim.** Until about the end of the 19c., the verb aim in the figurative senses 'have as an object, endeavour' was normally followed by at + gerund (they aimed at being on the successful side—Froude, 1879) or, much less commonly, by a to-infinitive (aiming to arrive about the breakfast hour—Hardy, 1891), except that the construction with a to-infinitive was markedly favoured in AmE. A recent (1986) paper by Åge Lind demonstrated that the type aim + to-infinitive (I don't know if you're dumb or just stupid, but by God I aim to find out—T. O'Brien, 1978; Much imagination has gone into the project, which aims to attract half a million visitors a year—Times, 1983) has largely replaced aim at + gerund (His prophecy aimed at harming me—Neh. (NEB) 6: 12; The directive aims at ensuring open passage through borders—Financial Times, 1984), except in passive constructions (The Alliance's strategy ... will clearly be aimed at persuading Labour supporters that a party led by Mr Foot has no chance of ousting Mrs Thatcher—Observer, 1983). The preference for the aim + to-infinitive construction is probably the result of analogy with verbs used synonymously, e.g. intend, mean, plan, but American influence may have contributed.

**ain't.** 'Do you hear? Don't say "ain't" or "dang" or "son of a buck" ... You're not a pair of hicks!' scolds a mother to her children in a New Yorker short story. ' "Fritz That's It" ain't' was the headline to a news item about the closing of a salad bar with the wonderful name of 'Fritz That's It'. 'Some people are poor, some are unhappy, some are rich and selfish, ... there is famine and war—ain't it a rotten shame,' said the broadcaster Brian Walden in the Sunday Times.

Ain't is not much of a word to get excited about, one would have thought. But excitement and it are often directly related. In 1942 Eric Partridge remarked that using ain't for isn't is 'an error so illiterate that I blush to record it'. In 1961 Webster's Third New International Dictionary listed the word with only mild admonishment ('though disapproved by many and more common in less educated speech, used orally in most parts of the US by many cultivated speakers esp. in the phrase ain't I'), and found itself virtually placed in the stocks for this entry alone. Dictionaries of current English tend to hold the word, as it were, in a pair of tweezers. The label 'colloquial' is applied to it. None admits it to the sacred unqualified ranks of standard English.

Ain't is an undisputed element in Cockney speech, whether in Dickens ('You seem to have a good sister.' 'She ain't half bad.') or in the deliciously outrageous rantings of the TV actor Alf Garnett. Trendy articles on popular music tend to associate the word with American blacks; a Rap singer called LLCoolJ ( – Ladies Love Cool James) in a 1988 issue of the International Musician is reported as saying My music sounds kinda real to me—it ain't no fantasy. The Australian lawyer Malcolm Turnbull (of Spycatcher fame) in an article on the Australian bicentennial, said that The Queen ... is an Englishwoman, the greatest this century. But an Aussie she ain't.

How did the word come about? And why do some people use it naturally, while others regard its use as a sign of irretrievable vulgarity or ignorance?

Its spelling is odd. In constructions of the type am, is, are, have + not, the word not in unabbreviated form is written separately whether or not the first element is contracted: am not/ 'm not, is not/ 's not, are not/ 're not, has not/ 's not, have not/ 've

*not*. When the negative itself is contracted, it combines with the full form of
the preceding element: *isn't, aren't,
wasn't, weren't, hasn't, haven't* (not *'sn't,
'ren't*, etc.). The form *ain't* is the odd man
out. It is not a reduced form of any logical
ancestor. Properly speaking, the last element in a tag-question of the type 'I am
here, am I not?', if reduced, should be
*amn't I*, as it is in many modes of speech
in Scotland and Ireland. But standard
English has opted instead for the
puzzling *aren't I*, a stiffnecked Sassenach
use if ever there was one from the point
of view of the Scots.

So what are we dealing with here? The
word *ain't* has been recorded in the popular speech of London and elsewhere
since 1778. It has made its way into a host
of catchphrases and songs: 'ain't it grand
to be blooming well dead?'; 'ain't love
grand?'; 'there ain't no such animal';
'ain't that something?'; 'it ain't necessarily so'; 'if it ain't broke don't fix it'; 'Is
you is or is you ain't my baby?'

It flourishes in Black Vernacular English in the US, where it has also acquired
the meanings 'do not, does not, did not'
(*Ain't you know Felo ben stay'n wid me?*). It is
used by many Americans, white as well
as black, in a manner that they might describe as 'cute': *Phylicia Ayers-Allen who
plays Bill Cosby's wife on the Cosby TV show—is
she or ain't she married in real life?* Alexander Haig once said of George Bush:
*Anybody who has to spend all his time demon-
strating his manhood has somehow got to
know he ain't got it.*

Yale alumni presumably consider
themselves to be 'cultivated speakers'. A
1988 issue of the *Yale Alumni Magazine* recorded one Wilbur Woodland as saying
*Still working the Cape Cod and Florida cycle.
And it ain't too bad.*

*Ain't* is a necessary part of the vocabulary of comic strips. 'Hägar the Horrible',
for example, shows a Viking-type warrior
in bed and reluctant to get up. The bal-
looned words read: *It ain't a question of
'when', but can I get up?!*

For over 200 years the bar sinister word
*ain't* has been begging for admission to
standard English. In tag-questions it has
been thwarted by the equally bar sinister
form *aren't*. In other uses it leads a
shadowy existence in the language of
various underclasses. It stands, as it were,
at the door, out on the pavement, not yet
part of any standard paradigm in the
drawing-room, except of course in
catchphrases and in other contexts of
referential humour.

**air** (verb). In the sense 'to broadcast',
this modern use (first recorded 1943) is
frequent in AmE (*it aired a heartwarming
TV commercial on the importance of savings
institutions—Wall St. Jrnl*, 1989) but is not
often encountered in BrE (*The obligation
to keep records of all programmes aired—
Economist*, 1981).

**aisle.** From L *āla* 'wing', *aisle* first meant
'a wing or lateral division of a church,
the part on either side of the nave, usually divided from the latter by a row of
pillars'. In the 18c., by confusion with
*alley*, it developed its dominant sense 'a
passage in a church between the rows
of pews or seats'. By a further shift of
sense that began in the same century,
originally in English dialects and esp. in
AmE, it came to mean, as it still does, 'a
passage-way in a building (esp. a theatre,
cinema, or other place of public entertainment), a railway carriage, etc.' From
this third sense came the colloquial
phrases *to have, send* (people) *rolling in the
aisles* (first recorded 1940).

**ait** /eɪt/. The recommended spelling of
a term (also *eyot*) for an islet, esp. one
on a river and specifically on the R.
Thames. It is derived from OE *iggaþ*, from
*īeg* 'island' + *-aþ*, which may have been
a diminutive suffix.

**aitch. 1** The letter H was written as *ache*
by Shakespeare and his contemporaries,
and answers to ME and OF *ache*. The
spelling *aitch* is not recorded before 1887.

**2** English /h/ occurs only as an aspirated pre-vocalic fricative at the beginning of a syllable. The dropping of the
sound (aitch-dropping), and its converse,
the false insertion of initial /h/ ('arm' for
*harm*; 'hill' for *ill*), were already noted by
John Walker in his *Pronouncing Dictionary*
(1791) as characterizing many types of
popular regional speech. As the *OED* says
s. v. H, 'In recent times the correct treatment of initial *h* in speech has come to
be regarded as a kind of shibboleth of
social position.' It is perhaps worth
pointing out that loss of /h/ is also a
feature of RP in unaccented, non-initial
contexts in connected speech, e.g. *I*

pushed him back /aɪ 'pʊʃt ɪm 'bæk/, and I could have hated her /aɪ kəd əv 'heɪtɪd ə/; also at the beginning of a syllable after certain consonant groups, as in exhaust, exhortation, and in such suffixes as -ham in Chatham, Clapham, Durham, and -herd in shepherd.

**-al.** 1 As an adjectival suffix it is normally found in words of L or Gk origin: oral (L ōrālis, from ōr-, ōs 'mouth'); hebdomadal (late L hebdomadālis, from Gk ἑβδομάς (-αδ-), from ἑπτά 'seven') 'weekly'. Words of English origin (house, path, etc.) have never developed adjectives in -al. In L itself -ālis was added to existing adjectives, e.g. annuālis (from annuus), giving English annual. Fowler (1926) regarded coastal as a barbarism (properly costal, he thought, from L costa), but his hostility to the word has not been taken up by modern writers. See TIDAL.

On adjectives in -ic/-ical (e.g. comic/comical, geographic/geographical, historic/historical), see -IC(AL) and the individual words.

2 As a suffix to nouns. Fowler had no objection to older nouns ending in -al that had passed into common use, and that, for the most part, were not competing with virtually synonymous words ending in -ation, -ion, -ment, etc., e.g. acquittal (first recorded 1430), arrival (1384), proposal (1683), refusal (1474), retrieval (ante 1643). He expressed strong disapproval, however, of an upsurge of interest during the Victorian period (among historians and novelists) in -al nouns that did compete with synonymous words of different formation. Thus, he objected to accusal (used instead of accusation), appraisal (for appraisement), bequeathal (for bequest), refutal (for refutation), retiral (for retirement), etc. 'When there is need', he said, 'on an isolated occasion for a verbal noun that shall have a different shade of meaning ... or that shall serve when none already exists ... it is better to make shift with the gerund (the accusing, etc.).' His judgement seems to have been sound. Such words are rarely encountered now, except that appraisal (a markedly literary word in the 19c.) and retiral (in Scotland) are now everyday words in the language of personnel departments (e.g. promotion depends in large measure on the results of annual appraisals; to fill a vacancy due to retiral). See also OVER-THROWAL. A recent literary example of a

new formation in -al (Lear's addressal of Cordelia in the first scene—English, 1988) perhaps indicates that the Victorian habit of creating such words is not entirely extinct.

**à la.** Adopted from French ( = à la mode), it is used in English, frequently printed in italics, without regard to gender, as a compound preposition before nouns of French origin (à la carte), English nouns (à la country of origin), names (giant landscape photocollages à la David Hockney), and place-names, street-names, etc. (a regatta à la Henley; the BBC should give serious consideration to an autumn shuffle à la 10 Downing Street—C. Freud, 1989). It is especially common in the language of cookery for the names of particular dishes (à la broche, à la meunière, etc.), mostly adopted as phrases direct from French. By contrast, the corresponding masculine preposition au is not free-forming in English. It exists only in a few fixed phrases (au courant, au naturel, au pair, au pied de la lettre, etc.) taken over intact from French.

**alarm, alarum.** 'From the earliest period [sc. the 14c.] there was a variant alarum due to rolling the r in prolonging the final syllable of the call' (OED). By the end of the 19c., alarum was restricted to an alarm signal, e.g. the peal or chime of a warning bell or clock, or the mechanism producing it. These senses are now virtually obsolete, and the word survives mainly in the old stage direction alarums and excursions to describe off-stage skirmishings, sudden divagations, etc. Examples: Flemish lantern clock with striking-work and alarum—Antiquarian Horology, 1984 (advt); I now respond less briskly than I should to the alarums and excursions of everyday events—G. Priestland, 1983.

**albeit.** One of the most persistent archaic-sounding words in the language. This worn-down form of the conjunctive phrase all be it (that) ( = let it entirely be that) has been in continuous use as a conjunction meaning 'even though it be that, even though' (Their voices, too, albeit the accent was provincial, were soft and musical, 1878) since the time of Chaucer, and in quasi-adverbial use (When a certain (albeit uncertain) morrow is in view—Thackeray, 1847) since the late 18c. Some more recent examples: In Iatmul a man

*is a master in his own house, albeit he has
to fight for it against a wife practically as
strong as he is*—M. Mead, 1949; *It is an
unwelcome, albeit necessary, restraint*—A.
Storr, 1972; *'Jesus!' they said in Italian albeit
in a conversational manner*—H. Secombe,
1981; *A great line of poetry, albeit by a
mendacious fascist, will outlast ... the most
sanctified of good deeds*—*TLS*, 1988.

**albino.** Pl. *albinos*. See -O(E)S 6.

**Albion.** Partly because of its late appearance as an entry in the *OED*, speculation about the origin of the word *Albion*
and of the phrase *perfidious Albion* was
once rife. *Albion* was known already to
the OE translator of Bede's *Eccles. Hist.*,
and much earlier still to Pliny (in Latin)
and to Ptolemy (in Greek). Etymologically, it answers ultimately to a reconstructed IE form \*albho- (cf. L *albus* white),
the allusion being to the white cliffs of
Dover. The French phrase *la perfide Albion*
is believed to have been first used by the
Marquis de Ximenès (1726–1817).

**alcaics.** In prosody, a stanza written
after the manner of Alcaeus, a lyric poet
of Mytilene in Lesbos (*c*600 BC), consisting
of:

˘|–˘|–˘|–˘˘|–˘|–

twice repeated, followed by

˘|–˘|–˘|–˘˘|–˘˘
–˘˘|–˘˘|–˘|–˘˘

Horace used a modified alcaic stanza in
many of his odes:

*O matre pulchra filia pulchrior,
Quem criminosis cunque voles modum
Pones iambis, sive flamona
Sive mari libet Hadriano.*

(*Carm.* 1. 16)

The Horatian four-line form was imitated by Tennyson:

*O mighty-mouth'd inventor of harmonies,
O skill'd to sing of Time or Eternity,
God-gifted organ-voice of England,
Milton, a name to resound for ages.*

Both the Latin (four-line) and Greek (five-
line) types are illustrated in:

*Unblemished fresh life leafing with opened buds
The future's misty limits of thought, whose lids
Confuse our nights but bring to touching
Soft-scented, newly-born skin of dawn light;*

*Now eastern blue streaks stealthily press the
  grass
Imprinting morning after long months of rain,*

*Warm shadows darken into protecting shade
That heightens promise, deepens dreaming,
Glistening like lavender's folded leaf-drops,*

(F. Warner, 1989)

**alexandrine.** In prosody, a line (usu.
iambic) of twelve syllables (the Fr. *vers
héroïque*), used in English, for example,
as the last line of a Spenserian stanza
(*All things decay in time and to their end do
draw*) or as a variant in a poem of heroic
couplets, rarely in a whole work. In Fr.
the alexandrine is found, for example,
in Alexandre de Bernay's dodecasyllabic
version (late 12c.) of the great cyclic *Roman d'Alexandre* (e.g. *D'Alixandre vous vœil
l'istorie rafreschir*). The most famous English example is Alexander Pope's couplet,
*A needless Alexandrine ends the song, That,
like a wounded snake, drags its slow length
along.*

**alga.** Pl. *algae*, pronounced /ˈældʒiː/.

**alibi.** **1** Properly 'a plea that when an
alleged act took place one was elsewhere'
(from L *alibī* 'elsewhere, in another
place', old locative case of *alius* 'another'): (examples from fiction) *Just because the suspects had arranged a false alibi
to account for their whereabouts ... they were
not necessarily murderers*—T. Capote, 1966;
*Since you think I murdered him, I had better
produce my alibi*—S. Brett, 1979.

**2** Colloquially, and, it would seem,
inevitably, it has come to mean 'an excuse, a pretext; a plea of innocence' (*Don't
offer alibis for losing* [a game of tennis]—
W. T. Tilden, 1922; *Pip's devotion gave her
... a complete alibi in all charges of frustration and virginity*—W. Holtby, 1936; *The
power-loom provided both the State and the
employers with a cast iron alibi*—E. P. Thompson, 1980); and also 'a person providing
an excuse' (*Tom and Maureen are my
alibis*—C. Hare, 1949). The colloquial use
is first recorded in American sports and
then in detective fiction. The damage
done to the original legal sense now
seems to be irreversible.

**3** By the usual process known as CON-
VERSION, the now dominant sense has
generated a verb meaning 'to clear by
an excuse; to provide an alibi': *Aunt Primrose ... hadn't opportunity. She's alibi-ed by
Mrs Fitch*—J. Cannan, 1958; *There's got to
be somebody to alibi us*—L. Duncan, 1978.

**alien.** From the 14c. to the 19c. inclusive, in the sense 'of a nature or character different *from*', *alien* was construed with *from* (e.g. *This uncouth style, so alien from genuine English*—Henry Reed, 1855). As the *OED* says, this sense gradually passed imperceptibly into 'of a nature repugnant, adverse or opposed *to*'. The crossover came at some point near the turn of the century, and the construction with *to* is now routine (*Thinking, and certainly brooding, were quite alien to his character*—J. C. Oates, 1980; *The implied snobbery of the remark was quite alien to the whole way in which she had been brought up*—A. N. Wilson, 1982). The older construction still occurs occasionally: *a reflection upon how far man has come to feel himself alien from the animal kingdom of which he is a member*—A. Storr, 1968.

**alignment.** Fowler argued for this spelling in 1926. *OED* 2 says that 'The Eng. form *alinement* is preferable to *alignment*, a bad spelling of the Fr.', but lists the entry under *alignment*. The word is now universally spelt *alignment*.

**-(al)ist.** For such alternative forms as *educationist* and *educationalist* see -IST 2.

**alkali.** Final syllable now universally pronounced /-aɪ/ (formerly often /-ɪ/). The more usual plural form is *alkalis* but the by-form *alkalies* is found in many standard textbooks.

**all.** 1 *all of*. Except before certain personal pronouns (*all of it, all of us, all of its own*) and in certain idiomatic uses (e.g. *we had walked all of ten miles* = as many as; *all of a sudden*), *of* can normally be dispensed with in nominal phrases: e.g. *all those years ago; all the time; my father lights all his lamps; we sold all our stocks and shares; almost all his former colleagues*. The construction with *of* is comparatively modern (first recorded *c*.1800), and is probably due to form-association with *none of, some of, little of, much of, few of, many of* (*OED*). From the following set of examples it will be seen that *all of* is often used as a means of separating *all* from a proximate pronoun. It is also used in partitive *of*-phrases. Examples: *All of those activities whose very virtue is that they enable you to think* [etc.]—H. Jacobson, 1983; *I'd like all of that piece of meat*—CGEL, 1985; *He will have to be all of these things*—A.

Brookner, 1986; *'It's one of my limitations.' 'I haven't begun to list all of mine.'*—P. Roth, 1986; *all of the new company's profits had been used to salary him*—B. Ripley, 1987; *At each stop, all of us visitors were greeted by a hail of celebratory statistics*—New Yorker, 1989.

**2** Since the 1960s *all's* (or *alls*) has been noted in informal AmE in the sense 'all that' (it is probably a shortened form of *all as*): e.g. *I get calls for people who want eel, snake or alligator* [skin] , *but all's I've got is pigskin*—Chicago Sun-Times, 1987; *Actually alls we did was smear some guts and shit from a dead rat on his eyes*—E. L. Doctorow, 1989; *So all's I need, pretty much, is a tie*—New Yorker, 1990.

**3** When *all* is the subject of the verb *to be* followed by a plural complement, the linking verb is expressed in the singular: *All I saw was fields*—N. Williams, 1985; *In some sense, all we have is the scores—incomplete and corrupted as they often are*—New Yorker, 1989.

**4** See also ALL RIGHT; ALL THAT; ALL THE; ALL TOGETHER; ALL TOLD; ALREADY; BUT 9c. For *is all*, see IS 8. For *at all*, see AT 2.

**all-around.** An optional AmE variant (*everything on this all-around Italian menu is good; the best all-around American city school; a good all-around player*) of *all-round* adj., the latter being the only form used in BrE.

**allay.** For inflexions, see VERBS IN -IE, etc. 1.

**alleged.** Routinely applied to actions, events, and things that have been asserted but not proved (*the alleged medical breakthrough, the alleged crime, the alleged illness that prevented him from coming to the wedding*) or that need defending (*her alleged innocence*). The related adverb *allegedly* is pronounced with four syllables: see -EDLY.

**allegory.** 'A narrative of which the true meaning is to be got by translating its persons and events into others that they are understood to symbolize' (Fowler, 1926). 'A figurative narrative or description, conveying a veiled moral meaning, an extended metaphor' (*OCELit.*, 1985). The form flourished in medieval literature to the extent that 'every kind of serious realism was in danger of being choked to death by the vines of allegory' (E. Auerbach, *Mimesis*, 1953, ch. 10). Later allegorical works of signal importance included Spenser's *Faerie Queene* (1590–6), Bunyan's *Pilgrim's Progress* (1678–84), and Dryden's *Absalom and Achitophel* (1681). Allegorical elements are present in much modern writing, e.g. Virginia Woolf's *Between the Acts* (1941), which by means of a village pageant presents 'a communal image of rural England, past and present', but the noble tradition of poetic allegory exemplified by works like the *Faerie Queene* seems to be in abeyance.

**allegro.** Pl. *allegros*. See -o(E)s 6.

**alleluia.** L form (from Greek), the Septuagint representation of Heb. *hallēlū-yāh*, i.e. 'praise ye Jah or Jehovah'. In English, an exclamation meaning 'Praise the Lord', which occurs in many psalms and anthems. In AV spelt *Alleluia*, by Handel spelt *Hallelujah*. Both forms are acceptable in context. Pronounce /ˌælɪˈluːjə/; but with initial /ˌhæl-/ for the form *Hallelujah*.

**allergy.** An early 20c. formation (first in German as *Allergie*) from Gk ἄλλοσ 'other, different' + ἔργ-ον 'work', meaning 'sensitiveness to pollen, certain foods, antibiotics, etc.' (the normal use). It acquired the figurative sense 'a feeling of antipathy to a person or thing' by the 1940s; slightly earlier the corresponding adj. *allergic* had come into general use, with varying degrees of informality. It is a standard example of a popularized technical term. Allergies to anything disliked range from the understandable to the far-fetched: to colonels (*Punch*, 1942), maidenheads (Auden, 1944), all kinds of rationing (the European public, according to Koestler, 1951), opera on television (*Observer*, 1958), negotiating with terrorists (*Times*, 1986), etc.

**alley.** Pl. *alleys*.

**alliteration.** A consonance or agreement of sounds (not necessarily of letters) usually at the beginning of a word or syllable, as in *b*ig, *b*old, and am*b*er; *k*nife, *gn*arled, and *n*ote. In the literature of the Germanic languages, alliteration could exist between all vowels indiscriminately: in OE, for example, not only does *e*llen alliterate with *e*nde and *ē*ac

with *ē*age, but *e*cg alliterates with *ī*ren and *ī*sig with *ū*tfūs.

Throughout the ages, in the literature of many countries, alliteration has been used for various special effects, whether for humour, as in Voltaire's Non, il n'est rien que Nanine n'honore; for emphasis, as in Victor Hugo's La Marseillaise ailée et volant dans les balles, Les bombes, les obus, les tambours, les cymbales; or for delicate effects of sound, as in Shakespeare's Full fathom five thy father lies. It is not infrequent in Latin verse, as in Ennius' at tuba terribili sonitu taratantara dixit.

In English literature, the systematic use of alliterative verse is associated with the medieval period, most memorably in poems like *Beowulf*, *Piers Plowman* (In a somere seyson whan softe was þe sonne), and *Sir Gawain and the Green Knight*. The great cycle of English alliterative verse ended with the poem *Flodden Field* (1513). In modern times the tradition has been revived in some of Auden's verse (With labelled luggage we alight at last, Joining joking at the junction on the moor, 'The Exiles', 1968).

Alliterative collocations abound in the language itself, e.g. *to aid and abet*, *as good as gold*, *by fair means or foul*, *might and main*, *part and parcel*, *as right as rain*, *wild and woolly*.

**allot.** Use *allotted*, *allotting*, but *allotment*.

**allow.** **1** This verb matches *admit* in having a wide range of common uses, transitive and intransitive, with subordinate *that*-clauses, with an infinitive complement (*Poetry is allowed to border upon the horizon of mysticism*—J. B. Mozley, 1877), and so on, but also for several centuries alternating in many senses with the phrasal verb *allow of*: e.g. *His condition would not allow of* ( = permit the occurrence of) *his talking earlier*, 1732; *Jortin is willing to allow of* ( = to accept as valid) *other miracles*—J. R. Lowell, 1849. Constructions with *of* now seem somewhat old-fashioned but are still part of the normal grammatical apparatus of the present day, esp. in the sense 'to admit the realization or possibility of, permit'.

**2** In the sense 'to acknowledge, concede', *allow* followed by a subordinate clause has been in continuous use since the 17c.: *I suppose it will be allowed us that marriage is a human society*—Milton, 1643; *He allowed that the old Pre-Raphaelites had . . . exquisite merits*—Hawthorne, 1858; *What was their civilisation? Vast, I allow: but vile*—J. Joyce, 1922; *'You know best, Captain,' Hugh Macroon allowed with grave courtesy*—C. Mackenzie, 1947.

**3** In the AmE colloquial phrase *allow as how*, the meaning is 'to state as an opinion, have to admit that': *She allowed as how my old friend J.J. was flying on Monday morning*—N. Thornburg, 1976; *He allowed as how she was faithful*—T. Morrison, 1981. This is not a standard construction in Britain (see *as how* s. v. AS 8), but it is probably to be found in regional use.

**allowedly.** Pronounce as four syllables. See -EDLY.

**all right.** The use of *all right*, or inability to see that there is anything wrong with *alright*, reveals one's background, upbringing, education, etc., perhaps as much as any word in the language. *Alright*, first recorded in 1893 (*I think I shall pass alright*—Durham Univ. Jrnl) is the demotic form. It is preferred, to judge from the evidence I have assembled, by popular sources like the British magazines *The Face* (*Kool and The Gang are alright. They put on a good show*, 1986), the *New Musical Express*, and *Sounds*, the American magazine *Black World*, the Australian journal *Southerly*, the *Socialist Worker*, by popular singers (*Elvis sings five numbers including the memorable 'That's Alright Mama'*—*Oxford Times*, 1979), and hardly ever by writers of standing (*You'll be alright, love*—C. Achebe, 1987; *'Yes, they visit tombs and live in ashrams alright,' Farrokh sneered*—A. Desai, 1988). It is commonplace in private correspondence, esp. in that of the moderately educated young. Almost all other printed works in Britain and abroad use the more traditional form (first recorded as adj. 1837, as adv. 1844) *all right*, as adv. ('*O, all right,' she said, 'go and be damned.'*—G. Greene, 1980), predicative adj. (*Lisbon was all right*—K. Amis, 1958; *You seem to be all right*—S. Hill, 1971; *I wouldn't want anybody to think that everything is all right*—New Yorker, 1986), and attributive adj. (*A painting or two hung in an all right place*—T. Morrison, 1981). The sociological divide commands attention.

**all-round.** See ALL-AROUND.

**all that.** As a reinforced demonstrative adverb qualifying an adj. or adv. it is now found quite commonly in negative constructions: *The Spanish gypsies . . . hired to do the sweeping were not all that handy with a broom*—Harper's Bazaar, 1962; *I looked round the stock. It wasn't all that brilliant, I must admit*—J. Leasor, 1969. *All* is simply an intensifier in such constructions. Gowers (1965) cited examples of *all that* qualifying an adv. (e.g. *The figures show that even Lazards do not sell £2m. all that frequently*), and judged that the use was 'now well on its way to literary status'. It is now a standard construction. For the more debatable type *I was that angry*, see THAT demonstrative adv.

**all the.** Used with comparatives, *all* is an intensifier qualifying the adverb *the* (anciently a locative or instrumental case of the definite article): *she loved him all the more because he was a true scholar; he thought all the better of her for not coming late*. This construction, which does not seem to have been used before the 20c., is now standard.

**all together, altogether.** Confusion between the two is not uncommon. The distinction is shown in the following examples: (*a*) ( = everyone together) *Until at last, gathered all together again, they find their way down to the turf; A long pull, a strong pull, and a pull all together; Great efforts have been made . . . to bring the troops out all together in brigades, and even in divisions*. The OED s.v. *altogether* ¶ cites an array of examples (1765–1930) of *altogether* written 'where *all together* is logically preferable'. The opposite phenomenon is shown in *All together, IIASA received White House support through four administrations*—Bull. Amer. Acad. Arts & Sci., 1987. (*b*) *He didn't altogether* ( = entirely) *agree with me; you owe me £400 altogether* ( = in all, in total amount); *The weather was bad and the hotel overcrowded. Altogether* ( = considering everything) *it was a disappointing holiday*.

**all told.** This fixed phrase = 'when all are counted; in all' is first recorded in 1850 (*OED tell* v. 21c (*d*)), e.g. *All told, those Celtic teams won the NBA title 11 times in 13 years*—Basketball Scene Annual, 1988. Originally a development of *tell* in the sense 'to count', it has now made its natural way into contexts where enumeration is not present = 'taking all factors into account' (e.g. *All told, I enjoyed life in the army*).

**allusion, allude.** 1 For pronunciation see LU (pronunciation).

2 In normal circumstances, an *allusion* is 'a covert, implied, or indirect reference, a passing or incidental reference' (*OED*). The corresponding verb *allude* is normally used in the same way with a limitation placed on the manner of reference. In practice, however, the complexity of the language has ensured that indirectness is not always part of the sealed meaning of the two words, and the reference is sometimes ambiguously direct or indirect or just plainly direct. Examples showing a broad band of direct and indirect reference: (allude) *He would allude to her, and hear her discussed, but never mentioned her by name*—E. M. Forster, 1910; *In his surviving works, Aristotle never mentions Alexander nor alludes directly to his stay in Macedonia*—R. L. Fox, 1972; *He had star quality, an element often alluded to in Arlene's circle of show-biz friends*—G. Vidal, 1978; *Not just the words forming a sentence but also the meanings to which these words allude*—B. Bettelheim, 1983; (allusion) *She came across allusions to her family in the papers*—V. Sackville-West, 1931; *Midway in the questioning ... he'd begun to notice the number of allusions to a particular November weekend*—T. Capote, 1966; *There were hints and allusions about his troubles to his friends*—D. Halberstam, 1979.

3 See LITERARY ALLUSIONS.

**ally.** Originally stressed, both as noun and verb, on the second syllable. As a verb it still is, but as a noun is now normally stressed on the first. The adjective *allied* is stressed on the first syllable when it precedes a noun (*the Allied forces*), otherwise on the second.

**Alma Mater.** An affectionate title ( = bounteous mother), once given by the Romans to several goddesses, especially to Ceres and Cybele, now applied to one's university or school seen as a 'fostering mother'.

**almanac.** Normally so spelt except in traditional titles like 'The Oxford Almanack' and *Whitaker's Almanack*.

**almoner.** The *OED* (1884) and Daniel Jones (1917) gave preference to the pronunciation /ˈælmənə/, but /ˈɑːm-/ is now customary, doubtless after the analogy of *alms* /ɑːmz/. The word itself is passing out of use: (*a*) the official distribution of alms has no place in a welfare-state economy; (*b*) the word has given way to (*medical*) *social worker* as a title in British hospitals for an official with certain duties concerning the welfare of patients.

**almost.** 1 Its occasional use as a quasi-adj. meaning 'close to being, not quite' (*an almost Quaker*; *his almost impudence of manner*; *an almost Prime Minister*) seems slightly forced, but it is neither new (*OED* 1552– ) nor unparalleled, since many adverbs and adjectives have the same form (see ADJECTIVE 8).

2 See MOST, ALMOST.

**alone.** As an adv. phr. correlated with *but*, *not alone* has been vying with *not only* for some seven centuries, esp. in poetry (*Wisdom ... Which not alone had guided me, But served the seasons that may rise*—Tennyson, 1850). This use is now archaic or obsolete. On the other hand, sentences with *not* and *alone* separated are commonly found (*It is not young people alone who need to be taught to swim*), though the alternative construction with *not only* standing together (*It is not only young people who ...*) is more usual.

**along.** 1 *Along about* is an AmE regionalism meaning 'approximately': *In the one case you start out with a friend along about eleven o'clock on a summer's night*—M. Twain, 1879; *'Nettie Bill got married along about the same time I got married to Jack,' Aunt Lou said*—M. Grimm, 1989.

2 *along of.* Used to mean 'owing to, on account of' (*A present from the Gentlemen, along o' being good!*—Kipling, 1906; *'The trouble I've had along of that lady's crankiness,' he confided*—N. Marsh, 1963), and '(together) with' (*You come along a me, Ginger*—J. Cary, 1940), *along of* is 'common in London, and southern dialects generally' (*OED*), but is not standard.

**alongside.** 1 For some two centuries, *alongside* has vied with *alongside of* immediately before a noun, verbal noun, or pronoun (e.g. *a boat lying alongside the rock*/*alongside of a sheet of water*). Both constructions remain available: (along-side) *The transport* Stamboul ... *was along-side the harbour wall*—D. A. Thomas, 1988; (alongside of) *Many minds have been en-gaged alongside of mine in the same pur-suit*—R. Fry, 1920; *alongside of preaching the Gospel ... there are other ways in which we have to change the lives of these savages*—C. Hampton, 1974.

**2** Recently, *alongside* has been used conjunctively in the financial pages of newspapers with a number of meanings, e.g. together with, as a result of, as well as: *GEC is also fancied to be taking an interest in the shares, alongside buying back some of its own*—*Times*, 1984; *Connells Estate Agents rose 3p to 103p alongside a statement from the chairman*—*Times*, 1985; *The group has suffered alongside the rest of its sector in recent weeks*—*Times*, 1985.

**alot.** This illegitimate form of *a lot* is beginning to turn up in informal corres-pondence in AmE: e.g. (from some letters written in 1991) *My parents have been out alot the past 2 weeks; The recession has affected the advertising business alot; We still had alot of fun, just a different kind of fun.* Cf. AWHILE.

**already.** The adverb is so spelt (*I have lost a great deal of time already*) and of course is not be confused with the two separate words *all ready* (*We are all ready to start now*, i.e. all of us are ready).

**alright.** See ALL RIGHT.

**also.** The word is an adverb and is almost always so used: e.g. *Besides being an astronomer and mathematician, Grassi was also an architect; the gas can also be compressed by the blast wave.* Occasionally, in the speech of hesitant people, or as an afterthought, it strays into conjunc-tional territory: e.g. *He has made a good impression. He writes well, he keeps to dead-lines, and follows house rules. Also he's an agreeable person.* / *Remember your watch and money; also the tickets.*

It is sometimes used as an 'additive conjunct' within a sentence, where *and also*, *and*, or *but also* would be in place: e.g. *Great attention has been paid to the history of legislation, also* [ = and also] *to that of religion.*

It should perhaps be noted that con-junctional uses of *also* are a marked feature of uneducated speech: e.g. *Also Dolly May ain't no chicken neither*—E. Jolley, 1985.

**alternate.** **1** *As adjective.* In the regular senses '(said of things of two kinds) com-ing each one after one of the other kind' (e.g. *walls built of alternate layers of stone and timber*), and '(of things of the same kind) every second one' (e.g. *the congrega-tion sang alternate verses; sprinklers may be used on alternate days only*, i.e. on every other day), *alternate* is standard in all forms of English.

In the course of the 20c., in AmE, *alternate*, pronounced /ˈɔːltərnət/, has usurped some of the territory of *altern-ative* in its ordinary sense '(of one or more things) available in place of another'. The Book of the Month Club offers 'alternate selections'. A route, a material, a lyric, etc., can be described as 'alternate' rather than (as in the UK) 'alternative': *It obscures the possibility of another, complementary re-ading of* rode *as an alternate spelling for* rud—*Amer. Notes & Queries*, 1985; *An altern-ate way to make these rellenos is to stuff the meat mixture into whole green chiles*—*San Diego Union*, 1987. There are *alternate* (or *alternative*) *schools* in the US which offer a non-traditional curriculum.

**2** *As noun.* Frequently used in AmE in the sense 'an alternative, a reserve (player), a variant': *I was fourth alternate in the Miss Teenage South Carolina pageant*— William Boyd, 1984; *A recent rule change would allow the two alternates to compete in the two preliminary rounds* [of a relay race]—*Runner*, 1984; *Earthworm, the standard expression, is certainly no longer a rare alternate*—*Amer. Speech*, 1985; *the twelve jurors and six alternates in Room 318 of the United States Courthouse*—*New Yorker*, 1986.

**alternative.** **1** *As noun.* The traditional view that there can only be one of two (not more) alternatives because L *alter* means 'other (of two)' can no longer be maintained. The word can still be used in this manner (e.g. = a proposition con-taining two statements, the acceptance of one of which involves the rejection of the other; or = either of the two mem-bers of the alternative proposition (*no other alternative*); or, especially, the other or remaining choice (*a fate compared to which death would have been a joyful altern-ative*). The *OED* amply illustrates such traditional uses. But, beginning with an example from John Stuart Mill in 1848, it also shows the word in extended use meaning 'a choice between more than two things'. Gladstone is reported as saying in 1857 *My decided preference is for the fourth and last of these alternatives.* Recent examples: *No one would suggest that ... [she] should go through with a pregnancy and delivery ... when she could never care for the resulting baby. But what are the alternatives?* —Sunday Times, 1987; *Some alternatives will be screened out immediately because they would violate what can be viewed as boundary conditions on choices*—Bull. Amer. Acad. Arts & Sci., 1989.

**2** *As adjective.* Since the late 1960s the adjective has increasingly been used to mean 'purporting or claiming to represent an acceptable or preferable alternative to that in traditional use', e.g. *alternative* (i.e. non-nuclear, not using fossil fuels) *energy, alternative* (i.e. mainly homoeopathic or holistic) *medicine, alternative* (i.e. pirate) *radio, alternative* (i.e. rejecting the traditional way of life of) *society.*

**although, though.** In most concessive clauses, *although* and *though* are interchangeable with no change of meaning: (*Al*) *though he was only thirty he was bald; There was not a single black in the party,* (*al*)*though Lucinda had directed that this be otherwise.* (Of the two, *though*, which is perhaps slightly less formal, is much the commoner.) Except that *although* cannot be used (*a*) as an adverb in medial or final position: ( = nevertheless, however) *He did his duty, though*; (as an intensive after a question or emphatic statement: indeed, truly) *'What a sad story!' said Maria. 'Isn't it, though?'*—G. Vidal, 1948; *he knew, though, that his team would not win.* (*b*) after *as* or *even* as part of a conjunctive phrase: *We fell into step and talked as though eight months were minutes; free him even though he will die; even though he was a Tory, he was opposed to the privatization of the water industry.*

**alto.** Pl. *altos.* See -O(E)S 5.

**altogether.** See ALL TOGETHER.

**alto-relievo** (sculpture) /ˌæltəʊ rɪˈliːvəʊ/. Pl. -*vos*: see -O(E)S 6. The Italian form *alto-rilievo,* pronounced /ˌaltəʊ riˈljervəʊ/, is sometimes used. Cf. BAS-RELIEF.

**aluminium.** The BrE spelling harmonizes best with other names of elements, as *magnesium, potassium, sodium,* etc., whereas the AmE spelling *aluminum* is the one given to the word by its discoverer, Sir H. Davy, c1812. *Aluminium* is stressed on the third syllable, and *aluminum* on the second.

**alumnus.** From L ( = nursling), and meaning former pupil or student, *alumnus* is more frequently encountered in AmE than in BrE. It is pronounced /əˈlʌmnəs/ and its plural *alumni* as /əˈlʌmnaɪ/. A former female student is an *alumna* /əˈlʌmnə/, pl. *alumnae* /əˈlʌmniː/. The form *alumni* is normally used for a mixed gathering of former students.

In both AmE and BrE, rival views on the pronunciation of Latin sometimes lead to a reversal of the way in which the endings of the two plural forms are pronounced.

**alveolar.** See DENTAL.

**a.m.** As an abbreviation of L *ante meridiem* 'before noon', it is always pronounced as /ˌeɪˈrem/, and is normally written in the form *8.15 a.m.* (or *am*; in AmE *8:15 a.m.*). Note that 12.30 a.m. = half an hour after midnight, and 12.30 p.m. = half an hour after noon. The abbreviation is sometimes used as a noun = the period before noon: *I arrived here this a.m.*

**amateur.** The pronunciation dictionaries present an array of possibilities for this word, but /ˈæmətə/ is now standard and /ˈæmətjʊə/ and /-tʃ-/ have become minority pronunciations.

**amazedly.** Pronounce with four syllables. See -EDLY.

**ambidextrous.** Not -*erous*.

**ambience.** Derived from F *ambiance* (a form also occasionally used in English) only a century ago (first recorded 1889) in the sense 'environment, surroundings, atmosphere', the word is now firmly established in the language. Some people still affect the French pronunciation, but it is most often pronounced as /ˈæmbɪəns/. Its entrance into English was assisted by the fact that the corresponding adj. *ambient* had been a part of the language since the late 16c. and had also been used absolutely. *Ambience* filled an obvious gap.

**ambiguity.**    **1** The many-sidedness of language accounts for the kind of poetic ambiguity described by William Empson in his *Seven Types of Ambiguity* (1930). With this kind of pleasurable literary ambiguity we are not concerned here.

**2** Deliberate, and often agreeable, ambiguity is sometimes a feature of advertising slogans, the captions of cartoons, and general anecdotes, e.g. (doctor to overweight woman) *'What is the largest weight you have been?' 'Sixteen stone.' 'And the smallest?' 'Six pounds.'*

**3** The early grammarians and writers on usage placed great emphasis on misconceptions arising from the wrong ordering of words. For example, William Cobbett (1823), Letter XXI: 'Of all the faults to be found in writing this [*sc*. the wrong placing of words] is one of the most common, and, perhaps, it leads to the greatest number of misconceptions.' He claimed to have 'noted down about two hundred errors in Doctor Johnson's Lives of the Poets'. Henry Alford (1864) wrote at length on the matter and quoted as one of his examples, *I with my family reside in the parish of Stockton, which consists of my wife and daughter.* Fowler (1906) said that 'A captious critic might find examples [of false ordering of words] on almost every page of almost any writer.' There is an air of unreality and implausibility about these old precepts and about the examples given in support. Copy editors and proofreaders remove the great majority of such crudely ambiguous constructions at the pre-publication stage.

**4** Modern grammarians approach the problem more sympathetically. Easily targetable misconstructions are still attacked: e.g. (a newspaper report of road works causing a safety hazard to schoolchildren) *The council plans to notify parents whose children are affected by post.* The most eminent modern authority, *CGEL* (1985), gives prominence to ambiguity which arises through ellipsis: e.g. (§15.67) *He loves his dog more than his children* (who loves whom most?); (13.68) *He specializes in selling old and valuable books* (two kinds of books?); (13.70) *the meetings on Monday and Friday* (how many meetings on each

day?); (8.43) *The dog is not allowed to run outside* (directional or positional?).

**5** In written work, ambiguity can be reduced in several ways: by changing the order of clauses, by supplying ellipted elements, by restructuring the sentence altogether, or by the use of punctuation to mark the boundaries of clauses. In spoken English, potential ambiguity is often cancelled by a shift of intonation at the boundary of clauses.

**6** Nevertheless, vigilance is required, especially in contexts where backward reference (see ANAPHORA), ellipsis (see ELLIPSIS), and misrelated clauses (see UNATTACHED PARTICIPLES) are involved. If these matters are not attended to, ambiguity of various degrees of seriousness can certainly arise.

**7** See AMPHIBOLY, AMPHIBOLOGY.

**ambivalent.**    This 20c. (first recorded 1916) Jungian word meaning 'having either or both of two contrary or parallel values' quickly moved into literary and general use. C. S. Lewis called death ambivalent because it is 'Satan's great weapon and also God's great weapon'. At the core of its meaning is the idea of equivocation. Moods, characters, relationships, attitudes, behaviour—anything judged to contain contradictions—can be described as ambivalent. It is a striking addition to the language, though it is tending to be overused.

For the most part it is distinguished from *ambiguous*, which means rather '(of speech, words, etc.) having more than one possible meaning', '(of events, etc.) wavering or uncertain in direction or tendency'.

**amen.**    I was brought up to pronounce the word /ˌɑːˈmen/ and was puzzled to hear others saying /ˌeɪˈmen/. Speakers are probably equally divided in the matter.

**amend, emend.**    The first of these is much the more usual word. It is used of the making of (minor) adjustments to a document, a motion, a parliamentary law, etc., in the interests of clarity, equality, etc. It is also commonly used in contexts of personal behaviour. *Emend* is the property mainly of textual scholars who propose improvements to a word or phrase by the addition, deletion, or alteration of some linguistic element.

**amenity.** 1 The dominant pronunciation in RP seems to be /əˈmiːnɪtɪ/, and in regional and overseas varieties of English /əˈmenɪtɪ/.

2 In the 20c. the word, while retaining its older general meaning 'the quality of being pleasant or agreeable', has been extended to denote the more agreeable or pleasurable environmental aspects of a house, a village, etc. According to taste, a skating rink, a library, a discothèque, a meadow, a yachting marina, etc., can all be described now as amenities.

**America.** English speakers outside North America usually call the United States 'America' and its inhabitants 'Americans'. (So does everyone in the United States from the President downwards.) The term 'North America' is mostly used to mean the United States and Canada together. Countries to the south of the United States are described as being in Central America (Mexico, Nicaragua, etc.) or South America (Brazil, Argentina, etc.). The Spanish-speaking inhabitants of Central and South American countries are often referred to as Latin Americans. The Black inhabitants of North America used to be called Negroes, then Blacks (or blacks), and, more recently and increasingly, African-Americans. The varieties of English spoken in the United States are normally called 'American English', occas. just 'American'. The original Indian inhabitants of both North and South America are called 'American Indians', 'Indians', or (recently) 'Native Americans'.

**amid, amidst.** 1 *Amid*, recorded as a preposition and adverb before the Norman Conquest, developed a by-form *amides* (with the final -s which survives in such words as *always*, *backwards*) by the 14c., and, by form-association with superlatives, a further by-form *amidst* (cf. *against*, *amongst*) in the 16c. *Amides* dropped out of use and *amid* and *amidst* survive only as prepositions.

2 Both words have become somewhat restricted in currency, and they both have a slight air of formality, though the matter is disputed. In a note written in the 1880s, the *OED* says that 'There is a tendency to use *amidst* more distributively than *amid*, e.g. of things scattered about, or a thing moving, in the midst

of others.' Now, a century later, the distributional pattern of the words is not clearly ascertainable, though, from the fact that a major database (1989) contained 62,700 examples of *amid* and 4, 900 examples of *amidst*, it clear that *amid* is much the more common of the two. Typical examples: (amid) *I ... have often stood by the Frome at Woolbridge, enjoying the mellow manor house amid its watermeadows—Times*, 1987; Outside Days *is a bedside book which conjures up memories of happy days amid the high hills—Spectator*, 1989; 'we shall enjoy strong, sustained growth and prosperity into the 1980s,' he said amid Conservative cheers—*Daily Tel.*, 1989; (amidst) *he returned here for more tests amidst rumours that he had Parkinson's disease—Washington Post*, 1984; *when he did give up a day's work to sit amidst these men and women*—T. Keneally, 1985; *this woman, sitting with such modest dignity amidst my students and colleagues*—M. Frayn, 1989. In many contexts both words can be replaced by *among*, *amongst*, or *in the midst of* without change of meaning or of function.

**Amish.** A sub-dialect of AmE spoken by the Amish people, a strict sect of Mennonite followers of Jacob Amman or Amen (*fl.* 1693), the Swiss founder of the sect. Found mostly in Pennsylvania and Delaware, the Amish people maintain an older style of life: horse and buggies for transportation, no electricity in their homes, plain dress and so on. Old Order Amish speech is marginally distinctive in pronunciation (e.g. *house* is pronounced /haːs/ not /haʊs/), with many syntactic features that do not quite match those of neighbouring, non-Amish varieties of AmE. For example, reflecting German usage, they do not use—have not adopted—the English progressive tense: they say *he works very hard* rather than *he is working very hard*.

**amn't.** A frequent variant in Irish and Scottish English of *aren't* used as part of the tag question *amn't I?* See BE 4.

**amoeba.** The UK spelling, as against *ameba*, the customary US spelling. The plural form is *amoebas* (US *amebas*), less commonly *amoebae* /-iː/, US *amebae*.

**amok, amuck.** The spelling *amok* (as in the Malay original) and the pronunciation /əˈmɒk/ are recommended, though the variant pronunciation /əˈmʌk/ is frequently heard.

**among, amongst.** 1 For uses of *among* and *between*, see BETWEEN 2.

2 The OE antecedent *on gemang* yielded *onmang* before 1100, whence by regular phonetic gradation *amang*, *among*. By the 14c. the variant *amonges* had emerged (cf. AMID, AMIDST 1) and by the 16c., by form-association with superlatives, *among(e)st*. The surviving forms *among* and *amongst* have competed with each other for several centuries, with *among* now much the commoner of the two (in my collections in the proportion 10:1).

3 There is no demonstrable difference of sense or function between the two, and the distribution is puzzling except that *amongst* seems to be somewhat less common in AmE than in BrE.

Typical examples: (among) *the giants war among themselves*—J. M. Coetzee, 1977; *there were a lot of very young people among the temporary staff*—P. Fitzgerald, 1980; *Britain also has . . . the lowest level of welfare expenditure among the countries of the European Community*—*Times*, 1985; *she wove her way among the crowd*—M. Ramgobin, 1986; *a familiar, enchanting presence among the habitués of the city's film festivals*—*New Yorker*, 1987; *the arguing out of concepts can last for hours among her advisers*—*Daily Tel.*, 1987.

(amongst) *they fight amongst themselves*—W. Wharton, 1978; *he was grateful to the Kabbels . . . for taking account of her amongst their berserk schemes*—T. Keneally, 1985; *the other guests served themselves discreetly and talked amongst themselves*—B. Rubens, 1987; *they stood on the edges of the lamplight amongst the wattles by the creek*—P. Carey, 1988.

It is clear that *among* can be used with a collective noun (Fitzgerald 1980, Ramgobin 1986) or with the name of something that is widespread (e.g. hay, scrub, wreckage), but otherwise only with plural nouns or pronouns. An older view, favoured by Fowler, that *amongst* is commoner than *among* before a following vowel does not seem to be borne out by the evidence: *the 'insolence of office' which (among other things) drove Hamlet to contemplate suicide*—*Times*, 1985; *the depressive begins to appear as the one among*

us *who sees the terror of the game more clearly than the rest*—*TLS*, 1987.

**amoral.** Used to mean 'not within the sphere of moral principles', this word, first recorded in 1882, has largely ousted the slightly older words *non-moral* (a1866) and *unmoral* (1841). See A-[1]. Cf. IMMORAL.

**amount, number.** In most circumstances *amount* is used with non-count nouns to mean 'quantity' (e.g. *a reasonable amount of forgiveness, glue, resistance, straw*), i.e. nouns which normally have no plural; and *number* is used with plural nouns (e.g. *a certain number of boys, houses, jobs*, etc.). *Amount* is now fast breaking into the territory of *number*, sometimes, but by no means always, when the following plural noun is viewed as an aggregate or collection. A range of examples: *Fame had magnified the amount of the forces*—1849 in *OED*; *I have any amount of letters for you*—G. B. Shaw, 1893 ( = a great many); *I expect you get a fair amount of road accidents on these winding roads?*—R. Billington, 1988; *the grunts* (sc. GIs in Vietnam) *were all carrying those little automatic cameras . . . getting enough pictures because no amount of words was going to tell it*—M. Doane, 1988 (US, = not even the greatest amount (of)); *Billy's had a tremendous amount of problems*—T. McGuane, 1989 (US); *But booksellers have less and less space for the amount of books that are being published*—*The Author*, 1990; *fearing . . . that the next thing Sinead would treat them to was an account of her husband's suicide, of the amount of pills he took in that hotel*—F. O'Brien, 1990; *The amount of bulbs she would find between the stones next spring*—A. Huth, 1991. We may rail against the loss of a useful distinction—and I do—but can it be stopped now?

**ampersand.** The name of the symbol & used as a space-saving device. H. W. Fowler used it throughout *MEU* and in early editions of *COD* and *POD* as well as in his correspondence with the OUP. It is exceedingly common in handwritten work. In print it survives mainly in the names of jointly owned firms, e.g. Marks & Spencer, and not always then, e.g. Faber and Faber Ltd. In origin it is a 19c. corruption of '*and per se and*', the name of the character & as it appears at the end of the alphabet in primers and hornbooks, i.e. '& (standing) by itself, and'. The sign & itself seems to be a stylized version of L *et* 'and'.

**amphiboly** /æm'fɪbəlɪ/, **amphibology** /ˌæmfɪ'bɒlədʒɪ/. In rhetoric, a figure of speech signifying ambiguity that arises 'from the uncertain construction of a sentence or clause, of which the individual words are unequivocal' (*OED*). For example, the road sign *Slow Children*, meaning 'Slow down, Children in the vicinity', could perversely be taken to refer to the walking pace or the learning speed of children in the vicinity. A classic example occurs in Shakespeare's *Othello* III. i: Cassio. *Doest thou heare my honest friend?* Clowne. *No, I heare not your honest friend, I heare you.*

**amphimacer.** See CRETIC.

**ample.** Just over a century ago, when the relevant section (A-Ant, 1884) of *OED* was published, *ample* was 'legitimate only with nouns denoting immaterial or abstract things' (*MEU*, 1926): e.g. *ample opportunity, praise, provision, time.* Fowler did not accept that it could properly be attached to nouns like *butter, coal, oil, water* that denote substances of indefinite quantity. *There is ample coal to carry us through the winter* he regarded as 'wrong'. Time has moved on and such extended uses are now regarded as unexceptionable.

**amuck.** See AMOK.

**amusedly.** Pronounce as four syllables. See -EDLY.

**an** (indefinite article). See A, AN.

**an-.** See A-¹.

**anachronism.** Linguistic anachronism in historical fiction and drama is commonplace. At first sight it would seem entirely regrettable: no one would question the desirability of aiming at a broad restriction of language to the words, phrases, meanings, and constructions known to be current at the time of the actions or events being depicted. But there are boundaries of language which cannot be fixed exactly at any given point of time. It goes without saying that the terminology of fashion, science, warfare, technology, domestic life, and so on, should be rigorously verified from the largest dictionaries and grammars available in so far as they deal with the language of the period in question. Thus, a writer setting a novel in Britain in the 1920s would be ill-advised to use the slang expression *he went bananas* because it seems extremely unlikely that the expression was current then. But such a writer would find it difficult, for example, to establish the degree of obsolescence of the words *abolishment* and *appraisement* (which were in process of yielding to *abolition* and *appraisal* respectively), the currency in the UK of American words like *commuter* and *guy* ( = a person), and the frequency with which to use, say, the subjunctive mood after certain verbs. That said, writers of historical fiction and drama should try to achieve linguistic verisimilitude by a judicious use of the reference works that are available.

**anacoluthon.** (Gk ἀνακόλουθον 'wanting sequence'.) The name given to a change or break in the grammatical construction of a sentence or phrase, e.g. (a recapitulatory pronoun in casual speech) *put little bits of bacon on which the fatter they are the better*—The Victorian Kitchen, BBC2, 1989 (cook speaking); (*is* omitted) *She's had five husbands and ʌ on the lookout for the sixth*—Oxford University examination script, 1989; (with an intervening pause) *he did not see—was prevented by the brightness of the sun from seeing—the traffic lights.* Instances of anacoluthia occur at all periods from the Anglo-Saxon period onward.

**anacrusis.** (Gk ἀνάκρουσις 'prelude, upstriking'.) In prosody, a syllable or syllables preceding the point at which the reckoning of the normal measure begins. It is a particular feature of OE verse and that of the other Germanic peoples. In modern verse, where analysis is often disputed, anacrusis may account for the initial *Or* in

> *Or wás there a déarer one*
> *Still, and a nearer one.*
> (Hood)

**anaemia, anaesthesia.** Usually spelt *anemia* and *anesthesia* in AmE; also their derivatives *anemic, anesthetic,* etc.

**anagoge** /ˌænə'gəʊdʒiː/ and **anagogy** /'ænəgəʊdʒɪ/, both meaning 'mystical or spiritual interpretation', are pronounced in English with /-dʒ-/ in the final syllable, not, as Fowler wished, with /-g-/. See GREEK G.

**analogous.** The g in the final syllable is to be pronounced /-g-/ and not, as one hears all too often, /-dʒ-/.

**analogy.** 1 The only kind of analogy with which we are concerned here is linguistic analogy, defined by the OED as 'Similarity of formative or constructive processes; imitation of the inflexions, derivatives, or constructions of existing words, in forming inflexions, derivatives, or constructions of other words, without the intervention of the formative steps through which these at first arose'. Thus (an OED example) 'the new inflexion *bake*, *baked*, *baked* (instead of the historical *bake*, *book*, *baken*) is due to analogy with such words as *rake*, *raked*, *raked*, etc.'

2 The process of analogy can be further illustrated by the way in which (a) new words are formed from native bases on the analogy of a pattern taken over from a foreign language, e.g. *starvation* (first recorded in 1778) as the noun corresponding to the verb *starve* (OE *steorfan*) by analogy with, for example, *vexation* (c1400 from OF), the noun corresponding to the verb *vex* (also ME from OF from L *vexāre*, *-ātum*); and (b) the manner in which loanwords are made to fit into existing word patterns in English, e.g. *strive* (from OF *estriver*) taken into the native conjugation of *drive* (OE *drīfan*), the only verb of French origin to be so treated.

3 The same process can be seen at work in the emergence of new past participles and past tenses of some verbs, e.g. *dug* (16c., earlier *digged*) by analogy with *stuck*; and *snuck* (19c., chiefly US) beside the traditional *sneaked*.

4 False analogies frequently produce irregular analogical formations in the language of children, e.g. 'I couldn't *of*' from recognizing that unstressed *have* and unstressed *of* are both pronounced /əv/ in informal English; and 'I am *being* have' from a false analysis of the imperative 'Behave yourself!'

5 It is easy to assemble a list of unintended casual errors arising from false analogies: *Thou shalt not make unto thee*

*any* craven *image* (oral example, 1989, instead of *graven*); *The Most Eligible* Batchelors *of 1960* (Observer Mag., 1988, after *batch* n.); *the assumption that it* [sc. wire] *is being* payed out *from the saddle horn when it isn't* (London Rev. Bks, 1988, after *played out*); *'a few minutes from the sea' can be a gruelling* treck *over stony ground* (Times, 1988, after *deck*, etc.).

6 *Word-formation*. In the 20c., analogy has been at work in the identification and extraction of suitable final elements and regarding them as new suffixes for exploitation: e.g. *-(a)thon* (extracted from *marathon*) producing *talkathon*, *telethon*, *walkathon*, etc.; *-burger* (extracted from *hamburger*) producing *beefburger*, *cheeseburger*, *steakburger*, etc.; *-teria* (by an analysis of *cafeteria* as *café* + *-teria*) producing *healthateria*, *valeteria*, *washeteria*, etc.

7 The apparent abrasiveness of some 20c. processes of analogy should be judged against the performances of earlier centuries. It is salutary to bear in mind that a great many analogical formations can be said to be badly formed or at least 'irregularly formed', and that this has not stopped them from becoming part of the unquestioned core of the language, e.g. *chaotic* (from *chaos*, after the analogy of other Greek-derived words like *demotic*, *erotic*, *hypnotic*); *dilation* (from *dilate*; only *dilatation* is etymologically sound); *operatic* (from *opera*, after *dramatic*).

**analyse.** After a period of uncertainty (Dr Johnson, for example, used the form *analyze*), this verb has settled down as *analyse* in BrE and *analyze* in AmE. Both forms are etymologically defensible (see the OED).

**analysis.** 1 'The resolution or breaking up of anything complex into its various simple elements' (OED), in chemistry, grammar, etc. It is the opposite process to *synthesis*. Pl. *analyses* /-iːz/.

2 The Fr. phrase *en dernière analyse* has been drawn on in English to provide the common phrases *in the last* (or *final*) *analysis* (first recorded in 1877).

**anaphora.** 1 (First recorded in the 16c.) In rhetoric, the repetition of the same word or phrase in several successive clauses: e.g. *The voice of the Lord is powerfull; the voyce of the Lord is full of Maiestie; The* *voyce of the Lord breaketh the Cedars* (Ps. 29: 4–5).

**2** (First recorded in the 20c.) In grammar, the use of a word which refers to, or is a substitute for, a preceding word or group of words. In the sentence 'The city was deserted when it was overrun by the rebels', 'it' refers back to 'the city', i.e. is anaphoric. Cf. CATAPHORIC.

**anastrophe.** A term of rhetoric meaning 'inversion, or unusual arrangement, of the words or clauses of a sentence': e.g. *Day is done, gone the sun*—New Yorker, 1989; *Beats there a heart amongst us so jaded ... that it has failed to be touched ... by the sound of Roy Orbison?*—The Face, 1989.

**anathema.** **1** Derived from late L *anathema* 'an excommunicated person' and Gk ἀνάθεμα orig. 'a thing devoted', but in later usage 'a thing devoted to evil, an accursed thing', it came into English in the 16c. initially in the broad sense 'the formal act, or formula, of consigning to damnation'. With the passage of time, this sense weakened until the word became freely used as a general weapon of ecclesiastical, and then of lay, rancour. Its plural is most commonly *anathemas* (*he knew no curses except the day-to-day anathemas of the Webfeet*—J. Mark, 1982), but in the specialized sense 'a thing devoted or consecrated to divine use' the singular form is normally pronounced, with shift of stress, /ænəˈθiːmə/, plural *anathemata* /ænəˈθiːmətə/.

**2** Beginning in the 18c., *anathema* has also been used as a quasi-adj. meaning 'accursed, consigned to perdition' (*Saint Paul wished to become anathema himself, so he could thereby save his brethren*—Abraham Tucker, 1765). In the 20c. the word continues to hover on the edge of the domain of the adjective in such uses as *he is anathema to me, taxes are anathema* [NB not *anathemas*] *to most people*. Cf. ADAMANT.

**anchorite.** This word, derived from Gk ἀναχωρητής (cf. Gk ἀναχωρέ-ειν 'to retire, retreat'), has driven out the by-form *anchoret*. 'Anchorites were persons, most often women, who were ritually enclosed and permanently set apart both from lay society and from the regular religious life, whereas hermits retained freedom of movement' (*Dictionary of the Middle Ages*, 1982).

**anchovy.** Now usually stressed on the first syllable, /ˈæntʃəvɪ/. Walker (1791) lists only *anchóvy*, i.e. /ænˈtʃəʊvɪ/; this placing of the stress was given preference in the OED (1884) and by Daniel Jones (1917), but is now rarely heard.

**anchylosis.** See ANKYLOSIS.

***ancien régime.*** Print in italics and with acute *e*.

**and.** The simplest-looking words are often among the most complicated, and *and* is no exception. **1** The normal function of this connective conjunction is, of course, to join sentence elements of the same kind: e.g. *Dido and Aeneas*; *first and foremost*; *the compensations are generous*; *she served quickly and efficiently*; *for ever and ever*; *an acute and wary sense of the ordinary*. It can imply progression (*faster and faster*), causation (*misbehave and you'll not get your pocket money*), great duration (*she ran and ran*), a large number or a great quantity (*miles and miles, piles and piles*), and addition (*four and four are eight*).

**2** In practice *and* is often omitted for contextual effects of various kinds, especially when two or more adjectives occur in conjunction. Thus, from my files: (without *and* and using a comma or commas instead) *Czechs were marginal, remote, troublesome, peculiar Europeans with unpronounceable names*; *factories outlined against a still, sunless sky*; (without *and* and without commas) *the teeming jerry-built dun-coloured traffic-ridden deafening city*—P. Lively, 1987; *he envied Jenkin his simple uncluttered uncomplicated innocent life*—I. Murdoch, 1987.

**3** There is a persistent belief that it is improper to begin a sentence with *And*, but this prohibition has been cheerfully ignored by standard authors from Anglo-Saxon times onwards. An initial *And* is a useful aid to writers as the narrative continues. The OED provides examples from the 9c. to the 19c., including one from Shakespeare's *King John*: Arthur. *Must you with hot Irons, burne out both mine eyes?* Hubert. *Yong Boy, I must.* Arthur. *And will you?* Hubert. *And I will.* It is also used for other rhetorical purposes, and sometimes just to introduce an improvised afterthought: *Tibba still pined and slavered for the school lunches. And little other* care hath she.—A. N. Wilson 1982; *I'm going to swim. And don't you dare watch*—G. Butler, 1983. It is also used in expressing surprise at, or asking the truth of, what one has already heard: *O John! and you have seen him! And are you really going?* —1884 in *OED*.

**4** *and all*. Another well-established use of *and* is in the phrase *and all*. Wright's *English Dialect Dictionary* gives prominence to this use, meaning 'and everything; also, besides, in addition'. In some of the examples it seems to lack any perceptible lexical sense and to be just a rhythmical device to eke out a sentence. Wright's 19c. evidence is drawn from almost every county and he also lists examples from dialectal contexts in the works of Tennyson, Gissing, Kipling, and others: (Scottish) *Woo'd and married an' a'*; (Westmorland) *when she saw me she wept, I wept ano'*; (West Yorkshire) *Whoy, we'n been up an darn anole*; (Lincolnshire) *He wants sendin' to Ketton* [Kirton-in-Lindsey prison], *an a-cat-o'-nine-tails an'-all*. The use has seeped out into more general use in our own century: *When I held her in my arms she was like a dying bird, so thin and all*—M. Doane, 1988; *We had a hell of a job pushing it, what with the sarnie-boards and all*—Caris Davis, 1989; *Isn't it amazing? He has a Ph.D. and all.*—J. Shute, 1992.

**5** *there are kings and kings*. A cricket commentator on BBC TV said wearily at the end of a barren over by the West Indian bowler Curtly Ambrose, *There are maidens and there are maidens, but that wasn't one of his best.* He was using a construction first recorded in English in the 16c. 'expressing a difference of quality between things of the same name or class', as the *OED* expresses it. The use, the *Dictionary* says, is 'commonly called a French idiom' and refers to Molière's *il y a fagots et fagots* in *Le Médecin malgré lui* (1666), but the English evidence is earlier. The *OED* cites examples from the 16c. to the 19c., including *Alack, there be roses and roses, John!* (Browning, 1855). To which may be added the following 19c. and 20c. examples: *Well, as to that, of course there are kings and kings. When I say I detest kings, I mean I detest bad kings* — W. S. Gilbert, *The Gondoliers*, 1889; *There are Coloureds and Coloureds, just as there are whites and whites*—D. Matthee, 1986; *There are ways to steal and ways to steal*— New

Yorker, 1988; *There is homelessness and homelessness. The word has become a shibboleth for opposition politicians and the 'caring' media ... The sort of homelessness which means despair is quite different from the sort which means adventure*—*Times*, 1991.

**6** See also AGREEMENT 3; AND/OR; *and which* (WHICH 5); COMMA 3,4; GOOD AND; *nice and* (NICE 2); TRY AND.

**and/or.** A formula denoting that the items joined by it can be taken either together or as alternatives. First recorded in the mid-19c. in legal contexts, and still employed from time to time in legal documents, *and/or* verges on the inelegant when used in general writing: *The Press has rather plumped for the scholar as writer, and/or as bibliophile*—Cambridge Rev., 1959; *political signalling by such means can be dangerous and/or ambiguous*—Bull. Amer. Acad. Arts & Sci., 1987. The more comfortable way of expressing the same idea is to use 'X or Y or both', or, in many contexts, just 'or'.

**anemone.** The crass pronunciation of this word, not uncommonly heard, as */ə'nenəmi:/*, i.e. with transposition of the medial *m* and final *n*, is to be shunned.

**anent.** This ancient preposition (in origin a phrase, OE *on efn*, *on emn* 'in line or company with, on a level with') survives in Scottish law ('in respect of or reference to'). In general English it is used to mean 'with respect to', but it often carries an air of affectation or of faint jocularity. It is also frequently used, with a tinge of pomposity, in letters to the editor (of a newspaper, dictionary, etc.). Examples: *Their arguments are anent What nanny really meant*—W. H. Auden, 1952; *Adamant you'll find me anent 'aficionado'*—O. Nash, 1961; *the consensus view of the reading public anent poetry: they, too, dislike it*—TLS, 1984; *His Lordship had been much influenced by averments anent section 74*—Lord Jauncey (a Scottish law lord), 1988; *a Dr Malcolm Carruthers had delivered a lecture to Edinburgh's Lister Institute anent the astonishing cardiac benefits of the kilt*—A. Coren, 1989.

**aneurysm.** Correctly spelt thus (not *aneurism*), with the *y* answering to the second upsilon in Gk εὐρύς 'wide'. Form-association with other words ending in *-ism* has led to the adoption of *aneurism* by many writers, but *aneurysm* is still the better form.

**angina.** 'Progress in Plautine prosody', said Fowler (1926), led the *OED* (in 1884) to give precedence to the pronunciation /'ændʒmə/. But a knowledge of Plautine prosody was not shared by the general educated public, it would appear, as the word was recorded by Daniel Jones in 1917 as /æn'dʒamə/, and this pronunciation, with stress on the second syllable, is the only one given in major authorities since then.

**angle** (noun[1]). In the sense 'the point or direction from which one views or approaches a subject of inquiry, an event, etc.', *angle* has been in use since the 1870s. An event can be viewed from *every possible angle*. It is frequently used with a defining word: the *OED* (sense 1c) provides examples of *statistical angles*, *selling angles*, and *propaganda angles*. *Angle* also vies with *line* in the sense 'approach, line of argument'. Thus used it is a lightweight word best avoided in polished prose. *The Berlin Wall: the West German perspective* or . . . *the West German standpoint* are better than . . . *the West German angle*.

**angle** (noun[2]), **angler**, **angling**. An 'angle' was originally (in OE) a fishing-hook. Somewhat later (during the 15c.) it came to be used as a verb meaning 'to fish with a hook and bait'; and, in the 16c., *angler* emerged in the sense 'one who fishes with a hook and line'. These derivatives stood alongside *fish* (verb), *fisher*, and *fisherman*.

The original noun *angle* fell into disuse in the 19c., but the verb *angle*, the noun *angler*, and the verbal noun *angling* remain, though only in carefully designated uses.

The verb *angle* has survived mainly in transferred senses. Politicians *angle* for votes; most of us, at one time or another, *angle* for a present, an invitation, a compliment, etc. Yet, to show that the distinction is far from a fixed one, it is also idiomatic *to fish for compliments* and *to fish* (not *angle*) *in troubled waters*.

The distribution of the pair *angling/fishing* is far from straightforward. In certain contexts they can be used interchangeably, whilst in others they have their own restrictions. *Fishing*, which covers everything from jam-jars

through rod-and-line to trawlers, can sometimes be too general a term. *Angling* refers to rod and line only. No one says 'I'm going angling tomorrow'; in such a sentence 'fishing' would always be used. *Fly-fishing* is idiomatic; *fly-angling* is not.

Coarse fishing means fishing in a river or pond for roach, rudd, perch, pike, and other freshwater fish by anglers equipped with tackle, groundbait, maggots, and floats.

An *angler* is one who fishes with rod and line; *fisher* survives only in the biblical phrase (Matt. 4: 19) 'I will make you fishers of men'; *fisherman* is a generic term for a person who fishes for sport or one who goes out to sea in a fishing boat to earn a living.

Some examples: (angler) *There are reputed to be 3 million anglers in this country—Natural World*, 1988; *big rivers—the Test, Itchen, Kennet and Avon—which were the cradle of dry fly-fishing as a sport 100 years ago, and which many anglers believe are dying— Sunday Times*, 1990; *Irish anglers have defeated the Dublin government's plan to impose a Ir£15 trout and coarse fish licence throughout the Republic—Times*, 1990; *'Fishermen' is a sexist word, so we say 'anglers', because fishing includes men and women—Chicago Tribune Mag.*, 1993; (angling) *Angling is a sport that knows no social barriers—Clitheroe Advertiser & Times*, 1990; *Mullet angling is all about challenges—Times*, 1990; (fly-fishing) (title on cover, 1990) *Fly-fishing & fly-tying. The new magazine for the stillwater game angler*.

**anglice** /'æŋglɪsɪ/ = in English; in plain English. This adverb, first recorded in the 17c., is modelled on *latine* /'lætmɪ/ = in Latin. Similar adverbs are *celtice* in Celtic, *gallice* in French, *graece* in Greek, *hibernice* in Irish, and *scot(t)ice* in Scots, all with the final two letters pronounced as /-sɪ/. All these words are sometimes printed with final *-è* to show that the *-e* is sounded.

**Anglo** (noun). For two centuries this term has been used in Canada to designate English-speaking, as distinct from French-speaking, Canadians. Since the 1930s, and especially in the southwestern States of America, an *Anglo* is an American with an English-speaking background, as distinct from a person whose first language is Spanish.

**Anglo-.** People in Scotland and Wales understandably view this combining form (as in *Anglo-French*, *Anglo-Irish*, etc.) with some distaste, but it continues to be used in most standard works. The logical and more diplomatic element *Brito-* has not achieved widespread currency, except in the somewhat specialized terms *Brito-Arctic* (relating to British territory in the Arctic) and *Britocentric* (having Britain as a centre), *-centricity*, and a few other compounds.

**Anglo-Indian.** This term was first used in the early 19c. to denote a person of mixed British and Indian descent, i.e. a half-caste. At a slightly later date it was introduced as a term for a British civil servant, businessman, memsahib, etc., who had lived for many years in India. Since India became independent in 1947, both uses have tended to die out.

**ankylosis.** This form, answering correctly to Gk ἀγκύλωσις 'stiffening of the joints', has now virtually driven out the once dominant form *anchylosis*.

**annex(e).** The verb is *annex* and the noun *annexe*, except that in AmE the noun is usually spelt *annex*.

**annihilate, annihilation.** In the OED (1884) the medial *h* was fully pronounced. At some point since then it became silent in standard English and is now never pronounced. See VEHEMENT, VEHICLE.

**annual.** See PERENNIAL.

**annul.** So spelt, but with *-ll-* in *annulled* and *annulling*. The corresponding noun is *annulment*. See -LL-, -L-.

**Annunciation** /əˌnʌnsɪˈeɪʃən/, now restricted in meaning to 'the intimation of the incarnation, made by the angel Gabriel to the Virgin Mary' (OED), is to be distinguished from *enunciation* 'the uttering or pronouncing of articulate sounds; manner of utterance' (OED), pronounced with initial /ɪ-/.

**anniversaries.** The normal practice is to refer to the tenth, twentieth, thirtieth, etc., anniversary of an event, but some particular names have come to be applied to the more significant anniversaries. Among these are: (weddings) *silver*

(25 years), *ruby* (40), *golden* (50), and *diamond* (60, sometimes 75); (public events) *centenary* /sənˈtiːnərɪ/, /-ˈtenərɪ/ (100) (but in North America and some other English-speaking countries more usually *centennial*; so also in the relevant combinations that follow); *sesquicentenary* (150), *bicentenary* (200); (not recorded) *semiquincentenary* (250); *tercentenary* (300), *quatercentenary* (400), *quincentenary* (500), *sexcentenary* (600), *septcentenary* (700), *octocentenary* (800); (not recorded) *nonacentenary* (900); *millenary* (1,000).

**announcer.** This broadcasting term (first recorded in 1922) for a person who announces the subjects of a programme or reads the news has now largely given way to a group of somewhat more specific terms: *anchorman* (first recorded 1958), *anchorperson* (1973) (the compère of a radio or television programme); *newscaster* (1930), and *newsreader* (1925). In practice, however, the more usual formula adopted is *The news is read by* (name of the newsreader).

**anorexia.** This word meaning 'absence of appetite' and most frequently found in the expression *anorexia nervosa* 'a condition marked by emaciation, in which loss of appetite results from severe emotional disturbance', has yielded two main adjectives, *anorectic* and *anorexic*. Neither can be said to be the dominant form at present. Examples: *He became listless, anorexic, and increasingly sleepy, refusing to stand or crawl*—Lancet, 1961; *This condition, which almost exclusively affects females, on average slightly older than the anorectic patient, comprises chaotic eating patterns with alternate bouts of carbohydrate bingeing*—Listener, 1985.

**another.** For *one another* see EACH 2.

**antagonize.** In the course of the 20c. this verb has gradually but not entirely moved away from its traditional sense 'to contend with, to oppose' (*The Democrats on the committee have given notice of a determination to antagonize this and all other bills for the admission of Territories as States*—Boston (Mass.) newspaper, 1882; *Our first object must be to antagonize the poison and at the same time uphold his powers*—J. G. Farrell, 1973), and established its main current sense, 'to make (someone) antagonistic, to incur or provoke the hostility of (someone)': *They conducted their duties humbly and reticently ... and went to great lengths not to antagonize anyone*—J. Heller, 1961; *so antagonized a group of men at a Fort Hanna tavern that they turned upon him*—J. C. Oates, 1980.

**Antarctic, Antarctica.** A widespread and regrettable tendency to omit the medial *c* when pronouncing these words should be resisted.

**ante-, anti-.** The first means 'before, preceding, in front of' (*antenatal* 'happening or existing before birth', *antebellum* 'before a (specified) war', *antechamber* 'chamber or room leading to a more important one'). The second, which is much more frequent, means 'opposite; opposed to; against' (*anti-hero* 'a person totally unlike a conventional hero'; *anti-American, -Semitism; anti-aircraft (gun)*).

As the *OED* points out, the analogy for the limitless number of *anti-* formations seems to have been given by *Antichrist* and its adj. *antichristian*, which (with the analogous *antipope*) were almost the only examples in use before 1600. Shakespeare has no independent *anti-* compounds (though he did use 'fused' *anti-* words such as *antidote*, *antipathy*, and *antipodes*).

In standard English both prefixes are pronounced /ˈæntɪ/, but in AmE *anti-* is pronounced /ˈæntaɪ/.

**antenna.** In the sense 'a sensory organ found in pairs on heads of insects, crustaceans, etc.', the pl. is *antennae*. In the sense '(radio) aerial', the pl. is *antennas*.

**antepenult.** A term of prosody and phonology, 'preceding the penult; the last but two', as in 'áltitude' and 'heaviness'.

**anterior.** It is worth noting that *anterior* is in English comparative in sense but not in construction. We say *anterior to*, not *anterior than*: *The hysteric's deception of himself is anterior to his deception of others*—R. D. Laing, 1961; *There is a certain paradoxical logic to thinking of writing as anterior to speech*—*Paragraph*, 1986.

**antetype.** See -TYPE.

**Anthony.** In standard English pronounced /ˈæntənɪ/ with medial /t/ not /θ/. But the pronunciation with medial /θ/ is common in regional speech in the UK.

**anthropoid.** After a period of uncertainty (the word is first recorded in 1832), it is now always pronounced as /ˈænθrəpɔɪd/, not /æn'θrəʊpɔɪd/ or /-pəʊɪd/.

**anthropophagi.** This learned word, familiar esp. from Shakespeare's *Othello* (*And of the Cannibals, that each other eate, The Anthropophagie*), is plural in formation and means 'man-eaters, cannibals'. It answers to Gk ἀνθρωποφάγ-ος 'man-eating', from ἄνθρωπος 'man' + φαγεῖν 'to eat', and is pronounced in English as /ænθrə'pɒfəgaɪ/ or /-dʒaɪ/.

**anticipate.** The *OED* recorded nine senses of this verb, the chief two of which are rendered in *The Oxford Miniguide to English Usage* (1983) as (1) To be aware of (something) in advance and take suitable action, to deal with (a thing) or perform (an action) before someone else has had time to act so as to gain an advantage, to take action appropriate to (an event) before the due time, e.g. *His power to ... anticipate every change of volume and tempo* (C. Day Lewis); *I shall anticipate any such opposition by tendering my resignation now* (Angus Wilson); *She had anticipated execution by suicide* (R. Graves); *Some unknown writer in the second century ... suddenly anticipated the whole technique of modern ... narrative* (C. S. Lewis). (2) To take action before (another person) has had the opportunity to do so, e.g. *I'm sorry— do go on. I did not mean to anticipate you* (J. le Carré).

Fowler presented and scornfully rejected a third meaning, 'to expect, to foresee'. Educated people of his generation and mine opposed this encroachment on the core meaning of these two verbs, but the tide kept moving in the opposite direction, and this third sense now seems to have become widely adopted. Examples: *The programmes were certainly more popular with viewers than many had anticipated*—*Parliamentary Affairs*, 1986; *Another thing he ought to have anticipated was the plethora of Rolls-Royces on the streets*—T. Clancy, 1987; *It is anticipated that it will deal with perhaps 50 to 100 of the most serious ... frauds*—*Counsel*, 1987; *Wing mirrors were selling better than they had ever anticipated*—M. Drabble, 1987; *Patricia's knife and fork and plate worked out much as Rose had anticipated*—I. Murdoch, 1987; *when he noticed his safe had been tampered with in the Soviet Embassy, he defected earlier than anticipated*—Peter Wright, 1987; *They have every right to be there, and we don't anticipate any change in that status*—USA Today, 1988; *One would not expect Cleopatra to have suffered such a fate, nor did she herself anticipate it*—A. Fraser, 1988.

This third sense divides people across the Usage Wall. Insults about it will continue to be hurled. At least it should be pointed out that usurpation of part of the territory of the verb *expect* leaves it with considerable independence. Unlike *expect*, *anticipate* cannot be followed by infinitive constructions: *I anticipate to come to Cambridge next week* is an incorrect use. Nor can *anticipate* mean 'expect as one's due': *I expect* (not *anticipate*) *payment for this*. And *expect* is not replaceable in the informal statement *expect me when you see me*.

**antistrophe.** **1** The returning movement, from left to right, in Greek choruses and dances, answering to the previous movement of the strophe from right to left. (*OED*) Cf. STROPHE 1.

**2** In rhetoric and grammar, the repetition of words in inverse order, e.g. *You must say what you mean and mean what you say.*

**antithesis.** In rhetoric, 'an opposition or contrast of ideas, expressed by using as the corresponding members of two contiguous sentences or clauses, words which are the opposites of, or strongly contrasted with, each other: *as he must increase, but I must decrease', in newness of spirit, not in oldness of the letter'* (*OED*). Antithesis is a marked feature of e.g. Samuel Johnson's style: *The colours of life in youth and age appear different, as the face of nature in spring and winter; The old man pays regard to riches, and the youth reverences virtue; Marriage has many pains, but celibacy has no pleasures* (*Rasselas*, ch. 26).

**antitype.** See -TYPE.

**antonomasia.** In rhetoric, (*a*) the substitution of an epithet or descriptive phrase for a proper name, e.g. *the Iron Duke* for 'the Duke of Wellington', *the Iron Lady* for 'Mrs Margaret Thatcher'; (*b*) the use of a proper name to express a general idea, e.g. *a Solomon* for 'a wise man', *a*

*Cicero* for 'an orator'. The original Gk word meant literally 'name instead'.

**anxious.** Anxiety (first recorded in a *c*1525 work by Sir Thomas More) lies at the basis of the traditional meaning of the adj. *anxious*, namely 'the quality or state of being anxious, uneasiness of mind'. In the 20c., psychiatric terms like *anxiety neurosis* have strengthened the belief that a morbid state of mind lies behind the words *anxiety* and *anxious*. Meanwhile, in the 18c., the adjective began to turn on its axis and came also to mean 'full of desire and endeavour; earnestly desirous (to bring about some purpose)'. The phrase *anxious to please* appeared in Robert Blair's poem *The Grave* (1743), and Lord Nelson declared in 1794 that *The General seems as anxious as any of us to expedite the fall of the place*. Other writers adopted the new use: *Punch was always anxious to oblige everybody*—Kipling, 1888; *All seemed pleased with the performance, and anxious for another of the same sort*—K. Amis, 1954; *You must be anxious to see your folk*—S. King, 1979; *She's very anxious that you should like her*—A. N. Wilson, 1982.

**any.** **1** As *pronoun* (*a*) 'Any is distinguished from *either* in representing a choice between three or more, while *either* limits the choice to two' (*CGEL* 6.61). This primary distinction should be borne in mind.

*Any* is correctly used with both singular and plural nouns, and also with 'non-count' nouns like *homework* and *merriment*. Examples: (sing.) *the right of any journalist to refuse to reveal his sources; any promiscuous person is at risk from AIDS*; (pl.) *I haven't been keeping any secrets from you; the Whitehall advisers were refusing to answer any questions*; (non-count) *any food found in passengers' luggage will be confiscated; will there be any music at the party?*

These uses show *any* used as a pronominal determiner. It is also correctly used as a simple pronoun in the same kind of sentences: *these values will never solve any of the world's problems; if you keep ferrets don't let any escape; It's as good an excuse as any to buy a new car.*

(*b*) *Any*, *anybody*, and *anyone* (as well as other indefinite pronouns) are now frequently, though somewhat controversially, followed by the plural pronouns *they* or *their*: e.g. *Can any illegal migrants*

**any**

*entering the country be sure that* they *will not be deported?*; anybody/anyone *who wants to improve* their *writing may attend the course.* Popular usage and historical precedent favour the use of a plural pronoun in such contexts, but many writers prefer to use *he* (*him, himself*) or *he or she*. See AGREEMENT 6.

(c) *In comparisons.* A fine net of illogicality mars constructions of the types *this is the most brutal piece of legislation of any passed by this government* (read *this is a more brutal piece of legislation than any other passed by this government*), and *a better book than any written by this author* (read *any others*).

**2** *As adverb. Any* can be correctly used as an adverb to emphasize a comparative adjective or adverb: *they are not treated like schoolgirls any longer; he can't play any better; she refuses to go any further.* In AmE, and occasionally in BrE, in informal contexts, it can stand alone with the sense 'at all': *We're used to responsibility. Doesn't worry us any*—A. Christie, 1937; *But it's not going to help any with my exams*—New Yorker, 1988.

**3** *Compounds.* (a) The pronouns *anybody* and *anyone* are interchangeable in most contexts, e.g. *What they keep in their handbags is anybody's* (or *anyone's*) *guess; they were giving free beer to anyone* (or *anybody*) *who played; the little boy staring out of these pictures could be anyone* (or *anybody*).

(b) For occasional occurrences with a following plural pronoun, see 1b above.

(c) *Anyhow* and *anyway* are interchangeable as adverbs, though I have the impression that *anyway* is now much the more frequent of the two. Examples: *Anyhow I'm carving out a career there teaching the boss's daughter to read novels*—T. Keneally, 1985; *Home is not the place for charm anyway*—London Rev. Bks, 1987; *Anyway, I want to get on the N train because it has a more artistic crowd*—New Yorker, 1987; *I'll prepare the list anyway, she decided, and a tentative date*—B. Rubens, 1987.

These adverbial uses of *anyway* are to be distinguished from the nominal phrase *any way: someone approached me to ask if there was any way I could help.*

(d) *any more, anymore.* Logically it would seem sensible to reserve the separated form for contexts in which the sense is 'even the smallest amount' (*the boy had eaten two of the apples and refused to eat*

any more of them) and the one-word form for the sense 'any longer'. But the language does not work as neatly as that. By and large *any more* is used in all areas when the sense required is 'even the smallest amount'. When the required sense is 'any longer' there are sharp divisions. AmE and other forms of English outside the UK tend to favour *anymore*, and this form is now being adopted by some British writers and publishing houses. The majority of authors and printers in the UK, however, still print *any more* for this second sense. Examples (all in negative contexts): (any more) *There's nothing for me in London, John, not any more*—M. Wesley, 1983; *no one will know any more what history is*—T. Keneally 1985 (Aust.); *I don't look at myself any more*—B. Rubens, 1985; *Nobody was going to look for his potential any more*—A. L. Barker, 1987; *He is not lying there any more*—P. Lively, 1987; *Baumgartner could not look into his face any more*—A. Desai, 1988; *he won't say the Pledge of Allegiance any more until the words 'under God' are excised*—Chicago Tribune, 1989; (anymore) *he wasn't a schoolkid anymore*—M. du Plessis, 1983 (SAfr.); *God, no jokes even in this hut anymore!*—V. O'Sullivan, 1985 (NZ); *You seldom hear about love anymore*—S. Bellow, 1987; *We don't call them wigs or hairpieces anymore*—New Yorker, 1987; *But Britain is not that sort of country anymore*—Sunday Times, 1988.

Contexts remain in which the words must be kept separate, e.g. *Mrs Carbuncle can't buy coal any more than the Fordyces can*—New Yorker, 1987. But even in such circumstances some writers crudely obscure the meaning by allowing the two words to merge: *She was mysterious and wouldn't tell me anymore* ( = anything else)—J. Winterson, 1985; *You know, as parents we don't know anymore than these kids know*—Yale Alumni Mag., 1987.

In American regional use, *anymore* is also used in positive constructions in the sense 'nowadays, now'. The *Dictionary of American Regional English* cites examples (beginning with one of 1859) from virtually every State, e.g. *We all use night-crawlers anymore; We put up quite a bit of hay here anymore; He's hard of hearing anymore.* Harry S. Truman is cited as saying, It sometimes seems to me that all I do anymore is to go to funerals.

(e) *anyone* ( = anybody, any person). Its normal uses (*she wasn't scared of anyone; you are asked not to discuss the case with anyone*) are to be distinguished from *any one* = 'any single (person or thing)' (*members may vote for any one of the above candidates*).

(f) *anyplace*. A markedly American adverb ( = UK *anywhere*) first recorded in the second decade of the 20c. Examples: *They may have figured that since I could swim like a fish I was as safe in the water as anyplace else—New Yorker*, 1986; *Why didn't we ever get to go anyplace—New Yorker*, 1988; Even in AmE, however, it is much less common than *anywhere*.

(g) *anytime*. Another characteristically American adverb ( = at any time). Examples: *She said she would vote for him anytime—New Yorker*, 1987; *I wouldn't have wanted to know her as a child, but once a man, anytime—M. Doane*, 1988.

(h) *anyway*. See 3c above.

(i) *any ways, anyways*. As an adverb = 'in any way, in any respect, at all', used in the *Book of Common Prayer* (*All those who are any ways afflicted... in mind, body, or estate*), in AV (*And if the people of the land doe any wayes hide their eyes from the men*), and in many literary contexts during the last four centuries. It seems to have dropped out of standard UK use now, though it survives in regional use. It is also encountered (always written as one word, *anyways*) in informal AmE, e.g. *So who promised this guy anything anyways? —Reader* (Chicago), 1983.

**aorist.** (Gk ἀόριστος 'indefinite'.) 'One of the past tenses of the Greek verb, which takes its name from its denoting a simple past occurrence, with none of the limitations as to completion, continuance, etc., which belong to the other past tenses. It corresponds to the simple past tense in England, as "he died"' (*OED*).

**apart from.** Standard in the UK from the early 17c. onward, e.g. *Apart from her role in displaying ... the collection in the new galleries, she has made a number of significant scholarly contributions—Ashmolean*, 1987; *the raven, who apart from anything else was much stronger in the air than the dove—Julian Barnes*, 1989. The equivalent expression in AmE from the early 19c. has been *aside from*, e.g. *Aside from that, the church leadership had trouble figuring out exactly what to do about him—*

*New Yorker*, 1986. The indications are that *aside from* is now dominant in other English-speaking countries (*Aside from which we must have something to show for our pains—M. Shadbolt*, 1986 (NZ)), and is beginning to supplant *apart from* among younger people in the UK. *Apart from* is, however, by no means extinct in AmE, e.g. *Apart from him, I don't know of anybody who ever made any particular very important mark in the world—L. Kirstein*, 1986; *you need four things in order to have New York City—I mean, apart from a piece of real estate on which x million people happen to be receiving goods and services—New Yorker*, 1990.

**apartheid.** Correctly pronounced /ə'pɑːtheɪt/ in standard English, not /ə'pɑːtheɪd/.

**apex.** The preferred pl. form is *apexes*, though *apices*, pronounced /'eɪpɪsiːz/, is still sometimes used. The corresponding adj. is *apical*. See -EX, -IX; LATIN PLURALS.

**aphaeresis.** 'The taking away or suppression of a letter or syllable at the beginning of a word.' (*OED*), e.g. *coon, cute,* and *round* for *raccoon, acute,* and *around*. See next.

**aphesis.** (Gk ἄφεσις 'a letting go'.) J. A. H. Murray's term for 'the gradual and unintentional loss of a short unaccented vowel at the beginning of a word; as in *squire* for *esquire, down* for *adown, St Loy* for *St Eloy, limbeck* for *alimbeck* ... It is a special form of the phonetic process called *Aphæresis* for which, from its frequency in the history of the English language, a distinctive name is useful. Now also used in the sense of APHÆRESIS' (*OED*). The corresponding adj. is *aphetic*.

**apiece.** This adverb, 'which marks segregatory meaning' (*CGEL*), is normally placed immediately after a direct object: e.g. *after buying his brothers a pint apiece [he] had to be content with half for himself—M. Bragg*, 1969; *the actresses have one beautiful costume apiece—New Yorker*, 1987.

**aplomb.** Pronounce /ə'plɒm/. The pronunciation /ə'plʌm/ is not wrong but is now a minority one.

**apocope.** /ə'pɒkəpi/ (Gk ἀποκοπή 'a cutting off'.) The cutting off or omission of the last letter or syllable(s) of a word, e.g. *curio* for *curiosity, cinema* for *cinematograph*; and, in more ancient times, OE *rīdan* → early ME *rīden* (two syllables) → later ME *rīde* (still two syllables) → modE *ride* (one syllable).

**apodosis** /əˈpɒdəsɪs/. The concluding clause of a sentence, as contrasted with the introductory clause or *protasis* /ˈprɒtəsɪs/; now usually restricted to the consequent clause in a conditional sentence, as 'If thine enemy hunger, *feed him*' (OED). In *CGEL*, the *apodosis* is called a matrix clause and the *protasis* a subordinate clause.

**apo koinou.** (Gk, 'in common'.) Applied to a construction consisting of two clauses which have a word or phrase in common. A standard example (cited by Visser) is *I have an uncle is a myghty erle*, in which *an uncle* is the object of *I have* and also introduces the following clause. The construction, which is common in Shakespeare's work (*You are one of those Would haue him used againe—Winter's Tale* v.i), survived into the 19c. and to some extent in the 20c. Examples: *There was a judge's daughter at Demerara went almost mad about him—*Thackeray, 1847/8; *I had it all planned out to go there this summer with a friend of mine lives in Winnipeg—*Sinclair Lewis, 1926; *You're the guy was in the papers—*T. Clancy, 1987.

*The Oxf. Dict. Eng. Gramm.* (1994) correctly comments 'The term is not in general use today; such a construction would be called a *blend* or treated as deviant or as an example of *anacoluthon*.'

**apology.** The normal word for a regretful acknowledgement of fault or failure, an assurance that no offence was intended (*I owe you an apology*; *he made his apologies to the chairman and left the meeting early*). An *apologia* is a written defence or justification of the opinions or conduct of a writer, speaker, etc., the currency of the word being largely due to J. H. Newman's *Apologia pro Vita Sua*, 1864. An *apologue* is a moral fable, like Aesop's Fables or George Orwell's *Animal Farm* (1945).

**apophthegm.** A terse, pointed saying; a maxim. Pronounce /ˈæpəθem/, despite the fact that leaving the *ph* silent conceals the derivation of the word from Gk ἀπόφθεγμα. In AmE frequently spelt *apothegm*, a spelling that was the more

usual in England till preference was expressed for *apophthegm* in Johnson's Dictionary.

**aposiopesis** /ˌæpɒsɪəʊˈpiːsɪs/ or /ˌæpəʊsaɪəʊˈpiːsɪs/. (Gk, noun of action, corresponding to a verb meaning 'to keep silent'.) A rhetorical artifice, in which the speaker (or writer) comes to a sudden halt, as if unable or unwilling to proceed (OED). Such ellipses are often the result of an emotional state of mind, e.g. *'Well, I never'—said she—'what an audacious'—emotion prevented her from completing either sentence—*Thackeray, 1847/8. But there are many other kinds, e.g. *I haven't the foggiest* [notion]; *Of all the ... [implying that 'this is the worst']; What the ...!; She is in fact a—. But I refrain from saying the word.*

**a posteriori** /eɪ pɒˌsterɪˈɔːraɪ/. (L, = 'from what comes after'.) A phrase used to characterize reasoning or arguing from known facts to probable causes. 'The prisoners have weals on their backs, so they must have been whipped' is an example of a posteriori reasoning. Contrasted with A PRIORI.

**apostrophe**[1]. (Gk, 'a turning away'.) An exclamatory passage in which a speaker or writer pointedly addresses some person or thing, either present or absent. Examples: *O proud death What feast is toward in thine eternall cell ...? (Hamlet); Busie old foole, unruly Sunne, Why dost thou thus, Through windowes, and through curtaines call on us?* (Donne); *Milton! thou shouldst be living at this hour: England hath need of thee* (Wordsworth); *Red Rose, proud Rose, sad Rose of all my days! Come near me, while I sing the ancient ways* (Yeats).

**apostrophe**[2].

| |
|---|
| **A** History. |
| **B** Some golden rules. |
| **C** General. |
| **D** Possessives. |
| **E** Relinquishment of the apostrophe. |

**A** *History.* The mark ' was introduced in English in the 16c. to indicate that a letter or letters had been omitted. The apostrophe before *s* became regulated as an indication of the singular possessive case towards the end of the 17c., and the apostrophe after *s* was first recorded as an indication of the plural possessive case towards the end of the 18c. Since then gross disturbances of these basic patterns have occurred in written and printed work, as will be evident from what follows. Such instability suggests that further disturbances may be expected in the 21c.

**B** *Some golden rules.*

**1** An apostrophe is required before a possessive s in the singular (*the boy's hat, the water's edge*) and after a possessive s in the plural (*the boys' gymnasium, the ladies' maids, in four days' time*). Except that in the small group of words that do not end in -s in the plural, the plural possessive is indicated by 's (*children's shoes, men's boots, women's handbags, the oxen's hoofs*).

**2** *It's* = 'it is' (see **C2** below). *Its* (no apostrophe) is the possessive form of the pronoun *it* (see **D5** below).

**3** *Errant apostrophes.* From the 17c. onwards an apostrophe was often used in the plural number when the noun ended in a vowel, e.g. *grotto's, opera's, toga's*. Since the mid-19c., grammarians have condemned this use, but it continues to appear, to the amusement of educated people, in signs and notices, especially in shop windows (*potato's 10p a 1b, video's for rent*). Henry Alford (*The Queen's English*, 1864) reported, 'One not uncommonly sees outside an inn, that "fly's" and "gig's" are to be let.' This use is often called the greengrocers' (or grocers') apostrophe because of the frequency with which plural forms such as *apple's, cauli's*, and *orange's* appear in their shops.

**4** For proper names ending in -y, use s (not 's or -ies) in the plurals (*the two Germanys, two bitter Januarys, three Hail Marys*).

**5** 's is legitimately used as an informal shortened form of *is* (*the joke's on him*), *has* (*he's got a knife*), *as* (*I'm sore's hell*), *does* (*What's he do?*), and *us* (*let's go*); and (very informally) = *it is* ('*S that bloody comet*), and *that's* ('*By car?' 'Sright.'*). Each of these uses is fully treated in the *OED*.

**6** For the aberrant type *Who's turn to deliver?* see WHO'S.

**C** *General.*

**1** *Abbreviations.* Though once commonly used in the plural of abbreviations and numerals (*QC's, the 1960's*), the apostrophe is now best omitted in such circumstances: *MAs, MPs, the 1980s, the three Rs, in twos and threes*, Except that it is normally used in contexts where its

omission might possibly lead to confusion, e.g. *dot your i's and cross your t's; there are three i's in inimical; the class of '61* ( = 1961).

**2** *Contractions.* Apostrophes in contractions of the type *I'll* = 'I will' should be joined close up to the letters of either side of them: *don't, haven't, isn't, shan't, won't; I'll, he'll, we'll; I'd, he'd, she'd; I've, you've; we're, they're; he's, she's, it's* ( = it is).

Apostrophes are no longer needed in *cello, flu*, and *phone* (for *violoncello, influenza*, and *telephone*) as these are now established words in their own right.

**3** An apostrophe is correctly used in a variety of special formations to indicate that a letter has been omitted: *e'en* (even), *fo'c's'le* (forecastle), *ne'er-do-well, o'er* (over), *rock 'n' roll.*

**4** For the use of *'d* for -ed, see -ED AND 'D.

**5** *Miscellaneous.* No apostrophe in *Guy Fawkes Day*. 's in *St Elmo's fire, St George's Cross, St John's wort, (St) Valentine's Day*. s' in *All Saints' Day, April Fools' Day, Presidents' Day* (q.v.).

**D** *Possessives.*

**1** For the ordinary uses, see B1 above.

**2** *Personal names.* Use 's for the possessive case in English names and surnames whenever possible; i.e. in all monosyllables and disyllables, and in longer words accented on the penult, as *Burns's, Charles's, Cousins's, Dickens's, Hicks's, St James's Square, Thomas's, Zacharias's*. It is customary, however, to omit the 's when the last syllable of the name is pronounced /-ɪz/, as in *Bridges', Moses'. Jesus'* is an acceptable liturgical archaism.

**3** *Classical names.* In ancient classical names use s' (not s's): *Mars', Herodotus', Venus'*. Ancient names ending in *-es* are usually written *-es'* in the possessive: *Ceres' rites, Xerxes' fleet*. In longer words of this type, *-es'* should also be used: *Demosthenes', Euripides', Socrates', Themistocles'*.

**4** *French names.* Those ending in s or x should always be followed by 's when used possessively in English. It being assumed that readers know the pronunciation of the French names (in *Rabelais, le Roux*, and *Dumas*, for example, the final consonant is left unpronounced), the only correct way of writing these names in the possessive in English is *Rabelais's* /ˈræbəleɪz/, *le Roux's* /lə ˈruːz/, *Dumas's* /ˈdjuːmɑːz/.

**5** *Pronouns.* An apostrophe must not be used with the pronouns *hers, its, ours, theirs, yours.* But an apostrophe is required in possessive indefinite or impersonal pronouns: *anybody's game, each other's books, one's sister, somebody else's fault* (see ELSE 1), *someone's hat.*

**6** *In place-names.* Use an apostrophe before the *s* in *Arthur's Pass* (NZ), *Land's End, Lord's Cricket Ground, The Queen's College* (Oxford), *St John's* (Newfoundland), *St John's Wood* (London), *St Michael's Mount* (Cornwall). After the *s* in *Queens' College* (Cambridge). Do not use an apostrophe in *All Souls* (Oxford), *Bury St Edmunds, Earls Court, Golders Green, Johns Hopkins University* (Baltimore), *St Albans, St Andrews, St Kitts* (Caribbean). Recent standard editions of local maps are the best guide to the correct spelling of the hundreds of names of this type.

**7** *Group possessives.* These normally require an apostrophe only after the last element, e.g. *the Duke of Edinburgh's speech, Faber and Faber's address, Lee and Perrin's sauce, the Queen of Holland's crown, Lewis and Short's Latin Dictionary, my aunt and uncle's place, a quarter of an hour's chat.*

**8** There is no agreed solution to the problem of the types (a) *Hannah's [Jamie Lee Curtis] love interest,* in which the heroine Hannah in a TV film is played by an actress called Jamie Lee Curtis. The alternatives are (b) *Hannah [Jamie Lee Curtis]'s,* or (c) *Hannah's [Jamie Lee Curtis's].* My preference is for type (b), but type (a) is also legitimate, whereas type (c) seems over-fussy.

Some other examples (drawn from American sources of 1989–90) of multiple apostrophes (or their omission) which tend to be resolved in an *ad hoc* manner: *Wayne's daughter Kim's latex ear* (correct); *It [Burger King's] decision was not unexpected* (read *Its*); *Michael's mother's new boyfriend* (correct); *A former boxer ignores Gillespie's (Carroll O'Connor) advice* (read *Gillespie (Carroll O'Connor)'s*); *It's Ronny Finney, Fatty Finney's brother's son's second boy* (correct but clumsy).

**9** For the type *the sentence's structure,* see 'S AND OF-POSSESSIVE.

**10** For the type *a friend of my mother's,* see DOUBLE POSSESSIVE.

**11** For the type *for appearances' sake, for Jesus' sake,* etc., see SAKE.

**12** For the type *the Council's abolition* ( = the abolition of the Council), see OBJECTIVE GENITIVE.

**E** *Relinquishment of the apostrophe.* Since about 1900, many business firms, institutions, and journals have abandoned apostrophes in their titles, e.g. *Barclays Bank, Citizens Advice Bureau, Diners Club, Farmers Weekly, Harrods, Mothers Pride Bread,— Teachers Training College.* (It can be argued that in some of these the word ending in *s* is a plural word used attributively.) Some other names appear in various forms, with and without apostrophes, e.g. *F. W. Woolworth/ Woolworths/ Woolworth's.* This trend towards the dropping of the apostrophe in such names and titles seems certain to continue.

**appal.** The customary BrE spelling, whereas *appall* is more usual in AmE. The derivative forms are *appalled* and *appalling* in both countries. See -LL-, -L-.

**apparatus.** Pl. *apparatuses.* In standard English pronounced /æpə'reɪtəs/, but abroad (esp. in Australia and NZ) /-'rɑːtəs/ and in AmE /-'rædəs/. See -US 2.

**apparent.** The pronunciation /ə'pærənt/ is now the dominant one, with /ə'peərənt/ heard only occasionally. The reverse was the case when the relevant section of the *OED* was issued in 1885.

**appeal.** In the US commonly used in legal contexts as a transitive verb = 'to appeal against', e.g. *Curtis has announced that it will appeal the verdict—Publisher's Weekly,* 1963. The normal equivalent in Britain is to *appeal against* the verdict of a lower court. The transitive use appears occasionally in UK sources, e.g. *The mother appealed the court's decision—Internat. Jrnl of Law & Family,* 1988.

**appear(ed).** For its use (and also *seem(ed)*) with a perfect infinitive (e.g. *GEC appears to have taken a firm grip on the project),* see PERFECT INFINITIVE 1.

**appeasement.** For centuries used without discreditable or humiliating overtones in the broad sense 'the act or process of giving satisfaction, pacification'. Even in 1920 Winston Churchill could say, with reference to Turkey, and merely as an expression of opinion worth examining, *Here again I counsel prudence and appeasement. Try to secure a* *really representative Turkish governing authority, and come to terms with it.* Since 1938, in political contexts, the word has regularly been used disparagingly with allusion to the attempts at conciliation by concession made by Mr Neville Chamberlain, the British Prime Minister, before the outbreak of war with Germany in 1939. It cannot be used now other than disparagingly.

**appendix.** The recommended pl. forms are *appendices* (in books and documents) and *appendixes* (in surgery and zoology). But this distinction, though a useful one, cannot be said to have been universally adopted. See -EX, -IX.

**applicable.** Walker (1791), *OED* (1885), and Gimson (1977) all recommend placing the stress on the first syllable, but at every turn one now hears the word stressed on the second syllable. The passage of time will settle the matter one way or the other.

**apposition. 1** The placing of a noun or noun phrase beside, or in (exact) syntactic parallelism with, another noun or noun phrase. This is a major feature of the language, and there are many types. The parallel elements are known as appositives.

In the sentence *Sir James Murray, the lexicographer, was born in Hawick,* the second element, *the lexicographer,* is appositive to the first, *Sir James Murray.* Similarly, in the sentence *The highest mountain in New Zealand, Mount Cook, is called Aorangi by the Maoris,* the second element, *Mount Cook,* is appositive to the first, *The highest mountain in New Zealand.* In both cases the second element syntactically duplicates the first. This is the most straightforward type of apposition in English.

**2** Appositives may be either restrictive (i.e. defining) or non-restrictive (i.e. descriptive), though there is considerable overlap between the two types: (restrictive) the grammarian *Otto Jespersen; William* the Conqueror; *Blenheim Palace,* the Duke of Marlborough's house in Oxfordshire; (non-restrictive) *he picked up the goods at the warehouse,* a huge complex of brightly painted buildings; *she loved the paintings of Claude Monet,* one of the leading exponents of impressionism.

**3** The appositive element is placed first in the type 'title or descriptive label + personal name', e.g. Chancellor *Kohl of West Germany;* singing sensation *Bles Bridges;* British Rail lobbyist *Richard Faulkner;* Caledonian Society secretary *John McGregor.* Originally US, this type of construction is now rapidly becoming adopted in all English-speaking countries, esp. in newspapers.

**4** Appositives can be introduced by *and* or *or,* e.g. *he worked in Duke Humfrey, a section of the Bodleian Library, and* arguably one of the most elegant rooms in the whole library; *their political interests lay in the Gulf, or* the Persian Gulf as it is often called.

**5** Most of the above types show the appositives divided by a mark of punctuation, but this sentence from John Fuller's *Flying to Nowhere* (1983) shows a type not requiring punctuation: *But I do not find* a tipsy man a reliable witness.

For a full-scale treatment of numerous other classes of full and partial appositives, readers are referred to *CGEL* 17.65–93.

**appraisal.** This word has now nearly ousted *appraisement* in non-technical language, though the latter word was dominant in the 19c. See -AL. The derivative *reappraisal* (used esp. in the phrase *agonizing reappraisal*) is a 20c. formation.

**appraise, apprise.** Confusion of near-sounding words or 'minimal pairs' is an ever-present possibility. *Appraise* means 'to assess the value of (something or somebody)': *When a man is stripped of all worldly insignia, one can appraise him for what he is truly worth*—C. Chaplin, 1964; *The curtain of motion and colour had been momentarily lifted, so that the reality behind it could be appraised*—J. Fuller, 1983; *it was an interval at least long enough for him to appraise the situation*—Antonia Fraser, 1988. *Apprise* means 'to inform, to give notice to', and occurs normally in the construction *apprise (someone) of* (something): *Attlee asked the Chiefs of Staff to confirm that they were fully apprised of the shipping situation*—D. Fraser, 1982; *He was . . . annoyed that I had not bothered to apprise him of the upsetting news sooner*—P. Bailey, 1986.

**appreciate. 1** Pronounce the *-ci-* as /ʃɪ/ not /sɪ/.

**2** Its normal use in business letters to mean 'esteem at full value, acknowledge with gratitude' (*I appreciate everything that you have done to help us*) is unexceptionable. So too is its use to express polite requests, e.g. *it would be appreciated if you would send your invoice in triplicate*. There is some justification, however, in Sir Ernest Gowers' suggestion that writers of such letters should avoid using constructions of *appreciate* followed by a *how* or a *that* clause, e.g. *I appreciate* (better *understand*) *how hard it is for you to make ends meet without a housing allowance*; *I appreciate* (better *realize*) *that you are disappointed by our failure to promote you*.

**apprehend, comprehend.** In so far as the words are liable to confusion, i.e. as synonyms of *understand*, *apprehend* denotes the grasping or discerning (of some general idea), and *comprehend* the capacity to follow the broad outline (of an argument, a person's language, behaviour, etc.). Examples: (apprehend) *Neither could apprehend the nature of their relationship, and each was flattered by it*—P. White, 1957; *She drew a breath, long enough to apprehend that he was about to step from one world into another*—I. Murdoch, 1962; *There are some natures too coarse to apprehend the mysteries*—G. Vidal, 1962; (comprehend) *He wandered off, thinking I was round the bend, not comprehending my complete and luminous sanity*—W. Golding, 1959; *Speak more slowly so that we can comprehend everything you say*—B. Malamud, 1966; *To comprehend language fully, to assemble it correctly and to express it properly is a task that has no equal in human capability*—Anthony Smith, 1984.

**apprise.** In the sense 'to give notice to' always thus spelt, not -*ize*. See APPRAISE; -ISE 1.

**apricot.** Pronounced with initial /eɪ-/ in BrE but mostly as /ˈæprɪkɒt/ in AmE.

**a priori** /eɪ praɪˈɔːraɪ/. (L, = 'from what is before'.) A phrase used to characterize reasoning or arguing from causes to effects, deductively. *Because they were wearing handcuffs it was obvious that they had been taken into custody* and *Because I've failed I live on her money* are examples of a priori reasoning. Contrasted with A POSTERIORI.

**apropos.** /ˈæprəpəʊ/. Brought into English at the time of Dryden from French *à propos* 'with reference to', *apropos* is now always written as one word in English and without an accent. It can be used as an adj.; *Instead of answering me directly he said something (I didn't know at the time whether it was apropos or not), the significance of which I realised only later*—S. Themerson, 1951; but its main uses are (1) as a preposition, e.g. *Voltaire has something to say on English irregularity too, this time apropos Shakespeare*—N. Pevsner, 1956; *Her voice, as has been mentioned apropos that of Boudica, was not harsh*—Antonia Fraser, 1988; (2) with *of*, forming a compound preposition, e.g. *Apropos of nothing she declared that love must be wonderful*—G. Clare, 1981; (3) occasionally, under the influence of *appropriate*, followed by *to*, e.g. *Is there not a passage in John Stuart Mill apropos to this?* Type (3) is not recommended.

**apt, liable.** When used with a following *to*-infinitive these adjectives are frequently synonymous. In the type *be apt to*, however, *apt*, meaning 'to have a natural tendency', carries no necessary implication that the state or action expressed by the infinitive is undesirable from the point of view of its grammatical subject, e.g. *Her morning routine was as set as a single person's is apt to be*—S. King, 1979; *she . . . began to finger an old Christmas decoration that the cat was apt to play with*—M. Bracewell, 1989. It indicates simply that the subject of the sentence has often behaved in this way before.

The type *be liable to* overlaps with it, but often carries the implication that the action or experience expressed by the infinitive is not only likely to happen but is also undesirable, e.g. *The ambitious young male executive appears particularly liable to suffer from neglecting his home life*—Intercity, 1989. See LIABLE 2.

But the border between the two uses is far from absolute, and in many contexts the two adjectives are interchangeable, e.g. *The questions that the interviewers are liable/apt to ask include . . .*

For the type *be liable to* = 'be likely to', see LIABLE 3.

**aqu-, acqu-.** It is important to distinguish words beginning with *aqu-* (nearly all denoting water, from L *aqua* 'water', as *aqualung, aquarium, aquatic*, etc., but also *aquiline* 'like an eagle' from L *aquila* 'eagle') from those beginning with *acqu-* (as *acquaint, acquiesce, acquit* etc., most of which were adopted in ME from OF or L words beginning with *acqu-*).

**aquarium.** Pl. *aquaria* or *aquariums.* See -UM 3.

**Arab, Arabian, Arabic.** 1 The three adjectives are now broadly differentiated along the following lines: *Arab* means 'of the Arab people', *Arabian* 'of Arabia (the peninsula between the Red Sea and the Persian Gulf)', and *Arabic* 'of the language, literature, or script of Arabic-speaking people'. So we have *Arab businessman, horse, racehorse-owner, sheikh; Arab courage, entente, philosophy; the Arabian desert, fauna* (*Arabian baboon, camel, hyrex*), *flora* (*Arabian gum, jasmine*), *gulf,* and *nights; Arabic alphabet, literature, numerals, script.*

2 The expression *street Arab,* first recorded in 1859, and for about a century commonly applied to a homeless vagrant (esp. a child) living in the streets of a city, is now regarded as ethnically offensive.

3 The normal pronunciation of *Arab* is /'ærəb/, but the jocular and faintly derogatory pronunciation /'eɪræb/ is occasionally heard, esp. in AmE.

**arbiter.** This word meaning 'one whose opinions or decision is authoritative in a matter of debate', and, formerly, 'one chosen by the parties in a dispute to arrange or decide the differences between them', has almost entirely given way in standard English to *arbitrator* in the second sense. Examples: (arbiter) *she was not so much an arbiter of fashion as she was fashion itself*—D. Halberstam, 1979; *The great nineteenth-century critic and arbiter of taste, John Ruskin*—L. Hudson, 1985; (arbitrator) *The owners demanded £112 million. The government offered £75 million. The arbitrators awarded £66.5 million*—A. J. P. Taylor, 1965; *Either party may apply to have the dispute referred to arbitration by the judge or by an outside arbitrator*—R. C. A. White, 1985. In practice, however, when unresolved disputes arise between unions and management, the phrase usually employed is 'a call for arbitration, the need for arbitration, etc.' rather than 'a call for the matter to be referred to an arbitrator'.

**arbour, arbor.** An *arbour* (US *arbor*) is 'a bower, a shady retreat', a word of complex etymology received into the language in the 14c. from OF *erbier* (mod. *herbier*) herbarium. It is a separate word from the first element of the (orig. US) term *Arbor Day,* a day specially set aside for the planting of trees, which has been adopted (with the American spelling) in Canada, Australia, and New Zealand. This *Arbor* is from L *arbor* 'tree'. See -OUR AND -OR.

**arc** (verb). The derivative forms are spelt *arced* and *arcing,* pronounced /ɑːkt/ and /'ɑːkɪŋ/ respectively, i.e. with the medial *c* pronounced as /k/ despite the fact that it is followed by *e* and *i.*

**arch-, arche-, archi-.** Though the prefix *arch-* 'chief' (as in *archbishop, archdeacon, arch-enemy,* etc.) is pronounced /ɑːtʃ/ in all words except *archangel* and its derivatives, the longer forms are always pronounced /'ɑːkɪ/, e.g. *archiepiscopal, Archimedes, archipelago, architect.*

**archaeology.** In AmE, often *archeology.* See Æ, Œ.

**archaism.** Archaism seems to be a necessary component of the language. At any given time modes of expression and of word formation are retained long after they have been discarded as part of the 'natural' language. Archaic spellings that call to mind a sturdy antiquity include (in imitation of Izaak Walton's *The Compleat Angler*) *The Compleat Bachelor* (title of book by Oliver Onions, 1900); *She writes and sings and paints and dances and plays I don't know how many instruments. The compleat girl* (M. McCarthy, 1963); *The Compleat Hiker's Checklist* (*Modern Maturity,* an American magazine, 1988). Other examples: *the Canterbury Clerkes, Whitaker's Almanack,* the *Culham Fayre*; and a multitude of uses with *olde,* e.g. *A lot of olde realle beames in Amersham and a lot of olde phonie cookynge too* (*Good Food Guide,* 1959); *Charming stone built olde worlde cottage of immense character* (*Rhyl Journal & Advertiser,* 1976).

Long-abandoned verbal inflexions are used to add a contextual element of antiquity: *If Mimi's cup runneth over, it runneth over with decency rather than with anything more vital*—A. Brookner, 1985 (cf. Ps. 23: 5); *The whole creation groaneth and travaileth in pain together*—I. Murdoch, 1987 (cf. Rom. 8: 22.)

Normal word order is disturbed, often as a deliberately archaistic device: *I would rinse out somebody's mouth at Clairol loved I not honour more*—W. Safire, 1986.

The prefix *a-* with a present participle (a centuries-old type of word formation) is having a new lease of life: *a-basking, a-changing,* etc. See A-².

Words that might normally be paint-stripped from the fresh timber to which they cling in modern writing include ALBEIT, ANENT, *betimes,* CERTES, DERRING-DO, NAY, PERADVENTURE, PERCHANCE, QUOTH, SURCEASE, *twain,* UNBEKNOWN(ST), *yea, yesteryear.* For further examples, see WARDOUR STREET.

Everything depends on the skill with which such old words, spellings, and pieces of grammar are deposited in particular contexts. Like Marcel Proust's famous madeleine cakes, they can conjure up rich memories of an older age. But they can also be as out of place as a donkey jacket worn at the Cenotaph on Remembrance Day.

**archetype.** See -TYPE.

**arctic.** To be pronounced as /ˈɑːktɪk/ with the first *c* fully in place.

**ardour.** AmE *ardor.* See -OUR AND -OR.

**are, is.** When one of these is required between a subject and a complement that differ in number (e.g. *these things ... a scandal*) the verb should normally follow the number of the subject (*are,* not *is, a scandal*). Similarly *The only difficulty in Finnish is* (not *are*) *the changes undergone by the stem.* See AGREEMENT 7, BE 1.

**aren't I?** See BE 4.

**argot.** See JARGON.

**arguably.** First recorded only in 1890, this sentence adverb is now in very widespread use in the sense 'as may be shown by argument or made a matter of argument'. It is frequently followed immediately by a comparative or a superlative adjective. Examples: *Mozart's sinfonia concertante for violin and viola, arguably the greatest of his concertos*—Times, 1959; *Arguably, Pip's search for Estella's true identity ... can be seen as a displaced search for his own identity*—Essays & Studies, 1987; *Fibich was*

*arguably worse off even than Hartmann, for he knew no one*—A. Brookner, 1988.

**arguing in a circle.** (logic) The basing of two conclusions each upon the other. That the world is good follows from the known goodness of God; that God is good is known from the excellence of the world he has made.

*argumentum ad* —. *argumentum ad hominem,* one calculated to appeal to the individual addressed more than to impartial reasoning; *argumentum ad crumenam* (purse), one touching the hearer's pocket; *argumentum ad baculum* (stick) or *argumentum baculinum,* threat of force instead of argument; *argumentum ad ignorantiam,* one depending for its effect on the hearer's not knowing something essential; *argumentum ad populum,* one pandering to popular fashion; *argumentum ad verecundiam* (modesty), one to meet which requires the opponent to offend against decorum. Also *argumentum e* (or *ex*) *silentio,* an argument from silence: used of a conclusion based on lack of contrary evidence.

**-arily.** Under American influence, in the second half of the 20c., adverbs ending in *-arily* (e.g. *momentarily, necessarily, primarily, temporarily, voluntarily*) have come to be commonly pronounced by standard BrE speakers with the main stress on *-ar-* even in the formal reading of the news on the main broadcasting channels. This placing of the stress was opposed in my book *The Spoken Word* (1981), prepared for the guidance of BBC newsreaders, but without avail. All such words were traditionally pronounced in RP with the main stress on the first syllable.

**arise,** in the literal sense of getting up from bed, has given place except in poetic or archaic use to *rise.* In ordinary speech and writing it means merely to come into existence or notice or to originate from, and that usually of such abstract subjects as *question, difficulties, doubt, occasion, thoughts, result, effects.*

**aristocrat.** The pronunciation /əˈrɪstəkræt/ seems now to be virtually obsolete in Britain, having been ousted by /ˈærɪstəkræt/. The reverse is the case in AmE. The word itself (in the form *aristocrate*) was a popular formation of the French Revolution in 1789.

**Aristotelian.** The form with final *-ian* has almost entirely supplanted the once common form ending in *-ean*. The discarded form was pronounced /ˌærɪstɒtɪˈliː-ən/ and the current one /ˌærɪstəˈtiːlɪən/.

**arithmetical, geometrical progression.** *Arithmetical progression* is marked by an equal increase between the items in the progression, e.g. the numbers 1, 3, 5, 7, 9, etc., showing an increase of 2 in each case. A *geometrical increase* is larger: a typical example would be the sequence 1, 2, 4, 8, 16, 32, etc., in which each number is double that of the number before it. In other words, one involves growth (or decline) at an unchanging rate and the other at an increasing rate. Both expressions, but especially *geometrical progression*, in popular use tend to be employed to suggest a rapid rate of increase. But both terms are relative. If the rate of increase is very small both sequences can be used to indicate a relatively slow rate of increase, e.g. (arithmetical) 10,000, 10,001, 10,002, 10,003, etc; (geometrical) .00001, .00002, .00004, .00008, etc.

**armada.** The 19c. pronunciation /ɑːˈmeɪdə/ has entirely given way to the current one, /ɑːˈmɑːdə/.

**armadillo.** The pl. form is *armadillos*. See -O(E)S 7.

**armour.** US *armor*. See -OUR AND -OR.

**aroma.** It is a matter of curiosity that *aroma* (a) has lost its original primary sense ('spice'), a sense recorded from the 13c. to the 18c.; (b) has been most commonly used of an *agreeable* smell since it became restricted in the early 19c. to the broad sense 'the distinctive fragrance exhaled from a spice, plant, scent, etc., or from items of food'. Confirmation of the agreeable nature of aromas is shown by the emergence in the 20c. of the term *aromatherapy* ( = massage with fragrant oils as a form of therapy). *Aroma* is Greek in origin, but the Gk pl. form ἀρώματα (E *aromata*) was used only sporadically in former times in English and is now obsolete.

**around, round.** As adv. and prep. both words have a wide variety of uses, in some of which the longer form and the shorter one are interchangeable, and in others where they are not. *Around*: **1** Still the normal form in standard English in certain collocations, as *around and about, all around* (*are signs of decay*), *she's been around*, and esp. in phrasal verbs having the general sense 'to behave in an aimless manner', as *fool around, mess around, play around, wait around*.

**2** Can be used in place of *round* without loss of idiomatic quality in contexts referring to surrounding (something or someone), as *seated around the table, the cheerfulness around her*. But

**3** there are many circumstances in which, in BrE, *round* is more or less obligatory, as *all the year round, winter comes round, the wheels go round, he came round to see me, send the hat round, show one round, the Whitbread Round the World Race.*

In AmE, *around* is much more common in the third group, and also when used in the sense 'approximately', e.g. *around four o'clock, around 70% of the people*. See ABOUT 2.

The distribution of the two words is, however, subject to considerable variation, as can be seen from the following examples: (*around*, prep.) *Jesse ... moped around the house all summer*—Lee Smith, 1983 (US); *Coming around the farthest mark*—New Yorker, 1986; *around the time of his birth*—S. Bellow, 1987 (US); *the area around Waterloo*—R. Elms, 1988 (UK); *I wrapped a blanket around me*—L. Maynard, 1988 (US); *You know how to get all the people around the operating table*—K. Russell, 1988 (UK); *around that time*—Julian Barnes, 1989 (UK); *they stood grouped around their luggage*—M. Bracewell, 1989 (UK). (*round*, prep.) *it stood just round the corner from his father's house*—Van Wyck Brooks, a1961 (US); *a map rolled up round a broom handle*—J. Winterson, 1985 (UK); *He looked round the table as if daring anyone to smile*—D. Lodge, 1988 (UK). (*around*, adv.) *Sir William, whom I haven't gotten around to discussing*—R. Merton, 1985 (US); *Stay around till she gets back*—New Yorker, 1987; *Hartmann's sunny ... attitude was marvellous to have around*—A. Brookner, 1988 (UK); *The devices have been around awhile*—USA Today, 1988; *I went around to the front door*—New Yorker, 1989. (*round*, adv.) *in the end she talked me round*—N. Bawden, 1987 (UK); *the news had gotten round pretty fast*—New Yorker, 1988.

In the face of such evidence one can only suppose that contextual euphony is perhaps the strongest factor in determining the choice, except when the meaning required is 'approximately'.

**arouse.** The relation of this to *rouse* is much like that of *arise* to *rise*; that is, *rouse* is almost always preferred in the literal sense and with a person or animal as object. *Arouse* is chiefly used with the senses 'call into existence, occasion', and with such abstract nouns as *suspicion, fears, cupidity, passions*, as object of the active or subject of the passive: *This at once aroused my suspicions; cupidity is easily aroused. Rouse* would be more suitable in *I shook his arm, but failed to arouse him.*

**arpeggio.** Pl. *-os.* See -O(E)S 4.

**arquebus.** See HARQUEBUS.

**arrant** is historically the 'same word' as *errant*, i.e. a spelling variant, and OF *errant*, from which they are both derived, is itself a product of two branches, Vulgar L *iterāre* 'to journey, travel' and L *errāre* 'to wander'. These broad senses have produced in English, for *arrant*, 'notorious, manifest, downright' (*arrant fool, knave, liar, nonsense*, etc.), and for *errant*, 'roving in quest of adventure' (*knight errant*) and 'astray, straying from proper behaviour, erring in opinion' (*errant children, errant husband, an errant taste in dress*).

**arride,** recorded in the work of major writers from Ben Jonson (1599) to Rudyard Kipling (1937) meaning 'to please, gratify, delight', is a standard example of a word of Latin origin (L *arrīdēre*) that has now dropped entirely out of use.

**arrive.** The absolute sense 'to achieve success or recognition', which is first recorded in 1889, is a Gallicism.

**arrogate.** See ABROGATE.

**arsis.** (Gk ἄρσις 'lifting, raising'.) In classical prosody there has been much dispute as to the exact meaning of this word and readers should turn to *The Oxford Classical Dictionary* for elucidation. But in modern English prosody, it means 'a stressed syllable or part of a metrical foot' (and thus = L *ictus*), as in Tennyson's 'The spléndour fálls on cástle wálls'. Cf. THESIS 2.

**art.** When used attributively, *art* denotes imaginative skill applied to design, as in paintings, pottery, architecture, etc. The plural form *arts* is used attributively to mean 'other than scientific', as in *The Arts and Reference Division of the Oxford University Press; an arts degree*, i.e. one in English, History, Classics, etc.

**artefact.** This is now the normal spelling of the word in BrE. In AmE it varies with *artifact*, the latter being the more usual spelling (*Her recorder, a stylish artifact of high-density plastic produced in Japan*— New Yorker, 1988). Etymologically, *artefact* is the better form as the word is derived from *arte*, ablative of L *ars* 'art' + *factum*, neuter pa.pple of *facere* 'to make', but *artifact*, perhaps formed by analogy with *artifice* and *artificial*, has been recorded in British as well as American sources for more than a century.

**artiste,** rhyming with *feast*, is applicable to either sex (*a trapeze artiste*). In the sense 'professional dancer, entertainer, singer, etc.' it serves a useful purpose in that (as Fowler remarked) 'it conveys no judgement that the performance is in fact artistic'. But in most contexts nowadays it has been superseded by *artist*.

**as.**

1 Case after *as*.
2 *as* = 'in the capacity of'.
3 Omission of *as*.
4 Causal *as*.
5 *as*, relative pronoun.
6 *as* and *when*.
7 *as from, as of*.
8 *as how*.
9 *as if, as though*.
10 *as per*.
11 *as such*.
12 *as to, as for*.
13 *as or like*.
14 *equally as*.
15 *as a fact*.
16 *as concern*(s), etc.
17 *as far as*.
18 *as long as*.
19 *as well as*.

1 Case after *as*. Case is distinguishable after *as* only in a few pronouns. To escape censure, and also sometimes to avoid ambiguity, it is better in formal writing to adopt the style *he is as clever as I/he/she/they* rather than *he is as clever as me/him/her/them*. Examples: *It was obvious that he had been consulted as well as I*—G. Greene, 1965; *he started to encounter kids*

*as gifted as he, some even more so*—New Yorker, 1986; *Such as we are bred to face the artillery*—M. Shadbolt, 1986. Informally, in ordinary speech, the second pattern is much the more usual. The choice of pronoun hinges on whether *as* is regarded as a preposition like *after* or *before*—no one would write or say *after I, before she*, etc.—or whether it is judged to be a conjunction with ellipsis of the following verb ( = *he is as clever as I am*, etc.). It is important to keep in mind that if the word after *as* is a name or an ordinary noun, no difficulties of case arise, e.g. *Oxford is as famous as Heidelberg; John lives in the same village as James.*

The potential ambiguity of *he loves me as much as my sister* ( = either 'as much as my sister does' or 'as much as he loves my sister') is best avoided by including a verb in each part of the sentence (*he loves me as much as he loves my sister; he loves me as much as my sister does*).

**2** *as* = 'in the capacity of'. Its normal use is shown in these examples: *it is as a historian that he is best known; I hear you are employed as a nanny.* But danger lies in the placing of the *as*-clause. Care must be taken to avoid the creation of a false antecedent. Wrongly attached links occur in each of the following sentences: *As a medical student his call-up was deferred*—P. Fitzgerald, 1986; *As a voluntary, charitable project the Government is about to charge us some £30,000 VAT for extending our premises*—letter to The Times, 1988; *As a 32-year career law enforcement professional, you know that I do not like being forced to release prisoners from jail*—Chicago Tribune, 1988.

**3** Omission of *as*. There are circumstances in which debatable constructions arise from the omission of *as*: (*a*) when it collides with another *as*, e.g. *But it is not so much as a picture of the time as as a study of humanity that Starvecross Farm claims attention.* In such circumstances, it is better to reconstruct: ... *more as a picture of the time than as a study of humanity* ...

(*b*) In a group of words forming constructions of the type *Henry Balfour was then appointed (as) Curator of the Collections.* Ditransitive constructions, i.e. those in which the verb has a direct object and also a second complement (e.g. *they reckoned him a good man*), are now relatively uncommon when the omitted element is *as*. *CGEL* provides a long list of verbs, including *choose, consider, count, deem, elect, proclaim*, which can be and are so used. But they can also be used with *as* (e.g. *they chose him as leader*). For a similar number of verbs (*accept as, acknowledge as, class as, regard as, treat as*, etc.), *as* is now normally obligatory (e.g. *Neil Kinnock is still to be classed as a socialist, but a distinctly flexible one*). For this group, 19c. examples showing the verbs used ditransitively lie ready to hand in the *OED* and in Visser, e.g. *The nation everywhere acknowledges him master* (1855); *Whatever constitutes atonement, therefore, must ... be regarded a safe ... remedy* (1836). But the trend towards the retention or restoration of *as* in this group looks likely to continue.

Casual omission of *as* is common enough in informal contexts (*she used to come regular as clockwork; it was soft as butter*) and in proverbs (*he was good as gold*; (*as*) *old as the hills*).

**4** Causal *as* meaning 'since, because, seeing that', etc. Fowler (1926) strongly objected to constructions of the type *I gave it up, as he only laughed at my arguments*, i.e. in which the *as*-clause follows the main clause, while accepting constructions where the causal *as*-clause is placed first (*As he only laughed at my arguments, I gave it up*). His objection now seems very dated. The *OED* (s.v. *as* 18a) lists as unlabelled, i.e. acceptable, modern examples: (*as*-clause precedes) *As you are not ready, we must go without you*; and (*as*-clause follows) *He may have one, as he is a friend.* When its meaning is 'in consideration of, it being the case that', *as* can nowadays be used at the head of a subordinate clause standing either before or after the main clause. The placing of the clauses is governed by the degree of emphasis allotted to each part of the sentence

**5** *as*, relative pronoun. When preceded by *same* or *such*, constructions with *as* used as a relative pronoun are still for the most part standard: *We can expect the same number to turn up as came last year; such repairs as have been made to the house are most acceptable.* But it is not difficult to find contexts, esp. in the 19c., where the same construction has a strong tinge of archaism: *there was such a scuffling, and hugging ... as no pen can depict*—Thackeray, 1847/8.

In all other circumstances, the use of *as* as a relative pronoun is restricted to non-standard or regional speech both in BrE and in AmE: *it's only baronets as cares about farthings*—Thackeray, 1847/8; *I mean the singers, Ma'am—them as sang at the concert to-night*—Atlantic Monthly, 1865; *This is him as had a nasty cut over the eye*—Dickens, 1865; *I don't know as I expected to take part in this debate*—Harper's Mag., 1888; *Them as says there's no has me to fecht*—J. M. Barrie, 1891; *There's them on this island as would laugh at anything*—W. Golding, 1954; *There's plenty as would like this nice little flat, Mr. E.*—A. Burgess, 1963; *Not as I know of*—DARE questionnaire, 1966–9; *You're the kind as'll never know it*—R. Elms, 1988.

**6** *as and when*. This common phrase meaning 'whensoever' (introducing a future event or action whose occurrence or frequency remains in doubt) is of surprisingly recent origin (first recorded in 1945). It is also used elliptically in informal language to mean 'when possible, eventually'. Examples: *The correct procedure, as and when we win our case, is then to apply for a writ of mandamus*—Listener, 1979; (elliptical use) *They confirmed the existing main roads as future main traffic arteries to be widened 'as and when'*—Listener, 1965. Cf. IF AND WHEN.

**7** *as from, as of*. In the drawing up of agreements or contracts, it is customary to indicate that a particular item or items is to take effect *as from* a stated date. This use is reasonable if it is retrospective: e.g. *The rate of payment is increased by 5% as from the 1st of September last*. For present and future dates the *as* is superfluous. Thus *your redundancy takes effect from today* (rather than *as of today*); *your membership of the Club will become valid on 1 January next* (rather than *as from 1 January next*).

Phrases of the type *as of now, as of today*, etc., first recorded in the work of Mark Twain in 1900, are now well established in standard English in the UK and elsewhere. Examples: *I'm resigning from the committee as of now*—D. Karp, 1957; *As of last term, Oxford has a new centre for the performing arts*—Oxford Mag., 1986; *As of today, I do not believe Tebbit has enough votes to win*—J. Critchley, 1990.

**8** *as how*. Used for four centuries as a contracted interrogative sentence (e.g.

*We shall see sometime how heretikes come to turne the groundes of our faith wholy vpside downe. As how?*, 1579), this formulaic construction, in an extended form, seems to be enjoying a new lease of life in demotic American use: *I took these concerns for my own, as how could I not?*—E. L. Doctorow, 1989.

It is to be distinguished from the UK and US dialectal uses of *as how* as a relative pronoun meaning 'that': *Seeing as how the captain had been hauling him over the coals*—F. Marryat, 1833; *The doctors came and said as how it was caused all along of his way of life*—Bret Harte, 1871 (US); *Just across the border here into Texas, the folks figured as how it was the thing to do to join the Union*—1949 in DARE. See also SEEING.

Neither construction forms part of standard English.

**9** *as if, as though*. In the great majority of *as if* clauses, when the choice of verb is between *were* or *was*, the subjunctive form *were* is preferable. It indicates that something is hypothetical, uncertain, or not factually true. But when uncertainty or hypothesis is less obviously present or not present at all, the indicative form *was* should be used. Examples: (were) *It was rather as if the capital city were a vast pan*—A. N. Wilson, 1985; *It was as if Sally were disturbed in some way*—A. Brookner, 1986; *One can elect to focus instead on Freud's achievement as if it were . . . itself a domain*—Dædalus, 1986; *It's not as if the park were in disarray*—New Yorker, 1986; *Dolly made sympathetic comments to John Pickering, as if he were the chief mourner*—A. Brookner, 1993; (was) *There was a silence, as if he was searching for something to say*—A. Guinness, 1985; *I felt as if I was losing my grip on the facts*—C. K. Stead, 1986.

*As though* operates in a similar manner. Examples: (were) *He was looking at her as though she were for sale*—A. Thomas Ellis, 1985; *His body felt as though he were trembling, but he was not*—B. Moore, 1987; (was) *The Faithful had gone back to their chorus sheets as though nothing was happening at all*—J. Winterson, 1985; *He devoured all, exhausted, as though his life was in danger*—A. S. Byatt, 1987.

It is as well to keep in mind, though, that in a great many contexts the question of a *were/was* distinction does not arise. Examples: *We fell into step, and as usual, talked as though eight months were minutes*—V. Woolf, 1920; *'Don't nag me,' he said, as though they had been married a long time*—A. Carter, 1967; *as if he had an exceptionally high specific gravity*—I. Murdoch, 1974; *as though she had remembered there was something she must do*—Maurice Gee, 1983.

There is a further distinction. *As if* and *as though* are followed by the past tense when the verb refers to an unreal possibility, i.e. when the statement introduced by *as if* or *as though* is untrue or unlikely (e.g. *Every critic writes as if he were infallible*—C. Connolly, 1938; *It's not as though he lived like a Milord*—E. Waugh, 1945). Conversely, the present tense is used when the statement is true or likely to be true (e.g. *I suppose you get on pretty well with your parents. You look as if you do*—K. Amis, 1960; *He speaks as though even the rules which we freely invent are somehow suggested to us in virtue of their being right*—M. Warnock, 1965.

**10** *as per.* The L preposition *per* meaning 'through, by, by means of' has long been established in English in many set phrases (*per annum, per capita, per diem*, etc.). The most notable slang product of these uses is the phrase *as per usual*. Examples: *I shall accompany him, as per usual*—W. S. Gilbert, 1874; *As per usual somebody's nose was out of joint*—J. Joyce, 1922; *As per usual, I had not returned to England refreshed*—L. Ellmann, 1988; *Same old jolly camp-fire life went on as per usual*—Julian Barnes, 1989. Humorous variants are legion, e.g. *She knew better, didn't she. As per always*—P. Bailey, 1986.

The compound preposition *as per* is more or less restricted to business letters and to such publications as DIY manuals (e.g. *as per specifications*).

**11** *as such.* In the spoken language any possible ambiguity in the use of the phrase is normally removed by the placing of the emphasis. In *The black people of South Africa have no objection to sanctions as such provided that only business firms are made to suffer*, the intonation of the sentence would make it clear that *as such* belonged with *sanctions*, and not, say, with *The black people of South Africa*. But in most circumstances it is better to substitute a synonymous expression like 'in principle'. Thus instead of *There is no objection to the sale of houses as such* write *There is no objection in principle to the sale of houses*.

**12** *as to, as for.* Some older uses of *as to* meaning 'with respect to, with reference to' (e.g. *As to myself, I am not satisfied*—D. Hartley, 1748) have given way to *as for* (e.g. *As for you, son, your mother will hear of this*). *As to* survives, however, when the noun governed is non-personal (e.g. *As to the matter raised at the Governing Body, my view is* ...), and esp. when the sense required is 'according to, proportionate to' (e.g. *correct as to colour and shape; the rates of postage vary both as to distance and weight*).

*As to* is frequently used before subordinate questions, as in *The Politics Fellow left no instructions as to whether you should write a second examination paper or not*. It is also used after certain preceding passive clauses (e.g. *I am not much troubled as to its outward appearance*—G. Gissing, 1903; *I am also simultaneously bemused* ... *as to why people should sunbathe indoors*—The Face, 1986). In most other circumstances, though, *as to* is best left unused, especially when other constructions are available or when its presence is simply unnecessary: *He asked his mother (as to) when he would be regarded as old enough to go to discos; Ms Jones raised the question as to whether a similar conflict will arise between the urban elites and the peasantry* (better *the question of whether*); *I brooded all the time as to whether I had hit the right note* (better *on whether*). *He reminded me how to behave* is a better construction than *He reminded me as to how I should behave*.

*As for* sometimes implies a degree of scorn, e.g. *As for Smith minor, he can't even swim across the baths yet*, or a threat (see the second example above). But not necessarily: *As for me, I was more than content with the description of me as a map of low desires*—H. Jacobson, 1986.

**13** *as* or *like.* The two words are often interchanged like hockey sticks, and many mistaken transferences of role occur. In general, *as* should be used before adverbs (*there are times, as now, when I could hate him*) and prepositions (*he acted well again, as in 'Henry V' last year*). *Like* should be used before nouns, noun phrases, and pronouns, and occasionally as a conjunction. Sometimes the choice between the two words affects the meaning: *let me discuss this with you as your father* is obviously not the same as *let me discuss this with you like your father*.

**14** For the debatable construction *equally as* (*good*, etc.), see EQUALLY.

**15** *as a fact*. See FACT 1.

**16** *as regard(s)*. See REGARD 1.

**17** *as far as*. See FAR 4.

**18** *as long as*. See LONG.

**19** *as well as*. See WELL *adv.* 1.

See also AS FOLLOWS.

**as ... as, so ... as.** In simple comparisons, the normal construction in standard English is clearly *as ... as* (*as busy as a bee*, *as mad as a hatter*, *as soon as you can*). Quite commonly, however, in the 19c. and earlier, the antecedent could also be *so*, esp., but not only, in negative sentences: *You have never so much as answered me*—Scott, 1818; *No country suffered so much as England*—Macaulay, 1849; *When did a morning shine So rich in atonement as this?*—Tennyson, 1855. Nowadays *as ... as* is overwhelmingly the more common of the two, but *so ... as* is far from extinct. Examples from my database: (a) *variations ... of a star as small as three kilometres across could be detected*; *there is not quite as much text as was expected*; *his porno movies can't be as much fun as we're led to believe they are*; (b) *It's here. So long as nobody disturbs it*; *she had seldom known anything so luxurious as this steam room*.

**as bad or worse than.** Difficulties arise in this and in the contrasting construction *as good or better than* because both *bad* and *good* (as well as other adjectives) obviously require *as*, not *than*, in comparisons. The juxtaposition of *as* and *than* without intervening punctuation is not logically defensible. Thus the sentence *we're sure they can judge a novel just as well if not better than us* (*London Rev. Bks*, 1987) needs correcting to *just as well as, if not better than, us*. But a wiser course is to avoid the difficulty by placing the comparative at a later point in the sentence, e.g. *Bowie was as deranged as Osterberg, probably more so; he started to encounter kids as good as me, some even more so*.

**ascendancy, ascendant.** **1** The spellings preferred in OUP house style are *-ancy*, *-ant*, though the words are still printed as *ascendency* and *ascendent* by some publishing houses.

**2** *Have ascendancy over* and *be in the ascendant* are the normal phrases, and in

them *ascendancy* means 'dominant control' and *ascendant* 'supreme, dominant'. Occasionally and understandably the corresponding verb *ascend* 'to rise' leads to the derivatives being used in a progressive or upward sense, '(in the process of) gaining control', but these are not standard uses of the words.

**ascension.** Except in the technical language of astronomy, *Ascension* (always with a capital initial) is now virtually restricted to mean the ascent of Jesus Christ into heaven on the fortieth day after His resurrection.

**aseptic, asexual.** Both words contain the privative prefix *a-*. See A-[1].

**as follows.** The phrase *as follows* is naturally always used cataphorically, i.e. with forward reference, and is not replaced by *as follow* even when the subject of the sentence is plural: *His preferences are as follows ...*; *his view is as follows ...* The reason for its fixed form is that it was originally an impersonal construction = 'as it follows'.

**ashamedly.** Pronounce as four syllables, /əˈʃeɪmɪdlɪ/. See -EDLY.

**ashen.** See -EN ADJECTIVES 3.

**Asian.** **1** From about 1930, and esp. in the second half of the 20c., *Asian* has replaced *Asiatic* in official use because of the alleged depreciatory implication of the latter (traditional) term.

**2** In Daniel Jones (1917) the only pronunciation given for *Asia*, *Asiatic*, *Persia(n)*, *version*, etc., was with medial /-ʃ-/. During the 20c. /-ʃ-/ has gradually been overtaken by /-ʒ-/ in all such words, and the new pronunciation now looks likely to become the dominant one soon in BrE as it already has in AmE.

**aside, a side.** Written as one word, *aside* is an adverb meaning 'to or on one side' (*to put aside*, *to take aside*, etc.) or a noun meaning 'words spoken in a play for the audience to hear, but supposed not to be heard by the other characters' (*COD*). In the sense 'on each side' it must be written as two words, e.g. *they were playing seven a side*, i.e. with seven players in each team, *a seven-a-side game*.

**aside from.** See APART FROM.

**asocial.** See A-[1].

**as per.** See AS 10.

**assassinate.** The traditional restriction of this word to mean 'to kill an important person for political reasons' (e.g. the assassination of Archduke Francis Ferdinand at Sarajevo in 1914) has tended to be modified in more recent times. The word is now also often applied to the killing of any person who is regarded by the killer(s) as a legitimate political or sectarian target, for example in Northern Ireland, Israel, and Lebanon. In the course of the 20c., an older figurative meaning of the word, 'to destroy someone's reputation', has also been revived. Examples: (political) *If the NLF* [National Liberation Front in Vietnam] *felt his death would serve a political purpose, he would be assassinated*—F. Fitzgerald, 1972; *Palestinian guerrillas sought for a second time in three months to assassinate King Hussein*—H. Kissinger, 1979; (fig.) *Helping the Prime Minister in his political battle to assassinate Mr. George Brown*—*Guardian*, 1962.

The same considerations apply to the corresponding noun *assassination*. Examples: (political) *the point-blank assassination of an off-duty detective at the dog track* [in Belfast]—*New Yorker*, 1994; *Assassinations of individual foreigners later escalated into massive bombings*—*Bull. Amer. Acad. Arts & Sci.*, 1994; (fig.) *The effects of these 'character assassinations' have been disastrous ... on the willingness of scientists to work for the government*—*Listener*, 1958.

**assay, essay** (verbs). These two words are now for the most part satisfactorily separated in meaning. *Assay* is usually restricted to contexts of testing or evaluating, e.g. the quality of metals or of ore, the content of chemical substances, or a person's character. The slightly archaic word *essay*, on the other hand, usually means 'to attempt (a task, etc.); to attempt (to do something)'. In older literature, at least until the end of the 19c., the two words were often used interchangeably meaning 'to attempt, try to do (anything difficult); to make the attempt (to do something)'.

**asset,** said Fowler (1926), is a false form. He meant that *assets* (derived from late Anglo-Fr. *assets*, from L *ad satis* 'to sufficiency', cf. modF *assez* 'enough') was in origin the true form, and that until the 17c. it was regularly construed as a singular. By the 19c. *asset* had emerged as an ordinary singular form (e.g. *his ability to speak Chinese is an asset of the first order*), and it is now presented as a headword in all major dictionaries (except, oddly, *OED* 2).

**assignment.** An assignment is an allocation, and in law a legal transfer of a right or property, or the document that effects the transfer. It is also, orig. in AmE, a task or piece of work allotted to a person. The dominant sense of *assignation*, which is pronounced /æsɪg-'neɪʃən/, by contrast, is now 'an agreement to meet, esp. a secret meeting of lovers'. Its original meaning of 'apportionment' is now seldom encountered.

**assimilation.** In phonetics, 'the influence exercised by one sound segment upon the articulation of another, so that the sounds become more alike, or identical' (D. Crystal, 1980). Thus in connected speech there is a tendency, for example, for *lunch score* to be articulated as /'lʌntʃ ʃkɔə/, i.e. with the normal /s/ of *score* assimilated to the final sound of *lunch*; and esp. for a final *n* to become *m* in connected speech under the influence of a following labial. Examples from ITN news bulletins in January 1990 included 'om probation', 'a milliom pounds', and 'have beem put'. In ancient times assimilation accounts for such words as L *illumi natio* 'illumination' from *in-* + *lumen*, *luminis*, and L *irrationalis* 'irrational' from *in-* + *rationalis*. Cf. DISSIMILATION.

**assist.** 1 The sense 'to be present (at a ceremony, entertainment, etc.)', now uncommon and sounding affected, is a Gallicism: *And assisted—in the French sense—at the performance of two waltzes* (Dickens).

2 The sentence *May I assist you to potatoes?* in Mrs Gaskell's *Wives and Daughters* (1864) is a genteel way of saying 'May I help you to potatoes?' In many contexts *help* and *assist* are interchangeable, but in general *help* is the word to prefer. The two verbs share the same constructions. Examples: (followed by *in* and gerund)

They assisted him in receiving and entertaining his guests (Poutsma); (followed by object and *to*-infinitive) *Mr. A. is assisting his wife to show a book of photographic portraits to a girl on a visit* (Poutsma); (followed by object and complement) *a young man who assisted him with the management of the farm* (T. Capote, 1966). The type *assist* + object + plain infinitive is also technically possible, but I have found no examples.

**association.** See -CIATION 3.

**assonance.** As a term of prosody it means partial rhyming, either (*a*) a correspondence of vowels but not of consonants, as in the last line of Yeats's *Byzantium*, That dolphin-torn, that gong-tormented sea, or (*b*) a correspondence of consonants but not of vowels, as in Wilfred Owen's *Arms and the Boy* where *blade/blood, flash/flesh, heads/lads, teeth/death, apple/supple,* and *heels/curls* lie at the end of successive lines. Type *b* is also called *half-rhyme*.

**assume, presume.** In many simple contexts when the meaning is 'to suppose', the two words are interchangeable: e.g. *I assume/presume you are coming to the party.* Otherwise the choice of word depends on the degree of tentativeness behind the assumption or presumption. The *OED* definitions are very similar. *Assume* is 'to take for granted as the basis of argument or action'; *presume* is 'to take for granted, to presuppose, to count upon'. There is a faint suggestion of presumptuousness about *presume*.

Fowler expressed the semantic difference thus: 'in the sense *suppose*, the object-clause after *presume* expresses what the presumer really believes, till it is disproved, to be true; that after *assume*, what the assumer postulates, often as a confessed hypothesis.' This is reasonable. But he then went on to claim, less reasonably, that the *that* of the object-clause is usually expressed after *assume* but omitted after *presume*.

The constructions available after the two verbs are very similar. *Assume* can be followed by a *to*-infinitive, a *that*-clause, or a direct object; *presume* occurs in the same three constructions but also with a direct object and complement. Some examples (principally given to illustrate the various constructions): (assume) *William assumes the willingness of the Assembly*—E. A. Freeman, 1869; *He was writing 'Gerontion', a dramatic monologue in which he assumes the persona of the 'little old man'*—P. Ackroyd, 1984; *This was assumed to be because of their high amplitude resonances*—Working Papers, School of English and Linguistics, Macquarie Univ., 1985; *When you're young you assume everybody old knows what they're doing*—M. Amis, 1987; *He had assumed that the economic growth of the 1950s would continue unabated*—Daily Tel., 1987. (presume) *Death is presumed from the person not being heard of for seven years*—Law Rep., 1871; *Those who ... presumed that if he spent his time with me I must also ... be either rich or disingenuous*—L. Durrell, 1957; *I know that in law every man is presumed innocent until proved otherwise*—A. Maclean, 1971; *It is a reckless ambassador who would presume to preempt his chiefs*—H. Kissinger, 1979; *He looked surprised—almost annoyed—as if a servant had presumed too great a familiarity*—P. P. Read, 1981.

**assuming (that)** is sometimes used as a 'marginal subordinator' (*CGEL*) in the sense 'for the sake of argument, on the assumption that', e.g. *assuming that the museum is open on Monday, we shall leave at 10 a.m.* But a sensitive person would not accept the construction *assuming that the museum is open on Monday, the car will pick us up at 10 a.m.* (cars cannot assume).

**assure, assurance.** *Assure* and *assurance* have never found general acceptance in the sense of paying premiums to secure contingent payments, though they are used by some insurance offices and agents, and so occasionally by their customers, especially when death is the event insured against (*life assurance; assure one's life*). Apart from such technical use, *insure* and *insurance* hold the field.

**assure, ensure, insure.** These three words have intersecting paths in contexts involving aspects of certainty, assuredness, and security. The following sentences show the main lines of usage. *Assure* ( = give an assurance in order to remove doubt, etc.) *I assure you of my love; I assured him that he had not been overlooked;* ( = be certain) *rest assured that I will be at the station when the train arrives;* ( = place insurance) *those who assure with this Company.*

*Ensure* ( = make certain, guarantee) *Security checks at airports should ensure that no firearms are carried by passengers; that will ensure your success.*

*Insure* ( = protect oneself financially by insurance) *He was insured against theft or loss.* Also, in AmE only, = *ensure*, e.g. *He kept saying that he would take her out on the day the show was going to be broadcast, to insure that she didn't see it—New Yorker,* 1992; *The revolution has done enough to insure that there is no return to the decrepit imperial system—Dædalus,* 1993.

**assuredly.** Pronounce as four syllables, /əˈʃʊərɪdlɪ/. See -EDLY.

**asterisk** (*). Used conventionally: **1** As a guide to a footnote (placed at the end of the context requiring elucidation, and at the head of the footnote itself).

**2** Formerly, a group of three asterisks was placed thus (∴) to draw attention to a particular passage in a book or journal.

**3** Esp. in books written before the second half of the 20c., as a device to indicate omitted letters, esp. in coarse slang words, e.g. *c\*\*t, f\*\*k.*

**4** In etymologies, placed before a word or form not actually found, but of which the existence is inferred, e.g. *wander* f. OE *wandrian* = MLG, MDu *wanderen,* etc.,:- WGmc. *\*wandrōjan.*

**5** In modern linguistic writing, placed before unacceptable forms or constructions that are cited to draw attention to what are the correct ones, as *\*childs* (for *children*), *\*Leave the room, kindly* (for *Kindly leave the room*).

**asthma.** Pronounced /ˈæsmə/ in BrE, but /ˈæzmə/ in AmE.

**astronaut.** First recorded, in a forecasting manner, in 1929, it has settled down as the customary word for a person trained to travel in a spacecraft. *Cosmonaut* (1959) is the usual word for a Russian equivalent.

**astronomical.** Some restraint is called for in the figurative use of the word to mean 'immense', esp. of figures, distances, etc. First recorded in 1899, this use of the word turns up with great frequency in popular and journalistic

work, esp. applied to large sums of money, prices, wage increases, foreign debts, etc.

**asylum.** **1** Pl. -*ums.* See -UM 1.

**2** First established in the 19c. as the customary word for a hospital for psychiatrically disturbed patients (though much older in the general sense 'a sanctuary, a secure place of refuge'), *asylum* in this sense has gradually fallen into disuse in the 20c. Such places are now usually called 'hospitals', 'clinics', or 'units', preceded by a proper name (in Oxford, for example, *Ashurst Clinic, Highfield Adolescent Unit, Ley Clinic, Littlemore Hospital, Warneford Hospital*). The dominant sense of the word now occurs in *political asylum* (first recorded 1954), the condition of being, or permission to remain in a country as, a political refugee.

**asyndeton.** (Gk, = 'unconnected'.) Unlinked coordination, esp. words not joined by conjunctions, is an ordinary feature of the language, e.g. *his comfort, his happiness, his life depended on the goodwill of his friends; carefully, quietly, remorselessly, the cat stalked the bird.* Asyndeton is also a rhetorical device in literature: *Is this the Region, this the Soil, the Clime, ... this the seat That we must change for Heav'n (Paradise Lost* i. 242–4); *Come back in tears, O memory, hope, love of finished years* (C. Rossetti).

**at.**

| **1** *at about.*
| **2** *at all.*
| **3** *at or in.*
| **4** *where ... at.*

**1** *at about.* Occasional objections to this use must now be set aside. The *OED* (s.v. *at* prep., sense 40) illustrates the use (e.g. *at about seven o'clock in the evening; at about that rate*) from 1843 onward in the work of George Borrow, Virginia Woolf, and others.

**2** *at all.* This prepositional phrase meaning 'in every way, in any way' has a variety of uses in standard English, e.g. (with negative construction) *I did not speak at all*; (interrogative) *did you speak at all?* (conditional) *if you spoke at all.* In former standard use, the phrase was restricted to affirmative constructions with the meanings 'of all, altogether;

only'. This use survives in Ireland and in some dialects in Britain and in the US: *And what at all have you got there?*—J. Barlow, 1895; *John Cusack is the finest dancer at all*—P. W. Joyce, 1910; *He is the greatest man at all*—*Dialect Notes*, 1916; *Use one statement at all*—1976 in *DARE*; *Was he the right man at all?*—J. Leland, 1987. Two oral examples of 1990 from speakers in Dublin (in both examples = altogether): *He felt very awkward in this company at all*; *I had a great time at all*.

**3** *at* or *in*. With proper names of places: 'Particularly used of all towns, except the capital of our own country, and that in which the speaker dwells (if of any size), also of small and distant islands or parts of the world' (*OED*). This rule admits of many exceptions, some of them arising from the new perspectives afforded by air travel to 'small and distant islands or parts of the world'. In general terms, *in* has gained ground and *at* retreated somewhat.

The implication of *in* is that the subject has been, or is, physically in (the place referred to): *we stayed in Fiesole for two weeks*; *St Peter's College is in Oxford*. But the choice depends in part on the dimensions of the place referred to. Reference to a specific place normally requires *at* (*at the North Pole*; *the plane landed at Nadi in Fiji*); reference to an area, country, etc., requires *in* (*she grew up in Switzerland*; *in Ontario, in Acton*). Large cities are treated as areas (*Professor Miyake lives in Tokyo*) but can also be regarded as specific stopping-points on journeys (*the plane called at Tokyo on the way to Seoul*). A further distinction is provided by the pair of sentences *he is at Oxford* ( = is a member of the University of Oxford) and *he is in Oxford* ( = living in, visiting, the city of Oxford).

**4** *where … at*. The tautologous regional use of *at* in such sentences as *Where does he live at? This is where I get off at* does not belong in the standard language. On the other hand the colloquial phrase *where it's (he's, she's) at*, meaning 'the true or essential nature of a situation (or person); the true state of affairs; a place of central activity', swept into AmE in the 1960s and thence into other forms of English. For example, David Lodge used the title 'Where It's At: California Language' for an article in *The State of the Language* (1980) about the trendy language of modern California.

**-atable.** For the types *demonstrable* (rather than the rare form *demonstratable*) and *debatable* (with -*atable*) see -ABLE, -IBLE 2v. Some of the forms that Fowler recommended in 1926, *incubatable, inculcatable,* and *inculpatable*, seem to have been figments of his imagination as there is no record of them in the *OED* or other large general dictionaries. The shorter form *inculpable*, on the other hand, has been in continuous use since the 15c.

**ate** (past t. of *eat*). The standard pronunciation is /et/, but /eɪt/ is also common and is equally acceptable. In AmE /eɪt/ is customary.

**atelier.** This 19c. loanword from French ( = workshop, studio) is still usually pronounced in English in a manner approximating to Fr. /atəlje/. Of anglicized pronunciations perhaps the more common are /əˈtelɪeɪ/ and /ˈætəlˌjeɪ/.

**-athon.** This combining form, barbarously extracted from *marathon*, has generated a great number of formations, most of them temporary but some persistent, denoting something carried on for an abnormal length of time, usually as a fund-raising event, e.g. *talkathon, walkathon,* and (with reduction to -*thon*) *radiothon, telethon*. Among the other formations noted in the 1980s and 1990s (most of them AmE) are *bake-a-thon, dance-athon, envirothon, jobathon, operathon, read-athon,* and *snoozathon*. The earliest words formed in this manner date from the 1930s.

**-ative, -ive.** There has been a great deal of slipping and sliding in the use of these rival suffixes in some, but not all, of the relevant words. The earliest English words in -*ative* entered the language in the ME period, some directly from Fr. -*atif*, -*ative* and others from the participial stem -*āt*- in Latin verbs in -*āre*. Some others were simply formed in English on the model of existing adjs. in -*ative*. A number of them were joined by rival forms in -*ive* (i.e. with the -*at*- omitted), and in some cases the rivalry between the -*ative* and the -*ive* forms continues.

The more important of these pairs are treated at their alphabetical places. There follows a select list (from scores of words of this type) with an indication of the date of first record of each. A dash signifies that the *OED* has no record of the word in question.

| -ative | -ive |
|---|---|
| affirmative 1509 | — |
| anticipative 1664 | — |
| argumentative 1642 | argumentive (1668, once only) |
| assertative 1846 | assertive 1562 |
| authoritative 1605 | authoritive (1645, once only) |
| contemplative 1340 | — |
| demonstrative 1530 | — |
| elucidative 1822 | — |
| exploitative 1885 | exploitive 1921 |
| figurative 1398 | — |
| interpretative 1569 | interpretive 1680 |
| interrogative 1520 | — |
| investigative 1803 | — |
| preventative 1654–66 | preventive 1639 |
| qualitative 1607 | qualitive (1846, once only) |
| quantitative 1581 | quantitive 1656 |
| representative 1387–8 | — |
| retardative 1847 | retardive (1797, now rare) |
| talkative (q.v.) 1432–50 | — |
| vegetative 1398 | vegetive 1526 |

The dominance of the -*ative* forms is self-evident, but each pairing or non-pairing deserves separate investigation.

**-ato.** Musical terms ending in -*ato* retain the Italianate pronunciation /ɑːtəʊ/, e.g. *obbligato, pizzicato, staccato.* Two non-musical words (neither of them derived from Italian) ending in -*ato, potato* and *tomato,* have /-eɪ-/ and /-ɑː-/ respectively in BrE but /-eɪ-/ for both in AmE.

**atop.** Used since the 17c. as an adverb (also, less frequently, *atop of*) and preposition, it is now rare except as a preposition. As such, it is common in all levels of writing. Examples: *the half-drunk glass of the stuff that waited atop a pile of 'Smith's Weeklies'*—T. Keneally, 1980; *Now I am seated atop the piano, spinning*—Lee Smith, 1983; (after a hurricane) *the deckchair atop a bush in St James's Park*—*Times,* 1987; *her graying hair arranged into a crown atop her head*—M. Doane, 1988.

**atrium** (hall of Roman house, etc.). Pl. *atria* or *atriums.* See LATIN PLURALS 1.

**attaché.** In BrE always /əˈtæʃeɪ/, whether for a technical expert on the diplomatic staff of a country or in *attaché case,* a small case for business papers. In AmE the dominant (but not the only) pronunciation for both senses is /ˌætəˈʃeɪ/, in printed work the acute accent is often omitted (esp. in advertisements for the case), and *attaché* is often used by itself to mean 'attaché case'.

**attic, garret.** An attic is a room at the top of a house immediately below the roof or (also called a loft, a place inside the sloping roof of a house, used for storing suitcases or other not-often-used items. A garret is a small, usu. spartan, room at the top of a house, usu. one rented cheaply to a student, a painter, etc.

**attorney.** See BARRISTER, SOLICITOR.

**attraction** (grammar). See AGREEMENT 8.

**attributive.** 1 In grammatical work, 'that expresses an attribute': normally an adjective placed immediately before a noun (*brown shoes, daily paper*) or a noun similarly placed (*beauty contest, bedside lamp, end result*). Plural attributive nouns, once relatively rare, are now commonplace (*appointments book, customs duty, narcotics dealer, procedures manual*).

2 The language permits the placing of more than one noun in a series: (two attrib. nouns) *a museum conservation department; an Oxfordshire gentry family; a quality control manager;* (three or more attrib. nouns) *a dilapidated South Side low-income apartment complex; Oxford City Football Club president II; the undisputed Secret Service crossword king.*

3 For attributive adjectives, see ADJECTIVE 2.

4 *An historical note.* By 1400 a large class of compound nouns of the type *cherry-stone* and *hall-door* had come into being, and before long such two-unit expressions became attached to other nouns, resulting in noun phrases like *coffeehouse conversation* (Hume, 1752) and *fellow-workman* (Coverdale, 1535). It was an easy step to the formation of more complex assemblages like *whoreson malt-horse drudge* (*Taming of the Shrew,* 1596), *tortoise-shell memorandum book* (Smollett, 1771), and many others. What has happened in the 20c. is that this ancient process has gathered momentum, especially in the hands

of journalists and civil servants, to produce all too frequently phrases of the type *university block grant arrangements, rate support grant settlement,* and *Slav Bosnia Famine Witness Dr John Smith,* as well as eye-stretching comments of the kind (in a letter to *The Times* in July 1991) *the dilemma of trading off patient waiting time against treatment resource idle time.*

**au.** See À LA.

**au courant, au fait.** 1 These loanwords from French, both meaning 'acquainted or conversant with what is going on', joined the language in the mid-18c. and have been freely used by good writers since then. *Au fait* was often construed in the past with *of* (*I will put you* au fait *of all the circumstances of the case*—A. Granville, 1828), but is now almost always followed by *with* (*he was very keen to keep me* au fait *with his progress*—R. Cobb, 1985). Since it entered the language, *au courant* has been followed by either *of* or *with* (*They are quite eager to place me* au courant *of all their proceedings*—J. S. Mill, 1830; *It would have been wiser if the United States had been kept* au courant *of the negotiations from the very outset*—*Daily Tel.,* 1928; *keeping its public* au courant *with recent work*—D. Macdonald, a1961).

2 Both phrases should normally be printed in italics.

**audience, audition.** 1 In the sense 'a formal interview', *audience* has been used since the 16c. for one given by a monarch, the Pope, or other high personage; *audition* was adopted at the end of the 19c. for one given by an impresario to an aspiring performer. Since the 1930s *audition,* by the normal process of CONVERSION, has also become established as an ordinary verb in the entertainment industry.

2 Despite its etymological meaning, *audience* (ultimately from L *audīre* to hear), with the advent of motion pictures and of television, has gradually established itself as a normal word for those watching a cinema or TV film.

**audit** (verb). British readers of American books or newspapers are still likely to be baffled by the 'new' use (first recorded in 1933) of the verb to mean 'to attend (a course or other form of instruction) in order to participate without the need

to earn credits by writing papers' (*OED* 2), e.g. *She audited his undergraduate lectures; she waylaid him in the department office*—A. Lurie, 1974.

**au fait.** See AU COURANT.

**au fond.** See À FOND.

**auger, augur.** An *auger* (from OE *nafogār,* with metanalytic loss of initial *n*) is a tool for boring holes. *Augur* (from L *augur* a soothsayer) is used occasionally to mean a prophet, but occurs more frequently as a verb, esp. in the phrases *augur well* or *augur ill* (or *badly*) 'to have good or bad expectations *of, for*'. Examples: *Everything augured badly—they weren't meant to be together*—E. J. Howard, 1965; *The novel augured well for a successful career in fiction-writing*—J. Pope Hennessy, 1971.

**aught.** By about 1300 OE *āwiht* 'anything' had become *aught.* It remained in use for many centuries (e.g. *Excuse me, dear, if aught amiss was said*—Pope, 1702) but now survives mainly in the fixed expressions *for aught I know, for aught I care.* From about the same date it was also spelt *ought* (e.g. *Grieve not, my Swift, at ought our realm acquires*—Pope, 1728), and so became a homonym of the modal verb *ought* (which is from OE *āhte,* pa.t. of *āgan* 'to own'). The presence of the modal verb may have helped to drive *aught* into restricted use.

**augur.** See AUGER.

**aural, oral.** In standard English these are both pronounced /ˈɔːrəl/. In some other varieties of English they are distinguished, by virtue of the fact that the second of the pair is pronounced /ˈɒrəl/.

**autarchy, autarky.** The first (from Gk αὐταρχία) means 'absolute sovereignty'; the second (from Gk αὐτάρκεια) means 'self-sufficiency'. They share the same pronunciation in English, but should be carefully distinguished in spelling.

**authentic, genuine.** Both words mean 'entitled to acceptance or belief, as being in accordance with fact', but *genuine* commonly (and *authentic* less commonly) has the additional nuance 'not sham or feigned'. The distinction made by Fowler (1926) was that '*authentic* implies that the contents of a book, picture, account, or the like, correspond to facts and are not fictitious, and *genuine* implies that its reputed is its real author'. The sentence *The Holbein Henry VIII is both authentic and genuine*, he said, is valid because it really is a portrait of him and is by Holbein, not by another painter.

But such a fine distinction is not universally applicable: data, documents, a Chippendale chair, a signature, and much else can be described as either *authentic* or *genuine* without a perceptible shift of meaning. On the other hand, an authentic account of a series of events or of an unusual or complex experience is one that is convincing, one that can be believed (*Harrier was thrilled to detect genuine disappointment in her voice*—M. Bracewell, 1989). In such contexts *genuine* means 'real, not fake'. Hitler's alleged diaries, 'discovered' in the 1980s, sounded authentic but were found to be not genuine.

**author** (verb). By the process of CONVERSION this 16c. verb came into being as a companion to the earlier noun (*The last foul thing Thou ever author'dst*—Chapman's *Iliad*, 1596). It has had a chequered career, used both transitively and intransitively, in literary works since then, but has come into widespread use in America, especially in newspapers and often in transferred senses (of 'creative' acts or events in non-literary spheres, e.g. the cinema, sport), during the 20c. It does not find any kind of acceptance in the quality newspapers, or in literary works, in Britain. American examples: *Housman appears to have authored it the year before*—*Amer. N. & Q.*, 1983; *Some of them authored only one or two books*—J. Brodsky, 1984; *Punctuation and grammatical errors in a will drawn by a distinguished ... attorney could be used by a jury in determining whether or not he had indeed authored it*—*Verbatim*, 1986.

**authoress.** A word of long standing in the language (first recorded in 1478), *authoress* has never come to the fore meaning a female author, though it is still used from time to time. Examples: *Both authoresses, one early and the other late Victorian, were of nearly the same age*—*Country Life*, 1972; *The authoress tells us that 'breakthrough scientists ask silly questions'*—*Nature*, 1974; *Charlotte M. Yonge, the authoress*—K. M. E. Murray, 1977. The *OED*'s

comment is just: 'Now used only when sex is purposely emphasized; otherwise ... *author* is now used of both sexes.' See -ESS; FEMININE DESIGNATIONS.

**authorial.** The reluctance to use *author* (verb) and *authoress* does not extend, at any rate among scholars, to the word *authorial*. It is used frequently, for example, to distinguish the spelling or vocabulary of an author from those introduced to the text by a scribe or printer. Examples: *A mass of error both typographical and authorial*—J. Ritson, 1796; *The two final sonnets ... have often been felt to be irrelevant, substandard, and perhaps not authorial*—K. Duncan-Jones, 1983; *Through all his faction the intrusive authorial voice ... directs and comments*—*Music & Letters*, 1986; *Updike's memoirs bear the title*, Self-Consciousness, *to indicate the natural authorial awareness ... of hidden damage*—*NY Rev. Bks*, 1989.

**authoritarian, authoritative.** The two words are readily distinguishable: -*arian* means favourable to the principle of authority as opposed to that of individual freedom; -*ative* means possessing due or acknowledged authority; entitled to obedience or acceptance (*OED*).

**automaton.** The plural is *automata* when used collectively, otherwise (much less commonly) *automatons*. See LATIN PLURALS.

**automobile.** In Britain this word survives mainly in the title of the motorists' rescue and repair service, the Automobile Association. The customary words for the vehicle are *car* and *motor car*; *limousine* is also used as a rather pretentious word for a showy car. In the US the customary words are *automobile* and *car*; *limousine* (often abbreviated in speech to *limo*) is most often a large, usu. chauffeur-driven, car, or a small bus used for carrying passengers to and from an airport.

**avail.** 1 The noun poses no problems. It is used most frequently in the phrases *of no avail* (his efforts were of no avail), *to no avail/to little avail* (he tried the key but to no avail, etc.,), *without avail*.

2 Uncontroversial uses of the verb are common: (used intransitively) *words avail very little with him*; (transitively, with a

personal object) *his good works availed him nothing*; (with reflexive pronoun + *of*) *none of the English departments ... avail themselves of such opportunities*—F. Tuohy, 1964; *I availed myself of the invitation to move about*—D. Lessing, 1979.

**3** In the 19c. when the adjective *available* acquired the sense 'that may be availed of', the verb came to be used as an indirect passive: *Power ... must be availed of, and not by any means let off and wasted*—Emerson, 1861. This use has persisted, with limited currency, in the 20c., esp. in America: *the wonderful system of drainage is being availed of*—*Daily Tel.*, 1927; *individual contracts ... may not be availed of to defeat or delay the procedures*—*Legal Times* (US), 1982; *a personal service corporation will not be considered to be formed or availed of for the purpose of evading ... income tax*—ibid., 1982.

**4** Some other new (and avoidable) uses are shown in the following examples, all from sources outside the UK: (reflexive pronoun omitted) *I am keen ... to go and train under an international coach ... and also avail of the time to play as many league matches in the international circuit*—*Society* (Bombay), 1987; (passive uses) *The President explained that private firms and industrial enterprises will be availed loans to rehabilitate them* (i.e. will have loans made available to them)—a 1986 report from Uganda; *it is only in the conditions of peace that the possibilities of creating a democratic society and people's prosperity could be availed* (i.e. made available)—1986 report from Afghanistan; (*avail* used transitively with a double object) *The association with this country has availed Koreans many advantages*—*Washington Post*, 1986.

Clearly the verb *avail* (with or without *of* and with or without a reflexive pronoun) is on the move, but for the most part outside the UK.

**avant-garde.** From the 15c. onward the word had a single meaning in English, namely 'the foremost part of an army', but this use was taken over at some point in the 19c. by *vanguard*. The 20c. has witnessed the revival of *avant-garde* with the transferred meaning 'the pioneers or innovators in any art in a particular period'. It continues to be used with great frequency, but still retains its

quasi-French pronunciation /ˌavãˈgɑːd/, not as yet /ˈævənˌgɑːd/.

**avenge, revenge.** A note of 1885 in the *OED* observed that 'the restriction of *avenge* and its derivatives to the idea of just retribution, as distinguished from the malicious retaliation of *revenge*' is not 'absolutely observed, although it largely prevails'. *Webster's Third* (1986 version) says that while both verbs are used in the sense 'to punish a person who has wronged one or someone close to one ... *avenge* more often suggests punishing a person when one is vindicating someone else than oneself or is serving the ends of justice', whereas '*revenge* more often applies to vindicating oneself, and usually suggests an evening up of scores or a personal satisfaction more than an achievement of justice'.

The distinctions are possibly overstated and they often go unobserved. Examples: (avenge) *Edwy had the power to avenge himself upon Dunstan*—*OED*, 1861; *he avenged himself for Father's obstruction of all his efforts to nominate an heir*—P. Scott, 1962; *The ferocity and guile with which Absalom had avenged the rape of his sister*—D. Jacobson, 1970; *That brave god will leap down from his steed when he has to avenge his father's death*—K. Crossley-Holland, 1980; *Through characterization the novelist has the means to avenge himself on his enemies if, of course, he is willing to risk an action for libel*—P. D. James, 1993. (revenge) *He was father's partner, and father broke with him, and now he revenges himself*—Dickens, 1865; *It is likely that the gunmen were from a Protestant paramilitary organization revenging the shooting by the IRA earlier this week of three Protestants*—*Times*, 1989; *murmurings ... that Ham's wife ... had decided to revenge herself upon the animals*—Julian Barnes, 1989; *It wasn't just that I could never revenge myself on him ... I felt the first dim recollection that my own life had lost all purpose.*—Simon Mason, 1990; *If I were to revenge myself upon you ... that would be an act of despair*—I. Murdoch, 1993.

**aver.** This formal word means 'to assert as a fact; to state positively, to affirm'. Just as assertion means rather more than mere statement, so *aver* means more than the neutral word *say*. Examples: *The shopman averring that it was a most*

*uncommon fit*—Dickens, 1838; *It is passion-*
*ately averred in Tortilla Flat that Danny alone*
*drank three gallons of wine*—J. Steinbeck,
1935; *Belloc liked to aver that you belonged*
*to the flower of the bourgeoisie if you knew*
*the maiden names of your four great grand-*
*mothers*—A. N. Wilson, 1984; *He was, she*
*avers, 'a real Sweetie'*—Godfrey Smith, 1990.

**averse, aversion.** There is no further
point in affecting woundedness if one
encounters *averse to* rather than the construction favoured by Dr Johnson, *averse*
*from*. In a note of almost unprecedented
length, the *OED* states that *averse to* and
*aversion to* are justified by the consideration that these words express a mental
relation analogous to that indicated by
*hostile, contrary, repugnant, hostility, opposi-*
*tion, dislike,* and naturally take the same
construction. Historically, both constructions, with *to* and with *from*, have
been used, often by the same writers,
but *averse to* and *aversion to* are now
without question the more usual. Examples: (averse to) *Nor was he averse to*
*being reminded of Calcutta*—A. Desai, 1988;
*while I am in no way averse to the high life*
*... of north-eastern London, it does have its*
*limitations*—R. Elms, 1988; *Vic wasn't averse*
*to keeping Everthorpe guessing whether he*
*and Robyn Penrose were having an affair*—D.
Lodge, 1988; (aversion from) *that terrible*
*reality—passing morality, duty, common*
*sense—her aversion from him who had owned*
*her body*—J. Galsworthy, 1921; *Dr Mainwar-*
*ing's prescription had not cured her aversion*
*from the prospect of becoming hopelessly senile*
*in the company of people who knew her*—K.
Amis, 1974; (aversion to) *The Air Force's*
*aversion to operational testing goes back at*
*least to 1968*—New Yorker, 1988; *he had a*
*lifelong aversion to British officialdom*—J. le
Carré, 1989. See also ADVERSE.

**avid.** Loanwords seldom take over all
the meanings and constructions of the
word in the original language. *Avid*, first
recorded in English only in 1769, is no
exception. The *Oxford Latin Dictionary* lists
six main senses for L *auidus*: (1) greedy for
gain, covetous; (2) having an immoderate
appetite for food, greedy; (3) (of qualities,
actions, etc.) voracious, greedy, in-
satiable; (4) eager, ardent; (5) ardently
desirous (of), eager (for); (6) lustful, passionate. In English, *avid* has settled down
in most contexts as a kind of medley of
senses 3, 4, and 5 of the Latin word, that

is 'extremely, and usu. praiseworthily,
eager'. It is used attributively, or, if predicatively, is usu. followed by *for* or by a
*to*-infinitive. Examples: (attributively) *an*
*avid collector of old coins*; *an avid reader*; *an*
*avid interest in politics*; (predicatively), *The*
*Africans are avid for advancement* (D. Lessing, 1968); *people in Eastern Europe are now*
*avid to find out how democracy works*.

**avocation.** In the 17c. and 18c. (after L
*āuocātio*) *avocation* was used in English to
mean 'the calling away or withdrawal
(of a person) from an employment, a
distraction'. Since, however, the business
which called one away could be either
of minor or major importance, a nice
balance of meanings developed. Already
in the 17c. it came to mean (still its
main sense) 'the minor or less important
things a person devotes himself to', or
(with complete loss of the original Latin
meaning) 'one's usual occupation, vocation, or calling'.

**avoirdupois.** If the need should arise,
pronounce this awkward word /ˌævədə-
'pɔɪz/.

**avouch, avow, vouch.** *Avouch* was once
a proud word of multiple meanings
(eleven senses spread over a column and
a half in the *OED*) and wide currency, e.g.
*Thou hast auouched the Lord this day to be thy*
*God*—(Deut. (AV) 26: 17); *Then my account I*
*well may giue, And in the Stockes auouch it*
(Autolycus in *The Winter's Tale*, IV. iii. 22).
It is now slipping into obsolescence, but
when it is used, usu. in somewhat elevated circumstances, it means guarantee, solemnly assert, maintain the
truth of, vouch for (*a miracle avouched by*
*the testimony of ...*; *millions were ready to*
*avouch the exact opposite*). *Avow*, still in
common use, means declare (something)
openly, admit (*avow one's belief, faith*, etc.;
*avow oneself to be a vegetarian*; *the avowed*
*aim of the government is to bring inflation*
*down*). *Vouch* is restricted to the phrase
*vouch for* (somebody or something) (*I can*
*vouch for him, for his honesty*; *experts were*
*produced to vouch for the authenticity of the*
*painting*).

**avuncular.** A minor curiosity of the
language is that this word, meaning 'of
or resembling an uncle', lacks a feminine
equivalent. In 1982, readers of William
Safire's column in the *New York Times*

*Magazine* suggested (several of them facetious) *amital* (L *amita* aunt), *auntique*,
*auntsy*, *materteral* (L *matertera* maternal
aunt), *tantative* (F *tante* aunt), *tantular*,
and *tantoid*, but none of them has taken
hold.

**await, wait.** *Await* is a transitive verb
meaning 'to wait for' (*he awaited his fate*,
*I shall await your answer*) or '(of something)
to be in store for' (*a surprise awaits you*).
It cannot be used intransitively. *Wait*,
on the other hand, can be used both
transitively and intransitively in a wide
range of senses and idiomatic phrases
(*wait a minute*; *you must wait your turn*; *he
waited for an hour*; *wait till I come*; *wait
and see*; etc.). See WAIT (verb).

**awake, awaken, wake, waken.**
**1** *History.* The network of forms in OE
and ME for these four verbs amounts to
a philological nightmare. From earliest
times, they have been unstable and unpredictable in two main respects: (*a*) the
choice of form for the past tense and
past participle, depending on whether
the verb was felt to belong to the strong
conjugation (with pastness indicated by
change of stem vowel) or to the weak
conjugation (with past tense and past
participle ending in -*ed*, -*ened*); (*b*) the
likelihood that any of them could at
most times and in the work of any given
writer be used either intransitively to
mean 'to arise from or come out of sleep'
or causally (and transitively) 'to rouse
(someone) from sleep'. To make things
worse, the presence or absence of a following *up* seems to have been optional,
esp. since the 19c., after *wake* and *waken*.
The complicated chain of events is set
down under the respective entries in the
*OED*, to which readers must be referred.

**2** *Principal parts.* In modern English
the principal parts are still somewhat
fragile or fluid, with much blending of
forms and function. The customary
forms in use are *awake*, pa.t. *awoke* (rarely
*awaked*), pa.pple *awoken* (rarely *awaked*);
*awaken*, pa.t. *awakened*, pa.pple *awakened*;
*wake*, pa.t. *woke* (rarely *waked*), pa.pple
*woken* (rarely *waked*); *waken*, pa.t. *wakened*,
pa.pple *wakened*. *Awaked* and *waked* are
more likely to occur in English dialects
(and also in the language of children)
than in standard English.

**3** *Awake*, *awaken*, and *waken* often have
a tinge of formality not present in *wake*.

**4** Examples of the more common uses
of this quartet of verbs (set out in the
order *wake*, *awake*, *awaken*, *waken*, i.e. the
frequency order of the four verbs):
(*a*) *When do you usually wake (up) in the
morning?*; *I wake (up) at about seven every
morning*; *Wake up! It's morning*; *I woke (up)
early this morning*; *I had woken (up) at dawn*;
*I was woken (up) by the wind in the night*; *I
woke (her) up at nine o'clock.*
(*b*) *I awoke from a deep sleep*; (somewhat
literary) *She awoke to the sound of driving
rain*; *she awoke her sleeping child*; *the accident awoke old fears.*
(*c*) *We must awaken motorists to the
danger of speeding on foggy motorways*; *the
episode awakened her interest in impressionist
painting*; *enough noise to awaken the dead.*
(*d*) *We were wakened by the storm*; *when
she fell asleep nothing would waken her.*

**aware.** The *OED* classes *aware* as a predicative adjective, i.e. one that can normally stand only after a noun or the
verb *to be*. Its more usual constructions,
which are still in use, are illustrated
by the types (*a*) (followed by *of* + noun)
*Arnan loked, and was aware of Dauid* (Coverdale, 1535); (*b*) (followed by a *that*-clause)
*Are you aware that your friends are here?*
(1885). In both constructions *aware* could
be pre-modified by *very much* or by (*very*)
*well*. In the 20c. it has started to be used
attributively (e.g. *an aware person*, one
who is alert to the possibilities of a given
set of circumstances), or absolutely, preceded by an adverb in -*ly* (e.g. *he was
environmentally aware all right, rather too
much so*). Both uses are paralleled by a
widening of the applications of the corresponding noun *awareness*. Examples: *it
... succeeded in striking a démodé note
in that aware community*—M. Allingham,
1938; *the painfully aware state that seems
to have succeeded her earlier calm*—A.
Brookner, 1985; *Lord Scarman recognises
that the awareness campaign needs forcefully
to target the government*—City Limits, 1986.

**away.** Used with intensive force before
adverbs (*away back, down, up*, etc.), originally and chiefly in AmE, since the early
19c.: *Perhaps away up in Canada*—J. Palmer,
1818; *Manufacturers of all good cars are
away behind in their deliveries*—NY Even.
Post, 1906. It now seems to be yielding

in all major forms of English to the shorter form *way*: hence such common phrases as *way back* (*in the 1960s*), *way down south*, *way off course*, *way out here*, *way over there*. See WAY 1.

**awesome.** British visitors to the US are surprised to find this word used with great frequency as an enthusiastic term of commendation, a far cry from its traditional meaning 'full of awe, inspiring awe (usu. with profoundly reverential implications)'. *WDEU* (1989) says that the word 'has been part of the standard hyperbole of sports broadcasting and writing for several years', and cites supporting evidence, including the following: *The depth of quality on the Steeler squad is awesome—there is no apparent weakness anywhere*—A. Mount in *Playboy*, 1979. The word also turns up, for example, in the language of young 'preppy' people in the US: it is glossed as 'terrific, great' in the *Official Preppy Handbook* (1980), edited by L. Birnbach and others. The traditional reverential use is far from extinct, however, in the US, though it is more at risk there than in Britain. It is just that in certain social groups the word has made its way into areas where it can be used of public or domestic experiences that seem as important to them as the experiences of religion.

**awful.** For nearly two hundred years this adjective has been used as 'a mere intensive deriving its sense from the context = Exceedingly bad, great, long, etc.' (*OED*). Thus *an awful looking woman*; *an awful while since you have heard from me* (Keats); *an awful scrawl*; *what an awful duffer I am*, etc. For a similar period it has also served as an adverb (*it is awful hot, it is awful lonely here*—Trollope; *Aunt Polly's awful particular about this fence*—M. Twain), though this has a non-standard ring about it now. Side by side with these, and over a similar period, the adverb *awfully* has become established, esp. in Britain, as a simple intensive = very, exceedingly, extremely. Examples: *You'll be awfully glad to get rid of me*—William Black, 1877; *I'm awfully glad you didn't run away*—J. B. Priestley, 1929; *You are awfully good to look after Elizabeth for us*—M. Binchy, 1982; *I counted only four entries . . . (with awfully badly spaced headlines)*—*Creative Review*, 1988.

These are all natural semantic developments, but they have tended to leave the primary reverential sense stranded.

**awhile, a while.** 1 When used as a noun meaning 'a period of time' *while* is written as a separate word, separate, that is, from the indefinite article that often precedes it (and obviously also when an adjective intervenes, as in the type *It's a long while since I last saw you*): *I moved away and looked at the T'ang Dynasty horses for a while*—*Encounter*, 1990; *I'm going away for a while*—B. Neil, 1993. This is by far the normal (and recommended) practice.

2 From the 14c. onward, however, the word (originally a phrase) has also been used as an adverb: e.g. *Sailing awhile to the Southward*—Defoe, 1725; *Awhile she paused, no answer came*—Scott, 1810; *when he reached the street-sign he stopped awhile and stood beneath it*—C. Dexter, 1983; *They delayed awhile, but finally the dance was held*—*New Yorker*, 1990.

3 The *OED* uses ¶ as a beacon to warn readers of misuse. Under the entry for *awhile* (adverbial phrase), it says 'Improperly written together, when there is no unification of sense, and *while* is purely a [noun]'. Three examples are given, from Caxton (1489), John Yeats (1872), and Ouida (1882). It is easy to find further examples of this regrettable tendency: *For awhile explication was impossible*—G. Meredith, 1861; *You'll be in Canberra for awhile now*—M. Eldridge, 1984 (Aust.); *He took awhile to die*—B. Ripley, 1987 (US).

¶ The tendency towards writing the expression as one word in all circumstances is being assisted by the fact that in many contexts *awhile* could easily be replaced by *for a while* without loss of sense or rhythm: e.g. *Mrs Hardcastle was silent awhile, frowning to herself*—F. Kidman, 1988 (NZ); *I asked her one day, after she'd been home awhile, what she thought* [etc.]—M. Golden, 1989 (US); *She could wait awhile, it wouldn't kill her*—*New Yorker*, 1991.

**axe.** Thus spelt almost everywhere in English-speaking countries except the US, where it is often printed as *ax*.

**axis.** Pl. *axes*, pronounced /ˈæksiːz/.

**aye, ay.** 1 The word meaning 'yes' appeared suddenly about 1575 and is of disputed origin. It is now always pronounced /aɪ/ and is normally spelt *aye*, esp. in parliamentary language (*The ayes have it*), in nautical language in *Aye, aye, sir* (the correct reply on board ship on receipt of an order), and in some northern dialects in Britain, esp. in Scotland (*Are you coming? Aye, I'm on my way*).

**2** The word meaning 'ever' was first recorded *c*1200 in the *Ormulum*, and is derived from ON *ei*, *ey* (cognate with OE *ā*), used in the same sense. It is normally pronounced /eɪ/ and spelt *aye*.

**3** The spelling *ay* is occasionally used for both senses.

**azure.** This 14c. loanword from French has fluctuated in pronunciation in the last two centuries between /ˈæʒ(j)ʊə/ (almost certainly thus in Walker, 1791), /ˈæʒə/ (*OED*, 1885, Daniel Jones, 1917, but both cite other pronunciations as well), and /ˈeɪ-/ (given as a variant in *OED* and Jones). The initial sound is now usually /æ-/ not /eɪ-/, and the medial consonant [-z-] or /-ʒ-/, while the final sound varies markedly between /-ʒjʊə/ rhyming with *pure* /pjʊə/ (the dominant pronunciation) and /-ə/.

# Bb

**babe, baby.** Both words appeared about the same time (late 14c.) and are probably both derivatives of an infantile reduplicated *baba*. Though *babe* and not *baby* is used in the Bible (AV), *baby* is now the regular form for the primary sense 'young child' and *babe* mostly an affectionate or old-fashioned variant (*it's such a pretty babe!*). In the *NEB*, *babe* is normally replaced by *baby* (Luke 1: 41) or *infant* (Heb. 5: 13); and *babes* by *infants* (1 Pet. 2: 2), *children* (Ps. 17: 14), or some other words like *the immature* (Rom. 2: 20), though it is occasionally retained (Ps. 8: 2, Matt. 21: 16). Inexperienced or guileless persons are still often called *babes and sucklings*, or *babes in the wood*. A *babe in arms* (*his father ... picked him up in front of visitors to hold him like a babe-in-arms*—I. McEwan, 1987), is also an unknowledgeable person (*when it comes to computers, you're a babe in arms*). In general English, *baby* is often used as a term of affection among adults (*my poor baby*). It is also used to imply unmanliness (*at the sight of blood he is no end of a baby*). Slang uses have proliferated since Victorian times (*terrible environments inhabited by some tough babies*). In the US, *babe* is used as a term of friendly address by a man to another man or to a woman, especially among Black speakers (*You and me, babe, We're going places, just you watch* M. Doane, 1988). This use is beginning to appear in UK fiction: *Hello, Tina. Sorry, babe, I felt a bit rough.*—R. Elms, 1988; *He looked me up and down. 'Who you, babe?'* —A. Billson, 1993. The first name of an American baseball player (*Babe Ruth*) has added a suggestion of sportsmanlike skill to the word *babe*; and *baby doll* is applied to a young woman seen as exhibiting a doll-like type of ingenuous beauty.

**bacchant** /'bækənt/ is **1** a priest or worshipper of Bacchus, the Greek and Roman god of wine,

**2** a drunken reveller, a Bacchanalian (from L *Bacchānālia* a Roman festival in honour of Bacchus). The female equivalent, a priestess or female follower of Bacchus, is *bacchante*, pronounced /bə'kæntɪ/ (less commonly and more demotically /bə'kænt/), pl. *bacchantes* /bə'kæntɪz/ (sometimes /bə'kænts/).

**bacillus.** Pl. *bacilli* /bə'sɪlaɪ/. See LATIN PLURALS; -US.

**back.**

> **1** *back of.*
> **2** *in back of.*
> **3** *in back.*
> **4** *the back of beyond.*

**1** *back of.* As a prepositional phrase meaning 'behind, in the back of', *back of* has been used for three centuries, mostly in North America, and still persists there (*In the shade back of the johouse*—E. Pound, 1949; *His computer ... locates a spare space back of the plane* Keyboard Player, 1986), but also occurs occasionally in Britain (*No one could live there, back of the railways*—M. Laski, 1953.)

**2** By contrast, *in back of* (modelled on *in front of*) is markedly American (*Should I or should I not go out to the swimming pool in back of my sister's condominium?*—A. Beattie, 1980; *My house floated by in back of its belt of olive trees*—New Yorker, 1987).

**3** *in back.* Used absolutely in AmE in the phr. *to get in back* 'to get into (one of) the back seat(s) of a car' (*'What luck,' she muses, sliding back in [the car]. 'Get in back, Herman.'*—B. Ripley, 1987).

**4** The phrase *back o' behint* occurs in some English dialects, and *the back of beyond* throughout the English-speaking world, both of them in the sense 'a very remote or out-of-the-way place'.

**back-formation.** A word coined by J. A. H. Murray—whence Ger. *Rückbildung*—for the formation of a word from a longer word which has the appearance of being derived from it. So *burgle* (first recorded 1872) is a back-formation from the much older word *burglar* (15c.), *edit* (18c.) from *edition* (16c.) or *editor* (17c.), *lech* 'to behave lecherously' (1911) from *lecherous* (14c.)

and related words, *reminisce* (verb) (1829) from *reminiscence* (16c.), and *televise* (1927) from *television* (1907). In the 20c. the type *window-shopping* (1922), yielding the verb *window-shop* (1951), is a potent model. Some back-formations had, and some still have, more than a tinge of jocularity, e.g. *buttle* 'to act as a butler' (1867) from *butler* (13c.). Whereas back-formations such as *diagnose* (from *diagnosis*), *extradite* (from *extradition*), *grovel* (16c.) (from *grovelling* adv. 'in a prone position', 14c.), *legislate* (from *legislation*), *resurrect* (18c.) (from *resurrection*), and *sidle* (18c.) (from *sideling* adv. 14c., or *sidelong* adv. 16c.) sit comfortably and neutrally in the language, for many people *donate* (from *donation*), *enthuse* (from *enthusiasm*), and *liaise* (from *liaison*) are as tasteless as withered violets. Most Americans use *burglarize* and reject *burgle*; in Britain *burglarize* is rejected as a 'vulgar Americanism', and *burgle* is used without constraints. The process of back-formation has a long history, and it will doubtless continue to generate new words in the years ahead. In the space of a few decades the pair *television*/*televise* has established itself on the analogy of such a pair as *revision*/*revise*. Many others will follow.

**backlog.** The OED recorded the word (with a hyphen) in the sense 'a large log placed at the back of the fire' (first noted in the 17c.). From the slow-burning nature of such logs, figurative uses in the general sense 'a reserve supply, an accumulation' established themselves from the last quarter of the 19c. onward (*the contrast between the immediate emotion and the backlog of our considered view*—J. Bailey, 1976). In the 1930s the word also acquired the meaning 'arrears of unfulfilled orders, uncompleted cases, unfinished work, etc.' (*court cases of which there was a considerable back-log*—*Listener*, 1958; *the backlog of visitors granted temporary admission is now over 3,500*—*Daily Tel.*, 1986), and this third sense is now the dominant one. The word is never hyphened now.

**backslang.** A kind of slang in which words are pronounced backwards, e.g. *cool* for *look*, *ynnep* for *penny*. *Slop* in the sense 'policeman' is a modification of *ecilop*, backslang for *police*; *skoob* is a modern reversal of *books* and means 'the ceremonial burning of a book or books'

(*OED*); and *yob* (first recorded in 1859) is backslang for *boy* (but now means 'hooligan').

**backward(s).** **1** In most adverbial uses, *backward* and *backwards* are interchangeable, but usage varies subtly from person to person and from region to region. It is broadly true to say that in North America *backward* seems to be somewhat more usual than *backwards*, and in Britain the other way round. Examples: (*a*) *Talk ran backward from the events of the morning*—A. Munro, 1987 (Canad.); *Rowers not only face backward, they race backward*—*New Yorker*, 1988; *I walked backward to look at her in the sun*—E. L. Doctorow, 1989 (US); (*b*) *he'll ... bend over backwards to please a client*—M. Bail, 1975 (Aust.); *he hauled the cart out backwards*—J. M. Coetzee, 1983 (SAfr.); *the door kept swinging backwards and forwards*—A. Brookner, 1984 (UK); *not knowing where he was, and trying to work his way backwards*—R. Cobb, 1985 (UK); *Damson spurted out of the driveway backwards*—*New Yorker*, 1988.

**2** As an adjective the only form used is *backward*: *he watched her walking away without a backward glance*—R. Sutcliff, 1954; *Getting involved with the blind in any way seemed like a backward step*—V. Mehta, 1987.

See FORWARD(S).

**bacterium** is the singular form, and *bacteria* its plural. See -UM 2. Erroneous uses of *bacteria* as a singular are regrettably common in newspapers: e.g. *A common gut bacteria may be a major cause of rheumatoid arthritis*—*Independent*, 1991; *the author reports that the bacteria for Legionnaires' disease has been found*—*Chicago Tribune*, 1995.

**bad.** **1** Infiltrating uses of this ancient word include a slang, originally US Black English, reversal of sense so that it means 'excellent, marvellous': *Jazzmen often call a thing ... 'bad' when they like it very much*—*NY Times*, 1959; *You See Through is an original tune of theirs—it's B-A-D!*—*Cut*, 1988.

**2** It is also repeatedly used informally as an adverb (*there's one thing I want in this world, and want bad*—M. E. Ryan, 1901; *I only came cause she's so bad off*—L. Hellman, 1934; and in sentences like *he didn't do too bad* (in a game, etc.), *he needed* (something) *bad*, especially in AmE and in English dialects. As a quasi-adj. after the verb *feel*—*I feel bad* (about doing something) or *I feel bad* ( = my health is not good at the moment)—*bad* is unobjectionable even in quite formal English (*To be absolutely honest, what I feel really bad about is that I don't feel worse*—M. Frayn, 1965. But after most other verbs (*play, sing, work*, etc.) and when used with the auxiliary verb *to be*, *badly* is required in standard English: e.g. *he behaved badly*; *they badly wanted to see the game*; *my brother was badly wounded in Tunisia*; *the Smiths are not badly off* (i.e. are not poor).

**bade,** pa.t. of BID, is pronounced /bæd/.

**badinage.** Print in romans (though practice is divided at present), and pronounce /ˈbædmɑːʒ/. Examples: *All these years of half-measures, of flattery and badinage*—A. Brookner, 1985; *Jean whose comical taunts and restless badinage had always stirred up what might otherwise have proved too quiet a scene*—I. Murdoch, 1987.

**bafflegab.** A term, first recorded in 1952, applied to abstruse technical terminology used as a means of persuasion or obfuscation. The word was coined by Milton A. Smith, assistant general counsel for the American Chamber of Commerce, and was 'defined' by him as 'Multiloquence characterized by a consummate interfusion of circumlocution ... and other familiar manifestations of abstruse expatiation commonly utilized for promulgations implementing procrustean determinations by governmental bodies'.

**baggage.** To British ears the word *baggage*, in its primary sense, has a residual American flavour, perhaps as a legacy of the *baggage trains* and *baggage wagons* of pioneering days in America. With the arrival of mass air travel, however, *baggage* and *luggage* have acquired dual, and then multiple, nationality as words. We read of *baggage supervisors* at *luggage carousels* (*Daily Tel.*, 1987); luggage *disappears in the confusion* followed immediately by *his policy pays up to £500 per person for loss of* baggage (*Financial Times*, 1982); *passengers' luggage that came through the baggage claim area had been badly damaged* (*Washington Post*, 1987). Computer searches in large US and UK databases (1987) yielded

an equal number of instances of both words in each country.

A few collocations seem to be more or less fixed, e.g. *excess baggage, baggage area, baggage claim, baggage handler*. A few British railway stations still have *left-luggage offices*; and Intercity trains normally have the label LUGGAGE attached to the relevant storage shelves. But (in 1988) the main terminal at Gatwick Airport had a sign that read *Left Baggage*. In the US a person who looks after the checked baggage of passengers on a train is a *baggageman*. The same word is also used there for a baggage porter in a hotel. At air terminals the terms *hand luggage* and *hand baggage* seem to be freely used in both countries.

In figurative uses, *baggage* is the customary word everywhere: *a pragmatist who travels light, without cumbersome ideological baggage*—*Financial Times*, 1983; *some emotional baggage of my own*—*New Yorker*, 1984; *none of this ancient baggage of criminology is current in the thinking now*—*Dædalus*, 1988.

**bahuvrihi** /bahʊˈvriːhɪ/. A Sanskrit word (lit. 'having much rice', f. *bahú* much + *vrīhí* rice) taken over by grammarians in other languages and applied to compound words that differ in grammatical function in one or, both of their constituent elements : thus Gk ῥοδο-δάκτυλος 'rose-fingered', L. *magnanimus* 'great-minded' (both the original Gk and L words and also the translations are bahuvrihi compounds), *five-pence* (piece), *high-quality* (goods). Some linguistic scholars call them *exocentric compounds*. Cf. ENDOCENTRIC COMPOUNDS.

**bail out, bale out.** The spelling *bail* (ultimately from OF *bailler* 'to take charge of') is always used in contexts of securing the release of a person after an undertaking is given to guarantee his or her reappearance in court on an appointed day. Figuratively, too, a person or a company, etc., may be *bailed out*, or released, from a difficulty, for example by the repayment of his or its debts. In the senses 'to scoop water out of a boat', or 'to make an emergency descent by parachute from an aircraft' the spelling *bale* (*out*) is now usual, as if the action were that of letting a bundle through a trapdoor, even though the word is of different origin (from OF *baille* bucket).

**balance.** The numerous figurative senses of the word are drawn from the principal function of the apparatus thus called, namely equipoise or equilibrium (*the fragile balance between peace and war*) or harmony (as in a work of art). The word is also frequently used by accountants in relation to their *balance sheets*. Two centuries ago the word branched out in the US and came to be used to mean as well 'something left over, remainder' (*the balance of our things, the balance of the penalty*). This extension of meaning has faced two centuries of disapproval but has certainly not ended its run yet (*after the rain started the balance of the holiday was a disaster; soil types comprise 15 hectares of combination Warkworth clay and Mahurangi soil, with the bulk of the balance being mainly peaty loam—NZ Herald, 1986*). The fate of the extended sense is, as it were, permanently weighed in the balance by some, and no longer hangs in the balance to those who have not heard that there is a problem.

**balding.** This word, now universally used as an adjective meaning 'going bald' (*his balding head laureled with a wreath of sandy hair*—B. Moore, 1987), is surprisingly modern (first recorded in *Time* magazine in 1938). The finite verb *bald* was used in the 17c. in the sense 'to deprive of hair', and is encountered occasionally in modern English in the sense 'to go bald' (*it was noticeable that he was beginning to bald*).

**baleful.** For centuries (orig. from OE *b(e)alu* 'evil, mischief') *baleful* has been a mostly literary word meaning principally 'malignant, pernicious' (applied to the arts of sorcerers, birds of prey, etc.), or a more general one meaning 'ominous, foreboding trouble' (*the baleful presence of his father in the house was like a constant reproach*—R. Hayman, 1981; *foghorns boom in still longer and lower choruses of baleful warning*—I. Banks, 1986. It is used esp. of eyes or looks (*he fixed his baleful eye on us*). Cf. the corresponding adverb in: *the goat was still staring balefully at her*—Encounter, 1988. Formerly (14–19c.) *baleful* also had a secondary sense 'unhappy, miserable' (*baleful spirits barr'd from realms of bliss*, 1812 in *OED*), but this sense is now obsolete. Cf. BANEFUL.

**bale out.** See BAIL OUT, BALE OUT.

**balk.** See BAULK.

**ballad,** at first, a song intended as the accompaniment to a dance (cf. late L *ballare*, OProvençal *balar* 'to dance'), then, a light, simple song of any kind. Between the mid-16c. and the early 19c. a popular song printed as a broadsheet, celebrating or scurrilously attacking persons or institutions : Autolycus hawks such ballads in *As You Like It*. Now, either a short sentimental or romantic composition, often in slow tempo, with each verse sung to the same melody, a pop song; or, much more encouragingly, a simple, spirited poem in short stanzas, such as *Sir Patrick Spens*, Keats's *La Belle Dame Sans Merci*, and Oscar Wilde's *The Ballad of Reading Gaol*. Ballads, both ancient and modern, frequently made use of a refrain. The *Lyrical Ballads* of Wordsworth and Coleridge manifest (as Margaret Drabble has it) 'their own interpretation and development of the term'.

From *Sir Patrick Spens* (probably 17c.):

Late late yestreen I saw the new moone,
 Wi the auld moone in hir arme,
And I feir, I feir, my deir master,
 That we will cum to harme.

From *The Ballad of Reading Gaol* (1898):

I never saw a man who looked
 With such a wistful eye
Upon that little tent of blue
 Which prisoners call the sky,
And at every drifting cloud that went
 With sails of silver by.

**ballade** /bæˈlɑːd/ In the 14c. and 15c. a poem ('ful delectable to heryn and to see') of three or more stanzas in RHYME ROYAL, often contrasted (by Chaucer and his contemporaries) with a *rondel* and a *virelay*. A typical example is Chaucer's *Compleynt of Venus*. In general, a poem consisting of one or more triplets of seven- or (afterwards) eight-line stanzas each ending with the same line as refrain, and (usually) an envoi. François Villon (*fl. c*1460) wrote many French ballades, including the celebrated *Ballade des dames du temps jadis* with its refrain 'Mais où sont les neiges d'antan?' The form was revived in England in the Victorian period, especially in the works of Swinburne, W. E. Henley, and Austin Dobson.

A 20c. example by J. C. Squire:

Ballade of the Poetic Life
*The fat men go about the streets,*
  *The politicians play their game,*
*The prudent bishops sound retreats*
  *And think the martyrs much to blame;*
  *Honour and love are halt and lame*
*And Greed and Power are deified,*
  *The wild are harnessed by the tame;*
*For this the poets lived and died.*

*Shelley's a trademark used on sheets:*
  *Aloft the sky in words of flame*
*We read 'What porridge had John Keats?*
  *Why, Brown's! A hundred years the same!'*
*Arcadia's an umbrella frame,*
*Milton's a toothpaste; from the tide*
  *Sappho's been dredged to rouge my Dame—*
*For this the poets lived and died.*

*And yet, to launch ideal fleets*
  *Lost regions in the stars to claim,*
*To face all ruins and defeats,*
  *To sing a beaten world to shame,*
  *To hold each bright impossible aim*
*Deep in the heart; to starve in pride*
  *For fame, and never know their fame—*
*For this the poets lived and died.*

Envoi
  *Princess, inscribe beneath my name*
*'He never begged, he never sighed,*
  *He took his medicine as it came'—*
*For this the poets lived—and died.*

**ball game.** From the North American use of the term to mean 'baseball' has come a medley of phrases in which *ball game* means 'a state of affairs' : *a different/a new/a whole new/another*, etc., *ball game*. This figurative use, first recorded in 1968, is now in danger of overuse even in countries where baseball is hardly played at all.

***ballon d'essai*** /bæ'lõ deˈseɪ/, pl. ***ballons d'essai.*** Lit. 'trial balloon'. A French phrase adopted in English a century ago in the sense 'an experimental project or piece of policy put forward to test its reception' (*A good deal of Hume's theory of belief is rather a* ballon d'essai *than meant altogether seriously—Mind,* 1942; *Why launch* ɟm ballons d'essai *concerning financial 'savings' which turn out on closer examination to be of so unfeasible a character?—Financial Times,* 1982.

**ballyrag.** This verb, meaning 'to maltreat by jeering at or playing practical jokes on', is now of diminishing currency in standard English, but Gerard Manley Hopkins knew it (in the form *ballyragging*), and it seems to be widespread

(in the spellings *ballarag, ballerag, balrag,* etc.) in dialectal use (Ireland, Yorkshire, Devon, etc.), and (spelt *bullyrag*) in the US. An Irish example: *Owen had been out ballyragging the country, talking at fairs and meetings*—J. O'Faolain, 1980. Some dictionaries place it under *bu-* and others under *ba-*. Its etymology is unknown. Its regional currency, from the time that it was first recorded in the 18c., suggests that for the foreseeable future the word will lie just below the surface of the standard language, but will emerge from time to time in regional fiction, etc.

**balm** /bɑːm/, a fragrant resinous exudation from certain trees, once used as an unguent, is the same as *balsam*. It is mentioned in the Bible (Jer. 8: 22, *Is there no balme at Gilead? is there no Physition there?*) where other versions have *gumme* or *resyn*. The phrase *balm of Gilead*, rejected by Fowler (1926) as a 'battered ornament', is now hardly ever heard.

**balmy.** In Britain the only current senses of the word are 'yielding balm' (cf. BALM) and 'deliciously fragrant, soothing' (used esp. of the weather: *it was a warm, balmy afternoon*; cf. also *those happy and balmy days for fathers, when they and their wishes were immediately obeyed*—G. Clare, 1981). Until recently it was also used to mean 'weak-minded, insane' (*'I s'pose you're balmy on her,' he said resignedly*—R. Crompton, 1922; *people here must have gone balmy*—J. B. Priestley, 1929). This use is still current in the US, but in Britain it has given way to BARMY.

**baluster.** Etymologically *baluster* is an antecedent of BANISTER, and was once the customary word for the posts and handrail(s) guarding the side of a staircase. It is now restricted in meaning to 'one of a series of short moulded vaselike shafts supporting a coping or rail' (Fr. *balustre*, ult. from L *balaustium*, Gk βα-λαύστιον 'blossom of the wild pomegranate', one feature of the pillar or column resembling the double-curving calyx tube of this flower). A *balustrade* is a row of balusters topped by a rail and forming an ornamental parapet to a terrace, balcony, etc.

***bambino*** /bæmˈbiːnəʊ/. Pl. *bambini* /-ni/. A colloq. word for a young (esp. Italian) child. In origin it is a diminutive of It. *bambo* 'silly'.

**ban** (verb). From a presumed pre-OE verb *bannan* meaning 'to proclaim to be under a penalty' came two main strands of meaning: **1** to summon by proclamation (long since in abeyance, though it produced the pl. noun *banns*, notice of an intended marriage).

**2** to curse (in older texts : *ever she blessed the old and banned the new*—W. Morris), to interdict, prohibit. In our precarious age, many things have been or are *banned* or under threat of *ban*, e.g. (until 1984) in Poland the independent trade-union movement called *Solidarność* (Solidarity), and in 1987 in Britain P. Wright's book *Spycatcher*. Also, in 1987, English football teams were *banned* from playing against European ones because of the riotous behaviour of some of their fans at a stadium in Brussels in 1985. The slogan *Ban the bomb* (i.e. nuclear weapons) was common from the beginning of the 1960s onwards.

**banal.** Once pronounced /'beməl/ or /'bænəl/, and scorned by both Fowler and Gowers as an unnecessary word 'imported from France by a class of writers whose jaded taste relished novel or imposing jargon', this word, pronounced /bə'nɑːl/, is now part of the normal vocabulary of English. Conversations, jokes, postcards, questions, etc., can be banal. The development of sense in French, 'of or referring to compulsory feudal service'→'open to the use of all the community'→'commonplace, feeble, trite', was mirrored in English from the mid-18c. onward, but the only surviving sense in English is the last of these three. The derivative *banality* /·'ælɪtɪ/ is common (*particular banalities, like 'Let's face it', come and go in a few decades*—John Jones, 1983), the adverb *banally* /·'ɑːl,lɪ/ less so. In AmE the words are commonly pronounced /'beməl/, /beɪ'nælɪtiː/ and /'beməliː/, but there is much variation.

**bandeau.** Pl. *bandeaux*, with the final letter pronounced /z/. See -x.

**bandit.** The *OED* (section published in 1885) reported that 'the pl. *banditti* (for It. *banditi*) is more used than *bandits*, esp. in reference to an organized band of robbers'. By 1926 Fowler could say that '*bandits* tends to prevail over *banditti*, especially when the reference is to more

or less clearly realized individuals'. *Bandits* was much used in the 1939–45 war for hostile aircraft. Nowadays it never gives way to *banditti* when the reference is to home-grown robbers or gangsters, but occasionally shares its currency with *banditti* when the context is of hostile acts by bands of guerrillas abroad, e.g. Sardinia. Areas in Northern Ireland where the IRA were active were known as *bandit country*.

**baneful,** first recorded in the 16c. (ult. from OE *bana* 'slayer'), means 'life-destroying, poisonous' (applied to a serpent's breath, wolves, noxious herbs, etc.)—a use now rare or obsolete—or, figuratively, 'pernicious, injurious' (of superstitions, over-possessive love, harmful beliefs, etc.): *Defeat the scare of kidnappings, violence and other baneful evils that shake modern society*—Pope John Paul II, 1984; *the baneful memory of that night haunted her, sometimes tormented her*—I. Murdoch, 1987). It corresponds to the noun *bane* (*selfishness is the bane of my life*). Cf. BALEFUL.

**banister** (usu. in pl. **banisters**). Etymologically *banister* (earlier also *bannister*) is an alteration of †*barrister* (both 17c.) which in turn is an alteration of BALUSTER, partly by association with *bar*. In the early 19c. *banister* was regarded by writers on architecture as improper or vulgar long after it had passed into good literary usage (R. B. Sheridan, Wilkie Collins, and others). It is now, of course, the customary word for the posts and handrail(s) guarding the side of a staircase.

**banjo.** Recommended pl. -*os*; but see -O(E)S 1.

**baptist(e)ry.** Standard sources in Britain use *baptistery*. In the US *baptistry* appears to be the more frequently used of the two forms. Cf. L *baptistērium*.

**bar,** used as a slightly formal preposition meaning 'except', has been in use since the 18c. : e.g. *this sortie, bar miracles, has decided the fate of Paris*—1870 in *OED*; *My sister-in-law for whom I probably care more than I care for anyone in the world bar one other*—P. Lively, 1983. It is also used in the phrase *bar none* to add emphasis to a statement (*the best detection expert I know, bar none*—R. Rendell, 1983); in the fixed phrase *all over bar the shouting* (orig. 19c.), said when the result of a contest, etc., appears certain except for formalities; and in racing to mean 'except' (the horses indicated) : used in stating the odds (*33–1 bar the rest*). Cf. the similar use of *except* and *save*, and cf. also BARRING.

**barbarian, barbaric, barbarous.** Etymologically a *barbarian* is a foreigner (ult. from Gk βάρβαρος 'non-Hellenic'), that is a person whose language and customs differ from the speaker's; thus at various times and with various speakers and writers, non-Hellenic, non-Roman, non-Christian, etc., alien. Such barbarians plundered more civilized nations (*into that chaos came real barbarians like the Huns, who were ... destructively hostile to what they couldn't understand*— Kenneth Clark, 1969). By a normal process of sense-development the word came to be applied (16–17c.) to *any* person or group regarded as uncivilized or uncultivated. The corresponding adj. has had a matching history. It has been applied, principally, to slaves, foreign languages, foreigners, the Saracens, prehistoric man, etc., almost always with reference to past time. The noun *barbarian* is sometimes resurrected in anger when some outrage has been committed in modern times: *Ancient civilisations were destroyed by the imported barbarians; we breed our own*—Dean Inge, 1922; *Sir Keith Joseph ... appealed to the National Union of Students not to disrupt meetings, branding the protesting students 'the new barbarians'*— *Times*, 1986.

Since the 15c., *barbaric* has been applied to the customs, language, culture, etc., of foreign people, and in particular those regarded as backward or savage (*the noble savage ... turns out to be a barbaric creature with a club and scalping knife*—H. J. Laski, 1920). From the 17c. onward (Milton speaks of *barbaric pearl and gold*), the word has also been used of aesthetically attractive objects seen, or brought from, abroad: Lawrence of Arabia, for example, described the colourful garments of his Arabic companions as *splendid and barbaric*. In present-day use, however, the word is applied almost invariably to heinous crimes (e.g. rape, child abuse, kidnapping, murder) or unacceptable social behaviour (e.g. the hooliganism of football supporters : *if the Syrians are written off as barbaric, what of the Palestine Liberation Organisation?* —*Economist*, 1975; *until fairly recently Italians were happy to consider kidnapping a barbaric Sardinian custom*—*Economist*, 1975. *Barbaric* is fractionally stronger than *barbarous* in such contexts.

*Barbarous* is used of deeply reprehensible behaviour : *the barbarous activity of the SS; the coal industry's barbarous effort to keep its workers in utter servitude* [in the 19c.]—*NY Times*, 1987. But it is also pressed into service to describe works of art, badly formed words (see BARBARISMS), and other matters falling in the realm of taste rather than of physical violence : *formulating his phrases carefully in the barbarous French these people used*—D. Bagley, 1966.

**barbarism, barbarity, barbarousness.** The apparently primary present-day sense of *barbarism* is described in the next article. Historically, and still commonly, it denotes brutality (*the law-abiding nations of the world will put an end to terrorism and to this barbarism that threatens the very foundations of civilized life*—G. P. Shultz, 1984), absence of culture, extreme rudeness. *Barbarity* and *barbarousness* are synonymous words for savage cruelty or extremely uncivilized behaviour. *Barbarousness*, though not *barbarity*, can be used of a linguistically poor construction, formation, etc.

**barbarisms.** The word *barbarism* is commonly, and with strictly logical appropriateness, used to describe words that are badly formed, that is words that are formed in a manner that departs from the traditions of the language concerned. In the past, philologists applied the term with particular relish to heterogeneous combinations of Latin/Greek/English elements exhibited in such words as *breathalyser*, *coastal* (instead of \**costal*), *helipad*, *impedance* (instead of \**impedience*), *speedometer* (instead of \**speedmeter*), *television* (Greek and Latin), and *triphibious*. The 'barbarisms' of earlier centuries for the most part escaped censure: e.g. false word-division (*adder* venomous snake, from OE *nædre*); insertion of redundant letters (*bridegroom*, from OE *brȳdguma*, from which *guma* man was assimilated to *groom*; and *daffodil*, from medL *affodilus*, akin to *asphodel*). In the present blizzard of word-creation, it is difficult to see how older standards of word-formation can be maintained. Only a fraction of modern coinages are made after consultation with a philologist. See HYBRID FORMATIONS; METANALYSIS.

**barbecue.** This 17c. word (from Haitian *barbacòa* 'a framework of sticks set upon posts'), once used mainly of large social entertainments in the open air at which animals like pigs or oxen were roasted whole, has become suburbanized and domesticated. As likely or not, a modern barbecue is an outdoor feast of steak, fish, or chicken cooked rapidly over hot charcoal. From jokey respellings of the word, e.g. *Bar-B-Q*, the word now often appears, esp. in America, written as *barbeque*. The diminutive form *barbie* is esp. common in Aust. and NZ.

**barely.** Like HARDLY and SCARCELY, *barely*, in the sense 'only just', should normally be followed by *when* in any succeeding clause: *Chance had barely begun to sip his drink when dinner was announced*—J. Kosinski, 1983. Only in very informal contexts is the use of *than* permissible: *Barely had her spirits fallen, leaving her to brood over the sea, than the pinch was repeated*—L. Bromfield, 1928). The use with following *than* is condemned by the *OED* s.v. *than* conj. 3d.

**baritone.** Adopted in its musical sense from Italian *baritono* in the 17c. and therefore spelt with medial -*i*-, though many earlier writers (e.g. G. Eliot, Palgrave) preferred *barytone*. Contrast BARYTONE.

**bark** is sparingly used in literary contexts (by poets like William Cowper and Walter Scott and modern writers like Joyce Carol Oates) as a synonym for 'boat, ship'. The doublet *barque* (also derived, but through French, from L *barca* 'ship's boat') is a sailing vessel of specified rigging.

**barman.** A man serving alcoholic drinks behind a counter is normally called a *barman* in Britain. In the US he may variously be called a *barkeeper* (or *barkeep*), a *barman*, or, most commonly, a *bartender*. If such a person happens to be a woman, in Britain (less commonly elsewhere) she is called a *barmaid*, and in America a *bartender*.

**barmy,** now usually so spelt in Britain (formerly also BALMY), is a slang word applied to someone or something that is crazy (*probably gone barmy with shock, done his nut*—Maggie Gee, 1981; *barmy pressure groups like the League of Empire Loyalists*—A. Ryan, 1984). It can be regarded as a derivative of *barm*, the froth on fermenting malt liquor (cf. a now lost adj. *barmy-brained*, over-excited, flighty), but the evidence points rather to its being a respelt (and thus differentiated) form of BALMY. The two words are pronounced identically, i.e. /ˈbɑːmɪ/ in RP.

**baronage, barony, baronetage, baronetcy.** The forms in -*age* are collectives: the body of barons (*baronage*) or of baronets (*baronetage*) collectively. A *baronage* is an annotated list of barons or peers, and a *baronetage* is a similar list or book of baronets. Those in -*y* are abstract terms: the domain, rank, or tenure of a baron (*barony*) or of a baronet (*baronetcy*). A baronet is a member of the lowest hereditary British titled order, ranking next below a baron, but with precedence over all orders of knighthood, except that of the Garter. The title is abbreviated as *Bart.* or *Bt.*

**baroque** /bəˈrɒk/, US /bəˈrəʊk/, was first applied to a florid style of architectural decoration, e.g. that of Francesco Borromini, which arose in Italy in the late Renaissance and became prevalent in Europe during the 18c. Later the term was extended to refer to profusion, oddity of combination, or abnormal features generally in other arts, including the visual arts (e.g. the sculpture of Bernini and the painting of Rubens), creative writing, and music (e.g. some of the work of J. S. Bach and Handel): *at Blenheim the English Baroque* [style of architecture] *culminates*—J. N. Summerson, 1953; *the conjunction of Christian and classical imagery* [in *Lycidas*] *is in accord with a baroque taste which did not please the eighteenth century*—T. S. Eliot, 1957. This term and ROCOCO are frequently used without distinction.

**barque.** See BARK.

**barrage.** Pronounced /ˈbɑːrɪdʒ/ in the 19c., but now normally /ˈbærɑːʒ/. **1** At first just an artificial bar across a river to control the level of the water, later a massive structure built across a body of water to allow extensive irrigation or to permit the creation of hydroelectric power.

2 In the 1914–18 war it came to be applied to a barrier of continuous artillery fire (*artillery barrage*) concentrated on a given area, and, in the 1939–45 war, also (*balloon barrage*) to a set of inflated balloons (called *barrage balloons*) placed in the air as a defence against low-flying hostile aircraft. The word is frequently used in a figurative sense, especially in phrases of the type *a barrage of questions, cheers, complaints, statements*, etc.

**barring.** As a marginal preposition (in origin an absolute use of the present participle of the verb *bar*), *barring* has been in use since the late 15c. and is still frequent: '*we've finished here—barring accidents*'—*he held up crossed fingers*—G. Charles, 1971; *They're all fine, thank God. Barring himself, who's got the divil of a hangover this morning*—A. Lejeune, 1986. Cf. the similar use of *excepting* and *saving*, and cf. also BAR.

**barrister, solicitor.** In BrE a *barrister* (in full *barrister-at-law*) is a person called to the Bar and having the right of practising as an advocate in the higher courts; a *solicitor* is a member of the legal profession qualified to advise clients and instruct barristers but not to appear as advocates except in certain lower courts. The processes of law are governed differently elsewhere: in the US, for example, an *attorney* (in full *attorney-at-law*) advises clients on business and other matters and may also represent them in court; clients are usually represented in court by a *counselor* (in full *counselor-at-law*) or a *trial lawyer*. An American *District Attorney* (or *DA*) is the prosecuting officer of a district. The *Attorney-General* is the chief legal officer in England, the US, and some other countries.

**bar sinister** is a layman's term found e.g. in the works of Sir Walter Scott, and still widely used in contexts suggesting illegitimacy or, figuratively, a lack of concern for tradition or good taste (*slashing through the area like a bar sinister is Route 1, which is dotted with beer bars and one-night motels*—*Washington Post*, 1979). In heraldry itself the customary terms for

illegitimacy are *bend sinister* or *baton sinister* (from L *sinister* 'left, left-hand').

**bartender.** See BARMAN.

**barytone.** This doublet of BARITONE is now reserved for the sense in Greek grammar 'not having the acute accent on the last syllable'. It faithfully reproduces the medial υ of Gk βαρύτονος (from βαρύς 'heavy' + τόνος 'pitch, tone').

**basal** /'beɪsəl/ and **basic** /'beɪsɪk/ entered the language in the 19c. as slightly artificial formations from the noun *base* (cf. Fr. *basal, basique*). They proceeded to usurp the territory of the adjectives *essential, fundamental*, and *minimal* in varying degrees. *Basal* occurs mostly in technical and scientific use: *basal anaesthesia* (a light unconscious state forming a basis for deeper anaesthetization at a later stage), *conglomerate, ganglion, metabolism* (of an organism in a fasting and resting state); only occasionally (esp. US) in a general sense (*basal readers and early reading textbooks*—*Publishers' Weekly*, 1963). *Basic* is overwhelmingly the more common in general contexts (*basic allowances, argument, pay, requirements, wage*, etc.). It is also used to describe a minimum core of vocabulary and grammar needed for linguistic communication: see BASIC ENGLISH. A few collocations of *basic* are found in technical and scientific language: *basic dye* (chemistry), *refractory* (steel manufacture), *slag* (agriculture; in NZ often with pronunc. /'bæsɪk slæg/).

**basalt.** Now normally pronounced /'bæsɔːlt/ with stress on the first syllable, but before about 1920 more regularly /bə'sɔːlt/ with stress on the second syllable.

**bas bleu.** Since the 18c. some writers have affectedly used the French *bas bleu* (pl. *bas bleus*) instead of its English equivalent *bluestocking*, i.e. a literary woman, but it has never become fully naturalized in English.

**base** overlaps in meaning with *basis*, but *basis* is used only figuratively in the general senses 'that on which something depends; a main constituent; a determining principle' (*the basis of a discussion; a solution with glycerine as its basis; legally mandated on the basis of race; on a friendly basis*). *Base*, on the other hand, has mainly literal senses: the bottom, lower part (*the base of a pillar, column, or statue*); a foundation (*the bases used in cosmetics*; *a special concoction of poison, with an arsenic base*; *a ring-shaped nitrogenous base*); a philological root (*the word* cairn *is derived from a Celtic base*); one of the four baseball stations. Figurative uses exist: *that is my feeling, and there must be some base for it*—G. Murray, 1918; *expand its current 33,000-customer base up to a million*—Industry Week, 1986; *Maybe I'm off base but I get the feeling that you're ... unhappy*—T. O'Brien, 1986; *the winners' victories ... have divided their natural bases of support*—Nation, 1987. But they are becoming increasingly subordinate to similar uses of *basis*. The plural of both words is *bases*, that of *base* with pronunc. /ˈbeɪsɪz/, and that of *basis* with pronunc. /ˈbeɪsiːz/.

**based on.** One of the commonest of all phrasal verbs in the passive voice: *M's hypothesis is based on the known habits of bees*; *the typeface Baskerville is based on the typefaces of the eighteenth-century printer John Baskerville.* (The verb, of course, can also be used in the active voice: *M bases his hypothesis on ...*) This construction is to be distinguished from the participial use of *based* in *Cruise missiles are based on Sicilian soil* (where *based* stands by itself and is not part of a phrasal verb). It is as well to avoid using *Based on* as a kind of sentence-leading preposition: *Based on this assumption, the economy is not expected to improve before the autumn.* The relationship between *Based on* and *the economy* is not a direct one. In the 20c. the past participle *based* has become frequently used in combinations with preceding nouns: e.g. *carrier-based, land-based, Miami-based, rule-based, science-based, shore-based.*

**basha(w).** Occasionally encountered in older literature—the word first appeared in English in the 16c.—this is the earlier form of the Turkish title *Pasha*. It is as well, though, to keep *basha(w)* in mind when reading, for example, Henry Fielding's *Jonathan Wild* and George Eliot's *Middlemarch*. There were formerly three grades of pashas in Turkey, distinguished by the number of horse-tails carried on their personal standards, the highest grade having three. The Grand Vizier's standard had five, the Sultan's seven. Though the title *pasha* was abolished in

Turkey in 1934, and *basha* in Egypt in 1952, both forms survive in everyday Turkish and in Egyptian Arabic as informal titles of admirals and generals.

**basic.** See BASAL.

**BASIC.** A convenient acronym (from Beginner's All-purpose Symbolic Instruction Code) for a computer language using familiar English words.

**basically.** A 20c. adverb, at first used sparingly to mean 'essentially, fundamentally' (*the basically democratic quality which belongs to a hereditary despotism*— G. K. Chesterton, 1905), but now somewhat worn down by prodigious overuse, esp. as an adverb beginning and governing a sentence. Examples: *I know I'm attractive in a way, but basically I'm ugly*—L. Michaels, 1969; *Basically, I feel great, except for fatigue*—M. Ali, 1987; *Mr. Titterington says: 'Basically, it was going to be a very Conservative show until I stuck my oar in.'*— Daily Tel., 1987. *Basically I see myself as a frank individual*—S. Bellow, 1988; *Basically, decay is just a process*—K. Russell, 1988.

**Basic English** is the name given by C. K. Ogden in 1929 to a simplified form of the English language that he invented, comprising a select vocabulary of 850 words and simplified syntax. He intended it to be used as an international auxiliary language. Despite much experimentation it did not succeed, perhaps partly because, as H. G. Wells said, 'it was more difficult to train English speakers to restrict themselves to the forms and words selected than to teach outsiders the whole of Basic'.

**basis.** See BASE.

**bas-relief.** Evelyn, Dr Johnson, and others went this way and that in their treatment of this word and of its Italian equivalent, *basso-rilievo.* Virtually all possible spellings and pronunciations, some Anglicized, some not, have been recorded. An Englishman leaning towards French should write *bas-relief* (pl. *bas-reliefs*) and pronounce it /ˈbarəljef/, and one leaning towards Italian should write *basso-rilievo* (pl. *bassi-rilievi*) and pronounce it /basoriljevo/. In practice, however, *bas-relief* (pl. *bas-reliefs*) is most commonly printed in romans and pronounced /ˈbæs rɪˈliːf/; and the preferred form of the other word is *basso-relievo* (pl. *-vos*), printed in italic and with a semi-Anglicized pronunciation /ˌbæsəʊˈɪˈljeɪvəʊ/. Cf. ALTO-RELIEVO.

**bassinet** (occas. **bassinette**) /bæsɪˈnet/, a wicker cradle. The word is used more frequently in the US (*a baby was asleep in a bassinet at the foot of the bed—New Yorker,* 1988) than elsewhere. In folk etymology it is falsely connected with the current Fr. word *bercelonnette* (a diminutive of *berceau* 'cradle'); it is in fact a diminutive of Fr. *bassin* 'basin'.

**bath.** 1 In Britain the noun *bath* (pronounced /bɑːθ/, pl. /bɑːðz/) means 'immersion in water for cleansing or refreshment' (*to have, to run, a bath*), or the vessel in which one has a bath. One can also have *a bubble bath, a Turkish bath,* etc.; and a massacre can be described as *a blood-bath.* In phrases like *she baths the baby three times a day/ you can bath after me,* the verb is pronounced /bɑːθ(s)/. The verb *bathe* /beɪð/ means either 'to swim in a body of water, esp. the sea' (*he preferred to bathe rather than just sun himself on the beach*) or 'to apply liquid to (a wound, the skin, etc.) as a cleansing or soothing agent' (*the doctor bathed the wound with a saline solution*). There are various figurative extensions: *bathed in sunshine, perspiration, acclamation,* etc.

2 US usage includes most of the above, but shows a marked preference for *bathe* (not *bath*) in the sense 'to wash in a bath' (*he often bathes before he goes to bed*); or the same idea is expressed in a different way (*her daughter usually takes a bath after finishing her homework*).

**bathetic.** A 19c. word (first recorded in Coleridge) formed from *bathos* on the assumed analogy of *pathos/pathetic.* It continues to lead a slightly buried life (*many of these tales and vignettes are bathetic—London Rev. Bks,* 1980; *a bathetic contrast between the lengthy sentence ... and a mundane ... reality—Essays & Studies,* 1987), as does the adverb *bathetically* (*Bathetically, I found myself thinking of Groucho Marx—R. Adams,* 1980). It is no longer, as *OED* says, 'a favourite word of reviewers'. The unhappily formed variant *bathotic* (presumably modelled on *chaos/chaotic*) is vanishing from use, though there are examples (1863–1952) in *OED* 2.

**bathroom.** It is important to bear in mind that one of the primary senses of *bathroom* in the US is 'lavatory'. The word alternates there in this sense with *restroom, washroom,* (colloq.) *john,* and numerous other synonyms. Example: *The man elected last week as mayor ... grew up ... in Alabama, in a town where he was unable to ... use the same bathrooms as white residents—Chicago Tribune,* 1987. For a general description of the various synonyms of *lavatory,* see TOILET.

**baton sinister.** See BAR SINISTER.

**battels** (pl. noun). Since the 16c., one's college account at Oxford for board and provisions supplied: a word often misspelt. In 16c. L written *batellae, batilli,* perhaps derived from a Latinized form of a northern and Scottish verb *battle* 'to receive nourishment', or the corresponding adj. *battle* 'nutritious', akin to the verb *batten* 'to grow fat'.

**baulk, balk.** The pronunc. /bɔːlk/ is dominant, but /bɔːk/ is not uncommon, doubtless because of the analogy of *stalk, talk, walk.* For the noun in its main general senses 'stumbling-block, hindrance', *balk* now seems to be the dominant spelling in the US (*the series of small balks that were to delay the publication of* Dubliners—J. Updike, 1984), and *baulk* in the UK (*Bill was also in baulk, up-country somewhere—J.* le Carré, 1974). In its specialized senses, e.g. the *baulk line* in billiards and snooker, a length of sawn timber (*the gravediggers ... pulled the supporting baulks from beneath the coffin—J.* Wainwright, 1980), etc., *baulk* is the more usual spelling in Britain. For the verb, *balk* alternates with *baulk* in Britain (*the brash generalizations at which scholars balk—* TLS, 1986; *shareholders could have baulked at having any such restraint attached—Times,* 1987), but American writers prefer *balk.*

**bay.** See GULF.

**bay window, bow window, oriel.** A *bay window* is one that protrudes outward from the line of a room in a rectangular, polygonal, or curved form, thus forming an alcove in a room; a *bow window* is a bay window in the shape of a curve. Sometimes, but improperly, *bow window* is used as co-extensive with *bay window.* An *oriel* is a large polygonal bay window projecting from the upper face of a wall and supported by a corbel or bracket.

**-b-, -bb-.** Monosyllabic words consisting of a simple vowel (*a, e, i, o, u*) before *b* normally double the consonant before suffixes beginning with a vowel (*dabbed, webbing, glibbest, bobbing, shrubbery*) or before a final -*y* (*scabby, hobby*); but remain undoubled if the stem contains a diphthong (*dauber*) or a vowel + consonant (*entombed, numbed, barbed*). Words of more than one syllable ending in *b* (e.g. *Beelzebub, cherub, cobweb, hubbub, rhubarb, syllabub*) rarely have vowel-led suffixes; but *cherubic, cobwebby*, and *hobnobbed*, for example, are so spelt. See SPELLING 6.

**BC**, before Christ, should always be placed after the numerals and printed in small capitals (55 BC). The abbreviation is first recorded in English in John Blair's *Chronology and History of the Ancient World* (1756), but was not widely adopted until the 19c. Cf. AD.

**be.**

| 1 *be* joining sentence elements of different number.
| 2 *be* and *were* as subjunctives.
| 3 The case of the complement.
| 4 Paradigmatic forms.
| 5 Confusion of auxiliary and copulative uses.
| 6 Anxiety about placing an adverb or other word(s) between *to be* and a following participle.
| 7 Ellipsis of *be*.

**1** The verb *to be* frequently joins sentence elements of different number: *Gustave is other animals as well*—Julian Barnes, 1984; *the view it obscured* was *pipes, fire escapes, a sooty-walled well*—A. S. Byatt, 1985; *these huge biographies* are *usually a mistake nowadays*—N. Stone, 1985; *its odour* was *potatoes*—J. Creece, 1986. In such circumstances the verb *to be* should agree in number with the subject, not with the complement, except that the plural may be used when the subject is a collective noun of singular form (*his family* were *well-known puritans*—Bodl. Libr. Rec., 1987; *its prey* are *other small creatures*—D. Attenborough, 1987). The problem of congruence is neatly balanced by Christopher Fry, 1946: *And so what's my trouble? Demons* is *so much wind. Are so much wind.* A sentence headed by the pronoun *what* requires a singular form of the verb *to*

*be: what I'm really interested in, and what is most concrete about whatever I've done,* is *the objects in this house*—New Yorker, 1986; *what they needed* was *houses*. In accordance with the above rules, *aren't* should have replaced *isn't* in the following example: *whether certain ethnocentric attitudes in our own working class . . . isn't also a hampering factor*—Encounter, 1987. Cf. AGREEMENT 7; WHAT 1, 2.

**2** The irregular and defective verb *to be* now has two subjunctive forms, *be* (present subjunctive) and *were* (past subjunctive). They are the survivors of a system, too complex to be set down here, in OE and ME. Their uses now are limited by circumstances that are somewhat elusive, but hardly more elusive than the use of the subjunctive mood in other verbs (see SUBJUNCTIVE MOOD). The main types are illustrated in the following examples: (at head of clauses) *we would much prefer to support specific projects*, be *they in management school or in university laboratories*—Jrnl RSA, 1986; were *this done we would retain a separate Bar with skill*—Times, 1986; (after *if*) *if the truth* be *told, I never wanted to fly away with the sky-gods*—J. M. Coetzee, 1977; *if I* were *obliged to rough out a blueprint for the Church of the future, I would start with the need for good popular theology*—G. Priestland, 1982; (in dependent clauses after certain verbs, e.g. *demand, insist, suggest*) *the evidence supports, indeed demands, that it* be *ventilated before an independent tribunal*—Times, 1986; *the Admiralty insisted that the case* be *clarified*—P. Wright, 1987; *I suggest it* be *carried to the balneary*—tr. U. Eco, 1983; (in dependent clauses after certain nouns and adjectives, e.g. *demand, important*) *the demand that the invention* be *used to benefit both the people and the land*—Dædalus, 1986; *it is important in today's vote that the principle itself* be *accepted*—Times, 1985; in certain fixed phrases, e.g. *if I* were *you, as it* were; and in fossilized optative-type formulas, e.g. be *that as it may, evil* be *to him who evil thinks, far* be *it from me, the powers that* be, *so* be *it, God* be *thanked, be it on your head.*

In negative constructions of the type *be* + passive, it is customary to place *not* immediately before *be: he insisted that he* not *be followed*—Observer, 1987; *what is crucial is that such inequalities* not *be perceived as part of a 'class structure'*—Encounter, 1987.

The fading power of the subjunctive mood is underlined by the frequency with which *was* is used instead of *were* in sentences of the type *if I were/was you*: *I'd get out if I was you*—Maurice Gee, 1983; *if I was your husband I should view me with suspicion*—I. Murdoch, 1974.

If a reasonably high degree of formality is intended (see the examples above) the subjunctive forms of *be* are available. If formality is not essential, the ordinary indicative forms are acceptable. In some circumstances (see SUBJUNCTIVE MOOD) the option chosen genuinely reflects the presence, or alternatively the absence, of some degree of uncertainty, supposition, intention, etc.

**3** *The case of the complement.* Dean Alford (1864) joyfully rejected the famous old rule that a pronominal complement must be in the same case as the subject of the joining verb *to be. It's me*, like French *c'est moi*, had even then come to stay. 'We must show an equal disregard', he said, 'for the monks of Rheims in the Ingoldsby legend who saw the poor anathematised jackdaw appear: Regardless of grammar, they cried out, "That's him!"'

Alford cited *if you see on the platform an old party in a shovel, that will be me* as the only acceptable form of the final pronoun. In both speech and writing the type *it's me/that's him/it's her* is now virtually universal: *Don't be scared, it's only me*—New Yorker, 1986; *I suppose you knew it was me*—P. Wright, 1987; *Phone rings at 8.07; it must be her*—The Face, 1987. Except that if the pronominal complement is immediately (or soon afterwards) followed by *who* (or *that*), the nominative is still frequently used in formal styles: *it was he who would be waiting on the towpath*—P. D. James, 1986; *it was I who put up the hare*—B. Levin, 1987; *as though it was she and not the drug that had done it*—P. Roth, 1987. Contrast *it isn't me that Cheryl wants to see*—H. Kureishi, 1987: here *me* is effectively the object of *wants to see* not of *isn't*. Contrast also *But why, if it was him who arranged it all for their benefit?* —I. Murdoch, 1987.

Some older writers in the present century continued to use the nominative in a rule-abiding manner whether *who* was in the vicinity or not: *if I were he, I should keep an eye on that young man*—C. P. Snow, 1979. The formally correct nominative is also still found in fairly formal writing: *If anyone could write about the narcissistic personality, it was she*—NY Rev. Bks, 1987. But the objective case of the pronoun, which OED reports as 'common in colloquial lang. from end of 16th c.', has triumphed for the most part except when *who* or *that* is adjacent.

**4** *Paradigmatic forms.* No special difficulties arise with the normal indicative forms, *am, is, are*; p.a.t. *was, were*; participles *being, been*. In speech, reduced forms of *am, is,* and *are* (*'m, 's, 're*) are freely used, and these often make their way into written English (*I'm* = I am, *she's* = she is, *they're* = they are), though such reduction is impossible after a preceding sibilant consonant (*the moss is green*, not *\*the moss's green*) or affricate (*the church is cold*, not *\*the church's cold*). In constructions of the type *am, is, are + not*, the word *not* in unabbreviated form is written separately whether or not the first element is contracted: *am not/'m not, is not/'s not, are not/'re not*. When the negative itself is contracted, it combines with the full form of the preceding element: thus *isn't, aren't, wasn't*, and *weren't* (not *'sn't, 'ren't*, etc.).

There are two anomalous forms: *ain't* (see the article on it at its alphabetical place in this book); and *aren't* when used as a tag-question in the form *aren't I?*. The expected reduced form of *am I not?* is *amn't I?*, as it is in many modes of speech in Scotland and Ireland but not in standard southern BrE. It is possible (as a correspondent has pointed out to me) that, when followed by *n't, am* behaves exactly like *can* and *shall*, losing its final consonant and (in standard English) lengthening its vowel. The expected spelling would be *an't*, but in those forms of English which lose pre-consonantal *r* the short form of *am not* merges with that of *are not* in both speech and spelling.

Whatever the explanation, *aren't I?* is a regular and natural tag-question in standard BrE. An American scholar, John Algeo, said recently (1995) that 'the expression is now in widespread use in America with no consciousness of a British origin', though he acknowledges that it might indeed be of British origin.

In informal contexts, the reduced forms *i'n't* and *i'nt* were once (18c.) commonly written for *isn't, is not*: the evidence is to be found in the OED. One hears, or one imagines one hears, some people saying /ɪnt/ to this day for *isn't* in such a sentence as *It isn't fair* /ɪt ɪnt feə/. But this pronunciation, and also the vulgarism *innit?* /'ɪnɪt/ for *isn't it* (see *OED* 2), both lie outside the realm of standard English.

The charming forms *wast* (2nd pers. sing. past t.) and *wert* (2nd pers. sing. past subjunctive, and also, since the 16c., past indicative) are now used only as archaisms. Examples: *Art thou the first man that was borne? or wast thou made before the hilles?—*Job 15: 7; *I would thou wert cold or hot—*Rev. 3: 15; *Thou wer't borne a foole—* Shakespeare, *A Winter's Tale; Oh, wert thou in the cauld blast—*Burns; *Hail to thee, blithe spirit! Bird thou never wert—*Shelley.

5 *Confusion of auxiliary and copulative uses.* In *The Bill was overtaken by the 1964 election* the word *was*, as part of the passive construction *was overtaken*, is an auxiliary. *Its postponement was welcome* shows a copulative use of *was.* Ellipsis is a main feature of the language, but it would be less than polished to allow one instance of *was* to do double duty here: *\*The Bill was overtaken by the 1964 election and its postponement welcome.* It is essential to repeat *was* after *postponement.*

6 Anxiety about placing an adverb or other word(s) between *to be* and a following participle. Constructions of the type *she had to be humoured* and *they seemed to be looking in different directions* are, of course, common. In such constructions an adverb may be placed with perfect propriety between *to be* and the following (past or present) participle: *he turned out to be* secretly *engaged to someone else; mosquitoes seemed to be* actively *circulating on the furniture.* The placing of such adverbs before *to be*, because of a belief that by doing so one is avoiding a SPLIT INFINITIVE, is an unhappy example of HYPERCORRECTION. On the other hand, it is frequently necessary, for idiomatic reasons, to place the adverb after the participle in such constructions: *the injection always has to be given* slowly.

7 In certain contexts, forms of *to be* may be idiomatically omitted: e.g. *I will burn this and feed the ash to the fruit tree*, where there is obviously no question of inserting *will* before *feed.* More ambitious

ellipses are permissible in informal contexts like the following (which happen to be from US and Aust. sources but could be from anywhere): *John Benjamin Alfred Redman ... prefers to be called Johnny Redd. When you come equipped with Johnny Redd's range of talents, you're entitled; I'll come over and give you a hand with those thistles, Pat. Thursday suit you?* [sc. Would Thursday suit you?]

**bean.** From its use (early 19c.) as a slang term for a sovereign or a guinea has emerged a range of negative expressions (*they never had a bean, she never saved a bean*) meaning '(having) no money whatever'. Slang or very informal terms for 'money' keep slipping into and out of the language. Those that have gone include *chink* ('exceedingly common in the dramatists and in songs of the 17th c.'—*OED*), *dibs, oof* (Yiddish), *rhino,* and *tin.* Survivors (first used in the 20c. except where indicated) include *bread, dough* (1851), *lolly* (UK), *mazuma* (Yiddish), *the needful* (1774), *readies* (bank notes), and *spondulicks* (1857).

**bear** (verb). See BORN(E); FORMAL WORDS.

**beat.** The old pa.pple *beat* is widespread in non-standard and regional speech but has almost vanished from standard English except as the second element of *dead-beat.* It may lie somewhere in the background of the 1950s expressions *beat generation* and *beatnik,* under the influence of the noun *beat,* the strongly marked rhythm of jazz and popular music. Jack Kerouac (1922–69), the coiner of *beat generation,* at some point unpersuasively rationalized the first element of the phrase as a shortening of *beatitude.* Some examples: *The* [AIDS] *virus had the boffins beat—*London Rev. Bks, 1987; (informal context) *you hear on television nowadays about little children getting beat up or treated nasty—*New Yorker, 1988; *she could have screamed ... she could have beat the man off—*A. Munro, 1988 (Canad.).

**beau.** The pl. form is *beaux,* an old-fashioned form for an old-fashioned word, or, less commonly, *beaus.* Both are pronounced /bəʊz/. See -x.

**beau geste** /bəʊ 'ʒest/. Especially since the appearance of P. C. Wren's *Beau Geste* (1924), the first of his Foreign Legion novels, this French expression has been pressing at the door of standard English in the sense 'a display of magnanimity, a fine gesture', but with only partial success. The Australian novelist Xavier Herbert, in his *Capricornia* (1939), used the expression in the plural: *She would have dressed poor Gigney, who did not understand* beaux gestes, *as a Eunuch.*

**beau idéal** or **beau ideal.** Pronounced as in French or Anglicized as /ˌbəʊ aːˈdiːəl/. In origin an abstract conception, the ideal Beautiful, *beau* being the noun and *idéal* the adj. In English it is mainly used in the sense 'the highest conceived or conceivable type of beauty or excellence, the perfect type or model'. Most people who know the expression in English probably think of it as = *beau* (adj.) + *idéal* (noun). Examples: *His beau idéal was the English country gentleman*—J. Pope Hennessy, 1971; *'Want of elegance' ... was decidedly an obstacle in raising Carlyle to the level of a* beau idéal—P. Rose, 1984; *He was the very beau idéal that Emily Post commended to her genteel readers in her perdurable* Etiquette—*TLS*, 1984.

**beauteous.** Since the 15c. a literary and chiefly poetical word, used most memorably in Wordsworth's sonnet (xxx) *It is a beauteous evening, calm and free.* As a formation it is paralleled by *bounteous* and *plenteous.* Examples: *It was a beauteous calm evening, warm and balmy after the hot afternoon*—J. Rathbone, 1979; *Some were Ancient Crones and some were beauteous and young*—E. Jong, 1980.

**beautician.** Despite the existence of parallel forms like *academ/ician* (from *academ/y*) and *geometr/ician* (from *geometr/y*), this 20c. word (first recorded in the US in 1924) still retains a slight flavour of etymological illegitimacy. Most words ending in -ician correspond to subject names ending in -ic or -ics: *arithmetic/arithmetician*, *music/musician*, *mathematics/mathematician*.

**beautiful. 1** From 'the House Beautiful' in Bunyan's *Pilgrim's Progress* (where 'Beautiful' is to be regarded as a proper name) has flowed a host of imitative uses, in which the adj. is placed after the noun it qualifies: e.g. *God's Acre Beautiful or the Cemeteries of the Future* (W. Robinson, 1880); *The House Beautiful* (Oscar Wilde, title of lecture, 1883); *The Body*

*Beautiful* (first recorded in *Ladies' Home Jrnl*, 1917). Similar, somewhat uncongenial, phrases like *the home beautiful, the bed beautiful,* and *the orchard beautiful* lie strewn about in modern advertisements.

**2** *the beautiful people.* Applied with stunning irrelevance to two separate groups of people: (*a*) the 'flower' people of the 1960s, hippies; (*b*) wealthy and fashionable people, the smart set, esp. those of the period of John F. Kennedy's presidency of the US (1961–3) or immediately after.

**because** (orig. *by cause that* after OF *par cause de*).

> **A** Standard usage.
> **B** More questionable *because*-clauses.
> **C** *because* or *for*.

**A** *Because* as a conjunction normally introduces a dependent clause expressing the cause, reason, or motive of the content of the main clause: *she wept because she loved him* (the *because*-clause, here following the main clause, answers the question 'why?'); *because we were running short of petrol, we began to look for a garage* (here the *because*-clause precedes the main clause); *I know he committed suicide, because his wife told me* (answering the question 'how do you know?'); *she thinks I'm upset because I wanted Fred to spend the night* (answering the question 'why does she think that?'); (preceded by *just*) *'And all they are doing is sitting around on their high-priced butts drinking tea, just because they haven't had your scripts.'* 'My word,' said Henry. All these, including the last from Malcolm Bradbury's *Cuts* (1987), are well-regulated uses of the conjunction *because*.

It is also in order to use *because* after an introductory *it is, it's, that's, this is: It is only because he regarded it as absolutely necessary that he took such harsh measures; it doesn't hurt, and I can tell you why, it's because I've changed my work; 'That's because I'm so damned good at journalism,' she added; Mars looks especially good on this encounter ... This is because the atmosphere obscures light rays.* All these examples are drawn from reputable sources.

In some informal contexts, *because* can be safely omitted: *Tammy put a hymn book up in front of her face she was so embarrassed.* It can also be replaced by, or re-expressed in, another construction: *Being poor, he could not afford to buy books* ( = because he was poor); *An experienced teacher, Mr Walton solved the problem quickly* ( = because he is).

Also acceptable in informal use is the unadorned retort *Because*, with the implication that a fuller reply is being withheld for some reason: *'Why didn't you leave the bottle?' 'Because!' I said shortly. I wasn't going to explain my feelings on the matter.*—M. Carroll, 1968.

**B** More questionable *because*-clauses, with varying degrees of questionableness, are: **1** *I know he committed suicide because the girl did not love him and his wife told me* (with a second *because* implied between *and* and *his wife*: a mixture of two constructions).

**2** *Mr David Alton ... has told colleagues that he is very unlikely to accept a spokesmanship in the Social and Liberal Democrats because he profoundly disagrees with Mr Paddy Ashdown, the party's new leader; He did not go to South Africa because he loved the game of rugby.* Examine both sentences for their ambiguity. It is often unsafe to place *because* after a clause containing a negative or negative equivalent.

**3** In constructions with *why* and *because* placed in that order: *he was implying that why he knew that she had kept the promise was because he had been seeing Arnold; why I spoke sharply was because she was rude.* Such constructions are sometimes called pseudo-cleft sentences.

**4** *Because* (or just *because*) at the head of a dependent clause governing a main clause: *Just because a fellow calls on a girl is no sign that she likes him*—G. Ade, 1897; *Just because I'm here now doesn't mean I didn't go, does it?*—B. Tarkington, 1913; *Because we don't explicitly ask these questions doesn't mean they aren't answered*—New Yorker, 1986; *Just because someone does not agree with or is offended by Knechtle's beliefs is no reason to keep him off campus*—Daily Northwestern (Illinois), 1988. Such *because*-clauses demand too long an attention span before the onset of the main clause.

**5** Though often defended, the type *the reason ... is because* (instead of *the reason ... is that*) aches with redundancy, and is still as inadmissible in standard English as it was when H. W. Fowler objected to it in 1926. A sub-editor on *The Times* should not have allowed *because* to stand on the front page of the issue of 19 January 1988: *He* [sc. Dr David Owen] *had believed the reason for Mr MacLennan's visit was because he had doubts about the new policy.* Other flawed examples: *The reason for the continued success of this physical atomism was because it was consistent in explaining a wide variety of new experimental results*—C. Gilman, 1987; *One of the reasons ... that singers like Randy Travis and Dwight Yoakam have sold themselves so successfully in this country is because they are selling authenticity*—The Face, 1987. See also REASON.

For acceptable uses of the phrase *the reason why*, see REASON WHY 2.

**C** *because* or *for*. *Because* is a subordinating and *for* a coordinating conjunction. The main standard uses of *because* are illustrated in section A above. *For* as a conjunction means 'seeing that, since': *Those houses ... ought not to be called houses, for they were unfit to be lived in*—Daily Worker, 1963. It always follows, and, as it were, acts as a kind of gloss on, or appendage to, a main clause. See FOR 2, 3.

**because of,** as a prepositional phrase followed by a noun or noun phrase, is straightforwardly admissible in most of its uses: *Three schools ... have been forced to close because of structural faults in their roofs; Hardy's legacy is somewhat hard to define because of its essential ambiguity.* It normally operates as a paraphrased equivalent of a notionally longer clause: *... because the roofs were found to have structural faults; ... because it is essentially ambiguous.* It should not be used, however, in constructions in which the word *reason* (or equivalent) appears in the main clause: *the reason we have no light is because of a broken fuse* (read *is that the fuse is broken*). Another example of the need to avoid redundancy.

**because why?** This phrase, used interrogatively (and frequently written in the form *Cause why* or *Cos why*), is chiefly dialectal (the OED cites evidence beginning in 1887) or in informal use. Further examples: *'Cos why? 'Cos I'm going to German-eye?*—E. M. Forster, 1910; *I know a lot of people that rant on about their religion and it doesn't do any good. Because why? Because they're trying to convince themselves, maybe?*—S. Chaplin, 1961.

**beccafico,** a small migrant Italian bird; pl. -os. See -O(E)S 6.

**bed and breakfast.** Also **B & B, b & b.** From their appearance on signs outside places where such services are available, these terms have come to be used concretely of the accommodation building itself (*The Garth Woodside Mansion, where Mark Twain was a guest, now is a bed and breakfast—Chicago Tribune*, 1989), as well as of the services offered.

**bedizen.** A somewhat archaic verb ('to deck out gaudily'). Pronounce /bɪ-ˈdaɪz(ə)n/. Examples: *He likes to think he's living in a palace full of bedizened captives waiting for him to arrive*—I. Murdoch, 1989; *Isn't life bedizened with jaunty contradictions?*—Julian Barnes, 1991.

**Bedouin,** a desert Arab. This customary English spelling of the word is derived from French. It contains, but ignores, the Arabic plural ending *-īn*. Some 19c. writers used the sing. form *bedawy*, pl. *bedawin* (and variants) as approximate equivalents of the Arabic forms of the word, but these spellings are no longer extant. The form *Bedu* has come into widespread use in the same sense in the 20c. (*The Bedu had their own food with them in their saddle bags*—T. E. Lawrence, 1917).

**beef,** the flesh of an ox, a bull, or a cow, is normally a non-count noun; but a waiter might say 'two roast beefs', meaning 'two orders for roast beef'. The word is also used (pl. *beeves*, US also *beefs* or *beef*) in the sense 'ox(en), esp. when fattened, or their carcasses'. In the slang sense 'complaint, grievance', the only pl. form is *beefs*.

**been and gone and —.** First observed by Dickens in popular speech in 1836, this 'facetious expletive amplification of the pa.pple. of a verb' (*OED*) is still commonly used in light-hearted contexts in standard modE: *And what's more, he's been and gone and got it printed*—P. Bailey, 1986.

**beg.** Once commonly used in formal, esp. business, correspondence as a formula of goodwill (*Begging my best remembrances to Mrs. Thomson*—Dickens, 1836), this verb is now mainly restricted to its general senses, 'to ask as a favour, to ask supplicatingly, etc.' Phrases of the type *I beg your pardon, I beg (leave) to differ from you*, etc., are still, however, in standard use.

**begin.** See COMMENCE.

**beg the question.** 1 In strict use, the English equivalent of Latin *petitio principii*, used in logic to mean the 'fallacy of founding a conclusion on a basis that as much needs to be proved as the conclusion itself' (Fowler, 1926). Gowers (1965) cited as an example, *capital punishment is necessary because without it murders would increase.*

2 In general use, the meaning is much more likely to be 'to evade a difficulty' or 'to refrain from giving a straightforward answer'. Examples: *Let's . . . beg the question of just who was in love with whom*—H. Jacobson, 1986; *He simply begged the question by saying that the decisions he disapproved invented new rights*—NY Rev. Bks, 1987; *John Major's vision of Europe seems to me entirely correct. But it begs the question: why did the prime minister all but sacrifice his office ratifying the Maastricht treaty when [etc.]?* —*Economist*, 1993.

**behalf.** The only use current in standard BrE is *on behalf of*, with two main senses, 'in the interest of or for the benefit of (another person, cause, etc.)', and 'as the agent or representative of (another)'. In AmE *in behalf of* vies with *on behalf of* in the same two senses, though Garner (1987) insists that a distinction exists. *In behalf of*, he says, means 'in the interest or in defense of' (*he fought in behalf of a just man's reputation*); and *on behalf of* means 'as the agent of, as representative of' (*on behalf of the corporation, I would like to thank . . .*).

A new non-standard use (or rather a revival of an obsolete one), first noted in the 1980s, is a tendency to substitute *on behalf of* for *on the part of*, or just to use *behalf* instead of *part*: e.g. (a person interviewing a golfer) *That was an 11th-hour decision on your behalf*; (read *part*) *His death was largely due to panic on his behalf*; ( = on the part of) *The detail may be trivial but it betrays an astounding lack of appreciation on behalf of the author*—Times, 1994.

**behemoth.** The usual pronunciation now is /bɪˈhiːmɒθ/.

**behest.** An ancient, somewhat formal, word, used esp. in the phrase *at the behest (of)*: e.g. *Van Dyck painted all five of the surviving royal children at the behest of their* *father*—Antonia Fraser, 1979; *The time and place were at her behest*—P. Lively, 1991.

**beholden.** Once the normal pa.pple of the verb *behold* (*These are stars beholden By your eyes in Eden*—E. Barrett Browning, 1850), *beholden* survives only in the sense 'attached, or obliged (to a person); under personal obligation for favours or services': *In his post as editor ... he had shown himself beholden to no clique*—T. Keneally, 1980.

**behoof.** An ancient word, now verging on obsoleteness, meaning 'advantage': e.g. *it was surprising how many people would dredge up from their past some exotic Jewish ancestor for Sefton's behoof*—H. Jacobson, 1983.

**behove.** Now most often used as a quasi-impersonal verb led by *it*, in the sense 'it is incumbent upon or necessary for (a person) to do (something)': e.g. *What books does it behove me to read?*; *it behoves us to know as much as possible about local government*. In the UK it is pronounced to rhyme with *grove* and *rove*, and in the US it is spelt *behoove* and pronounced to rhyme with *move* and *prove*. It is also used to mean 'befit' (usu. in negative contexts): e.g. *It ill behoves him to protest*.

**being as (how), being (that).** Used by many regional speakers, esp. in the US, and jocularly by standard speakers, *being as* (*how*) and *being* (*that*) have widespread currency just below acceptance in unmarked standard speech: *being as it's holiday time ...*—Sue Cook, BBC Radio 4, 1977; *naturally they* [sc. children] *get up early on Saturday mornings, being as how that is one of the two days in the week they do not have to get up early*—G. V. Higgins, 1979; (a daily help speaking) *Being as how you can't be married, you'd better have him christened*—J. Bowen, 1986.

*Being* (*that*), in particular, was not always so restricted: the *OED* records numerous examples from literary sources, including: *I believe your newspapers ... tell you all, but being there is nothing newer, I would do it too*—Lady Russell, 1692; *With whom he himself had no delight in associating, 'being that he was addicted unto profane and scurrilous jests'*—Scott, 1815. Cf. AS 8.

**belabour** (US **belabor**). In the senses, (*a*) to thrash, (physically) assail, and (*b*) to assail with words, recorded from 1600

and 1596 respectively. Examples: (*a*) *I got very mad as expected, and tried to belabour both of them*—A. Burgess, 1971; *and how they* [sc. Harpies] *belabour us, squawking, with their horrible claws and flapping wings*—F. Weldon, 1988. (*b*) *In the face of such overwhelming beauty it is not necessary to belabor the faithful with logic*—H. L. Mencken, 1923; *It was in vain that the fiery little George Augustus, and his wife, belaboured Walpole with their arguments*—C. Chenevix Trench, 1973. In 20c. AmE these two senses have been joined by another: *belabor* is being used interchangeably with *labor* in the type *to belabor* (*a point, a question, a theme,* etc.).

**believe me,** a phrase used to strengthen an assertion, is sometimes extended to the somewhat condescending form *believe you me*. Crashaw used *Beleeve mee* in this manner in 1646: the *OED* records his use and numerous others. The extended form *believe you me* is first recorded in 1926. Modern examples: *And believe you me this aspect of the war effort is not being ignored*—N. Williams, 1985; *'Is that so?' 'It is, believe you me, young man.'*—P. Bailey, 1986.

**belike,** as adverb, in use since the 16c. in the sense 'in all likelihood', has fallen into occasional use as an archaism or as an element of regional speech: *I'll put that in your lug will belike warm your pocket*—H. Allen, 1933.

**belittle.** It will probably come as a surprise to most readers of this book that the currency and acceptability of the verb *belittle* in the UK was ever in doubt. The word is first recorded in the late 18c. in America, and in its ordinary sense, 'to make (a person or an action) seem unimportant or of little value', became widely accepted in the UK only at some point in the 20c. Fowler (1926) and Gowers (1965) both rejected *belittle* and expressed a marked preference for *decry, depreciate, disparage, poohpooh*, etc.

**belly.** The OE word *bælig* meant 'a bag, skin-bag, purse, pod, husk'. By the 14c. *belly* had come to mean 'that part of the human body which lies between the breast and the thighs, and contains the intestines', i.e. the ordinary modern sense. Over the centuries, in its ordinary anatomical sense, the word has had mixed fortunes in the scale of mentionable words, and has still largely failed to displace *abdomen* (formal), *stomach* (the organ itself but also widely used of the surface area of that part of the body), or *tummy* (children's language, used by adults as a genteelism) in standard contexts. See GUTS.

Transferred senses of the word abound: in archery, sailing, printing, music (front part of instruments of the violin and lute types), etc. The underside of an aeroplane is its belly, and Italy was known as the 'soft underbelly' of Europe when the Allied forces drew up plans for the reoccupation of Sicily and the rest of Italy in 1944.

**beloved,** when used as a pa.pple (*beloved by all, was much beloved*), is disyllabic, /-'lʌvd/; as an adj. (*dearly beloved brethren; the beloved wife of*), or as a n. (*my beloved*), it is normally trisyllabic, /-'lʌvɪd/. Cf. *aged, blessed, cursed*.

**below, under.** At first sight these two words appear to be exact synonyms, but in a wide range of contexts one or the other is to be preferred. *Below* tends to be regarded as an antonym of *above*, and *under* as an antonym of *over*. *Below the bridge* means with it higher up the stream; *under the bridge*, with it overhead. Contexts in which *below* is customary include *below par, to go below* (in a ship), *below the belt, the information below, the temperature is 20° below*. By contrast it is not idiomatic to say *man below 45, below one's breath, incomes below £10,000*: *under* is more natural in each case. Contexts in which *under* is customary include *under the sun, the table, the circumstances, examination, the Stuarts, the protection of, one's wing, one's thumb, separate cover, sentence of death, fought under Montgomery. Under* is the invasive and more frequently used word of the two. Cf. BENEATH.

**bend sinister.** See BAR SINISTER.

**beneath.** Over the centuries, *beneath, below,* and *under* have tended largely to overlap. It would appear that, by the end of the 19c., *beneath* had become somewhat restricted in use: 'a literary and slightly archaic equivalent of both *below* and *under*. The only senses in which *beneath* is preferred are 7 ("beneath contempt"), and fig. uses of 4 (e.g. "to fall

beneath the assaults of temptation").' (*OED*). Fowler (1926) judged that, apart from the 'beneath contempt' sense, 'it is now a poetic, rhetorical, or emotional substitute for *under* or *below*'.

Be that as it may, *beneath* has a wide range of idiomatic contextual uses now. Examples: **1** (in a lower position than) *Lowe dropped to his knees, as if to drive the knife upwards beneath Leiser's guard*—J. le Carré, 1965; *I watched a child drag a butterbox on wheels beneath the cold streaky sky*—T. Keneally, 1980; *drinking pre-lunch aperitifs beneath crystal chandeliers*—P. Lively, 1987; *his body was positively abloom beneath the riding mac*—T. Wolfe, 1987; *the pipes and conduits that jostle each other beneath the street*—New Yorker, 1988; (fig.) *The Dog Beneath the Skin*—W. H. Auden and C. Isherwood, 1935.

**2** (not worthy of) *he considers such work beneath him; she had married beneath her* (i.e. to a man of lower social status). Cf. UNDERNEATH.

**Benedick,** not *Benedict*, is Shakespeare's spelling (varying with *Benedicke*) in *Much Ado*, and should be used whenever the sense is a newly married man, esp. an apparently confirmed bachelor who marries. As a Christian name, *Benedict* (but not *Benedick*) is now in common use.

**benefited** is the recommended spelling of the pa.t. and pa.pple of the verb *benefit*: see DOUBLING OF CONSONANTS 2 and -T-, -TT-. Nevertheless the norm in some publications, e.g. *The New Yorker*, is *benefitted*.

**Bengali** /bɛnˈɡɔːli/. Like many other Indian words in the period of the Raj, once commonly spelt with final -*ee*, but now always with -*i*.

**benign, benignant.** The shorter form was adopted from French in the 14c., and is by far the more usual of the two. It is used principally in the senses 'gracious, kindly' and 'exhibiting kindly feeling in look, gesture, or action'; and esp. in transferred use of anything considered to be propitious (e.g. the aspects of the planets), salubrious (e.g. the weather, the air), or not life-threatening (e.g. a tumour). *Benignant* /brˈnɪɡnənt/, formed in the late 18c. after *malignant*, has remained at the margin of the language, unlike the word on which it was modelled. It means 'gracious, kindly (esp.

to inferiors)' (*a benignant monarch*), or 'salutary, beneficial' (*the benignant authority of the new regime*). It is not normally used of medical conditions. Cf. MALIGN, MALIGNANT.

**bereaved, bereft.** When the verb *bereave* is used transitively to mean 'to rob, dispossess (someone) of (usu. immaterial things, e.g. hope, ideas, senses)', its pa.t. and pa.pple are normally *bereft* (*I faltered helplessly, nearly bereft of speech*—W. Styron, 1979; *Without her, he felt bereft as a child at a boarding school*—A. N. Wilson, 1982). In contexts of death it normally has *bereaved* as its pa.t. and pa.pple. When an attributive adjective is called for, *bereaved* is normally used (*a bereaved wife*). Similarly when the context requires a collective noun (*the bereaved assembled for the funeral*).

See -T AND -ED.

**Berkeley.** There can be few people who do not know that Berkeley Square in London is pronounced with the sound of *lark*, and Berkeley in California with the sound of *lurk*.

**berserk.** At first a noun (written esp. as *berserker, -ar, -ir*), adopted in the early 19c. from the language of the Icelandic sagas, meaning 'a Norse warrior who fought with frenzied fury' (*Out of terrible Druids and Berserkers, come at last Alfred and Shakspeare*—Emerson, 1837; *as though they were expecting another onslaught of rapacious Danes or shield-biting berserkers*—P. Theroux, 1983). In Icelandic the word was probably derived from ON *bern-*, *bjǫrn* bear (the animal) + *serkr* coat, sark, but some scholars derive the first element from ON *berr* bare (thus 'fighting in his bare shirt'), whence the 19c. English spelling (Carlyle, Emerson, and Charles Kingsley) *baresark*. Henry Kingsley and Kipling popularized its use as a quasi-adverb in the phrase *to go berserk*, to act in a frenzied manner, and the word is now mainly used thus: *a maladjusted Jamaican who goes berserk sometimes and screams and spits and bites*—E. Huxley, 1964; *I can better understand people who go berserk . . . than people who just can't bother to keep a kid alive*—A. S. Byatt, 1985.

*Berserk* may be pronounced either as /bə'zɜːk/ or /bə'sɜːk/.

**beseech.** For the pa.t. and pa.pple. use either *beseeched*, thus following e.g.

Shakespeare and Iris Murdoch (*Why had he not wept, screamed, fallen to his knees, beseeched, raged, seized Jean by the throat?* —*The Book and the Brotherhood*, 1987); or *besought*, in the tradition of Milton, Penelope Lively (*The Lord is praised and besought and worshipped*—*Moon Tiger*, 1987), and the New Yorker (*Trying to keep abreast, we besought a chat*, 1986). See -T AND -ED.

**beside, besides.** 1 *As adverb.* In modE, *besides* is the only adverbial form of the two and as such it means 'in addition, as well' (*Besides, he is already married; he has a current account and two other accounts besides*).

2 *As preposition. Beside* is the customary form when the meaning required is 'alongside, at the side of' (*dangling his arms beside his hips; she knelt down beside him; she suggested that the other passengers wait beside the road for the bus; a cat is seated beside the fire*). *Beside* is also the normal form when the meaning is 'away from, wide of (a mark)' (*beside the purpose, beside the question, beside the mark*). *Beside* is also used to mean 'in comparison with' (*beside Locke modern logical positivists are shown to have made very modest contributions to philosophy*).

*Beside* was once commonly, and is now occasionally, used to mean 'in addition to' (*Beside his master Andrea Sacchi, he imitated Rafaelle*—Reynolds, 1774; (mod.) *other men beside him can solve the problem*). *Besides* is much the more common form when the meaning is 'in addition to, other than' (*there has to be one other besides myself; besides all those tanks they have 1,000 warplanes*). A somewhat more complex example: '*So what does a folklorist do besides ask nosy questions . . . ?' Kelly asked*—New Yorker, 1987.

**bespeak.** The finite verb survives (*This, to say the least, bespeaks a rare relationship between a river and adjacent terrain*—New Yorker, 1987), with past tense *bespoke* (*She had no confidence in the stupid fashion which bespoke mincing and vapidity*—P. Carey, 1988) and past participle *bespoken*. But the older participial variant *bespoke* lingers only as an attributive adjective meaning 'made to order' (*bespoke suit, goods*, etc.) in contrast with *ready-made*.

**bestir** is now always used reflexively (*I must bestir myself*) and never, as it was until the 19c., as an ordinary transitive verb (*No Mauell, you haue so bestir'd your valour*—Shakespeare, *King Lear*, 1605; *More need that heirs, His natural protectors, should assume The management, bestir their cousinship*—Browning, 1873).

**bet** (verb). When Kingsley Amis chose *betted* rather than *bet* as pa.pple of the verb *bet* in *Difficulties with Girls* (1988) (*I'd have betted you wouldn't be much good at taking somebody out*), he was perhaps following a slight 20c. tendency (in the UK) to prefer the longer form of the two. But the two forms remain in contention. In the US, for both the pa.t. and the pa.pple, *betted* is 'rather unusual' (*WDEU*, 1989) and '*bet* is usual'. When a sum of money is stated immediately after the verb (*he bet me £50 that the secret ballot would confirm his view*), *bet* is the customary form in Britain too. In other circumstances the two forms are nicely balanced, but *bet* is recommended.

**bête noire.** Whatever the gender or nature of the person or concept referred to, *bête noire* (in italic type) should be so printed, and not *bête noir*. The plural is *bêtes noires*, with both *s*'s silent (*TV advertisements were his* bêtes noires).

**bethink.** *The Longman Dictionary of Contemporary English* (1987) accurately describes the verb *bethink oneself of* as '(literary or old use) to think about, consider: *You should bethink yourself of your duty, my lord!*' Most other dictionaries for foreign learners of English have no entry for the word, but well-read native speakers will all be familiar with it from its frequent use down the ages: *e g. bethink yourself wherein you may haue offended*—Shakespeare, 1605; *If they shall bethinke themselues . . . and repent*—AV, 1611; *Rip bethought himself a moment*—W. Irving, 1820.

**better.** This word has numerous idiomatic uses (*to know better, no better than she should be, go one better*, etc.) and a few somewhat debatable ones. **1** *had better* (often written as *'d better*). This modal idiom or semi-auxiliary is correctly used to mean 'would find it wiser to' (*You had better come and have a talk; 'She'd better go home,' she heard Miss Braithwaite say*—B. Rubens, 1987). In negative and interrogative contexts, the normal types are shown in *You'd better not come any closer* and *Hadn't we better go home now?*

**2** In a wide range of informal circumstances (but never in formal contexts) the *had* or *'d* can be dispensed with. Thus the type *You had better come tomorrow* can informally be reduced to (*You) better come tomorrow*. This reduced construction (without preceding pronoun and without *had*) is shown in the following example: *When you're feeling censorious, better ask yourself which you'd choose*—P. D. James, 1986. (Further examples, from 1831 onward, in the *OED*.) In practice this use of an unsupported *better* is much more common in North America, Australia, and NZ than in Britain. Examples: *I think I better get the taxi . . . so as to catch the last bus back*—P. Roth, 1979 (US); *Come on! . . . We better go on*—E. Jolley, 1980 (Aust.); *'Now there's one thing you better get straight before we start'*—M. Pople, 1986 (Aust.); *'You better be right,' she said*—A. Munro, 1987 (Canad.); *'Oh, you better believe I remember that, Judge'*—T. Wolfe, 1987 (US); *'You better lay low, Glory'*—Rosie Scott, 1988 (NZ). Cf. NOT 12.

**3** *better than* = more than. *WDEU* (1989) cites several modern American examples of this use, including *We were whistling along at slightly better than Mach 2* (*Saturday Review*, 1979), but adds that 'it is not generally found in the more formal kinds of writing'. The *OED* cites this use from miscellaneous writers of the period 1587 to 1823 (Lamb), but it does not turn up very often in standard use in Britain now (*Better than 95 per cent of the nation's alcoholics are middle-class*—Listener, 1984).

**4** *better half, better part* = larger portion. The *OED* gives examples from literary sources (Sidney, Wordsworth, etc.) from 1580 to 1805, and both expressions are still current, e.g. *for the better half of the last decade; for the better part of a year*. (Quite distinct, of course, from *my better half* = 'my husband' or esp. 'my wife'.)

**better, bettor,** one who bets. If only to distinguish it from the preceding word, *bettor* seems preferable, but in practice *better* is the customary form in Britain and *bettor* in the US.

**between.**

| 1 | *between you and me.* |
| 2 | *between and among.* |
| 3 | *between each, between every.* |
| 4 | *between + and.* |
| 5 | Repeated *between.* |
| 6 | *between 1914–18.* |

**1** The nation is divided in its use of *between you and me* and *between you and I.* Let me begin by declaring that the only admissible construction of the two in standard use in the 20c. is *between you and me.* I know that Bassanio read out a letter to Portia containing the sentence *All debts are cleerd betweene you and I if I might but see you at my death,* and that Mistress Page said that *there is such a league betweene my goodman, and he.* So be it. Goneril, on the other hand, said *There is further complement of leaue-taking betweene France and him* [not *he*]. *I, he,* and other pronouns were frequently used in earlier centuries in ways now regarded as ungrammatical. Grammatical assumptions were different then, especially when the pronoun in question was separated from the governing word by other words.

**2** *between* and *among.* The boundaries are much less clearly drawn in the use of *between* and *among.* Many people cling to the idea that *between* is used of two and *among* of many, but the *OED* maintains that 'In all senses, *between* has been, from its earliest appearance, extended to more than two … It is still the only word available to express the relation of a thing to many surrounding things severally and individually, *among* expressing a relation to them collectively and vaguely.' The *OED* further divides the uses of *between* into four main branches: of simple position; of intervening space; of relation to things acting conjointly or participating in action; of separation. These distinctions are needed to account for the wide fluctuations of usage over the last eight centuries.

The main present-day patterns can be fairly clearly discerned, I think, in the following examples: (two persons or things) *things that had happened a long time since—between Isaac and myself*—N. Williams, 1985; *museums have become an uneasy cross between theatre and boutique*—New Yorker, 1987; (more than two persons, things, etc.) *there was one iron between fifteen of us*—D. Lodge, 1962; *dividing his time between engineering, mechanical inventions, and writing for periodicals*—G. S. Haight, 1968; *The death of his sister had changed things between Marcus, Ruth and Jacqueline*—A. S. Byatt, 1985; *For Sale and Rent … Situated between Florence, Siena, Perugia. Easy access Rome*—London Rev. Bks,

1986; *About £5,000 had been wagered, he said, divided almost equally between the three main candidates*—Daily Tel., 1987; (number of events, groups, etc., less clearly defined) *Does he sigh between the chimes of the clock?*—J. M. Coetzee, 1977; *Other courses may be devoted to forging links between research teams in universities … and schoolteachers*—TES, 1987; (borderline cases in which *among* should perhaps have been used, esp. in the 1987 example) *I want to walk between the trees and smell them too*—E. Jolley, 1980; *a company has £25 million of profit to distribute between 10,000 workers*—Times, 1987; (*among* correctly used) *there's only us … left among the wreckage*—S. Barstow, 1960; *there were a lot of very young people among the temporary staff*—P. Fitzgerald, 1980; *there were differences among Christians dating back to the Church Fathers*—Bull. Amer. Acad. Arts & Sci., 1987; *the UN … does have machinery designed to … keep the peace among nations*—Chr. Sci. Monitor, 1987; *the day has not yet come when the giants war among themselves*—J. M. Coetzee, 1977; *it seeks the furthest extension of the educationally valuable among the masses*—Encounter, 1987; (*among* now sounding somewhat forced) *a conversation among Richard Smith, Sir Anthony Grabham, and Professor Cyril Chantler*—Brit. Med. Jrnl, 1989.

**3** *between each, between every* followed by a singular noun. Traditionally, constructions of the type *twenty-two yards between each telegraph pole* and *he had a cup of coffee between each tutorial* have been regarded with suspicion or downright dislike (and have been replaced by, for example, *between each telegraph pole and the next* and *between tutorials*). *CGEL* 9.21 n., for example, remarks: 'It is common, but not accepted by all speakers, to say *between each house* instead of *between each house and the next.*' On the other hand, Jespersen (*Mod. Eng. Gram.* vii) and other grammarians have had no difficulty in finding evidence in support of the condemned type: *Between each kiss her oaths of true loue swearing*—The Passionate Pilgrime, 1599; *pausing between every sentence*—G. Eliot, 1859; *a row of flower-pots were ranged, with wide intervals between each pot*—W. Collins, 1860; *staring at her furtively between each mouthful of soup*—M. Kennedy, 1924. This evidence must be respected and the constructions tolerated.

**4** between + and. The natural conjunction linking elements introduced by between is and (see 2 above), but or, as opposed to, (as) against, etc., are occasionally encountered and are always regrettable. Fowler (1926) rightly objected to against in *It is the old contest between Justice and Charity, between the right to carry a weapon oneself against the power to shelter behind someone else's shield* (read and); and to or in *The choice is between payment in money or in kind* (again read and).

**5** Repeated between. In long sentences, the reader or listener can be kept waiting for the second term or element to be mentioned in a between-construction, e.g. *You need to decide between voting for a party which, against all advice, introduced the poll tax, a form of tax first used in the 14th century, and one that dislikes the rates system but has no alternative to offer.* The solution to the problem is not to insert a second between between and and one. Either let it stand, or recast the sentence entirely.

**6** The type between 1914–18 is wrong in that it treats a dash as if it were the word 'and'. Read *from 1914 to 1918* or *between 1914 and 1918*.

**betwixt** as preposition and adverb is now archaic or obsolete in all uses in standard English except in the alliterative collocation *betwixt and between*, first recorded in 1832, and still common. *Betwixt* lives on in some dialects in both Britain and America.

**beware.** Derived in ME from the verb be + the adj. ware cautious, beware now has no inflected forms. It is used in the imperative, or, when used with a modal auxiliary like must or had better, acquires quasi-imperative force: *Beware of the dog!; beware lest he overtake you; you must/had better beware or someone will attack you.* The type shown in Longfellow's *Beware the pine-tree's withered branch!* (1842), i.e. showing beware constructed with a direct object, would still be acceptable in poetry or in formal prose.

**bi-.** This prefix, which was first used in contexts of time (*biweekly, bimonthly, biyearly, etc.*) in the 19C., is a cause of endless confusion. Each compound can mean 'occurring or continuing for two ——', 'appearing every two ——', or 'occurring twice a ——'. Because ambiguity is

usually present, and cannot be resolved by the devising of rules, it is always better to use unambiguous equivalents, e.g. *twice a week, twice-weekly; every two weeks, fortnightly; twice-yearly; every two years*; etc. See BIANNUAL; BICENTENARY.

**biannual.** First recorded in 1877, it has mostly been used to mean 'half-yearly, twice a year'. By contrast, the much older word *biennial* (first recorded in 1621) has been traditionally used to mean 'existing or lasting for two years; changed every two years; occurring once in every two years'. The distinction is far from watertight, and it is as well to reinforce the meaning contextually, e.g. *This biennial conference, first held in 1989, took place in Paris in 1991.*

**bias.** The recommended inflected forms of the verb are biased, biasing, and of the noun, biases. See -s-, -ss-.

**Bible.** Use an initial capital when it refers to the holy scriptures themselves (but *three bibles* = three copies of the Bible); use a small initial when the word is used in a transferred sense (*Wisden is the cricketers' bible*). The corresponding adj. *biblical* is always written with a small initial letter.

**bicentenary** /-sm'ti:nərɪ/, less commonly /-'tenərɪ/, as adj. and noun, is the more usual term in Britain, and **bicentennial** /-sen'tenɪəl/ abroad. In all areas it bears the senses '(marking) the two-hundredth anniversary; (celebration) of the two-hundredth anniversary'. See CENTENARY 2.

**biceps.** Pl. the same.

**bid. 1** In the auction-room and playing-card senses the pa.t. and pa.pple are both bid (*In May the Trust bid successfully at auction for Rex Whistler's well-known portrait of Lady Caroline Paget*—Ann. Rep. National Trust, 1989; *two no-trumps were bid*).

**2** In other senses, the past tense is usually spelt bade (*we bade him adieu; it bade fair to be the best holiday ever*) and pronounced /bæd/; the past participle is bidden, except that bid is used in a few phrases, e.g. *Do as you are bid.*

**3** When it is used—not very often nowadays—in the sense 'command', it governs a bare infinitive (*Why should he*

# bide | black

*... keep me from my own And bid me sit in Canterbury, alone?*—T. S. Eliot, 1935). When used in a passive construction, a *to*-infinitive follows (*he was bidden to get on with it*). This use has largely been replaced by *tell* (someone) *to do* (something).

**bide.** Apart from archaistic, regional, and poetical uses, this long-standing word (in OE and until the 19c. it meant 'to remain') is now idiomatic only in *to bide one's time* (past t. *bided*). The Scottish expression *bide a wee* is widely recognized in other areas.

**biennial.** See BIANNUAL.

**billet-doux.** Pronounce /ˌbɪleɪˈduː/. The pl. is *billets-doux*, pronounced /ˌbɪleɪˈduːz/.

**billion.** 'Since 1951,' the *OED* says, 'the US value, a thousand millions, has been increasingly used in Britain, especially in technical writing and, more recently, in journalism; but the older sense "a million millions" is still common.' It is best now to work on the assumption that the word means 'a thousand millions' in all English-speaking areas, unless there is direct contextual evidence to the contrary. Cf. MILLIARD; TRILLION.

**bimonthly.** See BI-.

**binomial** (noun). There are two principal uses: **1** The two-part technical Latin name of a plant or animal species: *Primula vulgaris* primrose, *Equus caballus* horse, *Homo sapiens* man. Both parts should be printed in italic, with the generic name given an initial capital. After the first mention of a species, later references may be shortened, if there is no risk of confusion, by the abbreviation of the generic name to the initial capital alone, followed by a full point: *P. vulgaris, E. caballus.* (*Hart's Rules.*)

**2** The name given to two-part expressions joined by *and* (occas. *or*) which are normally presented in a fixed order: (nouns) *bread and butter, cup and saucer, gin and tonic, by hook or by crook, ladies and gentlemen, law and order, odds and ends; Tweedledum and Tweedledee;* (adjs.) *fast and furious; spick and span;* (verbs) *to have and to hold, tried and trusted.*

**bishopric.** See SEE.

**bite.** In BrE and AmE, *bit* is the normal pa.t. form, and *bitten* the normal pa.pple.

In the US (as reported in *DARE*), *bit* is recorded locally as past participle 'esp. among males and lesser educated speakers' (*These apples are wormy, I think you got bit,* i.e. cheated).

**bivouac** (verb). The inflected forms are *bivouacked* and *bivouacking.* See -C-, -CK-.

**biweekly.** See BI-.

**biyearly.** See BI-.

**black.** **1** Beginning in the mid-1920s, American people of ultimately African descent campaigned for the abandonment of the words *Negro, Negress,* and particularly *Nigger,* in favour of *black* (or *Black*). As time went on, major newspapers and publishing houses accepted the argument that the *Negro* group of words had indefensible racial overtones, and used *black* (or *Black*) instead. For a long time, *black* had also often been applied 'loosely, to non-European races, little darker than many Europeans' (*OED*). As a result of this tendency, migrant Asians in Britain and their families are now often called *blacks.* The black section of a given community in Britain, therefore, often now means a cross-section of people of African, Caribbean, Indian, Malaysian, etc., descent. Meanwhile in the US the term AFRICAN-AMERICAN has come into use beside *black* for people of Negroid origin.

This area of terminology is riddled with anomalies. Whatever view one takes is likely to be controversial. In general the word least likely to give offence to American and British black people, or to anyone else, is *black.* Uncontroversial ways of describing people of South Asian descent are to use the terms *Indian, Malay, Pakistani, Bangladeshi,* etc., as appropriate, or to use *Asian* as a blanket term. The words *coloured* and *darky* are regarded as offensive by those to whom they are applied. See COLOURED. A minor curiosity is that African-Americans frequently use the word *nigger* without giving offence when addressing other blacks.

**2** The 20c. has witnessed the banning of books containing in their titles or text words that are regarded as racially offensive. At least one major American dictionary omitted words of this kind. All such vocabulary remains highly sensitive and looks like remaining so well into the 21c.

**3** In the 20c. there has been a sharp downturn in the use of long-established phrases like *to work like a nigger* (first recorded 1836) and *nigger in the woodpile* (first recorded 1852).

**black, blacken.** The shorter form is used when the intentional laying on of colouring matter is meant (to black boots with blacking; to colour one's face black as camouflage, in order to play a role as a black person, etc.); and, in the trade-union movement, to declare (something) to be *black*, i.e. to boycott (something). One can also *black* someone's eye. The longer form is the only one of the two that can be used intransitively (*his mood/the sky blackened*). *Blacken* also has a wide range of literal and figurative uses: *the ceiling blackened with smoke; to blacken one's character, reputation*, etc. See -EN VERBS.

**Black English. 1** The form of English spoken by many blacks, esp. as an urban dialect of the United States. A chance occurrence of the term in 1734 is recorded in *OED* 2, but the recognition of Black English as a describable and distinctive form of American English did not emerge until the civil-rights movement in the 1960s.

**2** As I have remarked elsewhere (*The English Language*, 1985, p. 164), Black Vernacular English, as it is often called, makes many holes in the standard American syntactical cobweb: e.g. *all my black brother* (uninflected plural); *a novel base on ...* (loss of final consonant in *based*); *he a black bitch* (absence of the verb 'to be'); *God didn't make no two people alike* (double negation). These are not casualties of an imperfect learning of standard American but features of a creolized form of English, shaped orally by some deep ancestral memory of patterns of speech brought many generations earlier by African slaves. William Labov's well-known work *The Social Stratification of English in New York City* (1965) demonstrated that Black English is not a fractured form of standard American English, but a stridently alternative form of American speech, a variety that is richly imagistic and inventive.

**blame** (verb). First used *c.*1200 to mean 'to find fault with' (e.g. *Thow blamest crist,*

*and seist ful bitterly, He mysdeparteth richesse temporal*—Chaucer), *blame* came to be used with *for* in the 18c. (first recorded in Defoe, 1727) with the meaning 'to censure (a person) for (something)'. A century later a construction with *upon* or *on* came into use, meaning 'to hold (a person) responsible for (something)'. Gowers (1965) described it as 'a needless variant' of *to blame* (someone) *for*, but it is now in standard use beside the construction that it competes with.

**blameable** is the preferred spelling in OUP house style. See -ABLE, -IBLE 6.

**blanch, blench** (verbs). **1** *Blanch* is used in two main ways: = to become pale (from fear, shock, cold weather, etc.); and, transitively, 'to make (vegetables, almonds, etc.) white by dipping (them) in boiling water for a specified time'. It was first recorded in the 14c. The by-form *blench* has also been used in the sense 'to become pale' since the early 19c.

**2** A separate verb *blench*, which started out in OE meaning 'to deceive, cheat', developed the sense 'to quail, flinch' in ME, and this sense is still current (*many people blench when they enter a dentist's surgery*).

**blank verse,** verse without rhyme, 'esp. the iambic pentameter or unrhymed heroic, the regular measure of English dramatic and epic poetry, first used by the Earl of Surrey (died 1547)' (*OED*). *Paradise Lost* and the greater part of Shakespeare's plays are written in blank verse.

**blatant.** 'Apparently invented by Spenser, and used by him as an epithet of the thousand-tongued monster begotten of Cerberus and Chimaera, the "blatant" or "blattant beast", by which he symbolized calumny' (*OED*). Except when used with allusion to Spenser, the word has (since the late 19c.) come to mean 'obtrusive to the eye, glaringly or defiantly conspicuous' (*the blatant way in which he intruded, a blatant lie, a blatant piece of late tackling*). It overlaps in meaning to a considerable extent with FLAGRANT.

**blend** (noun). A blend or portmanteau word is one derived by combining portions of two or more separate words, e.g. *motel* ( = motor + hotel), *Oxbridge*

( = *Oxford + Cambridge*), *smog* ( = *smoke + fog*). Such formations are now exceedingly common as the need arises on an unprecedented scale for fresh words to denote new fashions, sports, methods of entertainment, etc. A few modern examples: *blaxploitation* ( = *blacks + exploitation*), *croissandwiches* ( = *croissant + sandwiches*), *ginormous* ( = *giant + enormous*), *infomercial* ( = *information + commercial*), *liger* (offspring of a lion and a tiger), *raggazine* ( = *rag + magazine*), *rockumentary* ( = *rock + documentary*), *sexcapade* ( = *sex + escapade*). See -AHOLIC; -ATHON; BURGER.

**blended, blent.** The unvarying pa.t. and pa.pple of the verb *blend* in ordinary use is *blended* (*the sky blended in the distance with the sea; blended tea, wines,* etc.). In the 20c. *blent* is encountered only in literary works: *It was the memory of Saturday morning, blent with another emotion too vague to name*—S. Gibbons, 1937; *A serious house on serious earth it is, In whose blent air all our compulsions meet*—P. Larkin, 1955.

**blessed, blest.** When used as an attributive adjective, always spelt *blessed* and pronounced as two syllables, /ˈblesɪd/: *the Blessed Virgin Mary, the Blessed Sacrament, every blessed night.* So too when used as a plural noun with *the* (*Isles of the blessed*), and in the biblical expression *blessed are the meek.* When used as a finite verb, the past tense and past participle are normally written as *blessed* and pronounced /blest/: *the bishop (had) blessed his wife and children before he died.* So too in the colloquial expression *blessed if I know.* The spelling *blest*, once common in all the above uses, is now mostly restricted to poetry and hymns (*Blest are the pure in heart*—J. Keble, Hymn 370 in *English Hymnal*). It is also used in the colloquial expression *Well I'm blest.*

**blink** (verb). **1** In the sense 'to shut the eyes, evade', originally (in the 18c.) a sporting word (*to blink a bird, a covey*). It is now used in many contexts of evasion (e.g. *it is no use blinking the facts*).

**2** The same meaning is now sometimes expressed by the phrasal verb *blink at*, though *blink at* also means 'to view with surprise' (*they blinked at the unemployment figures*).

**bloc.** **1** A 20c. loanword from French, it is used, esp. in politics and business, of 'a combination of persons, groups, parties, or nations formed to foster a particular interest' (*OED* 2). The phrase *bloc vote* is now challenging the traditional *block vote.*

**2** The Fr. phrase *en bloc* 'as a whole' was adopted in English at a somewhat earlier date (first recorded in English in 1861). It is usu. printed in italic in English contexts (*Nearby ... are villages whose inhabitants in summer migrate en bloc ... to the islands off the coast where they spend their time fishing*—Discovery, 1934).

**blond(e).** **1** The word retains a vestigial mark of the grammatical gender of the language (French) from which it was adopted. A *blonde* is 'a *woman* with blonde hair'. A *blonde* person is 'a fair-haired *woman*'. By contrast 'a tall blond person' usu. = 'a tall blond (young) *man*'. But the distinction is not always an absolute one in all English-speaking countries: e.g. (*blond*, of a woman) *'We've taken her chair,' said blond, garrulous Mrs. Tessler*—L. S. Schwartz, 1989 (US); *The little girls whispered to each other, their blond heads shining in the rather dark room*—New Yorker, 1990; (*blond* of a man) *Crews of tall, blond men who hardly ever spoke*—T. Findley, 1984 (Canad.); *His blond eye-lashes gave him a bemused look*—P. Fitzgerald, 1988 (UK); (*blonde*, of the hair of a woman, a doll) *her blonde plaits reaching half-way down her bony back*—C. Dexter, 1989 (UK); *lugging that doll of hers, a thing with blonde shiny hair*—A. Duff, 1990 (NZ). *Blond* is also occasionally applied to inanimate objects: e.g. *It contains a blond desk, two chairs, and a weight-lifting bench*—New Yorker, 1989.

**2** The German phrase *eine blonde Bestie* (Nietzsche) 'a person of the Nordic type' was rendered in English as 'blond(e) beast' during the period of the Aryan heresy (ending in 1945) in Germany, but is now only of historical interest.

**bloody.** **1** Lit. 'like blood, smeared with blood, attended with much bloodshed, etc.', in use in the language as an adj. from earliest times, but since the 18c. diminishing in currency in these literal senses as the word came to be used as 'a vague epithet expressing anger, resentment, detestation; but often as a mere intensive, esp. with a negative as "not a bloody *one*" ' (*OED*). The *OED* placed it in the realm of 'foul language', but as the 20c. proceeded the degree of foulness attributed to the word receded sharply. It has come to mean little more than 'unpleasant, perverse' in most contexts (*Why go out of your way to be bloody about Archie when I'm trying to help him?*—A. Heckstall-Smith, 1954).

2 As an adverb it has been in use as an intensive since the later part of the 17c., meaning 'very ... and no mistake, exceedingly; abominably' (*OED*). The *OED* rules out any connection with the oath *'s blood!*, and there is no foundation to the belief that it is a shortening of *by Our Lady*. The *OED* instead attributes its use to 'the prevalent craving for impressive or graphic intensives seen in the use of ... *awfully* ... *devilish* ... *damned* ... *rattling*, etc.' Shaw was entitled to expect a sharp reaction from the audience when he caused Eliza Doolittle to exclaim *Walk! Not bloody likely.* (Sensation). *I am going in a taxi*, in *Pygmalion* (1914). Since then, however, and particularly since the end of the 1939–45 war, almost all traces of horridness have been shed from the word. Characters in TV plays, for example, regularly use expressions like *serve you bloody right, you bloody well will do it or else* without attracting large quantities of hostile correspondence. The word has become a vaguely used indicator of theatrical or actual anger or frustration. Nevertheless it is still not the kind of word to use 'in polite society'.

3 A pleasing myth (supported by an extended entry for the word in *The Australian National Dictionary* (1988)) is that Australians use the word *bloody*, whether as adj. or adv., more freely and with more vigour than people in other English-speaking communities. Certainly a dogged Australianness, rooted in the language of pioneering adversity, can be seen in many of the contexts in which it is recorded: e.g. (adj.) *You must think yourself a damned clever bushman, talking about tracking a bloody dingo over bloody ground where a bloody regiment of newly-shod horses would scarcely leave a bloody track*—M. J. O'Reilly, 1944; (adv.) *Here* [sc. in Tobruk] *we bloody-well are; and here we bloody-well stay*—C. Wilmot, 1944.

**bloom, blossom.** *Bloom* is not extended like 'flower' to a whole 'flowering plant'. It expresses (as the *OED* puts it) 'a more delicate notion than "blossom", which is more commonly florescence bearing promise of fruit, while "bloom" is florescence thought of as the culminating beauty of the plant'. Cherry trees are said to be in *blossom*, roses in *bloom*. In figurative uses of these nouns and their corresponding verbs, the same broad distinction applies. Someone or something full of promise is appropriately described as *blossoming*. By contrast, the bloom or blooming-time of a specified culture, a person (*the bloom of perfect manhood*), etc., is 'the most flourishing stage or season, the prime'.

**blow** (verb). The regular pa.t. and pa.pple are, respectively, of course, *blew* and *blown*, but it is worth noting that the once common past form *blowed* survives in standard English in such expressions as *Well I'll be blowed, I'm blowed if I'm going to do that* (sense 29 of *OED blow* v.¹).

**bluebell.** In southern England this is the wild hyacinth, *Scilla nutans*; in the north, and esp. in Scotland, it is another name for the harebell, *Campanula rotundifolia*, with fewer, larger, and thinner-textured flowers than the other. Abroad, the word is applied to a number of blue flowers shaped somewhat like bells, including plants of the genera *Viorna, Veronica*, and *Wahlenbergia*.

**blue book.** (Often with initial capitals.) Spec. (*a*) a parliamentary or Privy Council report, issued in a blue cover; (*b*) US a printed book giving personal details of government officials. (*New SOED*, 1993). See GREEN PAPER; WHITE PAPER.

**bluish** is the preferred spelling in OUP house style (not *blueish*).

**blurb,** a brief descriptive paragraph or note of the contents or character of a book, printed as a commendatory advertisement on the jacket or wrapper of a newly published book. The word is said to have been invented in 1907 by Gelett Burgess, an American writer, who appended the name Miss Blinda Blurb to a comic book embellished with a drawing of a young lady.

The modes of commendation and the language used tend to be predictable and repetitive. For example, phrases applied to textbooks, reference works, etc., on my shelves, include *an accessible and stimulating* (textbook), *a challenging and stimulating* (book), *a critical* (account), *a definitive* (guide), *essential* (reading), *expert* (guidance), *informative* (entries), *a lively and up-to-date* (book), *a major new* (theory of grammar), *up-to-date* (information). Phrases applied to new novels or their authors or contents include *acerbic wit, brimming with vulgar vitality, a compelling story, no more original comic writer in Britain, an ominously tangible novel, an outrageously gifted writer, a richly sensual book, strung-out sensitivity, a sustained and sardonic fable, a technical tour de force, an unusual compelling debut.*

**blusher.** In the sense 'a cosmetic used to give an artificial colour to the face', first recorded in 1965 (*OED* 2), it has now effectively replaced *rouge* (first recorded in 1753).

**boat.** A *boat* is 'a small open vessel in which to traverse the surface of water' (*OED*), whatever the means of propulsion. The word is also used of small fishing vessels, small passenger or cargo vessels, and the like. A large sea-going vessel, and in particular a naval surface vessel, is called a *ship*; but a submarine is called a *boat*.

**boatswain.** The originally nautical pronunciation /ˈbəʊsən/ is now general whether the word is written in full or as *bosun*.

**bodeful.** A fashionable literary word in the 19c. (*OED* 'very frequent in modern poets and essayists'), meaning 'foreboding, ominous', but now archaic and out of favour.

**bog(e)y, bogie.** The latest editions of the Oxford dictionaries prefer *bogey* for the golfing term, *bogey* for the mischievous spirit, and *bogie* for the railway term. The golfing term is said to have originated in a name (*bogey-man*) given in 1890 on a golf-course at Great Yarmouth by a Major Wellman to his 'well-nigh invincible opponent', namely the 'ground score'.

**boggle** (verb). In origin a 16c. verb formed from the noun *boggle*, a variant of *bogle*, a spectre of the kind such as horses are reputed to see, it now has both intransitive (*the mind boggles at the extent of the damage*) and transitive uses (*the suddenness of the collapse of Communism boggles the imagination*). The transitive uses still seem somewhat raw. The adj. *mind-boggling*, first recorded in 1964 (*A lot of mind-boggling statistics—Punch*), seems to have come into existence about a decade later than transitive uses of the verb.

**bohea.** Pronounce /bəʊˈhiː/.

**bolt** (adv.). The noun *bolt* meaning 'a sudden breaking away' and the verb *bolt* meaning 'to dart off' used similatively (cf. *sand-blind, snow-white, stone-cold*) in *bolt upright* (first recorded in the 14c.), and occasionally in other phrases: *he sat bolt upright against the wall; The Countess went bolt down into a chair*—G. Meredith, 1861.

**bona fide(s).** **1** As an adverbial phrase, *bona fide*, of Latin origin, means 'in good faith' and, when used at all, is normally pronounced in English contexts as /ˈbəʊnə ˈfaɪdi/. It is first recorded in the mid-16c.

**2** As an adjectival phrase (normally pronounced like the advb. phr.) it means 'acting or done in good faith' (*bona fide poverty, bona fide traveller*) and was first recorded in the late 18c.

**3** The noun phrase *bona fides* is pronounced in English /ˈbəʊnə ˈfaɪdiːz/ and is properly construed as a singular (L *fidēs* faith): e.g. *Bona fides is therefore opposed to fraud, and is a necessary ingredient in contracts*—1845 in *OED*. Erroneously treating it as a plural form of *bona fide* (assumed to be a singular noun although in Latin *fide* was the ablative singular of nom. *fides*), a vernacular tide of opinion has construed the phrase as if it meant 'guarantees of good faith' or 'credentials'. From about the early 1940s onward the new construction has tended to become dominant: e.g. *his bona fides were* (not *was*) *questioned*. WDEU (1989) lists several examples of this construction, some of them from 'the intelligence and counterintelligence business'.

The matter is not yet resolved. Until it is, readers of this book are advised to construe *bona fides* with a singular verb.

**bon mot.** In the sing. pronounced /bɔ̃ məʊ/ and in the pl. written *bons mots* and pronounced the same, or as /bɔ̃ məʊz/.

**bonne bouche.** In French 'a pleasing taste to the mouth' but in English 'a dainty mouthful or morsel' (which would be rendered in French as 'morceau qui fait ou donne bonne bouche'). Fowler (1926) resignedly remarked that such 'variation of meaning or form is no valid objection to the use of a phrase now definitely established'.

**Book of Common Prayer, The.** Until recently the official service book of the Church of England, it was originally compiled by Thomas Cranmer and others to replace the Latin services of the medieval Church. It first appeared in 1549, but the most familiar form now is the revised version of 1662. Like the Bible and the works of Shakespeare, it has contributed a great many familiar phrases to the language at large: e.g. *We haue erred and strayed from thy wayes, lyke loste shepe; Not waiyng* [weighing] *our merits, but pardonyng our offences.*

Between 1965 and 1971, Series 1, 2, and 3 of a revised Prayer-Book were issued as experimental forms of service, and in 1980 the Alternative Service Book was published. The aim of the revisers was to present the service book in up-to-date language. In the process some of the most memorable words and phrases in the Book of Common Prayer were replaced by others, and ancient and venerated points of accidence and syntax were removed. *Our Father, which art in heaven* became *Our Father in heaven; He ascended into heaven, and sitteth* became *He ascended into heaven, and is seated; Thereto I plight thee my troth* became *This is my solemn vow; With all my worldly goods I thee endow* became *All that I have I share with you.*

Anglicans are deeply divided about the merits of the two versions, but the BCP seems to be becoming marginalized by being mostly used only at restricted times or by special request (at weddings, funerals, etc.).

**bored.** The normal constructions are with *with* or with *by: they were bored with being left alone in the country; he became bored with Patrick; they were bored by the party political broadcasts before the general election.* A regrettable tendency has emerged in recent years, esp. in non-standard English in Britain and abroad, to construe the verb with *of.* Examples: *She would bore of the game quite suddenly*—M. Bracewell, 1989 (UK); *Oh, he's around, worst luck. I'm so bored of him. He's lost his virility*—C. Phipps, 1989 (UK); *Surely she must be bored of seeing this same setting all the time*—M. Tlali, 1989 (SAfr.); *I was conscious of all the problems ... of getting bored of something the minute you get it*—N. Fairburn, 1992 (Scottish); *They* [sc. children] *use the preposition 'of' in an unorthodox way: 'I'm bored of this,' they say (taking the construction from 'tired of')*—I. Opie, 1993. See OF C. 4.

**born(e).** The pa.pple of *bear* in all senses except that of birth is *borne* (*I have borne with this too long; he was borne along by the wind*); *borne* is also used, when the reference is to birth, (*a*) in the active (*has borne no children*), and (*b*) in the passive when *by* follows (*of all the children borne by her only one survived*). The pa.pple in the sense of birth, when used passively without *by*, or adjectivally, is *born* (*he was born blind; a born fool; of all the children born to them; melancholy born of solitude; she was born in 1950*).

**bosom.** Recorded from OE onward in the general sense 'the breast of a human being' and 'the enclosure formed by the breast and the arms' (*he clasped the fugitive to his bosom*; by extension, *in Abraham's bosom* (cf. Luke 16: 22) 'in the abode of the blessed dead'). It has also come into use at various dates in technical and literary senses (of the sea, a sail, a recess round the eye of a millstone, etc.). It was not until the 20c. that the word came to be used colloquially in the plural to mean 'a woman's breasts': e.g. *She gave him a quick glimpse of fine bosoms as she bent to the door of the icebox*—I. Fleming, 1965.

**botanic(al).** Both forms are recorded from the mid-17c. onward, but *botanical* is now much the more common of the two in Britain except in traditional names like the *Botanic Garden* in Oxford.

**both.**

1 both ... as well as.
2 Redundant *both.*
3 Need for symmetry in parts of speech in both ... and phrases.
4 Used with more than two items.
5 on both of our behalfs.
6 the both.
7 we both/both of us.

**1** *both ... as well as.* The construction is not often encountered, probably because the awkwardness of *as well as* as a correlative is all too obvious. Fowler (1926) cited 'Which *differs from* who *in being used both as an adjective as well as a noun*', and rightly rejected it in favour of '*both ... and as a noun*'.

**2** Redundant *both.* A detectable element of redundancy is introduced when *both* is used in conjunction with *at once, between,* or *equal(ly).* In such cases either *both* or the correlate should be omitted. Examples (from Fowler, 1926): *If any great advance is to be made* at once both *intelligible and interesting*; *The International Society is not afraid to invite comparisons* between *masters* both *old and new*; *The currents shifted the mines, to the* equal *danger* both *of friend and foe*; *We find* both *Lord Morley and Lord Lansdowne* equally *anxious for a workable understanding.*

**3** Need for symmetry in parts of speech in *both ... and* phrases. In such coordinate constructions it is essential for the conjoined items to be presented in the same manner: *both the men's and women's classes* should be replaced by *both the men's and the women's classes*; *a postwar reconstruction of both the political and economic structure of Germany* should be replaced by ... *both the political and the economic ...*; *Her article is both detrimental to understanding and peace* should be replaced by ... *is both detrimental to understanding and to peace.* Acceptable constructions: *the imaginative and dangerous energy of the son leaves the cautious father both puzzled and helpless*; *Being speechless is both a symptom and a cause of depression*; *its capacity to address both the internal and the international demands of the 1980s.*

**4** Used with more than two items. If language behaved like a simple mathematical system, the illogicality of using *both* of more than two items would be immediately apparent. In practice, *both* is almost always used with two homogeneous words or phrases: *both the people and the land*; *both by day and by night*; *he both loves and hates his brother*; *both now and evermore*; etc. From the 14c. onward, however, it has also been used 'illogically' in conjunction with more than two objects: *both man and bird and beast* (Coleridge, 1798) and *both Chaucer and Shakespeare and Milton* (De Quincey, c1839)

form part of an array of examples presented in the *OED*. Cyril Connolly wrote *... is both a musician, an archaeologist, and an anti-Fascist* in Horizon (March 1946). To judge from the infrequency of the 'illogical' construction in the 20c., however, the advisability of limiting *both* to two homogeneous objects is being recognized by the vast majority of writers.

**5** *on both of our behalfs.* Peter Carey's *Oscar and Lucinda* (1988) contains the sentence *I should not address you like this, even if I do hurt on your behalf, on both of our behalfs.* It just passes muster, but a prudent copy editor would have emended it to *on behalf of us both.* In general an *of*-construction is less awkward and less likely to be ambiguous than one using a possessive pronoun.

**6** *the both.* In spoken English, the use of *both* preceded by *the* is not uncommon: *Good morning from the both of us*—BBC Radio 4, 1977. It is more frequently encountered in regional speech, as, for example, *the both of you* heard on *The Archers* (BBC Radio 4, 1976). *The both* should not be used in formal prose.

**7** *we both/both of us.* The various (subjective and objective) types that follow are all equally acceptable: (*a*) *you both look cross*; *we both felt happy*; *it suited them both*; (*b*) *there was not enough for both of them*; *everybody knows both of us*; *both of us had small flakes of snow clinging to us*; *we've both of us got standards.*

**bother.** See POTHER.

**bottleneck.** Since the late 19c., applied to 'a narrow or confined space where traffic may become congested', and then (first recorded in 1928) figuratively, 'anything obstructing an even flow of production, etc., or impeding activity, etc.' (*OED* 2). In the looser kind of journalism, such bottlenecks are sometimes unsuitably said to be *cured, ironed out, broken,* etc., showing that the literal sense of the word is not always remembered.

**bottom line.** From its use in accountancy as 'the last line of a profit-and-loss account, showing the final profit (or loss)' (*OED* 2), the expression came to be used, and then to be overworked, to mean 'the final analysis or determining factor; the point, the crux of the argument' (*OED* 2). Figurative uses are first recorded in the late 1960s.

**bounden** is still used in *bounden duty* though not in *in duty bound*. For centuries it was the regular pa.pple of *bind*. It was also commonly used as an adj. ( = 'made fast in bonds; also fig.), but only *bounden duty* survives.

**bounteous.** Like BEAUTEOUS and PLENTEOUS, a literary word. It means 'generously liberal, munificent' or '(of things), ample in size or amount, abundant'. The word is more likely to be encountered in the works of Johnson and Tennyson, for example, than in the work of any 20c. writer. But it is not extinct: *Old hounds patrolling the corridors, seeing that none of the condemned flee back to the air, the light, the bounteous world above—* J. M. Coetzee, 1990.

**bourgeois.** *When I was a boy—a bourgeois boy—it was applied to my social class by the class above it*; bourgeois *meant 'not aristocratic, therefore vulgar'. When I was in my twenties this changed. My class was now vilified by the class below it*; bourgeois *began to mean 'not proletarian, therefore parasitic,* reactionary'. (C. S. Lewis, *Studies in Words*, 2nd edn., 1967.)

**bourn(e).** There are two distinct words, each of them spelt in the past and still sometimes today either with or without a final -e. One, meaning 'a small stream, a brook' (first recorded in the 14c.), survives in the south of England, used esp. in the context of winter torrents of the chalk downs, and in the place-names *Bournemouth* and *Eastbourne*; it corresponds to the northern word *burn*, also = 'a small stream'. The other word (first recorded in the 16c.), which is a loanword from French, means 'the limit or terminus of a race, journey, or course; destination, goal'. In the well-known passage in Shakespeare's *Hamlet* (1602), *The dread of something after death, The vndiscouer'd country, from whose borne, No trauiler returnes*, the word probably means 'frontier, boundary'.

**bowsprit.** Pronounce /ˈbəʊsprɪt/.

**bow window.** See BAY WINDOW.

**brace** (noun). ( = a pair). A collective noun having the same form in the sing.

(*a brace of pheasants*) and the pl. (*two brace of pistols*). Pl. uses are now rare.

**brachylogy.** (Cf. Gk βραχύς 'short'.) 'A shortened or condensed and grammatically incomplete expression, used in colloquial speech or specialised jargons to reduce time and effort, e.g. the greeting *Morning!* or the traffic sign *Road Up*' (Hartmann and Stork, 1973). The term is also sometimes applied to constructions that are overtly ungrammatical, e.g. *A is as good or better than B*, where formal grammar requires the insertion of a second *as* after *good*.

**bracket.** Since about 1880, one of the primary senses of the word has been a social one, 'a group bracketed together as of equal standing in some graded system' (*OED*). Thus, frequently, *income bracket, social bracket, top bracket*, etc.

**brackets.** Two marks of the form ( ), [ ], { }, < >, used to enclose an explanation, an aside, a pronunciation, an etymology, etc. ( ) are often called 'parentheses' or 'round brackets', [ ] 'square brackets', { } 'curly' or 'hooked brackets', and < > 'angle brackets'.

**Brahman.** Also formerly *Brahmin*, a member of the highest or priestly caste in the Hindu caste system. In American use, and occas. elsewhere, *brahmin* (thus spelt) is often applied to 'a highly cultured or intellectually aloof person' (orig. such a person in New England).

**brain(s),** in the sense of wits, may often be either sing. or pl. In *pick* (a person's) *brain(s)*, *rack* (one's) *brain(s)*, the number is indifferent; *has no brains* is commoner than *has no brain*, but either is acceptable English. Some phrases, however, admit only one number or the other, e.g. *have* (something) *on the brain*, *blow out* (someone's) *brains*.

**brand-new.** Correctly thus spelt, being (in the 16c.) formed from *brand* 'burning (wooden) torch' + *new* (i.e. fresh as from the furnace). Because the *-d-* is frequently not pronounced, the spelling *bran-new* was a common variant almost from the beginning, e.g. *Mr. and Mrs. Veneering were bran-new people in a bran-new house* (Dickens, 1865), but *brand-new* is customary now.

**bravado** /brə'vɑːdəʊ/ is an ostentatious display of courage or boldness, often concealing a felt timidity. *Bravery* is daring, valour, fortitude (as a good quality). *Bravura* /brə'vjʊərə/ is now virtually restricted to its musical sense, 'a passage or piece of music requiring great skill and spirit in its execution, written to task the performer's powers' (*bravura songs, a bravura performance*).

**brave.** Apart from its ordinary sense ('courageous, daring'), the word has been used for some four centuries as a general epithet of admiration or praise (e.g. *O that's a braue man, hee writes braue verses, speakes braue words*—Shakespeare). Apparently this use began to fall out of currency towards the end of the 19c. (to judge from the *OED*), but it has swept back into use in the 20c. (*a brave attempt, a brave step*, etc.), and, most notably, in Aldous Huxley's revival of Miranda's *O braue new world* (*Tempest*, v.i.183) in the title of his satirical novel *Brave New World* (1932).

**bravo.** It would be a brave person who would follow Fowler's advice (1926) to use *bravo* when applauding a male singer in an operatic performance, *brava* for a female singer, and *bravi* for the company. Gender and number distinctions have been abandoned in such circumstances, and *bravo* is the only cry of the three heard in English theatres now.

**breach, breech.** *Breach* is 'a breaking' (*in breach of his contract, breach of the peace, breach of promise* 'breaking of promise to marry', *step into the breach* 'give help in a crisis, etc.). *Breech* is principally, (*a*) in pl. *breeches* 'short trousers', memorably in *Breeches Bible*, the Geneva Bible of 1560 with *breeches* for *aprons* in Gen. 3: 7; and in *breeches-buoy*, a lifebuoy on a rope with canvas breeches for the user's legs; when used in the sense 'short trousers', normally pronounced /'brɪtʃɪz/; (*b*) = buttocks, now used only with reference to a baby's position at or before birth (*breeches birth*, with the baby's buttocks foremost); (*c*) the back part of a rifle or field gun barrel (*breech-loading gun*, one loaded at the breech, not through the muzzle).

Confusion of the two words occurs occasionally: e.g. *National capital … has hardly moved in to fill the breech* [read *breach*]

*created by the flight of comprador capital*—*Pacific Rev.*, 1988.

**breakdown.** Beside its primary meaning, 'a collapse, a failure of mechanical action or of health or of mental power', *breakdown* has been used since the 1930s to mean 'an analysis or classification (of figures, statistics, etc.). It is obviously important not to use this transferred sense in contexts in which it might have a tinge of ambiguity (e.g. *a breakdown* (better *an analysis*) *of engine failures in long-haul aircraft has not revealed any one main cause; a complete breakdown of our exports to dollar countries is not available at present.*

**breakthrough.** First used in the 1914–18 war to mean 'an advance of troops penetrating a defensive line', *breakthrough* has come to be used (since about the middle of the 20c.) of any significant advance in knowledge, achievement, etc. For a time it was an immensely popular vogue word, but it seems now to have joined the ranks of ordinary foot-soldier words, both in its literal and in its transferred senses.

**breech.** See BREACH, BREECH.

**brethren.** This ancient pl. of the word *brother* (first recorded *c*.1175) survives only in restricted use. It means 'fellow-members of a Christian society' (*dearly beloved brethren*); in particular the *Plymouth Brethren* (who call themselves 'the Brethren'), a religious body recognizing no official order of ministers, and having no formal creed, which arose at Plymouth *c*1830. A member of this body is called a *Plymouth brother*, and occas. also a *Brethren* (*Uncle Bill was coming to the meeting as well, even though he wasn't a Brethren*—N. Virtue, 1988).

**briar.** See BRIER, BRIAR.

**bridegroom.** The vicissitudes of etymology are seen in the emergence of this word in the 16c. The OE word was *brȳdguma* = *brȳd* 'bride' + *guma* 'man'. Had ME and early modE *grome*, 'lad, groom' (itself of unknown etym.) not been substituted for the second element, the word would have come down to us as *bridegoom*.

**brier, briar.** There are two distinct words, the first (from OE *brǣr*) meaning 'a prickly bush, esp. of the wild rose', and the other 'the white heath, *Erica arborea*, of southern Europe or a tobacco pipe made from its root'. The heath word is a 19c. loanword from Fr. *bruyère*. There is widespread inconsistency in the spelling of both words: the OUP house style for each is *brier*.

**brindle(d)** (adj.). The earlier form (first recorded in the 14c.) of the word meaning 'marked with bars or streaks of a different hue' was *brinded*. By the 19c. it was being ousted by the variant *brindled*, which had probably been formed (in the 17c.) 'by assimilation to such words as *kindled*, *mingled*, perhaps with some feeling of a diminutive sense' (*OED*), and *brindled* is now dominant. The noun *brindle* ( = 'brindled colour; a brindled dog') is a back-formation from *brindled*.

**bring.** 1 Partially distinguished from *take* according to movement towards the speaker (*bring*), or away from or accompanying the speaker (*take*): *take your raincoat with you and bring me a newspaper from the corner shop*. There are many circumstances, however, in which this simple distinction does not apply: e.g. *if we are going to the zoo shall we bring/take the camera?*

2 In regional speech in many areas in Britain and the US, the verb is conjugated *bring/brang/brung* (like *sing/sang/sung*) or even *bring/brung/brung*, but *brought* remains rock-solid for the pa.t. and pa.pple in standard English.

**brinkmanship.** Journalists and politicians have found a use for this word whenever two countries, groups, etc., come to the brink of war but do not engage in it. It is one of the products of nuclear confrontation between the Western powers and the USSR bloc in the 1950s. The word is attributed to the American politician Adlai Stevenson, who used it of the foreign policy of John Foster Dulles in 1956 (*Notes & Queries*, May 1959). For the formation, cf. *seamanship*, *statesmanship*, etc., and also Stephen Potter's facetious formations, *gamesmanship*, *one-upmanship*, and related words. See -MANSHIP.

**Brit.** A colloquial shortening of *Briton* or *Britisher*, first recorded in 1901, and

usually employed with more than a suggestion of teasing or, quite commonly, of hostility. *The Brit is at his old game* (1901 in *OED*), and *Brits out* (slogan on wall in N. Irish town, shown on ITN news, 6 Mar. 1977), just about sum it all up. In Australia and New Zealand, *Brit* is now challenging *Pom* as an everyday word for a British person, and has similar connotations. Elsewhere usage varies: the word often has an edge to it, but it is also favoured simply as being shorter than *Briton* and *Britisher*. In Britain itself there is less need for the word, but it is occasionally used for its informal convenience. Examples: *the average Brit has the greatest difficulty locating ... vital organs*—Radio Times, 1985; *No sooner had we arrived in Kenya than the goddam Brits began to scuttle*—D. Caute, 1986; *Cale in fact is a Brit who has emigrated to New York*—Plays International, 1988; *The bumbling Brits quietly built a better air force*—Literary Rev., 1989.

**Britain, British, Briton.** For the relation of these to *England*, *English(man)*, see ENGLAND.

**Briticism.** This word for 'a phrase or idiom characteristic of Great Britain, but not used in the English of the United States or other countries' (*OED*) seems to have been modelled on *Gallicism*, *Scotticism*, etc. Some writers, including H. W. Fowler, favoured *Britishism*, but in scholarly work *Briticism* (or, more usually, *BrE*, *British English*) is now the more usual term of the two.

**Britisher.** A regular US word (first recorded in 1829) for a British subject. People in Britain often register surprise, or are even slightly affronted, when the word is used, since the regular word used in this country for 'a native or inhabitant of Great Britain' is *Briton*.

**Britishism.** See BRITICISM.

**Brito-.** See ANGLO-.

**broadcast** (verb). For a short time in the 1920s it was not clear whether the past forms of the verb *broadcast* (in its airwaves sense) were to be *broadcasted* or *broadcast*. Learned arguments were displayed in a tract of the Society for Pure English (1924) and elsewhere, bearing on the apprehension of *broadcast* as a compound of cast, and comparing and contrasting the past forms of e.g. *forecast* and *roughcast*. In the event the shorter form *broadcast* has prevailed almost everywhere, though *broadcasted*, which is encountered occasionally, cannot be said to be wrong.

**broccoli,** now the only spelling (formerly also *brocoli*), is an Italian pl. n. (sing. *broccolo*), but is treated in English as a sing., non-count n., like *spinach*, etc. This vegetable has now been joined in the US by the *brocco flower* or *broccoflower* (*Brocco Flower is part of the mustard family, but it has a milder aroma than broccoli when cooking and a slightly sweeter taste than cauliflower when eaten raw—Chicago Sun-Times,* 1990).

**brochure,** first recorded in English in 1765, means lit. 'a stitched work' (cf. Fr. *brocher* 'to stitch'). In the sense 'a short printed work, i.e. a few leaves merely stitched together', it was more or less synonymous with the much older word *pamphlet*. From about the 1920s, however, *brochure* has tended to be restricted to mean 'a small, often glossy, pamphlet or booklet describing the amenities of a tourist resort or setting out the details of a fund-raising appeal, etc.'. *Pamphlet,* by contrast, continues to be used to mean 'a small treatise occupying fewer pages than would make a book, and normally left unbound'. Pronounce /ˈbrəʊʃə/ or /brɒˈʃʊə/, but in AmE only with the stress on the second syllable.

**broke(n).** The regular pa.pple and adj. *broken* (*the window had been broken during the night; a broken heart*) stand cheek-by-jowl with the predicative adj. *broke* (also *stony-broke*) 'ruined, without money'. The (orig. US) phr. *to go for broke,* meaning 'to make strenuous efforts, to go "all out"' (*If he were to go for broke on behalf of the Negroes … the President would endanger the moral reform cause—Guardian,* 1963), is now also commonly used in the UK.

**brow.** See MISQUOTATIONS.

**brunch.** This portmanteau word formed from *br(eakfast + l)unch* has made its way from university slang into more general use in the last century or so. It was first recorded in *Punch* in 1896, and, for half a century or so, was frequently written

within inverted commas or followed immediately by a bracketed explanation, but no longer.

**brunet(te).** In Britain, *brunette* is 'a (white) girl or woman of a dark complexion or with brown hair'. The same word is used as an adjective, 'of dark complexion, brown-haired'. In the US, the Fr. masculine form *brunet* is occasionally applied without distinction to both men and women, but there too the person to whom the word is applied is most commonly a girl or a woman. Examples: (brunette) *A pregnant brunette walks in off the street wearing black shorts—*T. Wolfe, 1965; (brunet) *tucked her blond locks under a series of brunet wigs—*G. D. Garcia, 1985.

**Brythonic.** See GAELIC.

**buck, doe, hart, hind, roe, stag.** The *OED* definitions make the distinctions sufficiently clear:

> *buck,* the he-goat, *obs.* … The male of the fallow-deer. (In early use perh. the male of any kind of deer.) … The male of certain other animals resembling deer or goats, as the reindeer, chamois. In S. Africa (after Dutch *bok*) any animal of the antelope kind. Also, the male of the hare and the rabbit.

> *doe,* the female of the fallow deer; applied also to the female of allied animals, as the reindeer … The female of the hare or rabbit.

> *hart,* the male of the deer, esp. of the red deer; a stag; spec. a male deer after its fifth year.

> *hind,* the female of the deer, esp. of the red deer; spec. a female deer in and after its third year.

> *roe,* a small species of deer inhabiting various parts of Europe and Asia; a deer belonging to this species.

> *stag,* the male of a deer, esp. of the red deer; spec. a hart or male deer of the fifth year.

**buffalo.** Pl. *-oes.* See -O(E)S 1.

**buffet.** When it refers to refreshments, the word is pronounced in standard English /ˈbʊfeɪ/, but railway staff seem mostly to say /ˈbʌfeɪ/ in British trains when drawing attention to the whereabouts of the *buffet car.* When the meaning is 'a cupboard in a recess for china and glasses', the word is pronounced /ˈbʌfɪt/. In AmE, /bəˈfeɪ/, i.e. with the main stress on the second syllable, seems to be the standard pronunciation.

**bugger.** The word is used as noun and verb with varying degrees of coarseness or vulgarity. It can also be used quite light-heartedly. Senses: (noun) **1** A sodomite. **2** Something or someone unpleasant or undesirable (*Heard one old lady say, 'It's a bugger this dark!'* [sc. the blackout]—Harrisson and Madge, 1940; *cheeky little bugger*; (said with a sympathetic voice) *poor buggers!*; *Needs 'sussing' proper, not the way you silly buggers go about it*—Match Fishing, 1990; *let's not play silly buggers* [sc. act foolishly]). **3** *bugger-all* 'nothing' (*I used to go and get her pension and do her shopping for her and I can tell you there was bugger-all left by the end of the week*—P. Barker, 1986. **4** A damn (*I don't give a bugger whether you won't or will*—Dylan Thomas, 1939). (Senses 3 and 4 uncommon in AmE.)

(verb) **1** To commit buggery with. **2** Used as a swear-word (*Bugger!*; *Bugger me!*; *Buggered if I know!*; *Well, I'll be buggered!*). **3** (with *up*) To ruin, spoil (*The rain buggered up the weekend for us*); (in passive) To be tired out (*he was completely buggered after two nights without sleep*). **4** (with *off*) To go away (*he buggered off home after the lecture*; *bugger off!*). **5** (with *about*, *around*) To mess about (*it's not wise to bugger about with electricity*). (Senses 3, 4, and 5 uncommon in AmE.)

There are still many circumstances in which such uses should be ruled out altogether. Nevertheless, attitudes towards words once judged to be unacceptable have changed considerably during the 20c., and it is no longer unusual to hear any or all of the above expressions used on the stage, on the radio, on TV, and in private conversations. Apparently there is much greater reluctance in America to use the word *bugger* in most of the senses listed here.

**bulk** is a noun signifying magnitude or size. As such it can be used correctly with singular nouns (the bulk of paper or of a book or of a tree, etc., is its size), and, somewhat adventurously, with collections like a people, the state, the clergy, one's land, etc. *Bulk-buying* and *-selling* are established terms. *Bulk* should not be used followed by *of* + an ordinary noun in the plural: *the bulk of policemen*,

the *bulk of brewers*, etc., would momentarily appear to refer to weight or size, not to numbers, unless the context proceeded swiftly to clear up the matter. In such contexts it is better to use *most*, *the majority of*, or some other synonym.

**bullyrag.** See BALLYRAG.

**bumble-bee** and **humble-bee** are independently formed, alternative names for the familiar large bee, of the genus *Bombus*, which makes a loud humming sound. *Bumble-bee* is much the more common of the two words.

**bunch.** As a collective noun, it has been used since the 16c. to signify a quantity, a collection, or a cluster of things (*a bunch of flowers, grapes, keys*, etc.). It is also commonly used to mean 'a company or group of people' (*the best of a bad bunch*, *the pick of the bunch*). The sporadic use of the type 'a bunch of + persons' in former centuries (e.g. *a bunch of cherubs*) does not support the view that this construction has unlimited currency at the present time, at least in Britain. The type *a bunch of spectators ran on to the pitch* verges on slang; whereas if the pl. noun is qualified by an adjective or other qualifier that indicates a feature or features held in common by them (*a bunch of corrupt politicians held the reins of power* [i.e. they had corruption in common]; *a bunch of weary runners crossed the line together an hour after the other competitors had finished* [i.e. they had weariness in common]), the informality is much less evident.

**bunkum.** This word meaning 'clap-trap, humbug' is one of the best-known American words to have spread to all English-speaking countries. Its origin is less well known. It is a respelling of *Buncombe*, the name of a county in N. Carolina. The phrase arose in America in the 1820s when the member of congress for that county needlessly delayed a vote near the close of a debate on the 'Missouri question'. The speaker insisted, however, that he was bound *to make a speech for Buncombe* in order to impress his constituents. (*OED*.)

**bur, burr.** In OUP house style, *bur* is the preferred spelling for 'a clinging seed-vessel or catkin', and *burr* for 'a rough edge; a rough sounding of the letter *r*; a kind of limestone'.

**burden, burthen.** Except as a rank archaism in poetry, rhetorical prose, etc., the form *burthen*, which was the original form of the word (OE *byrðen*), is now obsolete. In the sense 'the refrain or chorus of a song', *burden* represents, slightly indirectly, Fr. *bourdon*, the continuous bass or 'drone' of the bagpipe. The two words merged in the late 16c. (a long explanation can be found in the *OED*).

**bureau.** The recommended pl. is *bureaux*, pronounced /ˈbjʊərəʊz/, (but *bureaus* is not uncommon, esp. in AmE). See -x.

**burger.** A familiar shortening of *hamburger* and a fertile formative element in the 20c. There are *burger bars*, *burger parlours*, etc., throughout the English-speaking world. As a terminal element, *-burger* (first recorded as such in 1939) has generated *beefburger*, *cheeseburger*, *lamburger*, *nutburger*, *porkburger*, *steakburger*, and numerous other words.

**-burg(h),** a common element in place-names. When spelt *-burgh*, as in *Edinburgh*, it is pronounced /ˈɛdmbərə/, except that Americans tend to say /-bɜːrəʊ/, and *Edinburgh* in Texas is locally pronounced /-bɜːɡ/. When spelt *-burg*, as in *Hamburg*, it is pronounced /-bɜːɡ/. *Burgher*, 'a freeman or citizen of a foreign town, etc.', is pronounced /ˈbɜːɡə/.

**burgle, burglarize.** The first of these is a back-formation (first recorded in 1870) from *burglary*. It was at first thought to be facetious but is now the regular word in Britain (and in other English-speaking areas except N. America). AmE, from about the same date, seems to have mostly preferred *burglarize* (*one of us got hurt when we were burglarizing a pharmacy—New Yorker*, 1988; *another apartment building was burglarized last week following a barrage of house and apartment break-ins throughout Evanston in July—Summer Northwestern* (Illinois), 1988).

**burnt, burned.** *Burnt* is the usual form in the p.p. (*a thatched cottage was burnt down last week*) and as adj. (*burnt almond, a burnt offering*). In the past t., *burned* is the dominant form (*she burned her hands while preparing the barbecue*), but *burnt* is also permissible in all English-speaking areas. Some writers detect a preference for one form or the other as between

transitive and intransitive uses, but the evidence for such a distinction is unconvincing. However spelt, the word is normally pronounced /bɜːnt/, but *burned* as pa.t. and ppl adj. may also be pronounced /bɜːnd/. See -T AND -ED for other verbs of this type.

**burst, bust** (verbs). *Burst* is the regular verb with numerous senses (20 are listed in the *OED*) derived from the basic one (already in *Beowulf*) of 'to break suddenly, to snap'. *Bust* is a dialectal variant of it, first recorded in 1806. In two centuries, it has extended its territory in the standard language in such expressions (ranging from the colloquial to the entirely neutral) as the following: *I shall ... bust you one on the jaw* (P. G. Wodehouse, 1919); *busted* (arrested, jailed); *bust a house* (break in); *he busted a gut to get it done in time; he busted his leg playing football*. Derivatives: *a busted flush* (Poker); *bronco-busting; blockbuster; a bust-up* (quarrel). See BUST (ppl adj.).

**bus.** The form *'bus* (with apostrophe) is now extinct. Inflected forms (in OUP house style): pl. *buses*; as vb, present *buses*, past *bused*, pres.pple *busing*. See -S-, -SS-.

**business.** The regular word *business*, pronounced /ˈbɪznɪs/ with just two syllables, in its various senses stands apart from *busyness*, pronounced /ˈbɪzmɪs/, the simple abstract noun corresponding to *busy* (the state, etc., of being busy).

**bust** (ppl adj.) is freely used in the phrase *to go bust* 'to become insolvent', a use first recorded in a letter written by Rupert Brooke in 1913: *The Blue Review has gone bust, through lack of support*. See BURST, BUST (verbs).

**but.**

1 Normal uses as an adversative conjunction and preposition.
2 Used at the beginning of a sentence.
3 Case after *but* = except.
4 *but that, but what.*
5 Two successive *but*-constructions.
6 *But ... however.*
7 *cannot but* + bare infinitive.
8 *Always—but always.*
9 Miscellaneous uses.

**1 Normal uses as an adversative conjunction and preposition.** *But* is an adversative conjunction, and the words, phrases, or sentences contrasted by it must always be clearly adversative. Normal uses: *naughty but nice; nature is cruel but tidy; it was cool outside but even cooler inside; the answer is not to remove the parish system but to put more resources into it; he had many gifts and interests, but perhaps music was the greatest.*

But frequently means 'except (for)' when used as a preposition and 'except (that)' when used as a conjunction: (preposition) *the aftermath of the last economic crisis but one; everyone was pleased but John; There was little to be seen but a forest of brick chimneys.* (conjunction) *Claudia's eyes are closed but once or twice her lips twitch; I was willing enough, but I was ill-equipped; What else can we do but talk as if it were true?*

The contrast must never be neutralized by the placing of an additional circumstance in one of the contrasted elements. Fowler cites numerous examples of such partially cancelled contrasts, e.g. *In vain the horse kicked and reared, but he could not unseat his rider* (if the kicking was in vain, the failure to unseat involves no contrast; either *in vain* or *but* must be dropped).

**2** Used at the beginning of a sentence. The widespread public belief that *But* should not be used at the beginning of a sentence seems to be unshakeable. Yet it has no foundation. In certain kinds of compound sentences, *but* is used to introduce a balancing statement 'of the nature of an exception, objection, limitation, or contrast to what has gone before; sometimes, in its weakest form, merely expressing disconnection, or emphasizing the introduction of a distinct or independent fact' (*OED*). In such circumstances, *but* is most commonly placed after a semi-colon, but it can legitimately be placed at the beginning of a following sentence, and frequently is. Examples: *And went againe into the iudgement hall, & saith to Iesus, Whence art thou? But Iesus gaue him no answere.*—John (AV) 19: 9; *All Animals have Sense. But a Dog is an Animal.*—Locke, 1690; *Fare ye well. But list! sweet youths, where'er you go, beware.*—J. Wilson, 1816; *Parkin's emphasis on the agency of classes is unusually strong . . . I think it is too strong. But he could not weaken it.*—London Rev. Bks, 1980; *Of course thy loved her, the two remaining ones, they hugged her, they had mingled their tears. But they could not converse with her.*—I. Murdoch, 1993.

It should be said, though, that unless contextual dislocation is being deliberately sought as a rhetorical device, it is not desirable to litter the pages with constructions like *He is tired. But he is happy.*

**3** Case after *but* = except. Because of the historical levelling of inflexions of nouns, the problem arises only with pronouns that show case: *Everyone but she can see the answer* vs. *Everyone but her can see the answer.* The best course would appear to be to use the subjective case when the *but*-construction lies within the subject area of a clause or sentence (*No one but she would dream of doing that*), and to use the objective case when the *but*-construction falls within the object area of a clause or sentence (*No one else may use my typewriter but her*). The formula is not watertight, however. For example, when a subject containing *but* is delayed, *but* is merely an emphatic repetition of the main subject, the case remains the same: *But no one understood it, no one but I*—J. M. Coetzee, 1977. When the clause contains the verb *to be*, it is nevertheless usual for a late-placed *but* to be followed by the objective case: *No one is fool enough to work the straights but me*—J. Fuller, 1983. After interrogatives the objective case is the more usual: *Who can have done that but him?*

Fowler's description of the problem is worth repeating: 'The question is whether *but* in this sense is a preposition, and should therefore always take an objective case (*No-one saw him but me, as well as I saw no-one but him*), or whether it is a conjunction, and the case after it is therefore variable (*I saw no-one but him, i.e. but I did see him; No-one saw him but I, i.e. but I did see him*).' He concluded that when the *but*-construction falls within the object area the objective case has prevailed (*No-one knows it but me*).

**4** *but that* has many undisputed formal or literary (though somewhat fading) uses: e.g. (a) introducing a consideration or reason to the contrary: except for the fact that, were it not that (*OED*), which adds that 'formerly *that* was occas. omitted': *And but she spoke it dying, I would not Beleeue her lips*—Shakespeare, 1611; *I too should be content to dwell in peace . . . But that my country calls*—Southey, 1795; *He would not have set out for France* *by road but that he knew all flights had been cancelled*; (formulaic use without *that*) *it never rains but it pours*. (*b*) after *doubt*: *I do not doubt but that you are surprised*—Ruskin, 1870. (*c*) after *tell*: *How could he tell but that Mildred might do the same?*—*Blackwood's Mag.*, 1847.

The danger in *negative* and *interrogative* constructions of this kind is that a redundant *not* can inadvertently (and erroneously) be placed in the dependent clause: (sentence *b* with an incorrectly added *not*) *I do not doubt but that you are not surprised*; (cited by Fowler) *Who knows but that the whole history of the Conference might not have been changed?*

In the past, *but what* was sometimes used in similar constructions, but these uses are now mainly found in informal or non-standard types of English: *Nor am I yet so old but what I can rough it still*—Trollope, 1862; *It's no telling but what I might have gone on to school like my own children have*—Lee Smith, 1983 (US); *I never bake a pan of brownies . . . but what I think of him*—ibid.

**5** Two successive *but*-constructions. It is more or less self-evident that it is not desirable to add a *but-* construction to an unrelated *but*-construction in the same sentence. An example (from Fowler, 1926) of the rejected construction: *I gazed upon him for some time, expecting that he might awake; but he did not, but kept on snoring.*

**6** *But . . . however.* It is advisable to avoid conjoining *but* with *however*, and with other words which themselves express a limitation or distinction, as *nevertheless*, *still*, and *yet*: (*But*) *one thing, however, had not changed, and that was . . .*; (*but*) *nevertheless they went on arguing.*

**7** *cannot but* + bare infinitive. This construction, which has been in standard use since the 16c., is now very common. Examples: *The frailty of man without thee cannot but fall*—Bk of Common Prayer, 1549; *I cannot but be gratified by the assurance*—Jefferson, 1812; *she could not help but follow him into the big department store*—B. Rubens, 1987; *yet he could not help but admire Miss Leplastrier for the way she looked after the details*—P. Carey, 1988. It should be noted, however, that the use with *help* inserted between *cannot* and *but* has not been found in print before the late 19c.:

*She could not help but plague the lad*—H. Caine, 1894.

**8** *Always—but always.* *But* is often used after a pause to introduce a word that is being repeated for emphasis: e.g. *Nothing, but nothing, is going to be allowed to prevent Martha from meeting her deadline*—V. Glendinning, 1989; *she was always—but always—on a diet*—New Yorker, 1989.

**9** Miscellaneous uses. (*a*) *but* at end of sentence. One of the most surprising and largely uncharted modern uses of *but* is its occurrence as a qualifying adverb at the end of sentences. Taking a lead from the Scots and the Irish, not-quite-standard speakers in Australia, in some parts of South Africa, and perhaps elsewhere provide evidence of this construction which has not yet entered the standard English of England: e.g. '*He should have left the key with me,' she said. 'I'm his wife.' 'I didn't ask for it, but.'*—M. Richler, 1980 (Canad.); *'I been waiting round for years and years and I still don't know what it is, but.'*—M. Eldridge, 1984 (Aust.); *'Yes, I told 'im. Not the whole of it, but.'*—D. Malouf, 1985 (Aust.); *'That was a lovely cat, but'* [ = *that was a truly lovely cat*]—R. Mesthrie, 1987 (SAfr.); *'She's lovely.' 'Isn't she but,' said Jimmy Sr.*—R. Doyle, 1991 (Ir.); *'I like your café,' I said truthfully, for something to say. 'I'm not staying but,' she said.*—R. Scott, 1993 (NZ).

(*b*) *not but eight* = only, merely eight. See NOT 10.

(*c*) *all but* (adverbially) = everything short of, almost. Examples: *Man . . . All but resembleth God . . . All but the picture of his maiestie*—J. Bastard, 1598; *These were all but unknown to Greeks and Romans*—A. P. Stanley, 1862; *Edwin had persuaded his father to all but cut out his oldest son*—S. Chitty, 1981; *by the end of the war this attitude had all but disappeared*—P. Wright, 1987. From this use has emerged the adjectival use of *all-but*: *Our all-but freedom*—W. Empson, 1935.

(*d*) Used after an exclamation (*Ah! but, My! but*, etc.) to express some degree of opposition, surprise, etc. (a use first recorded in 1846): *Ah, but who built it, that we tiny creatures can walk in its arcades?*—M. Drabble, 1987; *My, but he was obliging*—New Yorker, 1987.

**buy.** **1** As noun, in such uses as *the best buy*, the word has been current since the third quarter of the 19c. Its currency has been greatly assisted by the coming into being in the second half of the 20c. of numerous consumer journals like *Which?*

2 As verb, the orig. American sense 'to believe', first noted in 1926, is now well established in everyday speech (*I'm willing to buy that for what it's worth*) but hardly at all in good quality prose in the UK. The past t. and p.p. forms of *buy* are, of course, *bought*. See VERBS IN -IE, -Y, AND -YE.

**buzz.** See -Z-, -ZZ-.

**by** (prep.). Owing to the variety of its senses, *by* can occasionally acquire an unwanted ambiguity in certain constructions. The absurdity of *he was knocked down by the town hall*, or of *In Poets' Corner where he* [sc. Dryden] *has been buried by Chaucer and Cowley* (G. E. B. Saintsbury, 1881) can be lessened by a contextual change of intonation, but is better avoided altogether by choosing a different preposition, or by some other means. Fowler (1926) worried about the accidental, slovenly recurrence of *by* in the same sentence: *Palmerston wasted the strength derived by England by the great war by his brag*. Fortunately such gross impropriety is not often encountered in written English now.

**by and large.** This adverbial phr. is first found in the 17c. in nautical (sailing ship) language meaning (to sail) 'to the wind (within six points) and off it' (*OED*).

It very rapidly—the new use first recorded in 1706—slipped into general use in its current figurative sense, 'without entering into details, on the whole'.

**by, by-, bye.** Nearly all the words in this group are derived from *by* preposition or adverb, the main exception being *by-law* (a variant of the obsolete *byrlaw*, of Scandinavian origin, = local custom). Over the centuries, the main body of *by-* words has settled down into three groups: **1** *by and by* 'soon'; *by the by* 'by the way, incidentally'.

**2** *bye* (in cricket and other games); *bye-bye* (familiar form of 'goodbye'); *bye-byes* (sleep).

**3** *by-* (tending to form one word with the following noun, but a hyphen is fairly regularly printed in some of the words; the lists that follow show OUP house style): (with hyphen) *by-blow, by-election, by-form, by-lane, by-law, by-product, by-street*; (one word) *bygone, byline, byname* (a sobriquet), *bypass, bypath, biplay, by-road, bystander, byway, byword*. The spellings *bye law* and *bye-election* are preferred by some other publishing houses.

**Byzantine. 1** Spelt with an initial capital when used of the architecture, art, politics, etc., of ancient Byzantium; but usu. with a small initial when it means 'intricate, complicated'.

**2** Several different pronunciations are current: /bɪˈzæntaɪn/, /baɪ-/, /-iːn/; /ˈbɪzən-taɪn/, /-tiːn/. The one I use myself is /baɪˈzæntaɪn/.

# Cc

**cabbalist(ic), cabbala,** etc. In these, and also in the other derivatives of *cabbala* ( = Hebrew oral tradition), *-bb-*, which reflects a doubled consonant in Hebrew, is the better spelling in English (not *-b-*).

**cacao** (pl. *-os*), pronounced /kəˈkɑːəʊ/ or /-ˈkeɪəʊ/, and in origin a Spanish word derived from Nahuatl *cacauatl*, is 'a seed pod from which cocoa and chocolate are made, or the tree from which such seed pods are obtained'. Cf. COCOA.

**cachet.** Marked as an unnaturalized French loanword in the *OED*, and scorned by Fowler ('should be expelled [from the language] as an alien'), *cachet*, pronounced /ˈkæʃeɪ/, is now an acceptable member of the family, both in its general senses ('a distinguishing mark or seal; prestige') and as used in medicine ('a flat capsule enclosing a dose of unpleasant-tasting medicine').

**cachinnation,** laughter. See POLYSYLLABIC HUMOUR.

**cachou,** a lozenge to sweeten the breath, is to be distinguished from *cashew*, a bushy evergreen tree, *Anacardium occidentale*, native to Central and S. America, bearing edible kidney-shaped *cashew* nuts. They are both pronounced /ˈkæʃuː/.

**cacoethes,** a Latin (ultimately Greek) word meaning 'an urge to do something undesirable', was frequently used in elevated English prose until about the end of the 19c., esp. in the phrases *cacoethes scribendi* 'an unhealthy passion for writing' (based on Juvenal's *tenet insanabile multos scribendi cacoethes*) and *cacoethes loquendi* 'an itch for speaking'. Both phrases are still used, but much less commonly than hitherto.

**cactus.** Pl. (in general use) *cactuses*, in botany *cacti* /ˈkæktaɪ/, but the distinction is far from watertight.

**caddie, caddy.** The golf-attendant has *-ie*; so too the corresponding verb. The small container for holding tea has *-y*.

*Caddie* was originally Scottish (from Fr. *cadet*); *caddy* is from Malay *kātī*.

**caddis-fly,** any small, hairy-winged insect of the order Trichoptera, is now always so spelt, not as *caddice-fly*.

**cadi** /ˈkɑːdɪ/, a judge in a Muslim country. Pl. *cadis*. The spelling with initial *c-* (not *k-*) is recommended.

**cadre,** used in the armed forces to mean 'a nucleus or small group (of servicemen) formed to be ready for expansion when necessary', is pronounced as /ˈkɑːdə/ or, in imitation of French, /ˈka:drə/. When used to mean 'a group of activists in a communist or revolutionary party, or a member of such a group', it seems to be most commonly pronounced /ˈkeɪdə/.

**caecum, Caesar, caesura,** etc. Now always printed with *-ae-* as two separate letters, not ligatured. See Æ, Œ. Some of these words are regularly spelt with medial *-e-* in AmE (e.g. *cecum, Cesarian, cesium*).

**Caesarean, Caesarian.** The dominant spelling is with *-ean*, esp. in the medical term *Caesarean section* (US *Ces-*). The word is often written with a small initial *c*.

**caesura. 1** In Greek and Latin prosody, the division of a metrical foot between two words, esp. in certain recognized places near the middle of the line (*OED*).

**2** An obligatory feature of OE verse like *Beowulf*: the caesura is indicated by a space in printed versions of the poems: e.g. *under heofones hwealf    healsittendra*. In later English verse, chiefly noticeable in long metres such as that of Tennyson's *Locksley Hall*: *Till the war-drum throbb'd no longer, // and the battle-flags were furl'd*. In post-medieval English verse, 'the term does not refer to anything in the *structure* of most English verse, ... and there is no reason to prefer it to "pause" or "syntactic break" in describing a line' (D. Attridge, 1982).

**café.** Usu. printed in English with an acute accent but occas. without. In either case it is pronounced /ˈkæfeɪ/. In non-standard or jocular English it is now often pronounced /keɪf/ or written as *caff* and pronounced /kæf/.

**caffeine.** Now always pronounced /ˈkæf-iːn/, but formerly (e.g. in Daniel Jones's *English Pronouncing Dict.*, 1917) as three syllables, /ˈkæf-ɪ-iːn/.

**cagey,** 'cautious and uncommunicative', was first recorded in America as recently as 1909, was not common in the UK until the mid-century, and is of unknown etymology. Sometimes spelt *cagy*.

**calcareous, calcarious.** The 'erroneous' form with final *-eous* is now standard. First recorded about 1790, the spelling with *-eous* was influenced by words in *-eous* from L *-eus*. The etymological sense of *calcareous* would be 'of the nature of a spur', whereas the word actually means 'of the nature or, or composed of, lime(-stone)', from L *calx, calcis* lime + *-arius*.

**calculate.** 1 *Calculate* makes *calculable*; see -ABLE, IBLE 6.

2 The sense 'to suppose, reckon' is American in origin (first recorded in 1805) but has not at any stage become standard in AmE, let alone elsewhere: 'This use of the word ... is not sanctioned by English usage' (Webster, 1847); 'Formerly chiefly New England, now more widespread, somewhat old-fashioned' *DARE*, 1985). The illustrative examples cited in the large American dictionaries are nearly all taken from regional sources, e.g. *Transactions of the Michigan Agricultural Society*, 1857, and *Report of the Maine Board of Agriculture*, 1882. A typical sentence: *I calculate it's pretty difficult to git edication down at Charleston*—C. Gilman, 1836.

**calculus.** The medical word ('a stone or concretion formed within the body') usu. has pl. *-li* /-laɪ/; the mathematical, usu. *-luses*.

**caldron.** See CAULDRON.

**calendar,** an almanac, not *k-*.

**calends,** the first of the month in the ancient Roman calendar, not *k-*.

**calf.** For plural, etc., see -VE(D), -VES.

**calibre** (US *caliber*) is now always pronounced /ˈkælɪbə/; the variant /kəˈliːbə/ has been discarded.

**caliph.** The transliteration of words of Arabic origin that have entered English through another language (in this case medieval French) normally leads to the emergence of a number of variant spellings and pronunciations. *Caliph* is now the dominant spelling in English (not *ka-*, *kha-*, *-if*) and /ˈkeɪlɪf/ the dominant pronunciation, not /ˈkælɪf/.

**calk** (verb). See CAULK.

**callus** (pl. *calluses*) means 'a hard thick area of skin or tissue'; the corresponding adj. *callous* is used to mean '(of skin) hardened or hard', but is much more frequently used in the figurative senses 'unfeeling, insensitive'. The spelling *callous* should not be used for the noun.

**calmative.** A word with not much history behind it (first recorded in 1870), no etymological support from Latin, and lying at the crossroads of two distinct pronunciations, /ˈkælmətɪv/ and /kɑːˈmə-tɪv/. *Sedative*, an older word with sound etymological credentials, is to be preferred both as n. and as adj.

**caloric.** Once used as the name (corresponding to Fr. *calorique*) given by Lavoisier to 'a supposed elastic fluid, to which the phenomena of heat were formerly attributed' (*OED*); now a regular adj. (pronounced /kəˈlɒrɪk/ meaning 'of or pertaining to heat'.

**calorie.** A word (first used in the 1860s) that stepped right outside physics laboratories (where it means 'a unit of quantity of heat') into widespread general currency as the 20c. proceeded. The general public have adopted what physicists call the *large calorie*, i.e. the amount needed to raise the temperature of 1 kilogram of water through 1°C, and use it as a measure of the energy value of foods. The word was formed arbitrarily in French from L *calor* 'heat'.

**cambric.** Pronounce /ˈkeɪmbrɪk/.

**camellia.** The spelling with *-ll-* is standard, as is the pronunciation with medial /-iː-/.

**camelopard,** an archaic name for the giraffe, does not contain the word *leopard* and should not be spelt or pronounced as if it did. Pronounce /ˈkæmɪləʊpɑːd/ or /kəˈmeləpɑːd/.

**cameo.** Pl. -s. See -O(E)S 4.

**camomile,** the literary and popular form of the word, answers to medL *camomilla*. The initial *ch* of the form *chamomile* answers to Lat. *chamaemelon* (Pliny) and Gk. Χαμαίμηλον 'earth apple' (from Χαμαί 'on the ground' + μῆλον 'apple').

**campanile.** Pronounce /kæmpəˈniːlɪ/. Its pl. in Italian is in *-i*, in English *-es*.

**can** (noun). See TIN.

**can** (modal auxiliary) has a wide range of uses. It usually expresses (*a*) possibility: *the data that can be gathered; anyone can make a mistake; manned spacecraft can now link up with other spacecraft in outer space; the virus can lie dormant in apparently normal skin; he can be very trying.* (*b*) ability: *his four-year-old son can ride a bicycle; at his peak Murray could read more than forty languages.* (*c*) permission: In informal circumstances, since the second half of the 19c., *can* has often been used in contexts of permission where *may* had earlier been obligatory: *Can I speak with the Count?* —Tennyson, 1879; *Father says you can come*—T. B. Reed, 1894; *No one can play the organ during service time without the consent of the Vicar*—Church Times, 1905. In everyday life, such informal uses of *can* now occur all the time: e.g. *can I speak to your supervisor, please?* But in any context where politeness or formality are overriding considerations, *may* is the better word: *May I come and stay with you?; May I have another whisky, please?*

In pa.t. contexts, *could* (and not *might*) is more or less obligatory: e.g. *At that time only rectors could* ( = were entitled to) *receive tithes.* In the sentence *I'll drop in to see you tomorrow, if I can* (i.e. if I am able to) the substitution of *may* would change the meaning. Cf. MAY AND MIGHT.

**Canaan(ite).** Fowler (1926) regarded the present-day pronunciation /ˈkeɪmən-aɪt/ as an 'evasion', insisting that the pronunciation prevalent in his day was /ˈkeɪmjən-aɪt/. Daniel Jones (1917) gave precedence to /ˈkeɪmən-aɪt/. The problem has vanished: the pronunciation with medial

/-j-/, which reflects the way the place-name /kèna'an/ was pronounced in ancient Hebrew, is no longer extant.

**canard.** Except in the realm of cookery (*canard sauvage*, etc.), where the French pronunciation of *canard* is retained, the word in its main English sense, 'an unfounded rumour or story', is now pronounced either as /kəˈnɑːd/ or /ˈkænɑːd/.

**candelabrum.** Because of its Latin origin, the word should normally have *candelabra* as its plural. It has not always worked out like that, and English patterns have partially established themselves. *Candelabrums* is sometimes used in AmE; conversely *candelabra* has often been treated as a singular from the early 19c. onward (Walter Scott spoke of *four silver candelabras* in *Ivanhoe*). It does not seem likely that the original pattern, *candelabrum* sing./*candelabra* pl., will be restored as the only correct forms.

**canine.** The pronunciation /ˈkemam/ is dominant, stressed on the first syllable and with /-eɪ/ as in *cane*. The *OED* (1888) gave preference to /kəˈnam/, stressed on the second syllable, but also listed /ˈkænam/. Daniel Jones (1917) recommended /ˈkænam/, but also gave /ˈkem-/ as 'less frequent'.

**cannon.** 1 From the 16c. onward, but no longer, the regular word for a piece of ordnance, to the types of which numerous exotic-sounding names were applied (*aspic, basilisk, culverin, serpentine,* etc.). Now, in military language, normally restricted to a shell-firing gun in an aircraft (a use first recorded in 1919).

2 Historically the word was used as an ordinary noun, with pl. *cannons*; but also collectively (*Cannon to right of them, Cannon to left of them, Cannon in front of them Volley'd and thunder'd*—Tennyson, 1855).

**cannot.** 1 This is normally written as one word (rather than *can not*) and is often pronounced like the reduced form *can't*. One encounters *can not* occasionally in letters, examination scripts, etc.; the division seems to do more with custom ('I have always written it this way') than with emphasis. The reduced form *can't*, which now seems so natural, is relatively recent in origin. It does not occur in the works of Shakespeare, for example, and the earliest example of it given in the *OED* is one of 1706.

**2** *Cannot* (or *couldn't*, etc.) is correctly used before *but* (see BUT 7); before the combination *help but* + infinitive (see BUT 7); and before the verb *help* + gerund (*I couldn't help thinking that he wasn't listening either*—B. Rubens, 1985). However, standard English does not admit constructions of the type *he can't hardly walk*, i.e. where *can't* is qualified by a negative adverb.

**3** *can't seem* + infinitive = seem unable to. This construction is relatively recent (the first example in the *OED* is one of 1898) and still has a tinge of informality about it. Examples: *He couldn't seem to get the boy out of his head*—I. Baird, 1937; *Somehow I can't seem to get warm*—M. Pugh, 1969. It belongs more in spoken English than in formal writing. See HARDLY 5.

**cañon, canyon.** Both forms are pronounced /'kænjən/. The form with the tilde seems to be less common than it used to be except when used with direct reference to Spain. The *Grand Canyon* in Arizona is always so spelt.

**cant.** In the 18c. and 19c., one of its primary meanings was 'the secret language or jargon used by gypsies, thieves, professional beggars, etc.' (*They talk'd to one another in Cant*—J. Stevens, 1707). During the same period, it was also applied contemptuously to the special phraseology of particular classes of (non-criminal, non-vagrant) persons (*All love—bah! that I should use the cant of boys and girls—is fleeting enough*—Dickens, 1839). These senses have drifted away except in scholarly work or historical novels. Instead *cant* now usu. means 'insincere pious or moral talk, language implying the pretended assumption of goodness or piety' (e.g. *the speech by the member for —— was saturated with cant*). See JARGON.

**can't.** See CANNOT.

**cantatrice,** 'a female singer', which is an early 19c. loanword from Italian or French (spelt the same in both languages), is now pronounced in an Italian manner as /'kantatri:tʃeɪ/ or in a French manner as /kãtri:s/ at choice. The French pronunciation is the more usual of the two in English.

**canto.** Pl. *-os*. See -O(E)S.

**canton.** Used of a subdivision of a country, esp. Switzerland, pronounced /'kæntɒn/. As a verb in military use, 'to put (troops) into quarters', pronounced /kæn'tu:n/. *Cantonment*, 'a lodging assigned to troops; a permanent military station in India', is pronounced /ˌkæn'tu:nmənt/.

**canvas, canvass.** **1** *Canvas* 'coarse cloth' is so spelt, with pl. *canvases*. When used as a verb, 'to cover or line with canvas', it is conjugated as *canvases, canvased, canvasing* (but often *-s-* in AmE).

**2** *Canvass* 'to solicit votes' yields the forms *canvassed, canvasser, canvasses,* and *canvassing*. The corresponding noun is also spelt *canvass,* pl. *canvasses*. Historically both words come from the same French original, and it is only in the 20c. that *canvas* has become fairly consistently restricted to cloth, and *canvass* to voting.

**canyon.** See CAÑON, CANYON.

**caoutchouc.** This strange-looking word, adopted in the 18c. from a Quechua word via Spanish and French, is pronounced /'kaʊtʃʊk/.

**caper.** See SINGULAR -S.

**capercaillie, capercailzie.** A Scottish word of Gaelic origin meaning 'wood-grouse'. The *lz* for *lʒ* [i.e. *l* followed by yogh] is a 16th c. Sc. way of representing *l mouillé* ... and is properly represented by *ly*' (*OED*). In fact, however, the prevailing spelling is *capercaillie* (not, as formerly, *capercailzie*), and the dominant pronunciation is /ˌkæpə'keɪlɪ/.

**capita, caput.** See PER CAPITA.

**capital.** **1** adj. A headmaster pointed out to me in 1990 that he had frequently encountered illiterate confusion of *capital* and *corporal* in contexts of the type *My son could do with some capital punishment now and then*'.

**2** noun. *Capital*, the most important town or city of a country or region, is to be distinguished from *capitol*, which is a building in which an American legislative body meets, the best-known of which is the one in Washington, DC.

**capitalist.** Many old-fashioned socialists, including my late father, regularly stressed the word on the second syllable, i.e. as /kəˈpɪtəlɪst/, but nowadays the stress is normally placed on the first syllable of it and of *capitalism*, *capitalization*, and *capitalize*.

**capitals.** Apart from certain elementary rules that everyone knows and observes, such as that capitals are used to begin a new sentence after a full stop, for the initial letter of quoted matter (but see PUNCTUATION), and for proper names like *John Smith* (with rare exceptions like the idiosyncratic *e. e. cummings*) and those of the days and months, their present-day use shows wide variation from one publishing house to another, and even within the pages of the same book, newspaper, etc.

What follows is an abridged and slightly modified version of the relevant section in *Hart's Rules*, pp. 8–14.

**A** Capital initials should be used for:

**1** *Prefixes and titles forming part of a compound name*: Sir Roger Tichborne, the Bishop of Oxford, the Duke of Wellington. Also, *Her Majesty the Queen, the Prince of Wales, His Excellency the British Ambassador, His Holiness, Your Honour*—when the title of a particular person; but in a general sense lower case is correct: *every king of England from William I to Richard II*; for *king* is used here in a general sense, where *monarch* or *sovereign* would be equally correct.

**2** *Parts of recognized geographical names*: (of countries or regions) *Northern Ireland* (as a political entity), but *northern England*, a plain description in general terms; similarly, *Western Australia, West Africa, South Africa, New England*, etc.; (names of straits, estuaries, etc.) *Firth of Clyde, Norfolk Broads, Straits of Gibraltar, Plymouth Sound, Thames Estuary*; (names of rivers) *River Plate* (Rio de la Plata), *East River* (New York), but *the Thames*, or *the river Thames*; (topographical and urban names) *Trafalgar Square, Addison's Walk* (in Magdalen College, Oxford), *Regent Street, London Road* (if official name), but *the London road* (that leading to London).

**3** *Proper names of periods of time*: Bronze Age, Stone Age, Dark Ages, Middle Ages, Renaissance; First World War, Second World War, or World War I, II, but *the 1914–18 war, the 1939–45 war*.

**4** *Proper names of institutions, movements, etc.*: Christianity, Marxism, Buddhism, Islam, the Church of England, the (Roman) Catholic Church; but lower case for the building or for a church in a general sense. *Church* and *State*—both capitalized when viewed as comparable institutions, also *the State* as a concept of political philosophy; *the Crown, Parliament, Congress* (US), *House of Commons* (*of Representatives*, US), *House of Lords, Ministry of Finance*, etc. Also, *HM Government*, or *the Government*, in official parlance and meaning a particular body of persons, the Ministers of the Crown and their staffs; but *the government* (lower case) is correct in general uses.

**5** *Parties, denominations, and organizations, and their members*: Air Force, Army, Navy (as titles of particular organizations), *Conservative, Labour* (in British politics); *Socialist, Liberal Democrat, Christian Democrat* (European countries, etc.); *Republican, Democratic* (USA); and so on. (But *socialist, republican, conservative, democratic*, etc., as normal adjectives when not party titles.) Also, *Baptist, Congregationalist, Methodist, Presbyterian, Unitarian, Church of England, Anglican, Roman Catholic, Orthodox* (i.e. Eastern Orthodox), *Evangelical* (continental and US). (But *congregational* (singing, polity), *unitarian* views of God, *orthodox* belief, *catholic* sympathies in non-denominational sense.) The general rule is that capitalization makes a word more specific and limited in its reference: contrast *a Christian scientist* (man of science) and *a Christian Scientist* (member of the Church of Christ Scientist).

**6** *Titles of office-holders*. In certain cases and certain contexts these are virtually proper names of persons: *HM the Queen, the Prime Minister, the Archbishop of Canterbury*. The extension of this principle depends on the context: *the President* (of the USA, of Magdalen College, Oxford, etc.). Similarly, *the Bishop of Hereford, the Dean of Christ Church*; and in a particular diocese, *the Bishop*, or within a particular cathedral or college, *the Dean* (referring to a particular individual, or at least a holder of a particular office: *the Bishop is ex officio chairman of many committees*). (But in contexts like *when he became bishop, the bishops of the Church of England, appointment of bishops*—such cases are better printed in lower case, and so with other office-holders.)

**7** *Names of ships, aircraft types, railway engines, trade names, etc*: The *Cutty Sark*, HMS *Dreadnought*; the Königs, the fastest German battleships in 1916 (capitals but not italic for *types* of ships). The Spitfire, the Flying Fortress, the Dakotas of the 1939–45 war. These are types, since aircraft do not usually have individual names; but 'the US bomber Enola Gay which dropped the atom bomb over Hiroshima on 6 August 1945' (not italic as not official like a ship's name). *A Viscount, a Boeing, a Concorde* (airliners). *A Ford Orion, a Renault 5* (trade names). *Anadin, Cow & Gate, Kleenex, Persil*, etc. Capitals must be used for proprietary names. Pronouns referring to the Deity should begin with capitals only if specifically requested by an author: *He, Him, His, Me, Mine, My, Thee, Thine, Thou*; but even so it is better to print *who, whom*, and *whose*. In religious writings, capitals are now either old-fashioned or a personal preference of some writers. The main trend is to use lower case in such circumstances.

**B** *Words derived from proper names*

**1** *Adjectives.* (i) Use a capital initial when usage favours it, and when connection with the proper name is still felt to be alive: *Christian, Dantesque, Hellenic, Homeric, Machiavellian, Platonic, Roman* (Catholic, Empire), *Shakespearian.* (ii) Use a lower-case initial when connection with the proper name is remote or conventional: *arabic* (letters), *french* (chalk, cricket, polish), *italic* (script), *roman* (numerals), and when the sense is an attribute or quality suggested by the proper name: *chauvinistic, gargantuan, herculean, lilliputian, machiavellian* (intrigue), *quixotic, titanic.*

**2** *Verbs.* (i) Use a capital initial when the sense of the verb is historical or cultural and has a direct reference to the proper name: *Americanize, Christianize, Europeanize, Hellenize, Latinize, Romanize.* (ii) Use a lower-case initial when the sense is an activity associated with but not referring directly to the proper name: *bowdlerize, galvanize, macadamize, pasteurize.*

**3** *Nouns.* Use a lower-case initial (i) When reference to the proper name is remote or allusive: *boycott, jersey* (garment), *mackintosh, morocco* (leather), *quisling, sandwich, suede, wellington* (boot). (ii) in names of scientific units: *ampere, joule,*

*newton, volt, watt.* (iii) in names of metres: *alcaics, alexandrines, sapphics.*

See also BINOMIAL 1.

**C** *Medial capitals.* Worth noting is the newish, mostly commercial, habit of inserting a medial capital letter into the name of a product, a process, etc.: e.g. *CinemaScope, InterLink* (a device for a medical injection), *GeoSphere* (made from satellite photographs of the earth).

**capping.** As a second element in *rate-capping*, etc., a modern political term of great potency. *Rate-capping* (first recorded with its derivatives in 1983) is 'the imposition of upper limits on the amount of money which a local authority can spend and also levy through rates, intended as a disincentive to excessive spending on local services' (OED 2). With the replacement of rates by the community charge or poll tax in 1989 (Scotland) and 1990 (England and Wales), *rate-capping* has been replaced by *community-charge capping, charge-capping*, or simply by *capping.*

**caption.** Adopted in the 14c. from medieval Lat. *caption-em* 'taking, seizing', *caption* has had a continuous history since then in various legal senses. Towards the end of the 18c. it began to be used, chiefly in the US (corresponding to *heading* or *title* in the UK), to mean 'the heading of a chapter, section, or newspaper article'. From about the 1920s it has gradually come to be used in all English-speaking areas for the title below an illustration, and, more recently, for a subtitle in cinematography and television. (The word is not derived from L *caput* 'head'.)

**carafe.** Formerly in Scotland used as a normal term for the crystal jug or decanter from which water at the table was served. In Victorian England the word was commonly applied as a genteelism to a glass water-bottle, over which a tumbler was placed, for use in a bedroom. In the course of the 20c., *carafe* has come to be adopted as a normal English word for an open-necked container for serving wine at table.

**carat.** See CARET.

**caravanserai.** This word of Persian origin, for a kind of inn where companies of merchants or pilgrims travelling together in Middle Eastern countries or in

N. Africa put up, took a considerable time to settle down in its fixed present-day spelling. It was first noted by Hakluyt in 1599; the older spellings included *Carauan-sara, caravansery*, and *caravansary*. The standard pronunciation is /ˌkærə-ˈvænsəraɪ/.

**carburettor.** The standard spelling in BrE, as against *carburetor* (with one *t*) in AmE.

**carcass,** pl. *carcasses*, are the forms recommended, not *-ase(s)*.

**care** (verb). The modern colloquial phrase of resignation, (*I*, etc.) *couldn't care less*, '(I am, etc.) completely uninterested, utterly indifferent' (*OED*'s first example is one of 1946), has partially yielded, since the 1960s and principally in AmE, to the construction (*I*, etc.) *could care less* with the same meaning even though the negative is omitted. Thus in some quarters the alternative statements *If a bill doesn't get paid, he couldn't/could care less* are identical in meaning. No one has satisfactorily accounted for the synonymy of what would appear to be straightforwardly antonymous uses.

**careen** (verb). In origin a nautical word (first recorded in Hakluyt, 1600) meaning 'to turn (a ship) over on one side for cleaning, caulking, or repairing', or, intransitively, '(of a ship) to lean over, to tilt when sailing on wind', *careen* carries a residual notion in non-nautical contexts of leaning or tilting. In a separate modern development in AmE, since the 1920s, *careen* has rapidly become standard in the sense 'to rush headlong, to hurtle, esp. with an unsteady motion', i.e. the speed is more central to the meaning than any latent notion of leaning or tilting. This modern sense hardly occurs in BrE, the broad sense being satisfactorily covered by the verb *career*. Examples: *A lot of Russians careening along the road on liberated bicycles*—H. Roosenburg, 1957; *The van careened across the road, almost running into the ditch*—B. Moore, 1987; *A giant [of a man] careens down the corridor of a crummy hotel in East Los Angeles*—The Face, 1988; (figuratively) *With a rakishly tilted upstairs screen porch and staircases that careened like carnival rides*—Lee Smith, 1983.

**caret** is a mark (ʌ ⋏) indicating a proposed insertion in printing or writing; *carat* is a unit of weight for precious stones; *carat* (US *karat*) is a measure of purity of gold, pure gold being 24 carats.

**cargo.** Pl. *cargoes*. See -O(E)S 1.

**Caribbean.** In the US and in the Caribbean itself, the word is pronounced just as often with the stress on the second syllable, /kəˈrɪbɪən/, as on the third, /ˌkærɪˈbiːən/. In Britain the standard pronunciation has the main stress on the third syllable.

**caries** is a Latin sing. noun (*cariēs*) meaning decay. In its ordinary dental sense, it was formerly always trisyllabic, /ˈkeəriːz/, in English, and this was the only pronunciation entered in Daniel Jones/Gimson (up till the 14th edn., 1977). It is now normally disyllabic, i.e. /ˈkeəriːz/.

**carillon.** The word has been in a parlous state as to its pronunciation in the more than two centuries since it entered the language from French. Who knows but what Dickens said /kəˈrɪljən/ in an English way and Thackeray /karijɔ̃/ to imitate the French? Or whether one or both or neither of them placed the main stress on the first syllable, i.e. /ˈkærɪljɒn/? My strong impression is that /kəˈrɪljən/ is now dominant, but the latest editions of *SOED* and *COD* place the stress on the first syllable. Who can tell?

**caring.** Since the 1960s, and esp. as the monetarist policies of the Thatcher government began to bear on the quality of social services available to certain sections of the general public, i.e. the amount of public money made available for the needs of the sick, the elderly, single parents, etc., the word *caring* in the sense 'compassionate' has come to have strong political overtones. Socialists interpret caring as a political willingness to set less stringent limits on the funding of the social services; Conservatives, on the other hand, while defending their own record in such matters, insist that sensible limits must be set. Phrases like *the caring party, caring policies, the caring professions*, with their social and political overtones, will doubtless be attended by controversy for a long time to come.

**carnaptious** (quarrelsome). Part of the charm of the language is our relative unawareness of the boundaries within which particular words lie. A BBC political correspondent, John Cole, used *carnaptious* in *The Listener* in 1988 only to find that his readers did not know what was to him an everyday word. It seems to be restricted to Scotland and Northern Ireland (the latter being Mr Cole's place of upbringing). The earliest example in the *OED* is one of 1858 from the *Ulster Journal of Archaeology*. It is probably derived from *car-* (an intensive prefix corresponding to the American *ker-*, as in *ker-plunk*) + *knap* (a verb meaning 'to break or snap with the teeth') + connective *-t-* + *-ious*.

**carousel.** 1 Historically, a tournament (first recorded 1650); then, a merry-go-round (1673, now chiefly US, where it is usu. spelt, like the French original, *carrousel*); from about 1960, a moving conveyor system for delivering passengers' luggage at airports. It is pronounced /kærə'sel/, occas. /-zel/.

2 It is, of course, to be carefully distinguished from *carousal* (revelry in drinking), which is pronounced /kə'rauzəl/, with the main stress on the second syllable.

**carpet.** 1 The phr. *on the carpet* (i.e. of the council table) meaning 'under consideration or discussion' and corresponding to Fr. *sur le tapis*, was widely used from the early 18c. onward, but almost completely fell out of use when *carpet* lost its earliest primary sense 'thick fabric used to cover tables, beds, etc.; a table-cloth'. The phrase survives principally in its new 20c. sense, 'undergoing, or summoned to receive, a reprimand' where the underlying metaphor refers to the covering on the floor.

2 See RUG, CARPET.

**carrel.** A *carrel* or *carol* was the name given in medieval English monasteries to a small enclosure or 'study' in a cloister. The word dropped out of use with the dissolution of the monasteries in the 16c.; but it was revived in the early 20c. to denote a private cubicle provided in a library for use by a reader (e.g. *The study cubicles in the college library at Ampleforth are still called 'carrels'—Medium Aevum*, 1960).

**carte.** In fencing, a variant of *quart(e)*, the fourth of eight parrying positions. However spelt, it is pronounced /kɑːt/.

**cartel.** In its older senses, esp. 'a written agreement relating to the exchange or ransom of prisoners' (a use first recorded in 1692), it was pronounced /'kɑːtəl/, with the stress on the first syllable. About the end of the 19c., after Ger. *Kartell*, the stress was moved to the second syllable, i.e. /kɑː'tel/, in the new, orig. Ger., sense 'an agreement or association between two or more business houses for regulating output, fixing prices, etc.; also the businesses thus combined'.

**carven,** the original pa.pple of the verb *carve*. It dropped out of use in the 16c., was revived in poetry and in rhetorical prose in the 19c. (*a screen of carven ivory*—E. Barrett Browning, 1856), and now seems to have fallen into disuse again, except occas. in poetry: *The carven priest Gilded and small*—C. Aiken, 1929.

**case** (nouns).

**A** There are two distinct nouns:

1 First recorded in English in the 13c. (from OF *cas* and L. *cāsus*), *case*[1] was at first 'a thing that befalls or happens, an occurrence'. From this abstract base sprang numerous extended meanings, esp. (*a*) an instance or example of the occurrence or existence of something (*the most recent case of that kind of behaviour that he could recall*); (*b*) an infatuation (*Richard had a case on Joanna*); (*c*) numerous technical senses in law (e.g. *This concluded the case for the prosecution*); (*d*) an instance of disease (*seven cases of cholera*); (*e*) in grammar, one of the varied forms of a word expressing its relationship to some other word in the vicinity (*the case-ending of the accusative singular of Latin nouns ending in -iō is -iōnem*).

2 Quite separately, *case*[2] entered English in the 13c. (from Norman Fr. *casse* and L *capsa*), meaning a receptacle, a box, a bag, a covering, etc. The two words have stood side by side ever since, and perhaps the majority of speakers have always been unaware that the two are distinct words.

**B** In practice it is easy to distinguish them if one recalls their origins, and, particularly, if one keeps in mind that only *case*[1] has generated a range of idiomatic phrases, some of them prepositions, and others conjunctions, e.g. *in case (of)*, in the event (that), supposing (that); *in case that*, in the event that; *just in case*, in the event or contingency that; *in that case*, in that event, if that were true; *in any case*, at all events, whatever else is done; *to put the case for*, to present the arguments in favour of.

In AmE, *in case* can also mean 'if': *In case it rains I can't go* ( = If it should rain).

*Case*[2] almost always means a receptacle (*a case of claret, cases of weapons*), the covers in which a book is bound, or, (in traditional printing) the frame in which a compositor places his types.

**C** Naturally it is important to avoid using *case*[1] in any context where it might be taken to be *case*[2]. Thus, *In the case of champagne, it is better to opt for the traditional labels* should be replaced by *For champagne* ...

Fowler (1926) cited a pageful of examples in which removal of the phrase containing *case*, or rewording, would improve the run of the sentence: e.g. *Though this sort of thing proceeds from a genuine sentiment in the case of Burns* [omit *the case of*]; *His historical pictures were in many cases masterly* [*Many of his*]. Such aberrant uses perhaps abounded in the essays of boys at Sedburgh, and also in the various newspapers that the Fowler brothers read with such care. They are much harder to find now: perhaps they are edited out by vigilant copy editors.

**D** A range of natural uses of *case*[1] follows: (law) *he ... was happy ... to help investigate a case that has since been tried—New Yorker*, 1987; (in transferred use) *Neither of these figures illuminates the case against Trident—*David Steel, 1985; (medicine) *As the number of reported cases of AIDS mounts, the magnitude and severity of the problem becomes increasingly evident—Dædalus*, 1989; (phrases) *Take your umbrella in case it rains* ( = because of the possibility that it may)—*CGEL*, 1985; *In case you want me, I'll be in my office till lunchtime* ( = should)—ibid.; *Claudia, through clenched teeth, says in that case she can do without the Big Dipper—*P. Lively, 1987; *At first, I felt used ... But in any case it was due to my innocence and my ambition—New Yorker*, 1986.

**casein.** In my own experience, normally pronounced /ˈkeɪsiːn/, but the standard guides to pronunciation give precedence to the trisyllabic form /ˈkeɪsɪɪn/.

**cases.**

1 Remaining cases in English.
2 Case after the verb *to be*.
3 After *as* and *than*.
4 After *but*.
5 After *not*.
6 Case-switching.

**1** Before the Norman Conquest, English was characterized by its system of case-endings. Nouns, pronouns, and adjectives had a range of forms distinguishing the nominative singular from the accusative, genitive, and dative (sing. and pl.). The nouns in particular fell into distinct groups according to their grammatical gender. The simplest of these patterns is shown in the OE word for 'stone': *stān* nom. and acc., *stānes* gen., *stāne* dat., *stānas* nom. and acc. pl., *stāna* gen. pl., and *stānum* dat. pl. The main modern English prepositions existed, but for the most part had a reinforcing rather than a strictly semantic role. From the Conquest onward, the case-endings rapidly disappeared except as signs of the possessive (sing. and pl.) of nouns and of the plural of nouns. Adjectives gradually became invariable. Pronouns alone were left with forms that distinguish case:

| subjective | objective | possessive |
|---|---|---|
| I | me | my |
| he | him | his |
| she | her | her |
| we | us | our |
| they | them | their |
| who | whom | whose |
| whoever | whomever | |

The main casualty of this process is that, because nouns form such a dominant part of the language, and because they do not change endings in the old accusative and dative positions, English speakers have partially lost an instinctive power to recognize case distinctions.

**2** The verb *to be* can usefully be regarded as a part of speech linking elements that are in the same case. In nouns and some pronouns, the cases before and aft are normally indistinguishable: *Paris is the capital of France*; *what's yours is mine*. In constructions introduced by *It is* or *This* (or *That*) *is*, however, the objective forms of pronouns have infiltrated standard English, esp. in short declarative sentences: thus *It's me* rather than *It is I*; *This* (or *That*) *is him* rather than *This* (or *That*) *is he*. The choice of pronoun in such circumstances still needs to be judged with care: a great many writers tend to prefer the subjective forms, esp, when the pronoun is qualified by a following *who-* (or *that-*) clause. Examples: (subjective) *If I were he, I should keep an eye on that young man*—C. P. Snow, 1979; *This time it was I who took the initiative*—R. Cobb, 1985; *It is we who are inappropriate. The painting was here first*—P. Lively, 1987; *that might very well be he at this moment, causing the doorbell to chime*—K. Amis, 1988. (objective) *Too much of a bloody infidel, that's me*—T. Keneally, 1980; *Hugh stepped forward. 'It's me, don't be frightened'*—M. Wesley, 1983; *'So ... ' says Jasper. 'That's him, the old fraud.'*—P. Lively, 1987; *Can this be me? Driving a car?*—New Yorker, 1988.

**3** After *as* and *than*. There is considerable variation, but in broad terms when *as* or *than* are felt to be prepositions the objective case is used, and when they are felt to be conjunctions the objective case is used. The subjective case is more formal than the objective, but is still widely used. Examples: (*as* subjective) *I sensed that he was an apprehensive as I about our meeting*—J. Frame, 1985; *Numbed as she is, she's as alive as Amaral or you or I*—New Yorker, 1987; (*than* subjective) *I hope you had a more cheerful Christmas than we*—E. Waugh, 1955; *On the whole the men ... are more formal and authoritarian in tone than she*—M. Butler, 1987; *He was eight years older than I, and planned to be everything that I, too, hoped to become in life*—Ld Hailsham, 1990; *They were nothing like Violet ... if anything, even less human than she*—A. Billson, 1993. (*as* objective) *Jim would have run the farm as good as me*—M. Eldridge, 1984 (Aust.); *He seems to be as lonely as me, and to mind it more*—D. Lodge, 1991; (*than* objective) *I wanted you to be wiser than me, better than me*—M. Ramgobin, 1986 (SAfr.); *The 1,700 paintings in the Uffizi have been described ... by better qualified writers than me*—P. Hillmore, 1987; *we're sure they can judge a novel just as well* [as] *if not better than us*—Julian Barnes, 1987.

See AS 1, THAN 6.

**4** After *but*. The objective form of the pronoun is preferable, though in practice both types occur: (subjective) *But no one understands it, no one but I, who have sat in corners all my life watching him*—J. M. Coetzee, 1977; (objective) *No one is fool enough to work the straights but me*—J. Fuller, 1983.

For a fuller discussion of the problem, see BUT 3.

**5** After *not*. Except in the type 'Who made that mistake?' 'Not me!', i.e. in informal speech, the subjective forms of pronouns still (just) tend to be preferred: *it must be he who's made of india-rubber, not I*—A. Carter, 1984; *Who would be Scrooge enough to call such ideas humbug? Not I.*—New Yorker, 1986. But see NOT 9.

**6** Case-switching. Case-switching of pronouns in certain circumstances is a marked feature of modern English. The nation is divided in its attitude to some of the types. Among the unacceptable patterns are constructions of the type *This is strictly between you and I* (see BETWEEN 1) and *They asked Jim and I to do the job* (see I). Also unwelcome, except in the representation of the speech of poorly educated people, are the types *Me* (or *Myself*) *and Bill* (at the head of a sentence or clause).

A number of other migratory uses of objective forms are standard, though fairly informal. They possibly go unrecognized as visitors in their new role. Examples: (sentences led by *Me, I*) *Me, I don't trust cats*—G. Keillor, 1989; *Me, I'm thick-skinned, charming, vain and happy*—R. Elms, 1989; (*me* + pres. pple) *Me thinking I'd probably got some filthy fever in spite of the jabs*—Julian Barnes, 1989; (straightforwardly illogical, but seeming natural) *we sat down on either side of the radiogram, she with her tea, me with a pad and pencil*—J. Winterson, 1985; (with an exclamatory infinitive) *What! me fight a big chap like him?*; (*Me too* in response to another person's assertion) *'Let's talk about each other, that's all I am interested in at the moment.' 'Me too,' says Tom.*—P. Lively, 1987; (in reply to a question) *'What do you make of that, Tonio?' 'Me?' he said.*—B. Moore, 1987; (*Me neither*) *'Too bad I can't reach the curtain.' 'Me neither,' he said.*—New Yorker, 1987; (silly *me*) *'After Diana had told me what Irena was asking?' 'Of course. Silly me.'*—A. Lejeune, 1986. See also ME 2.

On this evidence, further migrations of the objective forms to the subjective area seem likely. See also WHO AND WHOM.

**cashew.** See CACHOU.

**casino.** Pl. *casinos*. See -O(E)S 6.

**casket.** In America, and in some other English-speaking countries outside Britain, undertakers tend to use *casket* rather than *coffin*. In newspaper accounts of funerals in these countries, the words *casket* and *coffin* are often used interchangeably in the same column. In Britain the normal word is *coffin*, and *casket* is normally reserved for 'a small, often ornamental, box or chest for jewels, etc.', and sometimes 'a small wooden box for cremated ashes'.

The exchangeability of the two words in US newspapers was shown, for example, in accounts of the funeral of Ferdinand Marcos, former President of the Philippines, in October 1989: e.g. (caption) *Imelda Marcos kisses the casket containing the body of her late husband, Ferdinand Marcos*; (text) *Mr. Marcos' coffin, draped with a Philippines flag, was borne by 10 pallbearers* (both from the 15 Oct. issue 1989 of the *Chicago Tribune*). Similarly, in an account in an April 1989 issue of the *Chicago Sun–Times* of the funeral of sailors killed in an explosion on the battleship *Iowa*: *Flag-draped coffins of the 47 sailors killed in the worst naval disaster in more than a decade ... arrived in the United States on Thursday*; *Upon finishing his remarks to the tearful gathering, he knelt before one casket as if in prayer*—ibid.

**cast** (verb). This late OE word of Scandinavian origin, which once looked likely to drive out the native word *throw*, tends now to be restricted to a range of familiar phrases, idioms, and proverbs: *cast an eye over, cast a shadow over, cast a spell, cast lots, cast a vote, cast ashore, cast aside, Cast not a clout till May be out, be cast down* (downhearted), *cast accounts*, etc. It would be absurd to speak about its demise— eighty-three distinguishable senses are listed in the *OED*—but it does seem to be losing some of its former power. In ordinary contexts of propelling with force, *throw* is the more natural word.

**caste.** From the 17c. to the 19c., the spelling of the word meaning 'one of the several hereditary classes into which society in India has from time immemorial been divided' (*OED*) wavered between *cast* and *caste*. The latter form has prevailed. It is derived from Sp. and Pg. *casta* 'race, lineage, breed'. *Caste* is to be carefully distinguished from the noun *cast* in its multifarious meanings: *reflections of a moral cast, the cast* ( = the actors) *of a play, a cast in dice*, and numerous senses in angling, hawking, sculpture, etc. *Cast* has the same etymology as *caste*.

**caster, castor.** A *caster* is 'one who casts'; in hot-metal printing, 'a machine for casting type'; and, often, the first element in *caster sugar* (see below). *Castor* is (a) a somewhat archaic name for a beaver; (b) the only spelling of the first element of *castor oil*; (c) (also spelt *caster*) a small swivelling wheel for furniture; (d) (also spelt *caster*) a small pot with a perforated top (hence *castor* or *caster sugar*). The name *Castor* in Greek mythology, one of the twin sons (Castor and Pollux) of Tyndareus and Leda, now represented in the name of the constellation Gemini, is unrelated.

**casualty.** In the fifth edition of *The Concise Oxford Dictionary* (1964), where senses are arranged in historical order, the sense 'accident, mishap, disaster' is correctly placed first. In the eighth edition (1990), where the current senses are arranged in order of comparative familiarity and importance, the meaning 'a person killed or injured in a war or accident' is placed first, and the earliest historical sense 'an accident, mishap, or disaster' is placed last. The change in the ordering of senses is partly a matter of lexical technique, but in this case it also reflects a change in the currency of the respective meanings. Figurative uses of the now dominant sense are common: *lower profits will become the first casualty of the new government's policy.*

**catachresis.** In grammar, 'improper use of words; application of a term to a thing which it does not properly denote'. In rhetoric, 'abuse or perversion of a trope or metaphor' (*OED*). The popular uses of *chronic* = habitual, inveterate (*a chronic liar*), *infer* = imply, suggest, and *refute* = deny, contradict (without argument), are examples of lexical catachresis. In the *OED*, catachrestic uses are preceded by the sign ¶. In poetry or highly formal prose, nonce-deviations from ordinary linguistic uses are examples of rhetorical catachresis: e.g. Dylan Thomas's phrase *Once below a time*, and W. H. Auden's coinage *metalogue* (modelled on *prologue* and *epilogue*), 'a speech delivered between the acts or scenes of a play', an artificial word not taken up by anyone else.

**catacomb.** The dominant pronunciation in standard BrE is now /ˈkætəkuːm/. The older one, /-kəʊm/, the only pronunciation given by Daniel Jones in 1917, is now seldom heard in Britain but is customary in AmE.

**catalectic.** In prosody, = lacking a syllable (or more than one) in the last foot, a device frequently found in classical verse. Thus Browning's dactylic pattern is broken in

Júst for a/ hándful of/ sílver he/ léft us ʌ ,
Júst for a/ ríband to/ stíck in his/ cóat ʌ .

Cf. ACATALECTIC.

**cataphoric** (adj.). First (19c.) used of the action of an electric current, it is now (since the 1970s) used by grammarians of a reference to a succeeding word or group of words. In the sentence *After his discovery of New Zealand, Captain Cook went on to discover several Pacific islands, his* refers forward to *Captain Cook*, i.e. is cataphoric. There are numerous more complicated types of grammatical cataphora. Cf. ANAPHORA 2.

**catch.** 1 (verb). *Catch you (later)*, as an illogical but popular colloquial phr. of farewell, now stands alongside *See you (later)* (which is first recorded in 1891): '*Yeah, catch you, mate,' said Stephen, and slid out the door*—F. Kidman, 1988; *Catch you later. Let's have a drink sometime* (*The Bill*, ITV, 19 Apr. 1990).

2 *catch-22*. Mainly used in the phrase *a catch-22 situation*, and meaning 'a dilemma or circumstance from which there is no escape because of mutually conflicting or dependent conditions' (*COD*), it is tending to be overused. It is the title of a novel by Joseph Heller (1961) featuring predicaments of this kind.

**catchphrase.** A phrase that catches on quickly and is repeatedly used with

direct or indirect allusion to its first occurrence. The word is first recorded in the mid-19c. The adoption of catchphrases from popular songs, films, slogans, advertisements, etc., has become a marked feature of the language in the 20c. Hundreds of catchphrases are gathered up in Eric Partridge's *Dictionary of Catch Phrases* (1977) and in Nigel Rees's *Dictionary of Popular Phrases* (1990). Examples: *Have a nice day* (origin disputed, but quickly established in the US in the early 1970s); *to laugh* (or *cry*) *all the way to the bank* (attributed to the American entertainer Liberace); *Nice one, Cyril!* (from a 1972 TV advertisement for Wonderloaf); *Not tonight, Josephine* ('probably arose through music-hall in Victorian times', N. Rees; not believed to have been said by Napoleon himself); *over-paid, over-sexed—and over here* (applied to American GIs in Britain during World War II).

**catchup.** This word (first recorded in 1690) for a spicy sauce or condiment is now entirely replaced by *ketchup* in Britain. Of the various spellings of the word in the US, *ketchup* and *catsup* are the most common. The word is believed to be of Chinese origin via Malay.

**category.** To begin with, restricted to its original philosophical meaning, 'one of a possibly exhaustive set of classes among which all things might be distributed' (*COD*); later (attributed to Kant) 'one of the a priori conceptions applied by the mind to sense-impressions' (*COD*). These philosophical senses made their way imprecisely into general use from the 17c. onward, and settled down with the broad meaning, 'a class, or division, in any general scheme of classification'. This more general sense, though objected to by Fowler (1926) and others, is now uncancellably established. Examples: *She placed them in two categories: the honest imbeciles and the intelligent delinquents*—O. Manning, 1960; *He had had the whole of creation divided into two great categories, the things he was for and the things he was against*—M. Frayn, 1965; *'I hope you aren't a clergyman.' She said it with real vehemence, as if it was the one category of person she was not prepared to have in the house*—A. N. Wilson, 1981.

**cater** (verb). The more usual construction is with *for*: to provide a supply of food for (*some takeaway shops cater for people who order food by telephone*); (in transferred use) to provide (entertainment, education, etc.) for, to make provision for the needs of a group, a minority, etc. (*cater for all age groups*; *cater for the needs of people made homeless by the earthquake*; *perhaps they only cater for an élite spiritual clientele*).

The verb is occasionally construed with *to* (perhaps after *pander to*): *catering to the national taste and vanity*—Thackeray, 1840; *Nine years afterwards we find him ... catering to the low tastes of James I*—C. Kingsley, 1860; *He feels cheated because society does not cater to his irrational wishes*—B. Bettelheim, 1960; *Unlike other corporate hospitals, ours is just a diagnostic centre which caters to everybody*—*Business India*, 1986.

In AmE, *cater to* is said to be more usual than *cater for*.

**cater-cornered.** This adj. and adv. meaning 'diagonal, diagonally', spelt in various ways (*cater-corner, catty-corner(ed), catacorner(ed), kitty-corner(ed)*, etc.), is in daily use in AmE, but is virtually unknown in BrE.

**Catholic.** When used with a small initial, *catholic* (like its Gk original καθολικός) means 'universal' (*science is truly catholic*) or 'of universal human interest; having sympathies with, or embracing, all'. Swift, for example, declared of one of his works, *All my Writings ... for universal Nature, and Mankind in general. And of such Catholick Use I esteem this present Disquisition*. In ecclesiastical use, care must be taken to make the specific meaning clear. In everyday language, it is commonplace to hear *Catholics* (i.e. Roman Catholics) distinguished from *Protestants* (or *Muslims*, etc.). But the history of the term is strewn with difficulties and complications, and these must be left for theologians to disentangle.

**catsup.** See CATCHUP.

**catty-cornered.** See CATER-CORNERED.

**Caucasian.** The normal word in AmE (but rarely elsewhere) for a white person (as distinct from an African-American, a Japanese, etc.). It is also, as adj. and noun, the normal word used in all English-speaking countries for the people, language, etc., of the Caucasus.

**cauldron.** The normal spelling in BrE, whereas *caldron* is the more usual form in AmE.

**caulk** (verb). This word, meaning 'to stop up (the seams of a boat, etc.) with oakum, etc.', is normally spelt thus in BrE, and as *calk* in AmE.

**'cause.** Spelt thus (or as *cos, 'cos*, etc.), it is still sometimes used in informal contexts as a conjunction in place of *because*: e.g. *'Right now,' she said, 'people are stashing their money 'cause they know the day is coming'*—*Chicago Tribune*, 1988.

See 'COS.

**causerie.** Adopted in the early 19c. (first recorded 1827) from French in the sense 'a chat', it later came to be used to mean 'an informal essay, article, or talk, esp. one on a literary subject and forming one of a series'. This later sense was taken from Sainte-Beuve's *Les Causeries du Lundi* (1851–62), which was published as a series of articles in two French newspapers. In the 20c., *causerie* has also been used in English for 'a discussion, an informal seminar'.

**causeway.** This ordinary word for 'a raised road or track across low or wet ground or a stretch of water' (*COD*) has effectively driven the once common variant form *causey* into archaic or dialectal use. *Causeway* is in origin a reduced form of *caucé-way*, from Old Norman Fr. *coucié*, ult. from L *calx, calcis* 'lime'. *Causey* was very widely current in standard English from the 14c. to the mid-19c.; it is still found in a few place-names, e.g. *Causey Park*, Northumberland, *Causey Pike*, Cumbria. A Scottish example: *A skin of frost on the causeys*—W. McIlvanney, 1975.

**cavalcade.** A standard example of a word that has partially discarded its original meaning. Ultimately derived from late L *caballicāre* 'to ride on horseback', from *caballus* 'horse', it was brought into English via French in the senses 'a march or raid on horseback' (1591, last recorded 1647) and 'a procession on horseback' (1644). Almost at once, in the 17c., *any* procession came to be called a *cavalcade*: the earliest examples are of theatrical devils and of cows. Noel Coward used the word as the title of a play in 1931: he might equally well have called it *Pageant*.

Its final element, *-cade*, has yielded a new irregular formation, *motorcade* 'a procession of motor vehicles', first recorded in America in 1913, and, since the 1970s, increasingly found in British newspapers and other publications.

**Cave, caveat.** The British public-school slang word *Cave!* 'Beware!' is the L imperative of *cavēre* 'to take care, beware'. It is pronounced /ˈkeɪviː/, despite the short *a* of the original. By contrast, *caveat*, both in legal use and in the general sense 'a warning, an admonition', which until the mid-20c. was regularly pronounced /ˈkeɪviːæt/, is now just as regularly pronounced with a short *a* in the first syllable.

**caviare.** This delicacy is usu. so spelt in BrE, with the main stress falling on the first syllable, but *caviar* is gaining ground. The spelling *caviar* is the more common of the two in AmE.

**cayman.** The OUP house-style spelling of this word ( = a S. American alligator-like reptile), not *caiman*.

**-c-, -ck-.** Until the 18c., words like *critic*, *epic*, and *music* were often written with final *-ck*, *-cke*, or *-que*. The present-day *-ic* forms became established in the 19c. In inflected forms that preserve the pronunciation /-k/ before the native suffixes *-ed*, *-er*, *-ing*, and *-y*, a *-k-* is added: thus *mimicked, picnicker, frolicking*, and *panicky*. Before the classical suffixes *-ian*, *-ism*, *-ist*, *-ity*, and *-ize*, the soft *c* is left by itself: thus *musician, criticism, publicist, electricity*, and *criticize*.

**cease.** This 14c. loanword from French is slowly yielding to *stop* except in a few set phrases, e.g. *without cease*, unending; *cease-fire*, an order to stop firing; a period of truce. The natural idiom is *to stop breathing* not *to cease breathing*, and *to stop work*, not *to cease work*. But the battle between the two words is far from over and they look like coexisting for a long time to come.

**-ce, -cy.** The hundreds of (mostly abstract) nouns ending in *-ce* or *-cy* fall into three main classes: (i) Those which now always end in *-ce*: *annoyance, beneficence,*

*credence, elegance, magnificence, obedience, silence,* etc. The majority of these have competed with matching forms in *-ancy* or *-ency* at some time in the past: e.g. *credency* (recorded once in 1648), *silency* (obsolete in 17c.). (ii) Those which now always end in *-cy*: *agency, diplomacy, infancy, militancy, obstinacy,* etc. Some of these too have had matching forms in *-ce* at some point in the (often quite recent) past: e.g. *obstinace*, marked 'rare' in the *OED* but probably obsolete. (iii) Those which exist in both forms, usually with clearly distinguishable meanings: *dependence*, the state of being dependent; *dependency*, (esp.) a country or province controlled by another; *emergence*, a coming to light; *emergency*, a sudden state of danger; *excellence*, the state of excelling; *Excellency* (in *His* (etc.) *Excellency*, used as a term of address, principally to ambassadors.

The facts for each word are set out in the *OED*: it is unwise to theorize about the distribution of the rival forms without reference to the relevant *OED* entries.

In the lists that follow, those which once had a competing form are marked with a dagger; and those which continue to have a competing form (often with slightly different meanings) are marked with an asterisk:

| -ance | -ence |
|---|---|
| admittance | †coincidence |
| avoidance | *innocence |
| *brilliance | †intelligence |
| forbearance | *permanence |
| *irrelevance | *persistence |
| | *prurience |

| -ancy | -ency |
|---|---|
| †buoyancy | †cogency |
| †constancy | †decency |
| †vacancy | †frequency |

**cedarn.** See -EN ADJECTIVES.

**ceiling.** In the mid-1930s, governments and other administrative bodies began to use the word *ceiling* figuratively to mean 'an upper limit (to quantity, prices, expenditure, etc.)' (opposed to *floor*). It is self-evident that, if catachresis is to be avoided, the true nature of the metaphor should be kept in mind. A ceiling figure can, for example, be stated, reached, or warned about; a level of unemployment or inflation can be said to have reached or fallen below its ceiling. Verbs that are appropriate to levels, but not to ceilings (e.g. *extend, lower, raise, waive*), should not be used in conjunction with the figurative sense of *ceiling*. Examples of correct use: *Floors and ceilings can be adjusted to reflect the amount of downside protection needed*—*Energy in the News*, 1988; *Deposit interest rates were liberalised* ... *allowing banks to set their own levels after years of being locked into a fixed spread with floors and ceilings set by the government*—*Banker*, 1989.

**celebrant.** In BrE a *celebrant* is a person who performs a rite, esp. the Eucharist, a use first recorded in 1839. In the ordinary sense 'one who celebrates', i.e. at a party, an anniversary, etc., the word normally used is *celebrator*. In AmE in the 1930s, *celebrant* also came to be applied to a person celebrating anything (e.g. a birthday, a new appointment, a wedding anniversary), and the word is used alongside *celebrator* in this general sense (also occas. in BrE). Examples: *The party had gone splendidly until just after midnight when the celebrant himself had been involved in a pathetic little bout of fisticuffs*—C. Dexter, 1989; [On 4 July] *The heritage and history are honored, claimed, no matter whether the celebrant is a newly arrived Asian immigrant,* [etc.]—*Newsweek*, 1991; *With temperatures in the upper 30s on Friday, area celebrants were greeting the new year warmly*—*Chicago Tribune*, 1994; (caption) *A St. Patrick's Day celebrant enjoys a Hydro-Bike Explorer ... on the Chicago River*—*Chicago Sun-Times*, 1994.

**cello,** an abbreviation of *violoncello*, is now by custom normally written as shown, without a preceding apostrophe. Its plural is *cellos*.

**Celsius,** the name now used in weather forecasts, designates a scale of temperature on which water freezes at 0° and boils at 100° under standard conditions. Named after a Swedish astronomer, Anders *Celsius* (1701–44), it contrasts with the Fahrenheit scale in which the freezing-point of water is 32° and the boiling-point 212°.

**Celt, Celtic.** These two words are no longer spelt with initial *K*-. Except for the football club *Celtic* (in Glasgow), which is pronounced /ˈseltɪk/, both *Celt* and *Celtic* are pronounced with initial /k/ in standard English.

**censer, censor.** A *censer* is a vessel in which incense is burnt. A *censor* is an official authorized to examine printed matter, films, news, etc., before public release, and to suppress any parts on the grounds of obscenity, a threat to security, etc. (*COD*).

**centenary** /senˈtiːnərɪ/, less commonly /-ˈtenərɪ/. **1** As adj. and noun, the more usual term in Britain, and *centennial* /senˈtenɪəl/ abroad. In all areas it means '(marking) the hundredth anniversary', '(celebration) of the hundredth anniversary'.

**2** The recommended forms for longer intervals of the *centenary* (or *centennial*) kind are as follows: 150: sesquicentenary; 200: bicentenary; 300: tercentenary (or tri-); 400: quatercentenary; 500: quincentenary; 600: sexcentenary; 700: septingentenary; 800: octocentenary.

**centigrade.** Formerly the normal term for a scale of temperature in which the freezing-point of water is 32° and the boiling-point 212°, it has now been replaced in weather forecasts by CELSIUS.

**centi-, hecto-.** In the metric system *centi-* denotes division, and *hecto-* multiplication, by 100. Cf. DECA-, DECI-; KILO-, MILLI-.

**cento,** a composition made up of quotations from other authors, is derived from L *cento* 'patchwork quilt'. It is pronounced /ˈsentəʊ/ and its plural is *centos*. See -O(E)S 6.

**centre** (AmE *center*). The use of the verb with (a)*round* to mean 'to have (something) as one's or its centre or focus; to move or revolve round (something) as a centre; to be mainly concerned with' (*OED*) was first recorded in the 1860s and has had a continuous history since then. It probably developed under the influence of the prepositional patterns associated with the verbs *gather* and *move*. These constructions have been attacked (as being illogical) for much of the 20c. but to no avail, to judge from the following examples: *The little cluster of huts where he and his gang lived centred round the tattered dwelling of a sea-priest*—Kipling, 1893; *The foremost problems in European politics ... will centre round the revision of the Treaty of Versailles*—A. L. Rowse, 1931;

*That strange figure around whom this account properly centres*—W. Sansom, 1950; *The speaker's imagery and ideas center around the empirical world*—G. Smitherman, 1980; *There is the added enticement of a plot centred around a real historical event*—Listener, 1983. For those seeking a safety net, other constructions are available, esp. *to centre on* (or *in*), *to revolve round*.

**centrifugal, centripetal.** The first of these was usu. pronounced with the main stress on the second syllable—thus /sen'trɪfjʊgəl/—until the first half of the 20c., but the stress is now frequently placed on the third syllable, i.e. /ˌsentrɪ-ˈfjuːgəl/. For *centripetal*, on the other hand, /sen'trɪpɪtəl/ is still more usual than /ˌsentrɪ'piːtəl/. It seems likely that the stress pattern will in due course become the same for both words.

**century.** 1 Each century as ordinally named (*the 5th, the 16th*, etc., *century*) contains only one year (500, 1600) beginning with the number that names it, and ninety-nine (401–499, 1501–1599) beginning with a number lower by one. Accordingly 763, 1111, 1300, 1990, belong to the 8th, 12th, 13th, and 20th centuries. For dates before the birth of Christ, 55 BC was in the first century BC, and 500–401 BC was in the fifth century BC. For a different method of reckoning in Italy, see TRECENTO.

2 Despite the above reckoning, in modern usage, 1 January 1800 is often counted as being the first day of the 19c., 1 January 1900 as the first day of the 20c., and so on.

**cephalo-.** In the range of combining forms containing the element *cephalo-* (from Gk κεφαλή 'head'), there is much uncertainty about the pronunciation of the initial letter. When BSE (bovine spongiform encephalopathy) broke out, the broadcasting media invariably pronounced the third word as /en'kef-/. Many scientists also pronounce words like *cephalosporin* with initial /k-/, but just as frequently one hears an Anglicized /s-/. The variation is harmless, but my personal preference is to settle for the Anglicized sound.

**cerement,** a literary word (usu. in pl.) for 'grave-clothes', is pronounced as two syllables, /'sɪəmənt/.

**ceremonial, ceremonious.** *Ceremonial* means 'with or concerning ritual or ceremony' (*ceremonial occasions, for ceremonial reasons, ceremonial robes, a ceremonial bow. Islamic ceremonial slaughtering of sheep*). It is much the more common of the two words. Examples: *We've recently ... been carrying out a ceremonial parade on the day of a royal visit*—T. Parker, 1985; *Prince George ... had been paying a ceremonial visit to Peru*—A. Pryce Jones, 1987.

*Ceremonious* means 'having or showing a fondness for ritualistic observance or formality, punctilious' (*why be so ceremonious?*; *ceremonious politeness, ceremonious diction*; *a ceremonious former ambassador*). Example (of the corresponding adv.): *He rose slowly and motioned Chance to join him at center stage, where he embraced him ceremoniously*—J. Kosinki, 1983.

In these examples the form not used could hardly be substituted. But with some words *ceremonial* and *ceremonious* are both possible, though not with the same meaning. A visitor may make *a ceremonious* [i.e. excessively fussy] *entry* into a room, but an army of occupation *a ceremonial* [i.e. one marked by normal ceremony] *entry* into the main square of a conquered city.

**certes** /sɜːtɪz/, = 'of a truth, certainly'. This old word (first recorded in the medieval poem *The Owl and the Nightingale*) is encountered regularly in poetry and drama down to the 19c., but no later.

**certitude, certainty.** *Certitude* is a subjective feeling of absolute conviction. Ideally, *certainty* should be restricted to situations, results, information, etc., that can be shown objectively to be true. In that both words imply an absence of doubt about the truth of something, however, they are not always distinguishable in general use. In practice, *certainty* is much the more common of the two. Examples (in non-philosophical contexts): (*certainty*) *He was filled with certainty, a deep, sure, clean conviction that engulfed him like a flood*—R. P. Warren, 1939; *He never had the absolute certainty that one day he'd get the boat*—R. Ingalis, 1987; (*certitude*) *He* [sc. Hilaire Belloc] *appears in print as a man of many certitudes*—A. N. Wilson, 1984; *What can be asserted with well-nigh certitude is that the Barbara Bray reached Atherstone Locks*—C.

Dexter, 1989; *We craved certitude and order, and Oxford gave us both*—V. Mehta, 1993.

**cerulean** is one of the few words in BrE (as distinct from AmE) to have entirely lost its original *-ae-* (cf. L *caeruleus* 'sky-blue', from *caelum* 'sky'). Cf. Æ, Œ.

**cervical.** With the advent of nationwide cervical screening and cervical smears, the word *cervical* has emerged from laboratories (where it was normally pronounced with the main stress on the second syllable, /sɜːˈvaɪkəl/), into wide public use. A counting of heads would probably show that the dominant pronunciation now is /ˈsɜːvɪkəl/.

**cesarean,** or **cesarian,** the normal AmE spellings of CAESAREAN.

**Cesarevitch.** 1 In pre-revolutionary Russia, the eldest son of the Tsar (or Czar). Sometimes in historical work spelt with *-witch*, but now normally *tsarevich* or *czarevich*, pronounced /ˈzɑːrɪvɪtʃ/.

2 The name of the annual horse-race (instituted in 1839) at Newmarket is spelt *Cesarewitch*, pronounced /sɪˈzærɪwɪtʃ/.

**cess.** See TAX.

***ceteris paribus.*** This Latin phrase meaning 'other things being equal', is normally printed in italics and pronounced, according to taste, /ˌsetərɪs ˈpærɪbəs/ or /ˌketərɪs ˈpærɪbəs/. For the first word, initial /siːt-/ or /ˈkeɪt-/ also occur. It is a stock example of the differing traditions of pronouncing Latin words in English. My own preference is for /ˌketərɪs ˈpærɪbəs/.

**ch.** Words of French origin beginning with *ch* fall into two main groups: (*a*) those adopted in the medieval period or Renaissance, and (*b*) later borrowings. The earlier group are almost all now pronounced with initial /tʃ-/, e.g. *chafe* (14c.), *chain* (14c.), *chamber* (14c.), *chapel* (13c.), *charity* (12c.), *cherish* (14c.), *chief* (14c.), *choice* (13c.); and the later loanwords retain the /ʃ-/ of the French original, e.g. *champagne* (17c.), *Chardonnay* (20c.), *château* (18c.), *chef* (19c.), *chic* (19c.), *chiffon* (18c.), *crêpe de chine* (19c.).

**chagrin.** The dominant standard pronunciation of the noun in BrE is /ˈʃægrɪn/ and in AmE /ʃəˈgrɪn/. The adj. *chagrined*

is thus spelt, and pronounced like the noun with the addition of final /-d/.

**chain reaction.** A scientific term (first recorded in 1926) meaning: 1 In physics, a self-sustaining nuclear reaction, esp. one in which a neutron from a fission reaction initiates a series of these reactions (*COD*).

2 In chemistry, a self-sustaining molecular reaction in which intermediate products initiate further reactions (*COD*). Sense 1 has been carried over into general use since 1945 to denote 'a series of events, each caused by the previous one'. Examples: *We set in motion a chain-reaction of events*—W. Soyinka, 1972; *Challenged by the guards, they were then provoked into a chain-reaction of violence and murder*—J. G. Ballard, 1988.

**chair, chairperson.** One of the more startling results of feminist endeavour in the 20c. is the emergence of the gender-neutral terms *chair* and *chairperson* to replace the traditional words *chairman* (used of either gender) and the almost equally old term *chairwoman*. The chain of events is evident from the date of first record of the words concerned: *chairman* 1654, *chairwoman* 1699 (but noted by the *OED* as 'hardly a recognized name until the 19c.'), *chairperson* (mainly applied to women) 1971, *chair* (ditto) 1976. In Britain in particular the resistance to the use of *chairperson* and *chair* is very marked, but many groups and organizations, not all of them militantly feminist, are now using the modern terms. The terminological skirmishing bids fair to continue for a long time yet.

Something of the nature of the argument can be perceived in the following examples, two American and one British: (caption) *(from left) chairmen Barbara Allen of Highland Park, Lois Morrison of Deerfield and Anne Livingston of Highland Park*—*Evanston (Illinois) Rev.*, 1988; *Cochairs Carole Stone (from left) and Nancy Hillman with Princess Yasmin Aga Khan*—*Chicago Tribune*, 1988; *I was recently challenged for using 'chairman' to describe my position. My accuser went on to assert that I was being insensitive to the work of the Equal Opportunities Commission by not using 'chairwoman', 'chairperson' or 'chair'*—Ann Scully, *Times*, 1988.

**chaise longue** /ʃeɪz ˈlɒŋg/, a sofa with a backrest at only one end (*COD*). Pl.

*chaises longues* (pronounced the same as the singular). In AmE the word sometimes turns up, esp. in trade advertisements, as *chaise lounge*.

**challenged.** Among the more notorious formations arising out of the political correctness movement of the 1980s and 1990s are expressions containing *challenged* as second element: e.g. *cerebrally challenged* ( = stupid), *mentally challenged* ( = mentally retarded), *orally challenged* ( = having a speech impediment), *physically challenged* ( = disabled), and *vertically challenged* ( = shorter than average).

**Cham** /kæm/, an obsolete form of *khan*, a term formerly applied to the rulers of the Tartars and Mongols, and to the emperor of China, is used in English mainly with reference to Smollett's phrase *that great Cham of literature, Samuel Johnson.*

**chamois.** The name of the agile goat antelope is pronounced /ˈʃæmwɑː/. Its plural is spelt the same but pronounced /-wɑːz/. In the phrase *chamois leather*, the word is normally pronounced /ˈʃæmɪ/.

**chamomile.** See CAMOMILE.

**champagne, champaign.** These two words ( = sparkling wine, open country, respectively) are usually differentiated in pronunciation in BrE, the first being /ʃæmˈpeɪn/ and the second /ˈtʃæmpeɪn/; but in AmE are both pronounced /ʃæmˈpeɪn/.

**chancellery, chancery.** In British diplomatic use, the official term for the offices holding the general political section of an embassy or consulate is a *chancery*, but journalists frequently refer, for example, to the *chancelleries* of the great powers, thereby using the customary AmE term.

**chant(e)y.** See SHANTY 2.

**chap** ( = man, boy). The word is a shortening, first recorded in the late 16c., of *chapman*. It was not until the 19c. that its primary sense, 'a buyer, purchaser, customer', retreated into dialectal use, and the familiar present-day colloquial sense emerged.

**chap, chop** ( = lower jaw or half of the cheek). In *lick one's chops* and in the *chops*

*of the Channel* (the entrance into the English Channel from the Atlantic), the original form *chops* (itself of unknown origin) is normally used. (It is to be distinguished from the unrelated word *chop*, a slice of boned meat, in *lamb chop, pork chop*, etc.) The variant *chap* is used to mean either half of the bill of a bird such as a pelican or a stork, or the lower half of the cheek of the pig or other animal as an article of food. The somewhat fading adj. *chap-fallen* (also *chop-fallen*), 'dejected, crest-fallen', widely used from the 16c. onward, is now a relic of the time when *chap* was commonly used in the sense 'the lower jaw or half of the cheek'.

**chaperon** is recommended, not *chaperone*. Pronounce /ˈʃæpərəʊn/.

**char.** 1 Short for *charwoman* or *charlady* (a woman employed as a cleaner in houses or offices), whose occupation is *charring* (so spelt). The element *char* has nothing to do with charcoal or burning; it is a descendant of OE *cierr* 'a turn', ME *char(e)* 'an occasional turn of work'. Until the 19c. it was often spelt *chare*.

2 Now seldom heard: those who work in offices are normally called *cleaners*, and those who work in private houses, *dailies* or *daily helps*.

**charabanc** /ˈʃærəˌbæŋ/ is now the only extant spelling of this 19c. loanword from French meaning 'an early form of (motor) coach'. The French word *char-à-banc* literally means 'benched carriage'. As the *Economist* remarked in 1953, *But now the charabancs have all turned into coaches.*

**character.** Fowler (1926) argued the case for the disuse of *character* in two sets of circumstances: (a) as a substitute for an abstract-noun termination, *-ness, -ty*, etc.: e.g. *on account of its light character, X's whisky is a whisky that will agree with you* (lightness); (b) in the construction *of a* [adjective] *character*: e.g. *There is no unemployment of a chronic character in Germany* (no chronic unemployment). The economies he recommended are desirable in formal written work.

**charge** (noun). 'The phr. *in charge (of)* is used both actively and passively; *e.g.* to leave children *in charge of* a nurse, or a nurse *in charge of* the children. The latter is the more recent use.' (*OED*) The first now seems to be less commonly used than it was at the beginning of the 20c. It has been widely replaced by *in the charge of*, i.e. with *the* inserted. A third type, not mentioned by the *OED*, is shown in the phr. *in a nurse's charge*, i.e. employs a possessive. The various types have been extensively discussed, notably by E. Jørgensen (*English Studies*, 1986). He confirmed that all the possible constructions are still in use.

Examples: (1) ( = under the supervision of) *She didn't think it unreasonable to put Sebastian in Rex's charge on the journey*—E. Waugh, 1945; *They found Edmund in charge of Mrs Beaver a little way back from the fighting line. He was covered with blood*—C. S. Lewis, 1950; *The young prince was doing lessons at Ludlow in charge of the Queen's brother, Lord Rivers*—J. Tey, 1951; *The defendant is then put in charge of the jury, which means that the clerk tells the jury that the determination of the guilt or innocence of the defendant is a decision for them*—R. C. A. White, 1985; (2) ( = having command of) *Whenever I went out of town I always left him in charge of ... important business ... knowing that it was in safe hands*—M. L. King, 1958; *Davies: But what did he do? He's supposed to be in charge of it here.*—H. Pinter, 1960.

**charisma.** From Gk Χάρισμα 'gift of grace', the word made its way from theology to sociology in 1922 when the German sociologist Max Weber (writing in German) used it to mean 'a gift or power of leadership or authority'. It was quickly adopted by other sociologists. Nowadays any person who seems to the speaker to be 'set apart', i.e. to have some special quality or qualities, is said to have charisma. Recent political leaders to whom the word has often been applied include J. F. Kennedy, Mikhail Gorbachev, and Nelson Mandela, but charisma is often used locally of much less eminent people.

**charivari,** a serenade of 'rough music', with kettles, pans, tea-trays, and the like, used in France, in mockery and derision of incongruous or unpopular marriages, and of unpopular persons generally (*OED*), is pronounced /ˈʃɑːrɪˈvɑːrɪ/. It came into English in the mid-18c. but is seldom encountered now.

**charlatan.** Pronounced with initial /ʃ-/ like *shark*, not /tʃ-/ like *cheap*.

**chastise.** A Latinate-looking word (first recorded in English in the 14c.) for which no demonstrable base has been found in classical Latin, medieval Latin, or medieval French. It is correctly so spelt, not with -*ize*. And, as the *OED* remarks, 'The word is too early to be a simple English formation from *chaste* adj.'

**château.** Properly always written with a circumflex accent (but many newspapers lack the sign in their fonts). The pl. is *châteaux*, with the final letter pronounced /-z/. See -x.

**chauvinism.** The original sense, 'exaggerated or aggressive patriotism' (from the surname of Nicolas Chauvin of Rochefort, a Napoleonic veteran depicted in the French vaudeville *La Cocarde tricolore* (1831), continues in use. But in English (though not in French) it is beginning to yield to a range of extended uses signifying excessive loyalty to or belief in the superiority of one's own kind of cause, and prejudice against others, including *economic chauvinism*, *white chauvinism*, *female chauvinism*, and esp. *male chauvinism* (first recorded in 1970). The last of these is frequently used in the same sense without the modifier *male* (so also *chauvinist*).

**cheap, cheaply** (adverbs). The first can only be used to mean 'at a low price' and regularly follows the word it qualifies, whereas *cheaply* has that sense and can also be used to mean 'in low esteem': (cheap) *you can get them cheap at Woolworths*; (cheaply) *it's advisable to acquire one's first house as cheaply as possible*; *a tendency to treat religion cheaply*. The phrases *feel cheap* and *hold cheap* lose their idiomatic flavour if *cheaply* is substituted for *cheap*.

**check** (draft on bank), **checkered** (e.g. *a checkered career*). The AmE spellings of CHEQUE, CHEQUERED.

**checkers.** The American name for the game that is called *draughts* in Britain.

**checkmate. 1** A word of Persian origin (lit. 'the king is dead') via OF *eschec mat*.

**2** *Mate* is the form normally used in chess, and *checkmate* in transferred and figurative contexts.

**cheerful, cheery.** For the ordinary notions 'full of cheer; cheering, gladdening', *cheerful* is the customary word. It can be applied to a person's appearance, disposition, etc., and also to concepts or things of pleasant aspect (*a cheerful conversation, group, time*, etc.). It can also mean 'not reluctant, not opposed' (*his cheerful acceptance of the inevitable*). *Cheery*, which Johnson called 'a ludicrous word', is more colloquial: it suggests that the person, mood, voice, etc., to whom or to which it is applied is in good spirits, lively.

**cheers.** The word has long been used in all English-speaking areas as a salutation before drinking. In a way that must be baffling to foreigners, it is also now popularly used in BrE by a wide section of the population as an equivalent to 'thank you', as an expression of good wishes before parting, and even as an apology (e.g. for accidentally brushing against someone).

**cheque** is the standard spelling in BrE for the banking sense of the word. Etymologically it is a variant of *check* meaning 'a device for checking the amount of an item'.

**chequered** is the standard spelling in BrE for (*a*) the literal sense 'bearing a pattern of squares often alternately coloured'; (*b*) in the phr. *a chequered career*, i.e. one with varied fortunes.

**cherub** has pl. *cherubim* when used to mean an angelic being; in transferred use, e.g. applied to a well-behaved small child, the natural pl. is *cherubs*. The adj. *cherubic* is pronounced /tʃɪˈruːbɪk/.

**chevy.** See CHIVY.

**chiaroscuro.** Pronounce /kɪˌɑːrəˈskʊərəʊ/.

**chiasmus** /kaɪˈæzməz/, inversion in the second of two parallel phrases of the order followed in the first. If the two phrases are written one below the other, and lines are drawn between the corresponding terms, those lines make the Greek letter chi, a diagonal cross:

**chic.** See CH.

**chicane** /ʃɪˈkeɪn/. Nowadays one is most likely to encounter the word in the game of bridge ('the condition of holding no trumps') or in motor racing ('an artificial barrier'). The word is now rarely used as a shorter form of *chicanery*.

**Chicano** /tʃɪˈkɑːnəʊ/, less commonly *chicano*, a name (first recorded in 1947) applied to an American of Mexican descent. Most Chicanos live in the states of California, Arizona, New Mexico, Colorado, and Texas. Pl. *Chicanos*; fem. (sing.) *Chicana*.

**chick(en).** *Chicken* is the original (OE) and still the ordinary word for a young bird of a domestic fowl or for a domestic fowl prepared as food (*roast chicken, chicken thighs*). *Chick* serves as a diminutive, being used chiefly of a young bird, esp. one newly hatched, or, colloquially, of a young woman or a child.

**chide.** A verb that is slowly passing out of use in favour of *scold*, and, as it recedes, continues to show much variation in the pa.t. and pa.pple. It started out in OE as a weak verb with unchanged stem vowel in the past tenses. But in the 15c. and 16c. the paradigm *chide/chode/chidden* became common, and later *chide/chid/chided*. George Meredith, for example, in *Evan Harrington* (1861), used *chid* pa.t.: *her aunt … chid Mrs. Fiske, and wished her … at the bottom of the sea*. At present in standard English the dominant pattern is probably *chide/chided/chided*, but the paradigm is unstable.

**chiefest.** A minor curiosity is that *chief*, used as a quasi-adj. from the 14c. onward (*chief town, chief difficulty*, etc.), developed a superlative form *chiefest* at the beginning of the 15c. as it weakened in force from the absolute meaning of 'head' or 'supreme' to that of 'leading'. *Chiefest* is found in Shakespeare (*Within their chiefest Temple Ile erect A Tombe*, 1591), the 1611 AV (*And whosoeuer of you will bee the chiefest, shalbe seruant of all*— Mark 10: 44), and in many other classic sources until the mid-19c. It now seems to have passed out of use.

**chiffon.** See CH.

**childish, childlike.** The first of these can be used neutrally to mean simply 'like, or proper to, a child', but is more frequently used censoriously of adults or their behaviour in the sense 'silly, immature'. By contrast, *childlike* is more or less restricted to the favourable sense 'having the good qualities of a child such as innocence, frankness, etc.' Examples: (childish) *The childish titles of aristocracy*—Coleridge, 1809–10; *I feel ... my father's hand ... Stroke out my childish curls*—E. Barrett Browning, 1856; *John observing his daughter, saw her now as more grown-up, less childish*—I. Murdoch, 1976; *She said suddenly with touching, childish fear, '... I don't want to be alone.'*—A. Newman, 1978; (childlike) *Mr. Fox was marked by a childlike simplicity*—R. Chambers, 1866; *His childlike curiosity about life was held in check by childish timidity*—M. Holroyd, 1974; *She habitually wore an expression of childlike wonder*—R. West, 1977.

### children's language.
David Crystal (1986) estimated that 'by around 18 months, most children come to use a spoken vocabulary of about 50 words', but the choice of words differs from child to child. The common core of words usually includes *allgone, baby, bee(p)-beep, bye-bye, daddy, mama, nigh(t)-night*. By the third year a tattered form of grammar has been acquired (*this is him's car, her gave me one, I happy, Me do it, want Teddy drink, I ride train? Wayne naughty*). Andrew Radford (1988) called such constructions 'small clauses' and compared them with similar constructions in informal adult speech. Children aged 2–3 do not use auxiliaries, inflexions, preposed *wh*-phrases, infinitival *to*, or numerous other obligatory elements of adult speech. In this form of reduced grammar *I can see a cow* becomes *see cow*, and a great many clauses are verbless (*Paula good girl*).

After the age of three a child's acquisition of language proceeds by leaps and bounds. Forms like *bringed* and *taked* are gradually discarded. An average child of three probably uses as many as 3,000 English words, including 'hard' words like *helicopter, calculator, penguin*, and *vacuum cleaner* (not perfectly pronounced but clear nevertheless) and 'difficult' concepts like *triangle*.

### chilli
is the right spelling in BrE (more usu. *chili* in AmE) for the dried red pod of a capsicum. Pl. *chillies*, AmE *chilies*. It is etymologically derived via Spanish from the Nahuatl word for the plant. It has no connection with the name of the S. American country Chile.

### chill(y)
(adjs.). The normal word now is *chilly*, but from the 16c. to the 19c. it vied with *chill*: (chill) *Noisom winds, and blasting vapours chill*—Milton, 1640; *They had a fire to warm them when chill*—Mrs Shelley, 1818; *The chill elevation of political philosophy*—Gladstone, 1877; (chilly) *A chilly fear congeal'd my vital blood*—Pope, 1725; *The chilly mists of eventide*—Southey, 1793; *The chilly cry of the poor sweep as he crept shivering to his early toil*—Dickens, 1839.

### chimera
/kaɪ'mɪərə/ is the recommended spelling, not *chimaera*. It was adopted in the 14c. from OF *chimere* (from L *chimaera*, Gk Χίμαιρα 'she-goat, monster').

### china.
See PORCELAIN.

### Chinaman.
In 1926 Fowler could say that the normal uses were *A Chinaman* (rarely *Chinese*); *three Chinamen* (sometimes *Chinese*); *50,000 Chinese* (sometimes *Chinamen*); *the Chinese* (rarely *Chinamen*). At some point between then and 1965 (Gowers), *Chinaman* acquired a derogatory edge, and had virtually dropped out of use in BrE (except as a term in cricket for a left-handed bowler's offbreak to a right-handed batsman). *Chink* (first recorded in 1901) is an offensive slang term for a Chinese.

### chiropodist,
as the *OED* says, is 'a factitious designation, apparently assumed in 1785', and immediately attacked for its 'absurdity and needless affectation'. It was judged to be a high-flying term for a 'corn-cutter'. It was derived either from Gk Χείρ, Χειρο- 'head' + ποῦσ, ποδ- 'foot' (to indicate that both hands and feet were the subject of attention), or from the Gk adj. Χειροπόδης 'having chapped feet'. Time has healed the etymological wounds and the term is now in ordinary use for one who treats the feet and their ailments. The related noun *chiropody*, the art of treating corns, defective nails, etc., first recorded a century later in 1886, is also now in standard use.

### chivalry.
Despite the fact that it first entered the language from French in the 14c., it and its derivatives are now always

pronounced with initial /ʃ-/ not /tʃ-/, unlike most early loanwords beginning with *ch-* from French: see CH. The corresponding adj. *chivalric* in my experience is most often pronounced by medievalists with the main stress on the second syllable, but /ˈʃɪvəlrɪk/ is also widely used.

**chive** /tʃaɪv/. It is worth noting that this word (from a word of the same spelling in medieval Northern French) for the smallest cultivated species of *Allium* (type of onion) vied with *cive*, pronounced /saɪv/, from the 14c. until the 19c., when *cive* dropped out of use. ME *cive* was adopted from OF *cive*, which was descended from L *cēpa* onion.

**chivy.** 1 From the name *Chevy Chase*, the scene of a famous Border skirmish, *chevy* or *chivy* came to mean 'a hunting cry' (18c.) and 'a chase, pursuit, hunt' (19c.). By rhyming slang, *Chevy chase* also became used for 'face'.

2 Simultaneously, the first element *chivy* (or *chivvy*) also came into use as rhyming slang for 'face' (*I can't keep that look of modest pride on my chivvy forever—* Angus Wilson, 1958).

3 The verb *chivvy* (normally with double -vv-) entered the language in the early 20c. (first recorded in 1918) meaning 'to harry, harass, trouble, worry' (*the government was chivvied into changing its mind over certain details of the poll tax*).

**chlorine.** Always pronounced /ˈklɔːriːn/ now (formerly also with final /-ɪn/).

**chloroform.** Pronounced either as /ˈklɒrəfɔːm/ or as /ˈklɔːr-/.

**chock-full.** From the 17c. onward it alternated with *choke-full* and *chuck-full*, and the choice of headword became a vexatious matter for 19c. lexicographers. *Chock-full* has triumphed, possibly by association with the 19c. formation *chock-a-block* (used in the same sense).

**choir.** The dominant spelling from the time of its adoption from OF in the 14c. until the close of the 17c. was *quire* (or *queere*, etc.), and it is still so spelt in the Book of Common Prayer. But *choir* (respelt on the model of L *chorus*) gained the ascendancy about 1700 and is now standard in all main senses, and in *choirboy*, *choir-stall*, etc. The paper term *quire*

is unrelated: it is from OF *quaer, quaier* (modF *cahier*), ult. from L *quaterni*, a set of four.

**choler.** As one of the 'four humours' (*sanguis, cholera, melancolia, phlegma*) of medieval physiology, it was supposed to cause irascibility of temper. It survives in poetic or archaistic use in the sense 'irascibility'. The corresponding adj. *choleric* 'irascible, angry' has fared slightly better, though it too has somewhat limited currency, being much more formal than either of its main synonyms. Examples: (choler) *His face was improved by the flush of this momentary choler—*J. R. Ackerley, 1960; *in our bodies they [sc. the four elements] combine to form the Humours. Hot and Moist make Blood; Hot and Dry, Choler; [etc.]—*C. S. Lewis, 1964; (choleric) *[in medieval science] the choleric man is hot and dry, after the nature of fire—*W. C. Curry, 1960; *Gold's demeanor at the Y was habitually unsociable, his countenance that of someone introverted and choleric—*J. Heller, 1979; *Her red face makes her look choleric, in a most unattractive way—*M. Forster, 1986.

**chop, cutlet.** A chop (first recorded in 1640) is a slice of meat (esp. pork or lamb) cut from the loin, usu. including a rib. A cutlet (1706, an adaptation of Fr. *côtelette*, not connected with the English word *cut*) is a neck-chop of mutton or lamb, or a small piece of boneless veal, etc., for frying or microwaving. Vegetarians often prepare *nut cutlets* instead. Cf. CHAP, CHOP.

**chorale.** 1 Pronounce /kɔːˈrɑːl/. The final -e was apparently added in the 19c. soon after the adoption of the word from Ger. *Choral* in order to indicate that the main stress falls (as in Ger.) on the second syllable. For this convention cf. *locale, morale*.

2 Since the 1940s used in America in the names of choirs that sing principally choral music (e.g. *the Collegiate Chorale, the Paul Hill Chorale*).

**chord, cord.** The musical *chord* and the spinal *cord* are no more than third cousins. Their relationship is as follows: 1 There are two distinct words spelt *chord*: (a) (in origin a shortening of *accord* respelt with initial *ch*) in music, a group of (usu. three or more) notes sounded together, as a basis of harmony. (b) A technical word in mathematics, engineering, etc. (a straight line joining the ends of an arc, the wings of an aeroplane, etc.). From (b) came the idiomatic expression *to strike a chord* 'to recall something to one's memory'. This second *chord* is a 16c. refashioning of *cord* after the initial *ch* of its Latin original (*chorda*).

2 *Cord* = string, rope, rib of cloth, and in *spinal cord, umbilical cord, vocal cord,* etc., is descended from ME *corde*, adopted from OF *corde* in the 14c., which is derived from L *chorda*, a string of a musical instrument. The anatomical expressions (*spinal cord*, etc.) are still occasionally spelt as *spinal chord*, etc., but this spelling is not recommended.

**Christian name.** Now that Britain is a multicultural society, this traditional term for a person's first name is gradually giving way, esp. on official forms, to *first name, forename,* or (as in AmE) *given name*. A second name is usually called a *middle name.*

**chronic** is used of a disease that is long-lasting (as opposed to *acute*). It has the same sense, implying severity, when used of non-medical states or conditions (unemployment, shortages, etc.). The word is also used loosely as a vague term of disapproval. Examples: *it's chronic 'ere of a Saturday sometimes*—Kipling, 1904; *There was a chronic inaccuracy in him which vexed his father's soul*—J. Buchan, 1925; *Traffic congestion has become so chronic in Britain's cities that vehicles travel at an average speed of just 8 mph*—Back Street Heroes, 1988.

The adjvl or advbl phrase *something chronic* is also used in fiction in the speech of the 'lower orders': *It's made my eyes water something chronic*—H. G. Wells, 1910; *The men in these parts are something chronic ... They won't do anything till some time after we're due to start*—L. A. G. Strong, 1942. Cf. also the imitative phr. *something rotten: John Gielgud camps it up something rotten as Minnie's arriviste father*—Time Out, 1989.

**chrysalis.** Pl. either *chrysalises* or (in learned work) *chrysalides* /krrˈsælɪˌdiːz/.

**chuffed.** In standard southern BrE it is now mostly used, somewhat slangily, to mean 'delighted, very pleased': *He was chuffed at this new monumental skive he had discovered*—A. Waugh, 1960; *You were*

*pleased at the time. Chuffed in fact* Paul Scott, 1977. In some other varieties of English in the UK the word sometimes has the opposite sense, 'displeased, disgruntled': *Don't let on they're after you, see, or she'll be dead chuffed, see? She don' like the law*—Celia Dale, 1964. The two senses seem to reflect separate uses of dialectal *chuff* (adj.) listed in Joseph Wright's *Eng. Dialect Dict.,* (*a*) = proud, conceited; pleased, elated (various northern and Midlands counties, but not recorded in southern ones); (*b*) = ill-tempered, surly, cross (Lancs., Lincs., Berks., Kent, Devon, Cornwall, etc.). The channel by which the dominant ('delighted') sense entered standard English cannot be ascertained with any certainty.

**chute.** 1 A sloping channel or slide for conveying things to a lower level.

2 A slide into a swimming pool.

3 Short for *parachute*. As to the etymology, 'there appears to be a mixture of the F. *chute* fall (of water, descent of a canal lock, etc.) and the English word *shoot*' (OED). In the 20c., however, the spelling *chute* has been almost always used in BrE, rather than *shute* or *shoot*, for the three senses given here.

**-ciation, -tiation.** 1 The way in which nouns ending in *-ciation* or *-tiation* are pronounced divides standard speakers into those who are prepared to tolerate two successive /ʃ/ sounds and those who choose not to. The matter does not stop there, however, as the patterns are not always the same in the corresponding verbs (where they exist) ending in *-ciate, -tiate.*

2 Everyone comes out in favour of /s/ in the *-nci-* part of *denunciation, pronunciation,* and *renunciation*. Equally, everyone favours /ʃ/ in the *-iti-* part of *initiation* (and *initiate*) and *propitiation* (and *propitiate*).

3 For other common words in this group there is considerable fluctuation. Perhaps the least debatable distribution is as follows: (/ʃ/ dominant) *appreciate/ -iation, associate, consubstantiation, differentiate/-iation, negotiate/-iation, satiate/-iation, substantiate/-iation;* (/s/ dominant) *annunciate, annunciation, association, consociation, emaciate/-iation, enunciate/-iation.*

Neither form is dominant in *dissociate/ -iation.*

**cicada.** Pronounce /sɪˈkɑːdə/, but /sɪˈkeɪdə/ is often encountered in various English-speaking countries including Britain and America.

**cicatrice** (from French) and **cicatrix** (from L), pl. *cicatrices* /sɪkəˈtraɪsiːz/, are both used for 'a mark or scar left by a healed wound; (in botany) a mark on a stem, etc., left when a leaf or other part becomes detached'.

**cicerone,** a guide who gives information to tourists about places of interest. This loanword from Italian can be pronounced, according to taste, either as /tʃɪtʃəˈrəʊni/ or /sɪsəˈrəʊni/.

**cigarette.** The normal pronunciation in BrE is with the main stress on the final syllable, but Walter Nash (1986) observed that younger people are tending to move the stress forward to suit the rhythm of a clause or sentence. Thus *Cígarettes are déar* but *I smóked a cigarétte.*

**cinema.** 1 Fowler (1926), writing at a time when the film industry was becoming established, needed half a column of argument to justify the spelling and pronunciation of the word compared with the rival form *kinema* /kaɪˈniːmə/ (falsely stressed; cf. Gk κίνημα). *Cinema* /ˈsɪmɪmə/, of course, has triumphed, and has generated a formative prefix *cine-* (*cine-camera*, *-film*, etc.). *Cinematograph*, the full form in English, has been virtually displaced by its abbreviation *cinema*.

2 In Britain one goes to the *cinema* or *the pictures* to see a *film*. In America one goes to *the movies* or to a *theater* to see a *movie*. In Australia and NZ one goes to *the pictures* or a *picture theatre* to see a *film*. These are the dominant forms, but in each English-speaking country other variants are known and often used, esp. the word *movie(s)*. The film industry is commercially called the *motion picture industry*.

Examples showing variation of terminology: *One meets that sort of character amongst the older generation of the motion picture and theatre world*—M. Spark, 1984; *Making a movie is not enough these days; you also have to make it sell*— *Times*, 1990.

**Cingalese** is an archaic form of *Sinhalese*, which remains as a language name, and also as the name of the majority of the population of Sri Lanka, but

has been partially replaced by *Sri Lankan* as the general name for a native or inhabitant of the republic of *Sri Lanka* (until 1972 called *Ceylon*).

**cinque.** 1 Also *cinq*. The five on dice or cards is pronounced /sɪŋk/. *Ace*, *deuce*, *trey* /treɪ/, *cater* /ˈkeɪtə/, and *sice* /saɪs/, are the others of the series.

2 *Cinque Ports* /sɪŋk/ is the name for a group of ports (orig. five: Dover, Hastings, Hythe, Romney, and Sandwich; Rye and Winchelsea were added later) on the SE coast of England with ancient trading privileges.

**Cinquecento.** Pronounce /ˌtʃɪŋkwɪˈtʃentəʊ/. For its meaning see TRECENTO.

**cinquefoil.** Pronounce /ˈsɪŋkfɔɪl/ both in its botanical and its architectural sense.

**cion.** See SCION.

**cipher.** The preferred spelling (not *cy-*).

**Circe.** Pronounce /ˈsɜːsi/.

**circuit** is pronounced /ˈsɜːkɪt/ and *circuitry* /ˈsɜːkɪtrɪ/, but *circuitous* is pronounced /sɜːˈkjuːɪtəs/ with the *-ui-* treated as a diphthong.

**circumbendibus.** See FACETIOUS FORMATIONS.

**circumcise.** Like *exercise*, *televise*, and a few other words, always spelt with *-ise*. See -ISE 1.

**circumlocution** lacks a well-established adj., though *-locutional* (first recorded 1865), *-locutionary* (1863), and *-locutory* (1659) are all used from time to time. Synonyms like *periphrastic* and *roundabout* tend to be preferred.

**circumstance.** The debate about the merits of *in the circumstances* and *under the circumstances* has continued for most of the 20c. Fowler (1926), though predisposed to give preference to arguments based on the etymological meaning of classical roots, regarded the opposition to *under the circumstances* as 'puerile'. He did not accept that an English word containing *circum*, meaning in Latin 'about, around', logically required the use of 'in' rather than of 'under'. He observed that 'under the same Circumstances' was recorded in 1665 in the *OED* and 'in (the) circumstances' not until much later (in fact 1856). On the other hand, some other usage guides (including *The Economist Pocket Style Book*, 1986) have continued to rule out *under the circumstances* altogether.

The *OED* distinction, 'Mere situation is expressed by "*in* the circumstances", action affected is performed "*under* the circumstances"', is a fine one, but useful as a general guide. And it is as well to bear in mind that the phrases are often varied by the omission, strengthening, or qualifying of the middle element: so *in ʌ circumstances that he could not have foreseen, in certain circumstances, in such circumstances, in present circumstances, in straitened circumstances; under these circumstances, under no circumstances, she won't agree under any circumstances* are all idiomatic constructions.

**cirrus** has pl. *cirri* /'sɪraɪ/. See -US.

**cithern, cittern.** See ZITHER.

**city.** In many English-speaking countries the word is applied broadly to any large town. The official use of the term also varies from country to country. In Britain it is properly used of a town that is declared to be a city by royal charter and that contains a cathedral. As a result cities vary greatly in size and population: the population of Oxford (1981) was 98,521 and that of Wells (1981) was 8,374. Both are cities. The City of London, a municipality governed by the Lord Mayor and Corporation, is the part of central London in which Britain's major financial transactions are conducted. But it is one of the smallest cities, with an area of one square mile and a resident population (1981) of only 5,893.

**clad.** See CLOTHE.

**claim.** There are five uses of the verb *to claim* that are beyond dispute: **1** Used transitively meaning 'to demand as one's due or right' (*the creditor claimed repayment*).

**2** + a *that*-clause (*he claimed that his offer was a generous one*) (but see 6 below).

**3** + certain kinds of *to*-infinitive phrases (*every townsman could claim to be tried by his fellow-townsmen; he claimed to be the next-of-kin*) (but see 6 below).

**4** In insurance, = to make a claim (for an indemnity) (*Before deciding to claim for a small amount of damage*, etc.).

**5** To assert and demand recognition of (an alleged right, attribute, etc.) (*the degree of accuracy that has been claimed for them; both sides claimed victory; the only living but distant relative claimed the title after the childless duke died*).

There are two areas of potential or actual dispute:

**6** H. W. and F. G. Fowler in *The King's English* (1906) said that *claim* should not be followed by an infinitive except when the subject of *claim* is also that of the infinitive. 'Thus, *I claim to be honest*, but not *I claim this to be honest*.' They objected to sentences of the type *Usage, therefore, is not, as it is often claimed to be, the absolute law of language* (R. G. White, 1871); *The constant failure to live up to what we claim to be our most serious convictions proves that we do not hold them at all* (*Daily Tel.*, a1906). Fowler (1926) also condemned the use of *claim* followed by (or implying) a *that*-clause, when *claim* means not *demand* but *assert*.

Time has moved on, and the quotational evidence presented in *OED* 2 (sense 2c) and in *WDEU* (1989) partially cancels that cited by the Fowlers. Examples: *Refet Bey ... was hopping mad at an attempt which he claimed that the British had made to kidnap him*—World's Work, 1922; *siding with the forces of White repression in southern Africa in defence of what they claim to be the 'free world'*—TLS, 1971 ; *bothering him right on mike about a threatening phone call she had received, claiming him to have been the nasty, horrible caller*—Rolling Stone (US), 1972; *This volume's dust jacket claims that the mighty Argentine fantasist 'has come into English in haphazard fashion'*—J. Updike, 1982; *some respondents claimed to pay attention to cholesterol at every meal*—NY Times, 1988. This evidence cannot be gainsaid, but many a copy editor in Britain would nevertheless make an adjustment in such circumstances, esp. by substituting *allege, contend, declare, maintain*, or *say*, whichever seemed the most appropriate in the context.

**7** A rebel group or band of terrorists is often described in newspapers and in news bulletins as *claiming responsibility for* (an outrage). There are more neutral ways of expressing the same idea, e.g.

*The —— say they were responsible; the ——
say* (or *admit*) *that they shot the policeman*
(Burchfield, 1981).

**clamant,** used to mean 'clamorous,
noisy', is either poetical (*This clamant word
Broke through the careful silence*—Keats,
1818) or pompously journalistic (*A deliri-
ously clamant crowd rightly thought a try
inevitable*—*Times*, 1963). In some Sc. writing it is used, quite acceptably, in the
figurative sense 'urgent' (*My appetite was
a clamant, instant annoyance*—R. L. Stevenson, 1878).

**clandestine.** Pronounce /klæn'destɪn/.

**clarinet.** Derived from the Fr. diminutive *clarinette*, this form of the word for
a woodwind instrument has entirely replaced the former variant *clarionet*. The
main stress falls on the third syllable.

**classic, classical.** *Classical* is the customary word when reference is made to
the arts and literature of ancient Greece
and Rome (*a classical scholar, classical Latin,
classical metres*). The works studied, and
also the subject itself and the study of the
subject, are called (*the*) *Classics. Classical* is
also applied to serious or conventional
music (i.e. that of Beethoven, Mozart,
etc.) as distinct from light or popular
music; and in physics to the concepts
which preceded relativity and quantum
theory.

*Classic* means 'of acknowledged excellence' (*the classic textbook on the subject*),
or 'remarkably typical' (*a classic case of
cerebral palsy*). *Classic races* (or *the Classics*)
in Britain are the five main flat races,
namely the Two Thousand and the One

Thousand Guineas, the Derby, the Oaks,
and the St Leger.

**clause.** In a book like this, that by its
nature cannot offer a full-scale presentation of English grammar, the terminology must needs be kept to basic
essentials. In simple terms a *clause* is one
level above a *phrase* which in turn is one
level above a *word*. A main or principal
clause is of the type *He arrived at Heathrow
airport at 12 noon*. The sentence could
continue with a subordinate clause, e.g.
*because the plane was scheduled to depart at
2 o'clock*. The essential nature of a clause
is that it should have a subject (*He*; *the
plane*) and a predicate (*arrived at Heathrow
airport*, etc.; *was scheduled*, etc.). The subject and predicate can be further divided
into their *constituents*. Thus *the plane* consists of the definite article or determiner
(*the*) and a noun (*plane*); *at Heathrow airport* consists of a preposition (*at*) and a
noun phrase (*Heathrow airport*); *was scheduled* consists of an auxiliary (*was*) and a
main verb (*scheduled*), and can also be
called a verb phrase. Modern grammars
are studded with algebraic-looking
strings of various types. Thus the subordinate clause *because the plane was scheduled to depart at 2 o'clock* can be set out
in a serial manner as

conj. + NP (det. + n.) + VP (aux. +
main verb) + *to*-inf. + prep. phr.

or in diagrammatic form as illustrated.
Analyses of this kind will not be attempted in this book, but readers should
be warned that they will find a forest of
tree-diagrams and elaborate bracketings
and abbreviations in most modern grammars.

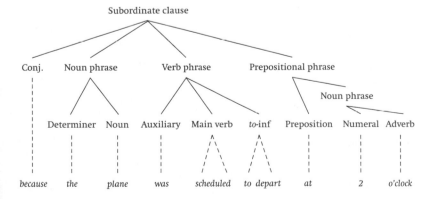

Those who quail before such analyses should at any rate be prepared to master such abbreviations as S (subject), V (verb), O (object), C (complement), and A (adverbial), e.g. *My son* [S] *considers* [V] *the price* [O] *quite reasonable* [C] *in the circumstances* [A].

**clean, cleanly.** *Clean*, spelt *clæne* (with a final -*e*), was already an adverb (as well as an adj.) in OE, and has never ceased to be one. Modern examples: *the bullet went clean through his shoulder-blade*; *it went clean out of my mind*; *the car plunged clean over the cliff*. (In a sentence like *the room must be swept clean*, 'clean' can be analysed as an adj. standing as complement of the predicate.) It would be absurd to add -*ly* to *clean* in any of these examples.

For the 'absent -*ly*' syndrome, see the next entry.

**clear, clearly. 1** Whenever a monosyllabic adverb is used in an obviously adverbial role some people suffer from what might be called the 'absent -*ly*' or something-is-missing syndrome. Because so many monosyllabic adjectives (*apt, brief, damp*, etc.) are never used as adverbs, it is an easy step to believing that none can be used. As so often, the historical explanation in the OED provides the answer: '*Clear* is not originally an adverb, and its adverbial use arose partly out of the predicative use of the adjective, as in "the sun shines clear"; partly out of the analogy of native English adverbs which by loss of final -*e* had become formally identical with their adjectives, esp. of *clean* (adverb), which it has largely supplanted.' It would be absurd to substitute *clearly* for *clear* in such phrases as *go clear, keep clear, stand clear, stay clear, steer clear*, or in sentences like *the thieves got clear away*.

**2** Nevertheless, it should be emphasized that the normal relationship between the two words is shown by the contrastive pair *a clear disappointment* (adj.)/*it is clearly a disappointment* (adv.).

**3** *Clearly* is overwhelmingly the more usual adverbial form of the two: (as simple adv.) *the author writes clearly and concisely*; (used as intensifier) *he was clearly mad*; (as a sentence adv.) *Clearly the Minister of State should resign*.

**cleave¹** (to split). **1** A separate verb from CLEAVE², *cleave¹* is a mostly literary word

with unfixed past tenses. In the pa.t., *clave* is archaic; *clove, cleft*, and *cleaved* are all permissible, but *cleaved* is the more usual of the three, esp. in scientific and technical contexts. The pa.pple is *cloven* or *cleaved* (esp. in many technical uses). The adj. is *cloven* in some fixed expressions (*cloven-footed, cloven hoof*) and *cleft* in some others (*cleft palate, in a cleft stick*).

**2** In grammar, a *cleft sentence* is one in which a particular element is highlighted. R. Huddleston (1984) cited three related sentences and classified them in the manner indicated: (i) *A faulty switch caused the trouble* (not highlighted); (ii) *It was a faulty switch that caused the trouble* (cleft construction); (iii) *What caused the trouble was a faulty switch* (pseudo-cleft construction). Such cleft and pseudo-cleft constructions abound in written work as well as in spoken English: (cleft) *It was I who was wrong to raise the issue*—P. Wright, 1987; (pseudo-cleft) *What he* [sc. Bernie Grant MP] *concedes the House has done is to increase his status, nationally and internationally*—Listener, 1988.

**cleave²** (to stick fast, adhere). A separate word from CLEAVE¹, *cleave²* is fairly regularly conjugated as *cleave/cleaved/cleaved*, except that the archaic variant *clave*, doubtless as an echo of the Bible, occurs occasionally. Biblical (AV) examples: *The Nobles held their peace, and their tongue cleaued to the roofe of their mouth* (Job 29: 10); *and Orpah kissed her mother in law, but Ruth claue vnto her* (Ruth 1: 14).

**cleft.** See CLEAVE¹.

**clematis.** The standard pronunciation in BrE is /'klemətɪs/, but the variant /klə'meɪtɪs/, with the main stress on the second syllable, is encountered often enough, and especially abroad.

**clench, clinch.** *Clinch* is a 16c. variant spelling of *clench* (from OE *(be)clencan*) and since then has been treated as a separate word. We *clench* our teeth, fingers, or fist. We *clinch* an argument, bargain, etc.; lovers *clinch* (embrace), and so do boxers and wrestlers (become too closely entangled). A remark or statement that settles an argument is a *clincher*. Usually nails are *clinched* (not *clenched*) to make a *clinker-built* (also *clincher-built*) boat.

**clerestory.** Pronounce /ˈklɪəˌstɔːrɪ/.

**clerk.** Pronounced /klɑːk/ in BrE and /klɜːrk/ in AmE.

**clever.** Fowler (1926) wrote a splendidly prejudiced piece about the misuse of *clever*, 'especially in feminine conversation' in the sense of 'learned, well read, bookish, or studious'. It is sufficient, perhaps just to recognize that *clever* is normally a term of approbation ( = skilful, talented; quick to understand and learn, as *COD* has it), but that it can be used contextually with severe limitations placed upon it, implying slickness or mere ingenuity (*clever Dick, too clever by half*, etc.).

**clew, clue.** They are 'the same word' (from OE *cliwen*), but *clew* has gradually become confined to two main uses, (*a*) a ball of thread in the legend of Theseus enabling him to find his way through a labyrinth; (*b*) as a nautical term = a lower or after corner of a sail; a set of small cords suspending a hammock. Since the 17c., *clue* has gone its own way and now has a wide range of uses in which its literal sense is obscured. It broadly means 'a piece of evidence' (in a treasure hunt, a detective novel, a crossword puzzle, etc.) putting one on the track of a discovery, a solution, etc.

**cliché.** *He was, however, on the whole, taking all things into consideration, by and large, not to put too fine a point on it, reasonably well self-sufficient.* Thus Anthony Burgess, in a classic mocking of clichés, in *Inside Mr Enderby* (1963). In the printing trade, French *cliché* was a stereotype block: it was the past participle of the verb *clicher* 'to stereotype', said to be imitative of the sound produced by the dropping of the matrix on the molten metal. Dislike of threadbare, stereotyped expressions is a relatively recent phenomenon: the word *cliché* is first recorded in English in 1892.

Despite the fact that Nicholas Bagnall wrote a spirited book called *A Defence of Clichés* in 1985, familiar overworked expressions are more often than not mercilessly pilloried. Among clichés continually ridiculed in Britain are: *at the end of the day, at this moment in time, explore every avenue, in this day and age, leave no stone unturned, over the moon, sick as a parrot*

(this and the previous phrase from the jargon of football), *situation* (preceded by a noun or phrase, e.g. *a crisis situation*), and *a level playing-field*. It should be noted that in the course of the 20c. the word *cliché* has come to be applied to commonplace things of other kinds—visual images, stock situations, remarks in radio or TV plays (*And now if you'll excuse me I've got work to do*), ideas, attitudes, etc. When reviewing Bagnall's book, I said that 'clichés are a sad subject, really, always reminding us of the repetitiveness of things, the humdrumness that lies beyond and within the doorstep if one's imagination should weaken or one's sense of humour runs out'. But perhaps the wisest comments on clichés were by Christopher Ricks (1980): 'A cliché begins as heartfelt, and then its heart sinks.' 'The only way to speak of a cliché is with a cliché.'

**client.** A person using the services of a professional person—a lawyer, an architect, an accountant, etc.—is a *client*. At the other end of the social scale, a person availing himself of the services of a prostitute is also a *client*. Social workers describe the people they assist as *clients* (not *cases* or *patients*). A person purchasing items from a shop is a *customer*; so too, probably, is someone who buys a car from a car showroom. Hairdressing salons call their customers *clients* rather than *customers*. A person on a doctor's list is a *patient*, an undergraduate is a *pupil*, literary agents have *authors* or *writers*, and theatres, restaurants, and hotels have *patrons*. A body of clients is a *clientele* /ˌkliːɒnˈtel/, no accent, not italic.

**climacteric. 1** The older pronunciation with the main stress on *-ter-*, which was favoured by Johnson, Walker, and some 19c. lexicographers, and was given precedence by the *OED*, has now been mostly replaced by /klaɪˈmæktərɪk/.

**2** It is used as adj. and noun of a critical stage in human life, variously taken to mean, as the *OED* expresses it, 'that period of life (usually between the ages of 45 and 60) at which the vital forces begin to decline (in women coinciding with the period of "change of life")'; or a person whose age is characterized by being a multiple of 7 (7, 14,

21, etc.); or by some other numerical factor.

**climactic** is an adj. formed in English in the 19c. from *climax* after the model of *syntax/syntactic*, but not on Greek analogies. Fowler (1926) disliked it: 'it may fairly be demanded of the literary critics who alone have occasion for the adjective that they should mend it or end it.' It closely resembles, but seldom seems to be confused with, *climatic*, of or relating to climate, also first recorded in the 19c. Examples: *Through subtle management of the narrative of this climactic moment, Kennedy is able to give equal weight to both mimetic and fabulistic readings—TLS*, 1988; *The audience wants a good climactic high C—Opera Now*, 1989.

**climate.** From its normal meteorological sense, *climate*, beginning in the 17c., developed a valuable figurative sense, 'the mental, moral, etc., environment or attitude of a body of people in respect of some aspect of life, policy, etc., esp. in *climate of opinion, of thought*' (OED): *To us he [sc. Freud] is no more a person Now but a whole climate of opinion*—W. H. Auden, 1940; *The whole climate of thought will be different*—G. Orwell, 1949.

**climax.** From a Greek word meaning 'ladder' the term was originally (first in the 16c.) used in English in rhetoric as 'a figure in which a number of propositions or ideas are set forth so as to form a series in which each rises above the preceding in force or effectiveness of expression' (OED). Thus (1) (1 Cor. 2: 9) *Eye hath not seene*, (2) *nor eare heard*, (3) *neither haue entred into the heart of man, /the things which God hath prepared for them that loue him*; three progressive stages of strangeness. Wales (1989) cites an example of 1553 from the work of Thomas Wilson; *Of sloth cometh pleasure, of pleasure cometh spending, of spending cometh whoring, of whoring cometh lack, of lack cometh theft, of theft cometh hanging, and here an end for this world.*

Popular use of this technical term in the 18c. led to the ordinary current meanings, (*a*) the event or point of greatest intensity or interest; a culmination or apex (COD); (*b*) 20c., the sexual sense.

**clime** is a literary word for a tract or region of the earth (also fig.) or for

climate: *This inclement clime of human life*—Edward Young, 1742; *Every man of every clime*—W. Blake, 1783–94; *Where a sweet clime was breathed from a land Of fragrance ... and flowers*—Keats, 1820. It can never be used to replace *climate* in ordinary discourse.

**cling.** The principal parts were once *cling/clang/clung*, but *clang* as pa.t. withdrew into regional use in the 19c. The pa.t. now in standard English is always *clung*.

**clique.** Pronounce /kliːk/, not /klɪk/.

**clitic.** See ENCLITIC.

**clone,** from Gk κλών 'twig', 'slip', came into English at the beginning of the 20c. in the technical language of botany (later in biology). Since about 1970 it has swept into popular use of a person or animal that strongly resembles its parent; and also of a person or thing whose behaviour, appearance, or activities are directly modelled on those of another (well-known) person or thing: e.g. —*is a typical Sloane Ranger clone; there were several Jaguar clones in the car park.*

**close** (verb). has a suggestion of formality or politeness when it is used in contexts in which *shut* might sound rather peremptory: *Close the door after you and we'll have a talk.* Private talks are usually held *behind closed doors.* But a child who has entered a room and carelessly left the door open will almost certainly be asked by its parent to *shut* (not *close*) *the door. Closed* has a life of its own. Some trade unions are *closed shops*; a gallery is often *closed* to the public on Mondays; the stock market *closed* one point up on the day's dealing; Albania is a *closed society*; Hebrew is a *closed book* (i.e. is unknown) to most people. In none of these examples would *shut* be possible.

**close** (adj.) is used in scores of collocations: *close contest, formation, friend, proximity, resemblance, shave*, etc. A *close season* (US *closed season*) is the season when something, esp. the killing of game, is illegal.

**close, closely. 1** No strict dividing line can be drawn between predicative uses of the adj. *close* (e.g. *the weather remained sultry and close*) and adverbial ones, e.g.

*to stand, sit, lie,* etc., *close* (i.e. in immediate contact or proximity); with verbs of motion, as *come, bring,* etc., *close*; or with *by, to, beside,* etc. (*close by the wall; he came close to solving the problem; she sat close beside him*).

**2** *Close* is a ME loanword from OF. It was not until the mid-16c. that it acquired the native suffix -*ly* in printed work. Then *closely* gradually became the dominant form.

**3** Despite its dominance, *closely* should not be allowed to oust *close* in such idiomatic uses as the following: *for close on* 200 years; *sticking close to his desk; the road runs close beside the river; stay close together.*

**4** Some grammarians treat *close to,* as in *he was standing close to the door,* as a complex preposition.

**closure.** Several names are used for the provision by which a debate in the House of Commons can be cut short in spite of the wish of a minority to continue it. The most important one is the *closure* (introduced in 1882), a decision by vote or under rules to put a question without further debate. The equivalent French word *clôture* (used in English without the circumflex) was occasionally used in the House of Commons instead of *closure,* but has fallen out of use there. It survives, however, in the procedures of some legislative bodies in the United States. A *gag rule* (or *gag law*) came into use in America at the end of the 18c. as a means of preventing or restricting discussion in a legislative body. The word *gag* was later adopted in the House of Commons, applied opprobriously to the action of a parliamentary majority in 'closuring' a debate. The *guillotine* (first recorded 1893), or closure by instalments, is a method used in the House of Commons for preventing obstruction or undue delays by fixing the dates and times at which specified stages of a bill must be voted on. The *kangaroo,* or *kangaroo closure,* is a UK term (first recorded in 1913) for a Parliamentary closure by which some amendments are selected for discussion and some excluded. The unselected (normally minor) items are then voted on without debate.

**cloth.** The Mitfordian pronunciation /klɔːθ/ lingers on, just, in the speech of elderly people, but has been almost entirely replaced in the standard language by /klɒθ/. The pl. *cloths* is normally pronounced /klɒθs/ but modern dictionaries of pronunciation also record /klɒðz/.

**clothe** has *clad* beside *clothed* both as pa.t. and pa.pple. While *clothed,* however, is suitable in most contexts (except when *dressed* is more suitable as less formal), *clad* is (1) either somewhat more formal or literary, or (2) never used absolutely, but always with some specification of the kind of clothing, or qualified by an adverb (e.g. *ill-clad, insufficiently clad*), or (3) in technical use. Accordingly, *clad* cannot be substituted for *clothed* in *You were fed and clothed at my expense* or *When he was dressed he let us in.* But it stands unchallenged by *clothed* (or *dressed*) in literary contexts like Coleridge's *I guess, 'twas frightful there to see A lady so richly clad as* she and Milton's *'Tis Chastity, my brother, Chastity: She that has that, is clad in complete steel;* also, qualified by an adverb, *They are lightly clad in summer;* and in metallurgy, etc. (*steel clad with Ni* [sc. nickel] *on one side; clad with stainless steel*).

As a second element -*clad* is frequent in such compound adjs. as *ice-clad, ivy-clad, olive-clad* (hills), *snow-clad.*

**clothes.** Until about the end of the 19c. the more usual standard pronunciation seems to have been /kləʊz/, with ample authority from rhymes in 17c. and 18c. poets, including Shakespeare (*Then vp he rose, and dond his close—Hamlet*). Daniel Jones (1917) described this pronunciation as 'old-fashioned'. The spelling-pronunciation /kləʊðz/ has now almost entirely replaced it.

**cloture.** See CLOSURE.

**cloud.** First recorded in the US in the 1950s, *on cloud seven* or esp. *on cloud nine,* meaning 'in a euphoric state', have become vogue words everywhere. Clouds are divided into classes. In Webster's Third (1961), *cloud nine* = a cumulonimbus cloud, a mountainous cumulus cloud, a thundercloud, the ninth highest in a ten-part classification (*stratus* is the tenth). According to William and Mary Morris (1988), the popularity of the phrase may be credited to an American radio show of the 1950s in which every

time the hero was knocked unconscious—which was often—he was trans-
ported to *cloud nine*. There he could speak
again.

**cloven.** See CLEAVE[1].

**clue.** See CLEW.

**co-.** Prefix of Latin origin. There are
three ways of writing *cooperate* (*cooperate,
co-operate, coöperate*), but the third, with
the diaeresis ¨, has been virtually abandoned. The form recommended is
*cooperate* (also *cooperation, cooperative, uncooperative*).

The spellings recommended for other
words containing the prefix *co-* may be
classified as follows: (*a*) Hyphen never
used: (*coagulate, coalition, coefficient,
cohabit, cohere, coincide, coition* (in most of
which *co-* is hardly felt to be a prefix in
English). (*b*) Hyphen rarely used: *coed,
coeducation, coequal, coeval, coexist, coexistence, coordinate* (also *uncoordinated*)
*copartner* (common words in which the
main element is transparent). (*c*) Scientific and technical words in scholarly
work: *coaxial, cosec, cosecant, coseismal,
cosine, cotangent* (clear to scientists, etc.).

Words commonly written with a hy-
phen include: (*a*) *co-respondent* (in divorce
case) (hyphen used to avoid clashing with
*correspondent*). (*b*) *co-latitude, co-pilot, co-sig-
natory, co-star, co-tidal, co-worker* (hyphen
used to avoid momentary perplexity on
the part of the reader). (*c*) *co-actor, co-
author* (noun and verb), *co-editor*; *co-
abound, co-ascend, co-enjoy, co-worship*, etc.
(relatively new formations; nonce formations). (*d*) Others: *co-op, co-opt* (clashing
vowels might invite pronunciation as
/kuːp(t)/).

**coastal.** Fowler's objection (1926) to this
formation as a 'barbarism' is a classic
example of a lost cause. The OF original
was *coste* (from L *costa* 'rib'), which yielded
a French adj. *costal* in the 16c. English
has pressed *costal* into service in sundry
technical uses (*costal respiration, vein*, etc.),
but developed a new form *coastal* in the
late 19c. for the sense 'of or pertaining
to the coast'. The distinction *coastal* (as
in *Coastal Command, coastal resort*) and
*costal* (in medical contexts) is a useful
one, and takes priority over etymological
considerations.

**coat-card.** Card-players may be surprised to learn that *coat-card* was the
original term (first recorded in 1563) for
'a playing card bearing a "coated" or
habited figure (king, queen, or knave)'
(*OED*). Its replacement, beginning in the
17c., by *court-card* was, of course, readily
suggested by the personages represented
on these cards.

**cobalt.** The pronunciation recommended is /ˈkəʊbɔːlt/ not /-bɒlt/.

**cobra.** The pronunciations /ˈkəʊbrə/ and
/ˈkɒbrə/ seem to be of about equal frequency in BrE.

**cocaine.** In our drug-ridden world, the
substance, pronounced /kəʊˈkeɪn/, is all
too familiar. The relevant entry in the
*OED* (1891) gave a three-syllabled pronunciation, equivalent in the IPA to /ˈkəʊkeɪ-
ˌæm/, as the regular one, and described
/kəʊˈkeɪn/ as 'vulgar'. Times have
changed.

**coccyx.** Bone at base of spine. Pronounce /ˈkɒksɪks/. The pl. is either *coccyges*
/-ɪdʒiːz/ or *coccyxes*.

**Cockaigne, Cockayne.** A medieval
name (first recorded in English in the
early 14c.), of obscure origin, of an imaginary land of abundance and bliss.
Since the early 19c. it has occasionally
been used as a humorous name for London, as the abode of Cockneys.

**cockle.** The common phr. *to warm the
cockles of one's heart* is placed in dictionaries under *cockle*, the name of a bivalve
mollusc. Two things need to be said: (1)
an alternative source is L *corculum*, dim.
of *cor* 'heart'; (2) the phr. (first recorded
in 1671) was formerly used with several
other verbs, *rejoice, please, delight, comfort*,
etc.

**cockscomb, coxcomb.** The first (formerly also spelt *cock's-comb*) is 'the crest
or comb of a cock' (*COD*); formerly also
'a jester's cap (shaped like a cock's
comb)'. The second, a respelt version of
the same word, means 'an ostentatiously
conceited man, a dandy'. Any of the
three spellings is likely to turn up in
earlier literature for any of the senses.

**cocky.** The adj. meaning 'conceited,
arrogant; saucy, impudent' has now entirely driven out the former variants
*cocksy* and *coxy*.

**cocoa,** originally an alteration, i.e. a corruption, of CACAO, is now the only spelling for 'a powder made from crushed cacao seeds, often with other ingredients; a drink made from this' (COD). So cocoa bean, butter. Cf. next.

**coconut.** Now the only spelling of 'a large ovate brown seed of the coco, Cocos nucifera, with a hard shell and edible white fleshy lining enclosing a milky juice' (COD). The first element coco is 'originally a playful use of Pg. and Sp. coco grinning face, grin, grimace, with allusion to the monkey-like appearance of the base of the shell of the nut' (ODEE).

**codeine.** Now always pronounced /ˈkəʊdiːn/, though formerly (19c.–early 20c.) it was trisyllabic, /ˈkəʊdɪˌiːn/ or /-əm/.

**codex.** Pl. codices /ˈkəʊdɪˌsiːz/. See -EX, -IX 2; LATIN PLURALS.

**codify.** Fowler (1926) believed that the diphthong (or 'long vowel') in the first syllable, /ˈkəʊdɪˌfaɪ/, rather than a short vowel /-ɒ/, placed this word in a category of its own among words ending in -ify. His belief was unsound: cf. gratify, verify, vilify, modify, mummify, all with short vowels in the first syllable; but ladify, deify, nitrify, glorify, notify, purify, stupefy, all with long vowels or diphthongs. The patterns are the result of various analogies, and are not governed by the ending -ify. In AmE the form of codify with short vowel in the first syllable is given precedence.

**codling, codlin.** (a) Codling, a young or small cod, is formed from cod + the diminutive suffix -ling, and should always be spelt thus. (b) A codling is a variety of apple (Kentish Codling, Keswick Codling, etc.), and the word is a 16c. respelling of quodling, earlier (14c.–15c.) querdling, of disputed origin. Codling-moth, a species of moth (Carpocapsa pomonella), the larva of which feeds on the codling and other apples, is frequently (by a kind of dissimilation) written codlin-moth and pronounced /ˈkɒdlɪn mɒθ/.

**coed, co-ed.** Its use as a noun meaning (a) a coeducational institution, (b) (esp. US) a female student at a coeducational institution, has become much less frequent as such institutions have become less uncommon. The distinction is much less noteworthy than it was, for example, when Auden wrote Enormous novels by co-eds Rain down on our defenceless heads in Nones (1951). But the word remains in frequent use as an adj., '(of an educational institution) open to or attended by both males and females'.

**coffin.** See CASKET.

**cognate.** 1 In philology, used (as adj. and noun) of a word in a given language having the same linguistic family or derivation as the corresponding word in another language, as Eng. father, Ger. Vater, L pater, all going back to the same Indo-European base *pətēr. The term is also applied to languages derived from the same original language, e.g. Spanish, French, Italian, and Portuguese, all descended from Latin, are cognate languages.

2 cognate object (or cognate accusative). In grammar, an object that is related in origin and sense to the verb governing it (as in live a good life, die a cruel death, he fought the good fight, as distinct from enjoy a good life, suffer a cruel death, win a hard fight). As CGEL remarks, 'Most cognate objects tend to convey a rather orotund style.'

**cognizance, cognizant, cognizable.** This group of words was frequently pronounced with the g silent (thus /ˈkɒnɪzəns/, etc.) until well into the 20c., and COD (1995) continues to list this mode of pronunciation as an acceptable variant. It is recommended that the g be sounded, if only to allow the words to conform with related words in which it has always been pronounced, as cognate, cognition, and cognitive.

**cohort.** In the Roman army a cohors (acc. cohortem) was a body of infantry consisting of 300–600 men (in some units 500 to 1,000 men). Ten cohorts made a legion. Taken into English in the 15c., the word was quickly used of similar divisions in other armies (e.g. the Cohort bright Of Watchfull Cherubim—Milton, 1667; The Assyrian came down like a wolf on the fold, And his cohorts were gleaming in purple and gold—Byron, 1815), and then of any band of persons united in defence of a common cause (My friends and my acquaintances ... I had a numerous cohort of the latter—Bolingbroke, 1719).

The word took a dramatic turn in the mid-20c. First it was applied in demography to a group of persons having a common statistical characteristic, esp. that of being born in the same year. As technical demographic jargon, the use lies out of reach of the general public and is doing no one any harm. But more or less simultaneously it began to be used in AmE to mean 'an assistant, a colleague, an accomplice': *The old poet had left, accompanied by two of his cohorts*—M. McCarthy, 1952; *Mr. Stratton consented, since it was four o'clock, to partake together with his cohort, of a sandwich and a glass of milk*—A. Cross, 1967. Before long the use turned up in the British press and in other contexts in Britain: *The impending trial of Bobby Seale, chairman of the Black Panther movement, and his eight cohorts in New Haven*—*Sunday Times*, 1970; in 1988 a BBC announcer introduced a dramatization of Dorothy Sayers's *Gaudy Night* by a few remarks about the heroine, Harriet Vane, and her *cohort*, Lord Peter Wimsey—P. Christophersen, 1989; *Brock and Emma had one wall, Bob, Johnny and their cohorts the other wall and centre aisle*—J. le Carré, 1989.

If, as seems likely, this new meaning becomes standard in due course, it will be a remarkable example of a descending scale of meaning (from 'a tenth of a Roman legion' to 'a single assistant'). Our language becomes vulnerable when the majority of its speakers forget major sections of Western history, in this case the fighting arrangements of Roman armies.

**coiffeur, coiffure.** *Coiffeur*, meaning a hairdresser, is pronounced /kwɑːˈfɜː/ (fem. *coiffeuse* /-ˈfɜːz/). *Coiffure*, meaning the way hair is arranged, is pronounced /kwɑːˈfjʊə/. All three are still sometimes printed in italics in English, even though they are relatively long-standing loanwords from French, *coiffeur* (1847), *coiffeuse* (1870), and *coiffure* (a1631).

**coinages.** Each century is characterized by a new burst of word-coinages. The 20c. is no exception: indeed, in terms of innovation, probably no century from the time of the first English records in the 8c. has been anything like as prolific. A brief, far from exhaustive indication of the main types of modern innovation must suffice here. **1** New words formed from existing prefixes: *apolitical* (1952), *decaffeinate* (1934), *metagalaxy* (1930), *microwave* (1931), *mini-series* (1972), *multistorey* (1918), *neoprene* (1937), *non-event* (1962), *retro-rocket* (1948), and *ultrasonic* (1923).

**2** New words formed from existing suffixes: *majorette* (1941), *sit-in* (1937), *voguish* (1927), *privatization* (1959), *carless* (1927), *pianola-like* (1945), *pushiness* (1920), *beatnik* (1958), *chillsome* (1927), and *jazzy* (1919).

**3** More complicated formations: *demythologize* (1950), *denazify* (1944), *paraphrasability* (1965), *post-doctoral* (1939), *prepsychotically* (1941), *renormalization* (1948), and *rubbernecker* (1934).

**4** New nominal compounds: *Cold War* (1945), *egghead* (1907), *lunatic fringe* (1913), *petrol pump* (1928), *plasma physics* (1958), *Rorschach test* (1927).

**5** Newly coined words, i.e. words not formed on existing elements (the smallest class): *ocker* (boorish person, Australian, 1971), *oi(c)k* (obnoxious person, 1925), *oomph* (sex appeal, 1937), *oracy* (fluency in speech, 1965).

**6** See also ACRONYM, BACK-FORMATION, BLEND, EPONYM, RHYMING SLANG.

The volume of innovation can be judged from a single issue of *The Barnhart Dictionary Companion* (fall–winter 1986), which lists more than 1,200 new words and meanings that by 1986 had not yet been included in the latest dictionaries. The words range from *adoptive immunotherapy* to *zero-zero option*. This quarterly collection, drawn from a selected range of printed sources in the US and Britain, suggests that the annual intake of new words at present is of the order of 5,000. The following examples give a broad indication of the kind of neologisms that are pouring into the language: *Cabbage Patch doll, chaos theory, dating parlor, Eurotunnel, Irangate, liposuction, megadebt, superkid, technospeak*, and *yuppie flu*. To which can be added some that I have collected myself: *differently abled, final club, francheezie sandwich, houby, lambada, mommy track, outplacement, sitnor, technonerd*, and *wordmeister*. But their permanence in the language can hardly be guaranteed.

**colander,** a word of long standing in the language (first recorded c1450), is normally pronounced /ˈkaləndə/. The variant form *cullender*, once common, is less often encountered now.

**Cold War.** This term for hostilities (threats, propaganda, building up of deadly weapons, espionage) short of armed conflict, esp. that between the USSR and the Western powers after the 1939–45 war, came into being in 1945 and came to an end, except as a historical term, in 1989–90, as various Eastern bloc countries overturned their communist governments, and the republics formerly making up the USSR itself began to seek ways of establishing a Western-style market economy.

**coleslaw,** a dressed salad of sliced raw cabbage, carrot, onion, etc., is so spelt, the first element being *cole* 'cabbage'. *Coldslaw* is a respectably old (first recorded in 1794), but erroneous, US variant. *Coleslaw* is an anglicized version of Dutch *koolsla*, from *kool* 'cabbage' + *sla*, a reduced colloquial form of *salade*.

**colic** yields the adj. *colicky*. For the spelling, see -C-, -CK-.

**coliseum, colosseum.** A *colosseum* is a large stadium or amphitheatre (from medL neut. of L *colosseus* 'gigantic'), esp. the one (with a capital *C*) in Rome. The variant *coliseum* is used, esp. in America, for a large public building for sporting events, exhibitions, etc., and, with capital *C*, as the name of music halls (including the well-known one in London), cinemas, and dance-halls.

**collaborate.** This verb, and the corresponding noun *collaborator*, acquired the sinister connotation of traitorous cooperation with the enemy during the 1939–45 war. Fortunately the original sense, namely, to work in conjunction with another or others, esp. in a literary or artistic production, can still be used without risk of ambiguity.

**collapsible.** See -ABLE, -IBLE 3.

**collation.** In the Roman Catholic Church a light meal allowed in lieu of supper on fasting days. Originally applied to the light repast or refection taken by the members of Benedictine monasteries at the close of the day, after the reading of Cassian's *Collationes Patrium* ( = Lives of the Fathers). In lay circles, the word is seldom used except in *cold collation*.

**colleague.** Primarily 'one chosen along with another, a partner in office' and (*OED*) 'not applied to partners in trade or manufacture', illustrated typically by a comment by Gladstone in 1851: *The treaty … upon … which, as a Colleague of your Lordship, I had the honour to be employed.* The word has been sharply democratized in the 20c. Representatives of trade unions speak of their *colleagues* (as *brothers* joins the ranks of less favoured words); fellow-strikers, fellow-prison officers, associates of an kind, now qualify for the term *colleagues*. The word is on a downward spiral.

**collectable, -ible.** See -ABLE, -IBLE 3.

**collective noun,** a noun that is normally singular in form and denotes a collection or number of individuals. Apart from the names of individual animals, birds, etc. (*deer, grouse, sheep,* etc.), terms for groups of animals, birds, etc. (*a pride of lions, a gaggle of geese,* etc.), and the names of institutions, firms, and teams (*BBC, Marks & Spencer, Tottenham Hotspur,* etc.), there are some 200 collective nouns in common use in English (*choir, flock, multitude,* etc.).

There are several main considerations. **1** In BrE it is in order to use *either* a plural verb *or* a singular verb after most collective nouns, so long as attendant pronouns are made to follow suit: *when the jury retires to consider its verdict; when the jury retire to consider their verdict.* The same principle applies to all the main collectives like *army, audience, clan, company, court, crew, folk, government, group, herd.* By contrast, in AmE the choice is much more restricted. For such words the following verb and any attendant pronouns are usually in the singular.

**2** The principle at work is that if the collective noun is thought of as a unit, a singular verb or pronoun follows naturally (*Mention the family and everyone is for it, however they define it*); but if the members of the group are thought of as individuals a plural verb or pronoun is appropriate (*I looked at my family and I knew that they did not know what I had seen*).

**3** Collective nouns that stand for a group of inanimate objects, e.g. *baggage,* *china, cutlery, earthenware, linen, luggage, stoneware,* are always treated as singular.

**4** When a collective noun is followed by *of* + a plural noun or pronoun, the choice between a singular and a plural verb remains open, but in practice a plural verb is somewhat more common: *a large number of conductors* want *to hear the great artists—Dædalus, 1986; a handful of their members* have *been agents of Moscow—London Rev. Bks, 1987; a handful of bathers* were *bobbing about in the waves—P. Roth, 1987; the current crop of bestsellers* include *a number of monuments to bad taste—The Face, 1987.* Contrast: *one group of disenchanted Levellers* puts *pen to paper—M. Lasky, 1985; a fleet of helicopters* was *flying low—New Yorker, 1986; a pool of sheep* is *assembled for each trials session—Daily Tel., 1987.*

**5** Particular cases: (*a*) *the United States, the United Nations, the Vatican, the Commons, the House of Lords, Congress* (US) are always treated as singular (e.g. *the United States* has *demanded a more open Japan—Dædalus, 1987.* (*b*) institutional names normally require a singular verb: *the CEGB* finds *it 25 per cent cheaper to buy in French electricity—Daily Tel., 1987.* But not invariably: *MI5* were *living in the past—P. Wright, 1987.* (*c*) animal, bird, etc., names that are singular in form but are being used in the plural naturally require a plural verb (and plural pronouns): *five bison* were *grazing in a shaded part of the valley; several silverfish* were *scuttling across the floor to* their *hiding-place.* (*d*) a number of abstract nouns, singular in form, are the equivalent of concrete plurals: *accommodation* ( = rooms), *kindling* ( = small pieces of wood), *royalty* ( = royal persons), *pottery* ( = earthenware pots).

See also AGREEMENT 5.

**college.** The word has many long-established meanings: *Apostolic college,* the body of Christ's Apostles; *college of cardinals,* the cardinals of the Roman Church, who constitute the Pope's council, and elect to the papacy from their own number; the *College of Arms, College of Physicians,* and so on. To be a member of such bodies is to be a person of considerable distinction. *College* was adopted at an early stage as a name for the independent societies within great universities, as the College of the Sorbonne in the ancient University of Paris, and the ancient colleges of Oxford and Cambridge, consisting normally of a *master* (*rector, provost, rector, president,* etc.), *fellows, scholars,* and *commoners.* Other universities have followed suit (e.g. *King's College, London,* a constituent member of the University of London), but have not adopted the Oxford and Cambridge terminology for the titles of teaching members of staff and their students. As time went on, esp. in the US and Scotland, the name sometimes became interchangeable with *university,* but was applied mostly to any institution for higher education affiliated to a university.

In wider educational circles, *college* has been traditionally used in the names of some of the ancient public schools (notably Eton and Winchester). In much more recent times it has come to be applied to a much broader group of educational and professional institutions—business colleges, training colleges for teachers, sixth-form colleges, secretarial colleges, military and naval colleges, colleges of agriculture, and so on. All this is part of the natural process of educational development in English-speaking countries. One consequence is that a general phrase like *at college* needs to be unravelled. It can mean that the speaker is at a famous university or at a very different kind of institution.

**collide, collision.** It is sometimes asserted that these two words are 'properly' restricted to circumstances involving a violent impact between two moving objects. There is no basis for such a belief. A car can *collide* with a tree, a bollard, or any other fixed object, as well as with another moving vehicle. Nevertheless there is a lot to be said for using *hit* instead of for factual reporting: [Two young men named] *died when their car hit a tree between Penzance and Land's End yesterday—Times, 1990.*

**collocation.** A term introduced in linguistics (though in much earlier use as a general word) by J. R. Firth to refer to the habitual co-occurrence of individual words. Thus *Tweedledum* goes with *Tweedledee,* and *spick* with *span.* A person killed in a traffic accident in Oxford in September 1988 was described by a witness not merely as a *bystander* but as an *innocent bystander. Bystanders* are almost always *innocent. Consequences* are often *far-reaching*; politicians are often *cautiously optimistic*; new medical discoveries often have *far-reaching implications*; *liars* are frequently called *habitual liars*. These are just elementary examples of an adhesive process which is part of the natural machinery of the language.

See SIAMESE TWINS.

**colloquial.** As a dictionary label, *colloquial* (usu. written as *colloq.*) indicates a use that is normally restricted to informal (esp. spoken) English. In dictionaries produced in the UK, the distinction has been taken as self-evident by users throughout the 20c. and it calls for no particular defence. Since the publication of Webster's Third in 1961, in which the label *informal* was used instead (though much more sparingly), some other dictionary houses at home and abroad have followed the lead, maintaining that the label *colloq.* is taken by users to mean 'derogatory'. It is hard to follow the reasoning, but it is nevertheless clear that the synonymous label *informal* seems now to be the more widely used of the two.

**collusion** /kə'l(j)uːʒ(ə)n/. The notion of fraud or underhandedness is essential to collusion. It is correctly used, for example, of insider dealing by two or more people acting in concert on the stock exchange. It would not be right to speak of joint authors of a book as being *in collusion* with each other. Such authors can be said to have *joined forces, collaborated*, etc., in the writing of the book, but not to have been *in collusion*, and not to have *colluded* in the writing of the book.

**colon.** The name of this punctuation mark (:) is derived from L *cōlon*, from Gk κωλον 'limb', member or clause of sentence, portion of strophe. As a mark of punctuation (as distinct from a term in rhetoric) it was adopted in English in the 16c. Since then its employment has been made subject to various (changing) rules. Henry Beadnell's *Spelling and Punctuation* (edn. of 1919), a standard work at the time and one of Fowler's sources in 1926, remarked that 'it is used to separate those parts of a sentence which have very little dependence on each other in construction, or which are only removed one degree from complete

independence'. He then stated five rules that covered, he believed, all the conventions. Fowler expressed his own view of the function of the colon more amusingly than helpfully: 'the colon ... has acquired a special function, that of delivering the goods that have been invoiced in the preceding words; it is a substitute for such verbal harbingers as *viz., scil., that is to say, i.e.,* etc.'

The most satisfactory guidance is given in *Hart's Rules*:

'Whereas the semicolon links equal or balanced clauses, the colon generally marks a step forward, from introduction to main theme, from cause to effect, premiss to conclusion., etc. e.g.:

> In business there is something more than barter, exchange, price, payment: there is a sacred faith of man in man. Study to acquire a habit of thinking: no study is more important.

It is regularly used to introduce examples, as:

> Always remember the ancient maxim: Know thyself.

A dash should not be added to a colon which is being used to introduce a list.'

To which it should be added that a colon is also used between numbers in a statement of proportion (as in 10:1) and in biblical references (as in Gen. 4: 3). It also appears on most digital clocks and watches to mark off the hours from the minutes (as in 21:30).

Americans use a colon after the initial greeting in a letter

> *Dear Mr Jones:*
> *Thank you for ...*

In BrE a comma is customary in such contexts (or no punctuation at all):

> *Dear Mr Jones(,)*
> *Thank you for ...*

**colossal.** In its ordinary sense 'like a colossus, gigantic', it is a relatively recent formation in English (first recorded in 1712 and not listed in Johnson's Dictionary, 1755). The use of the word as a mere intensive = 'remarkable, splendid' (*a colossal blunder, the novel is simply superb, colossal*) is a Germanism. Like all intensives, it can cover a wide spectrum of values, and should be sparingly used. The *OED*'s earliest example of its use as an intensive is one of 1892.

**colosseum.** See COLISEUM.

**colour.** In BrE the customary spellings for words related to *colour* are *colourable, colourant, coloured, colourful, colourist,* and *colourless,* but *coloration, colorific, colorimeter,* and *decolorize.* In AmE all have *-or-,* not *-our-.*

**coloured.** 1 In S. Africa, written with a capital C, it means a S. African of mixed descent; also, in pl., a racial group as officially defined under the former apartheid laws.

2 The word was formerly used euphemistically by white people in Britain and elsewhere as a term meaning 'nonwhite', but is much less commonly heard now.

**colour words.** 1 See PRIMARY COLOURS.

2 One of the fastest-growing areas in the language is the use of new colour names, esp. for commercial purposes. Such words as *beige* (first recorded as a colour word in 1879), *biscuit* (1884), *bisque* (1922), *Marina green* (1935), *marmalade* (1926), *oatmeal* (1927), *old gold* (1879), *oyster* (1922), *pillar-box red* (1916), *taupe* (1911), and *tea-rose* (1892) are typical of the names now applied to garments, motor vehicles, plants, etc. It is a far cry from the limited range of colour words available in Anglo-Saxon times and in ancient Greece and Rome.

**columnist.** Though the *n* in *column* is silent, the same letter is usually fully pronounced in *columnist.*

**combat** (noun). Pronounce /'kɒmbæt/, but be prepared to encounter the somewhat old-fashioned pronunciation with initial /'kʌm-/. Similarly with the derivatives *combatant* and *combative.*

**combat** (verb). 1 If pronounced with the stress on the first syllable, the same variation of the stem vowel is found as in the noun. The extended forms are *combated* and *combating.*

2 If pronounced with the stress on the second syllable (but this is not recommended exc. in AmE), the pronunciation is /kəm'bæt/ and the inflected forms are *combatted* and *combatting* (e.g. *It is obvious that covert action is only one instrument among many in combatting terrorism*—*Dædalus,* 1987).

**combining form.** A term, probably first used in the *OED* itself, for a linguistic form that normally occurs only in compounds or derivatives as a means of coining new words. Examples: *electro-,* formally representing Gk ἠλεκτρό-, combining form of ἤλεκτρον 'amber', as in *electrodynamic, electromagnet,* etc.; *kilo-,* an arbitrary derivative of Gk Χίλιοι 'a thousand', as in *kilogram, kilometre,* etc. Two combining forms can combine to form a new word: e.g. *leuco-* (Gk λευκός 'white') + *-cyte* (Gk κύτος 'receptacle') made *leucocyte* 'white blood cell'.

The connecting vowel in combining forms is normally *-o-* (Anglophile, barometer, philology, etc.) or *-i-* (altimeter, horticulture, etc.).

**come** (verb). 1 The slightly archaic use of *come* with a future date following as subject, as in *four years ago come Christmas* is compared by the *OED* (sense 36a) to a similar construction in French: *dix-huit ans vienne la Saint-Martin ... viennent les Pâques* 'eighteen years old come Martinmass ... come Easter'; i.e. = let Easter come, when Easter shall come. The construction is first recorded in the early 15c., and has been in continuous use since then.

2 In standard English *come* is often followed by *and* as an introduction to a verb indicating the purpose of coming. This construction has been in continuous use since the OE period: e.g. *Come, lovely nymph, and bless the silent hours*—Pope, 1704; *I wonder ... whether anyone will come and help me*—D. Potter, 1986. Formerly in standard BrE and still in AmE (occas. elsewhere) *come* is followed by a bare infinitive, i.e. divided from the following verb neither by *and* nor by *to:* e.g. *Quicke, quicke, wee'le come dresse you straight*—Shakespeare, 1598; *Come lie up close beside me*—E. Jolley, 1980 (Aust.); *We can sell this house, you can come live with us*—Lee Smith, 1983 (US); *Come let us say a prayer together and then we'll read some Tennyson*—J. Gardam, 1985 (UK); *Richie, come help me lift the turkey out*—J. Updike, 1986 (US); *'Do you think Dad knows one of his paintings is here?' 'No. If he knew, he'd come steal it off the wall.'*—New Yorker, 1992.

**come-at-able, get-at-able.** English tolerates the creation of many unwieldy formations. These two (the first of them first recorded in 1687 and the other in 1799) are notable examples. Also *come-at-ability* (1759). *get-at-ability* (1863), *un-come-at-able* (1694), *un-get-at-able* (1862), etc.

**comedian.** 1 Used first to mean a comic actor, a comic stage-player (first recorded in Shakespeare's *Twelfth Night*), it came, in the course of the 19c., to mean a professional entertainer first on the stage and later on radio and television. The term can be used both of men (Bob Hope, Danny Kaye) and of women (Victoria Wood).

2 From about 1860 onward a female comedian could also be called a *comédienne* (Elsie and Doris Waters, Barbra Streisand are modern examples), pronounced and written in a French manner. Both terms remain in common use.

3 The parallel terms *tragedian* (1592) and *tragédienne* (1851, also without acute accent) have had a similar history.

4 *Comedian* and *tragedian* have also for centuries been used to mean, respectively, 'a writer of comedies, a comic poet' and 'a writer of tragedies, a tragic poet'. In context, over several centuries, senses 1, 3, and 4 seem to have remained comfortably distinguishable, but neither word (esp. *comedian*) seems to have much currency in this fourth sense nowadays.

**comestibles** has been sparingly used as a humorous or affected word for 'food' since the early 19c.

**comic, comical.** The broad distinction is that *comic* is used of literary compositions, songs, dramatic performances, performers, etc., whose main aim is comedy, as *comic actor, opera, poet, relief, scene, strip*; whereas *comical* is principally used to mean 'laughable, mirth-provoking' in contexts of unintended humour, as *comical attempt, behaviour, effect, face, sight*, etc. The distinction is far from being a general rule, however. *Comic* is much the more frequent of the two words. Moreover, there are a great many contexts in which either word is serviceable: *his attempt to be musical is comic(al) in the extreme; there was something comic(al) in the sight of the archbishop lying flat on his back*. Both examples are listed in the *OED*, the first (1833) s.v. *comic* 3b and the second (1887) s.v. *comical* 4.

**comity** /'kɒmɪtɪ/. The primary sense of *comity* is 'courtesy, kindly and considerate behaviour towards others'. This sense, which was first recorded in the 16c., continues in use: *He was exposed and condemned, as though he could never re-enlist in the comity of the ward*—F. Tuohy, 1984; *In the Cuite era, these matters were usually worked out in comity with whoever was mayor*—New Yorker, 1986. It is derived from L *cōmitātem* 'courteousness', from *cōmis* 'courteous, friendly'.

The phrase *comity of nations* (or *peoples, states*, etc.) properly means 'the courteous and friendly understanding by which each nation respects the laws and usages of every other'. It is sometimes mistakenly taken to mean simply 'a company or group of nations or people who are friendly to one another' (perhaps originally by association with L *comes* 'companion'): e.g. *To show ... how much Iran had it at heart to obtain admittance into the comity of states*—tr. H. Vámbéry, 1864; *There were voices which spoke for Russia in the comity of civilised people other than those of ministers and Tsars*—P. Ustinov, 1983.

**comma.** The word *comma* is derived from L *comma*, Gk κόμμα 'piece cut off, short clause'. It was first used in English in the 16c., in imitation of the terminology of Greek rhetoric and prosody, to mean 'a phrase or group of words less than a colon (q.v.)' (*OED*). As a mark of punctuation it was also introduced into English in the 16c. The standard account of OUP practice in the use of the comma is that in *Hart's Rules*. The following description is based on Hart but with fresh examples taken from modern sources.
1 Generally, commas should be inserted between adjectives preceding and qualifying a noun, as: *in hoarse, melancholy tones; a strong-willed, intellectual woman; wild, surging, sealike landscapes*.

2 But, by convention, in many contexts where the last adjective is in closer relation to the noun than the preceding ones, or where the adjectives form a kind of unit, the comma is often omitted. Also frequently when the last adjective has no such priority. *abundant patriotic pageantry; a preternatural clerical clumsiness; a good little boy; a super-efficient liquid-cooled rear engine*.

**3** Where *and*, *but*, or *or* join two or more adjectives the comma is usually omitted: *a reluctant and limited step*; *the imaginative and dangerous energy of the son*; *solid but adventurous entrepreneurs*; *no extensive or protracted military intervention*.

**4** Where more than two words or phrases or groupings occur together in a sequence, a comma should precede the *and* (the omission of the second comma in the fourth example would render the sentence ambiguous). This is the so-called 'Oxford comma'. Examples: *an index of social, economic, and religious diversity*; *excesses of public, political, and intellectual opinion*; *areas of natural beauty, architectural monuments, and sites of historical interest*; *New premises were opened by Marks & Spencer, Jaeger, and Currys*.

The 'Oxford comma' is frequently, but in my view unwisely, omitted by many other publishers. Their preference is to omit it as a general rule (e.g. *tea, scones and cake*) but to insert it if there is a danger of misunderstanding (*tea, bread and butter, and cake*—examples from J. McDermott, 1990). A fuller example: The Mind of South Africa *is an extremely ambitious blend of personal memoir, ideological polemic and orthodox history*—R. Malan in London Rev. Bks, 1990.

**5** A comma is customary in sentences containing two main statements joined by a conjunction, or having some kind of complementary relationship. In such sentences, a comma is the equivalent of a slight pause. Examples: *Parliament is not dissolved, only prorogued*; *I believed, and therefore I spoke*; *The question is, can this be done?*; *My dear chap, please listen to me*; *One of the most densely populated countries in the world, it continually needs to reclaim land from the sea*; *Rents are high, and the shops are already going out of business*.

**6** Plainly parenthetic clauses, phrases, or single words require commas before and after them: *Yet there's a chance that, despite the razzmatazz, van Gogh may actually not be turning in his grave*; *But the general tenor of life, even in Amsterdam, is peaceful and well organized*; *There were, to be sure, occasional eccentrics in nineteenth-century novels*; *Be assured, then, that I will not let you down*; *He saw, a moment later, that the game was up*.

**7** A restrictive (or defining) relative clause does not require a comma. A non-restrictive (or non-defining) relative clause, i.e. one which by its nature supplies extra information, does. Thus (the examples are taken from CGEL 17.13): (restrictive) The woman who is approaching us *seems to be somebody I know*; (non-restrictive) The Bible, which has been retranslated, *remains a bestseller*.

**8** Such adverbs as *already* and *soon* when used as the first word of a sentence are usually followed by a comma. So too with *however* and *moreover*, and when these two words are used in mid-sentence they are preceded and followed by a comma. Examples: *Already, prints and posters have turned anguished, passionate paintings into mere features of the décor*; *Muggings, however, are still reported in the Vondelpark after dark*; *Moreover, you were late home after school*; *Soon, some inner compulsion erupts into the pretty pictures*.

**9** A comma is sometimes needed in order to avoid ambiguity: *In the valley below, the villages look very small* (so that *below* is not taken to be a preposition); *Mr Douglas Hogg said that he had shot, himself, as a small boy* (Mr Hogg is still alive).

**10** Omit the comma in such oppositive phrases as *my friend Judge Leonard, my son Jon*. But commas are needed in the type *His father, Humphrey V. Roe, was not so fortunate*.

**11** Omit the comma when giving house numbers in addresses: *44 High Street*. Omit the comma in dates; *27 July 1990* (not *July 27, 1990*).

**12** Warning. These rules apply now (1995), but wide variation can be seen in the work of many contemporary writers and, even more so, in that of earlier centuries. Thus: (a) We are all accustomed to the kind of endless comma-joined sentences that turn up from children or from our less literate friends, e.g. *We had a holiday in Florence, it was very hot, we could hardly bear it* ... This device is commonly called the 'comma splice'.

Curiously, this habit of writing comma-joined sentences is not uncommon in both older and present-day fiction. Modern examples: *I have the bed still, it is in every way suitable for the old house where I live now*—E. Jolley, 1980 (Aust.); *The audience tonight have been really sludgy, they were against us right from the start—Rosie Scott, 1988 (NZ); We saw Bob's cousins Denny and Donny, they live outside Las Vegas where they race cars on weekends—G. Keillor, 1989; Marcus ... was of course already quite a famous man, Ludens had even heard of him from friends at Cambridge—I. Murdoch, 1989. (b) The subject of a sentence should not be separated by a comma from the verb it governs: The charm in Nelson's history, is the unselfish greatness (read history is the); The primary reason that utilities are expanding their non-regulated activities, is the potential of higher returns (read activities is the).

**commando.** Pl. commandos. See -O(E)S 6.

**commence.** Families and neighbours are divided by the question of when to use commence and when to use begin. The main considerations are as follows: **1** Begin is a native word; commence, a French loanword, joined the language in the 14c. The two words have slugged it out for nearly seven centuries, and commence, in boxing parlance, is still well behind on points.

**2** It is a sound rule to use begin in all ordinary contexts unless start is customary (the engine started straight away; he starts work at 9 a.m.; the game started on time). Commence 'has more formal associations with law (to commence an action) and procedures, combat (hostilities commenced on 4 August), divine service, and ceremonial, in which it continues earlier Anglo-French use' (OED). As a general rule it should be reserved for such contexts.

**3** Until about the end of the 19c. both begin and commence were commonly used in constructions with a to-infinitive and with a gerund. In present-day English, begin is by far the more usual (she began to cry/she began crying). WDEU cites three 20c. examples of commence + to-infinitive, including a 1945 diary entry of Harry S. Truman: After discussing the day's prospects ... then commence to see the customers. My own files contain only one, and also only one example of commence + gerund: The boatman commenced smashing at the ice with the blade of his oar and then started tossing decoys out to the side and behind him—E. Hemingway, 1950. Worth noting (if only to condemn) is the mixed construction commence + to + gerund (both examples

are AmE): She commenced to following me around—A. Tyler, 1977; Then he commenced to coming by our place—M. Golden, 1989.

**4** The use of commence with a direct object (e.g. to commence author), recorded from the mid-17c. onward, is now archaic.

**commentate** is a predictable back-formation from the noun commentator. It was freely used from the late 18c. onward in the broad sense 'to furnish with comments', but such uses are now rare. The word needed to be revived in the 20c. in broadcasting contexts in the sense 'to deliver an oral commentary, esp. upon politics or sport, to act as a commentator'.

**commercialese.** The term refers to a range of stereotyped devices once commonly, and now much less commonly, used in business letters. Characteristic expressions (as listed by Gowers, 1965) included ult. (last month), inst. (this month), prox. (next month), even date (today), favour or esteemed favour (a customer's letter), to hand or duly to hand (received), same (it), your good self (you), beg (an obsequious verb widely used by tradesmen and by people writing letters to newspapers), advise (tell, inform), per (by), as per (in accordance with), re (about, concerning), be in receipt of (receive), enclosed please find (I enclose), and oblige (please), assuring you of our best attention at all times (a signing-off formula). A letter written in this kind of outdated commercialese might have begun: Your esteemed favour of even date to hand and we beg to thank your good self for same.

Concerted attacks on such language by committees of inquiry at various dates since about 1920, and especially by people like Sir Ernest Gowers, have largely driven out such language from business correspondence, though I have noticed the phrases has come to hand, interest as per attached statement, and we advise that in recent letters from a bank.

**commercial traveller.** This term for 'an agent for a manufacturer, wholesale trader, etc., who travels over a district, showing samples and soliciting orders' (OED), first recorded in 1807, has now been discarded in favour of the higher-sounding term sales representative (often shortened to representative or rep). The crossover point was approximately 1950.

**commiserate.** For about three centuries only a transitive verb (e.g. *She did not exult in her rival's fall, but, on the contrary, commiserated her*—H. Ainsworth, 1871), under the influence of *condole with* and *sympathize with*, etc., it is now always construed with *with* (*We commiserated with one another on our various hurts*—K. Hulme, 1984).

**commissionaire** is the recommended spelling for a uniformed door-attendant, not *-ionnaire*.

**commonality.** Since the early 1950s, this word has been chiefly used in the senses 'community of function, structure, or purpose'. It tends to be restricted to certain kinds of technical writing: e.g. *Commonality can be increased when subjects are asked to predict common normative responses*—Jrnl General Psychology, 1971; *One would therefore expect that the commonalities of the sciences, in particular their shared philosophical underpinnings, would now be more evident than ever*—Dædalus, 1977. But it is beginning to enter the common core of the language: e.g. *If he thought her attractive, and he thought them attractive, there had to be some commonality*—M. Gordon, 1985.

**communal.** Pronounce with the main stress on the first syllable. /ˈkɒmjʊn(ə)l/.

**commune.** The noun, pronounced /ˈkɒmjuːn/, and the verb, pronounced /kəˈmjuːn/, are regularly distinguished by the positioning of the main stress.

**community charge.** In the late 1980s until 1993 the official term in the UK for a tax levied locally on every adult in the community. It was called the *poll tax* by the majority of the population. Replaced in 1993 by the *council tax*, based on the band in which the estimated capital value of a property falls.

**commuter.** This long-established American word and the related verb *commute* became established in the UK about the middle of the 20c. with reference to a person who travels regularly by rail or road to and from his or her place of work, esp. a suburbanite travelling to and from a large city. A ticket used for this purpose on a train or bus is called a *commutation ticket* in the US and a *season ticket* in the UK.

**compact.** As a predicative adj. and as a verb it is normally pronounced /kəmˈpækt/, with the stress on the second syllable. When used attributively, as in *compact disc*, the stress moves to the first syllable: thus /ˈkɒmpækt/. As a noun, (a) a case for face-powder, (b) an agreement, it is always stressed on the first syllable. See NOUN AND ADJECTIVE ACCENT.

**comparable.** 1 The main stress falls on the first syllable, /ˈkɒmpərəbəl/. Nonstandard speakers frequently place it on the second syllable.

2 Its uses with *to* and *with* correspond to the senses at COMPARE.

**comparatively.** 1 This adverb, like *relatively*, has been regularly used with an immediately following adj. since at least the early 19c. (e.g. *A comparatively modern phraseology*—R. H. Barham, 1840). Direct comparison need not be in question: *comparatively* simply = 'fairly, somewhat'. Thus: *Emma ... was comparatively out-going*—Bodleian Library Record, 1987.

2 Uses with the adj. *few* have sometimes been seen as more debatable. It has been argued that if a comparison is directly made (e.g. *There were many casualties but comparatively few* [of them] *were fatal*) the use is clearly unobjectionable. Gowers (1965) strongly objected to the type *casualties were comparatively few*, i.e. there were not many casualties. Logically, however, it is difficult to separate such uses from those in (1), esp. as other qualifying adverbs are idiomatically used with *few* (*too few, very few*). As *CGEL* (5.24) expresses it, 'Being gradable, *many, much, few*, and *little* can be modified by intensifying adverbs: *too much, very few*, etc.'

**comparatives** (i.e. comparative adjectives). See ADJECTIVE 3, 4; -ER AND -EST.

**compare.** There are four main sets of circumstances. 1 When the meaning is 'to speak of or represent as similar, to liken', *compare* is normally construed with *to* (or in religious works sometimes *unto*): *Shall I compare thee to a Summers day?*—Shakespeare; *I haue compared thee, O my loue, to a company of horses in Pharaohs* *chariots*—S. of S. (AV) 1: 9; *For who in the heauen can be compared vnto the Lord?*—Ps. (AV) 89: 6; *it is hardly complimentary to Parrot, an undeniably handsome bird, to compare my grandson to him*—R. Graves, 1934. In these examples, stress is placed mainly on the likenesses, even though in Shakespeare's sonnet the loved one is declared to be *more louely and more temperate.* Fowler (1926), to bring out the distinctiveness of this group, gave the sentence *he compared me to Demosthenes* and said that it meant 'he suggested that I was comparable to Demosthenes or put me in the same class'.

**2** When the sense is 'to mark or point out the similarities and differences of two or more things', *compare* may be construed either with *with* or with *to.* Examples: (to) *This cramping tendency of town as compared to country*—G. C. Harlan, 1879; *to compare NIVEs* [certain types of non-standard English] *to other acquisitional phenomena is not to deny their legitimacy*—English World-Wide, 1987 ; *The gift of heaven, grace through virtue, is compared to the gift of nature*—English, 1987; (with) *To compare Great things with small*—Milton, 1667; *He did not individually compare other women with her, but because she was the first, she was equal in his memory to the sum of all the others*—J. Berger, 1972; *Isabella then compares Angelo's judgement with God's judgement*—English, 1987.

**3** When a subordinate clause or phrase is introduced by the past participial form *compared,* the preposition is either *to* or *with*: (to) *compared to Ward's witchhunters, Profumo is an almost blameless character in the story*—London Rev. Bks, 1987; *This was a modest sum compared to what other people spent*—T. Wolfe, 1987; *Even allowing for that, it's well down, Vic. Compared to last year*—D. Lodge, 1988; (with) *The church looked dimly mysterious compared with the glare of the passage*—P. D. James, 1986; *the same daily burden ... of everything being just slightly disappointing compared with what one knows one has inside oneself*—J. Updike, 1988.

**4** In BrE, the construction with *with* is obligatory when *compare* is used intransitively: *As athletes men cannot ... compare with horses or tigers or monkeys*—W. S. Jevons, 1876; *This compares favourably with the inertness of England*—1891 in OED; *New York does not for a moment compare with*

*Chicago*—W. Archer, 1904; *His achievements do not compare with those of A. J. Ayer*—Sunday Times, 1988. But *WDEU* cites AmE examples of *to*, e.g. *ham and bamboo shoots do not compare to those made at Ying's*—NY Times, 1977.

**comparison, grammatical.** See ADJECTIVE 3, 4; ABSOLUTE COMPARATIVE; ABSOLUTE SUPERLATIVE; DOUBLE COMPARISON.

**compendious** means 'containing the substance within small compass, concise, succinct'. Like many words indicating size, however, it is somewhat extendible in meaning, and is often misleadingly applied to works that are more marked by their comprehensiveness than by their conciseness. Jespersen's *Essentials of English Grammar* (1933) is a compendious work; *A Comprehensive Grammar of the English Language* (1985), edited by Randolph Quirk et al., is not.

**compendium.** For the pl., both *compendiums* and *compendia* are possible. The form preferred in OUP house style is *compendia.* See -UM 3.

**compensate.** **1** It provides a salutary reminder of one of the ways in which the language changes. The OED pointed out (in 1891) that /ˈkɒmpənseɪt/ 'is the pronunciation now usual in England, but appears to be quite recent; pronouncing dicts. had until *c*1850–70 only /kəmˈpɛnseɪt/, which is also that of the 18th c. poets. Tennyson has both.'

**2** The use of *compensate* in the broad sense 'to offset a disability or frustration by development in another direction' (COD) is a standard example (also *overcompensate*) of a scientific (Jungian) term making its way into popular use: e.g. *A young man afraid of marriage may compensate by specializing in dating and courting, becoming a 'man about town'*—J. E. Gordon, 1963.

**competence.** **1** Neither *competence* nor *competency* has any sense in which the other cannot be used; the first form is gaining on the second—so Fowler, 1926. The statement remains substantially true except that *competence* is now markedly more frequent than *competency.*

**2** *Competence,* as opposed to *performance,* has held linguistic scholars in thrall since this pair of terms was introduced by Noam Chomsky in 1962 to distinguish what a speaker of a language knows implicitly (*competence*) and the parts of it that the same person actually uses (*performance*). As David Crystal (1980) remarked, 'As a general conception, this distinction has been widely accepted, but increasing criticism has come from linguists who feel that the boundary between the two notions is not as clearcut as their definitions would lead one to believe. There are problems often in deciding whether a particular speech feature is a matter of competence or performance.'

**complacent, complaisant.** The first of these now has two main senses: (1) smugly self-satisfied. (2) calmly content. *Complaisant* also has two senses: (1) politely deferential. (2) willing to please, acquiescent. The two words form a minimal pair, in that they are distinguished in pronunciation by only one sound, medial /s/ in the first and medial /z/ in the second. They are etymological doublets in that they both ultimately answer to L *complacēre* to please, but they reached English by different medieval channels.

Historically, and even down to the 19c., the two words (and also their derivatives, *complacence* and *complacency* beside *complaisance* and *complaisancy*) became cross-linked and confused in some contexts. Burke, Scott, and Charlotte Brontë are cited in the *OED* as using *complacent* to mean 'obliging in manner, complaisant': e.g. *Mr. Moore ... was ... a complacent listener to her talk*—C. Brontë, 1849. But the distinctions indicated above, i.e. the two senses of each word, are now clearcut and should be observed.

**compleat.** A standard example of a 20c. revival of an older (16–18c.) spelling of *complete*. The title of Izaak Walton's *The Compleat Angler* (1653) was the principal model for the modern revival; *The Compleat Bachelor*—O. Onions (title of book), 1900; *She writes and sings and dances and plays I don't know how many instruments. The compleat girl.*—M. McCarthy, 1963.

**complected.** Used to mean 'complexioned, having a (specified) complexion', the word seems not to occur outside the US, where it was first recorded in 1806. (*Complexioned* is also used

in America.) Examples: *They told me the man they meant wasn't dark complected*—W. Faulkner, 1932; *George Gobbin was a tall dark-complected man in his late fifties*—J. D. MacDonald, 1977; *She was a good-looking, dark-complected lesbian—way out in the open about that*—M. Chabon, 1990.

**complement.** 1 In grammar, a relatively modern term (first recorded in 1874) for a word or phrase added to a verb to complete the predicate of a sentence. It is a term of wide application, and the way in which complements are classified varies from grammarian to grammarian. Some typical types (in each example the words in roman type form the complement): (adjectival) *the going looked* firm; *the task seemed* nice and easy; (adverbial) *she smiled at him* in a Mona Lisa way; *they remained* out of reach; *she spoke* clearly; (as subject complement) *he died* a happy *man* ( = he was a happy man when he died); *he is* a solicitor; (as object complement) *he called his brother* an idiot. It makes for simplicity if a direct object is not regarded as a complement, but there is no agreement among grammarians about this. Readers should be prepared to meet and accept various interpretations of which part of the predicate of a sentence is to be regarded as a complement.

2 As noun and verb, pronounced the same as *compliment*. The nouns are pronounced /ˈkɒmplɪmənt/ with a schwa (indistinct unstressed vowel sound) in the final syllable. The verbs are pronounced /ˈkɒmplɪment/ with a fully pronounced final syllable.

**complementary, complimentary.** The first means 'completing, forming a complement', and the second means 'expressing a compliment, praising'. They should not be confused, but sometimes are: (*complimentary* wrongly used) *There's a family of coated papers and a complimentary number of uncoated papers*—Paper Focus, 1989; (*complementary* wrongly used) *With a complementary glass of champagne*—advt in a US newspaper, 1991.

Similar confusion is also regrettably occurring (esp. in advts) between the corresponding verbs, *complement* and *compliment*.

**complex.** This Jungian psychological term (first recorded in English in 1907) for 'a group of emotionally charged ideas or mental factors, unconsciously associated by the individual with a particular subject, arising from repressed instincts, fears, or desires and often resulting in mental abnormality' (OED) is frequently used in technical literature with prefixed defining words, as *inferiority complex*, *Oedipus complex*, etc. The term rapidly made its way into popular use meaning no more than a fixed mental tendency of obsession: '*Muriel's losing her sex-complex.*' '*What on earth do you mean, boy?*' '*She's getting herself tangled up with some man.*'—A. Bennett, 1927; *A fond aunt with a commiseration complex*—Punch, 1928; *And both of them had a complex about economy and living within a budget*—M. McCarthy, 1954; *Governments should rid their minds of this foe complex which leads to so much trouble*—R. Macaulay, 1956.

**complexion.** Thus spelt, not *complection*. See -XION etc.

**complexioned.** See COMPLECTED.

**compliment.** See COMPLEMENT 2.

**complimentary.** See COMPLEMENTARY.

**compline.** The former variant spelling *complin* has now been abandoned. The pronunciation /ˈkɒmplɪn/ is now dominant, but the spelling-pronunciation /ˈkɒmplaɪn/ is heard often enough.

**compo,** a composition of plaster, etc., has pl. *compos*. See -O(E)S 5.

**compose.** See COMPRISE.

**composite.** Pronounced /ˈkɒmpəzɪt/. The former pronunciation (e.g. in Walker, 4th edn, 1806) /kəmˈpɒzɪt/ has been laid aside. The spelling-pronunciation /ˈkɒmpəˌzaɪt/, in which the final syllable rhymes with *light*, now frequently heard in the language of northern trade-union leaders (*composite decisions made by executive committees*), sounds harsh to standard speakers.

**compost.** Both syllables have a fully pronounced short *o*, thus /ˈkɒmpɒst/.

**compound.** See NOUN AND VERB ACCENT.

**compound** (verb). In law, the primary meaning is 'to condone (a liability or offence) in exchange for money; to settle by a payment of money'. *To compound a*

*felony* means 'to forbear from prosecuting (a felony) in exchange for some consideration: an offence at law'. In practice, it is often erroneously taken to mean 'to make worse, to aggravate', and the use of this extended meaning looks likely to persist.

**compound prepositions.** (Also called *prepositional phrases* and *complex prepositions*.) **1** *Simple prepositions*. At an elementary level, *at*, *by*, *from*, *to*, etc., are simple prepositions (*he was at home*, *she sat by the fire*, etc.). There are approximately seventy of these, about half of which have two syllables (*above*, *below*, *until*, etc.) or more (*notwithstanding*, *underneath*).

**2** *Marginal prepositions*. There is also a smaller group of marginal prepositions, i.e. words which operate in much the same way as simple prepositions, but which have affinities with some other word classes. These include *bar*, *barring*, *concerning*, *considering*, *given*, *granted*, *including*, and *pending* (e.g. *considering the difficulties at that time, you did very well*; *given the chance, he would like to earn more money*).

**3** *Compound prepositions*. These fall into two subsets, the first consisting of two words (e.g. *according to*, *apart from*, *because of*, *contrary to*, *next to*, *out of*, *previous to*, *thanks to*), and the second consisting of three (e.g. *in accordance with*, *in consequence of*, *with regard to*). Examples: *according to him, inflation should soon fall to single figures*; *the church stands next to the village hall*; *in consequence of the baggage-handlers' action, no planes will leave Gatwick today*.

Fowler (1926) excluded marginal prepositions from consideration and strongly objected to the overuse of compound prepositions: 'taken as a whole, they are almost the worst element in modern English, stuffing up the newspaper columns with a compost of nouny abstractions'. His colourful view no longer seems to be supported by the facts.

**comprehend.** See APPREHEND.

**comprise.** **1** See INCLUDE.

**2** *Distinguished from compose*. 'The special function of *comprise* . . . is to introduce a list of the parts making up the whole that is its subject; that is, it means *to consist of* or *be composed of*; all the parts compose the whole, the whole comprises (all) the parts. This close relation of the verbs is a convenience whenever euphony or some rhetorical need happens to be better served by a simple active verb (e.g. *comprise*) than by a compound passive (*is composed of*).'—Fowler, SPE Tract XXXVI, 1931 ( = part of an entry accidentally omitted from *MEU*, 1926). 'Unluckily,' Fowler added, 'close relation is mistaken for identity, and *comprise* takes the place of *compose*.' His statement is an excellent starting-point, but there is more to be said.

(a) Examples of correct uses: (*comprise*, used actively, 'consist of, be composed of, make up, contain') *A full pack comprises 52 cards*—cited in Burchfield, 1981; *Love comprises among other things a desire for the well-being and spiritual freedom of the one who is loved*—M. Spark, 1984; *the heritage of one nation often comprised works legitimately and fruitfully brought from another*—*Antique Collector*, 1990; (*compose*, used actively = to constitute, make up, in contexts where separate parts are stated to form a whole) *The five or six great tribes or tribe-leagues which composed the German nation*—J. Bryce, 1875; *A good society is a means to a good life for those who compose it*—B. Russell (*SOED* 1993); (*compose of*, used passively, in contexts that have the whole as subject and the constituents as object) *A government composed of princes and bishops*—H. T. Buckle, 1857; *the first British currency was composed of tin*—W. S. Jevons, 1875; *his face was. . .composed of little layers of flesh like pallid fungus*—I. Murdoch (*OMEU*).

(b) Examples of disputed or erroneous uses: (*comprise* used transitively, 'to constitute, make up, form') *All the most important documents, declarations and resolutions, which comprise or influence the American Constitution*—cited in Fowler, 1931; *the Government of the Federation and the three territories which comprise it*—cited in Gowers, 1965; *The four submarines comprising the nuclear deterrent*—cited in Gowers, 1965. It cannot be denied, however, that the sheer frequency of this construction seems likely to take it out of the disputed area before long. *WDEU* cites numerous examples of *comprise* used in contexts where the parts *comprise* the whole: e.g. *The buildings that comprise the Nunnery quadrangle*—*Saturday Review*, 1969; *Seven boys comprised the choir*—G. Keillor, 1985. (*comprise* used intransitively followed by *of*) [a government] *comprising of African nominees*—BBC (Rhodesian correspondent), 1976; *Up until the end of the August offensive food rations at Gallipoli were extremely poor. They comprised mainly of bully beef . . . and hard biscuits and jam*—*Empire & Desire: Gallipoli 1915* (National Library of NZ), 1990; (*comprise of*, used passively = be composed of, consist of, a relatively modern use, first recorded in 1874) *The stock is comprised mainly of small quantities of any one variety*—cited by Fowler, 1931; *Many of these words are comprised of monemes*—E. Palmer, 1964; *Internally, the chloroplast is comprised of a system of flattened membrane sacs*—*Nature*, 1970; *It is incorrect to say 'It was comprised of 20 students'*—Burchfield, 1981; *The Saxe-Coburg inheritance is comprised of the ducal palace and three castles*—*Daily Tel.*, 1991. (Opposition to this last construction is also weakening.)

**comptroller,** an erroneous but rather grand-looking spelling of *controller*, introduced about 1500 ('especially affected by official scribes', *OED*). The first syllable was mistakenly supposed to be *count*, etymologically *compt* (L *computāre*). *Comptroller* is still retained in certain official designations, e.g. *Comptroller of the* (Royal) *Household, Comptroller of the Lord Chamberlain's Office*, but *controller* is now the more common term of the two, e.g. *Controller of Audit* (Audit Commission for Local Authorities in England and Wales), *Controller Radio 3*, *Financial Controller* (of the Crown Suppliers).

**computerese.** Since the 1960s or so a significant proportion of people in Britain, America, and elsewhere have become rapidly familiar with the everyday terminology of computer systems. The language has been stretched and strained to a remarkable degree and has absorbed an entirely new layer of words and meanings. Examples of the more familiar terms follow: (extension of the meaning of existing words) *access, address, hacker, memory, menu, mouse, network, quit, retrieval, virus*; (new words, including blends) *database, hardware, mainframe, modem, software, spreadsheet, system crash, user-friendly, workstation*; (acronyms) *CD-ROM* (Compact Disk—Read-Only Memory), *MIS* (management information system), *RAM* (random access memory), *wysiwyg* (what you see is what you get), etc.; (commercial names of spelling checkers, etc.) *Appletalk, Compuware, Correctext, Macterminal, Wordperfect, Wordstar,* etc. As time goes on, such vocabulary is bound to become part of the day-to-day language of the whole community, as will the corresponding terminology in France, Germany, and everywhere else.

**comrade. 1** Pronounced in some contexts as /ˈkɒmreɪd/ and in others as /-rɪd/.

**2** As a form of address by a socialist or a communist to another, or as a prefix to a surname of such a person (a use first recorded in English in 1884), *comrade* suffered a mortal blow during the revolutions in Eastern bloc countries in 1989–90, and, one hopes, the word will simply become monosemous again, meaning just 'a close companion'.

**concave** means 'having an outline or surface curved like the *interior* of a circle or sphere'. *Convex* means 'having an outline or surface curved like the *exterior* of a circle or sphere'.

**conceal.** See FORMAL WORDS.

**concede.** In electoral contests it is appropriate for the defeated candidate to *concede defeat.* But, as a graceful gesture to the winner, some candidates have been known to *concede victory,* thus shifting the emphasis to the success of the other person.

**[concensus.** It hurts me to set down this common misspelling of *consensus,* but how else can the miscreants who use the wrong form find it in this book?]

**concentrate.** For the verb, Walker (4th edn, 1806) placed the main stress on *-cen-.* The *OED* (1891) presented this as an acceptable variant pronunciation, and as the only stressing for the adj. and n. *concentrate.* For the v., the adj., and the n. the stress is now always placed on the first syllable (but *concentrátion*).

**concept.** The philosophical sense of the word ('the product of the faculty of conception, an idea of a class of objects, a general notion'), first formulated in the 17c., remains in use. In non-philosophical circles, the word is widely used

in a weakened sense of a general notion or idea, esp. in contexts of marketing and design. Examples: *He was the man who invented the concept of a weekly news magazine*—D. Halberstam, 1979; *We aim to sell a total furnishing concept based on the 'one pair of eyes' principle*—*Sunday Tel.,* 1985. There is room in the language for both sets of uses.

**concerned.** The idiomatic expression *as far as ... concerned,* e.g. *as* (or *so*) *far as I am concerned* = 'so far as I have any say in the matter, for all I care' (*you may go to the devil as far as I am concerned*), is common enough. Gowers (1965), however, cited examples where the idiom was inappropriate and should not have been used: *The punishment does not seem to have any effect so far as the prisoners are concerned* (read *on* the prisoners); *The months of January, February, and part of March 1963 were disastrous as far as the building industry of this country was concerned* (for the building industry); *The girl is entirely unknown as far as the larger cinema audiences are concerned* (to the larger audiences). Such economies are worth making. Cf. FAR 4.

**concernment.** First recorded in the 17c., this word is now rare except in very formal contexts. It is being replaced in its several senses by the shorter word *concern.* A poetical example: *I preached one Sabbath on a text from Kings; He showed concernment for his soul*—R. Lowell, a1972.

**concert** (noun and verb). See NOUN AND VERB ACCENT.

**concerto.** The pronunciation /kənˈtʃɜː-təʊ/ and pl. *concertos* are recommended, but /kənˈtʃeətəʊ/ and pl. *concerti* (naturally in *concerti grossi*) are equally common in standard English.

**concessionaire** is the recommended spelling (not *-ionnaire*) for the holder of a (trading) concession.

**concessive. 1** In grammar, a concessive phrase or clause indicates 'that the situation in the matrix [*sc.* main] clause is contrary to expectation in the light of what is said in the concessive [phrase or] clause' (*CGEL* 15.40). The principal words introducing concessive constructions are *although* and *though*; others include *if, even if, even though, when,* and *while.*

Examples; (clauses containing verbs) *Although she was only 14, she was admitted to the University of Oxford; He was awarded a First even though he failed to complete one of the papers; While I respect your right to absent yourself, I think you should have left a note saying where you were; He was solemnly discussing peace terms when he had no intention of keeping his promises;* (introducing -ing, -ed, and verbless clauses) *While not seeming to mind, she suffered remorse in private; Even though given every chance, the filly came last in the Oaks; Though past the crawling stage, the baby could not yet stand up unsupported.*

2 Prepositions of concession include *in spite of, despite,* and *for all.* Examples: *In spite of his sprained ankle he could still ride a bicycle; despite this setback he had a successful career; for all his talent, he was unable to settle down.*

**conch.** Though Sir William Golding consistently pronounced the word as /kɒntʃ/ in a lecture that he gave on *The Lord of the Flies* and other matters at the University of Oxford in February 1990, the more usual standard pronunciation is /kɒŋk/. The word is derived from L *concha,* Gk κόγχη 'shell'.

**conciseness, concision.** To anyone familiar with the earliest senses of the word *concision,* namely 'the action of cutting to pieces; mutilation' (first recorded in 1382) and 'circumcision' (*Beware of dogs, beware of euill workers: beware of the concision*—Phil. (AV) 3: 2), where it is applied contemptuously to Judaizing Christians, the natural word for brevity of speech or writing is *conciseness.* Fowler (1926) regarded *concision* as a literary critics' word used by writers under French influence, 'and often requiring the reader to stop and think whether he knows its meaning'. *Conciseness,* first recorded in 1659, is now perhaps the more usual of the two words in the sense 'brevity (of style)', but both words are freely used.

**conclude.** Used with a subordinate *that*-clause in the sense 'come to a conclusion, deduce' (*he concluded that his life was in danger*) and absolutely or transitively in the sense 'to end' (*he concluded by making a few remarks about the umpiring; it took nearly an hour to conclude their business*). The once common construction with a *to*-infinitive in the sense 'to decide' (*My wife concluded to hire a balcony*—N. Hawthorne, 1858) is now archaic.

**concord.** For grammatical *concord* see AGREEMENT.

**concordat.** Pronounce in a fully Anglicized way, /kən'kɔːdæt/.

**concupiscence.** The main stress falls on the second syllable, /kən'kjuːpɪsəns/.

**concur.** The inflected forms *concurred* and *concurring* retain the stem vowel /ɜː/ of *concur,* but *concurrent* has /ʌ/, i.e. is pronounced /kən'kʌrənt/.

**condemn.** 1 A word that is baffling to foreigners in that the final *n* is silent in the infinitive and in the inflected forms *condemned* and *condemning,* but is sounded in its derivatives (*condemnable, condemnation, condemnatory*).

2 See CONTEMN.

**condensed sentence.** A term sometimes used for SYLLEPSIS or ZEUGMA.

**condign.** Originally used to mean 'worthy, deserving' (ult. from L *con-* 'together, altogether' + *dignus* 'worthy'), since the end of the 17c. it has come to be used only in the sense 'well-deserved (of punishment)': *He had been brought to condign punishment as a traitor*—Macaulay, 1848; *His reason has to do with deterrence and the exaction of condign punishment*—*Times,* 1983.

**condom.** J. C. Wells (1990) points out that the established pronunciation in Britain is /'kɒndəm/, but that the spelling pronunciation /'kɒndɒm/, with the same vowel sound in each syllable, became common with the spread of the HIV infection in the 1980s.

**conduct** (noun and verb). See NOUN AND VERB ACCENT.

**conductress.** See -ESS 5.

**conduit.** The dominant pronunciation in the UK is /'kɒndɪt/, but variants with /-djʊɪt/ or /'kʌn-/ are used by some standard speakers.

**confederacy, confederation.** See FEDERATION.

**confer.** The inflected forms are *conferred, conferring.* See DOUBLING OF CONSONANTS etc. 2; -R-, -RR-.

**conferrable.** See -ABLE, -IBLE 6.

**confidant, confidante.** *Confident* (stressed on the first syllable) was in use as a noun meaning 'a person entrusted with secrets' from the 17c. to the 19c. (*As he had neither friend nor confident, hardly even an acquaintance*—Scott, 1815). *Confidant*, with its feminine equivalent *confidante*, made its way into English from French after 1700. The English spelling in *-ant(e)* is usually taken to be an attempt to represent the pronunciation of the French *-ent, -ente*. Walker (4th edn, 1806) commented: 'so universal is its use [sc. the use of *confidant* pronounced in a French way] at present, that a greater mark of rusticity cannot be given than to place the accent on the first syllable, and to pronounce the last *dent* instead of *dant*.' *Confidant* is now used of both men and women, and *confidante* properly only of women.

**confine** (noun and verb). See NOUN AND VERB ACCENT.

**confiscate.** 'As in other words of the same form, *compensate, concentrate, contemplate*, etc., the stress is now usually on the first syllable, but till *circa* 1864 the dictionaries had only *confiscate*' (*OED*).

**conflict** (noun and verb). See NOUN AND VERB ACCENT.

**confusable words.** For various types of words that are commonly confused (*hoards/hordes, reek/wreak, reign/rein*, etc.), see DIFFERENTIATION.

**congeries.** Pl. unchanged. A collective name for 'a disorderly collection; a mass or heap' (from L *congeries*, from *congerere* 'to carry together'). Pronounce /kɒn-ˈdʒɪəriːz/; the pronunciation with four syllables as /kɒnˈdʒɛrɪˌiːz/ is old-fashioned. Example: *Whitehall, that congeries of government offices that takes in, for example, the Liverpool regional office of the Department of the Environment*—*Times*, 1985.

**conifer.** The pronunciations /ˈkɒnɪfə/ and /ˈkəʊn-/ are both common, but the first of these seems now to have the edge in BrE.

**conjugal.** The pronunciation recommended is /ˈkɒndʒʊɡəl/, not /kənˈdʒuːɡəl/.

**conjugation.** The inflection of verbs, or any class of verbs inflected in a specified way (*first conjugation, second conjugation, strong conjugation*, etc.). The word (from L *conjugātiō*) means 'yoking together'. Cf. DECLENSION.

**conjunction.** In grammar, a word used to connect clauses or sentences or words in the same clause. Conjunctions usually join like with like, i.e. a noun or its equivalent with another noun or its equivalent (*a rake* and *a hoe*; *she* and *Mr Ramsay*; *Nothing stirred in the drawing-room* or *in the dining-room* or *on the staircase*); an adjective with another adjective (*bright* and *breezy*; *crazy* but *nice*); an adverb with another adverb (*slowly* but *surely*); and so on.

In compound sentences, conjunctions fall into two main groups, *subordinating conjunctions* and *coordinating conjunctions*. Examples: (subordinating) as *the dinner isn't quite ready, let us have another drink*; *he went back to university* so that *he could complete his degree*; *I shan't go* if *you won't come with me*; though *she could not hear what was said, she could see that they were happy*; (coordinating, esp. *and, but, for, nor, or*) *She would go back* and *look for it*; *He should have been a great philosopher*, but *he had made an unfortunate marriage*; *it did her husband good to be three thousand*, or if *she must be accurate, three hundred miles from his library*; *For in one moment if there were no breeze, his father would slap the pages of his book together*.

John Algeo (*Internat. Jrnl Lexicography*, 1988) pointed out that *whilst, now, and nor, but nor*, and one or two other coordinating conjunctions are found in BrE, but not in AmE. Examples (cited by Algeo): *I would like to thank many friends and colleagues for their encouragement* whilst *I was writing this book*—Rosemary Jackson, 1981; *Now the socialists have adopted the expensive red rose as their emblem , may I suggest to Mrs Thatcher the simple blue forget-me-not*—*Times*, 1986; *But doctors prescribing the pills were not told of the maker's doubts, and nor was the Government's drug watchdog*—*Guardian*, 1986; *I am not a trained architect, but nor was Sir Edwin Lutyens*—*Sunday Times*, 1985.

Many of the more important conjunctions are treated in separate articles: see AND; BECAUSE; BUT; FOR; etc.

**conjunctive** (mood). This term does not apply to any aspect of the English verb. 'The forms denoted by *conjunctive* and *subjunctive* are the same, and *subjunctive* is the much better known name . . . *Conjunctive* should be given up, *subjunctive* be used as the name of the forms whatever their use, and the differences of function be conveyed by other words (*dependent, conditional, optative,* etc.)' (H. W. Fowler, 1926). The matter is mentioned only because anyone turning to 18c. or 19c. grammars will find the term *conjunctive* used occasionally. It is not used in any articles in this book.

**conjure.** In the dominant sense 'to perform tricks which are seemingly magical', *conjure* is pronounced /ˈkʌndʒə/; but in the sense 'to beseech, to appeal solemnly to (a person)' it is pronounced /kən'dʒʊə/.

**conjuror** is the recommended spelling, not *conjurer.*

**connection.** 1 At some point between 1964 (when the 5th edn of *COD* was published) and 1976 (6th edn), OUP gave up its long-standing preference for *connexion* and adopted *connection* as its house-style form. See -XION etc.

2 Fowler (1926) accepted the phr. *in connection with* only when it had 'a real meaning', e.g. *Buses ran in connection with the trains; The isolated phrase may sound offensive, but taken in connection with the context it was not so.* He intensely disliked its use as a compound preposition: 'vagueness and pliability [are] its only merits.' He then set up a coconut shy of examples (unattributed, but presumably from newspapers of his time) and dislodged them one by one. As a broad rule, if a simple preposition like *by* or *about* will do instead, use it.

**connoisseur** /kɒnə'sɜː/ was adopted in the 18c. from a now obsolete form of French *connaisseur* (cf. *connaître*).

**connote, denote.** Both words mean broadly 'to signify' but that is where the correspondence ends. A word *denotes* its primary meaning; it *connotes* attributes associated with the broad primary meaning. The primary meaning of the seasonal word *spring* is 'the season in which vegetation begins to appear, the first

season of the year (in the northern hemisphere from March to May)'. But spring *connotes* fresh growth, renewal, freshness, and so on. The primary meaning of *dog* is 'any four-legged flesh-eating animal of the genus *Canis*, kept as pets or for work or sport'. But the connotations of the word can include fidelity, obedience (esp. a sheep-dog), aggressiveness (esp. a Rottweiler), etc.

**conscience.** Write *for conscience' sake.* But see SAKE.

**conscientious.** Pronounce /kɒnʃɪ'enʃəs/, not /kɒnsɪ-/.

**consensus.** 1 It means 'general agreement' and should be carefully distinguished from *census* 'the official count of a population'. (One of the commonest misspellings at present is the writing of *concensus* for *consensus*.) The second element of *consensus* ultimately answers to L *sentīre, sens-* 'to feel', while *census* goes back to L *censēre* 'to assess'.

2 Since the mid-19c., *consensus* has been commonly used with a following *of* and noun (*consensus of opinion, evidence, authority,* etc.): some scholars have shied away from these on the ground that they are somewhat tautologous. A century later, beginning in the 1960s, attributive uses (*consensus view, consensus politics*) came to the fore, and these uses are now, perhaps, dominant.

**consequent, consequential.** 1 Where doubt can arise betweeen *consequent* and *consequential,* the shorter word should normally be used when the sense is the simple and common one of 'resulting, following in time': *in the consequent confusion he vanished; The very rapid increase of trade, and the consequent influx of wealth. Consequential* is usu. more appropriate when the context calls for a word meaning 'of the nature of a consequence or sequel', as in *two consequential amendments were passed; the widespread use of fossil fuels, and the consequential pollution of the atmosphere; Conservative MPs are hoping that she will take the opportunity of the consequential changes in the junior ranks to bring in some new faces—Times,* 1986.

2 Fowler's view (1926) that *consequential* does not mean 'of consequence' needs modification. It has meant 'self-important (of persons)' since the 18c.: *He*

*here consider'd it essential To shew he could be consequential*—1816 in *OED*; *Pampered and consequential freedmen*—F. W. Farrar, 1875; *It is very odd to look at all these poor consequential idiots and remember that war might at any moment make real mincemeat of them*—S. T. Warner, 1942.

The *OED* also gives five examples between 1728 (Fielding) and 1821 (*He must withhold no consequential fact*—T. Campbell) used in the sense 'pregnant with consequence, important', and marks the sense '*Obs.*'. But it is alive and well: *He considered his notebooks far more consequential than his published prose*—*TLS*, 1989.

**3** *Consequential* has various technical meanings in law and insurance, esp. *consequential damages, consequential loss*: the precise meaning of these terms may be found in specialized textbooks.

**conservative.** One of the hazards of commenting on English usage is that the passing of time invalidates some of one's judgements. Fowler (1926) forcefully rejected ('perhaps the most ridiculous of slipshod extensions') 'the rapidly spreading use of this word as an epithet, in the sense of moderate, safe, or low, with *estimates, figure*, etc.', e.g. *8,000 killed is considered a conservative estimate*; *Based upon the price of* $4\frac{1}{2}$*d. per ft., and with reasonable care this should be a conservative figure*. This use, first recorded in America at the beginning of the 20c., is now widespread and unopposable in all English-speaking areas.

**conservatory.** In the sense 'a school or academy of music' the English word is a 19c. Anglicized version of Ital. *conservatorio*. In Italy the word originated in hospitals for the rearing of foundlings and orphans in which musical education was given. It was adopted in French as *conservatoire* (a form also used in English), and in Germany as *Konservatorium*. *Conservatory*, in the sense 'a school of advanced standing specializing in one of the fine arts (as music, drama, or dance)' (Webster's Third), is esp. common in the US, e.g. the New England Conservatory of Music (Boston), the Oberlin Conservatory of Music, the American Conservatory Theater (San Francisco), and the Conservatory of American Dance (Evanston, Illinois).

**consider. 1** For most of the 20c., various writers on usage, including Gowers (1965), have expressed reservations about the use of *consider as* to mean 'regard as being' in sentences like *They considered him as a bad influence* (read *him a bad influence*); *I consider it as essential* (read *it essential*). On the other hand, *CGEL* 16.47 sets down all three types *We consider him a genius/as a genius/to be a genius* as mere (presumably acceptable) variants. Outside the confines of such simple sentences what is the position? There seems little doubt that all three constructions, with and without *of* or using *to be*, are in use. The choice does not seem to be based on particular rules, but rather on the nature of the surrounding words. Economy points in one direction and the rhythm of the sentence in another. Constructions with *as* seem to be the least favoured of the three and the least common.

Examples: (without *as*) *He is considered a rich man*—F. Marryat, 1842; *Gall considered it a gratuitous interference*—Mrs Henry Wood, 1867; *[This] was considered the most solemn of all oaths*—J. Bryce, 1875; *They were quite a trio, they decided, though each of the other two secretly considered herself the star*—F. Raphael, 1960 ; *The village boys ... considered it a privilege to enjoy a stroll with him in the evenings*—M. Das, 1987; (with *as*) *This species is not considered with us as migratory*—T. Bewick, 1797–1804; *Tungay considered the whole establishment, master and boys, as his natural enemies*—Dickens, 1850; *I fear that our lover will henceforth be considered by such a one as being but a weak, wishy-washy man*—Trollope, 1867; *The patient improved considerably but could not be considered as cured*—M. Balint, 1968; (with *to be*) *he ... considered himself to be a good sound Royalist*—R. D. Blackmore, 1869; *A determination to show people what she considered to be their proper place*—N. Mitford, 1949; *so progressives must consider all women to be clever!*—I. Murdoch, 1957.

**2** *consider of*, 'think attentively or carefully of', first recorded in the 16c., is now distinctly archaic. Examples: *Let us consider of the Hesperides themselves*—Ruskin, 1860; *A General Court will be held at The Bank on Thursday ... to consider of a Dividend*—*Times*, 1891.

**considerable.** The word continues to spread its grammatical wings more fully

in America than in Britain. **1** As *adj*. In
the UK it is applied only to immaterial
things, to abstractions: *I have given the
matter considerable attention*; *he was a man
of great energy and considerable ability*; *a
considerable sum of money*; *he suffered considerable pain*; *a considerable number of
people were queuing at the taxi-rank*. In
America this is also the natural use of
the adj.; but it is also used in AmE (*a*) =
great in size [qualifying non-count
nouns] (examples from American dictionaries): *silk fabric containing considerable gold or silver thread*; *certain
pharmaceutical preparations similar to cer-
ates, but containing considerable tallow*; *a
house with a considerable barn in back*; *the
product contains considerable salt*. Also, *One
of them housed a dark, pungent hill of milo
grain worth considerable money—a hundred
thousand dollars—T*. Capote, 1966; and (*b*)
absolutely = 'much, a great deal' (*During
the last two years considerable has been written on the subject*).

**2** As *noun*. Used informally in AmE =
a good deal; a fair amount or quantity
*of* (something). Examples: *Jim did considerable toward stirring up the farmers—Atlantic
Monthly*, 1932; *It was a kind of mixed hound,
with a little bird dog and some collie and
maybe a considerable of almost anything
else—W*. Faulkner, 1943; *he has done considerable for the community—Random House
Dict.*, 1987.

**3** As *adverb*. At first (17c. onward) in
use in Britain, but now obsolete or dialectal in this country. According to American dictionaries it is still current in
some regional varieties of AmE. Examples: *I presume I do have considerable
more time for writing than you do—O*. W.
Norton, 1862; *By-and-by she let it out that
Moses had been dead a considerable long
time—M*. Twain, 1884; *that speeded things
up considerable—*1984 in *WDEU*.

**considerateness, consideration.** The
first of these is not much used and then
only when an abstract noun corresponding to *considerate* is called for contextually and *considerateness*, as the more
'obvious' of the pair, comes to mind.
In practice *considerateness* has only one
current sense, 'thoughtfulness for
others' (*your considerateness over the years
has been greatly appreciated*), and it shares
this sense with *consideration* (*consideration
for others is one of his good points*). By

contrast, *consideration* has a number of
current senses, including (*a*) a circumstance taken into account, a reason considered (*To have a doctor living nearby is a
consideration of some weight with me*; *the
topography of the village of Fetherhoughton
may repay consideration—H*. Mantel, 1989);
(*b*) a reward, remuneration (*pigs were sold
to me for a trifling consideration*); (*c*) in
several idiomatic phrases, esp. *to take into
consideration, under* (*active*) *consideration,*
and *in consideration of. Consideration* is
a blossoming word; *considerateness* is a
wayside flower.

**considering.** **1** In origin an absolute
use of the present participle or verbal
substantive, it has been used for centuries as (*a*) (first recorded in the 14c.)
a preposition, with a simple object =
taking into account (*Considering the conditions, the horses galloped well*); (*b*) (15c.)
with an object clause (*I should think you
have every right to be cross, considering how
much you have done for her*); (*c*) (18c.)
elliptically = taking everything into account (*We came out of it not too badly,
considering*). Modern examples: *It's odd
that one boasts considering that no one is
ever taken in by it—V*. Woolf, 1921; *Considering the stories I had heard of a maze of
forgotten galleries, I had thought it advisable
to imitate Theseus and Ariadne and take with
me a large ball of string . . . to mark my
trail—G*. Household, 1981; *these were years
of disappointment, even grief, for* [Ogden]
*Nash, in what was, considering, a remarkably
successful career in writing—NY Times Bk Rev.*,
1990.

**2** See UNATTACHED PARTICIPLES.

**consist.** In most contexts, *consist*, when
followed by *of*, means 'be composed; have
some specified ingredients or elements'
(*an ordinary fence, consisting of a ditch and
a bank*; *Newton imagined light to consist of
particles*). By contrast, when followed by
*in*, it most often means 'have its essential
features as specified' (*not everyone can tell
in what the beauty of a figure consists*; *his
term of office consisted in a series of ill-judged
decisions*). Modern examples: (with *of*) *The
stage itself . . . consisted of a number of timber
boards set on concrete blocks—W*. Trevor,
1976; *The group consisted of five string players and two singers—J*. Simms, 1982; (with
*in*) *the new monastery . . . consists in a series*

*of buildings, arranged hexagonally*—L. Mumford, 1944; *Part of my unease about my cousin consisted in a fear that he would succeed in life and I would fail*—I. Murdoch, 1978.

**consistency, consistence.** Theoretically the two words are interchangeable in any given context, but in practice *consistency* is now almost always used. Fowler (1926) judged the *-cy* form to be invariable when the noun means being consistent, i.e. not inconsistent (*consistency is an overrated virtue*). In the noun meaning degree of viscosity in liquids, he said, usage varies (*Red-hot streams which generally present a consistence something like that of treacle*, and *Eruptions of mud, varying considerably in consistency and temperature*— both from T. H. Huxley, 1878). He hoped that the two forms would settle down in separate meanings, but it has not worked out like that. Examples: (*consistence*) *Judd began to eat his soup, which was of a milky, potatoey consistence, speckled with sweet herbs*—P. White, 1957; (*consistency*) *The hair lacquered until it had the consistency of wire-wool*—M. Leitch, 1981; *For some defenders of slavery, logical consistency required a defense of the slave trade as well*—J. M. McPherson, 1988.

**consistory.** In its normal ecclesiastical senses at first pronounced /kɒnsɪˈstɔːrɪ/, then /ˈkɒnsɪstərɪ/ until about the end of the 19c., with the main stress on the first syllable, but now always /kənˈsɪstərɪ/.

**consociation.** See -CIATION.

**consolatory.** Pronounce /kənˈsɒlətərɪ/.

**console.** The verb is pronounced /kənˈsəʊl/ and the noun ( = panel, cabinet, etc.) /ˈkɒnsəʊl/.

**consols** ( = government securities). Pronounce /ˈkɒnsəlz/.

**consommé.** Pronounce /kənˈsɒmeɪ/ and print in romans.

**consort.** The noun is stressed on the first syllable and the verb on the second. See NOUN AND VERB ACCENT 1.

**consortium.** Pl. *consortia* or *consortiums*. See -UM 3.

**conspectus.** Pl. *conspectuses*. See -US 2.

**conspicuity.** Marked 'rare' in the *OED* (the usual word, of course, being *conspicuousness*), it is having a new lease of life in road-safety contexts: *The word 'conspicuity' ... has become fashionable in road-safety circles ... in connection with the visibility of pedestrians after dark*—Walk, 1984; *Among the most important are points which deal with ... encouraging the wearing of conspicuity aids by all two-wheeled road users*—Today's Guide, 1986.

**conspicuous,** in the sense 'attracting notice or attention', is commonly used (some think overused) in the phr. *conspicuous by its absence* (first recorded in 1859): *Pinkie Aurangzeb was conspicuous by her absence; and Rani was queen for a day*—S. Rushdie, 1983.

**constable.** Pronounce /ˈkʌnstəbəl/, but don't be surprised if you hear some standard speakers saying /ˈkɒn-/.

**constitution(al)ist.** See -IST, -ALIST.

**construct, construe.** 1 In the *OED*, 'const.' is an abbreviation of 'construed with', usu. in formulas like 'const. *of* or *in*', meaning that the verb in question may be used either with a following *of* or with *in*. The word has been used since the 14c., as the *OED* says, in a range of broader senses, to mean 'to analyse or trace the grammatical construction of a sentence; to take its words in such an order as to show the meaning of a sentence; *spec.* to do this in the study of a foreign and especially a classical language, adding a word for word translation; hence, loosely, to translate orally a passage in an ancient or foreign author'. It is thus often virtually a synonym of *translate* or *interpret*.

2 *Construct* alternates with *construe* when used passively. Thus, '*different* is usually construed with/constructed with *from*'. In such contexts *constructed* is now the more common of the two.

3 A curiosity worth noting is that at an early date *construe* (from L *construere* 'to pile together, build up') was stressed on the first syllable, and the final syllable was reduced to *-stre*, *-ster*. As a result, /ˈkɒnstə/ continued to be the pronunciation down to the 19c., even after *conster* had disappeared as a written form.

Walker (1791) called this 'a scandal to seminaries of learning'. The main stress was moved to the second syllable at some indeterminable point in the first half of the 20c.

**4** Examples of the commonest uses of *construe*: *He could not construe the simplest German poem without the help of a translation*—M. Baring, 1924; *His life could be construed ... as a series of delinquent approaches to virtue*—V. S. Pritchett, 1980; *She said nothing to me which could not be construed as loyal admiration*—P. Ackroyd, 1983; *He asked his interrogators to specify anything he had written or said which could be construed as anti-Soviet*—R. Owen, 1985.

**constructive.** **1** In lay language and in politics used esp. in the sense 'having the quality of contributing helpfully': *a constructive approach*; *constructive criticism*, etc. It is opposed to *destructive*.

**2** In legal language it is often applied 'to what in the eye of the law amounts to the act or condition specified' (*OED*). Hence *constructive contempt, fraud, notice, trust*, etc. (for the precise meaning of which a good legal dictionary should be consulted).

**construe.** See CONSTRUCT.

**consubstantiation.** See -CIATION.

**consuetude.** Pronounce /ˈkɒnswɪtjuːd/. The adj. *consuetudinal* (commonly used in philology in the phr. *consuetudinal mood*, a mood indicative of a customary or habitual action or actions) is pronounced /kɒnswɪˈtjuːdməl/.

**consummate.** Pronounce the verb /ˈkɒnsəmeɪt/ and the adj. /kənˈsʌmət/.

**consummation.** For *a consummation devoutly to be wished*, see HACKNEYED PHRASES.

**consumption.** This long-standing term (first recorded in the 14c.) for 'wasting of the body by disease' (from 17c., 'a pulmonary disease, phthisis'), has been replaced in the course of the 20c. by *tuberculosis*, abbrev. *TB*, as part of the tendency to replace the names of dreaded diseases by more clinically sounding names.

**contact.** The verb, meaning 'to get in touch with', came into being in the US

in the 1920s and was greeted with open hostility by purists for several decades. By 1990 it had settled down as a natural and essential part of the language, though many of those who remember the controversy continue to avoid using it. The word is beginning to fall into the traditional stressing pattern of same-spelling nouns and verbs—first syllable for the noun, second syllable for the verb—but initial stressing for both the verb and the noun is still much the more common pattern. See NOUN AND VERB ACCENT.

**contact clauses.** Jespersen's term (*Mod. Eng. Gram.*, III, 1927), now widely used, for dependent clauses that are not formally joined by a relative pronoun or conjunction, e.g. *the man I saw last week*; *he has found the key you lost yesterday*; *there is a man below wants to speak to you*. As Jespersen pointed out, these have always been frequent in colloquial English but in the 18c. were opposed by Latinate scholars as such constructions do not occur in Latin. These constructions are often spoken of as clauses 'with omission of the relative pronoun', i.e. in the above examples, with omission of 'that/whom', 'that', and 'who'. Contact clauses are near relations of *apo koinou* (q.v.) constructions.

Jespersen cited a series of examples of contact clauses from the OE period onward, e.g. *Where is the thousand markes ʌ I gaue thee villaine*—Shakespeare, 1590; *Here she set up the same trade ʌ she had followed in Ireland*—I. Walton, 1653; *There are lots of vulgar people ʌ live in Grosvenor Square*—Wilde, 1893; *It was haste ʌ killed the yellow snake*—Kipling, 1895; *There isn't a boy in your stables ʌ would give me up like that*—G. B. Shaw, 1901.

Further examples are listed in the *OED* s.v. *that* relative pronoun 10, e.g. *What is it ʌ makes me beat so low?*—Tennyson, 1850; and s.v. *that* conjunction 10, e.g. *It may bee ʌ they will reuerence him*—(AV), 1611; *I fear ʌ They will not*—Tennyson, 1847. Grammarians are agreed that contact clauses are probably the product of a paratactic construction of two independent clauses, i.e. the loose running together of two independent clauses, rather than examples of the simple omission of a relative pronoun or conjunction.

See OMISSION OF RELATIVES.

**contagious.** A *contagious* disease is one transmitted by physical contact. It contrasts with *infectious* which, when used of an illness, means one that is liable to be transmitted by micro-organisms by air or water. In figurative uses, *contagious* tends to be used of both pleasant and unpleasant things (in the *OED* and *OED* files of corruption, folly, guilt, panic, and suffering, but also of laughter, shyness, and vigour), whereas *infectious* is mainly restricted to pleasant things (enthusiasm, good humour, laughter, sense of fun, simple delight, virtue, and zeal).

**contango.** Pl. *contangos*. See -O(E)S 6.

**contemn.** This literary word ( = treat with contempt, scorn) is pronounced with the final *n* silent: thus /kən'tem/. The inflected forms of the verb are similarly pronounced with silent *n* (thus /kən'temɪŋ/ and /kən'temd/, but *contemner* may be pronounced with or without sounding the *n*. See CONDEMN.

**contemplate.** Formerly there was much variation in the placing of the stress, but the orthoepists generally stressed on the second syllable down to the third quarter of the 19c.; since that time initial stressing has more and more prevailed, and *contémplate* begins to have a flavour of age (thus the *OED* in 1893). Now always /'kɒntempleɪt/.

**contemporaneous.** See next.

**contemporary.** 1 It is worth noting that though *contemporary* is the original (first recorded in 1655) form of the word, the by-form *co-temporary* (with or without hyphen) was used by many writers, including Dr Johnson, from the 17c. to the 19c. though characterized by Richard Bentley in 1699 as 'a downright barbarism': e.g. *Prudentius a Christian poet, cotemporary with Theodosius*—Fanny Burney, a1789; *Supported by cotemporary scholars*—Max Müller, 1861.

2 The word inevitably has two main senses: (*a*) living or occurring at the same time (*writers contemporary with the events they write of*); similarly with the corresponding noun (*the comic poet Alexis, a younger contemporary of Plato*). (*b*) modern, of or characteristic of the present day,

up-to-date (*Contemporary curtains for contemporary colour schemes*—M. Drabble, 1963; *Before I always wanted something up to date, what they call contemporary*—J. Fowles, 1963; *This classic is suddenly contemporary. The boat shoe is done in cotton canvas with a new shape*—Sears Catalog *1985*, 1984). Some writers use parentheses to cancel any possible ambiguity, e.g. *a contemporary report (from the 17th century)*.

3 Perhaps because of the occasional risk of ambiguity (Gowers (1965), for example, spoke of the ambiguity of the sentence Twelfth Night *is to* be *produced with contemporary incidental music*), *contemporaneous* (first recorded in 1656), which has been used from the beginning only in sense (*a*), is now more common than it used to be: *The high rate of interest, which is generally contemporaneous with a drain of specie*—G. J. Goschen, 1861; *His book is contemporaneous with our own ideas rather than with the thoughts of his generation*—H. J. Laski, 1920; *The recommendations of the Royal Commission that note-taking at interviews be improved ... encourage the taking of contemporaneous verbatim notes*—R. C. A. White, 1985.

**contemptible, contemptuous.** 1 The first means 'to be despised, deserving contempt, despicable' (*there are few things in life more contemptible than child abuse*), the second (often followed by *of*) 'showing contempt, scornful' (*Saddam Hussein showed a contemptuous disregard for the feelings of hostages held in Baghdad; he is contemptuous of the laws governing human rights*).

2 In historical contexts, e.g. in the works of Shakespeare, *contemptible* will be encountered in the sense 'exhibiting or expressing contempt' (e.g. *The contemptible idea I always entertained of Cellarius*—Gibbon, 1762), but this use is no longer extant.

**content** (noun and adj.). See NOUN AND ADJECTIVE ACCENT.

**content** (verb). *Content oneself with* (not *by*) is the right form of the phrase that means not go beyond some course of action: e.g. *You should not content yourself with being a mere collector of insects; He contents himself with reporting the results of other scholars.*

**content(ment).** The two forms now mean almost the same, *contentment* having almost lost its verbal use (*The contentment of his wishes left him unhappy*) and meaning, like *content*, contented state. *Contentment* is the usual word, *content* surviving chiefly in *to one's heart's content* and as a poetic or rhetorical variant.

**content(s),** what is contained. The OED (1893) said 'The stress *contént* is historical, and still common among the educated, but *cóntent* is now used by many, esp. by young people'. Fowler (1926) took the same view. But *content* (singular, as in *the cubic content*) and *contents* (plural, as in *the contents of the house will be sold at auction*) are now invariably pronounced with initial stressing, doubtless owing to a wish to differentiate the word from *content* = contentment.

**contest.** The noun (*a close contest*) is stressed on the first syllable, the verb (*he will not contest the seat at the next election*) on the second. See NOUN AND VERB ACCENT 1.

**Continent.** In the UK, despite the existence of the EC, of which Britain is a full member, *the Continent* still invariably means 'the mainland of Europe as distinct from the British Isles'. A *continental breakfast* is a light breakfast of coffee, rolls, etc., seen as typical of frugal breakfasts served in France, Italy, etc., as distinct from a fuller one. The phrase *Continental Sunday* is still occasionally heard, meaning one observed more as a day of public entertainment (as held to be customary on the continent of Europe) than as a day of rest and religious observance (as held to be customary in Britain) (*OED*).

**continual, continuous.** Only since the mid-19c. has it been customary to regard *continual* as being applicable to events that occur frequently but with intervals between, and *continuous* to anything that happens or proceeds in an unbroken or continuous manner. In practice the distinction is not as neat as that.

In its long history, since it was first recorded in the 14c., *continual* has been used at all times to mean 'incessant, perpetual', and also, less strictly (*OED*), to mean 'repeated with brief intermissions'. *Continuous* is a more modern word (first recorded in 1673) than *continual*. It

was first used in technical contexts in botany (of plants having their parts in immediate connection) and optics, and was not brought into general use until the 18c. In modern technical combinations, *continuous* (not *continual*) is the preferred term: e.g. *continuous assessment* (1959), *continuous creation* (1941), *continuous process* (1909, in industry, opp. *batch process*), *continuous stationery* (1942), *continuous tenses* (1887, in grammar), and numerous others.

Fowler's judgement (1926), though somewhat cryptically expressed, is about right: 'That is -*al* which either is always going on or occurs at short intervals and never comes (or is regarded as never coming) to an end. That is -*ous* in which no break occurs between the beginning and the (not necessarily or even presumably long-deferred) end.' In non-technical contexts, because of the lack of symmetry between the two words, it is often better to fall back on virtual synonyms: e.g. *recurrent* or *intermittent* instead of *continual*, and *unbroken*, *uninterrupted*, or *incessant* instead of *continuous*. Examples: (continual) *there was a grave ... that fascinated her, ornate and Jacobean, with four angles, an urn ... all worn away by the continual weather that had beaten it for three hundred years*—P. Larkin, 1947; *The house and garden had seen their best days, and the decline was now continual, from season to season*—R. Frame, 1986; (continuous) *Weldon, excellent comedian that he was, kept the audience in continuous laughter from the moment he came on*—C. Chaplin, 1964; *The correspondence between the two men was continuous throughout the next few months*—V. Brome, 1978; *The 1840s were years of continuous self-education for Philip Henry Gosse*—A. Thwaite, 1984.

**continually, continuously.** In historical terms, *continually* (first recorded in the 14c.) was the earlier by far: *continuously* was not recorded until 1678. During these three centuries, *continually* had been used to mean both 'incessantly, constantly, all the time' and less strictly (*OED*) 'with frequent repetition, very frequently'. The arrival of *continuously* signalled a fairly general restriction of *continually* to the less strict meaning. From the 18c. onward, *continuously* has tended to be the more dominant word of the two, esp. when the unbroken nature of a process (in space or time) or set of events (e.g. unbroken terms of office) is being stressed. But there are circumstances in which good writers are moved to use the two words more or less interchangeably. The choice is largely governed by the degree with which continuity is being emphasized. Many events or processes, like theatrical performances and long-haul flights, can properly be regarded as continuous if one disregards minor interruptions like an intermission and a refuelling stop. Examples: (continually) *Among all our talks, we returned continually to an argument about science*—C. P. Snow, 1934; *You can't do that at ease with a man continually on the go to the cellar for another litre*—G. Greene, 1967; *He said that the business of the court . . . was being continually held up by irrelevancies*—J. B. Morton, 1974; *The black coat lost its warmth and he shivered continually*—J. M. Coetzee, 1983; (continuously) *You're a masochist about it . . . you want to feel that your inability to create is continuously significant*—I. Murdoch, 1973; *This idea, plus a continuously variable output voltage position, is featured*—*Giant Bk Electronics Projects*, 1982.

**continuance, continuation, continuity.** *Continuance*, first recorded in the 14c., has nine main senses in the OED, four of them now obsolete. It has stood across the path of *continuation*, also first recorded in the 14c., in a wide range of applications, as if to say 'use me, not it', and the merging of meanings at all periods has been more noticeable than their separation.

At the present time, *continuance* is more often used when the context requires the sense 'a state of continuing in existence or operation; the duration of an event or action'; and *continuation* when the sense required is 'the act or an instance of continuing, esp. the resumption of an interrupted action or course'; or 'that by which anything is continued'. *Continuation* is by far the more frequent of the two. Some examples of each: (continuance) *Continuance of 1948–73 real income trends would make the average Japanese between three and four times richer than the average Briton*—*Economist*, 1975; *Confusion has arisen about their desperate continuance of the struggle which was manifestly lost*—Antonia Fraser,

1988; (continuation) *To impose a continuation of marriage on people who have ceased to desire to be married*—G. B. Shaw (cited in SOED); *The page four continuation of a front-page story*—J. Gardner (cited in SOED).

*Continuity* means 'an unbroken succession (of a set of events)' (*The political stability of most countries depends on the continuity of subsidies for agriculture*). In broadcasting, *continuity* is the process by which the separate shots or recordings of a programme are joined together in such a manner that they form a convincing sequence. The person responsible for this is called a *continuity girl* or a *continuity man*.

**continuant.** In phonetics, a speech sound which can be continued or prolonged, esp. all the vowels and the consonants *f*, *s*, and *th* /θ/ and their voiced equivalents, *v*, *z*, and *th* /ð/, as distinct from the stops, *b*, *d*, *p*, *t*.

**continue.** The comparatively modern (not in the OED) use of *continue* with *on*, esp. in contexts of travel (*After leaving Nice continue on to Cannes*), persists despite its palpable degree of tautology. (The 1719 example in OED, [It] *is likely to continue on to do so*, is an isolated and unrelated construction, with the meaning 'to go on doing'.) It is easily avoided if the tautology offends (e.g. *The main road to Turin goes on past the remote village of Vieve*—*Blue Guide to the South of France*, 1966).

**continuous.** See CONTINUAL.

**continuously.** See CONTINUALLY.

**continuum.** Pl. *continua*. See -UM 2.

**contract.** The noun has initial stressing and the verb is stressed on the second syllable. See NOUN AND VERB ACCENT 1.

**contractable, contractible.** See -ABLE, -IBLE 4.

**contractions.** 1 Fowler (1926) used the word to mean an abbreviation consisting of the first and last letters of a word, e.g. *Mr* for *Mister*, *St* for *Saint*, as distinct from abbreviations like *Jun.* for *Junior*, *Capt.* for *Captain*. A full point, he argued, was not needed for contractions but was desirable for abbreviations. This useful convention has been widely adopted, but (*a*) is occasionally awkward as when, for example, *Rev.* (with full point) and *Revd* (without) (both = *Reverend*) are distinguished in this way, (b) has been widely undermined by the abandonment of the full point in many publications for both abbreviations and contractions.

**2** Contractions of the type *I'm* ( = *I am*) and *don't* ( = *do not*) are exceedingly common in informal writing and increasingly found in various kinds of fairly formal contexts (e.g. in book reviews). The conventional forms of contractions have been subject to considerable fluctuation at various periods. Shakespeare, for example, used *Ile/I'le* for 'I will', *hee'l, shee'l, wee'l, you'l* for 'he, etc., will', and *I'ld, hee'ld*, etc., for 'I, etc., would'. But the *OED* records some of the most frequent modern colloquial contractions only from later periods: *ain't* 1778, *can't* 1706, *couldn't* 1882, *don't* 1670, *I've* 1885, *shan't* 1850, *shouldn't* 1848, *won't* 1660, and *wouldn't* 1836. Fluctuation in the use of the reduced forms is subject to all manner of social and stylistic assumptions which vary from one century to another.

**contractual.** The erroneous form *contractural* is surprisingly common. The existence of the suffixes *-ual* (*mutual, spiritual*, etc.) and *-ural* ( *procedural, structural*, etc.) accounts for the seepage of one type to another, but does not excuse it.

**contralto.** The pl. is now almost invariably *contraltos*, and not (as in Italian) *contralti*. See -O(E)S.

**contrary.** **1** The placing of the main stress has fluctuated over the centuries: 'the poets, from Chaucer to Spenser and Shakespeare, use both *contráry* and *cóntrary*.' (*OED*) In standard English, initial stressing is now obligatory for the adj. and the noun, except that /kən'treəri/, rhyming with *Mary*, is customary for the adj. in the sense 'perverse, obstinately self-willed', probably under the influence of the nursery rhyme *Mary, Mary, quite contrary, How does your garden grow?* Second-syllable stressing is also usual in *contráriwise*.

**2** The phr. *on the contrary* is properly used only in a statement intensifying a denial of what has just been stated or implied (e.g. *You say that war is inevitable; on the contrary I think the outstanding differences between the two countries can be settled by negotiation*). By contrast, the phr. *on the other hand* ( = from another point of view) is usually directly or impliedly paired with *on the one hand* ( = from one point of view) to indicate two contrasted sides of a subject, circumstances, considerations, points of view, etc. (*OED*): *If men of eminence are exposed to censure on the one hand, they are as much liable to flattery on the other*—Addison, 1711.

**3** *to the contrary*. In BrE the type *to the contrary* (e.g. *there is plenty of evidence to support this view and none to the contrary*) is normal, as it is in AmE and other forms of English. The use of *to the contrary* at the head of a sentence (e.g. *To the contrary, we have granted the press extraordinary protection for extraordinary reasons*— Bull. Amer. Acad. Arts & Sci., 1987) is restricted to AmE. In BrE such a sentence would begin *On the contrary*.

**contrast.** **1** The noun is pronounced with initial stress, and the verb with the stress on the second syllable. See NOUN AND VERB ACCENT 1.

**2** Be prepared to encounter certain unfamiliar transitive uses of the verb, and also uses of the verb construed with *by*, in older literature (e.g. *The brown and sun-tinged hermit and the pale decrepit elder contrast each other*—H. Fuseli, 1801–15; *The smooth slopes . . . are contrasted by the aspect of the country on the opposite bank*—A. W. Kinglake, 1863). Nowadays the normal constructions are with *with* and with *and*: *Her sudden energy contrasted with Henry's sudden exhaustion*—J. Frame, 1970; *My preoccupation with the poverty . . . I saw . . . would not have been so deep if I had not been able to contrast it with the relative affluence of the white world*—Angela Davis, 1975; *He is contrasting Adam in his state of innocence and us in our state of knowledge*—A. N. Wilson, 1983.

**3** The noun can be followed by *to, with*, or *between* : *Captain Massingham's picture . . . was in provocative contrast to all the soft and pretty comforts of his wife's bedroom*—M. Keane, 1981; *The light of intellectual processes in contrast with the darkness of the unconscious*—J. Hawkes, 1951; *Doing such a thing roused Wendy to angry thoughts of the contrast between the life she had once known and the one she was living now*—M. Laski, 1952; *Colour contrasts between brown stock* *bricks, red brick dressings, and white wood-work*—M. Girouard, 1977.

**contribute.** The standard pronunciation is with the main stress on the second syllable, /kən'trɪbjuːt/, but the non-standard pronunciation with initial stressing is beginning to encroach and is included in *COD* (1990) as a 'disputed use'.

**controversy.** The mood of the moment is to challenge orthodoxy by placing the main stress on the second syllable. This stressing is often used by newsreaders and also, in my experience, by many scholars and lexicographers, not to go any further. My verdict is that the traditional pronunciation with initial stressing is at risk, but is still, just, dominant among RP speakers in the UK. In AmE the stress is always placed on the first syllable in this word.

**contumely.** Stress on the first syllable, /'kɒntjuːmlɪ/, and give the word three syllables. Some RP speakers prefer to pronounce it as four syllables, /-tjuːmɪlɪ/.

**conundrum.** A 16c. word of unknown origin (not L, but possibly originating in some now-lost university joke). Pl. *conundrums*. See -UM 1.

**convenance, convenience.** The first is used mainly in the French phr. *mariage de convenance*, but *marriage of convenience* is much the more usual term in English.

**conversable, -ible.** See -ABLE, -IBLE 4.

**conversance, conversancy.** Both terms may be used for 'the state or quality of being conversant', but neither is common.

**conversant.** The placing of the stress has changed over the centuries from the third syllable to the first and, most recently, to the second. Second-syllable stressing is recommended, /kən'vɜːsənt/.

**conversationalist, conversationist.** Both terms have been in use since the early 19c., and neither can be faulted as formations. In practice, *conversationalist* is overwhelmingly the more common of the two.

**conversazione.** This word of Italian origin for 'a social gathering held by a learned or art society' is pronounced /ˌkɒnvəsætsɪ'əʊnɪ/. Its pl. is either *conversaziones* (the more usual) or *conversazioni* (pronounced like the singular).

**converse.** Pronounce the verb ( = engage in conversation) with the stress on the second syllable, and the noun, in its archaic sense 'conversation', on the first. The adj. and noun *converse* = (something) opposite, which are both pronounced with initial stressing, are unrelated words. See NOUN AND ADJECTIVE ACCENT.

**conversion,** in grammar, the deliberate use of one part of speech as another, is a modern term (first recorded in 1928) for an ancient process. It is also called *zero-derivation.* **1** In historical terms, conversion, and esp. the employment of nouns as hitherto unrecorded verbs, is exceedingly common. Examples (with the date of first use of the noun given first, and that of the verb second): *chair* (a1300, 1552), *distance* (c1290, 1578), *elbow* (OE, 1605, Shakespeare), *inconvenience* (c1400, a1656), *magic* (c1384, 1906), *napalm* (1942, 1950), *save* (c1250, 1906), *service* (c1230, 1893), *silence* (a1225, 1603, Shakespeare), *telephone* (1835, 1877). In these examples the maximum interval is one of over 600 years (*service*) and the minimum 8 years (*napalm*), but doubtless more extreme examples exist.

**2** Examples of the conversion of other parts of speech are less common, but they do exist: e.g. (adj. → verb) *dirty* (15.., 1591), *empty* (OE, 1526), *lower* (c1200, 1659), *ready* (c1205, a1340), *total* (c1386, 1716); (noun or verb → adj.) *go, no go* (OE, 1910), *touch and go* (1655, 1812); (verb to noun) *call* (OE, a1300), *interrupt* (1413, 1957); (miscellaneous) *but me no buts, the hereafter, a fun party, a must.*

**3** In the 20c., public resistance to such processes has hardened, and examples such as the following are almost invariably regarded with alarm: *Another rival ... candidate ... audibles his allegations*—Chicago Sun-Times, 1988; *somebody had blowtorched into it the names of all the people who spoke for it*—New Yorker, 1989; *it is difficult to example one aircraft accident statistic where the cause has been reliably attributed to cigarette smoking*—letter to Times, 1989.

Among the most fiercely resisted conversion-verbs that have come into prominence in new senses in the 20c. are *author, contact, critique, host, impact,* and *parent* ( = to be a parent).

**convert.** The noun has initial stress, and the verb is stressed on the second syllable. See NOUN AND VERB ACCENT 1.

**convertible.** Always thus spelt, not -*able.* See -ABLE, -IBLE 7.

**convex.** See CONCAVE.

**convict.** The verb is stressed on the second syllable, and the noun on the first. See NOUN AND VERB ACCENT 1.

**convince.** This verb, which has been in use in English in various senses and constructions since the 16c., began to be controversially used in the 1950s (with *to* + infinitive) to mean 'persuade'. Examples: *He worked very hard personally to convince Ike to run*—D. Halberstam, 1979; *Barril's overtures failed to convince him to come out of hiding*—Observer, 1983; *The miners tried to convince their colleagues to join them*—BBC World Service, 1991. It is a classic example of a new construction that is acceptable or at least unexceptionable to some and repugnant to others.

**convolvulus.** Pl. *convolvuluses.* See -US 1. (If you wish to use *convolvuli* instead, pronounce the final syllable /-laɪ/.)

**convoy.** Until the early part of the 20c. the noun (stressed on the first syllable) and the verb (on the second syllable) were regularly distinguished. Now both words are stressed on the first syllable.

**cony, coney** (a rabbit). OUP house style is *cony* (pl. *conies*), not *coney* (pl. *coneys*).

**cooee,** a sound used to attract attention, is the universal spelling now for the interjection, the noun (*within cooee (of)*, Aust. and NZ, = within hailing distance), and the verb.

**cookie** has settled down in this spelling (not *cooky*) in its various senses, (Sc.) a plain bun, (US) a small sweet biscuit, (slang) a person. Also in the colloq. phr. *that's the way the cookie crumbles,* that's how things turn out, the unalterable state of affairs.

**coolie.** 'The usual modern form' (Fowler, 1926), not *cooly* or *coolee*. Mercifully, the world has changed to the point that the word is much less frequently encountered now.

**coomb.** The customary form (rather than *combe*) in Britain for a valley or hollow on the side of a hill or a short valley running up from the coast. In place-names, however, -*combe* is the more usual form (*Ilfracombe, Winchcombe,* etc.).

**cooperate, co-opt, coordinate.** See CO-.

**cope.** The traditional construction with *with* has been in common use since the 16c.: e.g. *His being too unwell to cope with Dr. Johnson*—Mme D'Arblay, 1782; *Let the scholar measure his valour by his power to cope with intellectual giants*—Emerson, 1875; *There was pain later on but nothing his stubborn soul could not cope with*—W. Golding, 1979; *Like many religious professionals, I cope with festivals, but I can't really enjoy them*—L. Blue, 1985.

It now stands alongside absolute uses ( = to manage) which were first recorded in the 1930s: e.g. *Angela rang the bell wildly for someone to come and cope*—E. Bowen, 1934; *it wasn't as if Marcia was an invalid or unable to cope, even if she was a bit eccentric*—B. Pym, 1977.

**copular** (or **copulative**). In grammar, used of a word that connects words or clauses linked in sense or connecting a subject and predicate. 'The term *copula* refers to the verb *be*, and *copular* verbs are those verbs (including *be* and *become*) which are functionally equivalent to the copula' (*CGEL* 2.16). Typical copular constructions are: *he* felt *annoyed, she* sounded *surprised, he never* became *Prime Minister, the outcome* remained *uncertain, it all* came *true, the weather* turned *cold*; and esp. the type *Canberra* is the capital *of Australia.*

**coquette,** a woman who flirts. Pronounced /kəʊˈket/ (preferably), or /kɒˈket/. The related abstract noun *coquetry* is best pronounced /ˈkəʊkɪtrɪ/, and the verb *coquet* as /kəʊˈket/.

**corbel** (verb), to support or project on corbels, gives *corbelled, corbelling* (US *corbeled, corbeling*). See -LL-, -L-.

**cord.** See CHORD, CORD.

**cordelier,** a Franciscan friar of a strict order (wearing a knotted cord round the waist). Pronounce /kɔːdr'lɪə/.

**cordon.** When it stands by itself, *cordon* is always pronounced /'kɔːdən/, but when used with *bleu* or *sanitaire* it is commonly (?almost always) pronounced in a manner suggestive of its French origin: thus /kɔː(r)dɔ̃/.

**core.** *Rotten at the core (heart)* is a MIS-QUOTATION.

**co-respondent.** See CO-.

**corn.** In BrE *wheat* or *oats*, in AmE *maize* (also called *Indian corn* or *Indian maize*).

**cornelian,** a variety of chalcedony of a deep dull red, flesh, or reddish white colour, is in origin a refashioning of 15c. Fr. *corneline* (modF *cornaline*). Later in the same century a by-form *carnelian* appeared in English, influenced by L *carno*, *carnem* flesh, *carneus* flesh-coloured. On etymological grounds, *cornelian* is marginally the better form.

**cornucopia.** From a late L modification of L *cornū* 'horn' + *cōpiae* 'of plenty', its pl. in English is *cornucopias* (not *cornucopiae*).

**corolla,** a whorl or whorls of petals forming the inner envelope of a flower, has *corollas* as its pl. in English. In origin it is a L diminutive of *corōna* 'crown'.

**corona.** In its several English meanings (in astronomy, anatomy, botany, etc.) based on L *corōna* 'crown', its pl. is *coronae* /-niː/, rarely *coronas*. The Cuban cigar, the *Corona*, always has its pl. in *-as*.

**coronal.** Pronounce the noun ( = a circlet, a garland) /'kɒrənəl/, and the adj. (of the skull; in astronomy and botany, of a corona) /kə'rəʊnəl/. (Under the influence of the next word, the adj. *coronal* is now sometimes stressed on the first syllable.)

**coronary.** With the increasing frequency of the condition called (*a*) *coronary thrombosis* (often called merely *a coronary*) during the 20c., the word has become very familiar to lay people. It is pronounced /'kɒrənərɪ/ in BrE; in AmE the third syllable is fully pronounced.

**coronate** (verb). In the sense 'to crown', *coronate* (formed regularly from the pa. pple of L *corōnāre*) has been current in technical and semi-technical language, esp. in botany and zoology, since the 17c. In Britain the word is never used as a synonym of *to crown* (a monarch). Sporadic instances of such a use have been noted in the US.

**coroneted.** Spelt with one *-t-*, in accordance with the rule that unstressed syllables in verbs ending in *-t* are not doubled in their inflected forms (thus *benefited*, *targeted*, etc.). See -T-, -TT-.

**corporal, corporeal.** Neither is a common word except in particular phrases. *Corporal* means 'of or relating to the human body', and is common in the phr. *corporal punishment* (though this type of punishment is now effectively banned in EC schools). In other possible contexts (used of beauty, deformity, etc.) it is now almost entirely replaced by *bodily* or *personal*. *Corporeal* means principally 'bodily, physical, material, as distinct from spiritual' (*that which is created is of necessity corporeal and visible and tangible*—Jowett, 1875; *the corporeal presence of Christ in the Sacrament*).

**corps.** Pronounce in sing. /kɔː/, but in pl. (though the spelling is the same) /kɔːz/.

**corpulence, corpulency.** The first of these is by far the more frequent of the two words, though they both acquired the sense 'obesity' at a similar date and seemed to be variants of equal standing between the 16c. and the late 19c. In Hilary Mantel's *Fludd* (1989) the phr. *His Corpulence* is used jocularly as a term of address to a bishop (cf. *His Eminence*, *His Excellency*). See -CE, -CY.

**corpus.** The regular pl. is *corpora*, but *corpuses* is gaining ground. One of its most frequent uses in English language contexts now is for a representative group of (English) texts, spoken and (esp.) written, running to many millions of words, and used as measurable evidence, esp. for the currency and frequency of individual words and constructions. Among the best-known pioneering corpora of this kind are the Brown Corpus (1963–4), edited by W. Nelson Francis and H. Kučera, and updated in 1982; the corpus of the Survey of English Usage (University College London) (originated in the 1960s); and the COBUILD corpus (Collins–Birmingham University International Language Database), established by 1987. New computer-tagged corpora are being created all the time.

**corpuscle.** Pronounce /ˈkɔːpʌsəl/. The variant /kɔːˈpʌsəl/, used, I recall, by my father, seems to be dying out.

*corpus delicti.* In law, the facts and circumstances constituting a breach of the law, literally (in L) 'the body of a crime'. In lay use, the concrete evidence of a crime, esp. the body of a murdered person (*OED*).

**corral.** Pronounce this word of Sp. (and Pg.) origin /kɒˈrɑːl/. It is related to the S. Afr. word *kraal* (ult. from the same Pg. base) = a fenced village of huts.

**correctitude,** formed after *rectitude* in the late 19c., means 'correctness of conduct or behaviour'. It is not widely used.

**correctness.** Linguistic correctness is perhaps the dominant theme of this book. One of my concerns is the infringement of linguistic laws through ignorance or inadvertence. Obviously there are errors and errors, ranging from the illiteracy of *We can't possibly know everybodies position* (from an advertisement in a computer journal called *DEC User*, 1988) to the inadvertent hilarity of *For sale—four-poster bed, 101 years old. Perfect for antique lover* (from a report in the American magazine *People*, 1988). Schoolteachers derive a great deal of pleasure from the howlers of their pupils' essays. The type is well known: *Socrates died of an overdose of* wedlock; *The First World War, caused by the* assignation *of the Arch-Duck by an anarchist, ushered in a new error in the* anals *of human history.* It is only when one moves into the world of fully adult writing that infringements become reprehensible. Perhaps T. S. Eliot should have been clapped in irons for allowing *staid* (for *stayed*) to appear in the first edition (1939) of *The Family Reunion* (*You have staid in England, yet you seem Like someone who comes from a very long distance*). The same cell could be made ready for the person who wrote to the *London Review of Books* in 1988 about barbed wire being *payed out* from the saddle horn.

These might be said to belong to the sophisticated school of error. Easily the most common type is that shown in a Malapropistic setting down of words that are nearly right but not quite right. Phrases like *baited breath* (for *bated breath*), *wet the appetitite* (*whet*), and *hare's breath* (*hair's breadth*) lie scattered about in newspapers like broken bottles. Keep the words *snook, intransitive, disquieted,* and *recoup* in mind when reading the sentences that follow: *I'm in no position to cock a snoot at these new acts* (*The Face,* 1986); *One, a head of English, could not explain the function of an intransigent verb* (letter to *Sunday Times,* 1988); *Our man came away profoundly disappointed. And not a little disquietened* (*Auckland* (NZ) *Star,* 1988); *No explanation has been given on how investors in the expedition will recuperate the $2.5 million cost of the adventure* (Associated Press report on salvaging relics from the Titanic, 1987). Such are typical casualties of the fast-moving newsprint world. It is much more difficult to find errors of the same kind, and in such numbers, in copy-edited books put out by reputable publishers.

**corrector.** Thus spelt, not *-er*. See *-OR*.

**correlative** (adj.). 'Applied to words corresponding to each other, and regularly used together, each in one member of a compound or complex sentence: e.g. *either—or; so—as.*' (*OED*). A relatively modern term (first recorded in 1871), it is applied to clause-connecting words in English after the manner of similar words in Latin (*tantus—quantus, tam—quam, sic—ut,* etc.). Other common correlatives are *both—and, neither—nor, not only—but,* and *whether—or.* Correct and incorrect uses of correlatives are discussed under the relevant words.

**correspond.** In the senses (a) 'to be analogous or similar' and (b) 'to agree in amount, position, etc.', *correspond* is usually followed by *to*: (a) *Our nature corresponds to our external condition*—J. Butler, 1736; *We don't stop to consider much how far the pictured past corresponds to any reality*—R. Fry, 1920. (b) *The degrees of condensation of the air correspond to the degrees of cold*—R. J. Sullivan, 1794; *The peaks in such a curve correspond to the scattering of the electron wave*—*Sci. Amer.,* 1976. When it means (c) 'to be in harmony or agree-ment' or (d) 'to communicate by inter-change of letters', it is usually followed by with: (c) *The prudence of the execution should correspond with the wisdom ... of the design—Junius Letters, 1769; There were two bedrooms to correspond with the rooms down-stairs*—D. H. Lawrence, 1921. (d) *Those who have a mind to correspond with me, may direct their Letters to the Spectator*—Addison, 1711; *Sir Thomas Brews corresponded with John Paston II in 1477.* In practice, though, sense (c) is often followed by *to*. Had the *Junius* sentence been written now, it would probably appear as *correspond to the wisdom*. Note also *At least she had work that corresponded to her intentions when she left Germany*—D. May, 1986.

**corrigendum.** Pl. (much commoner than the sing.) *corrigenda*; see -UM 2.

**corrupter.** Thus spelt, not *-or*. See -OR.

**corsage** is still pronounced /kɔːˈsɑːʒ/ in a broadly French manner.

**corset.** In its ordinary sense 'a closely-fitting undergarment worn by women to give shape and support to the figure', it is properly used in the sing. But Somer-set Maugham (*Cakes and Ale*, 1930) is not the only person to have spoken of the garment in the pl. (*She did not put on her corsets again, but rolled them up*). The corresponding ppl adj. *corseted* and vbl sb. *corseting* have only one *-t-* in OUP house style. See -T-, -TT-.

**cortège.** Printed in roman type with a grave accent. Pronounce /kɔːˈteɪʒ/.

**Cortes,** the national assembly of Spain or Portugal is usu. pronounced /ˈkɔːtez/ in English. So too is *Cortés*, the 16c. Spanish conquistador.

**cortex,** the outer part of a bodily organ (*cerebral cortex, renal cortex*, etc.) has as its pl. *cortices* /ˈkɔːtɪsiːz/. Cf. L. *cortex, corticis* bark. See -EX, -IX.

**Corybant** /ˈkɒrɪbænt/, in classical myth-ology, a wild attendant of the goddess Cybele, has as pl. either *Corybants* or *Corybantes* /kɒrɪˈbæntiːz/.

**coryphaeus** /kɒrɪˈfiːəs/, (in ancient Greek drama) the leader of the chorus, has as pl. *coryphaei* /kɒrɪˈfiːaɪ/. See -US 1.

**'cos.** A frequent reduced form of *because*, first recorded in 1828, and used only in the representation of informal or re-gional speech: *They'll be good if I tell them, Mister.' 'Then why aren't they?' ''Cos I tell 'em to be bad.'*—E. Waugh, 1942; *that's why I seldom ever talk about paintings, 'cos you always sound like a jerk*—Steve Martin (an American) in *The Face*, 1988. See 'CAUSE.

**cosmonaut.** See ASTRONAUT.

**cosset** (verb). The inflected forms are *cosseted, cosseting*, with a single *-t-*. See -T-, -TT-.

**coster(monger).** See ABBREVIATIONS 1.

**costume.** The *OED* (1893) expressed a preference for second-syllable stressing for both the noun and the verb (some-what uncommon, except as ppl adj.), and Fowler (1926) similarly for the verb alone. Fowler said that first-syllable stressing of the noun 'is often heard'. First-syllable stressing for both noun and verb is now standard.

**cosy** as adj. meaning 'comfortable, etc.' and as noun meaning 'teapot-warmer' is always spelt thus in Britain and in most other English-speaking countries except the US, where the standard (but not invariable) spelling is *cozy*.

**cot.** 1 In the senses 'small bed for a baby' and 'hospital, naval, etc., bed', the word is Anglo-Indian in origin (Hindi *khāt* 'bedstead, hammock'). A *cot death* (US *crib death*) is a modern term (first recorded in 1970) for the unexplained death of a baby in its sleep.

2 In the sense 'a small house', it is a native word, answering to OE *cot* in the same sense. It survived as a poetical and literary word until about the end of the 19c. (e.g. *A few humble fishermen's cots*—E. E. Napier, 1849). As a second element in *bell-cot, sheep-cot* it means a small struc-ture for shelter or protection.

3 See COTE.

**cote** /kəʊt/. A native word (OE *cote*) = COT 2 and for many centuries more widely used than it, e.g. *bell-cote, sheep-cote, swine-cote.* The forms *dovecote* and *dovecot* have both been used since the 15c.; according to *COD* 1995, the spelling *dovecote*, pronounced /ˈdʌvkɒt/, is now dominant.

**cotemporary.** See CONTEMPORARY 1.

**co-tidal.** See CO-.

**cotillion** (a dance). Thus spelt in English, not quite reflecting the spelling of its French original (*cotillon*). Pronounce /kə-ˈtɪljən/.

**cottar, cotter, cottier.** Three forms of a word used at various times in the past, principally in Scotland (usu. spelt *cottar*, but also *cotter*) for 'a tenant occupying a cottage with or (late 18c.–19c.) without land attached to it; a married farm-worker who has a cottage as part of his contract' (*Conc. Scots Dict.*, 1985); and in Ireland (usu. *cottier*) for 'a peasant renting and cultivating a small holding under specified conditions of tenure. The word is best remembered, perhaps, in the title of Burns's poem 'The Cotter's Saturday Night'.

**cotyledon.** Pl. *cotyledons*; see -ON 3.

**couch**[1]. A *couch* is a near-synonym of *sofa* except that a sofa normally has a back and two arms, and a couch has often just one raised end and is designed in such a way that it is suitable for sleeping or reclining on. *Couch* is also the word used for 'a long padded seat with a headrest at one end, esp. one on which a psychiatrist's or doctor's patient reclines during examination' (*COD*). *Couch* is also a poetic or literary word for a bed. The distinction between a *couch* and a *bed* is brought out in these examples: *I've made a bed up for you on the couch*—M. Amis, 1973; *My brother and I share the verandah in the summer—what were two couches by day became our beds at night*—B. Mason, 1980. A *settee* is an upholstered seat with a back and usu. with arms, designed so that two or more people can sit on it. An *ottoman* is 'an upholstered seat, usu. square and without a back or arms, sometimes a box with a padded top' (*COD*). A *divan* is 'a long, low, padded seat set against a room-wall; a backless sofa' or 'a bed consisting of a base and mattress, usu. with no board at either end' (*COD*).

**couch**[2], any of several weed-like grasses. Usu. pronounced /kuːtʃ/, but /kaʊtʃ/ is also common. In origin it is a southern-counties variant of *quitch* (OE *cwice*), which has the same meaning.

**couchant.** In heraldry pronounced /ˈkaʊ-tʃənt/. It is usu. placed after the noun it governs (*two lions couchant*).

**could** is a modal auxiliary, formally the past tense indicative of CAN. In contexts of possibility, ability, and permission, it operates for the most part in the same way as *can*: (possibility, opportunity, risk, etc.) *the tide could swing decisively against him*; *it is difficult to imagine that a jury could find in his favour*; *pains that he thinks could be cancer*; *could he be making this up?*; *he's pleased, but his mood could change at any time*; (ability) *she could [was able to] remember it as if it were yesterday*; *all I could do was offer him a drink*; *she could not walk and think deeply at the same time*; *if only he could settle down and live like a normal person*; (permission, instead of *may*) *Could I come and see you next week?*; *Could I have breakfast now, please?* Cf. MAY AND MIGHT. *Could* is occasionally used as a past subjunctive, 'expressing an inclination in a conditional form' (*OED*): *I could wish our Royal Society would compile a Body of Natural History*—Addison, 1711; *I could not think of leaving you so soon*—Mrs Inchbald, 1786; *'I could hang up sheets, I guess.' But she's too busy to.*—New Yorker, 1987. The negative form is, of course, *could not*, frequently written in reduced form (*couldn't*). For *I could(n't) care less*, see CARE (verb).

**could of.** An illiterate alteration of *could 've* = *could have*, found depressingly often in children's letters and essays, and in the written work of poorly educated adults.

**coulomb** (unit of electric charge). Now pronounced /ˈkuːlɒm/.

**coulter** (a plough blade). Not *col-*.

**council, counsel. 1** A *council* is an advisory or administrative body of people formally constituted and meeting regularly (*parish council*, *city council*, *county council*, etc.) or a meeting of such a body. A member of one is a *councillor* (in AmE sometimes spelt *councilor*). *Council* is often used attributively of housing, etc., provided by a local council (*council flat*, *council housing*). Special uses of the word include *Privy Council*, a body of advisers appointed by the sovereign; and *the Queen* (or *King*) *in Council*, the Privy Council as issuing Orders in Council or receiving petitions, etc.

**2** *Counsel* is advice (e.g. in set phrases, *counsel of despair, counsel of perfection, to keep one's own counsel*), esp. that given formally; also, a barrister or other legal adviser (pl. unchanged: *K. began the defence of his client with public sympathy but under immense difficulties against five prosecuting counsel*). A person trained to give advice on personal, social, or psychological problems is a *counsellor* (AmE freq. *counselor*), e.g. *marriage guidance counsellor*. A *Queen's* (or *King's*) *Counsel* in Britain is a counsel to the Crown, taking precedence over other barristers. A *counselor* (or *counsellor*) in America is a barrister; also called a *counselor-at-law*.

**3** *Counsel* is used as a verb ( = to give counsel), but *council* cannot be.

**council tax.** See COMMUNITY CHARGE.

**countable nouns.** See COUNT NOUNS.

**countenance, face, physiognomy, visage.** *Face* is the ordinary name for the part; *countenance* is the face with reference to its expression; *physiognomy*, to the cast or form of a person's features, expression, etc. *Visage* is a literary word, used ornamentally for *face* without special significance.

**counterpart** means 'a person or thing exactly like another' or 'a person or thing forming a natural complement or equivalent to another': *The German firms have improved their performance compared to their British counterparts—Oxford Rev. Economic Policy, 1988; The party met the counterparts of their own high-level characters ... on this world—Dragon (US), 1989.*

**counterpoint.** See HARMONY.

**counter-productive.** A modern word (first recorded in 1959) which has swept into widespread use describing any action or series of actions having the opposite of the desired effect: e.g. *These tactics were not only useless but counter-productive—Ann. Register 1963.*

**count nouns** (or **countable nouns** or **countables**) are those that can form a plural or be used with an indefinite article (e.g. *ship, crisis, fellow-traveller, foot, kindness*), as distinct from *non-count nouns* (or *uncountable nouns*) like *adolescence, richness, scaffolding, warfare* which normally cannot. The distinction is esp. useful

for foreign learners of English. See MASS NOUN.

**countrified, countrify** are the recommended spellings (not *countryfied*, etc.). See -FY.

**coup** is pronounced /kuː/. The pl. is *coups* /kuːz/.

**couple. 1** See AGREEMENT 5.

**2** The type 'a couple weeks later', i.e. with no *of* between *couple* and *weeks*, is well established in AmE (first recorded in 1925): *the universe would have collapsed back onto itself in a coupl̃e million years*—J. Updike, 1986; *in the next couple months we got to know each other like real buddies*—G. Keillor, 1989. This use sounds alien to British ears.

**3** The type 'a couple more + pl. n.', i.e. where *couple* operates as a quasi-adj., is common in all English-speaking countries: *We can end this chapter by looking at a couple more examples of Middle English writing*—C. L. Barber, 1965; *It's going to be a couple more months ... before we decide what to do*—Washington Post, 1985; *We'd get a couple more snowstorms in Old Hadham, I suspected*—New Yorker, 1986; *while a couple more effeminate types toyed with him*—B. Ripley, 1987; *'How's your work?' 'Nearly done. A couple more days.'*—Maurice Gee, 1992 (NZ).

**4** *A couple of* = about two is used informally (*I'll just be a couple of minutes* = in approximately two minutes).

**couplet** (prosody), two successive lines of verse, usu. rhyming and of the same metrical type, e.g.

> But most by Numbers judge a Poet's song,
> And smooth or rough, with them, is right or wrong;
>
> —Pope.

**coupon.** Now always pronounced /ˈkuːpɒn/, but until about the middle of the 20c. it was also commonly pronounced /kuːpɔ̃/ in a quasi-French manner.

**course.** The adverbial phr. *of course* is a term of insistence. Contextually it can mean 'naturally, as is or was to be expected' (*And you were present? Of course; why not?*), or 'admittedly' (*as there were patches of fog about the driving conditions were not, of course, ideal, but there was no*

*obligation on the driver to stay at the depot).*
Fowler (1926) rightly urged caution in
the use of the phr. 'as the herald of
an out-of-the-way fact that one has just
unearthed from a book or reference, e.g.
*Milton of course had the idea of his line from
Tacitus; The House being in Committee, the
Speaker would not, of course, under ordinary
circumstances, have been present.'*

**court-card.** See COAT-CARD.

**courteous, courtesy.** Pronounce with
initial /'kɜːt-/.

**courtesan.** Pronounce with initial /kɔːt/
and place the main stress on the final syllable.

**courtier.** Pronounce /'kɔːtɪə/.

**court martial.** Pl. *courts martial.* For the
verb, insert a hyphen.

**Covent Garden.** The first element is
pronounced /'kʊvənt/.

**Coventry.** Pronounce /'kʊvəntrɪ/.

**covert** *adj.* The traditional pronunciation /'kʌvət/ is still favoured by most
people in Britain, but the current AmE
pronunciation /'kəʊvɜːt/ is gaining
ground outside the US all the time. The
American pronunciation is perhaps derived from the spelling (cf. *over*) but,
more likely, is influenced by its antonym
*overt.*

**cowardly.** Fowler (1926) pointed out
that acts of violence and persons who
committed them were often described
as *cowardly* (reflecting the view of the
person or persons attacked) even though
courage and determination might well
have been needed by the assailant(s).
International terrorism in the period
since the late 1960s has tended to draw
the word *cowardly* into ever more frequent use in exactly the manner described by Fowler, that is in contexts in
which innocent and unsuspecting persons are killed in acts of terrorism. Cf.
DASTARD, DASTARDLY.

**cowrie,** the shell of a mollusc, is spelt
thus, not *cowry.*

**cowslip.** As the word is derived from
*cow* + *slip* ( = viscous substance) and not
from *cow's* + *lip,* the correct pronunciation is with medial /s/ not /z/.

**coxcomb.** See COCKSCOMB, COXCOMB.

**coxswain.** Despite its spelling, the word
is normally pronounced /'kɒksən/, with
/'kɒkswem/ available as a pedantic alternative. In practice, however, the reduced
form *cox* /kɒks/ is commoner than *coxswain* noun and verb.

**coxy.** See COCKY.

**coyote.** Pronounce /'kɔɪəʊt/ with two
syllables or /kɔɪ'əʊtɪ/ with three.

**cozy.** See COSY.

**crabbed** (adj.), irritable, hard to de-
cipher, etc. Pronounce /'kræbɪd/.

**cracker.** In Britain, apart from its
Christmas, parties, and fireworks senses,
*cracker* is used in *cream cracker,* a thin
dry biscuit often eaten with cheese. In
America, *cracker* is used in all these
senses but the biscuits are *crackers* not
*cream crackers.* In America an implement
for crushing nuts is called a *nutcracker*
or *crackers,* in Britain usu. *nutcrackers.*

**craft** (verb). The verb (and its ppl adj.
*crafted*) had a thin and fugitive life from
its first recorded use in the 14c. until
the second half of the 20c. Since then it
has come into widespread use, esp. in
the language of advertising (*a beautifully
crafted antique-pine replica*) and in literary
and other criticism (*performances crafted
out of a shared language*).

**cranesbill** is the recommended spelling, not *crane's-bill.*

**cranium.** The recommended pl. form is
*crania,* not *craniums,* but both forms are
used. See -UM 3.

**crape, crêpe.** The first is used for a
band of black silk or ornamental silk
worn as a sign of mourning, and the
second for other gauzelike fabrics having
a wrinkled surface. There are some other
special distinctions: (*a*) *crape fern,* a New
Zealand fern; *crape hair,* artificial hair
used in stage make-up; but (*b*) *crêpe de
Chine, crêpe paper* (thin crinkled paper),
and *crêpe Suzette* (dessert pancake).

**crash.** Since the 1950s, *crash* has been
increasingly used attributively of something done urgently or in a speeded-
up manner (*a crash course in first aid,
in Russian,* etc.). In practice it seems to

remain free from the ordinary associations of the word—violent collision, sudden diving (of a submarine), sudden failure (of a computer system, of the stock exchange, etc.).

**crasis.** In phonology, the running of two vowel sounds into one, as when *cocaine* (q.v.) and *codeine* (q.v.) became reduced, in the course of the 20c., from three syllables to two. The phenomenon is also shown in the medieval and Renaissance practice of reducing *to* to *t* or *t'* before an infinitive beginning with a vowel (*t'adorn, tamend, t'enjoy*, etc.) and *the* to *th* or *th'* before a word beginning with a vowel (*thage, themperor, th'infection, thother*, etc.). In Greek phonology it is 'the combination of the vowels of two syllables, *esp.* at the end of one word and beginning of the next, into one long vowel or diphthong, as in κἀγώ for καί ἐγώ' (*OED*).

**crayfish** is the customary word in Britain for a small lobster-like freshwater crustacean. Americans call them *crawfish* as well as *crayfish*, and have also come to use *crawfish* as a verb = 'to retreat from a position, to back out' (*he crawfished out of the issue by claiming that he didn't drink*). Australians and New Zealanders tend to abbreviate the word to *cray* (so *cray-fishing, cray-pot*, etc.). *Crayfish* is a 16c. alteration of earlier *crevis(se)* (cf. modF *écrevisse*).

**crease.** See KRIS.

**crèche** is properly /kreʃ/ in English, but /kreɪʃ/ is a common enough variant.

**credence, credit, credibility.** 1 Apart from the phr. *letter of credence* (a letter of introduction, esp. of an ambassador) and the ecclesiastical sense (in full *credence table*) 'a small side-table, shelf, or niche which holds the elements of the Eucharist before they are consecrated', *credence* has only one meaning, namely 'belief, trustful acceptance'. It occurs chiefly in the phr. *give* (or *lend*) *credence to*, to believe (e.g. *the Western powers gave credence to Mikhail Gorbachev's views long before the Berlin Wall was demolished; one can give no credence to his promises; to lend credence to the conspiracy theory is to show how paranoid a person can be*). Further examples: *The scale of departures gave superficial credence to such a view*—M. Edwardes, 1983; *The author of chivalresque fictions gained the*

*credence of his readers . . . by his fabulations*—*TLS*, 1985.

2 The phr. *to give credit to* was also once (16–19c.) frequently used to mean 'to believe, put faith in': e.g. *Charges like these may seem to deserve some degree of credit*—Gibbon, 1781; *To give entire credit to whatever he shall state*—Ld Brougham, 1862. Nowadays the phr. is more likely to be used in the form *to give a person credit for* (something) = to attribute merit to a person for (something): *You chaps do tend to give the rest of us credit for perceptions about your work that we don't . . . always have*—J. Wain, 1953.

3 *Credibility* crosses paths with both words in that belief lies at the heart of its meaning. But it is used rather to mean 'the condition of being credible or believable' (*Christianity rests on the credibility of the Gospel story*) and 'reputation, status' (*because of his crackdown on crime, he has gained credibility as a Chief Constable*). Further examples: *The Scots . . . give scant credibility to Walters' promise to move the exploration office to Glasgow*—*Sunday Times*, 1987; *Let's look at . . . this idea of selecting artists who have some sort of credibility*—*Art Line*, 1989. A disparity between statements about policy and the outcome of that policy has come to be called a *credibility gap* (first recorded 1966).

**credible, creditable, credulous.** *Credible* means (*a*) (of a person or statement) believable, worthy of belief (*at a certain point his assertions ceased to be credible; it is not credible that he could have saved so much money on his earnings*); (*b*) (of a threat, deterrent, etc.) convincing (*nuclear weapons were judged by the Western powers to be a credible deterrent*). *Creditable* means 'bringing credit or honour, deserving praise' (*a creditable performance*; (const. *to*) *The places . . . were filled in a manner creditable to the government*). *Credulous* means 'over-ready to believe, gullible' (*a credulous public, a credulous age*). Examples: *They had laughed at him for a credulous fool*—H. Read, 1935; *There is a strong streak of the credulous in most Russians*—R. Owen, 1985.

**credo.** Pronounce /ˈkriːdəʊ/ or /ˈkreɪdəʊ/. Pl. *credos*. See -O(E)s 6.

**creese.** See KRIS.

**cremona.** Used from the 17c. to the 19c. for 'an organ reed-stop of 8-foot tone'

(*OED*), it is a corruption of Ger. *Krummhorn* (q.v.). The only acceptable meaning of *cremona* now is 'a violin made at Cremona in Lombardy (where the art of violin-making reached its highest perfection in the 17c. and early 18c.)'.

**crenel,** an indentation in the parapet of a tower. If so spelt it is pronounced /ˈkrenəl/. If written as *crenelle*, it is to be pronounced /krɪˈnel/. The corresponding verb is written with *-ll-* (*crenellated castles*); in AmE often with *-l-* (*crenelated*).

**creole.** What earlier generations described as *broken English, bastard Portuguese, nigger French,* and so on, are now seen as new languages (*pidgins* and *creoles*) which came into being because of the need for people with no language in common to communicate with each other in these adverse linguistic circumstances. A creole is 'a language formed from the contact of a European language (esp. English, French, or Portuguese) with another (esp. African) language' (*COD*, 1990). Detailed scholarly work on creoles, esp. since the mid-20c., has led to the discovery that they are rule-based and systematic, however opaque or dislocated they may seem to be to the uninitiated. John Holm (1988), for example, cites a text from an English-based creole called Ndjuka, spoken in the interior of Suriname in northern South America:

Mi be go a onti anga wan dagu fu mi. A be wan bun onti dagu.

'I had gone hunting with a dog of mine. He was a good hunting dog.'

The term *creole* is applied to a wide range of languages, including the original languages spoken by black slaves in the US (and partially surviving in the BEV speech of lower-class blacks in urban communities in the US), English-based creoles like Krio (in Sierra Leone), Guyanese, and Gullah, French-based creoles in Haiti and the Ivory Coast, and Portuguese-based creoles in Brazil and Cape Verde. Intensive study of creoles continues: the bibliography of John Holm's *Pidgins and Creoles,* i (1988), for example, runs to twenty closely printed pages. The groundwork for it all was laid by the German scholar Hugo Schuchardt in a series of studies published in the 1880s and 1890s.

**crêpe.** See CRAPE, CRÊPE.

**crescendo.** Pl. *crescendos.* See -O(E)S 3. In origin the word is the present participle of L *crescere* 'to grow'. It follows that its essential meaning implies a gradual increase in loudness or intensity towards some unreached peak. As a musical direction the word has been in use in English since the late 18c.: from there it became applied to other progressive increases in force and effect. Beginning in the 1920s, it would seem first in AmE, *crescendo* has become widely used to mean 'the *peak of* an increase in volume, force, or intensity; a climax'. Caterwauling car-horns, bar talk, the noise made by massed RAF bombers, noisy evangelism, and large wage claims are among the noises or topics said to have *reached* or *risen to a crescendo* in the *OED*'s illustrative examples. The new use still lies rawly in disputed territory, eschewed by anyone knowing the meaning of the Latin participial ending.

**cretic** /ˈkriːtɪk/. In prosody, a Greek (occas. English) metrical foot containing one short or unstressed syllable between two long or stressed ones. The words *otiose* and *multitude* would neatly form such a foot. Blake's 'Spring' (*Songs of Innocence*) makes use of the measure: *Sound the Flute!/Now it's mute./Birds delight/ Day and Night;/Nightingale/In the dale,/Lark in Sky,/Merrily.* It is also called *amphimacer* /æmˈfɪməsə/.

**cretin. 1** From its first use in English in the late 18c. until about 1900 it seems to have been regularly pronounced /ˈkriː-tɪn/, but /ˈkretɪn/ is now the normal pronunciation.

**2** It was originally used to mean 'one of a class of dwarfed and specially deformed idiots found in certain valleys of the Alps and elsewhere' (*OED*), or, as *COD* (1990) expresses it, 'a person who is deformed and mentally retarded as the result of a thyroid deficiency'. Its dominant meaning in lay use now (first recorded in the 1930s) is 'a fool, one who behaves stupidly'.

**crevasse, crevice.** A *crevasse* is a *deep* open crack or fissure in a glacier; in AmE it is also used to mean a breach in a river embankment. A *crevice* is a *narrow* cleft or opening, esp. one in the surface of anything solid, e.g. a rock or a building. Etymologically the words are doublets from the same Latin base (*crepātia*), from L *crepāre* 'to creak, rattle, crack'.

**crick, rick, wrick.** The earliest of these to appear was *crick* (as n., 15c.); the corresponding verb meaning 'to produce a crick (in the neck, etc.)' is not recorded until the 19c. The noun is probably of onomatopoeic origin. *Rick* (as noun first recorded 1854, as verb 1798) and *wrick* (noun 1831, verb 1886) are much later words: they are apparently of dialect origin and have Low German affinities. *Crick* and *rick* are commonly used of strains or sprains of the neck, back, joints, etc., but *wrick* has effectively dropped out of use.

**cringe** makes *cringing*. See MUTE E.

**crisis.** Pl. *crises* /'kraɪsiːz/. See LATIN PLURALS 1.

**criterion.** Pl. *criteria*: see -ON 1. Lamentably, the pl. form *criteria* is now frequently but erroneously used as a sing. My evidence happens to be from US newspapers, but the erroneous use is widespread: *However, this criteria is relative—Chicago Sun-Times* (weather forecasting section), 1989; *The major criteria for evaluating the wines ... was cleanliness, freshness and complexity—Chicago Tribune*, 1990.

**criticism.** In everyday use, *criticism* means 'finding fault, censure', but *literary criticism* and *textual criticism* are more in the realm of assessment or evaluation. The two senses, one adverse and the other evaluative, coexist without risk of ambiguity or cross-infection.

**critique.** The noun, meaning 'a critical essay or analysis; an instance or the process of formal criticism' (*COD*), is first recorded in a work of 1702–1 by Addison, and the verb, meaning 'to write a critique upon', in 1751. Fowler (1926) said of the noun, 'there is some hope of its dying out', but it continues in use: e.g. *This melange of stroboscopic graphics, jingles and one-liner critiques—Art Line*, 1989.

The verb has been controversially extended in meaning in the US to mean more generally 'to judge critically, to make a critical assessment of or comment on (an action, person, etc.), not necessarily in writing' (*OED* 2). Examples: *A football coach critiquing a fumble on a film of the game—New Yorker*, 1969; *I would not try to critique what Governor Rockefeller said to his legislature—NY Times*, 1973; *Critiquing the film would be a good way to review the issues—Amer. Ethnologist*, 1987.

**crochet,** in its ordinary handicraft sense, is pronounced /'krəʊʃeɪ/ as noun and verb. The *t* remains silent in the other parts of the verb: *crocheted* /'krəʊʃeɪd/, *crocheting* /'krəʊʃeɪŋ/.

**crocus** (the flower). As pl., *crocuses, croci* /-kaɪ/, and (used collectively) *crocus* are all in standard use, the most common being *crocuses*.

**croquet.** The noun and verb operate like CROCHET, with a silent *t* in the extended forms of the verb.

**crosier** is the recommended spelling, not *crozier*.

**cross.** Daniel Jones (1917) gave priority to the now mostly discarded pronunciation /krɔːs/. At some point in the first half of the 20c. the tide turned in favour of /krɒs/, and this is now by far the dominant pronunciation. Similarly with *frost, loss, lost*, etc.

**croup.** Both the throat trouble and the rump (of a horse) are now regularly spelt *croup* and pronounced /kruːp/, though they are distinct words from unrelated etymological bases.

**croupier.** The pronunciation /'kruːpɪə/ is recommended, but /-pɪeɪ/, in imitation of the French original, is still widely used. Cf. DOSSIER.

**crow.** The past tense varies between *crowed* and (chiefly in the UK) *crew*, but the pa.pple is always *crowed*: *Roosters, still in the shed, crowed—E. Jolley*, 1985; *He burst into her bedroom as the cock crew, or would have crowed, had there been any cocks in Surbiton—B. Rubens*, 1987. The pa.t. *crew* is normally used only of the noise made by cocks, and not, for example, when the sense is 'expressed gleeful satisfaction'.

**crown.** *The Crown* is often used in the sense 'the office of the monarch as head of State'. In such contexts the pronouns appropriate to *queen* or *king* should not be used. Thus *the Crown can act only on the advice of its* (not *her*, not *his*) *Ministers; the country is unlikely to return to government by the Crown, for which* (not *whom*) *such an experiment would be fraught with peril.*

**crucifixion.** Thus spelt, not *crucifiction*. See -XION etc.

**cruel** makes *crueller, cruellest* (in AmE often *crueler, cruelest*). See -LL-, -L-.

**crumby, crummy.** When the reference is to actual crumbs, as in *a crumby loaf* or *tablecloth*, use *crumby*. When the meaning is 'dirty, squalid, inferior, worthless', use *crummy*.

**crystalline.** The normal modern pronunciation is /ˈkrɪstəlam/, but the OED reports that some older writers (Milton, Gray, Shelley, and Palgrave), after Latin, used the pronunciation /krɪsˈtælm/.

**cubic, cubical.** *Cubic* is the form used in all senses, *cubical* only in that of 'shaped like a cube'. So *cubic content, equation, measure, metre*, etc.; but a *cubical box* or *stone*. *Cubic*, however, is sometimes used in the sense 'cube-shaped', for example of minerals crystallizing in cubes, as *cubic alum, saltpetre*. See -IC(AL).

*cui bono?* This Latin phrase literally means 'to whom (is it) a benefit?' In English it can be defined as 'who stands to gain? (with the implication that this person is responsible)' (COD). As Fowler (1926) pointed out, it should not be used to mean 'To what end?; What is the good?'

**cuirass.** Pronounce /kwɪˈræs/.

**cul-de-sac.** Pronounce /ˈkʌldə,sæk/ and print in romans. Pl. *culs-de-sac* (pronounced the same as the sing.).

**culinary.** Pronounced /ˈkʌlmərɪ/ in BrE and /ˈkjuːlm,eərɪ/ or /ˈkʌl-/ in AmE.

**cullender.** See COLANDER.

**cult.** In the course of the 19c., *cult* (sometimes written as *culte* in the French manner) began to be used by archaeologists and by writers on comparative religion in the sense 'a system of religious worship, esp. as expressed in

ritual'. This use continues, but it is now accompanied by the weakened sense 'a devotion or homage to a person or thing' (*the cult of Karl Marx, of post-modernism*), and by an even more weakened sense, 'a popular fashion followed by a specific section of society' (*the cult of popular music, the earring cult*). Throughout the 20c. a succession of combinations of *cult* came into being, *cult drama, figure, image, object*, etc.

**cultivable** and **cultivatable** are both used in the sense 'capable of being cultivated', with the former perhaps having the edge. For the type, see -ABLE, -IBLE 2 (v).

**cultivated, cultured.** Both words are used to mean 'having refined taste and manners and a good education', but they part company in some other circumstances. *Cultivated* distinguishes a crop raised in a garden or gardens (*cultivated blackberries*) from one grown in a wild state. *Cultured* is used of pearls (also called *culture pearls*) grown under controlled conditions.

**culture.** The word has had mixed fortunes in the 20c. It has flourished in the language of sociologists and anthropologists in the broad sense 'the civilization, customs, artistic achievements, etc., of a people, esp. at a certain stage of its development or history' (OED). From 1914 onward it came into contact with the German word *Kultur*, which had developed a derogatory sense in the eyes of British people (involving notions of racial and cultural arrogance, militarism, and imperialism). In 1939 the German writer H. Johst (not Goering, as is commonly supposed) wrote (in German): 'When I hear the word "culture" I slip back the safety-catch of my revolver.' It was mocked by some who tended to distort the spelling (*culchah*, etc.) to indicate that the acquisition of cultured ways implied a risible degree of affectation or vulgarity. Perhaps the most revealing 20c. combinations of the word were *culture shock* 'the feeling of disorientation experienced by a person suddenly subjected to an unfamiliar culture or way of life' (COD); and *culture vulture* 'a person eager to acquire culture'. In 1956 C. P. Snow brought about a decades-long discussion by employing the phrase *the two cultures* to mean respectively the arts and science. Opponents of Snow regarded the distinction as imprecise and unproven, but the phrase lingers on.

**cum.** The Latin preposition meaning 'with, together with' has been used in English for several centuries in some fixed phrases (*cum grano salis*, with a grain or salt; *cum privilegio*, with privilege (of sole printing); etc.) and in many English place-names, e.g. *Horton-cum-Studley*. In the 19c. it took on a fresh lease of life as a combining word used to indicate a dual nature or function: *He greatly preferred coffee cum chicory to coffee pure and simply*—J. C. Young, 1871; *Three short … dinner-cum-cocktail dresses*—*Manchester Guardian*, 1959; *'Do you work there?' 'Yes, as a sort of administrator cum priest.'*—J. Higgins, 1985; *The essentialist emancipatory aesthetic-cum-linguistic philosophy of Benedetto Croce*—*Br. Jrnl Aesthetics*, 1986.

**cum(m)in** /ˈkʌmɪn/. A case can be made out for the spelling *cummin* in that the word has two *m*s in its most familiar context, Matt. (AV) 23: 23 (*for yee pay tithe of mint, and annise, and cummine*). But etymology (OE *cymen*, from L *cumīnum*) tugs the word in another direction, and *cumin* seems to be the dominant spelling now.

**cumulative.** See ACCUMULATIVE.

**cumulus.** Pl. *cumuli* /-laɪ/. See -US.

**cuneiform.** After more than two centuries of use as a four-syllabled word (thus /ˈkjuːˈniːɪtɔːm/ or /ˈkjuːniːɪ-/), the word, while retaining its older spelling (cf. L *cuneus* wedge, medL *cuneiformis*), is now normally pronounced in BrE as three syllables, i.e. /ˈkjuːnɪfɔːm/. In AmE the customary pronunciation is still with four syllables, i.e. /kjuːˈnɪəfɔːrm/.

**cup.** For *cups that cheer*, see HACKNEYED PHRASES.

**cupful.** 'A *cupful* is a measure, and so *three cupfuls* is a quantity regarded in terms of a cup; *three cups full* denotes the actual cups, as in *three cups full of water*' (COD, 1990).

**cupola.** Pronounce /ˈkjuːpələ/.

**curaçao,** a liqueur. Spell thus (not *-çoa*) and pronounce /kjuərəˈsəʊ/.

**curare,** a poisonous substance. Spell thus (not *-ara*) and pronounce /kjʊˈrɑːri/.

**curator.** In the sense 'keeper or custodian of a museum or other collection' pronounced /kjuəˈreɪtə/, but in Scottish law (a ward's guardian) pronounced /ˈkjuərətər/.

**curb, kerb.** In BrE, *curb* as n. = a check or restraint, and as v. = to restrain. As noun, it also means a strap fastened to a bit (on a horse), and a fender round a hearth. In AmE, it also means a stone edging to a pavement (*sidewalk*) *or raised path* (BrE *kerb*).

**curio** is a 19c. familiar abbreviation of *curiosity*. Pl. *curios*. See -O(E)S 4.

**curriculum.** Pl. *curricula*. See -UM 2.

**cursed, curst.** The customary adjectival form is *cursed*, pronounced as two syllables (by contrast with the pa.t. and pa.pple of the verb *curse*, which are always monosyllabic). *Curst* is chiefly used (esp. in verse) to show that the rare monosyllabic pronunciation is meant, or, archaically, to denote the old sense 'cantankerous, shrewish'.

**curtailed words.** See ABBREVIATIONS 1.

**curtsy** = a woman's or girl's formal greeting, should be spelt thus, not *curtsey*. Pl. *curtsies*. The verb is *curtsy, curtsied, curtsying*. Etymologically it is a variant (first recorded in the 16c.) of *courtesy*.

**curvet** = a horse's leap. Pronounce /kɜːˈvet/. The extended forms of the corresponding verb should, according to rule, show doubling of *t* (*curvetted, curvetting*). See -T-, -TT-.

**customer.** See CLIENT.

**customs.** See TAX.

**cute.** Starting out in the 18c. as an abbreviation of *acute* in the sense 'quick-witted, keen-witted, clever', it was for a while often written with an apostrophe as *'cute* (*He will be a 'cute man yet,' resumed the locksmith*—Dickens, 1840). This sense ran into the sands in BrE but not in AmE. In AmE, beginning in the 1830s, a new informal use emerged, 'attractive, charming, pretty (often in a mannered or amusing way)'. The word lies spread-eagled across AmE and is exceedingly

common as a general term of approbation. It has much more restricted currency in BrE, except as applied to babies
when other adjectives seem inadequate.
In other BrE contexts, *cute* ranges from
tepid approval to latent disapproval. A
typical modern AmE example: *Yes, Lisa
runs a little gym in West Palm. We all go
there to work out. Isn't that cute?*—P. Booth,
1986.

**-cy.** See -CE, -CY.

**cybernetics.** A term introduced in 1948
by Norbert Wiener meaning 'the theory
or study of communication and control
in living organisms or machines' (cf. Gk
κυβερνήτης 'steersman'). It quickly established itself in our biotechnological
age, with reference to organisms treated
as if they were machines, to observed
similarities between neural activity and
the electronic devices of modern communications, and so on. A by-product of
the word is the use of the first element
*cyber-* in a wide range of computer
terms such as *cyberphobia* (an excessive
fear of computers), *cyberpunk* (a style of
science fiction), and *cyberspace*, or *virtual
reality* (the hallucinatory illusion of
moving through exotic locations that
are so devised on a computer screen
as to seem as 'real' as those of the real
world).

**cycle.** For *cycle* as a time-word, see
TIME.

**cyclopaedia, -paedic.** This shortened
form of *encyclopaedia* (q.v.), meaning
'circle of learning' (from Gk κύκλος
'circle' + παιδεία 'education, learning'),
has been in use since the 17c., though the
first book so called was that of Ephraim
Chambers (1728). It and the corresponding adjective *encyclopaedic* were almost
always spelt with -*ae*- (or the digraph
-*æ*-) in BrE until some point in the 20c.,

when forms with medial -*e*- (which are
customary in AmE) came along to challenge them, and look likely to supplant
them.

**Cyclopean.** *COD* gives preference to
/saɪklə'piːən/ rather than /saɪ'kləʊpɪən/,
but both pronunciations are admissible.

**Cyclops** (in Gk mythology, a member
of a race of one-eyed giants). The forms
recommended are: for the sing. *Cyclops*
/'saɪklɒps/; for the pl. *Cyclopes* /saɪ'kləʊpiːz/,
or, in non-specialized or jocular contexts,
*Cyclops* or the Anglicized form *Cyclopses*.
The use of *Cyclop* as a sing. is highly
irregular.

**Cymric,** Welsh. Pronounce /'kɪmrɪk/. Cf.
Welsh *Cymru* 'Wales'.

**cynic, cynical.** Except when there is
direct reference to the sect of philosophers in ancient Greece called *Cynics*,
and in the medical term *cynic spasm*,
the adj. form is always *cynical*. *Cynic* is
ultimately derived from Gk κυνικός 'dog-
like, currish'.

**cynosure.** The phrase *the Cynosure of
neighbouring eyes* from Milton's poem
'L'Allegro' (1632) has ensured that it and
variants of it (*the cynosure of all eyes*, etc.)
have remained in widespread use ever
since.

**cypher.** See CIPHER.

**Cyrenaic.** See HEDONIST.

**Czar.** See TSAR.

**Czarewitch,** etc. See CESAREVITCH.

**Czech.** After much fluctuation (*Tshekh,
Tsekh*, etc.) in the period since the
1840s, when the word was first recorded, the word has settled down
in the 20c. in its Polish form *Czech*,
pronounced /tʃek/.

# Dd

**dactyl,** a metrical foot (—⌣⌣) consisting of one long (or stressed) syllable followed by two short (or unstressed) syllables, as in the words *pottery* and *Julia*. Dactylic metres are common in Greek and Latin verse, but comparatively rare in English: *Cannon to right of them Cannon to left of them, Cannon in front of them Volley'd and thunder'd*—Tennyson, 1854.

**dado** /ˈdeɪdəʊ/ (the lower part of the wall of a room; the plinth of a column). Pl. *dados.* See -O(E)S 6.

**daemon.** This spelling is permissible only when the sense is '(in ancient Greece) an inner or attendant spirit; a supernatural being'. All the main senses of the word require *demon.*

**dahlia.** Pronounced /ˈdeɪlɪə/ in BrE and usu. /ˈdæljə/ or /ˈdɑːlɪə/ in AmE.

**Dáil** (in full *Dáil Éireann*), the lower house of parliament in the Republic of Ireland, is pronounced /dɔɪl/.

**dais.** Monosyllabic (/deɪs/) in English until the beginning of the 20c. but /ˈdeɪɪs/, i.e. two-syllabled, in Daniel Jones's *English Pronouncing Dict.* (1917) and in most authorities since then.

**Dame.** See SIR.

**damn.** In all uses as n., adj., v., and adv. the *n* is silent; also in the oblique forms *damned* (but often disyllabic in verse), *damning*. The *n* is sounded in *damnable, damnation, damnatory, damnification,* and *damnify.*

**dampen.** This verb was regarded as 'chiefly US' by the *OED* (1894) and by *COD* (7th edn, 1982), but it has now settled down as a fairly frequent variant of *damp* (v.) in the UK: *everyone ignored the snow that had failed to dampen the impact of John F. Kennedy's brilliant oratory*—J. Archer, 1979; *Rothschild dampens down the starburst*—*Times* (headline), 1988; *Rain fails to dampen day of fun*—*Oxford Times* (headline), 1988. See -EN VERBS FROM ADJECTIVES.

**dangling participles.** See UNATTACHED PARTICIPLES.

**dare.**

| | |
|---|---|
| **1** | With bare infinitive or *to*-infinitive. |
| **2** | *Dare* is a marginal modal. |
| **3** | The paradigm of *dare.* |
| **4** | Constructions with bare infinitive. |
| **5** | As a main verb + a *to*-infinitive. |
| **6** | *Dare* used as verb of past time. |
| **7** | *Dare* ( = challenge, defy) + object + *to*-infinitive. |
| **8** | *I dare say.* |

**1** With bare infinitive or *to*-infinitive. *Shall I part my hair behind? Do I dare to eat a peach?*—T. S. Eliot, 1915; *They dare not break in. They cannot break in*—T. S. Eliot, 1935. These two contexts neatly show two quite different ways of using the verb *dare*. In the first it is an ordinary verb followed by a *to*-infinitive, and in the second it is a marginal modal (see below) followed by a negative and a bare (or plain) infinitive. They bring us to the heart of one of the subtlest and most variegated verbs in the language.

**2** *Dare* is a marginal modal. The term 'modal auxiliary' is used to describe the verbs *can/could, may/might, shall/should, will/would,* their abbreviated forms *'ll* and *'d,* and their contracted negative forms *can't, won't,* etc. One of the characteristic features of these is that none has a final *s* in the third-person present indicative: we do not say *he cans, he mays, he shalls,* and so on. Four other verbs, *dare, need, ought to,* and *used to,* are called 'marginal modals' because their status is to some degree intermediate between that of modal auxiliaries and ordinary verbs. In Lord Alfred Douglas's famous line *I am the Love that dare not speak its name, dare,* with no final *s,* is a marginal modal.

**3** The paradigm of *dare.* In the standard language only four elements survive in strength: *dare, dares, dared,* and *daring.* Technically a fifth, namely *durst,* exists, but it no longer occurs in natural standard speech. In my recent reading I have encountered a few examples of *durst,* but only in historical, uneducated, dialectal, comic, or other special circumstances: *He'll come all right. He ain't got nowhere to sleep, see. And I durstn't face him.*—Miss Read, 1977; *Though fain to avoid eavesdropping, he yet had durst not betray his presence*—P. Jennings, 1983; *'Zounds! sirrah, how durst the Manhattan Punch Line attempt Sheridan's "Rivals"?'*—NY Times, 1984. Each time I felt I had unearthed some old or quaint linguistic bones.

**4** Constructions with bare infinitive. At the present time, there seem to be five main constructions in which *dare*, whether as a main verb or as a marginal modal, is used when followed by a bare infinitive. Most of them are negative or interrogative: (i) subject + *dare* + negative: *He hates only because he dare not love*—J. M. Coetzee, 1977; *I dare not speak these dreams to any person*—G. Keillor, 1986. (ii) subject + auxiliary + negative + *dare*: *Kennedy did not dare ask Congress for such a treaty*—Dædalus, 1986. *I once saw a herd get through the mud up a slope where I never thought they would dare go*—D. Matthee, 1986. (iii) subject + *dared* + negative: *she dared not stir until eleven was safely past*—A. Schlee, 1981; *No one dared defy the group by going out at the last moment*—I. McEwan, 1986; *She dared not even ask him how it was he had revised his opinions*—P. Carey, 1988. (iv) interrogative (or semi-imperative) + *dare* + subject: *How dare you come in without knocking!*—R. Dahl, 1984; *'Who is it?' 'It's me, Mudey. Gabriel. Open up.' . . . 'How dare you?'*—P. Bailey, 1986. (v) negative imperative auxiliary + *dare*: *Don't you dare put that light on!*—S. Delaney, 1959; *I'm going to swim. And don't you dare watch*—G. Butler, 1983; *Now you sit down there and don't you dare even look at anybody till I get back*—K. Amis, 1988.

Such uses together, according to my evidence, are three times as frequent as those in the following group. It should be stressed that some of the examples in this section show *dare* as first and third person present indicative and some as infinitive.

**5** As a main verb + a *to*-infinitive. *Dare* as a main verb followed by a *to*-infinitive is found in various negative, affirmative, and interrogative constructions: *I did not dare to look down*—B. Rubens, 1985; *How do they dare to be different?*—New Yorker, 1987; *No one dared to giggle, not even*

*Soraya*—C. Rumens, 1987; *she no longer dared to go into these shadowy apartments*—A. Brookner, 1988.

**6** *Dare* used as a verb of past time. Nowadays, because of its double status as an ordinary verb and as a marginal modal, *dare* can be used without inflection for past as well as present time: (present) *a person hardly dare think in this house*; (past) *She turned round. She dare not look at his face.* The OED (1894) condemned the past tense as 'careless', while citing examples from the work of Frederick Marryat and Charles Kingsley among others. Modern grammarians regard it as unobjectionable. Further example: *'Yes, yes,' she stuttered, then 'thank you', as an afterthought. She dare not look at his face now*—M. Duckworth, 1960.

**7** *Dare* ( = challenge, defy) + object + *to*-infinitive. In the following examples *dare* is used as a main verb in the sense 'to challenge, to defy': *She dared me to believe and take up her cause*—D. Malouf, 1985; *Deutsch confidently dared Kreskin to find the key*—Chicago Tribune, 1988; *He looked round the table as if daring anyone to smile*—D. Lodge, 1988.

**8** *I dare say.* This phrase (occas. with ellipsis of the pronoun) means (*a*) I make bold to say, I venture to assert; (*b*) I grant that much (*I dare say, but you are still wrong*). It is being increasingly written as *I daresay*, and said with the main stress on *dare*. Examples: *Dare say you'd like a last look at uncle Caleb's old cottage*—Ted Walker, 1981; *'From time to time I clean. Mrs Pollypot she don't like cleaning.' 'I dare say not.'*—M. Wesley, 1983; *I daresay I was shocked again*—N. Bawden, 1987; *I daresay I'll come back to it, in the fullness of time*—P. Lively, 1987; *'With your head, I daresay you could have worked out our story from the name of the shop*—P. Gilliatt, 1987.

**darkle** (verb), to grow dark, etc. Formed in the early 19c. as a (false) back-formation from DARKLING, and used by Byron, Thackeray, and William Morris, among others, *darkle* seems to have dropped out of use.

**darkling.** Formed as an adverb in the 15c. from *dark* adj. + *-ling* adverbial formative, it meant 'in the dark, in darkness', e.g. *O wilt thou darkling leaue me?*—Shakespeare, 1590. By confusion with the *-ing* of present participles, it later itself became a present participle (e.g. *His honest brows darkling as he looked towards me—*Thackeray, 1855), and, more commonly, an adjective, 'being in the dark, showing itself darkly, etc.': e.g. *Here like darkling nightingales they sit—*G. Meredith, 1859; *And we are here as on a darkling plain ... Where ignorant armies clash by night—*M. Arnold, 1867; *By the light of a darkling moon, I went in the greatest peace and satisfaction to my rest—*P. Matthiessen, 1979; *Jesus makes a primary act of conversion ... in seeing this darkling venture as the stumbling of ignorance away from reprobation and towards truth—*Geoffrey Hill, 1989.

**darky.** The word still has some currency as a term of racial opprobrium (e.g. *Was it something about not taking on the darkies as conductors?—*J. le Carré, 1983), but has largely been supplanted by other more neutral words.

**dash.** *Hart's Rules* is an excellent guide to the use of this mark of punctuation. What follows is a revised and extended version of Hart.

The en-rule (–) is used (a) to denote a span, e.g. 'folios 23–94'; (b) to specify a period of time, e.g. 'the 1939–45 war'; (c) between separate places or areas linked, for example, in a political context, e.g. 'the Rome–Berlin axis'; (d) between the names of joint authors to avoid confusion with the hyphen of a single double-barrelled name, e.g. 'the Temple–Hardcastle project' (that is, the joint project of Mr Temple and Mr Hardcastle).

Em-rules or dashes (which are longer than en-rules) are often used to show that words enclosed between them are to be read parenthetically: e.g. *What he has imagined—he contends—is theatrically stronger than what actually happened; Some critics thought—and some still think—that the Congress failed because* [etc.]; *His 1776 drawing of Les Combats de Diomède—a most important document in his stylistic evolution and in the realisation of his aims—suggests that chariots held a particular appeal for him.*

Just as commonly a dash is used to precede an explanation of or elaboration on something just said. In such cases the explanatory statement is usually followed by a full stop: *By this time I had become personally close to Michael Josselson—indeed he was to the end one of my closest friends; It is a kind of irony of history that I should write about the French Revolution in the very country where it has had the least impact—I mean England, of course.* (The five examples above are taken from *Encounter*, Sept. 1990.)

The dash should not be added to a colon; use a simple colon, not :—, as a mark introducing a list or the like. Fowler (1926) insisted that after a second dash 'any stop that would have been used if the ... dashes and their contents had not been there should still be used'. He was right, but such a circumstance seldom arises.

**dastard, dastardly.** The noun, in its traditional sense 'one who does malicious acts in a cowardly, skulking way, so as not to expose himself to risk' (*OED*) is now not often used. A typical example: *Though I hold no brief for Mr Neary ... yet I am loath to think he is the dastard you describe—*S. Beckett, 1938. By contrast, *dastardly*, like COWARDLY, is now frequently used in the manner described by Fowler (1926), applied to acts of terrorism in which unsuspecting, usually innocent, people are killed or injured by a terrorist group's bomb, or the like. Cowardice is no longer a necessary component of the meaning of the word, but underhandedness and public outrage are.

**data** /ˈdeɪtə/ is in origin a L plural (of *datum*) and is properly so used in English: *The data are* (not *is*) *insufficient.* In modern times usage varies as shown below **1** In philosophical (*sense-data*) and general use, *data* is usually considered as a number of items and construed as a plural: e.g. *It is no wonder if some authors have gone so far as to think that the sense-data have no spatial worth at all—*William James, 1890; *Most of the data concerning shock and vibration on airplanes are classified—*Macduff and Curreri, 1958. The corresponding singular is *datum*: e.g. *Can a doubleness, so easily neutralized by our knowledge, ever be a datum at all?—*William James, 1890; *What ... is immediately given in perception is an evanescent object called an idea, or an impression, or a presentation, or a sense-datum—*A. J. Ayer, 1956.

**2** In computing and allied disciplines it is treated as a singular noun and used with words like *it, this,* and *much,* and with a singular verb: e.g. *Data is stored on a disk—whether floppy or hard—as minute patches of magnetism—*P. Laurie, 1985; *Computers create and maintain the data distributed throughout an organization while communications moves the data to where it can earn the most—Computerworld,* 1989; *The problem . . . is that the raw data for the solution . . . lies in two vast collections of information—NY Rev. Bks,* 1989.

See -UM 2.

**datable** is spelt thus, not *dateable.* See -ABLE, -IBLE 6.

**date. 1** For *date, epoch,* etc., see TIME.

**2** For dates, OUP house style requires the type *25 June 1990,* with no comma between month and year. Many newspapers, however, and most Americans prefer the style *June 25 1990* or *June 25, 1990.* The conventional way of setting down a date in numerals differs in Britain and America: thus *5/7/90* means 5 July 1990 in Britain, but 7 May 1990 in America.

**3** See AD; BC.

**davits.** Being in origin an application of the Christian name *David* it was often in the past pronounced /ˈdeɪvɪts/, but the preferred pronunciation now is /ˈdævɪts/.

**day and age.** The phr. *in this day and age* slid into the language in the 1940s (a film called *This Day and Age* was issued in 1933 but is not certainly the source of the cliché), and is now used remorselessly by people who display little feeling for the language. It means no more than *nowadays* or *at the present time,* and the language would not be the poorer if it were to drop out of use.

**days.** See ADVERB 4.

**D-Day, D-day.** The military code-name (first recorded in 1918) for a particular day fixed for the beginning of an operation, specifically, and most famously, the day (6 June 1944) of the invasion of the Atlantic coast of German-occupied France by Allied forces. Also in transferred use, e.g. the day (15 Feb. 1971) on which decimal currency came into official use in Britain.

**-d-, -dd-.** Monosyllables ending in *d* double it before suffixes beginning with vowels if the sound preceding it is that of a single short vowel, but not if it is a diphthong or double vowel or a vowel + *r: caddish, redden, bidding, trodden, muddy;* but *deaden, breeder, goodish, braided, harden.* Words of more than one syllable follow the rule for monosyllables if their last syllable is stressed or is itself a word in combination (*forbidding, bedridden*), but otherwise do not double the *d* (*nomadic, wickedest, rigidity, periodical*).

**de-.** The prefix *de-* continues to be repeatedly called on to form new verbs (and derivatives). Twentieth-century examples (with date of first record indicated) include: *de-accession* (1972), *debark* (operate on dog to stop it from barking, 1943), *de-beak* (1937), *debrief* (1945), *decaffeinate* (1927), *decertify* (1918), *decriminalization* (1945), *de-emotionalize* (1942), *de-emphasize* (1938), *de-escalate* (1964), *deglamorize* (1938), etc. Cf. DIS-.

**dead letter,** apart from its theological and post-office uses, is a phrase for a regulation that still has a nominal existence but is no longer observed or enforced. Capital punishment can properly be called a *dead letter* in any country where the penalty remains on the statutes even though it may not have been acted on for many years. The term cannot properly be applied to aspects of life that are simply passing out of fashion, e.g. quill pens, steam locomotives, hot-metal printing, since they were not brought into being by regulation in the first place.

**deal** (noun). **1** The phr. *a deal* used pregnantly for *a good* or *great deal* in the sense 'an undefined, but considerable or large quantity' was much favoured by such writers as Shakespeare, Richardson, and Johnson, but is now mainly found in informal or dialectal writing or speech (*the decision saved him a deal of trouble*). It should not be used to mean 'a large number' (*a great deal of people have complained*).

**2** For Fowler (1926), *deal,* used to mean 'a piece of bargaining or give-and-take', was 'still slang'. In most of its applications (*a fair deal, a pay deal,* etc.), this use is now part of the normal fabric of the language.

**dean, doyen,** though originally the same word (cf. Fr. *doyen* from L *decānus*), meaning the senior member of a group, have become differentiated. *Dean* is the title of an ecclesiastical or academic officer; *doyen*, pronounced either /ˈdɔɪən/ or /ˈdwɑːjæ̃/, is a title of respect for 'the most senior or most prominent of a particular category or body of people' (*COD*).

**Dear** (as part of a formal greeting at the beginning of a letter). First introduced in the 15c., the beginning formula varied in the amount of elaboration: *Right dere and welbeloved; My most dere lorde and fader; Dearest broder*, etc. By the end of the 18c. it had settled down as part of the ordinary polite form of addressing an equal, ranging in increasing levels of informality and affection from *Dear Sir* and *Dear Madam* to *Dear Smith* (old-fashioned), to *Dear Mr/Miss/Mrs/Ms Jones*, to *Dear Ken/Mary*, etc., to *My dear Bill/Elizabeth*, etc., to *Dearest Margaret/Cedric*, etc. Increasingly now one receives circular letters beginning *Dear Customer, Dear Colleague*, etc., and formulas containing both the first and last names of the person addressed (*Dear Robert Burchfield*). The essential point is that the word *Dear* in itself is simply part of a letter-beginning formula, and no longer implies any particular or clear-cut level of intimacy or friendship.

**dear, dearly** (advs.). With the verb *love* and in some other contexts where the meaning is 'very much' (*he loved her dearly; I would dearly like to join you on Friday*), and *dear* merely poetic (*The dear-loved peaceful seat*—Byron, 1807); but with *buy, cost, pay, sell*, etc., when the sense is 'at a high price, or great cost', *dearly* (*they paid dearly for their crimes*) and *dear* (*the recession has cost Britain dear*) are both available, though usually one or the other feels contextually more idiomatic than the other.

**dearth.** From the 13c. onward, widely used to mean 'a condition in which food is scarce and dear', but as time went on more frequently used to mean a scarcity of anything (*a dearth of news, of supporters*, etc.). In general this noun seems to have mostly yielded to words of similar meaning (*shortage, scarcity*, etc.), or the idea is expressed in another way (e.g. *there isn't much news tonight; not many people turned up to give support*, etc.).

**débâcle.** Frequently now written and printed without accents, partly no doubt because of the shortage of special signs on modern daisy wheels and computer keyboards. Pronounce /deɪˈbɑːkəl/.

**debar** means 'to exclude from admission or from a right; prohibit from an action (*he was debarred from entering*)', by contrast with *disbar*, which can only mean 'to deprive (a barrister) of the right to practise; to expel from the Bar'.

**debatable.** Spelt thus. See -ABLE, -IBLE 6.

**debauchee.** Pronounce /dɪbɔːˈtʃiː/, with the main stress on the final syllable.

**debouch.** Pronounce /dɪˈbaʊtʃ/ to rhyme with 'pouch'.

**debris.** Spelt thus (no accent). The recommended pronunciation is /ˈdebriː/, but /ˈdeɪbriː/ is an acceptable alternative. In AmE it is often pronounced /dəˈbriː/.

**début** is usu. spelt thus in BrE and pronounced /ˈdeɪbjuː/. The accent is often dispensed with in printed work.

**debutant** (or **débutant**) /ˈdebjuːtɒ̃/ or /ˈdeɪ-/ is a male performer making his first public appearance.

**debutante** (or **débutante**) /ˈdebjʊtɑːnt/ or /ˈdeɪ-/ or /-tɒnt/ is (*a*) a young woman making her social début; (*b*) a female performer making her first public appearance. The shortened form *deb* is freely used for sense (*b*), esp. in spoken English.

**decade.** Until the 19c. frequently spelt *decad* (pronounced /ˈdekəd/), but now always *decade*. The preferred pronunciation is /ˈdekeɪd/, but increasingly, and in unexpected quarters, one hears /dɪˈkeɪd/, rhyming with *decayed*.

**deca-, deci-.** In the metric system, *deca-* means multiplied, and *deci-* divided, by ten; *decalitre*, 10 litres, *decilitre*, $\frac{1}{10}$ litre; so with *gramme, metre*, etc.

**decided, decisive** (adjs.). *Decided* means (*a*) (of something) definite, unquestionable (*a decided advantage*); (*b*) (of a person) having clear opinions, not vacillating (*he found them vacillating, he left them decided*). *Decisive* means (*a*) that decides an issue, conclusive (*a decisive superiority over its rival*); (*b*) (of a person) able to decide quickly and effectively (*the constable, a* *decisive man, arrested the pickpocket immediately).* However, for some two centuries the words have tended to be interchangeable in many contexts. *Decisive* has always been the more common of the two. Fowler (1926) distinguished the words in the following manner: 'A *decided* victory of superiority is one the reality of which is unquestionable; a *decisive* one is one that decides or goes far towards deciding some issue; a *decided* person is one who knows his own mind, and a *decided* manner that which comes of such knowledge; a *decisive* person, so far as the phrase is correctly possible at all, is one who has a way of getting his policy or purpose carried through.' His distinctions, though clear, do not exactly accord with the distributional spread of the two words (and of the corresponding adverbs *decidedly* 'unquestionably' and *decisively* 'conclusively') at the present time.

**decimate.** A key word in the continuing battle between prescriptive and descriptive linguists. L *decimāre* meant '(mil.) to punish every tenth man chosen by lot'. 'Punish' sometimes meant 'put to death'. In strict terms, therefore, *decimate* should mean 'to kill, destroy, or remove *one in every ten* of (something)', and it has been so used in English since the 17c. *Rhetorically* and *loosely*, as the *OED* expresses it, it has also been used from about the same time to mean 'to destroy or remove *a large proportion* of; to subject to severe loss, slaughter, or mortality'. The dispute between those who will and those who will not use *decimate* in its 'rhetorical or loose' sense is unresolved. An example of the disputed meaning: *But the forest has largely gone, decimated by a forest industry that is just now assaulting the final remains—Dædalus,* 1988. The word is even used sometimes just to mean 'kill': *He's a petty criminal but his long-lost brother bails him out of jail and protects him from the thugs intent on physically decimating him—Sunday Times,* 1989. This use is not recommended.

**declarant, declaredly, declarative, declaratory.** Pronounce /-'kleərənt/, /-'kleərɪdlɪ/, /-'klærətɪv/, and /-'klærətərɪ/ respectively. For the second, see -EDLY.

*déclassé.* Spelt thus (with accents) and printed in italics (fem. *déclassée*). The

meaning is 'that has fallen in social status'.

**declension.** In grammar, (a) the variation of the form of a noun, pronoun, or adjective, by which its grammatical case, number, and gender are identified; (b) the class in which a noun etc. is put according to the exact form of this variation. (*COD*) Cf. CONJUGATION.

**declinal, declination, declinature.** These 19c. attempts at forming a noun corresponding to the verb *decline* ( = to refuse in a courteous manner) have all failed: e.g. *The declinals were grounded upon reasons neither unkind nor uncomplimentary—*F. Palgrave, 1837; *The author must excuse our declination to accept as possible characters in any possible social system, people so unnatural—Pall Mall Gaz.,* 1884; *The reported declinature of office by the Marquis of Salisbury—Manchester Examiner,* 1885. The only available noun with the same stem is *declining* (*I took no offence at your declining the invitation to join us*); but the better course is to use the noun *refusal* modified by a suitable adj. (*courteous,* etc.). See -AL 2.

*décolletage, décolleté.* The first means 'a low neckline of a woman's dress, etc.' and the second '(of a dress etc.) having a low neckline' and '(of a woman) wearing a dress with a low neckline'. Both should be pronounced in the French manner and printed in italics.

**décor.** Spelt thus (with accent) and pronounced /'deɪkɔː/. The spelling *decor* (without accent) and pronunciation /'dekɔː/ are, however, equally common.

**decorous.** Dictionaries of the 18c. and 19c. mainly indicated second-syllable stressing, but the standard pronunciation now is /'dekərəs/.

**decoy.** By rule the noun should be pronounced /'diːkɔɪ/ and the verb /dɪ'kɔɪ/, but opposite stressing also occurs.

**decrease.** The noun is pronounced /'diːkriːs/ and the verb /dɪ'kriːs/. See NOUN AND VERB ACCENT 1.

**deducible, deductible** See -ABLE, -IBLE 7.

**deduction** is the inferring of particular instances from a general law, as opposed to *induction*, the inference of a general law from particular instances.

**deem** is a fairly formal word (frequently used in legal language) for 'judge, consider' (*He deemed it his duty to abstain from voting; because of your failure to reply you are deemed to have withdrawn from the project*).

**deep, deeply.** *Deep* as adverb is deeply entrenched in the language. It is used literally (*the boat was stuck deep in the mud*), and in many transferred and figurative senses (*the card-players sat up deep into the night; it doesn't go deep enough into the problem; still waters run deep*; etc.). In such types of sentences it would not be idiomatic to use *deeply*. *Deeply* is used esp. before past participles (as in the first sentence of this entry) and also when the meaning is 'profoundly, thoroughly' (*I should like to consider the matter a little more deeply*) or 'intensely' (*he was deeply conscious of his shortcomings*). See UNIDIOMATIC -LY.

**deer.** See COLLECTIVE NOUN.

**defect.** The pronunciations recommended are /ˈdiːfekt/ for the noun ( = imperfection, blemish) and /dɪˈfekt/ for the verb ( = abandon one's country or cause in favour of another). See NOUN AND VERB ACCENT 1.

**defective, deficient.** Some useful distinctions can be made. *Defective* tends mostly to mean 'faulty in quality, showing some definite fault, damage, etc., that impairs or cancels the efficiency of something' (*defective eyesight, hearing, measures, pronunciation, translation, valve, vehicle*, etc.); while *deficient* vies with *insufficient* to mean 'present in less than the quantity needed for effectiveness' (*deficient courage, diet, funds, water supply*, etc.). But there are many contexts in which the words converge in meaning, or are replaced by *insufficient, inadequate, faulty*, etc. Neither word is used now by professional people dealing with cases of mental abnormality.

**defence.** The spelling in BrE corresponding to AmE *defense*.

**defer.** The extended forms and derivatives are *deferred, deferring; deference* (respect), *deferral* (postponement), *deferment* (postponement).

**deficient.** See DEFECTIVE.

**deficit.** The only current pronunciation is /ˈdefɪsɪt/, though stressing on the second syllable (thus /dɪˈfɪsɪt/) and pronunciation with initial /ˈdiː-/ were common enough at the beginning of the 20c.

**definite article.** The use of the definite article *the* causes few difficulties to native speakers but is a major problem for foreign learners of the language. This is partly because many foreign languages, e.g. Russian, Chinese, Japanese, lack this feature, and partly because the distribution of the definite article in English is a matter of some complexity. Some of the main types are outlined below.
**1** *Names*. (*a*) In general, no definite article: *John Smith, High Street, Sutton Courtenay, Oxfordshire, England* (cf. *la France*); (religious festivals) *Christmas, Easter, Good Friday*; (months, days of the week) *January, Monday*; (continents) *Africa, Asia*; (lakes, mountains) *Lake Windermere, Mt Everest*); (locative names) *Buckingham Palace, Magdalen College, Paddington Station, Windsor Castle*. (*b*) But there are many exceptions: (unmodified names) *the Bible, the Pentagon, the Parthenon, The Times*; (names with premodification) *the Bodleian Library, the British Broadcasting Corporation, the English Channel, the National Gallery, the North/the South Island* (of New Zealand); (with postmodification) *the House of Commons, the President of the United States*.

**2** *Words*. In its ordinary use, the definite article *the* precedes a noun and implies a specific instance: e.g. *the cushion on the chair; he opened the door; the story is apocryphal*. It is also used as a guide to a particular place, person, etc, known by the speaker/hearer/reader to exist: e.g. *the earth, the Equator, the Church, the sky; the aristocracy, the working class; the French, the Chinese*. It is frequently used in a nonspecific way to mean 'whichever one of its kind is or was convenient, open, etc.': *I go to the cinema once a year; he heard it on the radio; she took the train to London. The* is often used of a part of the body following a preposition (instead of a possessive pronoun): *he took me by the shoulder; a pat on the back; they pulled her by the hair; he kicked his opponent on the shin*. The definite article is not repeated before the second of a pair of nouns joined by and: *the laws and ordinances of the ancient Hebrews.* Names of illnesses (*flu, measles, mumps,* etc.) are normally used without the definite article, but examples with *the* also occur in standard sources: *Annie has got the mumps*—B. Trapido, 1982; *It was that hot summer you had the measles*—N. Bawden, 1987.

**3** *Non-count nouns.* Mass or non-count nouns are normally construed without the definite article: *compelled by* guilt *to offer* atonement; *he couldn't stand* waste; *the important distinction between* narrative *and* dialogue; *the parents behaved with* dignity *and* grace. But many nouns can be used both as non-count nouns and count nouns: *instances of several claims to divine* illumination; *the Blackpool* illuminations.

**4** *Omitted.* In a number of other circumstances the definite article is regularly omitted: (subject names) *his main interest was* philology; (meal-names) *he had* breakfast *before catching the train;* (seasons) *snow on the ground and it is only* autumn; (places habitually visited) *he was taken to* hospital *for the operation; he goes to* church *every Sunday.*

**5** *Emphasis.* The definite article is also used for emphasis = the actual (person): *Not the Jeffrey Archer.*
See also THE.

**definite, definitely** are sometimes used as emphatic qualifiers, and simple deletion is often the best course: *there is a* (*definite*) *need for more accommodation for single parents*; *it has* (*definitely*) *been shown that there is a link between smoking and lung cancer.* Since the 1930s, *definitely* has come into widespread use colloquially as an emphatic affirmative meaning 'most certainly, indeed': e.g. *'Did they get more work done than at home?' 'Definitely.'* Would they recommend that the experiment is repeated another year? 'Oh, definitely.'—Sunday Times, 1959. It can be tiresome if overused.

**definite, definitive.** In most of their uses the distinction is clear-cut. One should bear in mind that *definite* means (*a*) having exact and discernible limits; (*b*) clear and distinct, not vague; and that *definitive* normally means (*a*) (of an answer, treaty, verdict, etc.) decisive, unconditional, final; (*b*) (of an edition of a book, etc.) most authoritative. Unfortunately, as *COD* (1995) notes, *definitive* (sense a) 'is often confused with *definite*,

which does not have connotations of authority and conclusiveness: *a definite no* is a firm refusal, whereas *a definitive no* is an authoritative judgement or decision that something is not the case'. Examples of the wrong use of *definitive*: *it is thought that any definitive news will be announced this afternoon; he promised Farmer Boldwood a definitive answer* (both from BBC Radio 4, 1977).

**deflection, deflexion.** From the 17c., when the word was first recorded, until the end of the 19c. the spelling with -*x*- was the more common of the two in all senses. *Deflection* is now the usual form in both Britain and America. See -XION etc.

**defrayal** continues to exist (somewhat uneasily) alongside *defrayment.* See -AL 2.

**defuse, diffuse.** A blatant example of confusion is the frequent use of *diffuse* (which properly means 'disperse') instead of *defuse* (which means 'remove a fuse (from an explosive device); remove tension or potential danger in (a crisis)'). Examples: *An early cut in base rates, which would ... diffuse the dispute between the Chancellor and the Prime Minister ... is now seen as a strong possibility in the City—Times,* 1988; *He turned up to find a gang of Hell's Angels terrorising the staff ... He was busy diffusing the situation (by suggesting they try a neighbouring café within another police district), when reinforcements arrived—Sunday Times Mag.,* 1989; *The Scott report is a time-bomb stealthy politicians and officials are trying to diffuse—Guardian,* 1995.

***dégagé*** /deɪˈgɑːʒeɪ/. Print in italics with accents. It means 'without constraints, unconstrained (in manner or address), unconcerned (about fashion or dress)'.

**degree.** The phrase *to a degree* originally (in the 18c., e.g. in Sheridan's *The Rivals: Assuredly, sir, your father is wrath to a degree*) meant 'to an undefined, but considerable, extent; extremely'. This sense survives (e.g. *the plot and situations improbable to a degree*—G. Saintsbury, 1916; *The first ones ... were primitive to a degree*—Times, 1984) alongside the frequent phrase *to the nth degree.* It is now also used in a very different way to mean 'to some extent': *WDEU* cites, for example, *To a degree this attitude is held by many in the Peace Now movement* (*NY Times Mag.*, 1980).

In practice any possible ambiguity is avoided by placing a suitable adjective between *a* and *degree* (*to a large degree, to a certain degree*, etc.: *All art is an abstraction to some degree*—Henry Moore, 1966).

**deity.** The traditional pronunciation /ˈdiːɪtɪ/ is now regrettably yielding to /ˈdeɪɪtɪ/. Similarly with the related words *deify, deification, deism*, etc. See -EITY.

**déjà vu.** Against the odds, this idiomatic French phrase (lit. 'already seen') has made remarkable progress in English since it was first recorded in the technical language of psychology in 1903 in the sense 'an illusory feeling of having already experienced a present situation'. Outside psychology its normal sense now is 'a clear impression that something has been previously seen or experienced', esp. in sentences of the type *he had a distinct feeling of* déjà vu. Once established in English, it tugged in two similar French phrases in its wake: *déjà entendu* (first recorded 1965) and *déjà lu* (1960), a feeling that one may have heard a passage of music, etc./ read a passage, a poem, etc., elsewhere.

**delectable.** A somewhat literary word meaning 'delightful, delicious'. Examples: *Delectable as life in a Cambridge college can be, there comes a time when a man can grow weary of the deliberations of the window box committee at King's*—N. Annan, 1981; *Every biting insect saw her as a delectable target*—William Boyd, 1982.

**deliberate, deliberative** (adjs.). These are readily distinguishable. The second now means only 'of, or appointed for the purpose of deliberation or debate' (*a deliberative assembly*) or, much less commonly, 'characterized by deliberation' (*in his cooler and more deliberative moments*). *Deliberate*, on the other hand, has several senses: intentional (*a deliberate foul*), slow in deciding (*a ponderous and deliberate mind*); (of movement, etc.) leisurely and unhurried (*he is very deliberate in his movements*).

**delimit** (verb). See LIMIT (verb) 2.

**delusion, illusion.** As is pointed out in *OMEU* (1983), some general distinctions can be made, though they are not absolute. *Delusion* is much used in psychiatric contexts: e.g. *delusions of grandeur*, a false belief concerning one's personality or status, which is thought to be more important than it is. Such uses have influenced the way in which the word is treated in the general language, signifying various kinds of misconception that arise from the internal workings of the mind. *Illusion* denotes a false impression derived either from the external world (e.g. *an optical illusion; a partition giving at least the illusion of privacy*), or from faulty thinking (*he thinks he will finish the book by the end of the year, but it may be an illusion*). The overlap of meaning between 'a false belief' and 'faulty thinking' is obvious, and it follows that there are circumstances in which the two words are interchangeable, at least technically.

It should be borne in mind, as a general guide, that *delusion* has a corresponding verb (*delude*) whereas *illusion* has not. On the other hand, the verb *disillusion* exists, and also the adj. *disillusioned*, whereas *delusion* has no such derivatives in *dis-*. Evidently such relationships (or the absence of them) partially govern the choice of noun when, on a strict interpretation of their lexical meanings, either *delusion* or *illusion* would fit. Examples: (delusion) *He suffered from the delusion that everything smelled of cats*—A. Koestler, 1947; *One of her recurrent delusions is that she is a mistress of dialects*—K. Tynan, 1961; *That was the way delusions started, thinking there was anti-Jewish feeling when there wasn't*—P. H. Newby, 1968; (illusion) *in the world as we know it … freedom is largely an illusion*—J. M. Roberts, 1975; *The illusion must be maintained that this was a purely Polish debate with no intrusion being made by the Soviet Union*—J. A. Michener, 1983; *Alfred Crowther loved his first-born child, but he had no illusions about him*—B. T. Bradford, 1986.

**de luxe** is regularly so written and printed in Britain, and pronounced /də ˈlʌks/, less commonly /də ˈluks/; and as one word, *deluxe*, in America, variously pronounced.

**demagogic.** This word and its derivative *demagogy* have resisted the general movement towards Anglicization: the second g remains /g/ not /dʒ/. Doubtless the main word of the group, *demagogue*, has had a bearing on the matter. See GREEK G.

**demagogue.** Sometimes spelt *demagog* in AmE.

**demand.** One demands something *of* or (more commonly) *from* someone (*demanded ten pounds from him*); one makes a demand *on* someone *for* something (*never-ending demands on him for his comments*).

**demean.** There are two verbs. The first, which was first recorded in the 14c. but is now somewhat archaic, is normally reflexive and means 'to behave, to conduct oneself (in a specified way)' (*The Prince Borghese certainly demeans himself like a kind and liberal gentleman*—N. Hawthorne, 1858). Cf. *demeanour*. The second, first recorded in 1601 and probably formed on the adj. *mean* after *debase*, is now dominant. Used transitively it means 'to lower in status, reputation, etc.' (*he simplified the argument without in any way demeaning it*); and, reflexively, 'to lower or humble oneself' (*he would not demean himself to accept such menial work*). Fowler's reservations (1926) about the second verb must now be set aside.

**demesne.** 1 The pronunciation recommended is /dɪˈmiːn/, but /-ˈmeɪn/ is also common.

2 *Demesne* means 'landed property' (usu. that occupied by the owner—*held in demesne*—not by tenants). It is sometimes also used (followed by *of*) to mean 'a region or sphere', where its doublet *domain* is more appropriate.

**demi-** ( = half-). The prefix is somewhat in retreat (as against *half-* and *semi-*, and, for words of Greek origin, *hemi-*), but survives in a fair number of English words (some of them loanwords from French), e.g. *demigod, demi-monde, demi-pension, demirep, demisemiquaver, demitasse. Demijohn* is probably a corruption of Fr. *dame-jeanne* Lady Jane; and *demiurge*, 'the creator of the universe' in the philosophy of Plato, represents Gk δημιουργός, lit. 'public or skilled worker'. See SEMI-.

**demise.** Outside legal contexts used, somewhat euphemistically, as a synonym of death (of a person), failure (of a political movement, etc.), cessation (the shutting down of a newspaper, business premises, etc.). Example of an extended

sense: *We were on our way to a funeral when we heard about Margaret Thatcher's demise* [i.e. her resignation]—V. Grove, 1990.

**demo,** abbrev. of *demonstration*. A 20c. word (first recorded 1936). Pl. *demos*. See -O(E)S 5.

**demon.** See DAEMON.

**demonetize.** Pronounce /diːˈmʌnɪtaɪz/, not /-ˈmɒn-/.

**demoniac** /dɪˈməʊnɪæk/, **demoniacal** /diː-məˈnaɪəkəl/ (adjs.) are not clearly differentiated in meaning. The choice between them is largely governed by the rhythm of the sentence. It is probable also that proximity of the nouns *demon* or *demoniac* would help to govern the choice of adjective. A third adj., *demonic* /dɪˈmɒnɪk/, is also often used = *demoniac*.

**demonstrate** and **demonstrator** have their main stress on the first syllable, **demonstrable, demonstrably,** and **demonstrative** on the second, and **demonstration** on the third.

**demote.** First formed in the late 19c. from *de-* + (*pro*)*mote*, this verb is now firmly established in the language, together with the corresponding noun *demotion*.

**demotic English.** There has been a marked increase in the 20c. in the writing down of semi-phonetic versions of (esp. unstressed) elements in the language. A typical example (from Tom Wolfe's *Bonfire of the Vanities*, 1987): '*But you gotta.*' ... '*You shoulda thoughta that, Irene.*' Here and elsewhere *a* represents 'to', 'have', and 'of', but the representation of informal speech elements goes much further than that. This demoticization of the written version of the spoken language is sometimes, but not always, used as an indicator of class, education (or lack of it), or race, but some of it (esp. the use of forms ending in *-a*) simply reflects the phonetic facts of relaxed or unstressed elements in the language.

Some examples: (reduction of *of*) *kinda, loadsa, lotsa, lotta*; (reduction of negatives) *dint* (didn't), *innit* (isn't it?); (reduction of *to*) *gonna* (going to), *hasta* (has to), *wanna* (want to); (reduction of *have*) *coulda, shoulda, woulda*; (miscellaneous) *doncha* (don't you?), DUNNO, *orl* (all), *termorrer* (tomorrow), *veddy* (very), *velly* (very), *whassa* or *what'sa matter*, *w(h)atcha* (what do, or are, or have you?). Most novelists and short-story-writers now make use of such devices in their works, but seldom in any systematic way.

**demur.** The pronunciations recommended are: for *demur, demurred, demurring* /-'mɜː-/; and for *demurrable* (Law), *demurrage* (shipping, railways), *demurral, demurrant* /-'mʌ-/. In law, *demurrer* has medial /-'mʌ-/, and in the ordinary language ( = one who demurs) /-'mɜː-/.

**dengue** (a tropical disease). Pronounce /'deŋgi/.

**denier.** Pronounce /'denɪə/.

**Denmark.** Marcellus's remark to Horatio after Hamlet has 'exited' with the Ghost (*Hamlet* I.iv.90), *Something is rotten in the state of Denmarke*, has suffered the fate of many such lines by being used and over-used in other contexts of rottenness. See IRRELEVANT ALLUSION.

**denote.** See CONNOTE.

**denouement,** the final unravelling of a plot or complicated situation in a play, novel, etc. In OUP house style, romans, no accent. Pronounce /der'nuːmɑ̃/.

**dental.** In phonetics, a consonant made by the tip of the tongue against the upper teeth, e.g. /θ/, /ð/, the speech sound represented by English *th*. They are closely related to *alveolar* consonants made by the tip or blade of the tongue against the teeth-ridge (or alveolar ridge), e.g. the speech sounds represented by English *t,d,n,s,z*.

**dent, dint,** are variants of the same word (*dent* first recorded in ME, *dint* in OE). At one time they both meant a blow or the mark or hollow made by such a blow. Now *dent* means **1** a slight mark or hollow in a surface, as one made by a blow, a collision, etc.; **2** a noticeable effect (*an overlong lunch made a dent in the afternoon*). *Dint* is mostly restricted to the phrase *by dint of*, by means of, but is sometimes used as a variant of *dent* 1.

**dentifrice** is a generic word (first recorded in the 16c.) for preparations used to clean the teeth, esp. toothpaste.

**denture(s)** is a genteelism (first recorded in 1874) for *false teeth*.

**denunciation.** Pronounce /-sɪ'eɪʃən/, and see -CIATION.

**deontic modality.** See EPISTEMIC MODALITY.

**depart** has been used transitively since the 14c. (e.g. *The clergy were ordered to depart the kingdom*, 1839 in OED; *Mrs. J. Gargery had departed this life on Monday last*—Dickens, 1861). WDEU (1989) cites numerous recent US examples of transitive use, e.g. *one boiling morning in July, we departed my father's house*—Harper's, 1971. But, except in the formal or literary phrase *departed this life*, the construction no longer forms part of the standard language in Britain. One *departs from* (a place) or *departs for* (a destination).

**departed.** For *the departed, the dear departed*, etc., see STOCK PATHOS.

**department.** See FIELD.

**depend.** In its main senses, *depend* is construed with *on* or *upon*: e.g. *He is the kind of man a girl may depend on to be a gentleman in whatever circumstances*—K. A. Porter, 1962; *It was quite wrong to come to depend too much upon one's children*—P. Fitzgerald, 1979; *I'll have a damn good try ... You can depend on that*—A. Price, 1982; *As grandfather grew older ... he seemed to depend increasingly on my company*—J. Simms, 1982. The slightly archaic sense 'to hang down' continues to be called on in literary contexts: *The leather sheath, from which two metal chains depended*—I. Murdoch, 1962; *From a beam crossing the low ceiling depended a mobile, the property of Parker*—E. Bowen, 1968. When followed by an indirect question, *on* or *upon* is usually not considered necessary in spoken English: e.g. *it depends what you mean by a minimum wage*; *it depends how much it costs*. Fowler (1926) judged such constructions to be 'slovenly' and 'indefensible', but his view now seems outmoded.

**dependant, dependent.** The noun has *-ant* in BrE (but sometimes *-ent* in AmE); the adj. *-ent* in both varieties.

**dependence, dependency.** See -CE, -CY (iii).

**deponent.** In Latin grammar, applied to verbs that are passive in form but active in meaning, as *cōnor, cōnātus sum, cōnārī* ('to try'), *morior, mortuus sum, morī* ('to die').

**depositary, depository.** A *depositary* is a person or authority to whom something is entrusted, a trustee. A *depository* is a storage place (for furniture, books, etc.); a store (of wisdom, knowledge, etc.); and (less frequently) a person or authority to whom something is entrusted.

**depot.** Romans, no accents. Pronounce /ˈdepəʊ/.

**deprecate, depreciate.** 1 It is important to bear in mind that *deprecate* (from L *dēprecārī* 'to try to avert by prayer') is in origin an antonym of *pray* (from L *precārī* 'to pray'); and that *depreciate* (from L *dēpretiāre* 'to lower in value', f. *pretium* price) is in origin an antonym of *appreciate, increase in value*. Both words entered the language in the 17c. and were for long kept apart. More recently a partial convergence of sense has become a matter of concern to those who are aware of the initial distinction between the two words.

2 *Deprecate* has come to mean (a) (trans.) (with a plan, proceeding, purpose, etc., as object) to express a wish against or disapproval of (*He deprecated an interim settlement, claiming that Egypt would reject it*—H. Kissinger, 1979); (b) (with a person or personal attribute as the object) to express disapproval of, to disparage, to deplore (*They continued in that vein, the one deprecating Mr K., the other praising him; When the news of this 'record' multiple birth emerged last weekend, few dared to deprecate it*—Sunday Times, 1987).

3 *Depreciate* means (a) (intrans.) to become lower in value or price (*the value of the paper money had depreciated almost to zero*—Edmund Wilson, 1940; *Experience has shown me that their cars are more reliable and depreciate less*—Mail on Sunday, 1985); (b) (trans.) to undervalue, to disparage (*When a patient depreciates a dream, I have found that it usually means his dream is particularly important*—R. Lindner, 1955; *Before this Wilde depreciated pity as a motive in art; now he embraced it*—R. Ellmann, 1969).

4 Senses (a) and (b) of *deprecate* and sense (b) of *depreciate* substantially overlap in meaning. The overlapping is emphasized when the adj. *deprecating* is preceded by *self-* (*He had the nervous, self-deprecating air of a man with enough money to have a bad conscience about it*—P. Hansford Johnson, 1959; *Barton ... smiled, and then his face changed again, the old, self-deprecating expression over it*—Susan Hill, 1971; *my relentlessly self-deprecating dinner conversation when we met at a restaurant on Broadway*—D. Menaker, 1990). Similarly, when the nouns *deprecation* and *depreciation* are preceded by *self-* (*self-aggression ... extends from relatively minor forms such as self-deprecation to self-destruction*—A. Giddens, 1977; *self-depreciation is not humility*—1827 in *OED*).

5 In general *depreciate* is tending to become restricted to financial contexts (sense 3(a) above), while *deprecate* is tending to crowd out *depreciate* in other senses and uses. But it is still desirable to keep in mind the distinction set down in the first paragraph of this article.

**deprival.** First recorded in 1611, it is still occasionally used for the act or state of depriving. See AL-. Example: *There are areas of need and social deprival here*—Oxford Diocesan Mag., 1984.

**deprivation.** Pronounce /ˌdeprɪˈveɪʃən/.

**Derby(shire).** Pronounced /ˈdɑːbɪ(ʃə)/ in standard BrE. In AmE, /dɜːrbɪ-/ is customary for the several senses (e.g. horserace [*Kentucky Derby*], a bowler hat) of the word.

**de rigueur.** Print in italics, and pronounce /də rɪˈɡɜː/.

**derisive, derisory.** Both words entered the language in the 17c. in the sense 'characterized by derision, scoffing, mocking'. *Derisive* also came to mean 'that causes derision, ridiculous' in the 1890s, and *derisory* followed suit in the 1920s. The two words are now for the most part kept separate, *derisive* being mostly used in the first sense (*derisive laughter, derisive remarks*, etc.), and *derisory* in the second (*a derisory pay offer, a derisory fee*, etc.).

**derring-do.** A literary word meaning 'heroic courage, feats of daring'. Spenser's Glosse to *The Shepheardes Calender*,

October (published in 1579) has *In derring doe, In manhoode and cheualrie*. He had taken the word from 16c. prints of Lydgate's *Chronicle of Troy*, where *derrynge do* is misprinted for original *dorryng do*, which echoes *In* dorrynge don *that longeth to a knyght* ( = in *daring to do* what appertains to a knight) of Chaucer's *Troilus and Criseyde* v.837. Lydgate also used the phrase as a fully developed noun (e.g. *Chron. Troy* v.136) (*Notes & Queries*, 1962, 369f.). Its currency in modern writers is due to Scott's use of *deeds of such derring-do* (*Ivanhoe*, ch. xxix). Example: *like birds that out of some wrongheaded derring-do had nested too high*—New Yorker, 1989.

**descant.** The noun is pronounced /ˈdeskænt/ and the verb /dɪsˈkænt/. See NOUN AND VERB ACCENT 1.

**descendant, descendent.** In BrE the noun is spelt *descendant* and the adj. *descendent*, In AmE each word may be spelt *-ant* or *-ent*.

**description.** Fowler (1926) deprecated the use of the word as a mere substitute for *kind* or *sort* (*no food of any description; crimes of this description*). As the use is not dealt with in any detail in the *OED* it is not easy to determine how old or how widespread it is. The use has been listed without comment in most smaller dictionaries since at least the 1960s. Examples: *I tried hard to keep talking, but I could not think of a single word of any description*—K. Waterhouse, 1959; *That's the first flying machine of any description that has ever landed on Muck*—Brian Moore, 1972.

**descriptive grammar.** A term properly used only of scientifically prepared grammars that describe the state of the standard language or regional forms of a given society at a given time, as revealed by a thorough examination and analysis of large quantities of relevant linguistic data. In practice, grammars written before the 20c. were systematically prescriptive or normative, and those written in the manner of *A Comprehensive Grammar of the English Language* (1985), by Randolph Quirk et al., descriptive. The two methods converge, esp. as shown in grammars prepared for schoolchildren and for foreigners. The key to the success of the descriptive method is

that it must be based on substantial amounts of real (not invented) data. See PRESCRIPTIVISM.

**desert, dessert.** Only *desert* in the sense 'an arid waste' is stressed on the first syllable: thus /ˈdezət/. The unrelated noun *desert* (usu. in the pl. *deserts*) 'what one deserves', the noun *dessert* meaning 'the sweet course of a meal', and the verb *desert* meaning 'to abandon' are all stressed on the second syllable: thus /dɪˈzɜːt/.

**deserter.** Thus spelt, not *-or*. See -OR.

**deservedly.** Four syllables. See -EDLY.

**déshabillé.** This French word, meaning 'a state of being only partly or carelessly dressed', is variously pronounced in English. The most usual pronunciation is probably /dezaˈbiːjeɪ/, but initial /deɪz-/ is also common. It is also commonly written without accents and pronounced /dezəˈbiːl/. The Anglicized variant *dishabille* /dɪsəˈbiːl/ is also used. As the word came into English in the 17c. it has had plenty of time to settle down, but has not.

**desiccated.** So spelt (not *dessicated*).

**desiderate.** First used in the 17c. in the sense 'to feel the need of, to regret the absense of', this verb is now archaic and seems to be falling by the wayside. See DESIDERATUM.

**desiderative.** 1 'From some Greek and Latin verbs secondary verbs are formed with special suffixes expressing the wish to do, or the being on the point of doing, the action. Thus the Latin for *be hungry, be in labour* is *esurio* from *edo* I eat and *parturio* from *pario* I give birth to; these are desiderative verbs.' (Fowler, 1926). In other words the desiderative verbs were used for the expression of wants, needs, and desires.

2 In modern linguistics the term *desiderative utterances* (Lyons, 1977) is applied to statements of the kind *I want the book; Have a good time!; Get well soon!; etc.*, i.e. statements of wishing or wanting containing an element of command.

**desideratum,** something lacking but needed or desired. Pronounced /dɪˌsɪdə-ˈreɪtəm/ or /-ˈrɑːtəm/. Also /dɪˌz-/. Pl. *desiderata*; see -UM 2.

**designedly.** Four syllables; see -EDLY.

**designer.** Since the 1960s a vogue word used first in fashion, etc., to denote goods bearing the name or label of a famous designer, with the implication that they are expensive or prestigious (*designer scarves, labels, jeans*, etc.). The word soon spread into much wider use, of which the following examples may be taken to be typical: *A still-growing line of what are known as 'designer drugs', new compounds that are synthetic 'analogues' of existing illegal mind-altering substances—Sunday Times*, 1985; *a giant slogan* [at the 1988 TUC conference in Bournemouth], *painted in the sort of designer colours (pink, grey and orange) which are nowadays associated ... with New Realism—Spectator*, 1988; *Mr Douglas Hurd, the Home Secretary, ... turned on 'designer violence' on the* [TV] *screen—Times*, 1988; *he remembered thinking to himself: so it's finally happened—designer industrial action—D.* Lodge, 1988. The short bristly growth on a man's unshaven face, something now much in favour, is known as *designer stubble*. *Designer* is a key word in the hyped-up world of the last three decades of the 20c.

**desist.** Pronounce /dɪˈzɪst/. The corresponding noun is spelt either *desistance* or *desistence*.

**desk.** We have all received letters *from the desk of* ——. The aim, one imagines, is to give an impression of personal attention. The reply to such letters, according to William Safire (1979), should begin *Dear Desk*.

**desolated,** as a polite exaggeration for *very sorry*, etc., is a GALLICISM, rendering Fr. *désolé*.

**despatch.** See DISPATCH.

**desperado.** Pronounce /-ˈɑːdəʊ/. Pl. *desperadoes* (AmE also *-dos*). See -O(E)S.

**despicable.** In *The Spoken Word* (1981) I recommended /ˈdespɪkəbəl/, with the stress on the first syllable, i.e. the traditional pronunciation, but the tide seems now to have turned in favour of /dɪˈspɪkəbəl/.

**despise.** One of the group of English words (*advertise, devise, surprise*, etc.) that must not be spelt with *-ize*. See -ISE 1.

**dessert.** See DESERT, DESSERT.

**destruct.** A back-formation from *destruction*, this verb is first recorded in 1958 in its modern sense connected with space missiles. It means 'to destroy (a missile that is malfunctioning or is off course)'. The normal past tense is *destructed*. It has also been used as a noun, esp. attributively (*destruct equipment, system*, etc.). In the 1960s the extended form *self-destruct* appeared, used intransitively to mean '(of a thing) to destroy itself automatically', both lit. and fig.: *Watergate came from within. The system itself has begun to self-destruct—Guardian*, 1973; *The tape would automatically self-destruct after twenty minutes—R. Perry*, 1979. Like the simplex, *self-destruct* is also used as a noun, esp. attributively (*self-destruct device, mechanism*, etc.).

**destructible.** See -ABLE, -IBLE 5, 7.

**detail.** In BrE the stress falls on the first syllable for both the noun and the verb. In AmE, second-syllable stressing is in the ascendant for both noun and verb.

**détente.** Retention of the French accent is desirable. Print in italics.

**deter.** 1 The pres.pple *deterring* is pronounced /dɪˈtɜːrɪŋ/, but the related noun *deterrent* has medial /-ˈter-/.

2 The *r* is doubled in the inflected forms (*deterred, deterring*) and in the derivatives *deterrence* and *deterrent*.

**deteriorate.** Pronounce all five syllables. The pronunciation with omission of the medial *-or-* is a vulgarism.

**determinately, determinedly.** *Determinately* should not be used to mean 'in a determined or resolute manner': the word required for that sense is *determinedly* (pronounced as five syllables: see -EDLY). *Determinately*, if used at all, should have the sense 'conclusively, finally'.

**detestation.** Pronounce /diːteˈsteɪʃən/, with the main stress on the third syllable.

**detour.** Print without accent and in romans, and pronounce /ˈdiːtʊə/.

**detract, distract.** *Detract* is usually followed by *from* and means 'take away (a part of something), diminish' (*this minor* *incident did not detract from their general performance). Distract*, also usually followed by an object and *from*, means 'to draw away the attention of (a person, the mind, etc.)' (*the excitement of the election distracted attention from the sharp increase in traffic accidents*). It is true that *detract* was (correctly) used between the 16c. and the early 19c. in the sense 'to divert' (e.g. *To detract their attention from every thing foreign—*1802 in *OED*), but this use is now obsolete in Britain. On BBC (TV) 1 (*Nation-wide*), on 14 September 1978, the Rochdale politician Cyril Smith said that Jeremy Thorpe *detracts attention* (from the real issues of the Liberal Party Assembly), but this can be taken to be a *lapsus linguae*.

**deuteragonist,** the person second in importance to the protagonist in a (Greek) play. The pronunciation recommended is /djuːtəˈrægənɪst/. Fowler (1926), by contrast, recommended /djuːtərəˈgəʊnɪst/.

**Deuteronomy.** Main stress on the third syllable, not the first: thus /djuːtəˈrɒnəmɪ/.

**develop, developable, development.** Not -*ope*.

**devil's advocate.** A translation (first recorded in 1760) of L *advocatus diaboli*, it means 'one who urges the devil's plea against the canonization of a saint, or in opposition to the honouring of any one; hence, one who advocates the contrary or wrong side' (*OED*). In simpler terms it means 'a person who tests a proposition by arguing against it'. It does not mean one who pleads for a wicked person.

**devise.** Not -*ize*. See -ISE.

**deviser, devisor.** *Devisor* is a person who devises (i.e. assigns or gives by will) property, and is in legal use only; *deviser* is the agent-noun of *devise* in all other senses. See -OR.

**devoid.** An adjective that from its first recorded use in the 15c. has been used only predicatively and followed by *of*. Examples: *A very simple style of dress, devoid of ornament or pretension—*W. G. Palgrave, 1865; *the Council is devoid of any powers—*BBC Radio 4, 1976.

**devolute.** In the sense 'to pass or transfer by devolution' this verb was freely used in the 16c. and then seems to have lapsed until it was revived (probably as a back-formation from *devolution*) in the late 19c: *The House will devise means of devoluting some of its work to more leisured bodies—Pall Mall Gaz.*, 1891. It has now largely given way again to *devolve*. Examples: (devolute) *Mr Tom King . . . is being urged to bring negotiations towards finding a suitable form of devoluted government for the province—Times*, 1985; *It will be interesting to see which country makes most progress in reducing teacher shortages—centralizing Britain or devoluting Poland—TES*, 1990; (devolve) *Our increasing dependence on tourism calls for highly devolved decision making—Glasgow Herald*, 1986; *Public support for the national question is at an historic high, with 50% of Scots favouring a devolved assembly and 30% of Scots wanting a separate parliament—Marxism Today*, 1986.

**dexter.** See SINISTER.

**dexterous.** The spelling given preference in *COD* rather than *dextrous*. However spelt it is pronounced /ˈdekstrəs/, i.e. with the second *e* silent.

**dhow,** a lateen-rigged ship used on the Arabian sea and adjacent waters. Now the accepted spelling (not *dow, daou, daw*). Pronounce /daʊ/.

**diabolic, diabolical.** In general, *diabolic* is used with direct reference to attributes of the devil (*Satan . . . merely bent his diabolic brow An instant* Byron, 1822), while *diabolical* is mainly used in the weakened senses 'disgraceful, disgracefully bad' (*Asked our postman about communications between Tunisia and England. He said they were 'diabolical'—*S. Townsend, 1982; *From my point of view that pitch was dangerous. In fact it was diabolical—Observer*, 1986.) But during the last four centuries this distinction has been by no means clearcut, and the rhythm of the sentence often governs the choice of form. Compare these examples: (diabolic) *Human technology, misused to serve the diabolic purposes of human egoism and wickedness, is a more deadly danger than earthquakes—*Toynbee and Ikeda, 1976; (diabolical) *Mephistopheles, a diabolical figure who played a great part in my imaginative life—*A. Powell, 1976.

**diachronic.** In linguistics, concerned with the historical development of a language, dialect, etc., as opposed to SYN-CHRONIC.

**diaeresis** (AmE **dieresis**). Pl. *diaereses* /-siːz/. Two dots placed over the second of two vowels (as in *naïve, Chloë, Eloïse*), and occas. in other circumstances (e.g. *Brontë*), to indicate that it is sounded separately. Since the sign is often not on modern keyboards it is often omitted in printed work; and it has also usually been dropped from such familiar words as *aërate, coöperate* (now *aerate, cooperate*). Occasional examples still occur, e.g. *I reëntered the chestnut tunnel—New Yorker*, 1987.

**diagnose.** Properly used to mean 'to make a diagnosis of (a disease, a mechanical fault, etc.)' *(he was able to diagnose the fault at once)*, but now often used with a person as object *(a baby who was incorrectly diagnosed as having died before birth, only to be delivered alive, but paralysed, 17 hours later—T. Stuttaford, 1990).*

**diagnosis.** Pl. *diagnoses* /-siːz/.

**diagram.** The extended and derivative forms have *-mm-* (*diagrammed, diagramming; diagrammatic*), but often *-m-* in AmE for the verbal forms.

**dial.** In BrE usu. *dialled, dialling* for the inflected forms, but *-l-* in AmE.

**dialect.** 'Within the British Isles, now as in the past, the English language exists and persists in an uncountable number of forms. Only one form—that taught to foreigners—is "standard" ' (Burchfield, 1985, 124). The words and meanings of many of these dialects are recorded in Joseph Wright's *English Dialect Dictionary* (1896–1905) and in a *Survey of English Dialects* (1962–8), edited by Harold Orton and others. There is a *Linguistic Atlas of England* (1978), edited by Harold Orton and others, and numerous monographs, articles, etc., on or about dialects in general or particular dialects. Such atlases and studies usually contain dotted and shaded maps showing where one linguistic feature ends and a competing one begins. For the most part dialectal words, meanings, and pronunciations remain in their own geographical area (subject only to the kind of internal linguistic change to which all forms of language are prone), and do not enter the standard language. Twentieth-century exceptions to this rule include *elevenses, to give* (someone) *an old-fashioned look*, and *gobsmacked*, all of which are now part of the standard language.

Most English people can recognize regional forms of pronunciation without having any form of professional training, though only in broad terms (Scottish, West Country, Geordie, Scouse, etc.). The ability to pin down the speech of a person to a particular city or locality is the preserve of fully trained dialect scholars and phoneticians. Work is proceeding in an encouraging manner on the analysis of particular linguistic features in selected places (Norwich, Reading, etc.), but full-scale dictionaries and grammars, both diachronic and synchronic, of the major dialects of all regions of the UK remain primary desiderata. It should constantly be borne in mind that non-standard features of grammar are in widespread use in the UK, and that they are not considered to be erroneous in these regions. They are also socially and ethnically oriented and class-bound: e.g. (working-class dialect of Reading) *I wants it; I don't eat none of that; he done it hisself* (J. Cheshire, 1982); *my hair needs washed* (Scotland and some areas immediately to the south of the border—P. Trudgill, 1983); *I don't want it but* (Scotland, Northern Ireland, parts of north-east England—P. Trudgill, 1983).

**dialectal, dialectic, dialectical** (adjs.). Throughout the 19c. all three words were freely used in the sense 'belonging to or of the nature of a dialect', but the only one now used in this sense in professional work is *dialectal. Dialectic* and *dialectical* are now confined to the language of logical disputation, and the phrase *dialectical materialism* to a specific, now discredited, theory propagated by Karl Marx and Friedrich Engels.

**dialogue** is neither necessarily, nor necessarily not, the talk of two persons. It is 'conversation as opposed to monologue, to preaching, lecturing, speeches, narrative, or description' (Fowler, 1926), and is ultimately derived from Gk διάλογος 'conversation' (διά 'through, across'). It has nothing to do with Gk δι- 'twice'.

Cf. DUOLOGUE. In AmE, *dialogue* is some-times spelt *dialog*. *Dialogue* is now very commonly used of discussions taking place between opposing groups (e.g. trade unions and management) or na-tions. In such contexts it is more or less synonymous with *negotiations*. *Dialogue* has a long history as a verb, beginning with Shakespeare (e.g. *Dost Dialogue with thy shadow?*, 1607), but is at present out of favour as part of the present-day hos-tility of many standard speakers to verbs believed to have been recently 'coined' from nouns.

**diamond.** Normally three syllables in BrE, /ˈdaɪəmənd/, but frequently just two, /ˈdaɪmənd/, in AmE.

**diaphanous.** See TRANSPARENT.

**diarrhoea.** Spelt thus in BrE (with *oe* still sometimes printed in the ligatured form *œ*) but as *diarrhea* in AmE.

**diastole,** in physiology, is pronounced as four syllables, /daɪˈæstəlɪ/.

**dice** (noun). In origin (14c.) the pl. of *die*, (noun) (as in *the die is cast*, now fig., 'the decisive step is taken'). The small cubes with faces bearing 1–6 spots used in games of chance are the *dice* (pl.); and one of them is also called a *dice*.

**dichotomy,** a division into two (ult. from Gk διχοτομία, from διχο- 'asunder' + -τομία used to form abstract nouns from adjs. in -τομος 'cutting', as *anatomy*, *lobotomy*, etc.). Since the Greek iota is short, the usual English pronunci-ation with initial /daɪ-/ goes against the etymology. The word has long-estab-lished senses in logic, astronomy, botany, and zoology (essentially 'division into two, bifurcation'), but the most note-worthy 20c. development is its free use in general contexts to mean 'something paradoxical or ambivalent'. Examples: *By a dichotomy familiar to us all, a woman requires her own baby to be perfectly normal, and at the same time superior to all other babies*—J. Wyndham, 1957; *Their uncritical use of the 'Communist' versus 'free world' dichotomy*—Listener, 1966; *The coffee-table featured a couple of Shakespeare texts and a copy of* Time Out—*an intriguing dichotomy, perhaps*—M. Amis, 1973; *there has been per-petuated a dichotomy between the relatively modern writings that can be 'appreciated'*

(*these are called 'literature'*) *and the relatively early writings that cannot* (*and these are called 'language'*)—R. Quirk, 1974. In general, the traditional words used in such contexts, e.g. *paradox*, *division*, *split*, *conflict*, *dilemma*, are often contextually more comfortable.

**dictate.** The noun, usu. in pl. (*dictates of conscience*), is stressed on the first syllable, and the verb (*he dictated a letter to his secretary*) on the second. See NOUN AND VERB ACCENT 1.

**dictionary, encyclopaedia, glossary, lexicon.** A *dictionary* is a book that lists, usu. in alphabetical order, and explains the words of a language or gives equiva-lent words in another language (a *bilin-gual dictionary*). The term is also used for reference books on any subject, the items of which are arranged in alphabetical order: thus Fowler's 1926 book was called *A Dictionary of Modern English Usage*; cf. also *A Dictionary of Canadianisms*, *Cham-bers's Technical Dictionary*, etc. The largest English dictionary is the *OED* (2nd edn, 1989) in twenty volumes. General dic-tionaries are sharply divided into those that largely omit names (*Napoleon*, *Berlin*, *Buckingham Palace*, *Rembrandt*, etc.) and those that include them in moderate, or even substantial, numbers.

An *encyclopaedia* is a book, often in several volumes, giving information on many subjects, including personal and geographical names, or on many aspects of one subject, usually arranged alpha-betically. The best-known example is the *Encyclopaedia Britannica*. A well known one-volume encyclopaedia is *The Hutchin-son Encyclopedia* (9th edn, 1990).

A *glossary* is an alphabetically arranged list of the more difficult terms or words found in the text, story, technical work, etc., to which it is appended. Its explana-tions are usually much briefer than those in a dictionary.

A *lexicon* is a unidirectional bilingual dictionary of an ancient language, e.g. ancient Greek, Hebrew, Syriac, and Arabic.

**dictum.** Pl. *dicta* (rarely *dictums*). See -UM 3.

**didacticism.** Fowler (1926), in an ele-gant but now dated essay characteristic of his generation, objected to 'The Anglo-Indian who has discovered that the suttee he read of as a boy is called *sati* by those who know it best is not content to keep so important a piece of knowledge to himself; he must have the rest of us call it *sati*, like the Hindoos (ah, no—Hindus) and himself; at any rate, he will give us the chance of mending our ignorant ways by printing nothing but *sati* and forcing us to guess what word known to us it may stand for.' Other examples were cited, including the replacement of *Caliph* by *Khalif* (but see CALIPH). Such 'didacticism', i.e. a tendency to insist on the alteration or replacement of traditional spellings, meanings, pronunciations, etc., in the interests of social, ethnic, or international amity, has continued throughout the 20c. Examples include: Asiatic → Asian; Eskimo → Inuit; Kenya, pronunciation changed from /ˈkiːnjə/ to /ˈkenjə/; Llanelly → Llanelli; Maori → Māori (or even Maaori); Mahometan → Mohammedan → Muhammadan (but nearly always replaced by Muslim); Moslem → Muslim; saree → sari; veldt → veld. For details see these words, or most of them, at their alphabetical places. See also ETHNIC TERMS; FEMININE DESIGNATIONS; LINGUISTIC ENGINEERING; POLITICAL CORRECTNESS.

**didn't ought.** A remarkable combination of the marginal modal *ought* and the periphrastic negative auxiliary *didn't* (treated in the *OED* s.v. *ought* v. 8). Almost certainly of dialectal origin, it has made its way into novels of the 19c. and 20c. and into informal speech as a typical construction used by rustic or sparsely educated speakers (*I . . . told him he didn't ought to go*—C. M. Yonge, 1854; *And I hope that none here will say I did anything I didn't ought. For I only done my duty*—M. Innes, 1942; *You didn't ought to have let that fire out* (Piggy speaking)—W. Golding, 1954).

**die** (noun). See DICE.

**die** (verb). When used with a preposition, the normal constructions are *to die of* (a malady, hunger, old age, etc.) or *to die from* (a wound, inattention, etc.); formerly (14–17c.) also *to die with* (a disease, the sword, etc.). Numerous examples of each construction and also of *to die for* (a cause, object, reason, or purpose), are provided in the *OED*.

**diesel,** named after a German engineer (d. 1913) called R. Diesel, is so spelt.

**dietitian.** The OUP house-rule spelling, not *dietician*.

**differ.** In the sense 'to be unlike or distinguishable', *differ* is almost always followed by *from* (*my interpretation of the passage differs substantially from his*). When it means 'to disagree, to be at variance with' the customary preposition, at least until the end of the 19c., was *with* (*To unite with those who differ with us*—J. H. Newman, 1833; *I differed with him in the conclusion he drew*—J. T. Coleridge, 1869). The second of these is now in relatively restricted use (at least in BrE) except when there is an *on*- or *about*-phrase as complement (*I'm very sorry to differ with you on/about that*).

**different.** 1 The commonly expressed view that *different* should only be followed by *from* and never by *to* or *than* is not supportable in the face of past and present evidence or of logic, though the distribution of the constructions is not straightforward.

(*a*) History. The *OED* lists examples of each of the three constructions from the dates indicated: *different from* 1590, *different to* 1526, *different than* 1644. In the 20c. a marked preference for *different from* has been shown in BrE; in the same period *different than* has flourished in AmE, but so too has *different from*. In both countries, in all kinds of circumstances, *different to* has been widely perceived as a credible alternative (cf. *indifferent to*, but not *indifferent from* or *than*). Modern examples of each construction: (from) *it presents a rather different view of abstraction from anything which one could find in the histories of modern art written 20 or 30 years ago*—Illustrated London News, 1980; *The Anglo-American approach to copyright was thought to be different from the approach taken by France and other European countries*—New Yorker, 1987; *He's no different from my other brother, in the end*—N. Gordimer, 1988; (to) *He looked no different at first to other boys Margaret had known*—M. Leland, 1986; *I found that a meadow seen against the light was an entirely different tone of green to the same meadow facing the light*—Scots Magazine, 1986; *It is readily possible . . . to argue for priorities different to those of the 'new vocationalism'*—Essays &

*Studies*, 1987; *They don't seem to be any different to us. They're real regular guys*—US sailor speaking of his Soviet counterparts, *Chicago Tribune*, 1989; (than) *It was no different than wolves cross-cutting on a deer*—T. Findley, 1984 (Canad.); *he wondered if people now made love ... in a different way than their grandparents*—T. Keneally, 1985 (Aust.); *This discrepancy is intriguing because most scallops have a very different mode of life than other species*—*Bull. Amer. Acad. Arts & Sci.*, 1987. The construction with *than* is widespread in AmE, but does not form part of the regular language in Britain.

A contrary view is sometimes expressed. I quote from *The Oxford Guide to English Usage* (1993): 'Both *different to* and *different than* are especially valuable as a means of avoiding the repetition and the relative construction required after *different from* in sentences like *I was a very different man in 1935 from what I was in 1916* (Joyce Cary). This could be recast as *I was a very different man in 1935 than I was in 1916* or *than in 1916*. Compare *The American theatre, which is suffering from a different malaise than ours*, which is greatly preferable to *suffering from a different malaise from that which ours is suffering from*.' A wiser course, perhaps, is simply to avoid *different than* if you happen to be writing (or speaking) in Britain. There are other ways of recasting sentences than by simply substituting equivalent American constructions.

(b) Logic. The principle upon which *different to* is questioned is based on the premiss that we do not say *differ to*. By this argument, all words in the same morphological family should be construed with the same prepositions; e.g. we ought to say *according with* (instead of *according to*) because we say *accords with*. Contrast also *full of* with *filled with*; *proud of* with *pride in*.

**2** *Different* has been informally used throughout the 20c. to mean 'out of the ordinary, special': *What a perfectly lovely couch ... Why, it is so beautifully different!*—D. F. Canfield, 1912; *They are always striving to write a piece of copy that will be 'different'*—*Publishers' Weekly*, 1930.

**3** In a weaker sense, *different* is sometimes used inadvisedly as a synonym of *various, separate, distinct*, etc. (*Different people told me I had done the right thing*;

*we went to different* [i.e. various] *places in Tasmania*).

**differentia.** Pl. *differentiae* /-ʃiiː/.

**differentiation. 1** See -CIATION, -TIATION.

**2** The term was applied by Fowler (1926) principally to pairs of words that at one time or another shared a common meaning in the standard language but have now become distinct. For example, among the *OED*'s 18c. quotations for *spiritual* and *spirituous* are these two: *It may not here be improper to take notice of a wise and spiritual saying of this young prince*—Isaac D'Israeli, 1791–1823; *The Greeks, who are a spirituous and wise People*—J. Savage, 1703. The linking of each with *wise* indicates that they are interchangeable. In present-day usage they have drawn apart. Differentiation has occurred. Similarly, *airship* when first used meant any type of aircraft, whether lighter or heavier than air; later, by differentiation from *aeroplane* (and *airplane*), *airship* became confined to the former kind. Differentiations become complete not by authoritative pronouncements or dictionary fiats (as Fowler correctly observed), but by being gradually adopted in speaking and writing. It is the business of a writer of a usage guide to give a clear indication of the stage reached by such differentiations in so far as the boundaries and limits can be determined from the available evidence.

Scores of pairs of words that have become partially or fully differentiated are dealt with at their alphabetical places in this book. Judgements are also offered about the rights and wrongs of the attitudes of writers and of members of the general public where controversial areas exist. For example: (fully differentiated in standard English but sometimes confounded) *affect/effect, defuse/diffuse, deprecate/depreciate, fortuitous/fortunate, imply/infer, masterful/masterly, prevaricate/procrastinate, refute/deny*; (partially differentiated, with a residual band of overlapping meaning) *administer/administrate, admission/admittance, continual/continuous, mutual/common*.

**differently. 1** Normally construed with *from* (*John Dunston had been brought up very differently from his father*—*Bodleian Libr. Rec.*, 1986), but historically, like *different*, sometimes construed with *to* or *than*.

**2** Occas. combined with *abled* (so *differently abled*) in AmE and used as a synonym of *disabled* (*the faculty* [at Oberlin College, Ohio] *voted to establish a committee for the differently abled—Observer* (Oberlin College), 1990. See POLITICAL CORRECTNESS.

**diffuse.** See DEFUSE.

**diffusible.** So spelt, not *-able*. See -ABLE, -IBLE 7.

**dig.** Of French origin in the 14c., it was conjugated as a weak verb (with *digged* as pa.t. and pa.pple) for some three centuries (*all the wels which his fathers seruants had digged in the dayes of Abraham—Gen.* (AV) 26: 15; *Also he built towers in the desert, and digged many welles—*2 Chr. (AV) 26: 10). Beginning in the 16c. it acquired a pa.t. *dug* and (in the 18c.) the same form began to be used for the pa.pple. The later forms are, of course, now invariable.

**digest.** The noun is pronounced /ˈdaɪdʒest/, with the stress on the first syllable (*Reader's Digest*), and the verb /darˈdʒest/, /dɪ-/, with the stress on the second. See NOUN AND VERB ACCENT 1.

**digraph.** Any combination of two letters used to represent a single speech sound, e.g. *dg* in *judge* /dʒʌdʒ/, *gh* in *tough* /tʌf/, *ng* in *long* /lɒŋ/, *ea* in *head* /hed/, *ee* in *fleet* /fliːt/, etc. The term is also sometimes applied to ligatures like *æ* as in *æsthetics* (if so printed) and *œ* as in *œuvre*.

**dike.** See DYKE.

**dilapidated.** Occas. thought to be correctly applied only to a *stone* structure that has fallen into disrepair, because its stem answers ultimately to L *lapis, lapidem* 'stone'. But even in Latin the verb *dīlapidāre* was used in extended figurative senses apart from the literal sense 'to bring (an edifice) into a state of partial ruin'.

**dilatation,** etc. This (first recorded c1400) and *dilatator* (1611) are the etymologically correct forms, since the medial element (as also in the verb *dilate*) answers to the L adj. *lātus* 'broad, wide'. But the shorter unetymological forms *dilation* (1598) and *dilator* (1605) also came into use at an early stage under the influence of patterns like *calculate* ∼ *calculation* (in which the stem simply represents the L first-conjugation pa.pple

termination *-ātum*). In surgery the forms *dilatation* and *dilator* are customary. In lay use (except when the reference is to an operation on a woman's cervix) the dominant forms are *dilation* and *dilator*.

**dilatory.** Pronounce /ˈdɪlətərɪ/. It should be noted that it is not formed like the words in the preceding entry, but is from L *dīlātum*, pa.pple of *differre* 'to defer, delay'.

**dilemma.** **1** The traditional pronunciation /dɪˈlemə/, with short *i* answering to the short iota in Gk δίλημμα, is now being threatened by initial /daɪ-/.

**2** The word is correctly used to mean 'a choice between two alternatives both of which are undesirable' (*They were in the dilemma of either violating the Constitution or losing a golden opportunity*). It is widely, but loosely, used to mean simply 'a problem, a difficult choice' (*What to do with one's spare time is a modern dilemma*). See POPULARIZED TECHNICALITIES. *Historical note.* The term was first used in the 16c. in rhetoric (and so in logic) to mean 'a form of argument involving an adversary in the choice of two (or, loosely, more) alternatives, either of which is (or appears) equally unfavourable to him' (*OED*). The alternatives are commonly spoken of as *the horns of a dilemma*, either of which would impale him.

There are many contexts in the past and at the present time when the line between the strict use and the 'loose' one is very fine, and even non-existent. But when words like *problem* or *difficult choice* fit neatly into such contexts they should be used in preference to 'dilemma'. A range of representative examples: *She was in the dilemma of those who must combine dignity with a bad cold*—M. Innes, 1942; *The dilemma is logically insoluble: we cannot sacrifice either freedom or the organization needed for its defence*—I. Berlin, 1949; *He was caught in a dilemma, a choice between doing a show or going on a much-needed vacation*—D. Halberstam, 1979; *Three corridors: one to the left, one ahead, one to the right . . . 'Dilemma. Left, right or centre?'*—D. Bogarde, 1980.

**dilettante.** Pl. either *dilettanti* (pronounced the same as the sing.) or *dilettantes.*

215

**dim.** For *dim religious light*, see IRRELEV-ANT ALLUSION.

**diminuendo.** Pl. *diminuendos* (see -O(E)s 6) or *diminuendi* (as in It.).

**dinghy.** The pronunciations /ˈdɪŋɪ/ and /ˈdɪŋɡɪ/ are both in common use.

**dingo** (wild Australian dog). Pl. *dingoes*. See -O(E)s 1.

**dint.** See DENT.

**diocese.** Pronounce /ˈdaɪəsɪs/. Pl. *dioceses* /ˈdaɪəsiːz/ or /ˈdaɪəsɪsɪz/. The spelling *diocess* was customary from the 16c. to the end of the 18c. and was the only form recognized by Dr Johnson and other 18c. lexicographers. See SEE.

**diphtheria.** Normally pronounced /dɪf-ˈθɪərɪə/, irregularly /dɪp-/.

**diphthong.** Pronounce /ˈdɪfθɒŋ/. The pronunciation with initial /ˈdɪp-/ is highly irregular. The word is properly applied only to a speech sound (very frequent in English) in one syllable in which the articulation begins as for one vowel and moves as for another (as in *coin, loud, fire, sure*, etc.). In professional work it should not be confused with DIGRAPH or LIGATURE.

**diploma.** The pl. is now always *diplomas*. Until the late 19c. the pl. form *diplomata* ( = state papers) was also used.

**diplomat, diplomatist.** Both words entered the language at the beginning of the 19c. *Diplomat* is now the more usual of the two, but the reverse was the case throughout the 19c.

**direct, directly.** Both have been used as adverbs for several centuries (*direct* first recorded c1450, *directly* c1400), but now *directly* is much the more common. *Direct* is usual when it means 'straight in direction or aspect, without deviation' (*Air NZ flights go direct from Heathrow to Los Angeles*; *the property is held direct from the Crown*; *purchasable only from the central office direct*); or 'without intermediaries' (*he appealed over the head of his line manager direct to the chief executive*). The natural form in most of the senses of the adjective is *directly* (*the beginning of a development which was to lead directly to some of his finest war poems*; *the wind is blowing directly on shore*; *directly opposite*; ( = immediately) *Directly after this, he was taken away*. Only *directly*, not *direct*, can occur immediately before an adjective (*directly responsible for the accident*). In informal contexts *directly* is used as a conjunction = 'as soon as, the moment after' (*Iodine and phosphorus combine directly they come in contact*). In this use *directly* is elliptical for *directly that/as/when*. In dialectal use and in America *directly* has also been used since about the mid-19c. to mean 'shortly, very soon' (*I'll come directly*). It is important to avoid ambiguity in the use of the two words: in *he went direct to Paris* the sense is 'without deviation'; but in *he went directly to Paris* the sense is 'immediately'.

**dirigible.** As noun and adj. spelt with -ible.

**dis-.** As a living prefix, it continues to form nouns, adjectives, adverbs, and verbs, expressing negation, indicating the reversal or absence of an action or state, etc. Twentieth-century accessions to the language in *dis-* include *disaffiliation* (1926), *disambiguate* (v., 1963), *disconfirm* (v., 1936), *disconfirmation* (1937), *diseconomy* (1937), *disincentive* (1946), *disinvestment* (1938), *dispersant* (1944), *disquietingly* (1901), and *dissimilatory* (adj., 1901). Cf. DE-.

**disassemble.** The customary modern sense 'to take to pieces, to take apart' is first recorded in 1922, and now forms part of the regular language, applied esp., but by no means only, to machinery that is being dismantled (*The reconstruction of disassembled paintings—TLS*, 1958). In dictionaries and in actual usage it stands well apart from *dissemble* 'to conceal'.

**disassociate.** A common but now widely condemned variant (first recorded in 1603) of DISSOCIATE. Examples: *it* [sc. an aeroplane] *appears disassociated from the weight, noise and violence of a huge jet shouldering its unnatural way through the pure air*—Janice Elliott, 1985; *Any other woman would have disassociated herself, gone where she wasn't known, changed her name*—A. L. Barker, 1987; *The Foreign Office at once issued a statement disassociating the Government from the idea*—Spectator, 1988.

**disastrous.** Be careful not to introduce a fourth syllable: it is not dis-ast-er-ous.

**disbar.** See DEBAR.

**disc.** *OED* (1896) gave *disk* as 'the earlier and better spelling' but in the second edition (1989) declared that '*disc* is now the more usual form in British English'. In the 20c. there has indeed been a marked preference for *disc* in Britain but for *disk* in America. The division has been lessened by the almost universal spelling of the computer storage device as *disk* (*magnetic disk, disk drive, floppy disk*, etc.). For the time being, however, *disc brake, the sun's disc, compact disc, disc harrow, disc jockey*, etc., remain the standard forms in BrE. For most of these *disk* is the more usual form in AmE.

**discernible.** Thus spelt in OUP house style, not *-able*. See -ABLE, -IBLE 7.

**disciplinary.** Pronounce /ˈdɪsɪplɪnərɪ/ or /dɪsɪˈplɪnərɪ/. The pronunciation with the stressed third syllable rhyming with *fine*, once occasionally heard, now seems to have vanished.

**discomfit.** Once (from the 13c. onward) mainly used in the primary sense 'to undo in battle' and 'to defeat or overthrow the plans or purpose of', *discomfit* is now mainly used in the weakened sense 'to disconcert'. It is also often qualified by *rather* (*Bell, conscious of past backslidings, seemed rather discomfited*—William Black, 1872). In this weakened sense it comes close to the etymologically unrelated verb *to discomfort* 'to make uneasy', and is possibly causing *discomfort* to be used less often as a verb. As a noun, *discomfort* 'lack of ease, slight pain' (*tight shoes, a slightly twisted ankle*, etc., *can cause discomfort*) remains common. Examples: *He seemed flustered, discomfited, and this amused me: Why should he feel embarrassed before a nigger preacher?*—W. Styron, 1966; *I should have corrected her, but, discomfited, missed the right moment*—A. Lurie, 1969.

**disconnection.** Now thus spelt in OUP house style, not *disconnexion*.

**discontent.** From Shakespeare's famous lines *Now is the winter of our discontent, Made glorious summer by this sonne of Yorke* (*Richard III* I.i.1) has been drawn perhaps the most prolific of modern political clichés, *a winter of discontent*.

**discord.** The noun is stressed on the first syllable and the verb on the second. See NOUN AND VERB ACCENT 1.

**discothèque.** Adopted from French (where it was coined on the model of Fr. *bibliothèque* 'library'), it could have come to mean in English as in French 'a record library, a library of discs'. Instead (first recorded in 1954) we adopted its second French sense, 'a club, party, etc., where (loud) music is played for dancing'.

**discount.** The noun is stressed on the first syllable and the verb on the second. See NOUN AND VERB ACCENT 1.

**discourse.** The noun is stressed on the first syllable and the verb on the second. See NOUN AND VERB ACCENT 1.

**discourse analysis.** A modern term (first recorded in 1952) for the linguistic analysis of stretches of language with the aim of disentangling the unspoken rituals, strategies, and implications built (unconsciously) into them and the inferences to be drawn from them. A standard textbook on the subject is by Gillian Brown and George Yule (1983).

**discover, invent.** The distinction between the two verbs was neatly made in 1783 by Hugh Blair in his *Lectures on Rhetoric and Belles Lettres*: *We invent things that are new; we discover what was before hidden. Galileo invented the telescope; Harvey discovered the circulation of the blood.* As *COD* has it, *discover* means 'to find out or become aware of, whether by research or searching or by chance (*discovered a new entrance; discovered that they had been overpaid*)'; and *invent* means 'to create by thought, devise; to originate (a new method, instrument, etc.); to concoct (a false story, etc.)'.

**discreet, discrete.** The two words are pronounced the same, and are etymologically 'the same word', i.e. are doublets (ME from L *discrētus*, pa.pple of *discernere* 'sift'), but differ widely in meaning. *Discreet* means 'circumspect in speech or action; unobtrusive (*a discreet touch of rouge*)'. *Discrete* means 'individually distinct, separate (*a telescope powerful enough to resolve galaxies into discrete stars*)'.

**discus.** Pl. *discuses*, not *disci*. See -US 1.

**discuss.** Anyone reading 19c. fiction should be alert to the use of *discuss*, when used in the context of an item of food or drink, to mean 'to try the quality of, to consume' (*A tall, stout, country-looking man ... busily discussing huge slices of cold boiled beef*—Scott, 1815; *Turner was always to be seen between ten and eleven at the Athenæum, discussing his half-pint of sherry*—G. W. Thornbury, 1861.

**discussible.** Thus spelt. See -ABLE, -IBLE 3.

**disenfranchise, disfranchise.** Both verbs have been in the language for several centuries. At present *disenfranchise* is the more common of the two.

Both verbs end in -*ise* not -*ize*. See -ISE 1.

**disgruntled** has settled into restricted use as an adj.—the older use of *disgruntle* as a finite verb belongs to the past—but it has brought into being in the 20c. the pleasing occasional back-formation *gruntled* (*I could see that, if not actually disgruntled, he was far from being gruntled*—P. G. Wodehouse, 1938) in the sense 'pleasing, satisfied, contented'.

**disguise.** Always spelt thus, never -*ize*. See -ISE 1.

**disgustful.** First recorded in a work by Beaumont and Fletcher in 1616 and commonly used from the 17c. to the 19c. to mean (in varying degrees) 'causing aversion or distaste', *disgustful* has gone into partial retreat in the 20c. and is now uncommon. Examples: *He brought out an English phrase with an air of handing it over to me between disgustful finger-tips*—P. Lubbock, 1923; *and now the time has come to enumerate a few of the reasons the whole event* [sc. the Albert Einstein Peace Prize] *was so disgustful to the present writer*—*Time*, 1985.

**dishabille.** See DÉSHABILLÉ.

**disillusion, disillusionize.** Both verbs came into being in the 1860s and both seem to have remained in respectable use at least until the end of the 19c. Since then it has been downhill all the way for the longer form, while *disillusion*, and esp. *disillusioned* as pa.t., pa.pple, and ppl adj., have flourished. *Disillusion*, formed from dis- 'deprived of, rid of' + noun, has numerous parallels (*discredit, disfigure, dismast*, etc.).

**disinformation,** first recorded in 1939 (applied to a German 'Disinformation Service'), is a keyword in the political vocabulary of the 20c. At the height of the Cold War (1945–89) the dissemination of concocted false information was one of the most potent and dangerous weapons of governments, and especially of the various intelligence groups, the CIA, the KGB, MI6, etc.

**disingenuous** means 'having secret motives, lacking in candour; insincere' (said of persons or their actions), whereas *ingenuous*, its antonym, means 'innocent, artless, frank'. Examples. (disingenuous) *I should be disingenuous if I pretended not to be flattered*—W. Golding, 1982; *Adam B. Ulam, the director of Harvard's Russian Research Center, likes to say 'Triple S' while flashing a disingenuous grin that's meant to make you wonder whether his Polish accent is perhaps thicker than you had supposed*—*New Yorker*, 1990; (ingenuous) *She took off the ingenuous little hat that Anna thought suitable for her years*—E. Bowen, 1938; *De Wilde's ingenuous imagination appeals strongly to Miss Lillie, who has a great deal of urchin in her*—K. Tynan, 1961.

**disinterest.** This noun has or has had three branches of meaning: **1** that which is contrary to interest or advantage, something against the interest of or disadvantageous to (a person or thing) (recorded from 1662 onward, but now rare or obsolete): e.g. *All gain, increase, interest ... to the lender of capital, is loss, decrease, and dis-interest to the borrower of capital*—Ruskin, 1876.

**2** Impartiality (recorded from 1658 onward and still current, but not in common use): e.g. *We here see Morris working, with entire disinterest, at his work, and caring above all things for fine workmanship*—*Sat. Rev.*, 1896.

**3** Absence of interest, unconcern (recorded from 1889 onward): e.g. *The days of service disinterest are over. The R.A.F. has begun to think seriously ... about the future of space*—*Guardian*, 1962; *He misread my quietude ... as either agreement or disagreement. It was neither. Pure, unadulterated disinterest*—C. Achebe, 1987; *because of his no-nonsense approach to art ... disinterest in modernistic rhetoric and formalist theory*—*Christian Sci. Monitor*, 1987. This third use is controversial; cf. DISINTERESTED.

It is worth noting that the notion of impartiality is more often conveyed by the word *disinterestedness* than it is by *disinterest*. Sense 3 is sometimes expressed by the word *uninterest*, but this word has tell-tale signs of being an example of 'error-avoidance', brought into being because of the controversial nature of *disinterest* 3.

The best course is to avoid using the noun *disinterest* altogether until it has reached safe shores. See UNINTEREST.

**disinterest** (verb), to rid of interest or concern. Labelled 'Now *rare*' in the *OED*, it seems to be having a new lease of life to judge from the following: *I try to disinterest myself from politics*—A. Huxley, 1923; *To dilute purposely the Canadian content of CN&Q is . . ., perhaps, to disinterest its Canadian subscribers as much as it might possibly interest and attract foreign subscribers*—Canad. N. & Q., 1984.

**disinterested.** Not recognized as a problem by Fowler (1926), the use of *disinterested* to mean 'uninterested' had become a matter of fierce controversy by the 1970s and 1980s. In *The Spoken Word* (1981) I reported that the use 'attracts more unfavourable comment from [BBC] listeners than any other word ... with the possible exception of *hopefully*'. The problem was neatly pointed up by the actor Julian Glover about his mother: *She's also a word pedant, and I hate having inherited this from her. She corrects me firmly about 'disinterested' and 'uninterested'— common but good examples. It's maddening, drives people bananas*—Sunday Times Mag., 1990.

The word can be approached in three ways: **1** *Etymology*. The prefix *dis-* has a primary function of expressing negation. On these grounds *dis-interested* would in a literal sense mean 'not interested, uninterested'. But *interested* can mean either 'having an interest in, being curious about' (*he was interested in philately*) or 'not impartial' (*an interested party*). As a result, *disinterested* can therefore mean 'not having an interest in' or 'not not impartial, i.e. impartial'. The etymological approach is too fundamental to be helpful.

**2** *Diachronic*. In the 17c. and 18c., from Donne to the Junius Letters (1767), *disinterested* meant 'not interested, unconcerned'. This sense then went into abeyance until it was revived in the 20c. (*Oxford, after three successive defeats, are almost entirely disinterested in the Boat Race*— Daily Tel., 1970). Meanwhile in its second sense, 'impartial, free from self-seeking', *disinterested* has been in unbroken use since 1659. The sense 'impartial' can therefore be said to have had a longer period of continuous use.

**3** *Synchronic*. Without doubt the sense 'uninterested, not interested' is making a strong challenge at the present time. The community is divided into those who believe that a genuine and useful distinction (*disinterested* = impartial/ *uninterested* = not interested) is being eroded, and others who, for whatever reason, have simply not heard the word being used to mean 'impartial', and just regard it as a routine antonym of *interested*.

A listing of examples underlines the strong presence of both meanings at the present time: ( = impartial, unbiased) *His disinterested, scholarly attitude*—K. M. Elisabeth Murray, 1977; *A disinterested approach to the literature of the past can liberate us from being dominated by the anxieties, problems, and even more by the clichés of today*—H. Gardner, 1981; *Many competent and disinterested experts on world poverty often stress the sterility of the East–West confrontation*— Encounter, 1981; *a tenacity and sheer steadfastness which must compel admiration ... from the disinterested observer. But of course none of the observers of twelfth-century England was disinterested*—Antonia Fraser, 1988. ( = not interested) *'Peanuts, lollies and chocolates,' he muttered in a supremely disinterested voice*—M. Pople, 1986 (Aust.); *A young girl* [sc. a nurse, summoned by a patient] *in a green smock came at a disinterested pace*—New Yorker, 1987; *Maclennan appeared disinterested in hearing any more about our conversations*—John Grant (former SDP MP), 1988; *Washington ensured that he would appear to be what in fact he was, a republican gentleman disinterested in power*—TLS, 1988; *she remains stubbornly neat and unadorned, disinterested in fashion*—Susan Johnson, 1990 (Aust.).

The varied pattern of meanings shown in other words in the same family is also bound to have had an effect on the distribution of the conflicting meanings of *disinterested*: thus *disinterestedly* (a) impartially 1711–, (b) without interest or concern 1941– (e.g. *Instead, she gazes disinterestedly and unblinkingly at her fawning inquisitor and removes a ring from her finger which she toys with absently*—Q, 1991); *disinterestedness* = impartiality a1682–. In these the 'impartial' sense has the edge.

My personal use and recommendation is to restrict *disinterested* to its sense of 'impartial', at any rate for the present. See UNINTERESTED.

**disjunctive.** In grammar, applied esp. to a conjunction (as *but*, *nor*, *or*) expressing a choice between two or more words, as *but* in *he is poor but happy*, and *or* in *she asked if he was going or staying*, as against one which has an additive function (e.g. *and*). The distinction is of some importance in determining the number of verbs after compound subjects: (disjunctive) *Neither the binary* nor *the synthetic approach is particularly satisfactory*; (copular) *Patience* and *humour* were *Bacon's best weapons*. Cf. COPULAR.

**disk.** See DISC.

**dislike** (verb). Constructions with a direct object (*he disliked Picasso's works*) and with a following gerund (*she disliked being absent at any time*) are normal. Occasional examples of a construction with a *to*-infinitive are non-standard: *She was hounded by a fear of imminent poverty that made her dislike to spend any money at all*—B. Guest, 1985; *she even disliked to be called 'Madame'*—P. P. Read, 1986.

**dislodgement.** The preferred spelling, not *dislodgment*.

**dismissible.** Thus spelt (*a dismissible offence*). See -ABLE, -IBLE 7.

**disorient, disorientate.** Both verbs have a long history (*disorient* first recorded in 1655, *disorientate* in 1704) and both are still in use (corresponding to the noun *disorientation*). In most contexts *disorient*, being shorter, is the better form.

**disparate** is in regular use in two main senses: 'essentially or markedly different or diverse in kind' (*disparate creeds, races, modes of thinking*), and 'unequal' (*people of disparate ages, countries of disparate size*). The first sense is the more usual of the two.

**dispatch.** The preferred form, not *despatch*.

**dispel** means to drive away in different directions, to disperse, and is used literally (*dispel clouds, fog*, etc.) and with generalized abstract nouns (*dispel fear, notions, suspicions*, etc.). It is not idiomatically used with an indivisible entity as object (*dispel an accusation, a rumour*, etc.); other verbs (*rebut, refute*, etc.) are available for such contexts.

**dispensable.** Thus spelt. See -ABLE, -IBLE 6.

**dispenser.** Thus spelt, not -*or*. See -OR.

**disposable.** Thus spelt. See -ABLE, -IBLE 6.

**disposal, disposition.** In some contexts there is no choice: *His disposition* [ = temperament, natural tendency] *is merciful; the disposal* [ = the getting rid of] *of the empty bottles is a problem*. The phr. *at one's disposal* no longer freely alternates with *at one's disposition* (as it did from the 17c. to the early 20c.). In most other contexts the choice depends upon the sense required. When doubt arises, it is worth remembering that *disposition* usually corresponds to *dispose*, and *disposal* to *dispose of*. So *the disposition of the books is excellent* (they are excellently disposed, i.e. arranged), but *the disposal of the books was soon managed* (they were soon disposed of, i.e. either sold or got out of the way); *the disposition of the shoulders reflected his parade-ground training* (the shoulders were disposed in a military manner), but *the disposal of the body proved impossible* (it could not be disposed of, i.e. buried or concealed).

**disproven.** It is a mild curiosity that *disproven* no longer alternates with *disproved* as pa.pple as it did until about the end of the 19c., despite the continuing currency in certain circumstances of *proven* beside *proved*. See PROVED, PROVEN.

**disputable.** Stressed either on the first or the second syllable. Second-syllable stressing now seems to be the more usual of the two.

**dispute.** Both the verb and the noun are stressed on the second syllable. In the 1970s and 1980s, first-syllable stressing of the noun came into prominence in the language of northern trade-union leaders, but this practice has not established itself in standard English.

**disremember.** A chiefly dialectal word both in Britain and in the US. The OED lists nine contexts in which it has been used instead of 'forget' in a recognizably regional manner (in the work of Maria Edgeworth, Mrs Gaskell, M. K. Rawlings, and others), and one, notably, in a poem by G. M. Hopkins (*Quite Disremembering, disremembering! all now*), from 1815 onward.

**dissect.** First syllable /dɪs-/, not as in *bisect, dice*.

**dissemble.** See DISASSEMBLE.

**dissimilar.** Construed with *to*, less frequently with *from*: (to) *the picture was dissimilar to all the others hanging in the gallery; He underwent a revelation not dissimilar to St Paul's on the road to Damascus—*Godfrey Smith, 1984; (from) *an entirely new style of coinage not dissimilar from the Roman.*

**dissimilation.** In phonetics, 'the influence exercised by one sound segment upon the articulation of another, so that the sounds become less alike, or different' (Crystal, 1980). In the nature of things this phenomenon is usually observable only after it has happened. It accounts for such words as *pilgrim* (from L *peregrīnum* 'one who comes from foreign parts', which in the Romance languages became, by dissimilation of *r* . . . *r, pelegrin(o)*, whence Fr. *pélerin*) (OED); and Ger. *Kartoffel* 'potato', earlier *tartoffel*, where the initial *k* replaced *t* in the 17c. under the influence of the following *t*. Cf. ASSIMILATION; FEBRUARY.

**dissociate.** First recorded in 1623, slightly later than its variant *disassociate*, it is now the more favoured form. Examples: *The fourth danger that a person who dissociates himself from the mass must confront is that he will become suspect to it—*Encounter, 1981; *The mother immediately dissociated herself from this conversation—*V. Glendinning, 1989; *He is at pains to dissociate Reagan's party from the one he helped steer to victory in 1968—*NY Rev. Bks, 1990.

**dissociation.** See -CIATION.

**dissoluble, dissolvable**. **1** Pronounce /dɪ'sɒljʊbəl/, /dɪ'zɒlvəbəl/.

**2** *Dissoluble* is the general word meaning 'capable of being separated into elements or atoms, decomposable'. *Dissolvable* is mainly reserved for contexts in which it means 'make a solution of in a liquid' (*sugar* is *dissolvable* or *dissoluble in water*), though it also occurs in other senses (*a Chamber dissolvable* or *dissoluble at the Minister's will*). See -ABLE, -IBLE 5.

**dissolute.** See -LU-.

**dissolve.** Pronounce /dɪ'zɒlv/.

**dissyllable.** See DISYLLABLE.

**distaff.** Originally a cleft stick on which wool or flax was wound in the process of spinning, the word *distaff* came to be used (from the 14c. onward) as a type of women's work or occupation, and hence, symbolically, for the female sex and the female branch (or line) of a family. The term *female line* seems to be preferred by scholarly genealogists (e.g. A. R. Wagner, 1960), but it is not unusual to hear the term *distaff side* used by gentlemen in after-dinner speeches or the like. Example: *I could tell you a thing or three. Always a great one for the distaff side, could never resist a bit—*G. W. Target, 1960.

**distendible, distensible.** For the sense 'capable of being distended', only *distensible* is now in current use. *Distendible* became obsolete in the 18c.

**distich.** In prosody, a pair of verse lines, a couplet. Pronounce /'dɪstɪk/. See -STICH.

**distil.** After much fluctuation since the word was first recorded in the late 14c., it has now settled down in BrE as *distil*, and in AmE mostly as *distill*. The oblique forms in both countries are *distilled, distilling*. See -LL-, -L-.

**distinct, distinctive.** *Distinct* normally means 'separate, different, not identical' (*the word has several distinct meanings*) and is often construed with *from* (*absolute as distinct from relative knowledge; holiness is quite distinct from goodness*). It also has a range of related senses, e.g. 'unmistakable, decided' (*she had a distinct impression of being watched*). By contrast, *distinctive* means 'distinguishing, characteristic' (*wearing a distinctive dress; the distinctive role of Anglo-Saxon bards*). Fowler (1926) claimed that the two words were 'often misused', but I have no evidence of such misuse. Examples of both words: (distinct) *'Ideal' Milk comes to you in a modern, raised rim tin—this is a distinct advantage which makes for easier opening with any kind of tin-opener—Woman's Illustrated,* 1960; *A very distinct . . . small tree known readily by the bark*—A. Mitchell, 1974; *I can still hear her distinct, rather emphatic, very self-assured speech*—R. Cobb, 1983; (distinctive) *The gods and goddesses symbolise specific beliefs and many of them have highly distinctive personalities*—K. Crossley-Holland, 1980; *Everyone who knew the Temple School will remember the distinctive smell of Freddie's office*—P. Fitzgerald, 1982; *Her main 'discovery' . . . was the distinctive way in which Marx had challenged all previous political traditions*—D. May, 1986.

**distinctly** followed immediately by an adj. or adjectival phr. was apparently a vogue word between the 1890s and the time of publication of *MEU* (1926). The OED cites an American source of 1893: *Now the favourite slang word of literature is 'distinctly'. Heroines are now 'distinctly regal' in their bearing, and there is about the heroes a manner that is 'distinctly fine'.* Fowler was scornful: *'Distinctly,* in the sense *really quite,* is the badge of the superior person indulgently recognizing unexpected merit in something that we are to understand is not quite worthy of his notice: *The effect as the procession careers through the streets of Berlin is described as distinctly interesting.'* This rough-hewn edge to the word seems to have disappeared. Examples: *That night the singing was distinctly husky and out of tune*—J. B. Morton, 1974; *Without an unembarrassable sense of humour she could not have completed a fifth witty book on a distinctly unglamorous and painful subject*—Sunday Times, 1988. Perhaps the older flamboyancy of the word has been transferred to other adverbs, e.g. on Radio 4 (6 Dec. 1990) the publisher Paul Hamlyn described publishing as *madly exciting.*

**distract.** See DETRACT

**distrait.** This adj., pronounced in a quasi-French manner as /dɪˈstreɪ/, is used of persons of either sex and of things anywhere along the band of meaning 'not paying attention, absent-minded' to 'distraught'. It entered the language in the 14c. (when it meant 'excessively perplexed or troubled') but it has always

resisted Anglicization, even to the extent that when used of females it sometimes retains its French form *distraite* (with the second *t* fully pronounced). Examples: *Mrs Arbuthnot's distrait expression made her look, as Miss Fisher said, quite above the world, but people did as she liked*—E. Bowen, 1935; *He was usually distrait, in pursuit of one of his creative dreams*—R. Church, 1955; *Did he feel pointless, feeble and distrait, Unwanted by everyone and in the way?*—Stevie Smith, 1975.

**distribute.** The main stress falls on the second syllable, though first-syllable stressing is now increasingly heard in the speech of otherwise standard speakers.

**distributive.** In grammar: 1 Numerous words have a distributive function when used with a following temporal noun: *once, twice, three times, four times,* etc., *a day, a week, a month,* etc. In such phrases *a* can usu. be replaced by *every* or *each,* and (not pleasingly) by *per.*

2 (Of adjs. and pronouns) 'having reference to each individual of a number or class, as distinguished from the whole number taken together' (OED), as *each, every, either, neither.* They normally govern a singular verb.

3 What is called the 'distributive plural' is shown in the following examples: *we watched the proceedings that led to her resignation with sadness uppermost in our minds* (each person has only one mind); *they wear gowns when the Vice-Chancellor is present* (each person wears only one gown). But in such contexts a singular noun is often idiomatic: *the parents were charged with causing the death* (or *deaths*) *of the two children; they all have a good appetite* (or *good appetites*); or even obligatory in order to avoid ambiguity: *children under seven must be accompanied by a parent* (only one parent is required).

**distributor.** Thus spelt in all uses, not -*er.* See -OR.

**disyllable.** The better spelling, not *dissyllable,* though the spelling with *diss*-'was universal in the 17–18th c.'. (OED) in imitation of Fr. *dissyllabe.* (In French the *s* was doubled in accordance with its pronunciation as /dis-/ not /diz-/.) It answers ultimately to L *disyllabus,* where the prefix *di-* means twice.

**ditransitive.** A ditransitive verb is one that appears to take two direct objects (without the intervention of a preposition): *he gave the baby a bottle* ( = a bottle to the baby); *he resented the author his talent if he had any*—H. Jacobson, 1983. See AS 3(*b*); FACTITIVE VERB.

**ditto.** Pl. *dittos*. See -O(E)S 3. Verbal forms *dittoed, dittoes, dittoing.*

**diurnal.** Its principal meanings are 'by day', as opposed to *nocturnal* 'by night'; or 'occupying a day', as opposed to *annual* ( = during a whole year, e.g. *annual rainfall), monthly*, etc. It is hardly comfortable as a substitute for *daily*: *daily paper, daily tasks*, etc., could only jocularly be called *diurnal paper, tasks*, etc.

**divan.** See COUCH. The recommended pronunciation is /dɪ'væn/.

**dive.** In Britain and in most other English-speaking areas the pa.t. is invariably *dived*, but *dove* /dəʊv/ has been noted in regional use in AmE since the mid-19c. and this form continues to turn up in fairly standard AmE sources today. Examples: *Straight into the river Kwasind Plunged as if he were an otter, Dove as if he were a beaver*—Longfellow, 1855; *He dove in and saved her life*—F. Scott Fitzgerald, a1940; *Forest Hill struck first when Mike Brown dove on a loose ball*—Toronto Daily Star, 1970; *She dove in and rolled onto her back*—New Yorker, 1988; *The plane ducked and dove, the lights went out*—New Yorker, 1989.

**divers, diverse.** The two words were not differentiated in ME (both being from OF, ult. from L *diversus* 'contrary', different, orig. 'turned different ways'; cf. *vertere* 'to turn'), but they gradually drew apart. *Divers*, pronounced /'daɪvəz/, means 'several, sundry', and is gradually falling into disuse (*divers colours*, i.e. several different colours; *divers workmen*, i.e. workmen of various kinds). There is a serial publication called *Essays by Divers Hands, being the Transactions of The Royal Society of Literature of the United Kingdom* (1921–   ). *Diverse* remains in frequent use as an everyday word, usu. pronounced /daɪ-'vɜːs/, meaning 'unlike in nature or qualities, varied' (*with habits so diverse; the diverse but not contradictory tenets of Christianity*).

**divest.** The traditional uses of *divest* (to undress, to disrobe; *to divest oneself of*, to dispossess oneself of) have been joined since the 1950s, first in AmE and then elsewhere, by the financial senses 'to sell off (a subsidiary company); to cease to hold (an investment)'. The word and also its corresponding noun *divestment* seem to fit neatly enough into the fast-moving vocabulary of modern finance: *A 1966 decree requiring Von's Grocery Stores to divest a certain number of required stores . . . resulted in divestment of its forty least profitable outlets*—NY Law Jrnl, 1973; *She advises American people to divest their investments in South Africa*—Christian Sci. Monitor, 1986.

**divisible.** Thus spelt, not -*able*. See -ABLE, -IBLE 7.

**divorcee.** First recorded in the early 19c., it is normally pronounced /dɪvɔː'siː/. This is the regular word in English, but on occasion the Fr. forms *divorcé* (divorced man) and *divorcée* (divorced woman) also occur in English.

**dj-.** This initial combination of letters was sometimes used in the past as a transliteration of foreign letters having the sound /dʒ/. Numerous words from Arabic, Turkish, and Berber acquired currency in English with initial *dj* (*djebel* 'mountain, hill'; *djinn* 'spirit power over people'; etc.), but these have all been respelt (normally with initial *j*) in modern English.

**djinn.** See JINNEE.

**do.**

| 1 As main verb. |
| 2 As auxiliary. |
| 3 *don't have* = *haven't got*. |
| 4 Contracted and uncontracted forms. |
| 5 *I don't think.* |
| 6 *did, does* at head of subordinate phrase. |
| 7 Non-standard uses. |

*Do* is one of the most complicated verbs in the language. One must first distinguish its use as a full verb with multiple meanings from its use as an auxiliary. (Unless otherwise specified, all the examples below are drawn from two novels, namely André Brink's *States of Emergency* (1988) and Anita Brookner's *Lewis Percy* (1989), but they could as easily have been drawn from almost anywhere.) **1** As main verb. In practice *do* is

used as a main verb much less frequently—the proportions vary from book
to book but are of the order 1:10—than
it is used as an auxiliary. Examples as a
main verb: *we used to do most things together* ( = carry out, perform); *anything
will do* ( = suffice); *I just brought you an
article I've done* ( = written); *I wanted to
bring them up to do you proud* (idiom); *it
can't do any harm* (idiom).

**2** As auxiliary. *Do* used as an auxiliary
is a much more complicated matter
though, paradoxically, it seems always
to fall into place without any particular
difficulty. Some of the main types: (a)
In interrogative and negative contexts
followed by a plain infinitive: *Did I somehow dread the prospect of returning?*; *Do you
realize this is the first time you've ever told
me anything about yourself?*; *I didn't know
(still don't) how to cope with the fact*; (with
uncontracted negative) *But does not the
fate of her small novella demonstrate the
futility of such an enterprise?* (b) In affirmative, including imperative, statements,
with varying degrees of emphasis: *It really
did seem as if a better world could still come
about*; *if you do choose me it is a different
me you'll have to contend with in future*; *But
do remember, won't you, that we live rather
a long way away.* (c) With inversion of
subject in various circumstances: *Only
after what seems to be hours ... do they
manage to shake off their pursuers*; *Such
arrested innocence affected him painfully, as
did her large eyes*; *So anxious did this make
him that he was determined never to court
such condemnation again.* (d) In tag questions: *And when you and I talk about history
we don't mean what actually happened, do
we?*—P. Lively, 1987; *You like Pen, don't you?*
—A. Brookner, 1989; *Yet I knew you at once,
didn't I?*—ibid. (e) Substitutive *do*; also, on
occasion, *do so*, i.e. contexts in which *do*
or *do so* refers back to the predicate of a
preceding clause: *who had married the
ungenerous Susan because that was what men
did*; *she might have spent pleasant harmless
days, as women of her generation were accustomed to do*; *So he gave her a coin and
playfully instructed her to telephone him the
day she turned eighteen. Which, against all
odds, she did*; *if that meant accepting George
... he was perfectly willing to do so*; *Yet this
marriage was by now so established that he
had no choice but to continue it. There was
no good reason not to do so.* (f) In BrE but
less commonly in AmE, *do* can also be

used intransitively (esp. after *have* or a
modal verb) to represent an earlier predicate: *We all get on enormously well and have
done for the last six weeks*—Susannah York,
1986; *Adults don't believe children but they
should do*—TV programme, 1986 (J. Algeo
in *Internat. Jrnl Lexicography*, 1988). (g)
Fowler (1926) rightly pointed out that *do*
must not be used as a substitute for a
copulative *be* and its complement: *It
ought to have been satisfying to the young
man, and so, in a manner of speaking, it did*
(read *was*).

**3** *don't have* = *haven't got.* 'American
writers seem to use the *do* construction
in many cases in which Britishers would
use *have got*' (Jespersen, 1940). After a
complex history in the 19c. (Visser, 1978,
pp. 1558 ff.), *do* + negative + *have* =
*haven't got* became firmly established as
American. It is now very commonly used
in most other English-speaking countries
as well, but still retains its American
flavour. Examples: (US) *We don't have any
beer. Just red wine. Sorry.*—New Yorker, 1986;
*But I don't have my pack with me*—J. Updike,
1986; *Well, I'm very sorry, but we don't have
time for that now*—T. Wolfe, 1987; (outside
America) *But you don't have a car*—M. Duckworth, 1986 (NZ); *Bathrooms in the bedrooms. Cocktail bars. Things we don't have*—A.
Lejeune, 1986; *we don't have that kind of
thing in my house, man*—A. Brink, 1988
(SAfr.); *'Don't you have central heating?' Clare
asked*—Francis King, 1988; *I don't have any
standards myself*—A. Brookner, 1989.

British visitors to America will have
encountered the following illogical use
of *do* as a substitutive verb: Q. *Have you
got a spare room?* A. *Yes, we do.* (It arises
because the question expected is 'Do
you have a spare room?'). Similarly (AmE
only), *'You've got to go.' 'No, I don't.'*

A different kind of 'illogicality' is
shown in a remark by John Updike on
BBC Radio 3 (18 June 1989) explaining
that he had always wanted to live in
New York but had only been able to do
so in the person of a character in his
novels: *When you can't be everywhere one
way is to invent a man who does.*

**4** Contracted and uncontracted
forms. The contracted forms *don't, didn't,*
and *doesn't,* none of which was recorded
in print before the later part of the 17c.,
are now customary in the representation
of speech, but by convention are still

hardly ever used in ordinary descriptive prose. Examples: (in conversations) *'Doesn't she ever go out, then?'* asked Lewis; *'Why don't you do that?'*; *'I didn't know what could have happened to you.'*; (uncontracted) *He did not think he could tell her about the stunning monotony of his everyday life*; *What she lived on Lewis never knew, but money did not appear to be a problem*; *The story I should like to write does not allow me to focus only on itself.*

**5** *I don't think.* Nowadays it passes almost without notice that in the phrase *I don't think* the negativity has often been 'illogically' transferred from where it 'properly' applies: *I don't think I've ever met anyone like you* ( = I think I have never, etc.); *The youth honestly didn't think Vuyani would survive his detention*; *I don't think he likes to be teased.* In each case the subject *is* thinking, and the negativity belongs in what follows. The monarchs of usage used to complain, but the migration of the negativity in such circumstances is now irreversible.

**6** *did, does* at head of subordinate phrase. This emphasizing use, which, according to Visser (1984), is not recorded before the 18c., is shown in the following examples: *He speaks uncommonly well, does Casaubon*—G. Eliot, 1872; *Never saved a cent, did old Don Juan*—H. A. Vachell, 1899; *It's interesting, she likes the old books, Dickens and Jane Austen, does my old lady*—K. Amis, 1988; *He's not that pathetic, you know, in that suit, and it only goes so far, does being pathetic*—ibid.; *She always knows better, does our Grace.*—Rosie Scott, 1988 (NZ); *He does have a sense of humour does Mr Marr*—Nigel Williams, 1992.

**7** Non-standard uses. (*a*) *done* as irregular pa.t. This use is common in regional and in uneducated speech in Britain and elsewhere. Examples: *I think it done him good*—M. Twain, 1873; *After what they've done to me, I never could forgive them. And I never done anybody any harm*—Listener, 1969; *'As for who done it,' he added, 'we have lots of suspects and lots of motives.'*—NY Times, 1987; *It can be seen that the past tense form* done *is reported by virtually all the schools in the south of the country* [sc. Britain], *but that it is less widespread elsewhere* [in Britain]'—J. Cheshire, 1989. (*b*) *don't* = *doesn't.* This illiterate use is illustrated in the following example: *I don't have to tell you, he's not the greatest orator in the*

*world, but that don't—doesn't—it was time to change gears in the third person singular— necessarily make any difference in a criminal trial*—T. Wolfe, 1987. Other examples: *It don't take ten thousand acres here to support one family*—O. W. Norton, 1862; *A man what don't profit from all a woman's telling and hiding the bottles ain't worth the trouble*—K. Tennant, 1946. (*c*) *done* in US dialect. First noted in US records in the early 19c., *done* as a perfective auxiliary or with adverbial force in the senses 'already, completely' is recognized as being a colourful regionalism but has failed to make its way into general use. Examples: *He had done gone three hours ago*—US source, 1836; *People have done forgot they had any Injun blood in 'em*—J. H. Beadle, 1873; *I don't know what you need with another boy. You done got four*—E. T. Wallace, 1945. (*d*) See DIDN'T OUGHT.

**do** (the musical note). Use *doh.*

**docile.** As recently as 1977, *Everyman's English Pronouncing Dict.* listed /ˈdɒs-/ as an alternative pronunciation for the first syllable, but the only standard pronunciation now in BrE is /ˈdəʊsaɪl/. Perhaps the most frequent pronunciation in AmE is /ˈdɑːsəl/. See -ILE.

**dock.** In BrE, an artificially enclosed body of water for the loading, unloading, and repair of ships (often called a *dry dock*), and also (in pl., also *dockyards*) the whole area concerned with the building, fitting, and repair of ships; but in AmE, a ship's berth, a wharf.

**doctor.** See PHYSICIAN.

**doctrinal.** The pronunciation /dɒkˈtraɪnəl/ is recommended, but in AmE /ˈdɑːktrənəl/.

**docu-.** Sliced from the word *documentary,* the element *docu-* has produced a crop of recently formed words in the entertainment business signifying entertainment that is a mixture of fact and fiction, e.g. *docudrama* (a dramatized documentary film, first recorded in 1961), *docu-fantasia* (1980), *documusical* (1974), *docusoap* (1990), *docutainment* (1983). Mainly confined to AmE.

**dodo.** Pl. *dodos.* See -O(E)S 1.

**doe.** See BUCK.

**do(e)st.** If for any reason this form of the verb *do* should be called on in a modern English context, it should be spelt *dost* as an auxiliary but *doest* as a finite verb.

**dogged.** As adj. ( = tenacious) always /'dɒgɪd/, but as the pa.t. and pa.pple of the verb *dog* (*he was dogged by misfortune*) always /dɒgd/, i.e. monosyllabic.

**doggie** is recommended for the noun (pet form for a dog) and *doggy* for the adj. (of or like a dog).

**dogma.** The old variant pl. *dogmata* (answering to the Gk and L pl. *dogmata*) was frequently used in English from the 17c. to the 19c. but is obsolete now. The ordinary pl. is *dogmas*.

**doh.** See DO (the musical note).

**doily,** a napkin, thus spelt, not *doiley, doyley, doyly, d'oyley, -ie*. In origin it is an eponymous word, named after a 17c. person having the surname Doiley or Doyley who, according to Samuel Pegge, kept a linen-draper's shop in the Strand, London.

***dolce far niente.*** This It. phrase (the individual words mean 'sweet do nothing') meaning 'pleasant idleness' was placed among 'Battered ornaments' as a 'foreign scrap' by Fowler (1926). It seems a slightly harsh judgement on a serviceable phrase used e.g. by Byron and Longfellow. Modern example: *The Austrians are not as efficient as the Germans, and they know it; they have too much charm, a concern with* dolce far niente—A. Burgess, 1986.

***dolce vita.*** This It. phrase (lit. = 'sweet life') meaning 'a life of pleasure or luxury' was first recorded in English in the 1960s and is particularly associated with people who, for one reason or another, reside abroad. Examples: *It didn't work,* la dolce vita *did not come packed with the detergent inside the new washing machine*—M. French, 1977; *Its expose of the* dolce vita *lifestyle of 30 years ago when it was written has lost its impact*—Stage, 1988.

**dole** (grief). A long-standing word in the language (adopted from OF in the 13c.), it is now seldom encountered except in poetry, and rarely then, perhaps forced out by the unrelated native word *dole* (payment made to an unemployed person). Examples: *To change Torment with*

*ease, & soonest recompence Dole with delight*—Milton, 1667; *Earth's warm-beating joy and dole*—E. Barrett Browning, 1850. On the other hand, the related adj. *doleful* is still in common use: *There are few more doleful sounds than the hearty laughter of a man without humour*—M. Holroyd, 1974; *The doleful figure of the Black Widow, as she haunted the macadam paths of the common*—R. Cobb, 1983.

**doll's house.** Thus in BrE, but *dollhouse* in AmE.

**dolour,** a literary word for 'sorrow, distress', is thus spelt in BrE, and as *dolor* in AmE. See -OUR AND -OR.

**domain.** For synonymy, see FIELD. See also DEMESNE.

**Domesday** (in full **Domesday Book**). Always pronounced /'du:mzdeɪ/ and therefore not distinguished in pronunciation from *doomsday*, the day of the Last Judgement.

**domestic.** 1 In Fanny Trollope's day in the first half of the 19c. and until more recent times *domestic servants* worked for their masters or their mistresses as part of a process called *domestic service*. Such people were also called *domestics* and *servants*. Changes of attitude in the 20c. have largely driven these terms out of use, and they have been replaced by a range of terms that are intended to reduce the social divisions once taken for granted: (*daily*) *help, au pair, domestic help, mother's help, resident companion/cook/ housekeeper*, etc.

2 *Domestic partners.* A term sometimes used (my evidence is American) for a living-together unmarried couple, whether heterosexual or homosexual.

**domesticity.** The pronunciation /ˌdɒme-'stɪsɪtɪ/ is recommended, not /ˌdəʊm-/.

**domicile.** Another word that is pronounced with final /-aɪl/ in BrE and either /-əl/ or /-aɪl/ in AmE. Cf. *docile, missile*, etc.

**domino.** Pl. *dominoes*. See -O(E)S 1.

**donate.** This back-formation from *donation*, which earlier in the 20c. was judged to be 'chiefly US', is now more or less restricted to contexts of contributing voluntarily to a fund or an institution: *It* [sc. a sparse scholarly room] *was*

*not without some pretty objects, mostly don-
ated by Gerard*—I. Murdoch, 1987; *In addition, it* [sc. General Electric] *donated a
million dollars for research on the river*—New
Yorker, 1987.

**donation.** First recorded in the 15c., it
is now mostly restricted to the sense 'a
voluntary contribution to a fund or an
institution'.

**done.** See DO (verb) 7.

**Don Juan.** J. C. Wells (1990) comments:
'In English literature, including Byron,
and usually when used metaphorically,
BrE prefers /ˌdɒn ˈdʒuːən/; in imitated
Spanish, and generally in AmE, /ˈwɑːn/ *or*
/ˈhwɑːn/.'

**Don Quixote.** Traditionally pronounced
/ˌdɒn ˈkwɪksəʊt/, but now frequently, in
imitation of Spanish, /ˈdɒn kɪˈhəʊti/.

**don't.** See DO (verb) 3, 4, 5, 7(*b*).

**doomsday.** See DOMESDAY.

**dossier.** Pronounced /ˈdɒsɪə/ and /ˈdɒsɪeɪ/
with about equal frequency. Both pronunciations are acceptable. Cf. CROUPIER.

**double.** When telephone (or other)
numbers or letters of the alphabet are
spoken in BrE the convention is to pronounce a sequence of two identical numbers or letters in the form *double four,
double r*: the number 848266 when read
out would end *double six*, and the name
*Forrest*, when read out, would be pronounced f-o-*double* r-e-s-t. In AmE the
equivalents would usu. be *six six* and
f-o-r-r-e-s-t.

**double comparison.** From the 14c. onward double comparatives and superlatives have been used occasionally for
emphasis, but since the 18c. they have
been regarded by standard speakers as
irregular, regional, or illiterate. Shakespeare used the convention repeatedly:
*a more larger list of sceptres; his more braver
daughter; this was the most unkindest cut of
all*; and Jespersen cites numerous examples from the 14c. (Chaucer) to the
early 20c. (Compton Mackenzie). The convention is still deeply embedded in local
dialects and in the language of children,
but not in standard adult speech.

**double construction.** Fowler's term
(1926) for a type of anacoluthon shown

in *They are also entitled to prevent the
smuggling of alcohol into the States, and to
reasonable assistance from other countries
to that end,* i.e. where 'to' is unjustifiably
used first to precede an infinitive and
then to precede a noun phrase. Fowler
called the process 'swapping horses'.

***double entendre.*** This expression,
adopted in the 17c. from a French phrase
that is no longer current in France,
means 'a word or phrase open to two
interpretations, one usually *risqué* or in-
delicate'. The equivalent modF phr.
*double entente* is also frequently used in
English (first recorded in 1895) in the
same sense.

**double genitive.** See DOUBLE POSSESS-
IVE.

**double meaning.** See OPPOSITE MEAN-
INGS.

**double negative.** A plain-speaking
shopkeeper explained to an interviewer
from the *Sunday Times Magazine* (12 July
1987): *I run a family business and I don't
want no hassle.* A black councillor in
Chicago, brought to trial on charges of
bribery and extortion, emerged from the
court saying: *I don't take no money from
no white folks* (*Chicago Tribune*, 19 Nov.
1990). A poorly educated person in a
novel called *Glory Days* (1988) by the NZ
writer Rosie Scott declares: *I don't have
nothing to do with them if I can help it.*
It is easy to find other examples from
various English-speaking areas: *'Clouds
come up,' she continued, ' but no rain never
falls when you want it.'*—E. Jolley, 1980
(Aust.); *He never did no harm to no one*—BBC
Radio 4, *The Archers*, 1987; *I don't give a
damn about nobody*—a black SAfr. speaker
in A. Fugard's *Tsotsi*, 1980. In all these
examples the double (or multiple) negative construction is obviously a comfortably natural way of expressing the idea.
The double negative emphasized the
negativity and did not cancel it. There
was no ambiguity, and communication
was not impeded in any way. But in the
20c. the construction is non-standard.

Repetition of uncancelling negatives
was the regular idiom in OE and ME in all
dialects. Thus, in Chaucer: *He nevere yet
no vileyne ne sayde In al his lyf unto no
maner wight*; and, later, in Shakespeare:
*And that no woman has, nor neuer none*

*Shall mistris be of it, saue I alone.* At some point between the 16c. and the 18c., for reasons no longer discoverable, double negatives became socially unacceptable in standard English. Playwrights placed them in the conversation of vulgar speakers, and 18c. grammarians like Lindley Murray roundly condemned them.

In present-day English, closely placed self-cancelling negatives are eminently acceptable if they are not overused or too intricate: e.g. *it has not gone unnoticed* = it has been noticed; *I don't feel inclined to disagree*; *a not unwelcome decision*; *I am not entirely dissatisfied.* On the other hand, the use of double or cumulative negation for emphasis is taken to be a certain indication of poor education and of linguistic deficit. But it was not always so in the past and attitudes can easily change again in the future.

**double passive.** See PASSIVE TERRITORY 3.

**double possessive. 1** *Currency.* The currency of the type *a friend of my father's* is not in question. It is called the 'postgenitive' in *CGEL*; Jespersen (*Mod. Eng. Gramm.* iii.19) says that *of*-phrases thus used should not be called partitive but 'appositional'; the *OED* describes the construction as '*of* followed by a possessive case or an absolute possessive pronoun: originally partitive, but subseq. used instead of the simple possessive (of the possessor or author) where this would be awkward or ambiguous, or as equivalent to an appositive phrase'. In practice one of its most useful functions is that it enables English speakers to distinguish between the simple types *a picture of the king* ( = an actual portrait of him) and *a picture of the king's* ( = one owned by him).

Examples: *Aunt Mary was a great admirer of hers*—G. Butler, 1983; *His aunt was a friend of my mother's*—M. Wesley, 1983; *But I was a child of hers, wasn't I?*—J. Frame, 1984; *It hardly seemed possible that the Harringtons had believed this story of Barney's*—P. Fitzgerald, 1986; *I was still undermanned ... but that was no fault of Arthur's*—P. Wright, 1987; *I am named after an old friend of my father's*—P. Carey, 1988; *Polly told me Gordon's father was an old friend of Mr. Logan's*—New Yorker, 1990. Contrast these examples in which the double possessive is not used: *Friedrich Wilhelm Keyl*

*... became a favourite pupil of Landseer*—*Bodleian Libr. Rec.*, 1987; *A shadowy girl, a niece of her late husband, was invited*—A. Brookner, 1988.

**2** *History.* Several types of double possessives emerged in the 14c. and 15c. and have been in use ever since: *A frend of his pat called was Pandare*—Chaucer, *c*1374; *A yong hors of the Quenes*—1502 in *OED*; *This was ... a false step of the ... general's*—Defoe, 1724; *I make it a rule of mine*—R. L. Stevenson, 1896.

**3** *Limitations.* It will be seen from the examples in (1) that the appositional *of*-phrase must be definite (i.e. not indefinite) and human: *a friend of my mother's* is idiomatic, but *a friend of the British Museum's* is not; *an admirer of hers* is idiomatic, but *an admirer of the furniture's* is not. It will also be observed that the phrase preceding *of* is normally indefinite (*a great admirer, a child of hers,* etc.). The only exceptions are those where the first noun phrase is preceded by the demonstratives *this* or *that* (*this story of Barney's*). It is not easy to explain why such constructions are idiomatic: one can only assert that they are.

See also OF A7, B.

**doublespeak.** (Also called **doubletalk.**) A dismaying feature of the 20c. is the steep increase in the devising of ambiguous or imprecise language used deliberately, esp. but not only in politics, as euphemisms or as a deliberate means of deceit or obfuscation. The term *doubletalk* emerged first (first recorded in 1938), followed by Orwell's *doublethink* (1949), and by *doublespeak* (1957). For similar concepts see BAFFLEGAB; GOBBLEDEGOOK.

Opposition to the use of doublespeak has noticeably increased since the 1970s. In America there is a newsletter called *Quarterly Review of Doublespeak.* Since 1974 a Committee on Public Doublespeak has made an annual award to a public utterance that is 'grossly unfactual, deceptive, evasive, euphemistic, confusing, or self-contradictory', esp. one that has 'pernicious social or political consequences' (William Lutz, 1987). The first award went to a press officer in Cambodia, who, after a US bombing raid, told reporters: *You always write it's bombing, bombing, bombing! It's not bombing! It's air support!* The nuclear power industry won the award in 1979 'for inventing a whole lexicon of jargon and euphemisms used before, during, and after the Three Mile Island accident and serving to downplay the dangers of nuclear accidents. An explosion is called "energetic disassembly" and a fire "rapid oxidation". A reactor accident is an "event", an "incident", an "abnormal evolution", etc.'

There is also an Orwell Award, given annually to a writer whose work has 'made an outstanding contribution to the critical analysis of public discourse'. Among the winners is Dwight Bolinger for his *Language: The Loaded Weapon* (1980), a work of considerable linguistic insight. Obfuscation of language continues to be a major feature of modern life. It is remorsely attacked and ridiculed by linguists and others, but the battle is far from over and looks like continuing well into the 21c.

**double subject.** The pleonastic type *The skipper he stood beside the helm* (Longfellow) has a long history in English. The *OED* presents examples from *c*1000 onward s.v. *he* 3 ('common in ballad style and now in illiterate speech'), from the 15c. to 1896 s.v. *she* 3 ('Now only *arch.* (*poet.*) and in uneducated use'), from the 15c. to the mid-19c. s.v. *it* 5 ('now esp. in ballad poetry'), and so on. This venerable construction continues to turn up occasionally but only in the special circumstances indicated by the *OED*. Examples: *The Liner she's a lady by the paint upon 'er face*—Kipling, 1896; *From time to time I clean. Mrs Pollypot she don't like cleaning*—M. Wesley, 1983; *My cousin he didn't go to college*—Jessica Williams (citing a second-language-learner), 1987.

**double superlative.** See DOUBLE COMPARISON.

**doublet.** A term of etymology meaning either of a pair of words of the same origin but which, for historical reasons, end up as two separate words. Examples: popL *cadentia (from *cadent-*, participial stem of *cadere* 'to fall' → OF *cadence* → Eng. *cadence*/ → OF *cheance* → Eng. *chance*/ medL *clocca* → OF *cloke* → Eng. *cloak*/ → OF *cloche* → Eng. *cloche*/ → MLG *klocke* → Eng. *clock*; L *factiōnem* → OF *faction* → Eng. *faction*/ → Eng. *fashion*; L *uncia* ('twelfth part') → OE *ynce* → Eng. *inch*/ → OF *unce* → Eng. *ounce*. There are a

great many doublets in English, e.g. *compute/count, dainty/dignity, fragile/frail, pauper/poor, radius/ray, secure/sure, tradition/treason.*

**doubling of consonants with suffixes.** The standard treatment of the problem is that in *Hart's Rules*, on which the present article is based. See also -B-, -BB-; -C-, -CK-; -D-, -DD-; -LL-, -L-; etc. **1** *Words of one syllable.* Those ending with one consonant preceded by one vowel (not counting *u* in *qu*) double that consonant on adding *-ed* or *-ing* unless it is *h, w, x,* or *y.*

| | | |
|---|---|---|
| beg | begged | begging |
| clap | clapped | clapping |
| dab | dabbed | dabbing |
| squat | squatted | squatting |

but

| | | |
|---|---|---|
| ah | ah-ed | ah-ing |
| tow | towed | towing |
| vex | vexed | vexing |
| toy | toyed | toying |

This rule also applies to the suffixes *-er* and *-est*:

| | | |
|---|---|---|
| fat | fatter | fattest |
| glad | gladder | gladdest |

Monosyllabic words not ending with one consonant preceded by one vowel generally do not double the final consonant (e.g. *clamp, clamped, clamping; squeal, squealed, squealing*). Exceptions are *bused, busing* (carrying people by bus).

**2** *Words of more than one syllable.* Those that end with one consonant preceded by one vowel double the consonant on adding *-ed, -ing,* or *-er, -est* if the last syllable is stressed (but not if the consonant is *w, x,* or *y*):

| | | |
|---|---|---|
| allot | allotted | allotting |
| begin | beginner | beginning |
| occur | occurred | occurring |
| prefer | preferred | preferring |

but

| | | |
|---|---|---|
| guffaw | guffawed | guffawing |
| relax | relaxed | relaxing |
| array | arrayed | arraying |

But words of this class *not* stressed on the last syllable *do not double* the last consonant on adding *-ed, -ing, -er, -est,* or *y* unless the consonant is *l* or *g* (see -G-, -GG-):

| | |
|---|---|
| audited, -ing | blanketed, -ing |
| balloted, -ing | bracketed, -ing |
| benefited, -ing | budgeted, -ing |
| biased, -ing | buffeted, -ing |
| bigoted | carpeted, -ing |

| | |
|---|---|
| chirruped, -ing | picketed, -ing |
| combated, -ing | pivoted, -ing |
| cosseted, -ing | plummeted, -ing |
| crocheted, -ing | profited, -ing |
| ferreted, -ing | rickety |
| fidgeted, -ing | ricocheted, -ing |
| filleted, -ing | riveted, -ing |
| focused, -ing | rocketed, -ing |
| galloped, -ing | targeted, -ing |
| gossiped, -ing | thickened, -ing |
| hiccuped, -ing | thicker, -est |
| leafleted, -ing | trellised, -ing |
| lettered, -ing | trumpeted, -ing |
| marketed, -ing | visited, -ing |
| offered, -ing | vomited, -ing |

Note: The words that most frequently break loose and appear with a doubled consonant are *benefitted*, *-ing* (under the influence of *fitted*, *-ing*), *leafletted*, *-ing*, *targetted*, *-ing*, and *focussed*, *-ing*; but it is better to keep to the basic rule in these words too. *Inputting*, *outputting*, and *worshipped*, *-ing*, *-er*, on the other hand, always show a doubled consonant in BrE, as do *kidnapped*, *-ing*, *-er*, but AmE often shows a single *p* in *kidnaped*, *worshiped*, and related forms.

In words ending in -*l* the last consonant is generally doubled whether stressed on the last syllable or not:

| | |
|---|---|
| annulled, -ing | initialled, -ing |
| appalled, -ing | instilled, -ing |
| bevelled, -ing | labelled, -ing |
| channelled, -ing | levelled, -ing |
| chiselled, -ing | libelled, -ing |
| compelled, -ing | marshalled, -ing |
| counselled, -ing | modelled, -ing |
| dialled, -ing | panelled, -ing |
| dishevelled, -ing | quarrelled, -ing |
| enrolled, -ing | revelled, -ing |
| extolled, -ing | rivalled, -ing |
| fulfilled, -ing | shovelled, -ing |
| grovelled, -ing | travelled, -ing |
| impelled, -ing | tunnelled, -ing |

Exceptions: *appealed*, *-ing*; *paralleled*, *-ing*.

Note: In AmE the -*l* is usually not doubled:

| | BrE (always) | AmE (usually) |
|---|---|---|
| cancel | cancelled | canceled |
| cruel | crueller | crueler |
| dial | dialled | dialed |
| duel | duelling | dueling |
| jewel | jeweller | jeweler |
| label | labelled | labeled |
| marvel | marvellous | marvelous |
| travel | travelling | traveling |

See also -LL-, -L-.

Note: It is not uncommon to find words like *crystallized*, *swivelled*, and *unravelling* spelt thus in standard AmE sources.

**doubt** (verb). A clause depending on *doubt* (which by definition expresses uncertainty) is normally led by *whether* or (less commonly) *if* when it is in the affirmative: *I doubt whether these portraits are genuine*; *he doubted if he could convince her*. In negative constructions *doubt* is normally followed by *that*: *he didn't doubt that they were coming*. These are overwhelmingly the dominant constructions. When the main clause is negative (denying uncertainty) or interrogative (leaving the question of certainty open), *doubt* is sometimes archaistically construed with *but that* (*not doubting but that he would find him faithful*; *do you doubt but that he will do it?*) and occasionally (also somewhat archaic) with *but* alone (*I do not doubt but England is at present as polite a Nation as any in the World*—Steele, 1711). Cf. BUT 4(*b*).

Increasingly, since the last quarter of the 19c. or so, *doubt* has also come to be construed in affirmative sentences in about equal measure either with a *that*-clause or with an objective clause not led by a conjunction, i.e. a contact clause. The sense is always 'think it unlikely'. Examples: (with *that*) *I doubt that the White House is responsible for this rash of tittle-tattle*—Alistair Cooke, 1981; *I doubted that I'd ever come across a Heddle anywhere*—P. Bailey, 1986; *he doubted that so abstruse a matter could ever be explained to those who were not already experts*—New Yorker, 1986; (with a direct objective clause) *in these two books . . . death is being served up nearly raw, and who can doubt they will be followed by others?*—London Rev. Bks, 1981; *he doubted Ferrari would sue him*—New Yorker, 1986; *But Mr [Bernie] Grant doubts the government will see it that way*—Listener, 1987; *I doubt there was anything really wrong with her*—A. Brookner, 1992.

Similar considerations govern the way in which the adj. *doubtful* is construed in contexts parallel to those described above.

**doubtfulness.** See DUBIETY.

**doubtlessly.** Despite its antiquity (first recorded *c*1440) this word has always led a precarious life alongside the more usual adverb *doubtless*. It seems to be having a new lease of life in the 20c., at any rate in the US, to judge from the following examples: *These mineral grains are doubtlessly heterogeneous—Bull. Geol. Soc. Amer.*, 1955; *But with one eye doubtlessly on the White House, Bentsen also moved quickly to establish himself more in the mainstream of his party—Dun's Review*, 1975; *The current argument . . . doubtlessly offers a cogent and easily understood explanation for the current deadlock in East–West relations—Washington Post*, 1984.

**doubtless, no doubt, undoubtedly.** Fowler's comment (1926) is still valid: '*Doubtless* and *no doubt* have been weakened in sense till they no longer convey certainty, but either probability (*You have doubtless* or *no doubt heard the news*) or concession (*No doubt he meant well enough*; *it is doubtless very unpleasant*). When real conviction or actual knowledge on the speaker's part is to be expressed, it must be by *undoubtedly, without* (a) *doubt,* or *beyond a* (or *any*) *doubt* (*he was undoubtedly,* etc., *guilty*).'

**dour.** The only standard pronunciation in Britain is /dʊə/, rhyming with *tour* not *sour*; but /daʊə(r)/ is common in AmE and Australia.

**douse, dowse.** Three verbs are in question and all three have been spelt with medial *-ou-* and also with medial *-ow-* at various times since the 17c. *Douse*[1] (first recorded 1559, probably from MDu.) is now pronounced /daʊs/ and means mainly 'to put off, doff (a cap, etc.); *Naut.* to strike (a sail)'. *Douse*[2] (first recorded in 1600, of unknown origin but probably onomatopoeic) is also pronounced /daʊs/, is commonly spelt *dowse*, and means 'to throw water over; to plunge into water'. (The *OED* recognized that these two may have been 'the same word', though 'the connection is not obvious'.) *Dowse*, usu. so spelt, is the third. First recorded in 1691, it seems to be of dialect origin; it is pronounced /daʊz/ and means 'to use a divining-rod to search for underground water or minerals'.

**dove** (noun). See PIGEON.

**dove** (pa.t.). See DIVE.

**dow.** See DHOW.

**dower, dowry.** The two words, in origin the same, are differentiated in ordinary use, *dower* being a widow's share for life of her husband's estate and *dowry* the property or money brought by a bride to her husband. In older, esp. poetic, use *dower* was often used in the sense of *dowry*. Both words are used in the general sense 'a talent, a natural gift'. In practice both words are now somewhat archaic, except that *dowry* is still used with direct reference to societies (esp. in Africa) where the custom of bringing a dowry is still part of ordinary life.

**down-.** Use of the adverb as a prefix to form new verbs has been a feature of the 20c., e.g. *downface* (first recorded 1909), *downgrade* (1930), *download* (computers, 1980), DOWNPLAY (1968), *downpoint* (1946), *downscale* (1945), *downsize* (1975), *downturn* (1909).

**down-market,** of or relating to the cheaper part of the market, is now so established in the language that it is surprising to see that its first example in the OED is one of as recently as 1970.

**downplay** (verb). Another word of recent origin (the earliest example noted is one of 1954) that has moved swiftly into standard use in the sense 'to play down, to minimize the importance of'. It is most frequently encountered in newspapers.

**Down's syndrome.** Named after J. L. H. Down, an English physician (1828–96), the word has moved swiftly into medical, and then into lay, use as a replacement for the traditional term *mongolism*. The earliest example of *Down's syndrome* in the *OED* is one of 1961.

**downstairs.** Used for both the adj. (*the downstairs loo*) and the adv. (*meet me downstairs when you are ready*). As adj. stressed on the first syllable, but as adv. on the second.

**down to.** Used in the sense 'attributable to', the phrase belongs to the second half of the 20c. Example: *The boom in Gucci and Pucci and . . . Lacoste 'names' on clothes, bags, and other ornamentation is all down to the Yuppies—Sunday Tel.*, 1985. Cf. *up to* (s.v. UP 3).

**downward(s).** The only form of the adj. is *downward* (*in a downward direction*), but *downward* and *downwards* are used without formal distinction for the adverb, *downwards* being the more common of the two in Britain. Examples: (downward) *And bats ... crawled head downward down a blackened wall*—T. S. Eliot; *After that, throwing herself face downward on her bed ... she would stay there sobbing*—E. Caldwell, 1973; *Every time he looked downward he grew dizzy*—J. M. Coetzee, 1983; (downwards) *She ferreted in her bug; then held it up mouth downwards*—V. Woolf, 1922; *His small, thick writing ... running downwards on the page*—R. Lehmann, 1936; *Any teletext improvements have to be downwards compatible so that older decoders will receive something recognizable*—*Television*, 1987; *The fact that commissioners' careers only seem to go downwards after they leave Brussels has a negative effect on morale*—*EuroBusiness*, 1989.

**dowse.** See DOUSE, DOWSE.

**doyen.** See DEAN.

**dozen.** A collective noun used in two ways: (a) when preceded by a numeral, always *dozen* (*two dozen eggs*); (b) in other plural uses, normally *dozens* (*there were dozens of birds on the lawn; he made dozens of mistakes*).

**drachm** is a British weight or measure formerly used by apothecaries, equivalent to 60 grains or one eighth of an ounce or (in full *fluid drachm*) 60 minims, one eighth of a fluid ounce; abbr. *dr.* See DRAM.

**drachma,** the chief monetary unit of Greece.

**draft, draught.** *Draft*, in origin a phonetic respelling of *draught*, is used for (a) a preliminary sketch or version (*to make a rough draft*); (b) a written order for payment by a bank; (c) a military detachment. A *draftsman* is one who drafts documents. *Draught* is used in all the other common senses (game of *draughts*, air-current, ship's displacement, beer on *draught*, a dose of liquid medicine, *draught-horse*). In AmE, *draft* is used for all senses and the game of *draughts* is called *checkers*.

**dragoman.** For this non-native word ( = an interpreter or guide, esp. in countries speaking Arabic, Turkish, or Persian) the normal plural is *-mans*, but inevitably *-men* is sometimes used instead.

**dram.** 1 A small drink of spirits. 2 = DRACHM.

**drank.** See DRINK.

**draught.** See DRAFT.

**drawing.** The use of an intrusive -r- in the pronouncing of this word, thus /ˈdrɔːrɪŋ/, is not defensible.

**drawing-room.** See SITTING-ROOM.

**dream.** For the pa.t. and pa.pple both *dreamt* and *dreamed* are used; *dreamed* is usu. pronounced /driːmd/ and *dreamt* /dremt/. *Dreamed*, esp. as the pa.t. form, tends to be used for emphasis and in poetry. No decisive evidence has been established about the distribution of the two forms, but *dreamt* appears to be somewhat more common in BrE than in AmE. See -T AND -ED.

**drier** ( = more dry; one who or that which dries), **driest**, **drily** are the preferred forms in OUP house style, but *dryish*, *dryness*.

**drink.** The only standard pa.t. is *drank* and for the pa.pple *drunk*. (In past centuries there has been a great deal of fluctuation in the choice of form, and considerable variation is still found in BrE and AmE dialects.)

**drink-driving.** A fairly recent term (first recorded in 1964) for the legal offence of driving a vehicle with an excess of alcohol in the blood; hence *drink-driver*. See DRUNK-DRIVING.

**dromedary.** Now usu. pronounced with initial /ˈdrɒm-/. The variant /ˈdrʌm-/ is now seldom encountered.

**drouth** /drauθ/, Sc. /druːθ/ is a widespread (Scotland, Ireland, USA) variant of the standard form *drought*, both in the sense 'dryness of climate' and meaning 'thirst'. A miscellaneous collection of examples: *You might take your death with drouth and none to heed you*—J. M. Synge; *Scarred hand, shut eyes, and silent mouth, Parched with the long day's bitter drouth*—W. de la Mare; *There are flood and drouth Over the eyes and mouth*—T. S. Eliot; *Once in my youth I gave, poor fool, A soldier apples and water; And may I die before you cool Such drouth as his, my daughter*—R. Graves; *while I sit here with a pestering Drouth for words*—S. Heaney; *My* *heart leaps up with streams of joy, My lips tell of drouth: Why should my heart be full of joy And not my mouth?*—Stevie Smith; *A friend, a poet, wrote me near the end Of summer there had been a drouth In Wales*—J. Dressel.

**drown. 1** The pa.t. is, of course, *drowned*. The form *drownded*, first recorded in the 16c., is now found only in dialectal and uneducated speech.

**2** The now customary intransitive sense 'to suffer death by submersion in water' was considered 'unusual' by the *OED* in 1897. The verb was more commonly used at the time in its transitive sense 'to suffocate (a person or animal) by submersion in water'.

**drunk-driving.** The AmE term (first recorded in 1937) for DRINK-DRIVING; hence *drunk-driver*. The synonym *drunken driving* is sometimes used both in AmE and in BrE.

**drunk, drunken. 1** In general *drunk* is used predicatively (*judged to be drunk and disorderly*; *he was drunk when he arrived at the party*; *she was as drunk as a lord*; *Napoleon was drunk with success*) and *drunken* attributively (*a drunken brawl*; *a drunken sleep*; *a drunken landlord*). See DRUNK-DRIVING.

**2** In many contexts there is a semantic distinction: *drunk* implies 'the worse for drink at present', whereas *drunken* is capable of meaning 'often the worse for drink, given to drink'.

**dry.** For the inflected forms (*drier*, etc.), see DRIER.

**dual.** The first element in a number of widely used specific collocations of the 20c.: *dual carriageway* (first recorded in 1933), *dual control* (of aeroplanes and motor vehicles, 1913), *dual nationality* (1961), *dual personality* (1905), *dual-purpose* adj. (1914), etc. Fowler (1926) warned against the use of *dual* in general contexts where *two*, *twofold*, *double*, etc., are adequate, but this looks to have been a warning about the approach of an imaginary enemy.

**dub.** Such small words tend to be of several different origins and are apt to have many unrelated meanings. The *OED* records seven distinct nouns and five

distinct verbs all written as *dub*. The oldest noun is Scottish and northern, was first recorded in the early 16c. (Gavin Douglas), and means 'a muddy or stagnant pool'. The most recent is Jamaican in origin, was first recorded in 1974, and means 'a re-mixed version of a piece of recorded popular music'. Perhaps the most widely known use of the verb is the earliest one (a1100), 'to confer the rank of knighthood by striking the shoulder with a sword'. The most recent verb spelt *dub* is a phonetic shortening of *double*, is first recorded in 1929, and means 'to provide (a film, etc.) with an alternative sound track, esp. in a different language'.

See -B-, -BB-.

**dubbin.** The now customary form of the word meaning 'a preparation of grease for softening leather and rendering it waterproof'. It is also used as a verb (pa.t. and pa.pple *dubbined*). The older variant *dubbing* (the form preferred by Fowler) seems to be dropping out of use.

**dubiety** /dju:ˈbaɪtɪ/ is an 18c. addition to the pre-existing words *doubtfulness* (16c.) and *dubiousness* (17c.). The three words remain in the language, with *doubtfulness* being the least used of the trio. Examples: *Jung liked looking at 'the mocking visage of the old cynic', who he said reminded him of the futility of his idealistic aspirations, the dubiousness of his morals* [etc.]—*Economist*, 1976; *Take, for example, the following figures, whose dubiety can scarcely be doubted, which Carter offered up to the people several weeks ago*—*Fortune*, 1980; *I had met at the opera. It had been 10 months since that meeting. I felt somewhat ashamed of my initial doubtfulness*—*Sports Illustrated*, 1985.

**duck.** When used collectively it is often unchanged in form (*a flock*, etc., *of duck*), but in ordinary plural contexts normally *ducks* (*several ducks came for the bread that we threw into the lake*). It is used without the indefinite article as an item of food (*we had duck for dinner*).

**dues.** See TAX.

**due to. 1** When considering the merits of this much-discussed phrase it is as well to bear in mind that there are various circumstances in which the sequence *due* + *to* is uncontroversial: ( = owing or payable to) *Then pay Caesar what is due to Caesar, and pay God what is due to God*— Matt. 22: 21 (*NEB*, 1961); ( + infinitive = likely to, supposed to, announced as) *Once ... he advised me to buy Caledonian Deferred, since they were due to rise*—F. M. Ford, 1915; *It was due to start at four o'clock, but didn't begin until twenty past*—W. Trevor, 1976; *the train is due to arrive at 12 noon*, ( = properly owed to) *So much is due to the wishes of your late husband*—Lytton, 1838; *Herr Wodenfeld hovered on the sidelines exhibiting that cunning deference due to one who had been deposed*—B. Rubens, 1987; (following the verb *to be* = ascribable to) *Death had come, and the doctor agreed that it was due to heart disease*—E. M. Forster, 1910; *Humanity can be due to laziness, as well as kindness*—V. Woolf, 1921; *This development was due directly ... to the power of advertisement*—E. Waugh, 1932; *Almost everything that distinguishes our age from its predecessors is due to science*—B. Russell, 1949; *part of her happiness, her unaltered sense of her own superiority, was due to a sense of virginity preserved*—A. Brookner, 1988. In all these uses *due* is construed as an adj. + complement (or + a *to*-infinitive).

2 Used as a prepositional phrase in verbless clauses = owing to, *due to* was described as 'erroneous' by W. A. Craigie (1940) and was said by Fowler (1926) to be 'often used by the illiterate as though it had passed, like *owing to*, into a mere compound preposition'. Hostility to the construction is an entirely 20c. phenomenon. Opinion remains sharply divided but it begins to look as if this use of *due to* will form part of the natural language of the 21c., as one more example of a forgotten battle. Examples: *Largely due to the defence efforts of the Western Powers, Europe was in a state of stalemate*—*Times*, 1955; *Michael ... hated mathematics at school, mainly due to the teacher*—*TES*, 1987; *The BBC serialisation of* The History Man *... caused something of a stir a few years back, partly due to the excellence of the book itself*—*London Rev. Bks*, 1987.

**due to the fact that.** Complex prepositional phrases ending in *the fact that*, e.g. *because of the fact that, due to the fact that, on account of the fact that*, can usually be replaced by the simple conjunction

*because*. But the use persists in writers of good quality: *Part of this* frisson *... is undoubtedly due to the fact that woman as a whole has been seen as a pacifying influence throughout history*—Antonia Fraser, 1988.

**duet, quartet,** etc. It now seems remarkable that this group of musical words were commonly spelt *duett, quartette* (also *-ett), quintette* (also *-ett), etc., from the 18c. onward to the point that the *OED* listed the main variants as joint headwords.

**dullness, fullness.** Use *-ll-*, as in all other words in which *-ness* follows a word ending in *-ll* (*drollness, illness, shrillness, smallness, stillness, tallness*, etc.). But be prepared to find forms with a single *-l-* in 19c. and earlier works: e.g. *which dulness in apprehension occasioned me much grief*—R. M. Ballantyne, 1858.

**dumbfound(ed).** Thus spelt, not *dumf-*.

**dump** (noun). See RUBBISH.

**dunno** /ˈdʌnəʊ/. First recorded in 1842 as a phonetic representation of a reduced form of *do not* (or *don't*) *know*, it is widely used in fiction and plays in the speech of vulgar or illiterate people: *He was fencing before we come here. Where? I dunno*—M. Eldridge, 1984; *'Now it's back the way it used to be.' 'Why? ...' 'Dunno, sweet. Do not know.'*—*New Yorker*, 1986. See DEMOTIC ENGLISH.

**duodecimo** (size of book). Pl. *duodecimos*. See -O(E)S 6.

**duologue,** a dramatic piece spoken by two actors. An irregular formation (first recorded in the 19c.) from L *duo* or Gk δύο 'two' after *monologue*. Cf. DIALOGUE.

**durance** /ˈdjʊərəns/. A distinctly archaic word (originally meaning 'duration') for imprisonment, surviving chiefly in the half-remembered set phr. *durance vile* 'forced confinement'.

**duress** /djʊəˈres/ is the only current form (not *duresse*). It originally meant 'hardness', but now survives mainly in the phr. *under duress* 'under compulsion'. Such compulsion to do something against one's will may be by actual imprisonment, but is commonly used more generally ( = under a threat of physical violence, etc.).

**durst.** See DARE 3.

**dustbin** (rubbish tin), **dustman** (rubbish collector) are normal terms in BrE, but are not used in AmE and are not common in other English-speaking areas. See RUBBISH.

**Dutch.** The word has a long and complicated history since it was first recorded in the 15c., but its primary reference now is to the inhabitants and the language of the Netherlands. Older attitudes or practices are reflected in various set phrases, ranging from the quaint to the disagreeable, as *Dutch cap* (a contraceptive pessary, first recorded in 1922), *Dutch courage* (bravery induced by drinking, 1826), *Dutch treat* (one in which each person present contributes his or her own share, 1887), *to go Dutch* (ditto, 1914), *double Dutch* (gibberish, 1789), etc. They are taken with good grace by the Dutch themselves, but the mood of the late 20c. is to discourage the use of derogatory expressions reflecting ancient national stereotypes.

**duteous, dutiful.** Both words were formed in the 16c. and of the two Shakespeare seems markedly to have preferred *duteous*, though he used both (*As duteous to the vices of thy Mistris, As badnesse would desire—Lear* IV.vi.253; *You know me dutifull, therefore deere sir, Let me not shame respect—Troilus* V.iii.72). As time went on, *duteous* (like *beauteous, plenteous*) fell into restricted literary use, while *dutiful* (like *beautiful, plentiful*) became an everyday word.

**dutiable.** See -ABLE, -IBLE 2.

**duty.** See TAX.

**dwarf.** The traditional pl. form in BrE is *dwarfs*, but *dwarves* is now increasingly being used, as it has been for a long time outside the UK. I have examples of *dwarfs* from BrE (C. S. Lewis, P. Lively, local and national newspapers) and AmE (R. Merton, P. Roth) sources; and of *dwarves* from BrE (A. Brookner, J. R. R. Tolkien, local and national newspapers), AmE (J. Updike, a Chicago newspaper), NZ (M. Gee, Rosie Scott), and S. African (J. M. Coetzee, D. Lessing) sources.

See -VE(D), -VES, etc.

**dwell** in the sense 'live, reside' (a use first recorded in the 13c. and frequent until the 19c.) is now more or less restricted to literary contexts. For the pa.t. and pa.pple *dwelt* (not *dwelled*) is recommended. See -T AND -ED.

**dye** (verb) has the inflected forms *dyed*, *dyes*, *dyeing* to avoid confusion with the conjugational forms of *die* (*died*, *dies*, *dying*).

**dyke, dike.** In the sense 'embankment', *dyke* is the preferred form (*COD*, 1995). The different word *dyke*, a 20c. word for a lesbian, is also usu. thus spelt.

**dynamic, dynamical.** Both words date only from the 19c., but since then *dynamic* has become the more usual word of the two in general contexts. *Dynamical* has tended to become restricted to certain technical expressions, esp. in the field of dynamics (a branch of physics).

**dynamo.** Pl. *dynamos*. See -O(E)S 5. In origin it is an abbreviation of *dynamoelectric machine*.

**dynast, dynasty.** Usu. for the first and overwhelmingly for the second the opening syllable is pronounced with /ˈdɪn-/ in BrE, but with /ˈdam-/ in AmE.

**dysentery.** In BrE it is so spelt and is stressed on the first syllable: thus /ˈdɪsəntrɪ/. In AmE stretched out as four syllables.

# Ee

**-e-.** See MUTE E.

## each.

1 Number of, and with, *each*.
2 *each other.*
3 *between each.*
4 *each's.*

**1** Number. *Each* as a pronominal subject is invariably singular, even when followed by *of* + pl. noun: *There are only two ties left: each has a thistle pattern; each of the three parties has a right to a confidential briefing* (not *have*). The same is true when *each* is an adj. qualifying the subject noun: *While each man* [of three] *on the bench has enjoyed his own many or infrequent passions, each remembers only the one*—N. Shakespeare, 1989. *Each* is wrongly used with a plural verb in the following sentence: *I only found out by sheer accident that you, Alan Lloyd and Milly Preston are all trustees, and each* have *a vote*—Jeffrey Archer, 1979 (read *each has* or *all of you have*).

When *each* immediately follows a plural pronominal subject the following verb is in the plural: *We each have our own priorities* (not *has* and not *our own priority*). In such contexts the combination *his or her* would produce an unacceptable sentence (*They each have his or her own priorities*). In such cases it is necessary to recast the sentence: e.g. *Each of them has his or her own priorities*. When *each* is not the subject, but is in apposition with a plural noun or pronoun as subject, the verb (and any complement) is invariably plural: *the three parties each have a right to a confidential briefing; lettuces cost 25 pence each; we have our own priorities*.

**2** *each other.* A belief that *each other* properly refers to two people and *one another* to more than two is reflected in many contexts: *But Mrs Bentley and Kitty were delighted with each other*—A. Brookner, 1982; *They* [sc. two people] *never left each other, or the flat*—A. Desai, 1988; *I mean, Stanley and Jessica and I could cry about it. We had one another*—Susan Hill, 1988. Frits Stuurman even wrote an article in *English Studies* (1978) entitled 'Each Other—

One Another: There Will Always Prove to Be a Difference'. But the belief is untenable, as can be seen from the following departures from the 'rule': (*each other* used to refer to more than two) *Everybody knew each other or about each other*—A. Brookner, 1983; *Let them* [sc. black people] *go ahead and kill each other, that's all they know*—N. Gordimer, 1987; *KHAD, the Russians themselves, foreign governments, the competing Afghan political parties in exile—all spy on each other*—New Yorker, 1987; *The* [several] *clerks had stayed at their desks, not looking at each other*—P. Carey, 1988; (*one another* used to refer to two) *He and Gussy were evidently very fond of one another*—A. N. Wilson, 1978; *There is no such thing as complete harmony between two people, however much they profess to love one another*—A. Brookner, 1984; *the two chairs ... slightly turned towards one another*—N. Gordimer, 1987; *as we* [sc. Felicia and I] *stood there, speechless for a moment, facing one another across the threshold*—M. du Plessis, 1983; *But after a day or two the* [two] *cats became accustomed to one another*—Julian Barnes, 1989. It is worth noting that the possessive of *each other* is *each other's*, not *each others'* (*they shared each other's possessions*).

**3** *between each.* See BETWEEN 3.

**4** *each's.* An uncommon but acceptable use: *The distressing conflict between Catholics and Jews ... is driven by each's belief that the other is attempting to* [etc.]—Chicago Tribune, 1989.

**each and every.** A tempting phrase, used for emphasis, as Edward VIII (when he was Prince of Wales) found when writing to Mrs Wallis Simpson in 1935: *I love you more & more each and every minute & miss you so terribly here.* It is not common, and it is not easily replaced.

**earl.** See TITLES.

**early on.** This phrase, a kind of back-formation from *earlier on* (itself modelled on *later on*), is first recorded in BrE in 1928 and only later in AmE. Examples:

'It might have been given him earlier.' ... 'Well—not too early on, Peter.'—D. L. Sayers, 1928; Early on, he puts a coin in a newspaper-vending machine—New Yorker, 1987; He was certain, early on, that Hitler was a world menace, not just a European threat—Bull. Amer. Acad. Arts & Sci., 1989; Earlier on, religion had supplied a drug which most of the clergy were quite ready to administer—V. G. Kiernan, 1990. An American correspondent (1989) remarked 'I'm constantly struck with the immense popularity in the U.S. of the once-exclusive-to-Britain expression "early on". Hardly anyone here can now use the word "early" by itself.' Doubtless an exaggeration, but it underlines the way in which the use of the phrase is gathering pace. The earliest example of later on in the OED is one of 1882.

**earn.** The normal inflected form for the pa.t., pa.pple, and ppl adj. (earned income) is earned. The pa.pple and pa.t. are occas. spelt earnt, perhaps because in fast speech this form is often pronounced /ɜːnt/. Examples: I earned my first wages ... in the winter of 1894—P. Bailey, 1986; Hugo Young's The Thatcher Factor (Channel 4, 8pm) has already earnt itself news space—Times, 1989; a statement ... showing all the transaction details and the interest earned—Lloyds Bank leaflet, 1989; The price of rubber rose again, not to the giddy heights of the boom, but to a level which earned a profit—N. Shakespeare, 1989; Ray and Alan Mitchell once worked gruelling hours and earnt good money as contract plumbers in London—Independent, 1992. See -T AND -ED.

**earth.** Freq. with initial capital (Earth) when considered as a planet of the solar system. In such contexts, like Mars, Venus, etc., it is normally used without the definite article (but the planet Earth).

**earthen, earthly, earthy.** Earthen is used only in the literal sense 'made of earth' (either soil, or clay used as pottery, hence earthenware). Earthly has two sense, (a) of the earth or human life on earth as opposed to heavenly, terrestrial, and (b), usu. in negative contexts, remotely possible or conceivable (is no earthly use; there wasn't an earthly reason). Earthy means (a) of or like earth or soil, and (b) somewhat coarse or crude. Typical collocations include earthen floor, earthen rampart; earthen jar, earthen vessel; the earthly paradise, an earthly pilgrimage; not an earthly ( = no chance); crystalline rocks occasionally occur in friable form and are then said to be earthy; an earthy taste; strong earthy expressions, earthy humour.

**easterly, northerly, southerly, westerly.** Chiefly used of wind, and then meaning blowing from an eastward, etc., direction; otherwise these four words are used to modify words implying either motion, or position conceived as attained by previous motion: an easterly wind; took a southerly course; the most easterly part of the constellation. But the East Indies (not easterly), East Side (of New York City); the South (not Southerly) Pole; the west (not westerly) end of the church; western (not westerly) ways of thought, etc.

**eastward.** As an adj. placed before a noun (in an eastward direction), always eastward. As an adv., either eastward or eastwards in BrE, but only eastward in AmE: thus to fly eastward(s).

**easy, easily.** I don't scare easy, said a contributor to the 5 Feb. 1990 issue of the New Yorker, thus drawing attention to the occasional informal use (the OED calls it 'colloq. or vulgar') of easy as an adverb. What it is that makes Shakespeare's As easie might I from my selfe depart, As from my soule which in thy brest doth lye (Sonnet 109, 1600) acceptable, even harmonious, and the New Yorker use distinctly informal is hard to say. Some set phrases containing easy used as an adverb, all of them first recorded in the 19C., are, however, firmly embedded in standard English: to take it easy (1867), to go easy (on or with, 1850), easy does it (Dickens, 1865), stand easy (1859), etc.

**eat.** The past is spelt ate (rarely eat) and pronounced /et/ in standard English, but /eɪt/ is common in regional speech and customary in AmE.

**eatable, edible.** Both words are used of food in the general sense 'that is in a condition to be eaten'. But edible is the only word of the pair used in contexts where the contrast is with poisonous or harmful (edible mushrooms, edible snails).

**ebullient.** The pronunciation recommended is /ɪˈbʌljənt/, not /ɪˈbʊljənt/: i.e. with the sound of pulse not that of bull.

**echelon** /ˈeʃəlɒn/ bears an acute accent in French (*échelon*) and mean primarily a rung of a ladder (cf. *échelle* 'ladder'). The word was adopted in English from French in the late 18c. to mean a formation (of various kinds) of troops and still has military applications. It was a natural extension (in the second half of the 20c.) to apply the word to grades or ranks in the civil service, and to speak of 'the higher (or upper) echelons' in any organization. Meanwhile the word had made step-by-step progress in French in the same way as it did in English. Gowers (1965) was surely wrong to object to the civilian meaning as a 'slipshod extension'.

**echo.** Pl. *echoes.* See -O(E)S 1.

**echoic.** In etymology, applied to words brought into being as imitations of the sounds they represent, as *plonk*, *plop*, *twang*, *tweet.* Other terms used for this class of words are *imitative* and *onomatopoeic.*

**ecology.** A word of comparatively recent origin (the earliest example in the *OED* is one of 1873), and at first, reflecting its Greek origin (Gk οἶκος 'house, dwelling'), normally spelt *oecology* or *oekology*, it has made spectacular progress in the language because of the growing consciousness in the last decades of the 20c. that the global environment of plants and animals (including human beings) is being placed seriously at risk by industrial pollution, by the destruction of rain forests, and by other factors. There can be few people left who are unfamiliar with what a century ago was a technical word ( = the branch of biology which deals with the relations of living organisms to their surroundings) used in tracts about the evolution of man and in proceedings of biological societies. The word has inevitably become politicized with the emergence of Green (political) parties and other environmental groups. The element *eco-* has been abstracted from the word and used to form numerous combinations of greater or less durability, e.g. *eco-activist* (1969), *eco-catastrophe* (1969), *ecofreak* (1970), *eco-friendly* (1989), *ecopolitics* (1973), *eco-sound* (1989).

**economic, economical. 1** The adj. normally answering to the subject of *economics* is *economic*, and that answering to *economy* in the sense 'frugality, sparing use (of money, language, etc.)' is *economical.* An *economic rent* is one in the fixing of which the laws of supply and demand have had free play; an *economical rent* is one that is not extravagant. In practice the first generally means a rent not too low (for the landlord), and the second one not too high (for the tenant). What is now being sent to third-world countries is economic aid, not economical aid.

**2** I have been unable to establish a consensus of any kind about the pronunciation of the first syllable of the two words. It seems to be a clear case of pleasing oneself whether to say /iːk-/ (my own preference) or /ek-/. Cf. UNECONOMIC, UNECONOMICAL.

**ecstasy** is thus spelt. It is among the most frequently misspelt words in the language.

**ecu, ECU,** European currency unit, is at present being pronounced by some economists and politicians as /ˈekjuː/ and by others as /ˈeɪkjuː/.

**ecumenical.** Now always thus spelt, not *oec-* (as it was from the 16c. to the 19c., representing L *oec-* and Gk οικ-). There is great diversity of opinion about the pronunciation of the first syllable, as between /iːk-/ (my personal preference) and /ek-/.

**-ed and 'd.** When occasion arises to append the ordinary adjectival ending *-ed* (as in *a decided difference, wedded bliss, a walled garden*) to words that have a fully pronounced vowel (esp. *a,i,o*) as the final letter of their natural form, it is best to avoid the somewhat bizarre appearance of *-aed*, etc., and write *'d* instead: *one-idea'd, mustachio'd, cupola'd arch, subpoena'd witness, a shanghai'd sailor.* Practice varies, however, from publisher to publisher, and it is worth noting that all three illustrative examples of *hennaed* (dyed or stained with henna) in the *OED* are so written. I also have examples on file of *antennaed, concertinaed, mascara-ed, pyjamaed, shampooed,* and *shanghaied.* See FEE (verb).

**-èd, -éd. 1** Some adjectives have a fully pronounced and uncancellable final -ed: e.g. *fair-minded*, *ragged*, *warmhearted*, *wicked*, *wretched*.

**2** In some other words the pronunciation of the -ed as a separate syllable, or the merging of it with the preceding syllable, can affect the sense, e.g. *the boy was aged* /eɪdʒd/ *12* but *an aged* /ˈeɪdʒɪd/ *man was sitting on the wall.* See also BELOVED; BLESSED; CURSED.

**3** In poetry, convention allows writers, *metri causa*, to pronounce as a separate syllable an -ed that would not be so pronounced in the ordinary language. In such cases the printers usually print the ending either as -èd or as -éd: *In curléd knots man's thoughts to hold*—Sidney (in D. Attridge, 1982); *Till the secret it gives Makes faint with too much sweet these heavy-wingéd thieves*—Shelley (in Palgrave, 1861).

**edema.** The AmE spelling of *oedema* (see OE, etc.).

**edible.** See EATABLE.

**editress.** See -ESS 2. The word is still current: e.g. ... *Alexandra Shulman, my editress at* Vogue, *for giving me a sabbatical* ...—acknowledgement by C. Lycett Green, 1994.

**-edly.** 'We accept that this was not the case and unreservedly apologize to ——': thus the normal formula when a newspaper offers apologies to a person whose views have been misrepresented. Count the syllables in *unreservedly* and everyone, I imagine, would say five. *Dickensian insolvency laws match their unembarrassedly Victorian values* (D. Jessel, 1990). Most people, I suppose, would pronounce *unembarrassedly* as six syllables to avoid the unseemliness of a final sequence /-sdlɪ/. '*Oh yes,' says the boy, shamefacedly, 'that's all right.'* (OED, 1881). Bearing in mind that *shamefaced* is pronounced /-feɪst/, what are we to do with the form in -edly? The going begins to look hard. The problem clearly amused Fowler (1926). He gave nearly three columns to -EDLY and inserted some seventy cross-references to the article itself, some of them highly idiosyncratic. Of Fowler's seventy, the ones that have stood in the language longest are *advisedly, assuredly,* and *unadvisedly,* all first recorded in the 14c. Further waves of such words came along

in the 16c. (*amazedly, ashamedly, deservedly, learnedly,* etc.), the 17c. (*avowedly, designedly, reservedly, resignedly, unexpectedly,* etc.), and the 18c. (*animatedly, detachedly, vexedly,* etc.). But the great age of -edly was the 19c.: no fewer than twenty-one of Fowler's seventy are first recorded then (*absorbedly, allegedly, presumedly, unabashedly,* etc.), and under *markedly* the OED adds the comment 'a favourite 19th c. adverb'. Two are not recorded in the OED before the 20c. (*painedly, unashamedly*), and *reportedly* (not mentioned by Fowler) is also first recorded in the 20c. Some of those listed by Fowler now seem very abstruse, e.g. *admiredly, discomposedly, harassedly,* and *labouredly.* Ten words that he lists are not recorded in the OED at all, *advancedly, annoyedly, ascertainedly, incensedly,* etc.

Certain patterns emerge. A fully pronounced -ed- is obligatory in the great majority, no matter how many syllables there are in the word. Thus, *learnedly* and *markedly* would always have three syllables, and *advisedly, assuredly, deservedly, reservedly,* and *resignedly* would always have four, while the great majority of words beginning with the prefix *un-* would normally have five (*unashamedly, unconcernedly, unreservedly,* etc.). In other words there is a predisposition, established over the centuries, to give syllabic value to the -ed- in words ending in -edly. Such words are frequently formed on the past tense or past participle of the corresponding verb. Thus *allege → alleged → allegedly.* In some cases such adverbs are extensions of adjectives, e.g. *belatedly, conceitedly, shamefacedly.* In still others the formations are parasynthetic (i.e. contain two different parts of speech), e.g. *cold-bloodedly, high-handedly, whole-heartedly,* all three, incidentally, 19c. formations, and all regularly pronounced as four syllables. The similar formations *good-naturedly* and *good-humouredly* entered the language a century earlier, but these are normally pronounced with four syllables, not five. A smaller group, words in which -ed- is preceded by *i* or *u* (as in *dissatisfiedly, frenziedly, hurriedly, studiedly, subduedly*), behaves differently. In these the vowel and -ed seemed normally to merge into one syllable: thus /-faɪdlɪ/, not /-faɪədlɪ/, etc.

There is a residue of -edly words in which the problem of pronouncing or not pronouncing the -ed- remains. One wonders, for example, how D. H. Lawrence would have pronounced *painedly* in his *England, My England* (1921); and how Browning would have pronounced *starchedly* (in *Red Cotton Night-Cap Country*, 1873). And how many syllables are to be pronounced in the following words, all listed from standard sources in the *OED*: *admiredly, depressedly, harassedly, labouredly, scatteredly,* and *veiledly*? I have a hankering to give full syllabic value to -ed- in some, but not all, of them. Some undiscovered deep-rooted principle seems to govern the matter in marginal cases. But in general the harmony of the sentence would probably not be seriously at risk in such cases, whatever one decided to do.

**educate** makes *educable*. See -ABLE, -IBLE 2(v), 6.

**educationalist, educationist.** See -IST 2.

**educible** is the customary adj. corresponding to the verb *educe* (not *educable*). See -ABLE, -IBLE 7.

**-ee¹. 1** A suffix drawn from OF words in -ee, -é, denoting the recipient of a grant or the like, e.g. *feoffee, grantee, lessee, patentee,* on the model of which many others were made, most of which cannot be construed as 'direct' passives, but denote the indirect object of the corresponding verbs; in *payee* (18c) 'one to whom something is payable', and *trustee* (17c.), there has been a further (uncontroversial) departure from the usual pattern.

**2** The common correspondence of agent-nouns in -or or -er, e.g. *lessor* and *lessee, obligor* and *obligee,* with nouns in -ee led to the general application of the suffix, sometimes risibly, as with *lover* and *lovee* (Richardson), *jester* and *jestee* (Sterne). Many such are nonce-words (I have noted *awakee, mentionee,* (AmE) *rumoree,* and *suggestee* in recent times); but others, like *abductee, addressee, amputee* (20c.), *employee,* (AmE) *enrollee, escapee, evacuee, examinee,* (AmE) *retiree,* (AmE) *standee* (a standing passenger), and *trainee,* look likely to be permanent. It should be noted that some words in -ee have the force of 'one who is ——ed' (*employee*) with parallel agent-nouns in -er (*employer*);

while others, remarkably, are added to intransitive verbs and mean 'one who —s' (*attendee, escapee, standee*).

**3** From the 14c. onwards -ee had also become the regular representation of Fr. *é* in adopted words like *debauchee* and *jubilee,* and also in similar words from other languages (e.g. *grandee,* Sp., 16c.). A great deal more about the suffix is set down by Laurie Bauer in his *English Word-Formation* (1983, 243 ff.).

**-ee².** A suffix (really several distinct suffixes) shown: **1** in diminutives like *bootee, coatee.*

**2** (now mostly abandoned in favour of -i) in words of Indian origin, *Bengalee, saree,* etc. (but retained in *dungaree, puttee*).

**3** irreg. in various unexplained circumstances, e.g. *bargee* (17c.), *goatee* (19c.), *jamboree* (19c.), *marquee* (17c.), and *settee* (18c.).

**-eer.** This suffix (the Anglicized form of Fr. -ier, L. -arius, -iarius) is **1** Appended to nouns meaning 'a person concerned with or engaged in', as *auctioneer* (first recorded in 1708), *mountaineer* (1610), *musketeer* (1590), *profiteer* (1912), *summiteer* (1957), *volunteer* (1618). A few formations denote inanimate objects, as *gazetteer* (1611). In a good many of the words so formed there is a more or less contemptuous implication, e.g. *pamphleteer* (1642), *profiteer, racketeer* (1928), *sonneteer* (1665).

**2** In verbal nouns with the meaning 'the being concerned with', as *auctioneering* (1733), *buccaneering* (1758), *electioneering* (1760). Most of these quickly produced finite verbs as back-formations.

**effect** (noun and verb). See AFFECT, EFFECT.

**effective, effectual, efficacious, efficient.** The words all mean having an effect, but with different applications and certain often disregarded shades of meaning. *Effective* means having a definite or desired effect, existing in fact rather than theoretically, coming into operation: *an effective speech, speaker, contrast; effective assistance, control; effective from 1 January. Effectual* applies to action apart from the agent, and means capable of producing the required result or effect: *effectual measures*; *his comments put an effectual end to the conversation*. A person cannot be described as *effectual*, but, paradoxically, a person lacking the ability to achieve results can be said to be *ineffectual*. *Efficacious* applies only to things (esp. medicines, treatment, etc.) used or adopted for a purpose, and means producing or sure to produce a desired effect: *an efficacious remedy*; thus *a drug of known efficacy*. *Efficient* applies to agents or their action or to instruments, etc., and means productive with minimum waste or effort, (of a person) capable: *an efficient cook, secretary*; *efficient work, organization*. *Efficient cause* in philosophy is an agent that brings a thing into being or initiates a change.

**effete** entered the language in the 17c. from L *effētus* ( = that has brought forth young (cf. *foetus*), hence worn out by bearing offspring), applied to animals that had ceased to bear offspring. It rapidly developed the transferred sense '(of a material substance) that has lost its special quality or virtue'; and by the late 18c. was being applied to persons or systems that had lost their effectiveness. In the 20c. it has also come to be applied to effeminate men, but, oddly, in view of its etymology, not to women who are or look as if they are past child-bearing age.

**effluvium.** The pl. is *effluvia*. See -UM 2.

**-efied, -efy.** See -FY 2.

**e.g.** is short for L *exempli grātia* and means 'for example'. Non-Latinists sometimes confuse it with i.e., which stands for L *id est* and means 'that is to say'.

**egis** is a spelling of *aegis* sometimes used in AmE.

**ego** /ˈiːɡəʊ/ *or* /ˈeɡəʊ/. The pl. is *egos*. See -O(E)S 6.

**egoism, egotism.** 1 The two words are modern (18c.) formations, the first correctly from L *ego* 'I' + the suffix *-ism*, and the second from the same elements together with an intrusive *-t-* (unexplained, but perhaps on the false analogy of such words as *idiotism*).

2 Philosophers keep the words apart. For example, in ethics, *egoism* is used for 'the theory which regards self-interest as the foundation of morality' (*OED*), and in metaphysics, 'the belief, on the part of an individual, that there is no proof that anything exists but his own mind' (*OED*). But since the mid-19c. *egoism* (followed by its derivatives) has been used in the broad sense 'excessive exaltation of one's own opinion, self-opinionatedness', thus making contact with *egotism* in its two senses, 'the obtrusive or too frequent use of the first person pronoun "I"; hence the vice of thinking too much of oneself, self-conceit, boastfulness' (*OED*). To the general educated public, at any rate those who are uninformed about the technical language of ethics and metaphysics, the net result is a residual and persistent belief that the words are more or less interchangeable. The same applies to the adjectives *egoistic(al)* and *egotistic(al)*. Here is a range of examples (the forms with medial *-t-* being the more common): (*egoism* and derivs.) *I have never gone out of my way for man, woman, or child. I am the complete egoist*—V. Sackville-West, 1931; *Rationalism in morals may persuade men in one moment that their selfishness is a peril to society and in the next moment it may condone their egoism as a necessary and inevitable element in the total social harmony.*—R. Niebuhr, 1932; *There were in man's nature not merely egoistic instincts concerned with self-preservation or the good of the Ego*—G. Murray, 1939; *The hero is filled only with himself; in his extreme egoism* [etc.]—E. Fromm, 1976; *I am . . . convinced that I can see . . . the crude wrecking of lives . . . —which they are egoistically . . . set on causing*—H. Brodkey, 1993; (*egotism* and derivatives) *Nothing so confirms an egotism as thinking well of oneself*—A. Huxley, 1939; *He was continually talking about himself and his relation to the world about, a quality which created the unfortunate impression that he was simply a blatant egotist*—H. Miller, 1957; *To justify or to condemn them in public is a squalid piece of egotism when it will hurt the living*—C. Day Lewis, 1960; *Samuel Taylor Coleridge . . . sees the avoidance of 'I' as an expression of concealed egotism and of the 'watchfulness of guilt'*—R. F. Hobson, 1985; *Jasper is sublimely egotistical, and the egotist of course sees himself as self-propagated*—P. Lively, 1987. The adjectives *egoistic* and *egotistic* are now under threat from the increasingly popular adjective *egocentric*.

**egregious.** The word was for a long time (16–19c.) used in two main opposing senses: (a) remarkably good, outstanding, distinguished (Marlowe's *egregious viceroys of these eastern parts* in *Tamburlaine* were held in great esteem, and so were many egregious heroes in later works); (b) remarkably bad, gross, flagrant, outrageous. Only sense (b) survives (*an egregious ass, an egregious waste of time, egregious errors*, etc.). The L adj. *ēgregius*, formed from *ē* 'out' + *grex, gregis* 'flock', meant 'towering above the flock' and was almost always used in a favourable sense.

**-ei-.** The following words are among those spelt with -*ei*- not -*ie*-: *apartheid, beige, ceiling, conceit, conceive, counterfeit, cuneiform, deceit, deceive, either, forfeit, geisha, heifer, heinous, inveigh, inveigle, kaleidoscope, leisure, meiosis, neigh, neighbour, neither, nonpareil, obeisance, perceive, plebeian, receipt, receive, seigneur, seise, seize, seizure, sleight, surfeit, their, weir, weird*. See also I BEFORE E; -IE-.

**eighties.** See NINETIES.

**-ein(e).** Until the early part of the 20c. this second element of a number of words was regularly pronounced as two syllables: see CAFFEINE, CASEIN, CODEINE, PROTEIN. They are all now simply pronounced as /-iːn/.

**either.**

1 Pronunciation.
2 Parts of speech.
3 Essential duality.
4 Number.
5 *either . . . or*.
6 Discarded uses.

**1** Pronunciation. Both /ˈaɪðə/ and /ˈiːðə/ are used with approximately equal frequency. In 1984 a correspondent informed the Oxford dictionaries that the Queen's private secretary had told him (a copy of the letter of 24 Oct. 1984 from Buckingham Palace to the correspondent was enclosed) that the Queen pronounces the word herself as /ˈaɪðə/ (in the letter written as 'eye-ther'). Doubtless the Queen, like the rest of us who use the same pronunciation, regards the alternative one with a degree of amiable tolerance.

**2** Parts of speech. *Either* is used as four distinct parts of speech: as adj. and pron.,

(a) = one or the other of two (*either book will serve the purpose; either of* [the two of] *you can go*); (b) = each of the two (*In the middest of the street of it, and of either side of the riuer, was there the tree of life*—Rev. (AV) 22: 2; *we sat down on either side of the radiogram, she with her tea, me with a pad and pencil*—J. Winterson, 1985; *These* [sc. two I-beams] *ran crosswise at either end of the trailer*—Tom Drury in *New Yorker*, 1992). And as adv. and conjunction, (a) = as one possibility (*she . . . had chosen not to call herself either Madeleine de Roujay or Madeleine le Freyne*—P. P. Read, 1986); (b) = as one choice or alternative (*either come in or go out but don't just stand there*); (c) with a negative, and usually placed at the end of a clause or sentence, = any more than the other, for that matter (*if she isn't the right material, she isn't one of that kind, either*—N. Gordimer, 1987; *At the time, I was not quite sure what to make of the Institute either*—New Yorker, 1987; *But then we realized that if this man could attend a television taping without his wife's being any the wiser, perhaps the show's broadcast wouldn't affect her, either*—New Yorker, 1992).

**3** Essential duality. *Either* means essentially 'one or other *of* two', or as *CGEL* inelegantly expresses it (13.39) 'According to the didactic tradition, the use of correlative coordinators is unacceptable when there are three or more conjoins.' A standard unopposable construction is shown in the following: *We can either rely on our children to translate for us or we can try to catch up*—*Illustr. London News*, 1980. If the number of coordinators is extended to more than two, opinions vary widely about the elegance or even the acceptability of the results. The principle of duality is broken in the following examples: (i) *Either France or Germany or Italy will break ranks in the matter*. (ii) *EITHER, (a) 'Dictionary Johnson'; did he deserve the title? OR, (b) Examine the achievement of any ONE English dictionary. OR, (c) What should be the aims and methods of an English lexicographer?*—Univ. of Oxford, Honour School of English, 1983. In most contexts in formal English, however, it is advisable to restrict *either* to contexts in which there are only two possibilities. In conversational English the type 'a narration of events, either past, present, or to come' is often unavoidable. Cf. Shakespeare's *they say there is Diuinity in odde Numbers, either in natiuity,* *chance, or death* (1598). Examples of the negative form *neither* followed by more than two alternatives are very common: *neither ABC nor CBS nor NBC has a permanent team in black Africa—New Yorker*, 1988.

**4** Number. In normal circumstances *either* governs a singular verb (*either candidate* [of two] *is acceptable*), but in the type '*either* + *of*-phr.' (esp. in interrogative clauses) the number of the verb often varies 'because of the fundamental plurality of the conception' (Jesperson, 1909–49, ii. 172): e.g. *either of them* [sc. the words *friend* and *mistress*] *are enough to drive any man to distraction*—Fielding, 1749; *Are either of you dining with Stewart to-night?*—E. F. Benson, 1911; *Have either of you two ladies received an anonymous letter?*—A. E. W. Mason, 1924. The same is true of the type '*neither* + *of*-phr.': (treated as pl.) *I daresay I . . . could have been a more loving mother to you and Jane, but as things are, neither of you require me*—E. Bowen, 1955; *Neither Anabel nor Mark are in London*—M. Wesley, 1983; *I have written about almost every subject except astrology and economics, neither of which are serious subjects*—P. Howard, 1985; (treated as sing.) *neither our mourning nor our rejoicing is as the world's is*—T. S. Eliot, 1935; *Neither of the two answers is satisfactory—Dædalus,* 1987. The informal type *Either John or Jane avert their eyes when I try to take a photograph of them* brings out the fundamental clash between notional and actual agreement. Similarly *If either James or Charles should turn up, offer them a drink.* See also AGREEMENT 6.

**5** *either . . . or.* Fowler (1926) correctly pointed out that in good writing *either* in *either . . . or* sentences should be placed with caution, and normally immediately before the first of the items being itemized: *teachers no longer feel that either they or the job they do is valued by society at large—TES,* 1986.

**6** Discarded uses. In past centuries various other uses of *either,* now discarded, were frequent: e.g. *either of both* 'either of the two' (*Wives were taken in Israel by bils of Dowry, and solemne espousals; but concubines without either of both*—Henry Ainsworth, 1621); *either other* 'one or the other of two' (*Let him take whether he liketh best, if either other of these words shall serue his turne*—John Jewel, 1567). Doubtless some of the uses now current will suffer

the same fate in due course. The central notion of duality is the one perhaps most at risk.

**-eity.** There has been a marked tendency in the second half of the 20c. to replace the traditional /-iːtɪ/ in the pronunciation of words containing this ending (*deity, homogeneity, spontaneity,* etc.) by /-eɪtɪ/.

**ejector.** Thus spelt, not *-er.* See -OR.

**eke** (adv.). Adverbs, like other words, come and go. This natural and ordinary adverb (OE *ēac*) meaning 'also' stood in the language more or less unthreatened until the 18c.: *but first, Mr. Ghuest, and M. Page, & eeke Caualeiro Slender, goe you through the Towne to Frogmore*—Shakespeare, 1598; *Supposing the wax good, and eke the thimble*—Sterne, 1759. But it has now become extinct, or at best is available only as an extreme archaism.

**eke out.** To begin with (16c.) the expression meant primarily 'to supplement, to supply the deficiencies of anything; esp. to make (resources, materials, etc.) last the required time by additions, by partial use of a substitute, or by economy' (*OED*). Defoe's Robinson Crusoe, for example, commented (1719), *My ink . . . had been gone . . . all but a very little, which I eked out with water.* This construction survives: *Her mother . . . edited 'Aunt Judy's Magazine' to eke out the clerical income*—H. Carpenter, 1985. Fowler (1926) declared this to be the only legitimate construction. 'The proper object is accordingly a word expressing not the result attained, but the original supply. You can eke out your income or a scanty subsistence with odd jobs or by fishing, but you cannot eke out a living or a miserable existence. You can eke out your facts, but not your article with quotations.' From the early 19c., however, *eke out* has also been used to mean 'to contrive to make (a livelihood), or to support (existence) by various makeshifts', i.e. the sense to which Fowler objected: *To eke out the existence of the people, every person . . . was called on for a weekly subscription*—Jefferson, 1825; *Some runaway slaves . . . contrived to eke out a subsistence*—Darwin, 1845; *He lived with his parents until their death, and thereafter eked out a marginal living as a messenger*—O. Sacks, 1985; *The elder Rossetti* eked out a precarious living as a teacher and translator—A. C. Amor, 1990. It is clear that both constructions are now standard.

**elaborateness, elaboration.** *Elaborateness* (first recorded in 1694) means 'the quality of being elaborate'. Since the early 19c., *elaboration* has borne the same meaning, but more commonly retains its earlier (17c.) meaning: 'the process of developing, perfecting, working out in detail, making more complex (a theory, a literary work, etc.)'. See -ION AND -NESS.

**elder.** As an adj., *elder* is now for the most part confined to the meaning '(of two indicated persons, esp. when related) senior, of a greater age (*his elder brother*)'. Hence the superlative form *eldest*. Outside this restricted use of family seniority, *elder* lingers in a few contexts such as *elders* meaning persons whose age is supposed to command the respect of the young, as the title of lay officers of the Presbyterian Church, the *elder brethren* of Trinity House, the *elder hand* at piquet, and *elder statesman*. In all ordinary contexts when the comparative and superlative forms of *old* are called for these are, of course, *older* and *oldest*.

**elector.** Thus spelt, not *-er*.

**electric, electrical.** In most contexts *electric* is the automatic choice (*electric blanket, chair, current, fan, heater, kettle, light, razor, shock*, etc.). *Electrical* is reserved for contexts in which the sense is more general, 'relating to or connected with electricity', e.g. *electrical engineer(ing), precipitation, storm*. See -IC(AL).

**electrocute** is (or once was considered to be) an irregular formation (first recorded in 1889) from the combining form *electro-* + *-cute* after *execute* (verb).

**eleemosinary.** Pronounce as seven syllables, /ˌelɪːˈmɒsmərɪ/ or /-ˈmɒzmərɪ/.

**elegant variation.** In a celebrated, leisurely essay, Fowler (1926) used this expression to describe the way in which 'second-rate writers' and 'young writers', in an effort to avoid using the same word in a sentence, fall into various stylistic traps. Among his examples of avoidable repetition he cited *A debate which took wider ground than that* actually *covered by*

the actual *amendment itself* (omit either *actually* or *actual*). And for a lapse of the opposite kind—a clumsy attempt at variation—he cited a context from Thackeray (*careering during the season from one great dinner of twenty covers* to *another of eighteen* guests) (use *guests* instead of *covers*, and allow the noun following *eighteen* to be understood). Nowadays, as typescripts make their way through the hands of copy editors and proofreaders, most clumsiness of the kind that stand in the original tend to be eradicated before the book, etc., appears, but vigilance is still required at every stage.

A commonplace method of avoiding repetition is to use a synonym or even a succession of identificatory phrases to refer to the same person or thing. Thus (from an American periodical) *K. has extraordinary hair; the man is nearly sixty years old, and look what a mane he's got!* And of the second type (a series of descriptive phrases) (cited from the NZ *English Newsletter*, 1991): *A few months ago she seemed near death. But* the world's most famous nun *continues her good work. Today* Mother Teresa *announced she is so moved by the plight of the Romanian children she is going to do something about it.* The Nobel Peace prize winner *will open a mission in Bucharest to care for the children.*

There are many other types of elegant, and of inelegant, variation.

**elegiac quatrain.** An alternative name given to the *heroic quatrain* owing to Gray's use of this metre in his *Elegy*. It has four iambic pentameters, rhyming alternately:

> The Curfew tolls the knell of parting day,
> The lowing herd wind slowly o'er the lea,
> The plowman homeward plods his weary way,
> And leaves the world to darkness and to me.

**elegiacs.** 'A scheme of metre in which a full hexameter is followed by a hexameter catalectic, i.e. one having a monosyllabic foot at the caesura and at the end of the line' (Egerton Smith, 1923). Thus the pattern is:

$$-\smile\smile\ |\ -\smile\smile\ |\ -_\wedge\ ||\ -\smile\smile\ |\ -\smile\smile\ |\ -_\wedge\ .$$

This well-known classical metre is followed, somewhat unconvincingly, in the following examples:

> Grant, O | regal in | bounty, || a | subtle and
> | delicate | largess;

Grant an e | thereal | alms, || out of the
| wealth of thy | soul:
(W. Watson (1858–1935), *Hymn to the Sea*)

Where, under | mulberry | branches || the
| diligent | rivulet | sparkles,
Or amid | cotton and | maize || peasants
their | waterworks | ply.
(A. H. Clough, 'Amours de Voyage', 1858)

**elegy.** At first (early 16c.) applied to 'a song of lamentation, esp. a funeral song or lament for the dead', and this has remained as the principal use of the word in English: notable examples are Milton's *Lycidas* (1637), Shelley's *Adonais* (1821), and Arnold's 'Thyrsis' (1867). But by 1600 the limitation of the word in this manner had partially given way to the sense 'a poem of melancholy reflection', and in particular to all the species of poetry for which Greek and Latin poets (and then English poets: see ELEGIACS) adopted the elegiac metre, or, as in Gray's *Elegy Written in a Country Church-Yard* (1751), wrote ELEGIAC QUATRAINS.

**elemental, elementary.** *Elemental* refers primarily to the medieval belief in the 'four elements' (earth, water, air, fire) or to the enormous observable power of the elements during storms: hence *elemental fire, forces, gods, spirits, substances*, etc. *Elementary*, by contrast, means 'rudimentary, introductory', as in *elementary book, school, subjects*. But in modern physics *elementary* means 'not decomposable', esp. in *elementary particle*, any of several subatomic particles supposedly not decomposable into simpler ones.

**elevator.** Principally in the US, but sometimes elsewhere, the normal word for what in Britain is called a *lift*. On the other hand, Americans as well as Britons use the word *ski-lift* for the device that carries skiers up a slope.

**elfish, elvish.** Both terms are used, and also *elfin*, without any discernible distinction. It is not that one is more established than the others, though *elvish* is first recorded in the 14c. and *elfish* and *elfin* not till the 16c. Spenser and Keats seem to have favoured *elfin*, Coleridge *elfish*, and Chaucer and the Gawain-poet *elvish*. Tolkien revived the obsolete term *elven* (*elven-kin, -tongue*, etc.) in the 20c. One can only suppose that the choice of word is largely a matter of taste, based on delicate associations of contextual appropriateness.

**elicit, illicit.** Though they are pronounced identically, confusion seldom arises since they are different parts of speech. *Elicit* (verb) means 'to draw out, evoke (an admission, response, etc.)'; and *illicit* (adj.) means 'unlawful, forbidden'. Letters to the editor in 1989 and 1992, however, did produce two examples of *illicit* used instead of *elicit*: *Your questionnaire targeted WASPs and illicited a typical reaction*—*Metro* (NZ), 1989. The second example occurred in a letter written to the *Guardian* on 6 Feb. 1992 by the broadcaster Sue Lawley. (In both cases let down by a copy-taker or proofreader?)

**eligible, illegible.** Only a very ill-informed person would confuse these two adjectives. *Eligible* means 'fit or entitled to be chosen (*eligible for a pension*)' or 'desirable, suitable' (*an eligible bachelor*). *Illegible* means '(of writing) not clear enough to read'.

**eliminate, elimination.** The essential meaning (etymologically 'turn out of doors') is the expulsion, getting rid of, or ignoring of elements that for any reason are *not* wanted. For example, police often say that they wish to interview a person in order to *eliminate* him or her from their inquiries. The *OED* stigmatizes as catachrestic a 19c. tendency to use *eliminate* to mean 'to extract or isolate for special consideration or treatment the elements that *are* wanted'. Examples: *He would ... eliminate the main fact from all the confusing circumstantials*—A. Bain, 1855; *[Hypotheses] of the utmost value in the elimination of truth*—Faraday, 1854. This rejected use appears not to have prospered in the 20c.

**ellipsis.**

1 Definition.
2 Axiomatically legitimate types.
3 Unacceptable types.
4 For discussion.
5 Omission of *that* (relative pronoun).
6 In non-standard speech.

1 Definition. Ellipsis is 'the omission of one or more words in a sentence, which would be needed to complete the grammatical construction or fully to express the sense' (*OED*). It is an important feature of the language, in its written

and especially in its spoken form. It can be seen in its most elementary form in utterances like *Told you so*, *Serves you right*, *Want some?*, and *Sounds fine to me*, in each of which a pronoun has been dispensed with. The primary nature of ellipsis is that it is a process by means of which certain notional elements of a sentence may be omitted if they are clearly predictable or recoverable in context.

**2** Axiomatically legitimate types. Ordinary English grammar normally requires the omission of certain elements: (definite article not repeated) *He heard the whirr and ∧ click of machinery*; (infinitive marker *to* not repeated before a second infinitive) *I was forced to leave and ∧ give up my work at the hospital*; (subject not repeated) *I just pick up wood in a leisurely way, ∧ stack it and ∧ slowly rake the bark into heaps*; (infinitive implied though omitted) *Knowledge didn't really advance, it only seemed to ∧.* In such circumstances insertion of the missing elements is not entirely ruled out if some degree of emphasis is required, but in ordinary declarative or narrative prose ellipsis is normal. Any substantial piece of writing, when analysed, yields examples of slightly more complicated but still legitimate types of ellipsis. From Nicholas Shakespeare's novel *The Vision of Elena Silves* (1989) I gathered the following among others: (various elements understood after the opening clause) *Henriques knew they would eat his tongue for wisdom, ∧ his heart for courage and for fertility ∧ make their women chew his genitals*; (noun replaced by pronoun and then left to be understood) *He also shared their resignation. He ascribed theirs* [ = their resignation] *to an ability to count to infinity; his* [resignation] *to the knowledge he was never made for the jungle*; (ellipsis of the adverbial phrase *of our stuff*) *'They file our stuff in Lima?' 'The most interesting.'* Cf. BE 7.

There are many other types of ellipsis. *CGEL* (1985) needed more than fifty pages to list, illustrate, and label them all (strict, standard, situational, structural, etc., ellipsis). The nature of what can safely be left to be understood is at the heart of the problem. The basic rules allow for a reasonable amount of flexibility in the imaginary wording of the understood parts: e.g. (adjustment of *thought* to *think* after the auxiliary *didn't*) ('*And you thought I was a virgin when I*

*married you?'* ... '*No, I didn't ∧'*; (adjustment of *felt* to *feel* in the ellipsis) *She hadn't felt fraudulent about it. Nor had there been any reason to ∧.* The adjustment factor is fractionally higher in T. S. Eliot's *Several girls have disappeared Unaccountably, and some not able to*, but one is predisposed to have to wrestle slightly more with a literary text than with a non-literary one.

**3** Unacceptable types. Unacceptable difficulties arise in various circumstances, e.g. if two auxiliary verbs that operate in different ways are placed together. One should not say or write *No state has or can adopt such measures*. Idiom requires *has adopted or can adopt such measures*. Fowler (1926) permitted the construction *He is dead, and I not*, but I would insist on the insertion of *am*. When a change of grammatical voice is involved, ellipsis spells danger. A reader cannot be expected to make the necessary adjustment from the active voice to the passive and supply an omitted part of the passive form. One of Fowler's examples (1926) will suffice to support the argument: *Mr Dennett foresees a bright future for Benin if our officials will manage matters conformably with its 'customs', as they ought to have been* (insert *managed*). Comparisons can also produce unacceptable sequences when unwise ellipses are attempted: *The paintings of Monet are as good or better than those of van Gogh* (read *as good as or better than*). One is again reminded that an understood element must be identical with the one it matches, or at any rate be very closely allied to it.

**4** For discussion. Some of Fowler's examples (1926) may perhaps be left for discussion. Some sound very antiquated; others lie in the no man's land between acceptability and rejection: e.g. *The ringleader was hanged and his followers ∧ imprisoned* (ellipsis of *were*); *Mr Balfour blurted out that his own view was ∧ the House of Lords was not strong enough* (ellipsis of *that*); *I assert ∧ the feeling in Canada today is such against annexation that* [etc.] (ellipsis of *that*); *The evil consequences of excess of these beverages is much greater than ∧ alcohol* (ellipsis of *with*). Only the first of these seems to me to allow the reader to proceed with minimal (or even no) discomfort. For the remainder, and for several other examples cited by Fowler, removal of the difficulty is best achieved by reconstruction, not merely by supplying the element that is intended to be 'understood'.

**5** Omission of *that* (relative pronoun). See THAT (relative pronoun) 2.

**6** In non-standard speech. Non-standard speech is characterized by the use of many unfamiliar types of ellipsis. For example, American writers such as Michael Doane and E. L. Doctorow underline the nature of a particular way of speaking among the underclasses of America by employing non-standard ellipses in their speech: (omission of *have*) *Watergate, man, Where you been?*—Doane, 1988; (omission of *do* or *can*) *Well how you expect to get anywhere, how you expect to learn anything?* —Doctorow, 1989. Such constructions are usually transparent to standard speakers but serve as sharp linguistic barriers nevertheless.

**else.** **1** When *else* is combined with an indefinite or an interrogative pronoun the usual possessive form is *anybody else's* (not *anybody's else*, as was the case until the mid-19c.), *who else's* (not *whose else*), *nobody else's* (not *nobody's else*), etc.: e.g. *They look to me like someone else's, to be frank*—P. Lively, 1987; *English feudalism was unlike anyone else's*—London Rev. Bks, 1988.

**2** When *else* was used with an interrogative pronoun it was possible until about the end of the first quarter of the 20c. to postpone it to the end of the sentence (*What did he say else?*). Now the only available order of words is shown in *What else did he say?*

**3** Since the Middle Ages, *else* has been used, sometimes as a quasi-conjunction, to mean 'otherwise, if not': *Strangle her, els she sure will strangle thee*—Spenser, 1596; *The land certainly had … vomited them out else*—Daniel Rogers, 1642; *and well was it for him that the outburst of his blind fury was over, else he had become an easy prey to his gigantic protagonist*—R. M. Ballantyne, 1858. The use seems now to be archaic, but it does occur occasionally: *Fortunately it [sc. a staircase] was not spiral, else I would have succumbed to vertigo*—B. Rubens, 1985; *My father was fortunate that my mother's nature was reserved, else he would have found himself out on the road that night*—H. Jacobson, 1986.

**elusive, illusory.** It is often convenient that one of a pair of near-synonyms drops out of use. This has happened to *elusory* (first recorded in 1646 and in common use until the early 20c.) and is beginning to happen to *illusive* (1679–). The beneficiaries are the words *elusive* (1725–) and *illusory* (1599–), which are sufficiently distinctive in sound and meaning never to be confused. Both words remain in standard use. *Elusive* is the adj. corresponding to the verb *elude*. That is elusive which we fail, in spite of efforts, to grasp physically or mentally: *elusive ball, person, weed*; *elusive image, pleasure, smell, spirit.* *Illusory* is the adj. corresponding to the noun *illusion*. That is illusory which we believe to be deceptive (esp. as regards value or content) or having the character of an illusion: *illusory comfort, distinction, dream, hopes, promise.*

**elvish.** See ELFISH.

**Elysium.** Pl. (if needed) *Elysiums.* See -UM 1.

**emaciate, emaciation.** For pronunciation, see -CIATION, -TIATION 3.

**em- and im-, en- and in-.** The words in which hesitation between initial *e-* and initial *i-* is possible are given in the form recommended: *embed, empanel, encage, encase, enclose* (and derivatives), *encrust* (but *incrustation*), *endorse, endorsement, endue, enfold, engraft, enmesh, ensure* (in general senses), *entrench, entrust, entwine, entwist, enwrap; insure* (in financial sense), *insurance, inure, inweave.* See also IM-; ENQUIRE.

**embargo.** Pl. *embargoes.* See -O(E)S 1.

**embarrass.** A commonly misspelt word.

**embarrassedly.** See -EDLY.

**embed.** See EM- AND IM-.

**embryo.** Pl. *embryos.* See -O(E)S 4.

**emend.** See AMEND.

**emergence, emergency.** See -CE, -CY.

**emergent.** In the sense 'urgent, pressing' recorded in the *OED* from Defoe (1706) until the late 19c. and marked as catachrestic. A professor from Hammersmith Hospital, speaking on Radio 4 on 5 Oct. 1978, used the word in this sense: *to say whether one patient is urgent or emergent.* The main American dictionaries list this meaning as standard, but its absence from some of the equivalent British dictionaries suggests that the sense is in restricted use in BrE.

**emigrant, immigrant, migrant.** An *emigrant* is one who leaves his or her own country to live in another. An *immigrant* is one who comes as a permanent resident to a country other than his or her own native land. It is all a question of whether a given person has just passed outward through the exit gate at a port or airport in one country, or has just been accepted as a resident member of another country. A *migrant* is (a) an animal or bird that migrates, (b) (in Australia and NZ) an immigrant. Turks who work in Germany, Pakistanis who work in Iraq, and similar groups of people working away from their native lands, are often called *migrant workers.*

**emolument.** See FORMAL WORDS.

**emote,** a back-formation from *emotion,* means 'to dramatize emotion; to act emotionally'. The word is first recorded in America in 1917, and it has been mostly restricted to the language of ballet and theatre critics and to photography: *The female sitter had to emote in some way, either by dressing up or by gazing with drooping head into a bowl of flowers*—Amateur Photographer, 1970; *Just that until a few years ago you could barely talk to strangers, so how are you going to get up and emote in front of an audience?*—L. S. Schwartz, 1989.

**emotional, emotive.** *Emotional* means (a) 'connected with, based upon, or appealing to the emotions', and (b) 'easily affected by emotion (*an emotional person*). Sense (a) is shared by *emotive,* but not (b). *Emotive* is more commonly used of language or behaviour that tends to arouse emotions, i.e. is not purely descriptive. Capital punishment is an *emotive* issue. The single word *home* is *emotive* to a sailor or to someone in prison. In literary criticism, I. A. Richards, writing in the 1920s and 1930s, established a broad distinction between the *emotive* meaning (i.e. one evoking an emotional response) of a poem or other literary work and its factual or referential meaning.

**empanel.** See EM- AND IM-. The inflected forms show a doubled consonant in BrE (*empanelled, empanelling*), but usu. a single -l- in AmE. See -LL-, -L-.

**empathy,** etymologically an adaptation of Gk ἐμπάθεια 'passionate affection', but introduced to English as a translation of Ger. *Einfühlung* at the beginning of the 20c., is a term used in psychology and aesthetics meaning 'the power of projecting one's personality into (and so fully comprehending) the object of contemplation' (*OED*). In lay use it tends occas. to replace *sympathy* or *feeling for* when the traditional words are sometimes more appropriate. Perhaps in most, though not all, of the following examples (all drawn from non-technical works), *empathy* is the *mot juste: Seeing our sadness, our empathy with the pain she was surely suffering, she said, 'What's wrong with you all?'*—Angela Davis, 1975; *The empathy of a true friend is what I have lived without for years*—G. Paley, 1980; *We develop empathy as a capacity to share in the experience of others, not just like our own but as our own*—P. Casement, 1985; *It was a hard life, and Byron recounts it with empathy and gusto*—A. Burgess, 1986; *Robyn has a tremendous empathy with the land and its relationship to structures*—More (NZ), 1988. The corresponding adj. is either *empathic* (probably the more usual form) or *empathetic.*

**employee.** During the second half of the 19c. *employé* (fem. *employée*) was more commonly used than *employee.* the *OED* in 1897 labelled *employee* as 'rare exc. U.S.'. Since then *employee* has become the dominant form in standard English. In AmE it is often spelt *employe,* but is till pronounced as three syllables, freq. with the main stress on the second syllable and with a final /-iː/ of intermediate length. See -EE[1] 2.

**emporium.** A formal word for a large retail store or a centre of commerce. The dominant pl. form is *emporia,* but *emporiums* is also in standard use. See -UM 3.

**-en adjectives.** (Reduced to -*n* after *r,* as in *cedarn, leathern,* not \**cedaren,* \**leatheren.*) **1** In earlier times, beginning in OE, a suffix added routinely to noun-stems to form adjectives: *brazen, golden*

(OE *gylden*), *leaden, leathern silken, silvern, waxen, wheaten,* and *woollen,* for example, can all be traced back to OE. Others are first recorded at a later date: *bricken* (19c.), *cedarn* (17c.), *earthen* (13c.), *flaxen,* (16c.) *hempen* (14c.), *oaken* (14c.), *oaten* (15c.), and *wooden* (16c.).

2 From the earliest time, however, and esp. from the 16c. onwards, there has been a marked tendency to use the corresponding noun attributively (thus *a flax* (not *flaxen*) *basket, a silver* (not *silvern*) *brooch*) and to allow the *-en* adjectives in their literal senses to become archaic or restricted to poetry (esp. *cedarn, leathern, silvern,* and *waxen*). In the same period the *-en* adjectives have moved in strength into metaphorical or figurative territory. We are now more likely to encounter e.g. *brazen impudence* than *brazen rods, a golden era* than *a golden crown, leaden skies* than *a leaden roof,* and *her silken touch* than *her silken curtains.*

3 *Ashen* is a special case, as two different words are involved: the ash-tree, and the ash from a fire. In its most common sense *ashen* = 'of or resembling (fire) ashes'; but it is also used to mean' of or relating to the ash-tree' and, archaically, 'made of ash wood'.

4 See EARTHEN.

**enamel.** The inflected forms show a doubled consonant in BrE (*enamelled, enamelling*), but usu. a single *-l-* in AmE. See *-LL-, -L-*.

**enamour** (US *enamor*). It is used chiefly in the passive, followed by *of* (occas. by *with*): *I am not so much enamoured of the first and third subjects*—Dickens, 1866; *He is, with good reason, not enamoured of the individual Communist guerrillas and commissars with whom his village had to deal*—TLS, 1984; *Dealers were not enamoured with the third quarterly figures from Reed International*—Daily Tel., 1984.

**en- and in-.** See EM- AND IM-.

**encage, encase.** See EM- AND IM-.

**enclave.** This 19c. French loanword was pronounced in a French manner, /ãklɑv/, in English until at least the end of the first quarter of the 20c., but is now routinely /ˈenkleɪv/.

**enclitic** (adj. and noun). (Derived from Gk ἐγκλιτικός, from ἐν 'on' + κλίνειν 'to lean'.) In grammar, (of a) word pronounced with little emphasis so that it 'leans on', i.e. merges with, the preceding word, e.g. *not* in *cannot,* and the final element in the demotic forms *coulda* ( = could have), *gonna* ( = going to), etc. In linguistics the reduced form *clitic* has became established in the second half of the 20c. to mean either an *enclitic* or a PROCLITIC.

**enclose, inclose.** See EM- AND IM-.

**encomium.** The recommended pl. form is *encomiums,* but *encomia* is also in standard use. See *-UM* 3.

**encrust, incrustation.** See EM- AND IM-.

**encyclopaedia.** The word, which was first recorded in English in the 16c., was adopted from late L *encyclopædia,* from a pseudo-Greek ἐγκυκλοπαιδείν, an erroneous form occurring in manuscripts of Quintilian, Pliny, and Galen, for ἐγκύκλιος παιδεία 'encyclical education', the circle of arts and sciences considered by the Greeks as essential to a liberal education (*OED*). It is now frequently spelt with final *-pedia,* but the form recommended in this book is *encyclopædia* (or *encyclopaedia*), though the form with *-ae* may not hold out much longer. Cf. CYCLOPAEDIA.

**endeavour** (AmE *endeavor*).1 See PASSIVE TERRITORY 3.

2 See FORMAL WORDS.

**ended, ending.** *Statistics for the six months ended* (or *ending*) *31 December.* If the terminal date is in the future *ending* is always used, and *ended* ordinarily when it is past. But it is pedantic to object to *ending* for a past date on the ground that a present participle cannot suitably be used of a past event. If the reference is to the initial date the word is always *beginning,* never *begun,* whatever the date.

**endemic, epidemic.** An endemic disease is one that is regularly or only found among a particular people or in a certain region; an epidemic disease is a temporary but widespread outbreak of a particular disease. Some scientists, but few laymen, use the terms *enzootic* and *epizootic* when such diseases occur in animals.

**endocentric compounds.** Compounds which have the same grammatical function as their main constituent element are called *endocentric compounds* (Gk ἔνδον 'within') : *armchair* = a kind of *chair*, *doorknob* = a kind of *knob*, *footstool* = a *stool* for resting the feet on when sitting. Cf. BAHUVRIHI.

**end of the day, at the.** One of the ignoble clichés introduced into the language in the 20c. (first recorded in 1974) and meaning no more than 'eventually; when all's said and done'.

**endorse, endorsement.** 1 See EM- AND IM-.

2 *Endorse* (lit. 'to write on the back of something') has long been used to mean: to confirm (a statement or opinion); to write on the back of a cheque or other document (for a specified commercial or banking purpose); and (in the UK) to enter details of a conviction for a motoring offence on (a driving licence). Since at least the early 20c. it has also been used with increasing frequency in the advertising world to mean 'to declare one's approval of (a named product, a policy, etc.)', a use that now seems irreversible.

**end-product** was at first (Rutherford, 1905) used in chemistry to mean 'a stable, non-radioactive nuclide that is the final product of a radioactive series'. It rapidly made its way into lay use in transferred and figurative senses, esp. in the concrete sense 'a finished article in a manufacturing process'. *End-product* and also the parallel term *end-result* have been criticized from time to time since they contain an element of redundancy, but they occur in good sources, including the work of J. S. Huxley and of Mary McCarthy, and in well-established journals such as *Nature*, *The Lancet*, and *The Spectator*.

**endue, indue.** See EM- AND IM-.

**enfold, infold.** See EM- AND IM-.

**enforce.** About the end of the 19c. a use of *enforce* to mean 'to compel, oblige', often construed with *to* + infinitive, began to be regarded as archaic, and it has now almost entirely dropped out of use. Examples: *the Law will inforce the Borrower to pay it*—Locke, 1691; *You would have been*

*enforced to compress your missive within … scanty bounds*—Palgrave, 1837; *They were prepared to take action with a view to enforcing this country into a premature and vanquished peace*—Fowler, 1926. A stray example heard on BBC Radio 4 on 22 May 1978 suggests perhaps that the use survives in the spoken language: *are the companies legally enforced to complete the forms?*—G. Clough. The most usual constructions now are to *enforce* (an action, conduct, one's will) on a person; and to *enforce* (a regulation). It is worth noting that the former variant spelling *inforce* survives in *reinforce*.

**enforceable.** Thus spelt; see MUTE E.

**enfranchise.** Always so spelt, not -*ize*. See -ISE 1.

**England, English.** England is of course the southern part of the island of Great Britain with the exception of Wales; but in practice, because the population of England is much greater than that of the other parts of the United Kingdom, because the seat of government is in London, and for broad historical reasons, the name is sometimes loosely used for the whole of Great Britain, a use that is understandably resented in Scotland, Wales, and Northern Ireland. The loose use is not restricted to English people: French *Angleterre*, It. *Inghilterra*, etc., are sometimes used in the same manner, and when a German speaks of *ein Engländer* he is not necessarily excluding the possibility that the person referred to is Scottish, Welsh, etc. It is a natural enough assumption that a person who speaks *English* as his or her native language is *ein Engländer*. For many purposes the wider words are the natural ones. We speak of the *British Commonwealth*, the *British Navy, Army*, and *Air Force*; we boast that *Britons* never never never shall be slaves; and *British English* is now the customary term used by all linguistic scholars to distinguish it from American, Australian, etc., English.

The loose use arises partly from the fact that people in the UK speak the *English* language, and are taught *English* history as one continuous set of events from Alfred to the present day; and there is Lord Nelson's famous signal at Trafalgar in 1805 that *England expects that every* man will do his duty. There are innumerable other patriotic or nostalgic lines of poetry or remarks of one kind or another in which it is not possible to be sure that the author had strict geographical boundaries in mind: e.g. *England is a nation of shopkeepers*—attributed to Napoleon; *England is the mother of parliaments*—John Bright, 1865; *The English have no respect for their language*—G. B. Shaw, 1916; *There'll always be an England*—popular song in the 1939–45 war. The loose use is doubtless regrettable, but it is persistent and the reasons for its use should be borne in mind by those who find it puzzling or objectionable.

**engraft, ingraft.** See EM- AND IM-.

**engrain.** See INGRAINED.

**enhance** means 'to heighten or intensify (qualities, powers, value, etc.); to improve (something already of good quality)' (*COD*, 1995). *Enhanced*, as ppl adj., and now frequently used of qualities, sound, radiation, and in spectroscopy of the lines of a metallic spectrum. In other words, *enhance* is correctly applied to things, values, reputations, etc., but cannot properly be used of people. *The book enhanced his reputation* is correct; *he was enhanced by the publication of his book* is not. Pronounce /mˈhɑːns/.

***enjambement*** /ɑ̃ʒɑ̃bmɑ̃/, if Anglicized as *enjambment*, is pronounced /mˈdʒæm(b)mənt/. It is used in prosody to mean the continuation of a sentence without a pause beyond the end of a line, couplet, or stanza.

**enjoin. 1** Used in the sense 'to prescribe authoritatively and with emphasis'. Fowler (1926) rejected 'the construction [of *enjoin*] with a personal object and an infinitive (*The advocates of compulsory service enjoin us to add a great army for home defence to . . .*)', and claimed that the OED examples had an 'archaic sound': e.g. *They injoined me to bring them something from London*—Steele, 1712; *The pope . . . advised and even enjoined him to return to his duties*—Froude, 1883. He recommended instead the type *to enjoin caution*, etc. *on* or *upon* (a person). Time has moved on, and both constructions now seem firmly established in standard use. Examples: *Nowadays analysts are enjoined to interpret*

everything that happens . . . in terms of transference*—M. Balint, 1968; *displayed the charity enjoined on Christians*—New Yorker, 1971 (WDEU); *The church had enjoined the faithful to say an Ave Maria*—B. Unsworth, 1985.

**2** Since the 16c., *enjoin* has also been used in the more or less opposite sense 'to prohibit, forbid (a thing); to prohibit (a person) *from* (a person or thing). This use survives principally in law, 'to prohibit or restrain by an injunction': *The Al-Fayed brothers . . . sought to enjoin* The Observer *from publishing the results of its continuing enquiries*—Observer, 1986; *because the* Times *by now had been enjoined from publication*—Bull. Amer. Acad. Arts & Sci., 1989.

**enjoy.** The natural and most common sense of the verb is 'to take delight or pleasure in'. From the 15c. onwards, however, it has also been used in a weaker sense to mean 'to have the use or benefit of (something which affords pleasure or some advantage)': e.g. *It* [sc. Allworthy's house] *stood . . . high enough to enjoy a most charming prospect*—Fielding, 1749; *Animals enjoying a much lower degree of intelligence*—W. B. Carpenter, 1874. From this weaker sense it is an easy step to what the OED calls the catachrestic use of *enjoy* with an object denoting something *not* pleasurable or advantageous. This use dates from the 16c. and is shown in expressions like *to enjoy poor health*, *to enjoy an indifferent reputation*. It is hard to see the logic of this sense but it seems to endure despite its inbuilt self-contradiction. Late 20c. examples of the original senses: *I enjoy very little leisure in the evenings*—P. Fitzgerald, 1978; *Select a sheltered spot which enjoys the sun for most of the day*—Practical Householder, 1990.

**enliven.** See LIVEN (verb).

**enmesh.** See EM- AND IM-.

**enormity, enormousness. 1** Origins. First let it be recalled that *enormous* is in origin a 16c. adaptation of L *ēnormis*, from *ē* 'out' + *norma* 'mason's square, pattern'; and that for some 400 years it stood alongside the shorter form *enorm* (now obs.) in the same range of senses. Both words were applied to anything deviating from the norm, whether in behaviour or nature ( = disorderly, irregular; wicked, monstrous, outrageous, etc.) or in size ( = abnormally large, vast, huge). By the end of the 19c. *enorm* had virtually gone, and *enormous* had but one sense, 'abnormally large'.

**2** Meanwhile the corresponding nominal forms *enormity* (15c., from Fr. *énormité*, = deviation from a normal standard of law, morality, etc., and from the late 18c. = hugeness, vastness) and *enormousness* (first recorded in the early 17c. = gross wickedness, and not until the early 19c. = vastness) competed with each other. It seems clear, however, that by the end of the 19c. a fairly clear distinction of meaning existed, or was believed to exist, between the two words. By then *enormousness* meant only 'hugeness of size', and *enormity* only 'extreme wickedness, etc.'. In other words one could speak of, say, *the enormousness of the pyramids*, and also of *the enormity of his crime*; but the two words were not interchangeable.

**3** Nowadays, in the majority of contexts, *enormousness* and *enormity* continue to be distinguished in most standard writings, or, to put it another way, *enormity* is usually restricted to contexts of crime, depravity, wickedness, etc., while large size is indicated by a range of synonymous nouns (*vast extent, hugeness, immensity, enormousness*, etc.). Because of its relative clumsiness, *enormousness* is not the most frequently used of these synonyms. But *enormity* seemingly cannot be kept within bounds and is encroaching on the territory set aside for *enormousness* and synonyms. There are also numerous circumstances in which the notions of wickedness and of hugeness coincide, and it is in these that the difficulty arises.

**4** Examples: (*enormity* used to mean 'some degree of wickedness') *I have got to the stage of disliking Randolph which is really more convenient than thinking I liked him & constantly trying to reconcile myself to his enormities*—E. Waugh, 1944; *I did not know how men so able, so humane, so uncorrupt as lawyers could stride on, turning a blind eye to the enormities of their profession*—J. Grimond, 1979; *He did not register the enormity of his crime*—H. Jacobson, 1983; (*enormity* = largeness (of immaterial concepts)) *Miss Witt ... frequently impressed upon George Osborne's mind the*

*enormity of the sacrifice he was making*—Thackeray, 1847/8; *But then habit commits its enormities quite casually*—W. McIlvanney, 1985; *As they lay in one another's arms ... the lovers were filled with a sense of the enormity of their love, and this took the form of deep sentimentality*—M. Bracewell, 1989; *I did not know then that one frequently fails to live up to the enormity of death*—A. Brookner, 1990; (*enormousness* = largeness) *The orange light of a lamp and the brown shadows around it have the manifold enormousness of a stared-at and altered dandelion*—New Yorker, 1987; *They [sc. the Great Hall of the People. etc., in Tiananmen Square] had become architectural dinosaurs, numbed in their own enormousness*—C. Thubron, 1987. It is not difficult to find examples of *enormity* used simply to denote great size or extent: *it was as though one had flown near enough to the sun to realize its monstrous enormity*—S. Sitwell, 1926 (WDEU); *A wide-angle lens captures the enormity of the Barbican Centre, London's new arts complex which has cost £153m and almost 11 years to build*—Times, 1982; *Frank ... wanting to protest that in spite of the enormity of the list* [of members of household staff] *he didn't live as grandly as all that*—P. Fitzgerald, 1988; *The enormity of such open spaces momentarily alarms her*—Susan Johnson, 1990 (Aust.).

**5** It is recommended that for the present *enormity* should not be used in plain contexts where the physical size of an object is the only feature involved: in other words, one should eschew the type *the enormity of the pyramids*. It is more difficult to find fault with *enormity* used of the size or immensity or overwhelmingness of abstract concepts, esp. when any element of departure from a legal, moral or social norm is present or is implied.

## enough, sufficient(ly).

**1** The distinction is often one of formality. As a broad rule the plain and vigorous word *enough* is to be preferred. Used as nouns, for example, in the sense 'an adequate amount', in most contexts *enough* is entirely adequate (*there was not enough for both of them*) and *sufficient* unnecessarily formal (*there was not sufficient for both of them*).

**2** Grammatical differences begin to emerge when they are used as adjectives (or, in modern parlance, as determiners).

In certain circumstances *enough* can be placed either before or after the noun it governs: *he had enough money to buy a car*; *he had money enough to buy a car*. The post-positioned use is fairly rare, but *sufficient* can never be used in this way. There is a further distinction. Whenever considerations of quality or kind are essential, *sufficient* is the better word. Compare *for want of sufficient investigation* with *there has been investigation enough*: the first implies that the investigation has not been thorough or skilful, the second that the time given to it has been excessive. Furthermore there are several non-count nouns with which *enough* cannot be used idiomatically: *a sufficient number* but not *\*an enough number*, *a sufficient supply* but not *\*an enough supply*.

**3** As adverbs, *enough* and *sufficiently* are mainly distinguished by the informality of the first and the relative formality of the second. The main grammatical difference between them is that *enough* is normally placed after the word it qualifies and *sufficiently* before: *he wasn't clever enough to understand the simplest thing*; *oddities in the judiciary are rare enough to be interesting* (but *sufficiently clever, sufficiently rare*).

**4** The best policy is always to use *enough*, not *sufficient* or *sufficiently*, unless it is obviously unidiomatic (see 2 above) to do so.

**enquire, enquiry, in-.** From the time that *enquire/inquire* came into English from OF *enquerre* (modF *enquérir*) in the 13c. until the early 20c., forms with initial *en-* or *in-* (also in the derivatives) were used with approximately equal frequency and with no difference of meaning. At present the *in-* forms are dominant for all senses in AmE, whereas in BrE the *en-* forms now tend to be restricted to the general sense 'ask a question' (*to enquire after her health*) and *inquire*, and esp. the noun *inquiry*, are used in contexts of formal investigations (*an inquiry agent, a police inquiry, the Council set up a special committee to inquire into the rumours about child abuse*). The distinction is, however, still far from absolute, as is borne out by the following examples: (article by a philosopher) *since this is a legitimate object of inquiry, it would be foolish to think that every assertion of a general*

*limitation on our understanding is an instance of prison theory—Encounter, 1987*; ( = request) *He parried her inquiry, 'I just thought I would ... have a longer look round.'—R. McCrum, 1991*. And it is worth noting that *CGEL* (1985) consistently uses *inquiry* to mean (in grammatical terms) a question: (§11.3) *I'd like to know the name of your last employer* [inquiry by statement].

In AmE the noun *inquiry* is often pronounced with the stress on the first syllable.

**enrol.** Thus spelt, with -*ll*- in the inflected forms (*enroller, enrolled, enrolling, enrolment*). In AmE the usual forms are *enroll, enroller, enrolled, enrolling, enrollment*).

**ensure, insure.** See ASSURE.

**entail.** Both the noun and the verb are stressed on the second syllable.

**enterprise.** Thus spelt, not -*ize*. See -ISE 1.

**enthral.** Thus spelt in BrE, with *enthralled, enthralling, enthralment* as the accompanying forms. In AmE the dominant pattern is *enthrall* (also *inthrall*), *enthralled, enthralling, enthrallment*.

**enthuse.** The credentials of this verb, which is an early 19c. AmE back-formation from *enthusiasm*, have been in question for a very long time and continue to be questioned. It serves the needs of journalists (*Edmund Blunden's sudden resignation, in mid-term, has enthused nobody—Guardian, 1968*) and of the racier kinds of novelists, but in general is held at arm's length by serious writers. It is more Edgar Wallace and Charles Chaplin (both cited by the *OED* as using it) than, say, Iris Murdoch (who is not). But Wilfred Owen used it in a letter that he wrote in 1912 (*I cannot enthuse over the things as Leslie does*), and one day it may be taken out of the drawer marked 'Use with caution' and form part of the unopposed vocabulary of the language.

**entitled** means (*a*) having a right (*to do something*) or a just claim (*to some advantage*); (*b*) (of a book, etc.) given the title of. It should not be used to mean bound (*to do*) or liable (*to a penalty*). Examples of misuse: *After the 1914–18* war Germany suffered bitterly and was entitled to *suffer for what she had done* (read *deserved to*); *because of the way the environment has been treated since the Industrial Revolution we are all* entitled to *certain penalties* (read *subject to*).

**entourage.** This early 19c. loanword from French is still pronounced in a French manner in English: /ɒntuˈrɑːʒ/ or /ˈɒntu-/

**entrée.** In cookery, an *entrée* is usually a dish served between the fish and meat courses, but also, esp. in America, the main dish of a meal.

**entrench, in-.** See EM- AND IM-.

**entrust, entwine, entwist.** See EM- AND IM-.

**enunciation.** See -CIATION, -TIATION.

**envelop** (verb). **1** The inflected forms are *enveloped, enveloping*.

**2** An older variant spelling with final -*e* (*envelope*), found as late as the 19c., is now obsolete.

**envelope** (for a letter). The Fr. spelling (-*ppe*) was ignored when the word was adopted in English in the early 18c. (first recorded in 1715), but since the adoption of the word the initial syllable has been pronounced either as /ˈen-/ or as /ˈɒn-/ with varying but now unestablishable degrees of frequency. The Anglicized form /ˈen-/ is now dominant and is recommended here.

**-en verbs from adjectives.** Fowler (1926) ventured to suggest that 'the average writer', if asked for an offhand opinion, would probably say that any adjective of one syllable had the power to produce a parallel verb ending in -*en*: thus *black* and *blacken*, *sad* and *sadden*, and so on. He then went on to show that the language did not work like that. So, for example, *moist* and *moisten* but no *wetten* to answer to *wet*. There are about fifty verbs ending in -en that have been formed from adjectives, including: *blacken, brighten, broaden, cheapen, coarsen, dampen, darken, deaden, deafen, deepen, fasten, fatten, flatten, freshen, gladden, harden, lessen, lighten, liken, loosen, madden, moisten, quicken, quieten, redden, ripen, roughen, sadden, sharpen, shorten, sicken,* *slacken, smarten, smoothen, soften, stiffen, stouten, straighten, sweeten, tauten, thicken, tighten, toughen, weaken, whiten, widen,* and *worsen*.

Examine the history of such pairs and it emerges that the process was at its most productive between 1200 and 1700. There is only one possible pair before the Conquest—the OE antecedents of *fast* and *fasten*—but the relationship of the two words is not at all straightforward. Sixteen of the words (*blacken, brighten, darken,* etc.) are first recorded from the ME period, i.e. effectively between 1200 and the late 15c. They turn up in official papers of the period, in private letters, in theological tracts, and in poems. Just over twenty more (*cheapen, dampen, deaden,* etc.) entered the language in the 16th and 17th centuries. *Flatten* is first recorded in a poem by John Donne (1630) and *shorten* in Thomas More's *History of Kyng Richard the Third* (1513). By 1700 the formative power of the suffix had largely disappeared. The 18c. seems to have yielded only *broaden, madden,* and *tighten*. Dr Johnson illustrated his entry for the verb *broaden* with an example from Thomson's *Seasons: Low walks the sun, and broadens by degrees, Just o'er the verge of day,* with the comment 'I know not whether the word occurs, but in the following passage'. The *OED* showed that it did occur elsewhere but could not find examples before Thomson. *Madden* is first found in a work (1735) by Pope, and *tighten* in one of Nathan Bailey's dictionaries (1727). Interest quickened in such formations in the bookish world of the 19c. (*smoothen*, first recorded in the 17c., was 'in frequent use *circa* 1820–30, esp. by Landor', according to the *OED*), but the dozen or so words of this type that are entered in the *OED* are nearly all marked 'rare' or else plainly were brought into existence to serve the metrical or other special needs of Victorian writers: e.g. *Eyes, large always, slowly largen*—C. Patmore, 1844; *When the child of God suffers his thoughts to wander, his affections to colden*—A. B. Grosart, 1863.

The suffix is no longer a living one. I have no record of any -*en* verb formed in the 20c. from a one-syllabled adjective. The most recent one noticed is *neaten*, which is first recorded in 1898. Technically, adjectives like *fab* (first recorded in 1961) and *naff* (1959) could produce *fabben* and *naffen* respectively as verbs, but it seems unlikely that they will.

Historically the emergence of *-en* verbs from adjectives is not as simple a process as I have so far presented it. It is not just a case of adjectives like *dead* yielding verbs like *deaden*. The word-forming patterns fall into several categories: (pairs in which the same form is used for both adjective and verb) *blind, foul, lame, still, wet*, etc.; (adjectives which have no corresponding verb in *-en* and no longer function as verbs themselves) *cold, good, grand, sore*, etc.; (a prefix added to form a verb rather than the suffix *-en*) *dense/condense, large/enlarge, new/renew, strange/estrange*, etc.; (no *-en* verb though the antonymic adjective has one) *cheapen* but not *dearen, deafen* but not *blinden, fatten* but not *thinnen* or *leanen, sweeten* but not *souren*, etc. There is another important factor at work. Nearly all the *-en* verbs that first came into existence in the 16c. and 17c. were preceded by words that were spelt the same as adjectives and as verbs. Thus *deep* already existed as a verb in the Anglo-Saxon period; *deepen* did not join it and compete with it until the 17c. *Moist* as a verb was used by Wyclif and Langland in the 14c. before being joined by *moisten* in the 16c. All these battles took place a long time ago. The victorious forms have emerged. Only the remnants of the old conflicts between *-en* verbs and their unextended forms remain, and this kind of conflict is not bothering anyone now.

**environs.** The pronunciation recommended is /ɪn'vaɪrənz/. An older pronunciation, /'envɪrənz/, has been driven out under the influence of the stress-pattern and long *i*, i.e. /-aɪ-/, of *environment*.

**Eocene.** See MIOCENE.

**envisage, envision.** *Envisage* is an early 19c. loanword from French meaning, at first, 'to look in the face of, to look straight at', and then (its current meaning) 'to view or regard under a particular aspect'. Fowler (1926) dismissed it as an 'undesirable Gallicism', and thought that *face, confront, contemplate, recognize, realize, view*, and *regard* seemed 'equal between them to all requirements'. Gowers (1965) added *imagine, intend*, and *visualize* to the list of words for which *envisage* was a 'pretentious substitute'.

Neither of them noticed the arrival, first in Britain (1921) and then strongly in America, of the more or less synonymous *envision* 'to see or foresee as in a vision'. From the evidence before me *envision* is now strongly favoured in AmE and *envisage* in BrE but the division is not complete. Examples: (envisage) *To envisage circumstance, all calm, That is the top of sovereignty*—Keats,1820; *Men continually envisaged the highest benefits which their souls could attain*—R. Chenevix Trench, 1845–6; *So mother envisaged us all here, gathered round staring down in this ghastly way*—P. Lively, 1989; *she continued to envisage various methods of killing Jack*—I. Murdoch, 1989; *Mr King ... spoke more diplomatically, emphasizing that he did not envisage any 'immediate' change in force levels*—Times, 1990; (envision) *His blackest hypochondria had never envisioned quite so miserable a Catastrophe*—L. Strachey, 1921; *And the more he envisioned this prospect, the more he was of two minds about it*—M. McCarthy, 1952; *He sat with his $650 New & Lingwood shoes pulled up ... envisioning Campbell, his eyes brimming with tears, leaving the marbled entry hall ... for the last time*—T. Wolfe, 1987; *he did not envision any basic change in the social structure or the standard of living*—Bull. Amer. Acad. Arts & Sci., 1989; *It may be only the stuff of newspaper editorials, of course, to envision a strategy in which the United Nations takes decisive action*—leader in Sunday Times, 1990.

**envoi.** Also *envoy*. In prosody, a somewhat archaic word for a short, and often metrically unmatching, stanza concluding a BALLADE.

**enwrap, inwrap.** See EM- AND IM-.

**eon.** A frequent AmE spelling of BrE *aeon*.

**epaulette** (AmE *epaulet*) is most often pronounced /'epəlet/ but variants, e.g. /'epɔːlet/ and /epə'let/, are still commonly heard.

**épée** (fencing-sword) is so spelt, i.e. with two accents.

**epenthesis.** In philology, applied to the phonetic process by which an unetymological sound is added to a given word when it is not present in its antecedent.

Examples: *chimney* pronounced as /'tʃɪm-blɪ/, with substitution of /bl/ for /mn/; *remembrance* pronounced as *rememb-er-ance*. Historical examples: *daffodil* from earlier *affodil* (current from the 15c. to the 17c.), from medL. *affodillus* (related to *asphodel*); *empty* from OE *æmtig* (no -*p*-). *Epenthesis* is derived from Gk ἐπένθεσις, from ἐπί 'in addition' + ἐν 'in' + θέσις 'placing'.

**epergne** /ɪ'pɜːn/, an ornament or dish, esp. in branched form, for the centre of a dinner-table. It has the appearance of being a French word, but is not one. First recorded in the 18c., it is sometimes said to be a corruption of Fr. *épargne* 'saving, economy', but the sense-connection is far from obvious.

**epexegesis** (lit. 'additional explanation'). In grammar, 'the addition of a word or words to convey more clearly the meaning implied, or the specific sense intended, in a preceding word or sentence; a word or words added for this purpose' (*OED*). English examples are *to find* in *This is very sad to find* and *to do* in *It is very difficult to do*. The word is derived from Gk ἐπεξήγησις (cf. ἐπί 'in addition', ἐξηγεῖσθαι 'to explain'). Cf. *exegesis*.

**epic.** The traditional use of this word is of a narrative poem celebrating the achievements of some heroic person of history or legend, such as the *Odyssey*, the *Iliad*, the *Æneid*, or Milton's *Paradise Lost*. The term has also frequently been applied to the OE elegiac poem *Beowulf*. Anything less ambitious, it was felt, did not deserve to be called an epic. Our more permissive age has witnessed the term being applied to any major literary work, theatrical performance, film, sporting event, etc, that stands out, or is claimed to stand out, from the ordinary; hence headlines like *Fabulous Hollywood Epic!* and *Epic Contest at Twickenham*. Examples: [Joyce's] Ulysses, *then, is a complete prose epic*—N. Frye, 1957; *The budget was supposed to have guaranteed an action-packed epic*—Movie, 1965.

**epicene.** **1** Some of the earliest English grammarians used the word *epicene* of one of several kinds of grammatical gender allegedly found in English. Ben Jonson (1640), for example, after describing the masculine, feminine, and neuter genders of English nouns went on to describe a fourth type: *the Promiscuous, or Epicene, which understands both kindes: especially, when we cannot make the difference; as, when we call them Horses, and Dogges, in the Masculine, though there be Bitches, and Mares amongst them* ... This use of the word has long since been abandoned. It reflected the use of the corresponding adjective in Greek and Latin to denote nouns which, without changing their grammatical gender, may denote either sex. English words like *author*, *doctor*, *teacher*, etc., are usually said to be of *common* gender.

**2** The dominant senses of the word in the 20c. are (*a*) having characteristics of both sexes; (*b*) having no characteristics of either sex; (*c*) (of a man) effete, effeminate.

**Epicurean.** See HEDONIST.

**epidemic.** See ENDEMIC.

**epidermis.** See POLYSYLLABIC HUMOUR.

**epigram** (lit. 'on-writing') has two main senses: **1** A short poem with a witty or ingenious ending. Baldick (1990) cites an epigram by Herrick which is adapted from Martial: *Lulls swears he is all heart, but you'll suppose By his proboscis that he is all nose.*

**2** A concise pungent saying, e.g. *To lose one parent, Mr Worthing, may be regarded as a misfortune; to lose both looks like carelessness*—O. Wilde, 1895. It also formerly meant an inscription on a building, tomb, coin, etc.

**epigraph.** **1** An inscription, esp. one placed upon a building, tomb, statue, coin, etc.

**2** A short quotation or pithy sentence placed at the beginning of a work, a chapter, etc., as a foretaste of the leading idea or sentiment in what follows.

**epigraphy.** The professional study of inscriptions. Cf. prec. (1).

**episcopalian,** belonging or referring to an episcopal church (i.e. one constituted on the principle of government by bishops), esp. (with initial capital) the Anglican Church in Scotland and the US, with elected bishops.

**epistemic modality.** From the branch of logic that deals with modality, modern grammarians have identified two kinds of grammatical possibility and necessity. Rodney Huddleston calls them *epistemic modality* and *deontic modality*, and, in his *Introduction to the Grammar of English* (1984), p. 166, he sets the two types down in this manner:

|            | Epistemic                               | Deontic                                |
|------------|-----------------------------------------|----------------------------------------|
| Possibility | i *You may be under a misapprehension* | ii *You may take as many as you like*  |
| Necessity   | iii *You must be out of your mind*     | iv *You must work harder*              |

In other words, epistemic modality signifies a qualifying or lessening or modification of certainty: cf. (*a*) *'Who is your favourite children's writer?' 'It has to be Beatrix Potter.'* with (*b*) *He had to finish writing the book by 31 December* = he was required (by some circumstance) to finish the book by then. Contrast also the epistemic modality of *There must be some mistake* (i.e. it is highly likely that there is a mistake) and *You must be home by midnight* (deontic obligation).

**epistle.** Its commonest use is of the letters of the New Testament (e.g. The Epistle of Paul the Apostle to the Romans). It is sometimes used playfully as a synonym for a letter of any kind.

**epithet** 1 An epithet is an adjective indicating some quality or attribute (good or bad) which the speaker or writer regards as characteristic of the person or thing described, as *Alfred the Great, Charles the Bold* (Holy Roman Emperor from 875). Dr Johnson cited the following example from Swift in his Dictionary (1755): *I affirm with phlegm, leaving the epithets of false, scandalous and villainous to the author.* It is also a noun used as a significant appellation, whether favourable or the reverse: *He assumed the proud Epithet of Sultan or Monarch of Tunis and all Barbary*—J. Morgan, 1728. Saddam Hussein earned the epithet 'Butcher of Baghdad'. These uses continue, particularly with unfavourable connotations. Examples: (favourable) *Geraldine recited*

*some of the traditional epithets used to describe Siva*—G. Vidal, 1978; *How often he would apply the epithet 'de la Marish' to some scene*—L. Whistler, 1985; (unfavourable) *The day the trial began, a group of young men in a provincially owned truck roared past the courthouse shouting epithets at Indian leaders*—NY Times, 1987; *The epithets of liar, racist and worse hurled at the vice president just won't stick*—Washington Times, 1988.

2 For the term *transferred epithet*, see ADJECTIVE 10.

3 The *Homeric epithet* is an adjective (in practice usually a compound one) added to the same thing or person. In Homer we find *the wine-dark sea*, while Hera is repeatedly described as *white-armed*, Athene as *bright-eyed*, and Zeus as *the cloud-gatherer*. From these it is an easy step to the kind of combinations so frequently used by English poets: *Rose-crown'd Zephyrus ... makes the green trees to buss*—Sylvester, 1598; *And still she slept an azure-lidded sleep*—Keats, 1820.

**epitome** /ɪ'pɪtəmi/. 1 Derived from a Greek word of the same spelling meaning 'an incision, an abridgement', it has two main meanings in English: (*a*) a person or thing embodying a quality, class, etc.; an embodiment; (*b*) a summary or abstract, esp. of a written work. Examples: *Some delight in abstracts and epitomes* (Dr Johnson); *Adolf Hitler was the epitome of evil; to win a Wimbledon title is regarded by every player as the epitome of success.* Some 20c. examples: *Local government was an epitome of national government*—W. Holtby, 1936; *The book ... is not intended to be popular. No doubt a lively epitome will one day be made for general reading*—E. Waugh, 1956; *The flat country of baobab, mopani and malala palm trees ... is the epitome of underdevelopment*—Times, 1990.

2 Surprisingly, it is occas. spelt *epitomy* even in quality newspapers: *Ventry, who was unmarried, was the epitomy of an English gentleman*—Times, 1987; *Mr Yeltsin was seen as the epitomy of the new breed of Soviet politicians*—Times, 1987. This spelling is non-standard.

**epoch.** 1 See TIME.

2 In 1816 Coleridge wrote of *the epoch-forming revolutions of the Christian world.* Later in the 19c. the adj. *epoch-making*

appeared, and the new word has been regularly and sometimes boastfully applied since then to any remarkable or sensational event, publication, etc., to the point that *epochs* must now be deemed to be occurring at very frequent intervals indeed. Examples: *I can hardly ever bring myself to make a decision at all. When I do manage it, it seems to me so epoch-making that I never choose the right time and place for an announcement*—P. H. Johnson, 1962; *This was an epoch-making moment in the history of Egypt, like the day a dam bursts*—N. Barber, 1984.

**epode** (from a Greek word meaning 'additional song'). In Greek choruses (and also in Milton's *Samson Agonistes*), the third part, distinct in metre, following the strophe and the antistrophe. The term was also used for a Greek metre invented by the lyric poet Archilochus (7c. BC) composed in couplets in which a long line was followed by a shorter one. Horace loosely followed this metre in his *Epodes*: the metre used in the book is one in which a full iambic line is followed by a shorter one regarded metrically as a mere appendage or 'added verse'.

**eponym,** a person after whom a discovery, invention, place, institution, etc., is named: thus Elias *Ashmole* (1617–92) after whom the Ashmolean Museum in Oxford was named, and Sir Thomas *Bodley* (1545–1613) who gave his name to the Bodleian Library in Oxford. Beowulf is the eponymous hero of the poem of that name, and Robinson Crusoe the eponymous hero of *The Life and Strange and Surprising Adventures of Robinson Crusoe* by Defoe. Legions of products, processes, etc., are named after or are believed to be named after particular people: in each case the person concerned, and also the product if it is identical in spelling or pronunciation, is an eponym. Examples: *Braille* (Louis Braille, 1809–52, French inventor), *clerihew* (E. Clerihew Bentley, 1875–1956), *diesel* (Rudolf Diesel, 1858–1913, German engineer), *mackintosh* (Charles Macintosh, 1766–1843), *Morse* (S. F. B. Morse, 1791–1872, American inventor), *sandwich* (4th Earl of Sandwich, 1718–92).

*e pur si muove* (also *eppur si muove*), a phr. attributed to Galileo (1564–1642) supposedly made after his recantation (in 1632) of belief in the Copernican system. It means only 'and yet it moves', 'but it does move', and the reference is to the Earth.

**equable.** The quality indicated is complex—not merely freedom from great changes, but also remoteness from either extreme, a compound of uniformity and moderation. A continuously cold climate or a consistently violent temper is not equable; nor on the other hand is a moderate but changeable climate or a pulse that varies frequently though within narrow limits.

**equal.** 1 The inflected forms have -ll- (*equalled, equalling*) in BrE but usu. -l- in AmE.

2 *The navy is not equal in numbers or in strength to* perform *the task it will be called upon to undertake*: perform *should be* performing. See GERUND 3.

3 *This work is the equal, if not better than anything its author has yet done. Equal* lends itself to this blunder: read *equal to, if not better than, anything ...*

4 Equality cannot normally be graded. It is idiomatic to say *almost equal, not equal, less than equal,* or *exactly equal,* but *more equal* is more questionable. It can mean 'more nearly equal': e.g. *The more to draw his Love, And render me more equal*—Milton, 1667; *the American woman is not the same as other women.... she is freer in her manners and ... she is freer because she is more equal* —H. Fairlie, 1976–7 (WDEU). George Orwell was memorably out of line when he wrote *All animals are equal but some animals are more equal than others* in *Animal Farm* (1945), but the use seems to have come to stay, and allusions to Orwell's context are very frequent: e.g. *All victims are equal. None are more equal than others*—J. le Carré, 1989.

**equally.** Over the centuries *equally* meaning 'to an equal degree or extent', has normally been construed with *with* but sometimes with *as*: e.g. *Being the inventor of the lyre, he* [sc. Hermes] *is patron of poets equally as Apollo*—F. W. Newman, 1853; *This work is equally one-sided and uncompromising with Wyclif's tracts*—T. Arnold, 18.. Fowler (1926) condemned the construction with *as* ('The use of *equally as* instead of either *equally* or *as* by itself is an illiterate tautology'), and the echoes of his condemnation rumble on. In practice, because of the presence of an element of redundancy in *equally as*, it is desirable in good writing to use one or other of the alternative constructions instead: *The labour crisis has furnished evidence just as* (not *equally as*) *striking*; *He was outplayed by a man with a game more original in tactics and as* (not *equally as*) *severe as his own*; *The expansion of the Canadian labour force will be as large* (or *as extensive* not *equally large*) *as in the last five years*.

**equerry.** At court the word is pronounced /ɪˈkwerɪ/, but in most other contexts the stress is placed on the first syllable, /ˈekwərɪ/, probably arising from the erroneous idea that the word is connected with L *equus* 'horse'. In fact it is derived from Fr. *écurie*, earlier *escurie*, from medL *scūria* 'stable'. *Equerry* is short for 'gentleman (or groom) of the equerry'.

**equilibrium.** Pl. either *equilibria* or *equilibriums*. See -UM 3.

**equivalence, equivalency.** Both forms are in use, the first of them being much the more common. Except in technical uses (in most of which *equivalence* is the only form used), the choice of form seems to be based more on the rhythm of the sentence than on any discernible semantic difference. See -CE, -CY.

**era.** See TIME.

**-er and -est, more and most.** The subject is dealt with briefly s.v. ADJECTIVE 3, 4, ABSOLUTE COMPARATIVE, and ABSOLUTE SUPERLATIVE. A more detailed treatment follows (owing a good deal to Fowler 1926).

> 1 The normal *-er* and *-est* adjectives.
> 2 Other common *-er* and *-est* adjectives.
> 3 *-er* and *-est* in adverbs.
> 4 Adjectives tolerating *-est* but not *-er*.
> 5 Stylistic extension of *-er* and *-est*.
> 6 Emotional *-est* without *the*.
> 7 Superlatives in comparisons of two.
> 8 Combinations.

**1** The normal *-er* and *-est* adjectives. The adjectives regularly taking *-er* and *-est* in preference to *more* and *most* are: (*a*) most monosyllables (*hard*, *rich*, *wise*, etc.; but not *like*, *own*, *real*, *right*, *wrong*, *French*, etc.); (*b*) disyllables in *-y* and *-ly* (*angry*, *early*, *holy*, *lazy*, *likely*, *lively*, etc.), in *-le* (*able*, *humble*, *noble*, *simple*, etc.), in *-er* (*bitter*, *clever*, *slender*, *tender*, etc.; but *more eager*), and in *-ow* (*mellow*, *narrow*, *shallow*, etc.); (*c*) many disyllables with accent on the last (*polite*, *profound*, etc.; but not *antique*, *bizarre*, *secure*, etc., nor the predicative adjectives *afraid*, *alive*, *alone*, *aware*, etc.); (*d*) trisyllabic negative forms of (*b*) and (*c*) words (*unholy*, *ignoble*, etc.). Adjectives of three or more syllables require the periphrastic forms with *more* or *most*: *more beautiful/the most beautiful*, *more interesting/most interesting*.

**2** Other common *-er* and *-est* adjectives. Some other disyllables in everyday use not classifiable under terminations, as *common*, *cruel*, *pleasant*, *quiet*, etc. (but not *constant*, *sudden*, etc.), prefer *-er* and *-est*. Some others, e.g. *awkward*, *buxom*, *crooked*, can take *-er* and *-est* without disagreeably challenging attention.

**3** *-er* and *-est* in adverbs. Adverbs not formed with *-ly* from adjectives, but identical in form with them, use *-er* and *-est* naturally (*runs faster*, *hits hardest*, *sleeps sounder*, *hold it tighter*); some independent adverbs, as *often*, *seldom*, *soon*, do the same; *-ly* adverbs, though comparatives in *-lier* are possible in archaic or poetic style (*softlier nurtured*, *wiselier said*), now normally appear as *more softly*, etc. The phrase *easier said than done* is a special case, in that there is no equivalent use of the positive *easy*.

**4** Adjectives tolerating *-est* but not *-er*. Many adjectives besides those described in (1) and (2) are capable in ordinary use, i.e. without the stylistic taint illustrated in (5) and (6), of forming a superlative in *-est*, used with *the* and serving as an emphatic form simply, while no one would think of making a comparative in *-er* for them: *in the cheerfullest*, *cunningest drunkenest manner*. It is a matter of deciding whether the *-est* form or the *most* — form feels more natural. Some of Fowler's 1926 examples (*in the brutalest*, *civilest*, *damnablest manner*) now sound distinctly unidiomatic. Adjectives ending in *-ic* (*comic*, *rustic*, etc.), *-ive* (*active*, *restive*, etc.), and *-ous* (*famous*, *odious*, *virtuous*, etc.) do not tolerate *-er* and *-est* forms except in special circumstances.

**5** Stylistic extension of *-er* and *-est*. As a stylistic device, it was once open to writers to extend the use of *-er* and *-est* to many adjectives of more than one syllable that normally take *more* and *most*. Historical examples (listed by Jespersen): *easiliest, freelier, proudlier, wiselier* (all from Shakespeare); *finelier, harshlier, kindlier, proudlier* (all from Lamb); *darklier, gladlier, looselier, plainlier* (all from Tennyson); *proudliest* (Carlyle); *neatliest* (G. Eliot); and many more. In the 20c., formations such as *admirablest, loathsomer, peacefullest, wholesomer*, etc., could only be used for comic effect or for some other special purpose.

**6** Emotional *-est* without *the*. For *In Darkest Africa* it may perhaps be pleaded that the title was deliberately chosen for its emotional content. But the use turns up from time to time in unemotional contexts: *Mlle Nau, an actress of considerable technical skill and a valuable power of exhibiting* deepest *emotion; An extraordinary announcement is made tonight, which is bound to stir* profoundest *interest among all civilized peoples; Mr Vanderlip is, therefore, in closest touch with the affairs of international finance*. Ways out of the difficulty include the replacement of ——est by (a) *very* ——, or *the most* ——, or just *the* ——est.

**7** Superlatives in comparisons of two. *They were forced to give an answer which was not their real answer but only the nearest to it of two alternatives*. This use of *-est* instead of *-er* where the persons or things are no more than two should normally be avoided; the *raison d'être* of the comparative is to compare two things, and it should be allowed to do its job without encroachment by the superlative. *Nearer* should have been used in the example given. But there are exceptions. Use of the superlative is idiomatic in such phrases as *Put your best foot foremost; May the best man win; Mother knows best*. And who would wish to introduce a comparative into Milton's *Whose God is strongest, thine or mine?* See also SUPERLATIVES 2.

**8** Combinations. Jespersen pointed out that an adjective used as the first element of a combination ending in *-ed* is very often in the superlative. Examples: *the sourest-natured dog* (Shakespeare), *noblest-minded* (Shakespeare), *the longest lived* (I. Walton), *the silliest, best-natured wretch* (Swift), *the mildest mannered man* (Byron), *the softest-hearted simpleton* (R. L. Stevenson).

**-er and -or.** **1** The agent termination *-er* can theoretically be joined to any existing English verb. In practice, many such words (and there are about 100 of them in common use) have *-er* as a termination and others have *-or*. Still others have both forms, often with a different meaning. The *OED* expresses the matter thus: 'The distinction between *-er* and *-or* as the ending of agent-nouns is purely historical and orthographical: in the present spoken language they are both pronounced /ə(r)/, except that in law terms and in certain Lat. words not fully naturalized, *-or* is still sounded /ɔː(r)/. In received spelling, the choice between the two forms is often capricious, or determined by other than historical reasons.' Scholarly attempts to account for the distribution of the *-er* and *-or* forms continue to be made (e.g. by L. Bauer in *Trans. Philol. Soc.*, 1990), but the problem remains unresolved. In this book, the most important of the individual words have been dealt with at their alphabetical places. See ACCEPTER; ADAPTER; ADVISER, etc.; -OR.

**2** When *-er* is added to verbs ending in *-y* following a consonant, *y* is normally changed to *i* (*carrier, occupier*, etc.), though *flyer* is now more usual than *flier*, and *drier* alternates with *dryer*. The *y* is retained between a vowel and *-er* (*buyer, employer, player*, etc.).

**-er and -re.** See -RE AND -ER.

**eraser.** Thus spelt, not *-or*. See -OR.

**ere.** Both as preposition and conjunction, *ere* has been clinging to the language with diminishing success for more than a century. It is an outstanding example of the near demise of a word that for centuries, from the OE period onward was part of the central equipment of the language. It is now only archaic or poetical. Examples: *And time seemed finished ere the ship passed by*—E. Muir, 1925; *I would give you a gift ere we go, at your own choosing*—J. R. R. Tolkien, 1954; *In that cluster of villages, London by name, Ere slabs are too tall and we Cockneys too few*, [etc.]—J. Betjeman, 1958.

**ergative.** In certain languages, including Eskimo and Basque, a grammatical case used in 'constructions where there is a formal parallel between the object of a transitive verb and the subject of an intransitive one (i.e. they display the same case)... In this approach, sentences such as *the window broke* and *the man broke the window* would be analysed "ergatively": the subject of the intransitive use of *broke* is the same as the object of its transitive use, and the agent of the action is thus said to appear as the "ergative subject"' (D. Crystal, 1980). Attempts have been made to identify examples of an ergative case in English, esp. in such contrasted types as *John opened the door* versus *the door opened*. Those wishing to pursue the matter further may do so by consulting John Lyons, *Introduction to Theoretical Linguistics* (1969) and, for example, an article in the 1989 issue of the linguistic journal *Dictionaries*.

*ergo.* The Latin word for 'therefore' is occasionally used in English, as an alphabetical equivalent of ∴, to precede a logical conclusion. Its most famous occurrence at a later date is that of Descartes in his *Discours de la méthode* (1637), *Cogito, ergo sum*, 'I think, therefore I am': this was facetiously echoed in a recent advertising slogan, 'I think, therefore IBM'. Examples: *The American Colony hotel is in the eastern, ergo Arab, part of the city*—C. James, 1983; *This wine reminds me of Chateau Latour, ergo it is fine wine*—M. Kramer, 1989; *He had not been told about her death; he was not saddened by it; ergo, he was not part of the Literary World*—S. Mackay, 1992.

**Eros.** In all of its uses in English, including (a) the name of the well-known statue in Piccadilly Circus, (b) = earthly or sexual love, contrasted with *agape*, (c) as the name of an asteroid discovered by a German astronomer in 1898, it is pronounced /ˈɪərɒs/, and thus in this respect has parted company with its Greek original, ἔρως, i.e. /ˈerəʊs/ or /ˈerəʊz/.

**erotica.** In practice this word, a neuter pl. noun in Greek meaning 'erotic literature or art', occurs mostly in contexts where its number does not arise (e.g. *literature classed under erotica; the erotica industry; the world of erotica*). If it governs a verb it should properly be regarded as a plural (*Erotica are much in evidence in the world of videos*) but is sometimes treated as a non-count (singular) noun: *Is all erotica male fantasy then?—Newsweek*, 1973 (WDEU).

**err.** Dialect speakers vary considerably in the way in which they pronounce *err*, *erring*, and *errant*, but in standard English the three words are regularly /ɜː/, /ˈɜːrɪŋ/, but /ˈerənt/.

**errant.** See ARRANT.

**erratum.** An *erratum slip* is one containing a single correction. A list headed *Errata* contains details of more than one correction. In other words *erratum* is the sing. and *errata* the pl. They are pronounced /eˈrɑːtəm/ and /eˈrɑːtə/ respectively. *Errata* is not comfortably used except as a heading. Sentences like the following sound very clumsy: *The Errata, if included, is a list of errors*. The pl. form is correctly used in this example: *These errata are of more than usual interest—Library*, 1988. See -UM 2.

**ersatz.** This German loanword meaning 'a substitute or imitation, i.e. something inferior', was first recorded in English in 1875 and became much used in English during the 1914–18 and 1939–45 wars. It is still liable to be printed in italic, and pronounced in a somewhat foreign manner as /ˈeərzats/; but for the most part it is now used attributively or as an adjective (*ersatz coffee, an ersatz culture*) and pronounced /ˈɜːzæts/ or /ˈɜːsæts/.

**erst, erstwhile.** It is always sad to see words of long standing passing into archaism, but these two words, the first of them found already in OE and the second a Spenserian coinage in 1569, both began to drop out of regular use at some point in the 19c., though *erstwhile* is still not all that infrequent when used as an adjective. *Erst* is an adverb meaning 'formerly, of old', and *erstwhile* is both an adjective ( = former, previous) and an adverb ( = erst). Modern examples of *erstwhile*: *Jude the Obscure in reverse; erstwhile scholar who transformed himself into a rustic swain*—A. Powell, 1976; *Many erstwhile Green Line travellers were doubtless driving their own cars*—K. Warren, 1980. See QUONDAM; WHILOM.

**escalate.** This 20c. back-formation from *escalator* looked for a time to be one of the most overused and therefore unwelcome words in the language. So too the corresponding noun *escalation*. *Escalate* was used briefly to mean 'to travel on an escalator' but that use seems to have dropped out. The primary sense is a figurative one, 'to increase or develop by successive stages', and as such is a useful addition to the language. *Escalator* continues to have both a literal sense (moving staircase) and figurative ones (*escalator clauses*, etc.). Examples: (escalate) *Peace as well as war can escalate*—Oxford Mail, 1963; *Only a tiny percentage of cannabis-smokers escalate to heroin*—Listener, 1967; *A dispute arose, initially involving one man, then 180 men, and then escalated until some 14,000 employees at both plants went out on indefinite strike*—M. Edwardes, 1983; *The raffle escalated into something much bigger*—Business Franchise, 1989; (escalation) *The risk of starting the process of escalation towards total war*—New Statesman, 1959; *Both leaders ... were anxious to check the increasingly painful escalation of defence costs*—Ann. Reg. 1963, 1964; '*Escalation' was a fashionable word last time I was in London*—B. W. Aldiss, 1980; (escalator, in fig. uses) *Labor leaders have never liked cost-of-living 'escalator' contracts, on the grounds that they tie the worker to a fixed standard of living*—Time, 1948; *Cattle prices are subject to escalator adjustment*—Farmer & Stockbreeder, 1960; *It is a ninety-nine-year lease, with an escalator clause that ties the annual payments to the cost of living index*—J. D. MacDonald, 1977.

**escape.** The verb is used transitively in a range of senses meaning broadly 'to elude; to avoid'. There are examples in the OED (e.g. *In a Work of this Nature it is impossible to escape Mistakes*, 1669; *He seems to have escaped suspicion*, 1751) and in modern dictionaries (*escape poverty, punishment, unhappiness*, etc.; *it escapes me for a moment why* [etc.]; *escape one's attention*). In the second half of the 20c. an older transitive use of the verb has been revived. It is used of astronauts escaping the earth's gravitational pull, and of a person escaping (instead of escaping *from*) a gaol, a mental home, etc. The following example happens to refer to a flight from a convent: *It transpires she may have escaped Santa Clara to look for a well-known terrorist*—N. Shakespeare, 1989. This modern transitive use still sounds somewhat unidiomatic (at least in the UK)—one's ears yearn for the insertion of *from*—but there is no mistaking its existence.

**escapee.** See -EE[1] 2. This remarkable use (instead of *escaper*; cf. *deserter*), which was first recorded in a work of 1875–6 by Walt Whitman, has come in for much adverse comment, but is now entered in all major dictionaries as part of the standard language.

**eschscholtzia.** The pronunciation recommended is /ɪˈʃɒltsɪə/, not /ɪsˈk-/ or /-ɒlʃə/; and the spelling recommended is as shown (named after J. F. von Eschscholtz, German botanist, d. 1831).

**escort** (noun and verb). The noun is stressed on the first syllable and the verb on the second. See NOUN AND VERB ACCENT 1.

**Eskimo.** Pl. *Eskimos* (but sometimes just *Eskimo*). See -O(E)S. The Fr. pl. *Esquimaux* was once also used in English travel books and the like. The Eskimo term *Inuit* ( = pl. of *inuk* person) is preferred by the people themselves.

**esophagus.** The AmE spelling of BrE *oesophagus*.

**esoteric.** See EXOTERIC.

**especial(ly), special(ly).** In the *Weekend Telegraph*, 11 Aug. 1990, I wrote *There are also dictionaries especially designed to help secretaries with spelling and word-division*. A reader reproved me and insisted that I should have written *specially* not *especially*. He was probably right, but the margin between the two words is a narrow and ragged one.

**1** History. The contest between the longer and the shorter forms began a long time ago. Both *especial* and *especially* have been in continuous use since about 1400, while *special* and *specially* have had a slightly longer life, both being recorded first in the 13c. As time went on the shorter forms (particularly the adjective) tended to force out the longer ones.

**2** Sense distinctions. As the OED pointed out, the primary meaning of L. *speciālis* is 'belonging to or concerned with a particular species, special as opposed to general'. OFr. *especial* (derived from it) developed the secondary sense 'pre-eminent, important'. Over the centuries, English *special* tended to be 'preferred in applications arising proximately from the primary sense, while *especial* is chiefly confined to the derivative sense. The distinction is still more marked in the adverbs *especially, specially'* (OED). Fowler (1926), more cryptically, expressed essentially the same view: 'The characteristic sense of the longer adjective and adverb is pre-eminence or the particular as opposed to the ordinary, that of the others being limitation or the particular as opposed to the general.' Fowler also noted, however, that 'there is a marked tendency in the adjectives for *especial* to disappear and for *special* to take its place'.

**3** In present-day English. Analysis of the evidence underlines the fact that (i) *special* has almost driven out *especial* in all senses; when *especial* is used at all it tends to mean 'notable, exceptional; attributed or belonging chiefly to one person or thing (*your especial charm*); (ii) *especially* is holding its ground, particularly in sentences such as the following: *loud music is now favoured at commemoration balls, especially by the younger set*; (immediately preceding an adjective) *Several points in his argument are especially deserving of notice*. Examples: (especial) *He loved Cathal ... and he had always known and knew it now with an especial terror, that Cathal was his vulnerable point*—I. Murdoch, 1965; *The impartial Law enrolled a name For my especial use*—R. Graves, 1975); (special) now used in a large number of fixed collocations: *special care, case, delivery, development, edition, education, effects, needs, pleading, -purpose* (attrib.), *stage, tribunal*, etc. (for the meaning, see dicts.). (especially) *This false proportion sometimes made it feel, especially at night, as if it were part of a ship*—I Murdoch, 1973; *The Sumo wrestlers are not especially tall, but they are especially big*—C. James, 1978; *We are especially grateful to you for filling in on such short notice for the Vice-President*—J. Kojinski, 1983; *I can't say anything especially wise about these paintings*—D. Donoghue, 1991; (specially) *I gathered these specially in bud, because I thought it would be nice to see them open out in the warmth of the house*—D. Madden, 1988; *It's a pretty anonymous mark. Not one I'm specially proud of, either*—P. Lively, 1991.

**espionage.** Now always thus spelt (cf. Fr. *espionnage*) and pronounced /ˈespɪənɑːʒ/. The word was first recorded in English in 1793 but the ending continues to resist Anglicization.

**esplanade.** The pronunciation with final /-eɪd/ is judged by the main authorities to be more common than that with /-ɑːd/. But *promenade* continues to tug it in the opposite direction. See -ADE.

**espresso.** This form (derived from It. *caffè espresso*, lit. 'pressed-out coffee'), first recorded in English in 1945, has entirely driven out the variant *expresso* (which was presumably invented under the impression that It. *espresso* meant 'fast, express').

**Esq.** Beginning in the 15c. an esquire was at first 'a young man of gentle birth, who as an aspirant to knighthood, attended upon a knight, carried his shield, and rendered him other services' (OED). From the same period onward, it was also 'a man belonging to the higher order of English gentry, ranking immediately below a knight' (OED). As time went on several classes of men became entitled to be called esquires—younger sons of peers, eldest sons of knights; judges, barristers-at-law, and many others. From the mid-16c. it became customary to refer to such people in writing in the form *Andrew Smith, Esquire. Esquire* in such circumstances soon became abbreviated to *Esq., Esqu.,* etc. (*Esq.* is now usual). By the mid-20c., *Esq.* had lost all sense of rank, and was being attached (in correspondence) to the name of any adult male. That state of affairs continues, except that other forms of address of a more informal kind are gradually replacing *Esq.* When *Esq.* is appended to a name, no prefixed title (such as *Mr, Dr, Capt.,* etc.) is used. It should be emphasized that except in one circumstance *Esq.* is hardly used outside Britain. The main exception is that in the US the term is often used by lawyers when referring to or addressing one another in writing. Curiously, in America, *Esq.* is often appended to the names of women lawyers as well as to men.

**-esque.** A suffix forming adjs., it represents Fr. -*esque* or It. -*esco* (from medL -*iscus*). It occurs in English in the sense 'resembling the style of', (*a*) in ordinary words, such as *grotesque* ( = resembling a grotto), *picaresque* (from a Sp. or It. word meaning 'rogue'), and *picturesque*; (*b*) appended to personal names, as *Audenesque* (20c.), *Browningesque* (19c.), *Disneyesque*

(20c.), *Giottesque* (19c.), *Reaganesque* (20c.), etc. See SUFFIXES ADDED TO PROPER NAMES.

**-ess,** suffix forming nouns denoting female persons or animals, adopted in ME from OFr. *-esse* (from late L *-issa*, ult. from Gk -ισσα as in βασσίλισσα 'queen', cf. βασιλεύς 'king'). **1** The first wave of *-ess* words in English (*countess, duchess, empress, hostess, lioness, mistress, princess,* etc.) all came into English from French. These words are all deeply embedded in the language.

**2** The suffix was rapidly judged to be a useful formative element in English and became attached to many established English words from the 14c. onwards, e.g. *Jewess* (14c., Wyclif), *patroness* (15c.), *poetess* (16c., Tyndale), etc.; and it supplanted the native female suffix *-ster* (which, used as a feminine ending, survives only in *spinster*). These new feminine-gender words were formed either by substitution of *-ess* for *-er* (*adulterer/adulteress, murderer/murderess,* etc.) or by the addition of *-ess* to the stem of the common-gender word (*author/authoress, giant/giantess,* etc.). There were some other refinements, e.g. the substitution of *governess* (already in Caxton) for the earlier *governeresse*; the emergence of *sorceress* (14c., Chaucer) from the common-gender *sorcer* (14c.–1549, now obs.) before the masc. form *sorcerer* (1526, Tyndale); and the establishment of the phonetically simplified forms *adventuress* (1754, Walpole; not \**adventureress*), *conqueress* (now obs.), etc. When *-ess* was added to a noun ending in *-der* or *-dor, -ter* or *-tor,* the vowel before the *r* was usually elided, e.g. *ambassadress, wardress; actress, doctress* (now obs.), *editress, protectress* (16c., J. Foxe), *waitress.*

**3** From ME until about 1850 the suffix retained its power as a more or less unrestricted means of indicating the femininity of the agent-noun. More than 100 words in *-ess* denoting female persons or animals are listed in the *OED*. They vary greatly in durability. (*a*) Except for *lioness* (ME), *tigress* (1611), and to a more limited extent *leopardess* (1567) and *pantheress* (1862), the suffix is not used in any routine way of female animals. (*b*) In *rectoress* (1729) and *vicaress* (1770), both now obsolete, the suffix means 'wife of a ——'. Both words have other meanings as well. One of the senses of *mayoress*

is 'the wife of a mayor'; similarly with *ambassadress.* (*c*) Words of small currency, e.g. *confectioness* (only 1640), *entertainess* (only 1709), *farmeress* (1672–), *preacheress* (1649–), *professoress* (c1740–), *saviouress* (1553–). Many words of this type never got beyond the stage of being merely fanciful or jocular.

**4** In the 20c. three main groups of *-ess* words have been threatened with the executioner's block: (*a*) *Jewess* and *Negress* are deliberately avoided by racially sensitive people who are not themselves Jewish or black, but continue to be used among themselves by the groups concerned. (*b*) With varying degrees of stridency, ill-defined groups of people do not favour the use of feminine-gender artistic terms like *authoress, poetess, paintress,* and *sculptress,* on the ground that there is no need for a separate term, any more than there is a continuing need for (say) *interpretress* (18c.), *philosopheress* (17c.), *tutoress* (17c.), all once common and all now obsolete. (*c*) Occupational terms, such as *air hostess, shepherdess, stewardess, waitress. Air hostess* and to some extent *stewardess* are gradually being replaced by *flight attendant; waitron* (already probably extinct) and *waitperson* have come along to nag at *waitress;* and *shepherdess* is more relevant to pastoral poetry than to modern farms.

**5** The suffix *-ess* is probably no longer a living formative suffix but there is a central core of *-ess* words (apart from those in (1) above) remaining in use more or less unchallenged, among them *abbess, actress, adulteress, adventuress, ambassadress, ancestress, benefactress, conductress, goddess, governess, heiress, huntress, instructress, manageress, mayoress, murderess, ogress, peeress, postmistress, priestess* (but only in non-Christian religions), *procuress, prophetess, proprietress, protectress, seductress, songstress, temptress, traitress, votaress.* All of these, and doubtless some others, are listed in their various senses and applications in *COD* 1995.

**essay. 1** As verb. See ASSAY.

**2** As noun ( = short composition) always stressed on the first syllable; as a rather dated formal word for 'an attempt' commonly stressed on the second syllable. As a somewhat dated formal verb, with the broad sense 'to attempt or endeavour to', it is always stressed on the second syllable.

**-est** in superlatives. See -ER AND -EST.

**Establishment.** *The Establishment*, as the journalist Henry Fairlie expressed it in a key article in *The Spectator*, 23 Sept., 1955, means 'not only the centres of official power—though they are certainly part of it—but rather the whole matrix of official and social relations within which power is exercised'. Others before Fairlie had used the expression in a more or less casual manner: Fairlie brought the term into widespread and continuing use. It is most often used as a blanket term of implied disapproval, denoting people or groups in positions of power about whom the writer or speaker has reservations. For something like two decades it was a much overused word, but it has now settled down as a useful word for any loosely defined influential or controlling group, not necessarily political (e.g. *the literary Establishment*).

**estate.** 1 *The three estates of the realm* are traditionally in the UK the Lords Spiritual, the Lords Temporal, and the Commons, but the term, according to the *OED*, 'has often been misused to denote the three powers whose concurrence is necessary for legislation, viz. the Crown, the House of Lords, and the House of Commons'. The standard use dates from the early 15c. and the misuse is first recorded in 1559. The *third estate* (Fr. *tiers état*) was commonly used of the French bourgeoisie before the Revolution, the other two *estates* (clergy, townsmen) being seldom spoken of numerically. The *fourth estate* has been applied to the newspaper press since the early 19c. Carlyle commented in 1841, *Burke said there were three Estates in Parliament, but in the Reporters' Gallery ... there sat a fourth Estate more important far than they all*. This statement has not been confirmed, but evidence of the use of the phrase in 1823 or 1824 is given in the *OED*.

2 As a curiosity, it should be noted that the older sense of *estate* as 'a landed property' (a use first recorded in the late 18c.) has been joined in the 20c. by *estate = housing estate*. As C. S. Lewis remarked in his *Studies in Words* (1960), *When I was a boy* estate *had as its dominant*

meaning 'land belonging to a large land-owner', but the meaning 'land covered with small houses' is dominant now.

**esteem.** The phrase *success of esteem* is a late 19c. calque on the Fr. phr. *succès d'estime* (itself first recorded in English earlier in the 19c.). It means 'a critical rather than a popular or commercial success'.

**esthetic.** An AmE variant spelling of *aesthetic*.

**estimable** as adj. corresponds to *esteem* and means 'worthy of esteem'. The notional adj. corresponding to *estimate* is *estimatable*, but it seems not to occur.

**estimation.** 1 Surprisingly, Fowler (1926) described the phr. *in my estimation* used as a 'mere substitute' for *in my opinion* as 'illiterate'. In its entry for *estimation* (sense 4) the *OED* lists a range of illustrative examples from Chaucer onwards in which the word is used to mean 'opinion, judgement of worth'. Examples: *The dearest of men in my estimation*—E. W. Lane, 1841; *It was about this time that Martin took a great slump in Maria's estimation*—J. London, 1909.

2 *Estimation* means (*a*) the process or result of estimating; (*b*) opinion, judgement of worth. By contrast, *estimate* (noun) means (*a*) an approximate judgement, esp. of cost, value, size, etc.; (*b*) a price specified as that likely to be charged for work to be undertaken (*my estimate for replacing the hot water cylinder is £150 plus VAT*).

**estrogen, estrus.** AmE spellings of BrE *oestrogen, oestrus*.

**Estuary English.** A fashionable term, coined in 1984 by a London scholar named David Rosewarne, for a kind of slightly non-standard English adopted as common currency by many young people at the present time as part of a minor reaction against the norms of standard English. 'The R.P. speaker accommodates "downwards" and the local accent speaker accommodates "upwards",' declared Rosewarne. Some of the features of this middle-ground form of speech are described (and are drawn on here) in Paul Coggle's *Do You Speak Estuary?* (1993). The estuary is the Thames estuary, and the term *Estuary English* refers to a variety of informal, allegedly classless, English spoken in London and some of the Home Counties, esp. North Kent and South Essex, and also, it is claimed, in some other southern counties. It is characterized in pronunciation by the use of glottal stops (instead of /t/) in such a sentence as *In Scotland the butter and the water are absolutely outstanding*; and the replacement of /l/ by /w/ in such words as *tall*, *ballpoint*, and *Maldon*. In *fault* both replacements occur. *Mouth* tends to come out as *mouf* and *anything* as *anythink*. The choice of vocabulary is important: e.g. *Cheers* instead of *Thank you*, and *mate* tends to replace *friend* (*Me and my mate went to the disco last night*: remember to pronounce the letter *t* as a glottal stop). These are mere specimens of a complex form of speech. It is enough, perhaps, to alert readers to the fact that some features of a new type of 'classless' English have been identified. Only time will tell whether Estuary English is just a passing fashion which will be rapidly replaced by some other equally graceless mode of speech. At the moment it is no more than a ripple in the great ocean of standard English.

**esurient** ( = hungry; impecunious and greedy) is a word of low frequency. See POLYSYLLABIC HUMOUR. Examples: *So should all women be accommodated, thought the esurient ladies in the audience*—E. Linklater, 1931; *As an esurient world power, the USSR maintains unchanged three interconnected foreign policy priorities*; [etc ]—*Christian Sci. Monitor*, 1980.

**-et** (suffix). **1** It occurs naturally as an originally diminutive ending in many words adopted from French at an early date, as *bullet*, *fillet*, *hatchet*, *pocket*, *tablet*, etc.; these words are now used without any consciousness of their original diminutive sense. *Bullet*, for example, is a 16c. adaptation of Fr. *boulette*, diminutive of *boule* 'ball'; *tablet* is a 14c. adaptation of OFr. *tablete* (mod. *tablette*), diminutive of L *tabula* 'table'.

**2** In more recent formations from French in the 16c. and 17c. *-et* represented Fr. *-ette* as well as *-et* (e.g. in *facet*, *islet*). In even more recent adoptions from French the spelling *-ette* is usually retained (see -ETTE); but shortened forms

are shown in BASSINET and in the AmE variant *cigaret* (BrE invariably *cigarette*).

**et al.** An abbreviation of L *et alia* (neut.), *et alii* (masc.), or *et aliæ* (fem.), it means 'and others', and is used esp. to avoid giving a list of authors, or other people, in full (*Lyly … read … and explicated innumerable passages of Ovid, Cicero, Virgil, et al.*—G. K. Hunter, 1962). Special care should be taken not to place a full stop after *et*. The phrase may be printed in romans or in italic type according to taste.

**etc.** Sometimes written as &c. **1** An abbreviation of *et cetera* (which in Latin means 'and others of the same kind'), it is now in English deemed to mean, (*a*) and the rest; and similar things *and people*; (*b*) or similar things *or people*; (*c*) and so on. Its use as an extension to a list of people must now be allowed, though it should be borne in mind that *et al.* is also available for this purpose. Fowler (1926) placed no restriction on the following types: ( = and other things) *His pockets contained an apple, a piece of string, etc.*; ( = or the like) *'Good', 'fair', 'excellent', etc., is appended to each name*; ( = and other persons) *The Duke of A, Lord B, Mr C, etc., are patrons*. There seems no reason to modify his view now.

**2** By the phonetic process of assimilation, *etc.*, which is properly pronounced /et ˈsetərə/, is quite frequently, but wrongly, pronounced as /ek ˈset-/. This comes about because no other common word in English begins with /ets-/, whereas words beginning with /eks-/ abound (*ecstasy*, *excellence*, *exclamation*, etc.).

**3** Since *et* means 'and', it hardly needs to be said that only an ignoramus would say or write 'and etc.'.

**-eteria.** See -TERIA.

**ethic, ethical** (adjs.). **1** The longer form is now dominant in all senses, and in particular, (*a*) relating to morals, esp. as concerning human conduct; (*b*) morally correct; (*c*) (of a medicine or drug) available only on a doctor's prescription.

**2** *ethical dative*. Originally in Greek and Latin, but also a feature found at all periods in English since the 14c., this is the name given to a use in which a person no more than indirectly interested in the fact or event described in the sentence is introduced into it. In practice it is most clearly shown in the presence of a pronoun which has no logical connection with the subject of the sentence: *The skilfull shepheard pil'd me certaine wands*—Shakespeare, 1596; *Prometheus once this chain [of gold] purloin'd, Then whips me on a chain of brass*—Swift, 1724; *Anti-social force that sweeps you down The world in one cascade of molecules*—G. Eliot, 1874. I have not found any 20c. examples, but feel sure that the use still exists even if only by the skin of its teeth.

**ethnic.** The original (15c. onward) meaning, 'pertaining to nations not Christian or Jewish; heathen, pagan', is now only in historical use. In the mid-19c., and at an accelerated rate since the mid-20c., it has come to be applied to any section of a community having common racial, cultural, religious, or linguistic characteristics that distinguish its members from the rest of the community. Ethnic minorities in Britain include Sikhs, Muslims, West Indian and other blacks, Cypriots, and many others. There are many ethnic Turks in Germany, and many ethnic Filipinos in the Middle East. Ethnic clothing means exotic clothing, i.e. conspicuous clothing seen as characteristic of a particular region abroad.

**ethnic terms.** The 20c. has been marked by an increasing awareness of the potential hurtfulness of unfavourable ethnic terminology and also by the realization that the problem cannot be legislated away. Some of the terms are too well established to cause any more than wry amusement though they are unmistakably opprobrious in origin, e.g. *Dutch courage* (first recorded 1826), *French leave* (1771), and *the English disease* (in economics, 1969). A second group of only playfully hurtful terms includes *Limey* (an Englishman, 1918) and *Pom/Pommy* (1912). A third large group of such terms continues to cause great offence to those to whom they are applied or in the contexts in which they occur: e.g. *dago* (1832), *kike* (1904), *Oreo* (an American black seen as having a 'white' outlook, 1968), *to welsh* (1857, though its connection with Wales is unfounded), *wog* (1929), *wop* (1914). The whole subject of such terminology is treated, for example,

in my book *Unlocking the English Language* (1989) and in all major dictionaries under the relevant words. See DIDACTICISM.

**et hoc genus omne.** A convenient occasional substitute for *etc.*, it means 'and the whole of that class or group': *The Herbert Warrens and the Sidney Lees*, et hoc genus omne—*Essays in Criticism*, 1953.

**-ette.**

1 Origin.
2 As diminutive suffix.
3 Imitation and other fabrics.
4 Slightly militant and decidedly unmilitant women.
5 Commercial food-names.

**1** *Origin.* The suffix was first used to form diminutive nouns, representing the feminine form corresponding to the masculine *-et* (see -ET). The main body of such words joined the language after the 17c., e.g. *chemisette* (1807), *cigarette* (1842), and *pipette* (1839). The basic idea is obviously a small chemise, a small cigar, and a small 'pipe'.

**2** *As diminutive suffix.* The equation of *-ette* with smallness gathered strength in the 19c. The literature of the period is studded with words, not all of them of long duration, like *bannerette* (1884), *essayette* (1877), *novelette* (1820), *sermonette* (1814), and *statuette* (1843). This branch of words has tended to weaken in the 20c., but is represented by such words as *dinette* (1930), a small dining-room, *diskette* (1973), a floppy disk, *kitchenette* (1910), a small kitchen, and *superette* (1938, chiefly US and Antipodean), a small supermarket. Smallness is not quite the meaning in a number of special cases. *Laund(e)rette* (1949), for example, is not a diminutive as such, but is just formed on *launder* (verb) or *laundry*. *Serviette* (15c. onwards) was for centuries exclusively Scottish, was reintroduced into standard English from French in the 19c., and is also not a diminutive. It also lies in the territory designated by A. S. C. Ross and Nancy Mitford as 'non-U'.

**3** *Imitation and other fabrics.* Beginning in the 18c., manufacturers began to place the ending on the names of materials intended as imitations of something else. *Muslinette* (1787), was among the first of such terms, followed by *leatherette* (1880), *flannel(l)ette* (1882), *cashmerette* (1886), and others. Some other names of fabrics, not in fact imitations, helped to establish a connection between -ette and 'name of a (dress) material'. These include *georgette* (1915), named after Mme Georgette de la Plante, a French dressmaker; *stockingette* (1824), now more commonly written as *stockinet*, and more likely to be, as the OED expresses it, 'a perversion of stocking-net'; and *winceyette* (1922), a suffixed extension of the older word *wincey*.

**4** *Slightly militant and decidedly unmilitant women.* The most spectacular development of the suffix came at the beginning of the 20c. when (in 1906) female supporters of the cause of women's political enfranchisement in Britain were called *suffragettes*. The suffix thus acquired a markedly feministic edge which has been partially inherited by later groups of terms applied to women. First came the decidedly unmilitant *majorettes* (1941) and *drum majorettes* (1938), with team-names (in N. America and the Antipodes) like *Mercurettes, Pantherettes, Rockettes,* and *Trojanettes.* In more recent times, especially in the 1980s, male chauvinistic writers in glossy and satirical journals began to make free use of the suffix in a pejorative manner: the language became peppered with antagonistic terms like *awarette, bimbette, editorette, hackette, hypette, punkette, reporterette, snoopette, whizzette,* and *womanette.* This last phase will doubtless pass, but the word *suffragette* will remain firmly in place as a term of considerable historical importance, and the marching girls keep marching on. Two other words designating classes of women, namely *undergraduette* (1919) and *usherette* (1925), for the most part passed out of use, except in historical contexts, somewhere about 1960.

**5** *Commercial food-names.* American newspapers are now (1990s) carrying advertisements for a host of such names, e.g. *Clubettes* ('bite-size' crackers), *Creamettes* (a kind of pasta), *Croutettes* (a stuffing mix), and *Toastettes* (a kind of tart). The model for these trade-names is possibly the word *croquette* (first recorded in 1706; Fr. *croquette*, prob. from *croquer* 'to crackle under the teeth, to crunch'), 'a fried breaded roll or ball of mashed potato or minced meat', etc.

**etymology.** **1** The OED sets out in a classical manner the stages through which the current English form of a given word emerged from its earliest recorded form or forms, together with details of its analogues in related languages; and for loanwords as much information about the spelling, meaning, etc., of the source word as is necessary to explain its passage into and its shape in English. Thus *quick* [Comm. Teut.: OE *cwicu*, *c(w)uc*, and *cwic*, *c(w)uc-*, = OFris. *quik*, *quek* (mod. Fris. *quick*, *qucck*), etc., etc.]; and *resin* [ME *recyn(e*, *reysen*, ad. F. *résine*, ad. L. *rēsīna* (Sp., Pg., and It. *resina*), cogn. with Gk ῥητίνη. See also ROSIN.]. The great majority of English words can be straightforwardly traced to their roots.

**2** A second group of words are of obscure origin and are so marked in standard dictionaries (or are said to be of unknown or uncertain origin). These include *boffin*, *burlap*, *coarse*, *dog* (occurs once in late OE as *docga*, but has no Germanic analogues; the normal OE word was *hund*, hound), *garbage*, *kibosh*, *nasty*, *to peter out*, *shandy*, *trash*, and scores of others.

**3** A third group of words have substantially changed their form in English under the influence of another word: e.g. *belfry*, a 15c. adaptation of earlier *berfrey* by association with *bell*; *bridegroom*, a 16c. alteration of *bridegoom* (from OE *brȳdguma*, lit. 'bride-man'); *shamefaced*, a 16c. alteration of *shamefast*.

**4** A fourth group of words have had fanciful explanations foisted upon them. Easily the best example of this phenomenon is *posh*, popularly, but erroneously, said to be derived from the initials of 'port outward starboard home', referring to the more expensive side for accommodation on ships formerly travelling between England and India.

See BACK-FORMATION; BLEND; FACETIOUS FORMATIONS; FOLK ETYMOLOGY.

**eulogy,** a speech or writing in praise of a person. See A, AN 3.

**euphemism,** a mild or vague or periphrastic expression substituted for one judged to be too harsh or direct, e.g. *to pass away* for *to die.* The employment of euphemisms can be viewed positively as the use of words of good omen, or negatively as the avoidance of unlucky or inauspicious words. Numerous examples of both kinds are listed in my essay

'An Outline History of Euphemisms in English' in D. J. Enright's *Fair of Speech* (1985).

Key words or concepts that have led to the proliferation of euphemisms include names for a privy (*bog(s)*, *little house, convenience, comfort station, rest room, toilet, loo*, etc.), prostitutes (*street-walker, fille de joie, fallen woman, scarlet woman, broad, call girl*, etc.), and the circumstances or results of war. In the 1939–45 war our fighter pilots *bought it* or *went for a Burton* as they crashed and were killed. In the Vietnam war the Americans discovered the calming usefulness of expressions like *pacification* (destruction of villages after evacuation of the inhabitants) and *defoliation* (destruction of forests used by the enemy as cover). In the Gulf war some soldiers were killed by what was chillingly described as *friendly fire*; and destruction brought about by misdirected bombs or missiles was called *collateral damage*. Euphemisms abound, and are frequently mocked, at places of employment. They include fair-sounding titles (e.g. *flight service director* for the chief steward on an aircraft); synonyms for the reduction of staff (*outplacement, restructuring, slimming down, downsizing*). Charwomen are now usually called *dailies* or *daily helps*; gaolers are called *prison officers*; and commercial travellers are called *sales representatives*.

The word *euphemism*, which is derived from the Greek words εὐφημισμός 'use of an auspicious word for an inauspicious one' and εὔφημος 'fair of speech', was first recorded in English in Thomas Blount's *Glossographia* (1656–81). The use of euphemisms can be traced back to the Middle Ages, but the prevalence of such evasive and concealing (or fair-sounding) terminology has never been more marked than in the cruel social and political areas of life in the 20c.

**euphuism,** an affected or high-flown style of writing or speaking, originally applied to work (esp. of the late 16c. and early 17c.) written in imitation of John Lyly's *Euphues* (1578–80). The name Euphues /ˈjuːfjuːˌiːz/ is derived from Gk εὔφυης 'well-endowed by nature'. The chief features of euphuism include 'the continual recurrence of antithetic clauses in which the antithesis is emphasized by means of alliteration; the frequent introduction of a long string of similes all relating to the same subject, often drawn from the fabulous qualities ascribed to plants, minerals, and animals; and the constant endeavour after subtle refinement of expression' (*OED*). It has no connection with EUPHEMISM.

**Eurasian.** A term first used in the 19c. for a person of mixed European and Asian (esp. Indian) parentage. Cf. ANGLO-INDIAN.

**Euro-.** One of the most prolific combining forms of the 20c. as various countries on the Continent edged into economic and political arrangements of various kinds among themselves and with the UK. Such formations include *Eurocentric* (first recorded in 1963), *Eurocheque* (1969), *Eurocrat* (1961), *Euro-dollar* (1960), *Euro-MP* (1975), and *Eurovision* (1951).

**evasion, evasiveness.** The latter is a quality only; in places where quality, and not practice or action, is the clear meaning, *evasion* should not be used instead of it: *his evasion of the issue is obvious*; *he is guilty of perpetual evasion*; but *the evasiveness* (not *evasion*) *of his answers is enough to condemn him*. See -ION AND -NESS.

**eve.** *On Christmas Eve*; *on the Eve of St Agnes*; *on the eve of the battle*; *on the eve of departure*; *on the eve of great developments*. The strict sense of *eve* being the evening or day before, the first two phrases are literal, the last is metaphorical, and the two others may be either, i.e. they may mean *before* either with an interval of days or weeks, or with a night intervening, or actually on the same day. Nevertheless, in spite even of the chance of ambiguity, they are all legitimate.

**even.** 1 *Placing.* As a general rule *even* should be placed next to the word or words that it governs or qualifies. These examples from my database show some of the main circumstances of its use: (before an adj.) *doctors must pursue costly and even dangerous investigations*; *they were disposed to be helpful, even solicitous*; (before a comparative adj. or adv.) *she is talking even louder*; (before a noun) *not a word addressed to me personally—not even a postcard*; (before an adv.) *He himself felt troubled, even slightly humiliated*; (before a finite verb) *he even enrolled in a business* studies course to get more organized; (between a modal or an auxiliary and an infinitive) *a changing field could even cause the magnetic liquid to flow*; *he had even managed to chuckle in a suggestive way*; *I did not even bother to read it*; (before a prepositional phr.) *My father was so polite, even in the family*; (before an advbl clause) *Annie Asra was the kind of girl to whom people give a job, even when they didn't originally intend to.*

**2** *Meaning.* COD (1990) says that it is 'used to invite comparison of the stated assertion, negation, etc., with an implied one that is less strong or remarkable (*never even opened* [let alone read] *the letter*; *does he even suspect* [not to say realize] *the danger?*; *ran even faster* [not just as fast as before]; *even if my watch is right we shall be late* [later if it is slow].'

**3** *An occasional use.* On occasion *even* can justifiably be placed before an adjacent verb rather than immediately before the word it logically qualifies: *it might even cost £100* (rather than *it might cost even £100*). In such cases the rhythm of the sentence comfortably brings out the sense. But care is needed.

**evenness.** So spelt.

**evensong.** See MORNING.

**event.** *In the event of* is a somewhat verbose cautionary phr. used to mean 'if such-and-such should occur': *The juridical and theological dilemma in the event of one Siamese twin predeceasing the other*—J. Joyce, 1922. In AmE the equivalent phr. usu. appears as either *in the event of* or *in the event that*: e.g. *next in line for the Presidency in the event that there is no Vice-President*—*Current Biography 1947* (WDEU); *Roll bar, a metal tubular structure over the cockpit which protects the driver in the event the car overturns*—*Publ. Amer. Dial. Soc.*, 1964.

**eventuality, eventuate.** *History.* These words, both first recorded in the 18c., have had a chequered history, esp. the latter. Derided by De Quincey (1834) as 'Yankeeish', by Dean Alford (1864) as 'another horrible word, which is fast getting into our language through the provincial press', and by Fowler (1926) as 'flabby journalese', *eventuate* has now settled into a fairly formal corner in the language. *Eventuality* is much less commonly used and has been much less frequently

attacked. Examples: (eventuality) *In certain eventualities this state of things might give rise to grave difficulties*—Lady Herbert, 1878; *Although he had been ordered not to destroy it, Harmel was prepared for the eventuality*—C. Ryan, 1974; *Of course, different situations demand different reactions and one cannot plan for every eventuality*—P. Mann, 1982. (eventuate) *He heard ... the discussions which eventuated in Acts of Parliament*—S. Smiles, 1873; *What would eventuate tomorrow or the day after or the year after*—W. Golding, 1979; *It had been intended to have educated Saudi women dealing with the public at the exhibition, but ... this had not eventuated*—*Times*, 1986; *Many of the things we worry about never eventuate*—D. Rowe, 1987.

In most circumstances *result* or *come about* are more suitable words than *eventuate*.

**ever.** **1** In informal conversation *ever* is sometimes used as an intensifier immediately after the interrogative pronouns *who*, *what*, etc.: *What ever did you do that for?*; *Why ever should you think that?* In the written form of such sentences *ever* should be written separately, i.e. should not be joined to the preceding pronoun: e.g. *Who ever thought I'd be in Paris?*—J. Frame, 1985.

**2** *Did you ever?* A colloquial phr., now mainly employed at the more demotic end of standard English, meaning 'Did you ever hear or see the like?' Victorian examples bring out the flavour of the expression very well: *He ... found that the surgery-boy had ... given calomel instead of ipecacuanha! Did you ever?*—G. E. Jewsbury, 1844; *'And where is she now?' 'In a studio.' ... 'Did you ever!' said Mrs. Fanshaw*—*Peel City Guardian*, 1892. A modern example: *Cody! Out here in the middle of nowhere, by sheer coincidence, Cody Tull! ... 'Well did you ever.'*—A. Tyler, 1982.

**3** *ever and anon* 'ever and again, every now and then'. Rather against the odds, this phr. (first recorded in Shakespeare's *Love's Labour's Lost, Ever and anon they made a doubt*) is still widely current in present-day English: *At dawn I had walked into the city, a little gingerly preceded by a sapper who thrust a bayonet ever and anon into the suspect soil*—M. Wheeler, 1955; *he ever and anon brought ... a cigarette-end to his lips*—A. Burgess, 1963.

**4** *ever so.* Prefixed in hypothetical sentences to adjs. or advbs., with the sense 'in any conceivable degree' or used elliptically to mean 'ever so much', the phr. has been in standard use since the late 17c.: e.g. *Though Sir Peter's ill humour may vex me ever so, it never shall provoke me to* [etc.]—Sheridan, 1777; *If ever so many queens are introduced into a hive*—an entomological work of 1816; *I couldn't lead the worship if it was ever so!*—C. M. Yonge, 1858. This use is beginning to slip away, however, and is now overshadowed by the same phr. used (since the mid-19c.) in affirmative contexts as a vague intensive, meaning 'vastly, immensely': *It's the greatest idea, and I'm ever so grateful.*—J. Leland, 1987; *It's ever so enjoyable,' Dawne remarked, with genuine enthusiasm*—New Yorker, 1987; *Ever so faintly the earth seemed to shake beneath our feet*—Kathy Page, 1989; *He's done ever so well out there. He's a fashion photographer, has his own studio.*—D. Lodge, 1991.

**5** See FOR EVER; NEVER 1, 2.

**every.** **1** Normal uses of *every* include the following: ( + noun) *they have every right to be there*; *it would be quite impossible to prosecute every motorist*; (adv. + *every* + noun) *virtually every aspect could be considered*; ( + noun of time) *every day, every morning, every Friday, every six months*, etc.; ( + number) *in every hundred*; *once every 1,000 molecules*; ( + adj. + noun) *she'd strained to establish every single word of the conversation*; (possessive pronoun + *every* + noun) *our every deed must help make us acceptable*; *she'll look after your every need*; ( + one) *she leaned forward and blew out every one of her candles*; ( + advbl phr.) *every now and then*; *every so often*; *every once in a while*.

**2** Difficulties arise in subject–verb agreement, esp. when the subject refers to people of both sexes. In the simplest case, *every* requires a singular verb: *not every book is worth reading.* A singular verb is also desirable in such a sentence as *Not every one of these enterprises shows a profit.* More complex uses of *every* are often felt to require a plural verb and/or a plural anaphoric pronoun: *every phrase, every line and every stanza are indissolubly welded*—M. Gullan, 1937 (*WDEU*); *Each and every one of my colleagues at the university will express their own opinion*—Lord Goodman, 1985. (In the latter sentence some

writers today would have used *his or her* or just *his* instead of *their.*) The matter is unresolved. When the subject is of the type *every —, every —, and every —,* a singular verb is usually defensible and even desirable. But the presence of *and* means that a plural verb could not be considered wrong. When a plural verb is followed by a backward-referring pronoun the unenviable choice is between *their* and *his or her.* Cf. AGREEMENT 6; EACH.

**everybody.** First (16–17c., and intermittently until the 19c.) written as two words (*every body*), this pronoun is now always written as one word in its ordinary meaning 'every person, everyone'. (Of course when *body* is a full noun the two words are separated: *Every body was recovered from no man's land by the stretcher-bearers.*) It alternates freely with *everyone* and both words normally require a singular verb. On the type *Has everybody got their first course?*, see AGREEMENT 6. The use with a singular (occas. pl.) verb, and in particular with a plural pronoun, has been a feature of the language for more than four centuries: *Everye bodye was in theyr lodgynges*—Lord Berners, c1530; *Now this king did keepe a great house, that euerie body might come and take their meat freely*—Sidney, 1580; *Every body else I meet with are full ready to go of themselves*—Bishop Warburton, 1759; *Every body does and says what they please*—Byron, 1820; *Everybody seems to recover their spirits*—Ruskin, 1866; *Everybody has a right to describe their own party machine as they choose*—W. Churchill, 1954; *There's a bus waiting outside the terminal to take everybody to their hotels,' said Linda*—D. Lodge, 1991. This is the way the language naturally operates, despite the obvious clash of number between the pronoun and what follows.

Cf. EVERYONE.

**everyday.** When used as an adj. (*an everyday event, everyday clothes,* etc.) meaning 'commonplace, usual; suitable for or used on ordinary days', *everyday* is written as one word. In contexts where it means 'each day' (*she went shopping almost every day*) two words (*every day*) are needed.

**everyone.** **1** In its regular use as an indefinite pronoun = everybody, it is now invariably written as one word (*living through hard times, as everyone was;*

*everyone was growing leaner and cleaner; you can't please everyone*), but this convention did not fall into place until the 20c. The *OED* (in 1894) gave precedence to *every one*, while Fowler (1926) presented a spirited argument in favour of the linked form.

**2** The pronoun referring to *everyone* is often pl., the absence of common gender rendering this violation of grammatical concord sometimes necessary (*OED*). Examples: *Every one Sacrifices a Cow or more, according to their different Degrees of Wealth or Devotion*—Dr Johnson, 1735; *Everyone then looked about them silently, in suspense and expectation*—W. H. Mallock, 1877; *Everyone was absorbed in their own business*—A. Motion, 1989; *the classical allegories look like surreal school outings in which everyone got to take their clothes off, and then was sorry*—M. Vaizey, 1991. Not everyone favours this practice, however: *Well, I'm afraid it's everyone for himself*—R. Hall, 1978 (NZ); *The idiom would be one of solid understatement with everyone turning his hand to what needed to be done*—N. Barley, 1983.

**3** When *one* is used as a full pronoun followed by *of* + noun or pronoun and happens to be preceded by *every*, the two words need to be written separately: *The drawings are academical in the worst sense of the word; almost every one of them deserves to be rejected.*

**4** Cf. AGREEMENT 6; EVERYBODY.

**everyplace,** a modern AmE synonym of *everywhere*: *Although, like everyplace else, the White Elephant had engaged a corps of college students*—*Saturday Rev.*, 1976 (WDEU). Cf. ANY 3(f); SOMEPLACE.

**everytime.** The phr. *every time* is commonplace. Regrettably it is now sometimes written as one word: *He passes it everytime he has to go to Party Headquarters*—R. Lindner, 1955 (US); *Everytime in his life a decision has had to be made, he's been able to follow his own priorities*—O. Marshall, 1984 (NZ).

**every which way.** This lively American phr. meaning (*a*) in all directions, and (*b*) in a disorderly manner, was first recorded in 1824 and is now fast gaining ground in BrE. Examples: *People running every which way*—D. Runyon, 1931; *Hemmed in every-which-way was Ross*—L. Deighton, 1962; *Like bouncing footballs …*

*words travel every which way*—*New Yorker*, 1987; *It was raining and she looked like hell, her hair all wet and her makeup every which way*—M. Doane, 1988.

**evidence** (verb). This uncomfortable verb has been in existence in several senses since the 17c. One of these is the verbal equivalent of the noun, namely 'to serve as evidence, to be evidence of' (*Occurrences evidencing the divinity of Christ*—J. O. Halliwell, 1859). Another is 'to attest' (*This is no reason for doubting their reality, when they are evidenced by Intuition*—F. Bowen, 1864). A third is 'to indicate, manifest' (*The courts eagerly seized on any expressions evidencing this intention*—K. E. Digby, 1876). It is obvious that all three senses occur in contexts well above the level of ordinary discourse. It is also apparent that such a verb is at risk of disapproval in any case as part of the traditional dislike of nouns (in this case a noun first recorded in the 14c.) being employed as verbs when other means are available. At all events, Fowler (1926) accepted the first sense as a 'right use' and rejected the third as a 'wrong use'. He left the sense 'to attest' without comment. His arguments in the matter no longer carry any weight. In practice the verb is not common in any of these senses, but it is legitimately available for all three. Examples: *I continued to reflect on social change as evidenced in infancy*—J. I. M. Stewart, 1977; *This is not … an ideal Universe, as was further evidenced by the eye-crossing patterns of the inlaid marble floor*—Douglas Adams, 1980; *This type of puss* [in basketball] *can be extremely devastating to the opposition, as evidenced by its use in providing the winning scoring opportunity in the final of the Munich Olympic Games*—Hoy and Carter, 1980; *the closer links with the London company were evidenced by the acquisition of LGOC-type buses and equipment*—K. Warren, 1980; *Everything he was and did evidenced distinction*—N. Gordimer, 1990.

**evilly** (adv.). So spelt.

**evince.** A learned word, first brought into English in the 17c. in several senses answering to L *ēvincere* (from *ē* 'out' + *vincere* 'to conquer'). Most of these senses (to overcome, prevail over; to convince; to confute; to constrain; to prove by argument; etc.) flourished for a time in learned work but were obsolete by the end of the 18c. The word is now used mainly in formal English (or, as Fowler put it, by 'those who like a full-dress word better than a plain one') to mean 'to show, make evident (a quality, a feeling), to show that one has (a quality, a feeling)'. Examples: *How is it our disruptive Price evinces such temperance and rectitude?* —G. Swift, 1983; *He had never been inside a church before, but neither then nor on any subsequent visit did he evince the least curiosity*—P. D. James, 1986; *Nobody he passed evinced the slightest interest in him nor seemed to constitute any kind of threat*—J. Leland, 1987; *They constantly evince a smug hermeticism that is graceless and slight*—*Times*, 1987.

**evolve.** See INVOLVE (verb) 1.

**ex-.** 1 No difficulties arise when *ex-* meaning 'former(ly)' is prefixed to a single word (*ex-convict, ex-president, ex-wife*); similarly when it is used to mean 'not listed' (*ex-directory*). Fowler (1926) scorned formations in which *ex-* is prefixed to a noun phrase, e.g. *ex-Lord Mayor, ex-Prime Minister, ex-Chief Whip*, on the ground that such persons are not ex-Lords, ex-Primes, etc. Who would dream of thinking so? The eye easily accommodates such expressions, the *ex-* being simply assumed to apply to the whole of the noun phrase. If, however, longer units are felt to be difficult (*ex-trade union leader, ex-Father of the House*, etc.), one can always fall back on *former* instead.

2 See LATE, ERST, etc.

**exceedingly, excessively.** 1 The first has been used with verbs since the 16c. to mean '(of degree) above measure, extremely', but by the 19c. was 'chiefly limited to verbs that indicate emotion, feeling, or the expression of them' (*OED*) (*I praised God ... and rejoiced exceedingly*— E. W. Lane, 1841). This use is now rare. With adjs. and advbs. it came to be restricted in use to the positive degree, whereas formerly it could be used with *more*, as in 'exceedingly more rare'. Examples: *She seems to me to be exceedingly pretty*—G. P. R. James, 1847; *This he found to answer exceedingly well*—*Medical Temperance Jrnl*, 1881; *In Persia this form of see-saw is exceedingly common*—J. B. Morton, 1974; *His room was exceedingly cold*—P. Fitzgerald, 1982.

2 From its first recorded use in the 15c., *excessively* has always meant 'in an excessive amount or degree; beyond measure, immoderately': e.g. *It makes them smart and burn excessively*—John Wesley, 1747; *The scenery seemed of an excessively rudimentary description*—Lady Brassey, 1877; *She may exercise excessively, spending hours each day in the gymnasium*—Abraham and Llewellyn-Jones, 1984; *One of those excessively sane people who react to everything in life with exasperating ... good sense*—T. Mallon, 1984.

3 There is a small area where 'exceedingness' and excessiveness overlap, but in practice no difficulty will arise if the base meanings of the related advbs. are kept in mind.

**excellence, -cy.** See -CE, -CY.

**excellent.** Properly used only in the positive degree (*an excellent essay, an excellent concert, an excellent talker*). Something can be *quite excellent* but not *more excellent* or *very excellent*. The only exception to the rule is that *most excellent* is permissible as an occasional conventional phrase of emphatic praise or approval (*most excellent brandy*, etc.).

**except.** 1 As a conjunction governing a conditional clause, *except* has been continuously in use in the sense 'unless' since the 15c., e.g. *Except the Lord build the house, they labour in vaine that build it*—Ps. (AV) 127: 1. This use has become archaic in the 20c., as *COD* (1990) noted. Fowler (1926) described it as 'either an archaism resorted to for one or other of the usual reasons, or else an illustration of the fact that old constructions often survive in uneducated talk when otherwise obsolete'.

2 As a conjunction 'introducing a predicative clause expressing a fact that forms an exception to the statement made' (*OED*), *except* is frequently used elliptically for 'except that', esp. by journalists in the UK and by every kind of writer abroad. Examples: *The day he turned 18, Trojan moved into his own council flat in Caledonian Road. Except he didn't like it*—*The Face*, 1987; *Except he died. Except I failed to carry and keep it*—E. Knox, 1987 (NZ); *'But it wouldn't be right to let the McCarthys pay for the wedding.' 'Except it might be cheaper,' said Vaun*—J. Leland, 1987

(Ir.); *the festivities passed off almost painlessly ... Except I forgot the Stilton, the port and the crackers—Spectator*, 1987; *They had created the Grand Concourse as the Park Avenue of the Bronx, except the new land of Canaan was going to do it better—*T. Wolfe, 1987 (US); *he could hardly bear to be there, except he must—*P. Carey, 1988 (Aust.); *There can't be any reason. Except she ruined them all, for everything, for the revolution, too—*N. Gordimer, 1988 (SAfr.). This construction is not appropriate in formal contexts in BrE.

**excepting** as a preposition has one normal use. When a possible exception is to be mentioned as not made, the form used is, instead of *not except*, either *not excepting* before the noun or *not excepted* after it: *All men are fallible except the Pope; All men are fallible, not excepting the Pope* or [less commonly] *the Pope not excepted*. Other prepositional uses of *excepting* are unidiomatic; but the word as a true participle or a gerund does not fall under this condemnation: *He would treble the tax on brandy excepting only, or without even excepting, that destined for medicine.* An example of the use deprecated is: *The cost of living throughout the world, excepting in countries where special causes operate, shows a tendency to keep level. Excepting* may also be used after *always. Excepted* is also sometimes used in post-position without *not*: e.g. *But in Tucson, visits to the stationery store excepted, the most cheering postponement I'd come up with was a walk out along the Old Spanish Trail—New Yorker*, 1987.

**exception.** The proverb *the exception proves the rule* is an abbreviated, commonly misconstrued version of the medL maxim *exceptio probat regulam in casibus non exceptis* 'the exception confirms the rule in cases not excepted'. In the context of the proverb, *proves* means 'tests the genuineness or qualities of', no more no less. There is an excellent comment on the use and misuse of the maxim in the *OED* (s.v. *exception*), which is too complicated to reproduce here, together with illustrative examples from 1640 onward.

**exceptionable, exceptional, unex-.** The *-able* and *-al* forms, especially the negatives, sometimes cause difficulty. *Exceptional* has to do with the ordinary sense of *exception*, and means out of the common; *exceptionable* involves the sense

of *exception* rarely seen except in *take exception to* and *open to exception*; it means the same as the latter phrase, and its negative form means offering no handle to criticism. The usual mistake is that shown in: *The picture is in unexceptional condition, and shows this master's qualities to a marked degree.*

**excessively.** See EXCEEDINGLY.

**exchangeable.** So spelt. See MUTE E.

**excise.** 1 Always so spelt, not with *-ize*, in all senses of the noun and of both verbs. See -ISE 1.

2 See TAX.

3 In the noun and verb in the 'tax' sense, the stress falls on the first syllable; and in the verb meaning 'cut, remove', on the second.

**excitable.** So spelt. See MUTE E.

**exclamation mark.** Except in poetry, the exclamation mark (!) should be used sparingly. Excessive use of exclamation marks in expository prose is a certain indication of an unpractised writer or of one who wants to add a spurious dash of sensation to something unsensational. There is a range of ordinary circumstances, including the following, in which the use of ! is customary: (sentences introduced by *how* or *what*) *How awful!*; *What a nuisance!*; (wishes) *God save the Queen!*; (alarm calls) *Help!*; *Every man for himself!*; (commands) *Stand still!*; (a call for attention) *Outside Edith's house, someone knocked. 'Edith!'*; (to indicate shouting) *'You're only shielding her. Shielding her!' His voice rose to a shriek.* Poetical and other literary uses are much more numerous. From *The Oxford Dictionary of Quotations* (1941) I quickly gathered a representative set of 19c. and 20c. examples which bring out some of the main tendencies: *Nearer, my God, to Thee, Nearer to Thee!* (S. F. Adams, 1805–48); *Mild o'er her grave, ye mountains, shine!* (M. Arnold, 1822–88); *It did not last: the Devil howling 'Ho! Let Einstein be!' restored the status quo.* (H. Belloc, 1870–1953); *Awake, my heart, to be loved, awake, awake!* (R. Bridges, 1844–1930); *Blow out, you bugles, over the rich Dead!* (R. Brooke, 1887–1915); *Oh, to be in England Now that April's there ... While the chaffinch sings in the orchard bough In England—now!* (R. Browning, 1812–89); *Talk about the pews and*

*steeples And the cash that goes therewith!*
(G. K. Chesterton, 1874–1936). They represent a wide range of commands, statements, assertions, and wishes.

**excusable.** So spelt. See MUTE E.

**executive.** Apart from its general use for one of the three branches of government, of which the others are the *legislative* and the *judicial, executive* is, in Britain, the name given to one of the three general classes of civil servants of which the others are the *administrative* and the *clerical*. In America it means a high officer with important duties in a business organization, and this meaning, outside the civil service, has now become common in Britain also. The word has also spread to other areas. Lucinda Lambton (*Listener*, 31 Aug. 1989) comments on its use in hotels: *You can quaff from 'executive bars' ... in 'executive suites' and top up from 'executive ice machines' on 'executive floors' ... We have ... luxuriated over 'executive menus' (smoked salmon is an extra with the 'executive breakfast'), and I once gratefully pocketed my 'complimentary executive gifts'.*

**executor.** In its ordinary sense (testator's posthumous agent) pronounced /ɪɡˈzekjʊtə/, stressed on the second syllable. It is to be distinguished, of course, from *executioner*, an official who carries out a sentence of death. For the feminine of *executor*, see -TRIX.

**exercise.** Always so spelt, not with -*ize*. See -ISE 1.

**exhaustible.** So spelt. See -ABLE, -IBLE 7.

**exigence, -cy.** *Exigency* is now the more usual and the more natural form of the two. Both forms entered the language in the late 16c. and seem to have been equally frequent in their main senses until about the end of the 19c., by which stage they no longer showed any clear functional differences. See -CE, -CY.

**exit** (verb). In stage directions the correct style is *Exit Macbeth* (when only one person leaves the stage) and *Exeunt Kings and Banquo* (more than one person). The two forms are respectively the 3rd person sing. pres. indicative and the 3rd person pl. pres. indicative of the L verb *exīre* 'to go out'.

**-ex, -ix.** 1 Naturalized Latin nouns in -*ex* and -*ix*, gen. -*icis*, vary in the form of the plural. The L plural is -*ices* /-ɪsiːz/, but there is considerable variation in the English forms. See LATIN PLURALS.

2 The simplest way to present the material is to set it out in tabular form (but see the separate entries for most of the words concerned).

| Singular | Plural |
| --- | --- |
| *apex* | *apexes** |
| *appendix* | *appendices** |
| *axis* | *axes* |
| (*calyx* | *calices* or *calyxes*) |
| *codex* | *codices* |
| *cortex* | *cortices* |
| *duplex* | *duplexes* |
| *helix* | *helices* |
| *ilex* | *ilexes* |
| *index* | *indexes** |
| *matrix* | *matrices* |
| *murex* | *murices* |
| *radix* | *radices* |
| *silex* | no pl. |
| *simplex* | *simplexes* |
| *vertex* | *vertexes** |
| *vortex* | *vortices* or *vortexes* |

* But usually *apices, appendixes* (medicine), *indices* (economics), *vertices* in scientific or technical work.

3 See -TRIX.

**exocentric compounds.** See BAHUVRIHI.

**ex officio.** A Latin phrase meaning 'by virtue of one's office or status' (*the Bursar was a member of the Governing Body* ex officio). Hyphened when used attrib. (*the* ex-officio *members of the committee*).

**exorcize, exorcise.** As the *OED* remarked in 1894, the spelling with final -*ize* had by then been almost driven out by that with -*ise*, perhaps under the influence of *exercise*. My impression is that *exorcise* is still the more usual of the two forms. Nevertheless it belongs etymologically with the classic -*ize* words (*baptize*, etc.), being from Gk ἐξορκίζειν (i.e. with zeta), and should be so spelt in works that retain the -*ize* ending in other words of the same kind.

**exordium.** The pl. is either *exordiums* (this form is recommended) or *exordia*. See -UM 3.

**exoteric and exotic,** of the same ultimate derivation, have entirely diverse applications. That is exoteric which is communicable to the outer circle of disciples (opp. *esoteric*); that is *exotic* which comes from outside the country (opp. *indigenous*); *exoteric doctrines*; *exotic plants*.

**expandable, expansible.** So spelt. See -ABLE, -IBLE 6, 7.

***ex parte.*** A legal Latin phrase meaning 'in the interests of one side only or of an interested outside party'. Hyphened when used attrib. (*an ex-parte affidavit*), as it most commonly is.

**expect.** Throughout the 19c. the use of *expect* to mean 'to suppose, surmise' came under attack, though the sense had been in use since the late 16c. The *OED* itself commented (in 1894): 'Now *rare* in literary use. The misuse of the word as a synonym of *suppose*, without any notion of "anticipating" or "looking for", is often cited as an Americanism, but is very common in dialectal, vulgar or carelessly colloquial speech in England.' Fowler (1926) took a different view: 'Exception is often taken to the sense *suppose*, be inclined to think, consider probable. This extension of meaning is, however, so natural that it seems needless purism to resist it. *Expect* by itself is used as short for *expect to find, expect that it will turn out that*, that is all.' Fowler's judgement has been borne out by the passage of time. The use is acceptable in standard speech, though it is far less suitable for use in expository prose.

**expectorate.** In the sense 'to eject, discharge (phlegm, etc.) from the chest or lungs, by coughing, hawking, or spitting', *expectorate* has been used, esp. in medical books, since the 17c. In modE it occurs in lay use from time to time as a genteelism for *spit*, but often only as a jocular substitute for it. Modern examples (the first of which is in a poem): *Your lips expectorate a stream Of self-igniting gasolene*—C. Raine, 1978; *Captain Anderson walked across the planking and expectorated into the sea*—W. Golding, 1980; *He was coughing less, expectorating less, and breathing more easily*—R. Hayman, 1981.

**expedience, -cy.** Both forms occur, *expediency* being the commoner of the two. The form chosen is governed, it would seem, by the rhythm of the sentence. See -CE, -CY. Examples: (expedience) *Burroughs said, 'I am against this board*

*settling for mediocrity and second or third best for the purposes of political expedience.'*—Chicago Tribune, 1990; *the Swansea college's department of philosophy has been riven by allegations of commercial expedience in the awarding of external MA degrees on the moral ethics of medical practices*—Independent, 1991; (expediency) *He did what needed to be done despite the political risk. He stood on principle rather than expediency*—Christian Sci. Monitor, 1990; *His [sc. F. W. De Klerk's] 'conversion' seems one of expediency, of cool-headed realism*—Independent, 1991.

**expert** (noun and adj.). See NOUN AND ADJECTIVE ACCENT.

**expertise,** a loanword (first recorded in 1868) from French, meaning 'expert skill, knowledge, or judgement', and pronounced with final /'-ti:z/. See -ISE 2. It is to be distinguished from the relatively uncommon verb *expertize* (or *-ise*) 'to give an expert opinion (concerning)', pronounced with final /'-taɪz/.

**expletive.** At the end of the 19c., when the relevant section of the *OED* was published, *expletive* was pronounced with the stress on either the first syllable (seemingly the dominant pattern) or the second. Thus /'eksplɪtɪv/ or /ek'spli:tɪv/. Fowler (1926) hoped that the two pronunciations would settle down in a manner that would distinguish the noun from the adj. This has not happened, and the form with second-syllable stressing (and long /i:/) is now standard for both parts of speech.

**exploit.** See NOUN AND VERB ACCENT 1.

**explore every avenue, to.** See CLICHÉ.

**export.** See NOUN AND VERB ACCENT 1.

**exposé** /ek'spəʊzeɪ/. First recorded in English in the sense 'a showing up of something discreditable' in 1809 (it had been used earlier in English in the sense 'a recital of facts or particulars'), *exposé* plays a useful rôle in the language. It is always written with an acute accent in BrE, but sometimes without in AmE. The *OED* notes that 'there is some 19th c. evidence for a disyllabic pronunc. corresponding to the spelling *expose*'.

**exposition.** Sharply distinguished in meaning from the preceding word. Its main senses are (*a*) an explanation or interpretation of something difficult; an expository article or treatise; (*b*) in imitation of Fr. *exposition*, a public exhibition.

**ex post facto** is a legal adverbial and adjectival phr. used to mean 'with retrospective action or force' (e.g. *increasing its guilt* ex post facto; ex post facto *laws*). As the *OED* comments, on etymological grounds 'the separation of *postfacto* in current spelling is erroneous'; but the phr. has been regularly written as three words since the 17c.

**expressible.** So spelt. See -ABLE, -IBLE 7.

**exquisite.** The placing of the main stress in this word has steadily changed in standard speech in the last two centuries from the first to the second syllable. Walker (1791) and Daniel Jones (1917) listed only first-syllable stressing. By 1963 Daniel Jones gave preference to first-syllable stressing but commented that the alternative pronunciation was 'becoming very common'. J. C. Wells (1990) reported that 69% of the members of his poll panel opted for /ɪk'skwɪzɪt/, i.e. for second-syllable stressing. My personal preference is to stay with the other 31%.

**extant.** Once meaning just 'existing' (heresy, fashion, roads, etc. could all be extant), *extant* now means 'continuing to exist, that has withstood the ravages of time' (ancient manuscripts, old churches, fossils, etc., are often described as 'still extant', 'the only extant examples', etc.

**extemporaneous(ly) and extemporary, -ily** are cumbersome words in contexts where the meaning required is along the lines of 'spoken or done without preparation'; *extempore* (four syllables) is seldom unequal to the need.

**extempore, impromptu.** Some people observe a distinction between an *extempore* speech, one that is made without notes (though perhaps prepared to some extent in advance), and an *impromptu* one, which is spontaneously given without preparations of any kind. But the distinction is a fine one. Many an 'impromptu' speech is only seemingly given off the cuff.

**extend.** Dire comments have been applied to some of the uses of this word. There is no difficulty about literal uses where the sense is 'to lengthen or make larger in space or time' (extend a chapter, a sermon, a railway line, a boundary, a term of office, etc.). But objections have been made to metaphorical applications of the word when *give*, *grant*, *accord*, etc., would serve as well. Fowler (1926) condemned *extend* used instead of *give* or *accord* as 'a piece of turgid journalese', and claimed that 'native English did not go that way'. '*Extend* in this sense', he said, 'has done its development in America, and come to us full-grown viâ the newspapers—a bad record.' Rightly or wrongly, such comments have cast a shadow over metaphorical uses of the word. It is perhaps advisable therefore to *offer* sympathy, hospitality, kindness, an invitation, etc., rather than to *extend* any of these.

**extendible, extensible** are the recommended spellings. See -ABLE, -IBLE 7.

**extenuate** (in practice appearing most frequently in its ppl adj. form *extenuating*) means 'to lessen the seeming seriousness of (guilt or an offence) by reference to some mitigating factor'. Etymologically the English word answers to L *extenuāre* 'to make thin, to whittle down, to lessen (in quantity, degree, etc.)'. In the 18c. and 19c. the verb was sometimes used of persons or circumstances in the sense 'to plead partial guilt for' (e.g. *extenuate the conduct of Madame La Motte; the purser's steward extenuated himself calmly enough*), but no longer.

**exterior, external, extraneous, extrinsic.** Etymologically the four words are closely related (from four Latin words meaning respectively 'outside', 'outward', 'external', and 'outward'), but the distinctions they now have reflect their history in English rather than their meanings in Latin. *Outside* is the fundamental meaning of all four; *outward* and *outer* are near-synonyms. Inevitably the description that follows is selective; for a fuller treatment one must turn to the *OED* and also to a good dictionary of synonyms.

1 *exterior* and *external*. That is exterior which encloses or is outermost, the enclosed or that which is inside being *interior*. These opposites are often applied to things of which there is a contrasting

pair, and with conscious or unconscious reference, when one is spoken of, to the other: an *exterior* door has another inside it; the *exterior* surface of a hollow ball, but not of a solid one, contrasts with the *interior* surface forming the boundary of hollowness within. Remedies, treatments, ointments, etc., applied to the outside of the body are *external* remedies, etc; such medicaments are usu. described as *for external use only*. Acts, appearances, qualities, etc., that are outwardly perceptible are *external*. *External debt* and *external relations* refer to spheres of operation outside the country concerned. *External evidence* is evidence derived from circumstances or considerations outside or independent of the thing discussed. An *external examiner* is one who tests students of a college, university, etc., of which he or she is not a member.

**2** *extraneous* and *extrinsic*. That is extraneous which is introduced or added from without, and is foreign to the object or entity in which it finds itself, or to which it is attached. A fly in amber, a bullet in one's shoulder, are extraneous objects. Extraneous points are irrelevant matters brought into a debate or discussion to which they do not properly belong. That is extrinsic which is not an essential and inherent part of something. It is applied, for example, to evidence beyond that afforded by a deed or document under consideration, or to a stimulus not provided by a body, a cell, or some other thing itself, but introduced from without. *Extrinsic* always contrasts with *intrinsic*; in various degrees it also overlaps in meaning with *inessential*, *irrelevant*, and *superfluous*.

**exterritorial.** See EXTRATERRITORIAL.

**extol** (verb). The only spelling in BrE (the inflected forms being *extolled*, *extolling*), and the dominant one (beside *extoll*) in AmE.

**extract.** See NOUN AND VERB ACCENT 1.

**extraneous.** See EXTERIOR.

**extraordinary.** The dominant pronunciation in standard English is with five syllables, not six, i.e. /ɪkˈstrɔːdmərɪ/, not /ekstrəˈɔːd-/.

**extraterritorial** (and derivatives). Now the dominant forms. They stood side by side with synonymous *exterritorial* and derivatives in the 19c., but the longer forms (the first element of which is from L *extrā* 'outside') have now prevailed.

**extrinsic.** See EXTERIOR.

**extrovert.** Now, regrettably, the established form in technical and lay use, supplanting the synonymous and better-formed *extravert* (which was used just as frequently for a time after its introduction by Jung c1915). The first element *extro-* is modelled on the termination of *intro-*.

**-ey and -y in adjectives. 1** The normal adjectival suffix is *-y*, not *-ey*. Doubt arises only (*a*) when the suffix needs to be appended to nouns ending in mute *e*; and (*b*) in a few special circumstances (see 3 below).

**2** Normal adj. formations in *-y*: *bony* (from *bone*), *breezy*, *chancy*, *crazy*, *easy*, *fluky*, *gory*, *greasy*, *grimy*, *hasty*, *icy*, *mangy*, *nervy*, *noisy*, *nosy*, *racy*, *rosy*, *scaly*, *shady*, *shaky*, *slimy*, *smoky*, *spicy*, *spiky*, *spongy*, *stony*, *viny*, *wavy*, *whiny*, *wiry*. Also recommended are *caky* (not *cakey*), *fluty*, *gamy*, *homy*, *horsy* (not *horsey*), *liny* (= marked with lines), *mousy*, *piny* (= like, or full of, pines), *pursy* (= short-winded; corpulent), *stagy*, *whity* (= whitish). The only word in this category that is usu. spelt with final *-ey* is *matey*.

**3** Special cases. (*a*) When an adj. is formed from a noun that itself ends in *-y*, the adjectival ending is *-ey*: *clayey* (not *clayy*), *skyey* (not *skyy*). (*b*) *cagey* and *phoney* (beside *cagy* and *phony*) are both of unknown etymology, and do not belong to this group. *Fiddly* is from *fiddle* (verb). (*c*) *holey* (= full of holes) has *-ey* to distinguish it from *holy* (= sacred). (*d*) Adjectives from nouns ending in *-ue* retain the *-e*: *bluey*, *gluey* (not *bluy*, *gluy*). (*e*) The rare adj. *treey* (= abounding in trees) was formed in the 19c.

**eye** (verb) makes *eyeing*, not *eying*. See VERBS IN -IE, etc. 7.

**eyeglasses.** See GLASSES.

**eyot** /eɪt/. A variant spelling of AIT.

**eyrie** (nest of bird of prey). This is the preferred spelling (not *aerie*, *aery*, or *eyry*), but *aerie* is also commonly used. It is derived from medL *airea*, *eyria*, etc., probably from OF *aire* 'lair of wild animals'. The spelling *eyerie* was favoured by Henry Spelman in his *Glossarium archaiologicum* (1664) by association with *ey* 'egg'. Milton has *Eyries* in *Paradise Lost* vii. 424. Pronounce /ˈɪərɪ/, /ˈaɪərɪ/, or /ˈeərɪ/.

# Ff

**-f.** For plurals and inflexions of words ending in *-f* (*dwarf, leaf,* etc.) and *-fe* (*knife, wife,* etc.), see -VE(D), -VES.

**fabulous** originally meant mythical, legendary, but was at an early stage (early 17c.) extended to do duty as an adj. qualifying something that is real but so astonishing or incredible that it has the appearance of being legendary. Round about 1960 the word, together with the abbreviated form *fab*, leapt into colloquial use in the much weaker sense 'marvellous, terrific', in other words became a term of general approbation. Examples: (fab) *She stretched her stockinged toes towards the blazing logs. 'Daddy, this fire's simply fab.'—Times*, 1963; *That's a fab idea. I think I will*—B. T. Bradford, 1983; *I have to say the mag is just fab—Video for You*, 1987; (fabulous) *I think it's* [sc. Salford is] *a fabulous place—Radio Times*, 1962; *Trueman puffed at a cigarette and said he felt fabulous*—A. Ross, 1963; *He took a tremendous gulp ... 'It's fabulous! It's beautiful!'*—R. Dahl, 1970; *He thought she looked fabulous, just like a dream*—R. Ingalis, 1987. The two words are used with abandon by the young and the moderately educated, but only with studied glee by those who know of many other ways of praising a pleasurable experience.

**façade.** The cedilla should always be retained.

**facetiae,** in booksellers' catalogues, is a euphemism for pornography. Etymologically it answers to L *facētiæ*, pl. of *facētia* 'a jest'.

**facetious formations.** Over the centuries many evanescent and some more durable words have been brought into existence as irregularly formed jocular formations. A broad selection from among the hundreds of such words follows, roughly grouped into categories. Containing a pun: *anecdotage* (*anecdote, dotage*), *correctitude* (*correct, rectitude*), †*gigmanity* (*gigman, humanity*), *judgematic*

(*judge, dogmatic*), *queuetopia* (*queue, Utopia*), *sacerdotage* (*sacerdotal, dotage*). English words made to resemble Latin ones: *absquatulate, circumbendibus, contraption, omnium gatherum.* Humorously long: *antigropelos, cantankerous, collywobbles, galligaskin, gobbledygook, panjandrum, skedaddle, spifflicate, splendiferous, spondulicks, supercalifragilisticexpialidocious, transmogrify.* Irreverent familiarity: *crikey* (Christ), *gorblimey* (God blind me), *gosh* (God). Literary nonce-words: *chortle, galumph,* etc. (L. Carroll), *grudgment* (Browning), *misogelastic* (hating laughter, Meredith), *nyumnyum, yogibogeybox,* etc. (J. Joyce). Mock errors: *Eyetalian, highstrikes* (hysterics), *trick-cyclist* (psychiatrist), *underconstumble.* Pseudo- or mock Latin: *bogus, bonus, cockalorum* (also *hey cockalorum*), *hocus-pocus, holus-bolus.*

**face up to.** This somewhat clumsy-looking phrasal verb was first noted in America and also in Britain in the early 1920s. Since then it has survived a barrage of criticism in Britain (e.g. 'A needless expression, the result of the tendency to add false props to words that can stand by themselves'—Eric Partridge, 1942), but the fury has subsided and the expression ( = to accept bravely; to confront; to stand up to) is listed in *A Supplement to the OED* (1972) and *COD* (1990) without any limiting labels. Examples: *He faced up to the paradox of man*—W. Raleigh, 1920; *It was our duty as guardians of the children to face up to the situation*—Punch, 1935; *He won't face up, can't face up, to them being gone*—K. Hulme, 1984; *Why don't you simply face up to the past?*—K. Ishiguro, 1986; *She has to face up to the fact that Harriet—this incongruously aspiring Harriet—is her own creation*—T. Tanner, 1986.

**facile. 1** For pronunciation, see -ILE.

**2** Its value as a synonym for *easy* or *fluent* or *dexterous* lies chiefly in its derogatory implication. A facile speaker or writer is one who needs to expend little pains (and whose product is of correspondingly little importance); a facile triumph or victory is easily won (and comes to little). Unless the implication in brackets is intended, the use of *facile* instead of its commoner synonyms (*a more economical and facile mode*; *with a facile turn of the wrist*) is inadvisable.

**facilitate.** *The officer was facilitated in his search by the occupants.* We facilitate an operation, not the operator. The *OED* lists three 17–19c. examples of the word used to mean 'to lessen the labour of, assist (a person)', but marks the use as catachrestic. Examples of correct use: *They often used hypnosis to facilitate recall*—A. Storr, 1979; *The door kept open between them to facilitate communication*—R. Cobb, 1983.

**facility.** The entry (sense 2b) in *OED* 2 'covers' the use of the word in the sense 'the physical means for doing something; freq. with qualifying word, e.g. *educational, postal, retail facilities*; also in *sing.* of a specified amenity, service, etc.', but it gives no indication of the extraordinary proliferation of the word in the second half of the 20c. Typical examples: *Other features include sound facilities*—Which Micro, 1984; *The Colombo Street Sports and Community Centre . . . had converted a derelict building into a thriving indoor sports and recreation facility*—Community Development Jrnl, 1988; *Major policy differences still exist between broadcasters and independent facilities*—Broadcast, 1989; *Your £4,000 overdraft facility has become a five-and-a-half thousand pound monster*—W. McIlvanney, 1989; *The four facilities* [sc. factories] *involved . . . would still be owned by the Ministry of Defence*—Times, 1989; *You don't need a generously proportioned tub to fit a spa or whirlpool bath facility*—Do It Yourself, 1990; *If you want credit, a bank facility is usually better value than even a good dealer can offer*—Opera Now, 1990.

**fact.** 1 The word is included in numerous unchallengeable idiomatic phrases, e.g. *in fact, in point of fact, as a matter of fact, the fact is, (and) that's a fact*, and *the fact of the matter is*. Fowler (1926) closed the door to one such, namely *as a fact*, and condemned it as a 'parvenu'. He cited several unattributed examples as evidence: e.g. *He says that a 'considerable part' of the 25 millions is spent on new officials like locusts devouring the land; as a fact, barely one-thirtieth of that figure is due to new officials* (read: *as a matter of fact*);

*as a fact, it is no more above serious blunders than are many other German institutions* (read: *the fact is*). Fowler was probably right, but the use must always have been uncommon as it is not recorded in the *OED*, and no recent examples have come to hand.

2 *the fact that.* But *the fact that Dallas schoolchildren applauded the news is recorded by Arthur M Schlesinger Jr*: the journalist Godfrey Smith wrote this sentence in 1988 and subsequently defended it when a correspondent criticized it and wanted it rewritten. On the other hand, *CGEL* (1985), a work not usually much given to condemnation, described the sentence *They convinced him of the fact that they needed more troops* as 'a rather clumsy expression'. Gowers (1965), after criticizing the periphrases *owing to the fact that* and *in spite of the fact that* (they 'seldom have any advantage over the simple conjunctions *because* and *although*'), seems to find no fault with *the fact that* in other circumstances, since he uses the phrase repeatedly in the 2nd edn of *MEU*, e.g. *Against this practice is the fact that it cannot be consistently followed without occasional ambiguity* (p. 73); *The fact is surely that hardly anyone uses the two words for different occasions* (p. 190). What are we to make of this? It must be recognized, I think, that *the fact that* ( = the circumstance that) can legitimately be used as a noun phrase in apposition to a following noun clause (*The fact—not a new one—that Eliot doesn't pull his punches*—D. J. Enright, 1957; *The fact that he wrote a letter to her suggests that he knew her*—CGEL, 1985). In such a sentence as *The fact that he is a millionaire has nothing to do with it*, it is possible to delete *The fact* and be left with a grammatical sentence, but a degree of emphasis or focusing is lost. On the other hand the phrases *owing to the fact that* and *despite the fact that* can normally, with advantage, be replaced by *because* and *although* respectively. It's all a question of holding the reins tight.

**factious, factitious, fictitious.** Though the words are not synonyms even of the looser kind, there is a certain danger of confusion between them because there are nouns with which two or all of them can be used, with meanings more or less wide apart. Thus *factious rancour* is the kind of rancour that can let the politics of a faction prevail over some nobler concept, e.g. patriotism; *factitious rancour* is a rancour that is not of natural growth, but has been specially contrived to serve someone's ends; and *fictitious rancour* is a rancour represented as existing but which is in fact imaginary.

**factitive verb.** 'A type of transitive verb such as *make, choose, elect, judge, name,* which may take two complements, e.g. *They elected Mr Miller chairman of the meeting* where *Mr Miller* is the direct object or noun complement and *chairman of the meeting* is the second complement' (Hartmann and Stork, 1973). Cf. DITRANS-ITIVE.

**factor.** The word passed through a phase in the mid-20c. when it became a fashionable substitute for such words as *circumstance, component, consideration, constituent, element, event, fact,* or *feature.* Its true meaning in lay language, 'a circumstance, fact, or influence which tends to produce a result', is first recorded in 1816 in a work by the poet Coleridge. Gladstone used it correctly in 1878 in *The first factor in the making of a nation is its religion.* It continues to play its part in the language *The two boys had little in common and there were many factors that prevented them from becoming friends: caste, upbringing and environment;* [etc.]— M. M. Kaye, 1978; *Because newsreels were becoming a bigger and bigger factor in American life, Roosevelt would then repeat vital parts of the speech for a newsreel camera*—D. Halberstam, 1979; *Associations have been sought between schizophrenic rates and such factors as maternal age, sibling size, birth order, and season of birth*—A. Clare, 1980), surrounded by hideously complex uses of the word in mathematics, genetics, medicine, etc. A new use of *factor* appeared in the 1980s showing that the word is far from spent: '(with defining word, as *Falklands factor*) designating an event, etc., which is judged to have a significant effect on the politics, economy, etc., of a given country or group'.

**factotum.** Pl. *factotums.* See -UM 1.

**faculty.** **1** In some British universities the term is used of a group of departments, etc., which together make up a major division of knowledge (e.g. in Oxford, *Faculty of Anthropology and Geography, Faculty of Biological Sciences, Faculty of Literae Humaniores*).

**2** In America it has come to mean 'the staff of an educational establishment', and, as such, is often used without the definite article and also as a pl. or a sing. noun, e.g. *Students and some faculty are compelling the American universities to painful self-scrutiny*—Listener, 1969; *They had all gone to Swarthmore, then they had all married highly placed faculty or executives of Chubb University*—C. McCullough, 1985; *Many young faculty send reprints to more established colleagues as a simple but powerful way of making professional contacts*—Nature, 1989; *A prestigious scholar ... would attract top faculty and really put the school on the map*—Boston Globe, 1989; *Faculty teaching the wind institute and flute master class ... include professor of horn Robert Fries,* [etc.]—*The Observer* (Oberlin College), 1991; *Faculty votes to require all incoming students to own a computer*—Daily Northwestern (Ill.), 1995.

**faecal, faeces.** So spelt in BrE (AmE *fecal, feces*).

**faerie, faery.** The realm or world of the fays or fairies, esp. the imaginary world depicted in Spenser's *Faerie Queene* (1590–6). As the *OED* comments, 'In present usage, it is practically a distinct word [from *fairy*], adopted either to express Spenser's peculiar modification of the sense, or to exclude various unpoetical or undignified associations connected with the current form *fairy.*'

**fag, faggot.** **1** In BrE a *fag* is a colloquial word for a piece of drudgery or a wearisome or unwelcome task; a cigarette; (at British public schools) a junior pupil who runs errands for a senior. In AmE it is a slang word for a male homosexual.

**2** In BrE a *faggot* is a ball or roll of seasoned chopped liver, etc., baked or fried. In BrE and also in AmE (in which it is usually spelt *fagot*) it is also a bundle of sticks or twigs bound together as fuel. In AmE *faggot* is also a slang word for a male homosexual.

**fain.** A long-standing word in the language (OE *fægen* 'glad'), it reached some kind of literary climax in the 19c. (*Such scuffling ... as no pen can depict, and as the tender heart would fain pass over*—

Thackeray, 1847–8; *I felt my heart grow sick at the sight of this bloody battle, and would fain have turned away*—R. M. Ballantyne, 1858; *Mr. Wegg was fain to devote his attention exclusively to holding on*—Dickens, 1865) and passed into archaism, both as predicative adj. and as adv., in the 20c.

**fair, fairly** (advbs). **1** The shorter form, in its ordinary sense 'in a fair way', is in common use in a few fixed collocations, e.g. *to bid fair, to play fair, fair between the eyes*. In dialect use and in some areas abroad (esp. in N. America and Australasia) it is used to mean 'completely, fully' (e.g. *It fair gets me down*). In both types *fairly* would not be idiomatic.

**2** It should be borne in mind that *fairly* has more than one sense: in a fair manner (*he treated me fairly*); moderately (*a fairly good translation*); to a noticeable degree (*the path is fairly narrow*); utterly, completely (*he was fairly beside himself*). In certain contexts it is sometimes difficult to know whether 'moderately' or 'utterly' is intended; but in speech this possible confusion is prevented by the marked difference in intonation (*OED*).

**fairy.** It is hardly necessary to point out that there are two distinct senses: **1** a small imaginary being with magical powers.

**2** (*derogatory slang*) a male homosexual. The context always makes it clear which sense is intended. In literary contexts the etymologically related word *fay* is occas. used in place of *fairy* in sense 1. Cf. FAERIE.

***fait accompli*** /feɪt ə'kɒmplɪ/, a thing that has been done and is past arguing against or altering.

**faithfully.** **1** For *Yours faithfully*, see LETTER FORMS.

**2** In *promise faithfully, faithfully* is a somewhat demotic substitute for *solemnly, truly, expressly*, etc. (*He promised faithfully to send the book the next day*—1894 in *OED*; *And I promised him yes faithfully Id let him block me*—J. Joyce, 1922).

**fakir** /'feɪkɪə/. Now the dominant form (beside *faquir*).

**fall.** The third season of the year was called the *autumn* from the 14c. onwards, and also, in the British Isles, *the fall of*

the leaf or simply *the fall* from the 16c. until about 1800. As time passed, *autumn* settled down as the regular term in Britain, whereas *the fall of the leaf* (less frequently *the fall of the year*) and then *fall* by itself gradually became standard in America from the late 17c. onwards. *Autumn* and *fall* are familiar names to everyone in each of the two countries, but in day-to-day speech *autumn* is the only standard form in BrE and *fall* is equally standard in AmE.

**fallacy.** In logic, a fallacy is a flaw that vitiates an argument. Some types of fallacy are of frequent enough occurrence to have earned names that have established themselves in the language. These include *arguing in a circle, argumentum ad hominem*, etc., *beg the question, ignoratio elenchi, non sequitur, petitio principii, post hoc ergo propter hoc*, and *undistributed middle*, all of which are treated in this book at their alphabetical place. The word is also widely used outside logic in the broad sense 'a mistaken belief, esp. one based on unsound argument'. Examples: *In pictorial art the fallacy that nature is the mistress instead of the servant seems almost ineradicable*—R. Fry, 1920; *The fallacy of equating size with quality*—R. Gittings, 1968.

**false analogy.** See ANALOGY 4, 5.

**falsehood, falseness, falsity.** The three words substantially overlap in meaning and are often interchangeable. The primary notion is 'departure from the truth or from the true state of affairs'. A *falsehood* is an uttered untruth, a lie; *falsehood*, as an abstract noun, is the intentional making of a lie. *Falseness* is essentially 'contrariety to fact, incorrectness, unreality'; or, more specifically, 'deceitfulness or faithlessness' or an instance of either. *Falsity* is almost always synonymous with *falseness* but hardly ever with *falsehood*. Typical examples: (falsehood) *But a half-truth was a falsehood, and it remained a falsehood even when you'd told it in the belief that it was the whole truth*—A. Huxley, 1939; *As it had always been, truth and falsehood were inextricably intertwined in that statement*—S. Naipaul, 1980; (falseness) *The falseness of its illusions*—OED; *His predecessor ... had recorded their falseness and cruelty*—OED; (falsity) *The engagement had probably not been a complete* *falsity, a piece of acting*—H. Carpenter, 1981.

**false quantity.** **1** English is a Germanic language and like the other members of the group has a primary system of vowels distinguished in length (or quantity). Thus *cat/arm/day* show respectively a short vowel, a long vowel, and a diphthong, all historically containing a variety of *a*-ness. Similarly *bed, sit, hot, run,* and *put* all have short vowels; *see, saw, her,* and *too* all have long vowels; and *my, boy, no, how, near,* etc., have diphthongs. From the time of the first records of English (8c.), fundamental changes to the original system have occurred and the details of such changes have been systematically mapped by scholars. Into this tripartite fluid system, at intervals over a period of more than 1,000 years, a great many words of classical origin have been adopted into English. Classical Greek and Latin had their own systems of vocalic quantities, and these quantities were not necessarily carried over into English. If the vowels and diphthongs of the incoming classical words matched those of similar-sounding words in the receiving language they were usually left unaltered, and were then subject only to the linguistic changes befalling native words themselves. If they did not match the English patterns the original classical quantities suffered a severe battering.

**2** The following English words regularly have short vowels by contrast with the long vowels of the corresponding forms in the donor languages. *cinema, clematis, echo, ethics, fastidious, ornithology, protagonist, semaphore, semi-, simian, Sirius, Socrates, vertigo.* English words have long vowels where the classical originals had short ones: *Eros, gladiolus, minor, saline, senile, status.*

**3** The rule is simple: the quantity (and often the placing of the main stress) of vowels in Latin and Greek loanwords is an unreliable guide to their pronunciation in English.

**falsetto.** Pl. *falsettos.* See -O(E)S 6.

**famed.** As ppl adj. in the sense 'celebrated, famous' it has had a patchy history since it was first recorded in the late 16c. (Shakespeare; contextually it is

perilously close to being just an ordinary pa.pple of the verb *fame*). Byron's use of it in *Childe Harold* (1812), *In famed Attica such lovely dales Are rarely seen,* is more or less forced on him by the contextual needs of the verse. The other examples are either literary or are more pa.ppl than adjectival. *Famed* cannot be regarded as a simple substitute for *famous.* In practice it appears most commonly in journalistic sources—*WDEU* cites *Ruskin's famed friend, Painter Sir John Millais* from *Time* 29 Dec. 1947—and in the type X, *famed for his accounts of fell-walking.* The best course at present is to avoid using it except in the manner of this second type.

**fandango.** Pl. *fandangos.* See -O(E)S 6.

**fan, fanatic.** **1** The adj. *fanatic* is tending to drop away in favour of *fanatical* but both forms are still found. (*Fanatic* is, of course, very common as a noun.) Examples: (fanatic) *What annoyed him was their fanatic sense of righteousness*—C. Potok, 1966; (fanatical) *He was a fanatical worker, often doing thirteen or fourteen hours a day*—Ann Thwaite, 1984; *The Rule of the Reformed Carmelites, which St John of the Cross joined, was severe but not fanatical*—L. Gordon, 1988.

**2** The abbreviated form *fan,* used to mean 'a keen and regular supporter (of a team, a person, etc.)' and recorded in AmE as early as 1682, became part of the ritual language of baseball in the 19c. and then passed into general use in AmE, BrE, and elsewhere in the 20c. in the senses 'a supporter', 'an admirer' (used of persons or subjects). Hence *fan letter* (first recorded 1932), *fan mail* (1924), *film fan* (1918), etc.

**fantasia.** (It. 'fantasy, fancy'.) Now regularly pronounced /fæn'teɪzɪə/, with the stress on the second syllable, though the Italianate pronunciation /fæntə'zi:ə/ is sometimes used when the word forms part of the title of a particular musical composition.

**fantastic** is one of the most popular 20c. (first recorded in 1938) colloquial terms for 'excellent, very enjoyable': e.g. *'Only ... very nice?' he asked woefully. 'Oh, it's great! I mean, it's fantastic!'*—*Los Angeles Times,* 1987. Cf. FABULOUS.

**fantasy, phantasy.** In its several senses the customary spelling now is *fantasy*, reflecting its proximate French, rather than its ultimate Greek, etymon. Until at least the end of the 19c. the two spellings occurred with approximately equal frequency. And as late as 1894 the *OED* commented 'In mod. use *fantasy* and *phantasy*, in spite of their identity in sound and in ultimate etymology, tend to be apprehended as separate words, the predominant sense of the former being "caprice, whim, fanciful invention", while that of the latter is "imagination, visionary notion".'

**faquir.** See FAKIR.

**far.**

| 1 *farther/further.*
| 2 *(so) far from —ing.*
| 3 *far-flung.*
| 4 *as far as/so far as.*
| 5 Some other uses of *far.*

1 For *farther/further*, see FARTHER.

2 *(so) far from ——ing* (used when something is denied and something opposite asserted). This construction headed by *so* and containing a verbal substantive in *-ing* is first recorded in the 17c. and was still current in 1926, when Fowler cited the example *So far from 'running' the Conciliation Bill, the Suffragettes only reluctantly consented to it.* The construction was also cited without any qualification by George O. Curme (1931): *So far from doing any good, the rain did a good deal of harm,* a blending of *Far from doing any good, the rain did a good deal of harm* and *The rain was so far from doing any good that it did a good deal of harm.* The *so* has tended to drop out of use since then, but the shortened form *Far from ——ing* is still widely current.

3 *far-flung.* This quite modern word (first recorded in 1895) immediately calls to mind Kipling's *Recessional* (1897) (*God of our fathers, known of old, Lord of our far-flung battle-line*). Fowler (1926) spoke of 'its emotional value ... as a vogue-word. The lands are distant; they are not far-flung; but what matter? *Far-flung* is a signal that our blood is to be stirred.' The *far-flung Empire* has been replaced by less stirring concepts, but the rich flavour of the adjective remains as a reminder of past grandeur.

4 *as far as/so far as.* (*a*) These phrases were once freely interchangeable as compound prepositions: *I'll take you as far as/so far as Banbury,* but the second of these is now obsolete or, at best, archaic. (*b*) In adverbial clauses, the natural constructions are virtually limited to *as far as——is concerned,* or *as far as——goes,* and the verbal extensions may not be omitted. Correct uses: *You can sit somewhere else as far as I'm concerned; as far as this theory goes, the outlook is promising.* Replacement of *as far as* by *so far as* is possible, but now seems to be less often encountered than it used to be. Doubtful or unacceptable examples (the words in brackets need to be supplied): *As far as getting the money he asked for [is concerned], Mr Churchill had little difficulty; As far as collecting statistics [goes], only industry is necessary.* See CONCERNED. *So* (or *in so*) *far as* may also be used to mean 'to the extent that' (*The results will be used in so far as they affect the classification of the group*).

5 Some other uses of *far.* As adv., *far* is used before comparatives (*It is a far, far, better thing that I do, than I have ever done*); before *too* + adj. (*far too busy*); in the adverbial phr. *so far* (*with whom I had so far had little to do*); = to a remote place (*he had wandered far and wide*); = at a distance from (*he felt fed up and far from home*); etc. As adj., it means principally 'situated at or extending over a great distance in space or time; more distant' (*a far cry; a far country; at the far end of the hall*). For a fuller range of uses of *far* (*far and away, far be it from me, far and near, go too far, so far so good,* etc.), it is essential to consult a large modern dictionary.

**farce.** The word in its sense 'a short dramatic work the sole object of which is to excite laughter' is a 16c. loanword from French (from OF *farce* 'stuffing'). 'The term, in the latinized forms *farsa, farcia,* was applied in the 13c. to phrases interpolated in the liturgical *kyrie eleison* ... and to passages in French inserted in the Latin text of the epistle at Mass (cf. medL *epistola farcita*); hence to impromptu amplifications of the text of religious plays, whence the transition to the present sense was easy' (*ODEE*).

**faro.** This gambling card-game is pronounced /ˈfeərəʊ/.

**farouche.** The English word, meaning 'sullen-mannered from shyness' is an 18c. loanword from French. The French word is an altered form of OF *faroche*, beside *forache* (cf. dial. *fourâche*, etc.) from medL *forasticus*, from L *foras* 'out of doors' (*ODEE*). It has no connection with *ferocious* (L *ferox, -ōcis*). See TRUE AND FALSE ETYMOLOGY.

**farrago** /fə'rɑːgəʊ/. Pl. *farragos*, (AmE) *farragoes*. See -O(E)S 6.

**farrow, litter.** *Farrow* is used of pigs only, *litter* of other quadrupeds producing several young at a birth.

**farther, further.** 1 *Etymology. Farther* is in origin a ME variant of the older form *further* (OE *furþor*, usu. taken to be the comparative form of *forþ* 'forth'). It is to be distinguished from OE *feorr* 'far', the comparative of which was *fierr* 'further'. From ME onward both *farther* and *further* came to be used as the comparative of *far*, beside the newly formed words *farrer* and *ferrer*, extant from the 12c. to the 17c. After that period the comparative forms *farrer* and *ferrer* remained only in dialects, being superseded in the standard language by *farther* and *further*. Wyclif (c1380) wrote *Sum ferrer and sum nerrer*, and a Scottish work of 1549 has *He vil see ane schip farrer* [ = *farther*] *on the seye*. Shakespeare used *farre* once instead of his customary forms *farther* and *further*: *Not hold thee of our blood, no not our Kin, Farre then Deucalion off* (*Winter's Tale*). But he used *further* 199 times and *farther* 47 times apparently without any functional distinction (as well as *farthest* 14 times and *furthest* twice). Curiously, the proportions are not dissimilar in a frequency analysis of a corpus of 1 million modern AmE words assembled in 1963–4: *further, furthest* (all uses) 165 examples; *farther, farthest* 29 examples.

2 *Distribution of farther and further.* (a) *Scholarly opinions.* The matter is not simple and opinions vary somewhat. The *OED* (1895): 'In standard Eng. the form *farther* is usually preferred where the word is intended to be the comparative of *far*, while *further* is used where the notion of *far* is altogether absent; there is a large intermediate class of instances in which the choice between the two forms is arbitrary.' Fowler (1926): 'The fact is surely that hardly anyone uses the two words for different occasions; most people prefer one or the other for all purposes, and the preference of the majority is for *further*; the most that should be said is perhaps that *farther* is not common except where distance is in question … On the whole … it is less likely that one [meaning or use] will be established for *farther* and *further* than that the latter will become universal.' Curme (1935): 'We use *farther* and *further* with the same local and temporal meaning, but *further* has also the meanings *additional, more extended, more*: "The cabin stands on the *farther* (or *further*) side of the brook." "I shall be back in three days at the *farthest*", or *at the furthest*. But: *further* details; without *further* delay. "After a *further* search I found her." "Have you anything *further* ( = *more*) to say?)". In adverbial function *farther* and *further* are used indiscriminately: " You may go *farther* (or *further*) and fare worse." There is, however, a decided tendency to employ *further* to express the idea of additional, more extended action: "I shall be glad to discuss the matter *further* with you." '

The most recent standard grammar, *CGEL* (1985), expresses the view that the *ar*-forms are chiefly restricted to expressions of physical distance: 'The two sets *farther/farthest* and *further/furthest*, which are both adjectives and adverbs, are used interchangeably by many speakers to express both physical and abstract relations. In fact, however, the use of *farther* and *farthest* is chiefly restricted to expressions of physical distance, and, in all senses, *further* and *furthest* are the usual forms found: "Nothing could be *further* from the truth"; "My house is *furthest* from the station." The most common uses of *further* are not as a comparative form of *far* but in the sense of "more", "additional", "later": "That's a *further* reason for deciding now." [but not: *a far reason*]; "Any *further* questions?"; "The school will be closed until *further* notice"; "We intend to stay for a *further* two months".'

(b) *A further view.* From the above it seems that there is a broad consensus that *farther* and *farthest*, whether as adjs. or advs., are recessive forms except in contexts where distance is in question. I would add a further consideration: in so far as the *a*-forms of the adv. or the adj. are used at all, they are now more likely to occur in AmE than in BrE, though the distinction is not clear-cut. In the early part of the 20c. *farther* was also common in the UK but less so now.

Examples: (i) Physical distance is involved. (*farther*) (Showing preference for *farther* in the UK in the first quarter of the 20c.) *The gulls rose in front of him and floated out and settled again a little farther on*—V. Woolf, 1922; *Oh, she would like to go a little farther. Another penny was it to the Strand?*—V. Woolf, 1925; (US) *And farther down the road was the state institution*—Marilynne Robinson, 1981; *I began to go farther and farther from the house*—New Yorker, 1986; *He pointed farther up the dirt road to a row of small . . . buildings*—P. Roth, 1987. An Aust. example: *Linda Craig was farther up the row*—M. Pople, 1986. (*further*) (UK) *I suppose you'll be catching the first boat back—but I shouldn't go further than France*—G. Greene, 1939; *Each day the snow grew a little whiter and spread further down the mountain-side*—O. Manning, 1960; *My God, what stairs! How much further? My legs ache*—M. Wesley, 1983; *As they climbed to the fourth, Veronica knew they could go no further*—B. Rubens, 1987. But *farther* is sometimes found in BrE: *Don't let's go any farther. Let's sit down here*—C. Rumens, 1987; *the motorway would make it seem no farther than the outer suburbs*—Stephen Wall, 1991.

(ii) The notion of actual distance is wholly or partially absent: (*farther*) (US) *'You get a lot farther using your nose than your palate,'* Patty says about wine-tasting—New Yorker, 1987; *However, it was so far and no farther*—S. Bellow, 1987; *Kasparov simply saw farther, 'much, much farther', than the machine*—NY Times Mag., 1990; (*further*) (UK) *He wrote for booklets containing further particulars of almost every device he saw advertised*—E. Bowen, 1949; *He . . . found English currency confusing and the* [taxi] *driver sought to confuse him further*—E. Waugh, 1961; *It doesn't do though to push the analogy much further*—P. Lively, 1987; *Denning's very subjective report . . . further vilified him*—London Rev. Bks, 1987. But *farther* is sometimes found in BrE: *Her 'old days' went back no farther than the late nineteen-twenties*—B. Pym, 1980. And *further* is sometimes found in AmE: *The trouble with the character was that he had to push it further and further*—New Yorker, 1986.

(c) *Special cases.* From about the beginning of the 20c. formal education in Britain for adults, and for young people who have left school, has regularly been called *further* (not *farther*) education. Used to introduce a sentence, i.e. to function as a sentence adverb, *further* is usual, not *farther*: *Further, shameful as it might be to admit it, the idea of the play had started to interest him rather*—K. Amis, 1958.

(d) *farthest, furthest.* For the superlative forms the semantic and geographical patterns of distribution seem to be the same as for the comparative forms: (*farthest*) *. . . and now the prince is scouring the farthest reaches of the globe for his bride*—J. M. Coetzee, 1977 (SAfr.); *Do you know my farthest journey has been away from suicide*—Peter Porter, 1981 (Aust./UK); *'Why, Lord, no honey!' I told her. 'It's the farthest thing from my mind.'*—Lee Smith, 1983 (US); *Coming round the farthest mark, they hauled down her big, billowing fisherman*—New Yorker, 1986; (*furthest*) *In our society, those who have the best knowledge of what is happening are also those who are furthest from seeing the world as it is*—G. Orwell, 1949; *he wanted nothing more . . . than to send her off with sixpence to spend at the furthest lolly shop*—M. Eldridge, 1984 (Aust.); *To the furthest place on earth from our imperial homeland*—V. O'Sullivan, 1985 (NZ); *This was the lower fountain, furthest from the house*—A. S. Byatt, 1987 (UK); *It seeks the furthest extension of the educationally valuable among the masses*—Encounter, 1987 (UK).

(e) *further and further* is commonly used for emphasis: *He hasn't given up the idea of such a return, though he tends to imagine it as taking place further and further in the future*—A. Lurie, 1974; *It always seemed possible to retreat further and further into the house to escape from intruders*—J. Fuller, 1983.

(f) The only current form of the verb is *further*; and *furthermore* (not †*farthermore*) is the only current form of the compound adverb.

**3** *Conclusion.* The views expressed in *CGEL* (1985) (see (2) above) satisfactorily account for the main patterns in BrE. They need to be reinforced by the prominence accorded to the *ar*-forms (see (2b) above) in AmE. But these are shifting sands, and further changes may well occur in the 21c., with the likelihood that *further* and *furthest* will continue to be the dominant forms in most varieties of English.

**Fascism, Fascist.** Understandably at the time—Benito Mussolini had not long been established in power—Fowler (1926) wondered about the degree of permanence of these words, and also about the degree of Anglicization with which they were to be pronounced. The pronunciation with medial /-s-/ was occasionally heard before and during the 1939–45 war, but the customary standard forms have long since settled down with a medial /-ʃ-/. The It. originals—*fascista*, pl. *-ti, fascismo*—dropped out of use in English long ago.

**fatal, fateful.** The words are etymologically closely related in that they both contained or contain the element *fate*. In times past they shared a number of senses, 'prophetic', 'fraught with destiny', etc., and even 'producing or resulting in death'. The facts, which are too complicated to summarize here, are all set down succinctly in the *OED*. But the two words have substantially drawn apart. Now *fatal* alone means 'causing or ending in death' (*a fatal accident*); it also means 'ruinous, ending in disaster' (*a fatal mistake, a fatal weakness*). By contrast, *fateful* usually means '(of utterances, decisions, encounters, etc.) having far-reaching (whether happy or adverse) consequences'. The shared territory of the two words is discernible only when the consequences are disastrous, and even then one's instinct normally leads us to prefer one of the two words rather than the other. Examples: (fatal) *There is some flaw in me—some fatal hesitancy*—V. Woolf, 1931; *Some of his suggestions were indeed harmless; but there was one which from our point of view could be fatal*—H. Macmillan, 1971; *Application forms and references can only provide a very rough and ready method of assessing an applicant's suitability—and it can prove fatal for employers to simply rely upon these measures*—Business Franchise Mag., 1989; (fateful) *The confusion of patriotism and personalities left behind by the fateful gathering*—J. Galsworthy, 1924; *In summing up 1934 we can see, in the light of what was to come, that it was a fateful year*—J. F. Kennedy, 1940; *He has had an intermittent run of bad luck, from spider bites and knee operations through to that fateful*

*puncture ... which cost him the race*—Bicycle Action, 1987.

**father.** Used in a number of SOBRIQUETS, e.g. *the Father of Lights* (God); *the father of faith, of the faithful* (Abraham); *the father of lies* (the Devil); *the Fathers of the Church* (early Christian writers); *the Holy Father* (the Pope). These and numerous others are listed and illustrated in the *OED*.

**father-in-law.** See -IN-LAW.

**fathom.** Pl. often *fathom* (i.e. unchanged) when preceded by a number: *six fathom deep*.

**fatten.** See -EN VERBS FROM ADJECTIVES.

**faucet.** Originally (15c.) 'a tap for drawing liquor from a barrel, etc.', *faucet* survives in technical use (combined with *spigot*) in BrE, but in domestic use has been entirely replaced by *tap*. In AmE *faucet* is in widespread use for an ordinary water tap but its distribution is uneven: '*faucet* is the usual New England name for the kitchen water tap, but *spigot* seems to be coming in from the southward, and *cock* and *tap* are also recorded' (R. I. McDavid, 1977). Examples: *Another counter ... was cluttered with test-tube racks, glass beakers ... and a sink whose faucet was dripping*—New Yorker, 1989; *The bathroom faucet dripped*—New Yorker, 1990.

**fault.** 1 (noun). *at fault*, guilty, to blame, began to oust the earlier *in fault* in the last quarter of the 19c. and is now standard. *The natural presumption is that it is not his already proved skill that is at fault but rather the nature of the theories*—Mulgan and Davin, 1947; *It will be obvious that the whole basis of the questionnaire ... is at fault*—H. J. Eysenck, 1957. An older sense of *at fault*, namely 'puzzled, at a loss', has dropped out of use.

2 (verb). Used in the sense 'to find fault with; to blame or censure', *fault* has been in use since the mid-16c. In 1850 Fanny Trollope hinted at the unfamiliarity of the use in Britain: *Her manner ... could not, to use an American phrase, be 'faulted'*. Some 20c. usage guides (but not MEU 1926, 1965) have questioned the validity of the use. In the event, it is not comfortably used to mean 'to blame or censure'; but in the sense 'to find fault with' it is unquestionably

standard now: e.g. *Martita wasn't too keen on Fay Compton I gathered (though she couldn't fault her perfect diction)*—A. Guinness, 1985.

**fauna, flora** are singular nouns used as collectives. Their plurals, rarely needed, are *faunas* and *floras* (less commonly *faunae* and *florae*). They are Roman goddess names pressed into service for the realms of animals and of plants respectively, especially those to be found in any given region or period.

**faun, satyr. 1** A *faun* is 'One of a class of [Roman] rural deities; at first represented like men with horns and the tail of a goat, afterwards with goats' legs like the Satyrs, to whom they were assimilated in lustful character' (*OED*).

**2** A *satyr* in Greek mythology is 'One of a class of woodland gods or demons, in form partly human and partly bestial, supposed to be the companions of Bacchus. In Greek art of the pre-Roman period the satyr was represented with the ears and tail of a horse. Roman sculptors assimilated it in some degree to the faun of their native mythology, giving to it the ears, tail, and legs of a goat, and budding horns' (*OED*).

**3** 'The confusion between the words *satiric* and *satyric* gave rise to the notion that the satyrs who formed the chorus of the Greek satyric drama had to deliver 'satiric' speeches. Hence, in the 16–17th c., the frequent attribution to the satyrs of censoriousness as a characteristic quality' (*OED*).

**faux amis.** Because English shares a great many words with French it is easy to fall into the error of assuming that the shared words operate in the same manner, and with the same meanings, in both languages. A publicity leaflet for a recent *Dictionnaire des faux amis, français ~ anglais* gave some excellent examples of the kind of pitfalls that can occur: 'Si vous demandez à l'un de vos invités anglais "Could you fetch me some wine from the cave?", ne vous étonnez pas de le voir revenir équipé de cordes et pitons pour cette expédition spéléologique, "cave" en anglais signifiant "grotte". De même, méfiez-vous de la réaction de la personne que vous complimenteriez sur sa "sparkling denture" car peu de gens aiment s'entendre dire qu'ils ont un

"dentier étincelant".' Another example: The word *fatigue* is spelt the same in English and in French, and has the same primary meaning. But *la fatigue/des fatigues du voyage* is best rendered in English by 'the tiring effects of the journey' or 'the tiring journey', not 'the fatigue of the journey'. Cf. also Eng. *Fidel Castro was wearing his usual green fatigues* and Fr. *Fidel Castro portait comme d'habitude son treillis vert*. Constant vigilance is needed if one is to avoid errors arising from the asymmetrical way in which pairs of words operate in the two languages.

**faux pas** /fəʊ ˈpɑː/. Pl. the same, but pronounced /fəʊ ˈpɑːz/.

**favour, favourable, favourite.** The normal spellings in BrE, as against AmE *favor, favorable, favorite*.

**fax.** First recorded in 1948 as a shortened form of *facsimile* (*process*, etc.) the word quickly established itself in the standard language as noun and verb, esp. during the 1980s when *fax machines* came into widespread use. The readiness to accept the word contrasts sharply with the continuing hostility to *pix* (pictures), *dix* (dictionaries), etc., and with the slow acceptance of *sox* (socks, pl. of sock) as the 20c. progressed.

**fay.** See FAIRY.

**fayre.** A curiously popular 'olde' or commercial spelling of *fair* in the sense of 'a gathering of stalls, amusements, etc., usu. outdoors' (*a medieval fayre*), or, worse, for *fare* 'food provided by a public house, etc.' (*delicious country fayre*). The spelling seems not to have spread to the names of exhibitions publicizing the products of a particular industry (*Frankfurt Book Fair, Audio Fair, world fair*, etc.).

**faze.** A 19c. AmE variant of the ancient (but now only dialectal) verb *feeze* 'to drive off, to frighten away'. *Faze* means 'to disconcert, disturb' and is used mainly in negative contexts (e.g. *he was not fazed by the unemployment figures or by other aspects of the recession*). It is now commonly used in informal contexts in BrE as well as in AmE. The word has no connection, etymological or otherwise, with the ordinary word *phase* (stage in a process, etc.), though it is sometimes so spelt in the US.

**-fe.** For plurals of words ending in -fe (*knife, wife*, etc.), see -VE(D), -VES.

**fearful, fearsome.** *Fearful* in the sense 'frightened, apprehensive' is usually const. with *of*, (archaic) *to* + infin., or with a clause introduced by *lest* or *that* (*fearful of what he would find; made him fearful to act; fearful lest he should fail; fearful that a full-scale attack would not succeed*). It is also used to mean 'unpleasant' (*a fearful row*). *Fearsome* means 'appalling or frightening', esp. in appearance' (*iron fencing with fearsome spikes at the top*). Some modern examples: (fearful) *As a result of my total misreading of Bradshaw arriving a whole day late at Rosslare ... I got into a fearful fluster*—C. Day Lewis, 1960; *Andrew was tolerantly aware that his mother was a fearful snob*—I. Murdoch, 1965; (fearsome) *My father treated me with a serious concern that was as fearsome as open anger*—W. Golding, 1967; *Ichiro continued to regard me with the most fearsome look*—K. Ishiguro, 1986; *Rhododendrons ... cast a deep shade over large areas and, as I learnt later, have become a fearsome forest weed, preventing the growth of other plants*—Outdoor Action, 1989.

**feasible. 1** Thus spelt, not -*able*. See -ABLE, -IBLE 7.

**2** There are three senses: (*a*) (of a design, project, etc.) capable of being done, practicable. (*b*) (of things in general, also of persons) capable of being dealt with successfully. (*c*) (of a proposition, theory, story, etc.) likely, probable. The first two are part of the standard language, but the third is more questionable: 'Hardly a justifiable sense etymologically, and (probably for that reason) recognized by no Dict. though supported by considerable literary authority' (*OED*, 1895). The 'literary authority' runs from Hobbes (1656) to Livingstone (1865). Fowler (1926) fell in with the *OED* in regarding sense (*c*) as questionable and cited several unattributed examples (from newspapers?) where either *probable* or *possible* should have been used: e.g. *Witness said it was quite feasible* [better *possible*] *that if he had had night binoculars he would have seen the iceberg earlier; We ourselves believe that this is the most feasible* [better *probable*] *explanation of the tradition*. Each case must be treated on its merits, but when the context requires the sense of likelihood or probability it is prudent to test first

whether *possible* or *probable* might not be the more satisfactory word, and to use *feasible* only if both of the other words seem unnatural or unidiomatic.

Examples of standard uses: *But changes become feasible over a period of time, and it was possible to create a better balance*—Harold Wilson, 1976; *There was a time when, with his friend Belinsky, he too had believed that a simple solution was feasible*—I. Berlin, 1978; *Christian unity is now a feasible objective for the first time since the Reformation*—D. Lodge, 1980.

**feature** (verb). Fowler tended to be nervous about new uses coming into widespread use in Britain from the general direction of Hollywood. The use of *feature*, he said, 'in cinema announcements instead of *represent* or *exhibit* is perhaps from America.... it is to be feared that from the cinema bills it will make its way into popular use, which would be a pity.' He cited an unattributed example of 1924: *Boys' school and college outfits, men's footwear and under-garments, as well as ..., are also featured*. His words went unheeded. The verb is now in standard use in the senses 'to give special prominence to; (trans. or intr.) to have as or be an important actor, participant, or topic in a film, broadcast, book, etc.' Examples: (trans.) *Balzac's 'The Eternal Flame', featuring Miss Norma Talmadge*—Westminster Gaz., 1923; *The Louvre, Oxford-circus, are featuring coats and skirts and top coats for Scotland in new designs*—Times, 1929; *They're featuring me* [sc. a newspaperman] *as a special service*—E. Waugh (*Scoop*), 1938; *I was to have my name featured for the first time at the top of the bill*—C. Chaplin, 1964; *Student concerts featuring him as composer and pianist*—R. Hayman, 1981; *A novel featuring several generations of a Rouen family*—Julian Barnes, 1984; (intr.) *The uncles and aunts who feature so largely in the memories of many orphan children*—R. Gittings, 1968; *Libraries and the youth service feature prominently in many of the local authority cuts*—Times, 1976.

**February.** The only standard pronunciation in BrE is /ˈfebrʊərɪ/, with both *r*s fully articulated. But dissenters are at the door: in numerous dells and vales and suburbs some people are content to say /ˈfebjʊərɪ/, i. e. to convert the first *r* into the initial sound of words like *yell*. The process is called DISSIMILATION. In America, the largest dictionaries report that, despite criticism, dissimilated pronunciations of *February* (i.e. with /-juəri:/ or /-jə,weəri:/) are used by educated speakers and are considered standard.

**fecal, feces.** See FAECAL, FAECES.

**fecund.** Two pronunciations are current, /'fekənd/ and /'fi:kənd/. For those who turn to etymology as a basis for the pronunciation of Eng. words, the L original (*fēcundus*) has a long *e*. My own impression is that /'fekənd/ is nevertheless the more usual pronunciation of the two among standard speakers.

**federation, confederation, confederacy.** It is unrealistic to think that these terms are mutually exclusive, but some indications of their historical applications can be given. The *OED* says: '*Confederacy* now [sc. in 1891] usually implies a looser or more temporary association than *confederation*, which is applied to a union of states organized on an intentionally permanent basis.' Specifically the *Southern Confederacy* was the term used for the Confederate States of America, the government established by the Southern US states in 1860–1 when they seceded from the Union, thus precipitating the American Civil War. *Confederation*, on the other hand, 'is usually limited to a permanent union of sovereign states for common action in relation to externals' (*OED*, 1891). For example, the *Confederation of the Rhine* was the union of all the German states except Prussia and Austria under the protection of Napoleon Bonaparte from 1806 to 1814; and 'the United States of America [is] commonly described as a *Confederation* (or confederacy) from 1777 to 1789, but from 1789, their closer union has been considered a "federation" or federal republic' (*OED*, 1891). A *federation* is 'now chiefly *spec.* the formation of a political unity out of a number of separate states, provinces, or colonies, so that each retains the management of its internal affairs' (*OED*, 1895). At present (1995) Switzerland, USA, Canada, Australia, and Malaysia are examples of federal government, each one differing markedly from the others in constitutional details. Some see the countries of Europe as moving towards the formation of a federal Europe.

**fee** (noun). For synonymy, see TAX.

**fee** (verb). The pa.t. and pa.pple are best written *fee'd*: see -ED AND 'D.

**feedback** is a 20c. technical term in electronics, defined by the *COD* (1995) as 'the return of a fraction of the output signal from one stage of a circuit, amplifier, etc., to the input of the same or a preceding stage; a signal so returned'. It is also a technical term in biology meaning 'the modification or control of a process or system by its results or effects, esp. in a biochemical pathway or behavioural response'. Round about 1960 the term was adopted by laymen and used very frequently to mean 'information about or response to an investigation of any kind' (e.g. *some useful feedback from journalists in the area*).

**feel** (verb). **1** Occasional, but misguided, opposition has been expressed to the use of *feel* in the intuitive sense 'to think, believe, consider', or, as the *OED* expresses it, 'to apprehend or recognize the truth of (something) on grounds not distinctly perceived; to have an emotional conviction of (a fact)'. The use goes back to Shakespeare (*Garlands ... which I feele I am not worthy yet to weare*—1613), and has been current ever since in standard use. Modern examples: *I feel you don't love him, dear. I'm almost sure you don't*— D. H. Lawrence 1920; *But perhaps it was a little flat somehow, Elizabeth felt. And really she would like to go*—V. Woolf, 1925; *Some were disturbed at the relation with the American government and felt that they were serving, in effect, as propaganda agents for the United States and the Royal Lao Government*—N. Chomsky, 1969; *If Pascal had been a novelist, we feel, this is the method and the tone he would have used*—G. Greene, 1969; *After two weeks in the Miller household, both Fenton and Amy felt they had lived there for ever*—C. Rayner, 1978.

**2** (*a*) *Feel* is used idiomatically in a quasi-passive sense with a wide range of adjs.: *to feel better, cold, good, happy*, etc. (*b*) On the phrases *to feel bad/badly*, see BAD 2.

**feet.** See FOOT (noun).

**feint.** Since the mid-19c. the usual spelling of the ordinary adj. *faint* in descriptions of stationery printed with light-coloured lines. *Faint* and *feint* are variant spellings of the same word, ultimately from OF *faint*, *feint* feigned, pa.pple of *faindre*, *feindre* 'to feign'.

**feldspar** /ˈfɛldspɑː/, rather than *felspar*. The first element is from Ger. *Feld* 'field', not Ger. *Fels* 'rock'.

**fellahin** /-əˈhiːn/ is the pl. of *fellah*, an Egyptian peasant. In the 19c. it was usually spelt *fellaheen*.

**fellow** (in compounds). **1** *Accentuation*. The simplex *fellow* is, of course, stressed on the first syllable, i.e. /ˈfɛləʊ/. In compounds of *fellow*, when the second element is a noun or verb, the pattern is normally of the type 'secondary stress followed by primary stress' (e.g. *fellow-* ˈtravel (verb), *fellow-*ˈtraveller). When the second element is a mere suffix this final element is unaccented (e.g. ˈfellowless, ˈfellowship).

**2** *Hyphenation*. There is no general agreement. Some authorities give, for example, *fellow-feeling*, *-sufferer*, *-traveller*; others print all three as separate words (*fellow feeling*, etc.). There is a noticeable tendency for printers in Britain to favour hyphenation and for American printers to print such words as two separate elements without hyphen. My own preference is to write *fellow citizen*, *fellow countryman*, *fellow creature*, *fellow man/men*, *fellow soldier*, *fellow student*, etc., i.e. the simplest types, as two words; to insert a hyphen in verbs (*fellow-travel*), agentive nouns (*fellow-commoner*, *fellow-sufferer*, *fellow-traveller*), and verbal nouns (*fellow-travelling*); and to write solid the type *fellow* + suffix (*fellowless*, *fellowship*). But one must be prepared to find the several patterns in printed matter. Such divergence cannot be legislated out of existence.

**fell swoop.** Usually in the phr. *at* (or *in*) *one fell swoop*, it occurs first in Shakespeare's *Macbeth* (1605): *Oh Hell-Kite! All? What, All my pretty Chickens, and their Damme At one fell swoope?* Except in this phrase, *fell*, an old word meaning 'deadly', is now restricted to poetical or

dialectal use; and *swoop* means, of course, a sudden descent, as of a bird of prey. The phrase now means simply 'in a single action', and is by no means restricted to deadly or dangerous actions.

***felo de se*** /ˌfɛləʊ də ˈseɪ/, /ˌfiːləʊ diː ˈsiː/. A 17c. Anglo-Latin formation meaning (a) suicide; (b) a person who commits suicide. The word is derived from L *felō* 'felon' + *dē* + *sē* 'oneself'. The pl. is properly *felones de se* /ˌfɛləʊniːz də ˈseɪ/, /ˌfiːləʊni diː ˈsiː/. It is less commonly written with hyphens (*felo-de-se*) and in the pl. as *felos de se*.

**felspar.** See FELDSPAR.

**female. 1** *Origin*. Adopted in the 14c. from OF *femelle*, it was almost at once refashioned as *female* to accord with its antonym *male*, an etymologically unrelated word. The OF word answers to L *fēmella*, a diminutive form of *fēmina* 'woman'. In popL *fēmella* seems to have been used to denote the female of any of the lower animals, as well as retaining its classical Latin (only Catullus) meaning of 'woman, girl'.

**2** *Uses*. From the 14c. onward *female* was used as both noun and adj., and of human beings and animals: e.g. (adj.) *twelve female children* 1634; *white female slaves* 1841; *female dragon* 1552, *a female scorpion* 1774, *a female salmon* 1870; (noun) *Saturne did onely eate up his male children, not his females* 1652; *The Danish and Swedish laws, harsh ... to all females* 1861 | *The females* [sc. elephants] *are of greater fiercenesse then the males* 1553; *The stag ... was ... acting as a sentinel for the females* 1847; *The abdomen of the females* [of ants, etc.] *sometimes increases in size* 1881.

**3** *Controversy*. Despite the long history of *female* in the sense 'a female person; a woman or girl', several 19c. usage guides advised against the use on the grounds that it was unsuitable to apply the same term to animals and to human beings. Even the *OED* (in 1895) said of *female* used as a mere synonym for 'woman' that 'The simple use is now commonly avoided by good writers, except with contemptuous implication.'

**4** *Present-day currency*. In the 20c. the phr. *the female of the species* has come into widespread use (e.g. *The female of the* species ... by the age of fifteen has a clearer sense of reality in these things–H. G. Wells, 1940). In other uses, woman is much the more usual word, except in contexts where female persons of varying age, including children, are being described, and, for example, on official forms where one is asked to indicate whether one is male or female. The use of female to mean 'a female animal' continues unabated when no separate word (mare, sow, etc.) exists.

**female, feminine; womanlike, womanly, womanish.** As adjectives, female and feminine stand widely apart in meaning and application. Female is used (a) of the sex that can bear offspring or produce eggs. (b) (of plants or their parts) fruit-bearing. (c) of or consisting of women or female animals or female plants. (d) (of a screw, socket, etc.) manufactured hollow to receive a corresponding inserted part. Feminine has more general applications and is usu. (but see (b) below) applied only to human beings and not to animals or plants: (a) of or characteristic of women or having qualities associated with women. (b) in grammar, of words in certain languages (e.g. OE, L, Gk) of a particular class, as distinct from masculine or neuter ones. So much can be gleaned from any standard dictionary. In broad terms female is principally used to indicate the sex of a person, animal, or plant, its 'non-maleness'; whereas feminine (apart from grammar) is used of qualities or characteristics regarded as typical of women— beauty, gentleness, delicacy, softness, resilience, etc. Feminine is more frequent than the synonymous words womanlike and womanly. 'Womanly is used only to describe qualities peculiar to (a) good women as opposed to men (womanly compassion, sympathy, intuition, etc.) or (b) developed women as opposed to girls (womanly beauty, figure, experience)' (Fowler, 1926). Womanish is chiefly applied in a derogatory manner to effeminate or unmanly men or their ways.

**feminine designations. 1** See ARTISTE; CONFIDANT(E); -ESS; -ETTE 4; -MAN; -MIS-TRESS; -PERSON; -TRIX; -WOMAN.

**2** For the great majority of occupational or agent nouns no distinctive feminine term exists: e.g. chef, chemist, clerk, cook, councillor, counsellor, cyclist, doctor, lecturer, martyr, motorist, nurse, oculist, palmist, president, pupil, secretary, singer, teacher, typist, etc. For a few of these the normal expectation at present is that the office, etc., will be held by a man (e.g. chef, president), and for others by a woman (e.g. nurse, secretary, typist), but the older patterns are fast breaking up. For example, in 1988, when I delivered the T. S. Eliot Memorial Lectures, the Master of Eliot College, University of Kent at Canterbury, was Dr Shirley Barlow. Cf. also It was a last resort for her–insisting that I be the servant and she the master–J. Kincaid, 1990.

**3** In practice, if the need arises to indicate the sex of an occupational or agent noun, woman, lady, or girl are sometimes used as prefixed modifiers: e.g. woman doctor, driver, student; lady barrister, doctor; girl Friday, (US) girl scout. The whole question of gender distinctions in occupational and related names is sensitive, verging on explosive. All possible 'solutions' introduce uglinesses or new inconsistencies or leave false expectations in their wake. Ours is an uneasy age linguistically.

**feminineness, feminism,** etc. A cluster of formations have for some time been vying with one another to designate the primary sense 'womanliness, feminine quality'. Some have gone to the wall while others remain. The OED lists the following (with date of first record indicated): feminacy (1847), feminality (1646), femineity (Coleridge, 1820), feminicity (1843), feminility (1838), feminineness (1849), femininism (1846), femininitude (nonce-word, 1878), femininity (14c. in Chaucer), feminism (1851), and feminity (14c. in Chaucer). The abundance of short-lived 19c. formations may simply bear witness to the enthusiasm of the editors of the OED and of their contributors for the coinages of their own century. Be that as it may, the existence of such forms throws light on the multiplicity of formative elements (-eity, -icity, -ility, -ism, -ity, etc.) that are available for the formation of new abstract nouns. And the disappearance or archaism of some of them (e.g. feminacy, femininism, femininitude) illustrates the way in which a language will tolerate only a reasonable number of similarly formed synonyms at a given time.

It is of interest to observe too that the strongest survivors of this group include a word of long standing (*femininity*) and two that first came into being in the 19c. (*feminineness, feminism*). It is also of interest to note that *feminism* stands apart from the others in having for its primary sense 'the advocacy of women's rights on the ground of the equality of the sexes'. It reflects Fr. *féminisme* used in the same sense, and is first recorded in the unsettled days of the last decade of the 19c. All the other words in this group are purely abstract nouns with only gentle social overtones. In 1868 Browning wrote *Put forth each charm And proper floweret of feminity*. At that time he and his contemporaries could hardly have foreseen the emergence, less than three decades ahead, of the stridency and hostility implicit in the concept of *feminism*.

**femur.** The pl. is either *femurs* /'fiːməz/ or *femora* /'femərə/. Cf. L *femur, -oris*.

**feoff** (and derivatives). Traditionally this feudalistic word (derived from medL *feudum, feodum*, a piece of land held under the feudal system) was spelt *fief* in English and pronounced /fiːf/; similarly *fiefdom* /fiːfdəm/. Some historians still spell and pronounce the two words in this manner. For the other derivatives, the customary spelling now is with *-eoff-* (so *feoffee, feoffment*, and *feoffor*), pronounced with initial /'fef-/. Inevitably some scholars have also adopted the spellings *feoff* and *feoffdom* and pronounce them /fef/ and /'fefdəm/ respectively.

***ferae naturae*** /'feraɪ nə'tjʊəraɪ/. 'The law applies only to animals ferae naturae; *Rabbits are* ferae naturae; *Rabbits are among the* ferae naturae. The first two sentences show the correct, and the third the wrong use of the phrase, and the three together reveal the genesis of the misuse. *Ferae naturae* is not a nominative plural, but a genitive singular, and means not '"wild kinds", but "of wild kind", and it must be used only as equivalent to a predicative adjective, and not as a plural noun' (Fowler, 1926). Nevertheless the *OED* defines the expression as 'animals living in a wild state, undomesticated

animals'. Among the illustrative examples is one where the phr. can be taken to be a genitive singular (*Women are not compris'd in our Laws of friend-ship; they are* feræ naturæ—Dryden, 1668) but may not be; and one (*He evidently viewed himself as the Underwood who alone could do his duty by the* feræ naturæ—C. Yonge, 1873) where it is certainly used as a nom. pl. It is a standard example of a mistaken interpretation of a Latin inflexion.

**feral** /'ferəl/, **ferial** /'fɪərɪəl/. **1** These are the dominant pronunciations, but the first elements may be /-e-/ or /-ɪə-/ for either word according to taste.

**2** *Feral* means 'wild' (cf. L *ferus* 'wild') and is commonly applied to animals in a wild state after escape from captivity, or (in NZ) to animals (esp. deer and goats) brought in from a wild state and farmed. *Ferial* (ult. from L *fēriae* 'holiday') is an ecclesiastical term meaning (*a*) (of a day) ordinary; not appointed for a festival or feast; (*b*) (of a service) for use on a ferial day.

**ferment.** See NOUN AND VERB ACCENT.

**ferrule, ferule.** The ring or cup strengthening the end of a walking-stick or umbrella is a *ferrule* (and is in origin an OF diminutive form, ult. from L *ferrum* 'iron'); it is also spelt *ferrel*. The flat ruler with a widened end formerly used for beating children is a *ferule* (and is derived from L *ferula* 'giant fennel'); it is also spelt *ferula*.

**fertile.** In 1895 the *OED* gave precedence to the pronunciation /'fɜːtɪl/ with a short vowel in the second syllable. The standard pronunciation in BrE now is /'fɜːtaɪl/ and in AmE /'fɜːtəl/. See -ILE.

**fervour.** Thus spelt in BrE; but *fervor* in AmE.

**-fest.** Beginning in the 19c., but increasingly in the 20c., the German word *Fest* 'festival' has made its way into English through the front door, so to speak, as the first element of the word *Festschrift*, beloved of scholars; and through the back door, principally in America, as the final element in a multitude of words denoting a festival of the kind indicated by the preceding qualifying word. Examples include *Beatlefest* (a festival celebrating the music of the Beatles), *discofest*, *filmfest, funfest, gabfest, rockfest,* and *talk-fest.* Such words have been formed within English, and are not adopted from German. At present words with final *-fest* have hardly penetrated beyond the periphery of standard English in Britain, but many of them, and the word-formation process itself, are strongly established in AmE.

**festal, festive.** Both words mean 'of or pertaining to a feast, festival, or other joyous occasion'. Collocations of these words listed in large modern dictionaries include *festal act, air, celebration, crowd, day, garment, mirth, occasion; festive air, atmosphere, board, champagne, decorations, drink, games* (in ancient Greece), *meal, mirth, mood, occasion, party, season* (Christmas period), *wear.* From which it can be seen that in many circumstances the two words are interchangeable. In practice *festive* is much the more common of the two.

**fetal.** See FOETUS.

**fête.** If the French abandon the use of the circumflex accent, as (in 1991) they were for a time threatening to do, it will be interesting to see how long it is preserved in this and other loanwords from French. Already *fete* (without accent) appears in much computer-set printed material.

**fetid** /ˈfetɪd/, less commonly /ˈfiːtɪd/. The spelling with *-e-* rather than *-oe-* seems now to be dominant. The L original is *fētidus,* with a long *e.* The refashioned form *foetidus,* which led to the Eng. form *foetid,* has no etymological justification.

**fetish.** 1 Until the early 20c. the first syllable was often pronounced like *feet,* but no longer.

2 The word is a 17c. adoption of Fr. *fétiche,* and was originally an African object or amulet having magical power. It is not an African word, though; the French word goes back ultimately to L *factīcius* 'factitious'.

**fetishes.** Fowler (1926) presented a list of grammatical and other linguistic features which, in his opinion, evoked irrational devotion, respect, or hostility, in other words had become *fetishes.* Among 'the more notable or harmful' were (the capitals indicate where in *MEU* the features were treated): SPLIT INFINITIVE; FALSE QUANTITY; avoidance of repetition (ELEGANT VARIATION); the rule of thumb for *and* WHICH; a craze for native words (SAXONISM); pedantry on the foreign spelling of foreign words (MORALE); the notion that AVERSE *to* and DIFFERENT *to* are marks of the uneducated; the dread of a PREPOSITION AT END; the idea that successive metaphors are mixed METAPHORS; the belief that common words lack dignity (FORMAL WORDS). It would seem that in this respect little has changed since 1926. At public and private functions, and in letters, when devotees of Fowler express their opinions to me, these are among the principal items mentioned, together with the erroneous use of classical plurals as singulars (*criteria, phenomena,* etc.), the use of *hopefully* as a sentence adverb, and a few other points. As perhaps at all times in previous generations, the more complex mechanisms of the language are left largely undiscussed except by scholars in learned journals and monographs.

**fetus.** See FOETUS.

**feverish, feverous.** Both words entered the language in the 14c. and down the centuries shared the same senses. They also seem to have had approximately equal currency. But *feverish,* used esp. to mean 'suffering from a fever' or 'intense, excited' (*feverish activity, a feverish race against time*), has now virtually driven *fevorous* out of the standard language.

**few.**

| 1 *few* and *less.*
| 2 As noun.

1 *few* and *less.* (a) The important starting-point is that in most circumstances (a) *few* and the comparative adj. *fewer* are used with countable nouns, i.e. with nouns that have both a singular and a plural form (*book/books;* so *fewer books, few books, a few books*); or with collective nouns (*fewer people, few people, a few people*). In such constructions *less* would be incorrect (see (c) below). By contrast *less,* which is the comparative of *little,* is properly used with uncountable or mass nouns: in other words *less* refers to quantity and is the opposite of *more* (*less affection, less power, less misery,* etc.). Neither *less* nor *few* should be used in conjunction

with the word *number. The number of think-ers … who have managed to say anything fundamental or new about translation is very few* (Roy Harris, 1986) is not correct. A number is *small, smaller,* or *very small,* not *less* or *few*(*er*) or *very few.*

(b) *Less* can be idiomatically used with plural nouns in certain circumstances, esp. distances (*it is less than seventy miles to London*), periods of time, sums of money (*costs less than £50*), or other statistical enumerations. The notion of quantity is present in each case. These examples are acceptable English: *unless he read him on consecutive days and for no less than three hours at a sitting*—P. Roth, 1979; *We have had reliable temperature records for less than 150 years*—NY Rev. Bks, 1988. A borderline case: *unashamedly rejoiced in having had in his house at one time no less than five Nobel Prize winners*—M. Drabble, 1987. See LESS 3. School examiners often invite candidates to write a précis of a passage of prose in *fifty words or less.*

(c) Regrettable, but prevalent among some standard as well as many non-standard speakers, is the use of *less* with an unprotected plural noun. Examples of wrong use: *I shall care about less things*—P. Fitzgerald, 1980; *there will be less 100% loans about*—Chairman of Halifax Building Society, BBC Radio 4, 1987; *We had been given less men … to perform a holding action*—Paintball Games, 1989. The incorrect use is very widespread and seems likely to be ineradicable, however regrettable that may be. As a character in Maurice Gee's *Prowlers* (1987) sadly remarks: *'Like', it seems, has taken the place of 'as if' 'Less' tips 'fewer' out. Less pedestrians, less immigrants.* See LESS 3 (historical note).

(d) See COMPARATIVELY 2.

(e) See SEVERAL.

**2** As noun. Used in a wide range of idiomatic expressions, e.g. *a few of those present, a good few* (colloq., = a fairly large number), *not a few* (a considerable number), *quite a few, many are called but few are chosen, he had had a few* (slang, = a few alcoholic drinks), *the Few* (the RAF pilots who took part in the Battle of Britain). None of these is in the least questionable (at the level of discourse indicated).

**fey.** A native word of great antiquity (OE *fæge* 'doomed to die') still survives regionally in its original sense. In the 19c., however, a weakened sense developed, 'disordered in mind like one about to die' (*Surely the man's fey about his entails and his properties*—J. Galt, 1823), and then it passed into meaning 'possessing or displaying magical, fairylike, or unearthly qualities' (*A gaze that was not at all fey, but … remarkably shrewd*—E. Coxhead, 1955; *she's got that fey look as though she's had breakfast with a leprechaun*—D. Burnham, 1969). Inevitably, it has made contact with FAY, and is sometimes thought to be 'the same word'.

**fez.** The inflected forms are (pl.) *fezzes* and (adj.) *fezzed.* See -Z-, -ZZ-.

**fiancé** /fɪˈɒnseɪ/, fem. **fiancée** (same pronunc.). See INTENDED (noun).

**fiasco.** Pl. *fiascos.* See -O(E)S 6.

**fiat.** The legal term is pronounced /ˈfaɪət/ and the car (*Fiat*) /ˈfiːət/.

**fibre** is the spelling in BrE and *fiber* in AmE.

**fibroma,** a fibrous tumour. The pl. is *fibromas* or, pedantically, *fibromata.*

**fibula.** Whether in anatomy ( = outer bone between knee and ankle) or in classical antiquity ( = brooch or clip), both *fibulae* /-iː/ and *fibulas* are acceptable as pl. forms. See LATIN PLURALS.

**fictitious.** See FACTIOUS.

**fiddle.** 'Of the verb the OED says "now only in familiar or contemptuous use". It is more in demand to convey the meaning of potter or cheat than that of play the violin, and it is a component of derisive words such as *fiddle-de-dee, fiddle-faddle,* and *fiddlesticks.* Perhaps the story of Nero's fiddling while Rome was burning started the decline. But the noun has escaped this taint, and is a partner in some most respectable expressions—*fiddle-back chair, fit as a fiddle.* So a violinist will still speak of his instrument as a *fiddle,* but not of his playing as *fiddling* or of himself as a *fiddler'* (Gowers, 1965). *Fiddler* survives memorably in the nursery rhyme 'Old King Cole' (*He called for his bowl, And he called for his fiddlers three*).

**fidget.** The inflected forms are *fidgeted, fidgeting.* See -T-, -TT-.

**fiducial, fiduciary.** The second is the ordinary form, *fiducial* being used only in some technical terms in surveying, astronomy, etc.

***fidus Achates*** /ˌfaɪdəs əˈkeɪtiːz/, a faithful friend, a devoted follower. In Virgil's *Aeneid*, *fidus Achates* is a companion of Aeneas.

**-fied.** See -FY.

**fief.** See FEOFF.

**field.** In the sense 'an area of operation or activity; a subject of study' (*each supreme in his own field*; *distinguished in many fields*), the word tends to be overused. This is all the more surprising in that numerous synonyms and nearsynonyms exist. A short selection of these follows, together with a brief illustrative example of each.

*Area, bailiwick, compass, department, domain, line, métier, province, realm, region, scope, specialty, sphere, subject, territory, theme. A debate covering a wide* area; *not in my* bailiwick; *expenses beyond my* compass; *fire drill is not my* department; *belongs to the* domain *of philosophy*; *casuistry is not in my* line; *Black-and-white was no more his* métier *than humour*; *presumably out of our* province [ = not our concern]; *in the whole* realm *of medicine*; *the wide* region *of metaphysical enquiry*; *find* scope *for one's abilities*; *he had selected Roman Law as his* specialty; *useful in his own* sphere; *wanders from the* subject; *turn from verified matters to some still unexplored* territory; *has chosen an ill-defined* theme.

It is important to keep in mind, however, that there are plenty of contexts in which *field* is the natural and legitimate word. A few examples: *As a Professor of English Literature she had no doubt that the whole field lay within her control*—Angus Wilson, 1952; *Working on him would give me an opportunity to learn something about politics—a field in which, thanks to* Commentary, *I was growing more and more interested*—N. Podhoretz, 1967; *He has ... made a number of breakthroughs in his field of labour and social welfare*—B. Castle, 1980; *The husband has been ... a figure of some renown in his field*—G. Swift, 1981. It is perhaps of interest to note that Alistair Cooke judged *field of study* to be an Americanism in the following context: *What was my 'field' of study?* (*I noticed then how*

*strange that he should use what in those days was an exotic Americanism.*)—A. Cooke, reporting a remark made by the Prince of Wales (i.e. the Duke of Windsor) *c*1932.

**fiery.** Two syllables, /ˈfaɪərɪ/.

**fifth(ly).** Both the *-f-* and the *-th* should be clearly sounded, /fɪfθ,-lɪ/.

**fifties.** See NINETIES.

**figure.** 1 Both the noun and the verb are pronounced /ˈfɪgə/ in BrE; similarly with the inflected form *figuring* /ˈfɪgərɪŋ/. The derivatives, however, mostly have /-gj-/: thus *figural* /ˈfɪgjʊrəl/, *figuration* /fɪgjʊˈreɪʃən/, *figurine* /fɪgjʊˈriːn/. But *figurative* is as often pronounced /ˈfɪgərətɪv/ as it is pronounced /ˈfɪgj-/. In AmE the standard pronunciation of the noun and verb is with medial /-gj-/, i.e. /ˈfɪgjə(r)/. The British pronunciation is sometimes heard in AmE but is usually condemned as substandard.

2 As verb, *figure* (often with *out*) is used, esp. in AmE, to mean 'to conclude, think, work out' (*He figured he could write the book in six months*), or with *on* in the sense 'plan' (*he figured on being home by six o'clock*). Also predominantly American is its use to mean 'seem likely or understandable' (*that figures*). Each of these uses is informal.

**figure of speech,** 'a recognized form of rhetorical expression giving variety, force, etc., esp. metaphor or hyperbole' (*COD*). In this book the figures of speech, or terms of rhetoric, are treated at their alphabetical place: see APOSIOPESIS; ASYNDETON; METAPHOR, etc.

**filial.** The only extant pronunciation in standard English is /ˈfɪlɪəl/, with a short initial vowel. Latinists, according to Fowler (1926), 'obsessed by the fear of false quantity (q.v.)', once often pronounced the word with initial /ˈfaɪl-/. Cf. L *fīliālis*, from *fīlius* 'son'.

**filigree.** The only current standard spelling (not *fila-* or *filla-*). The word is an abbreviated form of *filigrane*, a 17c. loanword from French, ultimately from L *fīlum* 'thread' + *grānum* 'grain'. In 17–19c. literature the spelling *filla-* was esp. common for both *filigrane* and *filigree*.

**fillers.** Meaningless fillers do not normally occur in scripted spoken (BBC) English but are not always avoidable in informal contexts, and constitute useful rhythmical aids to speakers who temporarily lose their fluency. The main fillers are: *actually, and er, and everything, and um, didn't I (I went into this shop, didn't I?), er, I mean, I mean to say, really, sort of, then (Nice to talk to you, then), um, well, well now, you know, you see.* It is obvious that too frequent a use of such fillers is undesirable, as in the following (invented) example: *'It's not really expensive. I mean, it only costs a pound.' 'It seems, sort of, a lot. I mean, when it used to cost only forty pence. I mean, where will it all end?' 'But, you know, forty pence then is like a pound now, and um, we must um make allowance for um inflation, and everything, if you know what I mean'* (Burchfield, 1981). The social levels that determine the use and frequency of such fillers are difficult to establish, but are being widely investigated by linguistic scholars.

**film.** See CINEMA 2.

***fils.*** See PÈRE.

**finable.** Not *fineable*: see MUTE E.

**final analysis.** See ANALYSIS 2.

**final clause,** in grammar, 'a clause that states the purpose or direct end of the action of the principal proposition' (G. O. Curme, 1931), esp. one introduced by *in order that, in order not to, in the hope that, to the end that, so (that); (expressing apprehension) that ... not, for fear (that), lest.* Such clauses are also called *clauses of purpose.* Examples: *He raised his hand* in order that *the bus might stop; They closed the border post* in order not to *exacerbate things further; He confessed everything* in the hope that *he would be forgiven; I have come to you now* so that *you will know that I have not left the country; She walked quickly* for fear that *she would be late for her appointment; Lord God of Hosts, be with us yet,* Lest we forget, lest we forget (Kipling).

**finale** has three syllables, /fɪˈnɑːlɪ/.

**finalize.** Hostility to this modern formation in *-ize* reached its peak in the 1940s and 1950s: 'As a synonym for *to complete* or *to conclude* it [sc. *finalize*] is superfluous and ugly' (E. Partridge, 1942); 'When I hear of ... things being adumbrated, or visualized, or finalized ... I think of that other aim of this [*English*] Association, "to uphold the standards of English writing or speech"' (N. Birkett, 1953). The word seems to have entered the language in the 1920s in Australia and New Zealand (though the Canadian scholar William J. Kirwin found an isolated example of 1901 in St John's, Newfoundland), but was for a long time regarded in Britain as an 'unwanted Americanism'. Time has moved on, and only elderly eyebrows are now raised when the word is used of rules, deals, etc., to mean 'to put into final form'.

**fine** (adj.). The phr. *not to put too fine a point upon it* is an apology for expressing a downright opinion, and means 'to put it bluntly'.

**fine** (noun). The phr. *in fine*, from the 15c. to the 19c. routinely used to mean 'to sum up; in short', is now seldom heard. Example: *We have, in fine, attained the power of going fast*—Ruskin, 1849.

**finger.** The fingers are now usually numbered exclusively of the thumb—*first* (or *fore* or *index*), *second* (or *middle*), *third* (or *ring*), and *fourth* (or *little*); but in the marriage service the third is called the fourth. It is also the fourth in the modern method of indicating the fingering of keyboard music, in which the thumb, formerly marked x, is now 1, and the fingers are numbered correspondingly.

**fingering,** as a name for stocking-wool is not from *finger*, but is an alteration of Fr. *fin grain* 'fine grain'. It was often written as *fingram, fingrim*, etc., in the 17c. and 18c. Cf. *grogram* from Fr. *gros grain*. See TRUE AND FALSE ETYMOLOGY.

**finical, finicking, finicky, finikin.** Of these, *finical* (first recorded in Nashe in 1592) is the oldest, and it seems to have been a rough and ready formation ('prob. academic slang', according to *ODEE*, 1966) from *fine* adj. + *-ical*. The others followed at intervals: *finikin* (1661, now obs.), *finicking* (1741), and *finicky* (1825). It is perhaps best to regard them all as unstable variants of the same word. By far the most commonly used of the group now is *finicky*, and it has two main meanings: (*a*) over-particular, fastidious; (*b*) needing much attention to detail.

**finish** (adj.). = rather fine. Best spelt thus (see MUTE E) despite the fact that it then becomes identical with the verb.

**fiord** is the recommended spelling, not *fjord*.

**fire** (verb). The progress of this word in the sense 'dismiss, sack' from AmE slang to everybody's slang is of interest in that it started out in AmE (it would seem from the *OED* in Dakota) as a phrasal verb, *to fire out*, first recorded in 1885. Fowler (1926) described *fire out* as 'still an Americanism'. Gowers (1965) gave the simple *fire* as the headword and said that it was 'still an American colloquialism, though making headway among us at the expense of the verb to *sack*'. *COD* (1995) calls it 'slang' and defines it as 'dismiss (an employee) from a job'. Its AmE origin seems to have been largely forgotten, and it stands alongside *sack* as a straightforwardly direct term for a disagreeable action.

**firm** (adv.) is used mainly in two fixed phrases, *stand firm* and *hold firm to*. In all other contexts the natural adverbial form *firmly* is used (*the bracket was firmly attached to the wall*).

**first.** 1 For *first* etc. *floor*, see FLOOR, STOREY.

2 *first thing* means 'before anything else; very early in the morning' (*he went for a run first thing*) and is normally used without the definite article; *the first thing* is used mainly in negative contexts to mean 'even the most elementary fact or principle' (*he does not know the first thing about croquet*).

3 *the first two*, etc. The earliest recorded examples (14c.) of *first* with a cardinal numeral follow the order *the two* (*three*, etc.) *first* ( = Fr. *les deux premiers*, Ger. *die zwei ersten*). By the 16c. 'the growing tendency to regard *first* as an ordinal led to the introduction of the form *the first two* (*three*, etc.), corresponding to "the second two (or three, etc.)". This is now the universal form in the case of high numbers [e.g. *the first sixty orders*]; but for numbers up to 3 or 4 many writers use it only when the number specified is viewed as a collective unity contrasted with the second or some succeeding 2, 3, or 4 in the series' (*OED*). Or as *CGEL* (5.22) expresses it: 'Ordinals cooccur with count nouns and usually precede any cardinal numbers in the noun phrase: *the first two days*; *another three weeks*.' Cf. LAST 1.

4 *first(ly)*, *secondly*, *lastly*. In such sequences where points or topics are enumerated the choice of form as between *first* and *firstly* is optional in modern English, as indeed it has been since at least the 17c. (examples of both forms are given in the *OED*). The absence of *firstly* from Johnson's Dictionary (1755), De Quincey's hostility to it in 1847—'First (for I detest your ridiculous and most pedantic neologism of *firstly*)'—,as well as the fact that the Prayer Book, in enumerating the purposes for which matrimony was ordained, introduces them with *First, Secondly, Thirdly*, and other considerations have led to the myth that *firstly* is actually wrong in a sequence *Firstly*, ... *secondly*, ... *thirdly*, ... etc.; or as *CGEL* (8.138n.) puts it 'for many people *firstly* is objectionable'. In practice many different patterns are used: *First*, ... *second*, ... *third*; *Firstly*, ... *secondly*, ... *thirdly*; *First*, ... *next*, ... *last*; (AmE) *First of all*, ... *second of all*, ... *thirdly*; and numerous others. Rightly or wrongly, my own instinct is to write *First*, ... *secondly*, ... *thirdly*, etc. But I am probably following a precept put before me by one of my schooldays teachers. Logic did not and does not come into it.

**first name.** See CHRISTIAN NAME.

**firth** (noun). 1 a narrow inlet of the sea.

2 an estuary. Originally (15c. from ON) a Scottish word, introduced into English literary use c1600, it was often in the past (in the work of e.g. Cowper and Tennyson) converted, by the process of METATHESIS, into *frith*. But *firth* is the only standard form now.

**fisc, fisk.** 1 (Usu. spelt *fisc*) in Roman antiquity 'the public treasury; the emperor's privy purse'.

2 In Sc. law, *fisk* (thus spelt) is 'the public treasury; the revenue falling to the Crown by escheat'. The word came into English via OF from L *fiscus* 'the imperial exchequer'.

**fish.** The pl. form is usu. *fish* (*he caught seven fish yesterday*; *nets full of fish*; *all the fish in the sea*; *a pretty kettle of fish*); but occas. *fishes* (*feed the fishes*; *food fishes like cod and flounder*). See COLLECTIVE NOUN.

**fisher, fisherman.** See ANGLE (noun²).

**fisk.** See FISC.

**fistic, fistical.** These words, both meaning 'pugilistic', are merely jocular formations, dismissed by the *OED* with the comment 'Not in dignified use'. *Fistic* was first recorded in 1806 and *fistical* in 1767. They are both listed in *COD* (1995), labelled 'jocular'.

**fistula.** The (rarely used) pl. is *fistulas*, but *fistulae* /-iː/, reflecting the pl. form in Latin, is also found in medical and zoological works.

**fit** (adj.). For *fit audience though few*, see HACKNEYED PHRASES.

**fit** (verb). The only pa.t., pa.pple, and ppl adj. form in BrE is *fitted* (*they fitted new locks to their doors*; *he fitted in well at his new school*; *fitted carpets*). In certain types of AmE, *fit* is used instead as an alternative, esp. in the senses 'to be of the right size and shape' (*Joan's dresses fit beautifully when she was nine*) and 'to be suitable' (*the name fit him to perfection*). Margaret M. Bryant (1962) commented: 'According to the Linguistic Atlas survey, in New England *fitted*, in the context of "His coat *fit/fitted* me", is more common in cultivated speech, whereas in the Upper Midwest *fit* is preferred by all classes of speakers, and used by 80% of the cultivated speakers.' She concluded that 'both *fit* and *fitted*, like knit/knitted, are standard [in the US]'. Examples of *fit* pa.t.: *the street looked like an improbable setting for such a modest vehicle, but when it pulled up before my grandfather's house it fit in very well*—tr. I. Allende, 1985; *His head fit snugly into his collar like a shell into a canister*—D. Pinckney, 1992; *Many questions were put; none fit.*—Bull. Amer. Acad. Arts & Sci., 1994.

**fix** (noun). Two originally American uses call for comment. **1** The noun is first recorded in any sense in an American military biography of 1816: *They are in a mighty good fix*, i.e. a position from which it would be difficult to escape. From that meaning it is an easy step to 'a difficulty, dilemma, predicament', and it was so used by Charlotte Brontë in a letter written in 1839: *It so happens that I can get no*

*conveyance ... so I am in a fix.* Further BrE evidence for this sense is presented in the *OED*, and its currency at the present time is not in doubt, though a slight American flavour remains. Examples: *Since she had vowed to remain celibate, she was in rather a fix when her father planned to marry her to the King of Sicily*—B. Cottle, 1983; *The patient will indeed be in a fix from which he may find it hard to extricate himself*—C. Rycroft, 1985.

**2** Beginning in the US in the 1930s, a dose of a narcotic drug came to be called a *fix-up* and then a *fix*. The word (only *fix*) reached Britain by the 1950s and with it the taking of hard drugs like heroin. Its currency has no doubt been assisted by investigative reports in the media of the prevalence of drug-taking, esp. among young people, in the last decades of the 20c. Examples: *He needed her as a drug addict needs his fix*—I. Murdoch, 1985; *What do we care where a junkie gets his fix?*—L. Cody, 1986. From drug-taking contexts figurative and transferred uses have emerged: *The addicts need their fix more often and less cut with trendiness than prime-time can give*—Guardian, 1984; *Many people seem addicted to exercise and get depressed if they don't get their daily fix*—Company, 1985; *The presidency alone had escaped then, and, in their beds at night, all the family members ... dreamed endlessly of this, the ultimate 'fix' of power*—P. Booth, 1986; *For them conventional war has been lived through—and they think nuclear weapons are a cheap fix to deter it*—Green Magazine, 1990.

**fix** (verb). This ancient verb (first recorded in the 15c.) has a multitude of senses beginning with the primary one of 'to fasten'. Out of all these, three chiefly AmE uses call for comment. **1** (Also *to fix up*.) To prepare (food or drink). *You must fix me a drink*, Fanny Trollope said in her *Domestic Manners of the Americans* (1839); and Bret Harte, an American writer, in a work of 1891, wrote *Mother'll fix you suthin' hot*. The use is familiar in BrE, but when used is still regarded as a conscious Americanism. Examples: (US) *He went to the store to get eggs ... She fixed his breakfast*—N. Mailer, 1979; *Fix some lemonade*—Anne Tyler, 1983; *We'd just bake a little something, fix up a bowl of punch*—G. Naylor, 1988; (UK) *When I am quite exhausted, go and cook a meal, fix a drink*—N. Bawden, 1981.

**2** (Also *to fix up*.) The sense 'to mend, repair', first recorded in AmE *c*1762, became established in Australia and New Zealand at some point in the second third of the 20c. *Mend* and *repair* remain the usual words in Britain, though *fix* is occasionally used, e.g. *My mum's 'other men' make dad's life a misery ... He is told ... 'Other men would have fixed that fuse in a few seconds'—News of the World*, 1990.

**3** The informal American expression *to be fixing to* meaning 'to prepare to, intend, be on the point of' which was first recorded in 1716, is still hardly ever encountered outside the US. Examples: *Aunt Lizy is just fixing to go to church—*1854–5 in *OED*; *If you're after Lily, she come in here while ago and tole me she was fixin' to git married—*E. Welty, *a*1983; *I was fixing to turn it down—Atlanta*, 1989.

**fixation.** Used since the 17c. in the general sense 'the action of fixing (in material and figurative senses)', it quickly made its way into popular use from the language of Freudian theory in the course of the 20c. From meaning 'the arresting of the development of a libidinal component at a pregenital stage, so that psychosexual emotions are "fixed" at that point' (*OED*), it has come to mean simply 'an obsession, an *idée fixe*' (e.g. *a young man with a fixation on boats*).

**fixedly.** Three syllables: see -EDLY.

**fixedness, fixity.** Both words entered the language in the 17c., and both notionally qualify as natural forms when an abstract noun is required for the quality or condition of being fixed (in various senses). There has been a sharp decline in the frequency of *fixedness* in the 20c. (possibly because of slight uncertainty about the number of syllables in the word (there should always be three), to the point that *fixity* has become the dominant form of the two. Examples showing various shades of meaning ('permanence, attentiveness, invariability, steadfast adherence to (a cause), etc.'): (fixedness) *Any impact from the hostage or debate episodes must work against a fixedness in public views about Reagan and Carter—Christian Sci. Monitor*, 1980; *Psychologists even have a special term to refer to the interference of prior knowledge on problem-solving activity: functional fixedness—Byte*, 1983; *A*

*film that seemed able to contemplate death ends up by denying even the fixedness of character—Independent*, 1990; (fixity) *Fine eyes, but rather disquieting, she found, in their intent, bright, watchful fixity—*A. Huxley, 1928; *Print-capitalism gave a new fixity to language—*B. Anderson, 1983; *Beaten into a fixity of revolutionary purpose, the peasants will have no more of it—TLS*, 1984; *Bhagawhardi's dreams had no such fixity, but presented ever-changing panoramas and dissolving landscapes to her eye—*O. Sacks, 1985; *I try to paint the objects and the people as though they are returning my gaze ... as though by looking at these things together they have a kind of hold, a fixity, which is like something that is being looked at—Landscape*, 1987; *What distinguishes perversion is its quality of desperation and fixity—NY Times*, 1991.

**fizz.** See -z-, -zz-.

**flaccid.** The pronunciation recommended is /ˈflæksɪd/, not /ˈflæsɪd/.

**flack, flak.** The first is a modern AmE slang word (first recorded in 1946 and of unknown origin) for a press agent. It has no currency in BrE. *Flak* meaning anti-aircraft fire (later also in fig. use, 'adverse criticism') is derived from Ger. *Fliegerabwehrkanone* 'pilot defence-gun'. The two are not confusible in Britain because *flack* is simply unknown, but there is a slight danger in AmE of their being thought to be 'the same word', with the result that the spellings are sometimes interchanged.

**flageolet.** Pronounce /flædʒəˈlet/.

**flagrant** is used of an offence or (formerly) an offender: glaring, notorious, scandalous (*a flagrant breach of a contractual agreement*; (archaic) *a flagrant misanthrope*). Its etymological meaning (and first meaning in English in the 16c.) is 'flaming, burning' (cf. L *flagrāre* 'to burn'). It is now more or less synonymous with BLATANT.

**flair** is a power of 'scent' (being originally from popL *flāgrāre*, an altered form of *frāgrāre* (cf. *fragrant*), 'an instinct for selecting or performing what is excellent, useful, etc.; a talent (*a flair for knowing what the public wants*; *a flair for languages*)' (*COD*, 1995). It was adopted from French in the late 19c. It is to be distinguished from the unrelated noun and verb FLARE.

**flak.** See FLACK.

**flambeau.** Pl. *-s*, or *-x* (pronounced /-z/); see -X.

**flamboyant.** Adopted in the early 19c. first by architects in the sense 'characterized by waved lines of contrary flexure in flame-like forms' (cf. Fr. *flambe* 'flame'), the word soon became used in transferred and somewhat weakened senses, 'flamingly or gorgeously coloured; ostentatious'. The transferred senses are now dominant.

**flamenco.** Pl. *flamencos*. See -O(E)S 6.

**flamingo.** Pl. *flamingos* or *-oes*. See -O(E)S 1.

**flammable.** The adj. and the corresponding noun *flammability* were revived in modern use (in BrE esp. by the British Standards Institution) and used in place of *inflammable* and *inflammability* to avoid the possible ambiguity of the *in-* forms where the *in-* might be taken for a negative. The modern form *flammable* contrasts with the negative form *non-flammable*. As with all such artificial revivals or alterations, the campaign for *flammable/non-flammable* has met with only partial success in lay use.

**flan.** See TART.

**flannel.** The inflected forms in BrE are *flannelled, flannelling* (AmE *flanneled, flanneling*); but *flannelette*, a napped cotton fabric imitating flannel. See -LL-, -L-.

**flare.** For all its familiarity, and its spread of senses, both as noun and verb (*flared trousers*; *the fire flared up*; *the shipwrecked crew sent up flares*; *the flare-path of an aerodrome*; etc.), the word is of unknown origin. The verb is first recorded in the 16c. and the noun in 1814.

**flat, flatly.** The dominant adverbial form is *flatly* (*flatly contradicting his own known views*; *he flatly refused to come down*; *rendered flatly into English*), but it is always used figuratively. The shorter form *flat* is used in fixed phrases (*he turned the offer down flat*; *flat broke*; etc.).

**flatulence, -cy.** The prevailing form is *flatulence* for both literal and transferred

senses. Examples: *The compressor will vibrate the gas past the cork over time. A week or two is OK. Beyond that, flatness starts to replace flatulence*—Washington Post, 1991; *It could also refer to the violent flatulence that occurs when an overblown epic is consumed by clichés*—Maclean's Mag., 1991. *Flatulency* is rare. Example: *He actually swooned at a ball, so violent was the stink from 'putrid gums, imposthumated lungs, sour flatulencies, rank armpits, [etc.]'*—B. Nightingale, 1985.

**flatways, flatwise,** with the flat side (instead of the edge) uppermost or foremost. Both forms have been in use from the 17c. onward, to judge from the OED, but *flatwise* is now the dominant form, at any rate in AmE: *flatwise tensile strength (400 psi) [of an adhesive] is 13% lower than standard FM 300*—Aerospace America, 1989; *Just cut in half flatwise, apply the topping and slip into a 450 F oven*—Washington Times, 1989.

**flaunt, flout.** The first means 'to display ostentatiously' and the second 'to display contempt for (the law, rules, etc.)', but *flaunt* is often wrongly used for *flout*. The wrong use, which has been particularly prevalent since the 1940s, has not been traced before the 20c. and was not mentioned by Fowler (1926). **1** Correct uses of *flaunt*: *He liked to see a woman flaunting her powers of attraction*—A. Brookner, 1988; *Americans, with their frank enjoyment of power and money, flaunt their luck*—J. le Carré, 1989.

**2** Correct uses of *flout*: *Countries engage in covert activities because they do not want to flout the rules openly*—Encounter, 1987; *Wilde achieved brilliant results at University (Dublin, then Oxford) while lazily flouting the rules*—Listener, 1987.

**3** Pointing the way to the correct use of both words: *'Your Excellency, let us not flaunt the wishes of the people.' 'Flout, you mean,' I said. 'The people?' asked His Excellency, ignoring my piece of pedantry*—C. Achebe, 1987.

**4** Incorrect uses of *flaunt*: *By flaunting these rules, Hongkong and Shanghai have challenged the Bank's authority*—Daily Tel., 1981; *Fanny Parkes, an intrepid type who flaunted convention by wearing trousers and sailing up the Jumna ... accompanied only by a native crew*—Listener, 1987; *The union *continued its campaign against Sunday trading yesterday, targeting shops which flaunted regulations*—Times, 1989.

**flautist, flutist.** In BrE, *flautist* is customary. In AmE both forms occur, the more usual one being *flutist*. As a matter of historical record, *flutist* (the more natural formation) was first recorded in 1603, and *flautist* (adapted from It. *flautista*) in 1860.

**flavour.** Thus spelt in BrE (also *flavourful, flavouring, flavourless, flavoursome,* but *flavorous*). In AmE, *flavor* is usual for the simplex and in all the derivatives.

**flaxen.** See -EN ADJECTIVES 1.

**fledgeling** is recommended, not *fledgling*.

**flee.** The verb is less frequently used now than it once was (but e.g. *Iraqi troops flee after the dawn attack by the Alliance forces; he decided to flee* [ = leave] *the country*), except in the pa.t. *fled*.

**fleshly, fleshy.** From medieval times onward the two words have been interchangeable in most of their possible senses, but in the 20c. the distinction is much the same as between *earthly* and *earthy. Fleshy* primarily means 'consisting of flesh' (*not in tables of stone, but in fleshy tables of the heart*—2 Cor. (AV) 3: 3), 'having a large proportion of flesh, plump' (*fleshy hands*), '(of fruit) pulpy', or 'like flesh' (*fleshy pink, softness,* etc.). *Fleshly* means 'sensual, carnal, lascivious' (*fleshly desire, lusts,* etc.), or 'unspiritual, worldly' (*not with fleshly wisedome, but by the grace of God*—2 Cor. (AV) 1: 12; *vainely puft vp by his fleshly minde*—Col. (AV) 2: 18).

**fleur-de-lis,** heraldic lily. This is the recommended spelling, not *fleur-de-lys* (a common variant) or *flower-de-luce* (as sometimes in AmE). Pronounce /ˌflɜːdə-ˈliː/. Pl. *fleurs-de-lis* (pronounced as the singular).

**flier, flyer.** In OUP house style, *flyer* is recommended for all senses. Perhaps *flier* is the more common of the two forms in AmE.

**flirtation.** Though superficially looking like a word of classical origin, *flirtation* was formed in the 18c., as was *flirtatious*

in the 19c., from the onomatopoeic word *flirt* (verb and noun).

**floatation.** See FLOTATION.

**floor.** See CEILING.

**floor, storey.** In Britain a *single-storey* house is one with a *ground floor* only; a *two-storey* house has a *ground floor* with a *first floor* above it; a *three-storey* house has a *second floor* above the first, and so on. (There may be an *entresol* or *mezzanine* floor between the ground and the first floors, but it does not count in the reckoning.) In America, more logically, our *ground floor* is the *first floor,* and our *first* is the *second,* and so on; thus in (say) a ten-storey building (US *story,* pl. *stories*) the top floor would be the tenth in America and the ninth in Britain. Cellars and basements are not counted in the reckoning of the number of floors or storeys.

**flora.** See FAUNA.

**florilegium,** an anthology. Pl. *florilegia.* See -UM 3.

**floruit** /ˈflɒrʊɪt/ is a L verb (3rd sing. perf. indic. of *flōrēre* 'to flourish') meaning 'he or she flourished', used with a date or dates when a writer, painter, etc., is believed to have been alive and working, i.e. when the exact dates are not accurately known. It is occasionally used as a noun: *The date of each Author's 'floruit' is added in the margin*—Liddell and Scott, 1843.

**flotation.** In the early 19c. *floatation* (thus spelt) emerged as a hybrid formation (from the native word *float* + *-ation,* a suffix of classical origin). It was soon joined by the etymologically unjustifiable form *flotation,* introduced to conform with *flotilla, flotsam,* etc., and *rotation.* The spelling *flotation* is now almost invariably used.

**flotsam and jetsam.** The traditional distinction is between goods found afloat in the sea (*flotsam*) and goods found on land after being cast ashore (*jetsam*). (But there are complications, which are set down in the *OED.*) *Flotsam* is a rationalized form of Anglo-Fr. *floteson* (modFr. *flottaison*); and *jetsam* is a contracted form of *jettison. Flotsam* is first recorded in the early 17c., and *jetsam* in

1570. The two words are almost always used in combination as a fixed phrase, and often figuratively = odds and ends; rejects of society, vagrants.

**flounder, founder.** The first verb means **1** to struggle in mud, or as if in mud, or when wading.

**2** (the more usual sense) to perform a task badly or without knowledge; to be out of one's depth. The primary meaning of *founder* is '(of a ship) to fill with water and sink'. It is commonly used too in the transferred sense '(of a plan, scheme, etc.) to come to nothing, to fail'. The two verbs are sometimes confused because *founder* can also mean '(of a horse or its rider) fall to the ground, fall from lameness, stick fast in the mud'. But, horses and their riders apart, it is really quite difficult to confuse sense 2 of *flounder* and the non-equine senses of *founder*.

**flour, meal.** Local applications in the various English-speaking countries mean that (temporary) misunderstandings can occur. In BrE *flour* is bolted meal, i.e. a cereal, esp. wheat, from which the husks have been sifted out when grinding. *Meal* is the edible part of any grain or pulse (usu. other than wheat) ground to powder. *Flour* used by itself means wheat-flour; applied to other kinds it is preceded by a qualifying word (*rye-flour*; *cornflour* (in AmE called *cornstarch*), a fine-ground maize flour used as a thickening agent). In Scotland *meal* is used as a contracted form of *oatmeal*. In America *meal* also = *cornmeal*, a meal made from white or yellow corn.

**flout.** See FLAUNT.

**flower-de-luce.** See FLEUR-DE-LIS.

**flu,** a 19c. abbreviation of *influenza*, now used as frequently as the fuller form except in quite formal contexts. The older variant spellings '*flu* and *flue* can now be discarded. Both *flu* (without definite article) and *the flu* would appear to be standard forms to judge from the following examples: *When the place is snowbound and the staff laid low with flu, the girls take over*—TLS, 1957; *While Bush was driven home and to bed with an apparent case of the flu, his wife remained at the dinner with Miyazawa*—Chicago Tribune, 1992; *It*

*was only the flu. They take it so much more seriously on the Continent*—A. Brookner, 1993.

**fluid, gas, liquid.** *Fluid* is the wide term including the two others; it denotes a substance that under pressure changes shape by rearrangement of its particles; water, steam, oil, air, oxygen, electricity, ether, are all fluids. Liquids and gases differ in that the first are virtually incompressible, and the second are elastic; water and oil are liquid and fluid, but not gaseous; steam and air and oxygen are gases and fluids, but not liquids. Oxygen in a liquid state is called *liquid oxygen*.

**fluky** (adj.) lucky. Thus spelt. See -EY AND -Y IN ADJECTIVES 2.

**flunkey,** liveried servant (usu. with implied contempt). This is the recommended spelling, not *flunky*. Pl. *flunkeys*.

**fluorine.** Pronounce /ˈfluəriːn/. See -IN AND -INE.

**flute.** The corresponding adj. is *fluty* (not *flutey*). See -EY AND -Y IN ADJECTIVES 2.

**flutist.** See FLAUTIST.

**fly.** **1** The noun is used as a collective of any of various small flies or aphids that infest garden or orchard plants, esp. *fruit fly*, *greenfly* (*the roses are infested with greenfly*).

**2** The verb makes is *flown* as well as *has flown*; see INTRANSITIVE PAST PARTICIPLES.

**3** The phr. *to fly a kite* means (*a*) raise money by an accommodation bill; (*b*) (more commonly) make an announcement or take a step with a view to testing public opinion. Cf. BALLON D'ESSAI.

**4** A *flyleaf* is a blank leaf in a printed document, esp. one between the cover and the title-page of a book or a similar blank leaf at the end of a book. It is not synonymous with *flysheet* (see 5(*a*)).

**5** A *flysheet* is (*a*) a leaflet, a two- or four-page circular; (*b*) a canvas cover pitched above a tent to give extra protection against rain or snow.

**6** A *flywheel* is a heavy wheel on a revolving shaft used to regulate machinery or accumulate power.

**flyer.** The recommended form, not *flier*. See FLIER.

**fob off** is used in two main constructions: **1** (followed by *with* a thing) to deceive into accepting something inferior (*customers have been fobbed off with battery-hen eggs described simply as 'farm eggs'*).

**2** (followed by *on* or *upon* a person) to palm or pass off (an inferior thing) (*a well-hyped novel can be fobbed off on readers even though it is poorly written*). See FOIST.

**fo'c'sle** is a much reduced form of *forecastle* (the forward part of a ship where the crew has quarters). The full form as well as the reduced one are both pronounced /ˈfəʊksəl/.

**focus.** **1** The pl. of the noun in general use is *focuses*, and in scientific use most often *foci* /ˈfəʊsaɪ/.

**2** The inflected forms of the verb are properly *focused*, *focuses*, *focusing*, but forms with -*ss*- are used by many printers and publishers. See DOUBLING OF CONSONANTS WITH SUFFIXES 2.

**foetid.** See FETID.

**foetus, fetus.** Despite its etymology (from L *fētus* 'offspring'; cf. *effete* from L *effētus* 'that has brought forth young', hence worn out by bearing) the form with -*oe*- has been consistently used in BrE for some four centuries, and should be retained. The AmE spelling is *fetus*, and this spelling has gained ground in medical writing outside the US, but has not yet established itself in lay use outside N. America. The corresponding adj. is *foetal* in BrE and *fetal* in AmE.

**fogey, fogy.** The second spelling was formerly the dominant one (Thackeray and Charles Kingsley, for example, wrote about old *fogies*, thereby implying a singular form *fogy*), but with the advent of the *young fogey* in the 20c. the spelling with -*ey* (pl. -*eys*) is now the more common of the two.

**föhn,** a hot southerly wind on the northern slopes of the Alps. Thus spelt (rather than *foehn*). The Ger. original is *Föhn*.

**foist** is mainly used in two constructions: **1** (followed by (*off*) *on*, (*off*) *upon*) to present (a thing) falsely as genuine or superior (*inferior articles foisted on the general public at exorbitant prices*). In this use it resembles sense (2) of FOB OFF.

**2** (Somewhat *archaic*) (followed by *in*, *into*) to introduce surreptitiously or unwarrantably (*he was eventually foisted into the see of Durham*). Foist should not be used in sense (1) of FOB OFF. Fowler (1926) cited an example of this erroneous use: *The general public is much too easily foisted off with the old cry of the shopman that 'there's no demand for that kind of thing'*.

**folio.** Pl. *folios*. See -O(E)S 4.

**folk** as an ordinary word for people in general is tending to fall out of use in BrE (except in northern parts) in the 20c. (swept away by the word *people* itself), but it survives strongly in certain specific senses: **1** as the last element in certain compounds and fixed expressions, or qualified by a preceding adj., e.g. *menfolk*, *north-country folk*, *townsfolk*, *womenfolk*.

**2** (in pl.) (usu. *folks*) one's parents or relatives.

**3** treated as sing., colloq. = traditional music.

**4** in attrib. combinations, some of which are loan-translations from German, e.g. *folk-dance*, -*dancing*, *memory*, -*singer*, -*song*, -*tale*, -*ways*; and esp. *folklore*. *Folks* is also used as a light-hearted form of address by entertainers, journalists, etc. *Folks* and the singular form *folk* are still strongly entrenched in AmE in private and public life.

Some typical examples: (UK) *The working folk of Lancashire have much in common, of course, with their Yorkshire neighbours*—J. B. Priestley, 1934; *Even folk who know little about Scotland have probably heard of the Trossachs*—*Scottish World*, 1989; *Yes folks, in 1990, 2,245 people were murdered in the city of New York*—B. Levin, 1991; *What Ursula brought home every week made all the difference to the old folk*—D. Lodge, 1991; (US) *that's all folks!* (the concluding line, written on the screen, of a Warner Bros. cartoon series called *Merry Melodies* from 1930 onwards, according to Nigel Rees, 1990); *You ever defend any coloured folks?* —R. Ingalls, 1970; *That really messes us up if my folks try to get hold of me*—L. Duncan, 1978; *If I was married and the head of my own house, then I could take my wife to church wherever the hell I felt like, I supposed. The folks wouldn't like it much*—R. J. Conley, 1986; *now that folks'd had time to go home and have supper, a few more customers had started to come in*—ibid.; *His folks will be back from work soon so he hared off taking short-cuts*—Young Americans, 1989.

**folk etymology,** 'the popular perversion of the form of words in order to render it apparently significant' (*OED*). Examples include: *hiccough* (a later spelling of *hiccup* under the mistaken impression that the second syllable was *cough*); *tawdry* (from a silk 'lace' or necktie associated with the East Anglian saint *St Audrey,* much worn by women in the 16c. and early 17c.; the final *t* of 'Saint' wrongly carried across to the name); *under weigh* (an alteration of original *under way*); *Welsh rarebit* (first recorded in 1785 as 'an etymologizing alteration' of the earlier *Welsh rabbit*). See ETYMOLOGY; FACETIOUS FORMATIONS.

**follow.** See AS FOLLOWS.

**following** is at present leading a slightly precarious life as a quasi-preposition meaning 'as a sequel to, in succession to (an event), consequent on': *The prologue was written by the company following an incident witnessed by them during anti-Jewish demonstrations following the hanging of two British soldiers in Palestine*—Evening News, 1947; *Used car prices are going up, following the Budget*—Observer, 1968; *Following the death of the actor who played Jethro Larkin, the character has been posthumously assassinated*—TES, 1987; *His wife, he suspected, following the discovery of four prophylactics in the pocket of her smartest jacket, was conducting an affair with the branch manager of the Banco de Crédito*—N. Shakespeare, 1989. A degree of uneasiness occurs (*a*) when *after* would have served as well (as perhaps in the first part of the 1947 example above), and (*b*) when *following* can possibly be taken to be either a present participle or a quasi-preposition (e.g. *Police have arrested a man following extensive inquiries*). There is little danger at present that *following* will oust *after* in ordinary prepositional uses (*he chased after* [not *following*] *me; please join us for coffee after* [not *following*] *dinner.* But note the following example from an announcement about the forty-sixth annual general meeting (1991) of the Friends of the Bodleian Library, Oxford:

*Members are invited to take tea in the Convocation Coffee House, University Church, Radcliffe Square, following the meeting.*

**foolscap,** a size of folio writing- or printing-paper of a kind that originally bore a watermark representing a fool's cap. The old spelling *fool's-cap* has long since been discarded.

**foot** (noun). The normal pl. form *feet* alternates with *foot* when used as a unit of length or height: *She is six feet/foot tall; a plank ten feet/foot long.* When such a phr. is used attrib. a hyphen is normally placed between the numeral and *foot; a 12-foot dinghy.* In contexts in which the number of inches is also given, *foot* is more common than *feet: C.A.* [a West Indian fast bowler] *is six foot eight.*

**foot** (verb). The traditional use of *foot* or *foot up* in contexts of reckoning the total of an account, a bill, etc., has passed out of standard use, but has left behind it the colloquial expression (first recorded in 1819) *to foot the bill.*

**for. 1** *For* is sometimes used as a coordinating conjunction, i.e. one that connects two independent sentences (see 3 below), but it should be said first that its most frequent use (apart from its employment as a preposition) is to join a sentence and a non-finite clause, esp. one of the type 'sentence + *for* + noun (phrase) + *to*-infinitive': *he waited for the lock to click; it is time for their legal rights to be clarified.* In such constructions the negative particle is normally placed before the *to* of the infinitive: *it would be more sensible for the code not to be enacted as law.*

**2** Unlike some other coordinating conjunctions its position in the sentence is sequentially fixed (though even a century ago it was not), i.e. it cannot normally be placed at the beginning of a sentence. Its function is to introduce the ground or reason for something previously stated: *he picked his way down carefully, step by step, for the steps were narrow*—G. Greene, 1988 ( *for the steps were narrow* could not have been placed before *he picked his way down,* etc.). In this respect it differs markedly from other coordinating conjunctions, e.g. *because, since,* which suffer from no such restriction.

**3** Used between sentences as a coordinating conjunction, *for* should normally be preceded by a comma or a semi-colon. Examples: *You had best spare her, sir, for she's your son's wife*—Thackeray, 1847/8; *My father was silently upset when I married a Campbell, for perhaps nowhere else as in Scotland are memories so deep*—H. M. Brown, 1986; *I hated walking through the woods; it was gloomy and damp, for the sun could hardly shine through the tops of trees*—New Yorker, 1989; *I wanted a setting for my own little life, for I did not think that I should know too many people*—A. Brookner, 1990; *That Yeltsin had called for Gorbachev's resignation only a few days before made the publication of the joint declaration all the more dramatic; for it said, in effect, that whatever their former differences, the signers* [etc.]—NY *Rev. Bks*, 1991. But this 'rule' is sometimes not followed when the linked sentences are fairly short: *The 1941 war had affected him little for he was over forty and his employers claimed that he was indispensable*—G. Greene, 1985; *he did not cry any more for it did not help*—D. Matthee, 1986.

**4** In AmE a number of verbs can be used with the conjunction *for* to introduce a non-finite clause: *No. I didn't intend for you to find out*—J. McInerny, 1985; *Let's face it, I can't afford for that bike to break down*—New Yorker, 1986; *I didn't like for Drew to be back there with him while I was driven like the baggage*—E. L. Doctorow, 1989. In BrE other constructions would be used instead (*I didn't intend that you should find out*, etc.).

**foramen** /fɒˈreɪmen/, (anatomy) an opening or hole, esp. in a bone. Pl. *foramina* /-ˈræmmə/.

### for- and fore-.

**A** Words with the prefix *for-* fall into two main groups:

**1** Those already formed by the end of the OE period: in these the prefix has at least four meanings: away, off, apart (*forget, forgive*), prohibition (*forbid, forfend*), abstention or neglect (*forbear, forbid, forgo, forsake, forswear*), excess or intensity (*forlorn*).

**2** A smaller group, now almost all obsolete, adopted from OF, where *for-* ultimately represents L *forīs* 'outside' (†*forcatch, forfeit*). *Foreclose* belongs in this group and was at first spelt *forclose*, but at a later date was respelt as *fore-*.

**B** Those with the prefix *fore-* contain a native prefix with the sense 'before'

(in time, position, order, or rank). Some of the main types show the following senses: in front (*foreshorten*), beforehand (*foretell, forewarn*), situated in front of (*forecourt*), the front part of (*forearm, forehead*), near the bow of a ship (*forecastle*), preceding (*forebear, forefather, forerunner*). It should be noted that *forbear* (stressed on the second syllable) is a verb meaning 'abstain or desist from', and that *forebear* (stressed on the first syllable) is a noun meaning 'ancestor' (except that all major dictionaries list *forbear* as a permissible spelling in the sense 'ancestor'); and that *forgo* 'abstain from, go without' has as pa.t. *forwent* and pa.pple *forgone*, while *forego* means 'precede in time or place' and has pa.t. *forewent* and pa.pple *foregone* (used esp. as adj. in *foregone conclusion* 'a predictable result'). The variants *forego* (for *forgo*) and *forgo* (for *forego*) are occasionally found but are ill-advised.

**forasmuch as.** If the need should arise to write or print this archaic conjunction, this is the form it should take.

**forbear.** See FOR- AND FORE- B.

**forbid.** **1** The pa.t. is most commonly *forbade*, pronounced /fəˈbæd/, but the spelling *forbad* and (for *forbade*) the pronunciation /fəˈbeɪd/ both occur and cannot be said to be wrong.

**2** One of the central uses of the verb is to construe it with *to* + infinitive (*he was forbidden to watch television after 9 p.m.*; *I forbid you to go*). A construction with *from* + a verbal form in -*ing* has also occasionally been used since the 16c.: *I forbede all syngular persones from the studyenge of this treatyse*—1526 in OED; *He forbade both men and women from entering them*—E. W. Lane, 1841; *the electric utilities are forbidden from burning the plentiful high sulphur Eastern coal*—Sunday Times, 1974 (WDEU). Visser also cites a use of *forbid* followed by an unprotected gerund: *The petition asked the king to forbid villeins sending their children to school*—S. J. Curtis, 1948. Fowler (1926) judged constructions with *from* + -*ing* to be 'unidiomatic' (he believed them to be based on analogical uses of *prevent* or *prohibit*), but the tide seems to be turning in favour of them. While the matter is unresolved, however, it is probably sensible to use alternative constructions or the verb *prohibit* instead.

307

# forceful, forcible | for ever, forever

**forceful, forcible.** **1** Fowler (1926) was worried about the tendency for the 'ordinary' word *forcible* to be overtaken by *forceful*, 'the word reserved for poetical or other abnormal use', with the result that 'we shall shortly find ourselves with a pair of exact synonyms either of which could well be spared instead of a pair serving different purposes'. His description does not square with the pattern of senses set down in the *OED*, and his forecast was wide of the mark. What we have is a pair of related words that overlap slightly in meaning, but are otherwise easily kept apart. *Forceful* means (*a*) (of a person, business firm, etc.) full of force, vigorous, powerful; (*b*) (of a speech, argument, etc.) compelling, impressive. *Forcible* means principally (*a*) done by or involving force (*forcible entry, forcible expulsion*); (*b*) (now not common) = forceful (sense *b*). *Forcible* is not normally used of persons, whether as groups or individuals, but mainly of actions, etc., brought about or carried out by the use of force. Examples: *it might be easier to ... start again from scratch, crystallizing a lifetime's experience into a hundred forceful pages?*—I. Murdoch, 1976; *He wanted to be forceful—and had an inclination to be yielding*—C. Whistler, 1985.

**2** For the distinction between *forceable* and *forcible*, see -ABLE, -IBLE 4.

**forceps.** The pl. is the same.

**fore.** *To the fore* was originally a Sc. and Anglo-Irish phrase meaning (*a*) (of a person) present, on the spot, within call; (*b*) still surviving, alive; (*c*) (of money) ready, available. The phr. came into English literary use during the 19c., but in standard English now is used only in the sense 'into view, to the front'. A person is said to have 'come to the fore' when he or she has come into prominence or has become conspicuous for any reason.

**forebear.** See FOR- AND FORE-. B.

**forecast** (verb). After much rivalry between *forecast* and *forecasted* as pa.t. and pa.pple, the first of these has more or less ousted the other.

**forecastle.** See FO'C'SLE.

**foregather.** See FORGATHER.

**forego.** See FOR- AND FORE- B.

**foregoing** meaning 'preceding, previously mentioned' is an adj. corresponding to the verb *forego* 'to precede' (see FOR- AND FORE- B).

**foregone.** See FOR- AND FORE- B.

**forehead.** The recommended pronunciation is /ˈfɒrɪd/ rhyming with *horrid*. The spelling pronunciation /ˈfɔːhed/ is common enough, though, among standard speakers.

**foreign danger.** **1** The adoption of foreign words and phrases can be a somewhat perilous exercise in that the meaning or spelling of the original expression is sometimes imperfectly recollected. People who use such expressions in English in a manner that is impossible in the donor language lay themselves open to varying degrees of derision. For specific examples of the dangers, readers are referred to, among others, the entries (at their alphabetical place) for *bête noire, bona fide(s), cui bono, e.g., et al., etc., galley, ibid., i.e., in petto, op. cit., pace,* and *qua*.

**2** There is a growing tendency to set fictional plots in far-away countries, and to make generous use of unglossed local expressions. The practice helps to produce verisimilitude but also stretches the patience of readers not familiar with the languages concerned. Recent examples include the repeated use of unglossed (though, it must be admitted, often guessable) Peruvian Spanish in Nicholas Shakespeare's *The Vision of Elena Silves* (1989); and unglossed or partially glossed Maori expressions in any number of recent New Zealand novels.

**forename.** See CHRISTIAN NAME.

**forenoon,** recorded since the early 16c. in the sense 'the portion of the day before noon', is now falling into disuse. *COD* (1995) labels it 'archaic exc. Law & Naut.'. Examples: *Whistling away to himself as though it were nine o'clock in the forenoon and the sun shining bravely*—L. G. Gibbon, 1932; *He employed her on a variety of minor jobs in the forenoon and at lunch time he dismissed her for the day*—N. Shute, 1955; *He had been going there every forenoon for as long as anyone could remember*—A. N. Wilson, 1985.

**for ever, forever.** **1** Written as two separate words when the sense is 'for all future time, in perpetuity' whether in liturgical phrases (*for ever and ever*) or in lay contexts (*he said he would love her for ever*). Also used (from the dates indicated) in various strengthened forms: *for ever and ay* (14c.), *for ever and a day* (mid-16c.), *for ever and ever* (early 17c.), *for evermore* (14c.).

2 Usually written as one word (*forever*) when used to mean 'continually, persistently, always' (*the children are forever asking for more pocket money*; *they are forever complaining about the rent*). In such sentences there is 'a subjective feeling of disapproval of the action described' (*CGEL*). In informal contexts, where the sense is 'a long time', either *forever* or *for ever* is appropriate (*it will take forever to wash all these dishes*; *it seemed as if she might walk for ever*).

3 In AmE, sense 1 is often written as one word: e.g. *on the morning they left Pittsburgh forever*—New Yorker, 1990.

**foreword, preface.** *Preface* is the traditional word (first recorded in the 14c.) for an introduction (by the author) to a literary work. *Foreword* is a late-coming word. It was brought into the language by professional philologists in the 19c. to mean simply 'a word or words said before something else' and then 'an introduction (to a literary work)'. Practice varies from book to book and some other terms are also used, e.g. Prefatory Note, Introduction. In the course of the 20c., however, especially for scholarly and technical works, publishers have sometimes favoured the inclusion of both a Foreword (usually written by an authoritative or distinguished person other than the author) and a Preface written by the author. In such cases the Foreword is always placed before the Preface in the preliminary pages.

**for free,** etc. Originally American, *for free* is now quite regularly used in BrE in quotation marks, as it were, as if to say 'I know it is pleonastic, but I am going to use it all the same'. In both countries the expression tends to be used only in fairly light-hearted contexts: *You don't expect to be ill for free*—Godfrey Smith, 1957; *I'd love a research assistant, but you have to pay for them. And most people want me to do things for free!*—C. Tickell, 1991. So too with the originally AmE expression *for real*: *patents have been applied for,*

*and it's for real*—New Yorker, 1957. A grammatically similar use, namely *for fair* 'completely, altogether', is restricted to AmE slang: *Then we danced and started on the beer for fair*—J. Kerouac, 1957.

**forgather.** The customary spelling with initial *for-* rather than *fore-* tends to suggest that it etymologically belongs with other native words in *for-* (see FOR- AND FORE-). In fact it is a 16c. Scottish loanword from Du. *vergaderen* used in the same sense, i.e. 'to assemble, meet together, associate', with accommodation to *for-* and *gather*. The spelling with initial *fore-* is also found but is not recommended.

**forge** makes *forgeable*. See MUTE E.

**forget** makes *forgettable*. See -T-, -TT-.

**forgive** makes *forgivable*. See MUTE E.

**forgo.** See FOR- AND FORE- B.

**forgot,** used in BrE as a past participle instead of *forgotten* is now, except in regional or uneducated speech, a deliberate archaism. It is sometimes used, beside *forgotten*, in AmE.

**forlorn hope.** Now surviving in English only in the sense 'a faint hope, an enterprise which has little hope of success', it has moved a long way from its original meaning. It is a 16c. adaptation of Du. *verloren hoop*, literally 'lost troop' (Du. *hoop* is cognate with English *heap*), and was in early use 'a picked body of men, detached to the front to begin the attack'. Later it meant 'a storming party'. The current figurative sense, which was first recorded in 1641, has driven out all memory of the original concrete sense.

**formalism, formality.** *Formality* is the customary non-technical abstract noun corresponding to the adj. *formal*: its ordinary senses include: 1 a formal or ceremonial act, requirement of etiquette, regulation, or custom (often with an implied lack of real significance);

2 the rigid observance of rules or conventions;

3 ceremony; elaborate procedure. (All three cited from *COD* 1995.) It has been in constant use since the 16c., and sometimes has depreciatory overtones, implying mere attention to externals.

*Formalism*, first recorded in 1840, has always been at least a semi-technical term, for example 'strict adherence to prescribed forms in theology'. In the 20c. it has spread its wings and has become widely used as a technical term for theories, movements, etc., in art, theatre, literature, linguistics, and mathematics, the complexities of which are described in the *OED* and other relevant reference works.

**formal words.** In Iris Murdoch's novel *The Message to the Planet* (1989) one of the characters is said to be *capable of solemnity, gravity, even gravitas*, the three nouns clearly representing a subtly ascending ladder of solemness. The example may suffice as an introduction to the notion that pairs or groups of nearly synonymous words, and the choices of them made by writers and speakers, are at the heart of the language. All modes of writing and speaking are marked by ascending, descending, or steady-state levels of vocabulary. At a surface level *peruse*, one can see, is a more formal word than *read*, *purchase* is more formal than *buy*, *luncheon* more formal than *lunch*, *purloin* more formal than *steal*, and *evince* more formal than *show*. The suitability of one of these rather than the other is a matter of discreet (and often delicate) contextual choice. Part of the distinction lies in the contrast between words of native origin and synonyms of (ultimately) classical origin: thus (with the native word placed first) the pairs *ask/enquire*, *bear/carry*, *begin/commence*, *drink/imbibe*, *gift/donation*, *hide/conceal*, etc. At other times it is just a matter of the contrast between a shorter word and a longer one: (none of them native words) *brave/valiant*, *pay/emolument*, *try/endeavour*, etc. And in practice the choice is rarely just between two words.

In his article 'The Five Clocks' (*Internat. Jrnl Amer. Linguistics*, 1962), Martin Joos identified five degrees or keys of style, which he called *frozen*, *formal*, *consultative*, *casual*, and *intimate*. His distinctions remain broadly true. The first may be observed, for example, in legal documents, the formal in (say) a philosophical monograph or in the pages of (say) the *London Review of Books*, and so on down to the intimate informality of the conversation of families and friends. The distinctions

in the end are infinitely complex and not properly definable, but, as a working hypothesis, and no more than that, Joos's classification of degrees of linguistic formality is a good starting-point. The subject is further developed in G. W. Turner's *Stylistics* (1973), who comments *inter alia* 'but five is an arbitrary number of steps to cut in a gradation from slang to ceremony'.

**format.** **1** The noun, in the 19c. and until about 1930 always /ˈfɔːmɑː/, is now always pronounced /ˈfɔːmæt/.

**2** In computing the inflected forms of the verb are *formatted*, *formatting*; see -T-, -TT-.

**former.** **1** Its natural senses as an adj. are: (*a*) of or occurring in the past or during an earlier period (*in former times*); (*b*) having been previously (*her former husband*); (*c*) (preceded by *the*; often used absolutely) the first or first mentioned of two (persons or things), as opposed to *the latter*. Several examples are cited in *WDEU* in which *former* and *latter* are used in contexts where there are three or more referents, e.g. *... there are three sorts of recruits ... The former of these probably joined with a view to an eventual captaincy*—*TLS*, 1949; *... though her bibliography includes Hecht, Snyder, and Daiches, she omits the latter's first name*—*Modern Language Notes*, 1957. Clearly the tendency exists and comes to the surface in respectable sources from time to time. But there are neater ways (esp. by using *first* or *last* in place of *former* and *latter*) of keeping to the logical path, and these should be sought out and used.

**2** Care should be taken to use *the former* (especially) and *the latter* only when they are close to their antecedents. It is undesirable to leave the reader with the task of rereading the passage or passages referred to in order to recall which is the former and which the latter of two persons, possibilities, etc.

**formidable.** The standard pronunciation is with the main stress on the first syllable. Second-syllable stressing, though increasingly heard (a limited opinion poll by J. C. Wells, 1990, actually revealed a slight preference for *formídable*), is not recommended.

**formula.** In general contexts the pl. is normally *formulas*, and in mathematics and chemistry *formulae* /-liː/. In AmE, *-as* and *-ae* are reported to be equally common in all senses.

**forte.** There are two separate (though etymologically related) words: **1** (noun) /ˈfɔːteɪ/ a person's strong point (from Fr., from L *fortis* 'strong').

**2** /ˈfɔːti/ (adj., adv., noun) (a passage to be performed) loudly (It., = strong, loud). In practice the pronunciation of *forte*[1] has been unstable for most of the 20c.: some still pronounce it as one syllable, /fɔːt/.

**forth.** The phr. *and so forth* is used as a convenient indication that an enumeration of items could be continued but, since the point has been made, need not be. It alternates with *and so on* and with the less stylish word *etc*. My impression is that *and so forth* is the option usually chosen by foreigners when they read their conference papers in English, reflecting their own formulas, *und so weiter, et ainsi de suite*, etc. For English speakers the choice is open, and a good many speakers, native or otherwise, use the extended formula *and so on and so forth*. At a *conversazione* arranged by the English Association in London in June 1991, I pointed out to Professor John Bayley that he had used *and so forth* (*Shakespeare was a playwright, actor, and so forth*) and that his wife, Iris Murdoch, on the same occasion had used *and so on*. Dame Iris felt that there was a slight and probably definable distinction, probably arising from the buried assertiveness of *forth* in phrases like *go forth, set forth*, and so on.

**forties.** See NINETIES.

**fortissimo.** When used as a noun ( = a passage of music to be played very loudly) the pl. is either *fortissimos* or (as in It.) *fortissimi*.

**fortuitous.** 'How sad it will be to lose "fortuitous" to the Visigoths,' wrote a friend to me in 1987, and I believe this to be the prevailing view, at least among standard speakers, in Britain. The word entered the language in the mid-17c., with the meaning 'that happens or is produced by chance, accidental'. The Cambridge Platonist Henry More wrote about *the fortuitous concourse of Atoms*, Addison of *fortuitous Events*, and Walter Scott of *a fortuitous rencontre*: for them, and for later writers, *fortuitous* meant 'by chance, by fortune'. The root or base of the word was L *fors* 'chance' and *forte* 'by chance'. That is how the matter stood until the beginning of the 20c. The derivatives *fortuitously, fortuitousness*, and *fortuity* had come into being and established themselves, always in contexts involving accident or chance: *Wiles, Trech'ry, Lies, Guilt, Flattery, Deceit, Like Atoms here fortuitously meet*, wrote Bishop Thomas Ken in 1711. Charles Reade, in *The Cloister and the Hearth* (1860), remarked that *one of the company, by some immense fortuity, could read*. No one stepped out of line.

Then something happened. Since about 1920, and with increasing frequency from about the middle of the century, the word has sometimes been used as a near-synonym for *fortunate*, that is, of something that happened by *good* fortune, not merely by chance or by accident. Fowler (1926) cited examples of the new sense but thought of it as a confusion brought about 'through mere sound'. He judged *fortuitous* used in place of *fortunate* to be a mere malapropism, and nothing to worry about. But the use has persisted. The Visigoths are at work among standard speakers: *Was it not very fortuitous for the government that this debate took place?* (BBC1 *Nationwide*, 1978); *Ben Elton could not have launched* Stark *at a more fortuitous time. In the midst of the present spate of bad news about pollution, he has brought out his first novel, a black SF comedy in which the world is hurtling to the brink of ecological disaster* (book review by Dorothy Wade, 1989) (*propitious* would have been a better word).

Of course the traditional use has not been driven out, as will be seen from the following examples (including two of *-ly*): *In this choice too I see something fortuitous, born of impulses which I am forced to regard as outside the range of my own nature*—L. Durrell, 1956; *His presence is not fortuitous. He has a role to play; and you will see him again*—A. Brink, 1979; *A prediction that London was to be destroyed on a certain date coincided fortuitously with a thunderstorm of exceptional severity*—E. P. Thompson, 1980; *Quite fortuitously, Morse lights upon a set of college rooms which he had no original intention of visiting*—C. Dexter,

1983. It is true, however, that some events that occur by chance may also seem to be fortunate or to have a favourable result. There is a dangerous zone along the edge of the two words where their meanings make contact or even merge. Possibly this is why some writers opt for *fortuitous* when others would choose *fortunate*, *lucky*, or *propitious*. The evidence that a zone of ambiguity is being created is fairly substantial. *WDEU* (1989) cites several examples: *This circumstance was a fortuitous one for Abraham Lockwood*—J. O'Hara, 1965; *But from a cost standpoint, the company's timing is fortuitous*—*Business Week*, 1982; *The opening of his firm had come at an extremely fortuitous time*—*Atlantic Monthly*, 1984.

Plainly the new meaning is knocking at the door. But readers of this book are urged meanwhile to restrict the word to its traditional ('accidental, by chance') sense. When an intrusive meaning contains a seed of ambiguity, it is advisable to stay with older ones.

**fortune.** The dominant standard pronunciation is with medial /-tʃ-/; that with /-tj-/ is a shade too precise.

**forum.** Pl. *forums*. See -UM 1.

**forward(s).** 1 As adj. the only possible spelling is *forward* ( *forward movement*, *forward play* [in cricket], *sufficiently forward in walking*; *a very forward child for his age*, *a rather forward* [ = presumptuous] *person*, etc.).

2 As adv., according to the *OED* (in a fascicle published in 1897), 'The present distinction in usage between *forward* and *forwards* is that the latter expresses a definite direction viewed in contrast with other directions. In some contexts either form may be used without perceptible difference of meaning; the following are examples in which only one of them can now be used: "The ratchet-wheel can move only *forwards*"; "the right side of the paper has the maker's name reading *forwards*"; "if you move at all it must be *forwards*"; "my companion has gone *forward*"; "to bring a matter *forward*"; "from this time *forward*" ... In U.S. *forward* is now generally used, to the exclusion of *forwards*.' Fowler (1926) was less sure: 'To this it must be added that there is a tendency, not yet exhausted, for *forward* to displace *forwards*, and that

even in the less than twenty years since the publication of that statement [*Fowler must have written this sentence before 1917*] there has been change. The reader will notice that, while he can heartily accept the banishment of *forwards* from the last three examples, it is quite doubtful whether *forward* is not possible in some or all of the first three.'

At present *forwards* as a directional adv. survives esp. in the phr. *backwards and forwards* (*the door kept swinging backwards and forwards*—A. Brookner, 1984; *It does not just go backwards and forwards*—R. McAlpine, 1985), but otherwise is not commonly used either in Britain or abroad. Examples: (forward) *Georgia Rose ... leaned forward and blew out every one of the candles*—Lee Smith, 1983 (US); *Hugh stepped forward. 'It's me, don't be frightened.'*—M. Wesley, 1983 (UK); *her mind refused to bring any such memory forward*—E. Jolley, 1985 (Aust.); *her head whips backward over the top of the seat and then forward again*—*Chicago Tribune*, 1988; *Then she leant forward, trying to find my face*—*Encounter*, 1988; (forwards) *This one, a small German monoplane, had smashed nose forwards into the field immediately behind the copse*—S. Hill, 1971; *Then he leaned forwards and touched Colin's forearm*—I. McEwan, 1981.

**foul** (adv.). From the ME period onward *foul* was used freely as adv. as well as as adj.: *Ye wil nat fro your wyf thus foule* [ = foully] *fleen?* (Chaucer, 1385); *Two of three her nephews are so fowle forlorne* (Spenser, 1590); *Carries his tail foul* (*London Gaz*, 1715); *Our Allies have ... played us foul* (Lord Nelson, 1799). It is still notionally available as an adv. but is not often used, exc. in quasi-adverbial circumstances, as in the combination *foul-mouthed* and in the idiomatic phr. *to fall foul of* 'to come into conflict with'. The normal adverbial form is, of course, *foully*.

**foulard,** a thin soft material of silk or silk and cotton, or an article made of this, is now usually pronounced /fuːˈlɑːd/, i.e. with the final consonant pronounced. (The *OED* gave precedence to a French-type pronunciation for this 19c. loanword from French, and, indeed, /ˈfuː-lɑː/ is still often used.)

**foully.** 1 Pronounce both *l*s.

2 See FOUL (adv.).

**founder.** See FLOUNDER.

**fount, fountain.** *Fount* (apart from the sense in typography, which is another word, related to *found* 'to melt (metal)') is a poetical and rhetorical back-formation from *fountain* (cf. *mount/mountain*), and normally means 'a source': *if streams did meander level with their founts* (Macaulay, 1830); *By Kedron's brook, or Siloa's holy fount* (W. M. Praed, a1839). Apart from such literary uses the word is also used in the popular phrases *the fount of all wisdom, knowledge*, etc.

**four.** In the 19c. the phr. *on* or *upon all fours* replaced the traditional (first recorded in the 16c.) *on* or *upon all four*, sc. the extremities (either the four legs of a quadruped or the legs and arms of a child). A homily of 1563 refers to *A bruit beast, creeping upon all foure*, and the Bible (Lev. (AV) 11: 42) has *Whatsoeuer goeth vpon all foure*. The added -*s* is shown in: *Edward ... could perceive him crawling on all-fours* (Scott, 1814). The phr. survives in its literal use ( = on hands and knees) but has also passed through various figurative stages, e.g. *to run on all fours*, evenly, not to limp like a lame dog; *to stand on all fours*, to present an exact analogy or comparison (with). Its most common form now is (often in negative contexts) *to be on all fours with*, or (preceded by various verbs) just *on fours with*, suggesting an exact analogy or equivalence with. Examples: *The railways maintain that conditions in Great Britain and America are not on all fours—Economist*, 1931; *It was impossible to make an agreement exactly on all fours with the Anglo-American bomber base agreement—Ann. Reg. 1960*, 1961.

**fowl.** The collective use of the sing. (*all the fish and fowl in the world*; see COLLECTIVE NOUN) still exists, but is not common except in compounds such as *guineafowl, wildfowl*.

**foyer.** Pronounce /'fɔɪeɪ/. The word is occas. pronounced /'fwɑːjeɪ/ in imitation of Fr.

**fracas.** Pronounce /'frækɑː/. The pl. is the same, but pronounced /-kɑːz/.

**fraction.** In arithmetic, 'a numerical quantity that is not a whole number, e.g. $\frac{1}{2}$, 0.5'. In general use, however, it means 'a small, esp. a very small, part, piece, or amount' (e.g. *the number of ospreys at large is now only a fraction of what it used to be*).

**fraenum,** in anatomy, a fold of mucous membrane or skin, esp. under the tongue, checking the motion of the part to which it is attached, is normally so spelt in BrE, but *frenum* is also found. Pronounce /'friːnəm/. Pl *fraena* /-nə/.

**fragile.** 1 Like many other adjectives ending in -*ile* (e.g. *docile, facile, fertile, futile*, etc.), *fragile* is now always pronounced in BrE with final /-aɪl/. It was not always so. In John Walker's *Pronouncing Dictionary* (1806) and in the *OED* (1897) only /-ɪl/ was given; but by 1932 A. Lloyd James in *Broadcast English* admitted only /-aɪl/. In AmE the standard pronunciation is with /-əl/.

2 *fragile/frail*. These two words, both being ultimately from L *fragilis*, are doublets, and are frequently interchangeable (e.g. *he felt frail/fragile after the late-night dinner party*). In general, however, a physical object (e.g. an old vase, a drinking glass, the Dead Sea scrolls) can be said to be *fragile* but not *frail*. Breakable objects sent through the post, or packed in containers for transport, are also frequently labelled *fragile* (not *frail*). Larger structures, e.g. a bridge, a building, are neither *frail* nor *fragile*, but, if in danger of collapse, are *unsound, unsafe*, etc. A physically weak or ill person could normally be called *frail* but not *fragile*. One's memory becomes increasingly *frail* (not *fragile*) with age. A truce or peace can be either *fragile* or *frail*. The area of meaning shared by the two words is considerable, but in the unshared areas, which are not precisely definable, careful navigation is required. Fowler's view (1926) that 'the root idea of break is more consciously present in *fragile* owing to its unobscured connexion with *fragment* and *fracture*' is unimpressive.

**fragmentary.** Pronounce /'frægməntərɪ/ with the stress on the first syllable.

**framboesia,** the medical condition of yaws. Thus spelt, and pronounced /fræm-'biːzɪə/. In AmE spelt *frambesia*.

**framework.** Gowers (1965), at his desk in the Civil Service, must have been afflicted by an overuse of the word in

official memoranda submitted to him:
'Few modern clichés have become more
pervasive than the phrase *within the
framework of* ... the very sight of it may
nauseate the sensitive reader.' His message was clear: look for a plainer alternative, e.g. *These scandals can only be dealt
with* within the framework of *the Trade
Union organization* (read *by the Trade Unions
themselves*). The advice is sound, but in the
everyday world evidence for an excessive
use of *within the framework of* is not easy
to find. It is hard to fault the following
examples: *He is contented and relaxed within
discipline, within a framework of obedience*—J. Gaskell, 1969; *The exercise of justice
is only possible within the framework of established institutions which command respect*—R.
Scruton, 1980; *The qualities that make a
good diplomat are often those associated with
the élite: discretion, tact, intelligence ... and
a willingness to work within the framework
of civilised courtesies*—M. Binyon in *Times*,
1991.

**Frankenstein,** the title of a Gothic tale
of terror by Mary Shelley (1818). It 'relates
the exploits of Frankenstein, an idealistic Genevan student of natural philosophy, who discovers ... the secret of
imparting life to inanimate matter. Collecting bones from charnel-houses, he
constructs the semblance of a human
being and gives it life' (*OCELit.*, 1985). The
monster thus created inspires loathing
in whoever sees it, and eventually turns
upon its creator before destroying itself.
Strictly, *Frankenstein* is the creator of the
monster, a person to be pitied. But since
the late 19c. the word has been used (by
those unfamiliar with the novel) as if it
meant Frankenstein's monster. This use
is so widespread and so embedded in the
language now that it looks unlikely to
be dislodged. Examples: *There are now
growing indications that the Nationalists in
South Africa have created a political Frankenstein which is pointing the way to a non-
White political revival*—*Daily Tel.*, 1971; *In-
doctrinated mass organizations answering to
the commands and agenda of an unselected
element of the bureaucracy could, critics
charged, easily prove a Frankenstein*—*Soldier
of Fortune*, 1990.

**frantically** (adv.) is now the usual form,
not *franticly*, even though the latter has
been in the language for two centuries

longer (1549 compared with 1749) than
the former.

**Frau, Fräulein.** Pronounce /frau/, /'frɔɪ-
lam/.

**free.** 1 *freeman/free man*. A *freeman* is (a)
a person who has the freedom of a city,
company, etc.; (b) a person who is not a
slave or serf. In the plur. *a free man* (e.g.
*at last I am a free man*, i.e. have retired
from work, have been released from
some binding restriction, etc.) the two
elements, *free* and *man*, are naturally
written separately.

2 *free will*. The expression should not
be hyphened except when used attributively (*his free-will hypothesis*; but *I did it
of my own free will*). Used attributively the
form *freewill* (one word, no hyphen) is
also commonly used (*freewill offering*).

**free gift,** an unattractively redundant
expression first recorded in 1909 (*Make
use of the Free Coupon printed here and you
will receive ... a free-gift parcel containing a
Bottle of Guy's Tonic*—*Daily Chron.*, 1909)
and repeatedly since then esp. of an
object given away without charge as part
of a sales promotion exercise: *Gimmicks
and the offering of free gifts to promote sales
was condemned*—*Guardian*, 1965.

**free rein.** Occasionally miswritten as
*free reign* as if the reference was to monarchs rather than to horseriders: *They say
that if they are given free reign to invest and
produce they will grow richer*—*New Yorker*,
1987. Correct use: *She gave free rein to her
feelings*.

**free verse.** See VERS LIBRE.

**French words.**

  1 Preliminary remarks.
  2 Phonological assimilation.
  3 The general process of assimilation.
  4 Degrees of Anglicization.

1 *Preliminary remarks*. French words
have flowed into English since the late
OE period. The process has never ceased,
but it is worth noting two periods of
intense intake: the Norman Conquest
(the result of cultural imperialism) and
the Age of Classicism, when movements
in science and philosophy led to the
appearance of lexical gaps in English
which were filled by French words,
rather in the same way as the computing

industry and the media in France are absorbing English terms at the moment. Of more recent imports, the rôle of French as the language of international diplomacy is clearly significant; and so, interestingly, are certain national stereotypes, particularly that of a perceived sophistication of the French in matters sensual and artistic (art, literature, food, wine, and sexuality). Part of the attraction of French seems to have been that it appears to fill lexical gaps in English, and also that the 'otherness' of French makes it possible to refer to facts and situations which English, reacting no doubt to the moral codes of the 19c., has preferred to leave unnamed, e.g. *affaire de cœur* (first recorded in English in 1809), *crime passionnel* (1910), and *ménage à trois* (1891).

**2** *Phonological assimilation.* The great majority of French loanwords are so firmly established that they form a natural part of the language and are pronounced as English words with no hint of foreignness: thus *button, glory, ounce, place, prime, uncle,* etc. It is a general truth, however, that words moving from one language to another tend for a time, sometimes a long time, to remain in a kind of no man's land, esp. when they contain sounds for which there is no exact equivalent in the receiving language. Uncertainty prevails until the foreignness of the adopted word is adjusted in order to fit into the phonetic arrangements and accentual system of English. Some words remain partially or permanently in a zone of incomplete adaptation.

**3** *The general process of assimilation.* Our main concern here is with those adoptions from French which lie within or near the borders of this unassimilated zone. Such words are usually pronounced in something like a French manner, and are normally printed in italic, or both. A typical example of the process of adoption is shown in the noun *abandon.* First recorded in English in 1822 it was pronounced /abã:'dɔ̃/, or thereabouts, throughout the remainder of the 19c. and at least as late as 1917 (the date of publication of the first edition of Daniel Jones's *English Pronouncing Dictionary*). It was italicized in Ruskin (1851) but printed in roman type in Joyce's

*Ulysses* (1922). By the 1930s it seems to have been treated as an ordinary English word by everyone. Hundreds of French loanwords had a similar history between the time of their adoption in English and their complete assimilation. Others are in the process of doing so.

**4** *Degrees of Anglicization.* At the present time, the degree of Anglicization of words and phrases adopted from French varies from person to person or from group to group, and any classification is likely to be vulnerable in some respect. With that proviso, the following groupings might be found to be of some value.

(*a*) Pronounced in a French manner (with the modifications listed below) and printed in italic. In such words the routine deviations from French include: introduction of diphthongs where French has none (*déjà, touché*); introduction of the English pronunciation of *r*—rolled between vowel sounds (*terrible*) and vestigial in word endings (*faire*); marked stressing of syllables where French has little or none; total or partial elimination of nasalization; introduction of the 'schwa' sound /ə/ for unaccented vowels (*apéritif* → /əp-/, etc.). Examples include *affaire de cœur, crime passionnel, déjà vu, enfant terrible, laissez-faire, ménage à trois, nom de plume, son et lumière, touché, tour de force.*

(*b*) Fully Anglicized and printed in roman type (the bracketed dates are those of the first record of each word in English): *baroque* (1765), *bizarre* (1648), *brunette* (1713), *clairvoyant* (1671), *éclair* (1861), *mayonnaise* (1841).

(*c*) Romanized in type but retaining partial elements of Fr. pronunciation (final letter not pronounced, etc.): *billet-doux, camembert* /-beə/, *escargot* /-gəʊ/, *sobriquet* /-keɪ/, *tournedos* /-dəʊ/.

(*d*) In literary and scholarly work, Gallicisms are scattered about like grain (the bracketed dates are those of the first record of each word in English): *à merveille* (1762), *arrière-pensée* (1824), *au fond* (1782), *au pied de la lettre* (1782), *esprit d'escalier* (1906), *point d'appui* (1819). But these expressions, and many others like them, are not in everyday use among standard speakers.

There is no doubt that unEnglish sounds will continue to be Anglicized in expressions like *tête-à-tête* and *tour de force*:

and that French words like *blasé*, *chic*, and *naïve*, for which there are no English synonyms with exactly the same shade of meaning, will continue to be used in the years ahead. And there is also no doubt that printers and publishers will disagree about which of them is to be printed in italic type and which not. See GALLICISMS.

**frenum.** See FRAENUM.

**frequency.** First recorded in the 16c., it stood alongside its alternative form *frequence* (also first recorded in the 16c.) for some three centuries; but the latter became obsolete at some point in the early 20c.

**frequentative.** Frequentative verbs are formed with certain suffixes to express repeated or continuous action of the kind denoted by the simple verb. The chief frequentative suffixes in English are *-er* (answering to OE *-rian*), e.g. *chatter*, *clamber*, *flicker*, *glitter*, *slumber*; and *-le* (answering to OE *-lian*), e.g. *crackle*, *dazzle*, *paddle*, *sparkle*, *wriggle*. Frequentative verbs are found in many other languages, e.g. Latin, Russian.

**fresco.** The recommended pl. is *frescos* (but *frescoes* is also admissible). See -O(E)S 1.

**friable** means 'easily crumbled'. If the need should arise for a word meaning 'able to be fried' it should be spelt *fryable*.

**friar, monk.** A *friar* is a member of any of certain religious orders of men, orig. in the 13c. and esp. the four Roman Catholic mendicant orders (Augustinians, Carmelites, Dominicans, and Franciscans), in former times living entirely on alms. A *monk* is a member of a religious community of men living under certain vows, esp. of poverty, chastity, and obedience. Monks normally live in monasteries or monastic houses, and seek salvation through their vows and their secluded way of life. A friar's sphere of work has traditionally been in the community, and his object is to do good work.

**fricandeau,** a cushion-shaped piece of meat, esp. veal. Pl. *fricandeaux* /-ˌdəʊz/. See -X.

**fricative.** In phonetics, (of) a consonant made by the friction of breath in a narrow opening (cf. L *fricāre* 'to rub'). In English, for example, /f/ and /v/ are labio-dental fricatives; /θ/ and /ð/ are dental fricatives; and /s/ and /z/ are alveolar fricatives. They contrast with *stops* (see STOP (noun)), such as /b/ and /p/.

**Friday.** The natural use shown in *He normally eats fish on Fridays* varies occas. with the type *He normally eats fish Fridays*, i.e. with omission of *on*, esp. in AmE. Only an American, I think, would say *I saw you Friday* (cf. *Noriega . . . said Monday the U.S. Southern Command in Panama . . . threatens the Central American nation*—USA Today, 1988), but there is a great deal of variation in such contexts in rapid speech.

**friendlily, friendly** (advs). Used to mean 'in a friendly manner, with friendship', *friendly* is recorded in the *OED* from *Beowulf* onward until the mid-19c. (e.g. *Some of the men marry three wives, who in general live friendly together*—James Cook, 1771–84), but has almost entirely dropped out of use. The alternative uneuphonious form *friendlily*, first recorded in the work of Rochester in the late 17c., is available but, because of its clumsiness, is not often used. Examples: *By the same token, went on Callie, friendlily smiling, 'I'm afraid I must ask you, Arthur, to take your boots off'*—E. Bowen, 1945; *The women . . . still addressed him friendlily*—W. Trevor, 1980; *All this was friendlily presented*—Roy Jenkins, 1991.

**frier.** See FRYER.

**frith.** See FIRTH.

**fritillary.** Now always pronounced with the main stress on the second syllable.

**frizz.** See -Z-, -ZZ-.

**frock** was originally a male garment, especially the mantle of a monk or priest (hence to *unfrock*), then the *smock-frock* that was the overall of an agricultural labourer, and finally the *frock-coat* that was for many years the uniform of the man-about-town. Discarded by men, the word came back into favour as a synonym of *gown* or *dress* for women or girls in the 19c. It remains in use, but *dress* or *gown* are the more usual terms for

formal garments, esp. those with designer labels.

**frolic** (verb). The inflected forms are
*frolicked, frolicking.* See -C-, -CK-.

**from.** Avoid the mixture of styles shown
in the type *he was chairman of the board
from 1979–1985* (read ... *from 1979 to
1985*).

**from whence.** *COD* 1990 baldly says,
'Use of *from whence* (as in *the place from
whence they came*), though common, is
generally considered incorrect.' The *OED*
(1899) also disapproved: '*from* ... used
more or less pleonastically before *hence,
thence, whence, henceforth,* etc.' The most
stubborn survivor of this set of phrases
is *from whence,* a phrase with a long and
distinguished history in direct and indirect questions and as a conjunctive
phrase introducing a relative clause.
Many historical examples are cited in
the *OED,* including the following:
(interrogative = from what place or
source?) *From whence these Murmurs, and
this change of Mind?—*Dryden, 1697; *My
wife, as I'm a Christian. From whence can she
come?—*Goldsmith, 1773; (indirect question) *Thys felowe, we knowe not from whence
he ys—*Tyndale, 1526; *No man can say from
whence the greater danger to order arises—*F.
Harrison, 1867; (relative or conjunctive
uses) *I will lift vp mine eyes vnto the hilles:
from whence commeth my helpe—*Ps. (AV)
121: 1 (cf. *NEB: If I lift up my eyes to the
hills, where shall I find help?*); *The quarter
from whence the following lucubration is
addressed—*Swinburne, 1887. Similar (though
fewer) examples are cited for the
other phrases (*from thence,* etc.).

How do matters stand now? Many authors clearly disown the expression altogether. But the conjunctive use persists
in the work of some standard authors:
*Above, from whence Jacob comes to greet me,
is a long narrow living room—*B. Trapido,
1982; *Father ... puts me excitingly on his
shoulder from whence I lord it over the
world—*P. Lively, 1987; *When they show the
captive a picture of the City of London, that
he may know from whence they come, he
displays no interest—*P. Lively, 1991; *Dark
clouds had gathered over the hills to the north,
from whence came the lucky changeling folk in
times long past—*S. Koea, 1994 (NZ). Philip
Howard criticized Penelope Lively for using the phrase in her *City of the Mind*

(1991) ('a solecism and tautology that jars
in such an elegant writer as Lively'). Who
is right? The best policy, perhaps, is to use
the simple words *hence, thence, whence* for
the present, not preceded by *from,* just as
a soldier in trench warfare is advised to
keep his head below the parapet. Wait for
a truce.

**frontier.** In BrE the standard pronunciation is /ˈfrʌntɪə/ (formerly freq. /ˈfrɒnt-/),
and in AmE the stress is commonly
placed on the second syllable.

**fruition.** Through false association with
*fruit,* the figurative sense 'the state or
process of bearing fruit' has become imposed on *fruition* as the 20c. proceeded,
and is now in standard use. Its original
meaning (first recorded in the 15c.) was
'enjoyment, pleasurable possession' (ult.
from L *fruitiōnem,* noun of action formed
from *fruī* 'to enjoy'). As the *OED* says, 'The
blunder was not countenanced by 19th-
cent. Dictionaries in this country, nor
by Webster or Worcester, though it was
somewhat common both in England and
in the U.S.' Examples: *A project for revealing the undiscovered burial chambers ...
is shortly to come to fruition—Times,* 1968;
*Cromwell's design on Dunkirk came to fruition
only in 1657—*C. Hill, 1970; *we are in the
grip of some evolutionary force ... which I
fear has already found its fruition in that
new race of young women I encountered in
the bus—*F. Weldon, 1978; *the style of the
Ensemble had perhaps not yet come fully to
fruition in Out of This Dream—Internat.
Musician,* 1988.

**frustum.** (In geometry) pl. *frusta* or, less
commonly, *frustums.*

**fryable.** See FRIABLE.

**fryer.** The better spelling (not *frier*) for
'a person who fries; a vessel for frying
(fish, etc.)'.

**frying-pan.** The usu. term in BrE,
whereas in AmE it alternates with *skillet*
and *frypan.* The distribution of these
three words, and the shape, weight, and
other features of this cooking utensil,
have varied considerably over the centuries.

**fuchsia.** Thus spelt (not *fuschia*), and
pronounced /ˈfjuːʃə/. The plant is named
after a 16c. German botanist called Leon-
hard Fuchs.

**fucus** (type of seaweed). Pronounce /ˈfjuː-kəs/; pl. *fuci* /ˈfjuːsaɪ/.

**fuel** (verb). The inflected forms are *fuelled, fuelling* (AmE *fueled, fueling*). See -LL-, -L-.

**fugacious,** apt to run away (a 17c. formation from L *fugax, fugācis* 'prone to run away', from *fugere* 'to flee'), has survived mainly in humorous literary use.

**fugue.** As verb the parts are *fugues, fugued, fuguing.* A composer of fugues is a *fuguist.* The corresponding adj. and adv. are *fugal* and *fugally* respectively.

**-ful.** The right plural for such nouns as *cupful, handful, mouthful, spoonful, teaspoonful,* etc., is *cupfuls,* etc., not *cupsful,* etc. See CUPFUL; PLURALS OF NOUNS 9.

**fulcrum.** Pronounce /ˈfʊlkrəm/ or /ˈfʌl-/. Pl. either *fulcra* or *fulcrums.* See -UM 3.

**fulfil.** Thus spelt in BrE with the inflected forms *fulfilled, fulfilling.* In AmE either *fulfill* or *fulfil* (inflected forms as in BrE). The corresponding noun is *fulfilment* in BrE and either *fulfillment* or *fulfilment* in AmE. See -LL-, -L-.

**fuliginous,** 'sooty, dusky'. In frequent use from the 16c. onward and still found in the work of good authors, even though it is doubtful if one person in a hundred is aware of its etymology (L *fūlīgō, -ginis* 'soot'). Examples: *Compared with Constable, whose colour is of the morning of the world, Delacroix is fuliginous and sultry* H. Read, 1931; *The fuliginous interior of a one-roomed timber house*—R. C. Hutchinson, 1952; *Gielgud . . . surrounded the play's fuliginous cruelties with settings of total black*—K. Tynan, 1961; *Two cartloads of books . . . Carlyle had torn out their vitals and fused them into his fuliginous masterpiece*—E. Johnson, 1977.

**full** (adv.). The commonest use is in the phr. *full well* (*you know full well that I told you*). Some other uses now sound rather forced: = quite, fully (*full twenty miles; full ripe*); = exactly (*hit him full in the face*); = more than sufficiently (*full early*). In the sense 'very', as in *full fain, full many a, full weary,* where *fully* cannot be substituted, it is poetical or literary. In Shakespeare's *Full fadom fiue thy Father lies,* the meaning is 'fully, quite', and *full* qualifies the numeral.

**ful(l)ness.** Spell as *fullness,* but be prepared to find *fulness* in 19c. and earlier works (e.g. *in the fulness of his heart*—R. M. Ballantyne, 1858). Cf. DULLNESS.

**full stop. 1** *Its ordinary use.* Traditionally defined as 'a punctuation mark used at the end of a sentence followed by a space and a capital letter', the full stop (.) or period or full point is, of course, so used: (examples from *Essays & Studies* 1991) *Fiction and history are kindred forms. Indeed, as late as the eighteenth century, history was regarded as a literary art.* For rhetorical or other effects a 'sentence' is frequently broken up into separately punctuated parts: *Culloden is Scott's watershed. And it is largely absent from his fiction.* Similarly, *He himself has no sense of history at all. But he and his discourse embody Joyce's wicked challenge to the historical imagination of others.* Also, *It was as though Hungary was not another place but another time, and therefore inaccessible. Which of course was not so*—P. Lively, 1987. Much more adventurous departures from the norm are now commonplace and have been for some time (examples from John McDermott's *Punctuation for Now,* 1990): *London. Michaelmas term lately over, and the Lord Chancellor sitting in Lincoln's Hall. Implacable November weather*—beginning of Dickens's *Bleak House,* 1852–3; *Oh, boy. What a week. Fourteen muggings, three rapes,* [etc.]—E. McBain, 1968. In *The King's English* (1906) the Fowler brothers wrote an impassioned essay (pp. 226 ff.) about the 'spot-plague', i.e. the tendency to make full stops do all the work. It is an interesting piece, built on the work of standard books of the time by Henry Beadnell and others, but is now only of historical interest.

**2** *Abbreviations and contractions.* The use of a full point in these is described in *Hart's Rules* (pp. 1–6), though some modifications are now needed. The distinction between abbreviations (e.g. I.o.W. = Isle of Wight) and contractions (e.g. Dr = Doctor, where the first and last letters are retained) is a useful one, but has been eroded in the 20c. by a widespread tendency to abandon the use of full points altogether for both types. As long as consistency is maintained, *a.m./p.m.* and *am/pm, St.* and *St, D.Phil.* and *DPhil,* and so on, i.e. types with or without full points, are both acceptable, unless ambiguity would arise by omission of the full point. Full stops are routinely used between units of money (£95.50, $27.50), before decimals (10.5%), and between hours and minutes (10.30 am; AmE 10:30 am). They are omitted in familiar abbreviations, e.g. BBC, OUP, TUC, or in acronyms pronounced as a word, e.g. Anzac, Aslib, NATO. Hart's Rules deals with numerous subtleties in the printing or omission of full points: e.g. 4to, 8vo, 12mo, etc. (sizes of books), points of the compass, names of well-known reference works (OED, DNB, etc.), names of books of the Bible, and so on. See CONTRACTIONS 1.

**fulsome.** (Pronounced /ˈfʌlsəm/ by the OED (1897) but now always /ˈfʊl-/.) First formed in the 13c., this word has acquired and then lost many senses, among them 'characterized by abundance' ( = full + -some); '(of food) tending to cloy or surfeit'; 'foul-smelling' (possibly from foul + -some). Its standard current meaning '(of language, style, behaviour, etc.) offensive to good taste by being excessively flattering, showing excessive flattery', first recorded in 1663, is a legacy of the older depreciatory senses of the word. This meaning seemed unthreateningly secure until the second half of the 20c., when some people began to use it in a favourable sense. For some, fulsome praise means 'high praise'. An age-old semantic process, in which a word loses its depreciatory element, is gaining a new recruit. The process seems to be proceeding more swiftly in AmE than in BrE, and to be more common among public speakers and journalists than in other quarters, to judge from the evidence I have seen.

Examples of the encroaching use (those of 1968 and 1987 from WDEU): I'm grateful for this very friendly and very fulsome introduction—B. Hays (US Congressman), 1968; I ... have a right to know that when I use 'fulsome praise' to mean great and sincere praise, many will construe it as an insult—Language, 1970; that very fulsome tribute to Mrs Shirley Williams by the PM—a reporter on BBC Radio 4, 4 May 1979, praising a speech by James Callaghan; I got a very fulsome apology from the President of Iraq—President Reagan, quoted on NBC News, 19 May 1987. Standard speakers

should brace themselves: the new meaning is bound to turn up on public occasions now and then. Fulsome is also occasionally being used to mean 'full-figured' (of a woman's figure) by fashion writers who analyse the word as consisting of full and -some as in handsome, wholesome, etc.: e.g. I am warned that these particular cassocks will only fit either the exceptionally petite or the handsomely fulsome—Daily Tel., 1985. It is unlikely that the two opposing senses will remain in the language permanently, but the outcome of the battle between them will doubtless not emerge until the 21c. Meanwhile everyone is advised to restrict the word to its 1663 meaning.

**fun.** This modernish noun (first recorded in 1700 and stigmatized by Johnson as 'a low cant word') has become an informal quasi-adj., esp. in the second half of the 20c. We had a fun time, exclaims many a young person after a party, an outing, a holiday, etc., or It was a fun thing to do, meaning 'an amusing or enjoyable thing'. But it has not yet gained admission to the standard class of adjectives in that, in serious writing, it (so far) lacks a comparative and a superlative. In ordinary attributive use fun is quite frequent, esp. in funfair, the American word funfest (a gathering for the purpose of amusement), and fun run (an invention of the 1970s).

**function. 1** That such and such a thing 'is a function of' such another or such others is a POPULARIZED TECHNICALITY: A man's fortitude under given painful conditions is a function of two variables—L. Tollemache, 1876. As not everyone can cope unaided with mathematical technicalities, the definition of the mathematical sense of function in COD 1995 may be found useful: 'a variable quantity regarded in relation to another or others in terms of which it may be expressed or on which its value depends ($x$ is a function of $y$ and $z$).'

**2** The sense 'a social or ceremonial occasion' (originally, in the 17c., a religious ceremony in the Roman Catholic Church) is suitable only for gatherings of some importance conducted with formality; for ordinary social occasions party is the right word. Example: It was not the kind of function to which Nat was accustomed to go, but his father's employer ... was a patron of the Appeal and pressed a ticket on him—F. Raphael, 1960.

**funebrial, funeral** (adj.), **funerary, funereal.** This is a tale of disappearing adjectives. The first of these, used of orations, verses, garlands, etc., associated with funerals, was first recorded in 1604 but was rare by the end of the 19c. and is now obsolete. *Funerary* (first recorded in 1693) vied with it as an adj. (*funerary bronze, festival, roll, urn,* etc.) but gradually became restricted to the rites of ancient, including classical, burials, and remains in use in such archaeological contexts. *Funereal* (first recorded in 1725) was and occasionally still is used to mean 'of or appropriate to a funeral', but it is not an everyday word. Its main meaning in day-to-day use is 'gloomy, dismal, mournful' (e.g. *funereal black, at a funereal pace*). The ordinary adj. from the 14c. onward was *funeral*. Its history as an adj. is one of retreat. The noun *funeral* (not recorded until the 16c.) gradually became the dominant and, as time went on, the only part of speech. It is a moot point at what stage from at least the 17c. onward that *funeral* came to be regarded, not as an adj., but as a noun used attributively in such expressions as *funeral pyre, funeral oration, funeral rites,* and *funeral expenses,* but of the general proposition there can be no doubt. *Funeral* must stand immediately before the word it governs. Social change in the last century or so brought the *funeral director* (1886), the *funeral home* (1936), and the *funeral parlour* (1927), in all of which *funeral* is perceived to be a noun used attributively, as also in the older expressions *funeral arrangements, funeral procession,* etc.

**fungus.** The recommended pl. is *fungi,* pronounced /ˈfʌŋgaɪ/ or /ˈfʌndʒaɪ/, but *funguses* is sometimes used.

**funnel** (verb). The inflected forms in BrE are *funnelled, funnelling,* but in AmE usu. *funneled, funneling.* See -LL-, -L-.

**funniment, funniosity.** Of these two 19c. words meaning 'drollery, (a) comicality', the first is now obsolete, and the second is used only jocularly.

**funny.** The word is mainly used in two contrasted senses: (a) amusing, comical (*a funny joke, puts on a funny voice*); (b) strange, perplexing, hard to account for (*a funny look, it's a funny old world, she's funny that way*). Since the 1930s (first recorded in a late novel by Ian Hay), sense (a) has come to be called *funny-ha-ha* and sense (b) *funny-peculiar.*

**furiously.** The use of the word in the following examples reflects an older idiomatic use of French phrases like *penser furieusement* 'to think hard', *donner furieusement à penser* 'to puzzle': *This gives one furiously to think*—W. J. Locke, 1910; *This attitude of his gave me furiously to think, and I was slowly forced to the conclusion that Alfred Inglethorp wanted to be arrested*—A. Christie, 1920; *That Jammy Hopkins should stay without moving for more than three consecutive minutes argued that he was being given furiously to think*—J. Tey, 1936. The use lingers on in English but apparently the original French expressions are no longer considered 'natural' and are therefore not used in day-to-day French. Phrases like *ça demande réflexion* or *ça donne à penser,* for 'think hard', and *ça me donne vraiment à penser,* for 'puzzle', are more commonly used.

**furore.** Pronounce as three syllables, /fjʊəˈrɔːri/. The word is spelt *furor* in AmE and pronounced /ˈfjʊərɔː/.

**furry.** Pronounce /ˈfɜːrɪ/.

**further.** See FARTHER, FURTHER.

**furze, gorse, whin.** All three words are synonymous (yellow-flowered shrub of the genus *Ulex*), but *furze* is chiefly restricted to BrE and *whin* is chiefly a Scottish and northern counties word. *Gorse* is used throughout the English-speaking world.

**fuse.** There are two distinct words: **1** (first recorded in 1644) a device for igniting a bomb, etc.; derived from It. *fuso,* from L *fūsus* 'spindle', (hence) a spindle-shaped tube orig. used for a bomb, etc. Usu. spelt *fuze* in AmE.

**2** (first recorded in 1884) a device or component for protecting an electric circuit; the verb *fuse* from L *fundere, fūs-* 'to melt', used as a noun.

**fused participle.** See POSSESSIVE WITH GERUND.

**fuselage.** Pronounce /ˈfjuːzəˌlɑːʒ/.

**-fy.** **1** English verbs in *-fy* are derived from or as from Fr. verbs in *-fier* (L *-ficāre*), e.g. *beautify* (16c.), *classify* (18c.), *countrify* (17c.), *dandify* (19c.), *horrify* (18c.), *pacify* (15c.), *speechify* (18c.). There has been some fluctuation in the spelling of the infinitival forms (*-ify/-yfy*) and also of the derivatives (*countryfied/countrified, dandyfied/dandified*, etc.). For these well-established words the recommended spelling is *-ify, -ified, -ifying*.

**2** A small group of words have *-efy* in the infinitive and therefore *-efied, -efying* in the inflected forms, e.g. *liquefied, -efying* from *liquefy*; *stupefied, -efying* from *stupefy*.

**3** The 'rule' outlined in section 1 has hardly affected the spelling of similar, but jocular or trivial, formations that have decorated the language since the late 16c., e.g. *bullify* (18c., = to make into a bully), *Frenchify* (16c.), *ladify/ladyfy* (17c.), *truthify* (17c.). The *-ee-* in *Yankeefied* (19c.) and the *-ey* in *cockneyfied* (19c.) naturally remain unchanged, i.e. the ending is not converted in them to *-ified*.

# Gg

**g.** See GREEK G.

**gabardine** (a durable cloth). This is the recommended spelling, not *gaberdine*. The spelling *gaberdine* is the one customarily used for the historical sense 'a loose long upper garment such as formerly worn by Jews, almsmen, beggars, and others'.

**Gaelic** usu. /'geɪlɪk/ but sometimes /'gælɪk/. Any of the Celtic languages spoken in Ireland, Scotland, and the Isle of Man. The term *Brythonic* (or *Brittonic*) is 'used [by comparative philologists] to describe the language brought to Britain by the bearers of that variety of primitive Celtic speech known as *P-Celtic*' (Kenneth Jackson, 1953), one of the branches of the primitive Celtic group of languages. The *Q-Celtic* group differed from them in that in them the Indo-European /kw-/ sound remained, whereas in the *P-Celtic* group this sound became /p/. Irish Gaelic, Scottish Gaelic, and Manx Gaelic belong to the Q-Celtic or Goidelic group, whereas Welsh (and also the Celtic of Cornwall and Brittany) belongs to the P-Celtic or Brythonic group. Cf. e.g. Irish Gaelic *ceathair* and Welsh *pedwar* 'four'.

**gag** (in the parliamentary sense). See CLOSURE.

**gage** is the spelling of the word meaning 'a pledge, a challenge, etc.' and of the abbreviated form of the *greengage* plum. In AmE it is also a variant of *gauge* (noun and verb), the 'measure' word, but *gauge* must be so spelt in BrE.

**gainsay.** This ancient verb (first recorded *c*1300) meaning 'to deny, to speak or act against' lay firmly in the standard language for nearly 600 years but seems to have fallen into relative disuse at some point in the 19c. and is now rarely encountered. It sounds appropriate in the text of the AV (*For I will giue you a mouth and wisedome, which all your aduersaries shall not be able to gainsay, nor resist*—Luke 21: 15; cf. the NEB ... *which no opponent will be able to resist or refute*), but slightly forced when used today, except in negative contexts. Examples: (positive) *One can gainsay de Gaulle's conclusion ...* —Antonia Fraser, 1988; *There are poets whose help we seek in our will-to-wane; we want them to encourage our gainsaying, to aggravate our stupor, our vice*—R. Howard, 1991; (negative) *No one ... will be admitted ... 'Doctor's orders,' they will be told. A phrase very convenient and one not to be gainsayed*—A. Christie, 1932; *There is no gainsaying that this kind of large orchestral statement ... has a feeling of* déjà entendu *about its formal structure*—Listener, 1965; *No reader of Whitman ... can gainsay ... his attitudes of unrelieved bluffness, heartiness, boosterism, boasting, [etc.]*—English Studies, 1966. The third-person present indicative form *gainsays* is usu. pronounced /-seɪz/, not /-sez/, unlike the simplex *says*.

**gal** ( = girl). First recorded in 1795, it is said by the OED (1899) to be a 'vulgar or dialectal pronunciation of *girl*'. Doubtless this was fair comment in 1899. Both the spelling *gal* and the pronunciation /gæl/ enjoyed quite wide currency in chic circles (as well as among Cockneys) both in Britain and America for much of the 20c.—partly brought about by the song *For me and my gal*—until feminists took against all uses of *girl*, however pronounced, to denote an adult female. See GIRL.

**gala** (a festive occasion). Pronounce /'gɑːlə/, though the traditional pronunciation, as shown in the OED (1899) and as used in the Durham Miners' Gala, is /'geɪlə/. The dominant pronunciation in AmE is /'geɪlə/, but /'gælə/ is also common.

**gallant.** The ordinary pronunciation, when the word means 'brave', is, of course, /'gælənt/. But when we use it in the sense '(of a male) markedly attentive to women' (also the related noun = a ladies' man, a paramour) we are mostly inclined to move the stress and say /gə'lænt/ as a harmless act of linguistic courtesy.

**galley.** 1 Molière's famous line (in his *Les Fourberies de Scapin* ii.xi) *Que diable allait-il faire dans cette galère?* 'What the devil is he doing in this *galère* (lit. 'galley')?' means 'What is he doing in this company?', i.e. mixed up with this (undesirable) set of people. Numerous allusions to the context are found in English writers from Chesterfield (1756) onward. In some of them the original French word *galère* has been retained, and in others it has been rendered as *galley*. What it does not mean is 'gallery'.

2 The pl. is *galleys*.

**gallice, -cè.** See ANGLICE.

**Gallic, Gallican, Gaulish, French.** *Gallican* is an ecclesiastical word, corresponding to *Anglican*. It is also used in palaeography of a certain kind of script. *Gaulish* means only 'of the (ancient) Gauls', and, even in that sense, is less usual than *Gallic*. *Gallic* is also much used as a synonym in some contexts for *French*. It means not simply 'French', but 'characteristically', 'delightfully', 'distressingly', or 'amusingly French', 'so French you know', etc.; or again not 'of France', but 'of a typical French person'. We do not normally speak of Gallic wines or trade or law or climate, but we do of Gallic wit, morals, politeness, and shrugs. So far as *Gallic* is used for *French* without any implication of the kinds suggested, it is merely an attempt at ELEGANT VARIATION.

**Gallicisms.** By *Gallicisms* are here meant French words or idioms that have been adapted to a larger or smaller extent in the process of adoption into English, or have been adopted element by element in a literal and often unidiomatic manner. The asymmetry of such linguistic borrowing is a well-known phenomenon in all languages. It is relatively rare for a word or phrase taken into language A from language B to retain *precisely* the same sense or range of senses as those of the original language and to maintain the equivalence as time goes on. Some examples of the various types of Gallicisms: 1 French words which have been adapted to suit the ordinary conventions of English, e.g. by dropping accents or by substituting English verbal endings, etc.: *actuality* (Fr. *actualité*), *redaction* (Fr. *rédaction*), ME *striven* (OF *estriver*) → modE *strive*.

2 Loan translation or calques, i.e. expressions adopted from French in a more or less literally translated form, e.g. *gilded youth* (Fr. *jeunesse dorée*); *give one furiously to think* (see FURIOUSLY); *jump or leap to the eye(s)* (Fr. *sauter aux yeux*); *knight of industry* (see KNIGHT); *marriage of convenience* (Fr. *mariage de convenance*); *success of esteem* (Fr. *succès d'estime*); *a suspicion (of)* = a hint (of) (Fr. *un soupçon*); *that goes without saying* (Fr. *cela va sans dire*).

3 *Mismatches.* These include (a) French-looking words for which there is no equivalent in French (e.g. *epergne*, dinner-table ornament), or have acquired a meaning in English not paralleled in French (e.g. *papier mâché*, lit. 'chewed paper'; the Fr. equivalent is *carton-pâte*). (b) 20c. mismatches: The now ubiquitous item of bedding which became available in the 1960s in Britain under the name of a *continental quilt* is now generally referred to as a *duvet* which in French means 'a sleeping-bag', the French term being *couette* (which has the same Latin root as *quilt*). A *cagoule* in French is either a monk's hood or a child's balaclava, and never 'a hooded thin windproof garment worn in mountaineering etc.' (*COD*) (the French for this garment being a *K-way*). *Fromage frais*, now available in British supermarkets, is in fact *fromage blanc*—*fromage frais* being a fresh, unmatured cheese. The term *mange-tout* denotes a type of green bean to French native speakers, who know the currently fashionable pea as *pois-gourmand*.

See FRENCH WORDS.

**gallop** (verb). The inflected forms are *galloped*, *galloping*. See -P-, -PP-.

**gallows.** Now usu. treated as a singular noun. In OE the singular *galga* and the plural *galgan* were both used for 'a gallows', the plural having reference presumably to the two posts making up the apparatus. From the 13c. onwards the pl. *galwes* and its later phonetic representatives have been the prevailing forms. Since the 16c. *gallows* has normally been treated grammatically as singular, with a new (and rarely used) pl. *gallowses* (*OED*).

**galop** /'gæləp/, the dance, is so spelt. As a verb, the inflected forms are *galoped*, *galoping*. See -P-, -PP-.

**galore.** A refreshingly informal word (adopted in the 17c. from Irish *go leór* to sufficiency, enough), which is always placed after the word it qualifies (*whisky galore*; *there is talent galore here*).

**galosh** (an overshoe), usu. in pl. *galoshes*. Thus spelt, not *golosh(es)*.

**galumph** (verb). One of Lewis Carroll's delicious inventions, perhaps with some reminiscence of *gallop* and *triumphant*, now usu. meaning 'to gallop heavily; to bound or move clumsily or noisily' (*OED*).

**gambade** (a horse's leap). If this French form of the word is used, the pl. is *gambades*. If the Spanish form *gambado* is used, the pl. is *gambados*.

**gambit.** A gambit is 'a chess opening in which a player sacrifices a piece or pawn to secure an advantage' (*COD*). In general contexts the idea of sacrifice has largely gone, and the word is used simply to mean 'an opening move in a conversation, meeting, set of negotiations, etc.'. It is a routine example of a POPULARIZED TECHNICALITY.

**gamboge** (a gum resin). The dominant pronunciation now is /gæm'bəʊʒ/, not, as formerly (e.g. in Daniel Jones, 1917), /-'buːʒ/.

**gambol** (verb) (skip or frolic playfully). The inflected forms are *gambolled*, *gambolling* (in AmE frequently with a single -l-). See -LL-, -L-.

**gamesmanship.** See BRINKMANSHIP; -MANSHIP.

**gammon** (ham). **1** The bottom piece of a flitch of bacon including a hind leg.

**2** The ham of a pig cured like bacon (*COD*, 1995).

**gamp.** In the UK a colloquial word for an umbrella, esp. a large unwieldy one. Named after Mrs Sarah Gamp, who habitually carried a large cotton umbrella, in Dickens's *Martin Chuzzlewit* (1844).

**gamut.** In music, one of its meanings is 'the whole series of notes used in medieval or modern music', and from this sense it has also come into general use to mean 'the whole series or range or scope of anything' (*he ran over the gamut of Latin metre*). The word is derived from medL *gamma*, taken as the name for a note one tone lower than A of the classical scale + *ut*, the first of six arbitrary names of notes forming the hexachord, *ut, re, mi, fa, sol, la*, said to be taken from the initial letters of a sequence of Latin words in the office hymn for St John Baptist's day.

**gamy** (having the flavour or scent of game left till high). Thus spelt, not *gamey*. See -EY AND -Y IN ADJECTIVES 2.

**gang agley.** A traditional Sc. idiomatic phrase meaning '(of a plan, etc.) to go wrong'. It is used in standard English as a half-remembered remnant from Burns's poem 'To a Mouse' (1785): *The best laid schemes o' mice an' men Gang aft a-gley*.

**ganglion.** The recommended pl. form is *ganglia*, not *ganglions*.

**gantlet** is a variant of GAUNTLET in AmE, esp. in the phr. *run the gantlet*.

**gantry** /'gæntrɪ/ in the modern engineering senses (structure supporting a crane, etc., or one supporting a space rocket prior to launching) is always so spelt. In the sense 'a wooden stand for barrels' it is also spelt *gauntry* and pronounced /'gɔːntrɪ/.

**gaol, gaoler,** the traditional spellings in the UK, are now under severe and probably unstoppable pressure from *jail*, *jailer*, which are dominant in most other parts of the English-speaking world. In practice the agent-noun is hardly used: it has been almost entirely replaced first by *warder* and at a later date by *prison officer*. But note: *Tiny radiator grille like a gaoler's spyhole*—Julian Barnes, 1991; *alone in a house which was empty except for parents who were now gaolers*—A. Brookner, 1991.

**gap.** This much-favoured word for 'a (usu. undesirable) difference in development, condition, understanding, etc.', now very frequently qualified by a preceding noun, shows no sign of weakening or passing out of fashion. Among the collocations listed in the *OED* are *credibility gap* (first recorded in 1966), *dollar gap* (1948), *export gap* (1952), *generation gap* (1967), *missile gap* (1959), and *technology gap* (1967).

**garage.** The pronunciation favoured by standard speakers is /'gærɑːʒ/. A minority of standard speakers say /'gærɪdʒ/ or transfer the main stress to the second syllable: thus /gəˈrɑːʒ/ or /gəˈrɑːdʒ/. The dominant pronunciation in AmE is /gəˈrɑːʒ/, followed by /gəˈrɑːdʒ/.

**garbage.** See RUBBISH.

**Garden.** For *the Garden* in Greek philosophy, see ACADEMY.

**gargoyle** is the only current spelling. The 19c. variant *gurgoyle* (perhaps modelled on medL *gurgulio*) is now obsolete.

**garret, attic.** See ATTIC.

**garrotte** (verb and noun). The customary spelling for the word to do with killing by strangulation. In AmE the dominant spelling is *garrote*, pronounced either /gəˈrɒt/ or /gəˈrəʊt/, but forms with one *r* or with two *t*s are also used.

**gas. 1** See FLUID.

**2** The pl. of the noun is *gases*, while the inflected forms of the verb are *gases*, *gassed*, *gassing*.

**3** It is the most usual word in AmE for petrol, gasoline.

**gaseous.** The dominant pronunciation now in standard English is /'gæsɪəs/. Daniel Jones (1917) recommended /'geɪzɪəs/ (which is now defunct), and gave /'geɪsɪəs/ as a variant, but the pronunciation with initial /'geɪs-/ is now not often heard.

**gasoline,** volatile liquid from petroleum, esp. (chiefly AmE) petrol; not *gasolene*.

**-gate.** A terminal element taken from the name *Watergate* (q.v.) from 1973 onwards and used to denote an actual or alleged scandal in some way comparable with the Watergate scandal of 1972. Among the more familiar formations of this kind is *Irangate* (1986), but the element has been a godsend to journalists wishing to bestow a potent name on short-lived real or alleged scandals, e.g. *Dallasgate* (1975), *Koreagate* (1976), *Muldergate* (SAfr., 1978), *Oilgate* (1978); *Billygate*

or *Cartergate* (Billy Carter's alleged Libyan connection, 1980).

**gaucho** (S. Amer. cowboy). Pl. *gauchos*. See -O(E)S 6.

**gauge.** Thus spelt, not *guage*. See GAGE.

**Gaulish.** See GALLIC.

**gauntlet.** The only spelling in BrE for the word meaning (*a*) 'a stout glove' and 'a challenge' (esp. in *throw down the gauntlet*); and (*b*) also for the separate word (of Swedish origin) in the phr. *run the gauntlet* 'pass between two rows of people and receive blows from them, as a punishment or ordeal'. Sense (*b*) is often spelt *gantlet* in AmE.

**gauntry.** See GANTRY.

**gay.** At some point in the mid-20c.—though occasional evidence exists from about 1935—homosexual men made it abundantly clear that they used the word *gay* of themselves, and wanted the public at large to use it too instead of the traditional word *homosexual*, and instead of all the derogatory terms such as *fag*, *faggot*, *fairy*, *homo*, *pansy*, and *queer*. Their choice of word arose, it would seem, in part at least, from the constant application of the word since the 17c. to a person, as the OED expresses it, 'addicted to social pleasures and dissipations' (esp. in *gay dog*, *gay Lothario*); and also to its use since the early 19c. to mean '(of a woman) leading an immoral life, living by prostitution'. In other words in some circumstances it was a customary word describing certain kinds of frowned-upon sexual activity. Victorian society must have known of these uses of the word *gay*, but detected in them no threat to the principal sense of the word, namely 'bright or lively-looking, esp. in colour; brilliant, showy' (OED), in regular use since the 14c.

Since the 1950s, attitudes have changed: widespread resentment has been expressed about the 'loss' of a treasured word and about the omnipresence of the sexual uses of *gay* to the point that the central older sense can now only be said with a slight change of intonation (signifying that the speaker is not yielding to the intrusive modern sense), except in contexts where the word can have no possible reference to a person's sexual preferences. Now, word has it, homosexual men in America are beginning to opt for the word *queer* instead of *gay*. Further developments can be expected.

Some examples to illustrate various attitudes: *my own dear mother, who has grey hairs, who at one time admitted to so little knowledge even of marital sex that I supposed I had been conceived in her sleep, now refers with familiarity and deference to gays*—H. Jacobson, 1983; *Gay should now only be used in the context of homosexuality, 88 per cent of the editors agreed*—*Righting Words* (US), 1987; (letters to the editors of *Newsweek*, 26 Nov. 1990) *Why all the fuss over "gay"? Could it be that, despite his protestations, Zorn* [sc. who wrote about the subject in a previous issue] *is offended by the homosexuals themselves?*—J. M. Morris; *As an elementary-school music teacher, I have deplored for years the homosexual adoption of the word gay . . . The beautiful "Have Yourself a Merry Little Christmas" declares, "make the Yuletide gay," which I changed to "it's a special day" to keep even first graders from nudging one another*—B. Sikes; *Thank you, Eric Zorn! I hope you get your word back, and I get my name back*—Jeffrey H. Gay, *In short, there is no historical case for homosexual ownership of 'gay'. So can we have our word back, please?* —Paul Johnson, 1995.

**gazebo** (structure designed to give a wide view). Pl. *gazebos*. The word was formed in the 18c., perhaps as a jocular derivative of *gaze* (verb) in imitation of L futures in -*ebo*.

**geezer,** a slang word for a person, now usu. an old man. It is a late-19c. adaptation of *guiser* (a mummer), showing a dialectal pronunciation.

**gelatin** /ˈdʒelətɪn/, the customary form in chemical use (including photography) and in AmE in all uses, but *gelatine* /-iːn/ is the more usual form in BrE in contexts of the preparation of food.

**gemma** (in certain plants). Pl. *gemmae* /-miː/.

**gender.** Since the 14c. the word has been primarily a grammatical term, applied to groups of nouns designated as masculine, feminine, or neuter. During all these centuries, however, as the *OED* shows, it has also been used as a term meaning 'the sex of a person' (e.g. *Of the fair sex . . . my only consolation for being of that gender has been the assurance it gave me of never being married to any one among them*—Lady M. W. Montagu, 1709). The *OED* (1899) labelled this sense 'Now only jocular'. Since the 1960s this secondary sense has come into much more frequent use, esp. among feminists, with the intention 'of emphasizing the social and cultural, as opposed to the biological, distinctions between the sexes' (*OED* 2). As a result the literature of the subject bristles with expressions such as *gender gap, gender identity, gender language, gender model, gender role*, and *gender-specific*; and fashionable courses at our universities abound in titles such as 'Literature and Gender in the English Restoration'.

**genealogy.** The existence of scores of words ending in -*ology* (*psychology, sociology*, etc.) traps some people into pronouncing and even spelling *genealogy* as if it too ended in -*ology*. It is derived ultimately from Gk γενεά 'race, generation' + -λόγος 'that treats of'. See -LOGY.

**general.** For the plurals of such compound nouns as *Attorney-General, Lieutenant-General*, see PLURALS OF NOUNS 9.

**generalissimo.** Pl. *generalissimos*. See -O(E)S 7.

**general practitioner.** See PHYSICIAN.

**generator.** Thus spelt, not -*er*.

**generic names and other allusive commonplaces.** When Shylock hailed Portia as *A Daniell come to iudgement*, he was using a generic name in the sense here intended; the *Historie of Susannah* (from the Apocrypha) was in his mind. We do the same when we talk of a *Croesus* or a *Jehu* or a *Hebe* (daughter of Zeus and Hera) or a *Nimrod* or of *Bruin* (name of the bear in *Reynard the Fox*), *Chaunticleer* (in Chaucer's *Nun's Priest's Tale*), and *Reynard*. When we talk of *a Barmecide feast*, of *Ithuriel's spear*, of *a Naboth's vineyard*, of *being between Scylla and Charybdis*, of *Procrustean beds*, or *Draconian measures*, or *an Achilles' heel*, we are using allusive commonplaces (to the *Arabian Nights, Paradise Lost*, the Bible, and classical antiquity). Some writers revel in such expressions, some eschew them, some are ill provided with them from lack of reading or imagination; some esteem them as decorations, others as aids to brevity. They are in fact an immense addition to the resources of speech, but they ask to be used with discretion. This article is not intended either to encourage or to deprecate their use; they are often in place, and often out of place; fitness is all. An allusion that strikes a light in one company will only darken counsel in another: most audiences are acquainted with the qualities of *a Samson*, *a Sancho Panza*, and *a Becky Sharp*, fewer with those of *a Count de Saldar* (Meredith's *Evan Harrington*) or *a Silas Wegg* (Dickens's *Our Mutual Friend*), and fewer still with those of *the Laputans* (in *Gulliver's Travels*) and *Ithuriel's spear*.

For examples of allusions of one kind or another see BENEDICK; DEVIL'S ADVOCATE; FRANKENSTEIN; ILK; IRRELEVANT ALLUSION; MISAPPREHENSIONS.

**genesis.** Thus spelt (cf. *genitive*).

**genie** /ˈdʒiːni/, a spirit of Arabian folklore. Pl. *genii* /ˈdʒiːnɪaɪ/.

**genitive.** See APOSTROPHE² B, D, E; OBJECTIVE GENITIVE; 'S AND OF-POSSESSIVE.

**genius,** (person of) consummate intellectual power. Pl. *geniuses* (not *genii*, which is the pl. of GENIE).

**gens** /dʒenz/ (in Roman history and in anthropology). Pl. *gentes* /ˈdʒentiːz/.

**gent** ( = gentleman). Apart from its use in commercial circles (*gents' outfitters*, etc.), and (in the UK) the colloquial euphemism the *Gents*, this abbreviation is now mainly used as a term indicating a man's social level (*he's a perfect gent*; *he was turned down for the post because he was not a gent*).

**genteel.** Its primary meaning is 'affectedly or ostentatiously refined or stylish', but it is often used ironically to mean 'of or appropriate to the upper classes' (COD).

**genteelism.** As Fowler (1926) expressed it, *genteelism* is 'the substituting, for the ordinary natural word that first suggests itself to the mind, of a synonym that is thought to be less soiled by the lips of the common herd, less familiar, less plebeian, less vulgar, less improper, less apt to come unhandsomely betwixt the wind and our nobility'. The Victorians, to go back no further, were familiar with the concept, though they did not use the word *genteelism* itself: '*We have made her a bow-pot* [sc. bough-pot].' '*Say a bouquet, sister Jemima—'tis more genteel.*' '*Well, a booky as big as a haystack.*'—Thackeray, 1847/8. The practice of employing fair-sounding words, or EUPHEMISMS, is probably as old as the language itself. It is only when it can be seen that a person is unknowingly placing such words in the wrong social context that a euphemism becomes a genteelism. Fowler's 1926 list of genteelisms has largely been overtaken by events. Indeed, any such list is bound to seem banal or uninstructive: e.g. *dentures* for *false teeth*; *expectorate* for *spit*; *lounge* for *sitting room*; *odour* for *smell*; *perspire* for *sweat*. The comedy of genteelism lies in its social incongruity rather than in the presence of degrees of innate genteelism in particular words.

**genteelly,** the correct spelling of the adverb corresponding to GENTEEL.

**gentle. 1** *The gentle art*. This phrase, long a favourite with anglers as an affectionate description of their pursuit, was cleverly used by the American painter James McNeill Whistler in his title *The Gentle Art of Making Enemies* (1890). The oxymoron was what made it effective. In the 20c. the phrase has thrived to the point that it stands on the brink of becoming a cliché. Examples: (in titles) *The Gentle Art of Bowling*—S. Aylwin, 1904; *The Gentle Art of Singing*—H. J. Wood, 1927; *The Gentle Art of Verbal Self-Defense*—advt in *Reason* (US), 1991; (general contexts) *Grant took full advantage of the lunchtime lull in traffic, and in derestricted areas excelled himself in the gentle art of speed with safety*—J. Tey, 1936; *Hype is an American word for the gentle art of getting a tune into the pop charts without actually selling any records*—*Sunday Times*, 1968.

**2** *Gentle*, as what the *OED* describes as 'used in polite or ingratiating address' (*Have patience, gentle friends*), lingered in writers' apostrophes to their *gentle readers* after it had disappeared from general use. Victorian novelists were much given to it. Authors have now invented other ways of creating a sense of intimacy with their readers; if the gentle reader is now addressed it will only be by way of a jocular or ironic archaism.

**gentleman.** Our use of *gentleman*, like that of ESQ., is being affected by our progress towards a less class-based society, but in more or less opposite directions. Almost any adult male these days in Britain (but not overseas) occasionally receives letters addressed to him in the style J. *Smith, Esq.* Meanwhile the word *gentleman* has largely fallen out of use as a recognizable indicator of social class. It has remained in use in the vocative plural: *Ladies and Gentlemen* (at the beginning of an address); *Gentlemen* (when only males are present). It has also been retained as a title of an office at court (*gentleman-at-arms, gentleman usher*, etc.) Agreements which are binding in honour but are not legally enforceable are still called *gentlemen's agreements*. Those old-fashioned cricket matches between *Gentlemen* and *Players* have long since been discontinued. And public lavatories no longer display the sign *Gentlemen.* The word is mostly used as a term of last resort, e.g. in auction rooms (*I am bid £60 by a gentleman in the second row*), and of (say) a person who has been asked to wait before entering (*There's a gentleman at the door who says he is an old friend of yours*). *Gentleman* was once a key word in social rankings: it now lies in the straitened circumstances of formulaic phrases, modes of address, and courtly titles.

**gentlewoman, lady.** Theoretically the first has no sense that does not belong to the second also, but *lady* has some for which *gentlewoman* will not serve—the Virgin (*Our Lady*), titled women (*Lady Thatcher*, a title she is said to dislike), a woman or girl described politely or sometimes as a genteelism or jocularly (*a perfect lady*), in the vocative plural (*Ladies and Gentlemen*), and in numerous compounds, e.g. *lady-in-waiting, Lady Mayoress.* In the lost worlds of Victorian and Edwardian society, *gentlewoman* was frequently used as a term for a woman of good birth or breeding. As late as 1932 Rosamund Lehmann could still speak of a *distressed gentlewoman* (meaning 'an impoverished woman of good birth'), but such people are now called *distressed gentlefolk.* In the complex world of social

distinctions, the word *gentlewoman* is now only rarely called on.

**genuflexion,** a bending of the knee. Thus spelt, but *genuflection* is also standard, esp. in AmE. See -XION etc.

**genuine.** 1 See AUTHENTIC.

2 The standard pronunciation in BrE is /'dʒɛnjʊm/. In America the pronunciation with final /-am/ is widespread but non-standard: 'the second [pronunciation], with the final syllable rhyming with *sign*, occurs chiefly among less educated speakers, esp. older ones. [It is also] sometimes used deliberately by educated speakers, as for emphasis or humorous effect' (*Random House Webster's College Dict.*, 1991).

**genus.** The standard pronunciation is /'dʒiːnəs/, though /'dʒɛnəs/ is frequently heard. Pl. *genera* /'dʒɛnərə/.

**geographic(al).** Both forms have a long history (*geographic* first recorded in 1630 and *geographical* in 1559). At the present time the longer form is dominant in all uses in the UK, but usage is more evenly divided in AmE. Webster's Third (1986 printing) lists the compounds *geographical biology, botany, coordinate, distribution, latitude, longitude, mile, point, position*, as headwords; and *geographic race, terrapin, tongue, tortoise, variation.* From this evidence it would appear that the choice is made according to the preferences expressed in different academic disciplines. But other indeterminable factors may be at work.

**geometric(al).** As with the preceding word, both forms have a long history (*geometric* first recorded in 1630 and *geometrical* in 1552). In this case the shorter form seems to be dominant in fixed collocations in BrE (*geometric mean, tracery*, etc.), but in general contexts the choice seems to depend on the rhythm of the sentence. In AmE the shorter form seems to be the more usual both in fixed collocations and in general contexts. Webster's Third (1986 printing) lists *geometrical clamp, construction, optics, pitch, radius* as headwords; but *geometric design, isomerism, lathe, mean, plane, progression, series, spider, stairs, tortoise, unit.* See also ARITHMETICAL PROGRESSION.

**German,** a member of the Germanic group of languages (which includes English, Dutch, Swedish, and Danish) of the Indo-European family, the national language of Germany and Austria and one of the official languages of Switzerland. Like English, German has many dialectal forms. The standard language is called *High German* (*Hochdeutsch*). *Low German* (*Plattdeutsch*) is a comprehensive name used by philologists for Dutch, Frisian, Flemish, and some dialects within Germany itself. The words *High* and *Low* are merely geographical, referring to the southern, or mountainous, and the northern, or low-lying, regions in which the two varieties developed.

English has adopted a fair number of words from German, among them (showing the first date of record in English) being *angst* (1944), *blitz* (1939), *edelweiss* (1862), *kindergarten* (1852), *poltergeist* (1848), *quartz* (1756), *rucksack* (1866), *schadenfreude* (1895), *waltz* (1781). In German, by custom, all nouns have a capital letter (so *Angst*, etc.), but the capital is not normally carried over into English except in proper names (*Gestapo*, *Nazi*, etc.). As is the way with loanwords, the sounds of the original language, except when they exactly accord, are often modified to a greater or lesser extent by the receiving language. But just as often the sounds of the original language are retained in English even though the spellings are visibly 'foreign'. Thus, because Ger. *w* is pronounced /v/ and Ger. *eu* is pronounced /ɔɪ/ we pronounce *Wagner* /'vɑːgnə/ and *Freud* /frɔɪd/. Further examples showing the preservation in English of the original German sounds: *Beethoven* (first syllable) /'beɪt-/, not /'biːt-/, *Junker* /'jʊŋkə/, *Mainz* /maɪnts/, *Mozart* /'məʊtsɑːt/, *poltergeist* /'pɒltəgaɪst/, *Riesling* /'riːslɪŋ/, *sauerkraut* /'saʊəˌkraʊt/, *Schiller* /'ʃɪlə/, *Volkslied* /'fɒlksliːt/.

**gerrymander** (orig. US), to manipulate election districts unfairly so as to secure disproportionate representation; formed from the name of Elbridge Gerry, governor of Massachusetts in 1812 when the word was coined. As his surname was pronounced with initial /g/ not /dʒ/, *gerrymander* was pronounced with a 'hard' g throughout the 19c. in Britain, and still by some people in America. But the only pronunciation in Britain now is with /dʒ/,

and this too is the dominant form in America at the present time. In both countries it is sometimes spelt *jerrymander*, thus emphasizing its pronunciation but cutting it off from its etymological roots.

**gerund.**

1  Gerund and gerundive.
2  Gerund and participle.
3  Gerund and infinitive.
4  Gerund and possessive.

*Preliminary note.* Readers should be warned at the outset that some modern grammarians have abandoned the term *gerund* altogether and speak instead of verbal nouns in *-ing*. In Sidney Greenbaum's *An Introduction to English Grammar* (1991), for example, the index lacks an entry for *gerund*, but has a long subset of entries under *-ing*. The adjective corresponding to *gerund* is *gerundial*.

**1** *Gerund and gerundive.* The distinction lies in Latin grammar not in English. The Latin gerund is a distinctive form of a verb functioning as a noun and is declinable: thus *amare* 'to love' has the gerunds *amandi* 'of loving', *amando* 'by loving', *amandum* 'the act of loving'. It corresponds to the English verb-noun *loving*. In Latin, *difficultas navigando* means 'the difficulty of sailing', *ars legendi* 'the art of reading', and *hic locus idoneus est dormiendo* 'this is a suitable place for sleeping'. From the same stem as *amandi*, etc., is formed in Latin an adjective *amandus* 'lovable', and in Latin grammar this is called the gerundive (or gerund-adjective). It is also often called a passive verbal adjective. Thus *vir laudandus* is 'a man to be praised, a laudable man'; *homo contemnendus* 'a person to be despised, a contemptible person'; *pugnandum est nobis* (lit.) 'fighting is to be done by us', i.e. 'we must fight'. The Latin gerundive usually contains or implies a sense of obligation or necessity.

**2** *Gerund and participle.* The English gerund is identical in form, but only in form, with the active present participle. *Jogging* is a present participle in *he was jogging when we last saw him*, and a gerund in *jogging is a popular form of exercise*. Examples: (present participle) *girls on bicycles with gowns billowing behind them*—A. S. Byatt, 1985; *So we turned away, left them guarding nothing in particular*—W.

Golding, 1985; (ppl adj.) *My mother and I sat and talked, in a* musing *way*—V. Ackland, 1985; (simple gerund) *Mr Justice Curlewis also ruled out* caning *as a penalty*—B. Levin, 1985; *The aborting of healthy and normal foetuses is generally agreed to be abhorrent*—*Daily Tel.*, 1987. Sometimes an -*ing* form can be interpreted as either a gerund or a participial adjective: a *flying machine* is either 'a machine for flying in' (gerund) or 'a machine that flies' (ppl adj.).

**3** *Gerund and infinitive.* The choice between the rival types 'word ( + preposition) + gerund' and 'word + *to*-infinitive or plain infinitive' is not always a simple one. For example (see AIM), *The directive aims at ensuring open passage through borders* is idiomatic; but so is *Much imagination has gone into the project, which aims to attract half a million visitors a year.* Also (see RATHER), both of the following examples are idiomatic: *She bestowed her activity, rather than letting it be harnessed to anyone else's needs*—A. Brookner, 1988; *Rather than try to improve our own appeal to those whose votes we must win, it is suggested we should do a deal of some sort with the Alliance*—*London Rev. Bks,* 1988. So too are both of the following: *Stella* began rolling her head *to and fro upon the pillow*—I. Murdoch, 1983; *Edith felt the hairs on the back of her neck* begin to crepitate—A. Brookner, 1986. In most circumstances, however, we have no choice in the matter: one or the other construction is obligatory. Examples. (gerund obligatory) *She* avoided marrying me *because she mistrusted her own power*—T. Keneally, 1980; *no man . . . has succeeded in persuading the courts that his right to* [etc.]—*Daily Tel.*, 1987; *Mr Kinnock is left holding the baby*—ibid., 1987; (infinitive obligatory) *I have decided to stick to my original intention*—P. P. Read, 1986; *he fails to take notice of those wanting university education*—*Times,* 1986; *he preferred not to torture himself with the specific knowledge* [etc.]—A. N. Wilson, 1986.

In practice, constructions of the second type (word + infinitive) occur far more frequently and after a much wider variety of words and phrases than the first type (word + gerund). Guidance in particular cases is given at many places in this book.

**4** *Gerund and possessive.* (a) See POSSESSIVE WITH GERUND for the types *I don't like his coming here* and *I don't like him coming here.* (b) Fowler (1926) had another conjunction of the two terms in mind: whether the presence of a possessive pronoun or possessive case of a noun before a gerund is, in various circumstances, desirable, necessary, or objectionable. Among the examples he cited are the following: *Jones won by* Smith's *missing a chance* (*Smith* would be possible; omission of *Smith's* would change the meaning); *He suffers somewhat, like the proverbial dog, from his having received a bad name* (*his* is 'a wasted word'); *This danger* [sc. the scorching of plants] *may be avoided* by whitewashing *the glass* (not desirable to have a possessive noun or pronoun before the gerund, i.e. leaving it vague who is to do the whitewashing); *Sure as she was of her never losing her filial hold of the beloved* (omit *her*, another wasted word).

**gesticulation, gesture.** Both words ultimately answer to L *gestus* 'action', related to *gerere* 'to carry'. *Gesture* entered English in the 15c. and at first meant 'manner of placing the body, posture, deportment'. By the 16c. it had come to mean 'movement of the body and limbs as an expression of thought or feeling' (its main current sense). The present-day sense 'an action to evoke *a friendly response*' first came into use as recently as the early 20c. From the time it entered the language in the 17c., *gesticulation* has had just one primary sense, 'the making of lively or energetic motions with the hands or body esp. as an accompaniment to or in lieu of speech'. In so far as the two words share this sense, it is the degree of animation that governs the choice, *gesticulation* being the one that indicates a more theatrical movement of the arms or the body.

**get. 1** When I was at school the teachers were hostile to this small and useful verb: we were constantly asked to rewrite sentences containing *get* by substituting another verb in its place (*fetch, obtain, earn,* and a multitude of others). The exercises were useful as a means of enlarging our vocabulary. On the other hand they left one wondering if *get* could ever be used in an acceptable manner. It needs no demonstration that *get* is indispensable in scores of idiomatic phrases: *get along with, get away with, get down, get down to, get even with*, etc.: *COD* 1995 devotes a complete column to such phrases and the *OED* several pages. *Get* also has a range of natural uses in which it passes virtually unnoticed: *get a job, get my book for me, get rich, get one's feet wet, flattery will get you nowhere, get going, get the upper hand*, etc.

2 The more controversial uses fall into the area that lexicographers mark as *colloquial* or *informal*. It is hard to see anything wrong with *he hadn't got any* (did not possess any) = AmE (and also commonly in other varieties of English) *he didn't have any* (see DO 3). Gowers (1965) suggested that the 'intrusion of *got* into a construction in which *have* alone is enough originated in our habit of eliding *have. I have it* and *he has it* are clear statements, but if we elide we must insert *got* to avoid the absurdity of *I've it* and the even greater absurdity of *he's it*.' Nevertheless the following examples give a hint of the kind of uses of *get* which, in Britain at least, have only restricted currency except in very informal writing: *He never got to write his volume on the Revolution—Dædalus,* 1987 ( = never got round to writing); *'Come on in, son,' he says ... 'I get you a beer or something?'—New Yorker,* 1989 ( = Can I get you a beer?); *A new song, a new movie comes out, you do it. You got to move with the times—ibid.,* 1989 ( = have got to, must). See GOT.

3 See GOTTEN.

4 For *have to be, have got to be,* see HAVE 4.

5 For the semi-passive type *get changed,* see PASSIVE TERRITORY 4.

6 See AHOLD.

7 See COME-AT-ABLE.

**geyser.** In the *OED* (1899) precedence was given to the pronunciation /ˈgeɪsə/ for both senses (hot spring; apparatus for heating water), with /ˈgaɪsə, -zə/ as variants. Time has moved on. J. C. Wells (1990) recommends /ˈgiːzə/ for both senses in standard English, but '/ˈgaɪzə/ (if at all) particularly for the meaning "hot spring"'. In America and NZ (the bathroom heater sense is not used in either country) the pronunciation is uniformly /ˈgaɪzə/.

**-g-, -gg-.** Words ending in g preceded by a single vowel double the g before a suffix beginning with a vowel, even in unstressed syllables: *wagging, sandbagged, zigzagged, shaggy; egged* (on), *bootlegging, nutmegged; digging, priggish; froggy, leapfrogged, logging; humbugged, mugging.*

**ghastlily.** The formal adverb corresponding to the adj. *ghastly,* but, for reasons of euphony, best avoided. See -LILY.

**ghat,** (in India) a mountain pass, steps to a river. Thus spelt, not *ghât, ghát, ghaut.* Pronounce /gɑːt/.

**ghetto.** Pl. *ghettos.* See -O(E)S 6.

**ghoul.** Pronounce /guːl/ rhyming with *fool,* not *owl.*

**giaour,** a derogatory word for a non-Muslim. Pronounce /ˈdʒaʊə/.

**gibber** (verb), **gibberish.** 1 Pronounce both with initial /dʒ/ not /g/. (It is worth noting that the *OED* (1899) gave only /g/ for the initial sound of *gibberish.*)

2 See JARGON.

**gibbous** (convex, protuberant, etc.). Pronounce /ˈgɪbəs/.

**gibe, jibe** (verb) (jeer, mock). The first spelling is recommended. See (the sailing term) GYBE. The American word *jibe* in the phr. *to jibe with* 'to agree, be in accord' is an unrelated word of unknown origin (in the 19c.).

**gift** (verb). Despite its antiquity (first recorded in the 16c.) and its frequent use, esp. by Scottish writers, since then, it has fallen out of favour among standard speakers in England, and is best avoided. On the other hand, *gifted* ppl adj. 'talented' (*a gifted violinist*) is standard. See also FREE GIFT.

**gigolo.** Pronounce this 20c. loanword from French with initial /ʒ/ or /dʒ/. Pl. *gigolos.* See -O(E)S 6.

**gild** (verb) (cover thinly with gold) has pa.pple *gilded* (*the porcelain is gilded by a magma of gold*); but as ppl adj. is sometimes *gilt* (*gilt-edged securities, gilt tooling*) and in other collocations is *gilded* (*gilded youth,* rendering Fr. *jeunesse dorée*). The word *guild* (medieval association of craftsmen or merchants) is always so spelt in standard English, not *gild*.

**gild the lily.** It has often been pointed out that this firmly established idiom ( = to try to improve what is already beautiful) is a not quite accurate rendering of Shakespeare's line in his *King John* (IV.ii.11): *To gilde refined Gold, to paint the Lilly; To throw a perfume on the Violet,* [etc.]. See MISQUOTATIONS.

**gill.** The separate words meaning 'the respiratory organ in fishes' and 'a ravine' are both pronounced /gɪl/. Those meaning 'a unit of liquid measure' and 'a female ferret' are pronounced /dʒɪl/. The female first name, whether spelt *Jill* or *Gill*, is pronounced /dʒɪl/.

**gillie** (Sc.), a man or boy attending a person hunting or fishing, has a hard initial /g/.

**gillyflower.** Pronounce with a soft initial /dʒ/.

**gimbals** (noun pl.). Pronounce with soft initial /dʒ/.

**gimmick,** a 20c. word (first recorded in an American *Wise-Crack Dict.* of 1926) of unknown origin. In our publicity-conscious and heavily politicized age the word surged into common use after about 1950, and now is widely applied to any trifling or ingenious device, gadget, idea, etc., that catches the public eye but is regarded as likely to be short-lived, unsound, etc., as time goes by. Seldom has such an unpromising word made its way into the central core of the language so swiftly and so successfully.

**gimp** (a trimming, a fishing-line). Thus spelt, not *guimp* or *gymp*. Pronounce with hard initial /g/.

**gingerly. 1** Its etymology remains obscure but the *OED* fairly plausibly suggests that it is a 16c. adaptation of an OF word *gensor, genzor* (variously spelt), properly comparative of *gent* 'noble', but used also as a positive, 'pretty, delicate', + *-ly*. The first uses of *gingerly* as an adverb had the meaning 'elegantly, daintily (chiefly with reference to walking or dancing)'.

**2** The word is still used both as adverb ( = in a careful or cautious manner) and as adjective ( = showing great care or caution). Examples of the adverb: *He descends gingerly from the cab*—New Yorker, 1990; *and they start gingerly to shift timbers and bricks*—P. Lively, 1991.

**Giotto, Giovanni.** See ITALIAN SOUNDS.

**gipsy.** See GYPSY.

**gird** (verb). The normal pa.t., pa.pple, and ppl adj. form is *girded*. But as recently as the 19c., and still as an archaism, the variant form *girt* occurs. Examples: *They girt her sons with the weapons of war*—A. Gallenga, 1848; *And how herself, with girt gown, carefully She went betwixt the heaps*—W. Morris, 1870; *The doctor ... girt on a cutlass ... and ... crossed the palisade*—R. L. Stevenson, 1883; *The water-girt promontory which is washed on the west by Lake Kilglass*—J. B. Bury, 1905; *Have you for the moment forgotten ... in which opera the chorus 'Upon our sea-girt land' occurs?*—TLS, 1972.

**girl** /gɜːl/. **1** In the standard language *girl* rhymes with *curl, pearl,* and *whirl,* but is sometimes pronounced /gel/ or /gæl/ (see GAL) with varying degrees of social affectation.

**2** Maud, the heroine in Anita Brookner's *Incidents in the Rue Laugier* (1995), had learnt all the lessons she was supposed to have learnt, knew that girls were no longer to be addressed as girls but as women. The passage reflected the fact that since the 1960s feminist writers have tried to drive out the word *girl* as applied to adult females, and have had some success. For example, women undergraduates at universities are now called *women*; and male managers who refer to the female members of staff as *girls* risk having the habit entered on the behaviour sheet as an indication of sexual harassment. But, in the way that language works, many established uses survive from an earlier period despite the intensity of the campaign. A young man still goes out with his *girl* or his *girlfriend*. The phrases *glamour girl, cover girl,* and *page three girl* remain in common use. There has been no move to censor (or ban from public libraries) Kingsley Amis's *Take a Girl like You* (1960) or Helen Gurley Brown's *Sex and the Single Girl* (1962). Throughout the century popular songs have contained the word in their titles or text, and the songs remain unaltered: *It's a long way to Tipperary, To the sweetest girl I know* (Jack Judge, 1912); *If you were the only girl in the world* (Clifford Grey, 1916); *A pretty girl is like a melody* (I. Berlin, 1919); *Poor little rich girl* (Noel Coward, 1925); *Diamonds are a girl's best friend* (Leo Robin, 1949).

**girlie.** Since the 1920s applied colloquially to publications, entertainment, etc., featuring young women, esp. scantily clad or in the nude (*girlie magazines*, etc.). Always so spelt, not *girly*.

**girt.** See GIRD (verb).

**given name.** See CHRISTIAN NAME.

**given (that).** Sense 32 of the *OED* entry for *give* reads 'The pa.pple. is used, esp. in an absolute clause, with the sense: Assigned or posited as a basis for calculation or reasoning.' Among the illustrative examples we find: *Given a reasonable amount of variety and quality in the exhibits, an exhibition ... is sure to attract large numbers—Manchester Examiner*, 1885. Jespersen (1940, v.54) provides further examples: *Given your Hero, is he to become Conqueror, King, Philosopher, Poet?—*Carlyle, 1840; *given that he lived through the danger, it was doubtful if he could ever live an active life again—*E. F. Benson, 1894. These provide the key to the use of *given* as a marginal preposition and subordinating conjunction, and of *given that* as a subordinating conjunction. They operate in a similar manner to *considering* (*that*), *granted* (*that*), and a few others. The crucial point is that, in such constructions, *given* and the other words like it are 'distinct from the participles in meaning and in not requiring subject identification, so that they cannot be viewed as the verb in a participle clause' (*CGEL* 14.14). In other words, they are not unattached participles.

These uses are part of standard English everywhere. Modern examples: (marginal preposition) *Given the more restrictive definition of being possessed by demons, he regrets its use in a religious context—National Geographic*, 1984; *He didn't think that, given her ambitions and temperament, she would enjoy it—*A. West, 1984; *Angel Clare's terrible rejection of Tess after the confession is wholly credible, given his preconceptions—*C. Whistler, 1985; *Given the world around us, that would be unhelpful, to say the least—*

*Times*, 1992; (*given* as subordinating conjunction) *Given what the boy's father was like, they had all been prepared for something roughish—*I. Murdoch, 1962; *This contrast is understandable, given what critics of the regime have labelled 'triumphalism'—New Yorker*, 1975; *Given how busy the Spanish monarchs were in the 1480s, it's a wonder they gave Columbus any notice at all—Chicago Tribune*, 1988; (*given that* as subordinating conjunction) *Given that not every art student can work as apprentice to a living artist, this is perhaps the nearest we can hope to get to the Renaissance system—Modern Painters*, 1989; *The biggest savings would come through rationalisation given that roughly two-thirds of the cost is attributable to teachers' salaries—TES*, 1990; *Given that the government shies away from a graduate tax, student loans are the next best way of making sure that students who benefit from higher education foot some of the bill—Economist*, 1991.

**glacial, glacier.** The standard pronunciations in BrE are /'gleɪʃəl/ and /'glæsɪə/ respectively; in AmE /'gleɪʃəl/ and /'gleɪʃər/.

**glacis** (bank sloping down from a fort). Pronounced in BrE /'glæsɪs/ or /-si/, in AmE /glæ'si:/ or /'glæsi:/.

**glad(den).** See -EN VERBS FROM ADJECTIVES.

**gladiolus.** 1 Pronounce /glædɪ'əʊləs/. The pronunciation recommended by the *OED* in 1899, namely /glæ'daɪələs/, is now obsolete. In classical Latin (Pliny) the word, which is a diminutive of *gladius* 'sword', had four short syllables, but this pattern has not been carried over into English. See FALSE QUANTITY 2.

2 The recommended plural is *gladioli* /-laɪ/. *Gladioluses* is too clumsy; *gladiolus*, i.e. the same form as the singular, is used in some plant catalogues, but has nothing to commend it.

**gladsome.** See -SOME. In continuous use from the 14c. to the 19c., it is now restricted to literary contexts or used as a deliberate archaism. Example: *She heads straight for him and gives him a gladsome embrace and a kiss on the cheek—*T. Stoppard, 1993 (stage direction).

**glamour** (AmE also *glamor*). **1** The corresponding adj. is *glamorous*. See -OUR AND -OR.

**2** The word, which was brought into general literary use by Walter Scott c1830, was originally Scottish, and etymologically was an alteration of the word *grammar* with the sense ('occult learning, magic, necromancy') of the old word *gramarye*. It then passed into standard English with the meaning 'a delusive or alluring charm', and, nearly a century later (in the 1930s), was applied to the charm or physical allure of a person, esp. a woman, first in AmE and then in BrE and elsewhere.

**glance, glimpse.** A *glance* (usu. followed by *at*, *into*, *over*, or *through*) is a brief look (e.g. *the women exchanged glances*; *have a quick glance at this*). A *glimpse* is what is seen by taking a glance, and not the glance itself (*he caught a glimpse of her in the crowd*). 'A *glance* at the map' is idiomatic, but 'a *glimpse* at the map' is not.

**glasses** is the primary term in both Britain and America for 'a pair of lenses in a frame resting on the nose and ears, used to correct defective eyesight or protect the eyes', but in Britain it alternates with *spectacles*. It is often qualified by a word specifying the particular purpose for which they are being worn: e.g. *reading glasses*, *sunglasses*. In AmE *eyeglasses* is commonly used instead of the simplex *glasses*, but, though the *OED* gives 19c. examples of *eyeglasses* in BrE use, this longer term is hardly ever encountered now in the UK.

**glassful.** The pl. is *glassfuls*. See -FUL.

**glimpse.** See GLANCE.

**glissade.** Pronounce /glɪˈsɑːd/.

**global.** From the 17c. to the 19c. it simply meant 'spherical, globular'. Beginning in the late 19c., in imitation of Fr. *global*, it acquired the meaning 'pertaining to or embracing the totality of a number of items, categories, etc.; all-inclusive' (e.g. *The global sum of £300 million looked like the result of bargaining with the Treasury—Ann. Reg. 1947*, 1948; *A 'global' picture can be obtained from the use of such techniques as Rorschach—Brit. Jrnl Psychology*, 1952). In the course of the 20c.

it has become widely employed in the sense 'world-wide, involving the whole world', esp. in military and environmental jargon: thus *global war(fare)*; *global village* (coined by Marshall McLuhan in 1960 in recognition of the fact that new technology and communications have effectively 'shrunk' world societies to the level of a single village); and esp. *global warming*, a term that became established in the 1980s to mean 'an increase in temperature at the surface of the earth supposedly caused by the greenhouse effect' (*SOED*).

**gloss.** There are two distinct words. *Gloss*[1] = an explanatory word or phrase clarifying the meaning of a word that might be unfamiliar to a reader, or a marginal note of explanation or comment. This word was written as *gloze* or *glose* from the 14c. onwards and was refashioned as *gloss* in the 16c. after L *glōssa*, Gk γλῶσσα. *Gloss*[2] = surface lustre or sheen; it entered English in the 16c., probably from a Low German language.

*Gloss*[1] (esp. when written as *gloze* or *glose*) had the secondary senses, 'flattery, deceit; a pretence, specious appearance'; and *gloss*[2] developed a secondary sense 'a deceptive appearance, a plausible pretext'. In other words the secondary senses of the two words substantially overlapped, with the result that in works written between the 16c. and the 19c. care must be taken to deduce which of the two words is meant. Thus, when Ben Jonson wrote *He . . Spurns back the gloses of a fawning spirit* in his *Poetaster* (1601), *gloses* meant 'flattering speeches'. When Clarendon in his *History of the Rebellion* (1647) wrote *Malicious Glosses made upon all he had said*, he meant that the glosses were capable of a sophistical or disingenuous interpretation. On the other hand, when Cardinal Newman wrote in 1834 of *the false gloss of a mere worldly refinement*, he was using *gloss*[2] figuratively.

It is easy to see that if we lacked the *OED*'s historical record, *gloss*[1] and *gloss*[2] could be taken to be 'the same word'. Indeed, in some modern dictionaries, notably some of those prepared for children and for foreign learners of English, etymological considerations are cast aside, and homonyms like *gloss*[1] and *gloss*² are presented as part of a single entry.

**glossary, dictionary, vocabulary.** A *glossary* is an alphabetical list of the hard words used in a specific subject or text, with explanations. Glossaries are usually of modest length, e.g. those of the British Standards Institution (of aeronautical terms, highway engineering terms, etc.), and those appended to the texts in series like those of the Early English Text Society and the Scottish Text Society. A DICTIONARY is usually a much more ambitious work, though sometimes, when restricted to the terminology of a specified subject, *dictionary* and *glossary* are virtually interchangeable terms. A *vocabulary* supplies the reader of a book in a foreign language (e.g. a school edition of Latin texts) with the English equivalents of the foreign words used in it. A *glossary* selects what is judged to be obscure; a *vocabulary* assumes that all is obscure. *Vocabulary* also means the whole stock of words used by those speaking a given language, by any set of persons, or by an individual.

**gloze.** See GLOSS.

**glue.** The inflected forms and derivatives are spelt *glues, glued, gluing*; and *gluey*.

**glycerine** /ˈglɪsəˌriːn/. So spelt in BrE, but usu. *glycerin* /-rm/ in AmE. In scientific writing the synonymous term *glycerol* is customary.

**gn-.** In nearly all current English words initial *gn* is pronounced /n/ with the *g* silent, though it is clear that in native words like *gnat* and *gnaw* the *g* was sounded until the 17c. The main exceptions are *gnocchi*, a 19c. loanword from Italian, which is sometimes pronounced with initial /nj/; and *gnu* (from a Bushman language in S. Africa), which is also sometimes pronounced with initial /nj/, beside the normal pronunciation /nuː/. All such words of Greek origin (*gnomic, gnostic*, etc.) are pronounced with the *g* silent.

**gnaw.** The pa.t. and pa.pple are *gnawed*. Some dictionaries continue to list *gnawn* as a variant form of the pa.pple but it has not been commonly used in the 20c. and now sounds distinctly archaic.

Examples: *From crock of bone-dry crusts and mouse-gnawn cheese*—W. de la Mare, 1921; *These shoes, particularly those of the children, are somewhat gnawn*—Agee and Evans, 1941.

**gneiss** (metamorphic rock). This 18c. loanword from German is now always pronounced /naɪs/, with the initial *g* silent.

**gnomic** /ˈnəʊmɪk/. Used of literary works, esp. Anglo-Saxon poetry, consisting of or containing aphoristic verses or maxims (sometimes called *gnomes*). The *gnomic aorist* in Greek (used 'to express what once happened, and has therefore established a precedent for all time') is a term used in the grammar of classical Greek, but has no precise counterpart in English grammar. The word itself is derived from Gk γνωμικός 'sententious'.

**go** (noun). Pl. *goes*. See -O(E)S 2.

**go** (verb). Five uses call for comment. **1** *Goes without saying* (first recorded in English in 1878) is a naturalized Gallicism (q.v., sense 2). For anyone opposed to the use of such loan-translations, other straightforwardly English equivalents are available, *needless to say, need hardly be said, of course*, etc.

**2** *go* + bare infinitive. This was the dominant construction in English until the 17c: e.g., in Shakespeare's *Winter's Tale, Ile* go see *if the Beare bee gone; I must* go buy *Spices for our sheepe-shearing*. It survives as a normal construction in AmE, but in BrE is used in only a few fixed expressions (*hide and* go seek, *he can* go hang *for all I care, to* go get (chiefly US)). In Britain the normal constructions are of the type *go and* + infinitive or *go* + *to*-infinitive. Examples of *go* + bare infinitive: *I'll* go put *your lovely flowers in water*—J. Updike, 1986 (US); *'I wish I could see a real Jew.' 'Go look in the mirror.'*—New Yorker, 1986; *I guess I'll* go finish *my shift*—Hill Street Blues (US TV series, 15 Nov. 1986). Occasional examples turn up outside N. America as late as the 19c. (natural uses) and in the 20c. (probably as conscious Americanisms): *'Very true,' said I, 'let us* go fetch *it.'*—R. M. Ballantyne, 1858; *'Sweetheart, I'm ravenous.* Go call us *a cab.'*—Maggie Gee, 1985 (UK); *Grace'd* [ = Grace had] *got the morning off from school to* go give *her brother some support* [at a magistrates court]—Alan Duff, 1990 (NZ).

**3** Related constructions. (a) *go and* followed by an infinitive: *She . . . said she would go and turn the sprinkler off herself*—New Yorker, 1986; *He packed the sacks with goods for Kaliel to go and sell*—D. Matthee, 1986 (SAfr.). (b) examples in which *go* and the accompanying verb are both imperatives: *It's late, child . . . Go and get some sleep*—J. M. Coetzee, 1977 (SAfr.); *Tell them, go and beg in the market-place*—A. Desai, 1988 (India). (c) *go + to-*infinitive: *Rozanov . . . had gone to live in the Ennistone Rooms*—I. Murdoch, 1983. (d) *go and* used colloquially to mean 'to be so foolish, unreasonable, or unlucky as to—': *and now all at once she remembered. Max Gill had actually gone and died*—New Yorker, 1988; *You herd cattle all day, you come to despise them, and pretty soon . . . you have gone and shot one*—G. Keillor, 1990. It should be emphasized that the constructions in (a) to (d) are standard in all forms of English.

**4** *go* = say. This use is seen by Sara Tulloch (1991) as occurring mainly 'in young people's speech . . . (usually in the present tense, reporting speech in the past)'. She sees it as 'an extension of the use of *go* to report a non-verbal sound of some kind expressed as an onomatopoeic word or phrase, as in "the bell went ding-dong" or "the gun went bang", perhaps with some influence from nursery talk (as in "ducks go quack, cows go moo")'. It is indeed common in children's speech and esp. (but not only) in America: *In school Friday our teacher held up a note and goes to Amy, 'Did you pass this note?' And Amy goes, 'Who, me?'*—dialogue in strip cartoon in Chicago Sun-Times, 1988. But it is also widespread in the language of the shell-suit brigade of adults in modern fictional contexts (my examples happen to be from AmE and NZ sources but the type occurs in the non-standard language of all English-speaking countries): *I go, if this is some kind of joke I'm like really not amused*—J. McInerny, 1988 (US); *He'd wait about three lulls . . . Then he'd go, 'Guess I'll have to have the circus tonight.'*—New Yorker, 1988; *They go, 'Are you laying a complaint?' I said not yet*—Rosie Scott, 1988 (NZ); *Butch and I were discussing this problem, and Butch goes, 'But you promised you'd do it.' Then I go, 'Well, I changed my mind.'*—Chicago Tribune, 1989.

Recent research (*OED Additions Series*, ii, 1993) has taken the use back to Dickens's *Pickwick Papers* (1836): *He was roused by a loud shouting of the post-boy on the leader. 'Yo-yo-yo-yo-yoe,' went the first boy. 'Yo-yo-yo-yoe!' went the second.* Somehow it seems more acceptable there.

**5** The (American) cox in the Oxford boat in the 1987 Boat Race had the slogan *Go for it* inscribed on the back of his rowing shirt. He thus drew attention to the arrival in Britain of this popular American phrase of the 1980s. Examples: *When you come up with a zingy locution that will astound your friends and confound your enemies—go for it!*—W. Safire, 1985; *I told her about Scott* [sc. a boyfriend]. *Eileen said, 'Go for it, Andrea.'*—New Yorker, 1986. The *OED* has a series of entries for various uses of *go + the* preposition *for*, including the very informal type *I could go for you in a big way, kid* ( = be enthusiastic about, be enamoured of), but has not yet dragged this nippy little fish into its net.

**gobbledegook.** Also spelt **gobbledygook.** This is an expressive term, first recorded in America in 1944, for official, professional, or pretentious verbiage or jargon. The word was almost certainly coined as a representation of a turkeycock's gobble. For some two decades or so the term was mainly restricted to AmE, the equivalent BrE expression being OFFICIALESE, but it is now in common use in BrE. In its most innocent form, gobbledegook lies in the proliferation of upgraded job titles, as, for example, when the chief steward on a long-haul passenger flight is called *the senior in-flight services manager*. Nowadays every head of a discernible section, however small, within an organization seems to be called some kind of manager.

Of much greater consequence is the use of obfuscatory or opaque statements by official groups. The following passage from an American policy document about transport plans (as reported in a Chicago newspaper of 1995) shows gobbledegook in its most potent form: *While EPA [the Environmental Protection Agency] will solicit comments on other options, the supplemental notice of proposed rulemaking on transportation conformity will propose to require conformity determinations only in the metropolitan planning areas (the urbanized* area and the contiguous area(s) likely to become urbanized within 20 years) of attainment areas which have exceeded 85 percent of the ozone, CO, NO2, PM-10 annual, or PM-10 24-hour NAAQS within the last three, two, one, three, and three years respectively. Doubtless the statement made good sense to members of the EPA. Its accuracy is not in question. The fault lies in its inability to make any more than laborious sense to the general public to whom it was addressed.

Dr James Le Fanu, the medical correspondent of *The Daily Telegraph*, reported (in 1995) a much more worrying case: 'A friend in her late thirties [received] a buff envelope from the Family Planning Clinic, where she had had a cervical smear six weeks earlier. The report read as follows: "The results of your test showed early cell changes (mild dyskaryosis suggesting CIN I) and wart virus changes." She was advised to have a repeat test in six months.' No further explanation was offered. She turned to Dr Le Fanu, and he 'translated' for her: 'There are some funny-looking cells ("dyskaryosis") which may or may not indicate the very earliest signs of pre-cancerous change ("suggesting CIN I") which almost always returns to normal with no treatment. However, when associated with evidence of infection with the wart virus, it is slightly more likely to progress up through grades CIN II and III—at which point something may need to be done, hence the need for a further test in six months' time.' Dr Le Fanu concluded that until those responsible for sending such reports to women without a translation of what they mean 'tens of thousands of women every year, just like my friend, will continue to be unduly and unforgivably frightened'.

This gap between the type of technical language used in the examples just cited, and also in taxation offices, insurance companies, local government departments, and the like, and the capacity of members of the general public to understand it lies at the heart of the problem.

Another central problem remains. Some people (especially the young) can understand the language of computer, video-recorder, and motor vehicle manuals. Others, quite simply, cannot. Moreover, the essential technical language of most professional subjects will always lie beyond the reach of someone unversed in the subject. For example, it is necessary to know the meaning of the words *lien* and *demurrage* if one is to understand the following passage (from a legal newsletter of 1988): *The Commercial Court has held that where a shipowner exercises a lien on cargo for demurrage due, he can claim demurrage for the further period of delay caused by his own exercise of the lien.* The words themselves are not replaceable by simpler terms.

Similarly, in the following passage by R. Jackendoff in the journal *Language* (1975) it is impossible to work out what is meant unless one is familiar with the terms *lexical insertion rules, paradigm,* and *deep structure*: [It has been suggested that] *paradigmatic information should be represented in the dictionary. As a consequence, the lexical insertion rules must enter partial or complete paradigms into deep structures, and the rules of concord must have the function of filtering out all but the correct forms, rather than that of inserting inflectional affixes.*

In everyday contexts, especially in welfare, tax, medical, and social-services documents, and in the drafting of our laws and charters, one hopes that the processes of simplification (and, where necessary, of translation) will continue. Numerous books have been written suggesting ways of reducing gobbledegook to a minimum, among them Gowers, *The Complete Plain Words* (1954 and later versions), Cutts and Maher, *Gobbledygook* (1984), M. Adler, *Clarity for Lawyers* (1990), and Cutts, *The Plain English Guide* (1995). But problems are bound to remain despite the best efforts of such writers.

See also BAFFLEGAB; JARGON; OFFICIALESE; PLAIN ENGLISH.

**gobemouche** /'gɒbmuːʃ/. If this engaging French word meaning 'a gullible listener, a credulous person' is to be used at all in English it should be printed in italics, and pronounced the same in the singular and the plural (*gobemouches*). The French, bearing in mind that the singular and the plural in French are both spelt *gobemouches*, will continue to wonder at the way in which we adopt individual French words with only partial accuracy. The French word literally means 'fly-catcher', from *gober* 'to swallow' + *mouches* 'flies'.

**godlily.** Technically the adverb answering to the adj. *godly*, but normally avoided (in favour of *in a godly manner*) because of its clumsiness. See -LILY.

**God's acre** (a churchyard) is modelled on Ger. *Gottesacker*. It is properly 'God's seed-field', in which the bodies of the departed are 'sown' (1 Cor. 15: 36–44) in hope of the resurrection (*OED*).

**Goidelic.** See GAELIC.

**gold(en).** See -EN ADJECTIVES. For objects made in part or whole of gold, the shorter form is now usual (*gold ring, gold watch*). *Golden* more commonly means 'coloured or shining like gold' (*golden hair, golden retriever, golden syrup*), and is the form preferred in figurative uses (*golden age*). But the distinction is far from complete. If not of gold itself, there are plain indications of wealth in the phrases *golden disc, Golden Fleece, golden goose*, and *golden handshake*, for example.

**golf.** The standard pronunciation is /gɒlf/. 'The Scottish pronunciation is /gəʊf/; the pronunciation /gɒf/, somewhat fashionable in England, is an attempt to imitate this' (*OED*, 1901). This imitative pronunciation is now seldom heard.

**golliwog.** Bear in mind that this word (first used as *Golliwogg* in 1895) for 'a black-faced brightly dressed soft doll with fuzzy hair' has been placed on the unfavoured list by those who disapprove of ethnic terms like *Negro* and *nigger*.

**golosh.** See GALOSH.

**good** (adv.). In continuous use from the 14c. to the 18c. it dropped out of standard use c1800, and now survives mainly in non-standard use, esp. in AmE. Examples: *But it didn't work too good, because you still managed to get a coffee break out of it—New Yorker*, 1991; *I didn't feel too good the next day—Stephen Wall*, 1991; *Like I said, the rent's all paid. I'm looking after the place good—Maurice Gee*, 1994 (NZ) (a remark by a young, uneducated single parent).

**good and.** Used colloquially to mean 'completely', it has been in common use in AmE since about the 1880s and elsewhere at a slightly later date. Examples: *When I was good and drunk I tried to start a conversation with the other people in the bar—Nigel Williams*, 1985; *Only I'm good and tired of watching high-quality people fuck*

*up in practical life—S.* Bellow, 1987. See HENDIADYS.

**goodbye.** The pl. of the noun is *goodbyes*. The AmE by-form *good-by* has pl. *good-bys*.

**good will, good-will, goodwill.** Except in the attributive use (*a good-will gesture*, that is, as a gesture of good will), the choice should be between the unhyphened forms, since the accent falls on the second syllable. *Good will* is required when the notion is virtuous intent, etc., and *goodwill* is customary when it means 'the established reputation of a business etc. as enhancing its value' (*COD*).

**gormandize** is the recommended form for the verb ( = to eat or devour voraciously) and *gourmandise* for the noun ( = the habits of a gourmand; indulgence in gluttony). But in practice there is much diversity in printed work, and the spellings are often regarded as interchangeable.

**gorse.** See FURZE.

**gossip.** The inflected forms are *gossiped, gossiping; gossiper; gossipy*.

**got.** It is interesting to see how new uses can be found for this small word, the pa.t. and pa.pple of GET. It is perhaps not surprising that most of them are informal verging on non-standard, and that they all seem to have begun their life in AmE. Some miscellaneous examples: (used with a *to*-infinitive = 'to secure an opportunity to') *This was considered a bonus for me, because I got to sit in the front*—F. Kidman, 1988 (NZ); *We got to see exactly what happens to the green when we were taken out on the floor of MRQ—New Yorker*, 1989; ( = have got to, must) *We just got to live. Isn't that so?*—(black speaker in) A. Fugard, 1980 (SAfr.); *It's* [sc. coffee is] *a drug, and we got to stay pure—New Yorker*, 1986: *He's one of 300 Americans who have recently arrived to teach English to Czechoslovaks. 'We got to help these people,' he says, 'any way we can.'—Newsweek*, 1990; (*got to be* = 'came round to being') *It got to be 11 p.m. We left the way we had come—New Yorker*, 1989; (*got to be* = has to be, inevitably is) *those who've heard the band ... pressurize their pop, R&B and free improv sources into what's got to be the most pleasurable prog-rock to come along in a decade—Musician*, 1991 (US); (*got* = possess, have)

*What you got in that jar, Alvie?*—M. Eldridge, 1984 (Aust.); *I can't get my head around it, Sharon. Suddenly I got three fathers*—at least—*Times*, 1987 (preview of the TV programme *EastEnders*); *Right now, we got nine cops in the Miami police department being tried for murder*—*The Face* (US speaker), 1987.
See also GET 2.

**gotten.** Nothing points more clearly to the North Americanness of a person than the ability to use the pa.pple forms *got* and *gotten* in a natural manner. *Gotten* forms no part of the paradigm of the verb *get* in the UK—though it once did—or in any other part of the English-speaking world, though I have observed it being used sporadically in New Zealand and Australia in recent years, esp. by young people, and its use may well be spreading elsewhere too. In AmE it is a fairly safe rule that *gotten* is used when the sense is 'obtained', but *got* when it means 'owned, possessed'. It has also been suggested that *gotten* is preferred to *got* when a notion of progression is involved. The matter cannot be neatly resolved since *get* has so many other senses and applications, and the prospect of examining the distribution of *gotten* and *got* in all of them would be very daunting indeed.

Some miscellaneous examples (NAmer. unless otherwise indicated): ( = obtained, received) *An Army friend . . . had gotten us tickets for a Tchaikovsky extravaganza*—P. Roth, 1979; *Have you gotten your paper the last couple of Sundays?*—*New Yorker*, 1986; (progression involved) *They've climbed fences in their sleep, gotten into gardens*—G. Keillor, 1985; *Has my reputation in town gotten that bad?*—T. Winton, 1985 (Aust.); *'You have gotten close to the whirlpool,' Fleda said*—J. Urquhart, 1986; *Prices in the Northeast have gotten so out of hand that buyers go to extraordinary lengths to come up with affordable housing*—*US News & World Rep.*, 1986; *People in the USA have gotten much healthier in the past 30 years*—*USA Today*, 1988; *It had gotten too quiet in the neighborhood*—T. McGuane, 1989; *It couldn't have gotten worse from cold air. You just felt worse*—J. Hecht, 1991; (gotten used to) *Been sewn up for a long time and the locals have gotten used to the idea*—T. Winton, 1985 (Aust.); *Joe had not gotten used to Astrid's friends*—T. McGuane, 1989; (other

uses with following pa.pples) *I just came out here for a drink of water, but I seem to have gotten turned around in the darkness*—*New Yorker*, 1988; *they bore the shame of having tempted Adam with the apple and had gotten them both evicted from the Garden of Eden*—P. Burke, 1989; *They may have finally gotten rid of Sir Robert [Muldoon] from under their feet, but they know there is still plenty of bark and bite to come*—*NZ Herald*, 1991; (unclassified) *Sir William, whom I haven't gotten around to discussing at all*—R. Merton, 1985; *They seem to have gotten together even without your help*—*New Yorker*, 1986; *this last year and a half I've gotten to fill out a lot of forms*—J. Updike, 1986.

With all this happening to *gotten* outside the UK it is curious that the word remains outside standard BrE, while *begotten*, *forgotten*, and *ill-gotten* remain such natural members of the everyday vocabulary of British people.

**gouge** (noun and verb). The *OED* (1901) gave /guːdʒ/ as an alternative pronunciation, but /gaʊdʒ/ is now the only standard one.

**gourd** (fleshy fruit or its shell). The *OED* gave /gɔəd/ and /gʊəd/ in that order in 1901, but the second is the only standard pronunciation now.

**gourmand, gourmet.** 1 The history of the two words is known but has left us with a curious legacy. *Gourmand* entered the language in the 15c. in the sense 'a glutton'. From the mid-18c. it also acquired the meaning 'one who is fond of delicate fare; a judge of good eating', and was so used (we learn from the *OED*) by Coleridge, Darwin, Charlotte Brontë, and some other writers of fame. *Gourmet* entered the language in the early 19c. and has been used only in the sense (as the *OED* defines it) 'a connoisseur in the delicacies of the table'. As the 20c. comes to an end, the preference of conservative speakers is to restrict the favourable sense (a judge of good eating) to *gourmet*, and to apply *gourmand* only to those for whom quantity is more important than delicate judgement. In practice the conservative view seems to be the prevailing one, but for how long? It has the advantage that it is frequently used attributively (*a gourmet meal*), whereas *gourmand* cannot be so used.

**2** *Gourmand* is either pronounced in a Gallic manner as /ˈgʊəmɑ̃/ or Anglicized as /ˈgʊəmənd/. *Gourmet* is always /ˈgʊəmeɪ/.

**gourmandise.** See GORMANDIZE.

**government.** While preparing my booklet *The Spoken Word* (1981) for the BBC, I found that this belonged to a small group of words that gave maximum offence to listeners if pronounced in a garbled manner, with the first *n* silent, i.e. as /ˈgʌvəmənt/ or even /ˈgʌvmənt/.

**governor-general.** The pl. form recommended is *governor-generals*, but *governors-general* is still current as an acceptable variant. See PLURALS OF NOUNS 9.

**gradable, gradability, gradience.** The useful concept of gradability in grammar and semantics was first extensively used by Quirk et al. (1972) and by Lyons (1977), and has passed into general currency. A particular set of words can be seen as forming a continuum of size, degree, vulnerability, etc.,: thus *few, little, much*, and *many* are gradable in that they do not represent a fixed number, quantity, size, etc. By contrast, words like *single* and *married*, or *male* and *female*, cannot be seen in terms of gradience: one is either single or married, male or female. Most adverbs and adjectives are gradable (*pretty, rich*; *slowly, patiently*) in that it is a matter of judgement what constitutes prettiness, richness, etc. Others are only controversially gradable, e.g. *perfect, unique(ly)*. Colour words show that there are considerable gradations between the primary colours. This is not the place for a full treatment of the concept of gradability. Readers who wish to pursue the matter further should consult Lyons (1977), CGEL, and other standard works.

**graduate** (verb). There is no problem about its ordinary intransitive sense (*he graduated from Yale in 1984*; *she graduated last year*). The newish AmE transitive type *he graduated Yale in 1984* is much more controversial and is best avoided. It should be noted that in AmE the word is not restricted to the completion of a university degree: it is also used for the completion of a high-school course.

**graece, -cè.** See ANGLICE.

**Graecism, Graecize, Graeco-.** Thus spelt in OUP house style, not *Grecism*, etc., but the forms with medial -*e*- rather than -*ae*- are customary in AmE. See also GRECIAN.

**graffiti,** which are such a prominent feature of our slogan-ridden age, is a plural word requiring a plural construction. Its singular is *graffito*. Before the 1960s it was mainly used by art historians and archaeologists of drawings or writing scratched on the walls of ancient classical ruins, such as those at Pompeii or Rome. In their work the (It.) singular form in -*o* and the pl. form in -*i* were unfailingly distinguished. The spray-can brigades of the period since the 1960s have brought the word dramatically to the attention of the general public, and *graffiti* is now often erroneously treated (cf. *confetti*) as a mass noun requiring a singular verb. There are encouraging signs that standard speakers are increasingly recognizing that one such slogan or message is 'a graffito', but it remains to be seen whether a clear number distinction between the two words can be established and maintained in English. The following examples from successive issues of the same American newspaper, the *Daily Northwestern* (Ill.), illustrate the competing types: (24 Oct. 1991) *A graffito in the fourth-floor women's bathroom in University library reads, 'I have been reading the same thing on these walls for three years and I'm bored!'*; (25 Oct. 1991) *The graffiti was left on a notepad hanging on the door of the unnamed professor's office.* Other examples: (graffiti treated as pl.) *May I add a word in praise of the London Library? I doubt whether any other lending library can claim that the graffiti in its books show such a concern for scholarship—TLS*, 1972; *The inscription on the fire engine in New College cloisters is written in Latin and so are some of the graffiti you find chalked on the walls of Oxford lavatories—J.* Morris, 1978; (graffiti treated as sing.) *That haunting graffiti inscribed on the approaches to Paddington station—Times*, 1980; *'I don't need drugs,' the T-shirt graffiti proclaims—Observer*, 1981; *Graffiti covers nearly every inch of the concrete barrier, its poignant messages seeming to fall on deaf ears—Gentleman* (Bombay), 1986; (graffito treated as sing.) *The sheer act of writing a graffito on a women's rest room wall means that the writer will put up with possible responses—Multilingua*, 1987; *A graffito on the Will*

*Rogers esplanade said: 'Welcome to the end of civilization.'*—R. Raynor, 1988.

**-gram.** 1 A long-standing suffix (ultimately representing Gk γράμμα 'something written') in many English words, including *cablegram*, *diagram*, *epigram*, *ideogram*, and *telegram*.

2 Since the late 1970s the suffix has combined with various words to produce a wide range of (mostly humorous) formations based on *telegram*, 'denoting a message delivered by a representative of a commercial greetings company, esp. one outrageously dressed to amuse or embarrass the recipient' (*OED* 2). Examples (some of which are respelt when used as proprietary terms) include *kissogram* (*Kissagram*), the best-known of them all; *Gorillagram*, *Mailgram*, *Rambogram*, and *strippergram* (*Strip-a-Gram*).

**grammar** is 'the study or rules of a language's inflections or other means of showing the relation between words, including its phonetic system; a body of forms and usages in a specified language (*Latin grammar*)' (*COD*, 1995). 'Words must be combined into larger units, and grammar encompasses the complex set of rules specifying such combination . . . We shall be using 'grammar' [in this book] to include both *syntax* and that aspect of *morphology* (the internal structure of words) that deals with *inflections* (or *accidence*)' (*CGEL*, 1985).

*Grammar*, in modern terms, is a branch of *philology* (the older term) or *linguistics* (the usual current term). Other components or branches of linguistics include: *etymology*: the origin, formation, and development of a word: OE *æcer* is the etymon of modE *acre*; *morphology*: deals with the internal structure of word-forms; the word *enrichment* consists of three *morphemes*, namely *en-* + *rich* + *-ment*; *orthography*: correct or conventional spelling; the study or science of spelling; *phonology*: the study of sounds in a language; *semantics*: the branch of linguistics concerned with meaning and change of meaning; *syntax*: the grammatical arrangement of words, phrases, and clauses, showing their connection and relation; *word-formation*: the way in which compound words are formed from simpler elements, e.g. *musicology* from *music* (noun) + the suffix *-ology*.

There is an immense gap between a simple presentation of grammar in a small book for children, in which, in the clearest possible manner, the parts of speech (noun, verb, adjective, adverb, preposition, etc.) are identified, and words, phrases, and sentences are differentiated from one another, and, on the other hand, the largest and most complex current work, *A Comprehensive Grammar of the English Language* (1985), a work bristling with difficult conceptual terminology and aimed at professional students of the language. Ideally every English-speaking person should begin to distinguish the several parts of speech at an early age, and continue to study the various aspects of the subject in a graduated manner throughout his or her time at school. An understanding of the way the language works is assisted by the learning of one or more foreign languages (e.g. Latin and French) and by the acquisition (an impossible dream!) of an outline knowledge of the history and nature of English from the eighth century to the present day.

**grammatical agreement.** See AGREEMENT.

**grammatical concord.** See AGREEMENT.

**grammatical gender.** In the earliest form of our language, called Old English or Anglo-Saxon (*c* 740 to 1066), nouns fell into three classes, masculine, feminine, and neuter, i.e. were distinguished by their grammatical gender. Thus *stān* 'stone' was masculine, *giefu* 'gift' was feminine, and *scip* 'ship' was neuter. The definite article and most adjectives varied to accord with the gender of the accompanying noun. This system was lost by the end of the 11c. by which time, among other factors, the distinctive unstressed vowel-endings, which were one of the mainstays of the system, had weakened to the neutral sound schwa, /ə/. Grammatical gender is an essential element of ancient languages like Latin and Greek, and also of numerous modern languages. The absence of grammatical gender in English is one of the reasons why English-speakers have considerable difficulty in mastering the major languages of Western Europe.

**gram(me).** The shorter form is now customary (abbreviated *g*, not *g.*, *gm*, or *gr*) for the sense 'fundamental unit of mass in the system of cgs (centimetre, gram, and second) units'. So *kilogram* ( = 1,000 grams), *milligram* (one-thousandth of a gram), *microgram* (one-millionth of a gram), etc.

**gramophone** (irregularly formed by inversion of *phonogram*), a late 19c. word, is now old-fashioned. The favoured current word is *record-player* (except that *phonograph* is still used in AmE for the same object).

**grand compounds.** It is recommended that the following current English compounds of family, etc., names be written thus (most of them in accord with *COD*, 1995): (ranks) *grand duchy, grand duchess, grand duke*; (family relationships) *grand aunt, grandchild, grandad, grand daughter, grandfather, grandma, grandmam(m)a, grand nephew, grand niece, grandpa, grandpapa, grandparent, grandsire* (an archaic word for *grandfather*), *grandson, grand uncle*; (miscellaneous) *grand jury, grand master* (chess), *Grand National, grand opera, grand piano, Grand Prix, grand slam* (sport, bridge), *grand stand, grand total, grand tour* (sightseeing tour). For some of the family terms a certain amount of latitude is permissible. e.g. *grand-aunt, grand dad, grand-uncle*. In all such terms an initial capital should be used at the beginning of letters (*Dear Grandpa*, etc.) and also in mid-letter (*Will you tell Grandma, please, that I loved her presents*). In chess circles *grandmaster* is very frequently written as one word.

**granny.** The more usual spelling (rather than *grannie*). See -IE, -Y.

**granted.** Like GIVEN, CONSIDERING, and one or two other words, *granted* can be used as a marginal preposition (*And, granted the initial assumptions . . . I think it stands the test*—A. White, 1965); absolutely (*Granted, it was not hard to interest a security man, who apart from a regular soldier had the most boring job on earth*—T. Keneally, 1985); and, usu. in the form *granted that*, as a subordinating conjunction (*Granted that he was somewhat motivated by a nonliterary consideration—the book is lengthily anti-Catholic—still I thought his deflation skillful and just*—Commentary (US), 1958).

**gratis.** The dominant standard pronunciation in BrE seems to be /ˈɡreɪtɪs/ and

in AmE /ˈɡræt-/. A third possibility, /ˈɡrɑː-tɪs/, is also used, esp. in the colloquial phr. *free, gratis, and for nothing*.

**gratuity.** When not used of a bounty given in certain circumstances to servicemen and servicewomen on their discharge from the armed forces, *gratuity* is a rather grand word for a 'tip'.

**gravamen** /ɡrəˈveɪmen/, a chief ground of complaint. Pl. *gravamina* /ɡrəˈvæmmə/.

**grave** (verb). (engrave, carve). Of the two pa.ppl forms *graved* and *graven* the second is much the commoner; but the verb itself is archaic except in so far as it has been kept alive in set phrases, esp. the biblical phrase *graven images*.

**gravel** (verb). The inflected forms are *gravelled, gravelling; gravelly* (in AmE the first two but not the third often written with a single -*l*-).

**gray.** A very common AmE variant of *grey*.

**greasy.** In its literal sense 'smeared with grease' always /ˈɡriːsɪ/; when applied to an objectionable person often /ˈɡriːzɪ/.

**great.** For *the great Cham, the Great Unwashed*, see SOBRIQUETS.

**greaten.** A stock example of a word that was part of the everyday language from the 17c. to the 19c. (see the substantial entry in the *OED*), but which has virtually passed out of use in the 20c. For the type, see -EN VERBS FROM ADJECTIVES.

**Grecian.** This adj. has steadily retreated before *Greek*, and is now idiomatically restricted to architecture, facial outline (esp. *Grecian nose*), and a soft low-cut slipper. Otherwise *Greek* is the natural word (*Greek alphabet, calends, chorus, fire*, etc.).

**Grecism,** etc. See GRAECISM.

**Greek g.** Words containing g derived from Greek gamma are of three kinds: **1** The small class in which g is silent in English (*gnome* aphorism, *gnomic, gnomon* (on sundial), *gnosis, gnostic; diaphragm, paradigm, phlegm*).

**2** A larger class (before a consonant or the vowels *a, o*) in which it is always 'hard', i.e. /ɡ/ (*glaucoma, Greek; galaxy, demagogue, gonorrhoea*).

**3** A large class of words in which the g happens to fall before *e*, *i*, or *y*. The English natural tendency to pronounce g 'soft', i.e. as /dʒ/, in such cases has been partially prevented by standard speakers who are familiar with the derivation of the words in question. In the following list the dominant current pronunciation is given when possible.

| | |
|---|---|
| *anthropophagi* | /g/ |
| *antiphlogistic* | /dʒ/ |
| *autogyro* | /dʒ/ |
| *demagogic* | /-ˈgɒgɪk/ |
| *demagogy* | /-ˈgɒgɪ/ |
| *geriatric(s)* | /dʒ/ |
| *gerontology* | initial /dʒ/, final /-dʒ-/ |
| *gymnastic(s)* | /dʒ/ |
| *gynaecology* | /g/ |
| *gypsophila* | /dʒ/ |
| *hegemony* | /g/ or /dʒ/ |
| *hemiplegia* | /dʒ/ |
| *isagogic* | first g hard, second g /dʒ/ |
| *laryngitis* | /dʒ/ |
| *meningitis* | /dʒ/ |
| *misogynist* | /dʒ/ |
| *misogyny* | /dʒ/ |
| *paraplegia* | /dʒ/ |
| *paraplegic* | /dʒ/ |
| *pedagogical* | /-ˈgɒdʒ-/ or /-ˈgɒg-/ |
| *pedagogy* | /-ˈgɒdʒ-/ or /-ˈgɒg-/ |
| *pharyngitis* | /dʒ/ |
| *phlogiston* | /dʒ/ |
| *syzygy* | /dʒ/ |

Note: *gynaecology* and its derivatives were very commonly pronounced with initial /dʒ/ in the first third of the 20c., but from about 1930 only with /g/. See also ANAGOGE.

**greenhouse effect.** This important but much misunderstood term is defined by the *New SOED* as follows: the heating of the surface and lower atmosphere of a planet due to the greater transparency of the atmosphere to visible radiation from the sun than to infrared radiation from the planet.
Cf. *global warming* (s.v. GLOBAL).

**greenness.** So spelt.

**Green Paper.** A publication issued by a British government department setting out various aspects of a matter on which legislation is contemplated, and inviting public discussion and suggestions. In due course it may be followed by a WHITE PAPER, giving details of proposed legislation. The first Green Paper was issued in 1967. (*Hutchinson Encycl.*, 1990).

**greenth.** Coined by Horace Walpole in 1753, it has never established itself in the language except in literary contexts or jocularly. See -TH NOUNS.

**grey.** In an unusually long note the *OED* (1901) says 'an enquiry by Dr. Murray in Nov. 1893 elicited a large number of replies, from which it appeared that in Great Britain the form *grey* is the more frequent in use, notwithstanding the authority of Johnson and later Eng. lexicographers who have all given the preference to *gray*'. In Britain a century later, *grey* is the only current form. See GRAY.

**greyhound.** The word goes back to Anglo-Saxon times, and it is known that the first element is unconnected with *grey*. Its ultimate etymology is uncertain, though it seems to answer to an ON *grøy* meaning 'bitch'. The word is often spelt *grayhound* in AmE.

**gridiron, griddle, grid.** It would be natural to assume, perhaps, that *grid* was the original word, *griddle* its diminutive, and *gridiron* a compound of it with *iron*. But the historical record in the *OED* reveals that *griddle* (13c., from OF, first in form *gredil(e)*) and *gridiron* (13c., first in form *gredire*) were the original terms; that *gridiron* was at an early date (14c.) formed by analogy with *fire-iron* and the simplex *iron*; and that *grid* (19c.) is a quite recent back-formation from *gridiron*.

**grievous** /ˈgriːvəs/. One occasionally hears the word lamentably pronounced as /ˈgriːvɪəs/, rhyming with *devious*, *previous*, etc. Those who say the word thus are also inclined to write it down as *grievious*.

**griffin, griffon, gryphon.** *Griffin* (also spelt *griffon*, and, e.g. by Lewis Carroll, *gryphon*) is 'a fabulous creature with an eagle's head and wings and a lion's body'. *Griffon* is (*a*) a dog of a small terrier-like breed, and (*b*) (in full *griffon vulture*) a large vulture. All three are variants of the same OF (ultimately Gk) word.

**grill, grille.** In BrE a *grill* is a device on a cooker for radiating heat downwards (those on the top of a cooker being called *rings*). The word also means a meal (e.g. *a mixed grill*) cooked over or under direct heat. A *grille* (less commonly spelt *grill*) is a metal grid protecting the radiator of a motor vehicle, or a latticed screen separating customers from staff (e.g. in a bank).

**grimace.** Most commonly stressed on the second syllable, esp. as verb, but with sections of the general public stressing both the noun and the verb on the first syllable. AmE seems to favour first-syllable stressing, but both stressings occur.

**grimy.** So spelt, not -*ey*. See -EY AND -Y IN ADJECTIVES 2.

**groin** is a physiological term, and *groyne* (AmE *groin*) is a low wall or timber framework built out from a sea-shore to prevent beach erosion.

**grosbeak** /ˈɡrəʊsbiːk/. Note that the *s* is fully pronounced.

**grotto.** Pl. *grottoes*. See -O(E)S 1.

**group names of animals** (a pride of lions, etc.). See PROPER TERMS.

**group possessives.** See APOSTROPHE[2] D7.

**grovel.** The inflected forms are *grovelled*, *grovelling*; in AmE often written with a single -*l*-. See -LL-, -L-.

**grown.** For a *grown man*, see INTRANSITIVE PAST PARTICIPLES.

**groyne.** See GROIN.

**gruelling** is the spelling in BrE, but usu. *grueling* in AmE.

**gruntled.** See DISGRUNTLED.

**Gruyère** (name of district in Switzerland). The cheese-name is thus spelt.

**gryphon.** See GRIFFIN.

**guano** (fertilizer). Pronounce /ˈɡwɑːnəʊ/. Pl. *guanos*. See -O(E)S 3.

**guarantee, guaranty.** 'Fears of choosing the wrong one of these two forms are natural, but needless. As things now are, -*ee* is never wrong where either is possible' (Fowler, 1926). The advice is still sound. For the verb the only form used is *guarantee*. For the noun the narrowness of the distinction between them may be judged from the definitions in *COD* 1995:

*guarantee*, a formal promise or assurance, esp. that an obligation will be fulfilled or that something is of a specified quality and durability; a document giving such an undertaking; *guaranty*, a written or other undertaking to answer for the payment of a debt or for the performance of an obligation by another person liable in the first instance; a thing serving as security for a guaranty. *Guarantee* is by far the more common of the two terms, *guaranty* being mostly restricted to legal and commercial contexts. See WARRANTY.

**guarantor,** a person, bank, etc., that gives a guarantee or guaranty.

**guerrilla** is the recommended spelling in English (thus, as it happens, according with that in Spanish), not *guerilla*. Pronounce /ɡəˈrɪlə/, the same as *gorilla*.

**guess.** In the late 18c. in America, the informal use of *I guess* meaning 'I think it likely, I suppose' became marginally detached from standard English uses of the phrase (which in the UK meant essentially 'it is my opinion or hypothesis (that)'. Throughout the 19c. and for much of the 20c. its Americanness was quite marked and was frequently commented on. But at some point in the 20c. the association of the colloquial use of the phrase with AmE alone ceased to be sustainable. Colloquial uses of *I guess* may now occur anywhere in the English-speaking world. Examples: *No, I guess it was Candy I told*—J. Steinbeck, 1937; *'You're too romantic,' he said … 'I guess I am.'*—D. Eden, 1955; *No, I guess I don't look at him very much*—G. Vidal, 1955; Martha. *You remember them now?* George. *Yes, I guess so,* Martha.—E. Albee, 1962; *I guess it takes a long time to grow up*—M. Sarton, 1978; *I guess you're supposed to think to yourself that you're in a garden*—R. Ingalls, 1985.

**guest.** The expanding circle of meaning round the word can be seen from the following uses: (no payment involved) a person invited to one's home to stay or for a meal, etc.; (payment involved) a paying guest at a hotel, motel, etc.; (wages paid) *guest worker* (e.g. Filipinos in Iraq, Turks in Germany); (payment occas. involved) acting as a guest (*guest speaker*); (payment due) a person who takes part by invitation in a radio or television programme (*guest artist*); and various other uses (for which see any standard dictionary of reasonable size).

**guild.** See GILD (verb).

**guillemot.** Pronounce /ˈgɪlɪmɒt/.

**guilloche** (architectural, etc., ornament). Pronounce /gɪˈlɒʃ/.

**guillotine** (in parliament). See CLOSURE.

**guimp.** See GIMP.

**guipure** (linen lace). Pronounce /gɪˈpʊə/.

**gulden** ( = Dutch guilder). Pronounce /ˈgʊldən/.

**gulf, bay.** Apart from the fact that each has some senses entirely foreign to the other, there are the differences (1) that *gulf* implies a deeper recess and narrower width of entrance, while *bay* may be used of the shallowest inward curve of the sea-line and excludes a landlocked expanse approached by a strait; and (2) that *bay* is the ordinary word, while *gulf* is chiefly reserved as a name for large or notable instances (*the Gulf, the Persian Gulf*).

**gullible** (easily cheated). Thus spelt, not *-able*.

**gumma** /ˈgʌmə/ (soft swelling, usu. of syphilitic origin, in connective tissue). Pl. either *gummas* or *gummata*.

**gunwale** (upper edge of ship's side). Pronounce /ˈgʌnəl/. Sometimes spelt *gunnel*.

**gurgoyle.** See GARGOYLE.

**Gurkha.** Thus spelt.

**gusseted** (adj.). Thus spelt. See -T-, -TT-.

**guts.** Not yet as neutral a word as 'courage' or 'determination', but frequently used, with an edge to the voice, in contexts where these synonyms, in the opinion of the speaker, would not quite have sufficed. The idiomatic phrases *to hate a person's guts* (dislike a person intensely) and *to sweat* (or *work*) *one's guts out* (work extremely hard) also attempt to double-underline the level of disliking and the intensity, length, etc., of the work (being) done. A well-known example of the word used in its main figurative sense: *I hate the idea of causes, and if I had to choose between betraying my* country and betraying my friend, I hope I should have the guts to betray my country— E. M. Forster, 1951.

**gutta-percha** /ˌgʌtəˈpɜːtʃə/. Note /-tʃ-/ not /-ʃ-/.

**guttural** (from L *guttur* 'throat'). **1** Thus spelt.

**2** In lay language, '(of a consonant) produced in the throat or by the back of the tongue and the soft palate', and applied particularly to certain throaty sounds in the German language. Phoneticians do not use the term, but divide these sounds into *pharyngeals* (e.g. the *h* in Arabic *Ahmed*) and *velars*. Thus Eng. *k* is a voiceless velar plosive and *g* is its voiced equivalent. The voiced velar fricative /ɣ/ is found in N. Ger. *sagen* /ˈzaɣən/ 'to say'; and the voiceless velar fricative /x/ is found in Ger. *ach* 'oh' and Sc. *loch* 'lake'. (Ger. *ich* 'I' has the voiceless palatal fricative /ç/.) The general public will doubtless continue to call the unfamiliar German and Scottish sounds 'gutturals', spelt, one hopes, correctly (not *gutterals*).

**guy** (noun). The long-standing use of the word to mean an effigy of Guy Fawkes in ragged clothing burnt on a bonfire on 5 November remains in force in Britain, and, for example, in Australia and New Zealand, though the custom itself is beginning to fade. As this use recedes, *guy* meaning 'a man, a schoolboy, a fellow' is now spreading rapidly from America to other English-speaking areas. In young people's parlance it has virtually become the customary term (replacing *chap* and other synonyms) for 'any male person', and it is sometimes also applied to females.

The use established itself in NAmer. towards the end of the 19c., and has made steady progress in Britain, and e.g. in Australia and New Zealand, in the course of the 20c. Some examples: (of males) *Take a look at them. All nice guys. They'll finish last*—L. Durocher, 1946 (US); *Half jokingly, he* [sc. Bing Crosby] *asked that his epitaph read, 'He was an average guy who could carry a tune'*—Newsweek, 1977; *I'm just as romantic as the next guy, and always was*—J. Lennon 1980; [He] *transferred his hands to his mouth before departing ... towards the door marked 'Guys'*—D. Francis, 1984; *You see guys from ... Westlake Boys* [School] *hassled by other guys*—North & South

(NZ), 1989; *I could see John by the bar talking to some guys—New Yorker*, 1989, (of females, esp. in pl. phr. *you guys*; chiefly NAmer.) *She was a regular guy, a good sport and a fine actress—*1982 in *American Speech, 1983*; *The new meaning … which still may evoke masculine images in some women, is nonetheless positive, especially since* guys *permits one to refer to numbers of a group without reference to their sex—Word*, 1985; [a man addressing his two daughters] *You guys are dwelling and brooding. A great thing right now would be exercise—New Yorker*, 1993.

**gybe** (noun and verb). The sailing term is spelt thus in BrE, but *jibe* in AmE.

**gymnasium. 1** Pl. (for preference) *gymnasiums*; but *gymnasia* is also widely used. See -UM 3.

**2** In Germany (pronounced /ɡɪmˈnɑːziəm/) and Scandinavia the word has an additional sense, 'a school that prepares pupils for university entrance'.

**gymp.** See GIMP.

**gyn(a)ecology. 1** See GREEK G (esp. the note).

**2** Spelt with -*ae*- in BrE and -*e*- in AmE.

**gyp,** as the name of a college servant, belongs to Cambridge and Durham, not Oxford (where the regular term is *scout*).

**gypsy, gipsy.** The recommended spelling now is *gypsy*, or, if one sees any magic in initial capitals in the names of formerly uncapitalled ethnic groups, *Gypsy*. The spelling gradually being ousted is *gipsy* (which was given priority in the *OED*, 1901). The derivatives are *gypsydom, gypsyfied, gypsyish*, etc. (or *G*-). A gypsy is 'a member of a wandering race (by themselves called *Romany*) of Hindu origin, which first appeared in England about the beginning of the 16th c. and was then believed to have come from Egypt' (*OED*). They are now dispersed throughout Europe, N. America, and elsewhere.

**gyre** (noun and verb). The standard pronunciation of the word is /dʒaɪə/, with an initial 'soft' g, but apparently gramophone records survive in which the poet W. B. Yeats recites his well-known lines with a 'hard' g in the word: *To watch a white gull take A bit of bread thrown up into the air; Now gyring down and perning there* (1920); *O sages standing in God's holy fire … Come from the holy fire, perne in a gyre* (1928).

**gyves** (fetters, shackles). This archaic word is pronounced /dʒaɪvz/ with a 'soft' initial g.

# H h

**h.** **1** The presence or absence of a fully pronounced voiceless glottal fricative /h/ (as it is technically called) in ordinary words like *have, house, ahead,* and *behave* 'has come to be regarded [in Britain] as a kind of shibboleth of social position' (*OED*). The Archbishop of Canterbury, George Carey, (*Time*, 2 Sept. 1991) described how he had been brought up in a working-class family in the East End of London. 'At the age of 17½,' he said, 'I discovered the letter *h* in the English language, which, you know, isn't much known among the English working class.' In practice large numbers of people in Britain go through life making no phonetic distinction between such pairs of words as *hedge/edge, hill/ill,* and *high/eye,* and are seemingly untroubled by those speakers who carefully distinguish them.

**2** Some phonetic curiosities. While it is true that /h/-lessness is characteristic of uneducated speech, standard speakers do not always notice that certain function words (e.g. *has, have, had*), pronouns, and pronominal adjectives lose /h/ in unstressed non-initial contexts in rapid speech: *she shoved him into her car* /ʃi: ʃʌvd m mtʊ ɜː kɑː/, *you should have heard him* /ju: ʃʊd əv hɜːd m/. Until about the beginning of the 20c., words containing *wh* (*where, whistle; nowhere,* etc.) were regularly pronounced with the /hw/ intact by a majority of RP speakers in England, and regularly too in Scotland, Ireland, and America. But *COD* 1995, by entering only /w/ for all such words, draws attention to the fact that the aspiration of *h* in such circumstances has largely disappeared from standard English in England. As a result, *whales* and *Wales* are both pronounced /weɪlz/, *whit* and *wit* /wɪt/, and so on. The *h* is silent in *heir* /eə/, *heiress, honest, honour,* and *hour,* and in AmE *herb* /ɜːrb/ (but in BrE /hɜːb/). On the use or non-use of aspiration in such cases as *a(n) habitual, a(n) historic,* see A, AN 3. See also AITCH; ANNIHILATE; HUMOUR; VEHEMENT, VEHICLE.

**habiliments.** See POLYSYLLABIC HUMOUR.

**habitude.** In continuous regular use since the 15c. in a variety of senses in which it competed with *habit* (manner of being or existing, a customary or usual mode of action, etc.), it has fallen on lean times and is now rarely used. Examples: *the slim figure of the orphan out of habitude from old times came close to the Virgin to whisper to her* [etc.]—H. Allen, 1933; *Shows them the main-spring of the play, The moral tragedy revealing, And country habitudes, country ways*—A. Cronin, 1979.

**habitué** (an habitual visitor or resident). Print in italics; but pronounce /həˈbɪtjʊeɪ/ in an Anglicized manner.

**háček** /ˈhætʃek/. The regular (but in Britain little-known) term for the diacritical mark ˇ placed over letters to modify the 'natural' sound in some Slavonic and Baltic languages. It is a Czech diminutive of *hák* 'hook'.

**hackneyed.** So spelt.

**hackneyed phrases.** It is impossible to draw a firm line between hackneyed phrases and clichés: they are genera of the same species. They are at the same time endearing and irreplaceable, and maddening and replaceable. Used sparingly they pass unnoticed, or even add a touch of quality, in novels, short stories, journalism, and some other kinds of writing. In passages that already lack any element of excitement or real interest they are unendurable. In obituaries they flourish like moss: *he did not suffer fools gladly; he had the defects of his qualities* (see below).

Many hackneyed phrases are mere allusions to famous passages in literature or politics: *to be or not to be* (Shakespeare); *not with a bang but a whimper* (T. S. Eliot); *the wind of change* (H. Macmillan). Many others, equally current, are drawn from possibly less well-known or half-remembered sources: *the cups that cheer but not inebriate* (Cowper); *a consummation devoutly to be wished* (Shakespeare's *Hamlet*);

a fit audience though few (cf. Milton's still govern thou my song, Urania, and fit audience find, though few); the feast of reason and the flow of soul (Pope); conspicuous by his absence (from a speech of 1859 by Lord John Russell, Among the defects of the Bill, which were numerous, one provision was conspicuous by its presence and another by its absence); he had the defects of his qualities (cf. Vauvenargues has the defects of his qualities, J Morley, 1878). Some are of Biblical origin: Is there no balme in Gilead? (Jer. 8: 22); For the king of Babylon stood at the parting of the way (Ezek. 21: 21); For ye suffer fooles gladly, seeing ye your selues are wise (2 Cor. 11: 19). A diminishing number are drawn from classical sources: O tempora, O mores! (O what times, O what habits!—Cicero); Hoc genus omne (All their kith and kin—Horace); in vino veritas (Truth comes out in wine—Pliny).

These are but a small selection from a virtually limitless list. Cf. CLICHÉ; LITERARY ALLUSIONS.

**had.**

| 1 had better.
| 2 had have.
| 3 had rather.

**1** had better. See BETTER 1, 2.

**2** had have. 'In the 15th and 16th c. (and later U.S. dialect) occur many instances of redundant have, had, in the compound tenses' (OED have 26; the use is labelled catachrestic): e.g. Had not he have be [ = been], we shold never have retorned—Malory, 1470–85; The sayd kyng had not so sone have returnyd—J. Style, 1509; 'If the fire hadn't have gone out,' he mused—J. F. Wilson, 1911 (US). The matter has been thrashed out in the journal English Today (1986), and the main conclusions are that the construction represents the type 'unreal conditions in the past', and that it is often written with either had or have in contracted form (thus If I'd have gone or If I had've gone). 'There is a problem with past unreal,' commented Professor Frank Palmer, 'because it needs to mark past tense twice, once for time and once for unreality. English does not have past-past forms, so instead introduces the verb "have".' The several contributors agreed in thinking that the construction was now common. One thought that it was moving from nonstandard speech into middle-class use in Britain. One contributor called it the

plupluperfect, another the pluplupast. Another pointed out the type 'had + schwa + past participle' as shown in the first part of a sentence in Galsworthy's Strife (1909), If we'd a-known that before, we'd not a-started out with you so early. There is a danger that this a- might be regarded as a shortening of of. Indeed, WDEU cites an example of just that from F. Scott Fitzgerald's The Great Gatsby (1925): It was four o'clock in the morning then, and if we'd of raised the blinds we'd of seen daylight. For those with access to DARE, numerous regional American examples of the various types are given s.v. have v[1] and v[2]. In view of all the warning signs, it must be sensible to avoid using constructions of the type 'had have + past participle' except in informal speech.

**3** had rather. The type I had rather (or in its reduced form I'd rather) is as idiomatic as I would rather/I'd rather. In historical terms it was formed on the analogy of the now archaic type I had liefer 'I should hold it dearer': I had rather err with Plato than be right with Horace—Shelley, 1819; I had rather gaze on a new ice age than these familiar things—J. Winterson, 1985.

**haemo-, hæmo-, hemo-,** a combining form from Gk αἷμα 'blood'. In BrE usu. written as haemo- (the digraph æ has now been virtually abandoned), and in AmE as hemo-. Hence haemoglobin/US hemoglobin. See Æ, Œ.

**haemorrhage.** Thus spelt in BrE (note the two rs), but as hemorrhage in AmE. See Æ, Œ.

**hagiography, hagiolatry, hagiology.** All are pronounced with initial /hæg-/, i.e. with a 'hard' g. See GREEK G.

**hail-fellow-well-met** (adj. phr.). Thus written (with three hyphens).

**hair-brained** is an erroneous variant (first recorded in 1581 and still occas. found) of hare-brained.

**hairdo.** Pl. hairdos.

**hale,** used esp. in the phr. hale and hearty, is not to be confused with hail in hail-fellow-well-met.

**half. 1** It needs to be borne in mind, first, that half is used as several parts of speech: as an adj. or predeterminer (immediately preceding another determiner), e.g. *half the audience, half that year, at least half his correspondence*; (also in partitive constructions in which *of* divides *half* and the determiner) *half of the Sikhs, half of the country*; as a noun, *the second half of the game, a year and a half, the first half of the year, one half of me,* (colloq.) *you don't know the half of it*; as a pronoun, *he'd eaten half of the cake*; and as an adverb (often in combinations), *he closes the curtains half way across, managed to half stammer the purpose of our visit, they were half asleep.*

2 When *half (of)* is used with a singular or non-count noun it is followed by a singular verb or pronoun: *half of the country is employed in agriculture; she took half a bag of mushroom compost and spread it under the roses.* When *half (of)* is followed by a plural noun a plural verb (or plural pronoun) is required: *half (of) the members of Nato have been deploying military aircraft; half the folks round here have never taken their children south of Nottingham.* Occasionally, by the principle of 'notional agreement', the type 'half *(of)* + non-count noun' can properly be used with a plural verb: *Nearly half (of) the population lose at least half their teeth before they reach the age of 40.*

3 The positioning of *half* in phrases concerned with quantities, measures of time, etc., is for the moment more or less fixed, though there has been considerable variation in such phrases in former centuries: e.g. *half an hour from now* (not *a half-hour*); *the next half an hour* (not *the next half-hour*); *a year and a half* (not *one and a half years*); *almost half a million dollars* (not *a half-million dollars*); *one and a half million pounds* (not *one million and a half pounds*); *about half a ton* (not *about a half-ton*). Exceptions occur: e.g. *A half-dozen seagulls cover the crown of a leafless Weeping Ash*—Maggie Gee, 1985.

4 *half seven.* Many young people in England and Wales are now freely using the type *half seven* ( = 7.30 a.m. or 7.30 p.m.) in imitation of Scottish and Irish usage. The standard type is still *half-past seven.* (The 17–19c. Sc. use of *half seven* = 6.30 is now almost obsolete, according to the latest Scottish dictionaries.)

5 The pl. of *half* is, of course, *halves* (e.g. *both halves of the game were played in poor weather*). But it is more usual to say e.g. *she cut the apple in half* (rather than *in halves*). The same 'rule' does not apply to *thirds, quarters,* etc.

6 A tendency in some quarters to place a redundant *a* in contexts of the type *a private meeting which lasted a half an hour* should be resisted; *half an hour* is all that is required.

7 *better half* = wife. See WORN-OUT HUMOUR.

8 *half-weekly, -yearly,* etc. These are to be preferred to the notionally equivalent *biweekly, biyearly,* etc. (which are ambiguous: see BI-).

9 *halfpennyworth.* Survives, since the disappearance of the coin, as a colloquial term for 'a negligible amount' (esp. after a negative: *doesn't make a halfpennyworth of difference*). Pronounced /ˈheɪpəθ/, and sometimes written as *ha'p'orth.*

10 For *half-breed, half-caste,* see MULATTO 2.

11 *half-brother, -sister.* See -IN-LAW 3.

**half-past.** It passes almost unnoticed that in ordinary conversation standard speakers frequently say /ˈhʌpɑːst/ when a numeral of time follows immediately, i.e. with reduction of /ˈhɑːf-/ to /ˈhʌ-/. See also HALF 4.

**half-rhyme.** See RHYME 2.

**hallelujah.** See ALLELUIA.

**halliard.** See HALYARD.

**hallo.** See HELLO.

**Hallowe'en** /hæləʊˈiːn/. Best spelt (with ') and pronounced (*w* omitted) thus.

**halm.** See HAULM.

**halo.** Pl. *haloes.* See -O(E)S 1.

**halyard** (nautical). Thus spelt, not *halliard, haulyard.*

**hamstrung, hamstringed.** The two forms competed with each other as the pa.t. and pa.pple of the verb *hamstring* between the 17c. and the 19c., but the only form now used is *hamstrung.*

**hand.** 1 *hand in glove.* To be *hand in glove* with someone means to be in collusion or association with someone. The phr. is mostly in predicative use (*The two of them* *were hand in glove from the beginning).* Earlier (17–19c.) *hand and glove* was perhaps the more common expression with the same meaning.

**2** *at close hand.* The idiomatic phr. *at hand* 'close by; about to happen' is sometimes expanded by the insertion of *close*, perhaps under the influence of *at close quarters:* thus *at close hand.* But *close at hand* is more idiomatic.

**handful.** Pl. *handfuls.* See -FUL.

**handicap.** The inflected forms are *handicapped, handicapping.* See -P-, -PP-.

**handkerchief.** The recommended pl. is *handkerchiefs,* not *-chieves.*

**handsel, hansel** (New Year gift). The first form is recommended. The inflected forms of the verb are *handselled, handselling:* see -LL-, -L-.

**handsome.** Applied equally to men and women of striking appearance ('beautiful with dignity', Dr Johnson, 1755) since the 16c., but now, when applied to women, tending to be used only of such as are middle-aged or elderly. Examples: *Today good-looking ... Tomorrow they'll be talking about what a* handsome *woman she is*—T. Wolfe, 1987; *Mrs Gorbachev is a stylish,* handsome *woman with an intense and intelligent manner*—Time, 1988; *Barbara Bush is what she is: 63. Sturdy. White-haired ... Handsome. (That's the word that reporters, struggling for an adjective to describe a woman who dares not to be glamorous, have unanimously settled on.)*—Chicago Tribune, 1988.

**hang** (verb). Use *hanged* as pa.t. and pa.pple of capital punishment and in imprecations; otherwise *hung.* The distinction goes back ultimately to the existence of two OE verbs (*hōn,* pa.t. *heng,* pa.pple *hangen;* and *hangian,* pa.t. *hangode,* pa.pple *hangod*) and an ON one (*hengjan,* a causal transitive verb). A full account of the subsequent history and entanglements of these three verbs in ME and in later centuries is set down in the *OED:* in brief, *hung* became established in literary English in the late 16c., with *hanged* largely restricted to the sense 'kill by hanging' and in imprecations of the kind *I'll be hanged if I'll do that.* In practice, a wide range of writers use *hung* in descriptions of executions.

This use is not erroneous but just less customary in standard English.

**hangar.** First recorded in Thackeray's *Henry Esmond* (1852) meaning 'a shed or shelter for carriages', and in 1902 in the sense 'building for housing aircraft', it was for long pronounced /ˈhaŋgar/, then /ˈhæŋgə/, i.e. with the medial *g* pronounced as in *finger* and *linger,* but at some point in the 20c. (presumably after 1926 because Fowler favoured the older pronunciation) only /ˈhæŋə/, i.e. as a homonym of *(coat-)hanger.*

**hanger-on.** The pl. is *hangers-on.*

**hanging participles.** See UNATTACHED PARTICIPLES.

**haply.** See WARDOUR STREET.

**hara-kiri** /ˌhærəˈkɪrɪ/. Thus spelt.

**harass.** **1** Nothing is more likely to displease traditional RP speakers in Britain than to hear *harass* pronounced with the main stress on the second syllable, i.e. as /həˈræs/. John Walker (1806), the *OED* (1899), Daniel Jones (1917), *COD* (7th edn, 1982) gave only first-syllable stressing, and that is the pronunciation I recommend. But second-syllable stressing (also in *harassment*) is dominant in AmE and, since about the 1970s, seems to be becoming the pattern favoured by the younger generation in Britain. It is hard to tell how long the traditional line can withstand the assault being made on it.

**2** Note that the word is spelt with one *r.*

**harassedly.** 'A bad form,' said Fowler in 1926. That is as may be. But it is hard to see whether it should be pronounced as four syllables or as three. See -EDLY. In the circumstances it is probably best to avoid using the word altogether.

**harbinger.** Pronounce /ˈhɑːbɪndʒə/.

**harbour.** **1** See PORT.

**2** Thus spelt in BrE, but as *harbor* in AmE. See -OUR AND -OR.

**hardly.**

**1** *hard* and *hardly.*
**2** *hardly ... than.*
**3** *without hardly.*
**4** *no — hardly.*
**5** *can't hardly.*

**1** *hard* and *hardly*. Except in the senses 'scarcely' (*we hardly knew them*) and 'only with difficulty' (*he could hardly speak*), the idiomatic adverb corresponding to the adj. *hard* is *hard*, not *hardly*. Examples: (hard) *How hard these actors worked*—Robertson Davies, 1986; *far from feeling relief at her death, my father took it hard*—A. Munro, 1987; (*hardly* = scarcely) *References to the pastoral convention can hardly excuse the general listlessness*—English, 1987; *captions are brief and hardly ever interpretative*—*TLS*, 1987; (used punningly) '*I hardly think*,' *was Digby's absolute favourite phrase and probably the most accurate thing he would say in his entire political life. Digby did hardly think, but this wasn't how he meant it*—B. Elton, 1991. Fowler (1926) cited some unattributed (?newspaper) examples in which *hardly* (used instead of *hard*) would reverse the sense: *For attendance on the workhouse he receives £105 a year, which, under the circumstances, is hardly earned; It must be remembered that Switzerland is not a rich country, and that she is hardly hit by the war; The history, methods, and hardly won success of the anti-submarine campaign.* Such traps should be sidestepped at all costs.

**2** *hardly . . . than*. This, and the parallel constructions *barely . . . than* and *scarcely . . . than*, which have arisen by analogy with *no sooner than*, are labelled by the *OED* (s.v. *than* conj. 3d) with the condemnatory sign ¶. The examples cited are: *He had scarcely won for himself the place which he deserved, than his health was found shattered*—Froude, 1864; *Hardly had the Council been re-opened at Trent . . . than Elizabeth was allying herself with the Huguenots*—F. W. Maitland, 1903. Jespersen (1909–49, vol. vii) added three more examples ('in vulgar or half-vulgar speech') from the works of minor 20c. writers. Fowler himself judged the construction to be 'surprisingly common' and cited several examples: *The crocuses had hardly come into bloom in the London parks than they were swooped upon by London children; Hardly has Midsummer passed than municipal rulers all over the country have to face the task of choosing new mayors.* In post-Fowlerian times both *hardly . . . than* and (esp.) *hardly . . . when* constructions have continued in use. Examples: (*hardly . . . than*) *Hardly were they past the carrier than two Corsairs 'scrambled' off the deck to 'intercept an enemy plane'*—Daily Tel., 1944; *Hardly had the chalky jet stream dissipated above the horizon*

*than it was time for another jetaway get-away to points west*—American Square Dance, 1991; (*hardly . . . when*) *We had hardly sat down in the lokanta when a gendarme captain swooped on me*—D. C. Hills, 1964; *Hardly had the two children been freed when they* [sc. a rescue team] *were on the spot, having covered the ground in a snow-tractor*—Country Life, 1971; *There was hardly a year when the winter ploughs did not turn up an old hunter . . . crouched still in his cold stone-kist*—D. K. Cameron, 1980; *but hardly a day goes by when you don't think about . . . bonfires, walks on the beach, and teaching the kids to swim*—New York Mag., 1990. From the evidence before me the standard construction of the two is *hardly . . . when*; but it can only be a matter of time before the *OED* needs to append a usage note to its catachrestic sign s.v. *than* conj. 3d. Meanwhile *hardly than* can easily be avoided.

**3** *without hardly*. See WITHOUT 4.

**4** *no — hardly*. An unacceptable construction: *There is no industry hardly which cannot be regarded as a key industry.* It should read *There is hardly any industry which* [etc.].

**5** *can't hardly*. Described by the *OED* as 'Formerly sometimes (as still in vulgar use) with superfluous negatives', and illustrated by a 'modern' (i.e. 1890s) example: *I couldn't hardly tell what he meant.* The vulgar use persists: *But now we really got a beast, though I can't hardly believe it*—W. Golding, 1954 (Piggy speaking); *Mom just loaded us up with fried chicken and I can't hardly walk*—New Yorker, 1988; *I'm so mad with him I can't hardly talk to him*—M. Golden (US); *No, Sweedon can't write anything. He can't hardly write his own pawn-tickets*—P. Fitzgerald, 1990.

**hard words.** 'Why does Anita Brookner use hard words like "rebarbative" and "nugatory"?', I heard someone ask recently at a cocktail party. One possible answer is that the famous novelist does not regard them as 'hard'. Candia McWilliam (author of *A Little Stranger*, etc.) explained to *Sunday Times* readers in 1989, answering the charge that her vocabulary was élitist, 'Well, yes, in the sense that they [sc. words] can be demanding, exigent and revealing.' In Peter De Vries's novel *Consenting Adults* (1981) a character declares 'I took piccolo lessons, and learned the meaning of words like putative and adumbrate and simulacrum.' Clearly there is an awareness that there are bands of vocabulary that lie outside a particular person's cognizance at a given moment.

For my own amusement I set down a few 'hard' words encountered in my recent reading and checked them in two desk dictionaries of current English, one British (*COD*) and the other American (*Random House Webster's College Dictionary*). The dictionaries had made very similar choices about inclusion:

|  | COD 1990 | RHWCD 1991 |
|---|---|---|
| *claustration* | × | × |
| *edulcoration* | × | × |
| *eidolic* | × | × |
| *idoneous* | × | × |
| *infraction* | √ | √ |
| *straticulate* | √ | × |
| *tergiversation* | √ | √ |
| *velleity* | √ | √ |

I have little doubt that Anita Brookner, Candia McWilliam, and Iris Murdoch would not hesitate to use them. Other writers would hold back. The great majority of us would be left stranded somewhere in the middle.

It is worth reflecting that the first English dictionary, Robert Cawdrey's *A Table Alphabeticall of Hard Usual English Words* (1604), contained a mixture of 'hard' and of 'usual' words. A typical sequence of twelve words chosen at random contains only 'usual' words: *halaluiah, hallucinate, harmonie, hautie* ( = haughty), *hazard, herault* ( = herald), *heathen, hebrew, hecticke, hemisphere, helmet, hereditorie* (Cawdrey was not concerned with strict alphabetical order). But there are some 'hard' words among the 2,500 words listed and glossed, e.g. *agnition* (acknowledgement), *carminate* (to card wool), *combure* (to burn up), *deambulation* (a walking abroad), *enarration* (exposition), *illiquinated* (unmelted), etc., all of them now lost from the language. Nothing changes, does it?

**harebrained.** Thus spelt (or with hyphen, *hare-brained*), not *hair-*.

**harem.** The established spelling in English of this Arabic word. In BrE the pronunciation varies between /'hɑːriːm/ and /hɑːˈriːm/, i.e. with variable placing of the stress; in AmE first-syllable stressing is usual but the pronunciation varies between /'heərəm/ and /'hærəm/.

**haricot.** Pronounce /'hærɪkəʊ/.

**harmonium.** Pl. *harmoniums*. See -UM 1.

**harmony, melody, counterpoint.** In music, *harmony* means 'a combination of simultaneously sounded musical notes to produce chords and chord progressions, esp. as having a pleasing effect' (*COD*); *melody* is, principally, 'an arrangement of single notes in a musically expressive succession' (*COD*); and *counterpoint* is, primarily, 'a melody played with another, according to fixed rules'.

**harness.** *Let not him that putteth on his harness* is a MISQUOTATION for *Let not him that girdeth on his harnesse, boast himselfe, as he that putteth it off* (1 Kings (AV) 20: 11).

**harquebus** /'hɑːkwɪbəs/ (early type of portable gun). Thus spelt now, but often as *arquebus*, and in other spellings, in earlier literature.

**hart.** See BUCK.

**hashish.** Pronounce /'hæʃiːʃ/.

**haulm** (a stalk or stem). Thus spelt, not *halm*.

**haulm, haunch, haunt.** All three words are now pronounced in RP with /-ɔː/, though at the time when the relevant section of the *OED* was issued (1899) /-ɑː/ was equally acceptable. In AmE both pronunciations are standard for *haunch* and *haunt*.

**haulyard.** See HALYARD.

**hautboy.** Standard authorities recommend /'əʊbɔɪ/ for this archaic variant of *oboe*.

**hauteur** (haughtiness). Pronounce /əʊˈtɜː/ and print in romans.

**Havana** /həˈvænə/. As city and cigar thus spelt. Cf. Sp. *La Habana* (the city), *habano* (the cigar).

**have.** 1 *No legislation ever* has *or ever will* affect *their conduct*. For this common mistake, see ELLIPSIS 3.

2 *Some Labour MPs would have preferred to have wound up the Session before*

*rising.* In such constructions the present infinitive is preferable to the perfect infinitive. Fowler (1926), s.v. *perfect infinitive* 2, gave numerous examples of the unnecessary use of the perfect infinitive, e.g. *If my point had not been this, I should not have endeavoured* to have shown [read *to show*] *the connection; Jim Scudamore would have been the first man* to have acknowledged [read *to acknowledge*] *the anomaly; it is doubtful whether they would have dared* to have gone [read *to go*] *further.* See also LIKE (verb) 1.

**3** For *don't have* = 'haven't got' see DO 3.

**4** (a) *have (got) to.* When the sense is 'must' a fairly clear distinction is made between habitual necessity (*I have to wear contact lenses*) and immediate necessity (*I've got to catch a plane at 3 p.m.*). In interrogative constructions the type *Do you have this book in stock?* is somewhat more common in AmE than in BrE; while the type *Have you got this book in stock?* is customary in BrE but unusual in AmE. UK examples: (have to) *He knew ... that in order not to lose control irretrievably of his life he had to hold on to his job*—W. Boyd, 1981; (have got to) *'I have got to go & see de Valera about it next month.' 'Got' is the operative word*—E. Waugh, 1952; *Don't you listen to anything I say? It's got to be away by tomorrow*—W. Boyd, 1981. (b) *have to be.* First recorded in 1967 in the sense 'must inevitably be', this colloquial phrase alternates with *have got to be* (which is also first recorded in the late 1960s). Examples: *The Rainbow Warrior tragedy notwithstanding, the success story of the decade would have to be Greenpeace*—Metro (NZ), 1989; *Bearing this in mind, the top trainers at the moment would have to be Hi-Tech Squash and Dunlop Green Flash*—The Face, 1991.

**5** For the use of a redundant or intrusive *have* in certain types of past clauses, see *had have* s.v. HAD 2. Laurie Bauer tested one of these types (counterfactual *have*) in a questionnaire put to some of his students in Wellington, NZ, and found fairly widespread acceptance of sentences such as *If I'd have been there, I could have shown them what to do*, and *Instead of opening your big mouth, I wish* you'd have *kept quiet.* Similar evidence has been found in Australia and N. America (e.g. *And now I bet she wishes it had've*

been—A. Munro, 1993), but the matter requires further investigation. Meanwhile it seems best to regard the construction as non-standard, or at any rate as not fully established in any form of standard English.

**haven.** See PORT.

**haver.** One of the pleasurable aspects of the language is the way in which some regional uses seep into standard English. This verb is a case in point. In Scotland it means (a) to talk garrulously and foolishly (first recorded in 1721), and (b) to hesitate, to be slow in deciding (1866). At some point in the mid-20c. sense (b) emerged in standard English use and is now well established. Sense (a) is still anchored in the north.

**havoc.** **1** The old phr. *cry havoc*, which originally meant to give to an army the order *havoc!* as the signal for the seizure of spoil, and so of general spoliation or pillage, is found in Shakespeare's *Julius Caesar* and in other historical works. As the centuries passed, the cry itself dropped out of use but the word *havoc* became more widely applied to other kinds of devastation and of disarray. It also passed into familiar use in combination with the verbs *make* (first recorded in Caxton and still is use), *work* (first recorded in 1900), *play* (1910), *create* (1961), and *wreak* (1926). And so we often (too often perhaps) read of storms playing havoc with crops, children creating havoc at school or at home, diseases wreaking havoc in Third World countries, and many another havoc. See WORK (verb).

**2** If the verb is ever called on its inflected forms are *havocked* and *havocking.* See -C-, -CK-.

**he.** **1** The long-drawn-out battle between *he* and *him* after the verb *to be* has been partly resolved in favour of *him*, esp. *that's him* (not *that's he*). But in slightly more extended constructions, esp. when the pronoun is qualified by a following *who-* (or *that-*) clause, *he* is still favoured by good writers (see CASES 2). The full vulgarity of using *him* instead of *he* as the leading item of the subject of the sentence is shown in: *Him and Ruihi were clearing out up north to live on a Maori marae*—N. Virtue, 1990.

**2** The use of *he* instead of *him* in an objective position, an indication of genuine illiteracy, is now depressingly common. Examples (in some of which vulgar speech is being shown): *Once Mrs Delaney stood over him and asked plaintively was there any chance of he and Gina reconciling*—T. Keneally, 1985; *Any contact with Flora would have to include he who was keeping an eye on her*—B. Rubens, 1987; *David Hirst wants he and other tenants to take over the management of their council block in south London*—Sunday Times, 1989 (caption to photograph); *The orderly precision of the custody arrangements between he and Sylvia*—M. Golden, 1989 (US); *To he who may conserve*—Sunday Tel., 1991 (headline); *To he that hath shall it be given*—letter to Independent, 1991 (ludicrously misquoting Matt. 13: 12).

**3** See HE OR SHE.

**head** (verb). The passive use shown in the following examples seems to be confined to AmE ( = BrE *is heading for*): *Where is he headed for, anyway?*—H. Crane, 1922; *They were all headed for Ralph's house, where they were going to store the scaffolding for the time being*—New Yorker, 1991; *It was pleasing to be a part of the herd, not to have the feeling that the rest of the world was headed in the opposite direction*—D. Pinckney, 1922.

**headline language.** Everyone knows that the need for brevity in headlines leads to uses that are not characteristic of the language at large. Prepositions, articles, auxiliary verbs, and some other parts of speech are often omitted altogether or used in modified ways. Nouns are run together attributively like verbal dominoes. Paradoxically, the damage to the language is probably non-existent. Readers simply regard headlines as blackboard pointers: the real message lies below. The more arresting types of headline are to be found in the tabloid newspapers. The examples quoted here are taken from the more sedate pages of *The Times*: (standard favourites) *Indian minister* quits *over Bofors*; *Baker* pledges *tougher action*; *Why God is more than mere talk*; *How hacks suffer on the stump*; (standard favourites in other sources) *ban, blast, bid, flay, oust, probe, quiz, rap, slash, swoop*; (two-word headlines) *Air deal*; *Airman*

*jailed*; *Cyclist killed*; *Holiday offer*; *Raid victim*; (attempted puns) *New law aims to dump fly-tippers*; *Off the rails* (a strike of rail workers forced a government minister to travel to Euro-Disneyland by car); *Fitting return* (otters have been seen on the banks of the river Otter in Devon for the first time in 20 years); *Crying for Tory* (Michael Heseltine acts as a town crier). Verbal uses are normal: (present tense) *UN imposes sanctions on Gadaffi*; (past tense) *UK agreed to sell Vulcans to Argentina*; (gerund) *Cresson on brink of resigning*; (passive voice) *Rodent is called to serve*; *Sinn Fein canvasser is murdered by Loyalists*; (future, often expressed by a *to*-infinitive) *Repairs will have to wait*; *Pesticide battle will hit tulips*; *Robbery case to be retried*; *Sanctions to fall on Libya this month*.

**headmaster,** etc. It is to be noted that nearly all compound words beginning with *head-* are stressed on the first syllable: *headache, headband, headland, headlights, headline, headphone, headroom, headset, headstone, headstrong*. But three stand apart in having the stress on the second syllable: *headmaster, headmistress*, and *headquarters*.

**headquarters** can legitimately be used either as a singular (*a large headquarters*) or as a plural (*the firm's headquarters are in London*).

**head up.** Someone can either *head* an inquiry, a committee, an investigating team, etc., or, esp. in N. America and in the enhanced language of personnel departments anywhere, *head up* a committee, a department, etc. In N. America, the haven of energetic phrasal verbs, *head up* first emerged in the 1940s, but its use in Britain is fairly limited.

**healthful, healthy.** 'TV anchormen, newspaper reporters, and even renowned authors use "healthy" when they should use "healthful" '—thus in the letters column of an American newspaper in 1990. The currency of the disliked use in America is not clear to me, but the problem hardly arises in Britain. *Healthful*, meaning 'conducive to good health', is rarely used. In the three main dictionaries prepared for foreign learners of English, it is labelled 'old-fashioned or literary' in one, 'formal' in another, and is omitted altogether in the third. *Healthy* is by far the commoner word of the two and is used to mean either 'possessing or enjoying good health' or, 'conducive to good health'. Normal idiom permits *a healthy child, animal, plant*; *a healthy bank balance*; *a healthy climate*; *a healthy appetite*; *a healthy curiosity*; *a healthy respect*. In none of these would *healthful* feel comfortable.

**heaps.** *There was heaps of time* (Mary Wesley, 1983) represents the normal colloquial idiom when the word following *heaps of* is a singular or a non-count noun. But when there is a plural noun in that position a plural verb is required: *there are heaps of papers to be marked*.

**heave.** There are two forms, *heaved* and *hove*, for the pa.t. and pa.pple, but in practice they are kept fairly strictly apart. In most ordinary senses of the verb ('lift, haul, utter (a sigh), throw, etc.') the normal form is *heaved* (*heaved a bag of coal on to the lorry*; *heaved a sigh*; *he heaved and was sick*). When the sense is 'come into view' the normal form is *hove*: *she hove around the Minister's flank with the effect of an apparition*—T. Keneally, 1980; *Other families' nurses ... quailed when we hove in sight*—P. Lively, 1987. In nautical language the dominant pa.t. form is *hove* (*the cable can be hove in*; *the ship hove to*; *the boat remained hove-to*; *the anchor was hove up*). In a moment of aberration the *Independent* (22 Aug. 1991) treated *hove* as an infinitive: *You never feel that Dr Mengele might suddenly hove into view.*

**Hebraism, Hebraist, Hebraize** are the usual forms, not *Hebrewism*, etc.

**Hebrew, Israeli, Israelite, Jew, Semite.** In current English, to judge from standard dictionaries like the *COD*, the distribution of these words is along the following lines: *Hebrew*, a member of a Semitic people orig. centred in ancient Palestine; the language of this people; (usu. preceded by *Modern*) the language of the (20c.) Israelis. *Israeli*, a native or national of modern Israel. *Israelite*, a native of ancient Israel. *Jew*, a person of Hebrew descent or whose religion is Judaism. *Semite*, a member of any of the peoples supposed to be descended from Shem, son of Noah, including esp. the Jews, Arabs, Assyrians, Babylonians, and Phoenicians. For a fuller treatment of some of the items, see JEW.

**hecatomb.** Pronounce /'hekətuːm/, but in AmE /-təʊm/. The word has no connection with *tomb*: it is derived from Gk ἑκατόμβη, properly 'an offering of a hundred oxen', from ἑκατον 'hundred' + βοῦς 'ox' (but even in Homer meaning simply 'a great public sacrifice' not necessarily confined to oxen).

**hectare.** This metric unit of land measurement is usu. pronounced /'hekteə/ when it is used at all in Britain (*acre* being still the standard unit for most people), but both it and /-ɑː/ are commonly heard in English-speaking countries that have officially adopted the term, with /-ɑː/ perhaps dominant.

**hectic.** The word entered the language in the 14c. (ult. from Gk ἑκτικός 'habitual, consumptive') in the sense 'characteristic of the condition resulting from the kind of fever which accompanies consumption'. It was often by extension applied to people with flushed cheeks and hot dry skin. It was not until the beginning of the 20c. (as medical treatment began to conquer consumption) that the adj. began to flourish, at first colloquially and more recently in standard contexts, in the senses 'stirring, exciting, disturbing' and esp. 'characterized by a state of feverish excitement or activity'. Examples of the modern sense: *concluding a hectic period of meetings*—news bulletin on BBC Radio 4, 1977; *In the meantime, during all this hectic and furious activity, Ham and Lucy were married*—T. Findley, 1984 (Canad.); *Both of us drank first shot at the same hectic pace*—T. Keneally, 1985 (Aust.); *the sole purpose of her slightly hectic rambling*—N. Bawden, 1987 (UK); *As targets of Vendler's most hectic attack ... we were at first horrified by the piece*—New York Rev. Bks, 1990. But the old sense of 'feverish-looking, flushed' is still extant: *While this was bound to be said and must have been expected, Uncle looked hectic, jerky*—S. Bellow, 1987; *the hectic face on the thin neck rose too sharply out of the collar of a silk blouse*—A. Brookner, 1990.

**hecto-.** See CENTI-.

**hedonist, Cyrenaic, Epicurean, utilitarian.** The first is a general name in philosophy since the 19c. for a person who adopts as a theory of ethics that the end or the *summum bonum* or proper aim of mankind is stated to be pleasure (in whatever sense). Hedonism is not a classical concept as such, but is derived from the Gk ἡδονή 'pleasure'. It is pronounced /ˈhiːdənɪzm/.

A *Cyrenaic* is a follower of a Greek philosopher called Aristippus of Cyrene, believed to have been the first to teach that immediate pleasure is the only end of action and that the present moment is the only reality (*Oxford Classical Dict.*). This 'minor Socratic' school of Aristippus, which became the forerunner of Epicureanism, seems to have come to an end *c*275 BC. *Cyrenaic*, which is pronounced /saɪrɪˈneɪɪk/, has been used in English since the 16c. but only with reference to classical antiquity.

An *Epicurean* is a follower of the doctrines of the Greek philosopher Epicurus (d. 270 BC). These doctrines are stated by the *OED* to be: (1) That the highest good is pleasure, which Epicurus identified with the practice of virtue. (2) That the gods do not concern themselves at all with men's affairs. (3) That the external world resulted from a fortuitous concourse of atoms. As a philosophical term the adj. *Epicurean* was first used in English in the 16c. From the 16c. to the 18c. the corresponding noun *epicure* had the secondary sense 'one who gives himself up to sensual pleasure, esp. to eating; a glutton'. But alongside this unfavourable sense, and now the only current one, emerged the meaning 'one who cultivates a refined taste for the pleasures of the table'.

A *utilitarian* is an adherent of the 19c. doctrines of Bentham and J. S. Mill known as *utilitarianism*. A utilitarian is defined by the *OED* as 'one who considers utility the standard of whatever is good for man'. As an adj. *utilitarian* means 'that regards the greatest good or happiness of the greatest number as the chief consideration or rule of morality'. As often happens to such concepts, some writers have used the adj. in a quasi-depreciative way, 'having regard to mere utility rather than beauty, amenity, etc.'.

**hegemony.** Pronounce /hɪˈgemənɪ/ or /ˈ-dʒemənɪ/. See GREEK G.

**hegira** (Muhammad's departure from Mecca in AD 622; the Muslim era). Pronounce /ˈhedʒɪrə/; but in AmE more commonly /həˈdʒaɪrə/. The spelling *hejira* is also used, but less commonly.

**heifer.** Thus spelt.

**height.** The only form in standard English. The form *highth* (formed on the analogy of *breadth, length, width,* and pronounced /haɪθ/), which was common in the 17c. and was the spelling used by Milton, is now used only by the poorly educated. The form *heighth,* pronounced /haɪtθ/, is also a vulgarism.

**heinous.** The recommended pronunciation is /ˈheməs/.

**heir.** 1 The first half of Tennyson's line from his *Locksley Hall* (1842), *I the heir of all the ages, in the foremost files of time,* was classified by Fowler (1926) as a hackneyed phrase, but the reference, if ever made now outside Eng. lit. circles, would probably pass unnoticed.

2 An *heir apparent* is an heir whose claim cannot be set aside by the birth of another heir. An *heir presumptive* is an heir whose claim may be set aside if an heir apparent is born. *Heir apparent* is often used now of a person seen likely to succeed the present incumbent as head of a political party, a business firm, etc.

**hel-.** This sequence of letters in words that have an immediately following vowel is pronounced /ˈhel-/ in *helical, helichrysum, helicon, helicopter, heliport,* etc.; but /hiːl-/ in *helianthus, heliograph, heliolithic, heliotrope, helium,* and *helix.* Those in the first group all answer to Gk ἕλιξ 'something in spiral form'; while all in the second group except *helix* answer to Gk ἥλιος 'sun'.

**helix** /ˈhiːlɪks/. See prec. Pl. *helices* /ˈhelɪsiːz/, occas. *helixes.*

**hellebore.** Pronounce /ˈhelɪbɔː/, with three syllables.

**Hellene** is pronounced /ˈheliːn/, but *Hellenic* most commonly /heˈlenɪk/ rather than /ˈ-liːnɪk/.

**hello,** etc. The multiplicity of forms and applications is somewhat bewildering and the distribution of the various pronunciations cannot be ascertained with any certainty. In my own speech, as far as I can tell, I tend to say /hʌˈləʊ/ as a greeting, which I would illogically write down as *hello!* rather than *hullo!*; or /ˌwel he'ləʊ/, which I would write down as *Well, hello!* (used when I cannot immediately recall the name of the person being greeted). I tend to answer the telephone with /he'ləʊ/, but the first syllable tends to degenerate into /hə-/ or almost to disappear altogether. Every standard speaker (to go no further) will have his or her own distributional pattern, and the pattern may vary from day to day and, esp. in intonation, from context to context. It is one of the clearest indicators of warmth of greeting, or of ascending or descending degrees of formality. Historically, many spellings preceded *hello*, which itself is first recorded in 1883, and, used as an answer to a telephone call, in 1892 (Kipling), e.g. *hollo, holloa, hollow; holler; hillo(a)*.

*Hallo* is first recorded (as a shout to call attention) in 1864. *Hullo* is first recorded in Thomas Hughes's *Tom Brown's School Days* (1857), and, used as an answer to a telephone call, in a book by one C. H. Chambers published in 1900. *Halloo* as noun and verb has mostly been restricted to hunting circles, = (to shout) a cry to urge on the dogs. No one seems to mind that there are so many possible variants. Meanwhile it should be noted that *COD* 1995 gives pride of place to *hello* as a term of greeting, with crossreferences at *hallo* and *hullo*. And novelists go on alleging that policemen say *'Allo, 'allo* as they come across something or someone suspicious. That will do for the present.

**helmet.** The corresponding adj. is *helmeted*. See -T-, -TT-.

**help** (noun). A genteelism (often in the form *daily help*) for a female domestic servant. The older terms *lady help* (assistant and companion to the mistress of a house) and *mother's help* (a women employed to help in the nursery) have mostly fallen out of use, replaced in various specified ways by *au pair, babysitter, child-minder, nanny*, and numerous other terms.

**help** (verb). Several main uses call for comment. **1** It should be borne in mind that the primary sense 'to assist' is far from being the only meaning of the verb. By the 16c., for example, *help* had come to be used in the sense (11 in the *OED*) 'to remedy, obviate, prevent; cause to be

otherwise'. The *OED* explains how this came about. As a result it had come to be used in such sentences as these (drawn from the *OED*) in which it is used with *can* or *cannot* + gerund: *He could not help thinking in verse; No man can help being a coward or a fool; Not one of us could help laughing*. Such sentences, being still idiomatic, can pass without further comment.

**2** In times past (and the debate is not quite over yet) concern has been expressed about sentences containing *cannot help but*, a construction first recorded in the work of the 19c. Manx writer Hall Caine. For the evidence, see BUT 7. Its critics say that it should give way to the type with a gerund and with *but* omitted: e.g. *Bertrand could not help venting his frustration upon Madeleine*—P. P. Read, 1986. The matter remains unresolved and both types are still in common use. Further examples of *cannot help but*: *She could not help but notice that all the passengers on the bus were pensioners*—S. Mackay, 1984; *He could not help but feel going home would put an end to this adventure*—Susan Johnson, 1993 (Aust.); *I couldn't help but wonder what a man had to do to get sawed longways fairly much in half*—T. R. Pearson, 1993 (US).

**3** Also much debated have been sentences in which, as the *OED* expresses it, '*cannot* is replaced idiomatically by *can* after a negative expressed or implied'. The *OED* lists four 19c. examples of this construction, including these: *Your name shall occur again as little as I can help, in the course of these pages*—J. H. Newman, 1864; *I did not trouble myself more than I could help*—C. H. Spurgeon, 1879. Fowler (1926) wrongly says that the *OED* stigmatizes the idiom as 'erroneous', but it does nothing of the kind. On this false premiss, Fowler then went on to 'correct' the types *Don't sneeze more than you can help* and *Sneeze as little as you can help* to *Don't sneeze more than you must* and *Sneeze as little as you can*. Such rewriting is unobjectionable as far as it goes, but does not demonstrate that the sentences of Cardinal Newman and the other 19c. writers are faulty.

**4** *help* and the use of the *to*-infinitive. I have set down elsewhere (*Points of View*, 1992) a description of the way in which the competing types 'he *helped me dig* out my driveway'/'he *helped me to dig* out my driveway' developed from the ME period onwards. Many writers, including Shakespeare and Marlowe, allowed the verb *help* to be followed by either a plain infinitive (one without the particle *to*) or by a *to*-infinitive. At the present time the distribution of the two types cannot be determined with any certainty but the general pattern seems to be something like this: (*a*) The construction with the *to*-infinitive appears to be the more usual one in Britain: *where he helped to look after German prisoners of war*—Brit. Med. Jrnl, 1986; *a well-designed phonics system helps most children to read well*—Daily Tel., 1987. But the construction with a plain infinitive also occurs: *Our every deed must help make us acceptable*—Times, 1986; *It helped silence critics on the party left*—K. O. Morgan, 1987. It is not altogether clear what governs the choice. (*b*) By contrast, in AmE and also in other forms of overseas English, the form with the plain infinitive predominates. Examples: *I had helped her carry it to her bedroom*—G. Keillor, 1986 (US); *in the hope that this may help provoke a transformation*—New Yorker, 1986 *Mandy helped him choose something for Claire*—C. K. Stead, 1986 (NZ); *the labourers' training school he helped create*—Highveld Style, 1986 (SAfr.); *When he is done he instructs Ria to help him pull the wire tight*—Susan Johnson, 1990 (Aust.). But the construction with a *to*-infinitive is also found in these areas often enough: *It may help us to conceive of their predicament if we imagine* ... —Dædalus, 1986 (US); *The levees were helping to aggravate the problem they were meant to solve*—New Yorker, 1987. (*c*) One governing factor, past and present, and in all present-day varieties of English, is a natural reluctance to adopt the sequence *to help* + a *to*-infinitive, that is, to repeat *to*. The construction does occur (*she allowed Pearl to help her to stack up her hair*—I. Murdoch, 1983), but it is not common. In this respect it is noteworthy that Shakespeare's examples of *help* + plain infinitive occur only when the verb *help* is itself preceded by the particle *to*. This reluctance to repeat *to* may partially account for some of the American and Antipodean occurrences of the construction with the plain infinitive: *One of my housemates ... offered to help me move in*—New Yorker, 1986; *she had moved heaven and earth to help me win*

*the Scholarship to Oxford*—the athlete Jack Lovelock, as reported in a recent biography. But this is not certain.

**helpmate, helpmeet** (a helpful companion or partner, usu. a husband or wife). The first is now the usual form, though *helpmeet* (first recorded in 1673) is the slightly older of the two words (*helpmate* 1715). Helpmeet is 'a compound absurdly formed by taking the two words *help meet* in Gen. 2: 18, 20 ("an help meet for him", i.e. a help suitable for him) as one word' (*OED*).

**hemi-.** See DEMI-.

**hemiplegia.** For pronunciation, see GREEK G.

**hemistich** (half a line of verse). Pronounce /-stɪk/.

**hemo-.** See HAEMO-; HAEMORRHAGE; Æ, Œ.

**hempen.** See -EN ADJECTIVES.

**hence.** Numerous historical examples of the pleonastic phr. *from hence* are listed in the *OED*, stretching from the 14c. to the 19c.: e.g. (of place) *Richard, not farre from hence, hath hid his head*—Shakespeare, 1593; *From hence I was conducted up a staircase to a suite of apartments*—W. Irving, 1820; (of time) *From hence Ile loue no Friend*—Shakespeare, 1604; ( = as a result) *From hence he has been accused, by historians, of avarice*—Goldsmith, 1771. But in the 20c. such uses have virtually dropped out of the language and are best avoided. See FROM WHENCE.

**hendecasyllable** (a line of eleven sylphr.lables; from Gk ἕνδεκα 'eleven'). A measure widely used in Gk and L verse, esp. by Catullus, and later in Italy and Spain. It has never been common in English poetry, but note, for example: *If mine eyes can speake to doe heartie errand* (Philip Sidney's *Arcadia*, 1580s); *I beholding the summer dead before me* (Swinburne's *Hendecasyllabics*, 1866).

**hendiadys** /hen'daɪədɪs/, a figure of speech in which a single complex idea is expressed by two words connected by a conjunction (in English always *and*) instead of subordinating one to the other. The word was formed in late L from the Gk phr. ἑν διὰ δυοῖν 'one by means of two'. English examples include *nice and cool* (instead of *nicely cool*), and *good and ready* (instead of *completely ready*; see GOOD AND). The term should not be applied to expressions like *might and main, spick and span, whisky and soda*, etc., in which the parts are of equal standing, and neither is subordinate in sense to the other.

**henna** (verb). The inflected forms are *hennas, hennaed,* and *hennaing.* See -ED AND 'D.

**he or she.** From earliest times until about the 1960s it was unquestionably acceptable to use the pronoun *he* (and *him, himself, his*) with indefinite reference to mean 'anyone, a person (of either sex), especially after indefinite pronouns such as *anybody, anyone, someone*, etc., gender-free words like *doctor, person*, etc., or in fixed phrases such as 'Every man for himself' or 'One man one vote'. One of the minor successes of the feminist campaign to dislodge sexist vocabulary from the language has been the virtual replacement of *he* after indefinite pronouns or with other backward reference by *he or she*; similarly with the objective and possessive forms *him, himself,* and *his.* Some solve the problem by using the pl. pronoun *they* (and *themselves*, etc.): see AGREEMENT 6. But the most favoured method is to 'double' the pronoun: *The potential user, if he or she already possesses hardware, may also be able to take the systems with him or her; If someone is ungainly, fat, ugly, unsure of his or her stage presence ... it can be difficult to be engaged in this country; The tragedy of a depressive's position is that he or she remains frozen in front of the mirror.* The problem is an old one, and various methods of avoiding the use of a backward-referring *he* have been in use over the centuries. The only change is that the process has been greatly accelerated in recent times.

**her.** **1** Case. For questions of *her* or *she*, see SHE 2, 3. Cf. HE 1, 2.

**2** For questions of *her* and *hers* (e.g. *her and his tastes differ*), see ABSOLUTE POSSESSIVES.

**3** For *her* and *she* used in personifications (*the Queen Elizabeth II made* her *stately way past the Needles; Switzerland has zealously guarded* her *neutrality in the 20c.;*

Britain stood alone in 1940 after she had lost her European allies during the blitzkrieg), see PERSONIFICATION. It is important not to use both *it/its* and *she/her* in the same passage, as in *The United States has given another proof of* its *determination to preserve* her *neutrality.*

**4** *Her* used as the subject of a clause is a certain indication that the speaker is illiterate: e.g. *He's slashing bracken along the boundary fence and* me *and* her *go there to get kindling*—M. Eldridge, 1984 (Aust.); Her *and Kitty didn't have much to do with each other anymore*—N. Virtue, 1990.

**herb.** For the AmE pronunciation, see H 2.

**Herculean.** **1** The dominant pronunciation now is /hɜːkjuˈliːən/, with the main stress on the third syllable, though as recently as 1899 the *OED* recognized only /hɜːˈkjuːlɪən/, stressed on the second syllable, a pronunciation confirmed by the metre in examples of the word cited from Marston, Milton, and Byron.

**2** Such proper name adjs. in *-ean* are evenly divided as to where the main stress falls. Thus *Ashmólean, Caesárean, Procrústean, Prométhean* are stressed on the antepenult; but *Cyclopéan* (beside *Cyclópean*), *Cytheréan, Européan, Pericléan, protéan, Sisyphéan, Sophocléan, Tacitéan* on the penult. Such adjs. ending in *-ian* always have the stress on the antepenult: *Aristotélian, Cicerónian, Darwínian, Shakespéarian*, etc. The *Bodleian* Library is /ˌbɒdˈliːən/.

**here.** **1** The normal agreement rules apply in most cases when *here's/here are* are used (*here's my ticket/here are my children*). But in rapid speech such a sentence as *Here's some flowers for you* is idiomatically acceptable.

**2** The types *this here friend of mine* and *these here bicycles*, with *here* in mid-position, are found only in vulgar or regional speech; but *here* in post-position is standard (*your friend here, these bicycles here*).

**hereby, herein, hereof, hereto, heretofore, hereunder, herewith.** The first and last are the strongest survivors in this group of threatened words (*I hereby promise never to smoke again; Herewith I enclose a cheque for my subscription*). The others, esp. as used in legal language, are in retreat before those who are seeking to replace 'archaic' jargon by plain

English expressions (substituting *by this means*, *in this matter*, *of this*, etc.).

**hereditary.** The type *an hereditary peer* is recommended, not *a*. Cf. A, AN 3.

**hero.** Pl. *heroes*. See -O(E)S 1.

**heroic.** 1 The type *an heroic act* is recommended, not *a*. See A, AN 3.

2 Some uses in literature and prosody call for comment: (*a*) *heroic poetry*, epic poetry. Such poetry normally uses a specified type of line: the hexameter in classical Greek and Latin, the alexandrine in French, the hendecasyllabic line in Italian, and the iambic pentameter in English. (*b*) The *heroic couplet* is a rhymed pair of iambic pentameters, a scheme favoured by Chaucer in the 14c. and characteristic of much late 17c. and 18c. poetry.

> And bathed every veyne in swich licour
> Of which vertu engendred is the flour
>                 Chaucer, *General Prologue*
> How happy is the blameless vestal's lot!
> The world forgetting, by the world forgot.
>                 Pope, *Eloisa to Abelard*

(*c*) The *heroic* (or *elegiac*) *quatrain* or *heroic stanza* is not used in epic poetry, but is so called because it uses the English heroic line: four pentameters rhyming *abab*, as in Gray's *Elegy written in a Country Church-Yard* (1751), or *aabb*.

> Can storied urn or animated bust
> Back to its mansion call the fleeting breath?
> Can honour's voice provoke the silent dust,
> Or flatt'ry soothe the dull, cold ear of death?

See ELEGIAC QUATRAIN.

**heroin, heroine** (both pronounced /'herəʊɪn/. One of the surprising things about our language is the way in which it tolerates the existence of homonyms like *heroin* (first recorded in 1898 and at first pronounced /hɪ'rəʊɪn/, stressed on the second syllable), and *heroine* (first recorded *c* 1659), with no suggestion that one word is threatening the other.

**Herr.** This German equivalent of *Mr* has been recorded in English works since the 17c. The *OED* provides examples of the types *Herr Hitler* (a quotation from Winston Churchill), *Herr Ulbricht*, and *Herr Alfred Krupp*, and also of the plural *Herren*. But, except for the relatively small number of English people who have learnt

German, the way in which the word *Herr* operates in German is not well known in Britain (e.g. *Herr Doktor Schmidt* is customary in German, but *Mr Dr Smith* impossible in English).

**herring pond.** From the 16c. to the 19c., and still occasionally, *the herring pond* has been used as a humorous term for an ocean, the sea, esp. the North Atlantic ocean. The illustrative examples in the *OED* are of the type *the other Side of the Herring-pond* (1689), *I'd send them over the herring-pond if I could* (1861). Examples from 20c. sources: *Even if the 'Herring Pond' is crossed previously by a single or dual control machine—Aeroplane*, 1914; *I hope you will pay us another visit when you next 'cross the herring pond'*—E. Waugh, 1945.

See WORN-OUT HUMOUR.

**hers.** See ABSOLUTE POSSESSIVES.

**hesitance, hesitancy, hesitation.** All three words are first recorded in English in the early 17c., and the story since then has been one of advancement for *hesitation* and of sharp retreat for the other two, esp. for *hesitance*. A not-always observed residual distinction is that *hesitancy* is used to mean 'a tendency to indecision, an unwillingness to do something, doubt about doing something' (*that perpetual hesitancy of people whose intelligence and temperament are at variance*), whereas *hesitation* is more often used to mean 'the fact or action of hesitating' (*we have no hesitation in determining what is right and wrong; she rejected it without hesitation*).

**heteronym,** a word having the same spelling as another, but a different sound and meaning, e.g. *agape* /ə'geɪp/ adv. and *agape* /'ægəpeɪ/ noun; *moped* /məʊpt/ verb pa.t. and *moped* /'məʊped/; *refuse* /'refju:s/ noun and *refuse* /rɪ'fju:z/ verb; *used* /ju:st/ accustomed and *used* /ju:zd/ made use of.

**hew.** The pa.pple is either *hewn* or *hewed*.

**hexameter,** a metrical verse line of six feet, used esp. in Gk and L epic poetry (*dactylic hexameters*), and imitated in modern languages, esp. in German, Russian, and Swedish verse (Baldick, 1990), and occas. in English, e.g. in A. H. Clough's *The Bothie of Tober-na-Vuolich* (1848): *This was the/ final re/tort from the/ eager, im/petu-ous/ Philip*.

**hiatus.** A term used in linguistics and prosody to mean a break between two adjacent fully pronounced vowels forming two successive syllables, e.g. *the eye*, *Noël*, *Eloïse*, i.e. expressions in which elision or diphthongization does not occur.

**Hibernian** is an archaic or poetic adj. and noun meaning 'of or concerning Ireland; a native of Ireland'. The related words *Hibernicism* (an Irish idiom or expression) and *Hiberno-* (a combining form, as in *Hiberno-English*), however, are in common use, esp. by linguistic scholars.

**hibernice, Hibernice.** See ANGLICE.

**hiccup.** 1 The inflected forms are *hiccuped*, *hiccuping*. See -P-, -PP-.

2 The variant *hiccough*, also pronounced /'hɪkʌp/, was formed in the 17c. 'under the erroneous impression that the second syllable was *cough*' (*OED*). It has nothing to recommend it.

**hide** (verb). The principal parts are now *hid* pa.t. and *hidden* pa.pple (*hid* being archaic or poetical only when used as pa.pple).

**hie** meaning 'go quickly, hasten' (*hie to your chamber*; *hied him to the chase*) is now archaic or poetical only. The inflected forms are *hies*, *hied*, and *hieing* (occas. *hying*). See VERBS IN -IE, etc. 3.

**hierarchic(al)** (adjs.). The choice of form seems to depend on the rhythm of the sentence rather than on any other factor. In practice the longer form is the more common.

**high, highly.** It should be borne in mind that *high* is both an adj. and an adv., and that in its role as an adv. it competes with *highly*. Examples: (*highly* not idiomatic) *a cart loaded high with the undergrowth he had cut out*; *most surfaces were piled high with magazines*; *party spirit ran high*; *the junior executive is aiming high*; (*high* not idiomatic) *it was highly amusing*; *she spoke highly of her tutor*; *a highly-strung person*; *a new wave of highly talented intellectuals*; *such interesting and highly placed people took him seriously*. In practice, *high* (adv.) is used when the sense required is (lit.) 'far up, aloft', and, fig., esp. with the verb *run* (*feelings were running high*) and in the phr. *high and low* (*he searched high and low for the lost ring*). *Highly* is the normal choice preceding adjs. or ppl adjs.; and, meaning 'with praise', with the verbs, *speak*, *think*, etc.

**hight.** The pa.pple (from the 14c.) of the OE verb *hātan* 'to call' survives as an archaism (often used jocularly) to mean 'called, named', but it is more likely to be encountered in older works than in modern ones: *Childe Harold was he hight*—Byron, 1812.

**hijacking.** See ABDUCTION.

**hike** (noun). Used to mean 'an increase (in prices, wages, etc.)' it is of fairly recent origin (first recorded in 1931) and is mainly restricted to N. America. It turns up occasionally in BrE use: e.g. on the BBC *World at One* programme (Radio 4) on 16 Feb. 1979 Brian Widlake spoke of *another hike in mortgage rates*.

**hillo(a).** See HELLO.

**him.** Further to the note s.v. HE 1, it is worth scrutinizing two examples, a century and a half apart, in which *him* is used (and, in the first, a grammatical statement made) in preference to *he*: *heedless of grammar, they all cried, 'THAT'S HIM!'*—R. H. Barham, 1840; *Yet too, it was not the old him that she had chosen to love*—I. Murdoch, 1989.

**Himalayas.** There seems no good reason to abandon the traditional English pronunciation /hɪmə'leɪəz/, stressed on the third syllable, or the traditional form (pl. in -s), in favour of the cultish form *Himalaya*, pronounced /hɪ'mɑːlɪə/, though it is true that the Nepalese word *Himālaya* is derived from Sanskrit *hima* 'snow' + *ālaya* 'abode', and has /ɑː/ not /ə/ in the second syllable.

**him or her.** See HE OR SHE.

**himself.** Normal uses of the pronoun (e.g. *he was talking about himself*; *Marcus hoped simply to avoid being touched himself*; *he made them himself*) are accompanied by various marginal uses where number- or gender-agreement or word-order gives slight pause for thought: *None of us was willing to commit himself to a clear-cut opinion*—R. Lindner, 1955; *Himself, he had worked with them often enough*—C. P. Snow, 1979; *Denis Healey has always been an aloof figure in the Labour movement, a cat that* *walks by himself*—K. O. Morgan, 1987; *It had often seemed as though the people in the poems were more real to the old man than himself and his brother Michael*—Encounter, 1989; *Himself and Bimbo were at the fryer*—R. Doyle, 1991 (Ir.). Cf. MYSELF.

**hind** (noun). See BUCK.

**hindmost.** See -MOST.

**hindsight.** At first used to mean 'the backsight of a rifle', since the last quarter of the 19c. it has become a pleasing abstract word meaning 'wisdom after the event' (*they realized with hindsight that their electioneering tactics had been wrong*).

**Hindu.** Now always thus spelt, replacing the dominant 19c. and Kiplingesque form *Hindoo*.

**hinge** (verb). The pres.pple is surely to be spelt *hingeing* (to bring out the pronunciation), even though *The Oxford Writers' Dict.* (1990) recommends *hinging*.

**hippo.** Pl. *hippos*. See -O(E)S 5.

**hippocampus** (zoology and anatomy). Pl. *hippocampi* /-paɪ/.

**hippogriff** (fabulous monster). The recommended spelling (but *hippogryph* is also found).

**hippopotamus.** The recommended pl. is *hippopotamuses*, not *-potami* /-maɪ/.

**his.** 1 From OE times onward, but esp. between about 1400 and 1750, and still occasionally (plainly as an archaism), *his* has sometimes been placed after a noun (esp. a proper noun) instead of the possessive *'s*; similarly *her* and *their*, where appropriate. Examples: *to such as will vouchsafe to reade the saide Sir Thomas More his aunswere*—Harpsfield's *Life of More*, 1558; *And this we beg for Jesus Christ his sake*—Book of Common Prayer, 1662; *In examining Æneus his Voyage by the Map*—Addison, 1712. Despite the distinguished history of such uses it was and is a mistake to interpret possessive *'s* as a reduced form of *his*.

2 *A graceful raising of* one's *hand to* his *hat*. For the use of *his* to refer back to *one's*, see ONE.

3 Modern practice tends to frown on male-oriented sentences like *Where your child gets his income determines how much*

*tax he owes* (USA Today, 1988); and to favour the type *Each player, alone on his or her part, was embarrassed whenever he or she became lost* (J. Updike, 1988).

**his or her.** See prec. 3; HE OR SHE.

**hisself.** Since the 14c. there has been a tendency to treat *self* as a noun ( = person) and to place possessive pronouns and adjectives in front: thus in the OED, *his owne self* (1406), *his very self* (1826), etc. A by-product of this process is the emergence of the dialectal (esp. northern and Sc.) form *hisself* (usu. written as one word). It is totally rejected by standard speakers as an element belonging only to uneducated speech or to the engaging vocabulary of northern counties and Scotland.

**historic.** 1 Its primary sense is 'famous or important in history or potentially so' (*a Norman castle on that historic spot; the historic figure of the Warrior Queen* [sc. Boadicea]; *we travelled to the historic breakthrough point* [of the Channel tunnel]). As William Safire noted in 1992, 'Any past event is *historical*, but only the most memorable ones are *historic*.' The traditional difference between the two words is ignored in the following examples: (*historic* used where *historical* would be more appropriate) *I wouldn't seek these churches out, even if they were in a guide-book and of historic significance*—M. Atwood, 1989; *Recent developments in Iran simply conform to the numerous historic examples of the compromises that successful revolutionaries always have to make*—Bull. Amer. Acad. Arts & Sci., 1993.

2 In grammar the term *historic present* is applied to 'the present tense when it is used instead of the past in vivid narration' (OED). Examples: *She had no notion of how welcome she would be. But Raymond opens the door before she can touch the bell, and he hugs her around the shoulders and kisses her twice* —A. Munro, 1989; *'Then she slams into me,' he continued, gesturing, 'and suddenly I'm in the world of things beyond myself.'*—New Yorker, 1991.

A short story called 'Green Winter' by Matthew Francis in *Encounter* (Nov. 1987) uses the historic present throughout, beginning *It is dark when the guards come to wake me* and ending *As we leave the hut ... the icy green of the surrounding fields looks suddenly foreign.* But the device is more often used as an occasional stylistic emphasizer. 'The speaker, as it were, forgets all about time and imagines, or recalls, what he is recounting, as vividly as if it were now present before his eyes' (Jespersen, 1909-49, part iv). The historic present has been used in English since at least the 13c. and, debatably, even in OE.

**3** When preceded by the indefinite article, the following pattern is recommended: *an historic, an historian; but a history*. See next 3, and A, AN 3.

**historical.** **1** A word of quite general meaning: of or concerning history; belonging to the past, not the present; from the historian's view; etc. The *OED* is a dictionary 'on historical principles'. Historical events are portrayed in several of Shakespeare's plays. *Roland's famous legendary death is a very easy perversion of his historical death* (E. A. Freeman, 1871). *Historical linguistics is the study of the developments in languages in the course of time* (R. H. Robins, 1964). *The movie ... is ... a great deal more than a decently-handled historical pageant* (*Guardian*, 1972).

**2** *Historical* is occasionally used (but mostly during the 19c.) in the same way as *historic* (sense 1) but this use is to be discouraged. Also a few linguistic scholars use *historical present* instead of *historic present*, but more perhaps by inadvertence than by conviction.

**3** The type *an historical* (not *a historical*) is recommended. See prec. 3; A, AN 3.

**hither.** This ancient word (used already in OE) and its correlative *thither* have been reserved for formal use or as archaisms during most of the 20c. It is curious, though, that the fixed phrases *hither and thither* (in this direction and that) and *hither and yon* (meaning the same) are still part of the ordinary language. Examples: *the European children who rush hither and thither, shrieking in French or English*—P. Lively, 1987; *No other great composer has so many unmusical admirers, who flock to Bayreuth on annual pilgrimages, and fly hither and yon if there is the prospect of a new and promising Tristan or Walküre production*—*Dædalus*, 1986. *Hither* is also present in the (mainly attrib.) expression *come-hither* 'enticing, flirtatious' (first recorded in 1900).

**HIV.** Since the abbreviation stands for 'human immunodeficiency *virus*', it is tautologous to speak of the *HIV virus*. But, tautologous or not, the phr. is now well established.

**hoard** (stock or store) and *horde* (large group) are homophones, and are sometimes mistakenly interchanged: e.g. *It does not involve hoards* [read *hordes*] *of customers parking in the streets*—G. Jones, *The Business of Freelancing*, 1987.

**hoarhound.** See HOREHOUND.

**hobbledehoy.** This colloquial word ( = a clumsy or awkward youth) of uncertain origin was first recorded in the 16c. and since then has been spelt in various ways (*hobbadehoy, hobbedehoy*, etc.). It has now settled down in the form shown in the headword.

**hobnob** (verb). The inflected forms are *hobnobbed, hobnobbing*. See -B-, -BB-.

**hobo.** Pl. *hoboes*. See -O(E)S.

**hock.** There are three separate words all spelt thus: **1** joint of a quadruped's hind leg; a cut of meat from this (also spelt *hough*).

**2** a German wine from the Rhineland.

**3** (verb) to pawn; (noun) a pawnbroker's pledge.

**hodgepodge,** a variant (esp. in AmE) of HOTCHPOTCH ( = a confused mixture, a jumble).

**hoe** (verb). The inflected forms are *hoes, hoed, hoeing*.

***hoi polloi.*** **1** Those who know that *hoi* is the (nominative masculine plural) definite article in Greek and that the phr. *hoi polloi* means 'the many; hence ordinary people, the man in the street' sometimes avoid saying or writing 'the hoi polloi'. Thus: *Alcibiades might have thought himself above* hoi polloi, *but none of the competitors* [in the Olympic Games] *in his day sought artificial aids to bolster their prowess*—B. Levin, 1992. Fowler (1926) thought that 'the best solution [was] to eschew the phrase altogether'. But the use with *the* is widespread and persistent. It is a standard example of an attempt to force Greek grammar on to the receiving language, or to let the loan-phrase operate in an English way. Dryden and Byron wrote 'the οἱ πολλοί', and who is to quarrel with them? And who is prepared to 'correct' W. S. Gilbert's lines from *Iolanthe*: '*Twould fill with joy, And madness stark the* οἱ πολλοί (*a Greek remark*.)?

**2** By an overturning of the true sense, *hoi polloi* has, since the 1950s, come to be used in parts of America to mean 'high society, the upper crust'. Substantial evidence of this unwelcome use is presented in vol. ii (1991) of *DARE*: e.g. *How can a night-club comedian go on Broadway?* ... *I'm a street-corner character, and Broadway audiences have a hoi-polloi attitude*—*New Yorker*, 1988. And not only in America it would seem: e.g. *I know our Terry's much too grand for the likes of us nowadays—too busy consorting with the hoi polloi at all those literary soirées*—S. Mackay, 1992.

**hoist** was originally a variant pa.pple of the now obsolete verb *hoise*. Both it and *hoised* were current at the turn of the 16c.: *For tis the sport to haue the enginer Hoist with his owne petar* [i.e. blown up by his own bomb]—Shakespeare, 1604; *they ... hoised vp the maine sail to the winde*—Acts (AV) 27: 40. At the same time *hoist* was already in use as a verb in its own right: *Let him take thee, And hoist thee vp to the shouting Plebeians*—Shakespeare, 1606.

**holey** (full of holes). The *e* is retained to distinguish it to the eye from *holy* (adj.).

**holily** (adv). See LILY 3.

**Holland.** See NETHERLANDS.

**hollo, holloa, holler, hollow.** See HELLO.

**home** (verb). (of a trained pigeon, return home; also for the phrasal verb *home in on*). The inflected forms are *homed*, *homing*.

**home** (adv.). Idiomatic uses are very numerous, esp. when *home* is accompanied by a verb of movement (*come, go home; I'll see you home; my son will be home soon; drive the nail home; pressed his advantage home*; etc.). When the sense is 'in his or her own home' the British preference is to use *at home* (*he stayed at home; is Jane at home?*) and the American *home* by itself.

**home, house.** The complexities of the distribution of these two words when used in the literal sense 'one's dwelling-place' cannot be set down here except by repeating what *OED* 2 states: 'In N. America and Australasia (and increasingly elsewhere) [*home* is] freq. used to designate a private house or residence merely as a building.' Examples (from *OED* 2): *More houses* (*or 'homes' as a house is kindly called here*) *are needed*—A. Ross, 1955 (Aust.); *Her three ... sons were shot to death in their home*—*Globe & Mail* (Toronto), 1968; *In Beverly Hills and Bel Air, we saw the homes* (*never called houses*) *of Jane Withers, Greer Garson, and Barbra Streisand*—*Guardian*, 1973. Estate agents and others tend to work on the assumption that *home* is a more personal and warmer term than *house*.

**homely.** The expectation in Britain if the word is applied to a woman is that she will turn out to be unpretentious, adept at housekeeping, warm and welcoming. In America, on the other hand, when used of a woman or her features, the word is much more likely to mean 'not attractive in appearance, plain'.

**Homeric.** See LAUGHTER.

**homey.** See HOMY.

**homo** (noun) /ˈhəʊməʊ/. An informal abbreviation of *homosexual*. Pl. *homos*. See -O(E)s 5. See also GAY.

**homo-,** etc. The prefixes *homoeo-* (as in *homocopath*), *homoio-* (as in *homoiousian*), and, much more common, *homo-* (as in *homosexual*) are all from the Greek word meaning 'same'; it is a vulgar error to suppose that in the last it comes from L *homo* 'man'. In these compounds the short *o* of the Gk originals is by no means uniformly retained in English. *COD* (1990) gives preference to initial /həʊm-/ in *homocentric, homoeopath, homoeopathy, homoeostasis, homoerotic, homogeneous, homomorphic, homophobia, homosexual*, etc.; and to initial /hɒm-/ in *homoeotherm, homograft, homograph, homoiousian, homologue, homonym, homophone*, etc. A pretty kettle of fish! My personal preference is to opt for /hɒm-/ in all of them, but it must be admitted that in practice there is a great deal of variation and uncertainty in public use of the various terms.

**homoeo-.** Words with this prefix, e.g. *homoeopathy*, are spelt with *homeo-* in AmE.

**homogeneity.** Pronounce the stem of the word as /-dʒɪˈniːɪtɪ/, not /-dʒɪˈneɪ-ɪtɪ/, i.e. with the stressed syllable having the sound of *bee* not of *bay*. But see -EITY.

**homogeneous** is a general word meaning 'of the same kind' (*a homogeneous community without foreign intermixture*). It is pronounced /hɒməʊˈdʒiːnɪəs/ or /həʊməʊ-/, with the main stress on the third syllable. It is to be distinguished from the scientific word *homogenous* /həˈmɒdʒənəs/ '(in Biology) having a common descent'; and should also not be confused in any way with *homogenized* (adj.) /həˈmɒdʒənaɪzd/, '(of milk) treated so that the fat droplets are emulsified and the cream does not separate'. An example showing *homogenous* misused for *homogeneous*: *In doing so, they produced a much more homogenous and orthodox form of narrative than that of the legendary tales themselves—Essays & Studies, 1991.*

**homograph,** a word spelt like another but of different meaning or origin, e.g. *pole* (piece of wood or metal), derived from OE *pāl*, and *pole* (as in *North Pole, South Pole*), derived from Gk πόλος. Homographs need not have the same pronunciation, e.g. *lead* (to guide) and *lead* (a heavy metal), respectively /liːd/ and /led/.

**homonym,** a word of the same spelling or sound as another but of different meaning. Homonyms identical in spelling include *calf* (young bovine animal), derived from OE *cælf*; and *calf* (fleshy hind part of the human leg), derived from ON *kálfi*. And those identical in sound but not in spelling include *tail* (of an animal), derived from OE *tægl*; and *tale* (narrative), derived from OE *talu*. Examples of the first type are called *homographs*, and those of the second *homophones*. Cf. HETERONYM. *Synonyms*, by contrast, are separate words that are equivalent in meaning, e.g. *furze* and *gorse*. Such exact equivalence is, however, rare, and the word *synonym* is applied much more frequently to pairs or sets in which the equivalence is partial only. See SYNONYMS.

**homophone,** a word having the same sound as another but of different meaning, origin, or spelling. Homophones are a major feature of the language. A short list (in which the examples are sound so long as each word is pronounced in RP): *air/heir, aloud/allowed, cue/queue, currant/current, heal/heel, hear/here, hoard/horde, literal/littoral, pair/pare/pear, peak/peek/pique, pore/pour, rains/reins/reigns, site/sight/cite, skewer/skua, son/sun, vain/vein.* Near-homonyms like *affect/effect, descent/dissent, eminent/imminent* also abound in the language. Robert Bridges wrote a classic paper on homophones (SPE Tract ii, 1919) and included lists of them divided into classes: the total number of words listed was 1,775. His main conclusion was that English was exceptionally burdened by them.

**homosexual.** Still the term used in most formal contexts (news bulletins, law courts, etc.) rather than GAY. There is no agreement about the pronunciation of the first *o*. Many people (including the present writer) use the same vowel in *homage* and *homosexual* just as many people, I suspect, use the sound of *home*. See HOMO-.

**homy** (adj.) (suggesting home; cosy). The dictionaries give conflicting advice about the spelling. Some give preference to *homey*, and others to *homy*. It is not a matter of great moment either way. My own preference is for *homey*.

**Hon.** Also *hon.* That this prefix may be an abbreviation of either *honourable* or *honorary* is a source of some confusion to people outside Britain. It stands for *honourable* in reports of debates in the House of Commons, where members may not be referred to by name, except by the Speaker, but must be called *the honourable member for ...* (*honourable and gallant* if a member of the Armed Forces, *honourable and learned* if a lawyer). It stands for *honorary* when prefixed to the holder of an office (*Hon. Secretary, Hon. Treasurer*, etc.) and indicates that he is unpaid. As an abbreviation of *honourable* it is also a courtesy title of the sons and daughters of viscounts and barons and of the younger sons of earls, as well as the holders of certain high offices, especially Puisne Judges in England and

Lords of Session in Scotland. (Privy Councillors and peers below the rank of marquess are *right honourable*; so are Lords Justices of Appeal, the Lord Mayor of London, the Lord provost of Edinburgh, and a few other civic dignitaries; marquesses are *most honourable* and a duke *his grace*.) *The Hon.*, when used as a courtesy title, requires the person's Christian name or initial, not his surname alone (*the Hon. James* or *J. Brown*, not *the Hon. Brown*); a common mistake is to suppose that the Christian name is unnecessary before a double-barrelled surname. The same rule applies to the prefixes REVEREND and SIR.

There is a great deal of variation in English-speaking countries outside Britain. For details readers will need to consult e.g. *Debrett's Correct Form* (1970 or later printing). The American practice is described by Bryan A. Garner (1987) as follows: 'Honorable is a title of respect given to judges, members of the U.S. Congress, ambassadors, and the like. It should be used not with surnames only, but with complete names (e.g. *The Honorable Antonin Scalia*) or with a title of courtesy (e.g. *The Honorable Mr. Scalia*). The abbreviation *Hon.* should be used only in mailing addresses.'

**honeyed, honied.** The first of the two is recommended.

**honorarium** (voluntary payment made for professional services rendered without the normal fee). **1** As pl. either *honorariums* or *honoraria* is acceptable. See UM 3.

**2** The word seems to have been pronounced with a fully aspirated initial /h/ until at least 1900, but now is regularly /ɒnəˈreərɪəm/.

**honorary.** See HON.

**honour. 1** Thus spelt in BrE, but *honor* in AmE.

**2** *it is a custome More honour'd in the breach, then the observance.* In context (*Hamlet* I.iv.16) it means a custom that one deserves more honour for breaking than for keeping; but it is sometimes quoted in the mistaken and very different sense of dead letter or rule more often broken than kept. See MISAPPREHENSIONS.

**honourable. 1** Thus spelt in BrE, but *honorable* in AmE.

**2** See HON.

**hoof.** For the pl. form, both *hoofs* and *hooves* are permissible, but *hooves* seems now to be the dominant form throughout the English-speaking world: *The sound of galloping hooves* (a stage direction)—B. Mason, 1963 (NZ); *the brave ... little donkey as she, on dainty hooves, picked her way down the uneven path*—E. Jolley, 1985 (Aust.); *even one* [sc. a tool] *for taking stones out of horses' hooves*—R. Rive, 1986 (SAfr.); *the felt-shod tittuping sound of a donkey's hooves*—P. Lively, 1987 (UK); *Act II began with the clopping of horses' hooves*—New Yorker, 1987; *This tragically malformed little beast* [sc. a toy] *also boasts detachable hooves*—The Face, 1987 (UK). Cf. ROOF.

**hope** (noun). The *OED* (sense 1c) gives examples of *hopes* used in a singular sense, esp. in the phr. *in hopes* (e.g. *I was in hopes you would have shown us our own nation*—Addison, 1702). This particular idiom, i.e. phrases containing *hopes* of the types *in the hopes of, in hopes of, in the hopes that, in hopes that*, seems now more characteristic of AmE (e.g. *He never said a kind word to them, and they worked like dogs in hopes of hearing one*—G. Keillor, 1989) than of BrE, though the *Oxford Advanced Learner's Dict.* (1989) cites a use with the verb *live* (*live in hopes of better things to come*). The customary equivalent in BrE is *in the hope of* (or *in the hope + that*-clause). Apart from the idiom just described, BrE freely admits the pl. form: e.g. *have high hopes of something or someone; raised one's hopes; our hopes for a peaceful outcome were not disappointed.*

**hope** (verb). **1** Fowler (1926) objected to double passives containing *hope* (e.g. *No greater thrill can be hoped to be enjoyed*) but such constructions seem now hardly ever to occur: see PASSIVE TERRITORY 3. (Jespersen, 1909–49, part iv, cites several 19c. examples of *had hoped* followed by the perfect infinitive: e.g. *I had hoped to have seen you and Clara pull together*—Swinburne. But this type, not being a double passive, is not parallel.)

**2** By analogy with legitimate constructions containing *expect*, the illegitimate type *I hope them to succeed* (in imitation of *I expect them to succeed*) could theoretically occur. This danger is more real in passive constructions of the type *A luncheon at which the Queen is hoped to be present*. One hopes never to encounter it; but, in commenting on the sinking of an oil-tanker, a speaker on the BBC TV programme *Nationwide* (13 Oct. 1978) did say *Work is hoped to begin on pumping out* ... The construction is clearly biding its time at the margins of standard English.

**hopefully.** **1** First, *hopefully* has been used since the 17c. as a routine adverb meaning 'in a hopeful manner'. Any number of examples can be found in support of the standard use: *We ... are that posterity to which you hopefully addressed yourself*—A. Burgess, 1963; *As lovers do, as lovers will, they travelled hopefully to Paris*—Maggie Gee, 1985; *Tam* [sc. a dog] *followed her, gazing hopefully upwards*—P. Lively, 1989; *Now she's looking hopefully around, as if she's eager to get burned again*—New Yorker, 1993.

**2** For the disputed modern use, see SENTENCE ADVERBS.

**horde.** See HOARD.

**horehound** (herbaceous plant). Thus spelt, not *hoarhound*.

**horrid,** etc. **1** The older (16–19c.) sense, 'bristling, shaggy, rough' (reflecting the primary sense of L *horridus*) is now extinct, but watch out for *horrid spears, mountain peaks, thickets*, etc., in earlier literature.

**2** An ascending order or unpleasantness might perhaps begin with *horrid* (the least emphatic of the 'disagreeable' group of words: *horrid little boy*), and go on with increasing degrees of severity to *horrible* (of accidents, weather, etc.) and *horrifying* (arousing horror). Some writers imagine that *horrendous* and *horrific* add a degree of disagreeableness to what is being described, but the power of these two words to capture the essence of horror has diminished in the 20c. through overuse in the more popular kinds of writing.

**horsy** (devoted to horses). Thus spelt rather than *horsey*. See -EY AND -Y etc. 2.

**hose** (collective pl. = stockings, socks, and tights). Now almost entirely restricted to trade use; but *hosiery* is perhaps the more usual term.

**hospitable.** Pronounce with the stress on the first syllable (but second-syllable stressing is very common and may prevail as time goes on).

**hospitalize** (send or admit a patient to hospital). A relatively recent word (first recorded in 1901), it is regarded with some suspicion in Britain, but is standard in AmE.

**hospitaller.** Thus spelt in BrE, but chiefly as *hospitaler* in AmE.

**host** (verb). Partly because it is assumed to be a modern use of the related noun as a verb (wrongly, as it happens: the verb is first recorded in the 15c.), and partly because of its omnipresence on TV shows and at Oscar awards (*hosted by* ——) and in other areas of entertainment, it is reviled by many standard speakers. If you are determined to use it (instead of 'act as host') go ahead, but do not be surprised if the use is greeted with a certain *froideur*.

**hostler.** The usual AmE spelling of OSTLER. In AmE, apart from its normal use ('a person who takes care of horses, esp. at an inn'), it also means 'a person who services trains, buses, etc., or maintains large machines'.

**hotchpot, hotchpotch.** The first is the usual spelling in law for 'the reunion and blending of properties for the purpose of securing equal division (esp. of the property of an intestate parent' (*COD*) Otherwise, in its main general senses, (*a*) a confused mixture, and (*b*) a dish of mixed ingredients, esp. a thick broth or stew made from meat and vegetables, the longer form *hotchpotch* is used. *Hotchpotch* (sense *b*), as a cookery dish, has been largely replaced in BrE by *hotpot*, a casserole of meat and vegetables, usu. with a layer of potatoes on top.

**hotel.** See A, AN 3.

**hough.** See HOCK 1.

**houri.** Pronounce /'hʊərɪ/.

**house.** See HOME, HOUSE.

**houseful.** Pl. *housefuls*. See -FUL.

**housewife.** The shortened pronunciation /'hʌzɪf/, which is customary for the sense 'a case for needles, thread, etc.', is now obsolete for the human housewife.

**hove.** See HEAVE.

**hovel.** Pronounce /ˈhɒvəl/, not /ˈhʌv-/.

**hover.** Pronounce /ˈhɒvə/, not /ˈhʌv-/.

**how.** 1 For *as how*, see AS 8.

2 *How come?* (in interrogative constructions = How did or does it come about (that)?). This colloquial American phr. was first noted in the mid-19c., and it still has strong NAmer. associations. Some examples: *How come you weren't there? Were you taking some kind of examination or something?*—New Yorker, 1988; *'How come you're still thin?' she asked with amusement*—A. Munro, 1989; *The night he got back I'd said, 'So what's the deal? How come you're home so soon? I thought this was the big breakout.'*—F. Barthelme, 1991.

**howbeit.** A falling-star word with a distinguished history spread over many centuries (as conj. from the 14c., as adv. from the 15c.). It is still listed in dictionaries of current English (marked *archaic*) but perhaps more as an old friend than as a reflection of actual usage. As an adv. ( = nevertheless, though) it is called upon from time to time for some contextual archaistic effect. Examples: *There is no authority in the phone book for us to call, And ask the why, the wherefore, and the howbeit*—C. Sandburg, a1967; *Howbeit, from the first I disliked the man*—J. Wainwright, 1980.

**however.** 1 When *ever* is used as an intensifier immediately after an interrogative pronoun such as *what, who, how*, as it often is in informal conversation (*Why ever did you do that?*), it should never be joined up to the pronoun in written English. So *How ever are you going to manage?* is right, and *However are you going to manage?* is wrong. See EVER 1.

2 *But ... however.* See BUT 6.

3 If the sense calls for *However* meaning 'nevertheless' to be placed at the beginning of a sentence, it should be followed immediately by a comma: *I should be angry if the situation were not so farcical. However, I had a certain delight in some of the talk*—W. Golding, 1980. This use is to be distinguished from *However* used at the beginning of a sentence as an ordinary adverb meaning 'no matter

how': *However carefully they seem to be pruned back ...*

4 The placement of *however* meaning 'nevertheless' is governed by the nature of the sentence in which it appears. It will be seen from the following examples (drawn from standard sources) that it is a matter of judgement, not of exceptionless rules, where the word is to be placed. The positioning of the word adds a degree of emphasis to what immediately precedes it. Note, too, that *however* in mid-sentence or at the end of a sentence is preceded and followed by a mark of punctuation, normally two commas. Examples: (early in sentence, after a phrase) *Susie is only one of five admitted last night. Two of these, however, are on medical wards; The man himself, however, was not greatly put out by the experience.* (mid-sentence, after a clause) *As time passed, however, I slowly began to see the originality of the resistance you offered; How complete a rescue operation can be mounted, however, is quite another matter.* (at end of sentence) *Lanzmann's tone over-generalizes, however; The luridity of her confessions isn't confined to self-revelation, however.*

5 Avoid at all costs the illiteracy of using *however* as a simple substitute for *but*, or of allowing a sentence to run on when *However* should have had a capital H at the start of a new sentence. Examples of wrong use: *Resources for doing so are not available, however, the matter will be reviewed at a later date; It was hoped that New HAG rents could be pooled to equalize rents, however a recent review has concluded that this will not be possible.*

**Huguenot.** Pronounce /ˈhjuːɡənəʊ/, not /-nɒt/.

**hullo.** See HELLO.

**human** (noun). When I was at school in the 1930s a rule was produced asserting that *human* was properly used only as an adjective, and that one must speak of *human beings*, not of *humans*. Presumably the rule had made its way into contemporary usage guides (though not into Fowler, 1926) from the *OED*. James Murray added a bracketed comment to the relevant *OED* entry in 1899: 'Formerly much used; now chiefly *humorous* and *affected*.' I tried to exorcise the ghost by causing a heap of 'Further examples' of *human* (noun) to be added to vol. ii of *OEDS* (1976). Some additional examples are offered below. But the rule embedded in my mind all those years ago still prevents me from using the word as a noun myself. Examples: *there rose before his inward sight the picture of a human at once heroic and sick*—W. Golding, 1954; *Behind the mental subterfuges ... by which humans escape from reality lies a deep sense of inadequacy*—English, 1989; *In mammals, including humans ... life is initiated ...*—Bull. Amer. Acad. Arts & Sci., 1990.

**humanist.** The word is used in a number of senses. The earliest humanists were the 14–16c. scholars of the Renaissance period who rediscovered the literature, philosophy, history, and oratory of the ancient classical world, and were called *humanists* as distinct from *divines*. The term was also applied to later disciples of the same culture, i.e. to classical scholars and teachers. Devotion to such pre-Christian culture led to the emergence of various systems of thought or action which are concerned with merely human interests (as distinguished from divine), or with those of the human race in general (as distinguished from the individual). For such humanists, humanism became the 'Religion of Humanity'. In the 20c. a new pragmatic system of thought introduced by F. C. S. Schiller and William James placed emphasis on the fact 'that man can only comprehend and investigate what is with the resources of the human mind, and discounts abstract theorizing; so, more generally, implying that technological advance must be guided by awareness of widely understood human needs' (OED 2). One is likely to encounter any of the three types of humanism in life or literature. And religious agnosticism, or as *COD* has it 'an outlook or system of thought concerned with human rather than divine or supernatural matters', is quite likely to form part of the way of life of the second and third type.

**humanities.** Study of *the humanities*, in Oxford called *Literae Humaniores*, i.e. classical language and literature, is a regular feature of the courses offered at many universities in Britain and elsewhere. The term *the humanities* is also used more broadly of arts subjects in

general, as opposed to science and technology.

**humankind.** First used in the 17c. as an occasional variant of *the human race* or of *mankind*, the term has gathered strength in the 20c., sometimes written as two words (*Human kind Cannot bear very much reality*—T. S. Eliot, 1935–42). The word is also favoured by those who judge *mankind* to be unacceptable in our politically correct times. Examples: *A venerable statement such as 'Let us make Man in our image' becomes 'Let us make Humankind in our image'*—Listener, 1985; *As the Ice Age waned in Europe 11,000 years ago, the lot of humankind remained harsh*—Sci. Amer., 1988; *Humankind was always interested in where we come from and where we go, the 'whence' and the 'whither'*—Bull. Amer. Acad. Arts & Sci., 1989; *One single species—humankind—is putting the Earth at risk*—BBC Wildlife, 1990.

**humble-bee.** See BUMBLE-BEE.

**humbug** (verb). The inflected forms are *humbugged, humbugging*. Also *humbuggery* (noun). See -G-, -GG-.

**humerus.** Pl. *humeri* /-raɪ/.

**hummus** /ˈhʊməs/, **humus** /ˈhjuːməs/. The first is a thick sauce made from ground chick-peas and sesame seeds; the second is the organic constituent of soil.

**humour.** 1 'The pronunciation of the *h* is only of recent date, and is sometimes omitted' (OED 1899). This is still largely true, though the pronunciation with silent *h*, i.e. /ˈjuːmə(r)/, is more likely to be encountered in AmE than in BrE.

2 The US spelling is *humor*. The derivatives are spelt *humoresque, humorist,* and *humorous* both in BrE and AmE; but contrast *humourless* (BrE) and *humorless* (AmE).

**hung.** See HANG (verb).

**huntress.** See -ESS 5.

**husband.** See WIDOW, WIDOWER.

**hussy,** a derogatory term for an impudent or immoral girl or woman, is now always thus spelt (not *huzzy*), and usu. pronounced /ˈhʌsɪ/. The OED (1899) gave only /ˈhʌzɪ/, but by 1926 Fowler was able to say that 'the traditional pronunciation (hūźĭ) is giving way before hŭśĭ, which, with the assistance of the spelling, will no doubt prevail'.

**hyaena.** See HYENA.

**hybrid formations** are words made up of elements belonging to different languages. In the last quarter of the 19c., as the study of etymology quickened, and as fascicles of the OED began to appear, it became widely known that the language contained a large number of words that had been 'irregularly' formed, i.e. not in accordance with the rules that were being established by the great linguistic scholars of the day. Among the irregulars were hybrid formations. As with any system of classification, borderline cases were found to exist and exemptions were made. Whole classes of words were held to lie outside the argument. For example, the medieval practice of clapping native prefixes like *be-* and native suffixes like *-ness* on to foreign loanwords was accepted as a 'natural' process: thus *bemuse, besiege, genuineness, nobleness,* and so on, all containing a native affix joined to a French stem, were not condemned as hybrids. Such words 'are hybrids technically, but not for practical purposes' (Fowler, 1926). Similarly exempt from criticism were words such as *readable, breakage, fishery,* and *disbelieve,* in which an English word has been joined by one of the foreign elements that have become living affixes in English: clearly *-able, -age, -ery,* and *dis-,* though of Latin-French origin, are all part of the family now, and their etymology is irrelevant in any discussion of hybrids.

When one moves into the main arena of hybrids it is noticeable at once that the arguments apply only to words formed in the 19c. and 20c. Typical examples include the following:

*amoral:* coined in the 19c. from Gk *á-* (privative prefix) and *moral,* a word of Latin origin.

*bureaucracy:* coined as Fr. *bureaucratie* in the 18c. by a French economist from *bureau* and Gk -κρατία 'rule', and adopted in English in the 19c.

*coastal:* coined in the 19c. from Eng. *coast* and the Latin-derived suffix *-al.*

*gullible:* coined in the 19c. from Eng. *gull* 'to deceive' and the Latin-derived suffix *-ible.*

*speedometer:* formed in the 20c. from Eng. *speed* and the combining form *-ometer,* ultimately from Gk μέτρον 'measure'.

*television:* formed in the 20c. from Gk τηλε- 'far', and *vision,* a word of L origin. If such words had been submitted for approval to an absolute monarch of etymology, some perhaps would not have been admitted to the language. But our language is governed not by an absolute monarch, nor by an academy, far less by a European Court of Human Rights, but by a stern reception committee, the users of the language themselves. Homogeneity of language origin comes low in their ranking of priorities; euphony, analogy, a sense of appropriateness, an instinctive belief that a word will settle in if there is a need for it and will disappear if there is not—these are the factors that operate when hybrids (like any other new words) are brought into the language. If the coupling of mixed-language elements seems too gross, some standard speakers write (now fax) severe letters to the newspapers. Attitudes are struck. This is all as it should be in a democratic country. But *amoral, bureaucracy,* and the other mixed-blood formations persist, and the language has suffered only invisible dents.

See BARBARISMS; METANALYSIS.

**hyena.** Thus spelt, not *hyaena,* i.e. the L. spelling with *-ae* has been abandoned.

**hygiene** /ˈhaɪdʒiːn/, **hygienic** /haɪˈdʒiː-nɪk/. Both words are pronounced with a long i in the second syllable. The (first) i was fully pronounced and the y was often pronounced as /ɪ/ when the relevant section of the OED appeared in 1899: thus /ˈhɪdʒiːn/ with three syllables and /hɪdʒɪˈiː-nɪk/ with four. As the form of *hygiene* does not strictly accord with its Greek original, it should perhaps be pointed out that the English word is an adaptation of Fr. *hygiène,* which in turn came from modL *hygieina,* from Gk ὑγιεινή (τέχνη) '(art) of health'.

**hypallage** /haɪˈpæclass="word">lədʒɪ/. A figure of speech for the transferring of an epithet from the more to the less natural noun. For examples, see ADJECTIVE 10.

**hyperbaton** /harˈpɜːbətɒn/. A figure of speech in which the logical order of words or phrases is inverted, esp. for the sake of emphasis, e.g. *this I must see*; (Shakespeare's *Othello*) *Nor scarre that whiter skin of hers, then snow.*

**hyperbole** /harˈpɜːbəlɪ/. A figure of speech for an exaggerated statement not meant to be taken literally, e.g. *a thousand apologies*. It is to be sharply distinguished from the geometrical term *hyperbola* /harˈpɜːbələ/, a particular kind of plane curve.

**hypercorrection.** A modern (20c.) term for a spelling, pronunciation, or construction falsely modelled on an apparently analogous standard form. Thus *geneology* (for *genealogy*, after numerous words in *-ology*); *miniscule* (for *minuscule*, after numerous words with the prefix *mini-*); *'ve* in *he would've* interpreted as 'of' instead of 'have'. See BE 6.

**hyper-, hypo-.** *Hyper-* means 'above; beyond normal' (thus *hyperactive*, *hypersensitive*); whereas *hypo-* means 'under; below normal' (thus *hypotension*, abnormally low blood pressure; *hypothermia*, the condition of having an abnormally low body-temperature). In most compounds the prefixes are pronounced the same, and the context needs to be carefully examined to make the meaning clear. Note that in *hypocrisy* and derivatives *hypo-* has merged with the following element and is pronounced with initial /hɪp-/: thus *hypocritical* is distinguished in pronunciation from *hypercritical*.

**hyphen** (verb). In the late 1950s I was persuaded by Dr C. T. Onions to use this form (and *hyphened*, *hyphening*) rather than *hyphenate*, etc., and I have done so throughout this book and in my other publications. I have little doubt, however, that *hyphenate* is the more common of the two forms.

**hyphens.** *The problem posed.* Some of the practical difficulties of hyphenation can be seen in the practice of the writers of solicitors' and accountants' newsletters sent out to their clients. In a newsletter of this type issued in 1988 I found the following examples: (hyphened) *oil-production platforms*, *on-the-spot assistance*, *our newly-issued Budget Tax Facts Cards*, *tax-free withdrawals*; (unhyphened) *dry dock installations*, *full income tax relief*, *a four year covenant*, *a tax efficient investment medium*, *a lump sum investment*, *a broadly based indirect tax*. The newsletter was clearly written by a person who was unaware of the need for basic consistency in the use of hyphens. Another difficulty is that practice varies from country to country and from publishing house to publishing house. Many American newspapers, for example, print *horse race* and *horse racing* as separate words, whereas the *New SOED* (1993) gives only *horse-race* and *horse-racing*. *The New Yorker* regularly places hyphens in *teen-age* and *teen-ager*; it is a long time since I last saw a hyphen in these words in works printed in Britain. Specialized journals often print *seaurchin* as one word; in more general work it is printed as *sea urchin* or *sea-urchin*. The familiarity of a term is important: I can still remember the occasion in the early 1970s when it was decided to adopt the spelling *radioisotope* (unhyphened) in the Oxford dictionaries.

*Authorities.* For those who have the time and patience to study the full implications of the stretchingly difficult subject of hyphenation, the following works (listed in chronological order), among others, offer valuable guidelines: K. Sisam, 'Word-Division' in SPE Tract xxxiii (1929); *Hart's Rules*; *The Oxford Writers' Dictionary* (1990); Ronald McIntosh, *Hyphenation* (1990). What follows is heavily dependent on the latest two of these.

*Main uses of the hyphen.* **1** To join two or more words so as to form a single expression, e.g. *ear-ring*, *get-at-able*, and words having a syntactical relationship which form a compound, as *weight-carrying* (objective), *punch-drunk* (instrumental); and in a compound used attributively, to clarify the unification of the sense, e.g. *an up-to-date record*, *the well-known man*, but *prettily furnished rooms*, *a stretchingly difficult subject*, and (predicative) *the records are not up to date*, *the man is well known.*

**2** To join a prefix to a proper name, e.g. *anti-Darwinian.*

**3** To prevent misconceptions: *thirty-odd people*, *extra-territorial rights*, and *more-important people* differ sharply in meaning from *thirty odd people*, *extra territorial rights*, and *more important people.*

**4** To avoid ambiguity by separating a prefix from the main word, e.g. *recover* (get better)/*re-cover* (an umbrella); (a footballer) *resigns*/*re-signs*.

**5** To separate two similar consonant or vowel sounds in a word, as a help to understanding and pronunciation, e.g. *sword-dance*, *Ross-shire*, *co-opt*.

**6** To represent a common second element in all but the last word of a list, e.g. *two-*, *three-*, *or fourfold*.

*Division of words at the end of a line of print*. The problems of hyphenation at the line-end are compounded in newspapers by the narrowness of the columns and by the customary assumption in most printing that the right-hand margin, like the left-hand one, should be straight (or 'justified'). Who has not encountered bad end-of-line breaks like *c-*/*hanging*, *berib-*/*boned*, *mans-*/*laughter*, *pain-*/*staking*, *rear-*/*ranged*? Ronald McIntosh includes *ch-*/*anges*, *ever-*/*yone*, *fru-*/*ity*, *occ-*/*urred*, and *unwat-*/*chable* among his own list of bizarre divisions. It is usually best to divide a word after a vowel, taking over the following consonant to the next line. In pres. pples take over *-ing*, e.g. *divid-*/*ing*, *sound-*/*ing*; but *chuck-*/*ling*, *trick-*/*ling*, and similar words. Generally, when two consonants or vowels come together divide between them, e.g. *splen-*/*dour*, *appreci-*/*ate*. Terminations such as *-cian*, *-sion*, *-tion* should not be divided when forming one sound: divide as *Gre-*/*cian*, *ascen-*/*sion*, *subtrac-*/*tion*. Hyphened words should be divided at the hyphen, and in dictionaries a second hyphen may be used to clarify their spelling, e.g. *second-*/*-hand*. This is not the end of the story: Ronald McIntosh lists thirty-three rules altogether for dividing words at the line-end.

*American printers*. 'Very broadly, [for hyphenation at line-ends] British practice has tended to emphasize morphological structure and word origin (as in *triumph-*/*ant*), and American practice has tended to give greater weight to pronunciation (compare *trium-*/*phant*) ... McIntosh warns us that American practice is likely to become more influential in British English as more and more software for word-processing programs is imported across the Atlantic' (R. E. Allen, Editor of the *COD*, 1990, writing in 1991). We have been warned.

*Envoi*. Whenever I am tempted to use the type *early-nineteenth-century poets* I find myself actually writing *poets of the early nineteenth century* instead. Similarly with many another multi-hyphened type. In 1989 Philip Howard said that he followed a similar course; instead of *a nuclear-weapon-free world* he would opt for *a world free of nuclear weapons*. Hyphening should not become burdensome. But that newsletter, with which I began this article, should have been tidied up, nevertheless.

**hypocoristic.** *Hypocoristic forms* or *hypocoristics* are pet names, nursery words, or diminutives. The majority of them have distinctive endings. Examples: *-er*: *rugger*, *fresher* ('freshman'); *-o*: *ammo* ('ammunition'), *arvo* ('afternoon', AustE); *-s*: *Babs* ('Barbara'), *preggers* ('pregnant'), *Twickers* ('Twickenham'); *-y*, *-ie*: *Aussie*, *baddy*, *bicky* ('biscuit'), *Katie*, *Susie*, *sweetie*. See also -IE, -Y.

**hyponym.** In linguistics, a specific term in relation to a more inclusive term: e.g. *daffodil*, *rose*, and *crocus* are hyponyms of *flower*; *robin*, *thrush*, and *finch* are hyponyms of *bird*.

**hypothecate, hypothesize.** If the meaning required is 'to frame a hypothesis' the right word to use is *hypothesize*, a word that has been in standard and widespread use since the 18c. In the 20c. the verb *hypothecate*, which means 'to give or pledge as security, to pawn', has been pressed into service as an occasional synonym of *hypothesize*, but has not attained the wide currency enjoyed by *hypothesize*.

**hypothesis.** Pl. *hypotheses* /-siːz/.

**hypothetic(al).** The shorter form *hypothetic*, first recorded in the 17c., has now given way to the longer form *hypothetical* to the point that it is not listed in the *COD* (1995). It remains in some other dictionaries of current English, however.

**hysteric(al).** Both adjs. have been in use since the 17c., but the dominant form now by far is *hysterical*.

**hysteron proteron** (from Gk ὕστερον πρότερον 'the latter (put as) the former'). A figure of speech in which what should come last is put first; an inversion of the natural order. Examples: *maisters, it is prooued alreadie that you are little better than false knaues, and it will go neere to be thought so shortly* (Dogberry in *Much Ado*); *Oh lift me from the grass! I die! I faint! I fail!* (Shelley).

# I i

**I. 1** The regrettable type *between you and I* (see BETWEEN 1) must be condemned at once. Anyone who uses it now lives in a grammarless cavern in which no distinction is recognized between a grammatical object and a subject. The same applies to the use of *I* as the second member of an objective phrase, e.g. *He drove Kirsten and I home; They asked Jim and I to do the job; I think she disapproved of Beth and I; our teacher wants you and I to work together on a science project; They disguise themselves like you and I.* The faulty *I* (instead of *me*) would not be used in any of these examples (one hopes) if it was not preceded by a first object and the conjunction *and*: *\*They asked I to do the job* is an impossible construction. The rejected type is a hypercorrection resulting from the slow migration of *me* (q.v.) to some constructions formerly requiring *I*, and from the widespread use of *I* after *as* (see AS 1) and *than* (q.v., 6). *I* is simply being swept this way and that by the tide. Some examples of *I* misused for *me* (drawn from otherwise standard sources): *Honestly if one day I should marry, I'll insist that my in-laws live in their own house far away from my husband and I*—M. Tlali, 1989 (SAfr.); *I think she disapproved of Beth and I, just quietly*—Susan Johnson, 1990 (Aust.); *'What is it?' asked Lemprière. 'Part of you and I,' said Septimus*—L. Norfolk, 1991; *after seeing Mary and I lingering over a late breakfast*—Chicago Tribune Mag., 1991.

**2** It is a minor curiosity that *I*, of all the pronouns, is the only one that must always be written with a capital. As the older forms *ic, ich, ik*, etc., gave way to the reduced form *i* during the Middle Ages, the new form was generally written as *i* or *y*, and was often merged by the scribes with the verb or auxiliary it governed. With the advent of printing in the late 15c. the new form *I* soon established itself as the only standard form, though instances of small *i* can be found as late as the 17c.

**-i¹. 1** For plurals with this ending, see LATIN PLURALS 3, 4 and -US.

**2** Words of Italian origin ending in *-e* or *-o* have plurals in *-i*. The *-i* is for the most part retained in English words adopted from Italian (e.g. *confetti, cognoscenti, dilettanti*); but not always (see GRAFFITI).

**-i². ** A termination used in the names of certain Near Eastern and South Asian peoples, e.g. *Bangladeshi, Bengali, Iraqi, Israeli, Pakistani.* See also -EE² 2.

**iambic, ** the adj. corresponding to *iambus* (pl. *iambuses*), in prosody, a foot consisting of one short (or unstressed) followed by one long (or stressed) syllable, the best-known measure in ancient Greek and modern English poetry. Examples: *They also serve who only stand and wait* (Milton); *The woods are lovely, dark, and deep, But I have promises to keep, And miles to go before I sleep* (R. Frost).

**-ian. ** See SUFFIXES ADDED TO PROPER NAMES.

**i before e. ** One of the traditional spelling rules in its simplest form is '*i* before *e* except after *c*' (so *believe, brief, fiend, hygiene, niece, priest, shield, siege*, etc.; but *ceiling, conceit, receipt, receive*, etc.). The rule can helpfully be extended by adding 'except when the word is pronounced with /eɪ/ (so *beige, freight, heinous, neighbour, sleigh, veil, vein, weigh*, etc.; and *apartheid*, if it is correctly pronounced)'. A further exception: a few words that are pronounced with /aɪ/ are regularly spelt with *ei*: *eiderdown, Eisteddfod, height, kaleidoscope, sleight.* So too are *either* and *neither* (however pronounced). But note *die, lie, piety*, all spelt with *ie*. Outside these familiar rules there lie groups of unresponding words. Among them are inflected forms, e.g. *holier; occupier; carried, defied*; and proper names, e.g. *Leicester, Leigh, Reith*; but *Piedmont*. There are also numerous familiar words that do not respond to the simple rules stated in the first paragraph. The spelling of these must simply be learnt by heart: e.g. (all with *ei*) *caffeine, codeine, counterfeit, foreign, forfeit, heifer, heir, leisure, protein,* *seize, their.* For longer, unanalysed lists, see -EI- and -IE-.

**ibidem** (in the same book or passage, etc.). Now more often pronounced /ˈɪb-ɪdem/ than /ɪˈbaɪdem/; but it is usually abbreviated as *ibid.* or *ib.* It is to be carefully distinguished in meaning from *idem.*

**-ible.** See -ABLE, -IBLE.

**-ic.** For *-ic* and *-ics,* see -ICS 1. See also SUFFIXES ADDED TO PROPER NAMES.

**-ic(al)** (adjvl suffix). This article attempts to assess in practical terms how complex the distribution is of adjectival forms ending in *-ic* and those ending in *-ical.* Since the mathematical unattractiveness of analysing the relevant evidence in large computerized corpora is self-evident—the number of adjs. ending in *-ic* and *-ical* is very large—I have searched my personal database instead and correlated the evidence with that presented in the *COD* (1990).

First, it should be borne in mind that we daily encounter many nouns (including proper names) and adjs. that happen to end in *-ic* but are not relevant to this article, e.g. *chic, Eric, logic, music, republic, topic, traffic.* Secondly, it would appear that a little more than half of the adjectives that fall within the sphere of this article always and only end in *-ic.* Thus *COD* lists *alcoholic* but not *alcoholical, basic* but not *basical, dramatic* but not *dramatical, patriotic* but not *patriotical, plastic* but not *plastical,* and so on. Thirdly, it would seem that about a quarter of all the relevant formations always and only end in *-ical.* Thus *chemical* but not *chemic, farcical* but not *farcic, practical* but not *practic, radical* but not *radic,* and so on. Fourthly, just on a fifth of all such words may end either with *-ic* or with *-ical,* sometimes with a difference of meaning and sometimes with no discernible difference. The more important of such words are dealt with at their alphabetical place, e.g. *classic/classical, economic/economical, historic/historical.*

Fifthly, for those pairs where there appears to be no difference of sense, I have formed a broad impression that the *-ic* forms are favoured in AmE and *-ical* ones in BrE; but the distribution is erratic, and is much influenced by the practice of particular publishing houses. Thus in American academic works the shorter forms *geographic, geologic, immunologic, lexicographic,* and *pedagogic* are more likely to occur than the longer forms *geographical,* etc., whereas in British publications the reverse is the case. On the other hand for many ordinary pairs, e.g. *comic/comical, ironic/ironical, problematic/problematical, symmetric/symmetrical,* the distribution is not governed by geography but by idiomatic or rhythmical considerations in a given context. Sixthly, it is a curiosity that all the above pairs of words have *-ically* as their adverbial equivalents, not *-ly.* The main exceptions are that *public* (which has no corresponding adj. form in *-ical*) has only *publicly* as the corresponding adverb; and there is no form *politicly.*

**-ics.** 1 Among the names of arts or branches of study are a few words in *-ic* that are invariable: the main ones are *logic, magic, music, physic* (the art of healing), and *rhetoric.* But the overwhelmingly normal form ends in *-ics: acoustics, classics, economics, ethics, mathematics, metaphysics, obstetrics, paediatrics, phonetics, physics, politics, statistics,* and scores of others. Of words in this group, *dialectic* and *dialectics* are used without change of sense when they mean 'the art of logical disputation'; *ethic* 'a set of moral precepts (the work ethic)' differs from *ethics,* which has a wider meaning; similarly *metaphysic* 'a system of metaphysics' has a much narrower application than *metaphysics*; a *statistic* is 'a statistical fact or item', not the name of the subject.

2 Grammatical number of *-ics.* See AGREEMENT 11. The primary rule is that *-ics* words used strictly for the name of a subject (*Economics, Ethics,* etc.) govern a singular verb. But when used more generally, and esp. when preceded by determiners such as *his, the, such,* etc., a plural verb is called for: *the economics of lending vast sums of money to Third World countries are difficult to understand; such ethics are abominable; the classics of the period include* Pride and Prejudice. Words in *-ics* that are not names of subjects always govern a plural verb: *heroics are out of place; hysterics leave me cold; optics were provided for measuring out spirits; antibiotics were not widely used until the mid-1940s; his tactics in the game were not easy to follow.*

**ictus** (in prosody). Rhythmical or metrical stress. Pl. same or *ictuses*.

**idea.** **1** It can be used both before *of* + a gerund (*it's not my idea of having a good time*) or before a *to*-infinitive (*the idea is to hit them where it hurts most*). The second type needs to be watched: Fowler (1926) cited an example (*Humperdinck had the happy idea one day to write a little fairy opera*) where *of writing* would certainly have been more idiomatic.

**2** *Idea'd* is preferable to *ideaed*; see -ED AND 'D.

**ideologue, ideology, ideological.** So spelt, not *ideal-*. The words are formed on Gk ἰδιο-, the combining form of Gk ἰδέα 'idea'. All three words are pronounced with initial /aɪ/ not /ɪ/, unlike *ideogram* and *ideograph*, which both have initial /ɪ/.

**ideology.** The first recorded use of the word was in 1798, when it meant 'the science of ideas'. Its main current meanings, 'the system of ideas at the basis of an economic or political theory (*Marxist ideology*)', and 'the manner of thinking characteristic of a class or individual (*bourgeois ideology*)' came into being at the beginning of the 20c. as a natural result of the decline of religious faith. We have had to find a word, free from the religious associations of *faith* and *creed*, for belief in those politico-social systems vaguely indicated by such words as *democracy, socialism, communism*, and *fascism*, which excite in their adherents a quasi-religious enthusiasm.

***id est.*** See I.E.

**idiom.** A classic paper on English idioms is that written in 1923 by Logan Pearsall Smith as SPE Tract xii. We use *idiom*, he said, for 'those forms of expression, of grammatical construction, or of phrasing, which are peculiar to a language, and approved by its usage, although the meanings they convey are often different from their grammatical or logical signification'. He neatly divides idioms into categories: (prepositions) *we tamper* with, but *we tinker* at; *we act* on *the spur of the moment* but at *a moment's notice*; (phrasal collocations) *odds and ends, part and parcel, neither here nor there*; (habitual phrases) *when all is said and done, how do you do?*; (ungrammatical at first sight) *it's me, how's tricks?, who did you see?*; (obsolete or archaic words preserved in familiar phrases) *hue and cry, chop and change, in a trice, at one fell swoop*; (familiar phrases from numerous subjects, e.g. from winds and the sky) *to blow hot and cold, a break in the clouds, once in a blue moon*; (from nautical language) *cut the painter, on the rocks*. The listing continues for 63 pages, but the nature of idioms has been sufficiently demonstrated from the examples just given. The language abounds with idioms that are grammatically or logically unpredictable or not always so, but are nevertheless a major factor in establishing its identity and its immense diversity. See also PHRASAL VERBS.

**idiosyncrasy.** Thus spelt, not *-cy*. The erroneous spelling, which arises from the existence of such pairs as *aristocracy/ aristocratic* and *democracy/ democratic*, therefore *idiosyncrasy/idiosyncratic*, fails on etymological grounds. The false models have as their etymon of the final element Gk κράτος 'power, rule', whereas the separate elements of *idiosyncrasy* come from Gk ἰδιο- 'own' + σύν 'together' + κρᾶσις 'mixture, combination'. It is not suggested that everyone should be a walking etymologist, but simply that people should learn to spell correctly. Lower the flag in sorrow whenever you see the word *idiosyncrasy* misspelt.

**idle** (verb). In various intransitive senses the verb has been an ordinary part of the standard language since the 16c. It was also first recorded in the late 18c. in the transitive sense 'to cause to be idle'. The *OED* provides examples of both types from good sources. In the 20c. the engine of a motor vehicle is said to *idle* (i.e. an intransitive use) when it is just 'ticking over' and the vehicle is not in motion. Since about the 1960s the transitive branch has been strengthened by being commonly used (a) of sportsplayers made idle by injury, etc. (*Gilchrist was … idled with a leg injury—Globe & Mail* (Toronto), 1968), and (b) in contexts describing the effects of industrial strikes (*Underground trains lined up at Neasden, north-west London, yesterday, idled by the one-day Tube stoppage—Times*, 1989).

**idolatry.** See -LATRY.

**idyll.** Probably in order to differentiate it from *idle* and *idol*, the word is now usually pronounced /'ɪdɪl/ with a short initial *i*. But there are some who say /'aɪd(ə)l/, and some (esp. in America) who spell the word *idyl*. The word answers to L *īdyllium*, Gk εἰδυλλιον, but it has gone its own way in English as far as the length of the first syllable is concerned. See FALSE QUANTITY.

**i.e.,** abbrev. of *id est*. **1** The full form *id est* is not usually either written or said, except as an affectation or facetiously.

**2** *i.e.* means 'that is to say', and introduces another way (more comprehensible to the reader, driving home the reader's point better, or otherwise preferable) of putting what has already been said. It does not introduce an illustrative example, for which E.G. is the proper formula.

**3** It is invariable in form, and should not be changed to *ea sunt*, etc., to 'agree' with the number of the explanatory word or phrase. In such a sentence as *which deals with persons* (ea sunt, *all present and future members*), *ea sunt* should be replaced by *i.e.* Cf. AS FOLLOWS; INTER ALIA.

**4** By its nature it is preceded by a mark of punctuation, usu. a comma. It is not normally followed by a comma unless the sense requires one, as when a parenthesis immediately follows: (no comma) *He attacked reactionaries, i.e. those whose opinion differed markedly from his own*; (comma needed) *He attacked reactionaries, i.e., it would seem, those whose opinions* [etc.].

**5** It is always written as lower case and in romans.

**-ie-.** The following words are among those spelt with *-ie* not *-ei-*: *achieve, adieu, aggrieve, Aries, befriend, belief, believe, besiege, bier, bombardier, brief, brigadier, cashier, cavalier, chandelier, chief, fief, field, fiend, fierce, friend, frieze, grenadier, grief, grievance, grieve, grievous, handkerchief, hygiene, lief, liege, mien, mischief, mischievous, niece, piece, piecemeal, pier, pierce, priest, relief, relieve, reprieve, retrieve, review, shield, shriek, siege, sieve, species, thief, thieve, tier, tierce, tiercel, wield, yield.* See also I BEFORE E; and -EI-.

**-ies.** A number of common nouns ending in *-ies* were regularly pronounced with /-riːz/ as the terminal element, i.e. two syllables, until some point in the 20C. See CARIES; RABIES; SCABIES; SERIES; SPECIES. Now all of them end in /-iːz/.

**-ie, -y** (in pet names and diminutives). See HYPOCORISTIC. It is often a matter of free choice between *-ie* and *-y*, except that *-y* is always used in *baby, daddy, granny,* and *mummy*; and in a number of words associated with Scotland (*beastie, kiltie, laddie, lassie,* etc.) *-ie* is the preferred form. In OUP house style (1990) preference is given to the following; *aunty, bobby* (policeman), *bookie* (bookmaker), *buddy* (friend), *bunny* (rabbit), *clippie* (bus conductress), *cookie* (AmE, biscuit), *darky* (dark person), *dearie, doggy, ducky* (darling), *fatty, frilly, Froggy* (Frenchman), *girlie* (as in *girlie magazine*), *goalie, goody, hanky* (handkerchief), *hottie, hubby, johnny, kewpie, kiddie, kitty* (kitten), *missy, Mountie* (in Canada), *nanny, nappy, nightie, piggy, pinny* (pinafore), *postie, puppy, rookie, sonny, sweetie, teddy, telly* (television), *Tommy* (British soldier), *tummy, walkie-talkie.* The pl. forms of all of these end in *-ies*. The familiar forms of personal names follow similar patterns, but there can be no question of insisting that one or another form is the 'preferred' one: the choice is simply a matter of agreement between the persons concerned. As a rough guide the following (of *-ie* and *-y* names) seem to be the more usual: *Amy, Annie, Betty, Billy* (but *Billie* if female), *Bobby* (but *Bobbie* if female), *Charlie* (or *Charley*), *Dicky, Dolly, Eddie, Elsie, Fanny, Florrie, Freddy, Georgie* (Georgina), *Ginny* (Virginia), *Jamie, Jeanie, Jenny, Jerry, Jessie, Johnny, Katie, Kitty, Milly, Molly, Nelly, Paddy, Patsy, Peggy, Polly, Reggie, Sally, Sandy, Sophie, Susie, Teddy, Tommy, Willie, Winnie.*

**if.** **1** The competing claims of *if* and *whether* at the beginning of a noun clause, and of *if* and *though* in the types *a cheap, if/though risky, procedure,* are matters for full discussion in large grammars. It must suffice here to say that in general *whether* is the better of the two in the first type of construction; and that *if* is as acceptable as *though* in the second, provided that no ambiguity results from its use.

**2** It is of interest to note that 'an *if*-clause without a continuation may ...

be simply a round-about way of saying something' (Jespersen, 1909–49, vii). Examples: *Well, if that isn't the best thing I've heard since I was home*—C. Mackenzie, 1919; (containing a mild request or command) *'I've got to leave this morning; so if you'll make out my bill, please?'*—C. Dexter, 1983; *'There's your tea. Drink it.'* ... *'If I could have another lump of sugar.'* G. Greene, 1988; (if dispensed with but implied; cf. *OED* sense 7¶) *In jail they keep accounts their own way, he was saying: you don't pay up, your throat gets slit*—New Yorker, 1989.

**if and when.** The Fowler brothers first attacked this phr. in *The King's English* (1906): 'This formula has enjoyed more popularity than it deserves: either "when" or "if" by itself would almost always give the meaning.' They supported their argument with six examples drawn from undated issues of *The Times* and one (which they find defensible) from Gladstone. 'Many writers', they said, 'seem to have persuaded themselves that neither "if" nor "when" is any longer capable of facing its responsibilities without the other word to keep it in countenance.' In *MEU* (1926) eleven more examples—they are unattributed but seem to be from newspapers—are cited. So there can be little doubt that the construction was in common use in the first quarter of the 20c. It was listed in vol. ii of *OEDS* (1976) and defined 'in reference to a future time but with a strong element of doubt', i.e. with nothing adverse said about the qualified nature of the when-ness. *CGEL* (1985) treated it as 'a stereotyped expression weakening the expectation (conveyed by *when* alone) that the condition in the clause will be realized', and mentioned *as and when* and *unless and until* as 'other institutionalized conjoinings of conjunctions'. One of our best modern novelists uses the expression: *Well, there hasn't been a fatality on a Docklands site yet, and if and when there is, it won't be the first sacrificial blood shed in this city*—P. Lively, 1991. Those who wish to uncouple the pair and use only *if* or only *when* are free to do so. But there are circumstances, it seems to me, when the conjoined pair has an independent value, and that is when its presence is required to reinforce the element of doubt. Cf. *as and when* s.v. AS 6.

**-ified, -ify.** See -FY.

**ignoramus.** Formerly a law term from the L word meaning 'we do not know', and used in law to mean 'we take no notice of (it)'; but from the 17c. onwards used simply as a noun to mean 'an ignorant person'. In view of the derivation of the word its pl. form can only be *ignoramuses* (not *ignorami*). See -US 5.

**ignoratio elenchi** /ɪgnə'reɪʃɪəʊ ɪ'leŋkaɪ/, 'a logical fallacy which consists in apparently refuting an opponent while actually disproving some statement different from that advanced by him' (*OED*), a scholastic Latin phr. first recorded in English in the 16c.

**I guess.** See GUESS.

**-ile.** Words ending in -*ile* (*docile, domicile, facile, fertile, fragile, missile, mobile, prehensile, servile, sterile, tactile, virile*, etc.) are usually pronounced with /-aɪl/ in BrE and with /-ɪl/ or /-əl/ in AmE. The division seems not to have become clear-cut until about 1900. Before that date the dominant pronunciation of these words in BrE as well as in AmE, was with a short *i*, that is /-ɪl/ or /-əl/. The main exceptions are *automobile* and *imbecile*, which are both usually pronounced with final /-iːl/ in BrE, and PROFILE. The statistics terms *decile, percentile,* and *quartile* show much variation, but the dominant pronunciation of these three words seems to be with /-aɪl/ in both BrE and AmE.

**ilex.** Pl. *ilexes*. See -EX, -IX 2. It means both the common holly and the holm-oak.

**ilk.** A word that arouses passions. In origin it is an adj. meaning 'same' (OE *ilca*), but in that sense in ordinary contexts it gradually retreated before *same* (of ON origin) during the ME period and vanished from standard English. In Scotland, beginning in the 15c., the phr. *of that ilk* emerged with the meaning 'of the same place, territorial designation, or name', chiefly in the names of landed families, e.g. *Guthrie of that ilk, Wemyss of that ilk* = Guthrie of Guthrie, Wemyss of Wemyss. In the late 18c., by analogy with the (misunderstood) Scottish use, *ilk* began to be used to mean 'family, class, set, or "lot"', and, by further extension 'kind, sort'. The *OED* has a string of examples of the new uses from 1790 to 1973, but described as 'erroneous'. Certainly many people are still irritated by the relatively new (and often derogatory) uses. But evidence is accumulating that many writers in Britain and abroad are prepared to press the word into service. Examples: *Her husband's employment was not of the ilk of the typical man on the job on the coast*—A. Kennedy, 1986 (NZ); *No, fifteen years a faithful husband, that was his ilk*—S. Bellow, 1987; *Rambo and Rocky and their ilk are the mere tip of a vast iceberg*—Encounter, 1987; *I'm being flippant. Irresponsible in the well-known propensity of my ilk*—K. Amis, 1988; *The image of the United States consorting with ... dope peddlers of General Noriega's ilk is not what Americans think is meant when our government says it is promoting freedom abroad*—New Yorker, 1991.

**ill.** There are one or two anomalies in the use of the word. As an adj. it is most commonly used predicatively and is followed by *with* (*he was ill with pneumonia*). But in the broad sense 'out of health' it can also be used attributively (*he was an ill man when I last saw him*; *mentally ill people*). It can also be used to mean 'faulty, unskilful' (*ill taste*; *ill management*); and it is found in many idioms (*do an ill turn to*; *ill at ease*), proverbs (*it's an ill wind that blows nobody any good*), and, as an adverb, is used to form many adjectival compounds (*ill-behaved*, *ill-considered*, *ill-mannered*). In many attributive contexts, *ill* is unidiomatic and is replaced by *sick*: *a sick child* is more usual than *an ill child*; and *sick leave* and *sick pay* cannot be replaced by *ill leave* and *ill pay*. By contrast with *ill*, *sick* is used primarily to mean 'vomiting or tending to vomit', in BrE that is. To underline the anomalies of the two words a person can *look ill* and then *go sick* (i.e. to report sick). The above comments apply mainly to BrE. There is considerable overlapping of meaning and idiom in the other English-speaking countries. In some of them *ill* and *sick* have no difference of primary meaning: *ill* is just the more formal word of the two. In N. America, if the meaning required is 'vomiting', the usual phrase is *sick to* (or *at*) *one's stomach*.

**illegal, illegitimate, illicit, unlawful.** The four words have similar workloads to their antonyms *legal*, *legitimate*, *licit*, and *lawful*, and their meanings and applications can be deduced from them: see LEGAL. The main difference is that *illicit* is much more commonly used than *licit*. In phrases like *illicit love affair* it 'carries moral overtones in addition to the basic sense "not in accordance with or sanctioned by law"' (Garner, 1987).

**illegible, unreadable.** In current use the *illegible* is not clear enough to be decipherable (*illegible handwriting*); and the *unreadable* is too dull or too difficult to be worth reading (*an unreadable book*).

**illicit.** See ELICIT; ILLEGAL.

**illiteracies. 1** 'There is a kind of offence against the literary idiom that is not easily named' (Fowler, 1926). He rejected *vulgar* as a word 'now so imbued on the one hand with social prejudices and on the other with moral condemnation as to be unsuitable'. Instead of *vulgarisms* he therefore described them as *illiteracies* ('their chief habitat is in the correspondence columns of the press'). Time has moved on and moods have changed. Some of the items marked by Fowler as illiterate are now simply labelled 'disputed' or 'colloquial' in the COD (1990): these are marked with an asterisk below. Others are analysed in a less peremptory manner in modern standard grammars of the language. Readers of the present book will find appropriate comments and advice on all the words, constructions, etc., that Fowler judged to be illiterate. His list (rearranged in broad alphabetical order) follows for its historical interest:

absolute adjectives qualified (*almost quite*, *\*rather unique*, *more preferable*; *\*aggravating* for annoying; present tense, etc., after *as if* and *as though* (*It looks as if we are winning* or *shall win*); *between ... or* used for *between and* (*The choice is between glorious death or shameful life*); *the hon. Smith*; *however*, *whatever*, *whoever*, etc., used interrogatively (*However did you find out?* for *How ever did you find out?*; etc.); bad hyphening (*the ruling-class*, *my wooden-leg*); *\*individual* for person; *\*like* as a conjunction; *me* etc. for *my* etc. in gerund construction (*instead of me being dismissed*); negative after *should not wonder* (*I shouldn't wonder if it didn't come true yet*); *\*re* in unsuitable contexts (*the author's arguments re predestination*); *Rev. Jones*;

same, *such*, and *various* used as pronouns (*have no dealings with such*; etc.); split infinitives used too frequently; *think to* = remember to (*I did not think to tell them when I was there*); *write with direct personal object (*though she had promised to write him soon*).

Gowers (1965) added three more examples of illiteracies: *I* for *me* when in company (*Between you and I*); *likewise* as a conjunction (*Its tendency to wobble ... likewise its limited powers of execution*); *neither* with a plural verb (*For two reasons neither of which are noticed by Plato*).

**2** Illiteracy (or *functional illiteracy*, as it is generally called) in the so-called underclasses is notoriously prevalent and deeply regrettable: its reduction can only be achieved by massive investment in education at a level which no country seems able to afford. What this book is concerned with, however, is to identify and demarcate the danger areas in standard English, that is in the language of the chattering classes, to explain how these danger areas came about, and to suggest ways of reducing them to a minimum. A general problem remains. In recent issues (1991) of *The Spectator* a column called 'Unlettered' presented each week 'a crass, illiterate, ignorant, irrelevant or embarrassing letter or notice from a company or public body'. Most of them contain, not so much examples of the items listed by Fowler and Gowers, but startling evidence of an inability to write clear and competent English sentences. Most of them are somewhat too long to reproduce here. Instead, as an example of poverty-stricken English, I have chosen an extract from a speech by a gubernatorial candidate in a primary election in Kentucky (as reported in *The Daily Telegraph*, 9 May 1991). The candidate stated that her first priority was stamping out illiteracy: *I would, the concept, yes, maybe if and when we can figure out, you know, the ability to handle it ... but I think that what we're doing now, you know, we need to let that see what's going to happen*. The correspondent who sent it to me remarked: 'If you have tears, prepare to shed them now.'

**illiterate** means 'unable to read' or 'poorly educated', by contrast with **innumerate**, which means 'having no knowledge of or aptitude for mathematical operations'.

**illogicalities.** For examples of unintended or idiomatic illogicalities, see the entries for BECAUSE B2; BUT 4; DO 5; MUCH 3(b); TOO 2. A stray example of an unintended illogicality on a notice in a public park: *Any person not putting litter in this basket will be liable to a fine of £5*. Those who have no litter must, it seems, go and find some or face a fine.

**illume, illuminate, illumine.** The first, a poetical shortening of *illumine*, which was used by Shakespeare and many later writers, is now entirely restricted to literary use (e.g. *The lamp illumed only the surface of the desk and one of Lucas's hands*—I. Murdoch, 1993). The third is also for the most part a literary word, and is used mainly in the senses 'to light up; to make bright' and 'to enlighten spiritually'. The normal word in these and other literal and figurative senses is *illuminate*.

**illusion.** See DELUSION.

**illusive, illusory.** See ELUSIVE.

**illustrate** (verb). The only current pronunciation in standard English is /ˈɪləstreɪt/, though the pronunciation with the main stress on the second syllable was given as a variant (in 1900) by the OED.

**illustrative.** The customary pronunciation in Britain is /ˈɪləstrətɪv/, but the form with second-syllable stressing, which is the only pronunciation given in the OED (1900), is still used often enough and is also the preferred form in the standard American dictionaries.

**im-.** See EM- AND IM-. The words, not there mentioned, that regularly have *im-*, not *em-*, are *imbrue* (a literary word = stain (one's sword, etc.)), *imbue* (saturate, etc.), *impale*, *impawn* (obs.?), and *imperil*.

**image.** The word is old (13c.) and over the centuries has developed a wide range of senses. In the late 1950s it began to be used in an entirely new way by advertising companies to mean 'a concept or impression, created in the minds of the public, of a particular person, institution, product, etc.; *spec.* a favourable impression' (OED 2). *Image merchants* (the title of a book on the subject by I. Ross published in 1959) created a public image of BBC programmes; Mr Gaitskell improved his *image* in a famous speech in which he said he and some of his colleagues would 'fight, fight, fight, and fight again to save the party we love'; popular singers, film stars, manufacturers of certain foods (hence *brand image*), indeed the great majority of people in public life, searched for and sometimes acquired an enhanced *image*. It is a word that sprang easily to the lips of the *image-makers*, and it has subsequently seeped into the vocabulary of every articulate person in the land.

**imaginary, imaginative.** The meanings of the two words are quite distinct and are not interchangeable. *Imaginary* means 'existing only in the imagination, not real' (*to have imaginary fears*); whereas *imaginative* means 'having or showing in a high degree the faculty of imagination, inventive, original'. An imaginary book is one that does not exist; an imaginative book is an exciting book that deserves high praise.

**imagine** makes *imaginable*. See MUTE E.

**imago** /ɪˈmeɪgəʊ/. Pl. (in lay use) *imagos*, or (chiefly in scientific work) *imagines* /ɪˈmædʒɪniːz/ or /ɪˈmeɪdʒ-/.

**imam** (leader of prayers in a mosque). So spelt. The older variant *imaum* is no longer in use.

**imbalance.** It is a minor curiosity that a noun meaning 'absence of balance' seems not to have been called for until the 19c. and that the corresponding verbs *disbalance* and *unbalance* are also 19c. words (except for an isolated 16c. example of *unbalance* v.). Ophthalmologists were the first to use the noun *imbalance* (in 1898) but from this technical use the word has branched out in the 20c. and is now used generally in many subjects and contexts. A typical example: *The imbalance in the world's financial system has become grotesque—Times*, 1969.

**imbecile.** See -ILE.

**imbed.** See EM- AND IM-.

**imbibe.** See FORMAL WORDS.

**imbroglio** /ɪmˈbrəʊlɪəʊ/. Pl. *imbroglios*. See -O(E)S 4.

**imbrue.** See IM-.

**imbue.** See IM-. For construction, see INFUSE.

**I mean.** See MEAN (verb) 1; MEANINGLESS FILLERS.

**imitate** makes *imitable, imitator*. See -ABLE, -IBLE 6; -OR.

**immanent** means 'indwelling, inherent; (of the supreme being) permanently pervading the universe (opp. *transcendent*)' (*COD*), and is derived from L *in* + *manēre* 'to remain'. It is naturally not to be confused with *imminent* 'about to happen', though the two words are pronounced in a very similar manner: *immanent* /ˈɪmənənt/; *imminent* /ˈɪmɪnənt/.

**immediately** has been occasionally used informally as a conjunction ( = immediately after), esp. in BrE, since the early 19c. Examples: *I started writing Jill immediately I left Oxford*—P. Larkin, 1983; *He vowed to call her immediately he had a story that would interest her*—L. Grant-Adamson, 1990; *Immediately I heard the front door I switched off his computer*—Nigel Williams, 1992.

**immense.** The very wide semantic area occupied by words meaning 'superlatively good, splendid' has always been shared by many, often short-lived, words (the current favourite in BrE seems to be *brilliant* or *brill*). Used in this manner, *immense* flourished from the 1760s until the end of the 19c. (e.g. *You look like a crown prince ... Perfectly immense*—F. M. Crawford, 1883), but Fowler declared in 1926 that it had 'lost its freshness and grown stale', and the meaning has since passed out of use.

**immigrant.** See EMIGRANT.

**imminent.** See IMMANENT.

**immoral** means 'not conforming to accepted standards of morality; morally wrong (esp. in sexual matters)' by contrast with *amoral*, which means 'not concerned with or outside the scope of morality; having no moral principles'. Cf. AMORAL.

**immovable.** 1 Thus spelt, not *immoveable*. See MUTE E.

2 It has a wide range of applications in the broad sense 'that cannot be moved'. When the sense required is 'that cannot be removed, esp. from office' the word to use is *irremovable*.

**immune.** A person is said to be (a) immune *to* an infection (i.e. resistant to, protected from or against); or (b) immune *from* some undesirable factor or circumstance (i.e. exempt from, not subject to). But the division is not clear-cut; in some contexts *from* is idiomatically used in type (a), and *to* in type (b).

**immunity** in non-medical contexts means 'freedom or exemption from an obligation, penalty, or unfavourable circumstance' (*COD*). By contrast, *impunity* has the much more limited sense 'exemption from punishment or from the injurious consequences of an action' (esp. in the phr. *with impunity*, without having to suffer the normal injurious consequences of an action).

**impact** (noun and verb). **1** (noun) This noun, which was first recorded in the late 18c., means lit. 'the action of one body coming forcibly into contact with another', and fig., from the early 19c. onward, 'the effective action of one thing or person upon another; an impression (esp. in the phr. *to make an impact* (*on*)'. The figurative sense seems to have been sparingly used until the mid-20c.; and at that point it passed into widespread use and ran into opposition ('why not use *effect*, *impression*, *ability to impress*, etc., instead?'). The wave of hostility has now receded, and all the more so as no permanent damage can be shown to have been done to the words that, for a time, it seemed to threaten.

**2** (verb) Used transitively ('to press closely into or in something'), the verb has had a distinguished history from about 1600 onward, i.e. it is older than the noun: e.g. *A stone-like mass ... had become impacted in the lower ilium* (from a standard medical work of 1897). Intransitive uses first appeared in the 20c.: e.g. *Something impacted with a soft thud against Lingard's temple*—Seamark (pseudonym), 1929. In 1962 it was reported in a work called *Basic Astronautics* that a Soviet space rocket had *impacted onto the Moon's surface*. A few years earlier figurative uses of the verb had begun to proliferate: e.g. *The Magazine ... is not the place for consideration of national or international events except in so far as they impact on Oxford*—Oxford Mag., 1956. Voices were raised against the use of the word as a

verb, and have continued to be raised. The Usage Panel of the *Harper Dict. of Contemporary Usage* (1985), for example, contemptuously rejected the use shown in the sentence *High school publishing is a tough market and is being impacted by declining enrollments*. Their verdicts were uncompromising: 'an ugly verb from a noun'; 'another barbarism'; 'this is bureaucratese and ought to be rejected'. They were wrong to describe it as a noun forcibly turned into a verb in the 20c. (see above). But there is no mistaking the hostility they and others have shown towards the verb. It therefore seems advisable to refrain from using the verb in ordinary non-scientific and non-medical contexts—at least for the present. It is very likely that it will pass into uncontested standard use as time goes on. At present it has something of the air of a guest who has turned up at a party uninvited.

**impale.** See IM-.

**impanel.** **1** See EM- AND IM-. *Empanel* is the usual form in BrE but *impanel* in AmE.

**2** However spelt at the beginning of the word, the pa.t. is usually *-panelled* in BrE and *-paneled* in AmE.

**impartable, impartible.** See -ABLE, -IBLE 4.

**impassable, impassible.** See -ABLE, -IBLE 4. The first means 'that cannot be traversed', is pronounced with /'pɑːs/ as its second syllable, and is derived from an unrecorded (late) L verb *passāre* (formed from the ppl stem of *pandere* 'to spread out'). The second means 'not subject to suffering', is pronounced with /'pæs/ in the second syllable, and is derived from *pass-*, the pa.ppl stem of L *patī* 'suffer'.

**impasse.** A word much used, but with no settled pronunciation in English. *COD* gives both /æm'pɑːs/ and /'æmpɑːs/, J. C. Wells (1990) gives pride of place to /æm'pɑːs/, *Collins Eng. Dict.* places four different pronunciations in line, a standard American dictionary gives /'ɪmpæs/ or /ɪm'pæs/ ... Some English speakers try to retain the French nasalized /æ̃/ of the first syllable. Oh dear! The word just must continue to endure the mangling it is receiving.

**impawn.** See IM-.

**impeach.** In BrE it means 'to charge with a crime against the State, esp. treason'. In AmE it means 'to charge (the holder of a public office) with misconduct'. What it does not mean is 'to dismiss from office'.

**impel.** The inflected forms are *impelled*, *impelling*. See -LL-, -L-.

**impenitence, impenitency.** Both words have been used for about four centuries with no perceptible difference of meaning; *impenitence* is recommended. See -CE, -CY.

**imperialism.** In the 19c. (when it came into being) the word simply reflected the British politics of the time. It meant essentially 'the principle or spirit of empire', seen for the most part as a benevolent process. In the 20c. it came to be used disparagingly, first by the Communist bloc of the outside forces which they believed to be ranged against them, i.e. the Western powers; and then, conversely, to refer to the imperial system or policies of the USSR in the countries over which it held sway. Imperialism now seems to be largely an historical concept, though the word is still used in political writing of the continuing or potential imperial policies of certain countries.

**imperil.** 1 See IM-.

2 The inflected forms are *imperilled*, *imperilling* (AmE *imperiled*, *imperiling*). See -LL-, -L-.

**impersonal verb.** Now, a verb used only in the third person singular without a particular subject, e.g. *it is snowing*. Once a rich class of verbs originally (in OE) governing either an accusative or a dative case. Examples of past uses: *What boots it with incessant care To tend the homely slighted shepherd's trade . . .?* (Milton) 'What good does it do . . .?'; *him listeth* 'he is pleased' (see LIST); *methinks* 'it seems to me'.

**impetigo.** Pronounce /ɪmpɪˈtaɪɡəʊ/. The corresponding adj. *impetiginous* is pronounced /ɪmpɪˈtɪdʒməs/.

**impetus.** The pl. (rarely needed) is *impetuses*.

**impinge.** The pres.pple must be *impinging*, with the *-e-* preserved, to stop it being thought of as related to the *pinging* of a bell.

**implement** (verb). The corresponding noun has been extant in its ordinary sense since the 15c. As a verb it came into use first in Scotland at the beginning of the 19c. in the now widespread sense 'to complete, carry into effect (a contract, agreement, etc.); to fulfil (an undertaking)'. The limitation to Scotland fell away in the mid-20c., and the verb came to be used, it would seem, by every committee, board, and governing body in the land. It has now passed through the customary stages that await any new verbs: a nervous welcome ('pedantry', said Fowler in 1926), widespread use ('why can't they use the traditional words *carry out, effect, fulfil*, etc.?'), followed by acceptance (the present attitude). Note that the noun has only one main stress, /ˈɪmplɪmənt/, while the verb has a secondary stress on the final syllable, /ˈɪmplɪˌment/.

**implicit comparative.** The term used in *CGEL* (1985) for what here is called the ABSOLUTE COMPARATIVE.

**impliedly.** Pronounce as four syllables, /ɪmˈplaɪɪdlɪ/. See -EDLY.

**imply.** See INFER.

**impolitic.** The corresponding adverb is *impolitically* rather than *impoliticly*.

**import** (noun and verb). The noun is stressed on the first syllable, /ˈɪmpɔːt/, and the verb on the second, /ɪmˈpɔːt/. See NOUN AND VERB ACCENT 1.

**important.** This adj. operates normally in most contexts (*an important test*; *the test was important*; *it is more important to cut the lawn than to spend too long in the garden*, etc.). But it also acts as a kind of sentence adjective when used parenthetically with *more* or *most*, a use first noted in the 1930s. In such contexts it can, but need not, be taken to be an elliptical way of saying 'what is more important'. Examples: *But, most important of all, it is now clear that reproducing a document which has been leaked in an unauthorised manner means* [etc.]—*Sunday Times*, 1972; *It can be readily synthesized from coal, oil or natural gas. More important, it can be produced simply by splitting molecules of water*—*Sci. Amer.*,

1973; *But, more important, a linked policy can be encashed—surrendered—before maturity date and the saver gets a high proportion of his savings returned to him—Daily Tel.*, 1973. See next.

**importantly.** In its ordinary sense 'in an important manner' it has stood in the language since the early 17c. (e.g. *To render this act importantly and lastingly useful*—A. Hamilton, 1796). Beginning in the 1930s it has also become commonly used as a kind of sentence adverb preceded by *more* or *most*. In some contexts it is interchangeable with *important* (see prec.) and so has the function of a quasi-adj. Examples: *Perhaps more importantly, income not applied to exclusively charitable purposes is not exempt from taxation—Times,* 1972; *It will of course be recognized as a great modern dictionary . . . but more importantly . . . for all the indications it gives of having registered the full impact of our so-called permissive age, is the way it preserves certain antique myths—THES,* 1972; *and, most importantly, soon after the birth of his daughter he is referred to as Honorary Secretary of the Donhead branch of the British Legion—Bodleian Libr. Rec.,* 1986. This, and the corresponding use of *important* (see prec.), must now be considered standard and useful acquisitions to the language, though it should be noted that strong objections were raised to both uses (esp. *more importantly*) in the 1970s and 1980s.

**importune.** Pronounce /ɪmpɔːˈtjuːn/.

**impossible.** This is sometimes thought to be an absolute adj., but see ADJECTIVE 4.

**impost.** See TAX.

**impostor.** The recommended spelling, not *imposter*.

**impostume, imposthume** /ɪmˈpɒstjuːm/. A now rarely used word meaning 'an abscess'; also fig. 'corruption', a 'festering sore'. It is worth mentioning as a word that 'has undergone unusual corruption both in prefix and radical part' (*OED*). The corruption happened partly in OF and partly in English. The starting-point was Gk ἀπόστημα 'abscess', and after various substitutions it ended up with a new Latinate prefix, an intrusive (though unpronounced) *h* by false analogy with *posthumous* (itself a false form),

and a changed ending. The *OED* lists more than a dozen distinct spellings for the word found in 15–19c. texts.

**impracticable, impractical.** *Impracticable* means 'that cannot be carried out, that is not feasible' (*it would be impracticable to place a ban on smoking in pubs; a manned mission to Mars is impracticable at present*). In general *impractical* (which is a relatively recent word, first recorded in the mid-19c.) means 'not practical, unpractical', i.e. (of an idea or course of action) not sensible or realistic; (of a person) lacking the ability to do practical things. Unfortunately *impractical* is tending to encroach on the territory of *impracticable*, and it is not easy to see how they can be permanently kept apart. Some examples: (impracticable) *In the end the scheme was abandoned as impracticable—* J. F. Lehmann, 1960; *As his arms were full of books it would have been impracticable for him to wave—*J. I. M. Stewart, 1974; *It was a memorable statement of the ascetic viewpoint at its most unpleasant and impracticable—*Peter Brown, 1988; (impractical) *I have always been ridiculously impractical . . . I cannot repair a fuse—*F. Howerd, 1976; *I ask my client to have a mirror to hand, as carrying one around with me has proved impractical—Practical Hairstyling & Beauty,* 1987; *Her plans were so impractical that someone like me was necessary to point this out—*A. Brookner, 1987.

**impregnable** is an alteration of ME *imprenable* (from OF) 'not able to be taken' (cf. Fr. *imprenable*). 'The *g* was evidently in imitation of the *g* mute in *reign, deign,* and the like' (*OED*).

**impregnate.** Both *impregnatable* and *impregnable* are notionally in use in the sense 'that can be impregnated'; but the awkwardness of *impregnable* in this sense is self-evident, and *impregnatable* is now extremely rare or obsolete.

**impresario** /ɪmprɪˈsɑːrɪəʊ/. Pl. *impresarios*. See -O(E)S 4.

**imprescriptible.** A legal term meaning '(of a right) that cannot be taken away by (negative) prescription, i.e. one that is not invalidated by lapse of time'.

**impressionable.** This 19c. loanword from French has now largely taken the place of the earlier (first recorded in

1626) and once dominant synonym *im-pressible*.

**imprint** (noun and verb). The noun is stressed on the first syllable, /ˈɪmprɪnt/, and the verb on the second, /ɪmˈprɪnt/. See NOUN AND VERB ACCENT 1.

**improbable.** See PROBABLE.

**impromptu.** 1 Pl. *impromptus*.

2 See EXTEMPORE.

**improve** makes *improvable*. See MUTE E.

**improvise** must always be so spelt, not *-ize*. See -ISE 1.

**impuissant.** Normally pronounced /ɪmˈpjuːsənt/ in English, not /ɪmˈpwiːsənt/.

**impunity.** See IMMUNITY.

**in.** 1 The combinations *inasmuch as, in order that* or *to, in so far*, and *in that* are dealt with separately at their alphabetical places.

2 A temporal use of *in* (e.g. *Although we have not spoken in more than a year . . .—New Yorker*, 1989) has been partially taken into BrE alongside the traditional preposition *for*. It is worth noting that temporal *in* = 'for' was used from the 15c. to the 19c. only in negative contexts (e.g. *To Westminster Hall, where I have not been . . . in some months*—Pepys, 1669). BrE 20c. examples in both negative and positive contexts: *Mark had never been near his house in a year*—C. Mackenzie, 1924; *The first bridge across the Bosphorus in 2,300 years . . . is now being built*—*Daily Tel.*, 1971.

3 There is a sharp division of usage between BrE *a bank* in *Threadneedle Street* and AmE *a store* on *Fifth Avenue*. Also distinct are BrE *she is* at *school* (i.e. not at home) and AmE *she is* in *school*. Cf. also AT 3.

**inacceptable** seems to be sharply retreating before its synonym *unacceptable* but, to judge from the evidence presented in *OED* 2, is not defunct yet.

**inadvertence, inadvertency.** Both forms are in use with no apparent difference of sense, but *inadvertence* is by far the more common. Typical examples: (inadvertence) *But whether the failure was due to inadvertence or to incompetence there could be no doubt that the penalty for it was a most*

severe one—*Times*, 1992; (inadvertency) *This is the moment of truth . . . our truth can no longer be hidden away in the dark recesses of inadvertency or neglect*—*Times*, 1991.

**-in and -ine.** The distinction in chemistry between the two terminations is outside the scope of this book. The general public have selected one or the other (normally *-ine*) for a wide range of substances, and have taken little note of whether they are neutral compounds, nitrogen compounds, or whatever. There has also been much variation during the last century or so in the pronunciation of the termination, as between /ɪn/, /iːn/, and /aɪn/ and a certain amount of variation still occurs.

The following spellings are now standard in lay language. Recommended pronunciations are also given: *casein, cocaine, codeine*, and *protein* (see as separate entries) all had three syllables until some point in the 20c. *albumin* /ɪn/; *Benzedrine* /iːn/; *casein* /iːn/; *chlorine* /iːn/; *cocaine* /em/; *codeine* /iːn/; *dioxin* /ɪn/; *fluorine* /iːn/; *gelatine* /iːn/; *glycerine* /iːn/; AmE *glycerin* /ɪn/; *histamine* /iːn/ or /ɪn/; *iodine* /iːn/ or /ɪn/, formerly /aɪn/; *margarine* /iːn/; *nicotine* /iːn/; *penicillin* /ɪn/; *protein* /iːn/; *quinine* /iːn/, AmE /ˈkwaɪnaɪn/; *strychnine* /iːn/; *vitamin* /ɪn/.

**in- and un-.** 'There is often a teasing uncertainty—or *in*certitude—whether the negative form of a word should be made with *in-* (including *il-, im-, ir-*), or with *un-*. The general principle that *un-* is English and belongs to English words, and *in-* is Latin and belongs to Latin words, does not take us far' (Fowler, 1926). Hundreds of words fall into each group and the historical pattern is untidy. A century ago, for example, *inarguable* and *unarguable* were fighting it out (they were both first recorded in the late 19c.) and both are still listed as current in the *COD*. By contrast, *inability* (first recorded in the 15c.) and *unability* (also 15c.) both remained in common use from the 15c. to the 18c., at which point *unability* inexplicably dropped out. But *unable* (*c*1380) was never challenged by an unrecorded *inable*, thus leaving us with the pair *inability/unable*. Close study of relevant pairs of words reveals that it is never safe to assume that there is a working model that will apply in all cases. Fortunately, for the most part, standard speakers instinctively acquire a working knowledge of the dominant patterns in current English, though they would be hard put to it to explain why they use a particular form and not another. Thus everyone knows that *instability* corresponds to *unstable*, and that *imbalance* corresponds to *unbalanced*. No banners are unfurled proclaiming the merits of *unstability/instable* or of *unbalance* (noun)/*imbalanced*. The number of words about which doubt exists (as with *inarguable* and *unarguable*) is not large. The passage of time will probably see the number further reduced, as e.g. between *inadvisable* and *unadvisable* and between *incommunicative* and *uncommunicative*. In other cases differences of meaning determine the choice of form: e.g., *inhuman* and *unhuman* do not mean exactly the same thing; there are contexts in which *unartistic* is more appropriate than *inartistic*; *immoral* and *unmoral* are not synonymous.

A working list of asymmetrical pairs: (i) adj./noun pairs: *unable, inability; uncivil, incivility; unequal, inequality; unjust, injustice; unquiet, inquietude*. (ii) *un-* words ending in *-ed*: *uncompleted, incomplete; unconquered, inconquerable; undigested, indigestible; undisciplined, indiscipline; unreconciled, irreconcilable; unredeemed, irredeemable; unseparated, inseparable*. (*Note*: The only indisputable *in- -ed* word is *inexperienced*.) (iii) *un-* words ending in *-ing*: *unceasing, incessant; uncomprehending, incomprehensible; undiscriminating, indiscriminate*. (iv) noun/noun pairs: *uncertainty, incertitude*.

*Some general observations*. (*a*) Negative *in-* seems to be no longer a living prefix in English though it was until the 19c.: thus *inacceptable, inadvisable, inappeasable, indecipherable*, and *insanitary* were all first recorded in the 19c. (*b*) The hard core of OE adjectives beginning with negative *un-* is represented by *unalike, unbound, unclean, uneven, unripe*, and *unwise*. This native prefix was later freely added to adjs. of foreign origin (e.g. those in the *unable* to *unquiet* group listed above); but it is worth noting that no adj. of native origin has ever acquired the negative prefix *in-* (thus no forms such as *inclean* or *inwise*). (*c*) Negative *in-* has never been prefixed to words that already begin with

*in-*: forms like *ininformed, inintelligible*, and *inintentional* are unacceptable. (*d*) A number of *un-* and *in-* forms are now being challenged by forms with the privative prefix *a-* (*amoral, apolitical*, etc.); and, in a continuing way, by NON-.

*A finding list*. The following list contains a representative selection of words that have occurred in the past, and are listed in the *OED*, in both *un-* and *in-* forms. An indication is given of which of the two forms now prevails or is likely to prevail at some time in the future; also an indication of the date of first record of the prevailing forms. *acceptable, un-* (15c.); *advisable, in-* (19c.); *alterable, un-* (17c.); *appeasable, un-* (16c.); *communicative, un-* (17c.); *completed, un-* (16c.); *consolable, in-* (16c.); *controllable, un-* (16c.); *decipherable, in-* (19c.); *describable, in-* (18c.); *digested, un-* (16c.); *discriminating, un-* (18c.); *distinguishable, in-* (17c.); *edited, un-* (19c.); *escapable, in-* (18c.); *essential, in-* (17c.); *frequent, in-* (16c.); *navigable, un-* (16c.); *sanitary, in-* (19c.); *substantial, in-* (17c.); *supportable, in-* (16c.); *surmountable, in-* (17c.); *susceptible, in-* (17c.). The puzzling distribution of the forms is confirmed by this list, in which thirteen words begin with *in-* and ten with *un-*. None of the words is of native origin; and it is to be noted that all centuries between the 15c. and the 19c. are represented as the time of first record of the recommended forms. Ours is indeed an unregulated and richly endowed language.

**inappeasable.** See IN- AND UN-.

**inapt** (and related words). The forms in standard use are *inapt, inaptitude, inaptly* (adv.), *inaptness; unapt, unaptly* (adv.), and *unaptness*. In each of them the primary sense reflects that of the adj., namely 'not apt or suitable, unlikely (to do something)'. *Inapt* contrasts with *inept* (and its derivatives *ineptitude, ineptness, ineptly* adv.) in which the primary sense is 'absurd, silly'. It needs to be noted, however, that there is an area of overlap: an unskilled or impractical person can be described as *inept* (an impolite use), *inapt* (reasonably polite), or even *unapt* (usu. followed by *for*; reasonably polite).

**inarguable.** See IN- AND UN-.

**inasmuch as.** This rather formal conjunction is conventionally printed as two words in OUP house style. It means 'to the extent that, in so far as'. Examples: *Inasmuch as she could be pleased, the idea of this marriage pleased her*—C. Blackwood, 1977; *Inasmuch as* The Fugitive *had any one coordinating editor it was he*—M. Seymour-Smith, 1982. It can also mean 'in view of the fact that, since': e.g. *I am unable to reply that I am much the better for seeing you, Pussy, inasmuch as I see nothing of you*—Dickens, 1870; *They are aware of their separateness, and of their difference, inasmuch as they belong to different sexes*—E. Fromm, 1957; *Inasmuch as Gray was Perdita's father, he was to be treated with a reasonable degree of respect*—B. Guest, 1985.

**in back of.** See BACK 2.

**in behalf of.** See BEHALF.

**incage, encage.** The second is the better form. See EM- AND IM-.

**incarnate.** 1 As adj. it is pronounced /ɪnˈkɑːnət/ and means (a) embodied in flesh, esp. in human form (*he is the devil incarnate*), and (b) represented in a recognizable or typical form, extreme (*stupidity incarnate, formality incarnate*).

2 As verb it is pronounced /ˈɪnkɑːneɪt/ or /ɪnˈkɑːneɪt/ and means (a) to embody in flesh, and (b) to put (an idea, etc.) into concrete form.

**incase, encase** (verb). The second is the standard form now. See EM- AND IM-.

**in case, in case of.** See CASE B.

**incentivize** (verb). As part of the seemingly endless coinage of new verbs ending in -ize in the 20c., this word meaning 'to give an incentive to' seems to have entered the fringe of the language, mostly in business contexts, somewhere about 1970. Examples: *We incentivized executives and gave them what the Americans call 'a piece of the action'*—J. Slater, speaking on BBC Radio 4, 1977; *Betamax owners had been 'incentivized to zip right past the commercial,' Sheinberg testified*—New Yorker, 1987. It seems likely to remain at the outer rim of the standard language for some time to come.

**inceptive, inchoative.** Alternative names given to verbs meaning 'to begin to do something'; in Gk -σκω and in L -scō are inceptive or inchoative terminations,

as γιγνώσκω 'learn' (i.e. 'come to know'), calescō 'grow warm'. The many English words in -esce, -escent, as recrudesce, iridescent, are from Latin verbs of this kind.

**inchoate** /ɪnˈkəʊeɪt/ means 'just begun, at an early stage' and, by a natural extension, 'not yet organized, rudimentary'. Under the influence of *incoherent* there is a slight danger that some people will be tempted to spell the word *incohate*. The English word is derived from L *inchoātus* pa.pple of *inchoāre* 'to begin'; the prefix *in-* is not a negative implying an antonym *choate*. Examples: *It was obviously necessary that we should continue our still inchoate discussion over a drink*—D. M. Davin, 1975; *The child ... too small and too weak, it cannot translate inchoate emotions into overt action*—P. Gay, 1985; *She is not allowed to express her real, if inchoate, feelings for Robert Marlin*—T. Tanner, 1986; *The first edition ... contained two sections that were in finished form, but much of the rest was tangled and inchoate*—R. Ellmann, 1986; (erroneously used to mean 'incoherent') *Syria and Iraq have proved almost inchoate with rage at the fearful imagery of their biggest single source of water being stemmed for a solid month*—Observer, 1990.

**incidentally.** 1 First, let it be emphasized that this is the only acceptable spelling of the word in current standard English. The obsolete form *incidently* (16–19c.) answered to the now obsolete use of *incident* as an adj. (see next). The occas. spelling of the word now as *incidently* probably reflects the virtual reduction of the word in rapid speech to four syllables, i.e. /ˈɪnsɪˈdent(ə)lɪ/. Cf. ACCIDENTLY. Example: *Viruses, incidently, remained* hors concours *throughout, it being doubtful whether they are living*—New Scientist, 1989.

2 Particularly, but not only, in the spoken language *incidentally* is used as a 'discoursal' sentence adverb meaning 'by the way; as an unconnected remark'. It is often used as a deft way of changing the subject. Examples: *Incidentally, in theory he must still owe me fifty pounds or so*—C. P. Snow, 1934; *Langmuir, incidentally, was the first scientist in private industry to win a Nobel Prize*—K. Vonnegut, 1981; *Incidentally, thanks to Tommy Hands, I nearly became a landlord myself*—A. J. P. Taylor, 1983; *... and incidentally to supply the name*

*by which the house was to be known thence-forth—Bodleian Libr. Rec.,* 1986.

**incident, incidental** (adjs.). The 20c. has witnessed the virtual demise of *incident* used as an ordinary adjective and its replacement by *incidental*. It is no longer idiomatic to write *incident* in such contexts as (shortened from examples in the *OED*): *Those in the highest station have their incident cares and troubles; The expedition and the incident aggressive steps taken in the campaign; The incident mistakes which he has run into; A bank note, more than sufficient to defray any incident charges.* In all these *incidental* (or some other word) would now be essential. *Incident* used with a following *to* and meaning 'apt or liable to happen; naturally attaching or dependent' remains in use, just, but carries more than a tinge of archaism: e.g. *the privileges incident to his promotion to Major-General; A State government is a natural growth, which* prima facie *possesses all the powers incident to any government whatever.* The freeing of *incident* from its traditional role as an adj. coincided with the greatly increased use of the noun *incident* in the 20c. to mean 'a particular episode (air raid, skirmish, etc.) in war; an unpleasant or violent argument, a fracas'. The term is widely used now of terrorist bombings, border raids, leakages of radiation from nuclear power stations—any serious events—right down to minor muggings, domestic arguments, affrays at public houses, etc. A somewhat delayed result of this use of the noun *incident* is the widespread use since the 1960s of *incident book, room,* etc., by the police. In such combinations *incident* is, of course, the noun used attributively.

**incise** (cut into, engrave). Always so spelt, not *-ize.* See -ISE 1.

**incisor.** Thus spelt, not *-er.*

**incline.** The verb is stressed on the second syllable, /ɪnˈklaɪn/, and the noun on the first, /ˈɪnklaɪn/. See NOUN AND VERB ACCENT 1.

**inclose.** *Enclose* is the usual form. See EM- AND IM-.

**include, comprise.** It should be recalled that the second of these seems to be prevailing in its battle with *compose* (see COMPRISE), regrettable though that is.

*Comprise* is simultaneously competing with *include.* When two words such as *include* and *comprise* have roughly the same meaning, examination will generally reveal a distinction; and the distinction between the present two seems to be that *comprise* is appropriate when the content of the whole is in question, and *include* only when the admission or presence of an item is in question: good writers say *comprise* when looking at the matter from the point of view of the whole, *include* from that of the part. With *include,* there is no presumption (though it is often the fact) that all or even most of the components are mentioned; with *comprise,* the whole of them are understood to be in the list. Thus the University of Oxford *includes* All Souls, Christ Church, Magdalen College, and Somerville College; but it *comprises* All Souls, Balliol College, Brasenose College, and more than 30 other colleges, as well as several permanent halls. This leads to the distinction that one cannot legitimately say that the University of Oxford *comprises* All Souls, Christ Church, Magdalen College, and Somerville College; in such a context only *includes* is correct. Similarly, *include,* not *comprise,* must be used in the following example: *The Serbian forces attacking Sarajevo are units of the former Federal army: they do not include regiments still stationed in Zagreb.*

**including.** A recent use of the type *including* ( = as well as) | a prepositional phrase, e.g. *including by reputable authors,* has been pointed out to me by a correspondent. Examples: (from a mid-1980s report of the Bodleian Library, Oxford) *copies of this notice are being distributed on a wide scale including to overseas establishments;* (from BBC Radio 4, Dec. 1990) *she is not so unpopular abroad, including in Europe;* (an American speaking) *Mrs Thatcher succeeded in imposing on the political class, including on the Labour Party, her …;* (from American sources) *we find free speech under assault throughout the United States, including on some college campuses—Internat. Herald Tribune,* 1991; *In the past 20 years, groups like. Médecins sans Frontières … have helped all over the world, including in Afghanistan, Ethiopia* [etc.]—ibid., 1991. Will the type become established? Time will tell.

**incognito** /ˌɪnkɒgˈniːtəʊ/. The pl. form of the noun (though rarely needed) is *incognitos*. The inflexions of the donor language, in this case Italian, have long since been abandoned in English: thus *a young incognito, this actress from London.* When the word is used adverbially (*they travelled incognito*) no change of form is called for. Contrast the fixed phr. (from Latin) *terra incognita* in which the L feminine ending is retained.

**incognizable,** etc. For pronunciation, see COGNIZANCE.

**incoherent.** See INCHOATE.

**incommunicado** /ˌɪnkəˌmjuːnɪˈkɑːdəʊ/. Always spelt with two *m*s in English, despite the spelling of the Sp. original (*incomunicado*).

**incommunicative.** The usual form now is *un-*. See IN- AND UN-.

**incomparable** /ɪnˈkɒmpərəbəl/. On the pronunciation, cf. COMPARABLE 1.

**incompetence, incompetency.** In lay language the two are not distinguished, except that *incompetence* is much the more frequently used of the two. Cf. COMPETENCE 1. In legal language, according to Garner (1987), 'The best advice is to reserve the *-cy* form to contexts involving sanity or ability to stand trial or to testify, and to use the *-ce* form when referring to less than acceptable levels of ability'. How closely such a distinction is observed in legal writing in Britain is a matter for others to determine. Once the choice of form is made, however, it is important to use it consistently, and not to change from one to another in a given passage.

**incompleted.** The usual form now is *uncompleted* (but *incomplete*). See IN- AND UN-.

**incongruous vocabulary.** To start with the ridiculous, I have before me a cartoon showing two zoologists having an alfresco meal in a tropical forest. One is saying to the other, *Hey! What's this Drosophila melanogaster doing in my soup?* The humour to the layman, of course, lies in the specification. Incongruity in the use of words stretches in a continuum from mirth to the deeply offensive, and few writers escape censure on one

count or another. I was myself attacked for using a French catchphrase in the Introduction to the first volume (1972) of *OEDS: Whereas in 1957, when we began our work, no general English-language dictionary contained the more notorious of the sexual words,* 'nous avons changé tout cela' ... I remain unrepentant, and so, presumably, does my critic. Strunk and White in their celebrated American book *The Elements of Style* (3rd edn, 1979) offer advice on specific matters of incongruity: e.g. 'Do not overwrite. Rich, ornate prose is hard to digest, generally unwholesome, and sometimes nauseating'; 'Avoid the use of qualifiers. *Rather, very, little, pretty*—these are leeches that infest the pond of prose, sucking the blood of words.' Sometimes the advice itself is presented in an over-ornate manner—it is the linguistic equivalent of shock therapy. Fowler (1926) attacked the sentence *Austria-Hungary was no longer in a position, an' she would, to shake off the German yoke.* He said that *be in a position to* is 'a phrase of the most pedestrian modernity; *shake off the yoke*, though a metaphor, is one so well worn that the incongruity is felt between it and the pedestrianism.' But his main scorn was reserved for *an'*: 'The gold-fish *an'* cannot live in this sentence-bowl unless we put some water in with it, and gasps pathetically at us from the mere dry air of *be in a position.*' The passage continues colourfully for several more lines, and it reminds us that 'minor lapses from congruity are common enough, and a tendency to them mars the effect of what a man writes more fatally than occasional faults of a more palpable kind, such as grammatical blunders.'

Two passages will illustrate the running wars concerning matters of linguistic congruity. First, Renford Bambrough on George Steiner in the 9 Sept. 1989 issue of *The Spectator:* 'Some of the English words and phrases used in the book are near the same precipice [as that approached by some Greek and Latin words and phrases just quoted]: phenomenality, paradoxality, councillary, monumentally atrophied, proof-values. There *are* such terms as 'truth-values' and 'truth-functions', but Professor Steiner would have done well to leave them in the logic books where they are doing no harm, and not bespatter the text with them in a way that leaves the reader unclear whether he knows what they mean or not.

'Then there are the words that certainly exist in the language, but whose use should be heavily taxed: seminal, cardinal, lineaments, talismanic. Nor is any value added by "very precisely", "of the essence", "of the utmost relevance", "in the full sense", "in the deepest sense", "quintessentially", "of the utmost pertinence". If we have heard but not understood the whisper, why should we be expected to understand the shout? Instead of "outside", Professor Steiner writes "extraterrestrial to" ...'

And D. J. Taylor on the Gentlewomanly High Style in *The Sunday Times*, 9 Dec. 1990: 'It will be familiar to anyone who has ever read a book by Lady Antonia Fraser or one of her daughters, a genteel style in which weather is never bad but always "inclement", in which a horse tends to be a "steed" or even, in very exalted moments, a "palfrey", a style pitched just this side of Wardour Street. [Katherine] Frank is a creditable exponent of this genre. Maria Brontë, Patrick's wife, doesn't write her tracts in support of the working classes. She can be found "plying her pen to solace the poor". Jack Sharp, Branwell's chum, is an "unsavoury rogue". And so on.'

All the sources quoted above illustrate levels of linguistic incongruity. The problem is to judge where the lines are to be drawn and the right levels established.

**in connection with.** See CONNECTION 2.

**inconsequent, inconsequential.** The two words overlap considerably. Both are used to mean 'not following naturally; lacking logical sequence', but when, for example, an argument is a disconnected one it would normally be described as *inconsequent*; and when the emphasis is on its unimportance the word to use is *inconsequential*.

**inconsiderateness, inconsideration.** Both words came into the language in the 16c. and have remained synonymous since then, chiefly in the senses 'thoughtlessness (resulting from lack of consideration)' and 'lack of consideration for others'. In practice *lack of consideration* is probably used more commonly than either of them.

**inconsolable, un-.** The first is now the favoured form. See IN- AND UN-.

**incontinently.** As adv. meaning 'straightway, immediately', *incontinently* and the corresponding adj. *incontinent* were commonly used from the 15c. to the 19c., but both are now obsolete. They ultimately answer to late L *in continēnte* (sc. *tempore*) 'in continuous time, without any interval'. They have been replaced by the etymologically unrelated adv. *incontinently* 'in a manner showing lack of restraint (esp. in regard to sexual desire)', and by the adj. *incontinent*, (a) unable to control movements of the bowels or bladder or both, and (b) lack of restraint (esp. in regard to sexual desire). The 'lack of restraint' words answer to L *continēntem*, pres. pple of *continēre* 'hold together'.

**incontrollable, un-.** The second is now the favoured form. See IN- AND UN-; UN-CONTROLLABLE.

**increase.** The noun, pronounced /ˈɪnkriːs/, and the verb, pronounced /ɪnˈkriːs/, fall into the normal pattern for such pairs. See NOUN AND VERB ACCENT.

**incrust, encrust.** *Encrust* is the usual form of the pair, and *encrustment*. But *incrustation*. See EM- AND IM-.

**incubator.** Thus spelt, not *-er*.

**incubus.** Pl. *incubuses* (preferred) or *incubi* /-baɪ/. Cf. SUCCUBUS.

**inculcate.** Derived from L *inculcāre* 'to stamp in with the heel' (cf. L *calc-*, *calx* 'heel'), 'tread in', 'cram in', this verb has traditionally been used to mean 'to endeavour to force (a view, a subject, a habit, etc.) into or impress (it) on the mind of another by emphatic admonition'. The usual prepositions governing the object are *upon*, *on*, and *in*. Examples: *These words deserve to be inculcated in our minds*; *An opinion difficult to inculcate upon the minds of others*; *All these teachers inculcate the duties of order, obedience and fidelity on the slaves*. A less standard construction, *to inculcate* (a person) *with* (an idea, habit, etc.), takes the verb further from its roots and its traditional meaning. It was roundly condemned by Fowler (1926) and, with varying degrees of hostility, by some later usage writers. Examples: *A passer-by saved him, formed a close friendship* *with him, and* inculcated him with *his own horrible ideas about murdering women; An admirable training-place wherein to* inculcate the young mind with *the whys and wherefores of everything which concerns personal safety.* The use probably arises by analogy with *to indoctrinate* (a person) *with*, which is the better construction of the two.

**incur.** The inflected forms are *incurred*, *incurring*. See -R-, -RR-.

**indecipherable.** See IN- AND UN-.

**indecorous.** Stress on the second syllable. It is worth noting that the *OED* (1900) gave priority to third-syllable stressing, as did Fowler (1926).

**indefeasible, indefectible.** The distinction between these two relatively uncommon words may perhaps best be kept in mind by associating them respectively with *defeat* and *deficit*. That is *indefeasible* which is not liable to defeat, i.e. (of claims, rights, etc.) that cannot be lost or annulled. That is *indefectible* which is not liable to deficit, defect, or decay, i.e. to failing for want of internal soundness: the word is therefore applied to qualities such as holiness, grace, vigour, resolution, affection, or abundance.

**indefinite article.** See A, AN.

**indent.** The verb, pronounced /ɪn'dent/, and the noun, pronounced /'ɪndent/, show the usual contrasting pattern for such pairs. See NOUN AND VERB ACCENT.

**independence, -cy.** The *-cy* form is now restricted, mostly in historical contexts, to the senses, (*a*) congregationalism (used since the 17c. of 'that system of ecclesiastical polity in which each local congregation of believers is held to be a church independent of any external authority', *OED*); (*b*) an independent or autonomous state. In all other contexts *independence* is the customary word.

**index.** In general use the normal pl. is *indexes*; but in scientific language and in mathematics *indices*. See LATIN PLURALS.

**Indian.** In BrE an Indian is first and foremost a native or national of India, or such a person born in and resident in the UK or elsewhere. What the Americans now call *native Americans* are for the most part called *American Indians* or (less commonly) *Red Indians* in BrE, except in the traditional phr. *cowboys and Indians.*

**indicative.** In all uses second-syllable stressing, i.e. /ɪn'dɪkətɪv/, is now customary.

**indict, -able, -ment.** Pronounce /-'daɪt(-)/. *Indict* means 'accuse', and the formal or jocular homophone *indite* 'compose, write'.

**indifference, -cy.** For all practical purposes the former has driven out the latter, though they were equally common, with more or less matching senses, from the 16c. to the 19c.

**indigestible** is the usual form for 'not digestible', while *undigested* is the customary form for 'not digested'. See IN- AND UN-.

**indirect object.** In grammar, the person or thing secondarily affected by the action stated in the verb, if expressed by a noun alone (i.e. without *to*, *for*, etc.), is called the indirect object. In Latin and Greek it is recognizable, as it once was in English, by being in the dative, while the direct object is in the accusative. The English dative now having no separate form, the indirect object must be otherwise identified, usu. by the fact that it stands between the verb and the direct object (e.g. *Hand* me *that book*), and, if it is to follow the direct object, must be replaced by a prepositional phrase (e.g. *Hand that book* to me). Variations are (1) when no direct object is expressed, as in *You told* me *yourself*; (2) when the direct object is a mere pronoun and is allowed to precede, as *I told* it *you before* (but not *I told the story you before*); (3) when the indirect object comes after a passive verb, as *It was told* me *in confidence. Note*: Variations (2) and (3) are now distinctly archaic. A speaker would now be more likely to say *I told you that* (or *about it*) *before*; *it was told to me in confidence.*

**indirect question.** **1** A passage from Iris Murdoch's *The Book and the Brotherhood* (1987) shows the type, or rather two forms of it: *Jean and Duncan then had a conversation. Jean said* where was his coat, *and he said* he thought he had left it in the car (my emphasis). In direct speech

Jean would have said *Where is your coat?*, and Duncan would have replied *I think I have left it in the car.* In such straightforward cases we make the necessary adjustments of pronouns and tenses without apparent difficulty.

**2** CGEL (1985) distinguishes two main types of indirect questions: (*a*) *wh*-questions, e.g. *I asked him when he would be 65* (equivalent to the direct question *When will you be 65?*); such clauses are introduced by interrogative words beginning with *wh* (*where, whether, what, who, why,* etc.; also *how*). (*b*) *yes-no* questions, e.g. *She asked me whether I was coming now* (equivalent to the direct question *Are you coming now? she asked me*). In both of these types the order of words in the indirect questions fractionally differs from that of the equivalent direct questions (*he would be* ↔ *will you be*; *I was* ↔ *Are you*; i.e. the order of the auxiliary verb and pronoun is reversed).

**3** Fowler (1926) noted that in some circumstances adjusting the order of such elements in indirect questions produces unidiomatic ('abnormal') constructions. Consider the following:

  A. *How old are you?*
  B. *Tell me how old you are*
    or *Tell me how old are you?*
  C. *He wondered how old she was*
    or *He wondered how old was she?*
  D. *He doesn't know how old I am*
    or *He doesn't know how old am I?*
  E. *How old I am is my affair*
    or *How old am I is my affair*

A is the direct question; in B, C, D, and E the first form contains the normal, and the second the abnormal, form of the indirect question. Fowler correctly discouraged the use of the 'abnormal' forms, and then went on to cite some examples (presumably from newspapers) of the undesirable types: e.g. (type D) *It shows inferentially how powerless is that body to carry out any scheme of its own* (normal order *how powerless that body is*); (type E) *How bold is this attack may be judged from the fact that …* (normal order *How bold this attack is*). Clearly it is important to have a 'feel' for the correct ordering of the elements in passages of indirect speech.

**indiscreet, indiscrete.** The first of these homophones (both pronounced /ˌɪndɪ-ˈskriːt/) is the common word meaning

'not discreet, lacking discretion'; and the second is the much less common word meaning 'not divided into distinct parts'.

**indiscriminate,** etc. The customary forms are *indiscriminate* (adj.), *indiscriminately* (adv.), and *indiscrimination*; but *undiscriminatory* (adj.) (in AmE also *indiscriminatory*), *undiscriminating* (adj.). See IN- AND UN-.

**indispensable.** Thus spelt, not -*ible*. See -ABLE, -IBLE 6.

**indisputable.** **1** Stressed on the third syllable.

**2** This is the standard form for the adj. *indisputable*; but cf. *undisputed* (adj.). See IN- AND UN-.

**indissoluble** (adj.). The pronunciation favoured by the OED (1900), namely with second-syllable stressing, has now given way to /ˌɪndɪˈsɒljʊbəl/, i.e. with the stress on the third syllable.

**indistinguishable.** Thus spelt (see -ABLE, -IBLE 6); but *undistinguished* (adj.). See IN- AND UN-.

**indite.** See INDICT.

**individual.** There is no particular problem with the adjective, but there is, or has been, a major one with the noun in some of its uses. When used in contrast with society at large (*the place of the individual in society*), the family, the people, etc., i.e. with a larger grouping, *individual* has no critics. Individuality as a special quality setting a person apart from other members of a group carries through to acceptable uses of the word *individual*. For other uses, a historical note is necessary. The objections to *individual* used 'without any notion of contrast or relation to a class or group: A human being, a person' (OED) apparently began c1870 and have continued. The OED itself (in a section issued in 1900) dismissed it: 'Now chiefly as a colloquial vulgarism, or as a term of disparagement', and provided supporting evidence beginning in 1742 (Johnson). Fowler (1926), in a mood of elegant contempt, said that '[we must like] this ghost of 19th-century jocularity as little as we enjoy the fragrance of a blown-out candle that just now gave us light, or of the smoking-room visited early next morning'. Thus the descriptive editors of the *OED*, and the great schoolmaster himself, legislating for polite society of the 1920s, joined forces in categorizing certain uses of *individual* as 'not to be imitated'.

The passion has not entirely disappeared, though the arguments supporting the labelling may be only half-remembered if at all. The present mood is perhaps best reflected in the way in which the noun is presented to foreign learners of the language. The *Oxford Advanced Learner's Dictionary of Current English* (4th edn, 1989) defines the noun as follows: **1** a single human being: *the rights of an/the individual compared with those of society as a whole.*

**2** (*informal*) a person of a specified sort: *a pleasant, unpleasant etc. individual. What a strange individual!*

**3** (*approving or derogatory*) an unusual or eccentric person: *He's quite an individual!* Foreigners are not in danger of being misled, except that they might have been alerted to the fact that standard speakers are much more likely to use *person* for sense 2, and *chap* (BrE) or *character* for sense 3.

Examples: (contrasted directly or indirectly with members of a larger group) *Individuals composing the crowd may be unanimous in hating the rules*—A. S. Neill, 1962; *She simply continues to treat them as individuals whom she knew*—J. Brodsky, 1982; *As individuals they resented her opinion of them but as members of the tribe they understood how she felt*—C. Hope, 1986; (examples showing varying degrees of informality) *There is confusion, too, over the criteria whereby an individual is deemed to be in need of psychiatric treatment*—A. Clare, 1980; *Individuals arrived there in their private cars with loads of medicines*—C. Hope, 1984; *You must find a diet which suits you. What works for one individual may not necessarily work for another*—Dumfries & Galloway Standard, 1986. In the 'informal' examples some writers would have used *a particular person, Some [of them]*, and *person* respectively.

**indivisible.** Thus spelt, not *-able*. See -ABLE, -IBLE 7.

**indoor, indoors.** The adj. is *indoor* (e.g. *indoor games*), a shortening of earlier *within-door*. The adv. is *indoors* (e.g. *he went indoors at sunset*), representing earlier *within doors*.

**indorse, indorsement.** Variants of the more usual ENDORSE and *endorsement*. See EM- AND IM-.

**induction.** See DEDUCTION.

**-ine.** For *glycerin(e)* etc., see -IN AND -INE.

**inedible** means 'not edible, esp. not suitable for eating'. It contrasts slightly in some contexts with *uneatable* 'not able to be eaten, esp. because of its condition'. *Inedible* is the more common of the two words. For the forms, see IN- AND UN-.

**inedited, un-.** The former is still used by some textual editors, but *unedited* is the more usual term in general literary use. See IN- AND UN-.

**ineffective, ineffectual, inefficacious, inefficient.** For distinctions, see EFFECTIVE.

**inept.** See INAPT.

**inescapable, inessential.** These are the customary forms now, not *un-*.

**inexactitude.** Now usu., and famously, in *terminological inexactitude*: *It* [sc. the employment of indentured Chinese labour on the Rand] *cannot in the opinion of His Majesty's Government be classified as slavery in the extreme acceptance of the word without some risk of terminological inexactitude.* Thus young Mr Winston Churchill, Parliamentary Under-Secretary of State for the Colonies, addressing the House of Commons on 22 February 1906 as spokesman of a government that had just won an overwhelming victory in an election in which denunciation of their predecessors for having sanctioned 'Chinese slavery' had played no small part. This piece of POLYSYLLABIC HUMOUR has worn better than most, thanks to the appeal of its sly whimsicality and the subsequent fame of its author.

**inexpressive** is now the more usual term rather than *unexpressive*. But *unexpressed* (adj.) is the established form. See IN- AND UN-.

**infectious.** See CONTAGIOUS.

**infer. 1** The inflected forms are *inferred*, *inferring*; but contrast *inferable, inference*.

**2** The only unopposed standard meaning of *infer* is 'deduce or conclude from facts and reasoning', by contrast with *imply*, which means 'strongly suggest the truth or existence of (something not expressly asserted)'. A. P. Herbert in *What a Word!* (1935) brought out the distinction in an anecdote: *If you see a man staggering along the road you may* infer *that he is drunk, without saying a word; but if you say 'Had one too many?' you do not* infer *but* imply *that he is drunk.* It is all a question of viewpoint and of logic. For the most part this distinction is carefully observed, but every now and then some writers (and speakers) make the wrong choice, almost always by using *infer* where *imply* is logically required.

Examples: (*imply* correctly used) *Vast stretches of abandoned concrete underfoot imply that someone once had plans for the land*—New Yorker, 1986; *I wouldn't want to imply that lead is the most important factor in children's intelligence*—a doctor interviewed on BBC Radio 4, 19 May 1987; *She's doing the bartending for money, her nonchalance implies*—M. Atwood, 1989; (*infer* correctly used) *You would have been able to infer from the room alone the nature of those who lived in it*—D. M. Davin, 1979; *We can infer, as Professor Ellmann suggests, that Ulysses is full of autobiographical material*—J. P. Donleavy, 1983; *No reference to any living person is intended or should be inferred*—S. Bellow, 1987; *Rose inferred from the letters that Sinclair was in love with Gerard*—I. Murdoch, 1987; (*infer* illogically used for *imply*) *I can't stand fellers who infer things about good clean-living Australian sheilas*—Private Eye, 1970; *I have seen references . . . to the watering of Ascot racecourse, inferring that the water had been taken from public mains at a time when economy is being urged on all consumers*—Daily Tel., 1973; *These were the ones who had made a slightly sulky entrance (inferring rebellion), and had then proceeded to sit on the floor*—M. Bracewell, 1989; *I'm one of those ex-Carrington patients. And I don't consider myself in that category inferred by that sentence*—Metro (NZ), 1989.

The clarity of the distinction between *imply* and *infer* is often questioned, and with a certain justification. It should be kept in mind that the *OED* (sense 4) gives unquestionable examples of *infer* used to mean 'to lead to (something) as a conclusion; to involve as a consequence: said of a fact or statement; sometimes, of the person who makes the statement', with excellent supporting evidence from the 16c. to the 19c. (and some less impressive 20c. examples). There is also abundant evidence that lawyers and judges sometimes use *imply* in contexts which seem plainly to call for *infer*. Garner (1987) cites three examples of *imply* in contexts where a layman would have expected *infer*: *When a possessory interest in property is conveyed, a court may* imply *from the circumstances that the parties also intended to grant or reserve an easement as well desite their failure to say so in the deed*; *In this case the defendant made an application for shares in the plaintiff's company under circumstances from which we must* imply *that he authorized the company, in the event of their allotting to him the shares applied for, to send the notice of allotment by post*; *The requirements of the rule are met if such an intention may be clearly* implied *from the language.* Linguistic attitudes tend to change as time goes on. *OED's* sense 4 may well become one of the natural uses of *infer* at some point in the 21c. The aberrant legal use of *imply* may also come to be regarded by all parties concerned as legitimate. Meanwhile they remain suspect and should be eschewed, in that they seem to be perverse overturnings of the primary senses of the two verbs.

**inferable.** See INFER 1.

**inferior.** Because it is not a true comparative it, like *superior*, cannot be used in comparative constructions with *than*: thus X *is inferior to* Y (not *than* Y). In a phrase like *inferior wine* the meaning is 'inadequate' rather than 'less good'.

**inferiority complex.** See COMPLEX.

**inferno.** Pl. *infernos.* See -O(E)S 3.

**inferrable.** This spelling is not recommended. See INFER 1.

**infinitely.** Used hyperbolically, *infinitely* means 'to an indefinitely great extent' (*he is infinitely cleverer than I am*). When size rather than degree is in question, *infinitesimally* is the better word (*the chances that a nuclear weapon will spontaneously explode are infinitesimally small*).

**infinitive.** **1** It is not practicable to set down here a full account of all the ways in which the infinitive form of verbs functions in English. First, though, some obvious points. The infinitive is the simple uninflected form of a verb, and is the form under which verbs are listed in dictionaries. An infinitive is frequently preceded by the particle *to* (*I am going* to cut *the hedge tomorrow*), but there are circumstances in which the form without *to* (called the bare or simple or plain infinitive) is either optional or obligatory. The split (or cleft) infinitive, i.e. an infinitive with an adverb or adverbial phrase inserted between *to* and the verbal part, as in *you have* to really watch *yourself*, has been a matter of intense public debate since at least the 1860s. See SPLIT INFINITIVE.

**2** The *to*-infinitive is related in a complex way to the verbal/nominal form in -*ing*, i.e. the gerund. For a basic account of some of the similarities and differences between the two constructions, see GERUND 3.

**3** The *to*-infinitive. Some of its features: (*a*) When a second infinitive is used after a *to*-infinitive, the second (and third, etc.) example is not necessarily preceded by *to*. Contrast *can be induced* to move *or* to change *its orientation* and *I prefer not* to live *and* work *in the same room*. (*b*) A number of formulaic phrases, the descendants of older and longer constructions, are in frequent use: e.g. *so to speak, to tell the truth, to put it mildly, to be honest, to say the least, come to think of it, to hazard a guess*. (*c*) Miscellaneous uses: (in which *to* = in order to) *She sat down to wait for him*; (at the head of a clause) *'To be frank with you,' she said, 'I don't believe it'*; (*to* at end of sentence with an infinitive implied but omitted) *Knowledge didn't really advance, it only seemed to*; *he has to make a decision but doesn't know how to*; (accusative + a *to*-infinitive) *he often took groups of students to see the work*. See also FOR 1.

**4** The bare infinitive. It is often optionally used after the verbs *dare, help*, and *need*, but often with a slight change of emphasis (see the entries for these verbs). But its use after modal verbs (*can, may, must, shall*, etc.) and after comparatives and superlatives (*better, had better, best, had best, rather than*, etc.: see these words) is much more significant; e.g. *Peter said he thought he* had better push

on; *Bertrand preferred to go to a restaurant rather than eat alone*. A select list of some other common uses of the bare infinitive: (at the head of a clause) *Try though I did, I couldn't open the bottle*; (after *let* + object) Let him enjoy *his ignorance*; (after *is* and *was*) *All* they want to do *is hide* in the kitchen; *All he had to do* was take *his seat*; (after *why* and *why not*) *Why suppress* it?; *Why not drop* in *at the surgery and let me write out a subscription for you?*. For the mainly AmE use of *come* and *go* + a bare infinitive, see these verbs.

**infinitude**   was gently dismissed by Fowler (1926) as a 'needless variant' of *infinity*, but he judged that 'Milton [the first recorded user of the word] and Sterne, however, will keep it in being for poets to fly to and stylists to play with when *infinity* palls on them'. In lay language (as distinct from that of mathematics and photography) the two words seem to have been more or less interchangeable till the late 19c. Standard modern dictionaries continue to list *infinitude* in its primary senses, and without any qualifying labels. English dictionaries for foreign learners, on the other hand, go their own way by labelling the word *formal*; and they bear out the formality in their illustrative examples: *the infinitude of space; the infinitude of God's mercy*. Stray 20c. examples: *As one did one's wonderful duty one could forget that one's legs were aching from the infinitude of the passages at Windsor*—L. Strachey, 1921; *'Well . . .' he said, looking with an infinitude of regret and reluctance at his newspapers*—J. Krantz, 1982; *There is an infinitude of incidental printed references to their work*—Rev. Eng. Studies, 1983.

**infirmity.**   *The last infirmity of noble minds* is a MISQUOTATION. The passage in Milton's *Lycidas* reads *Fame is the spur that the clear spirit doth raise (That last infirmity of noble mind) To scorn delights, and live laborious days*.

**inflame.**   *Inflam(e)able*, formed from the English verb, and used in the 17c. and 18c., has been displaced by *inflammable* (first recorded in 1605), adapted from French or Latin.

**inflammable. 1** See FLAMMABLE.

**2** It means (*a*) easily set on fire; (*b*) easily excited. It is to be distinguished from *inflammatory*, which means (a) (esp. of speeches, leaflets, etc.) tending to cause anger, bitterness, etc.; (b) of or tending to inflammation of the body.

**inflatable.** Thus spelt, not *-eable*. See MUTE E.

**inflection.** (Also spelt *inflexion*.) The general name for a suffix, etc., used to change the form of a word in order to express tense, number, comparison, etc., as *-ed*, *-ing*, *-s*, *-er* in *jumped, hunting, books, bigger*. Either spelling may also be used for the sense 'a modulation of the voice'. See -XION etc.

**inflexible.** Thus spelt, not *-able*. See -ABLE, -IBLE 7.

**inflict** is mostly construed with *on* or *upon* and means 'administer, deal (a stroke, wound, defeat, etc.)': *the Allied forces inflicted a humiliating defeat on Iraq in 1991*. At the beginning of the 20c., reflected in comments by Fowler in 1926, there was a tendency to use *inflict* when *afflict* was the word required: e.g. *At least the worst evils of the wage system would never have inflicted this or any other present-day community*; *The misconception and discussion in respect of the portraits of Shakespeare with which the world is in such generous measure inflicted are largely due to* ... *Afflict* means 'to inflict bodily or mental suffering on'. In the passive, a person is *afflicted with* or *by* (an illness, crippling debts, etc.).

**infold.** *Enfold* is the recommended form. See EM- AND IM-.

**inform** is a formal equivalent of *tell* and is in fairly restricted use: e.g. *the detective informed him of his rights*; *a station announcement informed us that the next train would arrive in ten minutes*. A customer, taxi, visitor, etc., might be *asked* or *told* to wait; it would not be idiomatic to say, for example, *Please inform Mr Jones to wait outside for five minutes*. The formulaic uses of *inform* in commercial, official, etc., letters (*I beg to inform you*; *I have the honour to inform you*, etc.) 'are generally unnecessary preludes to giving the information provided' (Gowers, 1965), but such phrases nevertheless persist.

**informant** is a neutral term, (a) one who gives information; (b) specifically, the regular term for a person from whom

a linguist, anthropologist, etc., elicits information about language, culture, etc. An *informer*, by contrast, is a person who informs against another person or persons, e.g. to the police, i.e. is a term with sinister overtones.

**infringe.** 1 The inflected forms are *infringed, infringing*; also *infringeable*.

2 'Latin scholars, aware that both *frango* and *infringo* are transitive only, will probably start with a prejudice against [*infringe*] *upon*; but Latin is not English, as some of them know' (Fowler, 1926). The earliest uses of *infringe* in English (16c.–early 18c.) showed the verb used only transitively. But from about the 1760s constructions of *infringe* with *on* or *upon* gradually became established (alongside transitive ones) and are now commonplace: the sense is 'encroach; trespass' (e.g. *the measure threatens to infringe upon and restrict our right to travel in certain countries*). The older use survives, of course: it is still possible to use *infringe* with *constitution, law, patent, privilege, promise, regulation, right, rule*, etc., as direct object, as it has been since the 16c.

**infusable, infusible.** The first is formed from *infuse* 'to pervade (e.g. *anger infused with resentment*); to steep (herbs, tea, etc.) in liquid to extract the content' + the suffix *-able*; it means 'able to be steeped, etc.'. The second is formed from the negative prefix *in-* + *fusible* 'that can be easily fused or melted'; *infusible* therefore means 'not able to be fused or melted'. It is a pity that the two words, which are far apart in meaning, are so narrowly distinguished in spelling.

**infuse.** The verb's oldest meaning (15c. onward) was 'to pour in'. Very soon afterwards it acquired the transferred and figurative senses 'to instil, insinuate (e.g. *to infuse new life into a community*); and also the sense 'to steep or drench (a plant, etc.) in a liquid, so as to extract its soluble properties'. Also, as the OED has it, 'with inverted construction', and by what Fowler (1926) called 'object-shuffling', from the 16c. onward it has been used to mean 'to affect or act upon (a liquid) by steeping some soluble substance in it; hence, to imbue or inspire (a person or thing) *with* some quality. In wider use, to impregnate, pervade, imbue (with some quality, opinion, etc.)'. These senses are all provided with good supporting evidence in the OED. It follows that *infuse with* and *imbue with* are alternative standard constructions in the types *the self-respect with which the troops had been infused* (or *imbued; he infused* (or *imbued*) *his troops with self-confidence*.

**-ing.** 1 *Picking up my Bible, the hill seemed the only place to go just then*—J. Winterson, 1985; *Packing to leave, her fingertips had felt numb on contact with her belongings*—M. Duckworth, 1986. For liberties of this kind, see UNATTACHED PARTICIPLES.

2 For the difference between participles in *-ing* and the gerund, see GERUND 2.

3 *Then we had our old conversation about the house being haunted*—C. Rumens, 1987. For such mixtures of participle and gerund, see POSSESSIVE WITH GERUND.

4 *In all probability he suffers somewhat, like the proverbial dog, from his having received a bad name*—cited in Fowler, 1926. For the need or no need of *his* and other possessives in such contexts, see GERUND 4.

5 *She bestowed her activity, rather than letting it be harnessed to anyone else's needs*—A. Brookner, 1988; *It seemed better to come and talk to someone here, rather than to continue to write letters*—Francis King, 1988. For the competing types *rather than + -ing* and *rather than + infinitive*, see RATHER 3.

6 *As well as closing the railway, it should make the Danube impracticable for traffic*—cited in Fowler, 1926; *Just like Dolly to usurp the mourning function as well as presuming to treat the evening as a normal evening party*—A. Brookner, 1985. On the use of *-ing* forms after *as well as*, see WELL 1.

**ingenious, ingenuous.** 1 There is substantial evidence in the OED that in the 17c. some writers (including Shakespeare) 'by confusion' used *ingenuous* for *ingenious* and vice versa. This is not the place for a full treatment of the history of the two words and of their intersections, but some details are essential. *Ingenious* came into English via French (from L *ingeniōsus* 'intellectual, talented') in the 15c. The meaning gradually weakened, and its current senses are

'clever, skilful (at doing something)' and '(of a machine, theory, etc.) cleverly contrived'. As the OED expresses it, 'Now usually somewhat light or sometimes even depreciative, expressing aptitude for curious device rather than solid inventiveness or skill'. *Ingenuous*, by contrast, means 'innocent, artless; open, frank'. It is sometimes difficult to distinguish it from DISINGENUOUS. It entered the language at the end of the 16c. directly from L *ingenuus* '(of occupations, studies, etc.) befitting a free man; hence (of character) honourable, frank, modest'. Later in the same century, as a by-product, it began to be used to mean 'innocently frank, guileless, artless'. For some two centuries, from c1600 to c1800, the two words intersected by sharing the broad senses 'befitting a well-born person' and 'open, frank, candid'. That is when 'the confusion' arose, and folk memories of this state of affairs remain. It has not helped that *ingenious* and *ingenuous* are virtually what linguists call a minimal pair, being divided only by the length of the vowel of the second syllable, and by a vowel in the penult.

2 It is a minor curiosity that the noun corresponding to *ingenious* is *ingenuity* (with a *u* in the antepenult), while that corresponding to *ingenuous* is *ingenuousness*.

**ingénue.** First recorded in English in Thackeray's *Vanity Fair* (1848), it is still often printed in italics as being not a fully naturalized word. It is also pronounced in Britain in a manner roughly resembling the French original, i.e. as /ˌæ3eɪ'njuː/. The accent is sometimes omitted. In other words it is a standard example of a loanword that is needed for its sense but has not yet acquired full British citizenship.

**ingraft.** *Engraft* is the recommended form. See EM- AND IM-.

**ingrained** is the normal spelling for the adj. meaning (a) deeply rooted, inveterate, (b) (of dirt, etc.) deeply embedded. And the finite verb *ingrain*, meaning (a) to implant (a habit, belief, etc.) ineradicably in a person, (b) to cause (a dye, etc.) to sink deeply into something, is usually so spelt.

**ingratiate.** This verb, yet another Latinate loanword of the 17c., is now normally used reflexively in the sense 'to render oneself agreeable to, to bring oneself into favour *with*'. It is sometimes improperly used transitively in the same sense: *Even if he does ingratiate the men, it will only be by alienating the women* (cited by Fowler, 1926, with disapproval); *He* [sc. Neil Kinnock] *refused to serve under Mr James Callaghan, is not tarred by loyalty to that Government, and thus ingratiates the Left—Daily Tel.*, 1983. But the following absolute use is beyond reproach: *'Good' behaviour designed to placate and to ingratiate—A*. Storr, 1972.

**inherent.** In standard English either /m'herənt/ or /m'hɪərənt/ is acceptable. J. C. Wells (1990) reports that his poll panel of RP speakers showed a 66% preference for the first of these.

**inherit.** A person who inherits is an *inheritor* (thus spelt, not -*er*). The feminine forms *inheritress* and (esp. in legal documents) *inheritrix* are also available if desired.

**in hopes of, in hopes that.** See HOPE (noun).

**initial** (verb). The inflected forms are *initialled, initialling* (AmE *initialed, initialing*). See -LL-, -L-.

**initialisms.** See ACRONYM 2.

**initiate.** Regular uses are (*a*) (usu. followed by *into*) to admit (a person) into a society, an office, etc.; (*b*) (usu. followed by *in* or *into*) to instruct (a person) in science, art, etc. It is not idiomatic (by what Fowler called 'object-shuffling') to *initiate* knowledge, a subject, a hobby, etc., *into* a person. In such a construction, *instil* is the better word.

**-in-law, step-, half-. 1** The plural of *in-law* formations, e.g. *brother-in-law*, is typically *brothers-in-law*, i.e. -*s* is added to the first word in the formation.

**2** -*in-law* 'is a phrase appended to names of relationship, as *father, mother, brother, sister, son*, to indicate that the relationship is not by nature, but in the eye of the canon law, with reference to the degrees of affinity within which marriage is prohibited. Formerly -*in-law* was also used to designate those relationships which are now expressed by *step-*,

e.g. *son-in-law* [formerly] = *step-son, father-in-law* [formerly] = *step-father*; this, though still [sc. in 1900] locally or vulgarly current, is now generally considered a misuse' (*OED*).

**3** A *half-brother* is a brother with only one parent in common; similarly *half-sister*.

**Inlay. 1** The noun is stressed on the first syllable and the verb on the second. See NOUN AND VERB ACCENT 1.

**2** The pa.t. and pa.pple of the verb is *inlaid*.

**inmesh.** *Enmesh* is the customary form. See EM- AND IM-.

**innavigable** has given way to *unnavigable*. See IN- AND UN-.

**innings.** In cricket, *innings* is both sing. (*the first innings*) and pl. (*the best of his many innings*). In baseball, the sing. form is *inning* and the pl. *innings*.

**Innocence, innocency.** Both words were more or less equally current from the 14c to the 19c., but *innocency* is now rare except in readings of Ps. (AV) 26: 6: *I will wash mine hands in innocencie: so will I compasse thine Altar, O Lord*.

**innocent of.** This standard expression (*entirely innocent of the crime with which he was charged*) has also been informally used since the early 18c. in a semi-humorous way to mean 'free from, devoid of': e.g. *The Sermon ... was quite innocent of meaning—*J. Wesley, 1743; *The windows are small apertures ... innocent of glass—*J. Colborne, 1884; *His skull-cap was innocent of decoration—*G. Durrell, 1954; *The eyes elaborately made up and the lips in contrast, innocent of any cosmetic at all—TLS*, 1977; *One editor's discussion of the Bowery B'hoy is innocent of reference to David Rinear's article—Theatre Research International*, 1986.

**innuendo. 1** The recommended pl. is *innuendoes*, not -*os*. See -O(E)S 1.

**2** The word was first (16–18c.) used as a formula to introduce a parenthetical remark commenting on or elucidating what had just been said or written, = 'meaning, to wit, that is to say'. This use was derived from L *innuendō* 'by nodding at, pointing to, meaning', ablative gerund of *innuere* 'to nod to, mean'; in medL used to introduce a parenthetic clause. Skeat's *Etymological Dict.* quotes an example from Thomas Blount's *Glossographia* (1674): *he* (innuendo, *the plaintiff*) *is a thief.* From this use, by extension, it came to signify the injurious meaning or signification alleged to be conveyed by words not *per se* injurious; and, by further extension, any disparaging allusive remark or hint.

**innumerate.** See ILLITERATE.

**inobservance** is the regular form now, not *unobservance*, but both are being overtaken by *non-observance*. The usual corresponding adj. is *unobserved*. See IN- AND UN-.

**inoculate.** 1 Thus spelt, not *inn-*.

2 See VACCINATE.

**in order that.** The *OED* (in a section published in 1904) gives only brief attention to this complex subordinator s.v. *order* sb. 29. The three illustrative examples show *in order that* followed in the dependent clause by the subjunctive *be* (1711), and by *may* + plain infinitive (1832, 1875). As it happens (see below), the apparent restriction of the following verb to the subjunctive or to the modal *may* (or *might*) seems to have been mainly justifiable at the time. Those grammarians who have treated the matter at all seem to agree that the verb in the dependent clause should not normally be any of the other modals (thus excluding *can* and *could*, *shall* and *should*, *will* and *would*, etc.). An examination of two new resources suggests that substantial changes have occurred in the course of the 20c. and are continuing. The changes are all in the direction of the removal of earlier constraints in what might appear in the dependent clause after *in order that*. The newly available resources are the electronic database of *OED* 2 and the electronic file of incoming quotations in the *OED* Department. They reveal that in the period 1901–26 (in which year *MEU* was published), which I shall call Phase 1, *may* and *might* were almost invariably used. In Phase 2, 1927–c1960, nearly one third of the examples of *in order that* in the *OED* 2 database showed verbs other than *may*, *might*, or a subjunctive standing in the dependent clause. In Phase 3, the period since about 1960,

the formerly excluded group (*can, could,* etc.) actually exceeded *may, might,* and the subjunctive in the proportion 10:7. This is not to say that *may, might,* and the subjunctive are being abandoned in these circumstances, but simply that older constraints upon other types of verb appearing in the dependent clause are gradually being withdrawn.

Some illustrative examples follow.

Phase 1 (may) *In order that the oiling may be confined to the road-bed only, the rails are kept free from spraying by guards on the sprinkling-car—Chambers's Jrnl*, 1902; (might) *Gater . . . had decided to sell his high-geared bicycle in order that he might become the possessor of a Brownie camera—A. Bennett*, 1906; *Stabilisation of wages is an urgent necessity in order that the industry might enjoy continued peace—World's Paper Trade Rev.*, 1922; (can) *The motor should be wound up fully for each record played, in order that the turntable can rotate at its normal and even speed—P. A. Scholes*, 1921.

Phase 2 (may) *A suitable block-and-tackle is essential in order that the boat may be hauled far enough up the shore to be safe from 'rafting' ice—Discovery*, 1935; (should) *What factor . . . must be present in order that the implicate should be dependent in being on the implicans—D. J. B. Hawkins*, 1937; (shall) *In order that he shall be said to make a moral judgement, his attitude must be 'universalisable'—A. E. Duncan-Jones*, 1952; (verb in subjunctive) *Since oxyanions have a large amount of resonance stabilization, it is necessary to overcome this stability in order that a chemical reaction take place—Chemical Reviews*, 1952.

Phase 3 (might) *he always insisted upon a certain reserve in order that the artist might give 'full measure' on the stage—Dancing Times*, 1990; (may) *the cutting of benefits to mothers who do not give the names of their children's father . . . in order that they may be made financially accountable—Rouge*, 1990; (can) *The Telematics Programme . . . looks at users' needs and requirements in order that entire networks can talk to each other ready for 1992—Practical Computing*, 1990; *Westlake alumnae cannot help but speculate that our beloved campus will become a teardown in order that a new coed school can be built—Vanity Fair* (US), 1990; (could) *In order that the oligomer could be deprotected easily, 'PAC amidites' were used in the synthesis—Nucleic Acids Research*, 1990; (verb in subjunctive) *in order that he be regularly* scared by Authority, he should present himself every six months to the Service's Legal Adviser—J. le Carré, 1989; Paulin vacillates in his claims in order that he not have to meet the responsibilities of arguing any of them out—London Rev. Bks, 1990; (does) I can only hope that such methodology will be adopted by teachers new to media work in order that learning about the media does not become a bookcover here and a story-board there with little attempt at a coherent conceptual context—TES, 1990.

This new evidence does seem to suggest that the old constraints are no longer valid. The models bringing it all about are clauses headed by so that, so, or that, where no such constraints applied.

**in order to.** From the 16c. to the 19c. used as a complex preposition meaning 'with a view to the bringing about of (something), for the purpose of (some prospective end)': e.g. A meeting ought ... to be called ... in order to a regular opposition in parliament—Burke, 1773; In order to the existence of love between two parties, there must be a secret affinity between them—E. M. Goulburn, 1869. The only standard use now is a construction (first recorded in 1711) with a following to-infinitive, with essentially the same meaning, i.e. 'with the purpose of doing, with a view to': e.g. Rozanov ... had taken a sharp right-hand turn in order to avoid going along the road—I. Murdoch, 1983; in order to make the material manageable, he divides it into three parts—Jrnl RSA, 1986; The SDP ... wants to reform the unions in order to bring them back in—Times, 1986. It is occasionally claimed that to, rather than in order to, is all that is needed, and it is true that to is much the more commonly used of the two in this kind of context: e.g. the voice used by the daughter to bully her mother—P. P. Read, 1986; the path takes an unscheduled turn to miss a big tree—C. K. Stead, 1986; over half the members of Nato have been deploying military aircraft to relieve the African drought—Times, 1986; We both came here to get other people out of trouble—A. Brookner, 1986. There is clearly room for both constructions. It is hard to pin down reasons for the choice of the longer form. Some have claimed that it is slightly more formal. The presence of a different kind of to-infinitive in the vicinity (see the first Times 1986 example above) may

sometimes be a factor, as will considerations of rhythm and emphasis.

**in petto** is an Italian phr. (petto, from L pectus 'breast') meaning 'in one's own heart, in contemplation, undisclosed'. It was adopted in this sense in English in the 17c. and still is so used by those familiar with Italian. But, by confusion with petty, it has been repeatedly and wrongly used in English since the early 19c. to mean 'in miniature, on a small scale'. When one sees the names of some of the malefactors (e.g. Kipling in his Kim (1901), He represents in petto India in transition—the monstrous hybridism of East and West; also Disraeli, T. E. Lawrence, and others), one must wonder if restricting the phrase to its Italian meaning is now a lost cause.

**input** (noun and verb). **1** Both the noun and the verb are now stressed on the first syllable. When the relevant section of the OED was issued in 1901, the noun, which was restricted to Scotland and meant 'a sum put in, a contribution', had initial stressing; but the verb (first recorded in the 14c. but by then obsolete) was stressed variably (infin. second syllable: inputted, inputter, inputting primary stress on first syllable, secondary stress on second). The secondary stress is still customary in the inflected forms.

**2** To the editors of the OED the word looked doomed except in Scotland. The advent of computers breathed new life into both the noun and the verb; and from the language of computers input made its spectacular entry into the everyday vocabulary of psychology, linguistics, and related subjects, and then into the language at large, used of data, information, effort, etc., 'put in or taken in or operated on by any process or system', as the COD has it.

**3** The pa.t. and pa.pple are either input or inputted; the pres.pple is inputting. See -T-, -TT-.

**inquire, enquire.** See ENQUIRE.

**in regard to.** See REGARD 1.

**in respect of, to.** See RESPECT.

**insalubrious, unsalubrious.** Both are well-established words meaning '(of climate or place) unhealthy', but the former is at present easily the dominant member of the two. See IN- AND UN-.

**insanitary, unsanitary.** Sanitation is a 19c. concept: the word itself was first recorded in 1848, and other words in the family also made their first appearance in that century (*insanitary* 1874, *unsanitary* 1871, *sanitary* 1842, *sanitate* 1882, *sanitize* 1836). Of the negative adjectival forms, *insanitary* is the more usual of the two. But there is something to be said for Fowler's view (1926) that *unsanitary* might be used 'of a place, etc., that neither had nor needed provisions for sanitation (e.g. *a primitive and unsanitary but entirely healthy life* or *village*; *insanitary* implies danger [to health]'. See IN- AND UN-.

**inside of.** Used of a period of time (*inside of a week* = less than a week), this colloquial expression is first recorded in an American work of 1839. From AmE it has gradually made its way into other forms of English, but not into standard BrE. The *OED* also records the closely related phr. *the inside of* 'the middle or main portion of a period of time'; the phr. is labelled 'colloq.' and is first recorded in 1890. Examples: (inside of) *Renny won a fiver off me because he made friends inside of the month*—M. de la Roche, 1940 (Canad.); *I'll be between the sheets inside of half an hour, old man*—A. Fullerton, 1954; *There too, faces were sometimes painted with 'any kind of paint that you could rub off inside of a few days'*—*Christmas Mumming in Newfoundland*, 1969; (the inside of) *Why, I can't even keep a man faithful to me for the inside of a month*—C. Isherwood, 1939; *At first Isabel had only meant to stay away for the inside of a week*—L. P. Hartley, 1955.

**insightful.** First recorded in a work by John Galsworthy in 1907, meaning, of course, 'characterized by insight', it has become an omnipresent word of praise in psychological, linguistic, literary, and other kinds of writing. Examples: *She created a film which was memorable, intriguing and moving, a warm and insightful reconstruction of a vanished age*—*Listener*, 1982; *The text typology is followed by an insightful outline of how the general knowledge ... was passed on*—*Internat. Jrnl Lexicography*, 1988; *It was a wonderful insightful exhibition*—*Modern Painters*, 1988. The frequency with which it has been used throughout the century has taken away much of its impact.

**insignia.** In origin it is the L pl. form of *insigne* 'mark, sign, badge of office', which is itself a substantival use of the neuter singular of the adj. *insignis* 'distinguished'. In two of its uses it has had an uncomfortable history since it entered the language in the mid-17c. (*a*) *Insignia* as pl. ( = badges or distinguishing marks of office) (*all the insignia of the Vice-Chancellor and proctors*). (*b*) Used erroneously (according to the *OED*) as sing. with *insignias* as its pl.: *Bells, ladle, and the fool's cap ... Insignias of their liking*—W. H. Ireland, 1807; *I saw not a single racer at Sestrière bearing an insignia that seemed out of place*—*Times*, 1971. (*c*) *insigne* (singular) 'badge, ensign, emblem', pronounced /ɪnˈsɪɡni/: *The men of Lord Louis Mountbatten's Southeastern Asia command wore it* [sc. the figure of a phoenix] *as an insigne in World War II*—W. R. Benét, 1948; *pieces here and there of old Wehrmacht and SS uniform, tattered civilian clothes, only one insigne in common, ... a painted steel device in red, white and blue*—T. Pynchon, 1973. Type (*a*) is the only one of the three in standard use in BrE. Type (*b*) is marked as 'fully standard' in American dictionaries. *Insigne* is rare and its use likely to cause bewilderment.

**insist** has three main uses: (*a*) *insist on*, demand, maintain (*he insisted on making an appointment*; *she insists on her suitability*); (*b*) followed by a *that*-clause, often one containing a verb visibly or notionally in the subjunctive (see below); (*c*) used absolutely (*since you insist, I will come to the party*). Examples of type (*b*) ( = to maintain persistently; always, when introducing an indirect statement, with indicative in the *that*-clause) *Protarchus ... insists that ... all pleasures are good*—B. Jowett, 1875; *She insists that he is older than he says he is*—mod.; ( = to make a demand with persistent urgency; with actual or notional subjunctive in the *that*-clause, and sometimes with omission of *that*): (with *that*) *Tony insisted that she accompany him to a meeting of the Literary Society*—A. S. Byatt, 1985; *Her father insisted that she stay at the comprehensive school until she had done A-levels*—P. Gilliatt, 1988; *He wants Western tourism to be Cuba's principal source of foreign exchange yet insists that Cubans not be attracted to the values of their guests*—*NY Rev. Bks*, 1989; (*that* omitted) *And I, maliciously, insisted he take the* *most comfortable chair*—P. Lively, 1987; *He insisted she handle the cutlet*—P. Carey, 1988; *That evening he insisted I share his opium pipe with him*—C. Phipps, 1989; *There were nights when she seemed so forlorn that he insisted she sleep over in the spare room at the back*—R. McCrum, 1991. In such contexts a putative *should* + infin. is often used instead of the subjunctive: e.g. the Carey 1988 sentence could be re-expressed as *he insisted that she should handle the cutlet.*

**in so far as.** This complex subordinator (as it is now called) has been in use since the 16c. (it has the rare distinction of being treated twice in the *OED*, s.v. *in* prep. 39 and *so* adv. & conj. 35c), and is still common. It is conventionally written as four words in OUP house style, but other standard authorities (including *CGEL*, 1985) use *insofar as* and others *in-so-far as*. It normally means 'to the extent that'. Examples: *Enforcement, insofar as salaries are concerned, is costing nothing*—Times, 1969; *The results of it will be used here in-so-far as they affect the classification of the group*—Watsonia, 1971; *The exercise of reviewing his life was proving monstrous in so far as it revealed the places in which it had gone irredeemably wrong*—A. Brookner, 1988. It is a somewhat formal phrase, and can sometimes be replaced without loss of meaning or dislocation of syntax by *so far as*, *though*, *since*, or some other simpler conjunction.

**insouciance, insouciant.** Opinion is tending towards a fully Anglicized pronunciation for these words: /ɪnˈsuːsɪəns/, /ɪnˈsuːsɪənt/. But some people still retain the French pronunciation in part or full, i.e. /ˌæsʊsjɑ̃ːs/, /ˌæsʊsjɑ̃/.

**inspan** /ɪnˈspæn/ (verb). A SAfr. word meaning 'to harness an animal or animals to (a wagon,etc.); to harness (people or resources) into service'. The inflected forms are *inspanned, inspanning*. See -N-, -NN-.

**inspect** makes *inspector* and the now-fading feminine equivalent *inspectress*.

**inst.** In business letters this abbreviation of *instant* (*your letter of the 6th inst.*) means 'of the current month'. Nowadays the name of the month is commonly used instead (*your letter of 6 June*).

**instability.** See IN- AND UN-.

**install** is now the recommended form, rather than *instal*. The inflected forms are *installed, installing*, but *instalment* (AmE normally *installment*). See -LL-, -L-.

**instance.** 'The abuse of this word in lazy periphrasis has gone far,' says Fowler (1926), 'though not as far as that of *case*.' Two examples are cited of which this is one: *The taxation of the unimproved values in any area, omitting altogether a tax on improvements, necessarily lightens the burden in the instance of improved properties.* Whether the battering by Fowler and others has brought it about or not, it is now quite hard to find examples of *in the instance of*. The best policy is to refrain from using this particular phrase, but to continue to use the word *instance* in its other dictionary-listed senses, e.g. 'an example or illustration of' (*just another instance of his lack of consideration*), 'a particular case' (that's not true in this instance); also the phrases *for instance*, and *at the instance of* 'at the request or suggestion of (a person)': *I was there at the instance of Mrs. Shuttlethwaite*—T. S. Eliot, 1950.

**instigate** properly means 'to spur, urge on (esp. to an antisocial or discreditable act)', but, as a correspondent has pointed out to me, it is occasionally wrongly used to mean 'institute, start, set up': e.g. (from a 1991 letter written by a prison governor) *Re improvement objectives—I have made a start on the above . . . The first change is to instigate a full induction programme*; (from a Treasury circular dated 16 May 1991) *Departments . . . should ensure that all staff are made aware of their obligations under the legislation, and instigate appropriate actions to ensure compliance.*

**instil. 1** Thus spelt in BrE, but predominantly as *instill* in AmE. The inflected forms are *instilled, instilling*. See -LL-, -L-.

**2** Fowler (1926) worried about what he called 'object-shuffling confusion', citing in support the sentence *The Tsar's words will undoubtedly instil the Christians of Macedonia with hope.* 'You can inspire men with hope,' he said, 'or hope in men; but you can only instil it into them, not them with it.' He was right. In such circumstances the word to use is *imbue* or *infuse*. There is a 1644 example (from

Milton) of *instil with* in the *OED* but it is sternly labelled *Obs. rare*⁻¹. Occasional passive uses of *instil with* in mod E merely show that the virus is not extinct : *I have prepared myself; my muscles are instilled with everything that must be done*—J. Updike, 1981; *During the war my mother and brother and I went to Norfolk, and there I was instilled with a love of the countryside*—Sunday Express Mag., 1986.

**instinct** (noun and adj.). see NOUN AND ADJECTIVE ACCENT.

**instinct, intuition.** See INTUITION.

**instinctual.** This adj., brought into being, it would seem, by psychologists in the 1920s, being not easily distinguishable from *instinctive*, is not often found outside learned monographs. It joins other sociolinguistic and archaeological formations (or revivals) of the 20c. like *aspectual* and *processual*. Such forms have the advantage that they are eminently translatable into the main continental languages of Europe; and, of course, have formal validity in that they join other well-established words ending in -*ual* like *conceptual*, *eventual*, *habitual*, and *textual*.

**institute, institution.** For concrete uses of the words, namely (*a*) a society or organization instituted to promote some literary, scientific, artistic, professional, or educational object, or (*b*) a building in which the work of such a society is carried on, there is no formal distinction between the two words. At the time of the formation of such a society one of the two terms was chosen and for the most part has remained in use. The earliest *institute* mentioned in the *OED* is the *Mechanics' Institute* (estab. 1823), and others listed include the *Royal Institute of British Architects* (founded 1834) and the *Royal Archaeological Institute* (1843). Of *institutions* the earliest mentioned in the *OED* is the *Royal Masonic Benevolent Institution* (founded 1798), but it is clear that various benevolent and charitable institutions existed from the beginning of the 18c. Other early institutions included the *Institution of Civil Engineers* (1818) and the *Smithsonian Institution* (Washington, DC, 1830). Typical examples of the two forms : *Taylor Institution* (Oxford, opened 1848); *Women's Institute*

(first in Canada 1897, then in various other countries in the early 20c.); *British Standards Institution* (UK national standards body, 20c.); *Institute for Advanced Studies* (at Princeton Univ., NJ, founded 1933). Since the early 19c. *institution* has also been widely used for well-established and much loved national customs, (sporting) events, constitutional concepts, etc. Thus cricket, the Grand National, the monarchy, the Boat Race, Wimbledon, the Trooping of the Colour, the (London) Lord Mayor's Show, the Chelsea Flower Show, the last night of the Proms, and a host of other aspects of life in Britain are said to be *national institutions*. The impermanence of such public recognition, however, is shown by the fact that in 1926 Fowler included the Workhouse and capital punishment in his list of such institutions.

**instruct** (verb). The derivatives are *instructor* (not -*er*) and (now rare) *instructress*.

**insubstantial** (first recorded in 1607) seems to have edged out *unsubstantial* (mid-15c.) from the language. There is no lexically definable difference between the two (they both mean 'lacking solidarity or substance; not real'); the occasional choice of *unsubstantial* seems to depend more than anything else upon the presence of the regular form *unsubstantiated* in the immediate vicinity. The main dictionaries for foreign learners of English give no houseroom to *unsubstantial*. See IN- AND UN-.

**insufficient** (adj.). Care should be taken in the choice of this word instead of *not enough*, *too little*, or some other synonym. A banker might stamp *insufficient funds* on an overdrawn account, and a tutor might write *insufficient evidence* on an unconvincing essay. But *inadequate* is often a better choice: a person, for instance, can be said to be *inadequate* but hardly *insufficient* (though Chaucer, Lydgate, Spenser, and others used it in this sense). And one can no longer say *He is insufficient in lands*, i.e. does not own enough land, as was possible from the 15c. to the 17c.

**insular** is a mild pejorative, bestowed normally with condescension. It is characteristic of the British that, although quick to discern intellectual and moral virtues in small self-sufficient indigenous communities, such as the Tibetans and the Eskimos, they affect to despise the mentality of sub-units of their own population. *Insular*, used of an attitude towards some aspects of international affairs, is merely *provincial* or *parochial* writ large.

**insupportable** (first recorded in 1530) is fighting it out with *unsupportable* (1586). In so far as a clearly definable difference exists between them, the first is usually chosen when the required meaning is 'unjustifiable', and the second when the required meaning is 'indefensible'. But the line is a thin one. The main dictionaries for foreign learners of English give space to *insupportable* but none to *unsupportable*. See IN- AND UN-.

**insure.** See ASSURE. See also EM- AND IM-.

**insusceptible** (first recorded in 1603) is now usually preferred to the synonymous *unsusceptible* (1692), though neither is in common use.

**intaglio** /ɪnˈtɑːlɪəʊ/ or /ɪnˈtɑːljəʊ/. Pl. *intaglios*. See -O(E)S 4. *Intaglio* is opposed to *relief* as a name for the kind of carving in which the design, instead of projecting from the surface, is sunk below it. Cf. It. *intagliare* 'to cut into'.

**integral.** As a noun it has little if any currency outside mathematics. As an adj. ( = necessary to the completeness of a whole; forming a whole) it, like the noun, should be pronounced with the stress on the first syllable, i.e. /ˈɪntɪɡrəl/, not /ɪnˈteɡrəl/.

**integrate** (verb). Since the late 1940s this 17c. word has been drawn into widespread use (similarly the associated words *integrated* and *integration*) to mean 'to bring (racially or culturally differentiated peoples) into equal membership of a society or system'. This welcome development is surprising only in the sense that it arrived so late in the day. The process of integration continues, more rapidly in some countries than in others: in other words *integrate*, in modern terminology, is, in this sense at least, a dynamic and not a stative verb.

**intelligent, intellectual.** While an intelligent person is one who is 'quick of mind; clever, brainy', an intellectual lives on a higher plane of abstract thought. The range of ability covered by the word *intelligent* is considerable, from that of a child seeming to have acquired skills ahead of the normal time to the kind of person, Fowler said in 1926, 'we most of us flatter ourselves that we can find in the looking-glass'. Wayward people in the dock are often described by prosecuting counsels as *intelligent* but .... A dog that performs a particular act, e.g. fetches a thrown stick, may be described as *intelligent*. An *intellectual* is a being apart, someone with rare special insights, such as, say, Isaiah Berlin, or, come to that, Karl Marx, the supreme example of an intellectual unmasked. *Intelligent people* are dispersed through the nation. They are hardly ever ham-fisted or impractical; they are simply too busy protecting society from anarchy to claim immunity from the acquisition of ordinary skills. *Intellectuals* are a distinguished but impermanent minority; they normally speak like archangels, or philosophers, or political scientists, usually in several languages, and have original views about the arts and about the diverse ways of mankind, but usually cannot cut a slice of bread straight or drive a car. Their reputations, like their opinions, come and go.

**intelligentsia.** A key word in 20c. political vocabulary, it was first formed in pre-revolutionary Russia to mean 'the part of a nation that aspires to intellectual activity; the class of society regarded as possessing culture and political initiative' (*OED* 2). The word survives (except with backward reference) mainly as a convenient appellation for any group of people judged to be more than ordinarily thoughtful or clever in a designated area of activity.

**intend** (verb). Normal constructions of this verb need not be dwelt on. They include the types ( + *to*-infinitive) *we intend to go*, ( + gerund) *we intend going*, and (in passive + *for*) *these are intended for children*. Non-standard constructions include the type *he didn't intend for Wales to lose* (read *that Wales should lose*); and the informal AmE type *Don't pick up a* *magazine unless you intend on buying it* (read *intend buying*).

**intended** (noun). In this age of broken marriages and of unblessed extramarital unions it is hardly surprising (though to Fowler in 1926 it was 'curious') that 'betrothed people should find it so difficult to hit upon a comfortable word to describe each other by'. *My intended* (a type first recorded in 1767), *my engaged* (unrecorded), and *my betrothed* (1588) are all notional possibilities but seem impossibly stilted or artificial. *My lover* is too explicit by far. Idiomatic substitutes are *my boyfriend/girlfriend*—though imminent marriage is not implied—*my future wife* (*husband*), and *my wife* (*husband*) *to be*. What it comes down to is that *my fiancé(e)*' is really the only appropriate expression, though various periphrases are available, e.g. *This is my friend X. We are to be married in November.* The fact that they are already living together is often left unsaid.

**intense, intensive.** They are both long-serving words (*intense* first recorded *c*1400, *intensive* in 1526) with many branches and sub-branches of meaning and frequent intersections. The broad strategy of the *OED* entry for *intense* was to place emphasis on its etymological sense, ('stretched, strained, high-strung') and therefore to highlight the following three senses: (*a*) (of a quality or condition) extreme, excessive, ardent; (*b*) (of a thing) having a characteristic quality in a very high degree; (*c*) (of a person or personal action) intent, eager, manifesting extreme emotion. *Intensive* to begin with (16–17c.) was used in senses (*a*) and (*c*), but settled down in the 17c. as the routine adj. corresponding to the noun *intensity* (of intrinsic strength, depth, or fullness). This broad semantic path often coincided with that of sense (*b*) of *intense*. In the 19c. and 20c. *intense* has retained its traditional senses, none of which has lost ground. *Intensive*, on the other hand, has moved into new territory. Since the early 19c. it has been applied to methods of cultivation, animal husbandry, fishing, etc., which increase productivity by the use of new techniques. Towards the end of the 19c. the notion of intensive medical treatment emerged, out of which came the term *intensive care* in the 1960s. And in the mid-20c. economists

introduced the distinction that an industrial process is either *capital-intensive* (intensively using capital) or *labour-intensive* (needing a large workforce). What we are looking at is a standstill adj. (*intense*) and a mobile one (*intensive*). But the limitations placed at the end of the 17c. upon certain uses of *intensive* (see above) remain as valid as ever.

**intensifier,** a class of adverbs that amplify or add emphasis to a gradable adjective, e.g. *greatly obliged, highly intelligent, perfectly reasonable*; or that have 'a general lowering effect' from an assumed norm, e.g. *barely intelligible, relatively small*. Also a class of adjectives that have emphasizing, amplifying, or downtoning force, e.g. *a sure sign, a true scholar, utter nonsense; a close friend, a complete fool; a feeble joke*. Some scholars use the term more widely, e.g. of *himself* in *he himself couldn't come*. The older term (still sometimes used) was *intensive* (noun and adj.).

**intensive.** See INTENSE and prec.

**intention** is construed with *of* + noun (*the noblest intentions of mankind*), *of* + gerund (*he acted with the intention of pleasing her*), or with a *to*-infinitive (*its main intention is to inform the public that any promiscuous person ... is at risk from AIDS—BMJ*, 1986). Some authorities (e.g. *CGEL*, 1985) assert that the dominant construction is with *of* + *-ing*, while others suggest that the construction with a *to*-infinitive occurs with approximately equal frequency. Examples: (followed by *of* + gerund) *I have no intention—no present intention—of standing for Parliament*—H. Macmillan, 1979; *He went to Cambridge to read Natural Science, with the intention of becoming a geologist*—H. Carpenter, 1981; (followed by a *to*-infinitive) *When the sun's warmth began to fade away, we had to reconsider our original intention to sleep out on the beach*—Angela Davis, 1975; *He had given notice of his intention to turn up this evening*—K. Amis, 1980; *I had no intention to upset Gillian*—Julian Barnes, 1991 (the speaker is half-French).

**inter** (verb). The inflected forms are *interred, interring*. See -R-, -RR-.

**inter-** is a combining form meaning 'between, among' (*intercity, interlinear*) or 'mutually, reciprocally' (*interbreed*); as opposed to *intra-*, which is a prefix forming adjectives, and has the meaning 'on the inside, within' (*intramural, intravenous*).

**inter alia** is Latin for 'among others' when 'others' are things. If the rules of Latin are to be carried over to English, when persons are referred to the correct form is *inter alios*. In practice, however, it is best to restrict *inter alia* to things (e.g. *he said*, inter alia, *that …* ), and to use *among others* when referring to people.

**intercalary.** The main authorities recommend second-syllable stressing, i.e. /ɪnˈtɜːkələrɪ/, but /ˌɪntəˈkælərɪ/ is also permitted. Only second-syllable stressing is permissible for the verb *intercalate*.

**intercept.** The agent-noun is *interceptor* (not *-er*).

**interchange.** The verb is stressed on the third syllable and the noun on the first.

**interchangeable.** Thus spelt, with the final *-e* of *interchange* retained.

**intercourse.** The sexual sense of the word (first recorded in 1798 but uncommon before the 20c.) is now threatening to drive out the traditional sense 'social communication or dealings between individuals, nations, etc.' (first recorded in 1494). The following passage in a letter written by Charles Dickens on 5 Aug. 1852 would now be subject to misinterpretation: *We looked forward to years of unchanged intercourse.*

**interdependence, -cy.** There is no difference of sense, but *interdependence* is the more usual of the two. See -CE, -CY.

**interdict.** The noun bears the main stress on the first syllable, and the verb on the third.

**interest.** It is worth noting that the dominant pronunciation in present-day standard BrE is /ˈɪntrɪst/, i.e. with the first *e* left unpronounced. The same is true of the derivatives *interested, interesting(ly)*.

**interestingly.** It is of interest to observe that its use as a SENTENCE ADVERB, first recorded in the 1960s, has escaped censure, unlike *hopefully*. It is often qualified

by *enough*. Example: *Interestingly, what exercises Lord Chalfont is not the existence of nuclear weapons, an existence which, he says, cannot be repealed*—M. Amis, 1987.

**interface.** In 1901, when the relevant section of the *OED* was issued, *interface* was virtually a new word. The earliest example was one of 1882, and the word had only the concrete sense 'a surface lying between two portions of matter or space, and forming their common boundary'. There were no figurative or transferred uses and there was no corresponding verb. In 1976, when volume ii of *OEDS* was published, a transformation had occurred. Computers had arrived, and so too had Marshall McLuhan. As noun and verb *interface* had become an instant vogue word in the 1960s, and the space devoted to it in the Supplement reflected that fact. The half-inch of the *OED* entry was supplemented by eleven inches of additional evidence. There were *interfaces* between governments and marketing systems, between lecturers and students, between unions and management, between medicine and ethics; concepts interfaced other concepts. All potentially adversarial systems were found to have *interfaces*. One of the key words of the 20c. had made its grand and noisy entrance, and was heckled and censured by those who had no taste for it. *Interface* continues to be used by physicists, computer specialists, and other technical writers, but in lay use has fallen into a recessive phase as vogue words tend to when they have trampled over the language for something like a generation.

**interior.** See EXTERIOR 1.

**interlocutor** /ˌɪntəˈlɒkjʊtə/, a person who takes part in a dialogue or conversation. Thus spelt, not *-er*.

**interlope** /ˌɪntəˈləʊp/, **interloper** /ˈɪntələʊpə/. Thus pronounced, i.e. with contrasting stressing.

**interlude.** The dominant pronunciation is with /-luːd/ rather than /-ljuːd/. See -LU-.

**interment,** the burial of a corpse, is, of course, to be distinguished from *internment*, confinement (of aliens, etc.) within the limits of a camp during a war or threat of war.

**intermezzo** /mtə'metsəʊ/. Pl. *intermezzi* /-tsi/ (preferred), or *intermezzos*.

**intermission,** an interval between parts of a play, film, concert, etc. Originally (19c.) an Americanism, but now probably as widely used in Britain as the traditional word *interval*.

**intermit.** The inflected forms are *intermitted, intermitting*. Also *intermittent* (adj.). See -T-, -TT-.

**in terms of.** This much-debated three-word sequence is now widely called a complex preposition, joining such long-established sequences as *in aid of, in front of, in place of; in common with, in compliance with; in exchange for; in relation to*. It has also been called 'a vague all-purpose connective' by one usage writer (H. P. Guth) in 1985, and 'it encounters a good deal of criticism when used to speak of one ill-defined thing *in terms of* another equally vague and ill-defined' (*Columbia Guide to Standard American English*, 1993). It was rather extravagantly described in 1993 by the Oxford philosopher Michael Dummett as representing 'the lowest point so far in the present degradation of the English language'. Examples (from 'broadcasters and military and governmental spokesmen') cited by Dummett include: *We have made great progress in terms of the balance of payments* (rewrite, says Dummett, as *The balance of payments has improved*); *Our troops have been highly successful in terms of advancing into enemy territory* (read *Our troops have advanced deep into enemy territory*).

How did this complex preposition come into being? The *OED* reveals that it has been in use since the mid-18c. as a mathematical expression 'said of a series ... stated in terms involving some *particular* (my emphasis) quantity', and illustrates this technical use by citing examples from the work of Herbert Spencer (1862), J. F. W. Herschel (1866), and other writers. From this technical use came at first a trickle and, after the 1940s, a flood of imitative uses by non-mathematicians. The *OED* lists the phr. *to think in terms of* (and labels it *colloq.*) 'to make (a particular consideration) the basis of one's attention, enquiries, plans, etc.': e.g. *The impact of Ibsen ... did much to revitalize the degenerate English theatre and force it to think in terms of living ideas*

and contemporary realities*—J. Mulgan and D. M. Davin, 1947. And it is a simple matter to collect examples of *in terms of* which have been written, like this example, by people who are not 'broadcasters or military or governmental spokesmen': e.g. *Dataquest pegs ESRI as the leading GIS company—in terms of both revenue and reputation—Computer Graphics World*, 1988; *He deals with the converso judaizing world in terms of its social and religious rituals, births, marriages, deaths, leading to the establishment of the Inquisition—Bull. Hispanic Studies*, 1990; *Rameau ... conceived his music precisely in terms of timbres, types of attack, degree of sostenuto—Country Life*, 1990; *Justifying space in terms of material wealth is as ridiculous as saying that man went to the Moon merely to be able to return with velcro zips and non-stick frying pans—New Scientist*, 1991; *The dating of his novels in terms of when they were written rather than when they were published is often uncertain, since in the upheavals of exile some were not published chronologically—NY Rev. Bks*, 1991. One can only conclude that there is a world outside the language of broadcasters and military and governmental spokesmen where the complex preposition *in terms of* is a useful particularizing device. But it should not be used merely as 'a vague all-purpose connective', as it seems to be in the following example: *When John Major emerged as a possible candidate to lead the Conservative party, one was struck by his engaging artlessness in terms of class—Daily Tel.*, 1991.

**intern.** In the US and Canada (on occasion elsewhere) the term (sometimes spelt *interne*) for a recent medical graduate, resident and working under supervision in a hospital as part of his or her training. The equivalent terms in Britain are *houseman, house physician*, and *house surgeon*.

**internecine.** Pronounced /mtə'ni:sam/ and used in English to mean 'mutually destructive'. Classicists will know that its 'true' meaning is 'characterized by great slaughter', a meaning found in some 17–19c. contexts, including Samuel Butler's *Hudibras* (1663). The word is an adaptation of L *internecinus*, from *interneciō* 'general slaughter, massacre', from *internecāre* 'to slaughter': the element *inter-* in Latin words meaning 'to kill' did not carry its usual sense of 'mutual, reciprocal'. Johnson's *Dictionary* (1755) interpreted the adj. to mean 'endeavouring *mutual* destruction', and thus set the word on its way to its only current meaning. It is used in this sense not only of battles between warring factions such as those in Bosnia in the 1990s, but also, more trivially, of boardroom or other internal battles in business circles or other walks of life.

**internist.** See PHYSICIAN.

**internment.** See INTERMENT.

**interpellate** /m'tɜːpɪleɪt/ (verb), '(in European parliaments) to interrupt the order of the day by demanding an explanation from (the Minister concerned)', is a virtual homonym of *interpolate* /m'tɜːpəleɪt/ 'to interject (a remark) in a conversation; to estimate (values) from known ones in the same range'. Contextually, however, the danger of confusion is minimal, and, in any case, the first of the two is now little used. The same considerations apply to the corresponding nouns *interpellation* and *interpolation*.

**interpersonal** (adj.). Recorded once in 1842, and reintroduced in 1938 by the psychologist H. S. Sullivan (1892–1949), *interpersonal* is mainly restricted to the language of psychology to describe behaviour between people in any encounter. From there it has seeped into the bureaucratic vocabulary of social workers, personnel departments, counsellors, head teachers, and so on—in other words people concerned with person-to-person relationships.

**interpretative, interpretive.** In standard BrE the traditional (and recommended) form is *interpretative* (formed on the ppl stem of L *interpretātum*), thus aligning itself e.g. with *authoritative, qualitative,* and *quantitative*. But it is coming under severe pressure from the shortened form *interpretive* (as, in some circles, *quantitative* is being threatened by *quantitive*). For further discussion of the matter, see -ATIVE, -IVE, PREVENTIVE.

Some examples will bring out the ubiquity of both forms in various kinds of writing: (interpretative) *An interpretative code is almost certainly going to result in a more compact compiled product—Personal*

*Computer World*, 1985; *Some of deconstruction's avowed interpretative misreadings of literature—Brit. Jrnl Aesthetics*, 1986; *Palaeobotany is a splendid interpretative science—Blitz* (Bombay), 1987; *Not all art, and not, perhaps, the greatest, need yield to such interpretative paraphrase—New Yorker*, 1987; *As Dr John Coiley, keeper of the museum ... says: 'If you're not interpretative, the general visitor won't learn anything.'—TES*, 1990; *You may be wondering why I am rabbiting on about interpretative processes when the theme of this article is how to build a bracket clock—Practical Woodworking*, 1990; (interpretive) *A converted wartime building, it will provide an interpretive base from which visitors can enjoy extensive views of the saltmarshes—Bird Watching*, 1986; *It is, however, confined to presenting a critique of ... the Interpretive approach—Brit. Jrnl Sociol.*, 1987; *He ... is working at present on a project to develop a network of marine wardens and interpretive centres—E. Wood et al.*, 1988; *Chinese culture ... has undergone major interpretive phases in recent decades—Dædalus*, 1991. For what it is worth, a search in a large computer database in November 1988 showed that the shorter form was the more common of the two in the proportion 5:1.

**interregnum.** The recommended pl. is *interregnums*, not *interregna*. See -UM 3.

**interstice** /m'tɜːstɪs/, an intervening space. A word that is most commonly found in the pl. form *interstices* /m'tɜːstɪsiːz/ or /-sɪz/.

**interval.** See INTERMISSION.

**intestinal.** Most authorities recommend placing the stress on the second syllable, thus /m'testɪnəl/. But third-syllable stressing, with lengthening of the stressed vowel, i.e. /mte'staɪnəl/, is gaining ground, at least in BrE.

**in that.** In the nature of things, because of the multiplicity of ways in which *in* and *that* are separately used, it is not easy to find the conjunction *in that* in the *OED*, but it is there, listed as sense 40 of *in* prep., defined as 'in the fact that; in its being the case that; in presence, view, or consequence of the fact that; seeing that; as, because', and illustrated by six quotations from the 15c. to the 19c., including this one from Shakespeare's *2 Henry VI* (1593): *Let him dye, in that he is a Fox*. It would seem to be a very flexible conjunction. Jespersen (1909–49, iii) provided further examples (16–19c.) and described *in that* as 'literary rather than colloquial'. He also reminded us that some scholars think 'that *that* in this group is the demonstrative pronoun *that* with omission of the conjunction'. Curme (1931) repeated the point by asserting that the type *He differed from his colleagues* in that *he devoted his spare time to reading* was the equivalent of *He differed* in that: *he devoted his spare time to reading*. He also believed that *Acts* (AV) 10: 33 *and thou hast well done*, that *thou art come* would perhaps be now rendered as '*in that you have come* or more simply *in coming*'. (In fact the corresponding passage in the *NEB* is crassly banal: *it was kind of you to come*.) *CGEL* (1985) paid little attention to *in that*: it simply called it a 'complex subordinator' and left it unillustrated.

In view of all this listing of uses of *in that*—its true force seems to baffle or not baffle the experts in turn—it is not easy to decide whether much value should be placed upon Fowler's insistence in 1926 that *in that* is wrongly used for simple *that* in several sentences that he cites, including *The legislative jury sat to try the indictment against Mr Justice Grantham* in that *during the Great Yarmouth election petition he displayed political bias*. Perhaps one should say that the jury is still out on that one. What one can say is that *in that* clauses are not uncommon, and that the meaning of *in that* is amply covered by sense 40 of *in* prep. in the *OED*. Witness, for example, this sentence by Henry Cecil Wyld in his *History of Modern Colloquial English* (3rd edn, 1936): *M.E. spelling . . . is to a certain extent phonetic*, in that *there is often a genuine attempt to express the sound as accurately as possible.* And also the following: *They work like disks* in that *they can be partially erased—Management Computing*, 1990; *it is an inescapable truth that the petroleum industry is subject to a rigid technical constraint*, in that *a large proportion of inventories are tied in the supply system—Oxford Institute for Energy Studies*, 1990; *The vessels . . . are unusual* in that *they have no engine room, being powered by ten diesel-driven alternator packs connected to the main switchboard—Ships Monthly*, 1991.

**in the circumstances.** See CIRCUM-STANCE.

**intimidate.** Fowler (1926) cited an incorrect use of *intimidate* followed by a direct object + *from*: *Similar threats were uttered in the endeavour* to intimidate Parliament from *disestablishing the Irish Episcopal Church*. As he pointed out, *from* is idiomatic after *deter* and *discourage*, but not after *intimidate* or *terrify*.

**into, in to.** 1 The two words should be written separately when the sense is separate, most commonly when *in* is a full adverb and *to* is an infinitive marker and means 'in order to'. Examples: (*in to*) *People dropped in to see him*; *The maid looked in to ask if they wanted coffee*; *the Secretary of State, George Shultz, slipped in to replace Reagan*; ( + noun) *he accompanied her in to dinner*; (*into*, esp. expressing motion or direction) *I was reduced to staring into the water*; *The desire to know can degenerate into mere trickery*; *The Highway Code can be taken into account*; *He wants to put his hand into hers*. In the first group it would be wrong to conjoin *in* and *to* (e.g. *People dropped into see him*); and in the third group to let them stand apart (e.g. *I was reduced to staring in to the water*).

2 The modern use (1960s onwards) of *into* to mean 'involved in, knowledgeable about' (*he is into Zen*) is still suitable only for distinctly informal contexts.

**intra-.** See INTER-.

**intransigent.** The recommended pronunciation is /ɪnˈtrænsɪdʒənt/, but /-ˈtrænz-/, /-ˈtrɑːns-/, and /-ˈtrɑːnz-/ are all permissible. The word came into English from French in the 1880s and was for some time pronounced in a French manner and often spelt (as in French) *intransigeant*.

**intransitive** (adj.). In sentences of the type *they lit a fire*, the verb *lit* is transitive in that it has an object (*a fire*). By contrast, in the sentence *they arrived at noon*, the verb *arrived* has no direct object and is called intransitive. Numerous verbs, e.g. *to lie*, *to seem*, are always used intransitively. Other verbs have both transitive and intransitive uses: e.g. (transitive) *he played the piano*; (intransitive) *she plays well*. Thus the presence or absence of a direct object determines whether a verb (or sense of a verb) is transitive or intransitive. Many verbs can appear to govern two objects, a direct one and an indirect one (*they gave her an apple*). These are called DITRANSITIVE verbs.

**intransitive past participles.** To begin with, it might be as well to remind readers that a past participle is the single-word participle that does not end in *-ing*, but most commonly in *-ed* (*polished*), *-d* (*heard*), *-t* (*lost*), or *-en* (*spoken*): thus *she had polished the silver*; *he had heard the news*; *she had lost the will to live*; *they had spoken to no one*. And that a transitive verb is one that takes a direct object (*he hits the ball*) and an intransitive verb one that does not take a direct object (*the lions roared*). It is well to remember, too, that many verbs are capable of being used either transitively or intransitively (cf. *he left the house an hour ago* and *I am going to leave now*). It should also be said at the outset that most of the pa.pples of transitive verbs may be freely used as participial adjectives: e.g. *the polished silver*; *heard melodies*; *lost property*; *the spoken word*.

There are various uncertainties about the use of the pa.pples of intransitive verbs as participial adjectives. First, a select list of phrases containing ppl adjs. where the corresponding verbs are in most circumstances used only intransitively: *absconded debtors* ( = debtors who have absconded), *a newly arrived guest* ( = a guest who has recently or has just arrived), *a collapsed lung*, *the departed guests*, *the deceased husband*, *the eloped pair*, *escaped prisoners*, *an expired licence*, *fallen angels*, *a grown man*, *the risen sun*, *a vanished tribe*. Secondly, when a verb is used both transitively and intransitively it is sometimes difficult to say whether the pa.pple is active or passive, and the answer may affect the sense. For example, *a capsized boat* may have capsized of its own accord (*the boat capsized*) or have been overturned by some agent, e.g. a high wave. A *deserted house* is one that is empty or has been abandoned; a *deserted wife* is one whose husband has left her. A *flooded meadow* reflects a passive pa.pple (the meadow became flooded when the river overflowed), but a *flooded river* is a stative one, i.e. one that is in flood. Thirdly, it is worth recalling that some 'participial' adjectives in *-ed* have no corresponding verbs: e.g. *the visit was unexpected* and *an unexpected visit*, but there is no verb *to unexpect*. Similarly, the

*players are downhearted* and *the downhearted players*, but there is no verb *to downheart*. Finally, it is worthy of note that some intransitive pa.pples have rather less than a straightforward relationship with the corresponding verbs: *an ill-advised action* is either one done without due consideration or one resulting from bad advice; *a well-read person* is one who has read widely, and by implication is continuing to read widely; *a well-behaved child* is one with a reputation for good behaviour or one who happens to be behaving itself well at the time of speaking. All of which reminds us that it is as well not to forget that the use of intransitive pa.pples and of the corresponding ppl adjs. calls for vigilance at all times.

**intrench.** See EM- AND IM-.

**intrigue** (verb). There are lessons to be learnt here. *Intrigue* is a standard example of a word of long standing in the language (first recorded in 1612) that was adopted in English first in the primary sense of its language of origin (French), namely 'to trick, deceive, cheat; to embarrass, puzzle, perplex', then branched out in various ways, esp. in the direction of amorous intimacy ('to carry on a secret amour') and of underhand plotting ('to employ secret means to bring about some desired end'), and finally in the 1890s drew further away towards the notion of fascinating curiosity ('to excite the curiosity or interest or fascination of'). The new sense flourished to the point that Fowler (1926) declared it to be a literary critics' word 'of no merit whatever except that of unfamiliarity to the English reader'. Fowler could not see why *fascinate*, *mystify*, *interest*, and *pique* were not adequate for the task: he might also have mentioned *absorb*, *captivate*, *charm*, *enchant*, *enthral*, *transfix* ... But the language shows an irrepressible liking for near-synonyms, and *intrigue* and *intriguing* in the new sense have found their way into acceptance, esp. in journalism and in domestic conversation. Examples: *She was a mine of intriguingly useless information—Daily Tel.*, 1970; *Even more intriguing than the sociology of fashion is its psychology—Observer*, 1974; *She created a film which was memorable, intriguing and moving—Listener*, 1982; *We are in turn sympathetic, intrigued, shocked, entertained—*

*but oh the yearning for the world she ma-gically conjured*—A. Huth, 1992.

**intrinsic.** See EXTERIOR 2. *Intrinsic* means 'inherent, essential, belonging naturally (*intrinsic merits, value*, etc.)' and is the opposite of *extrinsic.*

**introit.** The only pronunciation given in the *OED* (1901) (and the one I use myself) is /ɪnˈtrəʊɪt/. But /ˈɪntrɔɪt/ has forced its way into standard use as the 20c. progressed and is now dominant.

**introvert.** See EXTROVERT.

**intrusive r.** The insertion of an unhistorical *r* between one vowel sound and another as in the pronunciation of *drawing* and *India office* as *draw-r-ring* and *India-r-office*. The error is ridiculed by standard speakers. See LAW AND ORDER; LINKING R.

**intrust.** See EM- AND IM-.

**intuit** (verb). An 18c. back-formation from *intuition*, it seems to have been mostly reserved for literary use or the use of literary critics since then. *Intuit* turns up frequently, for example, in the works of Iris Murdoch and of A. S. Byatt. Examples: *Jenkin did not say any of this, but Gerard intuited it behind some clumsy expressions of sympathy and was irritated*—I. Murdoch, 1987; *As Browning was six years younger than his wife Elizabeth Barrett, he may well have feared that he would one day act as the wife intuits her husband will act*—N&Q, 1989; *Maud decided she intuited something terrible about Cropper's imagination from all this*—A. S. Byatt, 1990.

**intuition and instinct.** The word *intuition* being both in popular use and philosophically important, a slight statement of its meaning, adapted from the *OED*, may be welcome. The etymological but now obsolete sense is simply inspection (L *tuērī* 'to look'): *A looking-glass becomes spotted and stained from their only intuition* [i.e., if they so much as look in it]—O. Feltham, 1627–77. With the schoolmen it was 'The spiritual perception or immediate knowledge, ascribed to angelic and spiritual beings, with whom vision and knowledge are identical': *St Pauls faith did not come by hearing, but by intuition and revelation*—J. Taylor, 1660. In modern philosophy it is 'The immediate apprehension of an object by the mind without

the intervention of any reasoning process': *What we feel and what we do, we may be said to know by intuition*—J. Priestley, 1782; or again (with exclusion of one or other part of the mind) it is 'Immediate apprehension by the intellect alone', as in *The intuition by which we know what is right and what is wrong, is clearer than any chain of historic reasoning* (W. E. H. Lecky, 1865), or 'Immediate apprehension by sense', as in *All our intuition takes place by means of the senses only* (tr. of Kant, 1819). Finally, in general use, it means 'Direct or immediate insight', as in *Rashness if it fails is madness, and if it succeeds is the intuition of genius* (J. A. Froude, 1879). How closely this last sense borders on *instinct* is plain if we compare *A miraculous intuition of what ought to be done just at the time for action* (N. Hawthorne, 1851) with *It was by a sort of instinct that he guided this open boat through the channels* (William Black, 1873). One of the *OED*'s definitions of *instinct*, indeed, is: 'intuition; unconscious dexterity or skill'; and whether one word or the other will be used is often no more than a matter of chance.

How do matters stand now? At the philosophical level, I imagine, the differences between the two words remain subtle and continue to be topics for perpetual discussion. In general use it is hardly a matter of concern whether *intuition* is restricted to 'angelic and spiritual beings'. Rather, it is perceived as a valuable faculty possessed by human beings alone, and meaning 'immediate apprehension by the mind *without reasoning*'. *Instinct*, by contrast, is the pattern of behaviour *in most animals* in response to certain stimuli; and a similar property *in human beings* to act without conscious intention. Animals cannot reason, it is argued, but have only *instincts*; human beings have both *instincts* and *intuitions*.

**intwine, intwist.** See EM- AND IM-.

**inure, enure.** The second form of the word has some currency in law in the sense 'to take effect, to come into operation', but even in law the form *inure* is the more usual of the two (*the damages must inure to the exclusive benefit of the widow and children*—Garner, 1987). In its other senses, esp. in the passive followed by the preposition *to*, 'to accustom (a person) to something unpleasant', *inure* is the normal spelling.

**invalid.** The adj. in the sense 'not valid' is always pronounced /ɪn'vælɪd/. As noun, verb, and adj. in the 'ill' sense, the word has vacillated throughout the 20c. (indeed since the early 18c.) between /'ɪnvəliːd/ and /-ɪd/, with perhaps a slight preference for the former in BrE. The pronunciation with /-iːd/ reflects the influence of Fr. *invalide* (adj.).

**inveigle.** The recommended pronunciation is /ɪn'veɪgəl/ rather than /ɪn'viːgəl/, but both pronunciations are in standard use.

**invent.** See DISCOVER.

**inventor.** Thus spelt, not *-er*.

**inventory.** Pronounced /'ɪnvəntərɪ/ in BrE, and /ˌ-tɔːriː/ in AmE.

**inversion.** (Fowler's long entry, 1926, is a reprinted version of a paper he wrote as SPE Tract x in 1922. Readers may wish to consult this elaborate essay in the SPE Tracts themselves in one of our great institutional libraries.) **1** No one would doubt that the normal order of English sentences is subject/verb/object (SVO): *I found a black plastic bag*; or subject/verb/complement (SVC): *The caterers circulated with champagne and miniature quiches.* In practice, such sentences are usually somewhat more complex than the examples just given, with adverbial or other phrases inserted (e.g. *In the Lincoln College manuscript our poet has arranged words and ideas in chiastic and parallel patterns in the first half of his work—Bodl. Libr. Rec., 1992*); as well as subordinate clauses (e.g. *At the centre of his work the poet states his theme, that the traditional legend of St Brendan is heretical, writing in the twenty-sixth of fifty-two lines* [etc.]*—ibid., 1992*).

**2** In numerous circumstances, however, some of them grammatical and some stylistic, this normal order is disturbed and inversion occurs. By inversion is meant simply 'the abandonment of the usual English sentence order' or 'the deferring of the subject' (Fowler). It should be noted, however, that to native speakers such inversions are for the most part routine and pass virtually without notice. Some of the conditions in which inversion is usual, desirable, or permissible are as follows: (*a*) In the formulaic setting down of direct speech: *'Hey!'*

shouted Mrs. House, *who sat inside with her jumpsuit around her knees—New Yorker, 1992.* But this is optional: *'I was out in the orchards a while back,'* Milton said—W. Trevor, 1992. (*b*) To give prominence to a particular word or words: *Trusting she had been, she who had been reared in the bosom of suspicion—M. Drabble, 1987; Deprived all the rest of us are—D. Lessing, 1988; Dreams we did not discuss—G. Keillor, 1989; Most notable among the Bodleian's Gibson manuscripts are two volumes of collections for the edition of Camden's Britannia—Bodl. Libr. Rec., 1992.* (*c*) To obtain contrast or parallelism: *To Laura, Mrs. Laughlin said, 'Mr. House loves birds. He builds them houses in his spare time.' To him, she said, 'You'll have to cultivate a taste for squirrels.'—New Yorker, 1992.* (*d*) After negatives: *Yet never before had I seen anything so scarlet and so black—J. M. Coetzee, 1990; she said she wouldn't come to the party, nor did she.* (*e*) Routinely in questions: *Where have you been?; Who did you see?* (*f*) In sentences led by an adverb or an adverbial phrase: *Up stood Joe—M. Bail, 1975; Up and down the stoep struts Klein-Anna mastering the shoes—J. M. Coetzee, 1977; As Ernst Mayr has pointed out, seldom can biology boast the certainty of absolute laws—Dædalus, 1986.* (*g*) After *so*: *So didst thou travel on life's common way—Wordsworth; He had hardly been aware, so nervous was he, of what he had been saying—P. Carey, 1988.* (*h*) In a delayed anaphoric *do/did* or *is/was/were* clause: *One of life's battlers, was Raelene—M. Eldridge, 1984; She enjoyed a laugh, did Lilian—M. Drabble, 1987; She created it, and ruined it for me, did Irene—Maurice Gee, 1987.* Used mostly in informal contexts. (*i*) In conditional clauses without a conjunction: *Were this done, we would retain a separate Bar with skill—Times, 1986; Statistically, afterworlds—be they Christian, Greek, Pharaonic—must be populated almost entirely by children—P. Lively, 1987.* (*j*) In certain kinds of comparative (followed by *than*) clauses: *Poland's power structure included neither more nor fewer Jews than did the power structure in Rumania or in Hungary—Dædalus, 1987.* (This type was censured by Fowler. Visser (p. 170) traces the construction back through the centuries to Chaucer but admits that it has always been 'rare'.) (*k*) Poetical inversion: *His soul proud science never taught to stray—*

Pope; *But uglier yet* is the Hump *we get From having too little to do*—Kipling.

**3** Fowler (1926) provided a list of examples to show that inversion is often 'ugly, and that resort to it is the work of the unskilful writer'. Among these were the following: *Sufficient* is it *to terminate the brief introduction to this notice by stating* ... (It is sufficient); *Somewhat in the nature of a blow* is it, *therefore, to find that* ... (It is therefore somewhat); *Carrying far more than* can the steam-driven vessel (read *vessel can*). To which may be added some fashionable, exclamatory, rhetorical, or 'cleft' types of modern inversion: *Great literature it's not, but* ... *it's short, pithy* ...—*The Face*, 1987 (It's not great literature); *from Lord Rosebery to Ken Livingstone—ah, what a fall was there!*—*Encounter*, 1987 (what a fall this was!); (elderly woman:) *Spring it isn't*—*it's too bloody cold!*—oral example, 1987 (it isn't like spring); *He it was that scored more goals than anyone else in the World Cup*—S. Barnes, *Times*, 1987 (It was he who). Whether such inversions as these last ones serve any useful purpose is, of course, debatable.

**inverted commas.** See QUOTATION MARKS.

**investigator.** Thus spelt, not -*er*.

**investor.** Thus spelt, not -*er*.

**invincible, invisible.** See -ABLE, -IBLE 7.

**invite** (noun). A classic example of a word that seems to have been in continuous use beside *invitation* since the mid-17c. but has never quite made its way into uncriticized neutral use. It must have been known to Dr Johnson but he excluded it from his Dictionary (1755). It is also not listed in 'a new edition' (1863) of Charles Richardson's Dictionary. It was admitted to the OED (1901), labelled *colloq*. Fowler (1926) said that 'it has never, even as a colloquialism, attained to respectability; after 250 years of life, it is less recognized as an English word than *bike*'. Seven decades later the general verdict is still that it properly belongs to the informal (even on occasion comic) corners of the language. Examples: *Is it just an invite from the colonel for a working week-end?*—*Listener*, 1968; *The four detectives didn't await an invite into*

*the house*—G. F. Newman, 1970; *He scoffs, indicating the dodgier invites entreating his attendance at this or that launch*—*Sunday Express Mag.*, 1987; *He knows a particularly good printer who did the invites for his cousin's wedding*—*Precision Marketing*, 1989.

**invoke** (verb). The corresponding adj. is *invocable* (not *invokable*).

**involve** (verb). **1** Use a short *o*, i.e. /ɒ/, in words ending in -*olve* (*absolve, dissolve, evolve, resolve, revolve, revolver*, etc.).

**2** Before condemning journalistic uses of the verb, frequent though they are, readers are advised to bear in mind that in the 600 years since it entered the language, *involve* has enjoyed considerable currency in numerous reflections and extensions, including figurative ones, of the original meanings of L *involvere* 'to wrap up, envelop, entangle'. It is all a matter of the degree of envelopment or entanglement. Examples: ( = envelop) *I saw Fog only, the great tawny weltering fog, Involve the passive city*—E. Barrett Browning, 1856; ( = intertwine) *Our misfortunes were involved together*—L. Sterne, 1768; ( = beset) *The numerous difficulties in which this question is involved*—B. Jowett, 1875; ( = entangle in a difficulty of some kind) *Mr. Müller had been involved in financial difficulties*—A. W. W. Dale, 1898; ( = to contain explicitly) *Every argument involves some assumptions*—B. F. Westcott, 1892; ( = to include or affect in its operation) *It will be held a worthy subject of consideration what are the political interests involved in such accumulation*—Ruskin, 1857. Bearing in mind that few people carry round etymological baggage from day to day, and also keeping in mind the complex array of historical evidence set down in the OED, it is little wonder that 'precision' is hardly the right term to bring forward to describe the uses of such a many-branched word as *involve*. The idiomatic nature of *No other vehicle was involved*, as the police (and journalists) describe a one-vehicle accident, can hardly be questioned. Nor can that of *He was involved with* [i.e. was having an affair with] *the hotelier's wife*, or of *He became deeply involved in the conspiracy at an early stage*. Something of the primary sense of the word has survived in each of these examples.

Nevertheless a word of caution should be added. It is a notionally asterisked and tainted word in the sense that it is widely believed that other words are available to replace it. Some typical examples of uses rejected in other usage guides: *A collision took place involving a private motor vehicle and a lorry* (between); *There was no reduction last year in the number of cases involving cruelty to horses* (of); *Everyone involved in the demonstration* (taking part in); *The cost involved would be excessive* (omit *involved*). It is easy to become involved in an argument about such substitutions and rewritings. The best advice, perhaps, is to use the verb sparingly and with reasonable caution.

**invulnerable.** The word cannot normally be qualified by an adverb such as *fairly* or *very*, i.e. is an absolute adj., but there are some circumstances (illustrated in the *OED*) in which, for example, *rarely* and *increasingly*, and other adverbs of time, may properly be used with it.

**inweave.** The more usual spelling, not *en-*. See EM- AND IM-.

**inwrap.** The more usual form is *enwrap*. See EM- AND IM-.

**iodine.** Pronounced either as /ˈaɪədiːn/ or /-dm/. See -IN AND -INE.

**-ion and -ment.** As formative elements in English, the relationship is complex. At a surface level it is clear that some verbs form nouns of action in *-ion*: thus *abolish* → *abolition*, *admonish* → *admonition*, *inflict* → *infliction*, and *pollute* → *pollution*; while others form nouns of action in *-ment*: e.g. *achieve* → *achievement*, *assign* → *assignment*, *manage* → *management*, and *punish* → *punishment*. This is a useful distinction as far as it goes. The neatness of such a relationship, however, is misleading and will not stand up to historical examination. Some of the formations are not governed by arrangements in English but by the forms that already existed in OF. There are various chronological considerations (see below). And a small number of the contrasting forms stand side by side in the language but with differences of meaning. Note, first, that whereas *abolishment* and *abolition* and *admonishment* and *admonition* exist, they have unequal currency in current English. The lack of parallelism is brought out also by the non-existence of *inflictment* and *pollutement* on the one hand, and the non-existence now of *punition* on the other (though it was in use from the 15c. to the 19c.).

*Two main waves.* Nouns of action in *-ment* entered the language in two main periods: (i) (from Anglo-Fr.) *abridgement*, *accomplishment*, *commencement*, etc.; (ii) in the 16c., (from Fr. verbs) *banishment*, *enhancement*, etc.; (from native verbs) *acknowledgement*, *amazement*, *atonement*, etc. New formations in *-ment* after the 16c. are relatively rare except for words beginning with *en-* (e.g. *enlightenment*, 17c.) or *be-* (e.g. *bedevilment*, 19c.). The affix has scarcely been drawn on in the 20c.

*Differences of meaning.* A small number of verbs have corresponding nouns of action in both *-ion* and *-ment*. Thus *commit* corresponds both to *commission* (first 15c.) and *commitment* (17c.); and *excite* corresponds both to *excitation* (15c.) and *excitement* (17c., but its main current sense not until the 19c.). Both pairs of nouns have followed their own complicated tracks as the centuries passed, as a glance at the *OED* will show. All that one can say is that these paths have been asymmetrical.

**-ion and -ness.** Not surprisingly, the relationship of pairs of words in *-ion* and *-ness* differs from that discussed in the previous article in several respects. First, *-ness* is the suffix commonly attached to adjectives to form nouns expressing a state or condition: e.g. *bitter* → *bitterness*, *dark* → *darkness*, *dry* → *dryness*, *persuasive* → *persuasiveness*. In this function it vies with formations in *-ity*: e.g. *certain* → *certainty* (occas. *certainness*), *serviceable* → *serviceability* (occas. *serviceableness*). Secondly, *-ness*, unlike *-ment*, is still a living suffix, with at any rate the capability of being added to any new adjective. Thirdly, a reasonable number of the possible pairs have come to be used with different meanings, the forms in *-ion* likely to denote 'the act or process of —', and those in *-ness* more likely to denote 'the quality of being —'. Compare, for example, the following pairs: *abstraction* ~ *abstractness*, *consideration* ~ *considerateness*, *correction* ~ *correctness*, *indirection* ~ *indirectness*. The subtlety of the relationship between seemingly similar affixes is nowhere more clearly seen than in the groups briefly discussed in this and the preceding article. The wonder is that, by and large, the correct choice of the possible forms is made by standard speakers, for the most part naturally with unawareness of the lexical complexities involved. See CONCISENESS.

**Iran.** Pronounced /ɪˈrɑːn/ in BrE; but in AmE /ɪˈræn/ and sometimes /aɪˈræn/.

**Irangate.** See -GATE. The term was applied in the 1980s to a political scandal in America when it was alleged that funds obtained from the sale of American arms to Iran had been secretly transferred to the insurgent forces called the Contras who were opposing the Sandanista government in Nicaragua.

**Iranian.** The recommended pronunciation is /ɪˈreɪmɪən/, except in AmE in which /ɪˈrɑːnɪən/ seems to be more usual. A pronunciation with initial /aɪ-/ is also found in AmE.

**Iraq.** The standard pronunciation in BrE is /ɪˈrɑːk/; hence (a native or inhabitant) *Iraqi* /ɪˈrɑːkɪ/. In AmE /ɪˈræk/ and /ɪˈrækɪ/ seem to be more usual; also, less commonly, with initial /aɪ-/.

**irascible.** Thus spelt, not *-able*. See -ABLE, -IBLE 7.

**Irene.** The name of the Greek goddess of peace is always pronounced as three syllables, /aɪˈriːnɪ/: it is derived from Gk εἰρήνη 'peace'. Traditionally, when used as a Christian name in English, it has been pronounced in the same manner, that is with three syllables. But nowadays most of the Irenes one meets call themselves /ˈaɪriːn/, and this two-syllabled pronunciation seems likely to remain dominant in the foreseeable future.

**iridescent.** Thus spelt (being from L *īris*, *īrid*- 'rainbow'), not *irri*-.

**iron.** For *the Iron Duke, Lady, Maiden*, see SOBRIQUETS.

**iron curtain.** First recorded in 1794 of a curtain of iron which could be lowered as a safety device between the stage and the auditorium in a theatre, it soon (1819– ) began to be used of any impenetrable barrier. Its classic use in the 20c. was of 'a barrier to the passage of information, etc., at the limit of the sphere of influence of the Soviet Union'. The phrase was first used in this sense in 1920 by Mrs Philip Snowden (*We were behind the 'iron curtain' at last!*), but the *locus classicus* was a speech by Winston Churchill at Westminster College, Fulton, USA in 1946 (*From Stettin in the Baltic, to Trieste, in the Adriatic, an iron curtain has descended across the Continent*). With the dismantling of the Berlin Wall in 1989, and the breaking up of the Soviet Union, the phrase has lost its usefulness except as a potent reminder of a major confrontation between the great powers for nearly half a century.

**ironical** (adj.) alternates with **ironic**, the choice of form, it would seem, being governed by the rhythm of the sentence. Both mean (a) of the nature of irony, i.e. meaning the opposite of what is expressed: e.g. *Boyle ... paid, in his preface, a bitterly ironical compliment to Bentley's courtesy*—Macaulay, 1853; *She gave an ironical laugh as she looked at Guy*—O. Manning, 1977. (b) that uses or is addicted to irony: *Ostrowski was dignified, Lelewel ironical and inflexible*—W. H. Kelly, 1848.

Both words are now increasingly being used to mean simply 'odd, strange, paradoxical': *It is paradoxical, 'ironical' as people say today, that the constitution should bestow this power on someone who laments constitutionitis in others*—*Observer*, 1987; *Eurotunnel said yesterday that it was ironic that Flexilink should issue its document on the day that the European Investment Bank had demonstrated its confidence in the project*—*Times*, 1987; *It is savagely ironic that this [child] benefit should now be undermined by a Conservative government that has already pushed below the poverty line two million children in families in work*—*Times*, 1987; *It is ironic that such a beautiful orderly house should be the setting of our messy little farce*—Simon Mason, 1990. This weakened use looks as if it has come to stay.

**ironically.** This adverb, frequently followed by *enough*, is now being used daily as a sentence adverb (or disjunct, in the jargon), meaning not much more than 'strangely (enough), paradoxically (enough)'. Examples have been found from the 1940s onward, and it seems to be settling down into standard use, dragging *ironical* and *ironic* (see prec.) behind it. Examples: (ironically) *Ironically the bombing of London was a blessing to the* *youthful generations that followed*—I. and P. Opie, 1969; *Ironically, these may include assets within Texaco—if paying Pennzoil puts Texaco on the auction block*—Observer, 1987; *Less than 0.002 per cent of . . . muscle proteins comprise dystrophin, which it was why it was overlooked. Ironically, its discoverers believe it is the largest protein molecule in the body*—Times, 1987; '*We* [sc. two British tourists] *had heard he* [sc. an escaped prisoner] *was on the loose and ironically the night before we had camped outside the Grand Canyon to avoid any possible encounter with him*—Times, 1992; *(ironically enough) That caged rebuked question Occasionally let out at clambakes or College reunions, and which the smoke-room story Alone, ironically enough, stands up for*—W. H. Auden, 1945; *Ironically enough, the Israeli role in both remains crucial—thus underlining how relevant the idea of 'linkage' is, whatever cynical use Saddam Hussein has tried to make of it*—Sanity, 1991.

**ironist** (one whose works or conversation is characterized by an ironic tone) is the kind of casual formation that could have come into existence at any time since the 16c. (from Gk εἴρων 'dissembler, user of irony' + *-ist*) when *irony* itself came into the language. In fact it is first found in a work of 1727 by Alexander Pope and others.

**iron out.** The 17c. sense of *iron* (verb) 'to smooth or press with an iron' was occasionally represented as a phrasal verb (*iron out*) in the 19c., but *iron out* has been used much more frequently in figurative contexts in the 19c. and 20c., esp. of the smoothing out of differences of viewpoint. There are limits to its sensible use. It is not very stylish to speak of ironing out bugs in a new computer (*The new computer was delivered . . . last week . . . Ironing out the bugs will probably take until the new year*—Guardian, 1971), and incongruous to speak of *ironing out bottlenecks* (as Gowers pointed out in 1965).

**irony.** It should be borne in mind that there are several kinds and degrees of irony. The word was taken into English from L *īrōnīa* at the beginning of the 16c. in two main senses: (*a*) A figure of speech in which the intended meaning is the opposite of that expressed by the words used; usually in the form of laudatory expressions used to imply condemnation

or contempt; (*b*) (in its etymological, ult. Gk, sense) dissimulation, pretence, esp. with reference to the dissimulation of ignorance practised by the Greek philosopher Socrates as a means of confuting an adversary. Type (*b*) is also called *Socratic irony*. A third type, (*c*), is called *dramatic irony* in which the audience or reader knows more about the outcome of a play or an epic poem than the character or characters do because the author has, as it were, taken the audience or reader into his confidence at an early stage. *Dramatic irony* is first recorded in 1907 (in a work by Walter Raleigh, first holder of the chair of English literature at the University of Oxford). It was used as a literary device in Greek drama: as Fowler (1926) expressed it, '*all the spectators . . . were in the secret beforehand of what would happen. But the characters, Pentheus and Oedipus and the rest, were in the dark; one of them might utter words that to him and his companions on the stage were of trifling import, but to those who hearing could understand were pregnant with the coming doom*'. *Dramatic irony* is much used of certain episodes in *Beowulf: The well-known hymn on the loyalty and harmony reigning among the Danes which, in view of the poet's intimations of Hrothulf's treachery . . . may be considered as a fine piece of dramatic irony*—A. Bonjour, 1950.

These three types of irony are often further subdivided. Baldick (1990), for example, adduces *verbal irony* (a trivial example is the type *What a lovely day* said when it is actually raining), *structural irony*, *tragic irony*, and *cosmic irony* in addition to the three types mentioned above. Katie Wales (1989) cites various kinds of literary irony from the works of Shakespeare, Fielding, Hardy, Henry James, and Ford Madox Ford. These types and distinctions are matters for professional students of English literature. A typical example: *Irony is often said to be a figure in which the true sense contradicts the literal meaning. But in Swift's subtler irony the meaning need not be opposite exactly, and can be very elusive*—A. Fowler, 1987. The subtle weaving and unravelling of irony in literature and in life are matters of infinite gradations, and no amount of terminology will encompass them all or receive universal acceptance.

**irrecognizable, un-.** Both were first used in the early 19c. but *unrecognizable* is now the regular form.

**irreducible.** Thus spelt, not *-able*. See -ABLE, -IBLE 7.

**irrefragable.** Stress the second syllable, /ɪˈrefrəɡəbəl/.

**irrefutable.** The recommended pronunciation is /ɪˈrefjʊtəbəl/, i.e. stressed on the second syllable, but the form with third-syllable stressing, /ɪrɪˈfjuː-/, is gaining ground.

**irregardless.** In origin probably a blend of *irrespective* and *regardless*, it has been used for most of the 20c., chiefly in N. America, in non-standard or humorous contexts, to mean 'regardless'.

**irrelevance, -cy.** Both are used without distinction of meaning, but *irrelevance* is much the more usual of the two. See -CE, -CY.

**irrelevant.** The *OED* (1901) commented, 'A frequent blunder is *irrevalent*.' Fowler, in his jesting mode, commented that the form with consonants transposed 'does not get into print once for a hundred times that it is said; but it is not difficult, with a little fishing, to extract it from ladies'! Cf. ANEMONE. Obviously such metatheses are to be avoided.

**irrelevant allusion.** One of the frailties to which we are all prone from time to time is to make an unnecessary allusion to a familiar, often literary, phrase or line. Thus, a writer in the 29 July 1991 issue of the *New Yorker*: *When Switzerland talks about 'being Europe' instead of about 'servicing Europe', you know that there has been a sea change in the way Europeans think about who they are, and what they are, and where they are.* This is, of course, a half-remembered, gauche allusion to Shakespeare's *The Tempest, Nothing of him that doth fade, But doth suffer a Sea-change Into something rich, and strange: Sea-nimphs hourly ring his knell.* Sea changes happen in the context of sea nymphs, not in landlocked Switzerland. Instances of such irrelevance or semi-relevance are not difficult to find. Who has not heard someone say *There is method in my madness* when the madness is not such but merely a surface appearance of irregularity? The

allusion is, of course, to Polonius' *Though this be madness, yet there is method in't.* A trade-union strike in early winter automatically produces the headline *A winter of discontent?* in the tabloids, half-echoing Richard III's *Now is the winter of our discontent, Made glorious by this sonne of Yorke.* How many times has *light* been described as *religious* as well as *dim*, echoing the passage in Milton's *Il Penseroso, And storied windows richly dight, Casting a dim religious light?* How often has Marcellus' line in *Hamlet, Something is rotten in the state of Denmark*, been echoed down the centuries, when the context is merely something local (far from Denmark) that has a suggestion of moral or social decay? And how often have you said, when leaving a disappointing party or theatrical performance, *Home James, and don't spare the horses* when the vehicle is in fact a car? Our minds are half-filled with irrelevant allusions and with HACKNEYED PHRASES and LITERARY ALLUSIONS. We live in a world of reduced quotations with the quotation marks removed.

**irremovable.** See IMMOVABLE 2.

**irreparable** /ɪˈrepərəbəl/. A loss, harm, or a relationship that cannot be rectified or made good is *irreparable*: e.g. *The bridge-building spirit in which this series* [of cricket tests between Pakistan and England] *began has now collapsed, possibly irreparably*—*Times*, 1992. By contrast a material object that is badly damaged or badly worn is *irrepairable* (stressed on the third syllable), *not repairable*, or (most commonly) *beyond repair*.

**irreplaceable.** Thus spelt, not *-placable*.

**irrepressible, irresistible.** Thus spelt, not *-able*. See -ABLE, -IBLE 7.

**irrespective.** Here is a problem. The *COD* (1995) calls it an adjective followed by *of*, meaning 'not taking into account; regardless of', and lists *irrespectively* without definition as its corresponding adverb. The *OED* says that *irrespective* 'is now chiefly in adverbial construction, qualifying a verb expressed or understood; = *irrespectively*. Const. *of*.' By contrast, Jespersen, *CGEL*, and other grammars treat *irrespective of* as a group preposition or a complex preposition, or

as a 'stylistically clumsy' element introducing a concessive clause (e.g. *irrespective of cost, irrespective of the fact that*). The
difference of terminology reflects the
different perspectives of lexicographers
and grammarians. There is no quarrel
about usage, just a difference of opinion
about terminology. These examples may
suffice as guides to how the word is used:
*People sometimes judge actions to be right
irrespective of their consequences*—A. J. Ayer,
1972; *He is fed only to suit his mother's
convenience, irrespective of whether he is hungry or not*—A. Storr, 1972; *The Tudors set out
to employ talent at the centre of government,
irrespective of the social origins of their chosen
servants*—A. Briggs, 1983; *The beginner in
chess, who tries to follow his plans irrespective
of his partner's countermoves, will soon go
down in defeat*—B. Bettelheim, 1987; *The
security service must play a role of service to
the state and society, irrespectively of the
political alignment in Poland*—BBC Summary of World Broadcasts 13 Sept., 1989;
*In this case, they are unlikely to judge that
the competitor has this atypical attribute,
irrespectively of whether they rely on the explicit ad information* [etc.]—*Jrnl Consumer
Research,* 1991.

**irresponsible.** Thus spelt, not -*able*. See
-ABLE, -IBLE 7.

**irresponsive, un-.** Both words are current, but *unresponsive* is the more usual
of the two.

**irretentive, un-.** Both words came into
being in the mid-18c. Fowler (1926) recommended *irretentive*. I would use *unretentive* myself (of memory, etc.). But
neither word is widely current. See IN-
AND UN-. Examples: (irretentive) *A manli-
ness of tone, the direct opposite of the irretentive querulousness found in so great a number
of poets*—A. W. Ward, 1879; (unretentive)
*He says that he is seeded second even though
he claims that he has an 'unretentive memory'*—*Sports Illustrated* (US), 1985; *I* [sc. Mor-
decai Richler] *suffer from an unretentive
memory. A sieve. So, though I can recall devouring Stendhal, Dostoyevsky,* [etc.]—*NY
Times,* 1988.

**irreversible.** Thus spelt, not -*able*. See
-ABLE, -IBLE 7.

**irrevocable.** 1 Thus spelt, not -*ible*. See
-ABLE, -IBLE 6.

**2** Stressed on the second syllable, /ɪˈre-
vəkəbəl/.

**irridescent.** For this erroneous spelling,
see IRIDESCENT.

**is.**

| |
|---|
| **1** *is* and *are* between sentence elements of variant number. |
| **2** *is* and *are* in multiplication tables. |
| **3** *is* auxiliary and copulative. |
| **4** *is* after a compound subject. |
| **5** *is* or *has nothing to do with.* |
| **6** — *is* —. |
| **7** Postponed and repeated *is.* |
| **8** *is all.* |
| **9** The remarkable double *is.* |

**1** *is* and *are* between sentence elements of variant number. See AGREEMENT
7; ARE, IS; BE 1.

**2** *is* and *are* in multiplication tables.
In the type *five times five is/are twenty-five,*
either *is* or *are* is correct. There is just
about enough historical support in the
*OED* (s.v. *time* 19a) for both constructions
in such circumstances.

**3** *is* made to do double duty as an
auxiliary and as a copulative. See BE 5
(where the example given happens to be
of *was,* but the same principle applies).

**4** *is* after a compound subject. See
AGREEMENT 3 for examples of a composite
subject seen as a single concept (e.g.
*tarring and feathering*) followed by a singular verb. Within certain limits, and usually in informal contexts, *is how, is what*
may also acceptably follow a preceding
statement: (is how) *He is one of God's special
children, is how I look at it*—Lee Smith,
1983; (is what) *He's a fucking operator, is
what I think*—T. Wolfe, 1987; *One never
knows with these lefties, is what I always
say*—A. Brink, 1988; *You're wicked, is what
you are*—M. Doane, 1988; *You step up to
him and you cart him all over the park, is
what you do*—S. Fry, 1990.

**5** *is* or *has nothing to do with.* The *OED*
partially deals with the problem s.v. *nothing* 13 and *do* v. 33d. Between them these
entries make it clear that *to have to do
with* means 'to have dealings with, to
have relation to' and is an idiom of some
antiquity: e.g. *He wolde not haue to doo
with svche myscheuous men*—R. Eden, 1555;
*Away, I haue nothing to do with thee*—Shakespeare, 1605; *I neuer had any thing to doe
with the said Duke*—J. Wadsworth, 1630; *It
has nothing to do with the purpose*—Fraser's

*Mag.*, 1830; *All law has to do with pleasure and pain*—B. Jowett, 1875. The phrase *to be nothing to* is registered in the sense 'to be insignificant compared to some other person or thing' (*But all this is nothing to that which they both suffered for their conscience*), but there is no sign of *is nothing to do with*. It is perplexing to find that this use of *is* for *has* is neglected in some major works of reference, but help is at hand in Jespersen (1909-49, vii, § 15.83). First, he gives examples of *has*, including one from Shakespeare's *Venus and Adonis* (*Beautie hath naught to do with such foule fiends*) and another from Dickens's *Nicholas Nickleby* (*that has nothing to do with his blustering just now*). And then he goes on to say that 'in recent use we find also *is*, *was*, instead of *has*, *had*. Perhaps this may have started with the rapid pronunciation of *has* (*what's that to do with me?*) which was interpreted as *'s = is*; at any rate it is now extremely frequent in colloquial English.' Examples: *Besides, what I have had is nothing to do with it*—D. Jerrold, 1846; *This is nothing to do with your life*—H. S. Merriman, 1896. Fowler (1926) made out a stirring case for the recognition of the type with *is*: 'Most of us, when we have occasion to repel an impertinent question, and are not in the mood for weighing words in the scales of grammar, feel that *That is nothing to do with you* expresses our feelings better than *That has* etc.; that is to say, the instinctive word is *is*, not *has*.'

Whatever the strength of Fowler's feeling about the matter, it would appear that the type *X has nothing to do with Y* is the only established use at the present time. In July 1992 the electronic database of the *OED*, including a large corpus of examples from modern UK and NAmer. sources, yielded only this type and no examples of the type *X is/was nothing to do with Y*. A few examples: *Find out what Virginian Loyalist came here in 1785—which has nothing to do with me or you being Irish and Scots*—D. A. Richards, 1988 (Canad.); *Warhol's 'affectless' way of looking at objects in the world has nothing to do with any heightening of the powers of the senses*—*Modern Painters*, 1989 (UK); *The stark fact is that the Constitution has nothing whatever to do with issues of sexual morality*—R. Bork, 1990 (US); *The intellectual dishonesty that saturates the Whitney has nothing to do with lack of scholarship*—*Connoisseur*, 1990 (US); *Animal phobias have nothing to do with physical danger*—*Daily Tel.*, 1991; *I'm often told that anti-Semitism has nothing to do with racism*—*TES*, 1991.

**6** — *is* —. This is apparently a modern type used for emphasis with the force of 'really, after all'. Examples: *A job's a job, that was the thing*—Maggie Gee, 1985; *It wasn't exactly The Old Man and the Sea, but reading was reading and he was probably too old, at fifteen, to accept her censorship*—M. Leland, 1987; *She worried about Colin's wrist in the cast but a trip out was a trip out, and the day mustn't be spoiled*—N. Virtue, 1990. A different kind of emphasis or insistence is the double repetition of a noun in a formulaic type established by Gertrude Stein's *Rose is a rose is a rose is a rose*: *There is only one art form common to all sorts and conditions of people: the poster . . . a hoarding is a hoarding is a hoarding*—*Guardian*, 1970.

**7** Postponed and repeated *is*, (OED *be* 9d) This somewhat informal type has been in use since the early 19c. Examples: *He's a sad pickle is Sam!*—M. Mitford, 1828; *And Miss Rose got awfully angry, and she's clever, is my Miss Rose*—G. Meredith, 1861; *Yes, he is true to type, is Mr Heard*—R. Knox, 1932.

**8** *is all*. Used at the end of a sentence, often reinforcing the regret or sadness of the statement just made, this idiomatic expression sounds dialectal in origin, but is now found in standard (esp. American) fictional sources: *'Looks like you're into purple,' he says to her . . . 'It's my favorite color, is all,' she says*—*New Yorker*, 1987; *No one's interested, is all.*—M. Doane, 1988; *Sometimes,' says Tillie, 'God just wants to make sure He's got your attention, is all.*—*New Yorker*, 1989; *'I don't know why you always spite me by loving that house so.' 'I don't say I love it. I was born there, is all.'*—J. Updike, 1993.

**9** The remarkable double *is*. See REPETITION 1.

**isagogic.** For pronunciation, see GREEK G.

**-ise.** **1** This spelling is compulsory in the following words, most of which, let it be noted, are derived from French (esp. from the French pa.ppl ending *-is*). All are pronounced with final /-aɪz/. *advertise, advise, apprise, arise, chastise, circumcise,*

comprise, compromise, demise, despise, devise, dis(en)franchise, disguise, enfranchise,
enterprise, excise, exercise, franchise, improvise, incise, merchandise, prise (open), revise,
supervise, surmise, surprise, televise.

**2** -ise is also compulsory in a few
French loanwords that are pronounced
with final /-iːz/: chemise, expertise, reprise.
It is also compulsory in the noun premise,
and in promise (noun and verb).

**3** On the general question of the spelling of verbs ending in the sound /-aɪz/,
see -IZE, -ISE.

**4** For the coining of verbs in -ize, see
-IZE, -ISE.

**island, isle.** The two are etymologically
unconnected, the first being from OE
ī(e)gland, later īgland, from OE ī(e)g 'island'; and the second a reduced form of
L insula. The present spelling island was
adopted in the 16c. on the model of isle.

**-ism and -ity.** **1** -ism is a prolific, still
potent, suffix (representing Fr. -isme, L
-ismus, Gk -ισμός) in English, forming (a)
nouns of action from verbs; a typical
example is Gk βαπτίζειν 'to baptize', βαπ-
τισμός 'baptism': similarly English criticize ~ criticism, plagiarize ~ plagiarism, etc.;
(b) similarly formed, expressing the
action or conduct of a class of persons,
but with no accompanying verb, e.g. hero-
ism, patriotism; (c)) forming the name of
a system of theory or practice, religious,
political, etc., e.g. Buddhism, Judaism; (d)
(a subclass of (c) forming class-names for
doctrines or principles, e.g. agnosticism,
atheism, Communism, realism; (e) forming
terms denoting a characteristic feature,
esp. of language, e.g. Gallicism, Scotticism;
(f) in the second half of the 20c. forming
'politically correct' terms such as ableism
(prejudice against disabled people by
able-bodied people), speciesism (prejudice
or discrimination of one species over
another, esp. of the human species in the
exploitation of animals): see POLITICAL
CORRECTNESS. The suffix is so prevalent
in English words that, since the 17c., it
has often been used as a noun in its own
right: e.g. The proletarian Isms are very much
alike—G. B. Shaw, 1928.

**2** -ity is a rich formative element in
nouns expressing a state or condition,
(a) from adjectives of various kinds, e.g.

absurd ~ absurdity,        curious ~ curiosity,
modern ~ modernity; (b) specif. forming
abstract nouns from comparatives, e.g.
inferiority, majority, minority. It does not
follow that all adjs. have a corresponding
abstract noun in -ity: thus furious but no
*furiosity (the asterisk is conventionally
used here to indicate non-existent
formations), flat but flatness, not *flatity.
Also the relationship between pairs that
do exist is not, in word-formation terms,
always exactly the same: thus voracious
if it followed the model of curious would
produce not voracity but *voraciosity; grave
loses its -e and becomes gravity; divine
with a long second syllable produces
divinity with a short one, but both obese
and obesity have /iː/ in their second syllable.

**3** The absence of firm rules has meant
that over the centuries pairs of -ism and
-ity words have come into the language,
usually with different meanings. Thus
community and Communism share the
basic notion of L commūnis 'common',
but are widely different in meaning.
There are many such pairs in the language, some of which are treated at their
alphabetical place in this book, e.g. barbarism, barbarity; fatalism, fatality; humanism, humanity; Latinism, Latinity; legalism,
legality; liberalism, liberality; modernism,
modernity; realism, reality; spiritualism,
spirituality. The differences of meaning
are more or less self-evident.

**4** The only new formation in -ity in
the 20c. appears to be the controversial
word subsidiarity: it was formed as a rendering of Ger. Subsidiarität in 1936, not
as a direct formation on the adj. subsidiary. See also -TY AND -NESS.

**Israeli, Israelite.** See HEBREW.

**issue** (verb). Construed with with, and
having the meaning 'to supply, esp. officially' (issued them with passports), this use
of issue has passed from strong condemnation to acceptance (at least in BrE) in
the course of the 20c. It is interesting to
look back to the comment in Fowler
(1926): 'The military construction, to
issue a person with a thing (The Company
was issued with two gas-masks per man), on
the analogy of supply and provide, though
much popularized by the war, is not to
be recommended.' This use is said to be
not current in AmE, but it turns up

occasionally: e.g. *People in Russia's second city—hit by severe food shortages—are issued with coupons which entitle them to basic foodstuffs at subsidized state prices—Chicago Tribune*, 1991. In such circumstances the normal construction in AmE is *issue* + a direct object.

**issue** (noun and verb). The pronunciation /ˈɪʃuː/ is recommended, rather than /ˈɪsjuː/.

**-ist, -alist,** etc. **1** *-ist* forming agent-nouns from verbs in *-ize*, e.g. *baptist* ~ *baptize, dogmatist* ~ *dogmatize*, answers to Fr. *-iste*, L *-ista*, and Gk *-ιστής*, and is richly productive in the language. There are several types: (*a*) Forming a simple agent-noun and usu. having an accompanying verb in *-ize* (see above, also *antagonist, apologist, evangelist*, etc.) and an accompanying abstract noun in *-ism* (*antagonism*, etc.); (*b*) designating a person devoted to some art, science, etc., e.g. *archaeologist, economist, dramatist, philologist*; (*c*) designating an adherent of some creed, doctrine, etc., e.g. *atheist, Buddhist, Calvinist, hedonist*; (*d*) modern formations of various kinds, e.g. *balloonist, cyclist, fetishist, finalist*.

**2** Most of the *-ist* words have one undisputed form. The main exceptions (with the preferred form placed first) are: *accompanist, accompanyist* (?obs.); *agriculturalist, agriculturist; constitutionalist, constitutionist* (obs.); *conversationalist, conversationist* (?obs.); *diplomat, diplomatist* (?obsolesc.); *educationalist, educationist; egoist* (q.v.), *egotist; horticulturist, horticulturalist* (not in *OED* but current since at least 1863); *pacifist, pacificist* (obs.); *separatist, separationist* (obs.); *voluntarist, voluntaryist* (obs.). Note. *Tobacconist* is an irregular late-16c. formation from *tobacco* + *-ist* with insertion of an unetymological *-n-*.

**isthmus** /ˈɪsməs/. The pl. is *isthmuses*.

**it.**

| **1** Customary uses.
| **2** *it* used to highlight a sentence element.
| **3** Anticipatory *it*.
| **4** Pleonastic *it*.
| **5** *it* as a 'prop' in statements of time, etc.
| **6** *it's I/it's me*.

This article aims to record some of the central uses of the pronoun *it* on the assumption that they are mostly 'known'

to native speakers but not in any critically analysed manner. For further uses of *it*, which are very numerous, readers will need to consult the *OED* and *CGEL*.

**1** Customary uses. To state the obvious first, the pronoun *it* occurs most commonly as subject (*It was an unremarkable incident*), as object (*he took her hand and kissed it*), and as indirect object after a preposition (*she seemed to have forgotten about it*). Its functional forms are the possessive *its* (*the main thrust of any story is to be found in its incidentals*); the abbreviated form *it's* = 'it is' (*It's just for half an hour*); the interrogative *is it?* (*'Where the hell is it?' she said, all patience gone*); the negative interrogative *isn't it?* (*Sickening, isn't it, how we stick together?*); and *it isn't* as an ordinary negative (*It isn't here*).

**2** *It* is used to introduce an element that is being emphasized or highlighted (the so-called 'cleft' sentence). Various types are shown in the following examples: *Our lives are not worth living* → *It is we whose lives are not worth living; We laughed at you* → *It was we who laughed at you; I met him on Monday* → *It was on Monday that I met him; I've not come to discuss Henry but Poppet, your wife* → *It isn't Henry I've come to discuss. It's Poppet, your wife.*

**3** Anticipatory *it*. 'When the logical subject of a verb is an infinitive phrase, a clause, or sentence, this is usually placed after the verb, and its place before the verb is taken by *it* as "provisional" or "anticipatory" subject' (*OED*). (*a*) An infinitive phrase: *It was difficult to know to what level Moyra would be able to rise;* (*b*) With a clause introduced by *that* expressed or understood (esp. frequent with the passive voice in *it is said, believed*, etc., *that*): *It was said that it was a matter of attitude; Wasn't it true that we didn't hit it off at first?* When such a *that*-clause is introduced by *it is appropriate, crucial, essential*, etc., the *that*-clause has a *should* construction or a verb in the subjunctive: *it is crucial that you (should) be home by four.*

**4** Pleonastic *it*. In ballads and in rhetorical passages of prose *it* has traditionally been used pleonastically after a noun subject: *The raine it raineth euery day*—Shakespeare, 1601; *This piteous news so much it shocked her*—Wordsworth, 1798. Outside the higher forms of literature this type (e.g. *The bird it flapped its wings* *and flew off*) is now non-standard or dia-lectal.

**5** *It* as a 'prop' in statements of time. *It* is especially common in statements as to the time of day, season of the year, distance to or from a place, the state of the weather, and so on: *It was morning. Outside, it was cold and sunny; it's three miles from Abingdon.* This impersonal use of *it* is sometimes called a 'prop *it*'.

**6** *It's I/ It's me.* For the competing types *It's I* and *It's me*, see CASES 2.

It should perhaps be noted that the majority of the examples in this article have been taken for convenience from Simon Mason's novel *The Great English Nude* (1990) and from 1992 issues of *The New Yorker*, but they could have been drawn from almost any other standard source.

**Italian sounds.** Loanwords from Italian fall into the usual three broad categories: those still pronounced to resemble the original Italian (*pizzicato, scherzo*), those sometimes pronounced in a largely Anglicized manner (*intaglio, intermezzo*), and those that are fully acclimatized (*umbrella, volcano*). English speakers nor-mally recognize that the vowels of words in the first and second categories are to be pronounced in a 'continental' (i.e. tense, not lax or diphthongal) manner. There is rather less public knowledge of the consonantal sounds of Italian where these differ from English consonantal sounds. Phoneticians can detect subtle differences of aspiration, voicedness, and so on, between the way in which Italian and English speakers pronounce *b, d, f, g*, etc., but these need not concern us here. The consonants that cause diffi-culty are as follows (the IPA symbols represent the standard Italian pronunci-ations):

    *c* and *cc* before *e* and *i* = /tʃ/: *cicerone, Cinzano, arancia* 'orange', *capriccio*
    *cch* = /k/: *Ponte Vecchio*
    *ch* = /k/: *Chianti*
    *ci* before *a, o, u* normally = /tʃ/: *ciao!, cioccolata*
    *g* and *gg* before *e* and *i* = /dʒ/: *generalissimo, viaggio* 'journey'
    *gh* = /g/: *ghetto*
    *gi* before *a, o, u* normally = /dʒ/: *Giotto, Giovanni*
    *gl* = /lj/: *seraglio*
    *gn* = /nj/: *bagno* 'bath', *signor, gnocchi*
    *gu* = /gw/: *Guelph*

    *sc* before *e* and *i* = /ʃ/: *Fascisti* 'Fascists'
    *sch* = /sk/: *scherzo*
    *sci* before *a, o, u*, normally = /ʃ/: *sciolto* 'loose'
    *z* = /ts/: *scherzo*
    *zz* normally = /ts/: *pizzicato*
    *zz* occas. = /dz/: *mezzo*

**italics.** These are *a style of sloping type, like this*, indicated in handwritten or typewritten matter by a single underlin-ing. Italic type is conventionally used in English in a wide range of circum-stances, a reasonably full account of which is set down in *Hart's Rules*, pp. 23–8. The more important types are as follows:

> Book titles: *David Copperfield, To the Lighthouse.*
>
> Play and film titles: *Hamlet, Gone with the Wind.*
>
> Works of art: Leonardo da Vinci's *Last Supper*, Picasso's *Guernica.*
>
> Long poems which are virtually books in themselves: *The Faerie Queene, Paradise Lost*, and any other poems divided into books or cantos.
>
> Names of periodicals, newspapers, etc.: *London Review of Books, Dædalus.*
>
> Names of ships (preceded by 'the' in roman type): the *King George V*, the *Ark Royal*, HMS *Dreadnought.*
>
> Stage directions in plays: *Exeunt Don Pedro, Don Iohn, and Claudio.*
>
> Foreign words and phrases which are not fully naturalized in English: *amour propre, jeu d'esprit, ne plus ultra, Weltanschauung.*
>
> Words, phrases, or letters mentioned by name: 'The word *loyully* has three *l*s; the sentence adverb *frankly* is the equivalent of the phrase *to be frank*.'
>
> As a method of emphasizing (a device to be used sparingly): 'Oh come now, it can't be *that* bad.'
>
> As a method of distinguishing: 'The question is not whether indexing *can* be automated but whether and why it *should* be.'

*Note*: Fowler's elegant essay on the sub-ject in *MEU* (1926) may also be found in SPE Tract xxii (1925).

**itch** (verb). **1** Used transitively in the sense 'to cause to itch', *itch* is recorded from the 16c. onward is all manner of writing, but in BrE it is showing signs of becoming restricted to informal con-texts. It seems to be still standard in AmE. Examples all, as it happens, from standard sources: *The thick super-salty*

water of the Mediterranean, which tires and
itches the naked eye—Roy Campbell, 1951;
The dice already itch me in my pocket—L.
MacNeice, 1951; I heard those [fruit] flies
are coming to hide under your bed tonight—to
itch you!—New Yorker, 1990; the stone seat,
whose grittiness pressed through his trousers,
itching his thighs—A. Huth, 1991.

2 A more down-market use (restricted
to AmE?) is the sense 'to scratch (an itchy
part of the body)', as in: Don't itch your leg.
You can't do that on stage—Chicago Tribune,
1991.

**-ite.** As an ending of adjs it is derived
esp. from L pa.pples in -ītus, -itus, e.g.
ērudītus 'erudite', compositus 'composite'.
The length of the i in Latin has long
ceased to have any bearing on the way
in which the English derivatives are pronounced. The short i in apposite and opposite follows the Latin; that in definite
does not. The long i in bipartite and erudite
follows the Latin; that in recondite does
not. Composite (short i in Latin) has /ɪ/ in
RP but /aɪ/ in some regional forms of
English. The separate suffix -ite (originally from Gk -ίτης) seen in anthracite,
dynamite, Jacobite, etc., always has /aɪt/ in
English.

**its, it's.** Just a reminder that its is the
possessive form of it (the cat licked its paws)
and that it's is a shortened form of it is
(It's raining again) or it has (It's come). See
IT 1.

**-ity.** See -ION AND -NESS; -ISM AND -ITY.

**-ize, -ise in verbs.** 1 It is important
first to put aside the verbs which must
always be spelt with -ise: advertise, advise,
etc. A list of these is provided s.v. -ISE 1.
The AmE spelling of analyse, catalyse, etc.,
as analyze, catalyze, etc., is also a separate
matter: see -YSE, -YZE.

2 Spelling. In the vast majority of the
verbs that may in English be written
either with -ize or -ise and are pronounced
with /-aɪz/ the ultimate source is the
Greek infinitival ending -ίζειν (L -izāre),
whether the particular verb was an actual Greek one or was a Latin or French
or English imitation, and whether such
imitation was made by adding the termination to a Greek or another stem. A
key word showing the line of descent is
baptize, which answers to Gk βαπτίζειν
and L baptīzāre. But the French have

opted for baptiser, and a large proportion
of English writers and publishers have
followed suit by writing the word as
baptise; similarly with hundreds of other
formations of this type. In Britain the
Oxford University Press (and, until recently, The Times) presents all such words
with the termination spelt -ize. So do
all American writers and publishers. It
should be noted, however, that many
publishing houses in Britain, including
Cambridge University Press, now use -ise
in the relevant words. The matter remains delicately balanced but unresolved. The primary rule is that all words
of the type authorize/authorise, civilize/civil-
ise, legalize/legalise may legitimately be
spelt with either -ize or -ise throughout
the English-speaking world except in
America, where -ize is compulsory.

3 Status of such verbs. Quite apart from
the problem of the spelling of the
termination is the widespread current
belief that new formations of this kind
are crude, overused, or unnecessary.
('Within reason it [sc. the formation of
new verbs in -ize] is a useful and unexceptionable device, but it is now being employed with a freedom beyond reason',
Gowers 1965.) Objections to the word
FINALIZE, and to such words as permanent-
ize (first recorded in 1961 and judged by
myself in OEDS, 1982, to be 'a word of
little value and rarely found in serious
writing') and prioritize (first recorded in
1973 and described by me in 1982 as 'a
word that at present sits uneasily in the
language') are to be set against the long
history and distinctive usefulness of
such formations in English. The earliest
verb in -ize is baptize: it was first recorded
in English in the late 13c. Since then a
very large number of uncontested -ize
and -ization words have entered the language and are indispensable: e.g. authorize (first recorded in the 14c.), characterize
(16c.), civilize (17c.), fossilize (18c.), immor-
talize (16c.), memorize (16c.), patronize
(16c.), sterilize (17c.), terrorize (19c.).

There is always the danger of forming
misleading impressions about such
verbs. For several years I have taken to
making a note of what I felt were probably new or unrecorded formations in
-ization and -ize. They have been drawn
from a wide range of written and spoken
BrE and AmE sources. For example, at a
conversazione of the English Association

in June 1991 Iris Murdoch spoke of *writers who invisibleized a* [specified] *problem*. In 1987 a columnist in *Newsweek* carried a story of *an unmarried pregnant woman who said that she should have had her lover 'condomized'.* The exercise has been instructive. It turned out that about a quarter of the words I noted already had entries in the *OED*: e.g. *civilianization* (first recorded in 1946), *funeralize* (marked *Obs.*⁻¹, 1654), *strategize* (1943), *weaponization* (1969). In other words a proportion of the words that I felt to be possible intruders already had a respectable amount of currency (except for *funeralize*) and have entries in our largest dictionary. Other words in my files perhaps will have less permanent value: e.g. (in *-ization*) *marketization, peripheralization, securitization*; (in *-ize*) *couponize, disasterize, incentivize, keyboardize, marketize, minoritize, nuggetize*. But who is to tell? *Incentivize* and *marketization* are listed in the *Longman Register of New Words* (1990) and the others may follow. One must be careful not to give the thumbs down to words simply because one has not encountered them before.

As a further experiment I examined *The Oxford Dictionary of New Words: A Popular Guide to Words in the News* (1991). It lists only nine words in *-ization* or *-ize*: *debrezhnevization, decommunize, democratization, dynamize/ -ization, in vitro fertilization, marginalize/-ization, visualization*. In other words only a minute proportion of the 2,000 words presented ended in *-ization* or *-ize*, and of these six (*democratization, dynamize/-ization, fertilization, marginalize, visualization*) are 19c. words being used in new senses. Any feeling that the language is being swamped by new formations in *-ization* and *-ize* does not appear to be supported by the facts.

*Verdict.* The language has a large number of established words ending in *-ization* and *-ize*, but the arrival of new words of this type continues at a rate that, rightly or wrongly, makes many fastidious users of the language uneasy. Some of the 20c. newcomers will drop by the wayside. Others will survive into the 21c. and beyond. The creasing of brows about them will continue.

# Jj

**jabot.** Pronounce /ˈʒæbəʊ/.

**jackal.** Pronounce either /ˈdʒækɔːl/ or /ˈdʒækəl/.

**jacket.** Retain single *t* in the extended forms *jacketed, jacketing*. See -T-, -TT-.

**Jacobean, Jacobin, Jacobite.** These adjectival and nominal forms, ult. from L *Jacōbus* 'James', have been used as sobriquets of several different groups of people or things. *Jacobean* is used as adj. and noun with reference to the reign (1603–25) or times of James I of England, and, in particular, to an architectural style which prevailed in England in the early part of the 17c. The commonest use of *Jacobins* (the name earlier given in France to Dominican friars) is for the group of revolutionaries formed in Paris in 1789 (who used to meet in what was once a Dominican convent). The commonest use of *Jacobites* is for adherents of James II of England after his abdication in 1688, or of his son the Pretender. *Jacobites* is also sometimes used by writers for devotees of the works of Henry James (1843–1916).

**jaggedly.** Three syllables. See -EDLY.

**jail, jailer,** etc. OUP house style requires *gaol, gaoler*, etc., but the *jail*-forms are extremely common in BrE and obligatory in AmE.

**jamb** (side-post of doorway, etc.). Pronounce /dʒæm/ to rhyme with *lamb*.

**janizary,** rather than *janissary*, is the form given precedence in *COD* 1995.

**Jap,** a colloquial shortening of *Japanese*. As noun and adj. the word has had strong derogatory connotations since the beginning of the 20c., and is now falling into relative disuse in favour of the full form *Japanese*.

**jargon. 1** Over the centuries this word has been used in a number of ways, all connected with types of speech or writing rejected from normal standard English as being unintelligible, restricted in currency, inarticulate, or unfamiliar. It is perhaps the most suitable heading for an article describing the main distinctions between them. The words are *argot, cant, dialect, gibberish, idiom, jargon, lingo, lingua franca, parlance, patois, shop, slang,* and *vernacular*.

**2** It may be best, first, to consider the etymology of each of these terms before assigning meanings to them. The bracketed dates indicate the centuries in which the various words were first recorded. *Argot* (19c.), *jargon* (14c.; as a pejorative term 17c.), *parlance* (16c.), and *patois* (17c.) are French; *dialect* (16c.) and *idiom* (16c.) are immediately from French, but ultimately from Greek (διά-λεκτος 'discourse, way of speaking'; 'language of a district'; ἰδίωμα 'peculiar phraseology'); *cant* (16c.) and *vernacular* (17c.) are from Latin (cf. L *cantāre* 'to sing'; *vernāculus* 'domestic, indigenous', from *verna* 'home-born slave'); *lingo* (17c.) is probably from Portuguese *lingoa* but may be a corrupt form of *lingua (franca)*; *lingua franca* (17c.) is Italian and means 'Frankish tongue'; *gibberish* (16c.), *shop* (this sense 19c.), and *slang* (18c.) are English, the first from the verb meaning 'to chatter incoherently' + the ending *-ish* of English, Scottish, Danish, etc., the second a particular application of the ordinary word *shop*, and the third of unknown origin.

**3** The thirteen words and their present-day meanings: *argot*, the peculiar phraseology of a group or class, originally that of thieves and rogues. It is properly applied only to such groups in France.

CANT: see as a main entry.

DIALECT: see as a main entry.

*gibberish*: variously, unintelligible speech belonging to no known language; inarticulate chatter; blundering or ungrammatical language.

IDIOM: a fixed phrase established by usage and having a meaning not deducible from those of the individual words in the expression, e.g. *to keep a straight face*; *to sow one's wild oats*. See also as a main entry.

*jargon* has several meanings: the inarticulate utterance of birds; a term of contempt for something (including a foreign language) that the listener does not understand; and esp. any mode of speech abounding in unfamiliar terms, or peculiar to a particular set of persons, e.g. the specialized vocabulary of bureaucrats, scientists, or sociologists. See 4 below.

*lingo*: a contemptuous designation for foreign speech or language (*I can't speak their beastly lingo*); also, for language peculiar to some special subject and not intelligible to the listener or reader.

*lingua franca*: (a) a language adopted as a common language between speakers whose native languages are different, e.g. Latin in medieval Europe, Arabic in the Near East, Malay in South-East Asia, English in many parts of the world. (b) earlier, and now only in historical contexts, a mixture of Italian with French, Greek, Arabic, and Spanish, used in the Levant. See CREOLE; PIDGIN.

*parlance*: a particular way of speaking, esp. as regards choice of words, idiom, etc. Though originally from French, the word is no longer current in modern French. In practice, *parlance* must always be accompanied by a defining word or phrase, e.g. *in racing parlance, in the parlance of post-modernists, in common parlance*.

*patois*: the dialect of the ordinary people in a region, differing fundamentally from the literary language. It is not used of any form of speech in Britain, and is more or less restricted to designate popular modes of speech in France.

*shop*: occurs chiefly in the phr. *to talk shop*, to talk about one's work or business out of working hours.

SLANG: see as a main entry. See also BACKSLANG; RHYMING SLANG.

*vernacular*: the indigenous language of a particular country (*Latin gave way to the local vernaculars in France, Italy, Spain*, etc.). It is also sometimes applied to ordinary, as distinct from formal or literary, English. In America, Black English, a mode of speech widespread among black speakers, is now often called Black English Vernacular (BEV).

**4** The use and misuse of jargon. (a) No one has any quarrel with the use of *jargon* to mean 'the inarticulate utterance of

birds', esp. since it occurs mainly in literary sources from Chaucer to Longfellow.

(b) Nor do arguments break out when the term is applied, as, since the 17c., it often has been, to hybrid speech arising from a mixture of languages (e.g. *A mingled dialect, like the jargon which serves the traffickers on the Mediterranean and Indian coasts*—Dr Johnson, 1755).

(c) Academic jargon. It is generally accepted that writers of technical and scientific works and authors of other kinds of scholarly monographs are entitled to expect their readers to master the specialized terminology of their subjects. Scholarship has its own intricate rules and conventions, and user manuals theirs. In the passages that follow each of the writers doubtless hopes that what he has written will be understood, even though none of the passages is written in 'plain English'.

(The kind of 'Europeak' used at international linguistic conferences) *As an example of a common development in both genetically and areally closely related languages one could take the grammaticalisation of embraciation in declarative sentences in German, Dutch and Frisian* ... (A. Danchev, 1991). (The language of computer user manuals) *When the SET FORMAT TO SCREEN has been issued, an ERASE will clear the screen of all information that was previously on it, will release all the GETS* ... *and will reset the coordinates to 0,0. When the SET FORMAT TO PRINT has been issued, an EJECT will do a page feed and reset the coordinates to 0,0* (*DBASE II User Manual*, 1985). (Semiotics) *Moreover, nearly all the non-verbal signs usually rely on more than one parameter; a pointing finger has to be described by means of three-dimensional spatial parameters, vectorial or directional elements, and so on* (U. Eco, 1976).

(Literary criticism) *This view of the text ... has been seriously challenged in recent years, mainly by structuralist and semiological schools of criticism. According to these, the text has no within, beneath or behind where hidden meanings might be secreted. Attention is instead focused exclusively on the processes and structures of the text and on the ways in which these produce meanings, positions of intelligibility for the reader or the specific effects of realism, defamiliarisation or whatever* (T. Bennett, 1982). (Philosophy) *he suggests that someone with my realist views about the qualitative properties of experience ought to allow for the possibility of inverted spectra, thus pulling apart content and qualia. I think myself, in opposition to this, that inverted spectrum cases are precisely cases in which representational content is changed, so that it fits qualitative character* ... (C. McGinn, 1991).

But many scholars are dismayed by the employment of obscure terminology in academic writing. Thus, for example, the distinguished American philosopher John R. Searle (in *Bull. Amer. Acad. Arts & Sci.*, 1993): 'There is supposed to be a major debate ... going on at present concerning a crisis in the universities—specifically, a crisis in the teaching of the humanities ... Though the arguments are ostensibly about Western civilization itself, they are couched in a strange jargon that includes not only *multiculturalism* but also such terms as *the canon, political correctness, ethnicity, affirmative action*, and even more rebarbative expressions such as *hegemony, empowerment, poststructuralism, deconstruction*, and *patriarchalism*.' We have been (justifiably) warned.

(*d*) Repetitive jargon is a feature of live commentaries on sports and games: e.g. (Rugby football) *players left on the deck; hoisting one deep into Scotland's territory; the little chip and chase;* (Association football) *sick as a parrot; over the moon; take each match as it comes; that's what football is all about;* (Snooker) *the jaws of the pocket; get his cue arm going; the rub of the green; he failed to develop the black.* For some people, the sheer familiarity and predictability of such language add to the pleasure of the occasion. Others crave for more flexibility. At worst the linguistic sin of the playing field and the games room provides amusement for sophisticated couch potatoes.

(*e*) A steep downward path leads to the often mystifying jargon of various kinds of sociological writing. It is bearable (just) when Colin Renfrew, a distinguished archaeologist, calls the driving out of a people from their normal territory a *constrained population displacement*. A people fleeing in terror from an enemy (like the Kurds in Iraq in 1991) suffer this fate, and a card-index entry for the event can be made under a 'neutral' three-element heading. The Holroyds in Cyra McFadden's *The Serial* (1977) present

us with what has been called *psychobabble*—rosy enriched spontaneous pseudo-comfortable language reflecting the intricacies of their lives in an idyllic landscape known as Marin County in California (*Marin's this high-energy trip with all those happening people; Harvey and I are going through this dynamic right now, and it's kinda where I'm at. I haven't got a lot of psychic energy left over for social interaction.*).

(*f*) Worst of all is the kind of jargon employed as an obfuscating technique in bureaucratic and political contexts. One of its most distinguished critics, Sir Ernest Gowers, attacked it relentlessly in his book *Plain Words* (1948) and in the expanded version called *The Complete Plain Words* (1954). Everyone must be aware of the danger by now. The latest version of the book (ed. S. Greenbaum and J. Whitcut, 1986) sets down and analyses numerous passages of 'flat, tired, perfunctory, inconsiderate writing', of work that is 'verbose and stilted' or 'ambiguous', and so on. Every kind of pomposity and inelegance is surveyed, analysed, and 'translated'. Civil servants, politicians, lawyers, diplomats—indeed, anyone who has to address the general public in some official or business capacity—should have the latest version of the book at hand. Genuine communication in such areas of life has never been more important than in our inflammatory and dangerous times.

A typical example of immoderately obscure bureaucratic language is cited by the pseudonymous writer Theodore Dalrymple in a 1991 issue of *The Spectator*: 'Last week I received a circular entitled *Joint Care Planning to J.C.C.G.P.T. by Task Group for C.M.H.T.'s* ... For a reason which I now cannot recall, I opened this 20-page circular, of which someone was evidently very proud because it came in its own black folder. My eye fell on a section entitled *Service Sensitivity to Particular Groups*. Paraphrase will not do it justice, so I must quote *in extenso*. "Equal opportunities are not easily achieved. To have a policy and a strategy are insufficient. There has to be an internalisation of its fundamental philosophy which does not mean 'everyone is equal. I treat people the same whatever'. It is, however, to do with the differences between people, the uniqueness of each person. In terms of service delivery it is within

CMHT's that greater expression should be able to be given to the philosophy of equal opportunities in that by focusing on a given area with the benefit of a multi skill group together with users of the service a very clear strategy can be drawn up within the operation policy of each team that will lead not to a bland overall response to need but one that is tailored to and by the individual and community. In recruitment, selection and retention of staff, statutory and voluntary organisations need to continue to address the issue of achieving a shared understanding of what equal opportunities actually means and how it is to be put into practice." Insofar as this passage means anything—beyond its menacingly imperative tone—it means that you can't expect blacks to be the equal of whites, so they must be given jobs which, strictly speaking, they don't really deserve.' One's heart goes out to Theodore Dalrymple and other hapless recipients of such bureaucratic opacity.

**jarl.** Pronounce /jɑːl/, with the initial sound of *yew*.

**jasmine.** The three-syllabled by-form *jessamine* /ˈdʒesəmɪn/ was common in literary use in the 19c. (and earlier): *And the jessamine faint, and the sweet tuberose*—Shelley, 1820; *All night has the casement jessamine stirr'd*—Tennyson, 1855. But *jasmine* (first recorded in Lyte, 1578) continues to be the normal term in botanical parlance.

**jaundice, jaunt, jaunty.** The 19c.-early 20c. variant pronunciation of the stem vowel as /ɑː/ seems to have been given up by standard speakers at some point in the first half of the 20c. The only current pronunciations are /ˈdʒɔːndɪs/, /dʒɔːnt/, and /ˈdʒɔːntɪ/. Cf. LAUNCH.

**jazz.** See -Z-, -ZZ-.

**jehad.** See JIHAD.

**jejune.** 1 Now usu. pronounced /dʒɪˈdʒuːn/, with short first vowel, and stressed on the second syllable. Fowler (1926) commented 'now jĕ jŏŏn [i.e. /ˈdʒiːdʒuːn/] by recessive accent', but his view has been overtaken by events.

**2** Derived from L *jējūnus* 'fasting', the word first (17–18c.) meant 'without food' in English; then, in transferred use, '(of

concrete things, land, water, food, etc.) thin, meagre, unsatisfying'; and, more or less simultaneously, 'unsatisfying to the mind, dull, insipid, etc.: said of thought, feeling, action, etc., and esp. of speech or writing' (*OED*). Towards the end of the 19c., by a somewhat surprising association of ideas, *jejune* acquired the sense 'puerile, childish, naïve' as if it was connected with L *juvenis* 'a young person' or Fr. *jeune* 'young', or so some authorities say.

The *OED* lists examples of the new use from 1898 onward, including the following: *Is anybody ... now so jejune as not to realise that the state ownership of the deadweight of present nationalised industries must prevent Labour governments from being able to follow ... their social policies?*—*Economist*, 1975; *Mother seemed jejune, at times, with her enthusiasms and her sense of mission*—M. Howard, 1982. *WDEU* cites further examples of the new use from H. L. Mencken (1920) and from various journalistic sources. On the other hand, in a well-known essay in *The State of the Language* (1980), Kingsley Amis, railed against the use ('my favourite solecism of all time'). It seems unlikely that Fr. *jeune* or L *juvenis* had anything to do with the matter. A semantic shift from 'unsatisfying, insipid' to 'puerile' is not intrinsically improbable in a word that is in any case rather rare. Nevertheless, I do not feel comfortable with the new sense myself, and there are numerous synonyms of *puerile* (*childish, infantile, juvenile*, etc.) that can be used instead. In the circumstances, those who wish to use the word at all are advised to use it in its traditional senses, at least for the present.

**jerrymander.** See GERRYMANDER.

**jessamine.** See JASMINE.

**jetsam.** See FLOTSAM AND JETSAM.

**jettison.** In maritime law (since the 15c.) it means 'the action of throwing goods overboard, esp. in order to lighten a ship in distress'. Since the 19c. it (and also the corresponding verb) has been used of the action of throwing out or discarding any unwanted object (or idea, etc.).

***jeu d'esprit.*** Pl. *jeux d'esprit*, with the *x* left silent. See -X.

**Jew** (and related words). **1** A Jew is a person of Hebrew descent or one whose religion is Judaism. Jewish people live in many countries in the world. Citizens of modern Israel, many of whom are not Jewish, are called *Israelis*. The biblical *Israelites* were descendants of the Hebrew patriarch Jacob, whose alternative name was 'Israel'. The ancient language of the Israelites was *Hebrew*, which was spoken and written in ancient Palestine for more than 1,000 years. By *c*500 BC it had come greatly under the influence of Aramaic, which largely replaced it as a spoken language of the Jewish people. It was revised as a spoken language in the 19c., with the modern form having its roots in the ancient language but drawing words from the vocabularies of European languages, and is now the official language of the State of Israel (*Oxford Reference Dict.*, 1986).

**2** Since the 17c. the word *Jew* has sometimes been offensively applied to persons considered to be parsimonious or to drive hard bargains in trading. 'The stereotype, which is now deeply offensive, arose from historical associations of Jews as moneylenders in medieval England' (*COD*, 1990). The verb *jew* (or *jew down*) has also been recorded in use since the early 19c. to mean 'to cheat or overreach, in the way attributed to Jewish traders or usurers; to drive a hard bargain'. The need for the abandonment of such unenlightened language is obvious to all well-meaning people, and it is to be hoped that the 21c. will see these uses drop out altogether. An account of the controversial history of the word, as noun and as verb, may be found in my book *Unlocking the English Language* (1989).

**3** The normal adj. is *Jewish*. The simplex *Jew* used attributively or as an adj. (e.g. *Jew boy, girl, pedlar*, etc.) is now only used offensively, though it was not originally opprobrious (see the *OED*).

**jewel.** The inflected forms are *jewelled*, *jewelling*, and *jeweller* (in AmE usu. with -l-). See -LL-, -L-.

**jewellery.** Pronounce /'dʒuːəlrɪ/. The pronunciation /'dʒuːlərɪ/, with the last two syllables pronounced as in *foolery*, is vulgar. The spelling *jewelry* is sometimes used in BrE, and is the usual form in AmE.

**Jewess.** The late Marghanita Laski, novelist and broadcaster, told me round about 1970 that it was not acceptable for me to call her a Jewess, but that Jewish people were entitled to use the term among themselves. This habit of claiming exclusive rights to the use of particular words and denying them to outsiders is one of the paradoxes of the 20c. The word *Jewess* has in fact been in continuous use since the 14c., and, until the 20c., seems to have had no adverse connotations. The new state of the word is implied in the following example: *The antiquated language is even sharpened by a reference to an Ashley daughter-in-law as a 'Jewess'*—A. Hollander, 1990. See NEGRESS.

**jibe.** See GIBE; GYBE.

**jihad** /dʒɪˈhɑːd/. Now the normal spelling in English, rather than *jehad*, for this Arabic word meaning 'a holy war undertaken by Muslims against unbelievers in Islam'.

**jingles.** Fowler's term for 'the unintentional repetition of the same word or similar sounds'. His examples include: *The sport of the air is still far from free from danger*; *Mr Leon Dominian has amassed for us a valuable mass of statistics*; *Most of them get rid of them more or less completely*; *I awaited a belated train*; *Hard-working folk should participate in the pleasures of leisure in goodly measure*. Such jingles in the spoken language are likely to be light-heartedly commented on (e.g. *You're a poet but don't know it*). In the written word, a second reading of the text at drafting stage, or a reading by an independent person, normally ensures that such stylistic blemishes are noticed and removed before it is too late.

**jingo** (a supporter of any policy involving war). Pl. *jingoes*. See -O(E)S 1.

**jinnee** /'dʒɪniː/, 'a spirit in Muslim mythology, *not* djinn, ginn; pl. *jinn*, often used as singular' (*Oxford Writers' Dict.*, 1990).

**jiu-jitsu.** See JU-JITSU.

**jockey.** Pl. *jockeys*.

**jockey** (verb). Inflected forms are *jockeyed*, *jockeying*, *jockeys*.

**jollily, jolly** (advbs.). As a colloquial substitute for *very* (a *jolly good hiding*; *you* know jolly well) the adverb is *jolly*; in other uses (*he smiled jollily enough*) it is *jollily*, or, if this is felt to be too clumsy, *in a jolly* (*enough*) *way*. See -LILY.

**jonquil** /'dʒɒŋkwɪl/. The older pronunciation /'dʒʌŋkwɪl/, favoured by the *OED* (in 1901), is now obsolete.

**journalese.** *It is sad . . to find* [*him*] *guilty of such journalese as 'transpired'*—*Athenæum*, 1893. For more than a century the word *journalese* has been used to describe the more hackneyed or 'racy' language associated with newspaper writing, and esp. with the kind of colourful language used in the tabloids. The following examples of standard and virtually unnoticed journalistic words and formulas are taken from a single 1991 issue of *The Times*: *a sharp rise in retail prices*; *Major's initiative heralds drive for quality and wider choice*; *the cabinet big spenders* [ = the cabinet ministers responsible for the largest sectors of public expenditure]; *a North Korean woman terrorist* . . . *stands to earn $1.37 million in royalties* [from the sale of a book]. (headlines presented in the style of speech in a novel) *Problems self-inflicted, says Kinnock*; *Cut arms research, say Lib Dems*. (frequent use of the present tense in headlines) *Britain protests at American veto*; *Shamir takes a tentative step towards peace talks*. (frequent use of the infinitive with future time reference in headlines) *PLO to stay out of talks*; *Press to publish school leagues*; (but not always) *Late trains will entitle passengers to refund*.

**journey.** Pl. *journeys*.

**joust.** 1 *Spelling*. The historical spelling for this word from the 13c. to about 1800 was predominantly *just*: it is derived from OF *juster* from late L *iuxtāre* 'to approach, come together' (cf. L *iuxtā* 'near together'). Cf. the cognate word *adjust*. Nevertheless most early English dictionaries (including Johnson's, 1755) listed both forms, while Walter Scott and some other 19c. authors preferred *joust*. The only current spelling is *joust*.

2 *Pronunciation*. When written as *just* the word seems to have been pronounced /dʒʌst/ in the 19c. For the form *joust* the standard pronunciation (as given by Daniel Jones) in 1917 was /dʒuːst/. By 1932 A. Lloyd James (*Broadcast English*) recommended /dʒaʊst/, and this

pronunciation has now entirely supplanted the older /dʒuːst/ in BrE.

**jubilate** (or **J-**) (noun), the hundredth psalm; a call to rejoice. Pronounced (in an Anglicized manner) /dʒuːbɪˈleɪtɪ/ or (in a Latinate manner) /juːbɪˈlɑːteɪ/.

**judgematic.** See FACETIOUS FORMATIONS.

**judgement.** In BrE the more or less prevailing spelling is *judgement* (but *judgment* in legal works), whereas in AmE *judgment* is the dominant form in all contexts. The presence or absence of -*e*- is not a matter of correctness or the reverse, but just one of convention in various publishing houses.

**judging by, judging from,** used in the sense 'if we are to judge by', are now used at the head of a subordinate clause loosely connected to a preceding or following main clause. The examples that follow happen to be from American newspapers of 1988, but could as easily have turned up in British ones: (judging by) *Judging by what happened Monday, only the courts can now block the dodge engineered by politicians*; *the use of facsimile machines . . . is the fastest-growing transmission system, judging by the number of fax messages coming across this desk*; (judging from) *Judging from the review of the first performance . . . Wednesday, Solti's account of the Mozart Fifth Violin Concerto with Michael Ludwig had considerably more pizzazz than what was heard the evening before*; *Judging from the two hours available for previewing, one kind of propaganda is substituted for another*. The danger is that these will be perceived by some as UNATTACHED PARTICIPLES. If the threat seems real, fall back on *if we are to judge by*.

**judicial, judicious.** The first has to do with judges, law courts, and legal judgements, whereas *judicious*, in most of its uses, means simply 'sound in discernment and judgement, sensible, prudent'. A *judicial inquiry* is one that is conducted by properly qualified legal officers; a *judicial separation* is a separation of a man and his wife by the decision of a court. In Scotland a *judicial factor* is an official receiver. By contrast, *judicious* is usu. found in such sentences as these: *They made judicious use of the time available*; *a judicious plan of action*; *a tale should be* *judicious, clear, succinct.* Examples: (judicial) *Administrative and judicial authority still rested with the gentlemen Justices of the Peace, chosen from among the landowners—* G. M. Trevelyan, 1946; *The creation of judicial tribunals ... has probably been accelerated since the war—Conveyancer,* 1962; *The principle of hierarchy of law generally (but not always) leads to the possibility of judicial review of ordinary legislation.* (judicious) *Popularity had been cheaply purchased by the judicious distribution of dried apricots and jelly cubes—*M. Drabble, 1967; *The Lalpukur school had always believed in a judicious mixture of practical and theoretical knowledge—*A. Ghosh, 1986; (*judicial* incorrectly for *judicious*) *Many a country gentleman restored his depleted fortunes by a judicial alliance, marrying no doubt for love but prudently falling in love where money was—*C. Chenevix Trench, 1973. Overlap of meaning can legitimately occur in contexts where a judge is deemed to have made not merely a *judicial* decision (i.e. one in accordance with the law) but a *judicious* one, i.e. one that is wise and discerning when all factors, including legal, social, and political ones, have been taken into account. But the 1973 example (above) is beyond the pale.

**ju-jitsu.** Now the regular spelling in English (formerly also *jiu-jitsu* and *ju-jutsu*). The word is a 19c. loanword from Jap. *jūjutsu* (pronounced /dʒʊdʒitsʊ/, from *jū* (Chinese *jeu* 'soft, yielding') + *jutsu* (Chinese *shu, shut* 'science').

**jumbo.** Pl. *jumbos.*

**jump** (verb). The phr. *jump to the eye(s)* 'to be obvious or prominent' is a Gallicism (cf. Fr. *sauter aux yeux*), and, according to the *OED*, Fowler (1926) was the first to notice this. Examples: *the Banquo scene in 'Macbeth'—a scene which jumps to the eye—was overlooked—*G. Goodwin, 1929; *Things jump to the eyes of the reader of this passage which have yet been ignored—*M. D. George, 1931.

**junction, juncture.** A *junction* is a point at which two or more things are joined, esp. a place where two or more railway lines or roads meet, unite, or cross (*COD,* 1990). A *juncture* is a critical point of time, a convergence of events: since the 17c. it has always been a popular word (*at this juncture of time, different junctures,*

*sudden junctures, the present critical juncture of things;* and especially, since the later part of the 19c., *at this juncture*). Examples: *How idiotic to start an illness at this juncture, when she would get small help—*H. Green, 1948; *The United States came to Vietnam at a critical juncture of Vietnamese history—*F. Fitzgerald, 1972; *At this juncture, the opposing demands of two distinct worlds were visited upon her—*C. G. Wolff, 1977; *Liz hoped that at this juncture Shirley would go to bed—*M. Drabble, 1987; *The last thing she wanted at this juncture was to be under an obligation to Wilcox—*D. Lodge, 1988.

**junta.** This Spanish loanword is best now anglicized as /ˈdʒʌntə/, though the pronunciation with initial /ˈhʊ-/ is often heard in BrE and is more or less standard in AmE. The erroneous form *junto,* once common, is now obsolete.

**jurist.** 'In England, this word is reserved for those having made outstanding contributions to legal thought and legal literature. In the U.S., it is rather loosely applied to every judge of whatever level, and sometimes even to nonscholarly practitioners who are well respected ... The most common error in the U.S. is to suppose that *jurist* is merely an equivalent of *judge:* "We find no constitutional question concerning the validity of Charles Milton's conviction and sentence of death about which reasonable *jurists* [read *judges*] could differ," ' (Garner, 1987).

**juror.** A member of a jury is a *juror.* A male and a female juror are sometimes respectively called a *juryman* and a *jurywoman.*

**just** (verb and noun). See JOUST.

**just** (adv.). **1** Normal uses of this common adv. include the following: = exactly (*just what I need, that's just right*); = a little time ago (*I have just seen him getting into a car*); colloq. = simply, merely (*we are just good friends not lovers; it just doesn't make sense*); = barely, no more than (*he just managed to reach the airport in time; just a minute*); colloq. = positively (*it is just splendid*); = quite (*not just yet*); colloq. = really, indeed (*won't I just tell him!*); (in questions, seeking precise information) (*just how did you manage?*). All this (slightly adapted) from *COD* (1990). All these uses are current in the main

English-speaking countries. The last of them (*just* in questions) was originally American (where it was first recorded in 1884) but has now come into general use (*One wonders just how biased a view we develop of the human ecology of tropical Africa—TLS*, 1974).

**2** It is important to distinguish *not just* from *just not*: (*not just* = not only) *It was not just her that I disliked but all the people at the party*; (*just not* = simply not) *They are childish, they have just not grown up.*

**3** Examples of some other common uses, with comments on distribution, linguistic level, etc.: (used as emphasizer) *Not very polite just to run into a lady's place and run out again*—Roger Hall, 1978 (NZ); ( = only) *he said there was blood everywhere, and not just blood*—A. Munro, 1987; ( = simply) *this will just work the stain deeper into the skin and allow it to 'set'—New Yorker*, 1987; (a semi-conjunctional colloq. use =

it's just that) *Not that it mattered, just a fellow liked to know*—K. Amis, 1988; *I don't know why I'm telling you all this really. Just I wouldn't want to see you do something you might regret*—P. Lively, 1991; (*just now* = a short time ago) *What was it you were saying just now, child?*—E. Jolley, 1985; (*just now* = at present; the positive use is characteristically Scottish.) *Let's talk about them sometime. Not just now*—P. Lively, 1987; *But it's not just me; it's the whole of Scotland just now*—M. Gray, 1989; (*just now* = very soon, in a little while, a SAfr. use) *'Would you mind switching off the light after you lock up?' 'The men on cell duty will do that just now.'*—A. Sachs, 1966.

**juvenile.** The word, normally meaning a young person under an age (often 17) specified by law, is used neutrally in such phrases as *juvenile court, juvenile delinquency, a juvenile part* (in a play); but derogatorily in such contexts as *he's behaving in a very juvenile way.* Cf. TEENAGER.

# K k

**kadi.** See CADI.

**Kaffir.** Historically (and with no derogatory connotations), a member of the Xhosa-speaking peoples of S. Africa; also, the language of these peoples. Since about the middle of the 20c. it has come to be accepted that it is grossly offensive to use the word as a descriptive term for any black person. It is now an actionable offence in S. Africa to do so.

**Kaiser.** A term in almost daily use while H. W. Fowler was preparing *MEU*: HWF therefore commented on its pronunciation. It is now a fast-fading term, except in historical contexts concerning the Holy Roman Empire, etc. Whether the term is used of the head of ancient empires or in the name *Kaiser Bill* (Kaiser Wilhelm II, Emperor of Germany 1888–1918), the word is pronounced /ˈkaɪzə/.

**kale, kail.** Etymologically, these are Scottish and northern-counties variants of the southern word *cole*, a general name for various species of brassica, esp. of the curly variety, including colewort, borecole, and cabbage. Kale is grown as a crop (esp. for animal feed) throughout Britain. In the spelling *kail* (less commonly *kale*) the word is used in Scotland as a name for (*a*) certain kinds of curly brassica, (*b*) a broth or soup in which cabbage is a principal ingredient. Numerous combinations of the word are current in various parts of Scotland, e.g. *kail-bell*, the dinner bell or a call to dinner; *kail wife*, a woman who sells vegetables and herbs; and esp. *kail yaird*, a cabbage garden, a kitchen-garden. The *Kaleyard* (or *Kailyard*) *School* is a name applied to a group of 19c. fiction writers including J. M. Barrie (1860–1937) and S. R. Crockett (1860–1914), who described local town life in Scotland in a romantic vein and with much use of the vernacular.

**kalendar, kalends.** See CALENDER; CALENDS.

**kangaroo.** For the parliamentary sense, see CLOSURE.

**kaolin,** a fine soft white clay. Pronounce /ˈkeɪəlɪn/.

**karat,** AmE variant of *carat* (a measure of the purity of gold).

**kartell.** See CARTEL.

**kedgeree** /ˈkedʒərɪ, -iː/, now the usual spelling of a word spelt in many different ways (*kidgeree*, *khichri*, etc.) since it was first recorded in English in the 17c. It is a loanword from Hindi.

**keelson** /ˈkiːlsən/, a line of timber fastening a ship's floor-timbers to its keel. Now the preferred spelling and pronunciation (*COD*, 1995), not *kelson* /ˈkelsən/.

**keep** (verb). Used as a transitive verb, *keep* + object + *from* + *-ing* is normal English and means 'prevent from something': *Jimmie ... was glad that distance and duty kept Mr Neville from visiting him more than twice*—T. Keneally, 1972; *His hands held flat over his ears as if to keep his whole head from flying apart*—M. Amis, 1978; *I'd been wearing the old sunglasses all this time, to keep the dental headlights from blinding me*—New Yorker, 1989. The *OED* records the use of *keep from* in the intransitive sense 'to restrain or contain oneself from' + an *-ing* clause: e.g. *Nor was Louis able to keep from turning pale*—C. Yonge, 1877. This use now sounds distinctly archaic. On the other hand, the construction is flourishing in N. America: *Maria cut the wheel to the left, to keep from hitting the cans*—T. Wolfe, 1987; *Nathan pulled upward on the frazzled leg of his shorts and tried to keep from crying*—New Yorker, 1988; *He thinks we should all come ... to listen to some ... old university professor rabbit on about how to keep from going stale*—M. Atwood, 1990.

**Kelt(ic).** The standard forms now are CELT and CELTIC.

**kennel** (verb). The inflected forms are *kennelled, kennelling* (AmE *kenneled, kenneling*). See -LL-, -L-.

**kerb.** The standard spelling in BrE of the word meaning a stone edging to a pavement or raised path. In AmE spelt *curb*.

**kerosene, paraffin, petrol, petroleum.** *Kerosene* in the US (where it is sometimes spelt *-ine*), Australia, and NZ is a fuel oil suitable for use in domestic lamps and heating appliances. The corresponding term in Britain is *paraffin*. *Petroleum* (popularly shortened to *petrol*) is a hydrocarbon oil found in the upper strata of the earth, refined for use in internal-combustion engines, etc.; in the US called *gas* or *gasoline*.

**ketchup** is the established spelling in Britain. See CATCHUP.

**Khedive.** Pronounce /kɪˈdiːv/.

**Khrushchev,** Nikita Sergeyevich, Soviet statesman (1894–1971). Thus spelt, not *Kr-*.

**kibbutz** /kɪˈbʊts/, a communal farming settlement in Israel. Pl. *kibbutzim* /ˌkɪbʊtˈsiːm/.

**kid** (noun). Used to mean 'child', despite its long history (first clearly recorded in the 17c.), it is now markedly informal and should be restricted to such contexts (*he's only a kid*; *his wife and kids*; *spend the day with the kids*; *school kids*; *kids' stuff*; etc.).

**kidnap.** The inflected forms are *kidnapped, kidnapper, kidnapping*, except that in AmE forms with a single *-p-* are sometimes used. See -P-, -PP-.

**kidnapping.** See ABDUCTION.

**kidney.** Pl. *kidneys*.

**kiln.** Now normally pronounced /kɪln/, but the OED (1901) gave precedence to /kɪl/, with the final *n* silent, and Daniel Jones (1917) noted that 'The pronunciation /kɪl/ appears to be used only by those connected with the working of kilns'. The New SOED (1993) assigns /kɪl/ to Scotland.

**kilometre** (AmE *kilometer*). There is considerable variation in the placing of the stress. In BrE the stress is traditionally placed on the first syllable, /ˈkɪləmiːtə/, and this pronunciation is recommended. But /kɪˈlɒmɪtə/ is also common. In overseas English-speaking countries, including the US, second-syllable stressing

seems to be dominant, possibly under the influence of *barometer, speedometer*, etc. Other metrical units (e.g. *centimetre, millimetre, kilogram, kilolitre*) are always stressed on the first syllable.

**kilo-, milli-.** In the metric system, *kilo-* means multiplied, and *milli-* divided, by 1,000; *kilometre* 1,000 metres, *millimetre* 1/1,000 of a metre. Cf. CENTI-; DECA-.

**kilt,** as worn by Highland men. So spelt, not *kilts*.

**kiltie,** a wearer of the kilt; a soldier in a Highland regiment. So spelt, not *-y*.

**kimono.** Pl. *kimonos* (occas. without a pl. termination, as in Japanese itself). see -O(E)S 6.

**kin.** As noun = one's relatives or family (*he seems to have neither kin nor country; he's no kin of mine*), now a rather old-fashioned word except in the phrs. *next of kin, kith and kin*. It is also found in predicative use 'passing into *adj*.' (as the OED expresses it): *We are kin; we have the same blood in our veins*. See KITH AND KIN.

**kind** (noun).

| 1 | Ordinary uses. |
| 2 | *these kind of.* |
| 3 | Adverbial *kind of.* |
| 4 | *kind of a.* |

1 In the ordinary sense 'type, sort, variety', the normal use is shown in phrases such as *a new kind of soap powder; a nashi is a kind of pear; the rock formed a kind of arch; nothing of the kind; people of every kind*.

2 *these kind of.* Beginning in the 14c., the type *these kinds of trees*, though itself continuing in standard use, produced a strangely ungrammatical variant. As the OED expresses it, 'The feeling that *kind of* was equivalent to an adj. qualifying the following noun, led to the use of *all, many, other, these, those*, and the like, with a plural verb and pronoun, when the noun was plural, as in *these kind of men have their use*.' This illogical type is now exceedingly common in colloquial contexts: e.g. *She was used to these kind of smells in the night-time bedclothes*—M. Duckworth, 1960; *I memorized it for these kind of occasions*—J. McInerny, 1988.

3 Adverbial *kind of* turns up in informal contexts (now esp. in AmE) in the

types *I kind of thought you weren't coming*; *She kind of wasn't listening* (in which *kind of* is called a 'downtoner' in *CGEL*), meaning 'I rather thought ...', 'She wasn't listening carefully'. Examples: *I kind of want to choose my war since it's my life— Boston Sunday Herald Mag.*, 1967; *All these rich bastards driving up the property values have kind of made it impossible for everyone else—New Yorker*, 1987.

**4** *kind of a.* This slightly more complex type is shown in: *We're kind of a middle-aged Sonny and Cher—Washington Post*, 1973; *he took me to a place where there were pictures of naked women on the walls. It was kind of a club—New Yorker*, 1988. The BrE equivalent would read ... *a kind of* ... Historically the AmE construction is a reduction of a long-established type, ... *a kind of a* ... Examples: *I haue the wit to thinke my Master is a kinde of a knaue—*Shakespeare, 1591; *I ... thought myself a kind of a monarch—*Defoe, 1719; (a tautologous use) *Dash is a sort of a kind of a spaniel—*Miss Mitford, 1824.

The types described in 2 and 3 should not be used in standard language of reasonable formality, but both form a natural part of informal speech. Type 4 is restricted to AmE. See SORT.

**kinda.** As part of a widespread tendency in the 20c. to link reduced forms of *of* (and *have*, etc.) to the preceding word in the written form of the language, *kinda*, *shoulda*, etc., are often printed in place of *kind of*, *should have*, etc. Examples: *This some kinda gimmick?—Punch*, 1972; *That little chap must have been really desperate to take that kinda crap—*Caris Davis, 1989. The reduced forms are strictly excluded from formal writing.

**kindly.** **1** (adv.). There is a difficulty here. Fowler (1926) objected to the 'misplacement' of *kindly* in such a sentence as *Authors are kindly requested to note that Messrs——only accept MSS. on the understanding that ...*, when the writer should have said *Will authors please note ...*, since it is not Messrs—who are being kind. *COD* 1995 says that *kindly* is 'often ironically used in a polite request or command (*kindly acknowledge this letter*; *kindly leave me alone*)'. There is some evidence that the use to which Fowler objected is on the wane. In taxis now, for example, the printed message about smoking is commonly *Thank you for not smoking* rather than *You are kindly requested not to smoke*, *Will passengers kindly refrain from smoking*, or the very direct *No smoking*. In *CGEL* commands of either kind are called 'courtesy subjuncts': the term is applied to a small group (among them *cordially*, *graciously*, *kindly*, and *please*) of adverbs used in rather formulaic expressions of politeness and propriety. *Kindly* is undebatably used as an indication of politeness (*she kindly offered to baby-sit for us*), and in an admonitory fashion (*kindly behave yourself*). It is starting to drop out of use as a formula of polite request. And it is important to note that in admonitory uses it cannot be moved about in a sentence. *Kindly leave me alone* and *Kindly don't make a fuss* are idiomatic; *Leave me alone, kindly* and *Don't kindly make a fuss* are not.

**2** (adj.). Of course *kindly* is also often used as an adj. (*a kindly word*, *a kindly policeman*).

**3** *kindlily.* This adverb (first recorded in the 19c.) is recorded in dictionaries, but its clumsiness ensures that it is not often used: in practice it gives way to *in a kindly manner/way*, etc.

**kinema.** See CINEMA.

**king.** The types *King of Arms* and *King at Arms* have both been used in heraldry since the 16c. as a title of a chief herald of the College of Arms, but the former has now established itself in the designation of those who still hold such offices: Garter, Clarenceux, and Norroy and Ulster; in Scotland, Lyon.

**kinsfolk** (in AmE more commonly *kinfolk*) is pl. without the addition of -*s*.

**kith and kin.** This ancient phrase (first recorded in the 14c.), which originally meant 'country and kinsfolk' and then 'acquaintances and kinsfolk, friends and relations', has come to be used emotively of people of the same ethnic origin who are under a threat of some kind (driven out of power, invaded, forced to emigrate, etc.). It cropped up quite a lot, for example, at the time of the Falklands conflict.

**kitty-corner(ed).** See CATER-CORNERED.

**knee.** The adj. from *knock knees*, *broken knees*, etc., can now be safely spelt *kneed* (rather than *knee'd*); so *knock-kneed* (adj.). See -ED AND 'D.

**knee-jerk.** This popularized technical term ( = a sudden involuntary kick caused by a blow on the tendon just below the knee when the leg is hanging loose) has endeared itself to politicians, debaters, etc., used attributively in the figurative sense 'predictable, automatic, stereotyped', as in *knee-jerk reaction*. Example: *The* [motor] *industry's ... knee-jerk support for road construction and its opposition to tighter air pollution standards have not endeared it to the public—Times*, 1991.

**kneel** (verb). From the evidence before me, the pa.t. and pa.pple seem to be either *kneeled* or *knelt* in all English-speaking countries, but with *knelt* markedly in the ascendant: (kneeled) *And he kneeled down and cried with a loud voice*—T. S. Eliot, 1935; *His mother kneeled in the corner*—M. Leland, 1985; (knelt *US*) *George knelt beside the pool and drank from his hands with quick scoops*—J. Steinbeck, 1937; *He knelt comfortably, his heels tucked neatly under him*—J. Clavell, 1975; *Fauve knelt by the sofa on which Maggy sat*—J. Krantz, 1982; *They stood, and knelt, and sat splayed before us in splendid formation*—Atlantic, 1991; (knelt *UK*) *When he knelt by his bed to pray before sleeping, his angel returned*—W. Golding, 1964; *he was once remarked on as the only officer who knelt in church*—D. Cecil, 1978; *Some of the recruits knelt to pray before retiring, presumably for strength*—A. Burgess, 1987. The *OED* (1901) noted that 'the pa.t. and pple. *knelt* appear to be late (19th c.) and of southern origin'. See 'T AND -ED.

**knickers.** In BrE, a woman's or girl's undergarment covering the body from the waist or hips to the top of the thighs and having leg-holes or separate legs. In AmE, short for *knickerbockers*, loose-fitting breeches gathered at the knee or calf; also, a boy's short trousers. (Definitions from *COD* 1995.)

**knick-knack.** Thus spelt, not *nick-nack*.

**knife.** The pl. of the noun is *knives*, but the inflected forms of the verb *knife* are *knifes*, *knifed*, and *knifing*.

**knight.** 1 The obsolete phr. *knight of industry* is a rendering of the Fr. phr. *chevalier d'industrie*, an adventurer, a swindler ('personne qui se livre à des activités peu scrupuleuses, aventurier, escroc'—*Trésor de la langue française*).

2 For *knight of the road*, see SOBRIQUETS.

3 The plurals of *knight bachelor* and *knight errant* usually have -s after the first element (*knights bachelor*, etc.), but *Knight Hospitaller*, when used (they are also called *Knights of St John,—of Rhodes,—of Malta*), normally has -s after both elements. The pl. of *Knight Templar* is either *Knights Templars* or *Knights Templar*.

**knit.** The pa.t. and pa.pple form for the dominant sense of the verb, namely 'to make a garment, etc., with knitting-needles', is *knitted*. In other uses, *knitted* and *knit* occur with about equal frequency: *she knit(ted), or had knit(ted), her brows; a close-knit group; knitted pullovers*.

**knock-kneed.** See KNEE.

**knock up.** First-time visitors from Britain to the US need to be reminded that the BrE sense 'to arouse (a person) by a knock at the door' is not known in America, where the same phrasal verb is a slang way of saying 'to make pregnant'.

**knoll.** The standard pronunciation is /nəʊl/. Words with the diphthong /-əʊ-/ include *droll*, *poll* (voting), *roll*, *stroll*, *toll*; while those with the monophthong /-ɒ-/ include *doll*, *loll*, *moll*, *Poll* (parrot).

**knot.** This is a unit of a ship's or aircraft's speed equivalent to one nautical mile (approx. 2,025 yards or 1,852 metres) per hour. It is not a measure of distance, even though it is often loosely so called (e.g. *the ship went ten knots an hour*). Originally a knot was 'a piece of knotted string fastened to the log-line, one of a series fixed at such intervals that the number of them that run out while the sand-glass is running indicates the ship's speed in nautical miles per hour; hence, each of the divisions so marked on the log-line, as a measure of the rate of motion of the ship (or of a current, etc.)' (*OED*). As a matter of curiosity, the *OED* lists three examples of the loose use from unimpeachably (if somewhat old-fashioned) nautical sources, Anson's *Voyages* (1748), Cook's *Voyages* 1772–84), and Marryat's *Peter Simple* (1833).

**knout** (scourge used in imperial Russia). Pronounce /naʊt/, not /nuːt/ or /kn-/.

**know.** Now strikingly overused, esp. by non-standard speakers: **1** In the form of a rhetorical question, (*Do you) know what I mean?*. Examples: *Actually, I don't sing well. You know what I mean? I'm not sure what to do*—New Yorker, 1986; *it's not gonna get any better if you switch guys, you know what I mean?*—Musician (US), 1991. In rapid speech the question is often reduced to something like *know't I mean?*

**2** In *you know*, as a conversational filler. Examples: A. *They're supposed to be, you know, sexy.* B. *That's all right, but all men are the same, after one thing, but sometimes, you know, it can be wonderful*—Listener, 1965; *People get the wrong idea, thinking we might be, you know, glamorous or brilliant or something*—Sunday Times, 1974. It is a marked feature of informal conversation and unscripted speech. A Swedish scholar, Britt Erman, in her *Pragmatic Expressions in English* (1987), analysed the commonest of such expressions—*you know, you see*, and *I mean*—used in conversations between educated speakers. She concluded that *you know* has a number of functions. It is often used to introduce background information (such as a parenthesis), or extra clarification or exemplification. It is also often used to finish off an argument, or to mark the boundary between one topic or manner of speaking and another. All this may well be so, and may cause difficulty in Stockholm, but the fact remains that conversation that is studded with *you knows*, however closely the function of the expression is analysed, is also likely to be judged to be inept.

**know-how.** After a period in the mid-20c. when its usefulness was called into question in Britain, *that intangible but wonderful thing called American 'know-how'* (Encounter, 1953) has come to be accepted in all English-speaking areas as an indispensable term for technical expertness or practical knowledge. The word was first recorded in print in AmE in 1838, but did not come into widespread use until about a century later.

**knowing.** Used in the sense 'from what I know about (a person)' or short for 'knowing (a person) as I do', it is now frequently used in a formulaic way to emphasize the truth of an accompanying statement. Examples: *'What are you doing*

*today?' 'Lunching with Lalige. Some shopping. We won't lunch till late, knowing her*—M. Wesley, 1983; *Knowing Jeremy, it would be more a modus-operandi clash*—Gentlemen's Q, 1990. It is one of a number of admissible 'unattached participles'. Cf. JUDGING BY.

**knowledge.** Pronounce /ˈnɒl-/, not /ˈnəʊl-/. As the OED remarked (in 1901): 'The shortening of ... the first syllable is phonetically normal; cf. the 15–17th c. spelling *knoledge*; /ˈnəʊlɪdʒ/, used by some, is merely a recent analytical pronunciation after *know*.' Gowers (1965) noted that the pronunciation with *nōl* 'has disappeared even from those pulpits that were its last refuge'.

**knowledgeable.** So spelt.

**kopje.** This older spelling (from Du. *kopje*) has been entirely replaced in S. Africa by *koppie* /ˈkɒpɪ/. But *kopje* is still the customary spelling outside S. Africa.

**Koran.** Pronounce /kɔːˈrɑːn/. Sometimes written *Qur'an* or *Quran*, as a closer transliteration of the Arabic original.

**kosher.** Pronounce /ˈkəʊʃə/, not /ˈkɒ-/.

**kowtow,** pronounced /kaʊˈtaʊ/, entirely replaced *ko-tow*, pronounced /kəʊˈtaʊ/, at some point in the 20c.

**kraal.** The pronunciation is given as either /krɑːl/ or /krɔːl/ in SAfr. dictionaries., but in BrE it is pronounced /krɑːl/.

**kris** (Malay or Indonesian dagger). Now the preferred spelling. Formerly also *crease, creese, kreese*.

**krona, krone. 1** *krona* (pl. *kronor*) is the chief monetary unit of Sweden; and *krona* (pl. *kronur*) that of Iceland.

**2** *krone* (pl. *kroner*) is the chief monetary unit of Denmark and of Norway.

**krummhorn** (also **crumhorn**) is a medieval wind instrument with a double reed and a curved end. It is a German word, derived from *krumm* 'crooked' + *Horn* 'horn'. Cf. CREMONA.

**Krushchev.** See KHRUSHCHEV.

**kudos** /ˈkjuːdɒs/, glory, renown, is a 19c. adoption (orig. university slang) of Gk κῦδος in the same sense. A degenerate back-formation, *kudo*, has emerged from

about 1940, used to mean 'an honourable mention, praise for an achievement'. Examples: *This did not win Mr.
Eisenhower many kudos in the press—Wall St.
Jrnl*, 1961; *A kudo to* Life *for a fine story on
baseball's spring training—Life*, 1963. It is
true that back-formations resulting from
a misinterpretation of the function of a
final -s can be found: e.g. ASSET, *cherry*
(from OF *cherise*, taken as pl.), *pea* (from
*peuse*, taken as pl.). But this is an old

discarded process. No other word of Gk
origin (*bathos, chaos, pathos*, etc.) has suffered such an undignified fate.

**kukri**  (curved knife). Pronounce /ˈkʊkrɪ/.
Pl. *kukris*.

**kyrie eleison.**  Of the numerous variant
pronunciations, *COD* 1995 gives precedence to /ˈkɪərɪeɪ ɪˈleɪzɒn/. It is derived from
Gk Κύριε ἐλέησον and means 'Lord, have
mercy'.

# L l

**laager.** Pronounce /ˈlɑːɡə/.

**lab,** a late-19c. informal shortening of *laboratory*, now well established in everyday use.

**label.** The inflected forms are *labelled*, *labelling* (AmE *labeled*, *labeling*). See -LL-, -L-.

**labial.** Lit. 'of the lip', in phonetics it means '(of a consonant) requiring partial or complete closure of the lips (e.g. *p*, *b*, *f*, *v*, *m*, *w*; (of a vowel) requiring rounded lips (e.g. *oo* in *moon*)' (*COD*, 1995). Of these consonants, *f*, *v* are labio-dental (only one lip is used to form the sound), and *m*, *p*, *b* are bilabial; *w* is a labio-velar semi-vowel, involving lip-rounding, while the sound (i.e. /w/) is made at the velum.

**labium.** Pl. *labia*. See -UM 2.

**laboratory.** The standard pronunciation in BrE now is /ləˈbɒrətərɪ/, stressed on the second syllable. In 1902 the *OED* gave only first-syllable stressing for the word, with the last four syllables all unstressed. In AmE the word is stressed on the first syllable and the last two syllables are pronounced like *Tory*.

**labour.** The standard spelling in BrE (AmE *labor*). Hence *Labourite*, a member or follower of the Labour Party. In Australia, *labour* is the customary spelling except, curiously, in the official name of the Australian *Labor Party* (and *Laborite*). See BELABOUR.

**labouredly.** Probably best as three syllables. See -EDLY.

**lac.** See LAKH.

**lace.** The related forms are *lacier*, *laciest*, *lacily*, *laciness*.

**lacerate** (verb). The related adj. is *lacerable*. See -ABLE, -IBLE 2.

**laches** /ˈlætʃɪz/. In law, it means 'a delay in performing a legal duty, asserting a right, claiming a privilege, etc.' (*COD*). It is a sing. noun (derived from OF *laschesse*

(mod. *lâchesse* 'cowardice, laxness'), ult. from L *laxus* 'lax'): e.g. *Laches is* [not *are*] *pleaded as a defence.*

**lachrymose** (and related words). Irretrievably now always spelt with *lachry* though this group of words all answer to L *lacrima* 'tear' (and derivatives, e.g. *lacrimāre* 'to weep'). The *ch* of the prevailing spelling of this and the related words is due to the med. L. practice of writing *ch* for *c* before Latin *r*; cf. *anchor*, *pulchritude*, *sepulchre*. The *y*, in med. L. a mere graphic variant of *i*, has been retained in mod. Eng. orthography from the erroneous notion that *lacrima* is an adoption of Gr. δάκρυμα' (*OED*).

**lack** (verb). The use with *for* meaning 'to be short of something' (used only in negative constructions) seems to have originated in AmE and spread to BrE during the 20c. Examples: *Here's hoping he'll never lack for friends*—M. Twain, 1892; *The outward signs that she had marked upon him did not lack for inner causes*—E. Phillpotts, 1906; *He never lacked for friends*—R. Ellmann, 1987; *You get a lower standard of trim, but you don't lack for much in the way of essential equipment*—*Which? Car Buying Guide*, 1987.

**lackey.** Thus spelt, not *lacquey*.

**lacquer.** Thus spelt, not *lacker*.

**lacrim-.** See LACHRYMOSE.

**lacuna.** Pl. either *lacunae* /-niː/ or *lacunas*.

**lad.** A mostly affectionate word used in several senses: **1** a boy or youth (*he's only a lad*); a young son (*my lad is rather a shy boy*).

**2** (usu. in pl.) a man, esp. a workmate, drinking companion, etc. (*he's one of the lads; the lads will vote on the issue tomorrow*).

**3** a high-spirited fellow, a rogue (*he's a bit of a lad*).

**4** (in Britain) a stable-groom (regardless of age; sometimes applied to a female groom).

**laddie.** Thus spelt. See -IE, -Y.

**lade** (verb). Apart from the passive use of the pa. pple *laden*, this verb is now almost restricted to the *bill of lading* that the master of a ship gives to the consignor as a receipt for his cargo. In all of its remaining applications, *laden* has to vie with the much more natural word *loaded*. *Laden*, usu. followed by *with*, can be used of a vehicle, donkey, person, tree, table, etc., weighed down by goods, parcels, fruit, etc.: e.g. *the boughs of the tree were laden with apples*; *the cart was laden with sacks of corn*; *heavily laden buses*. It is also used in the sense '(of the conscience, spirit, etc.) painfully burdened with sin, sorrow, etc. (*a soul laden with the sin which he had committed*)'. But the word has all the hallmarks of impending archaism, and seems likely to fall into disuse in the 21c.

**ladleful.** Pl. *ladlefuls*. See -FUL.

**lady.** **1** In George Meredith's *Evan Harrington* (1861), a novel much concerned with the problems of social class, the heroine, Rose Jocelyn, niece of the British envoy in Lisbon, is rhetorically asked, *Would you rather be called a true English lady than a true English woman, Rose?*. The distinction mattered then, and, to judge from the first listed sense of *lady* in the *COD* (1995), 'a woman regarded as being of superior social status or as having the refined manners associated with this', still does. *Lady* stands alongside *woman*, sometimes more or less synonymously, more often not at all. It is used as a title (see 2 below), whereas *woman* is not. A natural use is seen in *the lady* (never *woman*) *of the house*. Other fixed collocations include the *Ladies* (or *Ladies'*), a women's public lavatory; the *Ladies' Gallery* (at the House of Commons); a *ladies'* (or *lady's*) *man*, one fond of female company; *ladies' night*, a function at a men's club, etc., to which women are invited; *ladies' room*, one of the terms for a women's lavatory in a hotel, office, etc.; *lady-in-waiting*, a lady attending a queen or princess; *lady-killer*, a practised seducer; and numerous others. *My lady wife* is sometimes jocularly used to mean 'my wife'; *old lady* is used rather than *old woman*, being thought to be more respectful. A speaker about to address a mixed adult audience conventionally begins by saying *Ladies and Gentlemen*. All these uses, and a good many more, are listed in the *COD* (1995). In practice none of them is challenged by *woman* or *women*.

**2** *Lady* is also used as of right in certain titles. The style *Lady Jones* is proper only for a peeress below the rank of duchess or a baronet's or knight's wife or widow. The style *Lady Mary Jones* is appropriate for a daughter of a duke, marquess, or earl. The style *Lady Henry Jones* is used for the wife or widow of the younger son of a duke or marquess. A *Lady Mayoress* is the wife or other chosen female consort, e.g. a daughter, of a Lord Mayor. Much fuller information about the correct use of such titles is available in the latest version of *Debrett's Correct Form*.

**3** *lady* by itself in the vocative is often loosely used for *madam* as a term of address (esp. in shops and businesses), as *governor* (or *guv*) or *squire* are for *sir*; but *madam* and *sir* are much the more usual terms.

**4** Prefixed to vocational words (*lady doctor*, *lady barrister*, etc.), *lady* is fast giving way to *woman* (*woman doctor*, etc.). See also FEMININE DESIGNATIONS; GENTLEWOMAN.

**ladyfy, ladyfied.** Thus spelt (but not often used).

**lady-in-waiting.** Pl. *ladies-in-waiting*.

**laid, lain.** See LAY AND LIE.

***laissez-aller, -faire, -passer.*** The preferred spellings (all three in italics), not *laisser-*.

**lakh.** In India, = 100,000 (not *lack*, *lac*); in sums of rupees, place a comma after the number of lakhs: thus 25,87,000 is 25 lakhs 87 thousand rupees.

**lam**, thrash, etc. So spelt (not, as sometimes until the 18c., *lamb*).

**lama, llama.** A Tibetan or Mongolian Buddhist monk is a *lama*. A *llama* is a SAmer. ruminant, used as a beast of burden.

**lamentable.** Pronounce /ˈlæməntəbəl/, with stress on the first syllable; in AmE also with the main stress on the second syllable (as in the noun and verb *lament*).

# lamina | large, largely

**lamina.** Pl. *laminae* /-niː/.

**lammergeyer,** a large vulture. Pronounce /ˈlæməɡaɪə/.

**lampoon, libel,** etc. There is often occasion to select a term for an intended-to-be hurtful or embarrassing attack (even if light-hearted) on a person, a group, etc. A *lampoon* is a satirical attack on a person, institution, etc. A *libel* (nowadays liable to be expensive if judged in a court of law to be one) is a published statement damaging to a person's reputation; contrasted with a *slander*, which is a false oral defamatory statement damaging to a person's reputation (but see the separate entry LIBEL (noun), SLANDER). Other terms for satirical or 'witty' statements, parodies, etc., that are potentially hurtful to those to whom they are addressed, but are mostly judged to be part of life's rich tapestry, include *pasquinade* (now rare), a lampoon or satire, orig. one displayed in a public place (It. *pasquinata*, from *Pasquino*, a statue in Rome on which abusive Latin verses were annually posted); *skit*, a light, usu. short, piece of satire or burlesque; and *squib* (now rare), a short satirical composition, a lampoon. (Definitions mainly from *COD*.) See SATIRE.

**lamprey.** Pl. *lampreys*.

**land** (noun). *The land of the leal* is a Scottish expression for 'the land of the faithful, heaven' (chiefly with allusion to Lady Nairne's song, 1798). *The Land of Cakes* (also with small initial letters) was a 17–19c. bantering sobriquet for Scotland, from the importance of oatcakes in the Scottish diet.

**landgrave,** an historical name for a German count having jurisdiction over a territory; fem. *landgravine* /-ɡrəviːn/. Cf. modGer. *Graf* count.

**landslide, landslip.** The original term (17c.) for 'the sliding down of a mass of land on a mountain or cliff side' was *landslip*. From the mid-19c. the preferred term in America was *landslide*; and before the end of the 19c. it was also being used, in the US and elsewhere, to mean a sweeping electoral victory. *Landslide* is probably the dominant term now in the literal sense in BrE as well as AmE. In NZ both terms are familiar, but warning

road signs about them usually just say DANGER. SLIP, or the like.

**landward.** The adjectival form everywhere is *landward* (*a landward breeze*); in Scotland it is also used specifically to mean 'rural, in or of the country as opposed to (a particular) town'. As an adv., *landwards* is the dominant form in BrE (*sail landwards*), and either *landward* or *landwards* is permissible in AmE. In Scotland, *landward* as an adv. is used to mean 'in, toward, or in the direction of the country as opposed to the town'.

**languor, languorous, languid, languish.** It is troublesome to foreigners that the first two are pronounced with medial /-ŋɡ-/ and the last two with medial /-ŋɡw-/. This clear-cut distinction is in fact quite recent: the *OED* (1902) listed both pronunciations for *languor*, i.e. those with /-ŋɡ-/ and with /-ŋɡw-/, and only /ˈlæŋɡwərəs/ for *languorous*. A similar contrast is shown in *liquor* /-k-/ compared with *liquid* /-kw-/, and *conquer* /-k-/ as against *conquest* /-kw-/.

**lansquenet** /ˈlænskənet/, a card-game of German origin. Derived from French (17c.). The French word is an adaptation of Ger. *Landsknecht*, lit. 'servant of the country'.

**lantern, lanthorn.** The second of these, commonly found in 16–19c. literature (e.g. *this lanthorn doth the horned moon present*—Shakespeare, 1590; *Fishing up a lanthorn he turned the light on her face*—G. C. Davies, 1873) is probably due to popular etymology, lanterns having formerly been almost always made of horn (*OED*).

**lapel.** Pronounce /ləˈpel/. The adj. is *lapelled* (also in AmE beside *lapeled*).

**lapis lazuli.** Pronounce /ˈlæpɪs ˈlæzjʊlaɪ/, though many pronounce the second word /ˈlæzjʊli/.

**lapsus calami, lapsus linguae,** a slip of the pen, of the tongue, respectively. If a pl. of *lapsus* is called for, it is *lapsus*, pronounced /ˈlæpsuːs/, not *lapsi*. See -US 2.

**larboard.** See PORT.

**large, largely** (advs.). In a restricted number of circumstances, *large* is the only idiomatic adverbial form of the two:

after the verbs *bulk* and *loom*; also *writ large*, and in the phr. *by and large*. In all other contexts *largely* ( = to a large extent) is the normal word (e.g.. *his failure was largely due to laziness*).

**largesse.** It goes against the grain to overturn one of Fowler's verdicts, but it is clear that *largesse* is now the natural form (not *largess*, as Fowler recommended in 1926).

**largo,** a musical instruction. Used as a noun ( = a largo passage or movement) the pl. is *largos*. See -O(E)S 6.

**larva** /ˈlɑːvə/. Pl. *larvae* /-viː/.

**laryngeal.** In this and other words in which -*ng*- is followed by *e* or *i* the g is 'soft': thus /ləˈrɪndʒɪəl/, *laryngitis* /lærɪn-ˈdʒaɪtɪs/. Otherwise it is 'hard': e.g. *laryngoscope* /ləˈrɪŋɡəskəʊp/.

**larynx,** organ forming air passage to lungs. Pl. *larynges* /ləˈrɪndʒiːz/.

**laser.** We are all so familiar with the word now—*laser beam, laser printer, laser-guided bomb*, etc.—that it is easy to forget that it is a recent coinage (1960) and that it is an acronym (formed from the initial letters of 'light amplification by the stimulated emission of radiation'). It was modelled on the slightly earlier word *maser* (1955), which itself is a combination of the initial letters of several words ('microwave amplification by stimulated emission of radiation'. Lasers are optical masers.

**lassie.** The diminutive form, commonly used in Scotland, of *lass* is so spelt (not -y). See -IE, -Y.

**lasso.** A word of Spanish origin (Sp. *lazo*, cognate with *lace*), it is pronounced /læˈsuː/. As noun its pl. is *lassos*. The verbal forms are *lassoed, lassoes, lassoing*. The noun and verb are often pronounced /ˈlæsəʊ/ in AmE.

**last.** 1 With a cardinal numeral. The more frequent order of words between the 14c. and the 17c. was *the two* (*three*, etc.) *last* ( = Fr. *les deux derniers*, Ger. *die zwei letzten*). But the form *the last two* (*three*, etc.) is now the more frequent of the two (*the last two volumes of Macaulay's History*), except where *last* is equivalent to 'last-mentioned' (*OED*). Cf. FIRST 3.

2 *last/lastly*. In enumerations there is considerable variation between the two. My preference is to use the sequence *First, ... secondly, ... thirdly, ... lastly* (or *finally*). See FIRST 4.

3 *at long last* (formerly also *at the long last*). Described in the *OED* (1902) as 'Now rare', the phr. has come back into common use: *At long last I am able to say a few words of my own*—King Edward VIII, 1936 (abdication speech); *Someone answers the phone at long last*—J. Aiken, 1971.

4 *last/latest*. In such a context as *In his latest book, Dr A ...*, it is clear that Dr A has written earlier books and that he is still alive and may well write others. If the statement runs *In his last book, Dr A ...*, the meaning could be the same, or it could also imply that this was the final book written by Dr A before he died. It is obvious, therefore, that if there is any danger of contextual ambiguity some word other than *last* should be used. In many idiomatic phrases *last* is still the only possible adj. of the two: = most recent; next before a specified time (*last Christmas; last week*); = preceding; previous in a series (*got on at the last station*); = only remaining (*the last biscuit; our last chance*); (preceded by *the*) = the least likely or suitable (*the last person I'd want to see; the last thing I'd have expected*); = the lowest in order (*the last name on the list*).

5 *the last analysis*. See ANALYSIS 2.

**late** makes *latish*. See MUTE E.

**late, erst, erstwhile, ex-, former(ly), quondam, sometime, whilom.** The choice of word among these is somewhat complicated. 1 See ERST, ERSTWHILE; WHILOM. None of the three is common in ordinary use: in most contexts they smack of WARDOUR STREET.

2 *ex-*: commonly prefixed to nouns, giving the meaning 'former(ly)'. See EX-.

3 *former*: for its three natural uses as adj., see FORMER.

4 *quondam*: in somewhat formal use as an adj. used attrib. (*his quondam friend* = his former friend), and occas. as an adv., it suffers from belonging to a vanishing tribe of Latin words in a Latinless age.

5 *sometime*: principally used in the sense 'former(ly)' of a person who once

held office (*sometime Lord Mayor of Oxford*) or a building which has changed its function (*the old Ashmolean Building, sometime the headquarters of the OED*).

**6** *late*: preceded by *the* or by a possessive pronoun (*my*, *his*, etc.) means (*a*) no longer alive, recently dead (*the late Nicolae Ceauşescu*; *her late father bequeathed the house to her*); (*b*) no longer having the specified status (*the late chairman of the Parish Council*). In ordinary, non-literary, English, *former(ly)* is the most serviceable word in this group, followed by *late* (but only in contexts requiring sense 6*a*), and *ex-* (but bear in mind that Fowler deplored its use with titles consisting of more than one word, e.g. *ex-Lord Mayor*).

**later on.** See EARLY ON. A recent example: *Later on I had worked in the* Post'*s editorial and circulation departments—Bull. Amer. Acad. Arts & Sci.*, 1989.

**lath** /lɑːθ/, a flat strip of wood, has pl. *laths* /lɑːθs/, less commonly /lɑːðz/.

**lathe** /leɪð/, a machine for shaping wood, has pl. *lathes* /leɪðz/.

**lather,** as noun and verb, is most commonly pronounced /ˈlɑːðə/ by standard speakers, but /ˈlæðə/ by a not insignificant minority. In AmE /ˈlæð-/ is the more common.

**lathi** /ˈlɑːtiː/, in India, a long stick used as a weapon, has pl. *lathis*.

**latine.** See ANGLICE.

**Latinism, Latinity.** The first is principally (*a*) an idiom or form characteristic of the Latin language, esp. one used by a writer in another language; (*b*) conformity in style or manner to Latin models. The second is the quality or the extent of Latin words in a person's writing (e.g. *the Latinity of Johnson's style is obvious*) or the extent of Latin, influence on a period's style. See -ISM AND -ITY.

**Latin plurals** (or Latinized Greek). Separate entries have been made for Latin words that are in regular use in English, and these are to be found at their alphabetical place together with their (often competing) plural forms. A few general features are worth noting. **1** No simple rule can be given for the distribution of the rival forms. Some common words

regularly retain the Latin pl., e.g. *bases*, *crises*, *oases*, *theses* (not *basises*, *crisises*, *oasises*, *thesises*). Some others exhibit both the original Latin form and the Anglicized one in a fairly random way, e.g. *atria/atriums*, *cacti/cactuses*, *lacunae/lacunas*. The context or individual taste governs the choice on most occasions. In an age when formal knowledge of Latin rules is fading fast, it is not surprising that there should be a general movement towards the use of English plurals like *crematoriums* (rather than *-oria*), *cruxes* (rather than *cruces*), *encomiums* (rather than *-mia*), *gymnasiums* (rather than *-sia*), *referendums* (rather than *-da*), but in such words a degree of self-satisfaction is certainly in order if a knowledgeable person chooses to retain the Latin plural form.

There is a further group for which the Latin plurals are more or less obligatory: not to use them represents a serious stepping out of line. The plurals of *alga*, *corrigendum*, *desideratum*, *nucleus*, *stratum*, are regularly *algae*, *corrigenda*, *desiderata*, *nuclei*, *strata*. A vulnerable group is that where the classical plural ended in *-ata*: thus *automaton*/pl. *automata*, *lemma*/pl. *lemmata*, *miasma*/pl. *miasmata*, *stigma*/pl. *stigmata*. It would not be surprising if the forms in *-ata* fell into disuse in the 21c. except in special circumstances. Some other 'irregular' Latin plurals are 'hanging on' in English, but standard speakers are finding it increasingly difficult to remember the morphological relationship between the singular and the plural forms: e.g. *apex/apices*, *codex/codices*, *corpus/corpora*, *cortex/cortices*, *genus/-genera*, *helix/helices*, *matrix/matrices*, *radix/radices*, *vortex/vortices*. The classical plural forms remain at risk. The choice of plural form sometimes depends on the subject area: e.g. *appendixes* in surgery and zoology but *appendices* in books. In scientific work *foci*, *formulae*, *indices*, and *vortices* are regularly used, but in general writing the ordinary plural forms in *-s*, *-es* are more usual.

**2** A good many originally plural nouns of classical origin have tended over the years to be wrongly treated as sing. nouns in English. See e.g. AGENDA; BONA FIDE(S); CRITERION; DATA; MEDIA; PHENOMENON.

**3** With a few exceptions too firmly rooted to be dislodged (e.g. *Adelphi*), Latin plurals in -*ī* should be pronounced /aɪ/, not /iː/: e.g. *bacilli, fungi, gladioli, narcissi, radii, stimuli*.

**4** Most Latin words in -*us* form their plural in -*i*, but some do not. See -US. It is a grievous mistake to write, for example, *hiati, ignorami, octopi*. See also -EX, -IX; -TRIX; -UM.

**Latin pronunciation.** Those who are interested in the details of standard English usage are very often curious to know how the Romans pronounced the Latin language in the classical period. The standard work on all these troublesome matters (how *ae, au, c, eu, g, oe, s, th, u* consonant, *y*, etc., were pronounced) is W. Sidney Allen's *Vox Latina* (2nd edn, 1978). His book is partly dependent on John Sargeaunt's paper 'The Pronunciation of English Words Derived from the Latin', SPE Tract iv (1920). The various types of pronunciation of Latin words—Classical (Ciceronian), Italian (Dantean), Continental (Chaucerian), and English (Shakespearean)—together with advice on the pronunciation of numerous legal phrases (*de jure, amicus curiae, ultra vires*, etc.), may be found in H. A. Kelly's paper 'Lawyer's Latin: *Loquenda ut Vulgus?*, in *Journal of Legal Education* 38 (1988), pp. 195–207. In the present book the recommended pronunciations of English words and phrases adopted from Latin are given under the individual words. The great majority of them differ markedly from the way in which the same words were pronounced in classical Latin.

**Latin quotations.** The spirit of our age is brought out in this extract from *The Spectator* of 26 September 1987: 'Last Friday, the *Independent* gave Nicholas Garland's cartoon the place of honour on its front page. His picture derived from the photograph of Mrs Thatcher standing among industrial desolation in Teesside. The caption was "If you seek for a monument, gaze around". This was not what Garland had written. His caption was "Si monumentum requiris, circumspice". It had been changed after 40 per cent of the *Independent*'s news conference had confessed that they did not know what the Latin meant (and so, presumably, did not know that it was Christopher Wren's epitaph). Does this mean that no Latin sentence or phrase can be used unexplained in a newspaper? Certainly not "Timeo Danaos et dona ferentes", neither "Dulce et decorum est pro patria mori"; perhaps "sic" and "RIP" may still survive. "O tempora, O mores", one might exclaim, if one thought that anyone was likely to know what one meant. In fact, however, the *Independent*'s caution was mistaken. If the Latin had appeared those who understood it would at once have felt flattered; those who did not know what it meant would either have kept their heads down and not owned up, or they would have engaged in the enjoyable pursuit of discovering its meaning. Broad-sheet newspapers are supposed to be for educated people: one of the features of being educated is that one is not totally flummoxed when one comes across something which one does not understand.'

**-latry,** representing Gk -λατρεία 'worship', is shown in *idolatry*, a 13c. loanword from OF (ult. a reduced form of Gk (NT) εἰδωλολατρεία). In English the formative element meaning 'worship of' is *-olatry*. It has been widely put to service in (*a*) *angelolatry* (first recorded 1847), *astrolatry* (1678), *bardolatry* (worship of the 'Bard of Avon', 1901), *bibliolatry* (1763), *demonolatry* (1668), *Mariolatry* (1612); (*b*) evanescent, low-currency, or facetious formations like *babyolatry* (1846), *crochetolatry* (1859), *lordolatry* (1846). The corresponding agent-nouns end in -*ater*: *bardolater, bibliolater, idolater*, etc.

**latter** survives almost solely in *the latter*, which provides with *the former* a pair of pronouns obviating undesirable repetition of one or both of previously mentioned names or nouns. **1** Like *the former*, it should only be used of the last *of a pair* (of persons, ideas, etc.), as in: *We can either rely on our children to translate for us or we can try to catch up. The National Air and Space Museum makes the latter choice easier*—*Illustrated London News*, 1980. The illogical use of *latter* to refer to the last of more than two antecedents is shown in: *His* [sc. Jonathan Kellerman's] *three previous novels are 'Blood Test', 'Over the Edge' and 'When the Bough Breaks', the latter of which won … the Edgar Allan Poe Award*—*Chicago Tribune*, 1988.

**2** Like *the former*, it should be placed close to the word or idea to which it refers so as not to mystify the reader. Fowler (1926) cites just such a mystifying passage: *The only people to gain will be the Tories and the principal losers will be the working-class voters whose interests the Labour Party is supposed to have at heart. It is a very poor compliment to the intelligence of the latter* [which, in heaven's name?] *to believe, as many Labour members seem to do, that their support of the Labour cause will be all the more ardent if their interests are thus disregarded.*

**laudable, laudatory.** The first means 'commendable, praiseworthy' (*he carried out his plan with laudable firmness*). *Laudatory* means 'expressing praise' (*a politician from one party is not likely to speak in a laudatory manner about a politician from another party*). The two words are never interchangeable. Examples: (laudable) *What was laudable about her politics was her passion to become involved*—M. Forster, 1988; (laudatory) *One wounding review is liable to be more memorable than ten laudatory notices*—R. Berthoud, 1987.

**laudanum.** Pronounce /ˈlɔːdnəm/, i.e. with two syllables.

**laughable** means 'ludicrous; highly amusing; that may be laughed at' (*he had acted in a way that was half pathetic, half laughable*).

**laughter.** In *Homeric laughter* the reference is to the *Iliad* i. 599 and the *Odyssey* xx.346. 'It is especially the laugh that runs round a circle of spectators when a ludicrous or otherwise pleasing incident surprises them' (Fowler, 1926). Even then Fowler doubted 'whether the frequent use of the phrase is justified by present-day familiarity with Homer'. Immediate familiarity with the detail of these two great works has further diminished in the interim, and the phrase *Homeric laughter* is now seldom heard.

**launch.** The *OED* (1902) gave preference to the pronunciation /lɔːnʃ/ but it also gave /lɑːnʃ/ as an alternative, and it is clear that the latter pronunciation was common then among standard speakers. Now the standard pronunciation in BrE is /lɔːntʃ/, with /-ɔː/ and also with palatalization of the final *ch*.

**laurustinus,** an evergreen winter-flowering shrub, *Viburnum tinus*, is usu. pronounced /ˌlɔːrəˈstaməs/, with the initial sound of *laurel* rather than that of *laurel*, i.e. /ˈlɒ-/. It is a 17c. modL formation from L *laurus* 'laurel' + *tīnus*, name of a plant (perh. the laurustinus). The spelling *laurestinus*, occas. found, is erroneous.

**lav,** a colloq. shortening (in BrE) of LAVATORY, is so written (no point).

**lavabo** /ləˈvɑːbəʊ/ or /-ˈveɪbəʊ/ has a range of senses from 'the ritual washing of the celebrant's hands at the offertory of the (Roman Catholic) Mass', to (sometimes pronounced /ˈlævəbəʊ/) 'a washing-trough used in some medieval monasteries'. The Roman Catholic rite (in its Latin form) is accompanied by the saying of Ps. 26: 6, beginning *Lavabo inter innocentes manus meas* (in which *Lavabo* = I will wash). The pl. is *lavabos*.

**lavatory.** In the early 20c. perhaps the dominant plain-speaking word for a WC (both the fixture itself and the room) was *lavatory*, but it has mostly given way now to other words, in BrE esp. *loo* (the usual middle-class word) and *toilet* (universal in working-class speech and upwards into the language of many white-collar speakers). For a description of the general distribution of the various terms, see TOILET.

**laver¹,** edible seaweed. Pronounce /ˈlɑːvə/; but /ˈleɪvə/ is also in standard use.

**laver²,** washing basin. Pronounce /ˈleɪvə/.

**law and order.** It is important not to insert an intrusive /r/ between *law* and *and*. Avoid the same fault in other cases: Say *drawing*, not *draw-ring*, *idea of*, not *idea-r-of*, *law-abiding*, not *law-r-abiding*. But be careful to make correct liaisons, as in *a pair of, for all I know*'.

**lawful.** See LEGAL.

**lawman.** Apart from various historical senses (a member of any of various groups of legal officials, e.g. in the five Danish boroughs), the word is now mainly used as an informal lay term for a law enforcement officer, whether in Wild West films or in crime novels. Examples: *Some lawmen took a delight in seeing the criminals squirm*—R. Barker, 1962; *A retired lawman, still sporting a tin* *star, demonstrated how he could kill with either hand—Radio Times,* 1972; *Lawman resigns. Kennebunkport—Richard Thornburgh, the US attorney general, has resigned, a White House official said—Times,* 1991.

**lawyer.** A generic term for a person whose profession is the law, or (informally) a person studying law at a university or law school. See BARRISTER, SOLICITOR.

**lay and lie. 1** Verbs. In intransitive uses, coinciding with or resembling those of *lie*, except in certain nautical expressions, *lay* 'is only dialectal or an illiterate substitute for *lie*' (*OED*). Its identity of form with the past tense of *lie* no doubt largely accounts for the confusion. In the 17c. and 18c. the alternation of *lay* and *lie* seems not to have been regarded as a solecism. Nowadays confusion of the two is taken to be certain evidence of imperfect education or is accepted in regional speech as being a deep-rooted survival from an earlier period. The principal parts of the two verbs are as follows: *lay* (transitive only, = put to rest), pa.t. *laid*, pa.pple *laid*; *lie* (intransitive only, = be at or come to rest), pa.t. *lay*, pa.pple *lain*. Correct uses: (lay) *please lay it on the floor* (present tense); *the teacher laid the book on the desk* (past tense); *they had laid it on the floor* (past participle); (lie) *go and lie down* (present tense); *she went and lay down* (past tense); *the dog is lying on the floor* (present participle); *the body had lain in the field for several days* (past participle).

The paradigm is merciless, admitting no exceptions in standard English. Incorrect uses: *She layed hands on people and everybody layed hands on each other and everybody spoke in tongues—*T. Parks, 1985 (read *laid*); *Laying back some distance from the road ... it was built in 1770 in the Italian style—Fulham Times,* 1987 (read *lying*); *Where's the glory in ... leaving people out with their medals all brushed up laying under the rubble, dying for the glory of the revolution?—New Musical Express,* 1988 (read *lying*); *We are going to lay under the stars by the sea—Sun,* 1990 (read *lie*); *They may have lost the last fight* [sc. the final of a rugby tournament], *but how narrow it was, and they can still be proud of all the battle honours that will lay for ever in their cathedral in Twickenham—*(sports column in) *Observer,* 1991 (read *lie*).

**2** Nouns. The various strands of meaning of the two nouns are clearly separated except in one case: BrE favours *the lie of the land* and AmE normally favours *the lay of the land*. The division is not absolute, however, as is shown in Muriel Spark's *Symposium* (1990): [a butler speaking] *'I am explaining to our young man,'* said *Charterhouse, 'the lay of the land for the forthcoming dinner.'*

The two forms of English share various other senses of the noun: e.g. both use *lay* for the sexual sense 'a partner (esp. female) who is readily available for sexual intercourse'; and both choose *lie* for 'the position of a golf ball when it is about to be struck'.

**lay-by,** an area at the side of an open road where vehicles may stop; a similar arrangement on a canal or railway. Pl. *lay-bys.*

**lay figure,** a dummy or jointed figure of a human body, has no connection with any of the English words *lay*, but is derived (18c.) from Dutch *led* (now *lid*) 'joint, limb'.

**lb.,** pound, pounds (weight), is an abbreviated form of L *libra* 'pound'.

**leaden.** See -EN ADJECTIVES.

**leadership** came into being in the 19c. in two main senses: (*a*) the dignity, office, or position of a leader (*Tartars under the leadership of their khan*). (*b*) ability to lead (*the person appointed will be expected to have outstanding leadership qualities*). These have been joined by a third sense, = leaders, in the 20c. (*a dinner for the heads of the Senate Committees and the Leadership on both sides and their wives—Mrs L. B. Johnson,* 1964; *They have refrained from making declarations that the union's policy is not in the best interest of the membership or that the leadership has failed to implement the policy—*cited in Gowers, 1965. Cf. MEMBERSHIP; READERSHIP.

**leading question.** In law, a *leading question* is a question which suggests to a witness the answer which he or she is to make. 'In Anglo-American law such questions are generally permissible only on cross-examination' (Garner, 1987). Laymen sometimes extend the term to mean a 'loaded' or 'searching' question, one requiring a guarded answer. The

expression has become a POPULARIZED
TECHNICALITY. In April 1978, on *Desert
Island Discs* (BBC Radio 4), the actress
Felicity Kendall, when asked by Roy
Plomley, *Are you musical?*, replied *That's a
leading question*. Her use of the expression
departed from Judges' Rules, but was
typical of the way in which lay people
now use it.

**lead, led.** The verb meaning 'to go in
front' is, of course, pronounced /liːd/ and
its pa.t. is *led*, pronounced /led/. Incorrect
use: *His idea was the one that lead to the
solution of the mascara mystery*—*Chicago
Sun-Times*, 1990. The chemical element
*lead*, pronounced /led/, is of quite separate
origin; the associated verb and its
parts (*leaded*, *leaden*, etc.) all have /led/ as
their first syllable.

**leaf.** As noun, the pl. is *leaves*. As verb,
the inflected forms are *leafs*, *leafed*, *leafing*
(*she was leafing through the book*), all pronounced with /f/ not /v/.

**-leafed, -leaved. 1** Dictionaries recommend /liːft/ for the first and /liːvd/ for the
second, but in practice the distinction is
perhaps less neatly observed.

**2** As the second element of combinations, -*leaved* is the customary word
for plants and trees: *broad-leaved* (trees),
*four-leaved* (clover) (but also *four-leaf clover*).
But not always: e.g. *a vigorous sword-leafed
fern*—A. S. Byatt, 1990. In some uses (*a
thick-leaved plant*, *a two-leaved door*), -*leaved*
is the more usual of the two forms. But
in older literature (down to that of the
19c.) there is much fluctuation between
*leafed* and *leaved* in contexts where the
meaning is simply 'having leaves or foli-
age' and esp. when there is no qualifying
prefixed word (*Bamboos ... sending from
every Joint sprouts of the same form, leafed
like Five-fingered Grass*—J. Fryer, 1698; *Three
lilies, slipped and leaved*—C. Boutell, 1864).
See -VE(D), -VES.

**leaflet** (verb). The inflected forms are
*leafleted*, *leafleting*; also *leafleteer*. See -T-,
-TT-.

**leak** (verb). For the unauthorized disclosure of secret or confidential information there are doubtless many historical
precedents. But it is of interest to note
that, apart from an isolated example of
1859, the verb *leak*, used transitively or

intransitively (*it had been deliberately
leaked to the press*; *almost all of them leak
to the press*) seems to have been used of
this practice only in the 20c. The phrasal
verb *leak out* (*the carefully guarded secret had
leaked out*), used intransitively, is found a
little earlier (first recorded in 1832), and
*leakage* (improper disclosure of information) is first recorded in 1863.

**lean** (verb). The pa.t. and pa.pple are
either *leaned* (normally pronounced
/liːnd/ but sometimes /lent/) or *leant* /lent/
in BrE, but usu. *leaned* in AmE. Examples:
(leaned) *Georgia Rose ... leaned forward and
blew out every one of her candles*—Lee Smith,
1983 (US); *Emira ... leaned against the
marble top of the work bench*—C. Else, 1985
(NZ); *I leaned sideways just a little, but
enough*—M. Pople, 1986 (Aust.); *Syl smiled
back at me and leaned across and took my
hand*—A. T. Ellis, 1987 (UK); *Costa Rica ...
had leaned heavily in favor of aid to the
rebels*—NY Rev. Bks, 1987; *the gentleman
leaned forward and asked ...*—A. S. Byatt,
1990 (UK); *Forbes ostentatiously leaned back
in the far corner*—E. Templeton, 1991 (US).
(leant) *Colin leant back in his chair, balancing
on the two rear legs*—A. Judd, 1981 (UK);
*Gareth slid the top of the scraper between
the box and lid and leant on it gently*—P.
Ferguson, 1985 (UK); *Then she leant forward, trying to find my face*—Encounter, 1988
(UK); *so she leant back on the pillow*—Susan
Johnson, 1990 (Aust.); *His tone was weary,
and he leant his head down on one hand*—I.
Murdoch, 1993 (UK). See -T AND -ED.

**leap** (verb). **1** The pa.t. and pa.pple are
either *leaped* (pronounced either /liːpt/ or,
in BrE, /lept/) or *leapt* /lept/ in BrE and also
in AmE. Examples: *Medicinal discovery, It
moves in mighty leaps, It leapt straight past
the common cold And gave it us for keeps*—P.
Ayres, 1976; *I can't say that wretch I leaped
in after was much of a loss to the human
race*—P. Bailey, 1986; *he leapt awkwardly
from the car*—New Yorker, 1988; *whole groups
of newly-shining young frogs which leaped up
in little showers of water under their feet*—A. S.
Byatt, 1990; *the canal had leaped a thousand
feet from one hillside to the next*—Atlantic,
1991 (US); *She had leapt on board the boat
like a boy*—New Yorker, 1994. See -T AND
-ED.

**2** The phr. *to leap to the eye* may, like
*jump to the eye(s)* (see JUMP), be a Gallicism.
Cf. Fr. *sauter aux yeux*.

**learn** (verb). **1** The pa.t. and pa.pple are either *learned* (pronounced /lɜːnt/ or /lɜːnd/) or *learnt* /lɜːnt/. In AmE *learned* is the more usual of the two forms but *learnt* is also found occasionally. Examples: (learned) *I learned ... that had I been free to dive and swim ... I should not have known the love of my mother and father*—P. Bailey, 1986 (UK); *The good chess player ... has learned to reconsider ... the overall situation after each step*—B. Bettelheim, 1987 (US); *So, what was learned from this experience?*—*Essays & Studies*, 1987 (UK); *Charlie learned ... that if he called for a vote he was bound to win*—*New Yorker*, 1987; *it was he who enquired, who invited confidences and information, who learned them,* [etc.]— A. S. Byatt, 1990 (UK). (learnt) *Trained as a sculptor, Perry learnt pottery at evening classes*—*The Face*, 1987 (UK); *a point that few of my bright young officers seem to have learnt at school*—B. L. Barder, 1987 (UK); *The wearers of contact lenses have, with difficulty I admit, learnt to master the blink*—*Br. Jrnl Philos. Sci.*, 1988; *The notion ... that because toddlers learn their own language in a 'natural' [way] ... therefore a foreign language should be learnt in the same way,* [etc.]— P. W. E. Bridle, 1991 (UK).

See -T AND -ED.

**2** In the sense 'to impart knowledge, to teach', *learn* has been in use since about 1200 and only gradually descended from acceptability. In 1902 the *OED* described this use as 'Now *vulgar*'. It is a classic example of a meaning that seemed ordinary and unexceptionable to writers like Caxton, Spenser, Bunyan, and Samuel Johnson (1755), but that fell into disfavour c1800 and is used now only by non-standard speakers and in representations of the speech of such people. Typical examples: *if she knows her letters it's the most she does—and them I learned her*—Dickens, 1865; *We asked whether he had learned the instrument at school ... 'No. He learned it himself and now he's learning me.'*—*Times*, 1974; *Children use 'learnt' for 'taught' ('I learnt the game in Germany, and when I came here I learnt it to Susan and Carol.')*—I. Opie, 1993.

**learned.** As pa.t. and pa.pple of LEARN (verb) it is always monosyllabic. Used as an adj. (*a learned journal*; *my learned friend*) it is always disyllabic, /ˈlɜːnɪd/.

**learnedly.** Three syllables. See -EDLY.

**lease** (noun). In BrE one speaks of taking *a new lease of life*, in AmE *a new lease on life*.

**least.** **1** For confusion between *much less* and *much more*, see MUCH 3. *Least of all* and *most of all* are sometimes confused in the same way. *Least of all* means 'especially not' and should be used only in negative constructions: *No cabinet minister, least of all the Right Honorable member for Loamshire, is beyond criticism*. *Most of all* means 'especially' and should be used only in positive constructions and in questions: *If that is the case, what justification exists for the sentences, and most of all for the way in which they were carried out?*; *Active politicians, most of all those burdened with heavy domestic responsibilities, are often little known abroad*.

**2** Use *less*, not *least*, when contrasting only two things: *I'm not sure which of the two of you is the less* (not *the least*) *irresponsible*.

**leastways, leastwise.** Both words mean 'at least, or rather' and both have a long literary history, the first from Chaucer onward, and the second from the 16c. (Thomas More) onward. Typical examples: *He was own brother to a brimstone magpie—leastways Mrs. Smallweed*—Dickens, 1852; *It was a sign that his money would come to light again, or leastwise that the robber would be made to answer for it*—G. Eliot, 1861. But clear signs of archaism or regionalism are noticeable in all the 19c. examples cited in the *OED*, and both words are now only in regional or colloquial use. Examples: *'Would you like ... to b'long to a secret serciety?' 'Dunno ... Never tried. Leastways, not as I can call to mind.'*— R. Crompton, 1923; *They don't have rows—leastways, not ones you can hear in the street*—J. Howker, 1985; *'What happened to them?' said Diung. 'Shot in the back,' said Foster. 'Leastwise, Johnson was ... He was the lucky one.'*—R. J. Conley, 1986 (US).

**leather.** **1** For *leather* and *leathern*, see -EN ADJECTIVES.

**2** In *leather or prunella* the meaning is not two equally worthless things, but the contrast between the rough leather apron of a cobbler and the fine gown of a parson. Pope's couplet (*Essay on Man*, 1734) is *Worth makes the man, and want of* it, the fellow; The rest is all but leather or prunella: what makes a man is his worth, not his clothes. The OED gives some 19C. examples of leather and prunella used inaccurately as an expression for something to which one is utterly indifferent: e.g. Then who shall say so good a fellow Was only 'leather and prunella?'—Byron, 1811.

**leave, let.** It is of interest that the two verbs are interchangeable in a few idiomatic phrases, but usually at a slightly different sociological level, leave sometimes being fractionally lower in the social register than let. **1** leave alone, let alone. The primary meanings of leave alone are (a) to refrain from disturbing, not interfere with (Leave me alone, please. You can see that I'm working); (b) not to have dealings with (we cannot but wish that Gladstone had left the matter alone). A line in the 19C. nursery rhyme 'Little Bo-Peep' (Leave them alone, and they'll come home) helped to establish the currency of sense (a) of leave alone. Let alone cannot be used in sense (b). Let alone usu. means 'not to mention, far less or more' (the Ministry, whose experts seem never to have entered a chemistry laboratory, let alone worked in one—ODCIE, 1983). But it can be used like sense (a) of leave alone: I can neither eat nor sleep. If he doesn't soon let me alone he'll be the death of me—ibid.

**2** The two verbs are also interchangeable in leave/let somebody or something be: e.g. Jerry: Good morning. (Cairy ignores him.) Jean: Say—. Cairy: No! Jean: Leave her be then—let her be—ibid. And also in leave or let well alone. Otherwise the two verbs go their own way.

**lectern.** see PODIUM.

**lecturership, lectureship.** The older form (17C.) of the two and the most common is lectureship, but the University of Oxford still retains the longer form (first recorded in the late 19C.): Grinfield Lecturership on the Septuagint; Jenkinson Memorial Lecturership in Embryology—Oxford University Calendar 1990–1.

**led.** See LEAD, LED.

**leeward.** Pronounced /'liːwəd/ by lay speakers, but /'luːəd/ in nautical circles. It means on or towards the side sheltered from the wind (as opposed to windward). It is invariable in form (not -wards) as adj., adv., and noun.

**leftward(s).** The only form of the adj. is leftward (a leftward glance; in a leftward direction). For the adverb, both leftward and leftwards are used both in BrE and in AmE. What contextually governs the choice of term is not clear.

**legalese,** a comparatively recent term (first recorded in 1914) for the complicated technical language of legal documents. Inevitably opinions differ about its usefulness. A correspondent to The Times in February 1990 claimed that 'in almost every walk of life language is used as much as a form of magic as a way of conveying meaning . . . The evocatively archaic cadences of the liturgy are . . . intended to endow syntax with arcane significance, nouns, verbs and adjectives with the binding power of a spell . . . Most people terminate their letters with adverbs which have little or no relation to those virtues of truthfulness and sincerity which they claim to possess, and the use of language in a basically irrational way to create feelings of awe and respect is as justifiable as the wig of a judge or the mitre of a bishop.'

Be that as it may, the 20C. has witnessed a broad movement towards the simplification of the language of legal documents. A useful short description of legalese is provided by Bryan A. Garner, an American scholar, in his Dict. Modern Legal Usage (1987): 'Legalese has, throughout the history of Anglo-American law, been a scourge of the legal profession. Thomas Jefferson railed against statutes "which from verbosity, their endless tautologies, their involutions of case within case, and parenthesis within parenthesis, and their multiplied efforts at certainty, by saids and aforesaids, by ors and ands, to make them more plain, are really rendered more perplexed and incomprehensible, not only to common readers, but to the lawyers themselves." (quoted in D. Mellinkoff, The Language of the Law (1963) 253).

'. . . For a humorous epitome of legalese, the following nineteenth-century example is nonpareil: "The declaration stated, that the plaintiff theretofore, and at the time of the committing of the grievance thereinafter mentioned, to wit, on, etc., was lawfully possessed of a certain donkey, which said donkey of the plaintiff was then lawfully in a certain highway, and the defendant was then possessed of a certain wagon and of certain horses drawing the same, which said wagon and horses of the defendant were then under the care, government, and direction of a certain then servant of the defendant, in and along the said highway; nevertheless the defendant ... then ran and struck with great violence against the said donkey of the plaintiff, and thereby then wounded, crushed, and killed the same,' [etc.]." *Davis v. Mann,* [Exch. 1842] 10 M. & W. 546, 152 Eng. Rep. 588.

'Other manifestations of legalese commonly appear. One aspect of it is its compressedness: "The question here is whether service of citation was proper in the face of a writ of error attack on a default judgment." Another is ceremoniousness, which arguably has a place in some legal instruments: "In testimony whereof, I have hereunto subscribed my name and affixed my seal, this 24th day of June, in the year of our Lord, one thousand nine hundred and eighty five."'

Garner's verdict is that simplification is desirable: 'The ... effect of the passage just quoted [the one about the donkey], and other passages throughout this work [sc. Garner's own book] should purge readers of any affection for or attraction to legalese.'

## legalism, legality.

The first means 'excessive adherence to law or formula', and the second, principally, 'lawfulness, conformity with the law'. Examples: (legalism) *Unbending legalism was the rock upon which both the ethics and politics of the colonist were built—Daily Chron.,* 1901; *We cannot advance United States interests if public officials ... resort to legalisms, word games—NY Times,* 1987; (legality) *Possibly the memory of some absurd legality or loophole tickled them—R. Narayan,* 1958; *She had never doubted the legality of the union between the lady and the gentleman—A. Brien,* 1987; *It does not follow that private legalities override public accountability—Guardian,* 1987.

See -ISM AND -ITY.

## legalisms.

Garner (1987) assembled a list of the circumlocutions, formal words, and archaisms that are characteristic of lawyers' speech and writing. 'Little can be said by way of advice except that generally lawyers ... are best advised to avoid them. It must be granted, however, that there may be those rare contexts in which the legalistic is preferable to the ordinary term.' Selected items from Garner's list (the legalistic term first the ordinary term following in brackets):

> *abutting* (next to); *anterior to* (before); *at the time when* (when); *be binding upon* (bind); *be empowered to* (may); *during such time as* (during); *for the reason that* (because); *in the event that* (if); *in the interest of* (for); *per annum* (a year); *per diem* (a day); *prior to* (before); *pursuant to* (under, in accordance with); *subsequent to* (after); *the reason being that* (because).

David Lodge catches the nature of certain kinds of legal language in the following passage from his *Paradise News* (1991): The text was written in typical legal jargon designed to cover every possible eventuality—*to purchase, sell, bargain, or contract for, encumber, hypothecate, or alienate any property, real, personal or mixed, tangible or intangible* ...

## legal, lawful, legitimate, licit.

'Legal is the broadest of these terms [sc. *legal, lawful, licit*], meaning either (1) "of or pertaining to law, falling within the province of law", or (2) "permitted, or not forbidden, by law". These two senses are used with about the same frequency. *Lawful* and *licit* share with *legal* sense (2), "according or not contrary to law, permitted by law". The least frequently used of these terms is *licit. Lawful* is quite common: "In March 1977, the company posted a notice on the bulletin board that contained a *lawful* statement on the solicitation and distribution of materials." *Lawful* should not be used in sense (1) of *legal,* however, as here: "The judgment must be affirmed if there is sufficient evidence to support it on any *lawful* [read *legal*] theory, and every fact issue sufficiently raised by the evidence must be resolved in support of the judgment"' (Garner, 1987). *Legitimate* belongs in the same group in that it frequently means 'lawful', but its meanings branch out in other directions as well, among them: (of a child) born of parents lawfully married to each other; (of an argument, viewpoint, etc.) logically sustainable; (of a sovereign) having a genuine claim to the throne; constituting or relating to serious drama as distinct from musical comedy, revue, etc.

**legend.** Until I investigated the matter I was surprised by Fowler's bald statement in 1926, 'Pronounce lĕ-', since I had never heard any other pronunciation of the word. Historically, it emerged, *legend* had two pronunciations, depending on whether it was placed with words adopted from Latin (in which case /'li:dʒənd/) or with those adopted from French (in which case /'ledʒənd/). John Walker (1791) strongly favoured /'li:dʒənd/, and Jespersen (1909) recognized the existence of both pronunciations in standard English. The arguments have been forgotten, and now the form with a short vowel in the first syllable is used without exception.

**legible, readable.** The first primarily means '(of handwriting, print, etc.) plain enough to be deciphered'. The second can be used in the same sense, but is much more frequently used to mean 'interesting or pleasant to read'. Cf. ILLEGIBLE, UNREADABLE.

**legislation, legislature.** The first is the process of making laws, or laws collectively. The second is the body that makes them. Both words entered the language in the 17c. and the distinction of sense has been consistently maintained since then.

**legitimate** (adj.). See LEGAL, LAWFUL, etc.

**legitimate** (verb). First recorded in 1533 and long without a rival in the sense 'to render lawful, legal, or legitimate', it gradually retreated before *legitimatize* (first recorded 1791), and, especially, *legitimize* (first recorded 1848). The oldest word of the three is now much less often encountered than it once was, and *legitimatize* has not greatly prospered. For better or worse, *legitimize*, despite the public antipathy to words ending in *-ize*, seems to be the more usual word in BrE, and, it would seem, is at least as frequent as *legitimate* in AmE. Examples: (legitimate) *My companion had up his sleeve something that would legitimate his employing my Christian name*—J. I. M. Stewart, 1974; *The museum ... tends to legitimate and hence fossilise the most challenging work*—*New Statesman*, 1988; *They support doctors' professional status by legitimating them as medical experts*—*Language in Society*, 1990; (legitimize) *Had the baby been a boy, he would have seriously considered legitimizing the union*—E. Pizzey, 1983; *You ... forget the very people who legitimize your authority*—C. Achebe, 1987; *Eliot ... goes out of his way to legitimize the 'objective', hence universal nature of Baudelaire's vision*—J. Berman, 1987 (US).

**leisure.** Pronounced /'leʒə/ in BrE. An older variant with a long first syllable, namely /'li:ʒ-/, is the one most frequently heard in AmE.

**leitmotif,** pronounced /'laɪtməʊti:f/, is now the more usual spelling in English, not *-motiv* (cf. Ger. *Leitmotiv*).

**lend. 1** See LOAN (verb).

**2** Used as a noun meaning 'a loan' in Scotland, in northern dialects of England, and, colloquially, in NZ. Examples: (16c. onwards examples from Sc. and northern sources cited in the OED); *Do ye think Mr. Awmrose could gie me the lend of a nichtcap?*—J. Wilson, 1826 (Sc.); *Could you give me the lend of a bob?*—F. Sargeson, 1946 (NZ); *Just ringing this feller to ask if I could have a lend of his gun*—J. Howker, 1985 (UK).

**length.** There is a widespread and regrettable tendency to pronounce *length* and *strength* to rhyme with *tenth*. The *ng* should be pronounced as in *sing*.

**lengthways, lengthwise.** From its first recorded use in the late 16c. until the present time the first of these has been used only as an adverb (*a hollow tube split lengthways*), never as an adjective. During the same period, *lengthwise* has been used as an adverb (*in a straight line lengthwise on the front of each seat*) interchangeably with *lengthways*; and also, more recently (first recorded 1871) as an adjective (*the driver was sleeping in a doubled-up lengthwise position*). Modern examples: (lengthways) *Fold in half lengthways*—*Woman & Home*, 1987; *Ropes were strung across the deck and lengthways, too*—*Sea Breezes*, 1989; (lengthwise) *They would eat these in the garden in the intervals between films, having left Danielle's boa draped lengthwise across their seats*—G. Adair, 1988; *He took another blank sheet of paper and laid it out lengthwise*—M. M. R. Khan, 1988.

Cf. LONGWAYS, LONGWISE.

**lengthy.** Before the 19c. used only by
American writers and therefore, it was
asserted, to be held at arm's length. In
1812 Southey remarked, *That, to borrow
a trans-atlantic term, may truly be called a
lengthy work.* And in 1827 Walter Scott
wrote, *The style of my grandsire ... was
rather lengthy, as our American friends say.*
It is now in everyday use throughout
the English-speaking world, meaning 'at
unusual length, often with reproachful
implication, prolix, tedious'. Not a person in a thousand would regard it as
anything other than an ordinary English
word.

**lenience, leniency.** Both words came
into the language in the late 18c. The
*OED* defines *lenience* as 'lenient action or
behaviour; indulgence', and *leniency* as
'the quality of being lenient', but in
practice there is little to divide them
except that *leniency* is much the more
commonly used. Modern examples: (lenience) *Sources close to the military say that
while Menem is urging maximum lenience,
Army Chief Gen. Isidro Caceres is anxious to
purge his force—Christian Sci. Monitor,* 1989;
*Crossing the line, Duxbury will head a new
department for the accountants, a revenue
investigations network, which advises people
who have already attracted the taxman's unwelcome attention. He will then negotiate for
lenience from his former colleagues—Daily Tel.,*
1989; (leniency) *They also accepted a
measure ... that gives years of regulatory
leniency to weakened savings and loans in
economically depressed areas—NY Times,*
1987; *The Chinese Republic released 144
prisoners ... as an expression of 'leniency'—R.
Deacon,* 1988. See -CE, -CY.

**lens.** Pl. *lenses.* See SINGULAR -S.

**lese-majesty.** 1 Now normally so written (without accents) and pronounced
/liːz ˈmædʒɪstɪ/. If written in the French
form *lèse-majesté* it should logically be
pronounced in a French way, but usually
comes out in a modified manner, approximately /leɪz ˈmæʒesteɪ/.

2 The term no longer has any legal
force in English (having been replaced
by *treason*). It is most commonly applied
to presumptuous or offensive behaviour
(esp. towards the sovereign). The Fr. word
is derived from L *laesa mājestās* 'hurt or

violated majesty', i.e. of the sovereign
people.

**less.**

1 *nothing less than.*
2 *much less, still less.*
3 *less, fewer; no less, no fewer.*
4 *less, lesser, smaller, lower, fewer.*

1 For the two meanings of *nothing less
than,* a possible source of ambiguity, see
NOTHING LESS THAN.

2 For the illogical use of *much less*
instead of *much more,* see MUCH 2. Similar
difficulties occur when *still less* is used
instead of *still more.* Incorrect use: *Of
course social considerations, still less considerations of mere wealth, must not in any
way be allowed to outweigh purely military
efficiency.* Here, if *still ... wealth* had been
placed later than *must not,* it would have
passed, since *still less* is appropriate in
negative constructions; placed before it,
it is inappropriate. It is reasonable to
understand *must* from a previous *must
not,* but not from a *must not* that is yet
to come. For a similar difficulty with
*least of all/most of all* see LEAST 1.

3 *less, fewer; no less, no fewer.* Modern
uncertainty about the use of *less* and
*fewer* has been dealt with briefly under
FEW 1(c). Here are some additional examples of incorrect uses found in standard sources: (heading) *A Million Less School
Leavers Won't Mean A Million Less Jobs* (but
*fewer* in the text) *School leavers. Over the
next few years you're going to see a lot fewer
of them—Independent Mag.,* 1988 (*you're
going to see a lot less of them* would have
changed the meaning); *a traffic expert who
believes that the answer, paradoxically, lies in
building less roads, not more—Listener,* 1988;
*There were less people about, as the weather
was chilly—N. Virtue,* 1988. In all of these
examples *fewer* should have been used.

A further example of an idiomatically
acceptable use (cf. FEW 1(b)): *She liked to
pray for not less than fifteen minutes—S.
Faulks,* 1989 (i.e. for a period of time
seen as a unit). Another borderline case:
*The police surgeon, and no less than three
outside doctors who were called in, confirmed
the opinion of Marzillian—I. Murdoch,* 1989.
This is the most vulnerable area—
contexts forcing writers or speakers to
choose between *no less than* and *no fewer
than*—and the choice often depends on
whether the notion of plurality or that

of quantity is dominant. Clearly Iris Murdoch here (and Margaret Drabble in the example given under FEW 1(b)) felt *no less than* to be preferable, and who is to contradict them? A routine correct use to end with, showing a clear distinction between the two words: *Indira Gandhi supported sterilisation in the belief that fewer children meant less poverty—Times*, 1991.

*Historical note.* The account given above (and under FEW 1 (*a, b, c*)) is an attempt to describe current attitudes towards the use of *less* and *fewer*. It should be borne in mind, however, that there is ample historical warrant for the type *less roads, less people*, etc. Such uses originate 'from the OE construction of *lǽs* adv. (quasi-*sb.*) with a partitive genitive' (*OED*). In OE, *lǽs worda* meant literally 'less *of* words'. When the genitive plural case vanished at the end of the OE period the type *less words* took its place, and this type has been employed ever since: e.g. *there are few Vniuersities that haue lesse faultes than Oxford*—Lyly, 1579. Hostility to the use emerged in the 18c., but 'folk memory' of the medieval type has ensured that there has been no break in the use of the type which I have branded as 'incorrect'. It is of interest to set down the way in which the problem of *less* and *fewer* is presented in *CGEL*, the standard descriptive grammar of current English: (§5.24) 'There is a tendency to use *less* (instead of *fewer*) and *least* (instead of *fewest*) also with count nouns: *You've made less mistakes than last time*. This usage is however often condemned. *No less than* is more generally accepted: No less than *fifty people were killed in the accident*.'

**4** *less, lesser, smaller, lower, fewer.* The entanglements become the more difficult to dismantle the more they are examined. As parts of speech, *less* and *lesser* were originally ordinary comparatives (*lesser* a double comparative) of *little*. But they are no longer clearly perceived simply as such, among an array of other uses. *Less* is now taken to mean 'a smaller amount of' (opposed to *more*), and to be the comparative of *a little*, not just of *little*. It cannot be preceded by the indefinite article: the type *a little butter* is acceptable, but *a less butter* is not. If the indefinite article is employed, some word other than *less* should follow it: thus *less noise* but *a lower level of noise; an item of less value* but *a lower value attaches*

to it; *I want to pay less rent* but *a lower rent is what I want; a smaller size*, not *a less size*. On the other hand, *lesser* can be preceded by *a: a lesser man than Churchill* (but not *a less man than Churchill*). For further ramifications of this cluster of comparatives it is essential to turn to the *OED* itself.

**-less. 1** This suffix forming adjectives was already strongly established in OE (spelt *-lēas*); it stood beside the separate adj. *lēas* governing a genitive, as *firena lēas* 'free from crimes'. The first element of such compounds was always a noun (e.g. *wīflēas* 'without a wife'). The separate adj. *lēas* did not survive in ME, but the suffix *-lēas* (which became our *-less*) attached to nouns created a virtually limitless class of words. Examples (with date of first record); *aimless* 1627, *endless* OE, *homeless* 1615, *landless* OE, *lawless* ME, *penniless* ME, *pitiless* ME, *restless* OE, *timeless* 1560. In many instances the nouns to which the suffix was attached was a noun of action, coincident in form with the stem of the corresponding verb, and some of the adjs. so formed had the sense 'not to be —ed', 'un—able' (rather than *devoid of*), e.g. *countless* (1588), *numberless* (1573). On the supposed analogy of these words, the suffix became appended to many verbs, e.g. *abashless* (1868), *dauntless* (1593), *resistless* (1586), *tireless* (1591), *weariless* (1430). But of all such formations only *countless, dauntless, numberless*, and *tireless* survive. As a living suffix *-less* can be appended now only to nouns and not, except fancifully, to verbs. Fowler's spirited crusade (1926) against the formation of new words in *-less* formed on verb stems is now only of historical interest.

**2** A hyphen is necessary when the suffix is added to a noun ending in *-ll*, e.g. *wall-less, will-less*, but not when appended to one ending in a single *-l*, e.g. *soulless, tailless*.

**lessee, lessor.** The *lessee* is the person who holds a property by lease, the *lessor* the person who lets a property by lease.

**lesser.** See LESS 4.

**lest. 1** Traditionally *lest* has been used in two ways, both paralleling uses of L *nē*: (*a*) As a negative particle of intention or purpose, introducing a clause expressive of something to be prevented or guarded against, = in order that not. Examples: *Forge your work as true as you can, least it cost you great pains at the Vice*—J. Moxon, 1677; *Look to the Purser well, lest he look to himself too well*—J. R. Leifchild, 1855; *Lord God of Hosts, be with us yet, Lest we forget, lest we forget*—Kipling, 1897; *Her head swings back and forth, lest she leave a listener untouched*—New Yorker, 1987; *I shall say nothing about alcohol lest I be pilloried by publicans*—J. Critchley, 1987; *it is a place where I don't go often, lest I be jostled*—M. Spark, 1988. (b) Used after verbs of fearing, or phrases indicating apprehension or danger, to introduce a clause expressing the event that is feared (often replaceable by *that*). Examples: *Lady Catherine grew frightened, lest her infanta should vex herself sick*—H. Walpole, 1750; *Fearing lest they should succumb*—Punch, 1881; *Knos and I became agonised with anxiety lest next time he would leave it too late*—C. Day Lewis, 1960; *Nervous lest an insanely jealous Bernard Levin should appear suddenly out of the clock, I tried to crash her on the subject of von Karajan*—Clive James, 1979.

**2** It will be observed that the subjunctive is (correctly) used in the examples in 1(a) (in the 1897 example the subjunctive form *forget* happens to be indistinguishable from the indicative one). In 1(b), *lest* is normally followed by *should*. Fowler (1926) pointed out that 'will and would after *lest* are merely a special form of the inability to distinguish between *shall* and *will*'. He cited these examples as mistakes: *We do not think Mr Lloyd George need be apprehensive lest the newspaper reader will interpret his little homily in Wales yesterday as . . .*; *The German force now lost no time in retreat, lest they would be cut off and surrounded.*

**3** *Lest* is occasionally used with the indicative but this construction is not recommended. Examples: *He was anxious lest Nunn was absent, or dead*—M. Frayn, 1965; *lest someone reminds me there are other denominations besides Methodists in this city . . .*—Bradford Tel. & Argus, 1985.

**let.** **1** Mistakes in the choice of case can occur in English only in the personal pronouns, since only these have distinctive subjective and objective forms. Such mistakes occur fairly often in unsophisticated speech in exhortations led by *let*: e.g. *And now, my dear, let you and I say a few words about this unfortunate affair* (read

*you and me*). A more serious error is to allow the subjective form of a pronoun to stand immediately after the exhortatory *let*: e.g. *let he who did this be severely punished* (read *him*).

*Historical note.* The *OED* marks this use as catachrestic and cites examples from the 17c. to the 19c. in minor literary works: e.g. *Let we* [1485 lete vs] *hold us together till it be day*—Malory's Arthur, 1634; *Awhile Let thou and I withdraw*—Southey, 1795. These cannot be used as evidence to support erroneous constructions of this type in present-day English.

**2** The type '*let us* (or *let's*) + infinitive', in which *let us/let's* is an imperative marker, is well established in the standard language (*let's hold more chat*—Shakespeare, 1588; *Let us begone from this place*—Dickens, 1840). Innovatory colloquial variants, irregular in form, have been brought into the language in America (and thence elsewhere) in the 20c.: e.g. *Let's you and I take 'em on for a set*—W. Faulkner, 1929; *Let's you and me duck out of here*—J. D. Macdonald, 1950. These, and also the absurd *let's us* (e.g. *Let's us go too*), are many distant suburbs away from formal standard English. In grammatical terms, *let's*, in these constructions, is simply being treated as a quasi-modal pronounced /lets/, with 's no longer perceived as a reduced form of *us*. When *let us* is used to introduce a firm request it must always be written as two words: *Please let us wash the dishes before we go.*

**3** The negative of *let's* is *let's not* (the most formal), *don't let's* (BrE), and, an unexpected type, *let's don't* (AmE).

**4** See LEAVE.

**let's.** See LET 2, 3.

**letter forms.** **1** For elevated modes of address at the beginning and end of letters—to dukes and duchesses, peers, ambassadors, etc.—readers should consult the latest printing of such a work as *Debrett's Correct Form.*

**2** *Letters to newspapers.* There is considerable variation. In *The Times* of 22 August 1991, for example, there were sixteen letters in the main section. All began with *Sir,*; the writers signed off in various ways (the order of frequency is shown): *Yours faithfully* (9), *Yours sincerely* (2), *I am etc.* (1), *Yours etc.* (1), *Yours truly*

(1), *Yours* (1), *I am, Sir, your obedient servant* (1). *Letters to business firms.* Normally *Dear Sir/or Dear Sir/Madam,* and *Yours faithfully.* But if the name of an individual is known to the writer the formulas are likely to be *Dear Mr Johnson/Dear Mrs Murray* and *Yours sincerely. Letters to private persons.* There is considerable variation and the borders are by no means clear. The following is no more than a general guide: (to a named person who is not a close friend) *Dear Mr/Mrs/Miss/Ms Fowler . . . Yours sincerely* (or since about 1960, optionally, *Dear Mary Fowler*); (to a person known slightly) as above . . . *Yours truly* (or, more cordially) *Yours very truly*; (to a close friend or colleague) *Dear John/Joan . . . Yours ever.* Frequently, however, no *Yours* formula is used at the end of letters. Some people prefer to put *Best regards, All good wishes,* etc., followed immediately by their name.

**3** In private correspondence, Americans tend to reverse the order of words used at the end of letters (*Sincerely yours, Cordially yours, Very truly yours*) or to write simply *Sincerely, Cordially,* etc.

**4** Between intimates there are limitless variations reflecting the degree of friendship or amorousness that is intended: *My dear(est) ——, My darling ——, Dearest ——,* etc./ *Best love, All love, Ever yours,* etc.

**5** *Historical note.* Epistolary style has varied considerably over the centuries. The broad pattern has been of movement away from extreme formality to relative informality. The following examples show typical opening and closing greetings from the 15c. to the 20c. Inevitably they give only a surface impression of the main styles used at various periods.

(*a*) From the Paston Letters, a collection of over 1,000 family letters written by members of a Norfolk family from about 1420 until soon after 1500: (1441) *Ryth worchipful hosbon/Yourrys, M. Paston*; (1445) *To myn welbelovid son/By yowre modre, Angneis Paston*; (1449) *Trusty and weel beloved cosyn/Be youre cosyn, Elisabeth Clere*; (1454) *To hys wurchypfull brodyr Jon Paston/Be yowre pore brodyr, Wyllyam Paston.*

(*b*) John Keats (1795–1821): (to Leigh Hunt, 1817) *My dear Hunt . . . Your sincere friend John Keats*; (to his brother George, 1819) *My dear George . . . Your most affectionate Brother John Keats*; (to Fanny Brawne,

1820) *My dearest Fanny . . . Ever yours affectionately my dearest—J.K.*

(*c*) Evelyn Waugh (1903–66): (to Tom Driberg, 1937) *My Dear Tom . . . Yours Evelyn (Waugh)*; (to his daughters Teresa and Margaret, 1953) *My Darling Daughters . . . Your loving papa E.W.*; (to Sir Maurice Bowra, 1959) *Dear Maurice . . . Yours ever Evelyn*; (to Constantine Fitzgibbon, 1964) *Dear Mr FitzGibbon . . . Yours sincerely E. Waugh.*

**6** *Envoi.* When I first became a member of the Senior Common Room in Magdalen College, Oxford, and then in Christ Church, Oxford, in the early 1950s, notes addressed to colleagues began with the style *Dear Lewis, Dear Dundas, Dear Stuart,* etc., i.e. with the surname only, occasionally (as a mark of close friendship) preceded by *My.* This simple mode of address has now vanished in favour of either *Dear* + Christian name or *Dear Mr/Dr* [etc.] *Williams.* A similar movement away from the use of surnames has occurred in many other spheres of professional and business life.

See RESPECTFULLY.

**leukaemia.** So spelt in BrE, but *leukemia* in AmE.

**levee.** There are two separate words:

**1** an assembly of various kinds; pronounced /ˈlevɪ/ or /ˈleviː/, occas. /ˈleveɪ/; derived from Fr. *levé*, var. of *lever* 'a rising', a substantival use of *lever* 'to rise'.

**2** (AmE) an embankment; pronounced /ˈlevɪ/ or /lɪˈviː/; from Fr. *levée* fem. pa.pple of *lever* 'to raise'.

**level.** **1** (noun). In the 20c. the noun has tended to be overused in the sense 'a plane or status in respect of rank or authority', usu. with a qualifying adj. or attributive noun: e.g. *discussions at Cabinet level; at the regional level; full consultation with users at the national level; the stories can be read and enjoyed on at least two levels.* The evidence in the OED suggests that this use became established in the 1930s.

**2** (verb). The inflected forms are *levelled, levelling* in BrE, and *leveled, leveling* in AmE.

**leverage.** The first syllable is pronounced /ˈliːv-/ in BrE but /ˈlev-/ in AmE.

**levy** (noun). See TAX.

**lexicon.** See DICTIONARY. The pl. is *lexicons*.

**Leyden jar.** The first word is pronounced /ˈlaɪdən/.

**liable, likely.** 1 See APT. Further to the comments made there, it should be borne in mind that *liable* is often used (not only in constructions with a *to*-infinitive) in contexts of taxation and of compensation: *I find that if I spend more than six months in this country I'm liable to pay income tax on everything I make in America as well as in England*—F. Donaldson, 1982; *Anybody who operates a business as a sole trader ... is liable for all of the debts of the business*—M. Brett, 1987; *They were liable for the distress and disappointment resulting from the breach of contract*—Holiday Which?, 1987.

2 Though the action or experience expressed by the *to*-infinitive after *liable* is often undesirable, it need not be so. It can indicate either the mere possibility, or the habituality, of what is expressed by the verb, e.g. *The cruellest question which a novelist is liable ... to be asked*—F. Raphael; *The kind of point that one is always liable to miss* –G. Orwell. (OMEU).

3 In some dialects in BrE and frequently in AmE, *liable* is used in place of standard English *likely*. Examples of the type *'Tis very liable he's* [sc. *a wounded pheasant*] *a-croped into one o' these here hovers* [sc. *shelters*] are listed in the *Eng. Dialect Dict.* US examples: *Norman Hunter's new record ... is liable to stand unmolested for many years*—NY Evening Post, 1903; *Boston is liable to be the ultimate place for holding the convention*—H. W. Horwill, 1935; *Without his glasses, he's liable to smash into a tree*—W. Wharton, 1982.

**liaise** /lɪˈeɪz/ (verb). Originally Services slang (first recorded in 1928), this back-formation ( = to establish cooperation, form a link) from the long-established noun *liaison* attracted much criticism from purists at first, but has survived and is now part of everyday (esp. spoken) BrE. It is reported to be much less common in AmE.

**liaison** /lɪˈeɪzən/ *or* /-ɒn/ became fully Anglicized at some point in the early 20c.; before that the final syllable was pronounced in a manner resembling Fr. /ɒ̃/. The word is variously pronounced in America: (in probable order of frequency) /ˈliːəzɒn/, /lɪˈeɪzɒn/. William Safire (1981) made fun of the pronunciation of the word by President Reagan and others as /ˈleɪəzɒn/. It is somewhat curious that the noun has been used since the early 19c. in the sense 'an illicit sexual relationship', but that the corresponding verb *liaise* is restricted to military and business contexts and cannot properly be used of an amorous relationship.

**liana** (a climbing plant) is pronounced /lɪˈɑːnə/, but if the variant *liane* is used pronounce it /lɪˈɑːn/.

**libel** (verb). The inflected forms are *libelled*, *libelling*, *libellous*, etc., in BrE, but usu. with a single medial -*l*- in AmE. See -LL-, -L-.

**libel** (noun), **slander.** The first is a published false statement damaging to a person's reputation, whereas *slander* is a malicious, false, and injurious statement spoken about a person (COD, 1995). In popular usage the terms are often used interchangeably. In law, what started out in the 17c. as a clear distinction has now become more complicated, depending on how the word *published* is interpreted. Anyone who is in doubt about the likelihood of infringing the law should not merely rely on dictionary definitions of the two words, but should take legal advice before proceeding to set down in writing, to broadcast, or simply to utter potentially hurtful statements about an identifiable person. In recent years courts have awarded very large sums of money to persons, esp. those in public life, who have successfully brought actions alleging defamation, libel, etc.

**liberal.** As an adj. it was originally (from the 14c. onwards) 'the distinctive epithet of those "arts" or "sciences" that were considered "worthy of a free man"; opposed to *servile* or *mechanical*' (OED). Cf. L *līber* 'free'. In medieval times such education was divided into two main groups—the quadrivium (arithmetic, geometry, astronomy, and music) and the trivium (grammar, rhetoric, and logic). The idea that a university course should be restricted to students of gentle birth, and that it should have a set range of subjects offered to all students, has long since been abandoned. Now entrance to universities and other places of higher education is open to all, admission being governed by academic merit, not by social level. But the word *liberal*, as applied to studies, still retains the residual sense 'directed to general intellectual enlargement and refinement; not narrowly restricted to the requirements of technical or professional training' (*OED*), despite the occasional illiberal acts of strongly politicized groups. In AmE the *liberal arts* are 'academic college courses providing general knowledge and comprising the arts, humanities, natural sciences, and social sciences' (*Random House Webster's College Dict.*, 1991).

**libertine.** To judge from Fowler (1926), the phr. *chartered libertine* ('one who follows his own inclinations') was at that time a hackneyed expression (? in newspapers). Now seldom encountered, it reads more like a scholarly allusion: it was first used of Henry V in the opening scene of Shakespeare's play: *When he speaks, The Ayre* [air] *a Charter'd Libertine, is still*. Dickens picked up Shakespeare's phr. in ch. 4 of his *Edwin Drood*: *He* [sc. Durdles, a stonemason] *is the chartered libertine of the place*.

**library.** Pronounced /'laɪbrərɪ/ (three syllables, two *r*s) by standard speakers, but often reduced to /'laɪbərɪ/ (one *r*) in rapid or careless speech. The awkwardness of pronouncing two *r*s in rapid succession lies behind the reduced form. In AmE, as J. C. Wells (1990) expresses it, 'where . . . the second syllable has a strong vowel, the reduction is more noticeable, hence less frequently heard and more strongly disapproved of'. Cf. FEBRUARY.

**libretto.** The pl. is either *libretti* or *librettos*.

**Libyan.** By some (including myself) pronounced as two syllables, /'lɪbjən/, but by the majority of standard speakers, it would appear, as three, /'lɪbɪən/.

**licence, license.** In BrE the first is the only spelling for the noun (AmE *license*), while the second is the normal form for the verb in both countries. Thus, in BrE, *motor vehicle licence, poetic licence*; *to license one's car, licensing hours*. Hence also *licensed*

*premises, licensed restaurants* (implying that they have a licence to serve alcoholic liquor), and (now a rather old-fashioned word) *licensed victuallers* (see VICTUAL). Occasionally one encounters *licenced* instead of *licensed* in such circumstances (rationalized, it is alleged, as being formed from the noun rather than from the verb), but this is a case of special pleading.

**lichee.** A variant spelling of the more usual LYCHEE.

**lichen.** The dominant pronunciation is /'laɪkən/ as in *liken*, but /'lɪtʃən/ is also common in BrE (though not in AmE). The word ultimately comes from Gk λειχήν.

**lich-gate** /'lɪtʃgeɪt/ (from OE *līc* 'body, corpse' + *gate*). So spelt, not *lych-*. It means 'a roofed gateway to a churchyard, formerly used at burials for sheltering a coffin until the clergyman's arrival' (*COD*).

**licit.** See LEGAL, LAWFUL, etc.

**lickerish** is the better form (rather than *liquorish*) of the adj. that means **1** lecherous.

**2** fond of fine food. It is a 16c. variant of the ME word *lickerous*, which is itself in origin an Anglo-Fr. variant of *lecherous*. The word is not etymologically connected with *liquor*.

**licorice** is a variant of LIQUORICE, a black root extract used as a sweet.

**lie** (verb), be prostrate. See LAY AND LIE.

**lie** (verb), speak falsely. The inflected forms are *lies, lying, lied*.

**lie** (noun), position. See LAY AND LIE.

**lien,** a right over another's property to protect a debt charged on that property (*COD*), is pronounced /'liːən/ or /liːn/.

**-lier.** For comparative-adverb forms, see -ER AND -EST 3.

**lieutenant.** Always pronounced /lefˈtenənt/ in BrE, and /luːˈtenənt/ in AmE. The British navy, on the other hand, omit the /f/ in the first syllable: thus /leˈtenənt/, /ləˈt-/.

**life.** The pl. is always *lives* (e.g. Johnson's *Lives of the English Poets*, 1779–81), except that the pl. of the artistic term *still life* is *still lifes.*

**life cycle.** First recorded in 1873 in the biological sense 'the series of developments which an organism undergoes in the course of its progress from the egg to the adult state' (*OED*), it came to be applied in the 1930s to the course of human, cultural, etc., existence from birth or beginning through various stages to decay and death and ending; and from the 1970s to the pattern of economic or business development shown in a country, a business firm, etc., spec. when the pattern is one of a distinguishable cycle of events from a beginning point to some later specified point. It is as well to check whether the simplex *life* is not adequate before drawing on either of these transferred uses of *life cycle*. But the extensions of meaning in themselves are natural ones, and *life cycle* in not in danger of becoming a tainted word except when it is used in non-cyclic contexts.

**lifelong, livelong.** The first, pronounced /ˈlaɪflɒŋ/, is a combination of *life* and *long*, and means 'lasting or continuing for a lifetime'. It was first recorded in this sense in the 19c. The second, pronounced /ˈlɪvlɒŋ/, is a combination of *lief* adj., 'dear, beloved', and *long*, and is an emotional intensive of *long*, used of periods of time (*the livelong day, night,* etc.). It was first recorded in the 15c., and is now in restricted (poetical and rhetorical) use.

**life-style.** Introduced in 1929 by the neurologist Alfred Adler (1870–1937) to denote 'a person's basic character as established early in childhood which governs his or her reactions and behaviour', the term quickly became extended to mean generally 'a way or style of living' (e.g. *Council of churches want freedom for students to create their own life-styles*—*Times*, 1973). The extended sense has often been assailed ('obnoxious', 'an unnecessary and clumsy excrescence on the language', etc.) but looks like surviving. Of course, if the simplex *life*, or the traditional phrase *way of life*, would in context adequately convey the required meaning, one of these should be used instead.

**ligature.** See DIGRAPH; DIPHTHONG.

**light** (noun). **1** For *dim religious light*, see IRRELEVANT ALLUSION.

**2** *in* (*the*) *light of.* There are two strands of meaning, complicated by different distributions of the forms with or without the definite article in BrE and AmE. From my own investigations and those of the American scholar John Algeo, the following patterns emerged. The form with the definite article is the older one (1749 onwards in the *OED*) and is the normal form in BrE (*We shall have to consider that in the light of the time limitations imposed on us; We have to read Chaucer's poetry in the light of the social changes going on in the fourteenth century*). It means 'from the point of view of; from the standpoint of'. In AmE this full form of the complex preposition is sometimes used in such contexts, but the shorter form *in light of* is just as frequent. *In light of* is an Americanism, with very limited currency in BrE. It is the dominant form in America, and is used almost exclusively when the meaning is 'in view of; because of' (*In light of what you've told us, we have decided to leave earlier; Their caution seems appropriate in light of the current stock-market fluctuations*). The question arose when on 6 Aug. 1989 I heard an American priest say on BBC Radio 4, *In light of the fact that ...*: it sounded American to me.

**light** (verb). In BrE the pa.t. is usually *lit* (*he lit the fire at 7 pm last night*), as also the pa.pple (*he had lit his pipe before he noticed the no smoking sign; frontiers and border checkpoints are always well lit*). When used attributively or as adj. the customary form is *lighted* (*a lighted cigarette*), except when it is qualified by an adv. (*a well-lit room, a badly lit cellar*). In AmE the same pattern exists, but one is more likely to encounter *lighted* in contexts where in BrE *lit* is more or less obligatory: *She lighted a candle and turned off the lamp*— New Yorker, 1987; *At night the parking lot is badly lighted*—New Yorker, 1987; *the explosion occurred immediately after he had lighted a cigarette*—Daily Northwestern (Evanston, Illinois), 1991.

**lightning** (electrical discharge in the sky) must always be spelt thus. *Lightening*, by contrast, is the ordinary pres.pple of the two verbs *lighten* (to reduce the weight of; to brighten up) or the gerund meaning 'the process of making or becoming light (*the task of lightening the burdens of taxpayers*).

## like.

1 As a conjunction.
2 As a preposition.
3 A hated parenthetic use.
4 Idiomatic phrases.

**1** *As a conjunction.* Fowler (1926) cites this sentence from Charles Darwin (1866): *Unfortunately few have observed like you have done.* The Great Schoolmaster's view of the construction was expressed with characteristic verve: 'Every illiterate person uses this construction daily; it is the established way of putting the thing among all who have not been taught to avoid it; the substitution of *as* for *like* in their sentences would sound artificial. But in good writing this particular *like* is very rare.' The *OED* (in a fascicle published in 1903) cited examples of *like* used as a conjunction from the works of Shakespeare, Southey, William Morris, Jerome K. Jerome, and others, and added the comment: 'Now generally condemned as vulgar or slovenly, though examples may be found in many recent writers of standing.'

The status of conjunctional *like* has been debated many times since then. The most recent authority, CGEL (1985), slightly disguises the problem by speaking of 'clausal adjuncts' and 'semantically equivalent phrasal adjuncts', but the verdict is nevertheless much as before. Constructions such as *Please try to write as I do* and *Please try to write like me* are standard; but *Please try to write like I do* is described as being only in informal use, and especially in AmE. Thus, throughout the 20c., the mood has been condemnatory. The use of *like* as a conjunction has been dismissed as 'illiterate', 'vulgar', 'sloppy', or, in the coded language of modern grammarians, 'informal'. Evelyn Waugh spoke for his generation of writers, and for many people still, when he said of Henry Green's *Pack My Bag* (1940): 'Only one thing disconcerted me . . . The proletarian grammar— the "likes" for "ases", the "bikes" for "bicycles", etc.'

I have reconsidered the matter by examining the works of 'many recent writers of standing', British, American, and from further afield, and the results are, I think, of interest. After I had set aside some really lazy sentences, four main conjunctional uses of *like* emerged. First, quite frequently, with repetition of the verb used in the main clause, and bearing the sense 'in the way that': *They didn't talk like other people talked*—M. Amis, 1981; *Gordon needs Sylvia like some people need to spend an hour or two every day simply staring out of the window*—P. Lively, 1987; *I'm afraid it might happen to my baby like it happened to Jefferson*—New Yorker, 1987. This use, which owes something to the song 'If you knew Susie like I know Susie', is common in all English-speaking countries, and must surely escape further censure or reproach. Naturally, though, we may continue to use other constructions if we wish to, and in good company: *She changed wallpapers and lampshades the way some women changed their underwear*—A. N. Wilson, 1986.

Secondly, it is frequently used in good AmE and Aust. sources (though much less commonly in BrE) to mean 'as if, as though': *It looks like it's still a fox*—New Yorker, 1986; *She acts like she can't help it*—Lee Smith, 1987 (US); *I wanted him born and now it feels like I don't want him*—E. Jolley, 1985 (Aust.). One of the few British examples in my files is one from BBC Radio 4 early in 1987: *It looks like Terry Waite will leave for London in two or three hours.*

Thirdly, it is interchangeable with *as* in all English-speaking countries in a range of fixed, somewhat jocular, phrases of saying and telling: *Send for your copy now. Like we said it's free*—Globe & Mail (Toronto), 1968; *Like you say, you're a dead woman*—M. Wesley, 1983; *Well, like I told you, I work with him upstairs*—P. Ackroyd, 1985; *Like I said, I haven't seen Rudi for weeks*—T. Keneally, 1985; *My whereabouts are in Merthyr Tydfil, like you said*—B. Rubens, 1985.

Fourthly, it is increasingly used, perhaps especially abroad, in contexts where a comparison is being made. In these it has the force of 'in the manner (that), in the way (that)': *You call us Mum and Dad like you always have*—M. Wesley, 1983; *How was I to know she'd turn out like she did?*—C. Burns, 1985 (NZ); *Like Jack and Jill came down the hill, Dilip also rolled down the box-office in 'Karma'*—Star & Style (Bombay), 1986; *The retsina flowed like the Arno did when it overflowed in 1966—Spectator,* 1987.

It would appear that in many kinds of written and spoken English *like* as a conjunction is struggling towards acceptable standard or neutral ground. It is not there yet. But the distributional patterns suggest that the long-standing resistance to this omnipresent little word is beginning to crumble.

**2** *As a preposition.* (a) Unquestioned uses: *drink like a fish; sell like hot cakes; it fits like a glove;* (in written sources) *one of those frilly little wooden stations like gingerbread houses*—A. Carter, 1984; *He saw the sunlight leave the grass like an eye suddenly closed*—P. Ackroyd, 1985; *The Pope was confined like a prisoner in the Vatican*—R. Strange, 1986. (b) Sometimes questioned is its use in place of *such as* meaning 'of the class of, for example', as in *a subject like philosophy; good writers like Dickens; When Scottish families came to York, like Mrs Eliza Fletcher's, Sydney [Smith] . . .* —A. Bell, 1980. The difficulty is that in such circumstances *like* can sometimes be ambiguous: for example, the title of Kingsley Amis's novel *Take a Girl Like You* (1960) could be taken to mean 'a girl, for example, you' or 'a girl resembling you'. Had the title been 'Take a Girl Such as You', there would have been no such ambiguity. For *as* used confusingly instead of *like*, see AS 13. Modern example: *New York, as most major cities, has found that the general public is very apathetic*—NY *Times*, 1970 (WDEU).

**3** *A hated parenthetic use.* Used parenthetically to qualify a preceding or following statement, and often no more than a dialectal or popular filler. This use was first noted in Fanny Burney's *Evelina* (1778). *Father grew quite uneasy, like, for fear of his Lordship's taking offence.* Further examples are cited in the OED from works by Scott, Lytton, De Quincey, Arnold Bennett, and some other well-known writers. As an occasional device it was unexceptionable. By the mid-20c., however, its use as an incoherent and prevalent filler had reached the proportions of an epidemic, and it is now scorned by standard speakers as a vulgarism of the first order. A range of typical examples (in most of which it is being used as a token of unsophisticated speech): *Naa, I was all into that last year,*

*but like I don't really think it's so relevant now*—M. du Plessis, 1983; *I'll say goodbye, like, and send you a message, like, somehow or other, when she turns up, like*—P. Bailey, 1986; *The Blitz? Like right in London?*—M. Pople, 1986; (waitress speaking) *The crowd here is hard to define. Like, they're pretty rich*—New Yorker, 1987; *Like, I just got this journal in the mail from this microtonal music society*—Melody Maker, 1988; *Hayley was pleased. 'That's him. He's, like, got her hypnotized.'*—Maurice Gee, 1990; *He was so young he was from the generation of human beings who use the word 'like' to mean 'said'. 'I'm, like, "You've got to be kidding"' was one of his expressions*—New Yorker, 1991.

**4** *Idiomatic phrases.* Used adverbially in a number of colloquial phrases, e.g. *like always* (20c.), *like anything, like blazes, like crazy* (20c., = like one who is crazy), *like fun, like mad.* These belong only in informal writing or speech. Examples: *Carsons' mill is blazing away like fun*—Mrs Gaskell, 1848; *The horse . . . went like blazes*—De Quincey, 1853; *They wept like anything to see Such quantities of sand*—L. Carroll, 1872; *We . . . heard our fellows cheering like mad* –W. Forbes-Mitchell, 1893; *There she was, beating them with her umbrella like crazy*—J. Osborne, 1957; *Skate was with him like always*—M. Doane, 1988.

**like** (noun). In 1988 I was upbraided by a Scotswoman for writing *Who has not seen the likes of the following?*. I turned to the OED, and noted with satisfaction that the use of *likes* as a (pl.) noun was listed there with illustrative examples from 1787 onward (alternative with *the like of*), including *2,500 [copies sold] in five months is a good sale for the likes of me* from Browning (1872). And I felt even more assured when I came across *that's a luxury for the likes of me* in Penelope Lively's *Moon Tiger* (1987).

**like** (verb). **1** In English-speaking countries where *will* regularly replaces *shall*, i.e. virtually everywhere except in England itself, *I would like* is used instead of *I should like*; in practice, the abbreviated form *I'd like* is widely used instead, esp. in the spoken language. The extended type *I would have liked* is now freely used everywhere (including England) in two distinct constructions: (a) *I/he*, etc., *would have liked* + *to*-infinitive: *I would have liked to pause there*—T. Keneally, 1980; *He would* have liked to run—I. McEwan, 1987; *I would have liked to keep it for myself*—Maurice Gee, 1987; *I'd have liked to know the outcome*—P. Lively, 1987. (*b*) *I/he.* etc., *would have liked + to + aux. + pa.pple.*: *I would have liked to have met James's mother*—A. Brookner, 1983; *I would have liked the children to have been interested in us*—M. Wesley, 1983; *She would have liked to have asked them*—B. Rubens, 1987. (*c*) In both such types *should* is sometimes used in England with a first-person pronoun. Examples: *I should have liked to spare you this*—T. S. Eliot, 1939; *I should have liked to talk to Maurice*—A. Brookner, 1982; *I should have liked to have enjoyed this more*—ibid. Type (*a*) is routinely acceptable. The examples in (*b*) and the last one in (*c*) could equally well or even with advantage have used the present infinitive in the subordinate clause without loss of meaning or time-reference.

2 *like for* + complement. This use (not in the *OED*) seems not to have spread its wings outside America. Examples: *He told Highridge he would like for the writer ... to be his guest for dinner*—T. Wolfe, 1987; *I'd like very much for you to meet him*—New Yorker, 1988. The *Amer. Dialect Dict.* (1944) lists examples from 1888 to 1943, mostly from southern States, but it is not so regionally restricted now.

3 *like to*, *liked to*. A home-kept American import from Britain. Formed on *like* (adj.), it was first recorded in the 15c. in the Paston Letters in the sense 'seem'. It survived in regional use in BrE until about 1800 before becoming archaic or obsolete, but survives in AmE, chiefly in compound tenses, with the meaning 'to look like or be near to doing (something), almost'. Examples from US sources: *The evening liked to have been a tedious evening*—J. A. Benton, 1853; *She liked to fainted just now*—J. E. Cooke, 1854; *Then when we got him abroad he liked to kick our brains out*—1938 in *Amer. Dialect Dict.*; *I like to had a fit*—M. K. Rawlings, 1939; *Well, the icing was rancid. I took a big bite and like to died*—M. Grimm, 1989.

**-like.** 'In formations intended as nonce-words, or not generally current, the hyphen is ordinarily used' (*OED*). Examples (most of them listed in the *OED*): (main element ends in -*l*) *cowl-like, eel-like, owl-like*; (main element ends in -*ll*) *cell-like, mill-like*); (others) *fir-like, friendly-like,*

*strange-like.* But established -*like* compounds are normally written as one word, e.g. *childlike, lifelike, statesmanlike.*

**likeable.** Thus spelt in OUP house style, not *likable.*

**likely.** 1 Freely used as an adj. : *it is not likely that they will come*; *a likely story*; (followed by a *to*-infinitive) *he is not likely to come now.*

2 As an adv. in standard English it is almost always qualified by another adv., esp. *more, most, quite,* or *very* but just as often stands without an adverbial prop in AmE: *It is possible to predict that within a few years the microfiche likely will move into the study and home*—Publishers' Weekly, 1971; *It was early morning, likely too early for anyone to have seen this happen*—New Yorker, 1989; *I ... caught myself trying to calculate how maybe he'd brought it on, why likely he'd had it coming*—T. R. Pearson, 1993.

**likewise.** Like *also*, it is not comfortably used as conjunction in standard English. Fowler (1926) cited *Its tendency to wobble and its uniformity of tone colour, likewise its restricted powers of execution* as an example of what to avoid (he actually called it an 'illiteracy'). On the other hand, it is correctly used when placed as an 'additive conjunct' (*CGEL*'s term) at the head of a sentence: e.g. *St Paul's Cathedral is one of the most easily recognizable sights of London. Likewise the Eiffel Tower is one of the most easily recognizable sights of Paris.*

**-lily.** 1 For reasons of euphony, adverbs in -*lily* formed from adjs. in -*ly* now seldom occur. For some examples, see FRIENDLILY. As Fowler (1926) remarked, 'It is always possible to say in *a masterly manner, at a timely moment,* and the like, instead of *masterlily, timelily.*'

2 Presumably for the same reason, a number of adjs. ending in -*ly* often or usu. remain unchanged when used as adverbs: GINGERLY, *jolly soon, a kindly thought* and *kindly said.*

3 A few -*lily* adverbs are listed as current English in *COD* (1995), e.g. *holily, jollily, sillily, wilily,* but all of these are formed from words in which -*ly* is part of the word-stem and not the usual adjectival ending.

**limb.** There are two distinct words:
**1** The ordinary word = a projecting part of a person's or animal's body (derived from OE *lim*).

**2** a specified edge of the sun, moon, etc. (from L *limbus* 'hem, border').

**limbo.** Pl. *limbos*, whether of the word meaning 'the supposed abode of the souls of unbaptized infants, etc.' or of the quite separate word meaning 'a West Indian dance in which a dancer bends low in order to pass under a bar'. See -o(E)s 6.

**lime.** The adj. corresponding to the white caustic alkaline substance *lime* is *limy*; that corresponding to the fruit called *lime* is *limey*. *Limey* as an AmE slang term for a person from Britain arose from the enforced consumption of lime-juice in the British navy.

**limit** (verb). **1** The inflected forms are *limited, limiting*. See -T-, -TT-.

**2** Its natural meaning is 'to restrict, to confine within bounds', e.g. *to limit one's research to the street-names of London*; whereas *delimit* means 'to settle or determine the limits or boundaries of (a frontier, etc.)'.

**limited** (adj.). A *limited* (or *limited liability*) *company* is one whose owners are legally responsible only to a limited amount for its debts. It does not imply that the number of members is limited.

**limn** (verb). Encountered in works about miniature portraits and in palaeography, the base-form is, like *solemn*, pronounced with the *n* silent. The inflected forms are *limns* /lɪmz/, *limned* /lɪmd/, *limner* /ˈlɪmnə/, and *limning* /ˈlɪmɪŋ/ or /ˈlɪmnɪŋ/.

**linage** /ˈlaɪnɪdʒ/ is the correct spelling for the word meaning 'the number of lines in printed or written matter'; contrasted with *lineage* /ˈlɪnɪdʒ/ 'lineal descent, ancestry'.

**line** (noun). For some synonyms in the sense 'occupation, etc.' (*What's My Line?*), see FIELD.

**lineage.** See LINAGE.

**lineament** /ˈlɪnɪəmənt/, usu. in the pl., is 'a distinctive feature or characteristic of the face'; *liniment* /ˈlɪnɪmənt/ is 'an embrocation'.

**lingerie.** In BrE pronounced in an approximately French manner as /ˈlæʒərɪ/ or /-riː/; in AmE most commonly as /ˌlɑːnʒəˈreɪ/ or /-ˈriː/. But the pronunciation of the word is still very unstable in all English-speaking countries.

**lingo.** The pl. is *lingos*. For some related words, see JARGON.

**lingua franca.** For some related words, see JARGON.

**linguistic engineering** is a term sometimes used for the process by which a group, an institution, etc., tries to impose an artificial set of new terms, on a largely reluctant public. It is what Fowler (1926) called DIDACTICISM. If you feel surprised or aggrieved when someone assails you for using *chairman* or *chairwoman* (rather than *chair* or *chairperson*), *history* (and never *herstory*, when, say, the subject-matter gives prominence to the role of women), *inflammable* (rather than *flammable*), *Peking* (rather than *Beijing*), in other words if you are being preached at by anyone insisting on the use of 'politically correct' terms, you are the victim of a process called linguistic engineering. This phenomenon is more marked and more strident in the 20c. than at any time in the past. Its more agreeable side is shown in the introduction of a medical term like *Down's syndrome* instead of *Mongolism*. But for the most part its effects are very divisive.

**linguistics.** First used in the early 19c. to mean 'the science of languages, linguistic science', it gradually overtook and then almost totally replaced the term *philology* in our universities in the second half of the 20c. The Karl Marx of linguistics is Ferdinand de Saussure (1857–1913), the French-speaking Swiss linguistic scholar, and the key work is his *Cours de linguistique générale* (first published in 1916, compiled from the lecture notes of his students). He established the primary distinction between a *synchronic* approach (seeing a language as a state at a particular point in time) and a *diachronic* (historical) one. He also perceived language as having two dimensions, which he called *langue* (the systems and totality of a given language) and *parole* (the act of speaking, individual speech). Linguistics has inevitably broken up into subdivisions, such as sociolinguistics, computational linguistics, psycholinguistics, and applied linguistics. The main outcome of the rise of linguistics as an academic subject has been the multiplication of opposing schools of thought, each with its own battery of complex terminology. The subject has propelled itself beyond the reach of ordinarily educated people (including graduates) who are not familiar with the works of de Saussure and those of his successors and adherents. The schisms and divisions are multiplying year by year, and the long-term outlook is very uncertain.

**liniment.** See LINEAMENT.

**linking r.** The term is used to describe the fully pronounced /r/ when a word like *pour*, which has a silent final r when pronounced in isolation (thus /pɔː/) or when followed by a word beginning with a consonant ( *pour cups of coffee* /'pɔː 'kʌps ... /), is followed by a word beginning with a vowel ( *pour out* /'pɔːr 'aʊt/). The linking r is always used by non-rhotic standard speakers in such expressions as *a pair_of gloves, wouldn't hear_of it, feather_edge.* Cf. INTRUSIVE R.

**links** (golf-course). The word has the same form as a singular noun and as a plural one: *a suburban links; there are numerous links within easy reach of the city.*

**Linnaean,** of the Swedish naturalist Carolus Linnaeus (1707–78). But spelt *Linnean* in the *Linnean Society* (London).

**liny** (marked with lines, wrinkled). Thus spelt, not *liney*.

**liquefy.** Thus spelt, not *liquify.* See -FY 2.

**liqueur.** Pronounced in BrE /lɪˈkjʊə/ and in AmE usu. /lɪˈkɜːr/.

**liquid.** 1 See FLUID.

2 A term used by some phoneticians in the classification of speech sounds for the apico-alveolar sounds /r/ and /l/. Also applied by some writers to the sounds denoted by the letters *m* and *n*.

**liquidize, liquidate.** The first is a relatively recent word (first recorded in 1827) meaning 'to make, or to become, liquid'.

In the 1950s it acquired its now dominant sense 'to purée, emulsify, or blend in a liquidizer'. The second is a much older word (first recorded in the 16c.), used in various obsolete and current senses: to render unambiguous, make plain (17–18c.) (obs.); to determine or apportion by agreement or by litigation (16–18c.) (obs.); to wind up (a business firm, etc.) (first recorded in 1870); and esp., as a calque on a Russian word, to put an end to, wipe out, kill, esp. one's political opponents (first recorded in 1924). The last sense is particularly associated with the brutality of totalitarian regimes.

**liquorice** /'lɪkərɪs, -rɪʃ/. In BrE the normal spelling of the word meaning a black root extract used as a sweet; AmE *licorice.* Cf. LICORICE.

**liquorish.** See LICKERISH.

**lira** /'lɪərə/, the chief monetary unit of Italy, has pl. *lire* (in English pronounced the same as the sing.).

**lissom.** The standard spelling, not *lissome.*

**list.** The impersonal verb *me* (etc.) *list* (also *me* (etc.) *listeth*) meaning 'I, etc., please, like, or desire' has only a shadowy existence in modE. So too the personal construction, used either with a following infinitive ( *if you list to taste our cheer*) or without one ( *the invaders harried where they listed*), in which it means 'to wish, desire (to do something)'. The verb is not listed in *COD* 1995 in any of its uses.

**lit.** See LIGHT (verb).

**litany, liturgy.** The first is 'a series of petitions for use in church services and processions, usu. recited by the clergy and responded to in a recurring formula by the people; specifically, that contained in the Book of Common Prayer' (*COD*). The second is 'public worship in accordance with a prescribed form'. It is tempting to suppose that the initial *lit-* in each word has the same derivation, but not so. *Litany* is ultimately from Gk λιτανεία 'prayer', from λίτη 'supplication'; whereas *liturgy* answers to Gk λειτουργία 'public service, public worship', from (recorded only in compounds) λεῖτος 'public'. At the present time, in transferred use, *litany* is frequently used for a simple enumeration seen as resembling a form of religious supplication (e.g. *a litany of curses, of woes, of lies*).

**litchi.** See LYCHEE.

**literally.** From the 16c. onward it has been used to indicate that the accompanying word or phrase must be taken in the literal sense: e.g. *It is found that the Act does not mean literally what it says*—*Law Times Rep.*, 1895. A word of this type, however, is liable to become weakened in sense, i.e. to be used in contexts where it does not exactly fit. Uses of this kind, displaying varying degrees of inexactness, have been recorded since at least the early 19c. The *OED* (1903) expressed it thus: 'Now often improperly used to indicate that some conventional or hypothetical phrase is to be taken in the strongest admissible sense.' Examples of the weakened sense and some showing a slight movement in that direction: *For the last four years ... I literally coined money*— F. A. Kemble, 1863; *And with his eyes he literally scoured the corners of the cell*—V. Nabokov, 1960; *You'd start in the morning and it would literally go on till late at night*—J. Gathorne-Hardy, 1984; *Many in this group think they can ignore depreciation as they are buying the car new and literally running it into the ground*—*Money Paper*, 1987; *Most of the buildings on the corniche have literally been face-lifted*—*Blitz*, 1989; *Zagreb believes that Europe is fiddling with face-saving diplomatic measures while Croatia—quite literally—burns—Spectator*, 1991; *They* [sc. supermarkets] *can literally play God, even to the point of sending food back to the genetic drawing board for a redesign*—*Guardian*, 1995.

It's a case of 'stop, look, and think' before using the word in any manner short of its exact sense.

**literary allusions.** In learned work cited passages are usually precisely identified, with author, title of work, date, page number, etc., as part of the normal scholarly apparatus. Thus in an essay on William Blake in *Essays & Studies 1986*, the writer cites from Milton's *Paradise Lost* in the following manner:

*He took the golden Compasses, prepar'd*
*In Gods Eternal store, to circumscribe*
*This Universe, and all created things:*
                                    vii.225–7

In a footnote he explained that the text is taken from the 20-volume edition of Milton's works published in New York in 1931. All this is as it should be.

There is a less precise, and therefore more demanding, level of reference, that of literary allusions. When using these, the writer simply assumes that the reader will recognize, or half-recognize, the phrase or passage alluded to. Public addresses and book reviews are garnished with such allusions, often accompanied by no more than a *as Virgil says*, *as the Bible tells us*, or by nothing at all. Thus, a minor example of the type, in the 22 September 1991 issue of the *NY Times Book Rev.*: *But, as Dr. Johnson's friend said, cheerfulness keeps breaking in* [a not quite accurate reference to Oliver Edwards speaking to Dr Johnson on 17 April 1778: *cheerfulness was always breaking in*]. One constantly encounters allusions, not always accurately set down, to well-known passages in English literature: e.g. (for the authors' names see at end) *To-morrow to fresh woods, and pastures new*; *Alas, regardless of their doom, The little victims play!*; *Oh, to be in England Now that April's there*; *And we are here as on a darkling plain*; *This is the way the world ends Not with a bang but a whimper*; *Sexual intercourse began In nineteen sixty-three* (*Which was rather late for me*). Sources: Milton, Gray, Browning, Matthew Arnold, T. S. Eliot, Philip Larkin.

**literary critics' words.** The work of literary critics in learned papers and monographs on aspects of deconstruction, post-modernism, and other recent theories has reached a point of complexity that is best marked by warning signals like 'Achtung' or 'Keep Out'. The arguments are studded with words like *intersubjectivity*, *alterity*, and *reifying*, and the inevitable pair *signifier* and *signified*. At lower, but more widespread, areas of activity, especially in the columns devoted to book reviews in newspapers, the word-choked terminology of the major priests of theory is mostly absent, but prevalencies of different kinds occur. From some recent issues of the top end of the newspaper world I have drawn some typical examples of recurring or formulaic words and phrases. Some words of praise are used with a view to their quotable value in blurbs: e.g. (of reference books), *definitive*; *epoch-making*; *meticulous*; *monumental*; (of novels) *an incredibly romantic story*; *a poetic and exhilarating novel*; *superb craftsmanship*; *an engrossing read*; *good, gripping reading*. Other popular words used by literary critics include *actuality, ambience, ambivalent, committed, compelling, dedicated, dichotomy, evocative, seminal*, and *sympathetic*. But unblurblike comments turn up as well: *disappointingly stodgy*; *a rambling account*; *raises more questions than it answers*; *a profoundly depressing book*; *dull stuff*. Dismissive sweeps are common: *of no more value to scholarship than Beefeaters are to the defence of the realm*; *Under the often glitteringly clever surface, there's little sense of the emotional depth the author wants to fathom*. Boldness of style or subject-matter is not seen as a fault: *a moving and anarchic novel*; *unflinching but unsensational*; *this dark and intelligent book*; *splendidly caustic*; *a volume as sound as it is daring*; *tough delicacy*.

**literary words.** The notion that certain words are more suitable for verse and literary prose than for the ordinary language held sway from the earliest times until the first part of the 20c. Fowler (1926) touched on the matter when he remarked '[A literary word] is one that cannot be called archaic, inasmuch as it is perfectly comprehensible still to all who hear it, but that has dropped out of [everyday] use and had its place taken by some other word except in writing of a poetical or a definitely literary cast.' He had in mind such words as *eve* (for *evening*), *gainsay* (for *deny*), and *visage* (for *face*). He might easily have added words like *ere*, *erstwhile*, *perchance*, and other WARDOUR STREET words. Such a view is no longer valid. Poetry and writing 'of a definitely literary cast' have become democratized and no longer necessarily contain an element of sectionally guarded vocabulary. As Katie Wales points out in her *Dictionary of Stylistics* (1989), 'literary language can be different and yet not different from "ordinary" or non-literary language; there is, as it were, a "prototype" of literary language, and also numerous variants. But it is the impossibility of defining it in any simple way that is its most defining feature.' Philip Larkin's chronologically arranged *Oxford Book of Twentieth-Century Verse* (1973) underlines the changes of attitude that have occurred as the century progressed. In the early part of the century Thomas Hardy typically wrote *Did her gifts and compassions* enray *and* enarch *her sweet ways With an aureate* nimb?' (my emphases). Many poets of the 1930s, including Auden and MacNeice, ventured into the plainer mysteries of the ordinary world. Kingsley Amis wrote lines such as these: *The journal of some bunch of architects Named this the worst town centre they could find*. Now there are no areas left designated 'for literary use only'. Witness (to go no further) lines like the following: *The butcher carves veal for two* (Hugo Williams); *The stars, the buggers, remained silent* (Brian Patten) The frontiers are down, and, as a result, the classification of literary words has largely passed out of the sphere of lexical enquiry into that of literary theory.

**literature.** Though the word itself is old—it was first recorded in the 14c. in the now-obsolete sense 'acquaintance with "letters" or books'—the usual modern sense, 'literary productions as a whole', is of quite recent emergence (first recorded in 1812). Before the 19c. was over it had already been pressed into service to mean printed matter of any kind, e.g. travel leaflets, political pamphlets, and, in the course of the 20c., the material in print on a given subject (*there is a considerable literature on anabolic steroids*). The extended senses have in no way diminished the primacy of the central sense, i.e. that shown in *English literature, French literature*, etc.

**lithesome,** pliant, supple. This 18–19c. word has largely dropped out of use in favour of its contracted variant LISSOM.

**litotes. 1** *Pronunciation*. Some standard dictionaries, including the *OED*, give /ˈlaɪtəti:z/ or the same form with a diphthongized second syllable /-əʊ-/ (cf. Gk λῑτότης with long iota); others, including *COD* 1995, give /laɪˈtəʊti:z/, i.e. with the main stress on the second syllable. When teaching OE (in which this figure of speech frequently occurs), I always pronounced the word (and heard my colleagues pronounce it) /lɪˈtəʊti:z/, i.e. with a short initial /-ɪ-/ and with the main stress on the second syllable. Clearly several pronunciations are current in standard English.

**2** *Meaning.* A rhetorical figure or trope meaning 'assertion by means of understatement or negation'. It has been a marked feature of English from earliest times. *Beowulf*, for example, has many instances: e.g. *gūðwērig*, lit. 'war-weary' but by litotes 'dead'; *nō þæt ȳðe byð tō beflēonne*, lit. 'that is not easy to escape from', by litotes 'that is impossible to escape from'. In 1 Cor. 11: 17, 22 *I praise you not* contextually means 'I blame you'. Typical modern examples include: *not a few*, a great number; *it was nothing* (used by a successful performer, examinee, etc., to downplay the achievement); *not unwelcome* (a polite way of saying 'very welcome'); *not uncommon*, quite common; *not bad, eh?*, (after an anecdote or mild boast), 'excellent'. *Litotes* is the opposite of HYPERBOLE or overstatement. Cf. also MEIOSIS.

**litre** is the BrE spelling, *liter* the AmE one.

**litter.** See FARROW.

**littoral.** The adj. means 'of or on the shore of the sea, a lake, etc.' The corresponding noun means 'a region lying along a shore'. They (esp. the adj.) are mostly used in oceanography, contrasted with ABYSSAL, *benthic*, and *pelagic*: *The littoral zone lies between the mean and low tide lines, also called the 'intertidal' zone. The sublittoral zone extends from below the mean low tide line to the edge of the continental shelf* ... —*Cambridge Encycl.*, 1990. The word is sometimes used in ordinary non-technical contexts as a formal substitute for *coastal*: *The Innuit of littoral Alaska*—W. J. Hoffman, 1895; *In a shallow, dry ford not far from this littoral road there were boulders in the sugar-cane*—V. S. Naipaul, 1987.

**liturgy.** See LITANY.

**liveable.** Spelt thus (in OUP house style). Cf. -ABLE, -IBLE 2.

**-lived.** In *long-lived, short-lived* the second element is pronounced /-lɪvd/, with a short stem vowel, in BrE, but usually with a long vowel in AmE. The pronunciation with a short stem vowel is presumably a product of the 20c. as the *OED* (entries of 1903 and 1914 respectively) recommends /-laɪvd/ for both words.

**livelong.** See LIFELONG.

**liven** (verb), to brighten, cheer, a late 19c. word (first recorded in 1884), and also the extended form *enliven* (17c.), are both formed on *life* noun + -*en* suffix, and therefore do not belong with the main group of -EN VERBS FROM ADJECTIVES.

**livid.** Dictionaries list the senses (first recorded in the 17c.) (*a*) of a bluish leaden colour, and (*b*) discoloured as by a bruise. But in practice the meaning most likely to be encountered now in everyday life (still marked *colloq.* or *informal* in some dictionaries) is 'furiously angry' (first recorded in 1912). *Livid* is derived from L *līvidus* 'slate-coloured; discoloured by bruises or chafing'. The corresponding L. verb *līvēre* was used in the literal sense 'to be discoloured', but also in the transferred sense 'to be envious or jealous'. Examples showing various 20c. senses: ( = furiously angry) *Father was livid then and shouted that he, Adrian, should think twice before being cheeky to his father*—T. Parks, 1985; *She was furious with me for ruining her tax-free investment. Livid.*—*Wall St. Jrnl*, 1987; *Further communications ... culminated in Shelley's most livid screech of rage against his father*—R. Christiansen, 1988; (discoloured, of a scar) *A huge, livid, recently healed scar ran along the right side of his face*—P. Abrahams, 1985; (applied to the colours violet and red) *I got some livid violet fruit juice*—D. Johnson, 1986; *two features of the book rather jar: a livid red is used for headings and underlinings, and the photographs are very dull*—*New Scientist*, 1989.

**living-room.** See SITTING-ROOM.

**llama.** See LAMA.

**-ll-, -l-. 1** Words in final -*l* are usu. treated differently from those ending in most other final consonants in that, in BrE, in inflected forms, the -*l* is doubled irrespective of the position of the accent. By contrast, in AmE, final -*l* is usually left undoubled in such circumstances. For details, see DOUBLING OF CONSONANTS WITH SUFFIXES 2. Exceptions: before -*ish*, -*ism*, and -*ist*, final *l* is not doubled (e.g. *devilish, liberalism, naturalist*); and before -*able*, the -*l* is usually doubled both in BrE and AmE (e.g. *annullable, controllable, distillable*).

**2** The 20c. has witnessed much shuffling and reshuffling of -*l* and -*ll* in the simple form of a good many verbs. A fairly safe guide is to use *annul, appal, befall, distil, enrol, enthral, extol, fulfil, install,* and *instil* in BrE, and *annul, appall, befall, distill, enroll, enthrall, extol, fulfill, install,* and *instill* in AmE. But, truth to tell, no firm rule can be applied. All of these words exc. *annul* and *befall* may turn up with either -*l* or -*ll* in good sources in either country.

**3** Before -*ment,* the usual spellings are *annulment, enrolment, enthralment, extolment, fulfilment, instalment,* and *instilment* in BrE; and *annulment, enrollment, enthrallment, extolment, fulfillment, installment,* and *instillment* in AmE. But forms with single -*l*- are often found in good AmE sources for the words listed here with -*ll*-.

**4** The doubling rule does not apply when the *l* is preceded by a double vowel or a vowel + consonant, e.g. *ai, ea, ee, oi, ow, ur,* as in *failed, squealed, peeled, boiled, howled, curled.* Another exception is *paralleled, -ing.*

**5** Derivatives and compounds of words ending in -*ll* sometimes drop one *l*: so *almighty, almost, already,* ALRIGHT, *altogether, always; chilblain, skilful, thraldom, wilful.* The AmE equivalents of the last four are *chilblain, skillful, thralldom, willful* (the last three alternating with forms in -*l*-).

**6** See DULLNESS, FULLNESS.

**Lloyd's,** society of underwriters in London (hence *Lloyd's list, Lloyd's Register*), not *Lloyds* or *Lloyds'.* But *Lloyds Bank* (no apostrophe).

**loadstar.** See LODESTAR.

**loadstone.** See LODESTONE.

**loaf.** The pl. of the noun is *loaves.* The third person sing. pres. indic. of the verb is *loafs.*

**loan** (verb). The types *we once loaned a Protestant lady a pamphlet by an eminent Catholic divine,* i.e. *loan* used as a ditransitive verb, and *the stalls are barrack chairs loaned for the occasion,* i.e. *loaned* used as pa.pple in a passive construction, were well established in standard English in the 19c. The *OED* placed no restriction on their currency. Fowler (1926), on the other hand, declared that the verb *loan* 'has been expelled from idiomatic southern English by *lend,* but was formerly current, and survives in U.S. and locally in U.K.' *OMEU* (1983) gave the verb limited acceptance: 'has some justification where a businesslike loan is in question, e.g. *The gas industry is using a major part of its profits to benefit the PSBR by loaning money to Government.* Otherwise it is a needless variant for *lend.*' *WDEU* (1989), after citing a miscellaneous set of eight examples mainly from American sources, asserts that '*loan* as a verb is entirely standard . . . Its use is predominantly American and includes literature but not the more elevated kinds of discourse.'

From the evidence available to me, it would appear that Fowler's verdict is the right one. The verb is principally used abroad. Examples: *Delaney told him he could loan him $50 a week*—T. Keneally, 1985 (Aust.); *I loaned her a copy of my Depression: The Way out of Your Prison*—D. Rowe, 1987 (Aust.); *the . . . problem was how to stretch the small amount of money he had been loaned by Herr Pfuehl*—A. Desai, 1988 (India); *friends to whom I had loaned Doctor Zhivago*—*Bull. Amer. Acad. Arts & Sci.,* 1990; *The Development Bank of Southern Africa . . . loaned R650-million for the purpose*—*Panorama,* 1991 (SAfr.). But it is occasionally used in BrE, sometimes in rather surprising sources (see 1988 example): *The coach was restored . . . and is loaned now to charity organisations to help with fund-raising*—*Lancaster Guardian,* 1987; *I wasn't sure that Peewee would be amenable to loaning me his violin again*—A. Lively, 1988; *It is part of the Christ Church plate in the treasury showcase, which also displays many items loaned by parishes within the diocese*—*Christ Church Oxford: A Pitkin Guide,* 1991. The 1991 example illustrates a particularly common use of the verb, that involving temporary lending of a work of art, etc.,, by one institution to another.

**loanword.** Now usu. written as one word (no hyphen, not two words).

**loath** /ləʊθ/, averse. Thus spelt in OUP house style, not *loth.* It is to be carefully distinguished from the verb *loathe* /ləʊð/, to hate, which, apart from its different spelling, has a fully voiced *th.* The *-oa-*

spelling is obligatory in *loathly* and *loath-some*. An erroneous example of *loathe* used for *loath*: *The young seventies family man ... was loathe to purchase the same car that his fifty-five-year-old father might own*—B. Elton, 1991.

**locale.** This noun meaning 'a scene or locality, esp. with reference to an event or occurrence taking place there' (*COD* 1995) was adopted in the 18c. from French in the form *local*, and respelt in the early 19c. (Walter Scott and others) as *locale* to indicate that the stress lay on the second syllable, /ləʊˈkɑːl/. Cf. MORALE.

**locate.** Recorded first in the American colonies in the early 17c., *locate* has retained a slight American flavour until the present day. The intransitive use to mean 'to establish oneself or itself in a place, to settle' (e.g. *numerous industries have located in the area*), for example, is more or less confined to N. America. In this and other uses in AmE the stress normally falls on the first syllable, /ˈləʊkeɪt/, as against second-syllable stressing in BrE.

The same geographical restriction applies to the intransitive use of *relocate* ( = to resettle), but there are signs that it is being used in BrE: '*We have received a threat, Dr Mallory, and I am advising your colleague,*' *he nodded contemptuously at Smith, 'to relocate'*—R. McCrum, 1991.

Various uses of the verb have become fairly well established in other English-speaking areas though usually in competition with more familiar synonyms. The central senses in Britain are (*a*) to discover the exact place or position of (*to locate the enemy's gun emplacements; She had located and could usefully excavate her Saharan highland emporium*—M. Drabble (*OMEU*)); and (*b*) in the passive, to be situated (*a supermarket located in the outskirts of the city*). Sense (*b*) is the less favoured of the two.

**locative.** In languages expressing relationships by means of inflections, the case of nouns and pronouns expressing location; cf. L *locus* 'place'. Many IE languages have or had such a case, typified by L *domi* 'at home', *rūrī* 'in the country', *Romae* 'in Rome', *Corinthi* 'at Corinth'. Modern English does not have a locative case, the locative idea being introduced by a preposition (*at, in, on*).

**loch, lough** (both pronounced with a guttural final sound as /lɒx/ or in Anglicized form as /lɒk/), respectively the Scottish and Irish words for 'lake'.

**locomote.** In origin (early 19c.) a back-formation from *locomotion*, meaning 'to move about from place to place', it has little currency in general use, but has been adopted or re-invented in biological use. Leucocytes, fibrolasts, and the like, in technical contexts, are often described as *locomoting*.

**loculus,** in zoology and botany 'a small separate cavity'. Pl. *loculi* /-laɪ/.

**locum tenens.** An example of the usual morass where the pronunciation of Latin words and phrases is concerned. All the locums I have ever met have pronounced the phr. /ˌləʊkəm ˈtenenz/, with a short /e/ in the first syllable of *tenens*. But standard dictionaries tend to give precedence to /ˈiːnenz/. In practice, though, such a person is usu. just called a *locum*. The pl., if it should be needed, is *locum tenentes*; and the position of such a deputy is a *locum tenency* (not -ancy).

**locus.** Usu. pronounced (e.g. in mathematics and biology, and in many legal phrases) as /ˈləʊkəs/, with pl. *loci* /ˈləʊsaɪ/. In the phr. *locus classicus* it is as frequently pronounced /ˈlɒk-/, i.e. with the short *o* of the Latin original. The pl. is *loci classici*, with both words ending in /-saɪ/.

**lodestar.** The recommended spelling, not *load-*.

**lodestone.** The recommended spelling, not *load-*.

**lodgement.** Thus spelt, not *lodgment*, except that the latter spelling is more usual in AmE. Cf. JUDGEMENT.

**logan.** Pronounced with a long *o* in *loganberry* /ˈləʊɡənbərɪ/, AmE /-ˌberɪ/, but with a short *o* in *logan-stone, loggan-stone* /ˈlɒɡənstəʊn/, a rocking-stone.

**logaoedic** /ˌlɒɡəˈiːdɪk/. 'A term applied by classical grammarians to lyric metres which apparently had a trochaic base, but contained also feet of dactylic and spondaic form, as in Alcaics and Sapphics' (Egerton Smith). The word is derived from Gk λόγος 'speech, prose' + ἀοιδή 'song' (as standing between the rhythm of prose and of poetry).

**loggia** (gallery, arcade). Pronounce /ˈlɒdʒ(ɪ)ə/. Pl. *loggias* or (like the It. original) *loggie* /ˈlɒdʒeɪ/.

**logion** /ˈlɒɡɪɒn/, a saying attributed to Christ. Pl. *logia* /lɒɡɪə/.

**-logue.** Principally **1** forming nouns denoting talk (*dialogue*, *prologue*) or compilation (*catalogue*).

**2** -*ologist* (*ideologue*). The equivalent AmE words are often, but not always, spelt with final -*og* (*analog*, *catalog*, *dialog*, etc.).

**-logy,** a combining form which normally has *o* as the combining element: thus *archaeology*, *tautology*, *zoology*, etc. The principal exceptions are *genealogy* (answering ult. to Gk γενεᾱλογίᾱ 'tracing of descent') and *mineralogy* (from *minera(l)* + -*logy*). The erroneous forms *geneology* and *minerology* are all too prevalent in the speech of otherwise standard speakers. Cf. -OLOGY.

**lollop.** The inflected forms are *lolloped*, *lolloping*. See -P-, -PP-.

**Lombard(y).** The standard pronunciation of the first element is now /ˈlɒm-/. The former common variant /ˈlʌm-/ seems to have become much less favoured.

**lone,** an adj. that can only be used attributively (*a lone man*, *a lone wolf*, etc., never *the man was lone*, etc.). It is often used poetically and rhetorically, sometimes, as in Browning's line *We stepped O'er the lone stone fence* (1864), with the sense 'lonely, unfrequented'. See ADJECTIVE 2.

**lonelily.** An awkward-sounding adv. formed on the adj. *lonely*. It was good enough for Matthew Arnold (*The weird chipping of the woodpecker Rang lonelily and sharp*, 1852), but in ordinary prose contexts and in speech it is best replaced by the type 'in a lonely fashion'. See -LILY.

**long.** The adv. *long*, preceded and followed by *as*, i.e. *as long as*, can mean **1** 'during the whole time that, for an extensive period of time' ('*Were they happy together?' I asked. 'Very,' he said. 'As long as it lasted.'*—B. Rubens, 1985; *You can come out and spend as long as you like there, any time*—N. Gordimer, 1987); or

**2** 'provided that, only if' (*she liked new clothes as long as they were well designed and functional*). *So long as* is also used in sense 2 (*What cared she so long as her husband was near her?*—Thackeray, 1847/8; *It will stay here. So long as nobody disturbs it*—N. Gordimer, 1987). In sense 2 the reduced form *long as* is starting to make inroads into the longer form, but only in very informal contexts: '*It's all right,' he said, 'long as you are here.'*—G. Greene, 1938; *I reckon they* [sc. black people] *figure long as they can't be buried here, why should they* [*water the place*]—*New Yorker*, 1990.

**longest word.** The innocent question 'What is the longest word in the English language?' can only be sensibly answered by saying that the longest words in the largest dictionary of English, namely the *OED*, are *pneumonoultramicroscopicsilicovolcanoconiosis* (a factitious word, first recorded in 1936, of 45 letters alleged to mean 'a lung disease caused by the inhalation of very fine silica dust'); *supercalifragilisticexpialidocious* (34 letters, 'fantastic, fabulous', 1949); *floccinaucinihilipilification* (a humorous word of 29 letters invented by William Shenstone in 1741 meaning 'the action or habit of estimating as worthless'); and *antidisestablishmentarianism* ('opposition to the disestablishment of the Church or England', 28 letters, early 20c.).

**longevity.** Pronounce /lɒnˈdʒevɪtɪ/, not /lɒŋˈɡ-/.

**longitude.** The pronunciation recommended is /ˈlɒndʒɪtjuːd/ rather than /ˈlɒŋɡɪ-/. The form *longtitude* (after *latitude*), with the first *t* fully pronounced, is illiterate.

**long-lived.** See -LIVED.

**long variants.** Fowler (1926) argued at length that 'The better the writer, the shorter his words', though he admitted that the 'statement needs many exceptions for individual persons and particular subjects'. He set down three main types of words in contrastive pairs: **1** where two forms or close relations of the same word are equally available and the longer form is chosen.

**2** where a short form exists and a longer form is made by the addition of a suffix such as *-atable, -en,* or *-ize.*

**3** where a longer form exists but bears a different meaning.

Examples of types 1 and 2 include (with the date of first record indicated): *administer* 14c./*administrate* a1639; *assertive* 1562/*assertative* 1846; *contumacy* 14c./*contumacity* 15c.; *cultivable* 1682/*cultivatable* 1847; *damp* (verb) 1564/*dampen* c1630; *doubt* 13c./*dubiety* c1750; *educationist* 1829/*educationalist* 1857; *epistolary* 1656/*epistolatory* 1715 ('erroneous formation', *OED*); *experiment* (verb) 1481/*experimentalize* 1800; *perfect* (verb) c1449/*perfectionize* 1839 ('rare', *OED*); *preventive* 1639/*preventative* 1654–66; *quiet* (verb) 1526/*quieten* 1828.

Most of these words are discussed at their alphabetical place. It is more or less self-evident that the first words of the pairs are usually to be preferred, though *dampen, dubiety, preventative,* and *quieten* have wide currency at the present time. What stands out is that the first words of the pairs are all recorded earlier than their contrasted forms, a fact that draws attention to the fluctuating fortunes of various suffixes over the centuries. What feels comfortable in one century does not necessarily feel as comfortable in another.

Examples of type 3 include: *advance* (noun)/*advancement*; *alternate/alternative*; *correctness/correctitude*; *definite/definitive*; *distinct/distinctive*; *estimate* (noun)/*estimation*; *intense/intensive*; *prudent/prudential*; *reverent/reverential*; *simple/simplistic.* The majority of these are discussed under the words in their dictionary place except those that are too familiar to need discussion.

Fowler listed a fourth type in which he preferred the longer form: *pacifist* 1906/*pacificist* 1907; *quantitive* 1656/*quantitative* 1581; *authoritive* (not listed in any dictionary)/*authoritative* 1605; *interpretive* 1680/*interpretative* 1569. Of these the longer forms are still recommended, except for *pacifist* (which is now the only customary form of the word).

**longways, longwise.** Both words have been in the language since the 16c., but both are now in retreat before the now commoner *lengthways, lengthwise,* both of which have also been in use since the 16c.

**loo.** See TOILET.

**look** (verb).

> 1 Non-standard uses.
> 2 *look + to*-infinitive.
> 3 *look + adverb.*
> 4 *look + adjective.*
> 5 *look like.*

Several uses call for comment.

**1** Non-standard uses. First, we should notice some regional and colloquial (chiefly AmE) uses. For example, *look you* (also variants), found in Fluellen's vocabulary in Shakespeare's *Henry V* and still a favourite way of bespeaking attention in Wales: *Look you! Plaid Cymru protested to the BBC yesterday over the timing of its only party political broadcast—Daily Mail,* 1974; *Well look now. It's all very well for you to say that I am not the Marion you are looking for* [the speaker is Welsh]*—B. Rubens,* 1985. Secondly, the predominantly AmE *looky here* and variants (the spelling reflects the level of speech being represented), a folksy extension of the imperative phr. *look here!* Examples: *Looky here! Ye'd oughtn't t' said that, Eben—E. O'Neill,* 1925; *Looka-here, Dr. Hare, I don't have a picture at this time—Black Scholar,* 1971; *'Hey, Lewis!' Bobby sits straight up. 'Looky there! Get it!'—Lee Smith,* 1983; *'Looky here, looky here,' he chanted. The tall SP man stared up at the photograph—Brian Moore,* 1987. Thirdly, the colloquial AmE form *lookit* (first recorded in 1917), used only in the imperative, with the meaning 'Listen!' or 'Look at (something)': *Oh, isn't that the classiest, darlingest little coat you ever saw! ... Lookit the collar. And the lining!—T. Dreiser,* 1925; *It's not even killing the lousy jungle, lookit. It's bringing it alive!—D. Bloodworth,* 1972; *And lo, look-it there ... a woman playing two machines at once—E. Leonard,* 1985.

None of these uses belongs in standard English; they are further evidence of the drip-by-drip way in which a common verb partly changes its nature in regions outside England.

**2** *look + to*-infinitive. This type, meaning 'to expect', has been in continuous use since the 16c. but is beginning to fall into disuse. Examples: *In these last wordes that euer I looke to speake with you—* Thomas More, c1513; *By whom we look* to be protected—Hobbes, 1651; *Two lovers looking to be wed*—A. E. Housman, 1896; *I shall hereafter look to be treated as a person of respectability*—T. Huxley, 1900. It is to be distinguished from the increasingly common type (though resisted by many standard speakers) *looking + to-*infinitive + a direct object. It is a logical extension of the earlier type, but it injects the notions of hope and intention. Examples: *I'm looking to the government to help to solve the problem*—oral source, 1987; *most people are looking to improve their living standards*—New Statesman, 1988; *The home team will be looking to get a result against the visitors next Saturday*—Times, 1988; [I] *am looking to finish this today*—COD, 1990. A third use of *look + to-*infinitive lies within the branch of meaning that the OED defines as 'to have a certain appearance'. The OED (sense 9c) illustrates it from works by Edmund Burke (1775) and other later writers. It means 'to seem to the view, to appear, to look as if'. Examples: *A little hat that looked to be made of beaver*—Clark Russell, 1890; *The Queen looked to be in good health*—Graphic, 1893; *The owl looked to be encircled by six cloaked hit-men*—J. E. Maslow, 1983. This type seems to be gaining ground, esp. in America.

**3** *look + adverb.* One must carefully distinguish what was formerly acceptable, indeed commonplace, from what is idiomatic now. The OED (senses 1c and 9a, b) lists numerous examples of the type (in which *look* is intransitive) used in the senses (a) 'to direct one's eyes in a manner indicative of a certain feeling', and (b) 'to have the appearance of being; to seem to the sight'. Examples: (a) *Where-fore looke ye so sadly to day?*—Gen. (AV) 11: 7, 1611; *The man look'd bloodily when he spoke*—R. Carpenter, 1642. (b) *The skies looke grimly*—Shakespeare, 1611; *The world looked awkwardly round me*—Defoe, 1719; *On the whole ... things as yet looked not unfavourably for James*—Macaulay, 1849. The only adverbs that can still idiomatically be used in sense (b) are *ill* and *well* (e.g. *you look well* = you appear to be in good health). In the ordinary sense 'to use one's sight', *look* is, of course, idiomatically used with adverbs: *he looked closely at the evidence; she looked longingly in his direction.*

**4** *look + adjective.* In the sense 'to seem to the sight', this is the normal type: e.g. *to look black, blue, cold, elderly, foolish, small, vain,* etc.

**5** *look like.* Time need not be spent on everyday constructions such as *he looks like a proud man,* where *look* means 'to have the appearance of being'. But *look like* used to introduce a clause (a construction first used in AmE) lies at the rim of standard English, whatever its credentials abroad. Examples: *Don't it look to you like she would of asked us to stay for supper?*—M. Mitchell, 1936; *Looks like your child's birthday is news again this year*—Guardian, 1973; *On top of the wooden beams were loops of thick metal cable that looked like what you would use if you were going to hang the Statue of Liberty*—New Yorker, 1989; *It makes you look like you're late*—H. Hamilton, 1900. Further examples, as well as of *act like* and *feel like* used in a similar manner, are given s.v. LIKE 1.

**loom** (verb). For *loom large,* see LARGE, LARGELY.

**loose, loosen** (verbs). The main distinctions between the two verbs, both of which came into the language in the ME period, are that *loose* means principally (a) to release, set free, (b) to untie or undo (something that constrains), and (c) (usu. with *off*) to let off (a gun, etc.); whereas *loosen* is mainly used to mean 'to make or become looser, less tight'. Standard speakers instinctively choose the right word: e.g. an arrow is always *loosed* (not *loosened*) by an archer; and knots, discipline, etc., are *loosened* (not *loosed*).

**loquitur** /ˈlɒkwɪtə/. This Latin word means 'he or she speaks'. With the speaker's name following, it is used as a stage direction or as a guide to the reader: e.g. *I drive out to my vineyard on weekends* (loquitur *a San Francisco lawyer*). If the pl. should be needed it is *loquuntur.* Cf. EXIT.

**Lord.** The younger sons of a duke or a marquess are known by the courtesy title of *Lord,* followed by their Christian name and surname, e.g. *Lord Edward Fitzgerald.* A letter addressed to such a person should begin *Dear Lord Edward,* but the surname should be added if the acquaintanceship is slight (Debrett's *Correct Form*). For *lord* used as an undress substitute for *marquess, earl, viscount,* see TITLES.

**Lord Bacon.** An incorrect mixture of styles bestowed on Francis Bacon (1561–1626) by Macaulay in an essay that appeared in the *Edinburgh Review* in July 1837. The permissible styles are *Bacon, Francis Bacon, Lord Verulam,* and *Viscount St Albans.* In literary criticism the only styles used are the first two.

**lordlily.** The technically available adverb corresponding to *lordly* (adj.), but, because of its clumsiness, it is hardly ever used. See -LILY.

**lordolatry.** See -LATRY.

**Lord's** is the name of a cricket ground in London. *Lords* (preceded by *the*) is a conventional shortening of *House of Lords,* the Upper House of Parliament in Britain.

**lose out.** From the mid-19c. used in AmE to mean 'to be unsuccessful, not to get a fair chance', this phrasal verb continues to be viewed with suspicion by standard speakers in Britain. It turns up typically in quickly written journalistic sources: e.g. *We are going to lose out unless the Government are prepared to do a tremendous public relations job for the tourist industry—Guardian,* 1971. But it is robustly rejected by many people. 'Why, in England, is *lose* now losing to *lose out?*' asked a correspondent in 1988. I could not find a convincing reason except to see it as part of a 19–20c. process of stretching the language by increasing the number of slightly risky phrasal verbs.

**loss, lost.** The pronunciation with /ɔː/, i.e. the sound used in *law,* which was standard in 1917 according to Daniel Jones, is now seldom heard. The standard pronunciation now is with /ɒ/, as in *lot.*

**loth.** See LOATH.

**Lothario** /ləˈθɑːrɪəʊ/. Pl. *Lotharios.* See -O(E)S 8.

**lot, lots.** 1 *A lot of people say so/Lots of paper is wanted.* See NUMBER 6(b).

2 While it is perfectly true that *a lot of* and *lots of* occur in the works of well-known writers such as Galsworthy and Winston Churchill, the main contexts in which one finds them (esp. *lots of*) are in plain workaday sentences or conversations. A selection of examples from my database (all from standard sources): (lot) *You'd start to see a lot of sense; a lot of people are concerned about land use; something worth a lot of money, that's for sure; Afterwards, there'll be a lot of clearing up to do;* (lots) *it was quite nice being telephoned by lots of young ladies; I would have lots of flowers and plants in pots.* The informal nature of both *a lot of* and *lots of* is underlined by the presence in many 20c. works of the contracted forms *lotta* and *lotsa*: (lotta) *No, I'm a lotta things, but I ain't crazy—Coast to Coast* (Aust.), 1969; *He did ... a lotta community theatre work in Houston—Ritz,* 1989; (lotsa) *The Notting Hill Carnival was lotsa fun for seven days and nights—It,* 1971.

3 See ALOT.

**lotus.** Pl. *lotuses.*

**loud, loudly.** In certain restricted circumstances *loud* may be used as an adverb: e.g. *he laughed loud and long; he screamed all the louder; she spoke loud enough for the whole class to hear;* (in radio, etc., communication) *I hear you loud and clear.* But the customary adv. is *loudly*: e.g. *complaining loudly about his treatment; loudly dressed.*

**lough.** See LOCH.

**lounge.** See SITTING-ROOM.

**lour** (of the sky, etc.), to look dark and threatening; (of a person) to frown, look sullen. This verb, which is sometimes spelt *lower,* is pronounced /ˈlaʊə/, like *tower.* It is first recorded in the 13c. and is probably of native origin but happens not to be found in OE. The verb *lower* (to cause to descend) is pronounced /ˈləʊə/, rhyming with *mower,* and is etymologically unrelated to *lour;* it is a derivative of the adj. *low* (ME *lāh,* from ON *lágr*).

**louvre.** Thus spelt in BrE (AmE *louver*).

**lovable.** Thus spelt, not *loveable,* in OUP's house style.

**love.** From its earliest use in the 16c. the phr. *to make love* (often followed by *to*) meant simply 'to pay amorous attention (to)'. At some point in the mid-20c. it came to have only the restricted and very precise meaning 'to have sexual intercourse', and the older meaning was forced out. When reading 16–19c. literature one must simply bear in mind that *making love* means no more than 'paying amorous attention': e.g. *Demetrius ... Made loue to Nedars daughter*—Shakespeare, 1590; *'who's had the* —*impudence to make love to my sister! cried Harry*—G. Meredith, 1861.

**lovelily.** Despite its antiquity (first recorded *c*1300) and its use by many excellent writers (e.g. *So lovelily the morning shone*—Byron, 1813), *lovelily*, the adverb corresponding to the adjective *lovely*, is now usually replaced by the type 'in a lovely manner', or by some other means. See -LILY.

**lovey.** A colloquial form of address to a loved one. So spelt (not *lovy*), thus conforming with the reduplicated form *lovey-dovey*.

**Low Countries.** See NETHERLANDS.

**lower, lour.** See LOUR.

**lowlily.** In use as an adverb (corresponding to the adj. *lowly*) since the 14c. (e.g. *Live and love*—*Doing both nobly, because lowlily*—E. B. Browning, 1844), but now comparatively rare. In ordinary prose it is better to express the same notion in another way, such as 'in a lowly manner', or the like. See -LILY.

**low, lowly.** The first can function either as an adj. (*a low neckline, a low moon*, etc.) or an adv. (*to aim low, to lie low*, etc.). The second is most commonly used as an adj. (*of lowly station, a lowly position in the firm*); as an adv. (= humbly, modestly; in a low voice; etc.) it has gradually retreated over the centuries into the safe havens of poetry and archaism.

**'low' words.** Dr Johnson occasionally used the word *low* in his Dictionary as an indication of the linguistic status of a word or meaning. For example, in a small section of the letter *S* we find the following entries: *To Sconce ...* 'To mulct; to fine. A low word which ought not to be retained.'; *Scrape ...* 'Difficulty; perplexity; distress. This is a low word.'; *To scuddle ...* 'To run with a kind of affected haste or precipitation. A low word.' *Sensibly* [four 'normal' senses, including 'perceptibly to the senses'] 5. 'In low language, judiciously; reasonably.' In broad terms, 'low' seems to have meant 'not used in the circles in which Johnson

moved'. Such uses are to be carefully distinguished from other unrecommended types, e.g. *Scelerat*, 'A villain; a wicked wretch. A word introduced unnecessarily from the French by a Scottish author.'; *Sciomachy ...* 'Battle with a shadow. This should be written *skiamachy*.'; *Scom ...* 'A buffoon. A word out of use, and unworthy of revival.'

**lu** (pronunciation). A long-drawn-out battle between /lu:/ and /lju:/ (or with /ʊə/ instead of /u:/ when followed by *r*) in the relevant words continues, and the present stage is a very untidy one. The untidiness is reflected in the judgements made in *COD* 1990 (some of which have been changed in the 1995 edn): **1** (only /lu:/) *alluvial, blue, clue, conclude, exclude, fluent, fluke, flute, glue, lubricate, lucerne, lucid, Lucifer, lucrative, lucre, ludicrous, lugubrious, lukewarm, lunatic, lupin, recluse, sluice.*

**2** (only /lju:/) *aluminium, celluloid, dilute, prelude, salutary* /-jʊ-/, *salutation, volume.*

**3** (/lju:/ or /lu:/; before *r*, /ljʊə/) *illuminate, lure, lurid, voluminous.*

**4** (/lu:/ or /lju:/) *absolute, allude, collusion, delude, delusion, dissolute, evolution, illusion, interlude, luminous, lunar, lute, Lutheran, salubrious, salute.* In general terms, it would seem from the words listed in 1 and 4 that /lu:/ is dominant but is far from universal. Standard practice in BrE is fairly represented in the above lists; but in AmE forms without /j/ are customary in all the above words except *celluloid, prelude, salutary*, and *volume.*

**lucre** (derogatory word for 'financial profit or gain'). Spelt thus both in BrE and in AmE (not *lucer*).

**luggage.** See BAGGAGE.

**luncheon** is a formal variant of *lunch*. It is of interest to note that the *OED* remarked in 1903: '*lunch* ... Now the usual word exc. in specially formal use, though formerly objected to as vulgar.' As a minor curiosity, however, *luncheon* is obligatory in the phrase *luncheon voucher*.

**lunging,** pres.pple of the ordinary verb *lunge.* Thus spelt, not *lungeing*. But for the separate verb *lunge* 'to train or exercise a horse with a lunge (a long rope)', use *lungeing* as the *-ing* form.

**lustre.** Thus spelt in BrE (AmE *luster*).

**lustrum** (period of five years). Pl. *lustra* (less commonly *lustrums*). See -UM 3.

***lusus naturae,*** a freak of nature. Pronounce /ˈluːsəs nəˈtjʊəriː/ or /-ˈtjʊəraɪ/.

**luxuriant, luxurious.** Because of their connection with *luxury* and with the verb *luxuriate*, these two adjectives have invaded each other's semantic space many times during and since the 17c. In strict terms *luxuriant* is primarily used of vegetation and means 'profuse in growth', as R. M. Ballantyne showed in his *Coral Island* (1858): *the trees and bushes were very luxuriant*. *Luxurious*, on the other hand, means 'supplied with luxuries, making a display of luxury', something that is brought out in the same book of Ballantyne's: *Altogether this was the most luxurious supper we had enjoyed for many a day*. Keeping the two words apart has proved to be difficult. The *OED* lists (a) 17–19c. examples of *luxuriant* 'misused' for *luxurious*, e.g. *It was a splendid apartment … luxuriant to a degree*—C. Gibbons, 1885; and (b) 17–19c. examples of *luxurious* used for *luxuriant*, e.g. *Their villages are situated in the midst of the most luxurious groves*—R. Southey, 1826. The best policy is to endeavour to keep the two words apart as Ballantyne did, but not to be too surprised if a particular context draws a writer away from strict conformity with this rule. It should be noted that in the attributive position, *luxury* is now tending to oust *luxurious* (e.g. *luxury apartment, coach, cruise, flat*, etc.).

**-ly. 1** For a tendency among some writers and speakers to suppose that all adverbs must end in -*ly*, and therefore to use *hardly, largely, strongly, tightly*, etc., see UNIDIOMATIC -LY.

**2** For participial adverbs like *determinedly*, see -EDLY.

**3** For *first(ly)*, see FIRST 4.

**4** Considerations of euphony and meaning make it desirable to avoid placing two -*ly* adverbs in succession when their function in the sentence is different. Thus *We are utterly, hopelessly, irretrievably, ruined* is acceptable because each of the -*ly* adverbs has the same relation to *ruined*. But the following show slightly uneuphonious contiguities: *Many of the manuscripts were until comparatively recently in the keeping of Owen's family English*, 1987; *he reverts to it (apparently disbelievingly) on several occasions*—Encounter, 1987; *Appearing relatively recently, Kyra represented change*—New Yorker, 1987. The cruder type *Soviet industry is at present practically completely crippled* is avoidable by substituting *almost* for *practically*.

**5** See also CLEAN, CLEANLY; CLEAR, CLEARLY; CLOSE, CLOSELY; QUICK.

**Lyceum.** For the meaning in Greek philosophy, see ACADEMY. Pl. *-eums*. See -UM 1.

**lychee** (a sweet fleshy fruit) /ˈlaɪtʃiː/. The dominant spelling in BrE at present for this word of Chinese origin is *lychee* rather than *lichee* or *litchi*. In AmE the forms *litchi* and *lichee*, both pronounced /ˈliːtʃiː/, seem to be dominant.

**lychgate.** See LICHGATE.

**lyric,** 'in the modern sense, any fairly short poem expressing the personal mood, feeling, or meditation of a single speaker (who may sometimes be an invented character, not the poet). In ancient Greece, a lyric was a song for accompaniment on the lyre and could be a choral lyric sung by a group … Among the common lyric forms are the sonnet, ode, elegy, … and the more personal kinds of hymns' (C. Baldick, 1990). From about the third quarter of the 19c., but more particularly since the 1920s, *lyric* has also been the customary term for the words of a popular song or a musical.

**lyric, lyrical** (adjs.). The longer form is often interchangeable with the shorter one (e.g. *lyric/lyrical poet, poetry, verse*), and is used in Wordsworth's title *Lyrical Ballads* (1798). But perhaps its most common use is in the colloquial sense 'highly enthusiastic' (*he waxed lyrical about the performance*).

# M m

**ma'am.** See MADAM.

**macabre.** Pronounced /mə'kɑ:brə/, but the final element is often realized as /-bə/ in rapid speech. Cf. CALIBRE.

**macaroni.** In current English it has only one sense (a tubular variety of pasta) and no plural form. In the historical (18c.) sense 'a British dandy affecting Continental fashions', its plural is *macaronies*.

**Machiavelli.** This is now the standard spelling in English of the name of the 16c. Florentine statesman and political writer (formerly also *Machiavel*, esp. in transferred use meaning a plotter or scheming person). The only form of the corresponding adj. is *Machiavellian*, and of the abstract principle, *Machiavellianism* (formerly also *Machiavellism*).

**machicolated.** Pronounce /mə'tʃɪkəleɪtɪd/, not /mə'kɪ-/.

**machination.** Pronounce /'mækmeɪʃən/, not /'mæʃ-/.

**machismo,** a show of masculinity, is pronounced /mə'tʃɪzməʊ/ or /mə'k-/. The word is still at an early stage of naturalization.

**macho** /'mætʃəʊ/, AmE /'mɑ:tʃ-/ may be used as a synonym of MACHISMO, but is more commonly used as an adj. meaning 'showily manly or virile'.

**mackerel.** Pronounce /'mækrəl/.

**mackintosh,** a waterproof coat. Now so spelt (patented by Charles Macintosh, Scottish chemist, 1766–1843). The shortened form *mac* (no point) is often used instead.

**Mac-, Mc-.** The major English dictionaries (and other works of reference) place all such names, whether they are spelt with *Mac-* or *Mc-*, at the notional alphabetical place of *Mac*: thus *Maccabees*, *McCarthyism*, *mace* and *macle*, *McNaughten rules*, *macramé*. By contrast, some American dictionaries list *Mac-* names s.v. *Mac*, but *Mc*. names between *Mb* and *Md*. In a standard American dictionary published in 1991, the entry for James Ramsay *MacDonald* is 26 pages away from that for *McCarthyism*.

**macula** /'mækjʊlə/. Pl. *maculae* /-li:/.

**mad** (adj.). The primary sense 'suffering from mental illness' has been in continuous use since OE (though it is now a lay term, not used by medical practitioners), but has shared its semantic space with other meanings from the dates indicated: in the phr. *like mad* 'with excessive enthusiasm, etc.' (17c.); 'foolish' (*a mad undertaking*) (OE); (const. *about*, *on*, AmE also *for*) 'carried away with enthusiasm' (*mad about opera*; AmE *mad for Egyptian artifacts*) (17c.); 'angry' (in standard, now colloquial, use since about 1300).

**mad** (verb). The verb came into general use in the 14c. in both transitive ('to make mad') and intransitive ('to be, become, or act mad') senses, and survived unchallenged until the 18c. when *madden* came into the language to join other verbs in *-en* formed from adjectives (see -EN VERBS FROM ADJECTIVES). Since then *mad* has been in full retreat in BrE except in the phr. *the madding crowd* (which echoes the line in Gray's *Elegy*, 1749, *Far from the madding crowd's ignoble strife*). It remains in use in AmE, however, esp. in the transitive sense 'to make mad, exasperate'.

**madam, madame.** 1 In the English word used as a polite or respectful form of address or mode of reference to a woman (*Dear Madam*; *Can I help you, Madam?*; *Madam Chair(man)*) the form without *-e* is appropriate; also (in BrE) a colloquial term for a precocious or conceited girl or young woman (*a right little madam*); or as a term for a woman brothel-keeper. When addressing royalty the shorter form *ma'am* /mæm/ should be used.

2 *Madame*, pronounced /mə'dɑ:m/, pl. *Mesdames*, is the right form of address to a woman belonging to any foreign nation

(substituted e.g. for Ger. *Frau* or Du. *Mevrouw*), if the correct local term is unknown to the speaker or writer, e.g. *Madame Mao. Modame* is also used as a term for a woman brothel-keeper.

**3** The pronunciation /'mɒdəm/ used by assistants in fashion shops at the beginning of the 20c. now seems to be obsolete (except in jocular or period use). Fowler (1926) called it 'odd', and suggested that it was 'perhaps due to a notion that French *Madame* is more in keeping with haunts of fashion than English *Madam*.'

**madding.** see MAD (verb).

**Madeira.** So spelt.

**madness.** See IRRELEVANT ALLUSION.

**maelstrom.** Pronounce /'meɪlstrəm/.

**maenad,** a female follower of Bacchus, not *me-*. Pronounce /'miːnæd/.

**maestoso,** (in music) majestic, stately. Pronounce /maɪ'stəʊzəʊ/.

**maestro** /'maɪstrəʊ/. Pl. *maestri* /'maɪstri/ or *maestros* /-trəʊz/.

**Mafia** /'mæfɪə/, /'mɑː-/. Its central meaning is 'an organized international body of criminals, orig. in Sicily, now also in mainland Italy and in the US'. But it is also widely used of any group regarded as having hidden powerful (not necessarily sinister) influence (*the literary mafia, the academic mafia,* etc.). In this transferred sense a small initial (thus *mafia*) is used. A member of the *Mafia* is a *Mafioso* /mæfɪ'əʊsəʊ, mɑː-/, pl. *Mafiosi* /-sɪ/.

**Magdalen(e).** **1** In the name of the Oxford (*-en*) and Cambridge (*-ene*) colleges, pronounced /'mɔːdlɪn/.

**2** In the use as a noun meaning 'a reformed prostitute' or 'a home for reformed prostitutes', use *magdalen*, pronounced /'mægdəlɪn/.

**3** In the full biblical name *Mary Magdalene,* of Magdala in Galilee (Luke 8: 2), *Magdalene* is pronounced /ˌmægdə'liːnɪ/ or /'mægdəliːn/, /-lɪn/. Here I am quoting the preferences of the main authorities. My personal preference is /'mægdəlɪn/.

**maggoty.** So spelt. See -T-, -TT-.

**Magi.** See MAGUS.

**magic, magical** (adjs.). Until recently *magic* was normally used only in the attributive position (*magic carpet, magic circle, magic spell*) when the sense was the normal one of 'of or pertaining to magic'. In this position, except in fixed phrases like *magic lantern* or *magic square*, it vied with *magical* (*the magical word 'democracy'; a magical effect on the economy*). The words were not freely interchangeable in this position but no clear-cut rule can be stated. Perhaps *magical* tended to be used mainly in contexts of high approbation, and *magic* in contexts of enchantment, but the line between them has always been untidily drawn. In the second half of the 20c. *magic* has come to be used colloquially in the attributive position or by itself (*Magic!*) as a term of enthusiastic commendation, simply meaning 'wonderful, exciting', e.g. *It's magic!; The food* (or *concert, exhibition,* etc.) *was magic*. Doubtless, except in fixed phrases, the two words will continue to vie with each other in the years ahead, esp. in figurative, extended, and weakened senses, both in the attributive and the predicative positions.

**magma.** The pl. is *magmas*, no longer *magmata*. The corresponding adj. is *magmatic*.

**Magna Carta, Magna Charta.** The name of the famous charter of 1215 is pronounced /ˌmægnə 'kɑːtə/ however it is spelt. The recommended spelling of the second word is *Carta*, except in AmE, where *Charta* is possibly the dominant form. It is worth noting, as reported by Gowers (1965), that 'In a Bill introduced in 1946 authorizing the Trustees of the British Museum to lend a copy to the Library of Congress, *Charta* was the spelling used. But when the Bill reached committee stage in the House of Lords, the Lord Chancellor (Lord Jowitt) moved to substitute *Carta* and produced conclusive evidence that that was traditionally the correct spelling. The amendment was carried without a division; so *Carta* has now unimpeachable authority.'

**magneto** /mæg'niːtəʊ/ is a shortened form of *magneto-electric machine*. Its pl. is *magnetos*. See -O(E)S 5.

**magnifico,** a Venetian grandee. Pl. *magnificos*. See -O(E)S 6.

**magus** /'meɪgəs/. Pl. (as in *the three Magi*) *magi* /'meɪdʒaɪ/.

**Magyar.** It is instructive to recall that the only pronunciation given in the OED (1906) was /ˈmadʲɑr/ and that Fowler 1926 gave only /ˈmɒdjɑr/ (in his system mŏ´dyar), both approximately reproducing the Hungarian pronunciation of the word. The word has been Anglicized in the course of the 20c. and the only current pronunciation is /ˈmæɡjɑː/.

**maharaja** (hist.), an Indian prince, not -jah. With capital initial when used as a title. The wife of a maharaja was a maharanee.

**mahlstick.** See MAULSTICK.

**Mahomet.** See MUHAMMAD.

**mahout,** an elephant-driver or keeper in India. Pronounce /məˈhaʊt/.

**maieutic** /meɪˈjuːtɪk/ (adj.) (of the Socratic mode of enquiry), serving to bring a person's latent ideas into clear consciousness. The word is a 17c. adaptation of the Gk adj. μαιευτικός, from the verb μαιεύεσθαι 'to act as a midwife', from μαῖα 'midwife'.

**maiolica** /məˈjɒlɪkə/. A variant spelling of MAJOLICA. Example: In Italy interest in porcelain often takes second place to that of maiolica and Renaissance and Baroque ceramics—Antique Collector, 1990.

**majolica** /məˈdʒɒlɪkə/, also /məˈjɒl-/, Renaissance Italian enamelled earthenware. Often spelt maiolica (see prec.) to accord with the It. form.

**major.** In origin the comparative of L magnus 'great', major is sometimes used in contrast with minor (e.g. a major road in contrast with a minor road; the two major political parties and several minor ones), and as a synonym of greater (the major portion of his speech = the greater portion), but it is not a true comparative in English in that it cannot be construed with than. Its occasional comparative role is brought out, however, in the British public-school custom of appending major (or minor) to the surname of the elder (younger) of two brothers at a school (Smith major, Smith minor). Voices have been raised in the past against using major as an absolute adj. equivalent to main, principal, important, etc., but it is difficult to see how any of these near-synonyms could be idiomatically substituted for major in most contexts where major is used in a natural way (as in the first sentence in this article). Major may not be a true comparative in English, but it is implicitly one in a way that main, principal, etc. are not.

**major-domo,** chief steward of a great household. Pl. major-domos. See -O(E)S 6.

**major-general,** an officer in the British army next below a lieutenant-general. Pl. major-generals.

**majority.** 1 Three allied uses need to be distinguished: (a) A superiority in numbers (The amendment was passed by a bare, small, great majority; the majority was small but sufficient to enable the Conservatives to form a government). (b) The one of two or more sets that form the more numerous group (Any future Prime Minister to whom the donnish majority was prepared to give an honorary doctorate would be one whose policies found favour with them—Times, 1985; the vast majority have now come to terms with their destiny—Encounter, 1987). (c) majority of + plural noun (The great majority of people are now apparently conditioned to accepting the trades unions as a virtual fifth estate of the realm—Times, 1985; The majority of school buildings are dilapidated and decaying—Encounter, 1987; The vast majority of microbes ... play a role that is not only valuable but essential—Bull. Amer. Acad. Arts & Sci., 1991).

2 Number. After majority in sense 1(a) the verb will always be singular. After majority in sense 1(b), as after other nouns of multitude, either a singular or a plural verb is possible, according as the body is, or its members are, chiefly in the speaker's thoughts. After majority of in sense 1(c) the verb is invariably plural.

3 Uncountable nouns. A or the majority of cannot properly be followed by an uncountable noun: thus a (or the) majority of experience, forgiveness, interest, tolerance, work, etc., are all unidiomatic.

4 With majority in sense 1(a), great, greater, greatest are freely used, and cause no difficulty. With majority in sense 1(b) they are not often used, but vast is common enough (see quot. 1987 above). With majority of in sense 1(c), great or vast are possible and common, but greater and greatest are unidiomatic.

**5** It should be noted finally that in British elections *majority* means the number by which the votes cast for one party, candidate, etc., exceed those of the next in rank (*he won by a majority of 2,400 votes*). In AmE such an excess of votes is called a *plurality*, as distinguished from a *majority*, which is such circumstances is applied only to an absolute majority of all the votes cast.

**make.** Contrast the normal active construction of *make* with an object and plain infinitive (*she made him feel unwanted*) and the equally normal passive construction with a *to*-infinitive (*he was made to feel unwanted*).

**make-belief, make-believe.** The verbal phr. *to make believe* 'to pretend to do something' (formed in the 14c. on the model of OF *faire croire*) yielded the nouns *make-believe* and *make-belief* in the early 19c. *Make-believe* happens to precede *make-belief* by some 20 years and remains the more usual of the two nouns.

**Malapropisms.** Lying at the heart of the language is a tendency to confuse like-sounding words. Mrs Malaprop, in Sheridan's *The Rivals* (1775), when challenged about her use of *hard words which she don't understand*, protested *Sure, if I reprehend anything in this world, it is the use of my oracular tongue, and a nice derangement of epitaphs*. She was producing misshapen recollections of the words *apprehend*, *vernacular*, *arrangement*, and *epithets*. Fowler (1926) believed that such Malapropisms 'pass the bounds of ordinary experience and of the credible'. In his view, Malapropisms normally come as 'single spies, not in battalions, one in an article, perhaps, instead of four in a sentence, and not marked by her bold originality, but monotonously following well beaten tracks'. By contrast, Kingsley Amis, in an interview in *The Times* in 1992, declared that 'Half the wives he knows are Mrs Malaprops, saying "courgettes" for "couchettes", and getting clichés wrong. Hilly [his first wife] does it all the time ... "If you can manage that, Adrian, the world's your lobster." "When you went down to Wales for your *ignaurial*."' Whether it is a case of 'single spies' or 'all the time', it is not difficult to make a small collection of Malapropisms without going to very much trouble: e.g.

*One, a head of English, could not explain the function of an* intransigent [intransitive] *verb and advised me to 'forget it'*—letter in the *Sunday Times*, 1988; *She's a child progeny* [progeny], *a vivacious* [voracious] *reader*—*gets through 15 books a week*—oral source, 1989; *When she* [sc. our daily] *heard our Gloucester house was haunted, she uttered the immortal line, 'You'll have to get the vicar in to* circumcise [exorcize] *it.'*—J. Cooper, 1991.

**male. 1** Both *male* and *masculine* entered the language from OF (ult. from L *masculus*) in the 14c. and gradually adopted different roles more or less parallel to those of FEMALE and FEMININE. *Male* (adj.) is used (*a*) of the sex that can beget offspring by fertilization or insemination; (*b*) (of plants or their parts) containing only fertilizing organs; (*c*) of or consisting of men or male animals, plants, etc. (*the male sex*; *a male-voice choir*; *male kennels*); (*d*) (of parts of machinery, etc.) designed to enter or fill the corresponding female part (*a male screw*). As noun, 'a male person or animal'. *Masculine* is (*a*) applied to qualities associated with men, manly, vigorous (not merely describing the fact of being male); (*b*) in grammar, of words in certain languages (e.g. OE, L, Gk) of a particular class, as distinct from feminine or neuter ones. In the 20c., the terms *male chauvinist* (first recorded in 1970) 'a man who is prejudiced against women', and *male menopause* (1949) 'a crisis of potency, confidence, etc., said to afflict men in middle life' have emerged as part of the vocabulary of women's movements aiming to identify attitudes that have produced notions of male supremacy in the past.

**malignancy, malignity.** The first formally corresponds to the adj. *malignant* and the second to the adj. *malign*, but there the parallelism ends. They are far from coeval words and have been dragged this way and that by the fortunes of war and of medicine. *Malignancy* is a Latinate word brought into the language in the 17c. It was used at first principally to mean 'envenomed hostility, the desire to inflict injury and suffering'. During the Civil War it came to mean 'disaffection to rightful authority', and was used as a hostile designation for sympathy with the royalist cause. As time went on, under the influence of *malignant*, it gradually became applied particularly to medical conditions (fevers, cancerous growths, etc.) judged to be life-threatening, though it is still used in the general sense 'the state or quality of being malignant'. *Malignity* is much the older word of the two: it was first used in the 14c. to denote the condition or quality of being malign, malevolent, or deadly. It has continued down the centuries to mean 'wicked and deeprooted ill will or hatred'; and from the 17c. to the 19c., like *malignancy*, it was applied to malignant or life-threatening medical conditions, but has now given way in this sense to *malignancy*.

**malign, malignant.** The shorter form was adopted from French in the 14c. and has gradually retreated before the later word (first in 16c.) *malignant*. *Malign* is used mostly of things that are evil in their nature or effects, and occasionally of diseases (fevers, syphilis, etc.). *Malignant* is the normal word applied to diseases or conditions (esp. cancerous tumours) that are life-threatening. It is also used to designate any aspect of life that is characterized by intense ill will, or to a person or persons keenly desirous of the suffering or misfortune of another person, or of others generally. Cf BENIGN, BENIGNANT.

**malinger.** Pronounce /mə'lɪŋgə/.

**mall.** In the senses 'a sheltered walk or promenade' and 'an enclosed shopping precinct' the pronunciation /mɔːl/ is increasingly being used in Britain now rather than /mæl/, as it has been for some time in America, Australia, etc. But /mæl/ is still obligatory in the London placenames *The Mall*, *Chiswick Mall*, and *Pall Mall*.

**malnutrition.** 'A word to be avoided,' said Fowler in 1926, 'as often as *underfeeding* will do the work.' But where is *underfeeding* now? *Underfed*, on the other hand, is still the customary word for 'insufficiently fed', alongside *malnourished*.

**mama** /'mæmə/, /mə'mɑː/, a colloquial term (used esp. by children) for *mother*.

**Mameluke,** a member of the military class that ruled Egypt 1254–1811. Pronounce /'mæmǝluːk/.

**mamilla** /mæ'mɪlə/ (AmE *mammilla*), the nipple of a woman's breast. Pl. *mamillae* /-liː/.

**mamma¹** /'mæmə/. = MAMA. Also *momma*.

**mamma²** /'mæmə/, a milk-secreting organ of female mammals. Pl. *mammae* /-miː/.

**man.** From the time of our earliest records until the second half of the 20c. *man* could be used without comment to mean 'a human being (irrespective of sex or age)'. (In OE the main words distinctive of sex were *wer* 'a man' and *wīf* 'a woman'.) The use is embedded in hundreds of traditional expressions (e.g. *Man cannot live by bread alone*; *Man proposes, God disposes*; *Every man for himself*; *Time and tide wait for no man*), and in the works of our greatest poets and philosophers. We must all tread cautiously now under the scrutiny of the more militant feminists, who judge this use of *man* to be an unacceptable outward sign of male dominance. For an outline account of the nature of the problem, SEE SEXIST LANGUAGE.

**-man.** Some of the traditional occupational terms ending in -*man* (e.g. *anchorman*, *chairman*, *craftsman*, *houseman* = male or female hospital doctor, *ombudsman*, *spokesman*) are now usu. restricted to mean 'a *male* anchorman, etc.'. When the holder of the office, etc., happens to be a woman, forms in -*woman*, or -*person*, or some other gender-free term are now commonly used instead. See FEMININE DESIGNATIONS; SEXIST LANGUAGE.

**manage.** The inflected forms are *managed*, *managing*, *manageable*.

**manageress.** See -ESS 5.

**mandamus** /mæn'deɪməs/. This term for a particular kind of judicial writ, which is a nominalization of L *mandamus* 'we command', has pl. *mandamuses*. See -US 5.

**mandarin.** The word meaning (a) (with capital initial) the official language of China, and (b) a high-ranking Chinese official (also in transferred use), is regarded by some authorities as a separate word from *mandarin*, a kind of orange, and they are therefore given separate

479

entries in some dictionaries. But it is possible that both are derived from the same Sanskrit original, the fruit being so named in the 18c. from the yellow of the mandarins' costumes.

**mandatary, mandatory.** The first is a 17–19c. legal term meaning 'one to whom a mandate is given'. The second is an adj. meaning (a) 'conveying a command; compulsory'; (b) spec. designating a state in receipt of a mandate from the League of Nations (1919-39) to govern a specified territory. It is also a noun meaning (a) one to whom a mandate is given; (b) spec. a power or state in receipt of a mandate from the League of Nations (1919–39) to administer and develop a specified territory.

**manes** /ˈmɑːneɪz/ or /ˈmemiːz/ (pl. noun). **1** The spirits of the dead, revered as minor deities.

**2** (construed as sing.) the revered ghost of a dead person.

**maneuver.** AmE variant of MANOEUVRE.

**mangel, mangold.** The alternative forms result from changes within the German language. The original form in German, *Mangold-wurzel*, is a combination of *Mangold* beet and *Wurzel* root. The altered form Ger. *Mangelwurzel*, due to association with *Mangel* 'want', was sometimes taken to mean 'root of scarcity' (hence Fr. *racine de disette*; but note that the current French words for this cattle food, *Beta vulgaris*, are *betterave champêtre* and *betterave fourragère*).

**mango.** The pl. is either *mangoes* or *mangos*. See -O(E)S 1.

**mangrove** /ˈmæŋɡrəʊv/. A 17c. word in English, obscurely related to Pg. *mangue*, Sp. *mangle* (which are said to be derived from an Arawak word in Haiti), and assimilated to the English word *grove*.

**mangy.** So spelt. See -EY AND -Y.

**-mania.** The *OED* reports that 'in the 19th c. it became somewhat common to invent nonce-words with this ending'. The process has continued in the 20c.: e.g. *Beatlemania* (The Beatles, Liverpool pop singers), *Olliemania* (Oliver North, central figure in the Iran–Contra scandal of the late 1980s), *Perotmania* (Ross Perot,

for a time a US presidential candidate in 1992), *turtlemania* (the craze for Teenage Mutant Ninja Turtles, fantasy characters, beginning in the late 1980s).

**Manichee.** Pronounce /ˈmænɪˈkiː/.

**manifesto.** Pl. *manifestos*. See -O(E)S 7.

**manifold.** Now always so spelt (formerly also *manyfold*) whether as adj. or noun. It is no longer felt to belong to the series *twofold*, *threefold*, etc.: it means 'diverse, various' rather than 'many times'.

**manikin,** a dwarf, an anatomical model. So spelt, not *mannikin* or *manakin*.

**Manila,** in the Philippine islands. So spelt. The recommended spelling for the hemp and the paper is *manila*, not *manilla* or *Manil(l)a*, but all of these forms are still found in standard sources.

**manipulate** (verb). The corresponding adj. is *manipulable* (rather than *manipulatable*). Cf. -ABLE, -IBLE 2(v).

**mankind.** **1** Stressed on the second syllable when it means 'the human species', and on the first when it means 'male people, as distinct from female'.

**2** In its generic sense 'the human species', it is now often replaced by HUMANKIND. See SEXIST LANGUAGE.

**manner.** The phr. *to the manner born* has two meanings: (a) destined by birth to follow a custom or way of life (Shakespeare, 1602); (b) colloq., naturally at ease in a specified job, situation, etc. It is sometimes erroneously or punningly written *to the manor born*.

**mannikin.** See MANIKIN.

**manoeuvre** /məˈnuːvə/. So spelt in BrE as noun and verb. The inflected forms are *manoeuvred*, *manoeuvring*. But AmE *maneuver* (inflected *maneuvered*, *maneuvering*).

**man-of-war.** Pl. *men-of-war*.

**manpower.** See SEXIST LANGUAGE.

**manqué.** In the sense 'that might have been but is not', *manqué* is printed in italics and is placed immediately after the noun it qualifies, e.g. a poet *manqué*.

**-manship.** This traditional terminal element (*craftsmanship*, *statemanship*, etc.)

was drawn on to form the political term BRINKMANSHIP (coined in 1956 by Adlai Stevenson), and Stephen Potter's facetious formations *gamesmanship* (1947), *lifemanship* (1950), and *one-upmanship* (1952), as well as many other fanciful imitations of Potter's words.

**mantelpiece, mantelshelf.** It cannot be said too often that these words are so spelt, not *mantle-*.

**manumit** (verb). The inflected forms are *manumitted, manumitting.* See -T-, -TT-.

**manuscript.** Abbreviated MS in singular (*MS Bodley 34*), and MSS in plural.

**many a.** The type *many a prophecy*, though notionally plural ( = many prophecies) always requires a singular verb (*Many a prophecy has turned out to be unfounded*). Similarly *many another.*

**Maori.** A number of things have happened within New Zealand to the word *Maori* since about 1980. As part of an ambitious race-relations policy, attempts have been made to persuade non-Maoris to adopt the pronunciation and morphology of the word in the Maori language itself, and to respell the word either as *Māori* (with macron) or as *Maaori*. In the Maori language the word is pronounced /ˈmɑːɔri/, i.e. as three syllables with the first vowel long. A small number of non-Maoris have adopted this pronunciation, but the majority of non-Maori speakers continue to say /ˈmaʊri/, as does everyone outside New Zealand. The spelling *Māori* and the use of the word as an uninflected plural in order to accord with the conventions of the Maori language are shown in *The Oxford History of New Zealand Literature* (1991). It is too early yet to see whether these new conventions will prevail. Examples from this book: *The 1980s . . . ended in a sesquicentennial year full of reminders, for most Māori and many Pākehā, of longstanding injustices still to be remedied*—T. Sturm; *There has been a written form of Māori since 1815*—J. McRae. From an earlier book (with an unmarked pl. but without the macron): *For years Maori have struggled to secure public recognition of rights based on their understanding of the treaty*—C. Orange, 1987.

**maraschino** /mærəˈskiːnəʊ/. Pl. *maraschinos*. See -O(E)S 6.

**marathon.** 1 This now familiar word for a long-distance race and, by extension, for any other long-drawn-out test of endurance, was introduced in the first revived Olympic Games at Athens in 1896. The modern race is one of 26 miles 385 yards (42 km 352 m), usu. through the streets of a large modern city. The circumstances of the run on which the modern event is based are disputed. The *OED*'s description is as follows: 'Herodotus records that Pheidippides ran from Athens to Sparta to secure aid before the battle [between the Greeks and the Persians], but the race instituted in 1896 was based on a later less sound tradition that Pheidippides ran from Marathon to Athens with news of the Persian defeat.'

2 For modern formations in *-athon* based on *marathon*, see -ATHON.

**marchioness.** (*a*) The wife or widow of a MARQUESS; (*b*) a woman holding the rank of marquess in her own right (cf. MARQUISE).

**margarine.** A battle royal has been fought throughout the 20c. between those who pronounce the word /mɑːˈdʒɑːriːn/, with a 'soft' *g*, and those who say /ˈmɑːgə-/ with a 'hard' *g*. The *OED* (1906) listed only the form with /g/, Daniel Jones (1917) gave precedence to /dʒ/, and Fowler (1926) said that the pronunciation with /dʒ/ 'is clearly wrong'. But in the last decade of the 20c. the form with /dʒ/ is overwhelmingly dominant, alongside the abbreviated form *marge* /mɑːdʒ/. It should be noted that perhaps the only other words in which *g* is 'soft' before *a* or *o* or *u* are *gaol* (and its derivatives) and *mortgagor.*

**marginal** is applied to what is so close to the dividing line between two opposing conditions or states that there is no saying which way it will go. *Marginal land* is land that is difficult to cultivate profitably; in Britain a *marginal* parliamentary seat is one held by a small majority and is therefore at risk in an election; a book, painting, etc., of *marginal* interest is one that has only marginal significance for the reader. From the late 1970s onwards, people who were judged by some to suffer from insufficient attention (e.g. ethnic minorities, gay rights groups) welcomed the use of the terms *marginalize* and *marginalization* to describe their condition. Marginalization ... *turn [ed] into one of the main social buzzwords of the eighties* (Tulloch, 1991). Figurative uses of *marginal* and its derivatives are clearly going through a period of greatly increased frequency resulting from the new social attitudes of our age.

Some modern figurative examples of *marginalize: until recently, children's books were regarded as marginal, less than serious as literature, and while feminist criticism was concerned to shift women and their writings away from the periphery, this scarcely looked a promising topic for exploration—indeed women writing for children seemed doubly marginalized—*J. Briggs, 1989; *This compromise ... managed to marginalize the traditional approach to the state—Dædalus, 1989; Gill does not disdain or marginalise Wordsworth's material concerns but rather shows how they relate ... to his sense of the Muse's just rewards—English, 1989.*

**marginalia** (pl. noun) meaning 'marginal notes' must obviously govern a plural verb (*the marginalia in this manuscript are of considerable interest*).

**marginal prepositions.** See COMPOUND PREPOSITIONS 2.

*mariage de convenance.* See GALLICISMS 2.

**marijuana.** Spell thus, not *marihuana.*

**marital.** The only current pronunciation is /ˈmærɪtəl/, but it is worth recalling that some authorities, including Daniel Jones (1917 and later edns), listed /məˈraɪtəl/ as a legitimate variant, probably because it accorded with the long *i* in L *marītus, marītālis.*

**markedly.** See -EDLY.

**Mark (or mark)** in the sense 'a particular design, model, etc., of a car, aircraft, etc., (*this is the Mark 2 model*)' earlier in the 20c. sometimes alternated with the French word *Marque* (or *marque*), but the latter properly means a make of motor car, as distinct from a specific model (*the Jaguar marque; the marque of Mercedes-Benz*).

**marquess** /ˈmɑːkwɪs/, a British nobleman ranking between a duke and an earl. 'The official spelling of this title is "Marquess", and this is adopted in the Roll of the House of Lords. Some newspapers ... spell the word "Marquis", as for the title in France. In the past, when spelling was not standardized, both forms were adopted in Britain. Some Scottish Marquesses, in memory of the "Auld Alliance" with France, prefer to use the French spelling' (*Debrett's Correct Form*, 1976).

**marquetry** /ˈmɑːkɪtrɪ/. Spelt thus, not *marqueterie.*

**marquis** /ˈmɑːkwɪs/, a foreign nobleman ranking between a duke and a count. But see MARQUESS.

**marquise** /mɑːˈkiːz/. (*a*) The wife or widow of a MARQUIS; (*b*) a woman holding the rank of marquis in her own right.

**marriage.** For *marriage of convenience*, see GALLICISMS.

**marriageable.** So spelt; see MUTE E.

**marshal** (verb). The inflected forms are *marshalled, marshalling*, but in AmE usu. *marshaled, marshaling*. See -LL-, -L-.

**Martello.** Pl. *Martellos*. See -O(E)S 8.

**marten, martin.** The first, a weasel-like carnivore, is to be distinguished from the second, a swallow of the family Hirundinidae.

**marvel** (verb). The inflected forms are *marvelled, marvelling*, but in AmE usu. *marveled, marveling*. See -LL-, -L-.

**marvellous.** Thus spelt in BrE, but usu. *marvelous* in AmE.

**Mary.** The plural is *Marys*. Ordinary nouns ending in consonant + *y* (*analogy, company*, etc.) have pl. in *-ies*, but this rule does not apply to names. See PLURALS OF NOUNS 6.

**masculine.** See MALE.

**mashie,** a golf-club. Thus spelt, not *-y*.

**Masorah,** a body of traditional information and comment on the text of the Hebrew Bible (*COD*). Thus spelt, not the many variants.

**massacre.** Thus spelt (noun and verb) in both BrE and AmE. The inflected forms are *massacres; massacred, massacring.*

**massage.** Pronounced /'mæsɑːʒ/ or /-ɑːdʒ/. The corresponding agent-nouns *masseur* and *masseuse*, by contrast, are stressed on the second syllable: thus /mə'sɜː/, /mə'sɜːz/. In AmE, *massage* is given second-syllable stressing; and the second syllable of *masseur* and of *masseuse* also bears the main stress.

**massive.** In its various extended senses (*massive achievement, assault, blow, overdose, retaliation*, etc.) the word has enjoyed considerable currency in the 20c., but this is not surprising in an age of scientific and technological discovery, and of fundamental changes in social attitudes. It hardly needs to be said that the word should not be overused in any given piece of writing.

**mass noun,** a noun which in common usage never takes the indefinite article *a* and normally has no plural, e.g. *bread, capitalism, leisure*. It is also called a *non-count noun* or an *uncountable noun*. See COUNT NOUNS.

**master.** See FEMININE DESIGNATIONS.

**masterful, masterly.** There is a problem here which remains unresolved. **1** *History*. From the 14c. to the present day *masterful* (*a*) has been used of persons (occas. of animals) who are imperious, self-willed, or domineering; or of actions that are high-handed, despotic, or arbitrary; (*b*) from the 17c. to the present day, but sternly attacked by Fowler in 1926 and by others since then, has also been used to mean 'characterized by the skill that constitutes a master'. From the 16c. to the 18c. *masterly* was used in sense *a* of *masterful*, i.e. 'imperious, domineering', but became obsolete in this sense at some point in the second half of the 18c. From Dryden (1666) (*When actions or persons are to be described ... how masterly are the strokes of Virgil!*) onward its only sense has been that of sense (*b*) of *masterful*, namely '(of persons, their qualities, actions, etc.) resembling or characteristic of a master or skilled workman (*a masterly piece of work*)'. A further historical consideration needs to be taken into account. The adverb corresponding to *masterful* in the natural formation *masterfully*, and it has been in regular use since the 14c. The *OED* cites instances of *masterly* used as the adverb since the 14c.

and it is still listed in standard dictionaries, but its awkwardness (or rather the potential awkwardness of the unused *masterlily*) is all too apparent. The asymmetry of the adverbs may have had some bearing on the asymmetry of the adjectives.

**2** In BrE more than in AmE, and in serious writing rather than in the sports pages of newspapers, *masterly* is the standard word for the 'skilful' sense, and *masterful* for the 'domineering' sense. It cannot be denied, however, that, much as one would wish it otherwise, the unconventional sense of *masterful* to mean 'skilful' turns up frequently and seems for the present ineradicable. Examples: *In the masterful rhetoric in the first part of his dissent, Justice Holmes seemed to be making the first attack—Bull. Amer. Acad. Arts & Sci..*, 1988; *Whatever disagreement one may have with* Battle Cry of Freedom, *it is a masterful work that cannot be ignored by anyone who wants to consider how the young country managed to ... wage suicidal war with itself—NY Rev. Bks,* 1988; *There are just enough such slippages in this generally masterful book to suggest that Fish still has room for further self-revision—TLS,* 1990; *Technically, Sher is in masterful control of his part—*theatre critic in *Sunday Times,* 1991; *Brave England denied by masterful Richards—*sports-page headline in *Times,* 1991; *Mackay's evocation of small community life in New Zealand almost a century ago is masterful—*fiction reviewer in *Spectator,* 1992.

But in good writing, *masterful* still emerges as the word to be used of a domineering person. Examples: '*Oh, I do like the way you talk to the waiters, so masterful,*' *sighed Esther—*S. Mackay, 1984; *Masterful was certainly what Pinky was: right from the moment when she had given him a smart tap on the nose, and shouted in his ear, 'Shall we get engaged?'—*J. Bayley, 1994.

**mat,** an AmE variant of MATT.

**materialize.** This verb, first recorded in 1710 (Addison), is still at a sense-forming stage. Its primary sense of 'to make material, to give a material existence to' was adopted in the 19c. by spiritualists in the sense 'to cause (a spirit, etc.) to appear in bodily form'. From there it is a small step to the transferred sense 'to come into perceptible existence'. First in

America in the 1880s and soon afterwards in Britain too (*Year after year passed and these promises failed to materialise—Blackwood's Mag.*, 1891) this new sense took hold. Opinions are divided about the level of writing in which this sense most comfortably sits, but there is no doubt of its existence in one of the bands of standard English. Modern examples: *When the partnership ... which had long been promised him did not materialise, he promptly resigned—*V. Brome, 1982; *Plans do not always materialise in the anticipated way—*B. Pym, 1982; *The alimony her ex-husband was supposed to pay her never materialized—*R. Deacon, 1988; *It's hard to predict how many alternative TV channels and services will materialise in the next few years—Which?*, 1988.

**matériel.** Used in English since the early 19c. but still not fully naturalized (hence the italics and the accent). It means 'the material and equipment used in warfare (as opposed to the personnel)'. In practice the accent is often dropped in Services' documents.

**matey** (adj.), sociable. So spelt (in BrE), not *maty*.

**math.** The AmE abbreviation of MATHEMATICS; in BrE always *maths*.

**mathematics.** As the name of a subject always construed with a singular verb (*Mathematics is a difficult subject*). But when used to mean 'the use of mathematics in calculations, etc.', a plural verb is often used (*The mathematics of the launching of the spacecraft are extremely complicated*). See -ICS 2.

**maths.** See MATH.

**matinée.** 1 Though this French word literally means 'morning', it now invariably means 'in the afternoon' (both in French and English) when used of a theatrical, etc., performance. The contrast is, of course, with an evening performance.

2 The accent is often dropped, esp. in AmE.

**matins.** Usu. const. with a singular verb. 1 The first of the seven canonical hours of the breviary, properly a midnight office, but sometimes 'anticipated' by

being recited in the afternoon or evening before, or recited instead at daybreak.

2 The order for public morning prayer in the Church of England since the Reformation. See MORNING. In both senses the word is sometimes spelt *mattins* (and 16–17c. *mattens*), and this spelling is given precedence in the *Oxford Dict. of the Christian Church*, ed. F. L. Cross (1957).

**matrix** /ˈmeɪtrɪks/. Pl. *matrices* /ˈmeɪtrɪsiːz/. See -EX, -IX 2.

**matt,** dull, without lustre. Thus spelt, not *mat*, in BrE. The more usual spellings in AmE are either *mat* or *matte*.

**matter.** The phr. *no matter* (followed by *when, how,* etc.) means 'regardless of' (*I will do it no matter what the consequences; he decided to go ahead no matter who had told him not to*).

**maty.** See MATEY.

**matzo** /ˈmætsəʊ/, unleavened bread. Pl. *matzos* or *matzoth* /-əʊt/.

**maulstick** /ˈmɔːlstɪk/ is the standard form, not *mahlstick*, for the word meaning 'a long stick used by artists to steady the hand holding the brush'.

**maunder** (verb). 'to talk in a dreamy or rambling manner; to move or act listlessly or idly' is a word of obscure origin (perhaps imitative?) first recorded in the 17c. It is not related to the verb *meander*, though the superficial resemblance of the two words has not escaped the notice of amateur etymologists.

**mausoleum.** Pl. *mausoleums* or *mausolea*. See -UM 3.

**maxilla.** Pl. *maxillae* /-iː/.

**maximum.** Pl. *maxima*. See -UM 2.

**may and might.** Since at least the 1960s, writers in Britain, America, and elsewhere, despite distress signals put up by linguistic scholars, have used *may have* for *might have* in contexts, each of them expressing a hypothetical possibility, like the following: *It was useless to try nailing reinforcing timbers at sea, since this may have upset whatever fragile forces were holding the planking together—Sunday Times,* 1977; *Had it not been for the media's willingness to report news of the charge in the* *sceptical terms they did, I may not have been acquitted*—P. Hain, 1981; *If some of the resources squandered this morning had been used more wisely, we may have been able to take steps to save his life*—Scotsman, 1989; *A mentally ill man may not have committed suicide had he been kept in hospital, rather than been discharged to be cared for in the community*—Guardian, 1990. How has this use arisen when the sequence of tenses seems to be clearly wrong? Is there no clear-cut rule?

It is not difficult to find contexts in which *may have* and *might have* are treated as virtual equivalents. A news item in a 1990 issue of the *Oxford Times* began: *Six prize-winning Shetland ponies stolen from a field near Chalgrove may have been sold at Reading market*. The headline immediately above it read *Stolen ponies might have been sold*. Perhaps the headline suggested a degree of tentativeness fractionally different from that in the text. But the context was not governed by presentness or pastness of tense. There are headlines and headlines. Had the *Guardian* placed the headline *Mentally ill man may not have committed suicide* above the sentence given above, the reader would not have been certain whether the person concerned had, or had not, committed suicide (he might, for argument's sake, have been murdered) until the text of the news item made it clear just what had happened. There are other considerations. As modal auxiliaries, *may* and *might* have more than one function. Examine this pair of sentences: *you may be wrong* ( = it is possible/possibly/perhaps you are wrong); *there might be a thunderstorm tonight* ( = it is possible that . . .). In such contexts both *may* and *might* can be idiomatically used: *might* usually carries with it a slightly more marked degree of tentativeness. What is beginning to emerge is that in certain circumstances *may* and *might* are interchangeable with only slight change of focus or emphasis. It is not always a rigorous matter of contrasting tenses, more a question of judging the degree of hypothetical possibility.

Consider further the interchangeability of *could* and *might* in the following pair of sentences: *There could be a storm when the council's decision becomes known*; and *Of course, I might be wrong*. Once more, *might* seems to provide a slightly greater

degree of uncertainty. *May* and *might*, as modal auxiliaries, have another function altogether, when they are used in contexts of permission. Examples: *You may have another piece of cake if you wish*; (a somewhat old-fashioned use) *Might I ask when the library opens?*. Used thus, *may* is more formal and less common than another modal, namely *can*, and is also distinctly more common than *might*. Cf. CAN. *CGEL* (1985) recognizes the problem described in my first paragraph, but relegates it to a footnote: 'There is a tendency for the difference between *may* and *might* (in a sense of tentative or hypothetical possibility) to become neutralized. Thus some speakers perceive little or no difference of meaning between *You may be wrong* and *You might be wrong*. This neutralization occasionally extends, analogically, to contexts in which only *might* would normally be considered appropriate: e.g. "An earlier launch of the lifeboat *may* [ = might] *have averted the tragedy*." The fact that sentences such as this occasionally occur is a symptom of a continuing tendency to erode the distinctions between real and unreal senses of the modals.'

Returning to the sentences cited in my first paragraph, it is my view that *may have* is wrong in all of them. Taking all considerations into account, reality and unreality, the degree of the hypothetical possibilities, and the sequence of tenses, *might have* is required in each case, though I can see how easy it is to confuse the roles of *may* and *might* when in some other circumstances they are more or less interchangeable.

**maybe.** Used as an adverb (in origin a shortening of *it may be* since the 15c. = 'possibly, perhaps', it had so dropped out of favour in Britain by the end of the 19c. that the *OED* (1906) labelled it 'arch.' and 'dial.' It has become fully restored to standard, though rather informal, use in BrE during the 20c., though, as a matter of curiosity, the great majority of the examples in my files happen to be from NAmer. sources. Examples: *He sat at the table munching, still thinking of Matilda. Despair? Self-hate? Maybe*—M. Wesley, 1983; *Maybe my chances had improved with separation*—S. Bellow, 1987; *If you quote the Liberal opposition saying that they did, aren't you implying that they didn't?*

*Maybe?—London Rev. Bks,* 1987; *Maybe a shotgun was all he had*—A. Munro, 1987; *Maybe if one of them had been sick that morning their father might not have gone to work. Maybe they should have ... said something at breakfast to delay him*—New Yorker, 1992.

**Mc-.** See MAC-, MC-.

**me.** 1 See I 1.

2 It may be sufficient if I list here, with indications of the degree of formality, the main circumstances in which *me* is now tending to usurp the territory that logically belongs to the subject-pronoun *I.* (*a*) Used informally at the head of clauses: *Me and the teacher are going to race tonight from the school to the store*—J. Crace, 1986; *Me and the lads have been wintering down here in Morocco*—London Rev. Bks, 1988; *Mrs Smeath and Aunt Mildred in the front [of the car], me and Grace and her two little sisters in the back*—M. Atwood, 1989; *Me and my deputies sometimes would be out riding, and we'd see a barn and think, That looks like a good place to hide*—Alec Wilkinson, 1992 (US). Cf. MYSELF. (*b*) *It's me*, in answer to the question *Who is it?*, is now standard. (*c*) In answer to the question *Who's there?*, the natural answer is *Me*, not *I*. (*d*) After *as* and *than* there is much diversity of usage. In such circumstance *I* is more formal than *me*. Examples: (as) *Jim would have run the farm as good as me*—M. Eldridge, 1984; *I sensed that he was as apprehensive as I about our meeting after seven years*—J. Frame, 1985; (than) *He was five years older than me*—E. Waugh, BBC TV, 1960; *I hit her as hard as I could. She was taller than me.*—M. Wesley, 1983; *He was taller than I, so that I had to stand a little to the rear*—B. Rubens, 1985; *He's some years older than me*—Chr. Sci. Monitor, 1987. (*e*) As a reply to another person's assertion, *me too* or *me neither* are normally used, not *I too* and *I neither*. Examples: *'Let's talk about each other, that's all I'm interested in at the moment.' 'Me too,' says Tom*—P. Lively, 1987; *'Oh no, I couldn't stand it!' 'Me neither!'*—R. McAlpine, 1985. (*f*) The types *Silly me!* and *Me, I go on about this all the time* are now commonly encountered in spoken English and in the dialogue of good novels.

3 In representations of popular speech, *me* frequently takes the place of the possessive pronoun *my*. Examples:

*That's him all right. He's got me coat and hat. That's him*—J. P. Donleavy, 1955; *I'll just go and clean me teeth*—Encounter, 1981; *These might do for me dad*—G. McGee, 1984.

Disturbance of some of the traditional pronominal roles is now a feature of modern English, but of the above only 2(*b*), (*c*), (*e*), and (*f*) are recommended, while 2(*d*) remains optional.

**meagre.** Thus spelt, but in AmE usu. *meager.*

**meal.** See FLOUR.

**mealie,** or (from Afrikaans) *mielie.* Used chiefly in the pl. form *mealies* (or *mielies*) as a SAfr. name for maize, Indian corn. The singular form occurs chiefly in compounds, e.g. *mealie-cob, -field, meal.*

**mean** (verb). 1 *I mean.* This is often legitimately used to introduce an explanation of what has just been stated: e.g. *But he was a marvellous butler. I mean, if you went there he'd welcome you in the most graceful and polite and proper way*—New Yorker, 1986; *He had got me a bit worried, though, I must admit. I mean, I'd only intended to put Danny Ram on his guard, not to provoke a major industrial dispute*—D. Lodge, 1988; *Courage to buy and courage to sell—I mean, maybe the real courage is in the keeping, before you sell*—R. Rofihe, 1990. But it is often used as a MEANINGLESS FILLER in unscripted speech, as I said in by BBC booklet *The Spoken Word* (1981): e.g. *'It's not really expensive. I mean, it only costs a pound.' 'It seems, sort of, a lot. I mean, when it used to cost only forty pence. I mean, where will it all end?'*

2 For (*Do you*) *know what I mean?*, see KNOW 1.

3 *A new passive use.* In the passive, the verb *mean* has been used since the 1890s to mean 'to be destined (by providence); to have special significance': e.g. *It couldn't have been* meant. *It was only the tide*—Kipling, 1897; *When I need you, you are here. You must see how meant it all is*—I. Murdoch, 1974. This use remains in force. It has been joined by another passive type in which *meant* followed by a *to*-infinitive draws part of its significance from the original sense of the verb, namely 'to purpose, to design', and now has come to mean little more than 'supposed, thought, intended': e.g. *For today he was meant to be having dinner with*

*Stephanie at the Dear Friends*—A. N. Wilson, 1986; *He was meant to give very good parties*—K. Amis, 1988; *I can't remember. Some woman who's meant to be absolutely marvellous—who is it, Scrope?*—C. Phipps, 1989; *'Just look at it* [sc. a poor-quality pink eiderdown], *just look at the colour of it.' ... 'Go and clean your teeth and drink plenty of water,' she said. 'But he's meant to be rich.'*—S. Mason, 1990.

**meander.** See MAUNDER.

**meaningful.** First recorded in 1852, it has enjoyed such widespread currency in the 20c. that voices have been heard to say 'No more!'. In the *OED* the words that *meaningful* qualify are *conclusions, expression, formula, gifts, part, result, smile, talks, tests, unit of sound, way, word,* and *a particular work.* They all seem appropriate in context, but I recall that the 20c. examples (which I selected myself) were drawn from 'respectable' sources (learned journals and scholarly books for the most part), and that the more ephemeral examples were left among the 'superfluous quotations'. Some more recent examples: *Ongoing situations and meaningful dialogues are two popular pieces of jargon ... at present*—P. Howard, 1978; *What would give you satisfaction, Emily? What would make your life seem more meaningful?*—R. Jaffe, 1979; *The profession must enjoy (or develop) a meaningful dialogue with the government*—Jrnl RSA, 1979; *Chris and Jayne turned to each other with raised eyebrows and meaningful looks*—J. Hennessy, 1983; *The fact that predictions last night suggested lengthy talks lasting a number of days ... gave rise to hopes of some meaningful negotiations*—Daily Tel., 1984; *To ensure a meaningful context for the matinées, they are placed at the centre of a programme*—TES, 1990.

**meaningless fillers.** An American newspaper of June 1992 reported the following conversation between two teenagers: *'Like wow, didja, y'know, see that new video?' 'Yeah, like, well, I did, yeah, but you know.' 'Didn't you think it was, well, like, you know, like wow, y'know?' 'Well, y'know I thought, like, well, y'know. Cool. Not!'* In another American newspaper of April 1992 a psychologist called Nicholas Christenfeld, who has studied speech patterns 'around the world', was reported as saying that 'filled pauses', such

as *um, er,* and *like* are universal, and that 'such utterances signal an individual's desire to speak, though the actual words may elude the tongue'. President Bush, he averred, averaged 1.7 *ums* a minute, and the TV personality Johnny Carson 1.5. A host of a talk show called Pat Sajak averaged 9.8 ('It's amazing he lasted as long as he did,' Christenfeld said.)

In *The Spoken Word* (1981) I said that such meaningless fillers 'do not normally occur in scripted [broadcast] material but are not always avoidable in informal contexts and constitute useful rhythmical aids to speakers who temporarily lose their fluency.' I certainly did not have in mind the mindless gibberish illustrated in the first paragraph above! A short list of meaningless fillers follows. It is obvious that too frequent a use of such fillers is undesirable. *actually, and er, and everything, and um, didn't I (I went into this shop, didn't I?), er, I mean, I mean to say, really, sort of, then (Nice to talk to you, then), um, well now, you know, you see.*

**means** is a plural noun in form, but in some circumstances may be construed either with a plural or a singular verb, thus joining a small group of nouns that may operate in this manner, e.g. *crossroads, gallows, headquarters, (golf) links, oats.* This state of affairs has come about because over the centuries the singular for *mean* was just as likely to be used as *means* when the sense was 'agency, method, way to an end': e.g. *Baptisme is necessarie as a meane to saluation* (C. Sclater, 1611); *Let us consider it as a mean, not as an end* (T. Balguy, 1785). The singular form has dropped out of use, but the folk memory of the construction with a singular verb has not. The matter is complicated, but Fowler's views (1926) are worth repeating: 'In the sense "income, etc.", *means* always takes a plural verb: *My means were* (never *was*) *much reduced.* In the sense "way to an end, etc.": *a means* takes a singular verb; *means,* and *the means,* can be treated as either singular or plural; *all means* (pl.) and *every means* (sing.) are equally correct; *the means do not,* or *does not, justify the end; the end is good, but the means are,* or *is, bad; such means are* (not *is*) *repugnant to me,* because *such* without *a* is necessarily plural: cf. *such a means is not to be discovered*; and similarly with other adjectives, as *secret means were found*, but *a secret means was found*.'

Some 20c. examples of various kinds: ( = the wherewithal) *Underprivileged kids may never have the means to drive*—P. Goodman, 1960; ( = method, mode) *Throughout my boyhood the normal means of passenger transport, if something more than a bicycle was required, was the cab*—J. Grimond, 1979; ( = way to an end) *Renunciation may be to them a means to attain peace*—R. Narayan, 1937; *They remained for her a means, and not an end, a bargaining power rather than a blessing*—M. Drabble, 1967; *One generation must not be condemned to the role of being a mere means to the welfare of its remote descendants*— I. Berlin, 1978; *Moreover, the means by which this end is achieved are themselves remarkable*—M. Foot, 1986.

**mean time, meantime. 1** From the 14c. to the 19c., whether in the temporal sense 'at the same time, all the while' or in the adversative or concessive sense 'while this is true, nevertheless', the phr. *in the mean time* was normally written thus, with *mean time* as two separate words. In all such uses now the phr. is written as *in the meantime*.

**2** From the 16c. to the 19c. *meantime* was as common as *meanwhile* as an adverb, e.g. *The ladies, meantime, were on the qui vive* (1842); *Meantime where was Lord Palmerston?* (1879). As adv., *meanwhile* is now the more usual of the two: e.g. *Holidays for the girls, meanwhile, were spent either at home or with relations*—Bodl. Libr. Rec., 1986; *The animals Mrs Murray cares for are always returned to the wild if possible. Meanwhile, they stay at her study centre*—Times, 1986; *Meanwhile, we had them call on the radio to the agents that had checked that house earlier in the day*—Alec Wilkinson, 1992 (US). But *meantime* is also widely used: *Meantime, with the industrial and educational revolutions, the numbers of people with the capacity and need to think for themselves increased immensely*—TLS, 1975; *Meantime, melt the remaining butter in a saucepan*—Delia Smith, 1978.

**mean while, meanwhile. 1** Like the preceding, the words *mean* and *while* were normally written separately (*in the mean while*) from the 14c. to the early 19c., but *meanwhile* has normally been written as one word since then in all uses. The phrases *in the meantime* and *in the meanwhile* are still to some extent interchangeable, though the former is the more usual: e.g. (in the meantime) *They say nothing to each other, waiting for someone to take their orders, in the meantime watching the women walk by*—P. Auster, 1986 (US); *The telephone will ... redial the number you last called, even if you've hung up in the meantime*—Which?, 1987; *We thought ... we would soon have to find another job. In the meantime we got on with the jobs we had*—M. Spark, 1988. (in the meanwhile) *In the meanwhile, the Government is effectively admitting that state spending is out of control*—Daily Tel., 1986; *In the meanwhile, I'll just lie here, flat on my back, fingering my perfect bones*—J. Shute, 1992.

**2** As adv., *meanwhile* is now the more usual of the two words. See MEAN TIME **2**.

**3** A loose resumptive use of *meanwhile* as an adverb is common in journalistic use, in TV commentaries, etc. It is the equivalent of phrases like *while we are on the subject* or *and another thing*. Examples: *When we [sc. journalists] begin our final paragraphs with 'meanwhile', we are usually (and a bit shamefacedly) trying to get away with a disgraceful non sequitur*—New Statesman, 1964; *Meanwhile, as we say in the trade, Motherwell [sc. a Scottish football team] go bottom [of the league table]*—D. Lynam, BBC TV, 14 March 1987; *(After a review of one TV programme) Meanwhile, Ask Dr Ruth (Channel 4) reminds one that the English and the Americans are divided by more than a common language*—Observer, 1987. *Meanwhile*, thus used, has been straining at the leash for some time. It was probably Hollywood's fault to begin with. The phrase *Meanwhile, back at the ranch* is one of the most memorable of captions used in silent Westerns of the 1930s, used to introduce a subsidiary plot. It made its way into later Westerns like *Champion the Wonder Horse* (1956), promoted from caption to voice-over. It has now spread in an astonishing manner to other walks of life. Examples (including some variant types): (Meanwhile, back) *Meanwhile, back on the American presidential campaign trail, Candidate Jimmy Carter repeatedly criticized Helsinki as a sellout*—Time, 1977; *Meanwhile, back at the ranch, I have been browsing through the* *press coverage of Death is Part of the Process and fancy I have discovered a phenomenon— the PMC, or Politically Motivated Critique— Guardian, 1986; Meanwhile back in the Commons, Mrs Thatcher tried to resist questions by saying the isues were sub judice because of the Australian case—Today, 1986; Meanwhile, back at the simulator, the question persists: 'What else could have gone wrong?* —Financial Times, 1987; (Meanwhile, down) Meanwhile, down the road in Janet Reger's Beauchamp Place shop, it's been a poison green Christmas for the men— Guardian, 1987; Meanwhile, down in prosperous Hampshire, Southampton City Council's own development company SEDCO has just launched a £500,000 fund for start-up and expansion finance—Today, 1987.*

**measles.** Most commonly construed as a singular (*Measles is decidedly infectious*) but occas. as a plural, and also, esp. in some regional forms of English, with *the* (e.g. *he fell ill with the measles*).

**measurements.** *A blank white eight-by-eleven sheet can look as big as Montana if the pen's not so hot*—G. Keillor, 1989. It is worth noting that in such double measurements the smaller number is usually placed first in AmE and the larger in BrE. Oxford lexicographers of my generation copied illustrative examples on to 6″×4″ slips, but our colleagues in America called them 4″×6″ slips. Cf. also *four-by-two* '(in Brit.) Mil. colloq., the cloth attached to a pull-through (for cleaning a rifle)'; and *two-by-four* '(orig. and chiefly US) a post or batten measuring 2 inches by 4 in cross-section'.

**meatus** /mɪ'eɪtəs/, (anat.) a channel or passage in the body or its opening. Pl. same (pronounced /mɪ'eɪt(j)uːs/ or *meatuses*. See -US 2.

**medal.** The derivatives are *medalled*, *medallist* (AmE usu. *medaled, medalist*). See -LL-, -L-.

**media** (pl. noun). In its sense '(usu. preceded by *the*) the main means of mass communication (esp. newspapers and broadcasting) regarded collectively', *media* (in origin the pl. of *medium*) is properly construed with a plural verb or pronoun. But it has swiftly followed *agenda* and *data*, and, like them, is often treated as a mass noun with a singular verb or pronoun. We are still at the debating table, however, on the question of *the media are/is having a field day in their/its reporting of the scandal.* When in doubt use the plural. Above all, never write *a media* or *the medias.*

**mediaeval.** Now fast passing out of use in favour of MEDIEVAL.

**medicine.** The recommended pronunciation is with two syllables, i.e. /'medsɪn/. But many RP speakers say /'medɪsɪn/; and in Scotland, the USA, and elsewhere the word is most commonly pronounced as three syllables.

**medieval.** The recommended spelling, not *mediaeval.*

**mediocre.** Pronounce /miːdɪ'əʊkə/.

**mediocrity.** Pronounce /miːdɪ'ɒkrɪtɪ/.

**Mediterranean.** Thus spelt.

**medium.** When the meaning is 'one who claims to be in contact with the spirits of the dead', the pl. is always *mediums*. In all other senses, *mediums* and *media* are both in use, except that MEDIA (newspaper/broadcasting sense) is invariable in form and has its own pattern of behaviour. See -UM 3.

**medlar,** a rosaceous tree. So spelt.

**meerschaum** /'mɪəʃəm/. So spelt.

**meet** (verb). For *meet* (*up*) *with*, see PHRASAL VERBS.

**mega-.** This prefix, representing Gk μεγα-, combining form of μέγας 'great', was drawn on repeatedly in the 19c. to form new words, both scientific (*megapode, megaspore*, etc.) and general (*megaphone*). It has lost none of its potency in the 20c.: (scientific and technical, usu. denoting 'a million (times)') *megabyte, megaparsec*; (in the language of possible nuclear warfare) *megacorpse, megadeath*; (casual formations) *megahit* (a major success), *megamerger* (in commerce), *megastar* (films, etc.), *mega-tyrant* (Stalin), etc.

**meiosis** /maɪ'əʊsɪs/, pl. *meioses* /-iːz/, derived from Gk μείωσις 'lessening', is a rhetorical device in classical and later literatures, and in ordinary speech and language, in which circumstances, a predicament, or any event is intentionally understated. An example is found in

*Romeo and Juliet* where Mercutio describes his *mortall hurt* as *a scratch, a scratch.* For examples of meiotic negative understatement, see LITOTES.

**-meister.** As a formative element, *meister* became familiar to English speakers in the 19c. in the word *Meistersinger* (from German), German lyric poets and musicians in the 14th to the 16th centuries. It has recently been repeatedly pressed into service, esp. in AmE, as a second element in various, mostly transparent, compounds, e.g. *dietmeister* (diet guru), *mediameister, spinmeister* (political spin doctor). Numerous other examples in my files include (all AmE) *angstmeister, chatmeister, dramameister, grungemeister, schlockmeister, talkmeister,* and *wordmeister.*

**melodeon.** Thus spelt, not *melodion, melodium.*

**melodrama.** 'In early 19th c. use, a stage-play (usually romantic and sensational in plot and incident) in which songs were interspersed, and in which the action was accompanied by orchestral music appropriate to the situations. In later use the musical element gradually ceased to be an essential feature of the "melodrama", and the name now denotes a dramatic piece characterized by sensational incident and violent appeals to the emotions, but with a happy ending' (*OED*, 1907). The term is also now applied to cinema and TV productions of a similar type; and to a series of incidents, or a story true of fictitious, resembling what is represented in a melodrama.

**melody.** See HARMONY.

**melted, molten.** As adjs., *melted* is the normal word (*melted butter, ice, snow,* etc.) of the two. *Molten* is used only of materials melted at extreme heat (*molten lava, lead, rocks,* etc.).

**membership** had two established uses before the 20c.: (*a*) (first recorded in 17c.) The condition or status of a member of a society or organization (*the oath of membership required fidelity*); (*b*) (19c.) The number of members in a particular organization (*a large membership is necessary*). A third, not widely favoured, sense, simply = 'members', has been added in the second half of the 20c., esp. in the language of the block votes of trade unions: *I hope our membership will listen to the advice of their elected representatives*; *There is growing restlessness among a section of its membership*—both examples cited in Gowers, 1965. Cf. LEADERSHIP; READERSHIP.

**membraneous** /mem'breɪnɪəs/, **membranous** /'membrənəs/. Either form may legitimately be used as adjectives corresponding to the noun *membrane.*

**memento,** a souvenir, should not be converted into the dubious formation *momento.* The recommended pl. is *mementoes.* See -O(E)S 1.

**memo** /'meməʊ/, a colloquial shortening of MEMORANDUM. Pl. *memos.* See -O(E)S 5.

**memorabilia** (pl. noun), souvenirs of memorable events, people, etc. It should always be treated as a plural. Incorrect examples (from Amer. newspapers 1994–5): *The memorabilia on the walls includes a magazine cover of an outlaw he killed in Miami*; *Older memorabilia, however, is more difficult to come by.*

**memorandum.** The recommended pl. is *memoranda,* but *memorandums* is also in standard use. See -UM 3. It is important never to use *memorandum* as a plural (*these memorandum*), and never to fall into the trap of regarding *memoranda* as a singular and consequently speaking of *these memorandas.*

**memsahib.** See SAHIB.

**ménage,** the members of a household, is written in romans, but normally retains its French accent and its not fully Anglicized pronunciation, i.e. /mei'nɑːʒ/. Contrast the next word.

**menagerie,** a collection of wild animals in a zoo. Though a comparatively recent (18c.) loanword from French it has fully settled into English: always written in romans, no acute accent, and pronounced /mə'nædʒərɪ/.

**mendacity** is habitual lying or deceiving, and **mendicity** is the practice or habit of begging.

**meningitis** /menɪn'dʒaɪtɪs/. See GREEK G.

**meninx** /'miːnɪŋks/, any of three membranes that line the skull. Pl. *meninges* /mɪ'nɪndʒiːz/.

**menstruum** /'menstrʊəm/, a solvent. Pl. *menstrua* /-strʊə/. See -UM 2.

**-ment.** For differences between this suffix and *-ion*, see -ION AND -MENT. Nouns in *-ment* are almost all formed on verbs: *merriment* is only an apparent exception (it was formed on an obsolete verb *merry* 'to make merry'); *oddment* is a genuine exception. In words like *garment, raiment*, and *testament*, the *-ment* is not a suffix but is part of the stem of the word.

**mentality.** First recorded in the late 17c. in the neutral sense 'that which is of the nature of mind or of mental action' (e.g. *There is a vast store of mentality even in the higher animals which has not yet been brought to perfection*, 1899), it was also used in the 19c. to mean 'intellectuality' (e.g. *Hudibras has the same hard mentality*—Emerson, 1856; *Pope . . . is too intellectual and has an excess of mentality*—G. Santayana, 1900). Praiseworthy applications of the word largely went out of the window as the 20c. proceeded, and now *mentality* is almost always used in disparaging contexts. Examples: *These Port managers, with their special knowledge and important position, tended to acquire the bureaucratic mentality: they said No automatically*—U. Le Guin, 1974; *I hate the triviality of journalism, you know, the sort of fluttering mentality that fills up the page*—H. Carpenter, 1978; *A kind of unacknowledged underground mentality had permeated all kinds of places*—A. Miller, 1987; *'Roberto being a Cuban understood my slight Third World mentality,' says Puttnam*—*Sunday Express Mag.*, 1987.

**Mephistopheles** /mefɪ'stɒfɪliːz/. The adjective is either *Mephistophelean* or *Mephistophelian* /ˌmefɪstə'fiːliən/.

**merchandise.** In BrE spelt thus as noun and verb, and usu. pronounced with final /-aɪz/ for both (sometimes with /-aɪs/ for the noun). In AmE the verb is often spelt with final *-ize*.

**mercy.** In at least three places in the Authorized Version, *mercy* (or *mercies*) is preceded by the word *tender* (e.g. *Through the tender mercy of our God*—Luke 1: 78; *turne vnto mee according to the multitude of thy tender mercies*—Ps. 69: 16). It seems harsh to see these as the source of the modern cliché *leave* (or *trust*, etc.) *something or someone to the tender mercies* (or *mercy*) *of* (a named person, etc.), but some

kind of connection cannot be ruled out: e.g. *Gaunt suddenly fell away from him . . . and left him naked to the tender mercies of his priestly enemies*—C. Clarke, 1863; *Too precious to trust to the tender mercies of a baggage pony*—Earl Dunmore, 1893; *To say nothing of your being at the tender mercy of others practically*—J. Joyce, 1922; *Smaller . . . traders and manufacturers . . . left to the tender mercies of the open property market*—*Listener*, 1965; *The baby left to the clumsy but tender mercies of a trio of bachelors was really only a designer toy*—*Highlife* (Br. Airways), 1991.

**merino.** Pl. *merinos*. See -O(E)S 6.

**merit** (verb). The inflected forms are *merited, meriting*. See -T-, -TT-.

**merriment.** See -MENT.

**mésalliance** normally means 'a marriage with a person of a lower social position', by contrast with *misalliance*, which is used of any unsuitable alliance, including an unsuitable marriage. Both words entered the language in the 18c.

**mesembryanthemum.** Despite the best efforts of Fowler (1926), this spelling with medial *-y-* has prevailed over *mesembrianthemum* in botanical works. It is a modL formation based on Gk μεσημβρία 'noon' and ἄνθεμον 'flower', i.e. midday-flower. Linnaeus used the irregular form and it has been slavishly followed by botanists since the 18c.

**Messrs.** It is worth setting down here the description of the term in *Debrett's Correct Form* (1976); 'The use of the prefix "Messrs." (a contraction of the French Messieurs), as the plural of "Mr.", is becoming archaic for commercial firms in Britain, Australia and New Zealand, but is still generally used by the professions, especially the law. It is not generally used in Canada or in the United States. For those who prefer to use the term "Messrs.", this is restricted to firms with personal names, e.g. Messrs. Berkeley, Stratton & Co. . . . It is never used in the following instances: (a) Limited Companies. (b) Firms which do not trade under a surname, or the surname does not form the complete name. e.g. The Devon Mechanical Toy Co., . . . John Baker's School of Motoring. (c) Firms whose name includes a title, e.g. Sir John Jones

& Partners. (d) Firms which bear a lady's name, e.g. Josephine Taylor & Associates.' In most modern work *Messrs* is printed without a full point.

Some examples: *Send in bill to Messrs. Grist*—G. Meredith, 1861; *Messrs Hodder & Stoughton should let me have all the Strand articles ... in slip proof*—W. S. Churchill, 1908; *Armour Glass ... similar to that made by Messrs. Pilkington Bros. in this country*—*Jrnl Society of Glass Technology*, 1932; *I am not surprised that Messrs. Auden and Isherwood came away with a sense of deep and humble respect for the people and the country who had hosted them*—*John o' London's*, 1939; *Messrs Wedgwood introduced a decorative type of marbled pink or purple 'gold' lustre called 'Moonlight Lustre'*—G. A. Godden, 1966; *Thanks to Messrs McCaskill and Fish with their TV weather forecasts most of us now have a good idea of how to recognise fronts and depressions on the satellite pictures*—*Motor Boat & Yachting*, 1991.

**messuage,** house with outbuildings (a legal term). Pronounce /'meswɪdʒ/.

**metal** (noun) and **mettle** are in origin the same word. From the late 16c. onward *mettle* began to move apart as a separate word used only in figurative senses, esp. '(of persons) ardent or spirited temperament; spirit, courage'. Meanwhile *metal*, while still current in figurative uses in the 16c. and 17c., esp. in the sense 'the "stuff" of which a person is made, with reference to character', gradually became more or less restricted to metallic substances (e.g. industrial metals; also, road metal).

**metal** (verb). The inflected forms are *metalled, metalling* (but usu. *metaled, metaling* in AmE). See -LL-, -L-.

**metallurgy.** The only pronunciation given in the OED (1907) showed the stress on the first syllable. COD (1990) gives precedence to second-syllable stressing. This tendency to adopt an antepenultimate stress pattern in four-syllable words is a marked feature of 20c. English. Cf. *despicable, disputable, pejorative*, etc. In AmE first-syllable stressing is still normal for *metallurgy*. The same patterns apply to *metallurgist*.

**metamorphosis.** The OED (1907) gave only /metə'mɔːfəsɪs/, i.e. stressed on the third syllable. COD (1990) added /ˌmetə-

mɔː'fəʊsɪs/ as a permissible variant. In either case the pl. form *metamorphoses* ends with /-iːz/.

**metanalysis.** (philology) Reinterpretation of the division between words or syntactic units: e.g. *adder* from OE *nædre* by analysis in ME of *a naddre* as *an adder*.

**metaphor. 1** Fowler's essay on the subject in the first edition of this book (1926) was first published as SPE Tract xi (1923): readers are referred to either of these. Readers may also wish to consult e.g. K. Wales, *A Dictionary of Stylistics* (1989), and the works mentioned in it s.v. *metaphor*; or the opening pages of W. Empson's *The Structure of Complex Words* (1951).

**2** Perhaps the two best-known figures of speech are the metaphor and the simile. They differ only in form: similes are normally introduced by *like* or *as* while metaphors tend to be imaginative substitutions. Straightforward examples: (metaphor) *True to the signal, by love's meteor led, Leander hasten'd to his Hero's bed*—G. White, 1769; (simile) *Th' Imperial Ensign ... Shon like a Meteor streaming to the Wind*—Milton, 1667.

**3** The ordinary language abounds in simple, often dead or invisible, metaphors: e.g. in this *neck* of the woods, the *mouth* of the river, no *smoke* without *fire*, throw out the *baby* with the *bathwater*.

**4** A type of metaphor that always arouses derision is the *mixed metaphor*, the application of two or more inconsistent metaphors to a given situation. A small collection drawn from *The Times* (mostly from the letter columns) in recent years: *Certain elements of the BMA leadership have gone over the top and taken fully entrenched positions; He has been made a sacrificial lamb for taking the lid off a can of worms; In coal mines, mice are used as human guinea pigs; Why do the dear old Labour party persist in burying their heads in the sand, parroting tired old formulae?; In a debate on shipping and the merchant fleet ... [he] called for a level playing field.*

**metaphysical** (adj.). See next.

**metaphysics. 1** As the name of a branch of philosophy, *metaphysics* has had an adventurous history and important consequences. Since at least the 1c. AD it was applied (as Gk τὰ μεταφυσικά) to the thirteen books of Aristotle dealing with questions of 'first philosophy' or ontology. They were in fact τὰ μετὰ τὰ φυσικά 'the things (works) after the Physics', in other words the works that in the received arrangement followed the treatise on natural science known as τὰ φυσικά 'the Physics'. In the 18c. the Greek word was falsely interpreted as 'beyond what is physical', and *metaphysics* came to be widely used to mean 'the science of things transcending what is physical or natural'. The adj. *metaphysical* followed suit. In more recent times the prefix *meta-* has come to be applied to subjects dealing with ulterior and more fundamental problems than the subject itself: e.g. *meta-ethics*, the study of the foundations of ethics; *meta-history*, inquiry into the principles governing historical events; *metamathematics*, the field of study concerned with the structure and formal properties of mathematics. There is no turning the clock back now.

2 As a further example of the use of the prefix *meta-* in an extended sense, *metaphysical* was adopted by Dr Johnson as the designation of certain 17c. poets (Donne, Cowley, and others) addicted to poetical conceits and far-fetched imagery.

3 *Metaphysics*, like other subject-names ending in *-ics*, is plural in form but is construed as a singular (see the first sentence above).

**metathesis.** (phonetics) The transposition of sounds or letters in a word. Examples include the transposition of the *p* and *s* in *hasp* (cf. OE *hæpse*); the *s* and *k* in non-standard *ax(e)* 'ask' (OE *āscian*); the *r* and *i* in *third* (OE *þridda*) and *thrill* (OE *þyrlian*); the *r* and *u* in *curled* (ME *crolled, crulled*). For a modern example, see IRRELEVANT.

**metempsychosis**  /ˌmetempsaɪˈkəʊsɪs/. Pl. *metempsychoses* /-iːz/.

**meter¹.** The normal spelling in both BrE and AmE for an instrument for recording the amount of gas, electricity, etc., used. Also *parking-meter*.

**meter².** AmE spelling of METRE¹.

**meter³.** AmE spelling of METRE².

**-meter,** combining form. 1 Forming nouns denoting measuring instruments (*barometer, speedometer*).

2 (prosody) Forming nouns denoting lines of poetry with a specified number of measures (*pentameter, hexameter*). Cf. METRE².

**method.** For *method in (his,* etc.) *madness*, see IRRELEVANT ALLUSION.

**meticulous** entered the language in the 16c. in the sense of L *meticulōsus* 'fearful, timid' (cf. L *metus* 'fear', *metuō* 'regard with awe'). But this use was short-lived and thinly evidenced, and has not been found since the 17c. For whatever reason, the word reappeared in the early 19c., but used to mean 'over-careful about minute details, over-scrupulous' (e.g. *The decadence of Italian prose composition into laboured mannerism and meticulous propriety*—J. A. Symonds, 1877), a use scorned by the Fowler brothers in *The King's English* (1906). They included it in a list of 'stiff, full-dress, literary, or out-of-the-way words'. At some point in the 20c. the over-carefulness element of the meaning dropped out, and the word is now routinely used simply to mean 'careful, punctilious, scrupulous, precise'. It is a useful word, together with its derivatives *meticulously* and *meticulousness*. Examples *He also set down with meticulous care an exact itinerary of Stillman's divagations*—P. Auster, 1985; *He'd read it all up beforehand, in the most meticulous manner*—J. Gloag, 1986; *Utz had planned his own funeral with meticulous care*—B. Chatwin, 1988; *Meticulous, obsessional scholars whose lives are dedicated to their work*—A. Storr, 1988. *Meticulosity* has occasionally been used, first (17c.) in the sense 'timorousness', and then (20c.) meaning 'the quality of being meticulous', but it has lost ground to *meticulousness* and is best left on one side.

**metonymy,** a figure of speech which consists in substituting for the name of a thing the name of an attribute of it or of something closely related. When we call Queen Elizabeth *the Crown*, we use metonymy. Similarly, *the White House* for the American presidency; *the stage* for the theatre; and *the pen is mightier than the sword* as a way of saying that the written word has more power than war. Cf. SYNECDOCHE.

**metope** (architecture, part of a Doric frieze). Pronounce as three syllables, /ˈmetəpi/, as in the etymon, Gk μετόπη.

**metre¹.** The spelling in BrE of the metric unit of length equal to about 39.4 inches. AmE *meter*.

**metre².** Any form of poetic rhythm, determined by the number and length of feet in a line. AmE *meter*. Cf. -METER 2.

**mettle.** See METAL (noun).

**mews.** In origin the pl. of *mew* 'a cage for hawks', it is now usu. construed as a singular (*a mews*) but may also (less commonly) be used as a plural (*The mews of London ... constitute a world of their own*—H. Mayhew, 1851). The royal stables at Charing Cross were called the mews from the 14c. onwards because built on the site where the royal hawks were formerly 'mewed' or moulted.

**mezzanine** /ˈmetsəniːn/ or /ˈmezə-/, a low storey between two others (usu. between the ground and first floors). See FLOOR.

**mezzo-relievo.** Italianate and semi-Anglicized pronunciations exist side by side at present just as they do for *basso-relievo* (see BAS-RELIEF). Perhaps the best compromise is to print it in italic and pronounce it in a semi-Anglicized manner as /ˌmetsəʊrɪˈljeɪvəʊ/.

**mezzotint.** Pronounce /ˈmetsəʊ-/, though the Italian original *mezzotinto* is pronounced /mədz-/.

**miasma** /mɪˈæzmə/, /maɪ-/, an archaic word for an infectious or noxious vapour. Pl. *miasmata* /-mətə/ or *miasmas*.

**mickle** and **muckle** are merely variants of the same word, and the not uncommon proverb *Many a mickle makes a muckle* is a blunder. The right form is *Many a little makes a mickle* (first recorded in the 17c.), in which *mickle* means 'a large sum or amount'. In the 20c., *pickle* has been used occas. in place of *little*. The aberrant form *Many a mickle makes a muckle* is first found (in 1793) in the work of George Washington.

**micro-** continues to be a combining form of limitless power. Among the *micro*-compounds that I have noted in the last few years, none of them so far registered in the *OED*, are: *microballoon* (silicone balloon used to fight aneurysms in the brain); *microburst* (intense local downfall of cool air); *microdiskectomy* (a back surgery technique); *microjolt* (minute electrical impulse used in cloning cattle); *microlaser* (in computer technology); *microsurfacing* (protective layer used on paved surfaces). But these are mere drops in the ocean. Hundreds of *micro-* compounds are already entered in the *OED*, including a great many first recorded in the 20c.: e.g. *microclimate, microdot, microfilm, miocroform, miocro-oven, microsurgery, microwave*. Many others will join them in the years ahead.

**mid** has superlative *midmost*.

**mid-air.** Fowler (1926) said flatly that '*mid air* should not be hyphened'. But times have changed and it always is hyphened now both in attributive and predicative uses.

**middle voice.** (grammar) Greek verbs may have, besides or without the active and passive voices, another called the middle voice, in most tenses identical in form with the passive, but expressing 'reflexive or reciprocal action, action viewed as affecting the subject, or intransitive conditions' (*OED*). The middle voice is not restricted to Greek. It is a feature, for example, of Old Norse. Here the middle voice consisted originally of the active forms with a reflexive pronoun suffixed: e.g. *verja* 'to defend'; (middle voice) *verjask* 'to defend oneself'; (as the equivalent of an intransitive verb) *sýna* 'to show', (middle voice) *sýnask* 'to seem'.

**middling, middlingly** (advbs.). There is a certain amount of historical and current evidence that *middling* does duty both as an adjective, where it is the natural form, and as an adverb, where it is not (*She was thin, and light, and middling tall*—H. James, 1880). But *middlingly*, current since Dr Johnson listed it in his Dictionary (1755), remains available. Examples: (middling) *He was middling tall and thin*—D. Francis, 1981; *It looks like a middling-good fastball that suddenly changes its mind and ducks under the batter's swing*—R. Angell, 1987 (US); *She lived on the edge of an upper-middle-class neighbourhood of ranch homes ... and middling-tall maple trees along the boulevards*—J. Welch, 1990 (US); (middlingly) *Middle-aged, middle-class, middlingly bright men*—Daily Tel., 1976; *Psyche ... fared middlingly at* *suggesting an efficient and considerate image for psychologists—New Scientist, 1987; Ivan Ilyich, who has a middlingly successful legal career and has become a judge, is confronted with the fact of his own mortality—A. N. Wilson, 1988.*

**midst** is most commonly used as a noun in fixed phrases, e.g. *in the midst of* 'among, in the middle of', *in our* (or *their*, etc.) *midst* 'among us (or them etc.)'. As a preposition = *amidst*, it is now rare. Example: *'Midst brightly perfumed water-flowing Eighteenth-century silks—E. Sitwell, 1924.*

**midwifery.** In standard use now mostly pronounced as four syllables, /ˈmɪd-ˌwɪfərɪ/, but it is only relatively recently that the second syllable has settled down as /-wɪf-/. The *OED* (1907) gave priority to /ˈmɪdwaɪfrɪ/, and also listed /ˈmɪdɪfrɪ/ (with the *w* silent). In AmE, usage is divided but /-waɪf-/ is commonly used for the second syllable.

**mielie.** See MEALIE.

**might.** See MAY AND MIGHT.

**mighty** (adv.) has been in use in English since about 1300 (*þair blisced lauerd ... þat ... was ... Sa mighti meke* 'Their blessed Lord who was so exceedingly meek'), and has continued to flourish in AmE at all levels except the most formal. In BrE it has receded somewhat in the course of the 20c., and examples are much harder to find: *As I say, I had mighty little to complain of—J. Buchan, 1926; Trolls are slow in the uptake, and mighty suspicious about anything new to them—J. R. R. Tolkien, 1937; Well, boys, before you move on out, I've got a mighty important announcement to make—P. Mann, 1982; Heating the greenhouse ... can be a mighty expensive proposition—Practical Gardening, 1986.*

**migraine.** In BrE the standard pronunciation is /ˈmiːɡreɪn/, but /ˈmaɪ-/ is also very commonly heard. In AmE, /ˈmaɪɡreɪn/ is standard.

**migrant.** See EMIGRANT.

**mikado.** Pl. *mikados*. See -O(E)S 6.

**mil** = one-thousandth of an inch, i.e. 0.0254 millimetre, as a unit of measurement for the diameter of wire, etc. 'The use is discouraged' (*Oxford Dict. for Science*

*Writers and Editors*, 1991). It is printed without a full point.

**milage.** See next.

**mileage.** Now the recommended form in OUP house style, not *milage*.

**milieu** /miːˈljɜː/. For the plural, *milieux* (pronounced the same as the sing.) is recommended, not *milieus* /-ljɜːz/. See -X.

**militate.** See MITIGATE.

**millenarian** /ˌmɪlɪˈneərɪən/, of or relating to, believing in, the millennium. The apparent inconsistency in spelling (-*n*-, -*nn*-) results from the fact that *millenarian*, like *millenary*, is derived from L *millēnī*, the distributive form of *mille* 'thousand'. See MILLENNIUM.

**millenary.** Now for the most part pronounced /mɪˈlenərɪ/ in BrE, though the *OED* (1907) gave only /ˈmɪlmərɪ/, and the *COD* did not admit the second-syllable stressing until 1982. AmE favours the form with first-syllable stressing. See CENTENARY 2.

**millennial.** See CENTENARY 2.

**millennium.** Note that it must be spelt with -*nn*-, not -*n*-. It is an 18c. formation from L *mille* 'thousand' + *annus* 'year', on the analogy of *biennium*, *triennium*, etc. Pl. *millennia* (or -*iums*). See -UM 3.

**milli-.** See KILO-.

**milliard.** In BrE = one thousand million. Now almost entirely superseded by BILLION.

**million.** 1 The spoken equivalent of £1½m is *one and a half million pounds* (not *one million and a half pounds*). See HALF 3.

2 When modified by a preceding quantitative word (a numeral, *many*, *several*, etc.) and followed immediately by a noun, *million* is unchanged in form (*two million people, several million pounds*). When an *of*-phrase follows, the plural form is normally used (*many millions of votes were lost*); but sometimes idiom allows the flexionless form to be used (*a few million of them*).

3 *Million* (not *millions*) is used in the type *Among the eight million are a few hundred to whom this does not apply.*

**4** In BrE *a million* is sometimes used elliptically for *a million pounds*, and in N. America (and other dollar-currency countries) for *a million dollars*.

**5** In printed work *million* is frequently abbreviated to 'm' (as in 1 above).

**mimic** (verb). The inflected forms are *mimics, mimicked, mimicking*. See -C-, -CK-.

**minacious** (first recorded in 1660) and **minatory** (1532) vied with each other in the sense 'menacing, threatening' until the end of the 19c. At some point in the 20c. *minacious* dropped out of use, but *minatory* /ˈmɪnətərɪ/ remains, though used only in fairly solemn contexts. Examples: *He talks in minatory terms to the ghost of the seventeen-year-old boy he had once been*—P. Ackroyd, 1984; *The minatory language from the Soviet Union*—New Yorker, 1988; *The assessment in London ... was that Iraq was engaged in minatory diplomacy designed to squeeze more cash out of its Kuwaiti neighbours*—Independent, 1990.

**mind** (verb). An absolute or intransitive use of *mind* or *mind you* in the imperative, calling attention to or emphasizing what the speaker is saying, is noted by the *OED* (sense 5c) from the early 19c. onward, with examples from Coleridge, Browning, and some other sources. The device is much used at the present time: e.g. (at beginning of statement) *Mind, he wasn't here these last years, scarcely*—J. Gardam, 1985; *Mind, you may feel that sort of thing is passing you by*—P. Lively, 1987; (at end of statement) *It's his Navy too, mind*—V. O'Sullivan, 1985; *Well, all right, but you aren't to do anything, mind*—D. Lessing, 1988; (in form *mind you*) *Mind you, if you think she behaved strangely, you should have seen me*—M. Amis, 1984; *Mind you, I don't know who the mother was*—A. L. Barker, 1987.

**mine.** **1** For *either my or your informant must have lied* or *either your informant or mine must have lied*, see ABSOLUTE POSSESSIVES.

**2** In present-day English, *mine* cannot be used immediately before the noun it qualifies except jocularly (*mine host*) or with a historical purpose. Historically it could be so used if the following noun began with a vowel or *h* (*mine ease, mine heart*). Before a noun the normal form of the possessive pronoun is, of course, *my*.

From the 13c. onward, and still archaistically, *mine* has been placed immediately after the noun it modifies (e.g. *O lady Venus myne!*—Lydgate, c1402; *For doting, not for louing pupill mine*—Shakespeare, 1592). Its natural modern use, however, is as a pronoun: e.g. *She held out her hand, which I clasped in both of mine; Irvan suddenly tilted his head toward mine; these qualities, even to myself, are indisputably mine*.

**mineralogy** is formed, probably after Fr. *minéralogie*, from *minera(l)* + -LOGY.

**mini-.** This combining form of *miniature* (reinforced by the first letters of *maximum*) has been greatly in demand in the 20c., esp. since the 1960s, to designate things, tendencies, etc., that are very small of their kind. A virtually unlimited number of *mini-* combinations has been brought into use in the 20c., e.g. *minibike, mini-budget, minicab, mini-crisis, mini-recession, miniskirt*. An unfortunate by-product of its popularity is the erroneous form *miniscule*, a misspelling of MINUSCULE: e.g. *with more than thirteen thousand entries spread across more than twenty-five hundred double-column pages of miniscule type*—New Yorker, 1986; *He ... peered at the miniscule handwriting*—F. King, 1988; *I was taken through events in miniscule detail*—C. Burns, 1989.

**minimum.** Pl. *minima*. See -UM 2.

**miniscule.** See MINI-; MINUSCULE.

**minister.** The *COD* (1990) definition 'a member of the clergy, esp. in the Presbyterian and Nonconformist Churches' is correct as far as it goes, but the matter is complex and requires fuller treatment. 'A person officially charged to perform spiritual functions in the Christian Church. As a general designation for any clergyman, it is used esp. in non-episcopal bodies. In the Book of Common Prayer it usually means the conductor of a service who may or may not be a priest. A Minister is also one who assists the higher orders in discharging their functions, and in this sense the Deacon and Subdeacon at High Mass are known as the "Sacred Ministers". In yet another sense, the word is used semi-technically of one who "administers" the outward and visible signs of a Sacrament. Thus should a layman baptize in a case of necessity, he would in that case be the "minister" of the Sacrament' (*Oxford Dict. of the Christian Church*, 1957). Cf. PARSON; PASTOR; PRIEST.

**minority.** For the most part the word stands at the opposite end of the field to MAJORITY or in relative contrast to it. Of a political party or other group, often followed by *of*, it means 'a smaller number or part (*a minority of MPs voted for the restoration of capital punishment*)'. *Minority government* is that in which the ruling party has received fewer votes than those given to the other parties together, a common state of affairs in 20c. Britain. A member of a committee or other body who has found little support for his or her views is said to be *in the minority*. But the most noticeable use of the word in the 20c. has been its application to any relatively small group of people differing from others in the society of which they are a part in race, religion, language, political persuasion, etc. Such minorities have always existed, and have often been commented on in earlier times, but they have not always been thought to need constitutional or social, etc., 'rights'. Ethnic minorities, be they blacks in a predominantly white community, disabled people requiring easier access to buildings, single parents needing financial assistance, homosexuals seeking legislation against discrimination, or any other group of disadvantaged people, have brought the word *minority* into unparalleled prominence. The seemingly paradoxical phr. *a growing minority* has come into being, meaning, of course, an increase, not a decrease, in the number of members of a given minority group. But this is not a startlingly new development: cf. *We are a minority; but then we are a very large minority*—Burke, 1789. Running parallel to this has been the 20c. use of *minority* as a quasi-adj. meaning 'of, for, or appealing to a minority of people, frequently with the association of "serious, intellectual" (as opposed to "lowbrow, mass")'. An early exponent of the idea, the Cambridge don F. R. Leavis, put out an influential book entitled *Mass Civilization and Minority Culture* in 1930. The BBC introduced special programmes, and later separate radio and television channels, for minority audiences seeking minority culture. The nature and boundaries of this culture

continue to form subjects of endless debate, but the existence of such a phenomenon is not disputed.

**Minotaur.** Now usu. pronounced /ˈmaɪnətɔː/, though the *OED* (1907) gave only /ˈmɪn-/.

**minus.** The informal use of *minus* as a marginal preposition = 'short of, having lost; without' (*they arrived back minus their tracksuits*) has been in existence since the mid-19c.: e.g. *We reached our munzil* [sc. halting-place] ... *about six in the evening, minus one horse*—J. B. Fraser, 1840; *The Englishman got back to civilization minus his left arm*—*Review of Reviews*, 1903; *I hope he comes minus his wife*—*CGEL*, 1985.

**minuscule** /ˈmɪnəskjuːl/ (but until about the mid-20c. almost always /mɪˈnʌskjuːl/).

**1** Spelt thus, not *miniscule*.

**2** In palaeography, 'a small letter, as opposed to a capital or uncial; the small cursive script which was developed from the uncial during the 7th–9th centuries; also, a manuscript in this writing' (*OED*).

**minute** (noun and adj.). See NOUN AND ADJECTIVE ACCENT.

**minutiae** /maɪˈnjuːʃiː/ (pl. noun), precise, trivial, or minor details. The singular form, rarely called for, is *minutia*.

**Miocene.** One of a family of geological epoch names ending in *-cene* (*Holocene, Pleistocene*, etc.), all irregularly formed from classical elements, but long since established and uncancellable. The periods covered by the terms are: *Holocene*, 0.01 (million years ago); *Pleistocene*, 2; *Pliocene*, 5; *Miocene*, 24; *Oligocene*, 38; *Eocene*, 55; *Palaeocene*, 65. The word *Miocene* was coined in 1831 by the distinguished scientist William Whewell (1794–1866) from Gk μείων 'less' + καινός 'new, recent'. So too was *Eocene* from Gk ἠώς 'dawn' + καινός; and, at various dates, several other scientific words. Fowler (1926) described these geological words as 'monstrosities' ('The elements of the word [sc. *Miocene*] are Greek, but not the way they are put together, nor the meaning demanded of the compound'), but the time has come for the hatchet to be buried. A simple *irreg.* suffices.

**mis-.** Words formed with this prefix do not require a hyphen, even when the stem of the word begins with *s*: thus *misbehave, miscarriage, miscount, mismanage, misshapen, misspelling, misspent,* etc.

**misalliance.** See MÉSALLIANCE.

**misandry** means 'hatred of men'. A *misanthrope* is a person who hates fellow human beings. A *misogynist* is a person who hates women, and *misogyny* means 'hatred of women'. A *misogamist* is a person who hates marriage, and *misogamy* means 'hatred of marriage'.

**misanthrope.** See prec.

**misapprehensions.** It may be convenient for readers to have a list of some of the misconceptions which are often accepted as being unquestionably true. Discussion of each will be found under the word printed in small capitals.

> That *the* COMITY *of nations* means the members of a sort of league.
> That CUI BONO? means What is the good or use?
> That a DEVIL's advocate, or *advocatus diaboli,* is one who pleads for a wicked person.
> That an EXCEPTION strengthens a rule.
> That FRANKENSTEIN was a monster.
> That *more* HONOURED *in the breach than the observance* means more often broken than kept.
> That ILK means clan or the like.
> That many a MICKLE *makes a muckle.*
> That a PERCENTAGE is a small part.
> That good writers do not end a sentence with a preposition (see PREPOSITIONS 2)
> That any order of words that avoids a SPLIT INFINITIVE is better than any that involves it.

**miscegenation,** mixture of races, is an irregular formation (in the 19c.) from L *miscēre* 'to mix' + *genus* 'race' + *-ation.* There is no classical precedent for a combining form *misce-* from *miscēre.*

**miscellany.** In BrE now always pronounced /mɪˈselənɪ/, though earlier in the 20c. (e.g. in the *OED,* 1907; Daniel Jones, 1917) first-syllable stressing was preferred. In AmE /ˈmɪsəˌlenɪ/.

**mischievous.** The spelling of the word as *mischievious* and the pronunciation /mɪsˈtʃiːvɪəs/ are both erroneous: the four-syllabled form is sometimes used by standard speakers in mocking imitation of uneducated speech.

**misdemeanour.** AmE *misdemeanor.*

**miserere, misericord** (hinged seat in choir-stall). The first is labelled 'An incorrect use' in the *OED*: it was first recorded at the end of the 18c. and is still used occasionally beside the regular form. The regular word in this use is *misericord.*

**mislead.** The pa.t. and pa.pple are both *misled.*

**misogamist, misogamy, misogynist, misogyny.** See MISANDRY.

**misogynist, misogyny.** See GREEK G.

**misprints to be guarded against.** These are of two main types: (*a*) choice of the wrong word of two similar-looking words, e.g. *adverse* and *averse, casual* and *causal, deprecate* and *depreciate, principal* and *principle, reality* and *realty;* (*b*) printing of forms that 'look' plausible but are in fact erroneous: e.g. *concensus* (for the correct form *consensus*), *indentify* for *identify, millenium* for *millennium, miniscule* for *minuscule, supercede* for *supersede.* Vigilance in proofreading is required to prevent such mistakes.

In his *Esprit de Corps: Sketches from Diplomatic Life* (1957), Lawrence Durrell describes the perils of printing the *Central Balkan Herald* in wartime Serbia. 'The reason for a marked disposition towards misprints was not far to seek; the composition room, where the paper was hand-set daily, was staffed by half a dozen hirsute Serbian peasants with greasy elf-locks and hands like shovels.' Among the headlines Durrell cites were: *Minister fined for kissing in pubic; Wedding bulls ring out for princess; Britain drops biggest ever boob on Berlin.* And various gems from the text of news reports: e.g. *In a last desperate spurt the Cambridge crew, urged on by their pox, overtook Oxford.*

*OCELang.* (p. 121) lists some traditional examples of misprints: 'On many occasions, editions of the AV have contained misprints which have led to their receiving special names, such as the *Wicked Bible* (1632), so called because the word "not" was omitted in the Seventh Commandment, making it read *Thou shalt commit adultery;* the *Vinegar Bible* (1717), so called because the Parable of the Vineyard became *The Parable of the Vinegar;* and the *Printer's Bible,* where a misprint makes the Psalmist complain that

*printers have persecuted me without cause,
as opposed to princes.'*

**misquotations.** In a perfect world,
familiar lines or passages from the great
classical works of English literature, or
from famous speeches, would never be
misquoted. But the evidence points elsewhere. Readers will judge for themselves
with what degree of success they normally cite the passages set out below.

In the sweat of thy *face* shalt thou eat
bread (not *brow*)—Gen. 3: 19.

I am escaped *with* the skin of my teeth
(not *by*)—Job 19: 20.

To gild refined gold, to *paint* the lily (not
*gild the lily*)—Shakespeare, *King John*.

A goodly apple rotten at the *heart* (not
*core*)—Shakespeare, *Merchant of Venice*.

An *ill-favoured* thing, sir, but mine own
(not *poor*)—Shakespeare, *As You Like It*.

I will a *round* unvarnished tale *deliver* (not
*plain . . . relate*)—Shakespeare, *Othello*.

But yet I'll make assurance *double* sure
(not *doubly*)—Shakespeare, *Macbeth*.

Tomorrow to fresh *woods* and pastures
new (not *fields*)—Milton, *Lycidas*.

*That* last infirmity of noble *mind* (not *The
. . . minds*)—ibid.

*Fine* by degrees, and beautifully less (not
*small*)—Prior, 'Henry and Emma'.

They kept the *noiseless* tenor of their way
(not *even*)—Gray, *Elegy Written in a
Country Church-Yard*.

A little *learning* is a dangerous thing (not
*knowledge*)—Pope, *Essay on Criticism*.

The best laid *schemes* o' mice an' men (not
*plans*)—Burns.

Water, water, every where, *Nor any* drop to
drink (not *and no a*)—Coleridge, *The
Rime of the Ancient Mariner*.

*Power tends to corrupt*, and absolute power
corrupts absolutely (not *Power
corrupts*)—Lord Acton, letter of 1857.

*My dear*, I don't give a damn—M. Mitchell,
*Gone with the Wind*, 1936. Compare Clark
Gable's words in the 1939 film version:
*Frankly, my dear*, I don't give a damn!

I have nothing to offer but *blood, toil, tears
and sweat* (not *blood, sweat, and tears*)—
W. S. Churchill, *Hansard*, 1940.

If she can stand it, I can. *Play it!* (not *Play
it again, Sam*)—H. Bogart, *Casablanca*
(1942 film).

**misrelated clauses, constructions,
etc.** In the last few years an American
correspondent has submitted numerous
examples of such infelicities, but it must
be emphasized that similar examples
occur in newspapers in all other parts

of the English-speaking world too. A
selection: *The Pope wore a full-length white
coat Saturday to protect against the biting
wind in his speech in Punta Arenas* (? a windy
speech); *Fluent in German, his position at
the university was professor in . . . Germanic
languages and literatures* (a position can
hardly be fluent); *Dean lived in Fairmount
with an aunt and uncle through his teens
after the death of his mother at age 9* (? the
youngest mother in history); *A Yugoslav
woman married to a Kuwaiti and about 20
Asians were among those fleeing* (a notable
example of polygamy). Despite the vigilance of sub-editors, correspondents, and
others involved, a small number of such
awkwardnesses are bound to occur from
time to time in the torrid world of newspapers. Cf. the more serious syntactic
blunders dealt with in the article UNAT-
TACHED PARTICIPLES.

**Miss.** 1 The title of an unmarried
woman or girl, or of a married woman
retaining her maiden name for professional purposes. Since the early 1950s,
challenged to some extent by MS, prefixed to the name of a woman, regardless
of her marital status.

2 *The Misses Jones* is the old-fashioned
plural, still used when formality is required, e.g. in printed lists of guests
present, etc.; otherwise the type *the Miss
Joneses* is now usual.

**misshapen, misspelling, misspent.**
See MIS-.

**missile.** In BrE /ˈmɪsaɪl/, in AmE /ˈmɪsəl/.
See -ILE.

**-mistress.** As a second element it occurs
in only a few words, all of them contrasting with words in *-master*: e.g. *headmistress*
(first recorded in 1872), *housemistress*
(1875), *paymistress* (1583), *postmistress*
(1697), *quartermistress* (1917), *schoolmistress*
(a1500), *stationmistress* (1897), *taskmistress*
(1603). The OED lists no feminine equivalents of *bandmaster, choirmaster, grand
master, harbour-master*.

**mitigate** is a usu. transitive verb meaning 'to make milder or less intense or
severe; to moderate': e.g. *The King's
eventual course of action did nothing to mitigate the conspirators' difficulties*—Antonia
Fraser, 1979; *The man was a kind of monster
of self, and nothing in his subsequent career*

had served to mitigate this harsh but considered judgment—L. Auchincloss, 1980; *The clear selection of a subordinate role still did not mitigate male fears of competition—* Dædalus, 1987; *Insurance companies ... are exercising their right of subrogation to mitigate their losses—*Times, 1991. For the last several decades *mitigate* has sometimes been confused with the intransitive verb *militate* (*against*) '(of facts or evidence) to have force or effect'. Correct uses of *militate against*: *all those evils which militate against nobility in life—*NY Rev. Bks, 1980; *His emphasis on tone militated against acceptance of French Impressionism—*F. Spalding, 1988. Examples of *mitigate against* erroneously used for *militate against*: *These factors mitigate against an adequate description of the material culture—*B. Cunliffe, 1974; *But the time factor may have mitigated against that course—*Times, 1977; *All of these factors might mitigate against cytologic detection and make diagnosis by punch biopsy less likely—*Cancer, 1979. The idea of countering (some condition or circumstance) is all that the two verbs have in common, apart from the fact that they are both three-syllabled words divided by only two consonants.

**mitre.** So spelt in BrE; in AmE usu. *miter*.

**mixed metaphor.** See METAPHOR 4. Further examples: *The chief economist of the Commonwealth Bank ... said it was an impressive and comprehensive package, although judgment would have to be withheld until the nuts and bolts of the reforms were digested—*Australian, 1991; *One day, when he's in charge of finance or foreign affairs, those lions and tigers will come home to roost. Stephen smiled. His boss was renowned for the mixture of his metaphors—*R. McCrum, 1991; *I became aware that very frequently important objects were not brought forward because the committee was seen to have no teeth—*letter in Independent, 1992; *David Blunkett, local government spokesman, deplored any attempt to 'steamroller and trigger people into making rapid decisions'—* Independent, 1992.

**mizen** (naut.). This spelling is used in OUP house style, but *mizzen* is also commonly used in all English-speaking countries.

**-m-, -mm-.** Monosyllabic words containing a simple vowel (*a, e, i, o, u*) before

*m* normally double the consonant before suffixes beginning with a vowel (*rammed, hemmed, dimmest, drummer*) or before a final *-y* (*Pommy, tummy*); but remain undoubled if the stem contains a diphthong (*claimed*), doubled vowel (*gloomy*), or a vowel + consonant (*alarming, balmy, squirmed, wormy*). Words of more than one syllable ending in *m* (e.g. *emblem, maximum, pilgrim, venom, victim*) rarely have vowel-led suffixes; but when they do the final *m* is normally left undoubled (*emblematic, maximal, pilgrimage, venomous, victimization*). Exceptions are words that end in what is perceived as a word (*bedimmed, hornrimmed, overcramming*) or that end in *-gram* (*diagram: diagrammatic, monogram: monogrammed*).

**mnemonic** /nɪˈmɒnɪk/. A device, such as a verse, to assist the memory, e.g. *Richard Of York Gave Battle In Vain* as a mnemonic of the order of the colours of the spectrum: red orange yellow green blue indigo violet; *All Cows Eat Grass* as a reminder of the bass clef ACEG: *No Plan Like Yours To Study History Wisely* as a reminder of the royal families of England: Norman, Plantagenet, Lancaster, Yorkists, Tudors, Stuarts, Hanover, Windsor; and *In fourteen hundred and ninety two Columbus sailed the ocean blue* as a reminder of a famous date.

The word was first recorded in English (from Gk μνημονικός 'mindful') in the 18c. (adj.) and the 19c. (noun).

**mobile.** Both the adj. ( — movable) and the noun ( — decorative structure) are pronounced /ˈməʊbaɪl/ in BrE. In AmE the adj. is usu. pronounced to rhyme with *noble*, and the noun is given the final sound of *automobile*.

**mobocracy** (colloq.). (*a*) rule by a mob; (*b*) a ruling mob. An excellent example of an irregular formation (first recorded in 1754) that has been admitted to the language without serious disapproval. It is formed from *mob*, itself a shortened form of L *mōbile vulgus* 'the movable or excitable crowd', after *democracy*. A rough-hewn word for a disorderly concept.

**moccasin.** So spelt, not the many variants.

**mocha** (variety of chalcedony; coffee). Now usu. pronounced /ˈmɒkə/ in BrE, though earlier in the century (*OED*, 1907; Daniel Jones, 1917) the pronunciation /'məʊkə/ was customary, as it still is in AmE.

**modal verb,** one of a group of auxiliary verbs, e.g. *shall, will, can, may, must*, used to express the mood of another verb. So *modality*, the expression of obligation, necessity, permission, etc., by means of modal verbs. See EPISTEMIC MODALITY.

**model** (verb). The inflected forms are *modelled, modelling* (AmE usu. *modeled, modeling*). See -LL-, -L-.

**modest.** Comparative and superlative forms are seldom called for: historically *modester* and *modestest* were used, esp. in the 17c. and 18c., but *more modest* and *most modest* are more idiomatic forms now.

**modulus.** (math.) Pl. *moduli* /-laɪ/. See -US 1.

***modus operandi.*** (L) a plan of working. Pl. *modi operandi* /'məʊdaɪ ˌɒpə'rændaɪ/. Print in italics.

***modus vivendi.*** (L) a way of living; (but usu.) an arrangement whereby parties in dispute can carry on pending a settlement. Pl. *modi vivendi* /'məʊdaɪ vɪ'vendaɪ/. Print in italics.

**mogul.** In its extended sense 'an important or influential person', always so spelt. In its historical sense 'any of the Muslim rulers of N. India in the 16c.–19c.', now usu. spelt *Mughal* after Arabic, Persian *muġul, muġal* 'Mongol'.

**Mohammed(an).** In the course of the 20c., MUHAMMAD(AN) became the customary spellings in English.

**moiety.** A legal and literary word for 'a half'. Pl. *moieties*.

**moire, moiré.** The first (in full *moire antique*) is the name of a watered silk material, and is pronounced /mwɑː(r)/. The second is (*a*) (adj.) (of silk) watered; (of metal) having a patterned appearance like watered silk; and (*b*) (noun) this patterned appearance; also, = *moire*. *Moiré* is pronounced /'mwɑːreɪ/.

**molasses,** though plural in form, is construed as a singular.

**mold.** See MOULD.

**mollusc.** Thus spelt in BrE, but also *mollusk* in AmE.

**molt.** See MOULT.

**molten.** See MELTED.

**moment.** For *at this moment in time*, see CLICHÉ. A modern example: *Well, love, you can wear it* [sc. *your hair*] *white but at this moment in time one can't, you know*—M. Wesley, 1983.

**momentarily.** In BrE it means only 'for a moment, briefly' (*He wondered momentarily if he had crossed into the wrong lane*). In AmE it may also mean 'at any moment, soon' (*This is your Captain speaking. Please fasten your seat belts as we shall be taking off momentarily*). The American sense is first recorded in the late 1920s. The AmE pronunciation, with the main stress falling on *-árily*, has made some inroads in BrE, where the word's main stress is traditionally placed on the first syllable. See next.

**momently** (adv.) has had a precarious hold on the language since, in its primary sense 'from moment to moment', it was first recorded in use in a work of 1676. It has also been used to mean 'at any moment, on the instant', and 'for the moment; for a single moment'. To judge from the available evidence this last sense seems to be the prevailing one now, and the word is largely restricted to literary sources. Examples: *Thoughts ... look at me With awful faces, from the vanishing haze That momently had hidden them*—George Eliot, 1868; *She believes that Alix will be there, and indeed momently she is*—M. Drabble, 1987; *You acknowledged, however momently (that infinite moment) that at least what I claim is true*—A. S. Byatt, 1990.

**momento.** An occasional, not recommended, variant of MEMENTO: e.g. *And the London frowsty Casino, a momento of which I enclose*—Dylan Thomas, 1951; *a satisfyingly real momento of my enjoyment*—M. Forster, 1968 (*WDEU*).

**momentum.** Pl. *momenta.* See -UM 2.

**monachal, monastic, monkish.** Each has a corresponding abstract noun: *monachism, monasticism*, and *monkery*. Of the three sets *monastic(ism)* is the one that is most often called on. *Monkish* and especially *monkery* sometimes have scarcely concealable derogatory connotations. And *monachal* and *monachism* have become progressively less well known since they entered the language in the 16c. to the point that they can safely be used now only in scholarly work.

**monarchal, monarchial, monarchic, monarchical.** All four adjectives entered the language within a generation of one another c1600, and all have been used in the broad senses 'of, belonging to, or characteristic of a monarch', 'ruled by a monarch', 'that advocates monarchy as a form of government'. All four are still listed in the major dictionaries of current English, except that neither *monarchal* nor *monarchial* is listed in the principal dictionaries for foreign learners of English. From the available evidence it looks as if *monarchial* and *monarchical* are now the preferred forms, and that *monarchal* and *monarchic* are drawn on rather less. Examples: (monarchial) *Monarchial displeasure is not lightly to be incurred—Observer,* 1981; *he* [sc. a resort developer in the Bahamas] *was sometimes criticized for operating in a monarchial style—NY Times,* 1988; *They are shown to have been ... a major piece of political propaganda to celebrate monarchial rule—TLS,* 1991; (monarchical) *He maintained a monarchical sensitivity in matters of protocol—M. Meyer,* 1967; *The United States came into being, having thrown off the monarchical and colonial principles of the British Crown—A. N. Wilson,* 1988; *the House of Commons, many of whose members have been agitating for months for a reduction in monarchical privilege—Times,* 1992.

**monastic(ism).** See MONACHAL.

**Monday.** See FRIDAY.

**monetarism, monetarist, monetary.** In these days when there are daily references to the market economy and to the concept of *monetarism,* all three words seem to have settled down in BrE with the first syllable pronounced /ˈmʌn-/ after a period when /ˈmɒn-/ and /ˈmʌn-/ stood side by side. In AmE the choice has gone the other way.

**moneyed** (adj.). So spelt, not *monied.*

**-monger.** Used as a second element in compounds, *-monger* has produced a few common words meaning simply 'one who trades in (the item specified in the first element)': thus *costermonger* (first recorded in 1514, from *costard* apple), *fishmonger* (1464), *ironmonger* (1343). Somehow these have remained untainted by the limitless class of *-monger* words which, from the mid-16c. onward, have come into being implying one who carries on a contemptible or discreditable 'trade' or 'traffic' in what is denoted by the first element (as the *OED* expresses it). Thus, for example, *gossip-monger* (1836), *scandal-monger* (1721), *scaremonger* (1888), *warmonger* (1590), *whoremonger* (1526).

**mongol.** In the first half of the 20c. regularly applied to a person suffering from *mongolism* or (what later came to be called) DOWN'S SYNDROME, a congenital disorder due to a chromosome defect.

**mongoose,** a small flesh-eating civetlike mammal: a word derived from Marathi *mangūs.* Its plural is *mongooses.*

**monies.** The natural plural of *money* when the sense is 'sums of money' is *moneys.* But in legal and accountancy parlance from at least the mid-19c. the irregular form *monies* has taken hold and now seems uncancellable. Examples (from three different countries): *Certain monies had been put aside for them—A. Brookner,* 1988; *The government has done just the opposite by giving Maori Affairs monies to the other departments—S. Jackson* in *Metro* (NZ), 1988; *Some of the bond monies are exempt because of their funding sources—Arizona Republic,* 1988.

**monk.** See FRIAR.

**monkery.** See MONACHAL.

**monkey** (noun). Pl. *monkeys.*

**monkey** (verb). The inflected forms are *monkeys, monkeyed, monkeying.*

**monkish.** See MONACHAL.

**monocle.** Adopted from French in the mid-19c. it means, as one modern dictionary defines it, 'a lens for correcting defective vision of one eye, held in position by the facial muscles', and another simply as 'an eyeglass for one eye'. It is derived ultimately from late L *monoculus* 'one-eyed'.

**monologue.** This and *soliloquy* are precisely parallel terms of Greek and Latin origin respectively; but usage tends to restrict *soliloquy* to talking to oneself or thinking aloud without consciousness of an audience whether one is in fact overheard or not, while *monologue*, though not conversely restricted to a single person's discourse that *is* meant to be heard, has that sense much more often than not and is especially used of a talker who monopolizes conversation, or of a dramatic performance or recitation in which there is one actor only.

**monotonic, monotonous.** The secondary sense of *monotonous* (same or tedious) has so nearly swallowed up its primary sense (of one pitch or tone) that it is worth remembering the existence of *monotonic*, which has the primary sense only. In *monotonic* the stress is on the third syllable, in *monotonous* on the second.

***Monseigneur*** /mɒnseˈnjɜː/, title given to an eminent French person, esp. a prince, cardinal, archbishop, or bishop. Pl. *Messeigneurs* /meseˈnjɜː/.

***Monsieur*** /məˈsjɜː/, title or form of address used of or to a French-speaking man, corresponding to *Mr* or *Sir*. Pl. *Messieurs* /meˈsjɜː/.

***Monsignor*** /mɒnˈsiːnjə/ or /-ˈnjɔː/, title of various RC prelates. Pl. *Monsignori* /-ˈnjɔːrɪ/ or *Monsignors*.

**mood.** It may save misconceptions to mention that the grammar word has nothing to do with the native word meaning frame of mind, etc.; it is merely a variant of *mode*, i.e. any one of the groups of forms in the conjugation of a verb that serve to show the mode or manner by which the action denoted by the verb is represented—*indicative*, *imperative*, and *subjunctive*.

**moot.** A *moot point* or *moot question* is a debatable or undecided one. The word, which is of native origin, should not be confused with *mute* 'silent'.

**moral, morale** (nouns). The history of these two words is complex and the

complexities cannot be set out here in the space available. It must suffice to say that Fowler (1926) treated the problem at length mainly because *morale* was being used in contexts of the type *the morale of the troops is excellent*, whereas the corresponding French word for this sense was *moral*. There is no longer a problem. *Moral* /ˈmɒrəl/ means 'a moral lesson of a fable, story, etc.; a moral maxim', and in the plural, 'moral behaviour, esp. in sexual conduct'; whereas *morale* /məˈrɑːl/ means only 'the mental attitude or bearing of a person or group, esp. as regards confidence, discipline, etc.'.

**moratorium.** The recommended pl. is *moratoriums*; but some use *moratoria*. See -UM 3.

**more.** 1 In origin from OE *māra* (adj.), comparative of *micel* 'big', its subsequent history is too complicated to be dealt with here. Suffice it to say that until about 1600 it was often contrasted with *mo* (from OE *mā*, comparative of *micle* 'much'): *mo* was used with pl. nouns and *more* with non-count nouns. Thus, in Shakespeare, *and let's first see moe ballads* (*Winter's Tale*), but *is there more toil?* (*Tempest*). The *mo/more* distinction dropped out during the 17c. and survives only in some regional forms of English.

2 Readers will hardly need to be reminded that *more* is now used before comparatives (*more hotter*) only by the illiterate, but that it was a standard use in earlier centuries (*and his more braver daughter could control thee—Tempest*). In comparative constructions that are legitimate, *more* is used with adjectives of three syllables or more (e.g. *difficult*, *memorable*), with many adjectives of two syllables (e.g. *afraid*, *awful*, *harmless*), and with adverbs ending in -*ly* (e.g. *highly*, *slowly*). In certain circumstances even monosyllabic adjectives can take *more*: (*a*) when two adjectives are compared with each other, e.g. *More dead than alive*; (*b*) for stylistic reasons, e.g. *This was never more true than at present*.

3 It is broadly true that *more* should not be used with absolute adjectives like *complete*, *equal*, *unique*, but that in practice they often are: see ADJECTIVE 4.

4 Number of noun and verb after *more than one*. Despite its plural appearance, the phrase is normally followed by a

singular noun and verb: e.g. *more than one journalist was killed* (not *journalists*, not *were*). But if *one* is replaced by a larger number, naturally a plural verb is required: *more than ten journalists were killed*. Any other slight disturbance of the phrase *more than one* (e.g. if it should be immediately followed by an *of*-phrase) leads to the use of a plural verb: e.g. *In the positions defined above in which more than one of these morphs occur—Language*, 1950.

**5** 'When one of the quantitative words which otherwise require *of* is accompanied by *more*, the latter word influences the construction so that no preposition is required' (Jespersen, 1909–49, vii). Compare *I had a great deal of money then* with *I had a great deal more money then*; and *a lot of power* with *a lot more power*.

**6** *More of* is sometimes used before a noun or noun phrase as an intensifier meaning merely 'greater in degree' (e.g. *he is more of an ass than I thought*); similarly, *more than* before an adj. (e.g. *You are more than welcome*).

**7** As adjective *more* was once (14–19c.) commonly used to qualify the designation of a person with the sense 'entitled to the designation in a greater degree': e.g. *A more heretike than either Faustus or Donatus*—Reginald Scot, 1584. This use survives only in the phr. *(the) more fool (you*, etc.).

**8** Historically (OE to the early 19c.) *(the) more* was used to mean 'greater in degree or extent': e.g. *for our more safety* Thomas Heywood, 1632; *to make the miracle the more*—Southey, 1829. This use survives only in the phr. *(the) more's the pity*.

**9** For difficulties arising from some uses of *much more* and *much less*, see MUCH 3(*b*).

**mores** /ˈmɔːriːz/ or /-eɪz/. In origin the plural of L *mōs* 'manner, custom', *mores* is normally construed as a plural in English (e.g. *within limits the mores of a morphological species* remain *the same throughout the geographic range*).

**morgue. 1** A place, usu. in a hospital, in which dead bodies are kept until burial or cremation, esp. bodies awaiting identification by relatives. Cf. MORTUARY.

**2** (In a newspaper office) a room or file of miscellaneous information, esp. for future obituaries.

**Mormon,** a member of a religious body calling itself The Church of Jesus Christ of Latter-Day Saints, a millenary religion founded in 1830 at Manchester, New York, by Joseph Smith (1805–44) on the basis of revelations in the 'Book of Mormon'.

**morning.** *Morning Service, Morning Prayer, Matins* are alternative terms (the first of them unofficial) for the order of public morning prayer in the Church of England. The corresponding service in the evening is officially called *Evensong*, less commonly *Evening Service*.

**morocco,** a fine, flexible material used in bookbinding and shoemaking. Pl. *moroccos*. See -O(E)S 6.

**morphia, morphine** are synonymous terms for a narcotic drug used medicinally to relieve pain, the former being a layman's term and the second the scientific term.

**morphology.** The (study of the) structure and form of words; e.g. the observation that *post-modernism* contains three elements (called *morphemes*), *post-*, a prefix meaning 'after', *modern*, adj., and *-ism*, a suffix used to form abstract nouns.

**mortal.** For *the mortal remains of*, see STOCK PATHOS.

**mortgagee,** the creditor in a mortgage, usu. a bank or building society. A *mortgager* (in law work *mortgagor*) is the debtor in a mortgage. Note that *mortgagor* is a rare case, paralleled only by *gaol* (and derivatives) and *margarine*, in which a *g* is 'soft' before *a*, *o*, or *u*.

**mortician** is an American genteelism (first recorded in 1895) for an *undertaker* (1698) or *funeral director* (1886).

**mortise,** hole for receiving the tenon in a lock. Thus spelt in OUP house style, not *mortice*.

**mortuary,** a room or building in which dead bodies may be kept until burial or cremation; a funeral parlour. Cf. MORGUE 1.

**Moslem** /'mɒzləm/. Now mostly abandoned in favour of MUSLIM.

**mosquito.** Pl. *mosquitoes.* See -O(E)S 1.

**most,** as noun, adj., and adv., has innumerable uses and many historical oddities: a few stray observations must
suffice. It is important to bear in mind
that it is 'the superlative of the three
adjectival notions now denoted by *great,
much,* and *many*' (*OED*). **1** As noun the
type *most of this* governs a singular verb
(*most of this is true*), whereas when the
noun following *of* is plural, a plural verb
is required (*most of those at the meeting
were silent throughout*).

**2** Avoid using *most* when a contrast
is being made between two people or
things: e.g. *of the two dictionaries this is the
more* (not *most*) *useful.*

**3** It is not necessary to repeat *most*
when two adjectives joined by *and* follow
(e.g. *one of the most brilliant and successful
people in America*).

**4** Instances of *most* + the superlative
form of adjectives (*most cleverest*) are now
a sign of illiteracy, but from the 15c. to
the 19c. went unopposed: e.g. *With the
most boldest, and best hearts of Rome*—Shakespeare, 1601; *One of the most wretchedst
Spectacles in the World*—W. Penn, 1683; *I
was always first in the most gallantest scrapes
in my younger days!*—T. Hardy, 1878. In
superlative constructions that are legitimate, *most* is used with adjectives of
three or more syllables (e.g. *notable, convenient*), with many adjectives of two syllables (e.g. *evil, dreadful, stylish*), and with
adverbs ending in *-ly* (e.g. *blatantly, nearly*).
It may also for stylistic reasons be used
with monosyllabic adjectives: e.g. *That
was the most cruel thing you could have said.*

**5** The two types of superlatives can cooccur without awkwardness: e.g. *Beyond
these tropes may lie the profoundest fears and
most appalling lusts; one of the deepest and
most sensitive studies I've yet read.*

**6** *Most* governing an adj. frequently
has an intensive rather than a superlative function: e.g. *a most remarkable
woman; this most fashionable garment.*

**7** Like *more, most* is not normally used
with absolute adjectives, but for stylistic
reasons it sometimes is: *That's the most
perfect thing I've ever seen*—V. O'Sullivan,
1985 (NZ). See ADJECTIVE 4.

**8** A rogue elephant, first recorded in
America in the mid-1880s, is the double
or strengthened superlative (*the*) *mostest.*
It is used only in jocular contexts. Examples: *Here's the hostess with the mostest*—
*Daily Herald,* 1958; *A great ambition to be
there firstest with the mostest*—*Times,* 1959; *I
reckon you admire the mostest in anything*—A.
Hunter, 1973.

**9** For *most important(ly),* see IMPORT-
ANT; IMPORTANTLY.

**-most.** Used to form adjs. in the superlative degree (e.g. *endmost, farthermost,
furthermost, headmost, hindmost, inmost, innermost, outermost, topmost, uppermost, utmost, uttermost*), *-most* is an altered form
of OE *-mest.* In late OE, adjs. in *-mest* came
to be regarded as compounds of *most*
and were repronounced to accord with
it, except that the absence of stress led to
the occasional retention of the spelling
*-mest* down to the 16c. and doubtless also
a relaxed pronunciation. The *OED* (1907)
remarked that in colloquial contexts the
pronunciation /-məst/ was 'usual', and
Fowler (1926) positively recommended
this pronunciation in all circumstances.
But no one has listened, and the standard
pronunciation now of all the relevant
words is with final /-məʊst/, i.e. exactly
as that for the separate word *most.*

**most, almost.** Treated in most major
dictionaries under *most* (adv.) but almost
certainly an aphetic form of *almost*: it is
often written *'most.* It has been in use
since the 16c. (first in Scotland) but now
is effectively limited to some UK dialects
and to AmE: e.g. *I 'most met my death
climbing up just now*—S. Merwin and W. K.
Webster, 1901; *He moved by jerks, and he
had most no tail*—Z. N. Hurston, 1935; *Most
anybody can play*—S. Greenbaum and J.
Whitcut, 1988; *Dewey knew no fear, would
just roar on into most any species of difficulty*—T. R. Pearson, 1993.

**mostest.** See MOST 8.

**mostly.** The ordinary sense 'for the most
part, in the main' causes no problems.
Examples: *The remaining letters are mostly
concerned with the octavo edition of Lorenzo;
An hour's drive along the arid, mostly barren
shore; O'Connor's days are spent mostly at
the Catholic Center on First Avenue and Fifty-
fifth Street.* Readers of texts from the 17c.

to the early 19c. will now and then encounter *mostly* used to mean 'in the greatest degree, to the greatest extent': e.g. *It* [sc. the epithet] *was applied to those Things which were mostly esteemed,* 1754; *the person whose society she mostly prized*—J. Austen, 1818 (*WDEU*). But this second sense is no longer current: *most* is used instead.

**mother.** For the *Mother of Parliaments,* see SOBRIQUETS.

**mother-in-law.** See -IN-LAW.

**Mother's Day.** In Britain, another name for *Mothering Sunday,* the fourth Sunday in Lent, traditionally a day for honouring mothers with gifts. In America, *Mother's Day* is an equivalent day on the second Sunday in May.

**moths** (pl. noun). In BrE the pl. is pronounced /mɒðs/; in AmE normally /mɔːðz/, sing. /mɔːθ/.

**motif** /məʊˈtiːf/. This mid-19c. loanword from French has drawn apart from the much older (14c.) French loanword *motive*. A *motif* is now chiefly a dominant theme or distinctive feature in a literary work; and in music a brief melodic or rhythmic formula out of which larger passages are developed. A *motive,* on the other hand, is a factor or circumstance that 'moves' or induces a person to act in a certain way.

**motivate, motivation.** These English formations of the 19c. (after Fr. *motiver, motivation,* Ger. *motivieren, Motivierung*), and their 20c. derivatives *motivational, motivator,* etc., have turned out to be perfectly suited to the needs and mood of those who are concerned with the psychological and social motives that influence people in their daily behaviour. The degree of *motivation* of people at work, at school, at war, etc., or its total absence, have been the subject of theses, monographs, memoranda, speeches, etc., throughout the century, and esp. in its second half. In 1967 George Steiner teasingly called *motivation researchers* 'gravediggers of literate speech'. Certainly the language used by such people and their analogues in commercial departments everywhere seems often to be characterized by verbosity, triteness, and banality.

**motive.** See MOTIF.

**mot juste.** Time has moved on since Fowler (1926) described *mot juste* as 'a pet literary critics' word, which readers would like to buy of them as one buys one's neighbour's bantam cock for the sake of hearing its voice no more'. He complained that it was to be found neither in French dictionaries nor in English ones. The lexicographers in Oxford and elsewhere have done their work, and have established that it entered English from French near the beginning of the 20c.: *Here and throughout we have conspicuously th: mot juste, not one too many and each where it will tell*—*Nation* (NY), 1912.

There does, however, appear to be a problem of usage between the two languages. French dictionaries mention the expression only in so far as it illustrates a precise sense of *juste. Mot juste* is thus not considered to be a compound in French and means nothing more than its literal translation—i.e. the right word, where right = correct. I am assured by the compilers of the Oxford–Hachette dictionaries that 'the French native speakers on the dictionary team are unanimous in opting for a purely literal use of *mot juste.* It would never be used in French in the same way or in the same context as in English.'

**motorcade.** See CAVALCADE.

**motto.** Plural *mottoes.* See -O(E)S 1.

**moujik,** Russian peasant. A variant of MUZHIK.

**mould** (AmE *mold*). The three common words so spelt (shape, fungous growth, loose friable earth, and three corresponding verbs) are etymologically unconnected: 14c. from OF, 15c. of uncertain origin, OE, respectively. They should at all times be treated as separate words, in the way that, for example, *calf* (bovine animal) and *calf* (hind part of the human leg) are.

**moult.** AmE *molt.*

**mouse** /maʊs/ (noun). The corresponding verb *mouse* is pronounced /maʊz/ and the agent-noun *mouser* /ˈmaʊzə/. The ordinary word *mousehole* is /ˈmaʊshəʊl/, but *Mousehole* (in Cornwall) is /ˈmaʊzəl/. The adj. *mousy* ( = shy, timid, etc.) is /ˈmaʊsɪ/.

**moustache** /məˈstɑːʃ/. Always so spelt in BrE, but *mustache* (stressed on the first syllable) in AmE.

**mouth** (noun). Pronounced /maʊθ/ in singular, but /maʊðz/ in plural. The verb is /maʊð/; and *-mouthed* in compounds (e.g. *foul-mouthed*) is /maʊðd/. See -TH(θ) etc.

**mouthful.** The pl. is *mouthfuls*. See -FUL.

**movable.** Thus spelt; except that *moveable* is preferred for legal work.

**movies.** See CINEMA.

**mow** (noun). The dialectal and AmE word for a stack of hay, corn, etc., is now usu. pronounced /məʊ/, though earlier in the century (e.g. in the *OED*, 1907) it was usu. said to be pronounced to rhyme with *cow*. The archaic noun and verb *mow* 'grimace' has long alternated between /maʊ/ and /məʊ/, but now, esp. in the phr. *mop and mow*, is pronounced /məʊ/.

**mow** (verb), to cut down grass, hay, etc. The pa.pple, when used as an adj., is *mown* (*the mown*, not *mowed*, *grass*; *new-mown*, etc.); as a true pa.pple both forms are permissible (*he had mowed*, or *mown*, *the lawn yesterday*).

**MP, M.P.,** Member of Parliament. Normally now printed without full points. So *MPs* (plural), *MP's* (possessive singular), *MPs'* (possessive plural).

**Mr, Mrs** Now usually printed without points: *Mr and Mrs J. Smith*.

**Ms, Ms.** /mɪz/. This title (now used with or without full point) was artificially formed in the early 1950s from *Mrs* and *Miss* and used before a woman's name in business correspondence in America when the marital status of the addressee was unknown or irrelevant to the matter in hand. It was enthusiastically taken up by feminists in the 1970s and was at first greeted with derision and hostility by the rest of the community. The heat has gone out of the controversy and *Ms* is now widely used, esp. in N. America, the rest of the English-speaking world tending to remain somewhat conservative. When *Ms* is used the style to be followed is: (on the envelope) *Ms Fiona Jones*; (at the beginning of a formal letter) *Dear Ms Jones*.

**much.** 1 Before setting down some debatable uses of *much* (adverb) it may be helpful if I list a few examples from standard sources of the 20c. in which *much* is used in an uncontestably idiomatic manner: *Prosperity was admittedly* much *easier to achieve on a small scale like that*; *That's something that happens* much *later*; *The nights were so* much *longer than the days*; *Her younger,* much *more aristocratic-looking brother*; *It doesn't do, though, to push the analogy* much *further*; *We would* much *prefer to support specific projects*; *The prospect of a stable, biracial, liberal—*much *less radical—coalition in the South*; *Frankly, I never cared* much *where I sat on a train*. A larger sample of evidence would confirm what is discernible here, namely that *much* frequently sets up an expectation that it will be followed by a comparative element or elements.

2 For the use of *much* rather than *very* with passive participles (*much obliged*, *much aggrieved*, etc.), see VERY 2.

3 *much more* and *much less*. (*a*) There is no problem with most uses of these: *University students may communicate* much *more freely with counterparts at research institutions*; *the ethos has* much *more to do with making a career*; *Frederica's recall of things seen was very* much *less lucid … than those of Stephanie*. In each of these examples the statement preceding *much more* or *much less* is straightforwardly and unambiguously positive. *Much less* is also correctly used after some negative statements: e.g. *I did not even see him,* much *less shake hands with him*. (*b*) Unintended difficulties arise, however, when *much less* is used (as it sometimes is) when *much more* is needed: e.g. *It is a full day's work even to open,* much *less to acknowledge, all the presents, the letters, and the telegrams, which arrive on these occasions*. (The underlying meaning is 'You could not even open them in under a day let alone acknowledge them.') *I confess myself altogether unable to formulate such a principle,* much *less to prove it*. (The underlying meaning is 'I confess that I am unable to formulate such a principle let alone prove it.') Each context must be considered on its merits, but if you are tempted to use a *much less* construction after a preceding statement that is plainly or by implication negative, consider whether *much more* is not more

appropriate or a fresh construction of the *let alone* type.

**muchly,** used to mean 'much, exceedingly', is surprisingly durable (it was first recorded in the 17c., though it is not listed in Johnson's *Dict.*, 1755), but it is now only a jocular variant of *much* (e.g. *thank you muchly for the chocolates*). The *OED* entry tracks its descent from seriousness to jocularity. Modern examples: *She stepped away from him as though evading her share in the pleasure. 'Thank you muchly,'* he said—M. Keane, 1988; *By this time I was muchly aware that most of the instructors just wanted to get me airborne so they could add a little more to their logs—Pilot*, 1988.

**mucous, mucus.** The first is the adjective (*mucous membrane*), the second the noun ( = a slimy substance).

**muezzin,** Muslim summoner to prayer. Pronounce /muːˈezɪn/.

**Mughal.** See MOGUL.

**Muhammad, Muhammadan.** The name of the founder of Islam is now spelt *Muhammad* in English, not, as formerly, *Mahomet, Mohammed*. For the name of a member of the faith of Islam, see MUSLIM.

**mulatto. 1** Pl. *mulattos*. See -O(E)S 6.

**2** With the abandonment of colonialist attitudes in the 20c., *mulatto*, once commonly used by the great seafarers and writers of the past (Drake, Dampier, etc.; Defoe, Thackeray, Stevenson, etc.) for a person of mixed white and black parentage, has virtually dropped out of use. The mood of the century has been to move towards the acceptance of whatever neutral terms are available. Thus *half-breed* is seldom heard or seen in print, whereas the more neutral term *half-caste* survives. See also ANGLO-INDIAN; EURASIAN. More specific terms, such as *quadroon* (a person of one quarter Negro blood) and *octoroon* (one-eighth) seldom occur outside serious sociological works.

**muleteer.** Pronounce as three syllables, /mjuːlɪˈtɪə/.

**mulish.** So spelt, not *muleish*.

**mullah,** a Muslim learned in Islamic theology and sacred law. So spelt, not *moll-, mool-, -a*.

**multitude.** For number with nouns of multitude or collective nouns, see AGREEMENT 5.

**mumbo-jumbo.** Pl. *mumbo-jumbos*. See -O(E)S 6.

**mumps.** Usually treated as a singular (*mumps is common in young children*). In some regions in Britain and abroad it is sometimes used with the definite article the (e.g. *When I was sick with the mumps . . . my mother found one* [sc. a balloon] *at the bottom of the steamer trunk*—M. Atwood, 1989).

**Munchausen.** Thus usu. spelt in English (cf. Ger. *Münchhausen*). The pronunciation recommended is /ˈmʌntʃaʊzən/.

**municipal.** Pronounce /mjuːˈnɪsɪpəl/, i.e. stressed on the second syllable.

**murderess.** See -ESS.

**murex.** Pl. *murices* /ˈmjʊərɪsiːz/ or *murexes*. See -EX, -IX 2.

**Muses.** 'Greek deities of poetry, literature, music, and dance; later also of astronomy, philosophy, and all intellectual pursuits. Throughout antiquity the prevailing conception of Muses follows Hesiod's *Theogony*. Muses approach the poet on Helicon and give him sceptre, voice, and knowledge. Hesiod is also responsible for the canonical number of nine . . . In late Roman times the Muses were differentiated according to their function . . . Calliope is Muse of the heroic epic, Clio of history, Euterpe of flutes, Terpsichore of lyric poetry (dance), Erato of lyric poetry or hymns, Melpomene of tragedy, Thalia of comedy, Polyhymnia of the mimic art, and Urania of astronomy. These functions and names vary considerably and names of other Muses are known. Daughters of Zeus and Mnemosyne, the Muses sing and dance at the festivities of Olympians and heroes, often led by Apollo' (*Oxford Classical Dict.*, 2nd edn, 1970).

**museum.** Pl. *museums*. See -UM 1.

**Muslim. 1** The recommended pronunciation is /ˈmʊzlɪm/ or /ˈmʊs-/, not /ˈmʌz-/.

**2** This is now the customary spelling of the name for a member of the faith of Islam, not *Moslem*, not *Muhammadan*.

**muslin** makes *muslined* (adj.). See -N-, -NN-.

**mussel,** bivalve. So spelt.

**Mussulman.** 1 In its primary sense 'a Muslim', now encountered only in older literature and in historical works. Pl. *Mussulmans*, the word being derived from Persian *musulmān*, and not a compound of English *man*.

2 A plural form *Mussulmen* was sometimes used earlier in the 20c. of inhabitants of Nazi concentration camps 'who had reached a state of physical and emotional exhaustion in which [they] displayed fatalism and loss of initiative' (*OED*). But the normal form *Mussulmans* was also used in this sense.

**must** (noun). 1 A quirkish use of the verb as a noun in obvious applications goes back to Dekker (1603): *Must is for kings, And low obedience for low underlings.* Another example: *In uttering these three terrible musts, Klesmer lifted up three long fingers in succession*—G. Eliot, 1876. It may still turn up at any time.

2 Beginning in America in the 1890s, but not becoming widely used until the 1940s, is its use to mean 'something that must be done, possessed, considered, etc.; a necessity'. It passed through a period of about four decades when it was usu. printed within quotation marks (e.g. *A film and a song made the Trevi Fountain a 'must' for tourists*—*Guardian*, 1973), and it is still regarded as being to some extent a not quite welcome guest in the language. But after a century of use, and half a century of concentrated use, it looks like becoming an indispensable and uncontested member of the family. It has also been used attributively (e.g. *this is a 'must' book*; *'must' legislation*) for most of the 20c., but this is less common than its use as a noun.

**mustache.** See MOUSTACHE.

**mustachio,** an archaic word for *moustache*. Pl. *mustachios*. See -O(E)S 4. The corresponding adj. is probably best written as *mustachio'd*, not *mustachioed*. See -ED AND 'D.

**mute e.** The problem of mute or silent *e*, and whether to retain it or not before suffixes, lies at the heart of the written

language. In its simplest form, the proposition can be stated as follows: if we wish to derive an adjective in -*ish* from the noun *mule*, do we write *muleish* or *mulish*? The answer is 'obviously' *mulish*. But if we wish to form a verbal noun from the verb *singe* the answer is 'obviously' *singeing*, not *singing*, because the latter is the verbal noun formed from the verb *sing*. Before setting down a general rule and its main exceptions it might be useful to have a list of the main types of words that are affected. Readers should bear in mind that some of the items stand for thousands, some for hundreds, and some for dozens, of similar cases. The base word is printed first, followed by notional forms of derivatives in brackets.

*excite (exciteable* or *excitable); gauge (gaugeable* or *gaugable); move (moveable* or *movable); notice (noticeable* or *noticable); rate (rateable* or *ratable); sale (saleable* or *salable); rare (rareer* or *rarer); blue (bluey* or *bluy); glue (gluey* or *gluy); mouse (mousey* or *mousy); age (ageing* or *aging); change (changeing* or *changing); dye (dyeing* or *dying); hie (hieing* or *hiing); hinge (hingeing* or *hinging); love (loveing* or *loving); route (routeing* or *routing); late (lateish* or *latish); pale (paleish* or *palish); change (changeling* or *changling); due (duely* or *duly); like (likely* or *likly); whole (wholely* or *wholly); judge (judgement* or *judgment).*

From a simple inspection of the list a basic rule begins to emerge: when a suffix is added to a word ending in a mute *e*, the mute *e* should be dropped before a vowel, but not before a consonant. The application of this rule narrows down the possibilities considerably. In accordance with it the normal forms are *excitable, movable, rarer, changing, loving, latish, palish, changeling, likely,* and *judgement.* The main exception is that the *e* should be retained even before a vowel if it is needed to indicate the 'soft' sound of a preceding *c* or *g*, or to distinguish a word from another with the same spelling: so (soft *c*) *noticeable;* (soft *g*) *gaugeable, hingeing;* (distinguishing two words) *singeing/singing.* Of the remainder, *bluey* and *gluey* are necessary because *bluy* and *gluy* would too closely resemble the monosyllabic words *buy* and *guy; hieing* is needed to avoid the awkwardness of the two *i*s in *hiing.* Individual exceptions are *duly, truly,* and *wholly;* so is *acreage.* For further information about particular cases, e.g. the use of both *ageing* and aging, *judgement* and *judgment*, *mileage* and *milage*, readers should consult them at their alphabetical places. For similar problems with adjectives ending in -*y* and verbs ending in -*ye*, see -EY AND -Y IN ADJECTIVES; VERBS IN -IE, -Y, AND -YE 7.

**mutual.** Evelyn Waugh classically drew attention to a two-centuries-old problem in a letter of 13 November 1950 to Louis Auchinloss, an American lawyer and writer: 'Apart from that, your misuse of 'mutual' (tricky word best left alone if you aren't happy about its precise meaning ... ) I have no criticisms.' What is the nature of the 'misuse'? And is it really one? The strength of feeling is unmistakable in the following examples from well-known usage manuals:

'What is "*mutual*"? Much the same as "*reciprocal*". It describes that which passes from each to each of two persons ... And *mutual* ought never to be used, unless the reciprocity exists. "The *mutual love* of husband and wife" is correct enough: but "a *mutual friend* of both husband and wife" is sheer nonsense' (Henry Alford, *The Queen's English*, 1864).

'... *mutual* for *common*, an error not infrequent now even among educated people ...' (Richard Grant White, *Words and Their Uses*, 1871).

'Every one knows by now that *our mutual friend* is a solecism. *Mutual* implies an action or relation between two or more persons or things, A doing or standing to B as B does or stands to A ... *Our mutual friend* is nonsense; *mutual friends*, though not nonsense, is bad English, because it is tautological. It takes two to make a friendship, as to make a quarrel, and therefore all friends are mutual friends, and *friends* alone means as much as *mutual friends* ...' (H. W. and F. G. Fowler, *The King's English*, 1906).

'... it follows that *our mutual friend Jones* (meaning Jones who is your friend as well as mine), & all similar phrases, are misuses of *mutual* ... In such places *common* is the right word, & the use of *mutual* betrays ignorance of its meaning' (*MEU*, 1926). From these statements it will be abundantly clear that in the usage manual tradition (reflected by Evelyn Waugh in 1950) *mutual* is correctly used only when it means 'reciprocal'.

There is another tradition, not to be ignored, and it is set down in the *OED*. It presents two broad strands of meaning, both of them in use for a very long time. This great dictionary shows that the meaning 'reciprocal', i.e. the favoured sense of *mutual*, has been in continuous use since 1477. But it also shows that the other sense of *mutual*, namely 'pertaining to both parties, common', has also been in unbroken use since the late 16c. (Shakespeare). In 1907, when the relevant fascicle of the *OED* was issued, the editors added the comment 'Now regarded as incorrect', almost as if the Fowler brothers had dropped in on them when the entry was being prepared. Abundant evidence was presented in the *OED* for the phrases *our mutual friend* (first in 1658, i.e. long before Dickens), *our mutual acquaintance* (1723), *our mutual opinion of Pope* (1820), and numerous others.

*Recommendations.* The contest between the two main meanings of the word remains unresolved, but one can say with reasonable certainty: (*a*) *Mutual* used to mean 'reciprocal' is, of course, acceptable: e.g. *Wilde and Yeats reviewed each other's work with mutual regard*—R. Ellmann, 1986; *That was my genuine feeling and until Lord Tonypandy's memoirs appeared last week I had every reason to believe that the feeling was mutual*—M. Foot, 1986; *In places we found mutual wariness between Grenadians and white tourists*—Holiday Which?, 1987. (*b*) So too are phrases of the type *a mutual friend, a mutual acquaintance*, in which *common* might be ambiguous, implying vulgarity: e.g. *We met ... at a dinner party at a mutual friend's apartment*—P. Monette, 1988. (*c*) *Mutual* is also acceptable in many other sentences with the meaning 'pertaining to both parties': e.g. *of mutual benefit to both the Scots and the English*—D. Stewart MP, BBC Radio 4, 1977; *No camaraderie exists between mother and daughter, no sympathy born out of mutual suffering*—J. Berman, 1987; *They could discuss mutual problems*—Information World, 1987. (*d*) But if it is possible idiomatically to use *common* or *in common* (in constructions of type (*c*)) this should be done: e.g. *They could discuss problems they had in common.*

See RECIPROCAL.

**muzhik.** Russian peasant. Now the preferred spelling in English, not *moujik*, etc.

**my.** 1 For *my or your informant*, etc. (not *mine*), see ABSOLUTE POSSESSIVES.

2 For the distribution of *my* and *mine*, see MINE.

3 The chiefly American (but, it would appear, originally Scottish) use of *my* as an exclamation, esp. at the beginning of a statement, is beginning to seep into other English-speaking regions, including Britain. The *OED* cites illustrative examples of the use from 1707 onwards. Modern examples: *Have you met him? Oh my!*—M. Wesley, 1983 (UK); *My, don't we look pretty*—H. Beaton, 1984 (NZ); *Oh my! My memory is shocking these days*—E. Jolley, 1985 (Aust.); *My, you're a deep one, thought Angela, not sufficiently taken in*—D. Potter, 1986 (UK); *My, but I was scared*—M. Grimm, 1989 (US); *I pointed to a pile of bird drawings. He went through them. 'My, you do this well.' I shook my head*—P. Saenger, 1991 (US).

**myriad** is almost always used of a great but indefinite number. It is worth remembering, however, that it goes back ultimately to a Greek word meaning 'ten thousand' (Gk μυριάς, μυριαδ-, from μύριος 'countless', μύριοι 'ten thousand'), and that the sense 'ten thousand' was to some extent current in English from the 16c. to the 19c. ('Chiefly in translations from Greek or Latin, or in reference to the Greek numerical system', *OED*.)

**myself.** The normal uses of *myself* as a pronoun call for no particular comment: as an emphatic form of *I* or *me* (*I began to feel guilty myself*; *I myself couldn't see him as a worker*); and as a reflexive form of *me* (*in a room by myself*; *I managed to restrain myself*). For broad stylistic reasons the following types are in varying degrees not recommended, though there is historical warrant for each of them. (Placed initially, chiefly in poetry, as a simple

subject; now obs.) *Myself when young did eagerly frequent Doctor and Saint*—E. FitzGerald, 1859; (substituted for *I* as part of a subject) *It wasn't that Peter and myself were being singled out*—F. Weldon, 1988; *Neither Cleo nor myself are naturally pushy types*—J. Dankworth, 1988; *My friends and myself do not find it a great problem*—*Paintball Games*, 1989; (substituted for *me* as the object of a verb or governed by a preposition) *Several of the ultra-popish bishops . . . had denounced the Bible, the Bible Society, and myself*—G. Borrow, 1842; *Palme Dutt's nervousness communicated itself to Isaac and myself*—Nigel Williams, 1985; *They've made myself and my wife very welcome*—professional footballer, BBC Radio 4, 1987; *'the rift between myself and Lord Hailsham is unseemly and ought to be ended,'* Judge Pickles said—*Times*, 1987.

Even so, the last group is stylistically questionable only when *myself* is placed before other elements in the object, i.e. as in the two 1987 examples. The other two examples are beyond reproach, as is the use shown in my booklet *The Spoken Word* in 1981: *This booklet results from a monitoring exercise of BBC radio in mid-1979 undertaken by Professor Denis Donoghue, Mr Andrew Timothy and myself at the invitation of* [etc.].

**myth.** It is perhaps rather surprising to discover from the *OED* that (*a*) the word came into English (from modL *mȳthus*) as recently as 1830; (*b*) that many 19c. authorities, including George Grote and Max Müller, spelt it *mythe*; and (*c*) that 'the pronunciation /maɪθ/, formerly prevalent, is still [*sc.* in 1907, when the relevant section of the *OED* was issued] sometimes heard'. It is, of course, now always /mɪθ/ in each of its senses, and always spelt *myth*.

**'n** (or **'n'**), colloq. shortening of *and* conj. Recorded occas. in the 19c. and early 20c. (the *Radio Times* mentions a one-step dance called *By 'n' Bye* in 1923), *'n* or *'n'* became common in the second half of the 20c. esp. under the influence of the phr. *rock 'n' roll*: e.g. *fish 'n' chips, good 'n' ready, sex 'n' violence*. They are sometimes written with hyphens (*sex-'n-violence*, etc.).

**nacreous** /ˈneɪkrɪəs/ is now the usual adj. corresponding to *nacre* 'mother-of-pearl', not *nacrous*. Both words entered the language in the early 19c.

**naff¹** (verb). This euphemistic substitution for *fuck* is most frequently used with *off* in the imperative phrase *naff off!* 'go away!' It seems to have first appeared in print in Keith Waterhouse's novel *Billy Liar* (1959), and Waterhouse himself insists that it was originally conscript talk as an acronym of '*n*asty, *a*wful, *f*uck it'. Other suggestions have been made, and the matter remains unresolved. What is not disputed, however, is that *naff off!* was brought into common currency by Princess Anne in 1982 when she told some persistent photographers to *naff off*.

**naff²** (adj.). Entirely unrelated to the previous word, it seems to have slipped into general use at the end of the 1960s, almost certainly as an adaptation of several regional words of similar sound meaning 'inconsequential, stupid; unpleasant, objectionable'. Since then it has been a vogue word, esp. among the young. In a column in the *Spectator* in 1989, Peregrine Worsthorne, after consulting two young girls about the meaning of the word, expressed it like this: 'As far as I can gather, anything pretentious or flashy is naff. Thus a pink Deux Chevaux is naff. Or a plastic Swatch watch. So, is the word a fashionable synonym for vulgar? Not quite, since London taxi drivers, it seems, can often be heard describing a piece of bad driving as naff. All classes use the word, shop girls as much as debutantes. In that

respect it is not at all a repeat of the U and non-U nonsense, although an element of snobbery is not altogether absent. For example, the girls definitely think that fish knives are naff, at any rate when used for fish cakes.'

**naiad** /ˈnaɪæd/. The pl. is either *naiads* or *naiades* /ˈnaɪədiːz/.

**naïf** /naːˈiːf/ or /naɪˈiːf/. Though recorded in English as an adj. from the 16c. onward and as a noun ('artless person') from the late 19c., it has gradually yielded to NAÏVE in all main uses.

**nail.** The phr. *to hit the nail on the head* 'to do or say exactly the right thing' is now the customary form. From the *OED* it seems that *to hit the right nail on the head* was the usual idiom in the 18c. and 19c., even though the form without *right* prevailed in the 16c. and 17c.

**naïve** (or **naive**) /naːˈiːv/ or /naɪˈiːv/. This most useful French word (in which language it is the feminine form of the adj. *naïf*) is still at an imperfectly naturalized stage in English, though it has been in use since the mid-17c. The grammatical gender of the word in French is ignored in English. See NAÏF. The corresponding English noun, in about equal measure, is *naïvety* /naːˈiːvtɪ, naɪ-/, *naivety*, or *naïveté* /naːˈiːvteɪ, naɪ-/. There seems to be no immediate sign that one or other of these forms is about to prevail.

**name** (noun). Of interest is the construction *name of X*, an informal shortening of *by the name of X*, now emerging in the best circles: e.g. *There was a sergeant in German Military Intelligence name of Jasper shot down in the Staroviche railway yards by partisans*—T. Keneally, 1985; *Out of here a black cat emerged, name of Frankie after some bygone entertainer*—K. Amis, 1988; *Keep your eyes peeled for a customer on his own, name of Sheldrake*—D. Lodge, 1991.

**name** (verb). **1** *you name it*. A persistent formula used after a short list of circumstances, etc., as an indication that further listing is unnecessary. It is first found in print in the 1960s. Later examples: *Whatever they choose to say, Directors, DG, Higher Command, War Cabinet, Prime Minister, you name it, I'm not sending my units back into Europe*—P. Fitzgerald, 1980; *Brunch minus the first hour, the one with just drinks in it, and minus too no doubt wine, beer, brandy, you name it and we're not going to get it*—K. Amis, 1988.

**2** *name(d) after, for, from* (someone). There is good evidence for each of these constructions in standard sources at intervals from the 15c. onwards; but at some point in the 20c. *name(d) for* settled down as the usual form in AmE, whereas *name(d) after* has long since been the standard form in BrE (only occas. *name(d) for*). Some examples: *At one time he* [sc. Peter VanRensselaer] *owned thousands of acres in this region and the city of Rensselaer is named for him*—G. Ade, 1930 (US); *each chapter is named for the element it recalls*—New Yorker, 1987; *In a city* [sc. Melbourne] *named for a British prime minister, in a state christened for a British queen*—Sunday Times Mag., 1988; *William Jack Mackenzie, named for his father and grandfather*—S. Mackay, 1992.

**names and appellations. 1** Names are the largest component in the language but paradoxically are the least well represented in dictionaries except specialized ones. In context we instinctively recognize the suitability or otherwise of the names of aeroplanes, racehorses, hurricanes, fashion colours, comets, dogs, cats, houses, pop groups (*Metal, The Beasties, The Hooters, Lindisfarne*, etc.), and so on, and accept their absence from general dictionaries because we 'know' that they belong to open-ended, virtually limitless classes. Names are not like 'words', esp. in the fixed manner in which they operate, usu. without articles (*Paris* is just Paris, not *a Paris* or *the Paris*), and in the fact that they normally have no plural. If a name does have a definite article, e.g. *the Kremlin*, it cannot idiomatically be used without it: we cannot say \*Boris Yeltsin is in Kremlin.

**2** Certain types of proper names, however, receive detailed treatment in specialized dictionaries: e.g. *English Place-Names* (Eilert Ekwall), the *BBC Pronouncing Dictionary of British Names* (G. E. Pointon),

*English River names* (Ekwall), *English Place-Name Elements* (A. H. Smith), *British Surnames* (P. H. Reaney), *English Christian Names* (E. G. Withycombe). There are also innumerable monographs on such matters and learned journals devoted to the collection and description of particular types of names.

**3** For a brief historical account of the use of personal names and designations, esp. from the 18c. to the early 20c., one may turn to R. W. Chapman's essay 'Names, Designations and Appellations' (SPE Tract xlvii, 1936). Brief descriptions of some modern conventions in the use of titles and forms of address may be found in the present book, for example s.v. ESQ.; MADAM; MR, MRS; MS. The general tendency in the 20c. is to settle for informality except in formal correspondence and on grand occasions (e.g. garden parties at Buckingham Palace, the opening of parliament). John Grigg neatly described the trend in *The Times* (2 Nov. 1991): 'In the last century and well into the present one, grown-up British people [he meant 'men'], with rare exceptions, addressed each other by their surnames. What we now call first names (then Christian names) were very little used outside the family. Men who became friends would drop the Mr and use their bare surnames as a mark of intimacy: e.g. Holmes and Watson. First names were only generally used for, and among, children. Today we have gone to the other extreme. People tend to be on first-name terms from the moment of introduction, and surnames are often hardly mentioned. Moreover, first names are relentlessly abbreviated, particularly in the media: Susan becomes Sue, Terence Terry and Robert Bob not only to friends and relations, but to millions who know these people only as faces and/or voices ... Most married women are still addressed, and wish to be addressed, as Mrs, while most unmarried women are still addressed as Miss. Ms is used by feminists to make their point, or by male chauvinists trying to make the opposite point, though of course it is also used by unprejudiced correspondents in a state of honest doubt.'

**4** For adjectival formations of the type *Aesopian, Audenesque, Borrovian, Platonic,* etc., see SUFFIXES ADDED TO PROPER NAMES.

**5** For names of groups of animals, etc. (*a pride of lions, a charm of finches,* etc.), see PROPER TERMS.

**6** For names of people after whom a discovery, invention, product, process, etc., is called, see EPONYM.

**7** Non-Americans continue to be mildly amused by the American habit of adding II, III, etc., to the names of descendants of persons of the same name. From a March 1992 American newspaper: *Other than her husband, she leaves three sons, the Rev. Raymond II, Silas and Mark; a daughter, Ann Semotan III, and 10 grandchildren.*

**8** Every now and then correspondence breaks out in newspapers about curiosities of nomenclature: e.g. the frequent use of certain types of first names (women's names from flowers: *Fleur, Flora, Lily, Marigold, May, Rose, Rosemary, Veronica,* etc.); the aptness of some surnames to the occupation of the owner of the name (e.g. *Mr Veale,* a butcher; *Mr Coffin,* an undertaker; *Mr Churchyard,* a vicar). Such correspondence is a paradisal garden for the amateur collector of unexpected connections.

**naphtha.** So spelt. Pronounce /ˈnæfθə/, not /ˈnæp-/.

**napkin,** according to Fowler (1926), Nancy Mitford, and others, 'should be preferred to *serviette*' (the latter word being judged to be a genteelism). As a totally separate matter, in Britain a baby wears a *nappy* (hypocoristic form of *napkin*) and in America a *diaper.*

**narcissus.** On grounds of euphony, the pl. form *narcissi* /-saɪ/ is recommended rather than *narcissuses.*

**narghile,** oriental tobacco-pipe. Pronounce /ˈnɑːɡɪleɪ/, i.e. as three syllables.

**narratress, narratrix,** female narrator. Both words have fallen out of use in the 20c., though some other words in *-tress* and *-trix* have survived. See FEMININE DESIGNATIONS; -TRIX.

**nary.** Markedly American from the mid-18c. onward until some point in the 20c., when it began to appear in British works. Except in regional AmE sources, it is almost always used in the form *nary a* + noun: e.g. *The computer that did all the dogwork gets nary a mention—Verbatim,* 1974. In origin it seems to be an extended respelling of *ne'er a* 'never a'.

**nasal.** In phonetics, (of) a letter or a sound pronounced with the breath passing through the nose, e.g. *m, n, ng,* or French *en, un.*

**nasturtium.** Pl. *nasturtiums.* See -UM 1.

**natheless, nathless** /ˈneɪθlɪs, ˈnæθ-/. A striking example of a word that was in regular use ( = nevertheless) from OE until the 19c. but which has now effectively dropped out of the language except in historical contexts.

**native.** It requires a certain amount of skill to avoid the seen and unseen tripwires stretched across uses of this word. *He is a native of Leeds, of Yorkshire, of England* are all acceptable as factual statements about where a person was born. *He speaks Italian like a native* and the phrase *native speaker* also bear no derogatory connotations—quite the reverse. There are scores of plants and animals in the former British colonies whose names are qualified by *native* on the grounds that they are similar to analogues in the British Isles but do not belong to the same species or genus, e.g. in Australia, *native bear* (koala), *native cat* (marsupial cat), *native oak* (casuarina). In New Zealand, *native bush* is the ordinary term for woods or forests made up of indigenous trees and shrubs. When, however, *native* is applied to the original or usual inhabitants of a (formerly) colonized country difficulties begin to arise. Throughout the 19c. and in the first decades of the 20c. it was acceptable, indeed customary, for British writers such as Mary Kingsley, R. M. Ballantyne, and E. M. Forster to speak of West Africans, Pacific Islanders, and Indians as *natives,* usu. implying social or cultural inferiority. In 1950 (I quote from the *OED*) a writer called J. C. Furnan spelt out what *native,* in its derogatory uses, means: *The meaning of 'Native' can be approximated ... Greedy for beads ... and alcoholic drinks. Suspect of cannibalism. Addicted to drumbeating and lewd dancing. More or less naked. Sporadically treacherous. Probably polygamous and simultaneously promiscuous. Picturesque.*

*Comic when trying to speak English or otherwise ape white ways.* Such attitudes have now been abandoned by all intelligent people in the former colonizing countries, but they tend to resurface whenever a particular country is faced with mass immigration from Third World countries.

In view of these opprobrious uses it is all the more remarkable that NAmer. Indians, from about the 1950s, have themselves adopted the appellation *Native American* (sometimes with a small initial *n*), and this term is now routinely used in much of the American press, and widely in day-to-day contexts by white, black, and Hispanic Americans.

**natter.** This everyday colloquial word in Britain, usu. meaning (as verb) 'to chatter idly' and (as noun) 'aimless chatter', is not widely used in AmE. It seems to be an alteration of the earlier dialectal verb *gnatter* (of uncertain origin) which, besides the sense 'to chatter', like *natter* also bore the sense 'to grumble'. The emergence of the word from restricted dialectal use to general UK use seems to have occurred during the 1939–45 war.

**nature.** In general, in most formal kinds of writing, it is better to use the type *an enigmatic theologian* rather than *a theologian of an enigmatic nature*. Similarly it is better to write *the dangerousness of the level crossing* (i.e. use an abstract noun) rather than the periphrasis *the dangerous nature of the level crossing*. But it does not follow that all periphrastic uses of *nature* are to be avoided: the *OED* disproves such blanket assumptions in its presentation of the phrases *of* (a certain) *nature*, and *of* or *in the nature of*. Examples: *With other Particulars of the like Nature*—Addison, 1711; *Your desires are to me in the nature of commands*—Fielding, 1749; *A plan of this nature*—Blackstone, 1765; *It was not in the nature of things that popularity such as he ... enjoyed should be permanent*—Macaulay, 1854; *Most of his public acts are of a ceremonial nature*—*London Calling* (cited in Webster's Third, 1961). In one's own writing, the best policy is to consider each potentially periphrastic use of *nature* on its merits and to use such constructions sparingly. Cf. CHARACTER; DESCRIPTION.

**naught, nought. 1** *naught*: now (in both BrE and AmE) an archaic or literary word meaning 'nothing'. It survives chiefly in the phrases *come to naught* 'to end in failure', *set at naught* 'place no value on', and *all for naught* 'all for nothing'. In AmE it is also used to mean the digit 0.

**2** *nought* (in AmE) means principally 'the digit 0' (as in the game called *noughts and crosses*; '0.5' pronounced as 'nought point five'). The variation in spelling is not a modern accident, but descends from OE.

**nausea.** The pronunciation recommended is /ˈnɔːzɪə/, not /-s-/ or /-ʃ-/.

**nauseated, nauseating, nauseous.** **1** In Britain *nauseated* occas. means 'affected with nausea', but much more frequently, in transferred use, means 'disgusted' (*he was nauseated by their behaviour* or, even more idiomatically, *he found their behaviour nauseating*). *Nauseous* is used of unpleasant smells, distressing circumstances (e.g. finding a dead body), anything offensive to taste or rational behaviour: (*a*) causing nausea; (*b*) loathsome, disgusting.

**2** In AmE, *nauseated* until recently only meant 'suffering from nausea, feeling sick to one's stomach'; and *nauseous* strictly meant 'causing nausea'. The distinction is no longer clearly observed, however, *nauseous* now being frequently used in the primary sense of *nauseated*. *WDEU* (1989), in a long article containing much printed evidence, concluded that *nauseous* 'is most often used as a predicate adjective meaning "nauseated" literally; it has some figurative use as well ... The older sense of *nauseous* meaning "nauseating", both literal and figurative, seems to be in decline, being replaced by *nauseating*. *Nauseated* is usually literal, but is less common than *nauseous*. Any handbook that tells you that *nauseous* cannot mean "nauseated" is out of touch with the contemporary language.'

Some examples, nearly all of them meaning 'affected with nausea': (nauseated) *The mere smell of the food, with its combination of onion and spices, made her slightly nauseated*—L. Duncan, 1978 (US); *Marian, a 40-year-old artist, woke up only to feel herself falling violently backwards ... 'I got out of bed, but felt dreadfully giddy, nauseated and then panic stricken'*—*Daily Tel.*, 1990; (nauseous) *The drug made him nauseous*—P. Monette, 1988 (US); *The shepherds, angels, wise men, pop-up camels, donkeys, fat-tailed sheep, etc. made her a little nauseous*—A. T. Ellis, 1990; *Guests felt unwell ... nauseous tribute to recollections of sea-legged fishermen*—*Independent*, 1991; *I started my new job, but I soon became nauseous and exhausted*—*Times*, 1991; *When a toadlicker licks the toad his mouth and lips will become numb and he will feel intensely nauseous*—*Times*, 1991.

**nautilus,** a shell. Pl. *nautiluses* or *nautili* /-laɪ/.

**naval,** pertaining to the navy, and **navel**, small depression in the centre of the belly, are both pronounced /ˈneɪvəl/. The first is derived from L *nāvālis* (from *nāvis* 'ship') and the second from OE *nafela* (cf. Ger. *Nabel*).

**nay. 1** As adv. it is curiously persistent (first recorded in this use in the mid-16c.) in the sense 'and even, and more than that': *What precisely is his* [sc. a bookseller's] *role when on the shop floor surrounded, nay probably inundated, by the goods which are his livelihood?*—*Logos*, 1990; *one could not but notice how theatrical, nay operatic, the whole adornment of the church was*—*Oxford Mag.*, 1991.

**2** As noun, *nay* means principally 'a negative vote': *they counted 20 ayes and 16 nays*.

**né** (or **ne**). Now beginning to be used in obituaries in American newspapers to indicate the original name of a man who changed his name at some point in his life: *Jethro Burns, ne Kenneth Burns* (1989); *Norman Charles (né Charles Norman Diggs)* (1990). Also occas. in UK sources in the sense 'formerly known as': *The Morning Star (né Daily Worker, as it was in May 1945) proved that some things have not changed*—*Pick of Punch*, 1985; ( = named at birth) *NWA's main man is Easy-E (né Eric Wright), considered the Pavarotti of the rap world*—*Daily Tel.*, 1990. The word is clearly embarking on a new career. I daily expect to find, for example *Michael Caine né Maurice Joseph Micklewhite* and *Cary Grant né Archibald Alexander Leach*. See NÉE.

**near** (adv.). In most of its adverbial uses *near* is completely natural (*the time drew near; near at hand; the bomb dropped near to their home; as near as one can say*). But it is becoming progressively more archaic

when used to mean 'almost, nearly': *not near as often; near dead with fright; near a century ago*. In such circumstances *nearly* is the natural form. An exception is *near-perfect* as used in the following example: *The capsule's attitude would have to be near-perfect when the rockets fired, or the angle of re-entry would be affected*—J. Glenn, 1962.

**near by, nearby.** As an adj. it should be written as one word (*a nearby hotel*), but as an adv. normally as two (*at a hospital near by*).

**neath,** beneath. Used only in dialects or in poetry since it first appeared in print in the 18c.

**nebula.** The recommended pl. form is *nebulae* /ˈnebjʊliː/, not *nebulas*.

**necessaries.** See NECESSITIES.

**necessarily.** In *The Spoken Word* (1981), I urged BBC broadcasters to avoid the American *-ar-*, and advised them to say *necessarily* not *necessarily*; and I commented '[The American pronunciation is] widely used in Britain, particularly by young people, but [is] best avoided for the present in formal contexts.' The advice still stands. But in informal contexts, additional weight given to the third syllable now passes almost unnoticed.

**necessities, necessaries.** At one stage earlier in the 20c. it seemed to some authorities that *necessaries*, not *necessities*, was the word to use when the sense required was 'any of the basic requirements of life, such as food, warmth, shelter, etc.'. This use of *necessaries* (sense B1 in the *OED*) had been current since the 14c. (*3e shal haue bred and clothes, And other necessaries i-nowe*—Langland, 1377) and was still favoured in the 19c. (*an island which was so prolific and so well stored with all the necessaries of life*—R. M. Ballantyne, 1858). But *necessities* has a similar history from the 15c. onward. A typical example: *Sufficient for many things more than the necessities of life*—E. Du Bois, 1799. After much locking of horns between the two words, *necessities* seems to have emerged as the stronger of the two in this concrete sense, though both words are still used: (necessaries) *I gathered up what few necessaries I could quickly lay my hands on—tobacco and papers, coffee, a can*

to cook in, and a couple of tin cups—
R. J. Conley, 1986; (necessities) *The possibil-*
*ities of bulk ordering of whole ranges of*
*hospital equipment and necessities, such as*
*blankets and linen, were realized early in the*
*development of the scheme*—A. Bevan, 1952.

**nectar.** The *OED* lists seven corresponding adjs. bearing the broad sense 'of the
nature of, sweet as, nectar': *nectareal* (first
recorded in 1652), *nectarean* (1624), *nectar-*
*eous* (1708), *nectarian* (1658), *nectarine*
(1611), *nectarious* (1771), and *nectarous*
(1667). Famous writers—among them
Crashaw, Milton, Gay, Pope, Smollett,
and Tennyson—made their choices, but
no one form emerged as the dominant
one. It is instructive in a way to see how
the editors of standard dictionaries of
current English react to such a situation.
I consulted ten desk dictionaries, British,
American, and Australian, six of them
prepared for the general market and four
for foreign learners of English. Three
general dictionaries listed *nectarean, nec-*
*tareous*, and *nectarous*; one listed only *nec-*
*tareous* and *nectarous*; two listed only
*nectarous*. None of the four dictionaries
for foreigners listed any. By this crude
test I suppose one could conclude that
the most usual adj. directly corresponding to *nectar* is *nectarous*. But who can say
with any certainty?

**née** (or, in modern printing, often **nee**).
From the French word meaning 'born',
the traditional use is to add it to a married woman's name after her surname
to indicate what her maiden name was,
e.g. *Mrs John Jones, née Smith*, or *Mrs Ann*
*Jones, née Smith*. This remains the standard use. The type *Mrs John Jones, née Ann*
*Smith* (i.e. with the Christian name added
after *née*) is not always acceptable. The
use of *née* as an explanation of a pseudonym or stage name (the type *Diana Dors*
*née Diana Mary Fluck*), and other irregular
types, suggest that *née* is a restless word
striving to find more room. Examples:
*On her marriage, then, Mrs. Maccomb née*
*Miss Yardes had gone to live at Seale . . . about*
*seventy miles from London*—E. Bowen, 1938;
*Mrs Agnes Childe was née O'Byrne, of Ratsey*
*and O'Byrne, ship's chandlers*—A. Price,
1982.
See NÉ.

**need** (verb). **1** The most important thing
to remember about the verb is that it

operates in two ways: (*a*) as an invariable
auxiliary verb or modal auxiliary followed by a bare infinitive; (*b*) as a finite
verb. Used as a modal auxiliary, it can
only be used in negative or interrogative
constructions, or in phrases with a negative implication, e.g. *he needn't leave*; *need*
*we leave?*; *needn't we leave now?*; *I need*
*hardly say*. As a finite verb it takes the
normal inflections and is followed by a
*to*-infinitive, e.g. *he needs to leave*; *he didn't*
*need to leave*; *did/didn't we need to leave?*

Some contextual examples: (*a*) *One need*
*interrupt the narrative no further than to say*
*this*—A. N. Wilson, 1984; *The Landlady need*
*never know*—J. Frame, 1985; *But need she*
*lie? Was he just a boy?*—M. Leland, 1985;
*Nothing that need embarrass you. Not at this*
*stage*—A. Lejeune, 1986; *You needn't bother*
*to let me know*—C. Rumens, 1987; *It need*
*not only be children who can enjoy guessing*
*games*—Spectator, 1988. (*b*) *Needing to be*
*with everybody else, I suppose*—M. du Plessis,
1983; *which was what he needed to do if his*
*feelings were to be relieved*—C. K. Stead,
1984; *the K2 tragedy shows that much more*
*needs to be done to bring home the lesson*—
Times, 1986; *the Bar and the judiciary need*
*to know what is being proposed*—Counsel,
1987; *she acted as if Strawberry needed to*
*be cuddled*—New Yorker, 1988.

**2** need + gerund. The type *this needs*
*washing* alternates with the type *this needs*
*to be washed*. Examples of the construction with a gerund: *The Commonwealth*
*Institute in London, whose copper-clad roof*
*alone needs £800,000 spending on it*—
Observer, 1988; *There's a lot of work needs*
*doing in there. You've got dead and diseased*
*trees need felling, you've got bramble and*
*stuff needs clearing out*—P. Lively, 1989; *It's*
*a place full of devils, it needs a mission*
*sending to it*—H. Mantel, 1989.

**3** need + pa.pple. The type *these clippers*
*need mended* is a worrying newcomer in
semi-educated speech in BrE and AmE.
Examples: *I walked round the cottage to see*
*what needed done*—C. Burns, 1989 (UK);
*When you see one* [sc. a dog] *drag its butt*
*on the ground like that, it needs wormed,*
Otis—T. McGuane, 1989 (US). The standard
alternatives are shown in 2.
See NEEDS; WANT (verb).

**needle.** The commonplace phr. *like look-*
*ing for a needle in a haystack* is a 19c.
variant of the much older phrases *like*

*looking for a needle in a meadow, in a bottle* (i.e. a *bundle* or *truss*) *of hay.*

**needleful.** Pl. *needlefuls.* See -FUL.

**needs.** In sentences containing *needs must* or *must needs* (e.g. *Needs must when the devil drives* [and variants]—old proverb; *The Squire must needs have something of the old ceremonies observed on the occasion*—W. Irving, 1822; *Now even my army, I fear, must needs Obey the conquering, never-conquered woman*—L. MacNeice, 1951; *The language of the bureaucrats and administrators must needs be recognized as an outgrowth of legal parlance*—NY Law Jrnl, 1973; *If needs must, I'll spin them a yarn about you being one of my long-lost sons*—P. Bailey, 1986), *needs* is in origin an adverb (OE *nȳdes, nēdes*) meaning 'of necessity, necessarily'.

**ne'er.** A literary (chiefly poetical) shortening of *never*, in regular use since the 13c., but now rarely used except in the compound *ne'er-do-well*, a good-for-nothing person.

**negation.** 1 For problems arising from double or multiple negation, see DOUBLE NEGATIVE.

2 Fowler (1926) classified various types of construction in which negative and affirmative statements occur in parallel clauses, and emphasized how important it was not to let the negativity of the first clause carry over into the second. Examples, followed by suggested improvements: *No lots will therefore be put on one side for another attempt to reach a better price*, but *must be sold on the day appointed* (read *but all must be sold*); (with negative inversion in first clause) *Nor does he refer to Hubrecht's or Gaskell's theories*, and dismisses *the palaeontological evidence in rather a cavalier fashion* (read *and he dismisses*); *It is not expected that tomorrow's speech will deal with peace*, but will be confined *to a general survey of …* (read *It is expected … will not deal*). To which may be added the following example in the proofs of a 1992 institutional report: *We greatly regret that —— decided that for personal reasons he would have to give up the chairmanship, but that he intended to remain an active member of the committee* (the final *that*-clause is in danger of being qualified by *We greatly regret*). The advice remains sound.

3 It is important to avoid ambiguity when the subject of a negative sentence contains the word *all*. For example, an American agony aunt tried to reassure a married woman, who was upset because her husband wanted to sleep in a separate bed, by saying *All married couples do not sleep together*. The advice should have been *Not all married couples sleep together*. It should be noted, however, that this unrecommended construction has many historical precedents. Jespersen (1909–49, v. 472) cites numerous examples, including the following: *I maintained that certainly all patriots are not scoundrels* (Boswell); *But all men are not born to reign* (Byron); *All Valentines are not foolish* (Lamb); *All our men aren't angels* (H. G. Wells). As Jespersen says, 'very often *all* is placed first for the sake of emphasis, and the negative is attracted to the verb'.

4 It is desirable not to multiply negatives to the extent that clarity is endangered. For two examples of over-negatived sentences see NOT 3.

5 For the ambiguity that can arise when a negative clause precedes one beginning with *because*, see BECAUSE B2.

6 For other problems concerning negation, see BUT 7; CANNOT 2, 3; HARDLY 4, 5.

**negative.** Orig. and chiefly in radio communication, for reasons of clarity, *negative* was widely adopted from about the 1950s to replace or reinforce *no*. Examples: '*Any result of my application for the return of my typist?' 'Negative,' said Mr Oates*—E. Waugh, 1961; *I shook my head. 'Negative,' I said*—P. Cleife, 1972; '*Do you see any friendlies around?' 'Negative.'*—T. Clancy, 1987. Cf. AFFIRMATIVE.

**négligé, negligée.** The spelling in English is fluid, and the accents are often dropped in printed work. In the sense 'a woman's informal or unceremonial attire' (e.g. *beautiful ladies in rather négligé raiment*) the 20c. seems to favour *négligé*. In the sense 'a woman's light dressing-gown', the preferred form is *negligée*, but the form without accent is equally common. However spelt, the word is usually pronounced /ˈneglɪʒeɪ/.

**negligible.** So spelt, not *negligeable* (cf. Fr. *négligeable*).

**negotiate.** One of Fowler's lost causes. In 1926 he strongly attacked its use in 'its improper sense of tackle success-fully', but this extended sense (orig. used in Hunting in the mid-19c.) is now routine in contexts of a person who succeeds 'in crossing, getting over, round, or through (an obstacle, etc.) by skill or dexterity' (OED). The word is pronounced /nɪˈgəʊʃɪeɪt/, not /-ˈgəʊsɪ-/.

**Negress.** Even if written with a capital initial, the word is no longer in favour. The standard terms now are black (or Black) and (in America) African-American. Negress apparently continues to be used in America among African-Americans themselves, but otherwise is judged to be derogatory (Blacks all over the world find the term 'Negress' offensive—TLS, 1974). See -ESS 4; JEWESS.

**Negro.** Pl. Negroes. see -O(E)S 1. In our racially sensitive age, a term to be used with great caution. See prec. The most favoured terms at present are African-American (in AmE) and Black (or black). See BLACK; NIGGER.

**neighbourhood.** It is inadvisable to use the cumbrous periphrasis in the neighbourhood of (a sum or figure) when roughly or about would serve as well.

**neither.**

1 Pronunciation.
2 Meaning.
3 Number of the pronoun and adj.
4 Number and person of verb after neither
   ... nor.
5 Position of neither ... nor.
6 Neither ... or.
7 Neither as quasi-conjunction.
8 Neither pleonastic (?) or strengthening a
   preceding negative.

**1 Pronunciation.** /ˈnaɪðə/ and /ˈniːðə/ are used with approximately equal frequency. See EITHER 1.

**2 Meaning.** The proper sense of the pronoun (or adjective) is 'not the one nor the other (of two things); not either'. Examples: Neither was exactly English—D. J. Enright, 1955; Neither of these figures illuminates the case against Trident—David Steel, 1985; Neither of the two answers is satisfactory—Dædalus, 1987. Like either it sometimes refers loosely to numbers greater than two: Heat, light, electricity, magnetism ... are all correlatives ... neither

... can be said to be the essential cause of the others—W. R Grove, 1846. But none or no should be preferred; cf. EITHER 3. This restriction to two does not hold for the adverb: buildings made of some translucent and subtly incandescent material, neither glass nor stone nor steel—P. Lively, 1991.

**3** Number of the pronoun and adj. The number of the pronoun and adj. is properly singular, and disregard of this fact is often taken to be a grammatical mistake, though, with the pronoun at least, very common. Examples: see EITHER 4.

**4** Number and person of verb after neither ... nor. If both subjects are singular and in the third person, the verb should normally be singular and not plural: e.g. Neither its chairman, Sir Frederick Dainton, nor its chief executive, Kenneth Cooper, is planning any dramatic gestures—Times, 1985. There are historical precedents for the use of a plural verb in such circumstances, but what was acceptable in the 18c. and 19c. may still become questionable in the 20c.: e.g. Neither search nor labour are necessary—Johnson, 1759; Neither painting nor fighting feed men—Ruskin, 1874; But neither Baker nor Bush are needed for that—Newsweek, 1991. A plural verb is required if either of the subjects (esp. the second one) is plural: e.g. Neither the Conservative figures nor the evidence of Labour's recovery since 1983 produce any sense of inexorable movement in political fortunes—Times, 1985.

Fowler (1926), in an unusually stretched-out passage, comments helpfully on the choice of verb form when the subject is complicated in various ways: 'Complications occur when, owing to a difference in number or person between the subject of the neither member and that of the nor member, the same verb-form or pronoun or possessive adjective does not fit both: Neither you nor I (was?, were?) chosen; Neither you nor I (is?, am?, are?) the right person; Neither eyes nor nose (does its?, do their?) work; Neither employer nor hands will say what (they want?, he wants?). The wise man, in writing, evades these problems by rejecting all the alternatives—any of which may set up friction between him and his reader—and putting the thing in some other shape; and in speaking, which does not allow time for paraphrase, he takes risks with equanimity and says what instinct dictates. But, as instinct is directed largely by habit, it is well to eschew habitually the clearly wrong forms (such as *Neither chapter nor verse* are *given*) and the clearly provocative ones (such as *Neither husband nor wife is competent to act without* his *consort*). About the following, which are actual newspaper extracts, neither grammarians nor laymen will be unanimous in approving or disapproving the preference of *is* to *are* or of *has* to *have*; but there will be a good majority for the opinion that both writers are grammatically more valorous than discreet:- *Neither apprenticeship systems nor technical education is likely to influence these occupations* (why not have omitted *systems?).| Neither Captain C. nor I has ever thought it necessary to* . . . (Neither to Captain C. nor to me has it ever seemed . . . ).'

It is important to bear in mind Jespersen's principle concerning 'the fundamental plurality of the conception [of *either* and *neither*]' (see EITHER 4) in certain circumstances; and also the principle of attraction exemplified in *Do you mean to say neither of you know your own numbers?* (H. G. Wells, 1905). *CGEL* also asserts (10.41) that the type *Neither he nor his wife have arrived* is more natural in speech than the type *Neither he nor his wife has arrived.*

**5** Position of *neither* . . . *nor.* In good writing *neither* in *neither* . . . *nor* sentences should be placed with caution, and normally immediately before the first of the items being listed. The type *Which neither suits one purpose nor the other* should be rewritten as *Which suits neither one purpose nor the other.* Cf. EITHER 5. In the following example *neither* should have followed *interview* not preceded it: *Students who neither interview on campus nor through SLAC still have a chance to attract the attention of businesses*—Oberlin College Observer, 1990.

**6** *neither* . . . *or.* When a negative has preceded, a question often arises between *nor* and *or* as the right continuation, and the answer to the question sometimes requires care. However, there is a problem in that the *OED* cites abundant literary evidence from the 16c. to the 19c. for *neither* . . . *or*: e.g. *Neither on the one side or on the other*—Burke, 1757; *Neither rabbits at Coniston, road-surveyors at Croydon, or mud in St. Giles's*—Ruskin, 1864;

*Wasn't it true that he neither knew anything or could do anything?*—G. W. Dasent, 1874. The matter must be left to the discretion of the reader, but I shall continue to use only *neither* . . . *nor* in my own writing.

**7** *Neither* as quasi-conjunction. This use, in which *neither* is preceded by another negative and means 'nor yet' or 'and moreover . . . not', is labelled 'Now *rare*' in the *OED*. Examples: *There was now no respite neither by day nor night for this devoted city*—Southey, 1827; *Christianity abrogated no duty . . . neither for Jew nor Gentile*—H. Coleridge, 1849. Fowler sagely commented 'This use is best reserved for contexts of formal tone.'

**8** *Neither* pleonastic (?) or strengthening a preceding negative. Fowler (1926): '*Neither* with the negative force pleonastic, as in *I don't know that neither* (instead of *either*), was formerly idiomatic though colloquial, but is now archaic and affected.' By contrast, the *OED* records the same phenomenon with an impressive array of illustrative examples from the 16c. to the 19c., and simply comments 'Used to strengthen a preceding negative: = either'. Examples: *There were no books neither*—Disraeli, 1844; *Lady Edbury would never see Roy-Richmond after that, nor the old lord neither*—Meredith, 1871. Fowler's advice seems about right, but it was a pity that his example was so basic.

**nem. con.** In my book *Unlocking the English Language* (1989) I described how Dr C. T. Onions (of the *OED*) and Fowler quarrelled about the definition of *nem. con.* Dr Onions said it meant 'with no one dissenting' ( = L *nemine contradicente*), and Fowler insisted that it meant 'unanimously'. Onions was right.

**neologism.** A neologism is a new word or new sense of a word. In broad terms the language tolerates the introduction of something of the order of 450 neologisms a year of a kind admissible to our largest dictionary, the *OED*, but many fall by the wayside as time goes on. The main nine types of neologism, together with examples, are listed in *OCELang.* (1992) together with typical examples of new words first recorded in the 1940s (*acronym, airlift,* etc.), the 1950s (*A-OK, beatnik,* etc.), down to the 1980s (*cash point, couch potato,* etc.). Nearly all neologisms are formed from elements already in the language. *OCELang.* mentions *googol* and *Kodak* as rare examples of words formed *ex nihilo*.

**nepenthes** /nɪˈpenθiːz/, a drink or drug 'capable of banishing grief or trouble from the mind' (*OED*), is mentioned in the Odyssey (Gk νηπενθές). The word was taken into English in the 16c. (Lyly) and was immediately joined by an unetymological *s*-less form *nepenthe* (Spenser), also trisyllabic, /nɪˈpenθiː/. Both forms are still in use, with *nepenthe* the more usual of the two in literary work and *nepenthes* the more usual in botanical circles.

**nephew.** The *OED* (1908) gave priority to the pronunciation with medial /-v-/, but then as now the word was often pronounced with medial /-f-/. In AmE, /-f-/ is standard.

**nereid** (mythology). Pronounce /ˈnɪəriɪd/.

**nerve-racking.** The preferred spelling in BrE. In AmE, *nerve-wracking* is a permissible variant.

**-ness.** For the distinction between the words *abstraction* and *abstractness*, and between *sensibility* and *sensibleness*, and similar pairs, see -ION AND -NESS and -TY AND -NESS.

It is worth noting that *-ness* is a strong living suffix (used esp. to form noncewords). In recent years I have noted numerous examples of such formations that are either not in the *OED* or are marked as rare: e.g. *accidentalness* (I. Murdoch, 1989), *energizedness* (*New Yorker*, 1986), *familyness* (*Newsweek*, 1992), *unimpressedness* (*New Yorker*, 1990).

**nestle, nestling.** The *-t-* is always silent in verbal senses. As a noun, *nestling* ('a bird that is too young to leave its nest') may also be pronounced with the *-t-* intact.

**net.** In the commercial sense ('not subject to deduction') the spelling is *net*, not *nett*.

**Netherlands, Low Countries, Holland, Dutch.** The *Netherlands* is now the official name for the Kingdom of Holland (*Queen of the Netherlands*, *The Netherlands Ambassador*, etc.), though formerly Holland was only part of *The Netherlands*, or *The Low Countries* (which included what is now Belgium and Luxembourg). The language spoken in The Netherlands is *Dutch*; and *Dutch* is the adjective used of the people there, and also in certain familiar phrases, e.g. *Dutch courage*, false courage gained from alcohol; *Dutch treat*, a party, outing, etc., to which each person present makes a contribution; *Dutch uncle*, a person giving advice with benevolent firmness. (The word for a costermonger's wife, immortalized in Albert Chevalier's *My Old Dutch*, is different: it is an abbreviation of *duchess*.)

**neuroma.** Pl. either *neuromas*, or (somewhat pedantic) *neuromata*.

**neurosis.** Pl. *neuroses* /-siːz/.

**never.** 1 The adverbial phrase *never so* prefixed to adjs. or advbs. in conditional clauses to denote an unlimited degree or amount is recorded from the 12c. onward, and is common in the Bible (*a showre of heauenly bread sufficient for a whole host, be it neuer so great*—Translators' Preface, para 4) and Shakespeare (*if thou dost intend Neuer so little shewe of loue to her*—1590) and until the late 19c. (*Were the critic never so much in the wrong, the author will have contrived to put him ... in the right*—Swinburne). It was joined in the late 17c. by *ever so* (substituted for *never so*, according to the *OED*, 'from a notion of logical propriety'), used in the same manner (see EVER 4), and for a time *ever so* seemed likely to replace *never so* in standard use. *Never so* is now deeply archaic. For the status of *ever so*, see EVER 4.

2 *never ever* (sometimes divided by a comma). This emphatic adverbial phr. is restricted to informal contexts, but is not uncommon. Examples: *She continued, a little too vehement. 'I've never, ever been bored.'*—Maggie Gee, 1985; *It seems odd that I never ever saw the meat truck on its way back to Huntington*—L. Maynard, 1988 (US); *How much that he never would have known he was capable of experiencing, never ever*—N. Gordimer, 1990 (SAfr.).

3 *CGEL* 1985 (8.112) lists a use of *never* as 'a negative minimizer': *You will never catch the train tonight* 'You will not under any circumstances catch the train tonight'. And it also points out that 'in nonassertive clauses *ever* (with some retention of temporal meaning) can replace *never* as minimizer, [especially] in

rhetorical questions: "Will they (n)ever stop talking? Won't they ever learn?" 'In some circumstances *never* may be used as a simple substitute for *not*: e.g. *I never knew you felt like that* ( = 'I didn't know'); *he never used to* ( = 'he didn't used to'). Uses of *never* = 'not (at all)', with little or no temporal sense, are particularly common in Irish English: e.g. *'That's never true,' exclaimed Mrs Riordan. 'How can you say such things on the anniversary of his going?'*—J. Leland, 1987; *'You're never going to cut hay in this weather?' asked Conor*—ibid.

**4** *Never!* (as an expression of denial and surprise). This use, which was first recorded in the 1830s in Dickens (e.g. *Could such things be tolerated in a Christian land? Never!*), is still commonly encountered (e.g. *There's a fellow ... a gun—a pistol.' 'Never!'*—N. Bentley, 1974).

**nevertheless, nonetheless.** These are mere stylistic variants as adverbs meaning 'in spite of that, notwithstanding, all the same'. The *COD* and *New SOED* favour the form *nonetheless* (one word), rather than *none the less*, for the second.

**news.** From the 16c. to the 18c. *news* 'tidings' was construed either as a plural or a singular noun, but since the early 19c. it has normally been treated as a singular (e.g. *News is what a chap who doesn't care much about anything wants to read*—E. Waugh, 1938).

**next.** **1** *the next three*, etc. In most contexts *next* precedes the noun it governs when it means 'the one (of two or more) immediately following thing, occasion, etc. (*the next three, the next time, the next chapter*, etc.). But there are circumstances in which it may legitimately follow the noun or stand in predicative use, e.g. *on Friday next, follow next in order, it would be next to impossible.*

**2** *next Friday, June*, etc. In normal use, if said or written early in a given week, *next Friday* means the immediately following Friday in the same week. Similarly *next June*, when referred to in any month during the preceding twelve months (at any rate in the preceding months June to December), means the first June following the date on which it is mentioned. In Scotland, and also in some northern dialects of England, *next* is employed to designate the days of

the following week: thus *next Friday*, the Friday of next week, is contrasted with *this Friday*, that of the present week' (*OED*). When making an arrangement, organizing a meeting, etc., for 'next Friday' it is advisable to add the date of the Friday that is meant, or clarify matters in some other way. For example, if the Friday of the week following is meant, you may say *Friday week, a week Friday, a week come Friday, a week on Friday*, or *a week next Friday*. As a separate matter, *next June, next Monday*, etc., can be used without a preposition (*let's make a start next June*); but if *next* is put after the noun, idiom requires a preposition (*the meeting will take place on Monday next*).

**nexus.** The English plural *nexuses* is intolerably sibilant, and the Latin *nexūs* /-uːs/, not *nexi* (see -US 2), sounds pedantic. The plural consequently is very rare. If one must be used, *-uses* is recommended.

**nice.** **1** *Nice* makes *niceish*: see MUTE E. (But *nicish* is also permissible.)

**2** *nice and* as a sort of adverb = 'satisfactorily' (*I hope it will be nice and fine; nice and early* = nicely early) is a kind of HENDIADYS that is an established colloquialism, but should be confined, in print, to dialogue.

**3** Meaning. Readers of medieval and Renaissance literature soon become familiar with the earlier central senses of the word (descended at various removes from L *nescius* 'ignorant'): (*a*) foolish, stupid, senseless; (*b*) wanton, loose-mannered; (*c*) numerous other applications. The editors of the *OED* were also moved to add: 'The precise development of the very divergent senses which this word has acquired in English is not altogether clear. In many examples from the 16th and 17th centuries it is difficult to say in what particular sense the writer intended it to be taken.' In the 20c. the various uses of the word have been easier to map but are still quite numerous. The primary and unavoidable sense now, 'pleasant, agreeable, satisfactory; (of a person) kind, good-natured', is so common that it has often been condemned: e.g. 'it has been too great a favourite with the ladies, who have charmed out of it all its individuality and converted it into a mere diffuser of vague and mild agreeableness' (Fowler, 1926). There is no sign, however, that the general public, or writers, are willing to give it up, and uses like *your nice long letter, it is nice to see you*, and *she has been so nice to me*, abound. Of course there are plenty of near-synonyms available, and it is often advisable to close the door on *nice* and admit one of these instead. Other senses of *nice* include 'bad, awkward' (*a nice mess you've made*); 'fine, subtle' (*a nice distinction*); 'efficiently done' (*nice work*).

**Nicene.** The Nicene Creed, a formal statement of Christian belief, is based on that adopted at the first Council of Nicaea held in Nicaea (now Iznik in Turkey) in AD 325.

**nicety,** unlike *niceness*, is trisyllabic, /ˈnaɪsɪtɪ/. Its primary sense is 'subtlety, intricate quality' (*a point of great nicety*), a meaning not shared by *niceness*.

**nickel,** (*a*) a malleable metallic element, and (*b*) a US five-cent coin, is so spelt, not *nickle*. The inflected forms of the corresponding verb are *nickelled, nickelling* in BrE, and usu. *nickeled* and *nickeling* in AmE.

**nick-nack.** Use *knick-knack*.

**nickname. 1** In origin an *eke-name* 'an additional name', with the initial *n* falsely carried over from the preceding indefinite article *an* (as also, for example, in *newt*).

**2** A nickname is a familiar or humorous name given to a person, place, thing, etc., instead of or as well as the real name. Examples are legion. A selected few: *Beantown* (Boston, Mass.); *chalky White* (generic); *Dicky Bird* (famous cricket umpire); *Dog Star* (the star Sirius); *Jolly Roger* (pirates' flag); *La Stupenda* (Joan Sutherland, soprano); *Pompey* (Portsmouth); *Sally Army* (Salvation Army); *Silicon Valley* (an area south of San Francisco); *spud Murphy* (generic); *tin Lizzie* (model T Ford); *'Tiny' Freyberg* (General Freyberg); *Tommy (Atkins)* (British soldier(s)); *Windy City* (Chicago). Some nicknames are regularly preceded by the definite article (e.g. *The Sally Army* but *Silicon Valley*). No simple rule governs the matter.

See SOBRIQUETS.

**nidus.** The pl. recommended is *nidi* /-aɪ/, rather than *niduses*.

**niece.** One of the most commonly misspelt words in the language.

**niello,** Italian metal work. Pl. *nielli* /-liː/, less commonly *niellos*.

**nigger.** The whole world knows now that the word is offensive when applied by a white person to a black, but that it may be used without offence by one black person of another. The traditional phrases *nigger in the woodpile* 'a hidden cause of trouble or inconvenience' (first recorded in 1852), *work like a nigger* 'to work exceptionally hard' (1836), and *nigger minstrels* 'entertainers, usu. white men with blackened faces' (1860) have all been laid aside except in historical contexts or when used with an obvious implication of ethnic abuse. Cf. NEGRO.

**nigh.** see WELL-NIGH.

**nightie, nighty,** a nightdress. Either spelling may be used.

**nihilism.** The word is usually pronounced with the *h* silent, i.e. as /ˈnaɪɪlɪzəm/. Similarly *nihilist*.

**-nik,** as a Yiddish (and Russian) suffix meaning 'a person or thing involved in or associated with the thing or quality specified', has made some headway, esp. in AmE, since the mid-20c., e.g. *beatnik, no-goodnik, peacenik*. Examples of these words and of some others are available in the *OED*. Further examples: *This is not the image of the Soviet Union entrancing detenteniks—Chicago Tribune*, 1989; *Being a 'compulsive neatnik' as my workmates describe me, I read with interest Mike Royko's column of April 7 in which he details the horrendous litter scene of Chicago Parks—* ibid., 1992.

**nimbus** (halo of a saint, etc.; a raincloud). Pl. *nimbuses* or *nimbi* /-baɪ/. See -US 1.

**nimby.** One of the best-known acronyms ('*not in my back yard*') in the later part of the 20c. (first used in the 1980s), 'expressing objection to the siting of something unpleasant, such as a nuclear waste dump, in one's own locality (although, by implication, not minding this elsewhere)' (S. Tulloch, 1991).

**nineties** (and **twenties, thirties,** etc.). These words do not require an apostrophe (*the 'nineties*, etc.) when used for the years 91–9 etc. of a century, and still less for those of a person's life. Nor is an apostrophe needed in decades expressed in arabic numerals, *the 1920s, the 1930s, the 1990s,* etc.

**ninish, ninth.** The final *-e* of *nine* is not carried over into these two forms.

**nite.** As a humorous or vulgar spelling of *night,* it turns up, for example, in the work of the 19c. American humorist Artemus Ward (C. F. Browne), but was not commonly used in advertisements and neon signs (*nite club, nite life,* etc.) until the 1930s (later in BrE). It should not be used in general writing.

**nitre.** Thus spelt in BrE, but as *niter* in AmE.

**-n-, -nn-.** The same rules apply as for other final consonants that are doubled or are left undoubled in the inflected forms of words. Monosyllabic words containing a simple vowel (*a, e, i, o, u*) before *n* normally double the consonant before suffixes beginning with a vowel (*tanned, penned, sinning, donnish, dunned*) or before a final *-y* (*sonny, funny*); but remain undoubled if the stem contains a diphthong (*raining*), doubled vowel (*soonest*), or a vowel + consonant (*scorned, damnable, burning*). Words of more than one syllable follow the rule for monosyllables if their last syllable is accented, but otherwise do not double the *n*: thus *japanned* and *beginner*; but *womanish, turbaned, awakening, spavined, tobogganist, cottony.*

**no.** A great many English-speakers are unaware of the complications lying behind the idiomatic use of this word, yet somehow manage to keep it under control. The 'secret' is to keep its use as an adjective (or, in modern terminology, as a 'negative determiner') apart from its uses as an adverb (see 2 below), and as a noun (see 3 below). **1** As adj. = 'not a, not any'. (*a*) Though hallowed by some eight centuries of usage, *no* accompanied by other negatives is now only dialectal or illiterate: e.g. *Don't try to con me with no such talk*—G. Ade, 1896; *He's not going to be put in no poorhouse*—Listener, 1968; *'Clouds come up,' she continued, 'but no rain never falls when you want it.'*—E. Jolley, 1980 (Aust.). (*b*) *No* is correctly used to qualify 'a noun and adj. in close connection,

implying that an adj. of an opposite meaning would be more correct or appropriate' (*OED*): e.g. *It makes no inconsiderable addition to the revenue of the crown*—Ann. Reg., 1772; *Anabel, so self-assured, no great animal lover she*—M. Wesley, 1983; *she was also an organizer of no mean talent* mod. (*c*) We normally have the choice of placing the negative *not* with a verb or of using *no* with another element in the sentence without change of meaning: e.g. *I can't see any footprints/I can see no footprints.* But, as CGEL 10.58 points out, such pairs of sentences can often show a difference of meaning: e.g. *He is not a teacher* denotes that his occupation is not teaching (i.e. he is a banker, accountant, etc.); but *He is no teacher* indicates that he lacks the skills needed to be a good teacher. (*d*) As a minor curiosity, it is worth noting that *no* may be used as an adj. before a singular noun (*no cup*), a plural noun (*no cups*), or an uncountable noun (*no homework, no salvation*), unlike the other determiners, such as *a, an, this,* and *that,* which cannot function in all three ways. (*e*) From the 16c. until quite recently *no* as adj. could be freely preceded by *the* or a personal pronoun (e.g. *his no niggardishe nature,* 1581; *the no-great privilege of Captivitie,* 1647). The construction survives—just—in *no small, no little* (e.g. *her no small achievement, the no little sum involved*), but seems to be moribund. (*f*) See NEGATION 2.

(*g*) Negative parentheses. The rule here to be insisted on concerns negative expressions in general, and is stated under *no* only because that word happens to be present in violations of it oftener perhaps than any other. The rule is that adverbial qualifications containing a negative must not be comma'd off from the words they belong to as though they were mere parentheses. The rule only needs stating to be accepted; but the habit of providing adverbial phrases with commas often gets the better of common sense. It is clear, however, that there is the same essential absurdity in writing *He will, under no circumstances, consent* as in writing *He will, never, consent,* or *He will, not, consent.* It is worth while to add, for the reader's consideration while he or she glances at the examples, that it would often be better in these negative adverbial phrases to resolve *no* into *not … any* etc.: *We are assured that* the Prime Minister will, in no circumstances and on no consideration whatever, consent (will not in any … or on any … Or omit the commas, at the least). / And Paley and Butler, no more than Voltaire, could give Bagehot one thousandth part of the confidence that he drew from … (could not, any more than … Or could no more than Voltaire give). /Proposals which, under no possible circumstances, would lead to any substantial, or indeed perceptible, protection for a home industry (which would not lead under any … Or which would under no possible circumstances lead). (h) Qualifying a gerund in the predicate, denoting the impossibility of the action specified. Historical examples: There is no healing of thy wounde—Bible (Geneva), Nahum, 1560; Do what they might, there was no keeping down the butcher—W. Irving, 1820; There's no accounting for tastes, sir—Thackeray, 1850. Modern examples: There was no denying that without Arabic I should not have been able to take charge of the boat—W. Golding, 1985; there was no knowing what a badly brought-up creature like that might not chop to pieces—K. Amis, 1988. (i) Writing of compounds. The forms recommended are: no-ball (cricket, noun and verb), nobody, no-claims bonus, no-go area, nohow, no man's land, no one.

2 As adv. To judge from the material in my database, no as an adv. is much less common than no as an adj. (the proportions are 1 : 4), but the distribution doubtless varies according to the type of writing. Uses of the adv. requiring comment include: (a) = 'not'. Now only a Scottish use (where it may be a reduced form of nocht 'nought' in origin): e.g. Oh, my dear, that'll no dae!—R. L. Stevenson, a1894; Who says the Scots are a dour lot? No' us anyway!—Inverness newspaper, 1973. (b) The commonest use by far is with comparatives. = 'not at all, not any': e.g. the facts are no longer in dispute; the Ivanovs were certainly no poorer than Tanya and I; Henry would no more accept imperfection in himself than in his work. (c) Expressing the negative in an alternative choice, possibility, etc., esp. in the phr. whether or no: e.g. A frog he would a-wooing go, Heigh ho! says Rowley, Whether his mother would let him or no—traditional nursery rhyme, c1809; It was a half-baked eloquence … But half-baked or no, David rose to it greedily—Mrs H. Ward, 1892; In January, whether the shop were open or no was a

matter of no great importance—J. Bowen, 1986. (d) In the representation of the speech of foreigners, no? is often used as a tag question: We all have families, no? We all need the family, also.—Maggie Gee, 1985 (Frenchman speaking); 'Funny,' she added in an afterthought, 'that they sent it to me, no?—A. Desai, 1988 (context in India); Can you not see, no?—M. du Plessis, 1989 (Afrikaner speaking). (e) For no less, see LESS 3.

3 As noun. (a) The recommended pl. is noes (The noes have it, the negative voters are in the majority). See -O(E)S 2. (b) When used to mean 'an utterance of the word no, a denial', the word no, by convention, is nowadays usu. printed in romans and without surrounding quotation marks: e.g. he would not take no for an answer; in answer to his question she gave a solemn no.

**noblesse** /nəʊˈblɛs/. Use italics, esp. in the phr. noblesse oblige /ɒˈbliːʒ/ 'privilege entails responsibility'.

**nobody, no one.** (a) Like anybody and anyone (see ANY 3), nobody and no one are interchangeable in most contexts: e.g. whether this was true or not nobody (or no one) can say for certain; No one (or Nobody) took a photograph. (b) Nobody and no one, like other indefinite pronouns (see ANY), are frequently, though not without a tinge of guilt, followed by the plural pronouns they, them, or their: e.g. Nobody wants to hear that their hero isn't a hero; No one likes to be turned out of quarters where they have lived comfortably for many years. This kind of construction, which is recorded in the OED from the 16c. onward, seems likely to pass into unquestioned use in the 21c.

**nodus** /ˈnəʊdəs/, a knotty point. Pl. nodi /-daɪ/.

**nomad.** 1 The corresponding adj. is nomadic. See -D-, -DD-.

2 It is worth noting that the OED (1908) gave precedence to the pronunciation /ˈnɒmæd/, i.e. with a short vowel in the first syllable, and that Daniel Jones (1917) gave this as the only pronunciation. The pronunciation /ˈnəʊm-/ did not become firmly established until some point in the second half of the 20c.

**nom de guerre, nom de plume, penname, pseudonym.** Here is a pretty kettle of fish. *Nom de guerre* (first recorded in English in 1679) is current French for an assumed name, but is far from universally intelligible to English-speakers, most of whom assume that, whatever else it may mean, it surely cannot mean 'pseudonym' or 'nom de plume', i.e. an assumed name such as 'George Orwell' for Eric Arthur Blair. *Nom de plume* was formed in English (first recorded in 1823) from three separate French words but as a compound has no currency in the French language itself. *Pen-name* is a 19c. English translation of the pseudo-French *nom de plume*. This leaves *pseudonym*, which is a 19c. adaptation of French *pseudonyme* (from a respectable Greek base); cf. *homonym*, *synonym*, etc. Of the four, *pseudonym* is the recommended word, though *pen-name* and *nom de plume*, despite the difficulties surrounding their origins, are also widely used.

**nomenclature.** Pronounced with the stress on the second syllable in BrE, but usu. on the first in AmE, with a secondary stress on the second syllable.

**nominal.** For this as the adjective of *noun*, see NOUN.

**nominative.** The grammatical word is always pronounced /ˈnɒmɪnətɪv/, with /ə/ in the third syllable. The adjective connected in sense with *nominate* and *nomination* (e.g. *partly elective and partly nominative*) normally has /-eɪt-/ for its third syllable.

***nominativus pendens*** /ˌnɒmɪnəˈtaɪvəs ˈpendenz/. An erudite term, now not often used by grammarians, for a 'dangling subject' or an 'anacoluthic subject' of the type *He, the chieftain of them all, his sword hangs rusting on the wall* (Scott, 1805). Fowler (1926) defined it as 'A form of anacoluthon in which a sentence is begun with what appears to be the subject, but before the verb is reached something else is substituted in word or in thought, and the supposed subject is left in the air.' Numerous examples from the OE period to the 19c. are given by Visser (1963, I.i. 61), but the construction is not often found in present-day English. An example from Shakespeare (1594): *That they which brought me in my Masters hate, I liue to looke vpon their tragedy.*

**non-.** Ours is in many respects a dissenting and disallowing age (consider how frequently one encounters a red circle with a stroke through it and an image of what is not allowed). It is therefore not surprising to see that all the main negative prefixes have flourished: see e.g. the entries for A-¹; DE-; DIS-; IN- AND UN-; and UN-. None has thrived as abundantly as *non-*. In vol. 2 of *OEDS* (1976) words with this prefix occupied something like sixteen pages of text, even though a headnote explained that 'The sections that follow contain [only] a selection of the more frequently occurring formations during the last hundred years or so'.

Some of the more notable *non*-formations of the 20c. (with an indication of the date of their first recorded use) are: *non-aggression* (1903); *non-belligerent* (1909); *non-destructive* adj. (1929); *non-event* (1962); *non-fiction* (1909); *non-flammable* (1961); *non-fraternization* (1945); *non-hero* (1940; cf. *anti-hero*); *non-iron* adj. (1957); *nonleaded* (petrol) (1955; cf. *unleaded*); *non-nuclear* adj. (1920; *non-person* (1959; cf. *unperson*); *non-proliferation* (of nuclear weapons, 1965); *non-returnable* (containers) (1903); *non-skid* adj. (1908); *non-standard* adj. (1924); *non-stick* (pan, etc., 1958); *non-stop* (1903); *non-U* (1954); *non-violence* (1920); *non-white* (1921). NB. In AmE such compounds are likely to be written solid, i.e. without a hyphen.

It should be borne in mind that many *non*-compounds contrast in meaning with other words having the same main element but a different negative prefix. Thus *non-moral* is not the same as *amoral* or *immoral*; *non-human* is close in meaning to *unhuman* but differs sharply from *inhuman*; *non-effective* is less condemnatory than *ineffective*; and so on. One of the most important 20c. groups of *non*-compounds is that in which *non*-means 'that does not' or 'that is not meant to (or to be)': e.g. *non-crease, non-iron, non-shrink, non-skid, non-stick*, all normally used adjectivally. The *OED* (*non*- 5a) lists some 17–19c. examples of *non*- used with infinitives (*to non-act, to non-answer*, etc.). This type has not flourished in the 20c., though some 20c. AmE examples of the finite verb *to non-concur* 'to refuse to concur, to disagree' are listed in the *OED*.

**nonce.** In origin from ME *for than anes* (unrecorded) = for the one, altered by wrong division (cf. *a newt* for *an ewt*), it survives only in the old-fashioned phr. *for the nonce* 'for the time being, for the present occasion'; and in the compound *nonce-word*, a word coined for one occasion, e.g. *finneganswaked* in Anthony Burgess's *Inside Mr Enderby* (1963), 'the intermittent drone was finneganswaked by lightly sleeping Enderby into a parachronic lullaby chronicle'.

**nonchalant, nonchalance.** Pronounce /ˈnɒnʃələnt/, /ˈnɒnʃələns/, i.e. as English words, but with /ʃ/ not /tʃ/.

**non-count nouns.** See COUNT NOUNS; MASS NOUN.

**none.** 1 It is a mistake to suppose that the pronoun is singular only and must at all costs be followed by singular verbs or pronouns. It should be borne in mind that *none* is not a shortening of *no one* but is the regular descendant of OE *nān* (pronoun) 'none, not one'. At all times since the reign of King Alfred the choice of plural or singular in the accompanying verbs, etc., has been governed by the surrounding words or by the notional sense. Some examples will clarify matters: (sing.) *a fear which we cannot know, which we cannot face, which none understands*—T. S. Eliot, 1935; *She is rather difficult to describe physically, for none of her features is particularly striking*—D. Lodge, 1962; *Except in the eye of God, none of these people was very interesting*—F. Tuohy, 1984; *A broken ring may be discovered where none was guessed at*—P. Carey, 1985; *None of the money has been spent on repairs*—CGEL, 1985; *So far none of Gorbachev's many pronouncements has changed things on the ground*—*Daily Tel.*, 1987; *I waited on, in the hope of further customers, but none was in the offing*—C. Rumens, 1987; *None of this was a matter of treachery*—P. Wright, 1987. (pl.) *we were forced to develop arts organizations in places where there were none*—*Dædalus*, 1986; *if you're talking about the aged and the unemployed and the sick, none of these are going to be affected*—R. Muldoon, 1986; *None of our fundamental problems have been solved*—*London Rev. Bks*, 1987; *She also says that though she had many affairs, none were lighthearted romances*—*New Yorker*, 1987; *I have seen and delighted in many marvellous gardens, but none have given me so much pleasure as hers*—A. N. Wilson, 1990.

Verdict: use a singular verb where possible but if the notion of plurality is present a plural verb has been optional since the OE period and in some circumstances is desirable. The type *None of them have finished their essays* is better than the clumsy ... *has finished his or her essay.*

2 As adv. First recorded as an adv. in the 12c. but the earliest types of use have not survived in modE. In the sense 'by no means, not at all', *none* (adv.) was common from the 17c. to the 19c. (e.g. *After some questioning, by which he saw that I was none informed regarding the page*—J. Galt, 1824); and remains in use in *none so* (though somewhat archaic), *none too* followed by an adjective (esp. forming compounds). Examples: (none so) *Your Italy, contains none so accomplish'd a Courtier*—Shakespeare, 1611; *The Horse Guards are none so fond of him*—*Spectator*, 1879; (none too) *I was none too sure of it*—A. Hope, 1896; *This none-too-accurate article on the DAE*—*Amer. Speech*, 1941; *Its class of none-too-hardy evergreens*—*Country Life*, 1974.

**non-entity, nonentity.** The invaluable hyphen should be used to distinguish the abstract sense 'something which does not exist, a non-existent thing' from the everyday sense 'a person or thing of no importance'. I imagine that most people would place level stress on *non-* and *ent-* in the first case; but would stress only *-ent-* in the second.

**nonesuch, nonsuch.** The first is the original form, but the second is now perhaps the more usual form both in the sense 'a person or thing that is unrivalled' and 'a leguminous plant with black pods'. However spelt it is usu. pronounced /ˈnʌnsʌtʃ/ in BrE, but when spelt *nonsuch* is sometimes 'thought of' as /ˈnɒn-/.

**nonet** /nəʊˈnet/, a composition for nine voices or instruments. Thus spelt, not *nonette*. Cf. DUET.

**nonetheless, none the less.** See NEVERTHELESS.

**non-flammable.** See FLAMMABLE.

**nongenary.** See CENTENARY 2.

**non-moral.** See AMORAL.

**nonpareil.** Pronounced in various ways, but the dominant standard pronunciation in BrE seems to be /nɒnpəˈreɪl/.

**nonplus** /nɒn'plʌs/ has *nonplussed* and *nonplussing* as its inflected forms in BrE, but the second *s* is often absent in AmE. See -s-, -ss-.

**non-rhotic.** See RHOTIC.

**nonsense.** Uses of *nonsense* as a countable word (i.e. used with the indefinite article and also in the plural) have become frequent, at least in BrE, in the 20c. Examples: *Everyone said, 'Lyne made a nonsense of the embarkation.'*—E. Waugh, 1942; *I knew you'd make a nonsense of it so I told Wallis to be ready to take over*—L. Cooper, 1960; *Ambitious nonsenses like* The Entertainer *and* Flame in the Streets—Movie, 1962; *To say that Mr Smith should attend the Geneva conference on Rhodesia as a member of the British delegation is a nonsense*—A. Crosland, on Radio 4, 1976. Its traditional role as an uncountable noun meaning 'absurd or meaningless words or ideas' remains intact.

**non sequitur** /nɒn 'sekwɪtə/, a conclusion that does not logically follow from the premiss or premisses (from L, lit. 'it does not follow'). Thus: *It will be a hard winter, for holly-berries* [which are meant as provision for birds in hard weather] *are abundant.* The reasoning called *post hoc, propter hoc* is a form of *non sequitur.*

**nonsuch.** See NONESUCH.

**non-U.** See U AND NON-U.

**no one,** no person, nobody. Thus spelt (no hyphen). Contrast the sense in *No one person was guilty.* See NOBODY.

**no place, noplace.** As adverb (also sometimes written with a hyphen) meaning 'nowhere', fairly firmly restricted to informal use in AmE: e.g. *You're going no place until Herb gets here*—M. Pugh, 1969. See -PLACE.

**no problem.** There is always room in the language for short idiomatic phrases expressing solace, condolence, comfort, and the like, e.g. *never mind, there there, no matter,* etc. One of the latest recruits, first recorded in the 1960s, is *no problem* meaning 'the question (of difficulty) does not arise'. John Osborne, in the 20 June 1992 issue of *The Spectator,* called it 'peoplespeak' and gave an example: *Last week, on doctor's orders, I telephoned a pathology factory to organise a blood test. 'No*

*problem.' How can they possibly know until I've had it? But I do hope they're right.*

**nor.** 1 The quite primary use of *nor* is as a coordinator after *neither* e.g. *neither my grandmother nor my grandfather had much formal education; his reassurance had in a way been neither here nor there at the time; unable to understand anything, neither what she had just seen nor the support of his expression*), or other negative or implied negative (e.g. *that is not why I want to marry Bertrand nor why he wants to marry me*). As the examples show, the words following *neither … nor* can be nouns, noun phrases, adverbial phrases, or clauses of various types.

2 On the type *Neither Anabel nor Mark are in London* (M. Wesley, 1983), see NEITHER 4.

3 The type *neither … nor … nor* clearly falls within the bounds of legitimacy: e.g. *Neither rain, nor cold, nor obscure polling places could keep voters from their civic duty yesterday* (referring to the US presidential election of Nov. 1992); *But the comment that receives the heartiest agreement concerns neither the war, nor the earthquake, nor the crime rate*—Observer Mag., 1992.

4 CGEL (13.36) points out that in non-correlative sentences containing a negative in the first element *neither* and *nor* are sometimes interchangeable: e.g. *They never forgave him for the insult, neither (or nor) could he rid himself of the feelings of guilt for having spoken in that way.*

5 *Nor* is occasionally used when there is no negative present or implied in the first clause. Such constructions were formerly more common (16–19c. in the OED: e.g. *His Age and Courage weigh: Nor those alone*—Dryden, 1697) than they are now. But the construction is still standard though rather formal: e.g. *This unique product carries a brand name that represents an entire country; nor is this an accident* (CGEL).

6 The correlative construction *nor … nor,* once common (e.g. *Nor Bits nor Bridles can his Rage restrain*—Dryden, 1697), is now only archaic or poetical.

7 Correlative constructions of the type *either … nor* and *neither … or,* being on the face of it illogical, should be avoided. But see NEITHER 6.

**normalcy.** 'Normalcy became a notorious word during the 1920 Presidential election, when Warren G. Harding proclaimed that what the country needed was a return to the "normalcy" of the days before World War I' (WDEU, 1989). Since then its notoriety has come and gone in AmE, but lingers on in BrE, where normality is the customary word. It may come as a surprise to many people that the competing abstract nouns normalcy, normality, and normalness all entered the language at approximately the same time, in the middle of the 19c. The surprise is perhaps reduced when it is noticed that the adj. normal itself, though recorded in the 17c. in the sense 'rectangular', did not acquire its modern everyday meaning until about 1840. So what we are dealing with here is a group of modern words that has hardly had time for the customary processes of assimilation or rejection to have taken their course. It would appear that normalness has hardly taken on at all. Normalcy and normality stand side by side in AmE as legitimate alternatives. In BrE normality is the customary term, and normalcy is widely scorned. The distribution of these two abstract nouns in the rest of the English-speaking world needs to be investigated further before any claims are made about their currency.

Examples of normalcy: partly in order to tidy up, tidy the room and restore it to normalcy—A. Desai, 1988 (Indian); such a surfeit of civility and normalcy from people who will pay with cash instead of cheques or credit cards in the morning—F. Kidman, 1988 (NZ); Every effort is made to suggest that a certain 'normalcy' prevails, that relations remain roughly as they were—Dædalus, 1992. And an example of the standard BrE form normality: The morning passed slowly, uneventfully, and with a beguiling normality—A. Brookner, 1989.

**north-.** Compounds of north- (north-east, North Pole, north-west, etc.) are pronounced with /θ/. Of the derivatives, norther (strong north wind in Texas, etc.), northerly, northern, and northerner have /ð/, but northward(s) has /θ/; and northing (distance travelled or measured) may be pronounced either with /θ/ or with /ð/.

**northerly.** For the special uses and meanings of this set of words, see EASTERLY.

**northward(s).** As adj. before a noun (in a northward direction) always northward. As adv. either northward (he travelled northward) or northwards (the latter form esp. in BrE).

**nosey,** a variant of NOSY.

**nostrilled.** So spelt in BrE (wide-nostrilled). In AmE often nostriled. See -LL-, -L-.

**nostrum.** Plural nostrums, not nostra. See -UM 1.

**nosy,** inquisitive, etc. Usu. so spelt.

**not.**

1 Normal uses.
2 not all/all ... not.
3 not in meiosis and periphrasis.
4 not in exclamations.
5 not pleonastic.
6 not ... but.
7 not only (correlative).
8 not splitting an infinitive.
9 Case of following pronoun.
10 not but.
11 Placing of not in subjunctive clauses.
12 better not, best not.
13 whether or not.

1 Normal uses. Before I set down some of the actual or potential fault lines in the use of the word, it seems desirable to list some of the normal uses of not. These include the type 'auxiliary verb + not + bare infinitive' (I do not believe that Shelley could have written these lines; one should not rule out the possibility; he could not diminish it); negation of an adj. or noun (appearances that are and are not identical; I'm not a tyrant); not followed by an adverb (not entirely, not merely, not really, not simply, not sufficiently, not surprisingly). It can negate a clause (Satan, not greatly concerned which way this particular battle goes) or one constituent of a clause (as if it were cast back, not unproblematically, upon the existing poem). It is frequently used in association with but (Shelley's sonnet is not a specific critique but at most an 'indirect response') and with even (No, that is arrogant and not even true); see also not only ... but also, 7 below.

In formal prose not is hardly ever reduced to n't (conjoined to a preceding auxiliary verb (thus we cannot hear in them the voice of mocking laughter). In conversations and other informal contexts in fiction the reduced forms are almost always printed: e.g. It wasn't true; I don't suppose it was difficult; he couldn't keep his eyes off it; so that she shouldn't see his small pity; you can't mourn for unborn grandchildren; They wouldn't want anyone to know. (In case the reader is wondering, the examples given in this paragraph are drawn from *Essays & Studies 1992* and P. D. James's *The Children of Men* (1992), but they are of general application.)

**2** *not all/all ... not.* Fowler admitted in 1926 that the proverb *All is not gold that glisters* (or, as Shakespeare had it in *The Merchant of Venice* II.vii.66, *All that glisters is not gold*) is not strictly logical, since the negative properly belongs with *all* not with *gold*, i.e. 'Not all that glisters is gold' (some things that glister are indeed gold). It would be futile to try to change the proverb now, but caution is desirable when in other contexts *not* and *all* are used in close proximity. Thus *Not all children of five can recite the alphabet* means that only some of them can. *All children of five cannot recite the alphabet* can (just) mean the same, but is in danger of being treated as meaning 'No children of five can recite the alphabet'. An American columnist, James J. Kilpatrick, in the *Chicago Sun-Times* (10 Mar. 1991) cited several crassly ambiguous uses of *all, everybody,* etc. . . . *not* from AmE sources, including the following: *All television Westerns haven't ridden off into the sunset; Everybody in America is not a good person and everybody in politics is not a good person; Every woman does not aspire to be a mother; All the people living around local courses are not golfers.*

**3** *not* in meiosis and periphrasis. It has long been the practice in standard English to use *not* with negative adjectives, adverbs, and past participles, implying the affirmative term: *at length Not unamaz'd, she thus in answer spake*—Milton, 1667; *We say well and elegantly, not ungrateful, for very grateful*—tr. Erasmus, 1671; *Not unclever but importunate*—Earl Malmesbury, 1794; *It is ... certainly true—and has not gone unnoticed—that the language* [etc.]—T. Tanner, 1986. The 18c. grammarian Lindley Murray was perhaps the first to distinguish such repeated negatives from DOUBLE NEGATIVES (which he roundly condemned): 'Two negatives, in English, destroy one another, or are equivalent to an affirmative ... "His language, though inelegant, is *not ungrammatical*"; that is, "it is grammatical"'

(*English Grammar*, 1795, cited from an edn of 1824, p. 172). Fowler (1926) was worried, not to say dismayed, by such uses ('It is by this time a faded or jaded inelegance, this replacing of a term by the negation of its opposite; jaded by general over-use'). But he went on to say in its defence that 'the very popularity of the idiom in English is proof enough that there is something in it congenial to the English temperament, and it is pleasant to believe that it owes its success with us to a stubborn national dislike of putting things too strongly'. Is it not also likely that standard users of the language have instinctively found a way of drawing attention to the non-standardness of double negatives by setting up a sequence of self-cancelling negative unions of their own devising?

In some closely related types, *not* is slightly (or startlingly) confusing, requiring one to do, as it were, some mental arithmetic to work out the intended sense: e.g. (slightly) *Not that watching the 'park' at work has not been every bit as interesting as watching the subjects which we have set off to investigate*—*Spectator*, 1988; (startlingly) *This is not to say that I do not think that some of the dissents are not preferrable* [sic] *and not to publish them would hurt every one*—*Bull. Amer. Acad. Arts & Sci.,* 1988.

**4** *not* in exclamations. Fowler (1926) pointed out 'the jumble of question and exclamation' in the sentence *But if you look at the story of that quadrilateral of land, what a complex of change and diversity do you not discover!* The type *what a complex you discover!* is a true exclamation; *what complexity do you not discover?* is a possible question. But *what a complex do you not discover!* illogically combines an exclamatory construction with an interrogative one.

**5** *not* pleonastic. *Not* is sometimes unnecessarily placed in a subordinate clause as a mere unmeaning echo of an actual or virtual negative in the main clause. Examples: *I shouldn't wonder if it didn't turn to snow now* (read *if it turned*); *Nobody can predict with confidence how much time may not be employed on the concluding stages of the Bill* (omit *not*); *He is unable to say how much of the portraiture of Christ may not be due to the idealization of His life and character* (omit *not*); *Who knows but* *what agreeing to differ may* not *be a form of agreement rather than a form of difference?* (omit *not*); *I shouldn't be surprised if the Secretary of State* didn't *agree with the trade union about this* (read *agreed*).

**6** not ... but. *Mrs Fraser's book, however, is* not *confined to filling up the gaps in Livingstone's life ... but it* deals most interestingly with her father's own early adventures. In such sentences the subject must not be repeated (or, as here, be taken up by a pronoun). The above sentence is easily made right by omitting *it*.

**7** not only *(correlative). Not only* out of place, said Fowler (1926), is 'like a tintack loose on the floor; it might have been most serviceable somewhere else, and is capable of giving acute pain where it is'. Or, as it is less colourfully expressed in *OMEU* (1983): '*Not only* should always be placed next to the item which it qualifies, and not in the position before the verb. This is a fairly common slip, e.g. *Katherine's marriage not only kept her away, but at least two of Mr. March's cousins* (C. P. Snow); *kept not only her* would be better. If placing it before the verb is inevitable, the verb should be repeated after *but* (*also*), e.g. *It not only brings the coal out but brings the roof down as well* (George Orwell).' Examples of correct placings (including cases where the second correlative *but* (*also*) is idiomatically omitted): *To deliberately change anything requires* not only *courage but the greatest motivation* — C. Burns, 1985; *He was proof that a breakthrough into the new world was* not only *possible, it was a fact*—D. Malouf, 1985; *men who* not only *rhymed Moon with June but also thought they were spelt the same way*—Sunday Times, 1987. *Rowers* not only *face backward, they race backward*—New Yorker, 1988; Examples of incorrect placings cited by Fowler (1926): Not only *had she now a right to speak, but to speak with authority* (she had now a right not only to speak); Not only *does the proportion of suicides vary with the season of the year,* but *with different races* (the proportion of suicides varies not only with).

**8** not *splitting an infinitive. Not* is routinely used before a *to*-infinitive, negating all that follows it: *Was I* not *brought up* not *to scream?*—M. Wesley, 1983; *But he preferred* not *to torture himself with the specific knowledge*—A. N. Wilson, 1986; *I used* not *to dream*—N. Bawden, 1987; *Lisa was*

enrolled in the Church of England in order not *to give offence*—P. Lively, 1987; *But she tried* not *to think about that*—B. Rubens, 1987; *She tried* not *to watch Oscar count out his farthings*—P. Carey, 1988.

But, not uncommonly, *not* is placed between ('splits') the *to* and its infinitive: *It was wrong to* not *love them*—P. Carey, 1985; *a perfect morning to* not *read 'Moby Dick'*—New Yorker, 1986; *I thought about how good it would be to* not *have Christmas but to be off in a room by myself*—D. Vaughan, 1986; *My advice to any woman who earns the reputation of being capable, is to* not *demonstrate her ability too much*—M. Spark, 1988. This 'unprincipled' placing of *not* is presumably intended to give additional emphasis to the negativity.

**9** Case of following pronoun. According to the *OED* (sense 6d), when emphasizing a pronoun after a negative statement, or in a reply, *not* is traditionally followed by the subjective case of a pronoun: *He is no Witch,* not he— Dekker et al., c1625; *He didn't care for himself—*not *he*—Thackeray, 1847–8; *They are not to be 'had' by a bit of worm on the end of a hook ...* not *they!*—J. K. Jerome, 1889. In most contexts the subjective case is still desirable: e.g. *Not I,* not *I, but the wind that blows through me!*—D. H. Lawrence, 1917; *And so it is my mother who's told the lie,* not *he*—N. Gordimer, 1990; *But I don't like it either; I just put up with it. Not he*—A. Brookner, 1991; *This woman is a skeleton,* not *I*—J. Shute, 1992.

But it cannot be denied that the objective case is increasingly being used and not just 'in special circumstances': *She should be here,* not *me, it's a good setting for her*—M. Wesley, 1983; *It wasn't that she hated living in Johannesburg. Many did of course. Not her*—M. du Plessis, 1989; *the underwear she takes off those places where he can touch her—he,* not *me.* (Not *I—he would correct me.*)—N. Gordimer, 1990; *He shook his head. 'you're wrong. Not her.'*—A. Billson, 1993.

See CASES 5.

**10** not but, *'only, merely, just'.* From the 14c. to the 16c. *not but* stood side by side with *nobbut* (from *no* adv. + *but* conj.). At that point *nobbut* gradually retreated into dialect use and is now so restricted: *I* nobbut *wanted to know if they'd gotten him cleared?*—Mrs Gaskell, 1855; *It's* nobbut *Thursday, isn't it? Well it seems like* months—J. B. Priestley, 1929. *Not but* as an adverbial phrase became obsolete in the 16c., but has re-emerged (? only in the US) in the 20c.: *When I was not but eight, I wrote a poem named 'God's Garden'*—Lee Smith, 1983 (US); *'She's not but twenty-three, Cobb said to his stepfather. 'You think she's into capital gains and tax shelters?—New Yorker*, 1988.

**11** Placing of *not* in subjunctive clauses. From about 1400 until the 19c. *not* was frequently placed immediately before a finite verb in the indicative mood: e.g. *I not doubt He came aliue to Land*—Shakespeare, 1610. This use has now lapsed but it survives in certain kinds of subjunctive clauses (esp. in AmE): e.g. *Aunt Clara preferred that I not ransack the memories of her mother and Uncle Eugene*—D. Pinckney, 1992 (US). See SUBJUNCTIVE 3(*b*).

**12** *better not, best not*. The negative equivalent of the use of an 'unsupported *better*' (see BETTER 2), is sometimes found in standard colloquial use in Britain, but is much more common in North America, Australia, and New Zealand. Examples: *better not to ask her at all*—E. Jolley, 1980 (Aust.); *I don't know what you're saying, but it better not be about me*—B. Oakley, 1985 (Aust.); *'Better not tell Willy what?' says Willy*—J. Elliott, 1985; *Well, I better not hear any complaints about how American tourists are staying away*—T. Clancy, 1987 (US); *Besides, I better not fight him, or he me*—New Yorker, 1988; *I was about to double up for a good laugh. 'Better not,' Arncz said. 'Better not.'*—D. Pinckney, 1992 (US); *'I better not right now, Louise,' said Dan*—New Yorker, 1992. In the UK, *best not* (a reduced form of the much older idiom *I/you* etc. *had best not*) is the more usual colloquial construction. Examples: *Charlotte said, 'Best not fall in the marsh in them.'*—J. Gardam, 1985; *'Ah,' says Claudia. 'Then best not to interfere.'*—P. Lively, 1987; *Best ... not to mention matters Japanese at all*—Listener, 1987.

**13** *whether or not*. See WHETHER 2.
See also BECAUSE B2. For *cannot* (*help*) *but*, see BUT 7, HELP (verb) 2.

**notary public.** Pl. *notaries public*.

**notepaper.** Nancy Mitford and her set preferred *writing-paper*, a much older word (1548– in *OED*; *notepaper* 1849–). Whether she knew the history of the two words, and whether she knew that *notepaper* is first recorded in the work of Jane Carlyle ('one of the best letter-writers in the English language', *OCELit.*), she did not say.

**nothing.** For *is* or *has nothing to do with*, see IS 5.

**nothing less than.** As the *OED* notes (s.v. *less*), the combination *nothing less than* has or had two quite contrary senses until the early 20c.: (*a*) (in which *less* is an adj.) 'quite equal to, the same thing as' (*But yet methinkes, my Fathers execution Was nothing lesse then bloody Tyranny*—Shakespeare, 1593; *His policy became nothing less than a series of gigantic blunders*—Bookman, 1895). This is the only current sense. (*b*) (in which *less* is an adv.) 'far from being, anything rather than' ( = Fr. *rien moins que*). Example: *Who, trusting to the laws ... expected nothing less than an attack*—Scott, 1827. The dictionary (in 1903) labelled sense (*b*) 'Now rare', but '*Obs.*' would be more appropriate now.

**noticeable.** Thus spelt, not *noticable*. See MUTE E.

**notional agreement.** (Also called *notional concord*.) Strict grammatical agreement (or concord) is essential in most circumstances in standard English, but there are some contexts in which this strictness is overridden and 'notional agreement' occurs: see AGREEMENT 5, 6, 7. Test sentences: *The jury reached* its/their *decision; mathematics* [i.e. a noun pl. in form] *is a difficult subject; the news* [noun pl. in form] *is bad; but* no *one hates it*, do *they?*

**not to worry.** Used as a quasi-imperative (reduced from the type *Laura told her not to worry*—E. Jolley, 1980; cf. *please not to mention that again*—G. Eliot, 1872), this idiom seems to have first emerged in BrE (and was later taken up in AmE and elsewhere) in the 1950s. Examples: *Not to worry. By the time he ... had finished with me ... I'd be doing long division*—Daily Mail, 1958; *Ah well, not to worry. Since Tom's untimely heart attack all that's over*—M. Wesley, 1983; *She obviously liked the looks on their faces. 'Not to worry. I'm not halakahically Jewish.'*—D. Pinckney, 1992 (US). Cf. the similar construction in: *'We'll send it to you.' 'Not to bother. I'm going down to the country this evening.'*—L. Meynell, 1965.

**nought.** See NAUGHT 2. Since there are alternative ways of pronouncing the digit 'o' according to the context in which it is used, an American lexicographical friend, at my request, sought the personal opinion of a few fellow-Americans (family, and people he dealt with in shops and banks). I also sought the views of an eminent American lexicographer who happened to be in London (he is Person 1). My preconception was that most would say *zero*, and that none would say *nought*. I was only half right, to judge from this small sample. Here are the results:

> **1** zip code 60201: (Person 1) *oh ... oh*; (Person 2) *oh ... oh*; (Person 3) *oh ... oh*; (Person 4) *oh ... oh*; (Person 5) *oh ... oh*; (Person 6) *oh ... oh*; (Person 7, a black bank teller) *oh ... oh\**; (Person 8, a barber) *oh ... oh\**; (Person 9, a white bank teller) *oh ... oh*.

> **2** in decimals, e.g. 0.5: (1) *oh*; (2) *zero*; (3) *zero*; (4) *zero*; (5) *zero*; (6) *zero*; (7) *zero*; (8) *zero*; (9) *oh*.

> **3** telephone 08804: (1) *oh ... oh*; (2) *oh ... oh*; (3) *oh ... oh*; (4) *zero ... zero*; (5) *zero ... oh*; (6) *zero ... oh*; (7) *zero ... zero*; (8) *zero ... oh*; (9) *oh ... oh*.

(\* On reflection changed to *zero ... zero*.)

Some of the informants suspected that the questions were designed to make a distinction between number and letter.

**4** The normal sequence in Britain is: **1** *oh ... oh*. **2** *nought point five*. **3** *oh ... oh*. But in football 5–0 becomes *five nil* or *five (to) nothing*; and in tennis the spoken equivalent of 40–0 is *forty love*. For temperature, the normal spoken form is *zero*: − 5° = *five degrees below zero*.

**noun** has two adjs.—*nominal* (first recorded in this sense *c*1430) and *nounal* (1871)—and is not entirely comfortable with either. The objection to the traditional form *nominal* is that it is a word much used in other senses. This has induced some modern grammarians to favour the ill-formed word *nounal*. In practice, however, *noun* by itself is sufficient in attributive use (*noun classes*, *noun clause*, *noun phrase*, etc.), and this form should be used whenever possible. If a 'real' adjectival form is called for (e.g. to contrast with *adjectival*, *adverbial*, etc.), then *nominal* is recommended. To the more traditional of us *nounal* sounds

repulsive. *Proper nouns* are names of particular people (*Darwin*), places (*Paris*), months (*January*), days (*Friday*), etc., and are always written with a capital initial letter. The remainder (*bottle*, *diary*, *hand*, etc.) are *common nouns*. See also COUNT NOUNS, NON-COUNT NOUNS.

**noun and adjective accent.** When a word of two syllables is in use both as a noun and as an adj., it might be expected that they would be distinguished in some manner in pronunciation, and in particular by the position in which the accent is placed. Such a view was encouraged by Fowler (1926): he firmly gave second-syllable stressing to *expert* in *He is an expert golfer*, and first-syllable stressing to the same word used as a noun in *He is an expert in handwriting*. The evidence in Daniel Jones's *English Pronouncing Dict.* (1917) points in a similar direction: first-syllable stressing is given for the noun *expert*, and second-syllable stressing for the adj.; but, he added, the adj. with first-syllable stressing 'is particularly common when the word is attributive'. This tendency towards stress differentiation seems to have lost ground in the 20c., but a few examples remain, especially when there is a difference of meaning. A table of some of the relevant forms shows no particular movement in the direction favoured by Fowler. The movement, though slight, is towards same-stressing. In the following table, 1 = stressed on the first syllable; 2 = stressed on the second syllable.

| | *c1920* | *1990s* |
|---|---|---|
| *adult* (adj.) | 2 | 1 or 2 |
| *adult* (n.) | 2 or 1 | 1 or 2 |
| *compact* (adj.) | 2 | 2[a] |
| *compact* (n.) | 1 | 1 |
| *content* (adj.) | 2 | 2 (only predic.) |
| *content* (n.) | 1[b] or 2 | 1 |
| *converse* (adj.) | 1 | 1 |
| *converse* (n.) | 1 | 1 |
| *expert* (adj.) | 2[c] | 1 |
| *expert* (n.) | 1 | 1 |
| *instinct* (adj.) | 2 | 2 (only predic.) |
| *instinct* (n.) | 1 | 1 |
| *minute* (adj.) | 2 | 2 (small) |
| *minute* (n.) | 1 | 1 (division of time; memorandum) |
| *supine* (adj.) | 2 | 1 |
| *supine* (n.) | 1 | 1 |
| *suspect* (adj.) | 1 or 2 | 1 |
| *suspect* (n.) | 1 or 2 | 1 |
| *upright* (adj.) | 1[d] or 2 | 1 |
| *upright* (n.) | 1 or 2 | 1 |

[a] But *cómpact disc.* [b] Jones (1917): 'The stress — ´ — [i.e. on the second syllable] is very frequent in the case of the plural *contents*, but is less common in the case of the singular.' [c] But see comment above. [d] Jones (1917): 'also [on second syllable in the sense 'erect'], under influence of sentence-stress.'

## noun and verb accent, pronunciation, and spelling.

When, for historical reasons, a word is made to do double duty as a noun and a verb, the language from almost the earliest times has shown a strong tendency to differentiate the two by pronunciation, as in *use* (noun with final /s/, verb with final /z/). Such a distinction is sometimes effected, as in the example just given, without change of spelling, but in other cases the two words are distinguished in spelling, as *bath* (noun) /*bathe* (verb), *calf* (noun) /*calve* (verb). A large number of disyllabic nouns and verbs are also distinguished by the placing of the stress, the normal pattern being shown in e.g. *áccent* (noun), *accént* (verb); *cómpound* (noun), *compóund* (verb). The following examples are given as illustrations of the main types and are not intended to be exhaustive.

**1** Disyllabic noun and verb distinguished by the placing of the stress (normally on the first syllable for the noun, and the second syllable for the verb): *accent, commune, compound, concert, conduct, confine(s), conflict, conscript, consort, contest, contract, contrast, convert, convict, decrease, defect, dictate, digest, discord, discount, discourse, escort, export, extract, ferment, import, imprint, incline, increase, indent, inlay, insult, produce, record, reject, suspect, transfer.*

**2** Such pairs are sufficiently numerous to make it seem that the distinction might be a regular one, and intrusive newcomers to the group are regarded by RP speakers with suspicion or downright hostility. For example, when North Country people speak of *industrial disputes* (while keeping the stress on the second syllable in the verb) they are following the pattern shown in (1) above but are not following the ('irregular') RP type *dispúte* (noun and verb). It should be noted that RP has any number of pairs in which the stress falls in the same place in both noun and verb (e.g. *cómpost* (noun and verb), *concérn, cóncrete, consént,*

*diréct, dislíke, displáy*), thus making it impossible to assume that the matter is governed by simple rules. There are other pairs in which the pattern is still changing: see e.g. PRESAGE.

**3** Pairs of nouns and verbs distinguished by the nature of the final consonant. (*a*) Spelt the same: *abuse, close* (noun, cathedral close), *excuse, grease* (verb sometimes /griːz/), *house, misuse, mouth, use.* (*b*) With change of spelling: *advice/advise, bath/bathe, belief/believe, breath/breathe, calf/calve, cloth/clothe, glass/glaze, grass/graze, grief/grieve, half/halve, life/live, loss/lose, proof/prove, relief/relieve, safe/save, sheath/sheathe, shelf/shelve, stife/strive, thief/thieve, teeth/ teethe, wreath/wreathe.*

**nouns as verbs.** See CONVERSION 1, 3.

***nouveau riche.*** Pl. *nouveaux riches* (pronounced the same as the singular).

**novecento.** See TRECENTO.

**novelty-hunting** is especially characteristic of our age, e.g. in the invention of children's dolls and fantasy characters (*My little Pony, Teenage Mutant Ninja Turtle*), video games (*Nintendo*, etc.), competitive games and activities (*boardsailing, rollerblading, T-ball* (kind of softball for children)), hair styles (*shelf, sorority bob*), and so on, or just in the invention of new words as such, esp. (usu. immediately transparent) BLENDS (e.g. *edutainment, infomercial, raggazine, sexcapade*), and VOGUE WORDS (e.g. *charismatic, couch potato, level playing field, photo opportunity, prestigious*).

**now** (adj.). Adjectival or attributive uses are recorded in standard sources from the 15c. onward (e.g. *The dreadful treatment of the now king*—Burke, 1793), but were falling into disuse by the end of the 19c. The adjectival use was revived in the 1960s in the sense 'modern, fashionable, up-to-date' with reference to the allegedly exciting innovations of that time (e.g. *Even a poet as now as Dylan has two kinds of female character in his imagery*— G. Greer, 1970). This modern use now sounds rather dated except with particular reference to that notorious decade and the immediately following years.

**no way.** See WAY 1.

**nowt.** This northern dialectal variant of *nought* 'nothing' has made inroads into colloquial standard English during the 20c., esp. in the phr. *There's nowt so* (or *as*) *queer as folk* (e.g. *Patrick went back to reflecting how there was nowt as queer as folk*—K. Amis, 1988). Genuinely dialectal uses of the word also turn up in speech and writing in national sources: *Getting owt for nowt, it's useless*—a Tynesider speaking on BBC Radio 2, 1977; *Well I tell you, there's nowt that's cheap if you want fruit*—J. Winterson, 1985 (Lancashire speaker in context).

**nth.** In mathematics, *n* is used to indicate an indefinite number, that is a number that is contextually determined (e.g. *Even an n-dimensional series of such terms ... is still denumerable*—B. Russell, 1903; *Relations ... are accordingly named dyadic, triadic, (or, for* n *terms,* n-*adic)*—E. Bach, 1964). From the mid-19c., by a natural extension of meaning, *to the nth power* or *degree* came to mean 'to any required power; hence fig. to any extent, to the utmost'. Fowler (1926), in his spirited way, rejected the extended use as a 'popularized technicality' and 'wrong', but his view has not prevailed, and the lay application of *nth* is now standard: e.g. *The Neapolitan ... is an Italian to the nth degree.*

**nubile.** Used in English in its classical Latin (*nūbilis*) sense '(of females) of an age suitable for marriage' since the 17c., the word has generated the sense '(of women) sexually attractive' in the second half of the 20c., and this later sense is now the dominant one.

**nuclear.** The only acceptable pronunciation in BrE is /ˈnjuːklɪə/. The spectacular blunder of pronouncing it as if it were spelt *nuc-u-lar* (cf. *circular, particular, secular,* etc.) is well documented in America and is particularly associated with President Eisenhower. *WDEU* (1989) says that speakers who pronounce the word as *nuc-u-lar* 'have succumbed to the gravitational tug of a far more prevalent pattern', since there is only one other reasonably common word that ends in unstressed /-klɪə(r)/, namely *cochlear*, and several that end in *-ular* (see above); but it admits that those who adopt President Eisenhower's form 'are likely to draw some unfriendly attention from those who consider it an error'.

**Nuclear English.** 'The term ... has been used in recent years for a proposed "core" language consisting of elements from natural English; in it, for example, such potentially ambiguous modal verbs as *can* and *may* would be replaced as appropriate by such paraphrases as *be able to* and *be allowed to*. Nuclear English would serve as the basis for purpose-built "restricted" international varieties comparable to Seaspeak, the medium of worldwide maritime communication' (*OCELang*). The concept is discussed at greater length, and rejected, in my book *The English Language* (1985): 'It is most unlikely that any such form of prescriptively reduced English will be regarded as acceptable either by the foreigners themselves (except the merest beginners) or by ELT teachers.'

**nucleus** /ˈnjuːklɪəs/. Pl. *nuclei* /-lɪaɪ/.

**nugae** (pl. noun), trifles. Pronounce this orig. L word as /ˈnjuːdʒiː/ or /ˈnuːgaɪ/.

**number** (as a grammatical concept). See AGREEMENT.

**number,** as a noun of multitude in the type '*a number of* + pl. noun', normally governs a plural verb both in BrE and AmE. Examples: *A number of American banks have been victims of a computer fraud known as 'the Salami'*—*Times*, 1985; *we must remember that a large number of conductors want to hear the great artists*—*Dædalus*, 1986; *a number of books in and on the Uzbek language were purchased for the Bodleian*—*Bodl. Libr. Rec.*, 1987; *A number of books by ballerinas have been published lately*—*New Yorker*, 1987. By contrast the type '*the number of* + pl. noun' normally governs a singular verb. Examples: *The number of MPs has increased*—*Daily Telegraph*, 1987; *With increasing experience, the number of features and aspects to be taken account of becomes overwhelming*—*Bull. Amer. Acad. Arts & Sci.*, 1987.

**numeracy** is a word coined by the Committee on Education presided over by Sir Geoffrey Crowther in 1959 as a term for that complement which is desirable in the sixth-form education of arts specialists in the same way as literacy is in that of science specialists. The committee defined it as 'not only the ability to reason quantitatively but also some understanding of scientific method and some acquaintance with the achievements of science'. Clearly there was a need for such a word, and it has settled into the standard language (also the corresponding adj. *numerate*) with particular reference to acquaintance with the basic principles of mathematics.

**numerals.** The writing and printing of figures and numerals is a highly specialized matter: for detailed guidance readers are referred to *Hart's Rules*. Some selected examples (drawn mostly from Hart) must suffice here. **1** *Words or figures.* Figures should be used when the matter consists of a sequence of stated quantities: *Figures for September show the supply to have been 85,690 tons, a decrease in the month of 57 tons. The past 12 months show a net increase of 5 tons; The smallest tenor suitable for ten bells is D flat, of 5 feet diameter and 42 cwt.* This applies generally to all units of measurement—tons, cwt., feet, as above, also of area, volume, time, force, electrical units, etc.

Separate objects, animals, ships, persons, etc., are not units of measurement unless they are treated statistically: *A four-cylinder engine of 48 b.h.p. compared with a six-cylinder engine of 65 b.h.p.; The peasant had only four cows; A farm with 40 head of cattle.*

In descriptive matter, numbers under 100 to be in words; but write *90 to 100*, not *ninety to 100*. Write as words in such instances as: *With God a thousand years are but as one day; I have said so a hundred times.*

**2** *Numerals generally.* Insert commas with four or more figures, as *7,642* and *525,500*; write dates without commas, as *1993*; omit commas in figures denoting pagination, column numbers, and line numbers in poetry; also in library shelfmarks, as *Harleian MS 24456.* In decimals use the full point, as *7.06*; and write *0.76*, not *.76*. Similarly in writing the time of day: *4.30 p.m.* and, with the 24-hour clock, *0.31, 22.15.* In references to pagination, dates, etc., use the least number of figures possible; for example, write *30–1, 42–3, 132–6, 1841–5, 1990–1*; but write e.g. *10–11*, not *10–1*; *16–18*,

not *16–8*; *116–18*, not *116–8*; *210–11*, not *210–1*; *314–15*, not *314–5* (i.e. for the group 10–19 in each hundred). The numbers from one to ten should always be written in full. Do not contract dates involving different centuries, e.g. *1798–1810* not *1798–810*.

In collective numbers: either *from 280 to 300*; or *280–300*; not *from 280–300* (which shows a mixture of styles). Do not begin a sentence with an arabic numeral. Thus write *Eighty-seven Coalition aircraft attacked targets near Baghdad on Monday*, not *87 Coalition aircraft . . .* It is also desirable to avoid placing arabic figures next to each other in this manner: *The theatre can accommodate 700, 450 of whom are in the stalls.* Reword as *. . . , of whom 450 are in the stalls.*

See also AD; BC; BILLION; DATE 2.

**numerous** is solely an adjective. Thus (*COD*): **1** (with pl.) great in number (*received numerous gifts*).

**2** consisting of many (*a numerous family*). Fowler (1926) cited an extract in which it is erroneously used as a pronoun: *These men have introduced no fewer than 107 amendments, which they know perfectly well cannot pass, and numerous of which are not meant to pass* (read *many of which*). See VARIOUS, which is used as a pronoun in AmE.

**nuncio.** Pl. *nuncios*. See -O(E)S 4.

**nuptial** /ˈnʌpʃəl/. The word is sometimes pronounced in non-standard AmE as /ˈnʌpʃʊəl/ or /-ptʃʊəl/, i.e. with the sound of -*ual* as in *actual, conceptual, gradual*, etc. Cf. the similar adoption of an erroneous final syllable in the 'nuc-u-lar' version of NUCLEAR.

**nursling.** Now the form favoured in the house style of OUP, rather than *nurseling*.

**nurturance,** emotional and physical nourishment and care. Introduced in 1838 in the technical language of psychologists, the word is still largely confined to technical and scientific books and journals. A recent example: *Typically, they [sc. Japanese women] saw devotion to the home as the creative nurturance of children—Dædalus,* 1992.

# O o

**-o** is a suffix of combining forms of words. On the analogy of Greek combinations in -o (e.g. λευκό-χλωρος 'pale green'; cf. λευκός 'white'), and their adaptations and imitations in Latin, late or medieval, like *Gallograeci, Anglosaxonicus*, etc., -o- has come to be the usual connecting vowel in English combinations of various kinds: in ethnic names, as *Anglo-Saxon, Indo-European*, etc.; in scientific terms generally, as *electromagnetic, psychotherapy*. Typical examples of -o appended to the stems of English words of classical origin include *chloro-, cirrho-, pneumo-; dramatico-, economico-, historico-, politico-*. Two main developments have occurred in English: **1** The creation of imitative (i.e. non-classical) formations within English, e.g. *cottonocracy, meritocracy; gasometer, speedometer* (a notorious example, *speed* being a word of native origin).

**2** The emergence of curtailed words in -o such as *chromo* (short for *chromolithograph*), *dynamo* (short for *dynamo-electric machine*), *electro* (short for *electro-plate, -plating*), *photo* (short for *photograph*).

**-o.** For the plural forms of nouns ending in -o, see -O(E)S.

**oaf.** Pl. *oafs*. The pl. form *oaves* occurred sporadically in the 19c.

**oaken.** See -EN ADJECTIVES 1.

**O and Oh.** There is no exceptionless rule in the matter, but the usual practice is to use *O* to form a vocative (esp. in poetry), and when it is not separated by punctuation from what follows, and to use *Oh* as an independent exclamation, followed by a comma or exclamation mark. Examples: (O) *O mighty Caesar!*; *O Death, where is thy sting?*; *O little town of Bethlehem*; *O that I had wings like a dove.* (Oh) *Oh, snatch'd away in beauty's bosom*; *Oh! the professions; oh! the gold; and oh! the French* (Darwin); *Oh, how do you know that?*.

**oarlock.** See ROWLOCK.

**oasis** /əʊ'eɪsɪs/. Pl. *oases* /-siːz/.

**oaten.** See -EN ADJECTIVES 1.

**oath** /əʊθ/. Pl. *oaths* /əʊðz/.

**obbligato.** Note that this musical term of Italian origin has -bb-. Pl. *obbligatos*. See -O(E)S 3.

**obdurate.** The standard pronunciation is /'ɒbdjʊrət/ with the obscure vowel /ə/ in the final syllable, not /-eɪt/. The stress falls on the first syllable.

**obedient.** For its (now old-fashioned) use as part of a signing-off formula at the end of letters, see LETTER FORMS.

**obeisance.** Pronounce /əʊ'beɪsəns/. The stressed syllable -beis- rhymes with *base*, i.e. it is not (as in my own use) a disyllable, /-'beɪɪs-/.

**obelus**, a dagger-shaped reference mark in printed work. Pl. *obeli* /-laɪ/.

***obiter dictum***, an incidental remark. Pl. *obiter dicta*.

**object** (verb). In its prevailing current sense 'to have an objection to, disapprove of', *object* is frequently followed by *to* + noun (phrase) (*he objected to capital punishment*) and *to* + gerund (see below). In the 19c. and early 20c. it was also commonly construed with *to* + infinitive. Examples: (with a *to*-infinitive) *whether I would object to give the ladies the benefit of my assistance*—W. Collins, 1860; *They objected to be actors in a farce*—Manchester Examiner, 1885; *we object to pass Sundays in a state of coma*—Gissing, 1892; *She objected to marry a Jew*—G. B. Shaw, 1903; (with *to* + gerund) *Would the lady object to my lighting the pair of candles?* —Dickens, 1865; *Battle was being given to the front door, which objected to opening on principle*—E. Bowen, 1969.

**objection.** The pattern is the same as that for OBJECT (verb). Examples: (with a *to*-infinitive) *he had no great objection, but* *also no great wish, to fight*—C. Kingsley, 1853; *I have no objection to join with you in the enquiry*—Jowett, 1875; *Cecil's objection to go through with it*—G. B. Shaw, 1911; (with *to* + gerund) *I have no objection to doing it myself*—G. B. Shaw, 1911.

**objective genitive.** The ordinary possessive genitive is shown in *John's dog* (i.e. John has a dog), *the world's oceans* (the world has oceans). A subjective genitive is shown in *Thackeray's novels* (i.e. those he wrote), *the girl's decision* (the girl has decided). An objective genitive is shown in *the boy's murder* ( = the murder of the boy). Further examples of the objective genitive: *a London cultural policy just before the* [Greater London] *Council's abolition*—Essays & Studies, 1987; *The Queen Mary's sailing had been postponed*—New Yorker, 1987.

**objective, object** (nouns). In the broad sense 'something sought or aimed at' it is not always possible to decide with assurance which one is to be used. Possibly this is the case because *objective* so used is a comparatively recent (late 19c.) shortening of the military expression *objective point* '(lit.) the point towards which the advance of troops is directed', while *object* in this sense is much older (16c.), and the semantic 'space' occupied by each is as yet unresolved. Some examples may be useful as guides: (a) (object) (showing a modern, somewhat overused, phrase) *The main object of the exercise is to merge Theatre Tickets and Messengers' turnover with ours*—Times, 1959; (other examples) *Mrs Newman said placidly that her main object in visiting Cologne was to purchase a crate of toilet water*—A. Schlee, 1981; *Kierkegaard left Copenhagen for Berlin. His professed object in doing so was to attend a course of lectures*—P. Gardiner, 1988; *How easily I become blind like this to everything except the immediate object of my attention*—P. Toynbee, 1988. (b) (objective) *Khrushchev and I declared that our common objective remained the ultimate prohibition of nuclear weapons*—H. Macmillan, 1971; *His main objective was to pile up a huge personal fortune*—S. Bellow, 1987; *There is no question about our objective that all pupils should study literature written in English from other countries and cultures*—TES, 1990.

**objector.** Thus spelt, not -*er*.

**objet d'art.** Pl. *objets d'art* (like the singular pronounced /ɒbʒeɪ ˈdɑː/).

**obligate.** The verb *obligate* in non-legal use is almost always used in the passive (*be obligated, feel obligated,* etc.) followed by *to* + a noun (phrase) or by a *to*-infinitive, and it is not now found in standard use in BrE, having been replaced in all main uses by *obliged*. It is a routine example of a word that was once (17–19c.) standard in BrE but has retreated into dialectal use, while remaining common (beside *obliged*) in AmE and to some extent elsewhere. Its use in law lies outside the scope of this book. Examples: (active) *The Truman Administration devised a policy that obligated the United States to come to the aid of all 'free peoples who are resisting attempted subjugation by armed minorities.'*—New Yorker, 1990; (passive) *President Ford is obligated early next month to report to Congress on the 'progress' of negotiations looking toward a Cyprus settlement*—NY Times, 1975; *After the divorce she felt obligated and guilty to both her parents and her child*—B. Ripley, 1987 (US); *it was the last thing I wanted, but I felt obligated*—Rosie Scott, 1988 (NZ); *we will be obligated also to recommend educational and communications strategies that can overcome* [etc.]—Bull. Amer. Acad. Arts & Sci., 1992.

**obligatory.** The *OED* (1904) gave only the pronunciation /ˈɒblɪɡətərɪ/, i.e. with the first syllable bearing the stress, and Daniel Jones (1917) gave preference to this pronunciation while also listing /ˈɒblɪɡeɪtərɪ/ and, the only current form in standard BrE, /əˈblɪɡətərɪ/ (stressed on the second syllable). In AmE the fourth syllable is usu. pronounced /-tɔːr-/.

**oblige** (verb). See OBLIGATE.

**obligee** /ɒblɪˈdʒiː/. In law, 'a person to whom another is bound by contract or other legal procedure' (*COD*).

**obligor** /ɒblɪˈɡɔː/. In law, 'a person who is bound to another by contract or other legal procedure' (*COD*).

**obliqueness, obliquity.** The first is the more usual of the two esp. in the literal sense 'the quality of being oblique or slanting'; *obliquity* is used particularly in figurative contexts, 'divergence from moral rectitude', 'indirectness'. But in the sense 'indirectness', the choice of word is often just a matter of personal preference. Examples: (obliqueness) *He* had mentioned it—with a mischievous obliqueness—to Gott—M. Innes, 1937. (obliquity) *A mild and subtle influence compounded of glancing opinions, smiling obliquities, tender and persuasive flatteries*—A. Brookner, 1985; *most critics were models of discretion and obliquity*—New Yorker, 1992.

**oblique stroke.** An extract from a letter (11 Oct. 1992) from a friend in the diplomatic service: 'FCO-speak, or rather FCO-write, is currently ... littered with the oblique stroke or slash, entailing constant ambiguity—is it an indolent substitute for "or", or "and", or "I don't know which and can't be bothered to decide between them"?' Well, which is it?

**oblivious.** 1 Derived from L *oblīviōsus* 'forgetful, unmindful', *oblivious* was used from the 15c. onward for just on four centuries only in its Latin sense (e.g. *Gods memory is not so oblivious, that it can so soone forgett this covenaunt*—J. Bell, 1581; *The slow formality of an oblivious and drowsy exchequer*—Burke, 1780). This sense survives (often followed by *of*), but is less prominent now than it was a century or so ago. Examples: *Never before ... has a great painter been completely oblivious of the style, or styles, of his time*—K. Clark, 1949; *Oblivious of any previous decisions not to stand together ... the three stood in a tight group*—D. Lessing, 1985.

2 In the mid-19c., by a normal process of semantic change, it came to be used in the sense 'unaware of, unconscious of' (construed either with *of* or *to*). The *OED* (1904) firmly labelled this new sense *erroneous*. Fowler (1926) dutifully accepted the Dictionary's verdict, and cited numerous examples of the 'erroneous' sense. Time has moved on, and the new sense now forms part of the orthodox vocabulary of every educated person. Examples: *I stayed indoors all day for several days, oblivious to the damp heat of Falmouth*—C. Day Lewis, 1960; *For a man who has lived here all his life Makinen is oddly oblivious to the city's history*—Daily Tel., 1970; *He would ... look up at the startled faces of his company entirely oblivious of any offence that might have derived from his choice of expressions*—A. N. Wilson, 1984; *Up till this point she had been as oblivious to her surroundings, she might as well*

have been alone—P. Farmer, 1987. In practice it is often difficult to judge which sense was dominant in the writer's mind.

**obnoxious.** Derived from L *obnoxiōsus* (from *obnoxius* 'exposed to harm, subject, liable'), *obnoxious* came into English in the late 16c. in its Latinate sense, and this sense was current in standard use until the end of the 19c. (e.g. *A similar case, and is obnoxious to similar criticism*—Law Times, 1891). By 1926 Fowler could remark that this *de jure* sense was 'now so restricted that it is puzzling to the uninitiated'. The prevailing current sense (or *de facto* sense, as Fowler described it), 'offensive, objectionable, odious', first recorded in 1675 (*A very obnoxious person; an ill neighbour, and given much to law sutes with any*—A. Wood), apparently arose by association with *noxious*.

**oboe.** See HAUTBOY. A player of the instrument is an *oboist*.

**obscene.** In a century that has tried repeatedly to define the nature and desirable limits of obscenity, that is of lewdness, in books, films, etc. (always by extending the limits), the word has gathered strength in its other main meaning, 'highly offensive or repugnant' (of violence, famine, accumulation of wealth, etc.). Both strands of meaning go back to the late 16c. Examples: ( = lewd) *Our later writers are saucy rather than obscene*—Athenæum, 1899; *Appeal Court judges ruled ... that not only sex, but drug addiction, made a book obscene and depraved*—Daily Tel., 1964; *The obscene is whatever touches nerves, whether of desire or loathing, which society, usually in the person of one of its most susceptible members, such as Mrs Mary Whitehouse, deems better left untouched*—A. Burgess, 1988. ( = highly offensive) *Something in the very robustness of Germany's economy seemed to the terrorists and their sympathizers profoundly obscene*—Time, 1977; *The obscene proposals of the Monday Club to initiate repatriation and reverse race relations legislation*—Times, 1981; *The idea of these old women being walled up and told what to do by a superstitious parson was* (Tibba allowed herself the modernism) *obscene*—A. N. Wilson, 1982.

**observance, observation.** Both words are of long standing in the language (*observance* first recorded in the 13c., and *observation* in the 14c.) and both have impeccable etymological origins in Latin, the first being originally a noun denoting a state or condition ('regard, reverence, etc.') and the second a noun of action from L *observāre* 'to observe'. The centuries have worked their magic, and the *OED* divides *observance* into five branches of meaning and *observation* into ten. In practice the words are seldom interchangeable. Full details of the differences of application cannot be set down here. It must suffice to say that *observance* is the word normally used for the attending to and carrying out of a duty, custom, ritual, etc. (*observance of the speed limit*; *observance of the local custom of removing one's shoes at the entrance to a holy place*). It has none of the customary senses of *observation* (seeing, noticing, speaking, etc.: *some observations on the use of the subjunctive*; *under observation in the casualty ward*), and *observation* in turn does not mean attending to or complying with. It is worth noting that *observance* has developed only one common compound, *the Sunday Observance Society*; whereas *observation* has a string of compounds in regular use, *observation balloon, car, platform, post, ward*, etc.

Some examples of faulty uses: (*observance* for *observation*) *From him Mr Torr inherits both his gift for exact observance and lively humour*—1926 in Fowler; *His early poetry, the product of exalted sensation rather than of careful observance*—1926 in Fowler; (*observation* for *observance*) *The observation of the Sabbath*—Macaulay, 1825\*; *The British Government has failed to secure the observation of law and has lost the confidence of all classes*—1926 in Fowler.

(\* The use was not faulty in 1825.)

**obsolete** (verb). In the 17–19c. in standard use in the sense 'to render obsolete; to discard', this verb has dropped out of use in BrE but remains in use (or was revived) in N. America. Examples: *Our precoated steel failed to obsolete the glass TLC plate*—Sci. Amer., 1975; *Wire is being obsoleted in communication today*—Islander (Vancouver, BC), 1977.

**obstetric, obstetrical** (adjs.). The shorter form is much the more usual of the two, and is recommended. Its formation is in fact irregular: a midwife is *obstetrix, -īcem*, so that the 'regular' form would be the uneuphonious *obstetricic*. But only pedantry would take exception to *obstetric* at this stage.

**obtain** (verb). There is a slight gain in formality by using *obtain* instead of *get*, as shown in Fowler's anecdotal example: '*Customer*—Can you get me some? *Shop assistant*—We can obtain some for you, madam.'

**occasion.** Thus spelt. For some reason it is one of the most frequently misspelt words in the language.

**occiput,** the back of the head. Pronounce /ˈɒksɪpʌt/.

**occur.** The inflected forms are *occurred*, *occurring*. See -R-, -RR-.

**occurrence** /əˈkʌrəns/, the stressed syllable having the stem vowel of *current*. In AmE the stressed syllable is the same as that for *occur*.

**ocellus** /ɒˈseləs/, simple (as opposed to compound) eye (of an insect). Pl. *ocelli* /-laɪ/.

**ochlocracy** /ɒkˈlɒkrəsɪ/, mob rule. Cf. Gk ὄχλος 'crowd'.

**ochre,** name of a yellow pigment. Spelt *ocher* in AmE. See -ER AND -RE. Hence *ochrous* (adj.) (sometimes *ochreous*), AmE *ocherous*.

**octaroon.** See OCTOROON.

**octavo** /ɒkˈteɪvəʊ/. Pl. *octavos*. See -O(E)S 3. Abbrev. *8vo*.

**octet.** For the form, see DUET; NONET; QUARTET.

**octingentenary, octocentenary, octocentennial.** Of these three forms for the eight-hundredth anniversary of an event, the first is now obsolete, the second is the preferred form in Britain, and the third the more usual form in English-speaking countries abroad. See CENTENARY 2.

**octodecimo,** a book based on 18 leaves, 36 pages, to the sheet. Also called *eighteenmo*. Abbrev. *18mo*. Pl. *octodecimos*: see -O(E)S 3.

**octopus.** The only acceptable pl. in English is *octopuses*. The Greek original is ὀκτώπους, -ποδ- (which would lead to a pedantic English pl. form *octopodes*). The pl. form *octopi*, which is occasionally heard (mostly in jocular use), though based on modL *octopus*, is misconceived. See LATIN PLURALS; -US 4.

**octoroon,** occas. *octaroon*, is a person of one-eighth Negro blood; the offspring of a quadroon and a white person. The unetymological -*r*- in *octoroon* is modelled on *quadroon*. See MULATTO 2.

**octosyllabics.** The usual name of the eight-syllable rhyming iambic metre used in *Hudibras, The Lay of the Last Minstrel*, etc.: *What ever sceptre could inquire for; For every why he had a wherefore./The way was long, the wind was cold, The Minstrel was infirm and old; . . .*

**oculist.** See OPHTHALMOLOGIST.

**-odd.** Appended to a number, sum, weight, etc., to mean 'somewhat more than' (e.g. *forty-odd members*) it should be written with a hyphen in order to avoid any possibility of ambiguity.

**oddment.** See -MENT.

**ode.** 'In reference to ancient literature (and in some early uses of the word in English): A poem intended or adapted to be sung; e.g. the Odes of Pindar, of Anacreon, of Horace. *Choral Odes*, the songs of the Chorus in a Greek play, etc.; *spec.* a short Old English poem, esp. *The Battle of Brunanburh*. In modern use: A rhymed (rarely unrhymed) lyric, often in the form of an address, generally dignified or exalted in subject, feeling, and style, but sometimes (in earlier use) simple and familiar (though less so than a song' (*OED*). *The Battle of Brunanburh* is no longer called an ode, but a 'panegyric poem'. Among the most famous odes in English literature are 'Ode on a Grecian Urn' (Keats, 1819), 'Ode to a Nightingale' (Keats, 1819), and 'Ode to the West Wind' (Shelley, 1819).

**odour, odorous.** The slightly genteel word *odour* (instead of *smell*) is so spelt in BrE but as *odor* in AmE. (In the phr. *in bad odour* 'out of favour', *odour* is idiomatically used.) In both countries the corresponding adj. is spelt *odorous*.

**odyssey.** Pl. *odysseys*.

**oe, œ, e.** See Æ, Œ. There is a broad tendency in AmE, and much less so in

BrE, to simplify *oe* to *e*. The following lists show the dominant spelling in each country of some common words in this group:

| BrE | AmE |
|---|---|
| oedema | edema[1] |
| Oedipus | Oedipus |
| oenology | enology[1] |
| oesophagus | esophagus |
| oestrogen | estrogen |
| oestrus | estrus[1] |

Also in medial position:

| | |
|---|---|
| amoeba | ameba[1] |
| fetid | fetid |
| foetus[2] | fetus[3] |
| homoeopathy[2] | homeopathy[3] |

1 Also *oe*-   2 Also -*e*-   3 Also -*oe*-

The earlier forms (esp. in the 19c.) *oecology, oeconomy*, and *oecumenical* have now been entirely replaced by *ecology, economy*, and *ecumenical* in both forms of English.

**-o(e)s.** At one time or another we are all in difficulties with the plural of words which in the singular end in -*o*. There is no simple rule declaring that all such plurals end in -*oes*: if there were, *heroes* would 'look' fine, and so would *potatoes*; but *egoes* and *Eskimoes*, for example, would not. Conversely, if the rule stated that all such words ended in -*os*, *curios* would 'look' fine, and so would *cantos, memos, ratios*, and *zeros*; but *heros* and *potatos* would not. When in doubt about particular words, all that one can do is to follow the practice of one's favourite dictionary. The broad categories provided by Fowler (1926) and confirmed by Gowers (1965), despite the fact that they overlap somewhat, are as useful as any. **1** Words used as freely in the plural as in the singular, and completely naturalized as English words, usually have -*oes*, though there are very few invariable examples: the names of some animals and plants, for example, fall naturally into this class. So *buffaloes, calicoes, cargoes, dominoes, echoes, embargoes, haloes, heroes, mosquitoes, mottoes, Negroes, potatoes, tomatoes, torpedoes, vetoes, volcanoes*. But *banjo, dodo, flamingo, fresco, grotto, innuendo, mango, memento, peccadillo, salvo* (discharge), and a large number of others have either -*os* or -*oes* as their plural forms. Recommendations for the majority of these are given at their alphabetical place in the present book.

**2** Some monosyllables take -*oes*: so *goes*, *noes*; but *mos* (*wait for two mos*), *pos* (chamber pots).

**3** Words of the kind whose plural is seldom wanted or is restricted to special uses have -*os*: so *bravados*, *crescendos*, *dittos*, *dos* (the musical note), *guanos*, *infernos*, *lumbagos*, *obbligatos*, *octavos*, *rectos*, *tobaccos*, *versos*, *vertigos*.

**4** When a vowel (whether letter or sound) precedes the -*o* in the singular, -*os* is the usual plural, probably because of the bizarre look of -*ioes*, etc.: so *adagios*, *arpeggios*, *bagnios*, *bamboos*, *cameos*, *curios*, *embryos*, *folios*, *imbroglios*, *impresarios*, *intaglios*, *kangaroos*, *mustachios*, *nuncios*, *oratorios*, *portfolios*, *radios*, *scenarios*, *seraglios*, *studios*, *trios*, *videos*.

**5** Curtailed words, usu. made by dropping the second element of a compound or the later syllables, always have -*os*: so *altos*, *demos*, *dynamos*, *hippos*, *homos*, *kilos*, *magnetos*, *memos*, *photos*, *pianos*, *pros* (professionals), *rhinos*, *stylos*, *typos*.

**6** Alien-looking words and comparatively recent loanwords have -*os*: so *albinos*, *allegros*, *boleros*, *caballeros*, *calypsos*, *casinos*, *commandos*, *contangos*, *credos*, *diminuendos*, *egos*, *fandangos*, *farragos*, *fiascos*, *flamencos*, *gauchos*, *gigolos*, *kimonos*, *limbos*, *magnificos*, *merinos*, *mikados*, *moroccos*, *mulattos*, *piccolos*, *placebos*, *provisos*, *quartos*, *silos*, *solos*, *stilettos*, *tiros*, *torsos*, *weirdos*. It should be borne in mind, however, that -*os*, rather than -*oes*, is the routine plural spelling for the majority of nouns ending in -*o* whether they are 'alien-looking' or recent loanwords or not.

**7** Polysyllabic words tend to have -*os*: so *armadillos*, *clavicembalos*, *generalissimos*, *manifestos*, *Punchinellos*.

**8** Proper names have -*os*: so *Gallios*, *Lotharios*, *Martellos*, *Neros*, *Romeos*; similarly with ethnic names, as *Eskimos*, *Filipinos*. Also *Tornados* (fighter aircraft).

Some other words ending in -*o* (e.g. *concerto*, *contralto*, *graffito*, *hairdo*, *intermezzo*, *libretto*, *soprano*) are treated at their alphabetical place.

## of.

| | A | Fowler's observations and pronouncements: |
|---|---|---|
| | 1 | Wrong patching. |
| | 2 | Patching the unpatchable. |
| | 3 | Side-slip. |
| | 4 | Irresolution. |
| | 5 | Needless repetition of *of*. |
| | 6 | Misleading omission of *of*. |
| | 7 | Some freaks of idiom. |
| | B | Jespersen's riposte to Fowler. |
| | C 1 | *of* for *have*, '*ve*. |
| | 2 | Splitting of the *of*-genitive. |
| | 3 | Type *of an evening*. |
| | 4 | A final word. |

**A** Fowler's observations and pronouncements. 'Straying' in the use of *of*, said Fowler in 1926, 'is perpetual, and the impression of amateurishness produced on an educated reader of the newspapers is discreditable to the English Press.' From newspapers of the time, but not identified and therefore no longer recoverable, Fowler classified seven uses of *of* which he called 'crimes of grammar'. They are repeated here for their historical interest.

**1** Wrong patching. In the ten examples to be given, the same thing has happened every time. The writer composes a sentence in which some other preposition than *of* occurs once but governs two nouns, one close after it and the other at some distance. Looking over his sentence, he feels that the second noun is out in the cold, and that he would make things clearer by expressing the preposition for the second time instead of leaving it to be understood. So far, so good; care even when uncalled for is meritorious; but his stock of it runs short, and instead of ascertaining what the preposition really was he hurriedly assumes that it was the last in sight, which happens to be an *of* that he has had occasion to insert for some other purpose; that *of* he now substitutes for the other preposition whose insertion or omission was a matter of indifference, and so ruins the whole structure. In the examples, the three prepositions concerned are in roman type; the reader will notice that the later of the two *ofs* can be either omitted or altered to the earlier preposition, and that one of these courses is necessary:

*An eloquent testimony* to *the limits of this kind* of *war, and* of *the efficiency of right defensive measures*; *Which clearly points the need* for *some measure of honesty and* of *at least an attempt at understanding of racial ambitions*; *He will be in the best possible position* for *getting the most out of the land and* of *using it to the best possible advantage*; *He would have recovered the power to*

*manœuvre his armies in mass, a power abso-*
*lutely necessary either to achieving a military*
*decision, or in case of necessity of retiring*
*in good order; The definite repudiation of*
*militarism as the governing factor in the*
*relation of States and of the future moulding*
*of the European world; The varying provisions*
*in the different States respecting the length*
*and nature of the voter's qualification, as*
*well as of the kind of persons excluded from*
*the suffrage; A candidate who ventured to*
*hint at the possible persistence of the laws of*
*economics, and even of the revival of the*
*normal common-sense instincts of trade; The*
*Ministry aims not merely at an equitable*
*division of existing stocks, but of building*
*up reserves against the lean months; It begins*
*with the early enthusiasm of St Petersburg*
*for the war, and of the anti-German feeling*
*which transformed the city into Petrograd;*
*The magistrate commented on the nuisance*
*of street-collections by means of boxes, and*
*of the scandal of a system under which a*
*large proportion of the money given goes for*
*the expenses of collection.*

2 Patching the unpatchable. These resemble the previous set so far as the
writers are concerned; they have done
the same thing as before; but for the
reader who wishes to correct them there
is the difference that only one course is
open; *of* must be simply omitted, and
*between* or *without* cannot be substituted.
We can say *for you and for me* instead of
*for you and me* if we choose, but not
*between you and between me* for *between you
and me*; *with cries and with tears* means
the same as *with cries and tears*, but *without
cries or without tears* does not mean the
same as *without cries or tears*. Examples:
*It could be done without unduly raising the
price of coal, or of jeopardizing new trade;
He will distinguish between the American
habit of concentrating upon the absolute essentials, of 'getting there' by the shortest path,
and of the elaboration in detail and the love
of refinements in workmanship which mark
the Latin mind; Without going into the vexed
question of the precise geographical limitations, or of pronouncing any opinion upon
the conflicting claims of Italy and of the Yugo-
Slavs, what may be said is that ...*

3 Side-slip. Besides the types given in
the previous sections, so beautifully systematic in irregularity as almost to appear regular, there are more casual
aberrations of which no more need be
said than that the sentence is diverted

from its track into an *of* construction
by the presence somewhere of an *of* ...
Examples: *Sub-section 3 prohibits the Irish
Parliament from making any law so as to
directly or indirectly establish or endow any
religion or prohibit the free exercise thereof or
of giving a preference or imposing a disability
on account of religious belief or ecclesiastical
status; The primary object was not the destruction of the mole forts, or of the aeroplane
shed, or of whatever military equipment was
there, or even of killing or capturing its
garrison; Lord Parmoor referred to the progress which had been made in the acceptance
of the principle of a League of Nations, mentioning especially its inclusion in the Coalition
programme, and of the appointment of Lord
Robert Cecil to take charge of this question
at the Peace Conference; Its whole policy was,
and is, simply to obstruct the improvement
of the workingman's tavern, and of turning
every house of refreshment and entertainment
in the land into that sort of coffee tavern
which ...*

4 Irresolution. Here again we have illustrated *Germany's utter contempt for her
pledged word and of her respect for nothing
but brute force; His view would be more
appropriate in reference to Hume's standpoint
than of the best thought of our own day.* The
results of having in mind two ways of
putting a thing and deciding first for one
and then for the other: *we have illustrated,*
and *we have an illustration of; to Hume's
standpoint (than to the thought),* and *to the
standpoint of Hume (than of the thought).*

5 Needless repetition of *of*. *There is a
classical tag about the pleasure of being on
shore and of watching other folk in a big sea.*
A matter not of grammar, but of style
and lucidity; in style the second *of* is
heavy, and in sense it obscures the fact
that the pleasure lies not in two separate
things but in their combination.

6 Misleading omission of *of*. *The prohibition of meetings and the printing and
distribution of flysheets stopped the Radicals'
agitation.* Unless an *of* is inserted before
*the printing,* the instinct of symmetry
compels us to start by assuming that *the
printing* etc. *of flysheets* is parallel to *the
prohibition of meetings* instead of, as it
must be, to *meetings* alone.

7 Some freaks of idiom. *You are the man
of all others that I did not suspect; He is
the worst liar of any man I know; A child
of ten years old; That long nose of his.* The
modern tendency is to rid speech of

patent illogicalities; and all of the above either are, or seem to persons ignorant of any justification that might be found in the history of the constructions, plainly illogical: the man of all *men*; the worst liar of *all liars*; a child *of ten years*, or a child *ten years old*; a *friend* of mine, i.e. among my friends, but surely not that *nose* of his, i.e. among his noses: so the logic-chopper is fain to correct or damn; but even he is likely in unguarded moments to let the forbidden phrases slip out. They will perhaps be disused in time; meanwhile they are recognized idioms ...

**B** Jespersen's riposte to Fowler (A(7) above). In SPE Tract XXIV (1926), Jespersen showed that the use of *of* in such constructions as *A child of ten years old* and *That long nose of his* is as old as Caxton, and argued that *of* is here not partitive but appositional. 'The historical treatment of the construction', he said, 'cannot be separated from that of "a friend of mine".' Like the *OED* (sense 44 of *of* and sense 4b of *old*), he cites numerous examples of this type which has 'for centuries been used by high and low'. Thus: *Any neighebore of myne* (Chaucer); *A yong hors of the Quenes* (1502); *Looke, here comes a Louer of mine, and a louer of hers* (Shakespeare); *Nay, but this dotage of our Generals* [sc. of Antony's] *Ore-flowes the measure* (Shakespeare); *A Steer of two Years old* (Dryden); *This child of twelve years old* (Shelley); *It is a positive rest to look into that garden of his* (J. R. Lowell).

In *CGEL* (1985) the construction is called the 'post-genitive', and is presented as a normal unopposable part of the language: *that irritating habit* of her father's; *an invention* of Gutenberg's; *several pupils* of his. It is pointed out, however, that the genitival phrase 'must be definite and human': e.g. *an opera* of Verdi's, but not *\*an opera* of a composer's; *an opera* of my friend's, but not *\*a funnel* of the ship's. For further discussion of this type, see DOUBLE POSSESSIVE.

**C 1** *of* for *have*, *'ve*. In the spoken language the auxiliary verb *have* when unstressed (in such phrases as *could have*, *might have*, *would have*) is normally pronounced /əv/ and so, often, is *of*. It needed only the keen ears of children and of other partially educated people, to misinterpret /əv/ = *have* as being 'really' one

of the ways of saying *of*. The erroneous use (first recorded in 1837) is found in all English-speaking countries. Typical examples (in each of which the speech represented is that of a child or of a poorly educated adult): *Well, you should of buyed some cigarettes for yourself so it's your own fault*—S. Mackay, 1984 (UK); *My friend Gladys wouldn't of wanted you to come here nosing*—N. Virtue, 1988 (NZ); *An' Edward might of been dying*—P. Lively, 1989 (UK); *Firecracker must of lifted a rock or something. Didn't see it hit*—C. Tilghman, 1991 (US).

**2** Splitting of the *of*-genitive. Elisabeth Wieser (in *English Studies* 1986) described a little-noticed phenomenon which she called 'the splitting of the *of*-genitive'. In such constructions the governing noun is divided from its *of*-genitive by any of various elements. The use appears to be widespread in all varieties of English. Examples: *The remnant that exists in my memory of a book read and admired years ago*—P. Lubbock, 1921; *The dubbing into British English of Fellini's 'La Dolce Vita' was a tour de force*—A. Burgess, 1980; *With the rise to power of the National Socialists*—D. M. Thomas, 1981; *There was a glint in his eye of the charming Continental rascal*—G. Mikes, 1983.

**3** Type *of an evening*. The chiefly AmE type *he works nights*; *she plays cards Thursdays* (see ADVERB 4) is a concise way of expressing the general English type *of an evening*, *of a Sunday afternoon*, though this latter type, which is still current, is now beginning to sound literary or archaic. A typical example: *All the intellect of the place assembled of an evening*—Carlyle, 1831. Until the 19c. the type *of nights*, *of mornings* was also possible: e.g. *Dice can be played of mornings as well as after dinner*—Thackeray, 1849.

**4** A final word. Dwight Bolinger (*World Englishes* vii, 1988) reminded us that as a function word like *of* 'occurs in increasingly cohesive and stereotyped larger assemblies [i.e. multiplies the ways in which it is idiomatically used], its nuclear sense begins to fade, and eventually it turns into a kind of grammatical particle'. It then begins to be replaced by other prepositions, e.g. *about*. Bolinger cited numerous examples from radio broadcasts of the period 1982–8 in which *about* is used where *of* would have been expected, e.g. *There was none of the aware-ness about drugs that we have now*; *I'm perhaps a little more conscious about that*; *The Vietnamese are downright disdainful about Chinese cookery*; *Pierre Salinger thinks that [President] Kennedy was wary about tele-vision*. The broad sense of the preposition in such contexts is 'on the subject of', and, at least in AmE, in uses like these *about* is beginning to sound the more natural of the two. Bolinger also cited examples of other replacements: *enam-oured of→enamoured with* (cf. *in love with*), *free of→free from* (e.g. *It's free of/from contam-inants*).

It begins to look as if preposition re-placement is becoming an occasional but significant feature of the language. Powerful analogies are strewn about everywhere waiting to be seized upon. It is perhaps not surprising, therefore, though at present still regrettable, that some people (esp. children) are now us-ing *bored of* instead of *bored with* (see BORED).

**of a.** Observed so far only in AmE sources is the colloquial type adj. + *of* a + noun, e.g. *wouldn't be that difficult of a shot*—L. Trevino, 1985 (during a TV in-terview). *WDEU* 1989 cites numerous ex-amples (including the one by Trevino) from informal oral sources, e.g. *How big of a carrier force?*—J. Lehrer, 1986 (TV news-cast). The type may possibly be a slowly evolving extension of a much older quasi-nominal use of the adj. *considerable*: e.g. (from *DARE*, 1985) *This morning about 6 o'clock considerable of a shock of an earthquake was felt in Boston*—1766 source. No BrE examples have been noted yet.

**of course.** See COURSE.

**off.** Used in the sense 'from the hands, charge, or possession of', *off* is (according to the *OED*, as preposition, sense 2) con-strued esp. with *take, buy, borrow, hire*, and the like. But there is no doubt that, despite the antiquity (mid-16c. onward) of the construction, it is now regarded as not quite standard. One no longer buys a car *off* a stated garage or person but *from* a garage or person. Typical ex-amples of the older use: *A villager had come ... to know whether Blincoe 'would take a goose off him'*—C. James, 1891; *She admitted borrowing the 1l. off the plaintiff*—*Daily News*, 1897. Cf. OFF OF.

**offence** (noun). Thus spelt in BrE, but as *offense* in AmE.

**offer** makes *offered*, *offering*. See -R-, -RR-.

**office** (noun). In AmE the ordinary word for a doctor's surgery (see PHYSICIAN): e.g. *The doctor's office, which occupied the two front rooms of his house, was cool and high-ceilinged*—*New Yorker*, 1990.

**office** (verb). The noun *office* has been used intermittently as a verb, in various senses, now mostly obsolete, since the 15c. Shakespeare, for example, used it to mean 'to appoint to an office' in *The Winter's Tale* (*So stands this Squire Offic'd with me*). The verb is no longer extant in BrE but is reasonably common in NAmer. use, in the sense 'to have an office' or, followed by *with*, 'to share an office'. Examples: (advt) *Chance for high grade realtor to office with lawyer*—*Atlantic Monthly*, 1936; *Mr. Mardian spoke of a man who 'officed in that same agency'*—*NY Times*, 1973.

**official** (adj.) means 'of or relating to an office or its tenure (*official duties*)' or 'properly authorized (*an official fellowship at Magdalen College*; *the official attendance was 47,000*)'. *Officious*, by contrast, means 'unduly forward in offering services, meddlesome' (*an officious fussy little clerk*; *an officious waiter*). See OFFICIOUS.

**officialese.** 1 The term, first recorded in 1884, is used, mostly pejoratively, to mean 'the language characteristic of officials or of official documents'. Adject-ives applied to it in the illustrative ex-amples in the *OED* entry for the word include *crabbed, jolting*, and *dry*. Wry com-ments about the unsuitability, unattract-iveness, or impenetrability of official memoranda and the terminology of groups of officials turn up all the time. The attitude is one of bemusement or of horror. A passage from a 1990 column by Theodore Dalrymple, a regular contri-butor to *The Spectator*, may be taken as representative of many such comments: 'I'm beginning to wonder whether I still speak modern English, or whether, approaching middle age, the language of my youth and education has become an archaic and somewhat quaint dialect. Last week, I read a review in the *British Medical Journal* of a book about Aids from which it became clear that a prostitute is no longer a prostitute: she is a 'sex worker'. From which it follows, I suppose, that gonorrhoea is no longer a venereal but an occupational disease. The pages of the *BMJ* are as Edward Gibbon, however, compared to the language used during meetings by the new managers of the NHS. To attend one of these meetings is to enter a world in which people are not busy: they are 'fully occupied operationally'. When these meetings have lasted more than three hours, the chairman does not say, 'It's late and I am bored', he says, 'I'm conscious of the time.' Up and down the country, in all our hospitals, members of committees are fully occupied operationally, identifying and then defining live issues, which of course can only be approached multidisciplinarily, utilising a pool of resources (that is to say, employees). It has to be taken on board that blocks of time can only be devoted to this if a ring-fenced tranche of money is made available to the accountable manager who, as we all know, is a very pivot person in a liaison situation. The key to the operation is resourcing the clinical input adequately. It goes without saying that the output must be rigorously audited. Only in this manner can a potential runner, out of all the options that are generated, be posited at headquarters.'

**2** A somewhat more ambiguous argument is that put forward by Gowers (1965): '*Officialese* is a pejorative term for a style of writing marked by peculiarities supposed to be characteristic of officials. If a single word were needed to describe those peculiarities, that chosen by Dickens, *circumlocution*, is still the most suitable. They may be ascribed to a combination of causes: a feeling that plain words sort ill with the dignity of office, a politeness that shrinks from blunt statement, and, above all, the knowledge that for those engaged in the perilous game of politics, and their servants, vagueness is safer than precision. The natural result is a stilted and verbose style, not readily intelligible—a habit of mind for instance that automatically rejects the adjective *unsightly* in favour of the periphrasis *detrimental to the visual amenities of the locality*. This reputation, though not altogether undeserved, is unfairly exaggerated by a confusion in the public mind between *officialese* and what

may be termed *legalese*. For instance a correspondent writes to *The Times* to show up what he calls this "flower of circumlocution" from the National Insurance Act 1959; it ought not, he says, to be allowed to waste its sweetness on the desert air. *For the purpose of this Part of this Schedule a person over pensionable age, not being an insured person, shall be treated as an insured person if he would be an insured person were he under pensionable age and would be an employed person were he an insured person.*

'This is certainly not pretty or luminous writing. But it is not officialese, nor is it circumlocution. It is legalese, and the reason why it is difficult to grasp is not that it wanders verbosely round the point but that it goes straight there with a baffling economy of words. It has the compactness of a mathematical formula. Legalese cannot be judged by literary standards. In it everything must be subordinated to one paramount purpose: that of ensuring that if words have to be interpreted by a Court they will be given the meaning the draftsman intended. Elegance cannot be expected from anyone so circumscribed. Indeed it is hardly an exaggeration to say that the more readily a legal document appears to yield its meaning the less likely it is to prove unambiguous. It is fair to assume that if the paragraph quoted were to be worked out, as one would work out an equation, it would be found to express the draftsman's meaning with perfect precision.

'If an official were to use those words in explaining the law to a "person over pensionable age not being an insured person" he would indeed deserve to be pilloried. The popular belief that officials use an esoteric language no doubt derives partly from the reluctance they used to feel to explain the law in their own words lest they should be accused of putting a gloss on it. But, now that the daily lives of all of us are affected by innumerable laws, officials have had to overcome this inhibition and act as interpreters; they could not get through their work if they had not learned to express themselves to ordinary people in a way that ordinary people can understand. Circumlocution is rife in present-day writing and speaking, but officials are not markedly worse than other people; they are probably better than most. But the following examples show that they still sometimes fall into the old bad habit of giving explanations in terms only fit for an Act of Parliament, if that: *"Appropriate weekly rate" means, in relation to any benefit, the weekly rate of personal benefit by way of benefit of that description which is appropriate in the case of the person in relation to whom the provision containing that expression is to be applied; Unemployment benefit is not payable in respect of 13.2.56 to 17.2.56 which cannot be treated as days of unemployment on the ground that the claimant notwithstanding that this employment has terminated received, by way of compensation for the loss of the remuneration which he would have received for that day each of those days if the unemployment had not been terminated, payment of an amount which exceeds the amount arrived at by deducting the standard daily rate of employment benefit from two thirds of the remuneration lost in respect of that day each of those days.*

'That circumlocution may occasionally be found even in the utterances of those official "spokesmen" who ought to know better, may be illustrated by this extract from a London evening newspaper: *Discussing Anglo-American talks on the Barnes Wallis folding-wing plane, a Ministry spokesman said: "The object of this visit is a pooling of knowledge to explore further the possibility of a joint research effort to discover the practicability of making use of this principle to meet a possible future NATO requirement, and should be viewed in the general context of interdependence." Or, (our version): "This visit is to find out whether we can, together, develop the folding wing for NATO."*'

**3** See BAFFLEGAB; GOBBLEDEGOOK; PLAIN ENGLISH.

**officinal** (adj.) (in medical contexts). The pronunciation /əˈfɪsməl/ is recommended. Cf. *medicinal* /mɪˈdɪsməl/.

**officious** formerly had a meaning in diplomacy oddly different from its ordinary one. A diplomatist until relatively recent times meant by an *officious* communication much what a lawyer still means by one without prejudice; it was to bind no one, and, unless acted upon by common consent, was to be as if it had not been. It was the antithesis of *official*, and the notion of meddlesomeness attached to it in ordinary use was entirely absent. But the risk of misunderstanding was obvious, and the use now seems to be obsolete. A distinguished British diplomat wrote to me as follows: 'I have to confess that I cannot recall ever having seen the use of "officious", in the sense that you describe, in nearly 30 years of exposure to diplomatic English. I *have* occasionally come across this usage in the French *officieux* ("unofficial, semi-official"—Harrap), and have thought it odd and unexpected enough probably to have noticed an equivalent English usage had it come my way. I see that the OED notes the French term as the basis for "officious" in this sense, and that seems obviously right ... The Foreign Office ... used to employ the term 'semi-official', sometimes 'demi-official' to describe diplomatic correspondence which was not as formal as a despatch or Third Person Note' (private letter, 17 Apr. 1993).

**offing, offish**, etc. The vowel sound /ɔː/ in *off* and its derivatives, which was standard at the beginning of the 20c., has now given way, except in the speech of elderly people, to /ɒ/.

**off of** is still strongly present in the language of the less well educated but is indisputably non-standard in Britain. The *OED* gives numerous examples of the use, beginning with one in Shakespeare's *2 Henry VI* (A [ = I] *fall off of a tree*), and also: *About a furlong off of the Porters Lodge—*Bunyan, 1678; *I'd borrow two or three dollars off of the judge for him—*M. Twain, 1884. But the 20c. evidence (as indeed in the 1884 example just cited from *Huckleberry Finn*) is all from sources representing non-standard speech. In AmE, *off of* 'is widespread in speech, including that of the educated ... but is rare in edited writing' (*Random House Webster's College Dict.*, 1991). US examples: *the night Wayne came at Randolph with a hammer to pull him off of Mary—*M. Golden, 1989; *the collection of virtually all older artifacts and most modern ones—pulled out of chapels, peeled off of church walls, removed from decayed houses,* [etc.]—S. Greenblatt in *Bull. Amer. Acad. Arts & Sci.,* 1990; *She had a way of moving her head to pitch it* [sc. her hair] *about, a way of sweeping it off of her face with her fingers—*T. R. Pearson, 1993.

**offspring.** From the 16c. to the early 19c. a plural form *offsprings* was current (e.g. *the widows and the offsprings of the poorer, the indigent clergy*, 1756), but since then this ancient word (found already in OE) has been invariable in form (*These are the offspring of Muslim parents*; *the son tried to become the worthy offspring of his famous father*).

**oft,** recorded in standard use from the OE period onward, survives only in special contexts or in literature: e.g. *Do this, as oft as ye shall drink it, in remembrance of me—Bk of Common Prayer*; *What oft was thought, but ne'er so well expressed—*Pope, 1711.

**often.** The *OED* (1904) gave only the pronunciation /ɔːfən/, but since then the vowel has almost universally been replaced by /ɒ/ as in *not* and a spelling pronunciation with medial /t/ has also emerged in standard speech. The *OED* commented that the pronunciation with medial /t/, 'which is not recognized in the dictionaries, is now frequent in the south of England, and is often used in singing'. Nowadays many standard speakers use both /'ɒfən/ and /'ɒftən/, but the former pronunciation is the more common of the two.

**oftener, -est.** Logically the comparative and superlative forms of *often* but in practice much less common than *more often* and *most often*.

**oh.** See O AND OH.

**OK.** Possibly the only English word universally recognized by foreigners throughout the world. Its origin has been established by the American scholar Allen Walker Read: it seems to have been an abbreviation in the late 1830s of *orl* (or *oll*) *korrect*, a jocular form of 'all correct'. Other suggestions, as I expressed it myself in *OEDS* (1982), 'e.g. that *O.K.* represents the Choctaw *oke* 'it is', or French *au quai*, or that it derives from a word in the West African language Wolof via slaves in the southern States of America, all lack any form of acceptable documentation'. The word is also written with full points as *O.K.*, and in spelt-out form as *okay*. As a verb, *OK's*, *OK'd*, and *OK'ing* are the commonest written forms, but *okays*, etc., are not far behind. *OK* has very wide currency in the spoken language but is rarely found in formal written work.

**okapi** (Zaïrean animal). Pl. unchanged or *okapis*.

**old.** 1 For the distinction between *older*, *oldest*, and *elder*, *eldest*, see ELDER.

2 For the type *a boy*, etc. *of ten* etc. *years old*, see OF A 7.

3 For the *Old Lady of Threadneedle Street*, see SOBRIQUETS.

**olde.** 'An affected form of *old* adj. supposed to be archaic and usually employed to suggest (spurious) antiquity, *esp.* in collocations often also archaistically spelt, as *olde English(e)*, *Englyshe*, *worlde*, *worldy*' (*OEDS*, 1982). Examples: *A lot of olde realle beames in Amersham and a lot of olde phonie cookynge too—Good Food Guide*, 1959; *The interior is old but not olde worlde—Guardian*, 1972; *Charming stone built olde worlde Cottage of immense character—*advt in a Rhyl (N. Wales) newspaper, 1976.

**olden.** First recorded in the 15c., this word has gradually fallen into disuse except in the expression *in olden times* (or *days*). Its origin is uncertain, but 'it has been suggested that the suffix may represent an early inflexion of *old*. Cf. Ger. *in der alten Zeit*' (*OED*). It is not to be reckoned among the numerous -EN ADJECTIVES.

**olfactory.** For *olfactory organ*, see POLYSYLLABIC HUMOUR.

**olio.** Pl. *olios*. See -O(E)S 4.

**-ology.** See -LOGY. New, usu. temporary, formations in *-ology* ('sportive noncewords', as the *OED* calls them) are still being coined all the time. Recent examples that have come to my notice include *coupology* (study of the plotting of coups), *crazeology* (title of a 1989 biography of a jazzman), *enterology* (procedure whereby a contortionist squeezes into a bottle of limited size), *Hamburgerology*, *spudology* (on headline of an article about a work called *The Amazing Potato Book*). The language seems to welcome such words as temporary passengers, so long as they get off at the next station.

**Olympiad, Olympian, Olympic.** 1 *Olympiad*. 'In its earliest use (Pindar,

Herodotus) it refers to the [Olympic] games themselves or to a victory in the games; only later does it come also to signify the period between celebrations; and only much later still, in Latin, does the second sense become the principal one' (W. Sidney Allen, 1992). In English the earliest recorded use (Trevisa, 1398) is the period between the ancient games, used by the Greeks for the computation of time. The sense 'a quadrennial celebration of the ancient Olympic Games' is first recorded in the late 15c. (Skelton). The Olympic Games were revived in Athens in 1896, and the term *Olympiad* was first recorded of these in 1907.

**2** *Olympian.* Shakespeare, Milton, and others used the adj. to apply to the ancient Greek games (*the Olympian games*), but now it is principally used of or referring to Mount Olympus in NE Greece, traditionally the home of the Greek gods, 'heavenly, celestial'. The noun *Olympian* is now sometimes used to mean 'a competitor in the modern Olympic Games'.

**3** *Olympic.* In modern use it is principally used of the games of ancient Greece or those of modern times. The games themselves are frequently called the *Olympics* (as well as the *Olympic Games*).

**omelette.** The customary spelling in BrE (rather than *omelet*), whereas *omelet* is the more usual form in AmE.

**ominous.** Now always pronounced /ˈɒm-məs/, with a short initial *o*, though Daniel Jones (1917) gave initial /ˈəʊm-/, as in *omen*, as an alternative.

**omission of ɩelatives,** etc. A characteristic feature of English speech and writing, current since at least the 13c., is the construction of relative clauses without the 'normal' introductory *that* (or *whom*, etc.). Some grammarians treat the matter as the omission of a relative pronoun or of the conjunction *that*; but see CONTACT CLAUSES. Examples: (omission of a relative pronoun) *I do loue a woman ... and shee's faire ∧ I loue*—Shakespeare, 1592; *In the day ∧ ye eate thereof, then your eyes shalbee opened*—Gen. 3: 5 (AV), 1611; *This is a spray ∧ the Bird clung to*—Browning, 1855; *She is engagingly grateful for the good luck ∧ she has had*—D. Davie, 1982; (omission of the conjunction *that*) *Direct mine Armes, ∧ I may embrace*

*his Neck*—Shakespeare, 1591; *It may be ∧ they will reuerence him*—Luke 20: 13 (AV), 1611; *We were sorry ∧ you couldn't come*—a1912; *I felt ∧ I should have gone to the tradesmen's entrance*—G. Greene, 1980.

The two 20c. examples just cited have a natural feel about them, i.e. the omission of the conjunction *that* passes unnoticed. But things do not always go so smoothly. Examples showing that omission can produce forced constructions: *Keating made it plain yesterday ∧ he did not accept the thesis ∧ swinging voters can be spooked into voting for Labor by [etc.]*—*Aust. Financial Review*, 1993; *A practical argument for Australian action is ∧ the British themselves may yet seek a future in Europe. Britain is not demanding ∧ Australia remain a constitutional monarchy—it accepts ∧ the choice rests with Australia*—*Weekend Aust.*, 1993.

**omit.** The inflected forms are *omitted*, *omitting*. Cf. -T-, -TT-.

**omnibus** (vehicle). Now almost entirely supplanted by the shortened form *bus*, but if it is used the pl. is *omnibuses*. Since the late 1920s the terms *omnibus book*, *volume*, etc., have been commonly used for a volume containing several novels, etc., previously published separately.

**omniscience, -scient.** The recommended pronunciation of *-sc-* in these words is /s/ rather than /ʃ/.

**omnium gatherum.** A mock Latin formation, first recorded in 1530, for 'a miscellany or strange mixture (of persons or things)'.

**on.** **1** For *on all fours*, see FOUR. For *on to*, *onto*, and *on*, see ON TO. For *on* and *upon*, see UPON. For *on* and *in* a street, see IN 3 (and below, 2).

**2** NAmer. (occas. elsewhere, but not in BrE) use of *on* where *in* is standard in Britain: (place) *My father, may he rest in peace, had a dry-goods store on Gesia Street*—I. B. Singer, 1983; *'I've lived on Eagle Street fifty-five years,' Mike said*—*New Yorker*, 1986; *I head north on University Avenue*—M. Atwood, 1989; *The street Logan lived on was tranquil, shaded by an impressive congregation of sturdy trees*—M. Golden, 1989; (time; BrE requires *at*) *On weekends she would play disk jockey like that for hours*—*New Yorker*, 1987; *At university we'd sometimes get together for a 'spree'. Especially on weekends*—A.

Brink, 1988 (SAfr.); *and on weekends having champagne and strawberries with the sons of the nobility*—J. Updike, 1988. But cf. BrE *on Monday, Tuesday, Boxing Day*, etc.

**3** Evidence is accumulating that *on* is beginning to invade some of the traditional territory of some other prepositions in BrE, esp. in newspaper headlines. A Norwegian scholar, Åge Lind, presented numerous examples in *Moderna Språk* No. 1, 1987: e.g. *Gilt-edged and equities drift lower on* [ = as a result of] *lack of buyers*—*Financial Times*, 1978; *Row on* [ = about] *massive cuts in defence*—*Observer*, 1981; *Pilot gets life on* [ = for] *wife killing*—*Times*, 1986 (the story that followed had *for*: *A pilot who stabbed his wife ... was jailed for life for murder*).

**-on.** Plural of words derived directly or indirectly from Greek or modelled on Greek words and having in English the termination *-on*: **1** Some may, and often or always do, form the plural in *-a*: so *asyndeton, automaton, criterion* (but see as main entry), *hyperbaton, noumenon, organon, oxymoron, parhelion, phenomenon* (but see as main entry).

**2** Others seldom or never use that form, but fall in with the normal English plurals in *-s*: so *cyclotron, electron, lexicon, neutron, proton, skeleton.*

**3** In others again, the substitution of *-a* for *-on* to form the plural would be a blunder, their Greek plurals being, if they are actual Greek words, of some quite different form, and *-s* is the only plural used in English. Such are *anion* (19 c.), *archon, canon, cation* (19c.), *cotyledon, demon, mastodon* (19c.), *nylon* (20c. invented word), *pylon, siphon.*

**on account of.** See ACCOUNT.

**on behalf of.** See BEHALF.

**once.** **1** Normally an adverb (*he goes to the theatre once a week; once bitten, twice shy; if we once lose sight of him we shall never set eyes on him again*), but also a conjunction (*once he's back, I'm sure things will settle down again*), and occasionally, in informal use, a noun (*No one but the broad-shouldered woman had spoken, and that only the once*).

**2** The old phr. *once and away* meaning 'once and for all' (*It is not enough to harrow once and away*—1759 in OED) seems to

have passed into meaning 'occasionally' in the 19c. and is now rare or obsolete. Meanwhile *once in a way* meaning 'occasionally' (1891– in the OED) has idiomatically acquired a preceding *for* when used to mean 'as a solitary or exceptional instance' (*I really think, father, you might for once in a way take some slight interest in the family*—G. B. Shaw, 1934). It in turn seems now to be threatened by the phrase (*every once in a while* 'every now and then, at long intervals' (1781 and later examples in the OED), which is now widely current.

**one.**

> **1** Writing of *anyone, no one*, etc.
> **2** The type *one and a half years*, etc./*a year* etc. *and a half.*
> **3** The type *one of, if not the best book(s).*
> **4** The type *one of those who need(s).*
> **5** Kinds of pronoun: numeral, impersonal, or generic, replacing *I.*
> **6** Possessive of the numeral and the impersonal.
> **7** Mixtures of *one* with *he, you, they, my*, etc.

**1** Writing of *anyone, no one*, etc. The written forms recommended are *anyone, everyone, no one, someone*: see these as main entries (*anyone* treated s.v. ANY).

**2** The type *one and a half years* etc./*a year* etc. *and a half.* See HALF 3.

**3** The type *one of, if not the best book(s).* The difficulty here is that one can so easily be trapped into writing a mixture of two constructions: *one of the best books* is normal and so is *the best book*, but they should not be used together in the manner indicated. Fowler (1926) cited a number of aberrant examples: e.g. *the Costume Hall—one of, if not the most, spacious of salons for dresses and costumes* (read *one of the most spacious, if not the most spacious, of salons*); *One of the finest, if not the finest, poem of an equal length produced of recent years* (read *One of the finest poems of an equal length produced [in] recent years, if not the finest*); *I think the stage is one of, if not the best of all, professions open to women* (read *I think the stage is one of the best professions open to women, if not the best of all*); *Fur was one of the greatest—perhaps the greatest—export articles of Norway* (read *Fur was one of the greatest export articles of Norway, perhaps the greatest*). It will be observed that acceptable sentences are produced in each case by a simple re-arrangement of the elements of the aberrant ones.

**4** The type *one of those who need(s)*. See AGREEMENT 9. Further examples of each type: (pl. verb in the subordinate clause) *She was one of those women who make an enchanted garden of their childhood memories*—A. Brookner, 1990; *Not only was Fenella one of those people who imagine ... that Roman Catholicism had more chic than other forms of Christianity,* [etc.]—A. N. Wilson, 1990; *Oh, he's one of those people who need a nurturing older woman to listen to their woes*—Antonya Nelson, 1992 (US). (sing. verb in the subordinate clause) *He's one of those Yanks who wants to be really English, you know*—A. Motion, 1989; *he is one of those people who has five minutes of fame and a place in history because he just happened to be in the right place at the right time*—J. G. Dunne, 1991 (US); *'Don't you think,' said Bernard, 'that Hawaii is one of those places that was always better in the past?'*—D. Lodge, 1991. A plural verb in the subordinate clause is recommended unless particular attention is being drawn to the uniqueness, individuality, etc., of the *one* in the opening clause.

**5** Kinds of pronoun; numeral, impersonal, or generic, replacing *I*. It is desirable at this point to distinguish the names given to certain uses of *one*. It is a pronoun of some sort whenever it stands not in agreement with a noun, but as a substitute for a noun preceded by *a* or *one*: in *I took one apple*, *one* is not a pronoun but an adjective; in *I want an apple; may I take one?*, *one* stands for *an apple* or *one apple*, and is a pronoun. In the following examples three different kinds of pronoun are exemplified: *One of them escaped*; *One is often forced to confess failure*; *One knew better than to swallow that*. In the first, *one* may be called a *numeral pronoun*, which description will cover also *I will take one*, *They saw one another*, *One is enough*, and so on. In the second, *one* stands for *a person*, i.e., the average person, or the sort of person we happened to be concerned with, or anyone of the class that includes the speaker. It does not mean a particular person; it might be called an indefinite, an impersonal, or a generic pronoun; in this book it will be called an *impersonal pronoun*. In the third, *one* is neither more or less than a substitute

*I*: in such contexts it is fully pronounced (/wʌn/ not /wən/), and is often used ironically or with a sense of social superiority (see below). The distinction between the second and third types is often a fine one—indeed there is much overlapping.

A range of examples illustrating types two or three, or in some cases both: *The Caterpillar murmured—'One doesn't pretend to be a Christian, but as a gentlemen one accepts a bit of bad luck without gnashing one's teeth*—H. A. Vachell, 1905; *Lady Seal ... had told Anderson it* [sc. the bombardment] *was probably only a practice. That was what one told servants*—E. Waugh, 1942; *How to persuade the Telegraph that ... one was a man of immense culture?* (saying 'one' when you mean 'I' would do for a start, I decided)—F. Johnson, 1982; *you must realize that there are risks that one doesn't take*—N. Gordimer, 1987; *at least one listener found it* [sc. a performance of a piano concerto] *more intriguing than a thousand more routine interpretations. This performance commanded attention; at times ... it brought one's blood to a boil*—Chicago Tribune, 1988; *One knew immediately that he was a bad lot*—C. Phipps, 1989; *a highly readable classic that nevertheless got into the Guinness Book of Records for having been rejected 69 times before publication. One wonders why, does not one?*—Fortune, 1990 (US).

**6** Possessive of the numeral and the impersonal. The question arises of what is the proper pronoun to be used when the pronoun *one* needs contextually to be followed by a reflexive or by a possessive. For example, when Caesar has been named, he can be afterwards called either *Caesar* or *he*, but when *one* has been used, does it matter whether it is repeated itself or is represented by *he*, etc.? There is no particular problem about the numeral pronoun *one*: its possessive, reflexive, and deputy pronoun is never *one's*, *oneself*, and *one*, but always the corresponding parts of *he*, *she*, or *it*. Thus *We all saw* one [sc. one athlete] *drop his baton*; *Certainly, if* one [of the seven women] *offers herself as a candidate*; *One* [sc. a detonator] *would not go off even when I hammered it*. Secondly, the impersonal *one* always can, and now usually does, provide its own possessive etc.—*one's*, *oneself*, and *one*. Thus *One does not like to have one's word doubted*; *If one fell*, *one would hurt oneself badly*.

The going is not always so smooth, however. At various times in the past, and still often in AmE, the above sentences would run *One does not like to have his* (or even 'their' *word doubted*; *If one fell, he would hurt himself badly*. This tendency is now fast disappearing in AmE because of concern over sexual bias, and the BrE pattern is tending to be used in AmE too.

**7** Mixtures of *one* with *he, you, they, my,* etc. These are all ill-advised, esp. when they occur in the same sentence. Examples (with remedies): *As* one *goes through the lists* he *is struck by the number of female candidates;* his *old lists were very different* (Replace *he* by *one* and replace *his* by *one's*); *As* one *who vainly warned* my *countrymen that Germany was preparing to attack her neighbours,* I *say that* . . . (Replace *my* by *his*, let *I* stand); (Description of a married couple) *One* has *the power because* they're *the* one *who's healthy, whatever the disability of the other* one *is, and the other* one *has power because* they *write the check* [from a 1992 American newspaper] (Reconstruct the whole sentence); *To listen to his strong likes and dislikes* one *sometimes thought that* you *were in the presence of a Quaker of the eighteenth century* (Replace *you were* by *one was*); *To be a good Imperialist* you *must assent to the impotence and decadence and backwardness of* one's *own Motherland* (Replace *you* by *one*, or *one's* by *your*).

Examples of pronoun mixture from modern sources: *As* one *walks . . . down any street in Nashville* one *can feel now and again that* he *has just glimpsed* [etc.] *New Yorker,* 1986, *so that if* one *wants the house and land* he *must compensate* his *brothers and sisters with money*—P. P. Read, 1986; *When looking back at Passau,* one *seems to see almost as many church spires as* he *does in Prague*—*Chicago Tribune,* 1988; *Yet* one *can now see that it was the usages of the newly literate which have prevailed, so that* you *cannot help but feel* [etc.]—*English Today,* 1988; *I like to believe* one *can be honest and sincere and committed in what* he's *doing*—*Chicago Sun–Times,* 1988.

An example of an 'unmixed' (i.e. standard) type: *If* one *has no base on which to formulate probing questions, can* one *actually give informed consent?*—*Dædalus,* 1986.

**one another.** See *each other* s.v. EACH 2.

**one word or two or more.** At any given time, the language contains elements that are written together (or with a hyphen) by some writers, and as separate elements by others. The editorial committees of publishing houses come and go and make their decisions in these matters as is their right. But, thank God, there is no superfamily of scholars and writers—such as an academy or linguistic politburo—with the power to impose uniformity on us all. As a result, the custom of this publishing house or that is to encourage *their* house-style: *to get under way* or *to get underway*; *straight away* or *straightaway*; *any more* or *anymore*; *common sense* or *common-sense* or *commonsense*; *loanword* or *loan-word* or *loan word*; *teenager* or *teen-ager*. In some cases a difference of meaning governs the way in which the parts are written (see e.g. EVERYONE and *every one*). More often the choice is just a matter of custom or fashion. An attempt has been made in the present book to make recommendations in all such matters, esp., but not only, for writers, printers, and the general public in the UK. Further joinings or severances of such word-elements are bound to occur in the future. Our language is a restless one: none of its components is static or wholly governable.

**ongoing** (adj.). First recorded in 1877, this adjective gained such widespread currency in the 1950s and beyond that it quickly attracted criticism as a vogue word and, in the phr. *an ongoing situation,* as a cliché on a par with *at the end of the day, in this day and age,* and *scenario.* 'The phrase *ongoing situation* should be avoided at all times [on the BBC],' I declared in *The Spoken Word* (1981), and restate now. It signals a person's linguistic impoverishment. The word is found in many acceptable collocations (examples of *ongoing movement, operation, process, relationship,* etc., are listed in *OEDS,* 1982, but *ongoing situation* is not one of the acceptable ones.

**only** (adv.): its placing and misplacing. **1** *A brief history of attitudes.* The first grammarian known to have commented on the placing of *only* is Robert Lowth (1762): 'The Adverb, as its name imports, is generally placed close or near to the word, which it modifies or affects; and its propriety and force depend on its position. [*footnote*] Thus it is commonly said, "I *only* spake three words": when the intention of the speaker manifestly requires, "I spake *only* three words." *Her body shaded with a slight cymarr, Her bosom to the view was* only *bare.* Dryden, *Cymon and Iphig.* The sense, necessarily requires this order, *Her bosom* only *to the view was bare.*' Numerous other grammarians followed suit and begged their readers to pay regard to the placing of *only.*

The *OED* (1904) took a middle view. It gave numerous examples of each of three different placings: (a) Preceding the word or phrase which it limits (illustrative examples from 1297 to 1899, e.g. *To distinguish ... that which is established because it is right, from that which is right only because it is established*—Johnson, 1751; *It is true, I have been only twice*—T. Harral, 1805); (b) Following the word or phrase which it limits (examples from 1340 to 1876, e.g. *What belongs to Nature only, Nature only can complete*—J. Brown, 1763; *In one only of the casements*—Lytton, 1838); (c) *Only* was formerly often placed away from the word or words which it limited; this is still frequent in speech where the stress and pauses prevent ambiguity, but is now avoided by perspicuous writers (examples from 1483 to 1875, e.g. *'Tis only noble to be good*—Tennyson, 1833; *I only asked the question from habit*—Jowett, 1875).

Fowler (1926), in a lengthy article, presented the case for the acceptance of the 'illogical' placing of *only* in such a sentence as *He only died a week ago*: since 'the risk of misunderstanding [is] chimerical, it is not worth while to depart from the natural'. He thought it reasonable that 'a reader should be supposed capable of supplying the decisive intonation' to bring out the meaning, and argued strongly that only 'one of the modern precisians who have more zeal than discretion' would wish to write *He died only a week ago*. His insistence was expressed in colourful language: 'the pedants who try to forward it [sc. a tendency shown by a language to eliminate illogicalities as time passes] when the illogicality is only apparent or the inaccuracy of no importance are turning English into an exact science or an automatic machine; if they are not quite botanizing upon their mother's grave, they are at least clapping a strait waistcoat upon their mother tongue, when

wiser physicians would refuse to certify the patient.'

He further argued that there were many longer sentences in which it is important to 'get [an] announcement of purport made by an advanced *only*. A precisian, he argued, is bound to insist on an orthodox placing of *only* in the following sentence: *It would be safe to prophesy success to this heroic enterprise only if reward and merit always corresponded.* Fowler claimed that the sentence 'positively cries out to have its *only* put early after *would*.' He concluded: 'there is an orthodox position for the adverb, easily determined in case of need; to choose another position that may spoil or obscure the meaning is bad; but a change of position that has no such effect except technically is both justified by historical and colloquial usage and often demanded by rhetorical needs.'

Jespersen (1909–49, vii) is quite explicit: 'Purists insist on placing *only* close to the word it qualifies, but as a matter of fact it is by most people placed between subject and verb, and stress and tone decide where it belongs.' One of his several examples neatly brings out the problem: [a journalist speaking] *I have only been married once. I mean I have been married only once*—G. B. Shaw, 1934.

Marghanita Laski (*Observer* 21 Apr. 1963) reported that her preparatory school, by means of a single sentence and in a single lesson, fixed for ever the ability of that class to use *only* correctly. The single sentence was *The peacocks are seen on the western hills.* The exercise was to place *only* in every possible place in the sentence (*The* only *peacocks; The peacocks are* only *seen*, etc.). Clearly the meaning changes with each new placing. *CGEL* (1985, 8.117) states that the natural intonation of a sentence containing an early-placed *only* will normally ensure that there is no ambiguity. In written English, however, it is quite another matter. In its written form the following sentence could be interpreted in three different ways: *John could only see his wife from the doorway* could imply (a) he could not talk to her; (b) he could not see her brother; or (c) he could not see her from further inside the room, depending on where the main sentence stress is placed ((a) on *see*, (b) on *wife*, (c) on *doorway*).

**2** The view that a reader can be expected to supply the necessary intonation or make immediate contextual adjustments when *only* is allowed to drift to a front position in a sentence is supported by the following examples, in all of which *only* is placed at a distance from the element(s) which it limits: *I was ... made to attend a Catholic businessmans luncheon* (*where I only got wine by roaring for it*)—E. Waugh, 1958; *Boris doesn't eat shanks so, of course, I only cook them when he is away on circuit*—E. Jolley, 1985 (Aust.); *Those days, you only applied to one college*—New Yorker, 1986; *garments which Hartmann thought should only be worn by students or those of indeterminate age*—A. Brookner, 1988; *I was glad I only did an occasional gig for the band*—Rosie Scott, 1988 (NZ); *the Soviet flag only flew on official buildings*—London Rev. Bks, 1989; *I don't like that food. I only like candy*—B. Nugent, 1990 (US); *Until I grew up, and she came to live with Uncle Roy and Aunt Deirdre, I only saw Granny at carefully spaced intervals*—A. N. Wilson, 1990; *He says he only took the job because the neon sign always cheers him up*—Julian Barnes, 1991; *I'm afraid I only seem to have five pounds*—P. Lively, 1991.

**3** The placing of *only* takes one to a front-line battle which has been taking place for more than 200 years. It would be perverse to ignore the evidence provided by so many of our best writers when they are prepared to allow *only* to make an early entrance in a sentence and leave common sense to work out the meaning. And yet the sentences analysed by Marghanita Laski and by *CGEL* are also persuasive. Fowler's view (1926), set out above, is the most flexible and the most moderate, and is therefore recommended here.

**onomatopoeia** (adj. **-poeic**): 'name-making'. The formation of names or words from sounds that resemble those associated with the object or action to be named, or that seem suggestive of its qualities: *atishoo, babble, cuckoo, croak, ping-pong, puff-puff*, and *sizzle* are probable examples. The word is also used of sequences of words whose sound suggests what it describes, as in Tennyson's *Myriads of rivulets hurrying thro' the lawn, The moan of doves in immemorial elms, And murmuring of innumerable bees.*

**on to, onto.** **1** *An historical note.* Written as two words, *on to* is first recorded as a (compound) preposition in 1581 (*I haue stept on to the stage ... contented to plaie a part*) and has been in continuous use since then (e.g. *He subsided on to the music-bench obediently*—Mrs H. Ward, 1888; *French windows opened from the breakfast-room on to the terrace and large walled garden*—P. Lively, 1981; *A tear trickled on to the pillow*—A. Brookner, 1991; *An important determining factor here was the need to ensure that the paint had dried completely on one side of the sheet before moving on to the other*—Bodl. Libr. Record, 1992; *my knees gave way and I collapsed on to the seat opposite*—A. Billson, 1993). Written as one word, *onto* is first recorded as a preposition in 1715 (*[Λ] place gutted away by the rain down onto Mr. Wiswells land*) and has been in continuous use since then (e.g. *He jumped down onto the rubbish*—C. S. Lewis, 1954; *There is nothing in the room to hold onto*—D. Hirson, 1986 (SAfr.); *He sank down once again onto his footstool*—A. Brookner, 1989; *He advanced onto the verandah*—I. Murdoch, 1989; *The blue sky threw its light down onto the fields below*—L. Norfolk, 1991; *Once again paper will disappear, this time onto optical discs like CD-ROM and CD-I*—Logos, 1992). In my own writing I have always written the preposition as two words, but the form *onto* is now, it would seem, in the ascendancy.

**2** When the *on*, however, is an independent adverb, not forming a compound preposition with a following preposition *to*, the sequence *on to* is obligatory: e.g. *Everybody has been on to* [ = 'wise to'] *that for some time*—J. C. Lincoln, 1911; *I can't help feeling that he's on to us ... That he knows about us*—J. Osborne, 1959. Also, in cases 'in which *on*, as the extension of a verb, is followed by *to* as a separate word, e.g. to *walk on to* the next station, to *flow on to* the sea, ... to *lead on to* another point; a ship *lies broadside on to* the waves' (*OED*).

**3** Fowler (1926) added further examples of *on* used as a full adverb before *to* and therefore written separately: *We must walk on to Keswick; Each passed it on to his neighbour; Struggling on to victory.* In *He played the ball on to his wicket*, he judged that 'as *He played on* could stand by itself, it is hard to deny *on* its independent status'. It should also be noted that *They* *drove on to the beach* would normally mean 'They continued the journey until they reached the beach' but could also mean 'They drove their vehicle to a position on the beach'; whereas *They drove onto the beach* could only mean 'They drove their vehicle to a position on the beach'.

See UPON.

**onward, onwards.** As an adj. the shorter form is obligatory (*resuming his onward journey*). As adv. both *onward* and *onwards* are used (*from the tenth century onward(s)*); the shorter form is perhaps the more common of the two in AmE.

**oolite** (a sedimentary rock). Pronounce as three syllables, /ˈəʊəlaɪt/.

**opacity, opaqueness.** Both mean 'the quality or state or an instance of being opaque' and both are used in concrete as well as in abstract senses. *Opacity* is the preferred form in various technical senses in physics, philosophy, etc.

**op. cit.** (L *opere citato*), an abbreviation meaning 'in the work already cited'. It is normally preceded by the name of the author and set out thus: Bloomfield, op. cit., pp. 54–5.

**operate** makes *operable*, *operator*: see -ABLE, -IBLE 2(v); -OR.

**operculum** (flaplike structure covering gills of fish). Pl. *opercula*. See -UM 2.

**ophthalmologist, optician, optometrist, oculist.** An *ophthalmologist* is a person who makes a scientific study of the eye, its structure, functions, and diseases; esp. a medical practitioner specializing in the diagnosis and treatment of the eye. An *optician* (also called an *ophthalmic optician* and nowadays often an *optometrist*) is a person qualified to test the eyes and prescribe and supply spectacles and contact lenses. A *dispensing optician* is a person who supplies and fits spectacle frames but is not qualified to prescribe lenses. An *oculist* is a somewhat old-fashioned term for an *ophthalmologist*. In America, one is more likely to go to an *optometrist* for the testing of eyesight and prescribing of 'eyeglasses', but the synonym *optician* is also used. The term *ophthalmologist* is used in the same way as in Britain.

**opine** (verb) (often followed by *that* + clause) means 'to hold or express as an opinion'. It is a somewhat stilted word, often used to introduce a view scorned, or at any rate not favoured, by the speaker.

**opinion.** For *climate of opinion*, see CLIMATE.

**opportunity.** One *takes the opportunity*, or *an opportunity*, or *every opportunity*, *to* do something, or, less commonly, *of* doing something (*the opportunity of going over the papers*). Constructions of *opportunity* followed by *for* + gerund are more difficult to find (*opportunities for procuring advantages to ourselves*), but examples of *opportunity* followed by *for* + noun (phrase) are reasonably frequent (*provided an opportunity for greater leisure*).

**oppose** makes *opposable*. See -ABLE, -IBLE 6.

**opposite. 1** *He can thwart him by applying it to the opposite purpose for which it was intended* (*he* is a pupil, *the* is the teacher, *it* the teaching): insert *from* (or *to*) *that* after *purpose*. Too much elision is undesirable.

**2** As adj., *opposite* 'having a position on the other or further side', is construed either with *to* (*two persons directly opposite to each other*) or with *from* (*on the opposite side of the river from that on which his house stood*). When used as a noun, *opposite* is construed with *of* (*Ariel is the extreme opposite of Caliban*). *Opposite* may also be used as a preposition (*they sat down opposite each other*), in which case it is a shortening of *opposite to* (also used).

**opposite meanings, words of.** A minor feature of the language is the coexistence at a given time of apparently opposite meanings of the same word. The phenomenon is shown, for example, in the verb *to head*, which means (*a*) to take off the head (historically a human head, now mainly applied to the lopping off of branches at the head of a tree or shrub), and (*b*) to put a head on, to form a head (*the fence is then headed with two feet of grass sods*). The verb *ravel* can mean 'entangle' (when used transitively), 'become entangled' (when used intransitively), and (often construed with *out*) 'disentangle' (examples in the *OED*), although nowadays *entangle* and *unravel* are the usual pair of words employed for the opposing meanings. A *seeded* field (one sown with seed) is to be distinguished from a *seeded* raisin (one from which the seeds have been removed). For further examples, see CARE (verb) (*I couldn't/could care less*); CHUFFED ( = pleased/displeased). Words of different origin account for CLEAVE[1] and CLEAVE[2] and for *let* 'allow' and *let* 'prevent'.

**optative. 1** Formerly often pronounced /ɒpˈteɪtɪv/, i.e. stressed on the second syllable, but now almost always on the first.

**2** Greek verbs have certain forms called the optative mood, used to express a wish or desire for the future, e.g. μὴ γένοιτο 'may it not happen!'. In modern English grammar the word is sometimes applied to the verbal form used in formulaic phrases expressing a wish, e.g. *So help me God!, Oh that I were young again!, God save the Queen!*.

**optic.** As a noun meaning 'the organ of sight, the eye' it was in good use ('the learned and elegant term', *OED*) from the 17c. to the 19c., but at some point it became pedantic, and by 1904 (when the relevant section of the *OED* was issued) humorous.

**optician.** See OPHTHALMOLOGIST.

**optimal, optimum.** For a little over a century (both words entered the language in the late 19c.) these words have been vying with each other as adjectives in the sense 'best or most favourable (under given conditions), most satisfactory'. In the *OED* the following collocations occur in the illustrative examples for the two words: *optimal decision, distribution, feedback, intensity, performance, sense-data, temperature; optimum behaviour, concentration, density, distribution, point, power output, production rate, size, temperature*. The overlap is obvious. Neither word is appropriately used as a simple substitute for *best* (*this is the best summer we have ever had*, not *the optimal summer* and not *the optimum summer*).

**optometrist.** See OPHTHALMOLOGIST.

**opus.** For a musical composition and in the phrase *magnum opus* the recommended pronunciation is /ˈəʊpəs/ with a long ō. The pl. is either *opera* /ˈɒpərə/ or *opuses* /ˈəʊpəsɪz/.

**opuscule** (minor musical or literary work) /əˈpʌskjuːl/. Pl. *opuscules*. *Opusculum*, pl. *opuscula*, is also used in the same sense: see -UM 2.

**or. 1** Number after *or*. When the subject is a set of alternatives each in the singular, however many the alternatives, and however long the sentence, the verb must be in the singular. Examples: *Costs ... will only be awarded when the behaviour or stance of one or other party is in some way unreasonable—Counsel*, 1987; *Constancy or predictability is their virtue—Encounter*, 1987; *One had to ... admit that a paint or steel company or a salt or coal mine was no place for the late Herr Baumgartner's widow—A. Desai*, 1988. An aberrant plural verb: *The effect of anti-racist casting may not be that which the playwright intended but ... it may not be that which the actor or director intend either—Listener*, 1988. If alternative members differ in number, the nearest prevails: e.g. *Were you or he there?; Was he or you there?* (A remedy, though, would be to give the verb twice: *Was I, or were you, there?*).

**2** Wrong repetition after *or*. 'With *and*, it does not matter whether we say *without falsehood and deceit* or *without falsehood and without deceit*, except that the latter conveys a certain sledge-hammer emphasis; but with *or* there is much difference between *without falsehood or deceit* (which implies that neither is present) and *without falsehood or without deceit* (which implies only that one of the two is not present)' (Fowler, 1926). Fowler then cited several examples in which either *or* must be changed to *and*, or the word or words repeated after *or* must be cut out: e.g. *No great economy or no high efficiency can be secured; We need something more before we can conclude that Germany is going to be democratized in any effective way, or before we can be sure that this move also is not a weapon in the war*. In practice this 'rule' is not to be literally applied. It would be hard to find fault with the words (*my, in, for, in*) immediately following *or* in the following examples: *Did my father or my grandfather perhaps simply gallop up ... to the farmhouse one day?—J. M. Coetzee*, 1977; *The 'insolence of office' which ... drove Hamlet to contemplate suicide is certainly not unknown in Whitehall or in British Embassies abroad—Times*, 1985; *He won't want to make trouble for her or for* *me*—P. P. Read, 1986; *We Europeans rode the streets in cars* or *in horse-drawn gharries*—P. Lively, 1987.

**3** See also AGREEMENT 4; EITHER 5; NEITHER 6; NOR 7.

**-or** is the Latin agent-noun ending corresponding to English *-er*: compare *doer* and *perpetrator*. English verbs derived from the supine stem of Latin ones—i.e. especially most verbs in *-ate*, but also many others such as *act, credit, invent, oppress, possess, prosecute, protect*—usually prefer this Latin ending to the English one in *-er*. Some other verbs, e.g. *conquer, govern,* and *purvey*, not corresponding to the above description have agent-nouns in *-or* owing to their passage through French or through some other circumstance. A select list of differences may be of interest: *corrupter* and *corrector*; *deserter* and *abductor*; *dispenser* and *distributor*; *eraser* and *ejector*. Some verbs generate alternative forms, generally preferring *-er* for the personal and *-or* for the mechanical agent (e.g. *adapt, convey, distribute, resist*), or *-or* for the lawyers and *-er* for ordinary use (*abet, devise, pay, settle, vend*). See also ACCEPTER; ADAPTER; ADVISER; CASTER; also -ER AND -OR.

**oral.** See AURAL; VERBAL.

**orate** (verb). The *OED* (1904) remarked: 'This word is occasionally instanced since *circa* 1600, but has only recently come into more common use, as a back-formation from *oration*, apparently first in U.S. *circa* 1860.' It means 'to hold forth, to "speechify", to make a speech pompously or at length.'

***oratio obliqua, recta*** ('bent speech, straight speech'). Latin names, the second for the actual words used by a speaker, without modification, and the first for the form taken by his or her words when they are reported and fitted into the reporter's framework. Thus *How are you? I am delighted to see you* (*recta*) becomes in *obliqua He asked how I was and said he was delighted to see me*; or, if the framework is invisible, *How was I? He was delighted to see me.* Most newspaper reports of speeches, and all third-person letters, are in *oratio obliqua* or reported speech.

**orator.** Thus spelt, not *-er*. See -OR.

**oratorio.** Pl. *oratorios*. See -O(E)S 4.

**oratress** (female public speaker). First recorded in the late 16c. and in continuous use since then until the 20c. Now moribund. See -ESS. The synonymous *oratrix* has also been used since the late 15c. and is also now obsolescent. See -TRIX.

**orbit** makes *orbital*; *orbited, orbiting*. See -T-, -TT-.

**orchid, orchis.** The first is the familiar (greenhouse) 'epiphytic plant of the family Orchidaceae, bearing flowers in fantastic shapes and brilliant colours, [etc.] (*COD*), grown outdoors in temperate and tropical regions. The *orchis* is 'any orchid of the genus *Orchis*, with a tuberous root and an erect flashy stem having a spike of usu. purple or red flowers; any of various wild orchids' (*COD*). Pl. *orchids* (for both words), but usu. *orchises* for *orchis*. Both words are originally from Gk ὄρχις 'testicle' (with reference to the shape of its tuber).

**ordeal.** Pronunciation: now always /ɔːˈdiːl/ or /ɔːˈdiːəl/, stressed on the second syllable. Johnson (1755) and Walker (1791) placed the stress on the first syllable, as did the *OED* (1904); and the last two indicated that the word was predominantly trisyllabic. The shift of stress to the second syllable must therefore have occurred in the 20c.: it is worth noting that Daniel Jones (1917) gave only second-syllable stressing.

**order** (noun). **1** See IN ORDER THAT; IN ORDER TO.

**2** *of the order of* is a COMPOUND PREPOSITION meaning 'in the region of, somewhere about' (*the average flow through the gorge is of the order of 2,000–3,000 cubic metres per second*). It is a popularized technicality from a more complex use of the term in mathematics.

**order** (verb). The construction *ordered* + object + pa.pple (often expressed in the passive) is first recorded in 1781 in AmE (*These things were ordered delivered to the army*—J. Witherspoon) and has been predominantly American ever since: BrE normally requires the insertion of *to be* before the pa.pple. Further examples: *My bill was introduced by Senator Williams of*

*Oregon, read by title, and ordered printed—*
J. H. Beadle, 1873; *the local military com-
mander was ordered removed—Time*, 1977;
*two gunmen burst into the cockpit and ordered
the plane flown to Algeria—Chr. Sci. Monitor,*
1987; *twelve other neighborhood people testified in court that they . . . would feel horrible
if the jury ordered them* [sc. church bells]
*unplugged—New Yorker*, 1990; *An 11-year-old
English girl allegedly left alone when her
mother went on a holiday in Spain was
ordered kept in the care of local authorities
yesterday—Dominion* (Wellington, NZ),
1993.

**orderly.** See -LILY. The word is used only
as an adj. (*behaved in an orderly manner*),
never as an adv. The notional adverbial
form *orderlily* does not exist.

**ordinance, ordnance, ordonnance.** An
*ordinance* is 'an authoritative order'; *ordnance* is 'a branch of government service
dealing esp. with military stores and
materials; *specif.* artillery'; and *ordonnance* is 'a plan or method of literary or
artistic composition; an order of architecture' (*The most conspicuous qualities of
the style are these: ordonnance, or arrangement and structure, precision in the use of
words, and relevant intensity—*T. S. Eliot,
1936; *the dead figure . . . is not exaggerated
and achieves a monumental 'ordonnance'
which is nearly classical—Burlington Mag.,*
1938). *Ordnance Survey* is an official UK
survey organization, orig. under the Master of the Ordnance, preparing large-
scale detailed maps of each region of the
country.

**oread** (mountain nymph in Greek and
Roman mythology). Pronounce /ˈɔːrɪæd/.

**organon** (Greek) or **organum** (Latin)
(system of logic, etc.). The pl. of *organon* is
*organa* /ˈɔːgənə/, less commonly *organons.*
The pl. of *organum* is *organa* /ˈɔːgənə/,
less commonly *organums. Organon* was
the title of Aristotle's logical writings,
and *Novum* (new) *Organum* that of Bacon's.

**oriel.** See BAY WINDOW.

**orient, orientate** (verbs). In the perverse
way in which such things often happen
these two verbs, one shorter than the
other but drawn from the same base
(French *orienter* 'to place facing the east')
have fallen into competition with one
another in the second half of the 20c.

The shorter form emerged in the 18c.
and the longer one, in the same sense
(as in the French original), in the 19c.
Both words then went in identical directions and developed the same extended
senses: 'to place in any particular way
with respect to the cardinal points of the
compass'; and, figuratively, 'to ascertain
one's "bearings"'. In particular both
words have become frequently used as
participial adjectives (*oriented, orientated*).
The symmetry of the two forms is
brought out clearly by the combinations
that happen to occur in the illustrative
examples given in the *OED*: (*a*) Preceded
by an adverb: *environmentally, psychologi-
cally, vertically oriented; politically orientated.* (*b*) Preceded by a noun (to which
it is hyphened): *adult-, art-, performance-,
user-oriented; degree-, performance-, user-
orientated.*

Readers may like to be reminded of
the nature of the contest by inspecting
the following examples: (orient(ed)) *Man
needs relations with other people in order to
orient himself—*R. May, 1953; *He can begin
the next big step . . . of becoming emotionally
more independent of his parents and oriented
instead towards the outside community and
its ways of doing things—*B. Spock, 1955; *In
a youth-oriented society for a woman to grow
old means to run the risk of being ignored—*
A. Hutschnecker, 1981; (orientate(d)) *In
a language like Malagasy it is also possible
to orientate the predicate with respect to what
in French would be a circumstantial complement—*E. Palmer, 1964; *We in Europe, says
Lévy Bruhl, have behind us many centuries of
rigorous intellectual speculation and analysis.
Consequently, we are logically orientated—*
E. E. Evans-Pritchard, 1965; *Kant's own
philosophy was undeniably orientated towards problems that lay at the heart of the
philosophical enterprise—*P. Gardiner, 1988;
*It was very much a London orientated maga-
zine—*N. Sherry, 1987.

In the face of the evidence, what is
one to do? In practice I have decided
to use the shorter form myself in all
contexts, but the saving is not great. And
one can have no fundamental quarrel
with anyone who decides to use the
longer of the two words.

**originator.** Thus spelt, not -*er.* See -OR.

**orison** (prayer). Pronounce /ˈɒrɪzən/.

**Orleans** (in *New O.*). Pronounce /ɔːˈliːənz/.
But some prefer to stress on the first
syllable: /ˈɔːliənz/.)

**ornament.** The noun is pronounced /'ɔːnəmənt/ with an obscure vowel in the final syllable. The verb, by contrast, normally has a clear /e/ in the final syllable.

**ornithology,** etc. In Greek the *i* is long, but in English words derived from Gk ὀρνῑθο- (*ornithophile, ornithorhyncus,* etc.) a short *i* is now usual. See FALSE QUANTITY.

**orography, oro-pharyngeal** etc. Two different formative elements yield words in *oro-*: (*a*) Gk ὄρος 'mountain' (e.g. *orogeny* or *orogenesis,* the formation of mountains; *orography,* the branch of physical geography dealing with the formation of mountains); (*b*) L *ōs, ōr-* 'mouth' (e.g. *oro-pharyngeal*). At least one word has strayed from the fold, namely *orinasal,* the usual spelling of the phonetic term meaning '(of French nasal vowels) pronounced with simultaneous oral and nasal articulation'.

**orotund.** An 18c. English, highly irregular, reduction of the L phr. *ōre rotundō* 'with round, well-turned speech' (lit. 'with round mouth').

**orthopaedic.** Thus spelt in BrE, but usu. as *-pedic* in AmE.

**osculate, osculatory.** (Cf. L *osculārī* 'to kiss'.) Since the 17c. (the verb) and the 18c. (the adj.) these words have been used in several technical senses (in geometry, anatomy, biology, etc.) with the central meaning of 'being, or coming into contact with', some specified thing'. As a noun an *osculatory* is 'a painted, stamped, or carved representation of Christ or the Virgin, formerly (18–19c.) kissed by the priest and people during Mass' (*OED*). During this period, and still (just), the verb and adj. have occasionally been called on in jocular contexts in place of the natural words *kiss* and *kissing* (e.g. *The two ladies went through the osculatory ceremony*—Thackeray, 1849).

**ostensibly, ostentatiously.** Both mean essentially 'by way of making a show', but the purpose in the first case is 'to profess in order to conceal a truth'; in the second it is merely display—'showing off, displaying something in a pretentious or vulgar manner'.

**ostler.** Pronounce /'ɒslə/. A term used 14–19c. (and still with historical reference) for a man who attended the horses

at an inn. The word is a phonetic respelling of *hosteler, hostler* (and thus etymologically related to *hostel* and *hotel*), representing the historical pronunciation with *h* mute. See HOSTLER.

**other.**

| | |
|---|---|
| 1 | *each other* and *one another.* |
| 2 | *on the other hand.* |
| 3 | *of all others.* |
| 4 | *other* or *others.* |
| 5 | *other* as adverb. |
| 6 | Two curiosities. |
| 7 | Archaic disturbance of word order. |

**1** *each other* and *one another.* See EACH 2.

**2** *on the other hand.* For the difference between this and *on the contrary* see CONTRARY 2.

**3** *of all others.* A fading use cited in the *OED* from Steele (1711) (*This Woman, says he, is of all others the most unintelligible*), and several other sources (e.g. *In Birmingham, the very place, of all others, where it is most likely to be of real service*—J. Morley, 1877). The phr. was called an 'illogicality' and a 'sturdy indefensible' by Fowler (1926), who cited as an example *You are the man of all others I wanted to see* (a mixture, according to Fowler, of *You are the man of all men* etc. and *You are the man I wanted to see beyond all others*). If you use the phr. you will be regarded as swimming against the tide. Further examples: *To the mind of the Jew, the man who of all others emphasized the holiness of God, the distinctive feature of this holiness was its separativeness*—G. Matheson, 1901: *But how could Israel, the nation which of all others understood the horror of mass murder, have allowed the Palestinians of Sabra and Chatila camps . . . to have been murdered*—R. Fisk, 1990.

**4** *other* or *others.* Used absolutely or as a noun, *other* was formerly often (and still occas.) an uninflected plural: *A body of men whom of all other a good man would be most careful not to violate*—Berkeley, 1713; *These writings, and all other of the same class*—R. H. Froude, 1826; *I know two other of his works*—J. H. Newman, 1844; *We find here, as in other of his novels, that he has no genius for . . .*—1926 in Fowler; *Petite teenage girls . . . often perform and record Paganini's caprices, concertos and other of his works with exceptional virtuosity*—Strad, 1990.

The now normal form *others* is recorded from the 16c. onward: e.g. *Loans from the citizens of London and others of her subjects*—H. Hallam, 1827; *In others of his sermons*—H. H. Milman, 1868; *As if the blame were with Madge and others of her generation for losing the words creek, paddock, footpath and others like them*—J. Frame, 1988; *They're being marketed as pheromones—smell substances secreted by animals that cause a response in others of the same species—Which?*, 1989; *I want to compare the will with others of the same period*—NY Rev. Bks, 1991; *Some typefaces look larger than others of the same size because their x-height, the height of the lowercase* x, *is higher*—PC Computing, 1992.

**5** *other* as adverb. The OED has a sprinkling of examples of *other* used as an adverb where *otherwise* might perhaps have been expected, e.g. *It is impossible to refer to them ... other than very cursorily*—Law Times, 1883. Fowler (1926) said that 'an adverbial use of *other* is recognized by the OED, but supported by very few quotations, and those from no authors whose names carry weight'. 'Its recent development may be heartily condemned as both ungrammatical and needless,' he continued. He then cited several newspaper extracts 'in which the only correction necessary is to insert the real adverb *otherwise* instead of the false adverb *other*': e.g. *Although the world at large and for long refused to treat it other than humorously; There was never a moment when it could less become Englishmen to speak other than respectfully and courteously of the Russian nation.* Fowler also attacked *other than* used 'tor what would naturally be expressed by some other negative form of speech': e.g. *Up to the very end no German field company would look with other than apprehension to meeting the 25th on even terms* (he suggested correcting to *without apprehension*); *Four years of war could not leave a people other than restless* (read *could not but leave restless*). COD (1990) describes this use of *other than* (e.g. *cannot react other than angrily*) as 'disputed', and adds 'In this sense *otherwise* is standard except in less formal use.'

As against these adverse remarks, the major American dictionaries just list the adverbial use of *other* in *other than* without comment and give routine illustrative examples (e.g. *not being able to sell the product other than by reducing the price*—Webster's Third, 1961; *We can't collect the rent other than by suing the tenant*—

Random House Dict., 1987). WDEU (1989) illustrates the use of *other than* followed by various parts of speech and in various constructions, and concludes that in some of them it is being treated as a compound preposition. 'All of these uses are standard English,' it declares.

The matter cannot be resolved. It would appear to be more or less true that adverbial uses of *other* are not widely accepted in BrE but pass without notice in AmE. But other adverbial phrases (*apart from, less than, in any other way than*, etc.) are available, and such sentences can easily be rephrased to avoid the problem altogether.

**6** Two curiosities. (*a*) An informal use of *a whole other* + noun seems to have begun life in AmE: *He had a family, a whole other life, in Florida*—J. Silber, 1991; *I thought she was going to come up with a whole other kind of animal, but this was just as good*—Julie Hecht, 1992; *You want a no-frills economy class pair of black hose? forget it. There's business sheer, silky sheer, ultra sheer and micro sheer (opaques are a whole other class).*—Globe & Mail (Toronto), 1993. (*b*) A pleonastic use of *other*. Examples (from American newspapers, both of 1991): *Drove a car ... westward along the Pyrenees and then north with my wife and two other friends, one of whom is an art historian;* (after a gas explosion) *A man in his early 40s was taken to Evanston Hospital ... Two other women at the scene were slightly injured.*

**7** Archaic disturbance of word order. *Other* is called a postdeterminer by modern grammarians in that it normally follows other determiners, including (sometimes) numerals (*our other neighbours; several other places; four other people; the other end;* but *the other two men in the room*). It is worth recalling that from the OE period until the 19c. (see *OED* adj. 5d) *other* was often placed in a manner that seems strange or impossible now: e.g. *The kynge of Fryse, & other his prysoners*—Caxton, c1489; *amonge other his good qualities*—N. Harpsfield, 1558; *hee stayed yet other seuen dayes*—Gen. 8: 10 (AV), 1611; *With other the great men of Scotland*—J. H. Burton, 1864.

**otherwise.** 'A definite outrage on grammatical principles' was Fowler's verdict in 1926 on the type *economic or otherwise*, i.e. the use of an adj. + *or* (or *and*) + *otherwise*. His examples included: *There are large tracts of the country*, agricultural and otherwise, *in which the Labour writ does not run*; *No further threats*, economic or otherwise, *have been made*; *No organizations*, religious or otherwise, *had troubled to take the matter up*; *Place a fair share of taxation on the owners*, ducal and otherwise, *holding land and not developing it*. He said that 'none of them would be less natural if the offending expressions were rewritten thus: *some agricultural and some not* / *No further economic or other threats* / *religious or non-religious* / *the ducal and other owners*.' But the language has moved on and the type condemned by Fowler is now in standard use, and has been extended to cover other parts of speech (preceding the *or* or *and*) as well. *OEDS* (1982) added this sense: 'Phr. *or* (occas. *and*) *otherwise*, following a noun, adjective, adverb, or verb, to signify a corresponding word of opposite or different meaning.' Examples: *Mrs. Lidderdale's dread ... was that her son would acquire a West country burr, and it was considered more prudent, economically and otherwise, to go on learning with his grandfather and herself—C. Mackenzie*, 1922; *I do not question the eruption at Santorin ... but the supposed connection of the underwater survey with the historicity or otherwise of the Atlantis myth—Listener*, 1966; *12,000 Cowley workers enjoyed (or otherwise) an enforced holiday because of a strike by plant attendants at the car assembly factory—Oxford Times*, 1973; *It's the balance of foods you eat that is healthy or otherwise—Which?*, 1989; *The blameworthiness or otherwise of another party bears no relation to the issue of the instant trader's diligent pursuit of safety—Statute Law Rev.*, 1991.

**ottava rima,** an Italian stanza of eight eleven-syllabled lines rhyming ababababcc, pioneered by Boccaccio in the 14c. and employed by Tasso, Ariosto, and other Italian poets. It was introduced into English by Thomas Wyatt (1503–42). The English version as used by Byron in *Don Juan*, as well as by Keats, Shelley, Yeats, and others, has iambic pentameters but in other respects follows the Italian model.

> Whate'er his youth had suffer'd, his old age
> With wealth and talking made him some
>    amends;
> Though Laura sometimes put him in a rage,

> I've heard the count and he were always
>    friends.
> My pen is at the bottom of a page,
> Which being finished here the story ends;
> 'Tis to be wish'd it had been sooner done,
> But stories somehow lengthen when begun.
>                 (Byron, 1818)

**ought** (noun). Since the mid-19c. *ought* has sometimes been used as a colloquial corruption of *nought*: e.g. *a half smeared-out game of oughts and crosses*—G. A. Sala, 1861; *'But did they find a rifle on Sutton?' 'Yep. Thirty-ought-six.'*—D. Anthony, 1972; *Strawberry Bill had played left field for Toronto in ought eight when Francis played third*—W. Kennedy, 1979; *Sir William Walton was born in nineteen ought two*—announcer on Radio WN1B (US), 29 Mar. 1993. It probably arose from a misdivision of *a nought* as *an ought*. Cf. NAUGHT 2; NOUGHT.

**ought** (verb). **1** This modal verb has a complicated history but in its normal uses now it is followed by a *to*-infinitive: *Two people advised me recently ... that I ought to see a doctor*—T. S. Eliot, 1950; *But oughtn't I first to tell you the circumstances?* —ibid, 1950; *You ought to have a cooked breakfast, these cold mornings*—D. Lodge, 1988; *If Canada should disintegrate ... what ought the U.S. to do?*—Wall St. Jrnl, 1990. *Ought* is peculiarly liable to be carelessly combined with auxiliary verbs that differ from it in taking the plain infinitive without *to*. *Can and ought to go* is right, but *Ought and can go* is wrong. *We should be sorry to see English critics suggesting that they ought or could have acted otherwise*: insert *to* after *ought*, or write *that they could or ought to have acted*. See ELLIPSIS for similar difficulties. The negative equivalent is *ought not to* (or, as a contracted form, *oughtn't to*). The *to* is optional in ellipses: *Yes, I think I ought (to)*. It is also optional in informal non-assertive contexts (*non-assertive* is the term used in CGEL): e.g. *They ought not (to) do that sort of thing*; *Oughtn't we (to) send for the police?* But the more natural standard expressions are *They oughtn't to do that* or *They shouldn't do that*; and *Shouldn't we send for the police?*

**2** See DIDN'T OUGHT.

**3** *hadn't ought*. Only found in dialectal use in parts of Britain and America: e.g. *Did you do that? You hadn't ought* ( = ought not to have done it).

## our.

1 *our, ours.*
2 *our* editorial and ordinary.
3 *our, his.*

**1** *our, ours. Ours and the Italian troops are now across the Piave.* The right alternatives are: *The Italian troops and ours; The Italian and our troops; Our and the Italian troops.* The wrong one is that in the quotation. See ABSOLUTE POSSESSIVES.

**2** The editorial *our,* like *we* and *us* of that kind, should not be allowed to appear in the same sentence, or in close proximity, with any non-editorial use of *we,* etc. In the following extract, *our* and the second *we* are editorial, while *us* and the first *we* are national: *For chaos it is now proposed to substitute law, law by which we must gain as neutrals, and which in our view, inflicts no material sacrifice on us as belligerents. We do not propose to argue that question again from the beginning, but …*

**3** *our, his. Which of us would wish to be ill in* our *kitchen, especially when it is also the family living-room?* If a possessive adj. were necessary, *his* and not *our* would be the right one, or, in our gender-conscious age, *his or her.* People of weak grammatical digestions, unable to stomach *his,* have a means of doing without the possessive altogether: why not simply *the kitchen,* here? It is undeniable that *which of us* requires a following possessive pronoun to be in the singular. It is no solution at all to replace *our* by *their.* Grammatical concord would still be breached by doing so.

**-our and -or.** Anyone who is really determined to pursue the history and distribution of such spellings as *colour/color* and *valour/valor* in BrE and AmE should turn to the *OED* entries for *-or* and for *-our* (as well as to the entries for the individual words themselves); and also to an article by C. M. Anson in *Internat. Jrnl Lexicography,* spring 1990. What follows is a heavily reduced description of the main patterns.

It is a simple matter, first, to set down a representative list of words that are now regularly distinguished in the two countries, BrE, followed by (AmE): *behaviour (behavior), candour (candor), colour (color), harbour (harbor), honour (honor), labour (labor), neighbour (neighbor), parlour (parlor), splendour (splendor), valour (valor).*

The list could be doubled (*clamo(u)r, glamo(u)r,* etc.) and one might be tempted to conclude that for every British *-our* the Americans write *-or.* Any such 'rule', however, is contradicted by a number of factors. For one thing, both countries use only *-or* as a terminal element in a wide range of abstract nouns (e.g. *error, horror, pallor, stupor, terror, torpor, tremor*), agent-nouns (e.g. *actor, governor, orator, sailor;* see also PAVIOR; SAVIOUR), and others of miscellaneous origin (e.g. *ambassador, anchor, bachelor, emperor, liquor, mirror, matador*). And for another, both countries regularly use *-our* in a number of words of diverse origin (e.g. *contour, paramour, tambour, vavasour* (but often *vavasor* in AmE)).

Three main historical factors govern the emergence of the various present-day spellings: (*a*) the adoption of Latin words ending in *-or* via Anglo-Norman (where spelt *-our*); (*b*) the later post-Renaissance adoption of *-or* words direct from classical Latin; (*c*) a steady adoption in America in the first half of the 19c., esp. by Noah Webster, of *-or* in all the main abstract nouns of the type *colo(u)r, hono(u)r,* etc. A radical solution lies close at hand, namely to adopt the spelling *-or* in all nouns listed or implied in the first list given above. There is some movement in this direction in Australia, and Canadians are for the most part free to choose whichever spelling they prefer. But everyone knows that such a change would be regarded by Britons as a kind of linguistic cleansing, something not to be contemplated in any circumstances. And, of course, there is always a great deal to be said for national pride concerning long-held traditions.

See also -ER AND -OR; -OR.

**-our- and -or-.** Even those nouns that in BrE regularly end in *-our* (see prec.), as opposed to AmE *-or,* e.g. *clamour, humour, odour, rigour, valour, vapour, vigour,* have adjectives ending in *-orous,* not *-ourous:* thus *clamorous, humorous,* etc. Derivatives in *-ist, -ite,* and *-able,* on the other hand, mostly retain the *-our-:* so *behaviourist, colourist, Labourite* (cf. *favourite,* of different formation), *colourable* and *honourable.* Derivatives in *-ation* and *-ize* are usually spelt, both in BrE and AmE, with *-or-:* so *coloration, invigoration; deodorize, glamorize, vaporize.*

**ours, our.** See OUR 1.

**ousel.** See OUZEL.

**out** (prep.). This use as a preposition instead of the customary *out of* has a long history: the *OED* lists examples from the mid-13c. to the 20c. (e.g. *When you haue pusht out your gates the very Defender of them*—Shakespeare, 1607). At the present time it is non-standard in the UK but is common (beside *out of*) in America, Australia, and NZ. It is in restricted use contextually in that it is usually employed in contexts of *looking, going,* etc. *out the door* or *the window.* Modern examples: *To drive with the left arm out the window—Amer. Speech,* 1962; *She looked out the window ... at all the other houses—Southerly* (Aust.), 1967; *We looked out the window at the snow—New Yorker,* 1986; *But Grandfather was out the door—M.* Pople, 1986 (Aust.); *I was looking out my side* [of the coal truck], *the way you do when you push out a curve—New Yorker,* 1987; *Now he looked past Bacon, out the bay window behind him—T.* Wolfe, 1987 (US); *I drove out the gates and left them open behind me, swinging in the wind—S.* Koea, 1994 (NZ).

**outcome** is a word that easily leads to tautologous constructions: e.g. *The outcome of such nationalization would undoubtedly lead to the loss of incentive and initiative in that trade.* The outcome of nationalization would be loss; *nationalization* would lead to loss.

**outdoor** is used only as an adj. (*outdoor games*); *outdoors* is an adverb (*the concert was held outdoors*) or a noun (*the great outdoors*).

**outermost.** For pronunciation, see -MOST.

**outfit.** The inflected forms are *outfitter; outfitted, outfitting* in both BrE and AmE.

**out-Herod** (verb). Shakespeare was responsible for introducing this type of verb: *I could haue such a Fellow whipt for o'er-doing Termagant: it out-Herod's Herod. Pray you auoid it—Hamlet.* It was no more than a casual expression, related to similar formations of his own in which the second element of the *out*-compound is not a proper name: *out-frowne false Fortunes frowne—King Lear; He hath out villain'd villanie—All's Well that Ends Well.* The type

produced a flood of imitative formations, esp. in the 19c. ('a few instances are found in the 17th c., esp. in Fuller, and in the 18th c. in Swift; but the vast development of this, as of so many other Shaksperian usages, belongs to the 19th c., in which such expressions have been used almost without limit' (*OED*).) Among the numerous examples listed in the *OED* are *to out-Alexander Alexander* (J. Wolcott, 1800), *to out-Milton Milton* (Lowell, 1870), and *to out-Zola-Zola* (*Literary World,* 1887); and the related type (ordinary nouns, not proper names) *to out-devil the devil* (B. H. Malkin, 1809), *to outmonster the monstrosities* (E. Blunden, 1930), and *to outrainbow the rainbow* (H. MacDiarmid, 1956). The type is shown in a weakened form when the object of the verb does not repeat or echo it, e.g. *Out-heroding the French cavaliers in compliment and in extravagance* (1809 in *OED*).

**out of.** See OUT (prep.).

**output** (verb). The pa.t. and pa.pple are either *output* or *outputted;* the pres.pple is *outputting.* See -T-, -TT-.

**outside of.** As a compound preposition *outside of* is used, esp. in AmE, in two main senses: (*a*) exterior to, outside (*These books are ... distributed outside of the U.S.A. and Canada by Academic Press—Nature,* 1975; *People in show business refer to those outside of it as 'civilians'—S.* MacLaine, 1987; *I remember an industrialist who, outside of presidential earshot, voiced stinging criticism of certain budget proposals—B.* and E. Dole, 1988). (*b*) with the exception of (*Outside of a slightly annoying tendency to call all female customers 'Hon', everything about Mr. Blume inspires confidence—New York,* 1972; *Outside of an unfortunate sermon in which he confused the words for charity and diarrhea, causing some tittering behind fans, he never put a foot wrong with his hosts—W.* Sheed, 1985; *Outside of the wound, I'd say primarily traumatic shock—R.* Ludlum, 1990. In most circumstances, however, uncompounded *outside* is sufficient, and is overwhelmingly the normal use in BrE.

**outspoken.** See -SPOKEN.

**outstanding.** The two main senses of the word need careful handling to avoid ambiguity: (*a*) conspicuous, remarkable (in a specified group); (*b*) not yet settled.

In an election *the other outstanding result* could mean either one of special interest or one not yet known (if the latter, it would be better to re-express as *the other result still outstanding*).

**outward, outwards.** As adj., the only legitimate form is *outward* (*the outward journey, his outward appearance*). As adv., *outward* is the customary form (*one eye is turned outward*), with *outwards* as a stylistic or personal variant, esp. in BrE.

**outwit.** The inflected forms are *outwitted, outwitting*.

**ouzel** (a bird). Usu. spelt thus, not *ousel*.

**over.** Since the later part of the 19c. there has been a strong tradition in American newspapers and in some American usage guides of hostility to the use of *over* with a following numeral to mean 'in excess of, more than' (the type *a little over £50, a distance of over 700 yards*). The anxiety continues: *The national view is a graphic composite of local reports across the country from over 50* (Oops! Make that 'more than' 50—I'm almost 'over' 50) reporting *stations*—columnist in *Chicago Sun-Times*, 1989; *Not perfect yet: 'over 150,000' AIDS deaths should have been* more than, used with all figures except ages—W. Safire, 1992. In Britain, *over* has been used with a following numeral without restriction or adverse comment throughout the same period. No voices have been raised in Britain or N. America against the type *over forty* (*years of age*). Examples: *These four sons were all over forty but they were treated as babies by their parents*—P. Kavanagh, 1948; *She was a little over twenty, very graceful and witty and cunning*—G. Greene, 1969.

**overall.** 1 Pronunciation. When used as a noun ( = two kinds of garment), it is stressed on the first syllable: thus /ˈəʊvərɔːl(z)/. As adj. and adv. the stress is variable: sometimes the main stress falls on the first syllable with a secondary stress on the final syllable, and sometimes it is the other way round: thus /ˈəʊvərˌɔːl/ or /ˌəʊvərˈɔːl/.

2 Parts of speech. As noun, an *overall* (in BrE) is a coat-like garment worn over one's clothes as a protection against stains, etc.; *overalls* are protective trousers, dungarees, or a combination suit worn by workmen, etc. (always construed with a pl. verb). As adj. it is always used in the attributive position, e.g. *the overall pattern, the overall effect*. It is also used as an adv., e.g. *Overall, the performance was excellent*.

3 Hostility to the use of the adj. (and by implication the adv. as well) surfaces from time to time, the argument being that *overall* is overused, and that a range of synonyms (*comprehensive, total, whole*, etc.) is available. Warnings about overuse seem to have had some effect and *overall* seems to be used much less commonly now than it was in the second and third quarters of the 20c. It is still advisable, however, to be on one's guard against using it when it contributes nothing to the sense: e.g. *It was not until the third ballot that Mr. Michael Foot secured an absolute overall majority*; *The overall growth of London should be restrained*.

**overflow** has pa.t. and pa.pple *overflowed*.

**overfly.** The inflected forms are *overflies* (pres. sing.), *overflew* (pa.t.), and *overflown* (pa.pple).

**overlay, overlie.** The pa.t. and pa.pple of *overlay* are respectively *overlaid* and *overlaid*, while those of *overlie* are respectively *overlay* and *overlain*. Confusion of the two sets of forms, and also of the present tense forms (respectively *overlays* and *overlies*) is even more likely to occur than in the inflected forms of *lay* and *lie* (see LAY AND LIE). There is abundant evidence in the *OED*, for example, for the use of either *overlay* or *overlie* for the sense 'to smother by lying upon' (a mother on her child, or a sow on a piglet). Both *overlay* and *overlie* also seem to be used in geology for the sense 'lie over or upon (e.g. of a stratum resting directly upon another)' (*the Palæozoic rocks do not appear to be overlain by recent marine deposits*—J. Ball, 1885; *At the edges of the cross section, we see a sequence of magnetized sedimentary blankets overlaying the crust*—P. J. Wyllie, 1976).

On the other hand, the sense 'to cover the surface (of something) with a coating, etc.' requires *overlay* (e.g. *wood overlaid with gold; they overlaid the walls with hessian*). And so do figurative extensions: e.g. *A habit of obedience overlaid the tumultuous desires and suppressions of her young daughter*—M. Keane, 1937; *Her anger was overlaid with bewilderment*—D. Welch, 1943; *This kind of insight overlays the patterns of his works, and the enjoyment of his art is as much the enjoyment of the overlay as it is of the patterns themselves*—*TLS*, 1975.

This is all tricky territory and the only safe rule is to consult a standard dictionary before making the choice of word that is traditionally used for a specified meaning. Having established that, the standard paradigms of the two verbs are as given above in the first paragraph.

**overlook, oversee.** At first sight these would seem to be synonymous, but in practice their meanings hardly overlap. *Overlook* means (*a*) to fail to notice; to condone (an offence, etc.); (*b*) to have a view from above (*their house overlooked the harbour*); less commonly, (*c*) to supervise, oversee (*Lord Hailsham, as Lord Privy Seal, is to overlook scientific development*—in Gowers, 1965). The primary sense of *oversee* is the same as that of sense (*c*) of *overlook*, i.e. to supervise (workers, work, etc.). It is the normal word in this sense.

Of the derivatives, *overseer* retains its meaning of 'supervisor', but *oversight* has tended to follow *overlook* instead of its parent verb and is now not often used except in the sense of a failure to notice something, or an inadvertent mistake. One must nevertheless not be surprised to find *oversight* used to mean 'supervision'. Examples: *There was no centralised mint and probably little centralised oversight*—F. H. Hayward, 1935; *The need to continue investigations on a broad front to keep an ecological oversight of the biogeodynamics of each metal*—*Nature*, 1971; *The broad answer is that there must be a representative from Scotland in the United Kingdom Cabinet*—*with a general oversight over the economy and the framing of Scotland's budget*—*to represent Scotland's interest on fuel policy and on defence*—Lord Home, 1976.

**overly** (adv.). Until about the 1970s, and in some quarters still, regarded as an Americanism, but the evidence shows that it is now widely used in BrE. Examples: *The Manitoba Minister of Agriculture is not overly impressed with the horsemen's woes*—*Globe & Mail* (Toronto), 1968; *Occasionally, an overly sensitive author will object strenuously*—*Surgery*, 1974; *That same novel is now with Macmillan. I am not 'overly' hopeful*—B. Pym, 1977; *Faith in their own judgment … will not make them overly*

*deferential about the wisdom of Ministers*—J. Cole in *Listener*, 1983; *Fitzpatrick's male adversary is an impassioned, overly emotional man*—*Times*, 1985; *she was so overly agreeable and pleasant that there had to be a reason for it*—J. Hecht, 1992 (US); *Since 1989, nearly 100 cases of overly lenient sentences have been referred by the Attorney General to the Court of Appeal*—*Times*, 1993; *She is not overly cheerful about the future of British drama*—M. Geare, 1993.

**overseas.** Now the customary word (rather than *oversea*) for both the adv. (*he was sent overseas for training*) and the adj. (*overseas postage rates*), though at least until the early 20c. *oversea* was the usual form for the adj.

**oversight.** See OVERLOOK.

**overthrowal.** The noun-forming suffix -*al* came into wide favour in the 19c. (see -AL 2) and *overthrowal* 'the act of overthrowing' (first recorded 1916) seems to have come into being as a kind of afterthought. It was first admitted to the *OED* in 1933, and it has found a place in Webster's Third, but smaller dictionaries find no place for it. Modern examples (both US): *I don't see his regime lasting more than 18 months before an overthrowal*—*Washington Post*, 1979; *The military says more than 2,000 people had been killed in insurgency-related incidents since the February overthrowal of President Ferdinand Marcos*—*Chicago Tribune*, 1986.

**overtone, undertone.** In their figurative uses both have the implication that there is more to a statement, undertaking, etc., than meets the eye. In music, an overtone is 'any of the tones above the lowest in a harmonic series' (*COD*). *Overtone* is therefore an apt metaphor for suggesting that a word, etc., has implications over and above the plain meaning (e.g. *'Artificial' cannot be used without an overtone of disparagement*). *Undertone* is not a technical term in music. It simply means 'a subdued tone of sound or colour' (*COD*) and is suitably used figuratively for something unexpressed but inferrable from evidence (e.g. *there was an undertone of optimism in the peace talks in Geneva*).

**ovum.** The pl. is *ova*. See -UM 2.

**owing to** is not inserted here because it is misused, but to give readers an assurance that it is as often as not a suitable substitute for *due to* (a phr. which often attracts adverse criticism in some of its uses). *Owing to* has two main meanings: (*a*) (as a predicative adj.) caused by; attributable to (*the cancellation was owing to ill health*); (*b*) (as a compound preposition) because of (*the trains were delayed owing to a signals failure*). Its use in sense (*b*) is particularly recommended. A literary example: *Owing to its length, my hair tends to fall forward in two curves on the temples*—S. Bellow, 1987. The wordy phr. *owing to the fact that* can usually be avoided: use a conjunction such as *because* instead.

**owl-like.** See -LIKE.

**OX.** The pl. is *oxen*. See PLURALS OF NOUNS 7(*b*).

**oxymoron.** A word of Greek origin (Gk ὀξύμωρον, formed from ὀξυ- 'sharp' + μωρός 'dull') meaning 'a figure of speech in which apparently contradictory terms appear in conjunction': e.g. *a cheerful pessimist; harmonious discord*. Longer examples: *And faith unfaithful kept him falsely true*—Tennyson, 1859; *And yet ninety-nine-point-nine per cent of the time Middlehope* [sc. a village] *is madly sane, if you'll permit the paradox*—E. Peters, 1978; *'Come on, my dear,' the woman said ... Mother never said it, only 'You useless lump.' Useless lump or my dear, the meaning was the same*—H. Mantel, 1986; *He gets out of it by saying that he used to think 'interesting Canadian' was an oxymoron, but that Eric was obviously an exception*—M. Atwood, 1990.

# Pp

**pace** (prep.). Derived from Latin, and first used in English in the 19c., it is the ablative singular of L *pax* 'peace', and means 'by the leave of (a person), with due respect to (someone or someone's opinion)'. L *pāce tuā* or *pāce vestra* mean 'with all due respect to you'; (with the noun in the genitive) *pāce Veneris* means 'if Venus will not be offended by my saying so'. In English *pace* is used chiefly as a courteous or ironical apology for a contradiction or difference of opinion' (*OED*). Though *COD* 1995 gives precedence to /ˈpɑːtʃeɪ/, in my experience the word is most commonly pronounced /ˈpeɪsɪ/. It is almost always printed in italic, and its use is restricted to works in which Latinisms are likely to be recognized. Examples: *Indeed, pace Chomsky and Halle, we would probably want it to be impossible for mid glides to exist at all*—A. H. Sommerstein, 1973; *I find (a) incredible (pace Herman Kahn)*—*Conservation News*, 1976; *Tolstoy . . . is not, pace Albert Sorel and Vogüé, in any sense a mystic*—Isaiah Berlin, 1978. Fowler and others warn against using the preposition to mean 'according to', and cite some unattributed examples; but evidence for such a use is lacking now.

**pachydermatous.** Brought into English in the 19c. in zoological works to mean 'thick-skinned' (of certain animals, including the elephant and the rhinoceros, from Gk παχύς 'thick' + δέρμα 'skin'), it was quickly applied, sometimes playfully, to persons regarded as 'thick-skinned, not sensitive to rebuff, ridicule, or abuse'. Modern examples: *Can I be sincere without wounding people less pachydermatous than myself?*—H. Nicolson, 1934; *Edward laughed. His happiness had made him pachydermatous*—L. P. Hartley, 1961.

**pacifier** is the customary word in AmE for a baby's dummy.

**pacifist, pacificist.** Both words came into being at the beginning of the 20c.

but the shorter form is now the only one in use. See -IST.

**package.** Used for 'the packing of goods' and for 'a bundle of things packed in a receptacle', the noun has been part of the standard language for many centuries. In the course of the 20c., first in AmE and then in all English-speaking areas, it developed figurative senses, and in particular 'a combination or collection of interdependent or related abstract activities', as in *package deal, offer, proposal*, etc. Any set of agreements, proposals, transactions, travelling arrangements, etc., can now be described as a *package*. The new sense is widespread and useful.

**paean** (a song of praise or triumph). Thus spelt in BrE (in AmE also *pean*).

**paed(o)-.** A number of words containing this element (from Gk παῖς, παιδ- 'child') are usu. spelt with *-ae-* in BrE and with *-e-* in AmE: thus *paediatrics/ped-, paedophile/pedo-*, etc.

**paid** pa.t. and pa.pple of PAY.

**pailful.** The pl. is *pailfuls*. See -FUL.

**paillasse.** See PALLIASSE.

**painedly.** See -EDLY.

**paintress.** See -ESS 4.

**pair.** 1 When used to mean (*a*) a set of two persons or things regarded as a unit (*a pair of eyes; a pair of gloves*), and (*b*) an article consisting of two equal parts which are joined together (*a pair of binoculars, clippers, jeans, pincers, pyjamas, scissors, shears, trousers*, etc.), the phr. is normally const. with a singular verb or pronoun (e.g. *pass me that pair of scissors; there's a pair of gloves in the drawer*). If *a pair of* is omitted a plural pronoun or verb is required: e.g. *those gloves, scissors, etc., need replacing.*

2 Used as a collective noun, *pair* can take a singular or plural verb according to notional agreement: either *a pair of*

*crocodiles were basking beside the river* or *a pair of crocodiles was basking* [etc.]. See AGREEMENT 5.

**3** Number contrast for *gloves, pyjamas, scissors,* etc., can only be made by using them with *a pair of:* thus *a pair of binoculars* = one object, and *two pairs of binoculars* (not *two binoculars*) − two objects.

**4** The phr. *a pair of twins* is common enough in speech ( = one set of two babies born at a single birth), but should be avoided if there is any risk of ambiguity (as at a convention of twins).

**5** The pl. form *pairs* is desirable after a numeral (e.g. *seven pairs of jeans*). The type *seven pair of jeans* is non-standard, at least in BrE.

**pajamas.** See PYJAMAS.

**palace.** The second syllable is pronounced /-ɪs/ according to *COD* (1995) and, as the preferred pronunciation, /-əs/ by J. C. Wells (1990). The same difference of opinion is found in two major American dictionaries, those of Random House and Merriam-Webster. Educated usage is obviously divided: my own preference is for /-əs/.

**palaeo-.** Thus spelt in BrE (sometimes printed with the digraph æ as *palæo-*), but *paleo-* in AmE. Some British publishers are beginning to use the AmE spelling. The first element should be pronounced /pæl-/ not /peɪl-/. Thus, for example, *palaeography* /ˌpæliˈɒɡrəfɪ/.

**Palaeocene** See MIOCENE.

**palaestra** (in classical antiquity, a gymnasium). Pronounce /pəˈlaɪstrə/ or /-ˈliːstrə/. If the variant *palestra* is encountered, presumably it is to be pronounced /-ˈlestrə/.

**palanquin, palankeen** /pælənˈkiːn/, a covered litter or conveyance. The first is now the customary spelling of this word, which was adopted in the 17c. from Portuguese (ultimately from an East Indian vernacular word).

**palatal.** A term used in phonetics of a sound made by placing the surface of the tongue against the hard palate (e.g. *y* in *yes*). *Palatalization* is the process of rendering palatal, e.g. of changing the

nature of a consonant in certain circumstances by advancing the point of contact between tongue and palate. This 'is an essential part of the [ʃ, ʒ] sounds in English words such as *she* and *measure*, being additional to an articulation made between the blade and the alveolar ridge; or again, it is the main feature of the [j] sound initially in *yield*' (A. C. Gimson, 1980).

**palindrome** (from Gk παλίνδρομος 'running back again'), a word, verse, or sentence that reads the same when the letters composing it are taken in the reverse order. Edward Phillips's dictionary (1706) cited as an example *Lewd did I live, and evil I did dwel* (which doesn't quite work today). Greek, Latin, and English examples are cited in A. J. Augarde's *The Oxford Guide to Word Games* (1984), e.g. *Roma tibi subito motibus ibit amor* (attributed to Sidonius); *Able was I ere I saw Elba* (attributed to Napoleon); and numerous others, including some that are much more complex than these.

**palladium,** a safeguard or source of protection. Pl. *palladia*. See -UM 2.

**palliasse.** Now the established spelling, pronounced /ˈpælɪæs/. Cf. Fr. *paillasse* (from *paille* 'straw').

**Pall Mall** (name of street in London). Now pronounced /ˌpælˈmæl/. Formerly also /ˌpelˈmel/.

**pallor.** Thus spelt in both BrE and AmE.

**palmetto,** small palm tree. Pl. *palmettos.* See -O(E)S 6.

**palpable** (adj.) corresponds to the verb *palpate,* to examine (esp. medically) by touch. It means (*a*) that can be touched or felt, and (*b*) (often fig.) readily perceived by the senses or mind. Sense (*a*) dates from the 14c. and is most memorably illustrated in Ostricke's verdict to Hamlet, *a hit, a very palpable hit,* during the fatal duel at the end of Shakespeare's *Hamlet.* Further example: *Two fifths of the head are palpable above the brim*—M. F. Myles, *Textbk Midwives,* 1985.

Sense (*b*) dates from the 15c., and is illustrated in the *OED* by examples from the works of Hooker, Pepys, Byron, and numerous others. Modern examples; *This problem (so individual in its origins) takes palpable public shape in the fiction*—C. G. Wolff, 1977; *Othello's crisis of identity becomes palpable when 'seeming' ceases to be synonymous with 'being'*—*Studies in Eng. Lit.* (University of Tokyo), 1990; *the tension-friendly tension*—*in the room was palpable*—*Atlantic*, 1991.

Sense (*b*) is now dominant.

**panacea** /pænə'siːə/, a universal remedy, a remedy reputed to heal all ailments. It cannot be used of a particular illness (*a panacea for measles* is not idiomatic), but is most frequently used in negative contexts or ironically (of another person's proposal) of wide-ranging suggestions for the solution of (esp. social or economic) problems. It is derived from Gk πανάκεια 'all-healing', from παν- 'all' + ἄκος 'cure'. Examples: *To ... deaden the pain of neuralgia, the early Victorian panacea of laudanum was prescribed*—D. Thomas, 1979; *Alongside travelling, change of diet was a popular panacea*—R. and D. Porter, 1988; *Kipling reminds us often that work is the only panacea for most of life's ills*—M. Pafford, 1989; *her approach to the high life ... is tempered with this decade's unrealistic just-say-no panacea*—*Details* (US), 1991.

**pandemonium.** Coined by Milton in the form *Pandæmonium* ( = place of all demons) in his *Paradise Lost* (1667), it is a striking example of a word that has made its way from the higher realms of literature into everyday use, always now meaning '(place of) uproar'.

**pander.** 1 (noun). (Originally *pandar* from *Pandaro, Pandare, Pandarus*, etc., the name of a character in the several classical, medieval, and later versions of the story of Troilus and Cressida, but now always *pander*.) A go-between in clandestine love affairs.

2 (verb) From the 16c. onward, most commonly intransitive and followed by *to*, it has been used to mean 'to gratify or indulge a person, a desire, a weakness, etc.'

**pandit.** See PUNDIT.

**panegyric, panegyrize, panegyrist.** Now pronounced /ˌpænɪ'dʒɪrɪk/, /ˌpænɪdʒɪ-ˌraɪz/, and /ˌpænɪ'dʒɪrɪst/ respectively. *Panegyric*, ultimately from the Gk adj. πανηγυρικος, means 'a public eulogy',

and was formed from παν- 'all' + ἄγυρις = ἀγορά 'assembly', i.e. a eulogy fit for a public assembly or festival.

**panel.** The inflected forms are *panelled*, *panelling* (AmE *paneled, paneling*). See -LL-, -L-.

**panful.** The pl. is *panfuls*. See -FUL.

**panic** (noun and verb). The inflected forms are *panics, panicked, panicking*, and *panicky*. See -C-, -CK-.

**pantaloons, pants,** etc. 1 Changes of fashion and the different choices of words made in Britain and America can often lead to considerable uncertainty and confusion in the use of these words. *Pantaloons* (first recorded in the 17c.) in the days of Evelyn and Samuel Butler was the name given to fashionable breeches word by men at the time. By the late 18c. the word was applied to 'a tight-fitting kind of trousers fitted with ribbons or buttons below the calf' (*OED*), and, in the course of the 19c. in America, became extended to close-fitting trousers in general. The word is now restricted (written with a capital *P*) to a foolish old man who is the butt and accomplice of a clown in a pantomime. *Pants* (first recorded in 1840 in America as an abbreviation of *pantaloons*) became in Britain a colloquial and 'shoppy' term for 'drawers' (*OED*), and then 'for underpants, panties, or shorts worn as an outer garment'. Now the word is mainly used to mean underpants or knickers in the UK, and trousers or slacks (for men or women) in America, but the distinction is not an absolute one. A number of idiomatic phrases containing the word have come into common currency in the 20c., e.g. *to bore the pants off*, to bore to an intolerable degree; *to be caught with one's pants down*, to be caught in an embarrassingly unprepared state; *to wear the pants*, to be the dominant member of the household. *Panties* (used with a pl. verb) is gender-restricted, meaning short-legged underpants worn by women and girls. See KNICKERS.

2 Numerous other terms for garments of the 'pants' or 'trousers' types are encountered in everyday life, and the terminology varies from decade to decade and from country to country. People living in a particular country, however,

have no special difficulty in recognizing the right term for each garment, as between *bloomers, boxer shorts, breeches, briefs, drawers, jeans, knickerbockers, knickers, plus-fours, shorts, slacks, trousers,* etc. Difficulties arise when one travels to another English-speaking country only to find that the choice of terms is sometimes radically different. One must just proceed with caution in such circumstances.

**3** In the attributive position *pant* is common enough in AmE, but *pants* is also found in all English-speaking countries. (*a*) *Then we went downtown and bought pant suits*—L. Ellmann, 1988; *The baby has got her foot caught in her own pant leg*—*New Yorker*, 1991. (*b*) *Floral print especially smart in this pants dress because it's done in navy and white*—*Sears Catalog*, 1969; *I took a jackknife out of my pants pocket*—R. B. Parker, 1974.

**paparazzo,** a freelance photographer who pursues celebrities to get photographs of them. The pl. (as in the It. original) is *paparazzi*.

**papier mâché.** The accents in *mâché* are essential.

**papilla, papula.** The pl. forms are *papillae, papulae.*

**papyrus** /pə'paɪərəs/. Pl. *papyri* /-raɪ/. See -US 1.

**para-.** Two prefixes of different origin are used in forming English words: (*a*) the Gk preposition παρά 'alongside of, beyond', as in *parable, paradigm, paradox, paragraph, parallel, paramedical, paramilitary,* and *paraphrase*; and (*b*) the L imperative of *parāre* meaning 'guard against', as in *parachute, parapet, parasol*; cf. Fr. *parapluie.*

**parable,** a narrative of imagined events used to illustrate a moral or spiritual lesson; an allegory (*COD* 1995). The forty allegorical parables attributed to Jesus of Nazareth provide the traditional models in Christian literature. Modern examples, as Baldick (1990) reminds us, include Wilfred Owen's poem 'The Parable of the Old Man and the Young' (1920), which relates a biblical story to the 1914–18 war; and a longer prose parable, John Steinbeck's *The Pearl* (1948).

**paradigm** /'pærədaɪm/, an example or pattern, esp. a representative set of the inflections of a noun or verb. The OE verbal set *bīdan* (to wait), *bītt, bād, bidon, biden* ( = infin., 3 pres. sing. indic., 3 sing. pa.t., 1, 2, 3 pl. pa.t., pa.pple), for example, make up a paradigm. Transferred uses, meaning 'a theoretical framework', have been particularly common in the 20c., e.g. *The unfolding of terror and duplicity which follows is easily seen as a paradigm of the suppression of Dubček's liberalizing administration*—*TLS*, 1973. The corresponding adj. *paradigmatic* is pronounced /ˌpærədɪg'mætɪk/, with the -g- fully in evidence.

**paradise.** As with *nectar*, a large number of adjectival forms have come into being for the word *paradise*: *paradisaic* (first recorded 1754), *paradisaical* (1623), *paradisal* (c1560), *paradisean* (1647), *paradisiac* (1632), *paradisiacal* (1649), *paradisial* (1800), *paradisian* (1657–83), *paradisic* (a1745), and *paradisical* (1649). Of these, *paradisal* and *paradisiacal* have the most life at present. Alongside these uses stand attributive uses of *paradise* itself in certain fixed expressions, e.g. *paradise crane, duck, fish, flycatcher.*

**paraffin.** A word that is commonly misspelt (usu. by giving it two *r*s and one *f*). See KEROSENE.

**paragoge,** the addition of a letter or syllable to a word in some contexts or as a language develops (e.g. *t* in *peasant*; cf. Fr. *paysan*). It is pronounced as four syllables, /ˌpærə'gəʊdʒiː/.

**Paraguay** is pronounced /'pærəgwaɪ/, but in AmE often /-gweɪ/.

**parakeet** is now the normal spelling in BrE, but *parrakeet* is also sometimes found in AmE.

**parallel.** Exceptionally among verbs ending in -*l* (see -LL-, -L-) *parallel* does not double the *l* in inflected forms (*paralleled,* etc.); the anomaly is due to the -*ll*- of the previous syllable. The same applies to the corresponding noun *parallelism.*

**parallelepiped,** one of the longest words that one is introduced to in geometry lessons at school, should be pronounced /ˌpærəlelə'piped/ or /-'paiped/.

**paralogism,** illogical reasoning. Pronounce /pəˈrælədʒɪzəm/.

**paralyse.** Thus spelt in BrE, but *paralyze* in AmE.

**paralysis.** Pl. *paralyses* /-siːz/.

**parameter.** A mathematical term of some complexity which, in the course of the 20c., has become perceived by the general public as having the broad meaning 'a constant element or factor, esp. serving as a limit or boundary'. This meaning is still at the controversial stage, the stage at which dictionaries and usage manuals attach the word 'loosely' to the popular meaning, while mathematicians smile knowingly and exclude the word from their social vocabulary. The mathematical and computer science uses are too technical to define and illustrate here, but examples of non-technical uses lie readily to hand: *There are parameters to these recollections which may not be immediately apparent: the world of learning ... and the war*—D. M. Davin, 1975; *Given a few early broadly defined parameters within which any reasonably sensitive adult works with an individual child (e.g. enthusiasm, patience [etc.])*—R. Cameron, 1986; *Lewis's refusal to accept her standards, her parameters, she regarded as threatening*—A. Brookner, 1989. In the sense 'boundary' *parameter* brushes against *perimeter* (e.g. of an airfield), but the boundary of an airfield is consistently its *perimeter*, marked by a *perimeter fence* or a *perimeter road*, never its *parameter*. It is worth noting, however, that *perimeter* was simultaneously becoming persistently used in figurative senses for the first time. See PERIMETER. Anyone feeling uneasy about *parameter* has a wide choice of near-synonyms to choose from: *border, boundary, criterion, factor, limit, scope*, etc; one or other of these is normally more suitable in context.

**paramo** (high treeless plateau in tropical S. America). Pl. *paramos*. See -O(E)S 6.

**paranoia.** It is strange to think that this familiar word (answering to Gk παρ-άνοια) was regularly spelt *paranœia* (with œ usu. printed as a ligature) for most of the 19c. The corresponding adj. is now usually *paranoid* (first recorded in 1904), rather than *paranoiac* (1928), though both are in use (also as nouns). *Paranoic* (adj.) is also in current use.

**paraphernalia.** In origin a pl. noun, being the neuter pl. of medL *parapher-nālis*, short for *paraphernālia bona*, the 'personal property' (*parapherna*) which a married woman was in law entitled to keep. By the early 18c. it had developed in English the more general sense 'miscellaneous belongings'. Since the late 18c. it has been construed either as an ordinary plural noun (*Hints in the Choice of Guns, Dogs, and Sporting Paraphernalia*, 1809), or as a collective singular (*A whole paraphernalia of plums*, 1845). Modern examples: *Edith received ... Samuel Cooper's sketchbooks, color boxes, and other artist's paraphernalia*—M. Mack, 1985; *Paraphernalia in his flat included Indian clubs and an adapted table to which boys were tied*—*Independent*, 1989; *A doctrine that aimed to restore religious practice to New Testament standard inevitably rejected the whole paraphernalia of medieval veneration of the saints*—*Rev. Eng. Studies*, 1990; *the executive position is today surrounded by often useless paraphernalia which does little more than ... reflect her ... standing in the company*—*Times*, 1992; *Painting paraphernalia ... abound*—*Observer Mag.*, 1993.

**paraplegic.** Pronounce with a 'soft' g, /pærəˈpliːdʒɪk/.

**paraselene** (bright spot on a lunar halo). Five syllables, /pærəsɪˈliːni/.

**parasitic, parasitical.** In technical writing the shorter form seems to be dominant. In general use the two terms occur with something like equal frequency. Examples: (parasitic) *This parasitic castle life had left my funds comparatively intact*—P. L. Fermor, 1986; *Budgies often scratch ... Sometimes it's because they have a parasitic problem*—*Fast Forward*, 1990; (parasitical) *For Freud, sociology and the other social sciences are parasitical on psychology*—P. Gay, 1985; *These works hold a parasitical relationship to existing sources of culture*—*Oxford Art Jrnl*, 1988.

**parataxis** (grammar, 'a placing side by side'). The placing of phrases or clauses one after another, with no linking word(s) to indicate coordination or subordination, e.g. *It's ten o'clock, I have to go home; Tell me, how are you?; I couldn't keep my eyes open, I was so tired* (examples from

Bloomfield 1933, *COD* 1995, and *CGEL* 1985, respectively). Such paratactic constructions have been a feature of the language from the OE period onward: readers who are interested in pursuing the matter should consult Mitchell's *Old English Syntax* (1985), ch. 5.

**paratroops** (pl. noun), troops equipped to be dropped by parachute from aircraft. A member of a *paratroop* regiment is called a *paratrooper.*

**parcel** (verb). The inflected forms are *parcelled, parcelling* (AmE usu. *parceled, parceling*). See -LL-, -L-.

**parcimony.** See PARSIMONY.

**pardon.** Used elliptically for *I beg your pardon*, it has been in use as a genteelism for about a century. A. S. C. Ross (1954) said that *Pardon!* is used by the non-U in three main ways: **1** if the hearer does not hear the speaker properly;

**2** as an apology (e.g. on brushing by someone in a passage);

**3** after hiccuping or belching.

**parenthesis. 1** Pl. *parentheses* /-siːz/.

**2** The most memorable use of the word *parenthesis* occurs in W. S. Gilbert's *The Gondoliers* (1889), Act II: 'Take a pair of sparkling eyes, Hidden, ever and anon ... Take a tender little hand, Fringed with dainty fingerettes, Press it—in parenthesis;—Take all these, you lucky man—take and keep them, if you can!' It shows one way in which a parenthetic remark is printed. But there are other ways. Typical examples taken at random from a single book (Tim Winton's novel *Shallows,* 1985): (explanatory phrase in round brackets) *he and Mara* (*then a milkman's pretty daughter*) *grovelled together long and effectively enough to cause the eventual birth of their son Rick*; (asides, explanatory remarks between dashes) *Hassa Staats lives in daily fear of cancer—he sweats in the small hours over it—and instead of a cigarette* [etc]; *It was a boom time for him—a Depression for everyone else—and the task was to buy up every skerrick of land available; The canning factory—for God-sake—is already working off a skeletal staff*; (he said-type formulas) *'Yes,' Staats agrees impatiently, 'the Onan's coming along nicely; What a day,* he thinks, *what a day for this town.* There are other types but these six examples must needs suffice to illustrate

the general nature of parenthetic comments. The essence of a parenthesis is that it interrupts the flow of a sentence, generally in order to explain or elaborate on something just written. Because they are interruptions, parentheses are best kept fairly short. It is important to bear in mind that a parenthesis may or may not have a grammatical relation to the sentence in which it is inserted. In *This is, as far as I know, the whole truth* there is such a relation, and in *This is, I swear, the whole truth* there is not; but one is as legitimate as the other.

**3** For the various shapes of printed parentheses, see BRACKETS.

**parenthetic, parenthetical.** The shorter form is the more usual of the two except when the sense is 'addicted to or using parentheses' (*many of the explanatory asides are needlessly parenthetical*) and also in certain technical uses (*Interpolated coordination may be distinguished from parenthetical coordination, where an unreduced coordinate clause is inserted parenthetically within another clause—CGEL* 13.96, 1985). Contrast the use of the shorter form in a less technical context: *When they end an included parenthetic sentence ... they have only a specifying function, not a separating function—CGEL* III.23, 1985.

**parget** (to plaster a wall, etc.). Pronounce /ˈpɑːdʒɪt/. The inflected forms are *pargeted, pargeting*; see -T-, -TT-.

**pariah.** For the pronunciation the *OED* (1905) gave precedence to /ˈpeərɪə/, Daniel Jones (1917) to /ˈpærɪə/, and *COD* (1990) to /pəˈraɪə/. There is no doubt that, after much experimentation and hesitation, the word has now settled down, as *COD* indicates, to rhyme with *Isaiah*, and no longer (except among older people) with *carrier.*

***pari passu.*** Frequently encountered in academic and legal work, this Latin adverbial phrase means 'at an equal rate of progress, simultaneously and equally'. It is derived from L *pār, pari-* 'equal' + *passu* abl. of *passus* 'step'. The dominant pronunciation in BrE is /ˌpɑːri ˈpæsuː/.

**Parkinson.** The medical condition known as *Parkinson's disease* was first thus called in 1877 (one year earlier in France as *maladie de Parkinson*) and was named after James Parkinson (1755–1824), English surgeon and palaeontologist. The whimsical 'law' known as *Parkinson's Law* ('work expands to fill the time available for its completion') first appeared in print in 1955 and was propounded by C. Northcote Parkinson (1909–93), English historian and political scientist.

**parlance.** See JARGON 3.

**parlay** is a 20c. AmE betting term (in origin a corruption of *paroli*, a term of It. origin used in faro and other card games since the beginning of the 18c.) for the leaving of the money staked and the money won as a further stake. Called an *accumulator* (*bet*) in BrE. Also used in AmE as a verb. Cf. next.

**parley,** a discussion of terms for an armistice or for settling a dispute. Pl. *parleys*. Also as verb, to hold such a discussion (inflected forms *parleys, parleyed, parleying*).

**parliament.** Pronounce /ˈpɑːləmənt/, not /ˈpɑːlɪəmənt/.

**parlour.** Thus spelt in BrE (AmE *parlor*). It has a wide range of uses: e.g. a sitting-room (a regional use for the 'best room'); in the combinations *beauty parlour, funeral parlour; ice-cream parlour; parlour game* (indoor game); also, a room or building equipped for milking cows. Over the centuries *parlour* has also been applied to a small official room set aside for private conversations (e.g. the *Mayor's Parlour* in a Town Hall); and an apartment in a monastery for conversation with visitors or among the monks themselves.

**parlous.** This syncopated form of *perilous* came into use in the 14c. and stood side by side with it in similar contexts and with much the same geographical distribution until the early 20c., when unmistakable signs of restricted use began to surface. Fowler (1926) described it as 'a word that wise men leave alone'. *COD* (1995) labels it *archaic* or *jocular*. It is a sad fate for a long-serving word. It can still be safely used, however, in fairly formal circumstances with *state* (*the economy is in a parlous state*) and *times* (*these are parlous times in Yugoslavia*). But *perilous*,

*dangerous, hazardous,* and other undisputed words are preferable in most other circumstances.

**Parmesan.** Pronounce the medial *s* as /z/.

**paronomasia.** From Gk παρονομασία 'a play upon words which sound alike', this rhetorical term has been used in English since the 16c. for serious (as opp. to banal or embarrassing) examples of word-play or punning. The best known of all (though concealed in English) is perhaps that of Matt. 16: 18 *thou art Peter, and vpon this rocke* [Gk πέτρα 'rock'] *I will build my Church.* See PUN.

**paroquet.** An older form of PARAKEET.

**parricide, patricide.** The first of these, which was first recorded in the 16c., has a wide range of meanings: one who murders either parent, or other near relative; also, the murderer of a person considered specially sacred as, for example, being the ruler of a country; also, one who commits the crime of treason. It is also used for any of these crimes themselves. In the same century *patricide* entered the language for the specific person or crime 'the murderer of/the murder of one's father'. The words are etymological doublets: *parricide* is from the L type *parricīda*, by Quintilian thought to be for *patricīda* (cf. L *patrem* 'father', *-cīda* 'killer'). *Patricide* neatly parallels *matricide, fratricide,* and *sororicide* to form a set of transparently distinguishable terms.

**parsimony.** Until the early 20c. *parcimony* was an acceptable variant spelling, reflecting the fact that classical Latin had both *parsimōnia* and *parcimōnia* (from *parcere,* ppl stem *pars-,* 'to spare, save').

**parsing.** **1** The resolution of a sentence into its component parts and the assigning of names (noun, verb, adjective, adverb, conjunction, etc.; subject, predicate, etc.) to each part of speech and to each component. See PARTS OF SPEECH.

**2** In extended use in computational linguistics, *parse* is 'to analyse (a string [of characters]) into syntactic components to test its conformability to a given grammar' (*OED* 2, 1989)—a simple definition of an extremely complex process.

**parson** is a colloquial term for a man of the cloth up to the level of a rector. In times past it had a long and distinguished history as a synonym (of equal formality) for a holder of a parochial benefice, a rector. The extension of the word in the 16c. to include a vicar, a chaplain, a curate, or any clergyman gave it an informal complexion, which now applies to all uses of the term.

**partake.** In origin a 16c. back-formation from the earlier *partaker* ( = *part-taker*, rendering L *parti-ceps*), it has come to be a somewhat formal, usu. intransitive, verb meaning 'to take a part in, to share in some action or condition' (*we need to partake in each other's joys*); and esp. (with *of*) 'to have a share or portion of (food or drink) in the company or others or of another person' (*Your papa invited Mr. R. to partake of our lowly fare*—Dickens, 1865). The notion of sharing with others, i.e. of receiving only a due part of the food being served, is crucial. To speak of *partaking of a boiled egg* if one is eating alone is to take the word beyond its proper limits.

**part and parcel.** A convenient alliterative collocation that has commended itself to English-speakers since the 15c. From the 15c. to the 19c. *parcel* was used to mean 'a constituent or component part', and, while it can no longer be used in that way, the sense survives in the alliterative phrase. The expression, let it be noted, means not just 'a part' but 'an essential part' (of something).

**Parthian shot** is now synonymous with *parting shot*, i.e. they both mean 'a remark or glance, etc., reserved for the moment of departure'. The use arose from a custom in Parthia, an ancient kingdom in W. Asia: Parthian horsemen would discharge their missiles into the ranks of the enemy while in real or pretended flight. The actual term *Parthian shot* (first recorded in 1902) and *parting shot* (1894) are comparatively recent, but the connection may just have escaped the attention of readers of earlier books that were 'read' for the *OED*.

**partially, partly.** 1 These two long-established adverbs—*partially* first recorded in the 15c. and *partly* in the 16c.—have shared the sense 'in part' from the beginning. The only clear semantic difference

between them is that from the 15c. until about 1800 *partially* (like Fr. *partialement*) also meant 'in a partial or biased manner, with partiality', a sense that *partly* could not share.

2 Fowler (1926) argued that *partially* was opposed to *completely*, and that *partly* was opposed to *wholly*. 'In other words,' he said, '*partly* is better in the sense "as regards a part and not the whole", and *partially* in the sense "to a limited degree".' There is something in this, but the distinction is not an absolute one. Fowler then cited two examples, both silently taken from the *OED*, which he claimed showed a 'wrong' use of *partially*: *The two feet, branching out into ten toes, are partially of iron, and partially of clay*—G. S. Faber, 1827; *As to whether* The Case is Altered *may be wholly or partially or not at all assignable to the hand of Jonson*—Swinburne, 1889. It is a straightforward example of a clash between two methods: the historical presentation of the evidence in the *OED* and the more judgemental presentation of the same evidence by Fowler.

From the examples that follow I do not think it reasonable to claim (as Fowler did) that *partially* and *partly* follow two distinct tracks in the language: (partially) *Nor can they sell glasses for children under 16 or to registered blind or partially-sighted people*—Which?, 1985; *I partially solved my money problems by being paid ten shillings to play regularly at the Black Horse*—A. Burgess, 1987; *The title partially preserved in the papyri specifies Delos as the place of performance*—Classical Q, 1988; *My body felt ... only partially under my control*—L. Ellmann, 1988; (partly) *Her dislike of him was of course ... partly based upon a sense that he disliked her*—I. Murdoch, 1980; *The tenth-century settlement at Manda was located on partly reclaimed land*—Antiquaries Jrnl, 1987; *Her untidy blonde fringe partly covered her eyes*—J. G. Ballard, 1988. Observe, for example, the grammatical equivalence of *partially-sighted people* and *partly reclaimed land*. Observe too that *partially* is opposed to 'completely' in Burgess and the *Classical Quarterly*; that the sense 'to a limited degree' is found in Ellmann's *partially*, and that 'in part' is the sense in Murdoch and in Ballard. This is not a two-lane pattern. The uses and meanings sometimes merge and sometimes stand apart.

Moving away from the examples given above, two main tendencies can be observed. *Partly* is used more often than *partially* to introduce an explanatory clause or phrase: *Partly because he needed the money* ...; *Partly for this reason* ... It should be noted too that the correlative construction *partly* ... *partly* is common (*partly in verse and partly in prose*), whereas *partially* ... *partially* is simply not idiomatic. The status of the two words is likely to change in the future, but it seems unlikely that they will fall into totally distinct grammatical and semantic lanes for a long time to come.

**participles.**  **1** Unattached participles (see as main entry).

**2** Absolute construction (see as main entry).

**3** Fused participles. See POSSESSIVE WITH GERUND.

**4** Initial participles, etc. Fowler (1926) identified a stylistic mannerism of newspapers, and rather fancifully described it as follows: 'In these paragraphs, before we are allowed to enter, we are challenged by the sentry, being a participle or some equivalent posted in advance to secure that our interview with the C.O. (or subject of the sentence) shall not take place without due ceremony.' His examples included: *Described as 'disciples of Tolstoi', two Frenchmen sentenced at Cheltenham to two months' imprisonment for false statements to the registration officer are not to be recommended for deportation; Winner of many rowing trophies, Mr. Robert George Dugdale, aged seventy-five, died at Eton; Found standing in play astride the live rail of the electric line at Willesden* ... *Walter Spentaford, twelve, was fined 12s. for trespass.* The type to which Fowler objected seems now to be fairly uncommon, at least in the quality press. Contrast these sentences (drawn from some Oct. 1991 issues of *The Times* and the *Sunday Times*) in (*a*) where the subject is delayed, and those in (*b*), the more common type, where the subject is highlighted by being placed first: (*a*) *Having filled the Tate with various enormous metal 'vessels', Sir Anthony Caro is moving still closer to nautical areas*; *Trailing Minnesota, two games to nil, the Braves were on the verge of extinction in game three.* (*b*) *Fred Overton, thought to be the last survivor of the Channel tunnel project halted

in 1923, has died aged 87; Neo-nazis and anarchists brawled in the streets of two east German towns, causing serious damage to property; John Casken, winner of the 1990 Britten Prize for composition, is to become the Northern Sinfonia's first composer in association.*

**5** Placing of the stress and changing of the pronunciation distinguishing a pa.pple (or adj. derived from the ppl. stem) from a verb having the same spelling. A standard example is *consummate* (adj.), usu. stressed on the second syllable, /kən'sʌmət/, as against *consummate* (verb), which is always stressed on the first, /'kɒnsəmeɪt/. Similarly *dílute* (adj.) but *dilúte* (verb); and the archaic adj. *frústrate* contrasted with the verb *frustráte*. A second group distinguishes the part of speech by placing the obscure vowel /ə/ in the final syllable of the adj. but /-eɪt/ in the final syllable of the verb. Examples: *animate* (adj.) /'ænmət/ but *animate* (verb) /'ænmeɪt/; similarly, *deliberate* (adj.) /-ət/ but *deliberate* (verb) /-eɪt/; and numerous other pairs of adjs. and verbs including *advocate, articulate, degenerate, designate, desolate, elaborate, legitimate, moderate, separate,* and *subordinate.* A third type is shown in *diffuse* (adj.) (spread out, not concise, etc.) /dɪ'fjuːs/ contrasted with *diffuse* (verb) (to disperse) /dɪ'fjuːz/, the pair being distinguished only by the contrast of the final sound, /s/ as against /z/. See also NOUN AND VERB ACCENT, and most of the above words at their alphabetical places.

**particoloured** (adj.). The first element appears to be a respelling of *party* (adj.), first recorded in the 14c. in the sense 'variegated' (*She gadereth floures, party white and rede*—Chaucer, *c*1386), ult. from L *partītus* 'divided', pa.pple of *partīre* 'to part, divide'. *Party* is still used as an adj. in heraldry, said of a shield divided into parts of different tinctures.

**particular.** Frequently used for emphasis, esp. after the demonstrative pronouns *this* and *that* (*he didn't like that particular tax; in this particular instance*) to the point that it has attracted adverse comment ('an unnecessary reinforcement', 'can often be left out, to the benefit of the sentence'). Up to a point such criticism is just, but it should be borne in mind that there are some contexts, esp. (but not only) after a negative, when the adjective supplies legitimate emphasis (e.g. *He had no particular reason for being there as far as I could tell*; *She didn't write that particular essay but many others just as good*).

**particularly.** Care should be taken to pronounce all five syllables. The pronunciation /pə'tɪkjʊlɪ/ = 'particu-ly' is a vulgarism.

**partisan, partizan.** There are two distinct words: **1** (Spelt with either *s* or *z*.) An obsolete word for an obsolete military weapon ( = a long-handled spear) used by foot soldiers in the 16c. and 17c. It is pronounced /'pɑːtɪzæn/, with the stress on the first syllable, and is etymologically of similar origin to the next.

**2** (The only current word, now usu. spelt *partisan*.) Generally, a zealous supporter of a party, cause, etc.; specifically, a guerrilla in wartime. It is usu. pronounced /,pɑːtɪ'zæn/ in BrE, with the main stress on the last syllable; but in AmE first-syllable stressing is favoured. It was adopted from French in the 16c., and the French word in turn was an alteration of Italian (Tuscan) dial. *partigiano*, from L *pars part-em* 'share, part' + a suffix answering to L *-iānus*.

**partitive.** As noun, a word, form, etc., denoting a part of a collective group or quantity (e.g. *any, some*; *half, portion*). The word is most familiar in *partitive genitive*, a genitive used to indicate a whole divided into or regarded as parts, expressed in modern English by *of* as in *most of us, half of the ground*. The partitive genitive is a common feature of inflected languages: e.g. L *Plato totius Graeciae doctorissimus fuit* 'Plato of all Greeks was the most learned'; OE *þæs landes sumne dæl* 'a part of the land'; *scipa fela* 'many (of) ships'.

**partly.** See PARTIALLY.

**parts of speech.** As a sign of the times, in 1990, an Oxford undergraduate, immediately before his final examinations, in which he was eventually placed in Class I, asked me what he called 'an embarrassing question': could I please explain to him the difference between an adjective and an adverb? I supplied

the essential facts and referred him to *CGEL* (1985). The episode startlingly reminded me that we live in an age when it can no longer be assumed that even a well-schooled person is capable of analysing sentences into their named parts and units. It is also an age when the linguistic analysis of sentences by professionals has reached a stage of complexity that puts their work beyond the reach of ordinary people.

In broad terms the central parts of speech named in the first modern English grammar, namely William Bullokar's *Bref Grammar for English* (1586), were those used from earliest times of Latin grammar. The same is true of other English grammars written between 1586 and the end of the 18c. A change of attitude came about in the later part of the 19c. and esp. in the 20c. Modern professional grammarians endeavour to analyse and describe English grammar in terms of its own features, with negligible reference to the grammar of Latin or of other languages. I have given an account of these changes of attitude and of substance in my book *Unlocking the English Language* (1989) and must refer readers to it. In the present book it seemed to me essential to retain the traditional terminology as far as possible, while trying to import into the relevant entries the discoveries and insights of modern linguistic scholars stripped of the opaque language in which such work is often written.

The main parts of speech used in this book are as follows: noun, verb, auxiliary verb, adjective, adverb, pronoun (including demonstrative and possessive pronouns), preposition, conjunction, article (definite and indefinite), interjection, and numeral. I have tried to avoid using specialized terms such as *approximator, deictic, deixis*, and *determiner* that have not yet made their way into the public domain. Most of the regular terms are treated at their alphabetical place, but the auxiliary verbs (i.e. *be, do*, and *have*) and the modals (*can/could, may/might, must, shall/will, should/would*) are treated as main entries.

**party.** Quite separate from various uses in law ( *guilty party, third party*, etc.), *party* preceded by the indefinite article (e.g. *a pious party, an aggrieved party*), has come to be used informally (the *OED* in 1905 labelled the use *low colloquial* or *slang*) to mean 'a person': and more vaguely 'the person (defined by some adj., pronoun, etc.)'. This second use (which seems to be commoner in AmE than in BrE) was described by the *OED* in 1905 as 'Formerly common and in serious use; now shoppy, vulgar, or jocular, the proper word being *person*'. The descent of the word into mercantile or other types of informality seems to have occurred during the 19C. Modern examples: *I don't know who the injured party is here*—R. Carver, 1986; *I've known this party for three years now and I hardly know him at all*—D. Goodis, 1986; *June had taken Imogen from her—'What a stout little party'—and settled down for the interview with Imogen on her knee*—J. Trollope, 1990; *She was staring right at the window, and the man's face was staring right at her. 'What party is he looking for, do you suppose?'*—New Yorker, 1992.

**party.** See PARTICOLOURED.

**pasha.** Now the normal spelling. See BASHA(W).

**Pashto** /ˈpʌʃtəʊ/. The official language of Afghanistan. Thus spelt, not *-u*, *Push-*.

**pasquinade.** See LAMPOON.

**pass.** (verb). The pa.t. and pa.pple are both *passed* (*the remark passed unnoticed; he had passed the afternoon reading*). The related adj., prep., and adv. are all spelt *past* (*past times; first past the post; she hurried past*), as is the noun (*memories of the past*).

**passable** /ˈpɑːsəbəl/ means 'barely satisfactory' (*a passable performance*) or 'able to be passed' (of a road, mountain pass, etc.). *Passible* /ˈpæsɪbəl/ is a theological term meaning 'capable of suffering', ult. from L *patī*, *pass-* 'to suffer'. See -ABLE, -IBLE 4.

**passed.** See PASS.

**passer-by.** The pl. is *passers-by*.

**passible.** See PASSABLE.

**passive territory.**

1 Constraints.
2 Scientifically passive.
3 The double passive.
4 Semi-passives.
5 *she was given a watch.*
6 Passive of *avail oneself of.*
7 The impersonal passive.

*Oxford defeated Cambridge* is an active expression and *Cambridge was defeated by Oxford* is its passive equivalent. In passive constructions the active subject has become the passive agent, and the agent is (in this case) preceded by *by*. In practice, however, in the majority of passives, the *by*-agent is left unexpressed (*peace was declared some time ago*), or the notional agent is introduced by a different preposition (*I am not disenchanted with Henry; a lot of people are concerned about land use*). There are also various restrictions on the use of both voices.

1 Constraints. Once pointed out, it is obvious that there is no natural passive equivalent of *she combed her hair* (the notional passive would be *her hair was combed by her*), since the pronoun *she* cannot idiomatically be converted into *by her* in such circumstances; or of *she had a nice laugh*, since *have* is one of several verbs (*to lack, to own*, etc.) representing a continuing state of affairs not a single act. On the other hand some verbs can be used only in the passive voice (e.g. *the creek was reputed to contain blackfish*—P. Carey, 1985) and have no active voice equivalent (we cannot say *the creek reputes to contain blackfish*). Similarly restricted verbs are *be said to be* (*he is said to be a good writer*) and *was born* (*she was born in London*).

In written English only about half of the notional forms of legitimate passives occur with any frequency (*is taken, was taken, will be taken, may be taken, has been taken, is being taken*, etc.); in spoken English other parts of the paradigm occur somewhat more frequently (*may have been taken, may be being taken*), and even the most extended forms (*has been being taken, may have been being taken*) are occasionally used in spoken English without causing undue inconvenience to the listener.

2 Scientifically passive. In scientific writing the passive voice is much more frequent than it is in ordinary expository or imaginative prose: e.g. *the cultures were fixed with 4% paraformaldehyde* rather than *I/we fixed the cultures with* [etc.]; similarly, *when DNA molecules are placed on a gel; an electron is scattered once every 1,000 molecules*. In ordinary prose true passives

are relatively uncommon—usually not more than two on an average page of a book. In scientific work they are a main constituent.

**3** The double passive. In constructions of the kind *members who are found to have taken cocaine; the race is thought to be won by those who travel lightest; the satellite is scheduled to be put into orbit in March; a vast natural garden, which has to be seen to be believed*, a passive verb is comfortably followed by a passive infinitive. In *a review tribunal is required to be reviewed itself after the first year* the double passive begins to obtrude, though the construction is still acceptable. Some grammarians (including Fowler) have condemned constructions in which passive uses of *attempt, begin, desire, endeavour, hope, intend, order, propose, purpose, seek, threaten,* and a few others, are immediately followed by a passive infinitive (e.g. *the order was attempted to be carried out/no greater thrill can be hoped to be enjoyed*). Such constructions should be avoided in favour of sentences written in the active voice. The *OED* provides examples of some of them: *it has been begun to be cleansed*—s.v. begin v. la¶, labelled obsolete (in 1887); *The evils that were intended to be remedied*—Bk of Common Prayer, 1662; *Was it thus intended and commanded by him to be drunken?*—Bentham, 1818; *all classes were threatened to be overwhelmed in one universal ruin*—Picture of Liverpool, 1834; *Persons who have any interest in lands which are sought to be registered can lodge a caution with the registering officer*—Law Times, 1891.

It is hardly necessary to speak out against such constructions now, as the fashion of writing them had virtually disappeared by 1900. The dictionary caught them at the point of near-extinction. Of course allowable examples of double passives still occur: *she couldn't be bothered to be interested*—D. Lessing, 1985; *it was clear that arms were allowed to be shipped to Iran*—New Yorker, 1986; *a young man who had saved his life in a Japanese prison camp after he had been sentenced to be beheaded*—M. Bradbury, 1987. And also stray examples of the condemned group of verbs: e.g. *Other records [of Nazi Germany] seized by the Soviet Army, were taken to Moscow and their contents have been begun to be made available to Western researchers only in the last five years*—Chicago Tribune, 1990. But they are relatively uncommon.

A related remnant of this kind of construction (but with ellipsis of *to be*) remains in standard use in AmE: *this is the first time in 30 years a person has been ordered deported for fascist activities*—NY Times, 1982. This use is first recorded in AmE in 1781: see ORDER (verb).

**4** Semi-passives. True passive constructions form part of the systematic world of finite verbs. Most of them have a direct active counterpart. Semi-passives (or false passives) occur when an apparent pa. pple is used immediately after a copular verb (e.g. *to be, to get, to seem*): *Phoebe was astonished; he was mistrusted in the village; I must get changed; he seems transformed*. Since in each case the pa. pple is formally interchangeable with an adjective (*happy; unpopular; ready; different*), such constructions are better regarded as consisting of a copular verb followed by an adjectival complement. There are other intermediate types ('the passive gradient', *CGEL* 3.74) between true passives and semi-passives: e.g. *I am prepared to take my oath; Fred was tired of the taste of hospital food, I'm just fed up with you; the wounds are all healed* are passively based constructions but hardly true passives, in that they cannot be straightforwardly converted into active equivalents.

**5** *she was given a watch*. In the 19c. some grammarians expressed reservations about such passives. Thus Henry Sweet (1898): 'we still hesitate over and try to evade such passive constructions as *she was given a watch/he was granted an audience* because we still feel that *she* and *he* are in the dative, not the accusative relation.' This view can now be safely discarded. The indirect passive construction (*Any Maltese who desired to free himself from his allegiance to the Grand-master was given a patent*—W. Porter, 1858; *Under such a charter the mayor is given power and opportunity to accomplish something*—J. Bryce, 1888) flourished throughout the 19c. and is now commonplace (*each subject was given a printed instruction sheet*—Working Papers, Macquarie Univ., 1985). Sweet's hesitation was based on the circumstance that OE *giefan* 'to give' governed a direct object (acc.) and an indirect one (with dat. inflection and/or preceded by *to*). By the 12c. the verb, as now, could govern two objects, direct and indirect (*te king iaf ðet abbotrice an* prior, 1154), but for some 600 years only the direct object could be placed at the head of a passive construction (e.g. *the rome of Gartier was never geven to no estraunger*, 1548). By about 1800 the constraint on the placing of an indirect object at the head of such constructions had disappeared.

**6** Passive of *avail oneself of*. See AVAIL 3, 4.

**7** The impersonal passive. Gowers (1965) advised against the use of *it is felt, it is thought, it is believed*, etc., in official and business letters ('it often amounts to a pusillanimous shrinking from responsibility'). He was probably right. The use or avoidance of the passive in such circumstances often depends on the level of formality being aimed at and often on the wisdom of accepting personal or group responsibility for the statement that follows. In general, however, it is better to begin by identifying the person or group who feel, think, believe, have decided, etc.: thus, *After due consideration the Finance Committee has decided not to* [etc.]; *I feel that the claim made in your letter is too optimistic.*

**past.** See PASS.

**pastel,** artist's crayon; light and subdued shade of a colour. Pronounce /ˈpæstəl/.

**pastiche** /pæˈstiːʃ/. Used in two main senses: **1** a medley, esp. a picture or a musical composition, made up from or imitating various sources.

**2** a literary or other work of art composed in the style of a well-known author. (COD 1990)

**pastille.** Spell thus, not *pastil*, and pronounce /ˈpæstɪl/.

**past master.** The OED says that the use of the expression to mean a person who is especially adept in an activity, subject, etc., 'apparently has arisen partly in allusion to the efficiency which results from having passed through such an office as that of master of a freemasons' lodge; sometimes it alludes to the efficiency from having "passed" the necessary training or examination to qualify as "master" in any art, science, or occupation'. It is no longer (as it once was) written as *passed master*.

**pastor** is now used, esp. in AmE, as the term for or title of a clergyman or clergywoman in charge of a nonconformist church, esp. a Lutheran or Methodist one (occas. also a Roman Catholic one). In 1989 an American correspondent collected a large number of clippings about services, funerals, etc., conducted by pastors in the state of Illinois. The word occurred as a title (*Fourth Presbyterian Church, John M. Buchan, Pastor*), and as a general descriptive term (*the Roman Catholic pastor of St Norbert parish; The Rev. Samuel Solomon was pastor of the African Methodist Episcopal Church in Gary; Trinity Lutheran Church, Evanston . . . Paul Christenson is the pastor*). The term was sometimes qualified by words indicating rank (*Pastor Emeritus, associate pastor*). The pastor of the Evanshore Presbyterian Church was a woman. A South Korean Presbyterian minister called Moon Ik Hwan was described as a *dissident pastor*. In news reports, obituaries, and other reasonably extended pieces, the word *minister* alternated freely with the word *pastor.*

**pastorale.** Now usu. pronounced as three syllables, /pæstəˈrɑːl/, but occas. as four /-li/. The pl. is *pastorales* (formerly alternating with *pastorali*).

**past tense.** It is of interest to note here (as CGEL 4.16 points out) that in certain circumstances the pa.t. may legitimately be used with reference to past and future time: **1** In indirect speech, e.g. *Did you say that you had* (or *have*) *a house to let?* | *How did you find out that I was* (or *am*) *the owner?* Such 'backshifts' are optional. Note also '*What did you say your name was?' 'Jones.'* (G. Greene, 1980). A pa.t. is used retrospectively with reference to future time in such a sentence as *My pupils will be annoyed that they mistook the time of your lecture and therefore missed it.*

**2** What CGEL calls the 'attitudinal past': e.g. *Do/Did you want to come in now?*

**3** The hypothetical past is used in sentences like the following: *It's time we left to catch the train; If you tried harder, you would probably win the game* (the implication being that you probably won't try harder).

**pasty** (noun) ( = a pastry case with a sweet or savoury filling) is pronounced

/'pæstɪ/. See TART. The adj. *pasty* (unhealthily pale), by contrast, is pronounced /'peɪstɪ/, thus conforming with *paste* /peɪst/.

**pâté** /'pæteɪ/. The accents are obviously needed to distinguish the word from *pate* ( = the head) and *pâte* ( = the paste of which porcelain is made).

**patella** /pə'telə/, the kneecap. Pl. *patellae* /-liː/.

**paten, patten.** The first is the now usual spelling of the word for the shallow dish on which the bread is laid at the celebration of the Eucharist; a *patten* is a (now disused) term for a shoe or clog specially shaped for walking in mud.

**patent 1** (noun) ( = authority giving a person the sole right to an invention). Pronounce /'pætənt/. So also in *Patent Office*, and in the phr. *letters patent* (in which *patent* is an adj.). The pronunciation /'peɪtənt/ in these words is not wrong, but is less usual in standard English. In *patent leather* my impression is that /'peɪtənt/ is the more usual of the two pronunciations in BrE, but /'pætənt/ in AmE.

**2** As adj. and adv. when *patent(ly)* means 'evident(ly), obvious(ly)' (*it was a patent lie; it was patently true*) the pronunciation is usu. with /'peɪt-/ in both BrE and AmE. Clearly the pronunciation of the word is unsettled, and resort to the Latin and French words lying behind the English ones does not help to resolve the problem.

**pathetic fallacy,** the attribution of human feelings and responses to inanimate things, esp. in art and literature. The term was introduced into the language by John Ruskin in his *Modern Painters* (1856). The phrase, which Fowler (1926) described as 'common though little recognized in dictionaries', received scant treatment in the OED but has a full entry in OED 2.

**patio.** Pl. *patios.* See -O(E)S 4.

**patois.** See JARGON 3.

**patricide.** See PARRICIDE.

**patriot.** The pronunciations /'pætrɪət/ and /'peɪt-/ have approximately equal currency in standard BrE at present, with the first perhaps slightly more common.

For the derivatives *patriotic* and *patriotism* the preferred pronunciation seems to be that with initial /'pæt-/. In AmE all three words have initial /'peɪt-/.

**patrol.** The inflected forms are *patrolled, patrolling.* See -LL-, -L-.

**patron. 1** *Patron* is pronounced /'peɪtrən/ and *patroness* /,peɪtrə'nes/, but the derivatives *patronage* and *patronize* both have initial /'pæt-/. In AmE initial /'peɪt-/ is usual for all four words.

**2** See CLIENT.

**patroness** was in regular use from the 15c. to the 19c., principally in the senses 'a woman who promotes social functions, as balls, bazaars, etc.' and 'a female patron saint', but is now seldom used. See -ESS.

**patten.** See PATEN.

**pavement.** One must keep in mind when travelling that the word means a paved way for pedestrians in Britain (corresponding to AmE *sidewalk*) and to some extent in the Atlantic States of America, and a paved road elsewhere in America.

**pavior, paviour,** one who paves. Both spellings are permissible in BrE (my own preference is for *paviour* on the model of *saviour*), but only *pavior* in AmE.

**pawky.** First recorded in the 17c. in Scotland and northern dialects in the sense 'artful, sly', from the 19c. onward the word has been applied to varieties of humour judged to be typical of the Scots: 'having a matter-of-fact, humorously critical outlook on life, characterized by a sly, quiet wit' (*Concise Scots Dict.*, 1985).

**pay. 1** In its ordinary senses the pa.t. and pa.pple are, of course, *paid.* The nautical verb *pay*, meaning 'to smear or cover with pitch, tar, etc., as a defence against wet', is of quite different origin and has *payed* as its pa.t. and pa.pple.

**2** Note that one *pays attention* (to something) but *takes note* or *notice* (of something). The verbs are not interchangeable in these expressions.

**pay off.** See PHRASAL VERBS.

**PC, pc.** Now (1996) just as likely to mean *political correctness* as *police constable, Privy Counsellor,* or *personal computer.*

**peaceable, peaceful.** Both words joined the language in the 14c. and there is a substantial overlap of meaning. In general, *peaceable* means (a) 'disposed to peace, not quarrelsome' (*the inhabitants are simple, peaceable, and inoffensive*), and (b) (*less commonly*) 'free from violence or disorder, characterized by peace' (*to do one's duty is not easy in the most peaceable times; peaceable, non-violent behaviour*). *Peaceful* is much the more common of the two words, and means (a) 'characterized by peace, tranquil' (*a peaceful country scene*), and (b) not violating or infringing peace (*peaceful coexistence*). Examples: (peaceable) *From the moment of the child's birth the unity with its mother can never be completely peaceable*—P. Roazen, 1985; (peaceably) *The baby lay peaceably in his carrycot, and was pleased to be joggling gently along*—J. Trollope, 1989; *here is proof that people with very different traditions can live peaceably together*—P. Ustinov, 1990; (peaceful) *The nights were peaceful and black*—L. Erdrich, 1988; (peacefully) *The water shone peacefully*—E. Jolley, 1981.

**peccadillo.** The recommended pl. is *peccadillos* rather than *peccadilloes.* See -O(E)S 7.

**pedagogue.** First recorded in the 14c. with the meaning 'teacher, instructor', by the end of the 19c. it had often come to be applied in a contemptuous way to teachers judged to be pedantic, dogmatic, or severe. It is sometimes spelt *pedagog* in AmE, and is in common use in the US usu. without any implication of fussiness or pedantry. Examples: *The master, a dryish Scotsman whose reputation as a pedagogue derived from a book that he had written*—Sci. Amer., 1955; *Georges de Beauvoir also thought teachers were low-minded pedagogues*—L. Appignanesi, 1988.

**pedagogy,** the science of teaching. The pronunciation recommended is /ˈpedə-ˌgɒdʒɪ/, with a 'soft' second g. Similarly in *pedagogical* /-ˈgɒdʒɪkəl/. But see GREEK G.

**pedal. 1** The inflected forms of the verb are *pedalled, pedalling* (but usu. *pedaled, pedaling* in AmE). See -LL-, -L-.

**2** The spelling *back-peddle* (noun and verb) shown in the following examples from 1991 issues of the *Independent* is erroneous: (headline) *A soft back-peddle by the Russian Communist Party; Croatia ... accepted the EC declaration with alacrity, while insisting that it would not back-peddle on the issue of full independence from Yugoslavia.* See PEDLAR.

**pedantic humour.** Children are surrounded by adults who, with varying degrees of unawareness and tactlessness, use words which lie outside the range of the very young—words such as *alleged, grandiose, immobile, incestuous, palpably, purloin, sartorial, vehement,* and *voluminous.* Wise children store them away like precious stones and bring them out for inspection or polishing and then for experimental display at some later date. In adulthood standard speakers find that for the most part they have settled for a boundaried choice of words and constructions, a language for life, modified only by important social, domestic, and political events as time goes on. The limits are not precisely drawn at any given time and they vary substantially from person to person. But each person's safe haven is instinctively known, and such havens require a protective zone round them.

Outside this zone lie not only the errors of the vulgar (which give rise to unlimited mirth), the soft rusticisms of country folk (always respected), and the barbarous jargon of specialists (unfailingly mocked), but also verbal tracts and territories that are unfamiliar and often therefore mirthful. Overuse of archaisms (*nay, peradventure, perchance, surcease,* etc.) can bring a secret smile to the lips of an audience or to readers. Moribund uses of the gerund, the subjunctive mood, and of the infinitive can have the same effect. There is a tendency to smile at those who lard their speech or writing with foreign expressions, *coûte que coûte, pari passu, penchant,* and so on. Words of high formality are also capable of eliciting a wry smile: e.g. (mostly drawn from the works of Anita Brookner, Peter de Vries, and William Golding) *crepitation, derogate, edulcoration, infraction, perdurable, pulchritude, putative, refulgent, sesquipedalian,* and *simulacrum.* Try to define them without consulting a dictionary and you will see what I mean. There is room for them in the language, of course. My purpose in listing them is not to condemn them but to show that all our linguistic territories are different. The choices we make are a source of wonderment and often, it has to be said, of rich humour. See POLYSYLLABIC HUMOUR.

**pedantry.** Fowler's classic statement of 1926 is still relevant: 'Pedantry may be defined, for the purpose of this book, as the saying of things in language so learned or so demonstratively accurate as to imply a slur upon the generality, who are not capable or not desirous of such displays. The term, then, is obviously a relative one; my pedantry is your scholarship, his reasonable accuracy, her irreducible minimum of education, and someone else's ignorance. It is therefore not very profitable to dogmatize here on the subject; an essay would establish not what pedantry is, but only the place in the scale occupied by the author; and that, so far as it is worth inquiring into, can be better ascertained from the treatment of details, to some of which accordingly, with a slight classification, reference is now made. The entries under each heading are the names of articles; and by referring to a few of these the reader who has views of his own will be able to place the book in the pedantry scale and judge what may be expected of it. There are certainly many accuracies that are not pedantries, as well as some that are; there are certainly some pedantries that are not accuracies, as well as many that are; and no book but that attempts, as this one does, to give hundreds of decisions on the matter will find many readers who will accept them all.'

Some of the main entries in which elements of pedantry are discussed in this book are as follows: Choice of words: FORMAL WORDS; LITERARY WORDS; PEDANTIC HUMOUR; POLYSYLLABIC HUMOUR; SAXONISM; WARDOUR STREET. Grammar: AGREEMENT; CASES; ELLIPSIS; PREPOSITION (at end); SPLIT INFINITIVE; UNATTACHED PARTICIPLES; VERBLESS SENTENCES. Pronunciation: ESTUARY ENGLISH; FALSE QUANTITY; GREEK G; PRONUNCIATION. Punctuation: AMPERSAND; APOSTROPHE[2];

BRACKETS; COLON; COMMA; DASH; EXCLAMATION MARK; FULL STOP; HYPHENS; ITALICS; QUESTION MARK; QUOTATION MARKS; SEMICOLON. Sensitivity: -ESS; ETHNIC TERMS; FEMININE DESIGNATIONS; LINGUISTIC ENGINEERING; POLITICAL CORRECTNESS; U AND NON-U. Spelling: DIDACTICISM; I BEFORE E; MUTE E; SPELLING. Style: ELEGANT VARIATION; U AND NON-U.

**peddler.** See PEDLAR.

**pedigree.** The corresponding adj. is pedigreed, not pedigree'd.

**pedlar,** the traditional term for an itinerant seller of small items, is losing ground rapidly before the AmE spelling peddler, esp. in the context of the peddling of drugs (in which it is virtually the only spelling). Examples of the contrasting spellings: a cafeteria called informally the Saigon, a place for poets, drug-pedlars and speculators, not professors' daughters—J. le Carré, 1989; We've been missing the independent Senegalese venders in mid-town lately: except for watch peddlers . . . the venders . . . seem to have vanished—New Yorker, 1989.

**ped(o)-.** See PAED(O)-.

**pee.** Since the introduction of decimal currency in Britain in 1971 the spelling pee has come into widespread use to represent the pronunciation of the initial letter of 'penny'. Example: May I trouble you for forty-two pee?—R. Rendell, 1974. The pronunciation as /piː/ is widely condemned by people who prefer /'penɪ/ as singular and /pens/ as plural.

**peewit** /'piːwɪt/. Thus spelt, rather than pewit (q.v).

**peignoir.** Pronounce /'peɪnwɑː/.

**pejorative.** Stressed on the (long) first syllable, thus /'piːdʒɒrɒtɪv/, in the OED (1905) and Daniel Jones (1917), the word is now normally stressed on the second syllable, thus /pɪ'dʒɒrətɪv/.

**pekoe** (black tea). Pronounce /'piːkəʊ/. (The OED, 1905, recommended /'pekəʊ/.)

**pellucid.** See TRANSPARENT.

**pelta,** a small light shield used by the ancient Greeks and Romans. Pl. peltae /-iː/.

**pelvis.** The 'natural' pl. now is *pelvises*, but in medical work *pelves* /-viːz/ is dominant.

**penal, penalize.** Both words are pronounced with a long first syllable, /ˈpiːn-/, but *penalty* with a short one, /ˈpen-/.

**penates** (Roman household gods). Pronounce /prˈnɑːtiːz/.

**penchant.** Still pronounced in a Gallic manner, /ˈpɑ̃ʃɑ̃/, in English.

**pencil** (verb). The inflected forms are *pencilled, pencilling* in BrE, but often *-iled, -iling* in AmE. See -LL-, -L-.

**pendant, pendent, pennant, pennon.** These are normally kept apart in the following manner: *pendant* is a noun meaning a hanging jewel or the shank and ring of a pocket-watch; also, in nautical language, a short length of rope fixed on the main- and foremasts of a square-rigged ship, used for attaching tackles; also, any length of rope used as a means of purchase to a distant object; *pendent* is an adj. meaning (over)hanging; *pennant* is a nautical word (sometimes written *pendant* but always pronounced 'pennant') for a narrow tapering flag used for signalling or for some other specified purpose; *pennon* is principally a long narrow flag, used esp. as the military ensign of lancer regiments. In the 15c. and 16c. *pennon* was also used for a long, coloured streamer flown from the mastheads or yardarms of warships on occasions of state or national importance, and they are said to have been 'on occasions as much as 60–80 feet in length' (*Oxford Companion to Ships and the Sea*, 1976).

**pendente lite,** 'during the progress of a lawsuit', is written in italic, and pronounced /penˌdenti ˈlɑːti/.

**pending** has been used as a preposition or quasi-preposition since the 17c. to mean 'during, throughout the continuance of' (*pending these negotiations*); and since the 19c. also to mean 'while awaiting, until' (*a final decision cannot be taken pending his trial*; *pending her return*). These prepositional uses are to be distinguished from *pending* used as a predicative adj., when it means 'awaiting decision or settlement' (*The printing of*

the first edition is still pending, because final production details have not been worked out), or 'soon to come into existence' (*patent pending*). A *pending tray* is, of course, a tray for documents, letters, etc., awaiting attention.

**pendulum.** Pl. *pendulums* (from the 17c. to the 19c. occas. *pendula*). See -UM 1.

**penetralia** (innermost shrines or recesses) is a pl. noun. It is in origin the neuter pl. of L *penetrāle*, from *penetrālis* 'interior, innermost'.

**peninsula** is the noun (*the Spanish Peninsula*) and *peninsular* the corresponding adj. (*the Peninsular War*). They are best kept strictly apart. The type *the Peninsula War* could be justified by regarding *Peninsula* as an attributive use of the noun; but the type *the Spanish Peninsular* is clearly wrong.

**penman** should be used only with reference to handwriting, not to the writing of books or articles. The second sense, 'an author, a writer', flourished from the 16c. to the 19c., but is now an affectation.

**pen-name.** See NOM DE GUERRE.

**pennant.** See PENDANT.

**pennon.** See PENDANT.

**penny.** The pl. for the separate coins is *pennies* (*he had four pennies in his pocket*), but for a sum of money is *pence* (*an increase of 50 pence*). See PEE. In N. Amer. a one-cent coin is often called a *penny*, pl. *pennies*.

**pension,** used in the sense 'a French boarding-house' has not been Anglicized (though it has been used in English since the 17c.) and is pronounced in a Gallic manner, /pɑ̃sjɔ̃/.

**pentameter,** a verse of five feet. One of the commonest metres in traditional English poetry is the *iambic pentameter*, i.e. a line consisting of five feet each containing an unstressed syllable followed by a stressed one: Ĕnfórcĕd tŏ séek sŏme cóvĕrt nígh ăt hánd. In classical Latin and Greek, *pentameter* is 'a form of dactylic verse composed of two halves each of two feet and a long syllable, used in elegiac verse' (*COD*, 1995).

**penult.** Frequently used in phonetics, = the last but one (syllable). Thus *referendum* /refə'rendəm/ has its main stress on the penult. The *pen* part of the word answers to L *paene* 'almost'.

**penultimate.** William Safire (*NY Times Bk Rev.*, 7 June 1981) reported instances in American newspapers of *penultimate* erroneously used to mean 'ultimate, final', as if the element *pen-* simply added emphasis to the adj. *ultimate*. I have another from the *Chicago Sun–Times*, 23 Nov. 1988: *'These are the penultimate in quality scarves,'* she said [to a customer]. *'Well, then, show me the better line,'* [the customer cruelly said]. *'This is the better line',* she responded. One can but hope that the word can maintain its true meaning in the decades ahead. See PENULT.

**people, persons. 1** Though questioned many times since the mid-19c., the types '*people* preceded by a numeral' and '*people* preceded by unspecific adjs. (*many, several, a few*) or the pl. pronouns *these* and *those*' have been part of the standard language for many centuries: *But right anon a thousand peple in thraste*—Chaucer, *c*1386; *These people saw the Chaine about his necke*—Shakespeare, 1590; *And many giddie people flock to him*—Shakespeare, 1593. Jespersen (1909–49, ii) cites *two people dying* from a 1722 work by Defoe, and *three thousand people* from Disraeli's *Lothair* (1870). Further examples: *Twenty million people suffer from rheumatism each year*—V. Bramwell, 1988; *Four out of five people thought that fresh fruit and vegetables should be labelled*—*Which?*, 1989; *a great many people feel that a hug can make their day*—*Chicago Tribune*, 1991; *People have been debating abortion for decades, and two people cannot resolve this issue*—*Daily Northwestern* (Illinois), 1991. Clearly these types are legitimate and have been for a long time, though there happens to be no specific entry for them in the *OED*.

**2** Competing with them, however, has been the specifically plural form *persons*. The *OED* cites *Fyftene persons* from a 14c. romance called *Richard Coer de Lion* and also a series of later examples, e.g. *more than ouer ninety and nine iust persons*—Luke (AV) 15: 7, 1611. In Shakespeare we find *Time trauels in diuers paces, with diuers persons* (1600); *Eight Wilde-Boares rosted whole at a breakfast and but twelue persons there* (1606).

**3** To judge from the following comment in the *Harper Dictionary of Contemporary Usage* (2nd edn, 1985) the plural form *persons* still has wide currency in AmE: 'The basic difference between *persons* and *people* is that *persons* is usually used when speaking of a number of people who can be counted and *people* is used when speaking of a large or uncounted number of individuals.' A fresh survey of the evidence, however, revealed that *persons*, whether preceded by a numeral or not, is tending in both BrE and AmE to yield to *people*, and to retreat into somewhat restricted, mostly (semi-)legalistic use: e.g. in notices in lifts, banks, police reports (e.g. *killed by a person or persons unknown*).

An American correspondent collected an array of evidence from American newspapers, brochures, etc., of the early 1990s. A selection of examples: (people) *Deluxe dinners for about 16 people cost about $75; Only $7.95 per person. Minimum 25 people; A tornado injured more than 20 people in downstate Illinois;* (in a book review) *the loss of these two good people;* (hotel advertisement) *Two beds, up to four people and a free hot buffet breakfast for all registered guests; About 3,000 people with reserve component status remain in the Gulf voluntarily, the Pentagon said; The airline bombing ... killed 270 people; Fewer than 40 people attended the banquet; It's only people who don't know, or don't care, who will remain at risk; a lot of people out there can do variety shows.* By comparison, examples of *persons* were much harder to find: (extract from a Federal Savings Bank notice) *It is also the Board of Directors' view that these provisions should not discourage persons from proposing a merger;* (from a Movie Ratings Guide) *R (Restricted). Persons under 17 not admitted unless accompanied by parent or adult guardian;* (brochure of a religious sect) *Open to all persons: prior membership in a twelve-step group is not required;* (advertisement) *the Chinese banquet dinner is only suited for a group of at least eight persons.* It looks as if BrE and AmE usage may not be very different in the choice of *people* and *persons*, but further sampling of the evidence is needed.

**per.** It is a sound general rule not to use this Latin word when an English

equivalent exists and is idiomatic: it is better, for example, to say that *the salary is £25,000 a year* rather than *£25,000 per year*. Office clerks can be forgiven if they say that a parcel is to travel *per rail* or *per* + name of delivery firm: they are simply following a long-established commercial convention. But in lay contexts *by* is better. In correlative statements of the type '—— head —— week', it is probably better (on grounds of euphony) to use *per* rather than the indefinite article: *£200 per head per week*, though there are notable exceptions, esp. *an apple a day keeps the doctor away*. In a number of fixed phrases in which the accompanying word is also Latin, *per annum, per capita, per diem, per se*, etc., and also in *per cent*, clearly *per* must be kept. An indication of the wide variety of uses of *per* can be seen in the following examples: *He fell to the ground at thirty-two feet per second*—D. Abse, 1954; *It contains between 100 and 1,000 atomic particles per cubic centimetre*—New Scientist, 1959; *The recipe will serve 4 to 6 portions and that works out at less than* $\frac{1}{2}$ *oz. of flour and* $\frac{1}{4}$ *oz. of sugar per portion*—Woman's Illustrated, 1960; *I think twelve tablets per calf then six every eight hours after that*—J. Herriot, 1974; *Lord and Lady Kilmaine . . . have arrived in London. Arrivals at Kingstown per Royal Mail steamers include* [etc.]—J. Johnston, 1979; *Fifteen minutes per baby, per breast, at specific intervals*—M. Roberts, 1983; *They can also bat in the happy knowledge that bouncers are restricted to one per over*—Willis and Lee, 1983; *This is my chore, per our agreement*—New Yorker, 1991.
See also *as per* s.v. AS 10.

**peradventure.** The word has been used in English for so long (first recorded as an adverb in the 13c. and as a noun in the 16c.) that it is somewhat surprising to find that it has stepped down a rung into archaism or even further into the world of jocularity. Examples of its use as adv. meaning 'perhaps' and as noun meaning 'uncertainty; conjecture' (esp. preceded by *beyond*) can be found, but for the most part only in somewhat atmospheric or humorous contexts or in some English-speaking countries abroad. Examples: *Where else* [except in Ghana] *in the English-speaking world will you still hear the word 'peradventure', meaning 'perhaps',*

*used in current speech?*—Times, 1982; *Peradventure they were the only two windows in the house*—B. Breytenbach, 1984 (SAfr.).
See WARDOUR STREET.

**per capita.** Properly meaning 'by heads' (in law), 'applied to succession when divided among a number of individuals in equal shares (opp. to *per stirpes*)' (OED), *per capita* has joined *per caput* as a normal English way of saying 'for each person or head (of population)'. Fowler (1926) regarded this use of *per capita* as 'a modern blunder, encouraged in some recent dictionaries', but attitudes have changed and it is now in standard use beside *per caput*. Examples: (per capita) *For the bulk of humanity per capita consumption remains the same*—New Statesman, 1965; *During the same period, per capita personal consumption rose 15 percent in terms of constant prices*—Dædalus, 1990; (per caput) *It may be argued that the per-caput cigarette consumption is not a good measure of the cigarette consumption in young women*—Lancet, 1976; *The Harbin Transistor Plant was achieving very high per caput sales*—Jrnl RSA, 1978.

**per cent. 1** Always so written in BrE, but usu. as *percent* in AmE. The formula 0.5% is normally rendered in the spoken form as 'half of one per cent', rather than 'o point 5 per cent'; but for other fractions of 1% we inevitably have to use, for example, 'point 7 per cent', or 'point 7 of one per cent' as the spoken equivalent.

**2** The type '30 per cent of + noun' normally governs a singular verb if the noun is a collective or a mass noun and a plural verb if the noun is an ordinary plural. Examples: *Fifteen per cent of the electorate has yet to make up its mind*—Daily Tel., 1987; *35 per cent of an officer trainee's time on the one year course is academic*—Daily Tel., 1987; *in practice in the six-to-fourteen age-group some 90% of children belong to the Pioneers in East Berlin*—Encounter, 1987. But the choice of concord is often governed by which element of the construction is felt by the writer or speaker to be dominant. Many people would have written the first example in the form 'Fifteen per cent of the electorate have yet to make up their mind(s)'.

**percentage.** 'The notion has gone abroad that a percentage is a small part' (Fowler, 1926), and he went on to remind readers that whereas 'a part is always less than the whole, a percentage may be the whole or more than the whole'. The advice is sound. It is desirable to modify the word by adjectives like *small*, *tiny*, *large*, or by the adverb *only*: e.g. *a large percentage of the professional books appearing in America or Britain represent small adjustments . . . of an immediately current idea—London Rev. Bks*, 1987; *out of reach of all but the tiny percentage of the people who are themselves involved in the new economic activity—New Yorker*, 1991. Cf. FRACTION.

**perchance** is the kind of word that immediately summons up a recollection of its most celebrated occurrence (*To die to sleepe, To sleepe, perchance to dreame*: I [ = Aye] *there's the rub*—Shakespeare, 1602) and, to the layman, of not much else. The word in fact has a rich literary history from the 14c. onward, and it began to fall into an archaistic or purely poetical mode only at some point in the 19c. It was often more or less interchangeable with *peradventure*, which has suffered a similar fate. See WARDOUR STREET.

**père.** Added to a surname to distinguish a father from a son, esp. when both are well known, e.g. *Amis* père ( = Sir Kingsley Amis) and *Amis* fils ( = Martin Amis).

**peremptory.** *COD* 1995 gives priority to /pə'remptəri/, with the main stress on the second syllable. *OED* (1907) gave only the form with first-syllable stressing. As a matter of historical interest, Daniel Jones (1917) commented; '/'pei əm-/ [stressed on the first syllable] is more usual when the word is used as a legal term, while /pə'rem-/ or /pr'rem-/ seems more usual in other circumstances.' J. C. Wells (1990) notes that 'both stressings are in use among English lawyers'.

**perennial.** Used of a plant it means 'lasting several years', by contrast with *annual* 'lasting for one growing season'.

**perfect.** 1 'There is a prescriptive tradition forbidding the use of *very* or the comparative with intensifying adjectives like *perfect*, *absolute*, *unique*, and also with the corresponding adverbs (*perfectly, absolutely, uniquely*)' (*CGEL*, 1985, 7.4 n.). The statement is in general true: some grammarians from Lindley Murray (1795) onward have ruled out *more perfect*, *most perfect*, etc., as 'improper', but the more

sophisticated authorities have noted that in certain circumstances rigid application of the rule is mere pedantry. The *OED* notes that the adj. *perfect* is 'often used of a near approach to such a state [of complete excellence], and hence is capable of comparison'. Numerous examples of comparative uses are cited in the *OED* from good 14–19c. sources, including one from Shakespeare (1597) (*Our men more perfect in the vse of armes*) and one from Leigh Hunt (1841): *The perfectest prose-fiction in the language.* Gowers (1965) cites G. M. Trevelyan on Lady Jane Grey: *As learned as any of the Tudor sovereigns, this gentle Grecian had a more perfect character than the best of them.*

The rule must therefore be modified to read something like this: in most circumstances *perfect* is used as an absolute adjective, but there are somewhat rare occasions when the speaker has in mind a near approach to such a state and a comparative adj. or the adverb *very* may be appropriately used with it. It can, of course, legitimately be governed by *almost* and *so*. Examples: *What figure is more perfect than the sphere?*—W. Golding, 1965; *She knew she had an almost perfect manner with subordinates*—R. West, 1977; *Maybe not purity but he seemed so perfect and so unreal, in a way*—C. Achebe, 1987.

2 As a curiosity, the phrase *a perfect stranger* is idiomatic, but it cannot be rephrased so that *perfect* is used predicatively (*as a stranger he is perfect*). In *a perfect stranger* the adj. is called a 'non-inherent' one, and is one of a class of adjectival phrases that cannot be re-expressed predicatively (e.g. *a firm friend*, but not *as a friend he is firm*).

**perfect** (verb). The stress is on the second syllable, thus distinguishing the verb from the adjective where it falls on the first.

**perfectible.** Thus spelt, not *-able*. See -ABLE, -IBLE 7.

**perfect infinitive,** i.e. the type *to have done*, etc. 1 In its simplest form it is used after the verbs *appear* and *seem*. Examples (from my files): *she appeared to have encouraged him*; *GEC appears to have taken a firm grip on the project*; *that seemed to have been an isolated event*; *Gorbachev seems to have* given eastern Europe the go-ahead to re-establish full diplomatic links with China; a fist-sized water bug that seemed to have fallen from the sky. In each case the reported event had occurred before the time of the statement itself. If the appearing and seeming occur at the same time as the event that is being reported a present infinitive is used instead: *Herr Schmidt appeared to be internationally important; they seemed to get to know one another quickly.*

2 For the types *should/would have liked to have met, should/would have liked to meet* see LIKE (verb) 1. See also HAVE 2.

3 The type *Jane was very sorry to miss you when you called* is preferable to *Jane was very sorry to have missed you when you called*, but both types occur frequently.

**perforce.** Used from the 14c. first as a phrase meaning 'by the application of physical force' and from the 16c. as an adverb meaning 'by moral constraint, of necessity', the word has largely fallen out of use and is now archaic. Examples: *We sat and listened, perforce, and the pure notes shivered the motes of dust from the kitchen shelves*—F. Weldon, 1988; *Since Harvey had perforce moved in with his mother he had not shaved*—I. Murdoch, 1993.

**perfume** (noun and verb). Both words are normally stressed on the first syllable. This appears to have been the result of a tidying-up process in the 20c., as the *OED* (1907) gave only second-syllable stressing for the verb.

**perhaps.** In everyday relatively formal speech this common adverb is normally pronounced /pə'hæps/. In relaxed or informal speech, and esp. in rapid speech, it is often pronounced /pᵊ'ræps/ or /præps/ in standard English, though many speakers would rather die than admit it.

**peril.** The inflected verbal forms (also for *imperil*) are *(im)perilled*, *(im)perilling* (AmE usu. with a single -l-), but the corresponding adj. is *perilous*.

**perimeter.** Except for an isolated 17c. (Ben Jonson) example, *perimeter* is not recorded in the *OED* in figurative use. Its extension from the traditional geometrical, military, and (civil) airport uses to denote the extremes or limits of a subject, topic, etc., seems to have occurred in

the last third of the 20c., thus coinciding with the period during which *parameter* has broken loose from its technical and scientific strait-jackets. The two events would seem to be connected. Modern figurative examples: *The perimeter of her own life was shrinking*—J. Urquhart, 1986 (Canad.); *Fourteenth Street determined the size of their lives, the perimeters of their hopes in ways that they honored yet despised*—M. Golden, 1989 (US).

**period.** For near-synonyms, see TIME. For the punctuation mark, see FULL STOP 1. For its use in abbreviations and contractions, see FULL STOP 2.

**periodic, periodical.** The first is only an adj. and is largely restricted to technical and scientific contexts, esp. in the expressions *periodic acid, decimal, function,* and *table* (for definitions of which see *COD* 1995). Also, in grammar, a *periodic sentence* is one consisting of a main clause, followed by 'several clauses, grammatically connected, and rhetorically constructed' (*OED*). I may as well cite a standard example given by Ogilvie and Albert in their classic *Practical Course in Secondary English* (1913): 'The various methods of propitiation and atonement which fear and folly have dictated, or artifice and interest tolerated in the different parts of the world, however they may sometimes reproach or degrade humanity, at least show the general consent of all ages and nations in the opinions of the placability of the Divine nature.' Note that the main statement is completed only at the end of the sentence.

*Periodical* is the more common of the two words in non-technical language (though in many contexts they are freely interchangeable in the sense 'appearing or occurring at regular intervals'), and is the only word of the two to be used as a noun ( = a newspaper, magazine, etc., issued at regular intervals).

**peripeteia.** An Aristotelian rhetorical term (from Gk περιπέτεια 'a falling round, a sudden change') for a sudden change of fortune or reverse of circumstances (in a tragedy, etc., or in everyday life), e.g. in *The Merchant of Venice*, the downfall of Shylock, when Gratiano repeats to him his own words, *O learned judge*, and Shylock's case collapses.

**periphrasis,** a rhetorical term for a roundabout way of saying something, a circumlocution, e.g. to refer to Shakespeare as *the Bard of Avon*. Circumlocutory phrases (e.g. *be of the opinion that* = think, believe; *in close proximity* = near(ly)) lie strewn about in the language and must be used with caution to avoid accusations of prolixity or windbagging. As Fowler remarked in 1926, '"The year's penultimate month" is not in truth a good way of saying "November".'

**periwig** is in origin a 16c. alteration of *peruke* (from Fr. *perruque*): the intermediate stages appear to have been *perwyke, perewyke, perewig, perrywig, periwig*, whence the abbreviation *wig* (*OED*). The corresponding verbal inflected forms are *periwigged* (used esp. as adj.), *periwigging*.

**perk.** See PERQUISITE.

**permanence, permanency.** 'One of the pairs (see -CE, -CY) in which the distinction is neither broad and generally recognized, nor yet quite non-existent or negligible' (Fowler, 1926). In practice *permanence* is much the more common of the two words. Both share the meaning 'the quality of being permanent'. When *permanency* is used it often means 'the state of being permanent' or 'something permanent'. But there is no clear-cut distinction between the two words. Examples: (permanence) *His determination, which he now estimated to have the size and permanence of an ice cube, began to grow even smaller*—B. Byars, 1981; *Its [sc. a memorial's] construction was touched with grace, its stark and solid geometry with permanence*—Newsday (US), 1991; *There is a sense of invulnerable permanence at General Motors*—Fortune (US), 1992; (permanency) *The new world view fundamentally altered long established notions about the permanency of the continents and the ocean basins*—Sci. Amer., 1974; *Permanency is also associated with marriage, and a person in a temporary position cannot marry and remain in the household without moving into a permanent position*—Man, 1983; *A stranger is not a permanency. One can easily shed a stranger*—G. Greene, 1988.

**permissible.** 1 Thus spelt, not *-able*. See -ABLE, -IBLE 7.

2 It means 'allowable', as distinct from *permissive* which means 'tolerant

(e.g. of grammatical or other linguistic deviations from the norm); liberal, esp. in sexual and drugs-related matters' (*the permissive society*).

**permit.** 1 The inflected forms of the verb are *permitted, permitting*.

2 When *permit* is used as a noun the stress falls on the first syllable, and as a verb on the second. See NOUN AND VERB ACCENT 1.

3 The verb is normally transitive (*permit the traffic to flow again*) but can also be used absolutely (*traffic permitting*), and also intransitively with *of*, when it means 'to admit, to allow for' (*the rule permits of certain exceptions*). Cf. ADMIT 1; ALLOW 1.

**pernickety** ( = fussy, fastidious) is a 19c. word of Scottish origin (its etymology beyond that cannot safely be postulated) which has spread to all English-speaking areas. In AmE, since the beginning of the 20c., it has alternated in standard colloquial use with the form with an intercalated *s*, namely *persnickety* (e.g. *That archetype of persnickety Yankee toolmakers*—R. Stein, 1967), but outside AmE *pernickety* is the only form used.

**perorate** (verb). A natural formation in the 17c. from the ppl stem of L *perōrāre* and not merely a back-formation from *peroration* (the noun was first recorded in the 15c.), it is nevertheless often bracketed with the likes of *enthuse* and *liaise* and is therefore avoided by many speakers and writers.

**perplexedly.** Four syllables: see -EDLY.

***per pro, per proc.*** See P.P.

**perquisite,** an incidental benefit attached to one's employment. This is the form used in business correspondence, notice-board announcements, etc., but the informal equivalent *perk* indicates how the word is most often presented in spoken English. It is not to be confused with *prerequisite* (noun and adj.), (something) required as a precondition.

**persiflage.** 1 After more than two centuries of use in English, the word is pronounced /ˈpɜːsɪflɑːʒ/, i.e. is still not fully Anglicized in pronunciation but it always appears in roman type.

2 It means 'light banter or raillery'. Hannah More defined it in 1799 as *The cold compound of irony, irreligion, selfishness, and sneer, which made up what the French ... so well express by the term* persiflage. Perhaps friendly teasing and levity are the main ingredients of the word as it is now used. In French the noun was formed from *persifler* 'to tease', from the intensive prefix *per-* + *siffler* 'to whistle' + *-age*.

**persistence, persistency.** Both words entered the language in the 16c. and are still interchangeable in the senses 'the action or fact of persisting' and 'the quality of persisting or being persistent'. Whatever the reason (? a preference for a three-syllabled word), *persistence* is by far the more common of the two words. Examples: (persistence) *The opening phrase continued to echo in my mind with an obstinate throbbing persistence*—L. Durrell, 1957; *By sheer persistence he'd achieved what at first seemed inaccessible*—E. North, 1987; (persistency) *agelessly silent, with a reptile's awful persistency*—D. H. Lawrence, 1921; (used of insurance policies) *Motorists Life said it has revised its commission schedule and added a persistency bonus program*—*Investment Dealers' Digest* (US), 1991. See -CE, -CY.

**persnickety.** See PERNICKETY.

**person** (noun). 1 For the plural *persons*, see PEOPLE.

2 *So now one person can start their own magazine*—*The Face*, 1991: an extreme (and easily avoidable) example of the increasingly popular types: anyone *who wants to improve* their *writing may attend the course* and *Everyone was absorbed in their own business*. See AGREEMENT 6; ANY 1(b).

**-person** (as a sex-neutral element). 1 At first to the puzzlement and then to the amusement, and now often to the despair, of many people (both men and women), feminists began to use, and then to insist that others use, *person* or, where appropriate as a second element, *-person* (e.g. *draughtsperson, salesperson, spokesperson*) instead of *man* or *-man*. The movement for the removal of *man* in its traditional sense of '(without regard to sex)' began in the 1970s, and had reached a point of maximum

intensity by the 1990s. See CHAIR, CHAIRPERSON.

As far as it is possible to judge, the phrases *the man in the street, one man one vote, man overboard*, and *every man for himself* remain for the most part intact. They are probably too deeply embedded in the language to be ousted, at any rate for the present. But depressing reformulations of old adages are beginning to appear, esp. in AmE: e.g. *The poor workperson blames the tools*—a comment by one Asst Prof. John Kupetz in *The Daily Northwestern* (Illinois), 1991, instead of the traditional proverb *A poor workman blames his tools*.

2 Associated with this development and reflecting the exasperation of people opposed to such radical manipulation of the language came facetious suggestions such as *gingerbread persons, personhandle* (for *manhandle*), *personhole* (for *manhole*), *personpower* (for *manpower*), *Personchester* (for *Manchester*), and *chairperdaughter* ( = a female *chairperson*), but none of these has taken root except as anecdotal items at the dinner table.

3 It is noteworthy that there is a marked tendency for the *-person* combinations to be applied mainly to women, leading to such pairs as *spokesman* (male)/*spokesperson* (female). Many of the occupational terms (*draughtsperson, salesperson*, etc.) occur in job advertisements as an indication that applications are acceptable from persons of either sex. Such disturbances to the language as have occurred since the 1970s can now be clearly seen as one aspect of the movement called POLITICAL CORRECTNESS. The arguments about them will continue into the 21c.

**persona.** The primary meaning of L *persona* was 'a mask, esp. as worn by actors'. It also meant 'a character in a play, a dramatic role', and '(with no idea of deception) the part played by a person in life, a role, character'. Ezra Pound, in a work of 1909 called *Personae*, introduced the term to literary criticism, where it has settled down to mean 'the assumed identity or fictional "I" ... assumed by a writer in a literary work: thus the speaker in a literary poem, or the narrator in a fictional narrative' (C. Baldick, 1990). In a separate 20c. development,

589

*persona* became employed in Jungian psychology to mean 'the set of attitudes adopted by an individual to fit himself or herself for the social role seen as his or hers; the personality an individual presents to the world' (*OED*, slightly adapted). This sense is first recorded in English in 1917. In general writing the danger exists of a merging of the two concepts. In scholarly literary work, the two strands of meaning are normally kept firmly apart. Examples: (in literary criticism) *To this extent, Lewis Eliot is, as it were, a convenient and comfortable persona for his author*—TLS, 1958; *Whether or not the actual Kipling persona is present, the narration is often forceful as well as vivid*—M. Pafford, 1989; *The outstanding example is Burns, whose ability to assume different poetic personae ... is one of his most individual characteristics*—TLS, 1989; (in Jungian sense) *But to satisfy our needs we have to become social persons, and every social person is a bundle of rôles or personæ*—Trans. Philol. Soc., 1935; *The stiff upper lip, the persona of the English gentleman, is a particularly appropriate mask for the schizoid person to adopt* Λ. Storr, 1968; *He can be a pompous, contentious man, yet his private persona sometimes contrasts sharply with his more abrasive public image*—Observer, 1972; *When he speaks on the subject, you can hear all three of his personas, as he answers with equal aplomb as fan, participant and TV exec*—Tennis, 1987; *Despite his assertion that he is merely a machine, Warhol constructs a carefully-crafted persona that he markets relentlessly*—Modern Painters, 1989.
The plural is usually *personae* /-naɪ/.

**personage, personality.** Since the 15c. a *personage* has been principally 'a person of high rank, distinction, or importance' (though sometimes applied ironically or laughingly to a person of self-importance). In the 20c. such a person has often come to be known more familiarly as a *VIP* ( = very important person). Since the late 19c. *personality* has come to be used for 'a person who stands out from others either by virtue of unusual character or because his or her position makes him or her a focus for some form of public interest (*OED*); now esp. a celebrity (*a TV personality)*'. Examples: (personage) *No longer was I a nondescript of the slums; now I was a personage of the theatre*—C. Chaplin,

1964; *Like some admiral of former days, he seemed a personage of importance*—Brian Moore, 1972; *Royal personages officially came of age at eighteen*—S. Weintraub, 1987; (personality) *He is a local councillor in a small town, and one of its prominent personalities*—Listener, 1962; *The advertisements will no doubt feature television comedians, showbusiness personalities and disc jockeys*—Listener, 1983; *'Tell us about some of the overseas personalities you have met,' said the interviewer*—J. M. Coetzee, 1990. Since the 1950s, *personality cult* has often been used to mean 'devotion to a leader that is deliberately fostered by the emphasis placed on certain aspects of his or her personality' (*OED*). Example: *The emphatic condemnation of the 'personality cult' at the 20th Congress of the CPSU by Mr. Khrushchev*—Encounter, 1959.

**persona grata,** a person, esp. a diplomat, acceptable to certain others. Print in italic and pronounce /pɜːˈsəʊnə ˈɡrɑːtə/. Its antonym is *persona non grata* /nɒn or nəʊn/.

**personally.** The unassailably legitimate uses of *personally* are (*a*) to signify that something was done by or to someone in person and not through an agent or deputy (*The writ was served on the defendant personally at his residence; The appointment was made by the Secretary of State personally*); and (*b*) to exclude considerations other than personal (*I welcome the decision though I was not personally involved*). Its use with personal pronouns (esp. *I*) as the equivalent of '(speaking) for myself, etc.' is more debatable and is best reserved for informal contexts and best avoided in formal correspondence, etc. Thus *Personally I don't approve* has no advantage over *For my part I don't approve* or over the simple *I don't approve*. But do not be surprised to find *personally* used in this seemingly redundant manner in conversational or informal passages in standard sources: Examples: *I am personally very gratified that the Committee has seen fit to make this recommendation*—C. P. Snow, 1934; *Well, you'd better get on, then ... Personally, I'm going to wait until Fiver and Pipkin are fit to tackle it*—R. Adams, 1973; *I personally would like to see fewer groups*—S. Biko, 1978.

**personalty.** Used only in law, one's personal property or estate (as opposed to *realty*). For other senses of the adj. *personal*, the corresponding noun is *personality*.

**personification** arises partly as a natural or rhetorical phenomenon and partly as a result of the loss of grammatical gender at the end of the Anglo-Saxon period. In OE a pronoun used in place of a masculine noun was invariably *he*, in place of a feminine noun *heo* ( = she), and in place of a neuter noun *hit* ( = it). When the system broke up and the old grammatical cases disappeared, the obvious result was the narrowing down of *he* to refer only to a male person or animal, *she* to a female person or animal, and *it* to nearly all remaining nouns. At the point of loss of grammatical gender, however, *he* began to be applied 'illogically' to some things personified as masculine (mountains, rivers, oak-trees, etc., as the *OED* has it), and *she* to some things personified as feminine (ships, boats, carriages, utensils, etc.). For example, the *OED* cites examples of *he* used of the world (14c.), the philosopher's stone (14c.), a fire (15c.), an argument (15c.), the sun (16c.), etc.; and examples of *she* used of a ship (14c.), a door (14c.), a fire (16c.), a cannon (17c.), a kettle (19c.), and so on. At the present time such personification is comparatively rare, but examples can still be found: e.g. *Great Britain is renowned for* her *stiff upper lip approach to adversity*; *I bought that yacht last year:* she *rides the water beautifully*; (in Australia and NZ) she's *right*; she's *jake*; she's *a big country*, etc.

Personification has long been used as a literary device, esp. in poetry: e.g.

> Ay, in the very temple of delight
> Veil'd Melancholy has her sovran shrine.
>                                     (Keats, 1819)

In the ordinary language, personified uses of words are widespread and so familiar that they pass virtually unnoticed. Proverbs are fertile territory for such uses (*Brevity is the* soul *of wit*; *the wish is* father *to the thought*); and such ordinary metaphors as *the* heart *of the matter*, *the* mouth *of the river*, and *to* eat *one's words* are used all the time.

**personnel.** Pronounce /pɜːsəˈnel/. First in French and, from about the 1830s, in English the word came to be used, esp. in the armed services, of the human as

distinct from the *matériel* or material equipment (of an institution, undertaking, etc.) (*OED*). It seems not to have been widely favoured outside the armed forces until the second half of the 20c. Now all self-respecting large firms have *personnel departments*, *personnel managers*, etc., and *Personnel* (by itself) is commonly used to mean 'personnel department'. Its use preceded by a numeral (e.g. *one copy to every twenty-five personnel*) has little to commend it, but occasionally occurs in reasonably standard sources.

**persons.** See PEOPLE.

**perspective.** In the type 'new perspectives — (a subject)', the customary preposition is *on* (not *of*). Thus 'Three Perspectives on Canada's Future' in *Dædalus*, 1988; *Perspectives on Thomas Hobbes*, ed. G. A. J. Rogers and Alan Ryan, 1989.

**perspic-.** The two groups of words beginning with these seven letters are mostly restricted to fairly formal contexts—the unlearned, as Fowler observed in 1926, are well advised to use simpler synonyms—and are mostly distinguishable along the following lines. *Perspicacious* means 'having mental penetration or discernment, discerning', and its corresponding noun is either *perspicaciousness* or *perspicacity*. *Perspicuous*, on the other hand, means 'easily understood, clearly expressed, (of a person) expressing things clearly', and its corresponding noun is either *perspicuity* or *perspicuousness*. The only snag is that since the 17c. *perspicuity* has often been improperly used to mean 'discernment, insight' by distinguished writers including Charles Dickens, and since the 16c. *perspicuous* has also been improperly used to mean 'discerning, perspicacious' by eminent writers including Swinburne. The best course, unless you are thoroughly conversant with the basic meaning of each group, is to avoid them all and use any of several synonyms instead. Examples of correct uses: (*perspicacious* and derivatives) *He went through the photographs. But it didn't take much perspicacity to tell that some ... were missing*—R. Rendell, 1983; *How much Edith longed for my wholehearted commitment to her I was not perspicacious enough to see*—R. Manning, 1987; *Ned ... was now envied for his perspicacity* *in avoiding the problem of giving houseroom to children who might have infested heads—M. Wesley, 1987; I was not overjoyed to recognise myself as the cuckold of the piece, an idiotic Hamlet betrayed ... by his own lack of perspicacity*—Simon Mason, 1990. (*perspicuous* and derivatives) *Its high speed was not a challenge but a courtesy; its structural intricacy not a dazzling pattern but a perspicuous design*—C. James, 1979; *What would any of us not give to be able to acknowledge in such manly terms, perspicuous and candid, not obsequious nor humbugging, the receipt of a modest competence?*—E. Powell, 1992.

**perspire.** See GENTEELISM.

**persuade.** See CONVINCE.

**persuasion.** When Matthew Arnold declared in 1879 that *Men of any religious persuasion might be appointed to teach anatomy or chemistry*, he was using *persuasion* in one of its traditional senses (first recorded in the early 17c. in a work by John Donne). The senses 'belief or conviction' or 'a religious belief, or the group or sect holding it (*of a different persuasion*)' are still central meanings of the word. Since the 1860s, however, the word has also acquired diverse senses that have little to do with 'belief or conviction', namely 'nationality; kind, sort, description'. The *OED* labels such uses *slang* or *burlesque*, and adds several illustrative examples, including these: *She said she thought it was ... a gentlemen of the haircutting persuasion*—F. Anstey, 1885; *No one of the male persuasion was present* (c1907). The verdict (*slang* or *burlesque*) seems just, but there are many borderline cases, and some that steer well clear of slanginess or the burlesque, e.g. *Sampson has even found a psychoanalyst of the sturdily Freudian persuasion*—TLS, 1990.

**pertinence, pertinency.** Both nouns continue to be used without any strong lexical distinction, except that *pertinence* perhaps more usually means 'the fact of being pertinent' and *pertinency* 'the quality of being pertinent'. The first is easily the more commonly used. Examples: (pertinence) *One essay, 'patriotism and Sport', has a particular pertinence today, in a world of fighting football fans*—M. Coren, 1989; *The ever-present question of whether Raven's titles are better than the*

*contents seems particularly pertinent ... The pertinence is to whether his roistering and hedonistic message is surviving in these less than rorty days*—Times, 1991; (pertinency) *If it is an essential part of scientific work to start one's work by acquainting oneself with all pertinent facts, it seems entirely arbitrary to exclude social implications from the realm of 'pertinency'*—Bull. Atomic Scientists, 1955.

**perturbedly.** Four syllables: see -EDLY.

**peruke.** See PERIWIG.

**peruse** is a formal word for 'read (through), study (with care)', or (somewhat archaic) 'examine (a person's face, etc.) carefully'. It has a long literary history with many fine shades of meaning, but is in much more restricted use now. See FORMAL WORDS.

**pervert.** The noun is stressed on the first syllable, and the verb on the second. See NOUN AND VERB ACCENT 1.

**peseta** (chief monetary unit of Spain). Pronounce /pə'seɪtə/.

**peso** (chief monetary unit of several Latin American countries). Pronounce /'peɪsəʊ/. Pl. *pesos*. See -O(E)S 6.

**pestle.** The *OED* (1907) gave the form /'pestəl/, with a fully pronounced -t-, as a legitimate variant, and other authorities of a later date (including the standard AmE dictionaries) have expressed the same view. But the word is always pronounced /'pesəl/ (like *nestle, wrestle*, etc.) in standard BrE.

**petal.** The inflected forms and combinations are *petalled* (adj.) but *petaloid* (adj.) and *petaliferous* (adj.). *Petalless* is normally written without a hyphen, and *petal-like* normally with one. See -LL-, -L-.

**Peter Principle.** A jocular term of profound significance, propounded in 1968 by the Canadian-born US scholar Laurence J. Peter (1919–90), for the principle that members of a hierarchy are promoted until they reach the level at which they are no longer competent.

***petite*** /pə'tiːt/. Though the adj. is in widespread use in English (normally only in its grammatically feminine form), and has been for more than two centuries, it is still felt to be quintessentially French, and more often than not is printed in italic type.

**petitio principii** /pɪˌtɪʃɪəʊ prɪn'kɪpɪaɪ/, a logical fallacy in which a conclusion is taken for granted in the premiss; begging the question (*COD*). Fowler cited as examples the assertion that foxhunting is not cruel, since the fox enjoys the fun; and that one must keep servants, since all respectable people do so. See BEG THE QUESTION.

**petrel** and *petrol* are homophones, both being pronounced /'petrəl/.

**petrol(eum).** See KEROSENE.

**pewit.** See PEEWIT. The form *pewit*, sometimes pronounced /'pju:ɪt/ rhyming with *cruet* (e.g. in Tennyson's 'Will Waterproof', 1842), was common from the 16c. to the 19c., and is still a recognized variant in some house styles.

**phaeton.** The *OED* (1907) gave priority to a three-syllabled pronunciation, i.e. /'feɪ-ɪ-tən/, and so did Fowler in 1926. This pronunciation is still favoured in AmE, but the word is normally pronounced as two syllables in BrE, i.e. /'feɪtən/.

**phalanx.** In its ordinary sense '(in Greek antiquity) a line of battle' its plural is normally *phalanxes*. In anatomy ('any of the bones of the fingers or toes') the pattern is usually *phalanx* sing., *phalanges* pl. /fə'lændʒi:z/. There is considerable variation in AmE in the placing of the main stress, and in the length of the first syllable (/'feɪ-/ is common) in both main senses.

**phallus.** The layman's pl. form is *phalluses* (less commonly *phalli*), but in the technical language of medicine and botany it is more usually *phalli* /-laɪ/.

**phantasmagoria.** Formed in the early 19c. from *phantasm* + a fanciful ending (but cf. Gk ἀγορά 'assembly, place of assembly'), it meant at first 'an exhibition of optical illusions produced by a magic lantern'. It soon acquired its main current sense, 'a shifting series of real or imaginary figures as seen in a dream or fever'. It governs a singular verb.

**phantasm, phantom.** The two were originally 'the same word', but they have drifted apart. Both words were formed in early ME from OF *fantasme* (orig. from Gk φάντασμα). For the complex process which led to the differentiation of the meaning of the two words, one must turn to the *OED*. Suffice it to say here that *phantom* is the more common word of the two, and is now used esp. to mean (*a*) a ghost, spectre; (*b*) a mental illusion; (*c*) as adj., in medical language, esp. of the continuing sensation of the presence of a limb that has been amputated. *Phantasm* is used to mean (*a*) an illusion; (*b*) a supposed vision of an absent (living or dead) person. *Phantasm* has two adjectival derivatives, *phantasmal*, *phantasmic*, whereas *phantom* is used both as noun and adjective.

**phantasy.** See FANTASY.

**Pharaoh.** So spelt. Pronounce /'feərəʊ/.

**Pharisee.** The customary adj. is *Pharisaic*, though *Pharisaical* is also used. The corresponding abstract noun in -*ism* is *Pharisaism* /'færɪseɪɪzm/ (less commonly *Phariseeism*).

**pharmacopoeia.** The final two syllables are pronounced /-'pi:ə/. Always so spelt in BrE, but commonly as -*peia* in AmE.

**pharyngitis.** The medial -*g*- is 'soft', /-dʒ-/.

**pharynx.** The recommended plural is *pharynges* /fæ'rɪndʒi:z/. The corresponding adj. is *pharyngeal* /færɪn'dʒi:əl/ or *pharyngal* /fæ'rɪŋɡəl/. In phonetics it means '(of a speech-sound) articulated in the pharynx', and, in my experience, is normally spelt *pharyngeal* and pronounced /fæ'rɪndʒəl/.

**phenomena.** See PHENOMENON.

**phenomenal.** In the early 19c. the word was first used to mean 'of the kind apprehended by (any of) the senses'. This sense, and the related noun *phenomenalism*, continued to be used in philosophy of the doctrine that human knowledge is confined to the appearances presented to the senses. In the somewhat maddening way in which the lay public like the sound of a new word but often fail to grasp its strict technical meaning, *phenomenal*, by the middle of the 19c., had come to be used as a general epithet of high praise, meaning 'of the nature of a remarkable phenomenon; extraordinary, remarkable, prodigious'. This sense has long since settled in as part of the standard language (*phenomenal growth, success, talent, etc.*).

**phenomenon.** Fowler's view, expressed in 1926, that the senses 'notable occurrence' and 'prodigy' were regrettable developments of the original 17c. specialized meaning of the word ('something perceived by any of the senses, or by the mind') is a by-product of his generation's anti-historicism. The specialized sense was first recorded in the 1630s, and the extended senses in 1771 (thing) and 1838 (person) respectively. What is to be regretted is the persistent failure by many writers (and speakers) since the 16c. to remember that *phenomena* is the plural of *phenomenon*, and is not itself a singular form. The *OED* lists examples of *phenomena* used as a singular from 1576 onward, to which may be added (the examples happen to be American, but the use is widespread): *This natural phenomena* [sc. the appearance of mirages on roads] *occurs more during the summer because* [etc.]*—Chicago Sun-Times,* 1990; *If this same phenomena occurs in people, it means that senior citizens should have greater difficulty working late at night—Daily Northwestern* (Illinois), 1991. In fact almost from the beginning the word has been distinctly unstable. The *OED* lists numerous examples of the plural form *phenomenons* and of the erroneous form *phenomenas* in addition to the erroneous singular in *-a*.

**philately.** First recorded in English in 1865 (one year earlier, as *philatélie*, in Fr.), it is one of the best-known of the numerous terms meaning 'the hobby of collecting (some specified objects)'. Cf. *cartophily* (first recorded in 1936) 'the collecting of cigarette-cards', and *deltiology* (1947) 'the collecting of postcards'. *Philately* is an artificial formation from Gk φιλ(ο)- 'loving' + ἀτελής 'free from tax or charge', the second element being regarded as a passable equivalent of *free* or *franco*, which were formerly stamped on prepaid letters. The plainer words *stamp-collecting* and *collector* are often preferred to the 'official' terms *philately* and *philatelist*.

**-phile** /faɪl/ is now the customary combining form (rather than *-phil* /fɪl/) forming nouns and adjectives denoting fondness for what is specified in the first element (*bibliophile, Francophile, etc.*).

**Philippines.** Thus spelt. The islands are inhabited by *Filipinos*.

**Philistine.** Originally a member of a people opposing the Israelites in ancient Palestine, the word (with lower-case *p*) has come to be applied to 'a person who is hostile or indifferent to culture, or one whose interests or tastes are commonplace or material' (*COD*). The extended sense is first recorded in 1827 in a work by Carlyle.

**philogynist,** a person who likes or admires women. Pronounced /fɪˈlɒdʒɪnɪst/ with a 'soft' *g*.

**philology.** First used in the 17c. in the sense 'love of learning and literature' it gradually became narrowed in sense in BrE to mean 'the study of the structure and development of language or of a particular language or language family'. Comparative philology, i.e. the study of parallel features in a related language family, became established in Britain in the course of the 19c. *Philology* in the linguistic sense never became established in the US. There, and since about the middle of the 20c. also in the UK, the dominant form for the study of the structure and development of language, etc., has been LINGUISTICS.

**philosophic(al).** Both forms of the word are in widespread use in many of their possible applications, except that most people would say *he took a philosophical* [i.e. sensibly calm or unperturbed] *view of* [some setback]' rather than *he took a philosophic view*, etc.' In the names of societies, transactions, proceedings, etc., the longer form is also dominant.

**philtre, -ter,** a love-potion. Spelt with *-tre* in BrE and with *-ter* in AmE.

**phiz.** See ABBREVIATIONS 1.

**phlegm.** The *g* is unsounded in *phlegm* and *phlegmy*, but is fully pronounced in *phlegmatic.*

**phlogiston.** The pronunciation recommended is /fləˈdʒɪstən/ with a 'soft' *g*.

**Phoebe, Phoenician, phoenix.** All three are regularly spelt with -oe- (not -e-) both in BrE and AmE.

**phone.** A well-established abbreviated form of *telephone* (noun and verb), chiefly used in the spoken language.

**phoney** (adj.). Also spelt *phony*. A now very familiar word, esp. since the *phoney war* (relative inaction before full-scale hostilities) of 1940, but not traced in print before 1900, and of uncertain origin.

**phonograph.** A now-disused term in BrE for an early form of gramophone, but still used in AmE for any type of gramophone or record player.

**phony.** See PHONEY.

**photo.** A reduced form of *photograph* (noun and, occas., verb). Pl. *photos*. Now esp. common in such attributive phrases as *photo call, finish, opportunity*.

**phrasal verbs.**

  1 Origin of the term.
  2 History and prevalence of the type.
  3 Frequent in AmE.
  4 Scholarly attitudes.
  5 Recommendation.

**1 Origin of the term.** Dr Johnson provided a classic description of the type in the Preface to his Dictionary (1755) but did not actually use the term *phrasal verb*: 'There is another kind of composition more frequent in our language than perhaps in any other, from which arises to foreigners the greatest difficulty. We modify the signification of many verbs by a particle subjoined; as to *come off*, to escape by a fetch; to *fall on*, to attack; to *fall off*, to apostatize; to *break off*, to stop abruptly; to *bear out*, to justify; to *fall in*, to comply; to *give over*, to cease; to *set off*, to embellish; to *set in*, to begin a continual tenour; to *set out*, to begin a course or journey; to *take off*, to copy; with innumerable expressions of the same kind, of which some appear wildly irregular, being so far distant from the sense of the simple words, that no sagacity will be able to trace the steps by which they arrived at the present use. These I have noted with great care; and though I cannot flatter myself that the collection is complete, I believe I have so far assisted the students of our language,

that this kind of phraseology will be no longer insuperable; and the combinations of verbs and particles, by chance omitted, will be easily explained by comparison with those that may be found.' The term *phrasal verb*, used in the manner that Johnson described, was said by Logan Pearsall Smith (1923) to have been suggested to him by the late Dr Bradley, i.e. the lexicographer Henry Bradley (d. 1923).

**2 History and prevalence of the type.** The earliest example known to me is *to give up* 'to surrender', which is recorded in 1154: [He] *sende efter him & dide him ȝyuen up ðe abbotrice of Burch—OE Chron.*, annal for 1132. The type thrived in the centuries that followed: *And downe fro his neke he it* [sc. a mantle] *lete, It covyrd ouer his kne—Ipomedon*, a 15c. romance; *They came forth out of all the townes ... and closed them in* [ = shut in, confined] 1 Macc. (AV) 7: 46, 1611; *Ye whiles sleep in* [ = sleep late] *on a morning*—a Scottish writer, 1827; *Bronco and I feel you're the logical one to head up a committee*—E. Lipsky, 1959.

**3 Frequent in AmE.** It is clear that the use of phrasal verbs began to increase in a noticeable manner in America from the early 19c. onward. From there, many have made their way to Britain during the 20c., to widespread expressions of regret and alarm. Some quite rightly object to the use of a hyphen in such a sentence as *So we can't* cover-up *any imperfections* (advt in *Sunday Times Mag.*, May 1989). The hyphen is appropriate only in the related nouns (*cover-up, hold-up, show-down, take-over*, etc.). But the main objection, expressed many times in the 20c. and not only in Britain, is to the use of phrasal verbs in senses not perceived as different from that of the simple verb. Gowers (1965), for example, objected to the combinations *meet up with* ( = meet), *visit with* ( = visit), *lose out* ( = lose), and *miss out on* ( = miss). Where is the gain, he asked, in the particle-supported phrases *pay off* 'prove profitable', *rest up* 'rest', *drop off* ( = drop, fall; 'It is expected that by that time the usual afternoon temperatures of about 90° will have started to *drop off* ')?

**4 Scholarly attitudes.** Bas Aarts (*Jrnl Linguistics*, 1989), by a subtle application of modern grammatical techniques, asserted that 'so-called phrasal verbs do not exist', and that all such are 'verb–preposition constructions in English'. But the article did not deal with the hostility to the use of a good many expressions of this type, and in particular those brought into being in recent times. Tom McArthur (*English Today*, April 1989) usefully analysed and classified the type. In particular he stressed that phrasal verbs were 'common, complex, colloquial and informal'. 'Revolutionaries', he said, 'can *bring down* or defeat a government, then *bring in* or introduce new laws, *bring off* or clinch deals with foreign countries, *bringing on* or creating new problems, while journalists *bring up* or raise awkward questions about the revolution and later *bring out* or publish books about it.'

**5** Recommendation. Guard against the use of phrasal verbs that one knows or suspects to be relatively recent formations or that seem too clever by half unless the one in question seems to add an element of intensification or a dimension of meaning not present in the simple verb. Do not place a hyphen in a phrasal verb. But keep in mind that phrasal verbs are a long-established and essential component of the language, useful, difficult for foreigners, but not cancellable or removable by edict or decree.

**phthisis.** The *OED* (1907) gave preference to the pronunciation /'θaɪsɪs/,. i.e. with silent initial *ph*, and so does Wells (1990). But /'fθaɪsɪs/ is recommended, as also by *COD* (1990). The stem vowel is long in English even though it is short in Gk φθίσις; see FALSE QUANTITY.

**phth, -phth-.** The *ph* must always be sounded (as /f/), whether initially, as in *phthallic acid*, *phthisis* (see prec.) or in other parts of the word, as in *diphtheria*, *diphthong*, *naphtha*, and *ophthalmology*. In the latter group the pronunciation with /-p-/ betrays gross insensitivity or carelessness or both.

**phylloxera.** The pronunciation recommended is /fɪ'lɒksərə/. In AmE /fɪlɒk'sɪərə/ seems to be more usual, but both pronunciations occur in both AmE and BrE.

**phylum** /'faɪləm/. The plural of this taxonomic term is *phyla*. See -UM 2.

**physic** is an archaic word for (*a*) a medicine (*a dose of physic*); (*b*) the art of healing, medicine (*Medicine has long been divided into two departments, Physic and Surgery*—J. Thomson, 1813). It was adopted in ME from OF *fisique*, ultimately from Gk φυσική (ἐπιστήμη) 'the knowledge of nature'.

**physician, doctor, surgeon,** etc. In the UK every medical practitioner is required to have a qualification as physician and also as surgeon, i.e. to have at least a Bachelor's degree in both medicine and surgery. Historically a *physician* is a person qualified in *physic*, the traditional term (14c. onwards) for the theory of diseases and their treatment, medical science. The healing function of the *physician* is preserved for ever in the biblical proverb in Luke 4: 23: *Yee will surely say vnto me this prouerbe, Physitian, heale thy selfe*. A *physician* now is (*a*) a person legally qualified to practise medicine and surgery; (*b*) a specialist in medical diagnosis and treatment; (*c*) any medical practitioner.

In general the terms *physician* and *doctor* (both first recorded in the 14c.) stand together and are frequently interchangeable. To the layman, however, the standard term is *doctor* (*I'm going to see my doctor*), abbreviated before the doctor's name as *Dr*. Since the late 19c. a doctor, including one who works in a group practice with other doctors, has been commonly known as a *general practitioner* or *GP* (in AmE, *family physician*, *internist*). A *doctor* (or *physician* or *general practitioner*) in Britain sees patients in his or her *surgery* (the AmE equivalent is *office*), or, by arrangement or in an emergency, at a patient's own home. In AmE the term *doctor* is also used for a qualified dentist or veterinary surgeon; and in all English-speaking countries the term *Doctor* is used of a person who has a doctoral degree in a non-medical subject, e.g. civil law, philosophy, etc. (hence DCL, D.Litt., D.Phil., etc.; such a person is formally addressed as *Dr* ——). A *surgeon* is a person qualified to practise *surgery*, i.e. the branch of medicine concerned with the treatment of injuries or disorders of the body by incision, manipulation, or alteration of organs, etc., with the hands or with instruments. An ear, nose, and throat specialist is a *surgeon*. A surgeon who specializes in the treatment of eyes is an *eye surgeon*. A *plastic surgeon* is a qualified practitioner of *plastic surgery*, i.e. the process of reconstructing or repairing malformed, injured (esp. burnt), or lost parts of the body chiefly by the transfer of living tissue. In Britain (except Scotland) a surgeon is normally addressed as *Mr* + surname, but in Scotland *Dr* is used for both physicians and surgeons.

**physicist** is, of course, a person skilled or qualified in physics, and is to be distinguished from PHYSICIAN.

**physics, physiology.** The two words at first (in the 16c.) shared the same broad meaning, 'natural science in general, natural philosophy'. They have moved apart. *Physics* is now 'the science dealing with the properties and interactions of matter and energy', whereas *physiology* is 'the science of the functions of living organisms and their parts'.

**physiognomy.** The *g* is silent. The primary sense is 'the cast or form of a person's features'. It was also once in widespread use to mean 'the art of judging character and disposition from the features of the face or the form and lineaments of the body generally'.

**physiology.** See PHYSICS. The corresponding adj. is *physiological* rather than *physiologic* in BrE, but the shorter form is also used in AmE.

**pianist.** Pronounced /ˈpɪənɪst/ in BrE, but predominantly /prˈænɪst/ in AmE.

**piano.** The musical instrument is pronounced /prˈænəʊ/, whereas the musical direction ( = performed softly) is /ˈpjɑː-nəʊ/. The pl. of the noun is *pianos*.

**piazza.** 1 Customarily pronounced in English in an Italianate manner, as /prˈætsə/. Formerly, as recorded in the OED (1907), pronounced /prˈæzə/, and so also still in AmE (beside /prˈazə/).

2 The pl. is either *piazze*, as in It., or *piazzas*. In AmE, when used to mean 'the veranda of a house', the pl. is always *piazzas*.

**pibroch,** an air on the bagpipe (*not* the bagpipe itself). Pronounced /ˈpiːbrɒx/,

with the final sound like that in *loch*, or with final /k/.

**picaresque.** (Ult. from Sp. *picaresco* 'roguish', from *picaro* 'a wily trickster'.) The earliest Spanish picaresque novels, that is fiction dealing with the adventures of an (amiable) rogue, date from the 16c. The type is represented in English by *Moll Flanders, Jonathan Wild, Tom Jones*, and other classic works. Curiously the term itself was apparently slow to come into use in English: the first record of its use is by Walter Scott in 1829.

**piccolo,** the smallest flute. Pl. *piccolos*. See -O(E)S 6.

**pickaxe.** For the spelling of the second element, see AXE.

**picket** (verb). The inflected forms are *picketed, picketing*. See -T-, -TT-.

**picket** (noun). A pointed stake; a person stationed at the site of a strike; (in military use) a soldier or group of soldiers watching for the enemy, a party of sentries. Use *picket* (not *picquet* or *piquet*) for all senses.

**picnic.** The inflected forms are *picnicked, -cking, -cker*. See -C-, -CK-.

**picture.** Pronounced /ˈpɪktʃə/, with the neutral vowel /ə/ in the second syllable, not /-tʃʊə/ or /-tjʊə/.

**pictures.** See CINEMA 2.

**pidgin,** a simplified language containing vocabulary and grammatical elements from two or more languages, used for communication between people who do not have a language in common. Cf. CREOLE. A pidgin differs from a creole in that it is essentially a trading language and is not the mother tongue of a given speech community. The word is probably a Chinese corruption of the English word *business*, and has no connection with *pigeon*, though both words are pronounced /ˈpɪdʒɪn/.

**pie.** See TART.

**piebald, skewbald.** A *piebald* animal (esp. a horse) is one having irregular patches of two colours, esp. black and white. A *skewbald* animal has irregular patches of white and another colour (properly not black).

**pièce de résistance.** The accents and the italic type are obligatory. Pl. *pièces de résistance*, pronounced the same as the sing.

**pied-à-terre** /pjeɪdɑːˈteə/. The hyphens and the italic type are essential. Pl. *pieds-à-terre* (pronunciation unchanged).

**pietà.** Print in italic and pronounce /pɪeɪˈtɑː/.

**pigeon, dove.** Used literally of a bird, the words are more or less coextensive in application, most doves being pigeons, and vice versa ('any of several large usu. grey and white birds of the family Columbidae', *COD*, 1995). But *pigeon* is the word in everyday use, and *dove* is mostly found in poetical or symbolical contexts. 'The dove has been, from the institution of Christianity, the type of gentleness and harmlessness, and occupies an important place in Christian symbolism' (*OED*), being equated with the Holy Spirit. Since the early 1960s the word *dove* has also been applied to a person who advocates negotiations as a means of terminating or preventing a military conflict, as opposed to a *hawk*, who advocates a hard-line or warlike policy. Species of dove found in Britain (but not only in Britain), with a prefixed word defining the species, include the *ring-dove* (also called *wood-pigeon*), *rock-dove* (also called *rock-pigeon*), and the *stock-dove*. The *turtle-dove* belongs to a separate genus (*Streptopelia turtur*). Other birds of the family Columbidae, with an appropriate prefix defining their role or appearance, are always called *pigeons*, e.g. *carrier pigeon, homing pigeon, pouter pigeon*. In English-speaking countries abroad, birds of the family Columbidae are much more likely to be called pigeons (with appropriate defining word prefixed) than doves; and the pigeonlike birds there are often members of a different family altogether, e.g. the *Cape pigeon*, a S. Afr. petrel (*Daption capense*). and the *New Zealand* (or *native*) *pigeon* (*Hemiphaga novaeseelandiae*).

**pigeon (English).** Now disused in favour of PIDGIN (*English*).

**pigmy.** See PYGMY.

**pigsty.** Pl. *pigsties*.

**pilau** /pɪˈlaʊ/. The prevalent form in BrE for this 'Middle Eastern or Indian dish of spiced rice or wheat with meat, fish, vegetables, etc.' (*COD*), beside *pilaff* /pɪˈlæf/ and *pilaw* /pɪˈlɔː/. AmE dictionaries give precedence to *pilaf* or *pilaff*. The word is of Turkish origin.

**pilfer** (verb). The inflected forms are *pilfered, pilfering*. See -R-, -RR-.

**pilot** (verb). The inflected forms are *piloted, piloting*. See -T-, -TT-.

**pilule,** a small pill (L *pilula*). This is the recommended spelling, not *pillule*.

**pimento** (small tropical tree), **pimiento** (sweet pepper). Pl. *pimentos, pimientos* respectively. See -O(E)S 6.

**pinch.** The idiomatic expression *at a pinch* 'in an emergency, if necessary', which is customary in BrE, answers to AmE *in a pinch*. Examples: *It could carry two passengers easily, three at a pinch*— M. R. D. Foot, 1966; *And then Danchkovsky looked in. In a pinch he could be called famous*—*New Yorker*, 1989. The polarization of the phrases is incomplete, and in the 19c. either phrase could be used in either country.

**Pindaric,** of or pertaining to the work of the Greek poet Pindar (Pindaros, 518–438 BC), a writer of choral odes. Pindar's odes 'are written in regular stanzas, either in a series of strophes on the same plan or in a series of triads, each consisting of strophe, antistrophe, and epode' (*Oxford Classical Dict.*), using three main classes of metre. The Pindaric odes of English imitators have 'an unfixed number of stanzas arranged in groups of three, in which a strophe and antistrophe sharing the same length and complex metrical pattern are followed by an epode of differing length and pattern' (C. Baldick, 1990). English writers of Pindaric odes include Thomas Gray ('The Progress of Poesy', 'The Bard', both 1747) and Abraham Cowley ('Pindarique Odes', 1656). Others, including Dryden and Pope, wrote poems that resemble the model set by Pindar, expressing 'exuberant heated ideas and passionate feelings in appropriately loose and (relatively) free rhythms' (Alastair Fowler, 1987).

**pinion** (verb). The inflected forms are *pinioned, pinioning*. See -N-, -NN-.

**pinkie** is the Sc. and AmE word for the little finger (sometimes spelt *pinky* in America). The adj. *pinky* 'tinged with pink, pinkish' must always be spelt with *-y*.

**pinna** (a part of the outer ear of mammals). Pl. *pinnae* /-ni:/.

**pinny** (colloquial shortening of *pinafore*). So spelt. Pl. *pinnies*.

**pi, pious.** See ABBREVIATIONS 1.

**pipy,** like, or having, pipes, not *pipey*.

**piquant.** Pronounce /'pi:kənt/.

**piquet.** See PICKET (noun).

***pis aller,*** a course of action followed as a last resort. Print in italics.

**piscina.** Pronounce /pɪ'si:nə/. For both the sense 'fish-pond' and 'stone basin for disposing of water used in washing the chalice, etc.', the pl. is either *piscinae* /-ni:/ or *piscinas*.

**pistachio** /pɪ'sta:ʃɪəʊ/. Pl. *pistachios*. See -O(E)S 4.

**pistil.** The inflected forms are *pistillary*, *pistillate*, *pistilliferous*, and *pistilline* (adjs.).

**pistol** (small firearm). The inflected verbal forms are *pistolled*, *pistolling*; also *-l-* in AmE.

**pistole.** The regular spelling for a former Spanish gold coin.

**piteous, pitiable, pitiful.** All three words share the broad sense 'arousing pity'; *piteous* and *pitiable* share the sense 'deserving pity'; and *pitiable* and *pitiful* share the sense 'evoking mingled pity and contempt'. But that is as far as the synonymy goes. *Piteous*, the least common of the three words, can also mean 'wretched'. The history of the three words, all of which were first recorded in ME, is very tangled, and readers of older literary works must be prepared to find that the historical senses are more diverse than the present-day ones. Examples: (piteous) *'What did I do this time?' Helen looked piteous*—M. Binchy, 1988; (pitiable) *How she had suffered for him, for her poor pitiable ridiculous father*—M. Drabble, 1987; *He wanted to say, 'I'm sorry . . .' as if to a pitiable innocent victim*—I. Murdoch, 1987;

*What was coming into being was a sort of pity. Owen seemed to me pitiable, unshriven*—A. Brookner, 1990; (pitiful) *In the centre of the profitless lawn a pitiful tube squirts water to a height of a couple of feet*—J. D. R. McConnell, 1970; *His blindness now struck her as utterly pitiful*—M. Forster, 1988.

**pituitary** /pɪ'tju:ɪtərɪ/. Another example showing differences of vowel quantity compared with the L original, *pītuītārius*. See FALSE QUANTITY.

**pity** (noun). The type *Pity you came home just then* is an acceptable (informal) shortening of *It is a pity that . . .*

**pixie** (small fairy). Thus spelt, not *pixy*.

**pizzazz.** Of all the variants, *pizazz, bezazz, pazzazz*, etc., this seems at present the most enduring. See -Z-, -ZZ-.

**pizzicato.** Pronounce /pɪtsɪ'ka:təʊ/. Pl. *pizzicatos*. See -O(E)S 6.

**placable** (easily placated). The OED (1907) gave precedence to /'pleɪk-/, but /'plæk-/ is now dominant.

**placard.** Both as noun and as verb pronounced /'plæka:d/.

**placate.** The pronunciation recommended is /plə'keɪt/, though the OED (1907) gave precedence to /'pleɪkeɪt/.

**-place.** See *anyplace* (s.v. ANY 3*f*); EVERYPLACE; NO PLACE; SOMEPLACE (AmE equivalents of *anywhere, everywhere, nowhere*, and *somewhere*).

**placebo.** Pl. *placebos*. See -O(E)S 6.

**plague.** The inflected forms are *plagues, plagued, plaguing*.

**plaice** (fish). So spelt.

**plaid** (/plæd/; variously pronounced in Scotland), a length of fabric worn over the shoulder as part of the ceremonial dress of members of the pipe bands of Sc. regiments. It is to be carefully distinguished from *tartan*, which is 'a woollen cloth with a pattern of stripes of different colours crossing at right angles; such a pattern, esp. one associated with a particular clan' (*Concise Scots Dict.*, 1985). A *plaid* can be made from *tartan* cloth.

**Plain English.** The current drive for plain English grew out of the consumer movement and the demand for fair dealing. It insists on clarity as well as accuracy and is opposed to convoluted, obfuscating language in official, legal, and commercial documents. It wages war on language typified by the use of such words as *aforesaid*, *heretofore*, and *thereto*, and argues that inflated announcements such as *Encashment of a foreign currency may incur a processing fee* may be stated more effectively as *We may charge you to change your foreign money*. Similarly, *He hails from a multi-delinquent family with a high incarceration index* is better expressed by *Other members of his family have been or are in jail*. Throughout the English-speaking world attempts are now being made, sometimes by the official groups themselves and sometimes by their linguistic opponents, to bring about the demise of bureaucratic language (see GOBBLEDEGOOK) and DOUBLE-SPEAK that characterize, discredit, and diminish much of the documentary language of our age. It will be a long struggle, following in the tradition of Sir Ernest Gowers (through his book *The Complete Plain Words*, 1954 and later versions), Rudolph Flesch, and Jeremy Bentham in earlier periods.

**plain sailing.** An early 19c. alteration of the original (late 17c.) phrase *plane sailing*. The *OED* says that *plane sailing* is 'a simple and easy method, approximately correct for short distances', and defines it as 'the art of determining a ship's place on the assumption that she is moving on a plane, or that the surface of the earth is plane instead of spherical'. The phr. *plain sailing* is now chiefly used figuratively for an uncomplicated situation or course of action.

**plait** (noun and verb). Pronounce /plæt/.

**plan** (verb). The more usual construction is with a *to*-infinitive (*The government plan to close them and redeploy their workers*—*Times*, 1986), but a construction with *on* + gerund also occurs with no change of meaning, esp. in AmE: *Do you plan on staying with Muriel forever?*—A. Tyler, 1985.

**planchette.** Pronounce /plɑːnˈʃet/.

**plane sailing.** See PLAIN SAILING.

**planetarium.** Pl. *planetariums* or *planetaria*. See -UM 3.

**plangent.** Pronounce /ˈplændʒənt/.

**plantain.** Both words are pronounced /ˈplæntɪn/, the shrub of the genus *Plantago* and the banana plant, *Musa paradisiaca*.

**plaster.** The inflected verbal forms are *plastered*, *plastering*. See -R-, -RR.

**plastic.** Now always /ˈplæstɪk/, though earlier in the 20c. it passed through a phase in some quarters of being pronounced /ˈplɑːstɪk/.

***plat du jour*** (a dish specially featured on a day's menu). Usually pronounced in a semi-Gallic fashion, /ˌplɑː dy ˈʒuːə/. Normally printed in italic in prose contexts.

**plateau.** Pl. *plateaux*, pronounced with final /z/, rather than *plateaus*. See -X.

**plateful.** Pl. *platefuls*. See -FUL.

**plate glass.** Write as two words unhyphened.

**platen** (roller in a typewriter, etc.). Thus spelt, not -tt-. Pronounce /ˈplætən/.

**platonic love.** In his *Symposium*, the Greek philosopher Plato (c429–347 BC) declared that love of a beautiful person can lead us to the love of wisdom and of the 'form' of beauty itself (*Oxford Classical Dict.*). As thus originally used, there was no specific reference to women. First recorded in English in 1631 (Ben Jonson), *Platonic love* from the beginning was 'applied to love or affection for one of the opposite sex, of a purely spiritual character, and free from sensual desire' (*OED*). The *OED* has a large number of quotations from Jonson onwards to illustrate this sense. It also provides evidence for the meaning 'applied to affection for one of the same sex'. Now normally printed with lower-case initial *p*.

**platypus.** Pl. *platypuses*, not *platypi*. See -US 4.

**playwright.** Thus spelt. A *wright* is a maker or builder; the word is not related to the verb *to write*. For the formation cf. *shipwright*.

**plc, PLC.** A modern abbreviation of *Public Limited Company*. The name was proposed in 1973 and brought into being in the Companies Act of 1980.

**plead.** The pa.t. and pa.pple in standard BrE are *pleaded*, but *pled* and *plead* /pled/ occur in America, Scotland, and some dialects in the UK beside *pleaded*.

In law courts a person charged with an offence can *plead guilty*, plead *not guilty*, or *plead insanity*, but in normal circumstances cannot *plead innocent*.

**please.** 1 The ordinary imperative or optative use of *please* by itself (*Will you come in, please?*) is a reduced form of *may it* (*so*) *please you*. It was first recorded in the 17c., but was not known to Shakespeare, whose shortest form is *please you*.

2 The somewhat antiquated type *Please to return the book soon* (i.e. *please* + a *to*-infinitive) has been in use since the 17c. but is now fast falling out of use. One of its earliest uses is in Milton's *Paradise Lost: Heav'nly stranger, please to taste These bounties which our Nourisher ... hath caus'd The Earth to yield.* A modern example: *Please to go over there and sit next to your friends*—T. J. Binyon, 1988 (a foreigner speaking).

**pleasure.** Normal idiom permits the use of *pleasure* to mean 'source of pleasure or gratification' (*it was a pleasure to talk to them*; *reading was his chief pleasure*); and also to be used in fairly formulaic constructions, such as *I have pleasure in declaring this building open*, and *I have already had the pleasure of meeting him*. The word crosses into more formal territory when it means 'a person's will or desire' (*what is your pleasure?*). Possessiveness can also pass into a command (*The accused was found guilty but insane and was ordered to be detained during her Majesty's pleasure*). Sometimes questioned is the introduction of a possessive pronoun into constructions of the type *it was a pleasure to*: thus *it is our pleasure to have you here*; *in the experiment which it was my pleasure to witness*. Readers should be on their guard if an objection is raised to this kind of construction. Fowler (1926) disliked it to the point of calling the type 'a mistake'.

**plebiscite.** 1 The pronunciation /ˈplebɪsɪt/ is recommended, not /-saɪt/, except that in AmE /-saɪt/ is dominant.

2 A *plebiscite* is most commonly used of a direct vote of the electors of a State (almost always one abroad) on a fundamental constitutional matter. For example, the word was applied in 1852 to

the ratification of the *coup d'état* in Dec. 1851 in France, conferring the imperial crown upon Napoleon III. A *referendum* is usually the referral of an important specific issue to the electorate for a direct decision by a general vote, e.g. the vote which led to the admission of the UK into the European Common Market in 1974.

**plectrum.** The recommended plural is *plectra*, not *plectrums*. See -UM 3.

**pleiad** /ˈplaɪəd/, figuratively, a brilliant cluster or group of persons or things (named after the *Pleiades*, a cluster of small stars in the constellation Taurus). In the French Renaissance a group of poets, including Ronsard and Du Bellay, adopted the name *La Pléiade*.

**Pleistocene.** See MIOCENE.

**plenitude.** First recorded in the 15c. ( = the condition of being full), and, since the 17c., in the sense 'plentifulness, abundance', it is now little used except in high-sounding prose. From the 17c. onward it has occasionally, under the influence of *plenty*, been written as *plenitude*, but this form is not recommended.

**plenteous, plentiful.** Several adjectives in *-eous*, e.g. *beauteous*, *bounteous*, and *plenteous*, are more likely to be encountered in 19c. and earlier literary works than in any 20c. context. The normal and natural word for 'abundant, copious, etc.' is *plentiful*.

**plenty** is historically and in essence a noun. It came into English from Latin (via OF) *plēnitātem* (cf. L *plēnus* 'full') in the 13c., and has had a continuous and varied presence in the language since then (*compliments in plenty*; *plenty of errors*; *in plenty of time*; *we have plenty*; etc.). Examples of its use as noun (esp. followed by *of*, and paralleled by *a lot of* and *lots of*) and as quasi-pronoun (*plenty more*): *There was plenty going on behind the scenes at the BBC in 1968*—BBC Year Bk *1968*, 1969; *There were parties every night and plenty of rooms for friends*—R. Whelan, 1985 (US); *There's plenty more at University Press Books/New York where that came from*—New Yorker, 1986; *There were still plenty of cars hurrying up and down, though no pedestrians but themselves*—K. Amis, 1988; *Money brings deceit ... as he had plenty of opportunity to*

*observe in Newport or in New York society*—J.
Bayley, 1988; *There were plenty of bright
ideas as to how the compass might be insu-
lated from the effects of ship's magnetism*—*Jrnl
Navigation*, 1988; *Plenty of excuses are
offered*—*Independent*, 1989; *The two orchestral Nocturnes have plenty going for them—
Gramophone*, 1990; *The very stylish, smoky-
toasty-vanilla signature of oak is all over this
wine from first sniff to finish, but there is
plenty more, built round a core of deep, black-
berry-raspberry fruit*—*Wine & Spirits* (US),
1991.
    Yet almost from the beginning it was
used predicatively as a quasi-adj. Examples: *If Reasons were as plentie as Blackberries, I would giue no man a Reason vpon
compulsion*—Shakespeare, 1596; *Mosques
are plenty, churches are plenty, graveyards are
plenty, but morals and whisky are scarce*—M.
Twain, 1869; *Both* [quartz and cinnabar]
*were plenty in our Silverado*—R. L. Stevenson, 1883. Such uses are hard to find in
standard 20c. sources: the natural word
to use in such circumstances is *plentiful*.
    In the 19c., attributive uses of *plenty*
came into widespread currency in regional forms of English, both in the UK
and overseas. The *OED* provides evidence
from an impressive range of dialectal
and non-standard sources, e.g. *Although
there are plenty other ideals that I should
prefer*—R. L. Stevenson, 1878; *The water
they brought was a little thick … but Dad
put plenty ashes in the cask to clear it—
*Australian source, 1899; *Leopard Society
in Sierra Leone. They kill plenty people*—G.
Greene, 1969; *When all dem fellas gambling
and heap up plenty money, we bawl out
'Police!'*—*Sunday Express* (Trinidad), 1973.
To which may be added: *There's plenty
places where people have died*—R. Anderson,
1982 (UK); *Watch where you walkin'! Plenty
snake hidin' in the grass*—B. Gilroy, 1986
(Guyana).
    From the 1840s onward, *plenty* has also
been much used, most frequently in non-
standard American contexts, as an adverb meaning 'abundantly, clearly, more
than adequately, etc.' Modern examples:
*I'll bet you got Irving Thalberg plenty wor-
ried*—B. Schulberg, 1941; *He'd suffer for it
plenty if he kept it up*—W. C. Williams, 1948;
*We were plenty lucky*—T. O'Brien, 1976; *Tell
her he's plenty mad*—B. Byars, 1981 (UK);
*The 'young daughter' here was seventeen and
plenty nubile*—*New Yorker*, 1984; *It's hard to
be completely sure just by looking. But he*

*seems plenty dead to me*—R. Silverberg,
1985; *I was plenty scared myself, not to
mention ticked off*—R. Banks, 1989; *I frowned
at my mother plenty*—*New Yorker*, 1990.
    *Conclusion*. Uses of *plenty* as an adj.
(whether predicative or attributive) and
as an adv. are rare in standard BrE, but
are well established in informal (and esp.
in non-standard) contexts in AmE. They
stand at the rim of acceptable standard
English and could well move into the
standard zone in the 21c.

**pleonasm** is a 16c. loanword in English
from L *pleonasmus* (ult. from Gk) meaning
'added superfluously, redundant'. As a
rhetorical device, 'the use of more words
in a sentence than are necessary to express the meaning' is a necessary feature
of poetry and of ornate prose. In this
book, however, we are not concerned
with rhetorical pleonasm, but rather
with expressions and grammatical constructions that contain an element or
elements of redundancy. They are
treated under many heads, e.g. DOUBLE
COMPARISON; DOUBLE SUBJECT; PREFER-
ABLE 2; REDUNDANCY; TAUTOLOGY.

**plesiosaurus**. Pl. *plesiosauri* /-raɪ/ or *ple-
siosauruses*. In practice more commonly
called a *plesiosaur* /ˈpliːsɪəsɔː/, with a normal English plural in *-s*.

**plethora,**    an oversupply, a glut. Pronounce /ˈpleθərə/. Contrast the vowel
quantities in the medL original *plēthōra*
(from Gk πληθώρη 'fullness, repletion').

**pleura**    (membrane lining the thorax
and enveloping the lungs). Pl. *pleurae*
/-riː/.

**plexus**. Pl. same (but pronounced /-uːs/)
or (more commonly) *plexuses*. See -US 2.

**Pliocene.**    See MIOCENE.

**plosive**    (noun). See STOP (noun).

**plough**    is the normal BrE spelling. AmE
*plow.*

**plumb-.**    The *b* is silent in *plumb* (noun,
verb, adv., and adj.), *plumber, plumbing,*
and *plumbless*, but sounded in *plumbago,
plumbeous, plumbic, plumbiferous, plumb-
ism*, and *plumbous.*

**plunder.**    The inflected verbal forms are
*plundered, plundering*. See -R-, -RR-.

**plurality.** See MAJORITY.

## plurals of nouns.

> 1 Words ending in *-ics.*
> 2 Words ending in *-f, -fe.*
> 3 Words ending in *-o.*
> 4 Words ending in *-s.*
> 5 Words ending in *-us.*
> 6 Names etc. ending in *-y.*
> 7 Irregular plurals.
> 8 Animal names.
> 9 Compounds.
> 10 Letters, figures, and abbreviations.
> 11 Plural though singular in form.
> 12 Plural forms only.
> 13 Foreign plurals.

1 Words ending in *-ics.* See AGREEMENT 11; -ICS 2.

2 Words ending in *-f, -fe.* For plurals of nouns ending in *-f* (*dwarf, hoof, leaf,* etc.) and *-fe* (*knife, wife,* etc.), see -VE(D), -VES. The pl. of *still-life* is *still-lifes.*

3 Words ending in *-o.* See -O(E)S.

4 Words ending in *-s.* (*a*) Names of illnesses (*AIDs, measles,* etc.): see AGREEMENT 12. *Chickenpox* and *smallpox,* originally pl., are now reckoned sing. (*b*) *corps, innings, mews, news* are treated as sing. nouns. See as main entries. (*c*) Names of games (*billiards, darts, deck quoits,* etc.) are normally treated as sing. nouns (but *billiard table, he threw a dart, a deck quoit,* etc.). (*d*) Proper names (*Athens, the Thames, the United States* etc.) are always treated as sing. nouns.

5 Words ending in *-us.* See -US.

6 Names etc. ending in *-y.* Proper names ending in *-y* have plurals in *-ys: the two Germanys; three Marys in the class; the Kennedys.* Note also *lay-bys, stand-bys, the 'if onlys', treasurys* ( = treasury bonds). For surnames ending in *-s,* the pl. is *-es* (*the Joneses, the Rogerses, the Simmses*).

7 Irregular plurals. (*a*) Mutation plurals: *foot/feet, goose/geese, louse/lice, man/men, mouse/mice, tooth/teeth, woman/women.* Also *brother/* (*occas.*) *brethren* (with irreg. pl. ending). (*b*) Irregular plurals: *child/children, ox/oxen.*

8 Animal names. These normally have the regular pl.: *cat/cats, elephant/elephants, hawk/hawks,* etc. When regarded collectively some animal names remain unchanged in the pl.: e.g. *they were in the country shooting duck; they had an excellent haul of fish.* The following names of animals and fish are unchanged in the pl.:

*bison, cod, deer, grouse, salmon, sheep, squid, swine.*

9 Compounds. Compound words formed from a noun and an adj., from two nouns connected by a preposition, or from a personal noun followed by an adverb, normally form their plurals by a change in the chief word: e.g. \*\**Attorneys-General,* \*\**courts martial,* \*\**Governors-General,* \*\**Poets Laureate; aides-de-camp, commanders-in-chief, fleurs-de-lis, men-of-war,* \*\**sons-in-law; listeners-in, passers-by.* In the words marked with a \*\* this rule is often ignored (so *Attorney-Generals,* etc.). They are being drawn towards such exceptionless pl. compounds as *forget-me-nots* and *ne'er-do-wells.* Compounds consisting of a verb + an adverb normally have *-s* at the end: *call-ups, close-downs, knock-outs, push-ups, stand-ins.* Words ending in *-ful* add *-s* at the end: *cupfuls, mouthfuls, spoonfuls* (see -FUL). Note also *gin-and -tonics, whisky-and-sodas* (not *gins-and-tonic,* etc.).

10 Letters, figures, and abbreviations. All types were once normally written with *'s: two VC's, the 1950's,* etc. Increasingly now the apostrophe is being dropped in the following types: *B.Litts, MAs, MPs, QCs, the sixties, the 1960s.* But after letters an apostrophe is obligatory: *dot your i's, mind your p's and q's.*

11 Plural though singular in form: *cattle, people* (the corresponding singular is *person*), *police, poultry, vermin.*

12 Plural forms only. (*a*) Names of instruments and tools: *bellows, binoculars, clippers, forceps, gallows, glasses, pincers, pliers, scissors, shears, tongs* (construed with a plural verb). (*b*) Names of articles of dress: *braces, briefs, flannels, jeans, knickers, pants, pyjamas* (US *pajamas*), *shorts, slacks, suspenders, tights, trousers* (all construed with a pl. verb).

13 Foreign plurals. Apart from those dealt with under LATIN PLURALS, -O(E)S, -UM, -US, and -X there are numerous common English words of foreign origin that regularly or frequently have a pl. form other than the ordinary English type in *-s.* These are dealt with at their alphabetical places: *concerto, criterion, libretto, phenomenon,* etc.

**plus.** First adopted in the 17c. from L *plūs* 'more' and used as a quasi-preposition, i.e. in a manner that did not exist in Latin. To begin with, and still primarily, it was used in English as the oral equivalent of + (as opposed to minus, or − ). In the 20c. it has gone from strength to strength as a quasi-prep. in the sense 'with the addition of, and also' (e.g. *A cup of Epp's cocoa and a shakedown for the night plus the use of a rug or two and overcoat doubled into a pillow*—J. Joyce, 1922; *There ... was a .. plane waiting plus an army scout car and a guard of soldiers*—E. Ambler, 1974). Even more striking, and causing widespread ripples of dismay among purists, is the use of *plus* from about the 1960s (first in America) as a conjunction meaning 'and furthermore, and in addition'. Examples: *You can fly an airplane ... and command a ship. Plus you ride horses*—T. Clancy, 1987; *I'm a pianist, but I feel all thumbs today. Plus which I've got a cold.*—*New Yorker*, 1987. This use persists in informal contexts, and is now joined by an adverbial use in which *Plus*, often followed by a comma, leads a new sentence which is only loosely connected to the previous one. Examples: *'So I'll quit romanticizing him. Plus, he never got to go on any road trips*—B. Ripley, 1987; *When you have to take a hot-water bottle to bed even after Memorial Day something is very wrong. Plus, the neighbor's dog was putting me over the top*—*New Yorker*, 1990. *Plus* used both as a quasi-prep. and as an adv. occurs frequently in advertisements (e.g. *20% off everything—plus no deposit; salary £50K—plus you will have a luxury company car*), and in informal contexts of various kinds, but should be excluded from contexts requiring any degree of formality.

**p.m.** As an abbreviation of L *post meridiem*, it is normally written in the form *3.15 p.m.* (or *pm*; in AmE *3:15 p.m.*). See A.M.

**pn-.** The *OED* (1908) gave only /nju:-/ for the initial sound of *pneumatic* (and derivatives) and *pneumonia* (and derivatives), but recommended the retention of initial /p-/ in other, more technical, words beginning with *pn-* (*pneum, pneumatology, pneumatophore*, etc.). It added a note (s.v. *Pn-*) 'The *p* is pronounced in French, Spanish, Italian, German, Dutch, and other European languages ... It is to be desired that it were sounded in English also, at least in scientific and learned words; since the reduction of *pneo-* to *nco-*. *pneu-* to *new-*, and *pnyx* to

*nix*, is a loss to etymology and intelligibility, and a weakening of the resources of the language.' Its recommendation has gone unheeded and /nj-/ is now invariable in all such words. Cf. PS-; PT-.

**pochard** (a duck of the genus *Aytha*). *COD* (1995) gives preference to /'pəʊtʃəd/. Wells (1990) recommends /'pɒtʃəd/ and gives /'pəʊtʃ-/ as a variant.

**pocket.** The inflected forms of the verb are *pocketed, pocketing*. See -T-, -TT-.

**pocketful.** Pl. *pocketfuls*. See -FUL.

**podagra** (gout of the foot). Pronounce /pɒ'dægrə/.

**podium** /'pəʊdɪəm/. Pl. *podia* or *podiums*. See -UM 3. A *podium* is a platform or rostrum (e.g. for a speaker or an orchestral conductor). A *lectern* is a stand for holding a book (usu. the Bible) in church, or a similar stand for a lecturer, etc.

**poetess.** See -ESS. The generic term *poet* is now overwhelmingly used of both male and female writers of verse, and examples of *poetess* are not easy to find (*poet* 476 examples and *poetess* only 12 in OUP's electronic corpus of examples from works written between 1987 and 1993).

In a standard textbook, Alistair Fowler (1987) discusses the work of the New England poet Emily Dickinson. *OCELit.* (1985) uses *poet* throughout for women writers of verse: e.g. *Hemans, Mrs Feliciu Dorothea ... a precocious and copious poet; Sappho ... a Greek lyric poet.* But cf. *Sappho ... 6th century B.C. Greek lyric poetess of Lesbos*—Collins Eng. Dict., 1986. Some examples of *poetess* from the Oxford corpus: *Amrita Pritam ... is a great poet of India (poetess, like authoress, is a term I am allergic to)*—Indian Bookworm's Jrnl, 1987; *It purports to be about a minor sentimental 'poetess' ... living in a small, raw, cowpat-strewn, treeless nineteenth-century town*—M. Atwood, 1989; *I aspired to be Gertrude Stein ... or some poetess tragically and forlornly trying to scrape some piece of misery off the sole of my soul*—R. Barr, 1989; *For Sylvia Plath, an American poetess living in England, it was sometimes her mother ... who sat on her typewriter and prevented her from believing in her work*—Raritan, 1990.

**poeticisms.** By these are meant modes of expression that are thought (or were once thought) to contribute to the emotional appeal of poetry but are unsuitable for plain prose: 'To most people nowadays, I imagine,' says T. S. Eliot, '*poetic diction* means an idiom and a choice of words which are out of date and which were never very good at their best.' Poeticisms are not favoured even by poets any more. The revolt against them advocated by Wordsworth in his preface to *Lyrical Ballads* has gone to lengths that would have surprised him. Nevertheless injudicious writers of prose are still occasionally tempted to use them as tinsel ornaments. See WARDOUR STREET.

**poetic, poetical.** In a great many contexts the choice of form is a personal matter, there being no clearly designated differences of meaning. Sentence rhythm is perhaps the main factor governing the choice, but, whatever the reason, *poetic* is the more common of the two words. There are, however, a few fixed collocations, e.g. *The poetical works of* —; *poetic diction*; *poetic justice*; *poetic licence*.

**Poet Laureate.** Pl. either *Poets Laureate* or *Poet Laureates*. See PLURALS OF NOUNS 9.

**pogrom.** Pronounced /ˈpɒgrəm, -rɒm/. A word of Russian origin ('devastation, destruction'), taken into English towards the end of the 19c., specifically referring to an organized massacre of any group or class of people, esp. applied to those directed against the Jews. It soon began to be applied to an organized attack on any community or group whether of the past (e.g. in the ancient Greek Empire) or the present (e.g. the Holocaust under the Nazi regime).

**poignant.** Pronounce /ˈpɔɪnjənt/.

**point in time, at this.** The SAfr. writer André Brink, in his *States of Emergency* (1988), drew attention to the unnecessary use of this popular 20c. phrase (*OED point sb.*[1] D.1g) when shorter and more direct synonyms are available: *if it could be amplified by a political dimension to involve what was happening in the country right now* (the actual phrase they used was '*at this point in time*'). Like *at this moment in time*, *at the end of the day*, and *in this day and age*, it is the kind of phrase that

calls out for the attention of copy editors and journalistic sub-editors.

**point of view** (first recorded in the early 18c.; cf. Fr. *point de vue*) is freely interchangeable with *standpoint* (19c., modelled on Ger. *Standpunkt*) and *viewpoint* (mid-19c.). The run of the context governs the choice of word.

**polemarch** (in ancient Greece, a civilian official). Pronounce /ˈpɒlmɑːk/.

**polemic, polemical.** The noun is *polemic* 'a controversial discussion, argument, or controversy, esp. over a doctrine, policy, etc.' *Polemics* (pl. in form but functioning as a singular) is 'the art or practice of controversial discussion'. The corresponding adj. is either *polemic* or *polemical*.

**policewoman,** a female member of a police force, esp. one holding the rank of constable: hence the abbreviation *WPC*, woman police constable (in the UK). Lay people are tending to use the gender-neutral term *police officer* (or *officer* for short) when addressing such a person.

**policy.** There are two separate words. The word meaning 'course of action, etc.' is from OFr. *policie*, from L *politīa* from Gk πολιτεία 'citizenship', government. The second *policy* meaning 'a contract of insurance' is from Fr. *police* bill of lading, contract, from medL *apodissa*, *apodixa*, from Gk ἀπόδειξις 'evidence, proof'.

**political correctness.** At some point in the 1980s, first in America and soon afterwards elsewhere, the term *politically correct* (abbreviated *PC*, *P.C.*) began to be used in the sense 'marked by or adhering to a typically progressive orthodoxy on issues involving esp. race, gender, sexual affinity, or ecology' (*Random House Webster's College Dictionary*, 1991). It, and the corresponding noun phrase *political correctness*, gathered up notions which had gradually evolved during the previous half-century or so. In crude terms the aim of the liberal-minded crusaders was to persuade the community at large to abandon earlier prejudices and suppositions in certain broadly designated areas of life, and to substitute a whole range of new vocabulary for expressions which were alleged to be false, hurtful, discriminatory, sociologically dangerous, or inept in some other way. The public reaction ranged from widespread acceptance of some of the attitudes and the attendant vocabulary to amusement and even to downright hostility to others. A leader in the *Sunday Times* (20 October 1991) attacked *PC* in the following manner: 'Something is rotten in the United States of America and it threatens the whole basis of that great society's role as protector of the free world and inspiration for those who yearn to be free. American politics is being corrupted and diminished by the doctrine of Political Correctness which demands rigid adherence to the political attitudes and social mores of the liberal-left, and which exhibits a malevolent intolerance to anybody who dares not comply with them.'

Among the first major works of reference to deal with the new terms *political correctness* and *politically correct* was *OCELang.* (1992): 'The phrase is applied, especially pejoratively by conservative academics and journalists in the US, to the views and attitudes of those who publicly object to: (1) The use of terms that they consider overtly or covertly *sexist* (especially as used by men against women), *racist* especially as used by whites against blacks), *ableist* (used against the physically or mentally impaired), *ageist* (used against any specific age group), *heightist* (especially as used against short people), etc. (2) Stereotyping, such as the assumption that women are generally less intelligent than men and blacks less intelligent than whites. (3) "Inappropriately directed laughter", such as jokes at the expense of the disabled, homosexuals, and ethnic minorities ... Both the full and abbreviated [i.e. *PC*] terms often imply an intolerance ... of [opposing] views and facts that conflict with their "progressive orthodoxy".'

Jocular aspects of the PC movement are set down in *The Official Politically Correct Dictionary and Handbook* by Henry Beard and Christopher Cerf (1992). Terms attracting derision included *differently abled* (of a person confined to a wheelchair), *nonwaged* (unemployed), *physically challenged* (disabled), *significant other* (lover, sexual partner, etc.), *vertically challenged* (shorter than average), *waitron* (waiter, waitress), *wimyn*, *wimmin* (women), *womyn* (woman); and a range of more or less transparent words in *-ism*:

*ableism, ageism, lookism, sizeism, weightism,* etc. Various aspects of this unforgiving politically correct attack on the values and terminology of an older generation are treated s.v. DIDACTICISM; FEMININE DESIGNATIONS; MS; PERSON; SEXIST LANGUAGE.

**politic, political.** As an adjective, *politic* now normally means (*a*) (of an action) judicious, expedient; (*b*) (of a person) prudent, sagacious. In the sense 'political, concerned with politics' it survives only in the phr. *body politic*. The corresponding adverb is *politicly*, but it is not often used. The adjective *political* is now the natural and only form to mean 'of, concerning, or engaged in politics'. The corresponding adverb is *politically*.

**politics** is construed as a singular noun when used to mean 'the art or science of government' (*Politics is a popular subject at many universities*), but normally as a plural when used to mean 'a particular set of ideas, principles, etc., in politics (*what are their politics?*). See -ICS 2.

**polity** is (*a*) a form or process of civil government or constitution; (*b*) a society or country as a political entity. It should not be confused with *policy* 'a course or principle of action'.

**poll,** the woman's name and an affectionate name for a parrot, is pronounced /pɒl/. *Poll* in voting, etc., and other uses, is pronounced /pəʊl/.

**polloi.** See HOI POLLOI.

**poll tax.** See COMMUNITY CHARGE.

**polypus** (a small benign growth, a polyp). Pl. *polypi* /-paɪ/ or *polypuses*. See -US 4.

**polysyllabic humour** is a marked characteristic of many a comedian and journalist. It was also resorted to by numerous popular writers of the past, typified by W. S. Gilbert: *Merely corroborative detail, intended to give artistic verisimilitude to an otherwise bald and unconvincing narrative* (1885).

There is no shortage of polysyllabic humour in the 20c: e.g. *£17.10s., the cost of their passage, may not be a healthy or proper contract, but it cannot in the opinion of His Majesty's Government be classified as* *slavery in the extreme acceptance of the word without some risk of terminological inexactitude*—W. S. Churchill, 1906.

> *I test my bath before I sit,*
> *And I'm always moved to wonderment*
> *That what chills the finger not a bit*
> *Is so frigid upon the fundament.*
>
> Ogden Nash, 1942

*Supercalifragilisticexpialidocious!*—title of a song in the Walt Disney film *Mary Poppins* (1964).

> *Words you've never used*
> *And have always wanted to—*
> *Get them in quickly.*
> *Dight in dimity*
> *Enlaced with lazy-daisy*
> *In fishnet fleshings . . .*
> *Jalousies muffle*
> *Criminal conversation—*
> *Discalced and unfrocked*
> *Ithyphallic, perforate—*
> *A case of jactitation.*
>
> (D. J. Enright, 1993)

Fowler (1926) compiled 'a short specimen list of long words or phrases that sensible people avoid'. The implication was that they were all examples of polysyllabic humour: (*a*) Proper names. *Batavian* Dutch; *Caledonian* Scottish (Fowler said 'Scotch'); *Celestial* Chinese (as in *Celestial Empire*); *Hibernian* and *Milesian*, both meaning 'Irish'. Of these only *Caledonian* and *Hibernian* are still in common use. (*b*) Ordinary nouns and adjectives: *cachinnation* 'laughter', *culinary* (adj.) 'kitchen', *diminutive* small, *epidermis* 'skin', *equitation* 'horse-riding', *esurient* 'hungry', *femoral habiliments* 'trousers (breeches)', *fuliginous* 'sooty', *matutinal* 'morning', *minacious* 'threatening', *olfactory organ* 'nose', *osculatory* (adj.) 'kissing', *pachydermatous* 'thick-skinned', *peregrinate* 'travel'. The clock has moved on. *Culinary* and *diminutive* are everyday words now. The humour has disappeared from most of the other expressions. The *COD* (1990) lists nearly all of them (*minacious* has dropped by the wayside) without any qualificatory labels. The only words labelled 'joc.' or 'archaic' or both are *esurient*, *habiliments*, and *peregrinate*. Nevertheless Fowler's list is still worth bearing in mind. The natural course is to choose the familiar rather than the unfamiliar word. Clarity is better than opacity, unless obscurity is the whole point of the context.
See PEDANTIC HUMOUR.

**pomade.** The *OED* (1908) listed both /pɒˈmeɪd/ and /pɒˈmɑːd/ and so do some modern desk dictionaries except that they give the initial unstressed syllable as /pə-/ rather than /pɒ-/. Other modern dictionaries give initial /pəʊ-/. Clearly the word, which is encountered more often in print than in speech, is still in process of becoming acclimatized, even though it has been in the language since the second third of the 16c. See -ADE.

**pomegranate.** The *OED* (1908) listed four pronunciations: /pɒm-, pʌmˈgrænɪt/; /ˈpɒm-,ˈpʌmgrænɪt/. *OEDS* (1982) added 'Now freq. with pronunc. /ˈpɒmɪgrænɪt/', and *OED* 2 (1989) changed 'freq.' to 'usu.'. The pronunciation with four syllables (and with a secondary stress on the third) now seems to be dominant (except in AmE, in which the three-syllabled pronunciation is more usual); and the initial syllable is always /pɒm-/, never /pʌm-/.

**pommel.** The noun means 'a knob, esp. at the end of a sword-hilt; also, the upward projecting front part of a saddle', and is pronounced /ˈpʌməl/ in BrE. The corresponding verb means 'to strike repeatedly esp. with the fist', and is normally spelt *pummel*, pronounced the same as the noun. The inflected forms of both spellings of the verb are *pummelled, pommelling* (or *pommelled*, etc.) in BrE, but usu. *pummeled, pummeling* (or *pommeled*, etc.) in AmE.

**poncho** (S. Amer. cloak). Pronounce /ˈpɒntʃəʊ/. Pl. *ponchos*. See -O(E)S 6.

**pond.** Used lightheartedly for the sea, esp. the N. Atlantic Ocean, since the mid-17c. Cf. HERRING POND. A modern example: *It is telling that Jackie Collins, born British, wrote thin, amoral novels when she was resident here . . . but huge, earnest tomes which even started to feature safe-sex warnings when she took up residence across the pond*—J. Burchill, 1993.

**pontifex.** Pl. *pontifices* /pɒnˈtɪfɪsiːz/. See -EX, -IX.

**pontificate,** to be pompously dogmatic (first recorded in this sense in 1922), has now replaced *pontify* (first recorded in the same sense in 1883).

**poof** /pʊf, puːf/ is a derogatory term for (*a*) an effeminate man; (*b*) a male homosexual. It is sometimes written as *poove*, pronounced /puːv/. The slang synonym *poofter* was first used in Australia at the beginning of the 20c. but has much wider currency now.

**poor.** The pronunciation is variable: either /pʊə/ or /pɔː/.

**poorly.** Apart from its routine uses as an adverb in the senses of *poor* (*he came out of the affair poorly*; *she performed poorly*), *poorly* is in widespread, mostly colloquial, use as a predicative adj. meaning 'unwell' (*she looked poorly after the game of squash*; *her husband had been poorly for months before he died*).

**poorness.** See POVERTY.

**popular etymology.** See ETYMOLOGY; FOLK ETYMOLOGY.

**popularized technicalities.** When Fowler was preparing *Modern English Usage* he remarked that 'the term of this sort most in vogue at the moment of writing (1920) [was] *acid test*'. In scientific use it meant 'the testing for gold by the use of nitric acid'. In transferred use it had acquired the broad sense 'a severe or conclusive test', a use perhaps popularized by Woodrow Wilson in 1918 (*The treatment accorded Russia by her sister nations in the months to come will be the acid test of their good will—Times*, 9 Jan. 1918), though this figurative use occurred earlier in the same decade. The emergence of such lay extensions of meaning may be designated as 'something lost, something gained'. And so it is with numerous other technical and scientific terms that have passed into lay use in the 20c., in the process losing something of the import of the original expression but nevertheless adding something of value to non-specialized language: see e.g. the entries for *allergy, chain reaction, climax* (sexual sense), *clone, complex, feedback, fixation, function, parameter, persona*, and *syndrome*. The process is as old as the language itself. The technical terminology of specialized subjects has at all times passed into general use. Lexicographers tend to look upon the emerging new uses as 'weakened' or even 'trivial' senses, but tend to remove the qualifications as time goes on. A handful of examples that have made their way from specialized subjects into general

use: law (*leading question, special pleading*); logic (*dilemma*); mathematics (*arithmetic/geometric progression*; *to the nth degree*); physics (*quantum jump* or *leap*).

**popular music,** songs, folk tunes, etc., appealing to popular tastes (*COD*, 1990). Slight indications of the heavy linguistic content of present-day popular music can be gleaned from any of the numerous journals devoted to the subject or from newspapers and magazines. A single issue (13 Feb. 1988) of the journal *Melody Maker* yielded the following terms: *funk chatter, a giant rollercoaster beat, indie, pomp-rock, raver, saw-toothed guitar drone, synths, uncool, weedybop*. The *Christian Science Monitor*, 22 Nov. 1989, had a short glossary helping to unravel 'the street-wise poetry' of rap music: *chill* (to cool out, hang loose), *crush* (good, great, terrific), *dis* (to put down, disrespect), *def* (good, great, terrific), *dope* (smart, wise), *fly* (good-looking, fine), *posse* (a group), and *stupid* (the best, tops). Inevitably such music is associated in the minds of many people with drugs, a subject which possesses its own endless lists of rapidly changing slang. The 28 Oct. 1991 issue of *Newsweek*, for example, provided a short list of buzzwords used by drug-dealing street gangs in American cities. Among them were *balling* (selling rock cocaine; *Mikey's balling over by the alley*); *gaffle up* (to arrest), *G-ride* (stolen car; from 'gangster ride'), *Jim Jones* (marijuana laced with cocaine and PCP), *kibbles and bits* (crumbs of cocaine). Such lists could be extended almost indefinitely. They are important because they partially define the terminological world with which young people are familiar at the present time, esp. those dressed in leather (or other informal gear) and with strange hair styles.

**porcelain** is 'a hard vitrified translucent ceramic' (*COD*), and *clay* is 'a stiff sticky earth, used for making bricks, pottery, ceramics, etc.' (*COD*), i.e. they have separate definitions. But as Fowler (1926) remarked, 'Porcelain is china, and china is porcelain; there is no recondite difference between the two things, which indeed are not two, but one.' Nevertheless there are differences of use; e.g. someone making preparations for an important tea party might well say *I must get out my best china* (not *my best porcelain*). But ornaments and vases, for example, may equally well be described as made of *china* or made of *porcelain*.

**Porch.** For *the Porch* in philosophy, see ACADEMY.

**porpoise.** Pronounce /ˈpɔːpəs/. The second syllable does not rhyme with *poise*.

**porridge** was formerly frequently treated in Scotland as a plural noun (*and put butter in* them).

**portcullis.** The corresponding adj. is *portcullised*. See -s-, -ss-.

**portfolios.** Pl. *portfolios*. See -o(E)s 4.

**port, harbour, haven.** The broad distinction is that a haven is thought of as a place where a ship may find shelter from a storm, a harbour as one offering accommodation (used or not) in which ships may remain in safety for any purpose, and a port as a town whose harbour is frequented by naval or merchant ships.

**portico.** Pl. *porticoes* or *porticos*. See -o(E)s 1.

**port, larboard.** The two words mean the same, but *port* has been substituted for *larboard* (the earlier opposite of *starboard*) because of the confusion resulting when orders were shouted from the too great similarity between *larboard* and *starboard*.

**portmanteau.** The recommended pl. is *portmanteaus* /-əʊz/, not *portmanteaux*. See -x.

**portmanteau words.** See BLEND (noun).

**Portugese,** a native or inhabitant of Portugal. Plural the same. The spurious 'singular' form *Portuguee* has had some currency in non-standard AmE since the early 19c.

**pose.** The verb meaning 'nonplus, puzzle (a person) with a question or problem' (with its noun *poser* 'a puzzling question') is shortened from an obsolete verb *appose*. It is a different word from that meaning 'to assume a certain attitude of body, etc.'

**posh.** See ETYMOLOGY 4.

**position** (noun). See -TION.

**position** (verb). The use of *position* as a verb has met with some criticism. But there are instances, going back for more than a century, of its use in the senses both of 'to place in a particular or appropriate position' and of 'to determine the position of', and if it can claim a useful role on the ground that neither *place* nor *pose* will always give quite the same meaning (which is at least arguable), it need not be rejected merely because it is a long-established noun used much later as a verb.

**position of adverbs.** See ADVERB 5. For split infinitives and false adjustments made by writers because of an irrational fear of the splitting of other sentence elements see SPLIT INFINITIVE (where a history of attitudes towards the splitting of the particle *to* and its following infinitive may also be found). It should be noted that Fowler's classic articles POSITION OF ADVERBS and SPLIT INFINITIVE, which are now mainly of historical interest, were first published in SPE Tract xv (1923) and simply reprinted in *MEU* in 1926.

**posse.** Pronounce /ˈpɒsɪ/, i.e. as two syllables.

**possessive.** See ABSOLUTE POSSESSIVES; APOSTROPHE[2] B, D, E; OBJECTIVE GENITIVE; 's AND OF-POSSESSIVE.

**possessive pronouns. 1** See the warning given under APOSTROPHE[2] D5, and the reminder s.v. ITS, IT'S.

**2** Distinguish *their* (possessive pronoun) from *they're*, a contracted form of *they are* (e.g. *They're off to Birmingham tomorrow*); and distinguish *your* (possessive pronoun) from *you're*, a contracted form of *you are* (e.g. *you're not my friend*).

**possessive with gerund. 1** *Historical note: the fused participle.* In a classic battle in SPE Tracts xxii (1925), xxv (1926), and xxvi (1927), Fowler and the grammarian Otto Jespersen disputed the merits of the type *Women having the vote reduces men's political power*. Fowler asserted that *women having* was 'a compound notion formed by fusion of the noun *women* with the participle *having*', called it a fused participle, and condemned it. He contrasted it with the legitimate types *Women having the vote share political power with men* (in which *having* is a true participle) and *Women's having the vote reduces men's political power* (in which the subject is the verbal noun or gerund *having* (*the vote*), and *women's* is a possessive case attached to a noun). He went on to cite a number of newspaper examples in which the subject is separated from its fused participle by a subordinate clause or clauses, e.g. *New subsections giving the Board of Trade power to make regulations for permitting* workmen *who are employed under the same employer, partly in an insured trade and partly not in an insured trade, being treated, with the consent of the employer, as if they were wholly employed in an insured trade.* 'It need hardly be said,' he added, 'that writers with any sense of style do not, even if they allow themselves the fused participle, make so bad a use of the bad thing as is shown above to be possible. But the tendency of the construction is towards that sort of cumbrousness, and the rapidity with which it is gaining ground is portentous.'

He went on to claim that 'a dozen years ago' (he meant when he was writing *The King's English*, 1906) it was not very easy to collect instances of the most elementary form of the fused participle, i.e. that in which the noun part is a single word, and that a pronoun or proper name. He then went on to cite a dozen examples, 'culled without any difficulty whatever from the columns of a single newspaper', including the following: *We need fear nothing from China developing her resources* (China's); *It is no longer thought to be the proper scientific attitude to deny the possibility of anything happening* (anything's); *I quite fail to see what relevance there is in* Mr Lloyd George *dragging in the misdeeds of* ... (George's); *The reasons which have led to them being given appointments in these departments* (their). All such sentences he regarded as 'grammatically indefensible' because 'the words defy grammatical analysis'. He then resorted to Latin grammar to support his view.

Jespersen (SPE Tract xxv) vigorously defended the construction condemned by Fowler. He gave numerous examples of its use by writers from Swift to Shaw, said that Fowler's 'article is a typical specimen of the method of what I call the instinctive grammatical moralizer', and criticized Fowler for believing in the validity of Latin parallels ('Each language surely has a right to be judged on its own merits'). Jespersen's examples were of two types: (i) the common-case of nouns denoting living beings (e.g. *what is the good of a man being honest in his worship of dishonesty*—G. K. Chesterton; *no chance of the* lady *coming back*—A. Bennett). (ii) pronouns (e.g. *that accounts for them bein' away all night*—J. Galsworthy; *I set my heart on you coming out to Spain*—C. Mackenzie). He thought that with an indefinite pronoun (*anybody, no one*, etc.) 'the common-case construction is now the general rule' (e.g. *He insists on no one knowing about the experiment*). He concluded that the construction represented 'the last step of a long line of development, the earlier steps of which ... have for centuries been accepted by everybody. Each step, including the last, has tended in the same direction, to provide the English language with a means of subordinating ideas which is often convenient and supple where clauses would be unidiomatic or negligible.' Fowler in his rejoinder (SPE Tract xxvi) admitted that he had underestimated the extent of its use, but was otherwise unshaken. He also poured scorn on Jespersen's use of the term *nexus* ('Nexus, you know, is a patent medicine for the grammatical rickets').

**2** Current practice. As the 20c. draws to a close the choice of construction is mostly resolved along the following lines: (i) The possessive with gerund is frequently used when the word before the -*ing* form is a proper name or personal noun (e.g. *Andrew, Reagan, sister, baby*): *this makes it difficult to deduce any motive for Cope's collecting the books*—Bodl. Libr. Rec. 1987; *One cannot say that Kafka's marvelling at mundane accomplishments was not genuine*—London Rev. Bks, 1987; *I was now counting on my father's being able to make some provision somehow*—V. Mehta, 1987; *There is always the problem of Reagan's saying that he doesn't remember when he approved the Israeli shipment*—New Yorker, 1987; *wondering if he should be angry ... about May's sleeping with him and then throwing him out, about his grandfather's having left no message or sign for him but a field of junked cars*—C. Tilghman, 1991.

But there are many exceptions and the choice is a highly personal one: e.g. *Preserving his reputation depended upon Housman disguising his real nature*—R. P. Graves, 1979; *how could she think of the baby being born in the house*—A. S. Byatt, 1985; *Sylvia stopped saying how exciting it was Gordon being so much in demand*—P. Lively, 1987; *These catastrophes were ascribed to Venice having become law*—NY Rev. Bks, 1987.

(ii) When the noun is non-personal, is part of a phrase, or is in the plural the possessive is normally not used: e.g. *They turned a blind eye to toffee apples going missing*—J. Winterson, 1985; *Then we had our old conversation about the house being haunted*—C. Rumens, 1987; *many will question the wisdom of government departments straying into competitive commercial areas*—Daily Tel., 1987; *Mrs Thatcher herself is not averse to this elegant bone being cast before her longstanding tormentor*—Daily Tel., 1987.

(iii) With personal pronouns, where there is a difference of form, usage is evenly divided: (possessive) *Then it became empty and the question had arisen of its being sold*—I. Murdoch, 1958; *Fancy his minding that you went to the Summer Exhibition*—A. N. Wilson, 1978; *Mr Luder wonders whether our digging up the road in St George's Square so soon after it was resurfaced is a world record*—Times, 1985; *'Is it all right?' he asked, needing reassurance. 'My coming to your party?'*—B. Rubens, 1987. (non-possessive) *You're talking rather loudly, if you don't mind me saying so, Claudia*—P. Lively, 1987; *He said something about you ruining his slippers*—C. Rumens, 1987; *There can be no question of you disturbing the clerks*—P. Carey, 1988; *There would be something so despicable about him blustering ahead with a palpably unsound argument*—C. Chambers, 1992.

(iv) With indefinite pronouns usage is divided, but the non-possessive form is now dominant: (possessive) *There are many sound reasons, then, for everyone's wanting to join in this new Gold Rush*—Encounter, 1988; *Mrs Longo has nothing against anyone's being Japanese, of course*—New Yorker, 1988. (non-possessive) *he didn't think for a time of anybody clawing at his back*—D. A. Richards, 1981 (Canad.); *should we not primarily be looking on Aids as a symptom of something having gone fundamentally wrong with our attitudes to sexuality?*—Daily Tel., 1987.

3 *Further outlook.* The possessive with gerund is on the retreat, but its use with proper names and personal nouns and pronouns persists in good writing. When the personal pronoun stands in the initial position it looks certain that the possessive form will be preferred for a long time to come: e.g. *His being so capable was the only pleasant thing about the whole dreadful day*—E. Jolley, 1985 (Aust.); *'My being here must embarrass you,' she says*—New Yorker, 1986. The substitution of *Him* and *Me* would take both sentences into a lower level of formality.

**possible.** 1 The phr. *to do one's possible* is a Gallicism (modelled on Fr. *faire son possible*) meaning 'to do one's utmost'. First recorded in English in 1792 it seems to have been much favoured for about half a century, but to have more or less fallen by the wayside since then except in overtly Gallic contexts. Example: *He did his possible, but old Turgid was neither to be led nor driven*—A. M. Bennett, 1797.

2 *Construction.* It was formerly permissible to construct *possible* with a following *to*-infinitive in the sense 'able, capable', i.e. the type 'the broken toy was possible to be mended': e.g. *Firm we subsist, yet possible to swerve*—Milton, 1667; *The only offence against him of which she could accuse herself, had been such as was scarcely possible to reach his knowledge*—J. Austen, a1817. This must once have seemed as natural as the type *it is/was possible + to*-infinitive (e.g. *it is not possible to distinguish one from the other*), in which *it* is used in an anticipatory or impersonal manner. Fowler (1926) expressed the matter succinctly: '*But no such questions are possible , as it seems to me, to arise between your nation and ours; No breath of honest fresh air is suffered to enter, wherever it is possible to be excluded.* These are wrong. Unlike *able*, which ordinarily requires to be completed by an infinitive (*able to be done, to exist*, etc.), *possible* is complete in itself and means without addition *able to be done* or *occur*. The English for *are possible to arise* and *is possible to be excluded* is *can arise, can be excluded.*'

**post hoc, ergo propter hoc,** 'after it, therefore due to it': the fallacy of confusing consequence with sequence.

On Sunday we prayed for rain; on Monday it rained; therefore the prayers caused the rain.

**posthumous** /ˈpɒstjʊməs/, i.e. the *h* is silent as well as being etymologically intrusive. The English word is derived from L *postumus* 'last, late-born', which in late Latin was erroneously attributed to *humus* 'the earth'.

**postilion.** The recommended spelling, rather than *postillion*, though both are used.

**postmaster general.** The traditional pl. is *postmasters general*. Cf. PLURALS OF NOUNS 9. The office was abolished in the UK in 1969.

**postmistress,** a woman in charge of a post office. See -ESS 5; -MISTRESS.

**postpone.** The word is often pronounced with the *t* silent.

**postprandial,** as in *a postprandial speech*, one given after dinner or lunch, is now mainly in formal or jocular use. Cf. L *prandium* 'a meal'. Examples: *She had asked Toby and Charlie and Leighton to join her and Reed for a post-prandial confabulation*—A. Cross, 1986 (US); *What do people prefer for their pre- and post-prandial drinks?* —*House & Garden*, 1990; *The best is for last*—*a postprandial stroll along the edge of Battery Park City*—*NY Woman*, 1990; (caption) *Rest for the wicked: the master of the house takes a post-prandial nap in the library*—*National Trust Mag.*, 1992.

**postscript.** The word is often pronounced with the medial *t* silent.

**potato.** Pl. *potatoes*. See -O(E)S 1.

**poteen** /pɒˈtiːn/, in Ireland, an illicit spirit, usu. distilled from potatoes. The variant *potheen* is pronounced /pɒˈtʃiːn/, esp. in Ireland.

**potence, potency.** In general contexts where the sense is 'power, the quality or state of being potent' the customary word is *potency*. In technical uses the two terms have gone their own way: *potence* is favoured in certain uses in watchmaking and as a particular movement of soldiers, etc., marching; while *potency* is used of the ability to achieve orgasm in sexual intercourse, and in certain specific

senses in homoeopathy, genetics, pharmacology, etc. See -CE, -CY.

**potentate.** Pronounce /ˈpəʊtənteɪt/.

**potful.** Pl. *potfuls*. See -FUL.

**pother.** 1 Pronunciation. Walker's *Critical Pronouncing Dict.* (1791) gave only /ˈpʌðər/, i.e. rhyming with *other, brother, mother, smother*; the *OED* (1908) gave both /ɒ/ and /ʌ/ forms in that order; while Daniel Jones (1917) gave only the form with /ɒ/. Later standard dictionaries give only /ɒ/. As the *OED* remarks, the current pronunciation 'appears to be a 19th c. literary innovation, after the spelling, and perh. influenced by association with *bother*.'

2 The word emerged in the late 16c. in the sense 'disturbance, commotion', but its etymology is unknown. The similar-sounding word *bother* was first recorded in the 18c. in the works of Irish writers (T. Sheridan, Swift, Sterne), is 'doubtless of Anglo-Ir. origin, but no plausible Irish source can be adduced; possibly an Irish pronunciation of *pother*' (*ODEE*).

**potter** (verb), to work or occupy oneself in a desultory but pleasurable way (*he likes pottering around in the garden*), is commonly used in BrE. The AmE equivalent verb is *putter*.

**poverty, poorness.** The dominant sense of *poor* is lacking adequate money or means to secure the necessities of life. The noun corresponding to this dominant sense is *poverty*, and *poorness* is not so used in modern English. In practice, the further the dominant sense is departed from, the more does *poverty* give way to *poorness*: e.g. *Poverty is no excuse for theft*; *the poverty* (not *the poorness*) *trap*; *the poverty* (or *poorness*) *of the soil*; *the poorness* (rather than *the poverty*) *of the harvest*; *the poorness of his performance*.

**p.p.** (or **pp**) is now widely regarded as an abbreviation of L *per pro* 'for and on behalf of', and is used against a principal's typed name when his or her secretary or other agent is signing on his behalf. By others it is regarded as an abbreviation of L *per procurationem* (hence the abbr. *per proc.*) 'through the agency of', in which case it should appear against the signature of the agent or proxy rather than the principal. Though this is without doubt the true origin of the abbreviation, now that Latin is not so widely understood as it was in Fowler's day, *per pro* is much more usually understood to be the phrase for which *p.p.* stands. I am told that the abbreviation is not used in either sense in the US, where *Signed by X in Y's absence* or *Dictated but not read* are usual. *Dictated by Y but signed in his* (or *her*) *absence* is a commoner variation on this theme in the UK.

**-p-, -pp-.** Monosyllabic words containing a single vowel (*a, e, i, o, u*) before *p* normally double the consonant before suffixes beginning with a vowel or before a final *y* (*trapped, stepped, ripped, dropped, cupped, puppy*); but remain undoubled if the stem contains two successive vowels (*leaper, sleepy*), or a vowel + a consonant (*carping, helped, chirping, romped, pulped*). Words of more than one syllable follow the rule for monosyllables if their last syllable bears the main stress (*entrapped, but outleaped*). They also double the *p* if, like *handicap, kidnap*, and *bebop*, the final syllable is fully pronounced, as opposed to the obscure vowel, i.e. /ə/, in *jalap* or *gallop*, or if, like *horsewhip* and *sideslip*, they are compounded with a monosyllable. The main exceptions are *worshipped, worshipper*, etc. (but often *worshiped*, etc., in AmE), and (occas. in AmE) *kidnaped, kidnaper*, etc.

Examples that follow the rules:
(*-p-*) *chirruped, enveloping, filliped, galloper, gossipy, hiccuped, jalaped, scalloped, syrupy, turnipy, walloping.*
(*-pp-*) *bebopper, equipped, handicapped, horsewhipping, kidnapper, kneecapped, shipping, sideslipped, worshipper.*

**practicable, practical. 1** The negative forms are *impracticable* (first recorded in the mid-17c.), *unpracticable* (1647, now rare), *impractical* (1865), and *unpractical* (1637). Of these, *impractical* is tending to encroach on the proper territory of *impracticable*: see IMPRACTICABLE, IMPRACTICAL.

**2** In current use *practicable*, when applied to things, policies, etc., means 'that can be done or used; possible in practice' (e.g. *Schemes which look very fine on paper, but which, in real life, are not practicable*). By contrast *practical* means 'concerned

with practice or use rather than theory; (of a person) inclined to action rather than speculation; able to make things function well' (e.g. *Having considered the problem he came up with several practical suggestions; He is a very practical person—able to paint his own house, to repair his car, etc.; Lexicography is more of a practical than a theoretical art*). In the following extracts the wrong word of the two is used in each case: *But to plunge into the military question without settling the Government question would not be good sense or practicable policy; We live in a low-pressure belt where cyclone follows cyclone; but the prediction of their arrival is at present not practical.*

**practically.** Since the early 17c. the primary meaning has been 'in a practical manner', as opposed to *theoretically* or *formally* (e.g. *Questions which are theoretically interesting to thoughtful people and practically interesting to every one—Manchester Examiner*, 1886; *Yet how was this undertaking to be practically achieved?*—B. Duffy, 1987; *'How much are those bamboo calligraphy brushes, then? asks one practically minded (and doubtless short-budgeted) course participant—TES*, 1990; *At tertiary level there are ... options for students: the colleges offer an educational programme of six semesters ... and are practically oriented—European Sociological Rev.*, 1991.

But in the course of the 18c. it developed a tributary sense 'virtually, almost', and now the tributary is at least as broad as the main stream. Examples of the second sense: *The application was supported by practically all the creditors—Law Times*, 1891; *I'm still at the point where my relationship to my own work is practically nonexistent—P. Roth*, 1979; *sitting through lessons with practically nothing to show for them afterwards—R. Sutcliff*, 1983; *When he returns to his department just before lunch he finds the building practically deserted—A. Brink*, 1988; *Her mother smiled practically all the time, for no understandable reason—P. Ustinov*, 1989.

**practice, practise.** In standard BrE, *practice* is invariably used for the noun and *practise* for the verb. In AmE, *practice* is the dominant spelling of both noun and verb; but *practise* is also used by some writers for both parts of speech.

**pram** (perambulator). See ABBREVIA-TIONS 1.

**pratique** (licence to have dealings with a port). Pronounce /ˈprætiːk/.

**pre-.** In compounds whose second element begins with *e* or *i* a hyphen is normally used: *pre-eminent, pre-empt, pre-ignition.* In others the hyphen is not necessary, but is freely used if the compound is one made for the occasion, or if any peculiarity in its form might prevent its elements from being instantly recognized: (no hyphen) *pre-arranged, predetermine, prenatal, preoccupy, preschool*; (with hyphen) *pre-position* (to distinguish from *preposition*, the part of speech), *Pre-Raphaelite, pre-tax, pre-war.*

**precede** is the spelling of the word meaning 'to come or go before in time, order, etc. (*in the word* money, y *is preceded by a vowel*); and *proceed* that for the word meaning 'to make one's way, to go on' (*he paused and then proceeded to demolish his opponent's arguments*).

**precedence, precedent.** The *OED* (1908) placed the stress on the second syllable of *precedence* and gave this vowel as /iː/. For *precedent* it gave /ˈpresɪdənt/ for the noun, i.e. stressed on the first syllable, and /prɪˈsiːdənt/ for the adj. *Precedence* is now always stressed on the first syllable and pronounced /ˈpresɪdəns/. The noun *precedent* is pronounced /ˈpresɪdənt/ and the adj. (now rare, having been virtually replaced by *preceding*) is usu. pronounced /prɪˈsiːdənt/, occas. /ˈpresɪdənt/. In AmE, *precedence* is sometimes pronounced in the *OED* (1908) manner.

**preciosity, preciousness.** The first is now virtually restricted to mean 'over-refinement in art or language, esp. in the choice of words', leaving *preciousness* as the general abstract noun corresponding to all the main general senses of the adj. *precious* (of memories, metals, stones, etc.).

**precipitance, precipitancy, precipitation.** All three nouns, corresponding in their senses to those of the adjs. *precipitant* and *precipitate*, were first used in English between 1500 and 1700 and came to be used in several overlapping

senses. They have failed to fall into conveniently separable semantic compartments since then. All that one can usefully say is that the first two have tended to give way to *precipitation*, esp. in technical senses in chemistry and meteorology (of rain or snow falling to the ground) and to *precipitateness* ( = hastiness, rashness).

**precipitate.** 1 The verb is pronounced /prɪˈsɪpɪˌteɪt/ (secondary stress on final syllable) and the adj. /prɪˈsɪpɪtət/ (obscure vowel in final syllable).

2 See next.

**precipitous.** In general it is advisable not to use *precipitous* (whose normal sense is 'of the nature of a precipice; (of a cliff) having a vertical or overhanging face') when *precipitate* ( = hasty, rash, inconsiderate; headlong; violently hurried) is called for, even though the two adjectives were freely interchangeable for most of the period between the 17c. and the 19c. Fowler's examples (1926) clearly show contexts in which *precipitate* would have been the better word: *Are the workers justified in taking the* precipitous *action suggested in the resolution?*; *The step seems a trifle rash and* precipitous *when one remembers the number of banking and commercial failures that* [etc.].

The only snag is that some actions, decisions, etc., are over-hasty, rash, etc., and at the same time represent a steep decline from some assumed or self-evident norm. The following examples show a spectrum of meanings: (literal) *We will start at the Pacific Coast's* precipitous *headlands and sea stacks, home of thousands of murres, cormorants, and puffins—Nature Conservancy*, 1988; *There was a* precipitous *wooden stair . . . to the ground floor—A. Craig*, 1990; *a half-buried beck where you can scramble past still pools . . . along* precipitous *ledges lined with wild yew trees—National Trust Mag.*, 1992; *(figurative) The . . . report also acknowledged that there were substantive issues at work in the* precipitous *decline—A. Wilentz*, 1989; *Mr Sununu counts as the first casualty of President Bush's* precipitous *decline in public opinion polls—Independent*, 1991; *A number of factors might be responsible for such a* precipitous *decline, including overzealous lepidopterists who have ignored signs asking them to leave the beleaguered butterfly alone—Discover*, 1992 (US); (encroaching on the territory of *precipitate(ly)*: all the examples are American) *I left precipitously because I didn't want to work there any longer—A.* Cross, 1986; *After the precipitous English victory over the Pequot, some three hundred Niantic warriors suspected of hiding Pequot refugees refused a challenge from forty of Mason's men—W. S.* Simmons, 1986; *According to surveys of parents, the middle class had precipitously abandoned its Watsonian regime* [etc.]—B. Ehrenreich, 1989; *Nor has he correctly addressed the problem of the black-on-black township violence that has brought the country precipitously to civil war—World Press Rev.*, 1991.

**précis.** The accent of the original French should be retained in the English word.

**preciseness, precision.** *Preciseness* is the natural abstract noun corresponding to the adj. *precise* in its main senses and was in use in the sense 'exactness, accuracy' in the 16c. nearly a century before *precision* came to be used in this way (its natural sense being 'the abrupt cutting off of one thing from another', from L *praecīsiō-nem* 'a cutting off abruptly'). The main distinction now is the common use of *precision* (but not *preciseness*) as an attributive noun in the sense 'marked by or having a high degree of accuracy (*precision instruments, precision timing*)'.

**predicate.** 1 The corresponding *-able* adj. is *predicable*: see -ABLE, -IBLE 2(v).

2 The verb is pronounced /ˈpredɪˌkeɪt/ (secondary stress on final syllable) and the noun /ˈpredɪkət/ (obscure vowel in final syllable).

3 The verb *predicate*, which is derived from L *praedicāre* 'to cry in public, to proclaim', means (*a*) to assert or affirm as true or existent; (*b*) (followed by *on*) to found or base (a statement, etc.) on (e.g. *a new conception of reality is ... predicated on dissatisfaction with formalist literature—TLS*, 1973). It should not be confused with the verb *predict* (derived from L *praedīcere*) 'to foretell', though it should be noted that the *OED* lists examples from 1623 to 1897 of *predicate* used erroneously for *predict*.

**predominantly, predominately.** Both words mean 'in a predominant manner'

and both have a long history: *predominantly* is first recorded in 1681 and *predominately* in 1594, but the latter was rare before the 19c. In the late 20c. *predominantly* is overwhelmingly the commoner of the two words and is the one recommended here. Examples (which happen to be N. Amer.): (*predominantly*) *Clinton had seized the opportunity to talk past the predominantly black audience in the room and deliver a message to white America—Seattle Times*, 1992; *the Enterprise Forum of the Massachusetts Institute of Technology, which meets monthly and deals predominantly with investor-backed high-tech startups—Boston Globe*, 1993; *The gutsy mayor won ... in the predominantly White city—Ebony*, 1993; (*predominately*) *Eastview, Ont., a predominately French-speaking, low-income suburb of Ottawa—Gaz.* (Montreal), 1992; *Jablanica, a predominately Muslim town, was shelled from the hills by Croats—USA Today*, 1993. It is worth noting that the old adj. *predominate* (first recorded in 1591, probably modelled on such adjs. as *moderate, temperate*) is now virtually extinct, having given way to *predominant* (first recorded in 1576).

**preface.** 1 For *preface* and FOREWORD, see the latter.

2 For *preface* and PREFIX (verbs), see the latter.

**prefectoral, prefectorial.** The first of these (from Fr. *préfectoral*), first recorded in English in 1872, seems to be the usual form (besides *prefectorial*) used of regions, jurisdictions, etc., in France, Japan, and elsewhere, known as prefectures. The customary adj. in schools that have prefects is *prefectorial* (formed from late L *praefectōri-us*), which is first recorded in English use in 1862. Examples: (*prefectoral*) [*He wanted to keep the van Gogh painting*] *close at hand for the time being but ... eventually he would like to display it at the Shizuoka Prefectoral Art Museum—Washington Post*, 1990; *This prefectoral decision reveals the fundamental importance of local circumstances—Fr. History*, 1991; *it has just drawn up a programme that covers Lome as well as all the regional capitals, all the prefectoral capitals* [*in West Africa*] [etc.]—*BBC Summary of World Broadcasts*, 1993; (*prefectorial*) *With these prefectorial words, Britain's environment secretary, Mr Michael Heseltine, waves his stick once again at his* unruly brood of local authorities—Economist, 1981; The Gumma Prefectorial Police also are targeting 16 Japanese citizens ... as being responsible for contributing to the cause of the accident—Business Insurance, 1988 (US); There is no prefectorial system in the college, just upper and lower school councils and something called the committee—TES, 1991.

**prefer.** 1 The inflected forms of the verb are preferred, preferring (see -R-, -RR-); but all other formations containing the base prefer (preferable, preference, preferential, etc.) have a single r.

2 prefer that. In AmE, a subordinate that-clause after prefer frequently has its verb in the subjunctive: e.g. He prefers that the shoes be ready—New Yorker, 1986; Aunt Clara preferred that I not ransack the memories of her mother and Uncle Eugene—D. Pinckney, 1992.

3 prefer + to-infinitive or with gerund. Both constructions are used but the first is much the more frequent: ( + to-infinitive) I prefer to be alone here—O. Manning, 1955; Most of them preferred to stand, to move about, gracefully—G. Vidal, 1967; I prefer not to live and work in the same room—C. K. Stead, 1986; But he preferred not to torture himself with the specific knowledge [etc.]—A. N. Wilson, 1986; ( + gerund) He [sc. a parrot] has never said anything but prefers watching TV—M. Kington, 1985.

4 Limits on prefer to. With a following direct object the normal construction of prefer is with to (she preferred honey to marmalade; he preferred walking to jogging). When the object is an infinitive a rigid adherence to this rule could sometimes be calamitous, e.g. producing the unacceptable construction I prefer to die to to pay blackmail (rewrite as I prefer to die rather than pay blackmail). Even if a gerund is used in such a construction (I prefer to die to paying) the construction is still unacceptable (rewrite as above). An example of prefer correctly followed by a clause led by rather than: Bertrand ... preferred to go to a restaurant ... rather than eat alone—P. P. Read, 1986. It should be added that prefer ... than without rather is not acceptable: We should greatly prefer to pay the doctors more than to limit the area of insurance (rewrite as We would much rather pay ... than limit); Many prefer to go bareheaded than to reassume the fez (rewrite

as Many go bareheaded rather than reassume the fez).

**preferable.** 1 The stress falls on the first syllable, /ˈprefərəbəl/, not on the second.

2 The double comparative form more preferable is an inexcusable pleonasm. Preferable by itself means 'more desirable (than)' and is therefore intensified, when necessary, by far, greatly, or much, but not by more: e.g. After a hundred and eighty [skips] an unclear head seemed much preferable to more skips—K. Amis (OGEU).

**prefigure.** For the pronunciation, see FIGURE 1.

**prefix.** 1 Both the noun and the verb are usu. stressed on the first syllable: thus /ˈpriːfɪks/. For its main senses, the OED (1908) gave /prɪˈfɪks/ as the pronunciation of the verb, and /priːˈfɪks/ is still sometimes heard. The shifting of the stress has presumably taken place in the course of the 20c.

2 In grammar, a prefix is an affix placed at the beginning of a word to adjust or qualify its meaning: e.g. be-, ex-, non-, re- in befog, ex-servicemen, non-smoking, reopen. Cf. SUFFIX.

3 Prefix (verb) is normally transitive in the sense 'to provide (something) as a beginning or introduction' (e.g. Contributors are requested to prefix a synopsis to their articles). But the type to —— their articles with a synopsis requires preface as the verb or needs to be re-expressed (e.g. to place a synopsis at the beginning of their articles).

**prejudgement.** For the spelling with medial -e-, see JUDGEMENT.

**prejudice.** In the senses 'bias' and 'partiality', it is followed by against and in favour of respectively. The type a marked prejudice to the eating of meat (probably modelled on objection to) is erroneous.

**preliminary.** See QUASI-ADVERBS.

**prelude.** Both the noun and the verb are now pronounced with the stress on the first syllable, /ˈpreljuːd/. But the OED reports that /prɪˈljuːd/ was indicated for the verb in the early part of the 19c. by 'all the verse quotations and the dictionaries down to c1830'. See NOUN AND VERB ACCENT.

**premature.** The pronunciation /'premə-,tjʊə/ is recommended, i.e. with the main stress on the first syllable and with the final syllable fully pronounced and not reduced to /-tʃə/. Some speakers reverse the order of the main and secondary stresses. In AmE the first syllable is pronounced with a long vowel, and the *tu* is usu. not palatalized: thus /,pri:mə'tʊər/.

**premier.** As noun its main current sense is 'a prime minister or other head of government'. It is particularly common in this sense in Canada. Examples: *New-foundland Premier Clyde Wells announced his support for Quebec's 'distinct society'—Globe & Mail* (Toronto), 1991; *She heard them talking about the same boring things: the new Japanese premier, factory explosions,* [etc.]—A. Tan, 1991 (US); *In 1980, they voted No to sovereignty-association ... and cast their ballots for Premier Robert Bourassa's Liberals in the ... provincial elections—Macleans Mag.,* 1992.
As adj. it is enjoying a period of great popularity in the senses (a) first in order of importance, order, or time: e.g. *In the Middle Ages Spain and Britain competed for the premier position as producers of high-quality wool—L. Alderson,* 1980; *It would be nice to chew the fat with Silver Falls's premier tourist attraction—W. Stewart,* 1990 (Canad.); *Claiming that the female of the breed is 'a cow for all systems' the organisers ... set about demonstrating their belief through the premier shop window of the Holstein breed—Field,* 1990; *Hypersonic flight has become the premier area for aerospace research in the United States—Mechanical Engineering,* 1991 (US); (b) specif. of credit cards, football leagues, etc.: *Should I seek an injunction against Barclays Bank for sending me—again—an invitation to avail of their Premier Card, accompanied by an application form?—Punch,* 1986; *Nobody should be in any doubt that the Premier League will need the trust's money—Economist,* 1992. (c) of earliest creation (*premier earl*).
Both noun and adj. are pronounced in BrE with a short vowel in the first syllable, i.e. /'premɪə/. Perhaps the dominant of several pronunciations in AmE is /prɪ'mɪər/, stressed on the second syllable.

**première.** Now fully established as a noun ( = the first performance or showing of a play or film) and as a verb ( = to give a première of). According to the OED, the noun is first recorded in English in 1889, and the verb in 1940. The pronunciation in BrE is /'premɪ,eə/, and the dominant one in AmE probably /prɪ'mɪər/.

**premise, premiss.** As noun, (a) In logic, a previous statement from which another is inferred (usu. spelt *premiss*). Pl. *premisses* or *premises*. (b) In the pl. form *premises*, a house or building with its grounds and appurtenances. As verb, *premise* is pronounced /prɪ'maɪz/, rhyming with *surmise*, or as /'premɪs/, and means (a) to say or write by way of introduction; (b) to assume from a premiss. The verb is always spelt with *-ise*, not with *-ize*.

**premium.** Pl. *premiums*. See -UM 1.

**prep.** See ABBREVIATIONS 1.

**preparatory.** For the use in *They were weighing it preparatory to sending it to town*, see QUASI-ADVERBS.

**prepared to.** Bombarded as he was by official documents of many kinds, Sir Ernest Gowers developed a distaste for formulaic phrases such as *is not prepared to, is not in a position to,* and *does not see his way to.* 'Such phrases as these are no doubt dictated by politeness, and therefore deserve respect. But they must be used with discretion' (cited from the 1986 revised edition of *The Complete Plain Words*.) He condemned such examples as *he was not prepared to disclose the source of his information; I am prepared to overlook the mistake* 'as wantonly blurring the meaning of *prepare*' (Gowers, 1965). '*Prepared to*', he added, 'should be reserved for cases in which there is some element of preparation, e.g. *I have read the papers and am now prepared to hear you state your case.*' The advice is perhaps being followed in Whitehall, but has not reached the rest of the world. Examples: *I came here to place it before a body of persons of European distinction. I am not prepared to discuss it with an irresponsible young woman—G. B. Shaw,* 1939; *The Government was not prepared to fight for realistic exclusive zones for British fishermen—Daily Tel.,* 1976; *Mr Baker stressed ... that the US and its allies were not prepared to compromise on 'full' membership of Nato for Germany after unification—Guardian,* 1990; *That suggests the rethink is motivated by the belief that more local authorities would be prepared to remove bad teachers—TES,* 1990; *If non-executives are to carry out their duties properly,* *they must be prepared to blow the whistle—Independent*, 1991.

The phr. *be prepared to* seems to have fully established itself when the sense required is simply 'to be willing or disposed (to do something)'. The same applies to the negative equivalent *be not prepared to*.

## prepositions.

> **1** Definition.
>
> **2** Preposition at end: (*a*) History of attitudes; (*b*) Circumstances in which a preposition may or even must appear at the end of a clause or sentence; (*c*) Anecdotal; (*d*) A select list of stranded prepositions in the works of writers of past centuries.
>
> **3** Final verdict.

**1** *Definition*. For those who have forgotten the elementary terminology of grammar, here is a standard definition of the term *preposition*: 'a word governing (and usu. preceding) a noun or pronoun and expressing a relation to another word or element, as in: "the man *on* the platform", "came *after* dinner", "what did you do it *for?*"' (*COD*, 1995). See also COMPOUND PREPOSITIONS. Just under a century earlier, J. C. Nesfield's famous *Manual of English Grammar* (1898) gave as his examples: *I put my hand on the table*; *A bird in the hand is worth two in the bush*; *He is opposed to severe measures*. And Nesfield reminded his readers that two prepositions are often needed: e.g. *The mouse crept out from under the floor*; *The rabbit escaped by running into* [clearly thought of by Nesfield as two words] *its hole*. Besides nouns and pronouns, the objects to prepositions can also be adverbs (*By* | *far the best*), phrases (*The question of* | *how to do this is difficult*), and clauses (*He told every one of* | *what we had heard*). (Examples from Nesfield.) After a preposition the objective form of a pronoun, where it differs from the subjective form, must always be used: *believe in* him; *between* us; *for* them. This is especially important when two pronouns are linked by *and* or *or*: *between you and* me (not *I*); *a gift from my brother and* me (not *I*); *asked if there was any chance of* him (not *he*) *and Gina reconciling*.

**2** *Preposition at end*. (*a*) *History of attitudes*. One of the most persistent myths about prepositions in English is that they properly belong before the word or words they govern and should not be placed at the end of a clause or sentence. Apparently Dryden set the myth going. In his *Defence of the Epilogue* (1672) he cited a line from Ben Jonson's *Catiline* (1611), *The bodies that those souls were frighted from*, and commented, 'The Preposition in the end of the sentence; a common fault with him, and which I have but lately observ'd in my own writings.' At some later date, it is believed, Dryden made a partial attempt to remove end-placed prepositions from his other prose works. In a note on this passage in the Everyman edition, George Watson comments that 'Each instance of the solecism was corrected by Dryden for the second edition (1684) of the essay *Of Dramatic Poesy*.' A Dryden scholar whom I consulted, Dr Derek Hughes, reported, however, that he hadn't found any evidence that Dryden conducted a more general purge of his early prose. 'He certainly did not tidy up the famous description in the Preface to *Annus Mirabilis* of "the faculty of imagination in the writer, which like a nimble Spaniel, beats over and ranges through the field of Memory, till it springs the Quarry it hunted after".'

The myth became entrenched, though the grammarians were inclined to treat the stranding of prepositions simply as a matter of informality rather than of error. Robert Lowth in his *Short Introduction to English Grammar* (1775), for example, after citing several examples from Shakespeare (e.g. *Who servest thou under?—Henry V*; *Who do you speak to? —As You like It*) and Pope (*The world is too well-bred, to shock authors with a truth, which generally their booksellers are the first that inform them of—*Preface to his poems), concluded: 'This is an idiom, which our language is strongly inclined to: it prevails in common conversation, and suits very well with the familiar style in writing: but the placing of the preposition before the relative, is more graceful, as well as more perspicuous; and agrees much better with the solemn and elevated style.'

Fowler turned to Ruskin's *Seven Lamps of Architecture* (1849) and reported that: 'In the text of the *Seven Lamps* there is a solitary final preposition to be found, and no more; but in the later footnotes they are not avoided (*Any more wasted* words ... *I never heard of.* | *Men whose occupation for the next fifty years would be the knocking down every beautiful building they could lay their hands on).*' The natural inference, Fowler decided, would be: 'you cannot put a preposition (roughly speaking) later than its word in Latin, and therefore you must not do so in English.'

Alford (1864) regarded the placing of prepositions at the end as something that 'is allowable in moderation, but must not be too often resorted to'. Henry Sweet (1891) set down various circumstances in which what he called 'detached prepositions' are 'liable to be disassociated from their noun-words not only in position, but also in grammatical construction, as in *he was thought of,* where the detached preposition is no longer able to govern the pronoun in the objective case because the passive construction necessitates putting the pronoun in the nominative. Prepositions are also detached in some constructions in connection with interrogative and dependent pronouns and adverbs, as in *who are you speaking of?, I do not know what he is thinking of, where is he going to?, I wonder where he came from*; such constructions as *of whom are you speaking?* being confined to the literary language ... Although detached prepositions approach very near to adverbs, yet they cannot be regarded as full adverbs for the simple reason that those prepositions which are otherwise never used as adverbs, such as *of,* can be detached with perfect freedom.'

The myth of the illegitimacy of deferring prepositions had clearly been destroyed by the end of the 19c. and modern grammars simply recognize that 'Normally a preposition must be followed by its complement, but there are some circumstances in which this does not happen' (*CGEL* 9.6).

(b) *Circumstances in which a preposition may or even must appear at the end of a clause or sentence.*

(i) Relative clauses: *I certainly don't know any that I'm attracted to*—A. Lurie, 1965; *He was certainly not a man she could lean upon*—G. S. Haight, 1968; *They must be entirely reliable and convinced of the commitment they are taking on*—TES, 1987; *The Falls on the Peregrine River was nothing like the waterfalls you see pictures of*—A. Munro,

1993. More formally: *I wanted a window seat from which I could watch the road she would come along*—G. Greene, 1980 [contains both an undeferred and a deferred preposition]. (ii) *Wh*-questions (i.e. questions in which *what, where, which, who,* and other pronouns and conjunctions beginning with *wh* are used): *What did Marion think she was up to?*—Julian Barnes, 1980; *What should we talk about?*—P. Lively, 1987; *Who is it you are smiling at so beautifully?*—P. Lively, 1987; *He said, 'What do you want an old dump like this for?'*—S. Bellow, 1987; *What do you want to go there for?*—J. Hecht, 1992.

(iii) *Wh*-clauses: *No one had said whom I reported to*—D. Davie, 1982; *Nowadays insolence is what you survive by*—London Rev. Bks, 1987; *She looks out the window and says a bit of poetry and they know who has gone by*—New Yorker, 1987; *Budget cuts themselves are not damaging: the damage depends on where the cuts are coming from*—Spectator, 1993. More formally: *We found an Italian restaurant, or rather we went in search of one of which I had already heard*—A. Brookner, 1990. (iv) Exclamations: *What a shocking state you are in!*

(v) Passives: *Even the dentist was paid for*—New Yorker, 1987; *Laszlo was made a fuss of by both teachers and fellow students*—P. Lively, 1987; *Believe me, it's not eyes they're interested in*—L. S. Schwartz, 1989. (vi) Infinitive clauses: *he still had quite enough work to live on*—P. Fitzgerald, 1980; *In our forays we spent our time looking for V-shaped branches to make catapults with*—Encounter, 1981; *the conflict would be hard to live with if* [etc.]—London Rev. Bks, 1981; *Daddy never had any savings to speak of*—D. Lodge, 1991.

(c) *Anecdotal.* (i) A correspondent in SPE Tract xv (1923) quoted a context in which deferred prepositions are used in an absurd manner: [Sick child] *I want to be read to.* [Nurse] *What book do you want to be read to out of?* [Sick child] *Robinson Crusoe.* [Nurse goes out and returns with *The Swiss Family Robinson.*] [Sick child] *What did you bring me that book to be read to out of for?* (ii) It is said that Mr Winston Churchill once made this marginal comment against a sentence that clumsily avoided a prepositional ending: *This is the sort of English up with which I will not put*—Gowers, 1948.

(d) *A select list of stranded prepositions in the works of writers of past centuries.* Anyone who is in any doubt about the frequency

of occurrence over the centuries of prepositions placed at the end of clauses or
sentences may wish to browse in the *OED*
entries for *about, by, for, from*, etc., used
as prepositions, or in Jespersen (1909–49,
iii. 10.2–6). The following examples are
drawn from these two sources: they represent only a fraction of the examples
provided in these two places alone. *Me
lihtide candles to æten bi (OE Chron.,* 1154);
*This thinge the whiche ye ben about* (Chaucer,
c1385); *The most of them ... attempt ...
vnlawfull meanes to liue by* (P. Stubbes,
1583); *I would haue told you of good
wrastling, which you haue lost the sight of*
(Shakespeare, 1600); *The worke which himselfe and Paul went about* (D. Rogers, 1642);
*Many stories of the lady, which he swore to
the truth of* (Fielding, 1749); *The grass it
fed upon* (H. Martineau, 1832); *What is a
clock good for?* (P. *Parley's Ann.*, 1840); *I
am far from saying that merit is sufficiently
looked out for* (A. Helps, 1847–9); *Which eye
can you see me upon?* (R. Hunt, 1865);
*Resolutions which perhaps no single member
in his heart approves of* (J. Bryce, 1888).

**3** *Final verdict.* In most circumstances,
esp. in formal writing, it is desirable to
avoid placing a preposition at the end
of a clause or sentence, where it has the
appearance of being stranded. But there
are many circumstances in which a preposition may or even must be placed late
(see 2(*b*) above), and others where the
degree of formality required governs the
placing. When formality is desired, *of
which I had already heard*, for example, as
Anita Brookner wrote (see 2(*h*) (iii) above),
has the advantage over *which I had already
heard of*.

**prerequisite.** Now occasionally misused for *perquisite*: e.g. *If she is not released,
perhaps she could be treated to the same
prerequisites enjoyed by the inmates of Park-
hurst prison—Independent*, 1991. A *perquisite*
(or *perk*) is an incidental benefit attached
to one's salary, etc., a customary extra
right or privilege; a *prerequisite* is something required as a precondition.

**presage.** The noun is stressed on the
first syllable: thus /ˈpresɪdʒ/. The verb may
be pronounced the same, or with the
stress on the second syllable, i.e. /prɪ-
ˈseɪdʒ/. In the *OED* (1909) the only pronunciation given for the verb was that with
second-syllable stressing. For the noun

the *OED* set down /ˈpresɪdʒ/ and /ˈpriː-/,
and added 'formerly /prɪˈseɪdʒ/'. It is remarkable that these old loanwords from
French (the noun in the 14c. and the verb
in the 16c.) have still not found a stable
place in the English stress-system. See
NOUN AND VERB ACCENT.

**prescience, prescient.** The customary
pronunciations now in BrE are /ˈpresɪəns/
and /ˈpresɪənt/, whereas the *OED* (1909)
gave only /ˈpriːʃɪəns/ and /ˈpriːʃɪənt/, with
a long first vowel and a palatalized
medial consonant.

**prescribe, proscribe.** One main meaning of *prescribe* is 'to lay down as a course
or rule to be followed'. Another main
meaning is 'to issue a (medical) prescription or recommend a specified (medical)
treatment'. It should be carefully distinguished from *proscribe* in its sense 'to
reject or denounce (a practice, etc.) as
dangerous, antisocial, etc.; to condemn,
outlaw'. An easy-to-remember distinction is that a prescribed book is one that
is set down for special study (e.g. *'Beowulf'
was among the prescribed texts for the final
examination*), whereas a proscribed book
is one that is prohibited or banned (e.g.
Salman Rushdie's *The Satanic Verses* (1988)
is *proscribed* by the authorities in many
Muslim communities).

**prescriptivism.** In the *OED* this linguistic term is defined as follows: 'The
practice or advocacy of prescriptive
grammar; the belief that the grammar
of a language should lay down rules to
which usage must conform.' It is first
found in 1954. The corresponding adj.
*prescriptive* is first recorded, in its linguistic sense, in a 1933 work by Otto
Jespersen: it is contrasted with *descriptive*
and equated with *normative*. Since the
terms *prescriptivism* and *descriptivism* now
occur in almost every kind of linguistic
work, it might be useful to readers if I
quote from my book *The English Language*
(1985) a short passage describing the
nature of each. 'In the present century,
starting more or less with the work of
Ferdinand de Saussure, emphasis has
been placed much more firmly than hitherto on language as it is used rather than
on how experts say that it should be
used. There is no clear boundary between
the doctrines of prescriptivism and those
of descriptivism, much more an attitude

of mind. Prescriptivists by and large regard innovation as dangerous or at any rate resistible; descriptivists, whether with resignation or merely with a shrug of the shoulders, quickly identify new linguistic habits and record them in dictionaries and grammars with [little or] no indication that they might be unwelcome or at any rate debatable. Prescriptivists frequently use restrictive expressions like *loosely, erroneously, sometimes used to mean, falsely, avoided by careful writers*, and the like ... Readers can easily distinguish between the two approaches by consulting the entries for expressions like *all right/alright, anticipate, decimate, dilemma, disinterested, due to, enormity, hopefully, imply/infer, minuscule*, and *refute* in the two types of reference book.' Much more could be said on the subject, and said better, but the above remarks will serve perhaps to introduce readers to one of the great linguistic battles of the 20c. See DESCRIPTIVE GRAMMAR.

**present** (noun and verb). The noun is stressed on the first syllable, and the verb on the second. See NOUN AND VERB ACCENT 1.

**presentiment, presentment, presentient.** There is some hesitation about the pronunciation of the medial *s* in these words, but the dominant sound is /z/ for the first and second (under the influence of *present*) and of /s/ in the third (because of the visible connection with *sentient*).

**presently.** The oldest sense (15c.) of those still current is 'at the present time, now'; it is now chiefly used in America and Scotland. What the OED calls its 'blunted sense', because it became weakened at an early date (by the 17c.) from 'immediately' to 'in a little while, soon', is not quite as old, but is now the main current sense in England and elsewhere. Examples: (a) ( = now) *Mr. William O'Brien, solicitor, Dumfries, for the accused, said Mr. Savage was presently unemployed, his last employment being a year ago—Dumfries Courier*, 1978; *Dr Otto von Habsburg abandoned claims to the monarchy in 1961 ... and is presently a member of the European Parliament—Times*, 1989; *'We are presently climbing through 30,000 feet on our way to our cruising altitude of 37,000 feet,' says the pilot—Chicago Tribune*, 1991 (the journalist who wrote the article regarded this as a misuse of *presently*); *After the war, women had equal access to university training but, even presently, many are not encouraged to attempt the most prestigious departments, such as economics, in the most prestigious schools—Dædalus*, 1992.

(b) ( = in a little while, soon) *'A very curious one, as you shall see presently,' replied Jack—R. M. Ballantyne*, 1858; *Presently, seeing Ralph under the palms, he came and sat by him—W. Golding*, 1954; *Muriel waited for an hour ... Her feet hurt and she was thirsty. Presently she set off to walk back to her lodgings—H. Mantel*, 1986. There is much criticism of the use of sense (a) both in America and in Britain because of the risk of ambiguity, but in practice the context normally makes it quite clear which sense is intended.

(c) ( = immediately). This sense flourished between the 15c. and the 18c. but is now obsolete. Historical examples: *the Master and the Boat-swaine Being awake, enforce them to this place; And presently, I pre'thee.—Shakespeare*, 1610; *The poor woman ... no sooner looked at the serjeant, than she presently recollected him—Fielding*, 1749.

**present tense.** The natural and most frequent use of the present tense is in contexts of present time, whether actual (*he wants to know now; the door is open*) or habitual (*Paris is the capital of France; he has his pride; the clock strikes every half-hour*). It is also used of past events (*A British writer wins the Nobel Prize*—newspaper headline) and of future events (*When do* [ = will] *you retire?; the Paddington train for Didcot leaves at 9.15 pm; term ends tomorrow*). For the historic present, see HISTORIC 2.

**present writer, the** (or **present author,** etc.). The phr. *the present writer* is used from time to time instead of *I* or *me* by writers whose name appears at the head or end of an article or other work. Fowler called the phrase a *Coa vestis* (transparent garment of fine silk made at Cos), and dismissed it as 'irritating to the reader'. But it is used by the great and the good and seems a harmless convention. Cf. *your columnist, your reviewer* (as used by journalists). Examples: *At last, the present writer, asked to say how he would go about it, replied in 'The ADS Dictionary—How Soon?' ... and in 1962 found himself appointed Editor—F. G. Cassidy, introd. to*

*DARE* i, 1985; *Like the present writer, Stevenson, Davenport, and Cousland had family connections with the Catholic Apostolics and hence greater ease of access to source material*—C. G. Flegg, 1992. But passions run high in the matter. In the 30 Nov. 1989 issue of the *Independent*, the journalist Miles Kington wrote *We agree that if we are writing in the first person, we will use the word 'I' and not nauseating expressions such as 'your humble author', 'my very good self', 'the present writer', 'a person not a million miles from myself,* [etc.]. And note also *the journalist feels no need to avoid using the first-person singular, 'I' or 'me', when necessary. He doesn't have to pretend he is someone else, with coy references to 'the present author'* (N. Bagnall, 1993). So perhaps caution is required after all.

**preserve.** Both the noun and the verb are stressed on the second syllable. See NOUN AND VERB ACCENT 2.

**Presidents' Day.** The third Monday in February, a legal holiday in most of the states of the US, commemorating the birthdays of George Washington and Abraham Lincoln. Often erroneously printed with an apostrophe before the *s* or without an apostrophe at all.

**prestidigitator, -tion.** Formal or humorous words for a juggler, a conjuror/juggling, conjuring. Adopted from Fr. *prestidigitateur/prestidigitation* ('quick-fingeredness') in the mid-19c.

**prestige. 1** Adopted in the 17c. from French in its original sense of 'illusion, conjuring trick', *prestige* (as also in Fr.) lost this sense in the course of the 19c. and gained the favourable meaning 'glamour, reputation derived from previous achievements, associations, etc.'. Etymologically the word answers to L *praestigium* 'a delusion, illusion', usu. in the feminine pl. form *praestigiae* 'illusions, juggler's tricks', for earlier *\*praestrigium*, from *praestringere* 'to bind fast' (*praestringere oculos* 'to blindfold', hence 'to dazzle the eyes').

**2** Surprisingly, the pronunciation /preˈstiːʒ/ has persisted though, according to the *OED* (1909), an Anglicized pronunciation /ˈprestɪdʒ/ also existed at the beginning of the 20c. Cf. *vestige*, the pronunciation of which has been totally Anglicized.

**prestigious.** A striking example of a word which has lost its original sense ('deceptive, illusory, in the manner of a conjuror's trick': see prec.) and has become the adj. corresponding to the noun *prestige* in the course of the 20c. The new, and now virtually the only current, meaning 'having or showing prestige' was first noted in Joseph Conrad's novel *Chance* (1913). As the century proceeded it appeared sporadically in journalistic sources, and then, against stiff opposition, in standard literary works. It also went through a period of wavering between the pronunciations /preˈstiːdʒəs/ and /preˈstɪdʒəs/ before the latter became dominant. *Prestigious* is challenged by *prestige* used attrib. (*prestige car, group, location, model, ware,* etc.), and to a minor extent by the less euphonious form *prestigeful* (first recorded in 1956), but it remains part of the day-to-day standard language.

**prestissimo, presto.** When used as nouns, both have *-os* as their plural. See -O(E)S.

**presume.** See ASSUME.

**presumedly.** Four syllables if used: see -EDLY. But it is better to use *presumably* or another synonym.

**presumptive.** For *heir presumptive*, see HEIR 2.

**presumptuous.** Beware against using the once valid (15c.–early 19c.) by-form *presumptious.* An example of the erroneous form: *I thought it would be very presumptious of me to quote from myself as an epigraph*—*Times Saturday Rev.,* 20 July 1991.

**pretence.** The customary AmE spelling is *pretense.*

**preterite.** In AmE often *preterit.* In English grammar, the better form is *past,* e.g. '*jumped* is the past tense of *jump*'.

**pretty** is in regular use as an ironical adjective: *a pretty mess you have made of it; things have come to a pretty pass.* It is also used, esp. in colloquial contexts, as an adverb meaning 'fairly, moderately' (*the performance was pretty good; he did pretty much what he liked*), but only when qualifying another adverb or an adjective. Otherwise the adverb is *prettily* (*she* *dresses prettily; she arranged the flowers very prettily).*

**prevaricate** means 'to speak or act evasively or misleadingly; to quibble, to equivocate'. It does not mean 'to delay' though it is sometimes so used by confusion with *procrastinate* (e.g. *After prevaricating for six months, the Council is now . . .*).

**prevent.** There are three competing constructions when the verb *prevent* is followed directly or indirectly by a gerund in *-ing*. All three are legitimate, but the third type is beginning to fall into disuse. (*a*) *prevent* + object + *from* + *-ing* (*OED* 1663– ): *Sleeman had tried to prevent a widow from committing suttee*—R. P. Jhabvala, 1975; *An unavoidable explosion was followed by an outbreak of fire, which . . . prevented a rescue party from reaching them*—R. Thomas, 1986; *tanks are being prevented from entering the center of the city*—New Yorker, 1989; *And yet this love was like a barrier which prevented me from helping her*—C. Burns, 1989; *Cushla was only just quick enough to grab Colin's arms to prevent him from belting Restel across the head*—N. Virtue, 1990. (*b*) *prevent* + object + *-ing* (*OED* 1688– ): *Two women climb up the iron bars, which are meant to prevent people or animals falling under the tram*—J. Berger, 1972; *Muscle relaxants could prevent an actual seizure recurring*—R. D. Laing, 1985; *the Government now has an enormous challenge on its hands if it is to prevent its new environmental awareness backfiring in its face*—J. Porritt, 1989. (*c*) *prevent* + possessive + *-ing* (*OED* 1841– ): *Either the fire-guard or his mother prevents his reaching the fire*—H. T. Lane, 1928; *his shoes were locked up to prevent his running away*—P. Fitzgerald, 1986.
See POSSESSIVE WITH GERUND 2(iii).

**preventable, -ible.** The first is recommended: see -ABLE, -IBLE 3.

**preventive, preventative.** Both words entered the language in the 17c. and they have been fighting it out ever since. Both are acceptable formations, and the most that can be said is that the shorter form is the more frequent of the two (approximately 5:1 if computer printouts can be taken to be certain guides) and is the one recommended here for most contexts. An extensive search in the electronic databases available to me showed that the distribution of the two words is not governed by geography or subject, except to say that the word that follows is fairly likely to be *medicine, care, maintenance, measure*, or *detention*, or words closely related to any of these.

Some examples: (preventive) *Inoculation, another cross-cutting response, was the first effective specific preventive measure to be deployed against smallpox*—Amer. *Jrnl Public Health*, 1988; *Two days later university students demonstrated outside the palatial mansion of the President for the release of Professor Okola from preventive detention*—B. Head, 1989; *Although many preventive measures do improve health, they are not without risks and costs and in fact seldom reduce medical expenditure*—BMJ, 1989; *Is that service backed up with driver evaluations and preventive maintenance to reduce accident losses and downtime?*—Industry Week (US), 1990; *Preventive medicine may be more effective if problem-based learning programmes are established in place of the traditional methods of education*—Physiotherapy, 1990; *In this double-blind trial, the preventive effects of 300 mg acetylsalicylic acid were compared with those of 30 mg acetylsalicylic acid*—Lancet, 1991; *The Government . . . is not attracted by the preventive function of licensing, viewing it as overburdensome on trade*—Statute Law Rev., 1991; *'We are being used, clearly, and that's unfortunate,' declares Arnold J. Schecter, professor of preventive medicine at the State University of New York at Binghamton*—Sci. Amer., 1991. (preventative) *When we hear talk of a 'preventative strike' we must translate that term into what it really means: a surprise attack*—V. Mollenkott, 1987 (US); *If we happen to live near an airport . . . apart from the usual preventative measures such as double glazing and noise insulation to our homes, there is nothing we can do but try our best to live with it*—Internat. Health & Efficiency Monthly, 1990; *Provides an overview of current knowledge on genetic predisposition to chemical toxicity, reviewing experimental . . . evidence and considering the implications for preventative action and future research*—Koeltz Scientific Bks Catalog, 1992 (US); *The All England Club had agreed to co-operate with the ban [of a spectator from major tennis tournaments]. In the end, no such preventative measures were necessary*—Times, 1993.

Both words are also used as nouns: (preventive) *The ingestion of atabrine, the wartime substitute for quinine as a malaria preventive, has caused ears to ring for a lifetime—P.* Fussell, 1989; *And ask about ... ivermectin—the heartworm disease preventive you give to your dog once-a-month—Outdoor Life* (US), 1990; *Best wrinkle preventive around: a good sunscreen wth an SPF of 15 or higher—Homemaker's Mag.* (Toronto), 1993; (preventative) *Merck's has been overflowing for more than a decade ... with preventatives, mainly enzyme inhibitors, which 'block' or thwart many kinds of diseases—Kiplinger's Personal Finance,* 1991 (US).

See -ATIVE, -IVE.

**previous.** For the construction in *will consult you previous to acting*, see QUASI-ADVERBS.

**pride.** The common proverb *Pride goeth* (or *goes*) *before a fall* is one of several reductive versions of Prov. 16 : 18, *Pride goeth before [ = precedes] destruction: and an hautie spirit before a fall.* Several other variants, e.g. *Pride must have a fall* (Johnson, 1784), are listed in standard dictionaries of proverbs.

**prie-dieu.** Pl. *prie-dieux,* pronounced the same as the singular. See -X.

**priest.** A priest is **1** an ordained minister of the Roman Catholic or Orthodox Church, or of the Anglican Church (above a deacon and below a bishop), authorized to perform certain rites and administer certain sacraments.

**2** an official minister of a non-Christian religion (*COD,* 1995). Women who are ordained ministers of the Anglican Church are also called *priests* (the word *priestess* being used only of female priests of non-Christian religions). Cf. MINISTER; PARSON; PASTOR.

**priestess.** See prec. and -ESS 5.

**prima donna.** **1** The first word is pronounced /'pri:mə/ but in AmE sometimes /'prɪmə/.

**2** The pl. is *prima donnas.*

**prima facie.** In BrE normally pronounced /ˌpraɪmə ˈfeɪʃiː/. In AmE there are several competing pronunciations: /'prɪmə/ or /'praɪmə/ or /'pri:mə/, /'feɪʃi:/ or /'feɪʃi:/.

**primarily.** In the *OED* (1909) stressed only on the first syllable: thus /'praɪmər-ɪlɪ/. But now, under AmE influence, increasingly stressed on the second syllable: thus /praɪ'meərɪlɪ/.

**primary colours.** 'Any of the colours red, green, and blue, or (for pigments) red, blue, and yellow, from which all other colours can be obtained by mixing' (*COD,* 1995). This definition, while in essence correct, inevitably fails to bring out the immense complication of the subject of colours and how we perceive them. 'The study of colour falls within the fields of physics, physiology, and psychology,' reports the *Oxford Companion to Art* (1970). And it is to such works that the reader must be referred. Experts in the field consider red, green, and blue (RGB) the true *primary colours* as defined originally by the physicist James Clerk Maxwell (1831–79). These are the colours used in colour television and other similar systems. The other set of colours are technically referred to as the complementary colours or, in non-technical fields, as artists', or printers', primaries. The specific shades of 'red', 'blue', and 'yellow' were known by various names such as *process red, process blue, blue-green, process yellow,* etc., but are defined in modern texts as *magenta, cyan,* and *yellow.*

**primate** /'praɪmeɪt/. **1** any animal of the order Primates /praɪ'meɪti:z/, the highest order of mammals, including tarsiers, lemurs, apes, monkeys, and man.

**2** an archbishop. The ordinary pl. for both senses is *primates* /'praɪmeɪts/, i.e. is not pronounced like the name of the zoological order.

**primer.** The word meaning 'an elementary or introductory book' is always pronounced /'praɪmə/ in Britain, but /'prɪmə/ in America, New Zealand, and sometimes in Australia. It is interesting to note that the *OED* (1909) gave priority to the form with a short stem vowel. The form with /ɪ/ is preserved in Britain when the word is used of a size of type.

**primeval.** Now the dominant, and recommended, form, not *primaeval.*

**princess.** As an independent word, usu. stressed on the second syllable: thus /prɪn'ses/. When used as a title (*Princess*

*Diana, Princess Margaret)* it has an initial capital and is stressed on the first syllable: thus /'prɪnses/.

**principal, principle.** Confusion of the two betrays inadequate instruction at an early age. The first is an adj. ( = first in rank or importance) or a noun ( = a head, ruler, or superior; also, numerous other senses including 'a capital sum as distinguished from interest'). The second is only a noun, and means 'a fundamental truth or law as the basis of reasoning or action (*arguing from first principles*); (in pl.) rules of conduct (*a person without principles*)'.

**prioritize.** 'A word that at present sits uneasily in the language', I commented in vol. iii of *OEDS* (1982). The comment remains valid (1996), except that perhaps I could have indicated that the word has remained locked in the jargon of business managers, politicians, and other officials, i.e. among people who sometimes like to dress up their documents and speeches with high-sounding words. The word has not been found in print before 1968.

**prior to.** For its adverbial use ( = before), see QUASI-ADVERBS. As a complex preposition used instead of *before*, it is best used, if at all, in formal writing 'except in contexts involving a connexion between the two events more essential than the simple time relation, as in *Candidates must deposit security prior to the ballot*' (Fowler, 1926). In such contexts *prior to* means 'as a necessary preliminary to'.

**Priscian.** The phr. *to break* (or *knock*) *Priscian's head* (or *pate*) means 'to violate the rules of grammar' (representing L *diminuere Prisciani caput*). It has a long history in English from Skelton (*c*1525) onward, but is now seldom used. Priscian's *Institutiones grammaticae* (early 6c. AD) was the most voluminous grammar of his time, and was influential throughout the Middle Ages.

**prise** (verb), to force (something) open or out by leverage, occasionally written as *prize*, is a standard word in Britain. It corresponds to AmE *pry* used in exactly the same way. (In AmE, *prize* and *prise* are also found, but the dominant form is *pry*.) Examples: (prise/prize) *Then, with a joint prizing, the lock finally snapped—C.*

Dexter, 1983; *Coupar felt as though the wind was prising open his ribs—*T. Winton, 1985 (Aust.); *The hoard of money was prised out of Blue Rabbit and hidden at the back of his football-boot locker—*J. Trollope, 1990; *We did not speak again or look at each other till the chest was prized open and the lid raised—New Yorker,* 1991. (pry) (All US) *Let them come unhouse me of this flesh, and pry this house apart—*Marilynne Robinson, 1981; *Marilyn pried her hand from her father's and slipped beneath the bed to board a little boat—*Phyllis Burke, 1989; *she had to pry Esther's tightly coiled fingers open—*M. Golden, 1989; *The girl pried the lid from the shoebox—*Tom Drury, 1992.

*Pry* is an early 19c. shortening of *prise, prize* (verb), 'apparently through confusing the final consonant with the -s of the 3rd pers. sing. present' (*OED*).

**pristine.** 1 The *OED* (1909) gave only /'prɪstm/, i.e. with a short /ɪ/ in the second syllable, but /'prɪstiːn/ is now usual. The stress is variable, and the second syllable is sometimes pronounced /-staɪn/.

2 Before the 20c. the sense of the word in English usu. reflected that of L *pristinus* (belonging to olden times, antique, ancient, not new), and at the turn of the century most uses of the word were commendatory. In the nature of things, what is early and primitive is often judged to belong to a golden age—an age when life was marked by simplicity, was unspoiled and uncorrupted by civilization. Such uses are fully recorded in the *OED*. From there it is only a short step to transferred and weakened uses, in which the emphasis is placed upon spotlessness and newness. Such transferred uses of the adjective began to appear in AmE in the 1920s and are now widespread. They have attracted criticism, particularly in British usage guides. In *The Spoken Word* (1981) I therefore advised BBC announcers and presenters to restrict *pristine* to contexts in which the meaning required was 'ancient, primitive, old and unspoiled, e.g. *a sage of some pristine era*'; and to avoid its use when the meaning required was 'fresh as if new, e.g. *the ground was covered with a pristine layer of snow*'. Since then, on the evidence before me, it is becoming increasingly difficult to find fault with many of the weakened uses. A cross-section of typical examples follows:

with oblong patches of pristine colour marking the erstwhile positions of the heavier furniture—C. Dexter, 1983; thinking of the sour blackened brick of the place (scoured clean to a pristine rust once more)—P. Lively, 1991; The most common subjects [of children's china] ... should be under £75 unless they appear on uncommonly fine pottery in pristine condition—Antique Collector, 1992; Laura pictured Mrs. House sitting here, looking out over her estate, using a pristine set of watercolors to render a sunset—Antonya Nelson, 1992 (US).

**privacy.** The OED (1909) gave only /'praɪvəsɪ/ as the pronunciation, i.e. with /aɪ/ as in private in the first syllable. Since then /'prɪvəsɪ/ has supplanted it as the dominant form in Britain (cf. privilege, etc.), but the older pronunciation is still the usual one in English-speaking countries abroad, including America.

**privative.** In grammar, designating particles or affixes that express privation or negation. The a- of amoral and aseptic (see A-[1]) and the in- of innocent (cf. L nocere 'to hurt') and inedible are privative, whereas the a- of arise and the in- of insist are not.

**privilege.** Thus spelt, not privelege or other erroneous spellings.

**prize.** See PRISE.

**pro** (professional). See ABBREVIATIONS 1. Pl. pros: see -O(E)S 5.

**pro and con.** Pl. pros and cons, reasons or considerations for and against a proposition, etc.

**probable.** It is worth noting that whereas the adjectives able, impossible, possible, and unable may be idiomatically used with a following to-infinitive, improbable and probable may not. Thus we've been able to verify that; it was impossible to tell; it's not possible to be sure; unable to sleep under the blanket on the divan are all idiomatic, but none of the adjs. could be replaced by improbable or probable. In such infinitival constructions, likely is normally used in place of probable and unlikely in place of improbable: e.g. they are just as likely to vote for Herr Kohl; it must be extremely unusual and therefore unlikely to happen again. The matter is scarcely worth mentioning except that constructions of improbable and probable with a to-infinitive as complement were formerly possible. They are listed in the OED with 17c. examples (e.g. Nor was the design improbable to succeed—Clarendon, 1647; 'Tis probable to be the truest test—S. Butler, a1680), and are marked rare and obsolete respectively.

Cf. POSSIBLE 2.

**probe** (noun). 1 If an adjectival form is called for it would need to be spelt probeable (despite the rules given under MUTE E) in order to distinguish it from probable.

2 The Russian probe was not able to measure the lower 25 kilometres of the Venusian atmosphere, reported The Times in 1968. It was a newish use of the word probe. The primary sense (from the 16c. onward) of probe is a blunt-ended surgical instrument for exploring the direction and depth of wounds and sinuses. The early 20c. extension of the word to mean a small device, esp. an electrode, used for measuring, testing, etc., and the mid-20c. extension of the word to mean an unmanned exploratory spacecraft for obtaining information about the nature of planets and other bodies in outer space are natural uses of the word, and entered the language unopposed. Less well received has been the widespread 20c., mainly journalistic, use of probe to mean a penetrating investigation. Its appeal to journalists lies largely in its brevity and therefore its suitability for use in headlines. See HEADLINE LANGUAGE.

The word usefully illustrates a way in which a change has been made in the presentation of definitions in desk dictionaries in the last quarter of the 20c. The traditional practice was to place the definitions in date order, i.e. the order was governed by the date in which particular senses entered the language. Now, as the 8th edn (1990) of the COD informs us, the definitions 'are listed in a numbered sequence in order of comparative familiarity and importance, with the most current and important senses first'. As a result the meaning 'a penetrating investigation' is now the first numbered sense in the COD. In earlier editions it appeared in mid-entry. The new technique is often questioned because of the difficulty of establishing which are in fact 'the most current and important senses' of modern words in our widespread and swiftly changing language.

**problematic, -atical.** The shorter form is the more common of the two in both BrE and AmE sources, and there is no clear difference in usage. See -IC(AL).

**proboscis** /prəʊ'bɒsɪs/: the medial *c* is silent. The pl. is *proboscises* /-ɪsɪz/ or (esp. in zoological work) *proboscides* /-ɪdiːz/. The forms reflect Gk προβοσκίς, -κιδ-, 'an elephant's trunk'. Its use to mean the human nose is merely jocular.

**proceed.** Thus spelt. Contrast *procedure* and PRECEDE.

**process** (noun). The *OED* (1909) gave precedence to /'prɒses/, i.e. with a short vowel in the first syllable, but /'prəʊses/ is now standard in BrE. The standard pronunciation in AmE is with /ɒ/.

**process** (intr. verb), to walk in procession, is a back-formation from *procession*. It is pronounced /prə'ses/. The unconnected verb *process* meaning 'to treat (food, esp. to prevent decay); to deal with (data, etc.)' is pronounced like the noun (see prec.).

**pro-choice.** See PRO-LIFE.

**proclitic.** A word pronounced with so little emphasis that it becomes merged with the stressed word that follows it, e.g. in some forms of regional English down to the first half of the 19c., *chill* from *ich* 'I' + *will*; also, in modern English, *at* in *at home*, pronounced /təʊm/.

**procrastinate.** See PREVARICATE.

**proctor.** Thus spelt, not *-er*.

**procuress,** a woman who makes it her trade to procure women for prostitution. The word was first recorded in this sense in the 18c. See -ESS 5.

**produce.** Another example of a stress-contrast between the noun (first syllable) and the verb (second syllable). See NOUN AND VERB ACCENT.

**proem,** a preface or prelude to a literary work, is pronounced /'prəʊem/. The corresponding adj. is *proemial* /prəʊ'iːmɪəl/. It came into ME from OFr. *pro(h)eme*, ult. from Gk προοίμιον, used in the same sense.

**professedly.** Four syllables: see -EDLY.

**professor.** In BrE, a holder of a university chair, a university academic of the highest rank. In AmE, a university teacher who has the rank (in ascending order of seniority) of *assistant professor*, *associate professor*, or *professor*.

**professorate** /prə'fesərət/ and **professoriate** /ˌprɒfɪ'sɔːrɪət/. Both words have been used since approximately the mid-19c. to mean either (a) the office of a professor; professorship, or (b) the professorial staff of a university. In AmE the second form is sometimes spelt without the final *e* (*professoriat*).

**professoress.** It is recorded in use in the 18c. and 19c. in the sense 'a female professor (also occas., the wife of a professor)', but has now fallen out of use. See -ESS 3.

**proffer** (verb). The inflected forms are *proffered*, *proffering*. See -R-, -RR-.

**profile.** The *OED* (1909) gives the pronunciations /'prəʊfiːl/ and /'prəʊfɪl/ in that order. The only standard pronunciation now, namely /'prəʊfaɪl/, was first entered in the *COD* in 1976. See -ILE.

**prognosis.** Pl. *prognoses* /-siːz/.

**prognosticate** (verb). The corresponding adj. is *prognosticable*. See -ABLE, -IBLE 2(v).

**program, programme.** 1 (noun). There is no doubt that the standard spelling in BrE, except in computer language, is *programme* and in AmE *program* (all senses). But it was not always so. The word, which is derived from Gk πρό-γραμμα 'a public written notice', was taken into English in the 17c. in the form *program*, and was the form regularly used by Walter Scott, Carlyle, and numerous other 19c. writers. One could reasonably have expected this spelling of the word to have survived in standard British use after the model of *anagram*, *cryptogram*, *diagram*, etc. Instead, from about the beginning of the 19c., the French spelling *programme* was adopted, and gradually established itself except in the US. In computer work in all English-speaking countries, the spelling *program* is routinely used for 'a series of coded instructions to control the operation of a computer'.

**2** (verb). The inflected forms in BrE for all uses (including those in computer work) are *programmed, programming*; in AmE either *programed, programing* or with medial *-mm-*.

**progress.** In BrE the noun is pronounced /'prəʊgres/ and the verb /prə'gres/. See NOUN AND VERB ACCENT 1. In AmE the dominant forms show the same stress-patterns, but the noun is normally pronounced with a short vowel in the first syllable. When used transitively to mean 'to cause (work, etc.) to make regular progress' (e.g. *Welders to be trained to make more tack items to allow them to progress their own work to completion—Observer*, 1978) the verb is usu. pronounced /'prəʊgres/ in Britain.

**progression.** See ARITHMETICAL, GEOMETRICAL PROGRESSION.

**prohibit.** In the sense 'to forbid, stop, or prevent (a person)' + a verbal construction, the historical sequence is as follows: (*a*) (first recorded in 1523) + *to do* a thing (e.g. *The patients . . . are peremptorily prohibited to bathe on Sundays*—C. Lucas, 1756; *Marshal Oyama prohibited his troops to take quarter wthin the walls*—cited, and disapproved of, by Fowler, 1926). (*b*) (first recorded in 1840) + *from doing* something (e.g. *There is no Act . . . prohibiting the Secretary of State for Foreign Affairs from being in the pay of continental powers*—Macaulay, 1840). Type (*a*) is now rare and probably obsolescent, and type (*b*) is the standard construction.

**prohibition.** The *h* is now normally silent (thus /ˌprəʊɪ'bɪʃən/) in this word, but is fully sounded in related words in which the main stress falls immediately after it, as in *prohibit, prohibitive*, etc.

**project.** In BrE the normal pattern is /'prɒdʒekt/ for the noun and /prə'dʒekt/ for the verb. See NOUN AND VERB ACCENT 1. In other English-speaking countries the noun is often pronounced /'prəʊdʒekt/, but the type with a short vowel in the first syllable is also found, esp. in America.

**prolegomena.** A plural noun meaning 'introductory observations on a subject'. The singular, which is rarely used, is *prolegomenon* (from Gk προλεγόμενον,

neuter of the pres. pple passive of προλέγειν 'to say beforehand').

**prolepsis.** Pl. *prolepses* /prəʊ'lepsiːz/ or /-'liːp-/. A rhetorical and grammatical term derived from Gk πρόληψις 'anticipation'. It is used in English in four main senses: **1** The anticipation and answering of possible objections in rhetorical speech, e.g. a passage beginning *I know it will be said that . . .*

**2** The representing of a thing as existing before it actually does or did so, as in Hamlet's exclamation when he was wounded, *I am dead Horatio*. Also in Keats's poem *Isabella* (1820):

> So the two brothers and their murdered man
> Rode past fair Florence.

i.e. the man who was afterwards their victim.

**3** The foreshadowing of events which take place at a later stage in the narrative, e.g. in the OE poem *Beowulf* the mention of the harmony at the court of Hrothgar is accompanied by a warning of treachery to come:

> Heorot innan wæs
> freondum afylled; nalles facenstafas
> þeod-Scyldingas þenden fremedon.
> (Heorot within was filled with friends; not
> yet then had the Scylding people
> resorted to treachery.)

**4** In informal English, contexts in which 'a noun phrase is positioned initially and a reinforcing pronoun stands "proxy" for it in the relevant position in the sentence' (*CGEL* 1985, 17.78, where the construction is actually called *anticipated identification*). Example: *That man you spoke about, I saw him again this morning*. Other sources use *prolepsis* of different constructions, e.g. 'the anticipatory use of adjectives, as in *to paint the town red*' (*COD*, 1990). Cf. also Shakespeare's *Timon of Athens* iv.iii:

>           when Ioue
> Will o'er some high-Vic'd City, hang his poyson
> In the sicke ayre:

The air is not 'sick' until Jove introduces his poison into it.

**pro-life.** A term widely applied, at present esp. in AmE, to those who are opposed to a range of practices including abortion, assisted suicide, euthanasia, and the withdrawal of medical care or nutrition to severely handicapped newborn babies and people in irreversible vegetative states. It is contrasted with the term *pro-choice*, applied particularly to those who support a woman's right to choose to have an abortion. *A pro-choice Catholic is an oxymoron*, declared a San Diego bishop in 1989. Others saw the matter in less clear-cut terms: e.g. *the oversimplified and cosmetized banners of pro-life and pro-choice*—C. Grobstein in *Bull. Amer. Acad. Arts & Sci.*, May 1990. Both terms came into common use in the 1970s, joining some slightly earlier terms of a similar kind, e.g. *anti-abortion/ pro-abortion, right-to-life/right-to-die*. The debates continue.

**prolific. 1** The adj. is in common use, but there is no consensus about the choice of a corresponding noun. The possibilities are *prolificacy* (first recorded, according to the *OED*, in 1796), *prolificalness* (1860), *prolificity* (1725), and *prolificness* (1698). Perhaps the first and the last are the least uncomfortable, but substitutes such as *fertility, fruitfulness*, and *productiveness* are usually better solutions to the problem.

**2** *Prolific* is properly applied to someone or something that produces (offspring, literary works, paintings, etc.) in great abundance. It is less safely applied to what is produced, as in *His works, which are prolific, include many which have been translated into English* (read *(very) numerous*).

**prologue.** Another example of a word showing a 20c. change of preference in the length of the stem vowel. The *OED* (1909) gave /ˈprɒlɒg/ as its first choice. Now all UK dictionaries give only /ˈprəʊlɒg/. The word is often spelt *prolog* in AmE. Whether written as *prologuize* or *prologize*, the (rare) corresponding verb ('to write or speak a prologue') retains its hard /g/ and its (long) diphthongal sound in the first syllable.

**promenade.** Pronounce /-ɑːd/, not /-eɪd/. In AmE, /-eɪd/ is the dominant pronunciation.

**Promethean.** Stressed on the second syllable: /prəˈmiːθɪən/. See also HERCULEAN.

**prominence** (first recorded, according to the *OED*, in 1598) has virtually driven out its rival *prominency* (1645). See -CE, -CY.

**promiscuous.** Its minor 19c. colloquial use to mean 'carelessly irregular, casual, random' (e.g. *I walked in ... just to say goodmornin', and went, in a promiscuous manner, up-stairs, and into the back room*—Dickens, 1837) is being forced out of the language by its dominant sense '(of a person) having frequent and diverse sexual relationships, esp. transient ones'.

**promise** (verb). In the sense 'assure, assert confidently' (i.e. not expressing a future undertaking, etc.), *promise* is used only in the spoken language, and only then in the phr. *I promise you*. Examples: *Why that's nothing more than a trick of the candlelight, Rosanna. I promise you, there's no blood on the crucifix.*—*Islands* (NZ), 1985; *They were some bozo individuals, I promise you*—*New Yorker*, 1986; '*Hey, you're making that up.' 'No, it's true, I promise you.'*—M. du Plessis, 1989 (SAfr.); (Hannah) *You mean the game books go back to Thomasina's time?* (Valentine) *Oh yes. Further ... really. I promise you. I promise you.*—T. Stoppard, 1993. The use is recorded in the *OED* with 15-19c. examples.

**promisor.** Thus spelt in legal language, but *-er* in ordinary use.

**promissory.** So spelt, not *-isory*. The stress falls on the first syllable.

**prone.** In the sense 'disposed or liable, esp. to a bad action, condition, etc. (*is prone to bite his nails*)' (*COD*, 1990), *prone* is possibly being ousted by any of the following adjectives: *apt, inclined, liable, likely*. The *OED* provides abundant historical evidence, however, from the 14c. onward for expressions such as *prone to lechery, prone to idolatry, prone to meditation*, (with a *to*-infinitive) *prone to worship false gods, prone to err, prone to receive the faith*, none of which sounds unnatural. And the 20c. has produced a crop of formations of the type *accident-prone, violence-prone, drought-prone*, etc. So the word may not after all be heading for extinction in the sense and constructions given above. Cf. SUPINE.

**pronounceable.** Thus spelt. See -ABLE, -IBLE 2.

**pronouncedly** has four syllables. See -EDLY.

**pronouncement** is kept separate from its morphological partner *pronunciation* because the senses of the two words do not overlap. The first means only 'declaration, decision' which the other never does.

**[pronounciation].** An all-too-common misspelling and mispronunciation of PRONUNCIATION.

**pronouns. 1** *Types of pronouns.* Once set down these are instantly recognizable. (*a*) personal pronouns: *I, me; we, us; you; he, him; she, her; it; they, them.* The subjective forms are given first, the objective second. Where only one form is given it does double duty. (*b*) possessive pronouns: *mine, yours, his, hers, its, ours, theirs; my, your, his, her, its, our, their.* The first group are principally used predicatively (*this is mine*), and the second attributively or (to use the modern term) as determiners (*my house*). (*c*) reflexive pronouns: *myself, yourself, himself, herself, itself, oneself; ourselves, yourselves, themselves.* (*d*) demonstrative pronouns: *this, that; these, those.* (*e*) indefinite pronouns: *all, any, both, each, either, none, one; everybody, nobody, somebody; everyone, no one, someone;* etc. (*f*) relative pronouns: *that, which, who, whom, whose.* (*g*) interrogative pronouns: *what, which, who, whom, whose.*

There are also a number of 'extended' pronouns and reciprocal pronouns (*whatever, whichever, whoever, whosoever,* etc.; *each other, one another*); and also a number of illiterate (or regional) forms, e.g. *his self, me* ( = my), *themself, theirselves, yous(e)*. Articles on the majority of the pronouns may be found at their alphabetical places, HE; I; ME; MYSELF; etc. See also AS; BUT; CASES. In these articles major points of disputed pronominal uses are presented, and in particular the widespread use of the wrong case of pronouns (e.g. *It's goodbye from Delia and I*) combined with wrong ordering (e.g. *her and John are responsible for this*). Most of the mistakes arise from the fact that, unlike nouns, most pronouns have both a subjective and an objective form. Defaulters are perhaps expressing a yearning for a normalization of the rules affecting both nouns and pronouns.

**2** *The role of pronouns.* A pronoun is a 'word used instead of and to indicate a noun already mentioned or known, esp. to avoid repetition' (*COD* 1990). The definition is sound for most ordinary pronominal uses, so long as 'noun' is taken to include 'noun phrase', and so long as it is understood that the antecedent does not have to be the exact semantic and morphological equivalent of the pronoun itself. *CGEL* (1985) cites examples illustrating these two points: *The man invited the little Swedish girl because he liked her* (2.44); *The clean towels are in the drawer if you need one* [ = 'a clean towel'] (12.15). Jacques Barzun, in a work published in 1985, claimed that 'there can be no proper link between a proper name in the possessive case and a personal pronoun: "Wellington's victory at Waterloo made him the greatest name in Europe" is all askew, because there is in fact no person named for the *him* to refer to.' But substitute *his* for *him* and the difficulty disappears. Even with *him* retained the reader (or listener) would have no difficulty in making the necessary morphological adjustment.

It is clearly desirable that an anaphoric (backward-looking) or cataphoric (forward-looking) pronoun should be placed as near as the construction allows to the noun or noun phrase to which it refers, and in such a manner that there is no risk of ambiguity. Fowler (1926) cited a number of newspaper examples in which the antecedent was too far away for safety or in which ambiguity was possible. A more recent example: *He knew something she didn't. He knew that two students were missing. A boy and a girl. Whether together or otherwise was not yet clear. They had last been seen standing by her car.* [Whose car?]—G. Butler, 1992. An example in which a character in a play is uncertain of the antecedent of the pronoun *he:* (Septimus) *Geometry, Hobbes assures us in the Leviathan, is the only science God has been pleased to bestow on mankind.* (Lady Croom) *And what does he mean by it?* (Septimus) *Mr Hobbes or God?*—T. Stoppard, 1993.

**pronunciamento.** An adaptation of Sp. *pronunciamiento.* Note that the second *i* in the Sp. original is not retained in Eng., nor the Sp. pronunciation of *ci* as /θj/ and *mi* as /mj/. Pl. *pronunciamentos:* see -O(F)S 7.

# pronunciation

## pronunciation.

1 Introductory.
2 Some general 20c. changes that have occurred or are happening now in RP in BrE.
3 Disputed pronunciations.
4 Silent letters.
5 Words containing -ough.
6 Words containing a short o in stressed syllables.
7 Proper names.

**1** *Introductory.* Guidance on pronunciation given in this book follows the system of the International Phonetic Alphabet (IPA), and, in particular, that version of it used in the *Concise Oxford Dictionary* (1990). It is based on the pronunciation of educated adults in the south of England, i.e. is the version of English usually called *Received Standard* (or *Received Pronunciation* (RP)). In most respects, Received Standard coincides with the version called *Modified Standard*, i.e. the version spoken by educated speakers who have modified the pronunciation system they used at an earlier date in some other region of Britain or in an overseas English-speaking country, so that the differences pass almost unnoticed by standard speakers. A great deal of information is provided where relevant about (*a*) pronunciations in standard AmE, and (*b*) about changes of stress and of particular sounds in the course of the 20c. in BrE. See e.g. the entries for (*a*) *leisure, lieutenant, liqueur, -lived;* (*b*) *armada, gynaecology, legend, margarine, metallurgy, myth, pariah, profile.*

**2** *Some general 20c. changes that have occurred or are happening now in (RP) in BrE.* (i) Under AmE influence, adverbs ending in *-arily* now mostly bear the main stress on the *-ar-* (*necessarily, primarily,* etc.). See -ARILY. (ii) A number of words ending in *-ies* (e.g. *rabies, scabies*), formerly pronounced as /-ri:z/, i.e. with a two-syllabled ending, are now pronounced with one, i.e. /-i:z/. (iii) Scientific words ending in *-ein(e)* (*codeine, protein,* etc.), formerly having a two-syllabled ending, now have a monosyllabic one, i.e. /-i:n/. (iv) The /ɔ:/ that was dominant at the beginning of the century in words such as *cross* and *loss* has almost entirely given way to /ɒ/. (v) In words ending in *-eity* (*deity, homogeneity, spontaneity,* etc.) /i:/ has been widely replaced by /eɪ/.

**3** *Disputed pronunciations.* These are numerous. I listed more than a hundred words with disputed pronunciations in *The Spoken Word* (1981) and made recommendations in each case. They are dealt with in this book at their alphabetical places. Examples include *apartheid, centrifugal, cervical, communal, contribute, controversy, decade, despicable, dilemma, dispute* (noun), *dissect, distribute, envelope, extraordinary, forehead, formidable, harass, homosexual, kilometre, medicine, municipal, privacy, recondite, sheikh, sonorous, subsidence, zoology.* It should be borne in mind that the number of such words probably does not differ in any significant way from that at any given time in the past from OE onwards. It is worth noting, for example, that one of the earliest dictionaries to provide pronunciations on a regular basis, John Walker's *Pronouncing Dictionary* (1791), added numerous comments in support of his recommendations for particular words, e.g. under *Envelope:* 'This word signifying the outer case of a letter is always pronounced in the French manner by those who can pronounce French, and by those who cannot the *e* is changed into an *o*. Sometimes a mere Englishman attempts to give the nasal vowel the French sound, and exposes himself to laughter by pronouncing *g* after it, as if written *ongvelope.*' And under *Gymnastick:* 'In this word and its relatives we not unfrequently hear the *g* hard, because forsooth they are derived from the Greek. For the very same reason we ought to pronounce the *g* in *Genesis, geography, geometry,* and a thousand other words, hard, which would essentially alter the sound of our language.'

Nevertheless it still happens that non-standard pronunciations occasionally turn up in the language of otherwise standard speakers: for instance the failure to pronounce the first *c* in *Arctic* and *Antarctic,* and the failure to sound the first *l* in *vulnerable* or the /ð/ in *clothes* /kləʊðz/.

**4** *Silent letters.* The word-initial consonants in the combinations *gn* (*gnash, gnome,* etc.; NZ *ngaio*), *kn* (*knife, know,* etc.), *mn* (*mnemonic*), *pn* (*pneumatic, pneumonia*), *ps* (*psalm, psychology,* etc.), and *wr* (*write, wrong,* etc.) have remained resolutely silent in the 20c. So has interconsonantal *t* in such words as *castle, epistle,* and *wrestle*, except that the *t* is widely sounded in *pestle*. The *t* is frequently sounded in *often*. Americans pronounce *herb* with the *h* silent, and a diminishing number of people in Britain pronounce *hotel* and *humour* with the *h* silent.

**5** *Words containing* -ough. These are difficult for foreigners and pose occasional problems even for native speakers. A short list of typical words with the standard pronunciation in BrE of the -*ough* in each case: /ʌf/ *enough, rough, slough* (cast off skin), *tough*; /əʊ/ *dough, though*; /uː/ *through*; /aʊ/ *bough, plough, slough* (swamp); /ɒf/ *cough, trough*; /ɒk/ *hough*; /ɔː/ *bought, ought*; /ɒx/ or /ɒk/ *lough*; /ə/ *borough, thorough*.

**6** *Words containing a short* o *in stressed syllables*. These are evenly divided between those always pronounced with /ʊ/ and those always pronounced with /ʌ/: /ɒ/ *brothel, column, compact, constant, honest, lozenge, mongoose*; /ʌ/ *brother, colour, company, constable* (sometimes /ɒ/), *honey, dozen, mongrel*. The combination *ov* is normally now pronounced /ɒv/ in \*Covent Garden, \*Coventry, \*hovel, \*hover, *novel*, and *sovereign*; but /ʌv/ is still sometimes heard in those marked with an asterisk.

**7** *Proper names*. Many proper names have a traditional pronunciation not easily inferred from their spellings. A few examples are given below, but the list merely skims the surface of a huge problem. For authoritative guidance about hundreds of others, readers are referred to J. C. Wells's *Longman Pronunciation Dictionary* (1990), and especially the *BBC Pronouncing Dictionary of British Names* (2nd edn by G. E. Pointon, 1983). *Althorp* (Northants) /ˈɔːltrəp/; *Beauchamp* /ˈbiːtʃəm/; *Beaulieu* /ˈbjuːli/; *Belvoir* (place in Leics.) /ˈbiːvə/; *Caius* (Cambridge college) /kiːz/; *Cherwell* /ˈtʃɑːwəl/; *Cholmondeley* /ˈtʃʌmli/; *Cockburn* /ˈkəʊbən/; *De'ath* (surname) /diˈæθ/; *Fiennes* /famz/; *Glamis* /ɡlɑːmz/; *Harewood* (Earl and House) /ˈhɑːwʊd/; *Home* (Earl) /hjuːm/; *Keynes* (family name) /kemz/; *Keynes* (in Milton Keynes) /kiːnz/; *Leveson-Gower* /ˈluːsən ˈɡaʊə/; *Magdalen(e)* (the colleges) /ˈmɔːdlɪn/; *Marjoribanks* /ˈmɑːtʃbæŋks/; *Ruthven* /ˈrɪvən/; *Sandys* /sændz/; *Walthamstow* /ˈwɔːlθəmstəʊ/; *Waugh* /wɔː/; *Wemyss* /wiːmz/; *Whewell* /ˈhjuːəl/; *Woburn* (Abbey) /ˈwuːbɜːn/; *Wrotham* /ˈruːtəm/.

**propaganda** is a 17c. loanword from Italian, derived from the modL title *Congregatio de propaganda fide* 'congregation for propagating the faith', at first a committee of Cardinals of the Roman Catholic Church having the care and oversight of foreign missions, and later (late 18c.) applied to any association or movement for the propagation of a particular doctrine. In this extended sense it was sometimes treated as a neuter pl. ( = efforts or schemes of propagation) with singular *propagandum*. The dominant current sense (usu. derogatory), 'the systematic propagation of selected information to give prominence to the views of a particular group; also, such information', first emerged at the beginning of the 20c.

**propel.** The inflected forms are *propelled, propelling*. See -LL-, -L-.

**propellant** (noun) is something that propels. It is reasonably familiar to the general public in the sense 'a substance used as a reagent in a rocket engine to provide thrust' (*COD*). The corresponding adj., which is much less frequent, is *propellent* 'propelling; capable of driving or thrusting (a space vehicle, etc.) forward'.

**propeller,** the only standard spelling in BrE for a revolving shaft with blades used to propel a ship or aircraft, etc. Sometimes spelt *propellor* in AmE.

**proper** (adj.). Its use following the noun it governs, with the sense 'strictly so called, actual' (e.g. *Before beginning the story proper . . . it is worth taking a moment to look upwards, high above the teeming masses of rush-hour London where most of the story is set*—B. Elton, 1991) is worth noting. The use of the word as an adv. following verbs of talking is 'dial., vulgar, or slang' (*OED*): e.g. *Perhaps she'll 'ave another go at teachin' me to speak proper, pore soul*—M. Allingham, 1952.

**proper terms.** The following list of terms for groups of animals, etc., is reprinted from *The Oxford Encyclopedic Dictionary* (1991), Appendix 22. Terms marked † belong to 15th-c. lists of 'proper terms', notably that in the *Book of St Albans* attributed to Dame Juliana Barnes (1486). Many of these are fanciful or humorous terms which probably never had any real currency, but have been taken up by Joseph Strutt in *Sports and Pastimes of England* (1801) and by other antiquarian writers.

a †shrewdness of apes
a herd or †pace of asses
a †cete of badgers
a †sloth or †sleuth of bears
a hive of bees; a swarm, drift, or bike of bees
a flock, flight, (*dial.*) parcel, pod ( = small flock), †fleet, or †dissimulation of (small) birds; a volary of birds in an aviary
a sounder of wild boar
a †blush of boys
a herd or gang of buffalo
a †clowder or †glaring of cats; a †dowt ( = ?do-out) or †destruction of wild cats
a herd, drove, (*dial.*) drift, or (*US & Austral.*) mob of cattle
a brood, (*dial.*) cletch or clutch, or †peep of chickens
a †chattering or †clattering of choughs
a †drunkship of cobblers
a †rag or †rake of colts
a †hastiness of cooks
a †covert of coots
a herd of cranes
a litter of cubs
a herd of curlew
a †cowardice of curs
a herd or mob of deer
a pack or kennel of dogs
a trip of dotterel
a flight, †dole, or †piteousness of doves
a raft, bunch, or †paddling of ducks on water; a team of wild ducks in flight
a fling of dunlins
a herd of elephants
a herd or (*US*) gang of elk
a †business of ferrets
a charm or †chirm of finches
a shoal of fish; a run of fish in motion
a cloud of flies
a †stalk of foresters
a †skulk of foxes
a gaggle or (in the air) a skein, team, or wedge of geese
a herd of giraffes
a flock, herd, or (*dial.*) trip of goats
a pack or covey of grouse
a †husk or †down of hares
a cast of hawks let fly
an †observance of hermits
a †siege of herons
a stud or †haras of (breeding) horses; (*dial.*) a team of horses
a kennel, pack, cry, or †mute of hounds
a flight or swarm of insects
a mob or troop of kangaroos
a kindle of kittens
a bevy of ladies
a †desert of lapwing

an †exaltation or bevy of larks
a †leap of leopards
a pride of lions
a †tiding of magpies
a †sord or †sute ( = suit) of mallard
a †richesse of martens
a †faith of merchants
a †labour of moles
a troop of monkeys
a †barren of mules
a †watch of nightingales
a †superfluity of nuns
a covey of partridges
a †muster of peacocks
a †malapertness ( = impertinence) of pedlars
a rookery of penguins
a head or (*dial.*) nye of pheasants
a kit of pigeons flying together
a herd of pigs
a stand, wing, or †congregation of plovers
a rush or flight of pochards
a herd, pod, or school of porpoises
a †pity of prisoners
a covey of ptarmigan
a litter of pups
a bevy or drift of quail
a string of racehorses
an †unkindness of ravens
a bevy of roes
a parliament or †building of rooks
a hill of ruffs
a herd or rookery of seals; a pod ( = small herd) of seals
a flock, herd, (*dial.*) drift or trip, or (*Austral.*) mob of sheep
a †dopping of sheldrake
a wisp or †walk of snipe
a †host of sparrows
a †murmuration of starlings
a flight of swallows
a game or herd of swans; a wedge of swans in the air
a herd of swine; a †sounder of tame swine, a †drift of wild swine
a †glozing ( = fawning) of taverners
a †spring of teal
a bunch or knob of waterfowl
a school, herd, or gam of whales; a pod ( = small school) of whales; a grind of bottle-nosed whales
a company or trip of widgeon
a bunch, trip, or plump of wildfowl; a knob (less than 30) of wildfowl
a pack or †rout of wolves
a gaggle of women (*derisive*)
a †fall of woodcock
a herd of wrens

**prophecy** /ˈprɒfɪsɪ/ shows the spelling and pronunciation of the noun. The corresponding verb is *prophesy* /ˈprɒfɪsaɪ/.

**prophetess.** See -ESS 5.

**proportionable, proportional, proportionate.** The first of these adjs. is virtually extinct. The second and third are synonyms, but (a) *proportional* is the only possible word in several fixed collocations, e.g. *proportional compasses* (US *dividers*), *proportional parts*, *proportional representation*; (b) *proportionate* tends to be preferred when the context requires the meaning 'proportioned, that is in due proportion', e.g. *heavy bombs cause damage proportionate to their weight.*

**proposition** (noun). It is worth recording that Fowler (1926), writing at a time when the sense 'a matter or problem which requires attention (e.g. *a tough proposition, an attractive proposition, a business proposition)*' was widely felt in Britain to be an intrusive Americanism, sternly remarked, 'it is resorted to partly because it combines the charms of novelty and length, and partly because it ministers to laziness.' He wanted *proposition* to 'be brought back to its former well defined functions in Logic and Mathematics, and relieved of its new status as Jack-of-all-trades'. His advice, his admonitions, and his numerous illustrative examples (more than twenty, in contexts where, it was claimed, *proposal, task, undertaking, possibility, prospect, enterprise*, etc., would have been more suitable) have all gone unheeded. The noun continues to be widely used in ordinary contexts in precisely the manner that Fowler found objectionable. Examples. *'Call this a store?' he would say. 'Call this a paying proposition?'*—A. Tyler, 1980; *For a quizzical, curious mind . . . adequate explanations are too thin to suffice and tinkering with the possibilities becomes an enticing proposition*—D. Shekerjian, 1990; *Altering an alloy composition calls for relatively large orders, but to adjust slightly the temper of a work-hardening sheet alloy is a more feasible proposition*—Professional Engineering (UK), 1992. The corresponding verb was first recorded in America in the 1920s in two main senses: (a) to make or present a proposition to (someone), e.g. *While being propositioned by Lord Beaverbrook about becoming the film critic of the* Evening Standard, *I nervously filled in a yawning silence by telling this anecdote*—Punch, 1967; (b) to request sexual favours from (a person), e.g. *In Hyde Park, that black whore had*

*propositioned him as he walked from work toward the Tube*—New Yorker, 1975. These senses, esp. (b), now form part of the ordinary fabric of the language.

**proprietor.** Thus spelt, not -er.

**proprietress.** See -ESS 5.

**propylaeum** (entrance to a temple; specif. the entrance to the Acropolis in Athens). Pl. *propylaea* /-'li:ə/.

**proscenium.** Pl. *prosceniums* (recommended) or *proscenia*. See -UM 3.

**proscribe.** See PRESCRIBE.

**prosecutor** (person who prosecutes, esp. in a criminal court). Thus spelt, not -er.

**prosecutrix.** A female prosecutor. For the pl., see -TRIX.

**proselyte.** The noun means 'a person converted, esp. recently, from one opinion, creed, party, etc., to another'. In BrE the corresponding verb ( = to convert (a person) from one belief, etc., to another) is *proselytize*. In AmE the verb is either *proselytize* or *proselyte*.

**prosody.** Traditionally, and still today, its primary sense is 'the theory and practice of versification' and, in particular, 'the laws of metre'. The corresponding adj. is *prosodic* /prə'sɒdɪk/; and 'one skilled or learned in prosody' is a *prosodist* /'prɒsədɪst/. In linguistics, since 1949 in the theories of J. R. Firth and his followers, *prosody* has established itself with the meaning 'a phonological feature having as its domain more than one segment'. The OED adds a note: 'Prosodies include the class of "suprasegmental" features such as intonation, stress, and juncture, but also some features which are regarded as "segmental" in phonemic theory, e.g. palatalization, lip-rounding, nasalization.'

**prosopopoeia** /prɒsəpə'pi:ə/. In use in rhetoric since the 16c. as the direct descendant of Gk προσοποποιία 'personification', it means principally 'a figure in which an inanimate or abstract thing is represented as being able to speak'. As a rhetorical device it was known already to the Anglo-Saxons (though not under that name). In the OE poem The *Dream of the Rood*, for example, the True Cross speaks:

þæt wæs geara iu,   (ic þæt gyta geman)
þæt ic wæs aheawen   holtes on ende,
astyrede of stefne minum ...
(It was long ago   (I still remember it)
that I was cut down   at the edge of the
forest,
moved from my trunk   ...)

The device is also employed in the OE riddles, and in many a poem in later centuries.

**prospect** (noun). Regularly stressed in all its senses on the first syllable. The corresponding verb ( = to explore a region for gold, oil, etc.) is stressed on the second syllable (thus /prəˈspekt/). See NOUN AND VERB ACCENT 1.

**prospectus.** The pl. is *prospectuses* (notwithstanding the L pl. *prospectūs*). See -US 2.

**prosper.** The inflected forms are *prospered, prospering*. See -R-, -RR-.

**prostate (gland).** See note s.v. PROSTRATE ad finem.

**prosthesis** (an artificial limb, etc.). The pl. is *prostheses* /ˌprɒsˈθiːsiːz/.

**prostrate** (adj.) means (strictly) 'lying face downwards', but also, more generally, 'lying horizontally', and, in transferred use, 'overcome, esp. by grief or exhaustion'. It is stressed on the first syllable: thus /ˈprɒstreɪt/. For the corresponding verb the stress moves to the second syllable: thus /prɒˈstreɪt/. Cf. SUPINE. Examples of the adj.: *The leader increasingly assumed divine and imperial attributes, and any recalcitrants were locked in church for days and days until they were all prostrate*—W. Weaver, 1986; *Among those often listed as alpines or rock plants are a number of prostrate or low-growing perennials or shrubs—Garden Answers* (UK), 1990; *Gavin Hastings had charged upfield and at a subsequent ruck Carminati clambered in and stamped twice on the prostrate John Jeffrey—Independent on Sunday*, 1990; *He was obsessed ... by the prostrate, naked, and callipygous corpse in the foreground—NY Rev. Bks*, 1991.
Care should be taken to distinguish *prostrate* from *prostate*. The second (in full *prostate gland*) is a gland surrounding the neck of the bladder in male mammals.

**protagonist.** 1 In Greek drama the protagonist (πρωταγωνιστής, from πρῶτος

'first' + ἀγωνιστής 'an actor') is the chief personage in a play, and this single sense is maintained in all scholarly work on Greek drama. The *deuteragonist* (δευτεραγωνιστής) and *tritagonist* (τριταγωνιστής) take parts of second and third importance.

2 Dryden seems to have been the first user of the term in English: in the preface to 'An Evening's Love' (1671) he wrote *'Tis charg'd upon me that I make debauch'd Persons ... my protagonists, or the chief persons of the drama*. The use became established and many writers since Dryden have used the word *protagonist* to mean (in the pl.) 'the leading characters in a play, story, etc.' Examples: *Living actors have to learn that they too must be invisible while the protagonists are conversing, and therefore must not move a muscle nor change their expression*—G. B. Shaw, 1950; *Both stories deal with protagonists who hit bottom, who have slumped beyond hope—Literary Rev.*, 1989; *For all its revolutionary dramaturgy, the Ring is still a nineteenth-century drama, honoring if not always living up to the requirement that consistent, coherent protagonists carry the action—Opera Q.* (US), 1991–2. Sometimes *protagonist* is used in the singular to mean no more than 'character' or 'actor' (in a novel, film, etc.); thus used it is unerringly and controversially preceded by *chief, leading*, or *main*: e.g. *Price's "Smiley" is Dr David Audley, the chief protagonist of 17 previous novels—Oxford Today*, 1990; *He chose Nicholson for lead primarily because he thought he possessed the on-screen presence that was needed to bring the main protagonist in this film, the part of Jonathan, alive—John Parker*, 1991.

3 From the first half of the 19c., the word has been applied, both in the singular and the plural, to 'a leading person in any contest or course of events; a prominent supporter or champion of any cause'. Examples: *If social equity is not a chimera, Marie Antoinette was the protagonist of the most ... execrable of causes—J. Morley*, 1877; *Strong opposition to more cuts in public expenditure were voiced at a meeting of the Cabinet on Tuesday. The protagonists were Mr Crosland ... Mr Shore ... and Mr Benn—Times*, 1976; *This year the season has been characterized by the duel between the McLaren and Ferrari teams whose protagonists are Senna and Prost—Ronda Iberia* (Iberia Airlines), 1990.

**4** Early in the 20c., speakers and writers, ignoring the etymology of the word, seem to have perceived *protagonist* as being formed on the Latin prefix *pro-* 'for, in favour of' and therefore to think of the word as meaning 'a proponent, advocate, supporter (of a cause, idea, etc.)', as opposed to an antagonist. A. P. Herbert caught the mood in his *What a Word!* (1935): *I have heard with horror … that the word 'protagonist' is being used as if it were pro-tagonist—one who is for something, and opposed to ant-agonist, one who is against it.* Examples of 20c. uses of *protagonist*, including examples which are not clearly distinguishable from those in para. 3 and others (in para. 2) preceded by *leading* or *main*: *A protagonist of and expert on the Added Value concept—Jrnl RSA*, 1979; *Hart, its leading protagonist in recent times, clearly saw [that] it sets a limit on the means allowed in pursuit of moral ideals—J.* Raz, 1986; *For the past year a campaign has been waged for my political liquidation. The main protagonist in this has been Slobodan Milosevic—Independent*, 1990; *The protagonists on both sides of the controversy over the future of London's archaeological service … have a case to answer—R.* Cramp, 1991.

**5** *Recommendation.* It is easy to see how the development of senses occurred once the word had come into the ordinary vocabulary of people who knew nothing of the nature of Greek drama. In English literature the protagonists are the chief actors in a play or the main characters in a work of fiction. In political life the protagonists are those who play a prominent part in the great debates and issues that divide peoples and nations. In such contexts, and also in the realms of business, trade, banking, etc., *protagonist* may properly be used of those who hold opposing views on important matters. It is understandable that such protagonists will sometimes be called *chief, leading, main,* or *principal,* though such uses have a whiff of pleonasm about them. Further than that one should not go. In contexts that call for a noun meaning 'one who is not antagonistic to, one who is *for* or *in favour of* a specified belief or set of beliefs', i.e. taking the element *pro-* as the equivalent of that used in, for example, *pro-choice* or *pro-life, protagonist* should not be used. For such contexts

several standard alternatives are available and should be preferred, e.g. *advocate, proponent,* and *supporter.*

**protasis** /ˈprɒtəsɪs/. The clause expressing the condition in a conditional sentence, e.g. the *if-*clause in *If I can come I will* or *I will come if I can.* Cf. APODOSIS. Pl. *protases* /-siːz/.

**protean** (adj.). Pronounce /prəʊˈtiːən/. The *OED* (1909) gave only first-syllable stressing (thus /ˈprəʊtɪən/), and this pronunciation is still preferred by many standard speakers. (The word is the adj. corresponding to *Proteus,* in Gk mythology a sea-god able to take various forms at will.) Cf. HERCULEAN.

**protector.** Thus spelt, not *-er.* See *-OR.*

**protectress.** A female protector. See *-ESS* 5.

**protégé** /ˈprɒteˌʒeɪ/. In this and the corresponding feminine form *protégée,* the accents should be retained in printed work.

**protein.** Now always pronounced /ˈprəʊtiːn/, but the *OED* (1909) gave it three syllables, /ˈprəʊtiːɪn/.

**protest.** **1** The noun is stressed on the first syllable and the verb on the second. See NOUN AND VERB ACCENT 1.

**2** The transitive use of the verb in the sense 'to protest against (an action or event); to make the subject of a protest' has been widely accepted in AmE throughout the 20c. and is beginning to make slow inroads in other English-speaking areas. But it is far from being a natural use in BrE. Some recent examples: *There were Maori customers toying with European attire and passionately protesting the cost—M.* Shadbolt, 1986 (NZ); *Anatoly Koryagin, who has been imprisoned for protesting the use of psychiatry for political purposes—New Yorker,* 1987; *But I must protest this latest twist from law—USA Today,* 1988; *The ruin of Belfast's Black Mountain protested by the local community—Independent,* 1991.

**protestant.** In its religious sense always written with an initial capital and pronounced /ˈprɒtɪstənt/. In its occasional non-religious use to mean simply 'a protesting person', it is sometimes pronounced with the stress on the second syllable. But *protester* is the normal word in this sense.

**protester.** Thus spelt, not *-or.*

**Proteus** (Gk sea-god; a changing or inconstant person or thing). The dominant standard pronunciation now is /ˈprəʊtɪəs/ rather than /ˈprəʊtjuːs/.

**prototype.** See *-TYPE.*

**protractor** (geometrical instrument). Thus spelt, not *-er.*

**protrude.** For the adj. meaning 'capable of being protruded', *protrusible* is perhaps preferable to *protrudable.* Related adjs. are *protrudent* and *protrusive* (protruding) and *protrusile* (of a limb, etc., capable of being thrust forward).

**provable.** Thus spelt, not *proveable.* See MUTE E.

**proved, proven.** Middle English developed two distinct infinitives from OF *prover* (ult. from L *probāre*), namely *proven* and *preven* (or *preoven*). In standard English *prove* alone survives with pa.t. and pa.pple *proved.* In Scotland and some northern dialects of England the pattern *preve,* pa.pple *proven* survived, the pa.pple being usually pronounced /ˈprəʊvən/. Cf. *cleave/cloven, weave/woven.* In Sc. law the verdict *not proven* is a central concept. In the rest of the UK the form *proven,* usu. pronounced /ˈpruːvən/, is occasionally used beside *proved* in non-legal contexts both as a pa.pple and as an attributive adj. In AmE *proven* is at least as common as *proved* in both parts of speech. The ordinary BrE form *proved* is shown in the following examples: *Four cheeses were chosen to spearhead the consumer packs, and initial results have proved positive—Grocer,* 1988; *Attempts at winning environmental controls . . . have so far proved of only limited effectiveness—Natural World,* 1988. Examples of *proven:* (attrib. adj.) *Anchor's proven track record with aerosol products led to the introduction of . . . a non-dairy dessert topping—Grocer,* 1988; *His love of precise dates and proven facts—N.* Shakespeare, 1989; (pa.pple) *That Uncle was a susceptible and sometimes hallucinated reader is proven by his enthusiasm for the books he urged on me—S.* Bellow, 1987 (US); *Medical science has utterly proven the case in this issue—Sunday Times,* 1987; *in the United States . . .*

*conducting necessary seroprevalence studies has proven politically problematic—Dædalus,* 1989 (US); *Some of our predictions were proven correct, but others just as obviously were not—H. S.* White, 1990 (US); *Traditionally those who aspire to the main board of ICI have proven themselves on the way there as successful chief executives—Independent,* 1991.

See DISPROVEN.

**provenance, provenience.** The first (which is stressed on the first syllable) is the BrE form for 'the place of origin of a manuscript, work of art, etc.', and the second (stressed on the second syllable) is its AmE equivalent.

**provided (that), providing (that).** In the type *Provided that/Providing that he had a good book to read, he did not mind how much it rained* the phrases *provided that/ providing that* are compound subordinate conjunctions introducing conditional clauses. In slightly less formal contexts *that* is often omitted, and then *provided/providing* may be regarded as quasi-conjunctions. All four types have a long history in English, as the OED shows: *provided that,* illustrative examples from c1460 onward; *provided* 1604– ; *providing that* 1423– ; *providing* 1632– . Some modern examples: (provided (that)) *Provided, only provided, that it not be that one, or anything like it—M.* Bradbury, 1987; *In summer he will show visitors around the chapel provided he likes their faces and they are not wearing shorts—Linguist,* 1992; (providing) *It works well enough providing I keep my blanket round me—J.* Winterson, 1987; *Kids are actually as tough as old boots. Providing they're fed and watered and have at least one primary attachment figure, they tend to survive—Metro* (NZ), 1988; *With your help, I would like to do just that* [sc. meet my father], *providing he's still alive—Chicago Sun-Times,* 1988. In such contexts, in those given in the OED, the meaning ranges from 'on condition that', 'on the undertaking that', or 'in case that' to 'if only'.

**province.** For synonyms, see FIELD.

**proviso.** Pl. *provisos.* See *-O(E)S* 6.

**provost.** In the names of the officers of the military police often pronounced /prəˈvəʊ/ (after Fr. *prévôt*), AmE /ˈprəʊvəʊ/. In all other senses /ˈprɒvəst/.

**prox.** Abbreviation of next.

**proximo** (abbrev. **prox.**) is still occasionally used in commercial letters = 'of next month' (*the seventh proximo*). It is a shortened form of L *proximo mense* 'in the next month'. See COMMERCIALESE.

**prude** makes *prudish* (not *prudeish*). See MUTE E.

**prudent, prudential.** While *prudent* means having or showing prudence, circumspect, judicious, *prudential* means pertaining to, or considered from the point of view of prudence, characterized by forethought (*prudential motives*). To call an act *prudent* is normally to commend it; to call it *prudential* is more often than not merely to say that it involved prudence. But the distinction is not always clear-cut.

**prunella.** For the meaning of *leather or prunella*, see LEATHER 2.

**prurience, pruriency.** Both mean an unhealthy interest in sexual matters, but *prurience* is the more common of the two. See -CE, -CY.

**pry.** The intransitive verb *pry*, meaning to inquire impertinently (into something secret or private), is first recorded in the 14c. and is of unknown origin. It is to be carefully distinguished from the dialectal and AmE verb (first found in print in the 19c.) of the same spelling, meaning to force open or out by leverage. See PRISE.

**PS,** abbrev. of *postscript*. A second postscript is *PPS*, and a third *PPPS*. No one, except in jest, writes more than three postscripts.

**ps-.** The *OED* (1909) described the dropping of the *p* sound as 'an unscholarly practice often leading to ambiguity or to a disguising of the composition of the word'. Fowler (1926) hoped that 'With the advance of literacy the pronunciation of the *p* in words beginning thus is likely to be restored except in *psalm* and its family, e.g. in the compounds of *pseud(o)*- and such important words as *psychical* and *psychology*.' Nothing of the kind has happened and all English words beginning with ps (e.g. *psalm*, *pseudonym*,

*psittacosis*, *psychiatry*, *psychopath*, *psychology*) are now pronounced with initial /s/, not /ps/.

**pseudonym.** See NOM DE GUERRE. The desire to remain anonymous when writing to newspapers is widespread, particularly in small, far-off provincial towns. I have before me the letters page of a 1989 issue of the daily newspaper in my birthplace in New Zealand. Of the eleven letters printed, only three carry the name of the writer. The remaining eight use mundane pseudonyms such as *Why Pay the Rates* (our streets are neglected), *Interested Observer* (complaining about the paucity of cricket pitches for women), and *Pussy Willow* (expressing shock at the number of cats brought to the vets to be put down). My impression is that this practice is far less widespread in Britain, though for letters about highly sensitive matters that might endanger the correspondent the letters editor sometimes permits the use of the formula *Name and address supplied*.

**psychic, psychical.** The second of these, used in the sense 'of or pertaining to the soul or the mind, as distinguished from *physical*', is much the earlier of the two adjs. The earliest example of *psychical* in the *OED* is one of 1642 from the work of the Cambridge Platonist Henry More (1614–87). *Psychic*, by contrast, has not been found in print before 1836. Both words survive and are interchangeable in many contexts, but, to judge from the length of their respective entries in standard modern dictionaries, *psychic* is much the more resilient of the two. The shorter form is preferred in matters to do with the occult (telepathy, clairvoyance, etc.), and is obligatory when applied to a bid in bridge 'that deliberately misrepresents the bidder's hand' (*COD*). On the other hand the collocation *psychical research* is somewhat more often used than *psychic research*, at least in Britain.

**psychological moment.** This familiar phrase ( = the most appropriate time for achieving a particular effect or purpose) arose in the French during the German siege of Paris in 1870. French journalists used the phr. *le moment psychologique* to render the German phr. *das psychologische Moment* through confusion of *der Moment* 'moment (of time)' with *das Moment* 'operative factor, momentum'. The phr. then passed into English journalistic use in its erroneous sense, and has tenaciously remained in the language, condemned, if at all, merely as an overused expression.

**psychosis.** Pl. *psychoses* /-siːz/.

**pt-.** The *OED* (1909) expressed a preference for the *p* to be fully pronounced in all such Greek-derived words (*pterodactyl*, *Ptolemaic, ptomaine*, etc.), but implied that a silent *p* was acceptable in the fanciful misspelling *ptarmigan* (derived from Gaelic *tàrmachan*). But, as in words beginning with *pn-* and *ps-*, the *p* is now always silent, despite the fact that it is pronounced in all the relevant words in French, German, and other European languages.

**ptomaine.** Now normally pronounced /ˈtəʊmeɪn/ or /təʊˈmeɪn/, but not always so in the past. The word was first blunderingly formed (as *ptomaina*) in It. by Professor Selmi of Bologna (*OED*): 'As the Greek combining stem is πτωματ-, the correct form of the word is *ptomatine*.' The *OED* (1909) gave a three-syllabled pronunciation for *ptomaine*: /pt-, ˈtəʊmeɪam/, and regretted 'the rise of the illiterate pronunciation /təʊˈmem/ like *domain*'.

**pub.** See ABBREVIATIONS 1.

**publicly** is the standard form of the adv., not *publically*.

**pucka.** See PUKKA.

**pucker.** The inflected forms are *puckered*, *puckering*. See -RR-, -R-.

**pudenda, pudendum** are both used with the same sense, but the first with a plural, the second with a singular, construction. See -UM 2.

**puggaree** (Indian turban; helmet scarf). The spelling recommended in OUP house style. The word is derived from Hindi *pagrī*.

**puisne.** Like *puny*, pronounced /ˈpjuːnɪ/. It is derived from Fr. *puis né* 'born afterwards', hence 'inferior'. A puisne judge is a judge of a superior court inferior in rank to chief justices (*COD*).

**puissant** (a literary or archaic word meaning 'mighty, powerful'). The pronunciation recommended is /ˈpwiːsənt/, but /ˈpjuːɪsənt/ is also used.

**pukka** (genuine). Spelt thus, not *pucka*, *pukkah*. See SAHIB.

**pulley** (noun). The pl. is *pulleys*. Used as a verb, the inflected forms are *pulleys*, *pulleyed, pulleying*.

**pullulate.** Pronounce /ˈpʌljʊleɪt/.

**pulque** (Mexican fermented drink). Pronounce /ˈpʊlkeɪ/ or /ˈpʊlkɪ/.

**pummel** (verb). See POMMEL.

**pun.** A standard account of puns—the good, the bad, and the indifferent—is provided in Tony Augarde's *Oxford Guide to Word Games* (1984), esp. pp. 204–15. Puns have been used in Western countries from earliest times. Aristotle approved of them in some kinds of writing. Pope Gregory I (c540–604) famously described English slaves as *Non Angli, sed angeli* 'not Angles, but angels'. Shakespeare is said to have used about 3,000 puns in his plays, among them Mercutio's dying words in *Romeo and Juliet*, *aske for me to morrow, and you shall finde me a graue man*. When Gen. Sir Charles Napier conquered the Indian province of Sind in 1843, he sent the British War Office a message consisting of one Latin word— *Peccavi* (I have sinned). These are celebrated examples. Each one of us has a store of private puns and the ability to make new ones. Rhetoricians like them and give them fancy names (see PARONOMASIA). Journalists like them, especially in headlines (see HEADLINE LANGUAGE). Schoolchildren adore the near-puns built into riddles: (Question) *What is an ig?* (Answer) *An Eskimo house without a loo.* Puns will doubtless continue to be a feature of the language in the centuries that lie ahead.

**Punchinello.** Pl. *Punchinellos*. See -O(E)S 7.

**punctilio.** Pl. *punctilios*. See -O(E)S 4.

**punctuation.** See AMPERSAND; APOSTROPHE²; BRACKETS; COLON; COMMA; DASH; EXCLAMATION MARK; FULL STOP; HYPHENS; ITALICS; QUESTION MARK; QUOTATION MARKS; SEMICOLON.

**punctum** (in biology). Pl. *puncta*. See -UM 2.

**pundit.** The normal transliteration of the Hindi original is *paṇḍit*, and this spelling (without the subscript dots) is used in printed work (with a capital initial) when prefixed to the name of a learned Hindu (*Pandit Nehru*). But *pundit* and *punditry* are used in general contexts, and are the only spellings admissible when the reference is to a (non-Indian) learned expert or teacher.

**pupa.** Pl. *pupae* /ˈpjuːpiː/.

**pupil.** For the derivatives the recommended spellings are *pupillage*, *pupillar*, *pupillary*, rather than the spellings with single -l-. See -LL-, -L-.

**purchase.** As a substitute for *buy* (goods for money), *purchase* is to be classed among FORMAL WORDS. But in figurative use (*the victory was purchased at great cost*) it is not open to the same objection.

**purée** (pulped vegetables, fruit, etc.). Thus spelt (with accent). Pronounce /ˈpjʊəreɪ/.

**puritanic, puritanical.** Both words came into use in the 17c. The shorter form is now hardly ever encountered, the normal words being *puritanical* or *puritan* (adj.). Examples: (puritanic) *There are other factors ... behind these puritanic landscapes of domestic bliss, and over-mathematicized architectures—Architectural Rev.*, 1934; (puritanical) *Both [D. H.] Lawrence and Catherine came from puritanical, conventional and smothering families—J. Meyers*, 1990 (US); *Gould's aesthetic was uncompromisingly puritanical and severe, yet his playing is never cold or dispassionate—London Rev. Bks*, 1992.

**purlieu.** Pl. *purlieus*, with the final letter pronounced /z/. See -X.

**purloin** is a jocular or formal word for 'steal, pilfer'.

**purple** makes *purplish*. See MUTE E.

**purport.** 1 The noun is stressed on the first syllable, /ˈpɜːpɔːt/, and the verb on the second, /pəˈpɔːt/. See NOUN AND VERB ACCENT 1.

2 *The noun.* Its normal uses, meaning 'the ostensible meaning of something; the sense or tenor (of a document, speech, etc.)', cause no particular problems. Its value as a word is that it is non-committal: it refrains from either endorsing or denying, but instead lightly questions (the truth of a statement, etc.).

3 *The verb.* Uses of the verb are more complicated, and opinions vary widely about the acceptability of some of the 20c. constructions. First, a word on its history. It first appeared in the 16c., used transitively in the senses 'have as its purpose; express' (e.g. *I ... enclose copies of letters ... purporting some of the above facts*, 1780). This construction is no longer idiomatic. A transitive use with a relative clause, however, first recorded in 1693, is still standard (e.g. *It purports that some one from Oxfordshire ... applied to the College of Arms to have his title recognised*, 1858). Constructions in which *purport* is followed by an infinitive, and meaning '(of a document, book, etc.) to profess or claim by its tenor', first appeared towards the end of the 18c. and are still valid (e.g. *This epistle purports to be written after St. Paul had been at Corinth*, 1790; *The Declaration which purported to give them entire freedom of conscience*, 1849; *A letter purporting to have been written by you*, 1870). Less idiomatic is such a construction headed by (the name of) a person (e.g. *Jack Downey ... who purported to accompany the presidential party and to chronicle its doings*, 1884). (Examples from the *OED*.)

Fowler (1926) watched these traditional constructions being further extended in the newspapers of his day, and did not like what he saw, esp. developments in the use of *purport* with a following passive infinitive: e.g. *Many extracts from speeches purported to have been made by Mr Redmond are pure fabrications; He had no information of a Treaty between Japan and Germany purported to have been made during the war*. He judged that *supposed* would have been the better word in each case. He also thought that *purport* itself should not be used in the passive. And he believed that the subject of *purport* should only in rare circumstances be a person. He then cited some 'normal' examples and some 'illegitimate' ones: (normal) *The story purports to be an autobiography; the Gibeonites sent men to Joshua purporting to be ambassadors from a far country*; (illegitimate) *Sir Henry is purported to have said 'The F.A. are responsible for everything inside the Stadium'*; *She purports to find a close parallel between the Aeschylean Trilogy and The Ring.*

**4** *Present-day constructions of the verb.* The main present-day types are shown in the following examples: *The revolutionary intellectual purports to believe ... that the utopian ideal of unalienated existence can only be realised through violent struggle*—P. Fuller, 1986; *That it purports to be contemporary seems to be given away by the fact Murphy has stickers like 'No Nukes' on the windshield of his boneshaker*—Punch, 1986; *He even hesitantly claimed a near-sighting of his own, recalling that he had seen 'what purported to be a saucer phenomenon'*—R. Ferguson, 1991; *Sonenscher deconstructs almost every concept, and the 'reality' to which it purports to refer*—French History, 1991; *The paper purporting to demonstrate cold fusion is a case in point*—New Scientist, 1991; *Those who purport to describe educational issues should not promulgate miseducative stereotypes*—TES, 1991; *More recently, the Family Court of Australia held in contempt a layman who falsely purported to be a lawyer*—D. Pannick, 1992. As it happened, the OED electronic database contained examples of the passive construction *be purported to* only in AmE sources: e.g. *They were purported to be massive particles with a unipolar magnetic charge*—T. Ferris, 1988; *Set in what was purported to be modern-day California, it was about a family of lively young nowsters who dressed in flowery jumpers*—J. and M. Stern, 1992. I have not found any examples of *purport* with a following passive infinitive, a construction greatly disliked by Fowler (see above).

**5** *Conclusion.* It would appear that the language is gradually admitting some uses of *purport* not recorded in the OED and others that were disliked by Fowler. It is a classic example of a verb continuing to vie for space with its near-synonyms, e.g. *allege(d)*, *claim*, *have the appearance of being*, *repute(d)*, and *suppose(d)*, and having a mixed reception from standard speakers as the linguistic battle takes its course.

**purpose** (noun). For *clauses of purpose*, see FINAL CLAUSE.

**purposefully, purposively, purposely.** A careful examination of dictionary definitions of these three words and of the corresponding adjectival forms of the first two will throw light on some of their differences of meaning and use. The sturdiest of them all is *purposely* 'on purpose, by design', which entered the language in the 15c. (e.g. *he purposely avoided all reference to his ex-wife*). Both *purposeful* 'having a definite purpose in mind, intentional' and *purposefully* are mid-19c. creations. So too is *purposive* 'serving or tending to serve some purpose', while the adverb *purposively* is a 20c. word (1908– ). *Purposive* was taken over by psychologists in the 1880s with the meaning 'relating to conscious or unconscious purpose as reflected in human and animal behaviour or mental activity' (OED), and it is in psychological work that the word principally turns up (e.g. *The behaviourists, with the exception of Tolman, rejected purposive explanation, because they avoided all reference to consciousness, subjectivity, ideas, or mind*—Oxford Companion to the Mind, 1987).

Examples of *purposeful, purposefully*: (purposeful) *During the hour ... they had transformed a restless bunch of chewing, pencil-tapping, fidgeting, chattering inattentive bundles of undirected energy with an attention span of about 10 seconds into a purposeful class*—TES, 1990; *Prussia's civil service and even that of France were models of purposeful efficiency compared with Britain's patronage recipients down to the Age of Reform*—W. D. Rubenstein, 1991; *She had discovered her purposeful, almost lithographic prose style*—NY Mag., 1992; (purposefully) *He proceeded ... to fashion for himself ... a course all his own, made up of Latin, French and English with a special view to following words from one language to another, until one saw dimly how language worked and how purposefully it drifted*—W. Sheed, 1985; *Beatrix was carrying a wicker basket over one arm and as he watched she set off purposefully down the drive*—D. Simpson, 1987; *Skeins of wild geese moving purposefully across the sky*—Gourmet (US), 1990.

**pur sang.** Adopted from Fr. in the mid-19c., this adjectival and adverbial phr. meaning 'of the full blood, without admixture, genuine' is often used after a noun: e.g. *It is in fact possible to be a* *sociologist* pur sang *and not a black* (*white, yellow, piebald, Scots, Croat, Methodist, Muslim, etc., etc.*) *sociologist—TLS*, 1975.

**pursuant.** The phrase *pursuant to* is regarded by modern grammarians as a complex preposition and by an older school (including the editors of the OED) as a QUASI-ADVERB. It is more or less restricted to law, and means 'under, in accordance with': e.g. *Appellant is a state prisoner incarcerated in the Louisiana State Penitentiary ... pursuant to a* 1964 aggravated rape conviction—B. A. Garner, 1987).

**pursuivant.** Now conventionally Anglicized to /'pɜːsɪvənt/, but at the beginning of the 20c. normally pronounced with a medial /w/, /'pɜːswɪvənt/.

**purulent.** Pronounce /'pjuːrʊlənt/.

**purveyor.** Thus spelt, not *-er.*

**Pushtu.** An occasional variant of PASHTO.

**put.** In *shot-put* (athletics), *shot-putter*, and *shot-putting*, the standard pronunciation of the medial *-u-* is /ʊ/, not /ʌ/. See PUTT.

**putrefy** (to go rotten). Thus spelt, not *-ify.*

**putt.** In golf always pronounced /pʌt/, both noun and verb. Cf. PUT.

**puttee** /'pʌti/, strip of cloth wrapped round each leg as part of army uniform. Thus spelt, though derived from Hindi *paṭṭī.*

**putter.** See POTTER.

**pyaemia** (type of blood-poisoning). Thus spelt, but AmE *pyemia.*

**pygmaean** *adj.* Thus spelt, not *pygmean.* Pronounce /pɪg'miːən/.

**pygmy.** For etymological reasons (cf. L *pygmaeus*, Gk πυγμαῖος (adjs.) 'dwarfish'), the better spelling, rather than *pigmy.*

**pyjamas.** The standard spelling in BrE, but *pajamas* in AmE.

**pyorrhoea** (dentistry). Thus spelt in BrE, but *pyorrhea* in AmE.

**pyramidal.** Pronounce /pɪ'ræmɪdəl/.

**pyrites.** Pronounce /paɪ'raɪtiːz/.

**pyrrhic¹,** of a victory won at too great a cost to be of use to the victor, is named after Pyrrhus of Epirus, who defeated the Romans at Asculum in 279 BC but sustained heavy losses.

**pyrrhic²** (noun and adj.). Used of a metrical foot of two syllables, both of which are short (in Gk) or light (in Eng.). Thus *tune and* in | *When in* | *disgrace* | *with For* | *tŭne aňd* | *men's eyes* (Shakespeare's Sonnet 29). Some prosodists, however, regard *pyrrhic* as an unsafe concept in the scanning of English verse. For example, the line just given could be interpreted as | *When in dis* | *grăce wĭth* | *Fórtŭne aňd* | *mén's* | *éyes* |, or as | *Whén in* | *disgráce* | *wĭth Fór* | *tŭne ánd* | *mén's éyes* |.

# Qq

**qua** /kweɪ/. This useful word is in origin the ablative feminine singular of L *qui* 'who'. One of its several meanings in Latin is 'in virtue of the fact that'. It was adopted in English in the 17c. as a useful link-word (it is variously described as an adverb, a preposition, and a conjunction in standard English dictionaries) meaning 'in the capacity or character or role as'. Fowler (1926) thought that it should be restricted to contexts in which 'a person or thing spoken of can be regarded from more than one point of view or as the holder of various coexistent functions, and a statement about him (or it) is to be limited to him in one of these aspects: *Qua lover he must be condemned for doing what qua citizen he would be condemned for not doing*; the lover aspect is distinguished from another aspect in which *he* may be regarded. The two nouns (or pronouns) must be present, one denoting the person or thing in all aspects (*he*), and the other singling out one of his or its aspects (*lover*, or *citizen*).' *WDEU* (1989), on the other hand, cites several examples in which *qua* is used between two identical nouns as 'a somewhat more emphatic synonym of *as*', e.g. *a key to any one poem* qua *poem*—L. MacNeice, 1941. It also cites examples of the less common type in which *qua* does not occur between repeated nouns, e.g. *It cannot, qua fiˈlm, have the scope of a large book*—J. Simon, 1974.

Both descriptions are sound, as the following string of examples from good 20c. sources will show: *Earth closets, too. Do they exist* qua *earth closets? No*—K. Mansfield, 1920; *the presence of actual words is apt to confuse any estimate of the evocative power of the music* qua *music*—C. Lambert, 1934; *Qua phonetician, de Saussure has no interest in making precise the notion of species*—R. S. Wells, 1947; *Look at the sky ... What is there so extraordinary about it? Qua sky?*—S. Beckett, 1956; *I don't think that 'Hard Times' is a particularly good novel* qua *novel, whatever it may be as a social document*—Broadcast, 1977; *Wayland Ogilvie's dissatisfaction with the pictures* qua

*pictures*—S. Brett, 1979; *James Kirkup's poem about Jesus ... is ... an indefensibly bad poem* qua *poem*—London Rev. Bks, 1981; *Dressed in an Armani suit ... and espadrilles, he plays a cop* qua *existential hero*—Literary Rev., 1989. As to assigning it to a part of speech, *qua* seems to coexist as a conjunction and an adverb but hardly as a preposition. And as to usage, *as* is often the better choice of word, *qua* word.

**quad.** See ABBREVIATIONS 1.

**quadrate.** Stressed and pronounced the same as adj. and noun but differently as verb: thus (adj. and noun) /ˈkwɒdrət/; (verb) /kwɒˈdreɪt/. See PARTICIPLES 5.

**quadrennium,** a period of four years. The L original is *quadriennium*, but the first *i* has been lost in English under the influence of other 'period of X years' words, such as *decennium, millennium, septennium*. So too *quadrennial* adj.; cf. the L type *\*quadriennialis* (not recorded).

**quadriga** (chariot drawn by four horses harnessed abreast). Now normally pronounced /kwɒˈdriːɡə/, but, earlier in the century (*OED*, 1904; Fowler, 1926) only /-ˈdraɪɡə/. Pl. *quadrigae* /-giː/.

**quadrille.** Now always pronounced /kwəˈdrɪl/, but, earlier in the century (*OED*, 1904; Fowler, 1926) and still in AmE, optionally also /kəˈdrɪl/.

**quadrillion,** a thousand raised to the fifth (or formerly, esp. in BrE, the eighth) power ($10^{15}$ and $10^{24}$ respectively) (*COD*).

**quadroon.** See MULATTO 2.

**quagmire.** The pronunciations /ˈkwɒɡmaɪə/ and /ˈkwæɡ-/ are equally acceptable.

**quality.** 1 For 'has the defects of his qualities' see HACKNEYED PHRASES.

2 The corresponding adj. is *qualitative* /ˈkwɒlɪtətɪv/, not *qualitive*. See -ATIVE, -IVE. Cf. QUANTITATIVE.

**qualm.** Pronounced /kwɑːm/, no longer optionally as /kwɔːm/.

**quandary.** Now normally pronounced /ˈkwɒndərɪ/, no longer also as /kwɒnˈdeərɪ/. In other words it has firmly joined the group of words in which *-ary* is unstressed (*aviary, boundary, burglary,* etc.), as opposed to those in which *-ary* is stressed (*canary,* AmE *ordinary,* etc.). Cf. VAGARY.

**quantitative** has been fighting it out with **quantitive** for more than three centuries, but there can be no doubt that the longer form has been, and still is, the dominant one of the two. See -ATIVE, -IVE.

**quantum.** In physics a *quantum* (pl. *quanta:* See -UM 2) is a 20c. term for 'a minimum amount of a physical quantity which can exist and by multiples of which changes in the quantity occur' (*OED*). A *quantum jump,* which can be small or large, is simply 'an abrupt transition between one stationary state of a quantum system and another'. In popular use, since the 1950s, and, from the 1970s, in the form *quantum leap,* the phr. has been gleefully and endlessly used in general contexts to mean 'a sudden large increase'. It is one of the most striking POPULARIZED TECHNICALITIES of the 20c.

**quarrel** (verb). The inflected forms are *quarrelled, quarrelling* (AmE usu. *quarreled, quarreling*). See -LL-, -L-.

**quart(e)** (fencing). See CARTE.

**quarter.** 1 *Hyphening.* The practice shown in COD 1990 is as follows: *quarter day, quarterdeck, quarter-final, quarter-hour* (but more commonly *quarter of an hour*), *quarter-light, quartermaster, quarter sessions.*

2 In phrases of time, value, etc; there is considerable flexibility of idiom: *a quarter of an hour, a quarter to eleven* (AmE *a quarter of eleven,* and, regionally, *a quarter till eleven*), *with nearly quarter of an hour to spare, quarter-past eleven, be there by quarter to* are all (except the bracketed AmE ones) standard in BrE. So are *for a quarter of the price, for quarter of the price, for a quarter the price,* and *for quarter the price.*

**quarter** (verb). The inflected forms are *quartered, quartering.* See -R-, -RR-.

**quartet, quartette.** The shorter form is now usual. See DUET.

**quarto.** Pl. *quartos.* See -O(E)S 6.

**quasi.** The recommended pronunciation is /ˈkweɪzaɪ/, not /ˈkwɑːzɪ/.

**quasi-adverbs.** A smallish group of adjectives are idiomatically used in such a manner as almost to fall into the broad class of adverbs. They are therefore called quasi-adverbs, and by some grammarians complex prepositions (because they are invariably followed by a simple preposition). Some of the main quasi-adverbs are *according to, contrary to, irrespective of, preliminary to, preparatory to, previous to, prior to, pursuant to, subsequent to.* The type is shown in *he was rolling up his sleeves preparatory* (not *preparatorily*) *to punching the other boy; He did it contrary* (not *contrarily*) *to my wishes* (note that neither *different from* nor *opposite to* can be used in this manner). Further comments may be found under some of the words at their alphabetical place. See also DUE TO; OWING TO.

**quassia** (S. Amer. evergreen tree). The pronunciation recommended is /ˈkwɒʃə/.

**quat.** This sequence of letters is not always pronounced the same. It is /kwɒt/ in *quatrain* and *squat.* In unstressed initial position (*quaternary, quaternion*) it is /kwət-/, except that in *quatorze* it is /kət-/. In *quatercentenary* and *quattrocento* it is /kwæt-/. In *aquatic* it is normally /-kwæt-/, (though my personal preference is / kwɒt-/). In *quatorzain* and *quatrefoil* it is usually /ˈkæt-/.

**quatercentenary.** See prec., and CENTENARY 2.

**quatorzain, quatrain, quatrefoil.** See QUAT.

**quattrocento** /ˌkwætrəʊˈtʃɒntəʊ/. (Also with a capital initial.) The style of It. art of the 15c., i.e. 1400–99. See QUAT; TRECENTO.

**queer.** See GAY.

**querist,** a questioner, a person who asks questions. This word, first recorded in the 17c., is now rapidly passing into restricted, mostly literary, use.

**question.** 1 For LEADING QUESTION, see that article.

**2** For the order of words in indirect questions (*He asked what he was to do*, etc.), see INDIRECT QUESTION.

**3** See BEG THE QUESTION.

**4** For *the question as to*, see AS 12.

**question mark.** The mark of interrogation in English, represented by the sign ? **1** Its ordinary use is shown in the following examples: *What time is it?*; *What does the word empathy mean?*; *Ah, did you once see Shelley plain, And did he stop and speak to you And did you speak to him again?*. The next word should normally begin with a capital letter.

**2** The question mark should not be used after indirect questions, e.g. *He asked why I was there*; *He asked whether I would come with him*; *I was asked if I would stay for dinner*.

**3** The question mark and the exclamation mark (!) are normally easily distinguishable in function, e.g. *How often does it happen?*; *How seldom it happens!* Sometimes, especially in popular writing, the two are combined for emphasis: *How often must I tell you?!*

**4** A question mark may be placed before a word, etc., whose accuracy is in doubt, e.g. *Thomas Tallis (?1505–85)*; *Phnom Penh is the capital of Cambodia (? Kampuchea)*.

**5** A question mark is not needed after certain types of requests, e.g. *Would passengers now on platform 2 please move to platform 3 if they wish to join the 8.50 train for Reading and intermediate stations*.

**6** A tag question (e.g. *He's much taller now, isn't he?*) must always be followed by a question mark.

**questionnaire.** Now almost invariably pronounced with initial /kw/, not /k/. The main stress falls on the last syllable.

**queue.** The inflected verbal forms are *queues, queued, queuing* (or *queueing*).

**quiche.** See TART.

**quick** has a long history in standard use as an adv. (first recorded in the 13c. and used, for example, by Shakespeare, Milton, Locke, and Dickens), but now has restricted currency ('This use is now usually avoided in educated speech and writing, though found in some standard

colloq. constructions'—*OEDS*, 1982). Examples of a thin line of current colloquial uses: *Come quick!*; *the 'get-rich-quick' society*; *quick-frozen food*; (formerly, message on makeshift beach hats) *Kiss Me Quick*.

**quid** (slang, = one pound sterling). The pl. is unchanged (*it cost me five quid*).

**quiescence, quiescency.** Both words entered the language in the 17c., and mean the same, but *quiescence* has always been the dominant word of the two. See -CE, -CY.

**quiet** (verb). In the sense 'to reduce to quietness, to soothe (a person, an emotion, a disturbance, etc.)', *quiet* has been used transitively since the 16c. (e.g. *The unexpectedness of this departure from the routine at first disquieted but then quieted us all*—M. Lindvall, 1991; *Bishop regularly quiets the butterflies by reassuring himself that his new job is no different, really, from the old one*—*NY Times Mag.*, 1992). From the late 18c., chiefly in N. America, it has also been used intransitively (e.g. *The effect of the drugs is often dramatic, with the children quieting down, paying attention to their schoolwork and in some cases doing better in school*—*Sci. Amer.*, 1974; *And sometimes, at night, when everything had quieted down ... he would lie supine and evoke the one and only image*—D. Nabokov, 1986; *I didn't last that long as a rowdy drunkard, and when I switched to opiates at least I quieted down*—*New Yorker*, 1992).

As one of the later -EN VERBS FROM ADJECTIVES, *quieten* emerged in the early 19c. as a transitive verb and in the late 19c. as an intransitive one (commonly with *down*). It has not had an easy journey: 'To "quieten" the children is not English,' said a writer in *The North British Review*, 1844; 'quieten, whether as transitive or as intransitive, is a superfluous word,' said Fowler in 1926. But, in BrE at any rate, *quieten* is now a commonplace verb (used either transitively or intransitively) and any stigma attached to it for something like a century has gently disappeared. Examples: *Bapaiji related the story in outraged tones, rocking the baby to make it quiet, while Dhunmai quietened down herself*—B. Desai, 1988; *Arnica also helps to calm and quieten the upset child*—*Health Shopper*, 1990; *By the time this issue of* The Linguist *reaches our Members,* business activity in the UK and other parts of Europe will have quietened down for the summer months—Linguist, 1993.

**quiet** (noun), **quietness, quietude.** The first is much more commonly used than the others, and quietude the least often of the three. Quiet means principally 'silence, stillness, tranquillity' (the quiet that sometimes precedes a storm; a period of peace and quiet). Quietness is 'the condition of being quiet or undisturbed' (the quietness of a congregation at prayer). Quietude, a formal synonym of quietness, means 'the state or condition of being quiet, peaceful, or calm' (the Coventry Canal (of more interest today to the industrial archaeological than to the lover of rural quietude)— C. Dexter, 1989; Their two and one-half acres retain a bucolic quietude—Angeles (US), 1991).

**quincentenary.** See CENTENARY 2.

**quintet, quintette.** The shorter form is now usual. See DUET.

**quire.** See CHOIR.

**quit** (verb). The pa.t. and pa.pple are either quit (esp. in AmE) or quitted.

**quite.** 1 A colloquial use that often puzzles or amuses visitors to Britain is the use of quite (or quite so) to express agreement ( = 'I quite agree') with a previous declarative statement: e.g. 'The Minister should have resigned.' 'Quite.' Other ways of expressing agreement exist (e.g. precisely, exactly, (yes,) indeed, rather (somewhat oldfashioned, BrE), right (AmE), absolutely, etc.), but quite, quite so, and rather are the ones that are likely to be regarded as distinctively British by visitors.

2 Used with a noun preceded by the indefinite article, quite traditionally meant 'completely, entirely': e.g. You are a humourist . . . Quite a humourist—J. Austen, 1816. This use has dropped out of the literary language, but survives in informal contexts in the 20c., often implying emphatic, and occasionally ironic, commendation: e.g. We had us a party last night—quite a party—D. Divine, 1950; Well now, that's quite something. Thank you—N. Blake, 1958; It had been quite a week and I wanted a day of relaxation—R. Crossman, a1974.

3 The most important modern development of quite, however, is its use in a weakened or 'down-toning' sense, 'rather, to a moderate degree, fairly'. This use surfaced in the 19c. but became established, esp. in BrE, in the 20c. The result is that quite can potentially mean either 'completely' or 'fairly'. For example, in Alun quite approves of my literary efforts (Nigel Williams, 1985), the contextual meaning is 'really does approve'. But in the social archaeologist quite often needs to use the techniques of the locational analyst (Geographical Jrnl, 1983) the contextual meaning is probably 'fairly often'. The sense 'somewhat, fairly' tends to occur most frequently with what are called gradable adjectives (those that may be modified by more, less, or very): e.g. the book is quite interesting; the music was at times quite loud; he was quite generous with his money. By contrast, when used with adjectives that are not normally gradable (e.g. different, enough, excellent, impossible), quite usually means 'completely, totally': e.g. he still had quite enough work to live on—P. Fitzgerald, 1980; it would be quite impossible for me to know you—E. Jolley, 1980; It's very hard to imagine a child growing up with quite different beliefs—P. Fitzgerald, 1986; You needn't feel at all awkward about approaching other publishers. In fact, I think you were quite right to—London Rev. Bks, 1988. But there are plenty of examples that do not clearly accord with this 'rule': e.g. I'm bound to own that it was quite nice being telephoned by lots of young ladies—Times, 1986. (How nice?) In the spoken language the intonation can usually clarify the difference between quite nice ( = reasonably nice, but I've known better) and quite nice ( = very nice). In the written language only the immediate context holds the key to the meaning of such a sentence as It was quite awkward for me to arrange to come here (P. Fitzgerald, 1986).

4 When quite is used with a preceding can't or don't and is followed by an infinitive, it is normally a 'downtoner': e.g. When you care enough, but can't quite say it; Many people don't quite know how to react when someone dies.

The difficulties lying behind the correct use of quite are plain to see, but, paradoxically, most educated people seem to have no particular problem in using or understanding the word in its various constructions and meanings.

**quiz.** See -z-, -zz-.

**quoin, quoit.** Pronounce /kɔɪn/, /kɔɪt/.

**quondam.** This Latin word meaning 'former(ly)' has been used in English as adv. and adj. since the 16c., but is hardly ever encountered now except as an attributive adj. applied to persons. Examples: *The memory of her quondam fiancé on his knees*—S. Mackay, 1984; *she saw all male members of her quondam department as persecutors*—A. S. Byatt, 1990. Cf. ERST, ERSTWHILE; LATE, etc.

**quorum,** the number of members whose presence is needed to make proceedings valid. Pl. *quorums.* See -UM 1.

**quota,** a share, the number allowed, etc. Pl. *quotas.*

**quotation marks** (also called *inverted commas*). The placing of quotation marks varies from publishing house to publishing house, and each system has its own validity. What follows is a slightly adapted version of the OUP house style as set out in *Hart's Rules.* **1** *General.* (*a*) Whenever a poetic quotation is given a line (or more) to itself, it is not to be placed within quotation marks; but when the line of poetry runs on with the prose, or when a number of quotations follow one another and it is necessary to distinguish them, then quotation marks are to be used. (*b*) Insert quotation marks in titles of essays: e.g. *Mr Brock read a paper on 'Description in Poetry'.* But omit quotation marks when the subject of the paper is an author: e.g. *Professor Bradley read a paper on Jane Austen.* (*c*) Quotation marks may be used to enclose slang and technical terms. They should not be used with house names or public houses: *Chequers, Cosicot, the Barley Mow.* (*d*) Single marks are to be used for a first quotation; then double for a quotation within a quotation. If there should be yet another quotation within the second quotation it is necessary to revert to single quotation marks.

**2** *Relative placing of quotation marks and punctuation.* All signs of punctuation used with words in quotation marks must be placed *according to the sense.* If

an extract ends with a point or exclamation or interrogation sign, let that point be included before the closing quotation mark; but not otherwise. When there is one quotation within another, and both end with the sentence, put the punctuation mark before the first of the closing quotation marks. Examples: 'The passing crowd' is a phrase coined in the spirit of indifference. Yet, to a man of what Plato calls 'universal sympathies', and even to the plain, ordinary denizens of this world, what can be more interesting than those who constitute 'the passing crowd'?; If the physician sees you eat anything that is not good for your body, to keep you from it he cries, 'It is poison!' If the divine sees you do anything that is hurtful for your soul, he cries, 'You are lost!'; 'Why does he use the word "poison"?'; But I boldly cried out, 'Woe unto this city!'; Alas, how few of them can say, 'I have striven to the very utmost'!

Thus, marks of exclamation and interrogation are sometimes included in and sometimes follow quotation marks, as in the sentences above, according to whether their application is merely to the words quoted or to the whole sentence of which they form a part. The sentence-stop must be omitted after ? or ! even when the ? or ! precedes the closing quotation marks. In regard to other marks, when a comma, full point, colon, or semicolon is required at the end of a quotation, there is no reason for perpetuating the bad practice of their undiscriminating inclusion within the quotation marks at the end of an extract. So place full points, commas, etc., according to the examples that follow.

(i) Example: *Our subject is the age of Latin literature known as 'Silver'.* The single word *Silver,* being very far from a complete sentence, cannot have a closing point belonging to it: the point belongs to the whole sentence and should go outside the quotation marks: ... *known as 'Silver'.*

(ii) If the quotation is intermediate between a single word and a complete sentence, or it is not clear whether it is a complete sentence or not, judgement must be used in placing the final point: *We need not 'follow a multitude to do evil'.* The words quoted are the greater part of a sentence—[Do not] *follow a multitude to do evil*—but not complete in themselves, so do not require their own closing point; the point therefore belongs to the main sentence, and is outside the quotation marks. Similarly in: *No one should 'follow a multitude to do evil', as the Scripture says; Do not 'follow a multitude to do evil'; on the contrary do what is right.* Here the comma and semicolon do not belong to the quoted words and are outside the quotation marks.

(iii) The quoted words may be a complete sentence but the closing point must be omitted because the main sentence is not complete: *You say 'It cannot be done': I say it can.* Here the colon clearly belongs to the main sentence, forming the punctuation between *You say* and *I say*, and is therefore outside the quotation marks.

(iv) When a quotation is broken off and resumed after such words as *he said*, if it would naturally have had any punctuation at the point where it is broken off, a comma is placed within the quotation marks to represent this. Example: The words to be quoted are: *It cannot be done; we must give up the task.* In quotation this might appear as: *'It cannot be done,' he said; 'we must give up the task.'* Note that the comma after *done* belongs to the quotation, which has a natural pause at this point, but the semicolon has to be placed after *said* and hence outside the quotation marks. On the other hand, if the quotation is continuous, without punctuation at the point where it is broken, the comma should be outside the quotation marks. Example: The words to be quoted are: *Go home to your father.* In quotation these appear as: *'Go home', he said, 'to your father.'* The comma after *home* does not belong to the quotation and therefore comes after the quotation marks. These rules, though somewhat lengthy to state in full, are simply instances of the maxim—place punctuation according to sense.

(v) The quoted words may be a complete sentence which ends at the same point as the main sentence: *He said curtly, 'It cannot be done.'* Logically, two full points would be required, one inside the quotation mark belonging to the quoted sentence, and one outside belonging to the main sentence. In such cases the point should be set *inside* the quotation marks (as ! or ? would be) and the point closing the main sentence omitted. In particular, when a long sentence is

quoted, introduced by quite a short phrase, it is better to attach the closing point to the long sentence: *Jesus said, 'Do not think that I have come to annul the Law and the Prophets; I have come to fulfil them.'* (Not *'... to fulfil them'*.)

(vi) Where more than one sentence is quoted, the first and intermediate sentences will naturally have their closing points within the quotation, and the last sentence should do so also: *Moses told you: 'Do not kill. Do not steal. Do not commit adultery.'* (Not *'... adultery'*.)

(vii) When a quotation is followed by a reference, giving its source, in parentheses, if it is a complete sentence, the closing quotation mark is placed according to the above rules, before the parenthesis, and there is another closing point inside the parentheses: *'If the writer of these pages shall chance to meet with any that shall only study to cavil and pick a quarrel with him, he is prepared beforehand to take no notice of it.'* (*Works of Charles and Mary Lamb*, Oxford edn, i.193.)

(viii) Where marks of omission (or, more rarely, *etc.*) are used, they should be placed within the quotation marks if it is clear that the omitted matter forms part of the quotation.

**3** *American English.* The most significant difference of practice is that in the types shown in 2(ii) above American publishers would normally place the final quotation mark *outside* the full point: *We need not 'follow a multitude to do evil.'* Similarly, in American style: *He believed in the proverb 'Dead men tell no tales.'* American publishers are also likely to reverse the procedure described in the first sentence of 1(*d*) above, i.e. to use double marks for a first quotation and single ones for a quotation within a quotation.

**quote** (verb) makes *quotable*. See MUTE E. The formulaic pair *quote ... unquote* used in dictation to introduce and terminate a quotation is first illustrated in the *OED* in 1935.

**quotes** (pl. noun). In copy-editing departments, printing houses, etc., *quotes* is used as the abbreviated form of 'quotation marks'. See ABBREVIATIONS 1(*b*).

**quoth.** See ARCHAISM. In origin it descends from OE *cwæð* he said (pa.t. of *cweðan* to say). When used, for example in historical novels, it is always placed before the subject, e.g. *Quoth an inquirer, 'Praise the Merciful!'*—Browning, 1884.

**Qur'an.** Now a frequent spelling in English of KORAN.

**q.v.** Abbreviation of (L) *quod vide* 'which see'.

# Rr

**r.** See INTRUSIVE R; LINKING R.

**rabbet** (carpentry). The better spelling, rather than *rebate*.

**rabbit** (noun). For *Welsh rabbit* (or *rarebit*), see WELSH RABBIT.

**rabbit** (verb). The inflected forms are *rabbited*, *rabbiting*. See -T-, -TT-.

**rabies.** Now always pronounced as two syllables, /ˈreɪbiːz/, but the *OED* (1904) and Fowler (1926) gave it as three, /ˈreɪbiːz/. See -IES; PRONUNCIATION 2 (ii).

**race.** In its ethnic sense 'a tribe, nation, or people, regarded as of common stock', *race* entered the language in the late 16c. and had considerable uncontroversial currency until the 20c. (*the British race, the German race, the Tartar race*, etc.). Now, as *OEDS* (1982) remarks, 'The term is often used imprecisely; even among anthropologists there is no generally accepted classification or terminology.' In practice, the word has retreated rapidly and is now largely replaced by such words as *nation*, *people(s)*, and *community*. As a minor curiosity, the ultimate origin of the word *race* is obscure. We know that it was adopted from French *race* (earlier *rasse*) c1570, and that it had entered French from earlier It. *razza*. Cognates exist in most European languages (Sp. *raza*, Ger. *Rasse*, etc.) but etymologists have found no pre-14c. evidence for the word in any European language.

**raceme, racemose.** Pronounce /rəˈsiːm/, /ˈræsɪˌməʊs/.

**rachitis** (rickets). Pronounce /rəˈkaɪtɪs/.

**racial** /ˈreɪʃəl/. Formed in English in the 1860s by tacking the suffix -*ial* on to *rac(e)*, the word swiftly gathered momentum in the 20c. as it was used to qualify such words as *prejudice, discrimination, equality, minority, conflict, segregation*, and *tension* in our bitterly divided age.

**racialism,** one of the key words of the 20c., entered the language at the beginning of the century, some thirty years before the now-dominant term *racism*. Both terms are used of discriminatory behaviour by authorities or groups (governments, police, employers, etc.) against people on grounds of colour, religion, or nationality. Linguistic racism is typified by the use of derogatory terms for ethnic groups, e.g. *Chink, Frog, honkie, Hun, Jap, kike, nigger, Oreo, wog, wop*. And the fecundity of abusive racial terminology is neatly underlined by a passage in the novel *High Cotton* (1992) by the black American writer Darryl Pinckney: *and predicted that I would come to no good among the no-accounts, burrheads, shines, smokes, charcoals, dinges, coons, monkeys, jungle bunnies, jigaboos, spagingy-spagades, moleskins, California rollers, Murphy dogs, and diamond switchers. He liked to be shocking.*

The phenomenon is by no means confined to white English-speaking people. As far as I can determine, name-calling is a universal phenomenon, but the offensive names are often squeamishly omitted from dictionaries by the lexicographers of many foreign countries. See ETHNIC TERMS.

**rack** in *rack and ruin* means 'destruction' and is normally so spelt in this phrase in BrE. One of nine homonymous nouns and seven homonymous verbs, it is a spelling variant of *wrack* (OE *wræc*) 'damage, disaster, destruction'. This *rack* is not etymologically related to the *rack* = framework, that = instrument of torture, that = an awkward gait of a horse, a *rack* of lamb, not to the verb *rack* in *rack one's brains*, and numerous others. The word of the same sound meaning 'seaweed' is spelt *wrack*. All the complexities of this exceedingly complicated word cannot be set down here: spare an hour (at least) to consult a large dictionary, esp. the *OED*.

**racket** is the recommended spelling for the bat used in tennis, squash, etc., rather than *racquet* (cf. Fr. *raquette*), but if your instinct leads you to prefer *racquet* no one is going to quarrel with you about it. The game played by two or four persons in a four-walled court is always written as *rackets* in BrE but usu. as *racquets* in AmE. The unrelated word *racket* meaning (*a*) a disturbance, (*b*) a fraudulent scheme, has no variant spelling.

**racoon** is the customary spelling in BrE for this furry NAmer. nocturnal carnivore, but *raccoon* in AmE. The word was first adopted in the 17c. from an American Indian language (Algonquian).

**raddle.** See RUDDLE.

**raddled,** the usual spelling of the adj. applied to a person, meaning 'unkempt, run-down in appearance'.

**radiance, radiancy.** The second is rare, but is occasionally drawn on for metrical or rhythmical reasons. See -CE, -CY.

**radiator.** Thus spelt, not *radiater*.

**radical** as noun and adjective is always spelt thus in its main senses. The noun *radicle* is (*a*) the part of a plant embryo that develops into the primary root; (*b*) a rootless subdivision of a nerve or vein (COD). The adjectival form of *radicle* in sense (*a*) is *radical*.

**radio.** Pl. *radios*. See -O(E)S 4. To the puzzlement of some people, *radio* gradually replaced *wireless* from about 1920 onwards, i.e. during the period when 'wireless' broadcasting turned into a world-wide industry. The word *wireless*, when used at all now, is normally uttered with a change of intonation ( = you know what I mean, don't you?) or by an explanatory disclaimer ('as we used to say', or some such phrase). The turning-point was about the middle of the 20c.

**radius.** The recommended pl. is *radii* /-dɪaɪ/, not *radiuses*.

**radix** /'reɪdɪks/ *or* /'rædɪks/. The pl. is *radices* /-dɪsiːz/ *or* -*ixes*. See -EX, -IX.

**railroad** (noun) is the customary term in AmE for BrE *railway*. In AmE, *railway* is sometimes use of the track on which trains run but not of the system of transportation itself. In both countries *railroad* is used as a verb to mean 'to pressurize or coerce into a premature decision', etc.

**raise, rear** (verbs). The first is esp. common in AmE of the cultivation of plants (*to raise corn*), the breeding and rearing of livestock (*to raise cattle*), and of the bringing up of children (*to raise children*); but *rear* is also used of livestock and children (*to rear a child*), and *bring up* of children (*to bring up a family of four*). In BrE, *raise* is sometimes used in all three senses, but the more usual terms are *to cultivate* or *to grow* (plants), *to rear* (animals), and *to bring up* (children).

**raise, rise** (nouns). An increase of salary is called a *raise* in AmE and a *rise* in BrE.

**raise, rise** (verbs). In almost all of their numerous senses, *raise* is transitive (*raise prices, raise money, raise an army, raise an objection, raise hopes*, etc.) and *rise* is intransitive (*he rose to his feet, prices have risen, he rose from the ranks, to rise in arms*, etc.). It is worth noting, however, that historically (15–18c.) *raise* was interchangeable with certain senses of *rise* in BrE, and that such uses persist in some regional forms of AmE (*the Mississippi is raising; the prices of rent have raised 10 per cent in the last two years*).

***raison d'être.*** This loanword from French means not merely 'reason' but 'a purpose or reason that accounts for or justifies or originally caused a thing's existence' (COD).

**Raj** (preceded by *the*), an historical term meaning 'British sovereignty in India', is pronounced /rɑːdʒ/.

**raja,** as Indian title or Malay or Javanese chief, is now normally written thus, not *rajah*.

**Rajput,** a member of a Hindu soldier caste, is now spelt thus, not *Rajpoot*. Pronounce /'rɑːdʒpʊt/.

**Raleigh.** Sir Walter *Raleigh* (or *Ralegh*; he never used the spelling *Raleigh* himself) is believed to have pronounced his surname as /'rɔːli/, but his name has been commonly pronounced either as /'ræli/ or /'rɑːli/ throughout the 20c. The place-name in N. Carolina is /'rɔːli/. *Raleigh* bicycles are /'rɔːli/ in AmE and /'ræli/ in BrE.

**rallentando** (musical direction). The recommended pl. is *rallentandos*, not *rallentandi*. See -O(E)S 7.

**Ralph.** The pronunciation /reıf/, now decreasingly heard, is a survival from the 17c. spelling *Rafe*. The usual pronunciation now is /rælf/.

**ramekin** /'ræmıkın/. This is the recommended spelling, not *ramequin*.

**rancour.** So spelt in BrE, but *rancor* in AmE. The corresponding adj. is *rancorous* in both varieties.

**ranee.** Still perhaps the more usual spelling (rather than *rani*) for the word meaning 'the wife of a raja'.

**ranunculus.** The pl. is either *ranunculuses* or *ranunculi* /-laı/.

**rapport.** An unusual example of a word that has become un-Anglicized in pronunciation in the 20c. after having the final *t* fully pronounced at an earlier stage (Walker, 1791; OED, preferred form, 1904). The only current pronunciation is /ræ'pɔː/.

**rapt.** In origin (15c.) from L *raptus*, pa. pple of *rapere* 'to seize', *rapt* first meant 'taken or carried up to or into heaven', then (16c., among other senses) 'fully absorbed, engrossed (*listen with rapt attention*)'. The homophone *wrapped*, pa.t. of *wrap*, which is first recorded in the 14c., is unrelated. It is a sheer coincidence that *wrapped up* in *He is wrapped up in his work* means 'engrossed'.

**rarefaction** (from L *rārēfacere* 'to make less solid, rarefy') has better etymological credentials than the less common noun *rarefication*, since there is no Latin verb *\*rārēficāre*.

**rarefaction, rarefy.** So spelt (in contrast with *rarity*), and pronounced /'rærı-/ (in contrast with *rarely* /'reəlı/.

**rarely.** 1 Used by itself, i.e. without an accompanying (*if*) *ever* or *or never*, this negative adverb ( = not often) has some interesting idiomatic uses, from the obvious (*He rarely leaves home now*) to others that are less obvious: (front-placing followed by subject–verb inversion) *Rarely had I seen such a mess*; (front-placing = 'on rare occasions', without causing subject–verb inversion) *Rarely, aggression is not premeditated*.

2 Uncontested uses with *ever* or *never* include *rarely if ever*, *rarely or never*. More debatable are *rarely or ever* ('by confusion of "rarely if ever" and "rarely or never"', OED), though illustrated by 1768 and 1811 examples in the OED; and *rarely ever*, though illustrated by 1694, 1709, 1728, and 1857 examples in the OED, and supported by a British and an American example in WDEU (1989): *and the thieves are rarely ever caught*—R. Blythe, 1969; *I rarely ever think about the past*—New Yorker, 1971. A later UK example: *These things are rarely ever explored outside the confines of the Catholic church*—Independent, 1989. In practice either *hardly ever* or *scarcely ever* is preferable.

**raring, to be.** The colloquial expression *to be raring to* (*go*, etc.) 'to be extremely eager to (do something)' made its way into the standard language from AmE or from English dialects at the beginning of the 20c. The infinitive *rare* is a variant of *rear* (verb) '(of a quadruped, esp. a horse) to rise on the hind feet'. Examples: *He's laid it on that the preacher makes some inflammatory remarks ... so that the congregation ... will be rarin' to go*—J. Tyndall, 1971; *We were at the starting-gate and raring to go*—Church Times, 1979; *John Patten, the education secretary, was described by his aides as 'raring to go' after throwing off the viral infection that put him in hospital last month*—Times, 1993.

**rase.** See RAZE.

**raspberry.** Pronounce /'rɑːzbərı/, despite the spelling -*sp*-

**rat-catcher.** A rat-catcher was originally a person employed to rid farm buildings, houses, etc., of rats and other rodents. Much later (early 20c.) the word was used to denote informal attire in the hunting-field rather than the black covert coat worn by riders to hounds who, although regular subscribers to the Hunt, do not aspire to pink. In its original sense the word is now largely obsolete. Most British rats now meet their doom at the hands of a local-authority official with the designation of *pest control officer* or *rodent operator*.

**rate** (verb). Despite the usual rules described s.v. MUTE E, *rate* (verb) makes *rateable* rather than *ratable* (lest the word should be momentarily connected with *rat*).

**rates** (pl. noun). See TAX.

**rather.**

1 Link inversion.
2 Emphatic use.
3 *rather than*.
4 With adjectives.
5 *a rather/rather a*.
6 *had rather*.
7 *would rather, 'd rather*.

**1 Link inversion.** In certain circumstances, some writers are moved to reverse the natural subject/verb order after comparatives, including *rather*. Fowler (1926) s.v. *inversion* judged that such link inversions are 'so little necessary as to give a noticeable formality or pomposity to the passage'. He cited as an example *His book is not a biography in the ordinary sense; rather is it a series of recollections culled from ...* (read *it is rather*). He gave further examples s.v. *rather*, including *I do not feel like one who after a day of storm and rain is glad to creep indoors, and crouch hopelessly over the fast-dying embers on the hearth; rather do I feel like one who ...* (read *rather I feel*). No example of such inversion has come to my notice: the practice may have died out.

**2 Emphatic use.** A particularly BrE use of *rather*, now somewhat old-fashioned, is as an emphatic response to a question, when it means 'Indeed, assuredly' (e.g. *Did you enjoy your holiday?—Rather!*) In such cases the main stress falls on the second syllable: thus /rɑːˈðɜː/.

**3** *rather than*. There are two main types of construction and some sub-types: (*a*) With a following *-ing* form. Examples: *Yet the insurgents themselves are subject to the accusation of pursuing ideological goals rather than seeking the truth—Bull. Amer. Acad. Arts & Sci.*, 1987; *the subject reveals its limitations, rather than providing a springboard for the author—English*, 1987; *At the same time society pays prematurely retired neighbours to do nothing rather than getting them to organise recreation, sport and leisure—Sunday Times*, 1987; *They walked to it that way, rather than descending from the moors—A. S. Byatt*, 1990. It is worth noting that the first example has an *-ing* form both before and after *rather than*. (*b*) With a following plain infinitive or *to*-infinitive. Examples: (plain infinitive) *Bertrand ... preferred to go to a restaurant ... rather than eat alone—P. P. Read*, 1986; *Better to part with what they must now, rather than*

*lose more later—M. Shadbolt*, 1986; *Rather than simply moralize against racism, the schools would aggressively recruit black students—NY Rev. Bks*, 1987; *he ... would probably straightaway certify her rather than delve into the years of confusion and mismanagement—B. Rubens*, 1987; *Rather than dwell on this shortcoming he went off into half-baked musings—K. Amis*, 1988. (*to*-infinitive) *I preferred to write, to explore the world of imagination, rather than to mix with others—J. Frame*, 1984; *The press ... has a responsibility to elevate rather than to degrade them—Bull. Amer. Acad. Arts & Sci.*, 1987. Note that some of the above examples have matching infinitives before and after *rather than*.

(*c*) Asymmetrical and therefore undesirable constructions with an infinitive in the first clause and an *-ing* form in the second. Examples: *It is a pity that Shandy père did not address himself to the problems of sociological methodology, rather than remaining an historical methodologist—R. Merton*, 1985; *What they are saying is that ... it is better to give way and let them have what they want rather than standing up for the rule of law—R. Muldoon*, 1986; *an employer, who had chosen to enter 'nil' in the contract rather than, say, striking out the clause altogether, could not expect* [etc.]—A London law firm's newsletter, 1987; (read *remain, stand up, to strike out, respectively*.) (*d*) Miscellaneous standard constructions containing *rather than*. Examples: *If Mimi's cup runneth over, it runneth over with decency rather than with anything more vital—A. Brookner*, 1985; *Beliefs about the female temperament promised that the pedagogical style of women teachers would be emotional and value-oriented rather than rational and critical—Dædalus*, 1987; *It had come as a jolt that life was something to be waged, rather than relied upon—New Yorker*, 1987; *a matter of calm and untroubled daily practice rather than of clandestine nocturnal comings-and-goings—English*, 1989. In general terms matching forms are best in the clauses preceding and following *rather than*. Simple examples (from *CGEL* 14.15): *They were screaming rather than singing* ( = *and not*); *She telephoned rather than wrote*; *He wanted to sunbathe rather than (to) swim*. The only snag is that many kinds of English, including literary English, are more complex, and call for greater subtlety, than can be gleaned from these short invented examples.

**4** With adjectives. Used with adjs., *rather* means 'somewhat, to some extent', i.e., to use the modern terminology, is a mild downtoner *(he became rather drunk; it was rather dark)*. At other times it acts as a mild intensifier *(he was driving rather fast; they seemed rather encouraged by the decision; rather better at cricket than at rugby)*. The degree of downtoning or intensification is usually clarified by the context.

**5** *a rather/rather a*. Both orderings are acceptable before a following adj. + noun: e.g. *a rather gruesome sight/rather a gruesome sight*. But an adj. must be present: *rather a sight* is idiomatic, but *\*a rather sight* is not.

**6** *had rather*. See HAD 3.

**7** *would rather*, *'d rather*. The full and the contracted forms both express preference and therefore may (but need not) be followed by a comparative construction beginning with *than*: *I'd rather read a book than watch television; he would rather not join the committee*.

**ratio.** Pl. *ratios*. See -O(E)S 4.

**ratiocinate.** The recommended pronunciation is /ˌrætɪˈɒsɪneɪt/ rather than /ˌræʃɪ-/. Similarly with its derivatives.

**ration.** Now always pronounced /ˈræʃən/ in BrE though the *OED* (1904) gave precedence to /ˈreɪʃən/, while commenting that the form with /æ/ was 'usual in the army'. Both pronunciations are common in standard AmE.

**rationale.** At the beginning of the 20c. this word (which is the neuter of the L adj. *ratiōnālis*) was pronounced as four syllables: /ˌræʃəˈneɪli/. At some point in the century, probably under the influence of the French loanwords *locale* and *morale* (there is no Fr. *rationale*), it became routinely pronounced as three syllables: /ˌræʃəˈnɑːl/.

**rationalize** in the sense 'to make (a business, etc.) more efficient by reorganizing it to reduce or eliminate waste of labour, time, or materials' (*COD*) is relatively recent: the first example in the *OED* is dated 1926. It is still a vogue

word in business language. A wartime example: *Ship-building yards were 'rationalised' out of existence and [coal] pits abandoned—Observer*, 11 June 1944.

**ratline** /ˈrætlɪn/, usu. in the plural (any of the small lines fastened across a sailing-ship's shrouds like ladder rungs, *COD*), is now the customary spelling, not *ratlin* or *ratling*.

**rattan** /ræˈtæn/, climbing palm, walking-stick, is the recommended spelling of this Malay loanword, not *ratan*.

**ravel** (verb). **1** The inflected forms are *ravelled*, *ravelling* (AmE often *raveled*, *raveling*).

**2** This is an example of a word capable of having two opposite meanings: 'to entangle or become entangled'; (often followed by *out*) 'to disentangle, to unravel'. Both senses have been extant for several centuries without causing confusion, though *unravel* is now the more usual word for the second sense. Two literary examples: ( = entangle) *Those wild, unhappy, self-defending Few, If not destroy'd in Time, will ravel all the Clew—Defoe*, 1706; ( = disentangle) *Must I rauell out My weau'd-vp follyes?—Shakespeare*, 1593.

**raze, rase.** In the sense 'to destroy, to tear down' the word is now normally spelt *raze* (esp. in *the building was razed to the ground*), not *rase*, notwithstanding that it entered the language c1400 from OF *raser* (from *rās-*, ppl stem of L *rādere* 'to scrape', etc.). Contrast the history of *erase*, formed in the early 17c. from *e-* (prefix form of *ex-* before some consonants) + the same Latin base as that of *raze*.

**re.** In origin the ablative of L *rēs* 'thing, affair', *re* (often in the form *in re*) has been used in legal documents before case names (e.g. *In re Rex v. Smithers*) for more than a century and, from there, in the 20c. has made its way into commercial correspondence and into office memoranda in the sense 'With regard to, with reference to, etc. (your enquiry, note, etc.)'. Some authorities have objected to it. In particular, A. P. Herbert in his *What a Word!* (1935) ridiculed *re* in this example: *We herewith enclose receipt for your cheque £4 on a/c re return of commission re Mr. Brown's cancelled agreement re*

*No 50 Box Street top flat.* Fowler (1926) was equally antagonistic and cited several examples with distaste: e.g. *Dear Sir,—I am glad to see that you have taken a strong line re the Irish railway situation./Why not agree to submit the decision of the Conference re the proposed readjustment to the people so that they alone can decide?* It will be observed that in each case *re* occurs in mid-sentence, where it is inappropriate. There can, however, be no objection to its use as an introductory preposition, which, with its following noun or noun phrase, announces the subject of the correspondence, memo, etc., that follows. 'Its conciseness makes it well-nigh irreplaceable' (Garner, 1987).

**re(-).** In words beginning with *re-*, the prefix of repetition, a hyphen is unnecessary (e.g. *reaffirm, recharge, regroup, reinterpret, reopen, reshape, reuse*) except (a) in words in which the second element begins with *e* (e.g. *re-edit, re-entry*); (b) to distinguish words containing this prefix from words spelt the same but having a different meaning (e.g. *re-collect* to collect again, but *recollect* to remember; *re-count* to count again, but *recount* to narrate; *re-cover* to cover again, but *recover* to return to health; *re-creation* creation over again, but *recreation* pleasurable activity; *re-form* to form again, but *reform* to make or become better, to improve).

**'re.** Used as a reduced form of *are*, pl. pres. indic. of *be*, it has been commonplace after personal pronouns in the representation of speech, or for metrical reasons in verse, from the time of Shakespeare onwards (thus *they're, we're, you're*). What seems to be a 20c. innovation, particularly outside Britain, is the use of *'re* after plural nouns and after *there* and *what*. Examples: *'Things're bad, though,' Pell said ... 'Recession.'*—T. Winton, 1985 (Aust.); *My cards're in my desk*—M. Cope, 1987 (SAfr.); *'There're oyster catchers too,' said Eric defensively*—A. T. Ellis, 1990; *What're you spying on him for?*—N. Gordimer, 1990 (SAfr.); *What're you doing?*—*New Yorker*, 1990; *When're we going to see Charles?*—Barbara Anderson, 1992 (NZ).

**reaction.** 1 *A brief history.* In chemistry since the early 19c. used to mean 'the action of one chemical agent on another, or the result of such action'. In physiology and pathology since the late 19c.

used to mean 'the response made by the system or an organ *to* an external stimulus'. Some authorities have therefore argued that *reaction* may only be used as a response *to* something that had acted *on* it. But the word did not begin its life in the language of chemists and medical practitioners.

The word had been used figuratively in general contexts at a much earlier date. Sir Thomas Browne, for example, used it in his *Religio Medici* (1643) to mean 'resistance exerted by something in opposition to the impact or pressure of something else': *It is the method of Charity to suffer without reaction.* Charles Wesley in vol. v of his *Works* (1771, edn of 1872) used it to mean 'reciprocal action': *A continual action of God upon the Soul, and a re-action of the Soul upon God.* Numerous other writers (cited in the *OED*) brought it about that *reaction* came to be used with various prepositions—e.g. *against, on,* and *upon*—particularly in the senses 'resistance; something exercised in return on something else'. Also, since 1792, the word had been employed in political language to mean 'a tendency to oppose change'.

This is the general picture before 1900. In the 20c. the word has continued to be used in various specialized ways by technical and scientific writers: these uses lie outside the scope of this book. In general use, two main branches of meaning have flourished: (a) 'a response (to an event, a statement, etc.); an action or feeling that expresses or constitutes a response' (*OED*), e.g. *They all hoot him down in a chorus of amused jeering. Hugo is not offended. This is evidently their customary reaction*—E. O'Neill, 1946; (b) 'a tendency to oppose change or to advocate a return to a former state of affairs, esp. in politics' (*COD*): e.g. *In spite of his achievements as a social reformer, he became a symbol of black reaction to organised labour*—*New Statesman*, 1965. There has also been a marked tendency, esp. by media interviewers, to use the word to mean 'immediate response', 'first impression': e.g. *What is your reaction to the statement that the government may extend VAT to domestic fuel?*

2 *Recommendation.* Leaving aside technical language, a simple enough rule is to restrict *reaction* to contexts in which the idea of a response to some previous

action, event, stimulus, etc., is present. But cross-questioners in general, and media interviewers in particular, will doubtless persist in using *reaction* in its weakened sense of 'immediate impression' when a handful of other words (*impression, judgement, opinion, response, view*, etc.) is available. See CHAIN RE-ACTION.

**readable.** See LEGIBLE.

**readership** is used in the world of 20c. newspapers and magazines to mean either 'the total number of (regular) readers of a periodical publication', or 'all such readers considered collectively'. Cf. LEADERSHIP; MEMBERSHIP.

**real** (adv.). Used only as an intensifier ( = very) with an immediately following adj. or adv. (*real nice, real slow; real soon*), it is non-standard or at best very informal in England, but more acceptable in Scotland and in America. The standard adv. in most contexts is, of course, *really* (e.g. *Are you really coming tomorrow?; I am really sorry*). Some US examples: *Stay around till she gets back, she'd be real sorry to miss you—New Yorker*, 1987; *You look real nice today, Carla —New Yorker*, 1987; *Lester is real alone—B. Ripley*, 1987.

**realm.** For synonymy, see FIELD.

**-re and -er.** One of the great dividers separating the spelling of BrE and AmE is that in many words BrE opts for *-re* and AmE for *-er*: e.g. *calibre, caliber; centre, center; fibre, fiber; goitre, goiter; litre, liter; louvre, louver; lustre, luster; manoeuvre, maneuver; mitre, miter; nitre, niter; ochre, ocher; reconnoitre, reconnoiter; spectre, specter; theatre, theater*. The contrast is, however, not totally systematic. For one thing, words such as *acre, involucre, lucre, massacre, mediocre, nacre*, and *ogre* are spelt thus in both forms of English: the presence of *-er* would turn the preceding *c* or *g* into a 'soft' variety. Moreover there is a group of words that are always spelt with final *-re* in America as in Britain: e.g. *antre* (cave, cavern), *cadre, double entendre, euchre, oeuvre, padre*. And some American publications opt for *-re* spellings in at least some words: e.g. *Through his sombre eyes, the pharmacist examined her face—New Yorker*, 24 Dec. 1990. One important distinction: BrE distinguishes poetic *metre* and *metre* used as the unit

of length from the electricity, gas, and parking *meter*; in AmE the only spelling used for each of these is *meter*.

**rear** (verb). See RAISE, REAR.

**rearward(s).** As adj. and noun always *rearward* (*a rearward view; to the rearward of the castle*). As adv. both forms are used (*to move obliquely rearward; the unfit were taken out of the front line and moved rearwards*).

**reason.** Various studies of the construction *the reason ... (is) because* leave no doubt about its frequency, especially in contexts where several words or a clause separate the two elements, and particularly in the work of 17c. and 18c. writers of good standing (Pepys, Wycherly, Pope, etc.). In the 20c. the construction appears with unquestionable frequency all over the English-speaking world in many kinds of writing (letters and novels by Robert Frost, P. G. Wodehouse, Ernest Hemingway, William Faulkner, etc.). Some further modern examples: *The only reason an individual is a suspect in a crime is because he's probably guilty—*letter in *Chicago Tribune*, 1988; (with *why* intervening) *the reason why she painted so many flower pieces in the early Thirties was because she was at the time having an affair with Constance Spry—J. R. Taylor, Times*, 1988; *Part of the reason that I don't feel any need to break out of romantic fiction, is because I know I can if I want to—More* (NZ), 1988; (with both *that* and *because*) *The reason I wrote about it then was that it was what was going on at the time, and the reason, I think, why I write about the war now is because it isn't what is going on at the time—F. Wyndham, London Rev. Bks*, 1988; (straightforward examples of *the reason ... is because*) *The reason I wrote the check is because I was the only one who could sign it, because the account was in my name—Daily Northwestern* (Evanston, Illinois), 1989; *The reason for this was because the director wished to create an atmosphere of gritty realism—B. Elton*, 1991; *The reason I like The Beatles is because they remind me of Chuck Berry—Q*, 1991. If you look for the construction in your morning newspaper or in the unguarded pages of popular magazines and light fiction, you will undoubtedly find it sooner or later. But, for the present at any rate, its absence from the works of our most talented writers and scholars is more significant than its presence in more informal printed work.

See also BECAUSE B5.

**reason why.** 1 As verb, *reason* 'to question, discuss' has frequently been followed by an object clause led by *what*, *why*, etc., or with *why* absolutely, since the 16c., most famously in Tennyson's poem 'The Charge of the Light Brigade' (1855), *Their's [sic] not to reason why, Their's but to do and die.*

2 As noun, *reason* has been idiomatically construed with *why* at the head of a subordinate clause since the 13c. (*Ancren Riwle*). Typical examples: *Aske me no reason why I loue you*—Shakespeare, 1598; *Is there any reason in you … why I sh^d respect you any more than the very Ethiopians?*—Bishop Hall, 1633. The construction remains valid in the 20c.: e.g. *but there is now no reason why lawyers can't do the same*—*Lawyer*, 1988; *The original reason why the MMB was first set up in the 1930s was to combat the 'monopoly power' over dairy farmers of United Dairies*—*Grocer*, 1988; *The reasons why I first adopted the practice are not the same as my reasons for continuing to follow it*—*Bull. Amer. Acad. Arts & Sci.*, 1988. But the border into redundancy is crossed if *reason*, *why*, and *because* all form part of the same sentence: e.g. (see prec., quot.[2] 1988); *I sincerely believe that the reason why First Secretary Mikhail Gorbachev is the first Soviet leader to promote such close cooperation with the U.S. is because he bears the map of the U.S. on his forehead*—letter in *Chicago Tribune*, 1989. Replace *because* by *that* in both cases.

**Réaumur.** Pronounce /'reɪəʊmjʊə/.

**rebate** (carpentry). See RABBET. If the spelling *rebate* is used, pronounce it /'riːbeɪt/. Otherwise /'ræbɪt/.

**rebec** (musical instrument) /'riːbek/. Thus spelt, not *rebeck*.

**rebel.** 1 The noun is stressed on the first syllable and the verb on the second. See NOUN AND VERB ACCENT 1.

2 The inflected forms of the verb are *rebelled*, *rebelling*. See -LL-, -L-.

**rebus** (a puzzle) /'riːbəs/. Pl. *rebuses*. See -US 5.

**rebut.** The inflected forms are *rebutted*, *rebutting*. See -T-, -TT-.

**receipt, recipe.** From the 14c. until shortly after the end of the 19c. *receipt* (from Anglo-Fr. *receite*, from L *recepta*, fem. pa.pple of *recipere* 'to receive') was a formula or prescription of the ingredients needed for some preparation, esp. in medicine and cookery (e.g. *I have put up two bottles of the gillyflower water for Mrs. Sedley, and the receipt for making it, in Amelia's box*—Thackeray, 1847/8). From about 1400, *receipt* was also used to mean 'an amount of money received', and from the beginning of the 17c. 'a written acknowledgement of money received (i.e. one of its main current senses). Meanwhile *recipe* (from L *recipe* 'take', 2nd sing. imperative of *recipere* 'to receive') from the late 16c. meant a formula for a medical prescription, and from the early 18c. a statement of the ingredients and procedures needed for the preparation of a dish in cookery. The overlapping of spellings and meanings came to an end in the 20c. At some point after the 1920s, *receipt* became almost entirely restricted to its two main current senses (the act or an instance of receiving something; an acknowledgement of money received), while *recipe* now means principally a formula for preparing food, and transferred applications (*a recipe for disaster*, etc.). Keep in mind when reading Victorian or earlier literature that *receipt* and *recipe* are liable to turn up in a confusing manner in any of the present-day senses of each word. *Receipt* is, of course, pronounced /rɪˈsiːt/ and *recipe* /'resɪpɪ/.

**receive.** A key word supporting the simple rule of spelling '*i* before *e* except after *c*'. See I BEFORE E.

**received pronunciation** (abbrev. **RP**), **received standard**. Names given to the form of speech associated with educated speakers in the southern counties of England and used as a model for teaching English to foreign learners. As Henry Sweet pointed out in *The Sounds of English* (1908), RP (which he called *Standard English*) is 'a vague and floating entity', and 'like all living languages, [it] changes from generation to generation'. Like Standard French, he asserted, '[it] is now a class-dialect more than a local dialect.

[Its] best speakers ... are those whose pronunciation, and language generally, least betray their locality.' The terms *received pronunciation* and *received standard* are not the only terms used: it is also known 'as the spoken embodiment of a variety or varieties known as *the King's English*, *the Queen's English*, *BBC English*, *Oxford English*, and *Public School English* ... It is often informally referred to by the British middle class as a *BBC accent* or a *public school accent*; and by the working classes as *talking proper* or *talking posh*. In England, it is also often referred to simply as *Standard English*' (*OCELang.*, 1992). It has always been a minority accent in Britain, and is at present spoken by an estimated 3–4% of the British population.

Except when explicitly giving the pronunciation of American or other overseas words, or words from other regions of the UK, my aim in this book is to present in the International Phonetic Alphabet the sounds and phonetic assumptions of RP speakers, while I am fully aware that there is no fixed circumference to this variety of speech and much fluctuation within it. The pronunciations given are normally those of the *COD* (1990), but instead of merely listing alternative pronunciations (where they exist within RP) I normally make specific recommendations. I have also heavily depended on the evidence presented in J. C. Wells's *Longman Pronunciation Dictionary* (1990), a standard work of great distinction. For older, discarded pronunciations I have frequently turned to John Walker's *A Critical Pronouncing Dictionary* (1791 and 1806 edns), the relevant sections of the first edition of the *OED* (1884–1928), and Daniel Jones's *An English Pronouncing Dictionary* (1917).

Among my assumptions is that RP speakers still favour /əu/ in *goat*, *show*, etc.; /aɪ/ in *fine*, *light*, etc.; /æ/ in *trap*, *bad*, etc.; /ɪ/ in the final syllable of *early*, *gently*, etc., though phoneticians have noticed a marked movement in RP towards /i/ (as in General American and some other overseas forms of English); and I regard it as axiomatic that /ɔ:/ is now a minority pronunciation in words such as *loss*, *off*, etc., having been replaced by /ɒ/. For convenience and simplicity I have not placed a bracketed (r) in the pronunciation of words such as *after*: that is, I have presented RP as fully non-rhotic. Again, for the sake of simplicity, I have treated words such as *little* /'lɪtəl/ and *rhythm* /'rɪðəm/ as having a fully syllabic final element, though the matter is debatable. I have no quarrel with those scholars who present such matters in a more flexible way.

The adjective most frequently applied to RP by speakers of General American is *clipped*, though this is not necessarily a term of disapprobation. Speakers of many other overseas varieties of English freely use the words *posh*, *la-di-dah*, and *very English* of RP accents, with varying degrees of admiration or the reverse. Regional UK speakers (depending on the region) notice with a shrug of the shoulders that RP speakers pronounce words such as *love*, *stumps*, *last*, *singer*, and numerous other words, in a different way from themselves, but are seldom moved to change their own manner of speech to approximate to RP unless they live for prolonged periods among RP speakers. There are numerous classic books, essays, and articles on the nature of received pronunciation, among them that by Henry Sweet (see above); in various works by H. C. Wyld including *A History of Modern Colloquial English* (1920; enlarged edn 1936); in A. C. Gimson's *An Introduction to the Pronunciation of English* (1962 and later edns); in J. C. Wells (see above); and in Tom McArthur's *Oxford Companion to the English Language* (1992). In this dictionary there are numerous articles, treated at their alphabetical place, on 20c. changes in RP, disputed pronunciations, and other matters to do with pronunciation. See PRONUNCIATION.

**recess** (noun and verb). The *OED* (1904) gave only /rɪ'ses/ as the pronunciation of both, i.e. with the stress on the second syllable, and this remains the dominant pattern, but /'ri:ses/ is becoming increasingly used for the noun.

**recessive accent.** Fowler's term (1926) for a process that is mostly of historical interest. **1** The term is defined by the *OED* as 'stress transferred towards or on to the first syllable of a word'. It should be borne in mind that English is a Germanic language, and that 'the primitive Germanic language developed a stress accent which fell upon the first syllable of all words ... Thus in Old English we find the stress on the first syllable of all simple words, and in most compound words: *wórd* word, *stánas* stones, *lúfiende* loving, *síþfæt* journey, *ándġiet* sense …' (A. Campbell, *OE Grammar*, 1959). This simple 'rule', however, has to be modified because OE nouns and verbs formed from prepositional adverbs prefixed to the stem came to have different stress patterns, the nouns being stressed on the first syllable and the corresponding verbs on the second. Examples of such doublets: *ándsaca* 'apostate', *onsácan* 'to deny'; *bígenga* 'inhabitant', *began* 'to occupy'. In practice the accentual patterns of OE words are more complicated than these two 'rules' would indicate. But the essential point about the *recessive accent* is that, from the Norman period onward, the large contingents of French, Latin, and Greek words that poured into the language became subject to the accentual rules of the receiving language. For example, the accent in French loanwords tended to move towards the first syllable. In Chaucer's General Prologue, the metre shows that *natúre* and *coráge* were stressed on the second syllable: *So priketh hem nature in hir corages.* But these two words, and many others (e.g. *château, garage, menu, plateau, tableau, village; charlatan, nonchalant*) are now stressed on the first syllable.

See also RECONDITE.

**2** *Stable patterns.* At the present time the stress patterns of the majority of English words drawn from French, Latin, or Greek, as well as those of native origin, are relatively stable, but they are not simple. Some two-syllabled words are stressed on the first syllable (*ínvoice, wíndow*) and others on the second (*alóne, arríve*). Similarly with three-syllabled words: (first syllable) *báchelor, quántity*; (second syllable) *eléven, impórtant*; (third syllable) *magazíne, understánd*. There are four types of four-syllabled words: (first syllable) *cáterpillar*; (second) *rhinóceros, unfórtunate*; (third) *circulátion, diplomátic*; (fourth) *aquamaríne, misunderstánd*. The complications in longer words is even more marked, with a goodly proportion of them having secondary stresses as well as primary stresses: e.g. *in‚feri'ority, ‚indis'tinguishable, im‚penetra'bility*. Moreover, many of these patterns are true only for the words when they are pronounced in isolation. The stress is often

moved in the rhythmic pattern of particular contexts: thus *‚thir'teen*, but *'thir‚teen steps*; *‚after'noon*, but *'after‚noon tea*.

**3** *Unstable accents.* In a number of words in modern times two conflicting tendencies have been or are being shown. In one group the accent has moved or is moving (sometimes only in qualified circumstances) in a recessive manner. In the second group the movement is the other way.

recessive

| | | |
|---|---|---|
| *capitalist* | $2 \to 1$ | |
| *cervical* | $2 \to 1$ | |
| *contribute* | $2 \to 1$ | (non-standard) |
| *distribute* | $2 \to 1$ | (non-standard) |
| *doctrinal* | $2 \to 1$ | (AmE) |
| *laboratory* | $2 \to 1$ | (AmE) |
| *subsidence* | $2 \to 1$ | |

progressive

| | | |
|---|---|---|
| *applicable* | $1 \to 2$ | |
| *centrifugal* | $2 \to 3$ | (young people) |
| *clematis* | $1 \to 2$ | (outside UK) |
| *communal* | $1 \to 2$ | (minority) |
| *controversy* | $1 \to 2$ | |
| *decade* | $1 \to 2$ | |
| *despicable* | $1 \to 2$ | |
| *disciplinary* | $1 \to 3$ | |
| *exquisite* | $1 \to 2$ | |
| *formidable* | $1 \to 2$ | |
| *harass* | $1 \to 2$ | |
| *hospitable* | $1 \to 2$ | |
| *integral* | $1 \to 2$ | |
| *lamentable* | $1 \to 2$ | |

The majority of these words are treated at their alphabetical place in the dictionary. The list is far from complete but from the examples listed it looks as if the recessive process is less common than the opposite one.

**Rechabite.** Pronounce /ˈrekəbaɪt/.

**réchauffé.** The French accents should be retained and the word printed in italics.

**recherché.** The word is more acclimatized than the last and is normally printed in roman type, but with the French accent retained.

**recidivist.** Pronounce /rɪˈsɪdɪvɪst/.

**recipe.** See RECEIPT.

**recipient.** Fowler (1926) cited four journalistic examples of the formal word *recipient* (e.g. *The Serjeant-at-Arms and Lady Horatia Erskine were yesterday the recipients of presentations from members of the Press*

*Gallery*), and scornfully asked 'Can any man say that sort of thing and retain a shred of self-respect?'. The language now mostly rejects the type *is/was the recipient of* in favour of just *receives/received* or *was presented with*, but there are exceptions. Examples of present-day uses: *The Shakespeare Prize is highly esteemed in Britain and, as a recipient, my stock has risen enormously*—A. Guinness, 1985; *she seems to have been the recipient of a small Poor Law dole throughout her life*—J. Halperin, 1990; *Britain is itself now the recipient of a huge influx of inward direct investment*—Economist, 1990.

**reciprocal** (grammar). *Each other* and *one another* are reciprocal pronouns, but they have limitations of use. Thus *Each could see the other in the distance* is idiomatic, and so is *They could see each other in the distance*, but the notional passive equivalent \**Each other could be seen in the distance* is not. Both reciprocal pronouns may be used in the genitive: e.g. *They got in each other's way*; *The prisoners stole one another's cigarettes*. See EACH 2.

**reciprocal, mutual.** To the difficulties presented by MUTUAL itself must be added that of the difference between it and *reciprocal*. *Mutual* regards the relation from both sides at once: *the mutual hatred of A and B*; never from one side only: not *B's mutual hatred of A*. Where *mutual* is correct, *reciprocal* would be so too: *the reciprocal hatred of A and B*; but *mutual* is usually preferred when it is possible. *Reciprocal* can also be applied to the second party's share alone: *B's reciprocal hatred of A*. *Reciprocal* is therefore often useful to supply the deficiencies of *mutual*. A, having served B, can say 'Now may I ask for a reciprocal [but not for a mutual] service?' Two parties can take mutual or reciprocal action, and the meaning is the same; one party can take reciprocal, but not mutual, action. *Mr Wilson said: 'I trust your Government saw in the warmth of the greetings accorded to his Royal Highness the manifestation of friendly goodwill which the people of the United States hold for those of Britain. Believing in the reciprocal friendship of the British people it will be my aim in the future to ...'* In this passage, *mutual* could not be substituted for the correct *reciprocal*; it, however, the words had been not 'of the British people', but 'of the two peoples',

*mutual* would have been as good as *reciprocal*, or indeed better. But it must be added that, since it takes two to make a friendship, which is essentially a mutual or reciprocal relation, to use either adjective is waste.

**recitative.** Pronounce /ˌresɪtəˈtiːv/.

**reckon.** 1 The inflected forms are *reckoned*, *reckoning*. See -N-, -NN-.

2 In 1992, in a conversation with an American, I used *reckon* to mean 'consider, think, be of the opinion' with *that* (in fact I omitted the *that*) before a relative clause (the sentence was of the type *I reckon it's time to go now*). He rejected the use as 'cowboy English', which I took to mean countrified or regional (American) English. He would have said *I guess* or *I think* himself. I have gone into the matter and it seems that the construction was in formal literary use in England from the early 16c. onward: e.g. *Men woulde not recon that hee coulde haue right to the realme*—More, 1513; *For I reckon, that the sufferings of this present time, are not worthy to be compared with the glory which shall be revealed in us*—Rom. (AV) 8: 18, 1611. Its literary use seems to have persisted in England until the late 19c.: e.g. *I reckon, said Socrates, that no one ... could accuse me of idle talking*—Jowett, 1875. With the omission of the relative pronoun *that*, it seems since then to have dropped a rung and become just part of the spoken language in England (though the standard dictionaries in England list the meaning without a restrictive label), in the Antipodes, and in certain parts of America (esp. southern and midland States). Some representative examples: *I don't reckon I've had a fare there for more than ten years*—I. Shaw, 1977; *I reckon he couldn't have been more than eighteen*—M. du Plessis, 1989 (SAfr.); *They* [sc. his students] *all reckon he's pretty amazing*—Susan Johnson, 1990 (Aust.); *D'y'reckon it's still breathing?*—Barbara Anderson, 1992 (NZ). The sharp drop in formality between the 16–19c. examples and the 20c. ones is very noticeable. *Reckon* in this sense seems to be a rollercoaster word with many ups and downs in status and distribution in the 20c. and more likely to come.

**reclaim** (verb). The main derivatives are *reclaimable* and *reclaimer*, but *reclamation*.

**recognizance.** Stressed on the second syllable, /rɪˈkɒgnɪzəns/.

**recognize.** The pronunciation with the g silent, i.e. as /ˈrekənaɪz/, is a vulgarism.

**recollect, remember.** 1 First, *recollect*, pronounced /ˌrekəˈlekt/, 'to remember' is to be distinguished from *re-collect* (with hyphen), pronounced /ˌriːkəˈlekt/ 'to collect again', though the two words have the same origin. See RE(-).

2 In most contexts *recollect* means 'to succeed in remembering', and implies a search in the memory or the recalling of something temporarily forgotten. But the distinction is not an absolute one, and it shares with *remember* much of the territory of the ordinary sense 'to call or bring back to one's mind'.

**recommend.** Traditional constructions of this transitive verb include those shown in the following types: (a) (with direct object) *He recommended Miss Jones for promotion*. (b) (followed by a *that*-clause) *I recommend that you stay at the George and Dragon*. (c) (with direct object + a *to*-infinitive) *I recommend you to control your temper*. (d) (with a subjunctive in the dependent clause) *the confidential report into Mr John Stalker ... recommends that he face a disciplinary tribunal on 10 separate counts*—Times, 1986; *One of the observers from the International Commission of Jurists ... had recommended she be approached*—N. Gordimer, 1990. Other less common types are shown in the following examples: (ditransitive) *Let me recommend you a little of this pike!*—Disraeli, 1826; *Can you recommend me a nice hotel?*—Times, 1985; (with direct object + plain infinitive) *If you go looking for her, I don't think I can recommend you attend*—N. Shakespeare, 1989.

**recondite.** The recommended pronunciation is that with the stress on the first syllable, i.e. /ˈrekəndaɪt/. But traces of an old dispute have been transmitted down the ages, leaving second-syllable stressing as a possible alternative. That sides were being taken as early as the 18c. are shown in the following extract from Walker's *Pronouncing Dict.* (1791): 'Dr. Johnson, Dr. Ash, Dr. Kenrick, Mr. Nares, Mr. Scott, Mr. Fry, and Entick, accent this word on the second syllable; Mr. Sheridan and Bailey on the last ... A few

words of three syllables from the Latin, when anglicised, without altering the number of syllables, have the accent on the same syllable as in the Latin, as *Opponent, Deponent*, &c.; but the general inclination of our language is to place the accent on the first syllable, as in *Manducate, Indagate*, &c.'

**reconnaissance.** Now completely Anglicized, i.e. /rɪˈkɒnɪsəns/.

**reconnoitre.** Thus spelt in BrE, but *reconnoiter* in AmE.

**record.** The noun is stressed on the first syllable and the verb on the second. See NOUN AND VERB ACCENT 1.

**record-player.** See GRAMOPHONE.

**recount** /rɪˈkaʊnt/, to narrate, is to be distinguished from *re-count* (noun) /ˈriːkaʊnt/, a counting again, and *re-count* (verb) /riːˈkaʊnt/, to count again.

**recourse.** See RESORT.

**recover, re-cover, recreation, re-creation.** See RE(-).

**recriminatory** (adj.). Pronounce /rɪˈkrɪmɪnətərɪ/.

**recrudescence** in medical use means 'the breaking out again of a disease, etc., after a dormant period'. In transferred uses it should be restricted to contexts in which some undesirable or unwanted setback or circumstance has occurred. It is not properly applicable to the reoccurrence of something agreeable or neutral. Fowler (1906) was already alarmed by the way in which journalists were misusing the word: '*Recrudescence* is becoming quite a fashionable journalistic word. It properly means the renewed inflammation of a wound, and so the breaking out again of an epidemic, &c. It may reasonably be used of revolutionary or silly opinions: to use it of persons or their histories is absurd.' He cited an unacceptable example from *The Times*: *A literary tour de force, a recrudescence, two or three generations later, of the very respectable William Lamb ..., his unhappy wife, Lady Caroline Lamb, and Lord Byron*.

**recto.** 1 Pl. *rectos*. See -O(E)S 3.

2 See VERSO, RECTO.

**rector.** In the Church of England the incumbent of a parish where all tithes formerly passed to the incumbent, as compared with a vicar, an incumbent of a parish where tithes formerly passed to a chapter or religious house or layman. Since the passing of the Tithe Act 1936 tithes have no longer been payable to any parish priest, but the designation *rector* is preserved where it previously existed; in all other parishes the incumbent is a *vicar*. *Rector* has a different meaning in some other churches; and the word is also used for the head of some schools, universities, and colleges.

**recur** /rɪˈkɜː/ (verb). 1 The inflected forms are *recurred*, *recurring*. See -R-, -RR-.

2 The stressed vowel in *recurred* and *recurring* is pronounced /ɜː/, but in *recurrence*, *recurrent* it is /ʌ/.

**recusancy, recusance.** Both forms are available, but *recusancy* is recommended (stressed on the first syllable).

**reddle.** See RUDDLE.

**Red Indian, redskin.** These terms are falling into disuse, being considered racially offensive, in favour of *North American Indian* and *Native American*.

**redingote** is an early 18c. French corruption of Eng. *riding-coat*, which was brought back into English before the end of the 18c. as a term for various fashionable coats.

**reduce** (verb). 1 The corresponding adj. is *reducible*. See -ABLE, -IBLE 7.

2 After *reduce to* and *be reduced to* the gerund, not the infinitive, is idiomatic: e.g. *He was reduced to retracting* (not *to retract*) *his statement.*

**reductio ad absurdum.** A method of proving the falsity of a premiss by showing that the logical consequence is absurd; an instance of this (*COD*). A *reductio ad absurdum* of the theory that the less one eats the healthier one is would be 'Consequently, to eat nothing at all gives one the best possible health'.

**redundancy.** Actual or concealed redundancy occurs with great frequency in the language. One type is shown in Muriel Spark's novel *The Only Problem* (1984): ' . . . she is lively and vital enough to be a member of a terrorist gang.' 'Lively and vital,' said Harvey, 'lively and vital—one of these words is redundant.' Here the redundancy is debatable. More understandable are examples arising from unfamiliarity with the extended form of an acronym or with the mechanics of a foreign language: e.g. *HIV virus* (*HIV* = human immunodeficiency virus); the *hoi polloi* (*hoi* is the Greek definite article); *LCD display* (*LCD* = liquid crystal display); *OPEC countries* (*OPEC* = Organization of Petroleum Exporting Countries); *SALT talks* (*SALT* = Strategic Arms Limitation Talks). Less defensible are phrases such as *sworn affidavits* (an affidavit is a written statement confirmed by oath); and *safe haven* (a haven is a place of refuge, implying safety).

Careless examples, for which it is hard to find an excuse, are phrases such as *10am in the morning, an armed gunman, a new recruit, RSVP requested,* and *a free gift.* Indefensible grammatical redundancy occurs in the following (the words within brackets should have been omitted): *We cannot dodge* (*out from under*) *the social consequences of our actions*—*Daily Tel.,* 1980; *One should not assume that because a critical mass of women on campus has been reached* (*that*) *their place is in any way guaranteed*—*Univ. of Chicago Mag.,* 1992.

See PLEONASM.

**reduplicated words.** The language is saturated with reduplicated words of various kinds. Most of them seem to represent a natural impulse to emphasize by repetition. They are of varying age and they illustrate various kinds of word-formation. For example, *ding-dong* and *helter-skelter* are first found in print in the 16c., *hocus-pocus* and *hoity-toity* in the 17c., *chin-chin* and *namby-pamby* in the 18c., *argy-bargy* and *bye-bye* in the 19c., and *arty-farty* and *heebie-jeebies* in the 20c. Some of them show reduplication by simple change of the initial consonant(s): e.g. *hanky-panky, harum-scarum, higgledy-piggledy, teeny-weeny.* Others result from a change of vowel in the second element: e.g. *dilly-dally, ding-dong, shilly-shally, tip-top.* Others again simply result from a repetition of the first element: e.g. *bye-bye, chop-chop, puff-puff.* The great majority of them are of self-evident meaning, but, for example, some of those that occur in Shakespeare's works now properly need to have glosses attached to them: e.g. *And such a deale of skimble scamble stuffe* (nonsensical) (*1 Henry IV*); *In hugger mugger to inter him* (secrecy) (*Hamlet*); *That hugs his kickie wickie heare at home* (beloved wife) (*All's Well*).

**re-enforce.** See REINFORCE.

**reeve** (a wading bird). See RUFF.

**reeve** (verb) (nautical). To thread or fasten a rope. Pa.t. and pa.pple either *rove* or *reeved*. The first of the two is preferred in OUP house style.

**refection** (meal). A literary or formal word.

**refectory.** The recommended pronunciation is with the stress on the second syllable, though in some monasteries the stress is placed on the first.

**refer** (verb). The inflected forms are *referred*, *referring*. See -R-, -RR-.

**referable.** Unlike the inflected forms of the verb *refer*, this adj. is spelt with a single -r-. Pronounce with the stress on the first syllable. Contrast CONFERRABLE (always so spelt). (If you prefer to pronounce it with the stress on the second syllable you must then spell it *referrable*.) See -ABLE, -IBLE 2.

**reference.** 1 Until quite recently, compendious works of reference were normally called *reference books*, *works*, etc. Now, esp. in publishing houses, the dominant term seems to be *reference* alone: e.g. *His work will long remain a basic reference for anyone working on Titian—TLS*, 1989.

2 Since about the 1920s the word *reference* has become established as the customary word for a (usually written) report produced by a referee; a testimonial.

**referendum,** referring to the electorate on a particular issue. Pl. *referendums* rather than *referenda*, though the latter form is widely used. See -UM 3.

**refill.** The noun is stressed on the first syllable, and the verb on the second. See NOUN AND VERB ACCENT 1.

**reflectible.** Use *reflexible*.

**reflection, reflexion.** The etymological spelling with *x* is the earliest (14c., from Fr. *réflexion*), but as the centuries passed it was gradually supplanted by that with *ct*, and *reflection* is now the dominant spelling in all uses of the word. See -XION etc.

**reflective, reflexive.** These are now not merely variants, but two distinct words. *Reflective* is used to mean principally '(of a surface) giving a reflection or image'; and '(of a person, disposition, etc.) thoughtful, given to meditation'. *Reflexive* is a grammatical term exhibited in such pronouns as *I wash myself in warm water, he made them himself, she herself is no angel, he came to the party by himself*, i.e. where the reflexive pronoun and the subject are the same entity.

**reflector.** Thus spelt, not *-er*.

**reflexible,** capable of being reflected. Not *reflectible*.

**reflexive.** See REFLECTIVE.

**reform, re-form.** See RE(-).

**refractor.** Thus spelt, not *-er*.

**refrangible,** that can be refracted, is now the current form, rather than *refractable*.

**refrigerator.** Thus spelt, not *-er*.

**refrigeratory** (adj. and (hist.) noun). If you have occasion to use it, pronounce it /rɪˈfrɪdʒərətərɪ/.

**refuse.** The noun is stressed on the first syllable, and the verb on the second. See NOUN AND VERB ACCENT 1.

**refutable.** First-syllable stressing is recommended, i.e. /ˈrefjʊtəbəl/, but second-syllable stressing, i.e. /rɪˈfjuː-/, is gaining ground. Cf. IRREFUTABLE.

**refutal.** A common word 17–19c., *refutal* seems now to be in full retreat. The customary word now is *refutation*.

**refute.** The traditional meaning (first recorded in the 16c.) is 'to prove (a statement, opinion, allegation, accusation, etc.) to be false; to disprove by argument'. When T. S. Eliot (in *Murder in the Cathedral*) wrote *If you make charges, Then in public I will refute them*, and when Rebecca West wrote *The case against most of them must have been so easily refuted that they could* hardly rank as suspects, both writers could be assured that their use of *refute* was beyond reproach. At some point in the second half of the 20c., however, traditionalists began to notice that people outside an educated social divide were beginning to use *refute* as a simple synonym of *deny*. In 1986 an enraged person wrote to the letters editor of the *Spectator*: 'In Mr Chancellor's day someone who didn't know the difference between "refute" and "deny" wouldn't have been employed by the *Spectator* as an office cleaner, let alone as a television critic.' In the 1980s the police, trade union leaders, and other sternly honest authorities were forever *refuting* (that is, denying) allegations of brutality, malpractice, dishonesty, and so on. The skirmishing continues. The likelihood that the new use represents a legitimate semantic shift is rejected by the traditionalists. Those who have no idea what a semantic shift might be, like the sound of *refute*, and will continue to use it in its partially standard new way. I have an uneasy feeling that the new sense will begin to sound normal in the 21c,–but not yet.

**regalia.** In its main sense 'the emblems or insignia of royalty; the crown, sceptre, and other ornaments used at the coronation of a king or queen', *regalia* must always be construed as a plural noun. Since the early 17c. the word has also come to be used (in the pl.) in the transferred sense 'the decorations or insignia of an order', i.e. of mayors, Freemasons, senior military officers, etc. Etymologically such transferred uses are inaccurate, but it is too late now to try to dislodge them. A typical example: *I ... answered a knock on my door only to find a storm-trooper dressed in RSPCA regalia standing there*—letter to the *Spectator*, 1993.

**regard.**

| 1 In compound prepositions.
| 2 *regard as*.

**1** Despite their obvious wordiness, the compound prepositions *with regard to*, *in regard to*, and *as regards*, are often used, as well as *regarding*, to introduce a statement (e.g. in business letters, beginning *With regard to your letter of 2 May ...*). In general writing, compound prepositions can be used at the beginning of a sentence or in mid-sentence: e.g. *As regards*

*function, these centuries are those in which the ancient patterns of cumulative negation last appear in the standard language*—B. M. H. Strang, 1970; *He also acted as almoner and adviser to Douglas of Cavers with regard to the annual distribution of alms to the poor*—K. M. E. Murray, 1977. The *OED* shows these phrases in older use: e.g. *I speak with regard to sensible things only*—Berkeley, 1713; *In regard to the matter ... he had, no doubt, been misled*—*Monthly Rev.*, 1792; *The world was believed fixed until ... it was found to change its place with regard to them*—W. R. Grove, 1842. They are all in standard use, but should be used sparingly and with discretion.

**2** *regard as*. In a range of constructions the verb *regard* in the sense 'to consider, judge, look upon' is used transitively followed by *as* and a noun or adjective. This is true whether the verb is used actively or passively. Examples: (active, + *as* + adj.) *it was equally irrational to regard George as simply accident prone*; (active, + *as* + noun) *he seems to regard him as a cross which just has to be borne*; (passive, + *as* + adj.) *urban commuters ... may nevertheless be regarded as foreign*; *practices which until now have been regarded as morally wrong*; (passive, + *as* + noun) *he is generally regarded as the foremost authority on the subject*. Ditransitive uses of the verb, though apparently common in Fowler's lifetime, are no longer idiomatic. Examples of such types (cited from Fowler 1926), with an indication of how they can be remedied: *Some County Associations regard it to be their first duty to accumulate large investment funds* (*regard it as*); *He regards it beyond question that Moses wrote practically the whole of the Pentateuch* (*regards it as beyond question*; or use *consider* instead); *It had regarded itself* [*as*] *as certainly out of the war as a great city could be* (use *consider* instead). To express the matter in a different way: do not use the type *He regarded them his friends*. Substitute *He considered them* (*to be*) *his friends*, or *He regarded them as his friends*.

See AS 3(b); CONSIDER 1.

**regardless** can be used both as an adjective (*A man who had been openly regardless of religious rites*—G. Eliot, 1863) and as a quasi-adverb (*although my voice was cracked and vanishing, ... I struggled on regardless*—S. Mason, 1990).

**regenerate.** As a verb pronounced /rɪ'dʒenəreɪt/, and as an adj. /rɪ'dʒenərət/.

**regime.** No accent. Print in romans.

**regimen.** Now mainly restricted to its medical sense, 'a prescribed course of exercise, way of life, and diet'. Some doctors use *regime* instead. Example: *a feeding regimen that takes no account of the baby's needs is likely to be experienced as frustrating*—Dædalus, 1993.

**region.** For synonymy, see FIELD.

**register.** 1 The agent-noun corresponding to the verb *register* is *registrar*, and in Cambridge University *registrary*.

2 In Britain, *register office* is the official term for 'a State office where civil marriages are conducted and births, marriages, and deaths are recorded with the issue of certificates' (*COD*). The unofficial term *registry office* is also in widespread use.

**regnal, regnant.** The -gn- as in *magnify*, etc., not as French, nor as in *poignant*.

**regress.** The noun is stressed on the first syllable, i.e. /'riːgres/, and the verb on the second, i.e. /rɪ'gres/. See NOUN AND VERB ACCENT 1.

**regret** (verb). The *t* is doubled in inflected forms (*regretted, regretting*) and in *regrettable*. See -T-, -TT-.

**regretful, regrettable.** 1 Fowler (1926) cited three examples in which *regretful*, properly meaning 'feeling or showing regret', is misused for *regrettable* 'causing regret': *The possession of those churches was unfortunately the reason of the regretful racial struggles in Macedonia*; *Sir Newton Moore's resignation of the Premiership of Western Australia was a regretful surprise to Australians in London*; *It was not surprising, however regretful, that Scotland had lagged behind*.

2 The adverbs *regretfully* and *regrettably* should be used only in senses corresponding to those of the correct uses of the adjectives. Both adverbs are now being commonly used as SENTENCE ADVERBS, and unfortunately, since the 1960s, *regretfully* has sometimes been used where *regrettably* ( = I regret to say, it is a pity that) properly belongs, e.g. *Regretfully, that is no ground for leniency*

*towards him*—New Statesman, 1976; *Regretfully, however, the editorial staff may justifiably be thought to stand criticized for what has been omitted*—Jrnl RSA, 1977.

**regularly.** All four syllables should be pronounced, i.e. /'regjʊləlɪ/. Reduction to three syllables, as if the word were spelt 'reguly', is a vulgarism.

**regulate** (verb). The recommended form of the corresponding adj. ( = capable of being regulated) is *regulable*, not *regulatable*. See -ABLE, -IBLE 2(v).

**regulus** (impure metallic product). Pl. either *reguli* /-'laɪ/ or *reguluses*.

**reign, rein.** The first is to do with royalty (the period during which a sovereign rules, etc.), and the second is a narrow strap used to guide or check a horse or child. The now widespread use in American publications of *reign* (instead of *rein*) in the phr. *give (free) reign to* is deeply regrettable. Examples: *They say that if they are given free reign to invest and produce they will grow richer*—New Yorker, 1987; *Under these measures, highly productive tenant farmers were given reign to new, independent operations*—Dædalus, 1990.

**reinforce, re-enforce.** The ordinary form (*rein-*) has been so far divorced from the simple verb (formerly *inforce* or *enforce*, now always the latter) that it no longer means to enforce again, as when a lapsed regulation is revived. For that sense *re-enforce* should be used; see RE(-).

**reject.** The noun is stressed on the first syllable, i.e. /'riːdʒekt/, and the verb on the second, i.e. /rɪ'dʒekt/. See NOUN AND VERB ACCENT 1.

**rejoin, re-join.** See RE(-). The hyphened form, if used at all, should be restricted to actual reuniting, and only if in the context it could be confused with *rejoin* 'to say in answer, retort'.

**relate** (verb). This long-established word (first recorded in Caxton) has enjoyed a period of great popularity in the second half of the 20c. in a sense derived from the jargon of social workers, namely 'to have an attitude of personal and sympathetic relation to'. Examples: *Group formation such as takes place in the classroom tends to be adult-centered and dependent upon the varying ways children relate to the*

665

**-related | reliable**

*teacher—Childhood Education*, 1950; *Candidates should ... be able to relate to senior officers of the University*—advt in *Globe & Mail* (Toronto), 1968; *Married people can still relate*—*Guardian*, 1971.

**-related.** This has flourished in the 20c. as the second element of compound adjectives: e.g. *AIDS-related disease*; *a smoking-related lung problem*; *oil-related employment*. A hyphen is customary but is not obligatory in all cases: e.g. *vivid sketches linking a child's inattention at school to the gang-and-drug related shoot-outs he witnesses at home*—*Dædalus*, 1993.

**relation, relationship, relative** (nouns). Their history as terms of kindred is set out in the *OED*. *Relation* 'a person related to one by blood or marriage' was the first to come on the scene (1502); as an abstract noun meaning modern English 'relationship' it is first recorded in the work of Jeremy Taylor in 1660. *Relative*, as a term of kinship, is first recorded in 1657. *Relationship* in its normal present-day sense 'the fact or state of being related' is first recorded in Pope's *Dunciad*, (a1744). The residue of several centuries of interaction is that *relation* and *relative* are now almost always interchangeable in the sense 'a kinsman or kinswoman', *relation* being the more usual of the two; while *relationship* continues to be used as an abstract noun in the sense given above. *Relation* and *relative* have a wide range of other senses: these are listed in standard dictionaries. As for *relationship*, one must be careful to distinguish the sense in e.g. *His relationship with his pupils is excellent* (working association), and *He was accused of having a relationship with one of his female pupils* (sexual harassment).

**relative clause,** a subordinate clause that is attached to an antecedent by a relative pronoun (*that*, *which*, *who*, etc.), e.g. *This is the house* that Jack built.

**relatively.** Like COMPARATIVELY, *relatively* has been in standard use before a following adjective since the early 19c.: e.g. *Justice ... denied to the relatively poor* (1825); *worth the relatively small sum ... paid for them* (1884); *The natural question to pursue is whether the Chinese state has been able to maintain control in this relatively open geopolitical region*—*Dædalus*, 1993.

**relative pronouns.** See the separate words: THAT; WHAT; WHICH; WHO; WHOM; WHOSE. See also the article OMISSION OF RELATIVES.

**relay, re-lay.** One must distinguish *relay* /ˈriːleɪ/, /rɪˈleɪ/ 'to receive (a message, broadcast, etc.) and transmit it to others' (pa.t. and pa.pple *relayed*) from *re-lay* /riːˈleɪ/ 'to lay again' (pa.t. and pa.pple *relaid*). See RE(-).

**relevance,** first recorded in the 18c., has almost completely taken precedence over *relevancy* (16c.) as the 20c. proceeded. Examples: (relevance) *... al* [sc. a novel] —*while laudable in its social intentions—is little more than a piecing together of stock responses to the current demand for 'relevance'*—*Times*, 1975; *Though the facts of Rembrandt's education and training are well-known, their relevance to his approach to self-portraiture has not been considered*—H. P. Chapman, 1990; *My primary criticism is that the relevance of the volumes to L2 acquisition and language pedagogy is not emphasized*—*Language in Society*, 1991. (relevancy) *A tendency to confuse relevancy with recency*—*TLS*, 1980.

See -CE, -CY.

**relevant.** In the 1960s and 1970s university campuses reverberated with the cries of students demanding that their academic studies be made *relevant* to the social and political aspirations of young people at the time, that is should be socially, morally, and politically *relevant*, as opposed to being merely traditional, evaluative, and aesthetic. This movement, and also the word *relevant* itself used in this way, have now largely disappeared to be replaced by notions of POLITICAL CORRECTNESS.

**reliable.** It will come as a surprise to many that, according to the *OED* (1908), this word (and by inference also *reliability*), though first recorded in 1569, came into common use only from about 1850, and at first perhaps occurred more frequently in American works than in British ones. It is now, of course, in standard use everywhere, as is also *reliability*. Earlier objection to the word was based on the belief that *reliable* ought to mean 'able to rely' rather than 'able to be relied on'. As Alford (1864) expressed it: '*Reliable* is hardly legitimate. We do not *rely u* *man, we rely upon a man*; so that reliable does duty for *rely-upon-able*. "Trustworthy" does all the work required.'

**relict** /'rɛlɪkt/ now has only technical senses in geology and ecology and (followed by *of*), archaically, 'widow', by contrast with *relic*, which has a wide range of meanings (a surviving custom, part of a deceased saint's body, etc.).

**relievo** /rɪ'liːvəʊ/. Also in the It. spelling *rilievo* /riː'ljeɪvəʊ/. Pl. *-os*. See -O(E)S 6. Cf. ALTO-RELIEVO; BAS-RELIEF; MEZZO-RELIEVO.

**religious.** For *dim religious light*, see IRRELEVANT ALLUSION.

**remain.** 1 Fowler (1926) rightly rejected as 'a ridiculous tautology' the phr. *continue to remain*. He judged it to be 'surprisingly common', and cited three examples, presumably from newspapers of his day: *The counsellors of the Sultan continue to remain sceptical; And yet through it all I continue to remain cheerful; It is expected that very soon order will be restored, although the people continue to remain restive.*

2 *I remain.* As part of the concluding formula of a letter, *I remain* was in frequent use from about 1600 to some point in the 20c. but now is rarely encountered. Examples: *I will ever remain Your assured friend Charles Percy* (1600); *I remain, my dear friend, Affectionately yours, W.C.*—Cowper, 1793; *Here is my letter done, and I remaining yours always sincerely, E.F.G.*—E. FitzGerald, 1873.

**remediable, remedial.** The first means 'capable of being remedied (*injustice is remediable*)', and the second 'affording or intended as a remedy (*remedial therapy*)'. There seems little danger of confusion at present, though this has not always been the case: the *OED* cites 15–16c. examples of *remediable* used to mean 're-medial, capable of remedying'.

**remember.** See RECOLLECT 2.

**reminisce** (verb) is a back-formation (first recorded in 1829) from the noun *reminiscence*.

**remise.** As verb and noun in law and in fencing, pronounced /rɪ'miːz/ according to *COD* and the *New SOED*, but as /rɪ'maɪz/ according to some other dictionaries, esp. American ones.

**remit.** 1 The inflected forms of the verb are *remitted, remitting*; see -T-, -TT-. The related adjective meaning 'able to be remitted' is either *remissible* or *remittable*.

2 *Remit* (noun), an item submitted for consideration at a conference, etc., is pronounced either as /'riːmɪt/ or /rɪ'mɪt/ in Britain, but always as /'riːmɪt/ in New Zealand. See NOUN AND VERB ACCENT.

3 Of the related nouns, *remittance* is right for the sending of money, *remittal* for referring a case from one court to another, and *remission* for all other senses.

**remonetize.** Pronounce /riː'mʌnɪtaɪz/.

**remonstrate** (verb). The only pronunciation given in the *OED* (1908) was that showing the stress on the second syllable, but it is now always pronounced /'remənstreɪt/, i.e. stressed on the first syllable. The corresponding noun *remonstrance* is stressed on the second syllable.

**remunerate, remuneration** are more formal words than *pay* and *payment* and are chiefly used of the higher salaries in business firms. To be avoided at all costs is the metathesized form *renumeration* (for *remuneration*): it occurred, for example, of all unfortunate places, in the 25 Aug. 1992 issue of *The Times* in an advertisement for the vacant post of General Secretary of the Association of University Teachers.

**Renaissance** (with a capital *R*) /rɪ'neɪsəns/, less commonly /rəner'sɑ̃s/ is the customary word for '(the period of) the revival of art and literature under the influence of classical models in the 14c.-16c.' With a small *r* it is used of any similar revival. The Anglicized form *Renascence* (or *r-*) is sometimes used in the same senses, but more commonly just means 'rebirth; renewal'.

**rendezvous** /'rɒndeɪvuː/, /-dɪvuː/. The pl. is the same, but pronounced /-vuːz/. As verb the inflected forms are *rendezvouses* /-vuːz/, *rendezvoused* /-vuːd/, and *rendezvousing* /-vuːɪŋ/.

**renegade** is now the only current form of the word, but *renegado* (the Sp. original) will still be encountered in historical novels.

**renouncement, renunciation.** From the beginning (*renouncement* 15c., *renunciation* 14c.) the second of the two words has been dominant, and *renouncement*, not having developed an independent meaning (cf. *pronouncement* and *pronunciation*), is now rarely used.

**renunciation** /rmʌnsɪ'eɪʃən/. See prec.

**rep** (textile fabric). This spelling is recommended, not *repp* or *reps*: the French original is spelt *reps*, pronounced /reps/.

**repa(i)rable.** *Reparable* /'repərəbəl/ is used almost only of abstracts such as *loss*, *injury*, *mistake*, which are to be made up for or to have their effects neutralized; *repairable* sometimes is used in that way also, but chiefly of material things that need mending. The negatives are *irreparable*, but *unrepairable*. See IN- AND UN-.

**repeat** (noun). The use of *repeat* to mean 'the repetition of a radio or television programme which has already been broadcast', and its attributive use in compounds such as *repeat fee, repeat performance*, have tended to give additional prominence to the word *repeat* in the 20c. and slightly less prominence to *repetition*. As a result there has emerged a marked tendency to use *repeat* where formerly *repetition* would have been the automatic choice: e.g. *members of the powerful 1922 Committee executive told Mrs Thatcher there must be no repeat of the dispute between Downing Street and Mr Lawson over the Exchange Rate Mechanism—Times*, 1989.

**repel** (verb). The inflected forms are *repelled, repelling*. See -LL-, -L-.

**repellent, repellant.** The first of these is recommended for all senses, as noun or adjective.

**repellent, repulsive.** That is *repellent* which keeps one at arm's length; that is *repulsive* from which one recoils. That is, the second is a much stronger word.

**repertoire, repertory** are essentially the same word, being the French and English equivalents of Latin *repertōrium* 'an inventory, a catalogue'. A *repertoire* is essentially a stock of dramatic or musical pieces which a company or player is prepared to perform; in transferred use, a collection (of habits, jokes, skills, languages, etc.) forming part of a person's or animal's special gifts or accomplishments. *Repertory*, besides having these meanings, is most particularly a type of theatrical presentation in which the plays performed by a company are changed at regular short intervals: hence *repertory company, player, theatre*, etc.

**repetition.** 1 Chance repetition of words is a natural feature of the language. Sometimes it happens because the repeated words are just part of the ordinary way in which verbs work: e.g. *Of course he too had had a choice and still had one*—I. Murdoch, 1989; *The way in which we do do such things*—BBC Radio 4, 1990. At other times it occurs because the same word is used twice with different functions: e.g. *The heart wasn't beating ... Whoever he was, the chap had had it*—M. Innes, 1956; *She brings with her her daughter Elizabeth-Jane*—M. Drabble, 1985; *They're all married, said her mother. Not that that would bother most people nowadays*—P. Lively, 1989; *such publicity as there was was left to the chairman and senior editors*—P. Howard, 1990; *He's out back in the darkroom. He'll be in in a minute*—N. Virtue, 1990. The phenomenon is not new: *Harry could forgive her her birth*—G. Meredith, 1861. Nor, of course, is it restricted to BrE sources: *the front page [of the newspaper] was missing and all there was was columnists and the life-style section*—G. Keillor, 1990 (US). Any awkwardness residing in such repetitions is normally passed over as something that is inevitable and unavoidable. If there is any question of loss of clarity a comma is inserted between the repeated words: e.g. *And while we're at it, it wasn't me with the fedora at Bea's sweet-sixteen*—M. Richler, 1980. In spoken English, in such slightly convoluted circumstances, a pause is often inserted between the repeated words: e.g. *We are getting people who had a right to be out* [pause] *out*—W. Waldegrave, BBC 1 News, 1990. Jespersen's *Modern English Grammar* draws attention to a spectacular example of word repetition in Addison's *Spectator* (May 1711). In reply to an article about the increasing use of *that* for *which* in the early eighteenth century, a pseudonymous writer called 'That' submitted an ironic reply: 'My lords! with humble submission, *That that*

I say is this: *that that that that* gentleman has advanced, is not *that, that* he should have said to your Lordships.' It takes a little working out but the meaning comes through.

Back to the twentieth century. Common or garden repetition has been joined in quite recent times by a remarkable domino-type repetition of *is*. It occurs only in spoken English, and principally in sentences beginning with *The problem is, The question is,* and similar phrases. An American scholar, Dwight Bolinger, first consciously noticed the reduplicated use of *is* in a speech by a former president of the Linguistic Society of America in 1971: *My real feeling is, is that there is* . . . This example could, one can suppose, have been the result of a momentary hesitation. But Bolinger, by then alerted to the construction, noted that it was breaking out everywhere, mostly in radio or television broadcasts, often with no perceptible pause between the first and second uses of *is*: e.g. *The problem is, is that* . . . (a Californian radio station, 1978); *The other problem is is on the demand side* (as against the supply side) (ditto, 1985). Sometimes the repetition is disguised or deflected by a change of tense: *The strange thing was, is that* . . . (1981); *Some of the problems in loading the structure was is that* . . . (1985). I was beginning to dismiss it as an idiosyncratic American use, when a correspondent from West Yorkshire, Mr James A. Porter, wrote to say that he had encountered this weird type of construction on the BBC 'scores of times': e.g. *The question is, is if the merger goes ahead, will* . . .? (David Owen, BBC, 30 August 1987); *What is clear is this, is what the Labour Party intend to do will lead to* . . . (N. Lawson, BBC, 26 May 1987); *But isn't that the problem, is that* . . . (BBC Radio 4 on a radio call-in with Nick Ross, 6 Nov. 1990). I watched out for examples myself and was soon rewarded: e.g. *The curious thing about it is that, is* . . . (Peter Barnes, BBC 1, 29 July 1990); *My message is to the government, is* . . . (G. Kaufman, BBC 1, 25 Aug. 1990). Somehow constructions of this type had crossed the Atlantic in the mysterious way in which such things happen.

The pleonastic doubling or repetition of *is* is clearly a marked feature of modern spoken English. A copytaker setting down an example of the type would normally, I think, insert a comma between the two occurrences or silently omit the second *is* as being otiose or 'a mistake'. In the examples I heard myself there was no perceptible pause before the second occurrence. Contextually the construction seems to be just some kind of unguarded syntactic stuttering. Or perhaps it is an example (to use modern terminology) of some deep-structured linguistic domino effect?

**2** Repetition of words and sounds is entirely legitimate in some other circumstances, e.g. (*a*) When done for rhetorical effect: *Looking at the far sandhills, Wiliam Bankes thought of Ramsay: thought of a road in Westmorland, thought of Ramsay striding along a road by himself hung round with that solitude which seemed to be his natural air*—V. Woolf, 1927; *For she could regard me without strong emotion—a familiar shape, a familiar face, a familiar silence*—Marilynne Robinson, 1981; *Long grey iron trains, the compartments jammed with people all the way, long long waits in grey steel-vaulted stations*—S. Bedford, 1989. (*b*) The type *there are kings and kings*. See AND 5.

**3** Certain other kinds of (mostly unintentional) repetition are undesirable. For example: (*a*) (awkward repetition of small words such as *of, with* or of terminations such as *-ly*): *The founders of the study of human nature; Taken up with warfare with an enemy; He lived practically exclusively on milk*. (*b*) Accidental repetition of the same or a similar-sounding word: e.g. *The Japanese democracy are affronted at what they regard as an affront to their national dignity; The cure for that is clearly the alternative vote or the second ballot, the former alternative being the more preferable; Anonymity seems to be a peculiar delight to writers on naval matters, though perhaps necessity has something to do with the matter; The features which the present Government in this country presents in common with representative and responsible government are few and formal*. The moral is simple: read through what you have written before it is committed to print.

**repetitional, repetitionary, repetitious, repetitive.** These four adjectives, first recorded in 1720, 1720, 1675, and 1839 respectively, are all still available, but the last two are easily the more common of the quartet at present. *Repetitious* (which the OED 1908 described as 'common in recent American use') is used particularly of speech or writing that is 'characterized by repetition, esp. when unnecessary or tiresome' (*COD*); and *repetitive* of work, rhythm, etc., that is characterized by repetition, not necessarily of an avoidable or tiresome nature. Contrast the neutral use of *repetitive* in *repetitive DNA*, a form of DNA that contains multiple copies of a particular gene in each cell; and *repetitive strain* (or *stress*) *injury* or *RSI*, the name given to a recently diagnosed medical condition affecting some individuals after prolonged sessions of keyboarding.

**replace.** 1 The corresponding adj. in *-able* is spelt *replaceable*. See -ABLE, -IBLE 2 (iii).

2 Note its two main meanings: (*a*) to put back in place (*He replaced the papers in the drawer after photocopying them*); (*b*) to take the place of, to succeed (*Atherton replaced Gooch as the England cricket captain at the end of the 1993 season*).

**replenishment, repletion.** The first means 'a fresh supply' or 'the act or process of replenishing'; whereas *repletion* means 'the action or fact or condition of being replete, i.e. filled up, stuffed full, or crowded'. See -ION AND -MENT.

**replete.** The primary sense is 'supplied with in abundance, stuffed with' (*a purse replete with £10 notes*); but, beginning in the 17c. and still, it is sometimes used where *complete* would seem to be more appropriate (*a medieval library replete with a chained-Bible table and a few manuscripts*). Quite full and abundantly full (*replete*) are not the same as characteristically containing (*complete*).

**replicate.** The word has been in use in various senses (e.g. to reply, to repeat, to fold back) from the 16c. onwards, but it has spread its wings in the 20c. esp. in various technical senses in biology, computer science, etc., in the broad senses 'to reproduce or give rise to a copy of itself', 'to imitate', and 'to copy exactly'. It is best avoided in non-scientific writing, when so many general words (*to be modelled on, to imitate, to duplicate*, etc.) are available to express the same broad ideas.

**reportage.** This French loanword is still only semi-Anglicized in pronunciation, /repɔːˈtɑːʒ/, though /rɪˈpɔːtɪdʒ/ is also used. Its older senses 'repute; gossip' have been dropped and now it means 'the reporting of events for the press or for broadcasting, esp. with reference to reporting style' (*SOED*), or an instance of this. Examples: *Németh's book was planned as a long psychological novel, Kuncz's book as a straightforward reportage*—A. Koestler, 1954; *His study of the Hyde Park orators might have been taken as a masterly piece of reportage*—London Rev. Bks, 1979.

**reported speech.** Direct speech in its printed form is conventionally placed within quotation marks: *'Here I have a suggestion which you might think about, Jane,' said Mr Pickering*—A. Brookner, 1993. Had the same statement been expressed in indirect or reported speech it would have been printed as *Mr Pickering said to Jane that he had a suggestion that she might think about*. A change of tense is often involved in reported speech as well as other minor changes. Thus, the direct speech statement *'I believe you should sell this place and make a move across the river. You will be quite comfortable there.'* (ibid.) would become in reported speech *Mr Pickering said that he believed she should sell her place and make a move across the river. She would be quite comfortable there.*

See also INDIRECT QUESTION.

**repp.** See REP.

**reprehensible.** Thus spelt. See -ABLE, -IBLE 7.

**reprieve.** Thus spelt, not *-ei-*.

**reprimand.** The noun is stressed on the first syllable. Some dictionaries, including the *SOED* (1993), give third-syllable stressing for the verb, but /ˈreprɪmɑːnd/ is recommended here.

**reproducible.** Thus spelt. See -ABLE, -IBLE 7.

**reprove** makes *reprovable*. See MUTE E.

**reps.** See REP.

**repugn** (verb) is pronounced /rɪˈpjuːn/, but in *repugnance* and *repugnant* the *g* and the *n* are both fully pronounced: thus /rɪˈpʌɡnəns/, /rɪˈpʌɡnənt/.

**repulsive.** See REPELLENT.

**reputable.** Stressed on the first syllable, /ˈrepjʊtəbəl/.

**request.** The noun is frequently followed by the preposition *for* (*a request for more time to come to a decision*), and the verb *ask* is often followed by a direct personal object plus *for* preceding the object requested (*the beggar asked her for some money*). But *request* (verb) cannot idiomatically be so used: *the beggar requested some money from her* is idiomatic, but *the beggar requested her for some money* is not.

**require** (verb). Two uses of the verb that, though presumably idiomatic in the countries concerned, are unidiomatic in Britain: *I require to know your surnames*—R. Rive, 1986 (SAfr.; BrE *need*); *Zoning ordinances required it be a certain distance from the street*—T. McGuane, 1989 (US; BrE *to be*).

**reredos.** Two syllables, /ˈrɪədɒs/.

**rescind** (verb). The corresponding noun is *rescission*, pronounced /rɪˈsɪʒən/.

**research** (noun). Normally stressed on the second syllable in BrE and on the first in AmE. But /ˈriːsɜːtʃ/ is now increasingly being used in BrE, esp. by the young.

**resentment.** *May I, as one in complete sympathy with the general policy of the Government, give expression to the strong resentment I feel to the proposed Bill. Resentment of, at, against,* never *to*.

**reservedly.** Four syllables. See -EDLY.

**residue, residuum.** In lay use the preferred form is *residue* (adj. *residual*). In law, *residue* (rarely *residuum*, pl. *residua*) is that which remains of an estate after the payment of all charges, debts, and bequests. In chemistry, a *residue* (less commonly a *residuum*) is a substance left after combustion, evaporation, etc. There are uses of both words in other technical and scientific subjects: for these, readers are advised to consult the appropriate dictionaries or textbooks. The adj. most commonly used in law is *residuary* (*a bequest of his residuary estate to charity*), but there are legal contexts, too complicated to be treated here, in which *residual* is appropriate.

**resignedly.** Four syllables. See -EDLY.

**resile** (verb) /rɪˈzaɪl/. An uncommon word meaning 'to withdraw from a course of action': e.g. *Bill Cash, unofficial leader of the rebels, told the Commons, 'I will not in any way resile from the objections that I have to the* [Maastricht] *treaty*—newspaper report, 24 July 1993.

**resilience, resiliency.** Pronounce with /rɪˈzɪl-/. The two words are not clearly distinguished in sense, but the first is much the more common of the two. Cf. -CE, -CY.

**resin.** See ROSIN.

**resist** (verb) makes *resistible*. See -ABLE, -IBLE 7.

**resistance.** *Resistance to* governs an *-ing* participle, not an infinitive. Example: *You have likened the resistance of Ulster Unionists to being* [not *to be*] *driven out of the Constitution … to the opposition …* See GERUND 3.

**resister,** one who resists, contrasted with **resistor,** a device having resistance to the passage of an electrical current.

**resoluble, resolvable.** Both are in use without distinction of meaning, the second being much the more common. Both are stressed on the second syllable. The corresponding negatives are *irresoluble* /ɪrɪˈzɒljʊbəl/ or *irresolvable* /ɪrɪˈzɒlvəbəl/, and *unresolvable*. See IN- AND UN-. See also DISSOLUBLE; SOLUBLE.

**resolution** (in prosody). The substitution in a metrical foot of two shorts for a normal long. Thus a dactyl $-\smile\smile$ or anapaest $\smile\smile-$ by resolution becomes a spondee $--$.

**resolution, motion.** As names for a proposition that is passed or is to be passed by the votes of an assembly, the two differ in that the passing of a motion results in action, and a motion is that something is done; while a resolution is not necessarily more than an expression of the opinion that something is true or desirable. Since, however, opinion often becomes operative, and since resolutions as well as motions are moved, i.e. are at least in one sense motions, the distinction is elusive. It is nevertheless of some value if not too rigidly applied.

**resolve** (verb). See INVOLVE 1.

**resolvedly.** Four syllables. See -EDLY.

**resort, re-sort.** The first has a wide range of meanings as noun and verb, and is pronounced /rɪˈzɔːt/. The second means 'to sort again or differently' and is pronounced /riːˈsɔːt/. See RE(-) (b).

**resort, resource, recourse.** Fowler (1926) cited a number of examples showing confusion in the use of these three: e.g. *She will not be able to do so ... without resource to the sword* (recourse, resorting, resort); *Surely he was better employed in plying the trades of tinker and smith than in having resource to vice* (recourse); *... should an autonomous régime for Macedonia have been agreed to by Turkey without resource to war* (recourse, resort); *... binding all Powers to apply an economic boycott, or, in the last resource, international force, against any Power which ...* (resort). These examples underline Fowler's resourcefulness in seeking out such evidence, but if such confusion still exists I have failed to find it. The three words are chiefly used in certain established phrases: (resort) *as a last resort, in the last resort* (after Fr. *en dernier ressort*); *without resorting to; to resort to; a holiday resort*; (resource) *without resources; at the end of his resources; one's own resources*, one's personal capabilities; *a person of many resources*; (recourse) *to have recourse to; without recourse*, a formula used by the endorser of a bill, etc., to disclaim responsibility for non-payment. These and others are all treated in the *OED* and the *SOED* with much greater detail than can be presented here.

**resource.** Pronounce either as /rɪˈzɔːs/ or /rɪˈsɔːs/. In BrE the stress always falls on the second syllable, but in AmE both first-syllable and second-syllable stressing are current.

**respect.** The compound prepositions *in respect of, in respect to, with respect to* and the simple preposition (or pres.pple) *respecting* are all listed in the *OED* with extensive historical evidence from literary works: e.g. *Item, shee is not to be fasting in respect of her breath*—Shakespeare, 1591; *After this, the Colony enjoy'd a perfect Tranquillity with Respect to the Savages*—Defoe, 1719; *He could not agree with him respecting the price*—M. Edgeworth, 1802; *She had struck him, in respect to the beautiful world,*

*as one of the beautiful, the most beautiful things*—H. James, 1904. Despite a certain amount of hostility or discouragement shown by some usage guides, all four expressions have continued in use. Examples: *Any allowance for relief at the higher rate in respect of interest on loans within MIRAS will be withdrawn*—*Financial Times*, 1991; *There are two pillars to this account ... much differentiated as well as distinct from a traditional petite bourgeoisie in respect of its structured 'overconsumptionism'*—*Oxford Art Jrnl*, 1991; *Miami of today ... cannot compare to the Milwaukee of 100 years ago in respect to the proportionate numbers of immigrants or their apartness in the larger culture*—*NY Times Bk Rev.*, 1991; *All stressed the need to reaffirm a position within the Labour Party against revisionism, particularly with respect to public ownership*—*Twentieth Century Brit. History*, 1991; *We were also convinced that consonance—meaning agreement with binding Roman Catholic teaching—had not been reached on certain essential points, even points respecting the eucharist and the ministerial priesthood*—*Church Times*, 1992.

**respectfully.** *Yours respectfully*, as a signing-off formula in letters (including those written to the editors of newspapers), is now rarely used. See LETTER FORMS.

**respectively** is correctly used when the sense required is 'each separately or in turn, and in the order mentioned' (e.g. *Such a scaffold is fully determined by the parameters R and c specifying respectively the equatorial radius and the widening of the barrel*—*Protein Engineering*, 1990; *Iraq and Syria have been ruled by small minorities (Sunni Arabs and Alawite officers respectively)*—*Internat. Affairs*, 1991; *The product is a joint venture by Rudy Rucker, a science fiction writer, and John Walker, who are employed respectively as 'mathenaut' and 'virtual programmer' by the computer-aided design specialist Autodesk Inc.*—*New Scientist*, 1991.

Fowler (1926) cited numerous examples in which *respectively* is used tautologically (e.g. *He wants the Secretary for War to tell the House in what countries they are at present stationed, and the numbers in each country respectively* (*each* is adequate) or unintelligibly (e.g. *The writing-room, silence-room, and recreation-room, have respectively blue and red arm-chairs*). He added a comment: 'The simple fact is that *respective* and *respectively* are words seldom needed, but that pretentious writers drag them in at every opportunity for the air of thoroughness and precision they are supposed to give to a sentence.' His comments are sound but their present-day relevance is uncertain. I have found no evidence of such misuses of *respectively* in the 1990s.

**respirable, respiratory.** The pronunciations recommended are respectively /'respərəbəl/ and /'respərət(ə)rɪ/. But second-syllable stressing, i.e. /rɪ'spaɪr-/ and /rɪ'spɪr-/, is common in each of them.

**respite** (noun). The *SOED* (1993) gives precedence to the pronunciation /'respaɪt/, but /'respɪt/ is at least as common.

**resplendence, resplendency.** The first is recommended. See -CE, -CY.

**responsible.** Thus spelt, not *-able*. See -ABLE, -IBLE 7.

**restaurant.** In standard BrE the final *t* is not sounded.

**restaurateur.** Thus spelt. The spelling *restauranteur* (with internal *n*), though increasingly common, is erroneous.

**restive** is used particularly of a horse that stands still or moves backwards or sideways and refuses to go forward; of a refractory child; or of a discontented army; or any other group of people harbouring a grievance. The word makes contact with *restless* when the meaning required is no more than agitated or fidgety, because of loss of sleep (*a restless night*), boredom (*unemployment left him listless and restless*), a poor performance of a play, etc. (*the audience was becoming increasingly restless*), or some other unsettling circumstance. *Restive* most often suggests that potentially hostile or disruptive action seems likely to occur, while *restless* is more often a kind of explanatory word, implying that there is a reason (lack of sleep, an unhappy relationship, a lacklustre concert, etc.) leading to the behaviour in question.

**restrainedly.** Four syllables. See -EDLY.

**restrain, re-strain.** See RE(-).

**restrictive** and **non-restrictive appositives.** See APPOSITION 2.

**restrictive clauses.** The main discussion of these and of **non-restrictive clauses** will be found s.v. THAT (relative pronoun), but it may be helpful, as an aperitif, if I set down here the definitions of the terms in standard sources.

Restrictive clauses are relative clauses 'delimiting the meaning or reference of a modified noun phrase or other element' (*SOED*, 1993). Leech and Svartvik (1975) cite the following pair of sentences to bring out the distinction: [1] *Children* who learn easily *should start school as early as possible*; [2] *Children*, who learn easily, *should start school as early as possible*. 'In [1] the relative clause is restrictive and tells us *what kind of* children ought to start school early. In [2], where the relative clause is non-restrictive, the speaker is talking about all children in general ... The clause does not in any way limit the reference of *children*.'

S. Greenbaum offers the following definition (*OCELang.*, 1992): 'A *restrictive relative clause* (also *defining relative clause*) is a relative clause with the semantic function of defining more closely what the noun modified by the clause is referring to. In the sentence *My uncle who lives in Brazil is coming to see us*, the relative clause *who lives in Brazil* restricts the reference of *my uncle*. The restrictive modification would distinguish this uncle from any others who might have been included. A *non-restrictive relative clause* (also *non-defining relative clause*) adds information not needed for identifying what a modified noun is referring to. The sentence *My uncle, who lives in Brazil, is coming to see us* contains the non-restrictive relative clause *who lives in Brazil*. This clause provides information about the uncle, but his identity is presumed to be known and not to need further specification. Non-restrictive relative clauses are usually separated from the noun phrases they modify by parenthetical punctuation (usually commas, but sometimes dashes or brackets). In speech, there may be a pause that serves the same function as the parenthesis.' When these are taken together it emerges that a non-restrictive relative clause can normally be recognized because of its parenthetical punctuation or its spoken equivalent—pauses in the flow of speech.

**rest room.** See TOILET.

**result.** The word is pushing into the territory of not just an outcome (of some action, game, etc.) but a favourable or desirable outcome. The earliest examples found so far are from the 1920s and show the word in the plural: e.g. *take some of those pamphlets with you to distribute aboard ship. They may bring results*—E. O'Neill, 1922. More recently, *result* has come to be used in this manner in the singular (? only in the language of footballers and their managers): e.g. *We needed a result ... Perhaps we should have done better than win 1–0*—*Observer*, 1976.

**résumé** /'rezjʊmeɪ/. In BrE thus written (with accents, in romans) and normally meaning 'a summary'. In AmE, often without the first accent or with no accents (*resumé, resume*), used to mean 'a summary', or specifically 'a curriculum vitae'.

**resurrect.** See BACK-FORMATION.

**retable** (frame at back of altar). Pronounce /rɪ'teɪbəl/.

**reticent.** Since it entered the language in the 1830s, along with the related noun *reticence* ('Not in common use until after 1830', OED), its standard meaning has been 'reserved; disinclined to speak freely'. Evidence is accumulating that it is now also being used to mean 'reluctant to act', and dragging the noun along in the same direction. Examples: *Not everyone is as reticent as London's civil servants about moving to Docklands*—*Times*, 1992; *Zhang Liang says he understands Western reticence about introducing genetically altered organisms into the environment*—*New Scientist*, 1993. The new use is non-standard at present, but has an air of inevitability about it.

**retina.** In technical work the pl. is usually *retinae* /-niː/, but in non-technical writing and speaking *retinas*. See LATIN PLURALS.

**retiral,** meaning 'retirement from office', is hardly used outside Scotland. Example: *Young person required for civil engineering stores to fill vacancy due to retiral*—advt in *Lochaber News*, 1978. See -AL 2.

**retractation, retraction.** The second is the customary word as the noun corresponding to all the main senses (withdraw a statement, etc.) of the verb *retract*, except that *retractation* is properly used with direct or indirect reference to the title of a book written by St Augustine 'containing further treatment and corrections of matters treated in his former writings' (OED); and also (beside *retraction*) to mean 'recantation of an opinion, statement, etc., with admission of error' (OED).

**retrieve** (noun). Apart from special uses in the training of gun-dogs, and in some American games (e.g. volleyball), hardly used now except in the collocations *beyond retrieve, past retrieve*. The customary noun for all the main senses is *retrieval*. See -AL 2.

**retro-.** For long pronounced either as /'retrəʊ/ or /'riːtrəʊ/. At some point in the 20c. the pronunciation with a short /e/ virtually drove out that with /iː/ in all the relevant words, and now /'riːtrəʊ/ is rarely heard.

**retrograde, retrogression** (and derivatives). There are two series: 1 *retrograde* adj. and (esp. in astronomy) verb, *retrogradation*; 2 *retrogress* verb, *retrogression* noun, *retrogressive* adj. In practice, for the adj. *retrograde* is the usual choice, and for the noun *retrogression* (except in astronomy and physical geography where *retrogradation* is customary). The verbs are seldom used (except for *retrograde* in astronomy and physical geography) There seems to be no prospect that one series will oust the other.

**retroussé** (adj.). (Of the nose) turned up at the tip. Print in romans, retain the acute accent.

**rev** (verb). The inflected forms are *revs, revved, revving*.

**reveille** /rɪ'vælɪ/. Thus spelt.

**revel.** The inflected forms of the verb are *revelled, revelling*; also *reveller*. In AmE usu. written with a single *l*. See -LL-, -L-.

**Revelation.** The last book in the New Testament was called *The Revelation of S. Iohn the Diuine* in 1611, and the singular form *Revelation* remains in standard use to this day. In popular use the pl. form *Revelations* is regrettably common.

**revenge** (verb). See AVENGE.

**Reverend** is abbreviated *Revd* (no point) or, with increasing frequency, *Rev.* (with point). It means 'deserving reverence', as contrasted with *reverent* 'feeling or showing reverence'. Archbishops are *Most Reverend*, bishops are *Right Reverend*, and deans *Very Reverend*; archdeacons are not *Reverend* but *Venerable*. The correct form of address to a member of the clergy is *The Revd* (or *The Rev.*) *J. Smith* (or with the Christian name in full). To describe such a person as *Rev. Smith* or to speak of such a person as *Reverend Smith* or *the Reverend* is incorrect, at least in standard use in Britain. The *Shorter Oxford* (1993) expresses it as follows: 'It is commonly considered unacceptable to use *Reverend* ... with a surname alone (rather than with a forename, initial, other title, or some combination of these) or without preceding the: thus *the Reverend Joseph Brown*, *the Reverend J. B. Brown*, *the Reverend Dr Brown*, but not *(the) Reverend Brown*, and not *Reverend* as a form of address, either spoken (*Hello, Reverend* (*Brown*)) or to begin a letter (*Dear Reverend* (*Dr Brown*)).' Actual usage is variable, but members of the clergy are quite rightly adamant about the need to keep to the older conventions while recognizing that in our age of receding Christian belief those who break with tradition are usually deeply respectful but quite simply ignorant of protocol.

In AmE scant regard is paid to the conventions of BrE in the matter, to judge from the following examples: *There really was something between her and Reverend Propper?*—J. Updike, 1988; *Reverend Samuels led the congregation in a prayer for Jackie Robinson*—M. Golden, 1989; *Reverend Knox decided he could join the staff on those terms*—G. Keillor, 1991; *The Reverend Cutcheon gave an address before the celebration began*—New Yorker, 1992. If a person is at the same time a minister or chaplain and (say) a professor, the correct style is *The Revd Professor James Jones*. The complications of how to address ordained members of religious orders in the Church of England, archbishops of the Church of Ireland, bishops of the Episcopal Church in Scotland, and so on, lie outside the scope of this book. For the

procedures, *Debrett's Correct Form* (1970 or later printing) is an excellent guide.

**reverent, reverential.** Between *reverent* and *reverential* the difference is much the same as that between PRUDENT and *prudential*, *reverential* being as applicable to what simulates reverence as to what is truly instinct with it, while *reverent* has only the laudatory sense. But *reverential* is often chosen merely as a LONG VARIANT; when *reverent* would not be out of place, *reverential* is a substitute as much weaker as it is longer.

**reversal.** See REVERSION.

**reversible.** Thus spelt (also *irreversible*), not -*able*. See -ABLE, -IBLE 7.

**reversion** has various senses, chiefly legal or biological, to be found in any good dictionary, and not needing to be set down here. It suffices to say that they all correspond to the verb *revert*, and not to the verb *reverse*, whose noun is *reversal*. In the following examples it has been wrongly given the meaning of *reversal*: *The reversion of our Free Trade policy would, we are convinced, be a great reverse for the working class*; *But to undertake a complete reversion of the Bolshevik policy is beyond their powers*.

**reviewal** had wide currency in the 19c. when nouns in -*al* were popular, but in most contexts now *review* (noun) is preferable.

**review, revue.** The former has numerous senses including the following: a general survey or assessment of a subject, policy, etc.; a display and formal inspection of troops; a published assessment of the value, interest, etc., of a book, play, etc.; part of the title of a regular journal (e.g. *The Review of English Studies*). A *revue* is simply a theatrical entertainment consisting of a series of short sketches and songs, i.e. has only one sense. It should be recorded here that *review* is occasionally used instead of *revue*, but the two words are best kept apart in the manner indicated above.

**revisal** had substantial currency from the 17c. to the 19c. but has now virtually dropped out of use in favour of *revision*.

**revivals.** See CARREL; CARVEN.

**revolve, revolver.** See INVOLVE 1.

**Rev., Revd.** See REVEREND.

**revue.** See REVIEW.

**rhapsodic, rhapsodical.** In classical literature a rhapsody (L *rhapsōdia*) was an epic poem or part of one, e.g. a book of the Iliad or Odyssey, suitable for recitation at one time. From this came the use of the word in English (16–19c.) to mean a literary work consisting of miscellaneous or disconnected pieces; and (from the late 19c.) a free musical composition in one extended movement, usu. emotional or exuberant in character. From the mid-17c. onward the word also stepped outside literary forms to be used of an exaggeratedly enthusiastic or ecstatic expression of feeling. All this from the OED. In past times *rhapsodic* and *rhapsodical* were used as adjs. corresponding to the literary senses of *rhapsody*. Now they are increasingly restricted to mean exaggeratedly enthusiastic, or to musical contexts, with *rhapsodic* being much the more common of the two. Examples: (rhapsodic) *By the same token, critics wax rhapsodic about* The Cook, the Thief [etc.] *because its images* [etc.]—M. Medved, 1992; *Their interpretation has a freedom of tempo and rubato that brings a rhapsodic quality to the music*—Strad, 1992; *When Barry Diller ... speaks of his Apple PowerBook, a laptop computer, he grows rhapsodic*—New Yorker, 1993. (rhapsodical) *People write rhapsodical volumes about the glories awash on Chesapeake Bay—the skipjacks, the oyster tonging*—V. Tanzer, 1989.

**Rhenish** (of the River Rhine). Pronounce /ˈrenɪʃ/ rather than /ˈriːn-/.

**rhetorical question.** A question is often put not to elicit information, but as a more striking substitute for a statement. The assumption is that only one answer is possible, and that if the hearer is compelled to make it mentally himself it will impress him more than the speaker's statement. So *Who does not know ... ?* for *Everyone knows*; *Was ever such nonsense written?* for *Never was* etc.

**rhino.** Pl. the same or *rhinos*. See -O(E)s 5. See also ABBREVIATIONS 1.

**rhinoceros.** Pl. the same (*a herd of rhinoceros*) or *rhinoceroses*.

**rhombus.** Pl. *rhombuses* (preferred) or *rhombi* /-baɪ/.

**rhotic.** Of a form of English, esp. Scots and AmE, that retains historical /r/ in medial and final position (Arthur, for fear, harder). The state or condition of being rhotic is *rhoticity* or *rhotacism*. RP is notable for its lack of rhoticity, i.e. is a non-rhotic form of English.

**rhyme,** identity of sound between words or the endings of words, a marked feature of verse. **1** The word entered English in the 13c. in the form *rime* (from OF *rime*) and this spelling (or *ryme*) remained dominant until the early 17c. when, through etymological association with the ultimate source, L *rhythmus*, the spelling *rhyme* began to be used. Gradually the new spelling established itself and *rhyme* is now the prevailing form. In Coleridge's *The Rime of the Ancyent Marinere* (1798) *Rime* means a rhyming poem.

**2** The rhyming element may be a monosyllable (*feet/seat, love/above*), known as a 'masculine rhyme', two syllables (*rabble/babble, guessing/blessing*), known as a 'feminine rhyme' or 'double rhyme', three syllables (*glamorous/amorous*), known as a 'triple rhyme', or more. *Half-rhyme* is much favoured by certain poets, e.g. Hopkins and Owen: *shell/shall, fronds/friends, hitting/hurting*. More complex types also occur: e.g. *sisterhood/good, Christ/sacrificed, room there/loom there/doom there*.

**rhyme royale.** A term in prosody meaning a metre in stanzas of seven iambic pentameters rhyming ababbcc. Chaucer's *Clerk's Tale* is a typical example:

> *This sergeant cam unto his lord ageyn,*
>     *And of Grisildis wordes and hire cheere*
> *He tolde hym point for point, in short and pleyn,*
>     *And hym presenteth with his doghter deere.*
> *Somwhat this lord hadde routhe in his manere,*
>     *But nathelees his purpos heeld he stille,*
> *As lordes doon, whan they wol han hir wille; ...*

**rhyming slang,** a type of slang (orig. Cockney) in which a word is replaced by words or phrases which rhyme with the word substituted, often with the rhyming element omitted (*New SOED*). Examples are first found in Cockney in the early 19c. They include: *apples and pears* (stairs, first recorded in 1857), *elephant's trunk* (drunk, 1859), *plates of meat* (feet, 1887), *rock of ages* (wages, 1937), *round the houses* (trousers, 1857), *shovel and broom* (room, 1928), *Sweeney Todd* (Flying Squad, 1938), *trouble and strife* (wife, 1929). When such phrases become familiar they are often used in a reduced form, e.g. *He's elephants* ( = He's drunk). In practice they are seldom explained, e.g. *Goethe never spoke ... All through the afternoon not a dickybird, Barley said*—J. le Carré, 1989. It is simply assumed that the reader knows that *dickybird* (first recorded in this sense in 1932) is rhyming slang for 'word'.

**rhythmic, rhythmical.** The two forms have stood side by side for nearly four centuries, and in the majority of contexts (except in physical geography, where the shorter form is favoured) are interchangeable. In practice the choice of form is made instinctively and arbitrarily. In William Thomson's classic work *The Rhythm of Speech* (1923) *rhythmical* seems to be used throughout (*Each of the series is organically rhythmical*; *Quintilian's total unfitness for dealing with any rhythmical question*). By contrast, Fowler (1926) uses both forms in his article on *rhythm*: *Rhythmic speech or writing is like waves of the sea; A sentence ... is rhythmical if, when said aloud,* [etc.]. D. Attridge, in *The Rhythms of English Poetry* (1982), keeps to *rhythmic* throughout his book (*Part of the rhythmic character ... of any line stems from the placing of word and phrase boundaries; the same unconscious immediacy with which we respond to some rhythmic patterns*). Recommendation: use whichever form seems natural, but be consistent.

See -IC(AL).

**ribbon, riband.** The first is overwhelmingly the customary form. The second occurs mostly in the phr. *blue riband* (also with capital initials), the name given to the record for the fastest sea journey between Southampton and New York, esp. in the 1920s and 1930s. Etymologically the two words are doublets, from ME *reban* (later, with parasitic *d*, *ryband*) from OF *riban* (still dialectal), modF *ruban*.

**rick** (noun and verb), a sprain; to sprain. Thus usu. spelt, rather than *wrick*. Both words are relatively recent (verb late 18c., noun mid-19c.). They seem to be of dialectal origin.

**rickety.** Thus spelt, not *-etty*. See -T-, -TT-.

**ricochet.** The spelling, accent, and pronunciation recommended are: *ricochet* /'rɪkəʃeɪ/; *ricocheted* /'rɪkəʃeɪd/; and *ricocheting* /'rɪkəʃeɪɪŋ/. Cf. CROCHET; CROQUET.

**rid.** The pa.t. and pa.pple are now normally *rid* rather than *ridded*. The active type *He ridded the stable of cluster flies*, however, occurs from time to time, as does the participial type *He has ridded the stable of cluster flies*. But when the pa.pple is passive the only possible form is *rid*: e.g. *I thought myself well rid of him*.

**Riesling.** Pronounce /'riːslɪŋ/ or /'riːz-/, but not /'raɪ-/ as in *rye*.

**right.** 1 The distribution of adverbial *right* used as an intensive meaning 'very' is not fully known. The *OED* lists examples from the 13c. to the 20c., including the following: (with adverbs) *The portrait of him she loved right dearly*—Disraeli, 1826; *They conquered it right royally*—Mag. Art, 1885; *The 'proofs' of the Blake book are coming in ... The illustrations look right well*—W. B. Yeats, 1891; *Cale was doing right well for himself*—P. Mallory, 1981; (with adjectives) *I should be right sorry To have the means so to be venged on you*—B. Jonson, 1611; *I was right glad ... to see your writing again*—Coleridge, 1800; *I did not feel right comfortable for some time afterwards*—M. Twain, 1869; *Miz Wilkes is right sensible, for a woman*—M. Mitchell, 1936. From these examples, and from other evidence, it is possible to surmise that, used in this way, *right* was at one time in good literary use in England; that it is much more informal now in BrE; and that it is mainly in regional (chiefly Southern) use in AmE. It remains in standard use in BrE in titles and forms of address, e.g. *Right Honourable, Right Reverend, Right Worshipful*.

2 In some other senses, adverbial *right* is standard everywhere, e.g. informally = straight (with temporal connotation) (e.g. *right away; right now; I'll be right down; I'll be right with you*); = quite, completely (e.g. *turn it right off;*

*he has come right round to the idea).* The emphatic *Right on!*, used since the 1920s as an expression of enthusiastic agreement or encouragement, is still mostly restricted to AmE.

**3** *right, rightly.* In uses where the meaning required is 'properly, in a fitting manner', *right* is idiomatically acceptable in a few fixed phrases, e.g. *treat them right; he guessed right; nothing went right.* But in this sense, and when the meaning is 'correctly', the natural form is *rightly,* e.g. *If I am rightly informed; Rightly, he was sent to prison; incapable of acting rightly; he thought, rightly or wrongly, that his manager was less than sympathetic.*

**righteous.** Pronounce /ˈraɪtʃəs/.

**rightward(s).** As adj. always *rightward* (*a rightward movement*). As adv. either *rightward* or, esp. *rightwards* (*the car drifted rightwards after hitting a puddle*), in BrE.

**rigour.** Thus spelt in BrE, but *rigor* in AmE. The corresponding adj. is *rigorous.* In the L phr. *rigor mortis* the spelling *rigor* is obligatory in both countries. See -OUR AND -OR; -OUR- AND -OR-.

**rilievo.** See RELIEVO.

**rime.** See RHYME 1.

**rinderpest** (disease of ruminants; Ger., cf. *Rinder* cattle). Pronounce /ˈrɪndəpest/.

**ring.** The conjugation of the verb (used of bells, telephones, etc.) is stable, the pa.t. being *rang* and the pa.pple *rung.* The unrelated verb *ring* 'to make or draw a circle round; to put a ring on (a bird, etc.)' has pa.t. and pa.pple *ringed.*

**ringleted.** Thus spelt, not *-etted.* See -T-, -TT-.

**riot** (verb). The inflected forms are *rioted, rioting.*

**rise.** See ARISE; RAISE, RISE (nouns and verbs).

**risky, risqué.** The first is the general word used whenever the meaning is 'involving risk'. The second, pronounced /ˈriːskeɪ/; less often /ˈrɪs-/ or stressed on second syllable, is used only to mean '(of a joke, etc.) slightly indecent or liable to shock'.

**ritardando** (musical direction). Pl. *ritardandos.* See -O(E)S 6.

**rival** (verb). The inflected forms are *rivalled, rivalling.* In AmE often *rivaled, rivaling.* See -LL-, -L-.

**rive** (verb) (archaic or poetic) /raɪv/, to split. Pa.t. *rived,* pa.pple *riven* /ˈrɪvən/.

**rivet.** The inflected forms have a single *t* (*riveted, riveting, riveter*). See -T-, -TT-.

**road, street,** etc. Of the many different names we give to thoroughfares for vehicles, *road* is the most comprehensive. The *New Shorter Oxford* (1993) defines it as 'A path or way between different places, usu. one wide enough for vehicles as well as pedestrians and with a specially prepared surface. Also, the part of such a way intended for vehicles.' It was the natural word for our long-distance highways such as the Great North Road and the Great West Road, and for what we used to call the Dover Road, the Portsmouth Road, and the Bath Road, until in the second half of the 20c. we resorted to a numerical system and called them respectively the A2, A3, and A4. A *street* is originally a paved way—from L *via strata*. According to the laws of Henry I a street 'was to be sufficiently broad for two loaded carts to meet and for sixteen armed knights to ride abreast' (Ekwall; *Street-Names of the City of London,* 1954). It was the name bestowed by the Anglo-Saxons on the great roads of the Roman occupation—Ermine Street, Watling Street, and others. Its current meaning is 'A public road in a city, town, or village usu. running between two lines of houses or other buildings; such a road along with the pavements and buildings on either side' (*New SOED*).

The history of the two words and of the words *lane, avenue, crescent, gate, row, place, terrace, rise, vale,* and some others is extremely complicated. Over the centuries thoroughfares have repeatedly changed their nature. For example, as Ekwall points out: 'In medieval London records a distinction between *street* and *lane* is fairly well kept up in street-names. But the difference between a comparatively narrow street and a comparatively wide lane might be slight, and there are cases of vacillation between *street* and *lane* in street-names ... [Thus]

Paternoster Row was *Paternosterstrete* 1307 ff., *Paternosterlane* 1321–35, finally Paternoster Row from 1334 on ... New Street (1185 ff.) was superseded by *Converslane* (1278 ff.) and by Chancery Lane.' A 'place' was originally 'an open space in a town, a market-place'. Now, it is 'a small square or a side-street, esp. a cul-de-sac, lined with houses' (*New SOED*). In London, *Grosvenor Place* and *Portland Place*, for example, might now be called 'streets' with perfect propriety. Similar 'anomalies' can be found in any major city in Britain. *Way*, as the name for a road, is as old as Fosse Way and Icknield Way. It has been preserved in *highway* and *railway*, and has been revived in the 20c. in *motorway*.

Certain fixed phrases have become established over the centuries: thus *No Through* Road, but *One Way* Street; *Cross*roads, but Street *corners*; *Rule of the* Road, but *High*way *Code*; *Keep in* Lane, but *Dual Carriage*way; *Major* Road *Ahead*, but *Clear*way.

No short article can do justice to the historical circumstances lying behind the names given to our streets and roads and avenues. There are no avenues in the village in which I live. But among the names given to the 'streets' of this quiet corner of the world are *All Saints Lane, Hobby Horse Lane, Bradstocks Way, Church Street, High Street, Tullis Close, The Green, Harwell Road, Drayton Road,* and *Milton Road*. Each name has an establishable history. And so it is in towns, villages, and cities up and down the country.

**roast.** **1** The distribution of *roast* and *roasted* as ppl adjs. is of interest: thus *roast beef, roast lamb,* etc., but *roasted peanuts, a whole roasted ox; a roast joint,* but *a well-roasted joint.* As a full pa.pple (*she had roasted a chicken for lunch*), *roasted,* is the only form used.

**2** For *rule the roast,* see RULE.

**rob** (verb) is used principally to mean 'to steal from' (*rob a bank, a jewellery shop*); also followed by a personal object + *of* + indirect object (*robbed her of her jewels*). The old use (13c. onwards) meaning 'to steal' (*he robbed money from the till*) is falling into disuse and should be avoided (even though it was used by

Joyce in *Finnegans Wake*: *Robbing leaves out of my taletold book*).

**robustious.** 'In common use during the 17th century. In the 18th it becomes rare, and is described by Johnson (1755) as "now only used in low language, and in a sense of contempt". During the 19th it has been considerably revived, esp. by archaizing writers.' (*OED*) Fowler (1926) described it as 'One of the words whose continued existence depends upon a quotation (*Hamlet* III. ii.10)'. In practice it has almost totally been replaced by *rumbustious* (first recorded in 1778), a corrupted form of itself.

**rock, stone.** In BrE a small piece of rock capable of being thrown is normally called a *stone* (or a *pebble*); in AmE the word used in this sense (as well as for a mass of stone forming a cliff, etc.) is *rock* (or a *pebble*), but not a *stone.*

**rococo.** See BAROQUE. A 19c. loanword from French (in which language it was a jocular alteration of *rocaille* shell-work) meaning ' **1** (of) a late baroque style of decoration prevalent in 18th c. continental Europe, with asymmetrical patterns involving scroll-work, shell motifs, etc.

**2** (of literature, music, architecture, and the decorative arts) highly ornamented, florid' (*COD*). Examples: *Haydn's symphonic music began as rococo entertainment*—Listener, 1959; *The drawing in nearly all Monticelli's pictures is reminiscent of the rococo*—Listener, 1965; *The painter to whom the epithet 'Rococo' has most often been loosely applied is ... Watteau, and in his rejection of the grand sujet and his fanciful and curvacious rhythms he does ... fit into the movement*—Oxford Companion to Art, 1970; *He produces rococo philosophical arguments for the conclusion that neither phrase coherently refers*—London Rev. Bks, 1987.

**rodent operator.** See RAT-CATCHER.

**rodeo.** Pl. *rodeos.* See -O(E)S 4.

**rodomontade.** Not *rho-*.

**roe.** See BUCK.

**roentgen** (unit of ionizing radiation). Thus spelt in English, though named after Wilhelm *Röntgen* (died 1923), German physicist. Pronunciation now normally Anglicized to /ˈrʌntjən/, though /ˈrɜːn-/ is equally acceptable.

**roguish.** Thus spelt. See MUTE E.

**role** (actor's part, etc.) is the preferred spelling in OUP's house style, but *rôle* is also valid.

**Rolls-Royce** (trade-mark). So spelt (with capitals and a hyphen).

**romance.** Both the noun and the verb bear the stress on the second syllable: thus /rəʊˈmæns/. The pronunciation of the noun with the stress on the first syllable, i.e. /ˈrəʊmæns/, is widespread but non-standard.

**Romanes.** The name of an annual lecture (the Romanes lecture) founded in the University of Oxford in 1891 by the eminent Victorian physiologist George John Romanes (1848–94). Pronounce /rəʊˈmɑːnɪz/.

**Romania.** Now the official spelling of the name of the country which was commonly spelt *Rumania* or *Roumania* in English-language contexts earlier in the 20c.

**Roman numerals** are used with decreasing frequency but are still customary in certain circumstances: e.g. for the names of monarchs (*George VI*); for the preliminary pages of books; for acts and scenes in plays (*Romeo and Juliet* II.iii); sometimes for the numbering of issues of learned journals; on old clocks and watches; and on some almanacs. The series runs as follows: (lower case) i, ii, iii, iv, v, vi, vii, viii, ix, x, etc. (note the positioning of i in iv and vi); (capitals) I, II, III, IV (or less often IIII), etc. Arabic 30 = xxx, 40 = xl (or less often xxxx), 42 = xlii, 49 = xlix; 50 = l (or L), 90 = xc, 99 = xcix: 100 = c (or C); 500 = D, 1,000 = M. The year 1995 is (according to the Oxford Almanack) 'For the Year of Our Lord God MDCCCCLXXXXV'; the year 2001 is MMI.

The main principle governing the system is that if a numeral is followed by another numeral of smaller value, the two are added together (l followed by i = 51); if it is preceded by one of lower value the smaller numeral is deducted from the larger one (v preceded by i = 4).

Multiples of a thousand are indicated by a superior bar: thus V̄ = 5,000, X̄ = 10,000, C̄ = 100,000.

**Romansh** /rəʊˈmænʃ/. Name given to a Rhaeto-Romanic dialect, esp. as spoken (by some 40,000 speakers) in south-east Switzerland. (Variant spellings are also used in some publications : *Ru-, Rou-, -ansch*.)

**Rome** makes *Romish*. See MUTE E.

**rondeau.** 1 'A medieval French verse form also used by some late 19th-century poets [e.g. Austin Dobson] in English. It normally consists of 13 octosyllabic lines, grouped in stanzas of five, three, and five lines. The whole poem uses only two rhymes, and the first word or phrase of the first line recurs twice as a refrain after the second and third stanzas. The standard rhyme scheme (with the unrhymed refrain indicated as R) is *aabba aabR aabbaR* (Baldick, 1990).

An example:

> *On London stones I sometimes sigh*
> *For wider green and bluer sky;—*
>   *Too oft the trembling note is drowned*
>   *In this huge city's varied sound;—*
> *'Pure song is countryborn'—I cry.*
>
> *Then comes the spring,—the months go by,*
> *The last stray swallows seaward fly;*
>   *And I—I too!—no more am found*
>     *On London stones!*
>
> *In vain!—the woods, the fields deny*
> *That clearer strain I fain would try;*
>   *Mine is an urban Muse, and bound*
>   *By some strange law to paven ground,*
> *Abroad she pouts;—she is not shy*
>     *On London stones!*

(Austin Dobson 'On London Stones', 1876)

**2** Pl. *rondeaux*, with the final letter pronounced /z/. See -x.

**rondel.** 'A medieval French verse form related to the triolet and the rondeau. In its usual modern form, it is a 13-line poem using only two rhymes in its three stanzas. It employs a two-line refrain which opens the poem and recurs at lines 7 and 8, the first line (or, in a 14-line variant, both opening lines) also completing the poem. The rhyme scheme—with the repeated lines given in capitals—is thus *ABba abAB abbaA(B)*. There is no fixed metre. This form was adopted by some poets in England in the late 19th century, including Austin Dobson and W.E. Henley' (Baldick, 1990).

An example:

*Love comes back to his vacant dwelling,—*
*The old, old Love that we knew of yore!*
*We see him stand by the open door,*
*With his great eyes sad, and his bosom*
*swelling.*

*He makes as though in our arms repelling*
*He fain would lie as he lay before;—*
*Love comes back to his vacant dwelling,—*
*The old, old Love that we knew of yore!*

*Ah, who shall help us from over-spelling*
*That sweet forgotten, forbidden lore!*
*E'en as we doubt in our heart once more,*
*With a rush of tears to our eyelids welling,*
*Love comes back to his vacant dwelling.*

(Austin Dobson 'The Wanderer', 1880)

**rondo** (music). Pl. *rondos*. See -O(E)s 6.

**Röntgen.** See ROENTGEN.

**roof.** The standard pl. form in BrE is *roofs*, but the minority form *rooves* seems to be gaining ground. In my own collection the ratio of *roofs* to *rooves* is 10:7. It is a classic example of a disputed plural, something that is brought out by a correspondent to *The Times* in September 1986: 'Almost daily now I am troubled by the sound of "rooves". Is there any hope of a cure?' Cf. HOOF.

**room.** Both pronunciations /ruːm/ and /rʊm/ are standard, the first being the dominant one of the two.

**roomful.** Pl. *roomfuls*. See -FUL.

**roost.** For *rule the roost*, see RULE.

**root** (noun). (philology) One of those ultimate elements of a language that cannot be further analysed, and form the base of its vocabulary (*OED*). For example, the word *unhistorically* can be shorn of its various affixes, *un-*, *-ic*, *-al*, and *-ly*, each of which modifies the root *-histor-*. Now *-histor-* answers to Gk ἵστωρ 'knowing, learned, wise man' (cf. Gk ἱστορία 'history'). The irreducible root is *histor*, which can still be accounted for by referring it to the hypothetical earlier Gk form *ϝίδτωρ, from *ϝιδ-, ἰδ- 'to know', which is the assumed base or root of the English words *idea*, *vision*, *wit*, etc. The etymological network is valid, though the task of constructing such networks is a relatively modern achievement, beginning in an acceptable form in the late 18c. and early 19c.

**root, rout** (verbs). Several separate verbs with these spellings are involved. The *OED* identifies two verbs spelt *root* and pronounced /ruːt/, and no fewer than ten verbs (most of them northern or Scottish and four of them obsolete) spelt *rout* and pronounced /raʊt/. For example, *root* (v.)[1] can mean '(of plants) to take or strike root', and *root* (v.)[2] can mean 'to poke about, to rummage'. The ten verbs spelt *rout* and pronounced /raʊt/ include those meaning 'to snore', '(of cattle) to bellow', and 'to cause (an army) to retreat in disorder, to defeat decisively'. There is coincidence of meaning, namely 'to poke about, to rummage', in *root* (v.)[2] and *rout* (v.)[8]. What all this amounts to is that it largely depends on where you live whether you say *root about* or *rout about* for the sense 'to poke about'. From the web of meanings and derivations for the other senses one can say with reasonable certainty that 'to root for (i.e. encourage by applause) a team' is largely restricted to American slang; that the sense '(of pigs) to turn up the ground with the snout while searching for food' is *root* in standard southern English and *rout* in many dialects of English; and that almost everywhere one can speak of *routing* a person out of bed (or a hiding place) without fear of being misunderstood.

**Roquefort** is pronounced /ˈrɒkfɔː/, i.e. is only partially Anglicized.

**roquet** (in the game of croquet) is pronounced /ˈrəʊkeɪ/. The final *t* remains silent in the verbal forms *roqueted* /ˈrəʊkeɪd/ and *roqueting* /ˈrəʊkeɪɪŋ/. In practice the sound in the last syllable of *roquet* is often /ɪ-/, hence /-ɪd/, /-ɪŋ/—rather than /eɪ/.

**rosary, rosery** (rose-garden). The first is the old word (from the 15c. in the *OED*), based on L *rosārium*; and *rosery* is a mid-19c. formation (after *nursery*, etc.). But in ordinary use *rose-garden*, *-bed*, *border* are now the more usual terms. (The most prominent sense of *rosary* is of course the Roman Catholic one.)

**rosin** is by origin merely an altered form (in the 14c.) of *resin* (also 14c., from OF *résine*, from L *rēsīna*) and can be used interchangeably with it as the name of the sticky secretion of certain trees, or for any synthetic material resembling a natural one. Most particularly, *rosin* is 'the solid amber residue obtained after the distillation of crude turpentine oleoresin (also *gum rosin*), or of naphtha extract from pine stumps (also *wood rosin*), used in adhesives, varnishes, inks, etc., and for treating the bows of stringed instruments' (*New SOED*).

**roster.** An 18c. loanword from Du. *rooster* 'grating, gridiron, table, list' (from the appearance of paper ruled with parallel lines). It was for long mainly a military word for 'a list or plan exhibiting an order of rotation', and in such circles was normally pronounced like *roaster*, i.e. as /ˈrəʊstə/. This pronunciation survives, but in ordinary use /ˈrɒstə/ is standard.

**rostrum** (platform for public speaking). Pl. (for preference) *rostra*; also *rostrums*. See -UM 3.

**rotary, rotatory.** Both are genuine adjectival forms, *rotary* (first recorded 1731 in Bailey) being slightly earlier than *rotatory* (1755 in Johnson). Each has a wide range of idiomatic uses, but only *Rotary* goes with *Club* (the first of which clubs was formed in Chicago in 1905). In technical (and some general) contexts *rotary* is by far the more common of the two: e.g. *rotary clothes-line*, *drill*, *engine*, *machine*, *rig*, *-wing* (*aircraft*). *Rotatory* is now more commonly stressed on the second syllable than on the first.

**rotten.** 1 The corresponding noun is *rottenness* (thus spelt).

2 For *something rotten in the state of Denmark*, see IRRELEVANT ALLUSION.

**rouble** (monetary unit of Russia). Thus spelt in BrE, but *ruble* in AmE.

**roué** (a debauchee) /ˈruːeɪ/. With accent, not italic.

**rough, roughen** (verbs). See -EN VERBS FROM ADJECTIVES. The two verbs are seldom interchangeable. The earlier of the two is *rough* (15c.; *roughen* 16c.), but *rough* has retained only a few idiomatic uses (e.g. *to rough* (someone) *up*, *to deal roughly with* (someone); *to rough it*, *to submit to hardships or inconvenience*; *to rough out*, *to plan or sketch out roughly*). For most other transitive and intransitive uses *roughen* is preferred

(*hands roughened by work*; *the scenery roughened as they pushed inland*).

**rouleau** (a roll of money). Pl. *rouleaux* (not italic), with the final letter pronounced /z/. See -x.

**round.** See AROUND.

**roundel.** 'An English version of the rondeau, devised by A. C. Swinburne for his collection *A Century of Roundels* (1883). It is a poem of eleven lines using only two rhymes in its three stanzas of 4, 3, and 4 lines. Lines 4 and 11 are formed by the repetition of the poem's opening word or phrase as a refrain, which may be rhymed with lines 2, 5, 7, and 9. The rhyme scheme (with the refrain represented as R) is thus *abaR bab abaR*, or, with a rhyming refrain, *abaB bab abaB*. The term was at one time a synonym for a rondeau or rondel' (Baldick, 1990).

An example (Swinburne's 'The Roundel'):

> A roundel is wrought as a ring or a starbright
>     sphere,
> With craft of delight and with cunning of
>     sound unsought,
> That the heart of the hearer may smile if to
>     pleasure his ear
>       A roundel is wrought.
>
> Its jewel of music is carven of all or of aught—
> Love, laughter, or mourning—remembrance of
>     rapture or fear—
> That fancy may fashion to hang in the ear of
>     thought.
>
> As a bird's quick song runs round, and the
>     hearts in us hear
> Pause answer to pause, and again the same
>     strain caught,
> So moves the device whence, round as a pearl
>     or tear,
>       A roundel is wrought.

**roundelay.** Not a precise term like RONDEAU, RONDEL, and ROUNDEL. It has several meanings, the main ones of which are defined by the *OED* as follows (with the dates of first recorded use): 1 A short simple song with a refrain (1573).

2 The music of a song of this type (1593–1600).

3 A kind of round dance (1589).

**rouse.** See AROUSE.

**rout** (to poke about). See ROOT, ROUT.

**route.** Normally pronounced /ruːt/, but sometimes in military use, esp. in *route march*, pronounced /raʊt/.

**rout, route** (verbs). The *-ing* forms of these two verbs are respectively *routing* and *routeing*. The pa.t. and pa.pple of each is *routed* (*the left-wing parties were routed in the election*; *the motor rally was routed through Coventry*).

**rowan** (mountain ash). Normally pronounced /ˈrəʊən/, but /ˈraʊən/ is also common, esp. in Scotland.

**rowel** (verb). The inflected forms are *rowelled, rowelling* in BrE, but usu. *roweled, roweling* in AmE. See -LL-, -L-.

**rowing boat** is the customary form in BrE, and *row-boat* (or *rowboat*) in AmE.

**rowlock.** Pronounced /ˈrɒlək/, occas. /ˈrʌlək/. The NAmer. equivalent word is *oarlock*.

**royal we.** See WE 4.

**-r-, -rr-.** Monosyllabic words containing a single vowel (*a, e, i, o, u*) before *r* normally double the consonant before suffixes beginning with a vowel or before a final *y* (*tarring, stirring, currish, furry*); but remain undoubled if the stem contains two successive vowels (*nearing, chairing, boorish*). Words of more than one syllable follow the rule for monosyllables if their last syllable bears the main stress (*preferred* but *proffered, interring* but *entering, abhorrent* but *motoring*. Exception: the group of verbs ending in *-fer* is not subject to strict rules in the corresponding adjectival forms ending in *-able*. See -ABLE, -IBLE 2. Those that transfer the accent in the adjective to the first syllable are regularly spelt with a single *r* (*preferable, (in)sufferable*). But when the adjective bears the stress on the second syllable, as it often does in *conferable, inferable, referable,* and *transferable,* application of the rule given above would seem to require *-rr-*. See INFER 1; REFERABLE.

**rubbish.** In BrE *rubbish* is the generic term for waste material placed in *dustbins*, plastic bags, wheelie bins, or other containers, and collected at regular intervals by *dustmen* for conveyance to a designated *tip*. At some tips, or other collecting points, *skips* (large metal containers) are placed in which householders may deposit separated loads of garden refuse, paper, cardboard, DIY waste, used clothing, etc., and a *bottle*

*bank* for recyclable bottles. In AmE, domestic rubbish is called *garbage* or, in certain categories, *trash*, and a dustbin is called a *garbage can* or *trash can*. In American newspaper reports, however, the terminology is somewhat fluid: several terms are sometimes used, seemingly synonymously, in the same report, e.g. (the emphasis is mine): *Coyotes ... at night scoot into town and raid the* garbage cans—*New Yorker*, 1988; *He said the amount represents the saving the city would realize by separating recyclable* trash *from* waste *trucked to area* dump *sites*—*Chicago Sun-Times*, 1988; (headline) *A Lot of* Rubbish. *With* dumps *filling up fast, America is finally embracing the new* garbage *ethic*—*US News & World Report*, 1990; *The inclusion of so many new materials in Chicago-area recycling programs means that less* trash *is being sent to garbage dumps. Usually, a separate truck comes for* yard waste *and a garbage truck makes a run to pick up non-recyclable* trash—*Chicago Tribune*, 1991.

In May 1988 a retired American lexicographer wrote to me to say that he liked to think that he did not use the various terms as synonyms. To him, 'garbage = orange and grapefruit rinds, egg shells, leftover food; what one puts in a garbage pail/can, usually covered, in the kitchen or outside the kitchen door. It has a bad odor. trash = what I throw into a wastebasket : letters, envelopes, newspapers, junk mail, broken toys; as in trash can. rubbish = stuff thrown out of the attic and garage; mown grass, tree limbs. Empty tin cans go out with the garbage, but I could also consider them trash. I'm sure there is much overlapping in the use of these words, even if I don't consider them synonyms.'

A casual collection of examples from other English-speaking countries suggests that *garbage* is becoming the main generic term outside BrE: *We all dashed to the bins with our* garbage—M. Pople, 1986 (Aust.); *I was replacing a wheel on our* garbage can—*Personality* (Durban), 1988; *the area being too poor for there to be any* garbage *that could be thrown in to enrich it*—A. Desai, 1988 (Ind.). But the distribution of terms for *rubbish* of various kinds (of the literal sort, that is, apart altogether from figurative and transferred uses as well as verbs) would be a profitable subject for a Ph.D. thesis or a section in a linguistic atlas.

**ruble.** See ROUBLE.

**rucksack.** The pronunciation /ˈrʌksæk/ is recommended, rather than /ˈrʊk-/.

**ruddle.** The usual spelling of the word for a kind of red ochre used to mark sheep; also for the corresponding verb. *Raddle* and *reddle* are regional variants.

**ruff** /rʌf/, a wading bird, *Philomachus pugnax*, has *reeve* /riːv/ as its feminine form.

**rug, carpet.** These two words, which are often used as synonyms (see below), are by no means always interchangeable. When the place of origin is given it usually seems to be a matter of taste whether one or the other word is used: e.g. *Kilim carpet/rug*, *Oriental carpet/rug*, *Persian carpet/rug*. *Carpet* is the normal, probably exceptionless, word used in *stair carpet*, *wall-to-wall carpet*, and *magic carpet*. Figurative extensions include *on the carpet* (see CARPET 1), *to sweep a thing under the carpet* (or *rug*); *to pull the rug out from under someone*: flexibility is possible only in the second of these. Of combinations, *carpet-bagger*, *carpet bombing*, and *carpet slippers* are always so called: *rug* could not be substituted. Similarly with numerous less well-known combinations in *rug*.

An American correspondent tells me that a shop selling rugs and carpets informed him that the primary difference was one of size: 'Anything 9′ × 12′ or larger is a carpet, anything smaller than that is a rug.' It seems unlikely that this is a general rule. One kind of rug is not a floor covering, namely a *travelling rug*, i.e. a kind of blanket used for warmth on a journey or as a ground cover at a roadside picnic. And such a rug is not

made of 'shaggy material or thick pile' as indoor rugs and carpets are.

**ruination,** the act of bringing to ruin, the state of being ruined, is a derivative (first recorded in 1664) of the once-common (16–19C.) but now obsolete verb *ruinate* (which answers to med L *ruinātus*, pa.pple of *ruināre*, i.e. is not, like *botheration*, *flirtation*, and *flotation*, a HYBRID FORMATION).

**rule.** 1 For *the exception proves the rule*, see EXCEPTION.

2 *to rule the roost*, to have full sway or authority. First recorded *c*1400 in the unexplained form *to rule the roast*, which lasted until the 19C. before being replaced at some point by *to rule the roost*. The later expression was first recorded in 1769 but does not seem to have become dominant until a later date, probably *c*1900.

**Rumania.** See ROMANIA.

**rumbustious.** See ROBUSTIOUS.

**rumour.** So spelt in BrE, but as *rumor* in AmE.

**rung** (pa.pple). See RING.

**runner-up.** Pl. *runners-up*.

**ruridecanal.** Pronounce /ˌrʊərɪdɪˈkeɪnəl/.

**russety.** So spelt, not *russetty*. See -T-, -TT-.

**Russian.** Since 1991 definitions of the word have had to be changed. The *New SOED* (1993) defines the primary senses of the noun as '1 A native or inhabitant of Russia or (more widely) its former empire, the former USSR, or the republics associated with Russia in the Commonwealth of Independent States.

2 The Slavonic language of Russia, the official language of the former USSR.'

# Ss

**-s.** **1** For the use of *-s* to form adverbs from nouns (e.g. *he works nights, we nip off to her place afternoons*), see ADVERB 4.

**2** The normal plural ending of nouns, namely *-s* (*boys, cars,* etc.), should never be preceded by an apostrophe (the type *video's for rent* is wrong, though such uses are often found in shops). See APOSTROPHE² B3.

**3** For the plurals of abbreviations and numerals (*MAs, the 1990s*), see APOSTROPHE² C1.

**'s.** **1** For *for conscience' sake*, etc., see SAKE.

**2** For *Achilles', Burns's*, etc., see APOSTROPHE² D2, 3, 4.

**3** For the types *the sentence's structure* and *the extent of the frame-up*, see 's AND OF-POSSESSIVE.

**4** For the type *a friend of my mother's*, see DOUBLE POSSESSIVE.

**Sabbatarian,** a strict sabbath-keeping Christian (for whom the sabbath is a Sunday) or a Jew (for whom the sabbath is a Saturday).

**sabbath,** (in full *sabbath day*) a day of rest and religious observance kept by Christians on Sunday, Jews on Saturday, and Muslims on Friday. In Britain the number of strict sabbath-keepers among Christians has sharply declined in recent years.

**sabbatical.** As adj. it is still sometimes used of observances, etc., appropriate to the sabbath. But its main use is of leave granted at agreed intervals to a university teacher for study or travel, originally every seventh year. Also as noun, a period of sabbatical leave.

**sabre.** So spelt in BrE, but usu. as *saber* in AmE.

**sac** is in English a biological and medical word, meaning a bag-like cavity, enclosed by a membrane, in an animal or plant; or, the distended membrane surrounding a hernia, cyst, tumour, etc. (*COD*). It answers to Fr. *sac* (in which language it is not thus restricted in meaning). See SACK, SAQUE.

**saccharin, saccharine.** The first is a noun meaning a sugar-substitute, and the second is an adj. meaning sugary or, in figurative use, unpleasantly overpolite.

**sacerdotage.** See FACETIOUS FORMATIONS.

**sachem,** chiefly in AmE, is pronounced /ˈseɪtʃəm/.

**sachet,** orig. a French diminutive of *sac*, is pronounced /ˈsæʃeɪ/.

**sack** (noun) dismissal and *sack* (verb) to dismiss have been used in English since the first half of the 19c. but are still only in colloquial use. The phr. *to give* (a person) *the sack* may be a calque on an earlier (but now obsolete) French phr. *donner son sac* (e.g. *on lui a donné son sac*). The modern French equivalent is (for the noun) *renvoi*, and for the phr. *renvoyer quelqu'un* or *mettre* (or *flanquer*) *quelqu'un à la porte*. Cf. FIRE (verb).

**sack, saque.** For the fashionable 17–19c. gown (or an appendage of silk attached to the shoulders of such a dress) various spellings were used: *sac, sack, sacke,* and *saque*. These will all be encountered in original-spelling editions of works of the period. Of them, *saque* is pseudo-French; and, while *sac* is a genuine French word with numerous senses (bag, sack, satchel, etc.), it is not used as the name of a garment. In the 1950s *sack* (*dress*) was, for a short time, revived as a term for 'a cut of dress, being short, unwaisted, and usu. narrowing at the hem' (*OED*).

**sacrarium** (sanctuary of a church). Pl. *sacraria*. See -UM 2.

**sacrifice.** For *the supreme sacrifice*, see STOCK PATHOS.

**sacrilegious,** the adj. formed from *sacrilege*. So spelt, and pronounced /ˌsækrɪˈlɪdʒəs/. It is sometimes misspelt by confusion with *religious*, but the words have different origins. *Sacrilegious* is ultimately from L *sacrilegus* 'one who steals sacred things', from *sacer, sacri-* 'sacred' + *legere* 'take possession of'; while *religious* is from L *religiōsus*, from *religiō* 'religious fear, religious feeling, a religious practice', etc.

**saga.** A word that has gone on acquiring new senses. In English it was first applied to 'any of the narrative compositions in prose that were written in Iceland or Norway during the Middle Ages' (*OED*, with its earliest example taken from a work by George Hickes, 1709). From the mid-19c. it began to be applied to any narrative having the real or supposed characteristics of the Icelandic sagas, esp. a novel or series of novels recounting the history of a family through several generations (Galsworthy's *Forsyte Saga* is a well-known example). From the Icelandic sense also came a spate of applications of the word to mean merely 'a legendary story, an orally transmitted story': in 1903, for example, an Anglo-Saxon scholar, L. F. Anderson, wrote anachronistically of the sagas known to the *Beowulf*-poet. This use was doubtless partly after Ger. *Sage*. In the course of the 20c. *saga* has also come to be applied (some would say 'loosely') to any long and complicated account of a series of (ordinary) events. I recall that my Old Norse tutor in the early 1950s was saddened by all the extended senses of the word: he felt that it properly belonged only to the Icelandic sagas. But it is clear that, for the moment at any rate, the boundaries between the various senses remain stable and unthreatened.

**sage.** Often applied playfully to a wise person. *Harrap's Book of Nicknames* (1990) lists ten writers upon whom the nickname *sage* has been bestowed, including *the sage of Baltimore* (H. L. Mencken), *the sage of Chelsea* (Thomas Carlyle, who moved from Scotland to Cheyne Walk, Chelsea in 1834), and *the sage of Monticello* (Thomas Jefferson). See SOBRIQUETS.

**sago.** Pl. (not often used, since the word is usually a non-count noun) *sagos*. See -O(E)S 3.

**Sahara desert.** One of the pleasant ironies of language is that exotic foreign place-names are not thought of as having a self-evident meaning, but are simply taken to be the names of particular places. In Arabic the word *Sahara* (actually, in transliterated form, *çahrā*) means 'desert'. It has been argued therefore that we should speak only of *the Sahara* and never of *the Sahara desert*. Arabists are certainly entitled to follow such a restriction, but the less well informed of us will doubtless continue to speak of *the Sahara desert* from time to time.

**sahib.** Pronounce /ˈsɑːhɪb/ or /sɑːb/. (In colonial India) a term of respect used by Indians when addressing an adult European male. A *pukka sahib* was a true gentleman; and *memsahib* was the equivalent term of respect for a married European woman.

**said. 1** *The said* ——. In legal documents phrases such as *the said witness* and *the said meeting place* are traditional and are beyond reproach. In non-legal contexts (e.g. *regaling themselves on half-pints of lager at the said village hostelry*) such uses verge on being semi- or fully jocular.

**2** *said he* (or *I*, etc.). This inversion of the natural order (*he said*) is often resorted to as a stylistic variation and is unobjectionable, e.g. *'Oh,' said a man to me, when the news had penetrated our circle of acquaintance, 'I hear they're actually giving you money for it.'*—H. Mantel, 1993. This is immediately followed by a passage in which the said formula precedes the direct statement: *Said another, 'Do you know—have you any idea—how many books are published in the course of a year?'* This construction is also acceptable: in the context one said formula balances the other. What is debatable, however, is the journalistic convention of using the second type as an eye-catching device, e.g. *Said a Minister: 'American interests are not large enough in Morocco to induce us to ...'* See INVERSION 2.

**3** Substitutes for *said*. These are innumerable, depending on the nature of the context. In Iris Murdoch's *The Green Knight* (1993), for example, direct speech is usually presented without any kind of *he said* formula, e.g. *'Once upon a time there were three little girls—'*; *'Oh look what he's doing now!'*; *'And their names were—'* (etc.). Here the dashes and the quotation marks make it clear that more than one speaker is involved. Much less common are the types *Clement said* and, esp. after direct speech, the type *said Lucas*. Substitutes for *said* are infrequent, but the following words are among those that occur: *called*. *continued, cried out, exclaimed, intervened, murmured, repeated, replied*, and *went on*. There are numerous examples of *said* + an adverbial phrase, e.g. *said in a low voice, said in a soft confidential tone*. The pattern is a little different in William Boyd's *The Blue Afternoon* (1993) in that the type used after the passage of direct speech always seems to be that shown in *Carriscant said* (rather than *said Carriscant*). His substitutes for *said* include *asked, began, declared, muttered*, and *shouted*. Boyd also uses numerous phrases of the type *said* + an adverbial phrase, e.g. *he said petulantly; he said almost lightheartedly*. Doubtless other authors, other patterns.

**sailor.** In the sense a seaman or mariner, always so spelt. But the normal agent-noun *sailer* exists for use in such contexts as *She* [sc. a ship] *is a slow sailer*. Cf. also the Australasian *trailer sailer*, a small sailing vessel, usu. one between 5 and 8 metres in length.

**Saint. St** (without point) or **S.** are now the customary abbreviations. Pl. *Sts* or *SS* (no points). In the alphabetical arrangement always place under *Saint*, not under *St*.

**sake.** *For appearances' sake, for Christ's sake, for God's sake, for Heaven's sake, for Pete's sake, for old times' sake* illustrate the obligatory use of the possessive apostrophe in such phrases. Practice varies widely in *for conscience' sake* and *for goodness' sake*, and the use of an apostrophe in them must be regarded as optional. In AmE, *sakes* is sometimes used in place of *sake*: e.g. *'Shush, for God's sakes!'* warned my mother—L. S. Schwartz, 1989; *I suppose that's hearsay, but, for heaven's sakes, it's hearsay from the guy who did the shooting*—New Yorker, 1993.

**salable.** See SALEABLE.

**salad days** (one's raw youth) is one of the phrases whose existence depends on a single literary passage (*My Sallad dayes, When I was greene in iudgement, cold in blood*—Shakespeare, 1606. It does not necessarily follow that present-day users

of the phr. are aware of its source. See LITERARY ALLUSIONS.

**saleable** seems to be the natural spelling, though *salable* is used by some printers and publishers.

**saline.** Pronounced /'seɪlaɪn/ not /'sæl-/, despite the fact that the corresponding Latin words (*salīnae* 'salt-pans', *salīnum* 'salt-cellar', etc.) all have a short *a* in the first syllable. See FALSE QUANTITY.

**salivary** (adj.). Now pronounced either as /sə'laɪvərɪ/ or /'sælɪvərɪ/. The corresponding Latin word was *salīvārius*, but see FALSE QUANTITY.

**Salonica** /sə'lɒnɪkə/ is the English name for *Thessaloniki*, the second largest city in Greece. Its Latin name was *Thessalonica*.

**saloon.** As the first element of compounds (*saloon car, saloon deck, saloon pistol, saloon rifle*) it is not normally joined to the second element by a hyphen; but *saloon-keeper*. A hyphen is normally required when *saloon* is the second element of compounds (*billiard-saloon*).

**salubrious, salutary, salutation, salute.** For pronunciation, see LU.

**salve** (noun and verb). The word for a healing ointment and also the corresponding verb are most commonly pronounced /sælv/ but may also be pronounced /saːv/. See next.

**salve** (verb). The transitive verb meaning to save (a ship or its cargo) from loss at sea or to save (property) from fire is a back-formation from *salvage*, and must always be pronounced /sælv/.

**salve.** The Latin greeting meaning 'hail!' may be pronounced /'sælveɪ/ or /'sælviː/.

**salvo**[1] (simultaneous firing of guns). Pl. *salvoes* or *salvos*. See -O(E)S 1.

**salvo**[2] (reservation, excuse). Pl. *salvos*. See -O(E)S 6.

**sal volatile.** Pronounce /ˌsæl və'lætɪlɪ/.

**Sambo** (applied offensively to a black person). Pl. *Sambos*. See -O(E)S 8.

**same.** *An historical note.* Its use as a substitute form preceded by *the* or *that* has been noted by the *OED* in literary sources from the 14c. onward: e.g. *Watermen haunt the waters, and fishes swim in the same*—P. Stubbes, 1583; *In the instant that I met with you, He had of me a Chaine, at fiue a clocke I shall receiue the money for the same*—Shakespeare, 1590; *But he that shall endure vnto the end, the same shall be saued*—Matt. (AV) 24: 13, 1611; *Her lute-string gave an echo of his name. She spoiled her half-done broidery with the same*—Keats, 1819. At some point, probably during the 19c., the use dropped out of serious literary work and into the realms of legal usage and commercialese, in the latter often with the omitted. Fowler (1926) cited numerous journalistic examples, presumably of his own day, in which he judged *same* or *the same* to be misused, e.g. *Having in mind the approaching General Election, it appears to me that the result of same is likely to be as much a farce as the last; I can only confirm the statement of the transfer, but the same will be made slowly.*

*Present status.* (The) *same* still abounds in commercialese: e.g. *We thank you for your order of 9 December for —— and we shall supply same as soon as fresh stock has come in.* Garner (1987), while admitting that (the) *same* was very frequent in legal writing (e.g. *The informer told the officer that a white male would usually load the buyer's car with marijuana at a residence and then deliver same to buyer*), nevertheless urged lawyers to use the appropriate pronoun (*it, him, them*, etc.) instead. *WDEU* (1989) cited several examples of (the) *same* from 20c. letters and periodicals and judged it to be 'often simply a mark of an informal style'. Opinion, it is clear, is divided. In such circumstances the best advice I can offer to readers is to fall back on ordinary pronouns to do the backward glancing, and leave (the) *same* to legal writers, to works of business, and to contributors of informal articles to such periodicals as *Esquire* and the *Saturday Review* (US). In spoken English, on the other hand, *the same* is very commonly used as a substitute form: e.g. *'I'll have the lemon sole and boiled potatoes, please.' 'I'll have the same.'* And phrases such as *all the same, just the same, the same to you!, the very same*, even though in some of them *the same* is a substitute form, are too well entrenched to invite attack.

**same as.** When used as an attributive adj. *same* is usu. construed with *as*. There

are several types of construction, among them: (*a*) where *as* introduces a clause, e.g. *Entering college at the same age as Fletcher had entered six years earlier*; (*b*) followed by a noun or noun phr., e.g. *Other rules ... point in the same direction as the first rule of the order*; (*c*) followed by an adverb, e.g. *I again consulted your magnificence, and you gave the same answer as before.* (Examples from the *OED*.) Less commonly *the same* is construed with *that* or without a relative pronoun, e.g. *It's the same textbook (that) I used when I was an undergraduate.*

**same like** used as an adverbial phr. meaning 'just like' or 'in the same way as' is first recorded in 1898 and has remained non-standard or jocular since then. Examples: '*See no evil, hear no evil, think no evil. Same like the monkeys,' observed Sergeant Percy Bond*—A. Christie, 1959; *I have rich friends, same like you*—I. Murdoch, 1980.

**samite** (medieval silk fabric). Pronounce /'sæmaɪt/ or /'sem-/.

**samurai.** Pronounce /'sæmʊraɪ/ or /'sæmjʊ-/. Pl. the same.

**sanatorium.** This is a modL word, first recorded in 1839, and formed from L *sānāre* 'to cure, heal'. It is the customary word in BrE for an establishment for the treatment of invalids, esp. of convalescents or the chronically ill; also (in BrE) a room or building for sick children at a boarding school, etc. (abbrev. *the san*). This form of the word is also current in AmE, but alternates there with *sanitarium* (first recorded in 1851), a quasi-Latin form derived from L *sānitās* 'health'. The occasional variant *sanatarium* is erroneous. The plural of both *sanatorium* and *sanitarium* is either *-iums* or *-ia*. See -UM 3.

**sanat-. sanit-.** The main words as they should be spelt are SANATORIUM; *sanative*, used occas. beside *curative* and in the same sense; *sanitary*, conducive to public health; *sanitation*, measures for the preservation of public health; *sanitarian* (adj.), of or relating to sanitation; as noun, a sanitation expert; and *sanitarium* (see SANATORIUM).

**sanction** (noun). The dominant late 20c. sense is (in the pl.) economic action taken by a state or alliance of states against another or others, usu. to enforce a violated law or treaty. It goes back to 16c. and 17c. technical uses in law and ethics concerned with specific penalties enacted in order to enforce obedience to the law or to rules of conduct. The main secondary sense, 'approval or encouragement given to an action, etc.', surfaced in the 18c. and still sits happily beside the other, despite the fact that the two senses are virtually antonymous. The emergence of the various meanings of *sanction(s)* is chronicled at considerable length in the *OED* to which the reader should turn for further clarification.

**sanction** (verb). By contrast with the noun, *sanction* (verb) from its first use in the late 18c. until the mid-20c. had only one primary set of senses, namely 'to authorize, countenance, or agree to (an action, etc.)'. Since the 1950s it has been joined by a secondary sense, 'to impose sanctions upon (a person, etc.); to attach a penalty or reward before making valid'. Examples are sparse, and this new use, though logical enough as a parallel to the main current sense of the noun, has only debatable currency and acceptability in the standard language. Examples: (heading) *Let Church sanction road killers—Universe*, 1956; *Georgina Dufoix, the only politician so far sanctioned for allowing the Palestinian guerrilla chief, George Habash, into France last week, said yesterday that she was resigning as president of the French Red Cross—Independent*, 1992.

**sandal** (verb). The pa.pple and ppl adj. are *sandalled* (but usu. *sandaled* in AmE.)

**sand-blind.** First recorded in the 15c., it is probably a corrupt form of unrecorded OE *samblind* half-blind (OE *sam*, corresponding to L *semi-*, meant 'half'). When the prefix lost its force, the word was perceived as being *sand-blind*, as famously shown in Shakespeare's *Merchant of Venice* (1596): *This is my true begotten Father, who being more then sand-blinde, high grauel blinde, knows me not.* Note also Dr Johnson's definition (1755): 'Having a defect in the eye, by which small particles appear to fly before them.' The word survives locally and in verse: e.g. *Hope ... Led sand-blind Despair To a clear babbling wellspring And laved his eyes there—W. de la Mare*, 1938.

**'s and of-possessive.** A friend of mine, an inspector of quarries, drew a questioning ring round the *'s* in a sentence he had found in an article about safety in quarries: 'It had been known for some time that the shovel's brakes were faulty.' Shouldn't it be 'the brakes of the shovel'? he asked. He had encountered, I explained, the problem of the type *the water's edge*: that is, whether to opt for an *of*-construction rather than a possessive apostrophe when the noun in question is an inanimate one.

The relationship of the two constructions is complex. In its elementary form it is sometimes stated that the *'s* construction should be employed only with the names of living things (e.g. *John's book, his mother's apartment, the neighbours' faces, her friend's apology, the dog's dinner*), and beyond that only with a few not easily classifiable and usually monosyllabic words (e.g. *the sun's rays, a day's work*). For inanimate nouns, and particularly for such nouns consisting of more than one syllable, the *of*-construction is customary (e.g. *the roof of the church*, not *the church's roof: the resolution of the problem*, not *the problem's resolution*). The rival constructions are the outcome of a morphological schism some 900 years ago. In Anglo-Saxon a wide variety of nouns referring to animates (human and animal), to materials, and to abstract qualities all had the power to 'possess', and this capacity was usually expressed by means of the genitive case (usually, but not always, *-es*). In the centuries that followed, some nouns, especially those denoting living things, continued to express the genitive by adding *'s* while most others came to be re-expressed as prepositional genitives. Fowler (1926) blamed newspaper headlines for what he called *'s incongruous*, namely for the increasing and (to him) unwelcome use of the shorter of the two genitival constructions. 'It begins to seem likely that *drink's victims*', he wrote, 'will before long be the natural and no longer the affected or rhetorical version of *the victims of drink*.' As a headline, *China's integrity*, he could see, was two words shorter than *the integrity of China*. But as a result of this shortening habit we are 'chastened', he wrote, by such constructions as *Ontario's Prime Minister* and *Uganda's possibilities*. For once, Fowler seemed to be off target. I turned to other sources for further elucidation.

There is general agreement that the non-personal genitive is frequently used with nouns of time (e.g. *the day's routine, an hour's drive*) and space (e.g. *the journey's end, a stone's throw, at arm's length*). It is also often used before *sake* (e.g. *for pity's sake, for old times' sake*), and in a number of fixed expressions (e.g. *at death's door, out of harm's way, in his mind's eye*). Jespersen noted the prevalence of *'s* genitives before the word *edge* (*the cliff's edge, the water's edge, the pavement's edge*, etc.). He also noted that *ship, boat*, and *vessel* tend to turn up with an *'s* genitive when we might expect *of* (*the ship's provisions, the boat's gangway*, etc.). In 1988 Noel Osselton demonstrated that the somewhat unexpected types *the soil's productivity* and *the painting's disappearance* (as well as others) represent a legitimate class of what he called thematic genitives. When a noun that cannot 'possess' is of central interest in a particular context, it tends to acquire the power to 'possess', and is therefore expressed as an *'s* genitive. One major genitival area remains virtually untransformable into *'s* genitives. Only the *of*-construction is appropriate for partitive genitives: e.g. *a glass of water* cannot be re-expressed as *a water's glass*, and try converting *a dose of salts*.

I tested these rules against my files and found them largely in accord with my own evidence. The great majority of *'s* genitives still occur with animate nouns (e.g. *my father's son, her friend's forearm, the boys' bicycles*). Non-personal nouns are usually followed by *of*-constructions (e.g. *the bottom of her glass, the turn of the screw, the edge of the table*). When non-personal nouns are used with an *'s* genitive, the categories set out above are the most prevalent: e.g. *I always seem to be sitting up in bed at the day's end*—Julian Barnes, 1980; *at the sea's edge its appearance was oily*—J. Fuller, 1983; *He jogged down-hill to the water's edge*—M. Wesley, 1983; *Ten days' holiday for you now, Betty*—M. Eldridge, 1984; *all scattered along the sandy mud by a high tide which had bored up the river's mouth*—C. Burns, 1989. It does seem from the evidence available to me that the *'s* genitive for inanimate nouns is commoner now than it was a century ago, though it and the *of*-construction

are not free alternatives. The reason for the shift in this direction lies deeply buried in a long-drawn-out historical process. Newspaper headlines, *pace* Fowler, have had little or nothing to do with it.

**sanguine.** In medieval and later physiology, descriptive of one of the four 'complexions', and supposed to 'be characterized by the predominance of the blood over the other three humours, and indicated by a ruddy countenance and a courageous, hopeful, and amorous disposition' (*OED*). In its later, more general senses 'disposed to hopefulness, optimistic', *sanguine* seems now to be losing ground, possibly because its derivatives *sanguinary, sanguineous, sanguineness*, etc., are less euphonious than the derivatives of *hopeful* and *optimistic*.

**Sanhedrin** /ˈsænɪdrɪn/ (highest court of justice in ancient Jerusalem) is the correct form (from late Hebrew *sanhedrīn*, from Gk συνέδριον 'council', from σύν 'together' + ἕδρα 'seat'), though a pseudo-Hebrew form in *-im* has been a rival spelling in English from the 16c. onward.

**sans.** 1 In phrases of French origin (e.g. *sans-pareil* 'not having its like'; *sans peur* 'fearless', *sans souci* 'without care or concern') *sans* must always be pronounced in a French manner, /sɑ̃/.

2 When used in ordinary English contexts, often with direct allusion to Shakespeare's famous context in *As You Like It* (1600) (*Second childishnesse, and meere obliuion, Sans teeth, sans eyes, sans taste, sans euery thing*), the Anglicized pronunciation /sænz/ is customary. Example: *The result was a high-quality recording, sans commercials*—A. Hailey, 1979.

**Sanskrit.** Thus spelt, not *Sanscrit*.

**Santa Claus.** First recorded in the *New York Gazette* in 1773 and for long mainly restricted to the US, the term is derived from a Dutch dialectal form *Sante Klaas* (cf. Du. *Sint Klaas*), Saint Nicholas. It is now virtually synonymous in English with *Father Christmas*.

**sapid.** Unlike its negative *insipid*, *sapid* ('palatable; not vapid or uninteresting') lives out its precarious life as a somewhat esoteric word. Examples: *The aromatic dish, a kind of thick red stew, was just as sapid* *as it smelled*—P. Roscoe, 1988 (Canad.); *The most prized wines were those that were ... infused with a sapid delicacy*—M. Kramer, 1989 (US).

**sapient** /'seɪpɪənt/, wise; of fancied sagacity. Mostly found in the higher realms of literature, i.e. it is not an everyday word. Examples: *Nor bring, to see me cease to live, Some doctor full of phrase and fame, To shake his sapient head and give The ill he cannot cure a name*—M. Arnold, 1867; *Polyphiloprogenitive The sapient sutlers of the Lord Drift across window-panes In the beginning was the Word*—T. S. Eliot, 1919.

**saponaceous,** lit. 'of, like, or containing soap; soapy'. It has sometimes been used jocularly to mean 'unctuous, flattering, ingratiating'. Examples: *He was undone by vulgar oranges, his saponaceous blues, his queasy purples, just as some men are undone by women and some by wine*—W. Lewis, 1937; *This was flat, as Hooker said; 'saponaceous samuel thought it was a fine opportunity for chaffing a savan' and he pitched into Huxley*—A. Desmond and J. R. Moore, 1992.

**sapor** /'seɪpə:/ or /'seɪpə/. First recorded in the 15c., it has gradually lost ground and is now rare. Its main sense is 'a quality perceptible by taste, e.g. flavour, sweetness'.

**Sapphics.** Lyric verses written in the sapphic stanza of four lines of the form —⌣——— | ⌣⌣–⌣–⌣ (3 times), —⌣⌣–⌣̱ (once), named after Sappho (7–6c. BC), the legendary woman poet of Lesbos. The stanzaic form has been imitated, sometimes not quite successfully, by a number of English poets including Sidney, Watts, Swinburne, and Ezra Pound. Examples:

> All the night sleep came not upon my eyelids,
> Shed not dew, nor shook nor unclosed a feather,
> Yet with lips shut close and with eyes of iron
> Stood and beheld me.
>
> (Swinburne)
>
> Breathing softly, wrapped in a shawl of daylight,
> Trusting blossom loveliness brought to being,
> Too small yet to lift up your head or turn round,
> I will stay with you.
>
> (Francis Warner)

**sarcoma** (tumour). Pl. *sarcomas* or *sarcomata* /sɑ:'kəʊmətə/.

**sarcophagus** (stone coffin). Pl. *sarcophagi* /-gaɪ/.

**sardine** (young pilchard) is pronounced /sɑː'diːn/. The unrelated word *sardine*, a precious stone mentioned in Rev. 4: 3, is pronounced /'sɑːdam/.

**sardonic.** A 17c. loanword from Fr. *sardonique*, it answers ultimately (with change of suffix) to Gk σαρδάνιος, used by Homer and others as the descriptive epithet of bitter or scornful laughter. (In late Greek, Σαρδόνιος 'Sardinian' was substituted for σαρδάνιος, the notion being that facial convulsions resembling horrible laughter, usually followed by death, resulted from eating a Sardinian plant. See the OED.)

**sari** seems to be emerging as the dominant form in English-language publications, though *saree* is still often found in English newspapers in India and elsewhere.

**sartorial** is derived from L *sartor* 'tailor', and means (a) of a tailor or tailoring; (b) of (esp.) men's clothes. It, and also the adverb *sartorially*, can be used facetiously (see PEDANTIC HUMOUR), but facetiousness is not a necessary ingredient of either word. Examples showing varying degrees of pedantic humour and of neutrality: *Sartorially speaking, men are at last catching up with the women*—*Daily Tel.*, 1970; *The club's most sartorially elegant member*—M. Underwood, 1974; *Palestinian sources disclosed that two suits had been packed, indicating that he* [sc. Yasser Arafat] *might yet use his sartorial style at the ceremony on the White House lawn to demonstrate his transformation to statesman*—*New Yorker*, 1993; (caption to a photograph showing a UK member of staff [of a publishing house] dressed in a medieval garment) *MD shows refreshing sartorial independence at the ELT Sales Conference*—*The Record* (OUP), 1993; *The true legacy of 1969–70 to 1993 is ... the courage, on the part of women, to write the narratives, sartorial and otherwise, of their daily lives*—*New Yorker*, 1993.

**Sassenach** /'sæsənæk/ or /-næx/. A derogatory term in Scotland and Ireland for an English person. Cf. Gaelic *Sasannach* and Ir. *Sasanach*, from L *Saxones* 'Saxons'. Cf. Gaelic *Sasunn*, Ir. *Sasana* 'England'.

**sat,** used as a present participle ( = standard English *sitting*), has been rediscovered by dialect scholars, who say that it is widely used in parts of the north and west of England. The following examples seem to confirm that its currency on the fringes of standard English is increasing: *I can't help thinking of that Tim sat there juddering his leg up and down*—K. Amis, 1988; *Now, I'm sat in a nice car, my husband at my side*—A. Duff, 1990 (NZ); *now, as a result of a conference débâcle, you are sat on the back benches with nobody wanting to sit next to you*—B. Elton, 1991. The use is exemplified in the *OED* s.v. *sit* v. 18b, with examples from OE down to 1864. It is firmly marked 'Now *dial*.' In other words it is an example of a use that was once standard but has gradually become regionally restricted over the centuries. Cf. STOOD.

**satanic.** Now the usual form rather than *satanical*, though *satanical* (first recorded in the mid-16c.) is the older of the two. *Satanic* is first recorded in Milton's *Paradise Lost* (1667).

**sati,** an occasional variant of *suttee*, the (historical) Hindu practice of a widow immolating herself on her husband's funeral pyre. The Sanskrit (Hindi, Urdu) original is *satī*.

**satiety.** Pronounce /sə'taɪtɪ/, rhyming with *variety*.

**satire.** The many-branched entry in the *OED* sets out the history of the word (first recorded in English in 1509) and is a primary guide to the ramifications of the noble art of satire in the ancient world of Aristophanes, Juvenal, Horace, and others and in the modern equivalent world of such writers as Pope, Swift, and Samuel Butler (*Hudibras*). The novel became a powerful vehicle for social satire in the 19c. (Dickens, Thackeray, et al.) and has continued in this role in the 20c. (Evelyn Waugh, Kingsley Amis, et al.). Beginning in the 1960s, satire took on a new dimension. Unprecedented levels of ignoble mockery became part of the world of popular entertainment. In Britain it began with the satirical revue *Beyond the Fringe* (1960) and a television programme called *That Was The Week That Was*. The programme ran its course, but has been succeeded by numerous others of a similar kind, though

not all of them are called satirical comedy shows. No institution, group, or individual lies beyond the reach of such entertainment in the form of ridicule. The nation survives. Cf. LAMPOON.

**satiric, satirical.** As Fowler (1926) remarked, 'the line of demarcation between the two [forms] is not always clear.' In BrE in most circumstances the longer form seems to be the more usual of the two; and in AmE it is just possible that the shorter form is the more likely of the two to turn up. Both words are of long standing in the language: *satiric* was first recorded (as an adj.) in 1509 (as a noun in 1387), and *satirical* some twenty years later.

**satiric, satyric.** The two spellings represent two different and unconnected words: *satyric*, which is in learned or literary use only, means of satyrs, especially, in *satyric drama* (a form of Greek play), having a satyr chorus. See prec.; FAUN, SATYR.

**satrap** (provincial governor, etc.). Pronounce /'sætræp/.

**Saturday.** For the adverbial use (*shall see you Saturday*), see FRIDAY.

**Saturnalia.** In Roman antiquity, *Saturnālia*, neuter pl. of the adj. *Sāturnālis* 'pertaining to Saturn', was always construed as a plural. In English, esp. in the transferred sense 'a period of unrestrained licence and revelry', it has often been perceived as a singular: e.g. *This was the beginning of a perfect saturnalia of tail-cutting and other operations [among the lambs]*—R. Haggard, 1899. Thus reinterpreted, a permissible pl. is *saturnalias*, but it looks strange to the trained eye. Modern examples: *The sexual revelries began, a continuous saturnalia*—A. Greeley, 1986 (US); *Mind and body are preparing for war. The body, during the waking hours, with its regimes, its saturnalias of self*—M. Amis, 1991.

**satyr.** See FAUN for distinctions.

**sauté** (lightly fried), *adj.* Accent to be retained. As verb, has pres.pple *sautéing* and pa.pple *sautéd*.

**savannah** (grazing plain in subtropical regions). Thus spelt (but frequently *savanna* in AmE). Also (with capital *S*) the seaport in E. Georgia and river of that name.

**save.** When used as a preposition (*no ornaments in the room save a crucifix*) and a conjunction (*I cannot remember anything about his appearance save that he had a morning coat*; *A small liqueur glass … empty save for a tiny drop*) it has an air of archaism or formality about it. (Examples from the *New SOED*.) It can often be replaced by *except* or *but*.

**saviour, savour, savoury.** Thus spelt in BrE, but as *savior, savor*, and *savory* in AmE. See -OUR AND -OR.

**saw** (verb) has pa.pple *sawn*, less commonly *sawed*. The pa.t. is invariably *sawed*. As an adj., of a shot-gun, *sawn-off* is the only form used.

**Saxonism.** 'A semi-technical term for: (1) The use of, and preference for, expressions of Anglo-Saxon origin. (2) A word or other expression of Anglo-Saxon origin or formed on an Anglo-Saxon or Germanic model, often contrasted with *classicism*, as in *foreword* with *preface*, *folkwain* with *omnibus*. Saxonisms are generally the outcome of a purist and nativist approach to the language. The aim behind many deliberately created forms has been to create compounds and derivatives to replace foreign borrowings; the device is rooted in the Old English practice of loan-translating words: *benevolentia* as *welwilledness* or *wellwillingness*; *trinitas* as *thriness* threeness … Saxonism has resurfaced only occasionally. In the 16c., it was a reaction to inkhorn terms; in his translation of the Bible, John Cheke used *hundreder* and *gainrising* instead of *centurion* and *resurrection* … The most enthusiastic 19c. Saxonizer was William Barnes [1801–86], English dialectologist and poet, who wished to turn English back into a properly Germanic language … His work is now largely forgotten and where remembered is usually seen as quaint and unrealistic' (*OCELang.*, 1992).

Most of Barnes' Saxonisms were not registered in the *OED*: such artificial words were mostly omitted on principle. Among these were *bodeword* commandment, *earthlore* geology, *gleecraft* music, *tastecraft* aesthetics, and *wondertoken* miracle. But *birdlore* was included, and dialectal words that he used such as *dew-bit* a

small meal before the regular breakfast, and *gil[t]-cup* buttercup. As Fowler (1926) remarked, 'The wisdom of this nationalism in language—at least in so thoroughly composite a language as English—is very questionable.'

**say.** 1 Except as a poeticism, the noun survives only in such phrases as *to have a say* (to have the right to be consulted) and *to have said one's say* (to have finished expressing one's opinion).

2 The use of the verb's imperative *to* introduce an hypothesis or an approximation (*Let us meet soon—say next Monday*; *You will need some cash—say £20*) is an established idiom.

3 The ordinary pronunciation of *says* (3rd pers. pres. indic.) is /sez/. It is therefore odd that from the mid-19c. *sez* should have been repeatedly used in representations of uneducated speech, esp. in the phr. *sez you* ( = so you say).

**sc.** See SCILICET.

**scabies.** Now always pronounced as two syllables, /ˈskeɪbiːz/, but the OED (1910) and Fowler (1926) gave it as three, /ˈskeɪbiːz/. See -IES; PRONUNCIATION 2 (ii).

**scalawag.** Probably the dominant AmE spelling of SCALLYWAG.

**scald** (medieval Icelandic poet). See SKALD.

**scallawag.** One of the AmE variant spellings of SCALLYWAG.

**scallop.** This is the recommended spelling (pronounced /ˈskɒləp/), not *scollop*. The verb makes *scalloped*, *scalloping*; see -P-, -PP-.

**scallywag.** The word emerged in American politics in the 1840s and is of unknown etymology. Its customary spelling in BrE is *scallywag*, but the word is spelt in various ways in AmE including *scalawag* and *scallawag*.

**scaly** (having many scales or flakes). Thus spelt, not -ey.

**scampi** (large prawns) is a plural noun. When used in the sense 'a dish of these' it is sometimes treated as a singular. The word is derived from It. *scampo* 'shrimp', pl. *scampi*.

**scandalum magnatum.** The second word is the genitive plural of L *magnas* 'a magnate', not a ppl adj. agreeing with *scandalum*. The phrase (now disused) means the utterance or publication of a malicious report against a dignitary. The plural is *scandala magnatum*.

**scant.** First taken into ME From ON in the mid-14c., *scant* flourished as noun, adj., adv., and verb for several centuries before it became restricted to literary, or, in some of its senses, to dialectal contexts. Such limitations happened first to the noun, and in time to the adverb and the verb. In the 20c. the retreat from ordinary currency has continued, and now even the adj. is limited to a few familiar collocations, e.g. *scant attention, scant courtesy, scant regard*, and (echoing Hamlet's mother Gertrude) *scant of breath*.

**-scape.** Based on *landscape* (first recorded in 1598) and *seascape* (1799), a number of formations in *-scape* have made their way into the language in the 19c. (e.g. *cityscape, cloud-scape*) and the 20c. (e.g. *lunarscape, moonscape*, and *roof-scape*). A number of nonce-formations of this type, e.g. *mindscape, moodscape, prison-scape*, and *winterscape*, are cited in the OED s.v. scape n.[3]

**scapula** (the shoulder-blade). Pl. *scapulae* /-iː/ or *scapulas*.

**scarce** (adv.) was often used instead of the fuller form *scarcely* in ordinary adverbial contexts until about the end of the 19c. (e.g. *I ran as I never ran before, scarce minding the direction of my flight, so long as it led me to the murderers*—R. L. Stevenson, 1886), but is hardly ever encountered now except in poetry. It has frequently been called on by 20c. poets to qualify a ppl adj. used attributively (e.g. *In the scarce-glimmering boles*—E. Blunden, 1922; *I match that child with this scarce-changed old man*—W. de la Mare, 1951).

**scarcely.**

| 1 *scarcely ... than*.
| 2 *scarcely* with negatives.

**1** *scarcely ... than*. The OED s.v. *than* cited an example of 1864 (see HARDLY 2), and Fowler (1926) cited another example (*Scarcely was the nice new drain finished* than several of the children sickened with diphtheria*), but the construction, which is modelled on the normal type *no sooner ... than*, is rare and does not form part of the standard language. *Scarcely ... when* or *Scarcely ... before* are preferable. Cf. BARELY. In common acceptable use, however, is the type *scarcely* + comparative adj. or adv. + *than*. Examples: *The deciduous conifers include the larches ... and, if you will allow it, the ginkgo, although the last is scarcely more closely related to the conifers than it is to the tree ferns*—Plants & Gardens, 1990; *The cystic stage is recognised by very small cavities, scarcely larger than a proto-scolex*—Lancet, 1991; *[The] bill encourages 'bare bone' policies—providing scarcely more coverage than Medicaid*—Village Voice (US), 1992.

**2** *scarcely* with negatives. *Scarcely* is not a negative as such as will be seen from the pair of sentences *he scarcely mentioned the subject/he didn't mention the subject*. But it is near enough to being one that a statement such as *I don't scarcely know* would be rejected out of hand by standard speakers as a 'double negative'. Caution is desirable whenever one is tempted to include a negative in the same sentence as *scarcely*, but the 'rule', if it is such, is not an absolute one. Idiomatic examples: *There is scarcely an aspect of the race that is not rife with meaning*—New Yorker, 1989; *Having been to other countries where those who run the game automatically rattle off gate money, betting 'handle' and TV figures as second nature, it seems scarcely possible that such statistics are not in constant use here*—Independent on Sunday, 1991; *There has scarcely been a time since the Russian Revolution when the American right was not fretting about the number of 'tenured radicals' ... installed at American universities*—R. Hughes, 1993 (US).

**scarf.** In its ordinary sense of a long strip of material worn round the neck, the plural is either *scarves* (the form recommended here) or *scarfs*. The pl. of the unrelated word *scarf*, meaning a joint or notch in timber, metal, etc., is *scarfs*.

**scarify.** There are two separate verbs. The first, pronounced /ˈskærɪfaɪ/, has been extant since the 15c. in various technical and figurative uses in the basic sense 'to scratch, to make incisions in'. The second, pronounced /'skeərɪfaɪ/, was irregularly formed in the late 18c. from *scare* (verb) + *-ify*, perhaps after *terrify*, and means 'to scare, frighten'. This second verb is classified as *colloq.* in the COD (1995) and *colloq.* (orig. *dial.*) in SOED (1993).

**scavenge** (verb), **scavenger** (verb). The noun *scavenger* is a 16c. altered form of earlier *scavager*, with an intrusive *n* as in *messenger*, *passenger*. It yielded the verb *scavenger* and (as a back-formation) *scavenge* in the 17c. Parallels to the first of these (i.e. verbal uses of nouns) are commonplace, e.g. *to bicycle, to mountaineer, to soldier.* At about the end of the 18c., *scavenger* (verb) more or less dropped out of use, leaving *scavenge* to hold the field.

*scena* (music). Pronounce /'ʃeɪnɑː/.

**scenario.** 1 Usually pronounced /sɪ'nɑːrɪəʊ/ in BrE and /-'neərɪəʊ/ in AmE, but there is much variation in both countries. Its plural is *scenarios*: see -O(E)S 4.

2 The word came into the language from Italian in its normal sense of 'an outline of the plot of a play, ballet, novel, etc.' in the 1870s, and thence in the sense 'a film script' in the first quarter of the 20c. when the world of the cinema became established. It could hardly have been foreseen that it would become immensely popular from the 1960s onwards in the broad sense 'a postulated sequence of (future) events'. Every kind of circumstance, situation, relationship, train of events, etc., came to be called a *scenario*, and there is no sign of any weakening of the grip that the word has on the language. Some typical examples: *With regard to the second scenario (redistributing libraries' costs to other entities, such as computer centers and research departments), there is little that one outside the academic community can say—Logos, 1990; Taking this scenario one step further, Hamlet himself becomes Edward de Vere, the seventeenth Earl of Oxford—Atlantic, 1991; Neither politicians nor military commanders would have been forgiven if the Iraqi war machine had been underestimated, so instead it was taken at face value and worst-case scenarios were projected—Britain's Gulf War, 1991; Road tolls are slowly edging their way on to the political agenda, and already two scenarios can be painted—Independent, 1992; At first, I* thought the stentorian snores were part of the script, then that one of the actors had accidentally fallen asleep (a highly enjoyable scenario)—Times, 1994. A wise writer uses the word sparingly.

**scene.** At first, in the 1950s, restricted to the language of beatniks and their informal followers in the senses 'an area of action or interest (*not my scene*)' and 'a way of life' (*well-known on the jazz scene*; *a bad scene*), the word gradually made its way, at first mockingly and later just colloquially, into the speech of the general public. These applications of *scene* remain at the informal rim of the standard language.

**scenic.** Pronounced /'siːnɪk/ in BrE and also usu. in AmE. But the variant /'senɪk/ also exists in AmE, and is dominant in some other English-speaking countries.

**sceptic.** Pronounced /'skeptɪk/. So spelt in BrE, but as *skeptic* in AmE. The word is of course to be distinguished from *septic* /'septɪk/, contaminated with bacteria.

**sceptre.** So spelt in BrE, but as *scepter* in AmE.

**schedule.** The traditional standard pronunciation /'ʃedjuːl/ is still dominant in BrE, but the AmE pronunciation with initial /'sk-/ is sometimes heard in BrE, esp. among young people.

**schema** (a synopsis, etc.). Pronounce /'skiːmə/. Plural *schemata* /-mətə/.

**scherzando, scherzo.** These musical terms, adopted from Italian, are pronounced respectively /skeə'tsændəʊ/ and /'skeətsəʊ/. Their plurals are *scherzandos* (less commonly *scherzandi*) and *scherzos*.

**schipperke** (Dutch breed of dog). Most commonly pronounced /'skɪpəki/ in English.

**schism.** The OED (1910) had only /'sɪz(ə)m/ but since then, except among the clergy, this pronunciation has been joined, if not quite replaced, in general use by the form with initial /'sk-/. Similarly in the derivatives *schismatic* and *schismatical*.

**schismatic, schismatical.** As adjs. the first is the more usual of the two. The noun ('a holder of schismatic opinions') is always *schismatic*.

**schist** (geology). Pronounce /ʃɪst/.

**schizoid, schizophrenia** are both pronounced in English with initial /ˈskɪts-/.

**schnapps** /ʃnæps/. Thus spelt.

**scholar** should be restricted to mean a learned person or the holder of a scholarship (at a school or university), and should not be used as a substitute for pupil, schoolchild.

**scholium** (marginal note in a manuscript) /ˈskəʊlɪəm/. Pl. *scholia* /-lɪə/. See -UM 2.

**school** (of fish, etc.), **shoal**. The two words are etymologically one (from a MDu. word) and are equally unconnected with the ordinary word *school* (which is derived from L *schola*). Both *school* and *shoal* are current and are used without difference of sense.

**sciagraphy** (art of the perspective of shadows). Spelt thus (rather than *skiagraphy*) and pronounced /saɪˈægrəfɪ/. The first element in this and the following word answers to Gk σκιά 'shadow'.

**sciamachy** (fighting with shadows). Spelt thus (rather than *skiamachy* or *skiomachy*) and pronounced /saɪˈæməkɪ/.

**scilicet** /ˈsaɪlɪset/, usually shortened to sc., is Latin (derived from *scīre licet* 'one may understand or know') for 'to wit, that is to say (introducing a word to be supplied or an explanation of an ambiguity)'. Examples: (explaining a term already used) *The policy of the NUT* (sc. *National Union of Teachers*); *The Holy Ghost as Paraclete* (sc. *advocate*); (introducing a word, etc., that was omitted in the original as unnecessary, but is thought to require specifying for the present audience) *Eye hath not seen nor ear heard* (sc. *the intent of God*). See also VIZ.

**scimitar** /ˈsɪmɪtə/. Thus spelt. Cf. Fr. *cimeterre*, It. *scimitarra*. The word is of unknown origin.

**scintilla.** Mostly used in the singular (*not a scintilla of doubt*), but if a plural is called for use *scintillas*.

**scion** /ˈsaɪən/. Thus spelt in BrE and AmE in the sense 'descendant of a (noble) family'. When used to mean 'a shoot of a plant, etc.,' often spelt *cion* in AmE.

**scirocco.** See SIROCCO.

**scission.** Pronounce /ˈsɪʃ(ə)n/, not /ˈsɪʒ-/.

**scissors** /ˈsɪzəz/ is construed as a plural noun (*the scissors are in the study drawer*), except in specialized senses in high-jumping (*the ordinary scissors is the least effective of the four styles*), wrestling, and rugby football. The primary sense can also be construed as (the sing. phr.) *a pair of scissors*.

**scleroma** /sklɪəˈrəʊmə/ (patch of hardened skin). Pl. *scleromata* /-mətə/.

**sclerosis** /sklɪəˈrəʊsɪs/ (hardening of body tissue). Pl. (if needed) *scleroses* /-siːz/.

**scone.** Now mostly /skɒn/ in BrE, but /skəʊn/ is also widely current. In AmE, *scone* most commonly rhymes with *tone*, but there is much variation. *Scone*, a village in central Scotland, the site of the coronation of Scottish kings, is pronounced /skuːn/.

**scope.** For synonyms, see FIELD.

**score** (noun) (− 20). The phrase *three score (years) and ten*, as a traditional way of describing one's allotted life-span, is still common. When followed by *of*, a *score* is normally construed as a plural: *a score of customers were waiting at the door.* Formerly *score* preceded by *a* was often treated as a numeral adj. (e.g. *I form'd a score different plans*—Sterne, 1768).

**scoria** is a singular noun, pl. *-iae*; but, as the meaning of the singular and of the plural is much the same (cf. *ash* and *ashes*, *clinker* and *clinkers*), it is no wonder that the singular is sometimes wrongly followed by a plural verb (*The scoria were still hot* etc.), or that a false singular *scorium* is on record.

**scotch** (verb). The existence of two etymologically unrelated verbs *scotch* complicates any discussion of two senses of one of them. One of these senses, 'to wound without killing, to inflict such hurt upon (something regarded as dangerous) that it is rendered harmless for the time being', is based upon Theobald's emendation of *Macbeth* III.ii.13, *We have scotch'd* [First Folio *scorch'd*] *the snake, not kill'd it.* In the 20c. this sense stands side by side with the sense 'to quash, destroy, bring to nothing' (e.g. *to scotch once and* *for all any lingering doubts or rumours that the pound is to be devalued by stealth—Times, 1955).* Both senses are valid. The second of them was probably influenced by *scotch* (verb)[2], which answers to a noun (first recorded in the 17c.) meaning 'a block placed under a wheel, etc., to prevent moving or slipping'; hence (for the verb) 'to render inoperative'. The evidence is complex and the word 'perhaps' appears more than once in the relevant entries in the *OED*.

**Scotch, Scots, Scottish.** It is not possible to set down here all the complications of this somewhat sensitive group of words. The adjective *Scotch*, in origin a contracted variant of *Scottish*, 'had been adopted into the northern vernacular before the end of the 18th c.; it [was] used regularly by Burns, and subsequently by Scott' (*OED*). But 'since the mid-19th c. there has been in Scotland a growing tendency to discard the form altogether, *Scottish*, or less frequently *Scots*, being substituted' (*OED*). *Scots* is also a long-standing variant of *Scottish*. The outcome is that all three adjectives are still current, but *Scotch* is the least frequent and survives mainly in certain collocations, e.g. *Scotch broth, Scotch egg, Scotch mist, Scotch terrier, Scotch tweed, Scotch whisky*, and a few others. *Scots* is the term regularly used of the form of English spoken in (esp. Lowlands) Scotland. It also occurs in the names of certain Scottish regiments. But the all-embracing general adjective meaning 'of or relating to Scotland, its history, its day-to-day life, or its inhabitants', is *Scottish*. These are middle-class preferences. 'Paradoxically,' A. J. Aitken reports in *OCELang*. (1992), 'for working class Scots the common form has long been *Scotch* ... and the native form *Scots* is sometimes regarded as an Anglicized affectation.' Outside Scotland, and esp. outside the UK, *Scottish* preferences are less well-known. *Scotch* is likely to occur, both as adj. and noun, in contexts which middle-class Scots would regard as either droll or improper.

**Scot, Scotsman, Scotchman,** etc. As nouns, *Scot* and *Scotsman/Scotswoman* are preferred by middle-class English-speakers in Scotland, but outside Scotland, and esp. outside the UK, *Scotchman* and *Scotchwoman* are still widely used. See prec.

**scottice, Scotticism, Scotticize.** The forms with -tt-, rather than -t-, are recommended. See ANGLICE.

**scoundrel** makes *scoundrelism*; *scoundrelly* (adj.). See -LL-, -L-.

**scout, gyp, skip.** College servants at Oxford, Cambridge, and Trinity College Dublin, respectively.

**scrannel** (thin, meagre). First recorded in Milton's *Lycidas* (1637), *Their lean and fleshy songs Grate on their scrannel Pipes of wretched straw*. Now used chiefly as a reminiscence of Milton's context, usually with the sense 'harsh, unmelodious', e.g. *His scrannel music-making*—W. H. Auden, 1951.

**scrimmage, scrummage.** The form with -u- is obligatory (or more usually just *scrum*) in rugby football, that with -i- in more general use ('a brawl, a confused struggle') and as a technical term for a particular sequence of play in American football.

**scriptorium.** Pl. *scriptoria* (for preference) or *scriptoriums*. See -UM 3.

**scrum(mage).** See SCRIMMAGE.

**scull, skull.** The single-handed oar has initial sc-, the cranium sk-. Both words entered the language in the ME period, but they are not etymologically related.

**sculpt** (verb). Adopted in the mid-19c. from Fr. *sculpter*, it is now apprehended as a back-formation from *sculptor*. The term is in regular use beside *sculpture* (verb) and the rather rare colloquial form *sculp*.

**sculptress.** The regular feminine equivalent of *sculptor*, in use since the 17c. (Evelyn), but now tending to be put aside in favour of the gender-free word *sculptor*. See -ESS 4.

**scutum.** (zoology) Pl. *scuta*. See -UM 2.

**sea change.** See IRRELEVANT ALLUSION.

**seamstress, sempstress.** The first of these is now the standard form. However spelt, the word is best pronounced /'semstris/.

**séance, seance.** The word is spelt as often with an accent as without. It is also pronounced with varying degrees of Anglicization, as /'seɪɑːns/, /-ɒs/, or /-ɒns/.

**sear, sere.** Several words are involved: *sear* for the verb meaning 'burn' or 'char', and the corresponding noun (mark produced by searing); *sere* for the noun meaning 'a catch of a gunlock'; for the ecological noun meaning 'a series of plant communities'; and for the adjective meaning 'dried up, withered'. In past centuries the adj. was often spelt *sear*: e.g. *I haue liu'd long enough, my way of life Is falne into the Seare, the yellow Leafe*—Shakespeare, 1605; *The rude materialities of life in this sear generation*—Cardinal Wiseman, 1837.

**seasonable, seasonal.** The first means suitable for the time of year, in keeping with the season, opportune, by contrast with *unseasonable*, occurring at the wrong time or season, e.g. *Hot weather is seasonable for August*; *he loved the seasonable mince pies that his aunt cooked at Christmas*; *You are apt to be pressed to drink a glass of vinegary port at an unseasonable hour* (S. Maugham); *the week-long unseasonable rain*. *Seasonal*, by contrast, means occurring at or associated with a particular season, e.g. *sheep-shearing is seasonal work*; *the seasonal migration of geese*.

**seated.** See SIT (verb).

**second.**

1 *second chamber.*
2 *second floor.*
3 *second-hand*, etc.
4 *second of all.*
5 *second* (verb).

1 The *second chamber* of some parliamentary democracies is the upper house in a bicameral parliament.

2 *second floor*. See FLOOR, STOREY.

3 *second-hand*. There is much variation in the use of the hyphen. *COD* (1995) opts for *second hand* for the hand recording seconds in a watch or clock; *second-hand* for the adj. meaning not new (*second-hand clothing*) and the adv. meaning at second hand, not directly (*she always buys second-hand*); and *second hand* in such phrases as *heard only at second hand*. But the adj. and adv. are very commonly written as one word (*secondhand*).

4 *second of all*: an Americanism based on the general English phrase *first of all*, e.g. *First of all he's not my Roosevelt, and second of all we don't have to discuss these things while we're eating*—L. S. Schwartz, 1989.

5 *second* (verb). When it means to transfer temporarily to other employment or another post, the stress falls on the second syllable: /sɪ'kɒnd/. But when it means formally support or endorse (a motion, nomination, etc.), the stress falls on the first syllable: /'sekənd/.

**secretary.** Pronounce only as /'sekrət(ə)rɪ/, not /'sekətriː/ as if it were spelt *seketry*, and not /'sekəteərɪ/ as if it were spelt *seketerry*. But in AmE, /'sekrəteərɪ/ is usual.

**secretive.** Now normally pronounced /'siːkrɪtɪv/, i.e. stressed on the first syllable, though the *OED* (1911) gave only the form with the stress on the second syllable, /sɪ'kriːtɪv/.

**sect.** Adopted in ME from OF *secte* or L *secta*, from *sect-* pa.ppl stem of *sequi* 'to follow', its dominant meaning is 'a body of people subscribing to views divergent from those of others within the same religion; a party or faction in a religious body' (*SOED*). Hence, over the centuries, often applied by Anglicans to various non-conformist groups (Quakers, Methodists, etc.), and by Roman Catholics to Protestant groups, and, outside religion, to the followers of a particular philosopher or school of thought.

**sectarian.** Now almost inevitably followed by the word *violence*, or other noun suggestive of killing or destruction, in Northern Ireland, the former Yugoslavia, or elsewhere in the world.

**sedile** /sɪ'daɪlɪ/, each of a series of usu. canopied and decorated stone seats, usu. three in number, placed on or recessed into the south side of the choir near the altar for use by the clergy. Pl. *sedilia* /sɪ'dɪlɪə/.

**see, bishopric, diocese.** A *bishopric* is the office belonging to a bishop; a *diocese* is the district administered by a bishop; a *see* is (the chair that symbolizes) a bishop's authority over a particular diocese. A *bishopric* is conferred on, a *diocese* is committed to, a *see* is filled by, such and such a person: *my predecessor in the* *see; all the clergy of the diocese; hoping for a bishopric.*

**seeing** is frequently used as a quasi-conjunction, usu. followed by *that* + a clause, to mean 'considering that, inasmuch as, because' (*seeing that you do not know it yourself*). Colloquial equivalents are *seeing as, seeing as how*: these are normally restricted to informal contexts, e.g. *Seeing as how you're always short of £sd, I thought you could maybe earn a bit*—S. Gulliver, 1974.

**seek** (verb). See PASSIVE TERRITORY 3.

**seem** (verb). **1** The pleonasm shown in *These conclusions, it seems to me, appear to be reached naturally* should be avoided.

**2** *can't seem* + infinitive. See CANNOT 3.

**3** See PERFECT INFINITIVE 1.

**Seidlitz.** Pronounce /ˈsedlɪts/.

**seigneur** etc. Spellings entered in the *COD* (1995): *seigneur, seignior; seigneurial, seignorial; seigniorage, seignorage; seigniory, seigneury*. In each pair the first is normally the preferred form. The pronunciation in all begins with /sem-/ followed by /j/ (i.e. the *y* sound). Differences in meaning or use exist but are too complicated to be given here: see the *OED*.

**seise, seisin.** Pronounce /siːz/, /ˈsiːzɪn/. The words are sometimes (but less often) spelt *-ze, -zin*, and belong etymologically to the ordinary verb *seize*. But in the legal phrases *to seise a person of*, i.e. put him in possession of, and *to be seised of*, i.e. to possess, the *-s-* spelling is usual.

**seize** (verb). See prec.

**seldom.** **1** (adj.). Used since the 15c., seemingly without break, but now more or less restricted to literary contexts: e.g. *My seldom night terrors*—W. Golding, 1959; *With her small seldom smile*—Edmund Wilson, 1961 (in Webster's Third).

**2** (adv.). Used in emphatic phrases, some now rare or obsolete: e.g. *seldom ever* (10–19c., now rare); *seldom or ever* (18–19c., obsolete); *seldom (,) if ever(,)* (17c., still current); *seldom or never* (14c., still current). Examples: (seldom ever) *I seldom ever dreamed of Lolita as I remembered her*—V. Nabokov, 1955; (seldom or ever) *The Players seldom or ever throw out the Voice with any Vehemence*—A. Murphy, 1752; *It was* *what they seldom or ever do*—D. Johnson, 1827; (seldom(,) if ever(,)) *The pettish Israelites (a people seldom if ever, pleased with God's present Providencies) who murmured under Moses*—W. Sclater, 1653; *The poison Sumach occurs in the western, but very seldom, if ever, in the eastern part of the State* [sc. Maine]—W. D. Williamson, 1832 (US); *Surrey backwoodsmen seldom, if ever, call much upon the rock-climbing or bog-defying wherewithal of their 4-w-d off-roaders*—Daily Tel., 1991; *A strong, confident exposition of the last movement completes an interpretation which has seldom, if ever, been surpassed on record*—Gramophone, 1992; *Sciacchetrà, which is seldom if ever exported, can be held for some years due to its high sugar content*—Gourmet (US), 1992; (seldom or never) *Also in wynter selden or neuer pytte water fresyth*—Trevisa, 1398; *those that doe dye of it, doe seldome or neuer recouer*—Shakespeare, 1606; *Microbes, that freely reach or parasitize normal individuals with intact antimicrobial defenses but seldom or never produce disease, are not considered to be 'true pathogens'*—Q. N. Myrvick and R. S. Weiser, 1988 (US).

**selector.** Thus spelt, not *-er*.

**self.** Acceptable in commercial use as a substitute for *myself*, and in particular when written on a cheque or counterfoil: e.g. *He drew a cheque for a hundred pounds to self on Friday*—G. Heyer, 1935; *He turned back the counterfoils … The uppermost … was marked 'Self', a withdrawal of four hundred pounds*—C. Watson, 1967. Though attested since the mid-18c., its use in non-commercial contexts is now merely jocular or fairly informal. Examples: *Mr. H. and self agreed at parting to take a gentle ride*—Dr Johnson, 1758; *As both self and wife were fond of seeing life … we decided a trip to Baden Baden would be a nice change for us*—J. Astley, 1894; *Drunks black out, remember nothing; A.A.* [sc. Alcoholics Anonymous] *requires memory, the acknowledgement of actions' effects on self and others, then apology and atonement*—B. Holm, 1985 (US).

See MYSELF.

**self-.** Such compounds are of unlimited number. The principal types are (a) those in which *self-* is in the objective relation to the second element (*self-abandonment, self-betrayal*, etc.); (b) used with verbal nouns (*self-searching, self-understanding*, etc.); (c) with agent-nouns (*self-educator,* *self-seeder*, etc.); (*d*) with nouns of state or condition (*self-awareness, self-mastery*, etc.); (*e*) with adjs. (*self-analytical, self-protective*, etc.); and numerous others. Occasionally such compounds are unnecessarily formed where the second element would suffice by itself. Writers should pause before writing *self-conceit(ed)*, for example, in contexts where *conceited* by itself would suffice, and *self-confessed* where *confessed* is adequate. Similar considerations apply to *self-assurance, self-complacent, self-confidence*, and some other *self-* compounds. The rule is quite simple: make a lightning decision about whether the *self-* adds anything essential to the proposed second element of a compound. It usually does (e.g. *self-addressed, self-defence, self-service*), but occasionally does not.

**self-deprecating, -deprecation, -depreciation.** See DEPRECATE, DEPRECIATE 4.

**selvage** (an edging of cloth). Thus spelt, rather than *selvedge*.

**semantics** (noun) (construed, like *politics*, etc., as a singular). Its primary sense, the branch of linguistics concerned with meaning, was adopted in English in the 1890s from French *sémantique* (M. Bréal, 1883), and has now largely replaced the slightly earlier term *semasiology*. The subject is now a highly sophisticated one beyond the reach of anyone but specialized linguistic scholars. But its essence can be gleaned from some elementary examples where fundamental changes of meaning have occurred over the centuries. The original meaning is given in brackets after each example; full details are set out in the OED. Examples: *buxom* (obedient, compliant); *deer* (an animal, a beast); *effete* (no longer fertile, past producing offspring; cf. *foetus*); *elaborate* (to produce by effort or labour); *horrid* (rough, bristling); *meat* (food); *meticulous* (timid, fearful); *treacle* (any of various medicinal salves). Since the 1940s, *semantics* and the corresponding adj. *semantic* have passed into general use in much weakened senses, as shown in these examples (esp. that of 1978); *We do ourselves and our Asian neighbors a distinct disservice when we insist on stretching them or shrinking them to fit our particular semantic bed—New Leader* (US), 1959; *Almost daily in the press briefing, whenever a newsman raises his hand*

*to ask for clarification of some mealy-mouthed statement: 'I am not going to debate semantics with you,' the spokesman replies*—K. Hudson, 1978.

**semaphore.** Pronounce /ˈseməfɔː/ with a short *e* in the first syllable, despite the fact that the first element answers to Gk σῆμα 'sign, signal'. See FALSE QUANTITY 2.

**semi-.** Compounds are innumerable, esp. with adjs. and participles as the second element. The earliest of these compounds in English is *semicircular* (1432–50) and *semi-mat. e* (c1440). In so far as a pattern is discernible at all, the prefix is extensively employed in technical terminology (e.g. *semiconductor, semidiameter, semigroup*), but it is preferred to *demi-, hemi-,* and *half-* in many ordinary words as well (e.g. *semi-automatic, semicircle, semi-conscious, semi-detached, semifinal, semi-professional, semi-skilled*). It is worth noting that the *e* in *semi-* is short, notwithstanding that it is long in the Latin original (sēmi-). See FALSE QUANTITY 2.

**semicolon.** This name of the punctuation mark consisting of a dot placed upon a comma (;) is first recorded in 1644. As the *SOED* (1993) notes, its primary function is to indicate 'a discontinuity of grammatical construction greater than that indicated by a comma but less than that indicated by a full stop'. The best account of its function is that provided in Hart's Rules (slightly abridged here): 'The semicolon separates two or more clauses which are of more or less equal importance and are linked as a pair or series: *Economy is no disgrace; for it is better to live on a little than to outlive a great deal. The temperate man's pleasures are always durable, because they are regular; and all his life is calm and serene, because it is innocent. To err is human; to forgive, divine. Never speak concerning what you are ignorant of; speak little of what you know; and whether you speak or say not a word, do it with judgement.'* To which it should be added that the semicolon is a useful device for separating a list of items set out in consecutive (as distinct from columnar) form: *Those present at the conference included Professor R. H. Robins, School of Oriental and African Studies; Dr M. K. C. MacMahon, University of Glasgow; Dr Rod McConchie,*

University of Helsinki; and Dr Brigitte Nerlich, University of Nottingham.

**Semite.** See HEBREW.

**semi-vowel.** In English phonetics the term applied to the sounds /j/ (normally spelt *y*, as in *young*) and /w/, i.e. to sounds intermediate between a vowel and a consonant.

**sempstress.** See SEAMSTRESS.

**senarius** /sɪ'neərɪəs/. A (Greek or Latin) verse consisting of six usu. iambic feet, esp. an iambic trimeter. Also called an *iambic senarius*. Pl. *senarii* /-rɪaɪ/.

**senator.** Thus spelt, not -*er*.

**senatus** /se'nɑːtəs/, the governing body or senate of certain universities, esp. in Scotland. The pl., if required, is -*tuses* or -*tus* /-tuːs/, not -*ti*. See -US 2.

**senhor, senhora, senhorita.** Portuguese and Brazilian titles. Pronounced in English /sen'jɔː/, /sen'jɔːrə/, /senjə'riːtə/.

**senior.** For *the Senior Service*, see SOBRIQUETS.

**sennight,** an archaic word for a week (from OE *seofon nihta* 'seven nights').

**señor, señora, señorita.** Spanish titles. Pronounced in English /sen'jɔː/, /sen'jɔːrə/, /senjə'riːtə/.

**sensational** entered the language in the literal sense 'of or pertaining to or dependent upon sensation or the senses' in the mid-19c. and almost immediately acquired the hyperbolic senses 'aiming at violently exciting effects; calculated to produce a startling impression'. Since then the extended senses have almost driven out the literal one, esp. in newspaper reportage (*a sensational upset at Wembley*), in advertisements (*sensational bargains in the New Year sales*), and in contexts of informal praise (*she looked sensational in her new dress*).

**sense** (noun). The phr. *sense of humour* means 'the faculty of perceiving and enjoying what is ludicrous or amusing'. It happens not to be recorded (in the *OED*) before the 1880s but it seems likely that it was in use at an earlier date. It is often, perhaps most often, used in negative contexts of someone who seems not to

recognize the humorous side of things (*he lacks a sense of humour; he has no sense of humour*).

**sense** (verb). It is worth noting that the current meaning 'to be or become vaguely aware (of something), to detect subconsciously (*I sensed a hardness in his tone*)' is first recorded (according to the *OED*) in 1872. The verb entered the language much earlier (late 16c.) with the meaning 'to perceive (an outward object) by the senses; also, to feel (pain)', and some other specific meanings (e.g. 'to expound the meaning of'). But instinct or vague awareness are contextually much more likely to be present in modern uses of the word. There are also technical uses in philosophy ('to have a sense-perception of') and in technology of a machine, instrument, etc., able to detect something (e.g. *In general particle detectors operate by sensing the ionization of atoms caused by the passage of a charged particle—Sci. Amer.*, 1978).

**sensibility.** Just as *ingenuity* is not *ingenuousness*, but *ingeniousness* (see INGENIOUS), so *sensibility* is not *sensibleness*, but *sensitiveness*. To the familiar contrasted pair *sense* and *sensibility* correspond the adjectives *sensible* and *sensitive*—an illogical arrangement, and one doubtless puzzling to foreigners, but beyond 'correcting'. See -TY AND -NESS.

**sensible, sensitive, susceptible.** In certain uses, in which the point is the effect produced or producible on the person or thing qualified, the three words are near, though not identical, in meaning. *I am sensible of* [sc. not unaware of] *your kindness, sensitive to ridicule, susceptible to beauty*. Formerly *sensible* could be used in all three types of sentence; but its popular meaning as the opposite of *foolish* has become so predominant that we are no longer intelligible if we say *a sensible person* as the equivalent of *a sensitive* or *a susceptible person*, and even *sensible of* is counted among LITERARY WORDS though surviving as a common phrase for the opening of formal speeches: *I am deeply sensible of the honour you have done me* ... The difference between *sensible of*, *sensitive to*, and *susceptible to* is roughly that *sensible of* expresses emotional consciousness, *sensitive to* acute feeling, and *susceptible to* quick reaction to stimulus: *gratefully, painfully, profoundly, regretfully,* sensible of; *absurdly, acutely, delicately, excessively,* sensitive to; *often, readily, scarcely,* susceptible to. With *of* the meaning of *susceptible* is different: it is equivalent to admitting or capable. *A passage susceptible of more than one interpretation; an assertion not susceptible of proof.*

**sensitize** is a word first made for the needs of Victorian photographers and made irregularly. It should have been *sensitivize*. One might as well have omitted the adjective ending of *fertile, immortal, liberal,* and *signal,* and say *fertize, immortize, liberize,* and *signize,* as omit the *-iv(e)-* of *sensitive.* The photographers, however, have made their bed, and must lie in it. At this distance of time we must just accept the shorter form. The OED (section issued in 1912) included only *sensitize* (e.g. *Education, while it sensitises a man's fibre, is incapable of turning weakness into strength,* 1880), not *sensitivize* (nor *sensize*). Nor does any other modern dictionary include the more regular form. There is no possibility of turning the etymological clock back.

**sensorium** (area of brain). Pl. *-ia* (recommended) or *-iums.* See -UM 3.

**sensual, sensuous.** The second of these is thought to have been expressly formed by Milton (in 1641) to convey what had originally been conveyed by the older *sensual* (first recorded *c*1450) (connection with the senses as opposed to the intellect) but had become associated in that word with the notion of undue indulgence in the grosser pleasures of sense. At any rate Milton's own phrase *simple, sensuous, and passionate* in describing great poetry as compared with logic and rhetoric has had much to do with ensuring that *sensuous* remains largely free from the condemnation that is more or less inseparable from *sensual.* In other words *sensuous* is the more neutral word of the two, while *sensual* more often than not has more than a hint of gratification, voluptuousness, or sexuality. It should be borne in mind, however, that in some contexts, when a faculty is being distinguished that is not rational or logical or intellectual, but one simply pertaining

to the senses, then either *sensual* or *sensuous* may be safely used.

**sentence.** Fowler (1926) had no doubts about the nature of a sentence: 'sentence, in grammar, means a set of words complete in itself, having either expressed or understood in it a subject and a predicate, and conveying a statement or question or command or exclamation. If it contains one or more *clauses,* it is a *complex sentence*; if its subject consists of more than one parallel noun etc., or its predicate of more than one verb etc., it is a *compound sentence*; if its subject or predicate or verb (or more) is understood, it is an *elliptical sentence* . . . Simple sentences: *I went* (statement); *Where is he?* (question); *Hear thou from heaven* (command); *How they run!* (exclamation). Complex sentence: *Where he bowed there he fell down dead.* Compound sentences: *You and I would rather see that angel; They hum'd and ha'd.* Elliptical sentences: *Listen; Well played; What?* Two sentences (not one): *You commanded and I obeyed.*'

Modern grammarians, by contrast, are nervous about defining the traditional terms of grammar ('Neither of these terms [sc. *sentence* and *grammar*] can be given a clear-cut definition', *CGEL*, 1985), but usually end up by providing working definitions: 'Sentences are units made up of one or more clauses. Sentences containing just one clause are called *simple,* and sentences containing more than one clause are called *complex*' (Leech and Svartvik, 1975). '*Sentence.* The largest unit of language structure treated in traditional grammar; usually having a subject and predicate, and (when written) beginning with a capital letter and ending with a full stop' (*Oxford Dict. Eng. Grammar,* 1994).

It may be found helpful if I set down here some of the main types of written English sentences (the first four drawn from *Essays & Studies* 1991). Similar examples can be found in any other well-written source.

Simple. *Culloden is Scott's watershed.*
Compound. *Fiction and history are kindred forms.*
Complex. *If history is about dates then Cousin Bette is self-evidently an historical novel.*
Long complex. *In 1854 and 1855, Dickens was also showing signs of overwork*

(the strain, not the work itself, was unusual) and was often angry, even to helplessness, at the turn of public events: progress in education, sanitation, social reconciliation, scant enough, was further impeded by the Crimean war (an excuse to defer home legislation), while the conduct of the war itself seemed increasingly criminal, with an army before Sebastopol virtually abandoned by the government and destroyed less by the enemy than by disease, lack of supplies, and general incompetence (it was another humiliation, felt deeply and not by Dickens alone, that the French ordered these things better, as was bitterly obvious to observers in the Crimea).

Verbless or otherwise incomplete sentences (stylistically acceptable in context). She only spoke to me once about her private life and that's what I've told you. As I remember it—G. Greene, 1980; They ate what was in front of them. While it was hot—M. Drabble, 1987; Max played ping-pong with the children. Then records—Encounter, 1987; Dionysiac release? I suppose so. A rehearsal for the release of seeds. An invisible embrace. The motion of the mountainside—H. Brodkey, 1993; Still, it had been a good party. An unforgettable party, actually. And still was—A. Huth, 1992.

**sentence adjective.** See IMPORTANT.

**sentence adverb.** In an August 1989 issue of The Times, coastguard Peter Legg, senior watch officer at Dover, was reported as saying Frankly we don't want them. He was referring to Channel swimmers and the hazard they present to shipping in the world's busiest waterway. His use of frankly, meaning 'if I may speak frankly', draws attention to one of the most bitterly contested of all the linguistic battles fought out in the last decades of the 20c. This unofficial war against certain uses of adverbs ending in -ly broke out in the late 1960s. Its chief focus was the adverb hopefully. The adverb was regarded as acceptable when it meant 'in a hopeful manner', as in to set to work hopefully, i.e. its traditional use since the 17c.; but not acceptable when used to mean 'it is hoped [that], let us hope', as in We asked her when she expected to move into her new apartment, and she answered, 'Hopefully on Tuesday'. The unofficial war rumbles on.

Let me state a general proposition: in the 20c. there has been a swift and immoderate increase in the currency of -ly adverbs used to qualify a predication or assertion as a whole. The -ly adverbs concerned include actually, basically, frankly, hopefully, regretfully, strictly, and thankfully. Suddenly, round about the end of the 1960s, and with unprecedented venom, a dunce's cap was placed on the head of anyone who used just one of them—hopefully—as a sentence adverb.

The simplest type of -ly sentence adverb is one that begins a sentence and is marked off from what follows by a comma: Unhappily, there are times when violence is the only way in which justice can be secured (in which Unhappily = it is an unhappy fact that); Agreeably, he asked me my name and where I lived ( = in a manner that was agreeable to me); Frankly, I do not wish to stop them ( = in all frankness, to speak frankly); Well, that won't happen at Pringle's, hopefully as they say. Hopefully ( = it is to be hoped). These examples are drawn, respectively, from T. S. Eliot's Murder in the Cathedral (1935), a 1987 issue of the New Yorker, Brian Moore's The Colour of Blood (1987), and David Lodge's Nice Work (1988). A reasonable cross-section of English writing, it would seem. Such sentence adverbs do not necessarily stand at the beginning of sentences: The investigators, who must regretfully remain anonymous—TLS, 1977; Aldabra Island in the Indian Ocean, where man 'has thankfully failed to establish himself'—Times, 1983.

Clearly, the question of the legitimacy of hopefully as a sentence adverb branches out far beyond the domain of the actual word itself. The second edition of the OED has entries for hopefully, regretfully, sadly, thankfully, and perhaps one or two others, used as sentence adverbs. It draws attention to their unpopularity among 'some writers'. Most of the illustrative examples given in the dictionary to support the constructions are drawn from works written since the late 1960s. What the 20c. evidence in these entries fails to bring out is that the present-day widespread use of sentence adverbs is no more than an acceleration of a much older process. The OED entry for seriously (sense 1) has an example of 1644 drawn from the diary of Richard Symonds, who marched with the royal army during the Civil War: Except here and there an officer (and seriously I saw not above three or four that looked like a gentleman). It is clear that seriously does not directly qualify I or saw or *gentleman*, but the whole of the sentence that follows it. In 1872 Ruskin, in *The Eagle's Nest*, used the same adverb in the same manner: *Quite seriously, all the vital functions . . . rise and set with the sun.* In both examples, *seriously* is an elliptical use of the phrase *to speak seriously*. The *OED* also cites a 1680 example of *strictly* from the work of the printer Joseph Moxon, qualifying (as it says) 'a predication or assertion as a whole' ( = strictly speaking: *This whole Member is called the Moving Collar, though the Collar strictly is only the round Hole at a.* Other pre-20c. adverbs used to qualify a sentence are not difficult to find: e.g. *Frankly, if you can like my niece, win her* (Lord Lytton, 1847).

The proposition, then, can be amended to read as follows: since at least the 17c., certain adverbs in -*ly* have acquired the ability to qualify a predication or assertion as a whole. Such adverbs are all elliptical uses of somewhat longer phrases. In the last third of the 20c., this little-used and scarcely observed mechanism of the language has broken loose. Any number of adverbs in -*ly* have come into common use as sentence adverbs. Conservative speakers, taken unawares by the sudden expansion of an unrecognized type of construction, have exploded with resentment that is unlikely to fade away before at least the end of the 20c.

**sentinel, sentry.** The first is the wider and literary word, and the one more drawn on for metaphorical use; the second is the customary military word. *Sentinel* makes -*elled*, -*elling* (AmE -*eled*, -*eling*); see -LL-, -L-.

**separate.** The adj. is pronounced /ˈsepərət/ and the verb /ˈsepəreɪt/. Poor spellers often incorrectly write the word as *seperate*.

**septenarius** (also **septenary**). In prosody, a verse of seven feet (often printed as two lines), esp. a trochaic or iambic tetrameter catalectic, commonly used in medieval works in verse such as the *Ormulum*, in ballads, and by Wordsworth (1800): *A slumber did my spirit seal; I had no human fears: She seemed a thing that could not feel The touch of earthly years.*

**septet, septette,** a composition for seven players; a stanza of seven lines.

The shorter form *septet* in now usual. See DUET.

**septic.** See SCEPTIC.

**septicaemia,** blood-poisoning. Thus spelt (but AmE *septicemia*).

**septillion,** the seventh power of a million ($10^{42}$); (orig. US) the eighth power of a thousand ($10^{24}$).

**septingentenary.** See CENTENARY 2.

**septum** (anatomy, botany, zoology). Pl. *septa*. See -UM 2.

**sepulchre.** Thus spelt, but AmE *sepulcher*.

**seq., seqq.** The sing. and pl. forms respectively of L *sequens* 'the following', and, in the pl., of *sequentes*, -*tia* 'the following (lines, etc.)', *sequentibus* 'in the following places'. Also, more fully, *et seq*. Now slowly giving way, as appropriate, to *f.*, *ff.*, and *etc.*

**sequelae** /sɪˈkwiːliː/, a medical word meaning 'a morbid condition or system following a disease' (cf. L *sequi* 'follow'). The singular *sequela* /-ə/ is rarely used.

**sequence of tenses.** Some of the regular patterns of sequences of tenses are given s.v. INDIRECT QUESTION and REPORTED SPEECH. It cannot be over-emphasized that in direct statements care should be taken to maintain the same tense throughout a sentence: e.g. *He undressed* and *got* into bed and *switched off* the light. (This and some of the examples that follow are taken from Iris Murdoch's *The Green Knight* (1993).) But writing and speech are also characterized by many types of variation. Consider: *This dog business* will end *in tears, said Jane. Louise who* also thought *that it* would end *in tears, said, '[etc.].'* Other normal patterns are shown in: (pa.t. followed by pres.t.) *She* thought, this *is the end of happiness, darkness* begins *here*; (pa.t. imperfect followed by pres.t.) *Bellamy was saying, 'Are you all* right?'; (pa.t. followed by a mixture of tenses) *She thought: something awful has happened and the children know it.*

Fowler (1926) distinguished normal sequence of tenses from what he called 'vivid sequence'. The first is shown in the type *He explained what relativity meant*;

and the second in *He explained what relativity means* (both being acceptable). But he added that there are traps for the unwary. Both the following types are acceptable: (normal) *One would imagine that these prices were beyond the reach of the poor*; (vivid) *One would imagine that these prices are beyond the reach of the poor*. But if *one would imagine* is parenthetic rather than at the head of the sentence it must be followed by *are*, not *were* if the intention is to refer to present time: (present time) *These prices, one would imagine, are beyond the reach of the poor*; (past time) *These prices, one would imagine, were beyond the reach of the poor*.

The waters are deep and muddy, however. The main point to be noticed is that a change of tense or mood is often obligatory, sometimes optional, and sometimes mistaken in sentences containing such verbs as *thought, believed* or *imagined*.

**seraglio** /sə'rɑːljəʊ/. Pl. *seraglios*. See ITAL-IAN SOUNDS; -O(E)S 4.

**serai** ( = caravanserai). Pronounce /se'raɪ/.

**seraph.** Pl. *seraphim* /-fɪm/ or *seraphs*.

**sere.** See SEAR, SERE.

**serendipity,** the faculty of making happy and unexpected discoveries by accident. Coined by Horace Walpole in 1754 after the title of a fairy tale, *Three Princes of Serendip*. (*Serendip* is said to be a former name of Sri Lanka (Ceylon).) Rarely used before the 20c., it now enjoys wide currency. It is sometimes loosely used to mean 'good luck, good fortune'; and the corresponding adj. *serendipitous* is sometimes used to mean 'occurring by (esp. fortunate) chance, fortuitous'.

**sergeant, serjeant** /'sɑːdʒənt/. Normally in military and police contexts spelt *sergeant* (hence *sergeant-major*), but in older use often written as *serjeant*. 'The spelling *serjeant* is now usually restricted to legal and ceremonial offices, exc. in historical and in certain official contexts' (*New SOED*).

**seri(ci)culture** (silkworm industry). The longer form (adopted in the 1890s from Fr. *sériciculture*; cf. L *sēricum* 'silk') has now been abandoned in favour of *sericulture*, pronounced /'serɪ-/.

**series. 1** Spelt the same as sing. and as pl.

**2** Regularly pronounced as three syllables until the early 20c. (the *OED* gave preference to the three-syllabled form in 1912), like several other words ending in *-ies* in the singular (see -IES), it is now always pronounced as two syllables, the second one being simply /-iːz/.

**serif** (typography), a slight projection finishing off a stroke of a letter as in T contrasted with T, is now always so spelt, not *cerif* or *seriph*.

**serjeant.** See SERGEANT.

**serpent.** See SNAKE, SERPENT.

**serum** (fluid that separates from clotted blood, antitoxin). Pl. *sera*. See -UM 3.

**servant.** Two phrases that are still sometimes used in BrE in certain kinds of formal correspondence: *your humble servant*, a formula preceding a signature or expressing ironical courtesy; *your obedient servant*, a formula preceding a signature sometimes used in letters to the editors of newspapers. See LETTER FORMS 2.

**service** (verb). A surprisingly late addition to the language (first recorded in Stevenson's *Catriona*, 1893), it was used at first only in the sense 'to be of service to, to provide with a service'. In the course of the 20c. it has become established in certain specific senses and now is a normal word for 'to perform routine maintenance on (a motor vehicle, etc.)'; 'to pay interest on (a debt)'; '(of a male animal) to copulate with (a female animal), to serve'; etc.

**serviceable.** So spelt. See -ABLE, -IBLE 2.

**serviette. 1** The main American dictionaries add the label 'chiefly Brit.' to this word, but the label presumably means 'not the customary word in AmE' as the word is common enough in several major English-speaking countries outside Britain.

**2** See NAPKIN.

**sesquicentenary.** See CENTENARY 2.

**sestet,** the last six lines of a sonnet. Cf. SEXTET.

**sestina,** a poem of six 6-line stanzas and a 3-line *envoi*, linked by an intricate pattern of repeated line-endings. Chiefly an Old Provençal, Italian, and French form, but occasionally copied in English, notably by Sidney in his *Arcadia* (1590), by Kipling in his 'Sestina of the Tramp Royal' (in *The Seven Seas*, 1896), and by W. H. Auden in his 'Paysage Moralisé' (1933).

**sett** is still a common variant spelling of *set* meaning (a) a badger's burrow; (b) a granite paving-block. Cf. the final *-tt* in MATT.

**seventies.** See NINETIES.

**several** as adj. is used only with plural count-nouns (*several shops* but not *several furniture*), and it indicates a number slightly greater than *a few* (cf. *several months later* and *a few months later*). By contrast with *a few* it cannot be qualified by *quite* or *only*. Contrast Several *people agreed with me*/Only a few *people agreed with me* (i.e. a small number); Quite a few *people agreed with me* (i.e. more perhaps than expected).

**Sèvres** (porcelain made in this suburb of Paris). Pronounce /ˈseɪvrə/.

**sew** (verb). Pa.t. *sewed*, pa.pple *sewn* or *sewed*.

**sewage.** In strict usage *sewage* is the waste matter conveyed in sewers, and *sewerage* is a system of or drainage by sewers.

**sexcentenary.** A six-hundredth anniversary or the celebration of one. Also as adj. See CENTENARY 2.

**sexist language. 1** As indicated in numerous articles in this book, e.g. -ESS; FEMININE DESIGNATIONS; HE OR SHE; HIS; MAN; -MAN; and -PERSON, feminists and others sympathetic to their views, from about the 1970s onwards, have attacked what they take to be male-favouring terminology of every kind and have scoured the language for suitable evidence and for gender-free substitutes. Their argument hinges on the belief that many traditional uses of the language discriminate against women or render them 'invisible' and for these reasons are unacceptable. The various types of alleged linguistic discrimination need not be repeated here. Perhaps the most obvious reference works on the subject are *The Handbook of Non-Sexist Writing for Writers, Editors and Speakers*, ed. C. Miller and K. Swift (1981) and *The Non-Sexist Word-Finder: A Dictionary of Gender-Free Usage* by R. Maggio (1988). Both books had mixed receptions.

**2** Some landmarks. As rough indicators of the development of feministic views on gender-free language it might be useful to set down in chronological order details of some decisions and discussions that have taken place in various English-speaking coutries since 1988:

In 1988 Cambridge University dons voted to eliminate sexist language from the university's *Statutes and Ordinances*. In 1988 the Style Manual of the Commonwealth of Australia (4th edn) included a detailed chapter, divided into 49 sections, setting out ways of avoiding sexist language. ('I find it quite nauseating,' commented the Queenslander (a man) who sent me a copy.) In 1989 the General Synod of the Church of England debated a report on the need to introduce non-sexist language into the liturgy. In 1989 a revised version of the Bible substituted *one* for *man* in such contexts as Ps. (AV) 1: 1 *Blessed is the* man *that walketh not in the counsell of the vngodly*. Cf. the *NEB* (1970) *Happy is the* man *who does not take the wicked for* his *guide*. In 1990 Radio New Zealand considered it unacceptable to say *fisherman, crewman, clergyman, actress, bridesmaid, maiden voyage, manhole, masterpiece, nobleman, mothering, mother tongue,* or *motherland*. In 1990 the mayor of Los Angeles, Tom Bradley, banned sexist terms from city reports. His mandated changes included *maintenance holes* for *manholes, people* or *humanity* for *mankind, staffed* for *manned,* and *chairperson* for *chairman*.

**3** Some revealing examples. When reviewing the *Handbook* of Miller and Swift, the Irish writer Brigid Brophy complained about the 'leaden literalness of mind' of M & S (as she called them) and 'their tin ear and insensibility to the metaphorical contents of language'. Other writers show in their works that they propose to ignore the shrillest of the advice of feminists. Witness, for example: *'It's every* man *for himself till teatime'—she raised her voice—*Elizabeth Jolley,

1985; *She was the* mastermind—mistressmind—*behind a deception that was going to rake in some considerable* ... *profit*—Colin Dexter, 1989; *In the master bedroom, Ruth announces that she will throw away all her mother-in-law's shoes*—Sarah Gaddis, 1990 (US).

**4** An academic view. In *English Today* (Jan. 1985) the dialect and sociolinguistic scholar Jenny Cheshire concluded: 'There is a built-in masculine bias in English, and this does have very serious implications for both the women and the men who use the language. And this bias will not disappear unless there is some measure of conscious reform in the language.' But where is the evidence that 'conscious reform' will be accepted by the English-speaking public? None of the significant changes to the language in past centuries has come about by 'conscious reform'. And none will in the future unless the whole community singly and collectively decides, not by edict or proclamation, and not even by a vote in the House of Commons, to allow new fashions to be regarded as standard, or at any rate irreversible.

**sextet.** An alteration of SESTET after L *sex* 'six'.

**sextillion.** Orig. (esp. in BrE), the sixth power of a million ($10^{36}$). Now usu. (orig. AmE), the seventh power of a thousand ($10^{21}$).

**sexto,** a size of book or page in which each leaf is one-sixth that of a printing-sheet. Pl. *sextos*: see -O(E)s 6.

**sez.** Curiously used as a conventional spelling of *says* (which, of course, is in fact pronounced thus) in slang contexts.

**sf-.** An un-English combination of initial letters, found only in loanwords from Italian, e.g. *sforzando* (music), *sfumato* (painting).

**sg-.** An un-English combination of initial letters, found only in loanwords from Italian, e.g. *sgraffito* (form of decoration on ceramic ware). Pl. *sgraffiti*/ sgræ'fi:ti/.

**shade, shadow.** **1** The difference in form is etymologically one of declension: *shade* descends from the nominative singular of OE *sceadu*, and *shadow* represents the oblique case (OE *sceadwe*) of the same noun.

**2** Details of the diversity of meaning in each of the two words are too many to be catalogued here: the *COD* has 12 numbered sections for *shade* and 10 for *shadow*. One primary distinction is that *shade* is a place or area sheltered from the sun, whereas *shadow* is a dark figure projected by a body intercepting rays of light (esp. from the sun). Whereas *shade* may be used to mean 'a slight amount' (*a shade better today*) *shadow* cannot. Whereas we have a *shadow Cabinet* (and a *shadow Home Secretary*, etc.) we cannot use *shade* in this way. We say *light and shade* (not *shadow*), *in the shade*, but *under a shadow, a shadow* (not *shade*) *of his former self*. There are any number of other unmatching uses.

**Shakespeare.** Now universally spelt thus but in the early decades of the 20c. *Shakspere* was the more usual form, and was recommended by the *OED* (1913) and by Fowler (1926). The corresponding adj. (and noun) may be written as *Shakespearian* (thus in the house style of OUP) or *Shakespearean* (in *The Times*).

**shako** /'ʃeɪkəʊ/, a cylindrical peaked military hat with a plume. Pl. *shakos*: see -O(E)s 6.

**shall and will.** (See also SHOULD AND WOULD.) The history of these auxiliary verbs and of their contracted forms is immensely complicated and cannot be satisfactorily summarized here. For such detailed information it is necessary to turn to such standard authorities as the *OED* and Jespersen (1909–49, iv). The long article in Fowler 1926, which closely follows his article 'Shall and Will, Should and Would in the Newspapers of To-day' in SPE Tract vi (1921), is a classic, but it too narrowly insists on the preservation of fast-fading traditional distinctions (though he did describe the time-honoured type I *will be drowned, no-one shall save me* as 'too good to be true'). It is broadly true that *shall* and *should* are slowly retreating in the standard language as used in England. In other English-speaking areas, *shall* and *should* have been almost totally replaced by *will* and *would*, or by the reduced forms *I'll/we'll*. What follows is mainly drawn from *The*

Oxford Guide to English Usage (1993), but with some fresh examples inserted.

Expressed simply, the traditional rules in the standard English of England are as follows: **1** (*a*) *I shall, we shall* express the simple future, e.g. *In the following pages we* shall *see good words . . . losing their edge* (C. S. Lewis); *I* shall *have to wear my old coat,'* she said apologetically (A. Brookner). This is especially true in questions, when the use of *will* would either change the meaning or would be unidiomatic: e.g. *I'll put the back rest up for you,* shall *I?* (P. Lively); '*Shall we have a cup of tea?'* he asked (A. Brookner). (*b*) *I will, we will* express determination or insistence on the part of the speaker, e.g. *'I don't think we* will *ask Mr. Fraser's opinion,'* she said coldly (V. S. Pritchett); *I* will *invite them to tea!* (A. Brookner).

**2** For the second and third persons, singular and plural, the rule is exactly the converse: (*a*) *You, he, she, it,* or *they will* express the simple future, e.g. *Seraphina* will *last much longer than a car. She'll probably last longer than you* will (G. Greene); *Will it disturb you if I keep the lamp on for a bit?* (Susan Hill); *Presently Claudia says, 'What* will *you do when the war's over?'* (P. Lively). (*b*) *You, he, she, it,* or *they shall* express intention or determination on the part of the speaker or someone other than the actual subject of the verb, especially a promise made by the speaker to or about the subject, e.g. *In future you* shall *have as many taxis as you want* (G. B. Shaw); *One day you* shall *know my full story* (E. Waugh); *Shall the common man be pushed back into the mud, or* shall *he not?* (G. Orwell). (Type 2(*b*) is no longer commonly found.)

The two uses of *will* and one of those of *shall* are well illustrated by: *'I* will *follow you to the ends of the earth,'* replied Susan, passionately. *'It* will *not be necessary,'* said George. *I am only going down to the coal-cellar. I* shall *spend the next half-hour or so there.'* (P. G. Wodehouse) The distinction between types 1 and 2 can also be seen in many well-known literary contexts, e.g. (type 1(*a*)) Shall *I compare thee to a summer's day?* (Shakespeare); (types 1(*b*) and 2(*b*)) *I* will *not cease from mental fight, Nor* shall *my sword sleep in my hand* (Blake).

**3** In informal usage *I will* and *we will* are quite often used for the simple future, e.g. *I* will *be a different person when I*

live *in England* (J. Rhys); *'Will I be there?'* I enquired. *'No, darling. You wouldn't enjoy it.'* (A. Brookner).

**4** Most frequently, however, the distinction is obscured by the contracted form *'ll*, e.g. *I'll book a table at Francesco's*; *'I'll come in after church and give you a hand,'* said Miss Lawlor; *'We'll be going on later for bridge,'* she explained (all three examples from A. Brookner, 1993).

**5** In the standard English of countries outside England, the absence of *shall* and the omnipresence of *'ll* and *will* are very marked, e.g. United States: *We'll be in touch*; *I'll cope with him in my office*; *'But I really must go to my class, Mr. Kohler. I will say . . .' 'Yessir?' 'I probably shouldn't say anything,'* I allowed (all three examples from J. Updike, 1986). New Zealand: *We'll get a Chinese take-away*; *I'll fill up while you get your keys*; *I hope you do it, Jack.' 'I will.'* (all three examples from Maurice Gee, 1992). South Africa: *I'll search to the south*; *Every hundred paces we'll call out*; *'We will not be going to the veld today . . .' she told them* (all three examples from D. Matthee, 1986).

**6** It is worth noting that in all English-speaking countries pure future uses are often expressed by *be going to* rather than by *shall, will, 'll,* e.g. *For myself, I'm going to become a sober citizen. A son of toil* (P. Lively); *Nobody is going to examine your coat* (A. Brookner); *'I'm going to teach him people are more important than money* (Maurice Gee).

**7** *Shall* survives outside England as well as in England itself in formulaic phrases, e.g. *one Abstract Expressionist, who* shall *remain nameless* (M. Bail, 1975 (Aust.)); *Shall we dance?* Also in legal drafting, where it means 'must, has a duty to', e.g. *The landlord* shall *maintain the premises.* (This use has been brought into question by some legal authorities in the US because of its alleged ambiguity.)

**shambles.** In general use now invariably treated as a singular noun, and used mainly as an informal word for 'a mess, a muddle' (*their marriage was a shambles*). The word has a colourful history. From its OE senses 'a stool, a footstool' and 'a table or counter for displaying goods for sale', it came to mean specifically 'a table or stall for the sale of meat'. The word was then applied to the slaughterhouse from which the meat came, and, from the 16c. onwards, any scene of blood and carnage. It was not until the 1920s that the weakened sense 'a scene of disorder, a mess' emerged. In some of its uses *shambles* is capable of being reduced to *shamble*, both now and in the past, but in the sense 'a scene of disorder, a muddle' (*its economy remains in a shambles*) it is always written with a final *-s*.

**shamefaced,** now the invariable form, is a 16c. alteration of the older word *shamefast* (cf. *steadfast*). The older word could now only be used as a poetical archaism.

**shamefacedly.** Four syllables. See -EDLY.

**shampoo.** As verb the inflected forms are *shampoos, shampooed, shampooing*.

**shanghai.** The recommended inflected forms are *shanghais, shanghaied, shanghaiing*.

**shan't.** From the 17c. onwards *shall not* has been frequently written in the contracted forms *sha'nt, shann't, sha'not, shan't*, etc. Of these, *shan't* survives in standard southern English as the regular contracted form, but is seldom used outside England. See SHALL AND WILL.

**shanty.** Two separate words are involved: **1** A crudely built hut (in Africa and elsewhere); hence *shanty town*, a poor or depressed area of a town, consisting of shanties. This word was first recorded in N. America in 1820 and seems to be from Canad. Fr. *chantier* 'the headquarters at which woodcutters assemble after their day's work'.

**2** A song, esp. one sung by sailors while hauling ropes, etc. Said to be a corruption of Fr. *chantez*, imperative of *chanter* 'to sing'. First recorded in 1867. Formerly in BrE and still usually in AmE spelt *chantey* or *chanty* (pl. *-eys* or *-ies*).

**shape.** The common phr. *in any shape or form* is plainly pleonastic but enjoys wide currency as a formulaic emphatic variant of *in any way, at all* (e.g. *not on offer in any shape or form*).

**shapely** is an adjective. The notional corresponding adverb *shapelily* should be avoided on grounds of euphony. See -LILY.

**shard** /ʃɑːd/. There are two separate nouns: **1** (Also *sherd* /ʃɜːd/.) A broken piece

of pottery, a potsherd. This is a native word derived from OE *sceard*.

**2** (*a*) Cow dung (now only dialectal); (*b*) An elytron or wing-case of a beetle. The *SOED* (1993) comments that sense 2(*b*) 'may be represented ... in *shard-born(e), sharded*, and *shards* in Shakespeare, the interpretation of which is disputed'. Sense 2(*a*) is apparently an alteration of *sharn* (OE *scearn*) 'dung, esp. of cattle'.

**sharif,** a Muslim leader, is a word of Arabic origin. It is etymologically unrelated to the English word *sheriff*, which is derived from OE *scīr-gerēfa*, lit. 'shire-reeve'.

**sharp** as an adverb is correctly used to mean punctually (*seven o'clock sharp*) or abruptly (*the van pulled up sharp*; *turn sharp right*). But in most other contexts it cannot replace *sharply* (e.g. *share prices dropped sharply during the afternoon*; *opinions were sharply divided about the matter*).

**shat** is a fairly frequent variant of *shit* or *shitted* as the pa.t. and pa.pple of the verb *shit* to defecate (*Cowface looked as though she'd shat herself*—H. Beaton, 1984; *the jay covered the square pedestal with twigs and leaves and shat on the head and arms*— New Yorker, 1990).

**shave** (verb). The pa.t. and pa.pple are *shaved*. But the ppl adj. form is usually *-shaven* (*a clean-shaven man in his twenties*).

**she.** **1** For *she* (and *her*) in personifications, see PERSONIFICATION. See also HE OR SHE; HER 3.

**2** The use of *she* rather than *her* after the verb *to be* follows the same pattern as that of *he* rather than *him* (see HE 1). Examples: (she) *And don't fear that it is she who might convert you*—P. P. Read, 1986; *If anyone could write about the narcissistic personality, it was she*—New York Rev. Bks, 1987; (her) *Phone rings at 8.07 ... it must be her, no one else would call at such a time*—The Face, 1987; *She comes down the basement stairs sideways, ... so you always know it's her*—New Yorker, 1987. In such circumstances *she* is the more formal of the two.

**3** After *than*, *she* is the more formal of the two pronouns, but both are in standard use. Examples: (she) *On the whole* the men are ... *more formal and authoritarian in tone than she*—Marilyn Butler, 1987; (her) *I'm not really fatter than her, am I?*—M. Duckworth, 1960 (NZ); *He's a million times better than her*—M. Bail, 1975 (Aust.). See CASES 3.

**4** Examples of the nominative *she* used wrongly for *her* after a preposition or in other respects in the objective position: *this uncombed stick-limbed fellow ... who had finally cracked the defences of she whom Harvey Fig had dubbed 'our pocket Venus'*—P. Carey, 1988; *Not even Patricia Johnson could bear to listen ... to the gruesome tale of death that led a Cook County judge to order she and her boyfriend be held without bail in the triple slaying of a Lakeview family in January*—*Chicago Tribune*, 1992.

**s/he** is a written representation of 'he or she', used by some writers since the 1970s as a nominative singular third person pronoun to include both genders. Example: *A child's sexual orientation is determined before s/he enters school*—*Amer. Educator*, 1978. Its self-evident artificiality suggests that its place in the language is far from secure.

**sheaf.** The noun has plural *sheaves*. The corresponding transitive verb meaning 'to make into sheaves' is *sheave*. See -VE(D), -VES.

**shear** (verb) has pa.t. *sheared* in its ordinary current senses (*we sheared our sheep last week; a machine sheared the bar into foot-lengths; the pressure from above sheared the rivets*). The normal pa.pple is *shorn* (*the sheep had been shorn; his shorn locks; shorn of one's authority*), except in contexts of metal-cutting (*a bar of wrought iron needed to be sheared across*). The more usual pa.t. in Aust. and NZ sheep-shearing sheds is *shore*.

**sheath.** Pronounced /ʃiːθ/, but in the pl. *sheaths* as /ʃiːðz/. The corresponding verb *sheathe* is /ʃiːð/. See -TH /θ/ AND -TH /ð/. See also NOUN AND VERB ACCENT, etc. 3.

**sheave** (verb). See SHEAF.

**sheep.** Pl. same.

**sheikh.** This is the spelling recommended in OUP house style. Pronounce /ʃeɪk/. Some other publishers prefer *shaikh* or *sheik*. The pronunciation /ʃiːk/ is also common.

**shelf** /ʃelf/. The pl. is *shelves* /ʃelvz/ and the corresponding verb is *shelve* /ʃelv/. The compound *shelf-ful* is best written with a hyphen; pl. *shelf-fuls*.

**shellac.** As verb has the inflected forms *shellacs, shellacked, shellacking*.

**shell-less, shell-like.** Thus spelt (with hyphens).

**Shelta, sheltie.** The first is the name of an ancient hybrid secret language used today by tinkers; *sheltie* (rather than *shelty*) is a Shetland pony or sheepdog.

**shereef, sherif** are variant spellings in English of SHARIF.

**sheriff.** Over the last several centuries a number of abstract nouns have been used for 'the office or post of sheriff': *sheriffalty* (first recorded in the 16c.); *sheriffdom* (16c. in this sense; 14c. in the sense 'district under the jurisdiction of a sheriff'); *sheriffry* (17c.; rare); *sheriffship* (15c.); *sheriffwick* (15c.; also 16c. in the sense 'district under the jurisdiction of a sheriff'); and *shrievalty* (16c.). The official term is the last of these, despite the disadvantage of not immediately announcing its (valid) etymological connection with *sheriff*.

**shew, show.** As the *Oxford Writers' Dict.* (1990) says 'use *show* except in Scottish law, and biblical and Prayer Book citations'. Pronounced /ʃəʊ/ (also in the first element of *shewbread*) however spelt.

**shibboleth.** **1** It is first and foremost 'the Hebrew word used by Jephthah as a test-word by which to distinguish the fleeing Ephraimites (who could not pronounce the *sh*) from his own men the Gileadites (Judges 12:4–6)' (*OED*). From this, from the 17c. onward, it has been used for any word or sound which a person is unable to pronounce correctly: hence a word or sound used to detect foreigners, or persons from another district.

**2** By extension, from about the same time, it has been used to mean 'a long-standing formula, idea, phrase, etc., held (esp. unreflectingly) by or associated with a group, party, class, etc.'; and, nowadays, also, 'an entrenched (political) belief or dogma'. Since there are numerous synonyms available to express these extended senses, the word is best left at present restricted to the uses defined in 1.

**shier.** See SHY.

**shillelagh.** Thus spelt. Proncunce /ʃɪˈleɪlə/ or /-li/.

**shine** (verb). The pa.t. and pa.pple forms are normally *shone* /ʃɒn/, AmE /ʃəʊn/, exc. that in the sense 'polished, made bright', and esp. in AmE, *shined* is used instead. Examples of *shined* (exc. in the 1989 and 1990[1] examples speakers of standard southern BrE would normally use *shone* instead): *Her shiny black paint shined in the sun—Black World* (US), 1974; *I asked him if it shined in the sunlight—*M. Doane, 1988 (US); *The car is a red Mercedes, newly shined—*S. North, 1989 (UK); *I have my shoes shined by a man who says he will pray for my wife and family—*J. Cheever, 1990 (US); *his hair was combed and flattened down with bay rum until it shined—*N. Virtue, 1990 (NZ); *It occurred to me that this was not a reflection from his glasses or his crown, no matter how much they shined—*D. Pinckney, 1992 (US).

**shingles** (acute viral inflammation), like other illnesses ending in *-s* (*measles, mumps,* etc.) is normally treated as a singular noun (*shingles is an unpleasant disease*), but if the emphasis is perhaps more on the resulting blisters than on the illness itself a plural verb may be used instead. See AGREEMENT 12.

**ship.** 1 In the proverb *Do not spoil the ship for a ha'porth of tar, ship* represents a dialectal pronunciation of *sheep.* The original literal sense was 'Do not allow a sheep to suffer for the lack of a trifling amount of tar'. Tar was used as a protection against flies and maggots.

2 See BOAT.

**-ship.** See BRINKMANSHIP; LEADERSHIP; -MANSHIP; MEMBERSHIP; READERSHIP; RELATION(SHIP).

**shire.** 1 A correspondent well versed in such matters comments: 'The *shires* traditionally refers to certain parts of the foxhunting areas of Leicestershire, Northamptonshire, and the former county of Rutland. Since local government reorganization in 1972 the expression *shire county* has also come into use

to denote those thirty-nine counties outside metropolitan areas which have county councils as distinct from the six metropolitan counties and London, which in England since 1986, do not. *Knight of the shire,* which in medieval times and later was a term used to denote each of the two representatives of every county in the House of Commons, now means any Conservative Member of Parliament who represents a constituency in a shire county and has been knighted or who, more loosely, has served for a long time in Parliament and holds traditional views.'

2 A partial explanation of the origin of the *-shire* ending of certain counties is that it helped to distinguish the name of the county (e.g. *Derbyshire, Yorkshire*) from the name of its county town (*Derby, York*). But the historical development of *-shire* names is extremely complicated and lies outside the scope of this book.

3 As the second element of county names, *-shire* is normally pronounced /ʃə/.

**shoal.** See SCHOOL.

**shoe** (verb). The inflected forms are *shoes, shoeing,* and *shod* /ʃɒd/.

**shoot, shute** (nouns). See CHUTE.

**shop** (noun). For the talk called *shop,* as compared with *cant, slang,* etc., see JARGON 3.

**shop** (verb). In AmE, since the 1950s, but not in BrE, the verb has been used transitively (as well as in the traditional intransitive uses) to mean 'to shop at (a store); to examine goods on sale (in a shop)': e.g. *One man who had shopped the entire store complained that he hadn't found what he was looking for—*S. Marcus, 1974.

**short-lived.** See -LIVED.

**shorts.** See PANTALOONS.

**short, shortly.** As adverbs the roles of the two words are fairly clearly separated. *Short* usually means 'before the expected time or place, abruptly' (*pulled up short; cut short the celebration; the statement stopped short of being offensive*). *Shortly* is used mostly to mean 'before long, soon' (*he is expected to arrive shortly; she spoke to me shortly after 5 pm*).

**should and would.** (See also SHALL AND WILL.) The most that one can offer in the small compass of an article here is to give a few ground rules with supporting examples. Some of what follows is drawn from *The Oxford Guide to English Usage* (1993). One feature that stands out is the frequency of *would* (or *'d*) compared with that of *should*: in my database the relative percentages are *should* 24%, *would* 66%, and *'d* 10%. Another striking feature is that *should* is very frequently used as a modal verb to mean 'ought to'.

**1** Examples of *'d* (in various constructions): *When you're feeling censorious, better ask yourself which* you'd *choose*—P. D. James, 1986; *I thought I'd drop by*—J. Updike, 1986 (US); *if you happened to be putting your kettle on ... I'd be delighted to join you*—K. Amis, 1988; *They'd end up in Japan at the rate they were going*—R. Billington, 1988; *That'd teach her mother to be selfish*—ibid; *A crowd of shoppers was roped off from the kind of buffet table you'd find on a cruise ship*—New Yorker, 1988 (US); *I'd sooner do it at home, thank you*—Maurice Gee, 1992 (NZ).

**2** Examples of *should* = 'ought to': *Mind you, if you think she behaved strangely, you* should *have seen me*—M. Amis, 1984; *'I probably* shouldn't *say anything,'* I allowed—J. Updike, 1986 (US); *small men* shouldn't *wear beards*—K. Amis, 1988; *You* should *wear glasses*—R. Billington, 1988; *We* should *have stayed in Wellington*—Maurice Gee, 1992 (NZ).

**3** Other uses of *should* (a) With the first person pronouns *I* and *we*, *should* is normally used in standard southern BrE as an auxiliary before verbs of liking, e.g. *be glad, be inclined, care, like, prefer*, e.g. *'Would you like a beer?' 'I* should *prefer a cup of coffee, if you don't mind'; 'I* should *like one of these, says Claudia*—P. Lively, 1987; *'Oh, and I* should *like you to know ... that I'm not a poet*—K. Amis, 1988; *I* should *like them to see the flat*—A. Brookner, 1993. For the types *should have liked to/would have liked to*, see LIKE (verb) 1(a–c). (b) *Should* is correct in standard southern BrE in tentative statements of opinion, with verbs such as *guess, imagine, say*, and *think*: *I* should *imagine that you are right; I* should *say so; I* should *have thought you'd got used to that principle by now.* (c) In other circumstances *would* is often used after *I/we*. Examples: *Well ...*

*I* wouldn't *exactly say*—P. Lively, 1987; *I* wouldn't *have minded that, but it wasn't that*—K. Amis, 1988; *If it hadn't been for dear Adèle ... I don't know what I* would *have done*—A. Brookner, 1993.

(d) *Would* is always correct with persons other than the first person singular and plural. Examples: *How they* would *respond at such a time is anyone's guess*—M. Amis, 1987; *She* would *put off thinking till she had the tea in her hand*—R. Billington, 1988; *Whenever anyone really important came along ... Simon* would *take the person to a restaurant*—K. Amis, 1988. (e) *Would* used in contexts where it expresses the future in the past: *I told you that* you would *find Russian difficult to learn; He was there. Later he* would *not be there.* The person's statement or thought at the time was *You* will *find Russian difficult to learn; He* will *not be there.* Further examples: *Dynamite* would *soon dominate the world's explosives market* [in 1867]—New Yorker, 1987; *Much later, Fox* would *be the intermediary in introducing Erasmus to Richard Bere*—J. B. Trapp, 1992.

**should of.** See OF C.1.

**shovel** (verb). The inflected forms are *shovelled, shovelling* in BrE but frequently *shoveled, shoveling* in AmE. See -LL-, -L-.

**show** (verb). For the spelling, see SHEW. The pa.t. is *showed* and the pa.pple normally *shown* (*the book was shown to him*) but occasionally *showed* (e.g. *Have you showed it to Thomas Blackhall?*—G. Butler, 1992; *Mr Marr hadn't showed up at any of the places where he should have been*—Nigel Williams, 1992; *I didn't see it happen but realised almost immediately the card was gone, must have showed panic by standing to shake out the pleats of my long skirt*—B. Neil, 1993.

**shred** (verb). For the pa.pple *shredded* and *shred* were used as acceptable alternatives until the 20c., but *shred* is now much less commonly used.

**shrievalty.** See SHERIFF.

**shrilly** (adv.). Pronounced /ˈʃrɪl-lɪ/, i.e. with both ls fully pronounced.

**shrink.** The pa.t. is *shrank* /ʃræŋk/, (nonstandard) *shrunk* /ʃrʌŋk/; pa.pple *shrunk* /ʃrʌŋk/ or (esp. as attrib. adj.) *shrunken* /ˈʃrʌŋkən/: *has shrunk, is shrunk* or *shrunken; her shrunken cheeks.* The contrast between the customary forms of the pa. pple and ppl adj. is neatly brought out by these examples from adjacent pages of the 27 Feb. 1990 issue of the *Guardian: For the entire race the margin between them has stretched and shrunk; Among the shrunken white community which has steadied at around 100,000 . . . there are few conspicuous signs of . . . racist attitudes.*

**shrivel.** The inflected forms are *shrivelled, shrivelling* in BrE but frequently *shriveled, shriveling* in AmE. See -LL-, -L-.

**shy.** The adj. makes *shyer, shyest, shyly, shyness, shyish.* The verb makes *shier* (shying horse). See DRIER; VERBS IN -IE, etc.

**Siamese twins.** This seems a suitable term for pairs of words which are traditionally linked by *and* or *or* and often have the same meaning as each unit in the pair (or a slightly strengthened one), or are related in other formulaic ways. Examples: (*a*) (used mostly for emphasis) *airs and graces, alas and alack, bag and baggage, betwixt and between, bits and pieces/bobs, fit and well, leaps and bounds, lo and behold, rant and rave, in any shape or form.* These are in their nature tautological but are not on that account made unidiomatic. (*b*) Others are fixed collocations, either because one of the components is used in an archaic sense and would not now be understood by itself, or because the combination has acquired a meaning different from that of either component alone. Examples: *at someone's beck and call, chop and change, fair and square, fast and furious, hue and cry, kith and kin, with all one's might and main, odds and ends, part and parcel, go to rack and ruin, no rhyme or reason, spick and span.* (*c*) Others again consist not of synonyms but of associated ideas. Examples: *bill and coo, bow and scrape, bright-eyed and bushy-tailed, flotsam and jetsam, huff and puff, hum and haw, a lick and a promise, loud and clear, nuts and bolts, spit and polish, thick and fast, ways and means.* Or consist of opposites or alternatives: *cut and thrust, fast and loose, hit and miss, hither and thither, by hook or by crook, through thick and thin, to and fro.*

There are various other types of 'Siamese twins' (or collocations), but the examples cited above will perhaps suffice to show that such linked phrases are a significant feature of the language. See COLLOCATION.

**sibilants** are FRICATIVES 'produced by forcing the air stream through a groove-shaped opening between the tongue and the roof of the mouth, e.g. [s] in *sin*; [z] in *zoo*; [ʃ] in *shop*; [ʒ] in *pleasure*' (Hartmann and Stork, 1972).

**sibling,** an ancient word for a relative (a sense obsolete by the end of the 15c.), was reintroduced by anthropologists in the first decade of the 20c. with the meaning 'each of two or more children having one or both parents in common'. It is for the most part restricted to scientific writing, but is occasionally brought into figurative or transferred use, e.g. *The line dividing the Kevin Street Sinn Fein organisation and its terrorist sibling, the Provisional IRA, is almost invisible—Daily Tel.,* 1972.

**sibyl(line).** The spelling (not *sybi-*) should be noted. *Sibyl* 'a prophetess' is to be distinguished from the modern Christian name *Sybil.*

**sic,** Latin for 'so, thus', is inserted (within brackets) after a quoted word or phrase to confirm its accuracy as a quotation, or occasionally after the writer's own word to emphasize it as giving his deliberate meaning. It amounts to 'Yes, he did say that', or 'Yes, I do mean that, in spite of your natural doubts'. It should be used only when doubt *is* natural; but reviewers and controversialists are tempted to pretend that it is, because *sic* provides them with a neat and compendious form of criticism. Examples: *I probably have a different sense of morality to* [sic] *most people—*Alan Clark as reported in the *Chicago Tribune,* 2 June 1994; *Could she take the* Tatler, Vanity Fair *and* Health *magazines that were laying* (sic) *around at home?—*V. Grove, 1994 (reporting a Desert Island Discs interview with the actress Britt Ekland); *his* [an inventor's] *crudely written notice declaring that it 'dose* [sic] *the work of a press that would cost £10,000'—Spectator,* 1994.

**sice, size, syce.** For the six on dice, *sice* is recommended (not *size*); for the Indian groom, *syce* is recommended (not *sice*); *size,* so spelt, is a gelatinous solution used in glazing paper, etc.

**sick.** See ILL.

**sickly** is an adjective. The notional but clumsy corresponding adverb *sicklily* is not used. See -LILY.

**sidle** is a back-formation from *sideling*, an obsolete form of *sidelong* 'inclining to one side; oblique'.

**siege.** So spelt; also *besiege*.

**sienna,** the pigment and its colour of yellowish-brown (*raw sienna*) or reddish-brown (*burnt sienna*), is so spelt despite its origin (from *Siena*, a city in Tuscany).

**sieve.** So spelt; pronounced /sɪv/.

**sight unseen,** without previous inspection. This phrase, of an ablative absolute type, first recorded in America in the 1890s, is now well established in the main forms of English everywhere. Examples: *I've always hated that phrase, haven't you, Paola, sight unseen, it's a tautology or something near, it simply means unseen, doesn't it?*—A. S. Byatt, 1990; *So much for my review sight unseen ... When the book arrived* [etc.]— D. J. Enright, 1990; *A friend writes: I acquired, sight unseen, a piece of land just outside Lynchburg.*—*New Yorker*, 1990.

**signal** (verb). The inflected forms are *signalled, signalling* in BrE but frequently *signaled, signaling* in AmE. See -LL-, -L-.

**signatory** is now the usual spelling (not *signatary*) for both the noun and the adjective.

**significant other.** One of a number of euphemisms, characteristic of our age, for an adult of the opposite sex with whom one is living in an unmarried state. Other terms include *live-in partner, lover, partner*, and *POSSLQ* (person of the opposite sex sharing living quarters). Example: *Then her Decembric significant other, the father of one of her biological children, has an affair with the May-esque if not Aprilian adopted daughter of her second marriage*—*New Yorker*, 1993.

**signifier** (and related words in linguistics). The Swiss scholar Ferdinand de Saussure (1857–1913) introduced the terms *signifiant* 'a physical medium (as a sound, symbol, image, etc.) expressing meaning, as distinct from the meaning expressed (the *signifié*)', and emphasized

the arbitrary relationship between the linguistic sign and that which it signifies. These French words are sometimes replaced in scholarly work by their Latin equivalents *significans* and *significatum*; and also, commonly, by the English words *signifier* and *signified*. The last two are also often used in the critical terminology of modern literary theory. Examples: *The bond between the signifier and the signified is arbitrary*—W. Baskin (tr. de Saussure), 1960; *He has only to offer the Signifier, 'Lie'—which could mean 'tell an untruth'—for Othello to snatch the Signified, 'lie with'*—*Dædalus*, 1979; *This novel understands that it is politics—torture, suffering, deprivation—which reminds us that our signifier-shaped existence is more corporeal than textual*—T. Eagleton, 1994. Meanwhile non-linguists, i.e. nearly everyone else in the English-speaking world, are content with the traditional terms *word* (or *phrase*, etc.) and *meaning* (or *sense*, etc.).

**signor, signora, signorina.** Italian titles. Pronounced in English /ˈsiːnjɔː/, /siːn-ˈjɔːrə/, /siːnjəˈriːnə/.

**Sikh.** Normally pronounced /siːk/ in English and /sɪk/ in Hindi.

**silex** (kind of glass). See -EX, -IX.

**silhouette.** Pronounce /sɪlʊˈet/.

**silken.** See -EN ADJECTIVES.

**sillabub.** See SYLLABUB.

**sillily.** One of the few current -*lily* adverbs, though it is often replaced by *in a silly way*. See -LILY.

**silo.** Pl. *silos*. See -O(E)S 6.

**silvan.** See SYLVAN.

**silvern.** See -EN ADJECTIVES.

**simian.** Pronounced /ˈsɪmɪən/ with a short /ɪ/ in the stem, despite the long stem vowel in L *sīmia*. See FALSE QUANTITY.

**similar.** It should be followed by *to*, not *as* (though it was formerly also construed with *with*). Examples: (obsolete use) *A legend of similar import with that of the death of Hercules* (1832 in *OED*); (standard use) *It seemed to me that she was acknowledging an emotion similar to my own*—C. Rumens, 1987; (non-standard use) *Wolverton Seconds showed similar form as their* seniors in their two home games—*Oxford Guide to Eng. Usage*, 1993.

**simile.** Part of the spiritual nourishment of writers and esp. of poets, similes are, of course, 'figure[s] of speech involving the comparison of one thing with another of a different kind, as an illustration or ornament' (*COD*, 1995). They turn up in profusion at all periods and are normally introduced by *as* or *like*. Some examples chosen at random: *Blue as the gendarmerie were the waves of the sea* (E. Sitwell); *Soft as rain slipping through rushes, the cattle Came* (E. Blunden); *She on the earth, with steadfast sight, Stood like an image of the Muse* (R. Pitter); *Then I turn the page To a girl who stands like a questioning iris By the waterside* (C. Day Lewis). Observe the spelling and weep for the thousands of students of English literature who persist in writing about 'similies'. Contrast METAPHOR.

**simony.** Now pronounced /ˈsaɪmənɪ/, though the *OED* (1911) gave only /ˈsɪmənɪ/. The modern pronunciation conforms with that of the name (Simon Magus) of the sorcerer who seems to have been the first to attempt this malpractice (Acts 8: 18).

**simple.** Note the post-positioning of the adjective in the legal term *fee simple*, an inherited estate, unlimited as to the category of heir (as against *fee tail*, which imposes limits).

**simpleness, simplicity.** See -TY AND -NESS.

**simplistic.** A word of surprising modernity (first recorded in its modern sense in the late 19c.; slightly earlier of medicinal plants or simples). It adds a connotation of excessive or misleading simplification to the ordinary adjective *simple*. Examples: *She's quite right ... It is simplistic to speak of malice*—T. Stoppard, 1976; *It was quite evident that these rather simplistic models were inadequate*—*Jrnl RSA*, 1980; *Unfortunately, Morris ... seems to dismiss his* [sc. Pope John XXIII's] *courageous attempts at reform as being naive and simplistic*—*Atlantic*, 1991. Care should be taken not to use *simplistic* in contexts where *simple* is adequate by itself.

**simulacrum.** Pronounce /sɪmjʊˈleɪkrəm/. Pl. -*cra*. See -UM 2.

**sin** (verb). The phr. *to sin one's mercies* 'to be ungrateful for one's blessings or good fortune' is found in 19c. literature (beginning with Scott's *Redgauntlet*, 1824), and was familiar to Fowler (1926) and Gowers (1965), but now seems largely to have dropped out of use. An isolated example: *Sometimes I speak hard things about my profession. I sin my mercies, as the Scots say*—E. Linklater, 1929.

**since.** 1 For the clumsy use of *since* after *ago*, see AGO.

2 For the type *P.S. Since writing this your issue of today has come to hand*, see UNATTACHED PARTICIPLES.

**sincerely.** For *Yours sincerely*, etc., see LETTER FORMS 2.

**'s incongruous.** Fowler's term for 's AND OF-POSSESSIVE.

**sinecure.** My impression is that /ˈsɪnɪkjʊə/ is now the dominant pronunciation, but the authorities differ, and /ˈsaɪnɪ-/ must therefore be regarded as having equal validity in standard BrE.

*sine die.* In common parlance usu. pronounced /ˌsaɪnɪ ˈdaɪiː/ or in relaxed speech /ˈdaɪ/, but Latinate pronunciations of both words, i.e. /ˌsɪneɪ ˈdiːeɪ/ are also current. Of course, 'adjourned indefinitely' or 'with no appointed date' are available as alternatives if you feel unsure about the pronunciation of the Latin phrase.

*sine qua non,* an indispensable condition or qualification, is pronounced /ˌsɪneɪ kwɑː ˈnəʊn/, though several variant pronunciations are also current, e.g. /ˌsaɪnɪ/ or /ˌsɪmi kweɪ ˈnəʊn/ or /ˈnɒn/. Uncertainty about the pronunciation probably causes many people to use English equivalent phrases instead.

**sing** (verb). The pa.t. is *sang* (formerly, esp. in the 18c. and 19c., often *sung*) and pa.pple *sung*.

**Singalese.** See SINHALESE.

**singeing** (burning lightly). So spelt to distinguish it from *singing* (uttering musical sounds). See MUTE E.

**single most.** A correspondent has pointed out to me that in strict terms *single* is tautologous when followed by *most* + an adj. and noun, as in the type *the single most valuable player in the team.* Example: *I've had mini-relationships and then gone for ... 12 months without sex ... and it has been the single most important growing time for me as a woman—NY Herald Tribune,* 1991.

**singular -s.** The instinctive feeling that a noun ending in *-s* is by nature plural has caused a good number of 'false' singular forms to come into being in the past, e.g. *caper* (the herb; earlier *capers*, from OF *caspres*, modF *câpre*, L *capparis*); *cherry* (from Old Norman Fr. *cherise*, taken as a pl., modF *cerise*); *pea* (from earlier *pease*, apprehended as a plural), *succour* (from OF *sucurs, socours*, modF *secours*). *Riches*, from Fr. *richesse*, is now always treated as a plural. Cf. also FORCEPS; GALLOWS; KUDOS; and LENS for aspects of the same phenomenon.

**Sinhalese.** A member of a people living chiefly in Sri Lanka or their Indic language. Formerly also *Sing(h)alese* and other variants.

**sinister** in heraldry means left (and *dexter* right), but with the contrary sense to what would naturally suggest itself, the left (and right) being that of the person bearing the shield, not of an observer facing it. For *baton, bend sinister,* see BAR SINISTER.

**sink** (verb). **1** The pa.t. is now overwhelmingly *sank* rather than *sunk.*

**2** For the pa.pple the longer form *sunken* is no longer used as part of a compound passive verb: *a ship would have been, will be, was, sunk,* not *sunken.* As adjectives both *sunk* and *sunken* are used and their distribution is not easy to determine. The most that can be said is that *sunken* is often used to mean 'submerged' (*a sunken ship*), 'fallen in, hollow' (*sunken cheeks, sunken eyes*), and often 'below the normal level' (*sunken garden, sunken porch*). *Sunk* is the form normally chosen for technical expressions such as *sunk fence, sunk initial, sunk key, sunk panel, sunk relief* (for the meanings consult an unabridged dictionary), but *sunken* is also used in some of these.

**3** Used predicatively with the verb *to be* and meaning 'in a hopeless position' the only form used is *sunk* (e.g. *'Hell!' thought Mr R., 'we're sunk!'*).

**sinus.** Pl. *sinuses* (despite the fact that the L pl. is *sinūs*). See -US 2.

**Sioux.** Pronounce /su:/. Pl. unchanged and pronounced the same or, optionally, /su:z/.

**siphon.** This spelling is recommended, not *syphon*. See Y AND I.

**Sir,** a titular prefix to the forename of a knight or baronet. Thus *Sir William Jones,* or, contextually, just *Sir William.* Foreigners, some of whom are unaware of the correct procedure, sometimes use the erroneous form *Sir Jones.* A further hazard: a double-barrelled surname should not be used instead of Christian name and surname: thus *Sir Douglas-Home* cannot be written for *Sir Alec Douglas-Home.* The same conventions apply to the corresponding female prefix *Dame.*

**siren.** Thus spelt, not *syren*. See Y AND I. Pronounce /'saɪərən/, not /saɪ'ri:n/.

**Sirius.** Pronounce /'sɪ-/, not /'saɪə-/. For departures from classical quantities (in this case L *Sīrius,* Gk Σείριος), see FALSE QUANTITY.

**sirloin.** A spurious but standard spelling of what should have been *\*surloin* (cf. *surname*) from an OF form *\*surloigne,* variant of *surlonge* [the modF word for 'sirloin' is in fact *aloyau*]. It is the upper (cf. Fr. *sur* 'above') and choicer part of a loin of beef. From the 17c. onward evidence has survived falsely attributing the 'knighting' of a loin of beef to Henry VIII, James I, and Charles II.

**sirocco.** Thus spelt in English, as it is in French. Pl. *siroccos:* see -O(E)S 6. The Italian spelling *scirocco* (the source of the English word) is now rarely used in English.

**sister,** a senior female nurse, esp. one in charge of a hospital ward. The equivalent AmE term is *head nurse.*

**sisterly.** For the notional adverbial form *\*sisterlily,* see -LILY.

**sit** (verb). The standard pa.t. is *sat*: see SAT (non-standard examples). The corresponding participial adj. is *seated: the seated part of the stadium; two-seated vehicles; please be seated.*

**sitting-room.** This or *living-room* is the term most often used of a room with sofa and (arm)chairs for relaxing in; in working-class parlance often called a *lounge*. A *drawing-room* is normally a (fairly) grand room used for the same purpose in a large house (*trooping into the draughty vestibule* [of an 18th-century mansion] *to deposit their coats, umbrellas and briefcases before taking the noisy baronial staircase to the drawing-room on the first floor*—R. McCrum, 1991). *Lounge* is also the normal word for a room set aside for sitting in in public places: *a hotel lounge; Gatwick's departure lounge.*

**situate.** 'The language of the law abounds in needless archaisms, and *situate* used as a past participle is one of them . . . Many wills contain phrases such as the following: "All the rest, residue, and remainder of my estate, real as well as personal and wheresoever situate, . . . I direct my executor to distribute [etc.]".' So in a standard American legal dictionary, and so also in Britain. The short participial form is also sometimes used in estate agents' advertisements. Examples: *The magistrates' court . . . will commit to the Crown Court centre designated by the Presiding judge as being the location to which cases should normally be committed from the petty sessions area in which it is situate.—Stones Justices' Manual* (edn 118), 1986; *Properties for sale . . . Land parcel, title No. LOC. 11/MARAGI/860, area 3.98 acres situate in Murang'a District.—Standard* (Nairobi), 1989.

**situation.** A useful noun for expressing the sense 'a set of circumstances, a state of affairs', esp. when preceded by a defining adjective, e.g. *the financial, industrial, military, political situation.* The placing of a noun immediately before *situation* should normally be resisted: *a crisis situation* adds nothing to *a crisis; a bankruptcy situation* adds nothing to *bankruptcy.* The widespread (often semi-jocular) habit of placing a nominal phrase before *situation* should also be resisted, e.g. *a premature baby situation; chairs suitable for multiple seating situations; the conditions had deteriorated to virtually a no-visibility situation.* See ONGOING; -TION.

**sixteenmo.** In printing, the English reading of the symbol 16mo, which also represents the Latin-derived term *sexto-decimo.*

**sixth.** Pronounce /sɪksθ/ with the /θ/ fully articulated.

**sixties.** See NINETIES.

**sizable** is the recommended spelling. See MUTE E.

**size.** See SICE.

**sjambok.** (In S. Africa) a rhinoceros-hide whip. Pronounce /ˈʃæmbɒk/.

**skald** (medieval Scandinavian poet), **skaldic** (adj.). Recommended spellings, not *sc-*.

**skeptic.** See SCEPTIC.

**skewbald.** See PIEBALD.

**ski.** In my boyhood (1930s) very commonly pronounced /ʃiː/, as in Norwegian, the language from which the word was adopted; but now always /skiː/. The pl. of the noun is *skis*; and the inflected forms of the verb are *skis, ski'd* or (for preference) *skied, skiing.* The related noun ('one who skis') is *skier.*

**skiagraphy.** See SCIAGRAPHY.

**skier, skyer.** For 'one who skis', use *skier.* For 'a high hit (in cricket)', use *skyer.*

**skilful.** So spelt in BrE, but usually *skillful* in AmE. See -LL-, -L- 5, and next.

**skilled.** The *skilled* and the *unskilled* are sheep and goats, distinguished by having had or not having had the requisite training or practice. Work is also classified as *skilled, semi-skilled,* or *unskilled. Skilful* by contrast describes the ability of a person (*a skilful climber*) or the ingenuity or dexterity of a thing (*a skilful painting*).

**skill-less.** Best hyphened as shown to avoid the ungainliness of three consecutive ls. Similarly *bell-less, hill-less, shell-less,* etc.

**skin.** *With the skin of my teeth.* See MIS-QUOTATIONS.

**skip** (college servant). See SCOUT.

**skip** (rubbish container). See RUBBISH.

**skull.** See SCULL.

**skyer.** See SKIER.

**slacken.** (See -EN VERBS FROM ADJECTIVES.) Since the 16c., when both words came into existence, *slacken* has been interchangeable with *slack* in many senses, esp. when used to mean 'to make or become slack', but *slacken* is now the more dominant word of the two. *Slacken* is now the ordinary word for 'to become slack' and 'to make (or let become) slacker': *a breeze, demand, one's energy, one's pace, a rope, the tide,* slackens; *we slacken our efforts, grip, opposition, the regulations. Slack,* not *slacken,* means to be slack or idle: *he accused me of slacking.*

**slake, slack** (verbs). The first of these two related verbs has been in existence since the OE period, while *slack* entered the language only in the 16c. In nontechnical use, *slake* now tends to be restricted to mean 'to assuage or satisfy (thirst, desire, etc.). For *slack,* see prec.

**slander.** See LIBEL (noun).

**slang** is 'language that is regarded as very informal or much below standard educated level. [It is also] the special vocabulary and usage of a particular period, profession, social group, etc.' (*New SOED,* 1993). 'To the ordinary man, of average intelligence and middle-class position, slang comes from every direction, from above, from below, and from all sides, as well as from the centre ... An antiquated example is *mob,* for *mobile vulgus.* That was once slang, and is now good English. A modern one is *bike,* which will very likely be good English also in time' (Fowler, *The King's English,* 1906). 'Seen as a sizeable distinguishable element in written English it can be plainly identified from the sixteenth century onward. Prior to that it lay about in unrecorded private speech. The first formal recording of slang in dictionary form came in B.E.'s *Dictionary of the Canting Crew* in 1699. As a phenomenon of modern times it can be pursued to many public and private lairs. It is stretched on class-structured racks. It crosses generation gaps but is sometimes mainly restricted for a period to a particular generation' (Burchfield, *The English Language,* 1985).

It is impossible to quantify the amount of slang extant at the present time, but everyone agrees that it is found at all levels of society, that it is largely excluded from formal writing (except in the representation of speech in literary

works), that the casualty rate is high (this year's favourite slang words are next year's abandoned ones), and that a small proportion of slang terms over a period of time join the uncontested central core of general unstigmatized vocabulary. Something of the flavour of modern slang can be seen in items so labelled in the range A–C in *The Oxford Dictionary of New Words* (ed. Sara Tulloch, 1991): *ace* (fantastic), *Adam* (a designer drug), *airhead* (stupid person), *angel dust* (an hallucinogenic drug), *awesome* (marvellous, great), *bad* (excellent), *bad-mouth* (to abuse someone verbally), *bimbo* (attractive but unintelligent young woman), *bonk* (an act of sex; to have sex), *bottle* (courage, guts), *brill(iant)* (fantastic, great), *chase the dragon* (to take heroin), *couch potato* (person who spends leisure time passively, mostly watching television and videos), *cowabunga* (an exclamation of exhilaration), *crack* (form of cocaine), *cred* (credibility, reputation), *crumblie* (an old or senile person). Some of these are more or less restricted to the US, others are widely used in the language of drug-taking, and some are restricted to the vocabulary of teenagers. These, and thousands more listed, in, for example, the latest edition of Eric Partridge's *Dictionary of Slang and Unconventional English,* will bump along in the language, some for a year or two, others for much longer. The survivors will be joined each year by thousands of other new slang words. Scholars collecting them have a task in perpetuity.

See RHYMING SLANG.

**slate** (verb). One of its dominant senses in BrE is 'to criticize severely, to scold' (*he was slated by his constituency party for voting against the amendment).* By contrast the word is used in AmE to mean 'to write or set down for nomination or appointment' or 'to schedule or designate (*slated a meeting of service chiefs for the following afternoon)'.* The context will normally make it clear which meaning is intended, but clearly care is called for, esp. in interpreting headlines such as *Summit meeting slated.*

**slave** makes *slavish.* See MUTE E.

**slay** (verb). Though poetic and rhetorical in BrE, it is still in ordinary use in AmE, for violent killing. The mainly

American use is sometimes carried over into UK newspapers (*Serial killer slays seven young men*). The pa.t. is *slew* and pa.pple *slain*. In the secondary slang sense 'to overwhelm with delight, to convulse with laughter' the pa.t. is usually *slayed* (*the comic scene simply slayed them*).

**sled, sledge, sleigh.** *Sled* is the mainly AmE word for 'a vehicle on runners for conveying loads or passengers esp. over snow, drawn by horses, dogs, or reindeer, or pushed or pulled by one or more persons' (*COD*), and *sledge* the BrE equivalent. A *sleigh* is 'a sledge, esp. one for riding on, rather than conveying loads, and drawn by horses or reindeer' (*COD*).

**sleigh.** See prec.

**sleight** (as in *sleight of hand*). Pronounce /slaɪt/. It is related to *sly* as *height* is to *high*.

**slew, slue** (verb). The first is now the normal spelling for this word in its sense 'to turn or swing forcibly from its ordinary position'.

**slier, slily.** See SLY.

**slime** makes *slimy*. See MUTE E.

**sling, slink.** Past tenses and past participles *slung* and *slunk*.

**sloe-worm.** See SLOW-WORM.

**slogan.** From the 16c. to the 19c. this word of Gaelic origin meant 'a war-cry or battle cry; *spec.* one of those formerly employed by the Scottish Highlanders or Borderers, or by the native Irish, usually consisting of a personal surname or the name of a gathering-place' (*OED*) (e.g. *The isle of Clareinch was the slogurn or call of war, proper to the family of Buchanan*—W. Buchanan, 1683). The less specific sense 'the distinctive note, phrase, cry, etc. of any person or body of persons' overlapped with the older sense in the 19c., and has now replaced it except in historical contexts. Throughout the 20c. slogans have become commonplace in politics of whatever kind: e.g. *Ban the bomb*; *Religion is the opium of the people*; *An equal opportunities employer*; *Business as usual*. See also ADVERTISING, LANGUAGE OF.

**slosh.** See SLUSH.

**slough.** The noun meaning 'bog, swamp' is pronounced /slaʊ/. The separate word *slough* meaning (as noun) 'a part that an animal casts' and (as verb) 'to cast or drop off' is pronounced /slʌf/. A cautionary note: many Americans pronounce the 'swamp' noun /sluː/ and spell it *slew*.

**slovenly** is used both as adj. and adv. The notional adverbial form *slovenlily* is, for reasons of euphony, not used. See -LILY.

**slow, slowly.** The primary distinction, namely that *slow* is an adjective (*take the slow train to Oxford*) and *slowly* is an adverb (*the yachts moved slowly out of the harbour*), is almost always followed in practice. But there are exceptions. In days gone by literary examples of *slow* used as an adverb abound: *But oh, me thinks, how slow This old Moon wanes*—Shakespeare, 1590; *I hear the far-off Curfeu sound ... Swinging slow with sullen roar*—Milton, 1632; *As the stately vessel glided slow Beneath the shadow*—Byron, 1812; *We drove very slow for the last two stages on the road*—Thackeray, 1848. Such evidence would perhaps lead us to expect that *slow* would have taken over many of the adverbial functions of *slowly* in the 20c., but this has not happened. Examples can be found (e.g. *where it was easy to drive slow and look into lighted uncurtained windows*—L. Ellmann, 1988), but only in carefully circumscribed circumstances. Further examples: (as a road sign) SLOW: this can be taken to be a shortened version of *go slow* or *drive slowly*; (industrial sense) *The union rep said we were to go slow from 2pm tomorrow*; (in comparatives and superlatives) *Thomas walked slower* (or *more slowly*); *James walked (the) slowest of all*; (in compounds) *slow-moving traffic*; *slow-paced lemur*. On the other hand there are circumstances when only *slowly* is permissible, e.g. immediately before a verb: *He slowly* (not *\*slow*) *closed the door*. The best course is to use *slow* as an adverb very sparingly indeed.

**slow-worm** (small legless lizard). Formerly also *sloe-worm*. It is not directly related to either *sloe* or *slow* but is derived from OE *slā-wyrm*, the first element of which is of uncertain origin.

**sludge.** See SLUSH.

**slue** (verb). See SLAY (verb); SLEW (verb).

**slumberous, slumbrous.** Now mainly poetic. Both forms occur and both can be paralleled: cf. *murderous, slanderous, thunderous* for the longer form, and *cumbrous, idolatrous, monstrous, wondrous* for the shorter. If I ever had cause to use the word I would spell it *slumberous* and pronounce it as three syllables.

**slur** (verb). The inflected forms are *slurred, slurring.* See -R-, -RR-.

**slush, sludge, slosh.** The differences are not very clear. There is the natural one, perhaps resulting from its sound, that *sludge* is usually applied to something less liquid or more solid than *slush* or *slosh,* e.g. to slimy deposits, clinging mud, or sewage, whereas thawing snow is typical *slush.* Of *slush* and *slosh,* the former is the more usual word for what is metaphorically watery stuff—twaddle or sentimentality; and *slosh* alone is a slang word for a violent blow or, as a verb, giving one.

**sly** makes *slyer, slyest, slyly, slyness, slyish* (not *slier,* etc.). Contrast DRIER; FLIER, FLYER.

**smell** (verb). **1** In BrE either *smelled* or (much less commonly) *smelt* is used for the pa.t. and pa.pple. In AmE *smelled* is the dominant form for both. See -T AND -ED. Examples: *Although he no longer smelled of sick his appetite was poor*—A. Brookner, 1988; *They* [sc. dogs] *smelt, especially when wet*—Good Housekeeping, 1989; *Barry has smelled it on winos after they've been on a three-week binge*—New Yorker, 1988; *And then he smelled something, and that wasn't his fault, was it?*—G. Keillor, 1991 (US).

**2** When used intransitively to describe the quality of a smell, the verb is normally followed by an adjective, not an adverb: *Hort's house always smelled so good*—New Yorker, 1988; *He smelled the cup of water and it smelled putrid*—Community Development Jrnl, 1988. When it is used to mean 'stink', however, an adverb of manner is the norm: *the prison cell smelled abominably. Smell* is also commonly followed by an *of* phrase: *I settled down on the leather seat that already smelled of orange peel*—M. Pople, 1986 (Aust.); *The wood smelled of beeswax*—A. S. Byatt, 1990. Had a qualifying word been inserted in either of these sentences it would undoubtedly have been an adverb, not an adjective (*smelled strongly of orange peel; smelled faintly of beeswax*).

**smite.** This ancient verb (pa.t. *smote,* pa.pple *smitten*), after centuries of use from Anglo-Saxon times onward, is now falling into disuse or is reserved for literary contexts (*When he saw Tissy moving about the quiet rooms of his house . . . his heart smote him*—A. Brookner, 1989). Possibly its most frequent sense at present is 'to infatuate, fascinate' (*he was smitten by her beauty*).

**smog.** This word, formed in 1905 from smoke + fog by Dr H. A. des Vœux, Hon. Treasurer of the Coal Smoke Abatement Society, is one of the most enduring blends of the century. Apart from its literal use it is also widely used in figurative contexts, of turgid speech, writing, etc.

**smoke** makes *smoky.* See MUTE E.

**smooth.** The noun, verb, and adverb (the last beside *smoothly*) are all so spelt now: the verb was also spelt *smoothe* from the 15c. to the 19c. For *smoothen* (verb), now much less common than it was a century ago, see -EN VERBS FROM ADJECTIVES.

**smoulder** (burn slowly, etc.). So spelt in BrE, but often as *smolder* in AmE.

**snail-like.** For hyphen, see -LIKE.

**snake** makes *snaky* (adj.). See MUTE E.

**snake, serpent.** *Snake* is the native and *serpent* the alien word. It is also true, though not a necessary consequence, that *snake* is the word ordinarily used, and *serpent* the exceptional. The OED's remark on *serpent* is 'now, in ordinary use, applied chiefly to the larger and more venomous species; otherwise only rhetorical . . . or with reference to serpent-worship'.

**snapshot** is 'a casual photograph taken quickly with a small hand camera' (COD). The corresponding verb *snapshot* is listed in the OED (with pa.t. *snapshotted,* ppl adj. *snapshotting*), with printed evidence from 1894 onward, but it leads a precarious life in the language, the more usual verb being *snap*. *Snapshooter* 'one skilled in snap-shooting birds or animals with a gun' has coexisted, since the last decade of the 19c., with the same word in its photographic sense. In photography now, the holiday-maker *snaps* or *takes snaps*, whereas the cameraman in the film or television studio *shoots*.

**sneak** (verb). Its origins are shrouded in mystery. First recorded in print in the 16c., it seems to have emerged from some uncharted dialectal area and made its way swiftly into the language of playwrights: *A poore vnminded Outlaw, sneaking home*—Shakespeare, 1596; *I hope he will not sneake away with all the money*—Dekker, 1604; *Where's Madrigall? Is he sneek'd hence?* —Jonson, 1625. From the beginning, and still in standard BrE, the pa.t. and pa.pple forms are *sneaked*. Modern UK examples: *I bet every father in the land sneaked out of the house to get an eyeful of that Miss Jane Russell*—P. Bailey, 1986; *I sneaked into his workroom*—Spectator, 1987.

Just as mysteriously, in a little more than a century, a new pa. t. form, *snuck*, has crept and then rushed out of dialectal use in America, first into the areas of use that lexicographers label as jocular or uneducated, and, more recently, has reached the point where it is a virtual rival of *sneaked* in many parts of the English-speaking world. But not in Britain, where it is unmistakably taken to be a jocular or non-standard form. The progress of *snuck* can be seen fairly clearly in the following chronologically arranged set of examples: *He grubbed ten dollars from de bums an den snuck home*—The Lantern (New Orleans), 1887; *They had all snuck in and were having a good time, making trouble*—J. T. Farrell, 1932 (US); *So I snuck off to the park and had a good read*—Oz, 1969; *This Crisp, he snuck over into the Blood's territory to make that mark*—New Yorker, 1986; *He had snuck through the orchards like a deserter*—J. Urquhart, 1986 (Canad.); *Happiness snuck up on her like a poacher in the night*—P. Carey, 1988 (Aust.); *It* [sc. bad luck] *had snuck up on him from behind as it always did*—Metro (NZ), 1988; *While I was groaning at my reflection in the bathroom mirror he snuck out of the flat*—A. Billson, 1993 (UK). What the future holds for *snuck* is unpredictable. Meanwhile no one has satisfactorily accounted for its

origin: there is no other verb in the language with infinitive -*eek* or -*eak* and pa.t. -*uck*. Consider the following verbs in /-iːk/: *creak, freak, leak, peak, peek, reek, seek, squeak, streak, wreak*, also *shriek*. None of them has shown any sign of following the path of *sneak* by acquiring a new pa.t. form.

**snivel** (verb). The inflected forms in BrE are *snivelled, snivelling*, but in AmE often *sniveled, sniveling*. See -LL-, -L-.

**snuck.** See SNEAK (verb).

**so.**

1　Phrases treated elsewhere.
2　*so long* (as formulaic farewell); *and so to—*.
3　The appealing *so*.
4　The didactic *so*.
5　Repetition of *so*.
6　*so* with past participle.
7　The explanatory *so*.
8　*so* with superlatives and absolutes.
9　*so* as conjunction, legitimate and illegitimate.

**1** For *do so*, see DO 2 (*e*); for *ever so*, see EVER 4; for *never so*, see NEVER 1; for *so far from, so far as*, see FAR; for *and so forth/and so on*, see FORTH; for *quite so*, see QUITE 1; for *so to speak*, see INFINITIVE 3(*b*); SUPERIORITY.

**2** *so long* used colloquially for 'goodbye' or '*au revoir*'. The popular view that this is an adaptation of *salaam* is without foundation. It is merely a special combination of *so* and *long*, and it has been in use (according to the *OED*) since 1865. English is richly endowed with farewelling formulas, varying greatly in their degree of formality: *adieu*; *bye(-bye)*; *see you*; *be seeing you*; *see ya*; *ta-ta*; etc. And *so to a division, and so to dinner*, etc. This formula for winding up the account of a debate or incident, borrowed directly or indirectly from Pepys, is apt to become addictive. 'It is wise to abstain from it altogether' (Fowler), or at any rate to use it sparingly.

**3** The appealing *so*. The type is *The English weather is so uncertain*. The speaker has a conviction borne in upon him, and, in stating it, appeals, with his or her *so*, to general experience as a means of confirmation; it means *as you*, or *as we all, know*. This is a natural use, but one more suitable for conversation, where the responsive nod of confirmation can be awaited, than for most kinds of writing. In print, outside dialogue, it has an air of unnecessary emphasis, even when the context is favourable, i.e. when the sentence is of the shortest and simplest kind, and the experience appealed to is really general. It is suggested that in the following extracts the context is not favourable: *For ophthalmology in the tropics a work of authority is so sadly overdue; But he does combine them ingeniously, though in instancing this very real power we feel that it might have been so much more satisfactorily expended; He was always kind, considerate, and courteous to his witnesses, this being so contrary to what we are led to expect from his successors; Slade would seem to have some of the philosophy of his kind, as well as the technique, which chiefly is the reason why one so hopes he will not be rushed on too rapidly.*

**4** The didactic *so*. This is a special form of the appealing *so*, much affected, for example, by Walter Pater and by many an enthusiastic lecturer. *In the midst of that aesthetically so brilliant world of Greater Greece* is an example. The *so* is deliberately inserted before a descriptive adjective, and is a way of saying, at once urbanely and concisely, Has it ever occurred to you how brilliant etc. it was? That is to say it differs from the *sos* in 3 in being not careless and natural, but didactic and highly artificial. Effective enough on occasion, it is among the idioms that should never be allowed to remind the reader, by being repeated, that he has already met them in the last hundred pages or so. Further examples: *Here an Englishman has set himself to follow in outline the very distinctive genius of Russia through the centuries of its difficult but always so attractive development; And still no one came to open that huge, contemptuous door with its so menacing, so hostile air.*

**5** Repetition of *so*. A change from the artificial to the artless. *So* is a much used word, but not indispensable enough to justify such repetitions of it as the following: *The pity is that for so many men who can so hardly keep pace with rising prices it should become so difficult to follow the sport; For ironically enough the very complexity of modern political life, which today makes it so necessary for the Government to improve their lines of communication with*

the people, has also done much to weaken the principal bridge that previously helped so much towards this end—the House of Commons; The situation was well in hand, but it had so far developed so little that nothing useful can be said about it, save that so far the Commander-in-Chief was satisfied.*

**6** *so* with past participle. The distinction usually recognized with VERY between a truly verbal and an adjectival past participle is not applicable to *so*; but it is well worth while, before writing plain *so*, to decide between it and *so much, so well*, etc. The insertion of *much* in the first and *well* in the second quotation after *so* would certainly be an improvement: *Admiral Farevelli reports that Tripoli batteries have been so damaged that Turkish soldiers have been forced to retire into town; Ireland being mainly an agricultural country, and England industrial, the Bill is not so suited to Ireland as to this country.*

**7** The explanatory *so*. Type: *He could not move, he was so cold.* The second clause is equivalent to a sentence beginning with *for* or *because*, and the idiom is mainly, but not solely, colloquial. In such sentences it is clear to the listener that the second clause provides an explanation of the first and no conjunction is needed.

**8** *so* with superlatives and absolutes. *So*, when it qualifies adjectives and adverbs, means to such a degree or extent; it is therefore not appropriately applied to a superlative, as in *The difficult and anxious negotiations in which he has taken so foremost a part in Paris*; or to an absolute, as in *It is indeed a privilege to be present on so unique an occasion.*

**9** *so* as conjunction, legitimate and illegitimate. The usual practice in standard English as late as the 19c. was to use *so that* to denote result or logical consequence, and sometimes = 'in order that': *[They] had conveighed their shippes in to the havens, so that he could not fight with them on the sea*—1548 in OED; *We will spend our evenings ... at our own lodgings, so that we may be found*—c1760 in OED; *The turf roof of it had fallen in: so that the hut was no use to me*—R. L. Stevenson, 1886. Constructions using *so* alone are recorded from medieval times, but are no more than sporadic. First in America in the 19c., and gradually elsewhere, *so* alone has gradually established itself in standard use, esp. in spoken English. When I began to collect such examples some years ago I placed them under the heading 'so as a loose connective'. In the face of the evidence it seems clear that many standard speakers now feel that the use without that is unobjectionable. Thus used, so normally follows a comma, but the punctuation is sometimes altered so that it leads a sentence.

Examples of so alone: *I'll knock ... and peep round the door When I come back, so you'll know who it is*—R. Frost, 1913 (US); *Shovelling coal up the back of the chimney, throwing it on so it would burn for hours*—C. P. Snow, 1951; *The Nigerian authorities asked for him to be returned under the Fugitive Offenders Act so he could stand trial on charges of treasonable felony*—1965 in Gowers; *My father had been a minor diplomat, so as a child I had lived in France, Turkey and Paraguay*—G. Greene, 1980; *The wind blew straight into his nose and mouth so he had to turn away for a moment to breathe*—Maggie Gee, 1985; *My husband ... kept his head down so he would not have to see the barn*—New Yorker, 1986. *I didn't like him, though, so I did nothing to alleviate it*—R. Elms, 1988; *He smashed the front windows first, so I just ran and hid*—N. Virtue, 1988 (NZ); *There is no clear record of the debates within the federal government, so the political dynamics of such delay are only conjecture*—Dædalus (US), 1989; (after a full stop) *But work is scarce, especially in winter. So he watches a lot of TV, soaps,* [etc.]—New Yorker, 1989.

**sobriquet** /ˈsəʊbrɪkeɪ/ is the recommended spelling, not *soubriquet*. Print in romans, not italics.

**sobriquets.** Under this heading are grouped together fifty or so out of the limitless number of NICKNAMES or secondary names that have become so firmly attached to particular persons, places, or things, as to be intelligible when used instead of the real or primary name. In the later part of the 20c., for example, who would fail to recognize the reference lying behind the sobriquet *the Iron Lady*? Or, for that matter, would anyone be baffled by *the Bard of Avon* or *the Maid of Orleans*? A heavily selective list follows:

Albion (England); *alma mater* (one's university); *Athens of the North* (Edinburgh); *Auld Reekie* (Edinburgh); *Auntie* (BBC); *Bard*

*of Avon* (Shakespeare); *Beefeater* (Yeoman of the Guard); *Big Apple* (New York); *Black Country* (industrial west Midlands); *Black Death* (14c. plague); *Black Maria* (prison van); *Black Prince* (eldest son of Edward III); *Blighty* (Britain, used by British servicemen abroad); *Buck House* (Buckingham Palace); *City of Dreaming Spires* (Oxford); *Cœur de Lion* (Richard I); *Dark Continent* (Africa); *Elia* (Charles Lamb); *Emerald Isle* (Ireland); *Garden of England* (Kent); *Great Cham* (Samuel Johnson); *Great Unwashed* (the lower orders); *Great Wen* (London); *Iron Duke* (Wellington); *Iron Lady, Iron Maiden* (Lady Thatcher); *Jack Tar* (sailor, RN) *John Bull* (Englishman); *Kiwi* (New Zealander); *knight of the road* (highwayman; tramp); *Latin Quarter* (hub of Paris's intellectual life); *Left Bank* (artistic district of Paris); *Maid of Orleans* (Joan of Arc); *Mother of Parliaments* (British Parliament); *Old Contemptibles* (British Expeditionary Force 1914); *Old Nick* (devil); *Old Lady of Threadneedle Street* (Bank of England); *Paddy* (Irishman); *Pom(my)* (British migrant to Australia or New Zealand); *Sage of Chelsea* (Carlyle); *Senior Service* (navy); *sport of kings* (horseracing); *staff of life* (bread); *Stars and Stripes* (US flag); *Swan of Avon* (Shakespeare); *The Thunderer* (The Times); *Tinsel Town, Tinseltown* (Hollywood); *Tommy (Atkins)* (British soldier(s)); *Uncle Sam* (USA); *Union Jack* (British flag) *Virgin Queen* (Queen Elizabeth I of England); *Warrior Queen* (Boadicea, Boudicca); *Wizard of the North* (Walter Scott).

For much fuller treatment, see e.g. B. Freestone, *Harrap's Book of Nicknames and Their Origins* (1990), and L. Urdang, *A Dictionary of Names and Nicknames* (1991).

**socage** (feudal tenure of land). Thus spelt, rather than *soccage*.

**so-called.** This much-used, often somewhat scornful, adjective frequently qualifies a word or phrase that is placed within quotation marks (e.g. *the so-called 'generation gap'*). Strictly speaking the quotation marks are not needed, but their use or omission is just a matter of taste.

**soccer** (formed irreg. from Association football + -er). Now the only spelling, though *socker* was often used in the last decade of the 19c. Most words having *cc* before *e, i, y* are pronounced with /-ks-/: *accept, eccentricity, success, accident, flaccid, coccyx*. The only exceptions, apart from *soccer*, seem to be *baccy* and *recce*, where the /-k-/ is taken over from *tobacco* and *reconnaissance* respectively.

**sociable, social.** For some confusion between pairs of adjectives in *-able* and *-al*, see EXCEPTIONABLE; PRACTICABLE. No such misuses occur with the present pair as with those; there is merely a tendency to use *social* not where it is indefensible, but where the other would be more appropriate. And this mostly in the past. Roughly, *social* means 'of, or (living) in, or belonging to, or interested in society'; and *sociable* 'fitted for or liking the society of other people, ready and willing to talk and act with others'. *Social* is rather a classifying, and *sociable* rather a descriptive adjective: man is *a social being*; Jones is *a sociable person*; people are invited to *a social evening*, and say afterwards (or do not say) that they had *a very sociable evening*. The *social services* are partly the concern of *social workers*. *Social security* is State assistance given to those lacking in economic security and welfare, e.g. the aged and the unemployed. The *OED*, under a definition of *social* that includes 'sociable' as an equivalent, gives two quotations in which *sociable* would now have been preferred: *His own friendly and social disposition—*J. Austen, 1816; *He was very happy and social—*Miss Braddon, 1878. Such interchanges of meaning seem unlikely now, but one does not know what the future holds.

**sociolinguistics.** A term introduced in the late 1930s for a subdivision of linguistics which may be defined briefly as 'the study of language in relation to social factors'. By studying such linguistic features as *h*-dropping (omitting the initial *h* when pronouncing *hay, horse*, etc.), the full pronunciation of postvocalic /r/ (as in *bar, heart*, etc.), the glottal stop (e.g. in the pronunciation of *butter* as /ˈbʌʔə/ and *pattern* as /ˈpæʔən/), kinship terms, certain grammatical constructions (e.g. the double negative in *I didn't see nothing*), and many other aspects of esp. non-standard pronunciation, grammar, and vocabulary, sociolinguists have established a broad stratification of types of English. Pioneering scholars in the subject include Peter Trudgill in Britain (e.g. his *Sociolinguistics: An Introduction*, 1974) and William Labov in the United States (e.g. in *The Social Stratification of English in New York City*, 1966). The subject is now a routine part of university courses in linguistics, and the annual

output of books and papers about it is formidable. Taken out of context some of the assertions of sociolinguists have been fiercely contested in that they seem to claim that all varieties of English, standard, regional, uneducated, inner city-based, are of equal value, aesthetically and linguistically (e.g. 'All varieties of a language are structured, complex, rule-governed systems which are wholly adequate for the needs of their speakers'—Trudgill, 1974). The great debate continues.

It hardly needs to be said that the aim of the present book is to identify the main elements of standard English in the UK, both those that are stable and those that are in process of changing, with numerous observations on standard American English, and with passing comments on differences of substance in the other main varieties of English in countries where English is spoken as a first language.

**sociologese.** 'We live in a scientific age, and like to show, by the words we use, that we think in a scientific way': thus Gowers in the second edition (1965) of this book. He then went on to attack what he called *sociologese*, i.e. a style of writing supposedly typical of sociologists 'which is over-complicated or jargonistic or abstruse' (as the *OED* has it). He built up his argument: 'Sociology is a new science concerning itself not with esoteric matters outside the comprehension of the layman, as the older sciences do, but with the ordinary affairs of ordinary people. This seems to engender in those who write about it a feeling that the lack of any abstruseness in their subject demands a compensatory abstruseness in their language. Thus, in the field of industrial relations, what the ordinary man would call an informal talk may be described as a *relatively unstructured conversational interaction*, and its purpose may be said to be *to build, so to speak, within the mass of demand and need, a framework of limitation recognized by both worker and client*. This seems to mean that the client must be persuaded that, beyond a certain point, he can only rely on what used to be called self-help; but that would not sound a bit scientific.'

Sociologese is an easy target and one that has been frequently targeted, not least by Gowers himself, while admitting that 'there are of course writers on sociological subjects who express themselves clearly and simply'. Gowers' examples of over-dense sociological language, even though they are regrettably cited without attribution, are worth preserving: (in the field of industrial relations) *The technique here reported resulted from the authors' continuing interest in human variables associated with organizational effectiveness. Specifically, this technique was developed to identify and analyse several types of interpersonal activities and relations, and to provide a method for expressing the degree of congruence between two or more of these activities and relations in indices which might be associated with available criteria of organizational effectiveness.* (on the reason why the 'middle class' speak differently from the 'lower working class') *The typical, dominant speech-mode of the middle class is one where speech becomes an object of perceptual activity, and a 'theoretical attitude' is developed towards the structural possibilities of sentence organization. This speech-mode facilitates the verbal elaboration of subjective intent, sensitivity to the implications of separateness and difference, and points to the possibilities inherent in a complex conceptual hierarchy for the organization of experience. [The lower working class] are limited to a form of language use which, though allowing for a vast range of possibilities, provides a speech form which discourages the speaker from verbally elaborating subjective intent, and progressively orients the user to descriptive rather than abstract concepts.* (on family life) *The home then is the specific zone of functional potency that grows about a live parenthood; a zone at the periphery of which is an active interfacial membrane or surface furthering exchange—from within outwards and from without inwards—a mutualising membrane between the family and the society in which it lives.*

For further comments on related matters, see GOBBLEDEGOOK; JARGON; OFFICIALESE. A useful book on this general field is Walter Nash's *Jargon: Its Uses and Abuses* (1993).

**socker.** See SOCCER.

**socketed.** Not *-tted*. See -T-, -TT-.

**socle** (plinth supporting a column, etc.). Pronounce /'səʊkəl/.

**Socrates.** Pronounced /'sɒkrəti:z/ in English despite the long *o* in Gk Σωκράτης. See FALSE QUANTITY.

**suddenness.** So spelt, with *-nn-*.

**soft** is primarily an adjective. After centuries of idiomatic use as an adverb (beside the 'regular' form *softly*) it had become mostly poet. and arch. by the end of the 19c.: *There is sweet music here that softer falls Than petals from blown roses on the grass*—Tennyson, 1833; *Sleep soft, beloved!*—E. B. Browning, 1850; *The wanderer ... Halts on the bridge to hearken How soft the poplars sigh*—Housman, 1896. It may still be used in a few combinations (*soft-spun, soft-tinted*), but for all practical purposes it has lost its primary power as an adverb.

**soften.** In the *OED* (1913) only /'sɔːfən/, but now regularly /'sɒfən/ with a short stem vowel.

***soi-disant.*** English is well supplied with words of similar meaning (of a person: *self-styled*; of a thing: *so-called, pretended*) and application (e.g. *ostensible, professed, professing, self-styled, so-called, supposed, would-be*) to the point that one begins to wonder about the semantic space left for another. Nevertheless *soi-disant* is in standard use.

**sojourn.** In standard BrE pronounced /'sɒdʒ(ə)n/, /-dʒɜːn/, or /'sʌdʒ-/ with equal acceptability. Note the following remark in David Lodge's novel *Nice Work* (1988): 'You seem to have acquired a very utilitarian view of universities, from your sojourn in Rummidge,' said Professor Penrose, who was one of the very few people Robyn knew who used the word sojourn in casual conversation.

**solarium.** Pl. *solaria*. See -UM 2.

**solder.** In BrE normally pronounced /'səʊldə/ or /'sɒl-/, but in AmE with the *l* silent.

**solemnness.** The recommended form, not *solemness*.

**Solicitor-General:** capitals and hyphen.

**soliloquy.** See MONOLOGUE.

**solo.** Pl. *solos*; see -O(E)S 6; in It. music *soli* /'səʊli/.

**so long** = goodbye. See SO 2.

**soluble. solvable,** make *insoluble, unsolvable:* see IN- AND UN-. Substances are *soluble* (or *dissolvable*), not *solvable*; problems are *soluble* or *solvable*. See also DISSOLUBLE and RESOLUBLE.

**solve.** Pronounce /sɒlv/, not /sɔʊlv/.

**sombre.** Thus spelt in BrE, but *somber* in AmE.

**sombrero** (broad-brimmed Mexican hat). Pl. *sombreros:* see -O(E)S 6.

**some.** As the central sense of *some* (adj.) is 'an unspecified amount or number of' (*some cheese, some apples; some of them*), the word has tended to be regarded with suspicion when it is used meiotically or ironically to suggest that something or someone is worthy of consideration, praise, or blame, as in the types *This is some party* ( = a fine party); *That's some example he sets us* ( = a 'fine' example, ironically); and *'They're some geologists,'* he added with unwilling admiration ( = remarkable). Such uses are recorded from 1808 onward in the *OED*, most famously one by Winston Churchill in a speech in 1941: *When I warned them* [sc. the French government] *that Britain would fight on alone whatever they did, their Generals told their Prime Minister and his divided Cabinet: 'In three weeks England will have her neck wrung like a chicken.' Some chicken! Some neck!*

But despite all the examples (including the Churchillian one), and the span of time involved, the use still seems to be informal and, as it were, on probation, in standard BrE. Perhaps, by cross-association it is linked in the public mind with the distinctly slangy adverbial phr. *to go some* 'to go well or fast; to do well' (e.g. *He'd known the girl for two months; for Daughtry that was going some*—H. Lieberman, 1982); or with the dialectal and American adverbial uses of *some* meaning 'somewhat', as in the following examples: *We had done been feeding it* [sc. a horse] *for two-three days now by forced draft . . . and it looked some better now than when we had brung it home*—W. Faulkner, 1940; *An old worker . . . turned the handle and tried it with a few roots. Asked what he thought of it he said with conviction: 'It's some stiff, maaster.'*—G. E. Evans, 1956. The more likely explanation, however, is that

multi-meaning words such as *some* tend to generate various informal, dialectal, and even substandard uses as necessary parts of their complex structure. Language is always in flux, and it is an essential requirement of such a system that informal candidates stand ready on the sidelines to join the rule-based standard game when, as so often happens, reinforcing elements are needed to replace vanishing or weakened ones.

**-some.** 1 It is a useful and revealing exercise to group adjectives ending in *-some* (ultimately representing OE *-sum*) according to the period in which they entered the language. Among the OE and ME words, those still in use include *buxom* (OE *buhsum*), *cumbersome, fulsome, handsome, irksome, loathsome, noisome, wearisome, wholesome,* and *winsome,* but the casualty rate has been high. In the 16c. appear *awesome, brightsome, darksome, gruesome* (later revived by Walter Scott), *quarrelsome, tiresome,* and *troublesome* among others; numerous others have subsequently fallen by the wayside. Of later date are *adventuresome* 18c., *bothersome* 19c., *cuddlesome* 19c., *fearsome* 18c., and *lonesome* 17c., all still current, but many others were just short-lived formations, e.g. *clipsome* fit to be clasped, 19c.; *dabblesome* given to dabbling, 19c. The general picture is of a suffix lying ready at hand waiting to be clapped on to a virtually limitless class of adjectives or nouns, but failing to produce words of long standing except in a small number of cases. See also AWESOME; GLADSOME.

2 OE *-sum* was also used after the smallest numerals in the genitive plural to form nouns meaning 'a group of (so many)': *twosome, threesome, foursome* (in all of which *-some* is pronounced /səm/). In recent times in AmE (occas. elsewhere) *-some* has been affixed to larger numbers, with full pronunciation as /-sʌm/, to mean 'about, approximately': e.g. *Me, I like the .45 Colt, but I been shooting that little baby for twenty-some years*—T. Clancy, 1987; *a silver-haired Catholic priest at the head of it* [sc. a coffin] *with a teen-age acolyte in a white smock, and forty-some mourners*—New Yorker, 1988; *Of our thirty-some days together so far, some have been better than others*—G. Keillor, 1989; *Forty-some hours since she had seen Charles, or heard from him*—New Yorker, 1989.

**somebody, someone.** 1 See AGREE-MENT 6; ANY 1(*b*), 3(*a, e*).

2 It is a matter of curiosity that the relevant section (1885) of the *OED* treats *anybody* as a main entry but merely subsumes *anyone* under that for *any*. This is not just a historical accident: from the ME period onward *anybody* (or *ani-body* etc.) was a fully fledged indefinite pronoun, whereas *anyone* (as distinct from the type 'any one of the two will suffice') does not appear until 1711. It is quite otherwise with *somebody* and *someone*, which have been in constant parallel use since the beginning of the 14c. Both stand as main entries in the *OED* (1913). The main controversial uses of *somebody* and *someone* have been dealt with elsewhere in this book, but it might be of interest to add here some additional examples from diverse sources of the late 20c. tendency to use a plural pronoun with backward reference to the by-nature-singular indefinite pronoun *someone*: *The extreme irritation one always feels when someone proves to be more percipient than one has decided they shall be*—A. S. Byatt, 1967; *For someone to dream about their own death is a ... sign that it is soon to come about*—Oxford student's essay, 1985; *The woman at the door shouted back at him, 'It's someone for Florrie.' 'Well, tell them to go upstairs.'* (only one caller)—R. Rive, 1990 (SAfr.). The trend seems irreversible.

**some day, someday.** Until the beginning of the 20c. normally written as two words. Now, in the types *let us meet some day soon* and *the debate will be held on some day to be agreed*, the two words continue to be written separately. But when the meaning is 'at some time in the indefinite future' usage favours the conjoined form *someday*, esp. in AmE: e.g. *someday we must talk about accreditation and the graduate school*—*Dædalus*, 1988; *he likes writers and wants to be one someday*—J. McInerney, 1988.

**someone.** See SOMEBODY.

**someplace.** 1 See ANY 3(*f*); EVERYPLACE.

2 The adverb *somewhere*, i.e. the normal form in BrE, is still widely used in AmE (e.g. *A television technician with a belt full of toys came out of the nave on his way somewhere*—*New Yorker*, 1988). In AmE (rarely elsewhere) *someplace* has surged into prominence in the second half of the 20c.: e.g. *she can get a good job herself someplace and they can get married*—Lee Smith, 1983; *they were truants from New Utrecht High School or someplace*—*New Yorker*, 1987; *If this discourse is not someplace in the culture, it is a rather tall order for us to try to invent it here*—*Dædalus*, 1988; *as if she had pushed off someplace else in her head*—N. Virtue, 1988 (NZ).

**somersault.** The several older variant spellings, *somersaut, somersalt, summersault*, etc., have now been discarded in favour of *somersault*.

**-something.** Recently in AmE (occasionally elsewhere) *-something* has been appended to round numbers, esp. *twenty* to *ninety* to indicate the approximate age of a person. Cf. -SOME 2. Examples: *In 'Ladykiller' ... a lonely thirty something woman who has stopped being a street cop* [etc.]—*Chicago Tribune*, 1992; *Yet in the scheme of things this year, these two fortysomething strivers* [sc. presidential candidates] *seem kind of old to me*—*Newsweek*, 1992.

**sometime, sometimes.** The second of these is invariably written as one word in all forms of English (e.g. *He could play Chinese checkers, and sometimes invited me to have a game*—A. Munro, 1987), but there is some variation in the use of *some time* and *sometime*. When used as an adj. (*a sometime Fellow of Balliol College, Oxford; his sometime colleague*) it must be written solid. Cf. LATE. Most people write the elements separately when the meaning is 'a certain amount of time' (e.g. *Some time after this interview, it happened that Mr. Cuff ... was in the neighbourhood of poor William Dobbin*—Thackeray, 1847/8; *We'll need some time to consider the matter*—mod.). When the contextual meaning is 'at some future time' the two words are normally run together (e.g. *Let's talk about them sometime. Not just now*—P. Lively, 1987; *Sometime in the 1990s Norway is short-odds favourite to become the 13th member of the EEC*—*Economist*, 1988. But this neat distinction is not always observed. Compare the following example ( = a certain amount of time) with the Thackerayan one above: *his* [sc. Kieslowski's] *last film No End, which caused the Polish authorities collective heartburn and took sometime to achieve distribution*—*Times* 19 May 1988.

Using a joined-up form to represent an adj. + noun has nothing to commend it.

**somewhen.** This potentially useful compound adverb has had a phantom-like existence for nearly two centuries, usu. coupled with *somehow* or *somewhere*, e.g. *I shall write out my thoughts more at length somewhere, and* some*when, probably soon*—J. S. Mill, 1833; *I cherished the belief that somehow and somewhen I should find my way to Oxford*—J. C. Masterman, 1975. At no time has it been in common use.

**somewhere.** See SOMEPLACE 2.

**sonant.** In phonetics, 'a consonant that can be either syllabic or non-syllabic; a continuant or nasal, as /l,m,n,ŋ,r/' (*SOED*).

**songstress.** See -ESS.

**son-in-law.** See -IN-LAW.

**sonnet.** A poem consisting of 14 lines (of 11 syllables in It., generally 12 in Fr., and 10 in Eng.), the English ones being arranged in three main patterns, the Petrarchan (or Italian), the Shake-spearian, and the Miltonic. Further details of the main types can be found e.g. in the *Oxford Companion to English Literature* and in Baldick's *Concise Oxford Dictionary of Literary Terms* (1990). Examples of each of the three types:

Petrarchan

*The world is too much with us; late and soon,*
  *Getting and spending, we lay waste our powers:*
*Little we see in Nature that is ours;*
*We have given our hearts away, a sordid boon!*
*This sea that bares her bosom to the moon;*
*The winds that will be howling at all hours,*
*And are up-gathered now like sleeping flowers;*
*For this, for everything, we are out of tune;*
  [octave]
*It moves us not.—Great God! I'd rather be*
*A Pagan suckled in a creed outworn;*
*So might I, standing on this pleasant lea,*
*Have glimpses that would make me less forlorn;*
*Have sight of Proteus rising from the sea;*
*Or hear old Triton blow his wreathèd horn.*

(Wordsworth)

*Note.* The Petrarchan sonnet has a break in sense between the octave and the sestet, two rhymes only in the octave, arranged abbaabba, and two, or three, other rhymes in the sestet variously arranged, but never so that the last two

lines form a rhymed couplet unless they also rhyme with the first line of the sestet.

Shakespearian

*Let me not to the marriage of true minds*
  *Admit impediments. Love is not love*
*Which alters when it alteration finds,*
  *Or bends with the remover to remove.*
*O, no! it is an everfixèd mark,*
  *That looks on tempests and is never shaken;*
*It is the star to every wandering bark,*
  *Whose worth's unknown, although his height be taken.* [octave]
*Love's not Time's fool, though rosy lips and cheeks*
  *Within his bending sickle's compass come;*
*Love alters not with his brief hours and weeks,*
  *But bears it out even to the edge of doom.*
*If this be error, and upon me proved,*
  *I never writ, nor no man ever loved.*

(Shakespeare)

*Note.* In the Shakespearian sonnet, though the pause between the octave and the sestet is present, the structure consists less of those two parts than of three quatrains, each with two independent rhymes, followed by a couplet again independently rhymed—seven rhymes as compared with the Petrarchan four or five.

Miltonic

*When I consider how my light is spent,*
  *Ere half my days, in this dark world and wide,*
*And that one talent which is death to hide*
  *Lodged with me useless, though my soul more bent*
*To serve therewith my Maker, and present*
  *My true account, lest he returning chide,*
*'Doth God exact day-labour, light denied?'*
  *I fondly ask. But Patience, to prevent* [octave]
*That murmur, soon replies: 'God does not need*
  *Either man's work or his own gifts; who best*
*Bear his mild yoke, they serve him best. His state*
*Is kingly: thousands at his bidding speed,*
  *And post o'er land and ocean without rest;*
*They also serve who only stand and wait.'*

(Milton)

*Note.* Of the Miltonic sonnet, which follows the Petrarchan in the arrangement of the octave, the peculiarity is that the octave and the sestet are worked into one whole without the break of sense elsewhere observed.

**sonorous.** The pronunciation /ˈsɒnərəs/ seems to have replaced /səˈnɔːrəs/ as the dominant form in standard BrE at some point in the 1980s. In other words the stress has moved from the second to the first syllable.

**soprano.** Pl. *sopranos* (see -O(E)S 6) or, less commonly, *soprani* /-ni/.

**sorceress.** See -ESS.

**sore** (adv.). In common use from Anglo-Saxon times until about the end of the 19c., this adverb has now retreated into the areas that lexicographers label 'now chiefly *arch.* and *dial.*'. Examples of faded uses: *Although it griev'd him sore*—Cowper, 1782; *For his blood was flowing; And he was sore in pain*—Macaulay, 1842; *The sore-pressed garrison which had retreated to its last defence*—G. Macdonald, 1866.

**sort** (noun). **1** In the ordinary sense 'a group of things etc. with common attributes; a kind', the normal use is shown in phrases such as *a fear of the right sort, the rosettes are of two sorts, nothing of the sort, he is some sort of doctor, all sorts of people watch football*.

**2** *these/those sort of*. From the 16c. onward, *sort* has been used collectively, preceded (illogically) by *these* or *those*: e.g. *Inchoatives* ... *are those sort of Verbs which express a gradual proceeding in any action*—E. Phillips, 1671; *'Those sort of rules are all gone by now,' said Mr. Arabin*—Trollope, 1857; *These sort of people are only interested in lining their pockets*—J. Leland, 1987. Not unexpectedly, the plural form *these/those sorts of* is also used: *He* ... *did an infinity of those sorts of things which were not professionally required of him*—T. Hook, 1825; *To afford her apartment in New York, she often took these sorts of library fellowships*—New Yorker, 1989. The type *these/those sort of* should now be used only in informal contexts. See KIND (noun) 2.

**3** Adverbial *sort of*. Since the late 18c., *sort of* has been used adverbially in somewhat informal contexts to mean 'in a way or manner; to some extent or degree; somewhat'. Examples: *I'll sort of borrow the money from my dad until I get on my own feet*—G. B. Shaw, 1903; *It just happened, sort of, and we couldn't either of us help it*—M. Laski, 1952; *I knew sort of underneath that I did want to do it*—J. Gathorne-Hardy, 1984; *The Shorewalkers looked sort of like mountain natives*—New Yorker, 1986; *I can't reproduce the way he talks* ... *but he just sort of zooms off*—Julian Barnes, 1991. See KIND (noun) 3.

**sotto voce.** Pronounce /ˌsɒtəʊ ˈvəʊtʃi/ in English.

**soubriquet.** See SOBRIQUET.

**Soudan, Soudanese.** Former spellings of *Sudan, Sudanese.*

**soufflé** /ˈsuːfleɪ/. The accent is obligatory.

**sough** (verb) (make a sound as of wind in trees, usu. encountered in the form *soughing*) and (noun). Both words may be pronounced as either /saʊ/ or /sʌf/. The first of these is recommended. The unrelated words *sough* noun ( = boggy or swampy place) and verb ( = face a ditch with stone) have only one pronunciation, /sʌf/.

**soulless.** Thus spelt, no hyphen.

**sound** (adv.). More or less restricted to the collocations *sound asleep* and *the sounder* (e.g. *he slept the sounder after a long walk*). Contrast *she slept soundly* (not *sound*) *for eight hours; the home team was soundly beaten.*

**sound bite,** a short pithy extract from a recorded interview, speech, etc. This term, and also *spin doctor* (a senior political spokesman or -woman presenting political views in a favourable light), and *photo opportunity* (an opportunity for media photographers to take pictures of well-known people) are among the unlovely media-political products of the 1980s. All three terms emerged in the US, but rapidly spread to other parts of the English-speaking world.

**soundly.** See SOUND (adv.).

**soupçon.** See GALLICISMS 2.

**south-.** Compounds (*south-east*, etc.) are pronounced with /θ/. Of the derivatives, *southerly, southern,* and *southerner* have /ð/, while *souther* (a south wind), *southing* (distance travelled or measured southwards), *southpaw,* and *southward* have /θ/ (except that in nautical language *southward* is pronounced /ˈsʌðəd/).

**southerly.** For some uses of this set of words, see EASTERLY.

**Soviet.** Since 1991, when the Union of Soviet Socialist Republics (USSR) broke up into independent republics, the word

729

Soviet has largely fallen into disuse, except with historical reference. In standard BrE the word is variously pronounced /'səʊvɪət/ and /'sɒvɪət/. Cf. RUSSIAN.

**sow** (verb) (scatter or plant seed, etc.). The pa.t. is *sowed* and the pa.pple either *sown* (the more usual form) or *sowed*.

**spadeful.** Pl. *spadefuls.* See -FUL.

**spastic** (adj.). This term, used since the 18c. of certain medical conditions (esp. cerebral palsy) characterized by spasmodic movements of the limbs, was ignorantly and offensively used by some people in the 1970s and 1980s as a term of abuse directed at anyone judged to be uncoordinated or incompetent. Also as noun.

**spats.** Short for *spatterdashes*: see ABBREVIATIONS 1(a).

**spavined** (adj.), not *-nned.* See -N-, -NN-.

**-speak.** Orwell's terms *Oldspeak* (standard English) and *Newspeak* (a sinister artificial language used for official communications), which he used in *Nineteen Eighty-Four* (1949), gave the English-speaking world a new formative element-*speak* denoting 'a particular variety of language or characteristic mode of speaking', as the OED has it. They were in due course joined by many others (some of them temporary formations), including *Haigspeak* (duplicitous talk), *airspeak* (unambiguous English used by air traffic controllers), and *seaspeak* (similarly unambiguous language being adopted by mariners). The combining form has now produced a virtually limitless class of contextually transparent formations. In the last few years I have noted the following examples, the great majority of them from AmE sources. (Several of them are taken from newspaper headlines and do not occur in the text of the articles.)

*archi-speak* (architects), *Britspeak, Bushspeak, catalogue-speak, Clintonspeak, collegespeak, criticspeak, diplo-speak* (diplomats), *fashionspeak, Fedspeak* (Federal government), *gutterspeak, idiotspeak, modelspeak, netspeak* (TV networks), *Nintendo-speak* (computer game), *taxspeak* (language used by politicians about their taxation policy). *The Barnhart Dict. New English* (1990) lists *artspeak, discospeak, Freudspeak, Olympspeak,* and *splitspeak* (the

**sow | specious**

vocabulary of broken relationships). Other scholars, other lists.

**spec,** = speculation. See ABBREVIATIONS 1(b).

**special.** See ESPECIAL(LY).

**speciality, specialty.** The two words, while they seem to cry out for different roles, have made little progress in that direction. The COD has it about right in listing two main senses for *speciality* in BrE: (a) a special pursuit, product, operation, etc., to which a company or a person gives special attention; (b) a special feature, characteristic, or skill. It then says that for both these senses *specialty* is also used, esp. in NAmer. It also lists a technical sense of *specialty* in law: an instrument under seal; a sealed contract. The following examples seem to support the distinctions set down in the COD: (*speciality*, all UK) *I like deceiving myself. It is comfortable. It is the House Speciality*—H. Mantel, 1986; *We had eaten nothing with the champagne except a small dish of potato crisps, a speciality from the island of Maui, thick and gnarled like tree bark*—D. Lodge, 1991; *The sons do a roaring trade with the footpads of Deptford fields, pewter a speciality*—L. Norfolk, 1991; (*specialty*, all NAmer.) *She considered dog issues her specialty*—T. Drury, 1991; *Big books came not only from the action-oriented surgical specialties, but the cerebral medical specialties*—Logos, 1992; *Al Roker, of Channel 4 News, m.c.'d, and since his specialty is weather, not music, he probably isn't to blame for the selection*—New Yorker, 1992; *The voice of Victor insisted that I order the Truite au Bleu, the specialty of the place.*—K. Weber, 1993.

**specially.** See ESPECIAL(LY).

**specialty.** See SPECIALITY.

**specie** /spiːʃi/ is coin money as opposed to paper money.

**species** /'spiːʃiːz/, prissily /'spiːs-/, is unchanged in the plural. The OED (1914) gave a three-syllabled pronunciation, thus /'spiːʃiiːz/, as a legitimate variant (cf. -IES), but this form dropped out of use at some later date.

**specious.** Like its Latin original (*speciōsus*), English *specious* began (in the 14c.) its life meaning 'fair or pleasing to the eye'. From then until about 1800 it is safe to assume that in most contexts *specious* persons, statues, buildings, flowers, birds, and so on, are being praised for their handsomeness, beauty, brilliance, etc. But caution is called for. Overlapping with the favourable sense, from about the beginning of the 17c., a qualification began to apply: having a fair or attractive appearance, yes, but in reality devoid of the qualities apparently possessed. From there it was an easy step to the general sense 'plausible, apparently convincing, but in reality wanting in genuineness, fallacious, etc.' Now it can only be used in a derogatory manner.

**specs,** = spectacles (for the eyes). First recorded as *specks* in 1807 and as *specs* in 1826. See ABBREVIATIONS 1(b).

**spectacles.** See GLASSES. BrE examples (one showing the normal reduced form in compounds): *She adjusted a pair of spectacles which hung on a chain round her neck*—A. Brookner, 1990; *his father sat there with his green spectacle-case in his hand*—P. Fitzgerald, 1990.

**spectator.** Thus spelt, not -*er*. Stressed on the second syllable.

**spectre.** Thus spelt in BrE, but usu. as *specter* in AmE. See -RE AND -ER.

**spectrum.** Pl. usu. *spectra*. See -UM 3.

**speculator.** Thus spelt, not -*er*.

**speculum.** Pl. usu. *specula*. See -UM 3.

**speed** (verb). The pa.t. and pa.pple forms are *sped* when the basic meaning is 'go fast' (*cars sped past*; *he had got into his car and had sped off down the road*). For other senses of the verb and for the phrasal verb *speed up* the pa.t. and pa.pple are normally *speeded* (*up*), e.g. in the intransitive sense '(of a motorist, etc.) travel at an illegal or dangerous speed' (*he speeded up and went through the traffic lights at 60 mph*; *the process of reform must be speeded up*).

**spell** (verb). Throughout this book and in my other work I have used *spelt* as the pa.t. and pa.pple of this verb, but the variant form *spelled* is widespread, esp. in AmE, and also when the meaning is 'explained in detail'. See -T AND -ED. Examples: (spelt) *men who not only rhymed*

*Moon with June but also thought they were spelt the same way*—*Sunday Times*, 1987; (spelled) *She was very handsome, but she spelled trouble*—S. Bellow, 1987; *Bech tried to … inscribe the names, spelled letter by letter*—*New Yorker*, 1987; *This is spelled out by Langland in two passages*—*English*, 1988; *Lil took it as an omen, sure that such an occurrence spelled bad luck*—S. Mackay, 1992.

The pa.t of the unrelated (mainly AmE) verb *spell* 'to relieve or take the place of (a person) in work, etc.' is always *spelled*: *That night Jake stayed on the sofa in the living room, spelling Olivia at her mother's side*—*New Yorker*, 1994.

**spel(l)icans.** See SPILLIKINS.

**spelling.**

1 An outline history.
2 Spelling reform.
3 Some notes on the relationship of English spelling and pronunciation.
4 Commonly misspelt words.
5 House style.
6 Double and single letters for consonantal sounds.
7 Cross-references.
8 Miscellaneous.

**1 An outline history.** From the time of the earliest records of English in the 8c. until the end of the 20c. the conventions governing the spelling of words in the language have changed radically several times. This sketch cannot do justice to what is in fact an absorbing story. Before 1066, and going back to the 8c., spelling was governed by the use of a combination of the roman alphabet and of a limited number of symbols from the earlier runic alphabet, including æ 'ash', þ 'thorn', ð 'eth', and ρ 'wynn' (w). Everything set down on parchment, etc., was written or (on wood or stone) carved by hand. Manuscripts containing the texts of poems, sermons, proverbs, and so on, were copied and recopied, often with striking illustrations, and with many of the letters having shapes characteristic of particular monasteries or particular areas. Scribal conventions were by any standard strict. Spelling variation tended to reflect dialectal differences of pronunciation, but in general terms spelling was rule-governed in a manner not matched in the next period, namely 1066 (Norman Conquest) until 1476 (first printed book). The OE texts that have come down to us can be said to show the closest relationship between sound and symbol of any period of the language. A typical passage:

Leofan men, ʒecnapað þæt soð is: ðeos þorold is on ofste, 7 hit nealæcð þam ende

(Beloved people, know what the truth is: this world is in haste, and it approaches its end. (Wulfstan (d. 1023), *Sermo Lupi ad Anglos*))

Note the use of some runic letters and the absence of French loanwords.

Norman scribes brought their conventions and their words with them and works written down in the period 1066–1476 reflect the habits of the new regime. Thus in Chaucer's *Anelida and Arcite* (14c.):

Singest with voice memorial in the shade,
Under the laurer which that may not fade
(Singest with memorial voice in the shade,
Under the laurel which may not fade)

Note the marked French influence on the vocabulary. Chaucer frequently used *qu-* in French loanwords (*quarter, querelc* 'quarrel', etc.) and also in native words that were spelt with *cw-* in OE (*quake* 'shake', *qualm* 'plague', *quelle* 'kill', etc.).

With the advent of printing, Caxton and the early printers in England adopted the spelling patterns of the late ME period and substantially rendered them immobile, though allowing more trivial variation than is permitted today. Such 'trivial variation' (see below) was permissible in all printed work of the period until the appearance of the increasingly sophisticated and influential dictionaries of the 17c. and 18c. The influence of 17c. and 18c. schoolmasters was also important in the emergence during this period of a national standard of spelling.

At this time spelling was not random or haphazard, but during the period 1476 to (say) 1755 (Johnson's Dictionary) moderate variation was permitted. Thus in Caxton's *Reynard the Fox* (1481) 'profit' is spelt *prouffyte* and *prouffyt*, 'peace' is spelt *pees* and *peas*, 'way' is spelt *waye* and *weye*, and 'opened' is spelt *opend* and *opened*. In Shakespeare's works a similar amount of relatively minor variation was admitted in the printed texts: 'Even a cursory glance will show that the spelling is by no means consistent, and a

word appearing in one form in a given line may appear with a different spelling only a few lines later—or even within the same line' (Vivian Salmon in Stanley Wells et al., *Shakespeare: The Complete Works*, original-spelling edn, 1986).

Variation is one thing; systematic spelling changes quite another. The period from 1476 to 1776 (American Independence) 'is the time when initial *fn* (ME. *fnēsen*, modE. *sneeze*) and *wl* (ME *wlatsom* loathsome) disappeared from the language; when *gh* or *f* took the place of earlier *h* or *ʒ* (yogh), pronounced /x/ in final position or before another consonant, as in *cough* (ME *coʒe*), *enough* (OE *genōh*), *fight* (OE *feoht*), and *plight* (OE *pliht*). The old runic letter *thorn* (þ) drifted in the way it was written until it so resembled the letter *y* that it had to be abandoned in favour of *th* ...' (these together with details of numerous other changes of spelling in this period are given in my book *The English Language*, 1985, esp. pp. 144–8).

Changes in pronunciation in this period are even more important. The 'Great Vowel Shift' was mostly at an end before Caxton's first book came off the press, but its reverberations are central to an understanding of what happened later. This is not the place to give details of all the sound changes that occurred during these three centuries. Suffice it to say that phonetic disturbances of a major kind occurred and most of them were not accompanied by changes in the spelling system. The result is that modern English spelling falls well short of being a reliable guide to the pronunciation of a sizeable number of English words. (Readers who wish to pursue the matter further should turn to any of several standard works on the subject. The most detailed, perhaps, is Barbara Strang's *A History of English*, 1970.)

**2** Spelling reform. The notorious inconsistencies of English spelling are described in numerous articles in this book (see below for a partial list). They are of such an order that it is tempting to think that they could be legislated away. All such attempts have so far failed for three main reasons. First, the absence of a single competent linguistic authority empowered to make such fundamental changes; secondly, reform, if radical, would automatically place millions of books, newspapers, etc., out of the reach of the general public until they were reprinted in the new spelling system (only a small proportion of the population are able to cope with the intricacies of the original-spelling version of the works of William Shakespeare, for example); and thirdly, the insuperable difficulty of the existence of divergent pronunciations throughout the English-speaking world. Whose standard English would qualify as the model for the respelling of the whole language: that of England, the USA, or (as a compromise) Canada or Australia?

There are numerous other difficulties. For example, spelling reform, if carried out at any more than a superficial level, would conceal the connectedness of word families that are divided only by the positioning of the stress. Consider the confusion that would arise if the spelling of the following pairs of words were adjusted to reflect the contrasting pronunciations of the key sounds in each: *adore/adoration, electric/electricity, fraternal/fraternize, history/historical, malign/malignant, mode/modular, nation/national, photograph/photography, sign/signature.* Furthermore, it is widely accepted that where systematic differences of spelling occur between two varieties of English, namely British and American (e.g. in obvious pairs of the type *centre/center, humour/humor, aesthetic/esthetic, catalogue/catalog,* etc.), the risk of misunderstanding is minimal.

My broad conclusion, after setting down considerably more evidence than I have here, is given on p. 145 of my book *The English Language* (1985): 'The English spelling-system is best left alone, except in minor particulars. Attempts to simplify or respell the language are likely to be unavailing for a long time to come.'

**3** Some notes on the relationship of English spelling and pronunciation. In general terms, written English has remained relatively static since the invention of printing in the late 15c., but spoken English, in its received form, has changed repeatedly since then. Loanwords have also been adopted from languages which have different spelling systems. As a result the sound /f/, for example, can be represented by a number of different spellings, e.g. *f* (*firm*), *gh*

(*rough, draught*), *ff* (*bluff, offer*), and *ph* (*philosophy*). The letter *h* can be silent (*honour*) or fully pronounced (*hand*). The sequence *-ough* can be pronounced /ʌf/ (*tough*), /aʊ/ (*slough*), /əʊ/ (*dough*), or /ə/ (*borough*). The sound /s/ can appear written as *c* (*cinder*), *s* (*send*), *sc* (*scent*), *ps* (*psalm*), *ss* (*assist*) or *sw* (*sword*); conversely the letter *s* can be pronounced as /s/ (*seven*) or /z/ (*is*). Occasionally it is silent (*aisle, demesne, island*). The digraph *ch* represents /k/ in *chasm*, /tʃ/ in *chain*, /ʃ/ in *charade*, or /x/ in *loch*. The trigraph *sch* is pronounced differently (in BrE) in *eschew, schedule, schism,* and *school*. All such variations can be accounted for historically.

**4** Commonly misspelt words. It is of interest to speculate about the amount of dislocation to the spelling system that would occur if English dictionaries were either out of reach or (as when Malory or Sir Philip Sidney were writing) did not exist. For instance, when writing private letters (in which spelling is thought not to 'matter' as much as it might in other circumstances) or in the stressful atmosphere of the examination room, the likelihood of misspellings increases at once. A select list of thirty words that are commonly misspelt in such circumstances: *accommodation, apartment, asinine, braggadocio, consensus, crucifixion, desiccated, ecstasy, guttural, idiosyncrasy, impresario, inoculate, liaise, liquefy, mayonnaise, mellifluous, millennium, minuscule, moccasin, necklace, pavilion, principal, principle, rarefy, resuscitate, sacrilegious, separate, supersede, verruca, withhold.*

**5** The 'house' style of printers and publishers varies from one to another. Where acceptable variants exist a choice must be made and should be adhered to throughout a given publication. The conventions of the Oxford University Press are set down in *The Oxford Writers' Dictionary* (1990). Thus it states: *absinth,* the plant; *absinthe,* the liqueur; *acknowledgement,* not *-ledgment; baptize,* not *-ise; Czar,* etc., use *Ts-; Djakarta,* Indonesia, use *J-; equinoctial,* not *-xial; feoff,* use *fief.* And so on for nearly 450 pages.

**6** Double and single letters for consonantal sounds. If a list were made of the many thousands of words whose spelling cannot be safely inferred from their sound, the doubtful point in perhaps nine-tenths of them would be whether some single consonantal sound was given by a single letter, as *m* or *t* or *c*, or by a double letter, as *mm* or *tt*, or two or more, as *sc* or *cq* or *sch*. Acquiesce and aqueduct, bivouac and bivouacking, Britain and Brittany, committee and comity, crystal and chrysalis, inoculate and innocuous, install and instil, harass and embarrass, levelled and unparalleled, personify and personnel, schedule and shed, science and silence, tic and tick, are examples enough. The use of double letters (*tt* etc.) or two letters (*ck* etc.) to give a single sound is due sometimes to the composition of a word, as when *in*not and *nocens* harmful are combined to make *innocent*, sometimes to the convention by which the sound of a previous vowel tends to be of one kind (long *a, e, i, o, u*) before a single letter and of another (short *a, e, i, o, u*) before two, and sometimes to other factors in word-formation, perhaps philologically explicable, but less obvious than in compounds like *in*nocent.

Among the rules are those that govern the doubling or not of a word's final consonant when suffixes are added in inflexions or word-formation. Directions are given for the various consonants under the articles -B-, -BB-, and so on, to be found in their alphabetical places.

Two more questions of single and double letters are of importance. In forming adverbs in -ly from adjectives in -l or -ll, neither a single nor a triple l is ever right; *full, purposeful, especial,* and *dull,* have adverbs *fully, purposefully, especially,* and *dully.* And in forming nouns in -ness from adjectives in -n both *ns* are retained—*commonness, rottenness, plainness,* etc. *Solemn,* with its silent *n,* needs hardly to be excepted: both *solemnness* and *solemness* are permissible, but the first of these is recommended as being the better visual form.

**7** Cross-references. Various points are discussed in short special articles throughout the book; and many words whose spelling is disputed will be found with or without discussion in their alphabetical places. The following collection of references may serve as a conspectus of likely mistakes and of readers' uncertainties.

For such words as *gaugeable, judg(e)ment, lik(e)able, mil(e)age, pal(e)ish, wholly,* see MUTE E.

For the plural of words in -o, see -O(E)S; many individual words are also given.

For the plural of words in -y, see PLURALS OF NOUNS 6.

For *cipher, sibyl, siphon, siren; gypsy, pygmy, syllabub; silvan sylvan, tire tyre, tiro tyro,* etc., see Y AND I, and the words themselves.

For *Aeschylus/Æschylus, Oedipus/Œdipus; archaeology/archeology, diarrhoea/diarrhea, homoeopathy/homeopathy, oenology/enology, paediatrics/pediatrics,* etc., see Æ, Œ.

For *tying, dyeing, denied, paid, buys, copied,* etc., see VERBS IN -IE, -Y, AND -YE.

For *one-idea'd/-eaed, her̩naed/-a'd, mustachio'd,* etc., see -ED AND 'D.

For the choice of -ize or -ise as the normal verb ending for the relevant class of verbs, see -IZE, -ISE IN VERBS. For a list of words in which -ise is obligatory, see -ISE.

For the plural of *handful, spoonful,* etc., see -FUL. The choice is often not between *handfuls* and *handsful,* but between *handfuls* and *hands full.*

For adjectives ending in -ble, see -ABLE, -IBLE.

For inflexions of verbs in -c like *mimic* and *picnic,* see -C-, -CK-.

For pairs like *enquiry* and *inquiry, inability* and *unable,* see EM- AND IM-; IN- AND UN-.

For adjectives like *bluey, clayey, holey, mousy, stagy,* see -EY AND -Y IN ADJECTIVES.

For *for(e)bear, for(e)go, forswear,* etc., see FOR- AND FORE-

For *cooperate co-op- coöp-,* etc., *pre-eminent* etc., *recollect* and *re-collect, recount* and *re-count,* etc., see CO-; PRE-; RE(-).

For *formulae -las, hippopotamuses -mi,* see LATIN PLURALS.

For *burnt -ned, leapt -ped,* etc., see -T AND -ED.

For *by and by, by the by, by-election,* etc., see BY, BY-, BYE.

For *classified, countrified,* etc., see -FY.

For *codeine, gelatine, glycerin(e),* etc., see -IN AND -INE.

For *into in to,* see INTO; for *onto on to,* see ON TO.

For *corrector, deserter,* etc., see -OR.

For *behavio(u)r, labo(u)r,* etc., see -OUR AND -OR.

For *clamorous, honourable, humorous,* etc., see -OUR- AND -OR-.

For *Dr. Dr, a.m. am,* etc., see FULL STOP 2.

For *Dickens's Dickens'* , *Venus' Venus's*, see APOSTROPHE[2] D2, 3.

For *preferred, referable*, etc., see -R-, -RR-.

For *the seventies*, etc., see NINETIES.

**8** Miscellaneous. For the rule '*i* before *e* except after *c*', see I BEFORE E.

The writing of the very common *anti-* 'against' instead of the rarer *ante-* 'before' (e.g. writing *antichamber, antidated* for *antechamber, antedated*) is to be carefully avoided.

Verbs in *-cede, -ceed,* are so many and so much used, and the causes of the difference are so far from obvious, that mistakes are frequent and a list will be helpful: *cede, accede, antecede, concede, intercede, precede, recede, retrocede, secede,* to which may be added *supersede;* but *exceed, proceed, succeed*. The commonest mistake is to write *preceeding* for *preceding*.

Adjectives and nouns in *-ble, -cle, -tle,* etc., make their adverbs and adjectives not by adding *-ly* or *-y*, but by changing *-le* to *-ly*: *humbly, singly, subtly, supply* (usu. not *supplely*), *tangly, treacly*.

Adjectives in *-ale, -ile, -ole,* add *-ly* for their adverbs: *stalely, docilely, vilely, solely;* but *whole* makes *wholly*.

For verbs ending in *-tre, -vre*, etc., the forms *centring, mitring, manoeuvring* are recommended in preference to *centering, manoeuvering* (except in the US where *centering* and *maneuvering* are customary). Similarly *ogrish* is recommended, not *ogreish*. See OCHRE.

Of adjectives in *-(e)rous* some never use the *e*, as *cumbrous, disastrous, idolatrous, leprous, lustrous, monstrous, wondrous;* some always have it, as *boisterous, murderous, obstreperous, slanderous, thunderous*. See DEXTEROUS; SLUMBEROUS.

Silent letters. One of the frustrations facing learners of English as well as native speakers is the number of silent letters, all arising from some quirk or other of the history of the relevant words. Thus *g* in *gnarled, gnaw, gnu*, etc.; *gh* in *might, sight*, etc.; *k* in *knight, knob, know*, etc.; *l* in *calm, palm*, etc.; *p* in *psalter, psychology,* etc.; *ptarmigan, pterodactyl*, etc.; *s* in *aisle, island*, etc.; *w* in *wrangle, write*, etc.

*q* without *u*. Surprise is often expressed that the language tolerates the presence of words containing an initial *q* without a following *u*. The few words involved fall into three classes: loanwords from Arabic (*qasida* 'elegiac poem', Qatar, State in Persian Gulf); loanwords from other 'exotic' languages (from Chinese, *qi* 'life-force'; from Eskimo, *qiviut* 'underwool of arctic musk-ox'); artificial formations (*Qantas*, Australian airline; *Qiana*, a US trade name for nylon).

**Spencerian,** of Herbert Spencer, philosopher, 1820–1903.

**Spenserian,** of Edmund Spenser, poet, ?1552–99.

**Spenserians** (prosody). The metre of *The Faerie Queene*, often used by later poets, especially by Byron in his *Childe Harold*. It consists of eight five-foot iambic lines, followed by an iambic line of six feet, rhyming ababbcbcc.

> *A Gentle Knight was pricking on the plaine,*
> *Y cladd in mightie armes and siluer shielde,*
> *Wherein old dints of deepe wounds did remaine,*
> *The cruell markes of many a bloudy fielde;*
> *Yet armes till that time did he neuer wield:*
> *His angry steede did chide his foming bitt,*
> *As much disdayning to the curbe to yield:*
> *Full iolly knight he seemd, and faire did sitt,*
> *As one for knightly giusts and fierce encounters fitt.*

**spew** (to vomit). Thus spelt, not *spue*.

**sphere.** See FIELD.

**sphinx.** Pl. *sphinxes*. The earlier (17–19c.) alternative pl. *sphinges* /ˈsfɪndʒiːz/ is now obsolete.

**spifflicate** (or *spiflicate*). A FACETIOUS FORMATION first recorded in the 18c.

**spill.** The pa.t. and pa.pple (also ppl adj.) are either *spilt* or *spilled*. The distribution of the forms is hard to pin down, but it would appear that *spilt* was the more favoured of the two until the end of the 19c. and that *spilled* is dominant in the 20c. *Spilt* seems secure in the expression *spilt milk*. There is also evidence to support the suggestion that *spilt* may be the more usual term in BrE in other uses and *spilled* the favoured form in AmE and elsewhere, but the evidence is not clear-cut. Examples of *spilled*: *The vision of spilled parcels hit her harder than she expected*—D. Malouf, 1985 (Aust.); *I was given the job because I never spilled any of the valuable stuff*—P. Bailey, 1986; *He nearly spilled his drink*—J. Updike, 1988; *which could equally be the reaction to hot coffee* *being spilled over somebody's knee*—Hugo Hamilton, 1990. Examples of *spilt: I gave her a cup of tea . . . and her hand . . . held it for a coincidental moment only before it was spilt on the carpet*—C. McWilliam, 1988; *Where the pubs were . . . forlorn establishments of spilt beer and Formica tables*—R. MacNeil, 1989 (Canad.); *The lounge boy . . . left too much change on the table and a puddle where he'd spilt the Coke*—R. Doyle, 1990 (Ir.); *Blacktop is unsuitable* [sc. for turning circles] *because spilt aircraft fuel will attack it*—Construction Weekly (UK), 1991; *She has never had much faith in therapy—all that crying over spilt milk*—Atlantic (US), 1992; *It spilt on his leg*—Independent, 1992; *Other multinationals doubtless polluted waterways or spilt their toxics*—Guardian, 1994. See -T AND -ED.

**spillikins, spel(l)icans** (pl. noun). For the game, *spillikins* is the recommended form. Also *spillikin* for the singular ( = splinter of wood, bone, etc.).

**spilth.** Used once by Shakespeare (*Our Vaults haue wept With drunken spilth of Wine*—1607) and occurs sporadically in literary works of the 19c., but now is archaic or obsolete.

**spin** (verb). The regular pa.t. and pa.pple forms are now *spun*. Before the 20c. *span* was commonly used for the pa.t. but it is now a deviant minority form. Examples: (span) *The other two . . . were swept half across the room before they span aside, one either way*—E. Peters, 1978; *He put the book back in the shelf and span the celestial globe gently*—C. Phipps, 1989; (spun) *my mother spun daydreams as easily as she mixed the turkey's mash*—M. Eldridge, 1984 (Aust.); *The other man spun towards the sound, gun extended, ready to fire*—A. Lejeune, 1986; *Beyond the lightly frosted glass, illuminated snowflakes spun and settled*—New Yorker, 1987; *The wheels spun up spray as we drove back to the farm*—Encounter, 1988; *I didn't get a chance to try it again because I was spun round, and dragged back*—A. Billson, 1993.

**spinach.** The pronunciation recommended is /'spɪnɪdʒ/, but /'spɪnɪtʃ/ is probably just as widely used in standard BrE and is the customary pronunciation in AmE.

**spin doctor.** See SOUND BITE.

**spindrift** (spray blown along the surface of the sea), a Scottish variant of *spoondrift* (from *spoon*, in sailing, to run before wind and sea + *drift*), has now replaced the 'correct' form.

**spinel** (red gem or precious stone). Pronounce /sprˈnəl/. In AmE it is sometimes spelt *spinelle*.

**spinet** (a small harpsichord). Pronounce /sprˈnet/ or /'spɪnɪt/. In AmE usu. stressed on the first syllable.

**spinney** (in BrE, a small wood, a copse). Pl. *spinneys*.

**Spinozism** (doctrine of the Dutch philosopher B. de Spinoza). Thus spelt.

**spiny,** not *spiney*. See MUTE E.

**spiraea** (rosaceous shrub). Thus spelt in BrE, but often as *spirea* in AmE.

**spiral** (verb). The inflected forms are *spiralled, spiralling* in BrE, but often *spiraled, spiraling* in AmE. See -LL-, -L-.

**spirit** (verb). The inflected forms are *spirited, spiriting*. See -T-, -TT-.

**spiritism.** Coined in the 1850s as an alternative to *spiritualism* (also first recorded in the 1850s in the sense 'the belief that the spirits of the dead can hold communication with the living'), it has lost ground since and now is virtually disused. *Spiritualism* has prevailed as the usual term for this sense, and also for the philosophical sense 'the doctrine that the spirit exists as distinct from matter, or that spirit is the only reality'.

**spiritual, spirituous.** The two words are now clearly divided in sense, the first meaning 'of or pertaining to the spirit or soul', and the second applied only to certain kinds of alcoholic drinks. From the 16c. to the end of the 19c. *spirituous* was frequently used as a synonym of *spiritual*, but no longer.

**spirituel,** (feminine) **spirituelle.** Since Dryden introduced this French adjective ( = of a highly refined character or nature) into English in his *Marriage à la Mode* (1673), 'the distinction between the masc. and fem. forms has not always been observed in English', as the *OED* remarks. A pity, but not surprising in that English rarely makes such distinctions in adjectives. Cf. *blond(e)* and *naïf/ naive.*

**spirt.** See SPURT.

**spit.** The pa.t. and pa.pple of the verb meaning 'to eject saliva from the mouth' are regularly *spat* in Britain, but in AmE either *spat* or *spit*. Examples: (spat) *'Kidnapped,' my aunt spat out*—G. Greene, 1988; *there were small chunks of foam spat out* [by mice] *in heaps of beige confetti*—*Encounter*, 1989; (spit) *I was so mad I could have spit*—*New Yorker*, 1989; *He spit. Then he jumped, from one rock to the next, leaping the boulders like a goat until he was gone*—M. Sumner, 1993 (US). The unrelated verb *spit* meaning 'to impale on a spit' has pa.t. and pa.pple *spitted*.

**spitting image.** A fashionable use of the phrases *spit and fetch, spit and image,* etc., in the late 19c. as an extension of the earlier 19c. phr. *the very spit of* produced the dialectal (and then standard) phrases *spitten image* (first recorded in 1910), *spittin' image* (1901), and *spitting image* (1929). The dominant phrase now is *spitting image*. Examples: *You are a queer fellow*—*the very spit of your father*—T. Hook, 1836; *He would be the very spit and fetch of Queen Cleopatra*—G. A. Sala, 1859; *He's jes' like his pa*—*the very spittin' image of him!*—A. H. Rice, 1901; *A nice-behaved young gentleman, and the spitten image of his poor mother*—A. Bennett, 1910; *In another twenty years ... she would be her mother's spitting image*—H. S. Walpole, 1929; *Spitten image of his dad, little Alf is, isn't he, Reg?*—W. Holtby, 1936; *The son's the dead spit of the old man*—A. Upfield, 1953; *Look at this, Father, appeared last Friday on Sister Philomena, the very spit and image of the nail marks in the palms of Our Blessed Lord*—H. Mantel, 1989; *And no man should grieve because he isn't the spit and image of Tom Cruise*—*Parade Mag.* (US), 1990.

**spiv.** A 20c. British colloquial expression for 'a man characterized by flashy dress, who makes a living by illicit or unscrupulous dealings' (*COD*). In inventing this word the English have emulated the American genius for coining monosyllabic words (e.g. *blurb, stunt*) whose sound is curiously suited to their meaning. The origin of *spiv* is uncertain, but it is probably connected with the late 19c. slang

or colloquial words *spiffed* 'made neat, smartly dressed' and *spiffy* 'smart, spruce'.

**splendiferous.** See FACETIOUS FORMATIONS.

**splendour.** Thus spelt, but *splendor* in AmE. See -OUR AND -OR.

**split infinitive.** No other grammatical issue has so divided the nation since the split infinitive was declared to be a solecism in the course of the 19c. First, it is essential to clarify what is and what is not a split infinitive. A brief history of the construction then follows. Finally, a description of the present state of the split infinitive is given with numerous illustrative examples showing the various types of split and unsplit infinitives.

1 *Definition.* The base form of an infinitive is shown in *to love*, in which the verbal part is preceded by the particle *to*. When such a combination is severed or 'split' by the insertion of an adverb or adverbial phrase (e.g. *to madly love, to really and truly love*) or other word or words the construction is called a *split infinitive*. In Latin such a construction could not arise because an infinitive (*amāre* 'to love', *crescere* 'to grow') is indivisible and is not preceded by a grammatical particle. In other words it is complete in itself. Keep in mind that the type *My mother taught me to be always prepared* is not an example of a split infinitive. It would became one only if *always* were placed between *to* and *be*: *My mother taught me to always be prepared.* Another type sometimes falsely taken to be a split infinitive is that containing *to* + insertion + verb in *-ing*. Examples: *I mean it's not as if I'm going to be actually risking my life*—K. Amis, 1988; *It was apparent she was very embarrassed when it came to openly deprecating a newly married girl to her mother-in-law*—M. Spark, 1990.

2 *A brief history.* The standard work on the history of English syntax, F. Th. Visser's *An Historical Syntax of the English Language* (4 vols., 1963–73), states that the earliest examples of split infinitives date from the 13c.; but the construction was not widely used between the 13c. and the 15c. (for example, there are only two examples in the works of Chaucer).

Typical examples: (adverb between *to* and the infinitive; note that *forto, for to* frequently = *to* in ME) *What movede the pape of Rome to thus accepte mennes persones*—Wyclif, c1380; *To enserche sciences, and to perfitly knowe alle manere of Naturals þinges*—*Secreta Secretorum*, c1400; *it longiþ forto not oonly bigynne ... but it longiþ* [etc.]—R. Pecock, c1443; (a noun, pronoun, or noun phrase inserted) *heo cleopode him to alle his wise for to him reade*—Layamon, c1250; *being moche redier forto suche writings lette and distroie þan* [etc.]—Pecock, c1445); (two or more words between *to* and its infinitive) *A kyng owith not to ... ouer oft haunte the company of his sugetis*—Prose version of *Secreta Secretorum*, c1425; *forto iustli and vertuoseli do a dede contrari to goddis comaundement*—Pecock, c1449. Visser goes on to say that 'From about the beginning of the sixteenth century to about the last decades of the eighteenth century the use of the split infinitive seems to have been as it were tabooed in the written language.' Nevertheless he cites four examples from the 16c. I had no difficulty in finding several examples of the avoidance of split infinitives in three 16c. lives of Sir Thomas More (e.g. *I am ready obediently to conforme my self to his graces commandements*—Roper, ?1557). There were no split infinitives in any of these three Lives.

The split infinitive seems to have come back into favour at the end of the 18c.: e.g. *I know not how I should be able to absolutely forbid him my sight*—F. Burney, 1778; *To sit on rocks to muse o'er flood and fell, To slowly trace the forest's shady scene*—Byron, 1812; *This jack-in-office had taken upon himself... to more than insinuate*—*Times*, 1839; *She wants to honestly and legally marry that man*—Hardy, 1895.

**3** *The present state.* There can be no doubt that there continues to be a noticeable reluctance to split infinitives both in the national press and in the work of many of our most respected writers. Thus in a 1987 issue of the *Daily Telegraph*: *there will be a further disposition seriously to underestimate the strength ... of the United States.* And in Peter Carey's *Oscar and Lucinda* (1988): *He was never ashamed publicly to bear witness.* Such placing of the adverb is overwhelmingly the norm at present. When Bernard Levin, the well-known columnist in The

*Times,* wrote (24 Oct. 1991) *he* [a former political prisoner] *was in Vilnius to formally close down the headquarters of the Lithuanian KGB,* the use called for special comment in the Diary of that newspaper two days later. In a leading article in the 18 May 1992 issue of *The Times* it was stated that 'The most diligent search can find no modern grammarian to pedantically, to dogmatically, to invariably condemn a split infinitive.' These light-hearted comments draw attention to the irrational nervousness that many people feel when they imagine that, by splitting an infinitive they are in danger of breaking a terrible taboo. What then are the present-day facts?

First, all the evidence points towards the reality of the feeling that it is 'wrong' to split infinitives. Examples showing an adverb placed immediately before the particle *to* are commonplace: e.g. *The threat of abolition enabled the Livingstone administration briefly to ride the inevitable wave of popular indignation it caused*—London Rev. Bks, 1987; *I had no wish* actually to read *it*—C. Rumens, 1987. Less commonly the adverb is placed after the verb: e.g. *it became urgent to demarcate accurately Alaska's eastern boundary*—Geogr. Jrnl., 1983; *little or no effort has been made to explicate clearly the mechanisms through which these needs* [etc.]—*European Sociol. Rev.,* 1986; *Party leaders have simply refused to attempt seriously to come to terms with the new situation*—Parl. Aff., 1986.

On the other hand it is clear that rigid adherence to a policy of non-splitting can sometimes lead to unnaturalness or ambiguity. Gradations of these can be observed in the following: (unnatural) *In not combining to forbid flatly hostilities,* (ambiguous) *In not combining flatly to forbid hostilities*; (unambiguous) *In not combining to flatly forbid hostilities.* Secondly, examples abound of most of the categories of split infinitives that Visser found in works of earlier centuries. The examples that follow represent only a small selection of the substantial file of examples that I have collected since 1987.

They turn up in all English-speaking areas: *That's when you have* to really watch *yourself*—Quarto, 1981 (UK); *David ... questing her attention, allowed one eyelid to minimally fall*—A. Brookner, 1984 (UK); *it led*

*Cheshires* to finally abandon *publishing fiction at all*—B. Oakley, 1985 (Aust.); *The goal is* to further exclude *Arafat*—*US News & World Rep.*, 1986; *For your safety and comfort we do ask you* to please stay *in your seats*—British Airways flight attendant, 1986; *It was no great achievement* to simply split *the third party*—R. Muldoon, 1986 (NZ); *Spring, the season she had been able* to utterly ignore—J. Urquhart, 1986 (Canad.); *And it allowed Fernanda Herford* to slightly more than double *her money*—Julian Barnes, 1993 (UK); *there is the question whether a nation burdened with guilt for the Holocaust can ever hope* to fully recover *national self-confidence*—*Dædalus*, 1994 (US).

They occur in popular sources: *Something had* to drastically change *in my life*—*The Face*, 1986; *Everything he had written seemed* to just deliberately and maliciously draw *attention to the fact that* [etc.]—B. Elton, 1991. And in the work of more serious writers: *We talked about ... how everything was going* to suddenly change—Nigel Williams, 1985; *I want to* really study, *I want to be a scholar*—I. Murdoch, 1987; *In face of all this Patrick managed* to quite like *him*—K. Amis, 1988.

The commonest type is that in which a simple adverb such as *even, ever, further, just,* or *quite* is inserted between *to* and the infinitive. Almost as common in this position are 'simple' adverbs in *-ly* (*actually, finally, fully, nearly, really, simply, utterly,* etc.). The negative adverbs *never* and *not* are often inserted in AmE, less commonly in BrE: e.g. *a perfect morning* to not read *'Moby Dick'*—*New Yorker*, 1986; *Many professional players hope* to never play *there again*—*American Way*, 1987; *The only unforgivable sin is* to not show up—G. Keillor, 1991.

Occasionally, for stylistic reasons or other special effects, adverbial phrases are inserted, not just a single adverb: e.g. *To suddenly, after all these years,* fire *them*—P. Carey, 1982 (Aust.); *a willingness* to not always, in every circumstance, think *the very best of us*—P. Roth, 1987; *You two shared a curious dry ability* to without actually saying anything make *me feel dirty*—J. Updike, 1988; [see above] B. Elton, 1991. Other types of split infinitives are rare in modern English, e.g. (with a pronoun inserted) *It was their nature* to all hurt *each other*—P. Carey, 1981 (Aust.).

**4** *Preference.* No absolute taboo should be placed on the use of simple adverbs between the particle *to* and the verbal part of the infinitive. 'Avoid splitting infinitives whenever possible, but do not suffer undue remorse if a split infinitive is unavoidable for the natural and unambiguous completion of a sentence already begun.' (Burchfield, *The Spoken Word*, 1981).

**splodge, splotch.** The second is two centuries older; the first perhaps now more usual and felt to be more descriptive.

**splutter, sputter.** Without any clear or constant difference of meaning, it may be said that in *sputter* the notion of spitting is more insistent, and that it tends on that account to be avoided when that notion is not essential.

**spoiled, spoilt.** Both forms are used for the pa.t., pa.pple, and ppl adj. of *spoil* (verb), though *a spoilt child* is more usual than *a spoiled child*. See -T AND -ED. In AmE *spoiled* is the more usual form of the two in all three parts of speech.

**-spoken.** For the curious use in *fairspoken, soft-spoken, outspoken,* etc. (instead of the more 'logical' *fair-speaking* etc.), see INTRANSITIVE PAST PARTICIPLES. As the *OED* remarks s.v. *outspoken*, 'the pa.pple has here a resultant force, as in "well spoken", "well read"'.

**spokesman** used to be the normal word for a person, male or female, who speaks on behalf of others, though *spokeswoman* was also used when relevant. A blitz by feminists has more or less succeeded in forcing us all into a corner unless we use *spokesperson* instead. But the corner is well populated. See PERSON (or -PERSON). Possibly transient substitutes to suit contextual needs have been noted in American sources, e.g. *spokesbody, spokescouturier, spokesmodel, spokesvoice.*

**sponge** makes *spongeable* (see -ABLE, -IBLE 2) and *spongeing*; but *spongy*.

**spontaneity.** See -TY AND -NESS. The traditional pronunciation with final /-iːɪtɪ/ is now being strongly challenged by one ending in /-eɪtɪ/. See -EITY.

**spook.** Pronounce /spuːk/.

**spoondrift.** See SPINDRIFT.

**Spoonerism.** Formed from the name of the Revd W. A. Spooner (1844–1930),

Dean and Warden of New College, Oxford. The type is illustrated by Anthony Burgess in his *Inside Mr Enderby* (1963): *But A Sale of Two Titties had struck Lady Fennimore as something like calculated insolence*. 'The eponymous Spooner was reputed to make errors of this type,' remarks *OCELang*; 'and a number of utterances are quoted as "original spoonerisms": *a well-boiled icicle* (a well-oiled bicycle), *a scoop of Boy Trouts* (a troop of Boy Scouts), and *You have hissed all my mystery lectures and tasted a whole worm*. It seems likely that these transpositions were exaggerated inventions by his students.' The term is now applied to normally accidental transpositions of any sounds, e.g. *a pea flit* for *a flea pit*. But inadvertence is not essential: e.g. W. H. Auden is said to have referred to Keats and Shelley dismissively as *Sheets and Kelly*.

**spoonful.** Pl. *spoonfuls*. See -FUL.

**spoony.** This archaic adj. ( − foolish, simple, esp. in amorous behaviour) and noun ( = a simpleton, esp. a person who behaves in a foolishly amorous manner) are thus spelt. The pl. of the noun is *spoonies*.

**sports.** Normally unchanged when in attributive use in BrE (*sports car*, *sports coat*, *sports jacket*, etc.) but often reduced to the singular in AmE when the sense is 'suitable for informal wear' (*sport clothes*, *sport shirt*, etc.). On the other hand the form in *-s* is usually retained in AmE in *sports car*, *sportscast*, and *sportsman*.

**s'pose.** In informal speech the first vowel is often elided. The *OED* lists eight examples of spellings representing this pronunciation, including the following: *I s'pose the poor must live somewheres,' said Mrs. Growler, the old maid-servant*—Trollope, 1873; *I s'pose we're not strictly his guests*—H. R. F. Keating, 1980.

**spouse,** pronounced either as /spaʊz/ or /spaʊs/, has over the centuries been used to mean either a married woman in relation to her husband or a married man in relation to his wife. A late 20c. development (only in AmE?) is to use the word of the live-in partner of a homosexual person.

**spring** (verb). The pa.t. is normally *sprang*, but occasionally *sprung* (esp. in AmE): *And then his hands sprung loose from the handle*—P. Carey, 1988 (Aust.). The pa.pple is *sprung*.

**springe** (a noose or snare). Pronounce /sprɪndʒ/.

**spry.** The *-y* is retained in its inflected forms and derivatives: *spryer*, *spryest*; *spryish*, *spryly*, *spryness*.

**spue.** See SPEW.

**spurt.** Use this spelling both for the verb ('gush out in a stream; make a sudden effort') and for the noun ('a sudden gushing out; an increase in pace'), not *spirt*.

**sputter.** See SPLUTTER.

**squalor.** Thus spelt, not *squalour*. See -OUR AND -OR.

**square** (adj. and noun) sprang into colloquial use in the 1940s applied to a person judged to be out of touch with the conventions of a popular contemporary movement (orig. jazz) or way of life. Now, it would seem, passing out of fashion to be replaced by more recent terms of generational abuse (*not with it*, *uncool*, etc.).

**squeeze.** The standard pa.t. and pa.pple are *squeezed* (*she squeezed out a tear; he was squeezed into a corner*). The non-standard or dialectal form *squoze* is used occasionally for the pa.t. Ronald Reagan used *squoze* at a press conference in August 1985 when commenting on a small skin cancer: *I picked at it and I squoze it and so forth and messed myself up a little*. *OED* 2 cites numerous examples of *squoze*, beginning with one of 1844 and including the following from Malcolm Lowry's *Ultramarine* (1933): *He just sort of squoze the rabbit*.

**squib.** See LAMPOON.

**squirearchy.** This spelling is recommended rather than *squirarchy*.

**-s-, -ss-.** 1 The general rules for the doubling or not doubling of final consonants are given s.v. DOUBLING OF CONSONANTS WITH SUFFIXES, and also, in so far as there are further circumstances to be taken into account, s.v. -B-, -BB-; -C-, -CK-;

-D-, -DD-, etc. Since very few monosyllables or words stressed on the last syllable end in a single *s*, it is hardly necessary here to do more than draw attention to specific cases. In the following words there is wide variation in practice among printers and publishers, but the recommended forms in OUP house style are in almost every case those with a single *s*: *atlases*; *biases*, *biased*, *biasing*; *boluses*; *bonuses*; *buses*, *bused*, *busing*; *canvas*: see CANVAS, CANVASS; *focuses*, *focused*, *focusing* (but see FOCUS); *gases*, but *gassed*, *gassing*; *incubuses*; *minuses*; *nimbuses*; *orchises*; *pluses*; *portcullised*; *trellised*, *trellising*; *yeses*.

2 Words ending in -*ss* naturally retain the *ss* in the plural: *abysses*, *busses* (kisses), *congresses*, *crosses*, *giantesses*, etc.

3 *Nonplus* makes *nonplussed*, *nonplussing* (all three words stressed on the second syllable).

4 A hyphen is needed to avoid a succession of three *s*'s: Inverness-shire, mistress-ship. For *misshapen*, etc., see MIS-.

**St.** For the question between *St Peter* and *St. Peter*, see FULL STOP 2; SAINT.

**stadium.** In contexts dealing with the ancient (classical) world, the pl. is *stadia*. For modern sports grounds, the pl. is *stadiums*. See -UM 3.

**staff.** 1 (*a*) a pole (pl. *staffs* or *staves*); (*b*) (music) a set of lines on which music is written (pl. *staffs* or *staves*); (*c*) a body of persons employed in a business, etc. (pl., if required, = such bodies of employees of various businesses, *staffs*). See STAVE (noun).

2 For *staff of life*, see SOBRIQUETS.

**stag.** See BUCK.

**stage** makes *stagily*, *staginess*, *stagy*.

**stalactites, stalagmites.** The first hang down from the roof of a cave, etc.; *stalagmites* are deposits rising from the floor of a cave, etc. Both words are stressed on the first syllable in BrE. In AmE the dominant pronunciation is with the stress on the second syllable.

**stamen.** Pl. *stamens*. The pl. of the Latin original, namely *stamina*, has moved into English in a different sense.

**stanch** /stɑːntʃ/ is the recommended spelling and pronunciation of the verb meaning 'to restrain the flow of blood (from)', not *staunch* /stɔːntʃ/. See STAUNCH.

**stand** (verb). In BrE candidates *stand* for office; in AmE they *run* for it.

**standard English.** The primary aim of this book is to identify and describe the principal elements of standard written English in Britain, often by contrasting them with non-standard, dialectal, overseas, archaic, or obsolescent features. An attempt has also been made to give prominence to the pronunciation of words (esp. the positioning of the stress) about which an educated person in the UK might conceivably have doubts. The standard Englishes of the other main varieties of the language, especially American English, are given as much prominence as space and circumstances permit, but inevitably standard speakers in the United States, Canada, Australia, New Zealand, South Africa, and elsewhere will need to interpret the recommendations made here in the light of their own experience.

The form of educated English used in their formal programmes by the broadcasting authorities based in London, by the London-based national newspapers, and by teachers of English to young people in this country and to foreigners is the variety presented here. Most other major English-speaking countries have guides which give prominence to the usage of their own people, for example *Webster's Dictionary of English Usage* (1989, mainly concerned with AmE), and Nicholas Hudson's *Modern Australian Usage* (1993). It is remarkable how often the problems they discuss are the same ones as those dealt with in this book; but the differences of detail and of emphasis are just as noteworthy.

**standpoint.** See POINT OF VIEW.

**stanza** in prosody is 'the basic metrical unit in a poem or verse consisting of a recurring group of lines (often four lines and usu. not more than twelve) which may or may not rhyme' (*COD*). In lay use sometimes called a VERSE.

**starchedly.** See -EDLY.

**starlight, starlit.** In *a starlight night*, *starlight* is a noun used attributively. In *a starlit night, a starlit sky, starlit* is a ppl adj. Both mean 'lit up or lighted by the stars'.

**state.** The initial letter is conventionally capitalized when it means 'an organized political community under one government' (*the United States of America, the State of Israel*), or a constituent unit of a federal nation (*the State of Virginia*); also when the word is used attributively in this sense (*State documents, a State visit*), and when it means 'civil government' (*Church and State, the Secretary of State for Northern Ireland*). In other contexts use a lowercase *s* (*a police state, the welfare state*).

**stately** can be used only as an adjective (*a stately home*). The notional corresponding adverb *statelily* is ruled out on grounds of euphony. See -LILY.

**static, statical** (adjs.). Of similar age (*static* first recorded in 1638 and *statical* in 1570), but the shorter form has prevailed in all uses. The corresponding adverb is *statically.*

**stationary, stationery.** The adj. ( = not moving) *-ary*; the noun ( = writing materials) *-ery*.

**statistic** (noun) in the sense 'a statistical fact or item' is a legitimate back-formation from *statistics*, the name of the science.

**statistic, statistical** (adjs.). Both adjectives are of similar age (late 18c.) when applied to statistics, but the longer form has for all practical purposes driven out the shorter one.

**statist, statistician.** Both entered the language in the sense 'one who deals in statistics' in the early 19c., but the longer form has prevailed and *statist* /'steɪtɪst/ is now rare.

**status.** 1 Pronounce /'steɪtəs/ despite the fact that the stem vowel was short in the Latin original (see FALSE QUANTITY). The pronunciation /'stætəs/ is sometimes used in AmE.

2 The Latin pl. *statūs* is not used in English, the regular form being *statuses*. See -US 2.

**statutable.** First used in the mid-17c., it now means (*a*) prescribed, authorized, or permitted by statute; (*b*) conformed

to the requirements of a statute or statutes; (*c*) of an offence: legally punishable. Its near-synonym *statutory*, first recorded in the early 18c., now means (*a*) enacted, appointed, created, or permitted by statute; (*b*) obligatory by custom; (*c*) required for the sake of appearances. Clearly there is some overlapping of meaning, but also some clear water between the ways in which the two words are used.

**staunch** /stɔ:ntʃ/. The recommended spelling and pronunciation of the adjective meaning 'trustworthy, loyal', not *stanch* /sta:ntʃ/. See STANCH.

**stave** (noun). (*a*) = STAFF 1(*b*); (*b*) each of the curved pieces of wood forming the sides of a cask, etc.; (*c*) a stanza of a poem; (*d*) an alliterating letter in a line of OE verse. Pl. of each, *staves.*

**stave** (verb). The distribution of the alternative forms *staved* and *stove* for the pa.t. and pa.pple is difficult to determine. Both are used in the various senses of the verb, with *staved off* probably dominant as the past tense of *stave off* 'to avert or defer (danger or misfortune)', and *stove in* for the past tense of *stave in* 'to crush by forcing inwards'.

**stay** (verb). See STOP (verb).

**stem** (noun). In etymology and word-formation, the root or main part of a noun, verb, etc., to which elements may be added to make new words. Thus *ship* forms *shipment, shipshape, shipwreck*, etc., by adding postfixed elements, and *airship, warship*, etc., by adding prefixed elements. The stem *ship* also forms *ships, shipped, shipping* by the addition of inflections.

**stem** (verb). From one of four distinct verbs (the one essentially meaning 'to remove the stem or stalk from') came the American phrasal verb *stem from* = to derive from or take origin from. First recorded in the 1930s it has now established itself in all the main English-speaking countries. Some UK and US examples: *Both literal and spiritual exposition stemmed from Origen—*B. Smalley, 1952 (UK); *Part of Pastor Spratt's charisma stemmed from his time spent as an advertising manager for Rathbone's Wrought Iron—*J. Winterson, 1985 (UK); *The chateau seems* to stem from the world of Disney rather than Louis Treize—P. Lively, 1987 (UK); *an apparent conflict stemming from his prior service as resident counsel to the Mayo Clinic—New Yorker*, 1987; *Investment-driven growth stemming from technological catch-up is an incomplete . . . explanation of Japanese economic history during the Showa era—Dædalus* (US), 1990.

**stemma** (a family tree; a line of descent). The pl. is *stemmata*.

**stencil.** The inflected forms are *stencilled, stencilling*, but often *stenciled, stenciling* in AmE. See -LL-, -L-.

**step** (verb). The phrases *Please step this way* and *step in* are rather formal equivalents of the same phrases using *come* instead of *step*.

**step-.** See -IN-LAW.

**stereo** /ˈsterɪəʊ/, a 20c. abbreviation of *stereophonic, stereophony*.

**sterile.** The *OED* (1916) admitted both /ˈsteraɪl/ and /-ɪl/ in that order. Now the only standard pronunciation in BrE is that with /-aɪl/, while AmE favours a short vowel in the second syllable (in practice a schwa). See -ILE.

**sternum.** Pl. (for preference) *sterna*, or *sternums*. See -UM 3.

**stevedore.** Three syllables, /ˈstiːvədɔː/.

**stewardess,** a female steward, esp. on a ship or aircraft. On English-speaking airlines, now giving way to *flight attendant*. See FEMININE DESIGNATIONS.

**stichomythia** /stɪkəʊˈmɪθɪə/. In verse plays, an interchange of short speeches consisting each of a single line. Common in Greek plays. Modern examples may be found in Molière's *Les Femmes savantes* III.v, and in Shakespeare's *Richard III*, IV.iv: Elizabeth: *But how long shall that title euer last? Richard: Sweetlie in force vnto her faire lyues end.* Elizabeth: *But how long farely shall her sweet life last? Richard: As long as heauen and nature lengthens it.* Elizabeth: *As long as hell and Richard likes of it.* [etc.] The word is Greek, from στίχος 'line' + μῦθος 'speech, talk'.

**sticking place, point.** In the context in Shakespeare's *Macbeth* I.vii (*But screw your courage to the sticking place, And wee'le*

*not fayle*) *place* is the word used. The phrase *sticking point* is first recorded in this sense in 1826, and in its common modern sense, 'the limit of progress, agreement, etc.', not until the 1960s. In *Macbeth* the reference (according to the *OED*) seems to be to the screwing-up of the peg of a musical instrument until it becomes tightly fixed in the hole.

**sticklebat, tittlebat.** The first is the correct and etymological form, the other being (*OED*) 'a variant, of childish origin'.

**stigma.** In the ecclesiastical, botanical, medical, etc., senses the plural is *stigmata* /ˈstɪgmətə/, sometimes (esp. in AmE) /stɪɡˈmɑːtə/. *Stigmas* is used only in the figurative sense of imputation or disgrace, in which a plural is rare. See LATIN PLURALS.

**stigmatize.** The mistake dealt with under REGARD 2 occurs sometimes with *stigmatize*: . . . *bravely suffering forfeiture and imprisonment rather than accept what in this same connexion Lord Morley stigmatized the 'bar sinister'* (Fowler, 1926). Things are not *stigmatized monstrous*, but *stigmatized as monstrous*.

**stile, style, stylus.** *Stile* (from OE *stigel*) is a native word for the means of passage across a fence, etc. A second *stile* (probably from Du. *stijl*) is 'a vertical piece in the frame of a panelled door, etc.'. *Style* (manner, fashion, etc.) is (via OF) derived from L *stilus* 'writing-tool' by confusion with Gk στῦλος 'column'. The English word *stylus* 'writing-tool' should also 'properly' be written as *stilus*. But it is obviously too late now to respell the words to accord with their roots.

**stiletto.** Pl. *stilettos*. See -O(E)S 6.

**still and all.** A resilient, rather casual adverbial phrase, first recorded in 1829, and used, much more commonly, since the 1920s in informal contexts. Examples: *Still and all, if you see something I haven't, let me know—R. Moore*, 1978; *Still and all, you might have worked out all right—T. Clancy*, 1987; *Still and all, something had undeniably changed—Encounter*, 1988.

**still life.** The correct pl. is *still lifes*. An example of the erroneous form *still lives*: *I think it* [sc. a fig tree] *would be a marvellous subject for one of your still-lives—C. Phipps*, 1988.

**stimulus.** Pl. *stimuli* /-laɪ/. See LATIN PLURALS.

**stimy, stimie.** See STYMIE.

**sting.** The inflected forms are regularly *stung* for both the pa.t. and the pa.pple.

**stink.** The pa.t. is *stank*, occasionally *stunk*. The pa.pple is *stunk*.

**stock pathos.** A number of words and phrases are frequently, but sometimes inadvisedly, put to use in letters of comfort or condolence. In the right circumstances some of these stock expressions can stand unchallenged, but each should be examined carefully before being pressed into service. A short list: *the land he loved so well; the supreme sacrifice; the pity of it; the mortal remains of; the departed; a lump in one's throat; tug at one's heart strings; stricken; but it was not to be.*

**stoep.** In South Africa, a terraced veranda in front of a house. Pronounce /stuːp/.

**stoic, stoical.** See -IC(AL). Both forms are used as adjectives, -ic being the commoner; but points of difference are discernible. In the predicative use *stoic* is rare: *his acceptance of the news was stoical; he was stoical in temper*, rather than *stoic*. In the attributive use, *stoic* preserves the original sense more definitely, while *stoical* departs from it. When we say *stoic indifference*, we mean such indifference as the Stoics taught or practised; when we say *stoical indifference* we think of it merely as resolute or composed. The *stoic virtues* are those actually taught by the Stoics, the *stoical virtues* simply those of the sterner kind. Lastly, while either epithet is applicable to abstracts, *stoical* is the word for persons: *with stoic or stoical composure; stoic or stoical life or temper or views; he is a stoical fellow; these stoical explorers; a stoical sufferer; my stoical young friend.*

**stokehold, stokehole.** The first is 'a compartment in a steamship, containing its boilers and furnace'. The second is 'a space [ashore] for stokers in front of a furnace' (COD).

**stomach.** See BELLY.

**stomacher** (formerly, an item of women's dress). Many 18c. and 19c. dictionaries list /ˈstʌmətʃə(r)/ as the pronunciation of this word, but it is now normally pronounced with /-kə/.

**stone** (noun). 1 See ROCK.

2 *to leave no stone unturned* is a tiresome cliché for 'to try all possible means'.

**stone** (adv.). Combinations in which *stone* has similative force have been used for centuries: e.g. *stone-blind*, as blind as a stone, 14c.; *stone-cold*, 16c.; *stone-dead*, 13c. A more recent development is the use of *stone* as a mere intensive = very, completely. It is not easy to pin down exactly when this happened, but such uses are now commonplace, esp. in AmE. Examples: *The Irish were stone courageous*—T. Wolfe, 1987; *He was stone angry, as if he had been brawling with some upstart stranger who got him with a lucky punch*—Rosie Scott, 1988 (NZ); *People ... got stone drunk and cruised through red lights*—G. Keillor, 1989.

**stood,** used as a present participle ( = standard English *standing*) presumably has a dialectal distribution in BrE similar to SAT, but if it has its history seems to have gone unrecorded in the *OED*. Its existence in modern regional use is not in question, but its precise distribution has not been established. J. Cheshire et al. (1989) conclude that it and the similar use of *sat* 'are now becoming characteristic of a general nonstandard or semistandard variety of English'. Examples: *She was stood in front of the mantelpiece trying to think of the name for the clock*—A. Bennett, 1981 (Yorkshire); *'But that's not the half of it.' Uncle Simon sat forward. 'Do you know what he did when he was stood there face to face with the priest, the man who positively identified him?*—G. Patterson, 1988 (NIr.); *And she'd pay the driver, and she'd be stood there, on the soiled concrete footpath*—A. Duff, 1990 (NZ, Maori speaker); *My husband was stood on the opposite side of the pits*—Cycling Weekly, 1993.

**stop** (noun). In phonetics, a word widely used for 'a speech sound which is the result of a complete closure tract' (Hartmann and Stork, 1972). Within the broad class of stops there are several subclasses, with much variation of terminology, e.g. *simple stop* (p in *pin*), *complex stop* (pʔ in French *pain* bread), *labial stops* (p, b), *alveolar stops* (t, d), *velar stops* (k, g), and *glottal stop* (ʔ). A different classification distinguishes *voiced stops* (b, d, g), in the production of which the vocal cords are brought into play, from *voiceless stops* (p, *t, k*) in which they are not. Some scholars use the term *plosive* instead of *stop*.

**stop** (verb). In 19c. fiction people frequently *stopped* with friends (i.e. remained overnight in the house of friends or relations), *stopped* at home, or *stopped* for dinner, where most of us would normally now say *stayed*. *Stay* is the much older of the two words in such contexts (first noted in use in the 16c.). What has happened is that the older term has more or less driven out the newer one. But not entirely: e.g. *'She's stopping with her daughter,' the woman said. 'She'll be back on Thursday.'*—H. Mantel, 1985. The corresponding noun for a place of temporary residence, a sojourn, is usually *stay*, though from the 17c. to the 19c. *stop* was regularly used for a stay or sojourn made at a place, esp. in the course of a journey. First in AmE and more recently in other branches of English, travellers were able to *stop off* or *stop over*, i.e. break their journey, at a designated place or places. And, from the late 19c. in NAmE, and from the beginning of the 20c. in BrE, travellers could make arrangements for a *stop-off* or a *stopover* at designated places on a long journey (esp. since mass travel by air became an established practice in the 1950s and 1960s).

**stops.** See APOSTROPHE²; BRACKETS; COLON; COMMA; DASH; EXCLAMATION MARK; FULL STOP; HYPHENS; ITALICS; QUESTION MARK; QUOTATION MARKS; SEMICOLON.

**storey.** 1 For the curious difference in sense between *storey* and *floor*, see FLOOR, STOREY.

2 Pl. *storeys*. In AmE the word is usu. spelt *story*, with pl. *stories*. Contrast BrE *three-storeyed house* and AmE *three-storied house*.

A *storied* window, urn, shrine, etc., is one ornamented with historical or legendary events: it is formed from the ordinary word *story* ( = narrative, etc.).

**stouten.** See -EN VERBS FROM ADJECTIVES.

**stove** (pa.t.). See STAVE (verb).

**Strad.** See ABBREVIATIONS 1(*b*).

**straight away.** *Straight* (which is in origin an adjectival use of the medieval past participle of *stretch* and is not etymologically related to its homonym *strait*) is first recorded in conjunction with *away*—as two separate words—in 1662: *some prisoners were hurri'd streight away to their Quarters*. As an adverbial form it stayed that way, as two separate words, until the beginning of the 20c. Then, perhaps under the influence of the fast-fading adverb *straightway*, it began to be written as one word. In 1923, for example, the *Daily Mail* reported that a horse called Evander had been badly hurt and was 'straightaway' withdrawn from a race. A character in Marganita Laski's *Tory Heaven* (1948) followed with *I said straightaway . . . that I'd like to be a land-agent. I'll buy some straightaway*, said Gina solemnly in Thomas Keneally's novel *A Family Madness* (1985). Bernice Rubens's crazy hero Mr Wakefield (1985) *rang the bell straightaway, to give himself no opportunity for second thoughts*. More examples turned up: *You could watch men see it in her, but never straightaway*—P. Carey, 1991; *He couldn't decide whether to make for home straightaway*—Stephen Wall, 1991. My instinct tells me to continue to follow the older practice of writing *straight away* as two words. But the evidence now points in the other direction, and I note that both the *COD* (1990) and the *New SOED* (1993) give only *straightaway* as headwords, relegating *straight away* under the combinations section under *straight*.

**straight, straightly.** The first is an adj. (*a straight line; keep a straight face*) and also an adverb (*came straight from Gatwick; in extreme heat he couldn't think straight; straight after breakfast*). The 'true' adverb *straightly* can hardly ever be used idiomatically: the most recent (19c.) examples cited in the *OED* now sound forced or archaic. But there are exceptions: *There was a kind of taut determination in the lumpy girl's straightly held shoulder*—A. McCaffrey, 1971.

**straight, strait.** *Straight* (not curved or bent) is from a medieval adjectival use of the pa.pple of ME *strecchen* 'to stretch'. *Strait* (tight, narrow) is from ME *streit*, from OF *estreit* 'tight, close, narrow' (also as noun, 'strait of the sea'), from L *strictus* 'drawn together, tight', pa.pple of *string-ere* 'to draw or bind tight'. In short, they are separate words with their own complex histories in English. *He straightened* up ( = he stood up straight) is also to be carefully distinguished from *straitened circumstances* (marked by poverty); and *straight-faced* (expressionless) is to be distinguished from *strait-laced* (puritanical). Also *strait-jacket* is better so spelt rather than *straight-jacket*.

**strappado** (old form of torture). Pl. *strappados*. See -O(E)S 3.

**strategic, strategical.** The *OED* (1917) gave precedence to a short vowel in the stressed syllable, but both words are now always pronounced with /-iː-/. It is worth noting that in the penult of adjectives and nouns in -*ic* (and the antepenult of -*ical* words), if -*ic* is preceded by a single consonant, there is an overwhelming majority of words with a short sound in the previous vowel (except *u*): so *barbaric, erratic, mechanic, tragic, academic, angelic, ethic, arthritic, prolific, chronic, exotic, historic, microscopic, spasmodic, lyric, paralytic,* and hundreds more; but with *u* we have *cubic, music, scorbutic* with /-uː-/. Nevertheless, *strategic* (with /iː/) has prevailed over *strategic* (with /e/). The most notable of other exceptions are *basic* and *scenic*, in which the long vowels are the natural result of familiarity with *base* and *scene*.

**strategy, tactics.** In war, as in politics and business, *strategy* tends to be the term used for a grand design or plan of action to be taken against an enemy, political opponents, or in order to ensure that a business enterprise is successful. *Tactics*, when there is a contrast with *strategy*, tends to be used of the detailed means adopted in carrying out such a grand design.

**strati-, strato-.** Such formations in which the main stress falls on the first element are normally pronounced in BrE with /æ/ not /eɪ/. Thus *stratiform, stratigraphic(al); stratocirrus, stratocumulus, stratosphere,* etc. In AmE, *stratocirrus, stratocumulus,* and other compounds of this kind usu. have /eɪ/ in the first element.

**stratum** /ˈstrɑːtəm/ or sometimes /ˈstreɪtəm/. Pl. *strata:* see -UM 2. Example: *The Reformation diffused the text of the Old Testament in previously untouched strata of the population—Bull. Amer. Acad. Arts & Sci.,* 1987. The *OED* lists three 18c. examples of *strata* used as a singular, with pl. *stratas,* and one 19c. example of *stratums* used as

a plural. Such erroneous forms are occasionally found in loosely edited modern books and newspapers. The traditional -*um* (sing.) and -*a* (pl.) distinction should be insisted on, at present at any rate, in all written work and also in the spoken word.

**stratus** /ˈstreɪtəs/, less often /ˈstrɑːtəs/, a continuous horizontal sheet of cloud. Pl. *strati* /-taɪ/.

**street.** See ROAD.

**strength.** The *ng* to be pronounced as in *sing* (not 'strenth' to rhyme with *tenth*).

**strew.** The pa.t. is *strewed* and the pa.pple *strewn* (for preference) or *strewed*.

**stricken.** Somewhat archaic as the pa. pple of *strike* (*he had become stricken with remorse*), it survives chiefly as the second element of compound adjectives, e.g. *grief-stricken, panic-stricken, poverty-stricken;* and as a participial adj. in certain contexts (e.g. *stricken deer; stricken hour* (of a clock); *stricken in years; a drawn, stricken face; the stricken city of Sarajevo*).

**stride** (verb). The pa.t. is *strode* and the (rare) pa.pple *stridden*. The rarity of occurrence of an actual pa.pple is underlined by the rather desperate use of *strode* in the popular magazine *The Face* (Nov. 1987): *Great strides are being strode in the cultivation of pre-teen female engineers.*

**strike** (verb). The pa.t. is *struck* (*the ship struck a rock*) and the pa.pple also *struck* (*the house was struck by lightning*) or, in certain contexts, and passing into archaism, *stricken* (q.v.).

**string** (verb). The normal pa.t. and pa. pple are *strung* (*they strung him up after a fair trial; his clothes were strung out on the line to dry*). A *stringed instrument* (in which *stringed* is formed from the noun *string*) is one having strings, but if the strings need renewing it will be *restrung*. Cf. HAMSTRING.

**strive** (verb). The regular pa.t. is *strove* and pa.pple *striven;* but *strived* is commonly used for both in AmE. Examples: *Her father ... had been a uniformed policeman ... who strived to make his beat a safe one for all—D. Koontz, 1984 (US); On the one hand, she'd never strived for celebrity— Musician (US),* 1991; *As surely as the painters* *of the nineteenth century strived to capture the city and people of Paris, so photographers in this century have been drawn to fururistic New York—Portfolio Mag.* (UK), 1991; *We've strived to lead the way in offering you the tools you need—Money* (US), 1993.

**stroma** (biology). Pl. *stromata*.

**strophe.** 1 'Part of a Greek choric ode chanted while the chorus proceeded in one direction, to be followed by a metrically exact counterpart as it returned' (Fowler, 1926). Cf. ANTISTROPHE 1.

2 A group of lines forming a section of a lyric poem (*COD*).

**strow** (verb). Formerly common, it is now only a by-form of *strew* (verb). Example: *Put it all back real neatly and strow it with old leaves and things—*A. S. Byatt, 1990.

**struma** (medicine). Pl. *strumae* /ˈstruːmiː/.

**strung.** See STRING (verb).

**strychnine** /ˈstrɪkniːn/. Thus spelt, not *-nin*.

**stubbornness.** So spelt, with *-nn-*.

**stucco.** Pl. *stuccoes* or *stuccos*. As verb, *stuccoes, stuccoed, stuccoing*.

**studding-sail.** Pronounce /ˈstʌnsəl/.

**studiedly.** Pronounce as three syllables. See -EDLY.

**studio.** Pl. *studios*. See -O(E)S 4.

**stupefy.** Thus spelt (cf. *liquefy, putrefy, rarefy*), not *-ify* (cf. *dignify, gratify, modify,* etc.).

**stupor.** Thus spelt, not *stupour*. See -OUR AND -OR.

**sty** (noun)[1] (pen for pigs). Pl. *sties*.

**sty** (noun)[2] (on eyelid). The OUP house style for the word is *sty* (pl. *sties*), not *stye* (pl. *styes*).

**style.** See INCONGRUOUS VOCABULARY.

**style, stile.** See STILE.

**stylo** ( = stylograph). 1 Pl. *stylos*. See -O(E)S 5.

2 See ABBREVIATIONS 1(*b*).

**stymie.** Now the regular spelling of the word in golf and in transferred senses, not *stimy* or *stimie*. As verb, the inflected forms are *stymies, stymied, stymieing*.

**suave.** Pronounce /swɑːv/, not /sweɪv/.

**subduedly.** Kipling used the word in his novel *The Light that Failed* (1891): *Maisie was crying more subduedly*. Pronounce as three syllables as *-ed* is part of the stem, not a separate syllable. See -EDLY.

**subject.** For synonyms in sense *theme* etc., see FIELD.

**subjective genitive.** See OBJECTIVE GENITIVE.

**subjunctive mood.** The subjunctive mood is one of the great shifting sands of English grammar. Its complexity over the centuries is such that the standard reference work on historical English syntax by F. Th. Visser (4 vols., 1963–73) devoted 156 pages to the subject (Visser called it the 'modally marked form') and listed more than 300 items in its bibliography. What follows here is by necessity merely an outline account of the present-day state of affairs, with a few backward glances at earlier conventions.

1 *Its nature*. The subjunctive is a verbal form or mood expressing hypothesis, usually denoting what is imagined, wished, demanded, proposed, exhorted, etc. Its main contrast is with the indicative mood. It is plainly recognizable in modE only in restricted circumstances: principally in the third person singular present tense by the absence of a final *-s* (*The confidential report into Mr John Stalker ... recommends that he* face *a disciplinary tribunal*; *If you want to irritate D., then suggest glibly that she* see *a sports psychologist*) and in the use in various circumstances of *be* and *were* instead of the indicative forms *am/is/are* and *was* (*Believing it to be fundamental that they* be *fully counselled by a professional of their own choice*; *Blanche almost wished that it* were *winter again*).

Contrast the use of the indicative mood in '*It is important that he* makes *friends,*' said Fibich; *I do wish he* was *coming too*. And contrast also contexts in which the verbal form is the same in the 'subjunctive' mood and in the 'indicative' mood: *We cannot talk as if the other parties* were *demons*; *She wondered if wild creatures* were *capable of envy*; *He asked that I do* him *the courtesy of never mentioning his former brokerage firm in his presence again*. Note. All the examples cited so far are from BrE sources of the 1980s and 1990s.

**2** *Its currency in the US and elsewhere.* The subjunctive mood was common in OE and until about 1600. Examples are harder to find in the period 1600–1900 but have become remarkably prevalent since then, first in AmE and then in other forms of English. Examples: *It would be as though the artist* were *not aware of the interim period*—Dædalus, 1986; *I was going to recommend that he* be *terminated*—New Yorker, 1987; *Vice President George Bush phoned Noriega . . . asking that Noriega* warn *Cuban leader Fidel Castro not to interfere in the operation*—USA Today 1988; *Gorbachev himself proposed that farms* be *leased back to farmers*—Bull. Amer. Acad. Arts & Sci., 1989.

The subjunctive is increasingly found in BrE: *It was as if Sally* were *disturbed in some way and was translating this disturbance into the habit of thought*—A. Brookner, 1986; *Each was required to undertake that if it* were *chosen it would place work here*—Times, 1986; *Better it* were *just Claudia and me, thinks Laszlo, but never mind*—P. Lively, 1987; *Fundamentalist Islam . . . decrees that men and women* be *strictly segregated*—Listener, 1988.

Note. In the Brookner example the change from *were* to *was* indicates a shift of mood after *and*. Visser (§839) cites parallel examples from OE onwards, e.g. *If Colonel Camperfelt* be *in Town, and his Abilities* are *not employ'd another Way* [etc.]—Steele, 1712.

The subjunctive is also increasingly found in other standard forms of English: *We suggest that a local hiring policy* be *put in place*—Globe & Mail (Toronto), 1985; *Now there* were *demarcations about where one lived and how one lived, especially if one* were *a black person*—M. Ramgobin, 1986 (SAfr.); *Your situation demands that either Kooti* be *nobbled or Whitmore nullified*—M. Shadbolt, 1986 (NZ); *She wished Sol Meyer would suddenly demand that she* act *her age and return to Parramatta*—P. Carey, 1988 (Aust.); *She declined a seat beside Charles on the sofa. She insisted Jane* sit *there*—Barbara Anderson, 1992 (NZ); *It was suggested he* wait *till the next morning*—M. Ondaatje, 1992 (Canad.).

Note. In many dependent clauses a putative *should* + infinitive is used instead: e.g. in the Shadbolt 1986 example *should be* could replace *be* without discernible change of meaning. The type with *should* is at least as common (e.g. *The report recommends that access to patent information* should *be widened and improved*). And so is the use of the indicative in similar constructions (e.g. *If Mr Ward is 'between airlines'* let *someone* suggest *that he* looks *towards Britain*—Weekend Guardian, 1990).

**3** *Various types.*

(*a*) *were* or *be* placed at the head of a clause (*rather formal*). Examples: *Were this done we would retain a separate Bar with skill*—Times, 1986; *Were I to get drunk, it would help me in the fight*—J. Updike, 1986; *Statistically, afterworlds—*be *they Christian, Greek, Pharaonic—must be populated almost entirely by children*—P. Lively, 1987; *All are under judgement,* be *they friend, or fellow, or bishop or journalist*—Times, 1987.

(*b*) Placing of *not* before a subjunctive *be* followed by a pa.pple: *Again he insisted that he* not *be followed*—Observer, 1987; *What is crucial is that such inequalities* not *be perceived as part of a 'classless structure' that stifles . . . individual initiative*—Encounter, 1987; *The register of residents requesting that their verges* not *be sprayed has become substantial*—Metro (NZ), 1989.

Note. This construction is routine in AmE, but less common elsewhere.

*Not* is also often placed immediately before a simple subjunctive: *I ask that he* not *track mud all over the floor*—New Yorker, 1988; *One essential quality for a holiday novel is that it* not *be too light*—F. Raphael, 1988; *What we shall be waiting for is the first success story in Eastern Europe. Pray that it* not *be too long in coming*—Christian Sci. Monitor, 1990.

(*c*) Subjunctives preserved in an array of fossilized clauses or sentences expressing a wish 'whose realization depends on conditions beyond the power or control of the speaker . . . and consist in short formulae of praise or prayer, in invocations, supplications, blessings, curses, oaths and imprecations' (Visser §841): *be that as it may; so be it; bless my soul; come what may; far be it from me to; God forbid (that); God bless you; God save the Queen, etc.; heaven forbid/forfend (that); heaven help us; So help me (God); Thy Kingdom come; long live the Queen, etc.; perish the thought; the powers that be; serve you right; suffice it to say that; woe betide.*

(d) The fixed phrase *as it were*. In the sense 'in a way, to a certain extent' the phrase is invariable: *Having to ask permission, as it were, to see her friends*— A. Lurie, 1985; *Suddenly, as it were overnight, the weather became hot and sultry*— A. Brookner, 1986.

**4** *General comments.* In BrE the subjunctive mood is most likely to be found in formal writing or speech (apart from some of the formulaic uses listed in 3(c) above), and particularly (the so-called *mandative subjunctive*) after verbs such as *demand, insist, pray, recommend, suggest,* and *wish*; nouns and adjectives such as *demand, essential, important, insistence, proposal, suggestion, vital,* and *wish*; and a number of conjunctions, such as *although, as if, as though, if, unless,* etc. But it is seldom obligatory, and indeed is commonly (?usually) invisible because the notionally subjunctive and the indicative forms are identical.

**submerge** (verb). The adj. ('capable of being submerged') corresponding to this verb is *submersible* rather than *submergible.*

**submissible** (adj.), able to be submitted. This is the better form (corresponding to the verb *submit*; cf. *permit, permissible*), rather than *submittable.*

**subpoena** /səb'piːnə/. Thus spelt (but occas. *subpena* in AmE). Pl. *subpoenas.* Pa.t. and pa.pple of verb, *subpoenaed:* see -ED AND 'D.

**subsidence.** The traditional pronunciation /səb'saɪdəns/, with the stress on the second syllable, is recommended. But the form with initial stress, namely /'sʌbsɪdəns/, under the influence of *residence* and *subsidy,* is also common in standard speech.

**substantial, substantive.** Both words mean 'of substance', but they have become differentiated to the extent that *-ial* is now the word in general use for real, of real importance, sizeable, solid, well-to-do, etc., and *-ive,* apart from its meaning in grammar, is chiefly used in special senses: in parliamentary procedure a *substantive motion* is one that deals expressly with a subject in due form; in

law *substantive law* (that which is to be enforced) is so called to distinguish it from *adjective law* (the procedure for enforcing it); in the Services *substantive* is used to distinguish rank or office that is permanent from one that is acting or temporary.

**substitute** (verb). The normal uncontroversial sense (construed with *for*) is 'to put (one or something) in place of another' (*in training for the ring they substituted rum for sherry; sacrifice could not be substituted for duty; the local priesthoods, who substituted their own favourite god for Re*—examples from the *OED*).

Beginning in the 17c., and running parallel to the normal sense, were transitive uses in which the sense is 'replace': *Double Pica* [a typeface] ... *was ... substituted by a new Letter* (1770); *Good brandy being substituted by vile whiskey* (1863, US); *Miss Hughes substituted Miss Oliver* (1867). The use went out of favour to the point that the *OED* (1917) commented 'Now regarded as incorrect'. The use has persistently returned in the 20c., however: *It* [sc. a debating club] *was frequented by the clerks and John Gilpins of the neighbourhood and presided over by a venerable greybeard citizen, known as the 'vice', who presumably substituted some more illustrious chairman*—J. R. Rodd, 1922; *The ecclesiastical principle was substituted by the national, the Empire and the Papacy by the Communes*— cited in Fowler, 1926; *If potatoes substitute bread, what is going to substitute potatoes? is a question every German will have to ask himself*—cited in Fowler, 1926; *The tribunal concludes that British Rail's proposal to compensate ... at rates of four, five and six per cent. are inadequate and substitutes them with levels of five, $7\frac{1}{2}$ and 10 per cent*—*Daily Tel.,* 1974. Repeatedly football commentators announce that a specified tired-looking player is about to be *substituted* (i.e. replaced) by a fresh-looking player on the bench (or, in the active voice, X substituted Y *just before half-time*).

This persistent transitive use of *substitute* still sounds non-standard to me, but it is pressing at the gate to be admitted into the standard territory inside.

**subtle.** The corresponding adverb is *subtly.* For a similar type of relationship, cf. *supple* (adj.)/*supply* (adv.).

749

**succedaneum** /sʌksɪ'deɪnɪəm/, a substitute, esp. for a medicine or drug. Pl. *succedanea.*

**succeed** (verb). In the sense 'to be successful' it cannot be followed by *to* + infinitive (*She succeeded to find an empty carriage*) but governs instead *in* + gerund (*She succeeded in finding an empty carriage*). See GERUND 3. Contrast the idiomatic use of *fail* in *She failed to find an empty carriage.*

**success of esteem.** See ESTEEM; GALLICISMS 2.

**succinct.** The *-cc-* is pronounced /ks/, as in *accent, success,* etc. Contrast *succour* /'sʌkə/.

**succour.** This rather formal word meaning 'aid, assistance, esp. in time of need' is spelt thus in BrE, but as *succor* in AmE.

**succuba, succubus.** Pl. *succubae* /-biː/, *succubi* /-baɪ/. For centuries equally acceptable terms for a female demon believed to have sexual intercourse with sleeping men (the male equivalent being INCUBUS), but *succubus* is the more usual word now.

**such.** Before embarking on this article it will be as well to bear in mind that this simple-looking word has had many branches and intricacies of meaning and construction from Anglo-Saxon times onward. Since the majority of these are not mentioned in what follows, let me begin by citing a few examples of incontestably legitimate uses drawn from my database, all from standard sources of the 1980s: *our hands are quite unsuited to such fragility; We laugh in this country for such strange reasons; Fortunately for Owen, such an event was to hand; there will be no such sermon given next Tuesday; her mind refused to bring any such memory forward; a very profitable company such as British Telecom; Its prey are other small creatures such as frogs and lizards; I was proud of his being accepted at such a good school; It should be of such a nature as to command credibility; Kennedy did not dare ask Congress for such a treaty; The one such marsupial we saw needed more ballast; shapeless dresses made of such materials as emerald silk moiré; One such case, indeed, has actually been reported in the press; How they would respond at such a time is anyone's guess.* From these alone one can see that *such, such a(n), such as, no such,* and so on, have a multitude of idiomatic uses that do not call for particular comment in a usage manual.

Another preliminary remark. Walter Nash in his *English Usage* (1986) pointed out that all of us have many preconceptions about language. 'You may decide, for instance, that *such a(n)*, as in *He was criticized for inventing such an unbelievable character*, is a dubious idiom, the preferred construction being *He was criticized for inventing so unbelievable a character* or *for inventing a character so unbelievable.*' The danger, he believes, is that we may 'seek out deviations that allegedly impair communication or reflect imprecision of thought.' In the following paragraphs every attempt will be made to elucidate rather than simply to condemn and also to place everything in a true historical perspective.

> 1 *such* followed by the relative pronouns *that, which, who.*
> 2 Emphatic *such.*
> 3 Agreement after *such as.*
> 4 *such as* or *like.*

1 *such* followed by the relative pronouns *that, which, who.* The OED (dem. adj. 12) gives a chain of examples from Anglo-Saxon times to 1888 but judged that by 1917 (when the relevant section of the OED was issued) the type was 'Now rare and regarded as incorrect'. Typical examples: *Such suffering Soules that welcome wrongs*—Shakespeare, 1601; *These seemed to him ... such which he never thought ... would be seriously opposed*—J. Strype, 1709; *Such of his friends that had not forsaken him*—Goldsmith, a1774; *Only such intellectual pursuits which are pleasant*—S. Grand, 1888. Fowler, in one of his moments of antagonism to the OED, objected that 'It is not in fact very rare'. Such constructions in the 20c., he said, 'are due either to writers' entire ignorance of idiom or to their finding themselves in a difficulty and not seeing how to get out of it'. He then cited ten examples from unassigned (but obviously journalistic) sources and rewrote several of them: e.g. *The Roumanian Government contends that it has only requisitioned such things of which there is abundance in the country* (such things as are abundant, or as there is abundance of); *It was proposed to grant to such casual employees of the Council who had been*

*continuously employed for three months and whose employment was likely to extend over twelve months, the privilege of additional leave* (read *those*, or *any*, for *such*). Whatever the reason, such uses seem to have disappeared from sight and I have found no recent examples. In most constructions of this type modern idiom prefers *such . . . as.*

The standard sources of grammatical evidence (Jespersen, Curme, *CGEL*, etc.) provide no parallel examples, either, though they give examples of idiomatic sentences in which *such* as either a predicative or an attributive adjective is followed by a *that*-clause of result: *There is such confusion that I can't collect my thoughts; The confusion is such that I can't collect my thoughts; Their condition was such that they could hardly speak.* And others with the insertion of the indefinite article (in which *such a* becomes adverbial): *She is such a good lecturer that all her courses are full; There was such a large crowd that we couldn't see a thing; These are such a heavy pair of shears that I can hardly use them for longer than half an hour.*

2 Emphatic *such*. The type *He's such a nice man* is widespread in spoken English, but less common in expository prose. Varying degrees of legitimate emphasis are shown in: *It was such a fine evening, His misbehaviour was such a bore, She has such beautiful manners.* A literary example from an American author: *Papa was, he still is, such a dude*—S. Bellow, 1987. Again, *such (a)* is adverbial.

3 Agreement after *such as*. Janice Elliott was following a long tradition (*OED* dem. adj. 10a) in her novel *The Italian Lesson* (1985) when she used the nominative form of the pronoun: *Such as he are never popular with hotels.* But in many constructions of this type the oblique form of the pronoun may be used with equal justification. Historical examples: (a) *Others such as he*—Shakespeare, 1611; *It is not fit for such as we to sit with the rulers*—Scott, 1819; *Death was not for such as I*—C. Brontë, 1847. (b) *while such as her die*—Swift, 1710–13; *Did ever man have such a bother with himself as me?*—H. G. Wells, 1900; *they were not bad, for such as her*—R. Macaulay, 1920. (Some of the examples are drawn from Jespersen.)

4 *such as* or *like*. (See LIKE 2(b).) Opinion is neatly divided about the merits of *like*

or *such as* used to introduce examples of a class. There is abundant evidence for *like* to be used when only one item, person, etc., is specified (*a writer like Tennyson*), and equally abundant evidence for *such as* to be used in the same way (*Many large gold coins, such as the doubloon*). The choice is often governed by the meaning: if the sense required is 'resembling' then *like* is preferable. And there is much to be said in favour of *such as* when more than one example of a class is mentioned: *All of the cat kind, such as the lion, the tiger, the leopard, and the ounce*—Goldsmith, 1774; *Writers such as Theophrastus and La Bruyere*—Mirror, 1779.

**suchlike.** 'Whether as adj. (*barley, oats, and suchlike cereals*) or as pronoun (*schoolmasters, plumbers, and suchlike*) [suchlike] is now usually left to the uneducated, *such* being used as the adjective and *the like* as the pronoun' (Fowler, 1926). Some voices are still raised, with diminishing levels of condemnation, but most dictionaries now simply list the two uses of the word as part of the standard language. The adj. and the pronoun have been in continuous use since the 15c.: it seems unlikely that they will be abandoned now or left to the mercy of any special group.

**Sudan, Sudanese.** Now the normal spellings in English, not *Soud-*.

**sudarium** (cloth for wiping the face), **sudatorium** (hot-air or steam bath), not being household words in English, retain their Latin plurals in *-ia*.

**suddenness.** So spelt, with *-nn-*.

**suede.** The accent of its original (Fr. *Suède* 'Sweden') is now usu. omitted when the word is used to mean 'leather of kidskin'.

**sufficient(ly).** See ENOUGH.

**suffix.** In grammar, a suffix is an affix placed at the end of a word to form a derivative (*-ation*, *-ing*, *-itis*, *-ment*, etc.). Cf. PREFIX.

**suffixes added to proper names.** *Aesopian, Betjemanesque, Borrovian, Heideggerian, Wittgensteinian*: such words crop up all the time in academic and general writing. Thus, from a 1988 encomium of a recently deceased cartoonist and writer in the London Review of Books: his best efforts took him from ... an aesthete's and illustrator's art to an account—more Proustian, or Powellian, than Swiftian—of the morals and manners of the Swinging London that ever was and isn't any more. The word Aristotelian apart, it was probably not until they reached the letter B that the editors of the OED began to face up to the problems presented by this kind of word. A policy of sorts began to emerge: in normal circumstances adjectives of this kind were not to be admitted to the OED until they were accompanied by a cluster of closely related forms. So Baconian and Baconic qualified for admission because of the existence as well of Baconianism and Baconist; Boswellian because of Boswellism and Boswellize; and Byronian and Byronic because of Byroniad, Byronist, Byronite, and Byronically. Throughout the OED the same policy of guarded admission was adhered to. Fame by itself was not enough, unless formations in -iana, -ism, -ite, -ize, and so on, were also found to exist. The editors were faced with infinite sets of adjectival formations if any adjectives formed on proper names were to be included. The nature of the inclusion/exclusion problem is neatly expressed in the OED entry for -esque: 'suffix, forming adjectives, represents French -esque, adapted from Italian -esco ... In Italian derivatives in -esco are formed ad libitum on names of artists, and French and English writers on art have imitated this practice. Examples of such formations, not calling for separate notice in the Dictionary, are Bramantesque, Claudesque, Turneresque.' See -ESQUE.

In the four-volume Supplement to the OED (1972–86) my colleagues and I, somewhat grudgingly, made good some of the omissions. For example, to take only writers, entries were made for Aeschylean, Alfredian, Anselmian, Anselmic, Arnoldian, Austenian, and a few others, in the letter A. The letter B contained entries for a few more such words—Blakeian (or -ean, -ian), Borrovian, Brontëan, Brontesque, Browningesque, Burnsian, and some other adjectives formed on the names of pre-twentieth-century writers, always with a small squad of related formations (Browningese, Browningite, Burnsiana, Burnsite, and so on). And so the pattern continued. The meaning was normally

expressed in a formulaic way: 'resembling the style of, partaking of the characteristics of', and the like. Such definitions, being easy to write, are like sweet spring water to lexicographers. Admittedly the OED had defined Johnsonian a little more specifically: 'applied especially to a style of English abounding in words derived or made up from Latin, such as that of Dr Johnson'. But such particularity was very rare. Those who consulted the OED were expected to know what the chief attributes of the writers, artists, etc., were, or else to seek them in other reference works.

The habit of appending -esque, -(i)an, or -ic to the names of authors to indicate resemblance appears not to have begun until the 16c. (Virgilian is first recorded in 1513, Platonic in 1533), and then only rarely. The names of a few well-known classical writers are first recorded in adjectival form in 17c. sources: Aristotelian in 1607, Ovidian 1617, Pindaric 1640, Plinian 1649, Ciceronian 1661, and Ptolemaic 1674. Of English names, Drydenian is even recorded during his own lifetime (1687). In the 18c. it was not unusual for such formations to be used very soon after the death of the writer concerned—thus Gibbonian 1794, Johnsonian (by Boswell) 1791, Richardsonian 1786—but, unless our records are faulty, most formations of this kind are not found until much later: Fieldingesque 1931; Ouidaesque ('marked by extravagance or lack of restraint') 1909. Shakespearian is not recorded before 1755. The choice of suffix seems to have been mostly governed by euphony, but names that end in -aw or -ow normally have the Latinate terminations -avian (Shavian) and -ovian (Borrovian) given to them. Thoreau becomes Thoreauvian. Some classical names have generated a cluster of adjectival forms since the 17c.—for example Ptolemy—but one of them (in this case Ptolemaic) usually settles down as the customary form.

There is one oddity. An unaccented final syllable of a name is normally lengthened and stressed when the adjectival form is made: thus Alfred but Alfredian (pronounced /-iːd-/ and bearing the main stress), and Dryden but Drydenian /-iːn-/. For reasons of euphony, a good many names—Amis, Beddoes, Burgess, Byatt, for example—probably have to get

by with fewer possible adjectival append-
ages than the likes of Byron.

**suffragette.** See -ETTE 4.

**sugar** makes *sugared* and *sugary.* See -R-,
-RR-.

**suggest** (verb). In his book *The Changing
English Language* (1968) Brian Foster cited
a sentence spoken by the actor Albert
Finney in a BBC *Face to Face* programme
in 1962: *The headmaster suggested I went
to drama school.* He might alternatively
have said *The headmaster suggested I go
to drama school,* or else *The headmaster
suggested* [that] *I should go to drama school.*
'Usage', Foster concluded, 'is in a somewhat fluid state.' The evidence I have
collected myself supports this view.
What can be said, however, with some
degree of certainty is that *suggest* (and a
few other words) now very commonly
generate a subjunctive form in a following *that*-clause (in so far as this form is
distinguished from the indicative equivalent: see SUBJUNCTIVE MOOD), esp. in
N. America but increasingly elsewhere.
*CGEL* (16.32) cites as alternatives *People
are* demanding *that she* should leave/
leave/leaves [esp. BrE] *the company.* It
would be the same if *suggesting* had been
used instead of *demanding.* Some of the
main patterns are shown in the following examples, all from standard sources:
  (a) *suggest* (or *suggestion*) + *that*-clause:
*We suggest that a local hiring policy be put
in place—Toronto Globe & Mail,* 1985; *Uncle
doesn't suggest that she* bring *a lamp from
the next room—S.* Bellow, 1987 (US); *the
suggestion that all HIV-positive individuals
be forcibly tattooed—Dædalus* (US), 1989; *If
you want to irritate D., then suggest glibly
that she* see *a sports psychologist—Times,*
1990; *including the suggestion that colleges
and the General Board be invited to consider
whether ways of reducing academic pressures
on students can be identified—Oxford Today,*
1993 (UK).
  (b) *suggested* + *that*-clause: *She suggested
... that Harry* turn *bisexual and get a
chauffeur as well—P.* Carey, 1981 (Aust.); *It
suggested that courts be allowed to detain
juvenile probation violators for up to a year—*
*New Yorker,* 1986; *I have also suggested to
Sir Percy that he* obtain *the services of a
young scientist—P.* Wright, 1987 (UK, but
book published in the US); *He suggested
that they* went *into the church at the end of*

*the road—A. S. Byatt,* 1990 (UK); *After tea
he suggested that she* have *a lie down until
dinner—R.* Rive, 1990 (SAfr.); *He even
suggested they* went *out—A.* Huth, 1991
(UK); *It was suggested he* wait *till the next
morning—M.* Ondaatje, 1992 (Canad.).
  In the BrE examples and in some of
the others a simple indicative verb or
the type *should* + infinitive would have
been equally idiomatic (see SUBJUNCTIVE
MOOD, note at end of (2)). As Foster remarked, 'Usage is in a somewhat fluid
state.'

**suggestible.** Thus spelt, not *-able.* See
-ABLE, -IBLE 7.

**suggestio falsi** /sə,dʒestɪəʊ 'fælsaɪ/. Pl.
*suggestiones falsi* /-ɪəʊniːz/. The making of
a statement from which, though it is not
actually false, the natural and intended
inference is a false one; an indirect lie.
Example: *It is rare to find a positively verifi-
able untruth in a school brochure: but it
is equally rare not to find a great many
suggestiones falsi, particularly as regards
the material comfort and facilities available—J.* Wilson, 1962. Cf. SUPPRESSIO VERI.

**suitor.** Thus spelt, not *-er.*

**suit, suite** (nouns). *Suite* is pronounced
/swiːt/ and *suit* /suːt/ or /sjuːt/. They are,
in terms of their origin, the 'same word',
and the differences of usage are the result of historical choices and driftings
since the 17c. The complex story is set
down in the relevant entries in the *OED.*
In modern English we speak of *a suit of
clothes, a suit of armour, a lawsuit, a suit of
playing cards* (any one of four), a man
*paying suit to* (courting) a woman; and
on the other hand *a suite of rooms* (in a
hotel), *a suite of furniture* (sofa and chairs
of the same pattern), a musical *suite,* etc.

**sullenness.** Thus spelt, with *-nn-.*

**sulphur.** The invariable BrE spelling for
it and its compounds (*sulphurate, sulphuric, sulphurous,* etc.), but normally *sul-
fur* (similarly with the compounds) in
AmE.

**sumac** (an ornamental tree). This spelling is recommended in OUP house style
rather than *sumach.* Pronounce /ˈʃuːmæk/.
But the spelling *sumach* and the pronunciation /ˈsuːmæk/ are also legitimate, esp.
in AmE.

**summersault.** See SOMERSAULT.

**summon, summons.** 1 A *summons* is a call to appear before a judge or magistrate. Pl. *summonses.*

2 As a transitive verb, *summons* means (in law) to serve with a summons. In ordinary use the normal form of the verb is *summon* (*the boy was summoned to the headmaster's study; the chairman summoned the members to a meeting*).

**Sunday.** For the adverbial use (*Sunday = on Sunday*), see FRIDAY.

**sung.** See SING (verb).

**sunk, sunken.** See SINK (verb).

**super-.** As a living prefix in English, '*super-* first appears about the middle of the 15th c.; it became frequent in Elizabethan times, and in the 17th c. it was very widely used' (*OED*). In the 20c. it has continued to be a prolific formative element. Among the formations of modern times (the dates are those of first record in the *OED* are: *supercharge* (verb) (1919), *supercharger* (1921), *supercluster* (astronomy, 1930), *superconductor* (1913), *supercontinent* (geology, 1963), *super-duper* (1940), *super-ego* (1924), *superfluorescence* (physics, 1966), *superman* (1913), *supermarket* (1933), *superpower* (1921), *supersonic* (adj.) (1919), *superstar* (1925), *superstore* (1965), and numerous other general, technical, and scientific words.

**[supercede.** For this misspelling, see SUPERSEDE.]

**superficies** (geometry). Five syllables, /ˌs(j)uːpəˈfɪʃɪˌiːz/. Pl. the same.

**superior.** Because it is not a true comparative it, like *inferior*, cannot be used in comparative constructions with *than*: thus *X is superior to its rivals* (not *superior than*). The equivalent comparative phrases are *better than, greater than*, etc. In the language of marketing, and sometimes in general social contexts, *superior* is used as a kind of blind or absolute comparative ( = above the average in quality) (*a superior bungalow; made of superior leather; a very superior box of chocolates*).

**superiority.** Fowler's term for the use of a slang expression or a socially divisive

remark preceded by a distancing or defensive comment implying that in normal circumstances the speaker would not deign to use such an expression himself or herself. Such distancing remarks include *as they say; if the word may be permitted; if the word is not too vulgar; in the jargon of today; in the vernacular phrase; not to mince matters; not to put too fine a point upon it; so to speak; to call a spade a spade; to put it vulgarly;* and also the use of depreciatory inverted commas.

Examples: *A grievance once redressed ceases to be an electoral asset (if we may use a piece of terminology which we confess we dislike); to put it vulgarly, that cock won't fight; Palmerston is to all appearance what would be vulgarly called 'out of the swim'; He seized my hand in what the lover of a cliché would call an 'iron grip'; For all I know the boy selling dusters coud have been, as they say, 'casing the joint'; 'Hopefully,' as people say, it will not rain on Sunday.*

Another form of 'superiority' is shown in the use of the parenthetic phr. *of course:* see COURSE.

**superlatives.** 1 For various uses of superlatives, see ADJECTIVE 3; ABSOLUTE SUPERLATIVE.

2 In general it is a sound rule that confines the use of comparative forms of an adjective to contexts in which two entities are being compared, and reserves superlative forms for comparison of three entities or more. But the English language is not a totally restrictive system. Jespersen (1909–49, vii.11.6₁) remarked that 'it is important to insist on the fact that in ordinary usage the superlative does not indicate a higher degree than the comparative, but really states the same degree, only looked at from a different point of view'. Whatever the explanation, seemingly illogical uses of the superlative occur from time to time in impeccably standard sources. Examples (the first three drawn from Jespersen ii.7.772): *to prove whose blood is reddest, his or mine*—Shakespeare, 1596; *whose God is strongest, thine or mine*—Defoe, 1720; *She was the youngest of the two daughters*—J. Austen, 1816; *dinghy, dingey. The first is best*—H. W. Fowler, 1926. Clearly there is a ragged edge at the rim of any strict rule, but the general pattern should normally be adhered to,

leaving exceptions only to the truly great or to literary or linguistic licence.

**supersede.** Under the influence of *accede, cede, intercede,* etc., *supersede* (ultimately from L *supersedēre* 'to sit above, be superior to') is frequently, but erroneously, spelt *supercede.* Examples: *he does not supercede his stature*—D. Hirson, 1986 (SAfr.); *Somewhere down the line, you are superceded by another load of players*—*Independent on Sunday* (headline), 1990; *and thoroughly nasty little conflict that illuminated our news bulletins for a while before it was superceded by that far greater conflagration in the Gulf*—*Spectator*, 1992. It is worth noting that the word appeared first as *superceder* in OF and only later as *-seder*; and that in English, forms with medial *c* have been recorded since the 15c. It was also often written as *supercedere* in medL, according to the *OED*.

**superstitions.** Among the most enduring of the superstitions or myths about our language are these: sentences should not begin with *and* or *but*; sentences should not end with a preposition; and infinitives should not be 'split'. For further examples of such beliefs, see FETISHES.

**supervise.** Thus spelt, not *supervize*. See -ISE.

**supervisor.** Thus spelt, not *-er*.

**supine.** 1 (adj.) lying face upwards, as opposed to *prone*, lying face downwards.

2 (noun) A Latin verbal noun used in the accusative case in *-um* with verbs of motion (*eo Romam servos emptum* 'I am going to Rome to buy slaves') or in the ablative in *-u* (*mirabile dictu* 'wonderful to relate'), esp. to express purpose.

3 *Pronunciation.* The grammatical term is pronounced with the stress on the first syllable, /ˈs(j)uːpam/. Similarly now with the adj., but it has not always been so: the *OED* (1917) gave preference to second-syllable stressing, and Daniel Jones (also 1917) gave only second-syllable stressing for it. Cf. CANINE; NOUN AND ADJECTIVE ACCENT; RECESSIVE ACCENT.

**supple** (adj.). The recommended form for the corresponding adverb is *supply* (two syllables) 'in a supple manner', rather than *supplely* (three syllables). Cf. *subtle* (adj.)/*subtly* (adv.).

**supplement.** The noun is stressed on the first syllable, /ˈsʌplɪmənt/. The verb differs only in that the final syllable often bears the main stress, /ˌsʌplɪˈment/, but there is a certain amount of variation and /ˈsʌplɪmənt/ is also found.

**supposal.** A long-lived (14–19c.) abstract noun ( = supposition; assumption) that seems to have dropped out of general use (see -AL 2 for the fate of such words) and to have survived mainly as a technical term in logic.

**suppose.** See S'POSE.

**supposedly.** Four syllables. See -EDLY.

**supposititous, supposititious.** These somewhat uncommon and often overlapping words 'may as well coexist, if there is work for two words, like *factious* and *factitious,*' commented Fowler in 1926. Both words entered the language in the 17c. and for a long period shared the sense 'fraudulently substituted for the real thing or person'. Nowadays the longer word is the one used in this sense (*Russia . . . is the supposititious child of necessity in the household of theory*—H. G. Wells, 1934), while *supposititious* is normally used to mean 'hypothetical, conjectural' (*We might take a small cottage outside Dublin . . . Not that I imagine that the atmosphere of our supposititous cottage could . . . become more unpleasant to you than the atmosphere you are at present breathing.*—J. Joyce, 1905). *Supposititious* is derived from L *supposītīcius* '(fraudulently) put in place of another', from *suppōnere,* a wide range of senses including 'to place under' and 'to suppose'. The shorter form is 'partly a shortened or illiterate form of *supposititious,* partly directly from *supposition*' (*OED*).

**suppressible.** Thus spelt, not *-able*. See -ABLE, -IBLE 7.

**suppressio veri** /səˌpreʃɪəʊ ˈviːəraɪ/. Pl. *suppressiones veri* /-ɪəʊniːz/. A misrepresentation of the truth by concealing a fact or facts which ought to be made known. Example: *It would not be easy to find a more flagrant case of* suppressio veri *than this omission . . . of any reference to the notorious Rohling scandal*—M. Hay, 1950. Cf. SUGGESTIO FALSI.

**suppressor.** Thus spelt, not -er. See -ER AND -OR; -OR.

**supreme.** 1 See ADJECTIVE 4.

2 For *the supreme sacrifice*, see STOCK PATHOS.

**surcease** (noun and verb). A literary word of some antiquity (noun 16c., verb 15c.) for 'cessation, respite' and 'to cease'. Now less commonly used (esp. the verb), except in AmE. Some examples of the noun possibly reflect the well-known context in Shakespeare's *Macbeth* (1605), *If th' Assassination Could trammell vp the Consequence, and catch With his surcease, Successe.* Modern examples of the noun: *The movies were low-cost balm and surcease for large audiences during the Depression*—A. Miller, 1987; *Spring training is meant to bring surcease to such dour notions*—New Yorker, 1987; *S. Rado coined the term 'pharmacothymia' to describe the disorder in which drugs are taken to find surcease from intolerable psychic pain*—E. A. Grollman, 1988; *The sculptured forms piled overhead, one into another without surcease*—A. Dillard, 1989.

**sure, surely** (adverbs). 1 The dominant adverb of the two by far is *surely*, and esp. in its use as an appeal to likelihood or reason. Examples (from my database) of standard uses of *surely*: *it was surely here, if anywhere, that he became the 'Poet'; the protester is surely honour-bound to show that he knows what he's talking about; those [sc. performance indicators] for dons will surely be easy to agree on; Tim was old enough, surely, to come and go as he pleased?; The only question to be asked . . ., surely, is what can he salvaged from the wreck; what was not wanted by its owner surely does not belong to anybody?; Not just for irony's sake, surely?* One of its commonest uses is in the phr. *slowly but surely*, when the meaning called for is 'so as to be certain to achieve or reach a result or end': *These things are slowly but surely coming about*—1912 in *OED*. All of the types cited above are part of general English, i.e. are not primarily associated with any specified region in the English-speaking world.

2 By contrast, *sure* used as an adverb is relatively uncommon in BrE. It exists in the time-honoured semi-proverbs *as sure as eggs is eggs* and *as sure as God made little apples*. It has also been used since the beginning of the 19c., first in Britain

and then elsewhere, as (part of) an affirmative response (19c. examples are given in the *OED*): *'Is that a fact?' 'Sure,'* murmured Archibald—P. G. Wodehouse, 1914; *If it had been a request to chop off one's right hand one would have said, 'Sure.'*—Mrs. L. B. Johnson, 1963; *I'm under arrest. I asked if you could finish your lunch, and they said sure, no hurry*—R. Stout, 1975. But most adverbial uses of *sure* are strongly associated with America, and to a lesser extent with Ireland. Witness the following examples: *Parts of it were pretty, sure*—A. Lurie, 1969 (US); *You sure were one sweet kid*—TV programme, Cagney & Lacey, 1987; *Sure how old are you, both of you?*—M. Leland, 1987 (Ir.); *A chemical fire. You worry about those, sure, said Clerk*—New Yorker, 1988.

3 Two useful sets of distinctions are given in *CGEL* (1985). The first group (7.56 n.):

> *Surely*, she's right. (Persuasive: 'you surely agree')
> *Sure*, she's right (AmE = agreement: 'of course')
> *Sure*, she's right. (IrE = asseveration: 'I assure you')

And the second (8.100), as responses to requests:

> (a) *Please get me the file on Robert Schultz.*
> *Certainly*
> *Sure* (esp. AmE informal)
> *Surely* (esp. AmE).

> (b) *Are you willing to help her?*
> (*Yes*) *certainly*
> *Sure* (esp. AmE informal)
> *Surely* (esp. AmE).

**surety.** The standard pronunciation now seems to me to be trisyllabic, /ˈʃʊərɪtɪ/, but the main dictionaries also list disyllabic /ˈʃʊətɪ/.

**surgeon.** See PHYSICIAN.

**surly.** A 16c. altered form of earlier (first recorded 14c.) *sirly*, from *sir* + *-ly*. The earliest sense was 'lordly, magnificent', but this sense gave way in the 17c. to the modern meaning, 'bad-tempered and unfriendly, morose'. The corresponding adv. is *surlily*, but it is usually avoided (in favour of 'in a surly manner' or the like) on grounds of euphony: see -LILY.

**surmise** (noun and verb). Thus spelt. See -ISE.

**surprisal.** From the 16c. until the end of the 19c. commonly used in various senses of *surprise* (noun), but now obsolete. See -AL 2.

**surprise.** 1 (noun and verb) Thus spelt. See -ISE.

2 For the erroneous type *I shouldn't be surprised if* + a negative in the subordinate clause, see NOT 5.

**surprisedly.** Four syllables. See -EDLY.

**surveillance.** The pronunciation recommended is /sɜː'veɪləns/, but /sə'veɪ(-j)əns/ is still valid though gradually falling into disuse.

**survey.** The noun is stressed on the first syllable and the verb on the second. See NOUN AND VERB ACCENT 1.

**surveyor.** Thus spelt, not *-er*. See -OR.

**survivor.** Thus spelt, not *-er*. See -OR.

**susceptible.** 1 For the spelling, see -ABLE, -IBLE 7.

2 See SENSIBLE.

**suspect.** The noun (*he is the chief suspect*) and the adj. (*a suspect package*) are stressed on the first syllable, and the verb (*he suspected nothing*) on the second. See NOUN AND ADJECTIVE ACCENT; NOUN AND VERB ACCENT.

**suspender.** In BrE a *suspender* is a device fastened to the top of a stocking or sock to hold it up; and a *suspender belt* is 'a woman's undergarment consisting of a belt and elastic suspenders to which the tops of the stockings are fastened' (*New SOED*). In NAmE, *suspenders* is the ordinary word for men's braces.

**suspense, suspension.** Formerly the two words were interchangeable in several senses, but no longer. *Suspense* in most of its uses is 'a state of anxious uncertainty or expectation', esp. in the phr. *to keep in suspense* 'to delay informing (someone) of urgent information', and, used attributively, as in *suspense novel*, *suspense thriller*. In bookkeeping a *suspense account* is one in which items are entered temporarily before allocation to the correct or final account. *Suspension* has numerous unshared senses, esp. (*a*) the act of suspending or the condition of being suspended; (*b*) the means by which a

(motor) vehicle is supported on its axles; (*c*) attrib. in *suspension bridge*, a bridge with a roadway suspended from cables supported by structures at each end (e.g. the Humber Bridge at Hull and the Golden Gate Bridge at San Francisco).

**suspicion** (noun). See GALLICISMS 2.

**suspicion** (verb). Apart from a stray 17c. example, this verb has been recorded since 1834, principally in AmE sources, in the representation of non-standard, regional, or facetious language instead of the synonymous *suspect*. Typical examples: *Our nineteen-year-old son, which he's home from Yale on his midyears and don't suspicion that his folks are rifting*—S. J. Perelman, 1946; *I suspicioned what she was, but I didn't have no proof*—D. Shannon, 1973. Presumably drawing on his Arkansas origins, President Clinton used the word in a speech in 1993 (cited by William Safire in the *NY Times Mag.*, 5 Sept. 1993): *The only thing that I can tell you is that everything I ever suspicioned about the way the Federal Government operates turned out to be true, plus some.*

**sustain** (verb). This old verb, taken into the language from OF in the 13c., has or had many senses (to succour, support; to uphold the validity of; to keep in being; etc.), some of which have survived and some not. The only controversial one is 'to undergo, experience, have to submit to (evil, hardship, or damage)'; and even then only the linking of *sustain* with a particular injury (a broken limb, or the like). Dean Alford (1864), as part of a lament about the deterioration of the language ('Its fine manly Saxon is getting diluted into long Latin words not carrying half the meaning') gave a series of examples from newspapers to support his view ('We never *eat* but always *partake*'; 'No man ever *shows* any feeling, but always *evinces* it'; etc.). His paragraph on *sustain* is filled with passion: '... In the papers, a man does not now *lose his mother*: he *sustains* (this I saw in a country paper) *bereavement of his maternal relative*.' Alford comments: 'to *sustain* a bereavement, does not properly mean merely to undergo or suffer a loss, but to behave bravely under it.' And then adds: 'Men never break their legs, but they always "*sustain a fracture*" of them.'

The *OED* (1918) seemed to take the same view ('In modern journalistic use (orig. U.S.) to suffer the injury of (a broken limb, or the like)'). Fowler (1926) was cautious about its use in the restricted sense: 'Nevertheless, *sustain* as a synonym for *suffer* or *receive* or *get* belongs to the class of *formal words*, and is better avoided ...' Such uses of *sustain* are now just listed without comment in modern dictionaries (*sustain injuries, bruises*, etc., as well as *sustain losses*, etc.). Perhaps the bluntest comment is that of *The Times Guide to English Style and Usage* (1992): '*sustain* a broken leg is pompous: just break a leg. Sustain injuries is acceptable, as is sustain losses.' I would add to that: the idea of brave resistance is inherent in the word, not merely the fact or occurrence of something adverse.

**sustainedly.** Four syllables. See -EDLY.

**suttee.** see SATI.

**svelte.** This French loanword has been used in English since the early 19C. to mean '(of persons, esp. women, occas. ironically of men, also occas. of animals) slim, slender, willowy'. Examples: *The Matron led the way, lovely, smiling, svelte, and graceful*—Miss Braddon, 1887; *There is a plush green carpet and a svelte grey cat with silky fur*—P. Carey, 1981; *Marlin Fitzwater* [President Bush's chief spokesman] ... *whose physique has ranged from svelte to portly in the last few years*—Chicago Sun-Times, 1991; *Foremost among these* [sc. animated short films] *is the story of a svelte, psychosexual, sci-fi assassin*—Aeon Flux—Wired (US), 1993. Since the early 20C. it has also been used to mean '(of edifices, cities, vehicles, etc.) elegant, graceful'. Examples: *And first the cities of north Italy I did behold, Each as a woman wonder-fair, And svelte Verona first I met at eve*—E. Pound, 1909; *Is our svelte hired limousine at the door?*—N. Marsh, 1974.

**Swan of Avon.** See SOBRIQUETS.

**swap** ( = exchange). In OUP house style the recommended form, not *swop*.

**swat** (verb), to crush (a fly) with a sharp blow. Thus spelt. *Swat* is in origin a 17C. regional variant of *squat*, to squash, flatten. Cf. SWOT.

**swathe** (noun) /sweɪð/, pl. *swathes*, one of a number of long narrow strips left

after the cutting of hay, the mowing of a lawn, etc. It is sometimes written as *swath* pronounced /swɔːθ/, pl. *swaths* /swɔːðz/.

**swathe** (verb and noun), to bind or enclose in bandages or garments; a bandage or wrapping. Always so spelt. Pronounced /sweɪð/.

**sweat.** See GENTEELISM.

**sweetie,** a term used in BrE for confectionery of various kinds ( = US *candy*). Also (in general English) *colloq.*, a likeable person; *sweetie-pie*, a term of endearment (esp. as a form of address).

**swell** (verb). In most of its uses the pa.t. is *swelled* (*after he tripped on a loose paving-stone his ankle swelled up*) and the pa.pple *swollen* (*his left ankle was swollen after the game*). The ppl adj. is normally *swollen* except in the phr. *swelled head* (beside *swollen head*). The *OED* has examples of *swelled* and *swollen* for the pa.pple and attributive adj. existing side by side in past centuries; and also of *swoll* as a by-form for the pa.t. until the 19C. before it became restricted to dialectal or illiterate use (*me ankles swoll so, and I 'ad dizzy spells*—illiterate speaker in A. S. Byatt's *Still Life*, 1985). It is broadly true that as pa.pple *swollen* implies harmful or undesirable swelling (*the river was swollen and running level with its banks*), whereas *swelled* has more neutral connotations (*the audience was swelled by the arrival of two coachloads of tourists*).

**swim** (verb). The pa.t. is now always *swam* and the pa.pple *swum*, but do not be surprised to see *swam* used for the pa.pple in earlier texts: *Who, being shipwrecked, had swam naked to land*—Dr Johnson, 1750. Also *swum* for the pa.t. as late as the 19C. (*Who turn'd half-round to Psyche as she sprang To meet it, with an eye that swum to thanks*—Tennyson, 1847).

**swine.** In AmE 'a pig' (*He was sweating like a swine*—New Yorker, 1993). In BrE, *swine* is used formally as a collective noun, but in the main the animals are called *pigs* rather than *swine*. In colloquial use, *swine* (pl. *swines*) also = (*a*) a term of contempt for a person; (*b*) a very difficult or unpleasant thing.

**swing.** Pa.t. and pa.pple now regularly *swung*, though *swang* was often used for the pa.t. by writers until the early 20c. (*and Silenus swang This way and that, with wild-flowers crowned*—Wordsworth, 1828; *His arms dangled rather than swang*—H. Belloc, 1912).

**swinging,** the ordinary pres. pple and adj. from *swing* is distinguished from *swingeing* by the retention of the final *-e* of *swinge* (verb) in the latter (*swingeing cuts in the armed forces*).

**swivel** (verb). The inflected forms are *swivelled*, *swivelling*, but usu. *swiveled*, *swiveling* in AmE. See -LL-, -L.

**swop.** See SWAP.

**swot** (verb and noun). In *colloq.* BrE, to study assiduously; a person who swots. Cf. SWAT. Both *swat* and *swot* are pronounced /swɒt/, and both are first recorded in the mid-19c. *Swot* is in origin a dialectal variant of *sweat*.

**sybil.** See SIBYL(LINE).

**syce** (formerly in Anglo-Indian), a groom. See SICE.

**syllabication, syllabification,** the division of words into syllables. Both forms are in common use by lexicographers and linguists.

**syllabub.** Thus spelt, not *sillabub*.

**syllabus.** The recommended pl. form is *syllabuses*, not *syllabi*.

**syllepsis** (Gk, = taking together), pl. *syllepses*. 'Gram. & Rhet. A figure of speech in which a word, or a particular form or inflection of a word, is made to cover two or more functions in the same sentence whilst agreeing grammatically with only one (e.g. a sing. verb serving as predicate to two subjects, sing. and pl.), or is made to apply to two words in different senses (e.g. literal and metaphorical)' (*New SOED*, 1993).

Examples: (*a*) *She has deceiv'd her father, and may thee*—Shakespeare, 1604 (*deceive* understood between *may* and *thee*); *He works his work, I mine*—Tennyson, (*work* (verb) understood between *I* and *mine*); *She's a lovely, intelligent, sensitive woman who has and continues to turn around my life in a wonderfully positive way*—Woody Allen, reported in *The Times*, 1992 (*turned* understood between *has* and *and*); ...

*his avowal of Christianity could have* (*and perhaps did*) *damage his career*—*Times* 1995 (*damaged* understood between *have* and (*and*). Constructions of this type are now regarded as ungrammatical.

(*b*) *Here, thou, great Anna! whom three realms obey Dost sometimes counsel take*—*and sometimes tea*—Pope, 1712–14; *She ... went home in a flood of tears and a sedan chair*—Dickens, 1836–7; *the small fire, over which toast could be made with the help of a long fork and much patience*—*London Rev. Bks*, 1987; *Sir Geoffrey Howe, who had arrived in a limousine, the editor of the Daily Telegraph, who had arrived in a motor-boat, and Dave Nellist, who had arrived in an anorak*—M. Parris, 1991.

Cf. ZEUGMA, but note that 'the term *syllepsis* is frequently used interchangeably with *zeugma*, attempts to distinguish the two terms having foundered in confusion' (Baldick, 1990).

**syllogism.** 'Logic. A form of reasoning in which a conclusion is deduced from two given or assumed propositions called the premises, which contain a common or middle term that is absent from the conclusion' (*New SOED*, 1993). A syllogism of the simplest form is: *All men are mortal; Socrates is a man; Therefore Socrates is mortal*. The predicate of the conclusion (here *mortal*) is called the *major term*, and the preliminary proposition containing it the *major premiss*. The subject of the conclusion (here *Socrates*) is called the *minor term*, and the preliminary proposition containing it the *minor premiss*. The term common to both premisses (here *man*) is called the *middle term*. There are several types of syllogism and only a small proportion of them are valid.

**sylvan,** rather than *silvan*, is by far the dominant spelling now, though *silvan* answers better to L *silua* 'wood'. Typical examples: *in sylvan surroundings, a million miles from Pattinson and Co and its sulphuric acid ... plants*—*Independent*, 1989; *amid scenes of sylvan beauty*—A. Wainwright, 1990.

**symbolic, symbolical** (adjs.). There is no difference of meaning but the shorter form is overwhelmingly the more frequent. See -IC(AL). Examples: (symbolical) *the occultism of symbolical Masonry*—J. McManners, 1990; (symbolic) *On the day* *before Easter, there is a symbolic burning of a cloth-draped wooden statue of Judas*—C. Hammerschlag, 1988; *Because they were symbolic of his psychoanalytic work, Freud was more than ready to share his 'archaeological' interests with patients*—B. Bettelheim, 1990; *The symbolic 'black rooster' (Gallo Nero) which is affixed to every Chianti bottle*—Q Rev. Wines, 1991.

**symbology,** the science or study of symbols. First recorded in the 19c., this word is irregularly formed, but the 'correct' form *symbolology* has not been used.

**symbols. 1** The symbol # is called a *number sign* by *Webster's Third New Internat. Dict.*, and a *hash sign* by the computer industry.

**2** For 0 in telephone numbers and postcodes (US zip codes), see NOUGHT. Air pilots and police normally read 0 as *zero* for clarity: 602 = *six zero two*.

**sympathetic.** Used to mean 'tending to elicit sympathy or to induce a feeling of rapport', this sense of *sympathetic* was too freshly minted for Fowler to approve of it in 1926 (the first recorded example of the sense in the OED is one of 1900). But it is now in unopposed and widespread use. Examples: *The true Don Juan ... is ... not a 'sympathetic' part*—M. Beerbohm, 1900; *Macbeth ... is not made sympathetic, however adequately his crime may be explained and palliated, by being the victim of a hallucination*—cited in Fowler, 1926; *Being a lover of the south, I personally found it* [sc. a novel] *more sympathetic*—Listener, 1965; *Despite the sympathetic portrayal of his father in these anecdotes, Lawrence turned against him after the death of his mother*—J. Meyers, 1990.

**symposium** /sɪm'pəʊzɪəm/. Pl. *symposia*. See -UM 3.

**synaeresis** /sɪ'nɪərɪsɪs/. (US *syneresis*). Pl. *synaereses*. The opposite of DIAERESIS, i.e. the making of two separate vowel sounds into one, as when *aerial* moved from /eɪˈɪərɪəl/ (four syllables) to /ˈeərɪəl/ (three syllables), and the ending *-ein* in some scientific words, e.g. *protein*, moved from /-ɪɪn/ (two syllables) to /-iːn/ (one syllable).

**synchronic** (adj.). Concerned with or describing the state of a language, culture, etc., at one particular time or period, past or present, without regard to historical development, as opposed to DIACHRONIC.

**syncope** /'sɪŋkəpɪ/. A grammatical term for the shortening of a word by the omission of a syllable or other part in the middle. Examples include *idolatry, pacifist*, and *symbology* for †*idololatry, pacificist*, and \**symbolology* as well as the careless pronunciation of *deteriorate* as four syllables instead of five.

**syndrome** /'sɪndrəʊm/, a medical word for a set of symptoms, and, in extended use, a characteristic combination of opinions, emotions, behaviour, etc., is in origin, like *syncope*, a trisyllabic word, i.e. /'sɪndrəmɪ/. English dictionaries, including the OED (1918), indicated trisyllabic pronunciation, either as the only one or as a variant, until well into the 20c. It was still listed as an alternative pronunciation in the seventh edition of the COD in 1982. Now universally disyllabic, on the model of *aerodrome, hippodrome, palindrome*, etc.

**synecdoche** /sɪ'nekdəkɪ/. A figure of speech in which a more inclusive term is used for a less inclusive one or vice versa, as a whole for a part (*England beat South Africa by four wickets*, i.e. the England and S.Afr. cricket XIs) or a part for a whole (*a fleet of fifty sail*, i.e. fifty yachts). Cf. METONYMY.

**synonymity, synonymy.** There is work for both words, the first (infrequently used) meaning synonymousness, and the second (the more general word) meaning the state of being synonymous (i.e. the same as the longer word), but also a system or collection of synonyms, and an article or treatise on synonyms.

**synonyms.** It is now universally recognized that no exact synonyms exist in a given language. In other words there is no pair of words of which one of the pair can be substituted for the other in all circumstances. Fowler himself made the point neatly in 1926: 'Synonyms in the widest sense are words either of which in one or other of its acceptations can sometimes be substituted for the other without affecting the meaning of a sentence; thus it does not matter (to take the nearest possible example) whether I say a word has "two senses" or "two meanings", and *sense* and *meaning* are therefore loose synonyms; but if "He is a man of sense" is rewritten as "He is a man of meaning", it becomes plain that *sense* and *meaning* are far from perfect synonyms.' The word *synonym* is therefore unavoidably used of words of approximately the same meaning 'but having a different emphasis or appropriate to a different context (as *serpent, snake; Greek, Hellene; happy, joyful; kill, slay*), or having a different range of other senses (as *ship, vessel; tube, pipe*)' (*New SOED*).

It is not practicable to provide a full list of all the articles in which the subject of synonymy or near-synonymy arises: they occur on almost every opening of this book. It must suffice to say that there are scores of relevant articles dealing with such matters as the differences, sometimes clear-cut and sometimes not, between such words as *feasible, possible,* and *probable; fixedness* and *fixity; hesitance, hesitancy,* and *hesitation; historic* and *historical; legible* and *readable;* and *masterful* and *masterly.* What is not attempted in this book is the listing of the very wide range of synonyms that are used for such commonplace notions as 'a fool' (*ass, bird-brain, blockhead, bonehead,* etc.), 'work' (*drudgery, effort, exertion,* etc.), 'obscure' (*abstruse, arcane, cryptic,* etc.), 'splendid' (*admirable, brilliant, exceptional,* etc.), and so on. For such information readers must turn to one or more of the many new or revised thesauruses and dictionaries of synonyms that have been issued in recent years.

**synopsis.** Pl. *synopses* /-iːz/.

**syntax.** See GRAMMAR.

**synthesis.** Pl. *syntheses* /-iːz/. Cf. ANALYSIS.

**synthesize** (verb), to make a synthesis of, has virtually replaced *synthetize,* used

in the same sense. Both words entered the language, according to the *OED,* c1830. The *OED* (1918) described *synthetize* as 'the correct form' (i.e. etymologically).

**syphon.** See SIPHON.

**syren.** See SIREN.

**Syriac, Syrian.** The first is the language of ancient Syria, western Aramaic, or, as adj., in or relating to this language. *Syrian* is a native or inhabitant of the modern State of Syria or of the region of Syria in ancient times; and, as adj., of or relating to the region or State of Syria.

**syringe.** Modern dictionaries offer conflicting advice about the positioning of the stress, but in my view second-syllable stressing is now dominant for both noun and verb.

**syrupy.** Thus spelt, not *syruppy.* See -P-, -PP-.

**systematic, systemic.** The ordinary word in lay language for 'methodical; done or conceived according to a plan or system' is *systematic.* The second word is reserved for certain technical uses in physiology (of or concerning the body as a whole), horticulture (of an insecticide entering a plant by its roots or shoots and passing through the tissues), and grammar (M. A. K. Halliday's term for a method of analysis based on the conception of language as a network of systems determining the options from which speakers choose in accordance with their communicative goals—*New SOED*).

**systole** (physiology). Pronounce as three syllables, /ˈsɪstəlɪ/.

**syzygy** (astronomy, etc.). Pronounce /ˈsɪzɪdʒɪ/. See GREEK G.

# Tt

**tabes** (emaciation). Pronounce /'teɪbiːz/.

**table** (verb). In BrE to table a motion means to place it on the agenda. In AmE it means 'to put aside (a bill, a motion) for an indefinite period'.

**tableau.** Pl. *tableaux* /-ləʊz/. See -x.

**table d'hôte** (circumflex accent, romans), a meal consisting of a set menu at a fixed price, as opposed to an *à la carte* (adv. and adj.), ordered as separately priced items from a menu, not as part of a set menu. See À LA.

**tablespoonful.** Pl. *tablespoonfuls*. See -FUL. But in practice the usual formula is of the type *take three tablespoons of . . .*

**taboo.** Stress on the second syllable whether as noun, adj., or verb. As part of the process of political correctness the word is now occas. being written as *tabu* to accord with the Tongan original. The inflected forms are *taboos, tabooed* (or *taboo'd*: see -ED AND 'D), and *tabooing*. The related Maori word *tapu* is pronounced /'tɑːpuː/ in NZE.

**tabula rasa.** Pl., if required, *tabulae rasae*.

**tactics.** See STRATEGY.

**tactile, tactual.** These two adjs. meaning 'of or relating to the sense of touch' have been rivals since they came into the language in the 17c. On the evidence before me it looks as if *tactual* is falling into disuse. In the primary sense 'perceptible by touch' *tactile* seems the natural adj. of the two to call on.

**taenia** /'tiːnɪə/ (US *tenia*). Pl. *taeniae* /-nɪiː/ or *taenias*.

**tag question,** a question converted from a statement by an appended interrogative formula; a formula so used (*New SOED*). The tag question is a prominent feature of English grammar and one that is difficult for foreign learners of English, who often resort to simplified formulas to achieve the same result. There are also regional variations within Great Britain and Ireland (see below), and degrees of emphasis and changes of intonation in certain kinds of confirmatory and peremptory tags. The scholarly literature on the whole subject is considerable: the examples that follow can only palely reflect the work done by numerous grammarians.

Some elementary examples: *You will do this for me,* won't you?; *She has been to South Korea,* hasn't she?; *I don't need an umbrella,* do I?; *We are expected for tea,* aren't we?. In each of these the verb in the main statement has been changed into its interrogative equivalent in the tag. If the statement is negative the tag is positive, and vice versa.

There follows a series of examples showing regional, social, and other variants on the standard type of tag question: *It's a glorious day,* isn't it?—CGEL, 1985 (confirmatory); *Students are human too,* aren't they? (as part of a possibly confrontational argument); *That's a bloody good reason,* innit?—J. Wainwright, 1973 (representing the speech of an uneducated person); *When will the taxi get here?—We'll know when it gets here,* won't we? (peremptory tag); *I work here, don't I?*—BBC TV *Eastenders,* 1987 (peremptory); *Do you know Lord Astor has made a statement to the police saying that these allegations of yours are totally untrue?* (M. Rice-Davies:) *He would, wouldn't he?* (satirical tag); *Surely a man must be kind to a man whose life he has saved,* isn't it?—Irish speaker in I. Murdoch, 1989 (formulaic tag used in some forms of Irish English); *You're going home now,* isn't it?—A. R. Thomas, 1994 (replacing person reference in some forms of Welsh English); *You are going tomorrow,* isn't it?—B. B. Kachru, 1994 (ditto, in S. Asian English); *You've just got everything,* don't you?—New Yorker, 1992 (change of verb in tag); *After all, the culprits had to be found,* not so?—A. Brink, 1988 (SAfr.) (verb not used in tag); *There must be thousands, no? said Don Miguel*—S. Amer. speaker in N. Shakespeare, 1989; *You're at university,* yes?—A. R. Thomas, 1994 (N. Wales).

The list of examples could be greatly extended, but it is clear from those given that the central core of permissible tag questions in standard BrE is by no means matched by the arrangements made elsewhere. Further instability in this area doubtless lies ahead.

**take** (verb). See BRING 1.

**talc** (verb). The inflected forms are *talcked, talcking,* or *talced, talcing.* The corresponding adj. *talcy* is normally so spelt.

**talented.** It is a chastening thought that the adj. *talented,* endowed with talent or talents, came in for severe treatment in the early 19c. Coleridge, in his *Table Talk* (1832), commented 'I regret to see that vile and barbarous vocable *talented* stealing out of the newspapers into the leading reviews and most respectable publications of the day.' And in the Jan. 1988 issue of the journal *English Today* it was reported that Carlyle, 'notorious for his eccentric style, had been criticized by his friend James Sterling . . . for inventing such unnecessary new words as "environment", "visualised", and "talented".' Carlyle accepted the criticisms: 'With unspeakable cheerfulness I give up *talented:* indeed, but for the plain statement you made, I could have sworn such a word had never, except for parodistic, ironical purposes risen from my inkhorn or passed my lips.' (In fact Sterling was not quite accurate: Carlyle seems to have invented the word *environment* but not the other two.) Sterling's objection to *talented* was based upon the belief that an adj. in *-ed* could not properly be formed from a noun (a verb *talent* in the sense required had existed in the 17c. and 18c. but was obsolete by the 1820s). But cf. (with the earliest recorded dates given in each case) *bigoted* (1645), *leisured* (1631), *moneyed* (1457), and *skilled* (1552), all formed from nouns + *-ed.*

**talent, genius.** In 1899, in the OED, Henry Bradley summed up the familiar contrast thus: 'It was by the German writers of the 18th c. that the distinction between "genius" and "talent", which had some foundation in Fr. usage, was sharpened into the strong antithesis which is now universally current, so that the one term is hardly ever defined without reference to the other. The difference between *genius* and *talent* has been formulated very variously by different writers, but there is general agreement in regarding the former as the higher of the two, as "creative" and "original", and as achieving its results by instinctive perception and spontaneous activity, rather than by processes which admit of being distinctly analysed.'

**talisman.** Since the word has nothing to do with the English word *man,* the pl. is *talismans,* not *talismen.* It came into English in the 17c. from any of three languages (Fr., Sp., or Pg.) which had derived it from an Arabic word connected with late Gk τέλεσμα 'a consecrated object'.

**talkative.** Fowler (1926) reminded us that this is the only word in *-ative* in which this element has been appended to a non-Latin, i.e. a native word. See -ATIVE, -IVE.

**talus**[1] /ˈteɪləs/, ankle-bone. Pl. *tali* /-laɪ/. Derived from L *tālus* 'ankle' in the 17c.

**talus**[2] /ˈteɪləs/, slope of a wall, etc. Pl. *taluses.* Adopted from Fr. in the 17c.

**-t and -ed.** Problems arise in a number of irregular verbs which have competing forms for the pa.t. and/or the pa.pple. They differ often in not only having rival forms in *-t* and *-ed* but also in some cases having a different vowel sound before the termination (e.g. *leapt* and *leaped*). Nor is it possible in every case to say in what circumstances the final sound is /t/ or /d/, even when the word is spelt with *-ed.* The distribution of the variant forms in Britain and in English-speaking countries abroad is for the most part not determinable in any precise way, though in most of the verbs the Americans show a marked preference for the forms in *-ed.* The main verbs affected are listed below. Further details of each are to be found under the individual words themselves at their alphabetical place.

> *bereave: bereft, bereaved; beseech: besought, beseeched; burn: burnt, burned; cleave*[1]*: cleft, cleaved; dream: dreamt, dreamed; dwell: dwelt, dwelled; earn: (earnt), earned; kneel: knelt, kneeled; lean: leant, leaned; leap: leapt, leaped; learn: learnt, learned; smell: smelt, smelled; spell: spelt, spelled; spill: spilt, spilled; spoil: spoilt, spoiled; toss: (tost), tossed.*

See also LEARNED.

**tangible.** Its primary sense is 'perceptible by touch' (from L *tangibilis*, from *tangere* 'to touch') (*some visible and tangible object of adoration*), but in practice figurative uses ('clearly intelligible, not elusive or visionary': *tangible evidence, tangible proof*) are probably more frequent. Cf. PALPABLE (adj.).

**tantalize.** An essential ingredient of this verb is the torment felt when something is unobtainable, or when something desirable is offered and then withdrawn. The verb is ultimately derived from the name of *Tantalos*, mythological king of Phrygia condemned to stand in water that receded when he tried to drink it and under branches that drew back when he tried to pick the fruit (*COD*). It should not be used as a simple synonym of *irritate, tease*, or *torment*.

**Taoiseach** (Prime Minister of the Irish Republic). Pronounce /ˈtiːʃəx/, with the final consonantal sound being that of Scottish *loch*. But both words are often pronounced in England with an Anglicized /-k/ instead of /-x/.

**Taoism.** Pronounce /ˈtaʊɪzəm/, rhyming with *Maoism*. (For an elaborate discussion of the way in which this pronunciation emerged from Chinese *dào* 'way', see Michael Carr's note in *Dictionaries* 12 (1990), 55–74.)

**tapu.** See TABOO.

**target** (noun). Gowers' article in 1965 has perhaps helped to reduce the number of illogical metaphorical uses of this noun: 'After the second world war this word was much used to express the quantitative result hoped for from some enterprise such as the output of a manufacturing concern or the amount subscribed for some public or charitable purpose. To an exceptional degree it has shared the experience of most popular metaphors of being "spoilt" by use in a way flagrantly incongruous with its literal meaning. Targets, it is said, must be "pursued vigorously"; to be "within sight" of one and to "keep fully abreast" of it are, it seems, positions that practically guarantee success, and when a target is "doubled" the implication that it will be twice as difficult to hit goes

unquestioned. But then, as Lord Conesford has remarked, "those who thus describe their ambitions never seem to entertain the faintest hope of actually hitting their targets, even when these are overall or even global ones; in their most optimistic moods they speak of 'reaching' or 'attaining' the target, an achievement which, since the bow and arrow went out of use, has never been rated very high."'

**target** (verb). The inflected forms are *targeted, targeting*. See -T-, -TT-.

**tarmac** (verb). The inflected forms are *tarmacked, tarmacking*. See -C-, -CK-.

**tarry** (verb), to linger, is pronounced /ˈtærɪ/; and the unconnected adj. *tarry*, smeared with tar, /ˈtɑːrɪ/.

**tarsus** (bones forming ankle and upper foot, etc.). Pl. *tarsi* /-saɪ/.

**tart.** What precisely is a *tart* as distinct from a *pie*? On both sides of the Atlantic a *tart* is more likely to be sweet than savoury, with a fruit, jam, custard, etc. filling. A *pie* can contain meat or fish, or it can have a fruit filling. A tart is generally an open pastry shell (though *Merriam Webster's Collegiate Dict.*, 10th edn, 1993, gives as one definition 'a small pie made of pastry folded over a filling'). A pie usually has a pastry crust covering the filling in its pastry or biscuit shell. In Britain a *mince pie* is a small individual round pie filled with mincemeat (i.e. a mixture of dried fruit and spices, etc.) rather than of minced meat. When reading books from earlier centuries it is well to keep in mind that various dishes, covered with a crust or not, and containing either a sweet filling or meat, fish, or vegetables, might turn up (*jam tart, strawberry tart; eel tart, veal tart*, etc., beside *goose pie, pigeon pie, eel pie; pumpkin pie*, etc.). A *pasty* is a pastry case more or less semicircular in shape with a savoury (or, less often, sweet) filling; a *turnover* is like it, but with a fruit or jam filling. A *flan* is a pastry or sponge base with a sweet or savoury topping, and a *quiche* is a tart or flan with a pastry base and a rich custard filling flavoured with bacon, onion, cheese, vegetables, etc.

In *The Times Cookery Book* by K. Stewart (1972) the chapter on pastries and pies gives savoury recipes first in this order:

steak and kidney pie, Cornish pasties, quiche Lorraine, bacon and mushroom flan, French onion tart, chicken and mushroom vol au vents, and cheese tartlets. The sweet recipes follow: lemon meringue pie (which has a meringue rather than a pastry top), apricot tart, plum flan with almonds, strawberry and cream cheese tarts, apple and blackberry pie, etc.

**tartan.** See PLAID.

**Tartar, Tatar.** A Tartar was originally 'a member of the combined forces of central Asian peoples, including Mongols and Turks, who under the leadership of Genghis Khan (1202–27) overran much of Asia and eastern Europe and later established a far-reaching and powerful empire in central Europe' (*New SOED*, 1993). *Tartar* was also the name of their language. The preferred ethnological term, ultimately derived from Turkish, is now *Tatar*. A Tatar is now defined as 'a member of a group of Turkic peoples probably originating in Manchuria and Mongolia and now found mainly in parts of Siberia, Crimea, the N. Caucasus, and districts along the River Volga' (*New SOED*). *Tatar* is the name of the Turkic language of these peoples.

**tassel** (verb) (provide with a tassel or tassels). The inflected forms are *tasselled*, *tasselling*; in AmE frequently *tasseled, tasseling*. See -LL-, -L-.

**tasty** is now restricted to mean '(of food) pleasing in flavour, appetizing', a meaning first recorded in the 17c. Its additional 18c. and 19c. sense 'characterized by or displaying good taste' (*My ... waistcoat ... is a much more tasty thing than these gaudy ready-made articles*—Thackeray, 1862) has now entirely given way to *tasteful* (itself first recorded in this sense in the 18c.). The antonym *tasteless* can be used in either of the senses '(of food) without taste or flavour' or 'devoid of good taste, lacking in discrimination'.

**Tatar.** See TARTAR.

**tattler,** a prattler, a gossip. Thus spelt, unlike the 18c. (1709–11) periodical *The Tatler*.

**tattoo** (verb). The inflected forms are *tattooed, tattooing; tattooer*. See -ED AND 'D.

**tautology.** Ultimately from Gk ταυτολο-γία 'the repeating of what has been said' (from ταυτό 'the same' + -λογία 'saying or speaking'), *tautology* has been used in English since the 16c. to mean 'the repetition (esp. in the immediate context) of the same word or phrase, or of the same idea or statement in other words; usually as a fault of style' (*OED*). Among the illustrative examples in the *OED* is one of 1686: *The Taedium of Tautology is odious to every Pen and Ear*. Odious, indeed, provided that the repetition has gone unnoticed or unintended by the writer or speaker. Tautological phrases such as *free gift, in this day and age, new innovation*, and *lonely isolation* can be condemned at once. So can the use shown in a report 'by Our Foreign Staff' in *The Times* (9 Sept. 1994): *The Cold War came to a final close in Germany yesterday with the withdrawal of the last British soldier from West Berlin.*' Other examples of tautologous expressions or clauses, borderline (the *hoi polloi*, etc.) or otherwise, are cited S.V. PLEONASM and REDUNDANCY. Fowler (1926) cited numerous examples of clumsily inbuilt tautology, including the following: *The wool profits were again made the subject of* another *attack by Mr Mackinder last night* (omit either *again* or *another*); *May I be permitted to state that the activities of the Club are not limited only to aeronautics?* (*limited* and *only* are tautological; read *limited to*, or *directed only to*).

These warnings given, *tautology*, as a rhetorical device, can be turned to advantage. Simple repetition is a feature of innumerable well-known literary contexts: e.g.

> Alone, alone, all, all alone,
>  Alone on a wide wide sea!
>
> (Coleridge, 1798)

> Keeping time, time, time,
>  In a sort of Runic rhyme,
> To the tintinabulation that so musically wells
> From the bells, bells, bells, bells.
>  Bells, bells, bells—
>  From the jingling and the tinkling of the bells.
>
> (Poe, 1849)

Whether the term *tautology* is nowadays ever employed to describe such skilful use of repetition is, however, extremely doubtful.

**tax, duty,** and some related words—*cess, customs, dues, excise, impost, levy, rates, toll,* *tribute, VAT.* With such sets of words it is sometimes convenient to have a conspectus of the distinctions and be saved the labour of turning them up for comparison in separate dictionary articles. Such convenience is all that is aimed at here, a brief definition of each word being given after some general remarks on the words *tax* and *duty.*

Historically a *tax* was a direct charge on a taxpayer which bore some relation to his or her ability to pay and was imposed for revenue only; a *duty* was an indirect charge, levied on transactions or commodities. But there is no longer any clear distinction between the two words in Britain. Few taxes are so called specifically: among the most notable are *income tax, inheritance tax,* and *council tax.* The rest are mainly *duties,* e.g. *customs, death, excise, estate,* and *stamp duties.* To keep the article within limits it is almost entirely concerned only with taxes and duties in the UK. The terminology abroad is only patchily similar to that of Britain, and in all English-speaking countries is subject to rapid change.

> *cess* (properly *sess* for obsolete *assess* (noun) assessment): a word formerly (16–19c.) used in England for a form of local rate; also formerly (and still locally) in Scotland, a land tax; in Ireland, formerly (16–19c.), the obligation to supply the soldiers and household of the Lord Deputy with provisions 'assessed' or fixed by government; and in India, a tax levied for a specific purpose (an irrigation cess, a road cess, etc.).

> *community charge:* see as main entry.

> *customs:* a duty levied on certain imported and exported goods.

> *dues:* any obligatory payment, the nature being normally specified by a prefixed noun, as harbour dues, trade-union dues.

> *excise:* a tax or duty levied on certain goods and commodities, esp. alcoholic liquor, produced for the home market.

> *impost:* a non-specific or general term for a compulsory payment exacted under statutory authority.

> *levy:* the act of imposing and collecting a tax or tariff.

> *rates:* a charge levied by a local authority based on the assessed value of a property (now replaced by the *council tax*).

*toll:* a fixed charge for passage over a bridge, ferry, etc., or (in some countries) at the entrance to a motorway.

*tribute:* historically, a payment made periodically by one State or ruler to another, esp. as a sign of dependence. Now usu. replaced by *reparations.*

*VAT* (value added tax): a tax, expressed in percentage terms, on the amount by which the the value of an article has been increased at each stage of its production and distribution.

**taxi** (noun). Pl. *taxis.*

**taxi** (verb). The inflected forms are *taxis, taxied, taxiing* (not *taxying*).

**teasel** (plant of the genus Dipsacus). Thus spelt, not *teazel* or *teazle.*

**teaspoonful.** Pl. *teaspoonfuls.* See -FUL. But in practice the usual formula is of the type *take two teaspoons of...*

**tec.** See ABBREVIATIONS 1.

**techno-.** The word *technology* (from Gk τεχνολογία, systematic treatment of grammar, etc., now the mechanical arts and applied sciences collectively), and also its main derivatives, *technological,* etc., joined the language in the 18c. But it was not until the 20c. that *techno-* became a free-flowing formative element, forming compounds mostly of self-evident meaning, e.g. *technobabble* (first recorded in 1987), *technochat, technocracy* (1919), *technocrat* (1919), *technofreak* (1973), *technomania* (1969), *technophobia* (1965), *technostress, technothriller,* most of them occurring in the technical language of computing and other high-technology areas.

**techy.** See TETCHY.

**Te Deum.** Most often pronounced /ˌtiː ˈdiːəm/ rather than /ˌteɪ ˈdeɪəm/.

**teem** (verb)[1], to be full of or swarming with (*the sea is teeming with fish*), is derived from OE *tēman* 'to give birth to'. It is not etymologically related to the next word.

**teem** (verb)[2], (of water, rain) to flow copiously, to pour (*it was teeming with rain*), came into English in the 14c. from ON *tœma* 'to empty'. Its earliest sense in English was 'to empty (a vessel, etc.)'. The application to water, rain, etc., seems to have come from dialectal use into the standard language in the early 19c.

**teenager,** a person from 13 to 19 years of age. The term seems to have been with us for ever, but the earliest recorded example (in an American source) is one of 1941. To begin with, and still in some publications, it is printed with a hyphen (*teen-ager*). The word is sometimes loosely used of any adolescent, without strict regard to age, particularly one judged to be behaving in an unsocial manner. The family of *teen-* words, all derived from the *-teen* of *thirteen*, etc., includes *teenage* (adj. and noun) and *teenaged* (adj.).

**teetotaller.** Thus spelt (but in AmE frequently *teetotaler*); but *teetotalism*. See -LL-, -L-.

**tele-.** From Gk τηλε, combining form of τῆλε 'afar, far off', *tele-* has become one of the great formative elements of the modern period. The notion of distance in its various senses is built into all words containing *tele-*. The words that have emerged have for the most part been formed without regard to etymological 'correctness': *tele-* has been freely joined to words of every kind. Thus *telecast* was based on *broadcast*; *telecommunication* was formed first in French in 1937 at a conference in Madrid (where the official language happened to be French); *telegenic* (adj.) was modelled on *photogenic*; *tele-lens* combines a Greek prefix with a word of Latin origin; *teleprinter*, *teleprompter*, and *teletext* were coined simply by clapping *tele-* on to *printer*, *prompter*, and *text*. Blends include *televangelist* (*tel(evision)* + *evangelist*) and *Telex* (*tele(printer)* + *ex(change)*).

The great developments in the core meaning of *tele-* as a formative element are shown in *telescope* (first recorded in 1648), *telegraph* (1794), *telegram* (1852), *telephone* (in the modern Alexander Graham Bell sense, 1876), *telepathy* (1882), and *television* (1907). One wonders what new central concept will be captured in new uses of *tele-* in the 21c. Meanwhile the limitless class of *tele-* words is joined, almost daily, it would seem, by more members, e.g., since 1980, by such words as *telebanking, telecounsellor, teledemocracy, telelection, telefraud, teleliteracy, telemarketing, telepresence, teleshopping, televent, teleworking.*

**televise** (verb). Thus spelt, with *-ise* not *-ize*, in BrE and AmE. It is a back-formation from *television*.

**temerarious.** First introduced into English by Thomas More in 1532 ('reckless, rash', from L *temerārius* in the same sense), it and also the adverb in *-ly* are now only in restricted use. Examples: *In temerariously excluding Wilfred Owen from the* Oxford Book of Modern Verse *... Yeats made clear that for him poetry ... cannot find its end in pity*—R. Ellmann, 1967; *One conversation ... involved my temerarious claim that if you focus attention on finding well-defined answers, then you're not doing research*—J. L. Casti, 1989.

**temperature.** From the normal medical use of the word ('the degree of internal heat of the body') a new and superficially illogical use developed at the end of the 19c.: e.g. *Has he got a temperature?* = a body temperature above the normal. The use is standard but is mostly encountered in spoken English.

**template.** Thus now spelt, not *templet*, though *templet* was the dominant spelling 17–19c., and *-plate* is 'pseudo-etymological after *plate*' (*OED*).

**tempo** (music). The pl. is either *tempos* or *tempi* /-piː/.

**temporal, temporary.** From the 14c. to the 19c. *temporal* shared with *temporary* the sense 'lasting only for a time'. Conversely, at various periods, *temporary* shared some of the senses of *temporal* (details in the *OED*). The central meaning of *temporal* now is 'of worldly as opposed to spiritual affairs, secular'. Hence the distinction between the *temporal* (civil) and the *spiritual* (ecclesiastical) authorities and the division of the House of Lords into *Lords Temporal* (lay peers) and *Lords Spiritual* (bishops).

**temporary, temporarily.** The first of these is pronounced as four syllables, /'tempərəri/; in AmE there is a secondary stress on the *-ar-*. The second word has five syllables and until recently was regularly pronounced in BrE with the stress on the first syllable, /'tempərərɪlɪ/, but is now almost as often /tempər'eərɪlɪ/, i.e. the regular pronunciation in AmE. See -ARILY.

**temptress.** see -ESS 5.

**tend** (verb)[1] (apheffic form of *attend*). A word of several meanings derived from *attend*. The main standard use in BrE is the transitive one, 'to take care of, look after (a person, esp. an invalid, animals esp. sheep, a machine)': *tend the sick, tended their flocks, tend a machine*. One use that has somewhat controversially drifted out of dialectal or Amer. English into occasional use in standard English is *tend to* meaning 'to give attention to, attend to'. Examples: (caption) *Police and ambulancemen tend to an injured woman outside Liberty's early yesterday—Times,* 1989; *Mr A.'s daughter ... believes she saw the thief watching from the end of the alley as she tended to her father—Times,* 1994.

**tend** (verb)[2] (from OF *tendre* 'to stretch', from L *tendere*). This common verb meaning 'to be apt or inclined, have a tendency' is used only intransitively: *he tends to lose his temper; she tended to do what her parents advised; the argument tends only in one direction*.

**tendinitis, tendonitis** (inflammation of a tendon). Both forms are acceptable. The source word medL *tendo* has acc. sing. *tendinem* or *tendōnem*.

**tenet** (dogma, doctrine). Daniel Jones (1917) gave /ˈtiːnet/ as his preferred pronunciation of the word. Fowler (1926) preferred the variant with a short /e/ in the first syllable. The dominant pronunciation now is the one recommended by Fowler, though some modern dictionaries still list both types.

**tenor.** 1 Thus now spelt, not *-our*, in all senses in both BrE and AmE.

2 See MISQUOTATIONS.

**tenses.** 1 Certain points requiring care are discussed under AS 9; HAD 2; HAVE 2; HISTORIC 2; LEST; LIKE (verb) 1; PERFECT INFINITIVE; SEQUENCE OF TENSES; SHALL AND WILL; SUBJUNCTIVE MOOD; and further articles cross-referenced in these.

2 To these may be added a few trifles which draw attention to some minor aspects of the English tense system: *Will that be all, Miss?*—T. S. Eliot, 1939 (servant to employer: future formula used of present time); *A chicken was roasting in the fireside oven, for it was Christmas Day*—J.

Gardam, 1985 ( = being roasted); '*They'll be rehearsing for the fête on Saturday,*' *Sylvia says*—M. Duckworth, 1986 (future form used to describe something actually taking place in the speaker's presence); *Next week she reveals the horrors of her recent experiences in a psychiatric ward*—*Truth* (NZ), 1986 (present tense of future time in an announcement).

**tenuis** /ˈtenjʊɪs/ (phonetics). A voiceless stop, e.g. *k, p, t.* Pl. *tenues* /-jʊiːz/.

**tercentenary.** See CENTENARY 2.

**tercet** /ˈtɜːsɪt/ (prosody). A set or group of three verse lines, esp. one of those in *terza rima* (see below), or half the sestet in some Petrarchan sonnets (see SONNET).

**teredo** /təˈriːdəʊ/ (bivalve mollusc). Pl. *teredos* /-dəʊz/, occas. *teredines* /təˈriːdiniːz/ (the word is derived from L *terēdō*, pl. *-ines*). See -O(E)S 6.

**-teria,** a suffix (orig. and chiefly US) (also *-eteria*) derived in the 1920s from *cafeteria* by analysis of its constituents as *café* + *-teria* and frequently in use since then. Examples are *danceteria* (night club), *washeteria* (self-service laundry).

**term.** 1 For *major, minor, middle, term* in logic, see SYLLOGISM.

2 See IN TERMS OF.

**terminable, terminal** (adjs.). The first means '(of an appointment, annuity, etc.) able to be terminated, cancellable'. *Terminal* as adj. means 'situated at or forming the extremity of (a crystal, a flower, etc.); of or situated at the end of a railway line; pertaining to a term (at a university, etc.); of or pertaining to the final stages of a fatal disease'. It is also used in the phr. *terminal building* (for passengers at an airport), though it is perhaps better analysed here as an attributive use of the noun (see next).

**terminal** (noun), **terminus.** The first has several meanings, among them 'a station at the end of a railway line, a terminus; a departure and arrival building at an airport; a device for entering data into a computer or receiving its output'. A *terminus* (pl. *termini* /-naɪ/ or *terminuses*) is the term often used in Britain for the station at the end of a railway or bus route (but not the passenger building at an airport), as well as being a technical term in mathematics and architecture. In AmE, and to an increasing extent in BrE, the terms *terminal* and *terminus* are used interchangeably for the final stop of any passenger journey by train, bus, or aeroplane.

**terminate.** 1 This has *terminable* (not *-atable*) as its corresponding adj.; and *terminator* (not *-er*; see -OR) as its corresponding agent-noun.

2 In the basic senses 'to bring to an end' (e.g. *terminated her pregnancy*; *terminate an agreement*) and 'to come to an end' (e.g. *this service terminates at Paddington, London*), *terminate* is sometimes (as shown in the examples) the natural idiomatic word. But in most contexts in which either of these senses is required, *end*, *finish*, or *stop* are more suitable words.

**terminological inexactitude.** See IN-EXACTITUDE.

**terminus.** See TERMINAL (noun).

**Terpsichore** (muse of dancing). Pronounce /tɜːpˈsɪkəri/. See MUSES.

**terrain** is best reserved to mean 'a tract of land esp. as regarded by the physical geographer or the military tactician' (*COD*), rather than used as a simple substitute for *area*, *ground*, *region*, or *tract*. If used figuratively the context should clearly reflect one or other of these two images.

**terrible, terribly.** Both words have tended down the centuries partially to lose touch with the notion of terror, and to come into use as mere intensives. A similar fate has befallen other words of originally similar meaning, such as *awful(ly)*, *dreadful(ly)*, *frightful(ly)*, *tremendous(ly)*. The remarkable thing about all of them is the way in which the original sense has co-existed with the weakened ones for long periods of time. The dominant senses of *terrible* now are the colloquial ones, 'very great or bad' (*a terrible bore*, *a terrible child*); and 'very incompetent' (*a terrible driver*); but 'causing terror or dread' (*a terrible thunderstorm*) is still a current meaning of the word.

**tertium quid.** A Latin phrase, apparently rendering Gk τρίτον τι 'some third thing', and meaning *sensu stricto* 'something indefinite or left undefined related in some way to two definite or known things, but distinct from both' (*New SOED*). An alloy of two metals or a chemical compound made from two elements might be called a *tertium quid*. But the phrase has been used to mean 'any third thing of a group of three' (the notion of indefiniteness being lost): in other words a third alternative or a middle course. Browning used it as the heading of book iv of *The Ring and the Book*, in which views are set forth about the culpability of the murderer Guido which are neither those of the 'Half-Rome' who have defended him in book ii nor those of 'The Other Half-Rome' who have condemned him in book iii. A new connotation was given to the expression when Kipling wrote a story called 'At the Pit's Mouth' (1888) beginning *Once upon a time there was a Man and his Wife and a Tertium Quid*; and the meaning 'the third party in an eternal triangle' is likely to be suggested by it to the popular mind. Some recent examples: Webster's Dictionary of English Usage *is a tertium quid, combining the virtues of the report and the pronouncement without their vices, and it is therefore a new breed of usage guide—English Today*, 1991; *Possibilities are opened up for another kind of Western, a secondary Western dealing with that New Man, the American tertium quid—Twenty Twenty*, 1991.

**terza rima** /ˌteətsə ˈriːmə/ (prosody), 'third rhyme'. A verse form invented by Dante Alighieri in his *Divina Commedia* (c1320) and followed by many English poets, among them Byron, Shelley, and Browning. It consists of a sequence of TERCETS rhyming aba bcb cdc ded, etc. Dante's lines consisted of five iambic feet with an extra syllable (hendecasyllabic lines). The English poets mostly wrote in iambic pentameters. The essential point of the rhyming scheme is that the second line of each tercet provides the rhyme for the first and third lines of the next, thus producing an effect of unending continuity. An example from Byron (1819), with three tercets and a line:

> *Oh! more than these illustrious far shall be*
> *The being—and even yet he may be born—*
> *The mortal saviour who shall set thee free,*
> *And see thy diadem, so changed and worn*
> *By fresh barbarians, on thy brow replaced;*
> *And the sweet sun replenishing thy morn,*

*Thy moral morn, too long with clouds defaced,*
*And noxious vapours from Avernus risen,*
*Such as all they must breathe who are*
*debased*
*By servitude, and have the mind in prison.*

**tessera** (square block in mosaics, etc.).
Pl. *tesserae* /-ri:/.

**testatrix** (feminine equivalent of *testator*). For pl., see -TRIX 3.

**test match.** Originally (examples in the *OED* from 1862 onwards) one of a series of cricket matches between England and Australia. In the 20c., at various points, the term came to be used of official representative cricket matches between any of the major cricketing countries; and, since the 1920s (first in S. Africa) of international rugby matches (except that in rugby football circles such matches are for the most part just called *internationals*).

**tetchy.** This spelling is more usual than *techy*. The word is more or less synonymous with *testy*, meaning 'irritable, peevish'. *Touchy*, on the other hand, most often means 'apt to take offence, oversensitive' rather than 'irritable', but potential or actual irritability lies at the heart of all three words.

**tête-à-tête.** The accents and hyphens are obligatory, but the word is normally printed in romans, not italics.

**tetralogy.** A group of four related plays, speeches, operas, etc.; *spec.* (Gk Antiq.) a series of four dramas, three tragic and one satyric, performed in Athens at the festival of Dionysus. (*New SOED*)

**tetrameter** (prosody). A line of four metrical feet, as in Coleridge's *Christabel* (1816): e.g. *A little child, a limber elf, Singing, dancing to itself.*

**tetrastich** (prosody). A less usual word for a stanza of four lines, a quatrain, as used e.g. in Gray's *Elegy Written in a Country Church-Yard* (1751).

**textual, textural.** The first means 'of, in, or concerning a text of any kind (*textual emendation*, *textual criticism*)'; whereas *textural* means 'of or belonging to texture (of paint, fibres, music, etc.)'. The writer of the following sentence chose the wrong word: *He seems to be*

*moving into popular journalism, and away from textural criticism.*

**thalamus.** Pl. *thalami* /-maɪ/.

**Thalia.** See MUSES.

**than.**

1  *than* and *prefer*.
2  Part of verb after *rather than*.
3  *different than*.
4  *other than*.
5  *barely . . . than*, *hardly . . . than*, *scarcely . . . than*.
6  *than* as conjunction and preposition.
7  Reflexive pronouns after *than*.
8  *than* and inversion.
9  *than* after non-comparatives (*inferior*, *superior*, etc.).
10  Miscellaneous standard uses of *than*.

1  For the solecistic use of *than* after *prefer*, see PREFER 4. Correct use of *prefer* followed by *rather than*: *I preferred to write, to explore the world of imagination, rather than to mix with others*—J. Frame, 1984.

2  Infinitive or gerund, etc., after *rather than*. See -ING 5; RATHER 3.

3  *different than*. See DIFFERENT 1.

4  *other than*. On the competing merits of *other than* and *otherwise than*, see OTHER 5. Evidence for the validity of various uses of *other than* (with *other* employed in more than one part of speech) continues to turn up. Examples: *it became clear that there was nothing other than social niceties to be shared* M. Sutherland, 1974 (NZ); *the United States . . . will have no other choice than to reduce the costs of its overseas commitments*—Dædalus, 1987; *We were seen as . . . unable to conceive of any interpretation other than Hollis' being guilty*—P. Wright, 1987; *there were now no people other than the tall, silent, courteous white-uniformed servants*—F. King, 1988.

5  *barely . . . than*, *hardly . . . than*, *scarcely . . . than*. See BARELY, HARDLY 2; SCARCELY 1.

6  *than* as conjunction and preposition. In the type *Diana has better manners than I*, *than* is a conjunction, implying the ellipsis of *have* after *I*. By contrast, in the type *Diana has better manners than me*, *than* is a preposition followed by the objective form *me*. The second type is much less formal than the first. In practice the types are inseparable when the

word after *than* is the same in the nominative case as in the objective, e.g. *He
was shorter than Bertrand; He's older than
you; Everything that lives in the sea had a
longer evolutionary history than anything on
land*. Sometimes, in order to bring out
that *than* is thought of as a conjunction,
the 'implied' clause is written out in
full: e.g. *I'll have to say this for them—they
all feel better than I do; She has long black
hair and is thinner than I am; He was a few
years older than I was; his attitude was kind
and generous to those less endowed than he
was*. See also CASES 3; ME 2(*d*).

**7** Reflexive pronouns after *than*. One
means of avoiding the problem of
whether to use a nominative or an objective pronoun after *than* (see preceding
paragraph) is to use a reflexive pronoun
instead, but the result is often rather
less than satisfactory. Examples: *One of the
most encouraging performances came from
Darren Cook ... against senior athletes with
far more experience than himself—Grimsby
Gaz.*, 1986; *Yes, I have one brother, five years,
less one month, younger than myself and fifty
times more successful—*oral conversation in
Oxford, 1987; *It had often seemed as though
the people in the poems were more real to
the old man than himself and his brother
Michael—Encounter*, 1989.

**8** *than* and inversion. *No tariff-armed
nation has got better entry for its potatoes to
the U.S.A. market than has Ireland; The
visit will be much more direct in its effect upon
the war than could be any indiscriminate
bombing of open towns*. Such inversions,
deprecated by both the previous editors,
are now rare. See INVERSION 2(*j*).

**9** *than* after non-comparatives. Such
originally Latin words as *inferior* and
*superior*, *junior*, and *senior*, are not true
comparatives and therefore cannot be
construed with *than*. See SUPERIOR.

**10** Miscellaneous standard uses of
*than*. The following examples from my
database illustrate uses that would pass
unnoticed as part of the ordinary standard language. They are drawn from a
wide range of standard sources of the
period since 1980. *More than 50 per cent
of the members of the entertainment-industry-
related unions are out of work; When no
more than a lad he acquired this socially
objectionable habit; Lanzman presses his
Polish peasants far harder than he does the
Nazis; He tended to look sideways, probably*
*because he could see more clearly with one
eye than with the other; They know that few
things are better for them than amity; The
extreme irritation one always feels when someone proves to be more percipient than one has
decided they shall be; He loved wrecking even
more than getting possession; It is possible to
be better off out of work than in work; One
need interrupt the narrative no further than
to say this; You know better than to be late;
Pat had been more involved than the others;
The recent amnesty should be seen more as
Solidarity's triumph than as the government's
free initiative; I think I spoke no more harshly
than was deserved; Girls are thought to need
less food than boys; As if I were more absolutely my particular self than ever; There was
nothing more important than that he get to
Epsom for the start of the card.*

They underline the virtual necessity
of the presence of a comparative (*more,
less, better, harder*, etc.) in the vicinity of
*than*. And they go some way to confirming that uncontroversial uses of *than* are
far more numerous and more significant
than constructions of the type *Diana has
better manners than me* (6 above).

**-th** /θ/ **and -th** /ð/. Monosyllabic nouns
ending in -*th* after a vowel sound sometimes differ in the pronunciation of the
plural forms, i.e. vary between /θs/ and
/ðz/. It is necessary to consider only those
words whose plurals are in regular use,
which excludes *dearth, ruth, sloth*, and
many others. The common words *mouth,
oath, path*, and *youth* all sound the plural
as /ðz/, not /θs/. The equally common
words *berth, birth, breath, death, fourth,
girth, growth, myth*, and *smith* always have
/θs/ in the plural form. Others again fluc-
tuate between /θs/ and /ðz/, e.g. (/θs/ recommended) *cloth, heath, hearth, lath,
wraith*; (/ðz/ recommended) *bath, sheath,
truth*. See also MOTH (pl. noun). It should
be added that the verbs or verbal nouns
connected with *bath, breath, cloth, mouth,
sheath, teeth*, and *wreath* have /ð/, namely
*bathe, breathe, clothe, mouthing, sheathe,
teething*, and *wreathe*. Cf. also *northern,
southern, smithy*, and *worthy*, all with /ð/.

**thankfully.** The ordinary adverbial use
( = with thankfulness; in a thankful
manner), which goes back to Anglo-
Saxon times, has continued in use over
the centuries: e.g. *He accepted thankfully
all my presents—*Defoe, 1725; *the Pacific
Islands, which are not members of ANZUS,*

*thankfully shelter under the ANZUS umbrella*—R. Muldoon, 1986; *Then, 'until Friday,' said Mrs Marsh, and shut the door thankfully behind her*—A. Brookner, 1992. Since the 1960s it has been joined by constructions in which *thankfully* is a SENTENCE ADVERB ( = let us be thankful (that); fortunately), as in *Thankfully his injuries were not serious*. Modern example: *Thankfully, however, the old style has not entirely disappeared*—Daily Tel., 1982. The new use has not attracted the same level of criticism as that given to similar uses of *hopefully*, but it is still too raw to be given general acceptance as part of the standard language.

**thanks to.** A correspondent wrote to me in 1988 complaining about the ironical use of *thanks to* in such contexts as these: *Thanks to the hurricane, millions of trees were destroyed*. 'What is wrong with *because of* or *due to*?' he asked. The OED supplies the answer. Since the 17c. the phrases *thanks to* and *no thanks to* have been used positively or adversely or ironically, the first to mean 'thanks are due to' or 'owing to, as a result of', and the second 'no credit to, not because or by reason of'. A positive use of *thanks to*: *Thanks to the researches of Prisco Bagni we know that in about 1681 the erotic mythologies by him ... had attracted the attention of 'Milady Devensier'*—Ashmolean, 1987. An ironical example was supplied (above) by the correspondent himself. Cf. the assertive use of *thank you very much* in: *I've got him now. Urhh! I know all I need to know about him, thank you very much*—K. Amis, 1988.

**thank you.** 1 In origin short for *I/we thank you*, it has inevitably produced a number of equivalent phrases showing varying degrees of formality and emphasis. In various circumstances it is lengthened to *thank you so much*, or *thank you very much*, sometimes with the addition of *indeed* for good measure. *Thanks* is a shade less cordial than *thank you*, and *many* and *best* and *a thousand thanks* and *thanks awfully* are frequent elaborations of it. *Much thanks* is archaic, surviving through our familiarity with *For this relief much thanks* (the sentinel Francisco in *Hamlet*), and now only used jocularly. The colloquial variant *Thanks a lot* has also become popular. If an acknowledgement of thanks is felt to be needed it

will be *Don't mention it*, or *Not at all*, or, in America, *You're welcome*.

2 A noticeable tendency to write *Thank you* as one word has emerged recently in private letters and also, for example, on bills at restaurants (*Thankyou for your lovely present; Thankyou for dining with us*). The standard form is still *Thank you* (two words), except when the formula is used as a noun (*We send our thank-yous; a thank-you letter*), in which case a hyphen is necessary.

**that** (demonstrative pron.). 1 It might be useful if I begin the treatment of the very complicated word *that* by setting down a series of examples of the word used pronominally in an entirely legitimate way. All are drawn from standard sources of the 1980s and 1990s.

*It goes deeper than that; Something worth a lot of money, that's for sure* (colloq.); *It's the first to the store, is that it?; 'You write clever stuff in Varsity and Granta ... don't you? 'Something like that,' said Tony; She had not meant it so, but it could have been read like that; Being Jewish is something you very occasionally apologize for and that's it; That's the way it ought to be; A different kind of excitement from that roused by the park; All right, that's what she meant; So Boston stopped asking questions, while he was sober, that is; What was all that about?; She was innocent: that is all anybody has ever been able to draw out of him; What benefactor will continue voluntarily to sponsor that?; He didn't say that, or anything like it; Upsy a bit, dear, that's a good girl* (colloq.); *He's a good officer ... It's just he talks too much, that's all* (colloq.); *She had a small, pretty face, I'll give you that; Hey, great work there, Toddie! Hey, look at that! That's my boy* (colloq.); *Too much of a bloody infidel, that's me* (colloq.); *Because if that happens, then I will be just like you—isn't that so?; Now how the hell did you manage that? You really are a one; The witnesses, if they could be called that, continued to repeat that they knew nothing; Their eyes tangled. 'I said "love", dammit.' 'Ah. That. I thought you said that.'; She cleared her throat to speak but left it at that; 'He wasn't a straightforward man, Francis'. That he wasn't.'*

What stands out from these examples is that *that* used as a demonstrative pronoun is overwhelmingly anaphoric, i.e. directs the attention back to a word, statement, or circumstance mentioned previously. It may also be used formulaically: e.g. *that's it, that is* (*to say*), *that's all*, etc.

A somewhat uncommon anaphoric use—postponement of *that* to the end of a statement—is worth noting: e.g. *a neat comeuppance for the Japanese, that*—NY Times, 1990; *It all seemed a bit* [*too*] *good to be true, that*—N. Virtue, 1990; '*Well, you poor girl,*' *said Margaret, and Wendy added,* '*Hard on a young person, that.*'—A. Brookner, 1993. This somewhat quirkish, but still standard, construction does not seem to be dealt with in the *OED*.

**2** It is well to remember that all these uncontestable uses of the pronoun *that* have a history of their own and are all likely to be encountered in pre-20c. contexts: e.g. *Look to me . . . That is look on me, and with all thine eyes*—Jonson, 1625; '*Very well,*' *cried I,* '*that's a good girl.*'— Goldsmith, 1760; *A man began to scream, and that so loud that my voice was quite drowned*—J. Wesley, 1772; *Do you, my dear K, have them sent to me, that's a darling*—T. Arnold, 1849; *What do you think I would give to be your age, and able to draw like that!*—Ruskin, 1884.

**3** *at that*. *The only seats available were at the back, and very uncomfortable at that.* This convenient use, described by the *OED* as 'orig. U.S. colloq. or slang', is first recorded in 1830, and is probably extended from such phrases as *cheap at that, dear at that* (*price*).

**that** (demonstrative adj.). The simple demonstrative adj. *that* is distinguished from the definite article *the* in that it *points out* something as distinct from merely *singling out* something. Some standard modern examples: *I don't feel that kind of attraction—lust you would call it*; *If I were he, I should keep an eye on that young man*; *What the painter intended to do was to introduce everything into that small space*; *That simple, honest, trusting marriage that had been theirs before Maria had appeared*; *Why did you take that picture of me?*; *It wasn't a nature reserve, that Ark of ours.* This type goes back to the earliest period of Middle English.

A subdivision of the same type, in which *that* is used to indicate quality or amount, and correlates with *that, as*, etc., was once common, but was archaic or unfashionable by the end of the 19c. Examples: *From me, whose loue was of*

*that dignity, That it went hand in hand, euen with the Vow I made to her in marriage*—Shakespeare, 1602; *With that cunning and dexterity as is almost imperceavable*—Milton, 1648; *An Error of that Magnitude, that I cannot but wonder*—I. Walton, 1678; *He . . . struck her . . . with that heaviness, that she tottered on the marble floor*— Dickens, 1848. Fowler (1926) cites an unattributed example: *He has that confidence in his theory that he would act on it tomorrow.* But he admits that the type was 'formerly normal English'.

**that** (demonstrative adv.). The type *I was that angry*, i.e. 'so angry, very angry', and its negative counterpart, *things aren't really that bad*, have been slipping into and out of standard use since similar uses were first recorded in the 15c. In 1912 the type was judged by the *OED* to be 'Now *dial.* and *Sc*[*ottish*]'. From the material I have collected, it would seem that both the positive and the negative types are common now but in the written language are normally used in plainly informal contexts, e.g. '*Shut up,*' *says Claudia . . .* '*It's not that funny.*'—P. Lively, 1987; *I've been that worried. I thought I'd lost you*—D. Lodge, 1988; *You and your brother, you're not really that alike, are you?* —Encounter, 1989. But it is only a short step away from reasonably formal territory, to judge from the following example: *The questioning attitude that comes naturally at student age is not that easily abolished*—Listener, 1987. To put it in modern military jargon, *that* used as a demonstrative adverb could be said to lie in a 'safe haven' at present, but it is open to criticism or attack or downgrading at any time.

Cf. ALL THAT.

**that** (conj.). **1** Kinds of clause attached by *that* (conj.). The conjunction *that* is legitimately used to join a subordinate clause to a preceding verb or its parts in the manner shown in the following examples: (preceded by a finite verb) *They understood that this was an errand of mercy*; (infinitive precedes) *I would hate to feel that we were corrupting you*; (pa.pple precedes) *He had taken it for granted that any compatriot of Louis Pasteur must have seen a needle or two*; (verb *to be* precedes) *It wasn't that Peter and myself were being singled out*; (*that* repeated when introducing a second subordinate clause) *He later* *discovered* that *it was St. Lucy's Day, and* that *this was a ceremony associated with its celebration;* (that (conj.)) followed by *that* (demonstrative adj.)) *he realized* that that *sprint was all that was left.* In every such *that*-clause the sentence out of which it is made or completed by prefixing *that* must be in the form of a statement, not a question, command, or exclamation. Sentences of the following type are unacceptable and need to be recast: *Crises, international or national, arise so rapidly in these days* that *who can say what a few years may bring forth?* (replace *who* by *none* and delete the question mark). A *that*-clause may sometimes, in a rather forced manner, occupy the front position in a sentence: e.g. *That any compatriot of Louis Pasteur must have seen a needle or two he had taken for granted; That England would win was hoped for.* But such reversals of the natural order are not common, and indeed are not possible in some of the examples cited in the first paragraph above.

2 Omission of the conjunction *that* is acceptable, not only in informal contexts, e.g. *They understood* (that) *this was an errand of mercy; I would hate to feel* (that) *we were corrupting you.* But self-evidently it must be retained in some of the examples cited in the first paragraph above. When the conjunction *that* is part of the correlative pairs *so ... that, such ... that, now ... that* (or *so that*, etc.), it is normally retained in formal writing but is sometimes omitted in informal contexts, e.g. *What would he do now* (that) *he had missed him in Toulouse?; The heat was up so high* (that) *almost everyone took off their coats. That* may not be omitted, however, if a phrase or clause intervenes between the verb and it, e.g. *I am saying, am I not,* that *I no longer loved Kioyoko.*

**that** (relative pronoun).

| 1 *that* or *who* with a human antecedent.
| 2 Omitted as relative pronoun.
| 3 Restrictive *that/which* and non-restrictive *which*.
| 4 Further complications.

1 *that* or *who* with a human antecedent. 'How many well-educated people have you heard say *Those are the people* that *attended the play?*'—thus a gloomy correspondent to a major newspaper, echoing a widely held belief that *that* as a relative pronoun should never be used

when the antecedent is personal. What in fact is the rule in such cases? A human antecedent is normally followed by *who* (rather than *that*) in a defining clause (*But it was I* who *got away to the steps up to the morning room*). An inanimate antecedent naturally calls for *that* (or *which*, as appropriate): *an electric blanket* that *knows how to warm various parts of the body; the bus* that *bore us away. That* is often used in contexts in which the antecedent is animate but not human (*a white poodle* that *sported a red hair bow*), and also in contexts where the antecedent is human but representative of a class (*a baby* that *cries in unsocial hours; a fellow* that *sells a bracelet is not necessarily interested in people*) or is an indefinite pronoun (*it was an obvious rebuff, and one* that *hurt Benn very deeply*).

Down through the centuries, *that* has often been used with a human antecedent. Chaucer, Langland, and Wyclif are all cited in the *OED* using *that* in this way, and examples are also given from writers in each of the later centuries. The 20c. abounds with writers who keep to the rule that only *who* is appropriate when the antecedent is human (whether a specified person or one representative of a class). Thus, from Penelope Fitzgerald's *The Golden Child* (1977): *And the vast, patient public ...* who *would soon be filing across the Museum courtyard to proffer their money at the entrance; The Russians have made a very considerable loan to Garamantia* [name of a fictitious country], who *is anxious to develop her agriculture; I know* that *there are some people in the service* who'd *rather not have them let in at all; a wonderfully good representation of the sea by a people* who *had never beheld it.*

*Summary.* Normally use *who* as the relative pronoun following a human antecedent and *that* (or *which*) following an inanimate antecedent. Either *who* or *that* may be used when the antecedent is animate but not human, or when the antecedent is human but representative of a class or is an indefinite pronoun. In contexts containing a double antecedent, of which the first is human and the second inanimate, *that* is naturally required (*he answered accusingly ... as though it was she and not the drug* that *had done it*).

2 Omitted as relative pronoun. One of the most capricious features of the

language is the way in which some seemingly necessary elements may be omitted without loss of meaning or dislocation of syntax. From the 13c. onward the relative *that* has been omissible in a variety of circumstances. Modern examples: *It reminded him of the Exhibition ∧ he was going back to*—P. Fitzgerald, 1977; *and that lipstick ∧ she used to put on*—H. Mantel, 1985; *it was your geography ∧ caused the doubt*—T. Stoppard, 1993. For further discussion of this phenomenon and additional examples, see CONTACT CLAUSES and OMISSION OF RELATIVES.

**3** Restrictive *that/which* and non-restrictive *which*. Take two sentences or parts of sentences from Anita Brookner's *A Family Romance* (1993): *with the ball-point pen which my father had bought for me in a curiously shaped department store*; *This sharpness of gaze gave her an air of vanity, which I dare say was justified*. The first contains a restrictive clause led by *which*. In it *which* could have been replaced by *that* without change of meaning and without giving offence to any rule of syntax. The second contains a non-restrictive use of *which* preceded by a comma. In the first sentence the *which*-clause defines and particularizes; and *that* would have done the same work just as well. In the second example, the *which*-clause provides additional information as a kind of parenthetic aside. In other words it is a non-restrictive clause.

The division between the two types is not an absolute one. In 1926 Fowler wisely observed: 'The relations between *that, who,* and *which,* have come to us from our forefathers as an odd jumble, and plainly show that the language has not been neatly constructed by a master-builder who could create each part to do the exact work required of it, neither overlapped nor overlapping; far from that, its parts have had to grow as they could.' He went on to stress that not all writers observe the distinction between restrictive clauses (which he called *defining clauses*) and non-restrictive clauses (which he called *non-defining clauses*): 'The two kinds of relative clauses, to one of which *that* and to the other of which *which* is appropriate, are the defining and the non-defining; and if writers would agree to regard *that* as the defining relative pronoun, and *which* as the non-defining, there would be much gain both

in lucidity and in ease. Some there are who follow this principle now; but it would be idle to pretend that it is the practice either of most or of the best writers.'

**4** Further complications. A wider spread of examples drawn from a range of standard modern sources will bring out the (partial) distinction between restrictive and non-restrictive clauses. But before setting out these examples, two main qualifications should be made. First, whereas the relative pronoun *that* cannot idiomatically be preceded by a preposition, there is no such constraint in the case of *which*. In practice therefore some relative clauses led by *that* end with a preposition. Secondly, another peculiarity of *that* is that it has no possessive of its own and therefore shares the possessive forms *of whom* and *whose* with *who*. The cases of *that* as a relative pronoun are therefore: subject *that*; object *that*; possessive *whose* (or *of whom*); prep.-preceded (*in, by, for, from,* etc.) *which*. It should also be borne in mind that whereas *that* leading a defining clause may often be 'understood' (i.e. omitted) without loss of meaning (*The house you saw* means the same as *The house that you saw*), *which* in the non-defining (or non-restrictive) clauses to which it is proper cannot be omitted (*This fact, which you admit, condemns you* cannot be changed to *This fact, you admit, condemns you* without altering the sense). A sentence in Penelope Fitzgerald's *The Golden Child* illustrates both the omission of the relative pronoun and the postponement of a preposition to the end of the clause: *It reminded him of the Exhibition he was going back to.*

Examples (all drawn from standard modern sources in my database) of three broadly distinctive types of *that* and *which* relative clauses:

(*a*) *that*-clauses: *Everything that lives in the sea has had a longer evolutionary history than anything on land*; *The financial resources that any opera production faces involves both private and government resources*; *it was Labour Party politics that captured his youth*; *It was precisely with rugged proletarians like these that Gaitskell felt personally most awkward*; *boundaries that one would like to see kept distinct have blurred*; *the great new glass tower that housed Eldorado Television.*

(b) *which*-clauses: *he is described frequently ... in terms which will set your teeth
on edge; the villa in Italy ... which she and
Arthur were being pressed to visit; She was
coming from the Rhodesian girls' school to
which ... she was sent because her father
had grown up in Salisbury; He sat there with
one foot on a fruit-box to support the leg on
which the guitar rested; the procedures ...
are those* into which *young ladies are to be
inducted; the legal and political anomalies
which exclusion of the Catholics has given
rise to.*

(c) *which*-clauses preceded by a comma:
*She held out her hand, which I clasped in
both of mine; Baxandall makes it sound not
just boring but unnecessary, which is severely
to misunderstand our intention; his company
plane, a Citation jet, which he uses to visit
his stores; my father's last illness, which she
persisted in seeing as a fantasy and betrayal;
You bought a Depression bargain, which I
tipped you off to; Besides him being a poodle,
which is not one of your heroic breeds of dog;
Dickinson's contract, which has four years
to run, would be paid in full; the concomitant
of ownership, which is the right to alienate
property.*

## the.

1 *The* in titles of books, etc.
2 With names of illnesses.
3 The type *Prime Minister Major.*
4 Pronunciation.
5 *by the hundred*, etc.
6 Omission of a second *the.*
7 *the* with two nouns and a singular verb.
Some other uses of *the* are treated s.v.
DEFINITE ARTICLE.

1 *The* in titles of books, etc. In ordinary
contexts *The* should always be retained
when it is part of the title of a book,
poem, journal, etc.: *The Times, The Old
Curiosity Shop, The Lay of the Last Minstrel,
The Doctor's Dilemma, The Merry Wives of
Windsor, The Oxford English Dictionary, The
Chicago Manual of Style.* It should be borne
in mind, however, that some familiar
titles are reduced forms of the original
ones: for example, Shakespeare's *Hamlet*
was originally *The Tragedie of Hamlet Prince
of Denmarke. The* is usually omitted or
reduced to lower-case romans 'the' in
abbreviated forms of titles: e.g. the *OED;
Merry W.* or even *M.W.W.; Lay of Last Minstrel.* Kingsley Amis (*Spectator*, 6 Sept.
1986) maintained that the type *Harold
Robbins, whose* The Storyteller *shows* was

'not English': he insisted that the language does not tolerate a possessive followed by the definite article. Write
*Robbins, whose novel* The Storyteller ...
instead, he suggested. Not everyone
agreed. In practice there is considerable
variation. In a single issue (autumn 1994)
of *Letters* (the journal of the Royal Society
of Literature), for example, I found: *Kenneth Tynan once wrote in the* New Yorker;
*J. R. R. Tolkien's* Lord of the Rings; *Trollope's*
The West Indies and the Spanish Main; *Dev-
lin's* The Judge; *Izaak Walton's* Compleat Angler.

All five examples either broke Amis's
Law or simply omitted the *The* of the
original title.

2 With names of illnesses. See FLU;
MEASLES; MUMPS.

3 The type *Prime Minister Major.* This
construction, with omission of *the* before
an occupational title followed by a person's name, is embedded in AmE but less
so in BrE. It is especially common in
journalistic work. Examples: *Chancellor
Helmut Kohl of West Germany announced
Bonn's conditional approval—Christian Sci.
Monitor, 1987; The notion of the 'enterprise
culture', well accounted for in Chancellor Lawson's 1986 budget speech—Essays & Studies,
1987; Vocalist Bjork, meanwhile, has taken
her first acting role in a television play—
Melody Maker, 1988; Two recent publications,
the report of the National Commission on
children ..., and a more popular work, by
economist Sylvia Ann Hewlitt, suggest* [etc.]
*Dædalus*, 1993

4 Pronunciation. The definite article
is the most common word in the language. Yet it is remarkable that the
majority of native speakers are hardly
aware that it is pronounced /ðə/ before
a word beginning with a consonant (*the
tree* /ðə triː/) and as /ðɪ/ before a word
beginning with a vowel (*the adverb* /ðɪ
ˈædvɜːb/) and /ðiː/ when stressed.

5 *by the hundred*, etc. *The mild revelations
of a gentle domestic existence which some
royal personages have given us command
readers by the hundreds of thousands.* The
idiomatic English is *by the hundred thousand; by hundreds of thousands* will also
pass, but with the plural *the* is not used.
So also with *dozen, score*, etc.

6 Omission of a second *the.* The definite article is normally dispensed with

before the second of two nouns joined by *and*: *The distortion and innuendo to which several of your correspondents have resorted*; *They throw valuable light on the ethos and attitudes of the ruling elite.* But it should be noted that a second *the* may be inserted (for emphasis or other contextual reasons) in such sentences. The same considerations do not apply to the second of two adjectives. For example, *the black and white penguins* has a different meaning from *the black and the white penguins*, and *the red and yellow tomatoes* (one group) is likely to differ in meaning from *the red and the yellow tomatoes* (two separate groups).

**7** *the* with two nouns and a singular verb. A composite subject of the type '*the* + noun + *and* + noun', if thought of as a single theme, may be followed by a singular verb. See AGREEMENT 3. Examples: *The innocence and purity of their singing comes entirely from their identification with the character*—B. Levin, 1985; *The respect and freedom that money can buy has made me not one jot happier than in my former years of penury*—B. Rubens, 1985; *The power and wealth of the United States is a natural source of envy*—*Daily Tel.*, 1987.

**theatre.** Thus spelt in BrE, but usually *theater* in AmE. See -RE AND -ER.

**thee,** objective case of the pronoun *thou*, has very little currency now (having been effectively replaced by *you*) except in verse surviving from older times (*Hail to thee, blithe Spirit!*—Shelley, 1820; *Dost thou love me, cousin? ... I have loved thee long*—Tennyson, 1842), in jocular or regional verse (*Dreamin' of thee! Dreamin' of thee*—E. Wallace, 1900, popularized in a 1930 broadcast by C. Fletcher), and in biblical contexts (*Get thee behinde me, Satan*—Bible (Geneva), Matt., 1560). It is also used as nominative case (a use unaccounted for) by Quakers (*Perhaps thee has noticed the point in our* Friends Journal *on February 15*—*Friend*, 1964).

**their.** **1** Fowler (1926) was among those who objected to the use of *their* in contexts that call 'logically' for *his* (though this use of the masculine gender to cover both has lately been called into question) or *his or her.* Examples: *But does anyone in their heart really believe that Ireland is only that?*; *no one can be easy in their minds about the present conditions of examination.*

The issue is unresolved, but it begins to look as if the use of an indefinite third person singular is now passing unnoticed by standard speakers (except those trained in traditional grammar) and is being left unaltered by copy editors. Examples: *I feel that if someone is not doing their job it should be called to their attention*—an American newspaper, 1984; *Everyone was absorbed in their own business*—A. Motion, 1989; *A mission statement that is sufficiently bland to encompass everyone's conception of their role in the university is of little use to anyone*—*Dædalus*, 1993.

**2** This possessive personal pronoun is to be carefully distinguished from *there* (adverb, noun, and interjection) and *they're* = 'they are'.

**theirs.** How many teachers at all levels need to point out daily that the correct spelling of *theirs* (and *ours*, *yours*) is not *their's* (etc.)? The misconception about the spelling is not all that surprising, perhaps, when one recalls the memorable lines from Tennyson's *Charge of the Light Brigade* (1855), *Their's not to make reply, Their's not to reason why, Their's but to do and die.*

**theirselves.** 'From the 14th c. there has been a tendency to treat *self* as a noun ( = person, personality), and substitute *their* for *them* ... This is prevalent dialectally, but in literary Eng. has place only where an adj. intervenes, as *their own, sweet, very selves*' (*OED*). Example of a demotic use of *theirselves*: *But they've only got theirselves to thank for it, I'm afraid*—P. Lively, 1991. See HISSELF.

**them.** **1** As the *OED* says, it is 'often used for "him or her", referring to a singular person whose sex is not stated, or to *anybody, nobody, somebody, whoever,* etc.'. Example: *Nobody else ... has so little to plague them*—C. Yonge, 1853.

**2** Its use as a demonstrative pronoun, permissible in standard sources from the 15c. onward, is now only dialectal or illiterate exc. in the humorous type *them's my sentiments.* Examples: *Them be my two children*—L. J. Jennings, 1877; *Them as says there's no has me to fecht*—J. M. Barrie, 1891; *We're out here to do justice and keep the peace. Them's my sentiments*—E. M. Forster, 1924.

**3** As a demonstrative adj., *them* has been downgraded in the 20c. (it is first recorded in the late 16c.) and now is only dialectal or illiterate. From the late 16c. to the 19c. it seems to have been in standard use. Examples: *I hope, then, the agent will give you encouragement about them mines*—M. Edgeworth, 1809–12; *Them ribbons of yours cost a trifle, Kitty*—S. Lover, 1842; *I've done all them skirting boards, but me back's not what it was*—J. Winterson, 1985; *'Ere, you're gonna look a right wally when you do them Skinny Puppy chaps aren't you?*—*Melody Maker*, 1988; *I didn't know much about planes in them days*—P. McCabe, 1992.

**themself.** The normal emphatic and reflexive form of *they* and *them* is, of course, *themselves*. The *OED* reports standard uses of the alternative form *themself* from the 14c. to the 16c., but then it seems to have lost its grip on the language and disappeared from sight. A remarkable by-product of the search for gender-neutral pronouns, *themself* re-emerged in the 1980s. It is a minority form, but one that turns up from time to time in Britain, N. America, and doubtless elsewhere. Examples: *It is not an actor pretending to be Reagan or Thatcher, it is, in grotesque form, the person themself*—I. Hislop, 1984; *I think somebody should immediately address themself to this problem*—A. T. Ellis, 1987; *After all, how could anyone defend themself against the soft-focus sleaze of a show like 'A Current Affair'* ...?—*Chicago Sun-Times*, 1991; *a principal who desires to acquire the tobacco for use or consumption by themself or other persons at their expense*—Canadian legal context in *English Today*, 1994. This new pronominal form can hardly be regarded as standard—yet.

**then.** Two uses, one adjectival the other adverbial, call for comment. **1** (adj.) Murmurs are occasionally heard against the type *the then Bishop of London* ( = the person who was the Bishop of London at the time), esp. if a hyphen has been inserted between *then* and the noun that follows. But the type has been in continuous use since the late 16c. and must surely pass without censure—in its simplest form, that is. It begins to look awkward when the immediately following word is an adjective (thus turning *then* into an adverb) or a complex title, and awkwarder still when a hyphen is used. Examples (all from recent American newspapers): *Four years ago great things were expected of Japan's then-embryonic biotechnology industry*; *Letterman countered that he was invited to pull such a stunt by the then-executive producer of 'Today'*; *a conclusion of then-White House Counsel Fred Fielding*; *once owned by the Whitewater business partners of then-Gov. Bill Clinton and Hillary Rodham Clinton.*

**2** Tail-end *then* (adv.). *Then* is frequently used, at least in BrE, 'as a particle of inference, often unemphatic or enclitic' (*OED*), sometimes merely as part of a formulaic farewell greeting: e.g. *As Lucas continued to stare at the window, Clement said, 'Well, goodbye then.'*—I. Murdoch, 1993; *I say. That time already? Good-night, then*—B. Neil, 1993; *He rose from his chair, patting his pockets, and said, 'So, well, I'll be off, then.'*—idem, 1993. It is also often used as the final element of an interrogative sentence: e.g. *'Not too much trouble getting here, then?' asked my father genially*—A. Brookner, 1993. It is also used as a stronger 'particle of inference' when the implied meaning is something like 'in that case': e.g. *What had she sat next to him for in the first place, then?*—M. Eldridge, 1984 (Aust.).

**thence.** Also *from thence*. Only in archaic or literary use. See FROM WHENCE.

**theoretic, theoretical.** As adjs. the two forms have coexisted in the language since the 17c., the longer form probably being the more frequent of the two in BrE at the present time. See -IC(AL).

**thereabout(s).** One of the medieval compound adverbs in *there*- that survive in common use without hint of archaism, the others being *thereby*, *therefor(e)*, and *thereupon*. The form with final -*s*, which is recorded from about 1400 onward, is now the more usual of the two (*it all changed in 1970 or thereabouts*; *there or thereabouts*). The casualty rate of adverbs in *there*- is high (see below), and one wonders how many of these words will still be in regular use in the centuries ahead.

**thereafter.** One of several traditional adverbs in *there*- now restricted to formal use, esp. in legal documents or ceremonial language. It was first recorded in OE (King Alfred). Other formal words in *there*-

include *therein* (first recorded in OE), *thereof* (OE), *thereto* (OE), and *theretofore* (c1350).

**thereat.** One of several adverbs in *there-* now falling or fallen into disuse after centuries of routine service in the language. It was first recorded in OE. Other archaistic words in *there-* include *therefrom* (first recorded a1250), *thereinto* (a1300), *thereon* (OE), *thereout* (OE), *therethrough* (c1175), *thereunto* (c1300), *therewith* (OE), and *therewithal* (a1300). *Thereanent* (c1340) is now restricted to Scottish use (see ANENT).

**therefor.** An adverb now mainly in legal use, stressed on the second syllable, and meaning 'for that object or purpose, for that reason, for it'. Examples (from the OED): *He shall supply a copy of such report ... on payment of the sum of one shilling therefor* (1885); *The ill-used crew promptly refused to do any more in her, and were, of course, clapped in jail therefor* (1899). Other examples (from Garner, 1987): *We are not unmindful of the sound and salutary rule, and of the obvious reasons therefor*; *The plaintiff discharged union cutters and shop foremen without just cause therefor.*

**therefore.** The most resilient of all the adverbs in *there-* and part of the central core of the language since the 12c. Unlike the preceding word it is always stressed on the first syllable. In longer explanatory sentences it is often preceded and followed by a comma, and when it occurs initially it is sometimes conventionally followed by a comma. In short sentences a comma is unnecessary. Examples: *Things obscure are not therefore sacred*; *The pound was falling against the Deutschmark. Therefore it was necessary to leave the ERM.*

Fowler (1926) pointed out that the presence of a comma before *therefore* tends to lead to additional stress being placed on the immediately preceding word. Some preceding words, e.g. *although*, 'are equal to that burden', while others, e.g. *and* and *it*, are not. Examples showing acceptable punctuation: (commas) *Although, therefore, the element of surprise could not come into play on this occasion, the enemy were forced to withdraw*; *I trust, therefore, that you will accede to this request*; (no commas) *The Franks were the stronger, and therefore the masters*; *It therefore behoves*

those who have made the passage of the Bill possible to attend once more.

**there is, there are.** Before launching into a sentence beginning with *there* + a part of the verb *to be*, one must decide whether the following subject is actually or notionally singular or plural. In most circumstances no difficulties arise. Thus: *There is a spider in the bath*; but *There were many who disagreed with the speaker.* The choice is more difficult when the subject is more complicated than in these two examples. Thus: *There were a table and some chairs in the room* (the subjects are transparently plural); *There was a plain deal table in the room and some wicker armchairs which Jorgenson had produced from somewhere in the depths of the ship* (*There was* because of the proximity of *a plain deal table* and the elaboration and 're-movedness' of the second subject).

Apart from such considerations there is a strong, not always resisted, temptation, found prominently but not only in uneducated speech, to introduce a plural subject with the reduced form *There's* (cf. Fr. *il y a*). Some informal examples from newspapers: *There's 35 locations* [sc. shops] *to serve you*; *But for every big, dumb move like this, there's half a dozen small, smart details.*

*Note.* In constructions of these kinds, *there* is described by some scholars as *introductory* and by others as an *existential*, an *anticipatory*, or a *dummy* subject.

**there you are.** This colloquial expression of regret, reluctant acceptance, resignation, etc., always preceded by *but*, has been in common use since the mid-19c. and its currency remains unquestioned. Examples: *It's a pity we have to shoot so many of them but there you are—* L. P. Hartley, 1953; *The conversation was a bit one-sided mind, devoted as it was to the morality of laying waste the African rainforests to make chairs for tourists, but there you are—*A. Coren, 1991; *I felt ridiculous of course, but there you are—*S. Wall, 1991.

**there you go.** An exceedingly common conventional phrase used by persons bringing a meal in a restaurant or an item or items bought in a shop, delivering the post, etc. preceded by *but* it is also sometimes used instead of *but there you are* (see prec.). Examples: *So we trooped off to a nearby public toilet, not the* *most salubrious of venues, but there you go*—R. Elms, 1988; *There yeh go, he said.*—*Keep the change*—R. Doyle, 1991; *Evan handed over her change. 'There you go then.'*—Barbara Anderson, 1993.

**thesaurus.** The pl. is either *thesauruses* or *thesauri* /-aɪ/. See -US 1.

**these.** See KIND (noun) 2; SORT (noun) 2.

**thesis.** 1 For the sense 'dissertation', the pl. is *theses* /'θiːsiːz/.

2 In Greek and Latin verse *thesis*, pronounced /'θiːsɪs/ or /'θesɪs/, is the term for an unstressed syllable or part of a metrical foot (as opposed to ARSIS).

**they, their, them.** 1 *one, anyone, everybody, nobody*, etc. followed by *they, their, them.* Over the centuries, writers of standing have used *they, their,* and *them* with anaphoric reference to a singular pronoun or noun, and the practice has continued in the 20c. to the point that, traditional grammarians aside, such constructions are hardly noticed any more or are not widely felt to lie in a prohibited zone. Fowler (1926) disliked the practice ('few good modern writers would flout the grammarians so conspicuously') and gave a number of unattributed 'faulty' examples, including the following: *The lecturer said that everybody loved their ideals; Nobody in their senses would give sixpence on the strength of the promissory note of that kind.*

The evidence presented in the OED points in another direction altogether. From the 16c. onward *they* has often been 'used in reference to a singular noun [or pronoun] made universal by *every, any, no,* etc., or applicable to one of either sex ( = "he or she")'. The examples cited by the OED include: *Every Body fell a laughing, as how could they help it*—Fielding, 1749; *If a person is born of a ... gloomy temper ... they cannot help it*—Chesterfield, 1759; *Nobody can deprive us of the Church, if they would*—W. Whewell, 1835; *Now, nobody does anything well that they cannot help doing*—Ruskin, 1866. Similar constructions are presented in the OED for *their* (from the 14c. onward) and *them* (1742– ). All such 'non-grammatical' constructions arise either because the notion of plurality resides in many of the indefinite pronouns or because of the absence in English of a common-gender third person singular pronoun (as distinct from *his* used to mean 'his or her' or the clumsy use of *his or her* itself).

Modern examples of *they, their,* and *them* used with singular reference may be found in this book in the articles for several of the indefinite pronouns, and also in the separate articles for THEIR and THEM. The process now seems irreversible.

2 Muddles with collective nouns. Consistency is essential. *The government is pressing ahead with* its *policy of privatization* is acceptable, and so (at least in BrE) is *The government are pressing ahead with* their *policy of privatization.* A mixture of numbers, e.g. *The government is pressing ahead with* their *policy or privatization,* should not be allowed to stand. Cf. AGREEMENT 5.

**thief.** Pl. *thieves.* See -VE(D), -VES.

**thimbleful.** Pl. *thimblefuls.* See -FUL.

**thin** makes *thinness.*

**thine.** See ABSOLUTE POSSESSIVES.

**thingumajig, thingumabob, thingummy** are the chief survivors of a large number of 18–19c. variants.

**think** (verb). 1 After *think, that* is usually omitted; see THAT (conj ) 2.

2 For the transfer of negativity from *I don't think* to the following clause, see DO 5.

3 *think to* + infin. This construction, meaning 'to expect', is an old one (e.g. *I neuer thought to heare you speake againe*—Shakespeare, 1597). It survives to this day, though in some contexts its meaning is 'remember' rather than 'expect' (*Did you think to ask him for his address?*).

4 *think up.* This colloquial phrasal verb meaning 'to devise, to produce by thought' first emerged in the mid-19c. in AmE, but is now in more widespread use.

5 *no thinking man.* One of the bluffing formulae, like *It stands to reason* and *with respect,* that are intended to put the reader's or listener's back up and incline

him or her to reject the view that is being presented.

**think** (noun). This late-coming word (not recorded until 1834, though the corresponding verb is OE) is labelled *colloq.* in the *New SOED* (1993). It is commonly used in such sentences as *Have a think about it*, and *You have another think coming* 'you are greatly mistaken'.

**thinkable** (adj.). This word, meaning 'able to be deemed real or actual; capable of being thought' seems not to have entered the language until the early 19c., some 450 years after its much more common antonym *unthinkable*.

**thirties.** See NINETIES.

**this** (demonstrative adj.). **1** The simple demonstrative adj. *this* is used 'to indicate a thing or person present or near (actually or in thought), esp. one just mentioned' (*OED*). Some standard modern examples of ordinary uncontested uses: *I wanted to enjoy this room; Only in this way can one hope to reconstruct the physical landscape of the time; The allies endorsed this hegemony because they stood to benefit from it; At this time it was Bodley's custom to remain in London; He wasn't particularly active in this respect; Both this house and No 45 will be converted.*

**2** There is a long tradition in English of using *this* instead of *these* in concord with 'a numeral expression denoting a period of time taken as a whole (... usually = 'just past or completed')' (*OED*). Examples: *Within this three houres will faire Iuliet wake*—Shakespeare, 1592; *The silence has kept my own heart heavy this many a day*—Ruskin, 1867; *This last six months*—L. Oliphant, 1883. It is a small example of the way in which a lack of formal agreement is justified by the concept of notional agreement.

**3** In narrative writing or speech, *this* is now modishly used to refer in a familiar manner to a person, place, object, etc., not previously mentioned (usu. taking the place of an indefinite article). The type seems to have emerged first in AmE in the 1920s. Examples: *He had this driver called Reg Whelan, father of seven*—T. Keneally, 1980; *He had this great little Nagra portable recorder for the birdcalls*—New Yorker, 1986; *She asked me to do this play for Icelandic television ... It's about this girl.*

*Her parents live in a huge house*—Melody Maker, 1988; *There was this man who trained his dog to go around the corner to Bud's Lounge with a dollar bill under his collar*—G. Keillor, 1989.

**this** (demonstrative adv.). Like *that* (see THAT (demonstrative adv.)), *this* is used, for the most part in informal contexts, in the sense 'to this extent or degree', very much in the manner of an intensive. Examples: *Perhaps this much of Plato is enough for one letter*—Ruskin, 1877; *I have a stack of telegrams this thick*—Boston Sunday Herald, 1967; *Yet the picture is usually not even this good*—Where, 1971; *Keep in mind, however, that no existing property is this typical*—Real Estate Rev., 1972.

**this** (demonstrative pron.). This common pronoun is used to indicate 'a thing or person present or near (actually in space or time, or ideally in thought, esp. as having just been mentioned and thus being present to the mind' (*OED*). Some typical modern examples from standard sources: *Am I going to undertake this just to oblige him?; What had I done to deserve this?; The result of all this was to re-open in Edith's mind the question ...; One need interrupt the narrative no further than to say this; I started to wonder—could I be making this up?; All college scarves are like this anyway; Should governments do more, or ought this to be left to the private sector?*. The test of the validity of *this* is that it must be crystal-clear to what it refers. It is comparatively easy to slip into the fault of leaving it unclear precisely what the antecedent is.

**thither.** See HITHER.

**-th nouns.** There are large numbers of well-established words ending in -*th* with corresponding adjectives, verbs, etc., as *breadth* (cf. *broad*), *dearth* (cf. *dear*), *depth* (cf. *deep*), *growth* (cf. *grow*), *health* (cf. *whole*), *length* (cf. *long*), *tilth* (cf. *till*), *truth* (cf. *true*), *warmth* (cf. *warm*), *wealth* (cf. *well* or *weal*), *width* (cf. *wide*). Long ago, however, -*th* ceased to be a significant living suffix for new nouns. Horace Walpole coined *blueth*, *gloomth*, and *greenth*, but to no avail. *Blowth* 'a blossoming', *mowth* 'a mowing', and *spilth* '(an instance of) spilling' came into the language in the 17c. but are now dialectal, archaic, or obsolete. Ruskin's coinage *illth* (antonym of *wealth*) has small currency except with direct reference to Ruskin himself.

**those.** See KIND (noun) 2; SORT (noun) 2.

**tho, tho',** informal (esp. in AmE) or poetic spellings of *though*. Examples: *Tho' much is taken, much abides*—Tennyson, 1842; *Tho all the good food and wine and reefer was gone now*—Black World, 1973; *Tho' the trip's less than a mile it's still a dreary bore*—NY Times, 1982.

**thou.** See THEE.

**though.** **1** For the choice of *though* or *although*, see ALTHOUGH. In conditional clauses led by *although* or *though*, the former is almost always used by conservative writers. For example, Anita Brookner in her novel *A Family Romance* (1993): (p. 119) *I thought I detected a change in her, although I had not seen her for some weeks*; (p. 120) *Quite possibly, although this seemed grotesque to me, Dolly was in love*; (p. 121) *I saw that Dolly despised us, although she would never have admitted as much, even to herself*; (p. 135) *and although I wanted to cry and sob my eyes were quite dry and my face composed*; (p. 140) *it was my mother who was her true friend, although she had the company of the church ladies on Monday evenings*. In the sense 'and yet' the shorter form is used: (p. 120) *These [glances] were indulged without mercy*, though *unaffectedly*. And the shorter form is compulsory after *even*: (p. 163) *but I should be quite content to watch, even though it might mean seeing the equivalent of one of those television programmes my mother so enjoyed*.

**2** For *as though*, see AS 9.

**thral(l)dom.** See -LL-, -L- 5.

**thrash, thresh.** Originally variants of the same word, now with pronunciations and spellings differentiated. To separate grain is *thresh*, to flog is *thrash*. *Thrash* is the usual spelling of figurative and transferred uses, e.g. to thrash one's opponents (in a game of cricket, etc.), to thrash out a problem. But both spellings are found for the senses 'to move one's limbs about violently' and 'to toss and turn (in bed)' (followed by *about*).

**Threadneedle Street, Old Lady of.** See SOBRIQUETS.

**threaten** (verb). For its use as a double passive, see PASSIVE TERRITORY 3.

**three-peat.** An unappealing neologism (an altered form of *repeat*) of the late 1980s in AmE: used both as a noun ( = three consecutive victories in a (championship) series, orig. in basketball), and a verb (to win a given contest three times in succession). Also *threepeat* (no hyphen).

**three-quarter(s).** The noun expressing a fraction is *three-quarters* (usu. hyphened). In a rugby football team a *three-quarter* is one of two, three, or four players (*three-quarters*) positioned between the half-back and the full-back. When used attrib. the form without final *-s* is obligatory: *the three-quarter line* (in rugby football), *three-quarter length* (coat, etc.), *three-quarter face* (in photography, the aspect between full face and profile), etc. Cf. *billiards* but *billiard-ball, -table*.

**threnody** /ˈθrenədɪ/, a song of lamentation. The standard form, rather than *threnode* /ˈθrenəʊd/ (altered on model of *ode*).

**threshold.** Pronounce /ˈθreʃəʊld/, or, with medial /h/ inserted, /ˈθreʃhəʊld/.

**thrice.** From about 1200 a regular word for 'three times', but now only archaic or poetical.

**thrive** (verb). The pa.t. is either *thrived* or *throve*, and the pa.pple either *thriven* or *thrived*. At present *thrived* seems to be the more usual form for both parts.

**throes.** So spelt, as in *to be in the throes of*, struggling with the task of. Keep it carefully apart from *throws* 'propels' and *throws* (pl. noun) (*allowed three throws with the javelin*).

**through.** **1** In the language of telephony *You are through* means in Britain 'You are connected'; but to an American *Are you through?* might be used by a switchboard operator to mean 'Have you finished your call?' Caution is therefore called for in transatlantic calls made through a switchboard. In ordinary language *Are you through with this?* is used idiomatically in both major forms of English to mean 'Have you finished with this?' But the simplex *through*, as shown in the following examples, seems characteristically N.Amer.: *Yaweh was through, now, with* *speech*—T. Findley, 1984; '*I'm through eating*,' *said my father, pushing his plate away*—L. S. Schwartz, 1989.

**2** The convenient AmE use of *through* to mean 'up to and including' (*Monday through Friday*) is familiar to standard speakers of BrE but is not often used by them except as a conscious Americanism. Examples: *they range from regimes of explicit rules and organizations at one end of the spectrum through private networks and informal conventions at the other*—Dædalus, 1991; *an eight-week summer program for disadvantaged children ages three through five*—ibid. 1993.

**thru,** an occasional non-standard AmE spelling of *through*. Examples: *I see the thing thru, alone*—E. E. Cummings, 1917; *When she wuz little and she had stuttered thru a sentence*—Black World, 1971; *Available for S types right thru to Mk 10s it retails for 26 notes*—Hot Car, 1977. The spelling has a history of its own in the 20c. in American spelling-reform circles, but the American public has rejected it except in informal circumstances (private letters, memos, etc.) and in the traffic term *thruway* ( = expressway).

**thunderer.** For *The Thunderer* = *The Times*, see SOBRIQUETS.

**Thursday.** For the adverbial use, see FRIDAY.

**thusly** (adv.). *Thusly—a word I picked up in Massachusetts and will give this single outing—if you are in that South Atlantic coaling station* [etc.]'—C. Freud in *The Times* (3 Aug. 1989). The word seemed out of order to me, since *thus* by itself would have sufficed. But its naturalness is more or less unquestioned in AmE (it was first recorded in the *OED* in an American source of 1865), and a string of examples lie in the Merriam Webster files. Their usage guide (1989) declares that '*thusly* is not now merely an ignorant or comic substitute for *thus*: it is a distinct adverb that is used in a distinct way in standard speech and writing.' Clement Freud's 'single outing' for the word suggests that it has not been washed ashore in the UK yet.

**thyme.** Though ultimately from Gk θύμον it seems to have had initial /t/ from the time of its adoption (15c.) in

English from French. It was frequently spelt *tyme* or *time* in former times (15–18c.). By contrast, some scientific words derived ultimately from Gk θύμον, as *thymene, thymidine, thymine, thymol*, always have initial /θ/ in English.

**tiara'd** is preferable to *tiaraed*. See -ED AND 'D.

**-tiation.** See -CIATION, -TIATION.

**tibia** /'tɪbɪə/. Pl. *tibiae* /-briː/. The stem vowel is short in English, despite the ī in its etymon, L *tībia*. See FALSE QUANTITY; LATIN PLURALS.

**tidal** (adj.). The formation was called into question by Fowler (1926) in his most puristic mood as being a modernish (1807) welding of the adjectival suffix *-al* to a noun of native origin (*tide*), a process not elsewhere paralleled. But the word is now, of course, essential and uncancellable. See -AL 1.

**tidbit.** See TITBIT.

**tigerish, tigress.** Thus spelt.

**tight, tightly.** The first is primarily an adjective, but it overlaps with *tightly* in some adverbial uses, e.g. in the imperative phr. *hold tight!*, and immediately after a few verbs. Thus, drawn from standard modern sources: *but, from what I know of them, they won't sit tight*; *Well, goodnight. Sleep tight*; *like something that had been wound tight and suddenly released*. *Tight* (adv.) also occurs in a few adjs., e.g. *tight-fisted, -fitting, -lipped*.

**tike.** See TYKE.

**tilde.** A mark (˜) put over a letter, e.g. over a Sp. *n* to indicate that it is to be pronounced /nj/, as in *señor*; and over a Pg. *a* or *o* to indicate nasalization, as in *São Paulo*.

**till, until.** The early history of the two words in medE, as set down with great thoroughness in the *OED*, is complicated. It is not true, for example, that *till* is a shortened form of *until*: in fact *till* is the earlier of the two. The present distribution is no less complicated. One can declare, with the *COD*, that '*Until* is used especially at the beginning of a sentence', but that doesn't take us far. One can also assert with reasonable certainty that *until* is fractionally more formal than *till*: this might account for the fact that *until* occurs much more frequently than *till* in edited prose (including fiction). In practice *until* is six times more likely than *till* to turn up in such work (both according to my own database and to a standard dictionary of word frequency).

Some typical examples of each word, all drawn from standard sources of the 1980s: (till) *Wait till you see what I've been doing; don't you dare even look at anybody till I get back; I could live here forever, he thought, or till I die*. (until) *The runic signature in the Fates was not found until 1888; This process will go on until election day; Many of the manuscripts were until comparatively recently in the keeping of Owen's family; This would continue for hours until he reached a more advanced stage; It used to be widely believed that in Britain one is innocent until proved guilty*. The evidence tends to confirm that *till* is sometimes the informal equivalent of *until*, but also that in many contexts the two words are simply interchangeable without affecting the stylistic level. Finally it should be noted that some authorities urge their readers not to use a third form, namely '*til*. In fact it is an informal 20c. contraction of *until*, and may be used with propriety in informal contexts.

See UNTIL 2.

**tilth.** An ordinary noun in OE (*tilþ*) corresponding to the verb *tilian* 'to labour', (by 1200) 'to cultivate land'. It has retained its place in agricultural and horticultural language ever since. Some recent examples: *Settle instead for a sprinkling of blood, bone and fishmeal ... once you've dug over the ground and are ready to break it down to a finer tilth prior to sowing—Amateur Gardening*, 1990; *As last season's mulch breaks down, it adds tilth to the soil—Harrowsmith*, 1993. Cf. -TH NOUNS for similar formations.

**timbre** /ˈtæmbə/ or, unacclimatized, /ˈtæbrə/. Adopted from Fr. in the mid-19c. in the sense 'the distinctive character or quality of a musical or vocal sound apart from its pitch and intensity', but still not fully Anglicized in pronunciation.

**time.** Under this word, as the most general term, it may be of use to some readers if I list some of its near-synonyms. Of the six following words each is given a simple definition (based partly on those in the *New SOED*) with a view merely to suggesting the natural relation between them. Though each is often used in senses here assigned not to it but to another (or not mentioned at all), the words *aeon, cycle, date, epoch, era*, and *period* form a broadly based series when they are used in their primary senses.

An *aeon* is an immeasurable period of time.

A *cycle* is a period in which a connected series of events or phenomena is completed, usu. as part of a repeating succession of similar periods.

A *date* is the identifiable or stated point of time (day, month, year, etc.) at which something occurs, has occurred, or will occur.

An *epoch* is the beginning of a distinctive period in the history of something or someone.

An *era* is a period of history characterized by a particular state of affairs, series of events, etc.

A *period* is a portion of time characterized by the same prevalent features or conditions.

A *time* and an *age* are often interchangeable with all or most of the above, but are less precise in meaning. Cf. also the terms *duration, juncture* (s v. JUNCTION), *moment, occasion, season, span, spell*, and *term*, all of which, in some of their uses, refer to divisions of time. Note that *in time* means, among other things, 'not (too) late', whereas *on time* means 'punctually, on the dot'.

**timeous** /ˈtaɪməs/ (adj.), in good time (Sc.). This is the better spelling, rather than *timous*.

**times.** Preceded by a cardinal numeral and followed by a number or expression of quantity (*6 times 4 = 24; an animal of ten times my strength*), *times* has been used as an agent of multiplication since at least the 14c. and is still, of course, in current use. Followed by an adj. or adv. in the comparative degree it is for the most part used only in an additive way (*a cathedral is usually at least five times larger than any parish church*). Though the OED gives a 19c. example of the type *ten* *times less* (*Men who had ten or twenty times less to remember*—Gladstone, 1876) and the type occurs from time to time in the 20c., it is better to assume that *times* normally implies multiplication, and to restrict its use to contexts in which multiplication is plainly intended.

**timpani** (pl. noun), a set of kettledrums. Thus spelt, rather than *tympani*. See TYM-PANUM.

**tin.** This ordinary word in BrE for 'an airtight sealed container made of tin plate or aluminium for preserving food' stands beside *can*, esp. when the container holds a drink. It is impossible to establish the currency of the two words except in an impressionistic way, but my own choice is, for example, *a tin of sardines, a tin of peaches, a can of beer; tinned fruit; a canning factory.*

**tinge** (verb). The ppl form *tingeing* is recommended (rather than *tinging*, which might be thought to represent the pronunciation /ˈtɪŋɪŋ/. Cf. HINGE (verb).

**tinker** (verb). The *OED* entry (sense 1(b)) shows that figurative uses of the verb ( = 'to occupy oneself about something in a desultory or aimless way') were normally construed with *at* in the 19c., but in the 20c. with *with*. Examples: *The public were tired of government which merely tinkered at legislation* (1880); *Nobody is prepared to tinker with a social structure that has withstood every kind of outside pressure*—Times, 1955; *Whatever moral doubts there may be about tinkering with nature, the biotechnology revolution will not be stopped in its tracks*—Oxfam News, 1990; *Some of the alterations suggest little more than ideological tinkering with the wording*—TES, 1991.

**Tinsel Town, Tinseltown.** See SOBRI-QUETS.

**-tion.** Fowler (1926) rightly ridiculed the excessive use of abstract nouns ending in *-tion*, and cited an unattributed but presumably journalistic example: *Speculation on the subject of the constitution of the British representation at the Washington inauguration of the League of Nations will, presumably, be satisfied when Parliament meets, but there is a certain nervousness at the suggestion that Mr Lloyd George will go over there as chief of the British delegation.* Anyone with half an ear would find such a succession of *-tion* words objectionable. But Fowler's general point is a sound one.

Two of the most overused words in *-tion* in the 20c. are *position* and *situation*. Examples (from the 2nd edn, 1965): *The situation in the industry has reached a tragic position; They based this opinion largely on the position of the company's financial situation; The Trades Union Congress should call a halt to the situation; We ought to be told the present position on this matter; The position in regard to unemployment has deteriorated; At the moment the political situation in Malta is in a strange position.* The sentences call out for recasting (e.g. *The industry is in a sorry state*, etc.). In fact relentless attacks on such turgid sentences seem to have had a beneficial effect. I quote from *The Times Guide to English Style and Usage* (1992):

> *position* like situation, an empty, overworked word as in "the unemployment position is unchanged". Try to rephrase such sentences. However, when preceded by an adjective ("a strong position") or a participial form ("a bargaining position"), it is acceptable. For "in a position to", "able to" or "can" are preferable.
>
> *situation* avoid. Even a quote using this vacuous word is probably waffle and not worth inclusion. The word situation cannot be rescued by being propped up by an adjective, as in "classroom situation", let alone the banned "crisis situation". As for ongoing situation, no-win situation and chicken-and-egg situation ...!

Doubtless recruits on other newspapers are given the same or similar advice.

**tip** (noun). See RUBBISH.

**tipstaff.** Pl. *tipstaffs* (not *tipstaves*).

**tiptoe** (verb). The inflected forms are *tiptoes, tiptoed, tiptoeing.*

**tirade.** Usu. pronounced /taɪˈreɪd/ in BrE, but commonly /ˈtaɪr-/ (i.e. stressed on first syllable) in AmE. Variants with a short /ɪ/ in the first syllable are also used by some standard speakers in both countries.

**tire** (of a wheel). See TYRE.

**tiro** (a novice). Thus spelt, rather than *tyro: see* Y AND I. Pl. *tiros: see* -O(E)S 6.

(American dictionaries give precedence to *tyro*.)

**tissue.** The pronunciation /'tɪʃuː/ is recommended, rather than /'tɪsjuː/.

**titbit, tidbit.** The first is the customary spelling in Britain, and the second in N. America. The word entered the language in the 17c., chiefly in the forms *tyd bit*, *tid-bit* (from Eng. dial. *tid* adj. tender, nice, special).

**titillate** means 'to excite pleasantly (*it titillated her fancy*)', whereas *titivate* means 'to adorn, smarten (*she titivated herself up for the party*)'.

**titles.** Some of the ground has already been covered, e.g. in the article NAMES AND APPELLATIONS, in other articles cross-referred to therein, and in THE 1. The complex difficulties of the correct designations of members of the nobility, bishops and archbishops, lord mayors, lady mayoresses, the Chief Rabbi, the Pope, ambassadors, and so on can only be safely resolved by consulting the current editions of *Debrett's Correct Form, Who's Who* (of the country concerned), *The Dictionary of National Biography*, and so on. Further detailed guidance is that provided to their journalists in *The Times Guide to English Style and Usage* (1992). Fowler (1926) wrote a 21-line piece deploring the emergence 'in the last twenty or thirty years' of the prefixes *Marquis, Earl*, and *Viscount*, and also of *Marchioness* and *Countess* in place of *Lady* ('that used to be good enough for ordinary wear'), and by this means neatly sidestepped most of the central issues. The holders of titles know how they should be addressed, on the envelope, at the head of a letter, and face to face. They are doubtless amused (or irritated) by some of the forms adopted by people who are unaware of the correct procedures. The best advice I can offer is to urge readers to consult the relevant guide of those listed above before addressing a person of rank.

**tmesis.** Pl. *tmeses* /'tmiːsiːz/. (grammar) The separation of parts of a compound word by an intervening word or words, as when *toward us* is expressed as *to usward*. The process is seen clearly in the colloquial types *hoo-bloody-ray, im-bloody-possible*, etc. The word is derived from Gk τμῆσις 'a cutting'.

**to.** The uses of the preposition *to* and of the infinitival particle *to* are discussed under several headwords in this book. For the erroneous types *she gave it to Elizabeth and I* (correctly *me*) and *to we* (correctly *us*) *lexicographers*, see CASES 6 and I 1. For the type *to fervently believe*, see SPLIT INFINITIVE. For the type in which *to* is placed in final position (e.g. *I learned what to put in my mouth and what not to*), see INFINITIVE 3. For the circumstances in which it is used between the verbs *dare, help*, and *need* and a following infinitive, see these verbs. A new use of the preposition *to*, replacing traditional *of*, now emerging in AmE, might be called the relational *to*: e.g. *interviewing a woman who neighbors said they believed was a girlfriend to the fired employee—Chicago Tribune*, 1987; *He's married and the father to a son—*ibid., 1989.

Finally, the preposition *to* used to mean 'at' is recorded in the *OED* from OE times onwards but is now only dialectal in Britain and only in non-standard use in AmE. Examples: *On Cantwarabyrig VII myneteres . . . to Hrofceastre—*a 10c. charter; *I haue heard say there is to Mountferrat . . . a deuoute & holy place—*a romance of c1500; *Lucy Passmore, the white witch to Welcombe—*C. Kingsley, 1855; *In Somerset . . . it is correct to say 'I bought this to Taunton'—*R. Jefferies, 1889; *We could do something in the afternoon. Were you ever to the Botanic Gardens?—New Yorker*, 1977.

**tobacco.** Pl. *tobaccos*. See -O(E)S 3.

**tobacconist.** See note s.v. -IST, -ALIST.

**toboggan** makes *tobogganer, tobogganing, tobogganist*. See -N-, -NN-.

**today, tomorrow, tonight.** Forms with a hyphen (*to-day*, etc.) were listed as alternatives in editions of the *COD* down to and including the 7th edn of 1982. They were dropped in the 8th edn of 1990. The lingering of the hyphen in these words in much printed work of the 20c. is a very singular piece of conservatism, but now it has virtually disappeared from sight in all three words.

**together.** See ALL TOGETHER, ALTOGETHER.

**together with.** See AGREEMENT 4.

**toilet.** *At least the children are told how to hold their knives properly and get walloped if they call the lavatory the toilet.* This example from Joanna Trollope's novel *A Passionate Man* (1990) neatly introduces the sociological problem about what was once called the *water closet*. There are many synonyms in all the main English-speaking countries, and the choice of word normally depends on what is judged to be the sociological level, age, etc., of the person addressed. A wide range of terms forms part of the passive vocabulary of most people, but not all of them are used in practice. On the evidence available to me, *toilet* is the first choice of the majority of people in Britain, while *lavatory* (now fading) and *loo* (especially) are the favourites of the chattering classes. There are many polite evasions (*the geography of the house, the facilities,* etc.), numerous dialect words (e.g. *netty* in the North Country), and many slang words (e.g. *the bog, karzy*). A public lavatory is often called a *public convenience* in BrE, with separate sections labelled *Ladies* and *Gents* or with signs indicating which is which.

In AmE the dominant word (in houses, public places, on aeroplanes, on long-distance buses, etc.) seems to be *restroom*, and, when the facilities are divided, *men's room* and *ladies room. Bathroom* and *washroom* are also common. *Toilet* is known but hardly ever occurs in print except when used attributively (*toilet attendant, toilet seat*) and in the phr. *public toilets*. There are also numerous slang words (*can, john,* etc.), and euphemisms (*comfort station, powder room,* etc.). Not uncommonly several of these synonyms are used in the text of the same newspaper article: e.g. (a 150-word report of the arrest of a man charged with smoking marijuana in the lavatory of a Continental Airlines flight from Chicago to Newark, NJ, in 1988) *bathroom, restroom, john,* and *lavatory. John* was the word used by the accused. *Water closet, WC,* and *privy* are known in both countries but are now used less and less. *Latrine* is a communal lavatory in camping sites, military barracks, etc.

**token.** The idiomatic phr. *by the same token* has been a convenient connective since the 15c. (Paston Letters) meaning (a) on the same ground; for the same reason; in the same way; and (b) ( = Fr. *à telles enseignes que*), the proof of this being that: introducing, as the OED expresses it, a corroborating circumstance, often weakened down to a mere associated fact that helps the memory or is recalled to mind by the main fact. It is important that there should be a clear connecting thread of meaning between the original statement and the consequent assumption. Some modern examples: *By the same token oil is an upstream resource in many Arab countries and we do not tell them how to use it—South,* 1991; *The writers might never quite get at what they believed was Giacometti's essence, but by the same token Giacometti was not going to go off on some completely different tangent and thereby get away from the writers—Mod. Painters,* 1992; *The MRC* [Medical Research Council] *looks out of step, but by the same token, the trial investigators are going to have to produce squeaky-clean results—New Scientist,* 1992.

**toll.** See TAX.

**tomato.** Pl. *tomatoes.* See -O(E)S 1.

**tome.** A large heavy book, not just a book of any size.

**Tommy (Atkins).** See SOBRIQUETS.

**tomorrow.** See TODAY.

***ton,*** prevailing mode or fashion. Though used in English since the 18c. it is still not fully naturalized: print in italics and pronounce /tɔ̃/.

**ton** (weight) and **tun** (the cask, vat, and wine-measure) are both pronounced /tʌn/. They were originally the same word (from OE *tunne,* from medL *tunna*), but became differentiated in the 17c.

**tondo** (a circular painting or relief). Pl. *tondi* /'tɒndi/.

**tonight.** See TODAY.

**tonne** (a metric ton equal to 1,000 kg.). In both BrE and AmE pronounced /tʌn/, i.e. the same as the ordinary word *ton* ( = 2,249 lb. avoirdupois).

**tonneau.** Pl. *tonneaus,* with the final letter pronounced /z/. See -X.

**tonsil.** The derivatives are spelt *tonsillectomy, tonsillitis, tonsillotomy.* See -LL-, -L-.

**tonsorial.** Used, only with facetious humour, of a hairdresser or hairdressing (cf. L *tonsor* 'a barber').

**too.**

| 1 With passive participle.
| 2 Illogical or ambiguous uses.
| 3 At the beginning of a clause.
| 4 As an intensifier.

**1 With passive participle.** With passive participles, when they are genuinely such and not adjectives ending in *-ed* such as *limited* or *tired*, the adverb *too* used by itself is slightly unidiomatic: *Swindon is too occupied with existing problems to be willing to take on any fresh adventurous projects* (read *too fully*); *But he was too engrossed in Northern Europe to realize his failure elsewhere* (read *too greatly* or *too fully*).

**2 Illogical or ambiguous uses.** These result from confusing two logical ways of making a statement, one with and one without *too*. *Praise which perhaps was scarcely meant to be taken too literally* means either 'which may easily be taken too literally' or 'which was not meant to be taken literally'. *We need not attach too much importance to the differences between Labour and Social Democrats* means either 'We may easily attach too much' or 'We need not attach much'. *It is yet far too early to generalize too widely as to origins and influences* means either 'If we generalize too early we may generalize too widely' or 'It is too early to generalize widely'.

**3 At the beginning of a clause.** Placed at the beginning of a clause and meaning 'in addition; moreover, furthermore', *too* had a long and untroubled history from OE until the 17c. (e.g. *Too, we profess our selves the Redeemed of the Lord*—J. Shute, 1641), at which point it became rare or obsolete, only to be revived in the 20c., chiefly in AmE. Examples: *She's had her novel published this year; but too, she's written some interesting articles on acupuncture*—cited in *CGEL*, 1985; *she was a veteran of the Friday shooting incident and, too, she had Fleurier, the patron, on her side*—T. Winton, 1985 (Aust.); *He told me a man needed sex. But I didn't want to. He got like that. But, too, he was a charmer*—New Yorker, 1988.

**4** As an intensifier, *too* has many standard (though informal) uses: e.g. *It's too good of you* ( = very good, extremely

good); *I'm not too sure that you're right* ( = not at all sure); *'This boy is not an athletic type.' 'Too true!'*. Note, too, the mainly AmE type *'I don't know what you're talking about.' 'You do too.'* ( = you certainly do).

**topmost.** See -MOST.

**tormentor.** Thus spelt, not *-er*.

**tornado.** Pl. *tornadoes*. The pl. of the fighter plane of that name is *Tornados*. See -O(E)S 1, 5.

**torpedo.** Pl. *torpedoes*. See -O(E)S 1. The inflected forms of the verb are *torpedoes, torpedoed, torpedoing.*

**torpor.** Thus spelt. See -OUR AND -OR.

**torso.** Pl. *torsos*. See -O(E)S 6.

**tortoise.** Pronounce /ˈtɔːtəs/. The spelling-pronunciation with final /-ɔɪz/ is nonstandard.

**tortuous, torturous.** The first means 'full of twists, turns, or bends' (*a tortuous route with many hairpin bends*) or '(fig.) indirect, circuitous, devious' (*a tortuous argument, mind, policy, etc.*). *Torturous,* a fairly rare word, is the adj. corresponding to *torture* and means 'involving or causing torture' (*a torturous execution, hours of torturous anguish*). The two sets of meanings almost come together in the following extract: *There was some debate about the pictures used to accompany this article, but it was nothing compared to, say, the torturous negotiations that preceded the Kylie Minogue story in the June issue of this magazine*—The Face, 1994. But I feel sure that *tortuous* fits the context, as indeed it (or *tortuously*) does in each of the following examples: *Sam did as he was instructed as Deborah torturously rolled her way back towards him*—B. Elton, 1991; *Turow spoke last night ... giving a detailed synopsis of the 'torturous path' that led him to be a practicing lawyer and author*—Daily Northwestern (Chicago), 1992.

**torus.** Pl. *toruses* or, less commonly, *tori* /ˈtɔːraɪ/. See -US 1.

**tost.** An archaic literary form of the pa.t. and pa.pple of *toss* (verb): *Wretch that long has tost On the thorny bed of Pain* (Gray); *these stormtost seas* (Carlyle). See -T AND -ED.

**total.** The adj. makes *totally* (adv.), *totality* (noun), *totalizator* (noun). The inflected forms of the verb are *totalled*, *totalling* (but usu. *totaled*, *totaling* in AmE): see -LL-, -L-.

**tother.** Formed in ME as *þe toþer* by a misdivision of earlier *þat oþer* 'the other', *tother* (also *'tother, the tother*) has stubbornly remained in the standard language as a colloquial (or sometimes jocular) variant of *the other* until the present day. The phr. *tell tother from which* 'to tell one from another' also surfaces from time to time.

**touchy.** See TETCHY.

**toupee** (wig) /'tuːpeɪ/. Thus spelt, not *toupée, toupet*.

**tourniquet.** Pronounce /'tʊənɪkeɪ/, not /-ket/.

**tow- and towing-.** The dominant forms in the more important compounds now are *tow-bar, tow-line, tow-net, tow-path,* and *tow-rope*, rather than *towing-bar*, etc. All five compounds are often written without a hyphen (thus *towbar*, etc.).

**toward, towards.** 1 As prepositions, *toward* is the more usual form in AmE, and *towards* the almost invariable form in BrE. But the distribution of the variants is subject to much variation. Routine examples: (UK) *Madeleine … led her guests towards Edmond*—P. P. Read, 1986; *Miles Harrier was making his way … towards their rendezvous*—M. Bracewell, 1989. (US) *'I'd hate to see you be so unforgiving toward me,' she said*—New Yorker, 1989; *There is a new policy toward risk that is at least as important to American society as tort*—Dædalus, 1980.

2 *Towards* is normally pronounced /tə'wɔːdz/ or /twɔːdz/ in BrE. The variant /tɔːdz/ also occurs in rapid speech. There is a similar wide spread of pronunciations in AmE.

**towel** (verb). The inflected forms are *towelled, towelling,* but usu. *toweled, toweling* in AmE. See -LL-, -L-.

**trachea.** In BrE usually pronounced /trə'kiːə/, but sometimes /'treɪkɪə/. Pl. *tracheae* /-iːiː/. In AmE, usu. /'reɪk-/. In the derivatives *tracheostomy* and *tracheotomy*, the first vowel is usu. /æ/ in BrE and /eɪ/ in AmE.

**trade marks.** Reference to proprietary terms in print—by journalists, novelists, or lexicographers—can lead to protests from the owners of the trade marks. They are concerned to defend their rights to the word registered as a trade mark; its use with a lower-case initial suggests that it is the generic name of the item. Every vacuum cleaner is not a *Hoover*, nor ball-point pen a *Biro*, not all sticky tape is *Sellotape*, nor jeans all *Levi's*, and if this is implied the trademark holder is in danger of losing the value, as measured in sales, of his trademarked product. Proprietary terms are of more than usual concern to lexicographers, not only because they are often the subject of protracted and complicated correspondence. Besides this, dictionary entries are consulted by the Registrar of Trade Marks at the Patent Office in London, and by his analogues in other countries, and the registration of such terms is sometimes delayed or brought into question because, among other factors, a given dictionary shows a term without indication of its proprietary status.

The preferred expression in the *OED* for the names of these marks is 'proprietary term'. It is used as a synonym of 'trade name', 'trade mark', and 'trade term', notwithstanding the fact that the four expressions are not synonymous in legal and business language. Proprietary terms should properly be entered in dictionaries with a capital letter in the lemma (e.g. *Bovril*). If in literary works or other sources systematically read for the *OED* such names are used with a lower-case initial letter, the entry also contains the uncapitalized form (e.g. *Bovril … Also bovril*). It should not be assumed that such a term, once registered, remains a trade mark indefinitely. *Cellophane, jeep,* and *linoleum* are examples of words which no longer have proprietary status. In 1962 *thermos* ceased to be a trade mark in the US, and Canada followed suit in 1967, but it remains a trade mark in Great Britain. An historical dictionary, even on the scale of the *OED*, cannot find room for more than a sprinkling of the many thousands of trade names on the current register, but I would argue for the inclusion of as many as possible of the commonest ones so that, for example, readers of C. Day

Lewis's *The Otterbury Incident* or Len Deighton's *Horse under Water* in future centuries should not be puzzled by references to *Mars bars*.

It is worth mentioning that the generic use of a trade mark in a dictionary is probably not in itself actionable in Britain at present, though the position may change at any time. Furthermore, membership of the EC might at some point complicate matters still more, since EC law in respect of the treatment of trade marks in dictionaries is unlikely to coincide with such laws in Britain at present. These are matters for lawyers.

**trade union** (not *trades union*). Pl. *trade unions*. But *Trades Union Congress*, abbrev. *TUC*.

**trade wind.** Write as two words.

**tradition(al)ism, -ist.** For the general question between such variants, see -IST. In this case the longer forms are usual, perhaps because the words are often opposed to *rationalism*, *-ist*, the form of which is fixed by *ration*'s not having the necessary meaning.

**traffic** (verb). The inflected forms are *trafficked*, *trafficking*; also *trafficker* (noun). See -C-, -CK-.

**tragedienne, tragédienne.** See COMEDIAN 3.

**tragedy.** Used in English since the Middle Ages of a play or other literary work, ancient or modern, in which the principal figure or figures fall from grace or are killed, *tragedy* soon became applied to any dreadful calamity or disaster in real life. In modern times it is often used hyperbolically in contexts where no deaths are involved, e.g. of an unwanted loss in a significant football or other match. For these lesser setbacks other terms are available: *tragedy* should be restricted to the emotional sports pages of the tabloid press.

**tragic, tragical.** The normal and natural word is *tragic*. *Tragical* is scarcely used at all except with direct or implied reference to the *very tragicall mirth* of Pyramus and Thisbe in *A Midsummer Night's Dream*.

**traipse** (verb), to tramp or trudge wearily. It emerged in the 16c. and is of unknown origin. The favoured spelling used to be *trapes*, but no longer. The word is apparently in widespread dialectal use, pronounced as two syllables, e.g. /ˈtreɪpəs/.

**trait.** Normally with the final *t* silent in BrE but sometimes /treɪt/. The final *t* is regularly sounded in AmE.

**trammel** (verb). The inflected forms are *trammelled, trammelling* in BrE, but usu. *trammeled, trammeling* in AmE. See -LL-, -L-.

**tranquil** (adj.). The derivatives have -*ll*- in BrE: *tranquillity, tranquillize, tranquilizer, tranquilly* adv. But AmE usu. has -*l*- in the first three of these. See -LL-, -L-.

**transcendence, transcendency.** The two words are usu. interchangeable, except in the sense 'of the deity: the attribute of being above and independent of the universe', for which *transcendence* is the usual term. See -CE, -CY.

**transcendent(al).** These words, with their many specialized applications in philosophy, are for the most part beyond the scope of this book; but there are popular uses in which the right form should be chosen. **1** The word that means surpassing, of supreme excellence or greatness, etc., is *transcendent*, and the following is wrong: *The matter is of transcendental importance, especially in the present disastrous state of the world.* See LONG VARIANTS for similar pairs.

**2** The word applied to God in contrast with IMMANENT is *transcendent*.

**3** The word that means visionary, idealistic, outside experience, etc., is *transcendental*.

**4** The word applied to Emerson and his 'religio-philosophical teaching' is *transcendental*. Note also the 20c. cult of *Transcendental Meditation*, a method of detaching oneself from problems, anxiety, etc., by silent meditation and repetition of a mantra (COD).

**transexual.** See TRANSSEXUAL.

**transfer.** **1** Noun *tránsfer*, verb *transfér*: see NOUN AND VERB ACCENT.

**2** The inflected forms of the verb are *transferred, transferring*. Derivatives include *transferrer*, but *transferable* (see

-ABLE, -IBLE 2), *transference, transferee,* and *transferor.* See -R-, -RR-. Of *transferrer* and *transferor,* the first is the general agent-noun, a person or mechanism that passes something on, and the second a legal term for the person who conveys his property to another, the *transferee.*

**transfixion,** the state of being transfixed. Thus spelt, not *transfiction.* See -XION etc.

**transgressor.** Thus spelt, not *-er.*

**tranship, transhipment.** Acceptable, but not recommended, variants of TRANSSHIP, TRANSSHIPMENT.

**transient, transitory.** The primary meanings (brief, fleeting) are the same, but *transient* is used with special senses in music, philosophy, electricity, and nuclear physics, and *transitory* in law.

**transistor.** Thus spelt, not *-er.*

**transitive verb.** See INTRANSITIVE (adj.); DITRANSITIVE.

**translator.** Thus spelt, not *-er.*

**transliterate,** to replace the letters or characters of a word or words from one language into those of another. Thus Gk πτερόν 'wing' is transliterated into roman letters as *pterón* (cf. *pterodactyl*).

**translucent.** See TRANSPARENT.

**transmit** (verb). The inflected forms are *transmitted, transmitting,* and the agent-noun *transmitter.* Also *transmittal* (noun). See -T-, -TT-. Related forms are *transmissible* and *transmittable:* see -ABLE, -IBLE.

**transmogrify.** See FACETIOUS FORMATIONS.

**transparence, transparency.** The words are interchangeable when the meaning is the quality or state of being transparent, but the form in *-ency* is much the more usual of the two in this sense. See -CE AND -CY. In the photographic sense ( = a slide), and in several other specialized senses, *transparency* is the only possible form.

**transparent,** and the near-synonyms *diaphanous, pellucid, translucent. Transparent* is the general word for describing what is penetrable by sight (lit. or fig.)

or by light, and it can replace any of the others unless there is some point of precision or of rhetoric to be gained. The three near-synonyms have the rhetorical value of being less common than *transparent,* and therefore appear more often in literary work. As regards precision, the following definitions of the words' narrower senses are offered, and to each are appended some appropriate nouns, and the adjective or participle that seems most directly opposed. The only one of the four capable of bearing the figurative sense 'obvious' is *transparent.*

That is *diaphanous* which is so insubstantial, thin, or gossamer-like that it does not preclude sight of what is behind it; *fabric, gown, veil, mist;* opp. *shrouding.* That is *pellucid* which does not distort images seen through it; *water, literary style;* opp. *turbid.* That is *translucent* which does not bar the passage of light but diffuses it; *quartz, tortoise-shell;* opp. *opaque.* That is *transparent* that does not obscure sight of what is behind it and transmits light without diffusion; *glass, candour, anger;* opp. *obscure, unclear.*

**transpire.** Let us first clear the ground by recalling its etymology and its sense-development in English. It was brought into English from French in the 16c. in the technical sense 'to emit as vapour'. The French word was derived from modL *tran(s)spīrāre,* from L *trans-* 'across' + *spīrāre* 'to breathe'. The English word was immediately applied to the process by which plants exhale watery vapour from the surface of the leaves; and, in the 17c., came also to mean '(of a liquid) to escape by evaporation; (of moisture) to give off through the skin (of an animal or human being)'. From these meanings it was just a short step (in the 18c.) to the figurative sense 'to become known by degrees, to leak out'. Dr Johnson did not like the new use: he defined it as 'to escape from secrecy to notice' and commented 'a sense lately innovated from France, without necessity'. This figurative sense of *transpire* has remained in the language since the 17c. though it is often said to be 'formal'. Examples: *The conditions of the contract were not allowed to transpire*—J. A. Froude, 1856; *But then, to our surprise, it transpires that he doesn't think much of our critics, either*—D. J. Enright, 1966; *Yaddo, it transpired, had been* *under FBI surveillance for some time*—I. Hamilton, 1982; *Unfortunately he died soon afterwards . . . Though the blow was softened when it transpired that he'd left her several million francs*—S. Faulks, 1989.

Meanwhile in the late 18c., first in America, *transpire* began to be used in the sense 'to occur, happen, take place': e.g. *There is nothing new transpired since I wrote you last*—A. Adams, 1775; *Transpire . . . 3. To happen or come to pass.*—*Webster's Dict.*, 1828; *Few changes—hardly any—have transpired among his ship's company*—Dickens, 1848. The use was condemned in vitriolic terms by the American writer Richard Grant White in his book *Words and their Uses* (1870). And he concluded: 'There is a very simple test of the correct use of *transpire*. If the phrase *take place* can be substituted for it, and the intended meaning of the sentence is preserved, its use is unquestionably wrong; if the other colloquial phrase, *leak out*, can be put in its place, its use is correct.' The *OED* (1914) firmly marked the sense 'to occur, happen, take place' as catachrestic, while presenting a string of 'misused' examples. It said that the misuse evidently arose 'from misunderstanding such a sentence as "What had transpired during his absence he did not know"'. Fowler (1926) spoke of 'the notorious misuse of this word . . . in making it mean happen or turn out or go on'. And so we have another example of a long drawn-out battle between competing senses, one seen as logical and etymologically sound, and the other has a 'monstrous perversion' (Richard Grant White, 1870) and a 'notorious misuse' (Fowler, 1926). It is still described as a disputed use by the *COD* (1995).

*Verdict.* If you are tempted to use *transpire* to mean 'to happen, occur' it will be greeted by many people with varying degrees of disapproval and by others with equanimity. The safest course is to leave the word to botanists, biologists, and physicists in their learned journals, or at most to restrict it to contexts in which it means 'to emerge, become clear'.

As a matter of interest, the modern French word for 'perspire' is *transpirer*. And this French verb is also used to mean '(*secret, projet, détails*) come to light, leak out'. The phr. *rien n'a transpiré* means

'nothing came to light, nothing leaked out *ou* transpired'.

**transsexual.** The better spelling, rather than *transexual*, despite the old models of *transcribe* (from L *transcrībere*; cf. Eng. *scribe*), *transept* (from L *trans-* + *septum*), etc.

**transship, transshipment.** The better forms, rather than those with a single medial *-s-*.

**trapes.** See TRAIPSE.

**trapezium.** Pl. either *trapezia* or *trapeziums*. See *-UM* 3. As a trap for the unwary, it means a quadrilateral with only one pair of sides parallel in BrE, while in AmE it means a quadrilateral with no two sides parallel. On the other hand, a *trapezoid* means a quadrilateral with no two sides parallel in BrE, while in AmE it means a quadrilateral with only one pair of sides parallel. Vigilance is called for.

**trash.** See RUBBISH.

**trauma.** The pronunciation recommended is /ˈtrɔːmə/, rather than /ˈtraʊmə/; but /ˈtraʊmə/ is more usual in AmE. Pl. (in medical use) *traumata*, (in lay use) *traumas*.

**travail, travel.** 1 The first is pronounced /ˈtræveɪl/ and the second /ˈtræv(ə)l/.

2 Used as a verb, *travail* has the inflected forms *travailed, travailing* in both BrE and AmE. By contrast, *travel* makes *travelled, travelling* (and also *traveller*) in BrE, but usually has forms with a single *-l-* in AmE.

**treachery** is a general term meaning a violation of faith or trust, a betrayal; a person who betrays something entrusted to him or her is *treacherous*. *Treason* is specifically violation by a subject of allegiance to the sovereign (or in countries without a monarch to the State); the corresponding adj. is *treasonable*, and the agent-noun is *traitor*.

**treble.** See TRIPLE.

**trecento** /treɪˈtʃentəʊ/. (Also with a capital initial.) The style of Italian art and literature in the 1300s (1300–99), i.e. what we call in English 'the fourteenth century'. Similarly *quattrocento* (the 1400s, i.e. our fifteenth century), *cinquecento* (the 1500s, i.e. our sixteenth century), and so on down to *novecento*, the style of Italian art and literature in the twentieth century (1900–1999). In Italian a capital initial is obligatory in all such uses, and this practice is sometimes followed in English-language studies of Italian art and literature of these periods. In all general contexts Italians follow the same rules as those in English: *il quattordicesimo secolo, il quindicesimo secolo, ... il ventesimo secolo*, the fourteenth century, the fifteenth century, ... the twentieth century.

**trefoil.** The *COD* (1995) and the *New SOED* (1993) give preference to the pronunciation with a short *e*, namely /ˈtrefɔɪl/, rather than /ˈtriːfɔɪl/. The standard dictionaries of AmE give their preferences in the opposite order.

**trek** (verb). **1** The inflected forms are *trekked, trekking*. The agent-noun is *trekker*.

**2** The verb normally retains an element of arduousness in its uses (*the boy had to trek two miles to school each day*), as a kind of folk-memory of the 19c. SAfr. use of the word to mean 'to migrate or make a long journey with one's belongings by ox-wagon'. It is not merely a synonym of *go, walk*. Similarly with the noun *trek*.

**trellis** (verb). The inflected forms are *trellised, trellising*. See -S-, -SS-.

**tremor.** Thus spelt. See -OUR AND -OR.

**trepan** /trɪˈpæn/ (verb). The inflected forms are *trepanned, trepanning*. See -N-, -NN-.

**trephine** /trɪˈfaɪn/ or /-ˈfiːn/ (noun). An improved form of the original instrument, the *trepan*. The word *trepan* is first recorded c1400, and *trephine* in 1628. *Trephine* was originally *trafine*, and, according to its inventor, John Woodall, was formed from L *três fínês* 'three ends'.

**tribunal.** Pronounce /traɪˈbjuːn(ə)l/ or /trɪ-/. The *i* is short in Latin; but see FALSE QUANTITY.

**tribute.** See TAX.

**tricentenary.** See CENTENARY 2.

**triceps.** Pl. the same.

**tricksy, tricky** (adjs.). Differentiation (see DIFFERENTIATION 2) is proceeding, in the direction of restricting *tricksy*, now much less used, to contexts in which the quality is regarded not with condemnation or dislike or apprehension ( = dishonest, cunning, difficult, etc.) but with amusement or interest ( = playful, frolicsome, etc.). It had formerly, to judge from the *OED* record, all the meanings to itself, being more than two centuries older than *tricky*. At the same time the gap is being widened by the increasing use of *tricky* to describe a task needing adroitness and care.

**triclinium** (Roman antiquity). Pl. *triclinia*. See -UM 2.

**tricolour** (a flag of three colours, esp. the French national flag of blue, white, and red). Thus spelt in BrE and pronounced /ˈtrɪkələ/. In AmE usu. spelt *tricolor* and pronounced /ˈtraɪˌkʌlər/.

**triforium** (gallery or arcade in a church). Pl. *triforia*. See -UM 2.

**trigger** (verb). The inflected forms are *triggered, triggering*. See -R-, -RR-.

**trill** (phonetics). Esp., in certain regional varieties of BrE, pronunciation of *r* with vibration of the tongue. The term is also applied to the rolled *r* in Italian and the uvular *r* in German.

**trillion.** It normally means now a million million (1,000,000,000,000 or $10^{12}$) both in AmE and BrE. Formerly, in BrE, it meant a million million million (1,000,000,000,000,000,000 or $10^{18}$). See BILLION.

**trilogy.** In ancient Athens there were dramatic competitions at which each dramatist presented three plays, originally giving successive parts of the same legend; the extant *Agamemnon, Choephoroe*, and *Eumenides*, of Aeschylus formed a trilogy, and, with the addition of the lost *Proteus*, a TETRALOGY. Later trilogies were connected not necessarily by a common subject, but by being works of the same author, presented on the same occasion. In modern use the word is applied to a work such as Shakespeare's *Henry VI*, comprising three separate plays, or to a novel, etc., with two sequels.

**trimeter** (prosody). A line of three measures, esp. preceded by a defining word (*iambic trimeters, trochaic trimeters*, etc.). |*Híghèr*| *still ånd*| *híghèr*| is a trochaic trimeter.

**trio.** Pl. *trios*. See -O(E)S 4.

**triolet.** A poem of eight (usu. eight-syllabled) lines rhyming abaaabab, the first line recurring as the fourth and seventh and the second as the eighth. An example (entitled 'Triolet') from Robert Bridges:

*All women born are so perverse*
*No man need boast their love possessing.*
*If nought seem better, nothing's worse:*
*All women born are so perverse.*
*From Adam's wife, that proved a curse*
*Though God had made her for a blessing,*
*All women born are so perverse*
*No man need boast their love possessing.*

And another, more light-hearted example (G. K. Chesterton) with a slightly different rhyming scheme:

*I wish I were a jelly fish*
*That cannot fall downstairs:*
*Of all the things I wish to wish*
*I wish I were a jelly fish*
*That hasn't any cares,*
*And doesn't even have to wish*
*'I wish I were a jelly fish*
*That cannot fall downstairs'.*

**triplet** (prosody). Applied to the occasional use, in rhymed-couplet metres, of three lines instead of two to a rhyme; common among the heroic couplets of Dryden and some 18c. poets. An example from Dryden's *Religio Laici* (1682):

*Or various Atom's, interfering Dance*
*Leapt into Form (the Noble work of Chance,)*
*Or this great All was from Eternity;*
*Not ev'n the Stagirite himself could see;*
*And Epicurus Guess'd as well as He.*
*As blindly grop'd they for a future State,*
*As rashly Judg'd of Providence and Fate:*

Triplets occurring among heroic couplets are sometimes marked by a brace, as, for example, in Pope's *Essay on Criticism* (1711):

*Musick resembles Poetry, in each*  ⎫
*Are nameless Graces which no Methods teach.*  ⎬
*And which a Master-Hand alone can reach.*  ⎭

**triple, treble.** If the musical sense of *treble* is put aside, and also specialized senses in betting, darts, and baseball, the two words are as often as not interchangeable. But there are diverging tendencies. First, though either can be adj., verb, or noun, *treble* for the moment is the more usual verb and noun, and *triple* the more usual adj. Secondly, in the adjectival use *treble* is perhaps increasingly being used to refer to amount ( = three times as much), and *triple* to plurality (consisting of three usu. equal parts or things), though the distinction is far from absolute. Thirdly, some believe that in AmE, *triple* is the commoner of the two words in their general uses; others believe that the same tendency is occurring in BrE.

A few examples will serve to point up some of the differences between the two words. Others will underline their interchangeability. (treble) *treble agent* (espionage); *a treble brandy*; *Treble the money would not buy it now*; *It sells for treble the price of whale oil*; *a treble difficulty* ( = three times the difficulty); *Think of a number and treble it*; *The newspaper has trebled its circulation.* (triple) *triple agent* (espionage); *triple alliance*; *triple century* (cricket); *triple crown* (in sporting events); *a triple difficulty* ( = a difficulty of three kinds); *a triple-layer sponge cake*; *her triple role as headmistress, gym instructor, and music teacher*; *The firm's income tripled last year.*

**tripod.** Walker (1791) listed a form with a short *i* as a variant pronunciation, i.e. /'trɪpɒd/, but preferred the form with a long *i*, i.e. /'traɪ-/. Fowler (1926) thought that the form with a short *i* was 'now certainly often heard, and is now not unlikely to prevail'. Walker's preferred form has prevailed, not Fowler's, and /'traɪpɒd/ is now the only standard form. Cf. *tripodal* adj. /'trɪpədəl/.

**triptych.** Pronounce with final /-k/.

**triumphal, triumphant.** *Triumphal* means only 'in honour of a victory' and is properly used only of a celebration (*a triumphal march, parade, procession*) or of a monument erected to celebrate a victory (*triumphal arch*). *Triumphant* answers to triumph in any of its senses, especially those of brilliant success or exultation. The *OED* entries show that the two words have sometimes been used interchangeably in the past (e.g. *triumphal* = 'victorious' by Gavin Douglas in 1513; and *triumphant* = 'in honour of a victory' by several writers from 1531 to 1876), but such crossovers are now rare in standard sources.

**triumvir** /traɪˈʌmvə/ or /ˈtraɪəmvə/. Pl. either *triumvirs* or, less commonly, *triumviri* /-raɪ/. See LATIN PLURALS. But *triumvirate* (set of triumvirs) /traɪˈʌmvɪrət/ is now more usual than either plural.

**trivia** (things of little consequence). A 20c. word (first recorded in Logan Pearsall Smith's title *Trivia* in 1902), adopted from the modern Latin plural of *trivium* ( = 'a place where three ways meet' in classical Latin), it is now used both as a singular and a plural noun: *Besides, trivia has its importance too. Or to put it another way, trivia have their importance too—Sunday Times*, 26 Feb. 1978. In practice it is most frequently used as an uncountable noun: *Henry's face—calm, almost seraphic, earlier when we had exchanged greetings and talked trivia at lunch—*M. Egremont, 1986; *Picture to yourself a monstrous skip crammed with trivia: singularly ununique childhood memories,* [etc.]—Julian Barnes, 1991; *Listeners bored with election trivia might prefer to hear music while watching the swingometers with the sound turned down—Classic CD*, 1991. In such contexts the pertinent pronouns and verbs used with *trivia* do not reveal its number.

**-trix.** 1 For words in *-trix* that are not agent-nouns with a male correlative in *-tor*, see CICATRIX; MATRIX. Cf. also -EX, -IX.

2 Since the 15c. a number of words in *-trix* have been formed to signify the female equivalent of agent-nouns in *-tor*. The most important of these have been *administratrix* (first recorded in the 17c.), *aviatrix* (20c.), *directrix* (17c.), *dominatrix* (16c., then obs., revived in the 20c.), *editrix* (20c.), *executrix* (16c.), *heritrix/heretrix* (16c.), *inheritrix* (16c.), *mediatrix* (15c.), *narratrix* (19c.), *oratrix* (15c.), *prosecutrix* (18c.), *testatrix* (16c.). The *OED* record of these feminine derivatives is uneven, and it is not possible to make safe statements about (a) their currency at a given time, (b) the currency of notional or

actual rival forms in *-ess*, (c) the distribution down the centuries of the 'regular' plural forms in *-ices* or the Anglicized ones in *-ixes*. It seems clear, however, that, except in legal language, *-trix* forms are not greatly in favour at the present time. Of the above (leaving one or two of the legal terms aside) only *dominatrix*, *editrix*, and *executrix* have much currency, the first because of the 20c. fascination with sexual practices, and the other two drawing attention to the slight unexpectedness of a woman's holding a hitherto male-dominated post.

3 *Plurals.* The notionally regular plural form of each of these *-trix* words, or at any rate those formed between the 15c. and the 18c., is *-ices*, and such forms are indeed found (*executrices*, *heretrices*, *mediatrices*, *prosecutrices*, etc.) at various times; but there has been a marked tendency over the centuries to use the Anglicized forms in *-ixes* instead, and this tendency seems likely to continue in so far as there is any future at all for words in *-trix*. When the pl. forms in *-ices* occur the main stressing pattern of the equivalent male word is sometimes retained and sometimes not. The pronunciations given in the *New SOED* (1993) are as follows:

(stress unchanged) *executor* /ɪgˈzekjʊtə/, *executrix* /ɪgˈzekjʊtrɪks/, and *executrices* /ɪgˈzekjʊtrɪsiːz/.

(stress changed) *mediator* /ˈmiːdɪeɪtə/, *mediatrix* /miːdɪˈeɪtrɪks/, and *mediatrices* /miːdɪˈetrɪsiːz/.

(stress variable) *prosecutor* /ˈprɒsɪkjuːtə/, *prosecutrix* /ˈprɒsɪkjuːtrɪks/ or /-ˈkjuː-/, and *prosecutrices* /prɒsɪˈkjuːtrɪsiːz/.

No wonder such plural forms tend to be avoided.

**troche** (small medicated lozenge). This altered version of the long obsolete word *trochisk* (from Fr. *trochisque*, ultimately from Gk τροχίσκος 'small wheel') is pronounced /trəʊʃ/ in BrE and /ˈtrəʊkiː/ in AmE.

**trochee** (prosody) /ˈtrəʊkiː/. A foot consisting of one long or stressed syllable followed by one short or unstressed syllable, i.e. — ◡, as in *manner* or *body*.

**trolley.** Thus spelt (not *trolly*). Pl. *trolleys*.

**troop** (verb). *Trooping the colour* is the orthodox modern term for a ceremonial mounting of the guard, but in the early 19c. examples in the *OED* it was *colours* that were trooped.

**troop, troupe.** In BrE a *troop* is an artillery or armoured unit of soldiers, or a group of Scouts; a *troupe* is a company of actors or acrobats. A *trooper* is a private soldier in a cavalry or armoured unit (whence the phr. *to swear like a trooper*), and a *trouper* is a member of a theatrical company, or, by extension, a reliable, uncomplaining person. In other English-speaking countries the terms *troop* and *trooper* overlap to some extent with those in the UK, but have other applications as well.

**troublous** (adj.). An archaic or literary word for 'full of troubles; disturbed (*troublous times*)'. In most contexts, *troubled* or *troublesome* are the more appropriate words.

**trough.** The only standard pronunciation in the UK now is /trɒf/, though Daniel Jones (1917) recommended /trɔːf/, gave /trʌf/ as a former variant, and added a note saying (somewhat mysteriously) 'Bakers often say trɒʊ'.

**trousers.** Always construed as a plural noun (*his new trousers were given to him by his sister*); but, when used attributively and in compounds, the word loses its final -*s*: e.g. *trouser leg, trouser pockets, trouser suit*.

**trousseau.** The pl. form recommended is *trousseaus*, but *trousseaux* is also admissible. See -x. Both plural forms are pronounced with final /-z/.

**trout.** Pl. usu. the same. But *old trouts* for the affectionately disrespectful slang term.

**trow.** Markedly archaic, but if used it should be pronounced /trəʊ/ both as noun ( = belief) and verb ( = believe). The noun is now rarely used.

**truculence, truculency.** The first entered the language in the 18c. and the second in the 16c. They have been used as synonyms since the 18c. ( = the condition or quality of being truculent), but *truculence* is now the dominant form of the two. See -CE, -CY. For pronunciation, see next.

**truculent.** The *OED* (1915) gave precedence to /'truːk-/ in the first syllable, and also in *truculence* and *truculency*, but the only standard pronunciation now is /'trʌkjʊlənt/.

**true and false etymology.** The English language, perhaps more than any other, has from its earliest times onward drawn loanwords from many other languages—Latin, Greek, French, Italian, Dutch, and so on—and has usually respelt the adopted words so that they 'fitted in' with' English conceptions of what constitutes properly established native models of spelling, prefixation, suffixation, conjugation, and so on. Some of the incoming words have undergone changes to the point that they seem to the amateur to be directly related to similar-sounding words of native origin (see e.g. *belfry* s.v. ETYMOLOGY). Many words of native origin were respelt at a later stage because of all kinds of spurious associations with similar-sounding words: e.g., from the list below, *crayfish, forlorn hope, greyhound, island, shamefaced, slow-worm.* Readers who wish to avoid some of the more obvious misconceptions about the origins of some of the commonest words in the language may find the following list a useful starting-point. Really determined readers should pursue them in *The Oxford Dictionary of English Etymology* (1966) or in its Concise version (1986), or indeed in any standard dictionary of moderate size. The words given are mere signposts to a huge and complicated subject. The words in small capitals are the few that happen to be treated at their alphabetical places. See also the articles ETYMOLOGY and FOLK ETYMOLOGY.

AMUCK, not Eng. *muck*
*andiron* and GRIDIRON, only by sound-association with Eng. *iron* (and *fire-iron*)
*apparel*, not L *parō* 'provide oneself with'
ARBOUR, not L *arbor* 'tree'
*barberry*, not Eng. *berry*
*belfry*, not Eng. *bell*
*blindfold*, not Eng. *fold*
*bliss*, not Eng. *bless*
*bound* (*homeward*, etc.), not Eng. *bind*
*Boxing Day*, not pugilistic
*bridal*, not an adjective in -*al*
*bridegroom*, not Eng. *groom*
BRIER pipe, not Eng. *brier* (prickly bush)
*bum* (buttocks), not a contraction of Eng.
   *bottom*
*buttonhole* (verb), not *hole* but *hold*

*catgut*, not made from the intestines of a cat

*cinders*, not L *cineres* 'ashes'

*cockroach*, not Eng. *cock* or Eng. *roach*

COCOA, COCONUT, unconnected

COMITY, not L *comes* 'companion'

*convey*, not L *vehō* 'carry'

*cookie* (biscuit, etc.), not Eng. *cook*

COT(E), separate words

COURT-CARD, a corruption

CRAYFISH, not Eng. *fish*

CURTAIL, not Eng. *tail*

CUTLET, not Eng. *cut*

DEMEAN (conduct oneself), not Eng. *mean*

*dispatch*, not Fr. *dépêcher*

*egg on*, not *egg* but *edge*

EQUERRY, not L *equus* 'horse'

*errand*, not L *errō* 'wander'

*fall asleep*, not Eng. *fall* but OE *fēolan* 'penetrate into (sleep)'

FAROUCHE, not L *ferox* 'fierce'

FOREBEARS, = fore-beërs

FORLORN HOPE, not *forlorn* nor *hope*

FUSE (for igniting explosive), from L *fūsus* 'spindle'

GINGERLY, not Eng. *ginger*

GREYHOUND, not Eng. *grey*

*humble pie*, a pie made from the *umbles* (intestines of a deer)

*incentive*, not L *incendō* 'set on fire'

*ingenuity*, modern sense by confusion of *ingenious* with *ingenuous*

*island*, respelt by association with *isle*

*Jerusalem artichoke*, not *Jerusalem* but *girasole* (sunflower)

LITANY, LITURGY, first syllables unconnected

MOOD (grammar), = *mode*, not *mood* (temper)

*old dutch*, not *Dutch* but *duchess* (see NETHERLANDS)

*pen, pencil*, unconnected

PIDGIN, not Eng. *pigeon*

POSH, not port outward starboard home

PROTAGONIST, Gk πρῶτος 'first', not L *pro-* 'for'

*recover*, not Eng. *cover*

*river*, not L *rivus* 'river'

*run the gauntlet*, not *gauntlet* (glove) but 17c. *gantlope* (see GAUNTLET)

SANDBLIND, not Eng. *sand*

*scissors*, not L *scindō sciss-* 'cleave, tear apart'

*shamefaced*, not Eng. *-faced* but *-fast*

SLOW-WORM, not Eng. *slow*

*sorrow, sorry*, unconnected

*vile, villain*, unconnected

*walnut*, unconnected with Eng. *wall*

WATERSHED, neither a store of water nor a place that sheds water

*Welsh rabbit*, not *rare bit* nor *rarebit*

**truffle.** Both the OED (1915) and Daniel Jones (1917) allow the vaguely French-sounding variant /'trʊf(ə)l/ (the French

equivalent is actually *truffe*), but it is now always pronounced /'trʌf(ə)l/, at least in standard BrE. American dictionaries list /'truː-/ as a permissible variant.

**truly.** See LETTER FORMS.

**trumpet** (verb). The inflected forms are *trumpeted, trumpeting*. See -T-, -TT-.

**trunkful.** Pl. *trunkfuls*. See -FUL.

**trustee, trusty.** The first, which is stressed on the second syllable, is 'a person or member of a board given control or powers of administration of property in trust with a legal obligation to administer it solely for the purposes specified' (COD). The second, which is stressed on the first syllable, is an archaic adj. meaning 'trustworthy' (*a trusty steed*) or 'loyal (to a sovereign)' (*my trusty subjects*); and also a noun (with pl. *trusties*) meaning 'a prisoner who is given special privileges for good behaviour'.

**truth.** The recommended pronunciation of the pl. *truths* is with final /ðz/, but /θs/ is also standard. See -TH (θ) and -th (ð).

**try and, try to.** Arguments continue to rage about the validity of *try and* followed by an infinitive instead of *try to*, 'To be used only in informal contexts', 'grammatically wrong' are among the verdicts of some writers on English usage. Fowler's judgement in 1926 was much more lenient. After briefly setting out the facts he concluded: '*try and* is an idiom that should not be discountenanced, but used when it comes natural.' He also made out a sort of case for the semantic distinctiveness of *try and* constructions. In 1983 a Scandinavian scholar, Åge Lind, examined a group of fifty modern English novels of the period 1960–70 and found that *try to* was likely to occur in certain syntactic conditions, *try and* in others, and that in some circumstances the choice seemed not to be governed by any particular reason. 'If a subtle semantic distinction exists it does not seem to be observed,' he concluded.

Over the last several years I have gathered a wide range of evidence, with the following results. Standard examples of *try to* occurred in many types of constructions: (preceded by an auxiliary verb) *I think we should try to help him as a family—*

I. Murdoch, 1983; (preceded by the infinitive marker to) *To try to forget is to try to conceal*—T. S. Eliot, 1950; *Mr Stratton's moods would always be a mystery, so much so that he had ceased to try to fathom them*—P. Carey, 1988 (Aust.); (preceded by an adverb) *I always try to travel light*—R. Elms, 1988; (separated from the infinitive it governs) *He's gone his own way, I go mine, or try to*—K. Page, 1989. Parallel examples of *try* and for all but the last type are not difficult to find: (preceded by an auxiliary verb) *We must try and find him at once*—J. R. R. Tolkien, 1954; *I will try and answer any questions you may have*—S. Hockey, 1981 (University of Oxford lecture); (preceded by the infinitive marker to) *he used to try and draw Dr De Wet out*—M. du Plessis, 1983 (SAfr.); (preceded by an adverb) *Frankly, even to try and make somebody happy is a gross and farcical mistake*—Simon Mason, 1990.

Some other examples are less easy to classify: *he glanced at her face to try and see if she was mollified*—P. P. Read, 1986; *I try and work on the assumption that they're all as smart as I am*—Clive James, 1987 (Aust.); *Let me try and set down the opposing points of view*—Julian Barnes, 1991. It should be noted that some of the examples of *try* and are drawn from the informal atmosphere of a lecture room or (James, 1987) a newspaper interview, or from non-British sources. *Try and* can also occur idiomatically in the imperative in such sentences as *Don't try and frighten me*, which, as it happens, is to be found in Thackeray's *Vanity Fair* (1847).

It is only when one turns to other parts of the verb (i.e. *tries, tried, trying*) that a gulf between the two expressions opens up. Try to substitute *tries and* (etc.) for *tries to* (etc.) in the following examples, and the impossibility of it all becomes apparent: *He tries to centre his mind on that sound*—C. K. Stead, 1984 (NZ); *I . . . paced around and tried to absorb all the details*—A. Brookner, 1986; *Einar tried to coach us in semaphore signals*—G. Keillor, 1986 (US); *us if trying to guess what her answer should be*—P. P. Read, 1986.

*Try and* gains a small amount of additional currency, perhaps, from the use of *and* to connect two verbs 'the latter of which would logically be in the infinitive' (as the *OED* expresses it). The commonest of such verbs are *come* (e.g. *You will come and see us sometimes, won't you?*)

and *go* (e.g. *Do go and thank him*). These two verbs, however, have no past or present tense restrictions. Clearly it is idiomatic to say *You came and saw me yesterday* and *He went and thanked him last week*. So the parallel with *try and* is far from exact.

**tryst.** This archaic word for a date, an assignation is normally pronounced /trɪst/ now, though the *OED* (1915) gave only /traɪst/. The main AmE dictionaries give preference to the form with a short vowel.

**tsar.** Treated under this spelling in the main British dictionaries and under *czar* in the main American ones. But both sets of dictionaries concede that there is much variation (including in the *OED*). The word, whichever way it is spelt, has an initial capital when used with a personal name.

**-t-, -tt-.** Monosyllabic words containing a simple vowel (*a, e, i, o, u*) before *t* normally double the consonant before suffixes beginning with a vowel (*batted, wetter, fitted, plottable, cutter*) or before a following *y* (*witty, nutty*); but remain undoubled if the stem contains a diphthong (*baiting, flouting*), doubled vowel (*sooty*), or a vowel + consonant (*fasted, nested, bolting*). Words of more than one syllable follow the rule for monosyllables if their last syllable is accented, but otherwise do not double the *t*: thus *regretted*; but *balloted, benefited, bonneted, buffeted, combatant, cosseted, discomfited, fidgeted, pilotage, trumpeter*. It is inconsistent to double the *-t-* in the past tenses of two- or three-syllabled words, though such forms as *benefitted, cosseted, plummetted*, and *targetted* are not infrequently found in standard sources.

The recommended forms for three special cases (esp. in the language of computers) are *formatting* (pres.pple), *formatted* (pa.t. and pa.pple); *inputting* (pres.pple), *inputted* or *input* (pa.t. and pa.pple); *outputting* (pres.pple), *outputted* or *output* (pa.t. and pa.pple).

**tub.** For the *Tub*, see ACADEMY.

**tubercular, tuberculous.** Before 1882, in which year the tubercle bacillus was discovered, the two words were used interchangeably to mean 'of the nature or form of a tubercle' and 'in reference to tuberculosis (or tubercular consumption)'. Since then, in medical use, *tuberculous* has been restricted to the second of these meanings and *tubercular* to the first. But in lay use the adjectives continue to be used without distinction of meaning, though *tubercular* is markedly the more common of the two words. *Tubercular* has also been used as a noun since the mid-20c. meaning 'a person having tuberculosis' (e.g. *In 1949 Orwell left the Isle of Jura to enter a convalescent home for tuberculars at Cranham in Gloucestershire*—I. Hunter, 1980.

**tuberose** (noun). (botany) This plant of the agave family, *Polianthes tuberosa*, is pronounced /ˈtjuːbərəʊz/. It is sometimes erroneously written as *tube-rose*, pronounced /ˈtjuːbrəʊz/.

**Tuesday.** For (*on*) *Tuesday*, see FRIDAY.

**tug.** For *tug at one's heartstrings*, see STOCK PATHOS.

**tulle.** Pronounce /tjuːl/.

**tumbrel, tumbril.** The first spelling is recommended.

**tumefy.** Thus spelt, not *tumify*.

**tumidity, tumidness.** See -TY AND -NESS.

**tumour.** Thus spelt in BrE, but *tumor* in AmE. See -OUR AND -OR.

**tumulus.** Pl. *tumuli* /-laɪ/.

**tun.** See TON.

**tunnel** (verb). The inflected forms are *tunnelling, tunnelled* in BrE (also *tunneller* noun), but usu. *tunneling, tunneled* (and *tunneler*) in AmE. See -LL-, -L-.

*tu quoque* (rhetoric; from L, lit. 'thou also'). An argument which consists in retorting a charge upon one's accuser. A standard example is the medical rejoinder *Physician heal thyself*.

**turban.** The corresponding adj. is *turbaned*. See -N-,-NN-.

**turbidity, turbidness.** See -TY AND -NESS.

**turbid, turgid.** The first means '(of a liquid) cloudy, unclear', and '(of a writer's style) confused, obscure'; whereas *turgid* means 'swollen, distended', and '(of language) inflated, bombastic'.

**turbine.** Most British dictionaries give /ˈtɜːbaɪn/ first and /ˈtɜːbɪn/ as a permissible

variant, whereas most standard American dictionaries reverse the order. (The *OED* (1915) gave only the form with /-ɪn/ and Daniel Jones (1917) gave preference to the same.)

**Turco-.** The normal combining form of *Turkish*, not *Turko-*. Thus *Turcocentric*, *Turcophilia*, etc.; *Turco-Russian discussions*.

**Turcoman.** A now mostly discarded variant of TURKOMAN.

**tureen.** Pronounce /tjʊˈriːn/ rather than /təˈriːn/ since the word has moved away from its earlier form *terrine* (ultimately from L *terra* 'earth'). The spelling *tureen* is possibly modelled on the place-name *Turin*. In AmE the first syllable is usually pronounced /tə-/.

**turf.** In the senses 'a piece of turf cut from the ground (used with other pieces to make a lawn)', and 'a piece of peat cut for fuel', the pl. is usu. *turves*. But *turfs* is also a standard form. See -VE(D), -VES, etc.

**turgid.** See TURBID.

**turkey.** Pl. *turkeys*.

**Turkish.** See TURCO-.

**Turkoman.** Pl. *Turkomans*.

**turnipy.** Thus spelt. See -P-, -PP-.

**turnover.** See TART.

**turps.** See ABBREVIATIONS 1.

**turquoise.** The recommended pronunciation in BrE is /ˈtɜːkwɔɪz/, but /-kwɑːz/ is sometimes heard (in imitation of the French equivalent word). In AmE, /ˈtɜː-kɔɪz/ seems to be the dominant form, i.e. without medial /w/.

**turret.** The corresponding adj. is *turreted*. See -T-, -TT-.

**tussore.** (In full *tussore-silk*) This form, pronounced /ˈtʌsɔː/, has more or less ousted *tusser* /ˈtʌsə/ in BrE. The dominant form in AmE is *tussah* /ˈtʌsə/. The word is a late 16c. loanword from Hindi (and Urdu) *tasar*.

**tutoress.** See -ESS 4.

**twenties.** See NINETIES.

**twilight.** The noun *twilight* is frequently used attributively or as an adj. (*twilight* *glow, shade, vision, years,* etc.) and is greatly to be preferred to the rather uncommon ppl adj. *twilit* except when the meaning is 'lit by or as by twilight' (*a twilit church*). In such contexts the choice is not restricted to *twilit*, as *twilighted* is also available (*warm twilighted evenings; twilighted ages*).

**-ty and -ness.** (Including some abstract nouns in *-ety* and *-ity*.) Though any adjective may be formed into a noun on occasion by the addition of *-ness*, the nouns of that pattern actually current are substantially fewer than those made from Latin adjectives with *-ty*, *-ety*, or *-ity* as their English ending. Thus from *one* and *loyal* and *various* we can make for special purposes *oneness*, *loyalness*, and *variousness*; but ordinarily we prefer *unity*, *loyalty*, and *variety*. Of the *-ty* words that exist, a very large majority are for all purposes commoner than the corresponding *-ness* words, usage and not anti-Latinism being the right arbiter. Scores of words could be cited, such as *ability*, *honesty*, *notoriety*, *prosperity*, *sanity*, *stupidity*, for which it is hard to imagine any good reason for substituting *ableness*, *honestness*, etc. On the other hand, words in *-ness* that are more usual than existing forms in *-ty* are rare, though *acuteness* and *conspicuousness* have the advantage of *acuity* and *conspicuity*. In general a *-ty* word that exists is to be preferred to its rival in *-ness*, unless total or partial differentiation has been established, or is contextually appropriate. Total differentiation has taken place between *ingenuity* and *ingenuousness*, *casualty* and *casualness*, *sensibility* and *sensibleness*. Partial differentiation results from the more frequent use made of the *-ty* words. Both terminations have, to start with, the abstract sense of the quality for which the adjective stands; but while most of the *-ness* words, being little used, remain abstract and still denote quality only, many of the *-ty* words acquire by much use various concrete meanings in addition. Thus *curiosity*, *humanity*, *variety*, beside the senses 'being curious, human, various', acquire those of 'a curious object', 'all human beings', and 'a sub-species'. Or again they may be habitually applied in a limited way so that the full sense of the adjective is no longer

naturally suggested by them. Thus *preciosity* is limited to literary or artistic style; *purity* can take a sexual tinge that *pureness* is without; *poverty* is often the abject state of being without material possessions, whereas *poorness* is often merely a substantial step in that direction.

Articles under which special remarks will be found are BARBARISM, BARBARITY, BARBAROUSNESS; BARBARISMS; ENORMITY, ENORMOUSNESS; INGENIOUS, INGENUOUS; OBLIQUENESS, OBLIQUITY; OPACITY, OPAQUENESS; PERSPIC-; PRECIOSITY, PRECIOUSNESS; SENSIBILITY. For similar distinctions between other nearly equivalent terminations, see -CE, -CY; -IC(AL); -ION AND -MENT; -ION AND -NESS; -ISM AND -ITY.

A few specimens may be added and classified that have not been cited above, but are notable in some way. (*a*) Some words in *-ty* for which there is no companion in *-ness*, the Latin adjective not having been taken into English: *celerity, cupidity, debility, fidelity, integrity, lenity, utility.* (*b*) Some more in which the *-ty* word has a concrete or other limited sense or senses not shared by the other: *ambiguity, capacity, commodity, fatality, monstrosity, nicety, specialty, subtlety.* (*c*) Some of the few words in *-ness* that are as much used as those in *-ty*, or more, though the *-ty* words exist: *clearness* (*clarity*), *crudeness, falseness, jocoseness, morbidness, ponderousness, positiveness, tenseness, unctuousness.*

(*d*) Some *-ness* words that have no corresponding word in *-ty* in common use, though the adjective is of Latin origin and might have been expected to produce one: *crispness, facetiousness, firmness, largeness, massiveness, naturalness, obsequiousness, pensiveness, proneness, robustness, rudeness, seriousness, tardiness, tediousness, tenderness, vastness, vileness.* (*e*) If there is also a *-tion* word, derived from the verb, this naturally signifies the process, and the *-ty* word, derived from the adjective, the result: e.g. *liberty* and *liberation*, *multiplicity* and *multiplication*, *profanity* and *profanation*, *satiety* and *satiation*, *variety* and *variation*. But sometimes these pairs develop by usage a sharper differentiation: e.g. *integrity* and *integration*, *sanity* and *sanitation*. (*f*) As the OED points out, such words as *difficulty, faculty, honesty, modesty,* and *puberty* represent Latin formations in which the suffix *-tās* is directly added to a consonantal stem. The number of these in English is very small.

**tyke, tike.** In its current senses ( = a Yorkshireman, etc.) the more usual spelling is *tyke*, not *tike*.

**tympanum** (the ear-drum). Pl. *tympana*: see -UM 3. Cf. TIMPANI (pl. noun).

**type.** Be careful to distinguish the modern AmE attributive use of the word ( = *type of*, with ellipsis of *of*), as in *The 110C systems may be used with virtually any type projector*, from the use of *-type* with a preceding defining word, as in *California-type barbecues*; *Fifties-type social realist films*. The first of these sounds forced and calls for the attention of a copy editor. The second is a natural use.

**-type.** Dictionary definitions of various technical terms in which *-type* is the second element should enable readers to work out the meaning of most of the contexts in which the word appears on the page before them. Thus, from the *New SOED* (1993):

*antetype*, a preceding type, an earlier example. (But this word is now rare and *prototype* can normally be used instead.)

*antitype*. **1** Something which a type or symbol represents. **2** A person or thing of the opposite type.

*archetype*. **1** The original pattern or model from which copies are made. **2** In Jungian psychoanalysis, a primordial mental concept inherited by all from the collective unconscious. **3** A pervasive or recurrent idea or symbol in legend, etc.

*prototype*. The first or primary type of something, the original of which a copy, imitation, representation, derivative, or improved form exists or is made.

**typo.** Whether used to mean a typographical error or a typographer, the pl. is *typos*. See -O(E)s 5. See also ABBREVIATIONS 1.

**typographic, typographical** (adjs.). Both forms are current and both are of respectable antiquity (*-ic* 18c., *-ical* 16c.). The longer form is much the more usual of the two in BrE, and the reverse seems to be the case in AmE. See -IC(AL).

**tyrannic, tyrannical** (adjs.). The shorter form, recorded in use from the 15c. to the 19c., has been set aside in the 20c. in favour of *tyrannical*. See -IC(AL).

**tyrannize** (verb). The transitive type *this attempt to coerce and tyrannize us* seems to be coming back into standard use since the first edition of this book (1926) appeared, when Fowler regarded it as a solecism. The traditional construction is *tyrannize over (us)*. Examples of the various types: *She tyrannised over the older woman in all their personal relations*—D. Lindsay, 1987; *the priests know nothing, but pretend to know much and tyrannise over the common people*—New Scientist, 1992; *The camps were largely tyrannized by a volunteer cadre of prisoners named kapos by the Nazis*—NY Times Bk Rev., 1992; *We can use it to tyrannize ourselves, to live in the future instead of in the present*, [etc.]—M. Williamson, 1992.

**tyrant.** The original Greek sense of the word is so far alive still that readers must be prepared to encounter it, esp. when the context is of an absolute ruler in a Greek city-state. Neither cruel nor despotic conduct was essential to the Greek notion of a tyrant, who was simply an absolute ruler owing his office to usurpation. The word connoted the manner in which power had been gained, not the manner in which it was exercised; despotic or 'tyrannical' use of the usurped position was natural and common, but incidental only.

**tyre, tire** (of a wheel). The first is the standard spelling in BrE, and the second in AmE.

**tyro** (a novice). See TIRO.

**Tyrrhene, Tyrrhenian.** Thus spelt.

# U u

**u.** Anyone who believes that there is a simple relationship between a given letter and the way in which it is pronounced should glance at what follows. It will be self-evident that the pronunciation of the letter *u* varies according to the circumstances in which it occurs. The list makes no claims to completeness, and the pronunciations given are limited to those of standard BrE. **1** /ʌ/ *ugly, under, undo*; /ʊ/ *butcher, Jungian, pulpit, put*; /uː/ *fruit, June, rule, ruminate*; /ə/ *circus, ukulele* (second *u*); /juː/ *amuse, unit, university*; /ju/ *fraudulent* (medial *u*), *uranium* (first *u*), *uvular* (second *u*); /ʊə/ *usual* (second *u* + *a*); /aʊə/ *flour, sour.* Note also: /gw/ *guano*; /kw/ *equal, quite*; /sw/ *suave*; (silent) *guarantee, fatigue, opaque.*

**2** Special cases. (*a*) In the 20c. the retreat of the palatalized /sj-/ in *Sue* (now usually /suː/, *suit, supreme, super,* etc.; but its retention (in BrE) in *assume*; also /-zjuː/ in *presume, resume.* (*b*) See LU.

**U and non-U.** First, it needs to be said, for the sake of accuracy, that *U*, as a simple abbreviation for 'upper class (esp. in linguistic usage)' was introduced by A. S. C. Ross, a professor of linguistics, in a 1954 issue of the learned journal *Neuphilologische Mitteilungen*, and, together with its antonym *non-U* ('not upper class'), was turned into a cult by Nancy Mitford when she included the essay in a book of essays that she edited in 1956 called *Noblesse Oblige*. The intention was to identify expressions and modes of behaviour which marked off the aristocracy (or upper class) from those deemed not to be so fortunate. This national game entertained the nation for decades and is still played, though perhaps with diminishing power, as the century draws to a close. The subject had fascinated Nancy Mitford long before Alan Ross wrote his article. For example, in ch. 4 of her largely autobiographical novel *The Pursuit of Love* (1945):

Uncle Matthew: 'I hope poor Fanny's school (the word school pronounced in tones of withering scorn) is doing her all the good you think it is. Certainly she picks up some dreadful expressions there.'
Aunt Emily, calmly, but on the defensive: 'Very likely she does. She also picks up a good deal of education.'
Uncle Matthew: 'Education! I was always led to suppose that no educated person ever spoke of notepaper, and yet I hear poor Fanny asking Sadie for notepaper. What is this education? Fanny talks about mirrors and mantelpieces, handbags and perfume, ...
Aunt Emily: ... (All the same, Fanny darling, it is called writing-paper you know—don't let's hear any more about note, please.)

Any bald list of U and non-U expressions tends to be misleading: the contextual nuances and qualifications need to be examined item by item. But a list (mostly drawn from Ross and/or Mitford) follows, nevertheless.

| U | non-U |
|---|---|
| *bike* | *cycle* |
| *false teeth* | *dentures* |
| — | *God Bless* (when saying goodbye) |
| *house* | *home* |
| *lavatory, loo* | *toilet* |
| *looking-glass* | *mirror* |
| *luncheon* (in middle of day) | *dinner* |
| *pudding* | *sweet, dessert* |
| *rich* | *wealthy* |
| *scent* | *perfume* |
| *be sick* (or *vomit*) | *be ill* |
| *Sorry?* (when something is not clearly heard) | *Pardon?* |
| *table napkin* | *serviette* |
| *vegetables* | *greens* |
| *writing-paper* | *notepaper* |

The mood of the language game did not escape the gentle satirical eye of John Betjeman. His poem 'How to Get On in Society' (1954) sets the scene to perfection:

*Phone for the fish-knives, Norman,*
*As Cook is a little unnerved;*
*You kiddies have crumpled the serviettes*
*And I must have things daintily served.*

*Are the requisites all in the toilet?*
*The frills round the cutlets can wait*
*Till the girl has replenished the cruets*
*And switched on the logs in the grate.*

*It's ever so close in the lounge, dear,*
*But the vestibule's comfy for tea*
*And Howard is out riding on horseback*
*So do come and take some with me.*

*Now here is a fork for your pastries*
*And do use the couch for your feet;*
*I know what I wanted to ask you—*
*Is trifle sufficient for sweet?*

*Milk and then just as it comes, dear?*
*I'm afraid the preserve's full of stones;*
*Beg pardon, I'm soiling the doileys*
*With afternoon tea-cakes and scones.*

The sociolinguistic diversion of U and non-U is tending to be pushed to one side by the comico-sinister rivalries of POLITICAL CORRECTNESS.

**uglily.** This formally correct adverb is usually avoided on grounds of euphony, and replaced by *in an ugly way* or the like. See -LILY.

**ukase** (arbitrary command). Pronounce /ju:ˈkeɪz/.

**-ular.** Such adjectives are technically made from diminutive nouns in *-ule* (e.g. *pustular* from *pustule*) but do not necessarily convey diminutive senses themselves. A *glandule* is a small gland, but *glandular* means 'of glands' not 'of small glands'. *Cellular* means 'of or consisting of cells', not necessarily of the small cells known to biologists as *cellules*. A selection of other *-ular* adjs. will suffice to show the incompleteness of possible *-ular* word families. In the following list, the adj. is followed by the corresponding *-ule* noun (if any), and then by any related adjs. *angular/–/angulated; circular/–/circulatory; globular/globule/globose; granular/granule/ granulated, granulose, granulous; modular/ module/–; molecular/molecule/–; tubular/tu-bule/tubulate(d), tubulous; valvular/valvule/–.*

**ulna.** Pl. *ulnae* /ˈʌlniː/.

**ult.** See ULTIMO.

**ultima** /ˈʌltɪmə/. In linguistics, the last syllable of a word, as *always*, *decorum*. Contrast ANTEPENULT; PENULT.

**ultimatum.** Pl. *ultimatums* (for preference), or *ultimata*. See -UM 3.

**ultimo.** Also abbrev. as *ult.* Formerly, but now rarely, used in commercial letters,

meaning 'of last month' (*your letter of the 12th ult.*) Cf. L *ultimō mense* 'in the last month'. See COMMERCIALESE.

**ultra,** orig. a Latin preposition and adverb meaning 'beyond', was adopted as an English noun in the early 19c. (pl. *-as*), meaning a person who goes beyond others in opinion or action of the kind in question. This was no doubt a development of its use as a prefix in such adjectives (and nouns) as *ultra-revolutionary* (*-ries*), *ultra-royalist(s)* (in France). Such compounds were curtailed into *ultra* (adj. and noun); but this was no longer felt to be, like *sub* when used for *subaltern* or *subscription*, an ABBREVIATION. It won independence of any second element, its own meaning being sufficient, and became a synonym of the now more usual *extremist*.

**ultramontane** (adj.) means (*a*) on the other side of a mountain range, esp. the Alps, from the speaker or writer (L *ultrā* 'beyond', *mons*, *mont-* 'mountain'); (*b*) strong adherent or supporter of the papal authority (the point of view being that of France or other countries north of the Alps). The history of ultramontanism, a movement which began in the early 19c., is complex: accounts of it must be sought elsewhere.

**ultra vires.** Pronounce /ʌltrə ˈvaɪəriːz/.

**ululate** (verb). Pronounce /ˈjuːljʊleɪt/ or (as I prefer) /ˈʌl-/.

**-um.** For general remarks on the plural of Latin nouns adopted in English, see LATIN PLURALS. Those in *-um* are numerous and call for special treatment. The Latin plural being *-a*, and the English *-ums*, three selections follow of nouns (1) that now always use *-ums*, either as having completed their naturalization, or for special reasons; (2) that show no signs at present of conversion, but always use *-a*; (3) that vacillate, sometimes with a differentiation of meaning, sometimes in harmony with the style of writing, and sometimes unaccountably. **1** Plurals in *-ums* only. Those marked † are not Latin nouns, and the *-a* plural for them would violate grammar as well as usage:

*albums, antirrhinums* (and other plant names), *asylums,* †*begums,* †*conundrums, delphiniums, Elysiums,* †*factotums, forums,*

803

†harmoniums, laburnums, lyceums, museums, nasturtiums, nostrums, †panjandrums, pendulums, premiums, †quorums, †Targums (also *Targumim*), †vade-mecums, †variorums, vellums.

**2** Plurals in *-a* only. Further comments may be found under those listed in small capitals:

ADDENDA, AGENDA, *bacteria* (and many other scientific terms), *continua, corrigenda, curricula*, DATA, *desiderata, effluvia, errata* (see ERRATUM), *fraena, labia, maxima, menstrua, minima, momenta, opercula, opuscula, ova, palladia, phyla, pudenda, puncta, quanta, sacraria, scholia* (and other learned words of this kind), *scuta, septa, simulacra, solaria, strata, triclinia, triforia, vela, vexilla, viatica, vivaria*.

**3** Words with either plural (with preferences, not always very marked, given in brackets):

*aquarium (-a), compendium (-a), consortium (-a), cranium (-a) (-ums* in jocular use for heads), *dictum (-a), emporium (-a), encomium (-ums), equilibrium (-ums), exordium (-ums), florilegium (-a), frustum (-a), fulcrum (-a), gymnasium (-ums), honorarium (-a), interregnum (-ums), lustrum (-a), mausoleum (-ums)*, MEDIUM *(-ums* in spiritualism; *-a* of the newspaper, etc., world), *memorandum (-a), millennium (-a), moratorium (-ums), planetarium (-ums), plectrum (-a), podium (-a), proscenium (-ums), referendum (-ums), rostrum (-a), sanatorium (-ums), scriptorium (-a), sensorium (-a), serum (-a), spectrum (-a), speculum (-a), stadium (-a* in ancient-world contexts; *-ums* for modern sports grounds), *sternum (-a), symposium (-a), trapezium (-a), tympanum (-a), ultimatum (-ums), vacuum (-ums)*.

See also CANDELABRUM.

**umbilical.** Pronounced /ʌmˈbɪlɪk(ə)l/, stressed on the second syllable, less commonly /ʌmbɪˈlaɪk(ə)l/, with stress on the third syllable and with the diphthong /aɪ/ in that syllable. The same double pattern is found in the noun *umbilicus*. The pl. of *umbilicus* is either *umbilici* /-saɪ/ or *umbilicuses*: see -US 1.

**umbo** (shield-boss, knob, etc.). Pl. either *umbones* /ʌmˈbəʊniːz/, as in Latin, or *umbos* (see -O(E)S 6).

**umbrae.** Pl. either *umbras* or *umbrae* /-briː/.

**umlaut.** Pronounce /ˈʊmlaʊt/.

**'un.** Sometimes written without the apostrophe, it represents a rendering in colloquial standard of dialectal forms of two different personal pronouns, *him* (*the ladies liked 'un*—P. Hill, 1977) and *one* (*those striped uns* [sc. roses] *have no smell*—G. Eliot, 1859).

**un-. 1** For the asymmetry of many *un-* and *in-* forms (e.g. *unstable* but *instability*), see IN- AND UN-.

**2** For the cancelling negation of the type *not uncommon* see NOT 3.

**3** Danger of ellipsis after *un-*words. *Untouched* means 'not touched', but with the difference that it is one word and not two, a difference that in some circumstances is important. In *I was not touched, and you were* the word *touched* is understood to be repeated, and does not carry the *not* with it. But in *I was untouched, and you were*, the ellipsis does not work in the same way: *un-* cannot be detached as *not* was in the other example. In the following example (cited by Fowler, 1926): *Dr Rashdall's scholarship is unquestioned; most of his writings and opinions on ecclesiastical matters are*, what is meant is that most of them are questioned, not unquestioned. Ellipses, like computer commands, work with mechanical logic, and must be treated with caution.

**unabashedly.** Five syllables. See -EDLY.

**unaccountable.** *Occurrences that are for the time being, and to the spiritualist, unaccountable by natural causes. Unaccountable* ( = not to be accounted for), like *reliable*, belongs to a legitimate class of words. But to use *by* after it, compelling the reader to resolve it into its elements, is asking too much. Read instead *not to be accounted for by*.

**unadornedly, unadvisedly.** Five syllables in each. See -EDLY.

**unanimous.** See NEM. CON.

**unapt, inapt, inept.** See INAPT.

**unarguable.** See IN- AND UN-.

**unartistic, inartistic.** The second is the usual word; but since it has acquired a sort of positive sense, 'outraging the canons of art', etc., the other has been introduced for contexts in which such condemnation is not desired; the *unartistic* are those who are not concerned with art. See IN- AND UN-.

**unashamedly.** Five syllables. See -EDLY.

**unattached participles.** Some grammarians call them dangling, hanging, or misrelated participles. Fowler (1926) called them unattached participles, and cited an example from a letter: *Dear Sir, We beg to enclose herewith statement of your account for goods supplied, and* being desirous *of clearing our Books to end May will you kindly favour us with cheque in settlement per return, and much oblige.* The reply ran, *Sirs, You have been misinformed. I have no wish to clear your books.* The mistake in the first letter was to attribute the desire to the wrong person. Such failures to look ahead and consider the grammatical compatibility of the following clause are exceedingly common, especially in unscripted speech. Lord Belstead, speaking on BBC Radio 4 in January 1988 about his own role after the resignation of Lord Whitelaw as Leader of the House of Lords, said, *Being unique, I am not going in any way to imitate him.* He did not intend to imply that he was himself unique. A commentator on *World at One* (BBC Radio 4) at the end of May the same year spoke of the Reagan/Gorbachev summit meeting: *After inspecting a guard of honour, President Reagan's motorcade moved into the centre of Moscow.* In December, 1987, Richard Ingrams wrote about the house in which he grew up: *Now demolished, I can call it to mind in almost perfect detail.* Obviously the entire motorcade did not inspect the Russian troops, and Mr Ingrams had not been demolished.

Naturally such misrelated clauses are not restricted to unscripted speech: *Picking up my Bible, the hill seemed the only place to go just then*—J. Winterson, 1985; *While serving ... as a sentry outside St James's Palace, on an extremely hot day in 1980, the Queen Mother sent an equerry down who instructed my partner and I* [sic] *to perform our duties on the opposite side of the road which was shaded and much cooler*—letter in *Independent*, 1991; *While serving in the RAF in North Africa the cockroaches and other creatures baked in the bread provided an interesting gauge as to how long the recipient had served out there*—letter in *Independent*, 1993.

Fowler went on to approve of sentences in which certain participles have acquired the character of (marginal) prepositions or subordinating conjunctions, and can stand before a noun or a clause without disturbing the logicality or grammatical soundness of the sentence: *Talking of test matches, who won the last?*; *Considering the circumstances, you were justified*; *Roughly speaking, they are identical*; *allowing for exceptions, the rule may stand.* His judgement was surely right. In such examples the participial form is now normally seen to be in harmony with what follows. Modern examples come easily to hand: *Assuming her memory isn't impaired, she's aware of the mix-up, and if she chooses to ignore it, so be it*; *Knowing my mother, this is her way of punishing us* (both from letters in the *Chicago Tribune*, 1987); *'Speaking of money,' said Beryl, 'do you mind my asking what you did with yours?'* (A. Munro, 1987). Somewhat more debatable is the use in scientific work of *using* as a semi-unattached participle: *Using carbon monoxide, his hiccups were cured for 30 minutes, but they came back again* (the writer, not his patient's hiccups, used carbon monoxide). See also the separate articles on BASED ON; CONSIDERING; GIVEN (THAT); GRANTED; JUDGING BY.

Some of the other key words in constructions of this type are *barring, excepting, owing to* (first recorded as a prepositional phrase in the work of Sir Walter Scott), *provided, providing, seeing,* and *supposing.* Some of them can also be used with *that*, with an added touch of slight formality: *Even assuming that the Socialist government was seriously committed ... all this could only be accomplished ...* (a 1986 issue of a sociological journal). Many of the constructions are traced back to earlier centuries by the OED. For example, *considering* construed as a preposition with a simple object is found as early as the 14c.: *And gentilly I preise wel thy wit ... considerynge thy yowthe* (Chaucer). *Provided (that)* and *providing (that)* construed as quasi-conjunctions have also been used for several centuries. Modern examples of the various key words of this type are strewn about in books, journals, and newspapers in every part of the English-speaking world: *Considering his gene pool, Sean Thomas Harmon is probably among the better-looking babies born Monday*

(*Chicago Tribune*, 1988); Given *bad light* ... *a nervous enemy firing this way and that may do most of our work for us* (M. Shadbolt, 1986 (NZ); Granted, *it was not hard to interest a security man, who apart from a regular soldier had the most boring job on earth* (T. Keneally, 1985 (Aust.)); Granted *that any interpretation is partial ... nevertheless there is a difference ... between* [etc.]' (D. Lodge, 1992).

It is a remarkable fact that criticism—and ridicule—of unattached participles did not begin until about a century ago. Historical grammarians—Jespersen, Visser, and others—and the *OED* cite clear examples from the Middle Ages onwards. A small selection must suffice: *Tis giuen out, that sleeping in mine Orchard, A Serpent stung me*—Shakespeare, 1602; *Having applied a smelling-bottle to her nose, the blood began to revisit her cheeks*—Smollett, 1748; *Meeting some friends and singing with them in a palace near the Hague his pen fails him to express his delight*—R. L. Stevenson, 1887. Scores of other examples have been dredged up from the past.

To conclude, here is an elementary example of a correctly attached participle: *Looking at Jim, I remembered the first time I had seen him*—Encounter, 1988. And here are further examples of incorrect or questionable uses: *Being a vegan bisexual who's into Nicaraguan coffee picking and boiler suits, you could safely assume that I vote Labour*—Private Eye, 1988; *By giving this youngster the chance to repair and race old cars, he's not tempted to steal new ones*—Home Office advt in *Times*, 1989; *Men-tioning* [ = when I mentioned] *a love of folk music, Mynors immediately produced a set of the most beautiful original broadside ballads for me to look at*—Times, 1989; *Driving near home recently, a thick pall of smoke turned out to be a bungalow well alight while the owner and neighbours watched helplessly*—Oxford Times, 1990; *Watching the President struggle with the crisis in Lithuania, the answer appears to be yes*—NY Times, 1990; *Yesterday, after conferring with my senior national security advisers and following extensive consultations with our coalition partners, Saddam Hussein was given one last chance*—statement by President Bush in *Chicago Tribune*, 1991. It must be admitted that unattached participles seldom lead to ambiguity. They just jar. Yet all those prepositional and conjunctional uses—*considering* (that), *given* (that), *speaking of*,

and so on—standing there quietly on the sideline, and also the centuries-old failure to fault overt examples of unattached participles, make the game of grammatical relatedness just that little bit harder to play.

See ABSOLUTE CONSTRUCTION; MISRE-LATED CLAUSES, etc.

**unavowedly.** Five syllables. See -EDLY.

**unaware(s).** *Unaware* is a predicative adj. usually followed by *of* (*he was unaware of her presence*) or *that* + clause (*I was unaware that you had moved to Brussels*); and, archaically in BrE, an adverb (*A Zephyr ... gathering round her unaware Fill'd with his breath her vesture and her veil*—R. Bridges, 1885–94). *Unawares* is only an adverb, esp. meaning 'unexpectedly' (*he stumbled on them unawares; the question took him unawares*) or 'inadvertently' (*dropped it unawares*).

**unbeknown(st)** (adjs). (in predicative use). Both the shorter and the longer form are current in standard English though their distribution is puzzling. *Unbeknown* entered the language in the 17c. and the *-st* form in the mid-19c. *Unbeknownst* seems to be the dominant form of the two at present, but the ordinary word for the notion is still by far *unknown*. Apparently *unbeknown* and *unbeknownst* were regarded for a period 'as out of use except in dialect or uneducated speech or in imitations of these' (Fowler, 1926), but these limitations no longer apply. Examples (both of which happen to be American): *where, unbeknownst to me, they had reached a deadlock about whether or not to publish*—Bull. Amer. Acad. Arts & Sci., 1989; *whose real father, unbeknownst to her, is her mother's one-time Jewish lover*—NY Rev. Bks, 1990.

**unbend** (verb) means 'to change from a bent position, to straighten', and also 'to relax from stress or severity' (*she likes to unbend with a gin and tonic after a visit from her hypochondriac friend*). By contrast, the adj. *unbending* normally means 'unyielding, inflexible; austere' (*she knew how unbending her father was*), though it can be used in the literal sense ('not stooping') as well (*the tall old foreigner stood erect and unbending*).

**unbias(s)ed.** The form with a single *s* is recommended, but that with *ss* is often found in standard sources. See -s-, -ss-.

**uncia** (copper coin in Roman antiquity). Pl. *unciae* /'ʌnsɪiː/.

**Uncle Sam.** See SOBRIQUETS.

**un-come-at-able.** This hangdog word, first recorded in 1694 and characterized by Dr Johnson as 'a low, corrupt word', has clung on in the language as a synonym of 'unattainable' and 'inaccessible', but is now rarely used. By contrast, *un-get-at-able*, first recorded in 1862 and equally a homespun formation, is in reasonably common use. An example: *The little untouchable, ungetatable thing, sitting so close to him and yet so completely removed*—E. von Arnim, 1925. Cf. *unputdownable* (of a book), so engrossing that one cannot stop reading it.

**uncommon.** The old slang or colloquial use as an adverb = remarkably (*an uncommon fine girl*), first recorded in 1794, has nearly died out, and is no longer in place in standard English outside the dialogue of period novels and in the representation of dialect speech.

**unconcernedly,     unconstrainedly.** Both words pronounced as five syllables. See -EDLY.

**uncontrollable.** Now the dominant form, rather than *incontrollable*. Both words entered the language in the 16c. See -IN- AND UN-.

**uncontrolledly.** Five syllables. See -EDLY.

**uncooperative, uncoordinated.** Both words are now written solid, i.e. without a hyphen or diaeresis. See CO-.

**uncountable nouns.** See COUNT NOUNS; MASS NOUN.

**under** (prep.). See BELOW; BENEATH; AND UNDERNEATH for distinctions.

**underlay, -lie.** The confusion described in the entry LAY AND LIE is worse confounded in the compounds. See the remarks on OVERLAY. The pa.t. and pa.pple of *underlay* are both *underlaid* (e.g. *it makes good sense when underlaid to the soprano part; the scent of lavender was underlaid with darker tones of meat*). The pa.t. and pa.pple

of *underlie*, on the other hand, are *underlay* and *underlain* respectively (e.g. *the principles of justice and retribution that underlay the beliefs of Anglo-Saxon rulers in the eighth century; The manners she had taught him were underlain by hostility to strangers*).

**underneath** (prep.). From earliest times in competition with *below*, *beneath*, and *under*, this word is tending to be restricted to its literal sense, 'directly beneath or covered by' (*underneath the arches of the bridge; he wore a bullet-proof vest underneath his uniform*). Figurative uses of the word have tended to drift into archaism (*Philosophy, thou canst not even Compel their causes underneath thy yoke—* Coleridge, 1822) and are now uncommon. As Fowler (1926) expressed it, 'its range is much narrower than that of *under*, being almost confined to the physical relation of material things (cf. 'underneath the bed' with 'under the stimulus of competition').' See BELOW; BENEATH.

**understatement.** See LITOTES; MEIOSIS.

**under the circumstances.** See CIRCUMSTANCE.

**undertone.** See OVERTONE.

**under way.** After the adoption of the Dutch word *onderweg* in the mid-18c., our ships, away from their moorings, were said to be *under way*. About forty years later some people connected with the sea, cleverly but erroneously, associated the phrase with the weighing of anchors and used *under weigh* instead. They were followed by myriads of writers, including Thackeray (*But though the steamer was under weigh, he might not be on board, Vanity Fair*, ch. 67). For nearly two centuries, writers, whether of the *way* or the *weigh* camp, regularly wrote the expression as two words, *under way* or *under weigh*: thus Captain Marryat, Washington Irving, Byron, Carlyle, and many others. Then something happened. The mysterious force that in earlier centuries had brought a great many other adverbs together (*any* + *way* → *anyway*) struck again. Ships, projects, experiments—almost anything—from the 1930s onward began in some circles to get *underway*. Uncle Sam was partly to blame but so was John Bull in the form of some of our most competent young writers—

Martin Amis and William Boyd, for instance. From the latter: *They walked arm-in-arm into the club where the dance was underway.* And from a 1987 issue of the *NY Rev.Bks: America's declared foreign policy of fostering stability ... in Central America might at last get underway.* The joined-up form is now entered in some dictionaries, and the *New SOED* (1993) even gives it precedence over the unjoined form. Most style books recommend *under way*, and so for the moment do I. The older variant *under weigh* seems to have virtually disappeared, though it was used by Anthony Burgess in his *Little Wilson and Big God* (1987): *arrangements were already under weigh.*

**underwhelm** (verb). An artificial, but successful, mid-20c. jocular adaptation of *overwhelm*. It means 'to leave unimpressed, to arouse little or no interest in', and has wide currency in standard English sources everywhere. Examples: *Both the prose and the play are underwhelming—TLS*, 1972; *The Sparks Street post office ran out of applications ... but a survey of other post offices ... showed the public was generally underwhelmed—Ottawa Citizen*, 1978.

**undeservedly, undesignedly.** Both words are pronounced as five syllables. See -EDLY.

**undigested, undiscriminating.** These are the customary forms now, not the corresponding forms in *in-* (*indigested, indiscriminating*), both of which were formerly in use. Cf. *indigestion, indiscriminate.* See IN- AND UN-.

**undisguisedly.** Five syllables. See -EDLY.

**undistinguishable.** Now less common than INDISTINGUISHABLE. See IN- AND UN-.

**undistributed middle.** In logic, a fallacy resulting from the failure of the middle term of a syllogism to refer to all the members of a class in at least one premiss. See SYLLOGISM.

**undisturbedly.** Five syllables. See -EDLY.

**undoubtedly.** See DOUBTLESS.

**uneatable.** See INEDIBLE.

**uneconomic, uneconomical.** The shorter form means 'not economic, incapable of being operated profitably',

and the longer one 'not economical, wasteful; not sparing or thrifty'. See ECONOMIC, ECONOMICAL.

**unedited, inedited.** See INEDITED.

**unequal.** In the sense 'inadequate in ability, resources, etc.', it is construed with *to* + a noun (*he was unequal to the task*) or *to* + gerund (*it was clear that he was unequal to completing his thesis*). (Formerly also with *to* + infinitive: *... to complete ...*)

**unequalled.** Thus spelt in BrE, but usu. *unequaled* in AmE. See -LL-, -L-.

**unequivocal.** The standard corresponding adverb is *unequivocally*, not *uniquivocably*.

**unescapable** (first recorded in 1614) has now retreated in favour of *inescapable* (first recorded in 1792). See IN- AND UN-.

**unessential** (first recorded in 1656) is still sometimes used interchangeably with *inessential* (first recorded in 1677). See IN- AND UN-.

**unexceptionable, unexceptional.** The first means 'with which no fault can be found; entirely satisfactory'; and *unexceptional* means 'not out of the ordinary; usual, normal'. See EXCEPTIONABLE.

**unfeignedly.** Four syllables. See -EDLY.

**un-get-at-able.** See UN-COME-AT-ABLE.

**ungula** (hoof, claw). Pl. *ungulae* /ˈʌŋɡjuliː/.

**unhuman, inhuman.** Both words have been in use for centuries (*unhuman* since the 16c., *inhuman* 15c.) in a range of similar meanings clustered round the broad sense 'not characteristic of the behaviour, appearance, etc., of a human being'. At present *inhuman* is much the more usual. See IN- AND UN-.

**unidiomatic -ly.** In a number of articles in this book, it is shown that there is often a fine idiomatic line to be drawn between adjectives and adverbs, and also between 'simple' adverbs and corresponding forms in *-ly*. See e.g. DEAR, DEARLY; DIRECT, DIRECTLY; HARDLY 1; HIGH, HIGHLY; IRRESPECTIVE; LARGE, LARGELY; PRETTY; RIGHT; SURE, SURELY; TIGHT, TIGHTLY; WIDE. Standard speakers for the most part instinctively know which form is appropriate in a given context. But for foreign learners of the language, young children, and dialect speakers (those, that is, who may try to adopt standard usage in particular circumstances), the going is much more difficult. For such groups the addition of -ly to an adjective must seem to be the obvious way of forming an adverb, since the majority of adverbs do indeed end in -ly and correspond to a clearly related simple adjective: *angry/angrily, gay/gaily, mere/merely, usual/usually*, etc. But the application of any such simple rule, that is to regard the addition of -ly as the only way of turning an adj. into a word meaning 'in the manner of, after the style of, etc.', is to fall far short of understanding how the language works.

It is not necessary to assemble here specimens of unidiomatic uses in which -ly is missing where it is wanted or is present when it is not needed, as numerous examples are to be found in the articles listed above, as well as in others. Standard speakers also need no special guidance about the difference of meaning between such types as *he had arrived late* and *he had arrived lately*. But there are always special cases which may surprise even the most articulate of readers, as for instance the jocular use of the double adverb MUCHLY and the transatlantic use of THUSLY. Three examples of adverbs lacking the notional -ly (taken from standard sources of the 1980s): *I burrow deep into my notebook; He took a great leap and landed square on the glossy back* [of a horse]; *Leo would be the first to concede that he had spread himself too thin.*

**uninterest.** This relatively uncommon word (it is not even listed in *COD* 1995) is first recorded in 1890 and is used only in the sense 'lack of interest, indifference'. Modern examples: *She had no idea ... whether all men went through periods of uninterest*—S. Faulks, 1989; *In undergraduates it tends to produce uninterest or rejection when compulsory*—*English*, 1991. Cf. DIS-INTEREST.

**uninterested** means 'not interested, indifferent' and has been in regular use since at least 1771 (*OED*). It is now being challenged in this sense by DISINTER-ESTED, but remains the standard and recommended form. Examples: *I wouldn't*

*say that—he was totally uninterested in both of us*—G. Greene, 1980; *Classical historiography was on the whole uninterested in local provincial history*—London Rev. *Bks*, 1981; *To viewers who are uninterested in politics, it was worse than the World Cup*—*Observer*, 1990.

**Union Jack.** As the *New SOED* expresses it, '(a) orig. & properly, a small British Union flag flown as the jack [sc. a small national flag usu. flown from the bow] of a ship; later & now usu., a Union flag of any size or adaptation, regarded as the national flag of the United Kingdom; (b) US (*union jack*) a jack consisting of the union from the national flag.'

**unique.** Some small modern dictionaries allow weakened uses of the word to go unchallenged. For example, the *Cambridge International Dict. of English* (1995), after stating that as an adj. *unique* is 'not gradable', defines it as follows: 'being the only existing one of its type or, [emphasis mine] *more generally, unusual or special in some way*'. (And so it is gradable after all?) The *COD* (1995) first lists its primary meaning ('of which there is only one; unequalled; having no like, equal, or parallel'), but adds what it calls a disputed use, 'unusual, remarkable (*a unique opportunity*)'. All around us, in print and in speech, is abundant evidence illustrating the 'correct' or 'proper' use, but also examples (esp. when *unique* is preceded by adverbs such as *comparatively, more, most, rather, somewhat, very*, and some others) of the secondary disputed (or informal) use. What are we to make of this?

*Origin.* L *ūnicus* yielded Fr. *unique* (and Sp., Pg., and It. *unico*) 'single, sole, alone of its kind', and in these Continental languages the word is still mainly restricted to contexts in which one-ness is implied (e.g. Fr. *sens unique* 'one-way road', *fils/fille unique* 'only son/daughter'). The French word made its way into English at the beginning of the 17c., at first narrowly restricted to mean 'single, sole, solitary' (e.g. *He hath lost ... his unic Son in the very flower of his age, c*1645). The *OED* points out that it was 'regarded as a foreign word down to the middle of the 19th c., from which date it has been in very common use, with a tendency to take the wider meaning of "uncommon, unusual, remarkable".'

809

Distribution now. It is a simple matter to lay one's hands on 20c. examples of the traditional, ungradable sense of *unique* (which is the one recommended in this book): *the unique and infrequently seen portrait of Sidney from the Upper Reading Room frieze*—Bodl. Libr. Rec., 1986; *Hopkins's inner ear is awash with an infinite and exquisite sense of unique vocal patterns*—TLS, 1987. It is almost equally easy to find examples of the weakened sense ('remarkable', etc.), esp. when *unique* is tautologically preceded by certain adverbs (*more, most, very*, etc.): *'Toad Hall,' said the Toad proudly, 'is an eligible self-contained gentleman's residence, very unique*—K. Grahame, 1908; *Almost the most unique residential site along the south coast*—advt in Country Life, 1939; *Our Institute is a very unique place, not only bridging the gap between Christians and Jews but also between academics and clergy*—New Yorker, 1993; *Some design choices become so unique that they border on the eccentric and make a property difficult to sell*—Chicago Tribune, 1995.

It must, I think, be conceded that *unique* is losing its quality of being 'not gradable' (or absolute), but copy editors are still advised to query such uses while the controversy about its acceptability continues.

**United Kingdom.** Now (1996), Great Britain and Northern Ireland; from 1801 until 1922 Great Britain and Ireland.

**unities.** The unities, or *dramatic unities*, are the unity of time, the unity of place, and the unity of action. In the past the terms were sometimes said to be derived from Aristotle's *Poetics*, but this is only partially true, as Baldick points out in his *Concise Oxford Dict. Literary Terms* (1990): 'In fact Aristotle in his discussion of tragedy insists only on unity of action, mentioning unity of time in passing, and says nothing about place.' These principles of dramatic composition were formulated by English and continental writers of the 16c. and 17c. According to Baldick, Jean Mairet's *Sophonisbe* (1634) was the first French tragedy to observe the unities. Mairet and others believed that a play should have a unified action, without subplots, and should represent the events of a single day within a single setting. But certain variations were permissible: 'The place the stage represented was allowed to shift from one

point to another within a larger area: a palace or even a city' (OCELit., 1985). The dramatic unities were not widely favoured by English dramatists of the period. For example, Shakespeare adopted them only in *The Tempest* and *The Two Gentlemen of Verona*.

**university.** At some point in the 20c. the word began to be construed without the definite article. 'After I went to school I went to the university,' wrote an elderly professor to me in 1986, 'not to university.' Example of the modern use: *Kolya, who resents the fact that I went to university instead of doing national service*—C. Rumens, 1987. In AmE, *to/at the university* is the normal construction when the reference is non-specific, i.e. the older convention is retained there.

**unlawful.** See ILLEGAL.

**unlearned, -nt.** See LEARN (verb).

**unless and until.** A modern 'strengthened' extension of *until* (the words *unless and* are often omissible without discernible loss of meaning). Fowler (1926) expressed doubts about its legitimacy, but, by citing numerous examples of its use, drew attention to its frequency. The OED provides further examples from standard sources beginning with the following: *We should as a rule stick to that pronunciation unless and until we find another native whose speech we have reason to think is more characteristic*—Daniel Jones, 1937. Recently the type has also appeared in the forms *unless or until* and *until and/or unless*: *Until and unless he discovered who he was, everything was without meaning*—D. Potter, 1986; *Iraq ... announced that it would honor the cease-fire ordered by the UN Security Council, until or unless the other belligerent, Iran, violated it*—Chr. Sci. Monitor, 1987; *But this is not now the case and unless or until it becomes so, the judgement of the Australian Supreme Court remains very questionable*—Times, 1987; *Tape 7 was not to be played unless or until you experienced 'Blockage'*—P. Carey, 1991; *Membership of the House of Commons is still the only legitimate qualification for real power in Britain and likely to remain so unless or until our national identity is totally submerged in Europe*—Spectator, 1991. As CGEL (13.38) points out, phrases of the type *as and when, if and when*, and *unless and until* 'weaken the expectation' and, as such, need no defence.

See *as and when* (AS 6); IF AND WHEN.

**unlike.** When used as an adj., *unlike* poses no special problems, whether used to mean 'not like' (*Utterly unlike in temper and tone*) or 'dissimilar' (*two animals as unlike as the bear and the lion*). So too when it is used as a preposition meaning 'dissimilar to' (*a man wholly unlike anyone she had met before*) or 'uncharacteristic of (a person)' (*it is unlike him to be late for work*). Fowler (1926), however, gave two special warnings: '(*a*) *I counted eighty-nine rows of men standing, and* unlike *in London, only occasionally could women be distinguished. Unlike is there treated as though it had developed … adverbial power…*; it has not, and something adverbial (*in contrast with London ways?*) must be substituted. (*b*) *M. Berger, however, does not appear to have—unlike his Russian masters—the gift of presenting female characters.* As with many negatives, the placing of *unlike* is important; standing where it does, it must be changed to *like*; *unlike* would be right if the phrase were shifted to before "does not appear".'

**unmaterial,** if chosen instead of the ordinary *immaterial*, confines the meaning to 'not consisting of matter', and excludes the other common meaning of *immaterial*, viz. 'that does not matter', 'not important or essential'. See IN- AND UN-.

**unmoral.** See AMORAL; IMMORAL.

**unparalleled.** Thus spelt (in all major forms of English).

**unparliamentary expressions.** From time to time Members of Parliament overstep the mark in the deeply ritualistic manner in which they accuse their political opponents of improprieties of various kinds and resort to expressions deemed by the Speaker (or Chairman of a Committee) to be 'unparliamentary' and therefore unacceptable. The authoritative guide in such matters is the most recent edition of Sir Thomas Erskine May's *A Treatise upon the Law, Privileges, Proceedings and Usage of Parliament* (first published in 1844). The 19th edn (1976), for example, after setting down a description of the complex procedures and practice of the House, includes an Appendix of Unparliamentary Expressions. It reads as follows (with footnote references deleted):

'From time to time the Chair has intervened to deal with the use of certain expressions in debate, which, in the context in which they were used, were abusive or insulting and of a nature to cause disorder. A list of some of these expressions is set out below. It must however be emphasized not only that the list is not exhaustive but also that the permissibility of some of them would depend upon the sense and temper in which they were used:

*Blackguard, Blether, Cad, caddishness, Calumny, gross, Corrupt, corruption, Coward, Criminal, Dishonest, Dog, Duplicity, Guttersnipe, Hooligan, Humbug, Hypocrites, Impertinence, Impudence, Jackass, behaving like a, Lousy, Malignant attack, Murderer, Personal honour, not consonant with, Pharisees, Prevaricating, Pup, cheeky young, Puppy, impertinent, Rat, Ruffianism, Slanderer, Slander, malignant, Stool-pigeons, Swine, Traitor, Treason, charges of, Vicious and vulgar, Villains, Wicked.'*

Later editions continue to treat the subject in great detail but do not include such lists. For lists of words and phrases allowed and disallowed in the Parliaments of the Commonwealth one must turn to the annual issues of *The Table*, an annual publication of the Society of Clerks-at-the-Table in Commonwealth Parliaments.

The 1994 issue of *The Table*, after noting that 'Straightforward accusations of dishonesty are not recorded, since they are universally unacceptable. Simple expletives and abuse are likewise omitted', gives systematic lists of words, phrases, and sentences disallowed, or objected to, in parliamentary assemblies throughout the Commonwealth in 1993. They include the following: (Canada) *lapdog*; *When it comes to Brians and hypocrisy, that Brian takes the cake*; *Judas*; *wounded hyena*; *rat, hypocrite, jackass*; (India, Rajha Sabah) *mad dog*; *horsetraders*; *terrorists*; (India, Lok Sabah) *Buffoonery*; *Then let half of them go to hell and half to heaven*; *You are all sinners*; (New South Wales) *he tried to massage the honourable member for Bligh*; *boofhead*; (NZ) *He's one of the laboratory rats*; *He is suffering from what looks to me like mad cow disease*; (UK House of Commons) *a friend and ally of Saddam Hussein; lickspittle; clever little sod; Baroness Bonkers;* (Victoria, Aust.) *Dirty little rat; Great big, flap-mouthed dork.* The normal procedure in such cases is for the Speaker to intervene and call upon the offending Member to withdraw the offending expressions, or make a sufficient apology for using them. More severe penalties follow if the Member refuses to retract or apologize.

**unpractical.** *Unpractical* and *impractical* are both used to mean 'not practical (of a person, a proposal, etc.); but *impractical* is also now quite widely used to mean 'not practicable'. See IMPRACTICABLE, IMPRACTICAL.

**unprecedented.** The second (stressed) syllable is pronounced /-pres-/, not /-priːs-/.

**unravel** (verb). The inflected forms are *unravelled* and *unravelling* in BrE, but often *unraveled* and *unraveling* in AmE. See RAVEL (verb).

**unreadable.** See ILLEGIBLE.

**unreligious,** chosen instead of the usual *irreligious,* excludes the latter's implications of sin, blasphemy, etc., and means 'outside the sphere of religion'. It is therefore a synonym of *non-religious.*

**unrepairable.** A rarish doublet of the more usual *irreparable* 'that cannot be rectified or made good'. On the differences of meaning and the positioning of the stress, see IRREPARABLE.

**unreservedly.** Five syllables. See -EDLY.

**unrestrainedly.** Five syllables. See -EDLY.

**unrivalled.** Thus spelt in BrE, but often *unrivaled* in AmE.

**unsanitary.** First recorded in the 19c., *unsanitary* has largely given way to *insanitary* (also a 19c. coinage) in BrE, but both forms are in common use in AmE.

**unseasonable.** See SEASONABLE, SEASONAL.

**unsolvable** differs from *insoluble* in having its reference limited to the sense of the English verb *solve*, and not covering, as *insoluble* does, various senses (dissolve as well as solve) of the Latin verb *solvere.* *Insoluble* has been the standard word for

'(of a difficulty, question, problem, etc.) not soluble' since the Middle Ages, and for 'incapable of being dissolved in a liquid' since the early 18c. On the other hand, *unsolvable* is a relative newcomer (early 19c., once also in a dictionary of 1775) in its primary sense of 'not able to be solved'.

**unstable** is the standard form of the adj. (rather than *instable*), but the corresponding noun is *instability.* See IN- AND UN-.

**unstringed, unstrung.** The verb *unstring* in all its senses now normally has *unstrung* (rather than *unstringed*) as p.a.t., pa.pple, and ppl adj. (*she unstrung the beads before restringing them; an unstrung harp; gave new life to his unstrung nerves*).

**unthinkable.** From the 15c. onwards, *unthinkable* has been in standard use in the philosophical sense 'unable to be imagined or grasped in the mind, unimaginable'. This sense is still current. By the beginning of the 20c., however, it was beginning to be used in various extended ways, applied, for example, to courses of action that were regarded by the speaker or writer as unacceptable or too horrible to contemplate (civil war, extreme action by terrorists, etc.) or just highly unlikely or undesirable. Fowler (1926) denounced all such extended uses: 'anything is now unthinkable from what reason declares impossible or what imagination is helpless to conceive down to what seems against the odds (as that Oxford should win the boat-race), or what is slightly distasteful to the speaker (as that the Labour Party should ever form a Government).' Fowler's condemnation of these extended or weakened uses has largely gone unheeded. Modern dictionaries simply record such uses as part of the standard language or at most part of standard colloquial English.

A wide range of modern examples follows: *In these circumstances the removal of British troops was unthinkable—*C. Allen, 1990; *In itself, the idea of representing 322 million Europeans by 518 elected Europarliamentarians is not unthinkable—EuroBusiness,* 1990; *What we would spend abroad without a qualm is unthinkable extravagance at home—*M. Duffy, 1991; *Margaret Thatcher give up? Unthinkable—*N. Wyn Ellis, 1991;

The ambition became a compulsion, failure literally unthinkable—R. Ferguson, 1991; Cameras were produced from tracksuit pockets, and souvenir snaps, of a form unthinkable only a year ago, were taken—Daily Tel., 1992; The notion that you should give a new [television] channel a particular qualitative charge and invite it to be different in particular ways is almost unthinkable today—Independent, 1992; In Romania communism and fascism are still on intimate terms, and genuine de-communization remains unthinkable without an exorcism of the authoritarian rightist past—New Republic (US), 1992; We have achieved in North America, Western Europe, and Japan a 'zone of peace' within which it is fair to say war is truly unthinkable—A. and H. Toffler, 1994 (US). An element of the unimaginable or the incogitable resides in all of them: the link with the traditional uses of the word is not entirely broken.

**until.** 1 until, till. See TILL.

2 until or till for before or when. In the following type, until (or till) is sometimes said to be unidiomatically used for before or when: In one of the city parks he was seated at one end of a bench, and had not been there long until a sparrow alighted at the other end. The OED (s.v. till conj. 1c) comments on this type: 'Formerly and still dial. and in U.S. used after a negative principal clause, where before (or when) is now substituted in Standard English.' The last example given is one of 1756: I was not long set till Margaret came to see me. But the OED gives no examples of this type s.v. until, and it would seem best to reserve judgement about the construction until more evidence of its existence is found.

3 unless and until (and variants). See UNLESS AND UNTIL.

4 until such time as has its uses, with an implication of uncertainty whether the event contemplated will ever happen. But often it is mere verbosity for until.

**unto.** The COD (1995) correctly labels it as an archaic variant of to prep. (in all uses except as the particle to forming the sign of the infinitive). It survives mainly in fixed phrases (do unto others; faithful unto death; take unto oneself), but

it is also occasionally used in other circumstances: e.g. Readers ... may well wonder ... whether each religious group dealt with in the separate chapters is unique unto itself—Bull. Amer. Acad. Arts & Sci., 1993.

**untoward.** Pronounced as three syllables, i.e. /ˌʌntəˈwɔːd/, but as /ʌnˈtəʊəd/ in certain circles, esp. in rapid speech. Normally as two syllables in AmE., i.e. /ʌnˈtɔːrd/ or /ʌnˈtəʊrd/, but also as three, i.e. /ˌʌntəˈwɔːrd/.

**unvoiced** (phonetics). See VOICED (adj.).

**unwashed.** For the Great Unwashed (or the g. u.) see SOBRIQUETS.

**up.**

| 1 Complexity.
| 2 up to date.
| 3 up to, down to.
| 4 up and.

1 Complexity. This most complicated word requires great subtlety of treatment in dictionaries. It is used as adverb, preposition, adjective, noun, and verb, and with great diversity within each part of speech. A syndicated column in some major American newspapers in 1994 (reproducing a letter to the Reader's Digest some 25 years earlier) provides an amusing introduction to some of the idiomatic uses of the word:

'We've got a two-letter word we use constantly that may have more meanings than any other. The word is up. It is easy to understand up, meaning toward the sky or toward the top of a list. But when we waken, why do we wake up? At a meeting, why does a topic come up? And why are participants said to speak up? Why are officers up for election? And why is it up to the secretary to write up a report? The little word is really not needed, but we use it anyway. We brighten up a room, light up a cigar, polish up the silver, lock up the house and fix up the old car. At other times, it has special meanings. People stir up trouble, line up for tickets, work up an appetite, think up excuses and get tied up in traffic. To be dressed is one thing, but to be dressed up is special. It may be confusing, but a drain must be opened up because it is stopped up. We open up a store in the morning, and close it up in the evening. We seem to be all mixed *up* about *up*. In order to be *up* on the proper use of *up*, look *up* the word in the dictionary. In one desk-sized dictionary, *up* takes *up* half a column; and the listed definitions add *up* to about 40. If you are *up* to it, you might try building *up* a list of the many ways in which *up* is used. It may take *up* a lot of your time, but if you don't give *up*, you may wind *up* with a thousand.'

All these and many more are treated in our standard dictionaries. For some discussion of a few of them (*give up*, etc.), see PHRASAL VERBS.

**2** *up to date*. This phrase should be written as three words unhyphened, except when it is used as an attributive adjective; then it is hyphened: *you are not up to date*; *bring the ledger up to date*; but *an up-to-date model*.

**3** *up to*, *down to*. These phrases often have distinct meanings: e.g. (*up to*) not more than (*you can have up to five*), less than or equal to (*sums up to £10*), occupied with or busy with (*what have you been up to ?*); (*down to*) having used up everything except (*down to their last tin of rations*). But both can mean 'until': cf. *up to the present*; *from Elizabethan times down to the nineteenth century*. In such contexts the choice of *up to* or *down to* depends on the vantage point of the speaker. More puzzling is the choice that is idiomatically available to us when either *down to* or *up to* may be used when the sense required is 'be incumbent on, be the responsibility of' (*it is down to/up to you to make the decision*). Cf. DOWN TO.

**4** *up and* ( + verb). Followed by a verb, *up and* means 'do (something) suddenly or unexpectedly' (*he just upped and vanished*; *she upped and married her cousin*; *suddenly the division ups and marches to Aldershot*). This use is first recorded in R. L. Stevenson's *Treasure Island* (1883): *And you have the Davy Jones's insolence to up and stand for cap'n over me!*

**up and down.** Gowers (1965) commented as follows: 'As geographical terms these words are ordinarily used in their natural senses. We speak of going *down* to the south and *up* to the north, *down* to the sea and *up* country from the coast. But two special uses are worth noting. **1** *In relation to London*. The use of

*up* for a journey to London and *down* for one from it preceded the adoption of those expressions by the railways. Perhaps the idea of accomplishment latent in *up* made it seem the right word for reaching the more important place. Geographical bearing was immaterial; going north from London was no less *down* than going south. "At Christmas I went down into Scotland," wrote Lord Chancellor Campbell in 1845, "and, crossing the Cheviots, was nearly lost in a snowstorm." "You don't mean to say," said Miss La Creevy, "that you are really going all the way down into Yorkshire this cold winter's weather, Mr. Nickleby."

'The railways conformed. They gave the name *up-line* to that on which their trains arrived at their London terminal and *down-line* to that on which they left. When a railway was built without any direct connexion with London the up-line was that on which trains ran to the more important terminus . . .' British Rail South West Region confirmed (1995) that the same terminology is still used, i.e. the trains arriving at mainline stations travel on the *up-line* and those leaving mainline stations travel on the *down-line*. Cross-country journeys sometimes require different terms.

It seems likely, but is difficult to establish, that passengers now speak of going *up* to London, *over* to London, *down* to London, as appropriate, depending on the location of the starting-point. But in the circles in which I move in Oxfordshire most people speak of going *up* to London. I do not recall anyone saying *I am going down to Edinburgh* (or even *to Birmingham*, *Lincoln*, etc.) in the half-century that I have lived in Oxfordshire.

**2** *In relation to Oxford and Cambridge*. To a member of these universities *up* means in residence. An undergraduate goes *up* at the beginning of term and remains *up* until he goes *down* at the end of it (unless he has the misfortune to be *sent down* earlier). But this special use relates only to the universities, not to the cities. A businessman (say) travelling to Oxford or Cambridge would use *up*, *down*, *over* to Oxford (or Cambridge), whichever is geographically appropriate.

**upcoming** (adj.). One of my most hard-line correspondents remarked (July 1994) that '*upcoming* seems to be ousting the perfectly serviceable *forthcoming* [in British English]'. It is true that this American use of the word (first recorded in the *OED* in this sense in 1959) is gradually making its way into English-speaking countries outside the US, but it is by no means ousting *forthcoming* yet. It just seems irresistible to some writers in certain contexts from time to time. Examples: *No change in the law was required, the existing statute was merely being clarified and there was no need to insert a special clause in the upcoming Finance Bill—Daily Tel.*, 1984; *Guinness Peat Aviation ... angrily rejected its merchant banks' advice to sell its shares at little more than $20 a share in its upcoming flotation—Economist*, 1992; *And in real life, Doucett will appear in an upcoming 'pictorial' in Playboy—Globe & Mail* (Toronto), 1993; *The MORC Level 30 would be the boat for the upcoming Cup—Canad. Yachting*, 1994: *he told last week of the spectator who telephoned Selhurst Park to enquire about Wimbledon's upcoming game—Spectator*, 1995.

**upon.** 1 For *(up)on all fours*, see FOUR.

2 According to the *OED* 'The use of the one form or the other [sc. *on* or *upon*] has been for the most part a matter of individual choice (on grounds of rhythm, emphasis, etc.) or of simple accident, although in certain contexts and phrases there may be a general tendency to prefer the one to the other.' The choice seems almost wholly arbitrary, and there is no saying why one has taken root in some phrases and the other in others: *once upon a time, row upon row of seats*; but *on no account, have it on good authority*; *Kingston upon Thames, Burton-upon-Trent*; but *Henley-on-Thames, Newark-on-Trent*.

**uppermost.** See -MOST.

**upright.** See NOUN AND ADJECTIVE ACCENT.

**upside** (adj.). *He was not a person looking for love in all the wrong places, but it was something that just slapped him upside the head—Chicago Tribune*, 1990 (of a space pilot in *Star Wars*); *No, he did not. Never did. And if he did, he would have gotten a frying pan upside his head—Newsweek*, 1995 (ex-wife of the American footballer O. J. Simpson, denying in a TV interview that he had ever slapped her). This thoroughly American use of the adjective *upside*,

which is almost always construed with a verb meaning 'hit' (*go, slap*, etc.) and *head*, seems to have been first noted in the late 1970s (*OED*) in the language of US blacks, and to have spread from there to other forms of informal AmE.

**upstairs.** As adverb (*go upstairs*) stressed on the second syllable, but as adj. (*the upstairs loo*) on the first. See NOUN AND ADJECTIVE ACCENT.

**upward(s).** 1 As adj. the standard form is *upward* (*an upward movement of the voice*; *upward mobility*). As adv. the normal but not invariable form in BrE is *upwards*, but in AmE mostly but not invariably *upward*.

2 *upwards of* (rarely *upward of*). First recorded in 1721 (*OED*) in the sense '(rather) more than', it remains in standard use (*there were upwards of a hundred people at the wedding reception*). Occas., according to the *OED* and to dialect sources, it is used to mean '(rather) less than' in some regions in England. Occas. also when *upwards of* is used of a large figure it means no more than 'approximately' (*upwards of £250,000*), but it is best to assume that the meaning is '(rather) more than' unless there is contextual evidence to the contrary.

**Uranus.** One hears this word stressed on the first syllable from time to time, i.e. /ˈjʊərənəs/, but more often, in my subjective judgement, on the second, i.e. /jʊˈreməs/. I am bound to report, however, that most of the standard desk dictionaries in Britain and America give priority to first-syllable stressing.

**urinal.** The dominant pronunciation in BrE is that with the second syllable bearing the main stress, i.e. /jʊˈraɪn(ə)l/, but the form with first-syllable stressing is also standard, and is the only one in standard use in AmE.

**us** (pron.). 1 Its normal uses as the objective form of the pronoun *we* (sometimes in its reduced form *'s*) are unremarkable: *They despise us; Let's look in at the party ... We needn't stay long; Both of us write books; Have you ever seen two giggling angels ... like us?; he looks down upon us country people; it's only us*.

2 It can legitimately be used instead of *we* in certain circumstances. Thus *you*

*are as well-informed as us* (more formally *as we (are)*). For other examples, see CASES and WE. In colloquial use it can also be used instead of *me* (*Done your homework, Oakley? Give us a look at it*).

**3** The main erroneous type is shown in *Us country boys have to stick together* (read *We country boys* ...).

**4** A minor optional type is shown in *So it's been strange, us being in England because of me* (*our being* would be equally acceptable). See OUR; WE.

**-us.** The plurals of nouns in *-us* are troublesome. **1** Most are from Latin second-declension words, whose Latin plural is *-i* (pronounced /aɪ/); but when that should be used, and when the English plural *-uses* is better, has to be decided for each separately; see LATIN PLURALS and the entries for individual words (CIRRUS, DISCUS, FUNGUS, LOCUS, UTERUS, etc.).

**2** Many are from Latin fourth-declension words, whose Latin plural is *-ūs* (pronounced / uːs/); but the English plural *-uses* is almost always preferred, as *prospectuses*. The Latin ending *-ūs* is occasionally seen in a few of the rarer words, e.g. *lusus, meatus, senatus*. Words of this class, which must never have plural in *-i*, are *afflatus, apparatus, conspectus, hiatus, impetus, lapsus, nexus, plexus, prospectus, sinus, status*.

**3** Some are from Latin third-declension neuters, whose plurals are of various forms in *-u*, so *corpus, genus, opus*, make *corpora, genera, opera*, which are almost always preferred in English to *-uses*.

**4** *Callus, octopus, platypus, polypus*, and *virus*, nouns variously abnormal in Latin, can all have pl. *-uses* and usually do; for any alternatives, see the words.

**5** Some English nouns in *-us* are in Latin not nouns, but verbs, etc.; so *ignoramus, mandamus, mittimus, non possumus*; for these, as for the dative pl. *omnibus* and the ablative plural *rebus*, the only possible plural is the English *-uses*.

**6** Some English nouns in *-us* are not Latin words at all: e.g. *caucus* (perh. Algonquin), *rumpus* (prob. fanciful).

**usable.** Thus spelt, not *useable*.

**use** (verb). **1** Pronounced /juːz/ when it means 'bring into use, bring into service'

(*use your discretion*), pa.t. *used* /juːzd/ (*she used him for her own ends*).

**2** The matter is much more complicated when *used to* (less commonly, and usu. in colloq. contexts, *use to*) is used in the sense 'did or had in the past (but no longer) as a customary practice'. Both *used to* and *use to* are pronounced the same, i.e. /juːstuː/, unstressed /-tə/. (The problem of whether *use(d) to* is an anomalous verb or a full verb is discussed by Eric Jørgensen in *English Studies*, 69 (1988); some of the examples that follow are taken from this article.) *Used to* can also, of course, be used with a following noun or pronoun in the meaning 'accustomed to, familiar with'.

The main types are: (*a*) + noun or pronoun: *she had got used to the sissy ... thin-blooded climate of Auckland*—D. M. Davin, 1986; *And you'll have to get used to it*—*New Yorker*, 1988. (*b*) + gerund: *Susan ... has accrued so much fame as a musher that she is used to being a curiosity*—*New Yorker*, 1987; *He still isn't used to her being old enough to drive*—ibid., 1988. (*c*) + infin. *I know what you're thinking, Patrick, and I used to think it too*—K. Amis, 1988; *'I used to wonder,' he went on now, 'if ever we'd meet again.'*—W. Trevor, 1988; *I used to joke bleakly that he had run off with my Muse*—*Guardian*, 1990. (*d*) had used + to-infin.: *She had used to squat with old Makata on the ground*—M. Spark, 1969; *As a teenager he'd used to wander up there and clamber around*—M. du Plessis, 1989 (SAfr.). (*e*) (Only in very informal contexts) With *do*-support in negative and/or interrogative constructions: *He didn't use to wear gloves*—P. Cheney, 1964; *What time did she used to return?*—L. Thomas, 1972; *'It didn't use to be that way,' Manuel said*—*New Yorker*, 1986; *Prostate cancer ... didn't used to be a problem*—*Times*, 1995. (*f*) (Now regarded as somewhat formal) Without *do*-support in negative and/or interrogative constructions: *You usen't to be like that*—A. Christie, 1964; *The Mistress usedn't to sleep well at night*—A. Christie, *Poirot Loses a Client* (n.d.); *He used not to sweat like that*—I. Fleming, 1964; *I used not to dream*—N. Bawden, 1987.

The negative/interrogative type *Use(d)n't people to ...* is also found, esp. in spoken English and in informal letters, and arguments rage as to whether it is 'better' than the type *Didn't people use(d) to* …? Restructuring of the sentence is often the way out. *People used to … didn't they?* is perhaps the best way to avoid the problem.

**usherette.** See -ETTE 4. A recent example: *What the hell are you holding that torch for as if you were a bloody usherette?* —A. N. Wilson, 1990.

**usual.** The standard pronunciation is /ˈjuːʒʊəl/, not the slipshod /ˈjuːʒ(ə)l/.

**usufruct.** Pronounce /ˈjuːzjʊfrʌkt/ or /ˈjuːs-/.

**uterus.** Pl. *uteri* /-raɪ/ rather than *uteruses.* See -US 1.

**utilitarian.** See HEDONIST.

**utilize.** This mid-19c. loanword from Fr. *utiliser* has led a precarious life for a

century and a half beside *use* (continuously in use since the 13c.). In virtually all contexts where one or the other word is needed to cover the sense 'to bring into service', *use* is the more satisfactory word. But a case can be made out for *utilize* when the required sense is 'to make practical use of, to turn to account'. The boundary is nevertheless a murky one and it is not all clear why *utilize* is preferred to *use* in the following examples: *utilizing the electric kettle and the little packets provided*—A. Thomas Ellis, 1990; *Levy utilizes eight Bodleian manuscripts; Katzenelenbogen utilizes,* inter alia, *Bodleian MS. Opp. 34, while Chavel relies heavily on Oxford Corpus Christi College Heb. MS. 165*—Bodl. Libr. Rec., 1992.

**utmost, uttermost.** See -MOST.

**uvula.** Pronounce /ˈjuːvjʊlə/. Pl. *uvulae* /-liː/ or *uvulas.*

# Vv

**vacation** is in America the ordinary word for what we call a *holiday*. In Britain it is not so used except (often abbreviated to *vac.*) for the interval between terms in the law courts and universities. The corresponding word for Parliament is *recess*. Used as an intransitive verb meaning 'to take a vacation' (*they always vacation in Miami*), *vacation* is mostly restricted to AmE.

**vaccinate** is technically synonymous with *inoculate*, but in practice *v.* tends to be restricted to mean to inoculate (someone) in order to procure immunity from smallpox, and *i.* for measures taken to procure immunity from other diseases (poliomyelitis, hepatitis B, etc.).

**vacillate.** Thus spelt, not *vacc-*. Pronounce /'væsɪleɪt/.

**vacuity, vacuousness.** The two forms have coexisted for more than three centuries, but *vacuity* has always been the dominant one of the two. See -TY AND -NESS.

**vacuum.** Pl. (in ordinary use) *vacuums*; (in scientific use) *vacua*. See -UM 3.

**vade-mecum** (modL., lit. 'go with me'), a handbook, guidebook, etc., carried constantly for ready reference. Pronounce /vɑːdɪ 'meɪkəm/. Pl. *vade-mecums*.

**vagary.** In the OED (1916) the only pronunciation given was that with the stress on the second syllable, i.e. /və'geərɪ/, but it is now normally stressed on the first, i.e. /'veɪgərɪ/. Cf. QUANDARY.

**vagina.** Invariably pronounced /və'dʒaɪnə/. The adj. *vaginal* is /və'dʒaɪn(ə)l/ in BrE but /'vædʒɪn(ə)l/ in AmE. Pl. (in ordinary use) *vaginas*, (in medical use) *vaginae* /-niː/.

**vainness.** So spelt, with *-nn-*.

**valance** (a short curtain round the frame or canopy of a bedstead, etc.). Pronounce /'vælens/, and spell with *-ance*, not *-ence*, thus distinguishing it from the

unconnected words *valence* and *valency*. See next.

**valence, valency.** In chemistry, in the form *valence* (e.g. in C. A. Coulson's 1952 book *Valence*), used for 'the power or capacity of an atom or group to combine with or displace other atoms or groups in the formation of compounds; a unit of this' (*New SOED*, slightly abridged). The dominant form in BrE for this phenomenon seems to be *valency*; in AmE *valence*. But scientists often cross linguistic barriers, and the geographical distribution of the two forms cannot be pinned down neatly.

**-valent.** In scientific use, practice varies widely in the pronunciation of compounds ending in *-valent*, such as *bivalent*, *divalent*, *trivalent*, and *polyvalent*. In broad terms, in chemistry the main stress is placed on the penultimate syllable, which is given a long vowel, /baɪ'veɪlənt/, /daɪ'veɪlənt/, /polɪ'veɪlənt/, etc. But in cytology, immunology, etc., it appears that these compounds have been influenced by *ambivalent* and *equivalent*, and are therefore stressed on the antepenult with the *a* short, i.e. /'baɪvələnt/, etc. Even in these subjects, however, the *i* in *divalent* is not shortened, i.e. the word is pronounced /daɪ'veɪlənt/.

**valet.** The noun is pronounced either /'væleɪ/ (my preference) or /'vælɪt/; in AmE also /væ'leɪ/. The verb, now also commonly used in BrE in the sense 'to clean, esp. the inside of, a motor vehicle', has pa.t. and pa.pple *valeted* /'væleɪd/ or /'vælɪtɪd/, in AmE also /væ'leɪd/, and pres.pple *valeting*.

**valise.** In BrE pronounced /və'liːz/, now archaic except as a term for a soldier's kitbag. In AmE pronounced /və'liːs/, a small piece of hand luggage.

**Valkyrie.** Pronounce with the stress either on the first syllable, i.e. /'vælkɪrɪ/ (my preference), or on the second, i.e. /væl'kɪərɪ/. Pl. *-s*.

**valley.** Pl. *valleys.*

**valour, valorous.** For spellings see -OUR AND -OR; and -OUR- AND -OR-. The noun is spelt *valor* in AmE.

**valve.** For the preference of *valvular* (not diminutive in meaning) over *valval* and *valvar*, except in technical senses in botany and medicine, see -ULAR.

**van Dyck, Van Dyck, Vandyke, vandyke.** The name of the Flemish painter Anthony *van Dyck* (1599–1641) was Anglicized to (Sir Anthony) *Vandyke* while he lived in England. The derived noun and adj. are usually spelt *Vandyke* (so *Vandyke beard, Vandyke brown*). But in the sense 'a painting by van Dyck' (*there are several van Dycks in Buckingham Palace*) it is perhaps best to preserve the original Flemish spelling of the name.

**vanguard.** See AVANT-GARDE.

**vanity.** 1 *Vanity* is occas. used in its old sense of futility, waste of time, without any of its modern implication of conceit (*I haue seene all the workes that are done vnder the Sunne, and behold, all is vanitie, and vexation of spirit*—Eccles. 1: 14).

2 The appurtenances of this wicked world that the candidate for Confirmation in the Church of England is required to renounce together with the devil and all his works, are its pomps and *vanity*, not, as often misquoted, *vanities*.

**vantage.** See ADVANTAGE.

**vapid.** Of its corresponding nouns, *vapidness* is usually better than *vapidity* (in strong contrast with the nouns of *rapid*), except when the sense is 'a vapid remark'; then -*ity* prevails, and still more the plural -*ities*. See -TY AND -NESS.

**vapour** and its belongings. For the spelling of the word itself, see -OUR AND -OR. Allied words are best spelt *vapourer, vapourings, vapourish, vapourless, vapoury*; but *vaporific, vaporize* (-*zation, -zer*), *vaporous* (-*osity*). For the principle, see -OUR- AND -OR-.

**variability, -bleness.** Both are in current use, without any clear difference of sense or application, though *variability* is the more common of the two. This is unusual (see -TY AND -NESS); but, while -*ity* would be expected to prevail, -*ness*

may owe some of its persistence to the Biblical 'with whom is no variableness, neither shadow of turning' (Jas. 1: 17).

***varia lectio.*** Pl. *variae lectiones.*

**variance.** *\*It is utterly at variance from the habit of Chaucer.* Idiom demands *with*, not *from*.

**variant** (noun), as compared with *variation* and *variety*, is the least ambiguous name for a thing that varies or differs from others of its kind; for it is concrete only, while the others are much more often abstract. *Variation* is seldom concrete except in the musical sense, and *variety* seldom except as the classifying name for a plant, animal, mineral, etc., that diverges from the characteristics of its species.

**varicose.** Pronounce /ˈværɪkəʊs/.

**variegated.** The *OED* (1916) gave it five syllables, i.e. /ˈveərɪəgeɪtɪd/, but it is now usu. pronounced as four, i.e. /ˈveərɪgeɪtɪd/, except in AmE, in which the pronunciation with five syllables is the dominant one.

**variorum,** when used as a noun, has pl. -*ums*; see -UM 1. The word is a compendious way of saying *editio cum notis variorum*, and means an edition of a work that contains the notes of various commentators on it. *Variorum* is a genitive plural in Latin, not the neuter nominative singular that a bookseller took it to be when he offered *a good variorum edition including variora from MSS. in the British Museum.*

**various.** Unlike *certain, few, many, several, some*, etc., the adj. *various* may not be used idiomatically as a pronoun followed by *of*, at least in BrE. The type *He gave me various of his books* has been recorded occasionally since the later part of the 19c. in AmE sources, but it is rare enough not to be treated in *CGEL* (1985) nor in the second edition of the *OED* (1989). Some recent Amer. examples: *In my suitcase was a range of prevention and education materials that had been developed by various of the programs I had visited*—M. Dorris, 1989; *Saddam Hussein and various of his Ba'athist confederates ... may be war criminals under definitions established by international law*—T. Dupuy, 1991; *Various of his* colleagues, including the oleaginous Stewart (James Spader), offer to go with him if he is dismissed—Amer. Spectator, 1994.

**varlet,** a menial or rascal (historically 'an attendant on a knight'), is now used only archaically or jocularly.

**varsity.** A word used with true idiomatic flavour only by those who have a direct connection with a particular university, and only then in a narrow range of contexts. From the OED: *I have such faith in the old University (never use the horrid word 'varsity, my lad; don't vulgarise the old place)*—H. Kingsley, 1872; *It seems to me that the real attractions of varsity life are reserved for the sportsman and the loafer*—E. H. W. Meyerstein, 1908. These give a hint of the barriers placed around the word. In the universities of Oxford and Cambridge the *Varsity* (or *'Varsity*) *match* is the annual rugby football match between teams from the two universities. Outside BrE the subtleties of usage are endless. In AmE the normal generic name is *college*, but in practice (as in BrE) the name of the institution generally stands alone (*He's at Princeton/at Cambridge/at the Sorbonne*, etc.). In New Zealand, in informal use, people are sometimes said to be *studying at varsity* or *playing in the varsity team*. But in Australia and New Zealand students *go to Uni* or *go to the Uni*. You have been warned. Sophistication and local knowledge are needed in all contexts when general or specific references to universities are called for. See UNIVERSITY.

**vasculum** (botanist's specimen-case). Pl. *vascula*. See -UM 2.

**vase.** The standard pronunciation in BrE now is /vɑːz/ and in AmE /veɪs/ (but also /veɪz/). The pronunciation /vɔːz/ also had 'some currency in England' at the beginning of the 20c. (OED, 1916).

**vastly.** In contexts of measure or comparison, where it means by much, by a great deal, as *is vastly improved, a vastly larger audience*, *vastly* is still in regular use. Where the notion of measure is wanting, and it means no more than 'exceedingly, extremely, very' as in *I should vastly like to know, is vastly popular*, it was fashionable in the 18c. (e.g. *The City … was vastly full of People*—Defoe, 1722; *This is all vastly true*—F. Burney, 1782;

*A'nt you come vastly late?*—Sheridan, 1799), but became less common as time went on, and is now in restricted use.

**vast majority.** See MAJORITY 1, 4.

**VAT.** See TAX.

**vaudeville.** The recommended pronunciation is /ˈvɔːdəvɪl/, rather than /ˈvəʊd-/.

**'ve.** Abbrev. form of *have*. Since the 18c. (according to the OED) often joined to a previous word, esp. a modal verb or a personal pronoun, in the representation of speech. Examples: *I've a good mind to take the tram*—F. A. Guthrie, 1885; *You would've thought at least she could've cut the bobbles off*—Margaret Forster, 1986; *Can't've been a nightmare, then, can it?*—Pat Barker, 1991.

**-ve(d), -ves,** etc. from words in -*f* and -*fe*. Corresponding to the change of sound discussed in -TH (θ) AND -TH (ð) that takes place in the plural, etc., of words ending in -*th*, like *truth*, there is one both of sound and of spelling in many words ending in -*f* or -*fe*, which become -*ves*, -*ved*, -*vish*, etc. As the change is far from regular, and sometimes in doubt, an alphabetical list follows of the chief words about which some doubt may exist, showing changes in the plural of the noun and in the parts of the verb and in some derivatives (d.). When alternatives are given the first is recommended.

*beef*. Pl. *beeves* oxen, *beefs* kinds of beef; d. *beefy*.
*calf*. Pl. *calves*; v. *calve*; d. *calfish*.
*dwarf*. Pl. *dwarfs* or *dwarves*; v. *dwarf*, pa.t. *dwarfed*; d. *dwarfish, dwarfism*.
*elf*. Pl. *elves*; d. *elfin; elvish, elfish*.
*half*. Pl. *halves*; v. *halve*.
*handkerchief*. Pl. *handkerchiefs*.
*hoof*. Pl. *hooves* or *hoofs*; v. *hoof, hoofed, hoofing*; d. *hoofy*.
*knife*. Pl. *knives*; v. *knife, knifed, knifing*.
*leaf*. Pl. *leaves*; v. *leaf, leafs, leafed, leafing; -leaved*; d. *leafy*.
*life*. Pl. *lives*; v. *live; -lived*; d. *lifer*.
*loaf*. Pl. *loaves*.
*oaf*. Pl. *oafs*; d. *oafish*.
*proof*. Pl. *proofs*; v. *prove*, pa.pple *proved/proven*.
*roof*. Pl. *roofs* or *rooves*.
*scarf*. Pl. *scarves* or *scarfs*.
*scurf*. d. *scurfy* having scurf; *scurvy* paltry, mean.
*self*. Pl. *selves*; d. *selfish*.
*sheaf*. Pl. *sheaves*; v. *sheave*; adj. *sheaved*.

shelf. Pl. *shelves*; v. *shelve*.
staff. Pl. *staffs*, (mus. etc.) *staves*; v. STAVE.
thief. Pl. *thieves*; v. *thieve*; d. *thievery, thievish*.
turf. Pl. *turves* or *turfs*; v. *turf*; d. *turfy*.
wharf. Pl. *wharves* or *wharfs*; d. *wharfage*, (Aust. and NZ) *wharfie*.
wife. Pl. *wives*.
wolf. Pl. *wolves*; v. *wolf, wolfs, wolfed*; d. *wolfish, wolvish*.

**vehement, vehicle.** The OED (1916) gave priority to an *h*-less pronunciation of both words, but included forms with the *h* fully pronounced as legitimate variants. In standard English now the *h* is never pronounced. But in *vehicular* the *h* is fully sounded. Cf. ANNIHILATE, ANNIHILATION.

**velamen** /vɪ'leɪmən/, an enveloping membrane of an aerial root of an orchid. Pl. *velamina* /-mmə/.

**velar** (adj.). (phonetics) Of a sound: articulated with the back of the tongue against or near the soft palate, esp. in the pronunciation of /k/, /g/, and /ŋ/. Cf. GUTTURAL; PALATAL.

**veld** (open country in southern Africa). Pronounced /velt/ in the UK, but /felt/ in S. Africa. Sometimes still written in the older form *veldt*.

**velleity, volition.** *Volition* in its widest sense means will-power. In a narrower but more usual sense it means an exercise of will-power for a specific purpose—a choice or resolution or determination. *Velleity* (now rare) is an abstract and passive preference. It is properly used either in direct opposition to *volition* or, when volition is understood in its widest sense, as equivalent to that inactive form of it which is sometimes called 'mere volition'. The man in Browning's 'Time's Revenges' (1845):

> —'And I think I rather ... woe is me!
> —Yes, rather would see him than not see,
> If lifting a hand could seat him there
> Before me in the empty chair
> To-night,'

is expressing a velleity, but not in the ordinary sense a volition. And the OED quotes from Bentham (1808): *In your Lordship will is volition, clothed and armed with power—in me, it is bare inert velleity.*

**vellum** (fine parchment). Pl. *vellums*. See -UM 1.

**velum** /'viːləm/, a membrane. Pl. *vela* /-lə/. See -UM 2.

**velvet** makes *velveted, velvety* (adjs.). See -T-, -TT-.

**venal, venial.** These words are so alike in appearance that they are sometimes confused in spite of their being so unlike in meaning: *venal* 'able to be bribed, influenced by bribery'; *venial* 'excusable, pardonable'.

**vendor, vender.** *Vendor* is the customary spelling in BrE in the legal sense 'the seller in a sale, esp. of property', and in the sense 'a vending-machine'. In AmE, *vender* is often used in both these senses.

**venery.** The word meaning 'sexual indulgence' is distinct in origin from that meaning 'hunting'. Both are pronounced /'venərɪ/. Both are now archaic.

**venison.** The OED (1916) and Fowler (1926) gave pride of place to the pronunciation /'venzn/ (two syllables). But this has now been supplanted by three-syllabled forms, either /'venɪs(ə)n/ or /-z(ə)n/.

**venturesome, venturous.** See ADVENTUROUS.

**venue.** Pronounce /'venjuː/. This term, once common in the sense 'a thrust or hit in fencing' (obs.—OED), and still used in law as the place appointed for a jury trial (esp. *lay*, and *change, the venue*) has largely taken over from *rendezvous, meeting-place, setting*, etc. as the place where an event (a concert, a race-meeting, etc.) is scheduled to take place.

**veranda.** Now usu. thus spelt, rather than *verandah*.

**verbal.** This common adj. has several established senses including: **1** Of the nature of a verb (*verbal noun*).

**2** Concerned with or involving words only rather than things or realities (*Opposition between these two modes of speaking is rather verbal than real*—B. Jowett, 1875).

**3** Consisting or composed of words (*the verbal wit and high-flown extravagance of thought and phrase which Euphues had made fashionable*—J. R. Green, 1874).

**4** Expressed or conveyed by speech instead of writing; oral (*The archbishop believed that a verbal agreement was all which* *would be demanded of him*—J. A. Froude, 1877).

All four senses have a long history of recorded use (1, 3, 16c.⁻ ; 2, 17c.⁻ ; 4, late 16c.⁻ ), but since the late 19c. some usage commentators have drawn attention to the possible ambiguity of sense 4, and have expressed a preference for *oral* in such contexts. Perhaps the best policy for the present is to restrict *verbal* in sense 4 to a few fixed phrases (e.g. *verbal agreement, contract, evidence*); but use *oral* in most other circumstances when a formal distinction is contextually called for between a spoken and a written statement. It is worth noting that the *oral tradition* believed to have preceded the writing down of ancient poetry (Greek, Old English, etc.) is always so called, never the *verbal tradition, verbal composition*, etc.

**verbal noun, verbal substantive.** A noun in *-ing* formed as the present participle of a verb and used as a noun (e.g. *smoking* in *smoking is forbidden*).

**verbatim.** Pronounce /vɜː'beɪtɪm/, not /-'bɑːt-/.

**verbless sentences.** The occasional use of verbless sentences in radio and television broadcasts is acceptable: e.g. *This report from Paul Reynolds*; *And so for the main points of the news again*. Sentences that lack a verb or are in other ways 'incomplete' are often found in good fiction, used as stylistic devices of various kinds, as afterthoughts or re-expressions, as a way of avoiding extensive listing, to represent broken thoughts, and so on. Examples of various types: *Friday morning. By tube to a lecture at the London School of Economics*—*Encounter*, 1981; *It asserts itself as impassive, impenetrable, enigmatic. Alluring*—M. Leland, 1987; *That way, they can work out their aggression. Once a year*—*New Yorker*, 1987; *'So,' my mother said. 'Maybe that's what this is. Just a coincidence.'*—*New Yorker*, 1987; *Made her want to weep sometimes. And not so much for her as for the kids. Their future. If you could call it that*—A. Duff, 1990 (NZ). For further examples, see SENTENCE.

**verb. sap.** Abbrev. of L *verbum sapienti sat est*, a word is enough for the wise person. Ostensibly an apology for not explaining at greater length, or a hint

that the less said the better, but more often in fact a way of soliciting attention to what has been said as weightier than it seems.

**verbs from nouns.** Throughout its history, English has made selective use of its power to press nouns into service as verbs, a process falling under the general name of CONVERSION (where examples are given). Despite the calmness with which Henry Alford greeted such new verbs in his book *The Queen's English* (1864) ('I do not see that we can object to this tendency in general, seeing that it has grown with the growth of our language, and under due regulation is one of the most obvious means of enriching it'), cries of anguish are almost invariably heard when a fresh example of conversion is noticed. Thus: 'The collapse of the tongue that Milton spoke into various mutant babbles is proceeding apace. We were told by one of the Post Office robots trained to handle enquiries that our request for further details of one of their services should be "attentioned" on the application form' (*Oldie*, 7 Jan. 1994). *Attention* (verb) is not registered in dictionaries yet and may well remain as part of the personal vocabulary of the 'office robot' in question. What cannot be questioned, however, is the legitimacy of the process of linguistic conversion.

**verbs in -ie, -y, and -ye** sometimes give trouble in the spelling of inflexions and derivatives. The following rules apply to the normally formed parts only, and are merely concerned with the question whether *-y-, -ie-*, or *-ye-* is to be used in the part wanted. **1** *-ay*: *plays, played, playing, player, playable* is the model for all except *lay, pay*, and *say*, and their compounds (*inlay, repay, gainsay*, etc.), which use *-aid* instead of *-ayed*. *Allay, assay, bay, belay, decay, delay, essay, flay, okay, relay, sway*, etc., follow *play* (thus *allayed, assayed*, etc.).

**2** *-ey*: *conveys, conveyed, conveying, conveyer* (one who conveys), *conveyable*. All follow this type, except that *purvey, survey*, have *purveyor, surveyor*, and *convey* has *conveyor* for the machine.

**3** *-ie*: *ties, tied, tying*. Other words in this group (*die, lie* (tell an untruth), *vie*) follow this type, and also for the most part have no *-er, -or*, or *-able* forms in common use. _Hie_ has either _hieing_ or _hying_ as its present participle.

**4** -oy: _destroys, destroyed, destroying, destroyer, destroyable_. This pattern is followed by _annoy, cloy, deploy, employ, enjoy, toy_, etc. There appear to be no exceptions.

**5** -uy: _buy, guyed, buying, buyer, buyable_. _Buy_ (and its compounds, e.g. _underbuy_) and _guy_ are the only two verbs in this group.

**6** -y after consonant: _tries, tried, trying, trier, triable; denies, denied, denying, denier, deniable; copies, copied, copying, copier, copiable_. Neither number of syllables, place of accent, nor difference between -y /ɪ/ and -y /aɪ/ , affects the spelling of the inflected forms. But see DRIER; FLYER; SHY.

**7** -ye: _dyes, dyed, dyeing, dyer, dyable; dyeing_ is so spelt to avoid confusion with _dying_ from _die_ (cf. _singeing_). _Eyeing_ is also recommended, rather than _eying_, because it more obviously preserves the connection with _eye_.

**verdigris.** Though adopted from OF _vert de Grece_ (cf. modF _vert-de-gris_) as early as the 14c., its pronunciation in English is still unsettled. The possibilities are (in order of preference) /ˈvɜːdɪɡriː/, /-ɡriːs/, and /-ɡrɪs/. All three pronunciations are listed in the standard BrE and AmE dictionaries, but not in the same order. The final element has no connection with the word _grease_.

**verger** (church official). Thus spelt except in St Paul's Cathedral in London and Winchester Cathedral, where the 17c. variant spelling _virger_ is still used.

**veridical** (adj.). Apart from its technical use in psychiatry, '(of visions, etc.) coinciding with or representing real events or people', _veridical_ has little currency now except in formal language, = truthful; true or faithful to an original.

**verify.** For inflexions, see VERBS IN -IE, etc. 6.

**verily** will not be forgotten as long as the Gospels are read, but outside biblical contexts it has rarely been used except rhetorically since the end of the 19c.

**veritable** had a strong presence in the language from the 15c. to the 17c., but then fell out of use until it was revived as a Gallicism in the 19c., esp. used intensively = 'deserving its name'. Examples: _They had a succession of governors who were veritable brigands_—A. Harwood, 1869; _At Rochefort there was ... a veritable hail of tiles, slates, etc. blown off the roofs—_Standard, 1897; _The book is a veritable encyclopedia of great sentences—Writer's Digest_ (US), 1988; _Tree ferns from the humid forests of Tasmania, Chilean firebushes, the dove tree from China ... these secret gardens are a veritable Noah's Ark of fabulous, rare plants—_Garden, 1991; _And yet the waves—all of the Atlantic—surged against mangrove, courida, concrete defences: a veritable sea wall—Globe & Mail_ (Toronto), 1991; _The baby ... idealizes the breast as though it were free of all frustration, as a veritable inexhaustible cornucopia of pleasure—_J. Sayers, 1991; _she lost no time in establishing a veritable spider's web of an old-girl network—Gourmet_ (US), 1992.

It is not used in English, as _véritable_ is in French, to mean 'authentic, genuine (of leather, pearls, tears, anger, etc.)'.

**vermeil.** Pronounce /ˈvɜːmeɪl/ rather than /-mɪl/.

**vermilion.** Thus spelt in OUP's house style, but _vermillion_ is also standard.

**vermin.** In each of its senses, esp. 'mammals and birds injurious to game, crops, etc.; vile or despicable persons', usu. treated as a plural (e.g. _these vermin infest everything; such vermin as them are a danger to the community_). But it is capable of being used as a singular: e.g. _You vermin!_ (addressed to one person). There is no plural form _vermins_.

**vermouth.** In BrE stressed on the first syllable, i.e. /ˈvɜːməθ/, and in AmE on the second, i.e. /vərˈmuːθ/.

**vernacular.** For _vernacular, idiom, slang_, etc., see JARGON.

**verruca** /vəˈruːkə/. Pl. (in ordinary use) _verrucas_; (in scientific use) _verrucae_ /-siː/.

**verse** has several meanings, including (_a_) a poetical composition, poetry (_wrote many pages of verse_); (_b_) a metrical line in a poem that conforms with the poem's rules of prosody (in ordinary use called a _line_); (_c_) a STANZA of a poem; (_d_) each of the short numbered divisions of a chapter in the Bible.

**versify.** For inflexions, see VERBS IN -IE, etc. 6.

**vers libre** /veər 'li:brə/. Also called *free verse*. Versification or verses in which different metres are mingled, or prosodic restrictions disregarded, or variable rhythm substituted for definite metre. Perhaps now the dominant form of verse. An example (taken at random from the *London Review of Books*, 6 Apr. 1995):

> *The table was a wreck.*
> *Bleared glasses stood*
> *Half-empty, bottoms stuck to wood.*
> *Cigarette stubs:*
> *Ashes:*
> *Bits of bread:*
> *Bottles leaning,*
> *Prostrate,*
> *Dead.*
> *A pink stocking: a corkscrew:*
> *A powder puff: a French-heeled shoe:*
> *Candle-grease.*
> *A dirty cup.*
> *An agate saucepan, bottom up.*

The reviewer of the poet's work commented: 'The ... description of the end of the party is full of details, but the short lines level them all into throwaways, as if only short stabs at speech were possible, as if syntax was beyond all human hope. The very colons are full of despair ...' There are better examples of *vers libre* in the work, for example, of Ezra Pound and T. S. Eliot.

**verso.** Pl. *versos*. See -O(E)S 3.

**verso, recto.** The *verso* is the left-hand page of an open book or the back of a printed sheet of paper or manuscript, as opposed to the *recto*.

**vertebra** /'v3:tɪbrə/. Pl. *vertebrae* /-bri:/.

**vertex,** highest point, angular point of triangle, etc. Pl. *vertices* /'v3:tɪsi:z/.

**vertigo.** After at least two centuries of fluctuation, *vertigo* seems to be always pronounced now as /'v3:tɪgəʊ/. Formerly also stressed on the second syllable, which was pronounced as either /-'ti:gəʊ/ or /-'taɪgəʊ/. Pl. -O(E)S 3

**vertu.** See VIRTU.

**very. 1** It should be borne in mind that *very* and *much* are complementary, each being suited to places in which the other is unnatural or wrong. Let me begin, as

I did under MUCH, by setting down some of the standard, unopposable uses of *very*. It is most often used with adverbs: *They are grasping the nettle* very *slowly*; *She fell asleep* very *soon*; *They admired each other* very *much*; and with adjectives: *She always wore* very *thin black silk stockings*; *Thomas was sent to a* very *select private school*; *We children led a* very *sheltered life*; *These are people who are* very *afraid*; *And a* very *good evening to you* (a conventional greeting used by public entertainers); less commonly with nouns: *At issue is the* very *nature of the system he shaped*; *He was ashamed of the* very *thing that gave him so much pleasure.*

In the 20c. it has also come to be used qualifying a noun or proper name used adjectivally for emphasis: *The total effect is* very *Kirov* ( = very much in the style of the Kirov ballet); *It's a band all dressed up in green-and-black—*very *art deco.* It also often occurs in conjunction with *much*: *During the reign of Matilda the kingdom was* very much *divided*; *Not seeing D. didn't matter* very much; *Neither management nor performers cared* very much *about who understood what.*

**2** *very* or *much* before a passive participle. The principle governing the choice of *very* or *much* before a passive participle is not easy to formulate. Some examples, all drawn from standard sources of the 20c.: (much) *Right now, a* much *enfeebled Soviet Union is receiving cheap grain from America*; *His father was* much *respected in that pretty town*; *The Stroves were* much *given to hanging their tenants*; *Kafka was* much *cosseted by the ladies in his office*; *The Dwarves are not* much *differentiated.* (very) *I'm* very *annoyed. Go away*; *Frank's* very *involved with squash this week*; *It got him* very *worried*; *You're acting* very *annoyed and upset.* (very much) *I believe your* [ = one's] *experience is* very much *connected to your* [ = one's] *memories.*

A partial principle emerges. If the word qualified is more adjectival than participial, the qualifier to use is *very* (or *very much*). Conversely, if the word governed is clearly a passive participle, *much* is normally required. As Fowler (1926) expressed it, 'It will at once be admitted that *I was much tired* is improved by the substitution of *very* for *much*, whereas, in *I was very inconvenienced, much* has undoubtedly to be substituted for *very*.' But it has to be admitted that there is a grey area where the choice is difficult. And there are many passive participles which by their nature are incapable of being qualified by either *much* or *very* (*defeated, finished, forced, located, undetected, unsolved,* etc.).

**-ves.** See -VE(D), etc.

**vesica** /ˈvesɪkə/. The bladder. Pl. *vesicae* /-siː/. The *OED* (1917) gave only /vɪˈsaɪkə/, a pronunciation still current, with others, in AmE. Cf. L *vēsīca.* See FALSE QUANTITY. The derivatives *vesical* adj., *vesicle* noun have a short /e/ in the (stressed) first syllable.

**vest.** In BrE a *vest* is an undergarment worn on the upper part of the body, equivalent to AmE *undershirt.* In AmE a *vest* is the term for what in BrE is called a *waistcoat.* In former times (17–19c.), *vest* was used for a variety of men's and women's garments. At the present time, in addition to the primary senses given above, an athlete's top garment is sometimes called a *vest*; and in AmE the term is also used for a short sleeveless woman's jacket.

**vet** (noun) is a well-established abbreviation (see ABBREVIATION 1) of *veterinary surgeon,* and in AmE of *veteran.* From the first of these has come the main two senses of *vet* (verb), namely to examine or treat (an animal); and to examine (a scheme, a preliminary draft, a candidate, etc.) carefully and critically for errors or deficiencies.

**veto.** Pl. *vetoes.* See -O(E)S 1. The inflected parts of the verb are *vetoes, vetoed, vetoing.*

**vexillum** (a military standard, etc.). Pl. *vexilla.* See -UM 2.

**via** /ˈvaɪə/. Used as a preposition = 'by way of' (*London to Sorrento via Naples*) it is in origin the ablative of L *via* 'way, road'. As such it was formerly printed with a circumflex accent and in italics (thus *viâ*), but no longer. *Via* is less comfortably used to mean 'by means of, through the agency of' in such a context as *He sent the parcels via rail* (read *by*). But the use is gaining ground: *It would in theory be possible to provide five more services with national coverage via satellite*—*Rep. of the Committee on the Future of Broadcasting,* 1977; *As a former Agony Aunt, I do know*

*that many people find comfort in discovering, via pages in newspapers and magazines, that others share their dilemmas*—C. Rayner, 1995; *By listening to Friends' opinions, we will be more successful at promoting membership of the organisation to other graduates via the mailing*—memo from the Development Office of the Univ. of Oxford, 1995.

**viable.** This 19c. loanword from French first had the sense '(of a foetus or newborn child) capable of maintaining life; cf. Fr. *vie*', but soon began to be used from time to time in figurative and extended ways, e.g. *What we have here is a romance in embryo; one, moreover, that never attained to a viable stature and constitution* (1883). It was not until the 1940s, however, that it became a vogue word applied to any concept, plan, project, proposition, etc., judged to be workable or practicable, esp. economically or financially. Like all vogue words it (and also its corresponding noun *viability*) has been repeatedly put to the sword by protectors of the language, but half a century on it still remains widely in use. Attitudes change slowly and it is probably advisable for the present to use other, well-established terms instead (e.g. *feasible, practicable, sustainable, tenable, workable*), at least until the stigmata of fashionableness and overuse cease to be attached to *viable.*

**viaticum** (Eucharist given to a person near death). Pl. *viatica.* See -UM 2.

**vibrato** (music). Pronounce /vɪˈbrɑːtəʊ/. Pl. *vibratos*: see -O(E)S 6.

**vibrator.** Thus spelt, not *-er.*

**vicar.** See RECTOR.

**vice** (noun) (clamping device). Thus spelt in BrE, but *vise* in AmE.

**vice** /ˈvaɪsɪ/ (prep.), in the place of, in succession to (*appointed Secretary vice Mr Jones deceased*). In origin the ablative of L *vix, vic-* 'change'.

**vice-** /vaɪs/ is the same word as the preceding one, but used as a combining form meaning 'acting as a deputy or substitute for' (*vice-chancellor, vice-president*) or next in rank to (*vice admiral*).

**vicegerent** /vaɪsˈdʒerənt/ is a wide-ranging term derived from medL *vicegerens* (from L *vix, vicem* 'stead, office, etc.' +

*gerens* pres.pple of *gerere* 'to carry, hold'), meaning a person appointed to discharge the office of another. Since the 16c. it has specifically been used to mean 'a ruler, priest, etc., regarded as an earthly representative of God; in particular, the Pope'. In Oxford, Pembroke College has a Vicegerent, corresponding to the Vice-Master, Vice-President, Vice-Principal, etc., of other colleges. *Vicegerent* should not be confused with *viceregent*, a rather rare term for the deputy of a regent (to a sovereign).

**viceregal** is the usual term for 'of or relating to a viceroy', not *viceroyal*, though the latter had some currency in the 18c. and 19c. In Australia and New Zealand, *viceregal* is used of a Governor-General.

**vice versa** /vaɪsə 'vɜːsə/. It is derived from a Latin phr. = the position being reversed, from the same word as *vice* (prep.) (see above) + *versa*, ablative feminine pa.pple of *vertere* 'to turn'.

**vicious circle.** In logic, *circle* and *vicious circle* mean the same—the basing of a conclusion on a premiss that is itself based on this conclusion. Or, as the *New SOED* (1993) defines it, the fallacy of proving a proposition from another that rests on it for proof. For an example, see ARGUING IN A CIRCLE. More generally, it is used to mean 'an unbroken sequence of reciprocal cause and effect; an action and reaction that intensify each other' (*New SOED*). The balancing acts of political parties in the 20c. have produced two related concepts: the *vicious spiral*, a process in which rises in wages are cancelled by rises in prices, and the *poverty trap*, a situation in which an increase of income is offset by a consequent loss of State benefits, making real improvement impossible (*New SOED*).

**victual.** Almost always in the pl. form *victuals*, pronounced /'vɪt(ə)lz/. A rather old-fashioned or rustic word for food, provisions. Hence *victualler* (AmE *victualer*), someone who supplies victuals; and, in BrE, *licensed victualler*, an innkeeper licensed to sell alcoholic liquor. The corresponding verb *victual* is also pronounced /'vɪt(ə)l/; its inflected forms are *victualled*, *victualling* (in AmE both usu. with -l-). See -LL-, -L-. *Victual*

(noun) entered the language in the 14c. as a modified form of OF *vitaille(s)* (modF *victuailles*) and was respelt in the 16c. to accord with late Lat. *victuālia*, neuter pl. of the adj. *victuālis*, from L *victus* 'nutriment, food'. The modern pronunciation of *victuals* reflects the earlier spelling of the word.

**vide.** Pronounce /'vaɪdɪ/ or /'viːdeɪ/; lit. 'see' (imperative). It is used in referring readers to a passage in which they will find a proof or illustration of what has been stated, and should be followed by something in the nature of chapter and verse, or by the name of a book or author. But for this purpose *see* will usually do as well as *vide*, and *see above* (or *below*) as well as *vide supra* (or *infra*), and *above* and *below* are superfluous if a page is mentioned. There is, however, no convenient English equivalent of *quod vide* (abbrev., q.v.).

**videlicet** /vɪ'deliset/ (adverb), lit. (in Latin) 'see (*vide*), it is permissible (*licet*)', is now rare, the abbreviation *viz.* being used instead (or the English equivalent word *namely*). See VIZ. for meaning. See also SCILICET.

**vie.** For inflexions, see VERBS IN -IE, etc. 3.

**view.** Lines of demarcation need to be drawn round two well-established phrases containing the word *view* and another which, though common until the early 20c., has now effectively dropped out of use. They are *in view of*, *with a view to*, and *with the* (or *a*) *view of*.

*In view of* means 'taking into account, or not forgetting, or in consideration of', and is usually followed by a noun expressing circumstances that exist or must be expected: *In view of the readiness she showed to help him with his research, all was forgiven; In view of these facts, we have no alternative but to recommend your dismissal.*

*With a view to* means 'with the aim or object of attaining, effecting, or accomplishing (something)', and is usu. followed by a gerund or verbal noun in -*ing*: *She accepted a post as secretary with a view to becoming a personal assistant.* (In the 18c. and the 19c. the more usual construction was with an infinitive (... *with a view to become* ...), but such constructions have now dropped out of favour.)

With the (or a) view of is a once-standard complex preposition (common in the 18c. and 19c.) meaning 'with the object or design of (doing something)': *With a view of ascertaining more accurately the nature of the sun; Power of taking possession ... with the view of carrying out the necessary work.* It has now effectively been replaced by *with a view to.*

Fowler (1926) cited examples in which the wrong phrase had been used. Among them: *I will ask your readers to accept a few further criticisms on matters of detail,* in view of *ultimately finding a workable solution* (read *with a view to*); *If Germany has anything to propose* in view of *the safeguarding of her own interests, it will certainly meet* with courteous consideration (read *with a view to*); *My company has been approached by several firms* with a view of *overcoming the difficulty* (read *to* for *of*); *They have been selected* with a view to illustrate *both the thought and action of the writer's life* (read *illustrating* for *illustrate*).

See also POINT OF VIEW.

**viewpoint.** See POINT OF VIEW.

**vigour.** Thus spelt in BrE but as *vigor* in AmE. The corresponding adj. is spelt *vigorous* in both countries. See -OUR AND -OR; -OUR- AND -OR-.

**vilify.** For inflexions, see VERBS IN -IE etc. 6.

**villain, villein.** 1 First it should be emphasized that neither *villain* nor *villein* is etymologically related to *vile* 'disgusting, base'. The latter is derived from L *vīlis* 'cheap, base'; while the other two are medieval forms of OF *villein*, ult. from L *villa* 'country-house, farm'.

2 Of the two branches of the medieval senses of the word *villain/-ein*, (*a*) a lowborn base-minded rustic, and (*b*) one of the class of serfs in the feudal system, the first by the late 16c. had worsened in meaning to 'an unprincipled scoundrel' (more or less regularly spelt *villain* from the 17c. onward), and the second gradually slipped into historical use as the feudal system was replaced by capitalism (and is now regularly spelt *villein*). It is a remarkable example of the severance of a single word into two separate words neatly distinguished in both meaning and spelling.

The adj. corresponding to *villain* was often spelt *villanous* from the 16c. to the 19c. but is now always *villainous*. For the pair, cf. *mountain/mountainous*.

**-ville.** Used as a terminal element since the 19c. (the *OED* includes an isolated 16c. example) to form the names of fictitious places or concepts denoting a particular quality suggested by the word to which it is appended. Recorded examples (some from the second half of the 20c.) include *Boneheadville, Deathville, Disasterville, Massmarketville, Mediaville, Nowheresville, Squaresville.*

**vinculum** (algebra and anatomy). Pl. *vincula* /ˈvɪŋkjʊlə/.

**viola.** The musical instrument is pronounced /vɪˈəʊlə/ and the flower /ˈvaɪələ/.

**violate** (verb). The corresponding adj. is *violable,* not *violatable.* Cf. *calculable, demonstrable, educable,* etc.; see -ABLE, -IBLE 2.

**violoncello.** So spelt (not -*lin*-; it is an Italian diminutive of *violone* a double-bass viol). The recommended pronunciation is /vaɪələnˈtʃeləʊ/. Its plural is *violoncellos*: see -O(E)S 6. In practice the instrument is usually just called a *cello.*

**virago.** Pronounce /vɪˈrɑːgəʊ/. Pl. *viragos.* See -O(E)S 1.

**virement,** a regulated process of transferring funds (esp. public funds) from one financial account to another. The word, taken from French in the early 20c., has not yet been fully Anglicized in pronunciation: /ˈvɪəmɑ̃/ is recommended for the present, but /ˈvaɪəmənt/ is probably equally current.

**virger.** See VERGER.

**Virgil** (Roman poet). Thus spelt in the house style of OUP, not *Vergil.* In Latin the poet's name was Publius Vergilius Maro.

**Virgin Queen.** See SOBRIQUETS.

**virile.** Now always pronounced /ˈvɪraɪl/ in standard speech in BrE, and /ˈvɪrəl/ or /ˈvɪraɪl/ in AmE. It is worth noting that the *OED* (1917) listed four pronunciations for the word, giving forms in which both *is* vary between /ɪ/ and /aɪ/.

**virtu** (as in *articles of virtu*). Thus spelt, not *ver-*; romans not italics. Pronounce /vɜːˈtuː/. It is a loanword from It. *virtù*.

**virtue.** The semi-proverbial phr. *to make a virtue of necessity* has a long history: it was first found in Chaucer as a rendering of OF *faire de necessité vertue* and of L *facere de necessitate virtutem*. The emphasis varies in its two main senses: (*a*) to derive benefit or advantage from performing an unwelcome obligation with apparent willingness; (*b*) to submit to unavoidable circumstances with a good grace. The sainted first editor of this book remarked that it is 'often applied to the simple doing of what one must, irrespective of the grace with which one does it'. This may well be true, but no printed evidence was provided. Over the centuries the indefinite article has sometimes been omitted from the phrase, and (15–16c.) *necessity* was sometimes replaced by *need*.

**virtuoso.** The recommended pl. is *virtuosi* /-siː/ or /-ziː/, rather than *virtuosos*.

**virulent.** Pronounce /ˈvɪrʊlənt/ rather than /ˈvɪrjʊ-/. See ʊ.

**virus.** Pl. *viruses*. See -US 4. A *computer virus* is 'an unauthorized self-replicating [computer] program that can interfere with or destroy other programs, and can transfer itself to other systems via disks or networks' (*New SOED*). It is the electronic equivalent of a microbiological virus.

**visa** (endorsement on a passport). This early 19c. loanword from Fr. *visa* (from L *vīsa*, neuter pl. of the pa.pple of *vidēre* 'to see') has replaced *visé*, the more common of the two words in the second half of the 19c. and early 20c.

**visage.** Pronounce /ˈvɪsɪdʒ/. See COUNTENANCE.

**vis-à-vis** (prep.). Pronounce /viːzɑːˈviː/. Though its primary meaning is 'face to face' in Fr. and Eng. (e.g. *His master dived down to him, leaving me vis-à-vis the ruffianly bitch*—E. Brontë, 1847), it is far more commonly used now to mean 'in relation to' (e.g. *British farmers' views vis-à-vis the Common Agricultural Policy*).

**viscount,** a British nobleman ranking between an earl and a baron, often contextually with a place-name attached (*Viscount Montgomery of Alamein*). The title, dignity, or rank of a viscount is a *viscountcy*. See TITLES.

**vise.** See VICE (noun).

**visé** See VISA.

**visibility, visibleness.** See -TY AND -NESS. The second was formerly in more frequent use than many *-ness* words with predominant partners in *-ty*. The 20c. prevalence of *visibility* in the sense 'the possibility of seeing (a motor vehicle, a vessel, etc.) in adverse weather conditions' has almost driven out *visibleness* in other senses as well. Since the 1950s *visibility* has acquired wide currency in the figurative sense 'the degree to which something impinges upon public awareness': e.g. *From a business standpoint, the visibility Carl receives during the Olympic Games can enhance his value to the companies—Observer*, 1984. It retains its currency and must be accounted one of the most prominent of VOGUE WORDS.

**visible, visual.** *Visible* means 'that can be seen by the eye; *visual* broadly means 'concerned with, or used in seeing' (*visual aid, visual display unit*). The visual arts (painting, sculpture, etc.) are concerned with the production of the beautiful or the startling in visible form. The differentiation (as far as it goes) of *visible* and *visual* is worth preserving. For instance, the wrong word is used in the descriptive phrase *Diagnosis by visual symptoms*; the method of diagnosis is *visual*, but the symptoms are *visible*.

**vision,** in the sense of statesmanlike foresight or political sagacity, enjoyed a noticeable vogue some years ago and still flourishes: *the vision thing* in the language of George Bush, President of the United States (1989–93). *Where there is no vision the people perish* (Prov. 29: 18) is perhaps what makes the word tempting to politicians who wish to be mysteriously impressive; at any rate they are much given to imputing lack of vision to their opponents and implying possession of it by themselves when they are at a loss for more definite matter.

**visit** (verb). Few expressions are more likely to bring hostility to American English to the surface than when an American speaks of visiting *with* someone (*We visited with our elder daughter at Easter*). This phrasal verb had some currency in the UK in the second half of the 19c. (the *OED* cites an 1850 example in a letter written by Euphemia ('Effie') Ruskin, and an 1872 one from George Eliot's *Middlemarch*), but seems to have become largely confined to AmE throughout the 20c. It should be borne in mind that in AmE *visit with* usu. means no more than 'call on (someone) for a friendly chat'. *Visit*, used by itself without *with*, has abundant currency in America.

**visitor.** Thus spelt, not *-er*, except in Daisy Ashford's small comic masterpiece *The Young Visiters* (1919).

**visit, visitation.** *Visitation*, once a formal word for visiting, as in the Prayer-Book Service for the Visitation of the Sick, is now little used except for official visits of inspection, esp. ecclesiastical, by someone in authority, and for an affliction attributed to divine or other supernatural agency. *Visitation* is also used in some parts of America (beside *viewing* and *visiting*) for the paying of last respects to the dead in the day or days between death and burial (*Visitation at Felician Sisters Provincial House Friday 10 a.m. to 7 p.m.*).

**visor.** Thus spelt in OUP house style (not *vizor*) in all its main senses (part of helmet, peak of cap, sun-shield in car, etc.).

**visual.** See VISIBLE.

**vitamin.** Pronounced /ˈvɪtəmɪn/ in BrE, but /ˈvaɪtə-/ in AmE (and also frequently in some Commonwealth varieties).

**vitellus** (yolk of egg). Pl. *vitelli* /-laɪ/. See -US 1.

**vitrify.** For inflexions, see VERBS IN -IE, etc. 6.

**vitta** (botany, zoology). Pl. *vittae* /ˈvɪtiː/.

**viva¹.** See VIVA VOCE.

**vivace** (music, in a lively, brisk manner). Pronounce /vɪˈvɑːtʃi/.

**vivarium** (enclosure for keeping animals in their natural state). Pronounce /vaɪˈveərɪəm/. Pl. *-ia*: see -UM 2.

**vivat.** See next.

**viva², vivat, vive.** Pronounced /ˈviːvə/, /ˈvaɪvæt/, and /viːv/, these are respectively the Italian, Latin, and French for 'long live —'. All may be expressed in the plural (*vivano, vivant, vivent*) when the object is plural: e.g. Fr. *vivent les vacances!* 'hurrah for the holidays!'. But some French authorities at any rate regard *vive* as invariable, e.g. *Vive les gens d'esprit*. All three verbs can be used as nouns in English, with pl. *-s* (e.g. *the repeated vivas of the vast crowd*).

**viva voce** /vaɪvə ˈvəʊtʃi/, an oral examination for an academic qualification. Often shortened to *viva*, and the shortened form is frequently used as a verb (*Most of the candidates vivaed are on the borderline between one class and another*).

**vive.** See VIVA².

**vivify.** For inflexions, see VERBS IN -IE, etc. 6.

**viz.** Abbreviated form of VIDELICET (*z* in its medieval shape being the usual symbol of contraction for *-et*). As is suggested by its usual spoken substitute *namely, viz.* introduces especially the items that compose what has been expressed as a whole (*For three good reasons, viz. 1 ..., 2 ..., 3 ...*) or a more particular statement of what has been vaguely described (*My only means of earning, viz. my fiddle*). It is sometimes printed without a full point, but *viz.* is the form recommended here. Care should be taken to distinguish *viz.* from *i.e.* and *sc.* (see these at their alphabetical places).

**vizor.** See VISOR.

**vocabulary.** See GLOSSARY.

**vocation.** See AVOCATION.

**vogue word. 1** The word *vogue* is a 16c. adoption of Fr. *vogue* 'rowing, course, success', from It. *voga* 'rowing, fashion', from *vogare* 'to row'. In the broad senses 'fashion, acceptance, currency, prevalence' it has had wide currency in English ever since. The *OED* examples show a range of concepts, activities, etc., that have been *in vogue* at various times and in sundry places since the 16c.: the study of medals, austere doctrines, travelling in a carriage, prodigality, mountaineering, tartan shawls, moustaches, waltzing, burlesque, etc. Curiously, except for an isolated 17c. example (*Pox on your Bourdeaux, Burgundie ... no more of these vogue names ... get me some ale*) the OED entry reveals that the application of *vogue* to words and names is predominantly a 20c. phenomenon. In the first edition (1926) of this book, Fowler introduced the concept of vogue words in the following manner: 'Every now and then a word emerges from obscurity, or even from nothingness or a merely potential and not actual existence, into sudden popularity ... Ready acceptance of vogue-words seems to some people the sign of an alert mind; to others it stands for the herd instinct and lack of individuality ... on the whole, the better the writer, at at any rate the sounder his style, the less will he be found to indulge in the vogue-word.'

**2** The attitude of writers of usage guides has been consistently hostile: e.g. '*Vogue words* wander in and out of journalism and can become instant clichés. They are as likely to deaden a story as to enliven it. The only safe rule is to beware of them. Current examples: backlash, bombshell, bonanza, brainchild, charisma, Cinderella of, consensus, crunch, escalate, facelift, lifestyle, mega-, persona, prestigious, quantum leap, rationale, trauma(tic), viable' (*Times Guide to English Style and Usage*, 1992). 'There are two reasons for avoiding a vogue word while the vogue lasts. First, any reader who has noticed the recent frequency of its occurrence will be irritated by it. Secondly, the usual result of its being used by many writers on every possible occasion is that its meaning becomes exceedingly vague' (M. Dummett, 1993).

**3** Usage guides have also stressed the ephemerality of vogue words: e.g. 'The borrowed Russian words *glasnost* and *perestroika* were vogue words a short time ago, and *infrastructure, ecosystem, caring, share,* and *senior citizen* are vogue words today' (K. G. Wilson, *Columbia Guide to Standard American English*, 1993).

**4** Vogue words come and go. They are as it were like baggage on an airport carousel, sometimes visible, sometimes not, appearing, disappearing, and reappearing. Many of them are treated at their alphabetical places in the book, and others in the entries for CLICHÉS, NOVELTY-HUNTING, and POPULARIZED TECHNICALITIES. See also -ATHON; -SPEAK; -VILLE.

A list of some of the vogue words and phrases of the 1990s follows: *at the end of the day; at this moment* (or *point*) *in time; ball game (a different,* etc.); *bottom line; brownie points; couch po: to; cutting edge; community (the Muslim,* etc.); *end result; flavour of the month, the; framework (in the—of); impact (v.); interface; in this day and age; kick-start (v.) (to—the economy); level playing-field, a; lifestyle; meaningful; name of the game, the; on a roll; ongoing; -oriented (e.g. market-oriented); parameter; persona; photo opportunity; pivotal; quantum jump/leap; scenario; situation; spin doctor; symbiotic relationship; syndrome; track record; visibility (e.g. of a product,* etc., *in the public awareness); window of opportunity; you name it.*

**voiced** (adj.) (phonetics). Uttered with voice or vibration or resonance of the vocal cords, as opposed to *unvoiced* or *voiceless*. Voiced consonants include *b, d, g, th /ð/, v,* and *z*. Their unvoiced or voiceless equivalents are *p, t, k, th /θ/, f,* and *s*.

**volcano.** Recommended pl. *volcanoes,* rather than *volcanos*. See -O(E)S 1.

**volition.** See VELLEITY.

**volley.** Pl. *volleys*. For verb inflexions, see VERBS IN -IE, etc. 2.

**volte-face.** Pronounce /vɒlt'fɑːs/.

**voluminous.** See LU 3.

**voluntarily.** Pronounced (in older and conservative use) /'vɒləntərɪlɪ/, but in AmE (and increasingly in BrE) /vɒlən'teərɪlɪ/, i.e. stressed on the third syllable. See -ARILY.

**vomit** (verb). The inflected forms are *vomits, vomited, vomiting*. See -T-, -TT-.

**vortex** (whirlpool, etc.). Pl. *vortexes*, but usu. *vortices* /-tɪsiːz/ in scientific and technical use. See -EX, -IX 2.

**votaress.** See -ESS 5.

**vouch.** See AVOUCH.

**vulgar, vulgarly.** For their use in apologies for slang, see SUPERIORITY.

**vulnerable.** The first *l* should be fully pronounced. Pronunciation with initial /ˈvʌn-/ is a vulgarism.

# W w

**wadi** (dry bed of a N. African torrent). This is the recommended spelling (pl. *wadis*), rather than *wady* (pl. *wadies*). Pronounce either /ˈwɒdi/ or /ˈwɑːdi/. (The Arabic word is transliterated as *wādī*.)

**wage, wages** (weekly pay). *Wages* is normally used in the plural (*Their wages are still too low*; *What are the wages for the job?*). The biblical *For the wages of sinne is death* (Rom. 6: 23) is the sole survivor of a once-common (14–early 18c.) construction with a singular verb. But *wage* (singular) is also used (*What sort of wage is he paid?*), and in some circumstances is obligatory (*wage earner*; *a wage cut*; *minimum wage*).

**wagon, waggon.** The form with a single *-g-* is recommended, but *waggon* is also used by some printing houses in Britain.

**wainscot.** Its derivatives are *wainscoted*, *wainscoting*. See -T-, -TT-.

**waistcoat.** Now always pronounced /ˈweɪs(t)kəʊt/, but until the early 20c. /ˈweskət/ was widespread. The *OED* (1921) called /ˈweskət/ 'colloq. or vulgar', but Fowler (1926) recommended it as standard. See VEST.

**wait** (verb). In addition to the transitive uses given s.v. AWAIT, *wait* has also acquired (late 18c.– ) a quasi-transitive colloquial use in the type *Don't wait lunch for me* ( = defer a meal).

**waitperson, waitron.** Two uncommendable coinages of the 1980s as 'common-gender' forms to replace *waiter* and *waitress*, i.e. a person, male or female, who serves at table in a hotel, restaurant, etc. The collective term *waitstaff* (*Waitstaff full and part time. Inquire within*—sign in the window of a restaurant in Chicago, 1992) has also been widely used in AmE since about the same time. *Waitron* is said to be formed on -*on*, as in *automaton*, seen as a contemptuous judgement of waiting at tables as a mindless, robotic activity: I wonder.

**waitress.** Under attack from feminists as being an unnecessary 'sexist' term, but it remains in normal standard use. See -ESS 4.

**waive** (verb) means 'refrain from insisting on or using (a right, claim, opportunity, legitimate plea, etc.)' (*COD*), and is derived from an OF verb meaning 'to allow to become a waif, to abandon'. Correct use: *Let us waive the formalities and proceed with the business of the meeting.* It has had a rich and varied history in English since the 13c. But in the early 19c. it began to be confused with the homonym *wave* (verb) and used to mean 'to put (a person or thing) aside, away, off with or as with a wave of the hand', e.g. *I cannot waive away all the teaching of history*. Fowler (1926) cited another example: *The problem of feeding the peoples of the Central Empires is a very serious and anxious one, and we cannot waive it aside as though it were no concern of ours.* Such confusions should be avoided.

**wake** (verb), **waken** (verb). See AWAKE.

**wale** (noun) has several senses, including 'a ridge on a woven fabric, e.g. corduroy', and (in the form *wale-knot*) 'a knot made at the end of a rope by intertwining strands to prevent unravelling or act as a stopper'. From the OE period until the 19c. it was also freely used to mean 'the mark or ridge raised on the flesh by the blow of a rod, strap, etc.'; but at some point in the 19c. *wale* was overtaken and supplanted in this sense by *weal* (by confusion with *wheal* 'pustule'). The etymological process was a complex one, but it simply meant that in practice in a 19c. book one is likely to encounter the word meaning 'a mark or ridge on the body caused by a caning or a whipping' to turn up in any one of three spellings (*wale, weal, wheal*). In the 20c. *weal* is the sole survivor.

**wallop** makes *walloped, walloping, walloper*. See -P-, -PP-.

**walnut.** In origin from OE *walhhnutu* 'nut of the (Roman) foreigner', the first element being the same as that in *Welsh* (OE *Welisc*, from *wealh, walh* 'foreigner' + *-isc*). It has nothing to do with *wall* (OE *weall*). See TRUE AND FALSE ETYMOLOGY.

**waltz, valse.** The first (from Ger. *Walzer*), described in a work of 1825 as 'the name of a riotous and indecent German dance', is the form that has established itself as the ordinary English, the other (Gallicized) form of the same word being largely confined to the music for the dance.

**wampum** (NAmer. Indian beads). Pronounce /ˈwɒmpəm/.

**want** (verb). The verb *want* has twenty-nine senses and sub-senses listed in the *OED*, divided into six main groups. Of these twenty-nine, thirteen are shown to have dropped out of use at various times since the word first entered the language from ON round about 1200, and one, recorded only in Shakespeare's *Macbeth* (*who cannot want the thought?* = who can help thinking?), is shown the red card by being labelled 'Confused use'. Such a rate of loss suggests that the language has been intolerant whenever this non-native verb looked like threatening the territory of other verbs of strong presence, especially *lack, need, desire*, and *wish*.

Lost constructions include *it wants of six* (*o'clock*) = it is not quite six (1709 in *OED*); *something wants* (followed by *to* + infinitive) = something is lacking (*Then, shall I see Laurentum in a flame, Which only wanted to compleat my shame*—Dryden, 1697); and *to want of* = to lack, not to have (*Unwrought gold and silver want considerable of that lustre and brightness they appear in at goldsmiths' shops*—George Smith, 1799). For some six centuries the verb *want* was welcomed into the whole of the linguistic space containing the *intransitive* notion of 'to be lacking or missing or being deficient in some respect'. Then the curtain began to come down. *Transitive* uses have fared better over the centuries: e.g. ( = not to have) *A purely optimistic creed always wants any real stamina*—L. Stephen, 1876; (in palaeography and bibliography) *folio 18 wants 1 leaf*—*Anglo-Saxon England*, 1976; ( = to come short of in telling the time of day) *It only wants five minutes to dinner*—Trollope, 1865. Even these uses, however, now sound distinctly antiquated, and some of them have retreated into dialectal

use. A Scottish use of *wanting* (present participle) meaning 'not having' is shown in Stevenson's *Kidnapped* (1886): *I would not go wanting* [ = without] *sword and gun*. *Want* still holds its own in the battle with *need* in the idiomatic phrase *to want for nothing*, in advertisements (e.g. *Wanted, a receptionist*; *Wanted, early books written on Fiji*), elliptically for 'Wanted by the police', and in a few other circumstances. But it is on the retreat.

The notion of lacking or needing leads easily enough to the notion of desiring or wishing to have, but such uses, though now routine—indeed dominant—are relatively recent. The *OED*'s earliest examples date from the early eighteenth century, just at the time when the verb was shedding its traditional sense of 'lacking'. Typical examples: *If every one of your clients is to force us to keep a clerk, whether we want to or not,* [etc.]—Dickens, 1840; *I want you to be a good boy*—S. Judd, 1845. This branch of meaning, the 'wishing' branch, firmly established itself in the nineteenth century and is now the dominant sense of the verb. Current idiomatic applications include 'to wish for the possession of' (e.g. *Tom wants a word processor for Christmas*), and 'to need a person sexually'. There are also two somewhat non-standard uses of the verb in place of *need*: (*a*) *want* + gerund (*David thought he wanted helping into bed*—M. Duckworth, 1960; *Well, this Perlmutter wants locking up for a start*—K. Amis, 1988); (*b*) *want* + infinitive (*you want to pull yourself together* [ = you need to, you must]).

New uses of the verb are arriving all the time. Modern uses, some of them not yet registered in the *OED* and other dictionaries, are shown in the following examples: ( + *that*-clause) *You want that I should lose both my lieutenants together?*—A. Lejeune, 1986; (with ellipsis of a verb of motion, originally 19c. Scottish) *The Federal Reserve chairman Mr Paul Volcker has reportedly told friends that he wants out*—*Guardian*, 1984; ( + *for* + object clause, US only) *My mother wanted so much for my sister to have the best animals*—*New Yorker*, 1989; (elliptical = do you want to?) '*Want to try one?' I asked, without thinking*—J. McGarry, 1990; (elliptical = want to) *It means he will be able to come in whenever he wants*—Nigel Williams, 1985; '*You can come in if you want,' she says*—B. Nugent, 1990; ( = wish) *I could do that if you want,* but it may mean replacing a few strings—P. Fitzgerald, 1980; So don't feel obliged to come along. Stay home if you want—F. Weldon, 1988; I'll meet you at Susan's party if you want—M. Bracewell, 1989. The long battles between want and lack or need seem to be almost over. Want is not often wanted in such circumstances. Now want is threatening want to and wish in certain conditions. In our day-to-day life we pick our way among all the uses of want with reasonable consistency and certainty. How we acquire that certainty is deeply mysterious.

**wantonness.** Thus spelt, with -nn-.

**wapiti** (a NAmer. deer). Pronounce /'wɒpɪtɪ/. Pl. wapitis.

**war.** And yee shall heare of warres, and rumors of warres (Matt. 24: 6) is the correct quotation (not war, not rumour). See MISQUOTATIONS.

**Wardour Street.** 'The name of a street in London, formerly occupied mainly by dealers in antique and imitation-antique furniture. Used attrib. in Wardour-street English, applied to the pseudo-archaic diction affected by some writers, esp. of historical novels' (OED). The term is first recorded in 1888, and it applies to certain uses (esp. those placed in the wrong century) of such expressions as the following. Some of these are treated at their alphabetical places: anent; a- as in aplenty; belike; betimes; eke ( = also); ere; erst (while); haply; hither; howbeit; nay; oft; peradventure; perchance; quoth; some there-compounds (e.g. -from); thither; to wit; trow; twain; ween; what time ( – when), whilom; wight ( = a person); withal; wot; yea; yon; yore (only in of yore).
See also ARCHAISM and INCONGRUOUS VOCABULARY.

**wardress,** a female prison warder. See -ESS 2. In practice both warders and wardresses are now called prison officers.

**-ward(s).** Most words ending in -ward(s) are used as adverbs, adjectives, and nouns. In BrE the -s is usually present in the adverb (backwards, downwards) but not in the adjective (a backward child, in a downward direction). In most circumstances AmE seems to prefer the forms without -s for the adverbs. The nouns, which are really absolute uses of the corresponding adjectives, tend to follow the adjectives in being without -s (looking to the eastward). See also the separate entries for many of these words: AFTERWARD(S); FORWARD(S); ONWARD(S); TOWARD(S); etc.

**warn.** Traditionally only a transitive verb (she warned them of the danger; I shall not warn you again; we warned her to be careful what she said; etc.), warn, since about the beginning of the 20c., has also come into common use in a narrow range of intransitive constructions. Examples: The Chancellor warned that more drastic measures might have to be taken; The BOAC warned that more flights will have to be cancelled if the strike goes on; (headline) Shadow minister warns against education cuts.

**warp** (noun). The warp is a set of parallel threads stretched out lengthwise in a loom. The threads woven across and between them are the woof or weft. The fabric that results is the web.

**warranty** is the term used in various branches of law (e.g. property law, contract law, insurance law) defining and delimiting certain kinds of covenants, contracts, and insurance risks. See GUARANTEE.

**washroom.** See TOILET.

**wash up** in BrE means to wash (crockery and cutlery) after use, to wash the dishes. In AmE it means to wash one's face and hands.

**wassail.** Pronounce /'wɒseɪl/.

**wast.** See BE 4.

**wastage** is partially differentiated from waste in that it tends to be restricted to mean (a) the amount wasted; (b) loss by use, wear, erosion, or leakage; (c) (preceded by natural) the loss of employees other than by forced redundancy (esp. by retirement or resignation). In most contexts wastage should not be used simply as a LONG VARIANT of waste: thus waste not wastage in to go to waste, a waste of time, a waste of words, a boggy waste; waste disposal unit, wasteland, waste-pipe, etc.

**waste-paper basket** is the normal word in BrE for a basket into which waste paper is thrown, as against *waste-basket* in AmE.

**Watergate.** The name of a building in Washington, DC, containing the headquarters of the Democratic Party, which was burgled on 17 June 1972 by persons connected with the Republican administration, an event that led to the resignation of President Richard M. Nixon in 1974. The terminal element *-gate* later became widely used in other formations (see -GATE) denoting actual or alleged scandals of other kinds.

**watershed.** At the beginning of the 19c. the word was formed in English (after the model of Ger. *Wasserscheide*, an old word in that language meaning 'water-parting') from *water + shed*, a now-dialectal word meaning 'separation, division, parting' (unrelated to the noun meaning 'a simple, usu. wooden, single-storey structure used for storage, etc.'). It was used by geologists to mean 'the line separating the waters flowing into different rivers or river basins; a narrow elevated tract of ground between two drainage areas'. Within a decade or two, *watershed* began to be used *loosely* (as the *OED* describes it) to mean (*a*) the slope down which the water flows from a water-parting; (*b*) the whole gathering ground of a river system.

We are not concerned here with the tangled history of the distribution of the words *watershed*, *water-parting*, and *divide* in geological works in the UK, America, and elsewhere, but with figurative uses of the word *watershed*. Since the last quarter of the 19c. the word has been used in the standard language to mean 'a turning-point in affairs, a crucial time or occurrence'. Examples: *A watershed of time between the Renaissance and the Counter-Reformation*—J. A. Symonds, 1886; *On the Town, which [Gene] Kelly himself describes as a watershed picture . . . which opened up the musical to location shooting*—*Listener*, 1980; *The publication of this book marked a watershed in their relationship*—M. Meyer (in the New SOED, 1993).

**wave** (verb). See WAIVE (verb).

**wax** (verb). Its primary meaning 'grow larger, increase' (as opposed to *wane*) leads naturally to the sense 'pass into a specified state or mood, begin to use a

specified tone'. In this meaning a following modifier must be an adj. not an adverb (*He waxed enthusiastic* [not *enthusiastically*] *about Australia*). Correct use: *When the Roman soldiers were asked to take part in the Claudian invasion of 43, they waxed indignant*—Antonia Fraser, 1988.

**waxen.** See -EN ADJECTIVES 2.

**way. 1** Among uses originating in AmE in the 20c. and achieving somewhat reluctant currency in BrE are *no way* (in colloquial contexts) = it is impossible, it can't be done (e.g. *He said he wouldn't start up a gang today—no way*—New Yorker, 1975); *way too* + adj. (e.g. *Stuart struggled with the suitcases, which were way too heavy for him, she thought*—New Yorker, 1987; *Jack? Nah, he's way too smart. You know Jack*—A. Billson, 1993); *the way* + clause, in which *the way* is treated as if it is a conjunction (e.g. *They'd assumed we'd run away to the army, the way three or four boys on our street had done*—Nigel Williams, 1985; *You help women the way people help dogs they're training to do tricks*—J. Hecht, 1991 (US)).

**2** *under way*, *underway*. See UNDER WAY.

**waylay.** For inflexions, see VERBS IN -IE, etc. 1.

**ways.** *We've come a ways in journalism, too*, wrote the American word maven William Safire in 1994, meaning (he said) 'we've come pretty far, but not too far', as distinct from 'a long way' if his meaning had been 'we had come really far'. In March 1994 the American Secretary of State Warren Christopher said that an overall peace agreement in Bosnia was *a ways down the road*. It was hard to say how far down the road peace was. Tom Wolfe in his *Bonfire of the Vanities* (1987) wrote *I was standing out in the street a little ways*; and T. R. Pearson in his *Cry Me a River* (1993) wrote *. . . concealed completely from me his pangs and his anguishments which surely went a ways towards confirming the wisdom of seeking Ellis's help.* Clearly *a ways* gives no more than an approximate indication of the distance a person, process, etc., has travelled along a particular route, and it may be qualified by an adj. (*good*, *great*, *little*, *long*, etc.). This plural form *ways*, thus used, was once standard in BrE: the *OED* (sense 23c) lists BrE examples from 1588 onward including the following: *Not that I hope . . . to live to any* *such Age as that neither—But if it be only to eighty or ninety: Heaven be praised, that is a great Ways off yet*—Fielding, 1749; *Falmouth . . . is no great ways from the sea*—Byron, 1809. But it is now only dialectal.

**-ways.** See -WISE.

**we.**

| 1 Normal uses.
| 2 Wrong case.
| 3 Indefinite *we*.
| 4 The royal *we*.
| 5 Addressing the sick and some others.

**1** Normal uses. Examples of normal uses of the first person plural *we* (including examples of *we* merged with reduced forms of accompanying auxiliary verbs): *We needn't stay long; We're going to a civic reception; We've still got plenty of time; we three girls; We Europeans rode the streets in cars; we neither of us; we both of us.*

**2** Wrong case. *We* is sometimes mistakenly used for *us* in the accusative case. Examples (drawn from standard sources): *that is good news for* we *who watch; a brassiere showing Jane Russell saying 'For* we *full-figured gals'; Australia's monarch-in-waiting has been sprayed with insecticide, like* we *mere mortals; Perhaps this product is best suited to* we *cloth-capped northerners.* Read *us* in each case. (It is worth noting that the *OED* (1921) cited similar examples from standard sources from *c*1500 to 1890, but commented 'now used only by the uneducated'.)

**3** Indefinite *we*. *We* is often used indefinitely in contexts in which the speaker or writer includes those whom he or she addresses, his or her contemporaries, family, fellow citizens, etc. Examples: *There is nothing we receive with so much reluctance as advice; What do we, as a nation, care about books?; In ordinary life we use a great many words with a total disregard of logical precision; We have to get our production and our earnings into balance.*

**4** The royal *we*. The *OED* gives examples from the OE period onward in which *we* is used by a single sovereign or ruler to refer to himself or herself. The custom seems to be dying out: in her formal speeches Queen Elizabeth II rarely if ever uses it now. (On royal tours when accompanied by the Duke of Edinburgh *we* is often used by the Queen to refer to them both: alternatively *My husband and I.*)

History of the term. The *OED* record begins with Lytton (1835): *Noticed you the* we—*the style royal?* Later examples: *The writer uses 'we' throughout—rather unfortunately, as one is sometimes in doubt whether it is a sort of 'royal' plural, indicating only himself, or denotes himself and companions*—N&Q, 1931; *'In the absence of the accused we will continue with the trial.' He used the royal 'we', but spoke for us all*—J. Rae, 1960. (The last two examples clearly overlap with those given in para 3.) It will be observed that the term 'the royal *we*' has come to be used in a weakened, transferred, or jocular manner. The best-known example came when Margaret Thatcher informed the world in 1989 that her daughter-in-law had given birth to a son: *We have become a grandmother,* the Prime Minister said. A less well-known American example: interviewed on a television programme in 1990 Vice-President Quayle, in reply to the interviewer's expression of hope that Quayle would join him again some time, replied *We will.*

**5** Addressing the sick and some others. *We* is often used instead of *you* by a nurse or other medical person when addressing a patient: e.g. *A young doctor came . . . 'Well, Mrs Orton, how are we?'.* A playful use of this convention is shown in the following example: *'We don't want to drop, do we?' 'I don't know what we're talking about,' Muriel said. 'Our head hurts.'*—H. Mantel, 1986. Used similarly in various other contexts: (a hairdresser speaking to a customer) *Do we have the hair parted on the left as usual, sir?*; (an army officer addressing a recruit) *Not quite professional soldier material, are we?*

**weal.** See WALE (noun).

**wear** (noun). See WEIR.

**wear, gear.** *Gear* in the sense of apparel, attire has been in use since the 14c. but in the 20c. has come to be regarded as an informal word in this sense (as opposed to *clothing, garment,* etc.). It remains, however, as a standard second element in *headgear* (what is worn on the head), though *-wear* is now the natural second element in *footwear, menswear, neckwear, sportswear,* and *underwear,* and also in the less well-known *eyewear, surfwear,* and *swimwear.*

**weave** (verb). Although some standard dictionaries treat them as a single verb, there are two distinct words, though admittedly they are tending to converge in one of their central senses. *Weave*[1] (pa.t. *wove*, pa.pple *woven*) is of native origin (OE *wefan*) and means to form fabric by interlacing long threads in two directions. *Weave*[2] (pa.t. *weaved*, irreg. *wove*, pa.pple *weaved*) entered the language in ME from ON *veifa* 'move from place to place', and means to move repeatedly from side to side, follow a devious course, (*and crucially*) thread one's way through obstructions. Any weaving manual will supply examples of *weave*[1]. The following examples of *weave*[2] will give some indication of the tendency of the two verbs to merge: (*weaved* pa.t.) *Then they got on to the little scooter and weaved down the lane*—J. Winterson, 1987; *Mrs. Gurney clambered over the driftwood and weaved across the wet sand toward the sea*—New Yorker, 1990; *Andrew found Ruth, and the two weaved their way to the car*—R. Rive, 1990 (SAfr.); (*wove* pa.t.) *They wove off* [on their bicycles] *through the theatre crowd, bending and bowing together*—A. S. Byatt, 1985; *She wove her way among the crowd, bumping into people, being bumped into*—M. Ramgobin, 1986 (SAfr.).

The tendency to converge is confirmed when one looks at figurative uses of *weave*[1]: e.g. ( = construct with elaborate care) *You stole the money, and you have woven a plot to lay the sin at my door*—G. Eliot, 1861; ( = to intermingle or unite closely or intimately as if by weaving) *Grave Dante weaving well his dark-eyed thought into a song divine*—R. Bridges, c1904.

**web.** See WARP (noun).

**wed.** Except in certain familiar collocations (e.g. *wedded bliss, wedding breakfast, With this ring I thee wed*), the traditional native verb *wed* is now fast falling into restricted use, though it is still much favoured by the popular press. It is also tending towards a single-form state, i.e. to be unchanged in its pa.t. and pa.pple. A range of typical modern examples: *He did wed one of his regular girlfriends ... in a Wicca ceremony*—Sky Mag., 1991; *The millionaire drummer has slapped a no-sex ban on their relationship until they wed*—Daily Mirror, 1992; *Supermodel Claudia Schiffer has ditched her boyfriend to wed Prince Albert*

*of Monaco, it was claimed last night*—Sun, 1992; *Matthew's graduate companion ... proposes a night's dissipation—'one last bachelorly bout before we wed the stern bride of the future'*—G. Swift, 1992; *How he met and wed Sweet Annie, a real 'piece of calico'*—Analog Sci. Fiction & Fact (US), 1994; *until a messy divorce allowed him to wed archaeologist Jacquetta Hawkes*—Guardian, 1994; *No one had to dwell on the ugliness of his policies if he were treated as a cartoon, sleepily wed to Cruella De Ville*—Spy (US), 1994. Also a common technological use: *Its power is wed to a patented suspension with 'unrivaled ride comfort'*—Fortune (US), 1991; *This power plant is wed to a double-pivot spring strut suspension*—Transpacific (US), 1992.

**Wedgwood.** So spelt, not *Wedge-*.

**Wednesday.** 1 Pronounce /'wenzdeɪ/ or /-dɪ/, i.e. as two syllables. The first *d* is silent.

2 For the type *Can you come Wednesday?*, see FRIDAY.

**ween** (verb) ( = think, suppose). A WARDOUR STREET word.

**weft.** See WARP (noun).

**weigh.** For *under weigh*, see UNDER WAY.

**weir** (dam built across river). Thus spelt. The older variant *wear* is now obsolete.

**weird.** Thus spelt. See -EI-.

**Welch.** See WELSH.

**well** (adv.).

1 *as well as.*
2 The preliminary or resumptive *well.*

1 *as well as.* Let us first set aside undisputed constructions in which the phrase simply means (*a*) 'in as good, satisfactory, etc., a way or manner as': e.g. *She affected ... to listen with civility, while the Hydes excused their recent conduct, as well as they could*—Macaulay, 1849; (*b*) with weakened force, passing into the sense of 'both ... and', 'not only ... but also': e.g. *Our churchmen have become wealthy, as well by the gifts of pious persons, as by ... bribes*—Scott, 1828. (Type (*b*) is probably still current but seems somewhat archaic.) Also, for the most part undisputed, is (*c*) the type in which *as well as* is used as a conjunction and means 'in addition to' (e.g. *He*

must irrevocably lose her as well as the inherit-ance—G. O. Curme, 1931) or 'and not only' (e.g. *It was obvious that he had been consulted as well as I*—G. Greene, 1965).

There are two main areas in which the grammatical status of *as well as* has been called into question. (*d*) Following verb singular or plural. Grammatical agree-ment calls for a singular verb to accord with the subject noun or noun phrase if it is in the singular. The standard construction is illustrated in the follow-ing examples, particularly in those in which commas are used fore and aft to bring out the parenthetical nature of the subordinating phrase or clause after *as well as*: *The back-ground as well as other parts is dotted or stippled*—W. M. Craig, 1821; *To talk as her thoughts came, as well as to wear her hair as it grew, was a special privilege of this young person*—G. Meredith, 1861; *This sheep, as well as the long-tailed Damara sheep and the small black-headed Persian, is very popular*—SAfr. Panorama, 1989; *He believes that tutor as well as pupil benefits from the arrangement*—Oxford Today, 1990. Some might have been tempted to use a plural verb in such constructions, i.e. to replace *is* by *are*, *benefits* by *benefit*, and so on, but they would have been ill-advised to do so. Nevertheless such plural constructions do occur from time to time: e.g. (cited by Jespersen) *When a man enlists in the army, his soul as well as his body belong to his commanding officer*—J. A. Froude, 1884.

(*e*) In most circumstances *as well as* may be idiomatically accompanied by a gerund: e.g. (the first two examples cited by Jespersen) *as you're a Trinity scholar as well as being captain of the eleven*—E. F. Benson, 1916; *As well as being unwashed, the girl seemed thickheaded*—W. B. Maxwell, 1919; *The protagonists are mercilessly guyed by the author, as well as being clobbered by the 'system'*—R. Yarrow, 1982; *Just like Dolly to usurp the mourning function as well as presuming to treat the evening as a normal evening party*—A. Brookner, 1985.

Fowler (1926) argued, however, that in certain circumstances the gerund is better replaced by another part of the verb in order to match the part of the verb used in the introductory clause. He cited several examples (with suggested improvements) including these: *The Terri-torial officer still has to put his hand into his pocket* as well as giving *his time*. (Read

give, which depends on *has to*, or else substitute *besides*.) *His death leaves a gap as well as creating a by-election in Ross and Cromarty*. (Read *creates*, which is parallel to *leaves*, or else substitute *besides*.) *A German control of the Baltic must vitally affect the lives of all the Scandinavian Powers* as well as influencing *the interests of a maritime country like England*. (Read *influ-ence*, which depends on *must*, or else substitute *besides*.) So there is room for disagreement in sentences containing *as well as* followed by a gerund. Each case must be judged on its merits.

**2** The preliminary or resumptive *well*. What the *OED* calls the preliminary or resumptive use of *well*, tantamount to a plea for a moment's grace to gather one's thoughts, is for most of us a reflex re-sponse to the stimulus of any question. Those taking part in broadcast dis-cussions would add to the pleasure of listeners if they would try to curb it, although it would be unreasonable to expect them to abandon altogether such a natural item of speech. It should also be borne in mind that the use of *well* in a resumptive manner has been a feature of the language from the Anglo-Saxon period onward.

**well and well-.** For combinations of *well* and a participle there is widespread uncertainty about whether the two parts should be hyphened or left separate. In fact the matter is easily resolved. If a participle with *well* is attributive (*a well-aimed stroke*) a hyphen is normally used in order to clarify the unity of the sense; but if the participle is predicative (*the stroke was well aimed*) a hyphen is not required. This is not an arbitrary rule; it follows from acceptance of the principle that hyphens should be used only when a reader is helped by their presence. Cf. *up to date* s.v. UP 2. See HYPHENS.

**well-nigh** (adv.) (very nearly, almost wholly or entirely). It has been in con-tinuous use since the OE period and is still reasonably common. But it looks as if it is drawing towards archaism or becoming restricted to literary use (at a slightly slower rate, it would appear, than its simplex *nigh*). Recent examples: *His journey toward anti-intellectualism was well-nigh complete by this time*—R. Ferguson,

1991; *He found that the craft of point-engraving on glass had almost vanished from twentieth century England, and wellnigh single-handed renewed it—Antique Dealer & Collectors Guide, 1992; If your country is the size of a postage stamp, your population is unsophisticated and your borders are well-nigh indefensible, you need luck—Economist, 1992.*

**Welsh.** In past times *Welch* was a frequent variant of *Welsh*, but in the 20c. *Welsh* has prevailed for all uses of the noun and the adjective except for the use of *Welch* in the name of the *Royal Welch Fusiliers*. Other Welsh units, including the *Welsh Guards*, are spelt with final *-sh* in the word *Welsh*.

**welsh** (verb). Used at first (since the mid-19c.) to mean 'to swindle (a person) out of money laid as a bet' (esp. in racing parlance), and later (since the 1930s), construed with *on*, 'to fail to carry out one's promise to (a person); to fail to honour (an obligation)', the word has understandably not gone down well in Wales and among Welsh people in America and elsewhere. Some writers have tried to take the sting out of the word by spelling it *welch*, but this practice is not widespread. However it is spelt or pronounced, etymologists have not been able to determine its derivation. It must not be assumed that the name of the people and the verb *welsh* are the same word. Our language abounds in homonyms (e.g. *light* not heavy, and *light* not dark; (Egyptian) *mummy* (Arabic origin) and *mummy* ( = mother)).

**Welsh rabbit.** This dish of cheese on toast emerged, with *rabbit* so spelt, in 1725. (In Mrs Hannah Glasse's *The Art of Cookery* (1747), on the same page, she apparently also called it a *Scotch rabbit*, but no one else seems to have followed suit.) In the same century the lexicographer Francis Grose defined *Welsh rabbit* in his *Classical Dict. of the Vulgar Tongue* (1785) as 'bread and cheese toasted' and added, 'i.e. a Welch rare bit'. To this day one encounters people who call the dish *Welsh rarebit* though there is no evidence of the independent use of *rarebit*. The origin of the name must remain a mystery: it is not thought to be Welsh in origin, there is no rabbit in it, and neither cheese nor toast is rare. A half-parallel to the name is *Bombay duck*, which was first exported from Bombay but is dried fish, not a duck.

**wen.** For *the Great Wen*, see SOBRIQUETS.

**were.** For the use of *were* as a subjunctive form (*as it were*, *wished that it were winter again*, *were this done*, etc.), see SUBJUNCTIVE MOOD.

**werewolf** /ˈweəwʊlf/. The recommended spelling (not *werwolf*) and pronunciation (not /ˈwɪə-/). Pl. *werewolves*.

**wert.** See BE 4.

**westerly.** See EASTERLY.

**westernmost.** See -MOST.

**westward(s).** See -WARD(S).

**wet.** The pa.t. and pa.pple forms are either *wet* or *wetted*. The shorter form seems to be on the retreat except in familiar contexts (e.g. *she's wet the bed; while coming ashore from the lifeboat they wet all their clothes; after they had wet their whistles*). Meanwhile *wetted* seems to be favoured in most 'free' uses of the verb and in many forms of English: e.g. *perspiration wetted those few strands of hair that fell against his broad forehead—D. A. Richards, 1981 (Canad.); In the square, the statues, roofs and monumental buildings were wetted slick—B. Moore, 1987 (Ir./US); Two weeks ago a heavy rain had leaked through the ceiling and wetted the box—NY Rev. Bks, 1987; with her clothes wetted and her pockets full of big round stones—A. S. Byatt, 1990 (UK).*

**wh.** In words beginning with the digraph *wh* the pronunciation depends on regional and social factors. In Received Pronunciation and some other accents in England and Wales it is usually /w/ with no aspiration or minimal aspiration, but in general American, in Canadian, and in Scottish and Irish English it is usually /hw/. Australian, New Zealand, and South African English mostly follow RP, but in these countries and sometimes in England one encounters varying (usually only faintly audible) degrees of aspiration, esp. in circumstances in which minimal pairs happen to occur (*whales/ Wales, white/Wight, Whig/wig*, etc.).

**wharf.** Pl. either *wharves* /wɔːvz/ or *wharfs* /wɔːfs/. See -VE(D), -VES.

**wharfinger** (owner or manager of a wharf). Pronounce /ˈwɔːfɪndʒə/.

## what.

1 Questionable number attraction.
2 *what* singular and *what* plural.
3 One *what* in two constructions.
4 *what* resumed by *and which* or *but which*.
5 *what* as affected expletive at end of sentence.
6 Pseudo-cleft sentences led by *what*.
7 *what* as relative pronoun.

**1 Questionable number attraction.** Fowler (1926) was in an unusually dogmatic mood when he discussed the type *what* (relative pronoun) + singular verb + a second verb followed by a plural complement. 'In each of the examples to be given it is beyond question that *what* starts as a singular pronoun ( = that which or a thing that), because a singular verb follows it; but in each also the next verb, belonging to the *that* of *that which*, or to the *a thing* of *a thing that*, is not singular but plural; this is due to the influence of a complement in the plural, and the grammatical name for such influence is *attraction*; all the quotations are on the pattern *What is said are words*, instead of *What is said is words*.' He went on to give a dozen quotations in each of which, he said, *are* should be replaced by *is*. Here are four of them: *What is required are houses or rents that the people can pay*; *What seems to be needed, and what, I believe, public opinion calls for, are stringent regulations to restrict the sale*; *What is wanted now are men who are Liberals today*; *What strikes the tourist most are the elegant Paris toilettes.*

Later studies of the matter have confirmed that Fowler's 'rule' is followed by most writers: e.g. *I know it's awful, the noise, keeping us all awake, but what really worries me is the numbers*—N. Bawden, 1987; *What bothered him was drivers who switched lanes without signalling*—New Yorker, 1989. But there is an understandable tendency to let 'attraction' take its course in certain circumstances. In the type *What upsets me most is his manners* attention is focused on the singular verb *upsets*. In the competing type *What upsets me most are his manners* the attention is focused on the plural noun 'manners'. Each context will need to be judged on its merits, but for the sake of simplicity it is better to adhere to Fowler's rule than to make it subject to qualifications.

**2** *what* singular and *what* plural. In each of the quotations in the prec. section, the writer made it plain, by giving *what* a singular verb, that he or she conceived *what* in that context as a singular pronoun. But the word itself can equally well be plural: *I have few books, and what there are do not help me.* So arises another problem concerning the number of verbs after *what*, and this second one naturally gets mixed up with the first, in which some of the examples could be mended as well by giving *what* a plural verb as by giving the complement a singular one. In dealing with this other problem, however, we will ignore the complication that the number of a verb may be affected by the 'attraction' of the complement, and consider only the question of when a verb governed by *what* should be treated as singular and when as plural.

First comes a particular form of sentence in which plural *what* is better than singular, or in other words in which its verb should be plural. These are sentences in which *what*, if resolved, comes out as the ——s that, —— standing for a plural noun actually present in the complement. After each quotation a correction is first given if it is desirable, and in any case the resolution that justifies the plural: *We have been invited to abandon what seems to us to be the most valuable parts of our Constitution* (read *seem*; abandon the parts of our Constitution that seem); *The Manchester City Council, for what was doubtless good and sufficient reasons, decided not to take any part* (read *were*; for reasons that were); *Confidence being inspired by the production of what appears to be bars or bricks of solid gold* (read *appear*; production of bars or bricks that appear).

But resolution of *what* often presents us not with a noun found in the complement, but with some other noun of wider meaning, or again with the still vaguer *that which*. A writer should make the resolution and act on it without allowing the number of the complement to force a plural verb on him or her if the most natural representative of *what* is *that which* or *the thing that*. In some of the following quotations the necessary courage has been lacking. Corrections and resolutions are given as before: *No other speaker has his peculiar power of bringing imagination to play on what seems, until he speaks, to be familiar platitudes* (read *seem*; on sayings that seem); *Instead of the stupid agitation now going on in South Wales, what are needed are regular working and higher outputs* (read *what is needed is*; the thing that is needed—rather than things, as opposed to *agitation*); *What provoke men's curiosity are mysteries, mysteries of motive or stratagem; astute or daring plots* (read *provokes . . . is*; that which provokes—rather than the things that provoke); *In order to reduce this material to utility and assimilate it, what are required are faith and confidence, and willingness to work* (read *what is required is*; but the qualities that are required justifies the plurals, though it does not make them idiomatic).

It will be observed that there is more room for difference of opinion on this set of examples than on either those in (1) or the previous set in (2), and probably many readers will refuse to accept the decisions given; but if it is realized that there are problems of number after *what*, and that solutions of them are possible, that is sufficient.

**3** One *what* in two constructions. It is relatively easy to come to believe that *what* can be used in one construction, and can then, though merely being 'understood', govern a second dissimilar construction. Fowler (1926) gave four examples of such anomalies and emphasized that there was no one way of correcting such misfits: *This is pure ignorance of what the House is* and its *work consists of* (and what its); *But it is not folly to give it what it had for centuries and was only artificially taken from it by force more than a hundred years ago* (what belonged to it for); *Mr —— tells us not to worry about Relativity or anything so brain-tangling, but to concentrate on what surrounds us*, and we *can weigh and measure* (and can be weighed and measured); *Impossible to separate later legend from original evidence as to what he was, and said, and how he said it* (and what he said).

**4** *what* resumed by *and which* or *but which*. Francis Turner Palgrave, whose name is inseparably connected with *what is probably the best*, and which *certainly has proved*

the most popular, of English anthologies (what is probably the best, and has certainly proved); *It is an instructive conspectus of views on what can hardly be described as a 'burning question'*, but which certainly *interests many Irishmen* (but certainly interests); *We are merely remembering what happened to our arboreal ancestors*, and which *has been stamped by cerebral changes into the heredity of the race* (and has been stamped).

A want of faith either in the staying power of *what* (which has a good second wind and can do the two laps without turning a hair) or in the reader's possession of common sense has led to this thrusting in of *which* as a sort of relay to take up the running. These sentences are not acceptable English; nothing can represent *what*—except indeed *what*. That is to say, it would be English, though hardly idiomatic English, to insert a second *what* in the place of the impossible *which* in each. If the reader will try the effect, he or she will find that the second *what*, though permissible, sometimes makes ambiguous what without it is plain; in the last example, for instance, 'what happened' and 'what has been stamped' might be different things, whereas 'what happened, and has been stamped' is clearly one and the same thing. The reason why *which* has been called 'impossible' is that *what* and *which* are of different grammatical values, *which* being a simple relative pronoun, while *what* ( = that which, or a thing that) is a combination of antecedent and relative. The second verb needs the antecedent-relative just as much as the first, if *but* or *and* is inserted; if neither *but* nor *and* is present, *which* will sometimes be possible, and so omission of *and* would be another remedy for the last example.

Two specimens are added in which the remedy of simply omitting *which* or substituting for it a repeated *what* is not possible without further change. The difficulty is due to the superstition against PREPOSITION AT END, and vanishes with it. *I can never be certain that I am receiving what I want and for which I am paying* (read *what I want and am paying for*); *But now we have a Privy Councillor and an ex-Minister engaged daily in saying and doing what he frankly admits is illegal*, and for which *he could be severely punished* (read *and what he could be severely punished for*). The repetition of *what* is required because the relative contained in the first *what* is subjective, and that in the second objective.

**5** *what* as affected expletive at end of sentence. This pompous use is listed in the *OED* with examples from 1785 (Fanny Burney) onward. Watch out for it and you will find it used from time to time, usu. in a mock-affected tone. Examples: *Can't say I've read it. It's a bit too literary for me. What? But they say it's jolly clever*—A. N. Lyons, 1914; *they do tend to insist when they can on dealing with the fellow they're used to, even if he's a poor old antique like me, what?*—K. Amis, 1988.

**6** Pseudo-cleft sentences led by *what*. This unattractive and inelegant type, called pseudo-cleft sentences by modern grammarians, is shown in the following examples: *Anyway, with this piece, what Gounod did was he went and put words to that silent part that was never meant to be exposed*—M. du Plessis, 1989 (SAfr.); *and she was bound to fall out so what Jimmy Sr did was, he went into the dunes and found a plank*—R. Doyle, 1991 (Ir.). It faintly reflects the type in which the actual words used by one person are reported by another: e.g. *What Dorothy said was 'My mother's on the phone.'* (*CGEL* 14.29).

**7** *what* as relative pronoun. From the OE period until the 19c. *what* was in standard use as a relative pronoun meaning 'that, who, or which', but the use is now in abeyance except among the uneducated or partially educated, typified by these two examples, one British and the other American: *I was the only boy in our school what had asthma*—W. Golding, 1954; *Boy, the guy what thought it up sure was a smart one*—G. Keillor, 1985. As for its distribution in Britain, make what you will of the following statement in a 1989 issue of a learned journal that deals with such matters: 'Nonstandard relative *what*, a feature of informal, vernacular speech, appears … to have been introduced at the top of the accessibility hierarchy, where relativization is less complex syntactically, and to have been introduced subsequently at successively lower positions on the accessibility hierarchy.'

**what ever, whatever.**

**1** The interrogative use. For the type *What ever did you do that for?*, see EVER 1. Despite the tendency to join the two words together (i.e. to write as *whatever*) in literary works from the 14c. to the 19c., they should no longer be joined, since the type has become restricted to colloquial contexts, with heavy emphasis placed on *ever*, intimating 'that the speaker has no notion what the answer will be' (*OED*).

**2** The antecedent-relative use. Constructions of the type *whatever* (or *whatsoever*) + noun (or noun phrase) + *that*-clause are incorrect in that *whatever* by itself means *all that* or *any/thing* etc.) *that*. To insert *that* would render the sentence ungrammatical. Omission of the roman-type *that* in the following examples would set matters right: *His cynical advice shows that whatever concession to Democracy that may seem to be involved in his words, may not be of permanent inconvenience; Keep close in touch with Him in whatsoever creed or form that brings you nearest to Him.*

**3** The concessive use. *Whatever one does, you are not satisfied; I am safe now, whatever happens; whatever you do, don't lie.* These are concessive clauses, short for 'Though one does A or B or C', 'Though this or that or the other happens', 'Though you do anything else'. They differ from the *whatever* clauses dealt with above in being adverbial, *whatever* meaning not *all* or *any that* (*that* beginning an adverbial clause), but *though all* or *any*. The difference is not a matter of hair-splitting. *Whatever he has done he repents* may mean (a) he is one of the irresolute people who always wish they had done something different, or (b) though he may be a great offender, repentance should count for something; *whatever* antecedent-relative gives (a) and *whatever* concessive gives (b). Punctuation can be used to distinguish the two: meaning (a) can be indicated by not having the two clauses parted by a comma, since *whatever* belongs to and is part of both; and meaning (b) can be indicated by the insertion of a comma between the two clauses. Admittedly this places a great weight on a simple comma, but there is no other way (short of rewriting) to make the meaning clear.

**wheal.** See WALE (noun).

**wheaten.** See -EN ADJECTIVES.

**when.** See IF AND WHEN.

**whence.** The word has a long (13–19c.) history in the language used in various routine ways, e.g. (in an indirect question) *He inquired whence the water came*—M. Edgeworth, 1802; (*from whence*—from what source) *From whence I did then conclude ... that Wine doth not inspire Politeness*—Swift, 1731–8; *From whence have we derived that spiritual profit?*—Dickens, 1853; (*from whence* = from which place) *Let him walke from whence he came*—Shakespeare, 1590. (Further examples are given s.v. FROM WHENCE.)

At some point in the 19c. attitudes changed: *whence* came to be more or less restricted to formal or literary use, and *from whence* to be regarded as an old-fashioned expression better replaced by *where ... from*. The outcome is that Tennyson's *O babbling brook ... whence come you?* (1855), for example, is now seen to be a quaint poetic way of saying '... where do you come from?' There is no going back. *Whence* (and to a greater extent *from whence*) is no longer a routine word in frequent use. But words come and go, and *whence* may well come back into routine use at some point in the future. Meanwhile here are two modern examples from standard British and American sources: *He has also, of course, a passport which nails him for who he is and whence he comes*—P. Lively, 1987; *On learning whence she hailed ... he was briefly enchanted to meet somebody he took to be an ideological bird of a feather*—New Yorker, 1987.

**when ever, whenever.** When used as an interrogative adverb in the type *When ever did you say you were sorry?*, *when ever* (always written as two words) is merely an emphatic extension of *when*. See EVER 1; WHAT EVER, WHATEVER 1. (*When did you ever say you were sorry?* is a variant of the same type.) *Whenever* is the right form for the ordinary conjunction.

**whereabouts** (noun). Towards the end of the 18c. *whereabouts*, plural in form, replaced the earlier noun *whereabout*, or rather the two forms stood side by side

until the mid-19c. at which point *whereabout*, singular in form, began to disappear. The only trouble is that standard authorities and the printing world at large do not agree about its number. One authority (1970) said bluntly '*Whereabouts* takes the singular'. So did another (1962): 'Even if the reference is to several persons, each with a different whereabouts, a singular verb is still used: "She has a brother and two sisters, but their whereabouts is unknown."' A few years ago a Polish professor of linguistics told me that she was taught to use a plural verb after all uses of the word. Standard dictionaries for the most part settle for a formula that recognizes that the noun is plural in form but is construed either as a plural or a singular noun. I have examined a large body of newspaper and other evidence, British and American, of the period 1988 to 1994. Leaving aside cases where the number is left undetermined (e.g. *In Scotland ... the erectors of signposts seem reluctant to reveal to the motorist the precise whereabouts of Dumfries*), my mini-corpus revealed that of the four possible constructions (of those in which the writer had to make a choice of number in the following verb) there was a clear preference for a plural verb. The examples that follow are drawn from this mini-corpus: **1** One person or thing + a singular verb: *Colleen's whereabouts was kept secret; Five months later, the whereabouts of the purse and its contents remains a mystery; Its whereabouts was never revealed; The exact whereabouts of its new home has yet to be determined.*

**2** One person or thing + a plural verb: *His current whereabouts were not disclosed; He did not call his wife, Felicidad, whose whereabouts are not known; His whereabouts are definitely known.*

**3** More than one person or thing + a singular verb: *the whereabouts of the remaining two paintings is unknown; the whereabouts of two, [names given], is unknown; the whereabouts of three others was unknown.*

**4** More than one person or thing + a plural verb: *by early Friday the whereabouts of the raiders still were unknown; The nuns left suddenly and a spokeswoman said their whereabouts were a mystery; Israel refuses to release any of its Arab prisoners until the whereabouts of six of its servicemen are* known; *the other rank of pipes mentioned were never a part of the organ at Pilgrim and their whereabouts are unknown.*

Types 2 and 4 were more frequent than types 1 and 3, but all four types must be regarded as standard.

**whereby.** One of the diminishing number of *where*-compounds still in common use. It means 'by what or which means'. Some examples: *'Acceptance in lieu', whereby heirs give a picture to a museum to pay off capital transfer tax, has also been kind to the National Gallery—Antique Collector,* 1990; *This latest approach to so-called gene therapy is known as 'biolistic' technology, whereby scientists shoot genes at very high velocities into animal cells as a coating on the surface of tungsten pellets—New Scientist,* 1990; *Cooperative arrangements whereby schools formed clusters to pool training, sport and music facilities, teaching ideas,* [etc.]— TES, 1990.

**where-compounds.** A small number of these are still in free general use, though chiefly in limited applications, with little or no taint of archaism. These are WHERE-ABOUTS, *whereas* (in contrasts), *whereby, wherever, wherefore* (as noun plural in *whys and wherefores*), *whereupon* (in narratives), and *wherewithal* (as noun). The many others—*whereabout, whereat, wherefore* (adv. and conj.), *wherefrom, whereof, whereon, wheresoever, wherethrough, whereto, wherewith,* and a few more—have given way, though to different degrees, in both the interrogative and the relative uses either to the preposition with *what* and *which* and *that* (*whereof* = of what?, what ... of?), or to some synonym (*wherefore* = why). Apart from their use in certain formulaic legal documents, words in this second group should be resorted to only in very formal contexts or as a matter of pedantic humour.

**where ever, wherever.** As for WHEN EVER, WHENEVER.

**wherefore.** It has been pointed out over and over again that when Juliet cries out from the balcony *O Romeo, Romeo, wherefore art thou Romeo?* she is asking not *where* he is but *why* or *for what purpose* or *on what account.* Shakespeare uses *wherefore* in such a manner more than a hundred times (and also as a noun: *in the why and the wherefore is neither rime*

*nor reason* (1590)). And so frequently did other writers from the 13c. to the 19c., But it did not deter a headline writer in a 1988 issue of the *Sunday Times* from placing the headline *Wherefore art thou, biographer* above a letter written by Anthony Holden who had published a new biography of the Prince of Wales and had allegedly gone to ground away from journalists.

**wherein.** As conjunction and adv. labelled 'formal' in some dictionaries, but surprisingly widespread nevertheless. Some examples (mostly from 'informal' sources): *Biamping. Method of powering speakers wherein each drive unit of a speaker is powered by a separate channel of an amplifier—What Hi-Fi?,* 1991; *Madison Avenue triangle: A cultural space wherein all forms of expression other than those used by baby boomers are eliminated from stores, airwaves and advertising—Playboy* (US), 1992; *the decorator showhouse—the phenomenon wherein a dozen or so interior designers each decorate a room in a single residence no one will ever inhabit—Spy* (US), 1992; *Balance: A condition wherein all proportions of a dog are in static and dynamic harmony— Dog World* (US), 1994; *Australia's cultural isolation constitutes a comedic time-capsule, wherein Britain's exported vaudeville tradition has been preserved long after its extinction back in Blighty—Guardian,* 1994.

**wherewithal.** As noun (always preceded by the definite article) recorded since the beginning of the 19c. and curiously persistent in the sense 'the means, the resources, esp. pecuniary resources'. Examples: *To supply him with the wherewithal to pay for the defence of the border—Eng. Historical Rev.,* 1917; *You don't need the intellectual wherewithal to function in society—S. Pinker,* 1994.

**whether.** 1 For *whether* and *that* after doubt, see DOUBT (verb).

2 *Whether or no(t). Whether he was there or was not there* easily yields by ellipsis *Whether he was there or not,* and that by transposition *Whether or not he was there. Whether or no he was there* is not so easily accounted for, since *no,* unlike *not,* is not ordinarily an adverb (see NO 2), but this type has been in use since the 17c. From the evidence before me, *whether or not* is commoner than *whether or no* (though the *OED's* verdict in 1923 was the reverse

of this). Some examples of each: (whether or not) *I brooded all the way whether or not I had hit the right note*—J. Gardam, 1985; *He had not made it plain whether or not I would be welcome on the bus*—F. Weldon, 1988; *his face seemed familiar in a way which made you forget to ask whether or not you judged it good-looking*—Julian Barnes, 1989; *What authors need to know is whether or not a term has more than one meaning within its subject field*—Internat. *Jrnl Lexicography*, 1989; *'Whether the report is true or not,' said Bush, 'I know I speak for all here when I try to express to the American people the sense of outrage that we all feel about this kind of brutality*—*Chicago Tribune*, 1989; (whether or no) (1809–1986 examples s.v. NO 2(c)). Whichever form is used, such a doubling of the alternative as the following should be avoided: *But clearly, whether or not peers will or will not have to be made depends upon the number of the Die-Hards.* Omit either *or not* or *will not*.

**3** *Whether* is often repeated as a clearer pointer than a bare *or* to an alternative that forms a separate sentence: *I cannot remember whether they were lowered into the street or whether there was a window opening out at the back.*

**4** *Whether* without an alternative. The type *Whether this ended the matter remained unclear*, i.e. leaving *or no(t)* to be 'understood', is legitimate in a limited range of circumstances. It is much less common than types with alternatives expressed (e.g. *I was not sure whether to admire her or feel sorry for her*).

**5** *the question (of) whether* + clause. Three slightly different types are shown in the following examples, all of them standard: *The whole* question *whether women actually are more pacific by nature is not the subject of the present book*—Antonia Fraser, 1988; *The* question *whether to be, or not to be, a career woman had never bothered her*—D. Lessing, 1988; *Ms. Frankel raised the* question *of whether, if the current pace of reform continues in China, a similar conflict will arise between the old urban elites and the peasantry*—*Bull. Amer. Acad. Arts & Sci.*, 1989.

**which. 1** For the use of *which* (normally preceded by a comma) as a relative pronoun at the head of restrictive and nonrestrictive clauses, see THAT (relative pronoun) 3, 4.

**2** As was pointed out s.v. THAT (relative pronoun) 4, whereas *that* as a relative pronoun may not be preceded by a preposition, no such limitation applies to *which*. Examples: *She then referred directly to my plastic hand* over which *I always wore a glove*—G. Greene, 1980; *the procedures . . . are those* into which *young ladies are to be inducted*—N. Gordimer, 1987.

**3** *Which* leading a clause or sentence of affirmation. This type, in which, sometimes almost as an afterthought, a new clause or sentence (often one of direct comment on a state of affairs just described) is introduced by *which*, is now increasingly common, esp. in spoken English. The type is not new (the OED gives examples from the 14c. onward), but its frequency and its informality are more marked in the 20c. than hitherto. Examples: *as if they had to explain that they had been whisked away by the fairies and had spent the evening under their hollow hill, listening to incantations.* Which *in a sense they had*—H. Jacobson, 1983; *He does Mr Rabinowitz's teeth* which *is super*—Nigel Williams, 1985; *It was as though Hungary was not another place but another time, and therefore inaccessible.* Which *of course was not so*—P. Lively, 1987. A more extreme example from a 1995 book review: *What is missing from* You Will Learn to Love Me *is the kind friend—a voice we can trust.* Which *brings us back to the original question of what binds our sympathies to a character.* These examples are perilously close to the type (18–19c.) condemned by the OED as 'in vulgar use, without any antecedent, as a mere connective or introductory particle' (e.g. *If anything 'appens to you*— which *God be between you and 'arm*—*I'll look after the kids*—*Daily Chron.*, 1905).

**4** With a personal antecedent. 'Now only dialectal except in speaking of people in a body, the ordinary word being *who* or (as relevant) *that*' (OED). Examples: *Yow* which *I haue loued specially*—Chaucer, c1386; *O God,* which *art author of peace, and louer of concorde*—*Book of Common Prayer*, 1548–9; *A couple of women . . . one* of which *. . . leaned on the other's shoulder*—Goldsmith, a1774; *His mother had ten children* of which *he was the oldest*—*Scribner's Mag.*, 1899.

**5** *and which.* 'And *which* or *but which* should not be used unless the coming *which*-clause has been preceded by a clause or expression of the same grammatical value as itself' (Fowler, 1926). This is a simple but safe rule. An example of a misfitted *and which* clause: *In contrast Peake's use of elevated language has a childlike quality which is appropriate given that the protagonist, Titus, is a boy, and which I found endearing*—Oxf. Univ. entrance candidate, 1989.

**6** Miscellaneous. When the antecedent to a relative pronoun is an indefinite pronoun (e.g. *anything, everything, nothing, something*) or contains a superlative adjective qualifying an impersonal antecedent, *that* is normally preferable to *which*. Examples (from the *Oxford Guide to Eng. Usage*, 1993): *Is there nothing small that the children could buy you for Christmas?*; *This is the most expensive hat that you could have bought*. But there are exceptions: e.g. *there was nothing... which distinguished him from any other intelligent young Frenchman studying law*—P. P. Read, 1986.

**while** (noun). See WORTH.

**while** (conj.). Also *whilst* (but not used in AmE). The principal use as a conjunction is in temporal contexts in which the meaning is 'during the time that (*They had begun drinking while he prepared to cook*); at the same time as (*He enjoyed drawing while he was being read to*)'. Such temporal uses have been dominant since the OE period, so much so that some writers on English usage have concluded that other uses are improper (see below). At least since the 16c., however, adversative or concessive senses have emerged, in which *while* means 'although' or 'whereas'. Thus CGEL (15.20) contrasts *He looked after my dog while I was on vacation* (temporal) and *My brother lives in Manchester, while my sister lives in Glasgow* (concessive).

Eric Partridge in his *Usage and Abusage* (1942) condemned the use of *while* to mean 'although' in no uncertain terms: '*while* for *although* is a perverted use of the correct sense of *while*, which properly means "at the same time as", "during the same time that".' But this indefensible remark was counterbalanced by a quotation from the work of A. P. Herbert which draws attention to the danger of confusion between the temporal and the concessive senses: *The Curate read the First*

*Lesson while the Rector read the Second.* Such rare (and amusing) examples apart, the temporal, concessive, and contrastive uses of *while* (or, in BrE, also of *whilst*) pose no threat to one another and are all part of the normal apparatus of the language.

Some examples of various types from standard sources of the 1980s: (while) *while domestic happiness is an admirable ideal, it is not easy to come by*—T. Tanner, 1986 (concessive); *Suffice it to say that while the judicial record ... has been one of appalling subservience to the Government ...* —Times, 1986 (concessive); *We are told that our institutions should concentrate on their business of education while being told also that they should do more to contribute to economic development and direct public service*—H. H. Gray, 1988 (US) (contrastive); (whilst) *Whilst it's in the computer it's stored*—S. Hockey, lecture in Univ. of Oxford, 1981 (temporal); *large numbers of users of the 'Volcano'* [kettle] *... tend to enjoy brewing up whilst on fishing expeditions on the other side of the Irish sea*—Times, 1986 (temporal).

**whilom** /ˈwaɪləm/. The last two letters represent a remarkable survival of the OE dative plural ending -*um* (OE *hwīlum* 'at times, at some past time'). *Hwilom* formed part of the normal vocabulary of writers down to the 19c. (e.g. Scott and Carlyle), but now belongs to the category of words known as WARDOUR STREET. Typical uses: (as adv.) *The wistful eyes which whilom glanced down ... upon the sweet clover fields*; (as adj.) *General Doppel, a whilom Medical man*. See also LATE. Twentieth century examples (the first of which is for a nonce-adj. *whilomst*): *He saw these things in the whilomst moment when the Navajos were making merry with the Choctaws*—Henry Miller, 1939; *Hidden in the tall marsh grass of the coastal lowland, the whilom seaport that once rivaled Philadelphia was remarkable*—W. Least, 1982.

**whilst.** See WHILE (conj.).

**whin.** See FURZE.

**whine** (noun and verb). The corresponding adj. is *whiny* (not -*ey*). The pres. pple and ppl adj. form is *whining*. See -EY AND -Y IN ADJECTIVES.

**whinge** (verb) has *whingeing* as its pres. pple and ppl adj.

**whirr,** (make) continuous buzzing sound. Thus spelt, not *whir* (exc. that *whir* is the more usual form in AmE). The inflected forms of the verb are *whirred*, *whirring* in all varieties of English.

**whisky, whiskey.** The first is the usual spelling in BrE (*Scotch whisky*) and in Canada (pl. *whiskies*), and the second the normal spelling in Ireland and USA (pl. *whiskeys*).

**whit** ( = a least possible amount: usu. in negative contexts). Said by Fowler (1926) and Gowers (1965) to belong to the vocabulary of WARDOUR STREET, *whit* seems to be having a new lease of life as an ordinary, untainted word to judge from the evidence in the *OED* files and my own: *When Pritchard allowed an encore of the sextet, it seemed no more than natural justice, and it spoiled the drama not a whit—Opera Now*, 1990; *A more natural, less inspired response to the Mother's mothering . . . would have mattered not a whit to the Mother, just as long as not one would ever have seen it—R. Goldstein*, 1991; *This much ballyhooed Andrew Lloyd Webber musical is fun—if you're not bothered by theatre that cares not a whit for words and contains not one ghost of an idea—New Yorker*, 1991; *She cared not a whit for clothes—A. V. Roberts*, 1992 (US); *Ashdown's latest repositioning changes Major's position not a whit—Sunday Times*, 1995; *the embodiment of a proprietorial dream no whit less obsessive than Gatsby's own—A. Coren*, 1995.

**Whit.** *Whit Sunday* (or *Pentecost*) is the seventh Sunday after Easter; hence *Whit Monday*. *Whitsun* and *Whit* are frequently used as informal shortenings of *Whitsuntide*, the weekend including Whit Sunday. *Whit Sunday* answers to OE *Hwīta Sunnandæg*, lit. 'white Sunday', probably from the white robes of the newly baptized at Pentecost.

**white** (adj.) yields both *whity* (adj.) 'whitish' and the derogatory *Whitey* (noun) 'a white person (as seen by Blacks)', white people collectively'. The pl. of the noun is *Whiteys*.

**white(n)** (verbs). The shorter form is first recorded in OE meaning 'to become white; to make white', and, by *c*1200, 'to cover or coat with white'. *Whiten* appeared in the early 14c. with the second and third of these meanings, and

acquired the sense 'to become white' in the 17c. This longer form is the more usual one now in the primary senses 'to make white' and 'to become white'. See -EN VERBS FROM ADJECTIVES. But *white out*, rather than *whiten out*, is the more usual phrasal verb when the sense required is 'to obliterate a typing or writing mistake with white correction fluid'.

**White Paper.** In the UK and some other countries, an official document that expresses government policy on an issue. It is usually preparatory to the introduction of a parliamentary bill. Its name derives from its having fewer pages than a government 'blue book', and therefore needing no blue paper cover (*Hutchinson Encycl.*, 1990).

**whither.** English dictionaries prepared for foreign learners of English are especially ruthless in their treatment of archaic words. There is a rich vein of such vocabulary that is familiar to adult native speakers, but is judged to be of minimal interest in ELT circles. Thus *whither*, which was in unbroken standard use as adverb and conjunction from OE times until the 19c. but then began to retreat into archaism, is presented in the *Cambridge Internat. Dict. Eng.* (1995) as follows: '*adv.* (not gradable) *old use* to where *Whither are they going?*'. *Collins Cobuild Eng. Lang. Dict.* (1987) is slightly more expansive: '*whither* means to where: a formal or old-fashioned word, e.g. *Traitor's Gate, whither so many came by water to their death.*' And in the margin it classifies *whither* as 'adv with vb, or conj subord'. The *Oxf. Advanced Learner's Dict.* (4th edn, 1989) added *rhet*[orical] to the various labels attached to the word.

Such limitations placed on a once-central word underline its decline in currency. It may therefore be useful to list some 20c. examples to show that *whither* has not disappeared yet: '*Whither bound?*' '*To Bromfield, friend. Am I going right?*'—E. Peters, 1982; (title of article) *Whither Islamic Economics*, in *Islamic Q*, 1988; *I write, now, from my bed, whither Dr Felton has banished me—M. Roberts*, 1990; *The United States . . . is having an increasingly disagreeable time in the Horn of Africa, whither it went, nine months ago, to facilitate the distribution of food for a few, perhaps two, months—Newsweek*, 1993. And of course churchgoers are accustomed to its use in the Bible, e.g. *The winde bloweth where it listeth, and thou hearest the sound thereof, but canst not tel whence it commeth, and whither it goeth*—John 3: 8.

**whitish.** Thus spelt, not *-eish*.

**whizz** (noun and verb). The recommended spelling, not *whiz*. Hence *whizz-bang*, *whizz-kid* (but the forms with *-z-* are also commonly used). The inflected forms of the verb are *whizzes*, *whizzed*, and *whizzing*. See *-z-*, *-zz-*.

## who and whom.

> 1 Introductory.
> 2 An historical note.
> 3 Modern practice.
> A Legitimate uses of *who*.
> B (Mostly) legitimate uses of *whom*.

**1** Introductory. *Who* is the only relative pronoun that is declinable (as against the indeclinable *that* and *which*): *who* subjective (or nominative), *whom* objective (or accusative), *whose* possessive.

As relative pronouns, *who* and *whom*, when used, must have a person or persons (not a thing) as antecedent. They are in many circumstances replaceable by *that* or are omissible: e.g. *Was there anyone now (who(m)/that) she could decently ask?* [Iris Murdoch, in her novel *The Green Knight* (1993), actually wrote *Was there anyone now* whom *she could decently ask?*] Despite its centuries-old use as an objective form (see section 2), *whom*, both as a relative pronoun and as an interrogative pronoun, is still frequently left aside in favour of *who*. *Whom* also continues to be illogically ('ungrammatically') used for *who* when the subjective case is properly called for: see sections 2 and 3B(e).

There are three approaches to the problem of the modern distribution of *who* and *whom*. First, that of those who see the use of *whom* as moribund or at best as socially divisive: ' "To whom do you wish to speak?" is usually regarded as formal (in some circles superformal or superpolite), indeed almost as something frozen, archaic, stifling, or artificial. Indeed, a conversation might be killed right there' (A. S. Kaye in *English Today*, 1991). '*Whom* has outlived *ye* but is clearly moribund; it now sounds pretentious in most spoken contexts ... If the language can bear the loss of *ye*, using *you* for both subjects and objects,

why insist on clinging to *whom*, when everyone uses *who* for both subjects and objects?' (S. Pinker, *The Language Instinct*, 1994).

Secondly, there are those who say or imply that the retreat of *whom* is regrettable (the examples happen to be from fiction but are none the less revealing): 'In *The Archers* on the other hand, no one leaves the "m" off "whom" ' (L. Ellman, 1988). ' "I don't know whom else to ask." The elder of the two policemen, Butterworth, noticed that she had said "whom" and decided that she was a credible witness' (A. Brookner, 1992).

Thirdly, there are the deadpan ('scientific') views of professional grammarians: '*Who* has an objective case *whom* ... for which, however, the uninflected *who* is substituted in the spoken language, as in *who(m) do you mean?*' (H. Sweet, *A New English Grammar*, 1891). 'In objective use, *who* is informal and *whom* is formal. The distinction is parallel to that between *who* and *whom* as relative pronouns. Similarly, interrogative *whom* functions like relative *whom*, except that interrogative *whom* strikes most people as even more formal than relative *whom*' (CGEL 6.38, 1985).

**2** An historical note. There is abundant evidence in standard sources of earlier centuries of departure from the normal nom./accus. relationship of *who*/*whom*. One need look no further than Shakespeare and the Bible: *who wouldst thou strike?*—Shakespeare, 1591; Albany. *Run, run, O run.* Edgar. *To who my Lord?*—Shakespeare, 1605; *but wayl his fall.* Who *I my selfe struck downe*—Shakespeare, 1605; *But whom say ye that I am?*—Matt. (AV) 16: 15. (Contrast the NEB's *who do you say I am?*)

The OED lists numerous other examples (15–20c.) of the breakdown of formal grammatical rules governing *who* and *whom*. Among them: (*who* used 'ungrammatically' for *whom*) *A great Prince who I forbeare to name*—Monmouth, 1641; *He has a right ... to choose who he will have for a teacher*—J. A. Froude, 1849; *Not being able to ask exactly* who *he liked*—R. S. Surtees, 1858; *Just over half ... of our sample* who *we assessed as working class concurred*—*Times*, 1984. See also section 3A(e) below. (*whom* used 'ungrammatically' for *who*) *I counsel ... all wise ... men, that they doo not* accompany wyth those whom they know are not secret—T. North, 1557; Comparing the ... humble epistles of S. Peter, S. James and S. John, whom we know were Fishers, with the glorious language ... of S. Paul, who [sic] we know was not—I. Walton, 1653; Are they yonder Knights whom you suppose will attack us?—Mrs Lennox, 1752; A strange unearthly figure, whom Gabriel felt at once, was no being of this world—Dickens, 1837; He saw the man whom he knew must be the King—R. H. Benson, 1906. See also section 3B(e) below.

**3** Modern practice.

**A** Legitimate uses of who:

(a) (who as relative pronoun with a person as antecedent) women who had once talked of love now talked only of sexual harassment—M. Bradbury, 1987; Arnold was a lawyer ... who could never practise law again—N. Gordimer, 1987; His father had had a brother, an older brother, who, he let it be known, had dominated him cruelly—J. Updike, 1987; I could not be close to anyone who I thought was rejected by God—V. Mehta, 1988. Cf. THAT (relative pronoun) 1.

(b) (who as interrogative pronoun) 'You mean Loseby?' 'Who do you think I mean?'—C. P. Snow, 1979; 'But who's to blame?' asked Madeleine.—P. P. Read, 1986; 'We've got Sir Luke Trimingham.' 'Who?' asked Henry. The acting knight, that's who,' said Gill—M. Bradbury, 1987; Who else did I know?—C. Rumens, 1987.

(c) (who introducing an indirect question) The sight of him raised the obvious question of who was going to restrap his leg—T. Keneally, 1985; She wanted to ask who he was—New Yorker, 1987; May I ask who's speaking?—C. Rumens, 1987.

(d) (who with preposition stranded) Who do you think you're speaking to?—W. McIlvanney, 1985; Who's she talking to? Has she got a visitor?—P. Lively, 1987; 'Who does he deal with?' Fred asked—New Yorker, 1987; After all, what did she know of his life, who he went to bed with—I. Murdoch, 1993.

(e) (who used where strict grammar calls for whom) not knowing who is sitting with who in the T.V. room—E. Jolley, 1985 (Aust.); Christ, who went for who first?—V. O'Sullivan, 1985 (NZ); 'The country is behind us.' 'Behind who? The unions do not want this demonstration.'—B. Moore, 1987; There's a woman called Kristin Johannesdottir who I really admire—Melody Maker, 1988; It

was an ambush, apparently. But it was unclear who had ambushed who—Times, 1988.

**B** (Mostly) legitimate uses of whom:

(a) (whom as object after a preposition) These were people to whom no civic monuments would ever be erected—W. McIlvanney, 1986; They ... argue about a man called Simpkins of whom the poet is jealous—Encounter, 1987; Bernard made useful contacts from whom he could hope to make the income he needed—NY Rev. Bks, 1987; besides which he can leave his property to whom he likes—Barbara Neil, 1993; The grand ramp, too, was crowded with pilgrims from all over India, many of whom had just arrived by special trains—V. Seth, 1993.

(b) (whom as object in relative clause) Was there anyone now whom she could decently ask?—I. Murdoch, 1993. Cf. section 3A(a) above. (c) (whom as interrogative pronoun) But whom should we support in the present fluid situation?—Bull. Amer. Acad. Arts & Sci., 1990. Cf. section 3A(b) above. (d) (whom with preposition stranded) There were other people whom I would have liked to speak to—G. Butler, 1983; I suppose I could say that I am a man whom life has passed indifferently by—B. Rubens, 1985; an astronomical fee for this young man—whom I guarantee three-quarters of you have never heard of—Dædalus, 1986. Cf. section 3A(d) above.

(e) (whom used 'ungrammatically' for who) ... further alienation of the people whom, whether he likes it or not, have his fate in their hands—Times, 1984; the baronet whom Golitsin claimed had been the target for homosexual blackmail—P. Wright, 1987; In late 1982, officers of the Royal Ulster Constabulary shot dead six people whom they said were armed members of the Irish Republican Army—Economist, 1988; Other times, because they are dealing with people whom they know might have recourse to the law, they hold back the gratuitously offensive nicknames—Guardian, 1988; The junior Civil Servant whom the Government has claimed is implicated in insider dealing has been allowed to return to work at the Office of Fair Trading—Times, 1988; Although married with three children, he is demanding £5,000 from the elderly woman whom he says has ruined his life—Sunday Times, 1990; Mr Irwin said it would not be fair to name the signatories whom he thought despised Mr Roache—Times, 1991; ... far more hostile to Diana whom she believes betrayed the Prince of Wales—Independent Mag., 1993. Despite the antiquity and prevalence of this construction (see section 2 above), it attracts severe criticism at the present time, and is best avoided. In all such constructions, *who* is still the better form.

4 See WHOSE.

**whodunit** /huː'dʌnɪt/ is a facetious formation, from a 'phonetic' respelling of *who done it?*, a non-standard form of *who did it?*. First recorded in 1930, it is still widely used as a colloquial term for a story, play, etc., about the detection of a crime, esp. a murder.

**who else's.** See ELSE 1.

**whoever** etc. 1 Forms. Subjective: *whoever, whosoever* (emphatic), *whoe'er* (poetic), *whoso* (archaic), *whose'er* (poetic). Objective: *whomever* (rare, chiefly literary), *whoever* (wrong case but common), *whomsoever* (chiefly literary), *whomsoe'er* (poetic), *whomso* (archaic or obsolete). Possessive: *whose ever/whosever* (formal and rare), *whoever's* (common in spoken English), *whosesoever* (archaic or obsolete). Typical examples of *whomever: To impose his will on whomever he sees comfortably settled*—M. Beerbohm, 1920; *ready at once to relax with whomever came to hand*—A. Brookner, 1992.

2 For the emphasizing function of *ever* in the type *Who ever can have done this?*, see EVER 1.

3 In its main uses, *whoever* (pronoun) means (a) whatever person or persons, anyone who (*Whoever finds him, shoot him dead!*—Scott, 1813); (b) used with conditional force: if anyone at all, no matter who (*Whoever it is, I won't see them tonight*—M. E. Braddon, 1863).

**whole other.** For the NAmer. type *a whole other*, see OTHER 6(a).

**wholly** (adv.). Thus spelt, not *wholely*.

**whom.** See WHO AND WHOM.

**who's** is a shortened form of *who is* (*Who's coming to the party?*) and *who has* (*Who's been reading the proofs?*) It is occasionally carelessly used instead of *whose*: e.g. *'Conor,' called Vaun, humping a churn of milk. 'Who's turn to deliver?'*—J. Leland, 1987.

**whose.** Human or inanimate antecedent. 1 No problems arise in questions:

in these *whose* always refers to a person: e.g. *Whose book is this?*; *'Can you give me his phone number please?' 'Whose?' The landlord's.'*—C. Rumens, 1987.

2 It is quite another matter in relative clauses. The twists and turns of grammatical teaching from the 18c. onward produced the folk-belief that while *whose* was the natural relative pronoun when the antecedent was human, or at a pinch was an animal, it should not be used with an inanimate antecedent. The OED (1923), by contrast, demonstrated that in all kinds of circumstances from medieval times onwards *whose* had been used as a simple relative pronoun with an inanimate antecedent. Fowler (1926) was at his most vehement in attacking the rigidity and the prevalence of the folk-belief: 'in the starch that stiffens English style one of the most effective ingredients is the rule that *whose* [as a relative pronoun] shall refer only to persons.' After citing some examples of *whose* used with an inanimate antecedent (including an example of *whose* as an objective genitive in the opening lines of Milton's *Paradise Lost: Of man's first disobedience, and the fruit Of that forbidden tree, whose mortal taste Brought death into the world …*), he concluded his article by declaring: 'Let us, in the name of common sense, prohibit the prohibition of *whose* inanimate; good writing is surely difficult enough without the forbidding of things that have historical grammar, and present intelligibility, and obvious convenience, on their side, and lack only—starch.'

3 Past and present usage. The second edition of the OED (1989) lists numerous examples of *whose* inanimate, including the following: *I could a Tale vnfold, whose lightest word Would harrow vp thy soule*—Shakespeare, 1602; *Mountains on whose barren brest The labouring clouds do often rest*—Milton, 1632; *The clock, whose huge bell … may be heard five leagues over the plain*—Southey, 1807; *There were pictures whose context she understood immediately*—I. McEwan, 1981.

To which may be added: (with person as antecedent) *She was not the only child whose parents were divorced*—N. Gordimer, 1987; (with animal as antecedent) *The exhaust from the car irritates the lion, whose head rolls from side to side*—New Yorker, 1987; (with inanimate antecedent) *An aged ferry* steamer ... whose *captain was a White Russian princeling*—R. Sutcliff, 1983; ... *going to live ... in a boardinghouse* whose *landlady was well known*—J. Frame, 1984; *we all bear the burden of making personal decisions* whose *soundness depends on our ability*—*Dædalus*, 1986; ... *a deeply disturbing book ... whose message will remain with you*—Susan Hill, 1986; *He looked up again at the tank* whose *huge cannon seemed to be pointing at him*—P. P. Read, 1986; ... *bought on an Access card* whose *credit limit has already been exceeded*—*Daily Tel.*, 1987; *It is a malady* whose *effects seem relatively independent of changes in the language in which they are experienced*—*TLS*, 1987; *Anne heaved her cases up to the front doors through* whose *glass panels she could make out a broad lobby*—S. Faulks, 1989.

Taken together they demolish the folk-belief that *whose* must always have a personal antecedent. But it would be equally wrong to suggest that equivalent *of which* constructions are being driven out. Readers may indeed prefer to read *of which* instead of *whose* in several of these 1983–9 examples (*the captain of which*; *the landlady of which*; *the soundness of which*; etc.). In fact it is likely that *of which* constructions outnumber those with *whose* when the antecedent is inanimate, esp. in essays and monographs. The *OED* (1923) was on target in its definition of *whose* 'in reference to a thing or things (inanimate or abstract). Originally the genitive of the neuter *what* ...; in later use serving as the genitive of *which* ..., and usually replaced by *of which*, except where the latter would produce an intolerably clumsy form'.

**why.** 1 Pl. *whys* (in the phr. *the whys and the wherefores*).

2 See REASON WHY.

**wide.** It should be borne in mind that there are a good many circumstances, mostly in fixed phrases, in which, though *widely* is theoretically the needed form, *wide* is the idiomatic form; see UNIDIOMATIC -LY for such uses. Thus *wide apart*, *wide awake*, *open one's eyes wide*, *wide open*, *is widespread*, are all idiomatically required (not *widely apart*, etc.); and there are many more.

**-wide.** This familiar formative element, found, for example, in *countrywide*, *nationwide*, *statewide*, and *worldwide*, shows signs of spreading its wings in AmE: e.g. *But there's much more involved in creating enterprisewide applications than mastering Notes*—*PC Week*, 1992; *We may be approaching a point where, on an economywide basis, layoffs may become counterproductive*—*Chicago Tribune*, 1993.

**wide(-)awake.** *He is wide awake*; but when used attributively, *a very wide-awake person.*

**widely.** See WIDE.

**wife.** Pl. *wives*. See -VE(D) etc.

**wig.** See PERIWIG.

**wight** (a person). A WARDOUR STREET word.

**wild.** 1 On the principle that hyphens should not be used except when needed to prevent ambiguity, *wild* in the sense of not domesticated or cultivated (*wild ass*, *wild rose*) should not ordinarily be given one; but when the two words are used together attributively they must be hyphened or consolidated (*wild-goose chase*, *wildcat strike*).

2 *Wildly* is the adverb. In such phrases as *run wild*, *went wild*, *wild* is a predicative adjective. See UNIDIOMATIC -LY.

**wilful.** Thus spelt in BrE, but as either *willful* or *wilful* in AmE.

**will** (noun). We owe to the German language a relatively small but important group of phrases of the type *the will to* + verb or noun. To judge from the *OED*, *the will to live* (reflecting Ger. *der Wille zum leben*) and *the will to power* (Ger. *der Wille zur Macht*) entered the language in the late 19c. These were joined by others in the 20c., e.g. *will to art* (Ger. *Wille zur Kunst*) an innate human drive towards artistic creation, *will-to-be*, *will to believe*. The expressions are mostly derived from Nietzschean philosophy and from analytic psychology as part of the description of instincts and internal drives that govern human behaviour.

**will** (verb). One needs an advanced course in Old English and the history of the language to understand the origin and development of what are now taken to be two separate verbs, namely the *will* that operates as a flexionless auxiliary with a bare infinitive (*the judge adjourned* *the case and* will give *his verdict tomorrow*) and the *will* that conjugates in the normal way (*wills, willing, willed;* formerly also *willest*) and corresponds to the noun *will* 'the faculty of conscious and deliberate choice of action (*he has a* will *of his own*)'. For the auxiliary *will,* see SHALL AND WILL. The full verb *will* has several meanings, including, as the *COD* reminds us, 'to have as the object of one's will; intend unconditionally (*what God* wills; willed *that we should succeed*)'.

**willful.** See WILFUL.

**wimmin.** This semi-phonetic spelling was used early in the 20c. as an ironical term for 'women' (e.g. *'Wimmin's a toss up,' says Uncle Penstemon*—H. G. Wells, 1910; *'Didn't I tell you?' sniffed Grumpy. 'She's crazy. Wimmin! Pah!'*—*Snow White and the Seven Dwarfs,* 1938), It was adopted by feminists by the 1980s as a form not containing the ending -*men.* In Britain the word was particularly applied to the women protesters camped outside an American airbase at Greenham Common in Berkshire. In America it has been widely used since the late 1970s in writing by or about feminists. A typical example (from a university campus newspaper of 1992): *The Women's Center and the Department of African-American Student Affairs are sponsoring a fireside on 'Black Wimmin's sexuality and its impact on sister-to-sister relationships'.* Whether it will stretch its wings in the 21c. or just gently drop out of use remains to be seen.

**wind** (verbs). The verb meaning to twist, coil, go on a circular course, etc. (pronounced /waɪnd/, pa.t. *wound* /waʊnd/) is unconnected with that meaning to exhaust the breath (*wind* /wɪnd/, pa.t. *winded* /'wɪndɪd/). The verb meaning to sound a wind instrument, esp. a bugle, by blowing is unstable: it is often pronounced /waɪnd/ rather than /wɪnd/, with pa.t. and pa.pple either *winded* or *wound* /waʊnd/. Neither verb is connected with the noun *wound* /wuːnd/ 'injury' or its corresponding verb *wound* /wuːnd/.

**windward** is invariable (not -*wards*) as adj., adv., and noun. It means (on) the side from which the wind is blowing (as opp. to LEEWARD).

**wine, wire.** The corresponding adjs. are respectively *winy* and *wiry.* See -EY AND -Y IN ADJECTIVES.

**-wise, -ways. 1** An historical note. As a terminal element of adverbs, -*ways* was originally a use of the genitive of the ordinary noun *way.* From the 12c. onward such adverbs were formed, sometimes with by-forms ending in -*way.* This class of words includes *alway(s), anyway(s), crossway(s), endway(s),* and *sideway(s),* not all of which have survived. Parallel to this group of words were many with the same first element but with -*wise* (from OE *wise* 'way, manner, fashion') as ending, e.g. *anywise, crosswise, edgewise,* and *sidewise.* The two types made contact, and came to be thought of by some as two forms of the same word. Dr Johnson (in his Dictionary s.v. *Way* 25) went further and said that '*Way* and *ways* are now often used corruptly for -*wise*', and this statement, as the *OED* points out, 'has probably led many to prefer -*wise* to -*ways* or -*way* on the ground of supposed correctness'.

As a result of these word-forming processes, spread out over many centuries, the use of -*ways* and -*way* as free-forming adverbial suffixes had come to an end by the 20c. But most of the well-established forms have survived, e.g. *anyway, breadthways, crossways, edgeways, lengthways, sideways,* and *widthways.* While -*way(s)* as a free-forming terminal element died away, -*wise,* meaning 'in the manner of', has continued to thrive as it has done since the 14c. Some examples: *Let us try once more to argue* Cardinalwise—W. Hughes, 1677; *Priests sitting with their legs tucked up,* tailor-wise, *in the attitude of Buddha*—*Cornhill Mag.,* 1885; *Her mass of chestnut hair parted* Rossetti-wise *in the middle*—R. Macaulay, 1923; *It … swerved out of control and came* crabwise *down the middle lane and hit my lorry*—*Times,* 1963; *dangling his arms beside his hips and rolling his head* idiotwise—J. McInerney, 1985.

**2** Distribution of forms in -*way, -ways,* and -*wise.* To judge from the *COD* (1990) the commoner adverbs of these types still in use are as follows: *anyway* (arch. *anywise*); *breadthways, breadthwise; clockwise; contrariwise; cornerwise; crabwise; crosswise,* var. *crossways; edgeways,* var. *edgewise; endways,* var. *endwise; lengthways,* var. *lengthwise; likewise; nowise,* var. *noway; otherwise; sideways,* var. *sidewise; slantwise; widthways, widthwise.*

**3** New use of *-wise*. Somewhere about 1940, chiefly in informal contexts, but also in more serious ones, *-wise* began to be tacked on to nouns (including verbal nouns in *-ing*) with the broad meaning 'as regards, in respect of'. The habit began in America and continues to be commoner in AmE than in other forms of English. Fastidious speakers treat it with mild disdain, or with a shrug of the shoulders as if to say that its use in this way is inevitable, painful or too clever by half though it is. Some examples: Plotwise, *it offers little more or little less of what-happens-next interest than may be found* [etc.]—*Saturday Rev.*, 1948; *John Robert Russell, 13th Duke of Bedford . . . in twelve TV performances, was the greatest,* successwise, *among the aristocrats*—*Spectator*, 1958; Acting-wise, *I like Katharine Hepburn, Joanne Woodward, Judy Garland and, of course, Marilyn*—*Gossip* (Holiday Special), 1981; *did I do the proper thing,* etiquettewise, *by just standing there . . ., watching and saying nothing?*—*Chicago Sun–Times*, 1992; *They all kept up with me,* drinkingwise—*New Yorker*, 1993.

**wish** (verb). Freely used with a direct object to mean 'to want, desire' from OE times onwards, the type was described by the *OED* (1928) as 'Now *dial.*; superseded in standard English by *wish for*, or colloq. in certain contexts by *want*'. Typical examples: *Would you* wish *a little more hot water, ma'am?*—Dickens, 1854; [*The maid*] *flew into a rage, and wanted to know if I* wished *a month's notice*—W. R. H. Trowbridge, 1901. My impression is that the *OED*'s judgement is the right one, at least for the UK. But *WDEU* (1989) regards the use as 'certainly standard' (in America), citing some insubstantial evidence in support: e.g. *a majority of employees* wished *a union shop*—*Current Biography*, 1948.

**wishful.** Passing out of use is the type *wishful* followed by a *to*-infinitive, e.g. *I was wishful to say a word to you, sir* (Dickens, 1852). The word is also on the retreat in the sense '(of a person) wishing, desirous' used in other constructions, e.g. *I am but Jason, who dwells here alone . . . Wishful for happy days* (W. Morris, 1867); *Wishful from my soul That truth should triumph* (R. Browning, 1875). The word's strength now is in the phr. *wishful thinking* and derivatives, used as disparaging terms for beliefs or expectations that go beyond

the facts and will almost certainly fail. *Wishful thinking* is a relatively new term: the earliest example in the *OED* is one of 1932.

**wisteria** /wɪsˈtɪərɪə/, any plant of the genus *Wisteria*, is by far the usual form, not *wistaria*, even though most authorities say that it was named after Casper Wistar (1761–1818), an American anatomist. (*The R.H.S. Gardeners' Encycl. of Plants & Flowers*, 1991 printing, lists only *Wisteria*.)

**wit** (verb). Except for the phr. *to wit* ( = that is to say; namely) and the derived adverbs *wittingly* and *unwittingly*, this verb ( = to know) survives chiefly in the 1st and 3rd singular present *wot* (esp. in T. E. Brown's often quoted remark (1892) *A garden is a lovesome thing, God wot!*). See WARDOUR STREET.

**witch-.** See WYCH-.

**witenagemot** /ˈwɪtənəgəˌməʊt/. The name of an Anglo-Saxon national council (lit. 'a meeting of wise men').

**with.** See AGREEMENT 4.

**withal** (adv.), moreover, as well, etc. A WARDOUR STREET word.

**withe** /wɪθ, wɪð, waɪð/, a flexible shoot of willow or osier used for tying a bundle of wood, etc. This form (pl. *withes*) is as acceptable as *withy* /ˈwɪðɪ/, pl. *withies*. (Note that the Bible (1611) has the pl. form *withs*: *And Samson said vnto her, If they binde mee with seuen greene* withs *. . . then shall I be weake*—Judg. 16: 7.)

**withhold.** Thus spelt, not *withold*.

**without.**

| 1 | = outside.
| 2 | As conj. = unless.
| 3 | Repetition after *or*.
| 4 | *without hardly*.

**1** = outside. Both as adverb (*listening to the wind* without; *clean within and* without), as preposition (*is* without *the pale of civilization*), the word retains this meaning; but it is no longer for all styles, having now a literary or WARDOUR STREET sound that may be very incongruous or even ambiguous. Awareness of this recessive meaning of *without* is essential, for example, in this passage written by the Irish poet Cecil Frances Alexander (1818–95): *There is a green hill far away, Without a city wall, Where the dear Lord was crucified, Who died to save us all.*

**2** As conj. = unless. Once (14–17c.) in standard literary use (e.g. *Such a one, as a man may not speake of*, without *he say sir reuerence*—Shakespeare, 1591), this type was described by Johnson in his Dictionary (1755) as 'not in use' (he cited two examples from the work of Sir Philip Sidney). It has subsequently become restricted to regional or demotic speech in the UK and Ireland, and is also liable to turn up in the sense 'without its being the case that' in regional or working-class speech in America and elsewhere. Examples: *Everything she looked at was that child ... She couldn't lie with that man without she saw it*—F. O'Connor, 1955; *Man can put up with only so much without he descends a rung or two on the old evolutionary ladder*—E. Albee, 1962; *the silly bitch can't even keep the fire going without I have to rouse on her*—M. Eldridge, 1984 (Aust.); *You see, Father, our parents won't let us outside without we put our uniforms on*—H. Mantel, 1989; *I can truthfully say he never sat an exam without he was bad with his asthma*—Pat Barker, 1991; *There's no way they can develop that site now without I sell them my properties, and I'm not doing that*—P. Lively, 1991; *The top copy for the customer, the carbon for ourselves. You couldn't do business without you keep a record of receipts*—W. Trevor, 1994.

**3** Repetition after *or. It can be done without any fear of his knowing it*, or *without other evil consequences.* The well-meant repetition of *without* is often needless, even wrong. See OR 2.

**4** *without hardly.* Example: *The introduction of the vast new refineries has been brought about quickly, silently, and effectively, and without the surrounding community hardly being aware of what was happening.* Again, like sense 2, this is non-standard, in that it represents a joining of a negative word and an approximate negative, i.e. is dangerously close to being a double negative. It is the kind of construction that should not appear outside the setting down of demotic speech. Trollope should not have used the phrase in *It seemed to her as though she had neglected some duty in allowing Crosbie's conduct to have passed away without hardly a word*

*of comment on it between her and Lily.* The English for *without hardly* is *almost without.* But the use continues to turn up now and then: e.g. *They manage to weave together without hardly ever touching*—New Musical Express, 1992. Vulgarity persists.

**witticism.** An ingenious hybrid formation coined by Dryden in 1677 from *witty* (adj.) after *criticism*.

**wizard.** For *Wizard of the North*, see SOBRIQUETS.

**wizened** (adj.). The survivor of three adjectival forms (the other two being *wizen* and *weazen*) meaning 'thin and shrivelled'. It was originally (16c.) restricted to Scottish texts but is now a general English word. All three words answer ultimately to an OE verb *wisnian* 'to shrivel'. *Wizen* (first recorded in the 18c.) is a clipped form of *wizened*, while *weazen* (19c.) is what etymologists call an 'altered form' of *wizen* (adj.).

**wolf.** See -VE(D) etc.

**wolverine** /ˈwʊlvəˌriːn/, a NAmer. and Eurasian animal. Thus spelt, not *-ene*. In AmE the main stress tends to fall on the final syllable.

**woman.** A great deal about the currency and distribution of this word has already been said s.v. FEMALE; FEMININE DESIGNATIONS; GENTLEWOMAN; WIMMIN. See also the next entry and WOMYN. For the sense 'women in general', *womankind* (rather than *womenkind*) or *womenfolk* seem to be the favoured words now. The 20c. phrase *women's liberation* (and the shortened form *women's lib*), meaning the liberation of women from inequalities and subservient status in relation to men, is less often heard in the 1990s as the old barriers come down one by one. There is no slackening, however, in the use of the term *women's rights* (coined in the 17c.), esp. with reference to the legal entitlement of women to the same terms as men in conditions of employment, in the selection of candidates as MPs, etc. And *women's studies*, academic studies concerning women's role in society, history, etc., has become a central option in the degree syllabuses of universities and in writing in general.

**-woman.** There has been a noticeable decline in the frequency of use of compounds in *-woman* in the 20c., mostly by the use of genderless equivalent terms such as *woman police constable* (abbrev. *WPC*) for *policewoman*. But many such terms remain in common use, including the following (*policewoman* among them) which are all listed in the *COD* (1990) without restrictive labels of any kind: *airwoman, chairwoman, charwoman, horsewoman, kinswoman, needlewoman, oarswoman, policewoman, saleswoman, servicewoman, spokeswoman, tradeswoman, washerwoman.*

**womanish, womanlike, womanly.** See FEMALE etc.

**womyn** (pl. noun). A factitious formation coined by feminists to replace *-men* as a sequence of letters in the word *women*. A headline in a 1991 issue of an American newspaper: *How to 'Do' the Michigan Womyn's Music Festival.*

**wonder** (verb). **1** For *I shouldn't wonder if it didn't rain*, see NOT 5.

**2** *wonder whether/if.* In formal style sometimes construed with a subjunctive in the dependent clause: e.g. *Hilliard wondered whether Barton were not right after all*—Susan Hill, 1971; *Watching her departing figure ... he wondered whether he were any happier*—A. Brookner, 1989; *they had never had a serious conversation, and she wondered if that were wrong*—idem, 1992.

**won't.** From a welter of contracted forms of *woll not = will not* (among them *wonnot, woonnot, wo'nt*, some of them recorded as early as the 15c.), *won't* emerged in the 17c. as the standard one, the others now being dialectal or obsolete. Cf DON'T; SHAN'T.

**wont(ed).** *Wont*, the pa.pple of an obsolete verb *won* (OE *gewunian*, pa.pple *gewunod*) 'to accustom oneself to', is still used as a predicative adjective (*as he is wont to do*). The participial origin of the word having been forgotten, *wont* was used as a verb (*Talbot is taken, whom we wont to feare*—Shakespeare, 1591) and acquired a pa.pple of its own, *wonted*. In its other parts the verb is no longer used, but *wonted* survives as an attributive adj. (*he showed his wonted skill*). There is also a noun *wont* (*as is his wont*). The traditional

pronunciation /wʌnt/, useful in differentiating this word from *won't* ( = will not), has been preserved in AmE (as one of at least three standard pronunciations), but not in BrE, in which the standard pronunciation is /wəʊnt/.

**woodbine** (wild honeysuckle) is the established form, not *woodbind*, though the word is derived from OE *wudubind(e)*.

**woodenness.** Thus spelt, not *woodeness*.

**woof.** See WARP (noun).

**wool** makes, in BrE, *woollen, woolly*, and in AmE *woolen, woolly* (also *wooly*).

**word-division.** See HYPHENS 6.

**word order.** **1** The legitimate inversion of subject/verb/complement order is treated in numerous articles in this book (see esp. INVERSION). Some typical examples follow: *Only in the dream-sequence, where he debates with the devil, is he triumphant*—Kenneth Muir, 1987; *Nor am I pretending that there can be only one legitimate interpretation*—idem, 1987; *Only in recent times have they insisted on and begun to receive justice*—Dædalus, 1988; *Rarely do they remind the stranger that they have the privilege of living in one of the world's most dramatic environments*—ibid., 1988; *Beats there a heart amongst us so jaded ... that it has failed to be touched ... by the sound of Roy Orbison ...?*—The Face, 1989.

**2** Unfortunately, however, unintended humour is often brought about by the careless ordering of words. Some typical examples follow (they happen to be from American newspapers, but similarly inept examples could doubtless be found in newspapers anywhere in the English-speaking world): *One of the biggest supporters of Schoolfest was Carolyn Blount, who donated the $21.5 million complex* along with her husband, Winton; *'His humility and his courage are something we will always remember,' said coach Bill McCartney, commenting publicly about Aunese's death for the first time*; *She strongly disapproves of our living together* for religious reasons; *'Sure enough, somebody at the San Diego hospital where she was born erroneously had typed in "male"*; *She did not want her last name* used to protect her daughter's identity.

**work** (verb). Perhaps the most important development to this polysemic verb (from OE *wyrcan*) is the virtual disappearance of its true pa.t. and pa.pple *wrought* (OE *worhte*; *geworht*). Richly documented in the *OED* from Anglo-Saxon times onward, *wrought* began to be challenged by *worked* in the 15c., and *worked* is 'now the normal form except in archaic usage (in which the older form *wrought* may appear in any sense), and in senses which denote fashioning, shaping, or decorating with the hand or an implement' (*OED*, 1928). Of course anyone who reads the Bible or pre-1900 literary works is familiar with *wrought*: e.g. *And by the hands of the Apostles, were many signes and wonders wrought among the people*—Acts (AV) 5: 12; *A splendid cover ... of tapestry richly wrought*—Cowper, 1784. One notable survivor is the participial adj. *wrought* in *wrought iron*. But not one person in a thousand would recognize the connection with the verb *work*. Another reasonably common residual use is the adj. *wrought-up*, agitated or excited: e.g. *I got so wrought up watching that trembly little fellow that I started cheering him on*—New Yorker, 1987.

The phrases *work havoc* and *wreak havoc* (in which *wreak* is not etymologically related to *work* or *wrought*) are still locked in battle: the indications are that *work havoc* but *wreaked havoc* will prevail as an uneasy pair. Logic does not necessarily play much of a part in linguistic development.

**workaday.** Since the Middle Ages, *workaday*, *workday*, and *working day* have all been used as nouns meaning 'a day on which work is done (as distinct from a holiday or a day off)'. The distribution of the three forms before the 20c. is not easily determinable, but at the present time it would seem that as nouns *workaday* is obsolete, while *workday* and *working day* are still very much in use in all forms of English. As an attributive adj. *workaday* means (*a*) ordinary, everyday; humdrum, routine; (*b*) characteristic of, used on days at work. In AmE, *workday* also means the period of time in a day during which work is performed (*an eight-hour workday*). In BrE, *working day* would be the preferred term in this sense.

**world.** *All the world and his wife*: first found in print in the early 18c. as a hyperbolical way of describing a large and miscellaneous gathering of people of both sexes present at a function of any kind (e.g. Miss. *Pray, Madam, who were the Company?* Lady Smart. *Why, there was all the World, and his Wife*—Swift, 1731–8), the type was harshly, it seems to me, described by Fowler (1926) as an example of 'worn-out humour'. The *OED* and other dictionaries simply list the phrase without condemnatory labels of any kind.

**worldly.** Take care not to write it as *wordly*, a common mistake.

**worn-out humour.** There is nothing so tired as a tired joke, nor so unamusing as an overused piece of facetious wordplay. However hilarious a pun or a deliberate misquotation was on first (or even second and third) hearing, it can outstay its welcome. To a new generation the jolly turns of phrase that amused our parents and grandparents seem laboured—and, what is worse, we have heard them before, and more than once. It is not the great comic writers of the past who now make us groan—*The Diary of a Nobody* and *Tristram Shandy*, Evelyn Waugh and P. G. Wodehouse, Damon Runyon and Ogden Nash, will surely never lose their savour. But how worn out are such jests as *May all your troubles be little ones* or *Don't do anything I wouldn't do*; such farewell formulas as *See you later, alligator* or *Hasta la vista*; such toasts as *Down the hatch*; such parodies as *to —— or not to ——*; such tired quotations as *single blessedness*; *no respecter of persons*; *Water, water, everywhere, Nor any drop to drink*; such oxymorons as the *gentle art* of doing something ungentle, or the *tender mercies* of a martinet; the insufferability of such truisms as *he does not suffer fools gladly*; such needless euphemisms as *in an interesting condition* or *pushing up daisies*; such meioses as *the herring pond* or *epithets the reverse of complimentary*; such playful archaisms as *hight* or *whilom* or *yclept*; such Gallicisms as *give one furiously to think* and *return to one's muttons*; such sobriquets as *his nibs* and *trick cyclist*; such arch appellations as *my better half, your lady wife*, or (the demotic equivalent) *the wife*; such amputated forms as *couth*, *kempt*, and *shevelled*; such blends (of hundreds) as *edutainment* and *shopportunity*; such threadbare circumlocutions as *well endowed with this world's goods*—all these, and many more, are overdue for a long rest, if not permanent banishment from the language. See also CLICHÉ; HACKNEYED PHRASES; IRRELEVANT ALLUSION; PEDANTIC HUMOUR; POLYSYLLABIC HUMOUR.

**worry.** For inflexions, see VERBS IN -IE, etc. 6.

**worsen.** See -EN VERBS FROM ADJECTIVES. The *OED* has an interesting note on the word's history: 'The word [first recorded c1225 but with no record of use between 1670 and 1806] is common in dialect (see *Eng. Dial. Dict.*) and was introduced to literature c1800–1830 (by writers like Southey and De Quincey) as a racy vernacular substitute for *deteriorate* and the like.' 'Racy' it no longer is: just a routine verb playing an ordinary part in the standard language.

**worsened words.** Changes in the meaning of words often reflect the value or values of what they stand for. Over the course of history a great many words have become commendatory or have lost their pejorative connotations: e.g. *boy* (fettered person → male servant → male child), *meticulous* (over-scrupulous → very careful, accurate), PRESTIGE, and PRESTIGIOUS. An even larger number have gone in the opposite direction, i.e. have worsened in meaning. *Colonialism* was once seen as a benevolent process bringing Western values and civilization to barbarous and backward countries and their populations. Events of the last century or so have led to the 'liberation' of most of the former colonies of Britain, France, Spain, and some other European countries, and *colonialism* and *imperialism* have become deeply suspect concepts. *Appeasement*, after five centuries of use in contexts denoting pleasure or satisfaction, moved into pejorative territory in the 1930s when Neville Chamberlain embarked on a foreign policy of conciliation by concession. *Collaboration* and *collaborator* took on sinister connotations during the war of 1939–45. *Democracy* has become a questionable term to apply to the systems of government adopted by many countries that are members of the United Nations. In several words a specific sexual sense has placed earlier general senses at risk, e.g. *erection* and *intercourse*. The word *gay* in the sense 'homosexual', to the dismay of many people, has largely displaced the traditional senses 'light-hearted, carefree'. *Sleaze* and *sleazy* are no longer applied to the flimsiness of textiles but to instances of moral turpitude, esp. in regard to financial transactions and sexual affairs of a scandalous nature. *Egregious*, once capable of meaning 'towering above the flock, remarkable' now normally means 'outstandingly bad, flagrant', i.e. remarkable in a bad sense.

The linguistic processes of amelioration and pejoration are immensely complicated, as the *OED* reveals in its treatment of thousands of upward-moving, downward-sliding, and sideways-moving words. The only certainty is that all three processes will continue inexorably and often for long periods unnoticed by the majority of a given community. Readers may wish to pursue the history of some ordinary words such as *buxom, candid, crafty, genteel, knave, lewd, sad, silly, specious,* and *villain* to see the way in which the language treats its own components. C. S. Lewis's *Studies in Words* is a mine of information about some of these words. Fuller treatment of them all is available in the *OED*.

**worser.** This double comparative has a long literary history beginning in the late 15c.: e.g. *Chang'd to a worser shape thou canst not be*—Shakespeare, 1591; *I find she is A diuell, worser then the worst in hell*—J. Ford, 1633; *For I, e'en I, the bondsman of a worser man was made*—W. Morris, 1887; *Some gals is better, some wusser than some*—E. Pound, 1959; *The people oh Lord Are sinful and sad Prenatally biassed Grow worser born bad*—Stevie Smith, 1975.

The *OED*'s verdict (1928) on the status of the word is still about right: 'The word was common in the 16th and 17th c. as a variant of 'worse', in all its applications. In modern use, it is partly a literary survival (esp. in phrases like *the worser part, sort, half*), partly dial. and vulgar.' Its vulgar use is illustrated by the following examples: *Your poor dear wife as you uses worser nor a dog*—Dickens, 1835; *That's my new word. Yeah, it's gonna be worser. You know, worse to worser*—Interview (US), 1990.

For the analogous *lesser*, see LESS 4.

**worship** makes *worshipped, worshipper, worshipping* in BrE, but often *worshiped* etc. in AmE. See -P-, -PP-.

**worst.** The idiomatic phr. *If the worst come(s) to the worst* has been in standard use in Britain since the late 16c. In AmE the usual form lacks the definite article, i.e. *if worst comes to worst*; and sometimes *worse* is used instead of the first *worst* or even instead of both.

**worsted.** Pronounce /ˈwʊstɪd/.

**worth, worth while, worth-while, worthwhile.** All four forms may be legitimately used but are set about with rules. To start with, *worth*, when used as an adjective, may only be used predicatively (*This Rembrandt is worth a million pounds*), and, as shown in this example, must always have what can best be described as an object (we cannot say *This Rembrandt is worth*). The object can be a verbal noun (*This is worth saying*). It can also be the word *while*, originally in this construction a plain noun, but now seen simply as an element that can be tacked on to *worth* without seeming unidiomatic: *this is worth while* literally means 'this is worth the necessary expenditure of *time*'. (For the sense 'time' we have only to recall phrases like *a long while ago*.) Another of the restrictions is that the type *This is worth while saying* is ungrammatical: simply omit *while*. By the same token, *while* should be omitted in the following examples: *A spare captain, to take charge of any prize that might be worth* while *turning into a raider; Was not that a line worth* while *pursuing?*

When used predicatively *worth while* is normally written as two separate words (*the experiment was worth while*). When the two words are used attributively (a use first emerging at the end of the 19c.) they should be hyphened (*a worth-while experiment*) or, now more commonly, written as one word (*a worthwhile experiment*). The use of the solid form *worthwhile* in the predicative position (*the experiment was worthwhile*) is undesirable, though now regrettably common.

**worthy.** Used to mean 'by reason of merit or excellence', *worthy* is usually constructed with *of* before a following noun or noun phrase (*thought such services worthy of some recognition*), but from the

14c. to the 19c. the *of* could also be idiomatically omitted (*Dame Polish is a figure well worthy the cordial and lavish commendation of Gifford*—Swinburne). The omission of *of* is no longer permissible except perhaps in highly stylized work ('in exalted contexts', as Fowler (1926) expressed it).

**-worthy.** Used as a combining form producing adjs. meaning (i) deserving (*blameworthy, noteworthy*), (ii) suitable or fit for (*airworthy, newsworthy, roadworthy*). They all seem so obvious. But a very large number of such compounds have fallen by the wayside since they came into the language. Among the casualties (with the date of first record indicated) are the following: *deathworthy* 13c., *faith-* 16c.; *fame-* 17c.; *fault-* 17c.; *honour-* 16c.; *labour-* 17c.; *laugh-* 17c.; *love-* 13c.; *mark-* 19c.; *name-* 16c.; *pains-* 17c.; *sale-* 16c.; *scorn-* 17c.; *song-* 19c.; *thank-* 14c.; *wonder-* 17c.; *worship-* 16c.

And among the survivors, some of them first used in this century or the last, are: *airworthy* 19c.; *battle-* 19c.; *blame-* 14c.; *hate-* 20c.; *note-* 16c.; *praise-* 16c.; *road-* 19c.; *sea-* 19c.; *trust-* 19c. Of these, *battleworthy* and *hateworthy* already look vulnerable. Others may follow as the language stumbles on into the 21c.

**would.** See SHOULD AND WOULD, where some of the main distinctions are set down, esp. the dominance of *would* or *'d* even with the first person pronouns *I* and *we*. For *would of* = *would have*, see OF C. 1.

**wove** (pa.pple and adj.). Instead of *woven* the usual form for these parts of *weave* (verb)[2] (of fabrics etc.), *wove* is used to mean 'Of paper: made on a wire-gauze mesh so as to have a uniform unlined surface' (*New SOED*). So *wove mould, paper*, etc.

**wove** (pa.t.). See WEAVE (verb).

**wrack.** See RACK; NERVE-RACKING.

**wraith.** Pl. *wraiths* /reɪðs/. See -TH (θ) etc.

**wrapt, wrapped, rapt.** See RAPT.

**wrath, wrathful, wroth.** The first two are literary words for extreme anger and extremely angry respectively, and are normally pronounced in BrE /rɒθ/ and /ˈrɒθfʊl/; but /rɔːθ/ is also standard in both words. *Wroth*, usually pronounced /rəʊθ/, is an archaic predicative adj., esp. in the phr. *wax wroth*, meaning angry (e.g. *Then the old man Was wroth, and doubled up his hands*—Tennyson). In AmE, /ræθ/, /ræθful/, and /rɔːθ/, respectively, are the usual standard pronunciations.

**wreak** is related to *wreck* at one remove. The difference is that *wreak* is the direct descendant of a standard OE verb *wrecan* 'to avenge, drive away, etc.', while *wreck* is an adoption in ME of an Anglo-Norman verb *wrec*, itself a descendant via ON of a Germanic verb *\*wrekan*. The same reconstructed form is the ancestor of OE *wrecan* (wreak). The subsequent history of the two verbs has been set down with great thoroughness in the *OED*. It is all very complicated. For the connection between *wreak havoc* and *work havoc*, see WORK (verb).

**wreath.** Pl. *wreaths* pronounced either /riːðz/ or /riːθs/. The corresponding verb *wreathe* is pronounced /riːð/. See -TH (θ), etc.

**wrestle.** The medial *t* is silent. Pronounce /ˈresəl/. See PRONUNCIATION 4.

**wrick.** See CRICK.

**write.** The transitive AmE use = 'to communicate with (a person) in writing' is shown in the following examples from short stories in the *New Yorker*: *I had written my mother about all this* (1987); *Liza, my dear, I have never written you yet to thank you for going out to our house* (1993). This construction was formerly standard in BrE ('frequent from *c* 1790' says the *OED*), but it is now in restricted use unless accompanied by a second (direct) object, as in *I shall write you a letter as soon as I land in Borneo*. In old-fashioned commercial correspondence the types *We wrote you yesterday*; *Please write us at your convenience* were often used, but nowadays *to* would normally be inserted before *you* and *us*.

**writ large.** In the ever-popular, usu. figurative, phr. *writ large*, first recorded in the 17c. (*New Presbyter is but Old Priest writ large*—Milton), *writ* is an otherwise hardly used pa.pple of *write*. Some examples of *writ large* and of other phrases containing *writ* (drawn from the *OED*): *The man was no more than the boy writ large, with an extensive commentary*—G. Eliot, 1866; *That my life ... Was but a tale Writ large by Zeus*—L. Morris, 1877; *This year's Defence White Paper ... is last year's writ quietly*—*Times*, 1959; *In a curious way he's* [sc. Sir Isaac Hayward's] *an amalgam, writ small, of Attlee, Morrison and Bevin*—*Observer*, 1961; *Every project has success writ large all over it*—Author, 1994.

**wrong** (adv.) occupies half as much space in the *OED* as *wrong* (adj.) and five times more space than the notionally correct adverbial form *wrongly*. In other words the subtleties attending the various uses are considerable. Adverbial *wrong* tends to be the choice when the sense required is 'in a direction different from the right or true one, astray' (e.g. *Tintoret ... may lead you wrong if you don't understand him*—Ruskin); or 'mistakenly, erroneously' (e.g. *In spite of her care and assiduity she guessed wrong*—Thackeray); in the phrases *go wrong* (e.g. *There are more ways of going wrong than of going right*—Herbert Spencer), or (somewhat slangily) *get wrong* (e.g. *Don't get me wrong; there's no offence meant*—N. Freeling, 1974); and in a limited number of compounds, e.g. *wrong-footed, -headed*. *Wrongly* (rather than *wrong*) is essential before the verb it governs (e.g. *he has been wrongly accused*); and also in the phrase *rightly or wrongly* (the adverbial phr. *right or wrong* does not exist in standard English).
   Cf. RIGHT.

**wroth.** See WRATH.

**wrought.** See WORK (verb).

**wry.** The various extended forms are *wryer, wryest, wryly*, rather than *wrier*, etc.

**wych(-).** The customary spellings of the two main trees involved are *wych elm* and *witch hazel*. (Sometimes with hyphens.) The first element *wych(-)* has nothing to do with witches, but is derived from OE *wic(e)-*, a formative element related to a verb meaning 'to bend'.

**Wykehamist.** Thus spelt.

# Xx

**-x,** as plural of loanwords from French. It is still customary to write *-x* instead of the English *-s* in many of the plurals of words in *-eau* and *-eu* adopted as loanwords from French, but to pronounce them with /z/, as in English plurals. Regularizing them all to a compulsory final *-s*, though theoretically desirable, seems likely to be a long drawn-out process. The main difficulty is that we like the look of final *-x* as a plural marker in words that are recognizably loanwords from French.

For words in the following lists (each of which is treated at its corrected alphabetical place), *-x* is recommended for the plural forms, except for *adieu, portmanteau, purlieu,* and *tonneau,* for each of which *-s* is dominant; and also except for *flambeau* and *trousseau,* which may have either *-s* or *-x*. The plural of all the words is pronounced /-z/, except that the *x* of *jeux d'esprit, prie-dieux,* and usu. of *milieux,* is left unpronounced.

(in *-eau*) *bandeau, beau, bureau, château, flambeau, fricandeau, plateau, portmanteau, rondeau, rouleau, tableau, tonneau, trousseau.* (in *-eu*) *adieu, jeu d'esprit, milieu, prie-dieu, purlieu.*

**xebec** /'zi:bek/, small Mediterranean boat. Thus spelt, not *zebec*.

**-xion, -xive.** In a small group of nouns drawn from Latin nouns (classical, ecclesiastical, medieval, etc., Latin) ending in *-iōn-em,* spellings with *-xion* and *-ction* have been in use in English at one time or another; similarly with some adjs. in *-xive* and *-ctive*. The main words concerned (with the related *-ive* adjs. given (in brackets) where they exist) are: *complexion; connection/connexion (connective); crucifixion; deflection (deflective); genuflexion/genuflection; inflexion/inflection; reflection (reflective/reflexive); transfixion.* The choice of *-ct-* in *connection* etc. has been governed partly by the spelling of the related verbs (*connect, deflect, genuflect, inflect, reflect*), and partly by the multitude of English nouns ending in *-tion*.

For more details about the above words, see each of them at their alphabetical place in the dictionary.

**Xmas.** In this abbreviated form of *Christmas,* which is first recorded in the 18c., X represents the first letter (chi) of the Greek word for Christ, namely Χριστός. Its convenience accounts for its common use in commercial printing and in personal letters and diaries, but in spoken English it is usually pronounced like 'Christmas' rather than as 'ex-mass', though the latter pronunciation is not uncommon.

**X-ray.** Thus spelt (with capital X and hyphen).

# Y y

**-y.** For the suffix used in making adjectives from nouns (*racy*, etc.), as it affects spelling, see -EY AND -Y IN ADJECTIVES. For the diminutive suffixes *-y* (*Johnny*) and *-ie* (*bookie, doggie,* etc.), see HYPOCORISTIC.

**y'all.** See YOU-ALL.

**y and i.** There are a few words in which *i* and *y* are interchangeable (for these, publishing houses usually adopt a house spelling), a few others in which BrE practice differs from that of AmE, and a handful containing the letters in successive syllables making it difficult to remember which is *i* and which is *y*. The main words involved are (those in small capitals are treated in separate articles):

> *cider*, CIPHER, GYPSY, *Libya(n)*, LICH-GATE, *Mytilene*, PYGMY, SIBYL, SIPHON, SIREN, STILE (over a fence), STYLE (manner), STYMIE, SYLLABUB, SYLVAN, TIRO, TYRE (of wheel), WYCH *elm*.

**Yankee.** The mistaken practice of applying this term, and also its shortened form *Yank*, to any citizen of the USA persists in Britain and in many other English-speaking countries. *Yankee* was originally the nickname of those who lived in New England; at its most comprehensive a hundred years later it was extended contemptuously by the confederates during the American Civil War to all soldiers of the federal armies. The derivation of the word has been the subject of much discussion and no agreement.

**yclept.** An archaic or jocular word meaning 'called (by the name of)', a direct descendant of OE *gecleopod*, pa. pple of *cleopian* 'to call, name'. See WORN-OUT HUMOUR.

**ye** (definite article). Sometimes written as *yᵉ*. A pseudo-archaic form of *the* (as in *Ye Olde Tea-Shoppe*), in which *y* is the descendant of a 14c. handwritten form of the OE and ME runic letter þ (thorn), which had come to resemble the letter *y*. It is pronounced /jiː/ like the pronoun *ye*, but is 'properly' /ðiː/, and in any case is just a variant of the definite article.

**yeah** /jeə/. The conventional spelling of a shortened form of *yes*. It occurs in the spoken word with greater frequency than is realized by many standard speakers. It is also used in the phr. *Oh yeah?* to express incredulity.

**year.** 1 For the use of *in* in the phrases *in a year* and *in years*, see IN 2.

2 A possessive apostrophe is needed in the types *five years' imprisonment* (contrast *a year's imprisonment*) and *in two years' time*.

**yellowhammer.** The latest view on the etymology of this bird-name is that the second element *-hammer* is perhaps from OE *amore* (name of an unidentified bird, but with cognates in some other Germanic languages), possibly conflated with OE *hama* 'covering, feathers'. The full name *yellowhammer* has not been found before the mid-16c., and then only in the form *yelambre*.

**yen** (Japanese currency unit). Pl. the same.

**yeoman.** *Yeoman service* and *yeoman's service* are both standard, but the greater ease of the former has made it the more usual form.

**yes** (noun). Pl. *yeses*. See -S-, -SS-.

**yet.** 1 (adv.). Most uses of *yet* (and of *as yet* and *but yet*) are uncontroversial, but it has an uneasy and complex relationship with some uses of *still*: witness the examples (from sources outside England): *You know you look ill yet, very ill*—J. S. Winter, 1888; *It says here in the newspaper that Mrs. Throckmorton was shot in her apartment last night, and the bullet is in her yet*—an Amer. newspaper, 1988. (In both cases *still*, correctly positioned, is required in England.) The type *Is it raining yet?* unambiguously means 'Has it started to rain yet?' in England, but in Scotland could be taken to mean 'Is it still raining?' if the *yet* is accented. *The Amer. Heritage Dict.* (3rd edn, 1992) in a usage note identifies an informal use of *yet* that is not current in England: 'In formal style *yet* in the sense "up to now" requires that the accompanying verb be in the present perfect, rather than in the simple past: *He hasn't started yet,* not *He didn't start yet.*' In other words, in informal AmE a *do*-support negative construction + infinitive is sometimes used instead of the BrE type of *have not* + pa.pple.

**2** (conj.). Formerly the presence of the conjunction *yet* often led to the reversal of word order in the immediately following words. Example: *Though ye haue lien among the pots, yet shall yee bee as the wings of a doue*—Ps. (AV) 68: 13. Such inversions would now be regarded as archaic.

**Yiddish** (adaptation of Ger. *jüdisch* 'Jewish') is 'a vernacular used by Jews in or from central and eastern Europe, based chiefly on High German with Hebrew and Slavonic borrowings, and written in Hebrew characters' (*New SOED*). English (esp. AmE) has adopted many words and phrases from Yiddish in the course of the 20c., e.g. *chutzpah* (shameless audacity); *kibitzer* (looker-on at cards, etc.); *klutz* (clumsy or inept person); *kvell* (to boast, gloat); *mazuma* (money); *need it like a hole in the head* (have no need for); *schlemiel* (clumsy person, fool); *schlep* (to haul, drag); *schmuck* (contemptible person); *shtoom* (silent); *What's with you?* (What's the matter?).

**yodel.** The inflected forms are *yodelled, yodelling,* but usu. *yodeled, yodeling* in AmE. See -LL-, -L-.

**yoghurt.** Thus spelt, or as *yogurt,* in BrE, and pronounced /ˈjɒgət/. In AmE (and also in Australia and New Zealand) the first syllable is pronounced /ˈjəʊg-/.

**yoke** is a wooden crosspiece fastened over the necks of two oxen, etc.; *yolk* is the yellow internal part of an egg.

**yon, yore.** See WARDOUR STREET.

**you.** For the mixture of *you* and *one* in the same sentence, see ONE 7.

**you-all.** (Also in abbrev. form *y'all.*) Used at all social levels in certain Southern states in the USA for *you* (either singular or pl., but predominantly pl.; 'always

with a plural meaning by those to whom the form is native, although often misunderstood as a singular by outlanders' (*CGEL*)). Examples: (y'all) *Unc' Toby won't feel right if y'all don't eat his lovely food*— E. J. Gaines, 1968; *Yes, Doctor. You'll be in the breakfast room. Y'all have a nice day*— J. S. Borthwick, 1982; *The only reason I can't do that is I would then be laundering y'all's money, which I can't do*—Chicago Tribune, 1988. (you-all) *You-all certainly is used as singular in the Ozarks—I have heard it daily for weeks at a time*—*Amer. Speech,* 1927; *In almost a score of years of residence in North Carolina I have never heard anyone say 'you all',* unless the plural was definitely and distinctly intended—ibid., 1944; *'Don't forget to get you-all some beer!' Ruth calls out to the men*—S. Gaddis in *New Yorker,* 1990.

**you know.** See MEANINGLESS FILLERS.

**you name it.** This popular phrase, implying that the list of entities just mentioned could be extended indefinitely, seems to have emerged in the 1960s and to have been in frequent use since then, almost to the point of having become a cliché. Typical examples: *Mallards, gadwall, partridge, quail—you name it—they're up here for the season every year*—Field & Stream, 1967; *'I've picked up rocks, glass, parts of beer cans—you name it,' she said*—New Yorker, 1988.

**you're** is a contracted form of *you are* (e.g. *You're a member of the Senior Common Room, aren't you?*).

**yours.** **1** The possessive pronoun *yours* is thus spelt (*this book is yours*), never as *your's.* So too, without apostrophe, *hers, its, ours,* and *theirs.* See ABSOLUTE POSSESSIVES.

**2** For epistolary uses, see LETTER FORMS.

**yous, youse.** Dialectal (esp. Liverpool and Glasgow according to *CGEL*) and low-prestige substitutes for the plural pronoun *you.* In printed work (as distinct from dialectal fieldwork) both *yous* and *youse* turn up most frequently in representations of the speech of poorly educated Amer., Aust., and NZ speakers. Some examples: *Youse kids makes me tired*—S. Crane, 1893 (US); *Say, yous guys, this is fellowworker McCreary*—J. Dos Passos, 1930 (US); *Look here, Ray, I fought them* *fellers when youse was all kids*—M. Eldridge, 1984 (Aust.); *It's the least I can do for youse*—E. Jolley, 1985 (Aust.); *What sort of caper yous buggers coming at now?*—V. O'Sullivan, 1985 (NZ); *You know, I thought you blokes—youse cobbers—were gunna beat me up when I first saw yers*—T. Winton, 1985 (Aust.).

**youth.** Pl. *youths*, pronounced /juːðz/. See -TH (θ) AND -TH (ð). Its primary sense is a young male person. It may also mean collectively 'young people (of both sexes)' (*the youth of the country*). In the pl. form *youths* it is used only of males, frequently disapprovingly (*the youths in the estate are well known to the police*).

**-yse, -yze.** In a small group of words with these endings the standard spelling in BrE is *-yse* (*analyse, catalyse, dialyse, paralyse*, etc.), and in AmE *-yze* (*analyze*, etc.). See -IZE AND -ISE IN VERBS 1.

# Zz

**Zarathustrian** /zærə'θʊstrɪən/. The adj. and noun corresponding to *Zarathustra*, the Old Iranian form of the name of the founder of the ancient Persian religion; = ZOROASTRIAN.

***Zeitgeist*** /'tsaɪtgaɪst/. A 19c. loanword from German meaning 'the spirit of the age'.

**zenith.** Pronounced /'zenɪθ/ in BrE, but with initial /'ziːn-/ in AmE.

**zero. 1** Pl. *zeros*, but *zeroes* is also permissible. See -O(E)s.

**2** For the circumstances in which the numeral 'o' is pronounced 'zero', see NOUGHT.

**zero-derivation.** See CONVERSION.

**zeugma** (Gk, = yoking). 'A rhetorical figure by which a single word is made to refer to two or more words in a sentence, esp. when applying to them in different senses.' (*New SOED*, 1993) In the past the term has sometimes been applied to the type of construction, now regarded as ungrammatical, that is illustrated s.v. SYLLEPSIS (*a*). But usage has tended to restrict *zeugma* to the type illustrated s.v. SYLLEPSIS (*b*). This second type is also known as a *condensed sentence*. Classical scholars continue to observe the distinctions between the terms *syllepsis* and *zeugma*, but in practice *syllepsis* is no longer applied to any English rhetorical device of the present day. The ancient distinctions have lost their usefulness in late 20c. English.

**zigzag.** The inflected forms are *zigzagged*, *zigzagging*; but *zigzags*. See -G-, -GG-.

**zinc.** The inflected forms of the verb hover between *zinced*/*zincing* and *zinked*/*zinking* (the danger of regarding *zinced* and *zincing* as rhyming with *minced* and *mincing* is obvious). In the derivatives of *zinc* there is considerable variation in practice between types having -k-, -c-, or -ck-: the headwords for these in the *New*

*SOED* (1993) are *zincian*, *zincic*, *zincite*, *zincification*, *zinco*, *zincography* (all pronounced with medial /k/), but it does not follow that everyone writing about the subject opts for the forms with -c-. See -C-, -CK-.

**zingaro** /'zɪŋgərəʊ, ts-/, an Italian gypsy. (Often written with a capital initial.) Pl. *zingari* /-ri/. Fem. *zingara*, pl. *zingare* /-reɪ/.

**zither,** a guitar-type musical instrument introduced into England from Austria in the mid-19c. The word is ultimately derived from L *cithara*, Gk κιθάρα. As the *OED* comments, 'Musical instruments are subject to great alteration of structure and shape, in process of time, and in different countries. Some of the resulting types become peculiar to one country, some to another ... Thus *cither*, *cithern* or *cittern*, *citole*, *gittern*, *guitar*, *zither*, are all found in English as names of extant or obsolete instruments developed from the *cithara*.'

**zoo.** A curtailed form of *zoological garden*, first recorded in the mid-19c. See ABBREVIATIONS 1.

**zoology.** Between 1921 when the *OED* gave only /zəʊ'ɒlədʒɪ/ and the late 20c. a battle has raged between those (including myself) who continue to pronounce the word thus and those who pronounce it as if it were spelt *zoo-ology*. As I remarked in *The Spoken Word* (1981): 'The practitioners of the subject, especially those more concerned with the animals themselves than with nomenclature and etymology, frequently use both this pronunciation [i.e. that with initial zō-ol-] and the form with initial zoo-with apparent indifference.'

**Zoroastrian** /zɒrəʊ'æstrɪən/. A follower of Zoroaster, the Latinized modification of *Zarathustra* (see ZARATHUSTRIAN).

**zwieback** (a kind of rusk). A word of German origin (in Ger. lit. 'twice-baked') for which no settled pronunciation has been achieved in English. In BrE, /'zwiːbæk/ is dominant, but /'tsviːbak/ is used by those who have some knowledge of German. In AmE any of several pronunciations are current, /ˈswiːbæk, ˈswaɪ-, ˈzwiː-, ˈzwaɪ-, -bɑːk/.

**-z-, -zz-.** Only a few words are affected, but no consistent pattern has emerged. Of the various possibilities, this book recommends:

buzz (n. and v.). So buzzes (n. pl.), 3rd pers. sing. of (v.); buzzer.

fizz (n. and v.). So fizzes (n. pl.), 3rd pers. sing. of v.; fizzy, etc.

frizz (n. and v.). So frizzes (n. pl.), 3rd pers. sing. of v.; frizzy, etc.

jazz (n. and v.). (No pl.: a non-count noun), jazzes 3rd pers. sing. of v.; jazzman, jazzy, etc.

pizzazz. (No pl.: a non-count noun). For variant spellings, see PIZZAZZ.

quiz (n.). Pl. quizzes. So quiz-master.

quiz (v.). So quizzes, quizzed, quizzing.

viz. (adv.). Invariable.

whizz (n. and v.). See WHIZZ.

zizz (n. and v.). So zizzes (n. pl.), 3rd pers. sing. of v.